The American Ceramic Society
735 Ceramic Place
Westerville, Ohio 43081-8720

AF229500

III-V Nitrides

MATERIALS RESEARCH SOCIETY
SYMPOSIUM PROCEEDINGS VOLUME 449

III-V Nitrides

Symposium held December 2-6, 1996, Boston, Massachusetts, U.S.A.

EDITORS:

F.A. Ponce
Xerox Palo Alto Research Center
Palo Alto, California, U.S.A.

T.D. Moustakas
Boston University
Boston, Massachusetts, U.S.A.

I. Akasaki
Meijo University
Nagoya, Japan

B.A. Monemar
Linköping University
Linköping, Sweden

MATERIALS
RESEARCH
SOCIETY

PITTSBURGH, PENNSYLVANIA

This work was supported in part by the Office of Naval Research under Grant Number ONR: N00014-97-1-0055. The United States Government has a royalty-free license throughout the world in all copyrightable material contained herein.

Single article reprints from this publication are available through
University Microfilms Inc., 300 North Zeeb Road, Ann Arbor, Michigan 48106

CODEN: MRSPDH

Copyright 1997 by Materials Research Society.
All rights reserved.

This book has been registered with Copyright Clearance Center, Inc. For further information, please contact the Copyright Clearance Center, Salem, Massachusetts.

Published by:

Materials Research Society
9800 McKnight Road
Pittsburgh, Pennsylvania   15237
Telephone (412) 367-3003
Fax (412) 367-4373
Website: http://www.mrs.org/

Library of Congress Cataloging in Publication Data

III–V nitrides : symposium held December 2–6, 1996, Boston, Massachusetts, U.S.A. / editors, F.A. Ponce, T.D. Moustakas, I. Akasaki, B. Monemar
    p. cm—(Materials Research Society symposium proceedings ; v. 449)
    Includes bibliographical references and index.
    ISBN 1-55899-353-3
    1. Nitrides—Congresses. 2. Thin films, Multilayered—Congresses.
  3. Crystal growth—Congresses. 4. Molecular beam epitaxy—Congresses.
  5. Photoelectronic devices—Congresses. I. Ponce, F.A. II. Moustakas, T.D.
  III. Akasaki, I. IV. Monemar, B. V. Series: Materials Research Society
  symposium proceedings ; v. 449.
QC176.9.84A14 1997   97-653
621.3815′2—dc21   CIP

Manufactured in the United States of America

CONTENTS

PART I: <u>CRYSTAL GROWTH - BULK AND MOCVD</u>

*Invited Paper

*Invited Paper

PART IV: <u>STRUCTURAL PROPERTIES</u>

*Invited Paper

PART V: <u>ELECTRONIC PROPERTIES</u>

*Invited Paper

*Invited Paper

PART VI: <u>LUMINESCENCE AND RECOMBINATION</u>

*Invited Paper

PART VII: <u>OPTICAL PROPERTIES</u>

*Invited Paper

PART VIII: <u>POINT DEFECTS</u>

*Invited Paper

*Invited Paper

PART XI: DEVICES

*Invited Paper

PREFACE

The symposium consisted of nine half-day oral sessions (25 invited and 54 contributed talks) and three poster sessions (192). The attendance was very high for all sessions, with an estimated peak of over 500 attendees.

The symposium reflected the large amount of work that has taken place in the last year, and much of the excitement that exists on this subject. The invited-talk program was designed to give a thorough review of the state-of-the-art in the field. The large number of contributions, in the form of talks and poster presentations, showed much progress in the growth and understanding of the III-V nitrides, and in the production of optoelectronic devices based on these materials. These proceedings represent the current state of the field, reflecting over 70% of the work presented at the symposium.

It is thus with much satisfaction on the advances reflected in these pages, and confidence in the future of III-V nitrides, that we present these proceedings for publication by the Materials Research Society.

F.A. Ponce
T.D. Moustakas
I. Akasaki
B.A. Monemar

December 1996

ACKNOWLEDGMENTS

We wish to thank the following organizations for their generous financial support of the symposium:

AIXTRON/MORITANI
Central Research Laboratories, Matsushita Electric Industrial Co., Ltd.
Cree Research
Fujitsu Limited
Hewlett-Packard Laboratories Japan, Inc.
Hitachi Ltd., Central Research Laboratory
JEOL USA, Incorporated
Office of Naval Research
Opto-Electronics Research Labs, NEC Corporation
Nichia Chemical Industries, Ltd.
Renishaw
Rockwell International Corporation
Sharp Corporation Central Research Laboratories
Sony Corporation Research Center
Toyoda Gosei Co., Ltd.
Xerox Palo Alto Research Center

We would also like to thank our invited speakers who with their talks and participation as session chairs, contributed greatly to the success of this symposium:

C.R. Abernathy
H. Amano
J.M. Baranowski
S. Binari
D.P. Bour
W.E. Carlos
R.F. Davis
S.P. DenBaars
V. Fiorentini
B. Gil
A. Hangleiter
T. Honda
M. Kamp

B-K. Meyer
S. Nakamura
A.V. Nurmikko
K. Onabe
H. Riechert
S. Sakai
F. Scholz
R.J. Shul
G.B. Stringfellow
C.G. Van de Walle
S. Yoshida
J.C. Zolper

We would especially like to thank all of the authors for the high quality of the oral and poster presentations, and for submitting manuscripts in a timely manner. In addition, we are very grateful to all the referees for their conscious review of the manuscripts. We are also very thankful to the staff of MRS for their efficient support.

MATERIALS RESEARCH SOCIETY SYMPOSIUM PROCEEDINGS

Volume 420— Amorphous Silicon Technology—1996, M. Hack, E.A. Schiff, S. Wagner,
R. Schropp, A. Matsuda 1996, ISBN: 1-55899-323-1

Volume 421— Compound Semiconductor Electronics and Photonics, R.J. Shul, S.J. Pearton,
F. Ren, C–S. Wu, 1996, ISBN: 1-55899-324-X

Volume 422— Rare-Earth Doped Semiconductors II, S. Coffa, A. Polman, R.N. Schwartz, 1996,
ISBN: 1-55899-325-8

Volume 423— III-Nitride, SiC, and Diamond Materials for Electronic Devices, D.K. Gaskill,
C.D. Brandt, R.J. Nemanich, 1996, ISBN: 1-55899-326-6

Volume 424— Flat Panel Display Materials II, M. Hatalis, J. Kanicki, C.J. Summers, F. Funada,
1997, ISBN: 1-55899-327-4

Volume 425— Liquid Crystals for Advanced Technologies, T.J. Bunning, S.H. Chen,
W. Hawthorne, T. Kajiyama, N. Koide, 1996, ISBN: 1-55899-328-2

Volume 426— Thin Films for Photovoltaic and Related Device Applications, D. Ginley,
A. Catalano, H.W. Schock, C. Eberspacher, T.M. Peterson, T. Wada, 1996,
ISBN: 1-55899-329-0

Volume 427— Advanced Metallization for Future ULSI, K.N. Tu, J.W. Mayer, J.M. Poate,
L.J. Chen, 1996, ISBN: 1-55899-330-4

Volume 428— Materials Reliability in Microelectronics VI, W.F. Filter, J.J. Clement,
A.S. Oates, R. Rosenberg, P.M. Lenahan, 1996, ISBN: 1-55899-331-2

Volume 429— Rapid Thermal and Integrated Processing V, J.C. Gelpey, M.C. Öztürk,
R.P.S. Thakur, A.T. Fiory, F. Roozeboom, 1996, ISBN: 1-55899-332-0

Volume 430— Microwave Processing of Materials V, M.F. Iskander, J.O. Kiggans, Jr.,
J.Ch. Bolomey, 1996, ISBN: 1-55899-333-9

Volume 431— Microporous and Macroporous Materials, R.F. Lobo, J.S. Beck, S.L. Suib,
D.R. Corbin, M.E. Davis, L.E. Iton, S.I. Zones, 1996, ISBN: 1-55899-334-7

Volume 432— Aqueous Chemistry and Geochemistry of Oxides, Oxyhydroxides, and
Related Materials, J.A. Voight, T.E. Wood, B.C. Bunker, W.H. Casey,
L.J. Crossey, 1997, ISBN: 1-55899-335-5

Volume 433— Ferroelectric Thin Films V, S.B. Desu, R. Ramesh, B.A. Tuttle, R.E. Jones,
I.K. Yoo, 1996, ISBN: 1-55899-336-3

Volume 434— Layered Materials for Structural Applications, J.J. Lewandowski, C.H. Ward,
M.R. Jackson, W.H. Hunt, Jr., 1996, ISBN: 1-55899-337-1

Volume 435— Better Ceramics Through Chemistry VII—Organic/Inorganic Hybrid Materials,
B.K. Coltrain, C. Sanchez, D.W. Schaefer, G.L. Wilkes, 1996, ISBN: 1-55899-338-X

Volume 436— Thin Films: Stresses and Mechanical Properties VI, W.W. Gerberich, H. Gao,
J–E. Sundgren, S.P. Baker 1997, ISBN: 1-55899-339-8

Volume 437— Applications of Synchrotron Radiation to Materials Science III, L. Terminello,
S. Mini, H. Ade, D.L. Perry, 1996, ISBN: 1-55899-340-1

Volume 438— Materials Modification and Synthesis by Ion Beam Processing, D.E. Alexander,
N.W. Cheung, B. Park, W. Skorupa, 1997, ISBN: 1-55899-342-8

Volume 439— Microstructure Evolution During Irradiation, I.M. Robertson, G.S. Was,
L.W. Hobbs, T. Diaz de la Rubia, 1997, ISBN: 1-55899-343-6

Volume 440— Structure and Evolution of Surfaces, R.C. Cammarata, E.H. Chason, T.L. Einstein,
E.D. Williams, 1997, ISBN: 1-55899-344-4

Volume 441— Thin Films—Structure and Morphology, R.C. Cammarata, E.H. Chason, S.C. Moss,
D. Ila, 1997, ISBN: 1-55899-345-2

Volume 442— Defects in Electronic Materials II, J. Michel, T.A. Kennedy, K. Wada, K. Thonke,
1997, ISBN: 1-55899-346-0

Volume 443— Low–Dielectric Constant Materials II, K. Uram, H. Treichel, A.C. Jones,
A. Lagendijk, 1997, ISBN: 1-55899-347-9

MATERIALS RESEARCH SOCIETY SYMPOSIUM PROCEEDINGS

Prior Materials Research Society Symposium Proceedings available by contacting Materials Research Society

Part I

Crystal Growth — Bulk and MOCVD

Metalorganic vapor phase epitaxial growth of GaInN/GaN hetero structures and quantum wells

F. SCHOLZ, V. HÄRLE, F. STEUBER, A. SOHMER, H. BOLAY, V. SYGANOW,
A. DÖRNEN, J.-S. IM, A. HANGLEITER, J.Y. DUBOZ[a], P. GALTIER[a], E. ROSENCHER[a],
O. AMBACHER[b], D. BRUNNER[b], H. LAKNER[c]
4. Physikalisches Institut, Universität Stuttgart, D-70550 Stuttgart, Germany
[a]Thomson CSF, Physics Lab, Lab. Central de Recherche, F-91404 Orsay, France
[b]Walter-Schottky-Institut, TU München, D-85748 München, Germany
[c]Werkstoffe der Elektrotechnik / FB9, Gerhard Mercator Universität, D-47048 Duisburg, Germany

ABSTRACT

GaInN/GaN heterostructures and quantum wells have been grown by low pressure metalorganic vapor phase epitaxy on sapphire using an AlN nucleation layer. We found a significant In incorporation only for growth temperatures of 700°C, although still very high In/Ga ratios in the gas phase had to be adjusted. The In content could be increased by reducing the H_2/N_2 flow ratio in the main carrier gas. GaInN layers typically show two lines in low temperature photoluminescence which are identified as excitonic-like (high energy peak) and impurity-related-like (low energy) by time-resolved spectroscopy. Quantum wells with a thickness between 8 and 0.5 nm showed only one emission line. The peak of the thinnest wells shows excitonic-like behaviour, whereas we found a smooth transition to an impurity-related-like type with increasing thickness. By scanning transmission electron microscopy studies we found indications for composition fluctuations in these thicker quantum wells which may cause localization effects for the excitons and thus be responsible for the observed optical spectra.

INTRODUCTION

The past several years have seen major advances in the epitaxial growth of group III-nitrides made possible by the dramatic development of modern epitaxial techniques like molecular beam epitaxy (MBE) and metalorganic vapor phase epitaxy (MOVPE). Consequently, highly efficient blue and green light emitting diodes (LEDs) have been realized and commercialized [1,2], and recently, blue and ultraviolet light emitting lasers have been reported [3,4]. Nearly all of these device structures have been grown by MOVPE. However, a lot of important problems concerning the growth and processing of these materials still remain unsolved. Therefore, we have concentrated our studies on this epitaxial technique particularly focusing on laser structures.

The MOVPE growth of high quality GaN requires high growth temperatures of typically 1000°C. This is mainly a consequence of the high chemical stability of the commonly used nitrogen precursor NH_3. Moreover, the higher thermal mobility of the adsorbed precursor atoms on the growing surface might improve the quality of the epitaxial layer.

For the adjustment of the band gap of the active region and thus the emitted wavelength of optoelectronic devices, In containing ternary layers ($Ga_{1-x}In_xN$) and quantum wells are needed. However, such layers require strongly different growth conditions. In order to achieve significant In concentrations, the epitaxy has to be performed at rather low growth temperatures of 700-800°C. In turn, extremely high V/III ratios to compensate for the low NH_3 cracking efficiency are needed. However, the In incorporation efficiency is still very low compared to the well-known phosphides and arsenides, where the Ga to In ratios in the gas phase and in the solid are identical [5]. For the growth of $Ga_xIn_{1-x}N$ with x ~ 6%, Nakamura et al. needed an In

Mat. Res. Soc. Symp. Proc. Vol. 449 © 1997 Materials Research Society

to Ga ratio in the gas phase of 17 at a growth temperature of 800°C [6]. Various reasons for this behaviour have been discussed. The In volatility at high temperatures [7,8] as a result of a weak In-N bond [9] and the high equilibrium vapor pressure of N over InN [10] may play an important role. Moreover, the presence of a miscibility gap in this ternary system [11] or the large disparity of the atomic radii of In and N [12] may limit the In incorporation. Simple thermodynamic calculations of the chemical reactions in the MOVPE reactor point to the large differences in the equilibrium constants for the formation of GaN and InN as a main reason [13]. Moreover, the large lattice mismatch between the growing GaInN and the underlying GaN has to be considered.

In order to investigate these problems, we have studied the growth of GaN/GaInN hetero and quantum well structures by low pressure MOVPE. We intended to increase the In content in the ternary layers while keeping their crystallographic and spectroscopic properties on a fair level. Moreover, we have investigated these layers by CW and time resolved photoluminescence (PL) and photothermal deflection spectroscopy (PDS) experiments. These data have been correlated to results obtained by transmission electron microscopy (TEM) and scanning TEM.

EPITAXIAL GROWTH

The epitaxial growth was performed in a home-made low-pressure MOVPE system using a horizontal reactor with rectangular cross-section. The SiC coated graphite susceptor was heated by halogen lamps. The standard alkyls triethyl-gallium and trimethyl-aluminum and -indium were used as Ga, Al and In precursors, respectively, whereas ammonia (NH_3) served as nitrogen source. Hydrogen was used as carrier gas for the alkyls. Both, hydrogen and nitrogen could be flown to the reactor as dilution gases.

All layers were grown on sapphire substrates which were precleaned by standard procedures and nitrided *in situ* in the MOVPE reactor at 1000°C immediately before the growth was initialized. As reported earlier [14, 15], excellent GaN layers could be grown at a reactor pressure of 100 hPa, a growth temperature of 1000°C and a V/III ratio of about 4000. A significant improvement was obtained by introducing an AlN nucleation layer (about 10 nm) grown at 800°C compared to a GaN nucleation layer grown at 500°C. These layers showed a mirror-like surface and exhibited extremely narrow linewidths of less than 50 arcsec in high resolution x-ray diffraction (HRXRD). Hall and C-V experiments revealed carrier concentrations around $1 \times 10^{17} cm^{-3}$ and mobilities of up to $600 cm^2/Vs$ at room temperature. In low temperature PL (T = 4.2K), an excitonic line of less than 3 meV full width at half maximum could be detected which was identified as a donor-bound excitonic transition by time-resolved measurements [16. The free-A-exciton line became dominant at about 20-30 K. The yellow luminescence was barely detectable at low temperature and fairly weak at room temperature [14].

GaInN layers have typically been grown on top of about 500 nm GaN grown under the optimized conditions mentioned above. At growth temperatures above 750°C, no significant In content in the layers could be measured. With decreasing temperature, an increasing In incorporation was found. However, the photoluminescence spectra of layers grown below 700°C became broad with rather low intensity. At temperatures of 650°C and below, we have observed the formation of In droplets. Therefore, we have chosen a growth temperature of 700°C as a good compromise of acceptable layer quality with still significant In content [14]. At the GaN-GaInN interface, the growth was interrupted in order to adjust the lower growth temperature. During this interruption, the NH_3 flow was increased to typically 4 slm while it was still flowing to the reactor. In order to establish an In to Ga ratio in the gasphase of about 5, the Ga flow was reduced. This resulted in a GaInN growth rate of about 50 nm/h and a V/III ratio of about 27,000. All layers for these studies have been grown intentionally undoped.

Most of these structures have been completed to double hetero structures by covering the ternary layer by another GaN layer of about 100 nm thickness. After the growth of the GaInN layer, the temperature was ramped back to 1000°C while keeping the TEG flow to the reactor. By this procedure, we intended to minimize the In desorption at higher temperatures by an instantaneous coverage of the GaInN film with GaN.

CHARACTERIZATION METHODS

Most of these epitaxial structures have been characterized by standard high resolution x-ray diffraction (Philips MRD). Photoluminescence at liquid helium and room temperature was excited with the ultraviolet line (300 nm) of an Ar ion laser. Moreover, optical gain spectra have been measured on some samples at room temperature using the stripe excitation method [17,18]. The PL decay was measured using a picosecond time-resolved PL setup, where the samples were excited with 5 ps pulses from a cavity-dumped frequency-doubled synchronously mode-locked dye laser. The luminescence was detected with a microchannel-plate photomultiplier and processed using single-photon-counting electronics [19].

Owing to the low thickness of most GaInN layers, direct absorption measurements could not be used to evaluate the band gap. Therefore, we have performed photothermal deflection spectroscopy (PDS) at room temperature. By this method, absorbance values αd as low as 10^{-5} (α: absorption coefficient, d: film thickness) can be determined. Details of this technique and our experimental setup are described in ref. 20.

In order to elucidate the microscopic structure of our ternary layers, high resolution TEM micrographs have been taken. Moreover, Z contrast images (Z being the atomic number) in a field emission scanning TEM (STEM) have been recorded yielding direct information about the local ternary composition of the probed GaInN films. For these experiments, cross section specimens have been prepared following standard procedures.

INFLUENCE OF CARRIER GAS

Similar to the MOVPE process of "conventional" III-V compounds, hydrogen (H_2) is used as carrier gas for the precursors and as main dilution gas to establish the desired high gas flow velocity of the reaction chamber. This is mainly due to the fact that H_2 is easily available in high purity and can be purified „in situ" just before entering the MOVPE machine by diffusion through a heated Pd wall. Only recently, very efficient purifiers for nitrogen (N_2) have also become available making this inert and non-explosive gas very attractive as the main carrier gas in MOVPE. Consequently, Hardtdegen et al. have optimized their growth processes for GaAs/AlGaAs and GaInAs/InP structures using exclusively N_2 as carrier gas. They have obtained excellent results [21]. Using N_2 does not only improve the process safety, but also results in strongly changed flow and thermal properties in the reaction chamber, which seems to be advantageous for some materials and structures.

For the growth of nitrides, the use of N_2 seems to be obvious. However, it is clear that this molecular nitrogen does not release enough atomic nitrogen even under the extreme conditions of the nitride growth process (temperatures around 1000°C) to act as a precursor gas. Therefore, H_2 is again the commonly used carrier gas for the growth of GaN and AlGaN, certainly because of the same reasons as mentioned above. Others used a mixture of H_2 and N_2 [22]. Only few groups reported about the exclusive use of N_2 [23]. Keller et al. switched their carrier flow from H_2 to N_2 when switching from GaN to GaInN growth [7]. Although the type of carrier gas certainly influences the growth process, we found only few reports about a systematic study of different carrier gas compositions. Yuan et al. observed improved GaInN quality when changing the shroud flow from H_2 to N_2 in their reactor [24].

In our standard growth procedure for the GaN layer growth, Pd-diffused H_2 is used as carrier gas for the metalorganic precursors and as main dilution flow, summing up to about 1300 sccm. The NH_3 flow is diluted with N_2 (about 1 slm). These conditions have also been applied to our first GaInN growth experiments. As described above, we have adjusted a growth temperature of 700°C in order to get a considerable In incorporation efficiency. For an In/Ga ratio in the gasphase $\gamma_{In/Ga}$ of about 5, we measured x ~ 7% in our $Ga_{1-x}In_xN$ layers by HRXRD.

In order to study the influence of a changed H_2/N_2 carrier gas composition, we have reduced our total H_2 flow from 1300 sccm to 300 sccm. This reduced H_2 flow was mainly used for the transport of the metalorganic precursors. Layers grown under otherwise unchanged conditions showed a higher In content of x ~ 12% as consistently determined by HRXRD, low temperature photoluminescence and room temperature gain spectra (fig. 1). However, the linewidths of the spectra of these samples are considerably broader than those of the layers grown with the larger H_2 flow. This is a direct consequence of the larger In content: Layers grown with lower $\gamma_{In/Ga}$ and low H_2 flow resulting in the same In content in the solid of about 7% show fairly the same spectral properties as the "large H_2 flow samples".

The increased In incorporation efficiency in a carrier gas with smaller H_2 content may be a result of the changed gas phase reactions of the precursors. Koukitu et al. performed simple thermodynamic calculations [13,25], from which they concluded that a large H_2 partial pressure may strongly reduce the In incorporation rate. Moreover, parasitic side reactions, sometimes blamed as possible reasons for In losses in the vapor, may be less severe in a N_2 atmosphere owing to the different thermal profile above the susceptor.

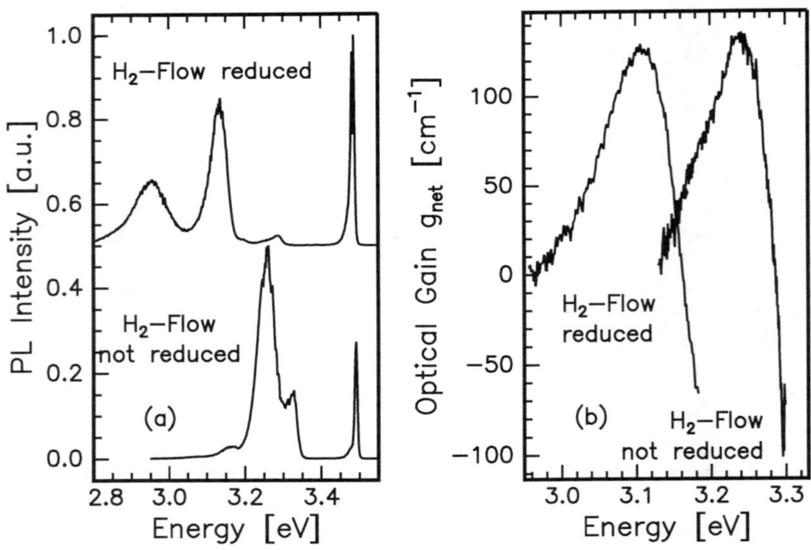

Fig. 1: Low temperature (4K) photoluminescence (a) and room temperature gain spectra (b) of GaN/GaInN double hetero structures grown with full and reduced hydrogen flow.

SPECTROSCOPIC RESULTS ON GaInN BULK LAYERS

As already shown in fig. 1, GaInN bulk layers frequently exhibit two peaks in low temperature photoluminescence. Fig. 2 shows PL spectra of a GaInN/GaN double hetero structure grown with reduced H_2 flow rate taken at various temperatures between 4K and room temperature. In the latter case, only the high energy peak I_1 remains. The simplest interpretation is that the high energy peak is a band-edge related emission, i.e. an excitonic emission at low temperature and a band to band emission at room temperature. Time-resolved PL experiments taken at low temperature further support this explanation: The high energy line I_1 shows a fast mono-exponential decay with a decay time of about 400 psec (fig. 3a). In contrast, the low energy peak I_2 shows a non-exponential decay on a very long time scale of some microseconds (fig. 3b). Moreover, this line shifts to higher energies with increasing excitation intensity, whereas I_1 remains at the same position. The maxima in the optical gain spectra recorded at 300K on double hetero structures (see fig. 1b) clearly are related to the high energy PL line I_1. However, the exact nature of the low-energy line is not necessarily related to extrinsic impurities, as will be discussed later in this paper.

GaInN QUANTUM WELLS

As described above, we have grown GaInN quantum wells embedded in GaN barriers. The well thickness L_z was varied between 0.5 and about 8 nm by changing the growth time accordingly. These structures showed mirror-like surfaces. In HRXRD, only the GaN peak with a similar half width as for single GaN layers could be detected, whereas the GaInN was not detectable owing to it's small thickness.

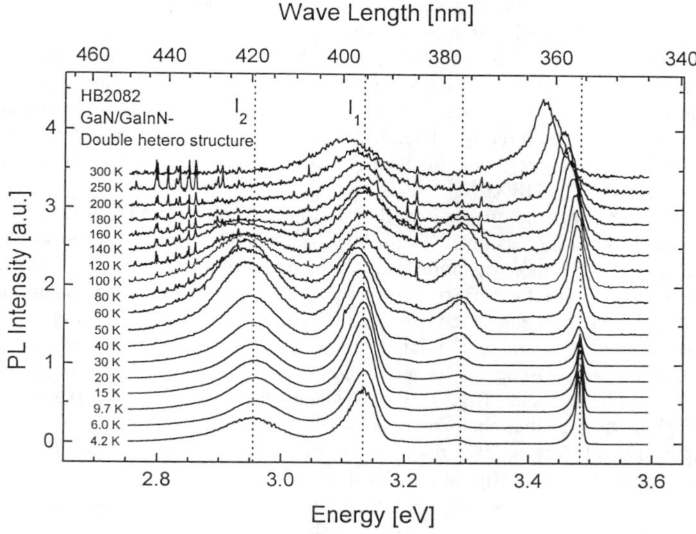

Fig. 2: Photoluminescence spectra of a GaN/GaInN double hetero structure taken at various temperatures between 4 and 300K.

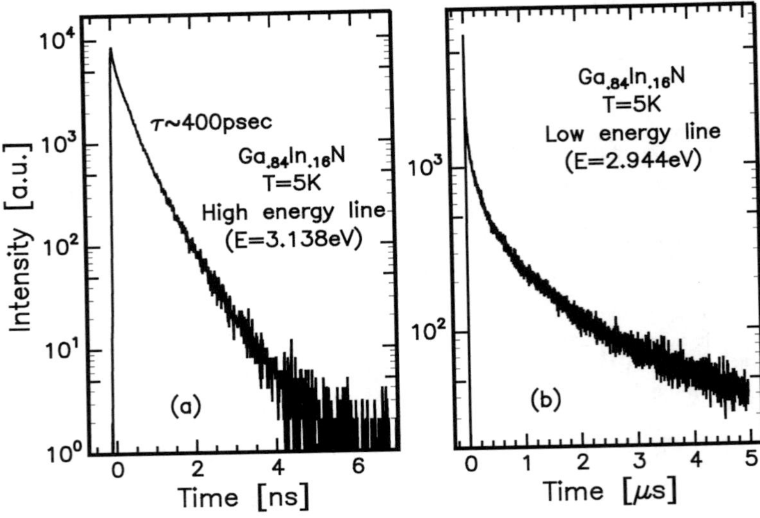

Fig. 3: Photoluminescence decay measured on the I_1 (a) and the I_2 line (b) of the GaN/GaInN double hetero structure.

Low temperature photoluminescence spectra of these structures are shown in fig. 4. The spectrum of the 15 nm sample is the same as already depicted in figs. 1 and 2. Due to the large effective carrier masses in GaInN, this sample can in fact be regarded as the „bulk limit".

At a first glance, the quantum well emission peaks show a shift to higher energies with decreasing thickness L_z, as might be expected from the quantum size effect. If we assumed that the PL line represents transitions directly related to the effective band-gap, the conclusion could be drawn [15] that such thin layers have a higher In content than thicker layers grown under the same condition and analyzed by HRXRD.

However, it is evident that the quantum well emission line ends up in the low energy peak I_2 of the 15 nm sample with increasing L_z. Moreover, the linewidths of these peaks decrease with decreasing L_z (fig. 5). For most quantum well systems, the opposite behaviour is observed, because the influence of interface fluctuations increases with decreasing quantum well thickness. Thus, it is not clear whether the PL peaks really show the effective band gaps of these structures or just some sub-band-gap related emission. Therefore, we have performed photothermal deflection spectroscopy in order to determine the effective band gap (fig. 6). Comparing the PDS and PL results (fig. 7) shows indeed that the PL peaks exhibit a much stronger shift to higher energies than the band edge values obtained by PDS.

Further insight in this behaviour was obtained by time-resolved PL experiments performed at 5 K (fig. 8). Similar to the I_2 line (fig. 3b), the „thick" quantum wells show a non-exponential decay on a very long time scale. However, with decreasing thickness, the decay becomes faster. The thinnest quantum wells ($L_z = 0.9$ nm and 0.5 nm) show a clear mono-exponential decay on a similar scale as the I_1 peak of the 15 nm sample. According to our explanation given above,

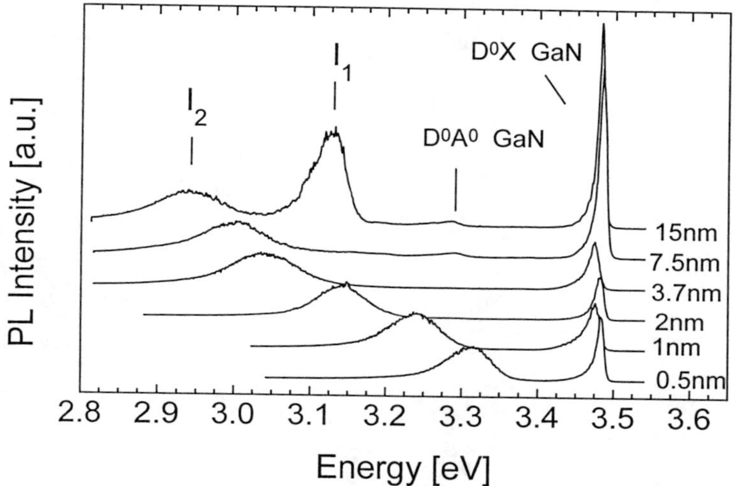

Fig. 4: Low temperature (4K) photoluminescence on GaInN quantum wells with various thickness L_z.

we attribute the emission line of the thicker quantum wells to an impurity-related like transition, whereas the thinnest quantum wells show an excitonic-like transition. Of course, the latter is expected for high quality quantum wells. Now, the variation in the PL linewidth can be easily understood, because excitonic lines usually show lower peak widths than impurity-related ones. Obviously, this is in good agreement with the results from fig. 7: The PDS data are caused by the effective quantum well band edge, whereas the distance between the PL line position and the effective band-gap increases with increasing L_z due to a smooth transition from an excitonic transition to an impurity-related-like one. This interpretation could be further verified by quasi-resonant PL measurements: Only for excitation energies above the effective band-gap (as estimated from the PDS results), the PL peak could be observed.

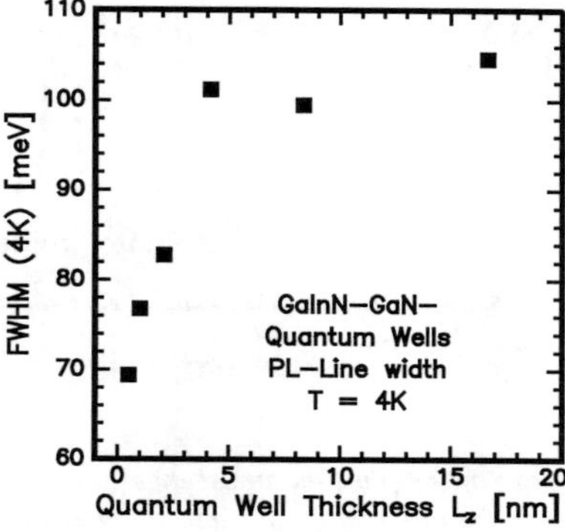

Fig. 5: Linewidth (FWHM) of the GaInN quantum well photoluminescence lines

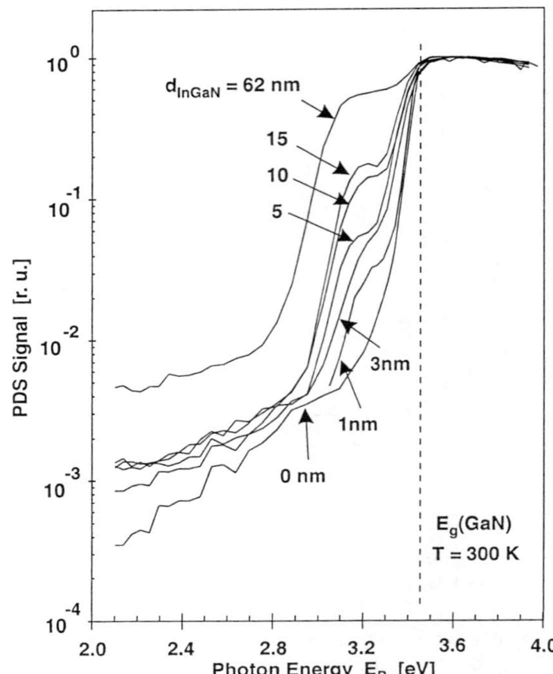

Fig. 6:
Photothermal deflection spectra of GaInN quantum wells (different samples than depicted in fig2, 4 and 5) taken at room temperature.

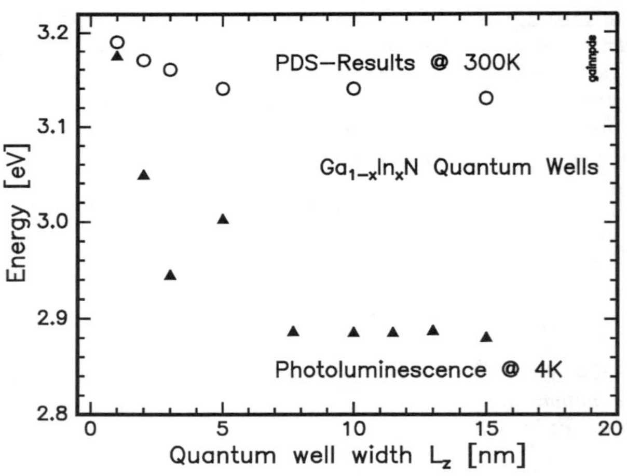

Fig. 7: Peak energy of the photoluminescence line of GaInN quantum wells at low temperature and effective band gaps of the same samples evaluated by PDS at room temperature.

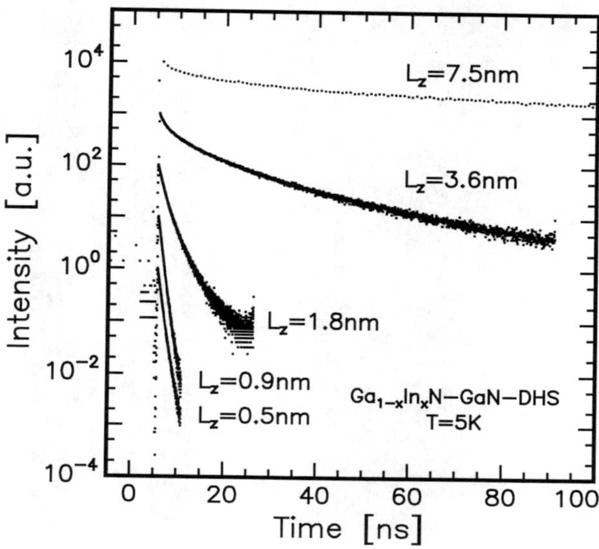

Fig. 8: *Photoluminescence decay curves obtained at 5 K on GaInN quantum wells of various thickness.*

TRANSMISSION ELECTRON MICROSCOPY STUDIES

This peculiar behaviour is certainly related to the microscopic structure of these quantum wells. Therefore, we have performed TEM experiments on these samples. Conventional TEM micrographs showed that the quantum well interfaces are abrupt on an atomic scale. There was no drastic roughness visible that might explain the increased PL linewidth of the thicker quantum wells. However, we observed some contrast modulation on these micrographs (not shown here). Therefore, we investigated the local chemical composition of the quantum wells by Z-contrast imaging in a STEM (fig. 9). Again, the quantum well with its abrupt interfaces is clearly visible. As expected from the conventional TEM figures, the quantum well shows some brightness modulations which have to be attributed to composition fluctuations in the ternary layer. Such modulations are also visible in the binary GaN barriers. However, a detailed analysis gave clear evidence that our interpretation about significant composition fluctuations is correct. Further experiments to determine the size of these fluctuations quantitatively are currently underway.

These findings may help to understand our spectroscopic results described above. As already implied in our interpretations, excitons may be created by the optical excitation in our experiments. In a perfect quantum well, these excitons can freely move around in the well plane. Thus, their recombination causes a rather narrow PL line. In these films with a considerable composition fluctuation on a scale comparable to or even smaller than the excitonic diffusion length, the decay process will be strongly influenced by the microscopic

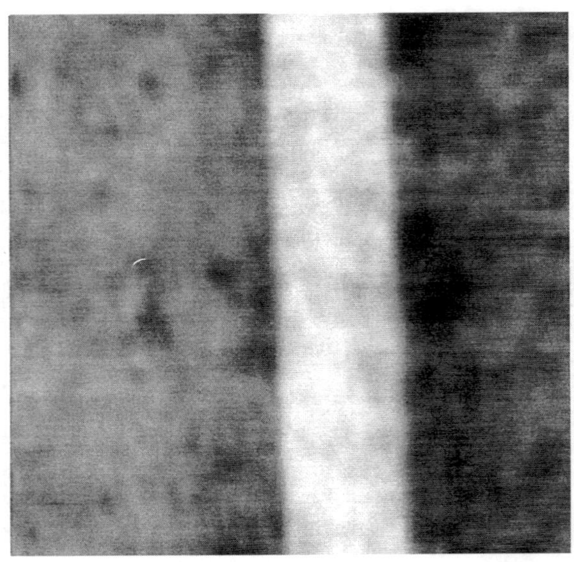

Fig. 9: Z-contrast scanning electron micrograph of a GaN/GaInN double hetero structure. The GaInN layer has a thickness of 17nm.

arrangement. This will result in line broadening or even localization effects. Indeed, we found further indications for such exciton localization in a closer analysis of our optical gain spectra, as described in more detail in [18]. This may even explain the appearance of the low energy peak I_2 in our PL spectra, analogous to the moving emission peak in ordered GaInP with a high density of domain boundaries [26].

It seems that such composition fluctuations are a universal problem in GaInN layers. As mentioned by Nakamura [27], this was also observed on the laser and LED samples grown recently by Nichia. Moreover, it is interesting to note that even in the Nichia quantum well LED structures the main electroluminescence wavelength is about 120 meV below the band gap expected from the composition of the nominally undoped GaInN active well [1]. This may be at least partly caused by the localization effects described above.

There are at least two possible reasons for such fluctuations: The large lattice mismatch between GaN and $Ga_{1-x}In_xN$ (about 1.2% for x~12%) may hinder a perfectly two-dimensional growth mode similar to the self-organized growth of quantum dots (Stranski-Krastanov growth mode) [28]. The existence of a miscibility gap [11] may further promote such fluctuations.

Taking into consideration the spectroscopic results explained above, we would expect less localization effects in very thin quantum wells. A preliminary STEM analysis of a 2.5 nm quantum well shows indications for a composition gradient in growth direction. Further work is needed to analyze such thin layers accurately.

SUMMARY

We have succeeded to grow high quality GaInN layers and GaInN/GaN hetero structures and quantum wells. The In incorporation could be drastically increased by increasing the nitrogen/hydrogen flow ratio in the carrier gas. We observed two emission lines in low temperature PL. The higher-energy-line has the characteristics of an excitonic peak, whereas the lower-energy line resembles an impurity-related transition. Quantum wells with thickness between 8 and 0.5 nm show only one PL peak. This peak smoothly changes it's character from impurity-related-like to excitonic-like with decreasing thickness, as observed by time-resolved PL, PDS experiments and linewidth analysis. A STEM analysis using Z-contrast imaging gives clear indications for composition fluctuations in GaInN layers. These fluctuations may cause localization effects for optically excited excitons thus explaining the observed spectroscopic data.

ACKNOWLEDGEMENT

The quasi-resonant PL experiments performed by F. Adler and the experimental assistance of J. Off are gratefully acknowledged. Moreover, we thank M. Pilkuhn for fruitful discussions. Parts of this work have been financially supported by the Volkswagenstiftung and by the Bundesministerium für Bildung, Wissenschaft, Forschung und Technologie.

REFERENCES

[1] S. Nakamura, M. Senoh, N. Iwasa, S. Nagahama, T. Yamada, T. Mukai, Jpn. J. Appl. Phys. **34** (1995) L1332.

[2] S. Nakamura, T. Mukai, M. Senoh, J. Appl. Phys. **76** (1994) 8189.

[3] S. Nakamura, M. Senoh, S. Nagahama, N. Iwasa, T. Yamada, T. Matsushita, H. Kiyoku, Y. Sugimoto, Jpn. J. Appl. Phys. **35** (1996) L74.

[4] S. Nakamura, M. Senoh, S. Nagahama, N. Iwasa, T. Yamada, T. Matsushita, H. Kiyoku, Y. Sugimoto, Jpn. J. Appl. Phys. **35** (1996) L217.

[5] G.B. Stringfellow, Organometallic vapor phase epitaxy (Academic Press, San Diego, 1989), p. 116.

[6] S. Nakamura, T. Mukai, M. Senoh, Appl. Phys. Lett. **64** (1994) 1687.

[7] S. Keller, B.P. Keller, D. Kapolnek, A.C. Abare, H. Masui, L.A. Coldren, U.K. Mishra, S.P. DenBaars, Appl. Phys. Lett. **68** (1996) 3147.

[8] C. Yuan, T. Salagaj, W. Kroll, R.A. Stall, M. Schurman, C.-Y. Hwang, Y. Li, W.E. Mayo, Y. Lu, S. Krishnankutty, R.M. Kolbas, J. Electron. Mat. **25** (1996) 749.

[9] F.G. McIntosh, K.S. Boutros, J.C. Roberts, S.M. Bedair, E.L. Piner, N.A. El-Masry, Appl. Phys. Lett. **68** (1996) 40.

[10] T. Matsuoka, H. Tanaka, T. Sasaki, A. Katsui, Inst. Phys. Conf. Ser. No. 106: Chapter 3 (1989), p. 141.

[11] I-hsiu Ho, G.B. Stringfellow, Appl. Phys. Lett. **69** (1996) 2701.

[12] S. Strite, H. Morkoc, J. Vac. Sci. Technol. **B 10** (1992) 1237.

[13] A. Koukitu, N. Takahashi, T. Taki, H. Seki, Jpn. J. Appl. Phys. **35** (1996) L673.

[14] F. Scholz, V. Härle, H. Bolay, F. Steuber, B. Kaufmann, G. Reyher, A. Dörnen, O. Gfrörer, S.-J. Im, A. Hangleiter, Solid State Electronics (1996), Proc. TWN '95, Nagoya, Japan, 1995 (in print).

[15] F. Scholz, V. Härle, F. Steuber, H. Bolay, A. Dörnen, B. Kaufmann, V. Syganow, A. Hangleiter, J. Crystal Growth (1996), Proc. ICMOVPE VIII, Cardiff, Wales, 1996 (in print).

[16] A. Hangleiter, J.S. Im, T. Forner, V. Härle, F. Scholz, Mat. Res. Soc. Symp. Proc. Vol. 395, p. 559 (1996).

[17] G. Frankowsky, F. Steuber, V. Härle, F. Scholz, A. Hangleiter, Appl. Phys. Lett. **68** (1996) 3746.

[18] A. Hangleiter, F. Scholz, V. Härle, J.S. Im, G. Frankowsky, these proceedings.

[19] J.S. Im, A. Moritz, F. Steuber, V. Härle, F. Scholz, A. Hangleiter, to be published in Appl. Phys. Lett.

[20] O. Ambacher, W. Rieger, P. Ansmann, H. Angerer, T.D. Moustakas, M. Stutzmann, Solid State Comm. **97** (1996) 365.

[21] see, e.g., H. Hardtdegen, P. Giannoules, III-Vs Review **8** (1995) 34 and references therein.

[22] S. Nakamura, T. Mukai, Jpn. J. Appl. Phys. **31** (1992) L1457.

[23] T. Matsuoka, N. Yoshimoto, T. Sasaki, A. Katsui, J. Electron. Mat. **21** (1992) 157.

[24] C. Yuan, T. Salagaj, W. Kroll, R.A. Stall, M. Schurman, C.-Y. Hwang, Y. Li, W.E. Mayo, Y. Lu, S. Krishnankutty, R.M. Kolbas, J. Electron. Mat. **25** (1996) 749.

[25] A. Koukitu, N. Takahashi, T. Taki, H. Seki, J. Crystal Growth (1996), Proc. ICMOVPE VIII, Cardiff, Wales, 1996 (in print).

[26] P. Ernst, C. Geng, G. Hahn, F. Scholz, H. Schweizer, F. Phillipp, A. Mascarenhas, J. Appl. Phys. **79** (1996) 2633.

[27] S. Nakamura, contribution to the Semiconductor Laser Conference, Haifa, Israel, Oct. 1996.

[28] see e.g. D. Leonard, K. Pond, P.M. Petroff, Phys. Rev. B **50** (1994) 11687.

GROWTH OF GaN BY SUBLIMATION TECHNIQUE AND HOMOEPITAXIAL GROWTH BY MOCVD

Shiro SAKAI, Satoshi KURAI, Katsushi NISHINO, Koichi WADA, Hisao SATO
and Yoshiki NAOI
Department of Electrical and Electronic Engineering, The University of Tokushima, Minami-josanjima, Tokushima 770, Japan

ABSTRACT

The growth of bulk GaN by sublimation method and a homoepitaxial growth by MOCVD are reported. A photo-pumped stimulated emission is obtained from a homoepitaxial layer. The source powder used as a source in the sublimation method is investigated in detail, and it is shown that the powder contains many kinds of compounds consisting mainly of gallium, nitrogen and hydrogen. Growth nucleation control is performed by partly covering an MOCVD-GaN or a scratched-sapphire (0001) by SiO_2. Hexagonal columns of the size of about 200 μm in diameter and about 200 μm in height are selectively and uniformly grown at the window sites. This technique enables the device processing of crystallites and also helps to increase crystal size by increasing growth time, since it prevents the nucleation of new crystallites which work as sink of the growing species.

INTRODUCTION

III-V nitride is one of the most promising material for fabricating light emitting devices for green, blue and ultraviolet region as well as high temperature devices. High quality GaN epitaxial films with flat surface were obtained by the two-step growth technique on sapphire substrates, using an AlN or a GaN buffer layer. Highly-efficient light emitting diodes (LEDs) and InGaN multi-quantum-well-structure laser diodes (LDs) were demonstrated. However, many substrate-related problems, such as the large differences in thermal expansion coefficient and lattice-mismatch between the epi-layer and the substrates, formation of the high density of dislocation, the difficulty in creating electrodes and cleavage, still remain as major problems. The sapphire substrate also adds some additional difficulties in most commonly used MOCVD technique. For example, substrate annealing and the two step growth conditions critically effect on the crystal quality.

The most suitable substrate for growing GaN is thought to be a bulk GaN which is synthesized under both high temperature and high pressure conditions and by the sublimation technique. The latter was studied in the early 1970's, [1,2] to grow GaN with the size ranging from several hundred microns to a few mm. We also reported the growth of small bulk GaN by sublimation method and a homoepitaxial growth by MOCVD. Photo-pumped stimulated emission from a homoepitaxial GaN was observed at room temperature [3,4], dislocation density was estimated to be less than $10^7/cm^2$ by TEM [5], and Raman spectra perfectly satisfied the selection rule. All these results indicate that the homoepitaxial layer has high crystal quality without strain.

However, the crystal size obtained so far is not large enough to replace commonly used sapphire substrates, and the reason for not being able to increase the crystal size is not clear. In

Mat. Res. Soc. Symp. Proc. Vol. 449 © 1997 Materials Research Society

addition, the homoepitaxial layer properties are not investigated in detail, and the nucleation control has not been tried.

In this paper, we review our work on the growth mechanism, the homoepitaxy and the nucleation control in the sublimation method.

EXPERIMENT

Sublimation growth

The system used in the sublimation growth is schematically shown in Fig. 1. The source powder was synthesized by heating 10 g of liquid Ga placed on quartz boat in NH_3 atmosphere at 1000 °C for 3 hours. The flow rate of NH_3 was 100 sccm. The source powder, the quartz substrate holder and the substrate facing to the powder were placed in the carbon crucible which was heated to 1050 °C. 50 sccm of NH_3 flow was introduced into the reactor. The distance between the powder and the substrate was 7 mm. The substrate was a sapphire (0001) unless otherwise mentioned.

Growth results

The hexagonal shaped cystallites, with the size ranging from several tens of microns to a little over a mm were obtained on the substrate or on the powder after 1to3 hours of sublimation growth. The size of the crystal did not increase with increasing growth time. The geometry of the substrate holder critically effected on the growth; a geometry that produced larger stagnant gas flow resulted in the higher density and the larger size of the crystals.

When the nucleation density is high enough for the crystallites to coalesce, a thick film with the thickness of several tens of micron to over 100 µm instead of the crystallites was obtained. A hexagonal pattern shown in Fig.2 was frequently observed on this surface.

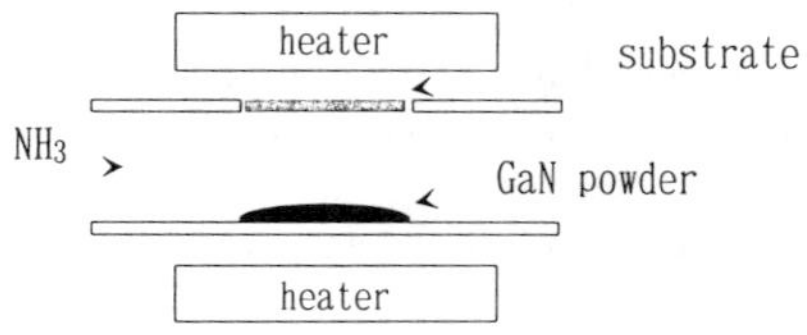

Fig.1 Reactor used in the sublimation growth.

Fig.2 Surface microphotograph of a thick GaN grown on a sapphire (0001) by sublimation method.

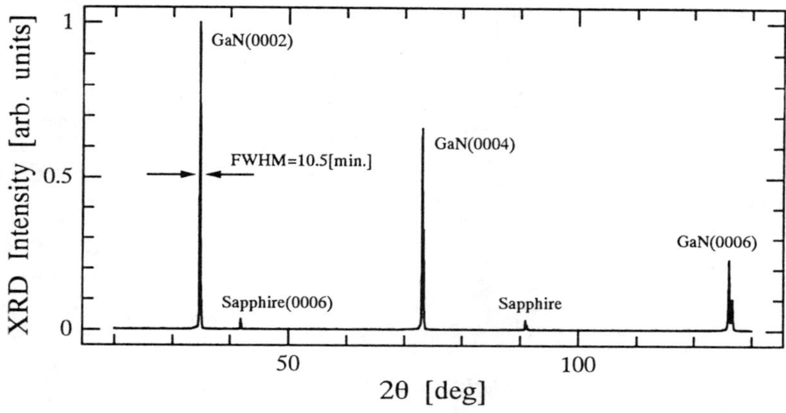

Fig.3 x-ray diffraction of the thick GaN grown on sapphire by sublimation method.

A thick film has the same crystal orientation as a substrate as indicated by the θ–2θ x-ray diffraction in Fig. 3. The dislocation density was measured by the cross-sectional TEM and was as high as 10^{11} cm^{-2}.

On the contrarily to a thick film, a small bulk crystal has very low dislocation density. No dislocation is seen in a TEM picture of Fig.4 in the observed area of 3.3 x 3 μm^2. The raman spectra of the bulk crystal (Fig.5 (a)) perfectly satisfies selection rule, while a prohibited peak at 745 cm^{-1} in $X(Z-)\bar{X}$ configuration is always observed in the heteroepitaxial films.

In spite of the high crystal perfection of the bulk crystal, its photoluminescence (PL) intensity is several orders of magnitude smaller compared to that of the heteroepitaxial film grown on sapphire substrate by atmospheric MOCVD, suggesting the existence of high density of point defect.

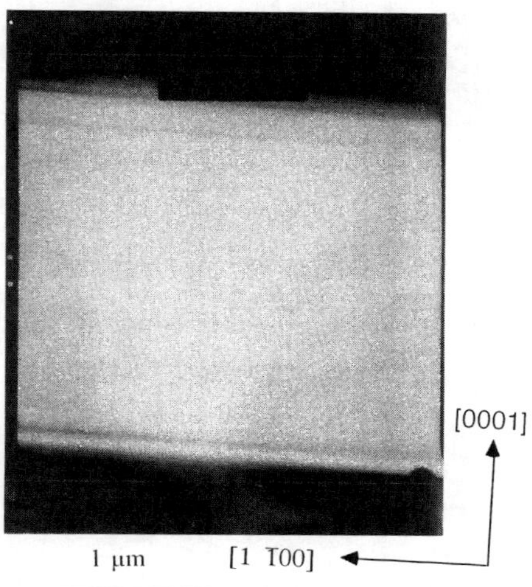

Fig.4 TEM picture of the bulk GaN.

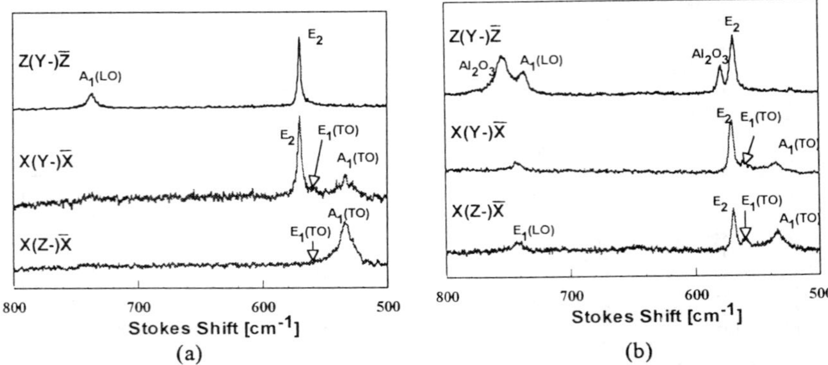

Fig.5 Raman spectra of a bulk GaN (a) and a heteroepitaxial GaN on sapphire grown by MOCVD (b).

Although it is difficult to measure electrical properties of the crystallites, a prober test reveals that the resistivity of the crystallite is much larger compared to the heteroepitaxial film which has the electron concentration of the order of 10^{17} cm^{-3} and the electron mobility of about 400 cm^2/vs.

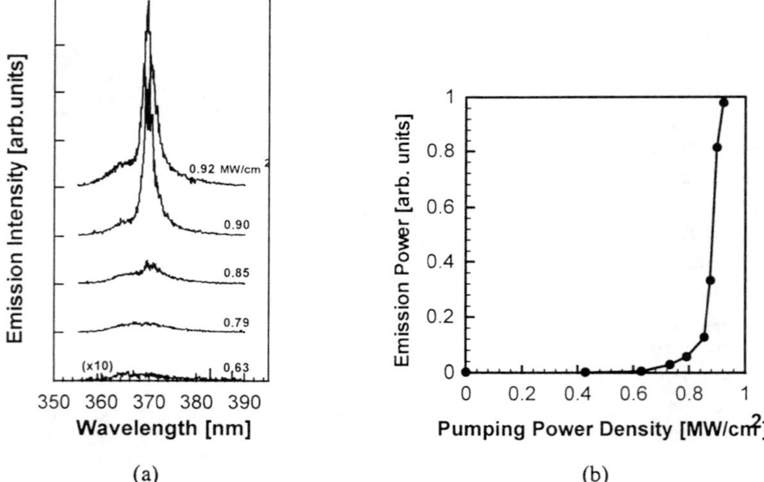

Fig.6 Emission spectra under high excitation (a), and the excitation power density-output power relation (b).

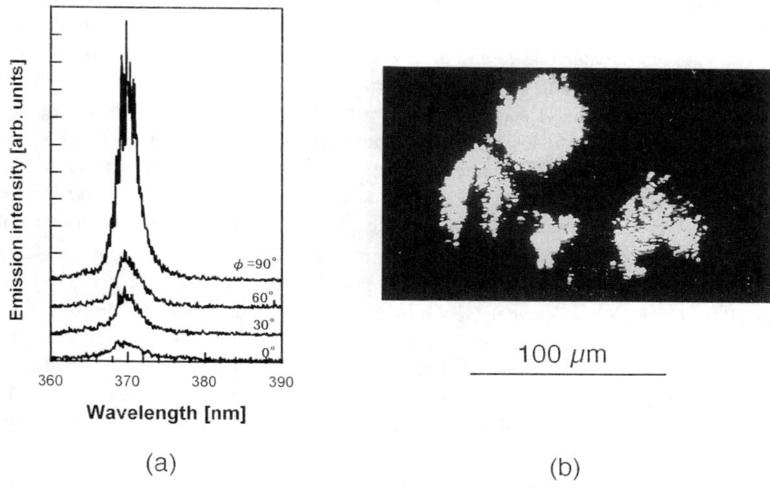

Fig.7 Polarization dependence of the emission above threshold (a), and a near field pattern of the stimulated emission from a homoepitaxial GaN grown on a bulk GaN.

Homoepitaxial growth by MOCVD

Since bulk crystallites prepared by the sublimation method is nucleated on a sapphire plate, it is easy to load them into MOCVD system to grow homoepitaxial GaN or heterostructures. The PL intensity of the homoepitaxial layer becomes several orders of magnitude stronger than the bulk crystallite.

A photo-pumping experiments are performed for the homoepitaxial layers using a nitrogen laser as a pumping source. The emission spectra and the excitation power density-output power relation of the homoepitaxial GaN are shown in Fig 6 (a) and (b), respectively. The output power superlinearly increases with excitation intensity, a spectrum line width suddenly narrows and the spectrum exhibits polarization above threshold, indicating that the stimulated emission is taking place. The threshold power density is almost the same as that of the heteroepitaxial GaN on sapphire reported elsewhere.[6] The polarization dependence and a near field pattern of the GaN above threshold are shown in Fig. 7 (a) and (b), respectively.

<u>Powder analysis</u>

The most significant problem of the sublimation-grown-crystal is its too small size. To investigate the growth mechanism of the sublimation technique, the source powder was analyzed, since the color of the powder before and after the use was different, and no growth took place if the growth was repeated using the same powder.

The quadrupole mass analysis (QMA) spectra of the powder are shown in Fig. 8. Many species other than GaN were detected. The powder was heated up to 1000 °C in vacuum, and the QMA signal intensity of these elements were measured as a function of time. Most of the peaks except that of nitrogen suddenly decreased after approximately 3 hours, indicating a depletion of volatile compounds. Therefore, the reason why crystal size saturated with the growth time was thought to be due to the depletion of the "active" species in the powder. θ–2θ x-ray diffraction measured for the powder after the annealing showed clear peaks associated with crystalline GaN, indicating that GaN still remained in the powder. Therefore, the active species that are contributing to the growth is not pure GaN, but the compounds which are mostly consisting of gallium, nitrogen and hydrogen atoms, such as GaNH or GaN_2H.

One way to overcome this problem is to repeat the growth using a new powder in each growth. The growth was repeated for 14 times, and indeed, the crystal size linearly increased with the growth.

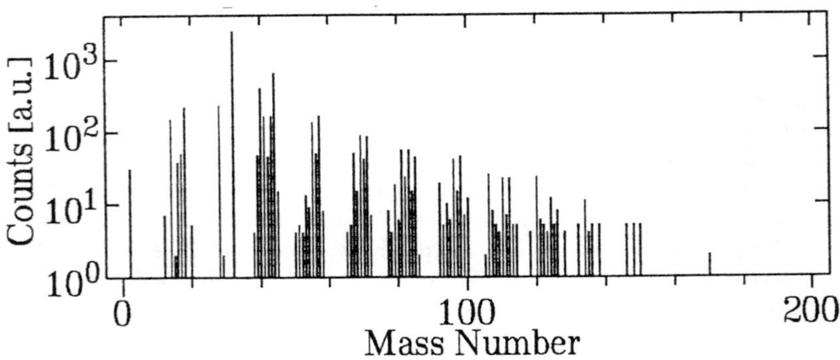

Fig.8 QMA spectrum from source powder used in the sublimation method.

Growth nucleation control

In the sublimation method, the growth nucleation takes place randomly on the substrate. Although the crystal size of several hundred micron is large enough to fabricate devices such as laser diodes, the random nucleation prevent to perform device processing. It is also important to control nucleation site and density in order to increase crystal size, because randomly nucleated crystallites consume growing species and prevent the growth of already existing crystals on the substrate.

Therefore, the nucleation density was investigated on the following materials: an electron-beam evaporated SiO_2 on a sapphire (0001), a Si (111), a sapphire (0001) which was scratched by 20-40 μm diamond powder in ethyl alcohol for 30 min in a ultrasonic bath, and a 0.3 μm-thick GaN grown on sapphire (0001) by atmospheric MOCVD. The growth time was fixed to 3 hours in this series of experiment.

The density of the nuclei was the largest on an MOCVD-GaN and on a scratched sapphire, and a continuous film rather than a discrete crystallites were obtained. On the other hand, the nucleation density both on a SiO_2 and on a Si (111) was much less. The nucleation density on a SiO_2 and on a Si (111) (about 6 x 10^3/cm^2) is almost two orders of magnitude smaller than those on a sapphire (1.41 x 10^5/cm^2). The density on an MOCVD-GaN and on a scratched-sapphire is difficult to estimate, but it is larger than 3.6 x 10^5/cm^2 assuming the size of the crystallite is the same as that on a sapphire. Therefore, a scratched-sapphire or an MOCVD-GaN can be used as a seeding material, and a SiO_2 or a Si (111) can prevent crystal nucleation.

The nucleation control was then tried by partly covering a seeding material (a scratched-sapphire or an MOCVD-GaN) by a masking material (SiO_2). 250 nm-thick-SiO_2 was deposited on a scratched-sapphire and an MOCVD-GaN by electron-beam evaporation, and squared windows of the size 60 x 60 μm^2 was opened by photolithography and etching within the area of 5 x 5 mm^2. The spacing between the square pattern was 200 μm x 500 μm.

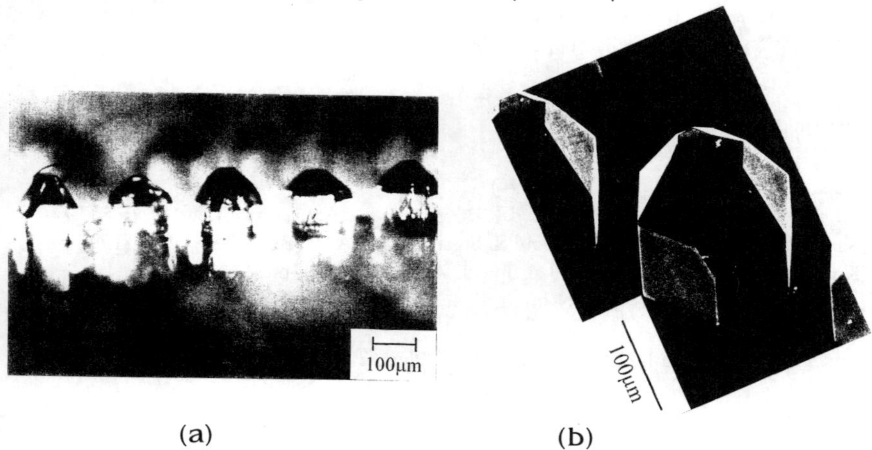

(a)   (b)

Fig.9 Microphotograh (a) and SEM picture (b) of GaN selectively grown by sublimation method.

Hexagonal columns of GaN were selectively and uniformly grown in the window opening on both samples as shown in Fig. 9 (a). The size of the crystal was about 200 μm in diameter and about 200 μm in height, and the top of the hexagonal columns was sharp. All the surrounding surfaces were very smooth as shown by SEM microphotograph in Fig.9 (b). The diameter of the column was much larger than the window width.

CONCLUSION

The growth of a bulk GaN by sublimation method, and a homoepitaxial growth by MOCVD are reported. A bulk GaN has very high crystal perfection and very low dislocation density, but its PL intensity is low, partly due to the existence of the high density of point defects. The PL intensity is much improved by growing GaN by MOCVD. A photo-pumped stimulated emission is obtained from a homoepitaxial layers.

A powder used as a source in sublimation method is analyzed in detail. The powder is shown to contain many compounds consisting mainly of gallium, nitrogen and hydrogen atoms, which are thought to be active species contributing to the growth.

Growth nucleation control was performed by partly covering an MOCVD-GaN or a scratched-sapphire (0001) by a SiO_2. Hexagonal columns of the size of about 200 μm in diameter and about 200 μm in height were selectively and uniformly grown at the window sites. This technique enables the device processing of the crystallites and also helps to increase crystal size by increasing growth time, since it prevents the nucleation of new crystallites which work as a sink of the growing species.

Acknowledgments

A part of this work was supported by the Satellite Venture Business Laboratory, "Nitride Photonic Semiconductor Laboratory" of the University of Tokushima and by a Grant-in-Aid for Scientific Research from the Ministry of Education, Science Sports and Culture. One of the authors (S.K.) was financially supported by JSPS.

REFERENCES

1. E.Ejder, J. Cryst. Growth **22**, p. 44 (1974).
2. R.Dingle, K.L.Shaklee, R.F.Leheny and R.B.Zetterstrom, Appl. Phys. Lett. **19**, p. 5 (1971) .
3. S. Kurai, Y. Naoi, T. Abe, S. Ohmi and S. Sakai, Jpn. J. Appl. Phys. **35**, p. L77 (1996).
4. S. Kurai, T. Abe Y. Naoi and S. Sakai, Jpn. J. Appl. Phys. **35**, p. 1637 (1996).
5. T. Okada, S. Kurai, Y.Naoi, K. Nishino, F. Inoko and S. Sakai, Jpn. J. Appl. Phys. **35**, p. L1318 (1996).
6. H.Amano, T. Asahi and I.Akasaki, Jpn. J. Appl. Phys. 29, p.L205 (1990).

MOVPE GROWTH AND OPTICAL CHARACTERIZATION OF GaPN METASTABLE ALLOY SEMICONDUCTOR

K. ONABE

Department of Applied Physics, The University of Tokyo
7-3-1 Hongo, Bunkyo-ku, Tokyo 113 Japan, onabe@photonics.rcast.u-tokyo.ac.jp

ABSTRACT

The $GaP_{1-x}N_x$ alloy semiconductor has been grown with the N concentration as high as 6.3% by metalorganic vapor phase epitaxy (MOVPE) using 1,1-dimethylhydrazine (DMHy) as the N source. The growth characteristics show the key role of the non-equilibrium circumstances during the growth, where the N desorption from the surface limits the N incorporaton. The band-edge states have been studied using time-resolved PL and PL excitation spectroscopies. The PL takes place via the tail states below the absorption edge. The band edge, which shifts to lower energies with increasing N concentration, originates from the A-line energy (*i.e.* isolated N states) rather than the indirect-gap energy of GaP in the limit $x=0$.

INTRODUCTION

The $GaP_{1-x}N_x$ alloy semiconductor is obtained when N is incorporated into GaP beyond the conventional doping. To do this, however, the extreme immiscibility of the alloy, due to the large atomic-distance mismatch ($\sim$20%) as well as the structural mismatch between GaP (zincblende) and GaN (wurtzite), must be overcomed by a far-from-thermal-equilibrium growth method. In fact, the N incorporation into GaP is expected to be as low as $x < 10^{-6}$ at 700°C thermal equilibrium[1]. Recently, the growth of the $GaP_{1-x}N_x$ alloy with a significant N concentration has been successful by molecular beam epitaxy (MBE) ($\sim$7%) [2] and metalorganic vapor phase epitaxy (MOVPE) ($\sim$6%) [1,3]. A latest work has reported the growth of the alloy with even 16% N [4].

A noteworthy fact about the optical properties of $GaP_{1-x}N_x$ alloy is the red-shift behavior of the fundamental absorption edge [5,6] and photoluminescence [1,2,4] with increasing N concentration. This can be interpreted as a manifestation of the huge bowing in the composition dependence of the energy gap [2,7,8], which is caused by the large electronegativity of N. A tight-binding calculation has given the N concentration dependence of the Γ-X and the Γ-Γ energy gaps, as shown in Fig. 1 [8]. It is known that with this alloy semiconductor, a very wide range of spectrum, not only between 2.3eV (GaP) and 3.4eV (GaN), but even much below as 1.5eV, may be covered.

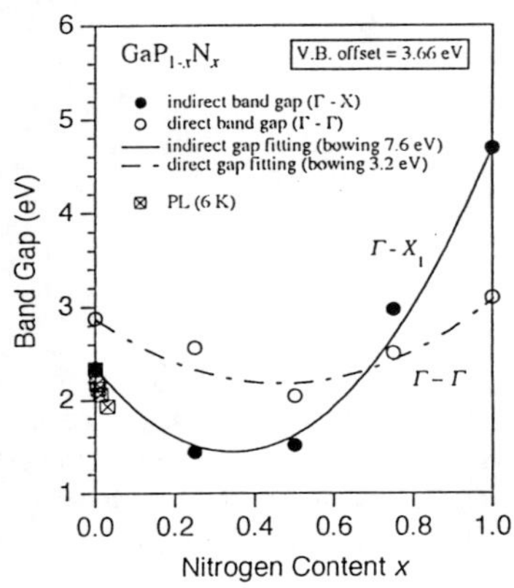

Fig. 1 N concentration dependence of the energy gaps (Γ-X and Γ-Γ) in the $GaP_{1-x}N_x$ alloy. A tight-binding calculation [8].

Mat. Res. Soc. Symp. Proc. Vol. 449 © 1997 Materials Research Society

GaP$_{1-x}$N$_x$ has revealed itself as an efficient luminescent material up to a few percent of N concentration [1,9]. In the doping regime ([N]$\leq 10^{19}$cm^{-3}, *i.e.* $x \leq 0.1\%$), it is well established that the N atoms form the deep efficient radiative centers, NN$_i$ (i=1,2,3,...10) [10]. Then, how do these isoelectronic N deep levels evolve into the band edge in the alloy regime? How can we define the alloy band edge at all?

In the present study, the GaP$_{1-x}$N$_x$ alloy with the N concentration (x) as high as 6% has been grown by metalorganic vapor phase epitaxy (MOVPE). The band-edge states, and hence the formation of the alloy energy gap, have been studied in detail using photoluminescence(PL), optical absorption, time-resolved photoluminescence, and photoluminescence excitation(PLE) spectroscopies.

MOVPE GROWTH

Growth Procedure

The GaP$_{1-x}$N$_x$ alloy films were grown on nominally on-axis (100)GaP substrates at 600-700°C by low-pressure (60 Torr) MOVPE with H$_2$ carrier using a horizontal reactor with an rf-heated susceptor [1]. Trimethylgallium (TMG), PH$_3$ and 1,1-dimethylhydrazine (DMHy, (CH$_3$)$_2$N$_2$H$_2$) were used as the Ga, P and N sources, respectively. DMHy is a liquid with an estimated vapor pressure of ~100 Torr at 16°C. DMHy is an efficient N source material which is easy to decompose at relatively low growth temperatures [11]. TMG and DMHy were bubbled at −10°C and +10°C, respectively. Prior to the growth of GaP$_{1-x}$N$_x$, a 0.3-μm-thick GaP buffer layer was grown at 750°C. Then the substrate temperature was lowered to the growth temperature for GaP$_{1-x}$N$_x$. The V/III ratio ranged from 20 to 200. The TMG flow rate was typically 4.5μmol/min (2.0 sccm). The epilayer thickness was 0.15-0.8μm.

Growth Results

Figure 2 shows the double-crystal x-ray rocking curves of the (400) reflection from GaP and GaP$_{1-x}$N$_x$ grown with different vapor phase compositions. The (511) reflection was also taken to estimate the lattice constant a of the alloy. The alloy compositions x were determined from the lattice constant a, assuming Vegard's law, in the way that $x = (a - a_{GaP})/(a_{c.GaN} - a_{GaP})$, where a_{GaP}(=5.541Å) and $a_{c.GaN}$(=4.51Å) were the lattice constants of GaP and cubic GaN[11], respectively.

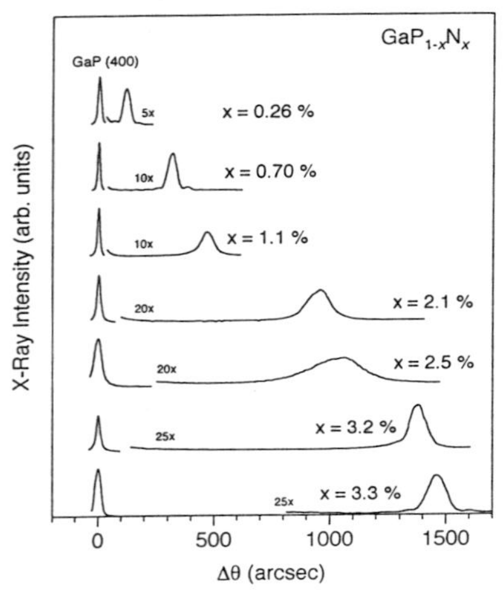

Fig. 2 Double-crystal X-ray rocking curves of the (400) reflections from GaP and GaP$_{1-x}$N$_x$ grown with different vapor phase compositions.

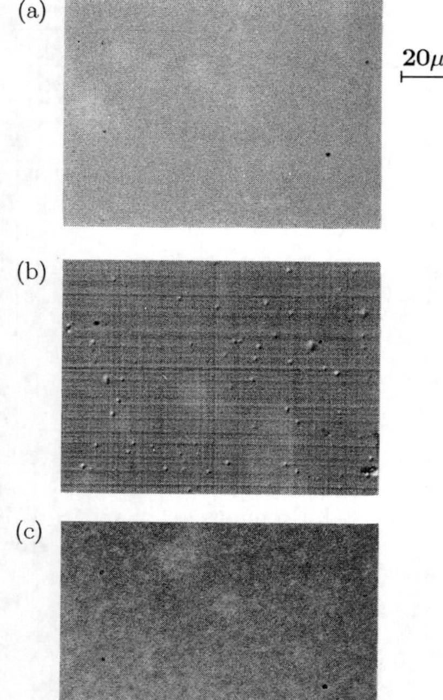

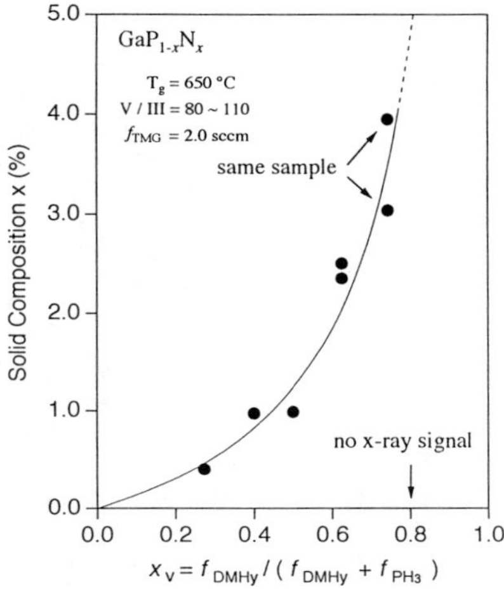

Fig. 3   Surface morphologies of GaP$_{1-x}$N$_x$ layers grown at 650°C with different vapor phase compositions x_v. (a) x_v=0.27, x=0.41%;   (b) x_v=0.63, x=2.4%;   (c) x_v=0.8, x is undetermined.

Fig. 4   Solid (x) versus vapor (x_v) compositions for GaP$_{1-x}$N$_x$ alloys at 650°C with V/III$\simeq$100, and f_{TMG}=2.0 sccm.

The growth rate of the GaP$_{1-x}$N$_x$ film was proportional to the TMG flow rate, f_{TMG}. It was 2.2μm/h for f_{TMG}=2.0 sccm. Figure 3(a)-(c) show the surface morphologies of the samples grown at 650°C with different vapor phase compositions x_v of the group V sources. x_v is defined as $x_v = f_{DMHy}/(f_{PH_3} + f_{DMHy})$, where f_{DMHy} and f_{PH_3} are the molar flows of the DMHy and PH$_3$, respectively. The corresponding alloy compositions x are also specified. For x_v=0.27 (x=0.41%, Fig.3(a)), the grown surface is fairly smooth and mirror-like. For x_v=0.63 (x=2.4%, Fig.3(b)), the surface is also mirror-like, but the cross-hatches due to the large lattice mismatch ($\sim$0.5%) are noticeable. When x_v is increased to 0.8 (Fig.3(c)), the surface becomes rough and powder-like, giving no x-ray diffraction signals except that from the substrate.

Figure 4 shows the solid (x) versus vapor (x_v) composition relationship at 650°C with the V/III ratio around 100, and f_{TMG} of 2.0 sccm. The epitaxial alloy layers with a mirror-like surface were obtained up to x=3% in the range x_v $\leq$0.63. When x_v exceeded 0.74, the crystalline quality was rapidly degraded, giving two separate compositions. No epitaxial layers were obtained at x_v $\geq$0.8.

At a fixed TMG flow rate and a vapor composition x_v, the solid composition x is much dependent on the growth temperature as shown in Fig. 5 for the case of f_{TMG}=2.0 sccm, V/III=80, and x_v=0.625. As the temperature was lowered from 700°C to 630°C, the solid composition x increased from 0.39% to 3.2%. When the temperature was as low as 600°C, the grown surface became rough, giving no x-ray diffraction signal from the alloy. The solid composition x is also dependent on the TMG flow rate as shown in Fig. 6 at the fixed growth temperature (660°C) and the group-V vapor composition (x_v=0.625). The solid composition increased to x=2.0% with increasing the TMG flow rate (which is proportional to the growth rate) up to 4.0 sccm, where the N incorporation became somewhat saturated.

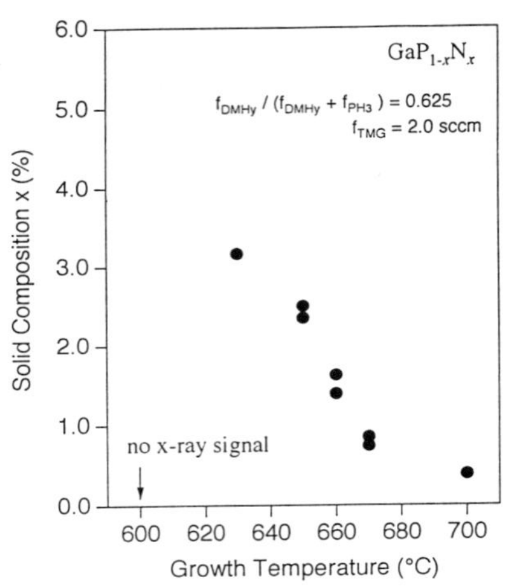

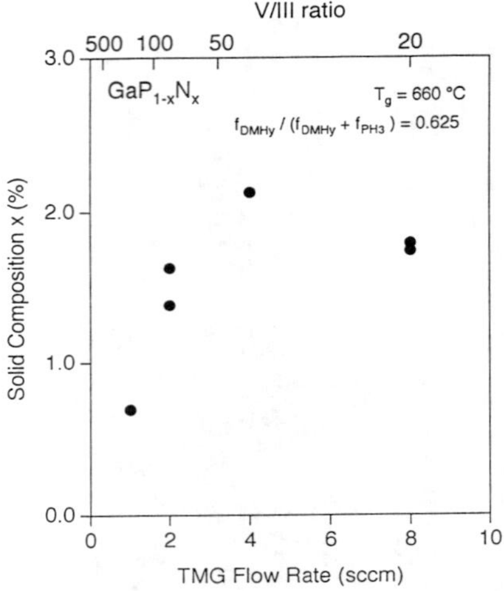

Fig. 5   Solid composition x versus growth temperature for GaP$_{1-x}$N$_x$ alloys at f_{TMG}=2.0 sccm, V/III=80, and x_v=0.625.

Fig. 6   Solid composition x versus TMG flow rate f_{TMG} for GaP$_{1-x}$N$_x$ alloys grown at 660°C with x_v=0.625.

Based on these growth characteristics, a much faster growth rate of 8.8μm/h was adopted with f_{TMG}=8.0 sccm, V/III=20, x_v=0.66 at 650°C, and the N concentration as high as 6.3% was obtained.

Discussion on Growth Kinetics

The strong dependence of the solid composition on the growth temperature and the growth rate suggests the key role of the nonequilibrium circumstances during the growth process. At higher growth temperatures, the N desorption from the surface will be substantially enhanced at the same time with the enhanced PH$_3$ decomposition, thus giving the low N incorporation. At higher growth rates (high TMG flow rates), the N desorption will be much suppressed due to the rapidly growing overlayers. In view of this growth knietics, we performed growth interruption experiments, in which the TMG supply was interrupted for a certain time after each one monolayer growth of GaP$_{1-x}$N$_x$.

Figure 7 shows the solid composition x as dependent on the growth interruption time t_i at the constant group-V vapor composition x_v=0.625 and the growth temperature of 670°C. The TMG supply duration was 0.4sec with the flow rate of 4.0 sccm in each cycle. It is found that the N incorporation is suppressed for longer interruption times. This behavior is modeled with a simple supply-and-desorption rate equation valid during the interruption,

$$\frac{dn}{dt} = G - \frac{n}{\tau}, \tag{1}$$

where n is the N concentration on the surface, G is the N supply, and τ is the N desorption time constant. Assuming small n and rapid adsorption of P to the sites where N is desorbed, we have

$$x \propto n = G\tau + A exp(-\frac{t}{\tau}). \tag{2}$$

The plots of Fig. 7 are fitted with τ=0.47sec. Thus the whole growth proccess is governed by the N desorption from the surface, which limits the N incorporation into the solid.

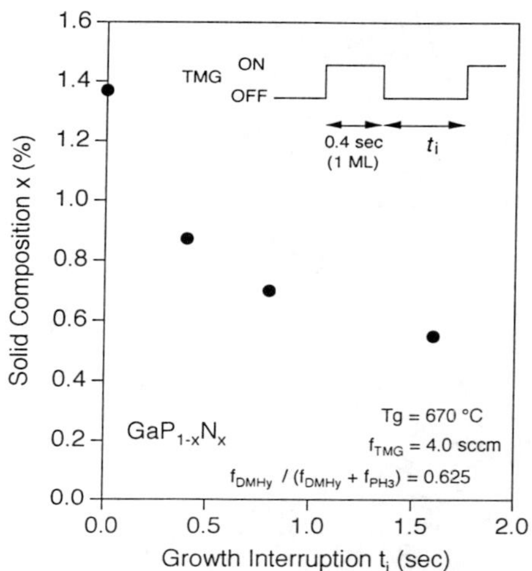

Fig. 7   Solid composition x versus growth interruption time t_i at 670°C with f_{TMG}=4.0 sccm and x_v=0.625.

OPTICAL CHARACTERIZATION

<u>Photoluminescence and Absorption</u>

Figure 8 shows the low-temperature (5.6K) PL spectra of $GaP_{1-x}N_x$ with various N concentrations [1]. The excitation source was a He-Cd laser (325nm). For GaP (x=0), the spectrum consists of isolated-N-bound A exciton (2.317eV) [10] and its phonon replicas. For low N concentrations, typically x=0.028% ([N]=7×10^{18} cm^{-3}), which represents the N-doping regime, the spectrum consists of the sharp emissions from the N-N pair bound excitons, NN_1 (2.18eV), NN_3 (2.26eV), NN_4 (2.29eV), NN_5 (2.30eV) and their phonon replicas. When x is increased to 0.36%, the deep NN_1 line and its phonon replicas become dominant, while NN_3 gets weaker and NN_4 disappears. For higher N concentrations, typically x=1.4%, which represents the alloy regime, the PL spectrum is dominated by a broad peak which shifts to lower energies with increasing x.

Such a red-shift behavior is also seen in the low-temperature (20K) optical absorption spectra shown in Fig. 9. The absorption edge shifts to lower energies with increasing N concentration. The A-line absorption peak is clearly seen at low N concentrations ($x \leq 0.75\%$), though it becomes broader and finally not discernable with increasing N concentration. The A-line peak is also observed in GaP (x=0) due to the unintentional N incorporation. The NN_3 peak behaves in the similar manner with the A line though not observed in GaP. The NN_1 peak is enhanced with increasing N concentration, and even observed at $x \sim 1.9\%$ until it merges into the large absorption band. It is noted that the emissions related to the NN_i pairs were not observed in the PL for $x \geq 1\%$. Thus, the PL does not reflect the density of states (DOS) of NN_i pairs. In fact, the NN_1 line dominates the other lines when $x > 0.2\%$, although the NN_1 (nearest-sites) pairs become most probable only when $x > 3\%$ [9]. Moreover, the PL is found to occur at the DOS tails which are lower in energy than the absorption edge. Namely, the broad emission of $GaP_{1-x}N_x$ alloys with high N concentrations does not originate from the band edge but from the tail states.

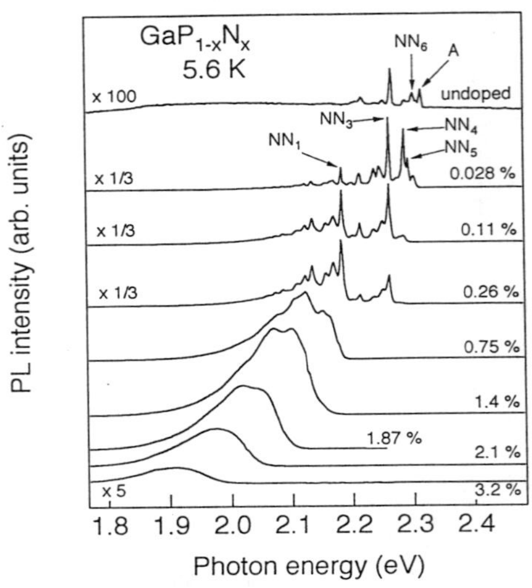

Fig. 8   Low-temperature (5.6K) PL spectra of $GaP_{1-x}N_x$ with various N concentrations.

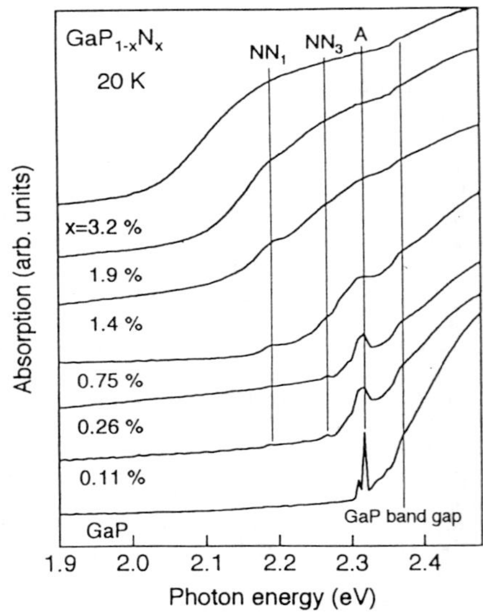

Fig. 9   Low-temperature (20K) absorption spectra $GaP_{1-x}N_x$ alloys. The energy positions of GaP band gap, A line, NN_3 and NN_1 lines are indicated.

<u>Time-Resolved Photoluminescence</u>

In order to examine the nature of the band edge states, time-resolved PL measurements were done at 4.2K [6]. The time-resolved PL intensity was measured with the time-correlated single photon counting method. The excitation source was the frequency-doubled light (315nm) obtained by a KH_2PO_4 crystal from a dye laser operating at 630nm with a repetition rate of 75.4MHz and a pulse duration of 2ps.

Figure 10 shows the PL intensity decay profiles of $GaP_{1-x}N_x$ alloys [6,12]. The decay profiles are much different between high ($x > 1\%$) and low ($x < 1\%$) N concentrations. For a low N concentration of $x=0.028\%$, for which the NN_5 line (540nm) was detected, the PL decay curve was fitted with a single-exponential with the decay time of $\tau=270$ns. This decay time is considered to be the radiative recombination lifetime of the excitons bound to NN_5 pairs, since the same order of magnitude of lifetimes has been reported for NN_1 and NN_3 bound excitons ($\sim$100ns) at 4.5K [13].

For a high N concentration of $x=1.4\%$, for which the lower-energy peak (600nm) of the broad emission was detected, the decay profile showed a double-exponential behavior which was fitted with

$$I(t) = C_{fast}exp(-\frac{t}{\tau_{fast}}) + C_{slow}exp(-\frac{t}{\tau_{slow}}), \tag{3}$$

where $\tau_{fast}=63$ns and $\tau_{slow}=1100$ns. The fast decay process is considered to be the relaxation to nonradiative centers. After the nonradiative centers are saturated, the radiative recombination or the relaxation between radiative levels, whose lifetimes amount to μs, will become dominant.

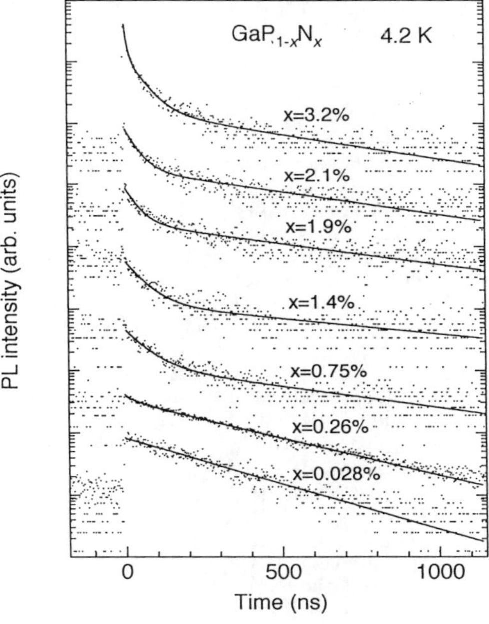

Fig. 10   PL decay profiles of $GaP_{1-x}N_x$ alloys at 4.2K. The solid lines are the best fits with eq. (3).

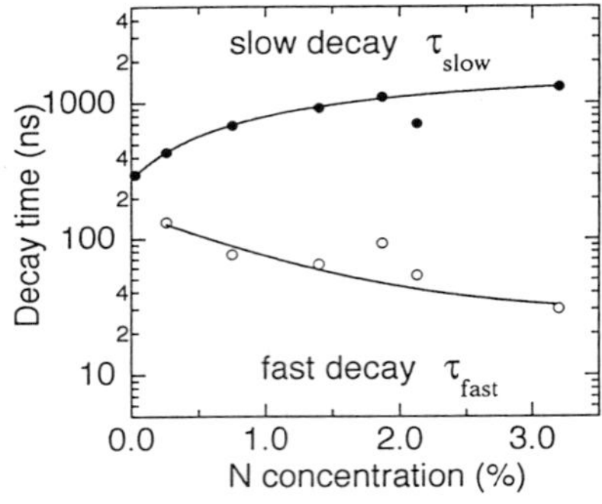

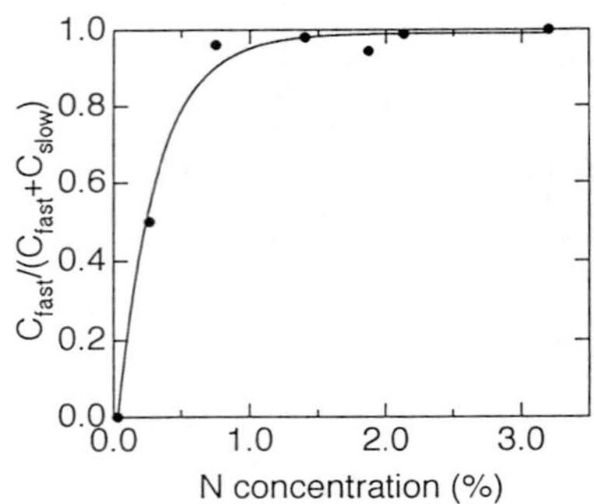

Fig. 11   N concentration dependences of the PL decay times τ_{fast} and τ_{slow}.

Fig. 12   Dominance of the fast decay process as dependent on the N concentration.

In Fig. 11, the N concentration dependences of the PL decay times τ_{fast} and τ_{slow} are shown. The slower decay time τ_{slow} increases with increasing N concentration, which indicates that the excitons become less localized for higher N concentrations. The relaxation between the radiative levels are probable due to the scattered spatial distribution of the DOS tail states. The whole process is more and more dominated by the fast decay process with increasing N concentration, as shown in Fig. 12, which indicates that the alloy becomes more defective.

Photoluminescence Excitation Spectra

Photoluminescence excitation (PLE) spectroscopy may give still additional information on the band edge states, or even the band edge formation itself. The PLE measurements were done using monochromatic light dispersed with a 0.5m monochromator from a 300W halogen lamp as the excitation source [14].

In Fig. 13, the PLE spectrum (20K) of $GaP_{1-x}N_x$ with x=0.028% (doping regime) is shown together with the corresponding PL spectrum. In the PLE spectrum, the A-line absorption peak is dominant, which implies the N atoms are almost isolated from each other in this low N concentration. Indeed, the average distance between the N atoms is estimated to be ~30Å for x=0.028%. Comparing the two spectra, it is obvious that the PL takes place after photoexcited carriers relax to the lower NN_i states. The GaP direct gap (Γ-Γ) edge, as well as the indirect gap (Γ-X) edge, is clearly observed.

In Fig.14, the PLE spectra (20K) of $GaP_{1-x}N_x$ with x=1.4% (alloy regime) is shown with the corresponding PL. In the PLE spectrum, a broad absorption peak centered at the A line is dominant. The absorptions due to NN_3 and NN_1 are also seen. It is very clear that the PL takes place at the DOS tail which is much lower in energy than the absorption edge. The average distance between the N atoms for x=1.4% is ~7Å, which is very close to the N-N distance of NN_3 (6.67Å). At the higher energies, the direct-gap edge became less clear.

Figure 15 shows the PLE spectra of $GaP_{1-x}N_x$ alloys with various N concentrations. With increasing N concentration, the absorption peak centering at the A-line energy becomes broader, and the absorption edge shifts to lower energies. The broadening of the A line is considered to arise from the interaction between the electrons trapped by the isolated N atoms. The average distance between N atoms becomes less and less with increasing N concentration. The GaP direct band edge becomes rounded for higher N concentrations. This is interpreted as that these band states are consumed to form the lower-energy states.

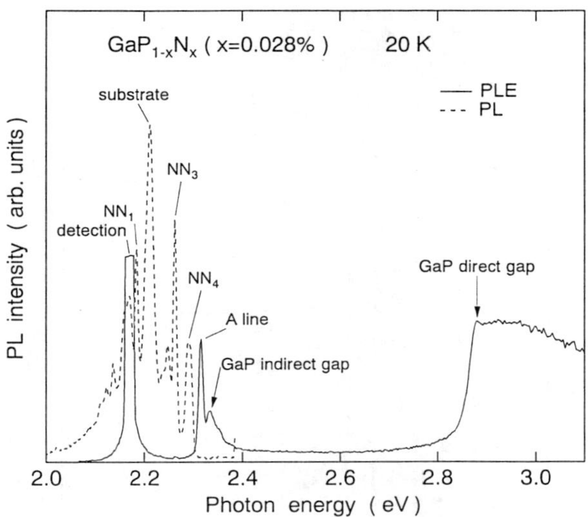

Fig. 13   PL and PLE spectrum (20K) of GaP$_{1-x}$N$_x$ (x=0.028%), which represents the low N concentration (doping regime).

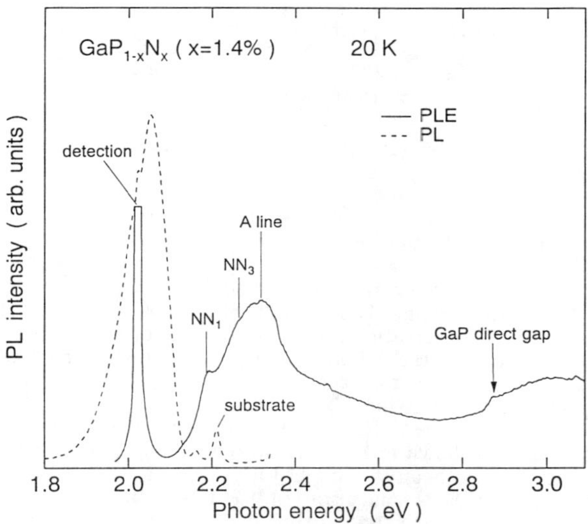

Fig. 14   PL and PLE spectra (20K) of GaP$_{1-x}$N$_x$ (x=1.4%), which represents the high N concentration (alloy regime).

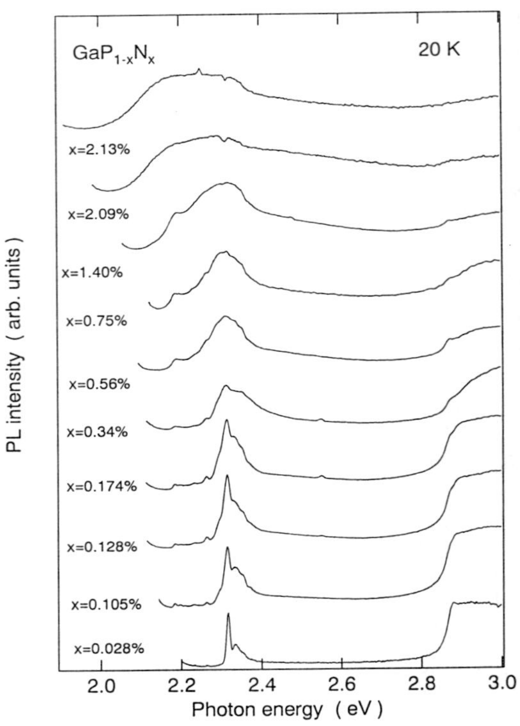

Fig. 15   PLE spectra (20K) of GaP$_{1-x}$N$_x$ alloys with various N concentrations.

Discussion on the Band Edge Formation

Figure 16 shows the N concentration dependence of the absorption edge energy determined from the PLE measurements. The red-shift of the absorption edge is very obvious as stated before. An important observation is that, in the limit of zero N concentration ($x \rightarrow 0$), the absorption edge approaches the A-line energy (2.32eV) rather than the indirect-gap energy (2.33eV) of GaP. This means that the conduction band-edge states originate from the isolated N states, which give the A-line absorption.

We can now depict the process of the alloy band edge formation with N incorporation into the GaP$_{1-x}$N$_x$ alloy as shown in Fig. 17. At low N concentrations, where the average distance between the N atoms is large enough, the A-line states formed by isolated N atoms are dominant. With increasing N concentration, the interaction among the N atoms makes the A line broader, which will give the red-shift of the absorption edge. At the same time, the higher energy side of the A-line states merges into the conduction band. At much higher N concentrations, the N states are almost delocalized, and we cannot distinguish the A-line states from the conduction band states. Such a process of the band edge formation is quite unique to GaP$_{1-x}$N$_x$ alloy, in which the isolated N states are formed deep in the GaP energy gap.

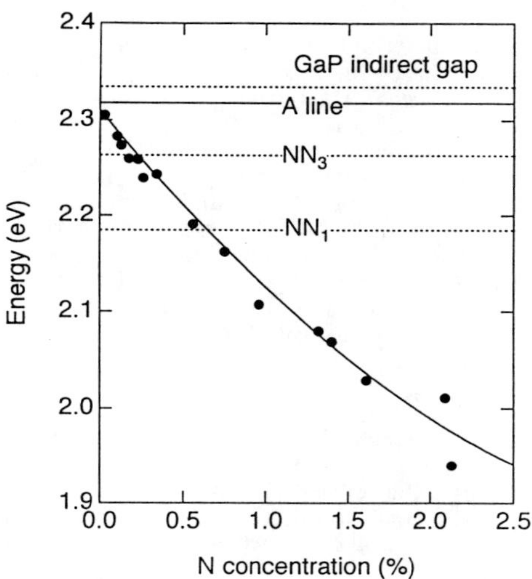

Fig. 16   N concentration dependence of the absorption-edge energy determined from the PLE measurements.

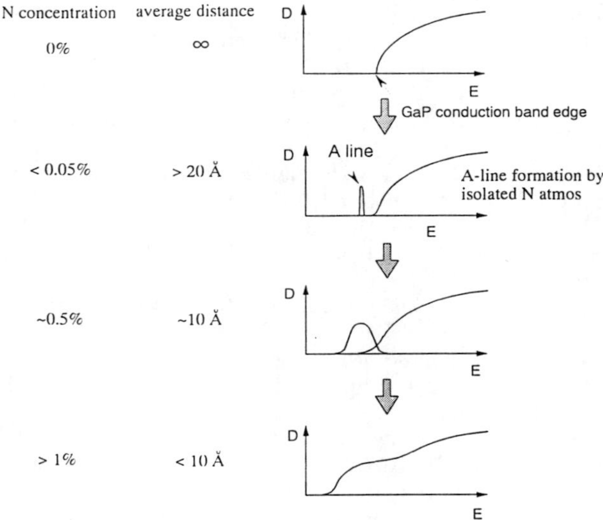

Fig. 17   Schematic representation of the band-edge formation of $GaP_{1-x}N_x$ alloy as dependent on N concentration.

CONCLUSIONS

The $GaP_{1-x}N_x$ metastable alloy semiconductor has been grown by MOVPE using DMHy as the N source. The growth characteristics show the key role of the non-equilibrium circumstances during the growth process, where the N desorption from the surface limits the N incorporation. An efficient N incorporation of 6.3% has been attained with a faster growth rate. When the N concentration is as low as 0.1% or less (doping regime), the PL is dominated by the NN_i-related lines, while the absorption edge is dominated by the A line (*i.e.* isolated N states). When the N concentration is higher than 1% (alloy regime), the PL is dominated by a broad emission from the DOS tail states, while the absorption edge is dominated by the NN_1 and the A line. The band edge originates from the A-line energy, which becomes less and less localized with increasing N incorporation, rather than the indirect-gap energy of GaP. The band edge shifts to lower energies with increasing N incorporation, which causes the red-shift of the PL and the optical absorption.

ACKNOWLEDGEMENTS

The author wishes to thank S. Miyoshi, H. Yaguchi and G. Biwa for their substantial contributions to this work. Thanks are also to Prof. Y. Shiraki and Prof. R. Ito for their continuous cooperation through this work. This work has been done using the facilities of Research Center for Advanced Science and Technology (RCAST), The University of Tokyo. This work is supported in part by a Grant-in-Aid (No. 06452107) from the Ministry of Education, Science, Sports and Culture.

REFERENCES

1. S. Miyoshi, H. Yaguchi, K. Onabe, R. Ito and Y. Shiraki, Appl. Phys. Lett. **63**, 3506 (1993).
2. J. N. Baillargeon, K. Y. Cheng, G. E. Hofler, P. J. Pearah and C. Hsieh, Appl. Phys. Lett. **60**, 2540 (1992).
3. S. Miyoshi, H. Yaguchi, K. Onabe, Y. Shiraki and R. Ito, Inst. Phys. Conf. Ser. No. 141, p. 97 (1995).
4. N.Y. Li and C.W. Tu, Mat. Res. Soc. Symp. Proc. **423**, p. 317 (1996).
5. X. Liu, S. G. Bishop, J. N. Baillargeon and K. Y. Cheng, Appl. Phys. Lett. **63**, 208 (1993).
6. H. Yaguchi, S. Miyoshi, H. Arimoto, S. Saito, H. Akiyama, K. Onabe, Y. Shiraki and R.Ito, Inst. Phys. Conf. Ser. No. 145, p. 307 (1996).
7. S. Sakai, Y. Ueta and Y. Terauchi, Jpn. J. Appl. Phys. **32**, 4413 (1993)
8. S. Miyoshi and K. Onabe, *Topical Workshop on III-V Nitrides*, Nagoya, 1995, P-1 (Solid State Electron., to be published).
9. S. Miyoshi, H. Yaguchi, K. Onabe, R. Ito and Y. Shiraki, J. Cryst. Growth **145**, 87 (1994).
10. D. G. Thomas and J. J. Hopfield, Phys. Rev. **150**, 680 (1966).
11. S. Miyoshi, K. Onabe, N. Ohkouchi, H. Yaguchi, R. Ito, S. Fukatsu and Y. Shiraki, J. Cryst. Growth **124**, 439 (1992).
12. H. Yaguchi, S. Miyoshi, H. Arimoto, S. Saito, H. Akiyama, K. Onabe, Y. Shiraki and R.Ito, *Topical Workshop on III-V Nitrides*, Nagoya, 1995, G-7 (Solid State Electron. to be published).
13. J. Zheng and W. M. Yen, J. Luminescence **39**, 233 (1988).
14. H. Yaguchi, S. Miyoshi, G. Biwa, M. Kibune, K. Onabe, Y. Shiraki and R.Ito, *Int. Conf. on MOVPE, Cardiff, 1996* (J. Cryst. Growth, to be published).

GaN CRYSTALS GROWN IN THE INCREASED VOLUME HIGH PRESSURE REACTORS

S. POROWSKI[*], M. BOCKOWSKI[*], B. LUCZNIK[*], M. WROBLEWSKI[*],
S. KRUKOWSKI[*], I. GRZEGORY[*], M. LESZCZYNSKI[*], G. NOWAK[*]
K. PAKULA[**] and J. BARANOWSKI[*, **]

[*]High Pressure Research Center Polish Academy of Sciences,
Sokolowska 29/37, 01-142 Warsaw, Poland
sylvek@iris.unipress.waw.pl

[**]Institute of the Experimental Physics University of Warsaw,
Hoza 69, Warsaw, Poland

ABSTRACT

GaN single crystals are grown from the solution of atomic nitrogen in liquid Ga at N_2 pressure up to 20 kbar. The crystals reaching dimensions of 1cm, with dislocation density of 10^3 - 10^5 cm^{-2} are currently obtained and successfully used for homoepitaxy by MOCVD and MBE.

The increase of the maximum size of GaN crystals with stable morphology requires the increase of the volume of the crucible. Beside the obvious geometric factors, the increase of the volume of the furnace and crucible allows to achieve much better control of supersaturation profiles in the solution and therefore better control of growth process. It helps to avoid morphological instabilities and leads to the growth of transparent, inclusion free crystals.

In the paper, the results of crystallization of substrate quality GaN crystals obtained with the use of large volume high pressure reactor will be presented. The crystals were characterized by High Resolution X-ray Diffraction and Atomic Force Microscopy. It will be shown that the quality of GaN crystals does not deteriorate with the increasing size and that epi-ready surfaces of GaN substrates can be obtained. Two-dimensional growth by propagation of monoatomic steps of GaN homoepitaxial layers will be presented as a verification of the quality of GaN substrates.

INTRODUCTION

In the recent years, the fundamental breakthrough has been achieved in the technology of GaN and GaN based optoelectronic devices. The most important results are: development of the technology of highly efficient blue and green Light Emitting Diods [1] and the creation of first GaN based lasers [2]. These devices have been constructed using heteropitaxial technology of GaN deposition on Al_2O_3 and SiC substrates. Due to the large lattice mismatch and the differences in thermal expansion coefficients between substrate materials and GaN, the layers have the dislocation density of order of 10^8 cm^{-2}. The most direct method to decrease the defect density in GaN layers is the use of GaN single crystalline substrates for the epitaxial growth.

Growth of GaN single crystals is extremely difficult due to high melting temperature - about 2800K [3] and very high dissociation N_2 pressure - higher than 40 kbar [4], at melting. Therefore for the growth of GaN bulk crystals the methods requiring lower temperatures and pressures have to be applied.

In this paper we report the recent results on bulk GaN crystallization from the solution in liquid Ga at high N_2 pressure. The crystals are grown at N_2 pressures up to 20 kbar and temperatures up to 1600°C, in temperature gradient.

Mat. Res. Soc. Symp. Proc. Vol. 449 © 1997 Materials Research Society

NITROGEN DISTRIBUTION IN LIQUID GALLIUM

For stable crystallization from the solution, the supersaturation in the growth zone has to be sufficiently small and uniform across the growing crystal faces. To create such conditions several factors specific for the considered crystallization method have to be taken into account. The most important features of the process are:

i/ the surface of liquid gallium is covered by the polycrystalline film which acts as an efficient sink for nitrogen near the crucible walls,

ii/ crystallization is carried out in temperature gradient to assure supersaturation in the growth zone,

iii/ the nucleation of GaN in the volume of the solution does not occur even at very high supersaturations.

The uncontrolled spatial variation of the nitrogen concentration is one of the principal factors disturbing the growth of GaN crystals. Therefore for the optimal growth, the maximization of the zone of uniform concentration of nitrogen in liquid Ga is required. The distribution of nitrogen in the solution is a function of N solubility [5], distribution of temperature, N_2 pressure, presence of growth centers and geometry of the growth system. The concentration distribution in liquid Ga has been estimated using the diffusive nitrogen transport approximation. The concentration field has been calculated using finite element code - Fluid Dynamics Analysis Package (FIDAP) [10].

The results of the calculations for crucibles of three different diameters are shown in Fig.1.

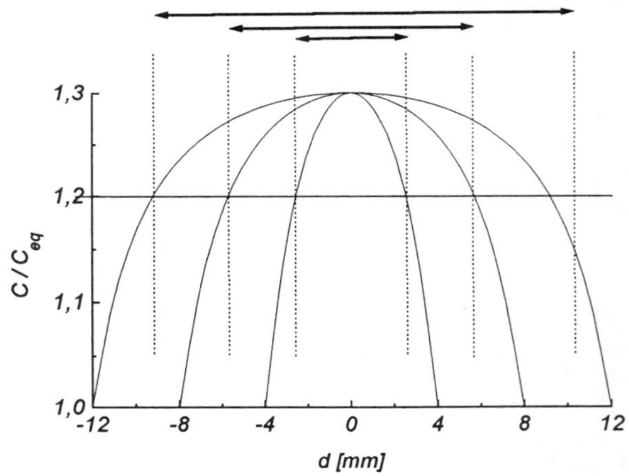

Fig.1 Radial distribution of nitrogen concentration in the growth zone; the crucible i. d. are: 8mm, 16 mm and 24 mm.

The widths of the zones for stable crystallization determined by the intersection of the horizontal line and the concentration profiles roughly correspond to the maximum size of crystals with stable morphology obtained in crucibles of 8 and 16mm inside. Above the horizontal line, the variation of N concentration does not exceed 5%. The zones with most uniform concentration of nitrogen allowing the best growth are located at the central part of the crucibles. At the sides, the concentration drops rapidly what is related to the growth of GaN on the crucible walls.

It follows from above that the size of the stable growth zone increases with the diameter of the crucible. The increase is faster than linear. For instance for the crucible of 24mm, the model predicts the stable growth zone approaching 17 mm. Therefore one can expect that, for the optimum experimental conditions, 17 mm crystals can be grown in 24 mm crucibles.

RESULTS OF CRYSTALLIZATION

<u>Size</u>

The increase of the crystal size obtained at present, has been achieved mainly by the increase of the solution volume where small and uniform supersaturation is easier to obtain than in smaller volumes.

The biggest crystals are obtained in crucibles with diameter of 18mm where the conditions are optimized Linear dimensions of these crystals are about 10mm whereas their surface area approaches 70mm^2. Some of these crystals are shown in Fig.2.

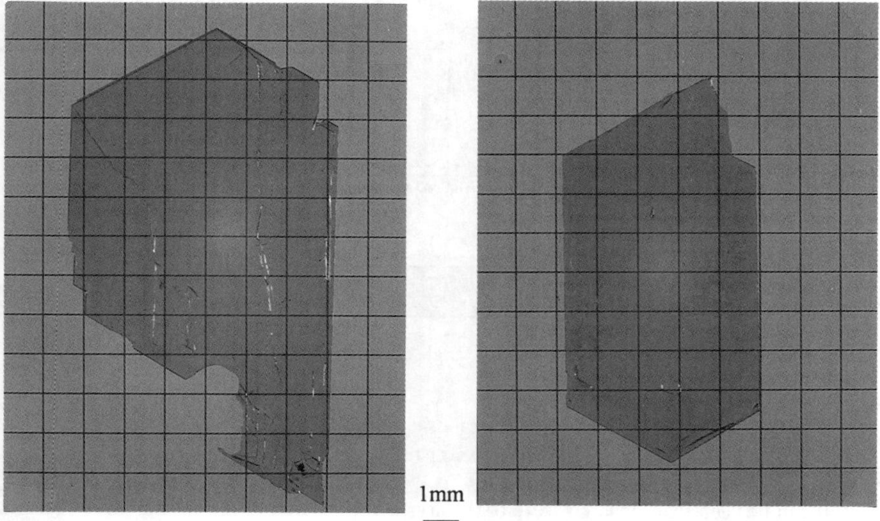

Fig. 2 GaN single crystals grown in crucibles of increased volume

Recently the experiments with high pressure reactors allowing crucibles of 30mm internal diameter have been started. However the conditions of the crystallization process are not yet optimized.

The growth rate in $\{10\underline{1}0\}$ directions, perpendicular to the c-axis of the crystal, is about 0.1mm/h that is the same as for smaller , earlier reported crystals i.e.[5]. This shows that the further increase of the crystal size is still possible without unreasonably long crystallization time.

<u>Structure</u>

The structural perfection of the crystals was examined using High Resolution X-ray Diffraction. ("real" rocking curve mode with the detector wide open). The beam was monochromatized by four 022 reflections grom Ge crystals and had 12 arcsec angular divergence.

The rocking curves were obtained for various positions of the samples with respect to the X-ray beam of a size of 3.5mm x 0.5mm. Typical result of the diffraction experiment for a symmetrical 00.4CuKα1 reflection is shown in Fig.3

sample SV23

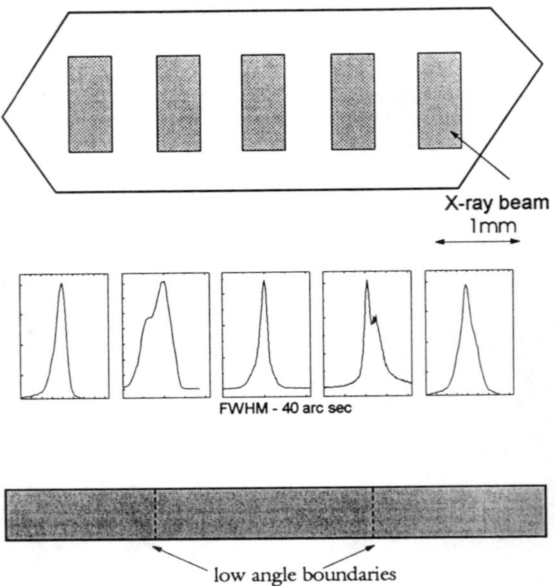

Fig.3 Mapping of GaN single crystal using X-ray diffraction rocking-curves,
00.4 CuKα1 reflection.

It can be seen that for large areas of the crystal (up to 2mm) the rocking curve consists of single narrow peak. Those areas are separated by low angle (about 1 arcmin) boundaries. When the X-ray beam comprises such a boundary splitted peaks can be observed.

Similar results were obtained using other reflections (including asymmetrical 11.3 and 11.4) what indicates that the samples do not contain directional unhomogeneities.

<u>Surface</u>

As it was shown by Z. Liliental-Weber et al. [6], the two polar (0001) surfaces of pressure grown GaN crystals are different in morphology. One of these surfaces is often atomically flat whereas the opposite one is rough and consists of 100-300Å regular pyramids. Both surfaces are partially covered by the macroscopic growth figures (usually small hexagons ~10μm in size) resulting from the cooling of the growth solution after crystallization in temperature gradient. Due to this the homoepitaxial layers grown on the "as grown" GaN substrates were not uniform although the step flow regime has been locally observed by AFM [7] for both hexagonal faces of the crystal. In particular, for the side which was initially rough, (with pyramids) it was shown that the surface becomes smooth within the range of a few monolayers, after annealing performed before the layer deposition by MOCVD [8]. In MBE [9] experiments the smoothing has been observed during the first stages of the epitaxial growth and has been checked by in situ measurements of the RHEED patterns. However in order to obtain large area uniform homoepitaxial layers, the substrates have to be polished and etched to remove the surface crystallites.

The quality of (0001) surfaces after mechano-chemical polishing has been estimated by RHEED measured at room temperature [9], by total X-ray reflection and by AFM examination of homoepitaxial layers grown on polished GaN substrates.

In Fig.4, the 2x2μm AFM scans for the MOCVD GaN layers grown on polished surfaces (the ones, which were initially rough - with pyramids) are shown.

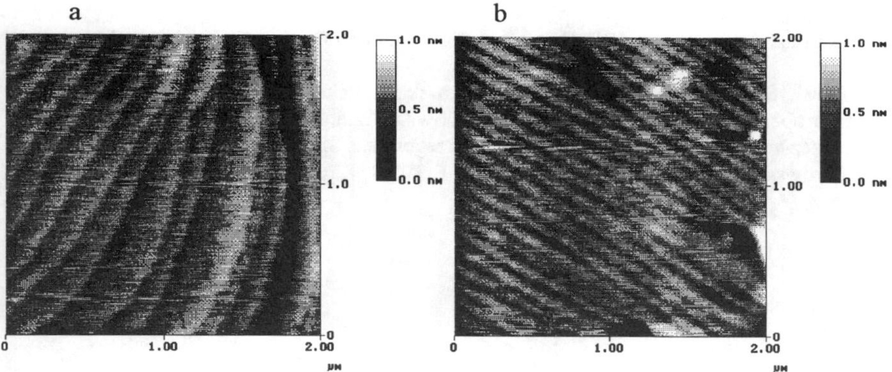

Fig.4   AFM scans of GaN homoepitaxial layers grown on slightly misoriented, (0001) surfaces
(100Å pyramids before polishing) of GaN substrates; a - misorientation about 7',
b - misorientation about 20'; the steps are ~2.6Å high

The diagram indicates the presence of monoatomic steps spreading across the substrate. The distance between steps depends on the misorientation angle between (0001) and a vicinal face obtained by polishing. The arrangement of steps allows to conclude that the growth plane was tilted by ~7 arc min from (0001) orientation for Fig.4a and by ~20 arc min for Fig.4b.

Also for the opposite side of the substrate (initially flat, with growth figures), the propagation of atomic steps has been observed for the polished samples.

CONCLUSIONS

The size of GaN single crystals of stable morphology can be increased by the application of bigger volume crucibles. In optimum conditions (uniform supersaturation) the rate of stable growth in the fastest growth direction, of 0.1mm/h has been achieved for crystals up to 10mm in size.

Structure quality of the crystals does not deteriorate with the increasing size. The quality of surfaces of the GaN substrates is sufficient to grow homoepitaxial layers by monoatomic step flow on both sides of the crystals.

REFERENCES

[1] S. Nakamura, T. Mukai and M. Senoh, J. Appl. Phys. 76 (1994), 8189

[2] S. Nakamura, M. Senoh, S. Nagahama, N. Iwasa, T. Yamada, T. Matsushita, H. Kiyoku and Y. Sugimoto, *Jap. J. Appl. Phys.* 35, (1996),L217

[3] Van Vechten J.A., *Phys. Rev.*B7, 1479 (1973)

[4] Karpinski J., Jun J. and Porowski S. , *J.Cryst. Growth*, 66, 1, (1984)

[5] I. Grzegory, J. Jun, M. Bockowski, St. Krukowski, M. Wroblewski, B. Lucznik and S. Porowski, *J. Phys. Chem. Solids* vol.56, No 3/4, pp 639-647, (1995)

[6] Z. Liliental-Weber, C. Kisielowski, S. Ruvimov, Y Chen, J. Washburn, I. Grzegory, M. Bockowski, J. Jun and S. Porowski, *Journal of Electronic Materials*, Vol.25, No9,1996

[7] S. Porowski, J. M. Baranowski, M. Leszczynski, J. Jun, M. Bockowski, I. Grzegory, S. Krukowski, M. Wróblewski, B. £ucznik, G. Nowak, K. Pakula, A. Wysmolek, K. P. Korona, and R. Stepniewski
International Symposium on Blue Laser and Light Emitting Diodes, Chiba, Japan March 5-7, 1996

[8] G. Nowak, S. Krukowski, I. Grzegory, S. Porowski, K. Pakula, J. M. Baranowski and J. Zak, First European GaN Workshop EGW -1, Rigi, Switzerland, June 1-3 1996

[9] M. Leszczynski, A. Barski, F. Wiedeman, I. Grzegory ans S. Porowski, to be published

[10] S. Krukowski, First European GaN Workshop EGW -1, Rigi, Switzerland, June 1-3 1996

GROWTH OF BULK AlN AND GaN SINGLE CRYSTALS BY SUBLIMATION

C. M. BALKAŞ*, Z. SITAR*, T. ZHELEVA*, L. BERGMAN**, I. K. SHMAGIN§, J. F. MUTH§, R. KOLBAS§, R. NEMANICH** and R. F. DAVIS*

* Department of Materials Science and Engineering, North Carolina State University, Raleigh, NC 27695-7907
** Department of Physics, North Carolina State University, Raleigh, NC
§ Department of Electrical Engineering, North Carolina State University, Raleigh, NC

ABSTRACT

Single crystals of AlN to 1 mm thickness were grown in the range 1950-2250°C on 10x10 mm^2 α(6H)-SiC(0001) substrates via sublimation-recondensation method. Hot pressed polycrystalline AlN was used as the source material. The color varied from transparent to dark green/blue. The crystal morphology varied with growth conditions. Most crystals were 0.3 mm - 1 mm thick transparent layers which completely covered the substrates. Raman, optical and transmission electron microscopy (TEM) results are presented. Single crystals of gallium nitride (GaN) were also grown by subliming powders of this material under an ammonia (NH_3) flow. Optical microscopy, Raman and photoluminescence results are shown.

INTRODUCTION

The current lack of easily produced single crystal wafers of III-V nitrides of moderate cost is an impediment to the rapid enhancement/development of epitaxial thin films and selected devices produced from these materials. The realization of blue and green light emitting diodes and blue lasers as well as prototypes of several microelectronic devices produced from GaN-based materials containing copious line and planar defects has been most fortunate. It also strongly indicates that the employment of substrates on which homoepitaxial films can be grown will result in marked improvements in the devices fabricated in these films..

The properties of the III-V nitrides which make growth of bulk crystals a challenge include (1) high melting temperatures, (2) low sublimation/decomposition temperatures relative to these melting temperatures, (3) very high equilibrium nitrogen vapor pressures at moderate temperatures and (4) low solubility in acids, bases and most other inorganic elements and compounds. Thus, novel bulk growth techniques must be employed which either take advantage of these inherent properties or surmount the difficulties presented by them.

Bulk growth of monocrystalline AlN by sublimation/recondensation [1-7], evaporation/ reaction [8] and solution routes [9-10] has been attempted in several laboratories. The results of the first two processes [1-8] have been small crystals. Prior to the present research, Slack and McNelly [1,2] had achieved the largest crystals (10 mm long by 3 mm diameter) using the sublimation/recondensation method at 2250°C. An extensive experimental study for producing unseeded single crystals has been conducted by Dugger [9-10] using AlN-Ca_3N_2 solutions contained in graphite crucibles. The solutions were cooled from 1550°C.at 2°/hr for 26 hours. Small (1.1 mm long x 0.3 wide) crystals were achieved via nucleation on the crucible walls. The important aspect of this work is that Ca_3N_2 is an effective solvent for AlN with the solubility exceeding 10% at 1200°C. However it is incompatible with graphite crucibles.

Relatively fewer attempts to grow bulk crystals of GaN have been reported. Elwell and Elwell [11] have reviewed this work to 1988. Two major approaches have been used: vapor phase transport [12-15] and solution techniques [16-21]. Success of these methods has been limited to small crystals. The high pressure solution approach has achieved approximately the same results. The initial research [19,20] in high pressure synthesis produced platelets limited in size to ≈1 mm. Recent publications by these authors (see, e.g., Ref. [21]) note that they have now achieved crystals 8 mm square x 0.2 mm thickness. The growth rate in this technique is reported to be ≈20 μm/hr. Recently, Sakai and co workers [22] have grown GaN needles by sublimation of GaN powder on sapphire substrates. However, crystal size was reported to be limited to small needles.

The recent results of research conducted by the present authors and described in the following sections represent important advances in the determination of the process parameters and the

Mat. Res. Soc. Symp. Proc. Vol. 449 © 1997 Materials Research Society

accomplishments regarding growth via sublimation of AlN and GaN bulk single crystals. This is also the first report of seeded growth of thick crystals of AlN. The crystals of both materials were of very high quality. Comparisons of our diffraction, photoluminescence and Raman data with those available in the literature indicate that our crystals are superior to any bulk GaN or AlN grown to date.

EXPERIMENTAL PROCEDURES

Sublimation Growth and Characterization of Single Crystals of AlN

The AlN sublimation experiments were conducted in a resistively heated graphite furnace having a maximum temperature of 2500°C. This furnace was also internally adaptable to allow the achievement of the temperature gradients required for material transfer from source-to-seed. An important feature of the growth system was the capability of changing the axial position of the seed crystal with respect to the source and heater.

A critical technical issue in the sublimation of AlN was the selection of the material for the crucible. Physical and chemical stability at very high temperatures and compatibility with the furnace atmosphere, source, seed and furnace construction materials and machinability were considered. Initial experiments with graphite resulted in severe degradation of the crucible components due to their reaction with Al vapor. Formation of aluminum carbide (Al_4C_3) was found in the crucibles below 1900°C. Above this temperature, large pits and loose carbon powder were found in the crucibles and believed to result from the formation and subsequent decomposition of Al_4C_3. To surmount these problems, the crucibles were coated with SiC via chemical vapor deposition. However, the Al vapor slowly diffused through the SiC barrier coating, reacted with the graphite at the interface and eventually caused the coating to peel. However, the lifetime of each coating was sufficient to conduct several 10-15 hr experiments.

Bulk 99% dense AlN produced via hot pressing without sintering additives was used as the source material to reduce the contamination that would come from the higher concentrations of surface oxide present on AlN powders. Using a solid form of the precursor allowed precise separation between the source and the seed and provided a more stable evaporation rate. The source was placed in the isothermal section of the furnace to ensure a constant evaporation rate. The source-to-seed distance was very important in the achievement of crystal growth, as explained below. The sublimation rate varied from approximately 10 mg/hr to 200 mg/hr as the temperature of the source was raised from 1950 to 2250°C at a background pressure of 500 torr.

The choice of seed material was very limited due to the high temperatures needed to both sublime AlN and to achieve rapid surface diffusion on the seed such as to produce bulk single crystal over relatively short growth times. Single crystal, 6H-SiC wafers (10 mm x 10 mm) were used as seeds in all experiments, because of their high temperature stability and close lattice matching to AlN (0.9% mismatch).

All experiments were performed under flowing ultra-high purity N_2 maintained at a background pressure of 500 torr. Two source temperature regimes were investigated: 1950-2050°C and 2050-2250°C. The lower temperature range was chosen primarily because of the degradation of the furnace and seed crystals. In both cases a temperature difference of 80-150°C was obtained between the source and the seed depending on the distance of separation (1 to 40 mm).

The as grown AlN bulk crystals were first characterized via optical and transmission electron microscopes. Color, size, transparency and macrostructural features were observed using the former at magnifications between 5X-40X. C-axis oriented TEM samples were prepared by grinding and polishing the bulk AlN crystals to a thickness of 100 μm, dimpling to a thickness of 20 μm and Ar^+-ion milling to electron transparency. The TEM studies were performed using a TOPCON EM-002B, operating at 200 kV. Micro-Raman spectra were obtained at RT from a back scattering geometry utilizing the 514.5 nm line of the Ar laser. The spot size and the spectral resolution were 4 μm in diameter and 2 cm^{-1} respectively.

Sublimation Growth and Characterization of Single Crystals of GaN

Initial experiments on the growth of GaN were conducted in a modified AlN sublimation system. High purity GaN powder produced in our laboratory [23] and compacted to a density of $\approx$ 60% was sublimed or evaporated in an ammonia atmosphere and condensed on a substrate. This powder possessed exceptional structural quality such as to be classified as the X-ray standard for this material [24]. The experimental parameters including source and seed temperatures, pressure, ammonia flow rate and source-to-seed distance were varied to optimize the growth conditions. A detailed description of the experimental procedures will be presented in a forthcoming publication.

The GaN crystals were characterized using optical microscopy, Raman spectroscopy and photoluminescence (PL). Raman spectra of the GaN crystals were acquired in a back scattering geometry from the face perpendicular to the c axis using same experimental conditions as in AlN. The incoming polarization state of the light was chosen to be perpendicular to the c axis. The PL studies employed an Ar-ion pumped, mode-locked, femtosecond (fs), titanium:sapphire laser with a frequency tripler as the excitation source. The duration of the excitation pulse was $\approx$250 fs. The spectra were acquired, collimated and focused onto the entrance slits of a 0.32 m spectrometer. Each sample emission was detected using a cooled GaAs photomultiplier.

RESULTS AND DISCUSSION

Aluminum Nitride

The highest growth rates and thickest crystals ($\leq$ 1 mm) were obtained with a source temperature of $\approx$2150°C. Although deposition was observed above 2150°C, the SiC seeds were sufficiently chemically unstable such that only individual crystals nucleated and grew from selected areas. The color of the crystals grown within the temperature interval of 2050-2250°C varied from dark blue to dark green. This is believed to be due to carbon and silicon incorporation, respectively, from the SiC substrate. A growth rate of 0.5mm/hr was achieved at a 4 mm separation and 2150°C source and 2080°C seed temperatures. The resulting bulk crystals were analyzed by XRD and Laue back reflection techniques and shown to be monocrystalline.

Growth of AlN in the source temperature range of 1950 to 2050°C was investigated both to prevent seed deterioration and increase crucible lifetime. Complete structural and chemical stability of the SiC seeds and substantial reduction in the deterioration of the SiC coated crucibles were attained. Crystals grown in this temperature range were always colorless. Typical growth rates were reduced to 30-50 µm/hr. Thus growth runs of 10-15 hrs yielded crystals having thicknesses of 0.3-0.5mm . An optical micrograph (OM) of a 0.4 mm thick transparent layer of AlN grown at 1975°C is shown in Fig. 1.

A bright field TEM micrograph taken in plan view along [0001] and associated selected area diffraction (SAD) pattern are shown in Fig.2. Uniform contrast density was observed throughout the specimen, indicating excellent quality single-crystalline AlN. No high angle boundaries, stacking faults or twinned regions were observed. The SAD pattern supports these results in that the diffraction spots are without arcs, which would indicate texture and/or grains with different orientation, and spikes, which would indicate stacking faults. Isolated cases of dislocations and low angle grain boundaries were observed at very large separations.

The Raman spectrum of the AlN bulk crystal is presented in Fig.3. The spectrum was acquired in a back scattering geometry from the c face. The selection rules predict the allowed modes for this geometry to be the $A_1(LO) \approx 893$ cm^{-1}, $E_2^{(1)} \approx 250$ cm^{-1}, $E_2^{(2)} \approx 660$ cm^{-1}. The spectrum exhibits the allowed modes with no detectable contribution from the forbidden modes [25]. Thus the AlN crystal has a well aligned c face of the WZ structure with no significant concentration of internal structural defects (which might relax the selection rules). The inset to the figure shows a high resolution spectrum of the $E_2^{(2)}$ mode.

Cracking and separation of the AlN crystal from the SiC seeds was commonly observed after cooling from within both temperature ranges. This is believed to be due primarily to the mismatch in the coefficients of thermal expansion; however the data for these materials are not available for the temperature range employed in this research. However, since AlN boules will ultimately be grown homoepitaxially on AlN seeds, such problems as cracking caused by foreign substrates will not be permanent.

Gallium Nitride

Colorless, hexagonal crystals were obtained by sublimation of the GaN powder. Figure 4 shows an OM of ≈1 mm long, well faceted, transparent GaN crystals with low aspect ratios which in contrast to the commonly reported needle-shaped crystals grown via vapor phase reaction.

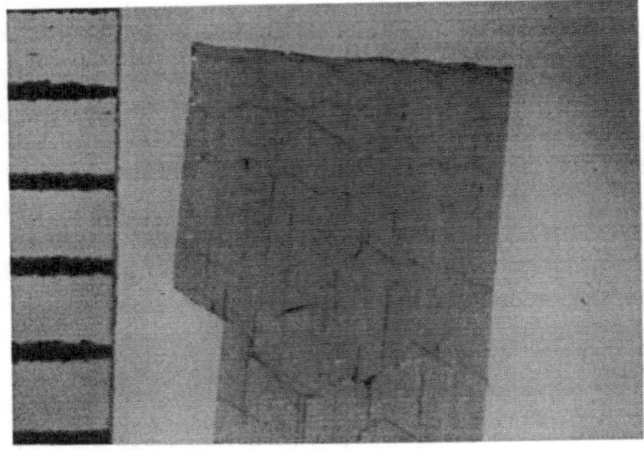

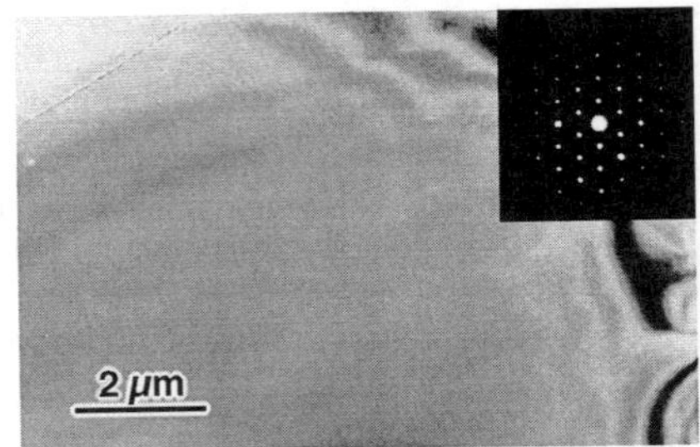

Figure 1: OM image of single crystal AlN grown at 1975°C source temperature. Space between ruled lines = 1 mm.

Figure 2: TEM image and selected area diffraction pattern of bulk AlN.

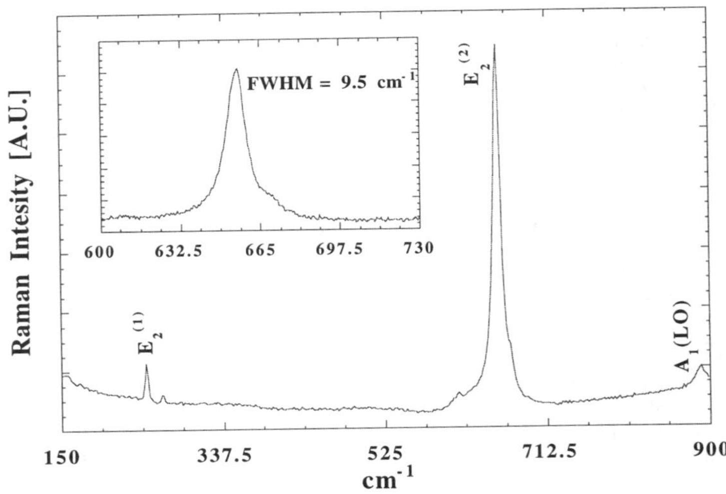

Figure 3: Raman spectrum of bulk AlN

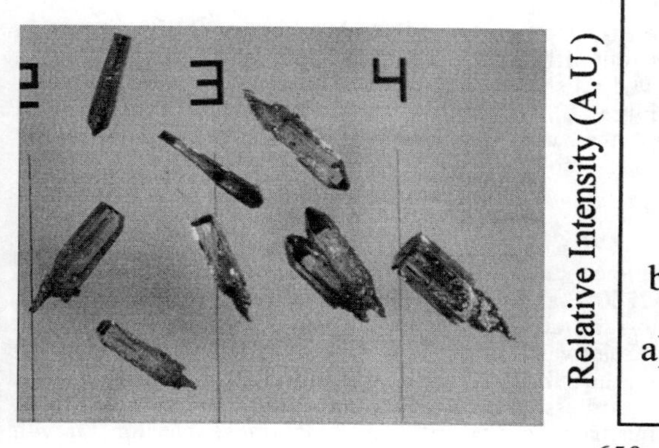

Figure 4: GaN crystals. Lines
are spaced 1 mm.

Figure 5: Photoluminescence spectrum
of bulk GaN

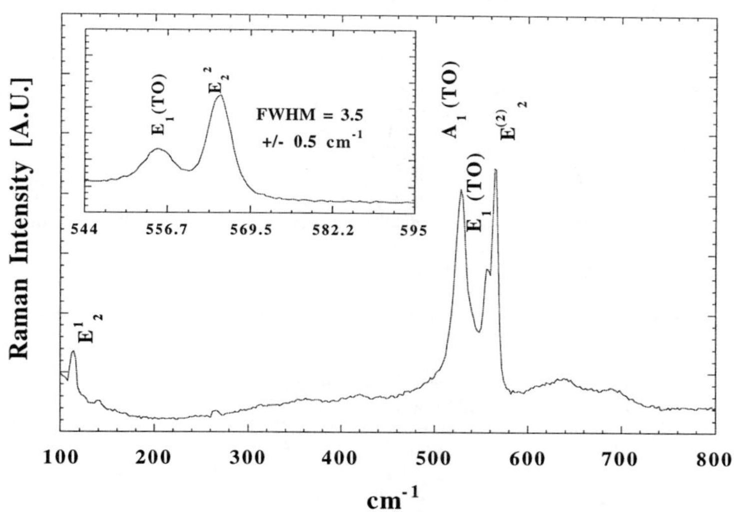

Figure 6: Raman spectrum of GaN

The direction of fastest growth and thus the crystal shape were controlled by changing the
Ga/NH$_3$ flux ratio and the growth temperature. This is a significance advance in that all previous
reports indicate that growth of bulk GaN from the vapor phase results primarily in long needles.

The allowed modes of the WZ structure are presented in Fig. 6. As in the case of AlN, the conservation of selection rules here is indicative of excellent crystallinity. The inset to figure shows that the $E_2^{(2)}$ mode is at 567 cm^{-1} and has FWHM $\approx$ 3.5 cm^{-1}. These values are indicative of a material of the highest quality reported to date.

A representative room temperature PL spectrum for bulk GaN taken at 300K is shown in Fig.5. Strong band edge related emission with a peak position at 365.0 nm (3.4 eV) and a FWHM of 9.0 nm (83meV) was observed, as shown in Fig.5 (a). The visible portion of the PL spectrum is expanded 500 times in Fig.5 (b). No yellow emission attributed to deep level transitions was detected on this scale or to the naked eye. The position of this peak obtained at 77 K was at 358.0 nm with a FWHM of 6.0 (53 meV).

CONCLUSIONS

The first seeded growth of bulk single crystals of AlN (0001) has been achieved on 6H-SiC (0001) substrates in the range of 1950-2250°C at 500 torr of N_2. The volume of these crystals is the largest ever reported. Close source-to-substrate distances and a 80-150°C temperature gradient were necessary to achieve these results. Color variations were observed above 2150°C and linked to the incorporation of Si and/or C from the substrate and the growth crucible. Cracks in these crystals were observed after cooling regardless of the growth temperature and believed to be caused primarily by stresses due to the mismatch in the coefficients of thermal expansion. Growth of bulk GaN crystals up to 1mm size was achieved via sublimation. Raman and PL spectrums of these crystals revealed excellent crystalline quality. No yellow emission was observed using the latter technique.

ACKNOWLEDGMENTS

The authors acknowledge support of the Office of Naval Research under contract N00014-92-J-1477 and Dr. W. Rafaniello and the Dow Chemical Company for the hot-pressed AlN.

REFERENCES

1. G. A. Slack and T. F. McNelly, J. Cryst. Growth., **34**, 263 (1976).
2. G. A. Slack and T. F. McNelly, J. Cryst. Growth. **42**,560 (1977).
3. W. F. Knippenberg and G. Verspui, US Pat, 3.634.149.
4. J. O. Huml, G. S. Layne, US. Pat., 3.607.014.
5. J. O. Huml, G. S. Layne, US. Pat., 3.598.526.
6. P. M. Dryburgh, The Ninth International Conference on Crystal Growth, **ICCG-9** (1989).
7. R. B. Campbell and H. C. Chang, Technical Report AFAPL-TR-67-23, Westinghouse Electrical Corporation, Pittsburgh, Pennsylvania (1967).
8. J. Pastrñák and L. Roskovcová, Phys. Stat. Sol., **7**, 331, (1964).
9. C. O. Dugger, Mat. Res. Bull., **9**, 331, (1974).
10. C. O. Dugger, Air Force Report, AFORL-TR-75-0486, Hanscom AFB Massachusetts, (1975)
11. D. Elwell and M. M. Elwell Prog. Crystal Growth and Charact., **17**, 53, (1988)
12. E. Ejder, J. Cryst. Growth, **22**, 44, (1974)
13. M. Gershenzon, ONR Rept. 243-004F, May (1981).
14. I. G. Pichugin and D. A. Yaskov, Neorg. Mat., **6**, 1973, (1970).
15. D. Elwell, R. S. Fiegelson, M. M. Simkins and W. A. Tiller, J. Cryst. Growth, **66**, 45, (1984).
16. C. J. Frosh, J. Phys. Chem., **66**, 877 (1962).
17. R. A. Logan and C. D. Thurmond, J. Electrochem. Soc., **119**, 1727, (1972).
18. T. Ogino and M. Aoki, Oyo Butsuri, **48**, 269, (1962).
19. J. Karpinski, J. Jun and S. Porowski, J. Cryst. Growth, **66**, 11, (1984).
20. J. Karpinski, and S. Porowski, J. Cryst. Growth, **66**, 1, (1984).
21. S. Porowski, et.al., Inst. Phys. Conf. Ser. no:137, **369**, (1994).
22. S. Sakai, S. Kurai, T. Abe and Y. Naoi, Jpn. J. Apl. Phys., **35**, L77, (1996)
23. C. M. Balkaş and R. F. Davis, J.Am. Ceram. Soc., **79**, [9], 2309, (1996).
24. C. M. Balkaş C. Basçeri and R. F. Davis, J. Powder. Difftraction, **10**, [4], 266, (1995).
25. L. Bergman and R. J. Nemanich, Annu. Rev. Mater. Sci., **26,** 551, (1996).

SYNTHESIS OF BULK, POLYCRYSTALLINE GALLIUM NITRIDE AT LOW PRESSURES

Alberto Argoitia[*], John C. Angus[*], Cliff C. Hayman[*], Long Wang[**],
Jeffrey S. Dyck[***] and Kathleen Kash[***]
[*]Chemical Engineering Dept., [**]Materials Science and Engineering Dept., [***]Physics Dept.
Case Western Reserve University, Cleveland, OH 44106

ABSTRACT

Bulk, polycrystalline gallium nitride was crystallized from gallium saturated with nitrogen obtained from a microwave electron cyclotron resonance source. The polycrystalline samples are wurtzitic and n-type. Well-faceted crystals give near-band-edge and yellow band photo-luminescence at both 10K and 300K. The results show that atomic nitrogen is an attractive alternative to high pressure molecular nitrogen for saturation of gallium with nitrogen for synthesis of bulk gallium nitride.

INTRODUCTION

The equilibrium pressures of N_2 over solid GaN in equilibrium with nitrogen saturated gallium is very high: approximately 1500 bar at 1500K and 25,000 bar at 2000K [1-3]. Efforts at bulk synthesis of GaN have primarily been at high pressures, and successful synthesis has been achieved [2]. However, because of the high pressures, the process is difficult to implement in most laboratories and, up to now, crystals are relatively small. Very recent studies have shown that the extremely high equilibrium pressures can be avoided by use of atomic nitrogen, N, rather than molecular nitrogen, N_2, to saturate gallium with nitrogen [4].

In this paper we describe the crystallization of bulk, polycrystalline gallium nitride from gallium saturated with nitrogen obtained from a microwave electron cyclotron resonance (ECR) source.

THERMODYNAMICS OF GALLIUM NITRIDE SYNTHESIS

The plasma is used to increase the thermodynamic activity of the nitrogen in order to raise the nitrogen concentration in the gallium to the solubility limit of GaN. The active species in the plasma include, N, N^+, N_2^+, and excited states of N_2. Each of these is sufficiently energetic to form GaN [5], but for simplicity we consider only atomic nitrogen, N. The plasma produces a sufficient partial pressure of N so the following overall reaction is thermodynamically favored.

$$Ga_{(l)} + N_{(gas)} = GaN_{(s)} \tag{1}$$

In the high pressure process, the pressure of N_2 is increased to the point where the following overall reaction is favored.

$$Ga_{(l)} + \frac{1}{2}N_{2(gas)} = GaN_{(s)} \tag{2}$$

Equilibrium thermodynamic calculations show the difference in required pressures is very large. For example, at 1500K, the required N_2 pressure for equilibrium of reaction (2) is approximately 1500 bar; for reaction (1) the equilibrium pressure of N is approximately 2×10^{-11} bar. This disparity arises from the extreme stability of N_2. Recombination of N to form N_2 is strongly

Mat. Res. Soc. Symp. Proc. Vol. 449 © 1997 Materials Research Society

favored thermodynamically; however, our results indicate that recombination is sufficiently slow within the gallium melt to permit the parallel formation of GaN from N by reaction (1).

EXPERIMENT

Synthesis

Synthesis of GaN was achieved by directing plasmas from an electron cyclotron resonance microwave source (Wavemat model MPDR 610) onto a pool of gallium. The gallium was held in a pyrolytic boron nitride crucible heated from below by a resistance heater. The ECR source was mounted directly above the crucible and gave an ion flux density of approximately 10^{16} $cm^{-2}s^{-1}$. The partial pressure of atomic nitrogen in the beam is approximately 0.05 mtorr.

A run was started by evacuating the reaction chamber to 10^{-9} torr. The gallium was then heated at 600C and 10^{-7} torr for 45 minutes to desorb dissolved gas. Next an argon plasma was employed for 10 minutes followed by a hydrogen plasma for 30 minutes to remove gallium oxide and other impurities from the gallium surface. The hydrogen flow to the ECR source was replaced by 10 sccm of nitrogen and the temperature raised slowly to approximately 1000C. During this step, the pressure was fixed at 0.5 mtorr by controlling the nitrogen flow rate. After about 15 minutes, at a temperature of approximately 700C, the reflectivity of the gallium surface changed. The surface became rougher and lost the specular appearance of metallic gallium, indicating formation of a crust of polycrystalline gallium nitride. The nitrogen plasma was maintained for 12 hours at the final temperature of 1000C. After removal from the chamber, the sample was etched with a mixture of hydrofluoric and nitric acids to remove the excess gallium.

At the end of a run, a polycrystalline GaN "dome" completely covered the liquid gallium. A typical "dome" was approximately 0.1 mm thick, weighed 40 mg, and had an upper surface area of 70 mm^2. The average linear growth rate was approximately 8 μm h^{-1}. However, the growth rate was significantly greater before the gallium was totally covered with solid GaN.

Characterization

Electron and X-ray diffraction confirmed the samples were wurtzitic GaN. A small region of cubic (zinc blende) GaN was observed by electron diffraction in one sample. Energy dispersive X-ray analysis showed a Ga/N ratio of 1 within the instrumental error of $\pm$ 5%.

Scanning electron micrographs of the polycrystalline product are shown in Figs. 1 through 4. A cross-section of one of the domes is shown in Fig. 1. Thin platelets of GaN aligned normal to the surface of the dome are evident. X-ray diffraction showed the crystals had a strong $[11\bar{2}0]$ texture in this region of the sample. This indicates that the fastest growth direction under these conditions is $[11\bar{2}0]$; however, in some regions the basal planes were parallel to the surface rather than normal to it. In the high pressure process, $[10\bar{1}0]$ is reported to be the fastest growth direction [6]. A detailed view of the oriented crystals is shown in Fig. 3. A dendritic morphology, typical of uncontrolled freezing from a supersaturated solution, is seen in Fig. 4.

A transmission electron micrograph is shown in Fig. 5. The insert is the electron diffraction pattern of a small hexagonal crystal similar to the one in the center of Fig. 5, tilted to the $[11\bar{2}0]$

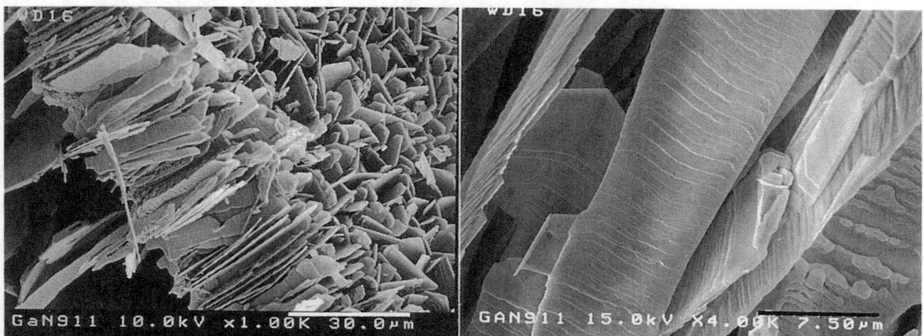

Figure 1. Scanning electron micrograph of cross-section of polycrystalline gallium nitride dome.

Figure 2. Detailed view of a portion of Figure 1.

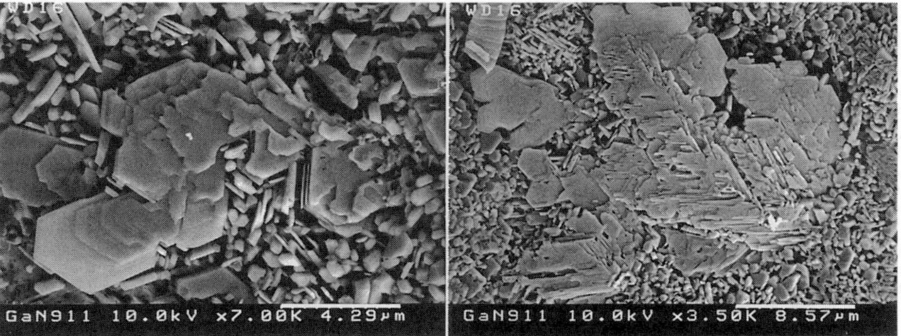

Figure 3. Scanning electron micrograph showing growth of hexagonal platelets of gallium nitride.

Figure 4. Scanning electron micrograph showing dendritic growth of gallium nitride.

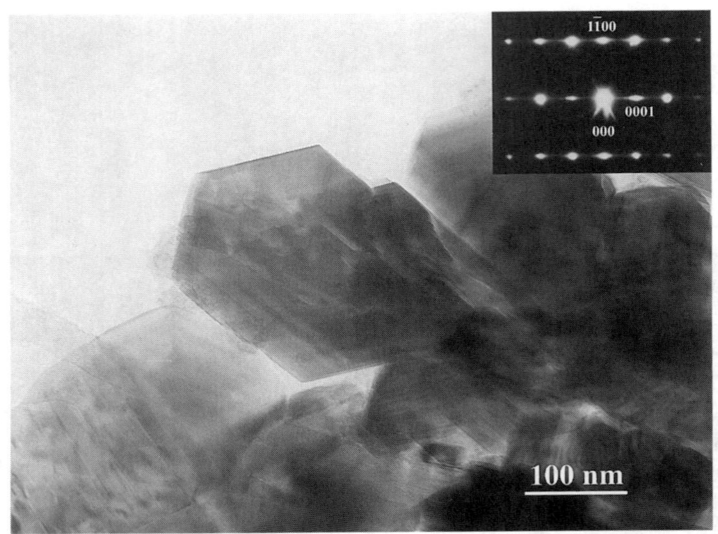

Figure 5. Transmission electron micrograph. The inset is the electron diffraction pattern of a small hexagonal crystal similar to the one in the center tilted to $[11\bar{2}0]$ zone axis.

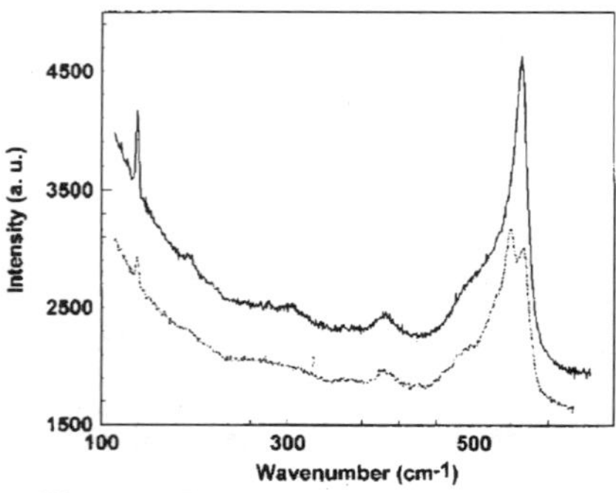

Figure 6. Raman spectra from two regions of polycrystalline gallium nitride dome.

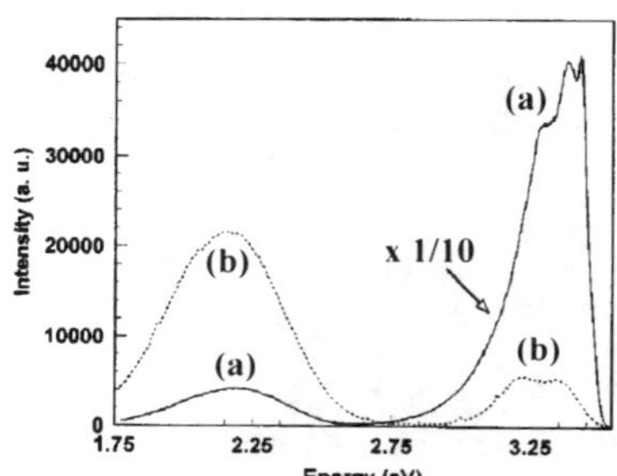

Figure 7. Photoluminescence spectra taken at (a) 10 K and (b) 300 K. Note the factor of 10 difference in scale factor for the two spectra.

zone axis. Streaks along the [0001] direction indicate the presence of planar defects. Diffraction from the $(1\bar{1}00)$, (0001), and $(10\bar{1}0)$ planes matched well with w-GaN. Dark bands in other micrographs, not shown, indicated the presence of planar defects.

Representative Raman spectra from two regions of a sample are shown in Fig. 6. Depolarized Raman spectra were taken in backscattering geometry, with spot size of approximately one micron. The low E_2 line of wurtzitic GaN at 144 cm^{-1} is clearly visible in both spectra. The lines in the 540 cm^{-1} to 560 cm^{-1} range have not been specifically assigned, but are consistent with several modes for wurtzitic GaN, *i.e.*, high E_2 (568 cm^{-1}), A_{1T} (531 cm^{-1}), and E_{1T} (560 cm^{-1}), or possibly from the TO (554 cm^{-1}) mode in the zinc-blende phase [7].

Photoluminescence spectra, taken at 10K and 300K, of the convex surface of a GaN dome are shown in Fig. 7. The excitation source was the 325 nm line of a HeCd laser, with 0.5 mW power focused to a 50 μm diameter spot. The detector was a photon counting photomultiplier tube with a 0.8 meter double monochromever, with spectral resolution of approximately 4 meV at a photon energy of 3 eV. The sample was mounted with colloidal graphite on the copper block tail of a liquid-helium-cooled cold finger dewar.

Multipeak near-band-edge luminescence and the broad "yellow band" luminescence, peaking near 2.2 eV, from well-faceted crystals are shown in Fig. 7. The shoulder in the luminescence spectrum at approximately 3.47 eV may be the A exciton or the exciton bound to a neutral donor [8]. A peak at 3.42 eV is not yet firmly identified. Peaks at 3.38 eV and 3.28 eV are in the energy range for impurity bound excitons and donor-acceptor pair recombination, as well as for the first and second LO phonon replicas of the free or neutral donor bound exciton [8]. Between 10K and 300K, the near-band-edge luminescence decreased by a factor of about 75, and the yellow band by a factor of 2. The integrated intensity decreased by a factor of about 7.5 over the same range, a rather weak temperature dependence in the presence of grain boundaries in such high density. However, samples with poorly defined faceting showed little or no luminescence.

The Seebeck coefficient was measured by by placing samples of the bulk, polycrystalline GaN on a water-cooled metal block and contacting them with a thin wire held at room temperature. The samples were n-type with a Seebeck coefficient of greater than 20 μV°C^{-1}. Since it is believed that the n-type conductivity arises from N vacancies, this result is consistent with GaN frozen from a gallium rich solution [9].

CONCLUSIONS AND DISCUSSION

Bulk, polycrystalline GaN was crystallized from gallium saturated with nitrogen derived from a microwave ECR source. The GaN is wurtzitic, n-type, and shows photoluminescence at both 10K and 300K. The use of atomic nitrogen circumvents the high equilibrium pressures required for the synthesis of bulk GaN from molecular nitrogen and gallium.

In the experiments reported here, the crystallization of the gallium nitride was rapid and essentially uncontrolled, leading to small crystals and, in some regions, to dendritic growth. We believe this occurred because the temperature was the coolest, and therefore the GaN solubility the lowest, at the top of the melt where the nitrogen concentration was the highest. Growth of larger bulk single crystals will require control of the temperature and composition gradients so local supersaturation in the bulk of the melt is avoided.

Growth of large GaN crystals by the method described here will have some of the limitations pointed out by Grzegory *et al.* for crystallization at high pressures. For example, the low solubility of GaN in Ga limits the concentration differences that can be employed and hence will limit growth rates. Grzegory *et al.* report growth rates of approximately 100 μm h^{-1} for the high pressure method. The average growth rate for the experiments reported here is approximately 8 μm h^{-1}, but this is limited by the formation of the solid dome.

ACKNOWLEDGMENTS

The support of a Materials Research Group grant from the National Science Foundation is gratefully acknowledged.

REFERENCES

1. J. Karpinski and S. Porowski, J. Crystal Growth **66**, p.11 (1984).

2. S. Porowski, J. Jun, P. Perlin, I. Grzegory, H. Teisseyre and T. Suski, Inst. Phys. Conf. Series, No. 137, Chapter 4, (5th Conf. on SiC and Related Materials, Washington, DC, 1993) p. 369.

3. P. Perlin, I. Gorszyca, N. E. Christensen, I. Grzegory, H. Teisseyre, and T. Suski, Phys. Rev. B **45,** p. 13,307 (1992).

4. A. Argoitia, C.C. Hayman, J.C. Angus, L. Wang, J.S. Dyck and K. Kash, to appear, Appl. Phys. Lett., January, 1997.

5. N. Newman in <u>III-V Nitride Materials and Processes</u>, edited by T.D. Moustakas, J.P. Dismukes, and S.J. Pearton, Proceedings Volume 96-11, (Electrochemical Society, Pennington, NJ, 1996) p. 1-19

6. I. Grzegory, M. Bockowski, B. Lucznik, S. Krukowski, M. Wroblewski and S. Porowski, MRS Internet Journal, Semiconductor Research, **1**, article 20, October, 1996.

7. L.E. McNiel, <u>Properties of Group III Nitrides</u>, edited by J.H. Edgar, EMIS Datareviews Series, No. 11, INSPEC, (Institution of Electrical Engineers, London, UK, 1994) p. 252-253.

8. I. Akasaki and H. Amano, <u>Properties of Group III Nitrides</u>, edited by J.H. Edgar, EMIS Datareviews Series, No. 11, INSPEC, (Institution of Electrical Engineers, London, UK, 1994) p. 222-230.

9. H.J. Scheel and C. Klemenz in <u>III-V Nitride Materials and Processes</u>, edited by T.D. Moustakas, J.P. Dismukes, and S.J. Pearton, Proceedings Volume 96-11, (Electrochemical Society, Pennington, NJ, 1996) p. 20-36

NITROGEN PLASMA PRETREATMENT OF SAPPHIRE SUBSTRATES FOR THE GaN BUFFER GROWTH BY REMOTE PLASMA ENHANCED MOCVD

Min Hong Kim*, Cheolsoo Sone*, Jae Hyung Yi*, Soun Ok Heur**, Euijoon Yoon*
* School of Materials Science and Engineering, Seoul National University, Seoul 151-742, Korea
** Inter-university Semiconductor Research Center, Seoul National University, Seoul 151-742, Korea

ABSTRACT

Low-temperature GaN buffer layers with smooth surfaces and high crystallinity could be prepared by a remote plasma enhanced metalorganic vapor deposition after the pretreatment of substrates with rf nitrogen plasma. Smooth AlN thin layer was formed on the (0001) sapphire substrate by the nitrogen plasma pretreatment for an hour. The AlN layer provided the nucleation sites for the subsequent buffer layer growth, thus highly preferred (0001) GaN buffer layers could be grown on the pretreated substrate. Formation of the AlN layer on sapphire and the surface smoothness were affected by pretreatment parameters such as exposure time, temperature, and rf power.

INTRODUCTION

III-V nitrides are most prospective materials for the optoelectric applications ranging from UV to visible region, judging from the recent success of the GaN-based laser diode[1]. Still, the major problem of GaN growth is the lack of a suitable sustrate. GaN has been successfully grown on (0001) sapphire substrates in spite of its large lattice mismatch with GaN (-13%). The successful GaN growth on sapphire comes from the 2-step growth using low-temperature buffer layers [2,3]. The initial nitridation of sapphire substrates is commonly practiced prior to the buffer growth by electron cyclotron resonance molecular beam epitaxy (ECR-MBE) [4-6] and by metalorganic chemical vapor deposition (MOCVD) [7,8], and the formation of AlN was reported after the nitridation [4-7]. Effects of the nitridation on the growth and optical properties of GaN were reported [7-9].

In this paper, we report the AlN formation by RF nitrogen plasma pretreatment of sapphire substrates. Effects of the nitrogen plasma pretreatment on the subsequent low temperature GaN buffer growth by remote plasma enhanced MOCVD are also reported.

EXPERIMENT

The MOCVD system with a vertical rf plasma column was used for the pretreatment of sapphire substrate surface and for the growth of low temperature GaN buffer layers. (0001) sapphire substrates were ultrasonically cleaned in trichloroethane, acetone, and methanol, and rinsed with deionized water. They were etched in a 3:1 H_2SO_4:H_3PO_4

Mat. Res. Soc. Symp. Proc. Vol. 449 © 1997 Materials Research Society

solution at 160℃ for 10 min. After the chamber evacuation to less than 1x10^{-7} Torr, the substrates were heated to 700℃ in a hydrogen ambient. The substrates were pretreated by hydrogen plasma at 700℃, 200 mTorr, rf power of 100W for half an hour in order to *in situ* clean the substrates. After the hydrogen plasma surface cleaning, the substrates were pretreated by a nitrogen plasma at various temperatures and rf powers. The low temperature GaN buffer layers were grown on the pretreated substrates at 500 mTorr and 500℃. Thicknesses of the GaN buffer layers were kept constant at 10 nm.

The pretreated substrates and GaN buffer layers were analyzed by *in situ* reflection high energy electron diffraction (RHEED). Compositional change in the pretreated substrates was analyzed by x-ray photoelectron spectroscopy (XPS). Surface morphology and roughness were measured by atomic force microscopy (AFM). The structural properties of GaN buffer layers were analyzed by x-ray diffraction (XRD) and RHEED.

RESULTS

RHEED patterns of the substrates pretreated at various conditions are shown in Fig. 1. Substrate temperature was fixed at 700℃. The azimuthal direction of the electron beam was $[11\overline{2}0]$ of the sapphire. After the hydrogen plasma treatment for half an hour, the RHEED pattern of the sapphire became streakier, as shown in Fig. 1 (a). Additional nitrogen plasma treatment was made at to nitridate the sapphire at 60W, however, no significant change was observed by RHEED, as shown in Fig. 1 (b). The RF power was further increased to 120W, and the streaky RHEED pattern of the AlN was observed after the pretreatment fo an hour (Fig. 1 (d)).

Compositional change of the surface after the nitrogen plasma treatment was analyzed by XPS using Mg K$_\alpha$ radiation. No Ga peaks were observed. N 1s peaks in XPS spectra are shown in Fig. 2. The O 1s peak was positioned at 532 eV for the calibration of binding energy shift due to sample charging. The intensity of the surface pretreated by the nitrogen plasma was much stronger than that of the surface pretreated by the hydrogen plasma. The N peak from the surface modified by hydrogen plasma resulted from the adsorption of nitrogen in air. The N 1s peak from the substrate

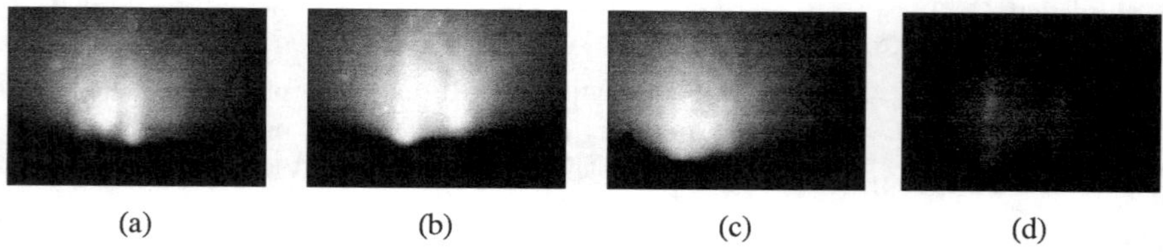

(a)   (b)   (c)   (d)

Fig. 1. RHEED patterns of the substrates pretreated by (a) hydrogen plasma at 100W for half of an hour, (b) hydrogen plasma at 100W for half an hour and nitrogen plasma at 60W for an hour, (c) hydrogen plasma at 100W for half an hour and nitrogen plasma at 120W for 20 min. and (d) hydrogen plasma at 100W for half an hour and nitrogen plasma at 120W for an hour.

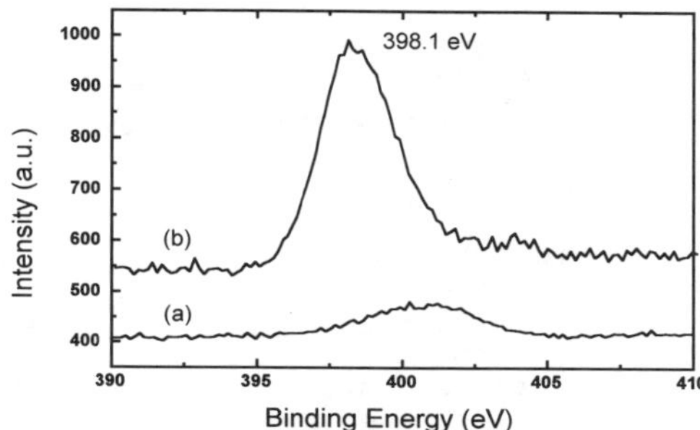

Fig. 2. XPS N 1s peaks after the plasma treatments. (a) hydrogen plasma at 100W for half an hour at 700℃, and (b) nitrogen plasma at 120W for an hour at 700℃.

pretreated by 120W nitrogen plasma shifted to a lower value of 398.1 eV. The peak shift to a lower value implies that the AlN layer was formed. The N 1s energy of AlN is reported 397.3 eV[10].

GaN buffer layers were grown on the pretreated substrates to investigate the effects of the surface pretreatment. The GaN buffer layers were not annealed at a higher temperature. The growth rate was about 10 nm/hr. RHEED patterns of the GaN buffer layers grown after various plasma pretreatment conditions are shown in Fig. 3. The electron beam was directed along the GaN $[12\bar{3}0]$, which is parallel to the sapphire $[11\bar{2}0]$. RHEED patterns of the GaN buffer layers on sapphire substrates pretreated by the hydrogen plasma alone and by a low power nitrogen plasma were mixtures of

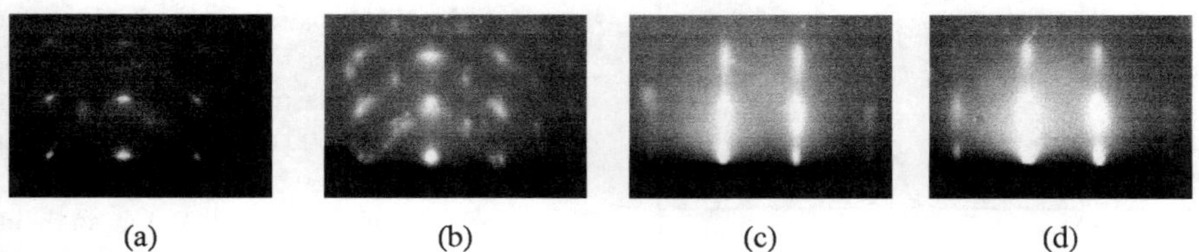

(a)   (b)   (c)   (d)

Fig. 3. RHEED patterns of the GaN buffer layers on the substrates pretreated by (a) H_2 plasma at 100W for half an hour at 700℃, (b) H_2 plasma at 100W for half an hour at 700℃ and N_2 plasma at 60W for an hour at 700℃, (c) H_2 plasma at 100W for half an hour at 700℃ and N_2 plasma at 120W for an hour at 700℃ and (d) H_2 plasma at 100W for half an hour at 700℃ and N_2 plasma at 120W for an hour at 500℃.

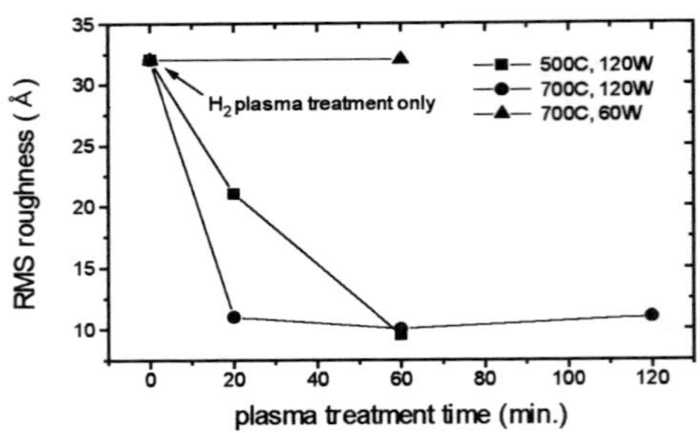

Fig. 4. The variation of rms roughnesses of the GaN buffer layers with pretreatment time.

rings and spots (Fig. 3 (a), (b)). On the other hand, smooth surface and good crystallinity were observed from the RHEED patterns of GaN buffer layers on the substrates pretreated by a high power (120W) nitrogen plasma after the hydrogen plasma treatment for half an hour at 100 W, 700 ℃(Fig. 3 (c), (d)). The AlN thin layer formed by exposure to the high power N_2 plasma could improve the nucleation of the GaN buffer layers. Presumably, a high density of nucleation sites promoted the 2-dimensional growth, therefore the highly textured GaN layers with smooth surfaces could be obtained in spite of the low growth temperature.

The surface roughness was characterized by AFM as shown in Fig. 4. The root-mean-squre (rms) roughnesses of the GaN buffer layers grown on the pretreated substrates decreased with the substrate pretreatment time at a high power (120W). The

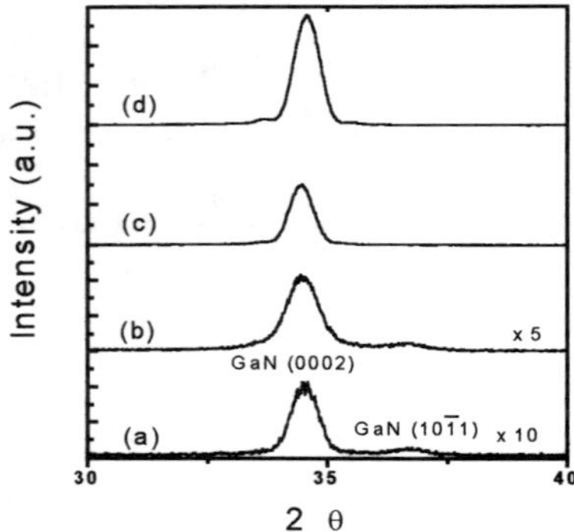

Fig. 5. XRD patterns of GaN buffer layers on the substrates pretreated by (a) H_2 plasma at 100W for half an hour at 700℃, (b) H_2 plasma at 100W for half an hour at 700℃ and N_2 plasma at 60W for an hour at 700℃, (c) H_2 plasma at 100W for half an hour at 700℃ and N_2 plasma at 120W for an hour and (d) H_2 plasma at 100W for half an hour at 700℃ and N_2 plasma at 120W for an hour at 500℃.

rms roughness of pretreatment at 700℃ was smaller than that of the pretreatment at 500 ℃ even after short pretreatments. The rms roughness at 60W did not change appreciably even after an hour pretreatment, and it is believed that the surface pretreatment is more effective at high plasma powers. The apparent grain size of GaN on the sapphire pretreated at 700℃ was larger than that of GaN when pretreated at 500 ℃ in AFM surface images. This implies that the nuclei formed on the surface pretreated by nitrogen plasma is larger at 700℃ due to the increase in surface diffusion.

The pretreatment of the sapphire substrates by nitrogen plasmas could improve the growth of the GaN buffer layers, resulting in the improved crystallinity. Fig. 5 shows the XRD patterns of the GaN buffer layers on the sapphire pretreated at various conditions. The GaN layers after pretreatment by the hydrogen plasma had both (0002) and (1011) peaks, as shown in Fig. 5 (a). However, highly (0002) textured GaN layers were obtained after pretreatment by the nitrogen plasma at 120W for an hour, as shown in Fig. 5 (c) and (d).

CONCLUSIONS

Highly (0002) textured GaN buffer layers were successfully grown on the sapphire substrates pretreated with nitrogen plasmas in conjunction with hydrogen plasma in a remote plasma enhanced MOCVD system. The smooth AlN layer was formed on the sapphire by the nitrogen plasma at 500 ~ 700 ℃ for an hour. The formation AlN by nitrogen plasma pretreatment resulted in better crystallinity of the GaN buffer layers. The low-temperature GaN buffer layers grown on nitrogen plasma treated sapphire substrates can be applied to the successful growth of III-V nitrides.

ACKNOWLEDGMENT

This work was supported by Ministry of Education through 1996 research fund for advanced materials and by Korean Science and Engineering Foundation through RETCAM in Seoul National University. Experimental support for XPS analysis from Mr. K. H. Cho and Ms. E. K. Lee in the Inter-University Center of Natural Science Research Facilities, SNU, is greatly acknowledged.

REFERENCES

1. S. Nakamura, M. Senoh, S. Nagahama, N. Iwasa, T. Yamada, T. Matsushita, H. Kiyoku, Y. Sugimoto, Appl. Phys. Lett. **68**, 2105 (1996)
2. I. Akasaki, H. Amano, Y. Koide, K. Hiramatsu, N. Sawaki, J. Cryst. Growth **98**, 209 (1989)
3. S. Nakamura, Jpn. J. Appl. Phys. **30**, L1705 (1991)
4. T. D. Moustakas and R. J. Molnar, Mat. Res. Soc. Symp. Proc. Vol. **281**, 753 (1993)
5. T. D. Moustakas, T. Lei and R. J. Molnar, Physica B **185**, 36 (1993)

6. M. E. Lin, B. N. Sverdlov, and H. Morkoç, J. Appl. Phys. **74**, 5038 (1993)
7. K. Uchida, A. Watanabe, F. Yano, M. Kouguchi, T. Tanaka, S. Minagawa, J. Appl. Phys. **79**, 3487 (1996)
8. C.-Y. Hwang, M. J. Schurman, W. E. Mayo, Y. Li, Y. Lu, H. Liu, T. Salagaj and R. A. Stall, J. Vac. Sci. Technol. A **13**, 672 (1995)
9. N. Grandjean, J. Massies and M. Leroux, Appl. Phys. Lett. **69**, 2071 (1996)
10. Z. Sitar, L. L. Smith, R. F. Davis, J. Cryst. Growth. **141**. 11 (1994)

EFFECT OF SAPPHIRE NITRIDATION ON GaN BY MOCVD

DONGJIN BYUN[1], JAESIK JEONG[1,2], JAE-INN LEE[1,3], BYONGHO KIM[2], JI-BEOM YOO[3]
AND DONG-WHA KUM[1]

[1] Division of Metals, Korea Institute of Science and Technology
P.O. Box 131, Cheongryang, Seoul 130-650, Korea
[2] Dept. of Materials Science and Engineering, Korea University
1 Anam-Dong 5-Ka, Sungbook-Ku, Seoul 136-701, Korea.
[3] Dept. of Materials Engineering, Sung Kyun Kwan University
300 Chungchun-Dong, Jangan-Ku, Suwon 440-746, Korea

ABSTRACT

Efficiency and lifetime of light emitting diodes and laser diodes inversely depend on defect density of the crystal. Reduction of defect density is accomplished by proper choice of the substrate or deliberate modification of substrate surface. Roughness of substrate surface for GaN deposition can be controlled by buffer growth and/or nitridation. Buffer layers or nitrided layers promote lateral growth of films due to decrease in interfacial free energy between the film and substrate. Optimum conditions for nitridation and GaN-buffer growth on $Al_2O_3(0001)$ were determined by means of atomic force microscopy (AFM). AFM analysis of nitridated sapphire surfaces was also carried out to find the optimum condition for nitridation of sapphire substrate before GaN-buffer layer deposition. Nitridation of sapphires was performed only with nitrogen. Based on the fact that GaN deposited on modified surface exhibited the better crystal quality and optical property, use of AFM roughness as a reliable criterion is suggested for process optimization of GaN film growth by metallorganic chemical vapor deposition.

INTRODUCTION

GaN is an attractive material with applications in blue and ultraviolet light emitting diodes and laser diodes, since the wurtzite GaN structure has a direct band gap of 3.4 eV at room temperature and forms continuous solid solution with InN and AlN which have the same wurtzite structure and direct band gaps of 1.9 eV and 6.2 eV, respectively[1-3].

Longer lifetime and better efficiency of LED's and LD's depend on the defect density in the film. In the case of GaN film grown by MOCVD, defect density is high (about $10^{10}cm^{-2}$). Dominant lattice defects are dislocations more in threading dislocations, much less in dislocation loops, and stacking faults. The threading dislocations is perpendicular to the c-plane in hexagonal structure. Stacking faults lie between the threading dislocations surround by partial dislocations and they form due to stacking disorder in c-planes. Most of the threading dislocations in GaN films starts from the interface between the substrate and the film, and continue to grow with the film. Threading dislocations with opposite vector can not meet each other and be annihilated. Therefore the density does not decrease with GaN film thickness. This is an inherent issue in the wurtzite type III-V nitrides which grow in c-direction. If the defect density needs to be reduced, it must be controlled to a minimum number from the beginning of the film growth.

On an atomically flat surface, there are three major sources of lattice defects. One is the defect introduced during deposition which is related with orientation selection during nucleation

Mat. Res. Soc. Symp. Proc. Vol. 449 © **1997 Materials Research Society**

and growth. The second is a results of misfit strain introduced by lattice mismatch between GaN and substrate. Strain due to thermal expansion/contraction difference also contributes lattice defects, and is the third source. Defect formation due to the mismatch is well characterized in the epitaxial films, and it may probably more so on an atomically rough surface. Sapphire and SiC are common substrate for GaN growth, and c-plane substrates are commonly used. Being very hard, the substrate surfaces are always rough in an atomic scale. The atomic scale steps serve as easy sources for growing defects during the early stage of deposition. Inversion domain boundary and double positioning boundary can be formed on steps in a c-plane substrate, which become threading dislocations in the wurtzite lattices. Therefore, the atomic scale morphology of substrate surfaces needs to be considered when defect density becomes major concern in an epitaxial GaN film. AFM analysis has been used to observe the nucleation layer and to optimize the process variables for GaN-buffer layer growth on SiC(0001)[4].

The AlN or GaN buffer layer on sapphire has improved the quality of GaN films grown by MOCVD at temperatures of about 1000°C[4-6]. High quality GaN can also be obtained by nitridation of a substrate before film growing[7-9]. Ammonia gas is used for nitridation of sapphire substrate. In both cases, there seem to be optimum conditions for best results. Oftentimes, the process window is reported as functions of temperature and time of a given CVD system. The low temperature buffer layer is called as a nucleation layer. However, the exact role of the buffer layer to improve the film quality is not fully understood. And the nature of nitridated surface needs further study. A possibility that 2-D nucleation and lateral growth are enhanced by improved surface morphology after buffer growth or nitridation is not considered. In this presentation, authors studied the effects of GaN-buffer layer and nitridation on optical behavior of GaN film quality and results are interpreted refer to surface roughness after these treatments in MOCVD. Use of AFM roughness is suggested as a materials variable for process optimization in MOCVD growth of GaN.

EXPERIMENT

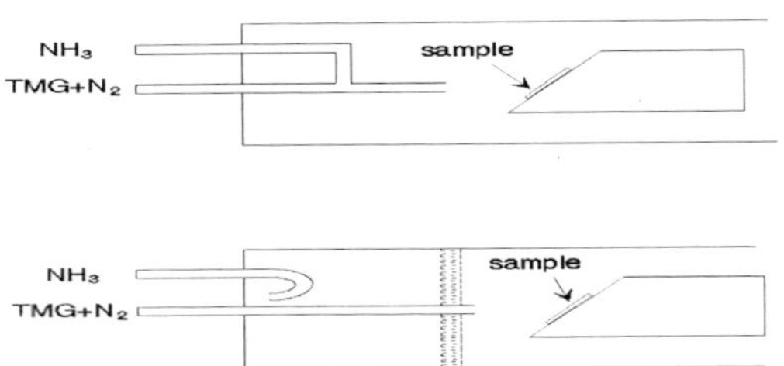

Figure 1. Schematic views of two types of gas delivery geometry. Type I and type II were used for nitridation study of sapphire, and only type I was used only for the buffer layer optimization.

Nitridation, buffer growth and GaN growth were carried out in a horizontal reactor with IR lamp heating at atmospheric pressure. Source gases for Ga and N were trimethylgallium (TMG) and ammonia (NH_3), respectively. N_2 was used as a dilutant and carrier gas for the TMG. Two types of gas delivery system were employed as seen in Figure 1. Type I and II were used for nitridation study with Al_2O_3(0001) substrate. Flow rates of TMG and NH_3 were 4.5 μ mole/min. and 1 slm, respectively. N_2 flow rate was 1 and 1.2 slm for type I and type II, respectively. Only type I was used for the optimization of buffer layer with and without nitridation layer. Flow rates were same except N_2 which was 1.2 slm. Substrates were degreased, dipped into 10% HF solution for 10 minutes, and rinsed in deionized water. Nitridation of the sapphire substrate was carried out at 1020°C with nitrogen flow only. GaN-buffer layer was grown at 550°C for 2 minutes after the nitridation for type I and II gas delivery system. The final step was to grow GaN film at 1020°C.

Optimization of GaN-buffer layer growth was carried out at 540°C for the specimens without nitridation and 520°C for the samples with nitridation. Buffer growth temperatures were chosen using the previous study ensuring the 2-dimensional nucleation and growth of GaN film[4]. GaN film depositions on the buffer layers were then carried out at 1020°C to give a GaN film thickness of about 1 μm.

Surface morphology after buffer growing or nitridation was observed by AFM (Park Scientific Instrument STM-SU2-210) with a scanning frequency of 4 Hz. XPS (SSI2803-S XPS with Mg) was performed for the characterization of nitrided layer. Optical properties were characterized by photoluminescence (PL) at 10 K using a He-Cd laser (λ=325nm). The power used for PL analysis was 40 mW and the power density was 6W/cm^2.

RESULTS

• GaN/Al$_2$O$_3$(0001) with N$_2$ Nitridation

After nitridation at 1020°C, samples were cooled down to room temperature, and RMS and average roughness were estimated by AFM. The AFM results are shown in Figure 2 for two types of gas delivery system, where roughness is plotted as a function of nitridation time. The initial roughness decreases to first minimum, reaches to a maximum and second minimum whose values are obtained at 3 and 4 minutes, respectively, and then increases gradually with nitridation time. The fact that the surface roughness reaches to a minimum value at intermediate time and increases gradually at longer times is similar to that of buffer layer.

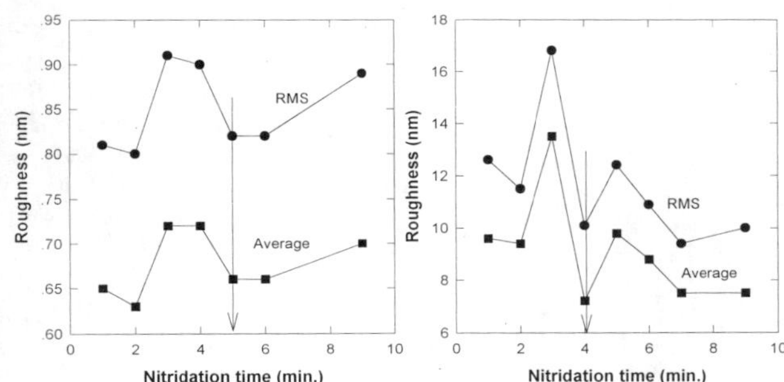

Figure 2. RMS and average roughness of nitrided surfaces are shown as a function of nitridation time using two types of gas delivery system. Note the roughness is minimum for the 5 and 4 minutes nitrided specimen for type I and type II gas delivery system, respectively.

The nature of the nitrided surface is analyzed by XPS. XPS spectra for Al$_{2p}$, N$_{1s}$ and O$_{1s}$ of the Al$_2$O$_3$(0001) surface without and with 4 minutes nitridation are shown in Figure 3. After the nitridation, N$_{1s}$ peak appeared and intensities for Al$_{2p}$ and O$_{1s}$ peaks decreased noticeably in the XPS spectrum. It is obvious that N has been incorporated on/into the Al$_2$O$_3$ surface during nitridation. Since surface roughness was changed with nitridation, one can see that nitridation process produce a new phase. The chemical nature of N is not clear. Uchida et al. reported the formation of an amorphous phase on sapphire by nitridation[8,9], while Masu et al. reported nanometer thick layer of AlN single crystal when NH$_3$ was used for the nitridation at 1050°C[8]. In the present study, the peak shift of the Al$_{2p}$ is not obvious. A possibility of the new phase is a mixture of AlN and nitrogen.

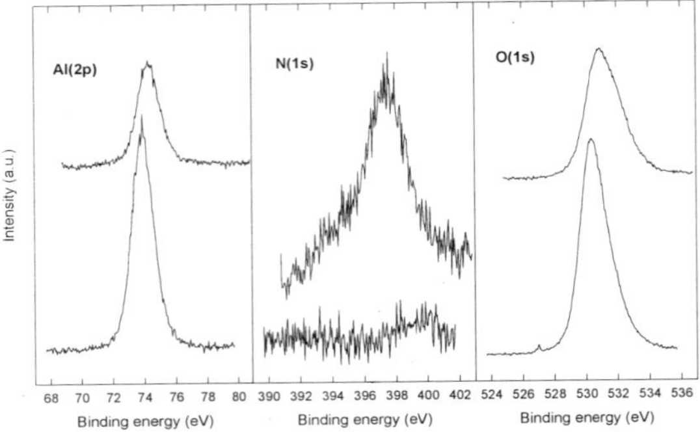

Figure 3. XPS spectra of specimen without (bottom spectra) and with 4 minutes nitrided layer (top spectra) are presented to confirm the formation of nitridation layer.

Low temperature (10 K) PL was performed to find the effect of nitridation on optical properties of GaN film. Low temperature (10 K) PL spectra are presented in Figure 4 for the two types of gas delivery system. All samples show the donor bound exciton peak around 3.47eV as reported previously[1,4,10-12], and showed no sign of yellow region emission. Minimum FWHM of PL spectra was observed from the samples whose nitridation time was 4 minutes and RMS was minimum. Minimum FWHM of sample from type I, whose roughness was about two orders lower that that from type II, exhibited about half the value from type II. This PL result implies that a GaN film grown on smoother surface exhibits better optical property. Current observation agrees with other works. Recent publications by Uchida et al.[7,8], Grandjean et al.[13] and Keller et al.[14] reported that a sample with minimum roughness obtained by nitridation gave the best GaN..

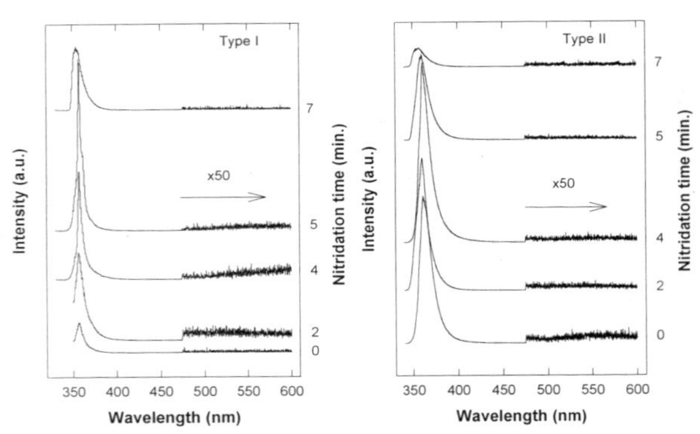

Figure 4. Low temperature (10K) PL spectra of GaN/Al$_2$O$_3$(0001) are presented. Minimum value of FWHM is obtained at 5 and 4 minutes of nitridation from type I and type II gas delivery system, respectively.

- GaN/Al$_2$O$_3$(0001) with GaN-buffer layer

To find the optimum buffer growth condition, the variation of surface roughness with growth time was monitored by AFM to correlate the roughness of the surface with the crystal quality of GaN. RMS roughness of the buffer layer without and with 5 minutes nitrided layer on sapphire surfaces were estimated by AFM and are shown in Figure 5 as a function of buffer growth time. Nitridations were carried out at 1020°C with nitrogen for 5 minutes.

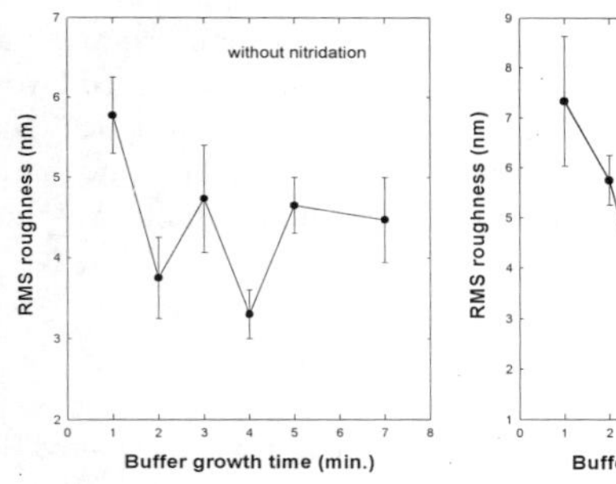

Figure 5. RMS (root mean square) roughnesses of the buffer layers on sapphire without and with 5 minutes nitrided layer are shown as a function of buffer growth time.

For both series of samples, they exhibit the same tendency in roughness variation. Surface roughness tends to decrease first, increases slightly and then decreases again to the second minimum. For the samples with 5 minutes nitrided layer, it is expected that the roughness will increase again after 7 minutes of buffer growth time even though no experiment was performed. Same type of fluctuation in roughness with buffer growth time was also observed from the SiC substrate[4]. Therefore, variation of roughness with growth time (buffer or nitridation layer) appears to be a common behavior and proper explanation for this behavior does not exist at present. The effects of nitridation on buffer growth behavior can be summarized as follows. As seen in Figure 5, the degree of fluctuation has increased and the times for minimum roughness have delayed with the presence of nitrided layer.

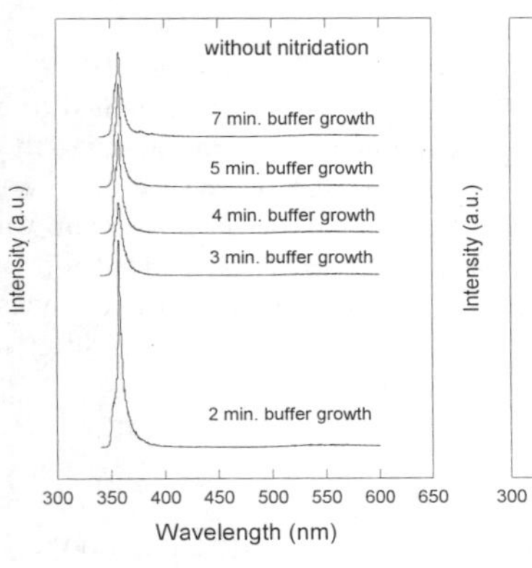

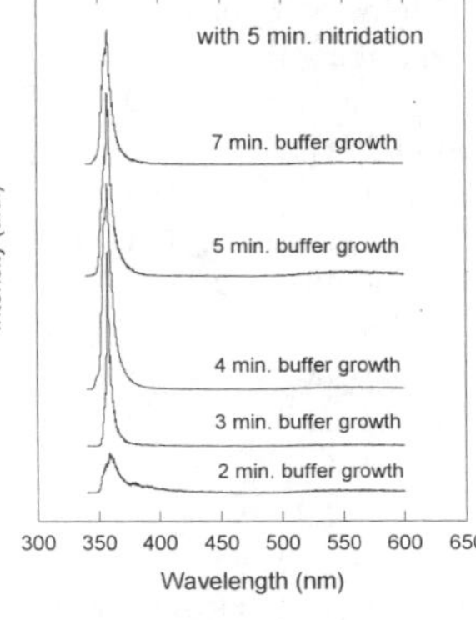

Figure 6. Low temperature (10K) PL spectra of GaN grown on different buffer layers on $Al_2O_3(0001)$ without and with 5 minutes nitrided layer are presented.

Low temp. (10 K) PL was performed to examine the optical properties of GaN. Low temperature (10K) PL spectra are presented in Figure 6 for the samples with and without nitridation layer. All samples showed the donor bound exciton peak around 3.47eV as reported previously[1,4,10-12], and a trace of yellow region emission was detected.

Minimum FWHM of PL spectra was observed from the samples grown on the first minimum roughness point, i.e., 2 minutes of buffer growth time without nitridation and 3 minutes of buffer growth time with 5 minutes of nitridation layer as seen in Figure 5. This clearly shows that the GaN films grown on a first minimum surface roughness in Figure 5 result in better optical property even though the surface chemistry of the substrate is different. 5

Optical properties and crystal qualities of GaN films showed that the optimum GaN-buffer growth condition is the first minimum point in Figure 6. That is, 2 minutes of buffer growth at 520°C without nitridation and 3 minutes of buffer growth at 540°C with 5 minutes nitrided layer. This result is consistent with vicinal 6H-SiC(0001) substrate case[4]. Therefore, both optical and crystal quality of GaN film is strongly dependent on the surface roughness of the substrate rather than the surface characteristics (with or without nitridation) or the substrate materials (Al_2O_3 or SiC).

CONCLUSIONS

The concept of surface morphology determined by AFM method was considered to understand the role of buffer layers and nitridation improving the quality of GaN film in MOCVD process. For the study, GaN/Al_2O_3(0001) using nitridation layer+GaN-buffer layer with two different gas delivery system, and GaN/Al_2O_3(0001) using GaN-buffer layer with and without nitridation layer were selected, and surface roughness after buffer growth and N_2 nitridation was determined by AFM.

Nitridation of sapphire resulted in N-incorporation into Al_2O_3 forming a new phase, and changes the roughness of the surface. The effect of nitridation improving GaN film quality has been interpreted as second minimum roughness after nitridation leads to better property.

AFM analysis on the GaN nucleation layers on sapphire with and without nitridation was employed to determine the optimal condition for the buffer layer growth. That is, 2 minutes of buffer growth at 520°C without nitridation and 3 minutes of buffer growth at 540°C with 5 minutes nitrided layer. GaN films with the optimized buffer layer show the best PL characteristics and better crystal quality. Results from other analytical methods including DCXRD and PL are consistent with the AFM analysis.

The effect of nitridation on the properties of GaN films can be deduced from the comparison of FWHM of low temperature PL and DCXRD. Smaller values were obtained from the specimen with 5 minutes nitrided layer. Since buffer growth conditions were optimized for both cases, it can be concluded that the nitridation layer improved the optical property and crystal quality of GaN films. It is also interesting to note that the value of the first minimum was decreased with nitridation, and smoother buffer layer surface may play a critical role in improving the quality of the GaN film. From this study, it may be suggested that the roughness of the substrate can serve as a reliable guideline in determining the optimum growth condition of GaN films, and for this purpose, AFM analysis proved to be a simple and effective tool.

ACKNOWLEDGMENTS

This work was supported by KIST-2000 Research Program (Grant No. BSV00020-002-4).

REFERENCES

1. S. Strite and H. Morkoc, J. Vac. Sci. Technol. **B10**, 1237 (1992).
2. I. Akasaki and H. Amano, J. Electrochem. Soc. **141**, 2266 (1994).

3. R. F. Davis, Physica **B185**, 1 (1993).

4. D. Byun, G. Kim, D. Lim, D. Lee, I.-H. Choi, D. Park and D.-W. Kum, Thin Solid Films, in print (1996).

5. H. Amano, N. Sawaki, I. Akasaki and Y. Toyoda, Appl. Phys. Lett. **48**, 353 (1986).

6. S. Nakamura, Jpn. J. Appl. Phys. **30**, L1705 (1991).

7. K. Masu, Y. Nakamura, T. Yamazaki, T. Shibata, M. Takashi and K. Tsubouchi, Jpn. J. Appl. Phys. **34**, L760 (1995).

8. K. Uchida, A. Watanabe, F. Yano, M. Kouguchi, T. Tanaka and S. Minagawa. J. Appl. Phys. **79**, 3487 (1996).

9. K. Maier, J. Schneider, I. Akasaki and H. Amano, Jpn. J. Appl. Phys. **32**, L846 (1993).

10. B-C. Chung and M. Gershenzon, J. Appl. Phys. **72**, 651 (1992).

11. R. Dingle, D. D. Sell, S. E. Stokowski and M. Ilegems, Phys. Rev. **B4**, 1211 (1971).

12. K. Uchida, A. Watanabe, F. Yano, M. Kouguchi, Proceedings of the International Symposium on Blue Laser and Light Emitting Diodes, p. 48 (1996).

13. N. Grandjean, J. Massies, and M. Leroux, Appl. Phys. Lett. **69**, 2071 (1996).

14. S. Keller, B. P. Keller, Y.-F Wu, B. Heying, J.S. Speck, U.K. Mishira, and S.P. DenBaars, Appl. Phys. Lett. **68**, 1525 (1966).

EFFECT OF THE NITRIDATION OF THE SAPPHIRE (0001) SUBSTRATE ON THE GaN GROWTH

N. GRANDJEAN, J. MASSIES, P. VENNÈGUES, M. LAUGT, M. LEROUX
Centre de Recherche sur l'Hétéro-Epitaxie et ses Applications,
Centre National de la Recherche Scientifique,
Rue Bernard Grégory, Sophia Antipolis, 06560 Valbonne, France.

ABSTRACT

The analysis of the sapphire surface nitridation by *in situ* reflection high-energy electron diffraction evidences the formation of a relaxed AlN layer. Its role on the early stage of the GaN growth is investigated by transmission electron microscopy (TEM). GaN crystallites of high structural quality, with the c axis perpendicular to the sapphire basal plane, are observed when the starting surface is nitridated. On the other hand, the growth of GaN on a bare substrate involves the formation of larger islands with numerous defects. TEM study reveals that the c axis of these latter crystallites is systematically tilted by about 19° with respect to the sapphire basal plane. Actually, this orientation corresponds to a particular epitaxial relationship between GaN and sapphire (0001) substrates. Finally, the optical properties of GaN thin layers are shown to be strongly dependent on the nitridation state of the sapphire surface.

INTRODUCTION

Group III-nitrides are now recognized as promising candidates for blue-violet light emitting devices. This is attested both by the marketing of high-efficiency GaN-based light emitting diodes [1] and the realization of a laser diode operating at room temperature at 417 nm under pulsed current injection [2]. This last success is quite astonishing when regarding the numerous difficulties for growing high-quality GaN epitaxial layers. Actually, the homoepitaxial growth of GaN cannot be achieved due to the lack of GaN substrates having sufficiently large dimensions. Sapphire is presently the standard substrate in spite of its large lattice-mismatch with GaN. Indeed, high-quality GaN materials have been obtained providing the growth of a buffer layer at low temperature [3]. It has been shown that a nitridation step of the sapphire substrate prior to the buffer layer influences the optical and structural properties of GaN thick layers [4-6]. In fact, this procedure is important for growing high-quality III-V nitrides on sapphire substrates, whatever the growth technique [5-8]. However, the role of the sapphire nitridation on the GaN growth is not yet well understood.

This paper describes the early stages of the growth of GaN on c-plane sapphire *versus* the nitridation state of the starting surface. The nitridation is followed by *in situ* reflection high-energy electron diffraction (RHEED). We conclude from the variation of the in-plane lattice parameter that an AlN layer is formed. Its consequence on the growth of GaN thin layers is investigated by transmission electron microscopy (TEM) and photoluminescence (PL) experiments.

Mat. Res. Soc. Symp. Proc. Vol. 449 ©1997 Materials Research Society

EXPERIMENTS

Growth of GaN was carried out in a gas-source molecular beam epitaxy system (GSMBE) equipped with a standard 10 kV RHEED electron gun and a CCD camera based video system. Ga and NH_3 were used as group III and V sources, respectively. The growth rates were determined *in situ* using the oscillations of the reflected intensity of a 670 nm laser diode. The growth temperature and the deposition rate were 780°C and 0.12 µm/h, respectively. The substrate nitridation is achieved by exposing the sapphire surface to a NH_3 flow of 20 sccm at high temperature (850-900°C). TEM investigations were performed with a JEOL 2010 FEG.

RESULTS

The evolution of the in-plane lattice parameter during the sapphire nitridation has been followed by measuring the RHEED streak distance variation with the electron beam along the $[1\bar{1}00]_{Al_2O_3}$ azimuth. While the surface morphology does not significantly change as indicated by the RHEED pattern remaining streaky-like, the in-plane lattice constant varies as a function of the exposition time of NH_3. After 10 minutes, the lattice mismatch reaches ~ 13 % and stabilizes (Fig. 1).

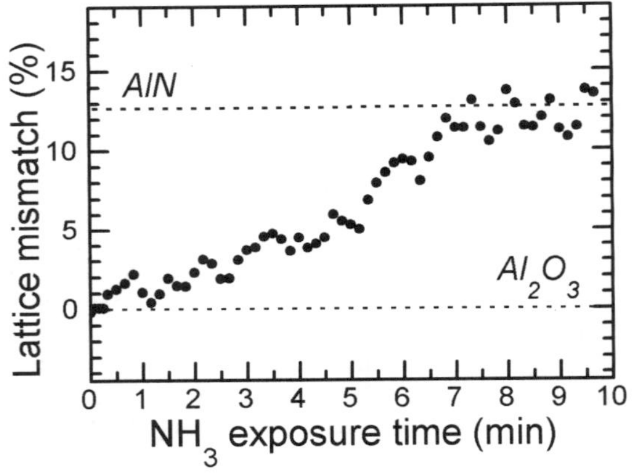

Figure 1: In-plane lattice mismatch variation measured from the RHEED pattern during the nitridation of the c-plane sapphire surface (electron beam along the $[1\bar{1}00]_{Al_2O_3}$ azimuth).

This value corresponds to the mismatch between AlN and sapphire (12.8 % at 780°C) with the orientation relationship $[11\bar{2}0]_{AlN}\|[1\bar{1}00]_{Al_2O_3}$. Remembering that Al atoms in the Al_2O_3 crystal form an hexagonal sublattice rotated by 30° about the c axis with respect to the Al_2O_3 unit cell, the conversion of octahedral Al sites in tetrahedral ones during the nitridation may result in an AlN layer with the previously quoted relationship. It can therefore be assumed that the in-plane lattice mismatch variation (Fig. 1) is due to the formation of a fully relaxed AlN layer at the surface of the Al_2O_3 (0001) substrate [7,8].

Figure 2: Bright field cross section transmission electron micrograph of 200 Å equivalent thickness of GaN deposited at 780°C on a nitridated sapphire substrate

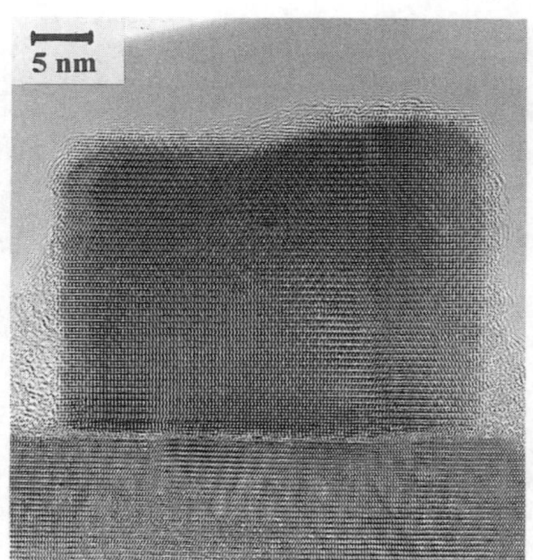

Figure 3: High resolution TEM image of a typical GaN crystallite observed in figure 2. The c axis of GaN is perpendicular to the sapphire basal plane (Al_2O_3 and GaN lattices are viewed in the $[1\bar{1}00]$ and $[11\bar{2}0]$ projection, respectively).

200 Å equivalent thickness of GaN have been deposited on nitridated sapphire substrates at a temperature of 780°C. A cross section TEM image of such a sample is displayed in figure 2. The GaN crystallites have a skyscraper-like shape and a high structural quality (Fig. 3). The epitaxial relationship between GaN and the c-plane sapphire deduced from TEM observations is $[0001]_{GaN}\|[0001]_{Al_2O_3}$ and $[11\bar{2}0]_{GaN}\|[1\bar{1}00]_{Al_2O_3}$, in agreement with previous reports [9-14]. In fact, the presence of an AlN interfacial layer (GaN/AlN/Al$_2$O$_3$) involves the following sequence of crystal orientation: $[0001]_{AlN}\|[0001]_{Al_2O_3}$ and $[11\bar{2}0]_{AlN}\|[1\bar{1}00]_{Al_2O_3}$ followed by $[0001]_{GaN}\|[0001]_{AlN}$ and $[11\bar{2}0]_{GaN}\|[11\bar{2}0]_{AlN}$. As a consequence, we cannot definitively conclude about the exact epitaxial relationship between GaN and sapphire when a nitridation step is performed. In this case, it can be tentatively assumed that GaN grows on a pseudo AlN (0001) substrate by keeping the same orientation. It should be noted that the lattice mismatch of GaN/AlN (2.5%) is considerably lower than that of GaN/Al$_2$O$_3$ (~16%). Therefore, the nitridation step strongly reduces the heteroepitaxial strain of GaN/Al$_2$O$_3$.

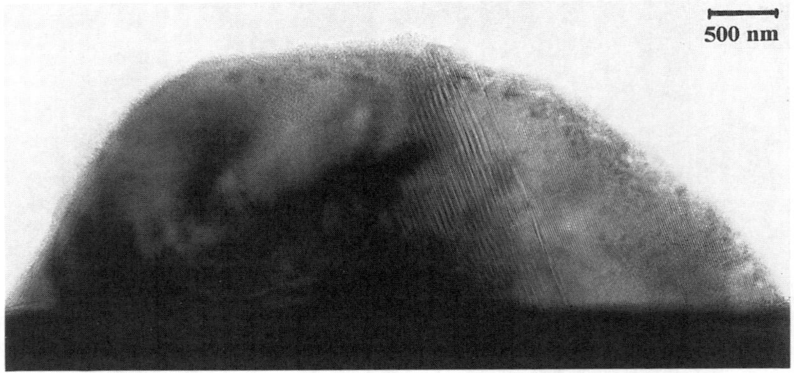

Figure 4: TEM image of a typical island resulting from the growth on a bare sapphire surface (Al$_2$O$_3$ and GaN lattices are viewed in the same projection as in figure 3).

In order to determine the actual epitaxial relationship between GaN and the c-plane sapphire substrates, no nitridation step was performed prior to the growth. This procedure results in three-dimensional growth as shown in figure 4 where is displayed a typical crystallite of such a growth. Besides the great density of structural defects, the main feature is that most of the islands (~75%) exhibit a large tilt of their c axis. Actually, the c axis and the sapphire (0001) plane have an angle of 19° (Fig. 4). Note that the 30° in-plane rotation with respect to the sapphire unit cell is conserved. The deduced epitaxial relationship is $[11\bar{2}0]_{GaN}\|[1\bar{1}00]_{Al_2O_3}$ and $[1\bar{1}03]_{GaN}\|[11\bar{2}0]_{Al_2O_3}$. The corresponding plane of GaN parallel to the sapphire (0001) surface is $(\bar{3}302)$. This orientation relationship leads to a theoretical tilt of the c axis of 19.5°, in good agreement with the value measured on the TEM images. The reason for the preferred tilted orientation (~75 %) when no nitridation is performed is probably due to the slightly better lattice matching condition compared to the standard one (c axis perpendicular to the sapphire basal plane) [15].

It is surprising that this epitaxial relationship has not been reported so far despite the numerous studies devoted to the growth of GaN on sapphire. The most probable explanation is the competition between the GaN deposition and the surface nitridation occurring simultaneously at the beginning of the growth. The relatively low growth temperature (780°C) used in the present experiments, compared to usual temperatures in metal-organic vapor phase epitaxy (MOVPE), can explain why the tilted orientation is preponderant: the nitridation rate can be slower than the growth rate. In contrast, for the MOVPE growth the nitridation is highly efficient due to growth temperatures of 1000°C or higher. AlN is thus rapidly formed on the substrate leading to the standard orientation for subsequent GaN deposition. Note that the tilted orientation has been very recently observed by another group also growing GaN by GSMBE [16].

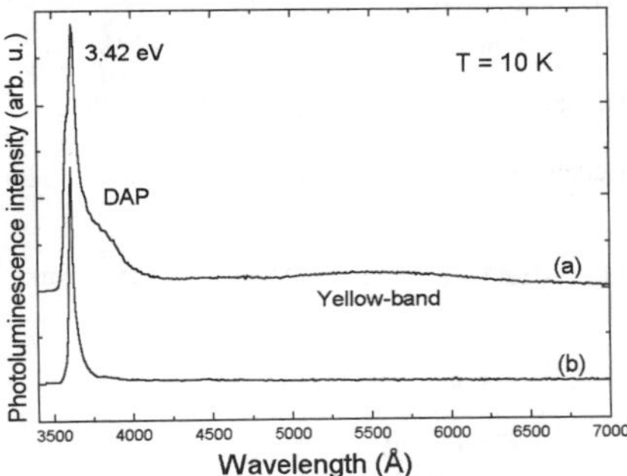

Figure 5: PL spectra at 10K of GaN thin layers grown on a bare c-plane sapphire surface (a) and a nitridated one (b).

Finally, the optical properties of GaN thin layers (200Å equivalent thickness) have been investigated as a function of the GaN crystal orientation (Fig. 5). Photoluminescence spectra at 10K exhibit a band-edge dominated by a peak at ~3.42 eV, whatever the nitridation state of the starting surface. Transitions in the 3.40-3.42 eV range are often reported, especially for MBE grown GaN. They have been assigned in the literature either to free hole - neutral donor transitions [17], or to donor acceptor pairs involving a shallow acceptor [18]. The main feature of interest here is that the intensities of both the donor-acceptor pair recombinations (~3800 Å) and the deep level related yellow band (~5500 Å) depend on the GaN orientation. The standard epitaxial relationship with the c axis perpendicular to the c-plane sapphire surface corresponds then to much purer optical spectra (Fig. 5b) than that corresponding to the tilted GaN layers (Fig. 5a). This sensitivity has been used to optimize the nitridation procedure [8].

CONCLUSIONS

This paper shows that two different epitaxial relationships can occur between GaN and c-plane sapphire substrates. The well known orientation with the c axis of GaN perpendicular to the (0001) sapphire plane is favored by the presence of a thin AlN layer formed by the nitridation of the sapphire surface. When GaN is deposited on a bare substrate, the c axis is tilted 19° with respect to the surface. In fact this orientation corresponds to a different epitaxial relationship between GaN and Al_2O_3 (0001) substrates which has not to our knowledge been previously reported. Finally, the optical properties of thin GaN layers are found to be strongly dependent on their orientation.

REFERENCES

1. S. Nakamura, T. Mukai, and M. Senoh, Appl. Phys. Lett. **64**, 1687 (1994)

2. S. Nakamura, M. Senoh, S. Nagahama, N. Iwasa, T. Yamada, T. Matsushita, H. Kiyoku, and Y. Sugimoto, Jpn. J. Appl. Phys. **35**, L74 (1996)

3. H. Amano, N. Sawaki, I. Akasaki, and Y. Toyoda, Appl. Phys. Lett. **48**, 353 (1986)

4. S. Keller, B.P. Kemmer, Y.-F. Wu, B. Heyring, D. Kapolnek, J.S. Speck, U.K. Mishra, and S.P. Denbaars, Appl. Phys. Lett. **68**, 1525 (1996)

5. O. Briot, J.P. Alexis, B. Gil, and R.L. Aulombard, MRS fall Meeting, Boston, November 1995

6. K. Uchida, A. Watanabe, F. Yano, M. Kouguchi, T. Tanaka, and S. Minagawa, J. Appl. Phys. **79**, 3487 (1996)

7. T.D. Moustakas, R.J. Molnar, T. Lei, G. Menon, and C.R. Eddy Jr., Mater. Res. Soc. Symp. Proc. **242**, 427 (1992)

8. N. Grandjean, J. Massies, and M. Leroux, Appl. Phys. Lett. **69**, 2071 (1996)

9. B.B. Kosicki and D. Kahng, J. Vac. Sci. Technol. **6**, 593 (1969)

10. D.K. Wickenden, K.R. Kaulkner, R.W. Brander, and B.J. Isherwood, J. Cryst. Growth **9**, 158 (1971)

11. H.M. Manasevit, F.M. Herdmann, and W.I. Simpson, J. Electrochem. Soc. **118**, 1864 (1971)

12. T. Sasaki and S. Zembutsu, J. Appl. Phys. **61**, 2533 (1987)

13. S. Yoshida, S. Misawa, and S. Gonda, J. Vac. Sci. Technol. B1, 250 (1983)

14. R.C. Powell, N.-E. Lee, Y.-M. Kim, and J.E. Greene, J. Appl. Phys. **73**, 189 (1993)

15. N. Grandjean, J. Massies, P. Vennéguès, M. Laûgt, and M. Leroux, to be published

16. S. Christiansen, M. Albrecht, W. Dorsch, H.P. Strunk, C. Zanotti-Fregonara, G. Salviati, A. Pelzmann, M. Mayer, M. Kamp, and K.J. Ebeling, MRS Internet Journal of Nitride Semiconductor Research, Vol. 1, Article 19

17. B.-C. Chung and M. Gershenzon, J. Appl. Phys. **72**, 651 (1992)

18. B.G. Ren, J.W. Orton, T.S. Cheng, D.J. Dewsnip, D.E. Lacklison, C.T. Foxon, C.H. Malloy, X. Chen, MRS Internet Journal of Nitride Semiconductor Research, Vol.1, Article 22

INITIAL GROWTH STAGE OF AlGaN
GROWN DIRECTLY ON (0001) 6H-SiC BY MOVPE

K. Horino, A. Kuramata, T.Tanahashi
Fujitsu Laboratories Ltd., 10-1 Morinosato-wakamiya, Atsugi 243-01, Japan

ABSTRACT

We investigated the growth process of AlGaN films grown directly on 6H-SiC (0001) substrates by metalorganic vapor phase epitaxy (MOVPE). We focused on the initial growth stage to clarify the mechanism of nitride growth on SiC. From Energy Dispersive X-ray (EDX) analysis we found that an Al-rich region generated naturally at the AlGaN/SiC interface. We also found that Al flux determined the density of grain which generated during the initial growth stage, and this grain density reflected the surface morphology.

INTRODUCTION

Wurtzite-type GaN and related materials are promising candidates for short-wavelength light emitting devices. Sapphire substrates are conventionally used for nitride growth, however, the cleavage is not so strong. On the other hand, (0001) 6H-SiC substrates have strong cleavage in the same direction as nitride films grown on them. SiC also has a small lattice mismatch to GaN (3.4%), high thermal conductivity (4.9 W/cmK)[1], and electrical conductivity. Therefore, SiC is expected to become material for nitride laser diode substrates.

Recently, the number of reports about nitrides grown on SiC substrates has increased. The uniform growth of GaN on SiC substrates have been reported both with[2-6] and without[7,8] AlN buffer layers. The growth of AlGaN on SiC without an AlN buffer layer has also reported[9,10] that $Al_xGa_{1-x}N$ alloys containing even low ($x \geq 0.05$) concentrations of AlN showed smooth surfaces and high crystal quality[10]. However, the growth mechanism of nitride on SiC has not been clarified.

In this study, we investigated the direct growth of AlGaN on SiC to clarify the cause of nitride growth on SiC.

EXPERIMENT

$Al_xGa_{1-x}N$ ($0 \leq x \leq 0.13$) films were grown on 6H-SiC (0001) $_{Si}$ by low-pressure MOVPE. Trimethylaluminium (TMA), trimethylgallium (TMG) and ammonia (NH_3) were used as precursors, and the respective flow rates were 0~1.6 µmol/min, 17 µmol/min, and 1.2 SLM. Hydrogen was used as the carrier gas, and its flow rate was 4.0 SLM. The total pressure of the system was 100 torr. The growth temperature, which was measured at the surface of the SiC-coated graphite susceptor, was 920°C. Growth rates were about 2.4 µm/h. In this study, the TMA flow rate was varied in order to examine the difference of growth process on Al flux, and growth time was varied for the purpose of observing the growth processes.

The $Al_xGa_{1-x}N$ films were characterized using several techniques. Surface morphologies were observed by scanning electron microscopy (SEM). Al composition of the thick films (> 1 µm) was calculated from an x-ray diffraction angle of (0004). It corresponded well to measurements made by an Energy Dispersive X-ray (EDX) Microanalyzer. The crystal quality was analyzed by an x-ray rocking curve (XRC) of (0004) diffraction. The depth profiles of Al, Ga, and Si, were analyzed by an EDX Microanalyzer with the probe scanning about 65 nm across the interface on the cross section of the sample. The variations in the peak intensity of each element was measured as the distribution of respective elements. The probe was focused to

73

about a 1 nm diameter, and the scanning time was 60 seconds. Width profile along the interface was also measured by the EDX Microanalyzer. The measurement was performed at seven points about every 166 nm. The measurement points were adjusted at the interface by the Si element in the SiC substrate. The probe was focused about 1 nm diameter, and measurement time was 100 seconds, respectively.

RESULTS AND DISCUSSION

Figure 1 shows the surface morphologies of $Al_xGa_{1-x}N$ crystals grown at different TMA flow rates. The crystal height (Fig. 1. (a)) and thickness of the films (Fig. 1. (b)~(e)) were about 2 μm and 1.2 μm, respectively. The respective AlN mole fraction 'x' of the films shown in Figure 1 (a) to (e) was 0, 0.06, 0.07, 0.09, and 0.13. Without TMA flow, crystals assumed an island-like formation. On the other hand, the crystals grown with TMA flow were film-like. However, we observed many hillocks on the films grown at lower TMA flow rates. The minimum AlN mole fraction where an $Al_xGa_{1-x}N$ film with smooth surface was obtained directly on SiC was estimated to be 0.08. This result was similar to that reported by Bremser [10].

Figure 2 shows the full width at half maximum (FWHM) of the XRCs of the films. FWHM narrowed as the AlN mole fraction increased. The FWHM of the films exhibiting smooth surfaces was about 250 arc seconds. These FWHMs are as small as those of GaN grown with AlN buffer layers in similar growth conditions. The crystal quality of these films was very high.

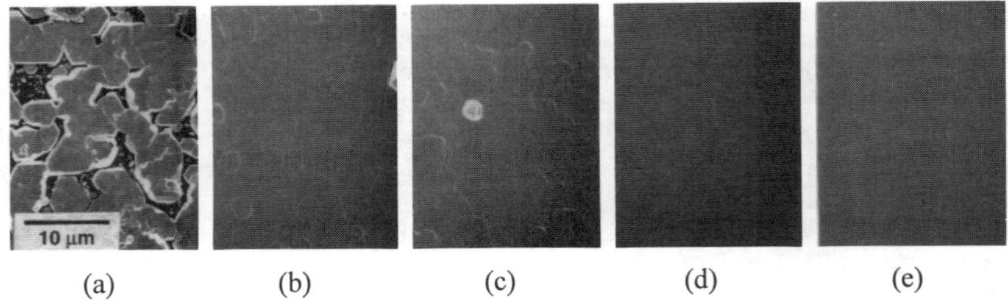

(a)   (b)   (c)   (d)   (e)

Fig. 1. Surface morphology of $Al_xGa_{1-x}N$/SiC grown at different TMA flow rates. Respective TMA flow rates are (a) 0, (b) 0.3, (c) 0.4, (d) 1.0, and (e) 1.6.

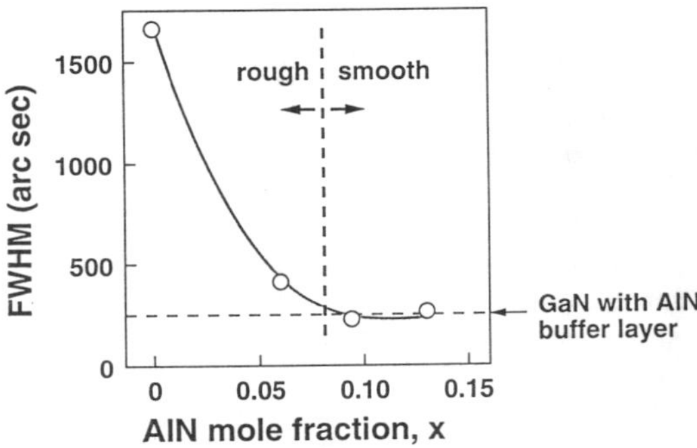

Fig. 2. Full width at half maximum of x-ray rocking curve of $Al_xGa_{1-x}N$/SiC

Depth profile of Al, Ga, and Si, near the $Al_{0.13}Ga_{0.87}N$/SiC interface is shown in Figure 3 (a). The scan area is shown in the cross-sectional image of scanning transparent electron microscopy (STEM) (Fig. 3 (b)). The point where Si decreases is considered to be the interface. At this point, we found that Al showed a peak. From a quantitative analysis, the AlN mole fraction at this point exceeded 50%. We found that an Al-rich region generated naturally at the interface of $Al_{0.13}Ga_{0.87}N$/SiC.

Figure 4 shows the surfaces of AlGaN grown at different growth times and with different TMA flow rates. We observed the grain on the surface of the SiC substrates in both flow rate conditions. The grain density as estimated by the photographs were about 4×10^8 cm^{-2} for lower flow rates and 3×10^9 cm^{-2} for higher flow rates. The surface morphologies of films grown for 1800 seconds were rough at a lower flow rate and smooth at higher flow rate. As the quantity of grain was observed not to increase over growth time, the surface morphology depended on the initial grain density.

From these results, we suggest a growth model in Figure 5. We consider that Al-rich nuclei generated first on SiC, and then AlGaN grain generated on the Al-rich nuclei, because the saturated vapor pressure of AlN was smaller than that of GaN, and/or AlN has a smaller lattice mismatch to SiC than GaN. As the nucleation probability of AlN is considered to depend on the Al flux, the density of both nuclei and grain appeared lower at lower TMA flow rates and higher at higher TMA flow rates. Finally, we obtained films both with hillocks and with smooth surfaces, as the grain grew laterally. We conclude that the roughness of the film grown at a lower TMA flow rate was caused from insufficient Al-rich nuclei.

Figures 6 (a) and 7 (a) show the width profile of the Al composition along the interface of the samples grown at lower and higher TMA flow rates, respectively. Figures 6 (b) and 7 (b) show the measurement points on the cross-sectional STEM image. The Al composition of the film grown at a higher flow rate was almost constant, however, it was fairly fluctuant at a lower flow rate. We believe this fluctuation is caused by insufficient nucleation, as shown in Figure 5.

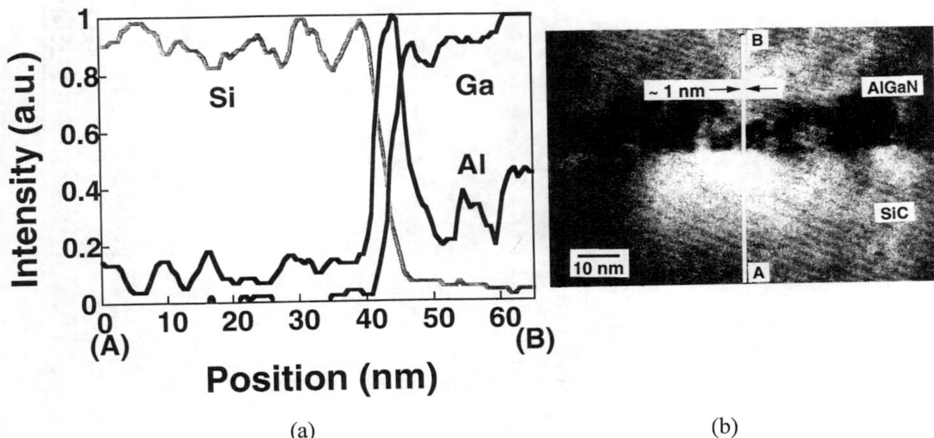

(a)   (b)

Fig. 3. Depth profile of composition near the $Al_{0.13}Ga_{0.87}N$/SiC interface: (a) profile of normalized peak intensity of Al, Ga, and Si, and (b) scanning area shown in a STEM photograph.

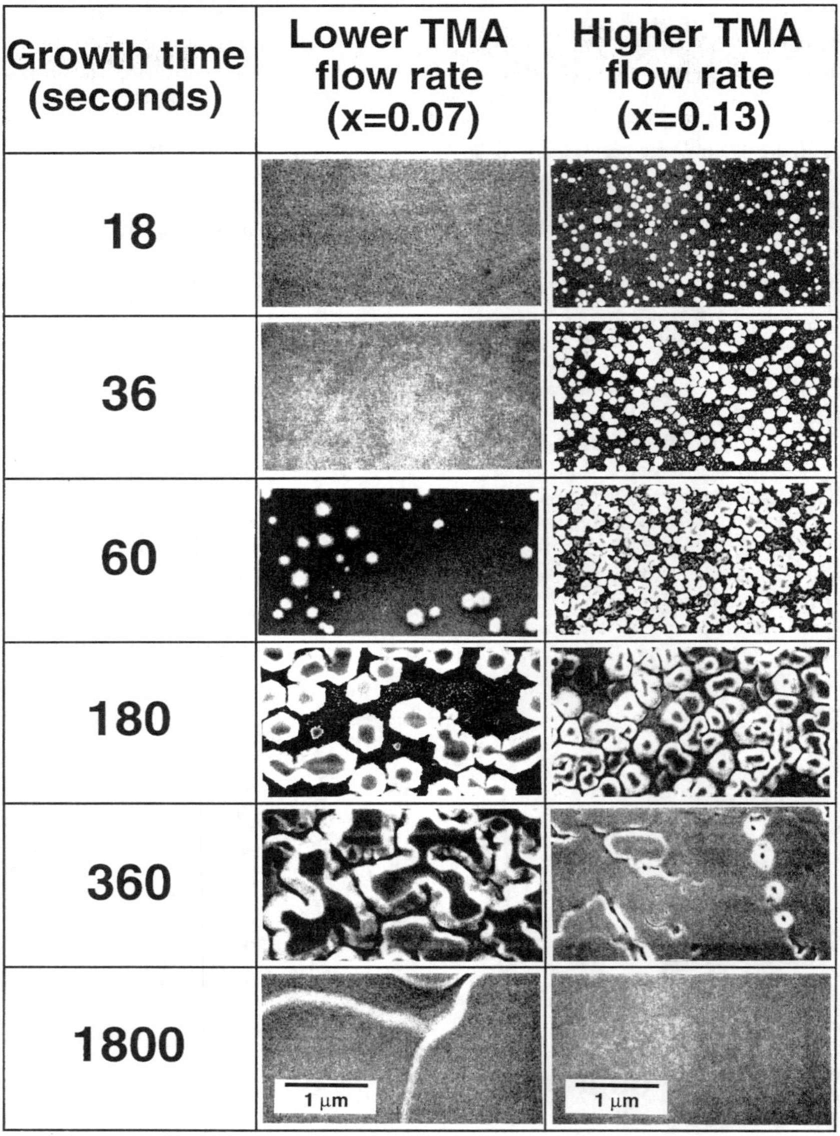

Growth time (seconds)	Lower TMA flow rate (x=0.07)	Higher TMA flow rate (x=0.13)
18		
36		
60		
180		
360		
1800	1 µm	1 µm

Fig. 4. SEM photographs of AlGaN grown with different growth times and TMA flow rates.

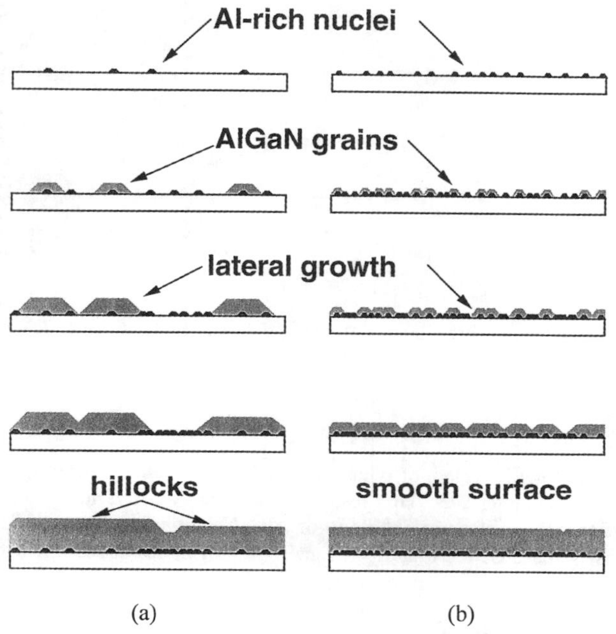

Fig. 5. Schematic diagrams of growth process of AlGaN grown directly on SiC at (a) lower and (b) higher TMA flow rates.

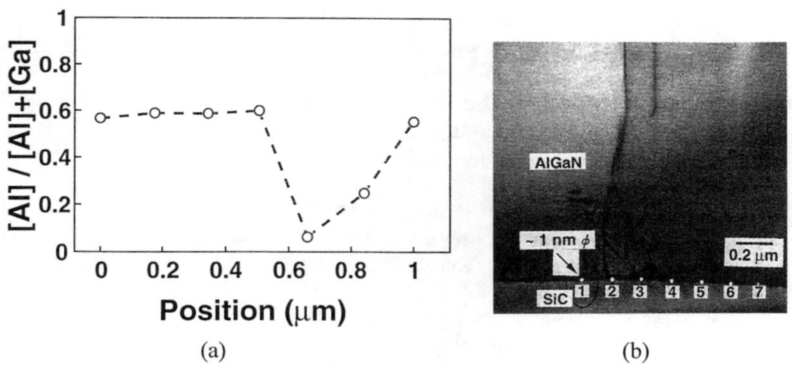

Fig. 6. Fluctuation of Al composition at the $Al_{0.07}Ga_{0.93}N$/SiC interface: (a) AlN mole fraction and (b) measurement points shown in a STEM photograph. Probe was focused to about a 1 nm diameter.

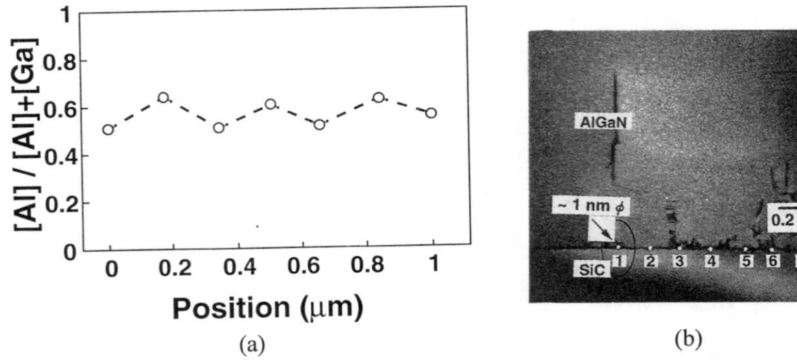

Fig. 7. Fluctuation of Al composition at the $Al_{0.13}Ga_{0.87}N$/SiC interface: (a) AlN mole fraction and (b) measurement points shown in a STEM photograph. Probe was focused to about a 1 nm diameter.

CONCLUSIONS

We found that an Al-rich region, of which the AlN mole fraction exceeded 50%, generated naturally at the AlGaN/SiC interface. This result suggests that Al-rich nuclei generated first on SiC. We also found that the surface morphology of AlGaN grown directly on SiC depended on the initial grain density, because the quantity of grain did not increase over growth time. From the width profile, we considered that an insufficient amount of Al-rich nuclei caused a low grain density. We conclude that a sufficient amount of Al-rich nuclei at the initial growth stage of AlGaN on SiC leads a uniform growth.

REFERENCES

1. *Landolt-Börnstein*, edited by O. Madelung (springer, New York, 1982), Vol. 17.
2. M. E. Lin, S. Strite, A. Agarwal, A. Salvador, G. L. Zhou, N. Teraguchi, A. Rockett, and H. Morkoç, Appl. Phys. Lett. **62**, 702 (1993)
3. T. Warren Weeks, Jr., Michael D. Bremser, K. Shawn Ailei, Eric Carlson, William G. Perry, and Robert F. Davis, Appl. Phys. Lett. **67**, 401 (1995)
4. F. A. Ponce, B. S. Krusor, J. S. Major, Jr., W. E. Plano, and D. F. Welch, Appl. Phys. Lett. **67**, 410 (1995)
5. K. Horino, A. Kuramata, K. Domen, R. Soejima, and T. Tanahashi, in *Proceedings of the International Symposium on Blue Laser and Light Emitting Diodes*, edited by A. Yoshikawa, K. Kishino, M. Kobayashi, amd T. Yasuda (University Convention Center, Chiba University, Japan 1996) p. 530-533
6. Wei Li and Wei-Xin Ni, Appl. Phys. Lett. **68**, 2705 (1996)
7. T. Sasaki and T. Matsuoka, J. Appl. Phys. **64**, 4531 (1988).
8. V. A. Dmitriev, K. Irvine, J. A. Edmond, C. H. Carter, Jr., A. S. Zubrilov, and D. V. Tsvetkov, Appl. Phys. Lett. **67**, 115 (1995)
9. Y. Kuga, T. Shirai, M. Haruyama, H. Kawanishi, and Y. Suematsu, Jpn. J. Appl. Phys. **34**, 4085 (1995)
10. M. D. Bremser, W. G. Perry, N. V. Edwards, T. Zheleva, N. Parikh, D. E. Aspnes, and R. F. Davis, Mat. Res. Soc. Symp. **395**, 195 (1996)

$Al_xGa_{1-x}N$ BASED MATERIALS AND HETEROSTRUCTURES

P. KUNG*, A. SAXLER*#, D. WALKER*, X. ZHANG*, R. LAVADO*, K.S. KIM*$,
M. RAZEGHI
* Center for Quantum Devices, Department of Electrical and Computer Engineering, Northwestern
University, Evanston, IL 60208.
Permanent address: Wright Laboratory, Materials Directorate, Wright Patterson AFB, OH
45433-7707
$ Permanent address: Department of Physics, Jeon Buk National University, Korea

ABSTRACT

We present the metalorganic chemical vapor deposition growth, n-type and p-type doping and
characterization of $Al_xGa_{1-x}N$ alloys on sapphire substrates. We report the fabrication of Bragg
reflectors and the demonstration of two dimensional electron gas structures using $Al_xGa_{1-x}N$ high
quality films. We report the structural characterization of the $Al_xGa_{1-x}N$ / GaN multilayer structures
and superlattices through X-ray diffraction and transmission electron microscopy. A density of
screw and mixed threading dislocations as low as 10^7 cm^{-2} was estimated in $Al_xGa_{1-x}N$ / GaN
structures. The realization of $Al_xGa_{1-x}N$ based UV photodetectors with tailored cut-off
wavelengths from 365 to 200 nm are presented.

INTRODUCTION

III-Nitride semiconductors, such as AlN, GaN, InN and their alloys, have become the leading
material system for wide bandgap, short wavelength, optoelectronics applications because of their
exceptional physical properties. III-Nitrides have a direct bandgap which is tunable in energy from
6.2 eV (AlN), 3.4 eV (GaN) to 1.9 eV (InN) making them ideal for high efficiency photonic
devices capable of operating from the ultraviolet (UV) to the red spectral regions. Such devices
include UV-visible light emitting diodes (LEDs), laser diodes and solar-blind UV photodetectors
for high density optical storage, high-brightness color displays, undersea and covert space-to-
space communications and the detection of spacecraft above the ozone layer where there is a strong
visible and infrared background.

We have already reported the growth and characterization of AlN and GaN epilayers with the
narrowest x-ray diffraction linewidths ever reported on sapphire substrates, as measured with an
X-ray diffractometer operated in the "open detector mode" [1,2]. The origin of the deep-level
associated yellow luminescence in GaN has been attributed to Ga vacancies [3]. High quality
metal-insulator-semiconductor devices have been fabricated for the first time using AlN as the
insulating layer [4]. GaN p-n homojunction based photovoltaic detectors have been fabricated and
modeled [5]. The kinetics of photoconductivity in n-type GaN has been studied [6].

In this paper, we report the growth and characterization of $Al_xGa_{1-x}N$ thin films on basal plane
sapphire substrates in the entire compositional range, the n-type and p-type doping of $Al_xGa_{1-x}N$ as
a function of alloy composition and dopant flow rate, the fabrication of Bragg reflectors using
$Al_xGa_{1-x}N$ alloys, the observation of a two-dimensional electron gas (2DEG) at $Al_xGa_{1-x}N$/GaN
interfaces, the reduction of dislocation densities in $Al_xGa_{1-x}N$/GaN heterostructures and the
fabrication of $Al_xGa_{1-x}N$ based ultraviolet photoconductors with cut-off wavelengths from 200 to
365 nm corresponding to the whole range of $Al_xGa_{1-x}N$ alloys.

EXPERIMENT

Material Growth

The thin films were grown using a horizontal low pressure metalorganic chemical vapor
deposition reactor (Aixtron, Inc.). The 4 inch graphite susceptor was RF heated and spun at about
100 rpm to enhance the uniformity of the films. Trimethylaluminum (TMAl), trimethyl- (TMGa)
and triethyl-gallium (TEGa) and ammonia (NH_3) were used as the source materials. Dilute silane
(35 ppm SiH_4 in H_2) and dilute germane (50 ppm GeH_4 in H_2) were used as the n-type dopant
sources, while biscyclopentadienylmagnesium (Cp_2Mg) was used as the p-type dopant source.
Organometallic and hydride sources were separated until just before reaching the susceptor to

Mat. Res. Soc. Symp. Proc. Vol. 449 © 1997 Materials Research Society

minimize parasitic reactions. The carrier gas was a mixture of hydrogen and nitrogen. The growth temperature was $1000°C$ and the growth pressure was 10 mbar.

The films were grown on basal plane sapphire (Al_2O_3) substrates. A thin AlN buffer layer was first grown to improve the initial nucleation and enhance two dimensional growth. The growth conditions of the buffer layer are described in detail in a previous report [2].

Material Characterization

The structural properties of the films were assessed through various techniques, including X-ray diffraction using a diffractometer operated in the "open detector mode", scanning electron microscopy (SEM) using a Hitachi S4500 field emission microscope and transmission electron microscopy (TEM) using a Hitachi HF2000 field emission microscope. The optical and electrical properties of the films were determined through optical transmission, photoluminescence using a 10 mW He-Cd laser and Hall effect measurements.

RESULTS

Undoped $Al_xGa_{1-x}N$ Thin Films on (00•1) Sapphire

The undoped $Al_xGa_{1-x}N$ thin films were grown on sapphire in the same growth conditions as for the GaN epilayers we reported earlier [2], except that the Al source partial pressure in the reactor was varied. The films were about 1 μm thick as determined by SEM, corresponding to a growth rate of about 1 μm/hr. The samples were transparent with smooth surface morphologies, and they were insulating. The alloy composition was determined both by optical transmission and X-ray diffraction. However, the X-ray diffraction peak positions were shifted from their values for bulk films due to significant residual strain in the thin (00•1) $Al_xGa_{1-x}N$ films which resulted from a combination of lattice and thermal mismatch between the film and the Al_2O_3 substrate [7]. Therefore, optical transmission offered a better measure of the Al concentration than X-ray diffraction and this is the method we used in the rest of the paper. Figure 1 shows typical optical transmission spectra of $Al_xGa_{1-x}N$, demonstrating that the entire ternary alloy composition is achieved. Figure 2 shows the X-ray rocking curve linewidths of the 00•2 diffraction peak for selected alloy compositions. The linewidths were lower than 4 arcmins for up to 50% Al, which represents the lowest values ever reported for such $Al_xGa_{1-x}N$ compounds. These linewidths are more than 10 times narrower than those reported in [8].

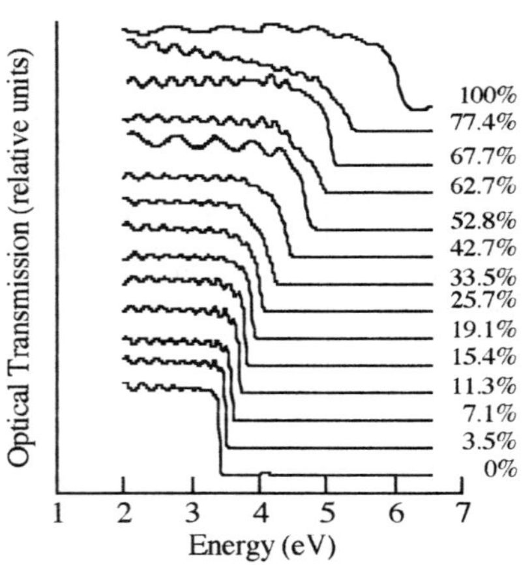

Figure 1. Room temperature optical transmission spectra of $Al_xGa_{1-x}N$ thin films on sapphire substrates.

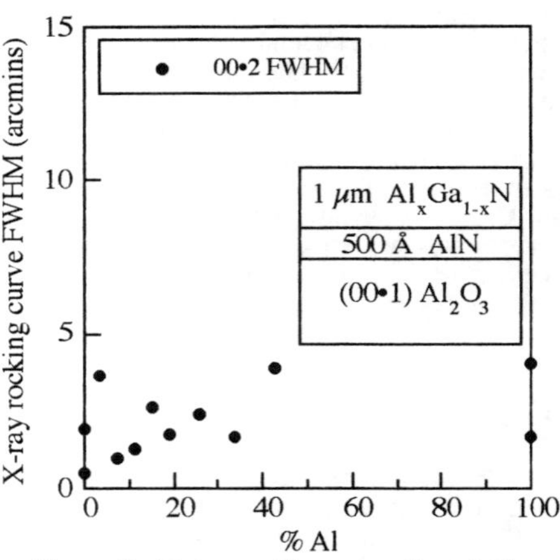

Figure 2. X-ray rocking curve linewidths of selected $Al_xGa_{1-x}N$ thin films on sapphire substrates.

<u>Doping of $Al_xGa_{1-x}N$</u>

N-type and p-type doping of $Al_xGa_{1-x}N$ compounds was achieved by incorporating the dopant during the epitaxial growth. Doping was studied as a function of Al concentration and dopant flow rate.

Figure 3 shows the resistivity, electron mobility and concentration in $Al_xGa_{1-x}N$ thin films doped with a fixed flow of SiH_4 as a function of alloy composition. It shows that the resistivity increases exponentially with Al concentration. The values for the resitivity were about 1 order of magnitude lower than those reported in [9] and several orders lower than those reported in [10]. Mobilities as high as 80 cm^2/Vs were measured on $Al_{0.2}Ga_{0.8}N$:Si, which is the highest value ever reported for such a high Al concentration. The graph also shows that $Al_xGa_{1-x}N$ compounds can be n-type doped for an Al concentration higher than 50%, which is very promising for the realization of device structures with large band offsets using III-Nitride materials. This is first time highly conductive $Al_xGa_{1-x}N$ films have been obtained for such high Al concentrations.

Figure 4 shows the resistivity of $Al_xGa_{1-x}N$ thin films doped with a fixed flow of Cp_2Mg as a function of alloy composition. In order to activate the incorporated Mg, the samples were subjected to thermal annealing prior to any electrical measurement. Here again, the plot shows that the resistivity increases exponentially with Al concentration, from ~ 2 Ω.cm for GaN up to ~ 10^5 Ω.cm for $Al_{0.3}Ga_{0.7}N$.

The n-type doping of $Al_{0.2}Ga_{0.8}N$ was studied as a function of SiH_4 and GeH_4 flow rates. The carrier concentrations increased linearly with dopant flow rate, up to 3×10^{19} cm^{-3}. The surface morphology of the films was not deteriorated by the doping in this range and the films were free from cracks.

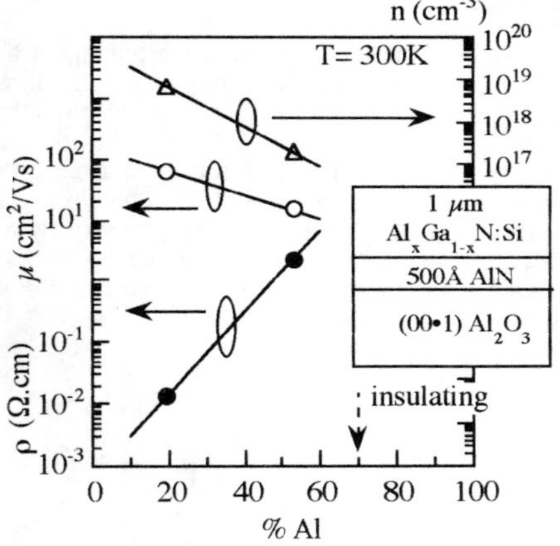

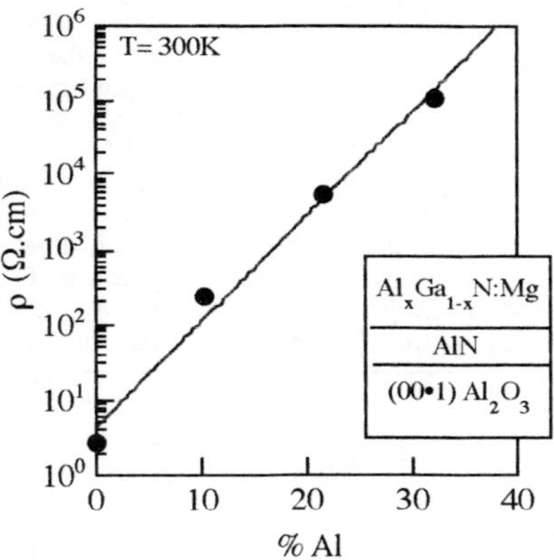

<u>Figure 3</u>. Room temperature resistivity, electron mobility and concentration of Si-doped $Al_xGa_{1-x}N$ thin films.

<u>Figure 4</u>. Room temperature resistivity of Mg-doped $Al_xGa_{1-x}N$ thin films.

<u>$Al_xGa_{1-x}N$ Based Bragg Reflectors</u>

Building on our success to grow high quality doped $Al_xGa_{1-x}N$ films, we realized Bragg reflectors on basal plane sapphire substrates. They consisted of 20 periods of Si-doped {$Al_{0.5}Ga_{0.5}N$ / $Al_{0.2}Ga_{0.8}N$} multilayers grown on an AlN buffer layer. The period thickness was on the order of 1000 Å, with each layer thickness on the order of 400-500 Å. Figure 5 shows the room temperature optical transmission for selected quarter-wave stacks. The dip in the transmission spectrum corresponds to the peak reflectivity of the structure. A peak reflectivity higher than 60% may be estimated from these curves. By varying the multilayer period thickness, we were able to tune the peak reflectivity from 330 nm to 456 nm. The structures were electrically conductive with high mobilities (~100 cm^2/Vs). These results are very promising for the realization of vertical cavity surface emitting blue lasers using III-Nitride materials.

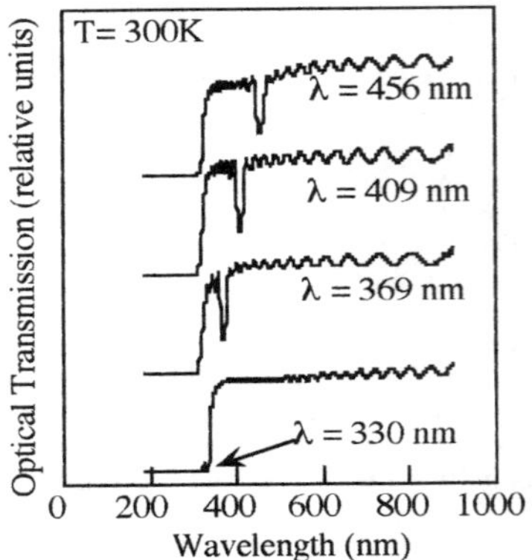

Figure 5. Room temperature optical transmission of selected Bragg reflectors using {Al$_{0.5}$Ga$_{0.5}$N / Al$_{0.2}$Ga$_{0.8}$N} multilayers

Two Dimensional Electron Gas in Al$_x$Ga$_{1-x}$N-GaN Heterostructures

In order to determine and improve the quality of our heterostructure interfaces, we attempted to achieve two dimensional electron gas in Al$_x$Ga$_{1-x}$N:Si / GaN structures on basal plane sapphire substrates. More precisely, a 0.7 μm thick insulating GaN was first grown on the AlN buffer layer. Then a ~30 Å thick undoped Al$_{0.25}$Ga$_{0.75}$N was grown, followed by a ~570Å thick Si-doped Al$_{0.25}$Ga$_{0.75}$N electron emitter layer. These thicknesses and the ternary composition were arbitrarily chosen. A typical X-ray diffraction spectrum of this structure is shown in Figure 6. The electrical properties, as determined by Hall measurements as a function of temperature, are shown in Figure 7. The sheet carrier density was constant near 10^{13} cm^{-3}, while the mobility increased up to 600 cm^2/Vs when the temperature was reduced. This behavior of the sheet carrier density and mobility clearly demonstrate the presence of a two dimensional electron gas in this structure.

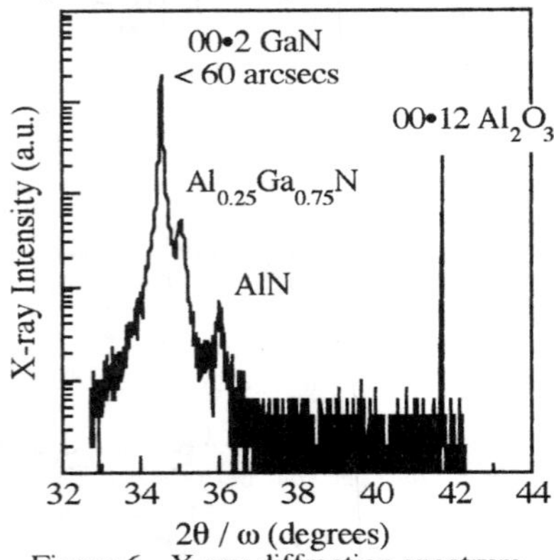

Figure 6. X-ray diffraction spectrum of a Al$_{0.25}$Ga$_{0.75}$N:Si / GaN structure exhibiting 2 DEG.

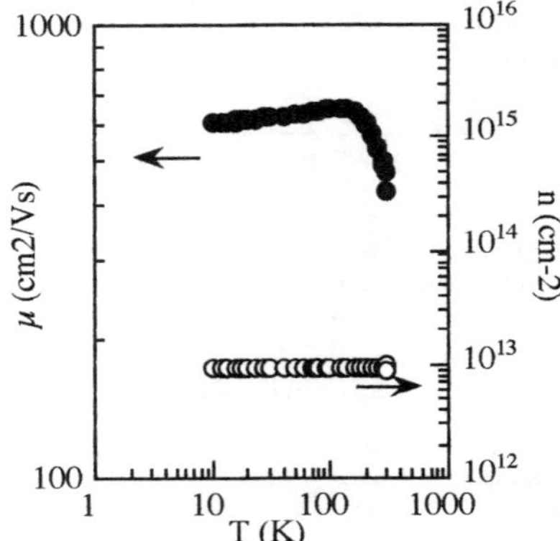

Figure 7. Temperature dependent Hall data for a Al$_{0.25}$Ga$_{0.75}$N:Si / GaN structure exhibiting 2DEG.

A more recent sample exhibited mobilities of ~1000 cm²/Vs and 2000 cm²/Vs at 300K and 77K respectively, but the temperature dependent electrical measurements were not measured. These mobilites are still lower than those reported recently [11], but the results presented here were obtained on unoptimized structures. More optimization is needed to further enhance the carrier mobility.

$Al_xGa_{1-x}N$ / GaN Superlattices

To further characterize our $Al_xGa_{1-x}N$ / GaN structures, we grew a few superlattices and structurally characterized them with X-ray diffraction and transmission electron microscopy. The X-ray diffraction spectrum of a superlattice consisting of 13 period {500Å $Al_{0.2}Ga_{0.8}N$ / 50Å GaN} structure grown on a sapphire substrate showed two main peaks corresponding to the bulk GaN layer and the $Al_{0.2}Ga_{0.8}N$ layers. The widths of these peaks were about 100 arcsecs. Satellite peaks can be clearly observed up to the tenth peak, confirming the excellent quality of the films. The TEM analysis will be discussed below.

Transmission Electron Microscopy of $Al_xGa_{1-x}N$ / GaN Heterostructures

The cross section TEM samples were prepared by first M-bonding two pieces face-to-face, then mechanically polishing down (~10 μm), followed by Ar^+ ion milling to reach electron transparency. The TEM observations were carried out with the microscope operated at 200 keV. The samples were viewed in bright field with $g=00\bullet2$.

Very low threading dislocation densities were achieved in $Al_xGa_{1-x}N$ / GaN structures. For example, a cross sectional TEM micrograph of a typical GaN / $Al_{0.33}Ga_{0.67}N$ / GaN heterostructure on sapphire substrate showed how threading screw and mixed dislocations were annihilated at each successive interface. A screw and mixed dislocation density of 10^7 cm^{-2} in the top GaN layer was obtained [12]. Further measurements are under way to confirm such low values.

The TEM cross sectional micrograph of a 15 period {100Å $Al_{0.33}Ga_{0.67}N$ / 50Å GaN} superlattice showed extremely sharp interfaces. High resolution micrographs showed that the interfaces were atomically sharp.

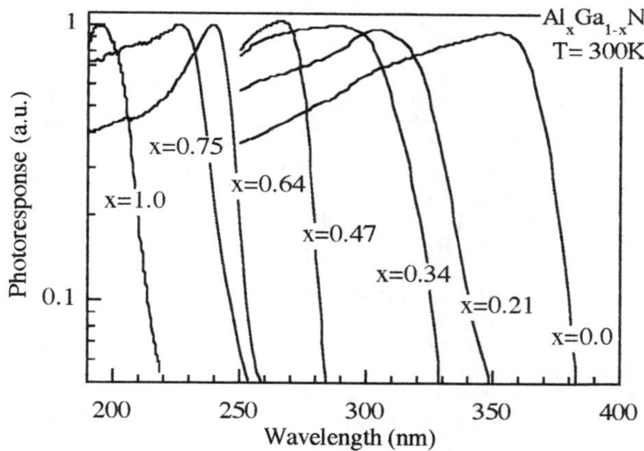

Figure 8. Room temperature normalized photoresponse of $Al_xGa_{1-x}N$ photodetectors.

$Al_xGa_{1-x}N$ Ultraviolet Photoconductors

Our success in achieving high quality $Al_xGa_{1-x}N$ alloys allowed us to realize ultraviolet photodetectors with various cut-off wavelengths. The photoconductors were fabricated using the

$Al_xGa_{1-x}N$ epilayers (0.5 to 1.5 μm) deposited as described previously. Metal contacts were deposited onto the film to realize low resistance ohmic contacts. The spectral responsivity of these devices was measured using a Xe arc lamp, focused into a monochromator after an optical chopper. The detector was placed at the exit of the monochromator and a standard synchronous detection scheme was used. The spectral power of the Xe lamp was calibrated using a Si ultraviolet photodetector.

Figure 8 shows the normalized photoresponse of these detectors. The cut-off wavelengths can be clearly tailored from 365 nm to 200 nm with the alloy composition. To the best of our knowledge, this is the only report of photodetectors with cut-off wavelengths as low as 200 nm, using III-Nitride materials.

CONCLUSIONS

In conclusion, we have reported the MOCVD growth, n-type and p-type doping and characterization of $Al_xGa_{1-x}N$ alloys on sapphire substrates. We have presented the fabrication of $Al_xGa_{1-x}N$ based Bragg reflectors with controlled peak reflectivity wavelength and the demonstration of two dimensional electron gas in III-Nitrides. $Al_xGa_{1-x}N$ / GaN heterostructures and superlattices exhibited sharp interfaces and clear satellite peaks in the X-ray diffraction spectrum. We have reported the reduction of the density of screw and mixed threading dislocations in $Al_xGa_{1-x}N$ / GaN structures to 10^7 cm^{-2} using high quality materials and interfaces. Finally, we have demonstrated $Al_xGa_{1-x}N$ based UV photodetectors with tailored cut-off wavelengths from 365 to 200 nm.

ACKNOWLEDGEMENTS

The authors wish to express their special thanks to M. Yoder, Y.S. Park and C. Wood at ONR and A. Husain at DARPA for their permanent support and interest. The authors would also like to thank M. Ahoujja and W. Mitchel at Wright Laboratory for the temperature dependent Hall effect measurements. K.S. Kim acknowledges the support from the Korea Science and Engineering Foundation. V.P. Dravid and W. Qian are thanked for the TEM microanalysis. This work is supported by the Ballistic Missile Defense Organization through ONR grant No. N00014-93-1-0235, and by DARPA/ONR through grant No. N00014-96-1-0714.

REFERENCES

1. A. Saxler, P. Kung, C.J. Sun, E. Bigan, and M. Razeghi, Appl. Phys. Lett. *64*, 339 (1994).
2. P. Kung, A. Saxler, X. Zhang, D. Walker, T.C. Wang, I. Ferguson, and M. Razeghi, Appl. Phys. Lett. *66*, 2958 (1995).
3. X. Zhang, P. Kung, D. Walker, A. Saxler, and M. Razeghi in <u>Gallium Nitride and Related Materials</u>, edited by F.A. Ponce, R.D. Dupuis, S. Nakamura, and J.A. Edmond (Mater. Res. Soc. Proc. *395*, Pittsburh, PA, 1996) pp. 625-629.
4. X. Zhang, D. Walker, A. Saxler, P. Kung, J. Xu, and M. Razeghi, Electron. Lett. *32*, 1622 (1996).
5. X. Zhang, P. Kung, D. Walker, J. Piotrowski, A. Rogalski, A. Saxler, and M. Razeghi, Appl. Phys. Lett. *67*, 2028 (1995).
6. P. Kung, X. Zhang, D. Walker, A. Saxler, J. Piotrowski, A. Rogalski, and M. Razeghi, Appl. Phys. Lett. *67*, 3792 (1995).
7. K. Hiramatsu, T. Detchprohm, and I. Akasaki, Jpn. J. Appl. Phys. *32*, 1528 (1993).
8. T.J. Kistenmacher, D.K. Wickenden, M.E. Hawley, and R.P. Leavitt, Appl. Phys. Lett. *67*, 3771 (1995).
9. H. Murakami, T. Asahi, H. Amano, K. Hiramatsu, N. Sawaki, and I. Akasaki, J. Cryst. Growth *115*, 648 (1991).
10. S. Yoshida, S. Mizawa, and S. Gonda, J. Appl. Phys. *53*, 6844 (1982).
11. M.A. Khan, Q. Chen, C.J. sun, M. Shur, and B. Gelmont, Appl. hys. Lett. *67*, 1429 (1995).
12. M. Razeghi, presented at the International Symposium on Nitrides, St. Malo, France, 1996 (unpublished).

A MODEL FOR INDIUM INCORPORATION IN THE GROWTH OF InGaN FILMS

E.L. PINER*, F.G. MCINTOSH**, J.C. ROBERTS**, K.S. BOUTROS***, M.E.
AUMER**, V.A. JOSHKIN**, N.A. EL-MASRY*, S.M. BEDAIR**, S.X. LIU**
*North Carolina State University, MS&E Department, Raleigh, NC 27695
**North Carolina State University, ECE Department, Raleigh, NC 27695
*** Philips Research Laboratories, Briarcliff Manor, NY
E-mail: piner@mat.mte.ncsu.edu

ABSTRACT

The development of high quality indium based III-nitride compounds is lagging behind the corresponding aluminum and gallium based compounds. Potential problems confronting the growth of epitaxial and double heterostructure InGaN will be discussed. A mass balance model is presented describing the competing reaction pathways occurring during the growth of indium containing compounds. Atomic layer epitaxy and metalorganic chemical vapor deposition grown InGaN films will be used to explain this model. Also, the growth parameters leading to the attainment of high InN percentages, reduced indium metal formation, and improved structural and optical properties of indium containing nitrides will be discussed.

INTRODUCTION

The development of high quality III-nitride compound semiconductor thin films has led to the commercial production of light emitting diodes (LEDs)[1] and the demonstration of laser diodes (LDs)[2] based on these materials. InGaN epitaxial films and their corresponding double heterostructures (DHs) play a critical role in the development of these nitride based devices. Research into the growth of InGaN films is hindered by several problems. These include: 1) The weak In-N bond necessitating the use of a high equilibrium vapor pressure of nitrogen to prevent In-N dissociation and the resulting desorption of the reactive indium species from the film surface. For comparison, the equilibrium vapor pressure of nitrogen over InN is several orders of magnitude higher than that of AlN and GaN.[3] 2) Indium metal droplet formation on the nitride film surface. Under certain growth regimes, indium metal can form leading to poor optical and electrical properties as well as low InN percentages in the growing film. 3) The influence of the reactive gas fluxes on the InN percent. The overall concentration of indium that is incorporated strongly depends on not only the indium precursor flow but also the gallium precursor and ammonia flows into the reactor.

This paper will review our current progress in indium based III-nitride compounds starting with a mass balance model that helps explain the reaction processes occurring during the deposition of indium containing nitride films.

EXPERIMENT

A versatile growth reactor has been designed for the growth of III-nitride compounds based on the rotating susceptor and has been described elsewhere.[4,5] The versatility of the system is due to its unique susceptor design, shown in Figure 1, which allows the system to be operated in three different growth modes: Metalorganic chemical vapor deposition (MOCVD), molecular stream epitaxy (MSE), and atomic layer epitaxy (ALE). For MOCVD and MSE, ammonia is injected into the reactor through Port 2 with Port 3 carrying only nitrogen. For MOCVD, the Rotating Susceptor, which holds the sample, remains stationary under the dual flow of organometallics and ammonia during the entire growth. For MSE, the rotating part rotates, thus intermittently exposing the sample to the organometallic/ammonia stream. This allows a thin layer of material to be deposited and effectively annealed during each rotation cycle. For ALE growth, nitrogen is injected through Port 2 and ammonia is injected through Port 3 thus separating the reactant gases and eliminating the possibility for gas phase reactions. The sample is first exposed to the organometallic gas stream then rotated under the ammonia gas stream. When the sample

Mat. Res. Soc. Symp. Proc. Vol. 449 ©1997 Materials Research Society

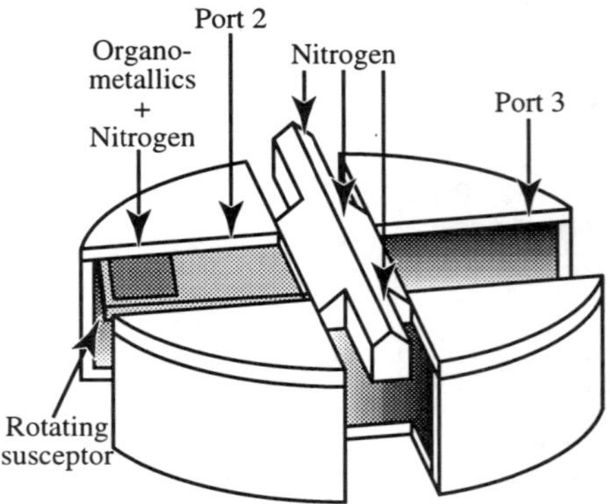

Figure 1.   Rotating susceptor design utilized for the growth of III-nitride compounds.

returns to the organometallic gas stream one cycle is complete. It should be emphasized that any of the three growth modes can be selected on demand at any time during a single experiment.

RESULTS

The incorporation of indium to form InGaN is controlled by several simultaneous competitive processes. The salient rate processes involved are:

1) Indium incorporation in the solid ternary alloy, F_s (atoms/cm^2/sec)
2) Atomic indium desorption from the growth surface, F_d (atoms/cm^2/sec)
3) Indium incorporation as indium metal droplets, F_m (atoms/cm^2/sec)

It should be noted that the rate process, F_d, can be due to the desorption of solitary adsorbed indium atoms or due to the breaking of the existing In-N bond. Thus, for an EDMIn incident flux, F_{in} (atoms/cm^2/sec), the following mass balance equation can be assumed:

$$F_{in} = F_s + F_d + F_m \tag{1}$$

For growth temperatures in the 600 - 800°C range it is expected that EDMIn is fully decomposed to indium atoms at the surface. Based on the In$_x$Ga$_{1-x}$N growth both by MOCVD (0 < x < 0.40) and ALE (0 < x < 0.25), we found that the growth temperature plays an important role as shown in Figure 2. In this figure the InN percent obtained by MOCVD critically depends on the growth temperature. This dependence has also been observed by others.[6]

The relative stability of gallium compared to indium on the nitride surface is supported by the observed behavior of gallium and indium atoms on arsenic terminated GaAs and InGaAs surfaces. The lifetimes of gallium and indium on an arsenic surface at 600°C are 10 and 1 second, respectively.[7] Thus it is safe to conclude that in the nitride system the desorption rate for indium atoms is greater than for gallium atoms and gallium atoms at the surface will have a very high chance of being incorporated in the growing surface. This explains the earlier MOCVD results [6,8] where a vapor pressure ratio of indium to gallium precursors as high as 12 was used to achieve an InN content of about 20% only. The reaction pathway, F_d, can also be used to explain the lower operating temperature range of ALE compared to MOCVD since ALE relies on growth interruption that can be much longer than the residence time of indium atoms on the growing surfaces. Thus, InN percents greater than about 25% were difficult to achieve.

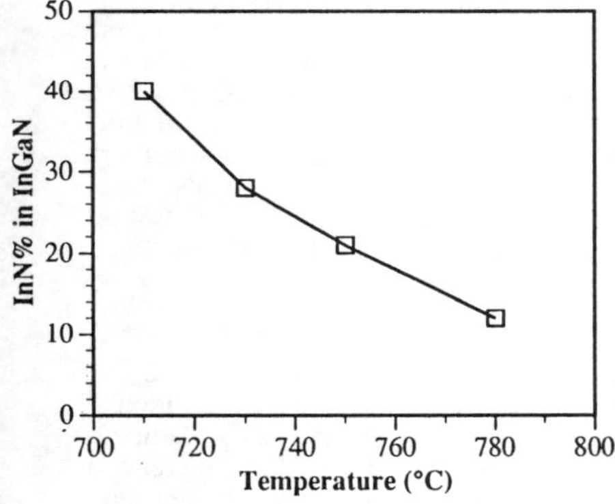

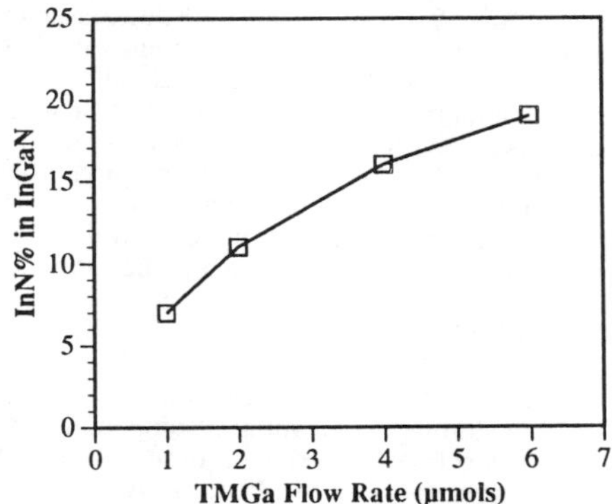

Figure 2. Temperature dependence of In incorporation by MOCVD.

Figure 3. Dependence of In incorporation on TMGa flow for the same EDMIn/TMGa ratio.

With reduced growth temperature F_d will be reduced and there will be a competition between F_m and F_s reaction pathways. Both indium and gallium atoms on the surface will react with ammonia independently or collectively to form $(InN)_x$ and $(GaN)_{1-x}$ in the ternary alloy according to the following reactions:[9]

$$In + NH_3 \longrightarrow (InN)_x + 3/2H_2 \qquad (2)$$
$$Ga + NH_3 \longrightarrow (GaN)_{1-x} + 3/2H_2 \qquad (3)$$

since the Ga-N bond is stronger than the In-N bond, the equilibrium constant for equation (3) is larger than equation (2). This will allow gallium atoms to be retained in the growing film more efficiently than indium atoms. This can result in surface stabilization by the Ga-N thereby aiding indium incorporation since the residence time for the indium atom is very short compared to the gallium atom. Therefore, in certain growth regimes, indium incorporation can increase with TMGa flow enhancing reaction pathway F_s and will result in higher values of InN as suggested by the data in Figure 3 for the ALE growth at 650°C for the same EDMIn/TMGa partial pressure ratio of 0.3 in the gas phase. The increase in TMGa flux in Figure 3 from 1 μmole/min to 6 μmole/min results in an increase in the InN percent from 7 to 20, respectively. A similar trend has been observed in the MOCVD growth of InGaN where the EDMIn flow was kept constant and the TMGa flow was increased. It should be mentioned that, both for ALE and MOCVD, increasing the TMGa flow rate will also increase the growth rate that can help in trapping indium atoms, however, InGaN films with poor optical and structural quality were obtained.

Intuitively, to incorporate higher percentages of InN in the growing film, lower growth temperatures and a high EDMIn flux, F_{in}, are needed, and that is where growth difficulties are encountered. The main problem will be the formation of indium metal droplets on the film surface, especially if the ammonia flow is fairly low, as detected by SEM, optical microscopy, and x-ray diffraction. The presence of sufficient densities of indium atoms diffusing across the film surface increases the probability of forming indium clusters. Once these clusters form a critical size, they become thermodynamically stable and can continue to grow. These indium droplets will act as sinks for the available indium surface atoms thus competing with and in some cases dominating the process of indium incorporation in the InGaN film. These indium droplets on the surface can explain our experimental observation that the percentage of InN in the film tends to saturate with increasing EDMIn flux in this growth regime showing almost no dependence on the EDMIn/(EDMIn + TMGa) partial pressure ratio in the gas phase, a dependence commonly observed in the growth of other III-V indium based compounds. Another observation from the x-ray diffraction spectra is that the amount of indium metal on the surface decreases with increasing ammonia flow. This can be due to the increased availability of nitrogen bonding sites for the indium as a result of the increased amount of nitrogen radicals from the ammonia. Indium metal can be completely etched away by HCl indicating its presence on the film surface. These droplets are mainly indium metal with traces of carbon and no gallium signal by EDS indicating they do not act as sinks for gallium atoms. This can be attributed to the stronger Ga-N bond versus the In-N bond and their effect on the surface diffusion properties, such as diffusion lengths of the gallium and indium atoms. Therefore, to achieve high quality, high percentages of InN in InGaN, the EDMIn, TMGa, and ammonia fluxes as well as the growth temperature and growth rate must be optimized for a given reactor to avoid the formation of indium droplets on the film surface.

CONCLUSIONS

The material quality of indium containing nitride alloys lags substantially behind that obtained in aluminum and gallium compounds. Potential problems with indium incorporation include the weak In-N bond, indium segregation at the growing surface, and the lack of lattice matched systems for heterostructures. A mass balance model is presented describing the reaction pathways available to the indium atom during deposition; desorption, incorporation in the solid phase, and metal formation.

REFERENCES

1. S. Nakamura, M. Senoh, N. Iwana, and S. Nagahama, Jpn. J. Appl. Phys. **34**, L797 (1995).

2. S. Nakamura, M. Senoh, S. Sagahama, N. Iwasa, T. Yamata, T. Matsushita, H. Kiyoku, and Y. Sugimoto, Jpn. J. Appl. Phys. **35**, L74 (1996).

3. T. Matsuoka, H. Tanaka, T. Sasaki, and A. Datsui, Inst. Phys. Conf. **106**, 141 (1989).

4. K. S. Boutros, F. G. McIntosh, J. C. Roberts, S. M. Bedair, and N. A. El-Masry, Appl. Phys. Lett. **67**, 1795 (1995).

5. N. Karam, T. Parados, W. Rowland, J. Schetzina, N. El-Masry, and S. M. Bedair, Appl. Phys. Lett. **67**, 94 (1995).

6. T. Matsuoka, N. Yoshimoto, T. Sakaki, and A. Katsui, J. Electron. Mater. **21**, 157 (1992).

7. J. Arthur, J. Appl. Phys. **39**, 4032 (1968).

8. S. Nakamura, Microelectron. J. **25**, 651 (1994).

9. H. Seki, and A. Koukitu, J. Cryst. Growth **98**, 118 (1989).

THE COMPOSITION PULLING EFFECT IN InGaN GROWTH ON THE GaN AND AlGaN EPITAXIAL LAYERS GROWN BY MOVPE

YASUTOSHI KAWAGUCHI, MASAYA SHIMIZU, KAZUMASA HIRAMATSU
AND NOBUHIKO SAWAKI
Nagoya University, Department of Electronics
Furo-cho, Chikusa-ku, Nagoya 464-01, Japan

ABSTRACT

InGaN has been grown on GaN and AlGaN epitaxial layers by metalorganic vapor phase epitaxy (MOVPE) and "the composition pulling effect" at the initial growth stage of InGaN has been studied in relation to the lattice mismatch between InGaN and the bottom epitaxial layers. Crystalline quality of InGaN is good near the interface of InGaN/GaN and the composition of InGaN is close to that of GaN. With increasing growth thickness, the crystalline quality becomes worse and the indium mole fraction is increased. The composition pulling effect becomes stronger with increasing lattice mismatch.

INTRODUCTION

High brightness blue and green light-emitting diodes (LEDs) used III-V nitride semiconductors have been already put to practical use [1, 2]. Recently, the realization of short wavelength laser diodes (LDs) has been reported [3, 4]. These optoelectronic devices consist of double heterostructures (DH) which consist of an InGaN active layer sandwiched between GaN and AlGaN clad layers. Therefore, the large lattice mismatch between InGaN/GaN or InGaN/AlGaN probably has a great influence on not only the characteristics of InGaN but also the crystal growth mechanism of InGaN.

Our previous papers [5, 6] reported "the composition pulling effect", namely, that the indium mole fraction is small at the initial growth stage of InGaN grown on the GaN epitaxial layer and increases with increasing growth thickness, and suggested that this effect is affected by the strain due to the lattice mismatch in InGaN/GaN.

In this paper, we fabricated InGaN/GaN and InGaN/AlGaN heterostructures with different lattice mismatches and investigated the composition pulling effect as a function of the lattice mismatch. For comparing with these heterostructures we grew InGaN directly on the low temperature (LT) buffer layer in which the lattice strain is relaxed near the LT buffer layer.

Mat. Res. Soc. Symp. Proc. Vol. 449 © 1997 Materials Research Society

EXPERIMENT

Crystal growth of InGaN and other III-V nitride semiconductors was performed by vertical-type metalorganic vapor phase epitaxy (MOVPE). The growth was carried out at atmospheric pressure. C-plane sapphire (α-Al_2O_3) was used as a substrate. Trimethylgallium (TMG), trimethylindium (TMI), trimethylaluminum (TMA) and ammonia (NH_3) were used as Ga, In, Al and N source gases, respectively. Hydrogen (H_2) gas was used as carrier gas during the growth process.

The substrate was heated to 1150°C in a stream of H_2 for 10 minutes for thermal cleaning of the substrate. Then, we fabricated two types of heterostructures; (1) InGaN grown on a GaN or AlGaN epitaxial layer and (2) InGaN grown on a LT buffer layer. The lattice deformation due to the lattice mismatch exists in InGaN of (1), while it is relaxed by the LT buffer layer in InGaN of (2). In the case of (1), the substrate temperature was lowered to 600°C to deposit the AlN LT buffer layer. Next, the substrate temperature was elevated to 1050°C to grow the GaN or $Al_{0.09}Ga_{0.91}N$ epitaxial layer (about 2.5μm thick). Again, the substrate temperature was lowered to grow InGaN. In the case of (2), the substrate temperature was lowered to 600°C to deposit the AlN, GaN or AlGaN LT buffer layer (about 50nm). Next, the substrate temperature was elevated to 1050°C for thermal annealing for a few seconds. Immediately, the substrate temperature was lowered to grow directly InGaN.

In order to change the lattice mismatch between InGaN and the bottom epitaxial layer, we changed the indium mole fraction of InGaN by controlling the growth temperature (800°C and 840°C) and also used the different bottom epitaxial layers of GaN and $Al_{0.09}Ga_{0.91}N$. During the InGaN growth, the flow rate of NH_3, TMG and TMI were maintained at 4.0 l/min, 7.33μmol/min and 18.7μmol/min. Details of the growth conditions and growth processes are described in other articles [5,6].

The indium mole fraction of InGaN was determined by Electron Probe Microanalysis (EPMA). Photoluminescence (PL) measurement was carried out at room temperature (R.T.) by using a He-Cd laser (325nm).

RESULTS AND DISCUSSION

Figure 1 shows surface and cross-sectional SEM images of InGaN grown on the GaN epitaxial layer at the growth temperature (Tg) = 800°C (type (1) sample) for different growth times of 5min, 10min, 15min and 60min. Crystalline defects begin to generate on the surface at the InGaN thickness of 0.2μm (fig. 1 (a)), the surface becomes gradually rough (fig. 1(b), (c)) and the surface of InGaN grown for 60min is very rough (fig. 1(d)). As seen in fig. 1(h), InGaN was divided into two types of areas : "a homogeneous, layer growth area" and "a rough, column growth area".

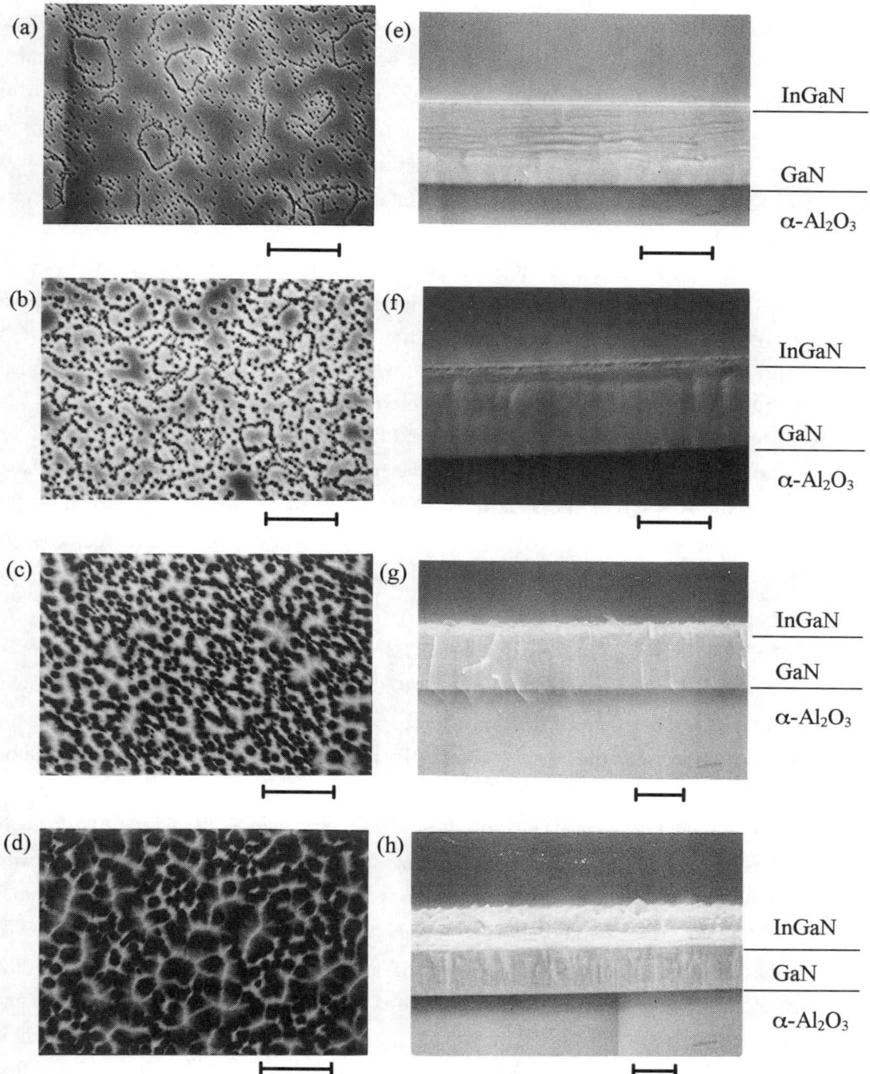

Fig.1. Surface ((a)-(d)) and cross-sectional ((e)-(h)) SEM images of InGaN grown on the GaN epitaxial layer at 800℃. InGaN were grown for (a),(e):5min,(b),(f):10min,(c),(g):15min and (d),(h):60min.Markers represent 2μm.

On the other hand, figure 2 shows a cross-sectional SEM image of InGaN grown directly on the AlN LT buffer layer at Tg=800℃ (type (2) sample). The indium mole fraction was 0.2 from EPMA. The surface is smooth and the homogeneous InGaN growth is obtained. Figure 3 shows PL spectra of InGaN grown directly on the AlN buffer layer at Tg=800℃ for different growth times of 15min, 30min and 60min (the thickness was varied from 0.5μm to 2μm). The spectra indicate the band-edge-emissions in which the peak wavelength was constant at 420nm. These results agreed with InGaN grown on another LT buffer layers (GaN and AlGaN).

Figure 4 shows PL spectra of InGaN grown on the GaN epitaxial layer at Tg = 800℃. The thin InGaN grown for the short growth time has a sharp PL spectrum at a peak wavelength of 387nm, and its indium mole fraction is 0.07 from EPMA. As the growth thickness increases, the peak wavelength shifts toward longer wavelength and it becomes constant at 420nm. The indium mole fraction determined by EPMA increased from 0.07 to 0.2, whose shift corresponds to the red shift of the PL peak wavelength. Figure 5 shows the growth thickness dependence of the PL peak wavelength for various InGaN layers that have been grown on the AlN LT buffer layer at Tg=800℃ (a) and Tg=840℃ (d), grown on the GaN epitaxial layer at Tg=800℃ (b) and Tg=840℃ (e), and grown on the $Al_{0.09}Ga_{0.91}N$ epitaxial layer at Tg=800℃ (c). Here, we can consider three different lattice mismatches between InGaN and the bottom epitaxial layers ((b), (e) and (c)).

In InGaN grown directly on the LT buffer layer, the peak wavelength was constant at 420nm for 800℃ growth (a) and 383nm for 840℃ growth (d) regardless of the growth thickness, suggesting the composition is not affected by the interface and is given simply by the thermal equilibrium between the gas phase and solid phase. For this reason the composition pulling effect does not occur in this case.

On the other hand, the thin InGaN grown at 800℃ on the GaN (b) and the $Al_{0.09}Ga_{0.91}N$ (c) epitaxial layers indicate shorter wavelength than that of (a) and with increasing growth thickness the peak wavelength shifts drastically toward longer wavelength and finally becomes the same value as (a). Furthermore, the thin InGaN grown at 840℃ on the GaN (e) epitaxial layer indicate a shorter wavelength than that of (d) and with increasing growth thickness the peak wavelength shifts toward longer wavelength drastically and finally becomes the same value as (d). The critical thickness at which the peak wavelength begins to switch is corresponding to that at which the surface becomes rough, as seen in fig. 1. It was found that the larger the lattice mismatch between InGaN and bottom epitaxial layers, the thinner the critical thickness, and that the larger the difference of lattice mismatch, the larger the compositional shift of InGaN. Thus, the composition pulling effect is strongly affected by the lattice mismatch between InGaN and the bottom layers. This implies that the indium

Fig.2. Cross-sectional SEM image of InGaN grown on the AlN LT buffer layer. InGaN were grown for 60min. Marker represents 2μm.

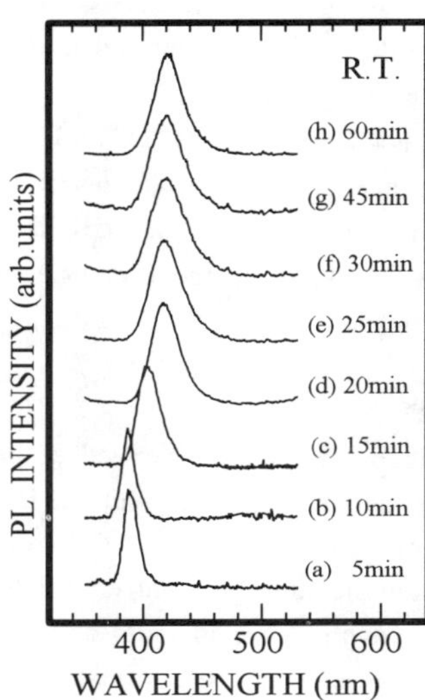

Fig.4. PL spectra at room temperature of InGaN grown on the GaN epitaxial layer. InGaN were grown for (a)5min, (b)10min, (c)15min,(d)20min, (e)25min, (f)30min, (g)45min and (h)60min.

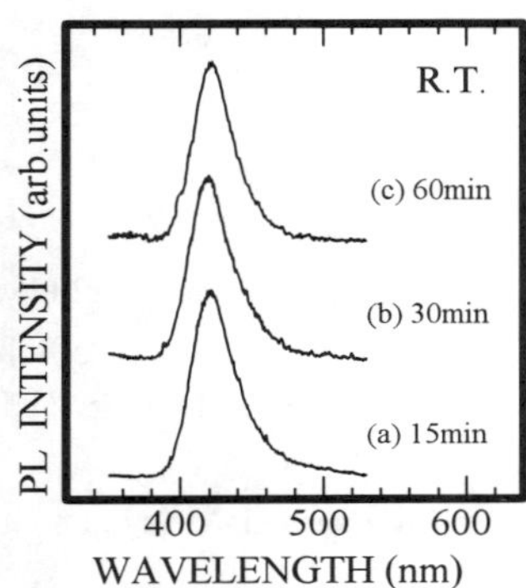

Fig.3. PL spectra at room temperature of InGaN grown on the AlN LT buffer layer. InGaN were grown for (a)15min,(b)30min and (c)60min.

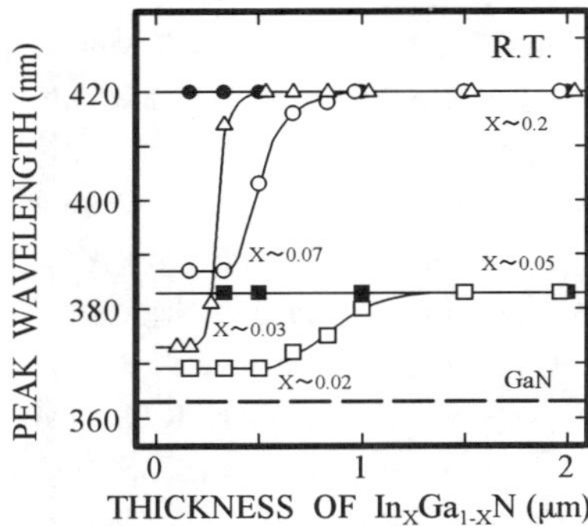

Fig.5. Growth thickness dependence of PL peak wavelength for InGaN.
(a)● InGaN on AlN LT buffer layer at Tg=800℃
(b)○ InGaN on GaN epitaxial layer at Tg=800℃
(c)△ InGaN on AlGaN epitaxial layer at Tg=800℃
(d)■ InGaN on AlN LT buffer layer at Tg=840℃
(e)□ InGaN on GaN epitaxial layer at Tg=840℃

distribution mechanism in InGaN is caused by the lattice deformation due to the lattice mismatch.   That is, indium atoms are excluded from InGaN to reduce the deformation energy during the InGaN growth.

With increasing thickness of the InGaN layer, the crystalline defects begin to generate and the surface becomes rough, hence the lattice strain is relaxed, which weakens the composition pulling effect. Finally, because the lattice strain is completely relaxed, the composition of InGaN has the same value as determined by the thermal equilibrium between the gas phase and solid phase.

SUMMARY

"The composition pulling effect" at the initial stage of InGaN grown on the GaN and the AlGaN epitaxial layers by MOVPE was investigated.   The lattice mismatch between InGaN and the bottom epitaxial layers had a great influence on this effect.   Indium atoms are excluded from the InGaN to reduce the deformation energy due to the lattice mismatch during the InGaN growth.   The composition pulling effect was weakened with increasing thickness of InGaN because the lattice deformation is relaxed by generation of crystalline defects.

ACKNOWLEDGEMENT

This work was supported in part by Mitsubishi Cable Industries Ltd.

REFERENCES

1. S.Nakamura, M.Senoh, and T.Mukai, Jpn.J.Appl.Phys. **32**, L8 (1993).
2. S.Nakamura, M.Senoh, N.Iwasa, S.Nagahara, T.Yamada and T.Mukai, Jpn.J.Appl.Phys. **34**, L1332 (1995).
3. S.Nakamura, M.Senoh, S.Nagahara, N.Iwasa, T.Yamada, T.Matsushita, H.Kiyoku and Y.Sugimoto, Jpn.J.Appl.Phys. **35**, L74 (1996).
4. S.Nakamura, M.Senoh, S.Nagahara, N.Iwasa, T.Yamada, T.Matsushita, Y.Sugimoto and H.Kiyoku, Appl.Phys.Lett. **69**, 1477 (1996).
5. M.Shimizu, Y.Kawaguchi, K.Hiramatsu and N.Sawaki, Topical Workshop on III-V Nitrides Abstracts, B-7 (Sept.21-23,1995).
6. M.Shimizu, Y.Kawaguchi, K.Hiramatsu and N.Sawaki, Solid State Electronics (to be published in 1997).

LOW-TEMPERATURE METAL-ORGANIC CHEMICAL VAPOR DEPOSITION OF GALLIUM NITRIDE ON (0001) SAPPHIRE SUBSTRATES USING A REMOTE RF NITROGEN PLASMA

CHEOLSOO SONE, MIN HONG KIM, JAE HYUNG YI, SOUN OK HEUR*, AND EUIJOON YOON
School of Materials Science and Engineering, Seoul National University, Seoul 151-742, Korea
*Inter-university Semiconductor Research Center, Seoul National University, Seoul 151-742, Korea

ABSTRACT

We report the low-temperature growth of GaN layers on (0001) sapphire substrates by a remote plasma enhanced metal-organic chemical vapor deposition in the temperature range of 500 - 800 °C. Effects of process parameters on the growth of GaN were studied. The structural quality of GaN improved as the growth temperature increased and the rf power decreased. Highly oriented GaN layers could be deposited at fairly low temperatures such as 500 °C under low rf power with low growth rate conditions.

INTRODUCTION

Low-temperature growth of III-V nitrides is attractive since it may increase the In incorporation in InGaN ternary layers and reduce thermal mismatch between the epilayers and the substrate [1]. Various growth techniques have been tried for the low-temperature growth of III-V nitrides [2-6]. Comparatively, a few works have been done in the growth of III-V nitrides by plasma assisted chemical vapor deposition [4-6], whereas many research results have been reported in the low-temperature growth of III-V nitrides by plasma assisted molecular beam epitaxy [2,3]. In this paper, we present the experimental results of the low-temperature growth of GaN on (0001) sapphire substrates by a remote plasma enhanced metalorganic chemical vapor deposition (RPE-MOCVD). In this technique triethylgallium (TEGa) and remote rf nitrogen plasma were used as reactant source gases at temperatures ranging from 500 - 800 °C.

EXPERIMENT

A schematic diagram of the RPE-MOCVD reactor is shown in Fig. 1. TEGa was chosen as a reactant source for Ga because the pyrolysis of TEGa was known to go on mainly via a low temperature β-elimination mechanism [7]. A remote rf nitrogen plasma was used to supply active nitrogen species. Plasma excitation of nitrogen was made in a quartz tube mounted on top of a vertical reactor. TEGa was injected separately into the reactor through a gas dispersal ring. Nitrogen was used as a carrier gas.

(0001) sapphire substrates were cleaned by a conventional procedure reported elsewhere [2,3]. Before the growth of GaN layers the substrates were pre-treated with a hydrogen plasma for half an hour at 700 °C for the removal of surface contaminants. Subsequently, they were pre-treated with a nitrogen plasma for an hour at 700 °C. GaN layers were grown either by a single-step process or by a two-step process. In the two-step process, 20 nm thick GaN buffer layers were grown at 500 °C, and an overgrowth was made at higher temperatures.

GaN layers were characterized *in situ* by reflection high energy electron diffraction (RHEED), scanning electron microscopy (SEM), and X-ray diffraction (XRD). Thicknesses of the GaN

Mat. Res. Soc. Symp. Proc. Vol. 449 © 1997 Materials Research Society

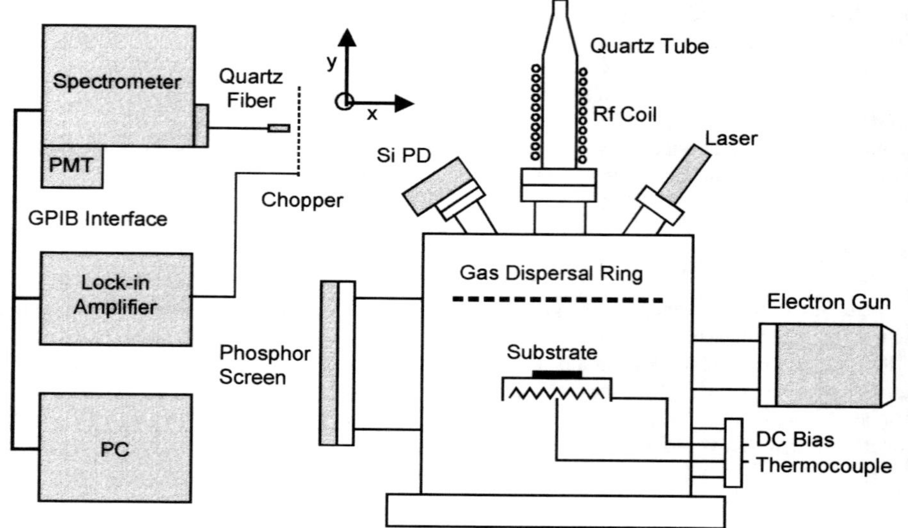

Fig. 1.   A schematic diagram of the RPE-MOCVD system with optical diagnostic tools.

layers were measured by *in situ* laser interferometry, cross-section SEM, α-step, UV-VIS interference fringes, and cross-section transmission electron microscopy (XTEM).   Optical emission spectroscopy (OES) and Langmuir probe measurements were used to diagnose the nitrogen plasmas at various process conditions.

RESULTS

An N_2^+ peak at 391.4 nm is observed in an electron cyclotron resonance (ECR) plasma by OES measurement, while atomic nitrogen emission peaks at 745, 821, and 869 nm were observed in rf plasmas at high rf powers [2,8,9].   No atomic nitrogen emission peaks were observed in our experimental conditions, as shown in Fig. 2.   In a quartz tube region the second positive nitrogen glow was dominant, but in the chamber region (2.5 cm above the substrate) relative intensities of the first positive nitrogen glow was increased.   Overall emission intensities of the nitrogen glow in the chamber region were greatly attenuated from those in the quartz tube region. It is well known that the second positive nitrogen glow is associated with the electron impact excitation of nitrogen molecules, while the first positive nitrogen glow is associated with the recombination of atomic nitrogens [8].   Thus, it may be deduced that the electron density reduces more quickly than the atomic nitrogen density in this system.   Substrates were placed outside the plasma generation region, but the glow region was observed to extend to the substrate as the rf power increased or the reactor pressure reduced.   An OES spectrum during growth in the chamber region is shown in Fig. 2(c).   New peaks related with Ga and hydrocarbons emerged, and detailed study is under way and reported elsewhere [10].   Nitrogen plasma is believed to crack TEGa in the gas phase.

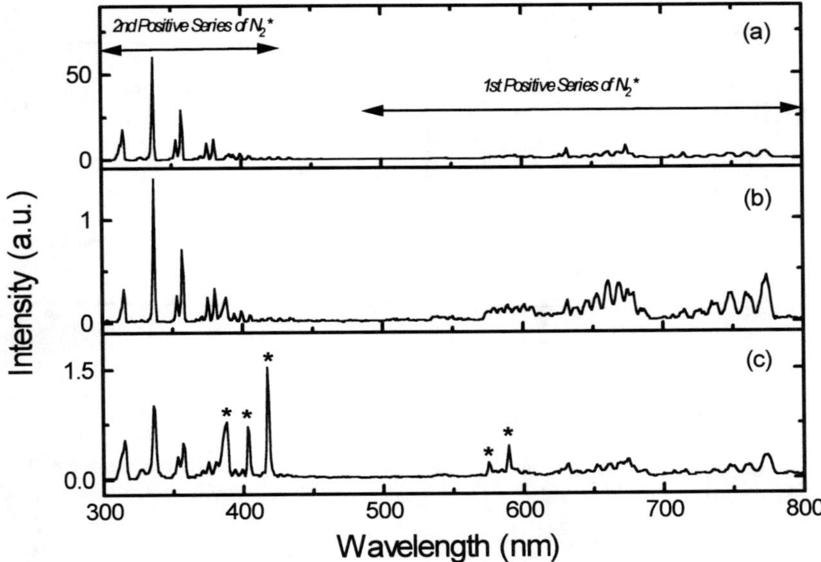

Fig. 2.   Optical emission spectra of an rf nitrogen plasma taken (a) at a quartz tube region and (b) at a chamber region 2.5 cm above the substrate,   (c) at the same chamber region during GaN growth.   New peaks from the dissociation of TEGa are marked.

Changes in GaN growth rate with growth temperature and total reactor pressure are shown in Fig. 3.   At a low rf power (60 W) with 100 sccm of nitrogen, the growth rate decreased abruptly as the growth temperatures increased above 700 °C.   However, at a higher rf power (140 W) with 250 sccm of nitrogen, the GaN growth rate could be maintain at its low temperature value up to 800 °C.   The flow rates of TEGa were similar in both cases.   It is known that TEGa is completely cracked above 300 °C [7], and the growth kinetics of GaN may be limited by mass transport process at low temperatures.   The growth rates began to decrease at high temperatures since a new limiting step, Ga desorption here, became important.   It is believed that the increased growth rate at high temperatures with the higher rf power and nitrogen flow rate results from the enhanced chemisorption of active nitrogen species on the growing GaN surface, resulting in the increased Ga incorporation in the surface [2,3].

The effect of the reactor pressure on the GaN growth rate was studied at a constant growth temperature of 700 °C.   The growth rate increased as the pressure decreased.   Homogeneous gas phase reaction may deplete the reactant species for the GaN growth as the reactor pressure increases, or the concentration of active reactants may change with pressure.

Structural properties of the GaN layers were studied by XRD measurements.   Structural quality of a single-step grown GaN improved with a lower growth rate and a lower plasma power at 500 °C as shown in Fig. 4(a) and 4(b).   Note the thickness differences of the two samples. Increased rf power may increase not only the density of active neutral nitrogen species but also the density of ions, resulting in poor layer quality.   To increase the growth rate of GaN layers with good structural properties, the growth temperatures were increased to 700 - 800 °C, and

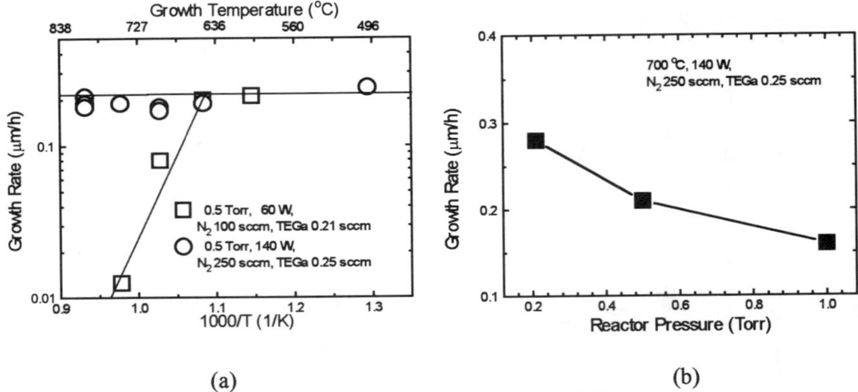

(a)

(b)

Fig. 3.   Changes in GaN growth rate with (a) temperature, and (b) reactor pressure.

their XRD patterns are shown in Fig. 4(c) and 4(d).   At high growth temperatures plasma damages could be dynamically annealed, so high quality GaN layers could be deposited at higher rf powers.

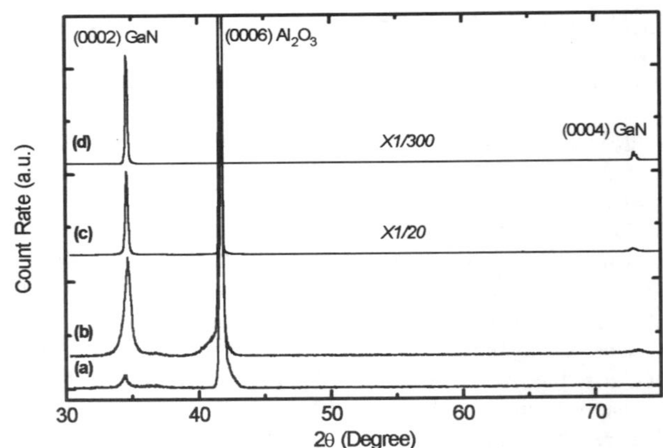

Fig. 4.   XRD patterns of GaN layers grown by a single-step process at various growth conditions.   Total pressure was fixed at 0.5 Torr.
(a) Temperature 500 °C,   plasma power 60 W, growth rate 0.28 μm/h, thickness 0.28 μm, (b) temperature 500 °C,   plasma power 40 W, growth rate 0.02 μm/h, thickness 0.02 μm, (c) temperature 700 °C,   plasma power 120 W, growth rate 0.14 μm/h, thickness 0.14 μm, (d) temperature 800 °C,   plasma power 140 W, growth rate 0.19 μm/h, thickness 0.95 μm.

98

Two-step growth of the GaN layers was made at 700 - 800 °C on a 20 nm thick buffer grown at a rf power of 40 W and 500 °C.   The surface morphology of the two-step grown GaN layers was mirror-like and had no surface structures such as hexagonal hillocks often observed in MOCVD samples at high growth temperatures.   The surface smoothness was also confirmed by the presence of optical interference fringes through UV-VIS transmission measurements. However, the surface became rough with increasing rf power and decreasing growth rate.   A spotty RHEED pattern was observed for this sample.   The XRD pattern and RHEED pattern of the the GaN with a mirror-like surface are shown in Fig. 5(a) and 5(b).   The streaky RHEED pattern shows that the surface of the GaN is smooth.   A strong (0002) GaN peak was observed in XRD pattern.   Note the logarithmic scale of the XRD pattern.

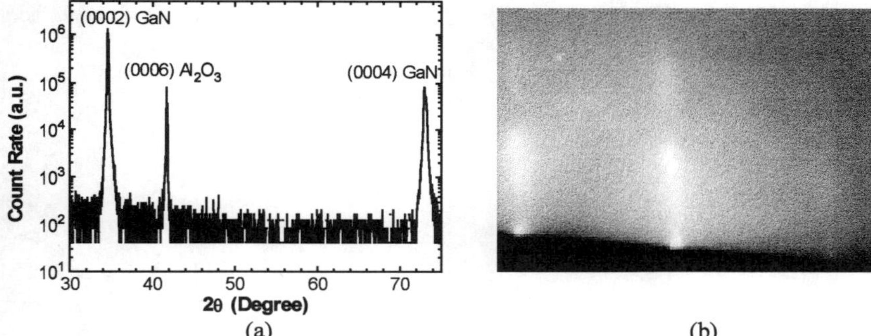

(a)   (b)

Fig. 5.   Structural and surface quality of the RPE-MOCVD grown GaN.   (a) XRD pattern of a two-step grown GaN layer by RPE-MOCVD method.   Total pressure, growth temperature, plasma power, growth rate, and total thickness was 0.5 Torr, 800 °C, 140 W, 0.2 μm/h, and 1.0 μm, respectively.   (b) RHEED pattern of the same GaN layer. The electron beam direction was along [12$\bar{3}$0] azimuth of the GaN.

CONCLUSIONS

Highly oriented GaN layers were successfully grown on (0001) sapphire substrates in the temperature range of 500 - 800 °C by RPE-MOCVD.   The concentrations of the active nitrogen species for the growth of GaN increased as the rf power increased   and the reactor pressure decreased.   At low temperatures such as 500 °C, growth rate and rf power should be reduced to obtain a good quality layer.   Good quality GaN layers were grown by proper optimization of process conditions such as rf power, total reactor pressure, and growth temperature.   A mirror-like surface morphology of the GaN grown by a two-step process was confirmed by a streaky RHEED pattern.

ACKNOWLEDGMENTS

This work was supported by Ministry of Education through 1996 research funds for advanced materials and by Korea Science and Engineering Foundation through the RETCAM in Seoul National University. We would like to acknowledge the help from System IC R&D Laboratories at Hyundai Electronics Industries Co. Ltd. and LG Electronics Research Center.

REFERENCES

1. H. Morkoç, S. Strite, G.B. Gao, M.E. Lin, B. Sverdlov, and M. Burns, J. Appl. Phys. **76**, 1363 (1994).
2. R.J. Molnar and T.D. Moustakas, J. Appl. Phys. **76**, 4587 (1994).
3. R.C. Powell, N.-E. Lee, Y.-W. Kim, and J.E. Greene, J Appl. Phys. **73**, 189 (1993).
4. T. Tokuda, A. Wakahara, and A. Sasaki, in Topical Workshop on III-V Nitrides, Nagoya, Japan, Sep. 21-23 (1995).
5. M. Sato, Appl. Phys. Lett. **68**, 935 (1996).
6. S. W. Choi, K. J. Bachmann, and G. Lucovsky, J. Mater. Res. **8**, 847 (1993).
7. G.B. Stringfellow, <u>Organometallic Vapor-Phase Epitaxy: Theory and Practice</u> (Academic Press: Boston, 1989), p.166-168.
8. A.N. Wright and C.A. Winkler, <u>Active Nitrogen</u> (Academic Press: New York and London, 1968), p.13-138.
9. R.P. Vaudo, Z. Yu, J.W. Cook, Jr., and J.F. Schetzina, Opt. Lett. **18**, 1843 (1993).
10. C. Sone, S. Han, E.S. Aydil, and E. Yoon, in preparation.

MOVPE GaN GAS PHASE CHEMISTRY FOR REACTOR DESIGN AND OPTIMIZATION

S.A. Safvi*, J.M. Redwing**, A Thon*, J.S. Flynn**, M.A. Tischler**, T.F. Kuech*
*Department of Chemical Engineering, University of Wisconsin, Madison, WI 53706
**Epitronics, 7 Commerce Dr., Danbury, CT 06810

The results of gas phase decomposition studies are used to construct a chemistry model which is compared to data obtained from an experimental MOVPE reactor. A flow tube reactor is used to study gas phase reactions between trimethylgallium (TMG) and ammonia at high temperatures, characteristic to the metalorganic vapor phase epitaxy (MOVPE) of GaN. Experiments were performed to determine the effect of the mixing of the Group III precursors and Group V precursors on the growth rate, growth uniformity and film properties. Growth rates are predicted for simple reaction mechanisms and compared to those obtained experimentally. Quantification of the loss of reacting species due to oligmerization is made based on experimentally observed growth rates. The model is used to obtain trends in growth rate and uniformity with the purpose of moving towards better operating conditions.

INTRODUCTION

The utilization of metal organic vapor phase epitaxy (MOVPE) as the major technique for the growth of GaN for device structures [1], has resulted in an increased effort towards understanding the growth process and designing suitable reactors. The growth of device-quality GaN is complicated by gas-phase interactions between trimethyl gallium (TMG) and ammonia. These interactions can lead to changes in the nature and gas-phase depletion of the growth nutrients. Such reactions can lead to a degradation in the growth uniformity, material quality and precursor efficiency [2]. The main gas phase reaction is the strong adduct reaction between NH_3 and TMG [3]. In this study, we have directly monitored this gas-phase reaction in order to better understand its impact on the growth process. The reactions have been monitored over the temperature range of 200-800 °C which is typically encountered in the gas phase environment of the MOVPE reactor, as the growth nutrients are transported to the growth front. From the information gained from the flow tube study [4] and earlier studies [3,5,6] we have developed an apparent chemistry that is incorporated into a vertical MOVPE reactor model. The two items of interest in GaN MOVPE reactor optimization are growth rate and growth uniformity. To illustrate the utility of this model in the optimization of reactor performance, reducing costly experimental trial and error, we determined the effect of altering a process parameter and a geometric parameter on the growth rate and uniformity. The trends obtained from such changes indicate how these variables could be used to improve reactor performance.

EXPERIMENTS OF GAS PHASE CHEMISTRY

We have studied the high temperature gas phase reactions between TMG and NH_3 by means of *in situ* mass spectrometry within a flow tube isothermal reactor. The schematic of the flow tube reactor system is presented in Figure 1. A two temperature reaction zone furnace was used. The two primary reactants, TMG and NH_3, were allowed to mix in the first hot zone of the reactor, that was kept at 150°C. This adduct and its initial reaction products are transported to the second temperature zone where they undergo decomposition over a temperature range of 200-800 °C. In order to label the products and to distinguish between possible reaction pathways, measurements of the co-pyrolysis of TMG and ammonia utilized both NH_3 as well as deuterated ammonia (ND_3). The TMG mole fraction at the inlet was 0.015-0.15, the reactor pressure was 76

Mat. Res. Soc. Symp. Proc. Vol. 449 © 1997 Materials Research Society

Torr and the residence time was ~1s. The reactor and the process conditions are discussed in more detail elsewhere [4]. The temperature dependence of decomposition of the TMG/H$_2$/NH$_3$ system has been presented earlier [4]. Several trends are readily noted. There is reaction between TMG and NH$_3$ resulting in the

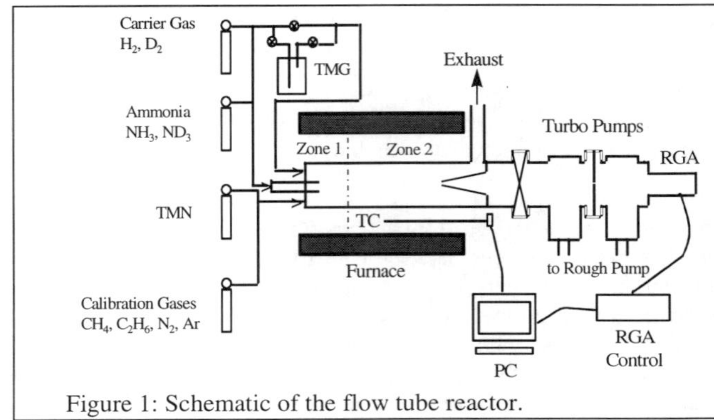

Figure 1: Schematic of the flow tube reactor.

elimination of a single CH$_4$ molecule over a temperature range of 200 - 500 ^{o}C. The constant peak height for larger m/e peaks over this temperature range indicates that there is a thermally stable product, most likely the adduct. No further reaction, beyond this initial release of methane, is noted until a temperature of ~ 500 ^{o}C. Also noted is the decrease in the TMG (adduct) - derived peaks that occur at about 50 ^{o}C higher than the TMG/H$_2$ mixture. A total of three CH$_4$ molecules per one TMG molecule are formed at high temperatures. The overall apparent activation energy and preexponential factor for the elimination of the last methyl group is 48.7±5.3 kcal/mol and 4x10^{13} s^{-1} respectively.

While a complete reaction model for the high temperature decomposition of TMG:NH$_3$ can not be made from these measurements, several conclusions can be made concerning the gas phase chemistry within the MOVPE growth environment. Table 1 lists several likely reactions. The formation of gas phase adduct (G1) occurs at an extremely high rate within the flow-tube reactor. The gas phase species, presumably TMG:NH$_3$ adduct, follows the same reaction path (G2) as adducts formed at lower temperatures, and immediately self-eliminates a CH$_4$ molecule at all temperatures monitored. The formation of a trimeric cyclic compound (G3) is predicted at low temperatures, however the experimental system used in this study was not capable of discriminating between these high molecular weight species. The observed species is stable over a temperature range of 200-500 ^{o}C. Decomposition (G4) proceeds until all three CH$_4$ molecules are removed per initial TMG molecule.

The implication of these studies for the design and operation of MOVPE reactors for GaN growth is several fold. In most or all MOVPE growth systems operating at conventional pressures (1-760 Torr), little TMG exists in the growth environment during NH$_3$-based GaN growth. The rapid adduct formation reaction, together with the immediate release of a single methane molecule, implies that the dominant gas phase species within the reactor is [(CH$_3$)$_2$GaNH$_2$]$_x$ with x=3 being most likely. The reaction of this species in the gas phase, through decomposition or further oligmerization, should be the principal mechanism by which the growth rate or uniformity is affected by the specific reactor design. Larger molecules, and eventually particles, will effect the growth rate through differences in

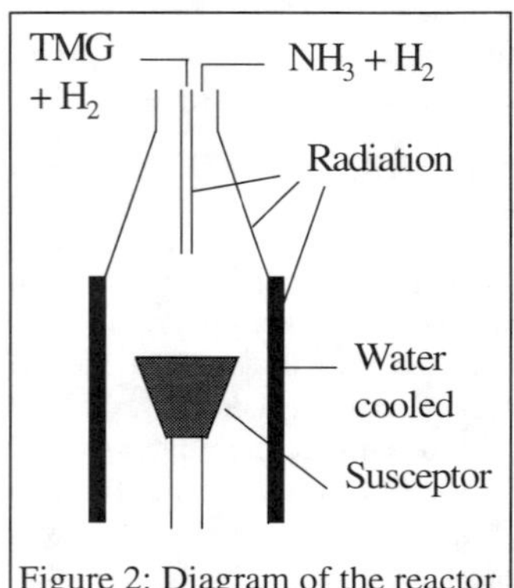

Figure 2: Diagram of the reactor

the transport rate to the surface, due to a difference in diffusion and thermal diffusion coefficients. At temperatures below 120 °C, the $Ga(CH_3)_3{:}NH_3$ adduct will condense due to its low vapor pressure. In typical MOVPE reactors, the temperature varies from room temperature to growth temperature. The growth rate could be improved by mixing precursors at temperatures zones higher than 120 °C, eliminating loss of reactants due to condensation. Alternately, the use of very high flow rates can improve the growth rate but typically at the expense of growth uniformity.

Reaction	Reactant → Products	Ref
G1	$Ga(CH_3)_3 + NH_3 \rightarrow Ga(CH_3)_3{:}NH_3$	4
G2	$Ga(CH_3)_3{:}NH_3 \rightarrow Ga(CH_3)_2{:}NH_2 + CH_4$	4
G3	$3Ga(CH_3)_2{:}NH_2 \rightarrow [Ga(CH_3)_2{:}NH_2]_3$	5,6
G4	$[Ga(CH_3)_2{:}NH_2]_x \rightarrow$ Product	this study

TABLE 1: List of gas phase reactions and their literature reference.

In our MOVPE model we have assumed reaction G3 and G4 to be competing. Reaction G3 is taken to represents loss of adduct to polymer formation, while reaction G4 represents decomposition of the adduct-species into a product which is assumed not to be able to participate in parasitic reactions. Due to lack of information on the nature of the product, its physical properties are taken to be that of $GaCH_3{:}NH$. Both adduct and this product are assumed to have unity sticking coefficients. The chemistry underlying the growth of gallium nitride from MOCVD is not well understood and hence kinetic parameters are not available. To get around this problem, an apparent kinetic parameter is estimated, based on experimental data for this reactor, to describe loss of reactants to parasitic reactions. This apparent parameter is obtained from growth rates under certain process conditions described in the next section. We assume that the apparent kinetic parameter would be insensitive to a slight perturbation of process conditions or reactor geometry.

MOVPE REACTOR

An idealized schematic of the quartz reactor is shown in Figure 2. TMG in hydrogen carrier gas is supplied in the inner tube while ammonia and hydrogen are supplied in the outer tube. The outer wall of the reactor is water cooled. A description of the experimental reactor has been presented elsewhere [7]. Cylindrical coordinates have been used in the model and the computational domain extends from 20 cm upstream of the substrate to 40 cm downstream. The fundamental equations of continuity, momentum and energy balances and species conservation are used to describe the system [8]. The assumption of no variation in circumferential direction reduces the problem to two dimensions. The physical and transport properties of the gaseous species used in this study are listed elsewhere [9]. The system of partial differential equations were solved using

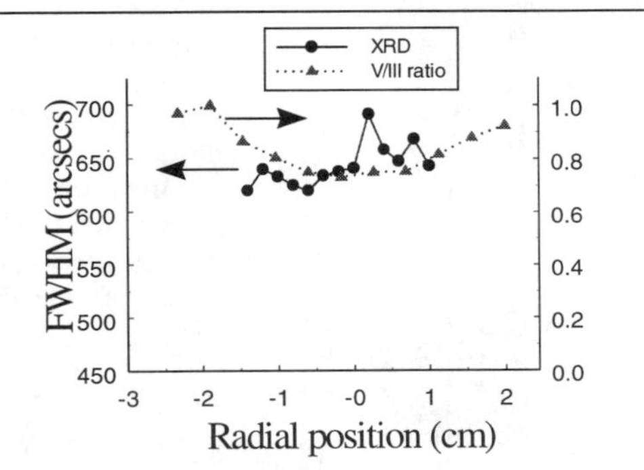

Figure 3: Radial distribution of the local V/III ratio, and x-ray FWHM for a film grown at a low Group III inlet velocity which promotes gas-mixing.

the finite element method. A typical mesh consisted of 1100 quadrilateral elements with the mesh being denser near the susceptor where the concentration and temperature gradients are the largest [9].

We have investigated the effect of local V/III ratio on the structural properties of the GaN films. One film was grown with 0.5 slm H_2 flow through the Group III inlet while the other with 2 slm. The TMG flowrate was adjusted to maintain a constant GaN growth rate of ~0.035 µ/hr at the center of the substrate. The GaN film thickness and double crystal x-ray rocking curve (ω-scan) full-width-at-half-max (FWHM) were measured as a function of distance across the substrate. A GaN layer with a very uniform thickness (< 10 %) and a uniform x-ray FWHM

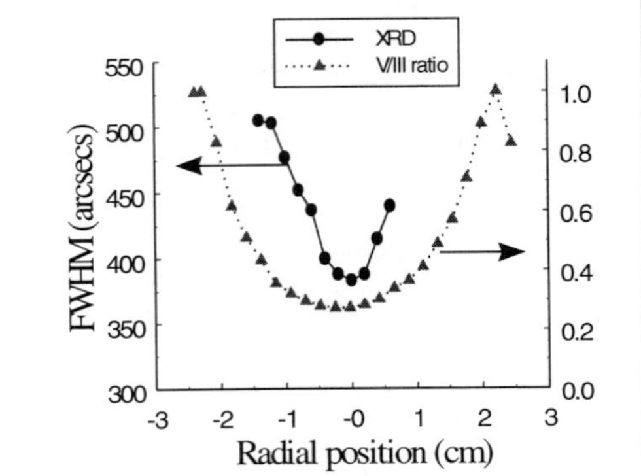

Figure 4: Radial distribution of the local V/III ratio, and x-ray FWHM for a film grown at a high Group III inlet velocity thats limits gas mixing.

across 2" was obtained using low Group III flowrate (0.5 slm) that permits sufficient gas mixing. For this film, figure 3 shows a plot of x-ray FWHM and the local V/III ratio, made dimensionless by its value at the center. Increasing the Group III inlet velocity (2 slm) significantly improves the material quality, decreasing the x-ray FWHM by a factor of two at the center of the film, but the variations in the film thickness of this sample approaches ~ 50%. Figure 4 shows the radial distribution of the X-ray FWHM and the local V/III for this film. These plots suggest that a high H_2 flowrate through the Group III inlet causes non-uniform distribution of the reactants resulting in non uniform growth rates and material properties. Typically, for a higher local V/III ratio the film quality is better. However, the quality also depends on the nature of the Group III source. Figure 4 shows that the quality of the film at the center of the wafer, where the reactants reach quicker, is better because the effect of parasitic reactions is minimized. As the Group III source is transported from the center to the edge, the residence time increases thereby changing the nature of the reactants. The result is that the material quality towards the edge approaches that obtained in the experiment which permitted sufficient gas mixing.

For the remaining experiments, the TMG flowrate was kept constant at 0.748 sccm. The percentage of NH_3 in the outer tube was kept at 16.7%. The runs were performed at a reactor pressure of 100 Torr and a susceptor temperature of 1000 °C. Since uniformity is an issue [7], we kept the flowrate of the inner jet constant at a low value of 200 sccm. We have assumed that the reaction (G3) leading to loss of adduct to parasitic reactions is of third order. To simplify the estimation we have neglected the temperature dependence. A value of 2.5×10^{17} $cm^6mol^{-2}s^{-1}$ best fit

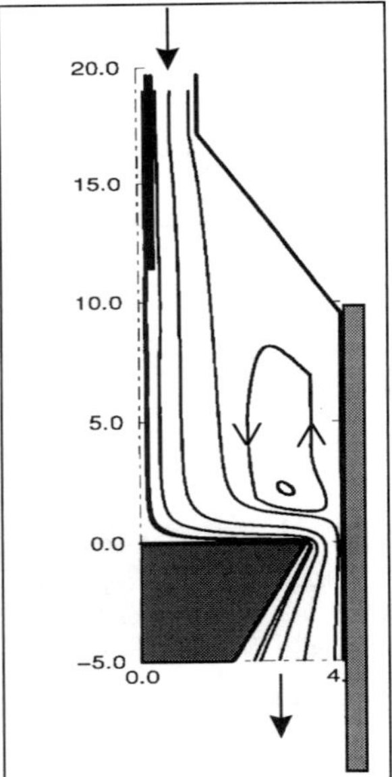

Figure 5: Flow profile for an outer feed flowrate of 12 slm and a gap of 11.4 cm.

the experimental data for the outer feed of 12 slm and a gap between the inner tube and the susceptor of 11.4 cm. Figure 5 shows the flow profile under these conditions while Figure 6 shows the growth rate data along with model predictions for this value of the apparent kinetic parameter.

In the first series of computation, we perturbed a process parameter, the outer jet flowrate from its original value of 12 slm keeping the composition constant. As the flowrate was decreased, the growth rate falls sharply as shown in Figure 7 for two reasons. The nutrients move slowly to the growth front leaving more time for parasitic reactions and the inner feed pathlines expand further away from the substrate reducing the concentration gradients to the surface.

In the second series of computations, the gap between the inner tube and the susceptor was lowered from 11.44 to 7.9 cm. One would expect that this would reduce the residence time and increase the growth rate. The results in Figure 8 indicate that as the gap is decreased the growth rate at the center does not change much but the growth uniformity improves slightly. Flow profiles indicate that as the gap is lowered the axial velocities near the center of the reactor, carrying the nutrients to the surface, decrease. This offsets the reduction in residence time, due to smaller gap, for the Ga containing species. This result reflects the advantages of detailed modeling in aiding the optimization procedure and understanding growth trends in the growth system.

In this study, two parameters were perturbed to demonstrate the kind of information that can be obtained by such a procedure. The same procedure can be applied to the other parameters with the purpose of moving to more optimal operating conditions. In a multiparameter system, a large number of computational runs aided with some experimental runs, are needed to obtain optimal operating conditions where perturbation of any parameter would lead to deterioration of reactor performance.

CONCLUSIONS

A flow tube apparatus was used to study adduct formation and decomposition for TMG and NH_3 at conditions encountered in MOVPE of GaN. We observed a very fast adduct formation and a simultaneous elimination of methane. The decomposition of this adduct seemed to be first order. The apparent preexponential factor and activation energy were found to be 4×10^{13} s^{-1} and 48.7 kcal/mole respectively. High Group III inlet velocities lead to higher growth rates and improved material properties at the expense of uniformity in growth rate and material properties. An apparent third order kinetic parameter is used to describe the loss of adduct due to parasitic reactions and was

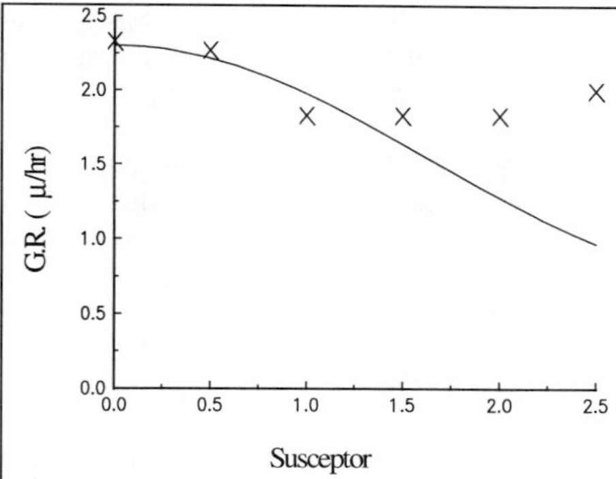

Figure 6: Growth rate data (x) for an outer feed flowrate of 12 slm. A third order apparent kinetic parameter for reaction G4 is used to fit the data.

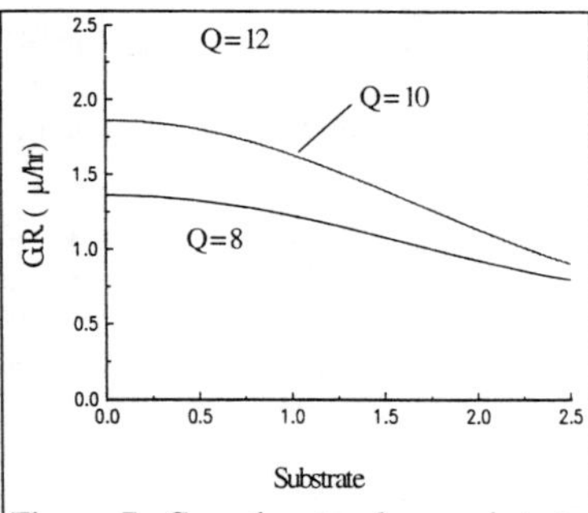

Figure 7: Growth rate along substrate for an outer feed flowrate of 12, 10 and 8 slm.

obtained from the growth data, and is specific to the growth environment. An apparent chemistry model is made based on the salient features of GaN MOVPE. This model can be applicable to any MOVPE GaN reactor but would require a rate parameter to quantify the loss of reactants due to parasitic reactions. We have demonstrated that for process optimization, such a numerical model can be used to reduce costly experimental trial and error.

ACKNOWLEDGMENTS

This work was supported by NSF-MRG on CVD and ARPA-URI on Visible Light Emitters. We thank N. Perkins at for many helpful discussions.

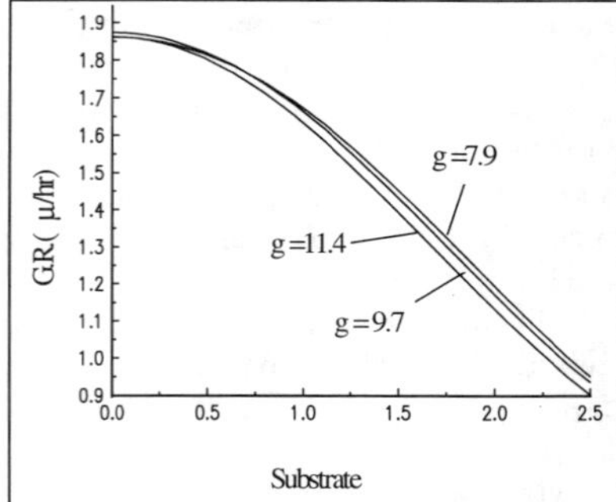

Figure 8: Growth rate along substrate for gap between inlet tube and susceptor of 7.9. 9.7 and 11.4 cm

REFERENCES

1. I. Akasaki and H. Amano, J. Crystal Growth **146**, 455 (1995).
2. R.H. Moss, J. Cryst. Growth **68**, 78 (1984).
3. B.S. Sywe, J.R. Schlup and J.H. Edgar, Chem. Mater. **3**(4), 737 (1991).
4. A. Thon and T.F. Kuech, Mat. Res. Soc. Symp. Proc., **395**, 97 (1995).
5. M.J. Almond, M.G.B. Drew, C.E. Jenkins, and D.A. Rice, J. Chem. Soc., Dalton Trans., 5-9 (1992).
6. M.J. Almond, C.E. Jenkins, and D.A. Rice and K. Hagen, J. Organomet. Chem. **439**, 251 (1992).
7. S.A. Safvi, J.M. Redwing, M.E. Tischler and T.F. Kuech, Mat. Res. Soc. Symp. Proc. Vol **395**, 255-260 (1995).
8. R.B. Bird, W.E. Stewart and E.N. Lightfoot, <u>Transport Phenomenon</u>, Wiley, New York, 1960.
9. S.A. Safvi and T.F. Kuech, submitted to the Journal of the Electrochemical Society.

SELECTIVE GROWTH OF GaN AND $Al_{0.2}Ga_{0.8}N$ ON GaN/AlN/6H-SiC(0001) MULTILAYER SUBSTRATES VIA ORGANOMETALLIC VAPOR PHASE EPITAXY

O.H. Nam, M.D. Bremser, B.L. Ward*, R.J. Nemanich*, R.F. Davis

Department of Materials Science and Engineering, North Carolina State University, Box 7907 Raleigh, NC 27695-7907

* Department of Physics, North Carolina State University, Raleigh, NC 27695-8202

ABSTRACT

The selective growth of GaN and $Al_{0.2}Ga_{0.8}N$ has been conducted on stripe and circular patterned GaN/AlN/6H-SiC(0001) multilayer substrates. Growth morphologies on stripe patterns changed with the widths of stripes and the flow rate of TEG. No ridge growth was observed along the edges of the stripe patterns, and the (0001) top facets were very smooth. Uniform hexagonal pyramid arrays of undoped GaN and Si-doped GaN were successfully grown on 5μm circular patterns. Field emission measurement of a Si-doped GaN hexagonal pyramid array exhibited a turn-on field of 25V/μm for an emission current of 10.8nA at an anode-to-sample distance of 27μm.

INTRODUCTION

Wide band gap III-nitrides are being extensively studied for application in optoelectronic and microelectronic devices, including field emitter. Gallium nitride is a promising material in this group for field emitters because of its low electron affinity (2.7-3.3eV) [1,2], high thermal, chemical and mechanical stability as well as the ability for controlled n-type doping. A recent report, in which AlN and Al-rich $Al_xGa_{1-x}N$ films have been shown to have a negative electron affinity, suggests these materials could be key elements of field emitters [2,3].

Selective growth techniques are available not only for the fabrication of semiconductor devices such as quantum well wire and dot structures, but also for field emitters. The selective growth of GaN and $Al_{0.1}Ga_{0.9}N$ on linear windows and GaN hexagonal pyramid fabrication on dot-patterned GaN/sapphire substrates by this technique have been reported [4,5]. The first field emission and the enhancement of this phenomenon from an undoped GaN hexagonal pyramid array on a GaN/sapphire substrate have been observed [6,7]. In a related area, the selective growth of GaN layers on sapphire substrates has been reported for low-loss optical wave guide structures for active and passive photonic devices [8]. To date, all research on the selective growth of GaN have used sapphire substrates. In this paper, we report the selective growth of GaN and $Al_{0.2}Ga_{0.8}N$ on stripe patterned GaN/AlN/6H-SiC(0001) multilayer substrates via organometallic vapor phase epitaxy. These substrates were produced using techniques described elsewhere [9,10]. We also report for the first time fabrication of GaN and Si-doped GaN hexagonal pyramid arrays on circular patterns and the field emission results from these arrays.

EXPERIMENTAL PROCEDURES

The selective growth of GaN and $Al_{0.2}Ga_{0.8}N$ was performed on stripe (window width = 3-80μm) and circular (diameter = 5μm) patterned GaN/AlN/6H-SiC(0001) multilayer substrates. To produce these substrates, undoped or Si-doped GaN films (n = $2E18cm^{-3}$) having a thickness of

Mat. Res. Soc. Symp. Proc. Vol. 449 © 1997 Materials Research Society

1.5µm were first grown on a high temperature AlN buffer layer on a 6H-SiC(0001) substrate in a cold-wall, vertical, pancake-style, RF inductively heated OMVPE system. The experimental growth parameters are described elsewhere [9,10]. A SiO_2 mask layer (thickness=1000Å) was subsequently deposited on each multilayer substrate by RF sputtering or LPCVD. Patterning of the mask layer was achieved using standard photolithography techniques and etching with a buffered HF solution. In the stripe patterned samples, the edges of the stripes were parallel to the <11-20> direction. Prior to selective growth, the patterned samples were dipped in a buffered HCl solution to remove the surface oxide of the underlying GaN layer.

Selective growth was conducted at 1000-1050°C and 45 Torr. Triethylgallium (TEG), triethylaluminium (TEA) and 1500 sccm NH_3 were used in combination with a 3000 sccm H_2 diluent. The ratio of NH_3 and TEG was varied from 960 to 2600. Silicon was incorporated into the GaN hexagonal pyramids during the selective growth using SiH_4 at a flow rate of 5.5nmol/min. The morphologies of the selectively grown samples were observed using scanning electron microscopy (SEM-JEOL 6400 FE). The final step was the deposition of Ti(200Å)/Au(1500Å) contacts via electron beam evaporation on Si-doped GaN layer.

The field emission measurements (FEM) were performed on the hexagonal pyramid arrays of Si-doped GaN in a UHV-FEM system having a working pressure of $2x10^{-8}$ Torr. During the measurement, samples were placed beneath a 5mm diameter movable Mo anode with a flat tip. The anode was controlled by a stepping motor such that one step yielded a translation of 0.44 µm. The current-voltage (I-V) measurements were taken at several distances ranging from 2 to 40 µm for anode voltages in the range of 0 to 1100 V.

RESULTS AND DISCUSSION

The characteristics of the stripe patterns of deposited GaN and $Al_{0.2}Ga_{0.8}N$ having various window widths were determined using SEM as shown in the micrographs in Figure 1. Both materials showed prismatic morphology with (1-101) side facets on the 3µm-wide stripes. Truncated prismatic growth with (0001) top facets and (1-101) side facets was observed on stripe patterns with widths greater than 5µm. Polycrystals of $Al_xGa_{1-x}N$ nucleate on the SiO_2 mask because of the chemical interaction between Al and SiO_2 [4], there is no, however, significant difference in the final growth morphology for the GaN and $Al_{0.2}Ga_{0.8}N$ patterns. Only a slight roughening of the (1-101) facets was observed for $Al_{0.2}Ga_{0.8}N$ selective growth. This is probably due to the poor selectivity of $Al_{0.2}Ga_{0.8}N$, that is, the deposition of polycrystalline material on the SiO_2 mask. No ridge growth was observed along the edge of the stripes, and the (0001) top facets were very smooth and flat regardless of the width of the stripes. This suggests that insignificant lateral vapor phase diffusion of the reactive species from the mask to the window area occurred during the selective growth. It is probably due to the large ratio (=0.5) of the window to the mask area, because lower values of this ratio induce more lateral vapor diffusion [11].

Figure 2 shows SEM images of GaN grown on 5µm-wide stripe patterns using different flow rates of TEG. The higher flow rate of TEG resulted in a smaller area of (0001) top facets and the development of (1-101) side facets. This behavior supports the model that GaN selective growth depends on the balance between the incoming flux on the (0001) top facets and the outgoing flux from the (0001) to the (1-101) side facets [5]. Increased TEG flow rate promotes two dimensional nucleation on the (0001) top facets, thereby reducing the surface diffusion length of the Ga adatoms. As a result, the growth rate of the (0001) facets becomes faster than that of the (1-101) facets.

The optimum conditions for selective growth of GaN hexagonal pyramids on circular patterns was based on the selective growth conditions on stripe patterns. The height of a hexagonal pyramid can be easily calculated by the relation of H=D·tan62°/2, where H and D are the height and the

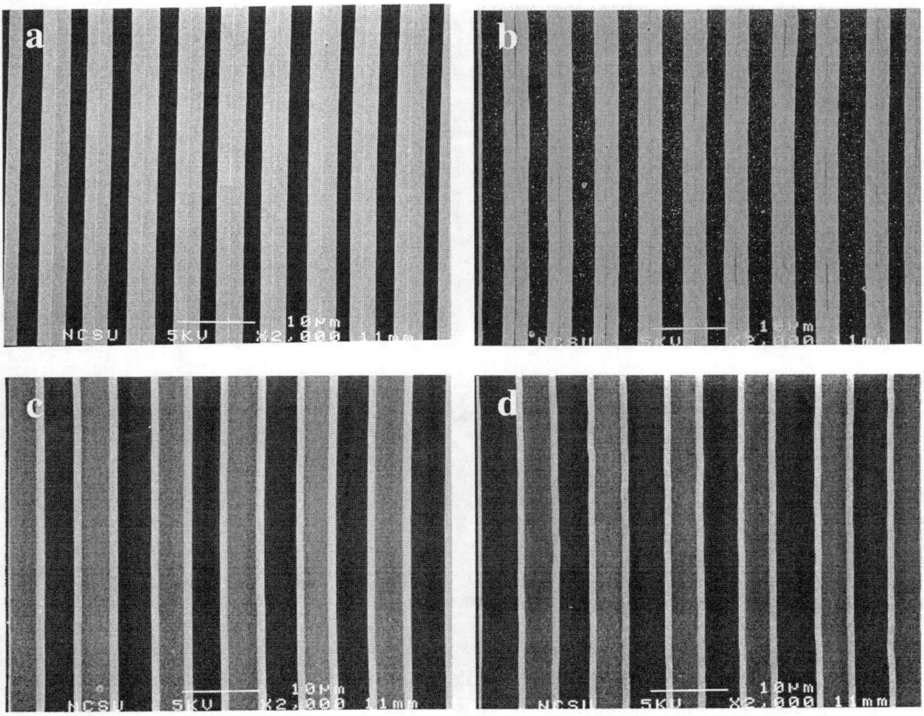

Figure 1. SEM micrographs of the selectively grown GaN(left) and $Al_{0.2}Ga_{0.8}N$(right) layers on different wide-stripe patterns. (a)(b) 3μm, (c)(d) 5μm

Figure 2. SEM micrographs of the selectively grown GaN layers on 5μm wide-stripe patterns at different flow rate of TEG. (a) 26μmol/min. (b) 70μmol/min.

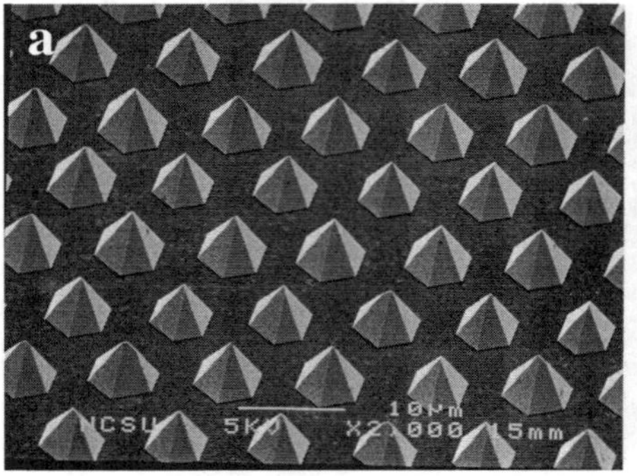
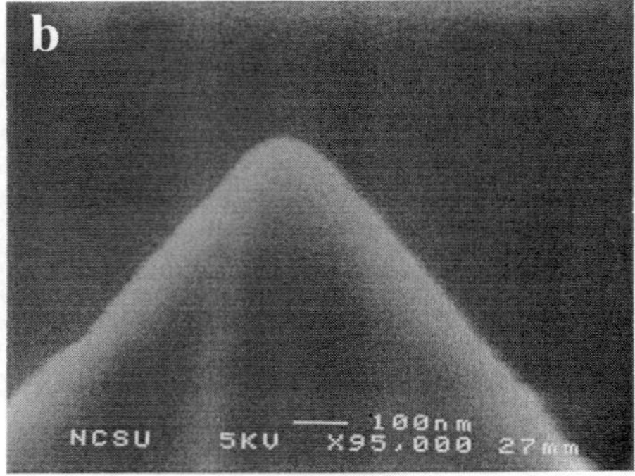

Figure 3(a). SEM micrograph of Si doped GaN hexagonal pyramid array. (b) High magnification SEM image of the apex of a hexagonal pyramid. Tip radius is less than 100nm.

diagonal width of the base of the pyramid. Each pyramid has six (1-101) side facets. The growth rate of these pyramids is strongly dependent upon the ratio of window-to-mask area in the patterned region as well as the selective growth conditions. The average diagonal width of the pyramids was 7.7μm using a ratio of 0.1. However, increasing the ratio to 0.23 resulted in an average diagonal width of 5.7μm for the same growth condition. These results indicate that the lateral diffusion of the reactive species from the mask to the window area is also an important factor for the fabrication of GaN hexagonal pyramids. As shown in Figure 3(a), the growth of an uniform array of Si-doped GaN hexagonal pyramid in an area of 0.5x0.5mm^2 was achieved for the first time. The high magnification SEM image shown in Figure 3(b) reveals that the tip radius of the pyramids is less than 100nm.

Field emission measurements (FEM) of the Si-doped GaN pyramid arrays were conducted using a UHV-FEM system. The emission current was measured as a function of the anode voltage applied to a Mo rod. The I-V curve in Figure 4(a) showes the turn-on voltage of 680V for a current of 10.8nA at a distance of 27μm between the pyramid array and the anode. This turn-on voltage corresponds to a turn-on field of 25V/μm. Using the same system, a polycrystalline p-type diamond film (p = 2.5E17cm^{-3}) grown on Si(100) exhibited a turn-on field intensity of 27V/μm. The Fowler-Nordheim (F-N) plot obtained from the I-V data as shown in Figure 4(b) reveals a linear relationship. This indicates that the emission from the Si-doped GaN array obeys F-N field emission theory. Research is in progress to reduce the turn-on field of the Si-doped GaN arrays to several volts per μm in order to be considered for practical applications.

CONCLUSIONS

The selective growth of GaN and Al$_{0.2}$Ga$_{0.8}$N has been conducted on stripe and circular patterned GaN/AlN/6H-SiC(0001) multilayer substrates. Prismatic morphology with (1-101) side facets was observed on 3μm wide stripes for both materials. Truncated prismatic growth with (0001) top facets and (1-101) side facets were obtained on stripe patterns with widths greater than 5μm. No ridge growth was observed along the edges of the stripe patterns and the (0001) top facets were very smooth and flat. Polycrystalline Al$_x$Ga$_{1-x}$N were deposited on the SiO$_2$ mask. For GaN selective growth on the stripe patterns, the increase of TEG flow rate resulted in the

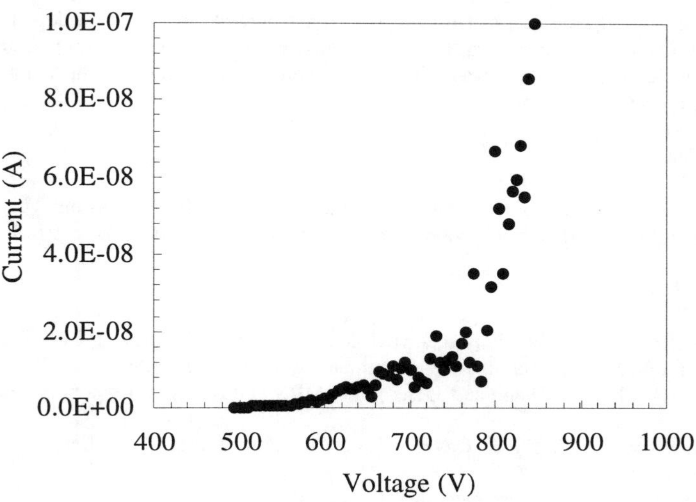

Figure 4(a). Emission current and anode voltage characteristic of Si-doped GaN hexagonal pyramid array.

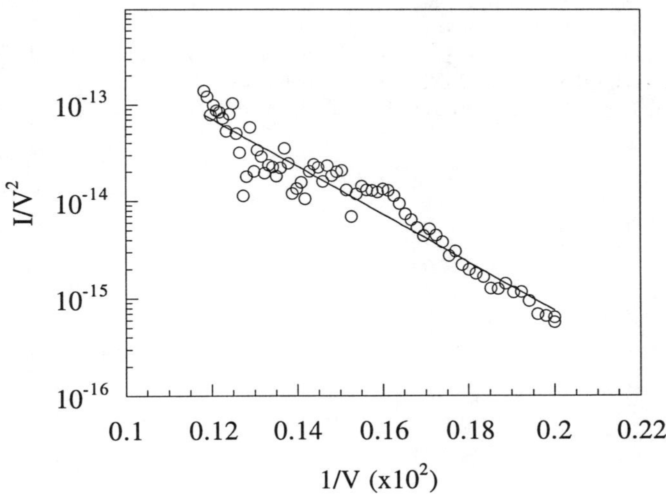

Figure 4(b). Fowler-Nordheim plot of the I-V data. The straight line indicates F-N tunneling.

development of (1-101) side facets. Uniform hexagonal pyramid arrays of Si-doped GaN were successfully grown on circular patterns having diameters of 5µm. Field emission measurements of these arrays showed a turn-on field was 25V/µm for an emission current of 10.8nA at an anode-to-pyramid array distance of 27µm.

ACKNOWLEDGMENTS

The authors express their appreciation to Dr. Hiramatsu for helpful discussion. This work was supported in part by the Office of Naval Research under contract # N00014-96-1-0765.

REFERENCES

1. J.L. Shaw, H.F. Gray, K.L. Jensen, J.M. Jung, J. Vac. Sci. Technol., **B14**, 2072 (1996)
2. R.J. Nemanich, M.C. Benjamin, S.P. Bozeman, M.B. Bremser, S.W. King, B.L. Ward, R.F. Davis, B. Chen, Z. Zhang and J. Bernholc, MRS Fall Meeting, Boston, November (1995)
3. M.C. Benjamin, C. Wang, R.F. Davis and R.J. Nemanich, Appl. Phys. Lett., **64**, 3288 (1994)
4. Y. Kato, S. Kitamura, K. Hiramatsu and N. Sawaki, J. Crystal. Growth, **144**, 133 (1994)
5. S. Kitamura, K. Hiramatsu and N. Sawaki, Jpn. J. Appl. Phys., **34**, 1184 (1995)
6. R.D. Underwood, D. Kapolnek, B.P. Keller, S. Keller, S.P. DenBaars and U.K. Mishra, Topical Workshop on Nitrides, Nagoya, Japan, September (1995)
7. R.D. Underwood, D. Kapolnek, B.P. Keller, S. Keller, S.P. DenBaars and U.K. Mishra, Solid State Electron., Submitted.
8. T. Tanaka, K. Uchida, A. Watanabe and S. Minagawa, Appl. Phys. Lett., **68**, 976 (1996)
9. T.W. Weeks Jr., M.D. Bremser, K.S. Ailey, E.P. Calson, W.G. Perry and R.F. Davis, Appl. Phys. Lett., **67**, 401 (1995)
10. M.D. Bremser, W.G. Perry, T. Zheleva, N.V. Edward, O.H. Nam, N. Parich, D.E. Aspnes and R.F. Davis, MRS Internet J. Nitride Semicond. Res. **1**, 8 (1996)
11. K. Yamaguchi and K. Okamoto, Jpn. J. Appl. Phys., **32** 1523 (1993)

A microstructural analysis of orientation variation in epitaxial AlN on Si, its probable origin, and effect on subsequent GaN growth

R. Beye*, T. George
Center for Space Microelectronics Technology, Jet Propulsion Laboratory, California Institute of Technology, Pasadena, CA 91109
*Robert.W.Beye@JPL.NASA.Gov

J.W. Yang, M.A. Khan
APA Optics Inc., 2950 N.E. 84th Lane, Blaine, MN 55434

Abstract

A structural examination of aluminum nitride growth on [111] silicon was carried out using transmission electron microscopy. Electron diffraction indicates that the basal planes of the wurtzitic overlayer mimic the orientation of the close-packed planes of the substrate. However, considerable, random rotation in the basal plane and random out of plane tilts of about ±3-4° are evident. The orientation variations were traced to the Si interface, where crystallites and an amorphous-like background were present. A strong relationship between these phenomena and substrates containing Si is established by comparing the present growth results with those reported elsewhere. Crystalline quality of the overgrown GaN on the AlN layer is described, with suggestions for the relation between surface pyramids or peaks and the mis-oriented buffer layer.

Introduction

Gallium nitride and aluminum nitride epitaxial films show great promise for use in optoelectronic and high power devices due to their large, direct band-gaps corresponding to blue and ultraviolet wavelength regime. Before such devices can be realized, reliable growth methods are needed that produce epitaxial layers of device quality. Thus far, heteroepitaxy is the only practical means, and much progress has been achieved with sapphire substrates [1]. Optimally, substrates are needed that are defect-free, cleave and etch well, are inexpensive, and are available with large surface areas. Silicon is thus an attractive candidate.

The direct deposition of GaN on silicon results in unacceptably high surface roughness and poor crystal quality. Aluminum nitride buffer layers have been employed to improve the GaN quality [2,3]. Epitaxial AlN is readily grown to have a flat surface on silicon, and has been reported to be able to grow as a single crystal [2,3]. The quality of GaN on AlN is known to be dependent upon the buffer layer growth conditions [3,4], but uncertainties remain about the nature of the AlN quality and how it influences subsequent GaN overgrowth. This article examines these issues with a structural examination of AlN and GaN/AlN on silicon and compares the findings to those reported in the literature.

Experimental

Single crystal AlN was first deposited over the Si(111) substrates using reactive MBE. The samples were analyzed using RHEED to confirm the single crystal nature of the sample surface. Subsequently, GaN was grown over the AlN coated Si(111) samples using conventional low pressure MOCVD. A growth temperature of 800°C and a pressure of 76 Torr were employed and TEGa and NH_3 were used as precursors. Cross-sectional samples for TEM were prepared for viewing by gluing the grown layers face-to-face with M-bond epoxy™[5]. The orientation of the material was such that a slice of each sandwich had a surface normal of $[11\bar{2}0]$ on one side of the epoxy and $[1\bar{1}00]$ on the other. Discs were ultrasonically cut from each slice with a diameter of 2.3 mm and subsequently glued into brass rings of 2.3 mm I.D. and 3.0 mm O.D to enhance

113

Mat. Res. Soc. Symp. Proc. Vol. 449 © 1997 Materials Research Society

mechanical stability. Discs were then thinned to 50 μm and dimpled on one side to about 10 μm before ion milling both sides to electron transparency. Plan view samples were prepared by first coring the material in a direction parallel to the substrate surface normal and gluing into rings as before. All thinning was performed as with the cross-sectioned samples, but taking place only on the silicon side. All TEM work was performed with a Topcon 002B using both high-resolution and high-tilt polepieces, depending on the need.

Results

AlN on Si- Figure 1 is a selected area electron diffraction pattern (SADP) obtained from a cross-sectional sample, illustrating the observed orientation relationship between substrate and AlN layer:

$$(111)_{Si} \,//\, (0002)_{AlN}$$
$$[\bar{1}10]_{Si} \,//\, [11\bar{2}0]_{AlN}$$

This relationship is the same as that between hexagonal and cubic structures that differ only in the stacking sequence of the close-packed planes. Extensive tilting experiments were performed to examine the possibility of the $(0002)_{AlN}$ assuming the orientation of one of the other $[111]_{Si}$ variants inclined to the interface, as well as the existence of cubic AlN. The basal plane normals of the wurtzite AlN were indeed found to lie parallel to the surface normal of the silicon throughout, but with ±3-4° tilt variation, witnessed by the arcs (not spots) of the AlN diffraction pattern of Figure 1. No SAD evidence of cubic AlN presence was observed in the cross-sectional samples, although cubic spots are masked in some orientations.

Figure 2 is a bright-field image taken from the same region as Figure 1, demonstrating the overall appearance of the AlN layer, the outer surface of which is essentially flat. Orientation variations throughout the layer result in the Moiré patterns being observed in high-resolution images. Figure 3 shows that the observed tilts extend to within a monolayer of the interface, with at least ±3-4° tilt variation as witnessed in the diffraction patterns. Figure 3 also shows that the basal plane tilting is not continuously varied, but rather confined to discrete crystallites, randomly mis-aligned with respect to the Si template. A thin, amorphous-like region at the Si interface is also seen in this and other images.

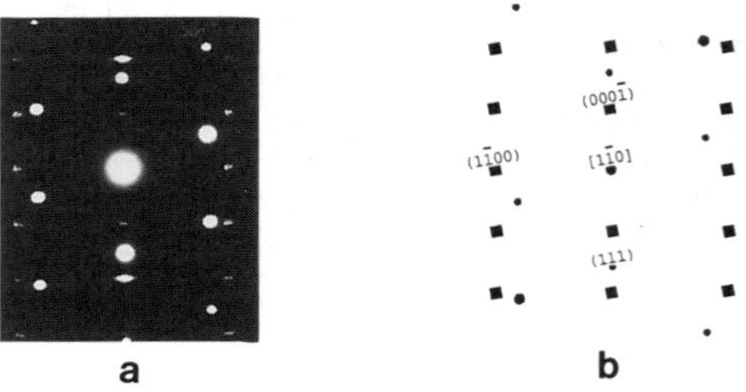

Figure 1: Experimental (a) and simulated (b) SADP of AlN on Si with the orientation relationship described in the text. Silicon reflections are shown as circles, AlN as squares.

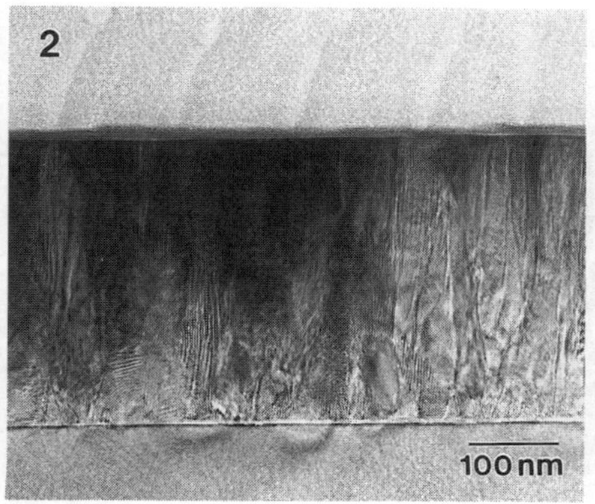

Figure 2:   TEM cross-section bright-field image of AlN on Si, showing smooth outer surface, and representing the area from which Figure 1 was obtained. The AlN band is seen between Si (below) and TEM specimen preparation epoxy (above).

Figure 3:   High-resolution image at the AlN/Si interface, showing tilted AlN crystallites (upper layer) corresponding to the arcs observed in Figure 1.

Plan-view samples yielded similar results. Figure 4a is a SADP of a plan-view sample at the [111] pole of silicon, confirming the above orientation relationship. In addition to the tilt of the AlN basal planes described above, this SADP shows extensive rotation in plane. Figure 4c was taken near the region represented in Figure 4a, but in an area thin enough to allow only AlN reflections. Rotation in the basal plane is seen to be much more extensive than tilting out of plane and nearly results in continuous rings connecting the reflections of lowest index. Figure 5 is a plan-view bright field image of the AlN layer. Here, the extensive in-plane rotation suggests a polycrystalline microstructure.

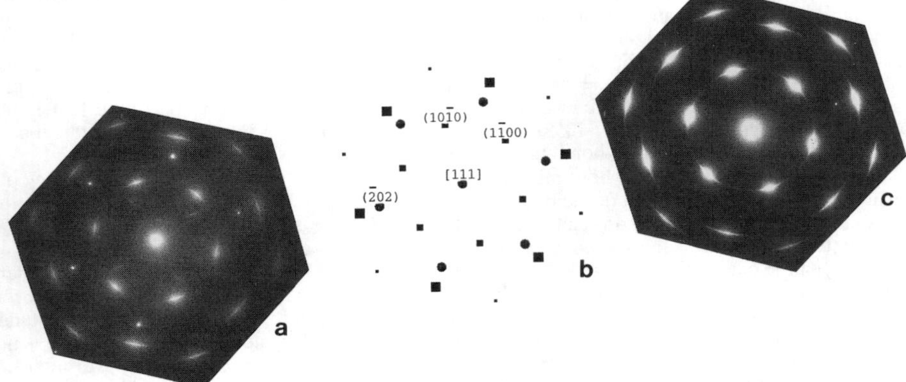

Figure 4:   Experimental (a) and simulated (b) SADP of AlN on Si with electron beam direction parallel to substrate surface normal, confirming the relationship seen in cross-section, and showing extensive in-plane rotation. Silicon reflections are shown as circles, AlN as squares. Experimental (c)SADP taken nearby, showing only AlN material with much streaking.

115

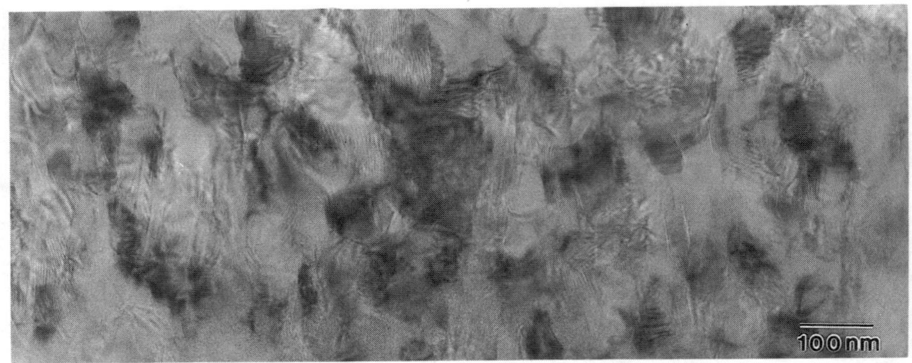

Figure 5: TEM plan-view image of AlN as in Figure 4. Features resembling grains are of similar scale as those seen in Figure 2.

That the angular range at the interface resembles that throughout the AlN suggests a correlation between the interfacial AlN domains and bulk mis-alignment. Basal d-spacings vary negligibly, while their orientation varies considerably, further substantiating this suggestion that the mis-oriented AlN nuclei are seeds for subsequent mis-orientation in the layer. This correlation has not been previously established, nor have the origins of the interfacial effects themselves. However, articles by other authors indicate similar interfaces, and similar diffraction patterns. Mis-alignment of GaN crystallites at the silicon interface and an amorphous-like region are seen between extended substrate surface steps in a study examining the effects of Si surface cleanliness on GaN growth. The authors minimized surface contamination by cleaning the substrates using an RCA cleaning procedure followed by an HF dip [6]. Another study [7] shows a SADP at the AlN/Si interface with much the same tilt, and describes an amorphous layer for AlN or GaN on Si, but not on sapphire. The authors report that analytical electron microscopy (AEM) found no oxide at the Si substrate. Ponce, et al.[8], describe similar interfacial AlN nuclei on SiC, whose mis-orientations become distributed throughout the AlN layer. It appears, then, that these orientation and amorphous-like domains are not the result of surface contamination or surface roughness, but rather of an inherent chemical or mechanical effect in the Al (or Ga)-N-Si system. The lattice mismatch between SiC and AlN is very small on the basal plane (about 1%) while that for a silicon substrate is significant (mismatch= 19%). Crystallite tilting occurs for both substrates suggesting that lattice mismatch is not the cause of the tilting. It should be noted that this effect is not observed on sapphire (mismatch >13%). Furthermore, as these effects are associated with silicon-bearing substrates, interfacial Si-bonding appears to be a likely cause.

The formation of amorphous silicon nitride during the initial stages of AlN growth would result in preferential alignment of the nuclei with the Si template, but with random tilt and rotation excursions proportional to the amount of Si_3N_4 formed. One reason AlN is preferred over GaN as an initial layer is due to the increased stability of AlN with respect to Si_3N_4 formation. Bulk thermodynamic calculations show that AlN is more stable than Si_3N_4 in the Al-N-Si environment [9], as opposed to Ga-N-Si system, in which Si_3N_4 is reportedly more stable [2]. However, interfacial tilts, rotations, and the amorphous-like region are seen for both systems. It then stands to reason that either silicon nitride formation is occurring in both systems, or silicon nitride is not responsible for the interfacial phenomena. In the latter case, silicon presence may still interfere with nucleation through chemical interactions at the interface such that the ideal AlN/Si orientation relationship is disrupted. Silicon has been reported to diffuse into GaN on Si at temperatures as low as 600°C [10]. Although diffusion is not expected to affect the orientation of already formed nuclei, such findings do indicate an affinity between Si and the AlN lattice. Interestingly, in the same study, similar interfaces are described, with Si concentrations on the order of several atomic percent at distances into the GaN similar to those of the disordered layer. One possible explanation for what happens at the interface involves a combination of both views. In this scenario, the surface Si and N combine, the product of which then attempts to mimic the Si template, but is restricted by the Si-N bonding. Subsequently, Al(or Ga)N is grown upon this best-fit arrangement. Such a situation would help to explain how the closed-packed plane and direction information is transmitted across the tilted, rotated, and amorphous-like interfacial region. Further investigation is needed to determine if the Si is indeed responsible for the rotated domains and, if so, whether it is from simple lattice distortion or a particular reaction with one of the epitaxial layer species.

GaN on AlN on Si-   With the texturing nature of the AlN layer established, growth of subsequent GaN is briefly described. Figure 6 shows a heavily faceted surface and polycrystalline microstructure belonging to the GaN. The peak surfaces are along $\{10\bar{1}1\}$ planes, corresponding to angles of 62° with respect to the basal planes. Such peaks may be related to Ga surface migration and evaporation from $\{10\bar{1}1\}$ planes [11]. In this case, when two-dimensional growth is interrupted, $\{10\bar{1}1\}$ surfaces may become established. Gallium may then easily evaporate and thus provide a slower-growing surface than that of the basal-plane. The basal plane, bounded by $\{10\bar{1}1\}$ surfaces grows until the surfaces converge, leaving a hexagonal pyramid. Initiation of a three-dimensional surface may arise from areas of GaN of opposite polarity [12], growing at a different rate; or from the intersection of grains or regions with slight orientation differences; or from dislocations at the surface. It is expected that the large tilts and rotations of the AlN basal planes of the buffer layer are largely responsible for the rough GaN surface. Efforts to mitigate this effect are underway.

Figure 6:   Cross-section TEM bright-field image of GaN on AlN on Si, showing a heavily faceted outer surface, corresponding to $\{10\bar{1}1\}$ planes.

Conclusions

In this work, the exact nature of the AlN buffer layer on silicon orientation variation has been established. It has been shown that, throughout the buffer layer, tilts of basal plane normals randomly deviate from the Si (111) surface normal by within approximately ±3-4°, while rotations in plane were shown to rotate randomly with the approximate range of ±10°. These deviations were shown to extent to the substrate surface, where a region of crystallites in a background of an amorphous-like appearance was present. It has been suggested that the interfacial phenomena are the responsibility of the chemical bonding from the Si-bearing substrate, and may encourage the formation of $\{10\bar{1}1\}$ peaks on subsequently grown GaN.

Acknowledgments

The authors would like to thank Amy Chang-Chien for TEM sample preparation. The work described in this paper has been performed at the Center for Space Microelectronics Technology, Jet Propulsion Laboratory, California Institute of Technology, and was sponsored by the Ballistic Missiles Defense Organization, Innovative Science and Technology Office, through an agreement with the National Aeronautics and Space Administration.

References

1	S. Nakamura, T. Mukai, M. Senoh, Appl. Phys. Lett., **64**, 13(1994).
2	K. S. Stevens, A. Ohtani, A. F. Schwartzman, R. Beresford, J. Vac. Sci. Technol. B **12**(2), 1186(1994).
3	A. Watanabe, T. Takeuchi, K. Hirosawa, H. Amano, K. Hiramatsu, I. Akasaki, J. Cryst. Growth, 128, 391(1993).
4	U. Rossner, J.-L. Rouviere, A. Bourret, A. Barski in GaN and related Materials, Edited by F. A. Ponce, R. D. Dupuis, S. Nakamura, J. A. Edmond, (Mater. Res. Soc. Proc. **395**, Pittsburgh, PA, 1996) pp. 145-150.
5	Measurement group, Inc., Raleigh, N. Carolina.
6	G.A. Martin, B. N. Sverdlov, A. Botchkarev, H. Morkoc, D. J. Smith, S.-C. Y. Tsen, W. H. Thompson, M. H. Nayfeh in GaN and related Materials, Edited by F. A. Ponce, R. D. Dupuis, S. Nakamura, J. A. Edmond, (Mater. Res. Soc. Proc. **395**, Pittsburgh, PA, 1996) pp. 67-72.
7	Ig-Hyeon Kim, Chan-Wook Jeon, Seon-Hyo Kim in GaN and related Materials, Edited by F. A. Ponce, R. D. Dupuis, S. Nakamura, J. A. Edmond, (Mater. Res. Soc. Proc. **395**, Pittsburgh, PA, 1996) pp. 313-318.
8	F. A. Ponce, B. S. Krusor, J. S. Major, Jr., W. E. Plano, D. F. Welch, Appl. Phys. Lett, **67**(3), 411(1995).
9	A. Ohtani, K. S. Stevens, R. Beresford, Appl. Phys. Lett. **65**(1), 61(1994).
10	S. N. Basu, T. Lei, T.D. Moustakas, J. Mater. Res. **9**(9), 2370(1994).
11	K. Hiramatsu, S. Kitamura, N. Sawaki, in GaN and related Materials, Edited by F. A. Ponce, R. D. Dupuis, S. Nakamura, J. A. Edmond, (Mater. Res. Soc. Proc. **395**, Pittsburgh, PA, 1996) pp. 267-271.
12	J.-L. Rouviere, M. Arlery, A. Bourret, R. Niebuhr, K.-H. Bachem, in GaN and related Materials, Edited by F. A. Ponce, R. D. Dupuis, S. Nakamura, J. A. Edmond, (Mater. Res. Soc. Proc. **395**, Pittsburgh, PA, 1996) pp. 393-398.

OBSERVATION OF NEAR BANDEDGE TRANSITION IN ALUMINUM NITRIDE THIN FILM GROWN BY MOCVD

XIAO TANG, FAZLA R. B. HOSSAIN, KOBCHAT WONGCHOTIGUL AND MICHAEL G. SPENCER

Materials Science Research Center of Excellence, Department of Electrical Engineering, Howard University, 2300 6th St., NW, Washington, D C 20059.
E-mail: spencer@msrce.howard.edu

ABSTRACT

The Cathodoluminescence (CL) measurements of undoped and carbon doped aluminum nitride (AlN) thin films near the band-edge region were performed at 300, 77 and 4.2 K, respectively. These films were grown on three different substrates: 6H-SiC, 4H-SiC and sapphire. A dominant peak was observed in undoped samples around 5.9 eV. This peak can be further resolved into three distinct peaks at 6.05, 5.85, and 5.69 eV for AlN on sapphire. The temperature dependence of the peak positions and line widths were investigated. These peaks are believed to be due to exciton recombination. Also, the absorption spectra of carbon doped AlN on sapphire were analyzed to study the Urbach tail parameters which play an important role in near band-edge transitions.

INTRODUCTION

III-V nitride semiconductors such as gallium nitride (GaN) and AlN have received extensive attention for their potential opto-electronic device applications including light emitting diodes (LEDs), UV detectors, wave guides and lasers. In the past, most semiconductor devices were operated in infrared and visible wavelengths.[1] Recently, a GaN laser has been demonstrated which emits in the blue[2]. AlN possesses a 6.2 eV direct bandgap at room temperature which makes it a very good candidate for optical devices operating in the ultra-violet (UV) and vacuum UV regions. There are very few reports about near band-edge emission in AlN so far. In this work, we report on transitions near the bandgap region of AlN.

EXPERIMENTAL

AlN samples were grown in a low pressure metal organic chemical vapor deposition (MOCVD) vertical reactor. The growth temperature and pressure were 1200^0C and 10 Torr, respectively. Trimethylaluminum (TMA), ammonia (NH_3) and hydrogen (H_2) were used as precursors. Carbon doping was done by flowing propane gas[3] during growth. The films were grown on three different substrates: 6H-SiC, 4H-SiC and sapphire.

AlN samples were placed in a ultra-high vacuum (UHV) chamber and evacuated to 10^{-10} torr for CL measurement. The typical film thickness is in the range of 0.35-0.4 0 μm. An electron beam (with an energy of 4 KeV and a current of 0.1 mA) was focused on the sample surface with a focal spot about 1 mm. The luminescence from this spot was collected through UV grade window and lenses, the luminescence was analyzed using a 0.39 m focal length monochromator followed by a photon counting photomultiplier tube. Data acquisition and plotting were performed using a computer.

RESULTS AND DISCUSSIONS

Figure 1 shows a CL spectrum of undoped AlN on 6H-SiC substrate at liquid nitrogen temperature. A strong luminescence peak at 5.92 eV dominates whole spectral range from 180 to

119

Mat. Res. Soc. Symp. Proc. Vol. 449 © 1997 Materials Research Society

580 nm. The other broad band occurring at 3.0 eV corresponds to a near bandgap transition in 6H-SiC substrate. The weaker peaks at 4.43 and 4.03 eV are related to oxygen impurity, as reported by Youngman and Harris[5].

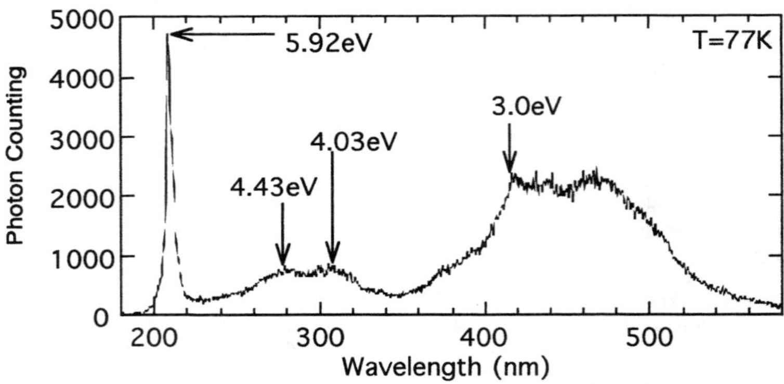

Figure 1. CL spectrum of undoped AlN on 6H-SiC at 300 K.

The luminescence from undoped AlN on sapphire substrate at room temperature exhibits a distinct peak at about 5.85 eV as shown in Figure 2. Using curve fitting software we were able to resolve this peak into three "Gaussian" peaks "A", "B", "C" corresponding to wavelengths of 6.05, 5.85 and 5.69 eV, respectively. A low intensity peak around 5.23 eV was observed for AlN on sapphire which is similar to that reported by Veselov et al[4].

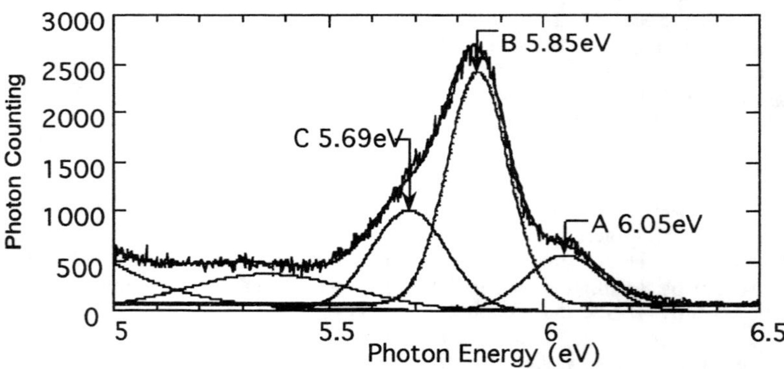

Figure 2. CL spectrum of undoped AlN on sapphire at 300 K.

When the temperature degreases from 300k down to 4.2 K, the position of peak "B" shows a gradual shift towards the higher energy side, and its full width at half maximum (FWHM) also decreases (the typical data is shown in Fig. 3). The energy position and linewidth of peak "B" are plotted in Fig 3 for AlN grown on sapphire and 6H-SiC substrates, It is observed that the linewidth of the film on 6H-SiC is narrower than those on sapphire. The narrowest line width recorded is about 90 meV for AlN on 6H-SiC at 4.2K.

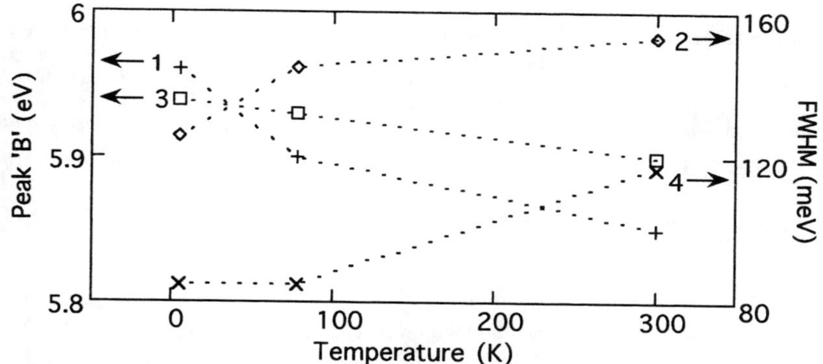

Figure 3. Temperature dependence of position and line width of peak "B". The peak position (1) and FWHM (2) are for undoped AlN on sapphire; The peak position (3) and FWHM (4) are for undoped AlN on 6H-SiC.

In addition to the CL study, an absorption measurement was also made for different carbon doped samples on sapphire substrates as shown in Fig. 4. The Urbach tail parameters[6] were calculated from the absorption curves (the insert of figure 4). A higher carbon doping causes a larger value of the Urbach tail parameter. The intensity of peaks "A", "B", and "C" decreases if a larger Urbach tail exists. As a result, no peaks are found near the band-edge of AlN samples with higher carbon concentrations.

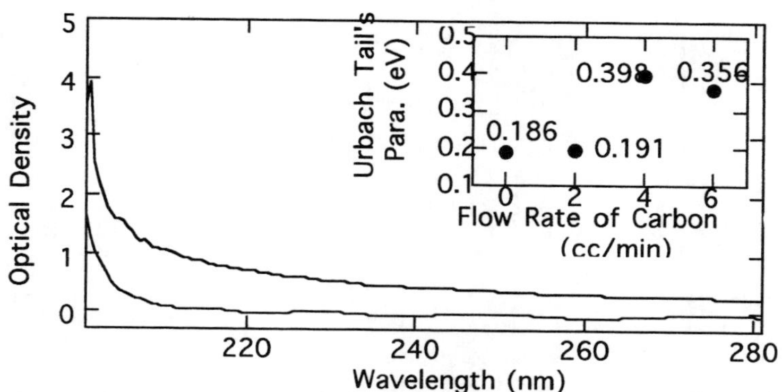

Figure 4. Absorption spectra of AlN on sapphire with different carbon concentrations. The insert shows the Urbach's tail parameters of AlN with different carbon flow rates in growth process

Recently, Davis et at[7]. have studied CL measurement of $Al_xGa_{1-x}N$ with X from 0 to 0.96 and found a bound exciton line (I_2-line emission) at 5.75 eV for X = 0.96. If this curve is extrapolated to X = 1, our peak at 5.92 eV for pure AlN fits very well. We believe that the UV peaks "A", "B" and "C" observed in this work are also to be due to exciton recombination. Further investigation is needed to confirm the origin of these peaks.

CONCLUSIONS

Near bandedge transition in films of AlN grown on 6H-SiC and sapphire are reported for the first time. The distinct peak around 5.9 eV is believed to be due to exciton recombination. This peak can be resolved into three peaks at wavelengths of 6.05, 5.85 and 5.69 eV for AlN on sapphire. As temperature decreases, the energy position of these peaks increases and their line widths become narrower. The Urbach's tail parameter were measured for AlN doped with carbon.

ACKNOWLEDGMENT

The financial support from the National Science Foundation (NSF) and the Office of Naval Research is greatly acknowledged.

REFERENCES

1. S. Strite and H. Morkoc; Journal of Vacuum Science and Technology B, **10**, 4, 1237-1266, July/August (1992)
2. S. Nakamura, M. Senoh, S. Nagahama and N. Iwasa; Appl. Phys. Lett., **68**, 15, 2105-2107, (1996)
3. K. Wongchotigul, N. Chen, D. P. Zhang, X. Tang and M. G. Spencer; Materials Letters, **26**, 223-226 (1996)
4. V. F. Veselov, A. V. Dobrynin, G. A. Naida and P. A. Pundur; Inorganic Materials, **25**, 1250, (1989)
5. R. A. Youngman, and J. H. Harris; Journal of American Ceramic Society, **73**, 11, 3238-46, (1990)
6. F. Urbach; Physics Review, **92**, 1324 (1953)
7. M. D. Bremser, W. G. Perry, T. Zheleva, N. V. Edwards, O. H. Nam, N. Parikh, D. E. Aspnes, and Robert F. Davis,.MRS Internet J. Nitride Semicond. Res. 1, 8(1996)

MOCVD GROWTH OF GaN FILMS ON LATTICE-MATCHED OXIDE SUBSTRATES

O. M. KRYLIOUK[*], T. W. DANN[*], T. J. ANDERSON[*], H. P. MARUSKA[**], L. D. ZHU[**], J. T. DALY[**], M. LIN[**], P. NORRIS[**], H. T. CHAI[***], D. W. KISKER[#], J. H. Li[##], K. S. JONES[##]

[*] University of Florida, Department of Chemical Engineering, Gainesville, FL 32611
[**] NZ Applied Technologies, 8A Gill Street, Woburn, MA 01801
[***] Crystal Photonics Inc., 2615 Westminster Ter., Oviedo, FL 32765
[#] IBM Research Division, PO Box 218, Yorktown Heights, NY 10598
[##] Department of Materials Engineering, University of Florida, Gainesville, FL 32611

ABSTRACT

The use of the nearly lattice-matched oxide substrates $LiGaO_2$ and $LiAlO_2$ has been explored for growth of GaN by MOCVD. As compared to the quality of films grown on sapphire, only growth on $LiGaO_2$ yielded good quality films, and required use of nitrogen as the carrier gas. Furthermore, high structural quality films were grown on $LiGaO_2$ at temperatures as low as 850°C. Dislocation densities estimated from cross-sectional TEM micrographs were found to be as low as 10^7 cm^{-2}. HRTEM studies revealed deformations at the surface of the $LiGaO_2$ adjacent to deposited GaN films, indicating possible interfacial reactions which may affect the film properties. The GaN film orientations corresponded directly to the substrate orientation, viz., ($\{0001\}/\{001\}$ and $\{1\bar{1}02\}/\{101\}$).

INTRODUCTION

The wide band gap nitride semiconductors have emerged as promising materials for a variety of electronic and opto-electronic devices such as high performance transistors, light-emitting and laser diodes, and field-emission displays.[1,2,3] These devices are providing new levels of functionality for both commercial and military systems and recent device demonstrations have motivated continued research. The Group-III nitride compounds and their alloys exhibit band gap energies throughout the visible and ultraviolet portions of the electromagnetic spectrum. In addition, these semiconductors show high breakdown field strengths, manifest high saturated drift velocities, and possess high thermal conductivity.[4] Specific devices of interest include (1) high power amplifiers, oscillators, and switches (bipolar transistors, MISFETs, HEMTs, IMPATT diodes), (2) thermistors, (3) solar-blind and other UV detectors, (4) LEDs of any color (including white), and (5) phosphors and cold cathodes for flat panel displays. The development of blue and green InGaN LEDs as commercial products,[5] the demonstration of a blue diode laser,[6] and the emergence of transistors with large breakdown voltages and high transconductance[7] have justified intense research in this materials system.

Although the majority of device demonstrations have used sapphire substrates to grow nitride films by metalorganic chemical vapor deposition (MOCVD), this substrate provides poor lattice and thermal expansion matching to GaN. It has been found that a low temperature GaN or AlN buffer layer significantly improves the structural quality of the subsequent nitride layers which are grown at high temperatures (e.g., > 1000°C for GaN). Even with this two temperature growth process, the dislocation density of GaN grown on sapphire ($\approx 10^{10}$ cm^{-2})[8] is considerably higher than obtained for other Group III-V systems for which lattice matched substrates are available.

Mat. Res. Soc. Symp. Proc. Vol. 449 © 1997 Materials Research Society

A difficulty with current growth procedures is achieving efficient indium incorporation. The high vapor pressure of In over In-rich solid solutions at typical nitride growth temperatures and the predicted miscibility gap in $In_xGa_{1-x}N$ and $In_xAl_{1-x}N$ alloys are responsible for this challenge. In this study, alternative substrates are explored which provide better lattice matching to GaN. This could lead to improved structural quality and increased indium incorporation efficiency through lowering the growth temperature.

ZnO is a potential substrate candidate for GaN, but this material is difficult to grow in bulk crystal form and exhibits a high sublimation pressure at typical GaN growth temperatures.[9] The structure of ZnO does suggest, however, that the substitution on Li on one Zn sublattice and Al or Ga on the other Zn sublattice might provide a more stable substrate material. This substitution yields the two oxide compounds $LiAlO_2$ and $LiGaO_2$, which congruently melt at 1700 and 1600°C, respectively. The substitution of Li and Al or Li and Ga on the two Zn sublattices results in a doubling of the unit cell and lowering of the crystal symmetry. The difference in ionic radii between Li and Al or Ga produces a slight distortion in the wurtzitic ZnO structure to yield stable tetragonal ($LiAlO_2$) and orthorhombic ($LiGaO_2$) structures.[10] Both $LiGaO_2$ and $LiAlO_2$ are closely lattice matched to GaN ($\approx 1\%$), provide a good thermal match, and have a high transparency in the visible and uv regions. Kung $et\ al$[11] recently reported the growth of reasonable quality GaN on $LiGaO_2$ by MOCVD. The promise of growing high quality GaN on these substrates provides the motivation for this study.

EXPERIMENTAL

Two independent MOCVD reactors were used in this study. A summary of typical growth conditions is given in Table 1.

Table 1. Summary of growth conditions

Growth Reactors	1. Water-cooled reactor with a tilted horizontal susceptor (UF) 2. Water-cooled reactor with a planar horizontal rotating susceptor (NZ)
Substrates	$LiGaO_2$ (100), $LiGaO_2$ (101), $LiAlO_2$ (100), and α-Al_2O_3
Precursors	TEGa, NH_3
Carrier Gases	Nitrogen, hydrogen; 2.5 to 3.0 slm
Flow Rates	TEGa: 50 sccm NH_3: 500 to 3000 sccm
V/III Inlet Molar Ratio	1000 to 5000
Growth Pressure	76 to 130 Torr
Growth Temperature	500 to 900°C
Growth Rate	0.3 to 0.8 μm/hr

Bulk single crystals of $LiGaO_2$ and $LiAlO_2$ were grown from the melt by the Czochralski technique at Crystal Photonics[12] and subsequently sliced on both sides. It was found that etching the surface of the $LiGaO_2$ or $LiAlO_2$ with common inorganic acids such as HCl degraded the surface quality, presumably due to anisotropic etching. Therefore, the wafers were degreased in acetone and methanol solutions and loaded into the reaction chamber without further processing.

RESULTS AND DISCUSSION

Initial attempts to grow GaN on either $LiAlO_2$ or $LiGaO_2$ using hydrogen as the carrier gas produced films of poor structural quality as determined by HRXRD. This was attributed to the reduction of the oxide surfaces by H_2 during substrate heating. It was also determined that a low temperature GaN buffer layer was not effective in producing high quality material. To avoid the surface degradation caused by the H_2, a high purity N_2 carrier gas was tested. Growth on $LiAlO_2$ was not improved. A significant improvement in surface roughness for growth on $LiGaO_2$, however, was obtained by growing in N_2 carrier gas with a pretreatment of the substrate with NH_3 at a temperature greater than 800°C. A simple thermodynamic stability analysis suggests the reaction of Li_2O, Ga_2O_3 and Al_2O_3 with NH_3 favors the formation of the GaN or AlN over Li_3N by more than 40 kJ/mol of nitride at 1000 K. It is thus believed that avoiding the use of H_2 and pretreating the substrate with NH_3 prevents significant reduction of the substrate surface and produces a nitride surface nucleation layer. Figure 1 shows an AFM (Atomic Force Microscopy) image of a 0.5 μm thick GaN film grown directly on $LiGaO_2$ at 850 °C.

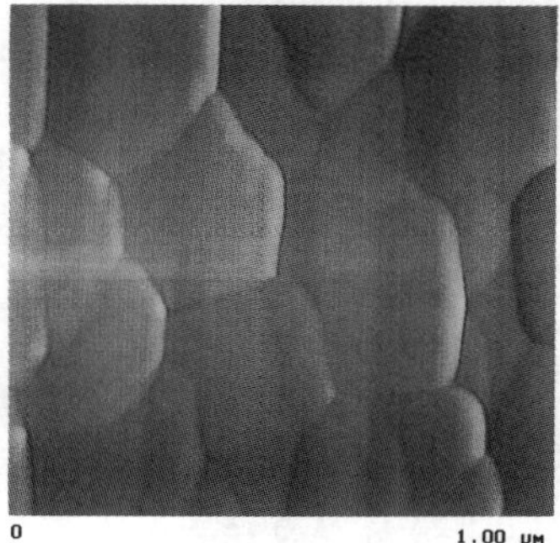

Figure 1. AFM image of the surface of a 0.5 μm thick GaN film on (001) $LiGaO_2$ at 850°C.

The quality of GaN films was investigated using cross-sectional TEM. Figure 2(A) shows a sample which has an estimated density of threading dislocations of only 10^7 cm^{-2} at a distance greater than 0.3 microns from the interface. The dislocation density was determined by estimating the thickness of the TEM foil and counting dislocations along a segment parallel to the interface. Densities were estimated to be 10^9 cm^{-2} at the interface. This low surface region defect density is comparable to results obtained for homoepitaxial growth of GaN films on small bulk GaN substrates.[13] SADP showed superposition of the GaN [2$\bar{1}\bar{1}$0] and the $LiGaO_2$ [010] diffraction pattern, suggesting that GaN layer is a single crystal and was grown epitaxially. Contrast analysis showed that there was residual strain at the interface between the GaN and $LiGaO_2$ substrate, despite the small mismatch.

Figure 2(B) shows a HRTEM photomicrograph of the interface between a GaN film and the $LiGaO_2$ substrate with lattice images aligned the GaN [2$\bar{1}\bar{1}$0] direction. Plane stacking faults were observed, but there was not a single polytype stacking order. These micrographs indicated that there was a nanocrystalline or amorphous interlayer between the GaN film and the $LiGaO_2$ substrate. It is speculated that either the substrate polishing procedure failed to completely remove sawing damage, or the surface region of the $LiGaO_2$ was altered during the initial stages of growth. A preliminary compositional study by SNMS (Spectroscopy of Neutrals Mass Spectrometry), indicated there exists an interfacial layer which is rich in Li between the GaN film and the substrate. It is possible that another phase such as Li_5GaO_4 could be formed by a surface reaction. We are actively studying this layer by various analytical procedures after annealing samples for various times and temperatures.

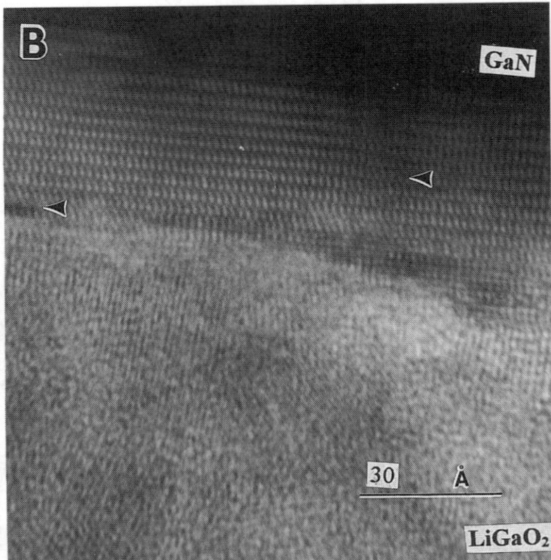

Figure 2. (A) XTEM photomicrograph of GaN film on LiGaO₂, bottom marker = 50 nm; (B) HRTEM image of the GaN/LiGaO₂ interface.

The interpretation of X-ray diffraction results has proven to difficult because the GaN and LiGaO₂ peaks are closely spaced. The thicknesses of the GaN films grown in this study are less than one μm, while the X-ray penetration depth was 3 to 5 μm. Thus signals were collected from the substrate as well as from the film. Figure 3(A) shows an X-ray diffraction pattern from a GaN film grown on (001) LiGaO₂. The three peaks occuring at 2Θ = 35.08°, 73.40°, and 126.35° may be indexed as the (0002), (0004), and (0006) diffraction peaks of GaN. Thus it is observed that (0001)-oriented hexagonal GaN grows on (001) LiGaO₂. With (101) LiGaO₂, however, it is noted that there is a strong peak at 2Θ = 47.96°, which may be indexed as the (1$\bar{1}$02) peak of GaN, while the second peak at 2Θ = 49.01° may correspond to the (202) diffraction peak of LiGaO₂. Clearly, the GaN film orientation is dependent on the orientation of the LiGaO₂ substrate. Although our TEM study indicates the presence of a disordered material at the film/substrate interface, the film is strongly influenced by the structure of the substrate below this interface. It is possible that the disordered regions occur as islands, with regions between the islands where there is full coherence. This issue is under study.

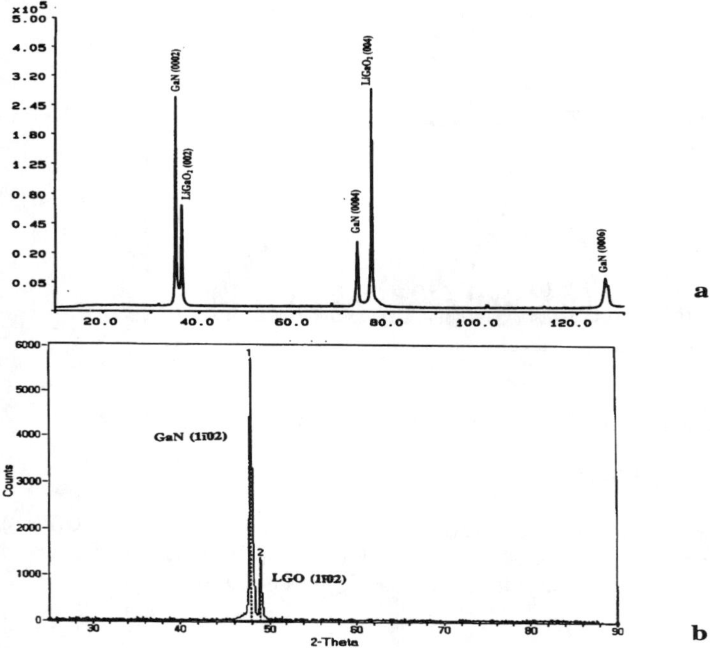

Figure 3. X-ray diffraction spectra: (A) GaN on (001) LiGaO₂; (B) GaN on (101) LiGaO₂

Finally, an in-depth analysis of the rocking curves produced from these samples has been pursued. Due to the close spacing of the substrate and film diffraction peaks, it has been difficult to interpret the results. The narrow peak shown in Figure 4 is tenatively assigned to the LiGaO₂ substrate, and the wider peak to the GaN film. Thicker films of GaN on LiGaO₂ are presently being prepared to assist in the interpretation of these data.

CONCLUSIONS

LiGaO₂ is a promising substrate for the growth of high quality GaN films. It was determined that the use of N₂ carrier gas is essential to achieve high structural quality when using LiGaO₂ as the substrate. XTEM micrographs indicate that the near surface region of the LiGaO₂ substrate was amorphsized, possibly due to a reaction product. These results also indicated that the first 300 nm of the GaN layer contained a high density of threading dislocations but that the density decreased to $\approx 10^7$ cm^{-2} at farther distances from the interface. Further study is in progress to better understand the initial stages of growth and to optimize the growth conditions.

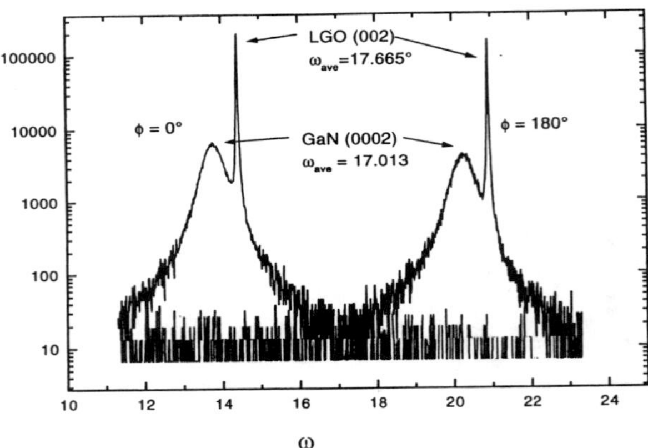

Figure 4. X-ray double crystal rocking curve of GaN on $LiGaO_2$.

ACKNOWLEDGEMENTS

The research at the University of Florida was sponsored by DARPA and the Office of Naval Research under contract N00014-92-J-1895. Work at NZ Applied Technologies was sponsored by BMDO and the Office of Naval Research under contract N00014-96-C-0253.

REFERENCES

1. H. Morkoc, S. Strite, G.B. Gao, M.E. Lin, B. Svederlov and M. Burns, *J. Appl. Phys.*, **76** 1394 (1994).
2. S. Strite and H Morkoc, *J. Vac. Sci. Technol.*, **B10**, 1237 (1992).
3. R.F. Davis, *J. Vac. Sci. Technol.*, **A11**, 829 (1992).
4. R.F. Davis, *Proc. Of the IEEE*, **79**, 702 (1991).
5. Nichia America Corporation, 1066 New Holland Avenue, Lancaster, PA 17601: catalog
6. S. Nakamura, M. Senoh, S. Nagahama, N. Iwasa, T. Yamada, T. Matsushita, H. Kiyoku, S. Sugimoto, *Japan. J. Appl. Phys.*, **35**, L74 (1996).
7. Y.F. Wu, B.P. Keller, S. Keller, D. Kapolnik, P. Kozodoy, S.P. Denpa, and U.K. Mishra, *Appl. Phys. Lett.*, **69**, 1438 (Sept. 1996).
8. W. Qian, M. Skowronski, and M. De Graef, *Appl.Phys. Lett.*, **66**, 1252 (March 1995).
9. David Bour, Xerox Palo Alto Research Center, unpublished results.
10. M. Marezio, *Acta Cryst*, **18**, 481 (1965).
11. P. Kung , A. Saxler, X. Zhang, D. Walker, R. Lavado, and M. Razeghi, *Appl. Phys. Lett.*, **69**, 2116 1996).
12. Bruce H.T. Chai, "Wurzite Structure Oxide Substrates for GaN Thin Film Growth", *J. Appl. Phys Lett.*, to be published.
13. A. Gassman, T. Suzuki, N. Newman, C. Kisielowski, E. Jones, and E. R. Weber, *J. Appl. Phys.*, **80**, 2195 (1996).

III-V nitrides growth by atmospheric-pressure MOVPE with a three layered flow channel

Kazuo Uchida, Hiroki Tokunaga, Yoshiaki Inaishi, Nakao Akutsu and Koh Matsumoto
Tsukuba Laboratories, Nippon Sanso Co., 10 Ohkubo, Tsukuba, Ibaraki, 300-26, Japan
Tsuyoshi Itoh, Takashi Egawa, Takashi Jimbo and Masayoshi Umeno
Deaprtment of Electrical & Computer Engineering, Nagoya Institute of Technology, Gokiso,
Showa-ku, Nagoya, Aichi, 466, Japan

ABSTRACT

We introduce III-V nitrides growth including GaN as well as InGaN by a newly developed atmospheric-pressure metal-organic vapor phase epitaxy system with a three layered flow channel which is a promising system for a large scale production. First, we have shown through computer simulation that a laminar flow of gases is maintained at 1000 °C in the three layered flow channel. Second, as a part of epitaxial results, we have found that the surface roughness of a low temperature-grown buffer layer on sapphire substrates, which can be measured by atomic force microscope, should be minimum in order to grow high quality GaN. We also report the growth of a double heterostructure of $In_{0.15}Ga_{0.85}N$/GaN which shows a strong near band-edge emission in room temperature photoluminescence.

INTRODUCTION

There has been great progress in the research on III-V nitrides growth as well as device fabrication[1-3]. To date, the commercial light emitting diodes which cover a spectral range from blue to green have already been available by metal-organic vapor phase epitaxy (MOVPE). Under these situations, we have developed an atmospheric-pressure MOVPE with a three layered flow channel which is a promising system for a large scale production. This three layered flow channel has been used in the large scale production (six-3 in. wafers) for ordinary III-V system because the large uniform diffusion boundary layer of source gases can be attained by adjusting the flow rate of each gases in the three layered flow channel[4]. The main purpose of this study is to verify the applicability of the three layered flow channel to the III-V nitrides growth in order to develop a large scale MOVPE for III-V nitrides.

In this paper we introduce the computer simulation of the gas flow and the growth results of low temperature-grown buffer layers, GaN and InGaN by a newly developed atmospheric-pressure MOVPE with a three layered flow channel. In particular, most of GaN films grown on c-plane sapphire substrates contain low temperature-grown buffer layers, which are either GaN or AlN. However, since the optimum growth temperature and its thickness have large variety in the literature[5-6], we first show the optimization method of buffer layers by atomic force microscopy (AFM) to produce high quality GaN films in MOVPE growth.

EXPERIMENT

I. Computer simulation of the gas flow in flow channels

A computer simulation of the gas flow in both the three layered and the center flow channel was performed using the control volume method on the commercial software (PHOENICS) for fluid dynamics. The biggest advantage in using the control volume method is that physical quantities such as mass, momentum and energy of fluid are preserved in a whole computing space. We used a 2-dimensional model with staggered mesh to simulate the gas flow in flow channels. Also, we employed following boundary conditions: the susceptor temperature was 1000 °C, the temperature of the center flow channel above the susceptor was 350 °C and the other surfaces of flow channels were heated up to 250 °C. N_2 gas at a flow rate of 10 l/min at

Mat. Res. Soc. Symp. Proc. Vol. 449 © 1997 Materials Research Society

atmospheric-pressure was flown in each layers in the three layered flow channel. Thermal expansion of gas was also taken into account.

II. Epitaxial growth

All epitaxial growth have been performed by a horizontal atmospheric-pressure MOVPE (Nippon Sanso Co., NM-2000) which is capable of handling a 2 in. wafer. Fig. 1 shows a schematic drawing of this MOVPE reactor. This reactor consists of two parts made of quartz. One is the three layered flow channel and the other is a center flow channel. In this three layered flow channel, source gases are separately flown in order to achieve very high gas velocity to avoid an adduct formation of reactive source gases without any entrance effects. The entrance effects cause an anomalous concentration distribution of source gases at an entrance in a reactor operating under high gas velocity. We usually flow purified N_2 at a rate of 15 l/min in the top layer, a mixture of metal-organic source and H_2

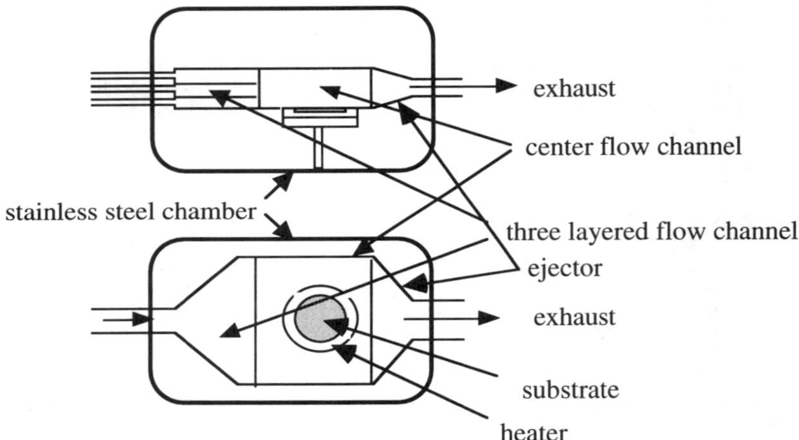

Fig. 1. Schematic drawing of atmospheric-pressure MOVPE reactor (NM-2000).

at a rate of 10 l/min in the middle layer and a mixture of specially purified 100 % NH_3 gas (Nippon Sanso Co., super ammonia) and H_2 at a rate of 10 l/min in the bottom layer. The center flow channel is rectangular shaped with very small height in order to achieve very high velocity of source gases. In this center flow channel a wafer susceptor with a carbon electric heater is located. These flow channels are installed in a stainless steel chamber. Using this MOVPE system, GaN and InGaN for this study were grown on c-plane sapphire (α-Al_2O_3) substrates. The substrates were loaded in the chamber without any chemical treatments, and then subject to an in-situ 1150 °C pretreatment in H_2 atmosphere prior to growth. Thin GaN or AlN buffer layers were grown at 450 °C or 300 °C, respectively. The temperature was then raised to 1050 °C for thick GaN growth with a typical growth rate of 2 µm/h at a V/III ratio of 5000. In the growth of double heterostructures of GaN/InGaN/GaN, the growth temperature of InGaN layer was either 700 °C or 725 °C. This was followed by a GaN cap at 1050 °C with a thickness of 0.1 µm. Photoluminescence (PL) system used in this study consists of a HeCd (325 nm) laser, a bi-alkali photomultiplier as a detector and a photon counter. The Hall measurement was used to characterize undoped GaN films using In contacts. AFM results were used to optimize the low temperature-grown buffer layer thickness for high quality GaN films.

RESULTS AND DISCUSSION

I. Computer simulation of the gas flow in flow channels

Fig. 2 shows our computer simulation result for the stream line of gas flow at a substrate temperature of 1000 °C. It was determined from this figure that a smooth laminar flow is achieved in the flow channel without any up-streams caused by the heat from the susceptor which is drawn as a dense line in this figure. This is because the height of these flow channels is very small. Consequently, it was also found that the gas stream goes to upward after the center flow channel.

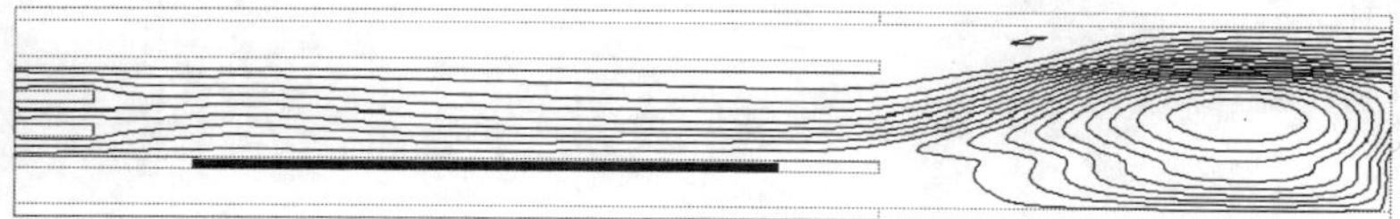

Fig. 2. The stream line of gas flow at a substrate temperature of 1000 °C at a total N$_2$ gas flow rate of 30 l/min .

II. Low temperature buffer layer and high quality GaN growth

In order to produce a mirror surface of GaN, an optimum thickness of both GaN and AlN low temperature buffer layers was characterized by the atomic roughness of their surfaces using AFM. The growth temperatures for GaN buffer layers and AlN buffer layers used in this characterization were 450 °C and 300 °C, respectively and these were grown on c-plane sapphire substrates after an in-situ 1150 °C pretreatment in H$_2$ atmosphere. Figs. 3 (a), (b) and (c) show the AFM surface images of GaN buffer layers grown with thickness of 10 nm, 20nm and 30 nm, respectively. It is found that the coverage of GaN layer on the sapphire substrate is not complete when the thickness is 10 nm (Fig. 3 (a)), and then becomes complete when the thickness reaches 20 nm (Fig. 3 (b)). However, further increase in thickness over 30 nm seems to cause a surface roughening as shown in Fig. 3 (c). These phenomena can be understood if the relation between the buffer layer thickness and the averaged atomic surface roughness is plotted as shown in Fig. 4. In this figure, open circles represent the data from GaN buffer layers, while solid circles represent those from AlN buffer layers. For GaN buffer layers, it was found that the minimum surface roughness occurs at the thickness around 20 nm. On the other hand, for AlN buffer layers, the minimum surface roughness occurs around 40 to 50 nm. Using GaN and AlN buffers with these optimum thickness, we have been successively obtaining high crystal quality with mirror surface GaN showing very high efficient band-edge PL emissions with very little deep emissions. Fig. 5 shows typical examples of the room temperature PL at a excitation power of 2 mW and the double crystal X-ray diffraction from 2µm-thick GaN grown at 1050 °C with 20 nm-thick GaN buffer layer grown at 450 °C. This x-ray diffraction shows a full width at half-maximum (FWHM) as small as 280 arc sec. The thickness uniformity of ±3 % over 2 in. wafer is obtained. The Hall measurement result on this sample indicates that the electron concentration is 3x10^{16} cm^{-3} with the electron mobility of 721 cm^2/V sec. The only difference in crystal quality caused by the different buffer layers is that GaN with an AlN buffer layer always shows a smaller electron mobility than that with a GaN buffer layer.

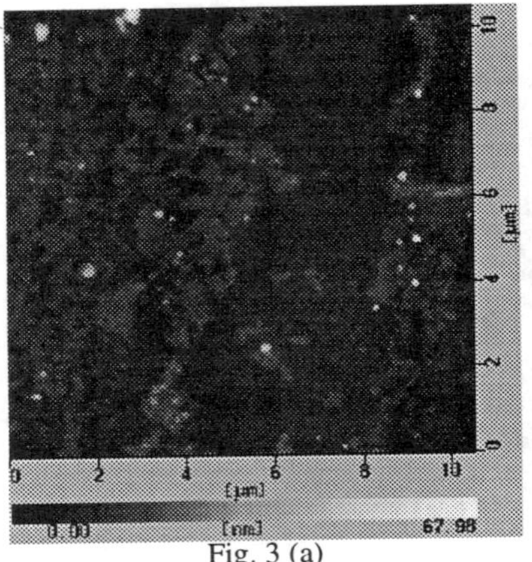

Fig. 3 (a)

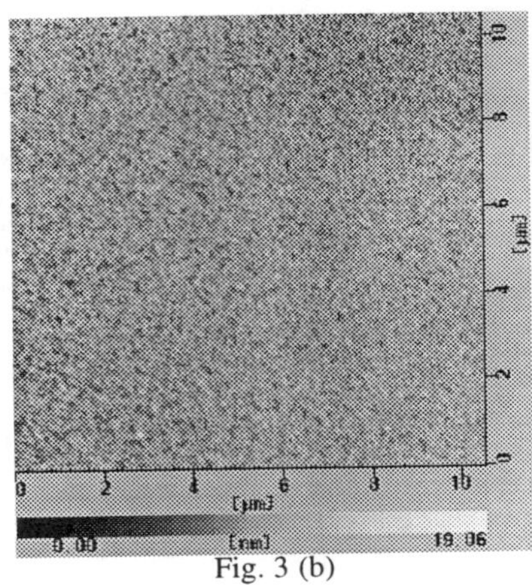

Fig. 3 (b)

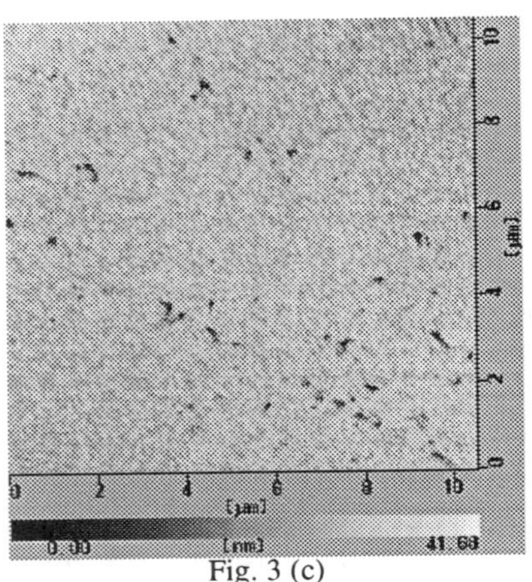

Fig. 3 (c)

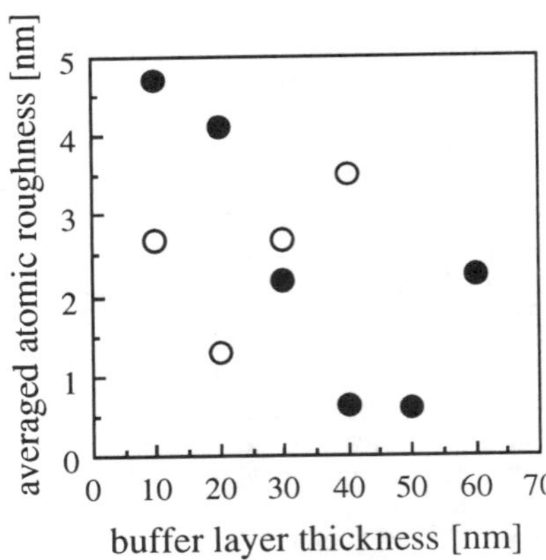

Fig. 3. The AFM surface images of GaN buffer layers grown with thickness of 10 nm (a), 20nm (b) and 30 nm (c).

Fig. 4. The relation between the buffer layer thickness and the averaged atomic surface roughness. Open circles represent the data from GaN buffer layers, while solid circles represent those from AlN buffer layers.

132

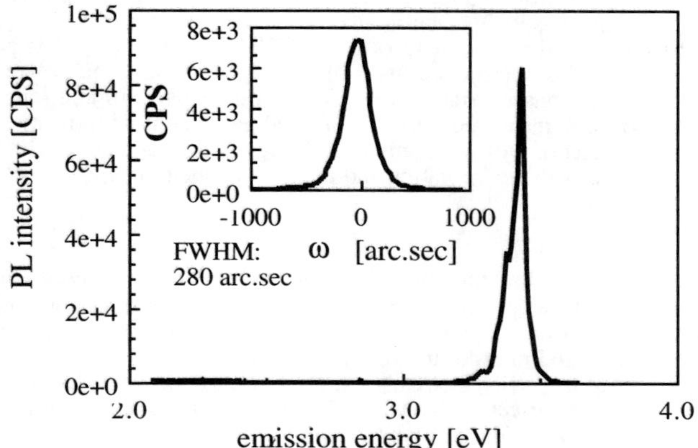

Fig. 5. The room temperature PL at the excitation power of 2 mW and the double crystal x-ray diffraction from 2μm-thick GaN grown at 1050 °C with 20 nm-thick GaN buffer layer grown at 450 °C.

III. Double heterostructure of GaN/InGaN/GaN

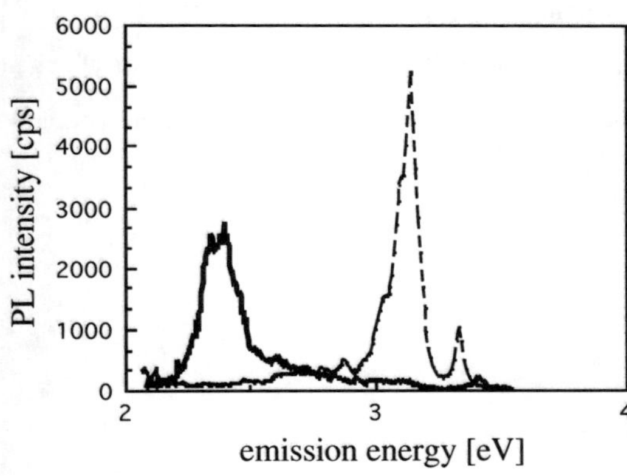

Fig. 6. The room temperature PL spectra at an excitation power of 2 mW from double heterostructures of GaN/InGaN/GaN. The solid curve represents the spectrum from 700 °C-grown InGaN while the dashed curve from 725 °C-grown InGaN.

We have also grown double heterostructures of GaN/InGaN/GaN. In these growth, the growth temperature of InGaN was changed. Fig. 6 shows the room temperature PL spectra collected from two different double heterostructures of GaN/InGaN/GaN at an excitation power of 2 mW. In these double heterostructures, the thickness of the InGaN layer and the input Ga/In ratio are 0.2μm and 0.24, respectively in both structures. However, the growth temperature of InGaN layers was different: One was grown at 700 °C (solid line) and the other at 725 °C (dashed line). In the PL spectrum from 700 °C-grown InGaN we can see the strong emission peak located around 2.4 eV, the broad tail toward higher energy and the small peak located around 3.4 eV which might originate from the cap GaN layer. The band-gap energy of 2.4 eV corresponds to the In content of 0.57 in the InGaN alloy. The double crystal X-ray data from this sample, however, indicates the In content of about 0.15. Further investigation is required to identify this emission peak at 2.4 eV. On the other hand, in the PL spectrum from 725 °C-grown InGaN, two dominant emission peaks are observed. One is at 3.14 eV and the other is at 3.33 eV. These emission energies correspond to the In contents of 0.15 and 0.07, respectively. The peak intensity ratio of these two emissions did not change upon increasing the excitation power. This result suggests that these two emissions originate from electron-hole pairs in continuum states. In addition, this In content of 0.07 measured by PL is in good agreement with the one obtained by the double crystal X-ray diffraction of this sample. Judging from these data, raising the growth temperature of InGaN decreases In incorporation and In inhomogeneity in epi-layers.

CONCLUSION

We have shown through computer simulation that a laminar flow of gases is maintained at 1000 °C in the three layered flow channel. We have found that the surface roughness of a low temperature-grown buffer layer on sapphire substrates, which can be measured by atomic force microscope, should be minimized in order to grow high quality GaN. We also report the growth of a double heterostructure of $In_{0.15}Ga_{0.85}N/GaN$ which shows a strong near band-edge emission in room temperature photoluminescence. Thus, we have verified the applicability of the three layered flow channel to the III-V nitrides growth through these results.

REFERENCES

[1] S. Nakamura in Gallium Nitride and Related Materials, edited by F. A. Ponce, R. D. Dupuis, S. Nakamura and J. A. Edmond (Mater. Res. Proc.**395**, Pittsburgh PA, 1996), pp.879-887.
[2] M.Koike, N.Shibata, S. Yamasaki, S. Nagai, S. Asami, H. Kato, N. Koide, H. Amano and I. Akasaki in Gallium Nitride and Related Materials, edited by F. A. Ponce, R. D. Dupuis, S. Nakamura and J. A. Edmond (Mater. Res. Proc.**395**, Pittsburgh PA, 1996), pp.889-895.
[3] H. Kong, M. Leonard, G. Bulman G. Negley and J. Edomond in Gallium Nitride and Related Materials, edited by F. A. Ponce, R. D. Dupuis, S. Nakamura and J. A. Edmond (Mater. Res. Proc.**395**, Pittsburgh PA, 1996), pp.903-907.
[4] J. Hidaka, T. Arai, H. Tokunaga and K. Matsumoto (Proc. of 8th Int. Conf. on Metal Organic Vapour Phase Epitaxy Cardiff, Wales, UK, 1996) , to be appeared in J. of Cryst. Growth.
[5] D. Kapolnek, X. Wu, B. Heying, S. Keller, B. Keller, U. Mishra, S. DenBaars and J. Speck, Appl. Phys. Lett. **67**, pp. 1541-1543 (1996)
[6] S. Nakamura, Jpn. J. Appl. Phys. **30**, pp. L1705-L1707 (1991).

GaN QUANTUM DOTS IN $Al_xGa_{1-x}N$ CONFINED LAYER STRUCTURES

Satoru Tanaka, Hideki Hirayama, Sohachi Iwai, and Yoshinobu Aoyagi
The Institute of Physical and Chemical Research (RIKEN), 2-1 Hirosawa, Wako, Saitama 351-01, Japan, E-mail: stanaka@postman.riken.go.jp

ABSTRACT

Nanoscale GaN quantum dots were fabricated in $Al_xGa_{1-x}N$ confined layer structures via metalorganic chemical vapor deposition (MOCVD), by using a 'surfactant' which can modify the GaN growth mode on AlGaN surfaces. A two dimensional growth mode (step-flow-like) of GaN films on $Al_xGa_{1-x}N$ (x=0~0.2) surfaces, that is energetically commenced under the conventional growth conditions, was intentionally changed into a three dimensional mode by adding tetraethyl-silane (TESi) used as a surfactant onto the AlGaN substrate surface prior to the GaN deposition. The surfactant is believed to inhibit the GaN film from wetting the AlGaN surface due to the change in surface free energy. The resulting morphological structures of GaN dots were found to be sensitive to; the doping rate of TESi, the Al content (x) of the $Al_xGa_{1-x}N$ layer, and the growth temperature. A very intense photoluminescence (PL) emission was observed from the GaN dots embedded in the AlGaN layers. The quantum size effect in terms of the blue-shift in a PL peak position was verified using the GaN dot samples having different dot sizes.

INTRODUCTION

Recently, GaN and other group III nitrides have been extensively investigated, largely in order to achieve high quality short wavelength light emitting and laser diodes (LED/LD). Laser action from InGaN multi-quantum well structures was recently achieved by Nakamura et al.[1,2] Among many problems needing to be solved for the future applications of these devices (especially short wavelength LDs), obtaining a low threshold current density (which is presently $8.7kA/cm^2$ high compared to that of III-V and II-VI LDs[2]) is essential. By reducing the dimension of the active layer structure, i.e. to that of a quantum wire or dot, if the size of structures is at least comparable to the effective Bohr radius of an exciton[3], then this value may be improved due to increased binding energies of excitons. Furthermore, a variety of physical phenomena due to many body effects of excitons is of great interest, for example, a biexcitonic transition[4] from the nanoscale GaN crystal is particularly interesting as a possible method to achieve new LD devices, quantum dot lasers.

Self-assembling dot structures have been constructed with InGaAs, InAs[5-7], AlInAs and different combination of phoshides.[8-10] A lattice mismatch in these can provide a Stranski-Krastanow (S-K) growth mode[11], which produces self-assembling nanoscale dots on a very thin wetting layer. Gallium nitride dot structures have so far only been reported from Dmitriev et al.[12] Various sizes of nanoscale GaN dots were fabricated *directly* on 6H-SiC substrates by metalorganic chemical vapor deposition (MOCVD) in which a three dimensional growth occurs. However, no investigation of GaN dot fabrication in the forms of both optical and electrical confined layer structures, which are essential for practical device applications, has been reported due to an inevitable two-dimensional growth mode in a GaN/AlGaN system.

In this letter a successful fabrication of nano-scale GaN quantum dots in AlGaN confined layers by MOCVD is reported. The role of a 'surfactant'[13] is discussed in terms of growth modes in the GaN/AlGaN system.

EXPERIMENTAL

The multilayer structures consisting of an $Al_xGa_{1-x}N$ capping layer, GaN dots, an $Al_xGa_{1-x}N$ cladding layer, and an AlN buffer layer[14] were fabricated on Si faces of on-axis

Mat. Res. Soc. Symp. Proc. Vol. 449 © 1997 Materials Research Society

6H-SiC(0001) surfaces by a conventional horizontal-type MOCVD system. Ammonia (NH3), trimethylaluminum (TMA) or trimethylgallium (TMG) with H_2 as a carrier gas and N_2 were independently supplied via a separate reactor tube and mixed just before the substrate susceptor for the growth of AlN, GaN dots, and $Al_xGa_{1-x}N$ solid solutions. The typical gas flow rate was 2 SLM, 2 SLM, and 0.5 SLM for NH_3, H_2, and N_2, respectively. The substrate temperature was measured with a thermocouple located at the substrate susceptor. After depositing the ~1.5 nm thick AlN buffer layer[14], the ~0.6 µm thick $Al_xGa_{1-x}N$ cladding layer was grown, with the Al content x being varied in the range of ~0.07-~0.2 by changing the gas flow ratio of TMA and TMG. The mole flux of TMA and TMG was 48 and 1.7-2.9 (for x = ~0.07-~0.2) µmol/min., respectively. The atomically smooth surfaces of the $Al_xGa_{1-x}N$ cladding layers were initially confirmed using an atomic force microscope (AFM). Tetraethylsilane ($Si(C_2H_5)_4$: TESi: 14 - 176 nmol) with the H_2 carrier gas was then supplied on the $Al_xGa_{1-x}N$ surface, followed by the short supply (5 second) of TMG and NH_3 gases for the growth of GaN dots. Finally, the GaN dots were embedded in the ~60 nm thick $Al_xGa_{1-x}N$ (x = ~0.2) capping layer. Samples without a capping layer were also fabricated to investigate the GaN dot morphology by AFM. The reactor pressure was maintained at 76 Torr, and the growth temperature was varied in the range of 1060°C and 1100°C but kept constant during a growth series. The resulting structures were examined by an AFM (SPI-3700), which was equipped with a Si_3N_4 sharpened pyramidal tip, and by photoluminescence (PL) using a He-Cd laser (325 nm) at 77K.

RESULTS AND DISCUSSIONS

(1) Two dimensional growth of GaN on AlGaN surfaces

A two-dimensional growth mode of GaN films on AlGaN surfaces was occurred, as demomstrated in the AFM micrograph in Fig. 1. GaN films were grown directly on to $Al_xGa_{1-x}N$ (x=0.2) surfaces without any surface treatment, resulting in a step-flow-like morphology. This feature resembles that of the AlGaN cladding layer which was also grown on the AlN buffer layer in a step flow mode as confirmed by AFM. The averaged step height was measured to be ~0.5 nm corresponding to 2 bi-monolayers of (0002) planes of GaN (see z-height analysis along the line A-B in Fig. 1). This was probably due to the fairy small lattice misfit (e.g. ~0.18% for x = 0.2 of $Al_xGa_{1-x}N$) and the larger surface free energy of the AlGaN surface as compared to that of GaN.

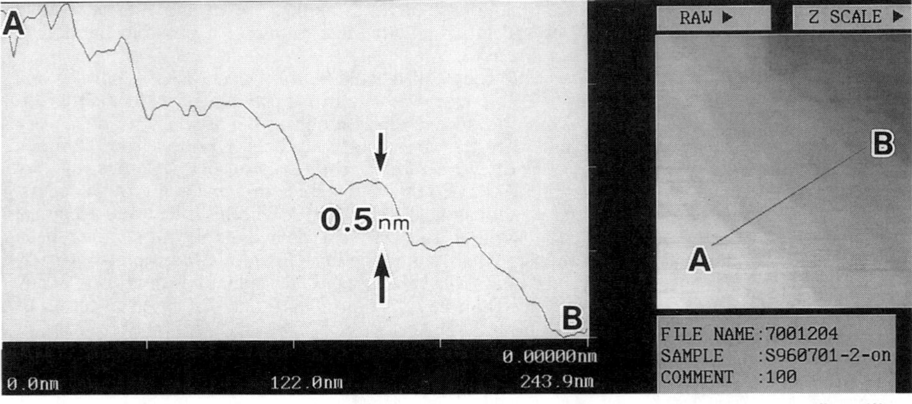

Fig. 1 AFM image of GaN grown on an AlGaN surface, showing a step-flow-like morphology. The z-height analysis along the line A-B indicates ~0.5 nm height steps on the GaN surface.

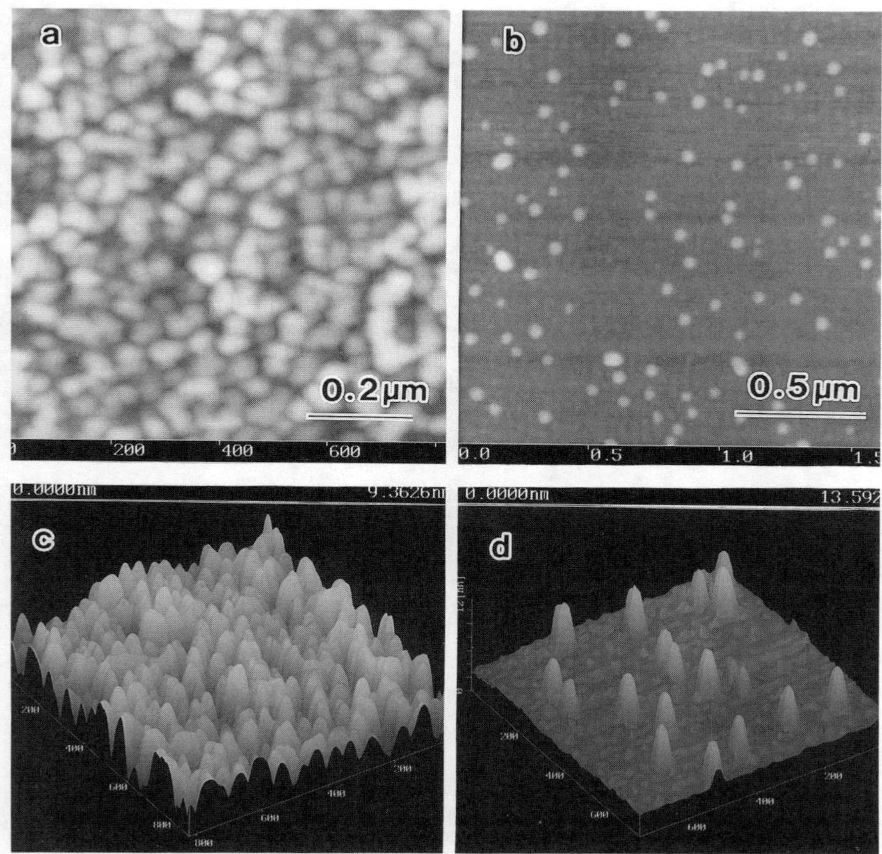

Fig. 2 AFM images of GaN quantum dots grown under the various conditions. The images of (a) plan-view and (c) bird-eye-view were taken of the sample grown at 1060°C and TESi dose of 14 nmol. The images of (b) plan-view and (d) bird-eye-view were taken of the sample grown at 1080°C and TESi dose of 41 nmol.

(2) Effect of TESi dose on GaN growth mode

The typical GaN quantum dot morphologies grown on the $Al_{0.2}Ga_{0.8}N$ surfaces are shown in Fig. 2(a) and (b), which were produced with growth temperatures of 1060°C and 1080°C, respectively. The GaN dots with an average width of ~40 nm and height of ~6 nm are uniformly distributed on the AlGaN surfaces. The dot density in Fig. (a) and (b) is ~2 x 10^9 cm^{-2} and ~4 x 10^{10} cm^{-2}, respectively. In the bird-eye-view AFM images of Fig. 2 (c) and (d), the detailed shapes of the dots corresponding to the plan-view images of (a) and (c) are clearly seen. It is of note that the GaN dot structures were only obtained when the AlGaN surface was exposed to the TESi gas before the GaN deposition procedure. The TESi doping rate, the Al and Ga content of $Al_xGa_{1-x}N$, and the growth temperature were all found to be important parameters which control the shape (aspect ratio = height/width) and the density of GaN dots. The dot density could be changed from ~10^7 to ~10^{11} cm^{-2} by various combinations of these parameters. The TESi doping rate and the growth temperature effectively controlled the density, for example, densities of ~5 x 10^9 and ~5 x 10^8 cm^{-2} were obtained with TESi doping rates of 44 and 176 nmol/min (both at 1080°C), respectively. The effect of the growth temperature on the density was rather drastic, varying by a factor of 10^3 between 1060°C and 1100°C. The dot size was also controllable by the TMG flux during GaN growth, i.e. growth time. By increasing the GaN growth time from 5 to 50 seconds the dot size was changed from 40 nm $\times$ 6 nm to 120 nm $\times$ 100 nm (width x height). The results of PL studies from various sizes of GaN dots are discussed in the following section.

(3) A model of GaN quantum dot formation mechanism

The GaN dot formation is probably explained by the surfactant effect. As has been studied in the Ge/Si system[13, 15-18], surfactants can be used to suppress island formation by modifying surface free energies.

In general, surface free energy balance at the growth front,

$$\sigma_s < \sigma_f + \sigma_i,$$

determines the growth mode, where σ_s, σ_f, and σ_i represent the surface free energies of the substrate, film, and interface free energy, respectively. The inequality shown here commences three dimensional growth in the system. Once a third element (surfactant), for example As[13], Sb[17], or Sn[18], is introduced onto the substrate surface, the sign of the inequality can be changed due to the alteration of substrate surface free energy. A similar reversed effect is expected to occur in the GaN/AlGaN system. It should be emphasized that a surfactant was introduced in order to enhance islanding in the present study, which was contrary to the inhibition it causes in the Ge/Si system. By adding the TESi gas to the substrate surface (AlGaN), the surface free energy may be decreased due to the termination of Al- and Ga- dangling bonds on AlGaN surfaces with Si. At this time, the presence or absence of Si-related bondings at the GaN dot/AlGaN interface, that is essential to clarify the exact role of TESi as a surfactant, is unknown. Moreover, in kinetic terms the surface diffusion length of adatoms may be enhanced by the presence of a surfactant, resulting in decreased dot densities under the higher TESi doping rate. The Si atoms may also be considered to play a role as a nucleation site. However, since the exact roles of TESi or Si are still unknown, a detailed study including surface chemistry needs to be carried out in the future.

(4) An optical property of GaN quantum dots

A photoluminescence study of the GaN dot structures was performed using the various GaN dot samples with different dot sizes discussed above. The GaN dots on all the samples were capped with an AlGaN layer in order to eliminate oxidation effects. Figure 3 shows the PL spectra of the samples with sizes in the range of 40 nm x 6 nm to 120 nm x 100 nm (width $\times$ height). Two peaks are apparent at ~3.5-3.55 eV and ~3.72 corresponding to emissions from the GaN dots and the AlGaN capping layer (x=~0.2), respectively. As the GaN dot size is increased, the peak position is shifted towards the lower energy side, indicating a quantum size effect. The PL peak of the sample with 40 nm x 6 nm size exhibits a blue shift of ~80 meV based on the standard position at 3.47 eV for a GaN film grown on the SiC substrate.[12]

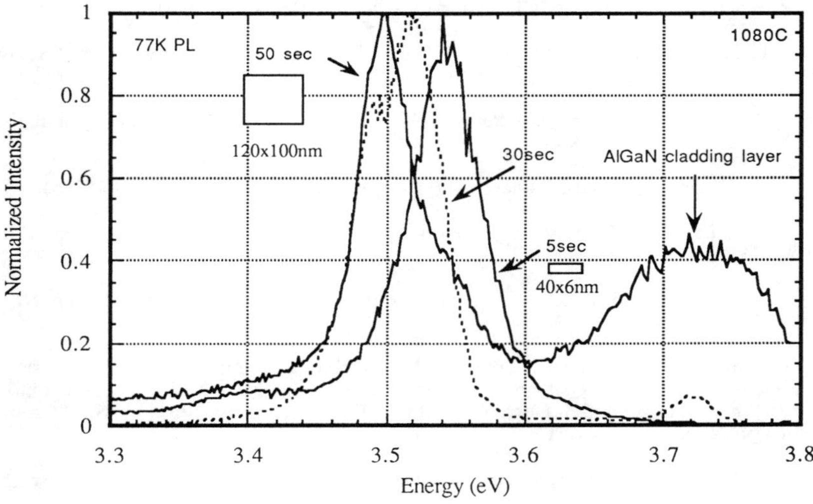

Fig. 3   77K PL spectra from different sizes of GaN quantum dots confined in $Al_xGa_{1-x}N$ (x = ~0.2) layers.

CONCLUSIONS

GaN quantum dots were fabricated on $Al_xGa_{1-x}N$ surfaces and in $Al_xGa_{1-x}N$ confined layer structures. The dose of TESi to the AlGaN surface was found to play a significant role as a surfactant and so establish dot structures, with no dot being formed without TESi. The size and density of the GaN dots were controlled by the TESi doping rate, the Al content of $Al_xGa_{1-x}N$ surface and the growth temperature. Finally, a very intense PL emission was observed from the GaN dots. The dot size dependence on the PL peak position indicated the quantum size effect.

REFERENCES

1.   S. Nakamura, M. Senoh, S. Nagahara, N. Iwasa, T. Yamada, T. Matsushita, H. Kiyoku, and Y. Sugimoto, Jpn. J. Appl. Phys. **35**, L74 (1996).

2.   S. Nakamura, M. Senoh, S. Nagahara, N. Iwasa, T. Yamada, T. Matsushita, H. Kiyoku, and Y. Sugimoto, Appl. Phys. Lett. **68**, 2105 (1996).

3.   B. Monemar, Phys. Rev. **B10**, 676 (1974).

4.   Y. Kawakami, Z. G. Peng, Y. Narukawa, Sz. Fujita, Sg. Fujita, and S. Nakamura, Appl. Phys. Lett. **69**, 1414 (1996).

5.   D. Leonard, M. Kishnamurthy, C. M. Reaves, S. P. Denbaars, and P. M. Petroff, Appl. Phys. Lett. **63**, 3203 (1993).

6.   R. Notzel, J. Temmyo, and T. Tamamura, Nature **369**, 131 (1994).

7.   J. Oshinowo, M. Nishioka, S. Ishida, and Y. Arakawa, Appl. Phys. Lett. **65**, 1421 (1994).

8.   N. Carlsson, W. Seifert, A. Petersson, P. Castrillo, M. -E. Pistol, and L. Samuelson, Appl. Phys. Lett. **65**, 3093 (1994).

9.   M. Sopanen, H. Lipsanen, and J. Ahopelto, Appl. Phys. Lett. **67**, 3768 (1995).

10.   A. Ponchet, A. L. Corre, H. L'Haridon, B. Lambert, and S. Salaün, Appl. Phys. Lett. **67**, 1850 (1995).

11.   I. N. Stranski and V. L. Krastanow, Akad. Wiss. Lit. Mainz Math.-Natur. Kl. Iib **146**, 797 (1939).

12.   V. Dmitriev, K. Irvine, A. Zubrilov, D. Tsvetkov, V. Nikolaev, M. Jakobson, D. Nelson, and A. Sitnikova, in <u>Gallium Nitride and Related Materials</u>, edited by R. D. Dupuis, J. A. Edmond, F. A. Ponce, and S. Nakamura (Mater. Res. Soc. Symp. Proc. **395**, Pittsburgh, PA, 1996) pp. 295.

13.   M. Copel, M. C. Reuter, E. Kaxiras, and R. M. Tromp, Phys. Rev. Lett. **63**, 632 (1989).

14.   S. Tanaka, S. Iwai, and Y. Aoyagi, J. Cryst. Growth (in press).

15.   A. Sakai and T. Tatsumi, Phys. Rev. Lett. **71**, 4007 (1993).

16.   H. J. Osten, J. Klatt, G. Lippert, B. Dietrich, and E. Bugiel, Phys. Rev. Lett. **69**, 450 (1992).

17.   D. J. Eaglesham, F. C. Unterwald, and D. C. Jacobson, Phys. Rev. Lett. **70**, 966 (1993).

18.   S. Iwanari and K. Takayanagi, Jpn. J. Appl. Phys. **30**, L1978 (1991).

GROWTH OF AlBN SOLID SOLUTION BY OMVPE

M. SHIN , A.Y. POLYAKOV, W. QIAN*, M. SKOWRONSKI, D.W. GREVE[1], and
R.G. WILSON[2]
Department of Materials Science & Engineering, Carnegie Mellon University, Pittsburgh, PA
15213-3890
[1]Department of Electrical & Computer Engineering, Carnegie Mellon University, PA 15213-3890
[2]Hughes Research Laboratory, Malibu, CA 90265
*currently with Department of Materials Science & Engineering, Northwestern University,
Evanston, IL 60208

ABSTRACT

Layers of AlBN were grown on sapphire by organometallic vapor phase epitaxy (OMVPE) at 1050^oC using triethylboron (TEB), trimethylaluminum (TMA) and ammonia as precursors. Boron is readily incorporated into the layers and its concentration in the solid phase can be made fairly high, up to at least 40%. However, single phase $Al_xB_{1-x}N$ films can only be grown for compositions not exceeding x=0.01. For higher B concentrations a second B-rich phase is formed. This phase is shown to be most probably wurtzite BN based on the results of transmission electron microscopy and x-ray diffraction. The growth of this phase occurs within the framework of wurtzite AlN islands providing the sites for lateral growth of wurtzite BN. This leads to formation of columnar structure of AlN and BN crystallites oriented in the basal plane and existing side by side. This is one of the first observation of purely thermal growth of sp^3 bonded BN.

INTRODUCTION

Group III nitrides are of much interest for a variety of applications in optoelectronics and electronics. However, great difficulties in bulk crystal growth and the lack of suitable lattice matched substrates for heteroepitaxial growth prevented fabrication of practical devices in this materials system. Only with low temperature AlN and GaN buffer layer, the crystalline quality of epitaxy layer improves up to certain degree. Recently controllable n- and p-type doping, growth of high quality heterostructures and quantum wells and fabrication of viable devices such as light emitting diodes (LED), various types of field effect transistors (FET) and, lately, of laser diodes (LD) [1,2] have been demonstrated. Still, even in the best nitrides films grown by OMVPE the lowest dislocation density achieved is in the high 10^8 cm^{-2} range [3], i.e. much higher than crystal perfection of more mature semiconductor systems, such as Si or GaAs.

4H and 6H polytypes of SiC offer a much better lattice match than does the sapphire. Also, the thermal conductivity of SiC is about an order of magnitude higher than that of sapphire which is a substantial advantage for high power microwave FETs now being developed in AlGaN. In addition, SiC substrates can be easily grown highly conductive n- or p-type, making it possible to form a good ohmic contact through the substrate in devices with vertical current flow such as LEDs or LDs for which many difficulties arise when growing on insulating sapphire substrates [1].

Even with the imperfect lattice match between AlGaN and SiC (about 1 % mismatch with AlN and about 3% mismatch with GaN), good quality epitaxial growth of AlN and GaN on SiC has been demonstrated. It is hoped that, could the exact lattice matching be achieved, the crystal perfection of AlGaN films would be determined mostly by the quality of SiC substrates [1,2].

It was proposed some time ago that lattice matching in the AlGaN/SiC system can be achieved by incorporation of boron, an atom with a smaller covalent radius than that of Ga or Al [4]. Although it has been demonstrated that a certain amount of B can be dissolved in AlN[5,6] the solubility limit and the phases which are formed when the solubility limit is exceeded have not been determined. In this paper we present such results for AlBN layers grown by OMVPE.

EXPERIMENTAL

AlBN layers were grown by OMVPE, on sapphire c-plane, in horizontal two-inlet reactor at 76 Torr. Hydrogen was used as a carrier gas and trimethylaluminum (TMA), triethylboron

Mat. Res. Soc. Symp. Proc. Vol. 449 © 1997 Materials Research Society

(TEB) and ammonia gas (NH3) were used for precursors of Al, B and N. TEB was used for growth of B containing compounds [7,8]. Sapphire substrates were cleaned in organic solvents and etched in hot mixture of H_2SO_4:H_3PO_4(3:1). These substrates were annealed in the reactor in H_2 at 1100°C before the growth. The buffer was AlN grown at 500°C and about 70 nm thick. Growth was done at 1050°C where good quality AlN layers could be deposited [9]. The typical sample consisted of two layers of same thickness. AlBN layer was grown on top of high temperature grown AlN. The flow of TMA and NH3 is fixed at 2.8 μmol / min and 2 l / min which gives the growth rate of 1 μm / hr. The flow of TEB was varied from 1.7 μmol / min to 34 μmol / min.

The thickness of the layer was determined by observing the interference fringes in optical transmission, and verified by cross-sectional transmission electron microscopy (TEM) measurement. The boron distribution was measured by secondary ion mass spectrometry (SIMS), and the signal was quantified using B implanted AlN standard sample [10]. The phases present in the samples were characterized by plan view and cross-sectional TEM and X-ray diffraction using Cu K_α radiation. The TEM samples were prepared by mechanical dimpling followed by ion milling.

RESULT AND DISCUSSION

SIMS profiles of two AlBN bilayer stacks grown with two different flow rates are shown in Fig. 1. The samples were double layer AlBN/AlN structure and each layer has a thickness of 0.5 μm. It clearly shows that TEB is decomposed at the growth temperature of 1050°C and B concentration in the solid phase is approximately proportional to TEB flow. The sharp drop of boron profile corresponds to the AlBN/AlN interface. If all of B is incorporated substitutionally in the AlN matrix, the lattice parameters of the layers would be changed accordingly. Assuming Vegard's law and published wurtzite BN lattice parameter [11], the lattice parameters of the AlBN layers shown in Fig. 1. should decreased by 1% and 6.4% compared to wurtzite AlN [12].

X-ray diffraction pattern for AlBN shows only (0 0 0 2) reflection from AlBN and (0 0 0 6) reflection from sapphire substrate. The sample with 10 sccm of TEB flow exhibits the change of peak position from 36.04° which is AlN basal plane position to 36.09° indicating that the c-lattice parameter decreased from 0.498 nm to 0.497 nm (Fig.2). This change corresponds to the composition of x=0.01 in $Al_{1-x}B_xN$. The remaining of B is apparently forming a second phase.

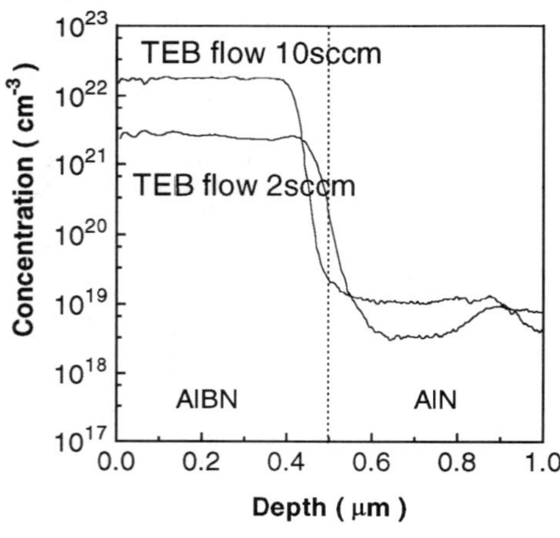

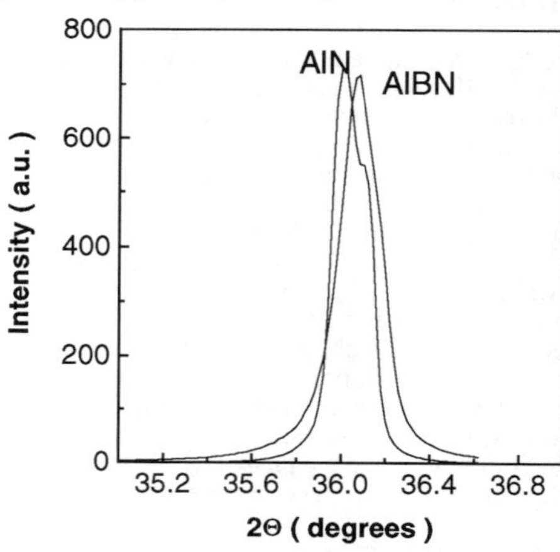

Fig.1. SIMS Profile of B in two double-layer grown with 2sccm and 10sccm of TEB flow

Fig.2. Comparison of X-ray peak position in (0 0 0 2) reflections of independently grown AlBN and AlN.

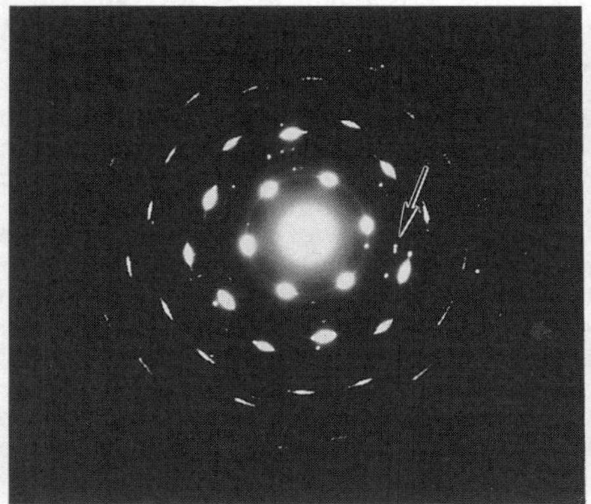 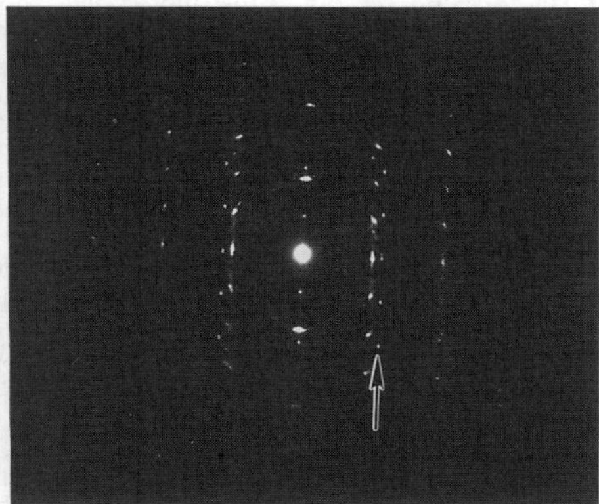

Fig.3. Selected area diffraction pattern of two layer sample along [0 0 0 1] zone axis

Fig.4. Selected area diffraction pattern of two layer sample along [$\bar{1}$ $\bar{1}$ 2 0] zone axis

Selected area electron diffraction (SAD) patterns of one of the two-layered AlBN/AlN films with 40 mol% B in the AlBN layer as measured by SIMS are shown in Fig. 3 from the plan-view sample along the [0 0 0 1] zone axis and Fig. 4 from cross-section sample along [$\bar{1}$ $\bar{1}$ 2 0] zone axis. SAD patterns of Fig. 3 and 4 clearly demonstrate the presence of extra set of diffraction spots belonging to second phase (marked by arrow). The weak intensity of this extra set of diffraction spots is believed to be due to low electron scattering by boron. The extra set of diffraction spots has 6-fold symmetry and is slightly rotated in respect to AlN diffraction (Fig.3). The calculated values of lattice parameters are 0.27 nm and 0.49 nm for a and c accordingly, and agree with values reported for wurtzite BN (w-BN). Bright field TEM micrographs of same sample are shown in Fig. 5 (plan-view) and Fig. 6 (cross-section). The second phase is marked with arrows. Since the boron atoms have a much lower electron scattering factor than that of Al atoms, the second phase should appear brighter than AlBN phase.

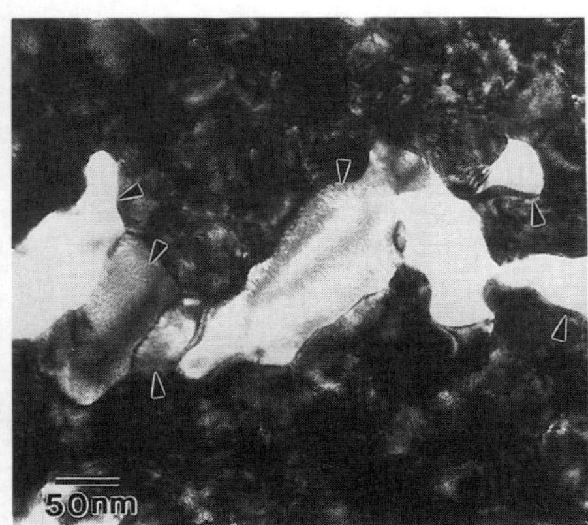

Fig.5. Plan-view bright field image of AlBN with B content of 40%. B-rich phase is brighter due to the low electron scattering of B atoms.

Fig.6. Cross-sectional image of same structure as Fig.5. B-rich phase is brighter. It can be seen above AlBN/AlN interface.

It can be seen from Fig. 5 that the second phase has characteristic dimension of $0.1\,\mu m$ and is uniformly distributed in the sample. We can see the columnar structure of bright area which represents the second phase in the AlBN layer and not in the AlN layer. From observations, it is clear that the second phase is some form of BN.

Among known BN polytypes, wurtzite BN (w-BN), hexagonal BN (h-BN), and rhombohedral BN (r-BN) have 6-fold symmetry. Table 1 shows the lattice parameters of these phases. The graphite-like phases have much larger c-parameter than AlN in contradiction with the measurement based on our SAD patterns (Fig. 5,6) The results of x-ray diffraction also support the conclusion of second phase being w-BN. From the lattice parameter of w-BN and crystallites alignment in the basal plane, the dominant reflection from (0 0 0 2) plane of w-BN should be found at the angle of about 43º for Cu K_α radiation [13-15]. However, this peak is too close to (0 0 0 6) peak of sapphire substrates at 41.7º to identify. On the other hand, the graphite-like phases have basal plane reflections at near 28º which should be easily observed if they exist. No such reflections were detected in x-ray diffraction pattern of AlBN layers. Two samples were grown with $1.5\,\mu m$ thick AlN layer under the AlBN layer which was enough to suppress the sapphire peak. The result is presented in Fig. 7. The 43.04º peak position agrees well with the c value of w-BN [11]. Due to some degree of uncertainty of lattice parameters of BN, it is difficult to say if this phase is pure BN or a solid solution which contains a small amount of Al.

The w-BN is not thermodynamically favorable and was previously formed only under the conditions of very high pressure and high temperature [16]. Only recently Haruyama *et al.* [5] claimed to have been able to grow w-BN on 6H-SiC substrates by OMVPE in the temperature range of 1100ºC to 1300ºC. The question which has to be asked is why this phase forms under OMVPE growth condition. One possibility is that the islands of AlN at the interface provide the nucleation sites facilitating the w-BN growth. The growth mode of this phase could be explained in the following way; first several monolayer-high islands of AlN are formed on the AlN underlayer and secondly the space between the islands is filled by laterally grown w-BN. AlN islands are restricted by this w-BN growth to growth laterally and grow upward with w-BN grown on the new step generated by growth of AlN. The evidence of BN growth on top of AlN or AlN growth on top of BN in the AlBN layer was not found (Fig.6).

Another possible mechanism is that the incorporation of very small amount of Al in BN stabilizes the w-BN. If this is the case, the growth rate of AlBN does not strongly depend on TMA flow as long as it is high enough to stabilize the BN growth. On the other hand, if the first hypothesis is the case, the growth rate of the layer should be the function of TMA flow rate since frame providing AlN should control the total growth rate. The growth with same TMA and varied TEB

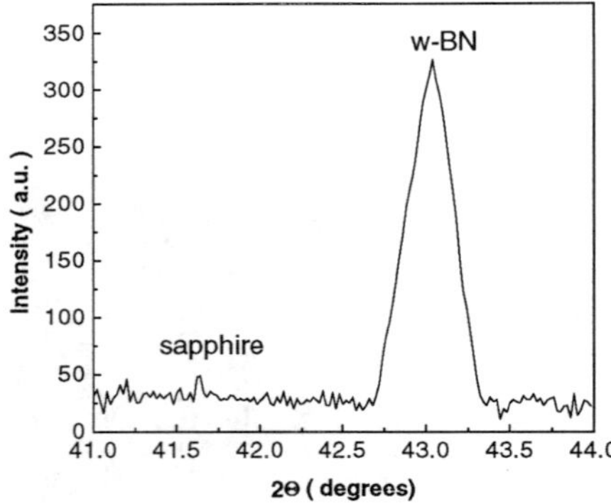

Poly type	a (nm)	c (nm)
h - BN	0.25	0.686
r - BN	0.25	1
c - BN	0.3616	
w - BN	0.2556	0.423

Fig.7. X-ray diffraction pattern of AlBN on top of thick AlN showing w-BN peak at 43.04º. Sapphire peak is barely seen due to thick AlN layer.

Table.1. Lattice parameters of some of BN polytypes.

flow from 0 sccm to 10 sccm which correspond to 40% total B concentration shows that there is only less than 25% growth rate difference. When TEB flow was constant and the TMA flow varied from $2.8\,\mu$mol/min to $0.7\,\mu$mol/min, the growth rate decreased accordingly. Hence the experiments seem to support the first hypothesis which is w-BN growth between AlN islands. Although the experiments seem to support our hypothesis, it is necessary to investigate further to get the positive proof of this growth mechanism.

Finally we tried to grow AlBN at 500°C where the grown material is either amorphous or very fine polycrystalline form and anneal the layer for recrystallization to incorporate more boron into the layer. This approach has been used very successfully in the case of GaBN [17]. For GaBN, we were able to incorporate up to 7% in the temperature range of 500°C-800°C. However, no x-ray peak shift has been detected in AlBN even in 10sccm of TEB flow. It's also noteworthy that in GaN-BN system, the second phase is most likely turbostratic BN and there's a growth poisoning of GaN at the onset of two phase growth.

CONCLUSIONS

In this presentation, we have shown that boron could be incorporated into the AlN layer substitutionally forming AlBN by OMVPE at 1050°C. For higher B concentration in the solid solution, the excess B starts to form the second phase which could be w-BN. Thermodynamically unfavorable w-BN could be facilitated by AlN islands at the interface between upper layer and AlN which induce the growth mode for w-BN in between.

It might be of interest to investigate the electrical properties of columnar w-BN and feasibility of this material being doped n- or p-type which will be our subsequent studies.

REFERENCES

1. S. Nakamura, M. Senoh, S. Nagahama, N. Iwasa, T. Yamada, T. Matsushita, H. Kioyoku and Y. Sugimoto, Jpn. J. Appl. Phys., **35**, L74 (1996)
2. S. Nakamura, M. Senoh, S.-I. Nagahama, N. Iwasa, T. Yamada, T. Matsushita, H. Kiyoku and Y. Sugimoto, Appl. Phys. Lett. , **68**, 2105 (1996)
3. B. Heying, X.H. Wu, S. Keller, Y. Li, D. Kapolnek, B.P. Keller, S.P. DenBaars and J.S. Speck, Appl. Phys. Lett., **68**, 643 (1996)
4. S. Sakai, Y. Ueta and Y. Terauchi, Jpn. J. Appl. Phys, **32**, 4413 (1993)
5. M. Haruyama, T. Shirai, H. Kawanishi and Y. Suematsu, $(B_xAl_yGa_{1-x-y})N$ quaternary system lattice matched to (0001) 6H-SiC substrate, in Proc. Int. Symp. on Blue Lasers and Light Emitting Diodes, Chiba Univ. Japan, March 1996, p. 106
6. A.J. Noreika and M.H. Francombe, J. Vac. Sci. Technol., **B2**, 722 (1969)
7. K. Nakamura, J. Electrochem. Soc., **133**, 1120 (1986)
8. H.M. Manasevit, W.B. Hewitt, A.J. Nelson and A.R. Mason, J. Electrochem. Soc., **136**, 3070 (1989)
9. M.Shin, A.Y.Polyakov, M.Skowronski, D.W.Greve, R.G.Wilson, and J.A.Freitas "Factors influencing the electrical and optical properties of AlGaN layers on sapphire" '96 Spring MRS. San Francisco, CA
10. R.G. Wilson, F.A. Stevie and C.W. Magee, <u>Secondary Ion Mass Spectrometry</u> (Wiley, New York, 1989)
11. J.H. Edgar, in <u>Properties of group III nitrides</u>, ed. J.H. Edgar (INSPEC Publications, London, 1994), pp. 7-21
12. W.J. Meng, in <u>Properties of group III nitrides</u>, ed. J.H. Edgar (INSPEC Publications, London, 1994), pp. 22-29
13. T. Akashi, H.-R. Pak, A.B. Sawaoka, J. Mater. Sci., **21**, 4060 (1986)
14. A. Onodera, K. Inoue, H. Yoshihara, H. Nakae, T. Matsuda and T. Hirai, J. Mater. Sci., **25**, 4279 (1990)
15. M. Ueno, K. Hasegawa, R. Oshima, A. Onodera, O. Shimomura, K. Takemura, H. Nakae, T. Matsuda and T. Hirai, Phys. Rev., **B18**, 10226 (1992)
16. V.L. Solozhenko, in <u>Properties of group III nitrides</u>, ed. J.H. Edgar (INSPEC Publications, London, 1994), pp. 43-70

[17] A.Y. Polyakov, M. Shin, M. Skowronski, D.W. Greve, R.G. Wilson, A.V. Govorkov and R.M. Desrosiers, in the abstracts of Electron Materials Conference, Santa Barbara, June 1996, paper CC9

Part II

Molecular–Beam–Growth Techniques

MBE GROWTH OF (In)GaN FOR LED APPLICATIONS

H. RIECHERT, R. AVERBECK, A. GRABER, M. SCHIENLE, U. STRAUß and H. TEWS
Siemens Corporate Research and Development
D-81730 Munich, Germany

ABSTRACT

We report on essential aspects of the growth of InGaN / GaN p-n junctions by gas-source molecular beam epitaxy (MBE) and present the first blue and green electroluminescence from such structures grown entirely by MBE.

A study of the growth conditions for a GaN nucleation layer on sapphire and for the subsequent growth of undoped GaN points out the necessity for Ga-stabilized growth. Unintentionally doped GaN grown at 1µm/h shows background doping levels below 10^{17} cm^{-3} and mobilities up to 320 $cm^2/Vsec$ (at 300K). Narrow photoluminescence with very low intensity in the yellow spectral range is observed. N- and p-type doping of GaN with Si and Mg yields layers with high mobilities (220 and 10 $cm^2/Vsec$, respectively at 300K) at carrier densities typical for devices.

Although incorporation of Indium is strongly temperature-dependent, InGaN-layers with In-contents of over 40% are obtained routinely. The optical properties of our InGaN layers typically exhibit the commonly observed, broad deep level luminescence. Finally, we present electroluminescence in the visible spectral range up to 540 nm from InGaN / GaN double-heterojunctions.

INTRODUCTION

For the growth of III-V nitrides, MBE offers the advantage of low growth temperatures by utilising plasma sources to generate highly reactive nitrogen in the form of atoms or excited molecular species. Thereby, growth proceeds directly through the reaction of elemental Ga and N and the substrate temperature is determined only by requirements intrinsic to epitaxial growth such as mobility of surface adatoms. Pioneering work [1] has shown that typically 700°C are sufficient to achieve this.

In contrast to this, metalorganic vapour phase epitaxy (MOVPE), which is at present widely used for the growth of these materials, relies on growth temperatures above 1000 °C due to the need to pyrolize the nitrogen precursor NH_3. These temperatures impede the incorporation of high vapor pressure elements such as Mg and In, which are needed for p-type doping and to shift the band-gap of GaN into the visible spectral range, i.e. for the growth of InGaN. This difficulty is made evident by the fact that to date there is only one group which has published results on green electroluminescence from InGaN [2]. The only reports on emission of visible light generated from MBE-grown heterostructures known to us are by Sakai et.al. [3] and Schetzina [4]. However in this case, the maximum wavelength attained is 420 nm (violet [3]) and the p-n junction was grown by MBE onto a GaN layer grown by MOVPE on sapphire [3] or SiC [4]. Thereby one of the main epitaxial difficulties, namely the growth of a starting layer of sufficiently high quality was not solved by MBE.

The goal of our work was therefore to exploit the possible advantages of MBE growth and to assess its usefulness, especially in the growth of complete InGaN / GaN heterostructures on sapphire for light-emission at wavelenghts in the blue and green.

Mat. Res. Soc. Symp. Proc. Vol. 449 © 1997 Materials Research Society

EXPERIMENTAL

All layers were grown in a standard MBE system (VG Semicon V 80H) equipped with conventional effusion cells for Ga, In, Al, Si and Mg. Nitrogen is supplied through a commercial RF-plasmasource with a pressure-controlled inlet line of N_2 gas. Conductances along the line into the growth chamber were optimized to attain a high efficiency for the generation of reactive nitrogen species (presumably N atoms [5]). Whereas previous work using a similar nitrogen source reported typical growth rates of about $0.3\,\mu m/h$ [4,6], we attained at least 1 $\mu m/h$. At present this is limited by the pumping speed of our MBE-system (about 1000 l/sec) and by the available RF power (600 W). Unless otherwise stated, all results reported here were achieved with a growth rate of 0.7 to 1 $\mu m/h$.

We use (0001)-oriented sapphire substrates with 2" diameter. To facilitate heating by radiation, the backside of the wafers is sputter-coated with a refractory metallisation. The stated growth temperatures were measured by pyrometer (with maximum sensitivity at 940 nm). A study of the effect of N generated from a RF-plasma source similar to ours has shown that no or little nitridation of the sapphire substrate takes place in the time before initiation of growth. [7]

To characterize our growth conditions we take advantage of the possibility to precisely measure the group-III-flux and the effective N-flux. Group III-fluxes (beam equivalent pressures) are measured by an ion gauge. For the effective N-flux which determines the growth rate in a Ga-stable growth mode (see below), one can establish a relation between N_2-pressure at the inlet side of the plasma source and the growth rate. For GaN growth, this relation is valid for all but the highest growth temperatures. The absolute calibrations are derived by measuring layer thicknesses from cleaved edges in secondary electron microscope (SEM). As described below, this routine allows the distinction between Ga-stable and N-stable growth regimes and the determination of the effective overpressures of Ga or N.

GaN and InGaN single layers were analyzed by room-temperature Hall, photoluminescence (PL), and X-ray diffraction (XRD). PL measurements were performed at 300K, 77K, and 4K. The luminescence was excited by a 20mW HeCd laser and detected by a cooled GaAs multiplier. Electroluminescence measurements (EL) were performed on processed wafers at room temperature using DC-currents. The EL was detected through the transparent sapphire substrate by use of an optical fibre.

GROWTH OF UNDOPED GaN

<u>Nucleation layer on sapphire</u>

We find empirically that the highest quality of GaN layers is achieved by initiating growth on (0001) sapphire with a nucleation layer of GaN, grown at similar temperatures as the subsequent layer, i.e. in the range of 650 to 750°C, but at a rate of about 200 nm/h. This was established by comparing structural, optical and electrical data of identically grown, 400nm thick GaN layers on various nucleation layers. We find that a good overall figure of merit is the electron mobility (at 300K) since even badly coherent layers may show good photoluminescence or narrow half widths in X-ray diffraction.

The most important parameter in the growth of GaN nucleation layers is found to be the N / Ga ratio. High quality GaN was only obtained for nucleation layers grown in the Ga-stable growth mode (see below). The amount of excess Ga as well as the growth temperature and thickness of the nucleation layer were found to be much less important.

Reflective high energy electron diffraction (RHEED) performed during growth shows that the layers are ordered almost from the beginning. The appearance of a "streaky" pattern indicates that large domains with a flat morphology are generated. This is confirmed by SEM-images of about 15 nm thick nucleation layers showing distinct islands with apparently flat surface. A high resolution transmission electron microscope (HR-TEM) image of the sapphire-GaN interface is shown in fig. 1. It reveals that large, apparently defect-free domains are formed at the beginning of growth. (Details will be published in a separate paper.) The low-magnification inset of this figure indicates that in the subsequent growth, extended defects form and evolve into groups propagating essentially along the growth direction. Thus the commonly observed domain structure is formed.

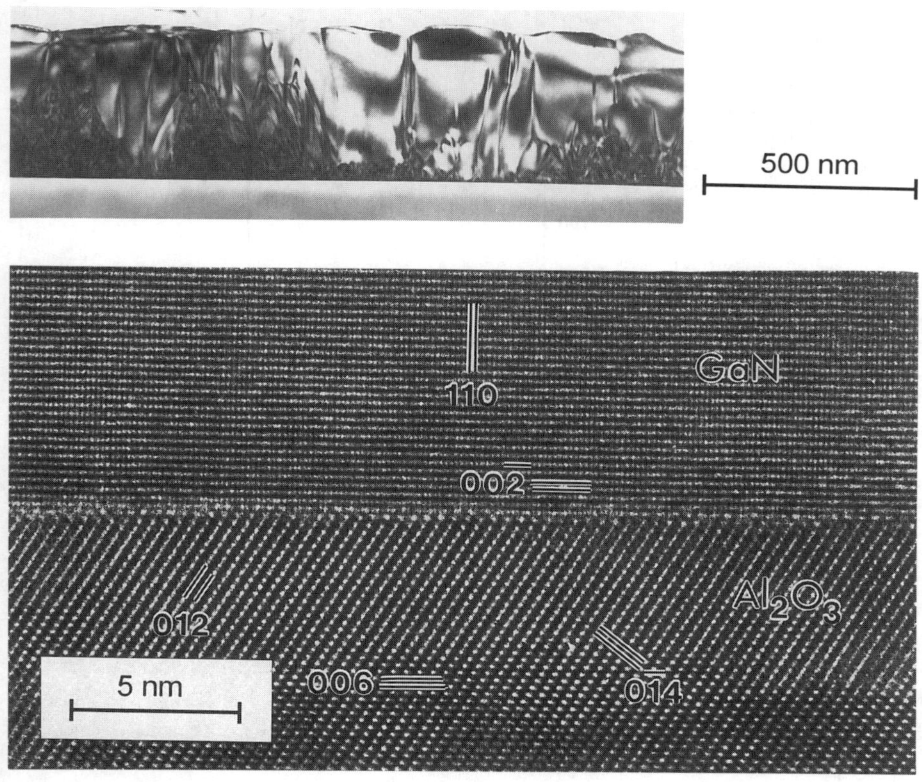

figure 1: HR-TEM of GaN [$\bar{1}$100] along sapphire [$2\bar{1}\bar{1}0$]

Basics of GaN growth in MBE

In the following section we illustrate how to establish the boundary between Ga- and N-stable growth and which influence the N / Ga ratio has on electrical properties as well as morphology. Fig. 2a shows the GaN growth rate R for a fixed Ga-flux as a function of N-flux (given by the N_2 pressure in the inlet line). The vertical line separates two fundamentally different regimes: to the left of the line, R increases with N-flux, i.e. the rate is N-limited and a certain excess Ga-flux is present. This is the regime of Ga-stable growth. Even further to the left of the line, with an onset depending on growth temperature, not all excess Ga can be desorbed and Ga droplets remain on the surface. To the right of the line, R saturates because all the available Ga is incorporated. Here, growth proceeds with excess N present, i.e. it is N-stable or Ga-limited. We find that the range of useful growth parameters is limited to the close vicinity of this transition line, quite unlike in the MBE of arsenides, where large variations of V / III-ratio are possible as long as group V stable growth is maintained. This is seen from figs. 2b and 2c, which show the dependence of background doping level and electron mobility on N / Ga ratio. It is obviously advantageous to grow GaN in a slightly Ga-stable mode to obtain a low background doping level and highest mobilities. (The mobilities in this series of layers are limited by the layer thickness of about 600 nm.)

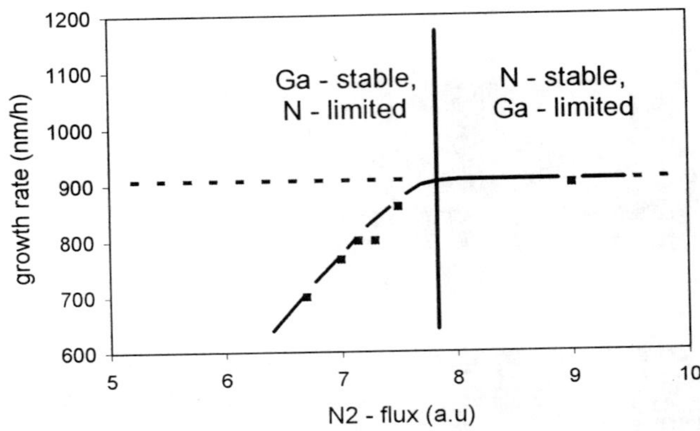

figure 2a: Growth rate of GaN as a function of N$_2$-flux

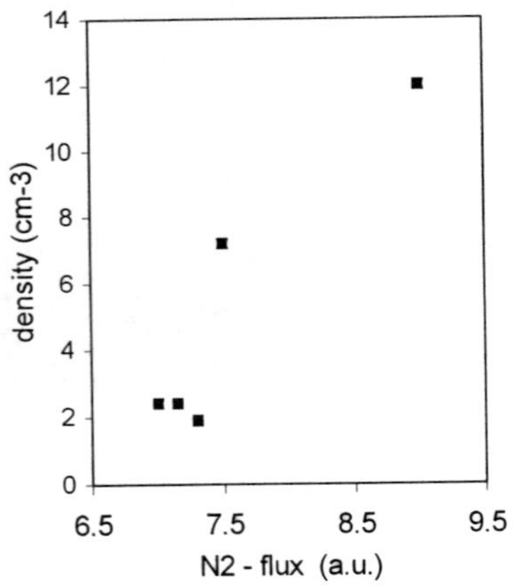

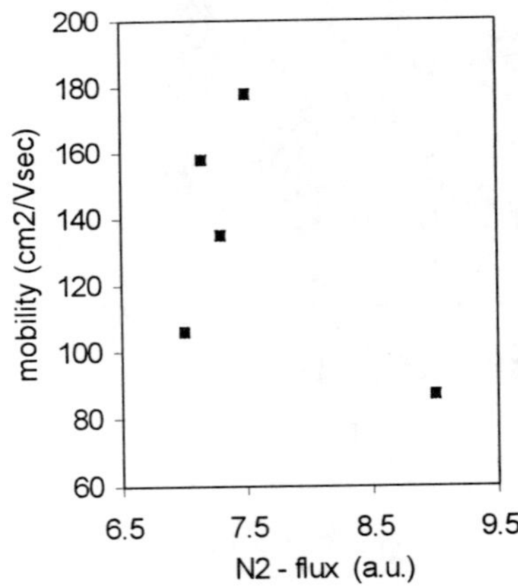

figure 2b: Electron concentration vs. N$_2$-flux

figure 2c: Electron mobility vs. N$_2$-flux

Fig. 3 shows the evolution of the surface morphology (on a microscopic scale with 50 000 fold magnification) with increasing N / Ga ratio. Again, the most favourable results are obtained with a slight excess of Ga. Apparently, this condition gives the optimum mobility of surface adatoms to allow large-scale two-dimensional growth.

<u>Structural, electrical and optical properties of unintentionally doped GaN</u>

GaN growth optimized as outlined above yields layers with the following properties:
The surface morphology is observed as shown in fig. 3b, but may be entirely flat. Since the characteristic dimensions are similar to those of "dislocation-free" domains observed in TEM images (fig. 1), the morphology probably images the boundaries of such domains.

XRD performed on layers with a thickness around or above 1μm yields a full-width at half-maximum (FWHM) in the symmetric (0002) reflection of 8 to 10 arcmin. In the asymmetric $(10\bar{1}2)$ reflection, the FWHM is about 50-100% broader as reported in [8] for high-quality, MOVPE-grown GaN. This means that the misalignment of the domains with respect to the substrate consists of a tilt out of the substrate normal and a rotation about this line. It is interesting to note that the authors of [8] also find this domain structure when little or no nitridation of the substrate is performed before growth.

The electrical properties systematically follow the behavior shown in fig. 2 (which contains data for growth at 680 °C). This must be taken as a strong argument against N-vacancies or Ga-interstitials causing the residual n-type doping. We can, however offer no alternative explanation. By growing at or above 700°C, the background doping level can be routinely reduced to below 10^{17} cm^{-3}, with lowest values around 5 x 10^{16} cm^{-3}. At such low carrier concentrations, the mobility tends to be below 200 cm^2/Vsec. Since the mobility increases

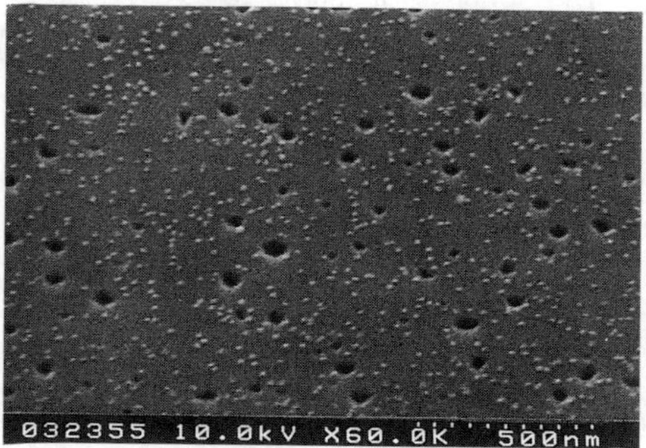

figure 3a: Scanning electron micrograph of a layer grown under Ga-rich conditions

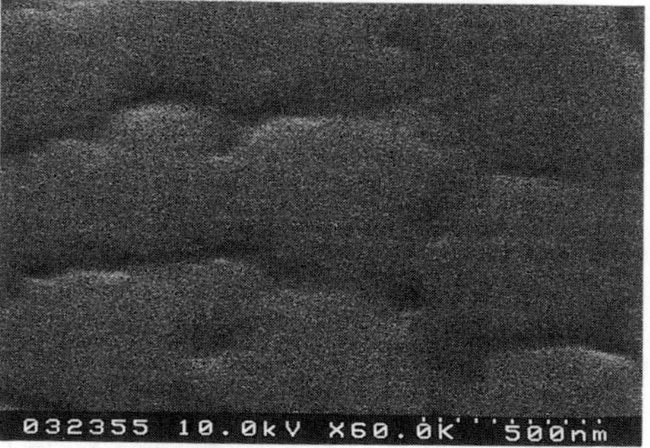

figure 3b: Layer grown at optimum N / Ga ratio

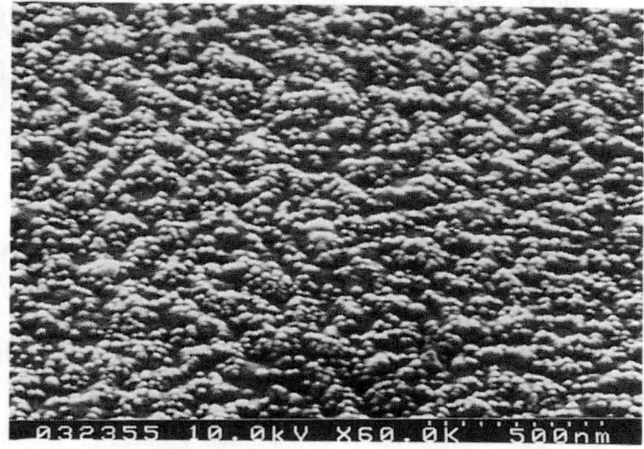

figure 3c: Layer grown under N-rich conditions

with slight Si doping, it appears to be at least partly limited by compensation. Highest mobilities are found at background doping levels of 3-5 x 10^{17} cm^{-3} with best values around 320 cm^2/Vsec for layers of 1.5 - 2 μm thickness. This is among the highest mobilities reported to date for MBE-grown GaN [9], [10].

The photoluminescence of our samples is clearly dominated by donor-bound excitons (D°X) as shown in fig. 4. We routinely find extremely low levels of yellow luminescence. At low excitation, the baseline in this spectral range is up to five orders of magnitude below the intensity of D°X [11]. At 4K (300K) the FWHM of the D°X peak is measured to be 8-9 (40) meV, which appears to be limited by nonuniform strain. The energy position of the luminescence line is about 15 meV lower than observed for MOVPE-grown GaN layers of similar thickness [12]. This may be explained by the lower growth temperature we use, which leads to lower strain between layer and substrate when cooling down the sample after growth.

GROWTH OF N- AND P-TYPE DOPED GaN

N-type doping is achieved with similar ease as in GaAs by the incorporation of Si. For electron concentrations of about 2-3 x 10^{18} cm^{-3} we find a mobility of 220 cm^2/Vsec. This is comparable with values achieved by MOVPE growth [13].

We have initially attempted to use Be as a p-type dopant, which according to Neugebauer [14] should be a shallow acceptor. However we found that at fluxes corresponding to a Be concentration of 1-5 x10^{18} cm^{-3} , growth is severely perturbed. This is indicated by an almost immediate change of the RHEED pattern from a two- to a three-dimensional one after opening of the Be-shutter. Such layers are highly resistive and microscopically exhibit a very rough, craterlike surface.

Using Mg as an acceptor does not show this problem, but due to its high vapor pressure at typical growth temperatures, Mg readily re-evaporates. We find that the Mg-concentration, measured by secondary ion mass spectrometry (SIMS), decreases by two orders of magnitude when the growth temperature is increased by 120°C. Consequently, we grow GaN:Mg at or below 700°C. The highest hole concentration measured by Hall effect is 6 x 10^{17} cm^{-3}, typically we obtain 3-5 x 10^{17} cm^{-3} with a mobility of 8 - 10 cm^2/Vsec. It should be

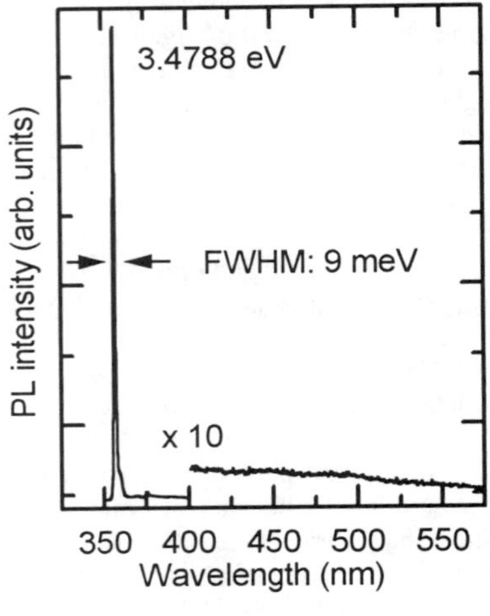

figure 4:Typical PL spectrum at 4K

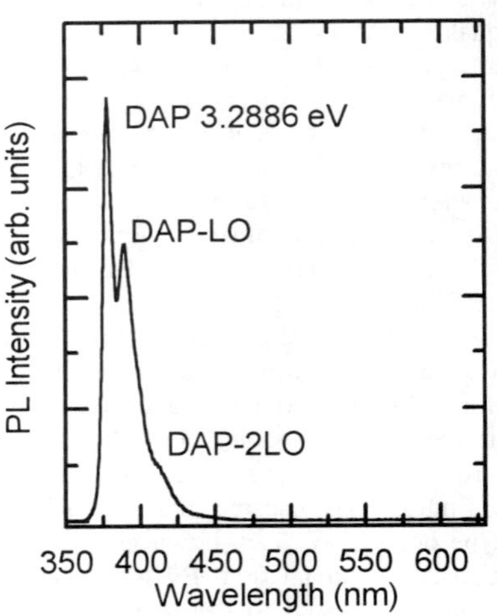

figure 5: PL spectrum of GaN:Mg at 77K

noted that these results are obtained without post-growth annealing in nitrogen as is necessary for MOVPE-grown material [15].

The PL of GaN:Mg at 77K is shown in fig. 5. It is dominated by the donor-acceptor pair-transition (DAP) at 380 nm, followed by clearly visible phonon replicas. In GaN:Mg with good electrical properties we do not observe PL features at or around 400 - 430 nm as reported by other authors [15].

GROWTH OF InGaN

The growth of InGaN is complicated by the relatively low thermal stability of the In-N-bond. Ambacher et. al. [16] have shown that the dissociation of InN begins at 650°C.

Consequently, loss of In from the growing layer not only occurs by immediate re-evaporation due to its high vapor pressure but also by thermal decomposition of already bonded In. This leads to a strong temperature dependence of the In-incorporation as shown in fig. 6. All data were taken from layers grown with the same In/Ga ratio. With this ratio In-contents around or above 50% can be easily reached. However, we find that growth at very low temperatures generally leads to poor layer morphology, such that at present the practical lower limit for growth is 650°C. At this temperature we obtain In-contents between 25 and 40% by variation of the In/Ga ratio. The growth rate chosen for InGaN is about 500 nm/h. A typical XRD rocking curve of InGaN is shown in fig. 7. The InGaN peak is much broader than that of GaN, which indicates that the composition is not uniform.

A PL spectrum of a 30nm thick $In_{0.25}Ga_{0.75}N$ layer on GaN, taken at 300K, is shown in fig 8. We observe the D^0X signal of the underlying GaN at 3.4228 eV and a strong photoluminescence line at 2.66 eV, which we attribute to InGaN. With respect to the

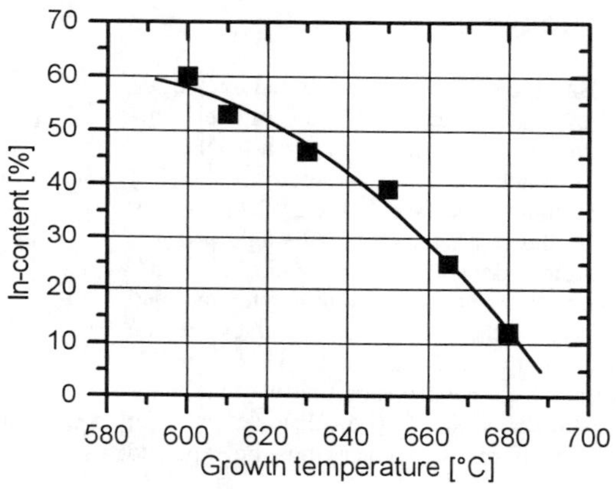

figure 6: In content as a function of temperature

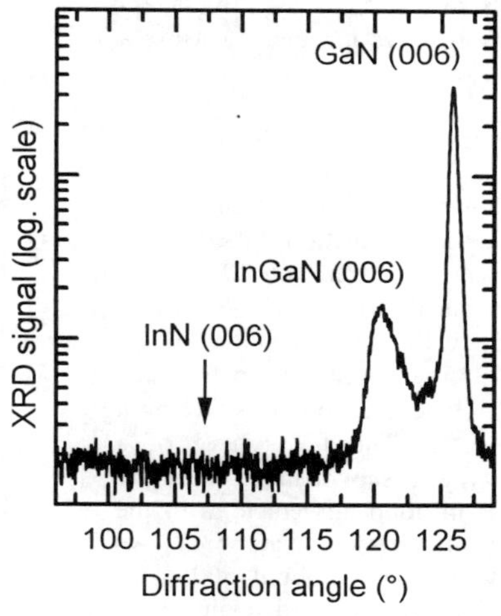

figure 7: θ-2θ scan of an $In_{0.25}Ga_{0.75}N$ layer

expected position of the band-gap [17], the InGaN-PL signal is some 160-200 meV lower in energy. PL emission via this deep level is typical for our samples up to an In-content of 40%. For higher In-contents, especially above 50%, the emission generally occurs about 400 meV below band-gap.

ELECTROLUMINESCENCE IN THE UV, BLUE AND GREEN

Based on the results presented above, we have grown p-n junctions to generate electro-luminescence. They consist of an n-GaN layer (1-1.5 μm thick, Si-doped to about 10^{18} cm^{-3}) and a 500nm thick layer of GaN:Mg, which, judged by the characterisation of single p-layers has a hole concentration of 3-5 10^{17} cm^{-3}. Heterojunctions have an additional, undoped InGaN-layer either 30 or 3 nm thick at the position of the p-n junction. These InGaN thicknesses were chosen such that the thin layers are elastically strained (i.e. dislocation free) whereas the 30 nm thick layers must be expected to be fully relaxed.
Light emitting diodes were fabricated from these layers. They have a transparent p-type Ohmic contact of 50μm or 200μm diameter, respectively. The Ohmic contacts to p-GaN are made by TiAu, those to n-GaN by TiAl. Mesa etching is done by reactive ion etching using a chlorine process.

GaN - homojunctions

Fig. 9 shows the electroluminescence spectrum of a GaN homojunction, taken at 5V forward voltage and a current of 20 mA. It looks almost identical to the PL-spectrum shown in fig. 5, with the exception of an additonal peak at 375 nm. Again, apart from the wide tail into the blue, no longer-wavelength features are observed. The I-V- curve of a structure of this type is shown in fig. 10. The forward voltage is in good agreement with the band-gap energy of GaN and at -5 V we measure a reverse current of about 500 μA . This attests to a junction of good quality with reasonable series resistance despite the thin n-layer.

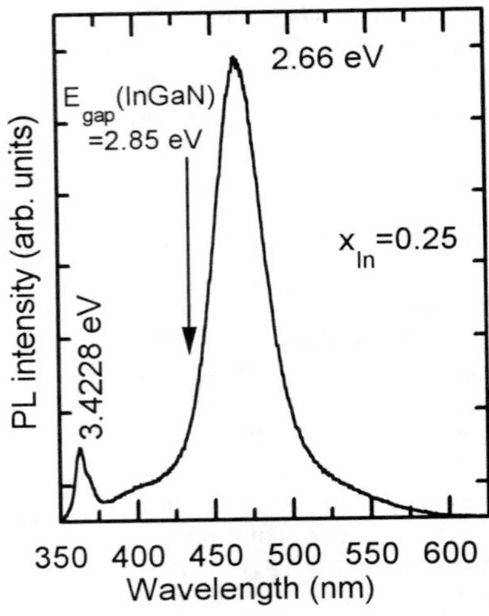

figure 8: 300 K PL spectrum of In$_{0.25}$Ga$_{0.75}$N

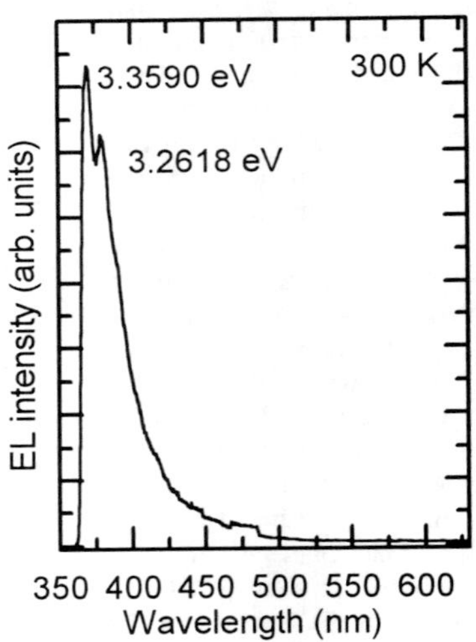

figure 9: 300K EL of a GaN pn homo-junction

<u>InGaN / GaN heterojunctions</u>

The electroluminescence spectrum from an InGaN / GaN heterojunction with a 3 nm thin InGaN layer is presented in Fig. 11. The spectrum was taken at a current of 20 mA and shows a green emission at 540 nm with a half-width of 50 nm. The shape of the line was found to be influenced by interference effects.

We have observed very similar results from structures with 30 nm thick InGaN [18]. In these relatively thick layers the In-content could be determined by XRD. For characterisation purposes, we have grown such structures with p-layers thin enough (below 200 nm) to allow optical excitation of the InGaN-layer. Thus, PL and EL can be observed in separate experiments on the same layer sequence. We found that EL and PL coincided for two samples with In-contents of 27% and 40%, respectively leading to emission in the blue and green. In both cases the optical emission was 160 meV below the expected position of the band-gap. The origin of the shift in EL is unclear. It could be due to the same deep level observed in the PL of InGaN single layers. However, in the case of p-n junctions it might as well originate from the unintended presence of Mg in the InGaN layer, which we detect by SIMS.

Some basic properties of these heterojunctions were examined after processing. Fig. 12 shows the shift of the EL signal to shorter wavelength with increasing injection current, indicating fast saturation in those parts of the InGaN layer with the highest In-content. As the width of the PL spectra suggests, the composition of our InGaN layers is apparently not uniform. A typical light-current characteristic is shown in fig. 13. It shows slight saturation above 20 mA. By measuring the light output through the substrate, we estimate that the luminescence efficiency of the investigated test-structures is about a factor of 50 to 100 below that of commercial devices. Still, at 20 mA their light output is clearly visible in bright daylight.

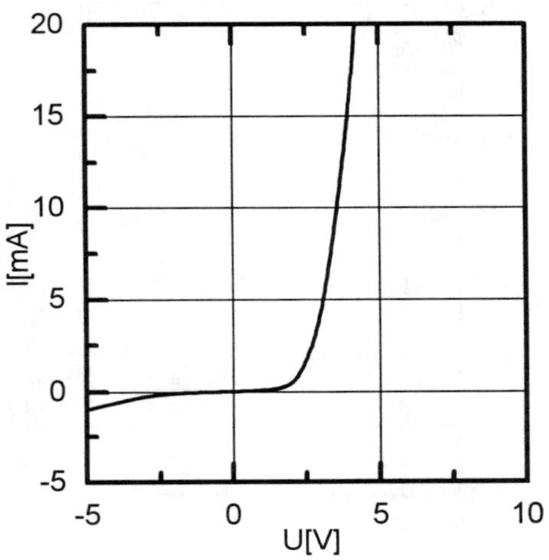

figure 10: I-V curve of a GaN pn structure

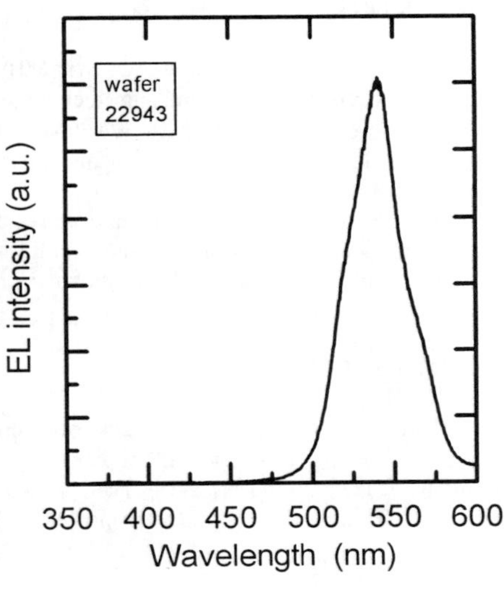

figure 11: 300K EL spectrum of an InGaN pn-structure

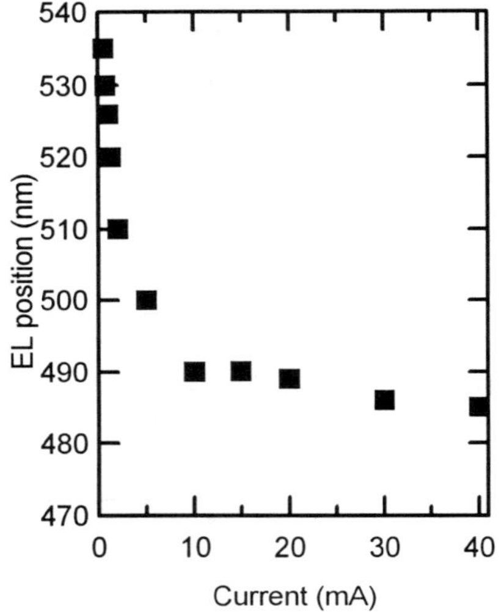

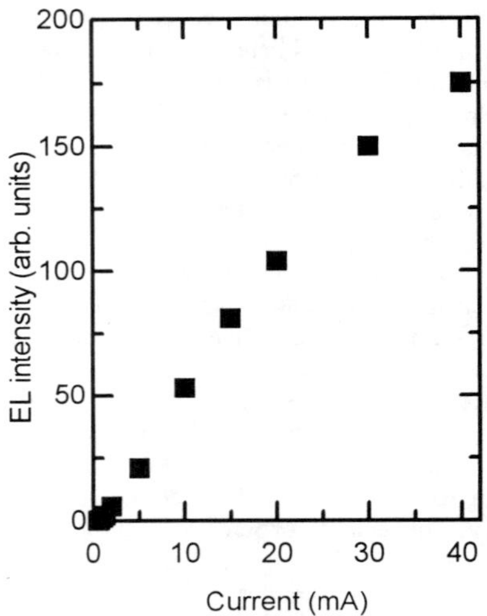

figure 12: EL position as a function of current

figure 13: EL intensity as a function of current

CONCLUSIONS

In conclusion, we have shown that MBE using a RF-plasmasource to generate reactive nitrogen is a viable and promising technique to grow InGaN / GaN heterostructures for optoelectronic device applications. We have achieved good quality in undoped GaN grown at 1μm/h, with electrical properties of n- and p-type doped GaN that are equal to those grown by other methods.

The relatively low growth temperatures accessible by MBE provide a clear advantage for both reproducible p-type doping with Mg and for the growth of InGaN with high In-contents. To demonstrate this potential, we have shown electroluminescence in the blue and green spectral range up to 540 nm.

ACKNOWLEDGMENTS

The authors would like to acknowledge the collaboration of H. Cerva, G. Franz, R. Hösler, C.-Y. Hung, B. Jobst, M. Schuster, R. Sedlmeier, S. Spruytte and R. Treichler as well as support by J. Heinen and N. Stath. H.R. would like to thank F. Ponce and W. Götz for helpful discussions. This work was partly supported by the German Ministry of Education and Research (BMBF).

REFERENCES

[1]   T. D. Moustakas, R. J. Molnar, T. Lei, G. Menon and C. R. Eddy Jr., Mat. Res. Soc. Symp. Proc. **242**, 427 (1992)

[2]   S. Nakamura, M. Senoh, N. Iwasa, S. Nagahama, T. Yamada and T. Mukai, Jpn. J. Appl. Phys. **34**, L 797 (1995)

[3]   H. Sakai, T. Koike, H. Suzuki, M. Yamaguchi, S. Yamasaki, M. Koike, H. Amano and I. Akasaki, Jpn. J. Appl. Phys. **34**, L 1431 (1995)

[4]   J. F. Schetzina, Mat. Res. Soc. Proc. **395**, 123 (1995)

[5]   R. P. Vaudo. J. W. Cook Jr. And J. F. Schetzina, J. Vac. Sci. Technol. B **12**, 1232 (1994)

[6]   H. Liu, A. C. Frenkel, J. G. Kim and R. M. Park, J. Appl. Phys. **74** (10), 6124 (1993)

[7]   C. Heinlein, J. Grepstad, H. Riechert and R. Averbeck, to appear in Materials Science and Engineering B (1996)

[8]   B. Heying, X. H. Wu, S. Keller, Y. Li, D. Kapolnek, B. P. Keller, S. P. DenBaars and J. S. Speck, Appl. Phys. Lett. **68** (5), 643 (1996)

[9]   R. J. Molnar, T. Lei and T. D. Moustakas, Appl. Phys. Lett. **62** (1), 72 (1993)

[10]   A. Botchkarev, A. Salvador, B. Sverdlov, J. Myoung and H. Morkoc, J. Appl. Phys. **77** (9), 4455 (1995)

[11]   O. Ambacher, unpublished results

[12]   D. Volm et al., Phys. Rev. B **53**, 16543 (1996)

[13]   D. L. Rode and D. K. Gaskill, Appl. Phys. Lett. **66** (15), 1972 (1995)

[14]   J. Neugebauer private communication, see also J. Neugebauer and C. G. van de Walle, in "Role of defects and impurities in doping of GaN", edited by M. Scheffler and R. Zimmermann (Proc. 23rd Int. Conf. on the Physics of Semiconductors, Singapore 1996) p. 2849-2856.

[15]   W. Götz, N. M. Johnson, J. Walker, D. P. Bour and R. A. Street, Appl. Phys. Lett. **68** (5), 667 (1996)

[16]   O. Ambacher, A. Bergmaier, M. S. Brandt, R. Dimitrov, G. Dollinger, R. A. Fischer, A. Miehr, T. Metzger and M. Stutzmann, to appear in J. Vac. Sci. Technol. (1996)

[17]   S. Nakamura, J. Vac. Sci. Technol. **A 13** (3), 705 (1995)

[18]   H. Tews, R. Averbeck, A. Graber and H. Riechert, Electron. Lett. **32** (21), 2004 (1996)

ON SURFACE CRACKING OF AMMONIA FOR MBE GROWTH OF GaN

M. KAMP, M. MAYER, A. PELZMANN, and K.J. EBELING
Department of Optoelectronics, University of Ulm, D-89069 Ulm, Germany

ABSTRACT

With the use of ammonia as nitrogen precursor for Group III-Nitrides in an on surface cracking (OSC) approach, MBE becomes a serious competitor to MOVPE. Homoepitaxial GaN exhibits record linewidths of 0.5 meV in PL (4 K), whereas GaN grown on c-plane sapphire also reveals reasonable material properties (PL linewidth $\approx$ 5 meV, n $\approx 10^{17}$ cm^{-3}, $\mu \approx 220$ cm^2/Vs). Beside results on hetero- and homoepitaxial growth of GaN, insights into the growth mechanisms are presented. The growth of ternary nitrides is discussed, p- and n-doping as well as first LED results are reported.

INTRODUCTION

Today, growth of GaN and its related compounds is still dominated by MOVPE [1, 2], which is mainly due to the lack of a suitable nitrogen source for MBE. Generally N_2 is used as nitrogen precursor, where the high binding energy of 9.5 eV requires the use of a plasma cracking source. Plasma sources such as DC, RF, ECR cannot fully avoid high energy ions to reach the surface which however, may cause crystal damage during growth. Self biased voltages, inherent to the plasma, produce sputtering effects leading to contamination of the epitaxial layers. Remaining challenges of those sources are reproducibility and homogeneity of the layers.

We report on the successful use of NH_3 as nitrogen precursor in MBE. Therefor we employ so-called on surface cracking (OSC), where uncracked ammonia is supplied to the surface for GaN growth.

For the first time we obtain MBE layer qualities that are comparable or even superior to layers grown by MOVPE, making MBE attractive for industrial applications. Advantages include excellent growth control, low growth temperature, low precursor consumption and therefore reduced pollution as well as various possibilities for in-situ analysis.

GROWTH KINETICS OF GaN USING NH$_3$

We briefly discuss the dissociation of ammonia under thermodynamic equilibrium, present experimental results on the catalytic dissociation and describe GaN growth studies.

NH$_3$ Dissociation under thermal equilibrium

To evaluate the suitability of ammonia for MBE growth for simplicity thermodynamic equilibrium conditions are assumed. Ammonia dissociation according to

$$2NH_3 \rightleftharpoons N_2 + 3H_2 \tag{1}$$

is determined for relevant pressures and temperatures. We assume that at $T = 0$ K a constant volume is filled with ammonia at the concentration $[NH_3]_0$. With increasing temperature the ammonia begins to dissociate into N_2 and H_2. The temperature and pressure dependent concentrations of $[NH_3]$, $[N_2]$ and $[H_2]$ are calculated by solving the three equations

$$2[N_2] + [NH_3] = [NH_3]_0 \quad (2) \qquad [H_2] = 3[N_2] \quad (3) \qquad K_C(T) = \frac{[NH_3]^2}{[N_2] \times [H_2]^3} \quad (4)$$

Mat. Res. Soc. Symp. Proc. Vol. 449 © 1997 Materials Research Society

Equation (2) describes the conservation of the number of N atoms, equation (3) the conservation of the H_2/N_2 ratio and equation (4) is the mass action law [3]. The equilibrium constant $K_P(T)$ can be taken from the literature [4], but has to be converted to K_C which is necessary for the constant volume and variable pressure conditions of our experiment. Together with the ideal gas equation and after simple calculus the results in Fig. 1 are obtained. NH_3 dissociation under thermal equilibrium is given for two different concentrations, the left hand solid curve holds for MBE conditions at the growth surface (1×10^{-5} mbar, temperature 1000 K, corresponding to a concentration of 3×10^{-11} mol/l). The thermal equilibrium calculations yield an almost complete dissociation at approximately 200 K.

The right hand solid curve is computed for MBE conditions in the gas injector (1×10^{-2} mbar, temperature 600 K, corresponding to a concentration of 2×10^{-7} mol/l). An almost complete dissociation is found at approximately 300 K.

The dashed curve represents experimental data obtained for ammonia dissociation in the MBE gas injector. Here, temperatures around 1100 K are required to achieve an almost complete NH_3 dissociation under non-equilibrium conditions.

The difference between thermal equilibrium calculations and experimental data for injector cracking reveals that the ammonia dissociation is not in thermal equilibrium, but limited by a kinetic barrier. This barrier can be reduced by a catalyist, which thereby pushs the dynamic equilibrium towards dissociation. However, a catalyst cannot shift the position of the thermal equilibrium since both forward and backward reaction are enhanced.

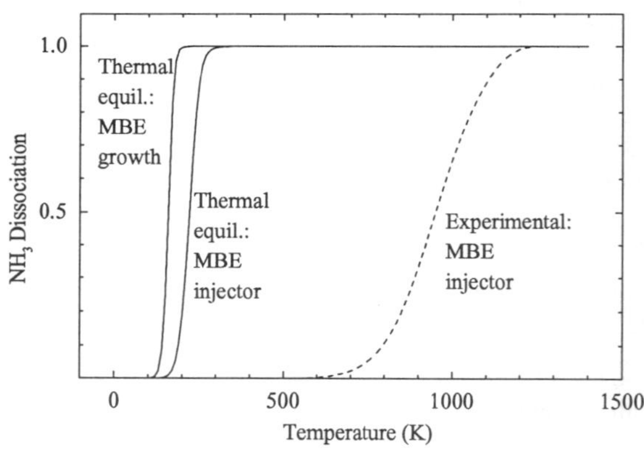

Figure 1: Temperature dependent NH_3 dissociation under thermal equilibrium and under real, non-equilibrium conditions (see text).

Catalytic NH$_3$ Dissociation

The catalytic dissociation of ammonia is investigated using the high temperature gas injector of the MBE system, a Riber HTI 432 injector. A quadrupole mass spectrometer (QMS) is employed to measure the NH_3 dissociation, all data are corrected for the cracking pattern.

The QMS has only a limited sensivity since it does not measure directly in the molecular beam. Within the accuracy of our setup we found no evidence for any desired reactive nitrogen species such as N, NH or NH_2 generated by the ammonia dissociation within the gas injector. This is in good agreement with our expectations since the free path length of a reactive species within the injector can be calculated to be approx. 10 mm, this would lead to fast neutralization of reactive species within the injector. The overall reaction inside the injector is therefor well described by NH_3 dissociation into N_2 and H_2.

Important insights into the mechanisms of ammonia dissociation are obtained by flux dependent studies. Fig. 2 shows the Arrhenius plot of the total amount of dissociated NH_3 which over a wide range of temperatures and fluxes is independent of the supplied amount of NH_3. In this regime, a kinetic limitation of the process is observed, whereas at higher temperatures the dissociation saturates and the concentration is determined by the flux. The dissociation cannot be pyrolytic only, since we obtain an upper limit in the kinetically

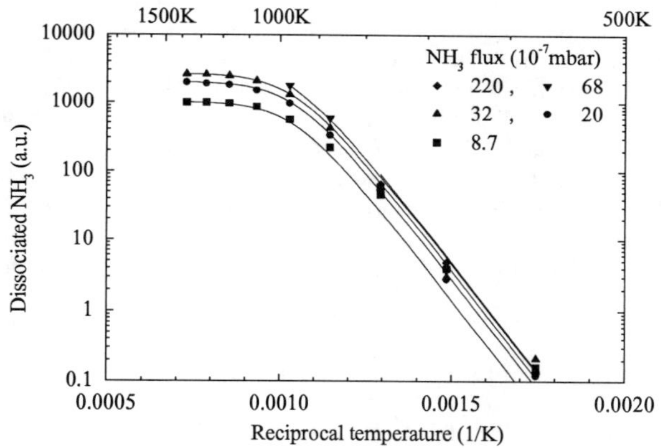

Figure 2: Arrhenius plot of the NH_3 dissociation. Symbols indicated experimental data, lines are calculated according to eq. (5)

limited region, where the total dissociation is independent on the flux. This saturation can be explained in terms of a catalytically enhanced dissociation model.

We assume that the number of catalytic surface states is limited. During the dissociation process each state is occupied for a certain time τ_0. At low fluxes all ammonia molecules find a vacant surface state to dissociate. The collision rate of ammonia with the surface is slower than the reaction rate of the dissociation process. With increasing flux the collision rate exceeds the reaction rate, excess ammonia does not find free surface states, the dissociation saturates (and is independent from the total flux, as observed). Higher temperatures enhance the reaction rate between the catalytic surface states and the ammonia. The rate increases, free surface states become available again for dissociation.

The overall process can be described by the transition function:

$$D = \Gamma(1 + \exp(\frac{E_a}{k(T - T_0)}))^{-1} \tag{5}$$

The coefficient Γ depends on the supplied NH_3 flux, the density of the surface states, the active surface area of the injector and a factor describing the efficiency of the adsorption process. Γ is used as a fitting factor for the flux dependence. T_0 is the transition temperature which is determined to be 950°C. The activation energy E_a of the kinetically controlled process is 1.2 meV. With this data we find an excellent agreement between experiment (symbols in Fig. 2 and our model (solid lines in Fig. 2)). The catalytic reaction is assumed to take place at the three molybdenum disks which act as baffle in the injector's orifice (molybdenum is a well-known catalyst for ammonia).

The above presented model is consistent with results obtained from dedicated surface science studies investigating the catalytic reaction between NH_3 and polycrystalline steel (Fe), powdered tungsten (W) and molybdenum (Mo), [5, 6, 7]. These studies reveal that NH_3, after physisorbtion to the surface, instantaneously chemiosorbs under release of a hydrogen atom at temperature above 200 K. At moderate temperatures NH_2 and NH are formed, H_2 is released from the surface. With increasing temperatures the remaining H is stripped off and the nitrogen forms a nitride with the surface. At high temperatures and sufficient ammonia fluxes such nitride layers may reach a thickness of several monolayers. Nitrogen molecules then evaporate by decomposition of the nitride as is concluded from the desorption energy which is the dissociation energy of the particular nitride rather than the binding energy of the nitrogen bonded to the surface. Differences in catalytic efficiency are due to the differences in the stability of the corresponding metal nitrides. Later is governed by the reaction between the unoccupied d-orbitals of the metal surface and the free electron pair of the ammonia molecule.
The experimental results have the following implications for epitaxial growth:

- NH_3 dissociation is sufficient under growth conditions.

- The surface is supposed to be NH_x terminated.

- The effective V/III ratio on the surface should be independent of the supplied V/III flux ratios over a wide range.

<u>Growth kinetics of GaN</u>

Above results suggest that the surface is covered with nitrogen under regular growth conditions, corresponding to a group V rich surface. This growth mode is common in conventional III-V growth, but is in distinct difference to the MBE growth of GaN with plasma MBE [8]. The observed linear dependence of our GaN growth rate on the supplied Ga flux is a further indication for a group V rich surface. The extrapolation of the growth rate to zero Ga flux (the intercept with the y-axis) has to be interpreted as the Ga desorption. At 750°C the intercept is -250 nm/h which is consistent with the observed Ga desorption at this temperature (see Fig. 3). A significant Ga desorption starts at temperatures above 500°C, the activation energy is surprisingly low with 0.65 eV. Ga desorption is significant at regular growth conditions ($\approx$ 30%).

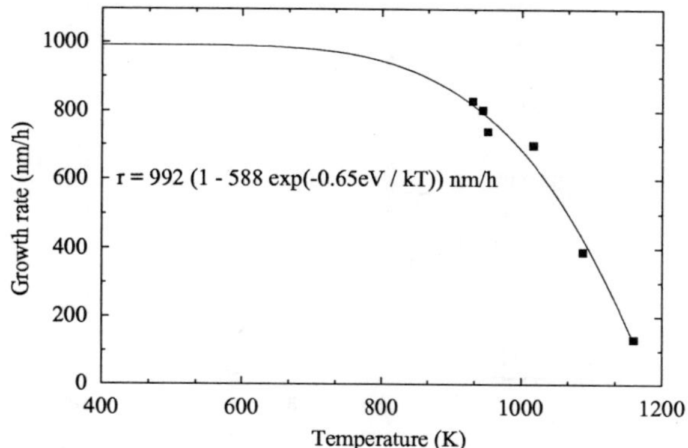

Figure 3: Dependence of GaN growth rate on temperature.

GROWTH OF GROUP III-NITRIDES

Using on surface cracking of ammonia for MBE growth we describe the relevant growth parameters and crystal properties for GaN, InGaN and AlGaN. First results on homoepitaxial growth of GaN will also be discussed.

Our otherwise standard MBE system (Riber 32) was adapted to group V gas sources. The system is turbo pumped, the attached gas control and handling system is home made. NH_3 is introduced into the system through a standard high temperature injector (Riber HTI 432). Effusion cells are used to supply group III species Al, Ga, In and dopants Si, Mg. Unless otherwise mentioned GaN layers are grown on c-plane oriented sapphire at growth rates of 750 nm/h to a thickness of approximately $1.8\,\mu$m.

GaN

The heteroepitaxial growth of GaN on c-plane sapphire requires decent preparation steps before subsequent growth of the GaN layer. Many authors have reported different approaches of sapphire preparation and nucleation layer growth [9, 10]. The following procedure provided the best results at our laboratory and therefor was used for our experiments.

1. The sapphire is heated up to approx. 850°C and kept at this temperature for 5 minutes under a certain NH_3 flux.

2. The substrate is ramped down to 700°C for the nucleation layer growth. At this temperature a 5 nm AlN layer is deposited, followed by a 15 nm GaN layer. Subsequently, the layers are annealed at 770°C for 5 minutes.

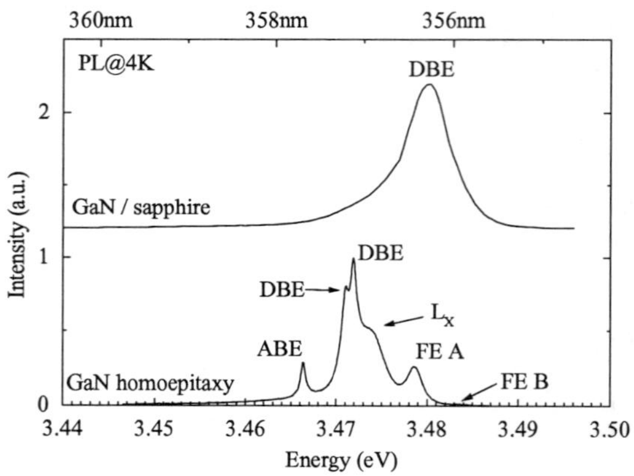

Figure 4: Photoluminescence of GaN on sapphire and homoepitaxial GaN.

3. The substrate is ramped to 750°C before GaN growth is initiated.

At a growth rate of 750 nm/h we achieve the following crystal properties for a GaN layer with a thickness of 1.8 μm.

- Photoluminescence dominated by donor bound excitions (DBE), linewidths of 5 meV at 4 K (Fig. 4),

- Carrier concentrations of 2×10^{17} cm^{-3} with carrier mobilities of $\mu = 220$ cm^2/Vs at 300 K. Free carrier concentrations down to 10^{16} cm^{-3} have been achieved. However, carrier mobilities of $\mu \leq 100$ cm^2/Vs indicate highly compensated material.

- X-ray rocking curves with linewidths of 400-450 arcsec in an ω-scan of the (0002) reflex.

AlGaN

AlGaN is grown by additional supply of aluminum under otherwise unchanged growth conditions. Depending on the Al-flux, AlN mol fractions up to 35% have been realized.
AlGaN layers with an aluminum content of 17 % reveal free carrier concentrations of n = 1×10^{18} cm^{-3} with carrier mobilities of 10 cm^2/Vs. The PL spectrum of an AlGaN layer is shown in Fig. 5, where AlGaN is used as barrier for a GaN quantum well. The aluminum incorporation turns out to be a nonlinear function of the Al/Ga flux ratio carefully considering corrections due to different ionization probabilities and thermal velocities. The experimental data displayed in Fig. 6 could be interpreted by competing chemical reactions between the two group III elements and ammonia.

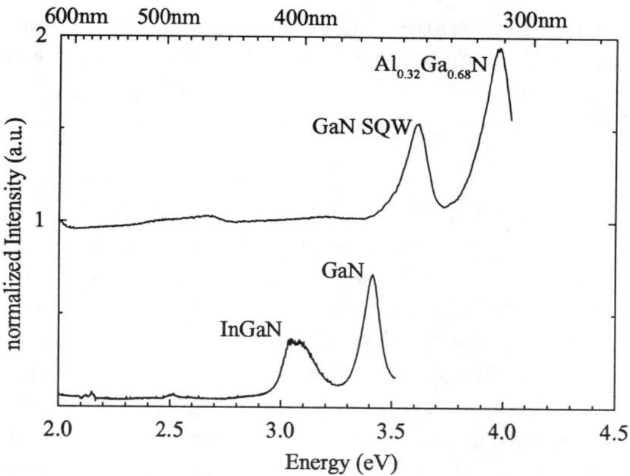

Figure 5: Photoluminescence of an AlGaN/GaN quantum well (77 K) and InGaN (300 K).

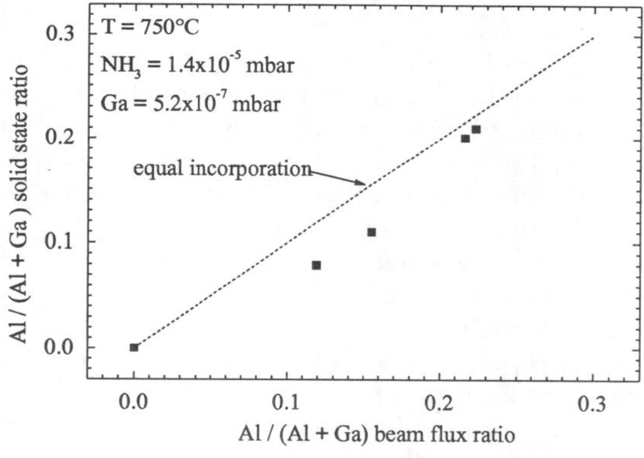

Figure 6: Aluminum/gallium ratio in AlGaN versus supplied aluminum/gallium ratio.

InGaN

The high vapor pressure of the elemental indium requires growth conditions different from those of GaN. The deposition temperature is reduced to 560°C to achieve measurable indium incorporation. The In/Ga ratio has to be increased to 15 to achieve an indium content of 4% according to XRD. Room temperature PL shown in Fig. 5 is still dominated by a rather broad transition at 3.06 eV indicating defect related luminescence. Further experiments have to be carried out to increase In concentration as well as to improve InGaN crystal quality.

GaN-homoepitaxy

GaN growth on sapphire has proven its ability to yield excellent LEDs as well as the GaN based lasers by the use of decent preparation techniques such as nitridation, low temperature nucleation layer, etc.. In spite of the amazing success of GaN based LEDs [11, 12] and lasers [13] both obtained by heteroepitaxial growth, homoepitaxy may overcome inherent problems of lattice mismatch, columnar growth, dislocations and thermal mismatch. Furthermore, GaN substrates are electrically conductive, allowing current injection through the substrate and have a sufficient thermal conductivity to facilitate high power applications.

The Polish Unipress Center for the first time has produced GaN substrates in useful sizes up to $9 \times 9\,\mathrm{mm}^2$ with thicknesses around 500μm [14]. Employing high pressure (15000 bar) and high temperature (1600°C) GaN platelets were obtained from seeded and unseeded Ga melts under N_2 pressure. Dislocation densities are in the $10^2\,\mathrm{cm}^{-2}$ range, whereas a high number of point defects still causes background carrier concentrations of $10^{19}\,\mathrm{cm}^{-3}$.

Such crystals seem the ideal test vehicle to investigate the potential of a certain growth technique since all problems related to nitridation and nucleation layers do not arise. For homoepitaxial growth, the as-grown (unpolished) GaN crystals were In-mounted to the Mo holder, heated up to 750°C under ammonia before growth was initiated under conventional growth conditions mentioned above [15, 16].

The lower curve in Fig. 4 shows the excitonic region of a PL spectrum for homoepitaxial GaN layer. The bound excitons transitions are much narrower than the transitions corresponding to free excitons. The transition of the acceptor bound exciton (ABE) at 3.4663 eV reveals a linewidth of 0.5 meV. The DBE lines at 3.4709 eV and 3.4718 eV have comparable linewidths, the separation is approximately 1 meV. The lines of the free excitons (FE) A (3.4785 eV) and B (3.4832 eV) and the yet unidentified line at 3.4735 eV show halfwidths of about 2 meV. Neither strong yellow luminescence nor luminescence near 3.2 eV due to donor-acceptor pair transitions are observed.

With increasing temperature the ABE transition becomes broader and weaker. The DBE transitions do not shift, but decrease in intensities. The line at 3.4735 eV undergoes significant changes. The intensity decreases with increasing temperature and nearly vanishes at 22 K. The intensity of the FE A strongly increases with temperature and becomes nearly equal to the DBE lines. The intensity of the FE B increases with temperature, too.

DOPING & DEVICE STRUCTURES

Excellent results on the homoepitaxial GaN growth have proven that the OSC approach makes MBE a serious competitor to MOVPE for GaN growth. MBE specific advantages such as a reduced growth temperature, in-situ analysis and low precursor consumption and low pollution may become important aspects. It remains, however, to demonstrate device structures of comparable quality. Therefore, we present our preliminary work on p- and n-doping, quantum wells and LEDs.

<u>P-doping</u>

P-doping is one of the major challenges on the way to fabricate light emitting devices in GaN. Early reports claimed that p-doping can only be achieved if a plasma source is used in addition to the ammonia [17]. Whereas Wang et al. could achieve free hole concentrations with simultaneous use of the plasma cell and ammonia, they could not achieve p-doping

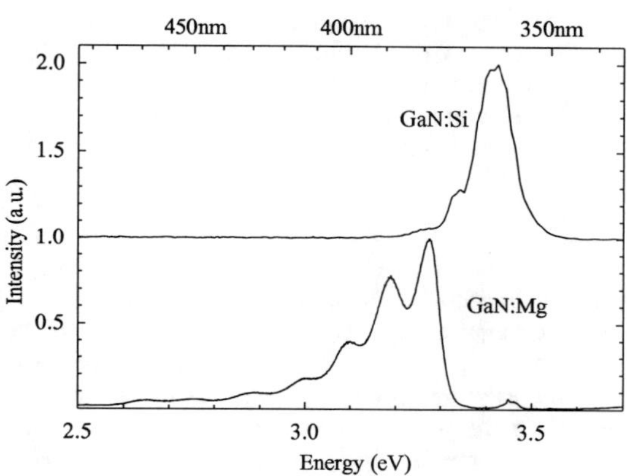

Figure 7: Photoluminescence of Mg-doped GaN (77 K) and Si-doped GaN (300 K).

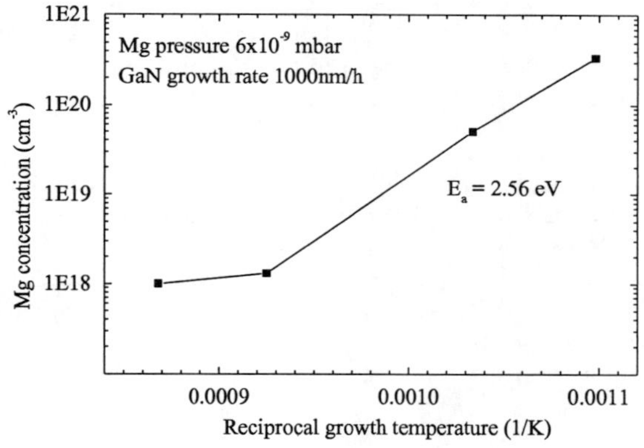

Figure 8: Temperature dependence of Mg incorporation into GaN.

when just ammonia was used.

We supplied Mg from an effusion cell for p-type doping under OSC conditions without any plasma source. P-doping was verified by several techniques, including current-voltage characteristics however, Hall measurements do not always yield reasonable results. In spite of the presence of hydrogen, p-type material is achieved without annealing, however tempering at 750°C for 10 min seems to improve p-conductivity.

Photoluminescence of Mg doped GaN is presented in Fig. 7. The spectrum is dominated by a rather narrow transition at 3.26 eV and the corresponding LO phonon replica at lower energies.

The Mg incorporation is governed by an almost exponential decrease with increasing temperature, the activation energy for the Mg desorption is 2.56 eV (Fig. 8). Mg desorption is the limiting mechanism under our experimental growth conditions. An effect of the increasing energy of formation which should yield a higher Mg incorporation at higher temperatures is not observed [18].

N-doping

Silicon from a standard effusion cell is used as donor in our experiments. Although, the number of experiments is still limited, up to now we have no indication for an eventual passivation of the source. Doping was investigated up to free electron concentration of $n=4.5\times10^{18}$ cm^{-3} with corresponding room temperature mobilities of 150 cm^2/Vs. The PL spectrum however is still dominated by a bound exciton (DBE) as can be taken from Fig. 7. Compared to undoped layers the photoluminescence intensity is strongly enhanced by a factor of 80. Microcracks as reported in literature have not been observed in the layers, probably due to the reduced growth temperature.

Quantum Wells

GaN single quantum well (SQW) structures have been grown with quantum well thicknesses of nominally 4 nm embedded in $Al_{0.32}Ga_{0.68}N$. Fig. 5 shows the PL spectrum of the SQW structure relative to the bulk GaN band edge. The quantum well luminescence is shifted by 150 meV. The photoluminescence of the 4 nm GaN layer is still broad at 128 meV. TEM and excitation spectroscopy can be used to determine if the broadening is due to interface fluctuations or higher sub-band emission.

Light Emitting Diodes

Homotype pn-junction were grown on sapphire. The layers were processed using CAIBE Cl_2 etching, Ti/Au n-contact metallization and Ni/Au p-contact formation. Fig. 9 depicts an I-V-characteristic of a MBE grown homotype LED.

The obtained electroluminescence is shown in Fig. 10. The luminescence is dominated by a rather steep increase at 380 nm, whereas the linewidth of appox. 85 nm is determined by the long tail towards lower energies.

Future work on electroluminescence will be focussed on heterostructures and the improvement of the p-doping.

SUMMARY

We investigated the on surface cracking of ammonia for molecular beam epitaxy of GaN based structures. Outstanding photoluminescence results with linewidths as low as 0.5 meV are achieved by homoepitaxial growth of GaN. Heteroepitaxial growth of GaN on sapphire

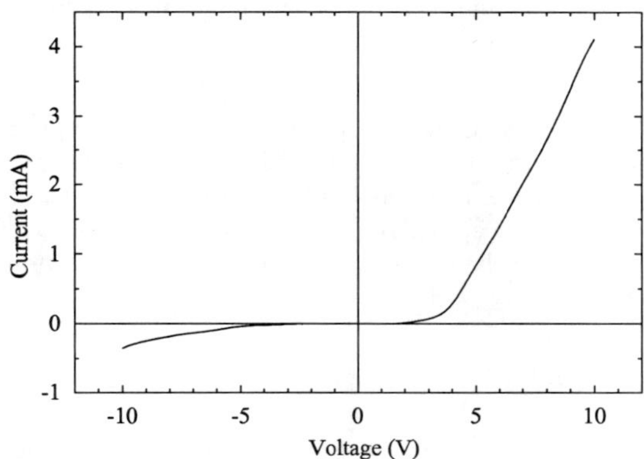

Figure 9: I-V-characteristic of a homotype GaN pn-junction.

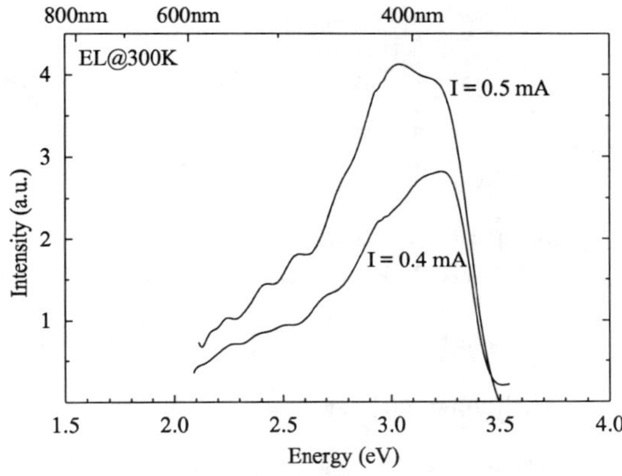

Figure 10: Electroluminescence from a homotype GaN pn-junction (300 K).

yields optical, electrical and structural properties comparable to MOVPE material. AlGaN has been grown with good material properties, whereas with InGaN further work is still necessary to achieve device quality material. P- and n-doping have been obtained using conventional Mg and Si effusion cells. First results on electroluminescence of homotype pn-junctions in GaN are encouraging but still leave room for further improvements.

Apart from technological aspects, fundamental insights into the growth mechanisms of the OSC process have been presented. Thermodynamic equilibrium calculations, mass spectroscopy and kinetic studies were employed to obtain a detailed understanding of the underlying processes. Important mechanisms of ammonia dissociation as well as of the kinetics of GaN growth have been determined.

ACKNOWLEDGEMENTS

We would like to acknowledge valuable contributions and active support from A.Thies, C.Kirchner, M.Schauler, F.Eberhard, P.Unger as well as financial support from the German Ministry of education, science, research and technology (BMBF) and the Volkswagen Foundation. The work on the homoepitaxial growth was done in close cooperation with the Polish Academy of Sciences namely H.Teisseyre, G.Karczewski, G.Nowak, M.Leszczyski, I.Grzegory, and S.Porowski.

REFERENCES

1. S.Nakamura, M.Senoh, N.Iwasa, and S.Nagahama, Jpn. J. Appl. Phys.Vol.34, pp. L797 (1995).
2. M.Koike, N.Sibata, S.Yamasaki, S.Nagai, S.Asami, H.Kato, N.Koide, H.Amano, and I.Akasaki, Mat. Res. Soc. Symp. Proc. Vol. 395, 889, (1996).
3. C.E.Mortimer, Chemie, G. Tieme Verlag Stuttgart (1987).
4. R.J.Meyer in Gmelins Handbuch der Anorganischen Chemie, Stickstoff, Verlag Chemie GmbH, Weinheim, (1936).
5. G.Ertl and M.Huber, Journal of Catalysis 61, 537, (1980).
6. K.Tamaru, Trans. Faraday Soc. 57, 1410 (1960).
7. M.Grunze in D.A.King The Chemical Physics of Solid Surfaces and Heterogeneous Catalysis, Synthesis and Decomposition of Ammonia, Vol. 4, Elsevier Scientific Publishing Company, Amsterdam (1982).
8. H.Riechert et al., to appear in : Mat. Res. Soc. Symp. Proc. of MRS Fall Meeting Boston 1996.
9. S.Nakamura, T.Mukai, and M.Senoh, J. Appl. Phys. 71, 11, 5543, (1992).
10. K.Uchida, A.Watanabe, F.Yano, M.Kouguchi, T.Tanaka, and S.Minagawa, J. Appl. Phys. 79, 7 (1996).
11. S.Nakamura, M.Senoh, N.Iwasa S.Nagahama, T.Yamada, and T.Mukai, Jpn. J. Appl. Phys.Vol.34, pp. L1332, (1995).
12. H.S.Kong, M.Leonard, G.Bulman, G.Negley, and J.Edmond, Mat. Res. Soc. Symp. Proc. Vol. 395, 903, (1996).
13. S.Nakamura, M.Senoh, N.Iwasa S.Nagahama, T.Yamada, and T.Matsushita, Jpn. J. Appl. Phys.Vol.35, part2. L74, (1996).
14. T.Suski, P.Perlin, M.Leszcynski, H.Teisseyre, I.Grezgory, J.Jun, M.Bockowski, S.Porowski, K.Pakula, A.Wysmolek, and J.M.Baranowski, Mat. Res. Soc. Symp. Proc. Vol. 395, 15, (1996).
15. H.Teisseyre, G.Nowak, M.Leszczynski, I.Grzegory, M.Bockowski, S.Krukowski, S.Porowski, M.Mayer, A.Pelzmann, M.Kamp, K.J.Ebeling, and G.Karczewski, MRS Internet J. Nitride Semicond. Res. 1, 13 (1996).
16. M.Mayer, A.Pelzmann, M.Kamp, K.J.Ebeling, H.Teisseyre, G.Nowak, M.Leszczynski, I.Grzegory, M.Bockowski, S.Krukowski, S.Porowski, and G.Karczewski, submitted to Appl. Phys.Lett. .
17. Z.Yang, L.K.Li, and W.I.Wang, Appl. Phys. Lett. 67, 12 (1995).
18. J.Neugebauer, and C.G.van de Walle, Appl. Phys. Lett. 68, 13, (1996).

GROWTH OF CUBIC GaN BY MBE AND ITS PROPERTIES

S. YOSHIDA*, H. OKUMURA*, G. FEUILLET**, P. HACKE*, K. BALAKRISHNAN*
*Electrotechnical Laboratory, 1-1-4 Umezono, Tsukuba, Ibaraki 305 Japan, syoshida@etl.go.jp
**Commissariat a l'Energie Atomique, Centre de Grenoble, 17 rue des Matryrs, 38054 Grenoble, Cedex 9, France

ABSTRACT

Molecular beam epitaxy (MBE) technique is a useful method to grow III-V nitrides, especially those having a metastable crystal structure, like cubic GaN (c-GaN), because of the capability of *in situ* observation of growing surfaces and its non-equilibrium growth mechanism. We have grown c-GaN on GaAs and 3C-SiC substrates by gas source MBE using dimethylhydrazine or activated nitrogen beam as an N source, and measured their luminescent and optical properties. This paper summarizes the MBE growth and properties of c-GaN, comparing with those of hexagonal one, and the control of the crystal structures is discussed in terms of growth method, orientation of substrate surfaces and growth conditions.

INTRODUCTION

GaN has been intensively investigated as a material for blue-light-emitting diodes (LEDs) because GaN is the direct wide bandgap semiconductor with band gap energy E_g=3.39eV at room temperature and forms a continuous range of solid solution with AlN and InN, which covers the band gap energy between 6.2 and 1.9eV. GaN has been also attracted much interests as a material for high-temperature and/or high-frequency, high-power devices because of its high electron saturation drift velocity (2.7×10^7cm/s) and high breakdown field (2×10^6V/cm).

The epitaxial growth of GaN was first reported by Maruska and Tietjen[1] in 1969 by halide vapor phase epitaxy (HVPE) method using the reaction of GaCl and NH_3. In 1971, Manasevit et al.[2] have succeeded in the epitaxial growth of GaN by metal organic chemical vapor deposition (MOCVD), and Pankove et al.[3] have fabricated MIS type GaN blue LEDs because p-type GaN could not be obtained even by the doping of acceptor impurities. The improvement of the crystal quality of MBE grown GaN by introducing AlN buffer layers on sapphire substrates has been reported in 1983[4]. The technique of depositing AlN buffer layers on sapphire has been introduced in the MOCVD growth of GaN epilayers in 1986[5], which resulted in the growth of epilayers with very low residual carrier densities. In 1989, Amano et al.[6] have succeeded in obtaining p-type GaN by using the low energy electron beam irradiation on the Mg-doped epilayers, and fabricated pn junctions for the first time. By using low temperature growth of GaN as buffer layers, Nakamura et al.[7] have obtained GaN films with good crystal quality. They have also found that the thermal annealing in N_2 brings about the p-type conduction of Mg-doped GaN in 1992[8]. These works on the heteroepitaxial growth of nitrides on sapphire substrates made possible to fabricate bright blue LEDs in 1993[9] and pulse operated laser diodes (LDs) in 1995[10].

Almost all the works on the fabrication of LEDs and LDs have been done by using MOCVD, not by MBE, because MBE-grown epilayers have poor crystal quality and high residual carrier density, compared with those grown by MOCVD. This can be attributed partly to the low growth temperature of MBE method. However, MBE technique is an attractive one for its important advantages in the capability of *in situ* observation or monitoring of crystal structures during growth and the capability of obtaining sharp interfaces by shutter control. Recently, the epilayers with high crystal quality, comparable with those by MOCVD, have been obtained by using a plasma-excited nitrogen beam. Most attractive point of MBE growth of nitrides is that as-deposited films doped with Mg reveal p-type conduction without post annealing. This is considered to be due to the absence of hydrogen, which is known as an acceptor killer, in the

Mat. Res. Soc. Symp. Proc. Vol. 449 © 1997 Materials Research Society

films grown by MBE using excited nitrogen sources. This capability has turned many researchers to MBE method. Vaudo *et al.*[11] have succeeded in the fabrication of GaN pn junction LEDs by plasma-assisted MBE without any post annealing..

It is well known that GaN usually crystallizes in the wurtzite hexagonal structure (h-GaN), which is quite different from other III-V compounds. However, in 1986, GaN films with the zincblende cubic structure (c-GaN) have been grown on GaAs by MOCVD[12]. Since then, the growth of c-GaN has been reported by using MBE as well as by MOCVD. One of the most attractive points of c-GaN is the growth on cubic substrates, like GaAs, MgO, Si and 3C-SiC, which can be easily cleaved, in contrast to sapphire substrates, being good for the fabrication of laser structures. These properties of c-GaN have generated much interest. It is supposed that the electrical properties of c-GaN should be superior to those of h-GaN, because of the prediction of a higher electron saturated drift velocity due to lower phonon scattering, a smaller electron effective mass and easier doping properties. Some physical properties , including luminescent and electrical properties, of c-GaN have been reported by several authors. However, the difference of the properties between c- and h-GaN have not been made clear fully because of still poor crystal quality of c-GaN compared with h-GaN.

In this report, the MBE growth of nitrides is summarized, and then, the growth of c-GaN is described from the view point of crystal structure control. Finally, the properties of c-GaN obtained are discussed. The existence of cubic rocksalt structure of GaN has been reported in the powder synthesized under high pressure[13]. However, the rocksalt structure has not been observed in the epilayers, and hence we will use here the term "c-GaN" for GaN having the zincblende structure.

MBE GROWTH OF NITRIDES

Group III nitrides (AlN, GaN and InN) have been grown by CVD (MOCVD and HVPE) and MBE. In the case of other III-V compounds, like GaAs, the solid sources of the constituents of the compounds, like Ga and As, are used to grow in MBE. In the case of nitrides, one of the constituents is nitrogen. However the nitride films can not be obtained by using nitrogen gas as a source. Ritter[14] pointed out that the chemisorption of reacting gas proceeds the reaction, *i.e.*, the reaction is preceded primarily by the combination of an active metal atoms with a chemisorbed gas molecule. Nitrogen is known to be chemisorbed on Ti, and thus Ti may react with nitrogen to form titanium nitride. On the contrary, nitrogen is not chemisorbed on Al, In and Ga, and thus nitridation of group III metals may not occur when nitrogen is used as a reactive gas[15]. Therefore, reaction gas with higher activity is required for an N source in the growth of nitrides, especially in case of MBE, where a low growth pressure is demanded. For that, the nitrogen must be excited, either chemically as in a reactive hydronitrogen or physically as in a plasma.

Reactive MBE

In 1974, we have grown single crystal films of AlN on sapphire using ammonia and Al molecular beams by MBE[16], which we called "reactive MBE" or RMBE, because AlN films grow through the reaction of ammonia with absorbed Al on the substrate surface. Recently, MBE using gas sources is called "gas source MBE (GSMBE)". GaN films were grown by MBE in 1981[17] and $Al_xGa_{1-x}N$ films with all the composition range x in 1982[18]. However, the crystal quality and the electrical properties of MBE grown GaN and AlGaN films were poor compared with those grown by MOCVD. In 1983, it was pointed out for the first time that the use of AlN buffer layers on sapphire substrates improves significantly the crystal quality of MBE grown GaN epilayers, resulting in high luminescence intensities and high carrier mobilities[4].

The reactivity of ammonia is not enough, resulting in the high residual carrier densities in the MBE grown nitrides, even with the large amount of ammonia supply up to the order of 10^{-4} Torr. Although hydrazine (N_2H_4)[12] and hydrogen azaide (HN_3)[19] have been used as alternate sources more active than ammonia, it is hard to treat these gases because of their explosive nature. We have proposed the use of an organic compound, 1,1-dimethylhydrazine (DMHy) as a N source for the growth of GaN on GaAs substrates[20]. First, we grew GaAs buffer layers on GaAs substrate by using Ga and diethylarsine (DEAsH) beam at 600°C, and then grew GaN by Ga and DMHy beams. We found that when DMHy and DMAsH beams are supplied simultaneously with a Ga beam onto the substrate surface, GaAs, not the ternary alloy GaNAs, grows. Therefore, we obtained GaAs/GaN multilayers only by the intermittent supply of a DMAsH under the continuous supply of Ga and DMHy beams. Figure 1 shows the depth profile of As and N compositions in a c-GaN/GaAs multilayer by SIMS measurements. However, fairly high density of carbon impurities was detected in the GaN films by electron probe micro analysis[21]. The carbon impurities in the films grown by DMHy is considered to originate from methyl radicals in DMHy. In the case of MOCVD, DMHy is reported to be a good precursor for the growth of GaN, probably due to the presence of hydrogen carrier gas in CVD.

The use of other precursors, like phenyl-hydrazine[22] and monomethyl-hydrazine[23], have also been reported. The MBE growth of GaN using group III metal organic, like TMG and TEG, called MOMBE, has been

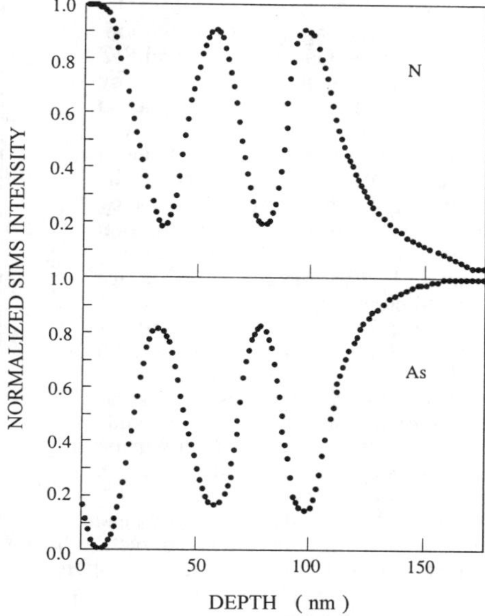

Fig.1 Depth profile of As and N compositions in a GaN/GaAs multilayer grown on GaAs by intermittent supply of DMAsH under the continuous supply of Ga and DMHy beams.

also reported[24]. It is noted that by using ammonia as an N source and at higher substrate temperatures around 800°C, the growth of good crystal quality GaN epilayers with high mobilities and the growth of p-type GaN have been reported, recently[25,26]. Moriyasu *et al.*[27] have reported first the reflection high energy electron diffraction (RHEED) intensity oscillation for the growth of nitrides by MBE, though it is hard to distinguish between the oscillation and noise, apparently. Very recently, Yang *et al.*[28] reported pronounced RHEED intensity oscillation for the GaN grown using ammonia as a nitrogen source.

MBE Using Excited Nitrogen Sources

Recently, MBE growth of nitrides by use of new types of N sources, plasma generated nitrogen radicals, instead of using ammonia and other compounds containing nitrogen, has been reported. Owing to the development of convenient nitrogen radical beam sources, like ion sources, and microwave-plasma, electron cyclotron resonance (ECR)-plasma and rf-activated grow discharge plasma sources, it becomes not so difficult to obtain activated nitrogen species , *e.g.*, atomic nitrogen, excited N_2^*, N_2^+ and N^+, efficiently. By using these radicals, many researchers have grown nitrides with good crystal quality, called plasma- assisted or plasma-enhanced MBE[29-31] and reactive ion MBE[33]. It has been pointed out that the growth rate of GaN by MBE (typically, several tens of nm/h) is much smaller than that by MOCVD (several μm/h). However, the development of high efficiency radical sources and high speed pumping system, which enable us to introduce more nitrogen source gas into the MBE chamber, bring about the increase in the growth rate of nitrides by MBE.

Many studies have been done for obtaining GaN films with good crystal quality by varying growth condition. We have grown GaN on GaAs and SiC as well as sapphire substrates by GSMBE using a microwave-plasma excited ammonia beam, or ECR plasma-excited nitrogen beam as an N source. We have proposed the monitoring of the growth of GaN by observing surface reconstruction patterns by use of RHEED during growth [34,35]. The changes of the surface reconstruction between (2×2) and (1×1) have been observed for h-GaN films by the changes of growth conditions, *i.e.*, substrate temperature, and Ga and nitrogen fluxes. GaN films were grown homoepitaxially on GaN epilayers, which have been grown on sapphire substrates by MOCVD, in order to avoid the influence of the

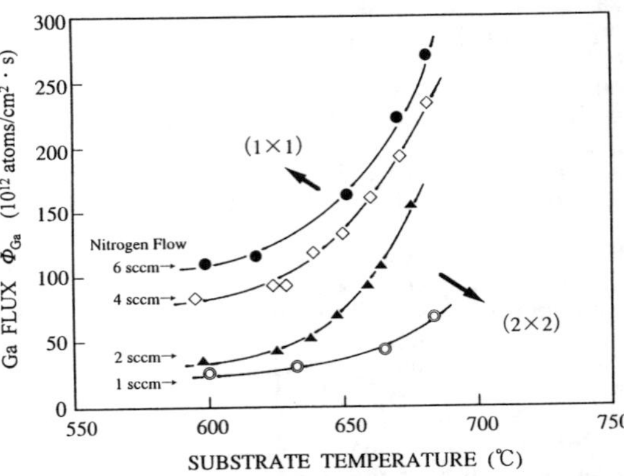

Fig.2 Mapping of the reconstruction patterns of a h-GaN (0001) surface during growth by GSMBE using a ECR plasma excited nitrogen beam at a constant nitrogen flow. The solid lines show the transition curves

defects and strain due to highly mismatched GaN/sapphire interfaces. ECR plasma-excited nitrogen was used as an N source. The Ga flux was calibrated using RHEED intensity oscillation of the growth of GaAs. Figure 2 shows the mapping of the reconstruction patterns on the substrate temperature vs. Ga flux plane at a constant nitrogen flow. The solid lines show the boundary between the regions of the reconstruction (2×2) and (1×1). When we change the substrate temperature, or the intensity of Ga flux near the boundary, the reconstruction pattern changed between (2×2) and (1×1), reversibly. Thus, we called the boundary as a transition curve. We grew h-GaN films under the condition that in the region of (2×2) and (1×1) and just on the transition curve, and found the photoluminescence peak width and the carrier mobilities of the films grown under the condition on the transition curve are superior to those grown under the condition deviated from the transition curve. Daweritz and Hey[36] obtained the phase diagram of surface reconstruction patterns for GaAs, and considered some surface

stoichiometry holds on the boundary between two phases. Following their idea of surface stoichiometry, we considered that the fluxes of Ga atoms and active nitrogen species are balanced on the transition curve, and Ga rich in (1×1) region and N rich in (2×2) region. Therefore we can monitor the surface stoichiometry by the observation of surface reconstruction. The transition curve shifts to higher Ga flux side with the increase of nitrogen flux. The transition curves have been explained based on a flux balance model between Ga and active nitrogen on the growing surfaces, taking the effect of re-evaporation of Ga atoms from the surface at high temperatures into account, as,

$$\alpha \Phi_{N*} = \Phi_{Ga} - A_0 \exp(-E_a/kT) \qquad (1)$$

where, $\alpha \Phi_{N*}$ and Φ_{Ga} are the fluxes of active nitrogen and Ga, respectively, and α the

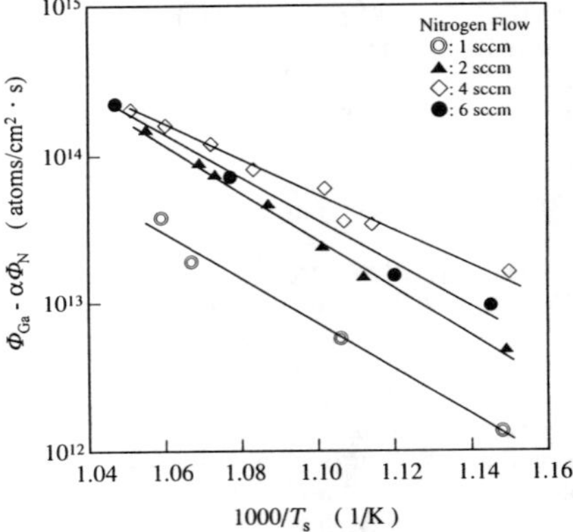

Fig.3 Plots of log($\Phi_{Ga} - \alpha \Phi_{N*}$) against $1/T$. The data points lie well on the straight lines. The activation energy of between 2.2-3.2eV are obtained from the slopes.

constant representing the ratio of Ga and active nitrogen to form a Ga-N bond.   The second term in the right hand side of Eq.(1) represents the amount of re-evaporated Ga atoms from the surface, where E_a is an activation energy for re-evaporation of Ga..   At low temperatures, re-evaporation is negligible, and we can obtain the value of $\alpha\Phi_{N*}$ from the saturated values of Φ_{Ga} at lower temperatures.   Figure 3 shows the plots of $\log(\Phi_{Ga} - \alpha\Phi_{N*})$ against $1/T$. The figure shows that the data points lie well on the straight lines, which suggests the experimental results can be explained well by Eq.(1), *i.e.*, flux balance model.   From the slope of the straight lines, the values of activation energy E_a between 2.2 and 3.2eV are obtained. These results suggest that the observation of the surface reconstruction patterns is a strong tool to monitor the optimum growth condition during growth.

GROWTH OF CUBIC GaN

GaN films with the zincblende cubic structure (c-GaN) have been grown on cubic substrates by MOCVD and GSMBE.   However, the grown c-GaN layers sometimes contain hexagonal phase.   And the crystal quality of c-GaN epilayers is still insufficient to make clear the properties of c-GaN, compared with h-GaN.   For example, the values of full width at half maximum, FWHM, for the XRD peaks of c-GaN epilayers (several tens of arcmin.) are one order of magnitude larger than those for h-GaN epilayers.   Here, we will discuss how to control the crystal structure of GaN in order to obtain high quality c-GaN epilayers without mixing with a hexagonal phase.

Growth Method

First report on the growth of c-GaN has been done by Seifert *et al.*[37], who found cubic phase dendric overgrowths on h-GaN grown by HVPE in 1974.   Mizuta *et al.*[12] have reported the existence of cubic phase in mosaic structure GaN films on GaAs substrates by MOCVD. The growth of c-GaN epilayers have been reported first by Paisley *et al.* in 1989[29], who have grown GaN on 3C-SiC (001) surfaces by using microwave excited nitrogen source in MBE.   Since then, many studies have been reported on the growth of c-GaN on (001) surfaces of MgO, Si, GaAs, GaP and 3C-SiC by using MOCVD and GSMBE.   Though the early studies on c-GaN have been done by CVD, many researchers have used MBE method to grow c-GaN, instead of using CVD[38].   MBE is known as a non-equilibrium growth method, *i.e.*, low substrate temperatures, compared with CVD, and therefore, MBE is believed to be a suitable method to grow metastable c-GaN.   As an N source, excited nitrogen or nitrogen radicals from ECR- and rf-plasma excited nitrogen beams and nitrogen ions have been used. However, there is no report giving the conclusion of the suitable nitrogen source for the growth of cubic phase GaN, though the effects of energetic ions on the film properties of h-GaN crystals have been reported[39,40].

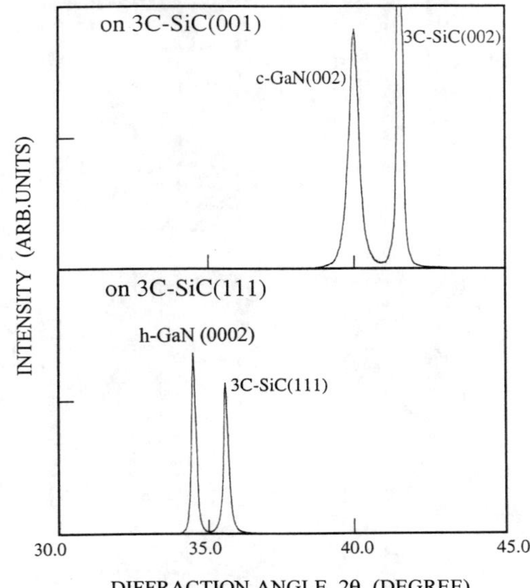

Fig.4   XRD patterns of GaN epilayers grown on 3C-SiC(001) and (111) substrates by GSMBE using an activated ammonia beam.

We have reported the growth of c-GaN on GaAs and 3C-SiC by using DMHy and micro-wave plasma-activated ammonia as an N source and successfully obtained c-GaN epilayers up to several micrometers thick[41]. By MOCVD, on the contrary, sometimes crystal structures turn from a metastable cubic phase to a stable hexagonal phase with the increase of film thickness[42]. This is considered to be due to the equilibrium growth condition in MOCVD technique, *i.e.*, high substrate temperatures, compared with those in MBE. The presence of hydrogen carrier gas in CVD, which is quite different situation from the case of MBE, may also affect the crystal growth.

Orientation of Substrate Surfaces

We have reported that the crystal structures of GaN epilayers grown on GaAs and 3C-SiC substrates depend on the crystal orientation of the substrate surfaces[41,43]. Figure 4 shows the XRD patterns of GaN films grown on 3C-SiC (001) and (111) substrates by using an activated ammonia beam as a N source. From the XRD patterns as well as RHEED observation, it was found that c-GaN grows on (001) surfaces of GaAs and 3C-SiC and h-GaN on (111) surfaces. Moustakas *et al.*[31] have reported the growth of c-GaN on (001) Si surfaces and h-GaN on (111) Si surfaces. Growth of c-GaN on (001) surfaces of cubic crystals, MgO[33] and GaP[44] has been also reported. These results suggest that c-GaN grows on the crystal surfaces having 4 fold symmetry, while h-GaN on the cubic crystal surfaces having 3 fold symmetry as well as on hexagonal crystals like sapphire.

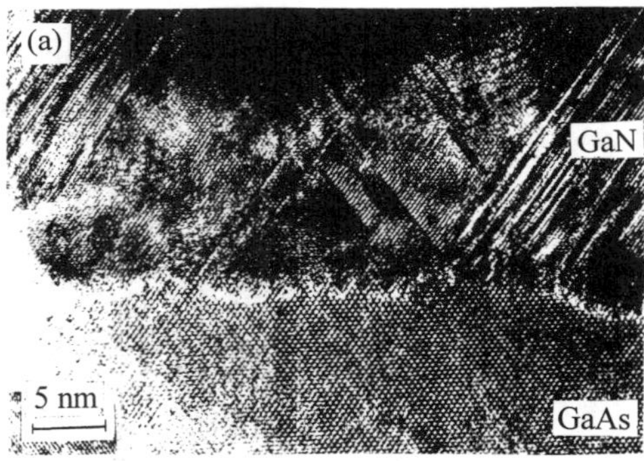

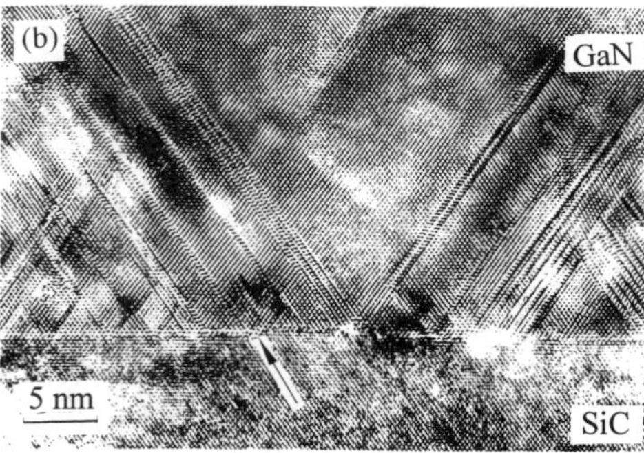

Fig.5 Cross sectional TEM images of (a) a c-GaN/GaAs, and (b) a c-GaN/3C-SiC interfaces.

In the case of GaAs substrates, this is not always true. The growth of c-GaN on (111) surfaces of GaAs[45], and h-GaN on GaAs (001)[46] have been reported. Inclusion of h-phase grains in c-GaN films grown on GaAs (001) has been often reported[30,47]. These results suggest that the growth conditions other than the crystal symmetry of the substrate surfaces also affect the selection of the growth of cubic and hexagonal crystal structures.

It is plausible that the initial step of the growth influences strongly on the crystal structure of the grown films. In fact, we have found that only when the GaAs (001) surfaces are nitrided fully by use of DMHy beam prior to the growth of GaN, c-GaN without mixing with a hexagonal phase can be obtained[43]. On partially nitrided GaAs surfaces, GaN films mixed with two phases have been grown. On the contrary, when GaAs surfaces were nitrided by a plasma-excited nitrogen beam, the GaN films grown were mixed with cubic and hexagonal phases[48]. This means that uniformly nitrided layer covered on GaAs surfaces, which is obtained, for example, by use of a DMHy beam, not by use of a plasma-excited nitrogen beam, is required to grow c-GaN without mixing of h-GaN. As seen by the cross sectional TEM image of a GaN/GaAs interface (Fig.5(a)), the interface is not flat but rough with (111) facets[49]. As h-GaN grows on (111) surfaces of GaAs, it is easy to suppose that mixed crystals may grow on rough surfaces caused by the damage of plasma. Sometimes, a lot of holes of GaAs are seen at the GaN/GaAs interfaces. These holes are considered to have formed due to the evaporation of As atoms from GaAs because of too high growth temperatures of GaN for Ga-As bonds, *i.e.*, 600-800℃.

In the case of SiC substrates, on the contrary, the nitridation of the surfaces by DMHy prevented the films from the epitaxial growth[50], which is considered to be due to the formation of amorphous silicon nitride layers on the SiC surfaces. We have done the epitaxial growth of GaN directly on SiC without nitridation process. Figure 5(b) shows the lattice image of a c-GaN/3C-

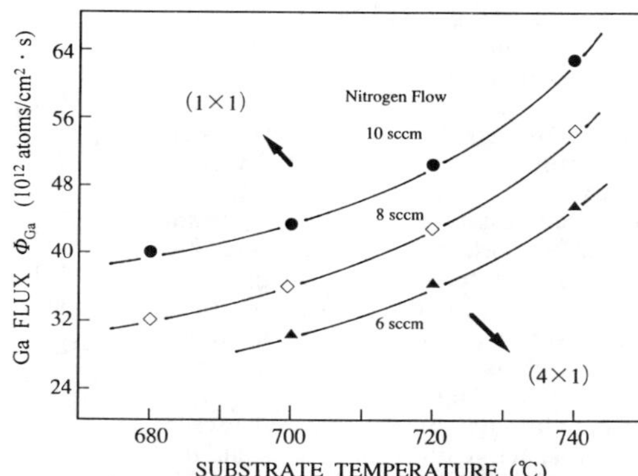

Fig.6 Mapping of the reconstruction patterns of a surface of the c-GaN epilayer grown on 3C-SiC (001) substrate by GSMBE using ECR plasma excited nitrogen beam.

SiC (001) interface. Comparing with the case of GaAs substrates, GaN/SiC interface is very flat with some atomic steps, though a lot of stacking faults can be seen in the c-GaN layer[49]. FWHMs of the XRD peaks are smaller than those for the GaN grown on GaAs, but still larger than those for h-GaN. This suggests that the c-GaN epilayers grown on 3C-SiC have higher crystal quality compared with those on GaAs, resulting from the better lattice matching at a GaN/SiC heterointerface (about 3 %) than that at a GaN/GaAs interface (about 20 %). In case of SiC substrates, no hole is seen at the interface, contrary to GaAs substrates, which is due to the refractory nature of SiC. We also observed no change of SiC surfaces by the supply of plasma excited nitrogen beam. From these points of view, 3C-SiC is a superior substrate material for the growth of c-GaN, compared with GaAs.

<u>Growth Condition</u>

It has been reported that the crystal structures depend on the growth conditions, *viz*, growth temperature, presence of As flux and V/III ratio on the growing surfaces. Yang *et al.*[45] have reported that c- and h-GaN can be selectively deposited on (111) GaAs substrates by varying the growth temperature in low pressure MOCVD, *i.e.*, c-GaN grows at 750 ℃, while h-GaN grows at 850℃. This result strongly suggests that the growth temperature is one of the key parameters to control the crystal structures, though these results have been obtained by MOCVD, not by MBE. The fact that, in the two step growth of h-GaN on sapphire by CVD, the low temperature grown buffer layers contain mainly cubic phase also

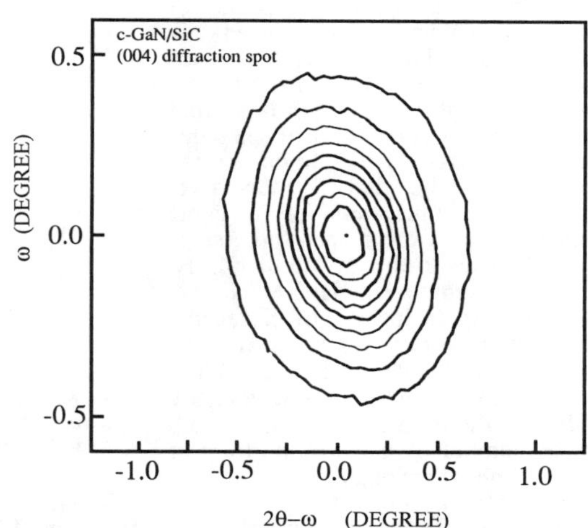

Fig.7 XRD reciprocal lattice mapping in the vicinity of (004) diffraction point for a c-GaN epilayer grown on 3C-SiC. The intensity contours of every 10% are shown.

supports these considerations. On the contrary, Cheng *et al.*[47] have reported the growth of mixed crystals of h- and c-GaN on GaAs and GaP (001) substrate surfaces by GSMBE using rf-activated plasma nitrogen source at 700℃ without As flux, and the growth of c-GaN with As flux. Their results suggest that As beam plays an important role for the selection of crystal structures. They proposed the kinetic growth model to explain their results. When arsenic is directed during growth, GaN is forced to adopt the crystal structure of the underlying cubic substrate through an interchange between the mobile active nitrogen species and the arsenic atoms occupying the group V lattices of the zincblende structure.

While, Brandt *at al.*[30] have reported the influence of the V/III ratio, *i.e.*, ratio of active nitrogen species and Ga atoms at the growing surfaces on the crystal structures. They grew GaN on GaAs (001) substrates by GSMBE using an active nitrogen beam generated by a high voltage dc plasma glow discharge. They monitored the crystal surfaces by RHEED during growth, and observed (1 × 1), (2×2) and c(2×2) reconstruction patterns, which is assigned to be the surfaces of Ga adatom coverage of 0, 0.5 and 1, respectively. Following the idea of surface stoichiometry introduced by Daweritz and Hey[36], Brandt *et al.* monitored surface stoichiometry by the observation of the intensity of half order reconstruction streak in RHEED patterns. They found N-rich condition during nucleation of the first 5 monolayers promotes the nucleation of a cubic template. Ga excess at this stage inevitably leads to the growth of h-GaN. The films grown under N excess condition, the inclusions of hexagonal phase with the c axis perpendicular to the (111) B of cubic GaN is detected While, those under near stoichiometric conditions, the films having very smooth surface morphology and with no trace of hexagonal phase are obtained.

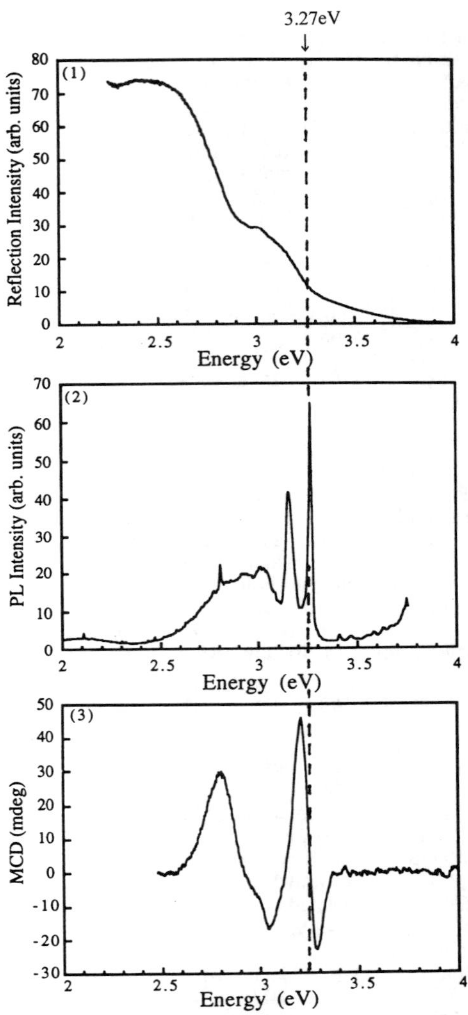

Fig.8 Optical and luminescent properties of a c-GaN grown on GaAs by GSMBE, measured at 4.2K: (1) optical reflection spectrum, (2) photoluminescence spectrum, and (3) magnetic circular dichroism (MCD) spectrum.

We have also reported (1×1), (2×2) and c(2×2) surface reconstruction patterns for c-GaN grown on GaAs (001) surfaces by GSMBE using ECR-plasma excited nitrogen source and (1×1) and (4×1) for that on 3C-SiC (001)[38,51]. Figure 6 shows the transition curve between (1×1) and (4×1) reconstruction regions on the temperature vs. Ga flux plane, with a constant nitrogen flux. This transition curve can be also explained by using Eq.(1). As in the case of h-GaN mentioned above, we monitored the optimum growth condition by the observation of the reconstruction patterns, and succeeded in obtaining c-GaN epilayers with good crystalline quality, as presented by Okumura *et al.* in this meeting[52]. Figure 7 shows the XRD reciprocal lattice mapping in the vicinity of (004) reflection for c-GaN epilayers on low pressure CVD grown 3C-SiC on Si. No spot from hexagonal phase can be observed in the mapping, which reveals no mixing of h-GaN in the epilayers. From the figure, $\Delta 2\theta = 18$min ($2\theta-\omega$ scan) and $\Delta\omega=30$min (ω scan) are obtained[38]. We have obtained the pole figure of cubic (002) Bragg diffraction for the c-GaN epilayers on GaAs and 3C-SiC (001) showing no peak except the center peak. This suggests that no misoriented cubic grain is included in the c-GaN epilayers we grew.

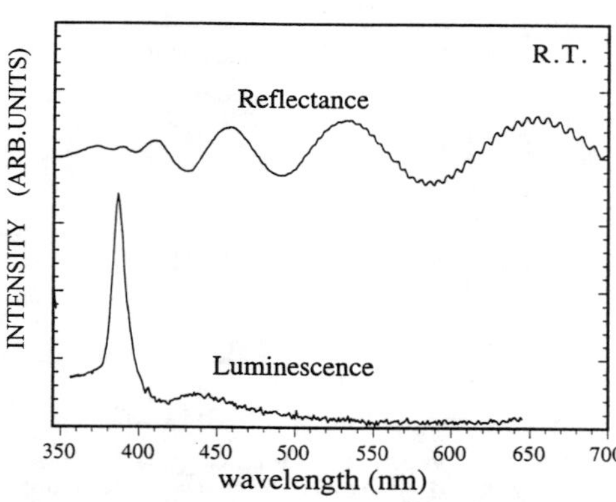

Fig.9 Optical reflection and photoluminescence spectra at room temperature for a c-GaN grown on 3C-SiC by GSMBE using an ECR plasma excited nitrogen beam.

The reconstruction patterns of (4×1) and (1×1) have been observed for c-GaN grown on and MgO (001) substrates[33,] as well as 3C-SiC (001) substrates[38,51,] which are different from those on GaAs (001). On the origin of this discrepancy, Feuillet *et al.* will present a paper in this meeting in detail[53].

PROPERTIES OF CUBIC GaN

Physical properties, including luminescent, optical and electrical properties, of c-GaN have been reported by several authors. However, the reported values scatter, and therefore, the differences of the properties of c-GaN from those of h-GaN have not been made clear. This can partly be attributed to the poor crystal quality of c-GaN obtained and the mixing of h-GaN phase. Recently, the growth of c-GaN epilayers with good crystalline quality have been reported, which makes possible to discuss the differences, for example, in the band gap energy. Electrical properties of c-GaN, *e.g.*, electron saturated drift velocity, have been theoretically predicted to be superior to those of h-GaN. However, there is no experimental report to support this prediction, so far. We observed the strong dependence of residual carrier densities on growth conditions, like ECR plasma power, and found high resistive ones at optimum conditions. However, it is hard to show the difference of the electrical properties from those of h-GaN. Therefore, only the bandgap energy, one of the most basic parameters of semiconductors, will be discussed here.

Comparing between hexagonal and cubic crystals of semiconductors other than GaN, the crystals with a hexagonal structure have larger bandgap energy E_g as in the cases of SiC, ZnS, and ZnSe, and those with a cubic structure have larger E_g in case of BN. For h-GaN, the values of 3.39eV at room temperature and 3.503eV at 1.6K for E_g, and the exciton binding energy E_{ex} of 28meV have been reported[54].

The values of E_g between 3.2-3.5eV have been reported by the measurements of photoluminescence (PL), cathodoluminescence (CL) and optical absorption for c-GaN epilayers grown by GSMBE and MOCVD[55-57]. While, theoretical calculation by empirical and model pseudopotential method have predicted a value of E_g 0.1-0.6eV smaller than that of h-GaN[58-60]. However, recently, the values reported have been concentrated around 3.2-3.3eV by the precise optical and luminescence measurements for c-GaN with good crystalline quality[33,61-63]. Powell *et al.*[33] obtained a value of $E_g(300K)=3.21\pm0.02$eV from transmission and/or reflection measurements carried out on c-GaN films grown on transparent MgO substrates. Ramirez-Flores *et al.*[61] have measured the temperature dependence of optical bandgap of the c-GaN grown on

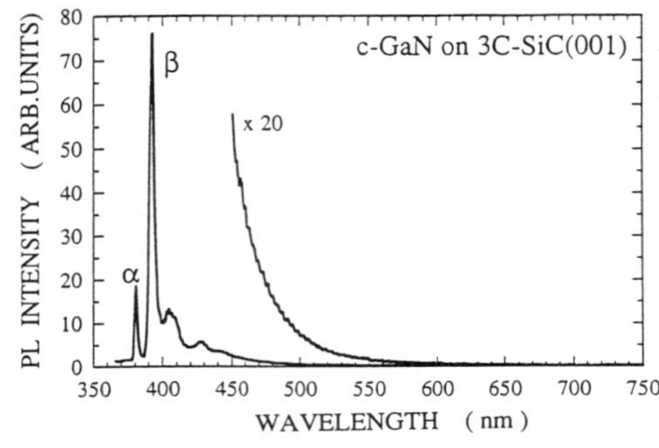

Fig.10 Photoluminescence spectrum of a c-GaN grown on 3C-SiC by GSMBE using an ECR plasma excited nitrogen beam.

MgO (001) by reactive-ion MBE by use of modulated photo-reflectance and photoluminescence measurements. The band gap was found to vary from 3.302 ± 0.004eV at 10K to 3.231 ± 0.008eV at 300K. Hwang et al.[62] have estimated the bandgap energy of c-GaN as 3.27 eV from the measure-ments of pressure dependence of PL spectra.

Figure 8 shows the optical reflection, PL and magnetic circular dichroism (MCD) spectra at 4.2K for a c-GaN epilayer grown on a 3C-SiC substrate by GSMBE using microwave-plasma excited ammonia beam as a N source. These spectra suggest the exciton band gap of c-GaN is around 3.27eV. Figure 9 shows the optical reflection and PL spectrum for a c-GaN epilayer grown on 3C-SiC by using ECR plasma-excited nitrogen beam, measured at room temperature. Even at room temperature, strong edge emission and pronounced oscillation of reflection below the bandgap energy are seen.

Figure 10 shows the PL spectrum for a c-GaN at 4.2K. Two strong emission peaks, named α and β, can be seen near 380nm. To know the origin of these peaks, we measured time dependent PL spectra for c-GaN on 3C-SiC. From the results shown in Fig.11, it is found that peak α decays with the time constant of several tens of psec, and the peak β remains after several ns. These values of time constant suggest that α peak relates to exciton, and β peak is assigned as D-A pair recombination, which also supports that the bandgap is around 3.26eV. In case of h-GaN, free-exciton and bound exciton peaks are well resolved[54]. However, in case of c-GaN, it is hard to resolve these peaks. The analysis in the time decay of α peak gives 2 kinds of components, i.e., the faster decay having the time constant of 15psec and the slow one of 395psec[64]. Klann et al.[65] measured time-resolved PL spectra of c-GaN epilayers on GaAs and also observed 2 kinds of components for the decay of the band edge emission, several tens of psec, and several hundreds of psec. They assigned that the faster component corresponds to the free exciton decay composed of both the radiative decay and the relaxation of the exciton towards localized state, and that the slower one corresponds to the radiative decay of localized exciton states. Recently, Menniger et al.[63] resolved two exciton peaks in CL spectra from micrometer-size single crystals and identified as free exciton peak at 3.272eV and bound exciton peak at 3.263eV with an exciton binding energy E_{ex} of 25meV (thus, fundamental energy

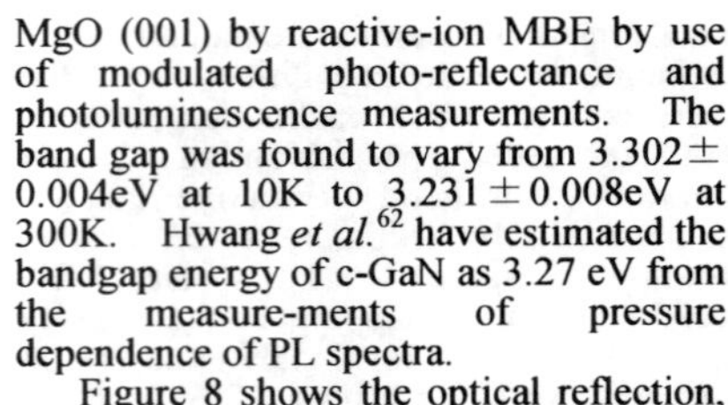

Fig.11 Time resolved photoluminescence spectra of a c-GaN grown on 3C-SiC, measured at 4.2K.

gap of c-GaN is estimated as 3.300eV). Hong *et al.*[66] reported a little larger values, 3.36eV for bound exciton and 3.375eV for free exciton gap, and the band gap of E_g=3.406eV at 6.5K.

SUMMARY

We have summarized the MBE growth of nitrides in comparison with the growth by CVD. The growth and properties of c-GaN have been described, comparing with h-GaN, including the control of crystal structures. Recently, the growth of c-GaN epilayers with good crystal quality have been reported, and some of the properties of c-GaN have been made clear.   However, it is still hard to find the merit of c-GaN compared with h-GaN.   For that, more investigations in the growth of high quality c-GaN and more wide measurements of their properties are longed for, and the growth mechanisms of c-GaN and h-GaN are required to be made clear.

AKNOWLEWDGMENTS

The authors are indebted to Y.Ishida for providing low pressure CVD grown 3C-SiC on Si, and to K.Ohta and H.Hamaguchi for their help in the MBE growth of c-GaN and XRD and luminescence measurements.

REFERENCES

1. H.P.Maruska and J.J.Tietjen, Appl.Phys.Let. **15**, p.327 (1969).
2. H.M.Manasevit, F.M.Erdman, and W.I.Simpson, J.Electrochem.Soc. **118**, 1864 (1971).
3. J.I.Pankove, E.A.Miller, and J.E.Berkeyheiser, RCA Rev. **32**, p.383 (1971).
4. S.Yoshida, S.Misawa and S.Gonda, Appl.Phys.Lett. **42**, p.427 (1983).
5. H.Amano, N.Sawaki, I.Akasaki and Y.Toyoda, Appl.Phys.Lett. **48**, p.353 (1986).
6. H.Amano, M.Kito, K.Hiramatsu and A.Akasaki, Jpn.J.Appl.Phys. **28**, p.L2112 (1989).
7. S.Nakamura, M.Senoh and T.Mukai, Jpn.J.Appl.Phys. **30**, p.L1708 (1991).
8. S.Nakamura, N.Iwasa, M.Senoh and T.Mukai, Jpn.J.Appl.Phys. **31**, p.1258 (1992).
9. S.Nakamura, T.Mukai, and M.Senoh, Appl.Phys.Lett. **64**, p.1687 (1994).
10. I.Akasaki, H.Amano, S.Sota, H.Sakai, T.Tanaka and M.Koike, Jpn.J.Appl.Phys. **34**, p.L1517 (1995), S.Nakamura, M.Senoh, S.Nagahara, N.Twasa, T.Yamada, T.Matsusita, H.Kiyoku and Y.Sugimoto, Jpn.J.Appl.Phys. **35**, p.L74 (1996).
11. R.P.Vaudo, I.D.Goepfert, T.D.Moustakas, D.M.Beyea, T.J.Frey and K.Meehan, J.Appl.Phys. **79**, p.2779 (1996).
12. M.Mizuta, S.Fujieda, Y.Matsumoto and T.Kawamura, Jpn.J.Appl.Phys. **25**, p.L943 (1986).
13. Y.Xie, Y.Qian, S.Zhang, W.Wang, X.Liu and Y.Zhang, Appl.Phys.Lett. **69**, p.334 (1996).
14. E.Ritter, J.Vac.Sci.Technol. **3**, p.225 (1966).
15. S.Yoshida, CRC Critical Rev. in Solid State and Material Science **11**, p.287 (1984).
16. S.Yoshida, S.Misawa and S.Gonda, Appl.Phys.Lett. **26**, p.461 (1975).
17. H.Gotoh, T.Suga, H.Suzuki and M.Kimata, Jpn.J.Appl.Phys. **20**, p.L545 (1981).
18. S.Yoshida, S.Misawa and S.Gonda, J.Appl.Phys.**53**, p.6844 (1982).
19. D.B.Oberman, H.Lee, W.K.Gotz and J.S.Harris,Jr., J.Crys.Growth **150**, p.912 (1995).
20. H.Okumura, S.Yoshida and E.Sakuma, J.Cryst.Growth **120**, p.114 (1992).
21. H.Okumura, S.Misawa, T.Okahisa and S.Yoshida, J.Cryst. Growth **136**, p.361 (1994).
22. C.H.Hong, Pavlidis, K.Hong, K.Wang and J.Singh, Inst.Phys.Conf.Ser.No137, p.413 (1993).
23. H.Tsuchiya, A.Takeuchi, M.Kurihara and F.Hasegawa, J.Cryst.Growth **150**, p.21 (1995).
24. S.Bharatan, K.S.Jones, C.R.Abernathy, S.J.Pearton, F.Ren, P.W.Wisk and J.R.Lothian, J.Vac.Sci.Technol. A **12**, p.1094 (1994), *ibid.* **13**, p.716 (1995).
25. R.C.Powell, N.E.Lee and J.E.Greene, Appl.Phys.Lett. **60**, p.2505 (1992).
26. W.Kim, A.Salvador, A.E.Botchkarev, O.Aktas, S.N.Mohammad and H.Morkoc, Appl. Phys.Lett. **69**, p.559 (1996).
27. Y.Moriyasu, H.Goto, N.Kuze and M.Matsui, J.Cryst.Growth **150**, p.916 (1995).
28. Z.Yang, L.K.Li and W.I.Wang, J.Vac.Sci.Technol. B **14**, p.2354 (1996).
29. M.J.Paisley, Z.Sitar, J.B.Posthill and R.F.Davis, J.Vac.Sci.Technol. A **7**, p.701 (1989).

30. O.Brandt, H.Yang, B.Jenichen, Y.Suzuki, L.Daweritz and K.H.Ploog, Phys.Rev. B **52**, p.R2253 (1995), Appl.Phys.Lett. **68**, p.244 (1996).
31. T.D.Moustakas, T.Lei and R.J.Molnar, Physica B **185**, p.36 (1993).
32. A.Botchkarev, A.Salvadlov, J.Myoung and H.Morkoc, J.Appl.Phys. **77**, p.4455 (1995).
33. R.C.Powell, N.-E.Lee, Y.-W.Kim and J.E.Greene, J.Appl.Phys. **73**, p.189 (1993).
34. G.Feuillet, P.Hacke, H.Okumura and S.Yoshida, <u>Proc. of Intern. Symp. on Blue Laser and Light Emitting Diodes</u>, (Chiba, March, 1996), Paper LN-4.
35. P.Hacke, G.Feuillet, H.Okumura and S.Yoshida, Appl.Phys.Lett. **69**, p.2507 (1996).
36. L.Daweritz and R.Hey, Surf.Sci. **236**, p.15 (1990).
37. W.Seifert and A.Tempel, Phys.Stat.Sol. (a) **23**, p.K39 (1974).
38. H.Okumura, K.Ohta, G.Feuillet, K.Barakrishnan, S.Chichibu, H.Hamaguchi, P.Hacke and S.Yoshida, J.Cryst.Growth, to be published.
39. J.M.Van Hove, G.J.Cosimini, E.Nelson, A.M.Wowchak and P.P.Chow, J.Cryst.Growth **150**, p.908 (1995).
40. R.J.Molnar and T.D.Moustakas, J.Appl.Phys. **76**, p.4587 (1994).
41. H.Okumura, S.Misawa and S.Yoshida, Appl.Phys.Lett. **59**, p.1058 (1991).
42. N.Kuwano, Y.Nagatomo, K.Kobayashi, K.Oki, S.Miyoshi, H.Yaguchi, K.Onabe and Y.Shiraki, Jpn.J.Appl.Phys. **33**, p.18 (1994).
43. S.Yoshida, H.Okumura, S.Misawa and E.Sakuma, Surf.Sci. **267**, p.50 (1992).
44. L.C.Jenkins, T.S.Cheng, C.T.Foxon, S.E.Hooper, J.W.Orton, S.V.Novikov and V.V.Tret'yakov, J.Vac.Sci.Technol. B **13**, p.1585 (1995).
45. J.W.Yang, J.N.Kuznia, Q.C.Chen, M.A.Khan, T.George, M.De Graef and S.Mahajan, Appl. Phys.Lett. **67**, p.3759 (1995).
46. A.A.Yamaguchi, T.Manako, A.Sakai, H.Sunakawa, A.Kimura, M.Nido, and A.Usui, Jpn.J. Appl.Phys. **35**, p.L873 (1996).
47. T.S.Cheng, L.C.Jenkins, S.E.Hooper, C.T.Foxon, J.W.Orton, and D.E.Lacklison, Appl.Phys. Lett. **66**, p.1509 (1995).
48. A.Kikuchi, H.Hoshi and K.Kishino, Jpn.J.Appl.Phys. **33**, p.688 (1994).
49. H.Okumura, K.Ohta, T.Nagatomo, and S.Yoshida, J.Cryst.Growth **164**, p.149 (1996).
50. S.Yoshida, H.Okumura, T.Okahisa and S.Misawa, Inst.Phys.Conf.Ser.No137, p.377 (1994).
51. G.Feuillet, H.Hamaguchi, K.Ohta, P.Hacke, H.Okumura, and S.Yoshida, to be published in Appl.Phys.Lett.
52. H.Okumura, K.Balakrishnan, G.Feuillet, Y.Ishida, and S.Yoshida, MRS 1996 Fall Meeting.
53. G.Feuillett, P.Hacke, H.Okumura, H.Hamaguchi, K.Ohta and S.Yoshida, MRS 1996 Fall Meeting.
54. R.Dingle, D.D.Sell, S.E.Stokowski and M.Ilegems, Phys.Rev.B **4**, p.1211 (1971), R.Dingle and M.Ilegemes, Solid State Commun. **9**, p.175 (1971).
55. S.Strite, J.Ruan, Z.Li, S.Savador, H.Chen, D.J.Smith, W.J.Choyke and H.Morkoc, J.Vac.Sci. Technol.B **9**, p.1924 (1991).
56. H.Okumura, S.Yoshida and T.Okahisa, Appl.Phys.Lett. **64**, p.2997 (1994).
57. H.Liu, A.C.Frenkel, J.G.Kim, and R.M.Park, J.Appl.Phys. **74**, p.6124 (1993).
58. W.R.L.Lambrecht and B.Segall, MRS Proc. Vol.242, p.367 (1992).
59. A.Rubio, J.C.Corkill, M.L.Cohen, E.L.Shirley and S.G.Louie,Phys.Rev.B **48**, p.11810 (1993).
60. W.J.Fan, M.F.Li, T.C.Chong and J.B.Xia, J.Appl.Phys. **79**, p.188 (1996).
61. Ramirez-Flores, H.Navarrow-Contreras, Lastras-Martinez, R.C.Powell and J.E.Greene, Phys.Rev.,B **50**, p.8433 (1994).
62. S.J.Hwang, W.Shan, R.J.Hauenstein, J.J.Song, M.-E,Lin, S.Strite, B.N.Sverdloy and H.Morkoc, Appl.Phys.Lett. **64**, p.2928 (1994).
63. J.Menninger, U.Jahn, O.Brandt, H.Yang, and K.Ploog, Phys.Rev.B **53**, p.1881 (1996).
64. H.Okumura, K.Ohta, K.Ando, W.W.Rule, T.Nagatomo and S.Yoshida, <u>Inst.Phys.Conf.Ser.</u> No.142, p.939 (1996).
65. R.Klann, O.Brandt, H.Yang, H.T.Grahn and K.Ploog, Phys.Rev.B **52**, p.R11615 (1995).
66. C.H.Hong, D.Pavlidis, S.W.Brown and S.C.Rand, J.Appl.Phys. **77**, p.1705 (1995).

MBE GROWTH AND OPTICAL CHARACTERIZATION OF InGaN/AlGaN MULTI-QUANTUM WELLS

R. SINGH *, W.D. HERZOG, D. DOPPALAPUDI **, M.S. ÜNLÜ, B.B. GOLDBERG***
AND T.D. MOUSTAKAS
ECE Department and Center for Photonics Research, Boston University, Boston MA 02215
* Email: raju@panda.bu.edu
** Also at the department of Manufacturing Engineering, Boston University MA 02215
*** Department of Physics, Boston University, Boston MA 02215

ABSTRACT

We report the growth of InGaN/AlGaN MQWs on c-plane sapphire by electron cyclotron resonance assisted molecular beam epitaxy (ECR-MBE). Two types of structures were investigated; one employing a GaN and the other a AlGaN barrier layer. The first structure consists of five periods of 80 Å thick $In_{0.09}Ga_{0.91}N$ wells separated by 90 Å thick GaN barriers. The second structure consists of seven periods of 120 Å thick $In_{0.35}Ga_{0.65}N$ wells and $Al_{0.1}Ga_{0.9}N$ barriers. The substrate temperature was kept constant during the growth of both the wells and the barriers, thus avoiding the need for any temperature cycling during the growth, which may lead to interfacial contamination. The films were characterized by cross sectional transmission electron microscopy (TEM), room temperature photoluminescence (PL) and sub-micron resolution luminescence microscopy. TEM images show sharp and abrupt interfaces, thus confirming the high interfacial quality of the MQW structures. Both structures exhibit strong RT luminescence emission peaking at 387 nm (FWHM=16nm) for the $In_{0.09}Ga_{0.91}N$/GaN structure and at 463 nm (FWHM=28nm) for the $In_{0.35}Ga_{0.65}N$/$Al_{0.1}Ga_{0.9}N$ structure. The high resolution luminescence microscopy studies reveal that the radiative recombination for the InGaN quantum wells is 60-70 times more efficient than for the underlying GaN film.

INTRODUCTION

InGaN alloy based active layers have been successfully used to fabricate high power blue-green light emitting diodes (LEDs) [1]. Multi-quantum wells (MQWs) of InGaN have also been used as the active layer for UV-blue laser diodes operating at room temperature [2]. In spite of the rapid progress by the Nichia group in the development of such devices, the growth and properties of the InGaN alloys are poorly investigated. We have recently reported on the phase separation in InGaN alloys having large indium concentrations, a result accounted for by both thermodynamic and kinetic factors [3,4]. However, such phase separation was not observed in thin InGaN films incorporated in quantum wells or double heterostructures, which suggests that devices based on the InGaN system will require the fabrication of the active regions in the form of quantum well structures [4,5].

There are only limited reports on the growth of InGaN SQW and MQW structures. The majority of the literature refers to such structures grown by the metalorganic chemical vapor deposition (MOCVD) method. Nakamura *et al.* reported quantum effects in $In_xGa_{1-x}N$/$In_yGa_{1-y}N$ superlattices grown by the MOCVD method [6]. Koike *et al.* reported on $In_{0.08}Ga_{0.92}N$/GaN MQWs grown also by the MOCVD method [7]. Finally, Keller *et al.* reported the growth and

Mat. Res. Soc. Symp. Proc. Vol. 449 © 1997 Materials Research Society

properties of $In_{0.16}Ga_{0.84}N$ /$In_xGa_{1-x}N$ (graded) SQWs also by the MOCVD method [8]. These authors observed quantum size effects in InGaN based QWs for thicknesses less than 30 Å. Our group has recently reported some studies on InGaN/AlGaN MQWs grown by the ECR-MBE method [5].

In this paper, we report on the growth and characterization of InGaN/GaN and InGaN/AlGaN MQWs deposited on (0001) sapphire substrates by ECR-MBE. Luminescence studies on these structures indicate that the radiative efficiency of these MQWs is 60-70 times higher than bulk GaN films.

DEPOSITION OF MQW STRUCTURES

The growth of the InGaN MQW structures was accomplished on thick GaN films grown on sapphire substrates. Details of the ECR-MBE growth method were reported previously and in the present paper only a brief summary is provided [9,10]. *C*-plane (0001) sapphire substrates were first subjected to ECR plasma nitridation which as we reported previously leads to a thin atomically smooth AlN film[10]. In the second stage these substrates were coated with approximately a 200-300 Å GaN buffer grown at 550 °C, which was also found to be atomically smooth [9]. Finally, the temperature was raised to 750 °C for the growth of a GaN film (approximately 0.5 μm or thicker).

The subsequent growth of the InGaN layers was conducted in the manner reported earlier [3, 5]. The first $(In_{0.09}Ga_{0.91}N/$ GaN) structure was grown at a substrate temperature of 670 °C, as measured by a thermocouple on the back side of the sapphire wafer. The gallium BEP was kept constant while the nitrogen plasma power employed during the InGaN layer growth was 100 W and reduced to 80 W for the GaN barrier layers. For the second $(In_{0.35}Ga_{0.65}N/$ $Al_{0.1}Ga_{0.9}N)$ multi-quantum well structure, a substrate temperature of 660 °C was used for the entire growth sequence. The gallium cell temperature was also kept constant, while the aluminum and indium cell shutters were opened alternately for the growth of the respective layers.

EXPERIMENTAL RESULTS AND DISCUSSION

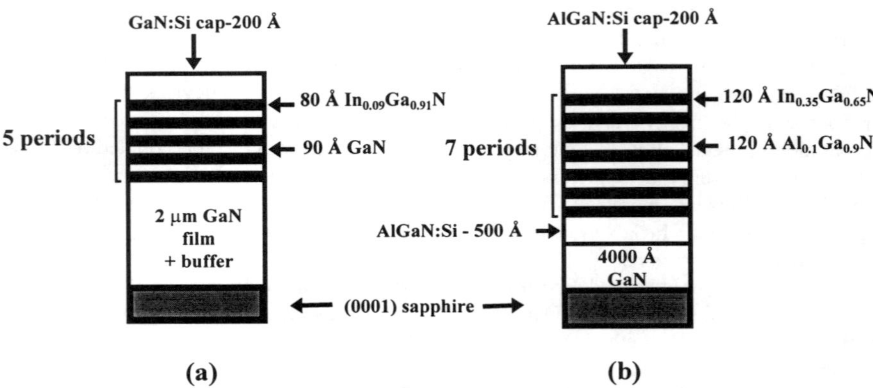

Fig. 1: Schematic of MQW structures (a) $In_{0.09}Ga_{0.91}N$/GaN with five periods and (b) $In_{0.35}Ga_{0.65}N$/$Al_{0.1}Ga_{0.9}N$ with seven periods

Fig. 1 show the schematic of the two InGaN based MQW structures investigated in this work. The sharpness of the interfaces was confirmed by cross sectional TEM studies as described previously [5]. The thicknesses of the wells and barriers were computed to be 130 Å each from the TEM micrographs, which is in close agreement with the thickness of 120 Å determined by beam flux measurements.

Room temperature luminescence was excited with the 325 nm line of a He-Cd laser. The power in the collimated beam was 10 mW with an incident beam diameter of 1 mm. The luminescence spectra were dispersed through a 0.5 m Acton Research spectrometer using a holographic grating and measured with a Hamamatsu R928 photomultiplier tube. The spectra were not corrected for the system response.

Fig. 2 shows the RT luminescence from the MQW structures illustrated in Fig. 1. The luminescence is dominated by emission from the InGaN wells, which peaks at 387 nm (3.2 eV) for the five period $In_{0.09}Ga_{0.91}N$/GaN structure (curve (a)) with a FWHM of 16 nm. The luminescence from the MQW structure composed of InGaN wells and AlGaN barriers is shown in curve (b). The luminescence peaks at 463 nm with a FWHM of 28 nm. Shown in the same figure are also small peaks to the GaN and AlGaN barrier layers. From these spectra, it is clear that the increase in the indium concentration leads to the broadening of the luminescence peak. However, these emission lines are significantly narrower than the corresponding luminescence spectra emitted by bulk InGaN films [4].

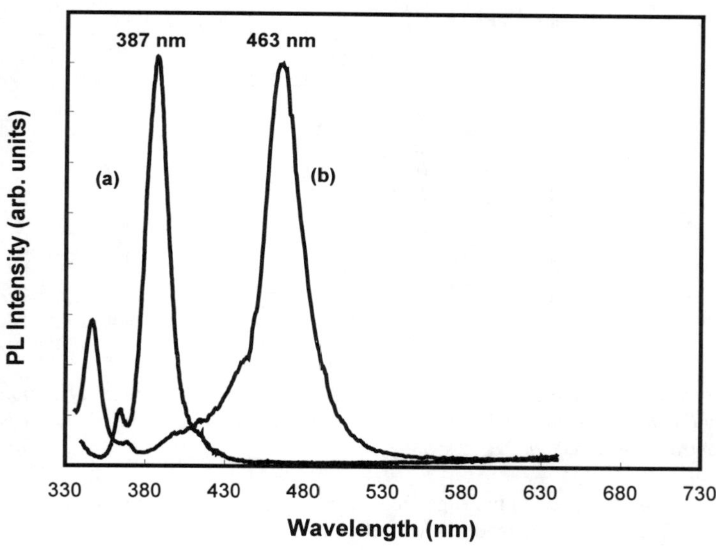

Fig. 2: RT luminescence spectra from the InGaN/AlGaN MQW structures shown in Fig. 1

Photoluminescence microscopy studies were conducted on the MQW structure by focusing the 335 nm line of an Ar$^+$ laser to an excitation spot of 0.6 μm FWHM by a reflecting objective. The spot size was measured with a scanning knife edge in the focal plane. The same objective collected the luminescence which was then dispersed through a 0.64 m spectrometer with a 300 line/mm grating onto a liquid nitrogen cooled CCD camera. The system provided 0.6 μm spatial resolution and 0.6 nm spectral resolution. The sample was scribed to expose a facet perpendicular to the growth direction. The InGaN heterostructure facet was scanned under the excitation optics using a piezo actuated stage under computer control. The spectral dependence of the luminescence was independent of the pump beam power densities ranging from 20 W/cm^2 to 1600 W/cm^2.

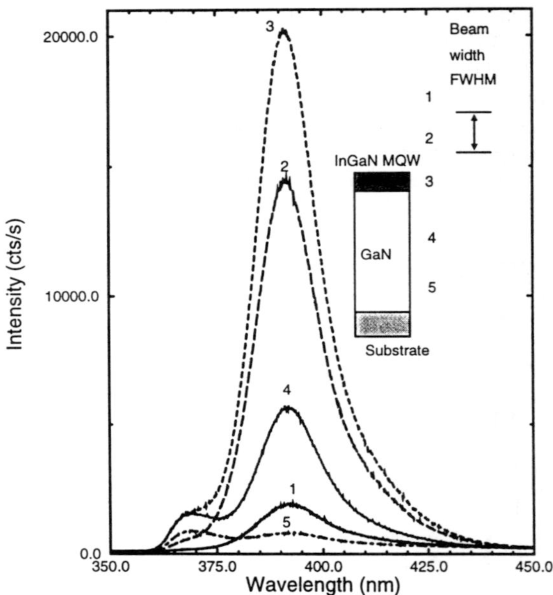

Fig. 3: Photoluminescence spectra of the In$_{0.09}$Ga$_{0.91}$N/GaN MQW structure as a function of position along the layer structure. The numbers in the inset correspond to the position where the respective spectra were collected.

A series of spectra taken on the edge facet of the layer structure as a function of position along the growth direction is displayed in Fig. 3. Starting from the surface and moving towards the substrate, the spectra are labeled with a number which corresponds to the position indicated in the schematic of the device structure (inset, Fig. 3). The InGaN PL intensity is highest when the beam is centered on the quantum wells (position 3). The luminescence intensity of the InGaN signal is 20 times larger than the GaN peak intensity. As the scan tip moves away from the wells and towards the substrate, the InGaN luminescence signal diminishes significantly (position 4) and the GaN PL increases relative to the InGaN PL.

Fig. 4 shows the peak amplitudes of the InGaN and GaN PL plotted as a function of position along the growth direction. The FWHM of the spatial dependence of the InGaN MQW PL intensity is 1.2 μm, which is larger than the laser beam spot size. The extra width is most likely due to the diffusion of photoexcited carriers from the GaN barriers and film into the MQW region.

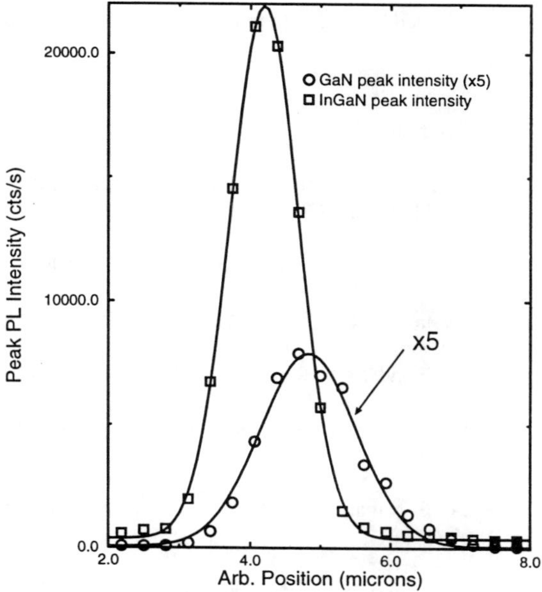

Fig. 4: Peak photoluminescence of InGaN and GaN as a function of position across the layer structure

The measured peak luminescence from the InGaN MQWs is 15 times stronger at peak than that from the 2 μm thick GaN beneath. Assuming that all the electron-hole pairs generated in the barriers diffuse to the QWs, and accounting for the difference in areas as well as relative collection and detection efficiency, we estimate that the InGaN MQWs are approximately 60-70 times more efficient for radiative recombination than the GaN film.

CONCLUSIONS

In conclusion, fabrication of multi-quantum wells of InGaN/GaN and InGaN/AlGaN has been accomplished by ECR-assisted MBE. Both wells and barriers were grown at the same substrate temperature (660-670 °C). The interfaces between the wells and barriers were found to be abrupt by TEM studies and the measured thicknesses matched well with those computed from beam flux measurements. Strong RT PL emission was observed with much narrower emission than that observed in bulk InGaN layers. Photoluminescence microscopy of the InGaN MQWs reveals that the recombination efficiency in InGaN layers is 60-70 times higher than in bulk GaN films. These studies clearly indicate the superiority of InGaN multi-quantum well active layers for the fabrication of high efficiency and power GaN based visible light emitting devices.

ACKNOWLEDGMENTS

This work was partly funded by ARPA (Grant. No. MDA972-95-3-0008).

REFERENCES

1.   S. Nakamura, Microelectronics Journal **25**, p. 651 (1994)

2.   S. Nakamura, M. Senoh, S. Nagahama, N. Iwasa, T. Yamada, T. Matsushita, H. Kiyoku and Y. Sugimoto, Appl. Phys. Lett. **68**, p. 3269 (1996)

3.   R. Singh and T.D. Moustakas, MRS Symp. Proc. **395**, p. 163 (1996)

4.   R. Singh and T.D. Moustakas, ECS Proc. **96-11**, p. 186 (1996)

5.   R. Singh, D. Doppalapudi and T.D. Moustakas, Appl. Phys. Lett. **69**, p. 2388 (1996)

6.   S. Nakamura, T. Mukai, M. Senoh, S. Nagahama and N. Iwasa, J. Appl. Phys. **74**, p. 3911 (1993)

7.   M. Koike, S. Yamasaki, S. Nagai, N. Koide, S. Asami, H. Amano and I. Akasaki, Appl. Phys. Lett. **68**, p. 1403 (1996)

8.   S. Keller, B.P. Keller, D. Kapolnek, A.C. Abare, H. Masui, L.A. Coldren, U.K. Mishra and S.P. Den Baars, Appl. Phys. Lett. **68**, p. 3147 (1996)

9.   T.D. Moustakas, T. Lei and R.J. Molnar, Physica B **185**, p. 36 (1993)

10.   T.D. Moustakas and R.J. Molnar, MRS Review Symp. Proc. **281**, p. 753 (1993)

IN-SITU REFLECTION HIGH ENERGY ELECTRON DIFFRACTION STUDY OF STRUCTURE AND MORPHOLOGY EVOLUTION OF AlN THIN FILMS DURING GROWTH

F.   JIN, G. W. AUNER, R. NAIK, P. ZATYKO, and U. RAO
Electrical Engineering Dept., and * Physics Dept., Wayne State Univ., Detroit, MI 48202

Abstract

AlN films were grown on Si (111) substrates using a PSMBE deposition system. Two types of growth methods were used: (i) a continuous growth at one temperature and (ii) a two stage growth where a buffer layer at low temperature is followed by deposition at higher temperature. The structure and morphology of the films grown at different temperatures are compared. The evolution of surface structure and morphology of the film during the growth were studied using in-situ RHEED, X-ray diffraction, and AFM.

Introduction

AlN is an important wide band gap material with many potential applications. Its fast Rayleigh velocity and high chemical and thermal stability make it an ideal material for acoustic wave sensor application. Furthermore, it can alloy with GaN and InN to form photonic devices with band gap ranging from 1.9 eV to 6.2 eV. Thus it is a promising system for semiconductor optoelectronic application.

High quality AlN thin films are hard to grow especially on Si substrates because of lattice mismatch and many other factors [1-8]. In this study we intend to gain some insights on the growth mechanism of AlN film as well as insights on what parameters are better to grow AlN thin films on Si substrates. We use Reflective High Energy Electron Diffraction (RHEED) to monitor the surface structure and morphology evolution of AlN thin films during growth and correlate the RHEED results with X-ray diffraction and AFM results. Two methods are used to grow AlN thin films on Si substrates in a PSMBE system, namely two stage growth and single stage growth. Unlike single stage growth, in two stage growth, a 200 Å buffer layer is grown first at low temperature (400 °C) followed by growth at higher temperature.

We show that high quality AlN films can be obtained with relative low growth temperature by controlling the energy of the ion flux to the substrate at a suitable level. Also, the RHEED images obtained at various stages during the film growth show the evolution of the structure and morphology of the films. RHEED results suggest that under various growth temperatures, high quality ultra thin films (less than 200 Å) probably formed on the substrates in the beginning of the growth. As the films become thicker and strains build up, the films crack and separate to many parts randomly oriented on the x-y plane (plane of substrate surface). The cracked films serve as templates for continued growth. The structure and morphology of the final films are highly depended on the growth temperature. The AFM and x-ray diffraction results also support our interpretation of the RHEED results.

Experimental Methods

A schematic diagram of the PSMBE deposition system is shown in Fig.1. The PSMBE (Plasma Source Molecular Beam Epitaxy) system consists of an ion pump and a throttled CTI cryopump. It maintains a base pressure typically around 1×10^{-9} torr. The system has a unique magnetically enhanced hollow cathode plasma deposition source lined with an Al cylindrical target. During growth, an argon/nitrogen plasma is formed within the hollow cathode. A low energy flux of   Al atoms/ions streams out of the source to the Si substrate together with nitrogen atoms/ions.The energy of the flux to the Si substrate is controlled by the DC voltage bias applied to the Si substrate. The Si substrate is rotated and is heated to the desired temperature, and the temperature is monitored by a IR pyrometer. The system also includes a RGA (Residual Gas Analyzer), an ellipsometry and a RHEED system which are used to monitor the evolution of the surface structure and the morphology during film growth. The Si substrates were cleaned in an ultrasonic bath of acetone followed by methanol. They were then dipped in 10% (by volume)

Mat. Res. Soc. Symp. Proc. Vol. 449 © 1997 Materials Research Society

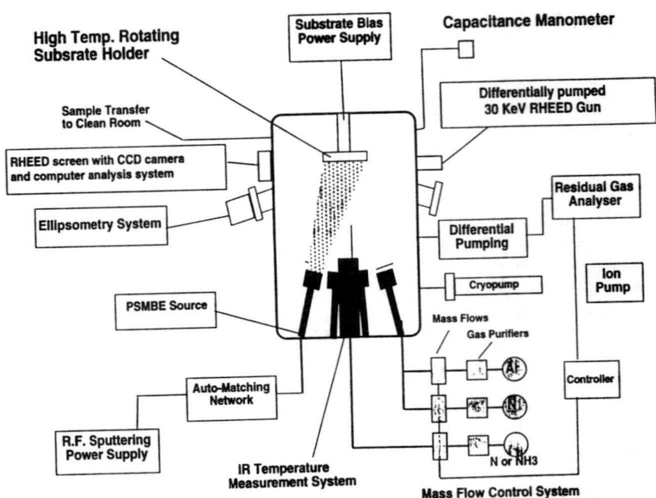

Fig. 1 A schematic diagram of the PSMBE deposition system.

hydrofluoric acid for 30 seconds to remove the oxides on the Si surface. The loaded samples were hydrogen terminated. Listed below is the growth parameters we used.

Sample	Substrate Temp.	DC bias	Dynamic Pressure
#1	400 °C	-15 V	1 m torr
#2	600 °C (with 200 Å buffer)	-15 V	1 m torr
#3	800 °C (with 200 Å buffer)	-15 V	1 m torr
#4	800 °C	-15 V	1 m torr

Results and discussions

Fig. 2 is a set of four RHEED images showing the surface structure evolution of the growth at 400 °C. The RHEED image of the Si substrate shows sharp streaks indicating a clean and smooth Si substrate. After deposition of a buffer layer (200 Å grown at 400 °C), the RHEED image shows diffusive spotty pattern. This suggests a relatively rough surface (diffusive image) and the existence of ordered crystal structures (RHEED spots displayed). The X-ray diffraction result on the buffer layer (see Fig. 5) further confirm the formation of a textured crystal structure (AlN (0002) peak observed) . The growth of a crystal structure at such low temperature is likely a result of the controlled energy of the ion flux to the substrate. As the growth continues, the spots in the RHEED images become sharper (see the other two images in Fig. 2), but the patterns remain the same. This suggests that the buffer layer served as a template for the continued growth, and the final film has a similar structure to that of the buffer. The x-ray diffraction of the final film grown at 400 °C (see Fig. 6) displays a stronger AlN(0002) peak which is consistent with RHEED observations. The two stage growth at 400 °C and 600 °C shows similar results, except that the RHEED images of the continued films are much sharper than that of the growth at 400 °C. This suggests that a smoother final surface is formed as a result of higher growth temperature. The results of continued growth at 600 °C shows that a certain level of the thermal motion on the growing surface is helpful in forming a smooth surface. Fig. 4 shows another set of RHEED images of a two stages growth at 400 °C and 800 °C. We can see that the RHEED pattern of the buffer layer is not followed by a continued growth in this case. The RHEED image of the final film shows diffusive spotty and ring patterns, indicating the existence of a polycrystalline structure and a relatively rough surface as well. Previous experiments have also shown that high growth

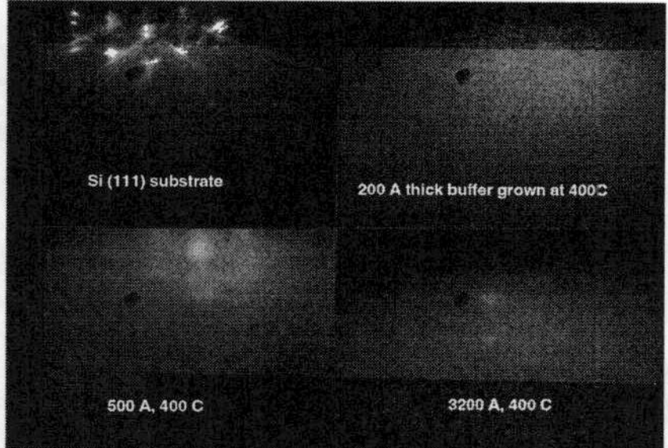

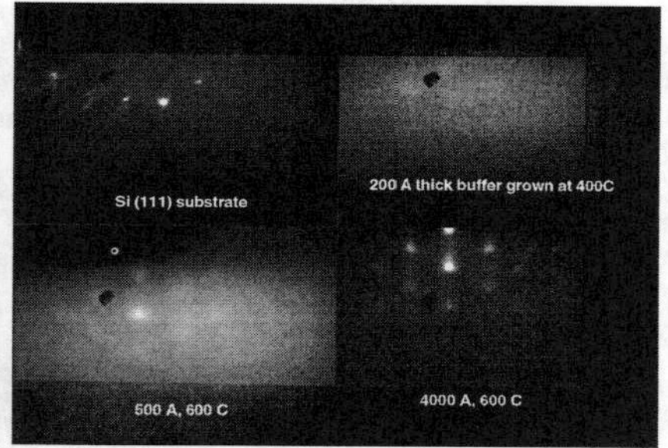

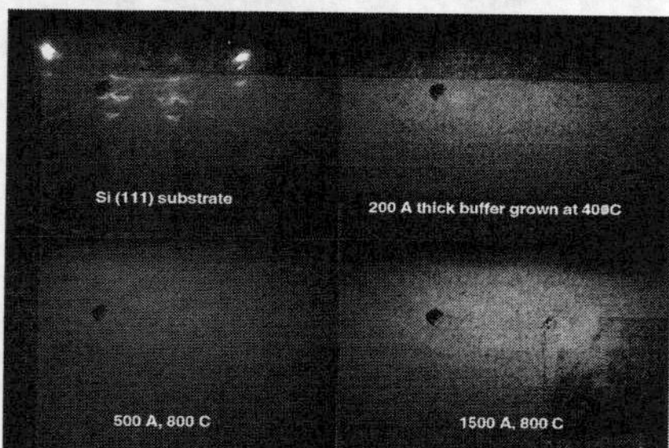

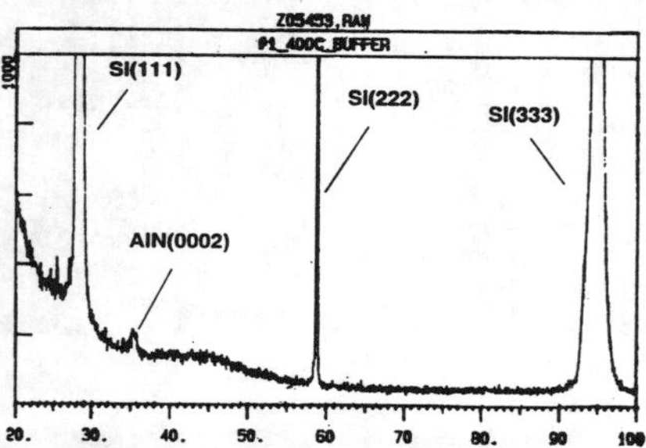

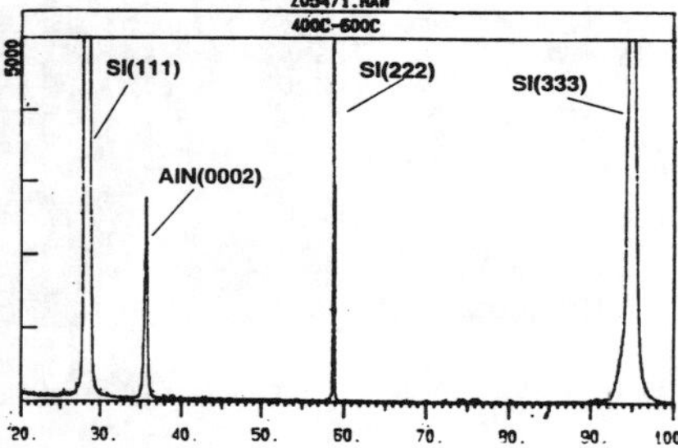

Fig. 2 (Upper left) RHEED patterns show the surface structure evolution of the growth at 400 °C.

Fig. 3 (Upper right) RHEED patterns show the surface structure evolution of the two stage growth at 400 °C and 600 °C.

Fig. 4 (Left) RHEED patterns show the surface structure evolution of the two stage growth at 400 °C and 800 °C.

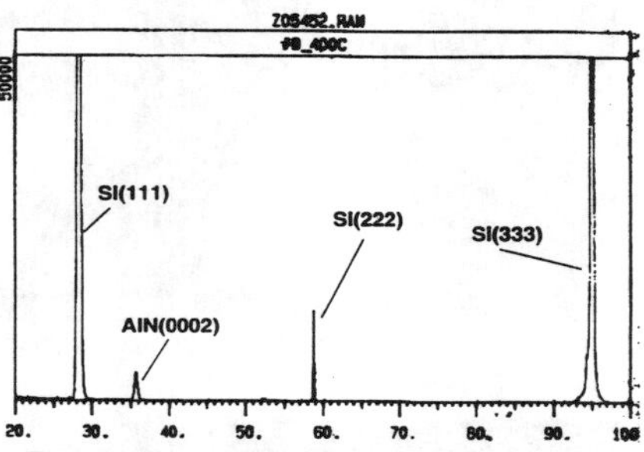

Fig. 5 X-ray diffraction of a 200 A thick buffer layer grown at 400° C.

Fig. 6 X-ray diffraction of the final film grown at 400° C.

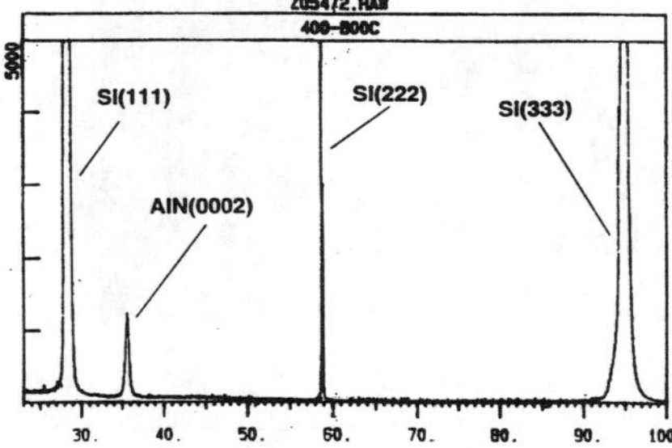

Fig. 7 X-ray diffraction of the final film grown at 400° C and 600° C.

Fig. 8 X-ray diffraction of the final film grown at 400° C and 800° C.

temperature tends to induce three dimension growth and therefore a more rough surface [9]. The AFM results in Fig. 9 also clearly indicates that the continued growth at 600 °C produce a smoother surface. Comparing the AFM image of the buffer layer and the AFM image of the continued growth at 600 °C, we see that the surface become much smoother as the growth continues.

The X-ray diffraction results shown in Fig. 5 - Fig. 8 show that only one AlN(0002) peak is observed on the final grown films. This suggest that these films have a highly preferred orientation along the Z direction (perpendicular to the surface). The RHEED results, on the other hand suggest that the films are probably fragmented into many parts which were randomly oriented in the X-Y plane (surface plane). Fig. 10 shows a set of RHEED images of two stages growth at

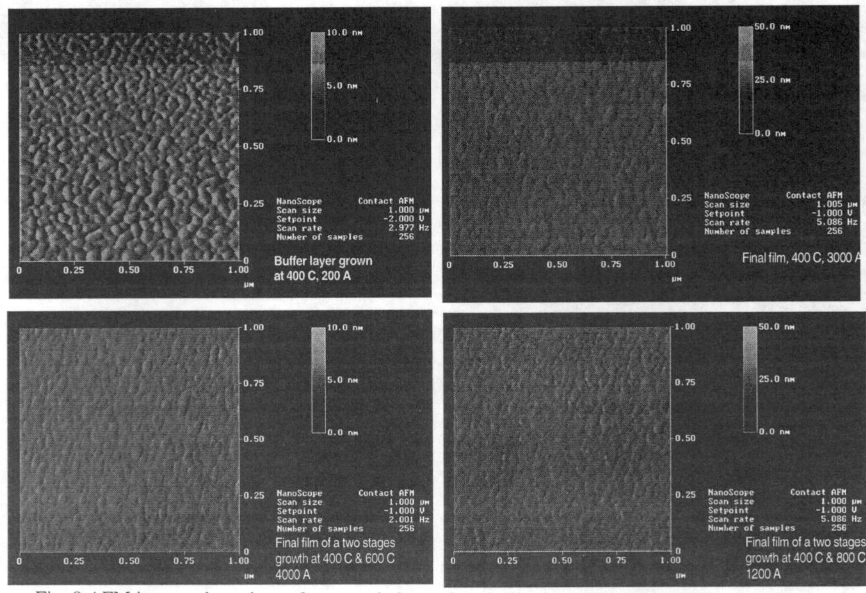

Fig. 9 AFM images show the surface morphology of the buffer layer and the final films grown on it.

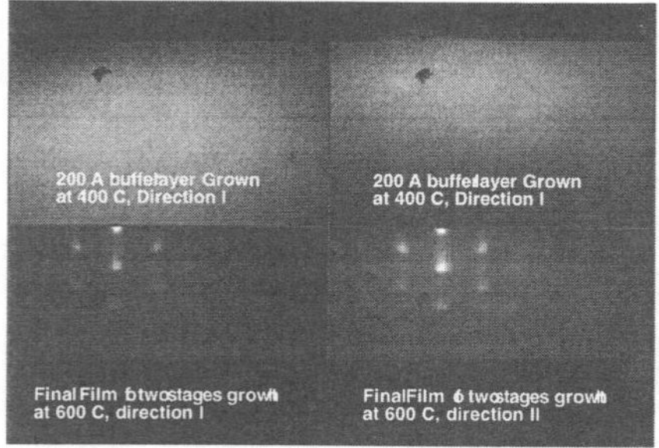

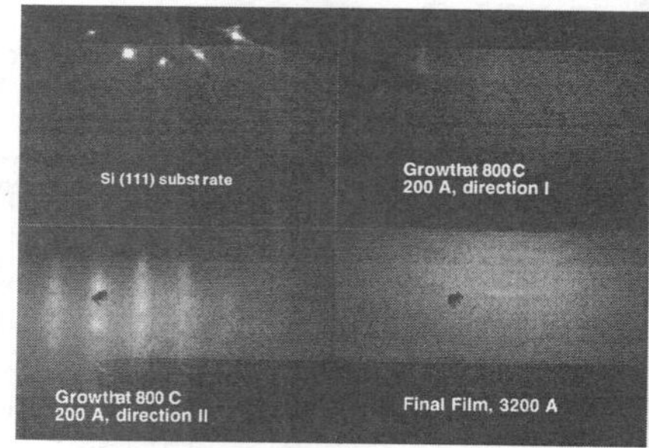

Fig. 10 RHEED images taken in different directions shows the random orientation of the crystals on the surface. The growth is the two stage growth at 400 °C and 600 °C.

Fig. 11 RHEED patterns show the surface structure evolution of the growth at 800 °C. Note that two RHEED images of the initial layer taken in different directions are shown.

400 °C and 600 °C taken at two randomly picked directions. The images show that the RHEED patterns remain the same as the direction of observation changes, indicating there's no preferred orientation of the film on the surface. We suspect that at the initial stage of each of the films discussed above, a highly ordered epitaxial layer (oriented both in Z-direction and the X-Y plane) is probably formed in the very beginning. As the growth continues and strain builds up, this layer breaks into many fragments. These fragments are randomly oriented and is the cause of the RHEED images in Fig. 10.

Fig. 11 is a set of RHEED images showing the surface structure evolution of growth at 800 °C . The RHEED streaks of the initial growth at 800 °C clearly indicate an epitaxial growth with a smooth surface. Although this initial layer is only 200 Å thick, X-ray diffraction shows a relatively strong AlN (0002) peak (see Fig. 12). The AFM image also shows a smooth surface (see Fig. 14). Two RHEED images of this initial layer taken in two different directions are presented here. We see that the film is oriented at certain directions both in the Z direction (shown by X-ray) and in the X-Y plane (see RHEED images of different directions). This evidence suggests that the initial layer is a high quality AlN thin film and an ideal template for further growth. But this template was not followed, the RHEED image of the final film shows a ring pattern, indicating a polycrystalline surface structure. Also a weak AlN (0002) peak in the x-ray diffraction of the final film confirms no continuation of epitaxial growth. Again, our interpretation of this dramatic change in structure is that the initial layer is highly strained due to lattice mismatch, as the film grows, the strain builds up and start to crack thus destroy the fine template. The high temperature

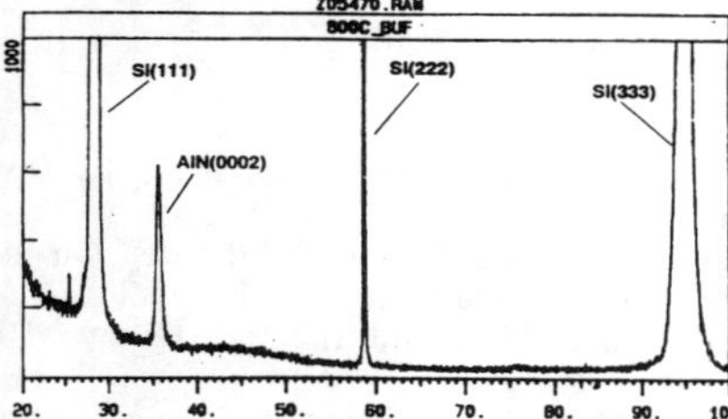

Fig. 12 X-ray diffraction of the initial layer grown at 800° C.

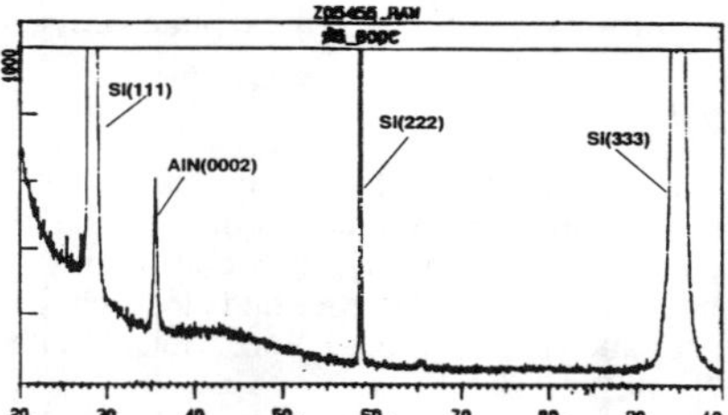

Fig. 13 X-ray diffraction of the final film grown at 800° C.

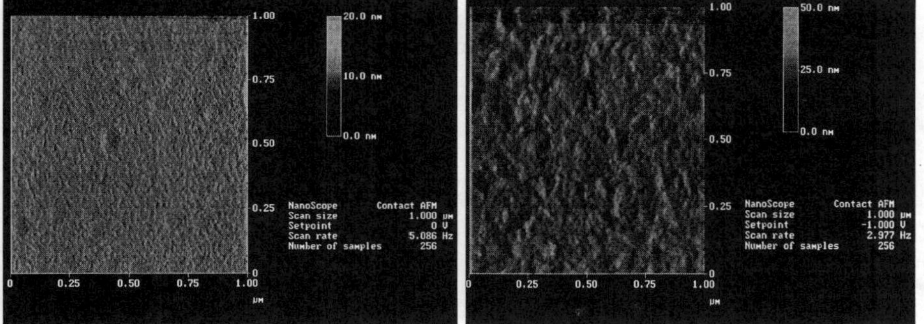

Fig. 14 AFM images show the surface morphology of the film grown at 800 °C.

growth (at 800 °C in this case) eventually lead to a polycrystalline structure. The AFM images in Fig. 14 also shows that the three dimensional growth at 800 °C leads to a much rougher surface compare to that of the initial layer.

Conclusions

We show that high quality AlN films can be obtained at relatively low temperatures as a result of controlled energy ion flux to the Si substrates. Two stage growth, growth at 400 °C and 600 °C gives the best quality AlN film. Also Growth at 800 °C shows that a high quality ultra thin (200 Å) epitaxial AlN was formed on the Si (111) substrate. This layer is probably highly strained. The in-situ RHEED study along with x-ray diffraction and AFM show that a typical growth sequence is probably as followed: i) Formation of a thin epitaxial initial layer, ii) at a certain film thickness, strains probably causes the film to break up to many fragments, and these fragments were randomly oriented, and iii) film continues to grow use the fragmented initial layer as a template. We also conclude that thermal motion help smoothing the surface. However high growth temperature (such as 800 °C) tend to induce a three dimensional growth and a polycrystalline structure.

Acknowledgment

This work is supported by the National Science Foundation (EEC-9420568) and Institute of Manufacturing Research at Wayne State University.

Reference

1. Roman and A.P.M. Adriaansen, Thin Solid Films, Aluminum nitride films made by low pressure chemical vapour deposition: preparation and properties 169. 241(1989).
2. Cox, D. C. Cummins, K. Kawabe, and T. Tredgold, J. Phys. Chem. Solids, 28, On the preparation, optical properties and electrical behavior of an aluminum nitride, 543(1992).
3. S.Strite and H.Morkoc, J.Vac. Sci. Technol. B, GaN, AlN, and InN: A review 10, 1237(1992).
4. Itoh, M. Kato and K. Sugiyama, Thin Solid Films, 146, Plasma-enhanced chemical vapor deposition of AlN coating s on graphite substrates, 255(1987).
5. K.Kubota et al., J. Appl. Phys. 66, Preparation and properties of III-V nitride thin films 2984(1989).
6. H.Windischmann, J. appl. Phys. 62, an intrinsic stress scaling law for polycrystalline thin films prepared by ion beam sputtering 1800(1987).
7. Hentzell, J.M.E. Haper, and J.J. Cuomo, Mater. Res. Symp. Proc. 27, Structure of Al-N films deposited by a dual ion beam process 519(1984).
8. J.S.Morgan, W.A. Bryden, RT.J. Kistenmacher, s.A. Ecelberger, and T.O. Poehler,J.Mater.Res. 5, Single-phase aluminum nitride films by dc-magnetron souttering 2677(1990).
9. Auner, T. D. Lenane, F. Ahmad, in Wide Band Gap Electronic Materials, 329-334(1995)

THE EFFECT OF HYDROGEN ON THE MOLECULAR BEAM EPITAXY GROWTH OF GaN ON SAPPHIRE UNDER Ga-RICH CONDITIONS

S. L. BUCZKOWSKI,[a] ZHONGHAI YU,[a] M. RICHARDS-BABB,[b] N.C. GILES,[a] L. T. ROMANO,[c] AND T. H. MYERS[a]

[a] Physics Department, [b] Chemistry Department, West Virginia University, Morgantown, WV 26506; [c] Xerox Palo Alto Research Center, Palo Alto, CA 94304; tmyers@wvu.edu

ABSTRACT

Nucleation and growth of GaN under Ga-rich conditions by molecular beam epitaxy using a nitrogen rf plasma source is shown to result in both a smoother GaN surface and a reduced inversion domain content. In addition, preliminary results of the dramatic effect of atomic hydrogen on growth kinetics for Ga-rich growth are presented.

INTRODUCTION

GaN typically nucleates and grows on sapphire by island formation. The use of low temperature buffer layers (450 - 600 $^\circ$C) of AlN [1] or GaN [2,3] results in a dramatic improvement in layer morphology and electrical properties. Annealing prior to high temperature growth causes coalescence of the nucleation islands, resulting in low angle grain boundaries which create the observed dislocation arrays [4,5]. It appears that this subgrain structure is stable during growth under most conditions. Thus, subsequent crystal quality is strongly dependent on the nucleation layer. The predominant growth mode is a further factor in dislocation reduction with increasing layer thickness. Two dimensional growth results in the highest degree of structural perfection in epitaxial layer growth. Typical molecular beam epitaxy (MBE) and metal-organic chemical vapor deposition (MOCVD) growth conditions appear to promote three dimensional growth of GaN. [6,7] This may lead to individual growth of the low angle grains, preventing dislocation recombination and annihilation. A recent study has reported MOCVD growth conditions resulting in two-dimensional step-flow growth [8], with a concomitant reduction in dislocation density to about 2×10^8 cm^{-2}. It is desirable to achieve conditions for MBE growth to allow such a growth mode to occur.

One of the obvious differences between MOCVD and MBE growth of GaN is the presence of hydrogen, primarily as a component of the molecule supplying the nitrogen. Most studies have either ignored the effect of hydrogen, or have only considered its effects in compensating p-type dopants such as Mg. For example, the use of ammonia in MBE growth is primarily to obtain a larger flux of active nitrogen through catalytic decomposition on the GaN surface, again with the hydrogen considered only as a reaction by-product. In this paper, we present evidence that: nucleation under Ga-rich conditions results in reduced dislocation and inversion domain content; continued growth under Ga-rich conditions promotes a smoother, two-dimensional growth front; and hydrogen can have a significant effect on the growth kinetics of GaN when the growth is limited by the amount of active nitrogen present.

EXPERIMENT

The GaN layers for this study were grown at West Virginia University by MBE in a system described elsewhere. [9] A standard MBE source provided the Ga flux. A cryogenically-cooled rf plasma source (Oxford Applied Research CARS-25) operating at 600W was used to

Mat. Res. Soc. Symp. Proc. Vol. 449 © 1997 Materials Research Society

produce active nitrogen flux. Atomic hydrogen was produced using a commercial thermal cracker (EPI-AHS). Our source-to-substrate distance was large, about 14 inches, leading to a lower total active-nitrogen flux than typically obtained from an rf plasma source. When scaled by the differences in source-to-substrate distance ($1/r^2$), the active nitrogen flux was comparable to that reported by others. The layers were characterized by Hall measurements, photoluminescence, x-ray diffraction, and atomic force microscopy (AFM) (Digital Instruments Nanoscope II). High resolution transmission electron microscopy was performed at Xerox Palo Alto Research Center using techniques detailed elsewhere [10]

RESULTS

All samples were grown on c-plane sapphire substrates (Union Carbide Crystal Products). Prior to growth, the substrates were degreased and etched in a phosphoric/sulfuric (1:3) acid mixture heated to 80°C. Based on our earlier study,[11] buffer layers were grown by heating the substrate to 730 °C under an atomic hydrogen flux for 20 minutes and then cooling to 630 °C for the growth of a 200 Å thick GaN buffer layer under a Ga flux of 5.0 x 10^{-7} Torr (BEP) with a 6 SCCM nitrogen flow at 600 W. This procedure results in nucleation island sizes around 0.3 to 0.5 μm, giving dislocations due to domain-wall coalescence of order 10^9 cm^{-2}. Buffer layer growth was initiated by simultaneous exposure to the Ga and N flux. The 20 nm nucleation layer was then annealed at 730 °C for 20 minutes under active nitrogen flux, cooled (or heated) to the growth temperature, and growth was resumed. Our conditions represented highly Ga-rich growth for the buffer layer. However, after the 730 °C anneal, examination of buffer layers by AFM indicated continuous films with no evidence of Ga condensation.

GaN layers were grown on the annealed buffer layers. In Fig. 1, the growth rate is plotted as a function of Ga beam equivalent pressure. For an active nitrogen flux in excess or equal to the Ga flux (nitrogen sufficient), one would expect a linear increase in growth rate with increasing Ga flux. This is represented by the solid line in the figure. When the surface ratio of Ga to active nitrogen is increased beyond unity, then either Ga condensation will occur if the desorption rate for excess Ga is low or the growth rate will become fairly constant with increasing Ga flux if the excess Ga desorption rate is large enough to prevent condensation. The surface morphology was distinctly different for layers grown under Ga-rich conditions as compared to layers grown closer to equal Ga and active nitrogen flux. Nitrogen sufficient conditions gave a highly textured, three-dimensional surface, as shown

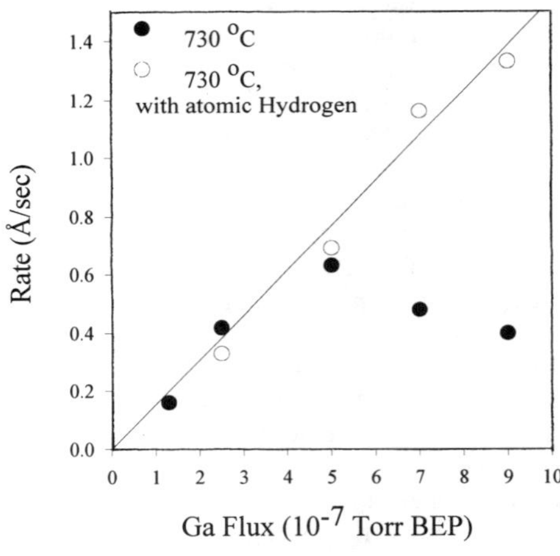

Figure 1. Growth rate v. Ga-flux.

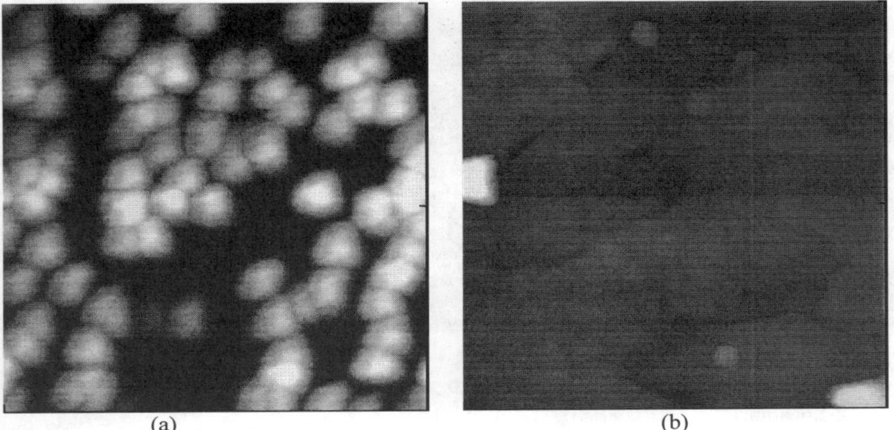
(a)   (b)

Figure 2. AFM micrographs of GaN grown under (a) nitrogen-sufficient and (b) Ga-rich conditions. The areas represented are 2 by 2 μm^2. The height scales are 135 nm in (a) and 50 nm in (b).

by the micrograph in Fig. 2(a). Such morphology is similar to that reported earlier [6] for the growth of GaN by MBE using an identical rf nitrogen source. Ga-rich growth close to, but below the Ga condensation point resulted in much smoother surfaces indicative of two-dimensional growth with an rms. surface roughness of 1.5 nm over a 15 μm span. The smoother texture can be seen in the AFM micrograph in Fig. 2(b) for a 250 nm thick layer. Sub-nanometer tall terraces representing atomic steps were present on the top of the "flat" regions of the sample.

This change in surface morphology is very similar to that reported earlier [12] for growth of GaN using an electron cyclotron resonance microwave (ECR) plasma source. In the ECR study, the amount of active nitrogen was changed by controlling the plasma density. Lower power led to smooth surfaces, while high powers resulted in textured surfaces. It is known that, in addition to increasing the active nitrogen flux, high power operation of an ECR source can lead to a significant amount of high energy ions in the nitrogen flux which could influence the growth morphology. In our study, the nitrogen source parameters were unchanged for the samples shown in Fig. 2 while the Ga flux was altered. Thus, the agreement between the two studies suggests that the change from smooth to textured growth is a universal phenomenon related to the ratio of active nitrogen to gallium at the growth front.

The sample whose AFM micrograph is shown in Fig. 2(b) was also measured by high resolution TEM. As shown in Fig. 3, the microstructure is characterized by a high density of threading dislocations at the GaN-sapphire interface. These quickly annihilate resulting in a line defect density in the low 10^9 cm^{-2} after growth of only about 250 nm of GaN. By comparison, samples grown under nitrogen-sufficient conditions had line defect densities in the mid-10^{10} cm^{-2}. This decrease can be attributed to the layer-by-layer growth mode indicated by the smooth surface morphology measured by AFM. Thicker layers should result in commensurably fewer dislocations. Perhaps of more significance, the sample exhibited a low concentration of inversion domain boundaries, where the GaN crystal structure is inverted along the c-axis due to nucleation of the opposite phase at the substrate. As reported previously, inversion domain

199

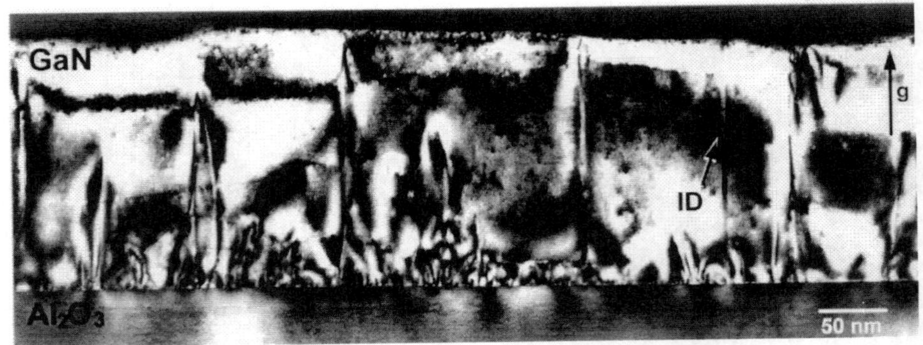

Figure 3. Dark Field TEM image taken with g = (0002) near the [11-20] zone axis of GaN. ID indicates the inversion domain in this micrograph. The remaining defects are nonedge-type dislocations.

boundaries are a significant source of defects in GaN grown by any technique. [10] Our direct nucleation of GaN on sapphire under Ga-rich conditions appears to suppress the formation of such inversion domains, resulting in significantly less than 10 % by volume. This is comparable to most MOCVD layers. In contrast, ECR-MBE layers grown on nitrided sapphire contain about an equal number of both orientations.[10]

Increasing the growth temperature should also increase the surface mobility of adatoms and promote better layer-by-layer growth, which may increase the rate of dislocation clean-up. Unfortunately, we [9] and others [13] have observed that the GaN growth rate decreases significantly at temperatures above about 730 ^{o}C. Further increase in temperature apparently either reduces the residence time of unreacted N on the surface or increases the Ga desorption rate, thereby reducing the growth rate. Prior studies of Ga desorption for GaN growth using ammonia [14] indicate that the Ga desorption rate increases rapidly above 730 ^{o}C. Thus, while this temperature can be increased somewhat by increasing the amount of active nitrogen, the regime between 750 and 800 ^{o}C may represent a practical limit to the growth temperature feasible by MBE using nitrogen plasma sources.

Introduction of atomic hydrogen resulted in a dramatic increase in the growth rate of the GaN under Ga-rich conditions, as also summarized in Fig. 1. The samples whose growth rates are indicated by the open symbols in Fig. 1 were grown under a total (atomic and molecular) hydrogen flux of 1 to 2 x 10^{-6} Torr BEP. The other growth parameters remained the same as for the corresponding filled-symbol case. Note that the growth rate was essentially brought up to the Ga-limit for each case of Ga-rich growth. In contrast, samples grown under nitrogen-sufficient conditions did not exhibit an enhanced growth rate. To see if the increased growth rate originated with molecular hydrogen, samples were grown under hydrogen flux with the cracker turned off. The resulting samples exhibited identical growth rates to the GaN grown without hydrogen, indicating that molecular hydrogen is not significantly affecting the growth kinetics. Also, to see if the atomic hydrogen was possibly forming active species with molecular nitrogen, an attempt was made to grow under an atomic hydrogen flux with the rf power turned off on the nitrogen source. The resulting GaN growth rate, if non-zero, was too small to be detected. Ammonia was not detected in an analysis of the background gases present during growth. This indicates that the enhanced growth rate is not due to the presence of ammonia formed due to gas-phase reactions.

The increase in growth rate for Ga-rich conditions is apparently related to the presence of atomic hydrogen at the surface of the growing layer. We propose that the atomic hydrogen becomes loosely bonded to the growing GaN surface. Nitrogen atoms adsorbed on the surface are then attracted by this hydrogen layer, resulting in an increased nitrogen residence time. The longer residence time increases the probability that a Ga atom will diffuse to within an interaction distance of the nitrogen, and thus enhance the growth rate of GaN. Thus, the atomic hydrogen could be increasing the effective active nitrogen concentration. In addition, the surface morphology for samples grown under atomic hydrogen more nearly resembled that shown as Fig. 2(a). This, along with the increased growth rate, is consistent with shifting the growth kinetics towards a more nitrogen-sufficient case. Of practical importance, this shifting of the growth kinetics may allow much higher growth temperatures for MBE of GaN. We are continuing to investigate this phenomenon.

PL measurements were made at liquid helium temperatures on the samples. All samples grown at or below 660 $^{\circ}$C exhibited significant luminescence centered at about 560 nm (2.2 eV). While the origin of this broad yellow luminescence band is still controversial, it is commonly attributed to deep states in the bandgap involving impurities or native defects. As reported earlier, [9] layers grown at 730 $^{\circ}$C did not exhibit detectable yellow luminescence. We take this as direct evidence that the higher growth temperatures are necessary to produce high quality material. The PL obtained from a GaN layer grown at 660°C under an atomic hydrogen flux is shown in Fig. 4. There is a single bound-exciton peak at about 3.48 eV and a small amount of yellow luminescence. As with the undoped layers, raising the growth temperature above 730 $^{\circ}$C eliminated the yellow PL for samples grown under hydrogen, [9] indicating that growth under hydrogen does not degrade layer quality.

The best GaN samples grown without atomic hydrogen were grown under Ga-rich conditions at 730 $^{\circ}$C and exhibited x-ray diffraction rocking curve full widths at half maximum (FWHM) between 2 and 3 arc minutes for samples with thickness about 1 μm. Hall measurements indicated n-type carrier concentrations as low as 1 x 10^{17} cm^{-3} with room temperature mobilities of 120 cm^{2}/V·s. The x-ray FWHM and Hall measurements of samples grown with atomic hydrogen were indistinguishable from those grown without atomic hydrogen. Although previously reported that hydrogen may introduce donor levels in GaN,[15] we have no direct evidence that this is the case here. As shown in Fig. 4 and in previously published spectra from samples grown at higher temperatures, we did not observe the 3.35 eV PL feature which has been associated with the hydrogen-related donor.[15] In addition, we did not

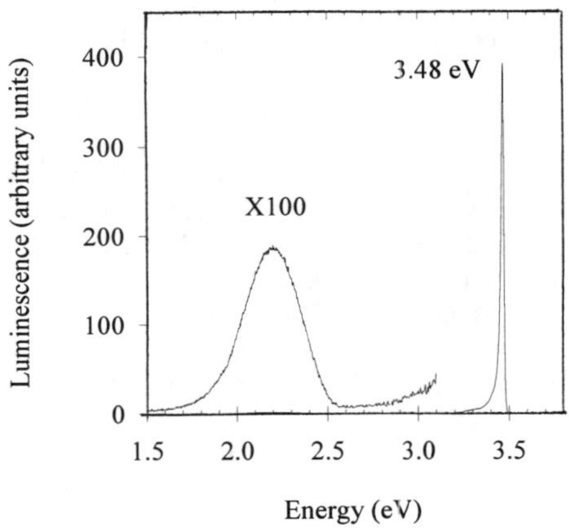

Figure 4. Photoluminescence of GaN grown at 660 $^{\circ}$C under an atomic hydrogen flux.

see any significant difference in background doping between layers grown with and without hydrogen.

CONCLUSIONS

In conclusion, we have demonstrated that the presence of atomic hydrogen can have a significant effect on the growth rate of GaN under Ga-rich conditions, resulting in an increase in growth rate limited by the total Ga flux. PL and Hall measurements indicate that layer quality is not degraded by growth under atomic hydrogen. The increased growth rate and change to a textured surface morphology are suggestive that the atomic hydrogen increases the effective surface concentration of nitrogen. In addition to the hydrogen-related growth rate enhancement, we have presented evidence that Ga-rich nucleation and growth results in a significant reduction of inversion domain boundaries, promotes dislocation reduction and promotes a smoother growth surface.

ACKNOWLEDGMENTS

This work was supported by DoD/ONR Grant N00014-94-1-1149, monitored by M. Yoder and ONR Grant N00014-96-1-1008 monitored by C. Wood.

REFERENCES

1. H. Amano, N. Sawaki, I. Asaki, and Y. Toyoda, Appl. Phys. Lett. **48**, 353 (1986).
2. S. Nakamura, Jpn. J. Appl. Phys. **30**, 1705 (1991).
3. J. N. Kuznia, M. Asif Khan, D. T. Olson, R. Kaplan, and J. Freitas, J. Appl. Phys. **73**, 4700 (1993).
4. W. Qian, M. Skowronski, M. DeGraef, K. Doverspike, L. B. Rowland, and D. K. Gaskill, Appl. Phys. Lett. **66**, 1252 (1995).
5. S. D. Lester, F. A. Ponce, M. G. Crafford, and D. A. Steigerwald, Appl. Phys. Lett. **66**, 1249 (1995).
6. C. Trager-Cowan, K.P. O'Donnell, S.E. Hooper and C.T. Foxon, Appl. Phys. Lett. **68**, 355 (1996).
7. Z. Sitar, L. L. Smith, and R. F. Davis, J. Cryst. Growth **141**, 11 (1994).
8. D. Kapolnek, X. H. Wu, B. Heying, S. Keller, B. P. Keller, U. K. Mishra, S. P. DenBaars, and J. S. Speck, Appl. Phys. Lett. **67**, 1541 (1995).
9. Zhonghai Yu, S.L. Buczkowski, N.C. Giles, T.H. Myers and M.R. Richards-Babb, Appl. Phys. Lett. **69**, 2731 (1996).
10. L.T. Romano, J.E. Northrup and M.A. O'Keefe, Appl. Phys. Lett. **69**, 2394 (1996).
11. M. Richards-Babb, S.L. Buczkowski, Zhonghai Yu, And T.H. Myers, Mater. Res. Soc. Symp. Proc. **395**, 237 (1996).
12. T.D. Moustakas and R.J. Molnar, Mat. Res. Symp. Proc. **281**, 753 (1993).
13. See, for example, S. Guha, N.A. Bojarczuk and D.W. Kisher, Appl. Phys. Lett. **69**, 2879 (1996).
14. C.R. Jones, T. Lei, R. Kaspi and K.R. Evans, Mater. Res. Soc. Symp. Proc. **395**, 141 (1996).
15. M.S. Brandt, N.M. Johnson, R.J. Molnar, R. Singh and T.D. Moustakas, Appl. Phys. Lett. **64**, 2264 (1996).

A STUDY OF MIXED GROUP-V NITRIDES GROWN BY GAS-SOURCE MOLECULAR BEAM EPITAXY USING A NITROGEN RADICAL BEAM SOURCE

W. G. BI and C. W. TU
Department of Electrical and Computer Engineering, University of California, San Diego, La Jolla, CA 92093-0407
D. MATHES and R. HULL
Department of Materials Science and Engineering, University of Virginia, Charlottesville, VA 22903-2442, U. S. A.

ABSTRACT

We report a study of N incorporation in GaAs and InP by gas-source molecular beam epitaxy using a N radical beam source. For GaNAs grown at high temperatures, phase separation was observed, as evidenced from the formation of cubic GaN aside from GaNAs. By lowering the growth temperature, however, GaNAs alloys with N as high as 14.8% have been obtained without showing any phase separation. For InNP, no phase separation was observed in the temperature range studied (310 ~ 420 °C). Contrary to GaNAs, incorporating N in InP is very difficult, with only less than 1% N being achieved. Optical absorption measurement reveals strong red shift of bandgap energy with direct-bandgap absorption. However, no semimetallic region seems to exist for GaNAs and a composition-dependent bowing parameter has been observed.

INTRODUCTION

Recently much attention has been paid to the study of a new material family, i.e., mixed group-V nitrides [1-11]. The main driving force for this is to investigate whether one can obtain direct-bandgap materials lattice-matched to Si for the purpose of integrating optical devices with the well established Si electronics. As the lattice constant of Si lies between those of nitrides and arsenides or phosphides, in principle it is possible to combine these nitrides with arsenides or phosphides to form materials lattice-matched to Si. For example, both $GaN_{0.2}As_{0.8}$ and $InN_{0.49}P_{0.51}$ can be lattice-matched to Si. Nevertheless, due to the large miscibility gap [11,12], these materials tend to phase-separate when the N composition is appreciable [13]. In addition, recent calculations [14] based on thermal equilibrium conditions show that the N solubility limit in III-Vs is very small (10^{13}~10^{20} cm^{-3} at 900 K). Experimental studies on GaNAs so far also indicate that only a small amount of N can be incorporated in GaAs. Weyers et al. [2,3] obtained 1.6% N in GaN_xAs_{1-x} using plasma-assisted metalorganic chemical vapor deposition (MOCVD) and NH_3 cracked in a microwave plasma as the N source. While Orton et al. [13] reported 20% N incorporation in GaAs grown by MBE using dissociated N_2 as the N source, the material is not an alloy but phase-separated. Theoretical calculations by Sakai et al. [15], based on the Van Vechten model [16,17], predict a negative bandgap region for GaNAs. Recent first-principle supercell calculations by Wei. et al. [18] and Bellaiche et al. [19] show the contrary, i.e., the alloy is a semiconductor, rather than a semimetal, in the whole composition range. It is our purpose in this work to investigate N incorporation in GaAs and InP, and the variation of the bandgap energy of these alloys.

Mat. Res. Soc. Symp. Proc. Vol. 449 © 1997 Materials Research Society

EXPERIMENTAL

Using a modified Varian Gen-II Molecular beam epitaxy (MBE) system equipped with two 2200 l/s cryopumps, the GaNAs samples were grown either on undoped (100) GaP substrates or on semi-insulating (100) GaAs substrates while the InNP samples were grown on semi-insulating (100) InP substrates. Ultra-high purity N_2 was injected through a N radical beam source (Oxford Applied Research Model MPD21) operated at radio frequency (*rf*) of 13.56 MHz and a *rf* power of 300 W to generate active N species. 7N elemental Ga, In and thermally cracked AsH_3, PH_3 were used. The film thickness was 0.5 µm. The growth temperature was varied from 500 to 650 °C for GaNAs and from 310 to 420 °C for InNP, respectively. After thermal cleaning of the substrates under a group-V overpressure (P_2 or As_2) to remove surface oxide layers, a 0.1 µm-thick buffer layer was grown first. Then the substrate temperature was lowered to the growth temperature, and a N plasma was ignited. The initial N_2 flow rate was set at 3 sccm in order to strike a plasma, and the background pressure was around 5×10^{-5} torr. During growth the background pressure was around $1.5 \sim 3\times10^{-5}$ torr depending on the N_2 flow rate used (0.54 ~ 2.1 sccm). High-resolution X-ray rocking curve measurement (XRC) was performed using a Phillips X-ray diffractometer. The N compositions in GaNAs and InNP were determined from (511) asymmetric reflections to take into account the strain-induced lattice constant change of the GaN_xAs_{1-x} and InNP films [20]. X-ray θ-2θ diffraction pattern was carried out using a Rigacu X-ray diffractometer. Optical absorption measurements were performed using a broadband halogen lamp, and the signal was detected at the exit of a monochromator by either a Si or a Ge photodiode depending on the wavelength studied.

RESULTS AND DISCUSSION

1. N incorporation in GaAs and InP

Because N-containing mixed group-V compounds tend to phase-separate, we first investigated this issue using X-ray θ-2θ diffraction measurement. The results of two GaNAs samples grown at 650 and 500 °C are shown in Figs. 1(a) and (b), respectively. As can be seen here, the X-ray θ-2θ curve of the GaNAs sample grown at 650 °C shows diffraction peaks due to GaN, indicating phase-separation of the film. This is similar to what has been observed by Orton et al.[13]. On the other hand, the sample grown at 500 °C shows no separation, as evidenced by the absence of X-ray diffraction peaks associated with cubic GaAs, cubic GaN, or hexagonal GaN. This might be due to the fact that although an extremely large miscibility gap exists for GaNAs, metastable alloys can be formed by using non-equilibrium growth techniques such as Gas-Source MBE. At low growth temperatures the thermal energy may not be large enough for the decomposition of the alloy while at high growth temperatures, the decomposition will happen, leading to the phase separation of GaNAs. The N concentration in GaNAs was found to increase with decreasing growth temperature. In fact, the sample grown at 500 °C has a N composition of 14.8%, as determined from X-ray (511) rocking curves. This value is the highest reported to date for a GaNAs alloy. The single crystallinity of the film was also confirmed by transmission electron microscopy (TEM), where no phase separation was observed. As the N solubility limit in GaAs is predicted to be very small by a theory based on

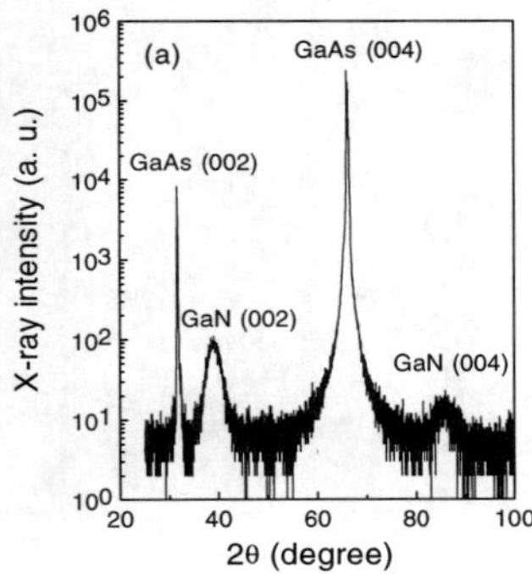
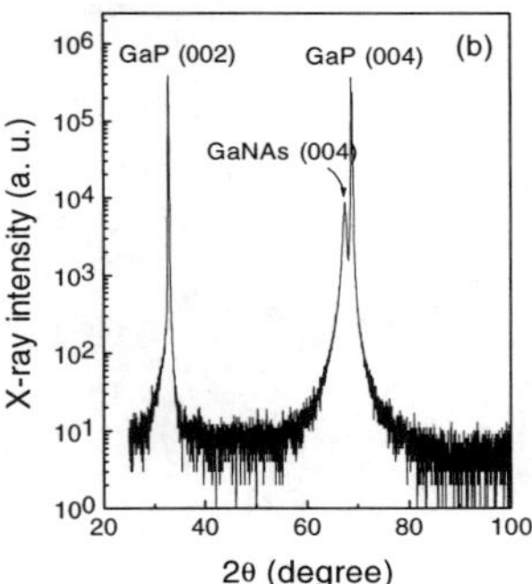

Fig. 1 X-ray θ-2θ curves for GaNAs (a) grown at 650 °C, showing phase separation, and (b) grown at 500 °C, showing no phase separation. The composition of (b) is 14.8%, the highest ever achieved so far.

equilibrium conditions and on an assumption that the lattice is relaxed around a nitrogen atom to the 6th neighbor in a zincblende nitride structure[22,23], while experimentally no phase separation was observed even for 14.8% N incorporation in GaNAs, we believe that by using non-equilibrium growth techniques, such as GSMBE, GaNAs alloys with N concentration far more than the thermal equilibrium solubility limit can be achieved.

To study the issue of phase-separation in InNP, we used the same method. The result of a typical X-ray θ-2θ scan profile for an InNP film is shown in Fig. 2 (a). Aside from the InP (002) and (004) peaks, no other X-ray peaks associated with cubic InN (002), (004), or hexagonal InN (0002), (0004) are present, indicating no phase separation. The reason why no InNP (002) and (004) peaks are seen in the θ-2θ scan is because of the relatively large scanning step used in the X-ray θ-2θ measurement. Using high-resolution XRC, the InNP epilayer peak is clearly seen, as shown in the inset of Fig. 2 (a). The N composition of the film determined from (511) XRCs is 0.53%. These results confirm the formation of an InNP alloy. Fig. 2 (b) is the (110) zone axis TEM diffraction pattern of the film. It shows single crystal. Besides, no extra spots are seen, suggesting single phase of the film.

One thing worth mentioning here is that although thermodynamic considerations[14] suggest that a much higher N incorporation in InP than in GaAs can be achieved, the highest N composition in InNP obtained so far is only 0.93%, while for GaNAs, as high as 14.8% N has been incorporated. This is due to the much higher equilibrium vapor pressure of N_2 over InN compared to that over GaN[23,24]. Therefore, the active N species needed for incorporating the

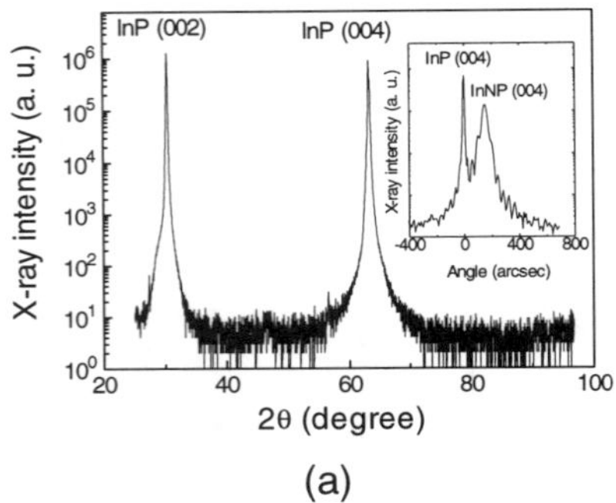

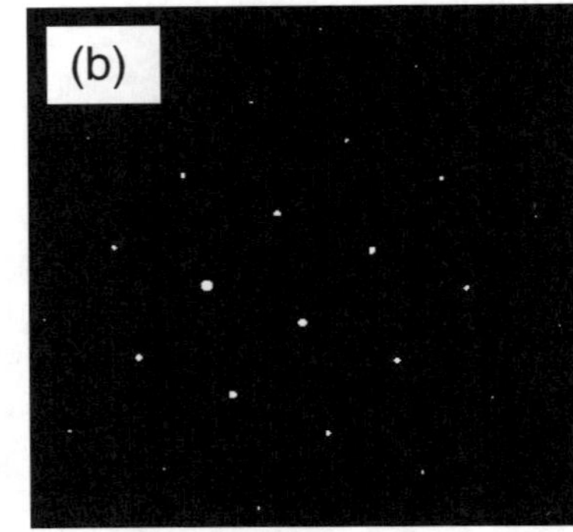

Fig. 2 (a) X-ray θ-2θ diffraction profile and (b) (110) zone axis TEM diffraction pattern of an InN$_x$P$_{1-x}$ (x=0.53%). The inset in (a) is the high-resolution (004) XRC result.

same N composition in InNP as that in GaNAs must be much larger for InNP than for GaNAs. In other words, with the same amount of active N species, less N can be incorporated in InNP than in GaNAs.

2. Bandgap bowing

Fig. 3 shows the bandgap of GaNAs as a function of N composition for our data (open triangles) and those of Kondow et al.[25] (solid squares). The dashed line is a best fit through these data and that of cubic GaN, and it shows that there may not be a semimetallic region. Recently Wei and Zunger[18] performed a first-principle supercell calculation of the bandgap energy of ordered GaNAs alloys. Owing to the large differences between atomic size and orbital energies of the As and N atoms, spatially separated and sharply localized band edge states are formed in the alloy, leading to the conclusion that for isovalent semiconductor alloys, the bandgap variation with composition can be divided into two regions: (i) an impurity-like region, at low concentrations, where the bowing coefficient is considerably larger and dependent on composition, and (ii) a band-like region, at high concentrations, where the bowing coefficient is relatively small and nearly constant. For materials like GaAsP, the transition concentration is very small and unobservable because of the similar size and energy levels of As and P, but for materials like GaNAs, the transition concentration can be a few percent. This can be seen more clearly in Fig. 4, where our data are plotted as solid squares, Wei and Zunger's calculation for ordered GaNAs alloys[18] as solid triangles, and Bellaiche et al.'s calculation for random alloys[19] as solid circles. The bowing parameter c(x) is determined from the equation

$$c(x) = [xE_g(GaN)+(1-x)E_g(GaAs)-E_g(GaN_xAs_{1-x})]/[x(1-x)]$$

Our data clearly show the impurity-like and band-like regions.

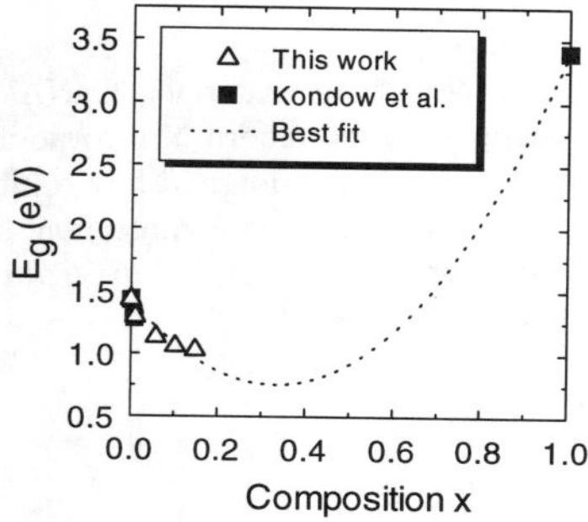

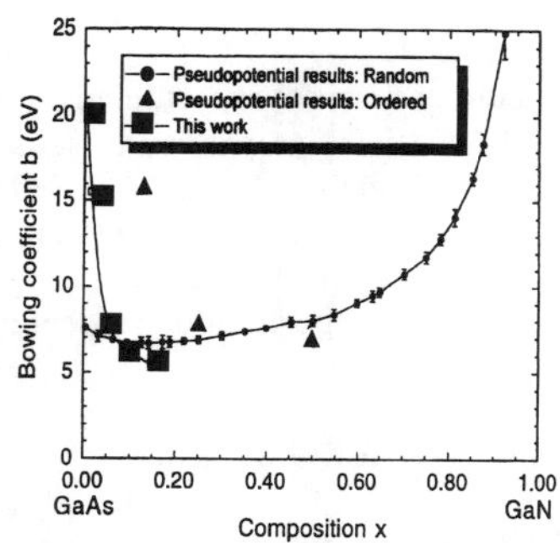

Fig. 3 The N-composition dependence of GaNAs bandgap energies. Our data are shown in open triangles and those of Kondow et al.[25] in solid squares. The bandgap of cubic GaN is also shown. The best fit indicates no semimetallic region.

Fig. 4 The bowing parameter of GaNAs as a function of the N composition. Our data are shown in solid squares, and the calculations of Zunger and his coworkers are shown in closed triangles for ordered GaNAs[18] and closed circles for random GaNAs[19].

The bandgap energy of InN_xP_{1-x} is shown in Fig. 5 as a function of the N composition. Our experimental data are shown as solid squares, and the solid line is a calculation using the Van Vechten's model of the dielectric theory of electronegtivity[16,17]. Similar to GaNAs, adding N in InNP pulls down the bandgap energy to longer wavelength, revealing a big bowing of the bandgap energy. Besides, our data follow the theory within the N composition range studied (x < 1%). However, as the N concentration in these InNP is very small, the agreement for larger N compositions still needs experimental verification, and further studies on samples with more N concentrations are needed to study the bowing parameter of this alloy.

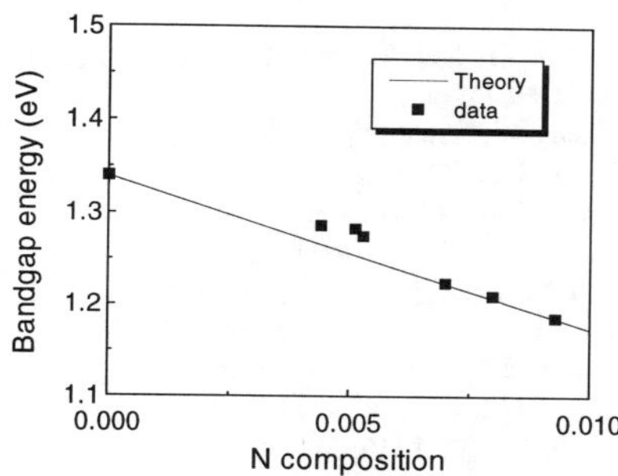

Fig. 5. Bandgap energy of InN_xP_{1-x} as a function of N composition x. The solid line is the result calculated from the Van Vechten model, and the solid squares are experimental data.

SUMMARY

In summary, we have studied N incorporation in two direct-bandgap materials, GaNAs and InNP, that in principle can be lattice-matched to Si, and have achieved record N incorporation, i.e., 14.8% in GaNAs, and 0.93% in InNP. The problems of phase-separation and low solubility limit of N in GaAs and InP are overcome by low-temperature gas-source molecular beam epitaxial growth. We also observed a big bowing coefficient for these mixed group-V nitrides. Besides, a composition-dependent bowing parameter is observed for GaNAs.

ACKNOWLEDGMENTS

We would like to thank Texas Instruments Systems Laboratory for the support of this work.

REFERENCES

[1] X. Liu, M. E. Pistol, L. Samuelson, S. Schwetlick, and W. Seifert, Appl. Phys. Lett. **56**, 1451 (1990).
[2] M. Weyers, M. Sato, and H. Ando, Jpn. J. Appl. Phys. **31**, L853 (1992).
[3] M. Weyers and M. Sato, Appl. Phys. Lett. **62**, 1396 (1993).
[4] M. Kondow, K. Uomi, T. Kitatani, S. Watahiki, and Y. Yazawa, J. Cryst. Growth **164**, 175 (1996).
[5] J. N. Baillargeon, K. Y. Cheng, G. E. Holfler, P. J. Pearah, and K. C. Hsieh, Appl. Phys. Lett. **60**, 2540 (1992).
[6] J. N. Baillargeon, P. J. Pearah, K. Y. Cheng, G. E. Holfler, and K. C. Hsieh, J. Vac. Sci. Technol. **B10**, 829 (1992).
[7] X. Liu, S. G. Bishop, J. N. Baillargeon, and K. Y. Cheng, Appl. Phys. Lett. **63**, 208 (1993).
[8] S. Miyoshi, H. Yaguchi, K. Onabe, and R. Ito, Appl. Phys. Lett. **63**, 3506 (1993).
[9] M. Kondow, K. Uomi, K. Hosomi, and T. Mozume, Jpn. J. Appl. Phys. **33**, L1056 (1994).
[10] M. Sato, J. Cryst. Growth **145**, 99 (1994).
[11] G. B. Stringfellow, J. Cryst. Growth **27**, 21 (1974).
[12] L. G. Ferreira, S. -H. Wei, and A. Zunger, Phys. Rev. B **40**, 3197 (1989).
[13] J. W. Orton, D. E. Lacklison, N. Baba-Ali, C. T. Foxon, T. S. Cheng, S. V. Novikov, D. F. C. Johnston, S. E. Hooper, L. C. Jenkins, L. J. Challis, and T. L. Tansley, J. Electron. Mat. **24**, 263 (1994).
[14] Y. C. Kao, T. P. E. Broekaet, H. Y. Liu, S. Tang, I. H. Ho, and G. B. Stringfellow, Mater. Res. Soc. Symp. Proc. **423**, 335 (1996).
[15] S. Sakai, Y. Ueta, and Y. Teauchi, Jpn. J. Appl. Phys. **32**, 4413 (1993).
[16] J. A. Van Vechten, Phys. Rev. **182**, 891 (1969).
[17] J. A. Van Vechten, Phys. Rev. **187**, 1007 (1969).
[18] S. H. Wei and A. Zunger, Phys. Rev. Lett. **76**, 664 (1996).
[19] L. Bellaiche, S. H. Wei, and Alex Zunger, Phys. Rev. B (to be published).
[20] W. G. Bi, F. Deng, S. S. Lau, and C. W. Tu, J. Vac. Sci. Technol. B **13**, 754 (1995).
[21] M. D. Sturge, E. Cohen, and K. F. Rodgers, Phys. Rev. B **15**, 3169 (1977).
[22] Q. X. Zhao and B. Monemar, Phys. Rev. B **38**, 1397 (1988).
[23] T. Matsuoka, J. Crsyt. Growth **124**, 433 (1992).
[24] J. B. Macchesney, P. M. Bridenbaugh, and P. B. O'Connor, Mater. Res. Bull. **5**, 783 (1970).
[25] M. Kondow, K. Uomi, K. Hosomi, and T. Mozume, 8th Int. MBE Conf., Paper A11-2, Osaka, Japan, August 29- September 2, 1994.

KINETICS OF NITROGEN IN GaAsN LAYERS DURING GaAs OVERGROWTH

Z. Z. BANDIĆ,* R. J. HAUENSTEIN,** M. L. O'STEEN,** and T. C. MCGILL*
* Watson Laboratories of Applied Physics 128-95
California Institute Of Technology, Pasadena, California 91125
** Department of Physics, Oklahoma State University, Stillwater, OK 74078

ABSTRACT

A set of GaAsN/GaAs strained layer superlattices was grown on GaAs (001) substrates by electron cyclotron resonance microwave plasma assisted molecular beam epitaxy. An *ex-situ* high resolution X-ray diffraction was employed to characterize interface quality and determine the effective nitrogen content, as a function of growth temperature, of buried GaAsN layers. The interface quality, which was assessed through FWHM's and intensities of superlattice peaks, and nitrogen content were found to change dramatically in the temperature range from 540 °C to 580 °C. A first order kinetic model was introduced to quantitatively explain this dependencies, in terms of energetically favorable N for As anion exchange and thermally activated N desorption and segregation processes. The strong nitrogen surface segregation process, which acts concurrently with nitrogen desorption is found to be responsible for both degradation of interface quality and reducing nitrogen content at elevated temperatures. The predicted nitrogen profile smearing of $(2-3)$nm, obtained from the model, is found to be in good agreement with cross-sectional transmission electron microscopy observations.

INTRODUCTION

Important technological applications of GaN, related alloys and heterostructures for blue and ultra violet (UV) light emitting diodes and lasers motivated intensive research and development activity. [1, 2, 3]. Most of the research concentrated on the growth of device structures,[4, 5] and epitaxial growth techniques of nitrides on various substrates.[4, 6]. GaAs, as one of the potentially interesting substrates, attracted some attention in the last few years,[7, 8, 9] mainly to circumvent some of the existing problems with Al_2O_3 and SiC. Besides as a substrate, GaAs can also be interesting for mixed anion nitride/arsenide systems, like GaAsN alloys and heterostructures.[10, 11, 12, 13, 14] Although limited by a small solubility[15], even small amounts of N in GaAs can cause large band gap bowing to infrared[15, 17]. There are several important microscopic processes which have to be understood in order to achieve control of the structural and chemical properties of the GaN/GaAs interface. Nitridation, or N for As exchange is usually the first step in growing GaN on GaAs substrates, [8, 9, 11] analogously to the growth of AlN buffer on Al_2O_3.[16] Another important phenomenon, nitrogen surface segregation occurs during GaAs overgrowth of GaAsN layer.[12, 13] We have recently shown[11, 12] that it is possible to produce high quality strained layer $GaAs_{1-y}N_y$/GaAs superlatices, grown on the (001) GaAs substrates. The initial high resolution X-ray diffraction (HRXRD) and reflection high energy electron diffraction (RHEED) characterization revealed the existence of several microscopic, thermally activated processes. However, full understanding of these processes and their influence on the interface properties has not been achieved yet. In this study, the interface quality and nitrogen content in $GaAs_{1-y}N_y$/GaAs superlattices were characterized by HRXRD. Dramatic changes of the nitrogen content and interface sharpness with temperature were observed. The quantitative model, based on the first order kinetic theory, was employed to explain our observation. The

Mat. Res. Soc. Symp. Proc. Vol. 449 © 1997 Materials Research Society

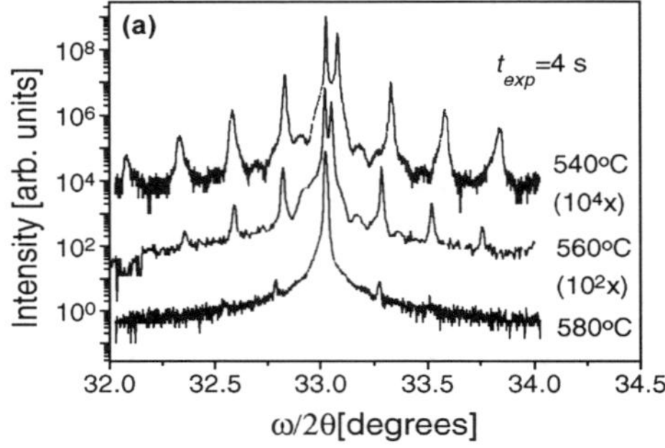
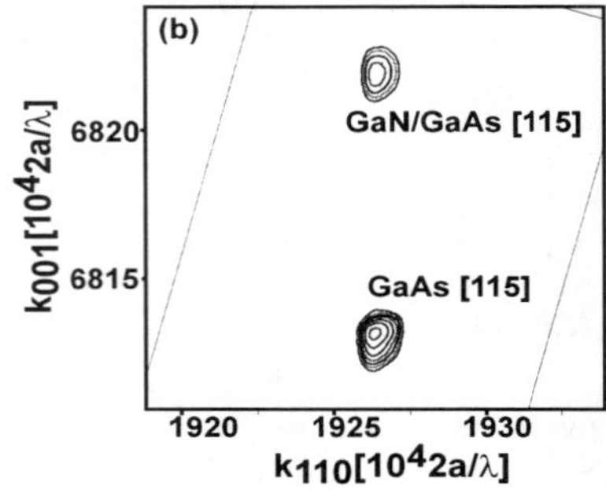

Figure 1: (a) The $\omega/2\theta$ rocking curves of the GaAsN/GaAs superlattices grown at temperatures between 540 °C and 580 °C, with 4 s nitrogen exposure time. The strong degradation of superlattice X-ray diffraction with temperature increase can be observed. (b) The reciprocal area scan for the superlattice grown at 550 °C. The superlattice is coherently strained and is of good structural quality, as observed from the circular shape of reciprocal space points. The contours are lines of equal intensities, which increase exponentially with base 2.

thermally activated nitrogen segregation process, acting concurrently with nitrogen desorption was identified to induce smearing of the interface. The nitrogen segregation length, derived from the model, is found to increase with the temperature, in agreement with HRXRD data. Transmission electron microscopy (TEM) experiments confirmed these observations.

EXPERIMENTAL DETAILS

A set of $GaAs_{1-y}N_y$/GaAs superlattices, with a superlattice period consisting of 75 monolayers of GaAs, followed by one monolayer of $GaAs_{1-y}N_y$ was grown on GaAs (100) substrates as a function of temperature (540–580 °C) in molecular beam epitaxy (MBE) system using an electron cyclotron resonance (ECR) nitrogen plasma source. $GaAs_{1-y}N_y$ monolayers were produced through brief (4–6 s) exposure to N* plasma of an As-stabilized GaAs surface, followed by GaAs overgrowth at a fixed growth rate (0.75 ML/s) for 100 s. The GaAs surface is then As-stabilized through exposure to As_2 flux for 30 s. Further details about growth procedures and ECR-MBE growth system can be found elsewhere. [9, 11] The entire sample set was characterized by *ex-situ* HRXRD, using Philips Materials Research Diffractometer. The (110) cross-section of one of the samples, grown at 550 °C, was observed by TEM.

RESULTS AND DISCUSSION

The $\omega/2\theta$ scans are performed in the 4-crystal mode using Ge(220) reflections, around the substrate (004) reflections. The resultant rocking curves are shown in Fig. 1(a). They reveal a strong dependence on the growth temperature, indicating a presence of thermally activated processes. The reciprocal space scans (area scans) around the substrate (115) reflections are done in the same mode, with the additional use of the Bonse-Hart collimator in front of the detector. The area scan for sample grown at 550 °C is shown in Fig. 1(b). It shows that superlattice is coherently strained and is of good structural quality. The width of the reciprocal space peaks parallel to the layer is small, indicating that tilting of the lattice layers is not present. The nitrogen contents y of the $GaAs_{1-y}N_y$ layers, normalized to one monolayer, are determined from superlattice peak positions. Fig. 2(a) shows the Arhennius

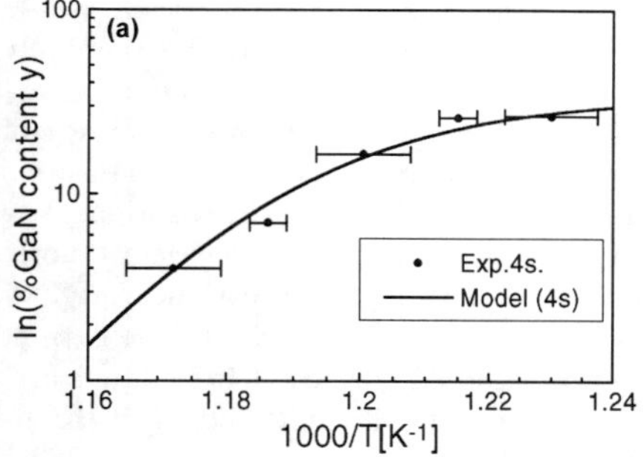
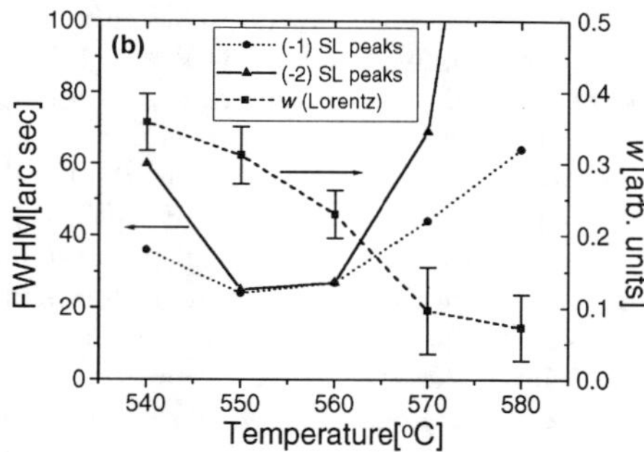

Figure 2: (a) Arhennius plot of the nitrogen content y observed in ECR-MBE grown GaAsN/GaAs superlattices. Experimental data points are obtained through HRXRD measurements for 4 s nitridations. The solid curve is fit from the model[Eq. (1)]. (b)The FWHM's of the -1 and -2 superlattice peaks (left vertical axis)as a function of temperature. The Lorentzian fit $Aw/(n^2 + w^2)$ of the dependence of the superlattice peak intensities as a function of the order n of superlattice diffraction gives parameter w (right vertical axis). They both show deterioration of structural properties and interface sharpness at high temperatures.

plot of the resultant compositional dependence on growth temperature, $y(T)$, for 4 s nitridations, including experimental uncertainties in temperature, which represent reproducibility from run to run within the sample set. The interface structural quality is assessed through the temperature dependence of the FWHM's and intensities of superlattice peaks, which are both shown schematically in Fig. 2(b). FWHM's of the -1 and -2 superlattice peaks as a function of temperature are shown in the Fig. 2(b) using the left vertical scale. The rise of FWHM's at 540 °C is attributed to a temperature too low for GaAs growth. The dependence of the superlattice peak intensities from the peak order is a measure of the sharpness of the interface. The sharp interfaces give rise to the superlattice rocking curves with a lot of peaks and slow peak decay with peak order, while superlattices with diffuse interfaces have few peaks which quickly become indistinguishable in the higher order, with the best example being alloys which only have a crystalline lattice peak. The intensity of the superlattice peaks as a function of the order of peak is fitted to the Lorentzian curve $Aw/(n^2 + w^2)$, where n is an order of the superlattice peak, w is the width of the Lorentzian curve and A the multiplicative constant. The reason why the Lorentzian curve is chosen is only to provide a parameter which could describe peak intensity as a function of peak order. The parameter w thus represent the quantitative measure of the smearing of the interface, and is plotted in Fig. 2(b) as a function of temperature, using the right vertical scale. From Fig. 2 we can notice rapid changes with temperature, where, at elevated temperatures N content is quite reduced, while interface quality is deteriorated.

For the purpose of understanding the physical origin of these rapid temperature dependencies, a first order kinetic model [12, 13] which incorporates three phases of our particular growth sequence: nitridation, GaAs overgrowth and As-soak, is employed. Previous modeling was concentrated on the effects of the temperature dependence of the nitrogen content.[12, 13] In this work, the focus is on the effects of thermally activated N surface segregation on the interface structural quality and sharpness, and therefore we outline only features of the model pertinent to this study. During the nitridation process, the energetically favorable N for As exchange occurs, due to a much stronger Ga-N than Ga-As bond (heats of formation 6.81 eV and 5.55 eV, respectively[18]). However, the thermally activated N desorption, with rate constant $\tau_d^{-1} = \tau_{0d}^{-1}\exp(-E_d/kT)$ and activation energy $E_d(\sim 2.1\,\text{eV})$, as determined from our RHEED measurement,[9, 11] limits the efficacy of the exchange. Additionally, during the GaAs overgrowth, the N surface segregation, with a rate constant $\tau_s^{-1} = \tau_{0s}^{-1}\exp(-E_s/kT)$, takes place concurrently with N desorption. The segregation process drives N atoms to the current growth surface where they can desorb, thus further limiting total N content and simultaneously producing interface smearing. After assuming that: (1) segregation is allowed only from the first subsurface to the current surface layer; (2) desorption occurs from all currently exposed surfaces (surface and first subsurface layer) and (3) desorption and segregation are statistically independent, it can be shown [12, 13] that final N content y in the nitrided layer[12, 13] is:

$$y(t_{\text{exp}}) = y_{\text{ss}}[1 - \exp(-t_{\text{exp}}/\tau)] \times \left[\frac{\tau_{\text{tot}}}{\tau_d + \tau_s e^{t_{\text{ML}}/\tau_\parallel}}\right] \quad . \tag{1}$$

where $y_{\text{ss}} \equiv r\tau$, $\tau \equiv (\tau_d^{-1} + r)^{-1}$, t_{exp} is the nitrogen plasma exposure time, r is the "dosing rate" estimated to be approximately 0.1 Hz from experimental values of y at 550 °C for 4 s and 6 s nitridations, $\tau_\parallel^{-1} \equiv (\tau_s^{-1} + \tau_d^{-1})$, $\tau_{\text{tot}} \equiv \tau_s + \tau_d$, and t_{ML} is one monolayer growth time. [12, 13] Equation (1) has been fitted to the experimentally obtained $y(T)$ and is plotted at Fig. 3. From this fit, the segregation activation energy has been estimated to be 0.9 eV±30%. [12, 13]

One of the important applications of the described kinetic model is the N profile smearing. To quantitatively estimate the extent of the smearing, we recall that the N contents S_n and S_{n+1} of the two consequent layer n and $n+1$ in superlattice period are related as [12, 13]:

$$S_{n+1} = S_n\left[1 - \exp(-t_{\text{ML}}/\tau_\parallel)\right]\frac{\tau_d}{\tau_{\text{tot}}} \quad . \tag{2}$$

From this equation we see that S_n form a geometric progression. By assuming continuous function $S_n(z)$ for the N profile, it is easy to show that $S_n(z) = S_n(0)\exp(-z/l)$ where l is the segregation length:

$$l = -\frac{a}{2\ln\{[1 - \exp(-t_{\text{ML}}/\tau_\parallel)]\frac{\tau_d}{\tau_{\text{tot}}}\}}. \tag{3}$$

The segregation length represents the decay length of the N profile and is therefore a useful parameter which quantitatively defines the interface quality. The plot of Eq. (3) for 4 s N exposures is given in Fig. 3, from which it can be seen that segregation length increases with temperature. This implies that interface structural quality and interface sharpness decrease with temperature, as observed in Fig. 2. To experimentally confirm our hypotheses we made TEM observations of our GaAs$_{1-y}$N$_y$/GaAs superlattices. Cross sections of the superlattices were prepared by polishing, dimpling and Ar ion-beam milling. Figure 4(a) shows a bright field TEM micrograph of the cross section of the GaAs$_{1-y}$N$_y$/GaAs superlattice grown at 550 °C, near the [110] zone axis, but normal to the surface. The (110) high resolution image of the superlattice is shown in Fig. 4(b). We did not observe GaN clustering, contrary to the results of a cross sectional Scanning Tunneling Microscopy study done on a similar structure.[14] Both of the micrographs in Fig. 4, indicate that the extent of the N profile smearing is indeed of the order of 3nm, in good agreement with the estimates obtained from the model.

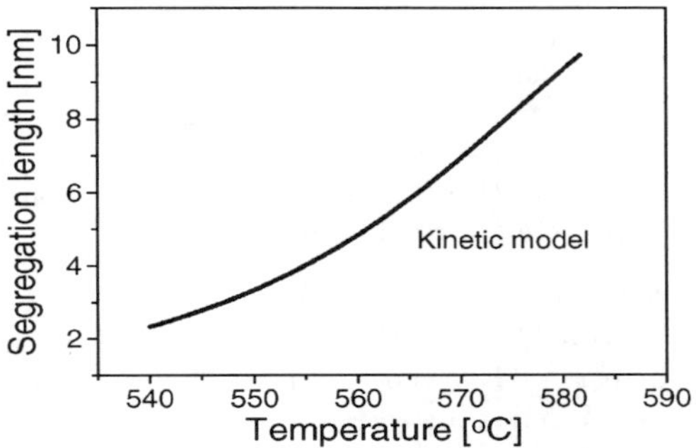

Figure 3: Nitrogen segregation length derived from the kinetic model [Eq. (3)]. It shows that the model predicts a large N segregation length at elevated temperatures, thus explaining the experimentally observed rapid changes with temperature.

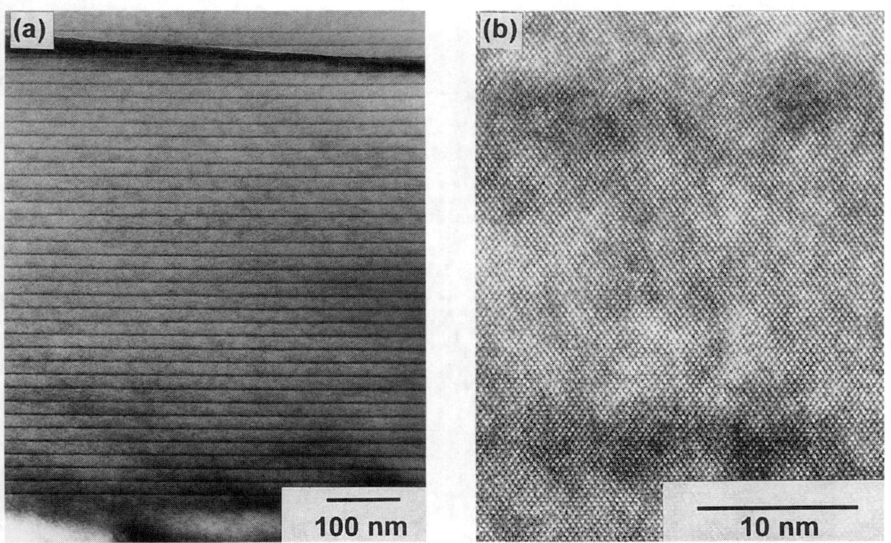

Figure 4: (a) The bright field TEM micrograph of the (110) cross section of the GaAsN/GaAs superlattice grown at 550 °C. The whole superlattice can be observed. The measured thickness of GaAsN layers is around 3nm, in good agreement with the model [Fig. 3] (b)The (110)high resolution micrograph of the GaAsN/GaAs superlattice. There are no lattice defects observed.

CONCLUSION

In conclusion, we studied the kinetics of N in the buried GaAsN layers in the $GaAs_{1-y}N_y$/GaAs superlattices. The thermally activated N surface segregation process was shown to be responsible for rapid changes of N content, and degradation of structural quality and sharpness of GaAsN/GaAs interfaces, observed in HRXRD experiments. The first order kinetic model quantitatively explained observed temperature dependencies, and predicted N profile smearing at the optimal growth temperatures to be on the order of $(2 - 3)$nm, in good agreement with our cross sectional TEM observations.

ACKNOWLEDGEMENTS

One of the autors (Z.Z.B) would like to thank Mrs. Carol Garland and Dr. Channing Ahn for their help related to the TEM experiments. This work was supported by Advanced Research Project Agency, and monitored by Office of Naval Research under Grant No. N00014-92-J-1845. Additionally, two of us (R.J.H. and M.L.O.) wish to acknowledge the support of Air Force Office of Scientific Research under Contract Nos. F49620-93-1-0211 and F49620-93-1-0389, and the support of the Advanced Research Projects Agency, monitored through the Army Research Office under Contract No. DAAH04-94-G-0393.

References

[1] S.Nakamura, M.Senoh, S.Nagahama, N.Iwasa, T.Yamada, T.Matsushita, Y.Sugimoto and H.Kiyoku, Appl. Phys. Lett. **69**, 1568 (1996).

[2] I.Akasaki, S.Sota, H.Sakai, T.Tanaka, M.Koike, and H.Amano, J.Electron. Lett. **32**, 1105 (1996).

[3] J.Edmond, H.S.Kong, M. Leonard, Inst. Phys. Conf. Ser. **142**, 991 (1996).

[4] I.Akasaki and H.Amano, Inst. Phys. Conf. Ser. **145**, 19 (1996).

[5] S. N. Mohammad, A. A. Salvador and H. Morkoç, Proc. IEEE **83**,1306 (1995)

[6] T.W.Weeks, M.D.Bremser, K.S.Ailey, E.Carlson, K.S.Ailey, W.G.Perry, and R.F.Davis, Appl. Phys. Lett. **67**, 401 (1995).

[7] O.Brandt, H.Yang, A.Trampert, and K.H.Ploog, Mat. Res. Soc. Symp. Proc. **395**, 27 (1995).

[8] M. Sato, J. Appl. Phys. **78**, 2123 (1995).

[9] R. J. Hauenstein, D. A. Collins, X. P. Cai, M. L. O'Steen, and T. C. McGill, Appl. Phys. Lett. **66**, 2861 (1995).

[10] M.Sato, Mat. Res. Soc. Symp. Proc., **395**, 285 (1995).

[11] R. J. Hauenstein, D. A. Collins, M. L. O'Steen, Z. Z. Bandić, and T. C. McGill, Mat. Res. Soc. Symp. Proc., **388**, 259 (1995).

[12] Z.Z.Bandić, R.J.Hauenstein, M.L.O'Steen, and T.C.McGill, Appl. Phys. Lett. **68**, 1510 (1996).

[13] Z.Z.Bandić, T.C.McGill, R.J.Hauenstein, and M.L.O'Steen, J. Vac. Sci. Technol. B **14**, 2948 (1996).

[14] R.S.Goldman, R.G.Briner, R.M.Feenstra, M.L.O'Steen, and R.J.Hauenstein, to be published

[15] J. Neugebauer and C.G. Van de Walle, Phys. Rev. B **51**, 10568 (1995.

[16] N. Grandjean, J. Massies, and M. Leroux, Appl. Phys. Lett. **69**, 2071 (1996).

[17] S. Sakai, Y. Ueta and Y. Terauchi, Jpn. J. Appl. Phys. **32**, 4413 (1993)

[18] CRC Handbook of Chemistry and Physics, edited by R. C. West (CRC, Boca Raton, FL, 1987), P. E102

REACTIVE MBE GROWTH OF GaN AND GaN:H ON GaN/SiC SUBSTRATES

M.A.L. Johnson*, Zhonghai Yu*, C. Boney*, W.C. Hughes*, J.W. Cook, Jr.*, J.F. Schetzina*, H. Zhao**, B.J. Skromme**, and J.A. Edmond***

*Department of Physics, N.C. State University, Raleigh, NC 27695, jan_schetzina@ncsu.edu
**Department of Electrical Engineering, Arizona State University, Tempe, AZ 85287-5706
***Cree Research, Inc., 2810 Meridian Parkway, Durham, NC 27713

ABSTRACT

Reactive N and H, created using rf plasma sources, were used to grow undoped GaN along with p-type GaN:Mg and p-type GaN:Mg:H thin films. By comparing the optical emission spectra from several rf sources with observed GaN grow rates, we deduce that nitrogen atoms and 1st-positive series nitrogen molecules (3.95 eV binding energy) are the reactive species responsible for GaN film growth. A Mg ground state acceptor binding energy of about 224 meV was determined from low temperature photoluminescence (PL) experiments for both p-type GaN:Mg and p-type GaN:Mg:H films.

INTRODUCTION

A key advance in the development of III-V nitrides for blue/green optoelectronic device applications has been the demonstration of p-type doping using Mg as a substitutional acceptor. However, fundamental aspects of p-type doping of GaN and related materials remains poorly understood at present, particularly with regards to MOVPE-grown materials which contain H. MBE provides a growth environment which contains no H and thus offers an opportunity to study p-type doping of GaN in detail. Alternatively, using reactive MBE, H can be introduced under controlled conditions by mixing it with N in an rf plasma source [1-5]. At NCSU, rf plasma sources were employed to generate sufficient "active" nitrogen for the growth of GaN by MBE at high temperatures (up to 900 °C) and growth rates up to 0.6 µm/hr on GaN/SiC substrates, and to selectively add H. This has resulted in device-grade GaN material with defect densities that replicate those in the underlying MOVPE-grown buffer layers. In the undoped GaN films, PL emission at 300 K consists of a single near-band-edge peak at 3.409 meV having a FWHM as narrow as 33 meV. Very little deep level emission is observed. A series of p-type GaN:Mg and GaN:Mg:H samples was prepared by varying the Mg MBE cell temperature from 200 °C (lightly doped) to 310 °C (heavily doped). Low temperature (1.7-80 K) PL experiments were completed at ASU on this set of samples from which the Mg ground state acceptor binding energy was obtained.

EXPERIMENTAL DETAILS

The GaN films were grown at NCSU using a modified EPI 930 three-chamber MBE system consisting of a main MBE film growth chamber, a second chamber for plasma cleaning of substrates using He/H plasmas, and a surface analysis chamber with provisions for Auger spectroscopy studies of substrates and epilayers at temperatures up to 800 °C. The three chambers are interconnected by an ultrahigh vacuum (UHV) sample transfer system.

The GaN films were grown by MBE on high-quality 2 to 3 µm thick GaN buffer layers prepared by MOVPE on basal plane 6H-SiC substrates at Cree Research, Inc. The substrates were cleaned prior to MBE film growth using trichloroethylene, acetone, and methanol followed by plasma-cleaning using a 1:1 hydrogen/helium gas mixture to remove carbon. After cleaning, the GaN/SiC substrate surface showed little or no carbon or oxygen contamination.

Three different rf plasma sources from different manufacturers were employed to grow undoped GaN. Each of the three rf sources were equipped with pyrolytic boron nitride (PBN) reaction chambers and exit apertures consisting of 37 small holes (~0.2 mm diameter). Optical emission spectra were obtained for each of these sources using a monochromator equipped with 150 g/mm and 1200 g/mm interchangeable gratings and correlated with measured GaN film growth rates. A series of p-type GaN:Mg samples was prepared by varying the Mg MBE cell temperature

Mat. Res. Soc. Symp. Proc. Vol. 449 © 1997 Materials Research Society

from 200 °C (lightly doped) to 310 °C (heavily doped). Low temperature (1.7-80 K) PL and reflectance experiments were performed at ASU on this set of samples as a function of temperature and excitation intensity.

RESULTS AND DISCUSSION

<u>Optical Emission Spectra From rf Plasma Sources.</u>

Optical emission spectra were obtained from each of the three rf plasma sources employed and correlated with film growth rates and quality. Fig. 1 shows emission spectra obtained for an SVT Associates (SVTA) rf source and an Oxford Applied Research MPD21 rf source, respectively. The sources were operated at a power of 400 watts and a pressure of 5 x 10^{-5} Torr. A number of nitrogen-plasma emission peaks associated with neutral nitrogen molecules are shown in the spectrum for each source which involve the 1st-positive and 2nd-positive series of molecular

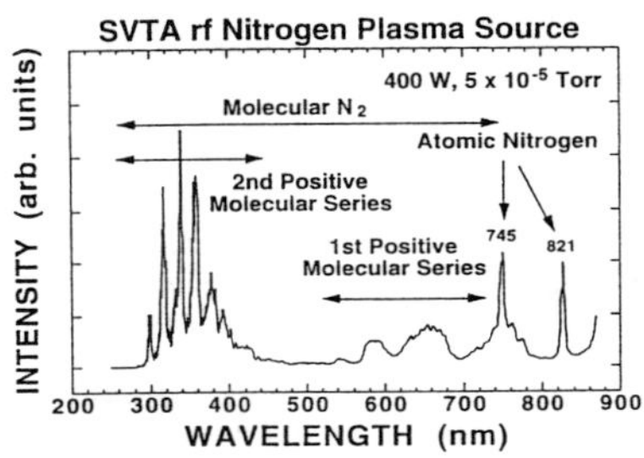

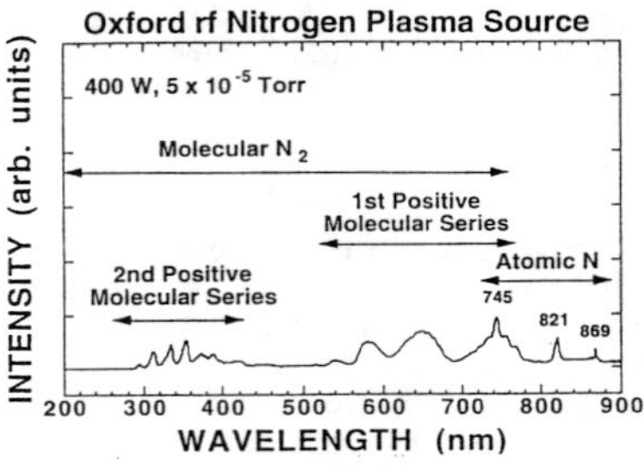

Figure 1. Optical emission spectra obtained for SVTA and Oxford rf nitrogen plasma sources.

transitions [1.2]. The 1st-positive series molecular nitrogen transitions ($B^3\Pi_g \rightarrow A^3\Sigma_u^+$ transitions) appear as five bands of regularly-spaced emission peaks in the visible and near-infrared (IR) spectral regions. The strongest 1st-positive emission peak of each band occurs at emission wavelengths of 540, 590, 660, 760, and 820, respectively, as shown in Fig. 1. The largest peaks of the 2nd-positive molecular nitrogen series ($C^3\Pi_u \rightarrow B^3\Pi_g$ transitions) occur at 316, 337, 357, 380, and 400 nm. The SVTA source produces large second-positive molecular emission peaks in the UV, and the emission spectrum appears whitish-violet to the eye. In contrast, the Oxford source emission spectrum shows diminished 1st-positive emission peaks and the emission appears whitish-orange to the eye. Both of these plasma sources produce high quality undoped GaN at comparable growth rates -- up to ~0.4 µm/hr using a GaN substrate temperature of 800 °C. This suggests that the principal reactive nitrogen species responsible for GaN film growth consist of N atoms and 1st-positive N$_2$ molecules. By using both of the above sources simultaneously, GaN growth rates of ~0.2 µm/hr at growth temperatures of ~900 °C were obtained. In the undoped GaN films, PL emission at 300 K consists of a single near-band-edge peak at 3.409 eV having a FWHM as narrow as 33 meV. No deep level emission is observed.

The EPI rf nitrogen plasma source employs a unibulb PBN reaction chamber of original design. This source produces a nitrogen emission spectrum which appears bright orange to the eye and which contains very strong 1st-positive molecular emission peaks and atomic emission lines as shown in Fig. 2. Note, in addition, that no second-positive emission peaks are present in the emission spectrum of the EPI source. Using the EPI rf plasma source, we have obtained high-quality GaN films at growth rates of 0.6 µm/hr at growth temperatures of 900 °C -- more that a three-fold increase in GaN growth rate compared to the other rf plasma sources investigated. This observed increase in film growth confirms that nitrogen atoms and 1st-positive N$_2$ molecules are the reactive plasma species responsible for GaN film growth.

Undoped GaN films grown at 900 °C using the EPI source display PL emission spectra at 300 K which consists of a narrow 49 meV near-band-edge peak at 3.409 eV and weak deep level emission as shown in Fig. 2.

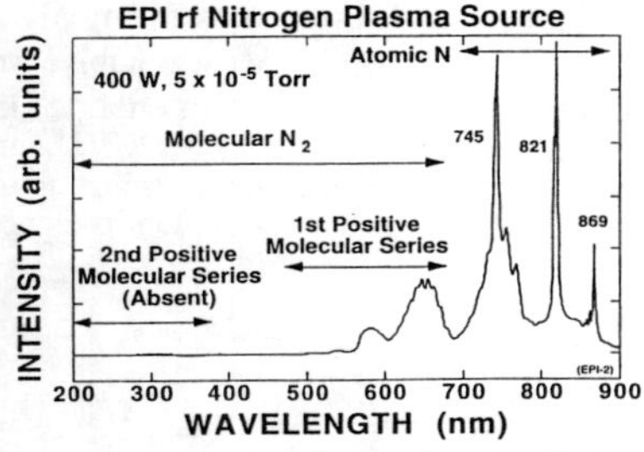

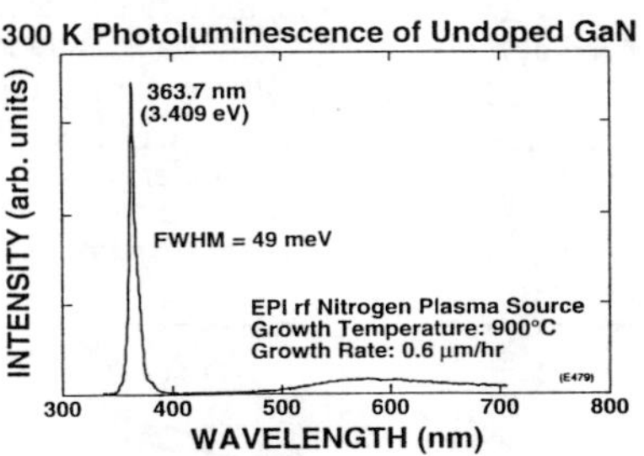

Figure 2. Optical emission from EPI nitrogen plasma source and PL from GaN grown at 900 °C.

P-type GaN:Mg and GaN:Mg:H Films

The optimum Mg cell temperature for p-type doping was found to be about 290 °C for GaN growth at 900 °C. C-V measurements yielded N_a-N_d ~ 2 x 10^{19} cm^{-3}, corresponding to a hole concentration of about 5 x 10^{17} cm^{-3}. The 300 K PL spectrum from such a layer is shown in Fig. 3. Note the strong emission at 3.26 eV and weaker emission at 2.95 eV that we associate with the Mg-acceptor, in addition to the band edge peak at 3.409 eV. The Mg-related features grow in magnitude as the Mg oven temperature increases. For $T_{Mg} \geq 320$ °C, however, C-V studies yielded n-type conduction

P-type GaN:Mg:H films were obtained using a 50% mixed H_2:N_2 plasma within the SVTA rf source. The mixed plasma is very effective for creating atomic H, as seen by the emission spectrum is shown at the bottom of Fig. 3. The emission appears bright red to the eye, due to the presence of very strong atomic H lines (Balmer Series). Note that no H_2 molecular emissions are present in the

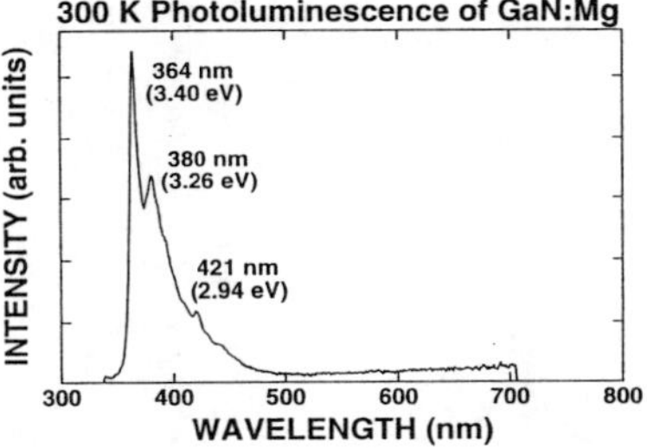

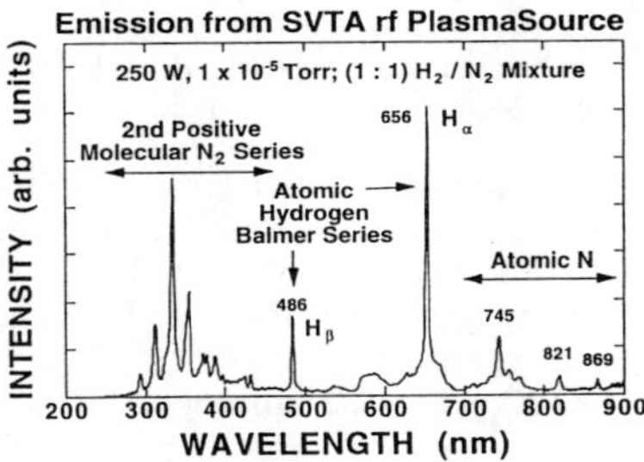

Figure 3. PL and H_2:N_2 plasma emission spectra.

spectrum. The 300 K PL or the GaN:Mg:H films were similar to the PL shown in Fig. 3 for GaN:Mg. C-V measurements showed p-type conductivity for Mg oven temperatures up to 320 °C.

Low Temperature PL of GaN:Mg and GaN:Mg:H

A total of eight Mg-doped and ten (Mg,H)-doped samples grown at various Mg cell temperatures were investigated at ASU to determine the spectroscopic properties of the Mg acceptor level and the effects of H incorporation on the doping efficiency. The excitonic luminescence of a series of lightly Mg-doped samples is shown in the lower three spectra in Fig. 4, together with a simple reflectance spectrum (uppermost) of the most heavily doped layer for comparison. (Similar reflectance spectra were recorded for all other samples but are omitted here for clarity.) At the lowest doping level, the neutral donor-bound exciton (D°,X) peak is dominant. It exhibits a very weak LO phonon replica with a Huang-Rhys coupling factor $S = 0.0021$. The A and B free exciton peaks are nearly superimposed because of the biaxial tensile strain in this sample, as is determined from the reflectance spectrum of this sample (not shown). The C exciton is resolved at higher energy in both PL and reflectance, whereas the A/B excitons form only a poorly-resolved shoulder in PL on the high energy side of the (D^o,X) peak. Most notably, we observe a weak two-electron replica of the (D^o,X) peak at about 20 meV lower in energy. In previous work, we observed a similar peak in a number of other samples grown by MOCVD and gas-source MBE, where we confirmed its identification using magnetospectroscopy in fields up to 12 T [6]. The donor binding energy is implied to be about 29 meV but its chemical identity is not yet established.

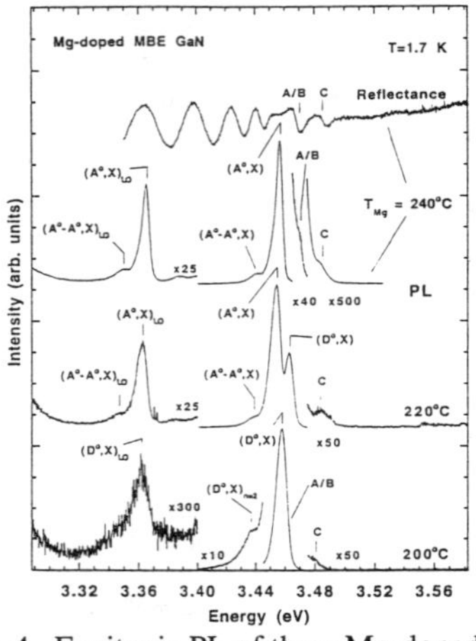

Fig. 4. Excitonic PL of three Mg-doped GaN layers and reflectance of one of them.

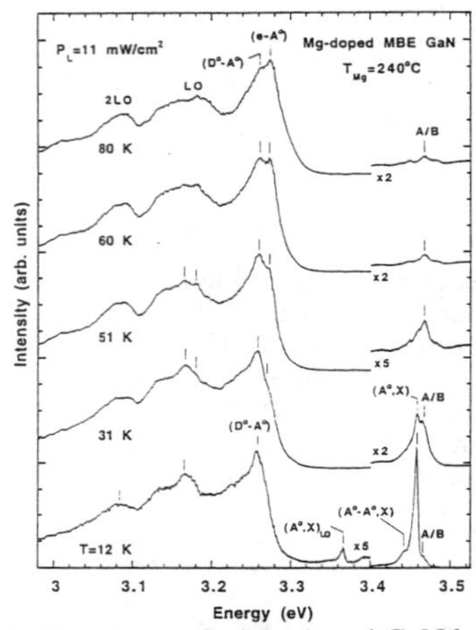

Fig. 5. PL spectra of a Mg-doped GaN layer as a function of temperature.

As the Mg doping level is increased, a new bound exciton peak appears about 8-10 meV below the (D^o,X) peak, which is still visible in the intermediate doping range. This peak exhibits much stronger LO phonon coupling with a Huang-Rhys parameter $S = 0.024$, over an order of magnitude stronger than that of the (D^o,X) peak. Its characteristically stronger phonon coupling and its association with Mg doping allow us to identify this peak as a neutral acceptor-bound exciton (A^o,X) peak. The stronger phonon coupling is due to the stronger localization of the holes in the (A^o,X) wave function as compared to (D^o,X). The peak identifications are further confirmed with reflectance (not shown). The peaks in these samples <u>cannot</u> be reliably identified on the basis of peak positions alone because of the substantial sample-to-sample random variations in strain, which shifts all the peaks to varying degrees [6]. Moreover, the $(D^o,X)/(A^o,X)$ peak separation is apparently a function of strain, since a slightly larger separation of about 12 meV was observed in samples with higher peak energies (3.467 and 3.455 eV, respectively) [7,8].

In the highest doped sample, the (A^o,X) peak dominates and the (D^o,X) peak is unobservable. The reflectance spectrum of this sample (upper spectrum) clearly shows structure related to the A/B and C free excitons, identifying them unambiguously. A low energy shoulder is observed on the (A^o,X) peak, which we attribute to an "undulation spectrum" involving excitons bound to closely-spaced <u>pairs</u> of neutral Mg acceptors. Similar features have been frequently observed in other *p*-type semiconductors [9,10]. This identification is supported by the phonon coupling strength of this peak ($S = 0.044$), which exceeds even that of (A^o,X). In still more heavily-doped samples (see below), this structure is broadened to low energies as a result of closer average pair spacing and grows to dominate the entire exciton spectrum.

As the doping level is increased, we also observed a nearly monotonic increase in the strength of a donor-acceptor pair (D^o-A^o) peak around 3.25-3.26 eV and its LO phonon replicas relative to the exciton emission. This peak shifts to higher energy at a rate of about 2-3 meV per decade of intensity as the excitation intensity is increased, identifying it as a $(D^o\text{-}A^o)$ transition [11]. (The peak sometimes shifts much faster, see below.) A shoulder extending from around 15 to 46 meV below the no-phonon peak is present, which we assign to acoustic phonon coupling rather than a second acceptor species as suggested previously [11]. We observe the same structure consistently with the same strength relative to the main peak in many different samples, making the latter interpretation quite unlikely. The $(D^o\text{-}A^o)$ peak position is not very useful to determine the acceptor binding energy, because it depends on the average strength of the Coulomb interaction between recombining pairs, which in turn is a function of doping level, nonradiative recombination rate, etc. We therefore performed variable-temperature PL measurements to observe the corresponding conduction band-to-acceptor $(e\text{-}A^o)$ transition, whose energy is a much better measure of the acceptor binding energy. These data are shown for a moderately-doped sample in Fig. 5 (similar data were obtained on other lightly-doped samples as well).

As the temperature is raised, the shallow donors $(E_D \approx 29$ meV) thermally ionize into the conduction band, quenching the $(D^o\text{-}A^o)$ peak in favor of a higher energy $(e\text{-}A^o)$ peak that becomes dominant in this sample above about 60 K. The (A^o,X) peak also quenches (more rapidly) as the temperature is raised in favor of a peak that probably involves mainly A/B free exciton recombination. Some anomalous structure is present in the LO phonon replicas of the acceptor-related peaks, which suggests a deeper $(D^o\text{-}A^o)$ or $(e\text{-}A^o)$ peak of some sort underlying the phonon replicas of the more usual peak. In this sample the A/B free exciton occurs at about 3.468 eV based on the reflectance spectrum. Using the free exciton binding energy of 26.4 meV determined in our recent magnetospectroscopy study [6], we deduce a (strained) band gap of about 3.494 eV at 1.7 K. The $(e\text{-}A^o)$ peak position at 60 K is 3.273 eV, and the thermal energy contribution $(k_B T/2)$ is 2.6 meV at this temperature. Precise measurements of (D^o,X) peak energies in samples with narrow $(\leq 2$ meV) exciton linewidths show a decrease in band gap of no more than 0.1-0.2 meV up to 49 K [1], so the gap should equal its low temperature value at 60 K to within 1 meV or better. We therefore deduce a Mg acceptor binding energy of 224 ± 4 meV. The accuracy is limited mainly by the relatively broad linewidths in the present samples.

In Fig. 6 we show the excitation intensity dependence of the near-band-edge PL spectrum of the sample of Fig. 5 at 1.7 K. As noted above, the $(D^o\text{-}A^o)$ peak shifts to higher energy with increasing intensity, supporting its identification. At the highest pump powers, however, we observe a new, relatively sharp peak at 3.267 eV, which is most pronounced when the excitonic peaks are strong compared to the $(D^o\text{-}A^o)$ peaks (note the indicated changes of scale). This peak is 189 meV below the (A^o,X) peak. It could be assigned either as a $(e\text{-}A^o)$ peak or as a "two-hole" replica of the (A^o,X) peak in which the acceptor is left in its first excited s-like state after the exciton recombines. Since the peak is consistently several meV lower in energy than the $(e\text{-}A^o)$ peak that appears at high temperature, we assign it as a two-hole transition. This identification implies a $1s\text{-}2s$ separation of 189 meV for Mg acceptors, which is a relatively large fraction (0.844) of their binding energy. While this value is larger than that for a purely hydrogenic case (0.75) and larger than typical values for semiconductors with large spin-orbit splitting, no quantitative calculations have been given in the literature for the $2s$ excited state of acceptors in the limit where the acceptor binding energy is much greater than the spin-orbit splitting of the valence band. Such calculations are now in progress to test the reasonableness of our assignment. The large ratio could also be due to a central cell correction, which will be about eight times larger for the $1s$ state than for the $2s$ state.

In samples doped with both Mg and H, we see several differences from the Mg-doped only case, as partially illustrated for two different doping levels in Fig. 7. New PL emissions that shift rapidly to higher energy with increasing excitation intensity are observed both in the exciton region and in the $(D^o\text{-}A^o)$ region of the spectra. This type of behavior is suggestive of highly compensated regions in the samples [12]. Such emissions are only rarely present without intentional H doping. Also, the doping seems to be less effective, as judged by the narrower $(D^o\text{-}A^o)$ and exciton peaks involving the Mg-doped material at high doping when H is used. Also, the absence of the exciton structure involving excitons bound to pairs of Mg acceptors $(A^o\text{-}A^o,X)$ in the H-doped case suggests that many of the Mg acceptors (if they incorporated at the same rate) have been de-activated by H. However, the Mg acceptor binding energy determined from temperature-dependent measurements of the (e-A°) peak position in the H-doped material is the same (within experimental accuracy) as that in material without added H. Those "shallow" acceptor levels that are present are therefore unaffected by the H. However, we also find evidence for deeper emissions around 3 eV. Further details of these results will be given elsewhere.

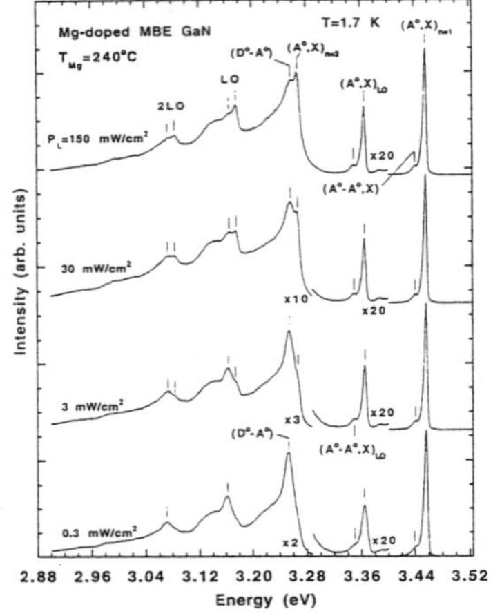

Fig. 6. Excitation intensity-dependent PL spectra for the sample of Fig. 5.

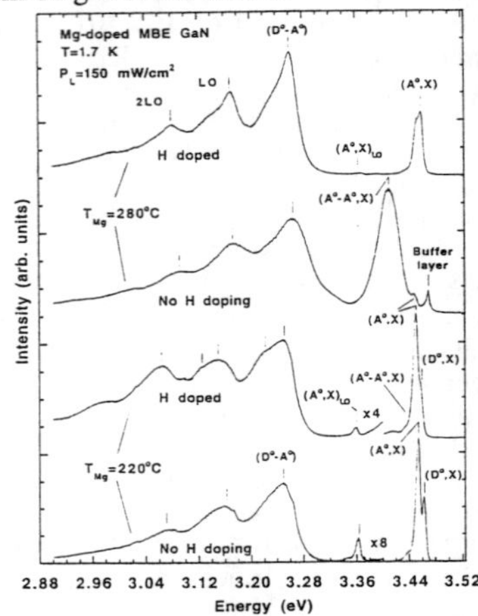

Fig. 7. Spectra of GaN layers with two different Mg doping levels, with and without H doping.

In conclusion, we determined the Mg acceptor binding energy in GaN both with and without H addition (it is the same). We also observe two-hole transitions in GaN for the first time, which involve the Mg acceptors. Use of reflectance to correct for variable strains and careful observation of phonon coupling strengths is found to be essential to identify the peaks reliably.

ACKNOWLEDGMENTS

Work at NCSU was supported by DARPA, ARO, and Cree Research. Work at ASU was supported by NSF through the ASU Materials Research Science and Engineering Center.

REFERENCES

[1] R. W. B. Pearse and A. G. Gaydon, <u>The Identification of Molecular Spectra</u>, Wiley, New York, 1963, pp. 209-220.
[2] A. N. Wright and C. A. Winkler, <u>Active Nitrogen</u>, Academic, New York, 1968, pp.15-220.
[3] A. R. Stringanow and N. S. Sventitskii, <u>Tables of Spectral Lines of Neutral and Ionized Atoms</u>, Plenum, New York, 1968, pp. 111-140b.
[4] R.P. Vaudo, X. Yu, J.W. Cook, Jr., and J.F. Schetzina, Optics Letters **18**, p. 1843 (1993).
[5] W. C. Hughes, J.W. Cook, Jr, J. F. Schetzina, J. Ren and J. A. Edmond, J. Vac. Sci. Technol.**B 13**, p. 1571 (1995).
[6] B.J. Skromme, J.W. Hutchins, H. Zhao, H.S. Kong, M.T. Leonard, G.E. Bulman, C.R. Abernathy, and S.J. Pearton, presented at the Electronic Materials Conference, Santa Barbara, Paper #W6, 1996.
[7] R. Dingle, D.D. Sell, S.E. Stokowski, and M. Ilegems, Phys. Rev. B **4**, p. 1211 (1971).
[8] M. Ilegems and R. Dingle, J. Appl. Phys. **44**, p. 4234 (1973).
[9] P.J. Dean and A.M. White, Solid State Electron. **21**, p. 1351 (1978).
[10] E. Molva and N. Magnea, Phys. Stat. Sol. (b) **102**, p. 475 (1980).
[11] R. Dingle and M. Ilegems, Solid State Commun. **9**, p. 175 (1971).
[12] P.W. Yu, J. Appl. Phys. **48**, p. 5043 (1977).

PRESSURE CONTROLLED GaN MBE GROWTH USING A HOLLOW ANODE NITROGEN ION SOURCE

M. S.H. Leung,*,** R. Klockenbrink,* C. Kisielowski,* H. Fujii,* J. Krüger,* Sudhir G.S.,*
A. Anders,** Z. Liliental-Weber,** M. Rubin,** and E. R. Weber*,**

* *Department of Materials Science and Mineral Engineering, University of California, Berkeley,CA 94720*
** *Materials Science Division, Lawrence Berkeley National Laboratory, Berkeley, CA 94720.*

ABSTRACT

GaN films were grown on sapphire substrates at temperatures below 1000 K utilizing a Hollow Anode nitrogen ion source. A Ga flux limited growth rate of ~ 0.5 µm/h is demonstrated. Active utilization of strain and the assistance of a nitrogen partial pressure during buffer layer growth are found to be crucial issues that can improve the film quality. The best films exhibit a full width at half maximum of the x-ray rocking curves of 80 arcsec and 1.8_5 meV for the excitonic photoluminescence measured at 4 K. A Volmer-Weber three dimensional growth mode and the spontaneous formation of cubic GaN inclusions in the hexagonal matrix are observed in the investigated growth temperature range. It is argued that this growth mode contributes to a limitation of the carrier mobility in these films that did not exceed 120 cm^2/Vs though a minimum carrier concentration of ~ 10^{15} cm^{-3} was achieved.

INTRODUCTION

In recent years, GaN thin film growth was developed in an unprecedented short time [1-3]. At present, films grown by metal-organic chemical-vapor deposition (MOCVD) or related techniques exhibit the best physical properties that allow for the fabrication of LED's, and for the development of power devices and laser diodes. In addition, growth of GaN thin films by molecular beam epitaxy (MBE) was successful [4, 5]. In contrast to MOCVD, a MBE growth process can exploit deviations from thermodynamic equilibrium that may help to increase doping levels or to grow AlN/InN/GaN quantum well structures of quality. However, GaN thin films are grown at temperatures that are low compared with the melting point of the material ($T_{growth} < 0.5$ T_m), though absolute growth temperatures are high (typical: MBE growth at 1000 K; MOCVD growth at 1300 K). This, together with the lack of lattice matched substrates makes the growth of GaN thin films a rather complex process. Additional complications arise for the MBE growth process. These range from technical limitations imposed by the development of reliable nitrogen sources that can give growth rates > 1µm/h to the lack of basic understanding of thin film growth at low temperatures.

In this paper we demonstrate that a hollow anode ion source (HA-CGD), utilizing a constricted dc glow discharge nitrogen plasma, allows for a reliable MBE growth of GaN thin films. The unprecedented low kinetic energy (~ 5 eV) of the activated nitrogen ions minimizes ion damage of the growing films. For the present study, a Ga flux limited growth rate of ~ 0.5 mm/h was used. We expect to realize growth rates > 1 µm/h in the near future. These properties distinguish the HA-CGD nitrogen source from commonly used ECR and RF sources. Those suffer in addition from a more complex design which makes them more susceptible to failure during operation. We demonstrate that our films compare in several aspects very well with state-of-the-art MOCVD grown films. In addition, we show that at growth temperatures below 1000 K hexagonal (2H) and cubic (3C) GaN can be grown and that a Volmer-Weber three dimensional (3D) growth mode results in films being composed of oriented sub-grains. It is argued that their shape, size and coalescence is limited by the diffusion of Ga ad-atoms, depending on strain and growth temperature.

Mat. Res. Soc. Symp. Proc. Vol. 449 © 1997 Materials Research Society

EXPERIMENTS

GaN is grown using a refurnished Riber 1000 MBE system. A Knudsen cell is used to evaporate pure Ga (99.9999%) while the activated nitrogen is produced by the HA-CGD source with pure nitrogen gas (99.9999%) along with a Millipore nitrogen purifier. Some details of the source design are given elsewhere [6]. A dc voltage generates a glow discharge in a hollow anode ion source that is constricted to an area in the plasma chamber close to the gas exit. It is the pressure difference between the plasma chamber and the MBE growth chamber that extracts the activated nitrogen species with energies around 5 eV. Liquid nitrogen cryopanels are used during growth to obtain a base pressure in the chamber of ~ 5 x 10^{-10} Torr. A thin Titanium (Ti) layer on the back of the substrate absorbs the heat radiated from the Tungsten (W) filament heater. The temperature of the substrate is monitored with a pyrometer. 10 x 11 mm^2 c-plane sapphire substrates are used. They are degreased by boiling in acetone and ethyle alcohol for 5 minutes each and blown dry with nitrogen gas and introduced in the growth chamber via a load lock. The substrates are heated up to 700 °C for thermal desorption of surface contaminants. At this temperature, they are exposed to activated nitrogen for 10 minutes. Subsequently, a thin low temperature GaN buffer layer (~ 250 Å) is deposited on the substrate. Finally the main epitaxial layer is grown on the buffer layer during 4 hours. Typical grown conditions are: Ga source temperature: 1210 K; nitrogen flow rate: 5 - 80 sccm; buffer layer growth temperature: 773 K; main layer growth temperature: 1000 K. The nitrogen partial pressure in the chamber during growth was varied in the range 10^{-5} - 10^{-2} Torr. The strain in the layers was engineered by using buffer layers of different thickness (0-30 nm) and by variation of the III/V flux ratio during main layer growth [7]. About 100 films were grown with the HA-CGD source and the source did not fail in a single event.

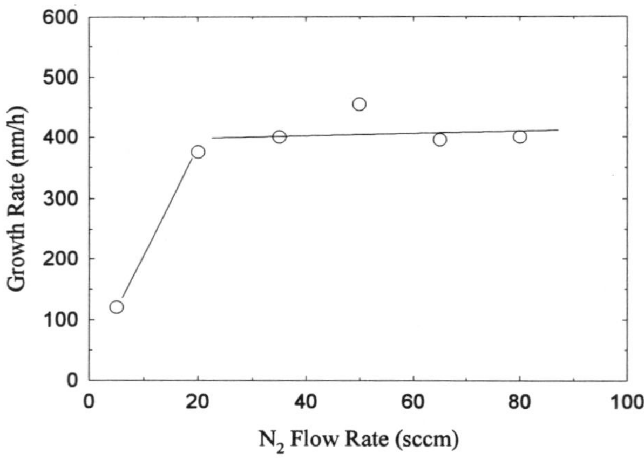

Fig. 1: Growth rate for different flows of nitrogen from the HA-CGD source. For a N-flux larger than 20 sccm the growth rate is limited by the Ga source temperature of 1210 K.

RESULTS AND DISCUSSION

In figure 1 we show that we achieved a growth rate of 400-500 nm/h that is Ga flux limited for nitrogen flows that exceed 20 sccm. The rate is determined by measuring the film thickness and for a growth time of four hours. The present design of the HA-CGD source allows for an increase of the nitrogen flow to 150 sccm. Consequently, we expect growth rates >1 µm/h to be realized in the near future.

The strain in the films was engineered by the buffer layer thickness and the composition of the film as described previously [7]. We note that in addition the nitrogen partial pressure in the growth chamber can be utilized to grow films with a smoother surface morphology. Figure 2 shows the dependence of the surface roughness on the nitrogen flow. Obviously, an increase of the nitrogen partial pressure during buffer layer growth reduces the roughness of the films considerably. A pressure of ~ 10^{-3} Torr decreases the mean free path of the activated species to source-substrate distance so that collisions with the background gas reduce the kinetic energy of the ions. This alters the structure of the buffer (nucleation) layer by forming a homogeneous layer with smaller nucleation sites.

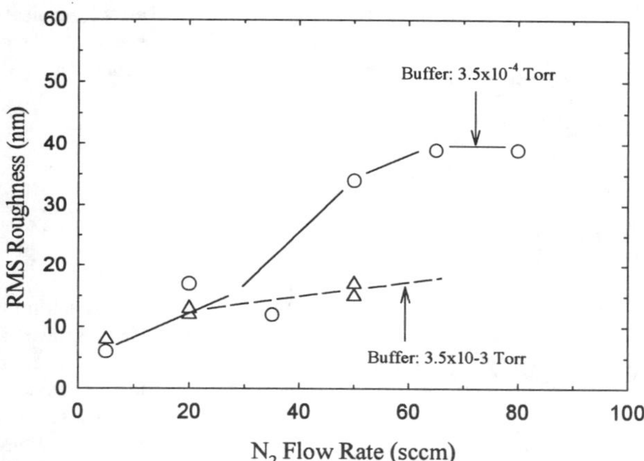

Fig. 2: Dependence of the surface roughness on the nitrogen flow realized with two different nitrogen partial pressures. Depicted are root mean square (RMS) AFM results obtained from 2 x 2 μm^2 areas.

The structural quality of the GaN films were characterized by x-ray diffraction measurements with a Siemens D5000 x-ray diffractometer containing a four bounce Ge monochromator. Typically, (0002) rocking curves exhibited a full width at half maximum (FWHM) ranging 2 to 10 arcmin depending on the growth condition. In the present study, a best value 80 arcsec was obtained. Fig. 3 shows the $\theta/2\theta$ scan of this particular GaN film grown on c-plane sapphire. The insert depicts the (0002) rocking curve. Such FWHM values are well comparable to the best heteroepitaxially grown MOCVD GaN samples.

The photoluminescence spectrum depicted in figure 4 was excited using the 325 nm line of a 50 mW HeCd laser. The luminescence light was then dispersed by a 0.85m double monochromator and detected with a photomultiplier via a lock-in technique. The sample temperature was 4 K. The spectrum is dominated by a donor bound excitonic transition at 3.468_8 eV with a FWHM of 1.8_5 meV, indicating that the layer is only slightly compressed [8]. To the best of our knowledge, an excitonic line as narrow as 1.8_5 meV has not yet been reported for heteroepitaxial GaN films. Such narrow PL lines can be obtained by engineering the strain in the material such that after the post-growth cooling the film is almost strain free.

The surface morphology of the films was investigated by a Park Scientific Atomic Force Microscope operated in air with tip force of 1 nN and a scan rate of 1 Hz. Details of the study are reported elsewhere in these proceedings [8]. No surface treatment is applied to the sample to preserve the surface features. Figure 5 depicts the surface morphology of a compressed GaN film measured by AFM as well as a cross-section transmission electron micrograph of the same sample. From the AFM image figure 5 it can be seen that the film is composed of individual large features of ~ 1 μm diameter. Such features can vary in

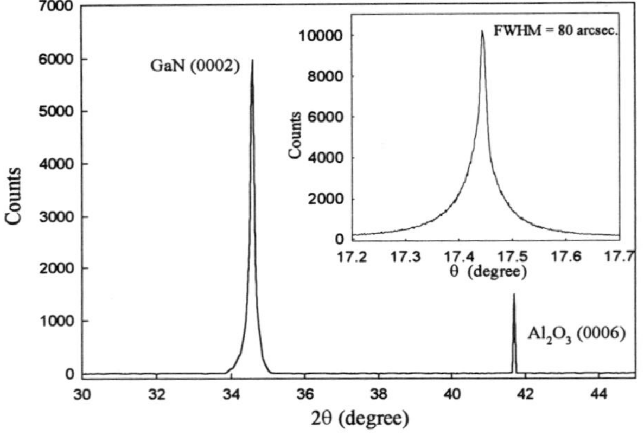

Fig. 3:
A θ/2θ scan of a GaN film grown on c-plane sapphire. The FWHM of the rocking curves is 80 arcsec, indicating an excellent structural quality.

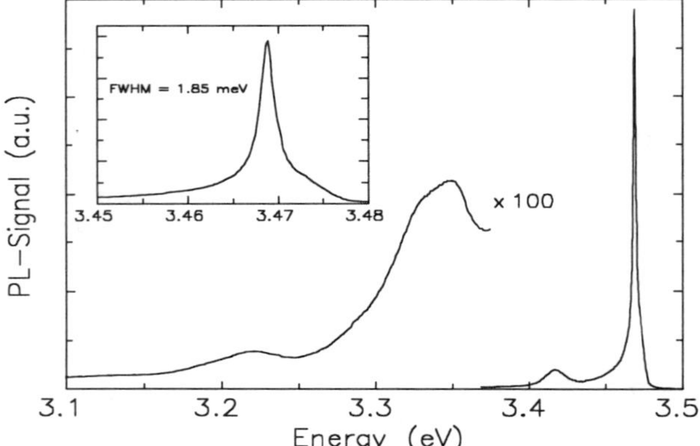

Fig. 4:
PL spectrum of a MBE grown sample, taken at 4 K. For the donor bound exciton we measured a line width (FWHM) of 1.85 meV. The line width at room temperature is 60 meV.

size, shape and coalescence depending on the growth temperature and on strain [8]. To the best of our knowledge, all MBE films grown in this temperature range exhibit similar surface structures. We argue that the formation of such structures is due to the low growth temperature. The corresponding TEM image reveals that the film is composed of individual but oriented column-like grains. In this particular example the sub-grains do not even coalesce. The grain size appears to be limited by surface diffusion of Ga adatoms [8]. These growth conditions lead to a very efficient reduction of the dislocation density close to the substrate/film interface. Consequently, individual sub-grains are almost dislocation free at the top of the layers. The surface is atomically flat as a result of the compressive strain in the sample. Closer inspection of the TEM image reveals inclusions with 60° and 120° angles, see figure 5. High resolution TEM (figure 6) confirms that such inclusions are cubic GaN crystals in the hexagonal matrix. The occurance of the cubic phase was found to decrease with increasing nitrogen flow, indicating that Ga-rich growth conditions favor the formation of cubic GaN. However, varying the composition of the films influences the strain in the layers so that it is not possible to exclude an impact of strain on the formation of the cubic phase. The

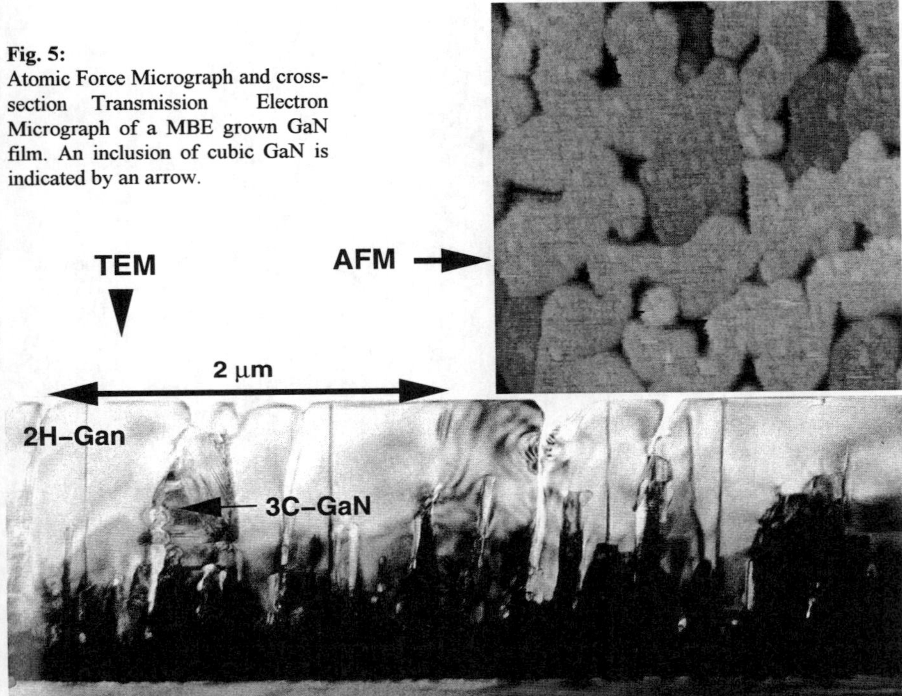

Fig. 5:
Atomic Force Micrograph and cross-section Transmission Electron Micrograph of a MBE grown GaN film. An inclusion of cubic GaN is indicated by an arrow.

observation of the cubic inclusions in areas close to the top of the layer also suggest their spontaneous formation.

Finally, we note that our films contain significant contamination with oxygen, carbon and hydrogen ($> 10^{18}$ cm^{-3}). In these films carrier concentrations down to 10^{15} cm^{-3} were obtained, however, the carrier mobility did not exceed 120 cm^2/Vs. Ongoing investigations indicate that the presence of the columnar structure in our films contributes to the reduction of the carrier mobility.

CONCLUSIONS

In conclusion, we demonstrate MBE growth of GaN thin films at a rate of ~ 0.5 µm/h by use of a Hollow Anode nitrogen ion source. A growth procedure has been developed that is pressure assisted and utilizes strain to engineer physical properties of GaN films. The obtained material compares well with MOCVD grown crystals as to its optical performance or its structural quality determined from the width of x-ray rocking curves. At MBE growth temperatures $\leq$ 1000 K a Volmer-Weber 3D growth mode is observed. As a result, the films are composed of columnar oriented sub-grains. The presence of such grains and the a large impurity content in our films may limit the carrier mobility. Spontaneous formation of the cubic GaN phase is observed if films are grown under Ga rich conditions.

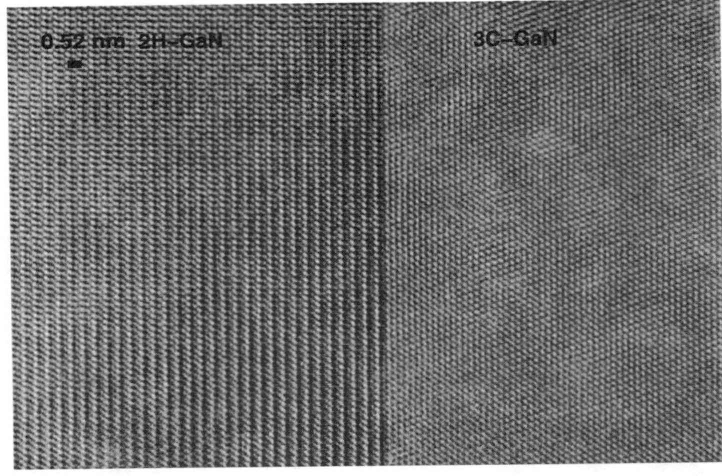

Fig. 6:
High resolution lat-
ice image of cubic
(3C) and hexagonal
(2H) GaN grown by
MBE. Cubic GaN
forms if the III/V
ratio is large. Zone
axes: [11-2]||[110]

ACKNOWLEDGMENTS

This work was supported by the Office of Energy Research, Office of Basic Energy Sciences, Division of Advanced Energy Projects (BES-AEP) and by the Laboratory Technology Transfer Program (ER-LTT) of the U.S. Department of Energy under Contract No. DE-AC03-76SF00098. A research scholarship provided by the German Science Foundation (DFG) to R. K. is gratefully acknowledged. This work benefits from the use of U.C. Berkeley's Integrated Materials Laboratory, which is supported by the National Science Foundation.

REFERENCES

1. S. Nakamura, SPIE, Proc. **2693**, 43 (1996).
2. M. Asif Khan, C. J. Sun, J. W. Yang, Q. Chen, Appl. Phys. Lett. **69**, 2418 (1996).
3. I. Akasaki, H. Amano, I. Suemune, Proceedings of International Conference on Silicon Carbide and related Materials, Kyoto, Japan, 18-21 Sept. 1995. IOP Publishing, 1996. p. 7.
4. R. Singh, D. Doppalapudi, T. D. Moustakas, Applied Physics Letters **69**, 2388 (1996).
5. W. Kim, A. Salvador, A. E. Botchkarev, O. Aktas, S. N. Mohammad, and H. Morcoç, Appl. Phys. Lett. **69**, 559 (1996).
6. A. Anders and S. Anders, Plasma Sources Sci. Technol. **4**, 571 (1995).
7. C. Kisielowski, J. Krüger, M. S.H. Leung, R. Klockenbrink, H. Fujii, T. Suski, Sudhir G.S., J.W. Ager III., M. Rubin and E.R. Weber, Proceedings of the International Conference on the Physics of Semiconductors (ICPS), Berlin 1996 (World Scientific, Singapore) 1996, p. 513.
8. H. Fujii, C. Kisielowski, J. Krüger, M. S.H. Leung, R. Klockenbrink, Sudhir G. S., M. Rubin, E. R. Weber, these proceedings.

IMPACT OF GROWTH TEMPERATURE, PRESSURE AND STRAIN ON THE MORPHOLOGY OF GaN FILMS

H. FUJII,[a]* C. KISIELOWSKI,* J. KRUEGER,* M. S. H. LEUNG,*,**R. KLOCKENBRINK,* M. RUBIN** and E. R. WEBER.*,**
*Department of Materials Science and Mineral Engineering, University of California, Berkeley, CA 94720, **Lawrence Berkeley National Laboratory, Berkeley, CA 94720
a) On leave from ULSI Device Development Laboratories, NEC Corporation, 2-9-1, Seiran, Otsu, Shiga, 520, Japan

ABSTRACT

GaN films grown on sapphire at different temperatures are investigated. A Volmer-Weber growth mode is observed at temperatures below 1000K that leads to thin films composed of oriented grains with finite size. Their size is temperature dependent and can actively be influenced by strain. Largest grains are observed in compressed films. It is argued that diffusing Ga ad-atoms dominate the observed effects with an activation energy of 2.3 ± 0.5 eV. Comparably large grain sizes are observed in films grown on off-axes sapphire substrates and on bulk GaN. This assures that the observed size limitation is a consequence of the 3D growth mode and not dependent on the choice of the substrate. In addition, the grain size and the surface roughness of the films depend on the nitrogen partial pressure in the molecular beam epitaxy (MBE) chamber, most likely due to collisions between the reactive species and the background gas molecules. This effect is utilized to grow improved nucleation layers on sapphire.

INTRODUCTION

GaN films are usually grown at temperatures that are low compared with the melting point (< 0.5 T_m) of the material. This, as well as the growth on lattice mismatched substrates (e.g. sapphire or SiC) with largely different thermal expansion coefficients, greatly affects the crystal quality and introduces strain into the layers. Amano et al. [1] and Nakamura [2] introduced the growth of AlN and GaN buffer layers to improve the film quality and to relax strain. A coexistence of hydrostatic and biaxial strain components in the GaN films was recently found [3] that allows to strain engineer the films. The design of an appropriate buffer layer structure is crucial for this purpose. However, the impact of growth parameters on the film morphology is not well understood. Wu et al. reported on the evolution of the morphology of GaN buffer and main layers [4]. However, its dependence on growth temperature, pressure and strain remained hidden. In this paper we report on the impact of growth temperature, strain, substrates and pressure on the morphology of GaN films.

EXPERIMENTAL

The GaN films were grown by ion-assisted MBE on c-plane and off-axes sapphire substrates. Homoepitaxial growth was performed on bulk GaN. Details of the growth procedure are reported elsewhere [5]. The films were homogeneous and well reproducible. Here, $\sim 1.5\mu m$ thick GaN layers were investigated. An n-type carrier concentration of $\sim 10^{19}$ cm^{-3} was present in

Mat. Res. Soc. Symp. Proc. Vol. 449 © 1997 Materials Research Society

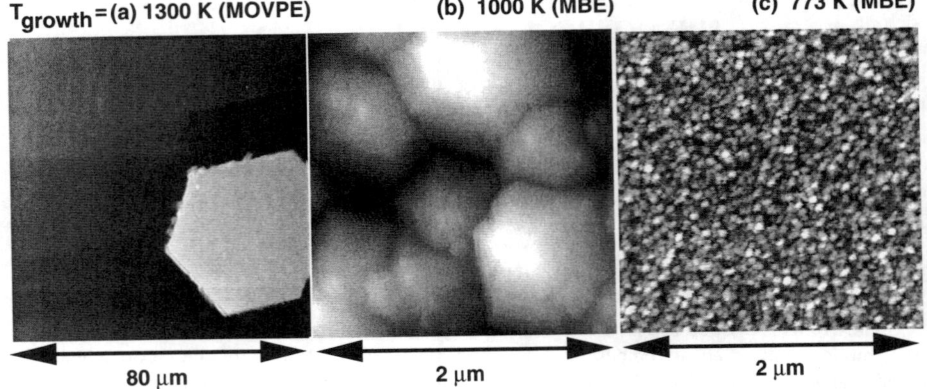

Figure 1: AFM plan-view images of GaN films on sapphire substrates grown at different temperatures by different methods as indicated.

all investigated MBE-grown films that did not change with the growth conditions. The n-type conductivity might be influenced by impurities. Oxygen, carbon and hydrogen were detected by SIMS at concentrations larger than 10^{18} cm^{-3}.

Here, we report on results obtained by AFM, by PL spectroscopy and by TEM. AFM measurements (contact mode) were done in air with an AFM tip force of 1 nN and a scan rate of 1 Hz. We used the root mean square (RMS) value of a 2 μm x 2 μm area to quantify the surface roughness. The feature sizes were evaluated statistically.

PL measurements were done at 4 K using the excitation of a 325 nm He-Cd laser line. The shift of the dominant donor bound exciton from its zero strain position at 3.467 eV [3] was evaluated to measure strain at 4 K. A narrow half width of the lines (best full width at half maximum: 1.8_5 meV at 3.468_8 for GaN film grown by MBE on sapphire) allowed to measure strains as small as 10^{-4}.

RESULTS AND DISCUSSION

Figure 1 shows AFM images of GaN films grown on sapphire at different growth temperatures. We compare the surface morphology of the MBE grown films (figure 1b,c)

Figure 2:
Feature size as a function of the reciprocal growth temperature. Open circles: Films grown by MOCVD and by MBE. Solid circles: strain engineered films grown by MBE.

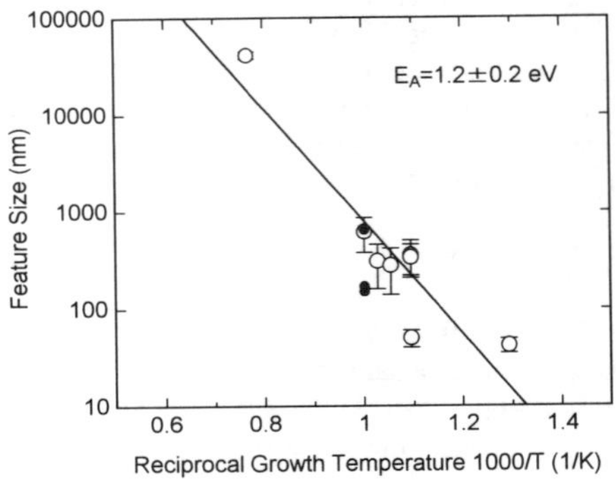

228

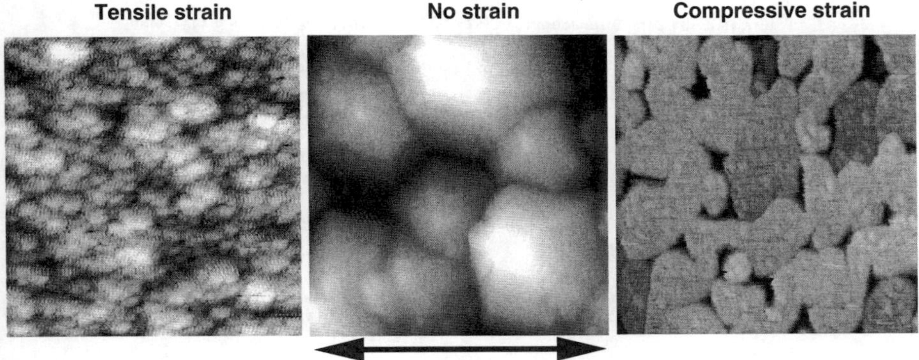

Figure 3 : AFM plan-view images of strain engineered GaN thin films grown by MBE at 1000K.

with the surface of a GaN film grown by MOVPE grown at 1300K (figure 1a). Details of the film morphology are given elsewhere [6]. Here, we stress that in figure 1a a Stranski-Krastanov [7] growth mode can be seen that causes two-dimensional (2D) plateaus and a three-dimensional (3D) grain. Step flow growth is observed in the flat areas. In contrast, figures 1b and 1c depict a Volmer-Weber [8] growth mode with 3D features only. Cross-section TEM reveals that the features seen in figures 1b and 1c are in fact oriented grains that form the GaN film [5]. The results shown in figure 1 are typical for all other investigated samples and demonstrate that the sizes of grains in the films depend on the growth temperature. Figure 2 depicts that the logarithm of the grain diameter is approximately a linear function of reciprocal growth temperature. Open circles show GaN films grown by MOVPE and by MBE. Solid circles are examples from a series of GaN films grown by MBE with different strains in the films. The grain size appears to be thermally activated. An activation energy $E_A = 1.1_5 \pm 0.2_5$ eV can be estimated from this Arrhenius plot. If we assume that the grain size is limited by a surface *diffusion length* $x \sim (Dt)^{1/2}$, where D is diffusion coefficient and t is the time, the activation energy of the diffusion coefficient is 2.3 eV. Brandt et al. reported on an activation energy of 2.4 eV for the *diffusion rate* of Ga ad-atoms on cubic GaN surfaces [9]. However, they could not specify whether or not it is the

Figure 4:

Dependence of feature size on the donor bound exciton position that measures strain. MBE growth. Open circles: variation of nitrogen flux. Open triangle (upwards): variation of buffer layer thickness. Solid square: Growth on MOCVD film. Open squares: pressure controlled buffer layer structure. Open triangle (downwards):Different main layer growth temperature.

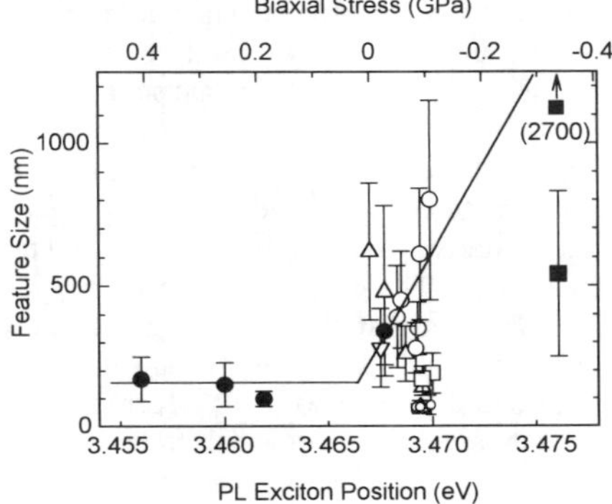

229

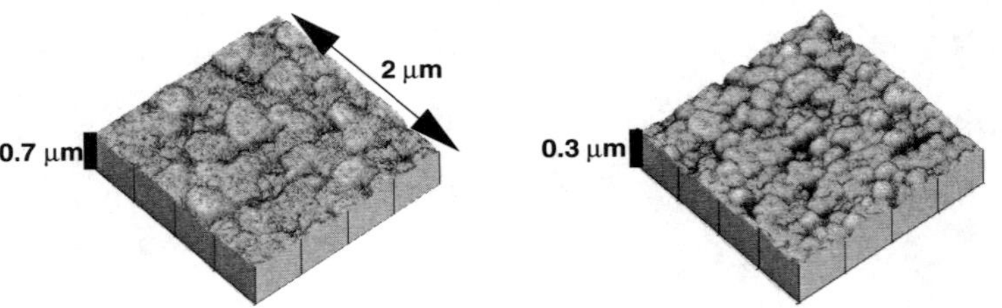

Figure 5: AFM 3D images of GaN films grown on sapphire substrates with a different buffer layer structure controlled by the nitrogen partial pressure in the MBE chamber.

diffusion length itself or the diffusion rate that depends on temperature. The similarity of the data suggests that it is the *diffusion length* of Ga ad-atoms which is temperature dependent and limits the grain sizes in our case. It is also seen in figure 2 that strain modifies the feature sizes as shown by the solid circles.

Next, we describe a correlation of the feature size with strain. The strain in the films was engineered by compositional changes and by the growth of buffer layers of different thickness [10]. Figures 3 shows that for a given growth temperature the feature sizes depend on strain that was extracted from the energetic position of donor bound excitons [3]. We investigated this correlation for a variety of films. From the caption of figure 4 it is clear that very differently grown samples were investigated. Yet, an analysis in terms of strain gives consistent results: the feature size in GaN films under compression can be larger than in films under tension. Thus, tensile and compressive strains affect the grain sizes in an opposite manner. Also, only films that are little strained or strain free exhibit a pyramidal surface structure while strained films are atomically flat [5]. We argue that it is the activation energy for the diffusion of Ga ad-atoms that is likely to be strain dependent. In fact, a tensile strain *widens* the in-plane lattice spacing of GaN and this may result in an increased trapping of the "large" Ga ad-atoms. However, compressed films may also exhibit smaller surface features at the same time. Ongoing investigations indicate that such small features can be related to the presence of extended defects that locally modify the strain. It should be noted that the differential thermal expansion of a GaN layer and its sapphire substrate will add a compressive component to the strain present at the growth temperature upon post-growth cooling. This was recently measured to occur up to 600K [11]. Thus, the strain measured at 4K is not equal to the strain state of the sample during growth.

Figure 5 shows AFM 3D images of GaN films grown on sapphire substrates with different buffer layer structures controlled by the nitrogen partial pressure during MBE growth. The nitrogen partial pressures during buffer layer growth was 3.5 x 10^{-4} Torr (left) and 3.5 x 10^{-3} Torr (right), respectively. The pressure during main layer growth is the same for both samples (3.5 x 10^{-3} Torr). It is seen that the film grown on the buffer layer with the larger nitrogen partial pressure is considerably smoother though it also exhibits smaller surface features. a nitrogen partial pressure in the range 10^{-3} - 10^{-4} Torr reduces the mean free path of the activated ions to a value comparable with the source-substrate distance. Thus, collisions with gas molecules reduce the kinetic energy of the adsorbed reactive species. Consequently, surface diffusion is reduced

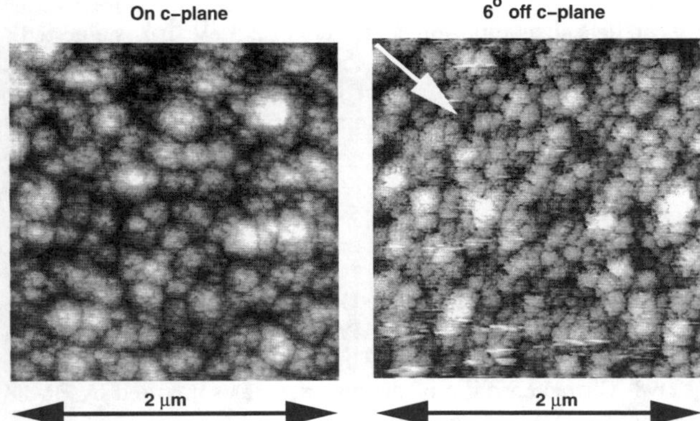

Figure 6:

AFM images of MBE grown GaN films on c-plane sapphire and on substrates 6° inclined towards the a-plane. The misorientation direction is indicated.

resulting in the formation of a uniform nucleation layer with grains reduced in size. This is favorable for dislocation annihilation during main layer growth. We exploit this effect to improve the quality of our MBE grown films [5].

Finally we evaluated a possible impact of the substrate material on the size of the grains in our films. Figure 6 shows AFM plan-view images of films grown by MBE at 1000K on c-plane sapphire and on 6° off c-plane sapphire towards [11-20](a-plane). The misorientation direction is indicated in figure 6. Similar small features are observed on all films grown on the other misoriented substrates. In the particular example of figure 6 the misorientation did not stimulate step flow growth but rather an alignment of the grains along terraces. The surface RMS roughness of the films increases with increasing off-angles from c-plane (figure 7). Using bulk GaN as a substrate material led to a morphology of the GaN thin film that is very simimilar to the one shown in figure 6. The results support our interpretation that the formation of grains in the GaN films grown by MBE at temperature below 1000K is caused by limited surface diffusion processes and not by the choice of substrates.

CONCLUSIONS

In conclusion, a three dimensional (3D) Volmer-Weber growth mode determines the morphology of GaN thin films grown by ion assisted MBE below 1000K. A transition to a step flow growth mode (2D/3D) occurs above this temperature. For the first time we show that the size of the grains in the films is temperature dependent and modulated by strain which induces an asymmetry with respect to

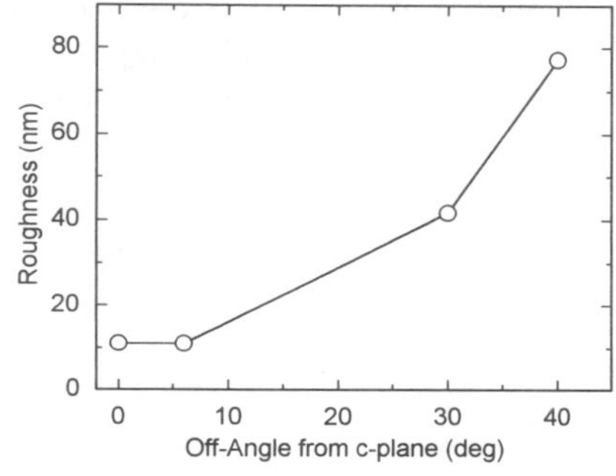

Figure 7:
Surface roughness as a function of the misorientation angle between the sapphire c- and a-plane.

compression or tension. If the formation of grains in the films is attributed to a limited surface diffusion length, we estimate a strain dependent diffusion coefficient of 2.3 eV. It is suggested to attribute this energy to the surface diffusion of Ga ad-atoms. The use of different substrates does not influence this growth mode significantly. An increase of the nitrogen partial pressure for buffer layer growth is found to be suitable for a better coverage of sapphire with small GaN nucleation sites.

ACKNOWLEDGMENTS

The MBE growth experiments were supported by the Office of Energy Research, Office of Basic Energy Sciences, Division of Advanced Energy Projects (BES-AEP) and by the Laboratory Technology Transfer Program (ER-LTT) of the U. S. Department of Energy under Contract No. DE-AC03-76SF00098. Characterization of the material was supported by BMDO administrated by ONR under Contract No. N00014-96-1-0901.

REFERENCES

1. H. Amano, N. Sawaki, I. Akasaki and Y. Toyoda, Appl. Phys. Lett. **48**, 353 (1986).
2. S. Nakamura, Jpn. J. Appl. Phys. **30**, L1705 (1991).
3. C. Kisielowski, J. Krueger, S. Ruvimov, T. Suski, J.W. Ager III, E. Jones, Z. Liliental-Weber, M. Rubin, E.R. Weber, M.D. Bremser, R.F. Davis, Phys. Rev. B **54**, Dec, 1996, in press.
4. X. H. Wu, D. Kapolnek, E. J. Tarsa, B. Heying, S. Keller, B. P. Keller, U. K. Mishra, S. P. DenBaars and J. S. Speck, Appl. Phys. Lett. **68**, 1371 (1996).
5. M. S. H. Leung, R. Klockenbrink, H. Fujii, J. Krueger, C. Kisielowski, Sudhir G. S., A. Anders, Z. Liliental-Weber, E. R. Weber and M. Rubin, this volume.
6. H. Fujii, C. Kisielowski, J. Kruger, M. S. H. Leung, R. Klockenbrink, Sudhir G. S., E. R. Weber and M. Rubin, submitted to J. Appl. Phys. (1996).
7. I. N. Stranski and Von Krastanov, Akad. Wiss. Mainz L. Math-Nat. K1 IIb **146** 797 (1939).
8. M. Volmer and A. Weber, Z. Physik. Chem. **119** 277 (1926).
9. O. Brandt, H. Yang, A. Trampert and K. H. Ploog, Materials Research Society Symposium Proceedings, **395**, 27 (1996).
10. C. Kisielowski, J. Krueger, M. S.H. Leung, R. Klockenbrink, H. Fujii, T. Suski, Sudhir G.S., J.W. Ager III., M. Rubin and E.R. Weber, Proceedings of International Conference on Physics of Semiconductors , Berlin 1996, World Scientific, Singapore 1996, p.513.
11. T. Suski, J. W. Ager III, G. Conti, Z. Liliental-Weber, J. Kruger, S. Ruvimov, C. Kisielowski, E. R. Weber, C. P. Kuo, I. Grzegory and S. Porowski, Proceedings of International Conference   on Physics of Semiconductors , Berlin 1996, World Scientific, Singapore 1996, p.2917.

DOPING STUDIES OF N- AND P-TYPE $Al_xGa_{1-x}N$ GROWN BY ECR ASSISTED MBE

D. KORAKAKIS [*], H.M. NG, K. F. LUDWIG, Jr.,[**] and T.D. MOUSTAKAS
Molecular Beam Epitaxy Laboratory, Dept. of Electrical and Computer Engineering and Center for Photonics Research, Boston University, 44 Cummington St., Boston, MA 02215, **Physics Department, Boston University, 590 Commonwealth Avenue, Boston MA 02215,*dimitris@engc.bu.edu

ABSTRACT

$Al_xGa_{1-x}N$ films ($x \leq 0.60$) were grown on c-plane sapphire and (0001) 6H-SiC substrates using ECR plasma assisted Molecular Beam Epitaxy. Evidence of long range ordering in the investigated $Al_xGa_{1-x}N$ films is presented. Without intentional dopants the films are semi-insulating with resistivities ranging from 10^3 to 10^5 Ω.cm. The films were doped n-type with Si and p-type with Mg. The carrier concentration in the Si doped films, as determined by Hall effect measurements, was between 10^{16} to 10^{19} cm^{-3}. At constant Si cell temperature, the carrier concentration was found to be reduced with AlN mole fraction, consistent with the observation that the donor ionization energy increases with Al content. Correspondingly, the electron mobility decreases with Al concentration, a result attributed to alloy scattering. The Mg doped films were found to exhibit p-type conductivity by thermoelectric power measurements with resistivities varying from 3 to 30 Ω-cm.

INTRODUCTION

The family of the III-V nitrides is gradually finding applications in optical devices (LED's, lasers and detectors) as well as electronic devices[1]. However, the majority of the scientific effort was focused on the growth and characterization of GaN films and very little effort has been dedicated to the study of its alloys with InN and AlN. In particular doping studies of such alloys are very limited[2,3].

In this paper, we report on n- and p-type doping studies of $Al_xGa_{1-x}N$ alloys with $x \leq$ 0.6. The films were grown by ECR-MBE and doped n-type with Si and p-type with Mg.

EXPERIMENTAL METHODS

The AlGaN films were deposited in a Varian Gen II MBE unit equipped with an ASTEX compact ECR microwave plasma source to activate the molecular nitrogen. The deposition method has been described in detail previously[4]. In this paper, we present only a brief description of the deposition method.

The AlGaN films were grown on (0001) sapphire and (0001) Si-terminated 6H-SiC substrates. The sapphire substrates were initially nitridated at $750^{\circ}C$[4] and subsequently a low temperature (550-600 $^{\circ}C$) GaN or $Al_xGa_{1-x}N$ buffer (approximately 300 Å thick) was grown. Following this step, the $Al_xGa_{1-x}N$ alloys (about 1 μm thick) were grown at 700-

Mat. Res. Soc. Symp. Proc. Vol. 449 © 1997 Materials Research Society

775 °C. Growth of $Al_xGa_{1-x}N$ alloys on 6H-SiC was done directly without employment of any buffer layer. The films were characterized by in-situ RHEED studies, XRD and SEM. Such structural studies were reported elsewhere[5].

The chemical composition of the investigated alloys was determined by accurate measurements of the c-lattice constant using XRD as well as energy dispersive spectroscopy (EDS) as reported previously[5].

The transport properties of the semi-insulating and lightly doped films were determined by four-probe resistivity measurements as a function of temperature. The moderately and heavily doped films were characterized by Hall effect measurements using the Van der Pauw geometry.

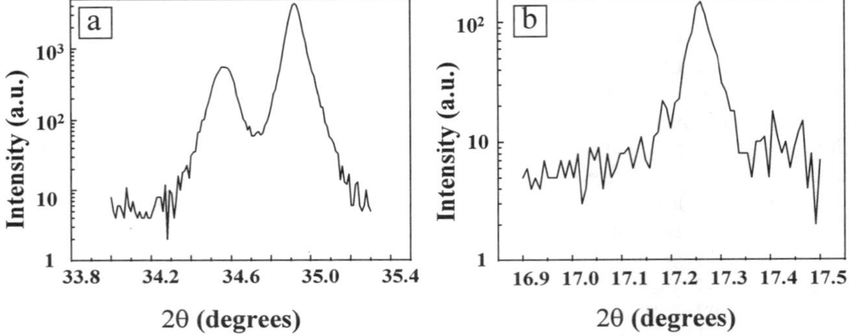

Figure 1: a) (0002) diffraction peaks GaN (0.2 μm) and $Al_{0.26}Ga_{0.74}N$ (1 μm) layers grown on c-plane sapphire b)(0001) diffraction peak of the same GaN and $Al_{0.26}Ga_{0.74}N$ layers indicating long range ordering of the $Al_{0.26}Ga_{0.74}N$ alloy.

EXPERIMENTAL RESULTS AND DISCUSSION

The films were characterized structurally by XRD using Cu $K\alpha_1$ radiation. Evidence of long range ordering in these $Al_xGa_{1-x}N$ films is provided by the existence of the (0001) diffraction peak, which is a forbidden reflection in the hexagonal close packed structures. Figure 1 shows the diffraction pattern at the (0002) peak (a) and the (0001) peak (b) of an $Al_{0.26}Ga_{0.74}N$, 1 μm film grown on a 0.2 μm thick GaN film on (0001) sapphire. Figure 1a shows both the GaN and $Al_{0.26}Ga_{0.74}N$ Bragg peaks. On the other hand, figure 1b shows only the diffraction peak which corresponds to $Al_{0.26}Ga_{0.74}N$. The existence of this diffraction peak can only be accounted by crystal ordering[6]. More detailed studies on this important observation is presented elsewhere[7].

Figure 2 shows the conductivity vs Al concentration for a number of undoped and Si-doped films grown on sapphire and 6H-SiC substrates. Some of the $Al_xGa_{1-x}N$ films on the sapphire substrates were grown using a GaN buffer and some an $Al_xGa_{1-x}N$ buffer. It appears the nature of the buffer did not effect the conductivity of the films. These data show that in both the undoped as well as the Si-doped films, the conductivity is reduced with the Al concentration. Similar observations on undoped $Al_xGa_{1-x}N$ films have

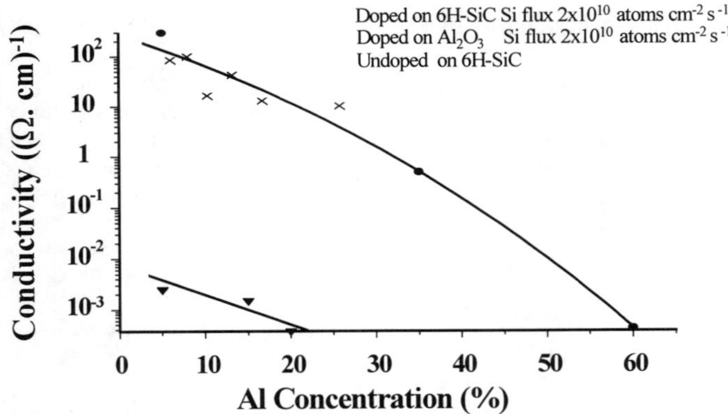

Figure 2: Dependence of conductivity on Al concentration of $Al_xGa_{1-x}N$ films a)Grown on sapphire (crosses) and 6H-SiC (circles) substrates and doped with Si (flux $2x10^{10}$ atoms cm^{-2} s^{-1}) b)Undoped on 6H-SiC substrates (triangles).

been reported by Yoshida *et al.*[8] and on Ge-doped films by Zhang *et al.*[2]. This transition from conducting to relatively insulating films with Al concentration is not well understood.

Figure 3 shows the dependence of conductivity on dopant vapor pressure for a number of Si- and Mg-doped $Al_xGa_{1-x}N$ films with practically constant Al concentration. These films were grown on both sapphire and 6H-SiC substrates. Thus the conductivity of the Si doped films was controlled systematically from 10^{-2} to 10^{2} (Ω cm)$^{-1}$ and the conductivity of the Mg-doped films was about 10^{-1} (Ω cm)$^{-1}$.

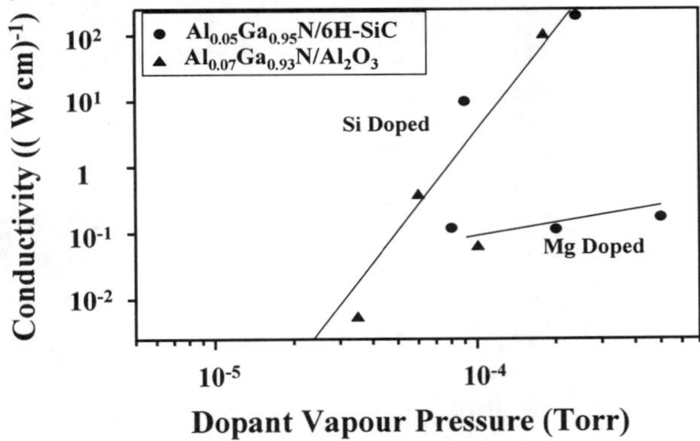

Figure 3: Dependence of $Al_xGa_{1-x}N$ films' conductivity on dopant vapor pressure

The type of conductivity in Mg-doped films was confirmed by thermoelectric power measurements.

Hall effect measurements were performed on Si-doped samples grown on sapphire substrates. Fig. 4 shows the carrier concentration vs AlN mole fraction for a number of films grown with Si-cell temperature varying from 1275-1375°C. Thus, the carrier concentration was controlled reproducibly from 10^{16} to 10^{19} cm^{-3}. The same data also show that the carrier concentration is reduced with increasing AlN mole fraction. The

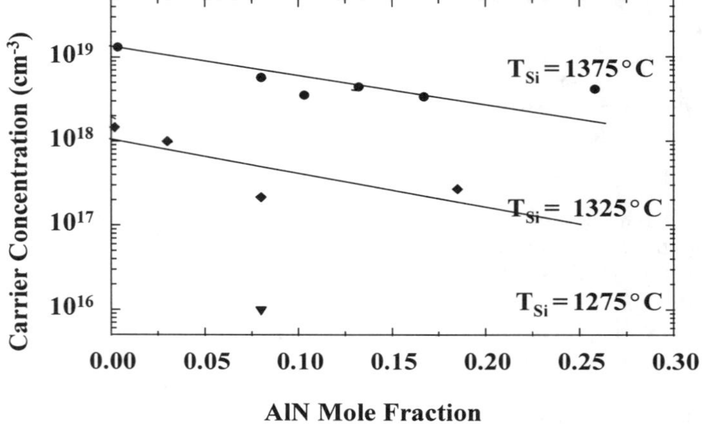

Figure 4: Carrier concentration dependence on AlN mole fraction

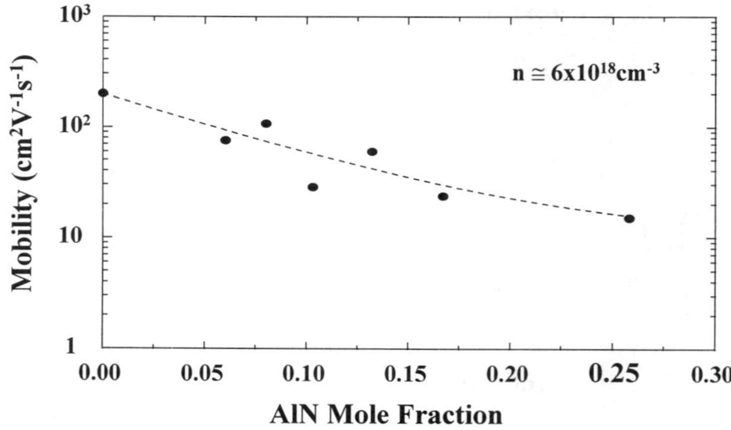

Figure 5:Mobility variation vs AlN mole fraction for heavily doped Al$_x$Ga$_{1-x}$N films

electron mobilities vs AlN mole fraction for the most heavily doped films (T_{Si}=1375°C) are shown in Fig. 5. The reduction of the electron mobility by almost an order of magnitude as the Al concentration increased from 0 to 25% is attributed to alloy scattering. In reference to the data of Fig. 2, it is apparent that the reduction of the conductivity of the Si-doped films by 6 orders of magnitude is the combined effect of reduction in carrier concentration and electron mobility with the Al concentration.

In order to understand the mechanism of carrier reduction with increasing AlN mole fraction for the Si-doped films, we studied the temperature dependence of the resistivity for the films grown with the Si-cell temperature at 1325°C. These data are shown in fig.6.

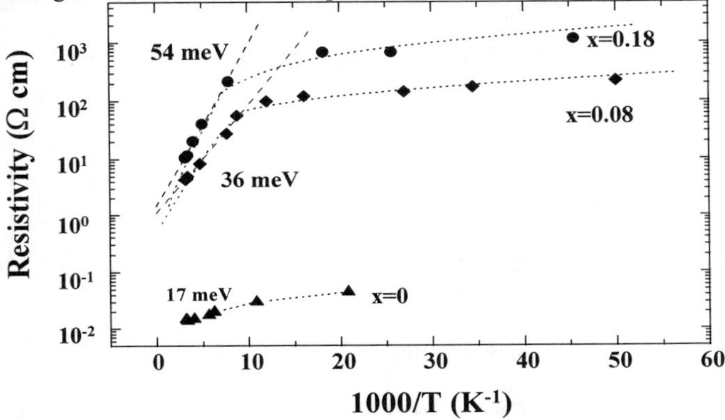

Figure 6: Temperature dependence of resistivity of $Al_xGa_{1-x}N$ films with different Al content

In all investigated samples, the resistivity becomes practically temperature independent at temperatures below 100K while it is thermally activated at higher temperatures. The temperature independent behavior at low temperatures was attributed by Molnar et al.[9] as due to impurity band conduction which is reasonable since these materials tend to be compensated[10]. The high temperature data indicate that the activation energy varies from 17meV for pure GaN to 54meV for $Al_{0.18}Ga_{0.82}N$. Thus, it appears that the Si donor states become deeper as the Al concentration increases. This result is intriguing but at this moment we do not have available a theoretical understanding.

CONCLUSIONS

In conclusion, a number of AlGaN films with Al concentration up to 60% were grown by plasma assisted MBE on sapphire and 6H-SiC substrates and doped n-type with Si and p-type with Mg. Long range ordering was observed in these AlGaN films by the existence of the (0001) diffraction peak. It was also observed that the conductivity of the Si-doped films as well as the unintentionally n-type doped films decreases with AlN mole fraction. Hall effect measurements on the same films indicate that this decrease is due to a combined effect in reduction in carrier concentration and electron mobility. The reduction of carrier concentration was correlated with the deeper ionization energies of Si as the AlN mole fraction increases. The reduction in mobility with AlN mole fraction was attributed to alloy scattering. The doping of AlGaN films with Mg led to material with resistivities ranging from 3 to 30 Ω-cm.

ACKNOWLEDGEMENTS

The authors wish to thank Mrs Anlee Krupp for assistance with SEM and EDS facilities.
The work was funded by DARPA (Grant No MDA972-95-3-0008)

REFERENCES

1) S. Strite and H. Morkoc, J. Vac. Sci. Technol. B <u>10</u>, 1237 (1992)

2) X. Zhang, P. Kung, A. Saxler, D. Walker, T.C. Wang and M. Razeghi, Appl. Phys. Lett. <u>67</u>, 1745 (1995)

3) M.D. Bremser *et al.,* MRS Internet Journal, Vol. 1, 8 (1996)

4) T.D. Moustakas and R.J. Molnar, Mat. Res. Soc. Symp. Proc., Vol 281, p. 753 (1993)

5) D.Korakakis *et al.,* MRS Internet Journal, Vol. 1, 8 (1996)

6) A. Guinier, "X-Ray Diffraction", W.H. Freeman and Co, (1963)

7) D. Korakakis, K. Ludwig and T.D.Moustakas, Appl. Phys. Lett.(to be published)

8) S. Yoshida, S Misawa and S. Gonda, J. Appl. Phys. <u>53</u>, 6844 (1982)

9) R.J. Molnar, T. Lei and T.D. Moustakas, 1993, Appl. Phys. Lett., 62, 72

10) Mott, Twose, 1961, Adv. Phys., 10, 107

MICROSTRUCTURES OF AlN BUFFER LAYERS FOR THE GROWTH OF GaN ON (0001) Al$_2$O$_3$

M. YEADON[*], W. KIM[**], A.E. BOTCHKAREV[**], S.N. MOHAMMAD[**], H. MORKOC[**], J.M. GIBSON[*]
[*]Materials Research Laboratory, myeadon@uiuc.edu
[**]Coordinated Science Laboratory, University of Illinois at Urbana-Champaign, Urbana, IL 61801

ABSTRACT

III-nitride semiconductors are emerging as highly promising candidates for the fabrication of wide band-gap electronic and opto-electronic devices. Sapphire (α-Al$_2$O$_3$) is currently one of the primary substrates of choice for the growth of GaN despite a large lattice mismatch. Significant improvements in the quality of III-nitride layers have been demonstrated by exposure of the substrate to reactive nitrogen species followed by deposition of a low temperature AlN or GaN buffer layer. In this paper we present a study of the evolution of the surface topography and defect microstructure of nitrided α-Al$_2$O$_3$ substrates and AlN buffer layers deposited by reactive molecular beam epitaxy (RMBE). Their influence on the morphology and properties of GaN layers is also discussed. Both nitridation time and AlN deposit thickness were varied systematically, at different temperatures and buffer growth rates. The microstructures were characterized using the atomic force microscope (AFM) and transmission electron microscope (TEM). Initial growth studies are ideally suited to *in-situ* experiments, and further investigations are also in progress using a unique UHV TEM with the facility for *in-situ* RMBE.

INTRODUCTION

Group III-nitride semiconductors, possessing wide bandgaps, have emerged as highly promising materials for use in electronic and optoelectronic device applications [1]. Significant progress has occurred recently in the development of high-efficiency blue light emitting diodes (LEDs) [2] and electronic devices [3-5]. Sapphire is frequently the substrate of choice for the growth of epitaxial III-nitride devices, despite large lattice and thermal mismatches. Thin, low temperature AlN [6,7] or GaN [8] buffer layers have been demonstrated to significantly enhance the quality of subsequent GaN layers grown by the MOCVD technique. Koide *et al.* [9] suggested that the low temperature AlN buffer layer improves the c-axis alignment of the epilayer by increasing the density of nucleation sites. The buffer layer is believed to promote lateral growth of the GaN film by decreasing the interfacial free energy between the film and the substrate [6]. Furthermore, it has been demonstrated that exposure of the sapphire surface to ammonia flow prior to growth of the buffer may lead to further improvement in electrical and optical properties despite an increased surface roughness [10]. Recent AFM investigations of nitrided sapphire substrates in MOCVD revealed that nitridation of sapphire for extended periods results in a high density of surface protrusions which favor 3D growth of the subsequently deposited GaN.

In this paper we present a study of the influence of different substrate nitridation conditions and thin film growth parameters on the surface morphologies and defect microstructures of RMBE-grown AlN buffer layers using AFM and TEM. Preliminary results

Mat. Res. Soc. Symp. Proc. Vol. 449 © 1997 Materials Research Society

of a study of the influence of the buffer layer on the surface morphology and electrical properties of the GaN active layer are also presented.

EXPERIMENT

The samples studied in the present work were grown in a conventional Riber 1000 MBE system modified for RMBE by addition of an ammonia gas injector. Deposition was performed onto acid-etched c-plane sapphire substrates mounted on a resistively heated Mo block. The substrates were heated to 850°C in the vacuum chamber for 3 minutes to desorb surface contaminants before exposure to ammonia (16sccm) at either 800 or 850°C. A series of samples were exposed to ammonia flow for various times between 1 and 30 minutes. The surface roughness was determined using a TopoMetrix Explorer AFM with oxide-sharpened silicon nitride tips. Nitrogen content at the surface of the samples was analyzed using a PHI 5400 XPS system with Mg target.

This experiment was then repeated, followed by deposition of AlN films with a systematic variation of film thickness at various deposition rates. The influence of nitridation time on the surface roughness and microstructure of these films was studied using AFM and TEM. GaN layers 2µm in thickness were then deposited on a series of AlN buffer layers of various thickness over the range 40-400nm. The surface roughness of these layers was again determined using AFM, and the carrier concentrations and mobilities determined from Hall measurements.

RESULTS

<u>Nitridation</u>

Examination of the surfaces of sapphire substrates nitrided for 1, 5, 15 and 30 minutes at 800°C revealed the continued presence of residual surface damage from the substrate fabrication process. The surface roughness was found to increase with exposure to ammonia, and after 30 minutes the presence of a high density ($\sim 10^8 \text{cm}^{-2}$)of surface outgrowths was observed. In figure 1 we present AFM images of the surfaces of sapphire substrates (a) without exposure to NH_3 and (b) after 30 minutes exposure at 800°C.

The surface morphology has clearly coarsened after exposure to NH_3 flow, in addition to the introduction of the surface protrusions. There is no clear correlation between the positions at which the protrusions have formed and the local surface topography. Similar protrusions were observed by Uchida *et al*. [11] after nitridation with ammonia at 1050°C in an MOCVD reactor, however the observed density after 20 minutes exposure was approximately three orders of magnitude greater than that observed here (figure 1(b)). It is likely that the combination of a higher substrate temperature and background ammonia pressure promotes a more rapid nitridation reaction leading to a higher density of protrusions in MOCVD-grown samples.

To further study the influence of temperature on surface topography during nitridation, the substrate temperature was increased to 850°C and fresh samples exposed to ammonia flow (16sccm) for periods of 1, 5 and 15 minutes. The surface roughness of these samples was found to be approximately 30% lower than those nitrided at 800°C and examination of the images

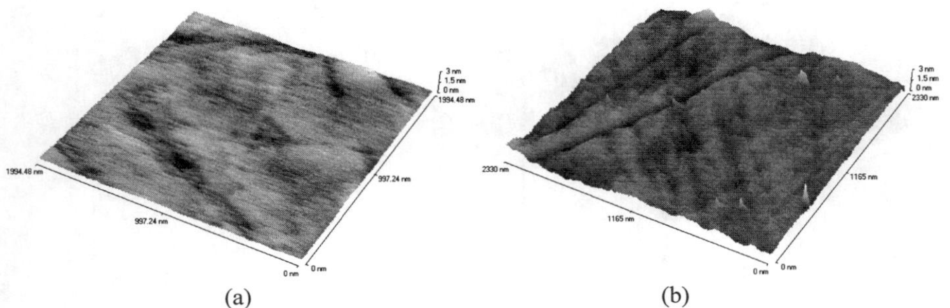

(a)   (b)

Figure 1   AFM images (2μm x 2μm) of the surface morphologies of (0001) α-Al₂O₃ (a) without exposure to NH₃ and (b) after exposure to NH₃ for 30 mins at 800°C (z markers represent 3nm).

suggests a smoothening of the surface damage. The nitridation process is believed to yield AlN_xO_{1-x} via the exchange of oxygen from the sapphire with nitrogen from dissociated ammonia. XPS analyses of our nitrided samples confirm the incorporation of nitrogen into the sapphire surface. The control sample, not exposed to NH₃, showed essentially no nitrogen 1s peak. Clear peaks were detectable, however, for the samples nitrided for 1 minute and 30 minutes, the intensity of the peak increasing with nitridation time confirming the presence of nitrogen close to the surface of our samples.

The position of the oxygen 1s line after 30 minutes ammonia exposure was found to be down-shifted by 0.25eV relative to the spectra obtained from both the samples nitrided for 0 and 1 minute. The origin of this shift may be attributed to the generation of a significant number of new bonding states between aluminum, oxygen and nitrogen atoms in the nitrided layer.

AlN Growth

The influence of nitridation on the surface morphology and defect microstructure of AlN layers subsequently deposited was investigated by comparing two samples, of similar thickness, deposited on sapphire surfaces nitrided for 1 and 10 minutes respectively. Although the substrate surface roughness increases with nitridation time, the roughness of the AlN layer was found to be inversely proportional to this nitridation time. An AFM image of the AlN layer deposited on a surface nitrided for 1 min is shown in figure 2(a). The morphology is composed of a high density ($\sim10^9 cm^{-2}$) of surface protrusions between 2.5 and 4.5nm in height. The surface of the film deposited after 10 minutes nitridation, however, exhibits a higher density of smaller protrusions on the surface, between 2 and 3.5nm in height (figure 2(b)). Plan-view TEM specimens fabricated from these samples showed both films to be highly oriented with the substrate, exhibiting the epitaxial orientation relationship $(0001)_{AlN}$ // $(0001)_{sub}$, $[1010]_{AlN}$ // $[1120]_{sub}$. Electron micrographs such as that presented in figure 2(c) confirmed the 3D nature of the microstructure. The dimensions of the crystallites observed by TEM corresponded to the dimensions of the surface features observed by AFM; however, the presence of cracks in the film

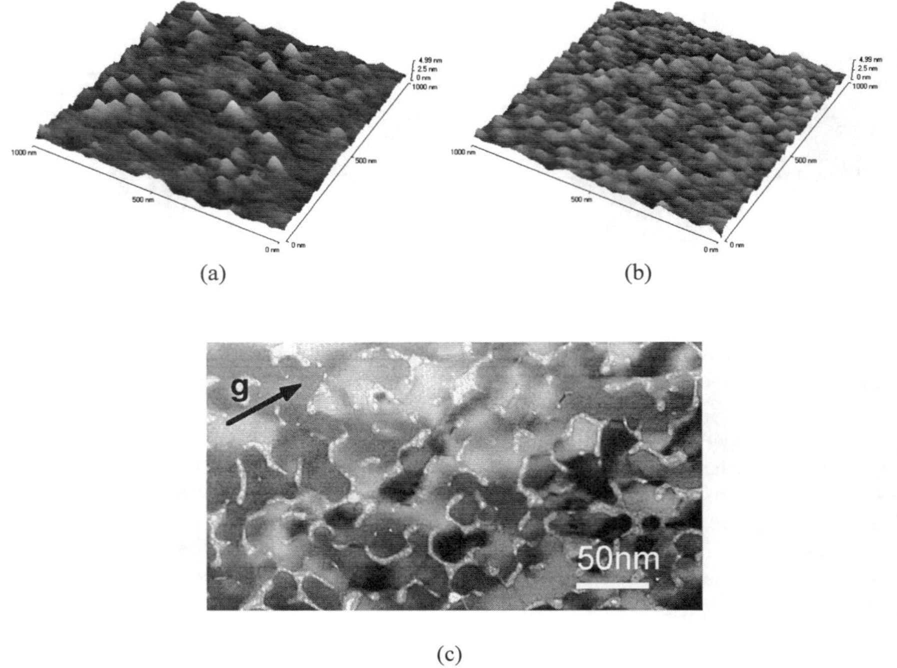

(a)　　　　　　　　　　(b)

(c)

Figure 2 AFM images (1μm x 1μm) of the surface morphology of 40nm thick AlN films deposited on (0001) α-Al$_2$O$_3$ after nitridation of the surface for (a) 1 minute and (b) 10 minutes (z markers represent 5nm). (c) Plan-view TEM micrograph of the sample shown in (a), g=1120.

not discernible by AFM are clearly evident in electron micrographs such as that presented in figure 2(c).

Our observations suggest an increase in the density of nucleation sites on the substrate surface with increasing nitridation time, leading to increased uniformity of the surface morphology and further investigations of these critical stages are planned using a UHV TEM with *in-situ* RMBE chamber.

The development of surface morphology was also investigated as a function of film thickness and growth rate. For a given substrate nitridation time, it was found that a significant increase in buffer layer surface roughness was obtained with increasing thickness. AFM images of the surface morphology of samples 40nm and 200nm in thickness are presented in figure 3(a) and (b), respectively, deposited on α-Al$_2$O$_3$ after 1 minute nitridation at 850°C. With increasing film thickness the surface undergoes a ripening transition leading to a coarse surface morphology. TEM investigations indicate that these samples do not become polycrystalline as has been reported on 150nm-thick MOVPE-grown AlN on sapphire [6]. Rather, the films remain epitaxial and exhibit a high density of defects close to the interface with the substrate, the density

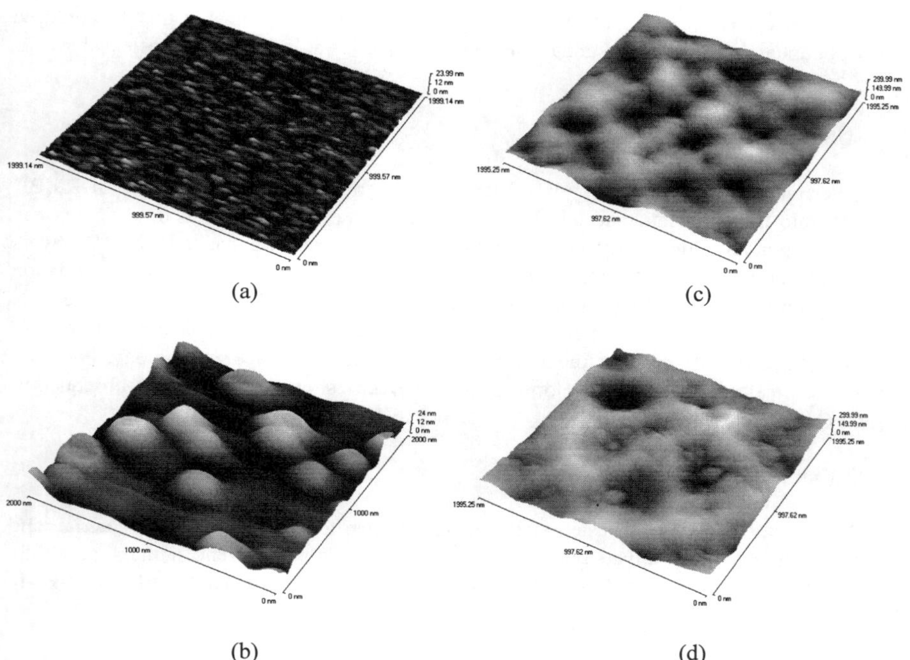

Figure 3 AFM images (2μm x 2μm) of the surface morphologies of (a) an AlN film 40nm thick and (b) an AlN film 200nm thick, (c) a 2μm GaN film grown on an AlN buffer identical to that shown in (a), and (d) a 2μm GaN film grown on an AlN buffer identical to that shown in (b). Z markers represent 24nm in (a) and (b), and 300nm in (c) and (d).

decreasing rapidly with distance from this interface as is typically observed in GaN films, e.g. [12]. Increasing the growth rate for a given layer thickness showed a dramatic decrease in the surface roughness of our films, however, indicating a classical surface migration rate-limiting mechanism.

Finally, the influence of the AlN buffer layer on the microstructure and electrical properties of GaN active layers subsequently deposited was investigated. A series of AlN layers of thickness between 40 and 200nm were grown under identical conditions to those shown in figure 3(a) and (b). The corresponding surface morphologies of the GaN layers, 2μm in thickness, are shown in figure 3(c) and (d), respectively. The GaN surfaces are composed of characteristic hexagonal domains whose average diameter increases with buffer layer thickness. Comparison of the images indicates that the morphology of the buffer layer has a direct influence on that of the active layer. It is apparent that the increase in AlN crystallite size with buffer thickness leads to a reduction in the density of nucleation sites for the active layer promoting the development of hexagonal GaN domains of increased diameter. Preliminary measurements of the electrical properties of these samples indicate an improvement in carrier mobility with buffer layer thickness; whilst the origin of this reduction is not clear it is speculated that the reduced density of boundaries between hexagonal domains in the GaN is at least partially responsible.

CONCLUSIONS

Investigations of the influence of substrate nitridation on the microstructure of RMBE grown AlN thin films indicate that the nitridation process in our system leads to an increase in the surface roughness of the substrates. AFM images of subsequently deposited thin AlN layers (40nm in thickness) indicate an improvement in surface uniformity with substrate nitridation time, suggesting that nitridation results in an increased density of nucleation sites for the AlN epilayer, which favors a three-dimensional nucleation and growth mechanism. Buffer layer thickness and growth rate have a more dramatic influence on the surface morphology, however, and exert a critical influence on the morphology and electrical properties of the subsequently deposited GaN layer.

ACKNOWLEDGEMENTS

This work was performed using facilities in the Center for Microanalysis of Materials and Beckman Institute at the University of Illinois, Urbana-Champaign. The authors gratefully acknowledge the Office of Naval Research for the funding of this work under grant N00014-95-1-0324.

REFERENCES

1. S.N. Mohammad, A. Salvador and H. Morkoc, Proc. IEEE **83**, p. 1306 (1995).
2. S. Nakamura, M. Senoh, N. Iwasa and S. Nagahama, Jpn. J. Appl. Phys. **34**, p. L797 (1995).
3. Z.-F. Fan, S.N. Mohammad, O. Aktas, A.E. Botchkarev, A. Salvador, H. Morkoc, Appl. Phys. Lett. (in press).
4. S.N. Mohammad, Z.-F. Fan, A. Salvador, O. Aktas, A.E. Botchkarev, W. Kim, and H. Morkoc, Appl. Phys. Lett., submitted.
5. O. Aktas, W. Kim, Z.-F. Fan, A.E. Botchkarev, A. Salvador, S.N. Mohammad, B. Sverdlov, H. Morkoc, Electron. Lett. **31**, p. 1389 (1995).
6. I. Akasaki, H. Amano, Y. Koide, K. Hiramatsu and N. Sawaki, J. Crystal Growth **98**, p. 209 (1989).
7. J.N. Kuznia, M. A. Khan, D.T. Olson, R. Kaplan and J. Freitas, J. Appl. Phys. **73**, p. 4700 (1993).
8. S. Nakamura, Jpn. J. Appl. Phys. **30**, p. L1705 (1991).
9. Y. Koide, N. Itoh, K. Itoh, N. Sawaki and I. Akasaki, Jpn. J. Appl. Phys. **27**, p. 1156 (1988).
10. C.-Y. Hwang, M. J. Schurman, and W. E. Mayo, Y. Li and Y. Lu, H. Liu, T. Salagaj, and R. A. Stall, J. Vac. Sci. Technol. A **13** (3), p. 672 (1995).
11. K. Uchida, A. Watanabe, F. Yano, M. Koguchi, T. Tanaka, and S. Minagawa, J. Appl. Phys. **79** (7), p. 3487 (1996).
12. W. Qian, M. Skowronski, M. De Graef, K. Doverspike, L.B. Rowland and D.K. Gaskill, Appl. Phys. Lett. **66** (10), p. 1252 (1995)

Growth and Characterization of AlN on 6H-SiC Substrates

M. Lekova*, G.W. Auner**, F. Jin**, R. Naik***, and V. Naik****,
*Dept. of Materials Science and Engineering, Wayne State University, Detroit, MI, **Dept. of Electrical and Computer Engineering, Wayne State University, Detroit, MI, ***Dept. of Physics, Wayne State University, Detroit, MI, **** Dept. of Physics, University of Michigan Dearborn, Dearborn, MI

ABSTRACT

AlN/SiC structures are of current importance for the development of high temperature electronic devices. A systematic study of the growth and structure of AlN on Lely grown 6H-SiC was performed. The SiC substrates were chemically etched followed by thermal desorption at 875 °C under UHV conditions. AFM and RHEED images show the relative improvement in surface morphology following the cleaning procedure. AlN films were grown on 6H-SiC substrates using plasma source molecular beam epitaxy (PSMBE). Substrate temperatures were varied from 400 °C to 800 °C and deposition energies from 1 eV to 25 eV during AlN growth. The AlN crystal structure and topology were analyzed by RHEED and AFM. The AlN/SiC film microstructure are correlated with the quality of the SiC surface and the parameters of the AlN deposition process. Epitaxial growth of AlN(0002) was strongly dependent on the SiC surface quality and deposition energy. The highest quality AlN was achieved at growth temperatures of approximately 600 °C and 15 eV ion energy. Higher ion energy resulted in amorphous film growth.

INTRODUCTION

AlN is currently being investigated for optoelectronic, SAW and high temperature electronic devices.[1-2] SiC is of interest as a substrate for AlN because of its close lattice match and similar thermal expansion characteristics [6]. The surface roughness and microstructure can influence the type of growth and hence the quality of the AlN film [5-6]. Furthermore, AlN/SiC MIS type structures are of importance for the development of high temperature sensors and power electronics. 6H-SiC also has a closer lattice match (mismatch under 1%) to AlN then other commonly used substrates such as silicon and sapphire. These qualities make AlN/SiC systems of particular technological interest. An understanding of the growth, interface structure and the development of optimal deposition parameters are critical to developing repeatable high quality devices. Therefore, as part of a larger study of AlN/SiC systems, a preliminary study on the growth and structure of AlN on Lely grown 6H-SiC substrates was performed.

EXPERIMENT

AlN films were grown on Lely 6H-SiC substrates using Plasma Source Molecular Beam Epitaxy (PSMBE). Details of the deposition system are described elsewhere [3]. The deposition species are generated in a hollow cathode plasma deposition source shown in Fig. 1. The source uses r.f. power to generate a nitrogen/argon (argon optional) plasma. The very intense plasma efficiently dissociates nitrogen and an internal sputter process provides a source of aluminum ions. The aluminum and nitrogen ions (both mono and diatomic) are accelerated to the substrate by a d.c.

Mat. Res. Soc. Symp. Proc. Vol. 449 © 1997 Materials Research Society

biasing system. The energy of the extracted ions is rather narrowly resolved. The ions coalesce at the heated rotating substrate holder where AlN crystalline films are formed.

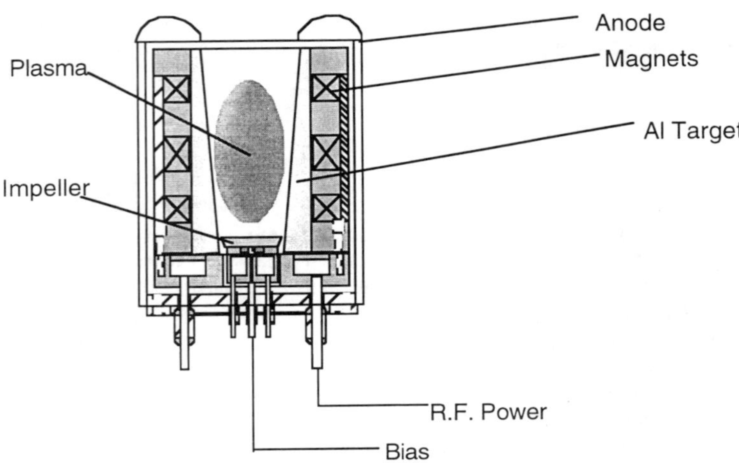

Fig. 1 Illustration of hollow cathode source for Plasma Source MBE system

The SiC substrates were pre-cleaned by standard degreasing followed by a chemical etching procedure. The chemical etching [4] was performed using H_2SO_4:HNO_3 (1:1) at 80 °C for 30 seconds followed by HCl:H_2O_2 (5:3) at 60 °C for 30 seconds followed by immersion in a 10% HF solution for 60 seconds to remove surface oxides and passivate the surface. The last two steps were repeated several times. The substrates were subsequently loaded into the UHV chamber under a typical base vacuum pressure of 6×10^{-10} Torr. The samples were then heated to 850 °C for 1 hour as a final cleaning step before deposition. Source power was set at 250 watts r.f., 20 sccm nitrogen flow, and a synamic pressure of 7×10^{-4} Torr. RHEED and AFM studies of the substrate were performed before and after cleaning to determine crystal quality. The samples were analyzed during growth by a differential pumped 35 keV Staib Instruments RHEED system with a K-space analysis system.

RESULTS AND DISCUSSION

As a part of a larger study of the growth of AlN on various types of SiC substrates, we performed preliminary work on the deposition of textured AlN films on Lely grown 6H-SiC substrates. Lely grown substrates have high crystalline quality but have rather scratched surfaces. X-ray diffraction analysis showed 6H-SiC were oriented slightly off c-axis. The AFM and RHEED images in Fig. 2 show the change in surface quality after chemical etching and after subsequent heating for 1 hour at 850 °C. The heating cycle clearly shows a better and smoother substrate surface and sharper streak RHEED patteren indicative of fairly high crystal quality surface.

Subsequent to the cleaning and preparation procedures, a series of experiments were performed in order to determine the effect of deposition energy and substrate temperature on growth morphology. Figure 3 shows the AFM and RHEED images for AlN deposition using aluminum and nitrogen ion energies from 1 eV to 25 eV. Deposition with ion energies of

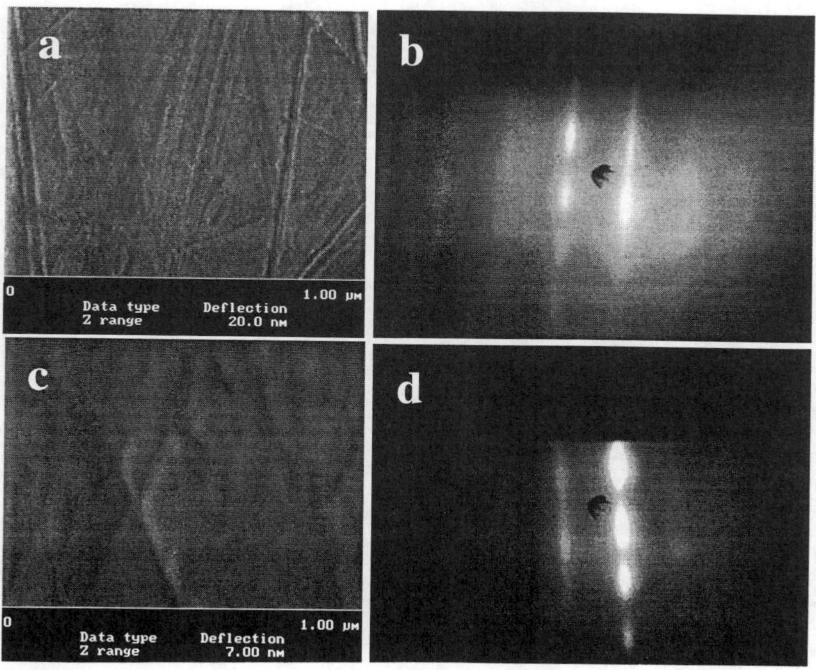

Fig. 2. AFM image (a) and the corresponding RHEED image (b) of the Lely grown 6H-SiC surface after chemical etching. AFM image (c) and the corresponding RHEED image (d) of the Lely grown 6H-SiC surface after heating for 1 hour above 850 °C.

approximately 1 eV at 600 °C result in a mostly textured, but partly polycrystalline, AlN film. Deposition at this energy on silicon or sapphire substrates have previously resulted in very polycrystalline films of relatively poor quality. However, the close lattice match of the 6H-SiC surface is more conducent to epitaxial growth. Subsequent increased ion energy results in higher quality and highly textured AlN films. The additional energy adding to the surface mobility of aluminum and nitrogen clearly results in better quality films and probably a higher degree of epitaxy. Figure 3b shows a streaky pattern of a rather high quality AlN film. Furthermore, the AFM image indicates a smooth surface. Growth between 12 eV and 15 eV on Si and sapphire substrates in previous studies have also found smooth highly textured AlN films. However, as the ion energy increases, a damage threshold (energy above approximately half the bulk displacement energy) is reached and the as deposited AlN is transformed to a mixture of crystaline and amorphous film. The highest deposition energy (25 eV) results in an apparently completely amorphous film.

Additional experiments were performed to determine the effect of substrate temperature on growth morphology. AlN was deposited on SiC substrates with a deposition energy of 15 eV.

The substrate temperature was varied from 400 °C to 800 °C. Growth at 400 °C is usually used as a buffer layer for higher temperature growth on silicon and sapphire substrates. At 400 °C a mildly rough surface can be seen by AFM in Figure 4a. RHEED images show a combination of streaky

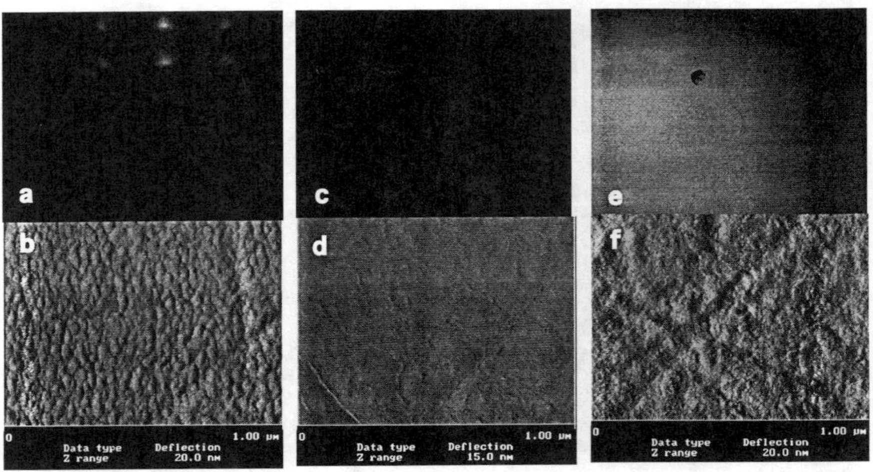

Fig. 3. a) RHEED image and b) AFM image of AlN film grown at 600 °C and 1 eV. c) RHEED image and d) AFM image of AlN film grown at 600 °C and 15 eV. e) RHEED image and f) AFM image of AlN film grown at 600 °C and 25 eV.

and spotty pattern. The AlN is textured with a AlN(0002) orientation. As the temperature increases, the added surface mobility gives a much smoother surface. AlN RHEED pattern indicates a textured growth. As the substrate temperature continues to increase, the AlN surface becomes more rough and the RHEED pattern more diffuse- possibly indicating a more 3-dimensional growth mode.

Figure 5 shows AFM images of Lely Grown 6H-SiC substrates and then subsequent growth of AlN. The films were grown with an ion energy of 15 eV. Snap shots of the RHEED images were taken of: (i) the SiC substrate, (ii) after a 200Å buffer layer was deposited at 400 °C, (iii) after a total of 500Å growth with 300Å grown at 600° C, and (iv) after 3400Å growth (3200Å at 600 °C). The RHEED images show a fairly well ordered mostly textured AlN buffer layer and subsequent 300Å. As the growth continues the AlN crystal improves and the final result is a high quality textured AlN(0002) orientation film. The evolution of the lattice mismatch (in terms of strain) is shown in the graph (Fig. 6). The graph shows a slight lattice mismatch (probably less than 1%) that gradually relaxes to the normal AlN lattice spacing. Most of the lattice change occurs during and immediately following the 400 °C buffer region.

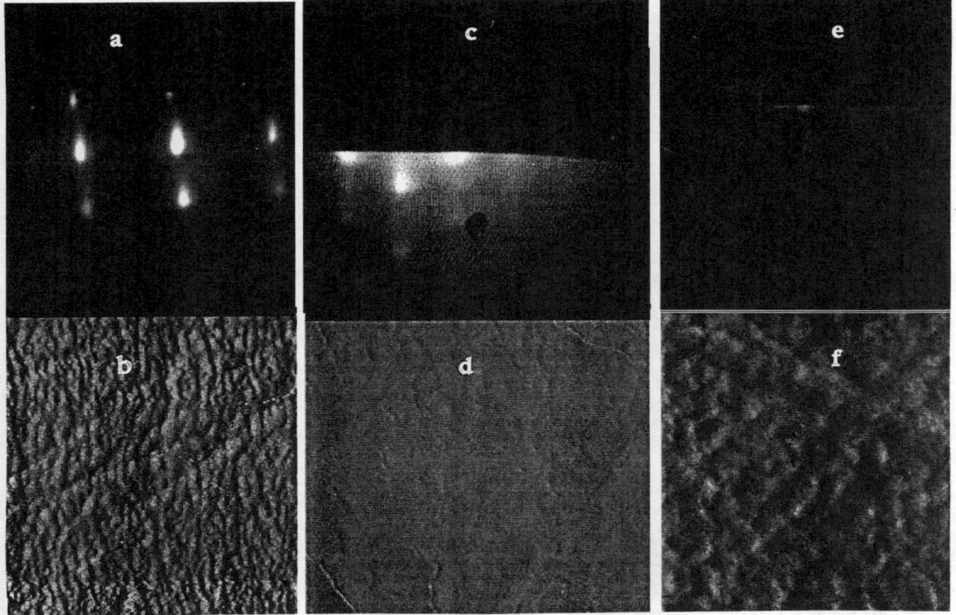

Fig. 4. a) RHEED image and b) AFM image of AlN film grown at 400 °C and 15 eV. c) RHEED image and d) AFM image of AlN film grown at 600 °C and 15 eV. e) RHEED image and f) AFM image of AlN film grown at 800 °C and 15 eV.

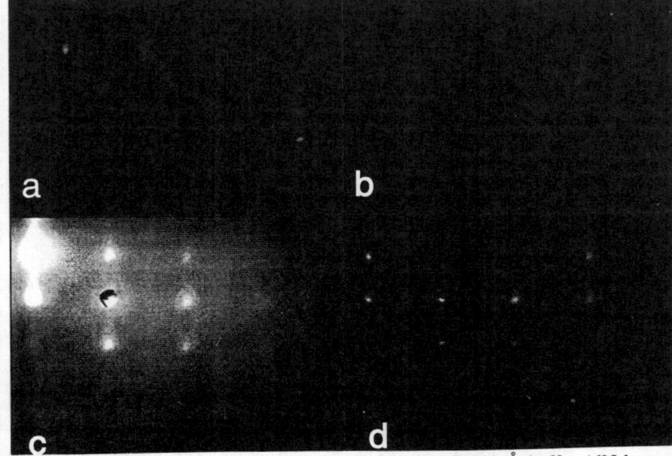

Fig. 5. RHEED images of a) the initial Lely grown 6H-SiC substrate; b) 200 Å buffer AlN layer grown at 400 °C and 15 eV; c) 500 Å AlN consisting of 200 Å buffer layer and 300 Å AlN grown at 600 °C and 15 eV; d) 3400 Å AlN consisting of 200 Å buffer layer and 3200 Å AlN grown at 600 °C and 15 eV.

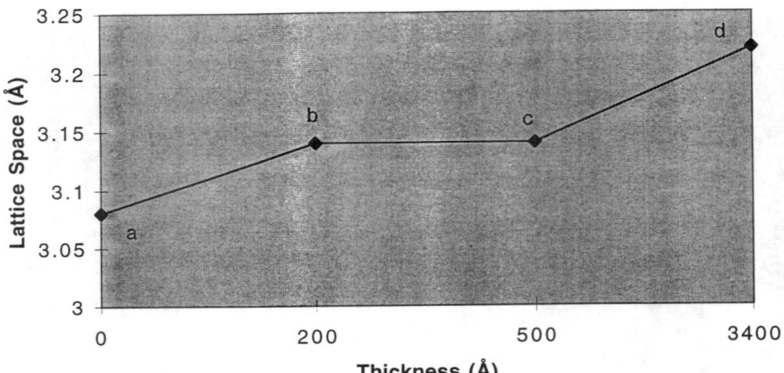

Fig. 6. Graph of the difference in lattice spacing vs. film thickness showing a) SiC/AlN interface; b) 200 Å buffer layer of AlN; c) 500 Å AlN (200 Å buffer layer); d) 3400 Å AlN (200 Å buffer layer).

CONCLUSIONS

AlN films 700-800 Å thick were grown on Lely grown 6H-SiC substrates by new energy controlled plasma source meolecular beam epitaxy system. Lely grown 6H-SiC substrates are noteably rough and scratched surfaces. AFM showed sugnificant improvement in surface roughness and RHEED images indicate good crystalinae quality after chemical etching followed by in-situ heating at 850 oC. Subsequent to sample cleaning and preparation, a systematic experiment varying substrate temperature (range 400 °C - 800 °C) and deposition energy (1 eV to 25 eV) during AlN deposition was performed. The best quality films were obtained at 600 °C temperature and 15 eV bias on the substrates. 3400 Å thick AlN film was grown on the SiC of substrate at 600 °C and 15 eV. RHEED images showed a thickness dependent increased quality AlN film coincident with a lattice mismatch that gradually relaxed, mostly in the buffer layer region, to near bulk AlN lattice spacing.

ACKNOWLEDGMENTS

This work was supported by the National Science Foundation (EEC-9420568) and the Institute for Manufacturing Research at Wayne State University.

REFERENCES

1. S.Strite and H.Morkoc, J.Vac.Sci.Technol.**B10(4)**,1237(1992).
2. J.H. Edgar, J.Mater.Res.**7(1)**,235(1992).
3. G.W.Auner, T.D.Lenane, F.Ahmad, R.Naik, P.K.Kuo, Z.Wu, Wide Band Gap Electronic Materials, 329, Academic Publishers 1995, Printed in Netherlands.
4. M.E. Lin, S.Strite, A.Agarwal, A.Salvador, G.L.Zhou, N.Teraguchi, A.Rockett, H.Morcoc, Appl.Phys.Lett. **62(7)**,702(1993).
5. F.A.Ponce, C.G.Van de Walle, J.E.Northrup, Phys. Rev. **B 53(11)**, 7473(1996).
6. L.B.Rowland, R.S.Kern, S.Tanaka, R.F.Davis, J.Mater.Res., **8(9)**, 2310(1993).

EFFECT OF STOICHIOMETRY ON DEFECT DISTRIBUTION IN CUBIC GaN GROWN ON GaAs BY PLASMA-ASSISTED MBE

S. RUVIMOV, Z. LILIENTAL-WEBER, and J. WASHBURN,
Lawrence Berkeley National Laboratory, Berkeley 94720, CA

T. J. DRUMMOND, M. HAFISH, and S. R. LEE,
Sandia National Laboratories, Albuquerque, NM 87185

ABSTRACT

High resolution electron microscopy has been applied to characterize the structure of β-GaN epilayers grown on (001) GaAs substrates by plasma-assisted molecular-beam-epitaxy. The rf plasma source was used to promote chemically active nitrogen. The layer quality was shown to depend on growth conditions (Ga flux and N_2 flow for fixed rf power). The best quality of GaN layers was achieved by "stoichiometric" growth; Ga-rich layers contain a certain amount of the wurtzite phase. GaN layers contain a high density of stacking faults which drastically decreases toward the GaN surface. Stacking faults are anisotropically distributed in the GaN layer; the majority intersect the interface along lines parallel to the "major flat" of the GaAs substrate. This correlates well with the observed anisotropy in the intensity distribution of x-ray reflexions. Formation of stacking faults are often associated with atomic steps at the GaN-GaAs interfaces.

INTRODUCTION

GaN currently attracts the interest of many researchers [1-16] because it is a promising material for device applications [1,2]. GaN crystallizes either in the stable wurtzite (α-GaN) or metastable zinc-blende (β-GaN) structure depending on growth conditions and substrate structure [5,8,11,12]. The metastable GaN polytype has some advantages with respect to the hexagonal one, namely, easy cleavage, smaller bandgap and higher carrier mobilities. β-GaN can be grown on various substrates including GaP [3,15], β-SiC [5,12,14], and Si [5,12], but the GaN/GaAs is of particular interest because GaAs is widely used for opto- and microelectronics. The growth of GaN on GaAs is complicated owing to the large lattice mismatch (~20%) and the substantially different chemistry of As and N [3,8,13-16]. On the other hand, the large mismatch in lattice parameters between the epilayer and the substrate results in full strain relaxation of the system after completion of the first or second GaN monolayer. The small size and great reactivity of nitrogen atoms relative to As atoms tends to cause deterioration of the GaAs surface: it is not stable under exposure to chemically active nitrogen. Formation of an islanded GaN layer with a highly facetted interface has been observed with exposure of the GaAs surface to chemically active nitrogen which is a significant fraction of atomic nitrogen produced by an rf plasma source. Although interfacial facetting was shown to decrease the defect density in the GaN layer, it is not suitable for device applications. To improve the structure of the GaN layer while maintaining a flat interface, optimization of the nucleation process is necessary.

The goal of the present study is to better understand how nitrogen reacts with a GaAs (001) surface and elucidate the correlation between interface structure and growth conditions. High resolution electron electron microscopy (HREM) was applied to several GaN/GaAs structures grown under different Ga fluxes for fixed nitrogen plasma conditions and fixed substrate temperature. The results show that there is an optimal Ga flux for an otherwise fixed set of nucleation parameters that results in a smooth GaN/GaAs interface and relatively low defect density in the GaN epilayer.

Mat. Res. Soc. Symp. Proc. Vol. 449 © 1997 Materials Research Society

EXPERIMENT

The growth of 50-100 nm thick GaN layers was carried out using a RIBER 32P molecular beam epitaxy (MBE) system. An Oxford Applied Research MPD21R rf plasma source operated at 300 W with a nitrogen flow of 2 sccm was employed to generate active nitrogen. Optical characterization of the plasma revealed the presence of N°, N^+, N^{2+}, and N^{2*}. The GaN was deposited at 620 °C on an n-/n+ GaAs:Sn buffer layer on an n+ GaAs:Si substrate. The buffer layer was grown at 580 °C on a GaAs (001) substrate using As_4 supplied through a valved cracking cell with the cracking zone run at 600 °C to avoid cracking the arsenic. Upon completion of the buffer layer growth was interrupted while the substrate temperature was adjusted to 620 °C under an As_4 flux. The upper n-GaAs layer doped at a level of 2×10^{16} cm^{-3} had a conventional As stabilized (2x4) surface reconstruction as observed by reflection high energy electron diffraction (RHEED). After stabilizing the substrate temperature the plasma was initiated behind a closed shutter. Nucleation proceeded by opening the plasma shutter immediately followed by opening the Ga shutter. The As flux remained on for an estimated eight monolayers of the GaN growth. Optimal nucleation was obtained by simultaneously applying equal fluxes of Ga and active nitrogen to the substrate at the highest possible growth rate. This minimizes nitrogen diffusion into the substrate by burying the interface as rapidly as possible. The Ga to nitrogen ratio was varied by changing the Ga flux delivered to the substrate.

GaN/GaAs epitaxial layers were characterized by conventional electron microscopy and high resolution electron microscopy using two cross-sections parallel and perpendicular to the "major" flat of the GaAs wafer. Electron microscopy was carried out on JEOL 200CX and Topcon 002B microscopes operated at 200 kV.

RESULTS AND DISCUSSION

1. "Stoichiometric" growth.

The best quality of GaN layers was achieved by "stoichiometric" growth in terms of Ga to N ratio. The term "stoichiometric" is used here only in the context of nucleation of a good GaN film under these growth conditions. Fig. 1 shows TEM cross-sectional images of near-stoichiometric GaN layer taken perpendicular (a) and parallel (b) to the major flat of the GaAs wafer.

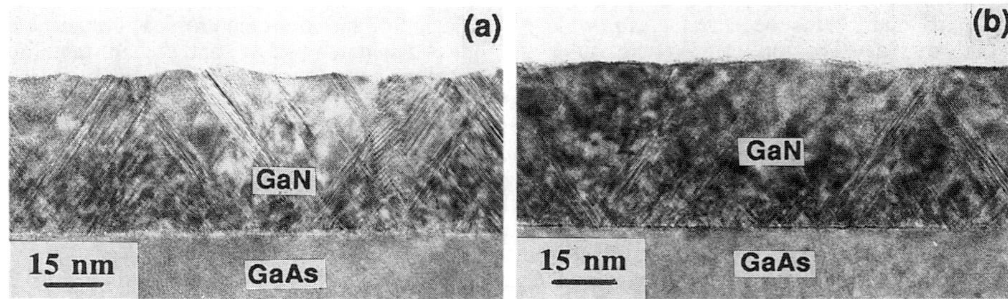

Fig.1, a-b.   Near-stoichiometric GaN layer grown on (001) GaAs. TEM cross-sectional images (a) and (b) were taken perpendicular and parallel to the major flat of the GaAs wafer, respectively.

Interfaces between the GaAs and the GaN were surprisingly abrupt for the samples under study as compared to the samples grown by other groups [12,13]. Interface roughness for the "stoichiometric" growth was in the range of 0.8-1.2 nm. Atomic structure of the GaN-GaAs interface is shown in Fig. 2. Because of the high misfit (f~20%) in lattice parameters between GaN and GaAs the interface contains a dense array of misfit dislocations extended in two perpendicular [110] and [-110] crystallographic directions. For 20% misfit f in lattice parameters almost each fifth (220) plane of the GaN lines up with each fourth (220) plane of the GaAs.

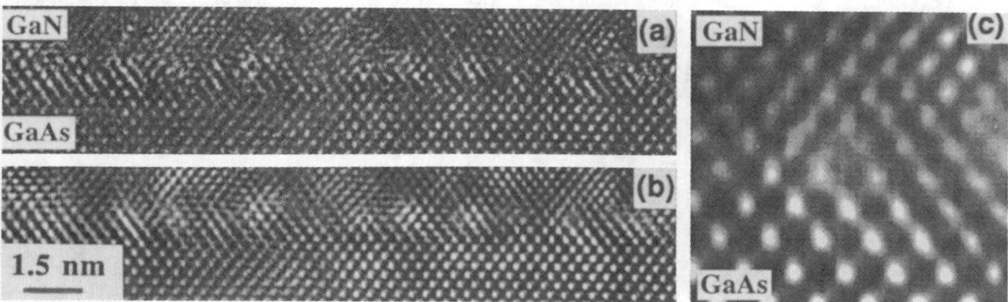

Fig.2, a-c.   Atomic structure of the GaN-GaAs interface: cross sectional HREM images before (a) and after (b) image filtering. 90° Lomer dislocation at GaN-GaAs interface is shown in (c).

For a fully relaxed GaN layer the equilibrium spacing between 90° misfit dislocations with Burgers vector **b** of 1/2[110] should be L=b/f~1.6 nm. Most observed dislocations are 60° dislocations that appear in pairs separated by a few atomic distances. These pairs of 60° misfit dislocations can be considered as dissociated 90° Lomer dislocations. About 10% of the misfit dislocations were undissociated 90° Lomer dislocations (see Fig.2 c). The average distance between misfit dislocations is approximately equal to the equilibrium value so that the GaN layer is almost fully relaxed. However, there is a local strain at the interface that is associated with inhomogenious distribution of the misfit dislocations (fluctuations in dislocation spacing).

While the Burgers vector of a 90° Lomer dislocation lies in the interface plane, 60° misfit dislocations have Burgers vectors of 1/2[110] which are inclined to the interface. Thus, 60° misfit dislocations have (111) planes as glide planes and can easily dissociate into 30° and 90° partials with formation of a stacking fault ribbon between them. Many of the 60° misfit dislocations are dissociated and, therefore, the GaN layer contains a high density of stacking faults (SFs). This results in streaking of the diffraction spots on electron diffraction patterns (Fig.3).

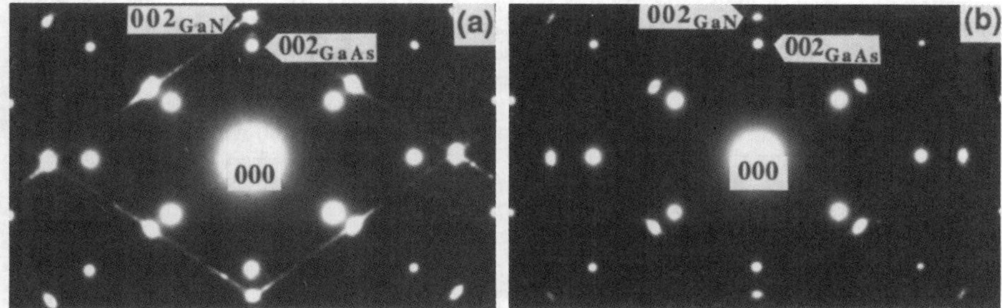

Fig.3, a-b.   Selected area diffraction patterns taken perpendicular (a) and parallel (b) to the major flat of the GaAs wafer. Note the different degree of GaN spot streaking in the figures (a) and (b).

The SF density drastically decreases away from the substrate toward the top surface in a 20 nm thick interfacial layer in agreement with previous observation. However, their density near the top of the GaN layer (Fig.1) is different by an order of magnitude in the two cross-sections the top of the GaN layer (Fig.1) is different by an order of magnitude in the two cross-sections taken perpendicular (a) and parallel (b) to the major flat of the GaAs wafer. It is lower in the cross sections taken parallel to the major flat of the GaAs wafer (Fig.1b). This asymmetry of SF distribution results in different degrees of GaN spot streaking in the selected area diffraction patterns taken perpendicular (Fig. 3a) and parallel (Fig.3b) to the major flat and in different shapes of (002) x-ray reflexions in reciprocal space (Fig.4).

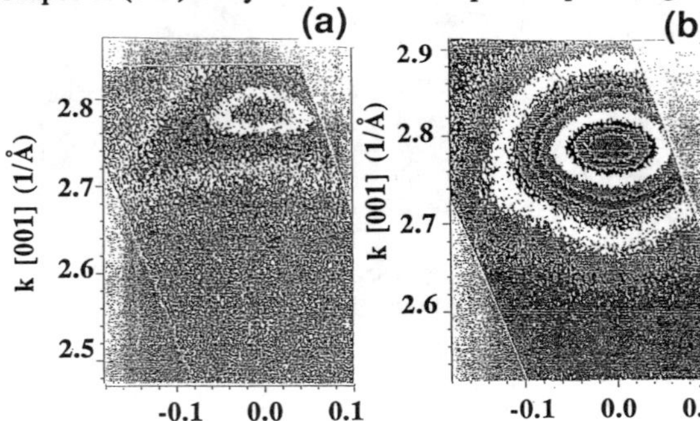

Fig.4, a-b.　x-ray KMAPs taken perpendicular (a) and parallel (b) to the major flat of the GaAs wafer. Note the different shape of (002) GaN reflexions in reciprocal space in figures (a) and (b).

The decrease of stacking fault density results from their annihilation and termination within the GaN layer. Formation of sets of stacking faults is often associated with steps at the GaN-GaAs interface so that their density correlates with interface roughness and local strain at the interface. This asymmetry of SF distribution might be associated either with the difference in the atomic structure of surface steps along [110] and [-110] crystallographic directions on the (001) GaAs or with different mobilities of α and β dislocations and their respective partials. SF density at the interface is only slightly less for the cross section parallel to the major flat while it is less by an order of magnitude for the top of GaN layer in comparison with that of the cross section perpendicular to the major flat. Similarly, an asymmetric defect distribution has been observed for other III-V compounds which was related to the asymmetry of the zinc-blende structure. Our observations suggest that the quality of the GaN layer will increase with thickness for stoichiometric growth conditions.

Near stoichiometric growth was shown to provide the best cubic GaN material [8,13]. However, GaN layers grown on (001) GaAs at 620 °C usually had very rough interfaces[8,13,16]. In order to avoid the deterioration of the GaAs surface and, hence, interface roughrning during GaN nucleation, As flux was exposed to the GaAs surface for the first eight monolayers of the GaN growth in the samples under study.

The role of this As flux during the GaN nucleation is not well understood at present, but it is reasonable to assume that As influences the behavior of Ga adatoms and, hence, their reactions with active nitrogen, resulting in changes of the density of nucleai.[13] Nether in-situ RHEED or ex-situ HREM showed any evidence that As incorporates in the GaN above the solubility limit (~1%) during nucleation. High arsenic pressure during the growth has earlier been shown to improve the quality of cubic GaN layers grown on GaAs and GaP substrutes [3,16]. On the other hand, As-doping of the GaN layer near the interface might decrease the stacking fault energy making a dissociation of 60° dislocations more likely.

2. Microstructure of "Ga-rich" and "Ga-poor" GaN layers.

Electron microscopy revealed dramatic changes in the structure of the GaN layers with variation of the Ga flux. Figure 5 shows typical cross-section electron microscopy images of GaN epilayers grown under Ga-rich conditions in terms of Ga to N ratio. The "Ga-rich" layer in Fig.5a has a columnar structure with the lateral size of the grains equal to 15-20 nm. It should not be inferred that Ga-rich growth conditions necessarily lead to the accumulation of metallic Ga on the surface. Ga-rich nucleation was expected [13] to promote a rough interface because of the high affinity of N to Ga.

Ga-rich growth leads to multiple twinning and formation of areas of the hexagonal phase during the initial stage of GaN growth in agreement with the results reported earlier [13]. Here, we observed that formation of the hexagonal phase is often associated with change in grain orientation from (001) to (111) due to the twinning (see Fig.5b).

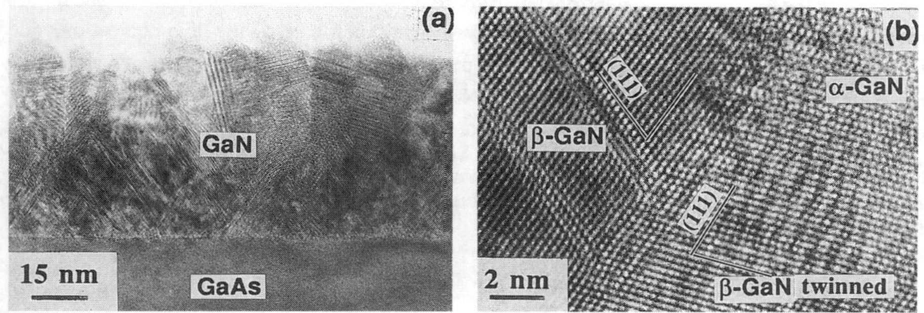

Fig.5. Structure of "Ga-rich" GaN layers grown on (001) GaAs substrate (a). Figure (b) is HREM image showing the formation of the hexagonal GaN phase on (111) planes of the GaN grain in twin orientation.

Figure 6 shows typical cross-section electron microscopy images of GaN epilayers grown under Ga-poor conditions. The GaN layer contains also contains slightly misotiented grains with amorphous like boundaries between them and a high density of stacking faults. SFs are generated at the interface and at grain boundaries as well.

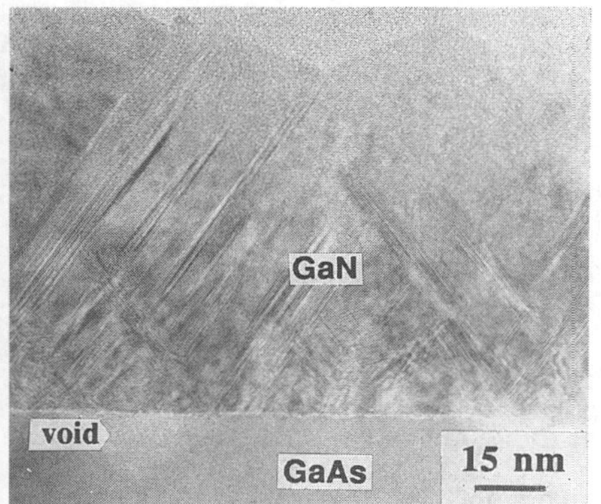

Fig.6. Structure of "Ga-poor" GaN layers grown on (001) GaAs substrate. Note the formation of pockets (voids) in the GaAs substrate below the interface

The twinned cubic and hexagonal grains were also observed at the interface with the substrate for nitrogen rich growth while the top of the GaN layer was cubic (Fig. 6). Formation of those grains was often associated with atomic steps at the interface and with amorphous pockets in the GaAs substrate.

In contrast to the "Ga-poor" layer, grains of α-GaN in the Ga-rich layers are formed at the interface and grow in columns extending through the layer (Fig.5a). Wurtzite grains are often crystallographically oriented with respect to the GaAs lattice. The predominant orientation relationship between GaN grains and the GaAs substrate is as follows: $\{0001\}_{GaN} \| \{001\}_{GaAs}$, $[1\text{-}210]_{GaN} \| [\text{-}110]_{GaAs}$, $[10\text{-}10]_{GaN} \| [110]_{GaAs}$ where the (-110) plane is parallel to the major flat of a GaAs wafer. Hexagonal grains in Fig.5a often tower over the (100) β-GaN surface by 7 nm resulting in a rough surface morphology. For nitrogen rich growth the GaN surface is planar to about ±4 nm and for near stoichiometric growth the surface is planar to about ±1.7 nm.

CONCLUSIONS

In conclusion, TEM shows that layer quality depends on the growth conditions particularly the Ga flux for fixed nitrogen plasma conditions and growth temperature. The highest crystalline quality β-GaN layer was obtained by "stoichiometric" growth while nucleation under Ga-rich conditions resulted in the formation of persistent grains of α-GaN and grown under very Ga-poor conditions showed grains of α-GaN near the interface which were subsequently overgrown by the surrounding β-GaN. Ga-poor samples also showed voids in the GaN and in the GaAs substrate. The GaN layers contain a high density of stacking faults which drastically decreases toward the GaN surface. Stacking faults are anisotropically distributed in the GaN layer: the majority intersect the interface along lines parallel to the major flat of the GaAs substrate. This anisotropy in SF distribution is likely to be associated with the different atomic structure of α and β dislocations in cubic GaN.

ACKNOWLEDGEMENT

The study at LBNL was supported by the Director, Office of Energy Research, U.S. Department of Energy under Contract No. DE-AC03-76SF00098. The use of facilities of the National Center of Electron Microscopy is greatly appreciated. The work at SNL was supported by the U.S. Department of Energy under contract No. DE-AC04-94AL85000.

REFERENCES

1. S. Nakamura, M. Senoh, S. Nagahama, N. Iwasa, T. Yamada, T. Matsushita, H. Kiyoku, and Y. Sugimoto, Jpn. J. Appl. Phys. **35**, L74 (1996).
2. S. N. Mohammad, A. A. Salvador, and H. Morkoç, Proceedings of the IEEE **83**, 1306 (1995).
3. L. C. Jenkins, T. S. Cheng, C. T. Foxon, S. E. Hooper, J. W. Orton, S. V. Novikov, and V. V. Tret'yakov, J. Vac. Sci. Technol. B **13**, 1585 (1995).
4. H. Liu, A. C. Frenkel, J. G. Kim, and R. M. Park, J. Appl. Phys. **74**, 6124 (1993).
5. T. D. Moustakas, T. Lei, and R. J. Molnar, Physica B **185**, 36 (1993).
6. R. C. Powell, N.-E. Lee, Y.-W. Kim, and J. E. Greene, J. Appl. Phys. **73**, 189 (1993).
7. M. Kasu, T. Makimoto, and N. Kobayashi, Appl. Phys. Lett. **68**, 955 (1996).
8. H. Yang, O. Brandt, M. Wassermeier, J. Behrend, H. P. Schönherr, and K. H. Ploog, Appl. Phys. Lett. **68**, 244 (1996).
9. J. Neugebauer, and C. G. van de Walle, In E. A. Fitzgerald, J. Hoyt, K.-Y. Cheng, and J. Bean (Eds.), Strained Layer Epitaxy - Materials, Devices, Processing and Device Applications, 379 (pp. 3). San Francisco, California, USA: Materials Research Society (1995).
10. R. J. Molnar, and T. D. Moustakas, J. Appl. Phys. **76**, 4587 (1994).
11. M. Nagahara, S. Miyoshi, H. Yaguchi, K. Onabe, Y. Shiraki, and R. Ito, Journal of Crystal Growth **145**, 197 (1994).
12. D. Chandrasekhar, D. J. Smith, S. Strite, M. E. Lin, and H. Morkoç, Journal of Crystal Growth **152**, 135 (1995).
13. H. Yang, O. Brandt, and K. Ploog, phys. stat. sol. (b) **194**, 109 (1996).
14. H. Okumura, K. Ohta, T. Nagatomo, and S. Yoshida, Journal of Crystal Growth **164**, 149 (1996)
15. T. S. Cheng, L. C. Jenkins, S. E. Hooper, C. T. Foxon, J. W. Orton, and D.E. Lacklison, Appl. Phys. Lett. **66**, 1509 (1995).
16. C. T. Foxon, T. S. Cheng, S. E. Hooper, L. C. Jenkins, J. W. Orton, D.E. Lacklison, S. V. Novikov, T.B. Popova, and V. V. Tret'yakov, J. Vac. Sci. Technol. B **14**, 2346 (1996).

SURFACE RECONSTRUCTIONS AND III-V STOICHIOMETRY: THE CASE OF CUBIC AND HEXAGONAL GAN

G.Feuillet*, P.Hacke, H.Okumura, H.Hamaguchi, K.Ohta, K.Balakrishnan, S.Yoshida

Electrotechnical laboratory, Tsukuba, Ibaraki, Japan
* Present Address: CEA/Grenoble, 38041 Grenoble Cedex 09 France . e-mail : feuillet@cea.fr

ABSTRACT

Surface reconstructions for MBE grown GaN are identified. Different cases are considered according to the type of substrate or crystal symmetry and surface phase diagrams are obtained . Through different examples, it is shown how growth monitoring can be efficiently achieved through the use of surface reconstructions. Finally, from the observation that a residual arsenic overpressure in the MBE chamber changes the surface reconstructions of cubic (001) GaN grown onto 3C-SiC (001) substrates to that commonly observed for GaN growth on (001) GaAs, it is proposed that arsenic might be a surfactant for nitride growth.

INTRODUCTION

Ga / N stoichiometry drastically influences the optical, structural and electronic properties of GaN layers grown by MBE [1-3] . But, if Ga fluxes are easy to determine, in plasma assisted MBE, evaluating the effective active nitrogen flux is somehow difficult. On the other hand, in MBE, in situ RHEED allows us to monitor the surface symmetry, namely surface reconstructions, which, if observed, evidence a stabilized growth front and are related to surface stoichiometry. It is one of the aims of this contribution to show that surface reconstructions can be obtained for GaN, whether cubic or hexagonal. Determining the stability conditions for the different surface structures yields surface phase diagrams which can be used as a tool to monitor the surface stoichiometry during growth. Epitaxial GaN layers grown for different conditions on the surface phase diagram are characterized and their optical, electrical and structural properties are shown to be influenced by the surface structure. The nature of the cubic GaN surface reconstructions is also shown to change if a residual arsenic pressure exists in the MBE chamber. Lower energy surfaces are obtained in this case, and a structural model is proposed which takes arsenic surface incorporation into account to explain the kinetics of cubic GaN growth. These considerations led us to believe that arsenic is a candidate for a surfactant effect on the GaN surface.

EXPERIMENTAL

GaN samples were grown by MBE with a Ga solid source and an ECR plasma source with N_2 gas. N_2 flow rates were varied from 1 to 10 ccm while GaAs fluxes were varied between 10^{12} to 10^{14} at. / cm2.s. Typical growth rates were low, in the range 1 to 2 ml / min.

Precise Ga flux measurements were carried out by measuring, through RHEED oscillations, the growth rate for GaAs grown with an As overpressure in the chamber, further allowing precise calibration of the Bayard-Alpert gauge.

For hexagonal c-oriented GaN, we used "pseudo-substrates", i.e. 3 to 4 microns thick GaN layers grown by MOCVD on c-plane sapphire. Growing homoepitaxially allows to circumvent the problem of heteroepitaxial nucleation and to achieve intrinsic hexagonal GaN surface reconstructions unaffected by strain or strain induced roughening.

For cubic (001) oriented GaN, two sorts of substrates were used, namely 3C-(001)SiC and (001) GaAs. In the case of SiC, because bulk cubic SiC crystal are not yet commercially available, here again we used "pseudo-substrates" in the form of 3 to 4 micron thick layers grown by Low-Pressure CVD. Although of the order of about 3.6%, the misfit between GaN and SiC is the smallest among commonly available substrates.

Mat. Res. Soc. Symp. Proc. Vol. 449 © 1997 Materials Research Society

GaN SURFACE RECONSTRUCTIONS

Hexagonal (0001) GaN

For hexagonal c-oriented GaN grown homoepitaxially on MOCVD GaN, in the low Ga flux - high substrate temperature growth regime, 2x2 surface reconstructions are observed. These surface reconstructions are stable when growth is interrupted by closing the Ga shutter, whether the ECR source is activated or not, down to room temperature and up to high temperatures, although above 800°C, all streaks are more diffuse. Upon increasing the Ga flux / lowering the substrate temperature during growth the 2x2 surface transforms reversibly into a 1x1 unreconstructed surface [4,5] .

Cubic (001) GaN on (001) SiC

Having worked out the conditions for the stability of reconstructions in the easiest case of homoepitaxial growth where no misfit strain can affect the growth front, we have extended the study to heteoepitaxial layers but for cubic GaN. In the case where GaN is grown on (001) SiC, 4x1 reconstructions are observed in the low Ga flux - high substrate growth regime. Similar to the hexagonal case, these surface reconstructions are stable when growth is interrupted, with the ECR on or off down to room temperature and up to at least 850 celcius, although above this temperature, surface degradation eventually occurs. Also, as for hexagonal growth, the 4x1 surface transforms in a reversible way into a 1x1 unreconstructed surface upon increasing the Ga flux / decreasing the substrate temperature [6,7] .

Cubic (001) GaN on cubic GaAs

When grown on (001) GaAs, 2x2 / c(2x2) surfaces have been observed(4). Indeed, we could observe (7) that the 2x2 surface transforms into a c(2x2) surface upon increasing the Ga flux or decreasing the substrate temperature. The transition was found reversible. The c(2x2) is stable up to high temperatures, although the strength of the reconstruction rods and of the integral order streaks decreases. No 1x1 growth surface could be detected even for high Ga flux even at rather low substrate temperature. When the Ga shutter if closed with the ECR source activated, the c(2x2) surface becomes 1x1 with the 2x2 as a transient state. This sequence of transitions is totally reversible.

GaN SURFACE PHASE DIAGRAMS

The conditions for the reversible surface phase transitions mentionned in the table above, have been examined [4,5] in the case of hexagonal c-oriented GaN. For cubic (001) GaN, similar determinations were carried out and are reported in this meeting [6,7] . This allowed us to draw surface phase diagrams, as represented in Figure 1a, b and c in a Ga flux - substrate temperature graph: a similar behaviour is clearly evidenced in the different cases. The exponential form of the transition curve between two different growth regimes is accounted for in terms of thermally activated Ga excess reevaporation from the growing surface [4,5] . The offset to the exponential curve, which moves upwards in the graph as the nitrogen flow rate is increased, is taken as being proportionnal to the available active nitrogen flux. According to this model a fixed surface stoichiometry is obtained for all points lying on the surface reconstruction transition line.

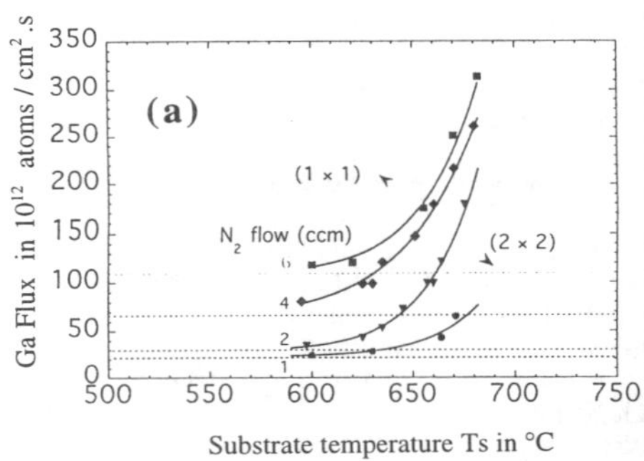

Figure 1(a): Surface phase diagrams for (0001) hexagonal GaN, for different N$_2$ flow rates

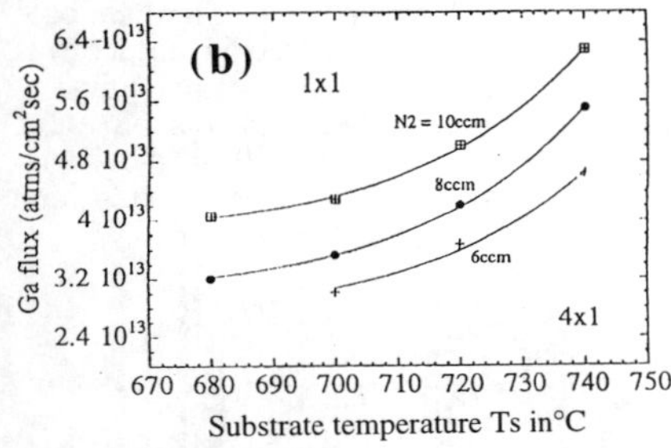

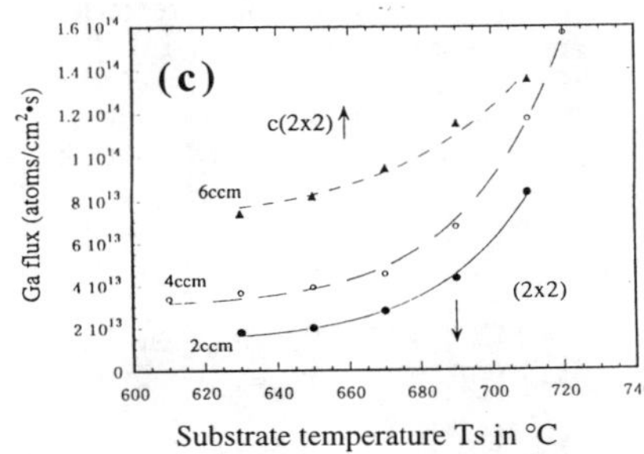

Figure 1: Surface phase diagrams for different N_2 flow rates : (b) (001) GaN grown on 3C-(001)SiC; (c) (001) GaN grown on (001) GaAs

MONITORING GROWTH WITH SURFACE RECONSTRUCTIONS

Ga droplets were consistently observed for growth conditions above, and only above, the transition curves: Ga rich conditions are thus achieved in this region of the surface phase diagram. It is suspected that, below the transition, all Ga atoms incident on the surface bond to N atoms on the surface and that excess nitrogen growth conditions will eventually be obtained in the lower part of the diagram.

GaN layers were grown in conditions on either side of the transition curve or strictly on the transition itself. Figure 2 represents the 4K PL half-width of hexagonal c-oriented GaN homoepitaxial layers together with their carrier mobility as a function of the Ga / N ratio normalised to that on the transition curve. PL half-width of about 3meV are obtained for the MOCVD layer, while it varies from 15 to 6 meV for the MBE top layer. The minimum half-width of 6meV is clearly obtained for growth conditions corresponding to the transition on the surface phase diagram with a clear optimum of 212 cm²/V.s for the carrier mobility[5] . For cubic GaN growth on (001) SiC, the same behaviour was found[6] . Here again the exciton related PL half-width is minimum for growth conditions on the 4x1 / 1x1 transition.

As to the structural properties, it was found out also that they are affected by the surface reconstructions conditions. For instance, in the case of cubic growth on (001) GaAs substrates, pole figures indicate that cubic only material is obtained for conditions on and above the transition (with Ga rich conditions however producing Ga droplets). However, for N rich conditions, i.e. below the 2x2 / c(2x2) transition, the overall diffraction intensity is weak: misoriented cubic grains exist in the cubic matrix, together with hexagonal grains with a c axis inclined to the (001) cubic axis[7] .

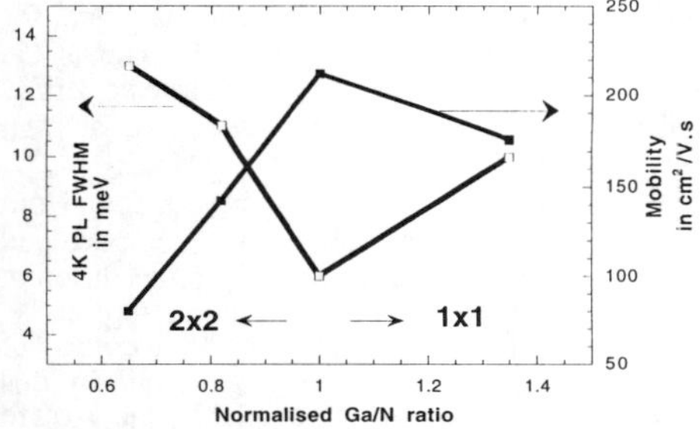

Figure 2: PL excitonic half width and carrier mobility for MBE homoepitaxial (0001) GaN with respect to Ga/N ratio normalised to 1 on the 2x2/1x1 transition

ARSENIC INDUCED SURFACE RECONSTRUCTIONS: A POSSIBLE SURFACTANT FOR
GaN EPITAXY

The 2x2 / c(2x2) surface reconstructions obtained when growth is carried out on (001)
GaAs are different from the 4x1 / 1x1 reconstructions as obtained when GaN is grown onto
(001)SiC. In order to understand the origin of this difference, the 4x1 reconstructed surface was
exposed to an As background pressure (As cell at 140°C with shutter closed, Beam Equivalent
Pressure : 10-8 Torr) during growth: the 4x1 surface rapidly and irreversibly transforms into a 2x2
surface (Figure 3). Similarly the 1x1 Ga rich surface transforms into a c(2x2) surface upon
bringing some As in the chamber. Subsequently, the usual 2x2 <---> c(2x2) reversible transition
could be observed.

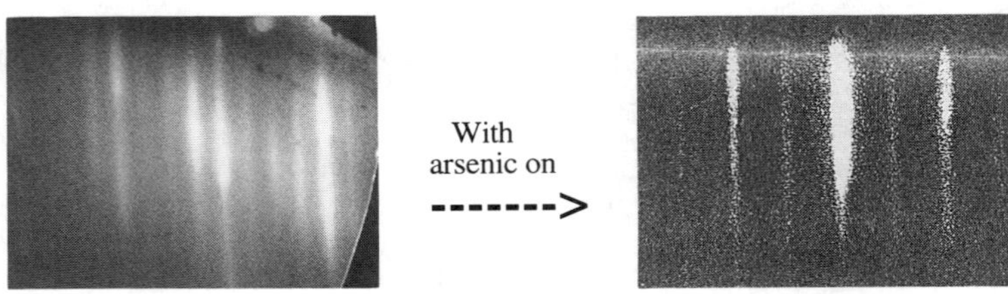

Figure 3: 4x1 to 2x2 surface reconstruction transition observed in the presence of As.

These observations lead us to believe that As is responsible for the 2x2 / c(2x2)
reconstructions when growth is carried out on (001) GaAs[8] . The presence of surface sitting As
atoms during growth on GaAs substrates could have different origins: arsenic desorption from the
underlying GaAs substrates at the usual GaN growth temperatures (600 to 800°C) or residual As
background pressure remaining in the chamber after deposition of the GaAs buffer. As segregation
from the substrate could also be involved.

The influence of As, or other group V atoms such as Sb, upon surface reconstructions is
well documented in the literature for group IV semiconductors. As intentionaly brought onto the
surface can form particular surface reconstructions involving dimers on the (001) oriented surfaces
of Si or Ge[9] . Unintentional As may also segregate: EELS and Auger investigations reveal for
instance that about half a monolayer of segregated As atoms covers the surface of epitaxial Ge
grown on GaAs even for growth temperatures of about 300°C with specific reconstructed LEED
patterns[10] . GaAs and GaN being almost immiscible materials, it is expected that, at high
temperature, As cannot be incorporated into GaN but rather will segregate onto the surface.

In a recent attempt to grow GaAsN ternary alloy layers by MBE[11] , it was found out
that, at low temperatures (low enough for GaAs growth), As is incorporated within the lattice and
GaAs is formed together with GaN. At high temperature however, and because of the different
bond strengths for GaN and GaAs (4.2eV vs 2.07eV), As is not incorporated, N atoms replace As
atoms on the surface and only GaN is formed. One might thus suspect that in the presence of
surface sitting As atoms, and at high enough temperatures, cubic GaN growth is governed by this
exchange process between As and N atoms: if sufficient Ga and N atoms are available, the growth
rate is thus limited by the As arrival rate and desorption. Because As to Ga bonds will be initially
formed in a transient state on the surface, the cubic GaAs surface structure will act as a template for
subsequent growth of cubic GaN. Cubic only material will be obtained if one monolayer of As
covers the surface.

We found that just a small amount of As is necessary to alter the surface symmetry.
However, the appearance of hexagonal grains within the cubic matrix after continued growth in
these conditions suggests that this amount is not enough to ensure growth of only cubic material,
in accordance with the observations in [11] and our discussion above.

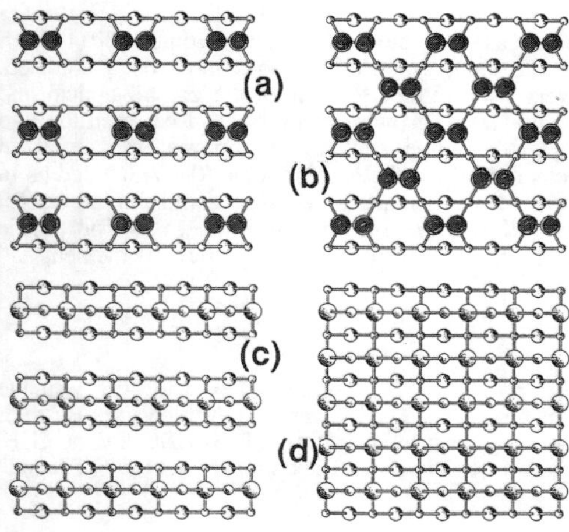

Figure 4: Arsenic (●) dimer model for (a) 2x2, (b) c(2x2) cubic (001) GaN surfaces in the presence of arsenic.(c) 0.5 monolayer Ga coverage starting from the 2x2 arsenic surface ; (d) 1 monolayer Ga coverage starting from the c(2x2) arsenic surface

These considerations and our experimental observations that arsenic changes the GaN surface symmetry indicate that some exchange between As and N atoms takes place on the growing surface, influencing the incorporation processes of Ga and N atoms. This exchange effect is at the basis of surfactant mediated growth as, for instance, in the case of As for group IV elements [11] and Te for GaAs [12,13] . It is thus proposed here that As could act as a surfactant for epitaxial growth of cubic (001) GaN.

From our observations, As is definitely taking part in the 2x2 and c(2x2) surface reconstructions. The model proposed in [3] , where Ga dimers are supposed to be involved, is reconsidered here by assuming As dimers on the surface. Because of the different bond stiffnesses between GaN and GaAs and contrary to the case when N is adsorbed on GaAs [15] it may be assumed that As induced surface strains will be mainly relaxed by out of plane deformations rather than in plane deformations: in this case 2x2 and c(2x2) reconstructions can be accounted for in the same way as in [3] but with As dimers (Figure 4 a and b). Interestingly, along the <1-1 0> direction parallel to the dimer row, after the exchange between As and N atoms, Ga will be incorporated along the dimer rows and the Ga surface coverage will be 0.5 monolayer (Figure 4 c). In the case of the c(2x2) surface on the other hand, the same exchange-incorporation process leads to a Ga coverage of 1 monolayer (Figure 4 d). This is in agreement with findings in [3] that the 2x2 (respectively c(2x2)) reconstructed surfaces are obtained for 0.5 (respectively 1) Ga monolayer on the surface.

The respective positions of the 4x1 / 1x1 and 2x2 / c(2x2) transition curves as presented in Figure 5 yield some information on the surface stoichiometry in the two cases. For the growth conditions corresponding to the reconstruction transition curve, a particular Ga/N surface stoichiometry is achieved: Ga rich conditions are achieved above the curve and nitrogen rich conditions below.

This is a strong evidence that the presence of arsenic on the growing GaN surface definitely changes the Ga/N surface stoichiometry. From Figure 5, the As induced 2x2 / c(2x2) transition occurs for a higher Ga flux and Ga excess conditions -with the eventual formation of droplets as found experimentally - will be obtained only above this new transition curve.

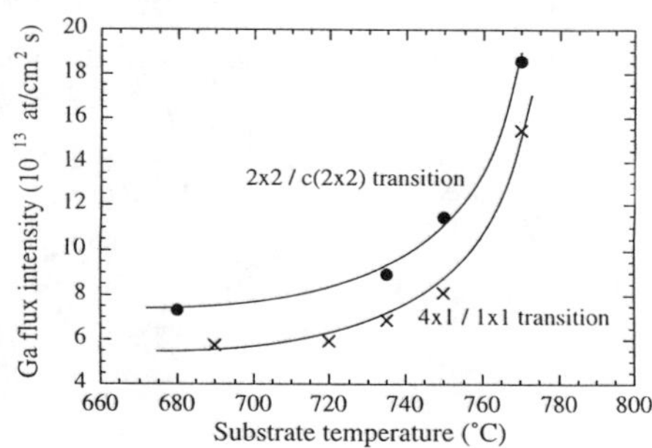

Figure 5: Respective positions of the 4x1 /1x1 and 2x2 / c(2x2) surface transitions

261

CONCLUSIONS

Surface reconstructions have been identified in the case of cubic (001) and hexagonal (0001) GaN. The growth conditions for the stability of the different types of surfaces have been clarified. Surface phase diagrams were shown to be a very useful guide to monitor the crystal optical and structural properties. As a general trend, it was shown that growing GaN right in between the two surface states corresponding to Ga and N rich growth regimes yields optimum quality material. In a second part, the detailed study of surface reconstructions in the case of cubic (001) GaN led us to conjecture that arsenic might be used as a surfactant for this material. Arsenic could thus be useful to produce cubic only material; it would also allow increased critical thicknesses for plastic relaxation and also for the 2D / 3D growth transition. Furthermore, As is a group III element and as such would not induce unintentionnal doping in GaN.

ACKNOWLEDGMENTS

The authors are indebted to Dr.Y. Iyechika from Sumitomo Chemicals Co. Ltd. for valuable discussions and providing the MOCVD grown hexagonal GaN layers. They also express their gratitude to Y Ishida from Electrotechnical Laboratory for providing the CVD 3C SiC layers. G.F. and P.H acknowledge financing from MITI through the Science and Technology Agency of Japan

REFERENCES

1. S.H.Cho, H.Sakamoto, K.Akimoto, Y.Okada , M.Kawabe, Jpn.J.Appl.Phys. **34**, 1995, p236.
2. T.Maruyama , S.H.Cho and K.Akimoto, Proc. of the 1995 Intern Conf on Silicon carbide and related Materials IOP publishing **142**, p851.
3. O.Brandt, H.Yang, B.Jenichen, Y.Suzuki, L.Daweritz, K.H.Ploog, Phys.Rev.B, **52**, N°4,1996, p.2253
4. G.Feuillet, P.Hacke, H.Okumura, S.Yoshida, Proc. of the 1st Int.Symposium on Blue Laser and Light Emitting Diodes, Chiba, March 1996, Paper LN4.
5. P.Hacke, G.Feuillet, H.Okumura, S.Yoshida Appl.Phys.Lett **69** (17) 2507,1996.
6. S.Yoshida, H.Okumura, G.Feuillet, P.Hacke, K.Balakrishnan, 1996 MRS Fall meeting, invited.
7. H.Okumura, K.Balakrishnan, G.Feuillet, K.Ohta, H.Hamaguchi, S.Chichibu, Y.Ishida, S.Yoshida, 1996 MRS fall meeting.
8. G.Feuillet, H.Hamaguchi, K.Ohta, P Hacke, H.Okumura, S.Yoshida Appl. Phys. Lett. to be published.
9. R.I.G.Uhrberg, R.D.Bringans, R.Z.Bachrach, J.E.Northrup, Phys. Rev. Let. **56**,1986,520
10. W.Monch, R.S.Bauer, H.Gant, R.Murschall, J.Vac.Sci.Technol. 21(1982), 498.
11. L.C. Jenkins, T.S.Cheng, C.T.Foxon, S.E. Hooper, J.W.Orton, S.V.Novikov, V.V.Tret'yakov, J.Vac.Sci.Technol., B **13**(4) 1995,p1585.
12. N.Grandjean, J.Massies, C.Delamarre, L.P.Wang, A.Dudon, J.Y.Laval, Appl. Phys. Lett. **63**(1) 1993 p66.
13. N Grandjean, J Massies, Phys.rev.B. **53**, 1996, p13-231.
14. J.Massies, P.Etienne, F.Dezaly, N.T.Linh, Surf.sci.**99**,121, 1980.
15. M.Kasu, T.Makimoto, N.Kobayashi, Appl.Phys.Lett.**68**, 955 (1996).

LARGE-AREA GROWTH OF InGaN/AlGaN USING IN-SITU MONITORING

E. WOELK**, D. SCHMITZ*, G. STRAUCH*, B. WACHTENDORF*, H. JUERGENSEN*

*AIXTRON GmbH, Kackertstr. 15-17, D-52072 Aachen, Germany

**AIXTRON, Inc., 1569 Barclay Bl., Buffalo Grove, IL 60089, U.S.A.

ABSTRACT

A newly developed Metalorganic Chemical Vapor Deposition (MOCVD) reactor for processing batches of seven 50mm wafers per run at deposition temperatures up to 1600°C is introduced. The substrates are individually rotated by means of gas bearings utilizing high purity gas for operation. In order to achieve the maximum uniformity on a wafer the uniformity of the growth temperature was maximized. Pyrometric measurements revealed a temperature uniformity across the susceptor of ±4°C at a temperature of 1300°C. Growth of GaN and GaInN produced uniform layers regarding composition and thickness. On a 50mm diameter wafer the standard deviation for the thickness of a GaN layer is 6% and the standard deviation for the composition of a GaInN layer as determined by photoluminescence is 2%.

INTRODUCTION

To satisfy the increasing demand for blue and green ultra-high brightness (UHB) light emitting diodes (LED) and to drive the price for these devices down, we developed a metalorganic chemical vapor deposition (MOCVD) reactor that is capable of depositing material films on seven wafers of 50mm diameter per run. Due to the high growth rates possible, MOCVD is the production method of choice for these films because the growth requires much less time compared to other techniques such as MBE. Starting point for the development was the Planetary Reactor that has been used in the production of ultra-high brightness (UHB) LEDs [1,2], lasers [3,4], high electron mobility transistors (HEMTs) [5,6] and photocathodes. So far all these devices have been made from films consisting of GaAlInAsP of various compositions. Film deposition for this material system requires low growth temperatures in the range from 600°c to 850°C. For the growth of single crystalline nitride films, process temperatures in excess of 1000°C are required. Such temperatures have not been encountered in CVD for mass production of electronic devices before. The extreme temperatures required a thermal optimization of the reaction chamber in order to deal with the low chemical stability of some of the precursor compounds.

One more particularity of the growth of GaN and related compounds has to be taken into account at the time of the design of a new reactor. For the growth of GaN the initiation of the growth has a great impact on the latter quality of the film and determines to a large degree whether the film is able to perform the desired functions. The initiation of the growth of GaN involves a growth step at a low temperature (500-600°C). After the growth of some 50Å at a low growth rate, the temperature is increased to 1000°C in order to allow the growth at high growth rates. It is desir-

Mat. Res. Soc. Symp. Proc. Vol. 449 © 1997 Materials Research Society

able to provide rapid heating and cooling of the susceptor to minimize the process time and to interrupt the growth for as short a time as possible.

MOCVD PRODUCTION REACTOR WITH OPTICAL ACCESS

Figure 1 shows the cross section of the reactor with a wafer capacity of 7 x 50mm wafers. The MOCVD of GaN and its relatives requires process temperatures that are well above those that are usually necessary for the well established MOCVD of GaAs and its relatives. In a cold wall reactor, at higher temperatures it becomes increasingly difficult to provide heating that yields a uniform surface temperature. In order to be able to tune the local RF power delivered to the coated graphite susceptor and provide a uniform temperature, the susceptor is heated by electrical currents induced by a coil system internal to the reactor. The distance of the members of the coil system and the susceptor can be adjusted which leads to a locally different heating of the susceptor. The maximum temperature of the susceptor is 1600°C which is also sufficient for the growth of SiC.

Figure 2 shows the temperature homogeneity across the susceptor. The measurement was taken using the pyrometer and the optical viewport of the reactor. The variation across a recess for a 50mm wafer is less than ±4°C at more than 1300°C. The variation between wafers is in the same order of magnitude. Also shown is the thermal image of a chip of SiC in one of the recesses. The temperature of the chip appears to be lower than that of the surface of the susceptor and less uni-

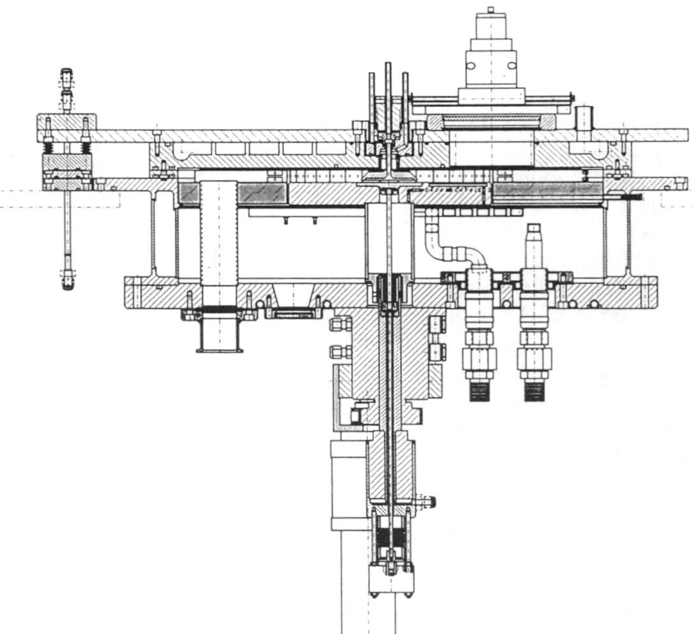

Figure 1: The High Temperature Planetary Reactor®

form. The apparent difference in temperature could be caused by the different emissivity of the coating of the susceptor and the SiC chip. We suspect that the apparent variation in temperature is real and caused by a lateral change of the thermal contact between the SiC chip and the susceptor.

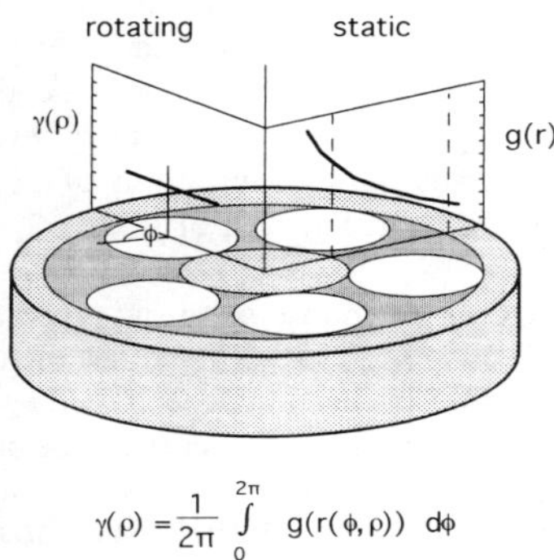

$$\gamma(\rho) = \frac{1}{2\pi} \int_0^{2\pi} g(r(\phi,\rho))\; d\phi$$

Fig.2: Growth rate over radius
  in the Planetary Reactor®

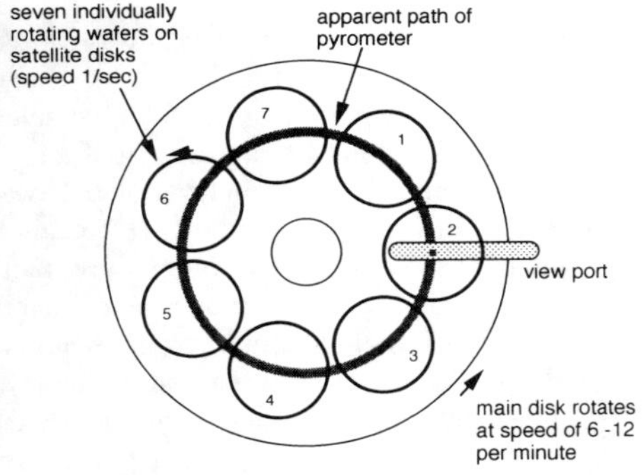

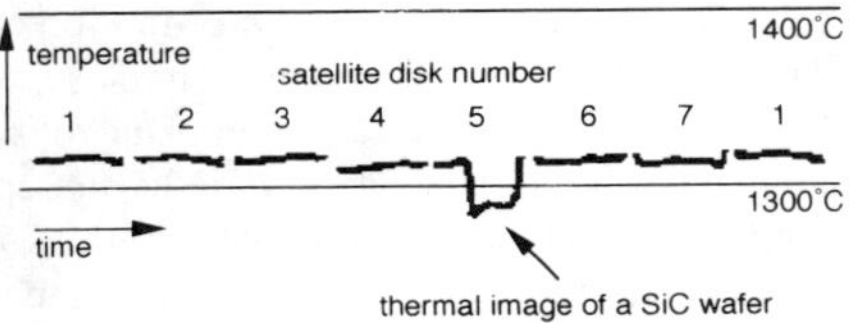

Fig.3: Pyrometric Measurement of the
  susceptor temperature

The two flow concept which has been used on the Planetary Reactor® separates the reagents in two carrier gas flows and prevents unwanted pre-reactions between the metalorganic compounds and the hydrides [8] which can lead to a complete deficiency of one chemical during the growth. The two carrier gas flows are mixed inside the reactor chamber where the gas phase is usually hot enough to inhibit parasitic reactions. Tight control of immediate process parameters such as pressure temperature and gas flow is essential for the growth of high quality electronic films. But also the secondary process parameters such as the wall temperature and the temperatures of the fins and baffles inside of the reactor have an impact on the properties of the layer. If optimally adjusted, however, they can increase the yield and reduce the maintenance [7]. The tools to control and monitor these parameters were transferred from the existing reactors to the new high temperature reactor.

Particularly important is the thermal management of the injection zone in the center of the reactor. In order to preserve the unstable trimethylindium, the injection zone is kept close to room temperature. Otherwise excessive amounts of trimethylindium and/or tri(m)ethylgallium would be required to compensate for the losses due to premature decomposition. Prior to the design and the construction of the reactor, numerical simulations suggested that the gas phase in the reactor was stable and laminar at the process conditions [9]. This prediction was verified in the actual growth runs.

NITRIDE GROWTH AND RESULTS

The laminar horizontal flow in the reactor during the process leads to a controlled depletion of the precursor compounds in the gas phase along the direction of the gas flow. This depletion in turn causes a well defined decrease of the growth rate along the radius. The rotation of each individual substrate leads to an integration of the growth rate between the leading and the trailing edge of the wafer. Figure 3 shows a generic profile of the growth rate along the radius and the integration obtained by rotation. It has been found that this profile holds true for the growth of most materials. The integration of an ideal growth rate profile leads to a uniformity of the layer of 0.5%. For real growth the uniformity is found to be in the range between ±1% and ±3%.

Even a simple GaN/GaInN LED structure grown by MOCVD requires growth and process steps that occur at different temperatures and gas flow conditions. A sequence of rapid heating and cooling and steady temperature steps is required. The rates for the heating depend on the heater power, the effectiveness of coupling that power into the susceptor and the thermal mass of the susceptor. The rate for cooling depends on the thermal mass of the susceptor and the mechanisms for the dissipation of the heat which are radiation and convection. A well designed thermal cycle can greatly reduce the stress in the material film and lead to a better performance of the device. The Planetary Reactor® used in this study offers flexibility in cooling and heating rates up to 6°C per second. Cooling rates were increased by a reduction of the thermal mass of the susceptor Heat-up rates were increased by the introduction of effective induction heating.

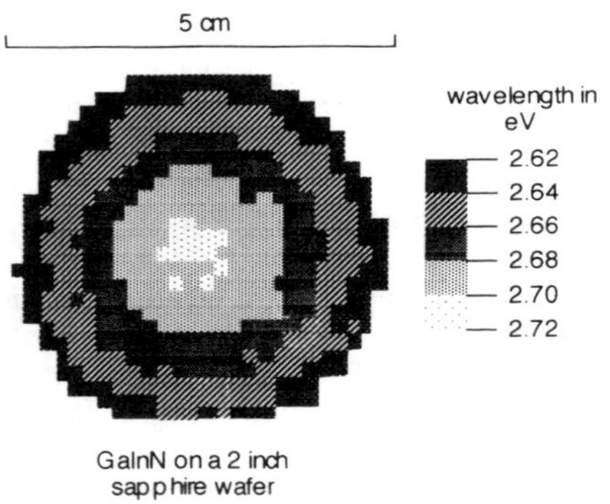

Fig.4: Photoluminescence mapping of a GaInN layer

The growth of the epitaxial layers was carried out at a temperature of 1000°C at a pressure of 100 mbar on the basal plane of 50mm diameter sapphire wafers. Prior to the growth of the layer, a buffer or nucleation layer of GaN was grown at a temperature of 500°C. To obtain high uniformity layers, the wafers were spun at a speed of about 50 rpm. In the reactor chamber ammonia, tri(m)ethylgallium, trimethylindium, and trimethylaluminum react to form the crystalline solid film. In all the growth runs, hydrogen was used as a carrier gas. The typical layer thickness was 1.3 µm which was obtained after an hour of growth. The growth rate was found to increase linearly with the concentration of TMGa in the gas phase and growth rates as high as 3 µm/hour were achieved. Due to the long exposure to the heat close to the susceptor the fraction of cracked ammonia is high at the wafers. This reduces the V to III ratio in the reactor to below 1000.

The finished layers were mapped using optical characterization. Figure 4 shows the wavelength mapping of a GaInN layer grown in a horizontal reactor using wafer rotation. The In-content of the layer appears to be around 40%. The actual In-content could be considerably lower than the photoluminescence may suggest due to a strong emission from energy levels in the band gap [12]. The circular symmetry is expected due to the rotation of the wafer during the deposition.

CONCLUSION

The two-flow Planetary Reactor® for III-Nitride growth on multiple wafers has been introduced. It was used to grow the nitride material films that are required for blue and green LEDs and lasers. The composition uniformity was found to be in the 1 % range. As a prove for the feasibility of MOCVD for production growth rates as high as 3 μm/hour were adjusted producing state of the art material. Today, this reactor represents the only MOCVD reactor with partially controlled secondary process parameters such as the wall temperature of the reaction chamber, that provides state-of-the-art GaN growth with the largest wafer capacity.

REFERENCES

[1]   R. Beccard, J. Knauf, G. Lengeling, D. Schmitz and H. Jürgensen, submitted to 22nd Int'l Symp. on Compound Semiconductors, Korea, August/September, 29-02, 1995

[2]   F. A. Kish, F. M. Steranka, D. C. De Fevere D. A. Vanderwater, K. G. Park, C. P. Kuo, T. D. Ostentowski, M. J. Peanasky, J. G. Yu, R. M. Fletcher, D. A. Steigerwald and M. G. raford, Appl. Phys. Lett. 64 (21), 23. May 1994

[3]   C. J. van der Poel, H. P. M. M. Ambrosius, R. W. M. Linders, N. J. Kiwiet, and J. Rijpers, J. Crystal Growth 124 (1992) 300-306

[4]   H. Jürgensen,. Symp. 'Laser Diode Technology and Applications IV', Los Angeles, January 19-25, 1992

[5]   P. M. Frijlink, J. L. Nicolas, H. P. M. M. Ambrosius, R. W. M. Linders, C. Waucquez, J. M. Marchal, J. Cryst. Growth 115 (1991) 203-210

[6]   H. Jürgensen and P. Frijlink, 'Gallium Arsenide Technology in Europe', ed. by J. Mun and A. A. M'Baye, Springer-Verlag (1994) 127-159

[7]   G. Strauch, J. Hergeth, B. Wachtendorf, M. Volk and E. Woelk, to be published in Proc. International Symposium on Blue Laser and Light Emitting Diodes, Chiba, 05.-07.03.1996

[8]   C. Chen, *MRS Fall Meeting, Symposium on Gallium Nitride and Related Materials*, Nov. 27, -Dec. 1, 1995, Boston, MA

[9]   E. Woelk, D. Schmitz, G. Strauch, C. Wachtendorf, J. Juergensen, M. Dauelsberg, L. Kandinski, Yu. Makarov, "MOVPE of III-Nitrides for blue and green light emitting diodes using in-situ monitoring", ICMOVPE VIII: 9th -13th June 1996, Cardiff.

MBE GROWTH OF III-V NITRIDE FILMS AND QUANTUM WELL STRUCTURES USING MULTIPLE RF PLASMA SOURCES

M.A.L. JOHNSON*, ZHONGHAI YU*, C. BONEY*, W.H. ROWLAND*, JR., W.C. HUGHES*, J.W. COOK*, JR., J.F. SCHETZINA*, N.A. EL-MASRY**, M.T. LEONARD***, H.S. KONG***, and J.A. EDMOND***
*Department of Physics, N.C. State University, Raleigh, NC 27695, jan_schetzina@ncsu.edu
**Department of Materials Science and Engineering, N.C. State University, Raleigh, NC 27695
***Cree Research, Inc., 2810 Meridian Parkway, Durham, NC 27713

ABSTRACT

MBE growth of III-V nitrides is being studied at NCSU using MOVPE grown GaN buffer layers on SiC as substrates. Rf plasma sources are being used for the generation of active nitrogen during MBE deposition. Through the use of multiple rf plasma sources, sufficient active nitrogen is generated in order to examine the properties of III-V nitride layers grown at higher substrate temperatures and growth rates. The resulting MBE-grown GaN films exhibit remarkably intense photoluminescence (PL) dominated by a sharp band-edge peak at 3.409 eV having a FWHM of 36 meV at 300K. No deep level emission is observed. AlGaN and InGaN films and quantum well structures have also been prepared using multiple sources. A modulated beam MBE approach is used in conjunction with the multiple rf plasma sources to grow InGaN. RHEED and TEM studies reveal flat 2D InGaN quantum well structures. Depending on the indium content, GaN/InGaN single-quantum-well structures exhibit electroluminescence at 300K peaked in the blue-violet to the green spectral region.

INTRODUCTION

Historically, both metal-organic vapor phase epitaxy (MOVPE) and molecular beam epitaxy (MBE) have played major roles in the development of III-V arsenide and phosphide materials and devices. However, in the recent rapid development of the III-V nitrides, MOVPE has played the major role [1-4]. Early efforts to employ MBE for growth of GaN and related materials using ECR plasma sources to general "active" nitrogen resulted in very slow film growth rates and poor quality material [5]. We have recently shown that high-quality GaN can be prepared by MBE by employing an rf nitrogen plasma source which produces a greater flux of active nitrogen and therefore permits MBE film growth at higher temperatures[6]. A homoepitaxial approach is used by performing the MBE growth on high-quality GaN buffer layers, prepared by MOVPE at Cree Research, Inc., on 6H-SiC substrates. This approach circumvents problems associated with GaN film nucleation on highly lattice-mismatched substrates such as sapphire or SiC and allows concentration on the issues associated with the MBE growth process itself [6]. In this paper, MBE growth of GaN, AlGaN, and InGaN films and quantum well (QW) structures are reported. By using up to three rf plasma sources for active nitrogen generation, MBE growth of nitrides at 1000 °C has been achieved for the first time. Use of multiple plasma sources has also been employed for growth of InGaN QWs using a modulated beam MBE approach at 800 °C.

EXPERIMENTAL DETAILS

The nitride samples were grown at NCSU using a modified EPI 930 three-chamber MBE system consisting of a main MBE film growth chamber, which accepts substrates up to 75 mm in diameter and which has provisions for up to ten MBE source ovens, a second chamber for plasma cleaning of substrates using He/H plasmas, and a surface analysis chamber with provisions for Auger spectroscopy studies of substrates and epilayers at temperatures up to 800 °C. The three chambers are interconnected by an ultrahigh vacuum (UHV) sample transfer system. Up to three inductively coupled RF plasma sources, made by Oxford, SVTA, and EPI, were used in the growth of these films. The nitrogen species generated by each of these sources has been previously studied by optical emission spectroscopy [6,8].

Mat. Res. Soc. Symp. Proc. Vol. 449 © 1997 Materials Research Society

The nitride films were grown on high-quality GaN buffer layers prepared by MOVPE on basal plane 6H-SiC substrates at Cree Research, Inc. The best MOVPE-grown GaN layers exhibit PL spectra at 298 K dominated by near-edge emission at 3.41 eV and double-crystal x-ray rocking curves as narrow as 85 arc sec FWHM (0002). These GaN/SiC substrates were cleaned prior to MBE film growth using trichloroethylene, acetone, and methanol followed by plasma-cleaning using a 1:1 hydrogen/helium gas mixture. This was followed by thermal annealing at 600°-1000°C. After cleaning, the GaN/SiC substrate surface showed little or no carbon or oxygen contamination.

The MBE film growth was monitored in-situ by reflection high-energy electron diffraction (RHEED). Selected films were also characterized by means of Nomarski interference-contrast microscopy, transmission electron microscopy (TEM), scanning electron microscopy (SEM), double-crystal x-ray diffraction, and variable-temperature photoluminescence (PL).

RESULTS AND DISCUSSION

<u>MBE Growth of GaN, AlGaN, and InGaN</u>

Figure 1 shows PL data obtained for an undoped GaN layer grown using multiple rf sources at 1000 °C. The 300 K spectrum on the left consists of a single sharp near-band-edge peak (FWHM = 36 meV) centered at 3.409 eV. Note that the 2.2 eV band of yellow-green deep level emission that is often present for GaN is completely absent from this PL spectrum, indicating the layer to be of very high quality. High quality is also manifested by the 4K PL spectrum shown at the right in Figure 1. The 4K spectrum consists of a bound exciton peak at 3.463 eV and a sharp (FWHM = 1.4 eV) free exciton peak at 3.471 eV. These peak assignments and energies are in excellent agreement with recent optical studies of MOVPE-grown GaN on SiC [7].

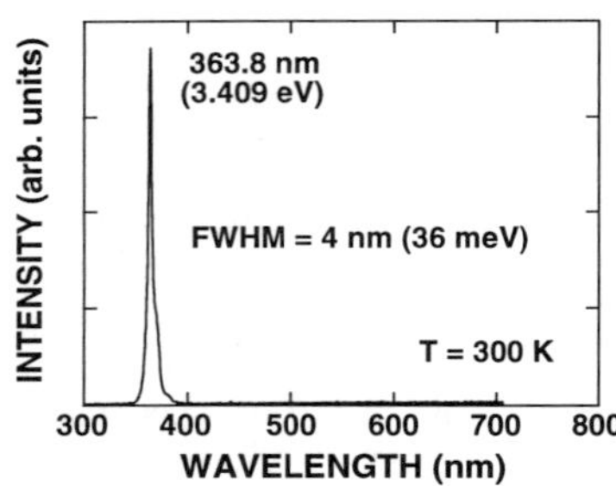

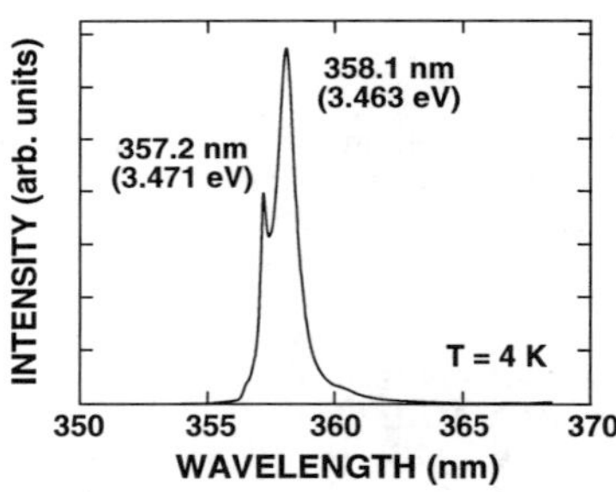

Figure 1:   Photoluminescence spectra for GaN film grown by MBE at 1000°C using multiple rf plasma sources.

N-type GaN has been grown using Si as the dopant. Mg has been used as the P-type dopant. $Al_xGa_{1-x}N$ films (x~0.06-0.08) grown by MBE on GaN/SiC substrates exhibit 2D nucleation as seen by RHEED and show strong PL edge emission at 3.62-3.69 eV at 295 K. $Al_xGa_{1-x}N$/GaN double heterostructures have also been grown. These p-on-n structures were processed into 250 μm x 250 μm mesa diodes using standard processing techniques. The LEDs emit bright violet light at temperatures ranging between 77 - 290 K.

We have also studied MBE growth of InGaN alloys. For visible light emission, InGaN is essential as the active recombination layer material in double-heterostructure quantum well devices such as laser diodes. The growth of high quality InGaN is complicated by thermodynamic limitations; InN is unstable and tends to dissociate at typical MBE growth temperatures of 600-800 °C [9]. Furthermore, the surface energies of InGaN are such that indium tends to cluster into metal droplets rather than to migrate freely to lattice incorporation sites. The formation of indium droplets results in a low indium incorporation rate in the growing film and a weak PL signal dominated by deep level emission. This kinetically induced phase segregation into indium droplets in the growth of high mole fraction InGaN layers has previously been reported for growth by MOVPE as well as MBE [10,11]. To overcome these difficulties, we have developed a modulated beam technique, based on atomic layer epitaxy and migration enhanced epitaxy, which employs continuous alternating layers of (In,Ga)N and (Ga)N. The intermittent deposition of a brief GaN layer stabilizes the indium containing layer before droplets can nucleate and results in high quality epitaxy. Factors influencing the composition of the resulting InGaN include the metal flux ratios, substrate temperature, and the relative lengths of the beam modulation periods. Multiple rf plasma sources have allowed us to grow InGaN by modulated beam MBE at substrate temperatures of 800 °C.

<u>RHEED Studies of InGaN Growth</u>

Figure 2 shows RHEED images obtained during growth of a GaN/InGaN/GaN single-quantum-well (SQW) double heterostructure. First a ~400 Å thick smoothing layer of GaN:Si was grown by MBE onto the GaN/SiC substrate. Figure 2(a) shows RHEED photos of the surface of the GaN:Si layer just prior to initiation of the InGaN QW growth. The streaky RHEED photos are indicative of 2D growth. We have referenced the observed reconstructions to the two perpendicular axes of the underlying substrate with the electron beam along the $[11\bar{2}0]$ and $[1\bar{1}00]$ axes respectively. A dim (2 x 2) surface reconstruction is observed on the GaN surface immediately prior to InGaN deposition. Figure 2(b) shows RHEED patterns of the InGaN surface during the QW growth which indicate that 2D growth is maintained. The QW consists of a total of twenty double-layers of InGaN/GaN which provide a total thickness of about 200 Å. Figure 2(c) shows RHEED patterns obtained during the growth of a GaN layer-component of the QW. Note that a strong (1 x 3) surface reconstruction is present and the streakiness of the patterns indicates that the layer is atomically flat. Twenty alternating RHEED patterns like those shown in Figures 2(b) and 2(c), respectively, are observed during deposition of the 20 double-layers of the entire QW structure. RHEED patterns obtained during the initiation of the top GaN:Mg layer also exhibit

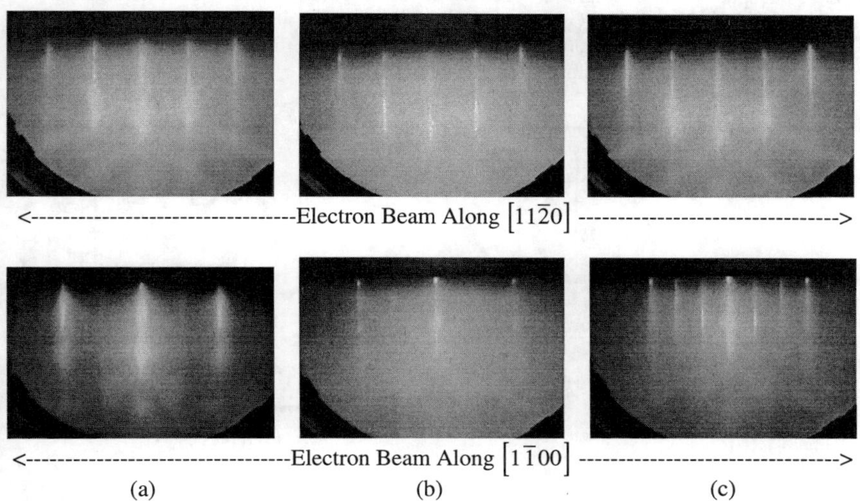

Figure 2:   RHEED photos of (a) GaN:Si layer just prior to growth of InGaN QW, (b) InGaN QW layer, and (c) GaN QW surface stabilization layer.

a (1 x 3) surface reconstruction like those shown in Figure 2(c). The RHEED patterns shown above provide direct experimental evidence that the MBE growth technique can be used to prepare very high quality InGaN QWs.

The (1 x 3) reconstruction has also been previously observed for GaN during the heating of GaN/SiC substrates under a nitrogen plasma flux prior to MBE growth [6]. These reconstruction patterns may alternatively be referenced to the hexagonal <11$\overline{2}$0> axes of the underlying substrate, separated 120° apart [12]. With this transformation of axes, the (1 x 3) reconstruction observed in this RHEED experiment corresponds to the ($\sqrt{3}$ x $\sqrt{3}$) 30° reconstruction. The real and reciprocal space lattices for this reconstruction are shown in Figure 3. The ($\sqrt{3}$ x $\sqrt{3}$) 30° reconstruction has previously been reported for group III adatoms on the close packed {111} planes of silicon [13]. The (2 x 2) reconstruction referenced to the hexagonal axes is shown in Figure 4.

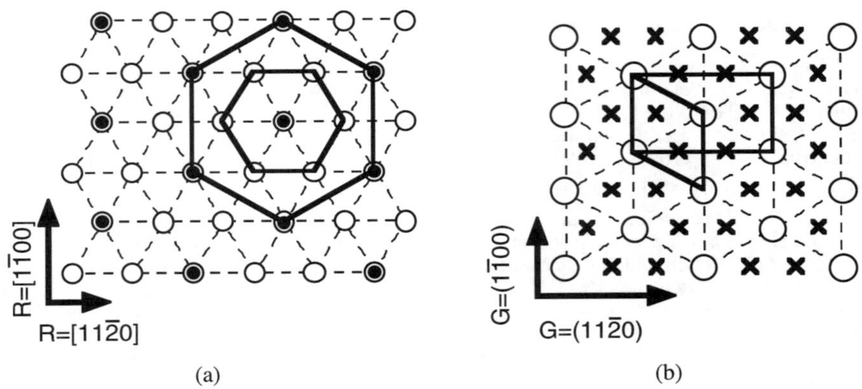

Figure 3:   Real space (a) and reciprocal space (b) net of the close packed {0001} hexagonal surface with equivalent (1 x 3) and ($\sqrt{3}$ x $\sqrt{3}$) 30° reconstruction lattice symmetries indicated. (Open circles (O)- bulk lattice sites. Filled circles (●)- surface lattice sites. Crosses (X)- surface reciprocal lattice sites.)

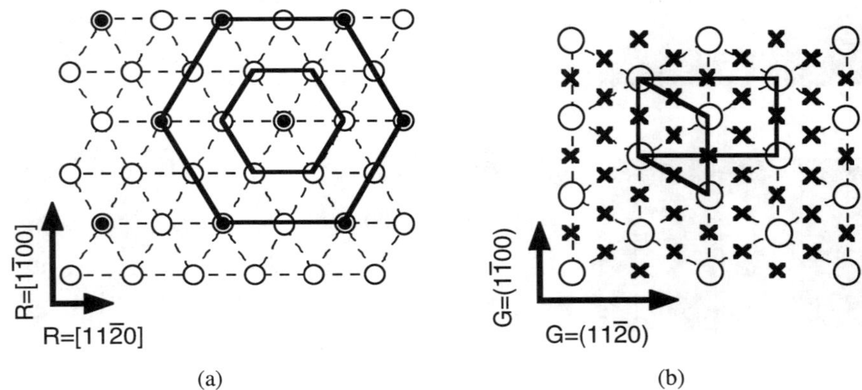

Figure 4:   Real space (a) and reciprocal space (b) of the close packed {0001} hexagonal surface with the (2 x 2) reconstruction lattice symmetries indicated. (Open circles (O)- bulk lattice sites. Filled circles (●)- surface lattice sites. Crosses (X)- surface reciprocal lattice sites.)

It is important to stress that the sites indicated for both the bulk and surface symmetries are lattice sites, not atomic positions. From these experiments alone, no conclusion may be drawn regarding either the type, number, or position of basis atoms associated with either the bulk or surface lattice sites. However, surface reconstruction during growth does imply that the lowest energy surface under these growth conditions displays long range ordering of the surface adatoms. This suggests that the thermodynamics and kinetics of MBE growth under these conditions exhibits a tendency to avoid adatom clustering. Thus, from the presence of the (1 x 3) reconstruction during the intermittent GaN stabilization layers we may conclude that this layer is effective in limiting the coalescence of indium into metal droplets during the growth of the entire QW structure layers.

Characterization of MBE Grown Heterostructures

TEM studies were also completed on selected InGaN SQW structures. Figure 5 shows a TEM vertical-cross-section obtained from a representative sample. Figure 5(a) shows the entire heterostructure. The SiC substrate is at the bottom of the image followed by a thin nitride nucleation/buffer layer and a ~0.5 µm thick layer of MOVPE-grown GaN:Si. The MBE-grown InGaN QW (~200 Å thick) is seen as the dark horizontal line near the center of the photo. Above this is shown a ~5000 Å thick layer of MBE-grown GaN:Mg to complete the double heterostructure. A higher resolution image of the InGaN QW is shown in Figure 5(b). Some of the individual InGaN (dark) and GaN (light) layers of the QW can be seen. Note that there appears

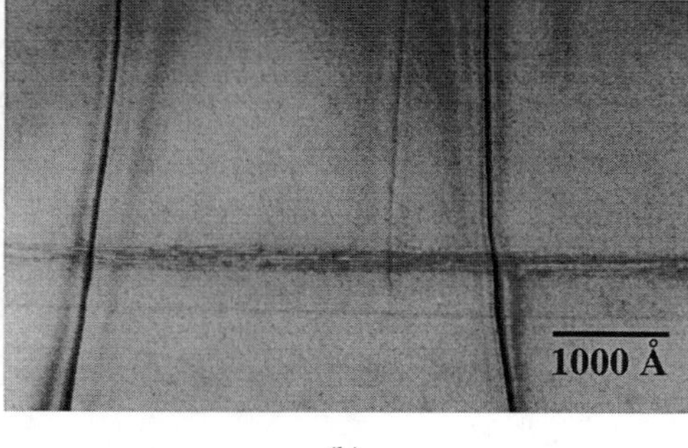

(a)   (b)

Figure 5: TEM photos of InGaN SQW sample showing (a) the entire heterostructure and (b) a high-magnification image of the InGaN QW and the MBE/MOVPE interface.

to be very little inter-diffusion between the very thin InGaN and GaN layers which comprise the QW. The MBE/MOVPE interface is seen as a thin flat line approximately 300 Å below the QW. It is important to note that there is no increase in dislocation density associated with the MBE growth process. The MBE technique replicates the quality of the underlying MOVPE-grown GaN layer, which corresponds to an average dislocation density of about 10^9 per cm^2 for the present samples.

The InGaN SQW LED samples exhibit bright electroluminescence (EL) at 300K as shown in Figure 6. Note that the emission from the MBE-grown samples consists of a single sharp peak in the blue/violet spectral region with FWHM comparable the best SQW LED emissions reported for MOCVD-grown samples. The peak luminescence at 425 nm corresponds to an indium mole fraction of 22% when Vegard's Law is used to calculate the composition from the band-gap energy and no shift due to strain is assumed. By increasing the indium content in the InGaN QW, we have also obtained EL peaked in the blue and green spectral regions.

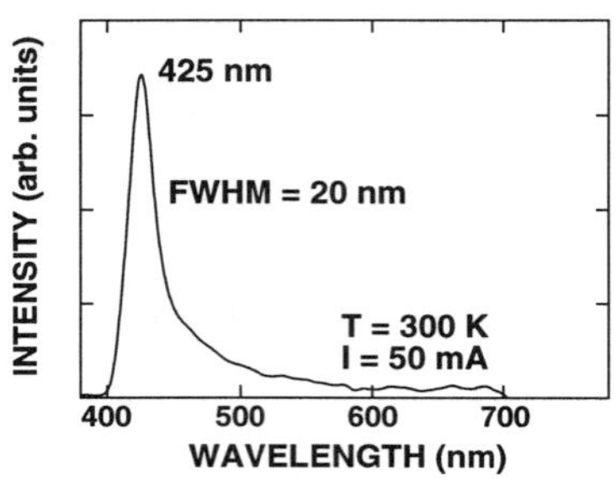

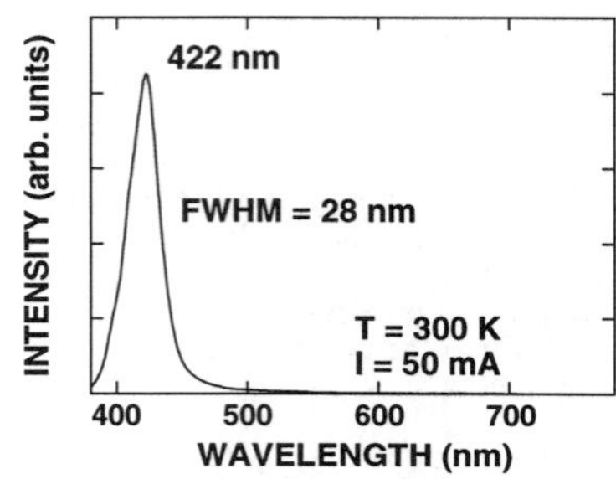

Figure 6:   300K EL from two different MBE-grown InGaN SQW heterostructures.

CONCLUSIONS

Multiple rf plasma sources have been used to generate the active nitrogen for the MBE growth of III-V nitride materials at high growth rates and substrate temperatures. Using this technique, very high quality material GaN has been produced at 1000°C. Combining the multiple sources with a modulated beam MBE approach, InGaN layers have been grown at 800°C. We have observed the (1 x 3) and (2 x 2) surface reconstructions by RHEED during modulated beam MBE growth of InGaN. The 2D reconstruction implies a flat surface, with very little indium phase segregation during the growth of these layers. This observation is confirmed by TEM analysis which reveals discrete InGaN layers with very little intermixing between the layers. Using this growth technique, InGaN SQW LED test structures have been made which emit in the blue-green visible light. The present work provides convincing evidence that MBE, along with MOVPE, is a powerful technique for the investigation and development of III-V nitride materials and devices.

ACKNOWLEDGMENT

The authors wish to thank J. Matthews and J. Brown of NCSU for their assistance with substrate preparation, x-ray diffraction measurements, and MBE system maintenance.

REFERENCES

[1]   H. Amano, M. Kito, K. Hiramatsu and I. Akasaki, Jpn. J. Appl. Phys. **28**, L2112 (1989).
[2]   S. Nakamura, T. Mukai and M. Senoh Appl. Phys. Lett. **64**, 1687 (1994).
[3]   S. Nakamura, M. Senoh, N. Iwagi and S. Nagahama, Jpn. J. Appl. Phys. **34**, L797 (1995).
[4]   S. Nakamura, M. Senoh, S. Nagahama, N. Iwasa, T. Yamada. T. Matsushita, H. Kiyoku and Y. Sugimoto, Jpn. J. Appl. Phys. **35**, L74. (1996).
[5]   S. Strite and H. Morkoç, J. Vac. Sci. Technol. B **10**, 1237 (1992).
[6]   W. Hughes, W. Rowland, M. Johnson, S. Fujita, J. Cook, J. Schetzina, J. Ren, and J. Edmond, J. Vac. Sci Technol. **B13**, 1571 (1995).
[7]   W. Shan, A. Fischer, J. Song, G. Bulman, H. Kong, M. Leonard, W. Perry, M. Bremser and R. Davis, Appl. Phys. Lett. **69**, 740 (1996).
[8]   M.Johnson, Z. Yu, C. Boney, W. Hughes, J. Cook, J. Schetzina, H. Zhao, B. Skromme and J. Edmond, Matl. Res. Soc. Symp. (Fall 1996).
[9]   R.D. Jones, PhD Thesis, Rensselaer Polytechnic Institute, (1986).
[10] M. Shimizu, K. Hiramatsu and N. Sawaki, J. Crystal Growth **145**, 209 (1994).
[11] J. Roberts, F. McIntosh, K. Boutros, S. Bedair, M. Moussa, E. Piner, Y. He and N. El-Masry, Mat. Res. Soc. Symp. Proc. **395**, 273 (1996).
[12] A. Kelley and G. Groves, <u>Crystallography and Crystal Defects,</u> (Tech Books, Herndon, VA, 1970), pp. 403-405.
[13] J. Lander and J. Morrison, Surf. Sci. **2**, 553 (1964).

Large Area Supersonic Jet Epitaxy of AlN, GaN, and SiC on Silicon

L.J. LAUHON, S.A. USTIN, W. HO
Department of Physics, Cornell University, Ithaca, N.Y., 14853

ABSTRACT

AlN, GaN, and SiC thin films were grown on 100 mm diameter Si(111) and Si(100) substrates using Supersonic Jet Epitaxy (SJE). Precursor gases were seeded in lighter mass carrier gases and free jets were formed using novel slit-jet apertures. The jet design, combined with substrate rotation, allowed for a uniform flux distribution over a large area of a 100 mm wafer at growth pressures of 1-20 mTorr. Triethylaluminum, triethylgallium, and ammonia were used for nitride growth, while disilane, acetylene, and methylsilane were used for SiC growth. The films were characterized by *in situ* optical reflectivity, x-ray diffraction (XRD), atomic force microscopy (AFM), and spectroscopic ellipsometry (SE).

INTRODUCTION

The wide band gap semiconductors AlN, GaN, and SiC are technologically important materials due to their desirable optical, electronic, thermal, and mechanical properties.[1] Much progress has been made recently in the heteroepitaxial growth of these compounds, work which is in part driven by the difficulty of growing bulk crystals of useful size and geometry. It is desirable to develop growth processes which are compatible with conventional device fabrication procedures, i.e., temperatures low enough to avoid damaging existing structures. The technique of supersonic jet epitaxy [2] allows films to be grown at temperatures lower than conventional chemical vapor deposition (CVD) by activating the reactant molecules with the kinetic energy of a seed gas in addition to the thermal energy provided by the substrate. The directed nature of the reactant flux reduces the amount of process gases used. The purpose of the research describe herein is to apply SJE to the growth of wide band gap semiconductors on 100 mm silicon substrates by using novel slit-jet nozzles to produce uniform thin films over a large area.

EXPERIMENT

A schematic overview of the apparatus is shown in Figure 1(a). The growth chamber is pumped by a diffusion pump with a cryobaffle to minimize hydrocarbon contamination. The base pressure before growth is typically 3×10^{-7} Torr. The substrate is held polished side down by a rotating sample manipulator with a boron nitride heater capable of heating the sample to $1000\,^{\circ}\mathrm{C}$ concurrent with rotation. Precursor gases are delivered to the substrate by one to four of the slit-jets shown in Figure 1(b). The tapered slits deliver more gas towards the edge of the wafer to compensate for the increase in differential area with increasing radius. This design, combined with substrate rotation, produces films which are uniform over a large area. The distance between the jets and the sample can be varied from 1.0-7.5 cm, with a typical value being 4 cm. The slit-jet nozzles routinely reach temperatures of $250\text{-}400\,^{\circ}\mathrm{C}$ during growth due to their proximity to the heated sample.

Both gas and liquid sources are used as precursors. In the case of a liquid precursor, a carrier gas is bubbled through the liquid to vaporize and transport the reactive substance. The gas delivery system allows the precursor gases to be seeded in carrier gases of lighter mass. When this gas mixture is sent through a small orifice, such as a slit-jet, into the vacuum chamber, a supersonic molecular beam is formed. The precursor gas attains a high

Mat. Res. Soc. Symp. Proc. Vol. 449 © 1997 Materials Research Society

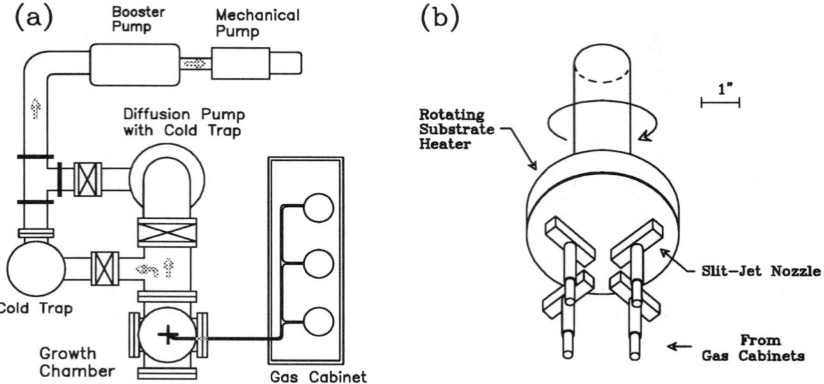

Figure 1: (a) The pumping configuration allows growth at pressures of 0.5-100 mTorr. The diffusion pump is bypassed for high pressure growth. (b) The slits on the gas nozzles are 1.25" long and tapered from 50-150 μm.

kinetic energy and becomes more reactive when seeded in this way. [3] The high kinetic energy attained by the precursor gases is necessary [2] to achieve growth at reduced temperatures via SJE.

100 mm Si(100) and Si(111) substrates were chemically treated *ex situ* to leave them hydrogen terminated by spin-etching [5] the wafers in hydroflouric acid diluted 10:1 in ethanol. The Shiraki [6] method was also used to clean and chemically oxidize some substrates. The silicon surfaces roughened when heated to temperatures necessary to desorb the oxide layers, probably as a consequence of the high base pressure in the growth chamber.[7] The hydrogen terminated layer, however, desorbs at temperatures lower than the growth temperature, and was therefore the preferred termination layer.

Three types of films were grown: AlN, GaN, and SiC. The Al and Ga sources were triethylaluminum (TEA) and triethylgallium (TEG) respectively, and the nitrogen source was ammonia (NH_3). Methylsilane, disilane (Si_2H_6), and acetylene (C_2H_2) were used for SiC growth. TEG and TEA are liquids at room temperature and therefore required the use of a carrier gas as mentioned above. Hydrogen, helium, and nitrogen were used as carrier gases; the kinetic energy of the reactive gas species increases as the mass of the carrier gas molecules is decreased according to the formula given in [3]. The flows of carrier gases were controlled by precision mass flow controllers and were in the range of 1-50 sccm for all films. The ratio of III/V species could be controlled by varying bubbler temperature and overpressure [8]. The growth temperatures were 550-800 °C for AlN and GaN, and 750-950° C for SiC.

The growth rate was monitored *in situ* by measuring the sample reflectance at a wavelength of 632 nm and an incident angle of 75°. The intensity of the light reflected from the growing film oscillates with a period dependent on the film thickness and index of refraction. The index of refraction and film thickness were measured *ex situ* using spectroscopic ellipsometry.[9] *Ex situ* characterization also included XRD and AFM.

RESULTS

The growth rate of AlN was studied as a function of three experimental variables: substrate temperature; precursor seed gas; and nozzle orientation.

Studies of growth rate versus temperature were not extensive enough to allow any conclusions to be drawn. With SJE as with CVD, an increase in growth rate with increasing substrate temperature can be attributed to the increasing thermal energy available to initiate reactions between the physisorbed precursor molecules, while a decrease in growth rate with increasing temperature can be attributed to a decrease in the sticking probability of a precursor molecule. In experiments more relevant to SJE, we sought to confirm that the slit-jet nozzles produce molecular beams which are kinetically activated, i.e., to confirm that the film growth was due in part to the SJE process and not just CVD. To this end, the kinetic energies of the precursor gases were varied by using different seed gases while other growth conditions were held constant. The resulting variations in growth rate are summarized in Figure 2(a). TEA molecules seeded in H_2 have 13 times as much kinetic energy as those seeded in N_2, which produced a 2.5 fold increase in growth rate.[10] No variation in growth rate was seen with NH_3 seeding variations, though it should be noted that the NH_3 molecules cannot, with the current system, be given as much kinetic energy as the TEA molecules due to their lesser mass. Previous SJE studies [2, 4] have also found that the growth rate of AlN does not depend strongly on the kinetic energy of the NH_3, whereas there is a strong dependence of growth rate on the TEA kinetic energy.

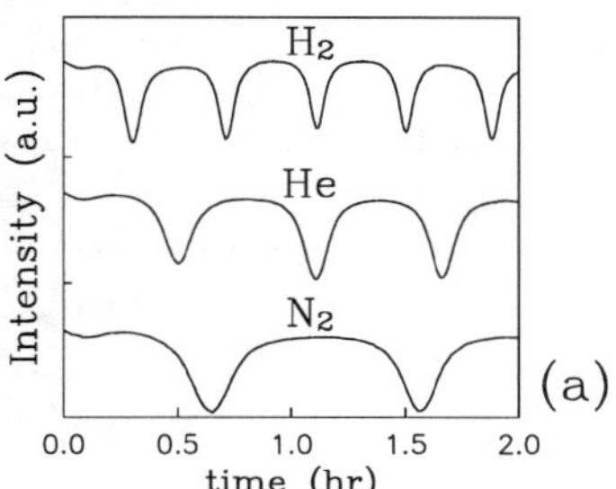

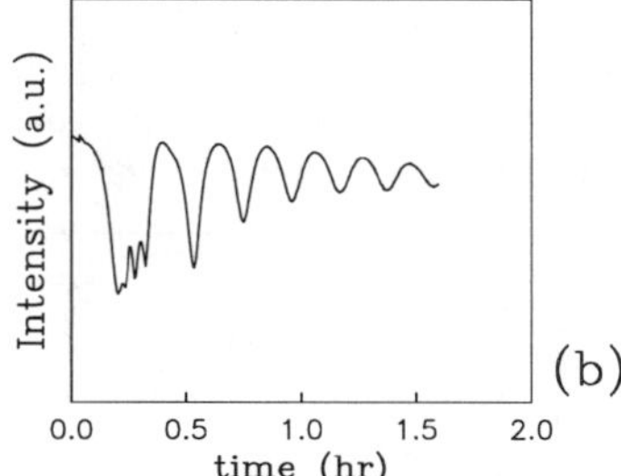

Figure 2: (a) Reflectance data from three AlN films grown with different seed gases show growth rates of 150-400 nm/hr. (b) Reflectance data from AlN film grown on silicon surface roughened by high temperature O_2 desorption.

Figure 2(b) shows reflectance data from an AlN film whose roughness is increasing with time. The roughness of the film decreases the maximum reflectance through scattering and increases the minimum due to the reduction in the coherence necessary for destructive interference. Similar behavior was seen with some methylsilane growths.

The instrument was initially run with the organometallic and nitrogen precursors delivered by separate nozzles oriented as shown in Figure 1(b). The maximum growth rate observed with this nozzle configuration was $0.15\,\mu m/hr$. One nozzle was then moved so that the flux from each of two nozzles impinged on the same area simultaneously and the growth rate increased to $0.6\,\mu m/hr$. The data shown in Figures 2 and 4 are from films grown with converging nozzles, whereas data shown in Figure 3 are from films grown in the original orientation.

One of the two converging jets was then moved to an angle of 75° from the surface normal. Growths were done with the NH_3 impinging at 75° and the organometallic impinging normally, and vice versa. The growth rate was not noticeably affected by the variation in NH_3 incident angle, but the growth rate dropped below $0.02\,\mu m/hr$ with the organometallic precursor impinging at 75°. From this it is concluded that the growth rate is sensitive to the normal component of the TEA molecular momentum.

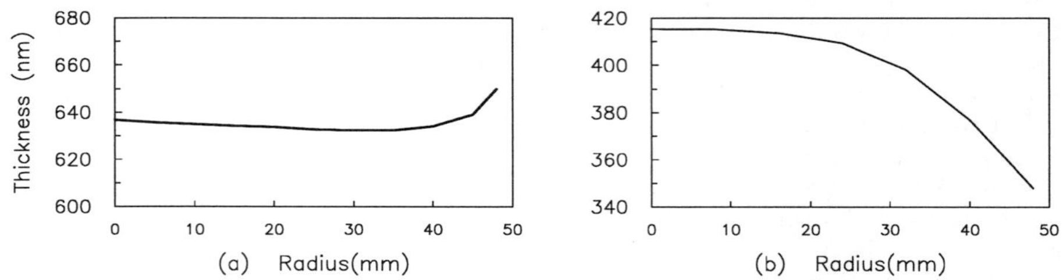

Figure 3: Thickness profiles of films on 100 mm Si(100) substrates. (a) Diffusion pumped-growth pressure of 1 mTorr. (b) Mechanically pumped- growth pressure of 20 mTorr.

Film thickness profiles were measured using spectroscopic ellipsometry [9]. The film thickness uniformity was affected by the total growth pressure in the chamber independent of the flow rates.(Fig. 3) The most uniform films showed a thickness variation of less than 1% over a 80 mm diameter. The index of refraction varied between films and across a single film by as much as 20% and showed a correlation with structural quality; films with better crystal quality tended to have indices of refraction closer to the value for single crystal AlN, which is 2.22 at visible wavelengths.

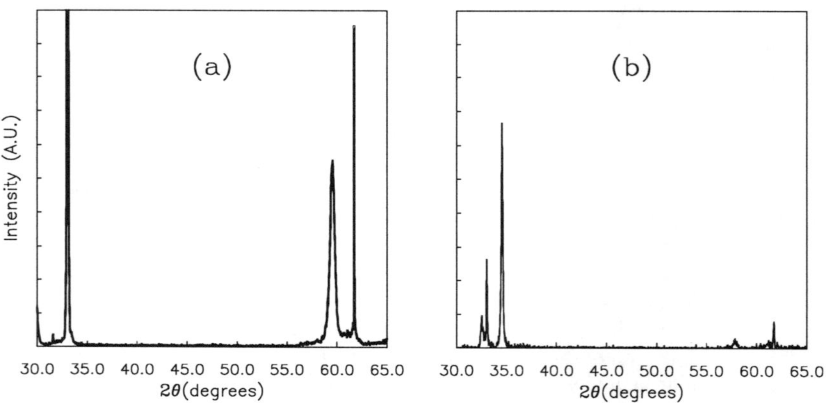

Figure 4: XRD θ-2θ scan of AlN/Si(100) and GaN/Si(100). (a) The peak at 59.3° is attributable to AlN(110) and all other peaks are from the substrate. (b) Three wurtzitic GaN peaks are visible: the (100) at 32.4°; the (002) at 34.6°; the (110) at 57.9°.

The films were in general polycrystalline, but AlN films grown with the convergent nozzles usually showed only one peak in $\theta - 2\theta$ scans.(Fig. 4(a)) An ω scan of the AlN(110) peak shown in Figure 4(a) has a rocking curve full-width at half-maximum (FWHM) of 9°, indicating the distribution in crystallite orientation with respect to the surface normal. For comparison, single phase AlN(0002)/Si(100) films grown using the supersonic jet apparatus mentioned previously have exhibited rocking curve FWHM as low as 3° [2]. The rocking curve FWHM values of AlN films grown on silicon tend to be larger than those for films grown on sapphire due to the larger lattice mismatch.

Films grown in the higher pressure environment, though of more uniform thickness, are not as structurally well ordered. Films grown at higher pressures show both the AlN(110) peak at 59.3° and the AlN(100) peak at 33.2° with varying intensity ratios. As the background chamber pressure increases, the mean free path of the molecules in the jet decreases, thereby reducing their kinetic energy prior to arrival at the surface. This can lead to more uniform films because more of the growth may be due to the background flux and less due to the directed jet. Growth from the uniform background is essentially low pressure chemical vapor deposition and not SJE, so it is not suprising that the crystal quality is adversely affected by elevated chamber pressures without a corresponding elevation to typical CVD temperatures.

GaN growths were carried out using SiC and AlN buffer layers on silicon, in part motivated by the success of other workers [11] in growing high-quality GaN using buffer layers to reduce the strain of lattice mismatch. Films were grown with the slit-jet nozzles in the non-convergent orientation. Figure 4(b) shows a $\theta - 2\theta$ scan of GaN/AlN/Si(100). Similar results were obtained when using SiC (grown with acetylene and disilane) as the buffer layer: GaN(002) was the predominant peak in the XRD spectrum.

The surfaces of AlN and GaN films were profiled using a Digital Instruments Nanoscope III AFM. The morphology of the GaN films grown on AlN buffer layers is similar that of AlN. The average surface roughness is in the range of 6-10 nm, which is 1-2% of the film thickness. For comparison, AlN(001)/Si(100) films grown in our lab with a different SJE system have an average surface roughness of 2 nm. The average grain size is approximately 50 nm which is consistent with a grain size that may be estimated from the FWHM of the $\theta - 2\theta$ XRD peak.[12]

XRD scans of films grown on Si(100) and Si(111) using jets of acetylene and disilane showed a mixture of the SiC(111) peak and various silicon peaks. The SiC(111) peak intesity grew with increasing growth temperature at the expense of the silicon peaks, but the signal intensity was weak in comparison to AlN and GaN peak intensities. The lack of SiC growth is partly due to the low reactivity of C_2H_2 compared with disilane under our growth conditions. Evidence for this includes the observation of stronger SiC(111) peaks on the back of the substrate; the back of the substrate is at a higher temperature than the front due to the proximity of the heater. This growth occured merely from the low pressure background gas concentration, again offering no advantage over conventional CVD, since the necessary reaction energy was coming primarily from the substrate and not the impinging molecules.

Preliminary studies suggest that methylsilane is a more suitable precursor for SiC growth in our apparatus. Quantitative growth rates could not be determined as the films grown to date have shown an increasing roughness with time, similar to that shown in Figure 2, which makes the index of refraction difficult to determine. XRD $\theta - 2\theta$ scans have shown cubic SiC(002)/Si(100) peaks for films grown at 850-900°C with 1 sccm methylsilane seeded in 19 sccm hydrogen.

CONCLUSIONS

Wurtzitic AlN and GaN films were grown on 100 mm silicon substrates and were of uniform thickness over 70% of the substrate. The growth conditions which produced the most uniform films did not necessarily produce the films with the highest degree of crystallinity. For AlN and GaN, the reactant gases needed to be delivered in separate nozzles to acheive a measurable growth rate. It was also determined that the flux delivered by the separate nozzles should converge to the same position on the substrate to maximize growth rate. It is possible to deposit single phase AlN at temperatures below 800 °C using this technique.

It was found that the apparatus is incapable of growing high quality SiC using C_2H_2 as a carbon source. Methylsilane was found to be a more suitable precursor for the apparatus as highly oriented cubic SiC(002)/Si(100) could be obtained at temperatures less than 900°C.

ACKNOWLEDGMENTS

Support of this research by the Defense University Research Instrumentation Program Grant No. N00014-95-01-0250 and the Ballistic Missile Defense Organization Grant No. N00014-93-0499 via the Office of Naval Research is gratefully acknowledged.

References

[1] S. Strite and H. Morkoc, J. Vac. Sci. Technol. B10(4), 1237 (1992).

[2] K.A. Brown, S.A. Ustin, L.J. Lauhon, and W. Ho, J. Appl. Phys. **79**, 7667 (1996).

[3] D.R. Miller, in *Atomic and Molecular Beam Methods*, edited by G. Scoles.

[4] S. A. Ustin, K. A. Brown, L.J. Lauhon, D.Q. Hu, and W. Ho, Mater. Res. Soc. Proc. **395**, 319 (1996).

[5] Spin-Etch processor manufactured by Laurell Technologies.

[6] A. Ishizazka, K. Nakagawa, and Y. Shiraki, Second Int. Symp. on MBE and Clean Surface Related Techniques, ed. by R. Ueda (Jap. Soc. Appl. Phys., Tokyo 1982) p. 183.

[7] B.S. Swartzentruber, Y.-W. Mo, M.B. Webb, and M.G. Lagally, J. Vac. Sci. Technol. A **7** (4), 2901 (1989).

[8] S.D. Hersee and J.M. Ballingall, J. Vac. Sci. Technol. A **8**, 800 (1990).

[9] Rudolph Research Auto El IV, at the Cornell Nanofabrication Facility.

[10] The films in Figure 2 were grown at 750°C, total pressure 1 mTorr, 18 sccm NH_3, 5 sccm carrier gas.

[11] I. Akasaki, H. Amano, Y. Koide, K. Hiramatsu, and N. Sawaki, J. Cryst. Growth **98**, 209 (1989).

[12] B.E. Warren, X-Ray Diffraction, (Dover Publications, Inc., New York, 1990), p. 253.

A Study of the Surface Morphological Features of the Polar Faces of ZnO by Atomic Force Microscopy (AFM) Methods and AlN Thin Films Deposited on ZnO Polar Faces by PLD

M.J. SUSCAVAGE*, D.F. RYDER, JR.**, P.W. YIP*
* Rome Laboratory, USAF
  80 Scott Dr, Hanscom AFB, Bedford , MA 01731
**Dept. of Chemical Engineering
  Tufts University, Medford, MA

ABSTRACT

The effects of both temperature and atmosphere on the resulting morphological features of the polar faces of single crystal ZnO were investigated and characterized by atomic force microscopy (AFM). In studies where ZnO was thermally processed in flowing oxygen at atmospheric conditions within the temperature range of 500°C to 900°C for 30 minutes, the Zn-surface (i.e., (0001)) showed a tendency to reconstruct with increasing temperature until terraces became evident at 900°C. Terrace heights were as small as 0.9 nm. In contrast, the O-surface (i.e., (000 $\bar{1}$)) was observe to change very little during the O_2-atmoshere, thermal treatment and remained comparatively rougher than the Zn-surface. ZnO samples which were thermally processed under high vacuum (i.e., 5 x 10^{-7} Torr) conditions exhibited a more dramatic contrast. The vacuum annealed Zn-surface was observed to develop very smooth surface features (Roughness = 0.09 nm) at annealing temperatures within the 700 - 800 °C range. In contrast, and as expected, the O-surface roughness increased due to surface reduction reactions. In addition to these findings, it is noted that AFM measurements may be utilized as a convenient method to distinguish between the two polar surfaces of ZnO.

Aluminum nitride was deposited on the Zn- and O- surfaces from 700 to 850°C by pulsed laser evaporation. X-ray diffraction indicated that the AlN was c-axis oriented with no interface reaction products detected between the ZnO substrate and AlN film.

INTRODUCTION

ZnO has the same crystal structure, and a reasonably close lattice match, to both GaN and AlN. Likewise, the noncentrosymmetric structure of ZnO imparts a crystallographic polarity that not only results in the formation of opposite polar surfaces (i.e., Zn- and O- surfaces), but also results in characteristic property differences between the surfaces. The energy, chemistry, and defect structure of each polar surface in ZnO will differ, and as such impact the resulting properties of the nitride film overgrowth. As previously noted, and from a crystallographic perspective, ZnO is a promising substrate candidate for the epitaxial growth of III-V nitride thin films. A study relating the changes of surface energy and chemistry on the two polar faces of single crystal ZnO subjected to the processing conditions of typical thin film deposition methods would contribute added insight into the use of ZnO as a III-V nitride substrate material.

It has previously been reported [1] that the polar faces of single crystal ZnO are both structurally and chemically distinct, with the ZnO (000 $\bar{1}$) being considered the O- surface and the ZnO (0001) being considered the Zn- surface. Given that ZnO is being considered as a substrate or buffer layer for the III-V nitrides [2-6], it is of importance to determine the contribution that ZnO surface structure and chemistry imparts to the epilayer's properties. While initials studies relating to these matters have recently been reported in the literature, the results

Mat. Res. Soc. Symp. Proc. Vol. 449 © 1997 Materials Research Society

are often contrary. For example, one noted problem with ZnO is that it is easily reduced at the processing conditions typically used for film deposition of GaN and AlN. In a recent article [4], GaN was deposited on the Zn- and O- surfaces by MBE. The results suggest that ZnO is not a suitable substrate for GaN films grown by MBE at 600°C due to reaction between Ga metal and ZnO to form Ga_2ZnO_4 at the interface. Other results [5] suggest that MBE grown GaN exhibits two-dimensional growth on ZnO when ZnO is used as a buffer layer on sapphire. No indication was given as to which ZnO face the GaN is being grown on. Given this controversy, there is a need to study, from a more fundamental perspective, the processing ramifications associated with the use of polar substrate materials for III-V nitrides.

In this study, we investigated the effects of both thermal and atmospheric conditions on the resulting surface microstructural features of the polar faces of ZnO. In addition, AlN films deposited by PLD methods on these polar surfaces is reported.

EXPERIMENTAL

ZnO (0001) single crystal substrates, polished on both faces, were obtained from Atomergic Chemetals Corp. (Farmingdale, NY). While optical microscopy did not reveal any differences in the microstructural features of the two substrate surfaces. Results obtained from a preliminary AFM (Digital NanoScope III, Santa Barbara, CA) investigation of as-delivered samples later showed that the noted differences in surface morphology were attributable to the microstructural features associated with the polar surfaces. This observation is consistent with the physical features reported by Mariano and Hanneman [1].

Samples of the oriented, single crystal ZnO substrates were thermally annealed at 1 atm in flowing oxygen for 30 min within the temperature range of 500 to 900°C. During thermal processing, the test sample was suspended by the edges to ensure that any changes to the bottom surface were not influenced by contact with a holder. Samples were allowed to furnace cool to room temperature under flowing oxygen before removal. In order to examine the effect of vacuum pressure (i.e., 5×10^{-7} Torr) and temperature on the stability of the polar ZnO surfaces, each polar face was annealed at 800°C for 15 min and then cooled to room temperature without deposition. Atomic force microscopy (AFM) operating in the tapping mode was performed on both ZnO faces in the as-received condition, after cleaning with trichloroethylene and methanol, and after each anneal. Samples were chemically etched in 20% HNO_3 for 90 sec. in order to determine the Zn- and O- surfaces via standard methods.

Aluminum nitride films were deposited on the Zn- and O- faces by pulsed laser evaporation. A KrF excimer laser operating at 248 nm, 25 Hz and an energy density of 2-4 J/cm^2 was used to evaporate an AlN target (99%, PureTech Inc., Carmel, NY). The custom-designed vacuum system consisted of a stainless steel chamber evacuated by a turbomolecular pump to a base pressure of 1×10^{-8} Torr. The films were deposited at temperatures from 700 to 850°C in 3 hours with a continuous flow of ammonia at a pressure of 3×10^{-4} Torr. Ammonia flow was started within five seconds after the start of deposition to reduce the possibility of reaction of ammonia with the ZnO substrate. Both the substrate and target were rotated while the laser was rastered across the target. Target to substrate distance was 11 cm with a resultant deposition rate of ~4 nm/min. Film thickness ranged between 700 and 800 nm. For more efficient utilization and to ensure deposition under the same conditions on the Zn- and O- face, each 1 cm^2 substrate was marked, sliced in half, and both halves were mounted on the heater block with one half flipped over.

RESULTS AND DISCUSSION

Figure 1 contains AFM images of the opposite polar faces of a ZnO substrate in the as-received polished condition, after a solvent cleaning step, and after the progressive O_2-atmosphere anneals from 500 to 900°C. The images clearly show that the surface morphology of each polar face are distinct, even in the initial as-received (i.e., uncleaned) and chemically cleaned state. The effects of thermal processing on surface microstructure are more dramatic

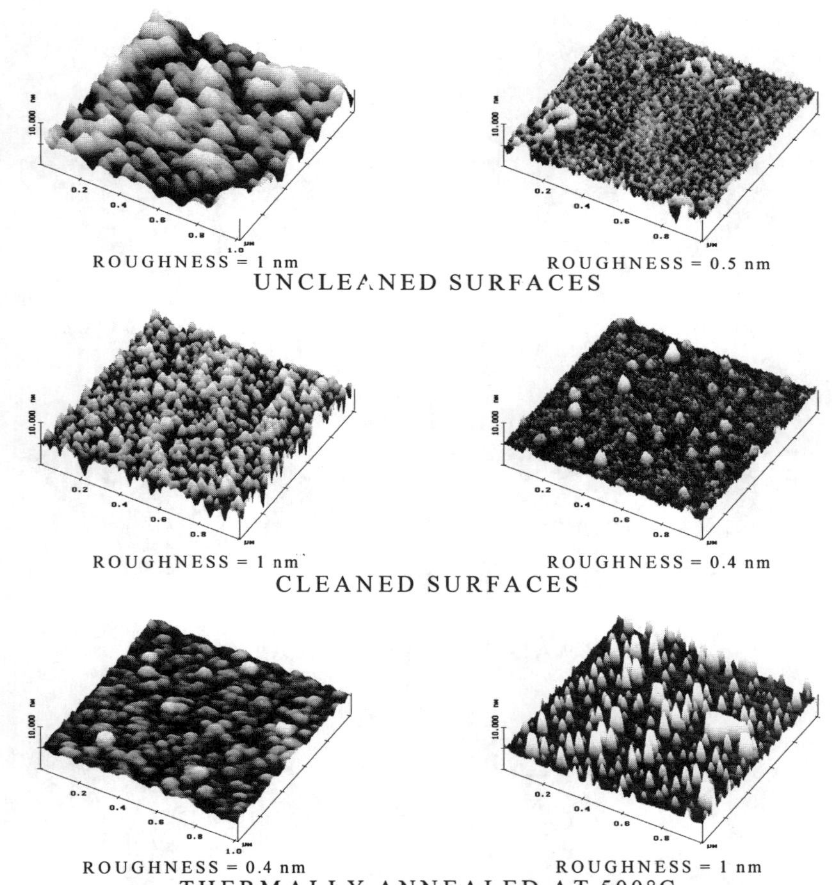

Figure 1. AFM images ZnO faces with the (000 $\bar{1}$) or O- surface on the left and the (0001) or Zn- surface on the right after annealing in oxygen.

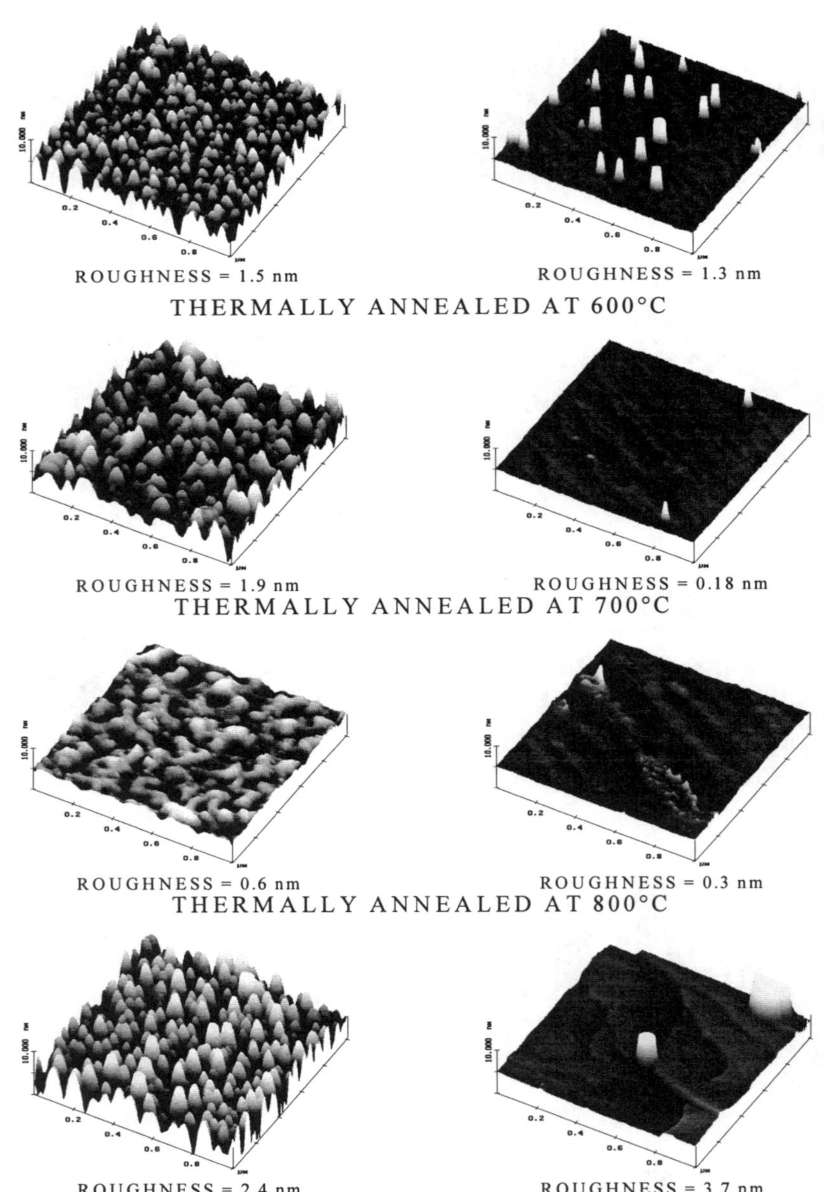

Figure 1 cont. AFM images of ZnO faces with the (000 $\bar{1}$) or O- surface on the left and the (0001) or Zn- surface on the right after annealing in oxygen.

for the Zn- surface than for the O- surface. The Zn- face reconstructs to form a flatter surface and eventually develops small terraces with a step height of ~ 1 nm after an anneal at 900°C. The O- surface does not appear to reconstruct in a similar fashion. It remains fairly stable throughout the course of anneals. To determine if the additive annealing history was influencing the morphological surface reconstruction, a fresh ZnO substrate was only annealed at 700°C in oxygen. The resulting surfaces were consistent with the 700°C image data from the progressive anneal trials. As such, it was assumed that there were minimal microstructural effects associated with thermal history. To examine the effects of high temperature and vacuum on the substrate, a new ZnO substrate was sliced in half and mounted in the vacuum system as described earlier and annealed at 800°C for 15 minutes. The AFM images in figure 2 show an extremely smooth Zn- surface and an unchanged O- surface. The roughness was 0.09 nm for the Zn- surface and 0.9 nm for the O- surface.

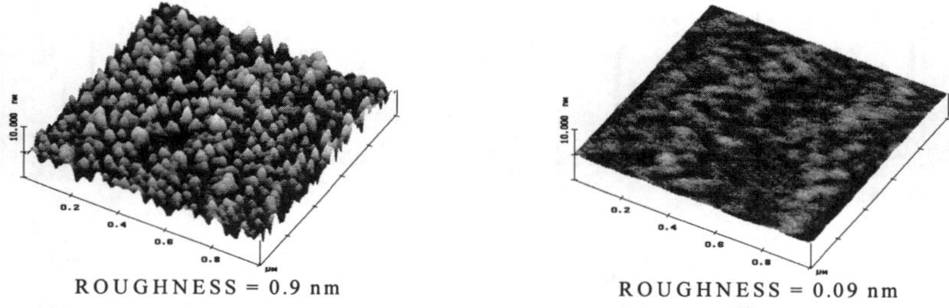

Figure 2. AFM images of ZnO faces with the $(000\,\bar{1})$ or O- surface on the left and the (0001) or Zn- surface on the right after an anneal at 800°C for 15 minutes at 5×10^{-7} Torr.

Figure 3 shows diffraction scans of AlN deposited on the Zn- and O- surfaces of ZnO at four different temperatures. The deposition time was three hours at all temperatures. The AlN appears to be well oriented on both surfaces up to 800°C with no evidence of interface reaction products. All films deposited at and below 800°C were smooth and transparent. Two point resistivity indicated the films were insulating when measured with an ohmmeter. At 850°C, no AlN is present in the diffraction scans, but no other peaks are present indicating the formation of reaction products. The films had a metallic luster but were transparent and smooth. Energy dispersive spectroscopy detected both aluminum and nitrogen on the sample along with zinc and oxygen.

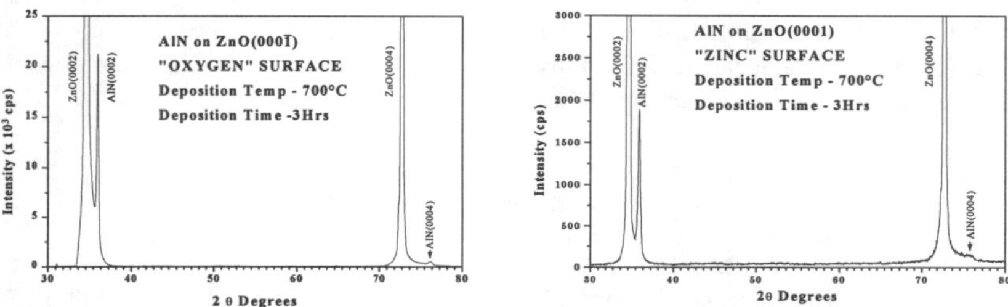

Figure 3. XRD scans of AlN deposited on the O- and Zn- surfaces of ZnO.

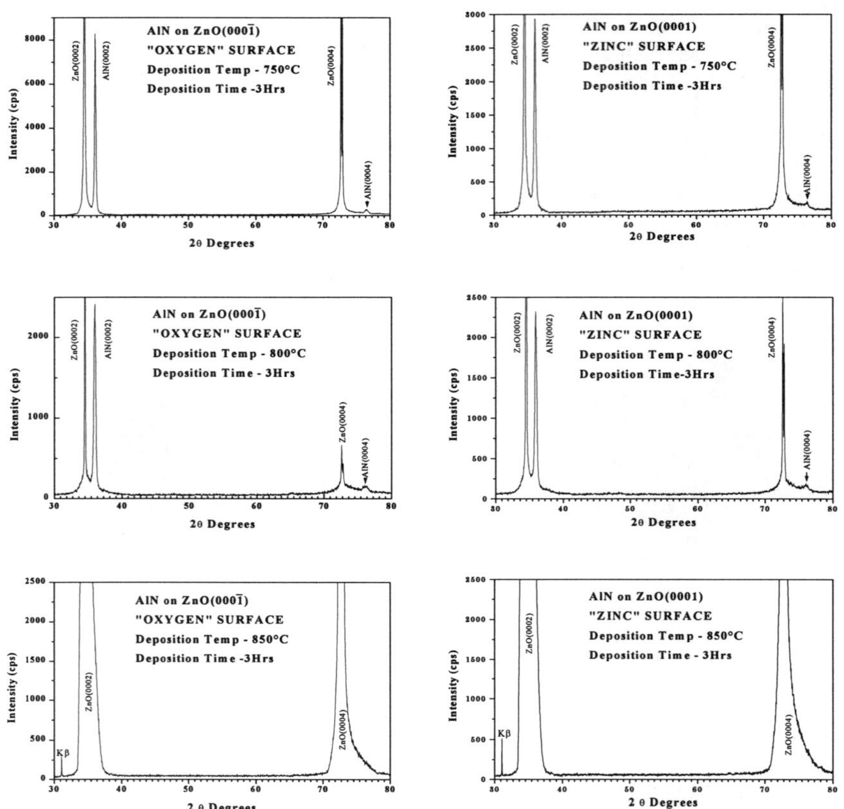

Figure 3 cont. XRD scans of AlN deposited on the O- and Zn- surfaces of ZnO.

CONCLUSIONS

It was determined from AFM analysis that thermal annealing of ZnO will result in extremely smooth surfaces on the Zn- face and stable but rough surfaces on the O- face. AFM is also a quick, nondestructive tool that can unambiguously differentiate between the Zn- and O- faces of ZnO. It may be possible that AFM can distinguish the faces of other III-V compounds (i.e. GaN and AlN) with polar faces. Pulsed laser deposition can deposit c-axis oriented AlN on ZnO with no apparent reactions occurring at the interface.

1. A.N. Mariano and R.E. Hanneman, *J. Appl. Phys.* **34**, 384, (1963).
2. T. Matsuoka, N. Yoshimoto, T. Sasaki, and A. Katsui, *J. Electron. Mater.,* **21**, 157, (1992).
3. M.A.L. Johnson, Shizuo Fujita, W.H. Rowland, Jr., W.C. Hughes, J.W. Cook, Jr., and J.F. Schetzina, *J. Electron. Mater.,* **25**, 855, (1996).
4. E.S. Hellman, D.N.E. Buchanan, D. Wiesmann, I. Brener, *MRS Internet Journal,* **1**, (16) (1996).
5. Z. Sitar,M.J.Paisley, B. Yan, and R.F. Davis, *Mater. Res. Symp. Proc.,* **162**, 537, (1990).
6. B. N. Sverdlov, G.A. Martin, H. Morkoç, D.J. Smith, *Appl. Phys., Lett.,* **67**, (14), 2063, (1995).

EFFECT OF GROWTH PARAMETERS AND LOCAL GAS PHASE CONCENTRATIONS ON THE UNIFORMITY AND MATERIAL PROPERTIES OF GaN/SAPPHIRE GROWN BY HYDRIDE VAPOR PHASE EPITAXY

S.A. Safvi*, N.R. Perkins**, M. N. Horton**, T.F. Kuech*'**
*Department of Chemical Engineering, University of Wisconsin, Madison, WI 53706.
**Materials Science Program, University of Wisconsin, Madison, WI 53706.

ABSTRACT

The effects of flowrate variation and geometry on the growth rate, growth uniformity and crystal quality were investigated in a horizontal Gallium Nitride vapor phase epitaxy reactor. To better understand the effects of these parameters, numerical model predictions are compared to experimentally observed values. Parasitic gas phase reactions between group III and group V sources and deposition of material on the wall are shown to lead to reduced overall growth rates and may be responsible for inferior crystal quality. A low ammonia concentration is correlated with the deposition of polycrystalline films. A low V/III ratio and an ammonia concentration lead to poor crystalline quality and increased yellow luminescence. An optimum HVPE growth process requires selection of reactor geometry and operating conditions to minimize these parasitic reactions and wall deposition while providing a uniform reactant distribution across the substrate.

INTRODUCTION

The nitrides of gallium, aluminum, and indium have great potential for applications in electronics and optoelectronic devices due to their wide bandgap range and stability at high temperatures. These nitrides have a direct bandgap ranging from 1.9 eV for InN to 6.3 eV for AlN. GaN with its bandgap of 3.4 eV is particularly suitable for making devices operating in blue to ultraviolet range [1]. The heteroepitaxial growth of thin GaN films on sapphire, the most common substrate, leads to defects arising from lattice mismatch and difference in thermal expansion coefficient. The development of GaN substrates is likely to be a key advance in nitride epitaxial technology, making it feasible to grow homoepitaxial thin GaN films [2]. A promising route for the development of GaN substrates is the heteroepitaxial growth of GaN films by rapid growth techniques, such as halide vapor phase epitaxy (HVPE), followed by the *in situ* etch removal of the initial substrate to leave a free-standing GaN film. The HVPE technique has been used previously to grow thick layers of GaAs [3], GaN [2, 4, 5], and InP [6]. Mochizuki et. al. [7] have studied the direct reaction between AsH_3 and surface adsorbed GaCl in order to understand the growth chemistry involved in HVPE of GaAs. However, no comparable GaN based studies have been reported. In an earlier work [8] we discussed that a convection-diffusion model of the reacting species was a good approximation of our HVPE process. The development of predictive models of the HVPE process can substantially reduce the time and cost associated with reactor optimization and scale-up by minimizing the required experimental trial and error. It would also aid development of an improved understanding of the HVPE process.

In this study we will describe a two-zone hot wall reactor used to grow thick HVPE GaN films. The emphasis will be on studying the effect of process and geometric parameter variation on film thickness, uniformity and material properties. Experimental results will be compared to computational predictions. The effect of local gas phase concentrations on film properties will be discussed.

Mat. Res. Soc. Symp. Proc. Vol. 449 © 1997 Materials Research Society

REACTOR MODEL & GROWTH STUDIES

The experiments were carried out in an atmospheric pressure quartz reactor which has been presented earlier[9]. The reactor has three separate concentric inlets for the reaction gases. A mixture of N_2 and HCl is introduced through the central tube, N_2 through the middle annular region, and NH_3 through the outer annular region. The reactor is divided into two separate temperature zones of 850 and 1050 °C. In the first reaction zone, operated at 850 °C, Ga metal is reacted with HCl gas (typical flow ~ 30 sccm) to yield GaCl and H_2 reaction products. The extent of reaction, obtained from decrease in the mass of Ga metal, was in the range of 50-70%. These reaction products are transported to the second zone through the central tube. In the second zone, typically maintained at 1050 °C, a high flowrate N_2 buffer and NH_3 were introduced from the middle and outer annular regions, respectively. The NH_3/HCl ratio was typically 30:1. Two inch

diameter (0001) sapphire substrates prepared with a standard solvent degrease are used for the HVPE deposition. For the base case, a N_2 buffer flowrate of 4.5 slm, a NH_3 flowrate of 882 sccm, and, in the central tube, a mixture of 30 sccm HCl and 150 sccm N_2 was employed.

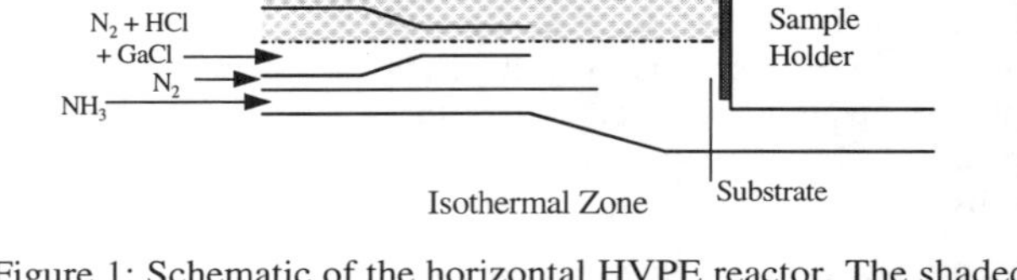

Figure 1: Schematic of the horizontal HVPE reactor. The shaded region defines the computational domain.

A schematic of the computational domain is shown in Fig 1. The variable Z shown in Fig 1 represents the relative axial distance of the sample holder from the inlet. Only the second temperature zone of the reactor, which is of interest for process optimization, was modeled. Cylindrical coordinates have been used. The fundamental equations of continuity, momentum, and species conservation are used to describe the system [10]. With the assumption of no variation in circumferential direction, the flow and concentrations are obtained in two dimensions.

<u>Physical and Transport Properties of Gaseous Species</u>: Experimental values of the viscosity of nitrogen [11] and ammonia [12] were fitted to equations as a function of temperature and are listed elsewhere [13]. Binary diffusion coefficients of gas phase species were either obtained from literature or estimated from their Lennard-Jones parameters [14, 15]. The values of the binary diffusion coefficient and Lennard-Jones parameters for all gaseous species used in this study are listed elsewhere [13]. Linear mixing rules were applied to determine local gas phase properties.

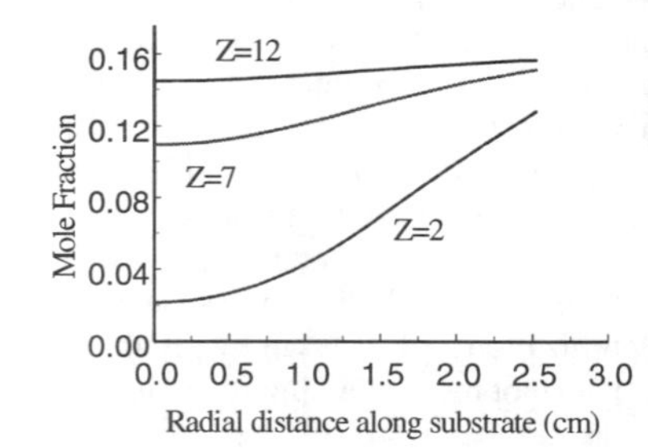

Figure 2: Ammonia concentration across the substrate for different substrate positions.

<u>Numerical Solution</u>: The system of partial differential equations describing flow and mass transfer was solved using Galerkin's finite-element method [16]. A typical mesh consisted of 1400 quadilateral elements with the mesh being denser in those parts of the domain where

steeper gradients existed. The system of nonlinear algebraic equations obtained after the application of Galerkin's technique was solved by using Newton's method. The system of equations and numerical methodology is presented in detail elsewhere [13]. The computations were performed on a Cray C90 supercomputer.

RESULTS AND DISCUSSION

In experiments where the sample holder was kept close to the inlet, the resulting samples have a dark polycrystalline patch in the center of the wafer, with a clear single crystalline film at the edges.

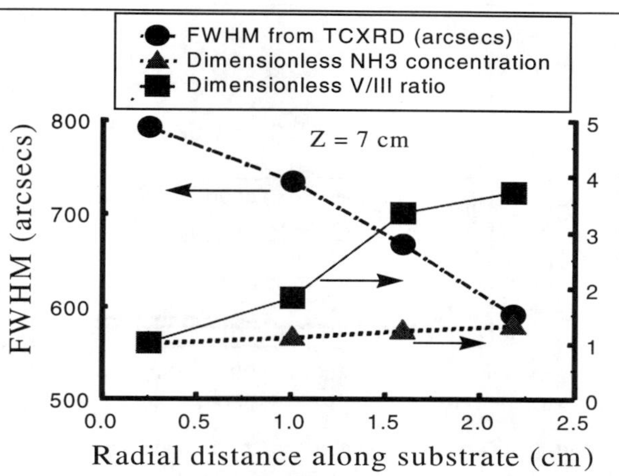

Figure 3: Radial distribution of the dimensionless NH_3 concentration, dimensionless V/III ratio, and FWHM from TCXRD (arcsecs).

This polycrystalline patch decreased in size and eventually disappeared as the substrate was moved further away from the inlet (Z increases in Fig 1). The patch could be caused by a deficiency of NH_3 (and hence low effective V/III ratio) near the center of the wafer as NH_3 is introduced from the outermost annular inlet to reduce prereaction. When the substrate is far from the inlet, the ammonia stream is well mixed with the other two inlet streams and there is a uniform concentration of ammonia over the substrate at the growth front. Fig 2 shows variation of NH_3 concentration over the substrate for different substrate positions.

At a distance of Z=7 cm, the film obtained was single crystalline across the wafer. Fig 3 shows a plot of NH_3 concentration and V/III ratio, across the substrate, made dimensionless with respect to their values at the center of the substrate. The corresponding full width of the triple crystal x-ray diffraction (TCXRD) rocking curve at half maximum ranged from 791 arcsecs at the center to 590 arcsecs at the edge of the wafer. The quality of the film

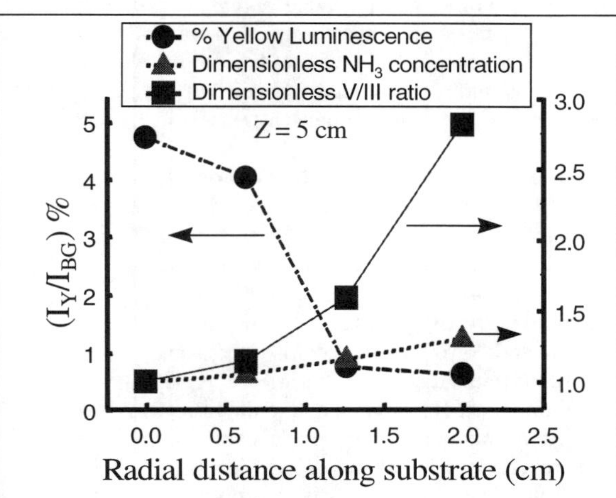

Figure 4: Radial distribution of the dimensionless NH_3 concentration, dimensionless V/III ratio, and percentage of yellow luminescence to band edge luminescence.

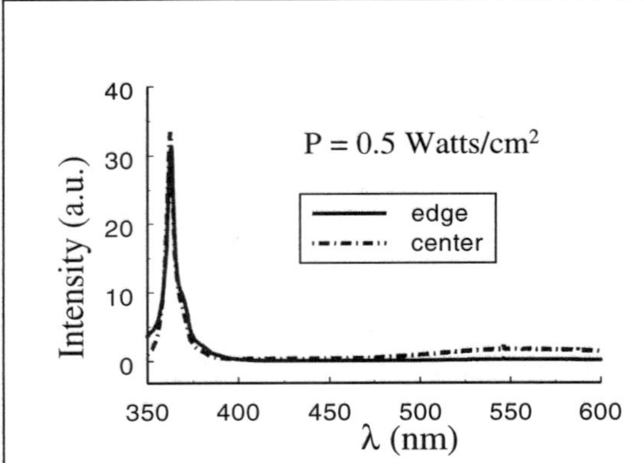

Figure 5: PL spectra taken at center and edge on a film grown at Z=5 cm. All other process conditions are as in the base case.

improves with increasing values of ammonia concentration, increasing the local V/III ratio.

GaN grown by any technique, suffers from defect luminescence in the 2.2 - 2.3 eV range [17], which limits its application to devices such as laser diodes. The reason for the optically detrimental yellow luminescence has been a subject of debate [18,19,20,21]. This yellow luminescence is either a result of native physical defects and/or impurities common to all growth systems. Fig 4 shows a plot of NH_3 concentration, V/III ratio, and the corresponding yellow luminescence along the radius for a film grown at Z=5 cm. The NH_3 concentration and the V/III ratio are made dimensionless with respect to their value at the center of the film, while the yellow luminescence peak intensity is given as a percentage of the band edge related peak. An Ar laser was used and the pump power incident on the wafer was 0.5 W/cm^2 and all measurements were done at room temperature. Fig 5 shows the PL observed at the center and the edge of the film. We observed that the ratio of yellow to band edge luminescence decreased with increase in local V/III ratio and improvement in crystalline quality (decreasing defect concentration) as determined by x-ray measurements. The quality of the films would be limited by the choice of reactor design and process conditions that enable a high and uniform V/III ratio at the substrate while having unmixed reactants upstream. A lower pressure process would be beneficial for the two constraints mentioned above, but it would involve substantial increase in equipment cost.

The predicted growth uniformity improved with increasing substrate-inlet distance as shown in Fig 6. This trend is expected since the reactant gases have more time to diffuse and become more uniformly distributed over the substrate. A lowering in the overall growth rate with increasing substrate-inlet distance, observed in our experiments [13], could be due to the loss of growth nutrients due to either deposition on the reactor walls or gas phase parasitic reactions.

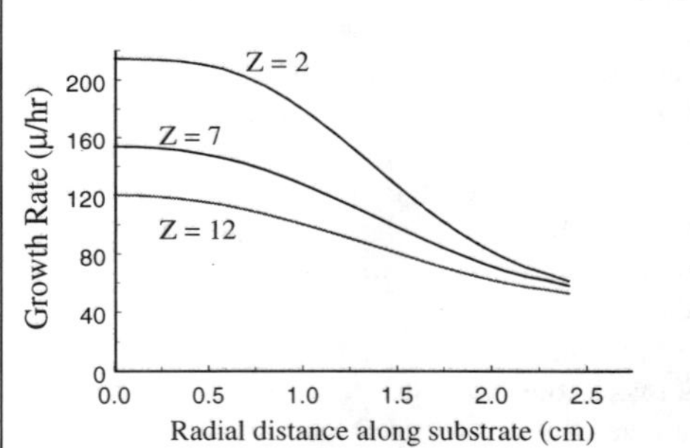

Figure 6: Predicted growth rate variation across the substrate for substrate position of Z =2, 7, and 12 cm.

The former effect is incorporated in the model but the latter effect is not, and this could explain the predicted growth rates being lower than those experimentally observed. This discrepancy between the predicted and observed growth rates increases with increasing substrate-inlet distance. Though the uniformity of the film grown at Z=12 cm was better than those grown at Z=2 cm and Z=7 cm, the quality of the films, as examined by TCXRD, was poorer. The TCXRD spectra exhibited double peaks across most of the sample, probably due to delamination of the film from the substrate upon cooling. Products of gas phase parasitic reactions or from reactions on the wall (e.g. HCl), interfere with the film growth and subsequent properties. Additionally, particles coming off the walls may be transported to the film in the absence of thermophoresis effects in the isothermal reactor leading to macroscopic defects on the wafer surface.

The coupled gas phase and surface reaction between GaCl and NH_3 has been studied by V. Ban using mass spectrometry [22]. The simplest gas phase reaction would lead to the formation of GaN monomer with eventual particulate formation and

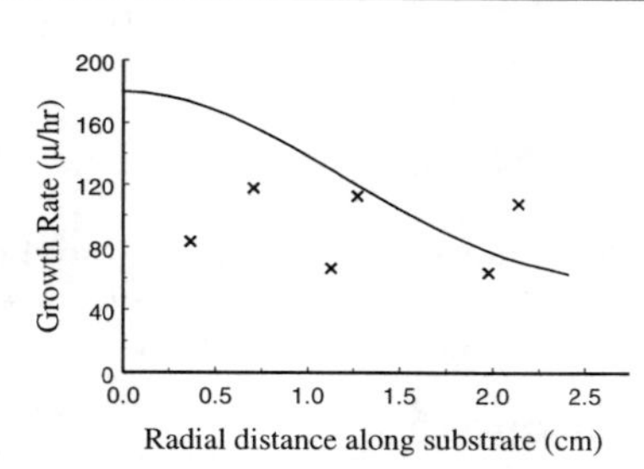

Figure 7: Comparison of computationally and experimentally (x) observed growth rates for a N_2 buffer flow of 2.25 slm, and Z=4.5 cm.

the resulting loss of reactants at the growth front. In the first zone (at 850 °C) formation of GaCl in the presence of excess HCl is favored. In the second zone (1050 °C), the presence of excess ammonia shifts the equilibrium such that the reduction of GaCl is favored. Both prereaction and wall effects could lead to the same results: lowered growth rates and inferior crystal quality. In order to isolate the effect of gas phase parasitic reactions, we performed some experiments at a sample holder held close to the inlet (Z= 4.5 cm) where the effects of the reactions at the wall on the substrate film growth are minimized. All process parameters are kept the same as in the base case, except for the N_2 buffer flow rate through the central annulus. For an N_2 buffer flow of 2.25 and 4.5 slm, computations show that the overall growth rate does not change appreciably, indicating that the walls have little effect on the growth process. Experiments, in contrast, reveal that the growth rates are significantly reduced upon lowering the N_2 buffer flowrate. Fig 7 and 8 show the predicted and experimentally obtained growth rates for a N_2 buffer flow of 2.25 and 4.5 slm, respectively. Lowering the buffer flowrate increases the residence time in the reactor and promotes the mixing of GaCl and NH_3. This gas phase mixing accelerates the pre-reaction between the two precursors. Increasing the residence time increases the extent of this reaction.

These results indicate that an optimum buffer flowrate of N_2 should be high enough to reduce any gas phase precursor prereactions, but low enough to allow diffusion of precursors so that they are uniformly distributed across the substrate.

CONCLUSIONS

We investigated the effect of process and geometric parameters on GaN grown by HVPE. Experiments in conjunction with modeling show that the crystalline quality of the film was found to improve and yellow luminescence decrease with increasing local concentration of ammonia and the V/III ratio at the substrate. A N_2 buffer between the group III and group V precursor was necessary

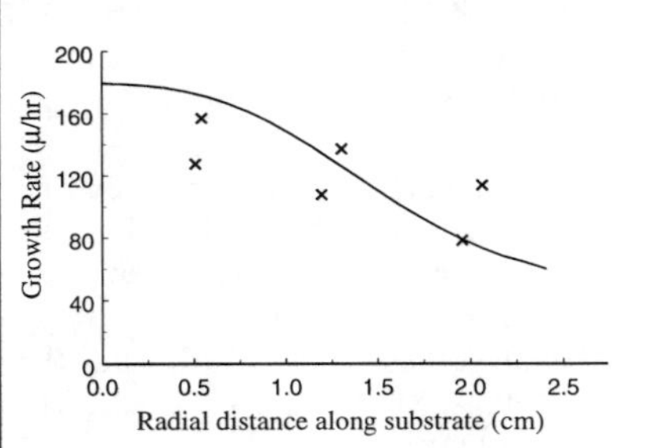

Figure 8: Comparison of computationally and experimentally (x) observed growth rates for a N_2 buffer flow of 4.5 slm, and Z=4.5 cm.

to limit parasitic prereactions. The quality of the films and the growth rate were sensitive to the buffer flowrates.

ACKNOWLEDGMENTS

The work was supported by the Naval Research Laboratories and ARPA - URI on Visible Light Emitters. The supercomputing time was provided by the DOD. J. Cederberg is acknowledged for assistance in the PL measurements.

REFERENCES

1. X. Zhang, P. Kung, A. Saxler, D. Walker, T.C. Wang, and M. Razeghi, Appl. Phys. Lett. **67,** p. 1745 (1995).
2. T. Detchprohm, K. Hiramatsu, N. Sawaki, and I. Akasaki, J. Crystal Growth **137,** p. 170 (1994).
3. D.W. Shaw, J.Crystal Growth **31,** p. 130 (1975).
4. H. Maruska, and J. Tietjen, Appl. Phys. Lett. **15,** p. 327 (1969).
5. O. Lagerstedt and B. Monemar, J. Appl. Phys., **45,** p. 2266 (1974).
6. R. Beccard, S. Beuven, K. Heime, R. Schmald, H. Jurgensen, P. Harde, and M. Schlak, J. Crystal Growth **121,** p. 373 (1992).
7. Y. Mochizuki, A. Usui, S. Handa and T. Takada, J. Crystal Growth **148,** p. 96 (1995).
8. S.A. Safvi, N.R. Perkins, M.N. Horton, A. Thon, D. Zhi, and TF Kuech, Mat. Res. Soc. Symp. Proc. **423,** p. 227 (1996).
9. N.R. Perkins, M.N. Horton and T.F. Kuech, Mat. Res. Soc. Symp. Proc. **395,** p. 243 (1995).
10. R.B. Bird et al., W.E. Stewart and E.N. Lightfoot, Transport Phenomenon, Wiley, New York 1960.
11. R.H. Perry & C.H. Chilton, Chemical Engineer's Handbook, 5th ed., McGraw-Hill, New York 1977.
12. R.C. Reid, J.M. Prausnitz and T.K. Sherwood, The Properties of Gases and Liquids, McGraw-Hill, New York, 1977.
13. S.A. Safvi and T.F. Kuech, Manuscript submitted to J. Cryst. Growth.
14. T.J. Mountziaris, S. Kalyanasundram and N.K. Ingle, J. Cryst. Growth, **131,** p. 283 (1993).
15. J.O. Hirschfelder C.F. Curtiss and R.B. Bird, Molecular Theory of gases and Liquids, Wiley, New York, 1967.
16. G. Strang & G. Fix, An Analysis of the Finite Element Method, Prentice Hall, Englewood Cliffs, NJ, 1973.
17. S. Strite and H. Morkoc, J. Vac. Sci. Technol. B, **10,** p. 1237 (1992).
18. J.I. Pankove and J.A. Hutchby, J. Appl. Phys.. **47,** p. 5387 (1976).
19. T. Ogino and M.Aoki, Jpn J. Appl. Phys. **19,** p. 2395 (1980).
20. E.R. Glaser, T.A. Kennedy, K. Doverspike, L.B. Rowland, D.K. Gaskill, J.A. Freitas, M.A. Khan, D.T. Olson, and J.N. Kuznia, Phys. Rev. B **51,** p. 13326 (1995).
21. P. Hack, A. Maekawa, N. Koide, K. Hiramatsu, and N. Sawaki, Jpn. J. Appl. Phys. **33,** p. 6443 (1994).
22. V. Ban, J. Electrochem Soc., **119,** 761 (1972).

MORPHOLOGY AND DIELECTRIC PROPERTIES OF REACTIVELY SPUTTERED ALUMINUM NITRIDE THIN FILMS

A.G. RANDOLPH, S.K. KURINEC*
Departments of Microelectronic Engineering and Materials Science and Engineering, Rochester Institute of Technology, Rochester, NY 14623, * skkemc@rit.edu

ABSTRACT

Aluminum nitride thin films ($\sim$ 100 nm) have been deposited on silicon substrate by reactive sputtering using Al target in 1:1 $Ar:N_2$ environment. The atomic force microscopy examination revealed continuous microcrystalline film structure. The Auger electron spectroscopic analysis show the presence of oxygen in the films. The annealing at 850 C in nitrogen is found to cause recrystallization and further oxidation of the films. The films can be characterized as lossy dielectrics with relative permittivity $\sim$ 10, higher than the bulk value of 8.9. Annealing the films is found to reduce anion vacancies and improve the dielectric strength within a range of a few MV/cm in these thin films.

INTRODUCTION

The high melting temperature (> 2800 C), chemical resistivity, and transparency in the visible spectrum make AlN an ideal passivation layer for microelectronic devices and a protective coating for electro-optical devices [1]. In its bulk state, AlN has a very high thermal conductivity, 320 $Wm^{-1}K^{-1}$, almost ten times that of alumina. Also, the thermal expansion coefficient of AlN is closer to that of silicon than alumina. These two factors make AlN an excellent thermal substrate for hybrid microelectronic devices.

Most of the deposition techniques of AlN require temperatures in the range of 1000 C to 1200 C for single crystal growth. But, some devices needing passivation and insulation, can not tolerate high temperatures. Considerable effort has been placed on lowering the deposition temperature and still get high resistivity, dielectric AlN films [2]. We have investigated AlN films deposited by reactive sputtering for their dielectric properties and morphology and the effect of annealing.

EXPERIMENTAL

The films were deposited by RF sputtering at a power of 500 W using Al target in 1:1 $Ar:N_2$ at 10 mTorr deposition pressure. The base pressure was 5X10-7 Torr. The deposition rate was estimated to be $\sim$ 5 nm/min. The films were deposited on n+ silicon wafers for dielectric measurements. To minimize the surface oxide layer, the Si wafers were cleaned in buffered hydrofluoric acid for 10 minutes and blown dried in nitrogen and immediately placed in the vacuum chamber. The films were annealed at 760 C, 800 C and 850 C for 30 min. under flowing nitrogen at 5 lpm flow rate.

The film thickness and refractive index was measured by an ellipsometer at a wavelength of 632.8 nm. The values of film thickness obtained by the ellipsometer were compared with those obtained by depth profiling using secondary ion mass spectroscopy (SIMS). The method involved sputter etching the films while tracking Si ion signal. An abrupt increase in Si signal indicated reaching the substrate. The etching was then halted and the depth of the crater formed was

Mat. Res. Soc. Symp. Proc. Vol. 449 © 1997 Materials Research Society

measured by a profilometer. We found that the ellipsometric value agreed with those given by SIMS technique within 5 nm in films of thickness ~ 100 nm.

The surface of each sample was scanned by a Digital Instruments scanning probe microscope (SPM) in the tapping atomic force microscope (AFM) mode. The Auger electron survey scans and depth profiles were taken to look for contaminations in the films. The FTIR spectra were obtained using PE 1770 infrared spectrometer.

For capacitance measurements, aluminum was evaporated and patterned by photolithographic process to form Al-AlN-n$^+$Si capacitors. The capacitors were square and circular shaped ranging in area from 0.1 to 2.0 mm^2. The capacitances were measured at 1 MHz and plotted against plate area to extract the out -of- plane dielectric constant.

RESULTS AND DISCUSSION

Film Composition and Structure

Figure 1(a) shows the AES survey scan of as deposited films after one minute of sputtering to remove the surface carbon impurity. The composition of the films could not be concluded from the AES data because the sensitive factors of constituent atoms can not known unless a standard AlN sample is available. The presence of oxygen is observed in annealed films also. Figure 1(b) shows the AES depth profiles for the film annealed at 850 C.

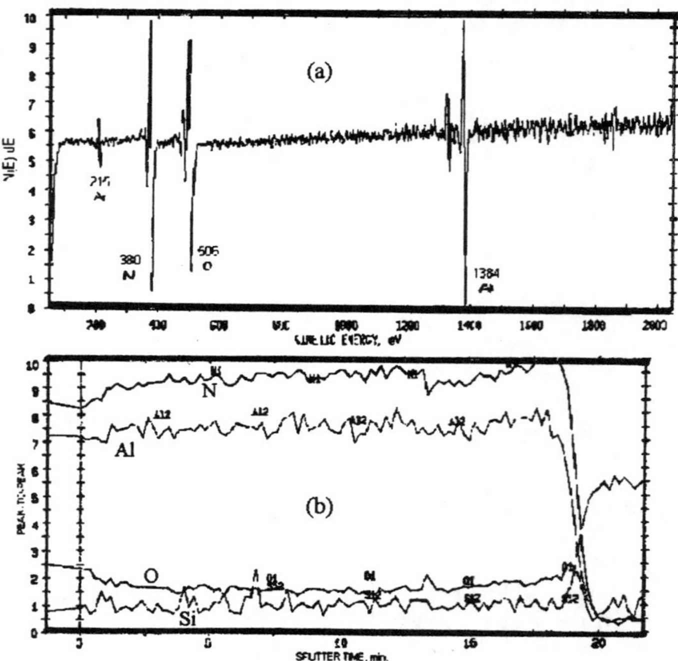

Figure 1. (a) AES survey scan of as-deposited AlN films; (b) AES depth profiles for N, Al, O, Si in AlN films annealed at 850° C in N$_2$.

It is observed that oxygen is present throughout the film and not just at the surface. It is believed that oxygen is incorporated during deposition since the rate of deposition is low (5 nm/min). The inert gases used in this study (Ar and N_2) contained 5 ppm of O_2 and 10 ppm of H_2O.

The infrared transmission of the films is shown in Figure 2. As deposited films reveal a sharp IR absorption at a frequency of 690 cm^{-1}, indicating that the films are predominately AlN [3]. On annealing, the films become progressively less transparent and the single absorption peak splits showing an additional absorption near 620 cm^{-1}. The infrared absorption in aluminum oxynitrides has been modeled by Ansart and Bernard [3] based on the molecular structure $(AlO_{3/2})_{1-z}(AlN)_z$ which is derived from $(AlO)_xN_y$. They have shown that increasing oxygen substitution for nitrogen in the AlN tetrahedron shifts the IR absorption to lower wavenumbers. However, they have argued that oxidation of AlN begins at 850 C. Our results suggest oxidation even at 760 C.

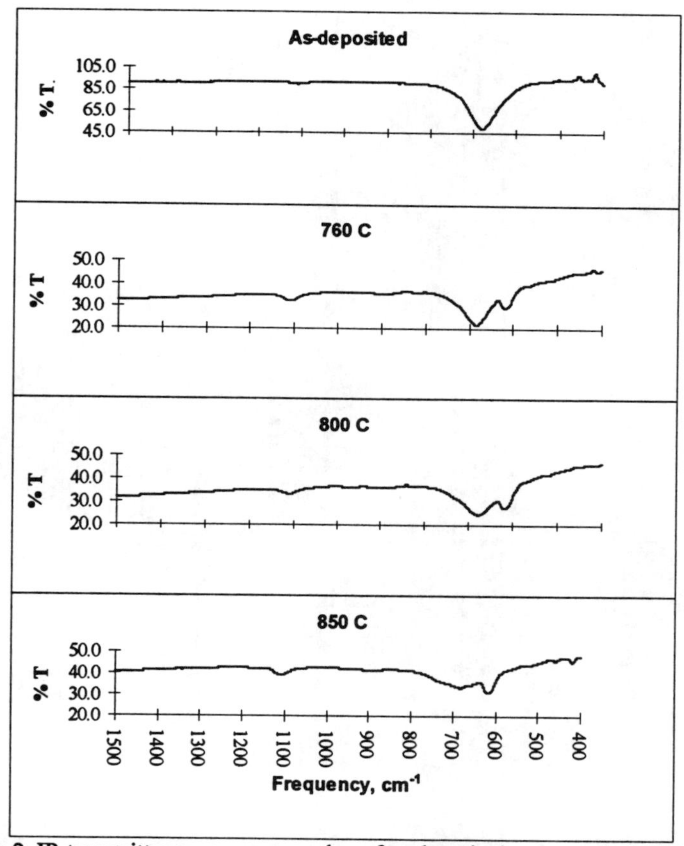

Figure 2. IR transmittance vs. wave number of as-deposited and annealed AlN films.

It is generally understood that O atoms substitutionally occupy N sites and are well situated near an Al vacancy. This is best explained by the following reaction where (O_N)

represents an O atom substituting on a N site and $(V_{Al})'''$ is an Al vacancy.

$$AlN + xAl_2O_3 \rightarrow Al_{1-1/3x}N(O_N)_x(V_{Al})'''_{1/3x} \qquad (1)$$

It tells us that for every three oxygen atoms incorporated on N sites, there exists one Al vacancy. These two conditions, O substitutions and Al vacancies, along with sintering temperatures, facilitate the conversion of tetrahedrally coordinated Al to octahedrally coordinated Al [4]. Figure 3 shows the AFM scans of as deposited and annealed films on silicon.

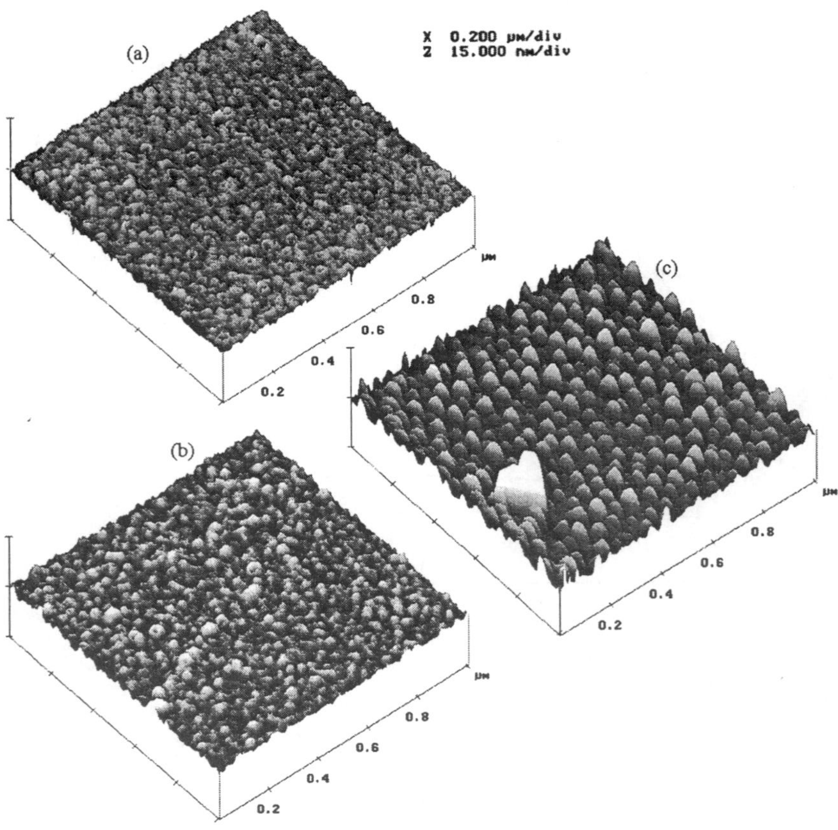

Figure 3. AFM surface scans of (a) as-deposited; (b) annealed at 800 C ; (c) annealed at 850 C AlN fims.

The as deposited films appear to be microcrystalline (Fig. 3 (a)). At 850 C, the film appears to recrystallize to a uniform grain size. It is known that pure AlN crystallizes in a wurtzite structure and pure AlON crystallizes in a spinel structure. Both are built on Al centered tetrahedral structures. The oxidation of AlN is therefore accompanied with crystallographic changes. It has been shown by Meng et. al. [5] that at room temperature, AlN deposited on Si have strong film texture with AlN(0001) in parallel with the growth direction. The film growth occurs via the island growth leading to compressive stresses.

<u>Dielectric Properties</u>

Figure 4 shows the capacitance measured on capacitors made with AlN films, as deposited and annealed, as dielectrics as a function of electrode area. The linear relationship in each case suggests that there are no gross defects such as pin holes present in the films. The values of the out-of-plane dielectric constant calculated from these plots are tabulated in Table 1.

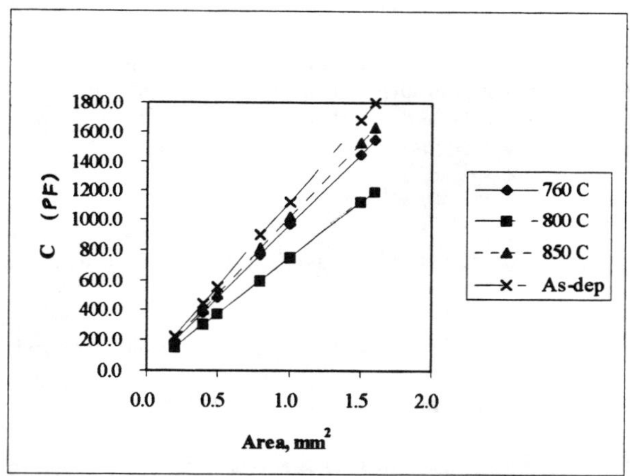

Figure 4. Capacitance vs. plate area of Al-AlN-Si capacitors.

Table 1. Summary of film dielectric properties.

Film	Refractive Index	Thickness (Å)	ε_r	Dielectric Strength (MV/cm)
as-dep	1.98	873	11.8	---
760	2.00	930	10.2	0.16 - 1.08
800	2.12	1075	9.1	0.11 - 3.24
850	2.05	927	10.7	0.3 - 1.08

It was rather surprising to obtain values larger than those reported for bulk AlN ($\varepsilon_r = 8.9$) and Al_2O_3 ($\varepsilon_r = 9.4$), even after taking into account the uncertainty in thickness measurement. The discrepancy may be due to the fact that the bulk values are average values, whereas those measured in thin films correspond to somewhat oriented films.

It was also observed that the capacitors with as deposited film as the dielectric showed excessive conductance.   Capacitors were leaky and no distinct breakdown was observed. Annealing the films, reduced the conductance and capacitors exhibited breakdown characteristics. However, the breakdown field strength showed a wide range of 0.1 - 3.24 MV/cm.

The leakage in GaN capacitors was interpreted by Zolper et. al. as an indication of vacancies due to N loss [6]. Nitrogen vacancies are believed to contribute to the n-type conductivity in GaN films. The same phenomena may explain the poor breakdown performance in sputtered AlN films. Nitrogen vacancies are suggested by the low reactivity of N_2 and further by the low N content from the AES estimates. On annealing, oxidation causes Al atoms to bond to an increasing number of oxygen atoms, reducing the anion vacancies and improving the breakdown strength.

CONCLUSIONS

Thin films of AlN deposited by reactive sputtering at room temperature in argon/nitrogen are nitrogen deficient and exhibit poor dielectric breakdown strength. Annealing the films causes oxidation of AlN and improves breakdown characteristics. The films show a value of relative permittivity of $\sim$ 10, and annealing does not seem to affect this value.

REFERENCES

1. A.U. Ahmed, A. Rys, N. Singh, J.H. Edger, and Z. J. Yu, J. Electrochem. Soc., **139**, p. 1146, (1992).

2. J.A. Kovacich, J. Kasperkiewicz, D. Lichtman, C.R. Aita, J. Appl. Phys., **55**, p. 2935 (1984).

3. F. Ansart and J. Bernard, Phys. Stat. Sol. (a), 134, pp. 467-73, (1992).

4. R.A. Youngman and J.H. Harris, J. Amer. Ceram. Soc., **73**, pp. 3238-46, (1990).

5. W. J. Meng, J.A. Sell, and T.A. Perry, J. Appl. Phys., **75**(7), pp. 3448-55, (1994)

6. J.C. Zolper, D.J. Rieger, A.G. Baca, S.J. Pearton, J.W. Lee, and R.A. Stall, Appl. Phy. Lett., **64**, pp. 538-40, (1996).

Low-Temperature Deposition and Characterization of $Al_xIn_{1-x}N$ Thin Films

GUOHUA QIU[*], J.O.OLOWOLAFE[*], TAO PENG[*], K.M.UNRUH[†], C.P.SWANN[†], and J.PIPREK[*]
[*]Department of Electrical Engineering, University of Delaware, Newark, DE19716
[†]Department of Physics and Astronomy, University of Delaware, Newark, DE19716
[*]Material Science Program, University of Delaware, Newark, DE19716

Abstract

Thin III-V nitride semiconductors films are commonly prepared using metal-organic chemical vapor deposition (MOCVD) and molecular beam epitaxy (MBE). These methods often require high temperatures (800-1000°C) for the films to grow epitaxially. In the present work, we deposited $Al_xIn_{1-x}N$ films on Si substrates by reactive magnetron sputttering method at low substrate temperature. The properties of the films have been studied by RBS, x-ray diffraction, and optical measurements. The $Al_xIn_{1-x}N$ films deposited at room temperature were confirmed to be crystalline by x-ray diffraction. Band gap energies of our $Al_xIn_{1-x}N$ alloys varies from 1.9 ev to 4.2 ev The bandgap energy vs. lattice constant curve was constructed and confirmed to bow downwards.

Introduction

The group III-nitrides are promising materials for semiconductor device applications in the blue and ultra-violet (UV) wave region. Impressive performance has been demonstrated in intense blue light emission and high-temperature operation. The major advantage is in their large and direct band-gaps ranging from 1.9ev to 6.2ev (E_g(GaN)=3.4ev, E_g(AlN)=6.2ev). Moreover, they are also promising for application as high-power microwave devices because of their high thermal conductivity and large electron saturation velocity. However, the nitride alloys AlGaN, InGaN and AlInN deserve significantly more attention than they have received thus far[1]. Nearly all the desirable nitride-based optical devices will require at least one of these alloys, as predicted theoretically by Jenkins and Dow[2] that $In_{0.85}Ga_{0.15}N$ and $Al_{0.6}In_{0.4}N$ are excellent candidates for blue-green wave-lengths emitters which can be doped either n- or p-type. Yet reliable measurements are scarce for all of these materials.

Experimental work on the $Al_xIn_{1-x}N$ alloy was extremely few[1] although this alloy system bears the largest band-gap range (from 1.9ev of InN to 6.2ev of AlN). Kubota et al [3] reported to prepare the $Al_xIn_{1-x}N$ alloy on sapphire substrates with AlN buffer layers by using a composite target composed of Al and In. It is , however, very difficult to stabilize the film composition because of the large difference of sputtering rate between Al and In. Qixin Guo et al[4] reported the growth of single crystal $Al_xIn_{1-x}N$ films by microwave-excited metalorganic vapor phase epitaxy, but with x up to 0.14 only. In this

Mat. Res. Soc. Symp. Proc. Vol. 449 © 1997 Materials Research Society

paper we report the growth of $Al_xIn_{1-x}N$ crystalline film in a wide composition range. The films were prepared using a multitarget sputtering technique. The structure and optical properties of the films are hereby reported.

Experimental

$Al_xIn_{1-x}N$ films were deposited on Si and glasses by magnetron sputtering Al and In targets simultaneously in a reactive nitrogen plasma. Al target is connected to DC power and In target is connected to RF power and they are from the equal distances to the substrate. The substrate was rotating during sputtering to ensure the uniformity of the deposited layer. Film thickness is measured by stylusprofilemeter after deposition. The growth rate were ranged from 1A/s to 3A/s. The crystal structures of deposited films were examined by X-ray diffraction using Cu Kα radiation. Alloy compositions were determined by Rutherford backscattering spectroscopy (RBS) using 2MV van deGraaff accelerator with 2MeV He^+ ions and the oxygen contamination was monitored by secondary ion mass spectroscopy (SIMS). Optical transmittance spectra were measured with a double beam spectrophotometer in the wavelength ranging from 200nm to 800nm. The optical bandgaps were then determined from the absorption spectra.

Results and discussion

A typical RBS of a $Al_xIn_{1-x}N$ film on Si substrate is shown in figure 1. The In peak is clearly seen with the leading edge at 1.8Mev. The Al and N peaks which overlapped with the Si signal can be seen with the leading edges at 1.1 and 0.63Mev, respectively. The composition and thickness of the $Al_xIn_{1-x}N$ film were determined by simulating the RBS spectrum with rump program. The result is shown with deposition conditions in Table 1.

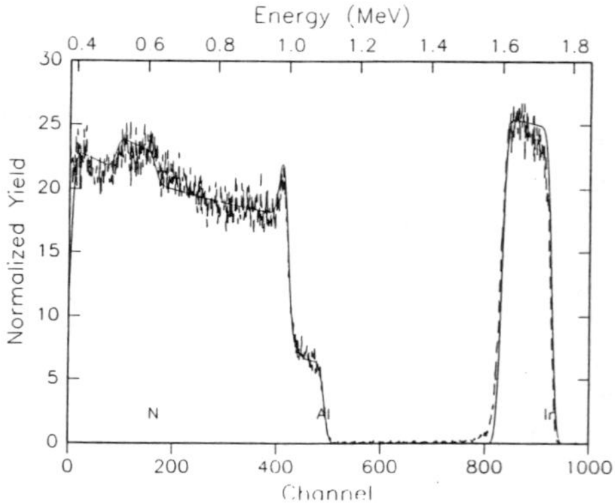

Figure 1. RBS spectrum of $Al_{.85}In_{.15}N$. The solid line is the rump fitting of the experimental curve.

Table1 Sputtering condition and film composition

Sample No.	N_2 flow rate(sccm)	Ar flow rate(sccm)	Al(x)	In(1-x)	Thickness(μm)
1	27	3	18	82	1.14
2	27	3	25	75	1.01
3	27	3	50	50	1.23
4	27	3	64	36	1.19
5	27	3	85	15	1.29

Low mass impurities in the film, particularly oxygen, were evaluated using SIMS profile. From the SIMS depth profile, we observed that the level of oxygen concentration is extremely low, below 10^{21} atoms/cm^3 throughout the film. Other impurities like H, C are hardly detectable.

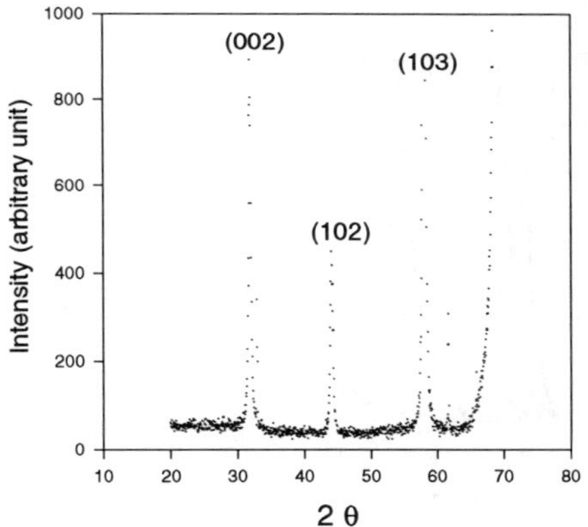

Figure 2. X-ray diffraction pattern of Al$_{.25}$In$_{.75}$N

The crystal structures of the Al$_x$In$_{1-x}$N films were examined by the X-ray diffraction technique. From the X-ray diffraction spectrum of alloy 2 shown in figure 2, it is clear that the Al$_x$In$_{1-x}$N film deposited at room temperature has wurtzite crystalline structure and the peaks are identified in the above spectrum with (002), (102) and (103) peaks located at 2θ = 31.95, 44.60, 58.75, in that order. The diffraction peaks shift towards higher angles with increasing Al concentration (i.e., in increasing value of x. This is because InN has larger lattice parameters than AlN . Lattice constants of the Al$_x$In$_{1-x}$N alloys are calculated from the position of (002) and (103) peaks and shown in Table 2.

Optical transmittance and reflectance spectra of the Al$_x$In$_{1-x}$N film were measured with a dual beam spectrophotemeter at room temperature in the wavelength range from

Optical transmittance and reflectance spectra of the $Al_xIn_{1-x}N$ film were measured with a dual beam spectrophotemeter at room temperature in the wavelength range from 200 nm to 800 nm and the optical absorption coefficient was calculated from these spectra.

Table 2. lattice constants of $Al_xIn_{1-x}N$

sample No.	comp. (x)	2θ of (102)	2θ of (103)	a (Å)	c (Å)
1	18	44.15	58.15	3.48	5.59
2	25	44.60	58.75	3.44	5.54
3	50	46.10	60.75	3.33	5.38
4	64	47.40	62.60	3.24	5.25
5	85	49.05	65.00	3.16	5.01

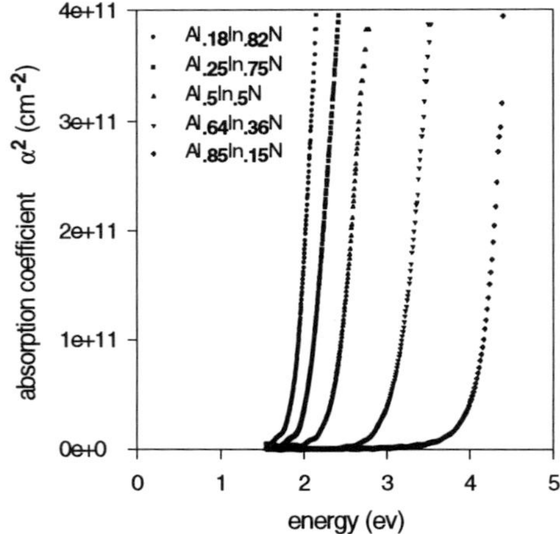

Figure 3. Variation of squared absorption coefficient α^2 of $Al_xIn_{1-x}N$ alloys

In figure 3 the square of the absorption coefficient α^2 versus photo energy of varying composition is plotted. Near the absorption edge the absorption coefficient is of the form $(hv-E_g)^{1/2}$, where hv is the photon energy and E_g is the band gap when direct transition is allowed. Linear dependence of α^2 versus photo energy was found for all the $Al_xIn_{1-x}N$ films. The $Al_xIn_{1-x}N$ alloys have direct band gaps as shown in figure 3. This is expected since AlN and InN are both direct band gap materials. Thus, E_g for the $Al_xIn_{1-x}N$ alloys is determined by the extrapolation of the linear part of the curves to the horizontal axis where $E_g = hv$ since $\alpha = 0$. The structure of group III-nitrides are mostly wurtzite and the heteroepitaxy of the heterojunction structure therefore, is expected along the c-axis. So, the relationship between the lattice constant a and the band gap energy, E_g, is very crucial and is shown in figure 4 for the $Al_xIn_{1-x}N$ films. The curve is highly non-linear, agrees with the results of Kubota et al[3]. However, Guo et al[4] found a linear relationship between band gap energy and molar fraction of $Al_xIn_{1-x}N$ in the In rich region, with x<0.14. The energy gap against composition curves of the alloy A_xB_{1-x} are conventionally fitted to a quadratic equation of the form[5]

$$E(x) = E_A + (E_B - E_A - b) x + bx^2$$

b, the bowing parameter is given by

$$b = 4 [E(x = 0.5) - (E_A + E_B)/2]$$

Two theoretical models were proposed for the calculation of b: the dielectric two band model (DM) introduced by Van Vechten and Bergstresser[6], and the empirical pseudopotential method (EPM) by Richardson[7]. In the EPM theory,

$$b = (\Delta r)^2 au$$

This means the bowing parameter b is proportional to the square of Δr, which is the difference of the covalence radii in the alloy. Δr for Ga-In 0.18Å, for Ga-Al is 0.005 Å and for Al-In is 0.165 Å[8]. Among the other two ternary nitrides alloys, InGaN bows downwards while AlGaN bows upwards. In general for many III-V ternary alloys bowing is downwards. Therefore, the downward bowing of the $Al_xIn_{1-x}N$ is not unexpected.

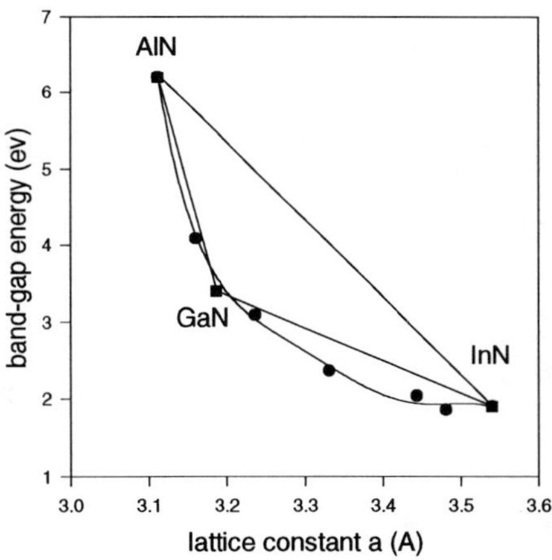

Figure 4. Band gap energy versus lattice constant curve for $Al_xIn_{1-x}N$ alloys compared with GaN.

It can be estimated from figure 4 that the $Al_xIn_{1-x}N$ alloy with $x \cong 0.78$ has the same lattice constant as GaN has the band gap energy slightly larger than that of GaN. This makes $Al_xIn_{1-x}N$ a possible candidate for a barrier material in the $Al_xIn_{1-x}N$ / GaN heterojunction system although the band gap margin is not so large.

Conclusion

$Al_xIn_{1-x}N$ alloy films were deposited by reactive magnetron sputtering at room temperature. The composition of the alloys were analyzed by RBS and the structures were determined to be crystalline by XRD. Band gap energies of our $Al_xIn_{1-x}N$ alloys varies from 1.9 ev to 4.2 ev. The bandgap energy vs. lattice constant curve was constructed and confirmed to be bowing downwards.

Reference

1. S.Strite and H.Morkoc, J.Vac.Sci.Technol., B,. **10**(1992),1237
2. D.W.Jenkins and J.D.Dow, Phys.Rev., **B34**(1989), 3317
3. K.Kubota, Y.Kobayashi, and K.Fujimoto, J.Appl.Phys., **66**(1989), 2984
4. Qixin Guo, Hiroshi Ogawa, and Akira Yoshida, J. Crystal Growth, **146**(1995), 462
5. R.Hill and D.Richardson, J.Phys., **C6**(1973), L115
6. J.A.Van Vechten and T.K.Bergstresser, Phys.Rev. **B1**(1970), 3351
7. D.Richardson and R.Hill, J.Phys., **C5**(1972), 821
8. J.C.Phillips, Bonds and Bands in semiconductors, Academic Press(1973), 22

LOW PRESSURE CVD OF GaN FROM GaCl$_3$ and NH$_3$

M. Topf*, S. Koynov**, S. Fischer, I. Dirnstorfer, W. Kriegseis, W. Burkhardt, and B.K. Meyer

I. Physics Institute, Justus-Liebig-University Giessen, D-35392 Giessen, Germany

Abstract

We report on the heteroepitaxial growth of GaN from GaCl$_3$ and NH$_3$ on (0001) Al$_2$O$_3$ and (0001) 6H-SiC substrates. In order to enable homogeneous growth within the entire deposition zone one has to use low process pressures in the 10^{-1} mbar range, where still a growth rate of ~ 2 μm/h can be achieved. We present a simple model to describe our process and explain our observations. A comparison of GaN deposited on different substrates and with GaN buffer layers is given by low temperature Photoluminescence (PL). Furthermore, impurities are traced by secondary ion mass spectroscopy (SIMS).

Introduction

Nowadays GaN is commonly grown from the gas phase by metalorganic chemical vapor deposition (MOCVD). No doubt MOCVD is the method of choice to produce GaN films of highest quality, like the previously reported CW lasers [1]. However due to the metal-organic (MO) Ga precursors a significant carbon and organic group contamination of the GaN films is observed. The Hydrogen is known to passivate acceptors in GaN, therefore a subsequent activation step is needed [2]. On the other hand the role of carbon in GaN is still an open question. There is some evidence for a shallow acceptor level [3,4], but also predictions for a self-compensating mechanism are reported [5].

Therefore growth of GaN with C free precursors and intentional C-doping would be very fortunate to answer such questions. One gas phase technique to grow without C is the hydride vapor phase epitaxy (HVPE) [6]. In this process GaCl is used as Ga precursor, which is synthesized *in situ* from HCl and Ga.

To overcome this additional synthesis we use directly the chemically stable GaCl$_3$ as Ga precursor. So far, only a few efforts have been reported with this kind of Ga source [7-10]. To our knowledge this is first time where a low pressure CVD process using GaCl$_3$ is taken for the synthesis of GaN. Here we will demonstrate the capability to grow GaN from GaCl$_3$ and NH$_3$.

Experimental

Fig. 1 displays a schematic of the LPCVD setup for GaN deposition. The substrates are placed on a quartz tray inside a horizontal quartz reactor. The reactor is resistive heated and the temperature is controlled by a thermocouple. The tray has an entire length of 35 mm which is used as the deposition zone. NH$_3$ is taken as N source and solid GaCl$_3$ is utilized as Ga source. During growth the GaCl$_3$ is transported via heated N$_2$ into the reactor. The GaN films were deposited either on 1x1 cm^2 specimens of (0001) Al$_2$O$_3$ or on (0001) 6H-SiC platelets approx. 1 cm in diameter. The substrates were degreased in organic liquids and dipped in HF prior

Mat. Res. Soc. Symp. Proc. Vol. 449 © 1997 Materials Research Society

loading them into the reactor. Depending on the duration of growth, the appropiate amount of $GaCl_3$ was destilled form the store vessel into the process ampoule.

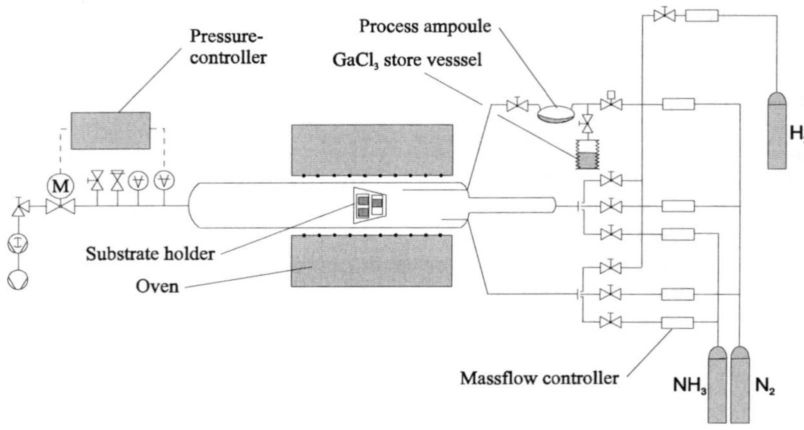

Fig. 1: Schematic of the LPCVD setup

Prior deposition the substrates were cleaned *in situ* by exposing them to a flow of 150 sccm H_2 at 1000°C at a system pressure of one mbar for 30 min. First the growth parameters were optimized for GaN deposition on Al_2O_3, whereas no buffer layer was used. Also a comparative study of the GaN deposition on Al_2O_3 and 6H-SiC was done under these conditions. The optimized growth parameters are summarized in Table 1. Typical film thicknesses range from 1 to 3 μm. Preliminary results are shown for films grown on Al_2O_3 with a low temperature buffer layer deposited at 600°C for 10 min resulting in an approx. 50 nm thick GaN buffer.

Table 1: Parameters for LPCVD of GaN

Growth pressure:	p	=	0.3 mbar
Substrate temperature:	T	=	970 °C
$GaCl_3$ flow	$F(GaCl_3)$	=	1.5 sccm
NH_3 flow	$F(NH_3)$	=	200 sccm
Carrier gas flow	$F(N_2)$	=	50 sccm
Growth rate	r	=	2 μm/h

The deposited films were investigated by standard x-ray diffraction measurements. The optical properties of these films were probed by low temperature photoluminescence (PL). The samples were mounted in a helium flow cryostat and excited by the 325 nm line of a HeCd laser. A comprehensive study of the structural properties of LPCVD grown GaN was done with transmission electron microscopy (TEM) and scannning tunneling microcopy (STM) and is reported elsewhere [11]. Impurities were traced by means of secondary ion mass spectroscopy (SIMS) using a Cameca/RIBER MIQ 56A system.

Results and Discussion

The first part of this section refers to the growth of GaN films and the investigation of the growth parameters. The second part deals with more detailed characterization of the GaN films themselves.

The growth parameters listed in Table 1 were found to enable two dimensional GaN film growth and homogeneous deposition over the entire deposition zone. For our setup the most significant parameter is the growth pressure. Starting from atmospheric pressure the growth rate decreased significantly towards lower pressure, resulting in 2 μm/h at 0.3 mbar. However above the mbar range we were not able to achieve homogeneous film growth in the entire deposition zone. Also the structural and optical properties of the GaN films inproved by lowering the deposition pressure as shown in Fig. 2.

The deposition process was found to be Ga limited by variation of the NH_3 and $GaCl_3$ flow rates. An almost linear dependence between the growth rate and the $GaCl_3$ flow is observed. In contrast the growth rate did not depend on the NH_3 flow as long as a certain NH_3 flow was established. Below this value no GaN deposition was observed.

To explain these results we propose the following growth model. The first step in our growth model is the successive thermal decomposition of $GaCl_3$ via the intermediate products $GaCl_2$ and $GaCl$ to Ga and Cl_2. At the low pressure in the mbar range diffusion is the most relevant mechanism for the transport of Ga vapours to the substrates and the reactor walls. Ga has a rather low vapour pressure even at elevated temperatures. Therefore it condenses on the exposed surfaces in the reactor and especially on the substrates. Together with gaseous ammonia (NH_3) the adsorbed Ga atoms react on the substrate surface and subsequently on the GaN surface to GaN as described by the following netto reaction:

$$2\,NH_3\,(g) + 2\,Ga\,(l) \longleftrightarrow 2\,GaN\,(s) + 3\,H_2\,(g)$$

Therefore the limiting factor within our growth model is the condensation rate of Ga since the partial pressure of NH_3 in our system is always higher than the corresponding equilibrium pressure of the reaction above [12]. In the frame of this model the linear dependency of the growth rate on the $GaCl_3$ is obvious. In addition the benefit of lowering the growth pressure resulting in a higher diffusion coefficent of Ga for a decreased total pressure is understood.

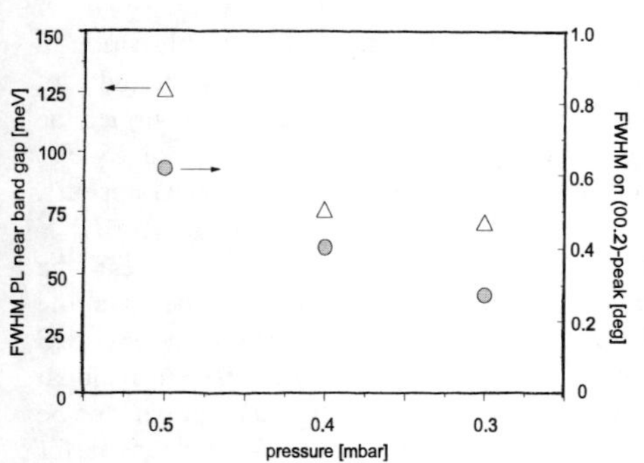

Fig. 2: The FWHM values of the (0002) diffraction peak and the nearband gap PL at T=2K are used to monitor the structural and optical properties of the GaN in dependence of the growth pressure.

The LPCVD grown GaN films appear transparent and exhibit a mirror-like surface. However so far all films show high n-type conduction with carrier concentrations in the 10^{18} to 10^{19} cm^{-3} range. SIMS analysis of these layers revealed a contamination predominantly with Si and O in the same concentration range. Possible sources for this contamination are believed to be the residual moisture of the NH_3 and the use of a quartz substrate holder.

To study the influence of the substrate materials on the optical properties of the GaN layers, GaN was deposited simultaneously on Al_2O_3 and 6H-SiC. Low temperature PL spectra of the near bandgap region of these layers are displayed in Fig. 3. All GaN films are clearly dominated by excitonic recombination. Films grown on Al_2O_3 exhibit the recombination of donor bound excitons (D^0X) at 3.480 eV (Fig. 3 a)). A contribution of an additional recombination at about 3.456 eV in the low energy shoulder of the D^0X can be observed. At lower energies the donor acceptor pair recombination (3.25 eV) followed by two LO phonon replicas appears. For GaN on 6H-SiC only a broader D^0X emission at 3.466 eV (Fig. 3 b)) can be observed. The energetic difference between the D^0X emission in GaN on Al_2O_3 and 6H-SiC can be well explained by the different residual strain in these films. It was shown to be compressive for GaN on Al_2O_3 and tensile for GaN on 6H-SiC leading to a blue and red shift, respectively, in respect to the bulk value of 3.472 eV for the prominent I_2 D^0X line [13]. Besides the significant smaller lattice mismatch between GaN and 6H-SiC (3.5%) in comparison to Al_2O_3 (16%) the optical properties of GaN on 6H-SiC are only equal or even lower compared to GaN on Al_2O_3. A detailed study of the structural properties revealed a rather undulated interface between GaN and 6H-SiC in contrast to an almost atomically flat one between GaN and Al_2O_3 which could account for these observations [11]

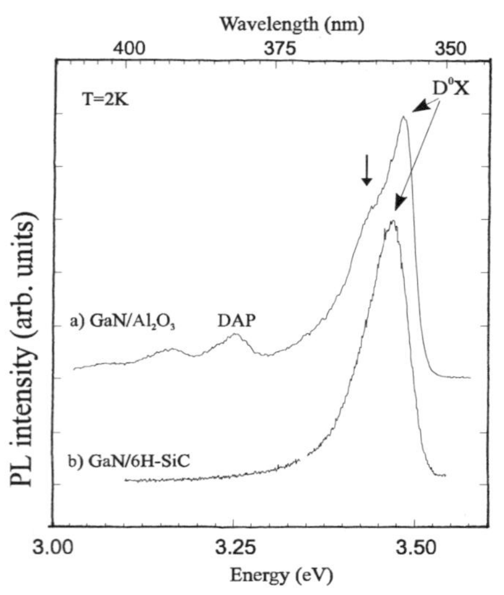

Fig. 3: PL spectra of GaN grown on Al₂O₃ and 6H-SiC. The spectra were normalized to the highest intensity and displaced vertically for clarity. For details see text.

An obvious improvement of the GaN properties can be achieved by the use of a low temperature buffer layer, as seen in Fig. 4 where the PL of a GaN film deposited without (A) and with GaN buffer (B) is compared. The GaN films themselves were grown under the same conditions. The GaN film grown without buffer layer is dominated by the same broad luminescence as discribed above, having its maximum at 3.483 eV. This changes dramatically for the film grown with GaN buffer, where the two luminescence lines are now resolved separately. The PL is dominated by the narrow recombination line of the D^0X at 3.470 eV. The different peak positions can be again explained within the picture of residual stress in these films [13], indicating much less strain in the film grown with buffer layer, as one would expect.

The second line appears around 3.42 eV. This line is believed to be related to O in GaN. It is only reported for GaN grown on Al_2O_3 and might be connected to the release of O from the Al_2O_3 substrate [14,15]. Our data could be explained within this szenario.

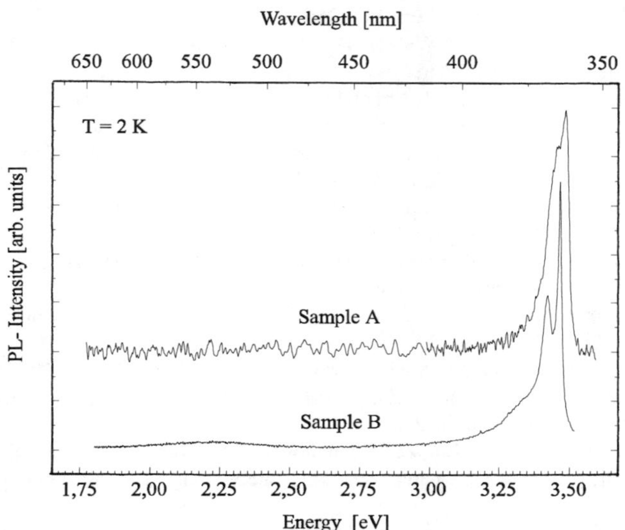

Fig. 4: Pl spectra of GaN films grown on Al_2O_3 without (A) and with a GaN buffer layer (B).

In summary we established a low pressure CVD process to grow GaN from $GaCl_3$ and NH_3. By using a GaN buffer layer we were able to grow high quality GaN on (0001) Al_2O_3 substrates with growth rates of 2 µm/h.

Acknowledgments

A. Graber is acknowledged for perfoming the Hall measurements. The authors would like to thank the Volkswagen-Foundation for financial support. S.F. gratefully acknowledges the Hanns-Seidel-Stiftung for a scholarship funded by the BMBF.

References

*	electronic mail: michael.k.topf@exp1.physik.uni-giessen.de
**	Permanent address: CLSENES Bulgarian Academy of Science, 1784 Sofia, Bulgaria

[1]   S. Nakamura, M. Senoh, S. Nagahama, N. Iwasa, T. Yamada, T. Matsiushita, Y. Sugimoto, and H Kiyoku, Appl. Phys. Lett. **69**, 4056 (1996).

[2]   W. Götz, N.M. Johnson, J. Walker, D.P. Bour, and R.A. Street, Appl. Phys. Lett. **68**, 667 (1996), and references therein.

[3]   S. Fischer, C. Wetzel, E.E. Haller, and B.K. Meyer, Appl. Phys. Lett. **67**, 1298 (1995).

[4]   M. Leroux, B. Beaumont, N. Grandjean, C. Golivet, P. Gibart, J Massies, J. Leymarie, A. Vasson, and A.M. Vasson, Proceedings of the EMRS Meeting, June 4-7, 1996, Strasbourg, France, to appear in Mat. Sci. Eng. B.

[5]   P. Boguslawski, E.L. Briffs, and J. Bernholc, Appl. Phys. Lett. **69**, 233 (1996).

[6]   H.P. Maruska, J.J. Tietjen, Appl. Phys. Lett. **15**, 327 (1969).

[7]   J.J. Nickl, W. Just, and R. Beringer, Mat. Res. Bull. **9**, 1413 (1974).

[8]   J.J. Nickl, W. Just, and R. Beringer, Mat. Res. Bull. **10**, 1097 (1975).

[9]   N.I. Chetverikov, I.F. Chetverikova, V.I. Chernyaev, V.B. Novikov, and others, Inorganic Materials **11**, 1792 (1975).

[10]   R. Lappa, G. Glowacki, and S. Galkowski, Thin Solid Films **32**, 73 (1976).

[11]   S. Koynov, M. Topf, S. Fischer, B.K. Meyer, P. Radojkovic, E. Hartmann, and Z. Liliental-Weber, unpublished.

[12]   N. Newman, T.C. Fu, X. Liu, Z. Liliental-Weber, M. Rubin, J.S. Chan, E. Jones, J.T. Ross, I. Tidswell, K.M. Yu, N. Cheung, and E.R. Weber, Mat. Res. Soc. Symp. Proc. **339**, 483 (1994).

[13]   D. Volm, K. Oettinger, T. Streibl, D. Kovalev, M. Ben-Chorin, J. Diener, B.K. Meyer, J. Majewski, L. Eckey, A. Hoffmann, H. Amano, I. Akasaki, K. Hiramatsu, T. Detchprohm, Phys. Rev. B **53**, 16543 (1996).

[14]   N. Grandjean, J. Massies, and M. Leroux, Appl. Phys. Lett. **69**, 2071 (1996).

[15]   S. Fischer, C. Wetzel, W.L. Hansen, E.D. Bourret-Courchesne, B.K. Meyer, and E.E. Haller, Appl. Phys. Lett. **69**, 2716 (1996).

NEW PATHWAYS TO HETEROEPITAXIAL GaN BY INORGANIC CVD   SYNTHESIS AND CHARACTERIZATION OF RELATED GA-C-N NOVEL SYSTEMS

J. KOUVETAKIS*, M. O'KEEFFE*, LOUIS BROUSEAU*, J. McMURRAN, DARRICK WILLIAMS*, D. J. SMITH**
*Department of Chemistry and Biochemistry, Arizona State University, Tempe Arizona 85287
**Center for Solid State Science, Arizona State University, Tempe Arizona 85287

ABSTRACT

We describe the development of a new deposition method for thin oriented films of GaN on basal plane sapphire using an exclusively inorganic single-source precursor free of carbon and hydrogen, Cl_2GaN_3. The films have been characterized by Rutherford backscattering spectroscopy (RBS) and cross sectional transmission electron microscopy (TEM) for composition morphology and structure. RBS analysis confirmed stoichiometric GaN and TEM observations of the highly conformal films revealed heteroepitaxial columnar growth of crystalline wurrtzite material on sapphire.   Auger and RBS oxygen and carbon resonance profiles indicated that the films were pure and highly homogeneous. We also report the reactions of Cl_2GaN_3 with organometallic nitriles to yield a crystalline, novel gallium carbon nitride of composition GaC_3N_3. Quantitative X-ray powder diffraction has been used to refine the cubic structure of this material which consists of Ga atoms octahedrally surrounded by on the average three C and three N atoms. The structurally analogous $LiGaC_4N_4$ phase has also been prepared and characterized.

INTRODUCTION

The potential microelectronic and optoelectronic applications of wide bandgap nitride semiconductors InN, AlN and GaN has resulted in considerable research associated with their growth and development. Gallium nitride, the most studied of the group III nitrides has a bandgap of 3.4 eV and forms solid solutions with InN and AlN from which heterostructures can be fabricated.[1] GaN-based heterostructures and quantum well light-emitting diodes have been developed and are commercially available. Recently Nakamura and co-workers have successfully demonstrated an InGaN-based multi-quantum-well laser diode.[2] Electronic devices ranging from field effect transistors to photodetectors have also been demonstrated.[3] The further development of these microelectronic and optoelectronic devices requires improved nitride material because despite the many advances, serious problems still hinder the synthesis of high- quality thin films. These problems include large background n-type carrier concentration due to nitrogen deficiencies, lack of suitable substrates, crystalline imperfections, and difficulties in p-doping.[4,5]

Alternative synthetic methods that provide stoichiometric nitride materials involve use of single-source precursors that incorporate strong Ga-N bonds. Particularly promising are gallium nitride precursors that contain the azide (N_3) group as the nitrogen source. Organometallic gallium azides, such as $(R_2GaN_3)_3$ ($R = CH_3$, C_2H_5), $[(CH_3)_2N]_2GaN_3$, and $(N_3)_2Ga[(CH_2)_3N(CH_3)_2]$ have been used to deposit stoichiometric GaN of reasonable crystal quality and chemical purity. [6-9] These methods permitted deposition at substantially lower temperatures than those required for traditional metal-organic chemical vapor deposition (MOCVD) processes.   They also have the potential of providing better deposition control at low pressures as well as eliminating the inefficient use of ammonia, thus leading to more effective and economical deposition processes.

A major obstacle to the realization of device-quality nitride material is the difficulty of achieving p-type doping when growth techniques such as (MOCVD) are used because the hydrogen impurities introduced during growth form complexes with Mg acceptors and severely diminish the doping efficiency of material.[10] Our approach to GaN involves development of

313

simple inorganic precursors that would allow growth in a carbon- and hydrogen-free deposition environment, that are compatible with p-doping processes, and that eliminate the possibility of carbon contamination in the films and use of NH_3 during growth. [11]

RESULTS AND DISCUSSION

1. GaN growth via decomposition of Cl_2GaN_3

We have succeeded in depositing heteroepitaxial GaN at very low pressures in an ultrahigh-vacuum chemical vapor deposition (UHV-CVD) chamber using, for the first time, an exclusively inorganic single-source precursor, Cl_2GaN_3. The remarkably stable Cl_2GaN_3 compound is prepared in nearly quantitative yields from a convenient synthetic route which is summarized below:

$$GaCl_3 + (CH_3)_3SiN_3 \rightarrow (CH_3)_3SiN_3 \cdot GaCl_3 \rightarrow (CH_3)_3SiCl + Cl_2 GaN_3$$

This route eventually gives a colorless air-sensitive solid that sublimes readily in vacuum, melts at 210°C, and it is not sensitive to shock. Electron impact mass spectrometry reveals $(Cl_2GaN_3)_3$-Cl as the highest mass peak at 511 amu and a fragmentation pattern consistent with a six-membered Ga-N ring structure. It is not as air-sensitive as the trialkyl gallium precursors currently used in GaN CVD processes and reacts rapidly but mildly with liquid water. En route to Cl_2GaN_3 the novel monomeric adduct $(CH_3)_3SiN_3GaCl_3$ is isolated and characterized by X-ray analysis as illustrated below in Figure 1.

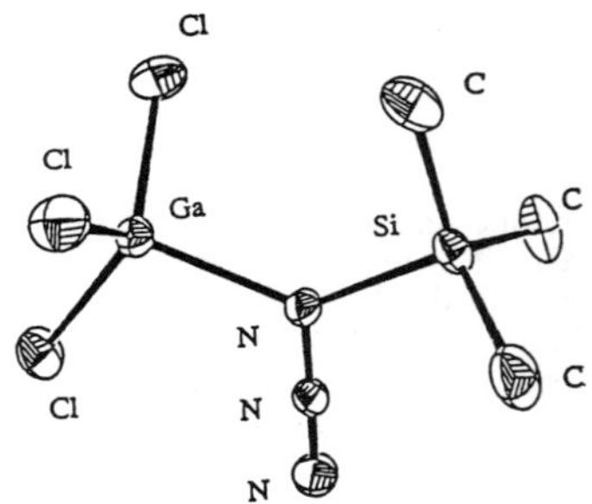

Figure 1. Single crystal molecular structure of $(CH_3)_3SiN_3 \cdot GaCl_3$.

Deposition studies of GaN films from $(CH_3)_3SiN_3 \cdot GaCl_3$, and Cl_2GaN_3, were carried out on sapphire and Si substrates at 600-700 °C. The Cl_2GaN_3 compound decomposes in the UHV CVD reactor via elimination of $GaCl_3$ and N_2 to produce GaN films as illustrated in the proposed decomposition reaction.

$$Cl_2GaN_3 \rightarrow 2GaCl_3 + 4N_2 + GaN$$

RBS and Auger analysis of of 200 to 1500 nm films deposited at 650 °C revealed highly stoichiometric GaN free of C or O impurities and with Cl contents less that 1 at.% . The Cl impurity content is lower and the crystallinity is better for films grown at higher temperatures, but the Ga-N composition is independent of deposition temperature and growth rate. High resolution cross sectional TEM examinations showed highly oriented heteroepitaxial growth on sapphire as illustrated in Figure 2. Our initial deposition results are promising and this material represents the first example of a potentially practical, totally inorganic precursor to grow good quality GaN by low-pressure methods. The deposition procedure was quite simple. It basically involves direct sublimation of the precursor at 70 °C into the hot zone of the reactor in the absence of any carrier gas or toxic ammonia. Current state-of the-art CVD processes of device-quality GaN require a large excess of NH_3 ranging from 1000 to 5000 fold. Our method and other similar precursor-related methods have the potential of eliminating the need of NH_3 for GaN growth. Other notable advantages of our method include high growth rates of 50-350 Å /min., low deposition temperatures of 650-700 °C highly stoichiometric GaN material, and a carbon- and hydrogen-free deposition environment that could be beneficial to p-doping processes.

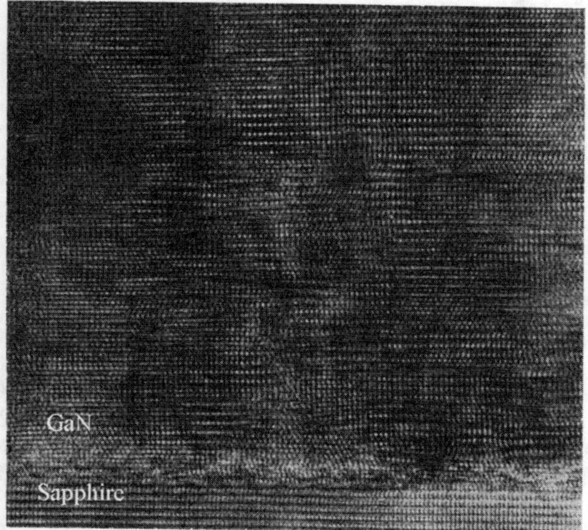

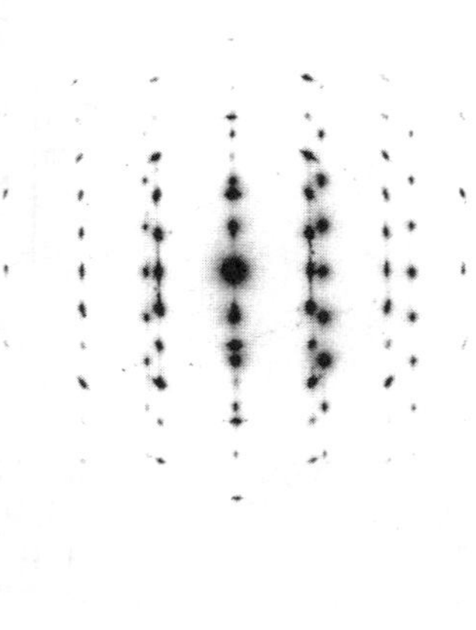

Figure 2. High resolution TEM image and corresponding diffraction pattern of wurtzite GaN on sapphire showing the heteroepitaxial growth

A potential disadvantage associated with the use of Cl_2GaN_3 as a practical source of GaN is that it does not have any significant vapor pressure at room temperature. This requires heating of the precursor container to 70 °C to obtain sufficient vapor pressure, in addition to turbo-pump processing during growth in order to achieve the low pressures necessary to transport the gaseous precursor. Although substantial growth rates have been achieved under these deposition conditions, higher volatility of the molecule is nevertheless desirable. We increased the volatility of Cl_2GaN_3 by reacting the polymeric material with a strong Lewis base trimethyl amine $N(CH_3)_3$ to form a monomeric adduct of composition $(Cl_2GaN_3)N(CH_3)_3$. The trimethyl amine adduct is a good GaN precursor and is readily prepared by direct interaction of $N(CH_3)_3$ and Cl_2GaN_3 .

$$N(CH_3)_3 + Cl_2GaN_3 \rightarrow (Cl_2GaN_3)N(CH_3)_3$$

The compound is a low melting (35 °C) white solid that sublimes readily at room temperature and thus easier to utilize in CVD experiments.

2. GaN growth via decomposition of $(Cl_2GaN_3)N(CH_3)_3$

The deposition process is similar to that developed for growth of GaN from Cl_2GaN_3. The CVD reactor used in this study is a cold-wall, inductively heated system described in detail elsewhere. The precursor is kept in a glass container equipped with a high-vacuum valve and it is directly attached to the reactor, which is constantly maintained at 2×10^{-8} Torr by a corrosion-resistant turbo pump. In a typical experiment, the source container valve is opened, resulting in a rapid rise of reactor pressure to approximately 5×10^{-5} Torr. Although film growth has been obtained at this pressure, we heat the precursor to its melting point in order to achieve reasonable growth rates of about 50 Å /min. at 2×10^{-4} Torr. Films ranging in thickness from 100 to 250 nm are normally deposited at 700 °C on sapphire and Si substrates with rates of 150-200 Å per minute. RBS analysis including carbon and oxygen resonance of these films reveals stoichiometric GaN with chlorine and carbon contamination of 1-2 at % as shown in Figure 3 (carbon and oxygen nuclear resonance reactions allow detection as low as 0.5 at % C and 1.5 at % O). Films grown at a slower rate of 50-80 Å/min. are pure as shown by RBS, and they display higher crystal quality, as revealed by TEM examinations.

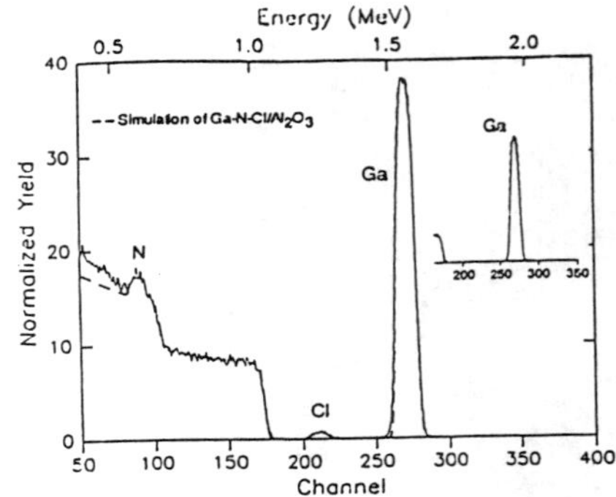

Figure 3. RBS spectrum of GaN films deposited on sapphire using (Cl$_2$GaN$_3$).N(CH$_3$)$_3$ at 650 °C (160 Å/min) and 700 °C (30-45 Å/min) inset. Films with composition Ga$_{0.95}$N$_{1.05}$ that contain 2-2.5 at % Cl contamination are grown at 650 °C as visible in the spectrum. Stoichiometric GaN with Cl contents lower than the detection limit (0.5 at.%) is deposited at 750 °C (inset).

3. Reactions of Cl$_2$GaN$_3$ with organometallic nitriles. Synthesis of GaC$_3$N$_3$ and LiGaC$_4$N$_4$

The novel GaC$_3$N$_3$ compound was prepared as a crystalline colorless solid by reaction of GaCl$_3$ with trimethylsilylcyanide.

$$Cl_2GaN_3 + 3SiMe_3CN \rightarrow 2\ Me_3SiCl + Me_3SiN_3 + GaC_3N_3$$

The composition of GaC$_3$N$_3$ is confirmed by spectroscopic and elemental analysis. Infrared spectra obtained of the solid are simple with strong absorptions at 2215 cm^{-1} (v(CN)) and 440 cm^{-1} v(M-CN). The absence of any absorption above 3000 cm^{-1} indicates that the solid does not contain N-H or O-H groups. Mass spectra were obtained at a source temperature of 225°C and revealed the molecular Ga(CN)$_3$ ion at 147 and 149 (m/z) corresponding to the 69 and 71 a.m.u. gallium isotopes of the compound. Carbon, hydrogen, and nitrogen elemental analysis confirmed that the carbon to nitrogen ratio is 1:1; expected for Ga(CN)$_3$:24.4% C, 28.4% N; found: 24.6% C, 26.2% N, <1% H. The slightly low N content found is due to incomplete combustion with formation of a GaN residue. X-ray powder diffraction data were collected with a Rigaku diffractometer using Cu Kα radiation. All diffraction peaks could be indexed using a primitive cubic cell with a = 5.295 Å. The only plausible possibility for cubic GaC$_3$N$_3$ is a disordered structure with Ga at the cell corners and octahedrally coordinated by (C,N) and joined by C-N bonds aligned along the cell edges as shown in Fig. 4. The structure was verified by quantitative Rietveld analysis. The C-N bond length is found to be 1.15 Å, a value normal for metal cyanides, and the Ga-(N,C) distance is 2.07 Å. Reactions of GaC$_3$N$_3$ with LiCN provided a new lithium-carbon-nitride of gallium which consists of LiN$_4$ tetrahedra and GaC$_4$ tetrahedra linked together at the corners in a simple cubic structure similar to that of Zn(CN)$_2$ as illustrated in figure 4 (bottom). The identity of this material has been determined by spectroscopic methods and the structure was checked by quantitative Rietveld analysis.

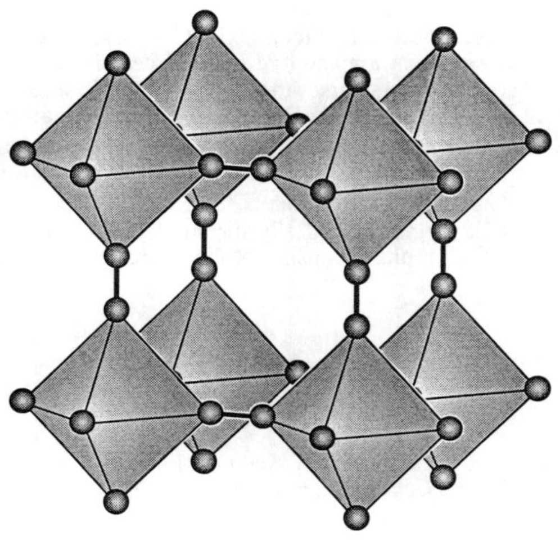

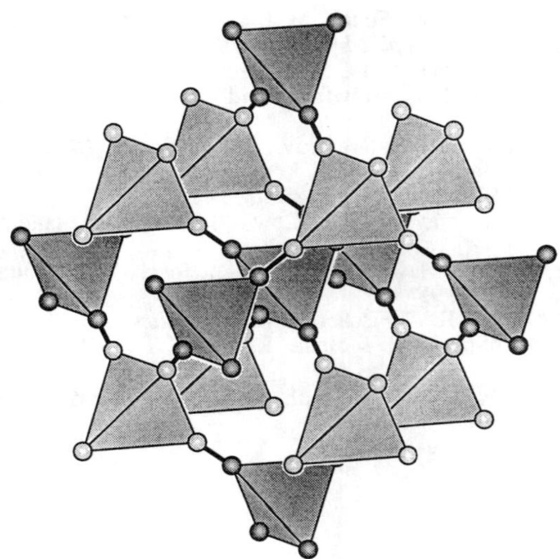

Figure 4. Top: the structure of GaC_3N_3 shown as $Ga(C,N)_6$ octahedra. Heavy lines are C-N bonds. Bottom: the structure of $LiGaC_4N_4$ shown as LiN_4 tetrahedra (lighter) and GaC_4 tetrahedra (darker). Heavy lines are C-N bonds.

CONCLUSION

A new low temperature route to heteroepitaxial GaN involving a new inorganic precursor, Cl_2GaN_3, that is entirely free of carbon and hydrogen, is described. The trimethylamine adduct $(Cl_2GaN_3)N(CH_3)_3$, is a convenient precursor to GaN because it is substantially more volatile than Cl_2GaN_3 and therefore easier to use in CVD processes. Good quality wurtzite GaN has been deposited by the thermal decomposition of $(Cl_2GaN_3)N(CH_3)_3$ at 650-700°C. Furthermore, reactions of Cl_2GaN_3 with $Si(CH_3)_3CN$ yield a crystalline gallium carbon nitride of composition GaC_3N_3. Quantitative X-ray powder diffraction has been used to refine the cubic structure of this material which consists of Ga atoms octahedrally surrounded by on the average three C and three N atoms. The related $LiGaC_4N_4$ phase consists of Ga atoms tetrahedrally surrounded by four C or N atoms.

ACKNOWLEDGMENTS

This work was supported by the National Science Foundation DMR 9458047. Electron microscopy was conducted at the Center for High Resolution Electron Microscopy at Arizona State University.

REFERENCES

[1] H. Morkoc, S. Strite, G. B. Gao, M. E. Lin, B. Sverdlov and M. Burns, *J. Appl. Phys. 76* , 1763, [1994]
[2] S. Nakamura, S. Masajuki, M. Senoh, M. Iwasa, T. Yamada, T. Matsuhita, H. Kiyotu, Y. Sugimoto, S. Nagayama, *Jap. J. Appl. Phys. 34*, L797, [1995]
[3] A.C. Jones, C.R. Whirehouse, and J.S. Roberts,*Chem. Vap. Deposition* , 1, *3* 65, [1995]
[4] S.D.Lester, F.A. Ponce, M.G. Crawford, and D.A. Steigerwald,*Appl. Phys. Lett.*, *66* , 1249, [1995]
[5] D.J. Smith, D. Chandrasakher, B. Sverdlov, A. Botchkarev, A. Salvador, and H. Morkoc, *Appl. Phys. Lett.* , *1830* [1995]
[6] J. Kouvetakis and D.B. Beach, *Chem. of Mat.* , 476, [1989]
[7] V. Lakhoita, D. A. Newmayer, A. H. Cowley, R. A. Jones, and J.G. Ekerdt, *Chem. of Mat.*, *3*, 441 [1995]
[8] D. A. Newmayer, A. H. Cowley, A. Decken, R. A. Jones, V. Lakhoita, and J. G. Ekerdt, *J. Am. Chem. Soc. 117*, 5893 [1995]
[9] A. Muehr, R. M. Mattner, R. A. Fischer, *Organometallics* , *15*, 2053, [1996]
[10] M.S. Brandt, N.M. Johnson,R.J. Molnar, R. Singh, T.D Moustakas, *App.Phys.Lett. 64*, 2264-67, [1994]
[11] J. M^C Murran, J. Kouvetakis, D. J. Smith, *Appl. Phys. Lett. 69*, 203-205, [1996]

TOF-LEIS CHARACTERIZATION AND GROWTH OF GaN THIN FILMS GROWN WITH ECR AND NH_3

E. KIM, A. BENSAOULA
Space Vacuum Epitaxy Center, University of Houston, Houston, TX 77204-5507
I. RUSAKOVA
Texas Center for Superconductivity, University of Houston, Houston, TX 77204
A. SHULTZ, K. WATERS
Ionwerks, 2472 Bolsover suite 255, Houston, TX 77005

ABSTRACT

GaN thin films were grown on various substrates by GSMBE using ECR nitrogen and ammonia. The growth of GaN was monitored by real time analysis, time of flight low energy ion scattering (TOF-LEIS) and RHEED.

Growth of GaN on GaAs, ZnO, Ge and Al_2O_3 was investigated. The substrates' surfaces were analyzed during pre-growth annealing and during GaN growth. The removal of surface contaminants and the modification of the surface stoichiometry from these surfaces are presented.

GaN films on sapphire (0001) grown under different conditions were examined real time by low energy ion scattering mass spectroscopy of recoiled ion (MSRI) and the relationship between the in-situ surface composition with ex-situ photoluminescence measurement results are discussed.

INTRODUCTION

Time of flight low energy ion scattering (TOF-LEIS) is a powerful technique which allows real time surface analysis during thin film growth without detectable damage to the epilayer [1]. Substrate annealing, nitridation and GaN growth are investigated on sapphire, silicon and gallium arsenide by time of flight direct recoil spectroscopy (DRS) and mass spectroscopy of recoiled ions (MSRI) [2,3]. In MSRI, only atomic ions recoiled or scattered from the sample surface can be deflected by a time refocusing analyzer when a voltage is applied on the sector. The time of flight of a given ion is

$$\text{TOF} = T_0 + k(M_{s,r}/e)^{1/2} \tag{1}$$

where T_0 is the time of flight of the primary ion traveling from the ion source to the analyzing sample, k is a constant related the sector voltage to the recoiled or scattered ion energy and $M_{s,r}/e$ is the corresponding mass per unit charge so that elements with low atomic weight, such as carbon, nitrogen and oxygen can be clearly separated on the time of flight scale which can be represented by a channel number.

For MBE nitride growth, electron cyclotron resonance (ECR), radio frequency (RF) plasma and reactive ion sources for nitrogen have been used by many research groups [4-6] in the growth of GaN thin films and alloys. Only recently has gas source MBE (GSMBE), with ammonia as the nitrogen precursor, been applied to these materials [7-9].

In this paper, growth of GaN on various substrates are examined by TOF-MSRI and effects of nitridation and buffer layer for GaN growth on sapphire using a combination of ECR and ammonia sources are discussed.

Mat. Res. Soc. Symp. Proc. Vol. 449 © 1997 Materials Research Society

EXPERIMENTAL

The GaN thin film deposition system is equipped with TOF-LEIS instrumentation, RHEED, an ASTeX compact ECR source, a gas source injector and conventional Knudsen cells. TOF-LEIS consists of a 5keV Ne^+ ion source mounted onto a series of optics designed to focus and pulse the ion beam for time-of-flight measurements.

After the samples were chemically cleaned outside of chamber, they were either indium mounted on the molybdenum block or directly mounted on the sample holder. For sapphire substrates, one micron thick titanium was evaporated on the back side to prevent heat dissipation and for the precise measurement of surface temperature with a pyrometer. Temperature was also measured by a thermocouple placed on the back side of the sample.

Substrates were annealed under vacuum prior to growth. Nitridation and buffer layer growth were performed by ECR or NH_3 and GaN final layer was grown with NH_3. The surface composition of the substrates and GaN epilayers were monitored by MSRI and RHEED. A 25mW He-Cd laser was used as the excitation source for photoluminescence measurements. PL measurements were taken at 10K.

RESULTS

Sample Annealing and Nitridation Studies

TOF-LEIS in conjunction with RHEED has been used to follow the surface reconstruction and surface composition during substrate vacuum annealing, nitridation and initial growth of GaN. We have used TOF-LEIS to optimize the annealing temperature and follow the surface nitridation for Sapphire, GaAs, ZnO, and Ge substrates. While sapphire, ZnO, and GaAs substrates are known to be important to GaN thin film growth, Ge was studied in the hope to achieve a better substitute to GaN/Si heteroepitaxy.

Detailed results for sapphire and GaAs have been published previously [2,3]. For sapphire and GaAs, we have demonstrated that a nitrogen terminated surface is easily achieved upon exposure to an ECR nitrogen (>40W) beam. Our recent studies using ammonia, as the nitrogen source, suggest that the nitridation layer is thinner and more difficult to achieve as compared to ECR experiments.

Figure 1 shows RHEED diffraction patterns and the corresponding MSRI spectrum for GaN/Sapphire substrates at different growth stages. It is clear that MSRI is a very sensitive tool for pre-growth surface treatment optimization and is uniquely sensitive to surface nitrogen. This is demonstrated by comparing a-1 and b-1 with a-3 and b-2 from figure 1. RHEED from an annealed Sapphire substrate and that from one exposed to ammonia are identical but MSRI shows a clear N shoulder. RHEED from a sapphire substrate exposed to a N-ECR plasma (b-1) shows faint streaks consistent with an AlN overlayer only after many minutes of exposure. In contrast the MSRI technique allows real time observation of the nitridation process from onset (a-3) to equilibrium (b-2).

In the GaAs case we have shown that depending on the nitridation procedure one can obtain the hexagonal or the cubic GaN phase. In the latter case the GaAs substrate was annealed in vacuum without an As overpressure. The Carbon impurity (mainly from hydrocarbons) on the GaAs (001) surface disappeared at about 480°C, and oxide desorption coupled with arsenic depletion was observed starting at 570°C. Arsenic was completely depleted from the surface at 640°C; at that point only Ga is detected by MSRI. Such a Ga

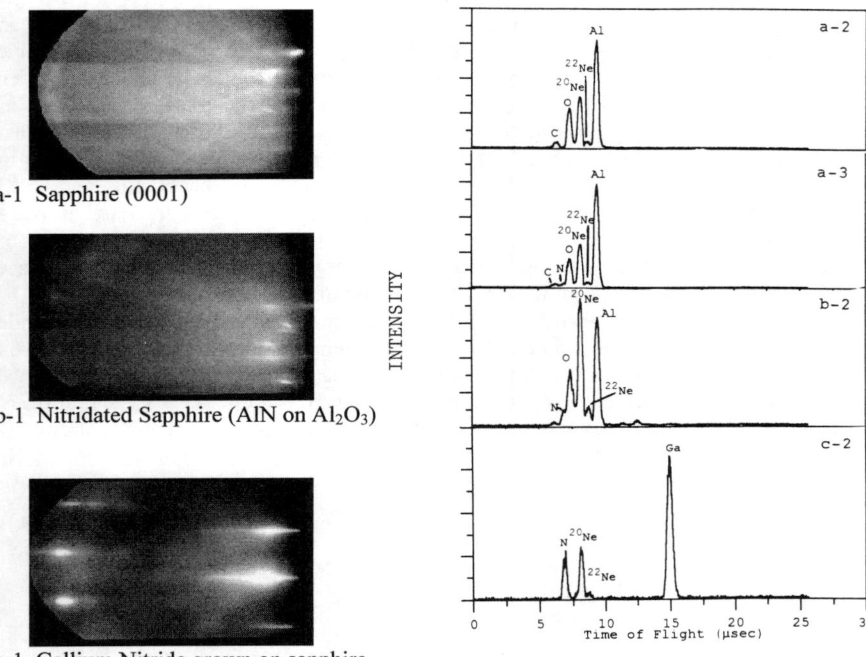

Figure 1. RHEED and MSRI data of GaN thin films grown on sapphire by GSMBE.
  a) a-1 shows RHEED from clean sapphire and the corresponding MSRI spectra (a-2). Fig. a-3 shows the MSRI spectrum after exposure to NH_3 beam where a nitrogen shoulder is apparent.
  b) RHEED from a sapphire sample exposed to nitrogen ECR showing a superposition of Al_2O_3 and AlN diffraction spots. The MSRI spectrum shows clearly a nitrogen surface peak on the Al_2O_3.
  c) RHEED and MSRI from GaN thin film on sapphire grown using NH3 and Ga.

terminated GaAs surface (at 590°C) when exposed to an ECR N beam, becomes nitridated as demonstrated by a surface nitrogen peak observed by MSRI. The growth of an ECR GaN buffer layer at the same temperature (590°C) followed with an ammonia grown GaN final layer, at 710°C, yields surfaces with 4 fold cubic symmetry RHEED patterns.

For ZnO (0001) substrates, carbon was found to be the major surface contaminant. The C surface composition was significantly reduced by a vacuum anneal at 550°C. The MSRI zinc signal was very weak at room temperature and did not vary until the surface carbon contamination was reduced (550°C). The MSRI peak intensity of O and Zn was seen to increase up to 660°C. Beyond that temperature the ZnO surface starts decomposing with surface O desorbtion. At this temperature the RHEED pattern becomes streaky. After the substrate was annealed, the sample temperature was lowered to 580°C. The ZnO was nitridated at this temperature for 20min. A clear nitrogen peak was observed in MSRI spectra and RHEED showed a chevron like pattern. First a buffer layer was grown at 535°C with an ECR source for 30min. MSRI detected prominent Ga and N peaks but C was still detectable on the surface. RHEED showed a polycrystalline GaN structure which did not change when the sample temperature was elevated to 650°C. After a thicker ECR grown buffer layer was grown (additional 30 minutes) at the same temperature the RHEED pattern remained that from a polycrystalline GaN surface. At this point the nitrogen source was switched to ammonia and a GaN layer was grown at the same temperature for another 30min. The MSRI Ga to nitrogen ratio increased but the RHEED showed a crystalline GaN structure but with a clear transmission pattern from a rough surface. Further studies are needed in the case of GaN/ZnO but it is clear from these preliminary results that nitrogen incorporation and surface kinetics in GSMBE are significantly different from those of ECR MBE growth.

In the Ge case, we found that the oxide layer desorption occurred at about 500°C, while the signal showed a decrease at around 300°C but remained constant up to a 700°C anneal (figure. 2b). This is consistent with previous studies that showed submolayer C coverage always present on Ge surfaces. We found however that exposure to an N-ECR beam very efficiently removed the surface carbon and resulted in nitrogen incorporation. The resulting film was however textured polycrystalline (from RHEED), thus the formation of a GeN overlayer could not be directly verified. To date only polycrystalline GaN thin films have been grown on Ge substrates.

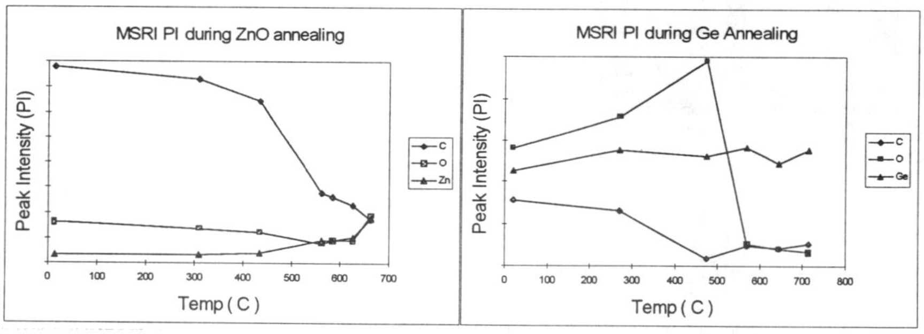

Figure 2. MSRI peak intensity change of surface elements during substrate pre-annealing.
a. zinc oxide b. germanium.

MSRI as a Real Time Characterization and Control Probe for GaN Growth

We have shown that MSRI can allow real time surface composition measurement for all elements including hydrogen. In the case of GaN, both the surface pretreatment (step one, annealing and nitridation), the buffer layer growth conditions(step two), and that of the final layer (step three) are found to impact the electrical and optical properties of the epilayers. In this study we have used a combination of N-ECR and ammonia in the growth of GaN thin films. The nitrogen species used in each of the above steps was found to be also important. As an example we show in figure 3 MSRI spectra and the corresponding PL FWHM for GaN films grown using either ECR or ammonia in step one and two and ammonia for the final GaN layer. We clearly see that for these conditions the use of ECR for all three steps gives the worst PL while that using an all ammonia process gives the best. Of importance the surface Ga/N ratio from the MSRI spectra, taken during growth, tracks with the PL FWHM.

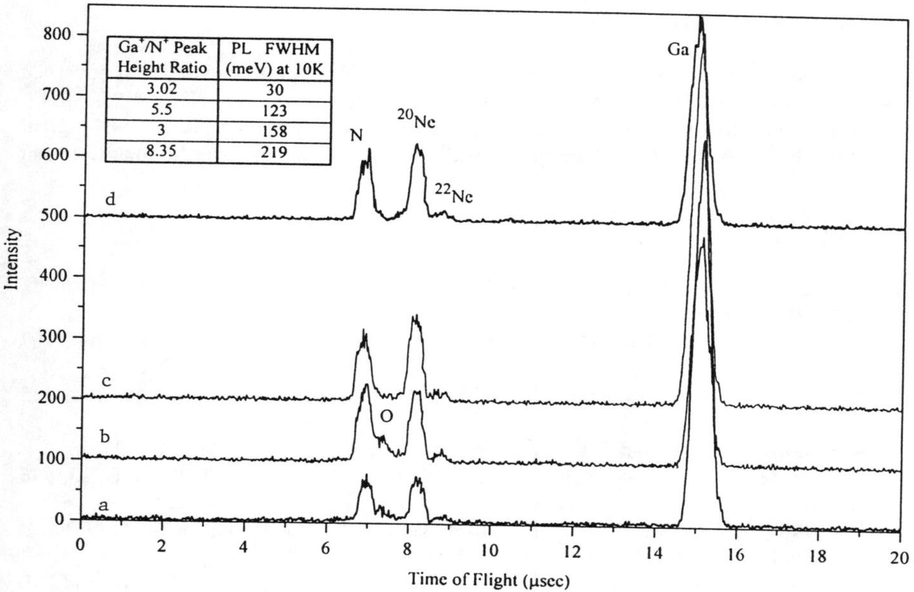

Figure 3. MSRI from GaN films grown under various growth conditions. The growth conditions and corresponding PL FWHM (meV) are described below. PL data all taken at 10K.
a. ECR nitridation, ECR buffer and NH_3 layer - Ga/N 8.35, PL FWHM 219meV
b. ECR nitridation, ECR buffer and NH_3 layer - Ga/N 3, PL FWHM 158meV
c. ECR nitridation, NH_3 buffer and NH_3 layer - Ga/N 5.5, PL FWHM 123meV
d. NH_3 nitridation, NH_3 buffer and NH_3 layer - Ga/N 3.02, PL FWHM 30meV

The closer to stoichiometry the surface is the better the PL result. While at first, spectrum b (Ga/N =3 but FWHM=158 meV) may seem not to follow the trend, a closer analysis of its MSRI spectrum shows a surface oxygen shoulder from either ammonia contamination or bad initial growth conditions (no buffer layer was used for this sample).

CONCLUSION

We have shown that MSRI is a unique in-situ technique which is compatible with epitaxial growth of optoelectronic thin films. Its unique sensitivity to nitrogen species makes it specifically adapted to nitride thin film growth optimization, composition control, and surface kinetics studies. We have also shown that the surface Ga/N ratio, as measured in real time by MSRI, correlates well with the PL characteristics of the thin films and thus can be used to optimize the optical properties of GaN materials.

ACKNOWLEDGMENTS

This work was supported by a NASA grant No. NAGW977, the state of Texas advanced research program (ARP No. 93-003652224), and an ONR grant No. N00014-95-1-0312. Ionwerks would like to acknowledge the support of USAF through a phase I SBIR grant monitored by WPAFB. The authors also thank Dr. I. Berishev and N. Badi for their technical input.

REFERENCES

1. Waters, A. Bensaoula, A. Schultz, K. Eipers-Smith, and A. Freundlich, J. Cryst. Growth, **127**, 972 (1993)
2. W. T. Taferner, A. Bensaoula, E. Kim, and A. Bousetta, J. Cryst. Growth, **164**, 167 (1996)
3. A. Bensaoula, W. T. Taferner, E. Kim, and A. Bousetta, J. Cryst. Growth, **164**, 185 (1996)
4. R. J. Molar, R. Singh, and T. D. Moustakas, J. Electronic Materials, **24**, 275 (1995)
5. K. Kim, M. C. Yoo. K. H. Shim, and J. T. Verdeyen, J. Vac. Sci. Technol. B, **13**, 796 (1995)
6. R. C. Powell, N. E. Lee, Y. W. Kim, and J. E. Greene, J. Appl. Phys., **73**, 189 (1993)
7. Z. Yang, L. K. Li, and W. I. Wang. Appl. Phys. Lett., **67**, 1686 (1996)
8. Wang, N. Y. Li, H. K. Dong, F. Deng, S. S. Lau, C. W. Tu, J. Hays, S. Bidnyk, and J. J. Song, J. Cryst. Growth, **164**, 159 (1996)
9. W. Kim, Ö. Aktas, A. E. Botchkarev, A. Salvador, S. N. Mohammed, and H. Morkoç, J. Appl. Phys., **79**, 7657 (1996)

THE GROWTH OF GaN FILMS BY MIGRATION ENHANCED EPITAXY

S.E.Hooper*, C.T.Foxon**, T.S.Cheng**, N.J.Jeffs**, G.B.Ren***, D.E.Lacklison***, J.W.Orton*** and G.Duggan*

*Sharp Laboratories of Europe Ltd, Edmund Halley Road, Oxford Science Park, Oxford OX4 4GA, United Kingdom

**Department of Physics, University of Nottingham, Nottingham NG7 2RD, United Kingdom

***Department of Electrical and Electronic engineering, University of Nottingham, Nottingham NG7 2RD, United Kingdom

ABSTRACT

Gallium Nitride epitaxial films were grown by migration enhanced epitaxy directly on sapphire (0001) without using any pre-growth substrate nitridation or low temperature buffer layers. In comparison with our material grown directly on sapphire by conventional molecular beam epitaxy, a significant improvement in the surface morphology and layer properties, measured by reflection high energy electron diffraction, X-ray diffraction, scanning electron microscopy, room temperature photoluminescence and the Hall effect, was observed for material grown by migration enhanced epitaxy.

INTRODUCTION

In order to overcome the large lattice mismatch and chemical dissimilarity between GaN and sapphire and therefore grow high quality GaN films, the majority of workers in both the metal organic vapor phase epitaxy (MOVPE) and molecular beam epitaxy (MBE) fields employ a combination of pre-growth substrate nitridation and low temperature GaN [1] or AlN [2] buffer layers. However, the suppression of 3-D growth using nitridation and low temperature buffers can be a lengthy and complicated process, particularly in MBE growth with its lower growth rates.
One way in which uniform 2-D nucleation has been achieved in the growth of III-V compounds such as GaAs and AlGaAs has been to employ the method of migration enhanced epitaxy (MEE). First reported by Horikoshi et al [3-5], MEE is believed to precisely control the stoichiometry of the growing crystal by alternating the supply of group III and V molecular beams to the growth surface. In this vein, MEE is potentially a promising technique for growing high quality GaN films directly on sapphire. Consequently, the purpose of the study presented in this paper was to investigate whether the properties of GaN flims grown directly on sapphire, without any pre-growth nitridation or low temperature buffer layers, could be improved by applying the MEE method.

Mat. Res. Soc. Symp. Proc. Vol. 449 © 1997 Materials Research Society

EXPERIMENT

GaN film growth

Details of the MBE apparatus are presented elsewhere [6]. A constant active nitrogen flux was supplied by a commercial rf plasma source and a Ga flux of 3×10^{14} atoms cm^{-2}s^{-1} was used. The GaN layers were grown on sapphire (0001) at a temperature of 750°C. The films were intentionally doped with Si and were 0.2-0.4μm thick. No substrate nitridation or low temperature buffer layers were used prior to GaN growth.

In order to study the MEE growth of GaN on sapphire, four independent layers were grown using various Ga and N beam shutter sequences. Schematic diagrams of the shutter sequences from growth onset at t=0 are shown in figure (1). Sample (A) was grown using a conventional MBE approach where both the Ga and N shutters were opened simultaneously at t=0 and remained open throughout growth. For sample (B) equal amounts of Ga and N were supplied alternatively at intervals of 2.0s, with growth being initiated by Ga. Sample (C) was initiated by 3.0s of Ga followed by a 3.0s anneal where no atoms were directed at the surface, before opening the N shutter for 2.0s. Similarly, sample (D) had 4.0s of initial Ga, a 4.0s anneal and 2.0s of N as its repeated growth cycle. Consequently, the effective V/III ratios of the four layers (A-D) were approximately 1.0, 1.0, 0.7, and 0.5 respectively. The Si cell shutter was open except during the anneal stages.

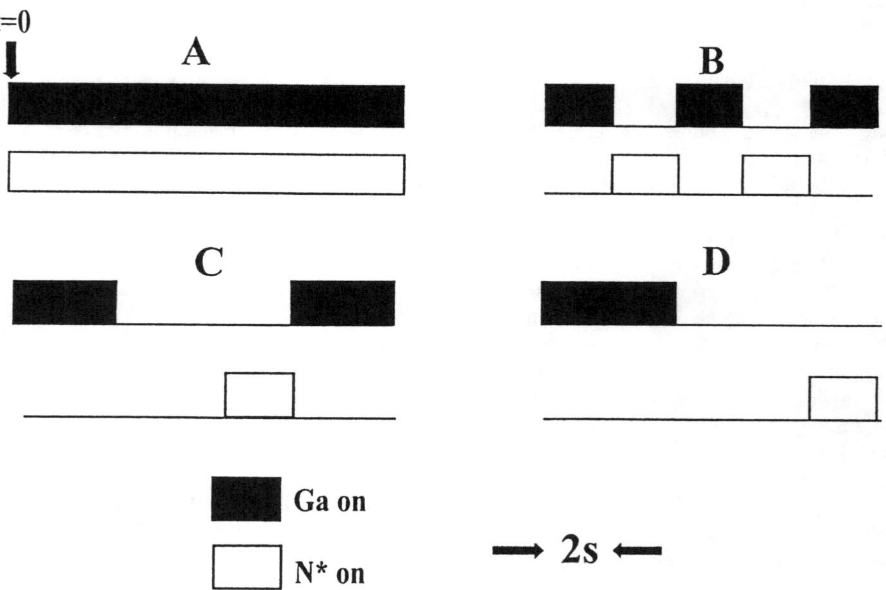

Fig. 1. Schematic diagrams of the shutter sequences used in the growth of samples (A), (B), (C), and (D).

RESULTS

<u>Scanning Electron Microscopy</u>

Scanning electron microscopy (SEM) observations of the cleaved edges of the four GaN samples (A-D), are shown in figure (2). Sample (A), grown using conventional MBE, showed the GaN layer to have poor lateral continuity and consist of oriented columnar grains. This columnar nature was consistent with both the spotty RHEED and large XRD FWHM exhibited by the sample. In contrast, SEM showed films (B), (C) and (D) to be more continuous, with sample (D) having a relatively flat surface morphology and the highest degree of continuous crystallinity.

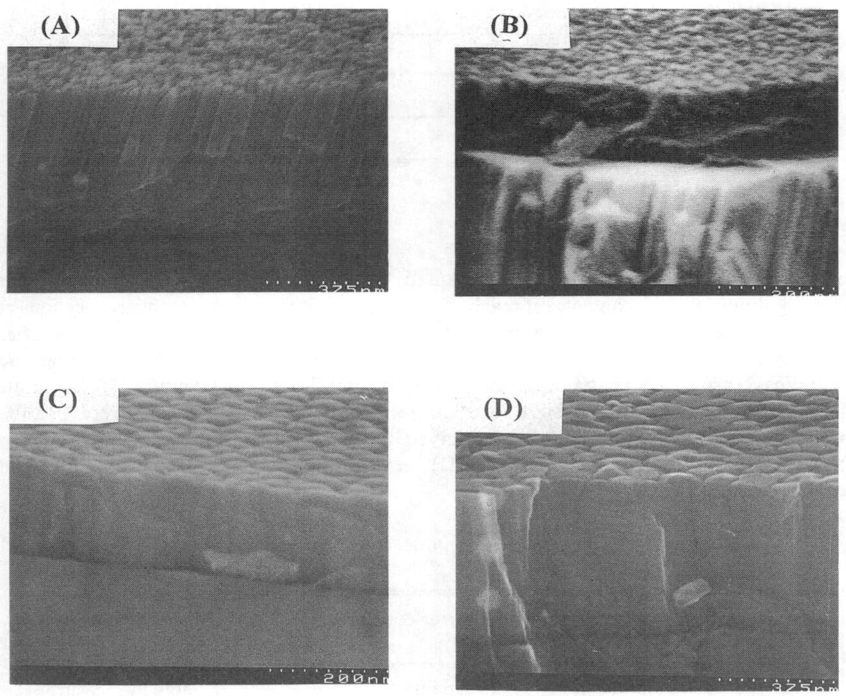

Fig 2. Scanning electron micrographs of the cleaved edges of the four GaN films (A-D).

<u>X-ray Diffraction</u>

The crystalline integrity of the GaN films was measured by x-ray diffraction (XRD) using a powder diffractometer equipped with a CuKα radiation source, with scans being made in the standard θ-2θ and ω modes. Table I lists the measured FWHM of the GaN(0002) reflection peak for each sample. A large reduction in the FWHM occured between (A) and (D), indicating a significant improvement in the crystal quality.

Table I: XRD measurements

Sample	GaN(0002) θ-2θ FWHM (arcmins)	GaN(0002) ω FWHM (arcmins)
A	6.12	60.00
B	6.96	38.40
C	5.76	30.48
D	4.32	22.14

<u>Electrical Measurements</u>

The GaN layer resistivities were measured by a four-point probe method using pressed indium ohmic contacts at each corner of a 5mm x 5mm square section cut from each sample. Room temperature Hall measurements were also made on these square sections using a 0.7T magnetic field. Table (II) lists the measured resistivities, carrier concentrations and electron mobilities for samples (A-D). Films (A) and (B) were highly resistive and no reliable Hall measurements were possible, this is consistent with the poor lateral crystallinity of these layers. Samples (C) and (D) exhibited n-type conduction with (D) having the highest electron mobility and lowest resistivity values.

Table II: Room temperature electrical measurements

Sample	Resistivity (Ωcm)	Carrier concentration (cm^{-3})	Electron mobility ($cm^2V^{-1}s^{-1}$)
A	Insulating	Unable to measure	Unable to measure
B	3.2	Unable to measure	Unable to measure
C	6.7×10^{-2}	7×10^{18}	13
D	5.0×10^{-2}	9×10^{17}	138

<u>Reflection High Energy Electron Diffraction</u>

Reflection high energy electron diffraction (RHEED) was used to monitor the growing surfaces throughout all GaN growth. No intensity oscillations of the RHEED specular spot could be detected from any of the four samples (A-D) during epitaxy. A (1x1) RHEED pattern with six-fold symmetry was observed from all samples during growth, with a significant improvement in the continuity of the RHEED streaks being recorded for samples (C) and (D) indicative of 2-D nucleation, faint Kikuchi lines were also visible for (C) and (D). Sample (A) showed highly broken RHEED streaks consisting of sharp spots signifying a 3-D nucleation process.

<u>Room Temperature Photoluminescence</u>

The samples (A-D) were optically probed by room temperature photoluminescence (PL) using a helium-cadmium laser. Figure (3) shows the PL from (A-D) measured under identical excitation conditions over the energy range 2.0 to 3.5eV. The MBE grown GaN layer (A) exhibited only broad, impurity related deep emission centered around 2.5eV. Conversely, intense band-edge luminescence at 3.4eV, with no deep emission, was recorded for the MEE grown film (D), characteristic of material with a high optical quality. Films (B) and (C) showed both a weak band-edge peak and some deep luminescence.

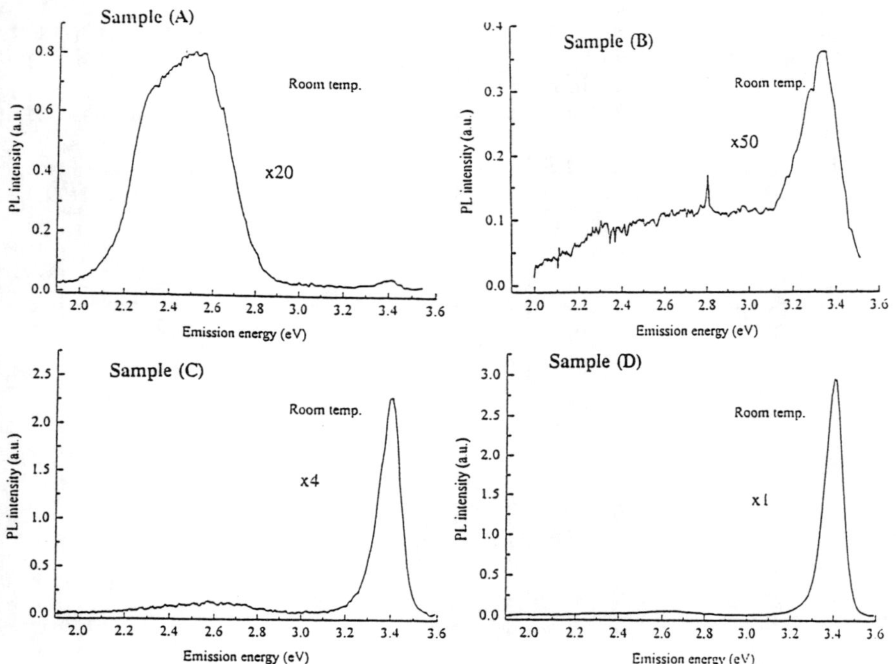

Fig.3. Room temperature photoluminescence spectra of the four GaN samples.

CONCLUSIONS

It is clear from the results presented in this paper that in comparison with our GaN films grown by MBE directly on sapphire without using any low temperature buffer layers, a significant improvement in the crystalline, electrical and optical properties of the material can be achieved using the MEE method of crystal growth.
The MEE technique has allowed highly Ga-rich growth conditions to be used (V/III~0.5) without the accumulation of surface Ga droplets which occur using conventional MBE under a similar V/III ratio and growth temperature. The reason why such Ga-rich conditions results in the best films is unclear at present, although it is well known that the MEE technique can provide precise control of the stoichiometry of the growing crystal. It should be noted that the best film presented in this limited study is not state-of-the-art quality, however, the MEE shutter sequence has by no means been optimised, and so more work is necessary to further improve the material. In summary, the potential of the MEE technique to grow GaN thin films directly onto sapphire has been demonstrated by this study.

REFERENCES

1. S.Nakamura, Jpn. J. Appl. Phys. 30, L1705 (1991)
2. H.Amano, N.Sawaki, I.Akasaki and Y.Toyoda, Appl. Phys. Lett. 48, 353 (1986)
3. Y.Horikoshi, M.Kawashima and H.Yamaguchi, Jpn. J. Appl. Phys. 27, 169 (1988)
4. Y.Horikoshi and M.Kawashima, Jpn. J. Appl. Phys. 28, 200 (1989)
5. Y.Horikoshi, H.Yamaguchi, F.Briones and M.Kawashima, J. Cryst. Growth. 105, 326 (1990)
6. S.E.Hooper, C.T.Foxon, T.S.Cheng, L.C.Jenkins, D.E.Lacklison, J.W.Orton, T.Bestwick, A.Kean, M.Dawson and G.Duggan, J. Cryst. Growth. 155, 157 (1995)

THIN FILM GROWTH OF GROUP III NITRIDES BY MASS SEPARATED ION BEAM DEPOSITION

C. RONNING, E. DREHER, H. FELDERMANN, M. SEBASTIAN, J. ZWECK*,
R. FISCHER*, H. HOFSÄSS
Universität Konstanz, Fakultät für Physik, Postfach 5560, 78434 Konstanz, Germany
*Universität Regensburg, NFW-2 Physik, Universitätsstr. 31, 93040 Regensburg, Germany

ABSTRACT

We have grown gallium nitride (GaN), aluminum nitride (AlN) and boron nitride (BN) thin films by mass separated ion beam deposition. All deposited films were found to be almost stoichiometric. AlN and GaN films are crystalline even after room temperature deposition whereas for the formation of crystalline boron nitride temperatures above 150 °C are necessary. The influence on the phase formation and the film structure of ion energy and substrate temperature on the one hand, and bond ionicity on the other hand, was investigated for these three systems.

INTRODUCTION

Direct deposition of low energy ions (20 - 1000 eV) is a thin film growth process far from thermodynamic equilibrium which can be used to nucleate and grow superhard materials like diamond-like amorphous carbon (DLC) [1], amorphous carbon nitride (CN) [2] or cubic boron nitride (c-BN) [3]. It is essential to use ion beam deposition or ion beam assisted deposition techniques to grow c-BN thin films.

Because the ion species, ion flux, ion energy and substrate temperature can be controlled precisely and independently, mass separated ion beam deposition (MSIBD) is an ideal technique for systematic studies of thin film nucleation and growth processes. Compared to ion-assisted deposition techniques for binary compound thin films, MSIBD has some unique features. In ion assisted methods (IBAD) the non-volatile component is usually deposited with thermal energies and the initial film growth starts from the surface. In contrast, with MSIBD both components are deposited as energetic and usually singly charged ions. No additional bombardment of the growing film with other ions, like Ar^+, or with neutral atoms or molecules occurs. The nucleation and growth process is therefore governed by bulk rather than surface processes.

In this paper we compare the MSIBD growth of thin nitride films of the group III elements gallium, aluminum and boron. Because of the high bond ionicity of group III nitrides we expect the formation of crystalline phases. There are only a few studies on GaN [4] and InN [5] growth by MSIBD described in the literature. According to these studies only a fraction of nitrogen ions were incorporated into the film and an excess of nitrogen was necessary to grow stoichiometric films. The structure of these films was not investigated in detail.

EXPERIMENTAL

AlN, GaN and BN thin films were grown by mass selected ion beam deposition using a UHV deposition system described in detail elsewhere [1]. Ga^+, Al^+, B^+ and N^+ ions were produced in a Sidenius type ion source by evaporation of liquid Ga, $AlCl_3$ and B_2O_3 in a N_2 gas stream, respectively. After acceleration to 30 keV and magnetic mass separation, the isotopically pure ion beam is guided into the deposition chamber, where the ions are decelerated down to energies E_{Ion}

Mat. Res. Soc. Symp. Proc. Vol. 449 © 1997 Materials Research Society

adjustable between 20 eV and 1 keV. These low energy ions are deposited onto Si (100) substrates at temperatures T_S between room temperature and 400 °C. The pressure during deposition was below 3×10^{-7} Pa. Prior to deposition the substrates were sputter-cleaned by 1 keV Ar^+ ions. The amount of deposited ions is precicely determined by an ion charge measurement, which is also used to switch the separation magnet periodically between the respective Group III element and $^{14}N^+$. During each cycle, 10^{15} ions with an ion ratio of $III^+/N^+ = 1$ were deposited. The typical ion current was 20-30 μA on an area of 1.8 cm^2. Films were characterized in-situ by Auger electron spectroscopy (AES) as well as ex-situ by infrared spectroscopy (FTIR) and Rutherford backscattering spectroscopy (RBS). Some samples were prepared for transmission electron microscopy (TEM) and diffraction (TED).

RESULTS AND DISCUSSION

<u>Aluminum nitride</u>

AlN thin films were grown at T_S = RT for E_{Ion} between 50 - 500 eV using $^{27}Al^+$ and $^{14}N_2^+$ or $^{14}N^+$ ions with an ion ratio of 2:1 or 1:1, respectively. The AES-measurements of these films show stoichiometric amounts of aluminum and nitrogen as well as the typical fine structure of the Al(LVV) line for AlN [6]. We have also detected a contamination with boron and oxygen, and in some samples also carbon. The boron and oxygen contamination is due to a contamination of the Al ion beam at 27 amu with $^{11}B^{16}O^+$ ions originating from the ion source insulators. A similar effect is responsible for the carbon contamination in films which were grown with N_2^+ ions. In this case the ion beam at 28 amu was contaminated with $^{12}C^{16}O^+$. Fig.1 shows the 700 keV $^4He^{2+}$ RBS spectrum of an AlN sample together with three RUMP simulations [7]. The simulations were done for different assumptions for the stoichiometry of the 80 nm thick AlN film. The spectrum is best fitted by Al:N = 1:1. Therefore, we conclude that the prepared films are almost stoichiometric AlN with a few percent contaminations of B, C and O.

A FTIR-spectrum of an AlN film which was deposited with E_{Ion} = 100 eV at T_S = RT is plotted in fig.2b. This spectrum is typical for all prepared AlN films, i.e. we found no dependencies as a function of E_{Ion}. The absorption line at 667 cm^{-1} can be attributed unambiguously to the optical phonon modes of crystalline AlN, but due to the broadness of the line it is not possible to distinguish between the zincblende and wurtzite structure [8]. The contamination described above and the presence of only nm-size crystallites might be responsible for the broadness of this line in comparison to single crystalline AlN [9]. This sample was also prepared for TEM and TED. With TEM we found AlN nanocrystallites (< 12 nm)

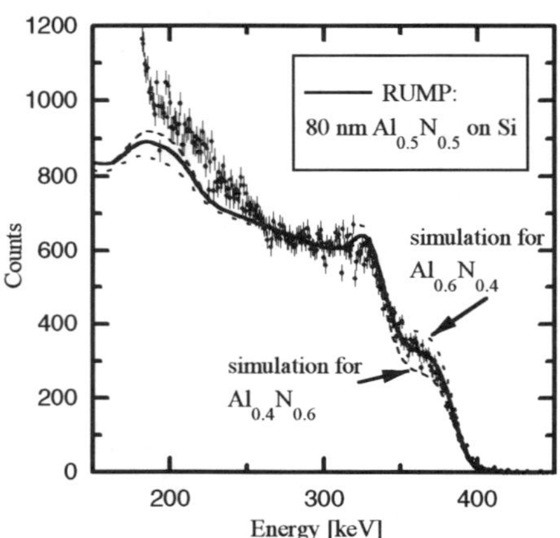

<u>Figure 1</u>: RBS-spectrum of an AlN film deposited with an ion energy of 50 eV at room temperature on Si. Lines: RUMP-simulations for different stoichiometries.

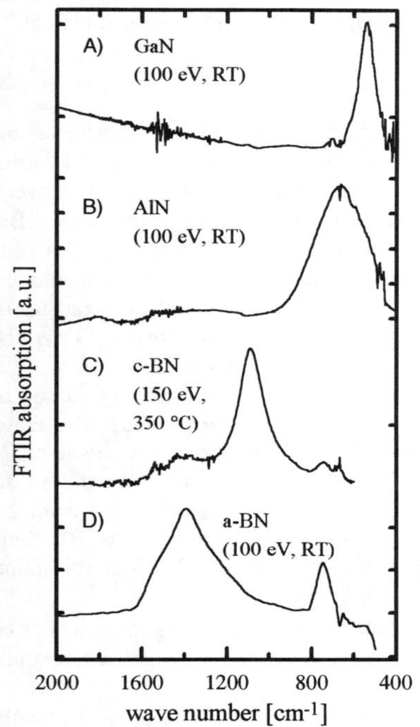

Figure 2: FTIR spectra of (a) GaN, (b) AlN and (d) BN thin films grown by MSIBD using E_{Ion}=100 eV at T_S=RT. (c) FTIR spectrum of a BN film deposited using E_{Ion}=150 eV at T_S=350 °C.

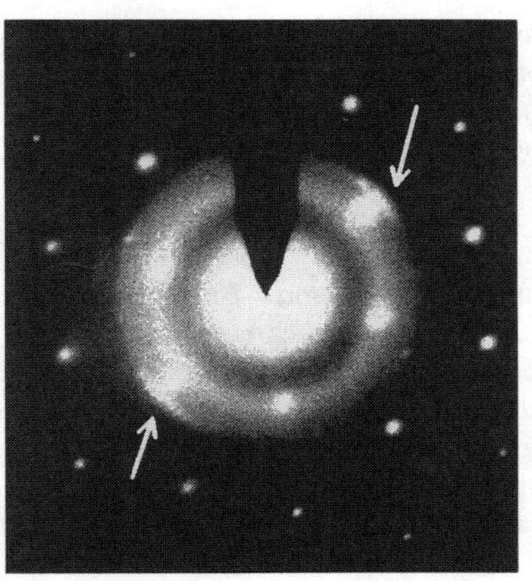

Figure 3: TED pattern of an AlN thin film deposited at room temperature. The arrows indicate diffraction spots from AlN crystallites.

embedded in an amorphous matrix. In the TED pattern shown in fig.3, bright spots originating from the single crystalline Si substrate as well as two long drawn-out spots from the AlN crystallites (arrows) and bright halos of the amorphous matrix are clearly visible. The existence of AlN-spots in the TED picture leads to the conclusion that the crystallites must have a preferential orientation.

One AlN sample was annealed under vacuum up to temperatures of 650 °C. No changes in the FTIR spectrum were observed as a function of T_A. For higher temperature the film disintegrated and evaporated from the Si substrates. This indicates that the crystalline structure of the films remains unchanged until they disintegrate at high T_A.

Gallium nitride

GaN thin films were grown using $^{69}Ga^+$ and $^{14}N^+$ ions with an ion ratio 1:1 at $T_S = 293$ K for different ion energies between 50 - 200 eV. In situ AES reveals only Auger electrons from Ga and N. We are therefore sure that the GaN films are free of contaminations like carbon and oxygen. Furthermore, the films should not contain any hydrogen. The AES spectra show a Ga:N ratio of 60:40, which is identical to ratios obtained by AES for single crystalline GaN layers grown by MOCVD [10].

A FTIR-spectrum of a GaN film which was deposited with $E_{Ion} = 100$ eV at $T_S = $ RT is plotted in fig.2a. This spectrum is typical for all prepared GaN films, i.e. we found again no dependencies as a function of E_{Ion}. The absorption line at 542 cm^{-1} is due to crystalline GaN [8] which is much narrower in comparison to the line of AlN (fig.2b). We attribute this to a better crystallinity of the film. To support this statement TEM and TED measurements are under preparation. One reason for a better crystallinity is of course the better purity of the film. Another reason may be the somewhat higher bond ionicity of 0.5 of GaN compared to 0.45 of AlN [11].

Boron nitride

Boron nitride exists similar to carbon in two main crystalline modifications, a hexagonal structure (h-BN) and a cubic zincblende structure (c-BN). The crystal structures and lattice parameters of h-BN and c-BN are almost identical to those of graphite and diamond, respectively. But in contrast to the carbon system, it was found in recent studies that the diamond-like c-BN phase is stable and h-BN is metastable [12]. The energy difference between these two phases is very low, but separated with a high energy barrier resulting in a high stability of both phases. Nevertheless, attempts to grow c-BN by chemical processes alone failed so far [13] and up to now the formation of c-BN thin films requires ion bombardment during growth. Thus the nucleation and growth of BN [14] is much more complex compared to AlN and GaN.

BN thin films were grown using $^{11}B^+$ and $^{14}N^+$ ions with an ion ratio 1:1 as a function of T_S and E_{Ion}. AES measurements indicate contamination-free and almost stoichiometric BN films. The c-BN and h-BN phases were identified by FTIR measurements, using the characteristic c-BN reststrahlen band around 1080 cm^{-1} and the absorption lines at 1380 cm^{-1} and 750 cm^{-1} due to h-BN stretching and bending vibrations (see fig.2c,d). The c-BN content was calculated from the ratio of areas of the c-BN and h-BN absorption lines. The results of these analyses of BN films grown for different values of E_{Ion} and T_S are summarized in the diagram for BN phase formation shown in fig.4. We observe c-BN formation above rather sharp threshold values of $T_S \geq 150$ °C and $E_{Ion} \geq 125$ eV. Except for $E_{Ion} > 500$ eV, these threshold values seem to be independent of each other. In the following we will discuss the details of the observed energy and temperature dependence.

At a threshold energy of 125 eV the c-BN content rises to about 85 % and remains nearly constant for higher ion energies. The observed threshold energy is much lower compared to published data for IBAD growth [15,16]. The sharp energy threshold indicates a sub-surface nucleation process requiring a certain ion range and a certain critical deposited energy density. Such a process is qualitatively described by the subplantation model introduced by Lifshitz et al. to explain the formation of tetrahedrally bonded amorphous carbon (ta-C) [17]. A quantitative description on an atomic scale is provided by the 'thermal spike' model of ion impact and rapid energy dissipation [18].

It is clearly seen that the temperature threshold for c-BN formation of 150 °C is also rather sharp. A similar sharp temperature threshold around 150 °C was observed for ion-assisted pulsed laser deposited BN films [19]. In contrast, Kester et al. reported an extended

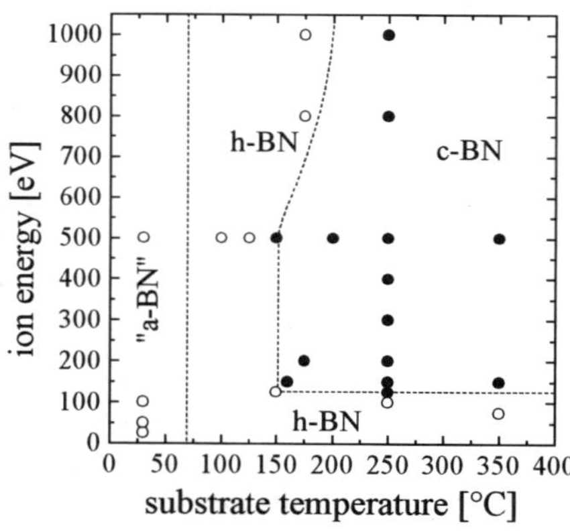

Figure 4: Phase diagram for ion beam deposited boron nitride films. The c-BN content is plotted as a function of the ion energy and the substrate temperature. Filled symbols represent samples with c-BN content exceeding 75 %, whereas open symbols represent h-BN or disordered a-BN samples.

temperature regime where mixed h-BN and c-BN phases were formed [16]. Another transition temperature lies between room temperature and about 100 °C, indicated by the dashed line in fig.4. Samples deposited at room temperature have an IR spectrum similar to h-BN; however, the h-BN peaks are broader compared to bulk h-BN and their positions are shifted to lower wave numbers (fig.2d). This indicates bond disorder and weaker bonding but not necessarily an amorphous phase. Nevertheless, we use the term 'a-BN' for these disordered h-BN films deposited at low temperatures. $T_S \approx 100$ °C is sufficient to obtain BN films exhibiting an IR absorption spectrum of crystalline h-BN. The fact that the crystalline phase is formed at higher temperatures can be attributed to the lower ionicity of 0.26 of the BN bond in comparison to AlN and GaN [11].

The observed temperature thresholds, indicated as dashed lines in fig.4, are closely related to the phase formation process during ion deposition, because they are not observed in annealing studies of BN films carried out for both c-BN and a-BN films [20]. In these studies, c-BN films were found to be stable up to 900 °C for annealing in air and up to 1200 °C in vacuum. a-BN films are stable up to 1100 °C for annealing in vacuum. Above these temperatures the films disintegrate and evaporate from the Si substrates. For a-BN films a gradual shift of the absorption peak towards the bulk h-BN value and a narrowing of the absorption line was observed. No phase transformation of the films from c-BN to h-BN or vice versa takes place.

Some samples were also prepared for TEM and TED analysis. Starting from the Si substrate, a disordered Si-B-N mixed layer is formed, followed by a textured h-BN layer with the c-axis parallel to the substrate surface. The thickness of the mixed interface layer is comparable to the ion range of the N^+ and B^+ ions used. On top of the textured h-BN the nanocrystalline cubic phase is formed. A texture of the c-BN is also seen in TED pictures, as they exhibit a structure in the c-BN diffraction rings [3]. The observed layered film structure is identical to the one obtained by IBAD [21]. The details of this complicated growth sequence are not well understood.

CONCLUSION

AlN, GaN and BN thin films were grown under identical deposition conditions using MSIBD. Therefore, we can compare the nucleation and growth processes for these different materials. AlN and GaN films deposited at room temperature are always crystalline, whereas BN films are disordered. We attribute this to the ionicity of the respective bonds. If the involved bonds have a sufficiently high ionicity, then crystallization is preferred (AlN and GaN), otherwise the phase will be amorphous or disordered (BN, C and CN) [18]. Another driving force for the formation of a crystalline phase is the substrate temperature. Whereas room temperature is sufficient to get crystalline AlN and GaN, temperatures above at least 100 °C are necessary to obtain crystalline c-BN or h-BN films.

All deposited films were grown with an ion ratio of $III^+/N^+ = 1$ and were found to be almost stoichiometric. Therefore, no excess of nitrogen was necessary and chemically-enhanced sputtering of nitrogen is negligible, e.g. compared to the deposition of CN [2]. In addition, the films are of high purity with the exception of the AlN films.

We have presented a phase diagram for boron nitride film growth. It is based on studies of the influence of the ion energy and substrate temperature on the phase formation. Above the temperature threshold of 150 °C and for low ion energies we observed the formation of h-BN. On the other hand, for ion energies above 125 eV c-BN is formed. The nucleation and growth of c-BN thin films is therefore a volume process occurring in the sub-surface region at a depth determined by the ion range. We interpret the energy threshold as the minimum energy needed to create a sufficiently high energy density to enable complete rearrangement of the atoms within the thermal spike volume [14]. This "molten" thermal spike volume recrystallizes during rapid dissipation of the deposited energy and the phase formation process is then triggered by the boundary conditions of the "solid" surrounding.

ACKNOWLEDGMENTS

Financial support by the Deutsche Forschungsgemeinschaft under grants # Ho1125/4-1 and Ho1125/8-1 under the auspices of the trinational ´D-A-CH´ German, Austrian and Swiss co-operation on the ´Synthesis of Superhard Materials´ is gratefully acknowledged.

REFERENCES

1.   H. Hofsäss, H. Binder, T. Klumpp and E. Recknagel, Diam. Relat. Mater. **3**, 137(1994).
2.   H. Hofsäss, C. Ronning, U. Griesmeier, M. Gross, Mater. Res. Soc. Proc. Vol. **354** (1995) 93; and this conference contribution A8.4.
3.   H. Hofsäss, C. Ronning, U. Griesmeier, M. Gross, S. Reinke, and M. Kuhr, Appl. Phys. Lett. **67**, 46 (1995); Nucl. Instr. & Meth. **B 106**, 153 (1995).
4.   F. Qin, X. Wang, Z. Liu, Z. Yao, Z. Ren, L. Lin, S. Su, W. Jiamg, and W.M. Lau, Rev. Sci. Instrum. **62**, 2322 (1991).
5.   I. Bello, W.M. Lau, R.P.W. Lawson, and K.K. Foo, J. Vac. Sci. Technol. **A 10**, 1642 (1992).
6.   R. Shinar, J. Vac. Sci. Technol. **A 10**, 137 (1992).
7.   L.R. Doolittle, Nucl. Instr. & Meth. **B 15**, 227 (1986).
8.   I. Corczyca, N.E. Christensen, E.L. Peltzer and C.O. Rodriguez, Phys. Rev. **B 51**, 11936 (1995).
9.   R.D. Vispute, J. Narayan, H. Wu and K. Jagannadham, J. Appl. Phys. **77**, 4724 (1995).
10.   T.W. Weeks, M.D. Bremser, K.S. Ailey, E. Carlson, W.G. Perry, and R.F. Davis, Appl. Phys. Lett. **67**, 401 (1995).
11.   J. C. Phillips, Rev. Mod. Phys. **42**, 317 (1970), *Bonds and Bands in Semiconductors* (Academic, New York, 1973).
12.   H. Sachdev, R. Haubner, B. Lux, and H. Nöth, Diam. & Rel. Mater. **6** (1997) to be published.
13.   S. Bohr, R. Haubner, and B. Lux, Diam. Relat. Mater. **4**, 714 (1995).
14.   H. Hofsäss, H. Feldermann, M. Sebastian, and C. Ronning, submitted to Phys. Rev. Lett..
15.   S. Reinke, M. Kuhr, and W. Kulisch, Diam. Relat. Mater. **4**, 272 (1995).
16.   D.J. Kester, and R. Messier, J. Appl. Phys. **72**, 504 (1992).
17.   Y. Lifshitz, S.R. Kasi, and J.W. Rabalais, Phys. Rev. Lett. **62**, 1290 (1989).
18.   H. Hofsäss and C. Ronning, in: *Proc. 2nd Int.Conf. on Beam Processing of Advanced Materials*, edited by J. Singh, S.M. Copley, J. Mazumder, (ASM International, Materials Park, 1996) p. 29-56.
19.   P.B. Mirkarimi, D.L. Medlin, K.F. McCarty, and J.C. Barbour, Appl. Phys. Lett. **66**, 2813 (1995).
20.   C. Ronning, E. Dreher, H. Feldermann, M. Gross, M. Sebastian, and H. Hofsäss, Diam. & Relat. Mater. **6** (1997), to be published.
21.   D.J. Kester, K.S. Ailey, R.F. Davis, and K.L. More, J. Mater. Res. **8**, 1213 (1993).

GROWTH OF (0001) ZnO THIN FILMS ON SAPPHIRE

A. J. DREHMAN and P. W. YIP
Rome Laboratory ,USAF, RL/ERX , 80 Scott Dr., Hanscom AFB, MA 01731

ABSTRACT

Using reactive *rf* sputtering, we have grown (0001) oriented ZnO films in situ on heated c-axis sapphire substrates, that are promising, particularly in terms of surface roughness, as buffer layers for the subsequent epitaxial growth of III-V nitride films. We compare the effects of on-axis and off-axis sputter geometries on the film epitaxy and smoothness. We also examined the effect of substrate temperature on the growth and smoothness and quality of the film. X-ray diffraction was used to verify the hexagonal ZnO phase, its c-axis orientation and, qualitatively, the degree of its epitaxy. Atomic Force Microscopy (AFM) was used to determine the ZnO growth morphology and roughness. Our best films, obtained by off-axis sputter deposition, have a surface roughness of less than 1 nm.

INTRODUCTION

Thin film GaN, as well as the mixed Ga-Al-In nitrides, are a promising new family of III-V semiconductors for large band-gap applications such as blue lasers and LED's. The growth of high quality (low defect density) GaN films requires finding a suitable substrate material. Due to having a better lattice match than sapphire, single crystal ZnO may be a good candidate [1-3], especially if it were available in large sizes. Currently, ZnO substrates are very expensive and are not available in large size. One approach to overcoming the difficulty of growing large single crystal ZnO substrates is to deposit a buffer layer of epitaxial ZnO upon another substrate material, such as sapphire, prior to the growth of the nitride film. Such a buffer layer could reduce the number of defects in the GaN film compared to growth directly on sapphire. To be of any value, a buffer layer must be both well oriented and very smooth in order to permit subsequent growth of high quality nitride films. A number of researchers have made ZnO films by sputter deposition on various substrates [4-7], but little attention was given to obtaining an appropriate surface roughness for subsequent epitaxial growth.

SPUTTER DEPOSITION OF ZnO FILMS

Using a 99.95% purity target, ZnO was sputter deposited onto heated (0001) sapphire substrates by two different methods. The first method was conventional on-axis *rf* diode sputtering using a 7.6 cm diameter sputter target which is located approximately 4 cm above the substrate.   The sapphire substrate was mounted on a plate which was radiantly heated by halogen lamps from below. Care was taken to calibrate the actual substrate temperature with that of the plate (which was directly measured by an imbedded thermocouple). Substrate temperatures of 400 to 650°C were used. The sputter gas was argon containing 5 to 20% oxygen to make a total pressure of between 20 and 60 mTorr. A relatively low sputter power was used, 75 to 100 Watts, with a corresponding average dc target bias of 1.2 to 1.4 kV. The resulting deposition rate was between 2 and 4 nm/min and the film thickness was 0.2 to 0.5 μm.

The second deposition method employed off-axis *rf* magnetron sputtering using a 2" sputter source. The sputter geometry is similar to that used for growing high temperature superconductor films and is shown in Figure 1. The substrate was heated from above using a radiant heater, and the sapphire

337

substrate was blackened on its backside (using $YBa_2Cu_3O_x$) in order to absorb the heat [8]. Substrate temperatures of 600 to 750°C were used, and again care was taken to calibrate the actual substrate temperature with that of the radiant heater block. A total pressure of between 80 and 120 mTorr was used with 10 to 20% oxygen in argon. A sputter power of 40 to 70 Watts resulted in an average dc target bias of 200 to 250 volts. Under these conditions we obtained deposition rates of 3 to 6 nm/min and grew films of 0.2 to 0.8 µm thickness.

ON-AXIS SPUTTERING RESULTS

For the on-axis diode sputtering a range of substrate temperatures, gas pressures and gas mixtures were used. It was found that below roughly 450°C one does not obtain pure (0001) oriented ZnO, but that other orientations, especially (1011), begin to be present. Although an exhaustive study was not performed of various sputter conditions, based on ten deposition runs, we found that the surface roughness is not strongly dependent on the sputter conditions. Typically we obtained a surface roughness of 18 to 25 nm (rms). The smoothest (0001) oriented film was obtained using a pressure of 35 mTorr Ar containing 6% O_2 , 75 Watts *rf* power and a substrate temperature of 600°C. Under these conditions a 0.4 micron thick film with a roughness of about 15 nm was obtained (Fig. 2.). Roughness on this scale is inappropriate for the subsequent growth of epitaxial nitride films.

OFF-AXIS SPUTTERING RESULTS

A more complete study of the effects of the sputter conditions was made for off-axis deposited films; over two dozen deposition runs were made using the off-axis geometry. We strove to maintain the (0001) orientation while minimizing the film roughness. We found that in order to obtain the (0001) orientation with off-axis deposition a higher substrate temperature is required than for on-axis deposition. This is probably due to the extra energy imparted to the growing film by the negative ion bombardment in the on-axis geometry, resulting in a form of ion assisted deposition.

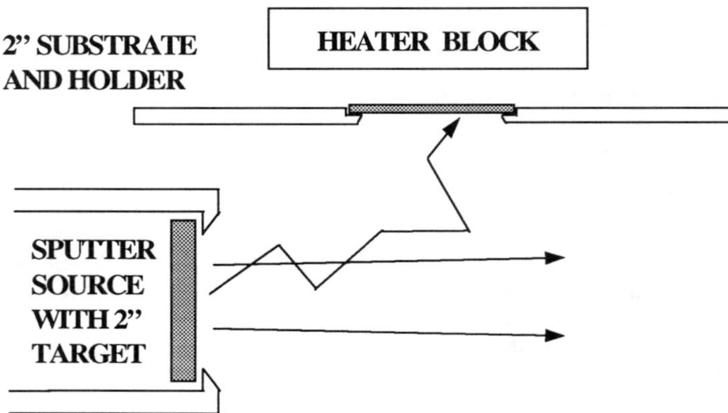

Fig. 1. Off-axis sputter geometry. Negatively charged ions, such as O⁻, are accelerated away from the negatively biased target, but in a direction that does not cause them to impact the substrate. Low energy atoms and molecules arrive at the substrate by diffusion in the gas.

At, and below, a substrate temperature of approximately 640°C we observed that the (0002) diffraction line became very weak (Fig. 4.), indicating a major change in the ZnO morphology; this is discussed later. Films deposited at these lower temperatures had a roughness of 2.5 to 3.5 µm. At temperatures above 700 to 750°C, depending on the oxygen partial pressure, one begins to see the (1011) orientation in addition to the (0001). The smoothest films were obtained using 8 to 15% O_2 in Ar, a total pressure of 90 to 125 mTorr, a substrate temperature of 630 to 660°C and an *rf* sputter power of 40 to 60 Watts. In this range of sputter conditions we were able to reproducibly grow 0.3 to 0.5 micron thick films with a roughness of less than 2 nm, the best being slightly less than 1 nm (rms), as seen in Fig. 2.

ATOMIC FORCE MICROSCOPY

The film morphology was examined using a Digital Instruments Dimension 5000 AFM. Figure 2 compares the roughness of the smoothest on-axis and smoothest off-axis deposited ZnO films. The software calculated the surface roughness from the image and we report the root-mean-square (rms) roughness. The smoothest on-axis deposited film had a roughness of approximately 15 nm while the smoothest off-axis film had a roughness of only 0.97 nm (rms).

Besides being significantly smoother, the off-axis deposited ZnO film has an apparent grain size of 50 to 100 nm compared to 150 to 300 nm for the on-axis film, despite being deposited at a higher temperature. Normally, higher temperature deposition results in a larger grain size, but the on-axis deposition must be enhancing grain growth, probably due to the bombardment of higher energy ions and atoms.

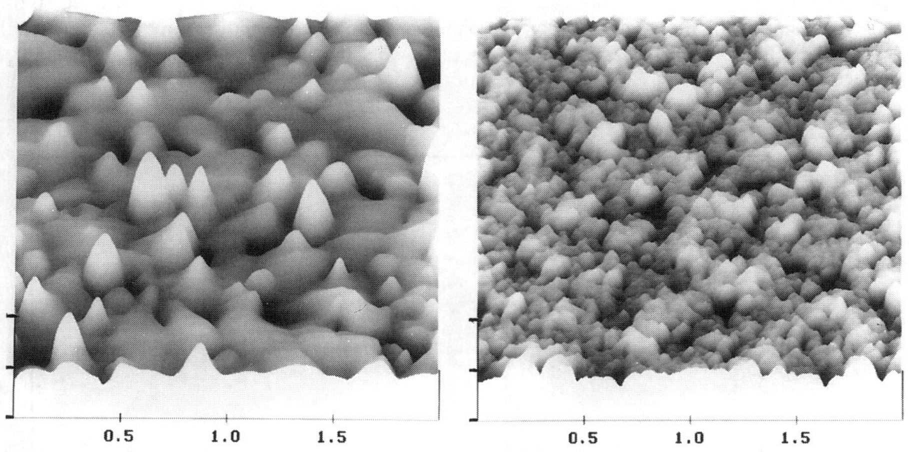

Fig. 2. Atomic force Microscope images of ZnO films. The image on the left is the smoothest on-axis sputter deposited film. The vertical scale is 100 nm per division, and the roughness is 15 nm. The right image is the smoothest off-axis deposited film, displayed with a vertical scale of only 10nm. It has an rms roughness of 0.97 nm. The horizontal scale for both is in µm.

X-RAY DIFFRACTION ANALYSIS

For most of our X-ray analysis we used a powder diffractometer and Cu Kα radiation. Since this instrument performs θ–2θ scans, only the orientations which have atomic planes that are parallel to the plane of the substrate are detected. For ZnO on (0001) oriented sapphire we detected only the (0002) and (0004) ZnO reflections and the (0006) sapphire reflection (range of 20 to 80°), even on a logarithmic plot (Fig. 4.). Several films were sent out for x-ray pole figure analysis to ensure that the observed (0001) orientation was the only orientation present and to determine how the ZnO was oriented in the plane with respect to the sapphire substrate. Scans were performed on both an on-axis and an off-axis ZnO film and no other orientations were detected. Then, using the (1012) reflection of ZnO in comparison with the (0112) reflection of sapphire, it was determined that there is a 30° in plane rotation between ZnO and sapphire about their mutual (0001) axis.

Figure 3. shows a section of a θ–2θ x-ray diffraction pattern which includes the (0002) reflection of (A) the smoothest off-axis ZnO film, and (B) the smoothest on-axis ZnO film. The patterns are typical of all of the (0001) oriented ZnO films: the on-axis films have a sharper peak, resolving the K$_\alpha$ split, than the off-axis deposited films. Although this measurement does not have the resolution of a double crystal rocking curve, it does clearly show that the on-axis films are relaxed and well oriented. In contrast, all of the off-axis deposited films, although smoother, have a more complex morphology, which we believe to be due to strain.

Figure 4. shows the effects of lower temperature off-axis deposition. Only the (0002) and the (0004) ZnO reflections can be seen. All three films are of nominally the same thickness (0.6 µm, within 10%), but the intensity of the (0002) reflection drops by over a factor of 100 going from a 650°C to a 625°C deposition temperature. [In the 600°C film there is a small peak at 36°, the first signs of a second orientation.] This factor of 100 decrease cannot be explained by polycrystallinity as there are no other reflections, nor by the film being partially amorphous, as there would be no reason for the fraction that is crystalline to be aligned in the (0001) direction. The most likely explanation is strain [5], as discussed later.

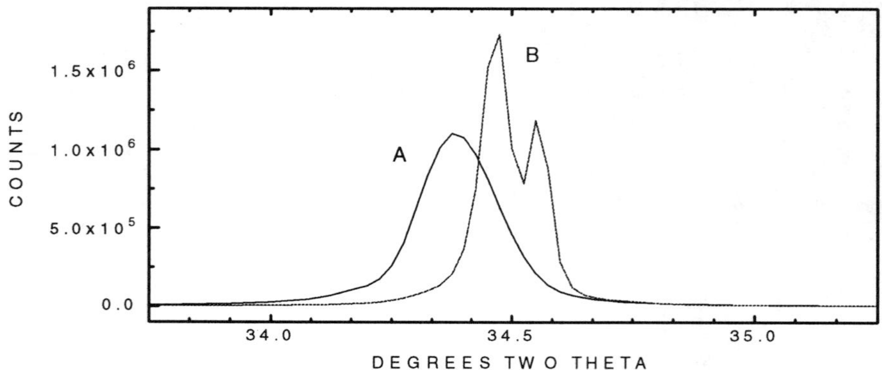

Fig. 3. Composite θ-2θ x-ray diffraction pattern showing the (0002) reflection of (A) off-axis deposited ZnO and (B) on-axis deposited ZnO. It is not certain that the peak shift is significant.

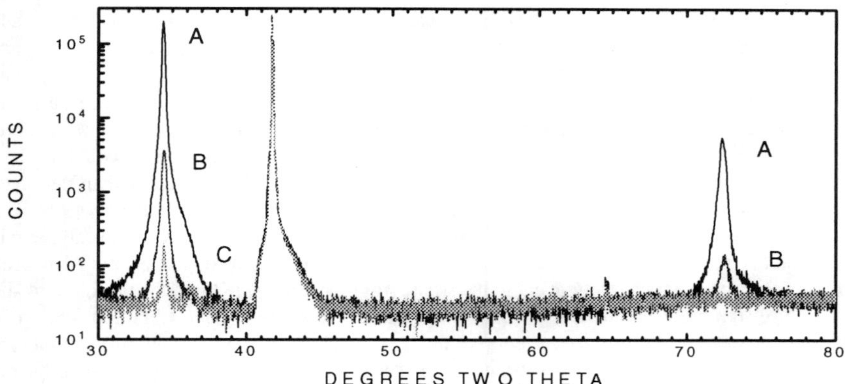

Fig. 4. Effect of substrate temperature on the (0002) and (0004) ZnO reflection of nominally equal thickness films. The scale used is logarithmic to show how large the change in intensity is for films deposited at (A) 650°C, (B) 625°C and (C) 600°C. The peak near 42° is the sapphire (0006) reflection. For the 600°C film there is evidence of another orientation with a peak near 36°.

DISCUSSION AND CONCLUSIONS

We demonstrate that very smooth ZnO films of (0001) orientation can be grown on c-sapphire. These films appear to be good candidates for a buffer layer on sapphire for the subsequent epitaxial growth of III-V nitrides. Although both on-axis and off-axis sputter deposition yielded (0001) oriented ZnO on (0001) sapphire substrates, the off-axis technique appears to be much better suited to obtaining smooth films for subsequent epitaxial growth. The primary difference between the two methods is that during on-axis deposition high kinetic energy atoms or ions are aimed at the depositing film, but in off-axis deposition these high energy particles are aimed away from the growing film (Fig. 1.).

In on-axis reactive sputter deposition of oxides, negative oxygen ions are formed near the target both as a result of the plasma and the oxygen being sputtered off of the ZnO target. As the target has a large negative potential, these ions are accelerated away from the target, roughly perpendicular to the target surface. Once beyond the dark space (the area next to the target which has an extremely weak plasma) these ions may easily become neutralized, but will retain their kinetic energy. Even multiple collisions will result in a bombardment of particles with a kinetic energy of over 10 eV, more than enough to break chemical bonds and enhance surface mobility. There will even be a degree of sputter etching due to these particles.

In contrast, during off-axis deposition, the high energy particles are directed perpendicular and away from the substrate. The higher sputter gas pressure means that the material which deposits on the substrate arrives by diffusion in the gas and therefore has a very low energy (possibly less than 0.1 eV). This results in a very gentle deposition, more like MBE than conventional sputtering. For example, in 1984 Nanto et al. [9] found that ZnO films made by *rf* magnetron sputtering onto unheated substrates had a much lower resistance (by several orders of magnitude) if they were deposited off-axis rather

341

than on-axis. Although their goal was a low resistance transparent film for electro-optic applications, rather than a smooth epitaxial film, they found that the much more gentle off-axis technique allowed the films to deposit without relaxation, and this resulted in significantly lower resistance films.

The broad (0002) ZnO x-ray peak that we observe in the off-axis films could be the result of several different structures. The film might be only highly oriented and therefore have a spread of (0002) orientations within a degree or two of the (0001) sapphire direction. This certainly would not be good for subsequent epitaxial growth of III-V nitrides. A grain size significantly smaller than that indicated by the AFM images could also account for a broad peak, but that would require the grains to be much thinner than the film thickness. Such a morphology would be inconsistent with typical epitaxial, or even columnar growth. The most likely explanation is that the off-axis deposited ZnO films retain a significant amount of strain. A gradient in the strain, from the Al_2O_3-ZnO interface to the film surface, would cause significant broadening of the (0002) peak as there would be a continuous change in the lattice parameter from the bottom to the top of the film. While we are not certain of the significance of the peak shift (roughly 0.1°), it is consistent with expanded ZnO c-axis, even up to the surface of the film, resulting from compressive strain in the <1010> directions at the Al_2O_3-ZnO interface.

The marked decrease in the intensity of the ZnO diffraction peaks for films deposited below 640°C is probably due to strain. A very large decrease (over a factor of 10) in the height of the (0002) ZnO peak has been attributed to internal stress (strain) in ZnO films deposited on Si versus GaAs substrates [5]. Similar strain may be the cause of the drastic (over a factor of 100) weakening of the (0002) and (0004) lines we observed for our ZnO films deposited at 625°C. If strain is the cause, then the off-axis deposition must truly be extremely gentle to preclude relaxation to such a large degree. There is another factor of 100 decrease for the film depositied at 600°C, but as there is a hint of another orientation present it is hard to draw a firm conclusion, but it certainly suggests that there is a further increase in the strain. Further x-ray analysis of the strain, both in the plane and perpendicular, will be the subject of a later report.

REFERENCES

1. T. Matsuoka, N. Yoshimoto, T. Sasaki, A. Katsui, J. Electron. Mater. 21, 157 (1992).

2. E. S. Hellman, D. N. E. Buchanan, D. Wiesmann, I. Brener, MRS Internet Journal, Vol. 1, Article 16 (1996), http://nsr.mij.mrs.org/1/16/complete.html

3. S. Strite and H. Morkoc, J. Vac. Sci. Technol. B. 10, 1237 (1992).

4. Syuichi Takada, J. Appl. Phys., 73, 4739 (1993).

5. Jwo-Huei Jou, Min-Yung Han, and Duen-Jen Cheng, J. Appl. Phys., 71, 4333 (1992).

6. Yasuhiro Igasaka and Hiromi Saito, J. Appl. Phys., 70, 3613 (1991).

7. H. Czternastek, A. Brudnik, M. Jachimowski and E. Kolawa, J. Phys. D: Appl. Phys. 25, 865 (1992).

8. Alvin .J. Drehman, John S. Derov, Jane A. Horrigan, Robert J. Andrews, and Derek S. Linden, IEEE Transactions on Applied Superconductivity, 5, 1793 (1995).

9. H. Nanto, T. Minami, S. Shooji, and S. Takata, J. Appl. Phys., 55, 1029 (1984).

GALLIUM NITRIDE THICK LAYERS :
EPITAXIAL GROWTH AND SEPARATION FROM SUBSTRATES

V.V. BEL'KOV, V.M. BOTNARYUK, L.M. FEDOROV, I.I.DIAKONU,
V.V.KRIVOLAPCHYUK, M.P.SCHEGLOV, Yu.V. ZHILYAEV
A.F.Ioffe Physical-Technical Institute, St.Petersburg 194021, Russia, bel@epi.ioffe.rssi.ru

ABSTRACT

We investigated the possibilities of vapour phase epitaxy in an open tube chloride system for thick GaN film deposition on sapphire substrates. The methodes of the buffer layer deposition were proposed and developed. The methods of fast (up to 100 microns / hour) was developed. Parameters of good quality gallium nitride epitaxy were obtained.

To determine the quality of fast grown epitaxial layers we used X-ray diffraction and photo-luminescence measurements. The halfwidth of the rocking curve for the best samples was equal to 4-6 minutes. Luminescense spectrum (T=77K) had a maximum near 3.46 eV. A signal in the visible wavelength range was hardly observed. Polished layers were transparent.

A special initial treatment of the substrates allowed us to separate thick (up to 300 micron) epitaxial gallium nitride layers from sapphire. It was shown that it is possible to use separated films for homoepitaxy of GaN.

GALLIUM NITRIDE GROWTH

The main aim of the work is to grow thick GaN layers of good quality, which could be used as substrates for fabrication of device structures.

To produce galium nitride layers from the gas phase in the open tube chloride growing system $Ga\text{-}HCl\text{-}NH_3\text{-}H_2$ has been implemented [1]. Liquid gallium of purity not worse than 5N was employed as the III-group element source. Ga was transported to the deposition zone with the help of gaseous HCl of 5N purity. Ammonia of 5N purity served as the nitrogen source. The transporting gas was hydrogen which was purified by passing through a doubled palladium filter. Sapphire wafers (mainly in basal plane orientation) were used as substrates for GaN growth.

Mechanical stresses which arise in substrates from mechanical treatment can affect severely properties of grown GaN layers. For this reason it is important to minimise mechanical stresses in them. With this aim in view the sapphire were treated thermally in a ceramic tube in air at 1800 ^{0}C for 8 hours.

Surface defects of the sapphire substrates were initially removed by chemical etching. A layer of 3-4 μm thickness was removed from the (0001) substrates. The etching rate of the $(10\overline{1}2)$ plane at such etching was extremely low, and it was very difficult to determine accurately the thickness of the removed layer.

GaN layers were grown both with and without a buffer layer. Gallium nitride buffer layers with thickness 20 nm were deposited at temperature 500-550 ^{0}C and were annealed at 900 ^{0}C. Due to variation of the HCl flow rate it was possible to change the gallium nitride growth rate in the range 30-100 μm/hour.

Initial sapphire substrate treatment by $GaCl_2$ vapour at 650 ^{0}C allowed us to separate GaN layers from the substrates. It was shown that separated gallium nitride layers could be successfully used as substrates for following GaN growth.

Mat. Res. Soc. Symp. Proc. Vol. 449 © 1997 Materials Research Society

X-RAY ANALYSIS

X-ray diffraction analysis of the grown GaN layers was carried out by means of a three-crystal spectrometer. Two types of measurements were performed: first, the halfwidth ω_Θ of the rocking curve (also denoted as FWHM) from crystallographic planes different for different directions of the layer growth was determined; second, values of the Bragg diffraction angle were determined with high accuracy (2″-3″). It follows from the Bragg equations $n\lambda=2d\sin\theta$ that $\Delta d/d = -\Delta\theta/\mathrm{tg}\theta$. Knowing the measured value of the angle θ and having calculated its value for GaN one can determine the relative crystal deformation $\Delta d/d$.

Table. Results of X-ray analysis

N	substrate	buffer	GaN thickness, μm	growth rate μm/h	reflection hk′l, GaN	ω_Θ	θ (measur.)
1	sapphire ($10\bar{1}2$)	no	30	30	$10\bar{1}2$	50′	$28^0 59′45″$
2	sapphire (0001)	no	30	30	0002	14′	$17^0 16′10″$
3	sapphire (0001)	no	100	30	0002	8′	$17^0 16′35″$
4	sapphire (0001)	+	30	30	0002	4′20″	
5	sapphire (0001)	no	100	100	0002	25′	
6	sapphire (0001)	+	100	100	0002	13′	$17^0 17′00″$

These experimental results (see Table) allow us to make the following conclusions:
- best results on the values of the rocking curve halfwidth ω_Θ for the (0001) plane orientation indicate that these layers are better ordered, and, therefore, growing GaN layers on such substrates is more advantageous
- lowest values of ω_Θ obtained for GaN / sapphire layers are related to the layers of about 100 microns thickness, which indicates increasing perfection of these layers with thickness
- deposition of low-temperature buffer layers leads to improvment of GaN crystal quality
- GaN grown on sapphire (0001) has (0001) orientation only, depending on the growth conditions the GaN layers / sapphire ($10\bar{1}2$) can be deposited with different orientation
- the epitaxial layer grown on sapphire with the orientation ($10\bar{1}2$) should be considerably thicker than that on sapphire (0001) to get the same crystal quality.

OPTICAL ANALYSIS

1. Raman spectra obtained from a bulk specimen (Fig. 1, curve 1) and a film of GaN (0001) (Fig. 1, curve 2) show that the spectrum of a film specimen repeats mainly the spectrum of a polycrystal, which indicates that the compound produced has a single phase. However this is a well pronounced dominating line in the film spectrum near frequency 570 cm^{-1}.

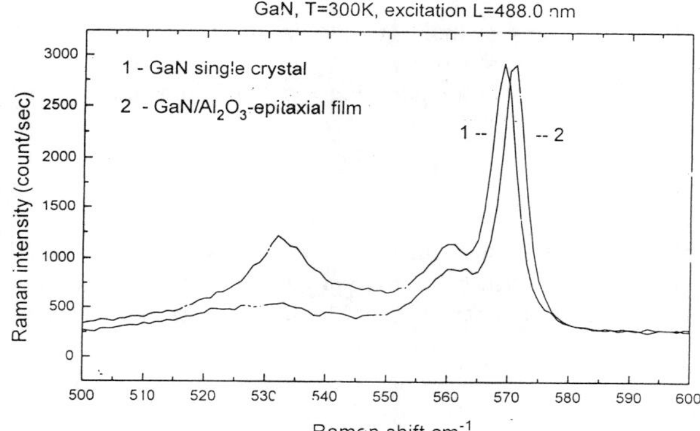

Fig. 1

In correspondence with the selection rules for the backscattering geometry in the parallel polarization case, when the direction of incident and scattering light coincides with the hexagonal optical axis, only one line with frequency near 569 cm^{-1} must be observed. It is known that shifts of the phonon lines with respect to their position for a bulk specimen are as a rule connected with the presence of deformation in the crystal, the low frequency shift being attributed to a stretching deformation and the high frequency shift to a compressing one. Thus, the observed shift by 1 cm^{-1} indicates that the resulting deformation in the specimens investigated has the character of compression.

A small value of the shift allows a conclusion that the hexagonal axis in the investigated film specimens is directed mainly perpendicular to the substrate plane. However these exist a small deviation of the C$_6$ axis from the perpendicular direction. This is indicated by the fact that in the Raman spectra one can see at frequency 533 cm^{-1} the rest of the TO phonon line.

2. To estimate the quality of fabricated epitaxial layers we studied also their photo-luminescence (PL). Investigations of the PL in reflection geometry were carried out at T=77K. Excitation was by the nitrogen laser pulses (λ=3371A). Attention was paid to investigation of the exciton lines. Existence of the exciton lines in the PL spectrum permits to estimate qualitatively the merits of a crystal, since at low temperature excitons are final states of the energy relaxation of electron excitations of a crystal.

The main line of the PL spectra is at the energy 3.462 eV which in correspondence with the data of many authors [2] is related to radiation of the exciton bound on shallow donors (Si). It should be noted that the line, which we attribute to a bound exciton, has a fine structure. In the majority of specimens the line maximum has a doublet structure and has well pronounced spectral peculiarities at both short- and long-wavelength sides from the main line. This indicates the nonuniform character of line widening and makes it impossible to determine correctly the radiation line halfwidth. However taking into account that the position of these peculiarities is the same for the majority of the specimens one can make a conclusion about fixed positions of impurities in a crystalline lattice. Of particular value in the well pronounced peculiarity from the short wavelength side of the bound exciton line. By the energy position and by the similarity with the spectrum shape of other well studied A^{3}B^5 and A^{2}B^6 materials this peculiarity can be ascribed to the free exciton radiation.

Besides, a band related to the donor-acceptor (DA) recombination (3.278 eV) and its LO-phonon replica (3.187 eV) are present in the PL spectra. The radiation in the energy range below

2.3 eV ("yellow band") corresponding to the recombination with participation of deep level impurities either is absent or is very weak.

It is necessary to note that after polishing the layers separated from sapphire substrate are transparent.

CONCLUSIONS

Based on the X-ray, PL and Raman scattering data one can state that the GaN/sapphire epitaxial films grown and studied by us have properties typical of good quality gallium nitride. It is possible to use these thick separated from sapphire layers as substrates for following gallium nitride deposition.

ACKNOWLEDGMENTS

We gratefully acknowledge financial support from AXT.

REFERENCES

1. H.P.Maruska and J.J.Tietjen, Appl. Phys. Lett., **15**, p.327 (1969).
2. S.Stribe and H.Morkoc, J. Vac. Sci. Technol. B, **10**, p.1237 (1992)

SURFACE ENERGY CONSTRAINTS FOR HETEROEPITAXIAL GROWTH ON COMPLIANT SUBSTRATES: MORPHOLOGY OF GaN GROWN ON Sc LAYERS

D.D. KOLESKE, A.E. WICKENDEN, J.A. FREITAS, JR, R. KAPLAN*, AND S.M. PROKES, Naval Research Laboratory, Electronic Science and Technology Division, Washington, D.C. 20375, *Sachs/Freeman Assoc. Inc., 1401 McCormick Dr., Landover, MD 20785

ABSTRACT

An empirical relationship linking surface energy to bulk modulus is presented which suggests that the thermodynamic growth mode for heteroepitaxy on compliant substrates should be 3-D. As an example of this behavior, GaN growth on Sc is shown for various growth conditions. 2-D growth is obtained when the GaN is grown on top of a low temperature GaN nucleation layer. Our results indicate that surface and interface energies, in addition to, lattice matching and thermal matching play an important role in determining the heteroexpitaxial growth morphology of GaN. There appears to be no net reduction in the dislocation density for GaN films grown on the Sc layers, because the GaN film has to be grown on a low temperature nucleation layer.

INTRODUCTION

Recently, GaN has shown promise as a material for blue emitting light devices, and high power and high temperature device applications [1]. For these device applications, smooth films of GaN are necessary and can be achieved on sapphire if the GaN film is grown on top of a thin AlN or GaN nucleation layer deposited at low temperature [2]. Despite the advantage that the low temperature nucleation layer provides in obtaining 2D morphology, the GaN films grown this way have many defects, primarily threading dislocations [1]. The exact role that these defects play in limiting device performance is not know at this time, however, it evident that their reduction should improve the material quality, ultimately leading to better device performance.

Since lattice matched substrates for GaN do not exist, growth on compliant substrates [3-7] have been suggested for GaN growth. Ideally, a compliant substrate would yield and adsorb the defects generated at the heteroepitaxial interface, in response to the strain at the interface. However, the merit of lattice matching between film and substrate has been questioned [8]. Bauer has suggested that the growth mode can be viewed thermodynamically, in terms of surface and interfacial energies [9]. Recently, interfacial bonding was found to play a greater role than lattice matching in determining the epitaxial growth [10] and crystal quality [11] of heteroepitaxial GaN films. Strain can also be generated at the heteroepitaxial interface due to the interfacial bonding between the substrate and thin film. Ideal heteroepitaxial substrates, therefore, should lattice match, thermal match, and bond so that no strain is generated between the substrate and the thin film.

BACKGROUND

The growth criteria based on surface and interfacial energies was initially used by Bauer to explain the growth morphology of metal epitaxy on insulators [9]. It has also been used to explain the growth modes for Si on Ge and Ge on Si surfaces [12], and Ag on MgO [13]. The growth morphology can be predicted by calculating the surface energy difference, $\Delta\gamma$, which is given by,

$$\Delta\gamma = \gamma_F + \gamma_I - \gamma_S \tag{1}$$

where γ_F is the surface energy of the film, γ_I is the interfacial energy and γ_S is the surface energy of the substrate. When $\Delta\gamma > 0$, 3-D growth or Volmer-Weber (VW) type growth is expected and when $\Delta\gamma < 0$, 2-D growth or Frank-van der Merwe (FM) should is expected [9]. This growth critera has been used to explain heteroepitaxial growth, however it ignores orientational ordering of the thin film to the substrate and the detailed bonding at the interface. If $\Delta\gamma \approx 0$, then the growth

Mat. Res. Soc. Symp. Proc. Vol. 449 © 1997 Materials Research Society

may initially start as FM type growth and then transition to VW type growth, which is called Stranski-Krastanov (SK) type growth. At an atomic level, van der Veen has suggested that FM growth is favored when strong bonds form between the 1st layer of the film and the substrate surface [14]. For the subsequent layers the bonding between layers will ultimately resemble the bulk which according to van der Veen will lead to SK type growth [14]. According to the Bauer criteria, film growth can be 3-D, even if the film and substrate are perfectly lattice matched.

To predict the growth mode by calculating $\Delta\gamma$, the values of γ_S, γ_F, γ_I, and need to be estimated. The surface energy and mechanical compliance of any material are directly related to the cohesive energy (i.e. bonding) of the solid. Here, an empirical relationship between the mechanical stiffness (i.e. inverse of the mechanical compliance) and surface energy for various elements and compounds is developed. In Figure 1, values for the surface energy, γ, (y-axis) and bulk modulus, B, (x-axis) are plotted. Values of γ

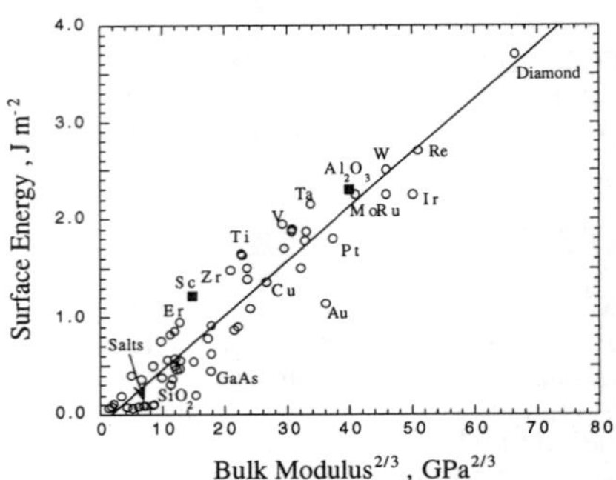

Figure 1. Surface energy, γ, plotted vs. bulk modulus, B, to the 2/3 power for elements and compounds (open circles). The values for Al2O3 and Sc are shown as filled squares. The slope of the line is 0.052 ± 0.01 Jm^{-2}Pa$^{-2/3}$.

and B for sapphire, Sc, and GaN are also listed in Table 1. The values of γ are given for each material at their melting temperature [15], while the values of B are given for the materials at room temperature [16,17]. Both the γ and B of most materials have a negative temperature dependence, i.e. as T increases both γ and B decrease [16-18]. Therefore, at heteroepitaxy growth temperatures below the melting point of the substrate and the thin film, γ will be larger and B will be smaller than shown in Figure 1 and Table 1. The power law relationship between γ and B was obtained by noting that γ is a 2-D quantity of the material while B is a 3-D quantity of the solid. A fit to the data in Figure 1 using the 2/3 exponent gives a prefactor of 0.052 ± 0.01 Jm^{-2}Pa$^{-2/3}$. The ratio of $\gamma/B^{2/3}$ to the prefactor 0.052 is listed in Table 1 for comparison. Note that relationship between γ and B presented here is empirical and is intended for predicting to first order the thermodynamic growth mode for heteroepitaxy on compliant substrates based on surface energy.

By definition a more compliant material should have a bulk modulus less than the heteroepitaxial film, i.e. $B_S < B_F$. For GaN on Sc this is true (see Table I), therefore according to the the relationship developed in Figure 1, $\gamma_S < \gamma_F$, should also be true. The interfacial energy, γ_I,

TABLE I. Values for the surface energy, γ, (column 2) and bulk modulus, B, (column 3) for sapphire, Sc and GaN are given. In column 4 the ratio $\gamma/(B)^{2/3}/0.052$ is calculated. In column 5 the value of the interfacial energy, γ_I, is calculated based on Ref 27, assuming $\gamma_I = (\gamma_S \gamma_{GaN})^{1/2}$ and $\gamma_{GaN} = 2.0$ Jm^{-2}. In column 6 the surface energy difference, $\Delta\gamma$, is calculated from Equation 1.

Substrate, S	γ^a, Jm^{-2}	B^e , GPa	$\gamma/(B)^{2/3}/0.052$	est. γ_I, Jm^{-2}	$\Delta\gamma$, Jm^{-2}
Al₂O₃ (0001)	1.76^b, 2.03^c	252	0.85, 0.98	1.85, 1.98	2.04, 1.90
Al₂O₃ (11_0)	1.86^b, 2.50^c	252	0.90, 1.21	1.91, 2.21	2.00, 1.66
Sc	1.22	56.7^f	1.59	1.54	2.27
GaN	1.95^d	245^g, 237^h	0.96, 0.98	0	0

^{a}From Ref. 15 unless otherwise noted ^{d}see Ref. 20 (2 atom cell for (10_0)) ^{g}see Ref. 22.
^{b}see Ref. 19, from first princples ^{e}From Ref. 16 and 17 ^{h}see Ref. 23.
^{c}see Ref. 19, from interaction model ^{f}see Ref. 21.

between the thin film and the compliant substrate should be positive and lie between γ_S and γ_F. To first order γ_I can be estimated as $(\gamma_S \gamma_F)^{1/2}$ [24], except for homoepitaxy, where $\gamma_I = 0$ by definition. γ_{GaN} has been calculated theoretically to be 1.95 Jm^{-2} for the (10_0) face of GaN [20]. Values for γ_I between various materials and GaN using $\gamma_{GaN} = 1.95$ Jm^{-2} are shown in the 5th column of Table 1. Also, in Table 1 are estimates of $\Delta\gamma$, based on the values for γ_S, γ_{GaN}, and γ_I. Note that the largest $\Delta\gamma$ occurs for Sc which has the lowest γ_S. Because γ is measured at the melting temperature, the magnitudes of $\Delta\gamma$ will be larger than listed in Table 1 at the growth temperature. With the relationships $\gamma_S < \gamma_F$ and $\gamma_I > 0$, the surface energy difference, $\Delta\gamma$, should be > 0, therefore, 3-D growth is expected. While this prediction is valid for heteroepitaxy on clean substrates, substrates with thin surface oxides or nitrides should have surface energies even lower than those shown in Figure 1 and Table 1, implying an even stronger tendency for 3-D growth.

RESULTS

GaN was grown on Sc metal surfaces using four different growth conditions. Sc has been proposed as a compliant substrate for GaN deposition [7].The GaN growth on a-plane sapphire [25] and the Sc surface preparation on this GaN film [7] have both been previously reported. Briefly, Sc was deposited to a thickness of 1000 Å on MOVPE grown GaN(0001) in UHV. The resulting Sc films gave sharp LEED diffraction peaks which indicate that the Sc film was in registry with the GaN film [7]. The Sc films where then transferred to the MOVPE reactor for GaN growth. The Sc surfaces were presumably had an oxide when placed in the MOVPE chamber.High

Figure 2. Resulting growth morphologies for GaN growth on Sc surfaces. A). GaN growth at 1040 °C without nucleation layer, B). GaN growth at 1040 °C with at 300 Å GaN nucleation layer grown at 450 °C, C). thick nucleation layer growth at 450 °C, and D). GaN growth at 1040 °C with at 900 Å GaN nucleation layer grown at 450 °C.

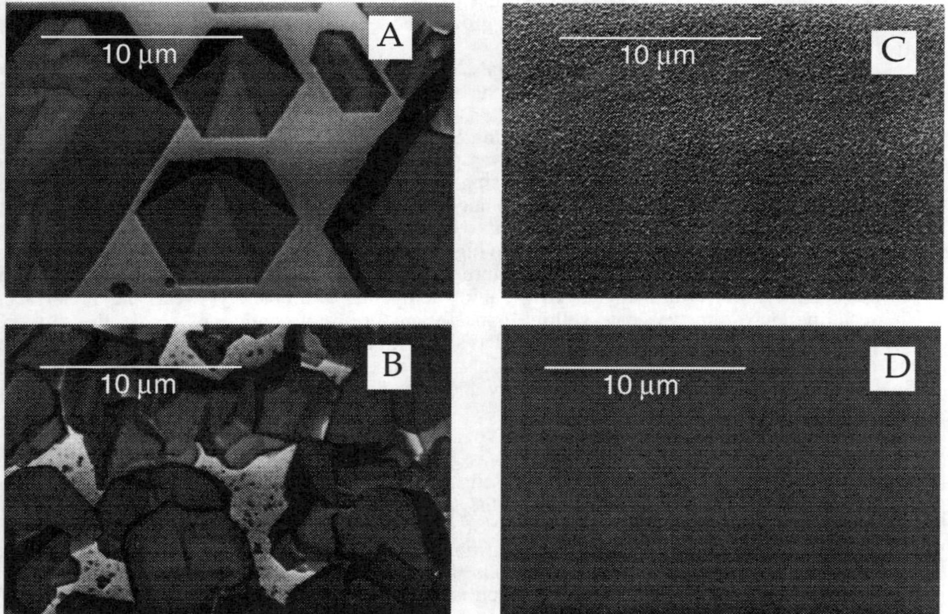

TABLE II. Run conditions that were changed for the growth runs A-D. In column 2, the nucleation layer thickness is listed and in column 3, the growth temperature is listed. Column 4 lists if growth occurred during the temperature ramp from low to high temperature. The morphology is reported in column 5 and the approximate coverage is reported in column 6.

run	nuc. layer d	T_G (°C)	G during ramp	morphology	coverage
A	none	1040	no	3-D hex. pyramids	50%
B	300 Å	1040	no	3-D hex. random	70%
C	5000 Å	450	no	amorphous	uniform
D	900 Å	1040	yes	smooth crystalline	uniform

temperature (1040 °C) GaN was grown on the Sc at reduced pressure (57 torr) using 53 µmole/min of trimethlygallium (TMG) under an NH_3 flow of 2.25 slm. For three of the films, low temperature (450 °C) GaN layers were used to nucleate growth on the Sc films using 13.3 µmole/min TMG and a 2.5 slm flow of NH_3 before growth of the high temperature GaN film. All high temperature grown films were uniformly doped using dilute Si_2H_6 in H_2 [26]. Four growth runs (A-D) where attempted to optimize the 2-D growth of GaN on Sc. Important growth conditions and results are listed in Table 2, and the growth morphologies are shown in Figure 2.

For run A, GaN was grown on the Sc by first heated the surface to 1040 °C followed by the introduction of NH_3 and 1 minute later the TMG. Despite the close lattice match between the Sc and GaN lattices, the GaN grows as isolated hexagonal pyramids with the c-axis oriented normal to the surface. This pyramidal growth is also observed on the c-plane sapphire pieces surrounding the Sc surface, however unlike the growth on Sc the GaN fully covered the sapphire surface. The GaN coverage on the Sc was estimated to be 50 % from SEM images. The Sc surface is easily nitrided at temperatures above 700 °C [7] and the formation of ScN evident by the yellowish appearance of the Sc layer after growth. This growth mode for GaN on Sc occurs because, $\gamma_{ScN} <$ γ_{GaN} and $\gamma_{ScN/GaN}$ is positive.

For film B shown in Figure 2B, GaN was grown at 1040 °C on a 300 Å thick GaN nucleation layer grown at 450 °C on the Sc surface. For GaN growth on a-plane sapphire, this growth sequence produces a GaN film with 2-D morphology, low carrier concentration and high mobility [24]. After the growth, the underlying Sc was yellowish in appearance, again indicating formation of ScN. In this growth the GaN was polycrystalline with a smaller crystallite size than the film A and a 70 % surface coverage. Using Raman spectroscopy the GaN crystallites were found to be wurtzite. While the 300 Å thick nucleation layer produces smooth specular films for GaN growth on sapphire, it does not appear to be thick enough to prevent exposing of the Sc surface to NH_3 during the heating from low to high temperature.

Film C was grown on Sc at a temperature of 450 °C to a thickness of 500 nm. This growth run was conducted to verify that GaN grown at low temperature completely "wets" the Sc surface. The resulting GaN film was pale yellow, optically smooth, and continuous across the surface. SEM images of the surface appeared textured as shown in Figure 2C. The low growth temperature and textured morphology suggest that the GaN is a mixture of amorphous and polycrystalline GaN. The underlying Sc film appeared shiny and metallic, suggesting that the Sc metal was not significantly nitrided during the growth.

Lastly, for film D, 2-D growth of GaN was achieved on Sc metal as shown in Figure 2D. For this growth, a 1000 Å thick GaN film was grown at 450 °C, growth was interrupted for 1 minute while the TMG flow rate was increased from 13.3 to 53 µmole/min., the TMG was then turned by on while continuously increasing the temperature from 450 to 1040 °C. This growth procedure was chosen to prevent exposing the Sc surface to the NH_3. The resultant GaN film was continuous and the Sc appeared metallic, signifying that the Sc was not substantially nitrided. A photoluminescence (PL) spectrum of this film is shown in Figure 3 along with a PL spectrum for an optimal Si doped GaN on a-sapphire which is shown in the inset of Figure 3. For the GaN grown on the Sc, the PL spectra show donor/acceptor pair recombination (near 3.2 eV) which

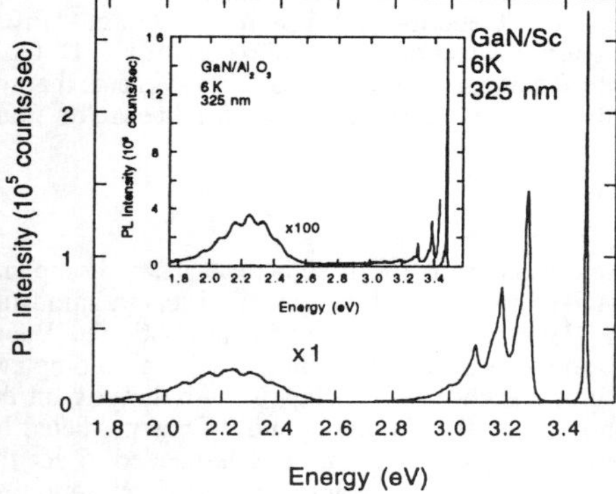

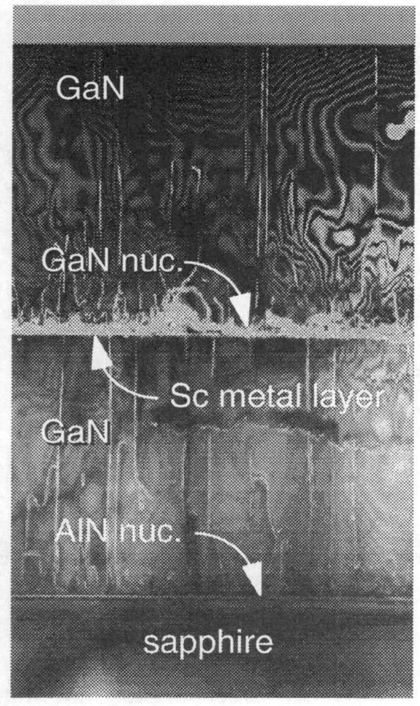

Figure 3. Photoluminescence (PL) spectrum of film D. In the inset a PL spectrum is shown for Si doped GaN on a-plane sapphire.

Figure 4. A cross section TEM of film D, showing from bottom to top the sapphire, AlN nucleation layer, Sc metal layer and GaN nucleation layer, and top GaN film grown at 1040 °C. Note that the dislocations do not propagate through the lower GaN film into the top GaN film. Instead a new network of dislocations is generated at the Sc/GaN nucleation layer interface.

suggests compensation of the GaN film. Because the probe depth of the laser is < 1000 Å, the states are due to the way the GaN was grown on top of the Sc and not due to Sc diffusion through the material during growth. From a planview TEM of film D, no significant reduction in the number is threading dislocations was found compared to typical growth of GaN on sapphire (10^8-10^{10} cm^{-2}). A cross section TEM of the GaN film on Sc is shown in Figure 4. The dislocations which propagate through the lower GaN film do not appear to propagate through the Sc layer into the top GaN film. Instead, a new network of dislocations is formed at the Sc/GaN interface, which then propagate up through the GaN film.

DISCUSSION

It is evident from the growth runs presented in the previous section that a nucleation layer is necessary to obtain 2-D morphology of GaN on the Sc surface. The nucleation layer thickness needed for the growth on Sc is approximately 3.3 times that for GaN growth on sapphire. It appears that if a thinner nucleation layer is used as in growth run B (see Figure 2B), the Sc surface becomes uncovered presumably during the ramp to growth temperature. During this annealing step to high temperature, increased ordering of the nucleation layer has been observed [27], and increased roughness has been observed in AFM images of the annealed nucleation layer [28], suggesting the GaN clusters migrate and coalesce on the surface. In addition, the interfacial bonding of the GaN clusters to the Sc metal may be less (1.56 Jm^{-2}, see Table 1) than for sapphire ($\approx$ 2 Jm^{-2}, see Table 1), suggesting that the clusters on Sc may be more mobile than on the sapphire surface. If the migration and coalescence of the GaN clusters is too large, open patches on the surface may occur if the nucleation layer is too thin. It is also known the GaN deposited in the nucleation layer is randomly oriented on the surface [29]. The growth morphology for film B is likely a result of having the GaN cluster density too low and the subsequent GaN growth is randomly oriented.

It appears from growth runs B and D that a nucleation layer thickness between 300 and 1000 Å is needed to growth 2-D morphology GaN films on Sc, suggesting that a correlation may exist between the nucleation layer thickness and the interfacial energy between the substrate and the GaN film. The surface energy for Sc from Figure 1 is 1.22 Jm^{-2}, while the surface energy for sapphire is 1.95 Jm^{-2}. From these two data points, it appears that the necessary nucleation layer thickness for 2-D GaN morphology decreases as the surface energy (or interfacial energy) increases. This suggests that the magnitude of $\Delta\gamma$ may be useful for estimating the needed nucleation layer thickness for GaN growth on other substrates.

CONCLUSIONS

In this paper, the thermodynamic growth mode for heteroepitaxy on any compliant substrate was demonstrated to be 3D (Stranski-Krastinov). For this, an empirical relationship between the bulk modulus and surface energy is developed and is used along with the Bauer growth criteria to predict the growth mode. An example of heteroepitaxial growth on a compliant substrate, namely MOVPE growth of GaN on Sc layers is shown. For GaN growth directly on the Sc surface at high temperatures, the growth morphology is discontinuous and 3-D as predicted by the Bauer criteria. A thicker (1000 Å) low temperature GaN nucleation layer is needed to get the GaN to completely "wet" the Sc surface compared to the thickness (300 Å) needed on sapphire. Photoluminescence spectra of the 2-D morphology GaN film with a thicker nucleation layer on Sc shows a substantial number of mid level compensation, which is a direct result of the growth conditions used to get the GaN film to grow 2-D on the Sc surface. From a TEM planview of the film no net reduction of the number of dislocations is observed. This is because the GaN films had to be grown on a low temperature GaN nucleation layer, which generates a new set of dislocations. Therefore, because of the constraints on the growth to get 2-D morphology, the compliant Sc layer does not apparently reduce the number of defects in GaN thin films.

ACKNOWLEDGEMENTS

Thanks to Loretta Shirey at the NRL NPF for SEM images and P. Pirouz at Case Western Reserve University for TEM results. Work sponsored by the Office of Naval Research (Colin Wood).

REFERENCES

1. For group III-nitride review articles see, R.F. Davis, Proc. IEEE 79, 702 (1991); S. Strite and H. Morkoc, J. Vac. Sci. Technol. B 10, 1237 (1992); S.N. Mohammad, A.A. Salvador, and H. Morkoc, Proc. IEEE 83, 1306 (1995).
2. H. Amano, N. Sawaki, I. Akasaki, and Y. Toyoda, Appl. Phys. Lett. 48, 353 (1986).
3. D. Teng and Y.H. Lo, Appl. Phys. Lett. 62, 43 (1993).
4. C.L. Chua, W.Y. Hsu, C.H. Lin, G. Christenson, and Y.H. Lo, Appl. Phys. Lett. 64, 3640 (1994).
5. Z. Yang, F. Guarin, I.W. Tao, W.I. Wang, S.S. Iyer, J. Vac. Sci. Technol. B 13, 789 (1995).
6. L.J. Schowalter, MRS Bulletin 21, 45 (1996).
7. R. Kaplan, S.M. Prokes, S.C. Binari, and G. Kelner, Appl. Phys. Lett. 68, 3248 (1996).
8. A.K. Green, J. Dancy, and E. Bauer, J. Vac. Sci. and Technol. 7, 159 (1970).
9. E. Bauer, Z. Kristallogr. 110, 372 (1958); Z. Kristallogr. 110, 395 (1958).
10. T. George, E. Jacobshon, W.T. Pike, P. Chang-Chien, M.A. Khan, J.W. Yang, and S. Mahajan, Appl. Phys. Lett. 68, 337 (1996).
11. T. George, W.T. Pike, M.A. Khan, J.N. Kuznia, and P. Chang-Chien, J. Electron. Mat. 24 241 (1995).
12. P.M.J. Maree, K. Nakagawa, F.M. Mulders, J.F. van der Veen, K.L. Kavanagh, Surf. Sci. 191, 305 (1987).
13. A. Trampert, F. Ernst, C.P. Flynn, H.F. Fischmeister, and M. Rühle, Acta Metall. Mater. 40, S227, (1992).
14. J.H. van der Merwe, "Recent Developments in the Theory of Epitaxy", in the Chemistry and

Physics of Solid Surface V, ed. R. Vanselow and R. Howe, (Springer-Verlag, Berlin, 1984), p. 365.
15. A. Zangwill, Physics at Surfaces (Cambridge, New York, 1988).
16. R. Bechmann and R.F.S. Hearmon, Group-III: Crystal and Solid State Physics, edited by K.-H. Hellwege and A.M. Hellwege, Landolt-Börnstein, New Series, Group III, vol. 1, (Springer-Verlag, Berlin, 1966).
17. R. Bechmann, R.F.S. Hearmon, and S.K. Kurtz, Group-III: Crystal and Solid State Physics, edited by K.-H. Hellwege and A.M. Hellwege, Landolt-Börnstein, New Series, Group III, vol. 2, (Springer-Verlag, Berlin, 1969).
18. H. Wawra, Z. Metallkde 66, 395 (1975), ibid. p. 492.
19. I. Manassidis and M.J. Gillan, J. Am. Ceram. Soc. 77, 335 (1994).
20. J.E. Northrup and J. Neugebauer, Phys. Rev. B, 53, R10477 (1996).
21. R.G. Leisure, R.B. Schwarz, A. Migliori, and M. Lei, Phys. Rev. B 48, 1276 (1993).
22. P. Perlin, C. Jauberthie-Carillon, J.P. Itie, A. San Miguel, I. Grzegory, and A. Polian, Phys. Rev. B 45, 83 (1992).
23. M. Ueno, M. Yoshida, A. Onodera, O. Shimomura, K. Takemura, Phys. Rev. B 49, 14 (1994).
24. J.N. Israelachvili, Intermolecular and Surface Forces, 2nd ed. (Academic, New York, 1992).
25. A.E. Wickenden, D.K. Gaskill, D.D. Koleske, K. Doverspike, D.S. Simons, and P.H. Chi, Mat. Res. Soc. Symp. Proc. Vol. 395, 679 (1996).
26. A.E. Wickenden, L.B. Rowland, K. Doverspike, D.K. Gaskill, J.A. Freitas, Jr., D.S. Simons, and P.H. Chi., J. Electron. Mat. 24, 1547 (1995).
27. A.E. Wickenden, D.K. Wickenden, and T.J. Kistenmacher, J. Appl. Phys. 75, 5367 (1994).
28. S. Keller, D. Kapolnek, B.P. Keller, Y. Wu, B. Heying, J.S. Speck, U.K. Mishra, S.P. DenBaars, Jap. J. Appl. Phys. Part 2, 35, L285 (1996).
29. X.H. Wu, D. Kapolnek, E.J. Tarsa, B. Heying, S. Keller, B.P. Keller, U.K. Mishra, S.P. DenBaars, J.S. Speck, Appl. Phys. Lett. 68, 1371 (1996).

SELECTED ENERGY EPITAXY OF GALLIUM NITRIDE

R. K. CHILUKURI*, SUIAN ZHANG**, E. CHEN*, R. F. DAVIS** AND H. H. LAMB*
*Department of Chemical Engineering, North Carolina State University, Raleigh, NC 27695-7905
**Department of Materials Science and Engineering, North Carolina State University, Raleigh, NC 27695-7905

ABSTRACT

A new apparatus for III-V nitride growth by selected energy epitaxy (SEE) is described. The multi-chamber system comprises a doubly differentially pumped molecular beam source, UHV-compatible growth chamber, x-ray photoelectron spectroscopy (XPS) chamber, UHV transfer line, and loadlock. The growth chamber is equipped for *in situ* quadrupole mass spectrometry and reflection high-energy electron diffraction (RHEED). Preliminary results of GaN SEE using hyperthermal beams of trimethylgallium (TMG) and ammonia (NH_3) are presented.

INTRODUCTION

State-of-the-art monocrystalline GaN films ($\leq 10^8$ defects/cm^2) have been used in the fabrication of blue LEDs and laser diodes. GaN forms a continuous range of solid solutions with AlN and InN allowing, in prospect, the fabrication of laser diodes with emission frequencies across the visible and ultraviolet regions. Heteroepitaxial growth of high-quality monocrystalline GaN thin films has proven difficult due to the lack of suitable lattice-matched substrates and the thermodynamic instability of GaN at typical chemical vapor deposition (CVD) temperatures. Sapphire, the most commonly employed substrate, exhibits a 16% lattice mismatch at the GaN(0001)/sapphire(0001) interface; moreover, the thermal expansion coefficient of sapphire is 25% greater that that of GaN. Only by employing a low-temperature AlN or GaN buffer layer can one obtain state-of-the-art monocrystalline GaN films on sapphire. In conventional CVD and MOCVD,[1,2] substrate temperatures of 1000-1100°C are required to achieve single-crystal growth; however, growth at these temperatures can lead to point defects, as GaN is thermally unstable above 600°C *in vacuo*.[3] Consequently, very large (≥ 1000) V/III precursor feed ratios typically are employed in CVD growth. Plasma-assisted molecular beam epitaxy (PAMBE) has been utilized to lower the monocrystalline GaN growth temperature to 600-700°C, but ion-induced damage and oxygen contamination are often incurred.[4,5]

An alternative approach to GaN heteroepitaxial growth, selected energy epitaxy (SEE), employs hyperthermal molecular beams with narrow translational energy distributions to achieve single-crystal growth at low temperatures.[7-9] Heavy precursor molecules (e.g., triethylgallium (TEG) and NH_3) are seeded in a supersonic free jet expansion of light molecules across an orifice. Ideally, the precursor molecules attain the terminal velocity of the light molecules and thereby acquire hyperthermal kinetic energies (0.5-10 eV). The acquired kinetic energy depends on the heavy-to-light mass ratio and the nozzle stagnation enthalpy; the energy distribution depends on the stagnation temperature, stagnation pressure and orifice diameter. Increasing the surface normal component of kinetic energy is expected to increase the sticking coefficient for precursors that undergo direct dissociative chemisorption;[10,11] whereas, increasing the parallel component may contribute to surface adatom diffusion. Moreover, SEE is better suited than MOCVD or PAMBE for fundamental studies of GaN growth using UHV techniques such as low-energy electron microscopy (LEEM) and photoemission electron microscopy (PEEM).

In this paper, preliminary GaN SEE results using hyperthermal molecular beams of trimethlygallium (TMG) and NH_3 are described. The GaN films were characterized by *in situ* RHEED, on-line x-ray photoelectron spectroscopy (XPS) and scanning electron microscopy

Mat. Res. Soc. Symp. Proc. Vol. 449 ©1997 Materials Research Society

(SEM). *In situ* quadrupole mass spectrometry (QMS) was used to identify the chemical species reaching the growth surface.

EXPERIMENTAL METHODS

<u>Apparatus</u>

The multichamber SEE facility (Figure 1) comprises a doubly differentially pumped molecular beam source, UHV-compatible growth chamber, x-ray photoelectron spectroscopy (XPS) chamber, UHV transfer line, and loadlock. The first differential pumping stage (nozzle chamber) is equipped with an 8000 L/s diffusion pump (Varian VHS-400) which is backed by a Roots blower (Tuthill 3206) and mechanical pump (Welch 1374) in series. The twin stainless steel nozzles feature replaceable laser-drilled apertures and 240-W cable heaters (Watlow) that allow heating to 800°C. Each nozzle is equipped with a *xyz* translation stage which is used for nozzle-skimmer alignment. The conical skimmers (Beam Dynamics) have 1-mm apertures. The base pressure of the nozzle chamber is 10^{-7} Torr.

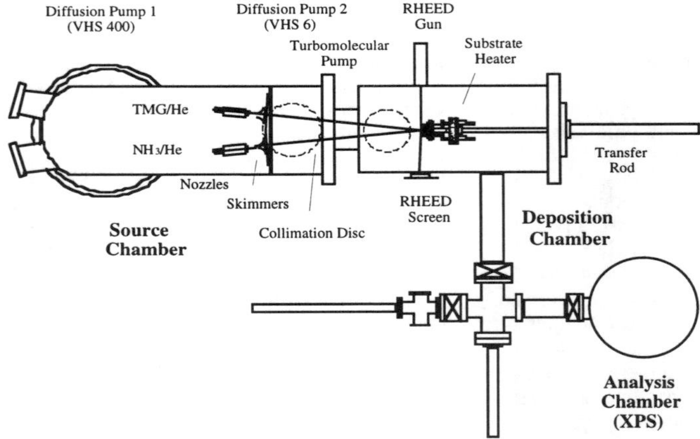

Figure 1.   Top-View Schematic Diagram of SEE facility.

The second differential pumping stage (chopper chamber) is equipped with a 2000 L/s diffusion pump (Varian VHS-6) which is fitted with a fluid-cooled baffle and backed by a mechanical pump (Varian SD700); the base pressure is 10^{-8} Torr. This chamber houses two rotating-disc choppers for time-of-flight (TOF) velocity measurements on seeded supersonic beams. The choppers are removed during film growth experiments. The chopper chamber is separated from the growth chamber by a plate containing two 5×5 mm^2 beam-defining apertures.

The UHV-compatible growth chamber is equipped with a hybrid turbomolecular/drag pump (Balzers 520M) and a PHI titanium sublimation pump (TSP); the base pressure is $<10^{-9}$ Torr. The growth chamber houses a triple-filter quadrupole mass sensor (Hiden HAL/3F 301 PIC) mounted on a rotatable table (MDC) and equipped with a *z*-motion stage for beam alignment. *In situ* reflection high-energy electron diffraction (RHEED) capability is provided by a Fisons LEG 110 15-kV electron gun and 100-mm Al-coated phosphor screen. The substrate heater assembly is fabricated from Mo and mounted on a magnetically coupled transfer rod. The heater, which is designed for temperatures to 1200°C, contains a PBN-coated graphite heating element. The Mo housing incorporates rear heat shields and is designed to absorb torque-induced stresses. The heating element is connected to a DC power supply (HP 6269B) via copper electrical feedthroughs and transfers heat via radiation to a solid Mo substrate holder (traveler). Sample introduction is via a small load-lock chamber that is evacuated using a Drytel 31 molecular drag/membrane pump. Subsequently, the sample is transferred to either the growth chamber or the UHV surface analysis chamber via a UHV transfer line; the transfer line base pressure (10^{-9} Torr) is established using a cryopump (APD Cryogenics, APD-4).

The surface analysis chamber is equipped with a PHI 3057 XPS system comprising a 10-360 spherical capacitor analyzer (SCA), Omni Focus III fixed-aperture lens, 16-element multichannel detector, and 257 DR11 PC interface card. A PHI 1248 x-ray generator with dual-anode (Al/Mg source is used. Non-destructive depth profiling is achieved by angle-resolved XPS (ARXPS): The sample is mounted on a tilt stage which is attached to a precision *xyz*-rotary manipulator (Thermionics) which is used to vary the photoelectron take-off angle. The analysis chamber is pumped by a Perkin-Elmer TNBX ion pump/TSP combination and has a base pressure of 2×10^{-10} Torr. The XP spectra reported herein were measured using Al Kα radiation using an input power of 300 W.

<u>Sample Preparation and GaN Growth</u>

Sapphire (0001) was used as the substrate for heteroepitaxial growth of GaN. Pieces (give size) from a X-in. sapphire(0001) wafer (Crystal Systems) were cleaned using the following procedure : a 15-min dip in an 80°C bath containing a 50:50 mixture of 85% H_3PO_4 and concentrated H_2SO_4, a 2-min DI water rinse followed by a 5-min dip in 10% HF solution and a final 2-min DI water rinse. Samples were blown dry with LN_2 boil-off before mounting on Mo travellers. Sapphire samples were cleaned *in situ* heating at 600°C *in vacuo* for 1 h followed by exposure to a NH_3/He seeded beam for 30 min at 600°C.

Table 1.   Source Conditions for GaN Growth Experiments

	TMG beam	NH_3 beam
Orifice Diameter (μm)	50	150
Stagnation Temperature (°C)	80	600
Stagnation Pressure (Torr)	660	515
Bubbler Temperature (°C)	-10	-
Total He Flow Rate (sccm)	40	100
Precursor Flow Rate (sccm)	0.53	14.5

After *in situ* cleaning, deposition was initiated by lowering the substrate temperature to 500°C and turning on the TMG/He seeded beam. Typical TMG and NH_3 source conditions are given in Table 1. These conditions were maintained for 1 h to facilitate nucleation of GaN on sapphire. Subsequently, the substrate temperature was increased to 600°C, and GaN was grown for 4 h. Growth was terminated by turning off the TMG/He beam and cooling the sample under the NH_3/He beam to ≤300°C.

RESULTS AND DISCUSSION

The RHEED patterns from sapphire (0001) substrates after *ex situ* cleaning and heating *in vacuo* for 1 h at 600°C typically showed some streaks against a relatively bright background. XPS confirmed the presence of residual carbon contamination. In contrast, a clear RHEED pattern with sharp streaks was observed after subsequent *in situ* exposure to a NH_3/He seeded beam for 30 min at 600°C. Moreover, XPS confirmed a reduction in the surface carbon concentration.

The XP spectrum of a GaN film deposited on sapphire (0001) using hyperthermal molecular beams of TMG and NH_3 is shown in Figure 2(a). For comparison, the XP spectrum of 5-μm thick GaN sample deposited on sapphire using seeded supersonic free jets of TEG and NH_3 is shown in Figure 2(b).[8] The spectra have been normalized with respect to the Ga($2p_{3/2}$) XPS peaks. The strong O(1s) and C(1s) peaks in the reference sample are explained by long-term exposure to the atmosphere. The XP spectra are closely similar which indicates that the film deposited by SEE is continuous with a thickness larger than the escape depth for substrate Al(2p) photoelectrons ($\approx$30 Å). Interference between the Ga(LMM) Auger peaks and the N(1s) photoelectron peak does not allow us to determine the surface Ga:N stoichiometry; however, both films appear to be Ga-rich, as a significant N(1s) shoulder is not observed. The C(1s) and O(1s) peaks for the SEE film indicate significant concentrations of these impurities.

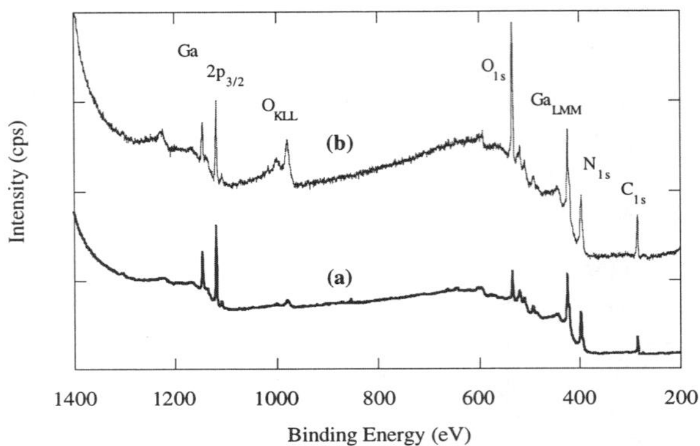

Figure 2.   X-ray photoelectron spectra of GaN films grown on sapphire (0001) using (a) hyperthermal beams of TMG and NH_3 (b) using seeded supersonic jets of TEG and NH_3.

Figure 3.   Inclined-view SEM image of GaN film grown using hyperthermal beams of TMG and NH_3.

An inclined-view SEM image of the GaN film deposited using hyperthermal beams of TMG and NH_3 is shown in Figure 3. The film thickness and the growth rate are estimated to be 300 nm and 60 nm/h, respectively. The film surface is covered by 3-dimensional structures that may be Ga clusters. In cross-section, the film appears to have a columnar growth morphology. The RHEED pattern of the SEE film contained streaks which were broader than those observed from the clean sapphire(0001) surface. X-ray diffraction patterns could not be obtained for the film due to its small thickness.

The quadrupole mass spectrum of a TMG-seeded He beam emanating from a 35°C nozzle contains peaks at mass-to-charge ratios of 115, 100, 85 and 70, which are consistent with the expected fragmentation pattern of TMG. This result indicates that TMG was transported to the growth surface without decomposition. In contrast, when the NH_3-seeded He beam was studied by QMS, significant NH_3 decomposition was observed at nozzle temperatures greater than 350°C. Figure 4 illustrates the dependence of the beam composition on nozzle temperature; the stagnation pressure was 535 Torr, and the NH_3 and He flow rates were 5 and 100 sccm, respectively. The QMS signals of the expected NH_3 decomposition products, N_2 and H_2, are seen to increase with temperature, exhibiting approximate Arrhenius relationships above 350°C. Conversely, the NH_3 and NH_2 signals decrease with increasing temperature over the same range. The NH and N QMS signals are very weak, and there is no evidence of NH_x fragments in the NH_3/He beam under any of these conditions.

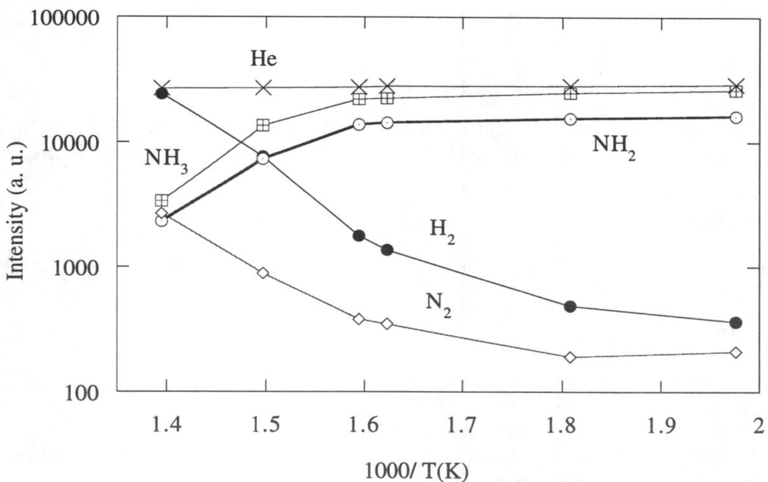

Figure 4.   Arrhenius plot of QMS signal intensity for a 5% NH_3 in He nozzle expansion; the nozzle stagnation pressure was 535 Torr.

The extent of NH_3 decomposition exhibits saturation behavior with respect to the NH_3 mole fraction (Figure 5) in the nozzle feed gas. The nozzle temperature and stagnation pressure were 410°C and 550 Torr, respectively, in this series of experiments. This result strongly suggests that NH_3 decomposition is catalyzed by the inner surfaces of the stainless steel nozzle. NH_3 decomposition in the nozzle is not expected to significantly impact the GaN growth chemistry, since ground-state molecular nitrogen is much less reactive than NH_3; however, NH_3

decomposition will obviously reduce NH_3 flux. We are currently seeking alternative materials of construction to inhibit surface-catalyzed NH_3 decomposition and examining a possible role of NH_3 decomposition in our previously reported GaN deposition results using seeded supersonic free jets.[7]

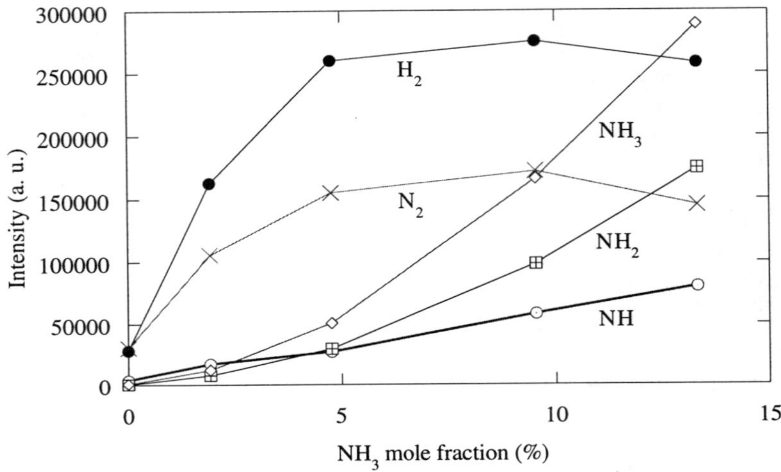

Figure 5.   Dependence of QMS signal intensities on NH_3 mole fraction for NH_3/He nozzle expansion. The nozzle temperature and stagnation pressure were 410°C and 550 Torr, respectively.

ACKNOWLEDGMENTS

The authors thank the Office of Naval Research for support under contract N00014-95-1-0319 and N00014-95-1-0122.

REFERENCES

1.   J. I. Pankove, Mater. Res. Soc. Proc. **162**, 515 (1990).
2.   S. Nakamura, Jpn. J. Appl. Phys. **30**, L1705 (1991).
3.   M. R. Lorenz and B. B. Binkowski, J. Electrochem. Soc. **109**, 24 (1962).
4.   C. R. Eddy, Jr., T. D. Moustakas and J. Scanlon, J. Appl. Phys. **73**, 448 (1993).
5.   C. Wang and R. F. Davis, Appl. Phys. Lett. **63**, 990 (1993).
6.   K. A. Pacheco, B. A . Ferguson, C. Li, S. John, S. Banerjee and C. B. Mullins, App. Phys. Lett. **67**, 2951 (1995).
7.   H. H. Lamb, K. K. Lai, V. Torres and R. F. Davis, in "Film Synthesis and Growth Using Energetic Beams", MRS Symp. Proc. **388,** 265 (1995).
8.   J. J. Sumakeris, R. K. Chilukuri, R. F. Davis and H. H. Lamb, in "Gallium Nitride and Related Materials", MRS Symp. Proc. **395** (1996).
9.   K. A. Brown, S. A. Ustin, L. Lauhon and W. Ho, J. Appl. Phys. **79**, 7667 (1996).
10.   S. T. Ceyer, Science, **249**, 133 (1990).
11.   J. R. Engstrom, D. A. Hansen, M. J. Furjanic and L. Q. Xia, J. Chem. Phys, **99**. 4051 (1993).

MATERIAL AND DEVICE CHARACTERISTICS OF MBE GROWN GaN USING A NEW RF PLASMA SOURCE

R. BERESFORD*, K. S. STEVENS*, Q. CUI*, A. SCHWARTZMAN*, AND H. CHENG**
*Division of Engineering and Center for Advanced Material Research, Box D, Brown University, Providence, RI 02912
**EPI MBE Products Group, 1290 Hammond Rd., St. Paul, MN 55110

ABSTRACT

A new rf plasma nitrogen source has been characterized for growth of GaN on basal-plane sapphire by molecular beam epitaxy. For rf power of 500 W and N_2 flow rate of 2 sccm, a maximum GaN growth rate of 0.80 µm/hr is obtained, implying a source efficiency greater than 5%. It is found that the GaN surface roughness is extremely sensitive to V:III ratio near unity and independent of growth rate in the range 0.3–0.8 µm/hr. Roughness as small as 1.0 nm (rms) is measured by atomic-force microscopy. Microstructure of the high-growth-rate films is similar to other GaN films, as observed in cross-section transmission electron microscope images. The electroluminescence spectra from homojunction light-emitting diodes exhibit a band of near-ultraviolet emissions corresponding to the energy separation of the intentional donor and acceptor levels on the two sides of the junction. The intensity of these emissions relative to the visible spectrum increases with drive current density, implying saturation of deep trap levels responsible for the visible light output.

INTRODUCTION

Epitaxial growth of the group III nitrides by molecular beam epitaxy (MBE) has been demonstrated by numerous investigators. However, the film properties as judged by device demonstrations [1, 2, 3] are generally inferior to what has been achieved via metalorganic chemical vapor deposition (MOCVD) methods [4], which are now in commercial use for light-emitting diode production and are yielding blue laser diodes of rapidly improving performance [5]. (see R. P. Vaudo *et al.* [6] for a recent exception to this trend.) It is not yet clear which issues are most important in understanding this discrepancy between MBE and MOCVD. The presence of energetic particle species from plasma nitrogen sources appears to interfere in the growth front [7, 8, 9]. On the other hand, MBE growth from thermally cracked ammonia is not dramatically improved compared to plasma-source growth [10, 11], except in that a high growth rate can be achieved. Therefore it may be considered that the growth temperature is a more important factor than the type of nitrogen precursor.

MOCVD growth is conducted at higher temperatures, typically 1050 °C versus 750 °C for MBE. Because of the sublimation of GaN in vacuum above 800 °C, large overpressures of active nitrogen would be required to conduct MBE growth at higher temperatures. For purposes of estimation, we will assume that to stabilize GaN against sublimation in vacuum, a pressure of active nitrogen at least equal to the measured vapor pressure is needed. Munir and Searcy [12] found the N_2 pressure above GaN to follow

$$\log P(atm) = 5.699 - \frac{15,923}{T(K)}.$$

For growth at 1050 °C, this pressure is 3.5×10^{-4} Torr. The available active nitrogen pressure in MBE is not known directly, since the plasma source is inefficient and most of the measured pressure is due to background N_2. However, we can make a comparison by assuming the thermal average velocity of N_2 molecules at room temperature, 4.75×10^4 cm/s. Then the sublimation rate corresponding to the vapor pressure 3.5×10^{-4} Torr works out to be 1.3×10^{17} cm^{-2} s^{-1}. Based on the known GaN lattice parameters, this sublimation rate removes the film at 110 µm/hr, more than

Mat. Res. Soc. Symp. Proc. Vol. 449 © 1997 Materials Research Society

two orders of magnitude greater than the growth rates that are achieved with existing sources. Therefore, we can estimate that growth at such high temperatures using existing source technology would require at least a hundred-fold increase in the active nitrogen pressure and therefore also a hundred-fold increase in the background N_2 pressure, into the milliTorr range, making it impossible to use thermal beams. This approximate analysis shows that plasma-source MBE at present may be inherently limited to a lower growth temperature.

It might be supposed that at such lower temperatures, the growth rate must be reduced in order to maintain film quality. Such reasoning is based on the reduction in surface diffusion currents with temperature. In this report, we show to the contrary that good MBE films can be grown at the relatively high rate of 0.8 µm/hr. This rate appears to be among the highest attained yet for plasma-source MBE grown films. The film quality is confirmed by fabricating *pn* junction diodes and obtaining the room-temperature electroluminescence spectra of the devices. Recombination radiation associated with the intentional donors and acceptors is observed. In addition, atomic-force microscopy and transmission-electron microscopy observations are presented to show that good film characteristics can be achieved, with rms surface roughness of 1.0 nm, and threading dislocation density of 6×10^9 cm^{-2}. These results are comparable to or better than other MBE grown films that were reported to require much slower growth rates. The achievement of a high growth rate will be helpful by making it feasible to optimize thick GaN buffer layers for improved crystallinity and therefore better optoelectronic device performance.

EXPERIMENT

The experiments were conducted in an EPI 930 MBE system fitted with the EPI UNI-Bulb rf nitrogen source in one of the effusion cell ports. In a series of characterization runs, the nitrogen plasma emission spectra were monitored via a Si mirror mounted on the transfer arm and a viewport on the source flange. The spectra exhibit the first positive series lines of molecular nitrogen and intense atomic nitrogen peaks, consistent with results from other rf plasma sources [13]. The intensity of the atomic nitrogen peaks was observed as a function of the rf power and the gas flow rate. For an rf power of 500 W, which is the maximum available in our present configuration, the N intensity has a broad maximum between 1–2 sccm and decreases slightly at higher flows. Above 3 sccm, the plasma cannot be maintained reliably. From pyrometer measurements of the pyrolytic BN (pBN) plasma bulb immediately after switching off the rf power, the temperature of the gas in the source is estimated to be 720 ± 30 °C. At this temperature, the viscosity and thermal average velocity of N_2 are taken to be 4.02×10^{-4} g/cm-s and 8.66×10^4 cm/s, respectively. Using this data and the computed conductance of the exit apertures of the pBN plasma bulb, the pressure in the source is approximately 45 mTorr per sccm of inlet flow.

With a 2 sccm flow rate, the chamber pressure is 2.2×10^{-5} Torr. For the maximum GaN growth rate of 0.80 µm/hr, the incorporation rate of nitrogen atoms is 9.7×10^{14} cm^{-2} s^{-1}. Assuming that the pressure measurement detects room-temperature nitrogen molecules, a pressure of 1.27×10^{-6} Torr accounts for the incorporated nitrogen. The remaining (inactive) nitrogen pressure corresponds to an impingement rate of background nitrogen (normalized to atoms) of 1.58×10^{16} cm^{-2} s^{-1}. Comparing the incorporation rate to the total impingement rate (active plus inactive) is a measure of source efficiency, which in this case is estimated to be 5.8%.

GaN growths were conducted on 1-cm^2 basal-plane sapphire substrates cut from 3-in diameter wafers. Before slicing the samples, the backside of the sapphire wafer is coated with approximately 4–6 µm of Mo by sputter deposition in order to provide an infrared-absorbing surface. The samples are solvent cleaned, etched in H_2SO_4:HPO_3 (3:1) at 100 °C, and then outgassed in vacuum at 850 °C or higher prior to the growth. In contrast to our experience with an electron-cyclotron resonance plasma source, the sapphire substrate is not nitridized upon exposure to the rf plasma, as the reflection high-energy electron diffraction (RHEED) patterns do not change. A thin buffer layer of AlN therefore is grown first. This layer is nominally 100 Å thick grown at 0.05 µm/hr and 700 °C. Then the temperature is reduced to 600 °C and a 200-Å thick GaN nucleation layer is deposited. After the temperature is ramped up to 750 °C at 20 °C/min, the

GaN growth is restarted. This procedure is similar to other nucleation approaches reported, but is not necessarily optimized.

For a range of Ga flux values, the GaN growth rate is controlled by the cation arrival rate and is simply proportional to Ga flux. As the Ga flux is increased further, the film growth rate does not increase and excess Ga appears condensed in droplets on the surface, indicating that the available nitrogen is exhausted. Scanning electron microscope (SEM) observations of cleaved samples from several growths enable determination of a maximum GaN growth rate, which occurs for a V:III ratio of unity (all activated nitrogen is consumed and no excess Ga remains). We assume that the sticking coefficient of activated nitrogen is unity in the presence of excess Ga. For 500 W rf power, the maximum rate is 0.80 μm/hr, while for 200 W rf power, the maximum rate is 0.35 μm/hr.

RESULTS

After carrying out the types of calibrations just described, growths at a specified rate and V:III ratio can be conducted by scaling the Ga beam equivalent pressure and adjusting the rf power. Such experiments have clearly shown that the surface morphology of GaN is extremely sensitive to V:III ratio in the range near unity but essentially independent of the growth rate in the range 0.30–0.80 μm/hr. For balanced V:III fluxes or slightly Ga-rich conditions, the GaN surface is very smooth and featureless (Figure 1). The measured surface roughness (rms) is 1.0 nm. For N-rich conditions, surface topography of a varied and extreme nature may develop with appearances that could be described as orange-peel, faceting, crystallites, domains, or sometimes whiskers. The typical rms surface roughness as shown in Figure 1b is 18 nm. Although we refer

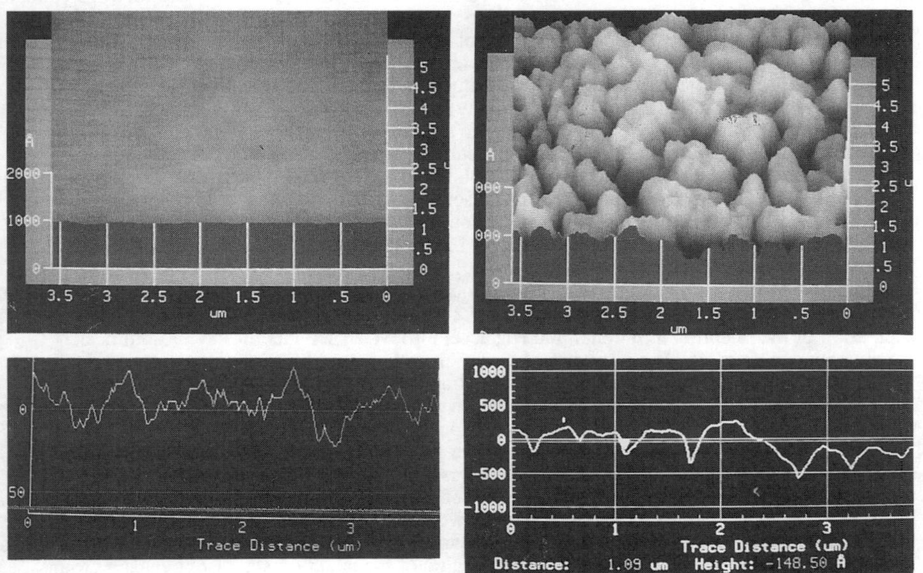

Figure 1. Atomic-force microscope observations of GaN surface morphology comparing Ga-rich (a) and N-rich (b) growth conditions. The corresponding V:III flux ratios are approximately 1.0 and 1.2, respectively. Representative line scans are given in (c) and (d). Note the different vertical axis scales. The rms roughness in (c) is 1.0 nm versus 18 nm in (d).

to the smooth surfaces as arising for V:III ratios of unity, the calibrations used are not very precise and the exact V:III ratio is not known. What is clear is that the transition in surface morphology occurs over a narrow range of values of the V:III ratio and that this range of Ga flux is also where the growth rate saturates. Therefore it appears that a high surface mobility of Ga is responsible for the surface smoothing.

Carrier type, density, and mobility were determined by room-temperature Hall effect measurements on 1-μm thick GaN films. The as-grown films are n type, with electron concentration that depends on the V:III ratio used in growth. The smooth and featureless films produced at unity flux ratio have a higher carrier density, typically 1×10^{18} cm^{-3}, which is consistent with excess Ga insofar as N vacancies are donors. For slightly larger V:III ratio, a background electron concentration as low as 8×10^{16} cm^{-3} has been observed. The measured mobility was 90 cm^2/V-s. This low value may represent the effects of microstructural disorder, which is especially prevalent in the initial 1 μm of film thickness. This residual electron density is nonetheless low enough to enable p type counter doping by Mg. Previous analysis of GaN growths identified an Mg flux that yields 2×10^{19} cm^{-3} atoms incorporated, as determined by secondary ion mass spectroscopy. Films grown with this Mg flux are p type as grown, with a hole density of 2×10^{17} cm^{-3} and a mobility of 1 cm^2/V-s. Intentional doping with Si has produced electron densities as high as 2×10^{18} cm^{-3}. Ti/Al was used to contact the n -type films; Ni/Au was used on the p -type films.

Large-area (10^{-3} cm^2) junction diodes were fabricated from these layers by growing 0.5-μm of p -GaN:Mg on top of 1.5 μm of n -GaN:Si. The top electrode metalization of Ni/Au/Ti was used as a mask for Ar ion milling of a mesa structure. Then a second shadow-mask metalization was completed to contact the exposed n -type electrode. The structure is shown in the inset of Figure 2a, which gives the room-temperature current-voltage characteristics. The forward voltage at 20 mA current is 4.0 V. The log-log plot also shown as an inset in Figure 2a makes it clear that the forward current characteristic is not exponential, nor is it a simple power law. There is no hard breakdown in reverse bias, instead the reverse current increases continuously. These characteristics are different from reports of commercial InGaN diodes, which showed an exponential characteristic with a large nonideality factor [14].

The room-temperature electroluminescence spectra for several pulse-mode drive current levels are shown in Figure 2b. For the highest current density that could be achieved in dc operation, the spectrum was similar to the one shown for 300 mA drive current. For consistent comparisons, all of the spectra are given for pulsed operating conditions. At higher current levels the device fails due to heating, probably at the contact to the p -type layer. Four main emission bands are found near 2.1, 2.4, 2.6, and 3.2 eV. Room-temperature observations of the 3.2-eV band have not been previously published, although they were referred to by Johnson et al. [3]. These earlier results involved an AlGaN/GaN double heterostructure, MOCVD-grown GaN buffer layers on SiC substrates, and a growth rate of 0.2 μm/hr, four times slower. The energy separation of the intentional Si donor and Mg acceptor levels accounts for the 390-nm peak wavelength of this band. The intensity relative to the other emissions increases very markedly as the drive current is increased, indicating that the trap levels responsible for the visible emissions can be saturated. The near-ultraviolet band has been observed in photoluminescence measurements previously, and is clearly associated with the Mg acceptor level in that context.

For a further point of comparison with other growth efforts, cross-section transmission-electron microscope observations of the GaN films were made as shown in Figure 3. The microstructure is characterized by a high density of threading dislocations close to the interface. As has been noted before, during the first μm of growth many of the line defects annihilate, leaving the film with a network of vertical domain boundaries that represent stacking sequence errors. The higher magnification view in Figure 3b from a region of the better crystal also reveals a group of horizontal stacking faults in one of these domains. By the usual estimation methods, the line defect areal density is about 6×10^9 cm^{-2}, similar to other reported GaN device layers.

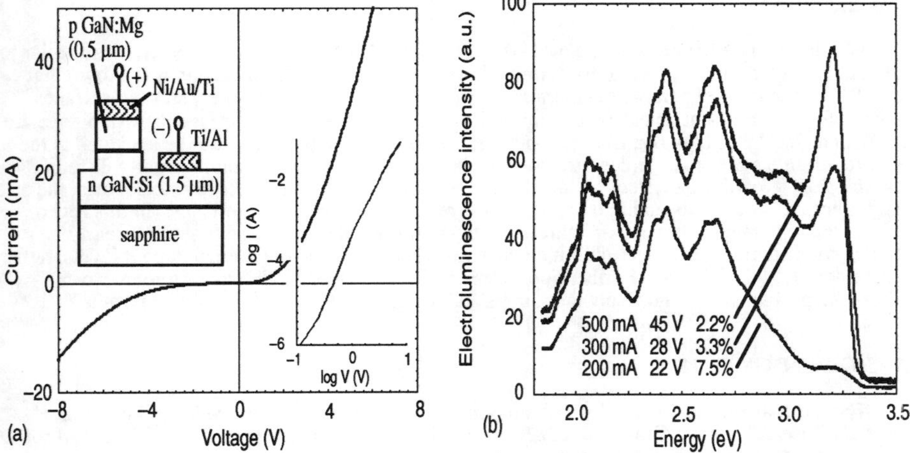

Figure 2. GaN homojunction diode current-voltage characteristic (a), with insets showing the device structure and the forward current on a log-log scale. The forward voltage at 20 mA is 4.0 V. Room-temperature electroluminescence spectra for several pulse-mode current densities are given in (b), labeled by the pulse current, voltage, and duty cycle. The device area is 10^{-3} cm^2. The pulse width is 400 ns. The emission band near 3.2 eV is ascribed to the intentional Si donor and Mg acceptor levels on either side of the junction.

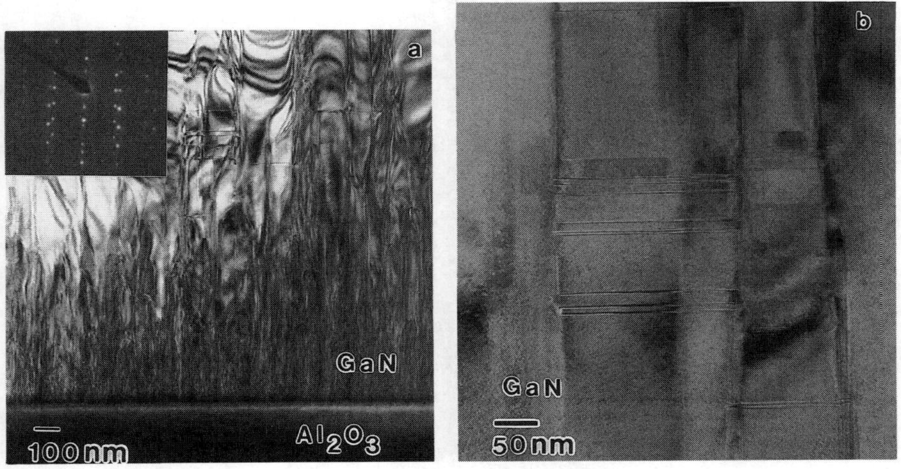

Figure 3. Bright-field cross-section transmission electron microscope image of GaN/sapphire (a) and a higher magnification view (b) showing a set of stacking faults. A "typical" GaN microstructure is evident, with a high density of dislocation lines threading normal to the interface.

CONCLUSIONS

We find that a high growth rate of 0.80 µm/hr yields GaN films with characteristics similar to those achieved by other investigators at much lower growth rates. Furthermore, it is shown that the V:III ratio is the key parameter determining surface morphology, with very smooth surfaces obtained for flux ratios at or just below unity. The conclusion is that Ga surface diffusion rates are sufficient at 750 °C to obtain surface smoothing. Doping of both polarities can be achieved at the high growth rate and the film microstructure as observed by TEM is not fundamentally different from the slow-growth case. Because the GaN layers are wanted for optoelectronic devices, the final judgment should reasonably be based on the resulting device characteristics. In this regard, we have demonstrated a "precursor" ultraviolet light-emitting diode, for the first time clearly showing room-temperature electroluminescence associated with the intentional dopant levels, rather than unidentified deep levels. Further optimization of buffer layers and device structures can address the problems of minimizing these traps and increasing the efficiency of the ultraviolet emissions.

ACKNOWLEDGMENTS

The work at Brown University was supported in part by the National Science Foundation Materials Research Group Grant No. DMR-9121747. The authors are grateful to Gün Ünsal for transmission electron microscope sample preparation.

REFERENCES

[1]   R.J. Molnar, R. Singh, and T.D. Moustakas, *Appl. Phys. Lett.* **66**, 268 (1995).

[2]   R. Beresford, A. Ohtani, K.S. Stevens, and M. Kinniburgh, *J. Vac. Sci. Technol. B* **13**, 792 (1995).

[3]   M.A.L. Johnson, S. Fujita, W.H. Rowland, Jr., K.A. Bowers, W.C. Hughes, Y.W. He, N.A. El-Masry, J.W. Cook, Jr., J.F. Schetzina, J. Ren, and J.A. Edmond, *J. Vac. Sci. Technol. B* **14**, 2349 (1996).

[4]   S. Nakamura, M. Senoh, N. Iwasa, and S. Nagahama, *Appl. Phys. Lett.* **67**, 1868 (1995).

[5]   S. Nakamura, M. Senoh, S. Nagahama, N. Iwasa, T. Yamada, T. Matsushita, H. Kiyoku, and Y. Sugimoto, *Appl. Phys. Lett.* **68**, 3269 (1996).

[6]   R. P. Vaudo, I. D. Goepfert, T. D. Moustakas, D. M. Beyea, T. J. Frey, and K. Meehan, *J. Appl. Phys.* **79**, 2779 (1996).

[7]   M.E. Lin, B. Sverdlov, G.L. Zhou, and H. Morkoç, *Appl. Phys. Lett.* **62**, 3479 (1993).

[8]   T.C. Fu, N. Newman, E. Jones, J.S. Chan, X. Liu, M.D. Rubin, N.W. Cheung, and E.R. Weber, *J. Electron. Mater.* **24**, 249 (1995).

[9]   A. Ohtani, K.S. Stevens, M. Kinniburgh, and R. Beresford, *J. Cryst. Growth* **150**, 902 (1995).

[10] Z. Yang, L.K. Li, and W.I. Wang, *Appl. Phys. Lett.* **67**, 1686 (1995).

[11] S. Yoshida, *Critical Rev. Solid State* **11**, 287.

[12]   Z.A. Munir and A.W. Searcy, *J. Chem. Phys.* **42**, 4223 (1965).

[13]   W.C. Hughes, W.H. Rowland, Jr., M.A.L. Johnson, S. Fujita, J.W. Cook, Jr., J.F. Schetzina, J. Ren, and J.A. Edmond, *J. Vac. Sci. Technol. B* **13**, 1571 (1995).

[14]   H.C. Casey, Jr., J. Muth, S. Krishnankutty, and J.M. Zavada, *Appl. Phys. Lett.* **68**, 2867 (1996).

GROWTH OF HEXAGONAL GALLIUM NITRIDE FILMS ON THE (111) SURFACES OF SILICON WITH ZINC OXIDE BUFFER LAYERS

Y. KIM,* C. G. KIM,* K.-W. LEE,* K.-S. YU,* J. T. PARK,** Y. KIM**

* Inorganic Materials Division, Korea Research Institute of Chemical Technology, Taejon 305-600, KOREA, yunsukim@pado.krict.re.kr
** Department of Chemistry, Korea Advanced Institute of Science and Technology, Taejon 305-701, KOREA

ABSTRACT

The growth of gallium nitride films on sapphire substrates has not been straightforward because of the large lattice mismatch between gallium nitride and sapphire. Zinc oxide is structurally the closest material to gallium nitride and therefore is finding use as the substrate for gallium nitride. Single crystal wafers of zinc oxide are hard to obtain and very expensive. However, a thin layer of zinc oxide on a suitable substrate might solve this problem. In this work, highly c-axis oriented zinc oxide buffer layers were grown on Si(111) substrates at temperatures 410-540 ^{0}C by chemical vapor deposition of bis(2,2,6,6-tetramethyl-3,5-heptanedionato)zinc, $Zn(tmhd)_2$, and the hexagonal GaN films were subsequently deposited on them at 500 ^{0}C using the single precursor tris(diethyl-μ-amido-gallium), $[(C_2H_5)_2GaNH_2]_3$. The compound $Zn(tmhd)_2$ was found to require oxygen for the deposition of zinc oxide. In the case of gallium nitride, low pressure chemical vapor deposition of tris(diethyl-μ-amido-gallium) worked reasonably well with or without a carrier gas. The buffer layers and the GaN films were characterized by X-ray photoelectron spectroscopy (XPS), X-ray diffraction (XRD), scanning electron microscopy (SEM), and reflection high energy elctron diffraction (RHEED).

INTRODUCTION

Since the successful fabrication of gallium nitride blue light-emitting diode was announced [1], numerous studies have been reported on the deposition of gallium nitride films on various substrates among which sapphire was the most popular material. Nevertheless, sapphire has many drawbacks as a substrate for gallium nitride. The lattice mismatch is rather large and therefore various buffer layers have been tested including aluminum nitride and gallium nitride. The fact that sapphire is an insulator makes fabrication of gallium nitride devices more complicated than necessary. Sapphire is also a difficult material to make into large area substrates. In this regard it is highly desirable to be able to use more readily available substrates that are not insulators.

Mat. Res. Soc. Symp. Proc. Vol. 449 © 1997 Materials Research Society

Recent application of 6H-SiC substrate for the deposition of gallium nitride [2] is somewhat successful in reducing the number of processes in the fabrication of light-emitting diodes. Silicon carbide substrate, however, is rather expensive. On the other hand, zinc oxide has the closest lattice parameters to gallium nitride [3] and is considered the most suitable substrate, but concerning its availability and price, it has no advantage over either sapphire or silicon carbide. If, however, zinc oxide films of good quality can be deposited on silicon wafers, they will become very useful substrates for the deposition of gallium nitride films. Zinc oxide buffer layers formed on sapphire have been utilized for the deposition of gallium nitride [4]. Since aluminum nitride has been shown to epitaxially grow on Si(111) [5], it can be thought that zinc oxide will have similar property. This research is the first attempt to grow zinc oxide buffer layer on Si(111) and deposit gallium nitride on this buffer layer. The single precursor tris(diethyl-μ-amido-gallium) has been employed for the deposition of gallium nitride for the first time.

EXPERIMENT

Syntheses of the Source Chemicals

The compound $Zn(tmhd)_2$ is now commercially available but we synthesized it according to the method reported by Hammond *et al.* [6] starting from zinc chloride and 2,2,6,6-tetramethyl-3,5-heptanedione (Htmhd). The product, white crystalline powders, was purified by sublimation and identified by elemental analysis and infrared spectroscopy. Thermogravimetric analysis (TGA) of this compound showed that it begins to evaporate around 110 ^{0}C. Tris(diethyl-μ-amido-gallium), the single source for gallium nitride, was synthesized using triethylgallium and ammonia similarly to the method reported by Almond *et al.* [7] for the synthesis of tris(dimethyl-μ-amido-gallium), but at room temperature. The synthesis of this compound was confirmed by ^{1}H NMR spectroscopy and mass spectrometry. It is a liquid at ambient temperature and has a reasonably high vapor pressure for chemical vapor deposition.

Chemical Vapor Deposition

CVD was carried out either under low pressure conditions in a simple apparatus made of quartz and Pyrex tubings or under high vacuum conditions in an all stainless steel ultrahigh vacuum chamber equipped with a RHEED apparatus and a residual gas analyzer. For the deposition of zinc oxide on Si(111) substrates, oxygen carrier gas was necessary to obtain signf11cant deposition with low carbon incorporation. $Zn(tmhd)_2$ were heated at 100-120 ^{0}C. The flow rate of the carrier gas oxygen was adjusted at 7-10 sccm and the total pressure of the low pressure CVD reactor was maintained at least above 500 mToor to obtain reasonable deposition rate. To avoid difficulties in distinguishing the most intense XRD peaks of ZnO and GaN, very thin ZnO buffer layers were grown only until the interference color started to develop when it was to be used as substrate for gallium nitride. Gallium nitride films were then deposited on the zinc oxide buffer layers with or without a carrier gas (Ar). Tris(diethyl-μ-amido-gallium) was used at room temperature or warmed around 40 ^{0}C. The substrate ZnO / Si(111) was kept at 500 ^{0}C.

<u>Characterization of the Buffer Layers and the GaN Films</u>

Both the zinc oxide buffer layers and the gallium nitride overlayers were examined by XPS, XRD, SEM, and RHEED. In the case of gallium nitride films, Auger depth profiling was carried out to ascertain the nature of carbon present in the films. *In situ* RHEED examination was almost impossible because of the extremely slow deposition rates of high vacuum chemical vapor deposition in this work. Therefore most of the RHEED studies have been carried out after the films were prepared by low pressure chemical vapor deposition.

RESULTS

Figure 1 is the X-ray photoelectron survey spectrum of a ZnO film prepared using Zn(tmhd)$_2$ and oxygen carrier gas at the substrate temperature of 450 ^{0}C. Deposition was carried out for 3 h. It clearly shows all the features of zinc oxide and is almost identical to the spectrum of a zinc oxide single crystal. Although not shown here, high resolution spectrum of Zn 2p region also indicated the formation of the oxide phase. The film was then examined by X-ray diffraction (Figure 2). The XRD pattern of the film shows only two peaks, Zn(0002) and Zn(0004) other than the substrate peak, Si(111). From this it is believed that the film is a highly *c*-axis oriented ZnO. However, the SEM image of the film (Figure 3) shows that very fine grains of zinc

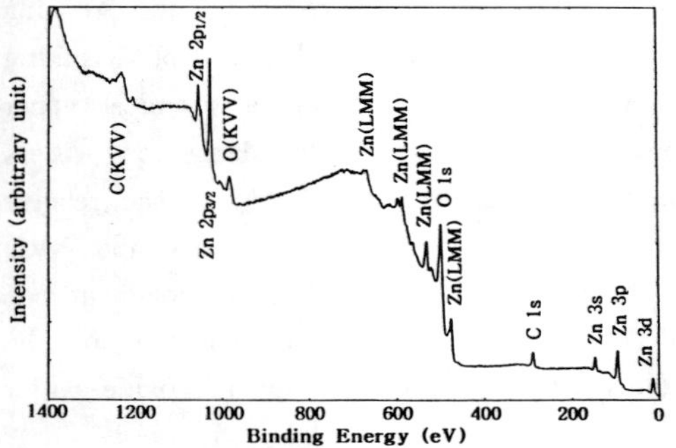

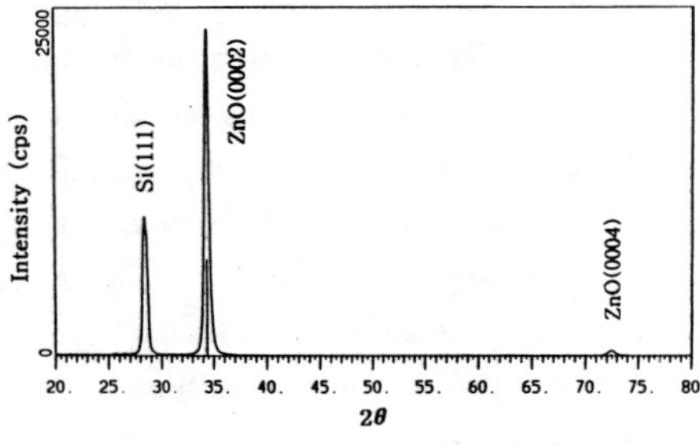

Fig. 1. XPS survey scan of the ZnO film.   Fig. 2. XRD pattern of the ZnO film.

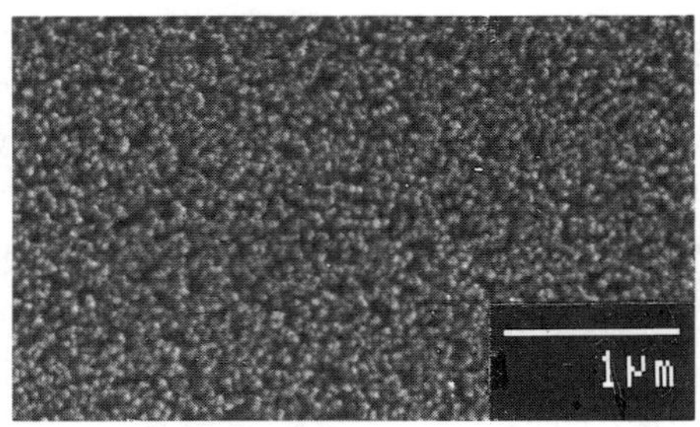 

Fig. 3. SEM image of the ZnO film.   Fig. 4. RHEED pattern of the ZnO film.

oxide are formed on the surface. This indicates that the film is predominantly oriented in c-axis but the grains are not very well aligned in the plane of the surface. This is also manifested in the RHEED pattern of the same film shown in Figure 4. The RHEED pattern has some spotty rings that are similar to those reported previously for a highly oriented ZnO films [8].

On the zinc oxide buffer layers thus produced, gallium nitride films were then deposited using the single precursor tris(diethyl-μ-amido-gallium) by low pressure chemical vapor deposition with the substrate kept at 500 ^{0}C. At higher temperatures, e.g., 650 ^{0}C, the thin zinc oxide buffer layers tend to sublime in a short time interval. The X-ray photoelectron survey spectrum of the film obtained from tris(diethyl-μ-amido-gallium) is shown in Figure 5. In this survey spectrum Ga $2p_{1/2}$ and Ga $2p_{3/2}$ peaks are prominent but the N 1s peak is very small. This is due to the fact that the atomic sensitivity factor of the N 1s orbital is much smaller than that of the Ga 2p orbital [9]. Even so, the film was found to contain much less nitrogen than gallium. Therefore the gallium nitride film is not stoichiometric. Since both Zn $2p_{1/2}$ and Zn $2p_{3/2}$ peaks do not appear in this spectrum, the buffer layer ZnO is thought to be completely covered by gallium nitride overlayer. Contamination due to carbon from the precursor can be seen in the spectrum. This is somewhat reduced after Ar^+ ion bombardment of the surface, but did not totally disappear in the Auger depth profiling analysis. Therefore carbon incorporation into the film occurs during chemical vapor deposition. It is also possible that the precursor decomposes in the deposition process and releases ammonia thereby reducing the nitrogen content in the film. The relative ratio of Ga:N of the film was found to be about 1:0.5. Nevertheless, the XRD pattern of the film (Figure 6) shows that the GaN(0002) peak is the predominant one with much smaller GaN(10$\bar{1}$0) and GaN(10$\bar{1}$1) peaks. This is indicative of the possibility that hexagonal gallium nitride films can be formed on the zinc oxide buffer layers deposited on Si(111) substrates.

The SEM image of the gallium nitride film (Figure 7) shows only a crude surface

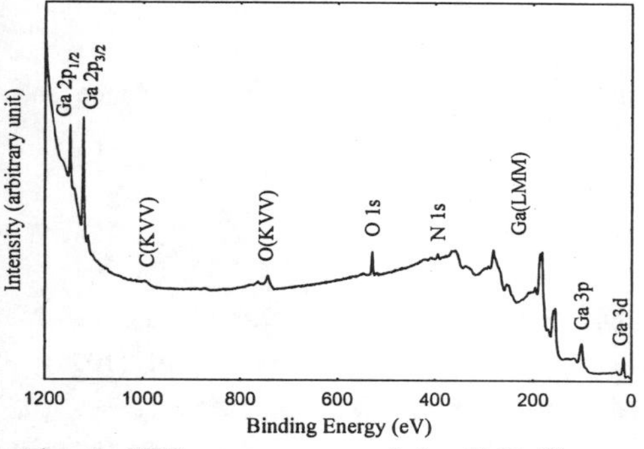
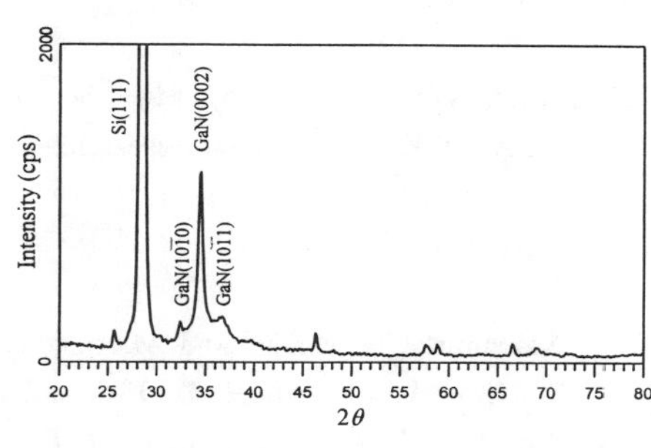

Fig. 5. XPS survey scan of the GaN film. Fig. 6. XRD pattern of the GaN film.

Fig. 7. SEM image of the GaN film.

Fig. 8. RHEED pattern of an *in situ* grown GaN film.

consisiting of small grains of gallium nitride and possibly different phases. However, the RHEED pattern (Figure 8) of the film obtained by *in situ* chemical vapor deposition in a UHV chamber (total pressure 0.14 mTorr, with substrate at 500 ^{0}C) does show faint but distinct development of a pattern indicating that the gallium nitride overlayer is forming an ordered structure. This phenomenon should be investigated more thoroughly.

CONCLUSIONS

From this research we can draw the following conclusions. Highly *c*-axis oriented zinc oxide films can be prepared by chemical vapor deposition of bis(2,2,6,6-tetramethyl-3,5-heptanedionato)zinc, $Zn(tmhd)_2$, on Si(111) substrates. This zinc oxide buffer layer can serve as the substrate for the deposition of hexagonal gallium nitride. The single precursor tris(diethyl-μ-amido-gallium), $[(C_2H_5)_2GaNH_2]_3$, has been found adequate for the deposition of hexagonal gallium nitride. However, epitaxial growth of hexagonal gallium nitride films has not been achieved yet probably due to unoptimized reaction parameters. Better controls of reaction parameters are necessary.

ACKNOWLEDGMENTS

The authors gratefully acknowledge the financial support by the Ministry of Science and Technology of Korea for their research.

REFERENCES

1. S. Nakamura, T. Mukai, and M. Senoh, Jpn. J. Appl. Phys. **30**, L1998 (1991).
2. J. G. Kim, A. C. Frenkel, H. Liu, and R. M. Park, Appl. Phys. Lett. **65**, 91 (1994).
3. D. Elwell and M. M. Elwell, Prog. Crystal Growth Charact. **17**, 53 (1988).
4. T. Detchprohm, H. Amano, K. Hiramatsu, and I. Akasaki, J. Crystal Growth **128**, 384 (1993).
5. H. M. Manasevit, F. M. Erdmann, and W. I. Simpson, J. Electrochem. Soc. **118**, 1864 (1971).
6. G. S. Hammond, D. C. Nonhebel, and C.-H. S. Wu, Inorg. Chem. **2,** 73 (1973).
7. M. J. Almond, M. G. B. Drew, C. E. Jenkins, and D. A. Rice, J. Chem. Soc. Dalton Trans. 5 (1992).
8. S. Oda, H. Tokunaga, N. Kitajima, J.-i. Hanna, I. Shimizu, and H. Kokado, Jpn. J. Appl. Phys. **24**, 1607 (1985).
9. C. D. Wagner, W. M. Riggs, L. E. Davis, and J. F. Moulder, " Handbook of X-ray Photoelectron Spectroscopy," G. E. Muilenberg (editor); Perkin-Elmer Corporation, Physical Electronics Division; Eden Prairie (1979).

ZnO BUFFER FORMED ON Si AND SAPPHIRE SUBSTRATES FOR GaN MOVPE

T. SHIRASAWA, T. HONDA, F. KOYAMA and K. IGA
P & I Lab., Tokyo Institute of Technology, 4259 Nagatsuta Midori-ku Yokohama 226, Japan. tshirasa@pi.titech.ac.jp

ABSTRACT

ZnO layers have been deposited by electron beam (EB) evaporation and laser ablation molecular beam epitaxy (MBE) as buffer layers to grow GaN by metal organic vapor phase epitaxy (MOVPE). The photoluminescence spectrum of the ZnO layer deposited by an EB evaporator shows an emission peaks of 367 nm. GaN was grown on ZnO/Si, Si and sapphire substrates under the same growth condition employing low-temperature-grown AlN buffers to prevent the dissociation of ZnO during the high GaN growth. The GaN on ZnO/Si shows sharp photoluminescence spectra at room temperature and 10 K. These results indicate a potential use of ZnO/Si substrates for GaN based blue-UV optical devices such as vertical-cavity surface-emitting lasers (VCSELs).

INTRODUCTION

Although the GaN system is very attractive for blue-UV light emitting devices[1], there is a large lattice mismatch (18%)between GaN and sapphire [2]. Also the sapphire is an insulator and hard to etch, that raises a problem for the fabrication of sophisticated devices such as VCSELs.

On the other hand, Si is the most popular semiconductor, and is advantageous due to its mature process, good conductivity and low cost. It also has a higher melting temperature than other substrates for compound semiconductors, so it can endure high temperature GaN growth. Therefore, the epitaxial growth of GaN on Si substrate is believed to be useful for various blue-UV light emitting devices. However, the direct growth of GaN on Si may form insulating SiN between GaN and Si substrate.

ZnO has a wurtzite structure and less lattice mismatch to GaN[3][4] and it is easily removable. It is also known that ZnO is an n-type semiconductor[4]. If ZnO is pre-deposited on Si, we can grow the GaN without SiN. In this paper, we report on the deposition of ZnO using EB evaporation and laser ablation MBE[6]. Also, the GaN growth is carried out on the ZnO/Si using MOVPE.

EXPERIMENTS AND RESULTS

Mat. Res. Soc. Symp. Proc. Vol. 449 © 1997 Materials Research Society

ZnO Deposition

At first, a ZnO layer was deposited by an electron beam evaporator on Si (111) and sapphire substrates for comparison. The substrate temperature was 300 °C and the O_2 was introduced into chamber with a gas pressure of 1×10^{-4} Torr to prevent oxygen dissociation from the ZnO layer The supply of O_2 gas during a cooling step after the deposition was found to be important to obtain a high quality ZnO layer, which was confirmed by photoluminescence describing in the following part.

Also, we used a laser ablasion MBE to deposit ZnO layers because of maintenance of EB evaporator. The substrate temperature was 400 °C and the O_2 was introduced into chamber with a gas pressure of 1×10^{6} Torr. The excimer laser whose frequency was 10 Hz was used to ablaze a ZnO target. Pre-ablation time and deposition time was 2 min. and 150 min., respectively. After the deposition, the temperature was kept at 350 °C for 35 min. to be annealed.

The ZnO layer obtained on a sapphire substrate by an EB evaporator has a sharp peak at 356.8 nm by photoluminescence measurement at 10 K. This result shows that the bandgap energy of this ZnO layer is around 3.47 eV and it is in a good agreement with previous data[5].

The sample obtained on Si (111) by laser ablation MBE shows a peak-to-peak roughness of about 30 Å, measured by an atomic force microscope (AFM).

The X-ray diffraction was measured on these samples. These show weak peaks near 34.5 ° originated by a ZnO layer which is highly oriented to the c-axis of substrates. However, a full width-at-half-maximum (FWHM) of the rocking curves of these sample are as wide as 20-30 °. They may be oriented to the c-axis with various rotation.

To grow GaN on these samples by MOVPE, it is necessary to prevent the evaporation of a ZnO layer at the growth temperature of GaN. These deposited ZnO layers disappeared at a temperature as low as ~ 500 °C. This indicates that the direct GaN epi-growth on ZnO by MOVPE is very difficult.

GaN Growth by MOVPE

GaN layers were grown on Si, Sapphire and ZnO/Si by MOVPE. This ZnO is prepared by laser ablation MBE. The vertical quartz reactor was used for MOVPE-growth of GaN in this study. The main gases including TMGa and NH_3 are introduced into a reactor by a H_2 carrier gas from a nozzle located at side wall. The top gases, H_2 and N_2, are introduced in order to press the main gas flow to the substrate. The growth is performed under atmospheric pressure. Each substrate is cleaned in organic solutions before the growth. The AlN layers were first deposited at low temperature in order to prevent the dissociation of ZnO at high temperature during GaN growth. The growth temperature of

AlN and GaN is 400 °C and 800 °C, respectively. The III/V ratio during the growth of GaN is about 3000 and the growth time is 120 min.

The surface morphologies of GaN on ZnO/Si are observed in detail by scanning electron microscope. It seems that the three dimensional GaN exists on very thin two dimensional GaN. The average height of loafs of GaN is about 3 μm. However, it is not clear that two dimensional growth occurred.

The results of X-ray diffraction measurements are shown in Fig. 1. It indicates that the GaN was grown on a ZnO/Si substrate with an AlN layer. The peak of a ZnO layer disappeared after the GaN growth, however, it is thought that the ZnO layer still exists because of its clear difference from that grown directly on a Si substrate. The diffraction peaks from GaN grown on Si and sapphire without a ZnO layer were not detected under the same condition.

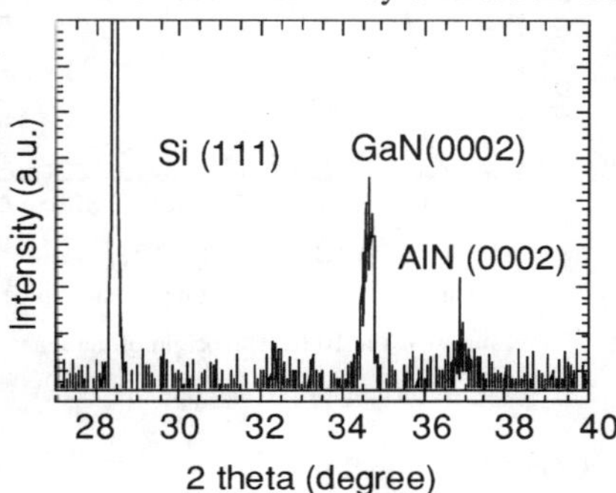

Fig.1 The X-ray diffraction of GaN/AlN/ZnO/Si(111). The srectrum from ZnO is included into the peak of GaN layer.

We have also carried out photoluminescence measurements at room temperature and low temperature (10 K). The optical source for photoluminescence is a He-Cd laser with a wavelength of 325 nm. The pumping power is 7 mW during the measurement. Figure 2 shows the room temperature photoluminescence spectra of GaN layers on ZnO/Si, sapphire and Si, respectively.

The emission from GaN on ZnO/Si near the bandedge is dominant rather than that from the deep level. The dominant peak has about five times stronger intensity than that of GaN grown on sapphire. As we mentioned above, an AlN buffer layer was not used as a nucleation layer but was employed to prevent the dissociation of the ZnO. On Si, the emission from GaN is very weak. The intensity is 300 times lower than the others. The reason of its low PL intensity is due to a large lattice mismatch with GaN and Si, resulting a difficulty of the direct GaN growth on Si.

The clear difference in low temperarure PL spectra between ZnO and GaN/ZnO/Si shows that the emission in Fig. 3. is originated from GaN. However, the GaN on ZnO/Si sometimes peels off the Si substrate. Thus, some techniques to conquer this problem are needed. For example the prior supply of Zn at the deposition of ZnO or using slightly tilted Si substrates to orient the c-plane rotation of ZnO on Si.

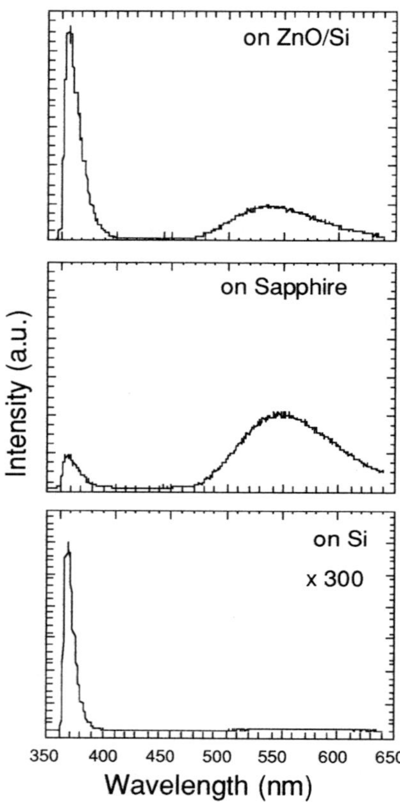

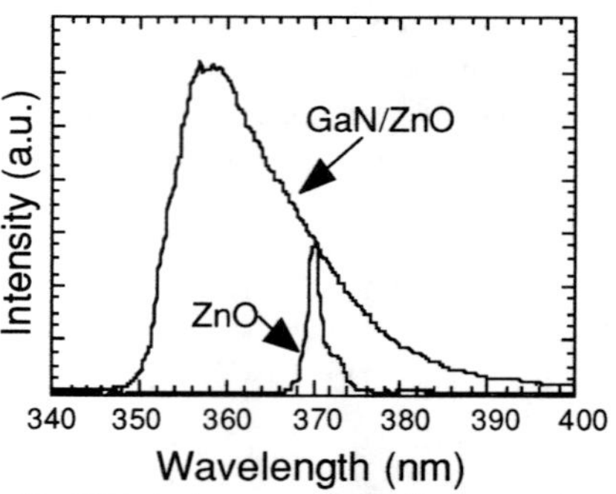

Fig.3 This is the spectra of photoluminescence measurements at 10 K. The origin of the GaN on ZnO is different from that one from ZnO.

Fig. 2 The photoluminescence spectra of GaN layers at room temperature. On ZnO/Si, on Sapphire and on Si, respectvly. The longutidudinal scale of on Si is 300 times smaller than the others.

CONCLUSIONS

The good quality of GaN grown on ZnO/Si were confirmed by X-ray diffraction and photoluminescence measurements. It indicates that ZnO/Si is a viable base material to grow materials of the GaN system for optical devices. We are planning to grow GaN based VCSELs by MOVPE on ZnO/Si substrates.

ACKNOWLEDGMENTS

The authors wish to thank to Mr. A. Ohtomo, Dr. M. Kawasaki and Prof. H. Koinuma of Materials and Structures Laboratory of Tokyo Institute of Technology for the deposition of ZnO layer by a laser ablation MBE and the AFM.

REFERENCES

1. S. Nakamura, M. Senoh, S. Nagahama, N. Iwasa, T. Yamada, T. Matsushita, H. Kiyoku and Y. Sugimoto, Jpn. J. Appl. Phys. **35**, L74 (1996).
2. H. Amano, N. Sawaki, I. Akasaki and Y. Toyoda, Appl. Phys. Lett. **48**, 353 (1986).

3. T. Matsuoka, N. Yoshimoto, T. Sasaki and A. Katsui, J. Electron. Mater. **21**, 157 (1992).

4. T. Detchprohm, H. Amano, K. Hiramatsu and I. Akasaki, J. Cryst. Growth **128**, 384 (1993).

5. H. K. Pulker, Applyed Optics **18**, 1969 (1979).

6. V. Cracium, J. Elders, J. D. E. Gardenieres and Ian W. Boyd, Appl. Phys. Lett. **65**, 2963 (1994).

Deposition of AlN on WS$_2$ (0001) substrate by Atomic Layer Growth Process

J-W. Chung and F.S. Ohuchi, Department of Materials Science and Engineering, University of Washington, BOX 352120, Seattle, WA 98195.

ABSTRACT

Close proximity of the lattice constant for tungsten disulfide and aluminum nitride has lead to an investigation to use WS$_2$ as a potential substrate for the growth of AlN. Metal organic chemical vapor deposition(MOCVD) has been develop to fabricated WS$_2$ thin films on Si(001) with their basal planes parallel to the substrate. AlN thin film was subsequently grown by atomic layer growth (ALG) process using dimethylamine-alane (DMEAA) and ammonia (NH$_3$). Deposition conditions for WS2 thin films by MOCVD, and AlN growth on WS2 by ALG are described.

I. INTRODUCTION

A need for electronic devices capable of operating at temperatures exceeding 300°C, as well as at short wavelengths for injection lasers, has recently motivated widespread research on III-V nitrides, including GaN, AlN, InN and their alloys[1~3]. Various starting substrate materials have been investigated for epitaxial growth of the III-V nitride materials[4~6], however, most of the conventional materials are poorly matched to the III-nitrides due different lattice structures, lattice constants and thermal expansion coefficients; hence the nitrides are subjected to large residual strains built in the lattice, leading to many defects.

Here we report an attempt to use alternate materials as the substrates for subsequent growth of III-V nitride thin films. These materials are tungsten disulfide (WS$_2$) and molybdenum disulfide (MoS$_2$), and crystallize into layered structures as shown in Fig. 1. In these materials, bonding within the layers is strongly covalent while each layer is held via weak van der Waals (VDW) interaction. Particular interest is that the lattice constants for WS$_2$ and MoS$_2$ (a = 3.15 Å and 3.16 Å, respectively) are remarkably close to those of AlN (a = 3.11 Å) and GaN (a = 3.18 Å). It is therefore conceivable that a properly oriented crystal surface of these materials may be used for the growth of AlN and/or GaN. To be considered for this application, it is necessary to grow highly oriented WS$_2$ or MoS$_2$ thin films such that the van der Waals layers are parallel to the plane of the substrate, exposing basal planes of the WS$_2$ or MoS$_2$ crystal structure in order to serve as substrates for subsequent growth of AlN or GaN, of which concept is illustrated in Fig. 2.

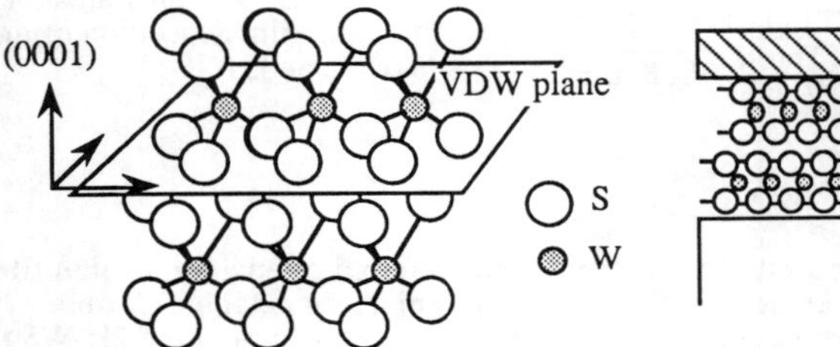

Fig. 1 Crystal structure of WS$_2$ Fig 2. Proposed AlN/WS$_2$/Si(100) structure

Mat. Res. Soc. Symp. Proc. Vol. 449 ©1997 Materials Research Society

In this paper, we describe our preliminary results of AlN thin film growth on the WS_2 thin films fabricated on Si(100) substrate. In the first part of our discussion, we present the results of WS_2 thin films fabricated by Metal Organic Chemical Vapor Deposition (MOCVD) using tungsten carbonyl ($W(CO)_6$) and hydrogen sulfide (H_2S) as precursors. Experimental conditions leading to deposition of highly oriented WS_2 thin films with their basal planes parallel to the substrate have been investigated. We then describe the growth of AlN films by a process, Atomic Layer Growth (ALG), using the novel precursor combination of dimethylamine-alane (DMEAA) and ammonia (NH_3).

II. WS_2 THIN FILM GROWTH

II.1 EXPERIMENTAL

MOCVD apparatus schematically illustrated in Fig.3 was specifically designed and constructed for the deposition of WS_2. Tungsten hexacarbonyl ($W(CO)_6$, Johnson Matthey, 99%) and hydrogen sulfide (H_2S, Matheson, 99.5%) were used as sources for tungsten and sulfur precursors, respectively. $W(CO)_6$ is evaporated in a bubbler immersed in a constant heat bath. The H_2S source flows to a gas line with its pressure regulated by the pressure gauge. Both sources are transported with dry Ar carrier gas, and their flow rates, $Ar(H_2S)$ and $Ar(W(CO)_6)$, are independently controlled by the mass flow meters so as to maintain the flow rate of source materials and the reactor pressure constant. Source materials are mixed just prior to being fed into the reactor. Experimental variables that determine the resultant products include: substrate temperature (T_s), source gas flow ratio ($R_{s/w}$) defined as $Ar(H_2S)/Ar(W(CO)_6)$, and reactor pressure (P_r), all of which must be optimized in order to obtain the desired thin film structure. In the deposition strategy, each parameter has been successively varied with other variables fixed. For example, the reactor pressure and the reactants flow ratio were fixed first, deposition was attempted by varying the substrate temperature.

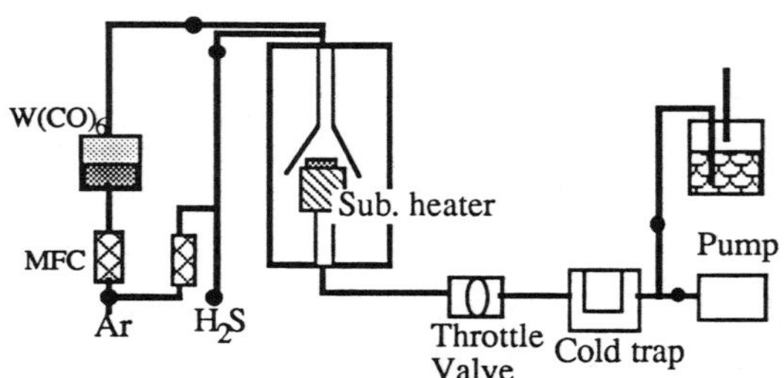

Fig. 3 MOCVD apparatus for WS_2 thin film deposition using $W(CO)_6$ and H_2S.

II.2 RESULTS

X-ray diffraction was primarily used to characterize the preferred orientation of thin films deposited under various experimental conditions. Thin films obtained exhibit a 2H-polytype hexagonal structure, where H denotes hexagonal symmetry. The spectrum of the 2H-WS_2 is characterized by a series of peaks, consisting of (00.2), (00.4), (00.6), and (00.8), which correspond to the reflections from the basal planes of WS_2. Appearance of these peaks indicates that thin films are grown with their basal planes parallel to the substrate. Here, we denote this crystal texture as C($\perp$). This nomenclature was adopted as originally suggested by

Galun et. al [7]. There are, however, a set of peaks, (10.0),(10.1) and (10.3), which originate from the crystallites with basal planes perpendicular to the substrate. This type of texture is denoted by C(ll). In thin film growth, both sets of peaks can co-exist, and their relative orientations are strictly determined by the deposition conditions.

Shown in Fig. 4 is a series of x-ray diffraction patterns for depositions made at the reactor pressures 100 torr with the substrate temperature varying from $T_S = 583$ °K to 723 °K. The deposition times ranged from 10 to 20 minutes, where the growth rates of 45-80nm/min were observed. It is apparent that there is a optimum substrate temperature regime where the most highly oriented $C(\perp)$-WS_2 films were obtained for fixed source gas flow rate ($R_{S/W}$) and reactor pressure (P_r) conditions, although crystallites with their basal planes non-parallel to the substrate, i.e. C(ll)-WS_2, are still present in the film.

Since the relative presence of $C(\perp)$ and C(ll)-WS_2 is strongly influenced by the deposition conditions, relative significance of three deposition parameters, T_S, P_r, and $R_{S/W}$, in relation to the preferred orientation for the growth of WS_2 thin films was studied. In x-ray diffraction spectrum, a set of reflected peaks by the basal planes ((00.2), (00.4), (00.6), (00.8)) arising from the $C(\perp)$, and a set of peaks consisting of the (10.0) and (10.1) reflections arising from the C(ll), are present. By calculating the intensity ratio of these peaks which is defined as $I_{(00.2)+(00.4)+(00.6)+(00.8)}/I_{(10.0)+(10.1)}$, the degree of preferred orientation for $C(\perp)$ and C(ll) in the film can be evaluated. Obviously, if all the basal planes are grown parallel to the substrate, this quantity becomes infinite. A plot of such quantities with varying deposition conditions is found to be useful in evaluating the effects of the experimental conditions on the structure of the deposited films.

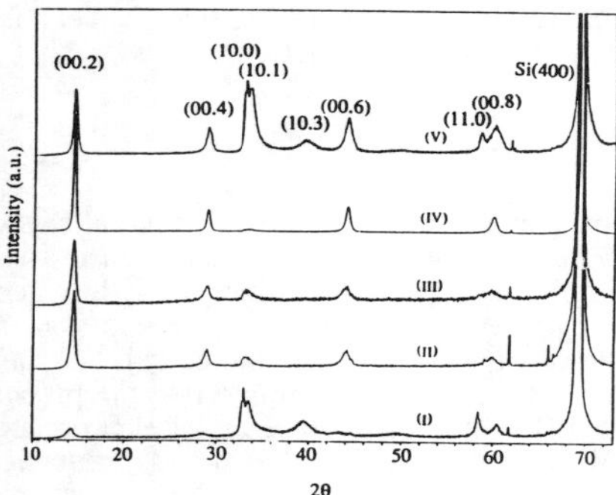

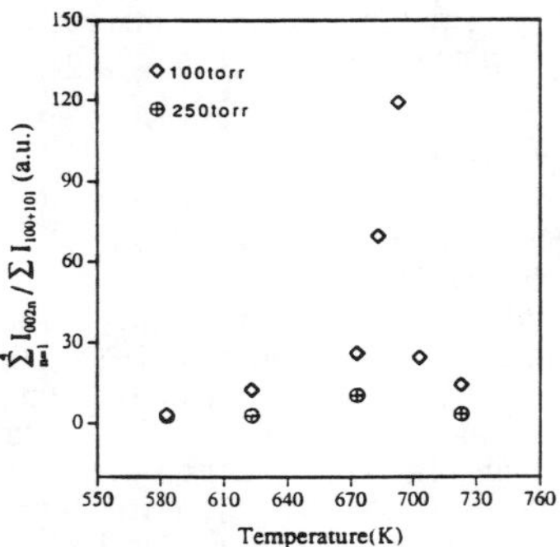

Fig. 4 X-ray diffraction patterns of tungsten disulfide thin films using Cu-Kα radiation Substrate temperatures varied at pressure 100 torr, at 583 °K, 623 °K, 673 °K 693 °K, and 723 ° K, with reactants ratio 0.66 fixed.

Fig. 5 The degree of preferred orientation of basal planes parallel to substrate with relation to substrate temperatures at 100 torr and 250 torr, respectively.

The plot of the intensity ratios described above as functions of substrate temperature at two different source gas flow ratios is shown in Fig. 5. It is found that choice of the reactor pressure is critical and strongly influences the overall quality of the deposited thin films. At the reactor pressure of 250 torr, a change in the substrate temperature does not significantly influence the overall thin film orientation, although the highly oriented $C(\perp)$ type thin films are hardly obtained. When the pressure is reduced to 100 torr, highly oriented C($\perp$)-WS$_2$ thin films are obtained only within a narrow range of deposition conditions. Using these deposition conditions, WS$_2$ thin film substrates for the subsequent deposition of AlN were fabricated on Si(100).

III. AlN THIN FILM GROWTH

III. 1 EXPERIMENTAL

AlN films were fabricated by Atomic Layer Growth (ALG) using novel precursor combination of dimethylamine-alane (DMEAA) and ammonia (NH$_3$) recently developed by Kidder [8]. ALG uses self-limiting heterogeneous reactions at the substrate surface to form a film one atomic layer at a time. Using this technique, Group III precursor, DMEAA, reacts with the surface in a self-limiting manner, i.e., one or less monolayers of the adsorbate is formed per deposition cycle. In the second step, Group V precursor, NH$_3$, reacts in a facile, site-selective manner with the adsorbate of the first precursor to from the desired film. This process leads to an ordered nucleation and growth process in which highly ordered crystalline or epitaxial films can be grown. By utilizing nucleation and growth strategies that promote an ordered, site-selective growth mechanism, high temperatures for enhancing surface mobility are not required. The second step is what gives ALG its larger advantage over MBE techniques which lack a reactive nitrogen source that can be produced efficiently in particular for nitride formation.

Deposition of AlN thin films on WS$_2$ thin film substrate fabricated on Si(100) has been carried out in a Crystal Specialities Inc. 425 MOCVD reactor equipped with a gas mixing manifold and three-say flush valves. In this apparatus, specially designed gas-mixing manifold enables for abrupt switching of the gas composition which is essential for the ALG process.

III.2 RESULTS

Development of the ALG process for AlN deposition was accomplished in several sets of growth experiments. The goals of these studies were to first identify a parameter space where AlN thin films would be deposited and to improve the flow conditions. It has been demonstrated the AlN growth on Si(100), Si(111) and Al$_2$O$_3$(00.1) by the ALG process using identical precursors, where a temperature regime of 613$\pm$40°K was obtained [8]. We have adopted similar conditions as the starting parameters, however, it is found that the process conditions for AlN deposition over WS$_2$ is quite different. From the series of experiments, the deposition procedure for the AlN growth on WS$_2$ was established as a 4-5-4-5 sequence, where DMEAA was flown for 4 seconds, then H$_2$ flush for 5 seconds, NH$_3$ flow for 4 seconds, followed by 5 seconds of H$_2$ flush. This sequence accomplishes one complete ALG cycle at 723°K with the reactor pressure ranging from 25 to 100 torr.

Shown in Fig. 6 is X-ray diffraction spectrum obtained from ALG grown AlN thin film on WS$_2$/Si(100) thin film substrate. Thickness of the AlN layers was not accurately measured, however, the estimated thickness based on the deposition time was in the order of several hundred Å. Appearance of a series of WS$_2$ peaks, consisting of (00.2), (00.4), (00.6), and (00.8), indicates that the WS$_2$ thin film is oriented with the basal plane parallel to the substrate, however, a small fraction of non-parallel component exists. An AlN(00.2) diffraction peak was detected at 2θ= 36.053° with the full-width-half-maximum (FWHM) of approximately 0.25°. Subset in Fig. 6 is the region corresponding to the AlN(00.2) reflection peak. The presence of only one AlN diffraction peak in data indicated that AlN thin film has grown with the AlN{00.1} planes aligned parallel to the substrate plane.

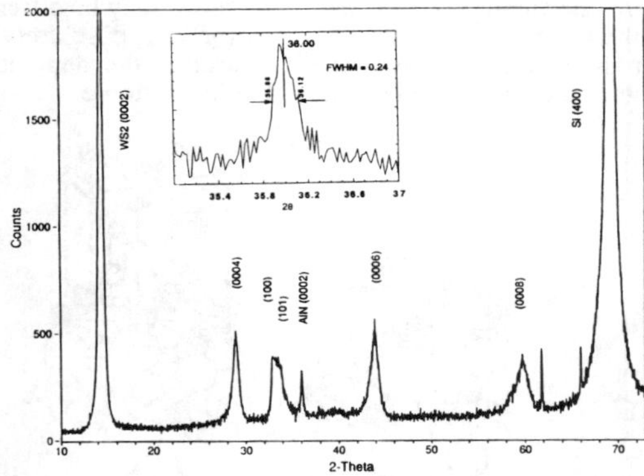

Fig. 6 X-ray diffraction spectrum obtained from ALG grown AlN thin film on WS$_2$/Si(100) thin film substrate.

X-ray photoelectron spectroscopy (XPS) and Auger electron spectroscopy (AES) were used to examined the chemical composition of the AlN films. While both spectra obtained from the "as-received" specimen show substantial amount of oxygen, indicating surface was oxidized by exposure to the ambient atmosphere, corresponding binding energies for N and Al were consistent to AlN.

IV. DISCUSSION

The present study has been motivated by the fact that the lattice constant of WS$_2$ is remarkably close to that of AlN, so that properly oriented WS$_2$ surface may be used as a template for epitaxial growth of AlN . To demonstrate this, we have developed MOCVD to fabricate WS$_2$ thin films with their basal planes oriented parallel to the substrate surface, on which AlN growth was attempted.

AlN can be produced via surface reaction using ammonia and aluminum-containing metal-organic precursor at elevated temperatures[9]. DMEAA precursor employed in the present study seems to be promising because of low dissociation energy of the aluminum-amine bond and stability of amine ligand, allowing the reaction to be carried out at temperatures much lower than those typically needed for AlN thin film deposition. For the atomic layer growth of AlN, the first step of the reaction involves the dissociation of DMEAA molecules by breaking aluminum-amine bonds, leaving AlH$_3$ species adsorbed on the WS$_2$ surface. Although the adsorption sites, nor chemical interaction between dissociated species with the substrate, are not known at present, it is likely to form two dimensional layer of DMEAA dissociated species, of which periodicity is determined by the WS$_2$ lattice constant. Subsequent exposure/reaction to NH$_3$ will form Al-N bonds, constructing AlN lattice structure. Because of close proximity of the lattice constant for WS$_2$ and AlN, the lattice constant for the first AlN layer may not be distorted, making subsequent growth of AlN unstrained. These steps are schematically illustrated in Fig. 7.

Relatively high deposition rate for AlN growth was found for sapphire substrates[9]; this is because the propensity of DMEAA molecules to form involatile solid with oxygen is relatively high. In the present case, however, the deposition rate was found to be a several factor slower than the sapphire case. This is partly understood that there is little or no oxygen on WS$_2$ thin film in our case, therefore the initial reaction that is necessary to initiate the subsequent reaction with NH$_3$ is not efficient as the sapphire case. Although a sulfur is

among the family including an oxygen, the chemistry of DMEAA to sulfur may be different
from that to oxygen.   This is why the deposition conditions found for WS_2 case are very
different from what was found for the sapphire case. At the present, the deposition
conditions are not yet optimized though, the details must be worked out in the future.

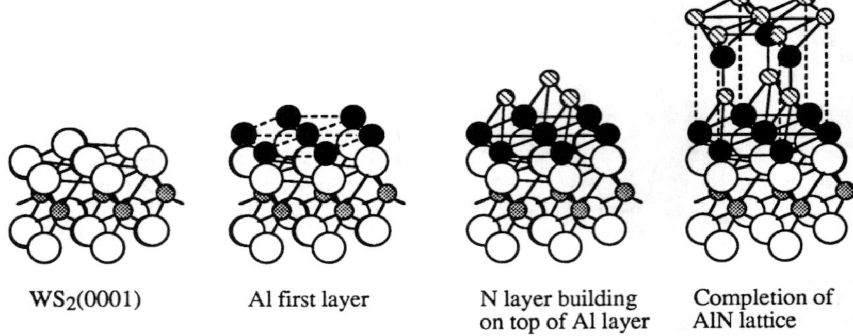

Fig. 7   Schematically illustrated model for AlN lattice growth on WS_2(0001) surface

V. CONCLUSION

Tungsten disulfide has been used as a substrate for the growth of aluminum nitride by atomic
layer growth mode using the novel precursor combination of dimethylamine-alane (DMEAA)
and ammonia (NH_3) has been demonstrated. Close proximity of the lattice constant of WS_2
to those of AlN may lead to a potential application of this materials as a substrate for
subsequent growth of AlN, however, the details chemistry for the reaction between first
DMEAA molecular layer with WS_2 surface must be worked out in order to develop this
process.

REFERENCES

1. R. F. Davis, Proc. IEEE 79, 702 (1991).
2. M. A. Khan, R. A. Skogman, J. M. V. Hove, S. Krishnankuttyand R. M. Kolbas,
    Appl. Phys. **56** L1257 (1990).
3. S. Nakamura, M. Senoh, S. I. Nagahama, N. Iwasa, T. Yamada, T. Matsushita,
    H. Kiyoku and Y. Sugimoto, Jpn. J. Appl. Phys. **35** 174 (1996).
4. Z. Sitar, M. J. Paisley, B. Yan, R. F. Davis, R Ruan and J. W. Choke, Thin
    Solid Films **200** 311 (1991).
5. R. G. Gordon, U. Riaz and D. M. Hoffman, J. Mater. Res. **7** 1679 (1992).
6. J. Chaudhuri, J. H. Edgar, B. S. Sywe and R. Thokala, J. Appl. Phys. **77**(12), 62
    (1995).
7. E. Galun, H. Cohen, L. Margulis, A. Vilan, T. Tsirlina, G. Hodes, M. Hershfinkel, W.
    Jaegermann, K. Ellmer and R. Tenne, Appl. Phys. **67**(23) L3474 (1995).
8. J. N. Kidder, "Synthesis of AlN thin films at low temeprature by MOCVD", Ph. D.
    Thesis, Univ. of Washington, Seattle, WA (1996).
9. M. Morita, N. Uesugi, S Isogai, K. Tsubouchi and N. Mikoshiba, Jpn. J. of Appl.
    Phys. **20**(1) 17 (1981).

MBE GROWTH AND CHARACTERIZATION OF ZNS/GAN HETEROSTRUCTURES

E. C. PIQUETTE, Z. Z. BANDIĆ, J. O. MCCALDIN,and T. C. MCGILL
Watson Laboratories of Applied Physics 128-95
California Institute Of Technology, Pasadena, California 91125

ABSTRACT

Heterostructures involving ZnS/GaN show promise for the injection of holes from p-GaN into n-ZnS. This combination could result in multi-color electroluminescent displays. We have grown single crystal ZnS on GaN and sapphire (0001) by MBE using elemental sources. The ZnS was grown at temperatures from 150 °C-400 °C, with beam flux equivalent pressures of $(0.3 - 2.0) \times 10^{-7}$ torr. Growth rates of up to 0.4 μm per hour were observed for the lower growth temperatures, with rapidly diminishing rates for temperatures above 350 °C. The GaN substrate consisted of a 3 μm epilayer grown on sapphire by MOCVD. XPS analysis revealed the presence of carbon surface contamination on the GaN, which was removed by *in situ* exposure to an RF nitrogen plasma. RHEED observations indicate that the zincblende ZnS layers commonly contain (111) twins, although twin free films may be grown at a high substrate temperature. The samples were characterized using photoluminescence and X-ray diffraction. X-ray peaks typically had FWHM of 400 arcsec for $\omega/2\theta$ scans, and somewhat worse for ω scans. Photoluminescence spectra of the ZnS films doped with Ag and Al demonstrated the well known blue donor acceptor transition at 440 nm.

INTRODUCTION

Zinc sulfide, with its wide direct bandgap, has long been known as a versatile and efficient light emitting compound. It has been extensively used as a phosphor, and continues to show promise in semiconductor light emitting devices, such as ZnS based flat panel displays [1, 2, 3], LEDs [4], and semiconductor lasers[5]. However, difficulties in producing high quality doped material, particularly p-type, have otherwise reduced the utility of the sulfide system.

Recently, high conductivity p-type GaN has become available [6, 7, 8]. With a valence band edge below that of ZnS [9], it has been proposed that GaN may be a good hole injector into ZnS and other light emitting semiconductor materials [10]. To this end, we have grown ZnS on GaN (0001) and sapphire (0001) substrates by Molecular Beam Epitaxy (MBE) in an attempt to characterize this potentially interesting system as well as gain insight into highly lattice mismatched systems in general.

While there have been many reports of ZnS grown by MBE on near lattice matched substrates such as Si [11, 12], GaP [13, 14, 15], and GaAs [16, 17, 18], little has been reported of epitaxial growth on the highly mismatched substrates GaN ($\sim$+20% mismatch) or Al_2O_3 ($\sim$-20% mismatch). These interfaces are very interesting because of the large mismatch and the unusual properties of the constituents. First, it is not expected that the ZnS epilayer will grow coherently strained for more than one monolayer. More likely relaxation in the layers closest to the interface will result in dangling bonds and dislocations other than the traditional misfit dislocations. In fact tilting and 3D relaxation are likely the most energetically preferred relaxation mechanisms[19]. Second, GaN and ZnS are known to be good light emitters even when the defect density is high. Hence, this interface provides

Mat. Res. Soc. Symp. Proc. Vol. 449 © 1997 Materials Research Society

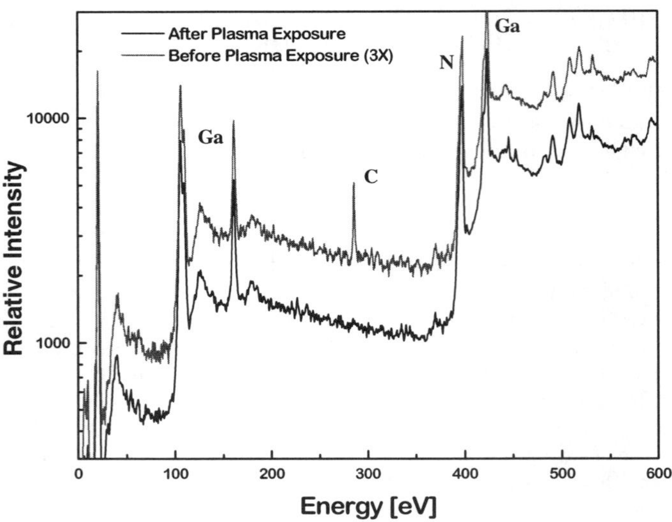

Figure 1: X-ray photoemission spectrum of GaN before and after exposure to *in situ* RF nitrogen plasma.

us with a potentially technologically useful interface even though it is likely to be heavily defected.

EXPERIMENT

ZnS epilayers were grown in a modified Perkin-Elmer 430 Molecular Beam Epitaxy system, equipped with a valved cracking source for sulfur (6N purity), Knudsen cell sources for elemental Zn (6N), Al (6N), Ag (6N), and an RF nitrogen (5N5) plasma source (Oxford Applied Research). Except where noted, the sulfur cracking zone temperature was held at 700 °C. The growth chamber maintains a base pressure during growth of 3×10^{-10} torr with cryopaneling fully cooled. Samples were either bonded with In to molybdenum holder blocks, or secured to similar blocks with Mo or Si clips. We found that In bonding provided better temperature uniformity and reproducibility than the use of clips. Temperatures were measured by thermocouple, and reflection high energy electron diffraction (RHEED) operating at 10 keV was used as a characterization tool during growth. Some ZnS films were doped with Ag and Al.

The GaN substrates consisted of a 3 micron thick epilayer grown by MOCVD on c-plane sapphire. Sapphire (0001) substrates were sourced from Union Carbide Crystal Products. Substrates were degreased in organic solvents and loaded into the vacuum system. Before epilayer growth, the samples were transferred *in situ* between the growth chamber and an XPS analysis chamber. XPS analysis on the GaN layers revealed carbon surface contamination of atmospheric origin. It was found that the carbon could be removed by exposing the sample to an RF nitrogen soak at a substrate temperature of 500 °C (Fig. 1). The plasma operated at a chamber pressure of 1×10^{-6} torr and 200 Watts RF power. This treatment also visibly improved the RHEED pattern of the GaN. Contamination could be reduced, but

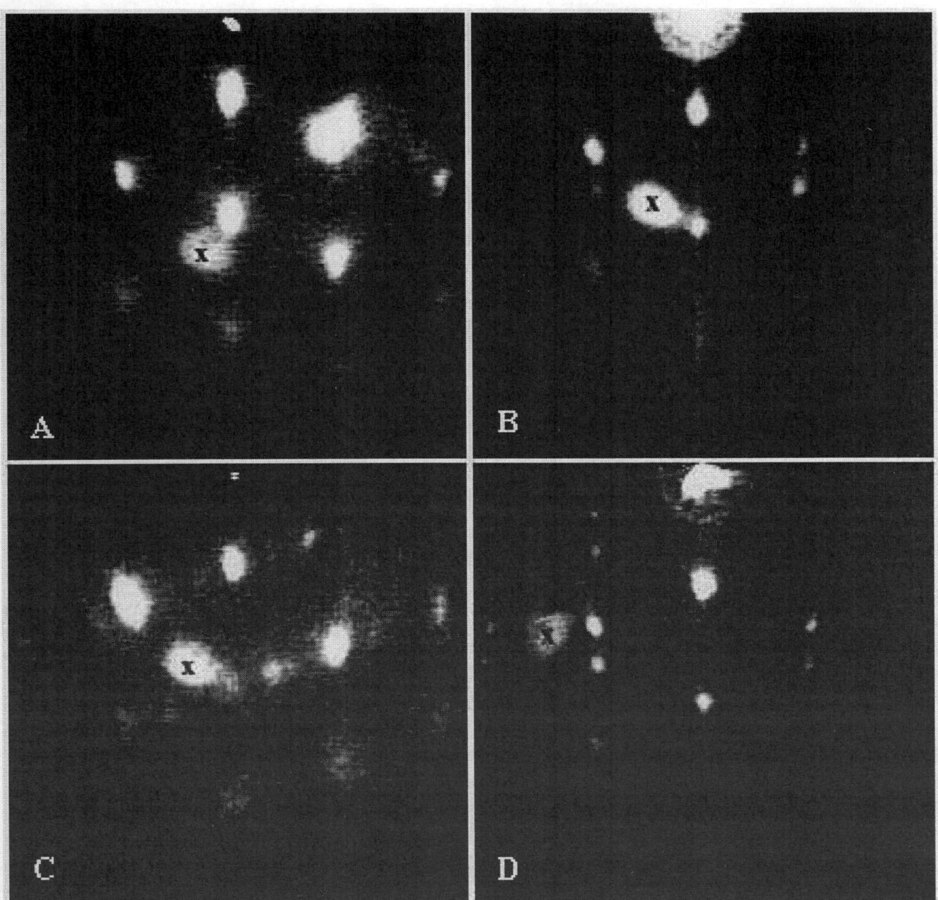

Figure 2: 10keV RHEED images of ZnS on sapphire (A) grown at 400 °C and showing no twins, (B) grown at 250 °C with sulfur cracking zone temperature of 700 °C and showing (111) twins, (C) grown at 250 °C with cracking zone at 300 °C showing (11$\bar{1}$) twins, and (D) ZnS:Al,Ag grown on GaN ((111) twinned). The smudge common to all pictures is due to a defect in the phosphor screen (x).

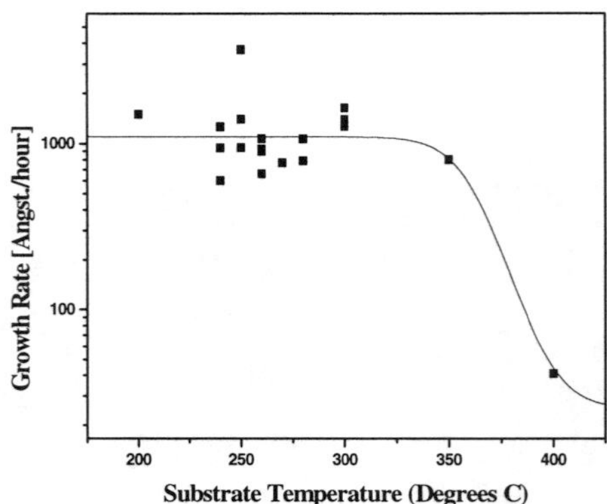

Figure 3: Temperature dependence of the growth rate of ZnS on sapphire (0001). The line is meant to guide the eye.

not eliminated, by thermal cleaning alone. The sapphire substrates were thermally treated at 500 °C for 30 minutes prior to growth, after which sharp RHEED streaks were manifest. We found that etching of substrates made no noticeable difference.

RESULTS AND DISCUSSION

Preceding MBE growth the GaN and sapphire substrates showed sharp RHEED streaks. After opening the source shutters the RHEED pattern quickly became spotty for most films, suggesting that growth proceeds in the 3D (Volmer-Weber) mode. The spotty pattern persisted throughout the growth period, and indicates that the ZnS films are of the zincblende structure with (111) orientation and are single crystalline or composed of mosaic crystallites which are aligned with the underlying substrate lattice. Most films displayed twinning in the basal plane (111) (Fig. 2(B)) which is commonly observed for growth on a (111) oriented substrate [18, 13], although this twinning was absent and replaced by (11$\bar{1}$) twins for some films grown with a lower sulfur cracking temperature of 300 °C (Fig. 2(C)). Films grown at the higher temperature of 400 °C showed no visible twinning (Fig. 2(A)) [20] and show superior surface morphology, although they suffer from extremely low growth rates. Antithetically, ring-like patters were observed for films grown at very low temperatures or with poor surface preparation. Streaky patterns were not observed for any of the samples grown at temperatures between 150 °C-400 °C and with source flux beam equivalent pressures (BEPs) of $(0.3 - 2.0) \times 10^{-7}$ torr.

Film thicknesses were measured by depth profiler and by spectroscopic ellipsometry, and growth rates were calculated assuming a constant rate throughout the growth period. Fig. 3 shows the calculated rates as a function of growth temperature. Rates of up to 0.4 μm per hour were observed for the lower growth temperatures, with rapidly diminishing rates for temperatures above 350 °C. This trend is in agreement with other work [13, 11, 18].

High resolution x-ray diffraction analysis was performed using a four crystal (Ge (220)) diffractometer and Cu Kα xrays. Results indicated that the films are composed of mosaic crystalline domains which are slightly tilted and rotated from one another. Such mosaic films are commonly observed in other materials systems where large lattice mismatch exists,

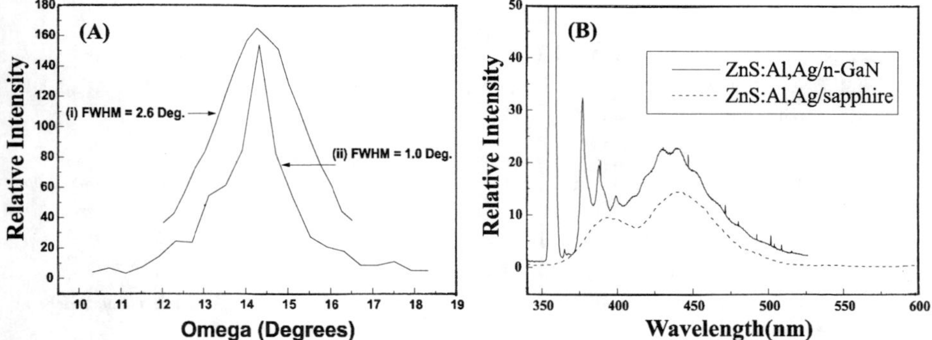

Figure 4: (A) X-ray rocking curves for ZnS grown on sapphire at (i) 250 °C and (ii) 400 °C. (B) Photo-luminescence spectra of ZnS doped with silver and aluminum on unintentionally doped GaN (solid), and sapphire (dashed).

such as Si on Al_2O_3, and GaAs on Si. The FWHM of (111) peaks were consistently close to 400 arcsec for $\omega/2\theta$ scans, invariable with growth conditions. X-ray rocking curves (ω scans) showed peak widths of up to several degrees, depending on growth temperature, film thickness, and doping (Fig. 4(A)). The FWHM of ω scans provides an indication of the degree of mosaic tilting that is present in the film. Undoped samples grown at 400 °C showed the best as grown rocking curve FWHM of approximately 60 arcminutes. Further improvement in crystalline quality was obtained by post-annealing at 1000 °C in a high pressure ($\sim$ 10 atm) sulfur environment [21].

Low temperature (5 K) photoluminescence measurements were made using the 325 nm wavelength of a HeCd laser. The samples doped with Ag and Al produced a bright blue luminescence, with a spectrum as shown in Fig. 4(B). Visible are the characteristic silver high (390 nm) and low (440 nm) energy emission bands [22]. Also present in the spectrum is the near band edge excitonic structure (350 nm - 400 nm) of the unintentionally doped n-type GaN substrate. The ZnS bandgap is too wide to probe near band edge features with this laser.

SUMMARY

Zinc sulfide thin films were grown on GaN and sapphire substrates by MBE, with the GaN surface being effectively cleaned prior to growth by exposure to an RF nitrogen plasma. RHEED patterns indicated that monocrystalline films could be grown on both substrates, but most contained microtwins in the (111) or (11$\bar{1}$) planes. No twins were visible in the films grown at 400 °C. Typical growth rates were measured at 0.1-0.2 μm/hour for temperatures below 350 °C and were much lower for higher temperatures. The samples were characterized using photoluminescence and X-ray diffraction. X-ray peaks typically had FWHM of 400 arcsec for $\omega/2\theta$ scans, and larger widths for ω scans, indicating a degree of mosaic tilts. The PL from the ZnS films doped with Ag and Al displayed bright blue luminescence.

ACKNOWLEDGEMENTS

This work was supported by Advanced Research Project Agency, and monitored by Office of Naval Research under Grant No. N00014-92-J-1845.

References

[1] C. N. King, J. Vac. Sci. Technol. A, **14**, 1729 (1996).

[2] Handbook of Display Technology, J. L. Castellano, (Academic Press, San Diego CA, 1992) Chapter 6.

[3] Y. Sato, N. Takahashi, and S. Sato, Jpn. J. Appl. Phys. **35**, L838 (1996).

[4] S. Yamaga, Physica B. **185**, 500 (1993).

[5] B. J. Wu, L. H. Kuo, J. M. Depuydt, G. M. Haugen, M. A. Haase, and L. Salamancariba, Appl. Phys. Lett. **68**, 379 (1996).

[6] S. Nakamura, N. Iwasa, M. Senoh, and T. Mukai, Jpn. J. Appl. Phys. **31**, 1258 (1992).

[7] H. Amano, M. Kito, K. Hiramatsu, and I. Akasaki, Jpn. J. Appl. Phys. **28**, L2112 (1989).

[8] C. Yuan, T. Salagaj, A. Gurary, P. Zawadzki, C. S. Chern, W. Kroll, and R. A. Stall, J. Electrochem. Soc. **142**, L163 (1995).

[9] M. W. Wang, J. O. McCaldin, J. F. Swenberg, T. C. McGill, and R. J. Hauenstein, Appl. Phys. Lett. **66**, 1974 (1995).

[10] J. O. McCaldin, M. W. Wang, and T. C. McGill, J. Crystal Growth **159**, 502 (1996).

[11] M. Yokoyama, K. Kashiro, and S. Ohta, J. Crystal Growth **81**, 73 (1987).

[12] I. P. McClean and C. B. Thomas, Semicon. Sci. and Technol. **7**, 1394 (1992).

[13] T. Yao and S. Maekawa, J. Crystal Growth **53**, 423 (1981).

[14] J. W. Cook, Jr., D. B. Eason, R. P. Vaudo, and J. F. Schetzina, J. Vac. Sci. Technol. B **10**(2), 901 (1992).

[15] K. B. Ozanyan, L. May, J. E. Nicholls, J. H. C. Hogg, W. E. Hagston, B. Lunn, and D. E. Ashenford, Solid State Commun. **97**, 345 (1996).

[16] O. Kanehisa, M. Shiiki, M. Migita, and H. Yamamoto, J. Crystal Growth **86**, 367 (1988).

[17] S. Ohta, K. Kashiro, and M. Yokoyama, J. Crystal Growth **87**, 217 (1988).

[18] K. Yoneda, T. Toda, Y. Hishida, and T. Niina, J. Crystal Growth **67**, 125 (1984).

[19] J. E. Ayers, S. K. Ghandhi, and L. J. Schowalter, J. Crystal Growth, **113**, 2156 (1991).

[20] K. Ichino, T. Onishi, Y. Kawakami, S. Fujita, and S. Fujita, J. Crystal Growth **138**, 28 (1994). See Discussion section.

[21] Z. Z. Bandić, E. C. Piquette, J. O. McCaldin, and T. C. McGill, in preparation.

[22] Physics and Chemistry of II-VI Compounds, Eds. M. Aven and J. S. Prener (North-Holland Amsterdam, 1967), Chapter 9.

Part IV

Structural Properties

STRUCTURAL AND OPTICAL PROPERTIES OF HOMOEPITAXIAL GaN LAYERS

J.M.BARANOWSKI[1,2], Z.LILIENTAL-WEBER[3], K.KORONA[1], K.PAKUŁA[1], R.STĘPNIEWSKI[1], A.WYSMOŁEK[1], I.GRZEGORY[2], G.NOWAK[2], S.POROWSKI[2], B.MONEMAR[4] and P.BERGMAN[4]

1. Institute of Experimental Physics, University of Warsaw, Hoża 69, 00-681 Warsaw, POLAND, <jacek@fuw.edu.pl>
2. High Pressure Research Centre Polish Academy of Sciences, Sokołowska 29/37, 01--142 Warsaw, POLAND,
3. Centre for Advanced Materials, Lawrence Berkeley National Laboratory, Berkeley, CA 94720, USA,
4. Department of Physics and Measurements Technology, Likoping University, S-581 83 Linkoping, SWEDEN

ABSTRACT

The review of structural and optical properties of homoepitaxial layers grown by MOVCD on single crystals GaN substrates is presented. The TEM technique is used to characterise the structural properties of epi-layers. It is found that the structural properties of GaN homoepitaxial layers are determined by the polarity of the substrate surface on which the growth takes place. It is shown that threading dislocations are present only in the layers grown on the [0001] "smooth" surface. On the other hand the layers grown on the [0001] "rough" surface are free from vertical defects. The characteristic feature of the growth on the "rough" surface are pinholes. The optical properties of homoepitaxial layers are predominantly determined by the growth polarity as well. It is shown also that the reflectivity measurement is the most precise way to determine the exciton energies and that emissions due to free excitons are strongly affected by polariton effects.

INTRODUCTION

Recent advances in heteroepitaxial growth of GaN by MOCVD have demonstrated that a high quality layers can be grown on sapphire substrates. The heteroepitaxial layers exhibit sufficiently good properties for production light emitting diodes[1] and injection diode lasers operating in blue wavelength region[2]. There is a lot of interest in understanding the growth mechanisms and defect structure. The investigation of homoepitaxial layers, which are of much higher quality than the heteroepitaxial ones can be very helpful to understand several growth problems and defect formation mechanisms. However, in contrary to many papers devoted to heteroepitaxy there is relatively a small number of papers about growth of homoepitaxial layers. This is because single crystals of GaN are rarely available. The first results on the homoepitaxial growth of GaN MOCVD epitaxial layers have been already reported.[3-6] In this paper a progress of structural and optical investigation of homoepitaxial layers is presented.

STRUCTURAL PROPERTIES OF HOMOEPITAXIAL LAYERS

Bulk GaN single crystals which have been used as substrates in the epitaxial growth have been grown from diluted solution of atomic nitrogen in the liquid gallium at temperature 1600°C

Mat. Res. Soc. Symp. Proc. Vol. 449 © 1997 Materials Research Society

and at nitrogen pressure of about 15-20 kbar by the method described previously[7]. GaN single crystals grown by this method have a plate-like shape of an area of 10 - 50 mm^2 and a thickness of about   100 - 200 μm.   These single crystal plates have the hexagonal [0001] c-axis perpendicular to the surface. Crystals grown by this method have electron concentration close to 5×10^{19} cm^{-3}. In spite of a high concentration of point defects the structural quality of GaN single crystals are very good indeed. The X-ray measurements have shown that the width of rocking curve for 00.4 Cu $K_{\alpha 1}$ reflection ranges between 30 and 50 arcsec.[7] The GaN platelets have relatively flat surfaces which may be used for the MOCVD growth. However, the preliminary results on mechanically polished surfaces will be presented in this paper. The GaN epitaxial layers have been grown in a horizontal atmospheric pressure MOCVD system adapted for nitrides growth. The trimethylgallium (TMG) and NH_3 have been used as sources of Ga and N respectively, in addition to H_2 as a carrier gas. The growth took place at temperatures ranging from   800°C to 1050°C and the flow of gases have been chosen in such a way that layers of about 1μm - 2μm thickness have been obtained. The growth of GaN layers was realised directly on single crystals substrates without deposition of a low temperature nucleation layer. It is known that the [0001] direction of GaN is polar, therefore, one surface should be terminated with Ga atoms and the other one with N atoms. The transmission electron microscopy (TEM) revealed that one of the surface is atomically flat with steps equal to 2 - 3 monolayers, and the other is rough, covered with pyramids of the 10-30 nm hight.[8-10]

Convergent - beam electron diffraction (CBED) was applied to determine polarity of the substrate bulk crystal and epilayers. It was shown earlier for bulk GaN samples that the bond along the c-axis between N and Ga atom is arranged in such a way that Ga was pointing towards the smooth surface and N towards the rough one[8-10]. This polarity assignment is opposite to the one obtained for polished GaN crystals. [6] Because the reason for this disagreement is not clear yet, a "smooth" and a "rough" surface will be used in this work.

TEM studies showed that the MOCVD epitaxy retains the polarity of the substrate. However, it has been found that two [0001] oriented GaN surfaces are not equivalent from point of view of properties of the grown homoepitaxial layer. In particularly the layer quality and type of defects present in the layer depend on the substrate polarity.

In epi-layer grown on the 'smooth' surface threading dislocations (Fig. 1a) and vertical defects (inversion domains) could be found. It was found that all threading defects started from the dislocation loops formed at the interface. In some cases threading dislocations were grouped into bunches. The estimated concentration of these threading defects was close to 5×10^6 cm^{-2}. As far as inversion domains are concerned, which are present at density 5×10^5 cm^{-2}, it was found that they originate from dislocation loops at the interface as well. In additional to these defects interstitial type dislocation loops and pinholes associated with inversion boundaries could be occasionally found in this layer. Beside described defects, a dark contrast clearly seen in Fig. 1a have been observed along the interface. Detailed high resolution TEM studies showed that this contrast is not related to any structural defect. Bright and dark field images taken with different diffraction conditions showed that some inhomogenity or segregation takes place within 35 nm thick layer at the interface. It should be pointed out that the interface between substrate and layer is very flat and relatively sharp. It seems that interface inhomogenities observed by TEM can be related with a some deviation from the stoichiometry which results from the decomposition of GaN at the growth temperature.

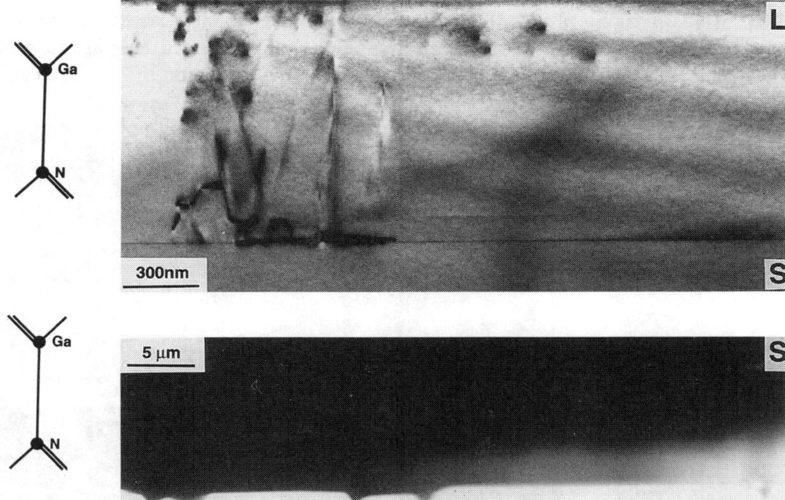

Fig. 1. (a) TEM micrograph of cross-section sample of the epitaxial GaN layer (L) grown on the „smooth" side of the substrate (S). Note formation of threading dislocations originating from the defect at the interface and dark contrast line at the interface. Detailed study of the interface region showed that this dark line represents inhomogeneities within 35 nm thick layer.
(b) Cross-section through the bulk GaN substrate (S). Only „rough" side of the bulk substrate is shown (on the opposite side from the layer).

Homoepitaxial growth has been done without buffer layers. Such growth requires that TMG is introduced into the MOCVD system at the temperature growth begins - close to 1050°C. However, at this temperature in the presence of H_2 and NH_3 the surface of the substrate is already affected by the decomposition of GaN. It has been found that the „smooth" surface of the substrate is more sensitive for the decomposition process of GaN than the „rough" one. In particular, it has been found that for homoepitaxy on the "smooth" surface the initiation of the growth have to be done at lower temperature than in the case of growth on the "rough" surface.

In GaN epi-layer grown on the „rough" surface no threading dislocations were observed. The cross section through the epi-layer grown on this surface is shown in Fig.2. No dark contrast at the interface between the bulk GaN substrate and the epi-layer could be found. The growth on this polarity of the surface is so good that the TEM could not conclusively determine the exact position of the interface. Only interstitial type dislocation loops mostly formed on c-planes with a density in the range 1×10^7 cm^{-2} were found. However, a high density of pinholes (in the range of 10^6 - 10^7 cm^{-2}) was found in the investigated epi-layer. The presence of pinholes of the hexagonal shape on the surface is a characteristic feature of the growth on this surface. There are not dislocations below and around the pinhole region as it is seen in Fig. 2a. It may be expected that some of pinholes originate from the interface. Pinholes seems to indicate places on the surface of the substrate on which due to contamination the nucleation of GaN is more difficult. The concentration of pinholes can be reduced when the growth on mechanically polish surface is taking place. It has been observed that the pinholes have tendency of closing themselves during the growth. However, a possibility that some of them are formed during the growth can not be excluded. A detailed discussion of the formation of pinholes can be found in this volume[11].

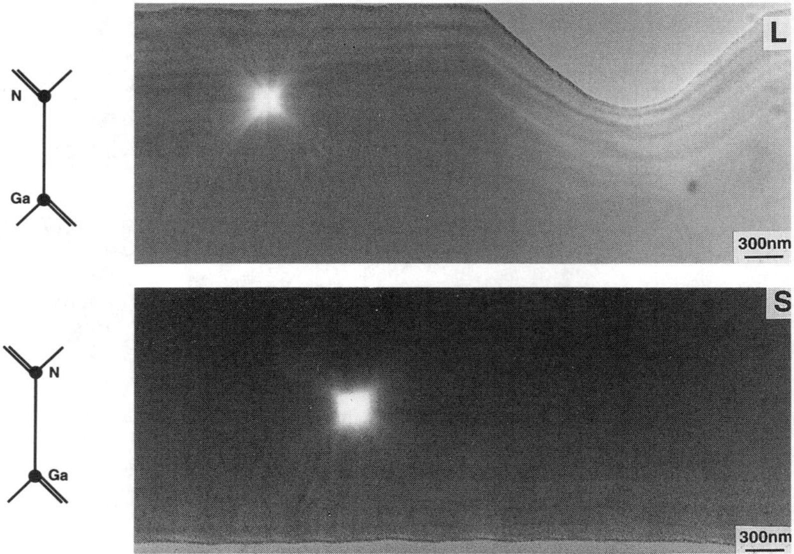

Fig. 2 (a) TEM micrograph of cross section sample of the epitaxial GaN layer (L) grown on the rough side of the substrate (S). Note high perfection of this layer and lack of any mark of the interface. Several pinholes shown in (a) were found in this cross-section sample.
(b) Cross-section through the bulk GaN substrate (S). Only opposite side from the layer (smooth side) is shown. White marks in the layer and in the substrate indicate the areas used for the particular CBED patterns.

The emission spectra are sensitive to the polarity of the surface on which growth takes place. This is shown in Fig.3 where spectra of layers grown on the "smooth" and the "rough" surface of the mechanically polished [0001] platelets are compared.

In spite that the two layers have been grown in the same conditions (with exception of the growth initiation temperature which is lower for the growth on the "smooth" surface) they differ in relative concentration of donors and acceptors. The strong emission lines due to neutral acceptor bound excitons are only present in the layer grown on the "rough" surface of the substrate, on which the interface is free from any sign of deviation from stoichiometry. On the other hand, the emission spectrum of the layer grown on the "smooth" surface is dominated by the neutral donor bound exciton. That indicates that in layers grown on the "sooth" surface predominant defects are of donor character. Deviation from stoichiometry connected for example with an excess of gallium could be consistent with this result.   Excess of gallium at the interface, created in the first stages of the growth, may diffuse into the layer and create

396

donors due to nitrogen vacancies or gallium interstitial. It is likely that deviation from stoichiometry leading to excess of Ga is visible in TEM in a form of dark contrast line at the interface. This hypothesis has a strong confirmation by the results of the emission spectra of two layers grown on unpolished surfaces shown in Fig.4. The growth and growth initiation temperature (the other growth conditions as well) were the same for layers of both polarity. The spectrum of the layer grown on the "smooth" surface is very broad and emission of light of energy above the energy gap of GaN is observed. This is the most likely due to Burnstain effect caused by a large concentration of donors.

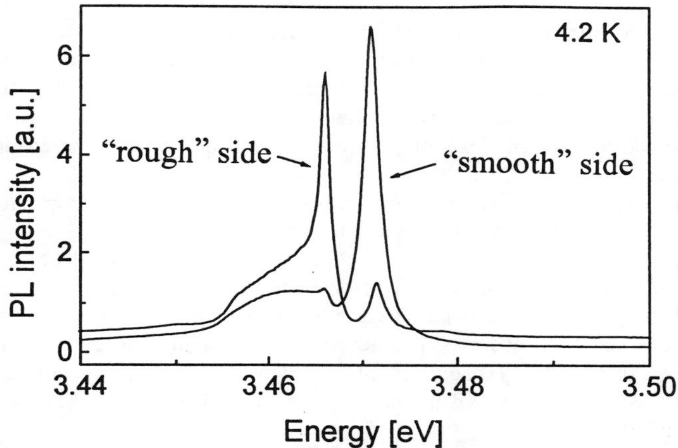

Fig.3. The emission spectra of two homoepitaxial GaN layers grown at the same temperature close to 1000°C on the "smooth" and "rough" surfaces of the GaN substrate (with a low temperature growth initiation done at the "smoth" side).

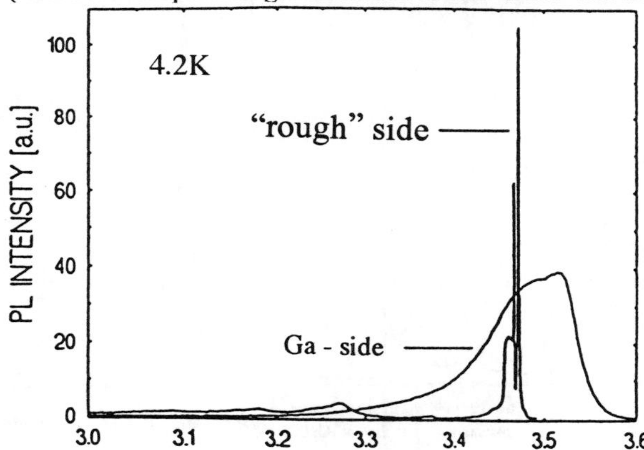

Fig. 4. The emission spectra of two homoepitaxial layers grown at 1000°C on the "smooth" and "rough" surfaces of the GaN substrate (without a low temperature growth initiation done at the "smooth" side).

The morphology of the surface of the epi-layer depends strongly on the state of preparation of the surface. This is in particularly important in the case of the growth on mechanically polished samples. It has been found that in the case of a well prepared surface of the "rough" surface, the step flow of steps of the same length and of the height equal to double atomic layer have been observed. A discussion of the surface morphology of homoepitaxial layers grown on polished GaN substrates can be found in this volume[12].

OPTICAL PROPERTIES OF HOMOEPITAXIAL LAYERS.

The optical properties of homoepitaxial layers are improved considerably comparing ones grown on sapphire. Therefore information obtained on homoepitaxial layers can be used to verify important parameters of GaN, such as free exciton binding energies and energies of excitons bound to acceptors and donors and photon - exciton coupling constants.

The luminescence within the exciton region of a good quality GaN homoepitaxial layer is shown in Fig. 5.

One can distinguish five different emission processes in this energy region. The first process is connected, with free exciton (FE) recombination which takes place in the range 3.475 - 3.490 eV and will be described later on. The emission line at 3.474 eV has been previously assigned as due to neutral donor bound exciton from deeper B valence band[4]. However, it is possible that it is connected with bound exciton to ionised donor. Very similar to the presented in Fig.4 spectrum has been observed in ZnO[13] and the binding energy of exciton to ionised donor has been found to be pronounce smaller than the one to the neutral donor. This is consider as the second process and the binding energy of 3 meV of exciton to ionised donor (D^+X) has been obtained.

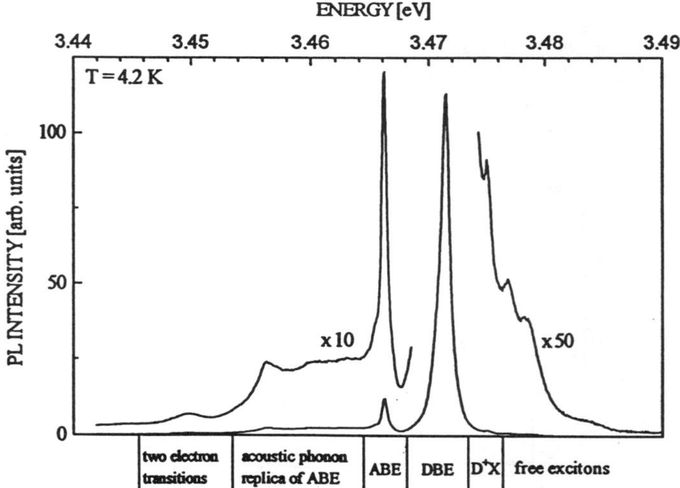

Fig. 5. The luminescence spectrum of undoped homoepitaxial GaN layer in the exciton region at 4.2K.

The third emission process is the well known recombination due to the exciton bound to the neutral donor (DBE), which takes place at about 3.472 eV. This process has been recently identify in Zeeman experiment[14]. The fourth process, also identify in Zeeman experiment[14] is connected with emission of excitons bound to neutral acceptors (ABE) observed close to 3.466 eV. The ABE emission line has a characteristic low energy shoulder, which is most likely due to coupling of this transition with acoustic phonons. A similar coupling of acoustic phonons with acceptor bound exciton emission have been observed in and CdS[15].   The fifth emission process responsible for the 3.45 eV line is the most likely connected with two electron transition. A similar process in which exciton bound to a neutral donor recombines leaving the donor in the first excited state has been observed in ZnO[13]. The energy difference between this line and the neutral donor bound exciton line is 22meV which corresponds to 30meV hydrogenic donor ionisation energy. It was found that decay kinetics of the 3.45eV line is close to the decay time of DBE and much faster than the decay time of ABE[16] acoustic phonon replica on which tail the measured 3.45 eV transition is superimposed.

Result of reflectivity on a good quality homoeptaxial layer grown on the N terminated unpolished surface is shown in Fig.6. The fit of calculated reflectivity to the experimental one is shown in Fig.6 as well. Details of the fit, which includes the polariton effects, are described elsewhere[17].

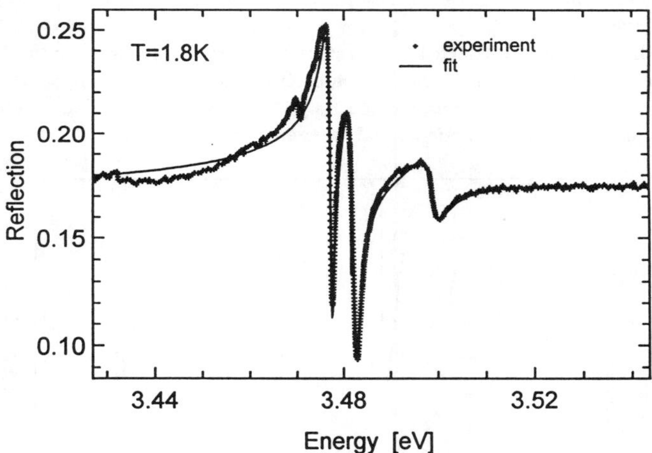

Fig.6. The experimental reflectance spectrum (dashed line) of homoepitaxial GaN layer and fitted theoretical curve (solid line).

The best fit has been obtained for following values of free exciton energies, the damping parameter G and longitudinal-tranversal splitting D_{LT} given in Table I.

Table I. The values of free exciton A, B and C energies, damping parameters G, longitudinal - transversal splitting D_{LT}, and polariability $4pa_0$ obtained from the fit to the reflectivity data presented in Fig.6.

Exciton	Energy (eV)	G (meV)	D_{LT} (meV)	$4pa_0$
A	3.4767	0.5	1.15	0.0039
B	3.4815	1.1	1.10	0.0037
C	3.4987	3.3	0.55	0.0018

A measure of coupling between exciton and photon given by D_{LT} is sometimes expressed by polarizability a_0 which is also included in the Table I. The values of the free exciton energies has been obtained from the best fit to reflectivity observed in one particular sample. However, one should remember that due to some residual strain present even in homoepitaxial layers, parallel shifts of excitonic lines even up to 1 meV have been observed.

The value of D_{LT} = 0.64 meV for GaN obtained on heteroepitaxial layer has been recently reported[18]. The values of $4pa_0$ obtained for GaN can be compared to $1.2x10^{-3}$, $14.0x10^{-3}$ and $7.7x10^{-3}$ obtained for GaAs[19], CdS[20] and ZnO[21] respectively. The longitudinal - transverse splitting D_{LIT} for these materials are equal 0.07 meV, 2.2 meV and 2.1 meV respectively. The exciton - photon coupling obtained for GaN is stronger than in GaAs but definitely weaker than in II-VI compounds and it seems to be connected with polarity of this material.

In addition to the strongly dispersed reflectivity lines due to the free excitons, there is also a weak structure connected with the neutral donor bound exciton at 3.472 eV. The DBE structure has been observed in heteroepitaxial GaN layers in calorimetric absorption measurements[18]. The DBE is usually not observed in reflectivity with one exceptional case for CdS[22]. The observation of DBE structure in reflectivity in GaN is an indication of a good quality of the homoepitaxial layer.

Energies of free excitons obtained from reflectivity are more precise than the ones obtained from luminescence measurements. The polariton effects which are usually neglected are also important in the interpretation of emission spectra. The emission and reflectivity spectra in the free exciton region of the same homoepitaxial layer are shown in Fig.7.

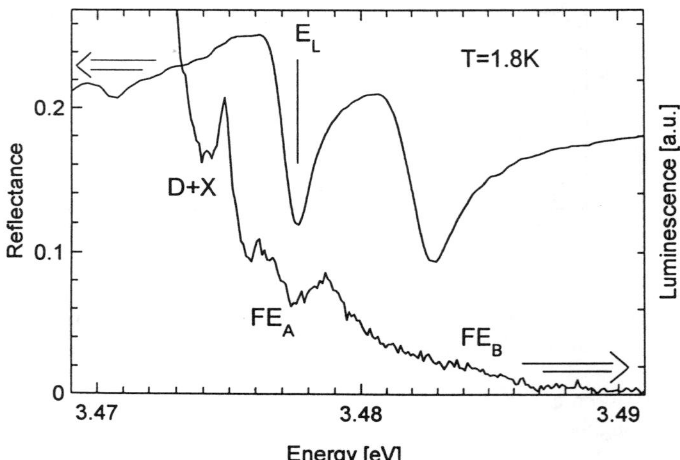

Fig.7. Reflectivity and luminescence within the free A and B exciton region of homoepitaxial GaN layer with a relatively high concentration of donors.

The emission spectrum shown in Fig.7 has clearly two peaks in the free exciton A region, and the reflectivity spectrum measured at the same sample shows the minimum, which falls between two emission peaks. The minimum of reflectivity corresponds to longitudinal exciton energy E_L. The intensity of emission connected with DBE in relation to the free exciton one is very high (ratio equal to about 100), what indicates a relatively large concentration of donors in this unintentionally doped sample. Therefore the scattering of polaritons by neutral donors may be expected to be important in this layer.

The emission spectrum of a relatively more pure layer is shown in Fig.8. In this layer the ration of intensity of DBE to free exciton one is close to 4 and therefore, a smaller concentration of donors is expected. It is also seen in Fig.8 that instead of the double peak a single one is connected with the free A exciton. Such effect has been also observed in GaAs, where   depending on the donor concentration a single or double free exciton peak was obtained[19].

It is known that in a direct band gap semiconductors, a strong coupling of free excitons with photons of the same energy exists. This coupling results in a mixed mode excitation called an exciton polariton which characterises in a two branch polariton dispersion curve. The existence of two polariton branches complicates description of interaction of light with a crystal. The Fresnel equations no longer provide sufficient boundary conditions and addition boundary conditions (ABC) have to be introduced containing the information about what fraction of the energy travels on which polariton branch. The polariton effects have been widely studied in GaAs[19] and CdS[20]. Clear reflectivity spectra obtained on homoepitaxial layers shown in Fig.6 enable us to study polariton structure of GaN.

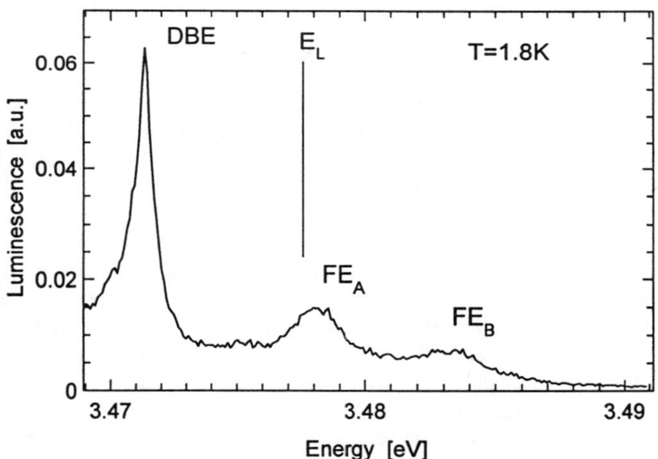

Fig.8. Luminescence connected with the free A and B excitons in a pure homoepitaxial
    GaN layer

The polariton dispersion curves for GaN have been calculated for parameters corresponding to the ones obtained from the reflectivity analysis and for ABC proposed by Pekar[23] and are shown in Fig.9. The positive values of real part of k^2 correspond to propagating modes and the negative ones give an imaginary values of the wavevector k and correspond to the evanescent mode. On the lower polariton branch (LPB), at large wavevectors, the polaritons are predominantly excitonlike. On the upper polariton branches (UPB) above the resonances connected with the A and B exciton energy the polaritons are photonlike with increasing wave vector.

It was found that in the GaN case a damping has significant influence on polariton disperse curves. The important difference between zero damping case (often use in the interpretation of polariton effects) and finite damping is, that polariton branches exist for all values of energies. Below the longitudinal exciton E_L energy (for polariton originating from A free exciton) there is one propagating and one evanescent mode.

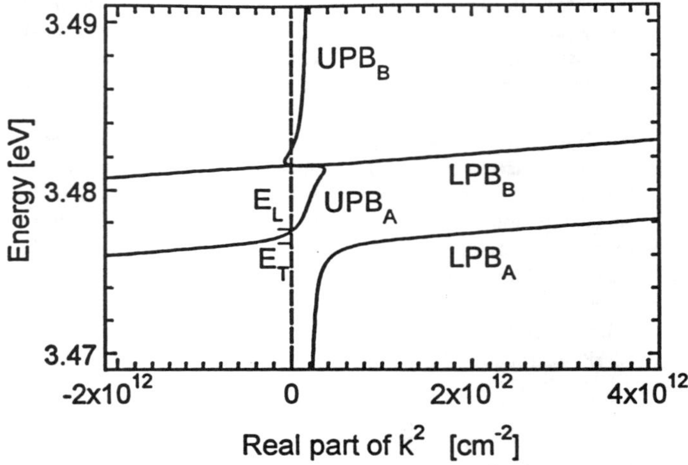

Fig.9. The calculated polariton energies versus real part of k^2 (where k is the wavevector of polariton) for GaN. Polariton branches presented in Fig. arise from photon coupling with A and B free excitons. The E_L and E_T indicated in Fig. correspond to longitudinal and transverse A free exciton energy.

The polariton dispersion has a profound effect on the luminescence mechanism due to polariton scattering process. Excitonlike polaritons can be inelastically scattered to the lower energy region of the dispersion curve which creates a "bottleneck" to further inelastic scattering due to a lack of suitable small wave vector phonons. In addition to this, the scattering from the LPB to the UPB photonlike region of the dispersion curve may takes place as well. Polariton luminescence results from those polaritons which upon reaching the surface, are converted into photons leaving the crystal. The polaritons from the UPB have more photonlike character and provide a path for polaritons to escape the crystal. Therefore the luminescence from the energy region above the E_L is primarily from the UPB. A single emission peak may be expected as it is seen in Fig.8.   However, much more common, not only in GaN but also in other semiconductors, it is a doublet structure such as shown in Fig. 7. Such doublet has been observed in n-type doped GaAs and it has been suggested that the elastic polariton impurity scattering is responsible for this effect[19].   It has been found also that neutral donors of wave function much more extended in space, scatter at least an order of magnitude stronger than neutral acceptors[20]. Therefore, a relatively high concentration of donors may create a path for polaritons to be efficiently scattered from the UPB to the LPB at the bottleneck energy, at which they finally reach the surface and convert themselves into photons. These explain qualitatively the results of luminescence shown in Fig.7   corresponding to a higher concentration of donors.

The polariton effects in free exciton emission are difficult to interpret quantitatively because they are connected with scattering of polaritons within the excited by light layer of the crystal.

It is relatively easy to include polariton effects in the analysis of reflectivity data, which are free from the scattering problem.

CONCLUSIONS

It has been shown that characteristic feature of the homoepitaxial layers grown on "smooth: surface of GaN substrate are threading dislocations of the density close to 5×10^6 cm^{-2} originating at the interface. Inhomogeneities observed by TEM at the interface in layers grown on the "smooth" GaN surface are the most likely connected with decomposition of GaN at the growth initiation temperature. The growth on the "rough" GaN surface with opposite polarity is more stable. Homoepitaxial layers grown on the "rough" side are of the best quality without threading dislocations and of very good interface between the bulk and the layer. However, growth on this side results in the formation of pinholes, whose origin is discussed in detail by Liliental-Weber et al [11] in this volume.

It has been shown that the reflectivity analysis gives the most accurate values of free exciton energies and proves the important role of polariton effects in GaN. It has been shown also that the impurity scattering of polaritons leads to emission from UPB and LPB which results in double peak of FEa luminescence. However, in more pure samples in which apparently the acoustic phonon scattering of polaritons is more than impurity scattering the emission is predominantly from the photonlike UPB.

ACKNOWLEDGMENTS

This work was partially supported by State (POLAND) Committee for Scientific Research Grant NO. 7 T08A 06110. ZLW work was supported by the Director, Office of Energy Research, Office of Basic Energy Sciences, Materials Science Division, of the U.S. Department of Energy under Contract No. DE-AC03-76SF00098.

REFERENCES

1. S.Nakamura, M.Senoh, N.Iwasa and S.Nagahama, Jpn. J. Appl. Phys. **34**, L979 (1995)

2. S.Nakamura, M.Senoh, S.Nagahama, N.Iwasa, T.Yamada, T.Matsushita, H.Kiyoku and Y.Sugimoto, Jpn. J. Appl. Phys. **35**, L74 (1996)

3. K.Pakuła, A.Wysmołek, K.P.Korona, J.M.Baranowski, R.Stępniewski, I.Grzegory, M.Boćkowski, J.Jun, S.Krukowski, M.Wróblewski and S.Porowski, Solid State Communications, **97** , 919, (1996)

4. J.M.Baranowski and S.Porowski, Proc. of the 23rd International Conference on Physics of Semiconductors, Berlin, ed. M.Scheffler and R.Zimmermann, World Scientific, Singapore 1996, in press

5. J.M.Baranowski and S.Porowski, Proc. of the 9th Conference on Semi-Insulating III-V Materials, Toulose (1996) in press.

6. F. A. Ponce, D. P. Bour, W. Götz, N. M. Johnson, H. I. Helava, I. Grzegory, J. Jun and S. Porowski, Appl. Phys. Lett. 68, 917 (1996); F. A. Ponce, D. P. Bour, W. T. Young, M. Saunders, and J. W. Steeds, Appl. Phys. Lett. 69, 337 (1996).

7. S.Porowski, J.Jun, M.Boćkowski, M.Leszczyński, S.Krukowski, M.Wróblewski, B.Łucznik, and I.Grzegory, Proc. 8th Conference on Semi-Insulating III-V Materials, Warsaw, ed M.Godlewski, World Scientific, p.61 (1994).

8. Z.Liliental-Weber, C.Kisielowski, S.Ruvimov, Y.Chen, and J.Wasburn, Journal of Electronic Materials, 25, 1545 (1996)

9. Z.Liliental-Weber, et al, Mat. Res. Soc. Symp., 395, 351 (1996).

10. Z.Liliental-Weber, C. Kisielowski, X. Liu, L. Schloss, J. Washburn, E.R. Weber, I. Grzegory, M. Bockowski, J. Jun, T. Suski, and S. Porowski, Topical Workshop on III-V Nitrides, Nagoya , September (1995)-Solid State Electr.-in press.

11. Z.Liliental-Weber, Y.Chen, S.Ruvimov, and J.Washburn, Mat. Res.Soc. Symp. Proc., this Symp. - submitted (1996).

12. S.Porowski, M.Boćkowski, B.Łucznik, S.Krukowski, I.Grzegory, M.Leszczyński, G.Nowak, K.Pakuła, J.M.Baranowski, Mat.Res.Soc.Symp.Proc., this Symp. - submitted (1996).

13. C.F.Klingshirn, Semiconductor Optics, Springer, p. 241 (1995).

14.R.Stępniewski, A.Wysmołek, K.Pakuła, J.M.Baranowski, M.Potemski, G.Martinez, I.Grzegory, M.Wróblewski and S.Porowski, Proc. of the 23rd International Conference on Physics of Semiconductors, Berlin, ed. M.Scheffler and R.Zimmermann, World Scientific, Singapore 1996, in press

15. D.C.Thomas and J.J.Hopfield, Phys. Rev., 128, 2135 (1962).

16. P.Bergman, B.Monemar and J.M.Baranowski, unpublished results.

17. R.Stępniewski, A.Wysmołek,K.Korona, J.M.Baranowski, K.Pakuła, M.Potemski, I.Grzegory and S.Porowski, Semiconductor Physics and Technology, submitted for publication.

18. D.Volm, K.Oettinger, T.Streibl, D.Kovalev, M.Ben-Chorin, J.Diener, B.K.Meyer, J.Majewski, L.Eckey, A.Hoffmann, H.Amano, and I.Akasaki, Phys.Rev. B53, 16543, (1996).

19. T.Steiner, M.L.W.Thewalt, E.S.Coteles and J.P.Salerno, Phys Rev. B34, 1006 (1986).

20. I.Broser, M.Rosenzweig, R.Broser, M.Richard and E.Birkicht, phys.stat.sol.(b), 90, 77, (1978)

21. J.Lagois, Phys.Rev. B16, 1699, (1977).

22. K.Bohnert, G.Schmieder, S.El-Dessouki and C.Klingshirn, Solid State Commun. 27, 295, (1978).

23. S.I.Pekar, J.Phys.Chem.Solids 5, 11, (1958).

NANOPIPES AND INVERSION DOMAINS IN HIGH QUALITY GaN EPITAXIAL LAYERS

F. A. PONCE*, D. CHERNS°, W. T. YOUNG°, J. W. STEEDS°, and S. NAKAMURA˜
* Xerox Palo Alto Research Center, Palo Alto, CA 94304
° University of Bristol, H. H. Wills Physics Laboratory, Bristol BS8 1TL, UK
˜ Nichia Chemical Industries, 491 Oka, Kaminaka, Anan, Tokushima 774, Japan

ABSTRACT

In this paper we report that, in addition to dislocations, two other types of defects are observed in high quality GaN thin films. These defects have a filamentary nature, are oriented along the <0001> direction. and may not be easily distinguished from the pure dislocations. Using a combination of conventional electron microscopy with convergent beam electron diffraction techniques we show that one of these types of dislocations consist of nanopipes, which are coreless dislocations with Burgers vectors <0001>. The other type of observed defects consist of inversion domains with $[000\bar{1}]$ orientation within the [0001] matrix. The origin of the inversion domains and nanopipes is discussed.

INTRODUCTION

High quality GaN thin films are characterized by high dislocation densities [1]. Dislocations are associated with a columnar array characterized by an angular distribution in the orientation of the columns of the order of 5 arcmin. Dislocations are generated at the GaN/sapphire interface and propagate in the growth direction; they are found typically in low angle boundaries [2]. The polarity of the lattice (i.e. whether Ga- or N-atoms are on the top of the basal planes) has been determined using convergent beam electron diffraction (CBED). Optoelectronic quality GaN films grow in the Ga-terminated [0001] direction [3]. Dislocations have been characterized by conventional transmission electron microscopy and by large angle convergent beam diffraction techniques, with results indicating that dislocations have Burgers vectors **c**, **a**, and **c+a**, where **c** and **a** are the unit cell vectors of the hexagonal wurzite structure [4,5].

While investigating the microstructure of high quality GaN films, we have identified some defects that tend to be invisible under typical observation conditions in transmission electron microscopy. In this paper we report the observation of filamentary defects with diameters of the order of nanometers, which occur either as hollow tubes or filled with GaN with polarity opposite to that of the matrix.

EXPERIMENT

GaN epitaxy was performed by the two-flow metalorganic chemical vapor deposition technique on (0001) sapphire substrates, following methods previous reported [6]. The material was not intentionally doped and exhibited resistive characteristics typical of high purity, undoped, GaN. Transmission electron microscopy (TEM) specimens were prepared in plan

Mat. Res. Soc. Symp. Proc. Vol. 449 © 1997 Materials Research Society

view, parallel to the (0001) basal plane of the sapphire substrate and the GaN epilayer. Electron transparency was obtained by Ar$^+$ ion beam milling at 4kV.

TEM was performed at 250 kV in a Philips EM430 microscope. Convergent beam electron diffraction patterns were obtained at various zone axes of GaN using a focused probe of 70nm, and a condenser aperture of 200 microns.

NANOPIPES

Fig. 1 shows a dark-field TEM image taken under two-beam diffraction conditions. The area exhibits three types of defects that propagate accurately in the [0001]$_{GaN}$ growth direction. Dislocations (D) are observed in the expected low-angle grain boundaries of the material. Other defects propagating in the same direction as the dislocations are observed. They are labeled as "T" and "I". It is not easy to distinguish these defects in standard TEM cross section views. Closer inspection shows that the "T" defects are actually empty tubes with dimensions of the order of 10nm, as shown in Fig. 2. These hollow tubes have generally constant cross-section in the range 5 to 25nm in diameter; they are faceted and exhibit an irregular hexagonal shape when viewed end-on, as observed in Fig. 2.

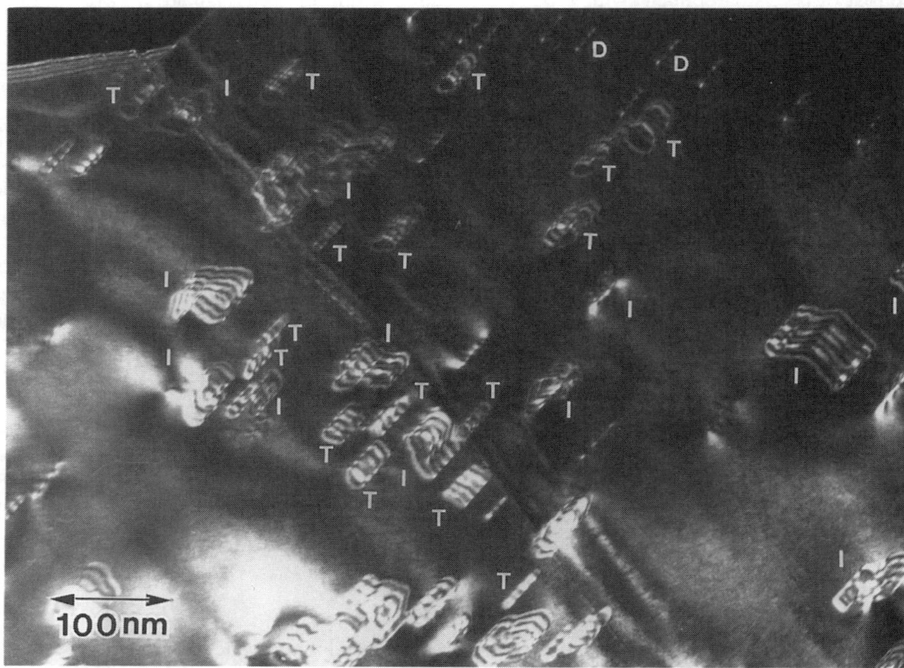

Fig. 1. Plan view transmission electron micrograph showing region with a large density of nanopipes (T) and inversion domains (I). Dislocations (D) belonging to low angle grain boundaries are also observed. This dark field image was taken under two-beam diffracting conditions in $\mathbf{g} = 2\bar{2}02$.

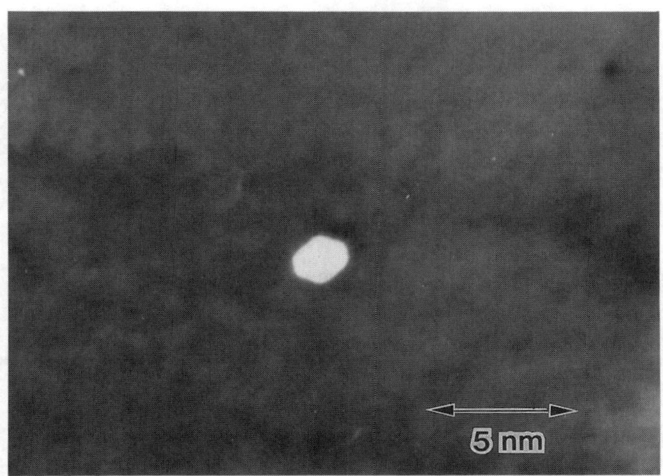

Fig. 2. Transmission electron micrograph of nanopipe observed along its axis. It presents an irregular hexagonal cross section.

Conventional TEM diffraction imaging analysis provides information for the direction of the Burgers vector. Large angle convergent beam electron diffraction (LACBED) imaging can accurately determine the magnitude, direction and sense of the Burgers vectors. LACBED imaging has been performed on the nanopipes, demonstrating that these hollow tubes contain a Burgers vectors in the [0001] direction with magnitude 0.52nm (equal to the c constant). Thus, with a few notable exceptions, nanopipes can be described as coreless dislocations with elementary Burgers vectors of the **c** type [7]. Frank predicted that a dislocation whose Burgers vector exceeded a critical value should have a hollow tube at the core, and predicted an equilibrium core radius of $r_{eq}=\mu b^2/8\pi^2\gamma$, where γ is the surface energy, μ is the shear modulus, and b is the Burgers vector [8]. Recent reports by Qian first noted that the contrast in two-beam bright field TEM images from a tilted nanopipe was similar to dislocation contrast, but no proof was advanced [9,10]. The difficulty lies in being able to measure accurately the magnitude of the Burgers vector, since the actual value for the nanopipe could be an integral fraction of the Burgers vector of a perfect dislocation. The LACBED technique used in this study is a new technique which has found an important application in solving this problem.

The observed tube diameters are too large to agree with the simple Frank equation. An upper limit using existing theoretical data gives 0.5nm for c-dislocations. Thus, coreless dislocations are not in an equilibrium state but may arise by the trapping of dislocations at pinholes at the early stages of growth as shown in Fig. 3a. Trapped dislocations arise if the displacement of the islands do not sum to zero. A mechanism that favors the generation of c-dislocations is presented in Fig. 3b, where it is shown that three-layer steps in the sapphire can couple to a two-layer step in the GaN lattice, thus giving rise to a **c**-type screw dislocation. Those dislocations with an **a**-component of Burgers vector eventually close out, and only the pure **c**-type dislocations can survive to generate nanopipes of constant cross-sections [7].

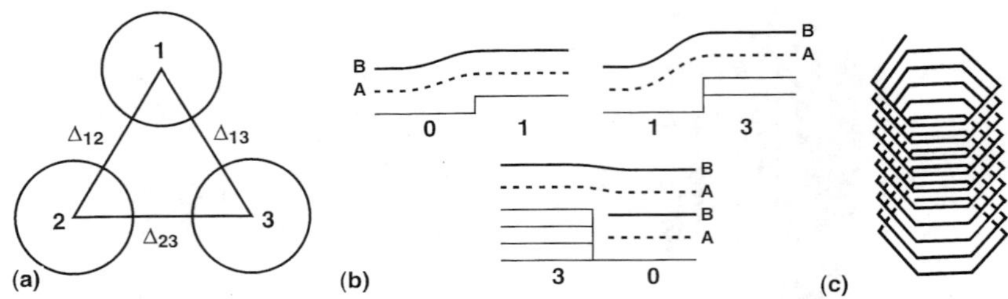

Fig. 3. Proposed mechanism for the generation of nanopipes in epitaxial GaN/Sapphire. (a) A dislocation is trapped if the island displacement $\Delta = \Delta_{12} + \Delta_{23} + \Delta_{31}$ do not sum to zero. (b) Steps on the sapphire surface in unit of $c_{sapphire}/6$ can join to trap a screw dislocation. (c) Addition of deposited material near a nanopipe containing a screw dislocation preserves the helicoidal nature of the defect about its central [0001] axis.

INVERSION DOMAINS

When the foil is tilted to excite a reflection with a c-component, a new set of defects appears which is not visible in reflections of type hki0 (i.e. in $\{10\bar{1}0\}$ or $\{11\bar{2}0\}$ reflections). Such case is shown in Fig. 1, where in addition to dislocations and nanopipes, a large density of defects labeled as "I" is noted. The regions of alternating black-white fringe contrast vary in size up to ~50nm across. The nature of the domains was clarified by comparing CBED patterns taken from the matrix and from regions within the domains which do not overlap the matrix.

Fig. 4 compares CBED patterns taken along the $[1\bar{1}02]$ axis from the matrix (a) and from an adjacent domain (b) in a region where the foil thickness was approximately constant. In each case, the patterns show an asymmetry between $\pm\mathbf{g}$ which is due to the GaN polarity [3]. The reversal of the asymmetry between matrix and domain clearly confirms that the new defects are inversion domains. Using the same orientation as with the CBED patterns shown before, dark field images can be obtained by using the asymmetric reflections. Fig. 4c shows a bright field image of the domain, taken at the $[1\bar{1}02]$ axis. The contrast of the domain is the same as that of the matrix, as expected for images formed using the central (0000) beam. Fig. 4d is a dark field image, using the $(\bar{1}101)$ disk (top central disk in the CBED patterns), and gives a brighter contrast for the domain than for the matrix. Correspondingly, Fig. 4e is a dark field image using the $(1\bar{1}0\bar{1})$ disk (lower central disk), giving a darker contrast for the domain than the matrix. These observations provide definitive evidence that the defects are inversion domains.

It is important to stress that the inversion domains are not easily seen under some commonly used conditions, namely with strong $\{10\bar{1}0\}$ or $\{11\bar{2}0\}$ reflections (see e.g. Ref. 11), and that there is no evidence of significant strain associated with the presence of these defects. These observations indicate that any lattice displacement should be parallel to the c-axis. Further TEM studies, shows that the Ga sublattice is displaced by 3c/8 or 7c/8 along [0001] between the matrix and inversion domains [12]. Two possible models consistent with these observations are shown in Fig. 5. Fig. 5b is an inversion boundary involving a c/2 shift of the lattice in addition to a pure inversion. The shift removes like-atom bonding but requires distortion of the bonds at the boundary itself. Theoretical and experimental evidence supports this model [12, 13].

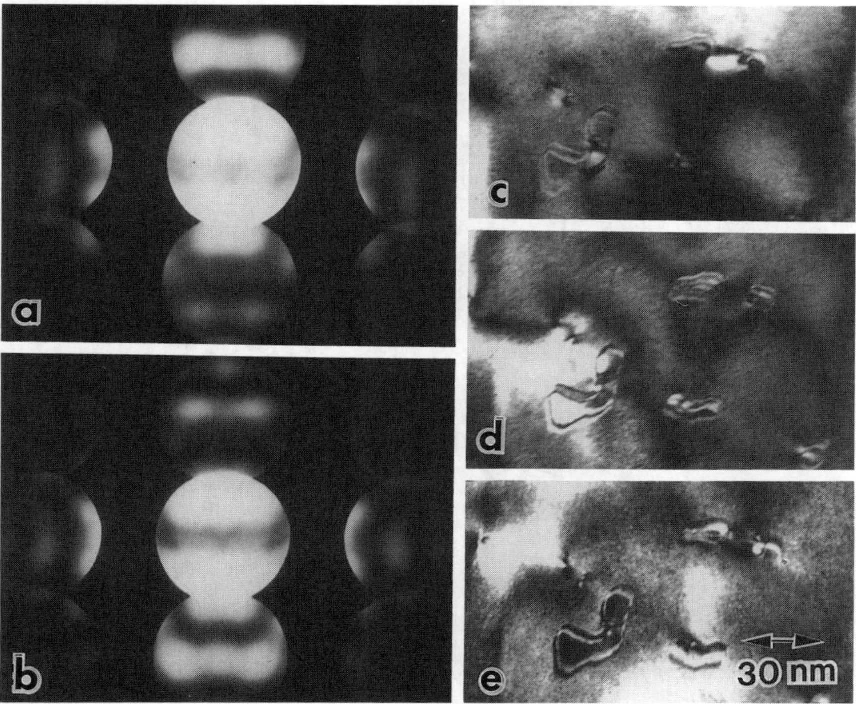

Fig. 4. CBED patterns taken at the [1$\bar{1}$02] zone axis from (a) inside an inversion domain, and (b) the surrounding film (matrix). Bright field (BF) and dark field (DF) TEM images taken at the [1$\bar{1}$02] zone axis of the same inversion domain: (c) BF image showing same contrast as the matrix, (d) DF image using the ($\bar{1}$101) disk resulting in brighter contrast, and (e) DF image using the (1$\bar{1}$01) disk producing a darker contrast for the domain in comparison with the matrix.

CONCLUSIONS

In addition to dislocations, the microstructure of high quality GaN thin films exhibits the presence of nanotubes which are coreless dislocations with Burgers vector c, and inversion domains with orientation $< 000\bar{1} >$ within the <0001> matrix. The nature of these faults has been unequivocally determined by a combination of conventional transmission electron microscopy and convergent beam electron diffraction techniques. Models for the generation of coreless dislocations and for the interface of the inversion domain boundary have been proposed.

ACKNOWLEDGEMENTS

This work was partially supported by Department of Commerce Advanced Technology Program (70NANB2H1241) and by ARPA (Agreement # MDA972-95-3-0008). The Bristol group gratefully acknowledges the use of NATO grant #CRG 960690 in the pursuit of this work.

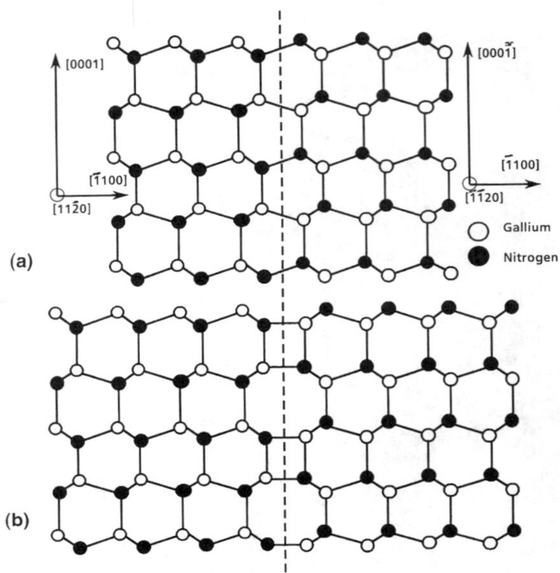

Fig. 5. Possible models for inversion domain boundaries along the prismatic plane of GaN. (a) Pure inversion boundary, (b) pure inversion plus c/2 shift boundary.

REFERENCES

1. S.D. Lester, F.A. Ponce, M.G. Craford, D.A. Steigerwald, Appl. Phys. Lett. **66**, 1249 (1995).
2. F. A. Ponce, J. S. Major, Jr., W. E. Plano, D. F. Welch, Appl. Phys. Lett., **65**, 2303 (1994).
3. F. A. Ponce, D. P. Bour, W. T. Young, M. Saunders, and J. W. Steeds, Appl. Phys. Lett. **69**, 337 (1996).
4. X. J. Ning, F. R. Chien, P. Pirouz, J. W. Yang, and M. A. Khan, J. Mater. Res. **11**, 580 (1996).
5. F. A. Ponce, D. Cherns, W. T. Young, and J. W. Steeds, Appl. Phys. Lett. **69**, 770 (1996).
6. S. Nakamura, Jpn. J. Appl. Phys. **30**, 1620 (1991).
7. D. Cherns, W. T. Young, J. W. Steeds, F. A. Ponce, and S. Nakamura, J. Crystal Growth (1997), in press.
8. F. C. Frank, Acta Cryst. **4**, 497 (1951).
9. W. Qian, M. Skowronski, K. Doverspike, L. B. Rowland, and D. K. Gaskill, J. Cryst. Growth **151**, 396 (1995).
10. W. Qian, G. S. Rohrer, M. Skowronski, K. Doverspike, L. B. Rowland, and D. K. Gaskill, Appl. Phys. Lett. **67**, 2284 (1995).
11. L. T. Romano and J. E. Northrup, MRS Proc. **449**, 423 (1996). (*These Proceedings*).
12. D. Cherns, W. T. Young, J. W. Steeds, F. A. Ponce, and S. Nakamura, (1997), submitted.
13. J. E. Northrup, J. Neugebauer, and L. T. Romano, Phys. Rev. Lett. **77**, 103 (1996).

DETERMINATION OF THE PERCENTAGE OF THE CUBIC AND HEXAGONAL PHASES IN GaN WITH NEXAFS

M.Katsikini[a,b], E.C.Paloura[a], T.D.Moustakas[c], E.Holub-Krappe[b], J. Antonopoulos[a]

[a] Aristotle University of Thessaloniki, Dept. of Physics, 54006 Thessaloniki, Greece
[b] Hahn-Meitner Institute (A.S), Glienicker Str. 100, D-14109 Berlin, Germany
[c] Boston University, College of Engineering, Boston, MA 02215, USA.

Abstract

Angle resolved near edge X-ray absorption measurements (NEXAFS) are used to access the existence and concentration of the two allotropic phases, cubic (β-GaN) and hexagonal (α-GaN), that coexist in a mixed-phase GaN sample grown by ECR-MBE. The resonance intensities in the NEXAFS spectra from a pure cubic GaN sample are independent of the angle of incidence ϑ, while they have a characteristic angular dependence on $\cos^2\vartheta$ for the hexagonal material, i.e. $I=A\pm B\cos^2\vartheta$. From the values of A and B the bond orientations with respect to the surface normal are calculated. Furthermore, the NEXAFS resonances in a pure α- or β-GaN sample appear at characteristic energies which are independent of the angle of incidence ϑ. Contrary to that the NEXAFS resonances in the spectra from a mixed-phase GaN show a characteristic ϑ-dependent shift. On the basis of this finding, a method is proposed and applied for the determination of the fractions of the co-existing polytypes in a mixed-phase sample.

I.   Introduction

Gallium nitride (GaN) has been a subject of intensive study because it finds applications in visible-UV light emitters and detectors as well as high-frequency, temperature and power devices[1,2]. GaN exists as the cubic (T_d) or the stable hexagonal (C_{6v}) polytype (hereafter called β-GaN and α-GaN, respectively). Because of the different thermal expansion coefficients and the lattice mismatch between GaN and the available substrates, stabilization of the β-GaN is difficult and growth often results in mixed phase crystals[3,4].

In crystalline solids, the near edge X ray absorption fine structure (NEXAFS) spectra are directly proportional to the partial DOS in the conduction band[5]. The NEXAFS electron dipole transitions are allowed from an initial s state to a final state with a p-component and their oscillator strengths depend on the material and the group symmetry[6]. The transitions to the continuum give rise to a step-like absorption while transitions to bound states give peaks (Gaussian-like absorption in the case of high experimental broadening) which are superimposed to the step-like absorption. Until recently NEXAFS has been extensively used to determine the symmetry of adsorbed molecules on surfaces[7]. However, it was recently shown that N-K-edge NEXAFS spectra from GaN can be used as a fingerprint of the cubic or hexagonal structure while from the angular dependence of the NEXAFS spectra conclusions can be drawn on the fractions of the α and β polytypes present in mixed phase samples[8,9,10].

Mat. Res. Soc. Symp. Proc. Vol. 449 © 1997 Materials Research Society

Here we propose a new method for the application of NEXAFS spectroscopy for the quantitative assessment of the percentages of the different polytypes in a mixed GaN crystal. Furthermore, the bond angles are calculated.

II. Experimental Details

The samples were grown at $600\,^{\circ}C$ on Si and Al_2O_3 substrates by microwave plasma electron cyclotron resonance-assisted molecular beam epitaxy (ECR-MBE). The α-GaN (GaN179) was grown on (0001) plane of Al_2O_3 (film thickness $d=1.67\mu m$), the β-GaN (GaN57) was grown on p-Si (100) ($d=1,60\mu m$) while a mixed phase sample (GaN67) was grown on n-Si (111) ($d=0,80\mu m$). Details on the growth conditions and structural information from high resolution transmission electron microscopy (HRTEM) and X-ray diffraction (XRD) measurements, has been reported previously[11 ,12 ,13 ,14].

The angular resolved NEXAFS spectra were recorded at room temperature, at the N-K-edge (390-440eV), using the SX-700-I plane grating monochromator at the electron storage ring BESSY in Berlin. Due to the high absorption cross section of the radiation in the soft X-Ray region the experiment was performed in a UHV chamber with base pressure $\leq 7x10^{-10}$ mbar. The spectra were recorded simultaneously in the electron (EY) and fluorescence (FLY) yield modes (using a high purity Ge detector), for different angles of incidence ϑ. Since the EY spectra are distorted due to energy dependent charging of the samples, only the FL spectra are used for the analysis. At the two limiting cases of normal ($\vartheta=90^{\circ}$) and grazing ($\vartheta=5^{\circ}$) incidence (the angle of incidence ϑ is defined by the incident beam and the sample surface), the electric field vector of the light is parallel and normal to the surface, respectively. Therefore, at normal and grazing incidence information from orbitals in the surface plane or normal to the surface is maximized, respectively.

III. Results and Discussion

The NEXAFS spectra were normalized to the flux of the monochromatised synchrotron radiation using the photocurrent from a gold grid with high transmission (90%). They were fitted using a step-function to simulate the absorption edge and a number of gaussians to fit the individual resonances. Prior to fitting the spectra were subjected to linear background subtraction and were normalized to the atomic limit (i.e. at a position where the spectra do not have any structure due to EXAFS oscillations or NEXAFS transitions). The energy position ΔE_i of the NEXAFS resonances is measured relative to the absorption edge, which is defined as the inflection point of the step function, and the ΔE_i values for the α- and β-GaN are listed in Table I.

Table I : Energy positions ΔE_i of the NEXAFS resonances for the cubic and the hexagonal samples. The errors due to the fitting are smaller than 100meV.

sample name	ΔE_1(eV)	ΔE_2(eV)	ΔE_3(eV)	ΔE_4(eV)	ΔE_5(eV)	ΔE_6(eV)
β-GaN57	1.80	4.70	7.20	8.70	11.40	-
α-GaN179	1.90	4.20	6.50	9.10	10.50	12.70

The spectra from the α-, β- and the mixed-GaN samples, recorded at $\vartheta=80°$ are shown in Fig.1a. The spectrum from the α-GaN is fitted and the gaussians are indexed following the notation using in the text. Contrary to the behavior of the α- and β-GaN, in the mixed-GaN the resonances shift with ϑ as shown in Fig.1b. The spectra shown in Fig.1 are normalized to the atomic limit (as described above).

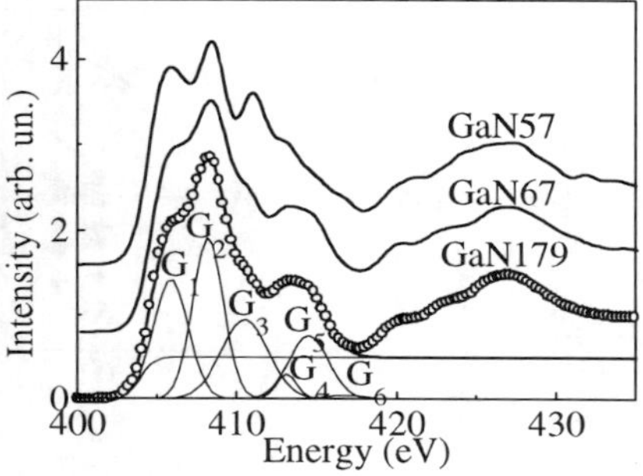
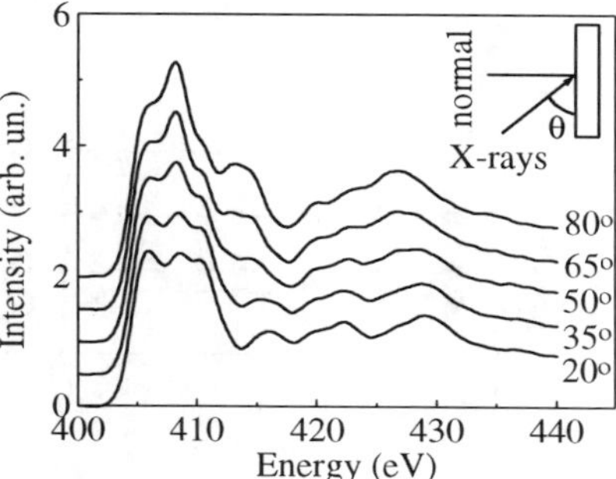

Fig.1a : Typical NEXAFS spectra from the α-, β- and mixed-GaN ($\vartheta=80°$)

Fig.1b : The ϑ-dependence of the NEXAFS spectra from the mixed-GaN.

III.1   Determination of bond angles

For the C_{6v} group symmetry the allowed transitions are $1a_1 \rightarrow a_1^*$, which will be strongest when the electric field vector **E** is parallel to the (0001) c-axis (z-axis), i.e. for grazing incidence, and the $1a_1 \rightarrow e_1^*$ which will be strongest when the **E** is parallel to the xy-plane (normal incidence). Therefore, from the angular dependence of the NEXAFS spectra we can get information about the p_x, p_y or p_z character of the orbitals. Moreover the intensities (I) of the NEXAFS resonances will depend linearly on $\cos^2\vartheta$, i.e. $I=A\pm B\cos^2\vartheta$, and from the values of A and B the bond angles can be determined. For the T_d group symmetry the allowed transitions are $1a_1 \rightarrow t_2^*$, which are observed with the same intensity for all angles of incidence (Fig.2a).

As reported previously the energy positions of the NEXAFS resonances are characteristic of the α and β polytypes[8-10], i.e. the NEXAFS spectra can be used as a fingerprint of the polytype. Furthermore, in a pure cubic or hexagonal sample the energy position and FWHM of the gaussians used for the fitting of the resonances are independent of the angle of incidence. The areas under the peaks as a function of ϑ and $\cos^2\vartheta$ for the cubic and hexagonal samples, are shown in Fig.2a and 2b, respectively. As shown in Fig.2a the NEXAFS spectra from β-GaN57 are independent of ϑ. In Fig.2b the areas under the resonances are normalized to the area corresponding to the magic angle (54.7°), i.e. the characteristic angle for which the NEXAFS spectra are independent of the angle between the final state vector or plane molecular orbital and the normal to the sample surface, for linearly polarized light (polarization plane is in the ring plane). The lines with positive slope correspond to transitions to final states resulting from mixing of s and p_z atomic orbitals and are strongest for grazing incidence (vector-like orbitals). The lines with the negative slope (strongest at normal incidence) correspond to transitions from the initial 1s state to a resulting from mixing of p_x and p_y atomic orbitals (plane like orbitals). The error in the area calculation induced by the pre-fitting treatment of the spectra can be estimated using the Thomas-Reihe-Kuhn sum rule[15] and is 2.5%[10].

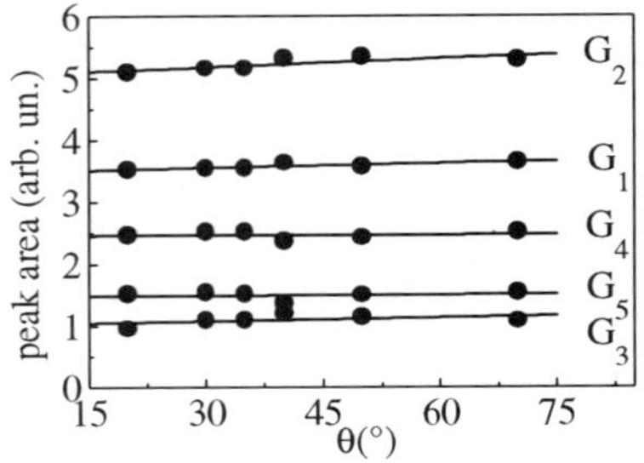

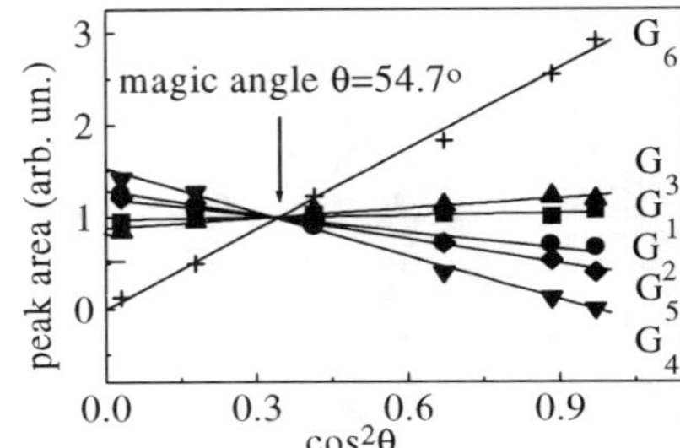

Fig.2a: Area under the NEXAFS resonances versus ϑ, for the β-GaN.

Fig.2b: Area under the NEXAFS resonances versus $\cos^2\vartheta$, for the α-GaN.

From the dependence of the areas under the peaks on ϑ the bond orientations with respect to the surface normal can be found. For the $1a_1 \rightarrow e_1^*$ transition the reduced intensity I (area under the resonance divided by the area at the magic angle) of the transition is given by $I = 1.6\ [1\text{-}0.5\sin^2\gamma - (1\text{-}1.5\sin^2\gamma)\cos^2\vartheta]$, were γ is the angle between the normal to the plane orbital and the normal to the surface. For $1a_1 \rightarrow a_1^*$ the reduced intensity I′ is given by $I' = 1.33\sin^2\alpha + (2.67\text{-}4\sin^2\alpha)\cos^2\vartheta$, where α is the angle of the vector orbital with the sample surface. Using the above equations the angles of the bonds can be determined with an error of about ± 5º. The transitions labeled G4, G3, G6 give angles which are in agreement with the expected values from the crystal structure. Among those, the transitions G4 and G6 correspond to plane orbitals parallel to the surface ($\gamma=0°$) and vector orbitals normal to the surface ($\alpha=0°$), respectively, i.e. bonds normal to the surface. G3 corresponds to vector orbitals (bonds) which are at an angle of $\alpha=51.9°$ with the c-axis (or 102.8º with each other). The transition G1 is independent of the angle of incidence and therefore corresponds to the magic angle (54.7º). The above calculated values agree well with those expected from the crystal structure[16], i.e. bonds at an angle of 54.5º with the c-axis or normal to the surface. However, the gaussians G2 and G5 give angles with a mean average of $\gamma=39°$. These correspond to bonds which have smaller than the expected angle with the normal to the surface and could probably be related with misoriented regions in the sample, i.e. regions where the c-axis deviates from the normal to the surface, or distorted tetrahedra.

III.2 Parametric curves

The ϑ-dependent shift of the NEXAFS resonances, shown in Fig.1b, is the signature of a mixed-phase sample. To quantify this result, i.e. in order to estimate the percentage of the coexisting α and β polytypes the following procedures can be followed : (1) the spectrum from the mixed sample can be fitted as a weighted average of the spectra from the calibrated pure α and β samples. This procedure yields very good results for the mixed sample GaN67 as shown in Fig.3a. Furthermore, it has yielded equally good results when applied in a different, independent set of calibrated GaN samples[9]. (2) A set of parametric curves can be constructed, which show the variation of ΔE_i as a function of the cubic fraction, with the angle of incidence as parameter. The parametric curves can be constructed for the ΔE_i of the gaussians G2, G3 and G4 (G1 and G6 cannot be used since G1 has the same characteristics for either polytype while G6 exists only in α-GaN), as well as for the FWHM or the resonance intensity. Here we shall discuss only the parametric curves constructed on the basis of the values ΔE_2, shown in Fig.3b.

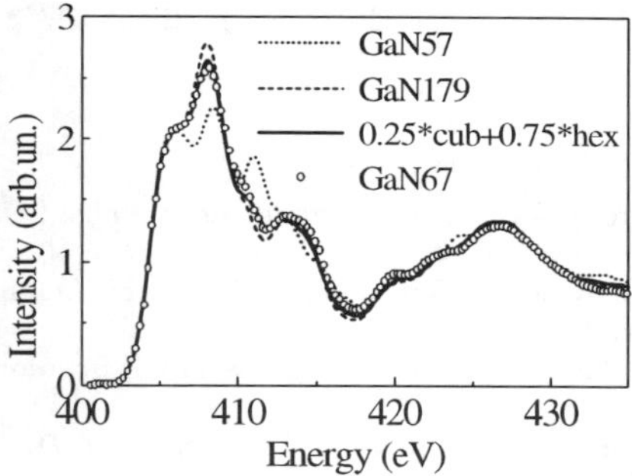

Fig.3a: Fitting of the NEXAFS spectrum from GaN67 as a weighted average.

Fig.3b: The parametric curves.

This diagram is constructed using spectra of mixed GaN with various fractions of α- and β-polytypes, calculated as weighted averages. The gaussians G_i^m for the mixed sample at a given ϑ are simulated by the weighted average of the corresponding gaussians G_i^c and G_i^h, where $i=2...5$ and the superscripts m, c and h stand for mixed, α- and β-GaN, respectively. The analysis presented here is based on G2 because the G_2^c and G_2^h have comparable FWHM and thus their average can be approximated in a satisfactory fashion with a gaussian.

The cubic fraction can be measured from the parametric curves when one or more experimental values of ΔE_2 are known from fitting of spectra recorded at one of more values of ϑ. The error-bar in the calculated cubic fraction is determined by the error in the value of ΔE_2, which can be calculated as described in Ref.10. Using the parametric curves the cubic fraction in GaN67 was found equal to $(25+\_5)\%$. The 5% error-bar was determined from the values of ΔE_2 for $\vartheta=20$, 50 and $80°$ and the corresponding error-bars. This result is in very good agreement with the results from the procedure based on the simulation of the whole spectrum, shown in Fig.3a, as well as with previously reported results from XRD and HRTEM[11-14].

IV.   Conclusions

It is demonstrated that the N-K-edge NEXAFS spectra from GaN can be used for the identification of the α- or β-GaN polytype as well as for the detection of the co-existence of cubic and hexagonal polytypes in a mixed-phase sample. Following the proposed analysis, quantitative results can be drawn on the fractions of the α- and β-GaN polytypes. One important advantage of NEXAFS is that it is non-destructive and it does not require any sample preparation procedure which might induce damage-related artifacts. From the angular dependence of the NEXAFS spectra the bond angles are determined. The results indicate that, besides the expected bond angles, some distorted bonds also exist. These distorted bonds form an angle of 40° with the normal to the surface instead of the expected angle of 54.5°. Possible origins of this distortion could be misoriented regions or the thermal stress which is not completely relaxed in thin films[17]. The identification of the origin of the distorted bond angles is a subject of further investigation.

Acknowledgments : This work was realized with financial support from the EC-HCM (CHGE-CT93-0027) program. The authors M.K. and E.C.P. wish to thank Dr. D. Arvanitis and Dr. J. Stoehr for fruitfull discussions.

References

1 . S. D. Lester, F. A. Ponce, M. G. Craford and D. A. Steigerwald, Appl. Phys. Lett. **66**, 1250 (1995).

2 . H. Sakai, T. Koide, H. Suzuki, M. Yamaguchi, S. Yamasaki, M. Koike, H. Amano and I. Akasaki, Jpn. J. Appl. Phys. **34**, L1429 (1995).

3 . T. Lei, M. Fanciulli, R. J. Molnar, T. D. Moustakas, R. J. Graham and J. Scanlon, Appl. Phys. Lett. **59**, 944 (1991).

4 . T. S. Cheng, L. C. Jenkins, S. E. Hooper, C. T. Foxon, J. W. Orton and D. E. Lacklison, Appl. Phys. Lett. **66**, 1509 (1995).

5 . J. C. Fuggle and J. E. Inglesfield, Unoccupied electronic states : Fundamentals for XANES, EELS, IPS and BLS (Springer Verlag, Berlin 1992) and references therein.

6 . W. R. L. Lambrecht, B. Segall, J. Rife, W. R. Hunter and D. K. Wickenden, Phys. Rev. B, **51**, 13516 (1995)

7 . J. Stoehr, NEXAFS Spectroscopy (Springer-Verlag, Berlin, 1992) and references therein.

8 . M. Katsikini, E. C. Paloura, J. Kalomiros, P. Bressler, T. D. Moustakas, Proc. 23$^\text{d}$ Int. Conf. on the Physics of Semiconductors, Berlin, July 1996 (in press).

9 . M. Katsikini, E. C. Paloura, T. S. Cheng and C. T. Foxon, Proc. 9$^\text{th}$ Int. Conf. on X-ray absorption fine structure, Grenoble, August 1996 (to be published in J. de Physique).

10 . M. Katsikini, E. C. Paloura, T. D. Moustakas, Appl. Phys. Lett. **69** (in press)

11 . T. D. Moustakas, T. Lei and R. J. Molnar, Physica B **185**, 36 (1993).

12 . T. Lei, T. D. Moustakas, R. J. Graham, Y. He, S. J. Berkowitz, J. Appl. Phys. **71**, 4933 (1992).

13 . S. N. Basu, T. Lei and T. D. Moustakas, J. Mater. Res. **9**, 2370 (1994).

14 . T. Lei, K. F. Ludwig, T. D. Moustakas, J. Appl. Phys. **74**, 4430 (1993).

15 . J. Berkowitz in Photoabsorption, Photoionization and Photoelectron Spectroscopy (Academic Press, New York 1979), p. 56.

16 . R. Zallen in Band theory and transport properties , Vol. 1, Ed. W. Paul (North-Holland, New York, 1982) p.19.

17 . K. Hiramastu, T. Detchprohm, I. Akasaki, Jpn. J. Appl. Phys. **32**, 1528 (1993).

NANO-TUBES IN GaN

Z. LILIENTAL-WEBER, Y. CHEN,* S. RUVIMOV, W. SWIDER, and J. WASHBURN,

Lawrence Berkeley National Laboratory 62/203, Berkeley, CA 94720,
*Hewlett Packard, Palo Alto, CA

ABSTRACT

Formation of vertical hollow nano-tubes in GaN grown on different substrates using different growth methods is described. These defects are shown to be of several different types, some related to threading dislocations, but others originating at tubular inversion domains, or crystal inhomogeneities. Kinetic mechanism based on slow growth rate on polar$\{0\bar{1}11\}$ surfaces is proposed to explain the origin of these defects.

I. INTRODUCTION

Gallium nitride is one of the promising semiconductors for laser diodes for the blue and UV wavelength regions [1-6]. Even though recent progress has greatly improved the crystal quality of GaN films grown on various substrates, its practical application is still hampered by the high defect density [7-11]. One defect which may hamper applications of this material is the nano-tube formed in this material during growth [12-14]. The origin of these defects is still not clear despite extensive research in this area.

Previous studies have suggested that the GaN layer surface is not uniform, and grown-in hillocks have an open-core screw dislocation forming a nano-pipe within the layer [12-14]. Their density was estimated in the range of 10^5-10^7cm.$^{-2}$ The radii of these nano-pipes are in the range of 3-50 nm and they appear to propagate along the c-axis of the film. Formation of such defects has been reported in a number of other crystals including SiC [15-17]. The micro-pipes in SiC have received great attention because they are known to be the defects that limit breakdown voltage of high power devices [18].

In this paper we try to understand the nature of vertical hollow nano-defects formed in GaN films grown on different substrates using different growth methods.

II. EXPERIMENTAL

Five different types of materials have been studied: three different GaN samples grown on Al_2O_3 with different methods-Molecular Beam Epitaxy (MBE), Metal-Organic Chemical Vapor Deposition (MOCVD), and Chemical Vapor Deposition (CVD), GaN grown by CVD on SiC, and homoepitaxial GaN grown by MOCVD on bulk GaN. Optical microscopy was used to give a general overview of the sample surface. More detailed studies of the surface morphology of the GaN epitaxial films grown by CVD on Al_2O_3 or SiC were done by UHV scanning tunneling microscopy and the details of these studies have been described elsewhere [19]. In particular, depending on the substrate material used, the dominant growth scenario and morphological manifestation of defects terminating at the surface of the respective GaN layers was deduced. Conventional and high resolution transmission electron microscopy (TEM) was then applied to plan-view and cross-section samples to obtain detailed information on defect distribution near the sample surface as well as near the interface. A Topcon 002B electron microscope at an acceleration voltage of 200 keV at point-to-point resolution of 0.18 nm was used for these studies.

III. RESULTS
III.1. Surface morphology

Optical microscopy suggests that the GaN films grow as a collection of hexagonal islands. This was true regardless of the growth method used. However, the island type, size and density varied

Mat. Res. Soc. Symp. Proc. Vol. 449 © 1997 Materials Research Society

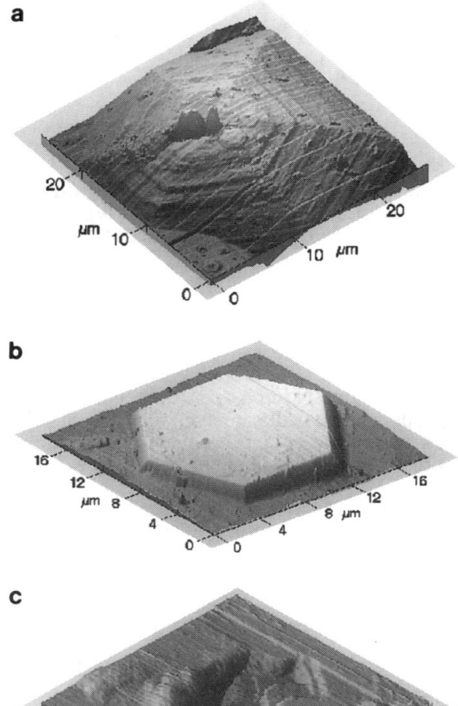

depending on the particular method used for growth as well as for the substrate used. Usually hexagonal islands are formed on GaN layer surfaces grown by any method (MOCVD, MBE or CVD) on any substrate used for growth. For the growth by MOCVD on the smooth side of bulk GaN crystals the island size was in the range of 15-60 μm. For the GaN grown by CVD on Al_2O_3 the diameters of islands scattered between 20-50 μm and for the samples grown on SiC was rather uniform in size (50 μm). Three types of islands were observed (Figs. 1a,b,c): a hillock type with either an inversion domain or twinned precipitate in the center (a), flat islands with small openings seen as dark dots distributed randomly (b), or with a large opening at the island center (c). This is in contradiction to the previous observations that all nano-tubes have empty cores [14]. It is not easy to obtain accurate information about the nature of these large openings in the hillock centers based on plan-view TEM studies, since the thickness at the center of the islands is much greater than the surrounding area. Therefore in most cases only the areas between the islands which get thinned earlier to electron transparency can be observed. In this paper we describe hollow nanodefects which were located mostly in flat areas between islands or in flat hexagonal islands.

Fig. 1. AFM micrographs showing different type of islands formed on the surface of GaN such as a hillock with a twin type precipitate in the island center (a), a large flat island with many small islands on top of it (b), with an opening in the island center (c). Note high density of small dark dots on each island representing nano-tubes.

III.2. Hollow nanodefects in GaN layers grown on Al_2O_3

For the GaN films grown on Al_2O_3 hollow vertical nanodefects were found for all three growth methods used. TEM in plan-view reveals that these defects had hexagonal shapes (very often elongated hexagons).

Some of these defects appeared to go through the entire epitaxial layer since with the right defocus the sapphire substrate can be seen (Fig. 2a). In these cases usually foreign particles could be seen at the substrate, which probably nucleated the vertical defect. About 70% of these hollow vertical defects were associated with a dislocation having an edge component (Fig. 2b). There were also many defects for which no edge dislocation displacement was found (2c). This would mean that either these hollow defects were associated with a pure screw dislocation, or they were not associated with any dislocation.

Cross-section electron microscopy shows that there are many hollow vertical defects which start in the subsurface area and terminate at the surface (Figs. 3a,b). In cross-section they have a V-shape with clear facets. Very often these V-shaped defects are filled with small crystalline particles. Energy-dispersive x-ray spectrometry (EDX) did not reveal any foreign atoms within the area, only Ga could be detected. Similar particles were seen in plan-view as well (Fig. 2b). From analysis of the lattice image and moire' fringes one can conclude that these particles are GaN, but they are not a part of the continuous layer. Such V-shaped defects are often associated with dislocations, usually more than one (Fig. 3). The dislocation which is attached to the tip of such a defect usually had a mixed character. There were also hollow vertical defects which are formed within the layer (Fig.4) that were clearly attached to screw dislocations. At the point where the hollow vertical defect starts the contrast associated with the screw dislocation disappears and two walls parallel to the c-axis start. In many cases these hollow vertical defects terminate within the layer; the segment length parallel to the c-axis varies from area to area (Fig. 4 a,b). There are also cases where one does not find any dislocation at the origin of such a defect, but a dislocation can be attracted to the vertical defect wall (Fig. 4c). All these defects start with the same V shape, they are also faceted in cross-section.

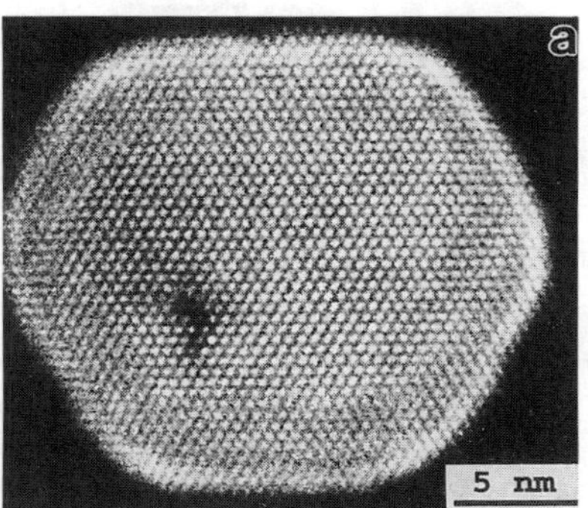
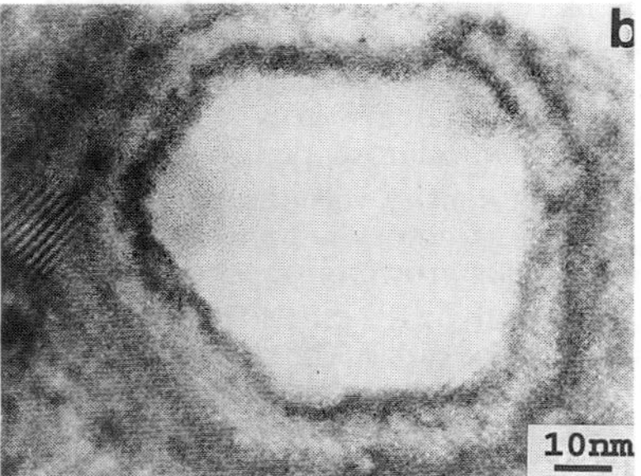

Fig. 2. Plan-view TEM micrographs showing different shapes of nanotubes formed in GaN grown on Al_2O_3; (a) a nanotube originated from a particle in the substrate, (b) a nanotube with an edge component of type of dislocations, (c) a nanotube for which no edge dislocation displacement was found.

III.3. Hollow nanodefects in GaN layers grown on SiC

For layers grown on SiC no substantial difference was observed in plan-view. However, a much higher fraction than was observed for the layers grown on sapphire had no displacement that would indicate a dislocation with any edge character. This means that they may be associated with pure screw dislocations, no dislocation at all or with tubular inversion domains. The density of vertical defects with an inversion domain character was negligible in the GaN films grown on SiC [7,19], therefore, association of vertical defects with inversion domains is unlikely for growth on this substrate.

III.4. Hollow nanodefects in homo-epitaxial GaN layers grown on bulk GaN

Homo-epitaxial layers were grown on bulk GaN either on the smooth surface or on the rough

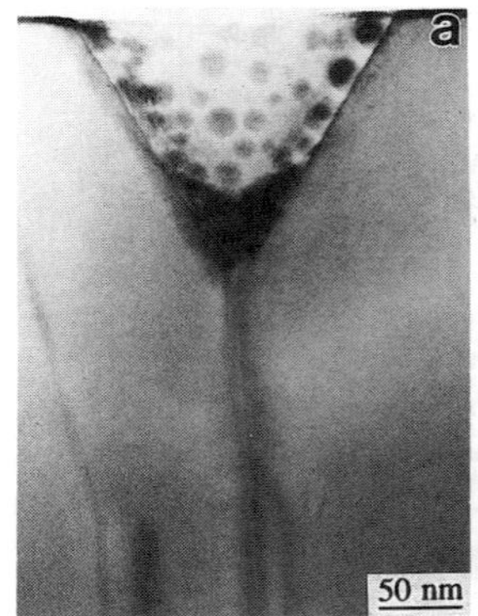
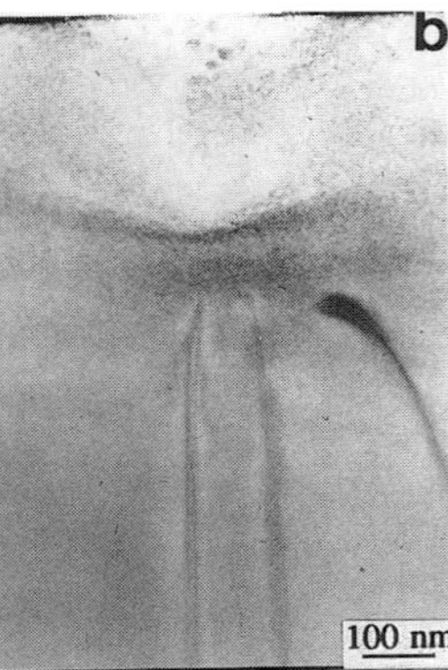

Fig. 3. Cross-section through the nanotubes formed in the GaN samples grown on Al_2O_3. More than one dislocation was connected with the hollow-type defects. At least one of them had a mixed type character. All remaining dislocations bend toward the hole. Note growth of GaN within the hollow in the form of spherical precipitates. (a) cross-section through the center of the faceted hollow; (b) a section closer to the hollow wall.

surface thereby in opposite polar directions. Based on Convergent Beam Electron Diffraction (CBED) studies we had previously concluded [8, 20-22] that the bonds parallel to the c-axis, in the direction N to Ga point toward the smooth surface. However, we are in the process of checking the possibility that this polarity arrangement may not be correct due to 180° rotations that may occur in the particular microscope used that are not described in the user instructions. A much higher hallow defect density was observed for growth on the rough surface ($1x10^7$ cm^{-2}). All of them had a V-shape in cross-section. Only two such V-shaped defects were found in the two cross-section samples grown on the smooth GaN surface (Fig. 5), which correspond to an estimated density of about $5x10^5$ cm^{-2}. The hollow vertical defects formed on the rough surface were never attached to dislocations. The two found in the layer grown on the smooth surface both started at the top of tubular shaped inversion domains originating at the interface with the substrate.

IV. DISCUSSION

These studies provide clear information that vertical defects formed in epitaxial GaN are of several different types. Hollow vertical defects are formed regardless of the type of substrate used or the growth method used. The explanation given by Quian et all [12-14] related these hollow vertical defects, referred to as nano-tubes, with open-core screw dislocations. This definitely applies to some of these vertical defects , but not to all. The original explanation for the formation of nano-tubes was given by Frank [24] who suggested the hollow-core screw dislocation. The explanation was given that a state of local equilibrium exists in which a dislocation would have lower total per unit length energy when its core is hollow compared to the core filled with a highly strained lattice. The size of the open-core, therefore, should be determined by a balance between the extra surface free energy and the decrease in strain energy near the dislocation core. This explanation probably would apply to a defect such as that shown in (Fig. 4 a,b).

However, the simple open-core dislocation model does not appear to apply to all hollow vertical defects observed. In many cases the defects appear to be formed in areas where there is a group of dislocations (Figs. 3a,b). Other vertical defects were clearly associated with tubular shaped inversion domains (Fig. 5).

One common feature which applies to all hollow vertical defects is the fact that they are faceted and start with a V-shape with a constant angle about 60° between "V-arms." These vertical defects are crystallographically oriented. For the nano-tubes taken in [01$\bar{1}$2] projection the walls of the tube are on (0$\bar{1}$11) planes, which are polar planes. From the measured growth rates for the homo-epitaxial films as well from the growth of bulk GaN plate-like crystals [8, 20-22] it is clear that the highest growth rates observed for these crystals are along the non-polar directions (about 100 times higher). The growth rate of crystals grown along [0001] polar direction is approximately 10-40% higher in one polarity (on the smooth surface of the bulk GaN) compared to the opposite [000$\bar{1}$] direction. The growth on (0$\bar{1}$11) planes is the slowest , however, not equal for Ga polarity and N

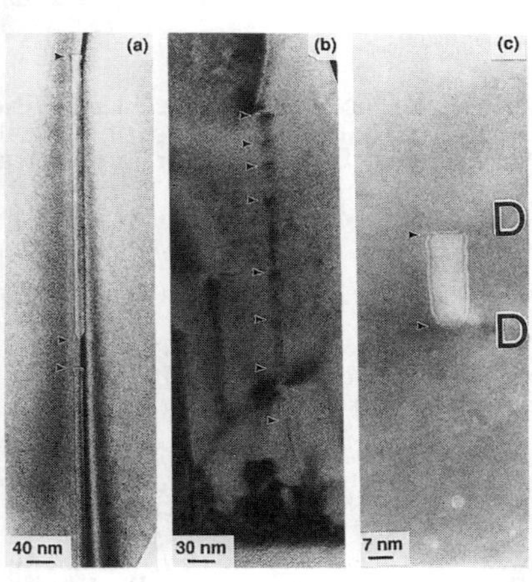
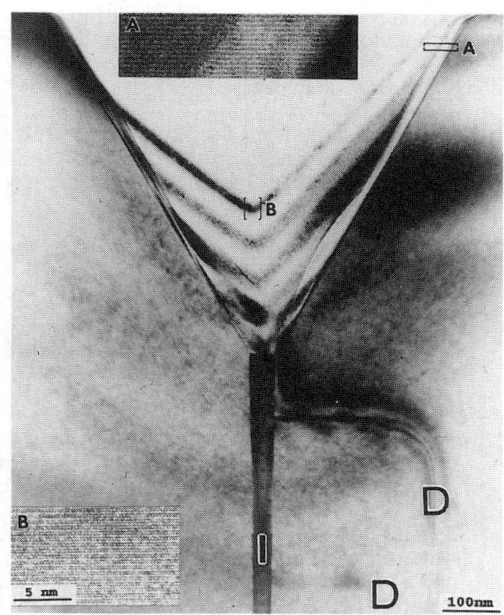

Fig. 4. (a,b) Different segments of nano-tubes related to screw dislocation, (c) bending of dislocation (D) toward the tube wall.

Fig. 5. A faceted hollow formed in the subsurface area of the GaN grown on the smooth side of a bulk GaN. This nano-tube was associated with an inversion domain (I) and two dislocations (D): upper with a mixed character and the lower with a screw character. A partial GaN growth was observed within the hollow. Note [A] and [B] inserts showing a structure of GaN in high resolution.

polarity. The polarity which applies for the growth on the rough surface would have even slower growth on $(0\bar{1}11)$ surfaces compared to the growth on planes with the reverse polarity. The slow growth rate on $(0\bar{1}11)$ polar planes is suggested by this study as the origin of "nano-tube" formation. This leads to faceting on these planes and formation of open tube vertical defects. This explains the observation that nano-tubes were formed with higher density for one polarity compared to the other. Since all these GaN crystals are not homogeneous, with a high density of small Ga clusters, they can locally slow-down growth. Therefore, especially for crystals which are growing on unfavorable crystal polarity (such one as for growth on the rough GaN surface) higher pin-hole density was observed. Once such a nano-tube (pin-hole) is formed dislocations which are not necessary at their nucleation could bend toward the hole to shorten their length and therefore reduce the total free energy of the system. For the nano-tubes with large diameter a large Burgers vector was determined in plan-view micrographs, for example the nano-tube shown in (Fig. 2b). Such a Burgers vector would be the total resulting Burgers vector for all the dislocations attracted to the hole.

The observation that during growth nano-tubes could close is interesting (Figs. 4a-c). The kinetic mechanism would explain this fact as well. Based on the observation of bulk plate-like crystals which grow in the "empty space" [8] from the cell-walls along $[1\bar{1}00]$ or $[01\bar{1}2]$ directions closing of these tubes might be possible by enhancing the growth in non-polar directions, since in this way islanding with facets on $(0\bar{1}11)$ planes can be avoided.

CONCLUSIONS

In conclusion, this paper describes nano-tubes which are formed in different type of GaN crystals. It was shown that the formation of these nano-tubes is independent from the substrate used or the growth method. Different type of extended defects can be associated with these nano-tubes making very difficult to explain the origin of these defects. Kinetic mechanism taking into account the slowest growth rate on the polar $(0\bar{1}11)$ surfaces was proposed to explain the origin of these defects. The high rate of growth in non-polar direction would explain how it is possible to close such nano-tubes within the layer and recover high crystal quality above the empty nano-tube.

ACKNOWLEDGMENTS

This work was supported by the Director, Office of Basic Science, Materials Science Division, U.S. Department of Energy under the Contract No. DE-AC03-76SF00098. The use of the facility at the National Center for Electron Microscopy at the Lawrence Berkeley Laboratory is greatly appreciated.

REFERENCES

1. S. Nakamura, M. Senoh, N. Iwasa, and S. Nagahama, Jpn. J. Appl. Phys. **34**, L979 (1995).
2. S. Nakamura, T. Mukai, and M. Senoh, Appl. Phys. Lett., **64**, 1687 (1994).
3. H. Amano, K. Hiramatsu, and I. Akasaki, Jpn. J. Appl. Phys. 27, L1384 (1988).
4. T.D. Moustakas and R.J. Molnar, Mat. Res. Soc. Conf. Proc. **281**, 753 (1993).
5. C. Wang and R.F. Davis, Appl.Phys. Lett. **63**, 990 (1993).
6. N. Newman, J. Ross and M. Rubin, Appl. Phys. Lett. **62**, 1242 (1993).
7. Z. Liliental-Weber, H. Sohn, N. Newman, and J. Washburn, "Electron Microscopy Characterization of GaN Grown by MBE on Sapphire and SiC", J. Vac. Sci. Technol **B 13** **(4)**, 1578 (1995).
8. Z. Liliental-Weber, S. Ruvimov, Ch. Kisielowski, Y. Chen, W. Swider, J. Washburn, N. Newman, A. Gassmann, X. Liu, L. Schloss, E.R. Weber, I. Grzegory, M. Bockowski, J. Jun, T. Suski, K. Pakula, J. Baranowski, S. Porowski, H. Amano, and I. Akasaki, Mat. Res. Soc.Symp. Proc. **vol. 395**, 351 (1996).
9. Z. Liliental-Weber, S. Ruvimov, T. Suski, J.w. AgerIII, W. Swider, J. Washburn, H. Amano, I. Akasaki, W. Imler, "Effect of Si-doping on the Structure of GaN," Mat. Res. Soc. Meeting, San Francisco, CA, **vol. 423**, (1996) in press.
10. F.A. Ponce, B.S. Frusor, J.S. Major, Jr., W.E. Plano, and D.F. Welch, Appl. Phys. Lett., **67**, 410 (1995).
11. S.D. Lester, F.A. Ponce, M.G. Craford, and D.A. Steigerwald, Appl. Phys. Lett., **66**, 1249 (1995).
12. W. Qian, G.S. Rohrer, M. Skowronski, K. Doverspike, L.B. Rowland, and D.K. Gaskill, Appl. Phys. Lett. **67**, 2284 (1995).
13. W. Qian, M. Skowronski, K. Doverspike, L.B. Rowland, and D.K. Gaskill, J. Cryst. Growth **151**, 396 (1995).
14. W. Qian, M. Skowronski, and G.S. Rohrer, Mat. Res. Soc. Mat. (1996).
15. V.G. Bhide, Physica **24**, 817 (1958).
16. H.M. Hobgod, D.L. Barret, J.P. McHugh, R.C. Clarke, S. Sriram, A.A. Burk, J. Greggi, C.D. Brandt, R.H. Hopkins, and W.J. Choyke, J. Crystals Growth, **137**, 181 (1994).
17. J. Heindl and H.P. Strunk, Phys. Stat. Sol. **193**, K1 (1996).
18. J.A. Powell, P.G. Neudeck, D.J. Larkin, J.W. Yang, and P. Pirouz, Inst. Phys. Conf. Ser., **137**, 161 (1994).
19. S. Koynov, M. Topf, S. Fiscer, B.K. Meyer, P. Radojkovic, E. Hartman, and Z. Liliental-Weber, to be published (1996).
20. Z Liliental-Weber, International School on Semiconductor Physics, May 1995, Jaszowiec, Poland (invited talk -no paper).
21. Z. Liliental-Weber, C. Kisielowski, X. Liu, L. Schloss, J. Washburn, E.R. Weber, I. Grzegory, M. Bockowski, J. Jun, T. Suski, and S. Porowski, Topical workshop on III-V Nitrides, Nagoya, September (1995)-Solid State Electr. in print
22. Zuzanna Liliental-Weber, C. Kisielowski, S. Ruvimov, Y. Chen, J. Washburn, I. Grzegory, M. Bockowski, J. Jun, and S. Porowski, J. Electr. Mat. vol. 25 #9, 1545-50 (1996).
23. F.A. Ponce, D.P. Bour, W.T. Young, M. Saunders, and J.W. Steeds, Appl. Phys. Lett. **69**, 337 (1996).
24. F.C. Frank, Acta Cryst. 4, 497 (1951).
25. B. Daudin. J.L. Rouviere, and M. Arlery, Appl. Phys. Lett., 69, 2480 (1996).

INVERSION DOMAIN BOUNDARIES IN GaN GROWN ON SAPPHIRE

L. T. ROMANO and J.E. NORTHRUP
Xerox Palo Alto Research Center, 3333 Coyote Hill Rd, Palo Alto, CA 94304

ABSTRACT

Inversion domain boundaries (IDBs) in GaN grown on sapphire (0001) were studied by a combination of high resolution transmission electron microscopy, multiple dark field imaging, and convergent beam diffraction. Films grown by molecular beam epitaxy (MBE), metalorganic vapor deposition (MOCVD), and hydride vapor phase epitaxy (HVPE) were investigated and all found to contain IDBs. Inversion domains (IDs) that extended from the surface to the interface were found to be columnar with facets on the $\{10\text{-}10\}$ and $\{11\text{-}20\}$ planes. Other domains ended within the film that formed IDBs on the (0001) and $\{1\text{-}102\}$ planes. The domains were found to grow in clusters and connect at points along the boundary.

INTRODUCTION

The role of microstructural defects in the III-V nitrides on device performance remains unclear. GaN alloys with defect densities of 10^{10} dislocations cm^{-2} have been successfully fabricated into blue laser diodes [1]. Identification of defects in these materials enables optimization of the growth processes used to make these materials and to understand their effect on the optoelectronic properties.

Threading edge dislocations with burgers vector $b = a/3\langle11\text{-}20\rangle$ are the most common line defects found for films grown on sapphire(0001) and 6H-SiC(0001). Recently planar defects that extend perpendicular to the GaN/sapphire interface have been identified as inversion domain boundaries (IDBs) [2-5]. In this paper we report further studies of IDBs on films grown by electron cyclotron resonance-molecular beam epitaxy (ECR-MBE), hydride vapor phase epitaxy (HVPE), and metal organic vapor deposition (MOCVD).

EXPERIMENT

Details of films grown by ECR-MBE [6], HVPE [7], and MOCVD [8] have been discussed elsewhere. All GaN films were grown on sapphire (0001) to thicknesses varying from 0.3μm for MBE films, 2-4 μm for MOCVD films, and 15-40μm films grown by HVPE. Films grown by MBE and MOCVD had a nitridation treatment before deposition of a low temperature GaN buffer layer. Films grown by HVPE either had a GaCl pretreatment at 1050°C or were grown on a sputter deposited ZnO buffer layer. The ZnO buffer layer could not be observed by TEM chemical analysis or XRD after film growth as discussed previously [8].

Plan view (PTEM) and cross sectional TEM (XTEM) samples were prepared by flat polishing to < 10 μm and Ar ion milling to electron transparency with a liquid nitrogen cold stage. Samples were studied in a JEOL 300 kV microscope using convergent beam electron diffraction, multiple dark field imaging (MDF), and conventional imaging.

Mat. Res. Soc. Symp. Proc. Vol. 449 © 1997 Materials Research Society

RESULTS AND DISCUSSION

Columnar inversion domains (IDs) discussed previously [2,4] were single domains with facets along either the {10-10} (M) or {11-20} (A) planes that extended from the interface to the film surface. PTEM studies of films grown by ECR-MBE that contain a high density of IDs (~ 50% of the film), show that the domains can be interconnected with facets on both the (10-10) and (11-20) planes. Figure 1 is a PTEM dark field image of an MBE film with a high density of IDs. In order to image the IDBs, a nonzero reflection along the inversion axis is required as demonstrated previously for imaging IDBs in AlN [9]. Fringe contrast occurs at the IDBs that are not visible when imaging with (hkil) reflections with l = 0. The image in Fig. 1 was taken using a reflection near the zone axis B = [1-216] with g = 2-201. This zone axis is ~ 15 degrees from the [0001] pole. Therefore the hexagonal domains enclosed by the IDBs in the image appear skewed.. Several isolated domains (~ 10-30nm wide) can be observed in the image, in addition to connected-domains. The connected-domain labeled T in the image are two very similar size domains with facets along the M planes. The domains are joined at the intersection of two different M planes. Domain E however, is an ID that extends along both the {10-10} and {11-20} planes. At point 'e', the domain E necks down and joins two smaller domains with facets along the M planes, which are connected to a larger domain with facets on both the M and A planes.

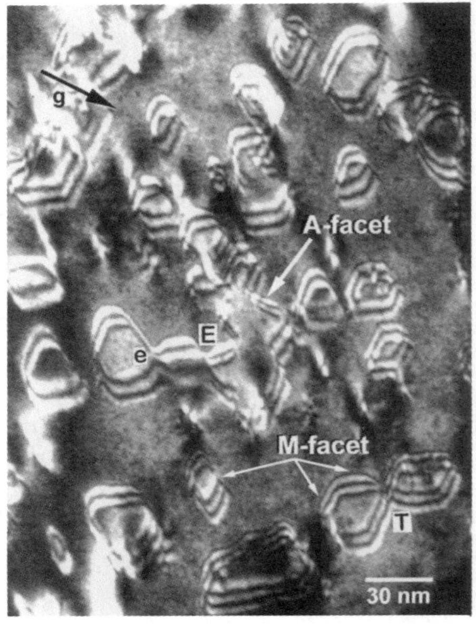

Figure 1. PTEM image of IDs in films grown by MBE taken n ear the [1-216] zone axis with g = 2-201

Columnar domains in the thick films grown by HVPE were found to end within the film and often form in clusters along the interface. Figure 2a is a XTEM low magnification image of several clusters of IDs in a GaN film grown by HVPE with the ZnO buffer layer. The domains visible in this image ranged in height from 3-7μm. A higher magnification image of the tallest domain (indicated by the pointer) is shown in Fig. 2b. It can be observed in Fig. 2b, that the tallest domain terminates by forming an IDB on the (0001) basal plane. Near this domain are several smaller domains that terminate either on the basal plane or the {1-102} planes. Given that these domains eventually end in the thicker films, indicate that the film would like to grow with one crystal polarity. The thickness of film that this occurs is probably dependent on the growth rate of the two crystal polarities and the growth conditions. From PTEM of the top surface of 15μm and 40μm films, no IDs were observed.

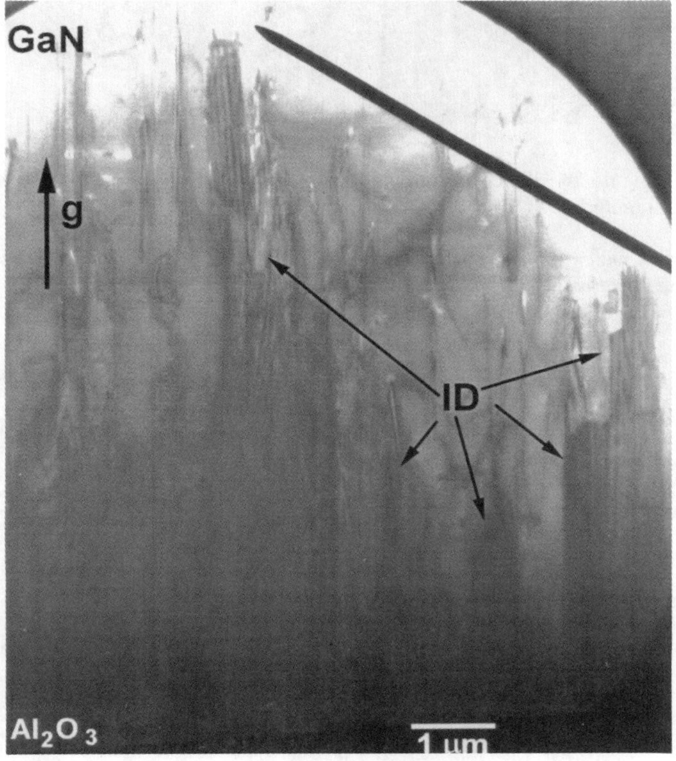

Figure 2a. Low magnification image of IDs in a film grown by HVPE with a ZnO buffer layer. The image is taken near B = (10-10) and g = 0002.

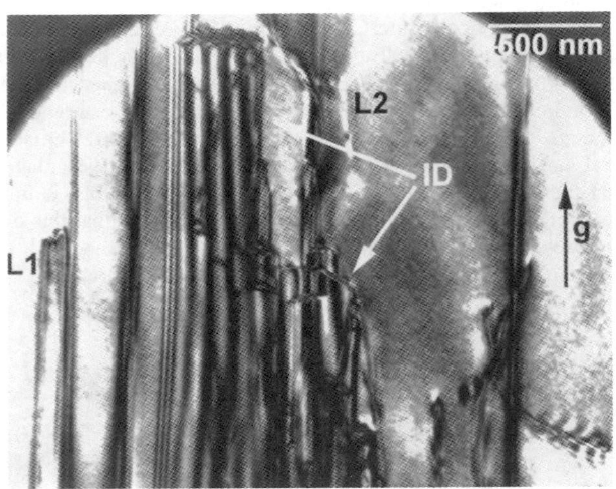

Figure 2b. Higher magnification image of an ID in Fig. 2a (indicated by the pointer) taken near B = [10-10] and g = 0002.

Several smaller "house" shaped domains found clustered near the interface of the same sample in Fig.2 are shown in Fig. 3. The domains in this image range in size from ~ 100-500nm. The facets of the smallest domains all appear to terminate on the {1-102} planes.

Figure 3. Dark field image of House shaped IDs in films grown by HVPE with the ZnO buffer layer taken near B = [10-10] and g = 0002.

An image of an isolated "house" domain of the same HVPE sample discussed above is shown in Fig. 4. Figure 4a and 4b are taken using MDF with g = 0002 along the [11-20] zone axis. A contrast reversal is obtained between the matrix and the domain which is expected for ID contrast as discussed previously [2]. In both of these images, the IDBs along the M-prism planes and the (1-102) plane is observed. However by imaging along either g = 10-10 shown in Fig. 4c or g = 11-20 (not shown), only the IDB along the (1-102) plane is visible. This indicates that the IDBs associated with the columnar portion of the domain has no basal plane component, which is similar to columnar domains discussed previously [2]. However the "rooftop" of the domain has a component parallel and perpendicular to the basal plane.

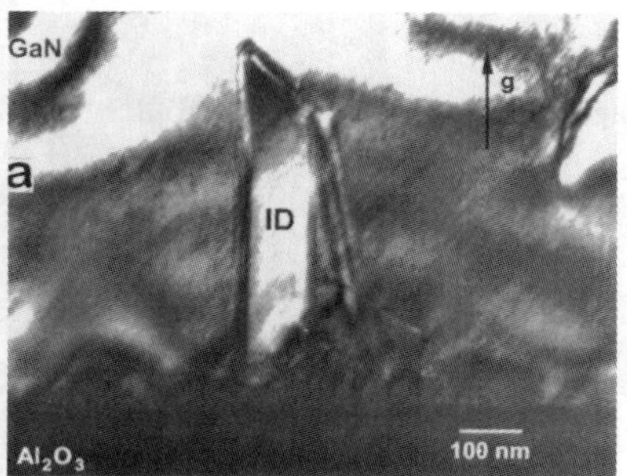

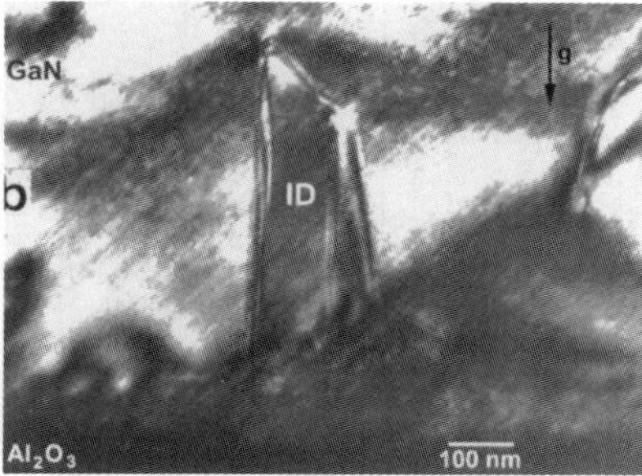

Figure 4. MDF image taken at B = [11-20] for a film grown by HVPE with a ZnO buffer layer with (a) g = +0002 and (b) g = -0002

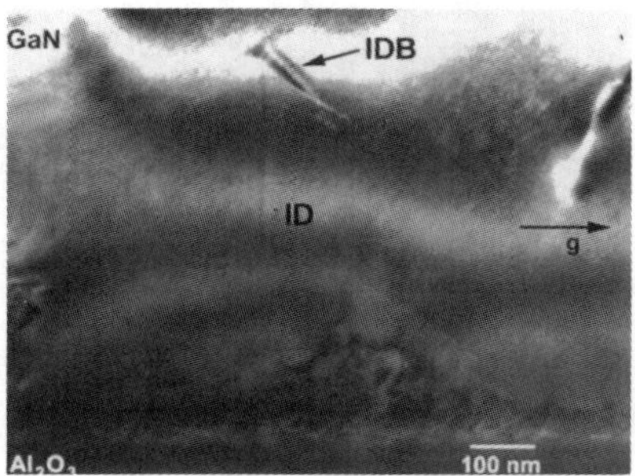

Figure 4c. Dark field image of the same region in Fig. 4a and 4b taken near B = [11-20] with g = 10-10

CONCLUSIONS

IDs were characterized in films grown by ECR-MBE, HVPE, and MOCVD. Films with a high density of columnar domains grown by MBE were found to be connected at the boundaries along both the {10-10} (M) and {11-20} (A) planes. The IDs in the thin MBE films all terminated at the film surface. For thicker films, IDs were found to terminate and form IDBs on the (0001) and {1-102} planes. The IDs formed both columnar and "house" like morphologies. They tended to cluster in the HVPE films grown with a ZnO buffer layer. Some of these domains grew several microns before terminating within the film. For MOCVD films and HVPE films grown with the GaCl pretreatment, only "house" shaped domains were found with a density of < 2% of the film. Therefore these domains tended to be isolated and grew to ~ 100-300nm high.

ACKNOWLEDGEMENTS

L.T.R acknowledges support by DARPA (agreement # MDA972-95-3-008), J.E.N acknowledges ONR Contract No. N00014-95-C-0169, and M.A.O'K acknowledges the DOE Contract No. DE-AC-03076SF00098. We would like to thank Raj Singh, Richard Molnar, and Dave Bour for providing films.

REFERENCES

1. S. Nakamura, M. Senoh, S. Nagahama, N. Iwasa, T. Yamada, T. Matsuhita, H. Kiyoku and Y. Sugimoto, Jpn. J. Appl. Phys. **35** (1996) L74

2. L.T. Romano, J. E. Northrup, M. A. O'Keefe, Appl. Phys. Lett., **69**, 2394 (1996)

3. X. H. Wu, L. M. Brown, D. Kapolnek, S. Keller, B. Keller, S. P. DenBaars, and J.S. Speck, J. Appl. Phys, **80**, 3228 (1996)

4. B. Daudin, J. L. Rouviere, and M. Arley, Appl. Phys. Lett. **69,** 2480, (1996)

5. F. A. Ponce, D. Cherns, W. T. Young, J. W. Steeds, S. Nakamura, *these proceedings*.

6. R. J. Molnar and T. D. Moustakas, J. Appl. Phys. **76,** 4587, (1994)

7. R. J. Molnar, R. Aggarwal, Z. L. Liau, E. Brown, I. Melngailis, W. Gotz, L.T. Romano, B. S. Krusor, N.M. Johnson, Mater Res. Soc. Symp. Proc. Vol. **395** (1996)

8. L. T. Romano, R. J. Molnar, B. S. Krusor, G. A. Anderson, D. P. Bour, and P. Maki, Mater. Res. Soc. Symp. Proc, (1996)

9. A. Berger, J. Am. Ceram. Soc., **74**, 1148 (1991)

MECHANISMS OF STRAIN REDUCTION
IN GaN AND AlGaN/GaN EPITAXIAL LAYERS

O. GFRÖRER, T. SCHLÜSENER, V. HÄRLE, F. SCHOLZ, and A. HANGLEITER
4. Physikalisches Institut, Universität Stuttgart, D-70550 Stuttgart, Germany
E-mail: o.gfroerer@physik.uni-stuttgart.de

ABSTRACT

We have investigated $Al_{0.12}Ga_{0.88}N$ layers with and without a 1 μm GaN buffer, grown on the c-face of $\alpha - Al_2O_3$ substrate with an intermediate AlN nucleation layer grown by LP-MOVPE. We used spatially resolved cathodoluminescence spectroscopy at a temperature of 8K to investigate the strain and the homogeneity of composition that can be determined from the energy of the luminescence peak. The larger thermal expansion coefficient of the sapphire in comparison to the nitrides leads to a biaxial compressive strain of the upper GaN layer when cooling down from growth temperature. For AlGaN layers directly grown on the nucleation layer this cannot be confirmed. The layer stays relaxed and fluctuations in the aluminium composition of 0.4% can be observed. When growing an intermediate GaN buffer, the AlGaN layer gets tensilely strained. This strain is of elastic nature and microcracks can be observed preferentially at the edges due to the smaller lattice constant of AlGaN in comparison to GaN. Even detaching of the AlGaN layers grown on the buffer can be observed. In the regions without cracks the layers are quite homogeneous. A deformation potential of (19±4)eV was estimated for $Al_{0.12}Ga_{0.88}N$.

INTRODUCTION

In addition to the luminescence wavelength in the visible and near UV spectral range and the direct energy gap, attributes like mechanical hardness and chemical resistance make the material system of the group III nitrides promising for realization of optical devices. Due to the lack of III-V nitride bulk crystals of adequate size, substrate materials are to be used that do not match in lattice constants nor in thermal expansion coefficient. Sapphire and SiC are the most commonly applied substrates at this time. In our work, we investigated samples grown by LP-MOVPE on sapphire substrates. Due to the large lattice mismatch between (Al)GaN and sapphire it is energetically favorable for the hexagonal (Al)GaN unit cells to grow $30°$ rotated on the substrate [1]. Nevertheless the lattice mismatch is more than 16% in $\vec{a}$ direction. For GaN bulk layers the build-up of strain is prevented by generation of dislocations during the growth under such conditions. The layers grow unstrained with a large resulting dislocation density [2, 3]. When cooling down from the growth temperature of about $1000°C$, the larger thermal expansion coefficient of the sapphire leads to a biaxial compressive strain of the upper GaN layer. This can be observed in a shift of the luminescence energy because a deformation of the Brillouin zone alters the band structure. For the AlGaN layers another behavior is observed that was investigated in this work by spatially resolved cathodoluminescence spectroscopy. Due to the larger energy gap of AlGaN in comparison to (In)GaN this alloy is of interest for heterostructure applications like quantum well structures. It is convenient for carrier as well as photon confinement. For this reason we focused our investigations on the influence of a GaN buffer layer on the AlGaN with the aim to get information about the homogeneity and the local dependencies of the luminescence energy and consequently of the strain. The origin of strain and especially the relaxation mechanisms

Mat. Res. Soc. Symp. Proc. Vol. 449 © 1997 Materials Research Society

as well as fundamental items like deformation potentials are not very well known for this material system.

EXPERIMENTAL

We investigated samples grown by low pressure metalorganic vapor phase epitaxy (LP-MOVPE) on the c-face of sapphire. The source materials were triethyl-gallium (TEG), trimethyl-aluminium (TMA) and ammonia (NH_3); H_2 and N_2 were used as carrier gases. The pressure in the reactor was in all cases 100 mbar. A nucleation layer of about 10 nm AlN grown at 800°C drastically improves the quality of the upper nitride layers grown at 1000 °C. Further details on the growth parameters are presented elsewhere [4].
The thickness of all $Al_{0.12}Ga_{0.88}N$ layers was 1.3 μm. In the first case a 1 μm GaN buffer layer is grown below the AlGaN, in the second case there is no buffer layer. To compare the layers with unstrained GaN and AlGaN, slivers were scraped off the samples using a diamond needle. Obviously at some slivers the AlGaN and the GaN are separated since no GaN luminescence can be observed. The aluminium content of the AlGaN layers was determined by room temperature X-ray diffractometry using Vergard's rule that is quite well observed for aluminium contents of less than 15% and by the luminescence energy of the relaxed slivers [5]. The layer thickness was measured by a scanning electron microscope (SEM). For our investigations we used low-temperature cathodoluminescence spectroscopy at 8 K to probe the local dependencies of the strain. The acceleration voltage of the primary electrons was 3 kV...10 kV according to the penetration depth needed for excitation of the regions of interest. For detection a 25 cm spectrometer was used followed by an optical multichannel analyzer intensified for the short wavelength region.

RESULTS

Let us first discuss our results for the case of the AlGaN layer grown on a buffer layer, in the second part the investigations without buffer will be presented. The left part of Fig. 1 shows the luminescence of a 1.3μm thick $Al_{0.12}Ga_{0.88}N$ layer grown on sapphire with an intermediate 1 μm GaN buffer layer and that of unstrained material. The luminescence energy of the AlGaN layer shifts to smaller values while that of the GaN layer shifts to larger values in comparison to the unstrained material. The origin of this behavior can be found in the strain of the layers. While the GaN is compressively strained due to the difference in thermal expansion coefficients between GaN and sapphire [2, 3], the AlGaN layer is tensilely strained. The lattice mismatch between GaN and AlGaN with an aluminium content of 12 % is much smaller than that between GaN and sapphire in $\vec{a}$ direction. The lattice constant of $Al_xGa_{1-x}N$ decreases with increasing aluminium content. For the AlGaN it is possible in this case to accommodate the lattice constant of the GaN at growth and is respectively tensile strained on the top. The influence of the difference in the thermal expansion coefficients [6], assuming a linear progression in composition, is smaller than the tensile elastic strain. So the resulting strain of the AlGaN layer is of elastic type induced during the growth. The right part of Fig. 1 shows the energetic position of the GaN and AlGaN luminescence lines scanned over a cleaved edge of the sample. As a result of smaller crystal forces at the edges, the material has the possibility to relax in this region. The direction of the relaxation indicates the type of strain. We observed an increase of the luminescence energy at the edge due to tensile, a decrease due to compressive strain for the AlGaN and the GaN layers, respectively. The tensile nature of the top layer can be observed also in the SEM

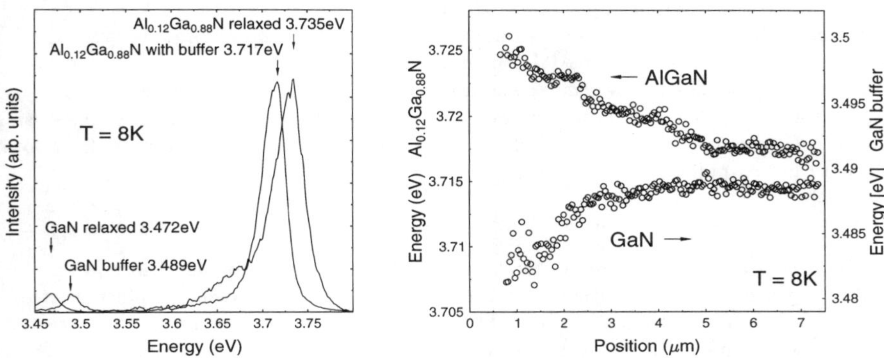

Figure 1: Characterization of a 1.3μm Al$_{0.12}$Ga$_{0.88}$N layer grown on a 1 μm GaN. The left picture shows the AlGaN and GaN luminescence lines compared with unstrained material, the right part shows the relaxation of strain at a cleaved edge of the sample.

image of Fig. 2. Cracks and even a lift off of the AlGaN can be seen in a few regions of the surface. The left part of Fig. 2 represents the energetic position of the luminescence

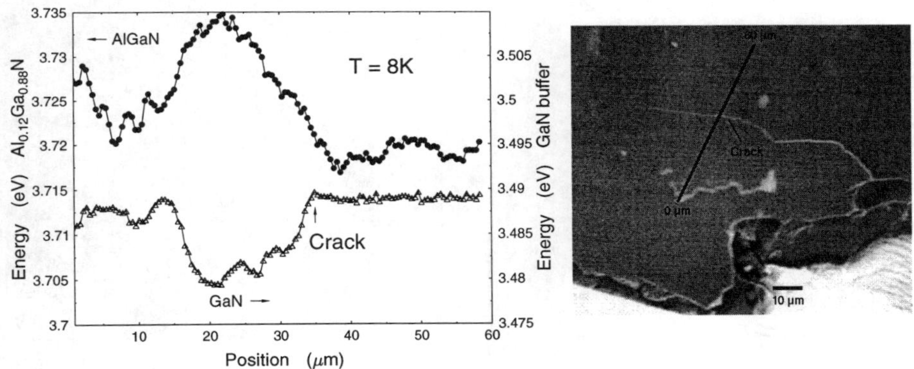

Figure 2: Position of the luminescence energy and SEM image of Al$_{0.12}$Ga$_{0.88}$N grown on a 1 μm GaN buffer layer. Due to tensile strain of the top layer detaching can be observed. The luminescence was scanned along the black line in the right image.

peak corresponding to the lateral position scanned along the line shown in the SEM image. Like in Fig. 1 it can be observed that the GaN layer is also strained, but in a compressive way. At 22 μm the AlGaN is completely lifted off the GaN. The luminescence energy of 3.735eV represents the value of relaxed material as it is shown in the left part of Fig. 1. The GaN layer does not relax totally, the contribution of the difference in thermal expansion

coefficient to sapphire remains as it was observed for GaN layers without AlGaN on top [7].
The energy shift between the fully strained region and the region of totally relaxed elastic
strain is 16meV and 10meV for AlGaN and GaN, respectively.

The relation between strain and energy shift is given by the deformation potential. In wurzite
structure there are four deformation potentials that can be combined to an effective single
value C. For the A exciton and in good approximation for the bound exciton [8] the following
relation can be applied:

$$\Delta E = 2[S_{13}(C_1 + C_3) + (S_{11} + S_{12})(C_2 + C_4)] \cdot \epsilon = C \cdot \epsilon \tag{1}$$

S_{ij} represent the coefficients of elasticity and C_i the components of the deformation potential. Using a deformation potential of 12eV for GaN [2, 7] the deformation potential for
$Al_{0.12}Ga_{0.88}N$ can be estimated to be (19 ± 4)eV. Most of the error is due to the uncertainty
of the exact composition and the intensity fluctuation of the optical transitions. For the
case of AlGaN grown directly on the AlN nucleation layer without a buffer the behavior
totally changes. In the SEM image no cracks can be observed. Scans over cleaved edges does
not show the typical bearing of relaxation, i.e. a shift of the luminescence energy to one
direction. The luminescence energy changes in a range of a few μm by more than 10 meV
with no recognizable tendency. The local dependency is shown in Fig. 3. In the cathodoluminescence image, dark regions of a few μm expansion can be seen representing areas of
higher energy. This is a quite amazing behavior. Growing a 1.3 μm GaN layer instead of

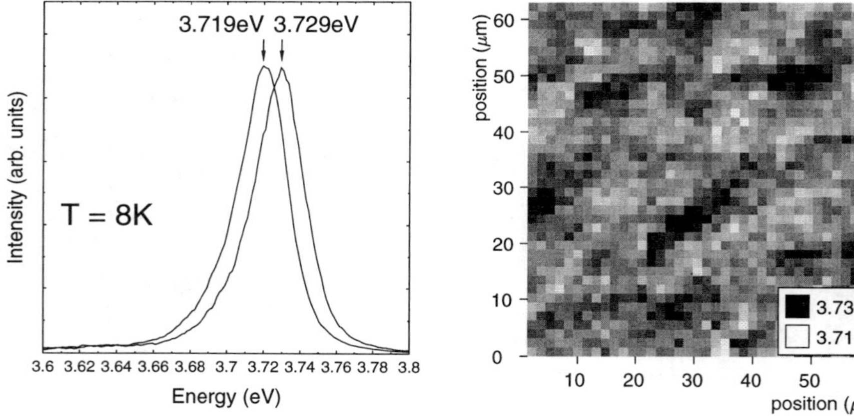

Figure 3: Luminescence of $Al_{0.12}Ga_{0.88}N$ grown directly on the nucleation layer. The left
picture shows two spectra at different position on the sample. In the right picture the
position of the luminescence was observed for a two dimensional area.

AlGaN, a compressive strain of more than 0.12% results. That can be definitely detected
looking at the relaxation at an edge as well as in X-ray diffractometry [7]. Fig. 4 represents
the rocking curves of the different samples. It can be observed that the rocking curve of the
AlGaN layer without buffer is obviously broadened in comparison to the one with a GaN
buffer layer. At least two reasons can be responsible for this behavior: a.) different layer

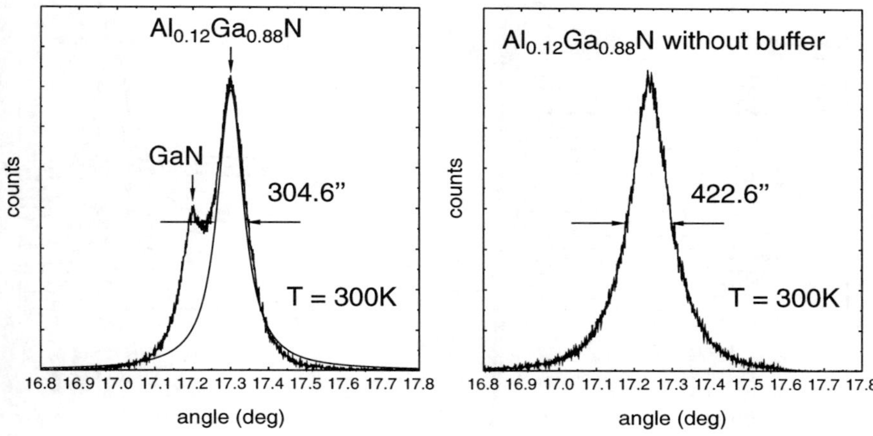

Figure 4: Rocking curve of Al$_{0.12}$Ga$_{0.88}$N grown with and without 1 μm GaN buffer layer. Angles are not corrected for misorientation.

qualities, i.e. dislocation densities, b.) fluctuations in composition. In cathodoluminescence spectra no strain of the layer could be observed. Investigations on GaN layers show that large dislocation densities broaden the luminescence line and shift some intensity to impurity related transitions but do not move the position of the main transition. This behavior can be found also for AlGaN as it is shown in Fig. 5. The luminescence is scanned perpendicular to an edge of the sample. At the interface to the sapphire, where most dislocations are expected, a low energy peak can be seen up to a height of about 500nm that suppresses the luminescence of the AlGaN. No influence on the energetic position of the AlGaN luminescence can be recognized. So the origin of the luminescence shift is due to fluctuation of the aluminium content depending on the position. Using the composition dependence of the energy [5], a fluctuation of the aluminium content of 0.4% can be calculated. The ultimate reason for this total different behavior of the AlGaN in comparison to GaN layers is not completely clear. An explanation might be a difference in adhesion between AlGaN and the nucleation layer and between GaN and the nucleation layer.

SUMMARY

We investigated the difference in strain and homogeneity of AlGaN layers grown by LP-MOVPE on the c-face of sapphire with regard to the influence of an intermediate GaN buffer layer. An AlN nucleation layer directly grown on the $\alpha - Al_2O_3$ is used to increase the quality of the upper nitride layers. It was found that a GaN buffer layer of 1μm thickness totally changes the behavior of the Al$_{0.12}$Ga$_{0.88}$N layer. With a buffer layer it can be observed that the strain of the AlGaN is tensile due to elastic stress and is induced during the growth. The GaN buffer layer on the other hand is, additionally to the stain induced by the mismatch in thermal expansion coefficients to the sapphire, compressively strained. Relaxation can be observed at the edges. Cracks are preferentially observed in regions close to the edges of

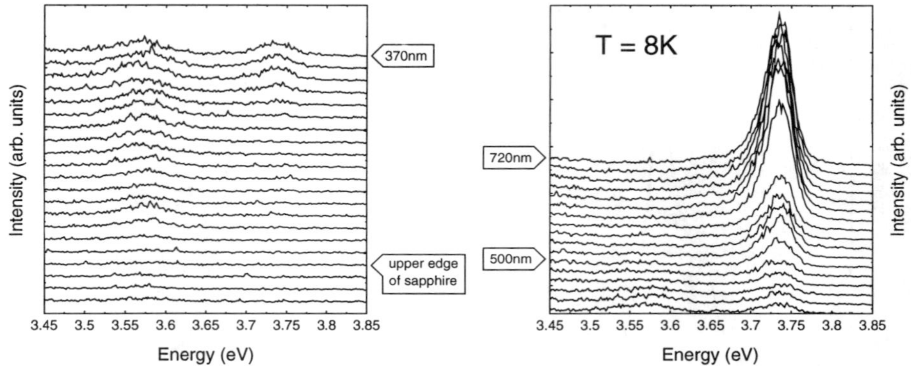

Figure 5: Luminescence of $Al_{0.12}Ga_{0.88}N$ without buffer layer scanned along the growth direction. The right part is the continuation of the left one in direction to the top of the sample. The low energetical peak has no influence on the energetical position of the AlGaN luminescence.

the sample but otherwise the top layer is quite homogeneous. From the energy shift in case of relaxation an effective deformation potential of (19 ± 4)eV can be estimated. Growing the AlGaN directly on the nucleation layer leads to a rough surface and broadened rocking curves. Spatially resolved cathodoluminescence spectroscopy indicates an absence of strain, but considerable composition fluctuations of 0.4%. At the interface to the substrate, where the most dislocations are expected, the luminescence of the AlGaN is suppressed.

REFERENCES

1. T. Sasaki and S. Zembutsu, J. of Appl. Phys. **61**, 2533 (1987).

2. H. Amano, K. Hiramatsu, and I. Akasaki, Jap. J. of Appl. Phys. **27**, 1384 (1988).

3. M. Leszczynski, T. Suski, H. Teisseyre, P. Perlin, I. Grzegory, J. Jun, , S. Porowsky, and T. D. Moustakas, J. of Appl. Phys. **76**, 4909 (1994).

4. F. Scholz, V. Härle, H. Boley, F. Stoiber, B. Kaufmann, G. Reyher, A. Dörnen, O. Gfrörer, S.-J. Im, and A. Hangleiter, in *Proceedings of the Topical Workshop on Nitrides 95, Nagoya, Japan* (1995). Solid State Electronics (1996), in press.

5. S. Yoshida, S. Misawa, and S. Gonda, J. of Appl. Phys. **53**, 6844 (1982).

6. *Numerical Data and Functional Relationships in Science and Technology, New Series*, edited by Landolt and Börnstein (Springer-Verlag Berlin, New York 1982), Vol. 17.

7. O. Gfrörer, T. Schlüsener, V. Härle, F. Scholz, and A. Hangleiter, in *Symposium C: UV, Blue and Green Light Emission from Semiconductor Materials, EMRS Spring Meeting, Strassbourg, France, June 1996*. Mat. Sci. Eng. B (1996), in press.

8. A. Gavini and M. Cardona, Phys. Rev. B **1**, 672 (1970).

STRUCTURAL AND OPTICAL CHARACTERIZATION OF HIGH-QUALITY CUBIC GaN EPILAYERS GROWN ON GaAs AND 3C-SiC SUBSTRATES BY GAS SOURCE MBE USING RHEED IN-SITU MONITORING

H. Okumura, K. Balakrishnan, G. Feuillet, K. Ohta*, H. Hamaguchi**, S. Chichibu**,
Y. Ishida, S. Yoshida
Electrotechnical Laboratory, 1-1-4, Umezono, Tsukuba, Ibaraki 305 JAPAN, okumura@etl.go.jp
*Shibaura Institute of Technology, 3-9-14, Shibaura, Minatoku, Tokyo 108, JAPAN
**Science University of Tokyo, 2641, Yamazaki, Noda, Chiba 278, JAPAN

ABSTRACT

By monitoring RHEED reconstruction patterns during gas source molecular beam epitaxy growth, the optimization of the growth for cubic GaN was carried out successfully. Cubic GaN epilayer having a X-ray diffraction width of 16min and a low temperature photoluminescence emission width of 19meV was obtained on a 3C-SiC substrate by adjusting the effective III/V ratio *in-situ* during the growth, which can be inferred from the surface reconstruction transitions. It was found that the surface reconstructions of cubic GaN surfaces are good indices for the optimization of growth parameters.

INTRODUCTION

Recently, much more interest has been aroused in III-V nitride material system owing to the achievements of blue or ultraviolet stimulated emission by current injection from semiconductor laser structures [1,2] as well as the commercialization of blue light emitting diodes (LED) [3,4] which are fabricated with GaN-based alloys and heterostructures. These GaN-based materials are heteroepitaxial layers having hexagonal (wurtzite) crystal structure grown on sapphire substrates by metalorganic chemical vapor deposition (MOCVD) technique.

On the other hand, it has been recently found that GaN can be grown in another crystal structure, *i.e.*, cubic (zincblende) structure [5, 6], which is usual for arsenides, phosphides and antimonides of III-V compound semiconductors. Energetically, cubic nitrides are less stable compared with corresponding hexagonal nitrides. Because cubic crystals have higher crystallographic symmetry than hexagonal ones, it is expected that cubic GaN would show lower phonon scattering, better doping characteristics, and so on. Moreover, large-area and conductive substrates, such as Si and GaAs, are available for the growth of cubic GaN. However, the number of studies on cubic GaN has been sparse compared with hexagonal GaN, and quality of cubic epilayers is still poor. One reason is the lack of an appropriate cubic substrate which is stable at the high growth temperatures and has a close lattice constant to GaN. The most commonly used substrates for GaN is sapphire, and the hexagonal structure of sapphire substrates results in the growth of hexagonal GaN. Another reason is the metastable nature of the cubic phase. Growth of the metastable cubic phase is unfavorable under the equilibrium growth condition such as that in MOCVD process. On the other hand, using molecular beam epitaxy (MBE) technique, which is well-known as a non-equilibrium growth method, it has been believed to be difficult to grow GaN with good crystalline quality due to the lack of sufficiently effective N sources, until recently. However, owing to the recent development of N radical sources, studies on cubic GaN have been increasing in number, and the interest in cubic-phase nitride has also been greatly increased.

MBE technique has a large advantage compared with MOCVD technique. It is the capability of *in-situ* monitoring using reflection high energy electron diffraction (RHEED) *etc*. Regarding the surface reconstruction of GaN surfaces, there have been several reports investigating the transition of reconstruction. Brandt *et al*. reported the reconstruction transition for cubic GaN (001) surface grown on GaAs (001) substrates [7]. They observed the transitions among (1x1), (2x2) and c(2x2) reconstructions, and attributed these transitions to the difference of surface coverage of Ga. For hexagonal GaN (0001) surface, transition of reconstruction between (2x2)

435

Mat. Res. Soc. Symp. Proc. Vol. 449 © 1997 Materials Research Society

and (1x1) has been also reported by Iwata *et al.* [8] and our group [9]. We also investigated the phase diagram of surface reconstructions as a function of substrate temperature and Ga flux intensity for the both types of GaN epilayers [9-11]. For the growth of III-V nitrides, it is well known that the accurate elucidation of the flux intensity of effective nitrogen species is difficult. The MBE growth of GaN has been so far carried out without any monitoring tools for III/V ratio. However, it was found that the conditions on the reconstruction transition curve in the surface phase diagram correspond to some effective III/V ratio [9, 10]. By monitoring the surface reconstruction transition, we can adjust the stoichiometric balance of reaction species on the surface *in-situ* for the growth of GaN.

We have grown cubic GaN epilayers by gas source MBE technique using a plasma source [11, 12]. The optimization of growth parameters using RHEED *in-situ* monitoring of the surface reconstructions enabled us to obtain high-quality cubic GaN epilayers on GaAs and 3C-SiC substrates. Also, flat surface morphology of 3C-SiC substrates much improved the structural quality of cubic GaN epilayers. In this paper, we report the structural and optical characterization of these cubic epilayers in relation to the growth conditions.

EXPERIMENTS

Cubic GaN epilayers were grown on GaAs(001) and 3C-SiC(001) substrates by gas source MBE. 3C-SiC substrates were grown on Si(001) surfaces by atmosphere chemical vapor deposition (APCVD) or low pressure chemical vapor deposition (LPCVD) technique. Ga was evaporated from a Knudsen-type effusion cell. As a nitrogen source, N_2 beam activated with an electron cyclotron resonance (ECR) plasma cell was used. After thermal cleaning of the substrates, followed by GaAs buffer layer growth in the case of GaAs substrates, streak RHEED patterns were obtained for the both types of substrates. On these substrate surfaces, at first, Ga was deposited under the activated N_2 beam at around 400°C. By the annealing of the substrates, spotty RHEED patterns of cubic GaN were obtained, and the epitaxial growth of cubic GaN was carried out subsequently. RHEED patterns in this sequence are shown in Fig.1 for the case of 3C-SiC substrates. The microwave power used for the ECR cell was 40-100W, the flow rate of N_2 was 1-8sccm, and the growth temperature was

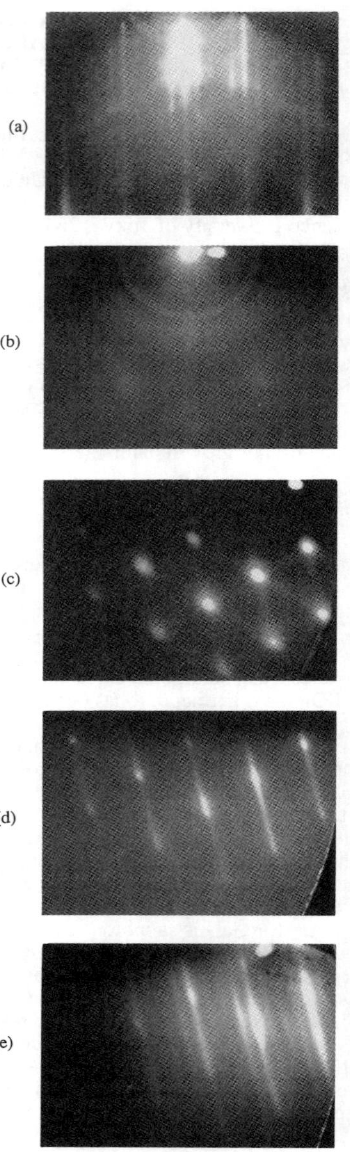

Fig.1 RHEED patterns in the sequence of cubic GaN growth on a 3C-SiC substrate, (a) initial 3C-SiC surface, (b) after the buffer layer deposition at 400°C, (c) after the subsequent annealing, (d) during GaN growth, and (e) after the subsequent annealing without Ga flux.

600-800°C. The obtained growth rate was 200-600Å/h. During the growth, RHEED patterns, *i.e.*, (2x2)/c(2x2) for GaAs and (4x1)/(1x1) for 3C-SiC substrates [10, 11, 13], were observed, and the Ga flux intensity and the substrate temperature were adjusted by monitoring the RHEED reconstruction transition.

PL measurements were carried out at 4.2K using a 325nm line from a He-Cd laser for excitation. In the XRD measurements, reciprocal lattice mapping and pole figure generation techniques were used in addition to the conventional ω - 2θ scanning mode.

RESULTS and DISCUSSION

Compared with hexagonal GaN epilayers on sapphire, the structural quality of cubic epilayers is still inferior. XRD peak full widths at half-maximum (FWHM) for (0002) diffraction of hexagonal layers are usually several min and the best one reported so far is around 30sec [14]. On the other hand, XRD peak widths of (002) diffraction for cubic layers are several tens of min. One of the causes of these large widths is the existence of many stacking faults in the cubic layers. From the cross-sectional transmission electron microscope observation, it was found that the region farther form the interface contains less stacking faults compared with the region near the interface, and that stacking faults originate from rough portion at the interface [15]. Figure 2 shows the thickness dependence of (002) diffraction FWHM of cubic layers grown on 3C-SiC. There is a tendency for the XRD width to decrease as the layer thickness increases, which is consistent with this TEM observation.

Considering the generation of stacking faults at the interface, it is well expected that smoother surfaces may results in the structural improvement of epilayers. Usually, 3C-SiC substrates are obtained by APCVD technique on Si (001) substrates, and the surface morphology of such SiC layers is not so good. However, LPCVD grown 3C-SiC layers exhibit quite flat and featureless morphology [16]. By using these 3C-SiC substrates, we have succeeded in achieving the XRD $\Delta 2\theta$ width as small as 16min in a ω-2θ scan for (002) diffraction spot. This result indicates that the initial surface morphology just before the nitride growth affects the structural quality of GaN epilayers.

In Fig. 3, the phase diagram of (1x1)/(4x1) surface reconstruction is shown as a function of

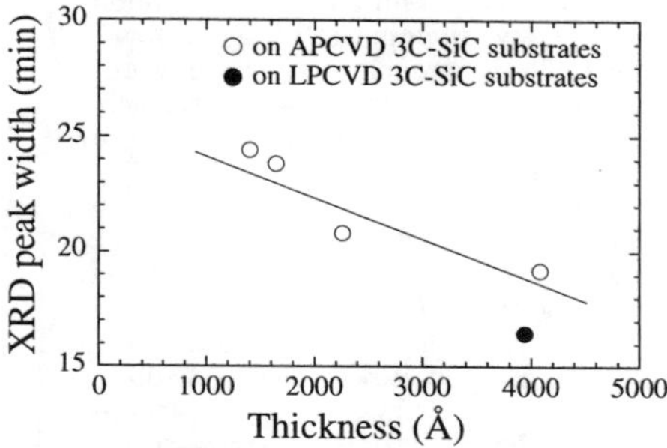

Fig.2 Thickness dependence of XRD (002) diffraction peak width of cubic GaN grown on 3C-SiC substrates.

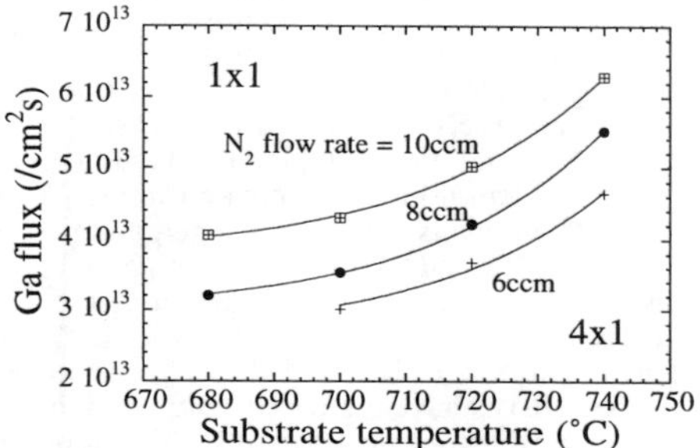

Fig.3 Phase diagram of cubic GaN(001) (1x1)/(4x1) surface on 3C-SiC substrates under the constant N$_2$ flow rate.

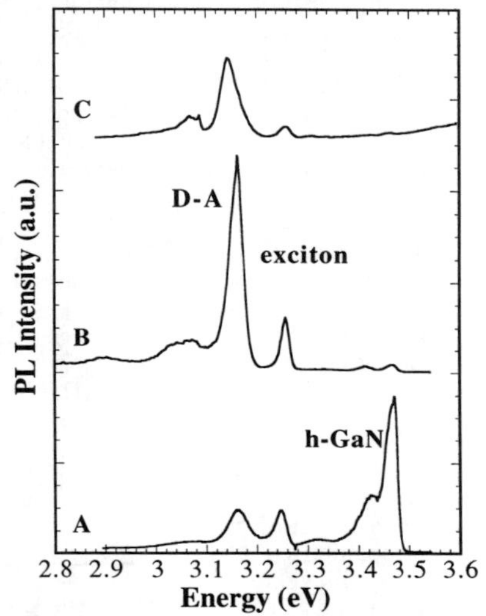

Fig.4 Low temperature PL spectra of cubic GaN epilayers grown under 3 types of growth conditions in Fig.3. Sample A was grown in (4x1) region, B on the transition curve, and C in (1x1) region.

437

substrate temperature and Ga flux intensity for several constant nitrogen flow rates. Phase boundary curves between (1x1) and (4x1) regions, where surface phase transitions occur, are expressed by the formula $\alpha\Phi_N = \Phi_{Ga} - A\exp(-E_a/kT)$, where Φ_N is the flux intensity of effective excited nitrogen species, Φ_{Ga} the Ga flux intensity, E_a the activation energy of Ga re-evaporation, k Boltzman constant, T the substrate temperature, respectively [9, 10]. For the growth on GaAs substrates showing (2x2)/c(2x2) transitions, similar phase diagram was obtained.

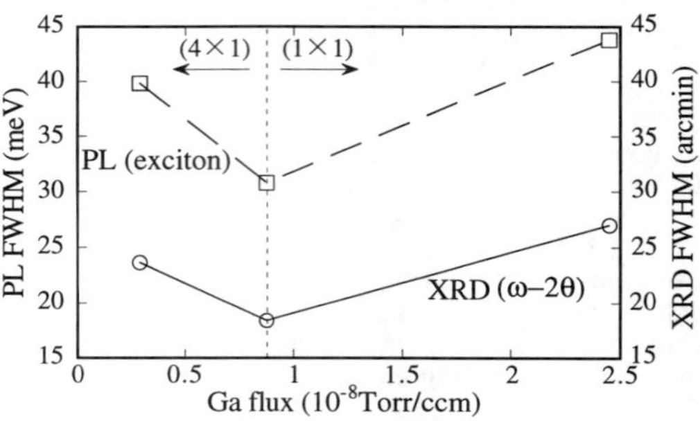

Fig.5 (002) XRD widths and PL exciton emission widths of cubic GaN epilayers grown under 3 types of conditions in Fig.3. ECR power during the growth was 40W.

Based on this phase diagram, we have grown 3 kinds of cubic samples at 3 different regions, i.e., (1x1) region, on the transition curve, and (1x4) region. Low temperature PL spectra for the 3 samples are shown in Fig. 4. The width of the exciton emission at 3.27eV is the narrowest for sample B grown just at the transition curve. Both samples A and C, grown in (1x1) and (4x1) regions apart from the transition curve respectively, exhibited wider widths. Moreover, samples A and C show emission attributed to impurity/defect-related levels at the lower energy region or that from mixed hexagonal phase at the higher energy side. The similar result was obtained for ω-2θ XRD measurement. Sample B also showed the narrowest XRD width among the 3 samples. In Fig.5, obtained widths of PL and XRD measurements are plotted against Ga flux intensity relative to N_2 flow rate. As far as the characterization by PL and XRD widths is concerned,, samples grown at the condition on the transition curves have the best quality, as demonstrated for hexagonal GaN[9].

Under the consideration of III/V stoichiometric balance during the growth [9, 10], transition curves in the surface phase diagram correspond to some constant effective III/V ratio, as proposed by Däweritz and Hey for GaAs [17]. Thus, the dependence of the quality of obtained layers on surface reconstruction during the growth is related to the effective III/V ratio on the surface. In the MBE growth using a plasma source, it is actually impossible to precisely measure N-related species (N_2^*, N_2^+, N^* etc.) which contribute to nitride growth on the growing surfaces. Moreover, III/V ratio should depend on the growth temperature strongly. The present results mean that the optimization of the effective III/V ratio for the growth of cubic GaN is carried out successfully by means of monitoring the surface reconstructions including the effect of growth temperature. So far, GaN epilayers have been grown under N-rich condition to avoid the

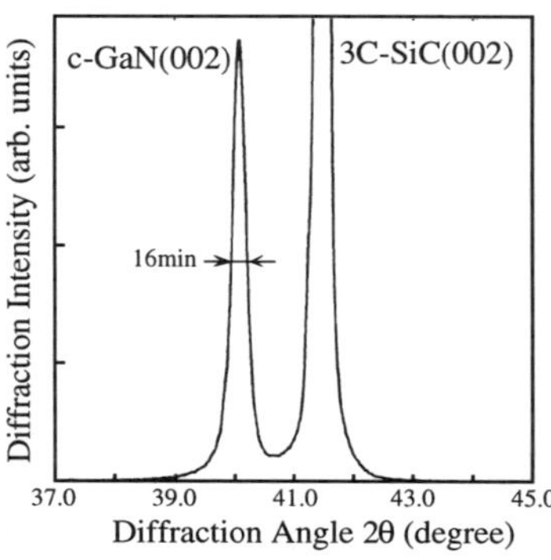

Fig.6 XRD pattern of a cubic GaN epilayer grown under the growth condition on the transition curve in Fig.3

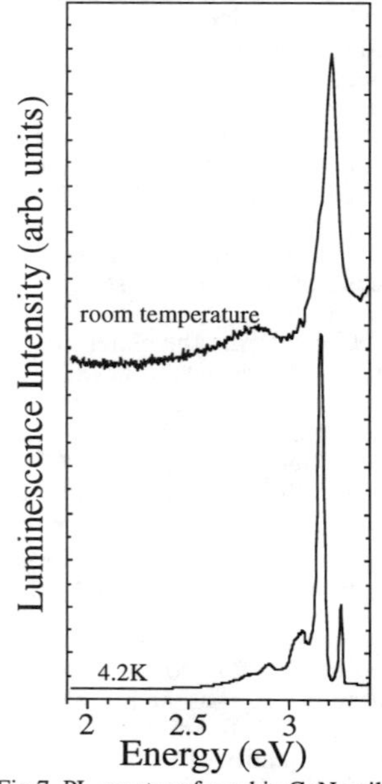

Fig.7 PL spectra of a cubic GaN epilayer grown under the growth condition on the transition curve in Fig.3.

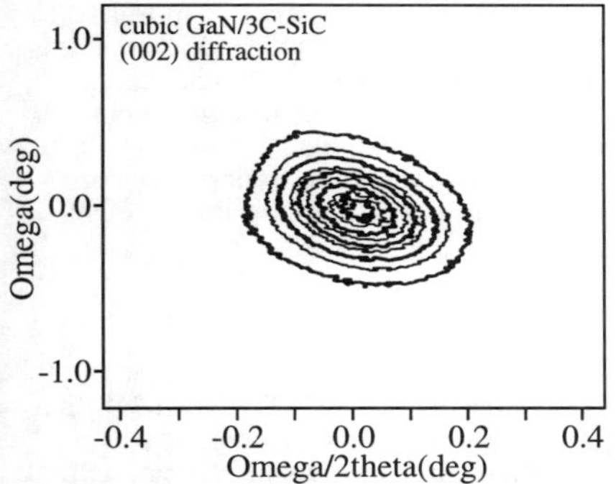

Fig.8 Reciprocal Lattice mapping of a cubic GaN epilayer grown on a 3C-SiC (001) substrate.

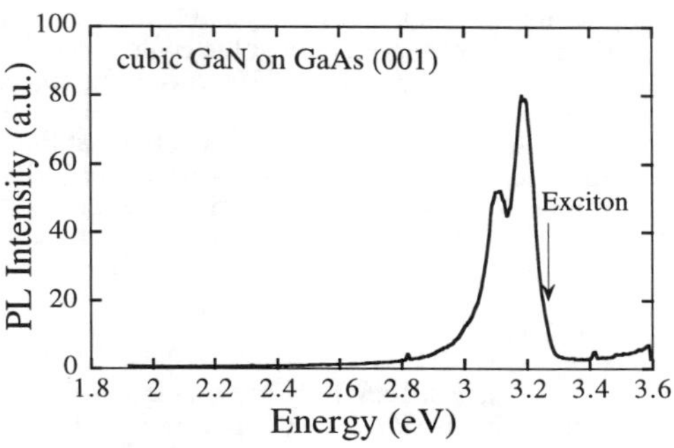

Fig.9 Low temperature PL spectrum of a cubic GaN epilayer grown on a GaAs substrate under the condition on the transition curve.

formation of Ga droplets. However, excess N-rich condition was found to result in the deterioration of the quality.

In Figs. 6 and 7, XRD patterns and PL spectra of the sample exhibiting the narrowest widths so far are shown, *i.e.*, 16min for XRD and 19meV for low temperature PL. This sample showed a pronounced emission band even at room temperature. In the lower energy region, there was no appreciable emission band. These results are the evidence of high quality of the sample.

The result of reciprocal space mapping of cubic GaN epilayers on a 3C-SiC substrate is shown in Fig.8. Compared with usual h-GaN epilayers, the shape of the reciprocal lattice point is isotropic. The value of $\Delta\omega$ is estimated to be 29min. Near this diffraction spot, there was no trace of other spots attributed either to cubic or to hexagonal GaN. Any streak lines caused by hexagonal stacking in cubic phase [18] are not observed either. The pole figure technique [11], which can detect any lattice planes inclined against the epilayer surfaces, also revealed the superior single phase feature of this cubic sample.

For the growth on GaAs(001) substrates, the effective III/V ratio was adjusted by observing (2x2)/c(2x2) surface reconstruction transitions. Compared with the case of 3C-SiC case, the streak feature of RHEED patterns and the reconstruction transition was not so obvious. The dependence of epilayer quality on the reconstruction region was not clear, also. However, the sample grown under the condition on the transition curve exhibited quite strong emission at the band edge region, as shown in Fig. 9.

In Fig.10, the results of XRD pole figure measurements for cubic GaN layers grown on GaAs (001) substrates are shown. Figs.10(a) and 10(b) were obtained under the detection con-

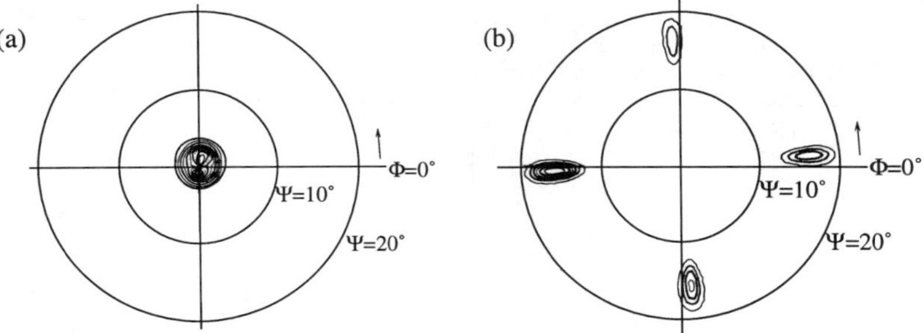

Fig.10 XRD pole figure measurements for a cubic GaN sample grown on a GaAs (001) substrate. Detecter angle 2θ was set for (a) cubic (002) and (b) hexagonal (0002) diffraction spots, respectively. Every 10 degrees of azimuthal angle ψ are shown by solid lines.

figuration for cubic GaN(002) and hexagonal (0002) diffraction spots, respectively. In Fig.10(a), diffraction is observed only at the center. This result shows that any misoriented cubic grain is not included in the epilayer. However, Fig.10(b) indicates that the sample contains hexagonal grains 16° inclined for 4 equivalent directions. From diffraction intensity, the hexagonal content is estimated to be several %. It is noted that this sample shows no detectable PL emission attributed to hexagonal phase, and any hexagonal XRD diffraction spots were not observed by conventional ω-2θ scan mode. In this sense, pole figure technique is a very useful tool for the characterization of phase mixing.

CONCLUSION

We have successfully grown high quality cubic GaN epilayers on LPCVD-grown 3C-SiC substrates by *in-situ* growth optimization using RHEED reconstruction monitoring. It was found that the growth condition on the reconstruction transition curves results in better structural and optical quality. The optimization by RHEED observation was a very useful tool especially for the improvement of the quality of cubic GaN epilayers.

REFERENCE
1. I.Akasaki, H. Amano, S. Sota, H. Sakai, T. Tanaka, M. Koike, Jpn. J. Appl. Phys. **34**, p.L151 (1995)
2. S. Nakamura, M. Senoh, S. Nagahara, N. Iwasa,T. Yamada, T. Matsushita, H. Kiyoku, Y Sugimoto, Jpn. J. Appl. Phys. **35**, p.L74 (1996)
3. S. Nakamura,T. Mukai, M. Senoh, Appl. Phys. Lett. **64**, p.1687 (1994)
4. S. Nakamura, M. Senoh, N. Iwata, S. Nagahama,T. Yamada, T. Mukai, Jpn. J. Appl. Phys. **34** p.L1332 (1995)
5. H. Okumura, S. Misawa, S.Yoshida, Appl. Phys. Lett. **59**, p.1058 (1991)
6. H. Okumura, S. Miwasa,T. Okahisa, S. Yoshida, J. Cryst. Growth **136**, p.361 (1994)
7. Brandt, H.Yang, B. Jenichen, Y. Suzuki, L. Däweritz, K. H. Ploog, Phys, Rev. **B52**, p.R225 (1995)
8. K. Iwata, H.Asahi, S. J. Yu, K. Asami, M. Fujita, S. Gonda, Jpn. J. Appl. Phys. **35**, p.L28 (1996)
9. P Hacke, G. Feuillet, H. Okumura, S. Yoshida, Appl . Phys. Lett. **69**, p.2507 (1996)
10. G. Feuillet, H. Hamaguchi, K. Ohta, P.L. Hacke, H. Okumura, S. Yoshida, to be published i Appl. Phys. Lett
11. H. Okumura, K. Ohta, G. Feuillet, K. Balakrishnan, S. Chichibu, H. Hamaguchi, P. Hacke and S. Yoshida, to be published in the special issue on nitrides in J. Cryst. Growth
12. H. Okumura, S. Misawa,T. Okahisa, S, Yoshida, J. Cryst. Growth, **136**, p.361 (1994)
13. G. Feuillet, P Hacke, H. Okumura, H. Hamaguchi, K. Ohta, S. Yoshida, presented at the 199 MRS Fall Meeting, Boston,1996 (unpublished)
14.W. Li, W.-X. Ni, Appl. Phys. Lett. **68**, p.2705 (1996)
15. H. Okumura, K. Ohta,T. Nagatomo, S. Yoshida, J. Cryst. Growth, **164**, p.149 (1996)
16.Y. Ishida, Private communication
17. L. Däweritz and R. Hey Surf. Sci. **236**, 15 (1990)
18.A. A. Yamaguchi, T. Manako, A. Sakai, H. Sunakawa, A. Kimura, M. Nido, A. Usui, Jpn. J Appl. Phys. **35**, p.L873 (1996)

STRUCTURAL ANALYSIS OF GaN AND GaN/InGaN/GaN DH STRUCTURES ON SAPPHIRE (0001) SUBSTRATE GROWN BY MOCVD

Hisao Sato, Yoshiki Naoi and Shiro Sakai
Department of Electrical and Electronic Engineering, The University of Tokushima
2-1 Minami-josanjima, Tokushima 770, Japan

ABSTRACT

We have investigated lattice structures of GaN and InGaN/GaN single-heterostructures (SH) and double-heterostructures (DH) by the reciprocal space mapping using X-ray diffraction technique from (0002) plane. For a single GaN layers, the transition of the film structure from grains with relatively independent orientation at about 0.3 µm followed by the coalescence at about 1 µm, to a uniform film with mosaic structure at 1.4 µm was clearly observed. For DH structure, ω-mode FWHMs both from InGaN (In composition approximately 7 %) and capping-GaN layers increased with increasing film thickness from 20 nm to 110 nm indicating the increased mosaic structure in the thick InGaN. We also observed the dislocation related to this increasing mosaic structure.

INTRODUCTION

The III-V nitrides, such as GaN, InN and their ternary alloys, have been widely investigated by numerous researchers with the purpose of constructing short wavelength optical devices as well as high temperature devices. That is because those materials posses peculiar characteristics: they have wider band-gap energy and are more stable than other materials even at the high temperatures. Nakamura has already demonstrated the blue-to-green light emitting diodes (LEDs) with high efficiency and the InGaN multi-quantum-well-structure laser diodes (LDs). [1,2] However, because of the lattice misfit between the nitride alloys and the sapphire substrate, the density of the dislocation in the III-V nitride layers in those devices remains to be as high as 10^9 cm^{-2}. [3]   In our research, we have investigated the structural properties of the III-V nitride layers, which we consider the most important elements to influence the effective device performance.

The reciprocal space mapping is a useful technique to investigate the mosaicity and the lattice coherence in the crystals. Several research groups have reported the structural analyses of the GaN films grown on the sapphire substrate using these techniques. [4,5] However, neither the results on GaN film thickness dependence nor the InGaN/GaN single-heterostructures (SH) and the double heterostructures (DH) have been investigated.

We have carried out the detailed investigation on the crystal structure of the nitride materials grown using the MOCVD technique. This paper presents the results of GaN single layers with different thickness of 0.35-1.4 µm and those of SH and DH structures.

EXPERIMENT

The samples used in our experiments were grown on the sapphire (0001) substrate, using the atmospheric MOCVD technique and also employing the two step growth method. TMG (trimethylgallium), TMI (trimethylindium) and NH$_3$ were used as source gases. A thin AlGaN buffer layer was grown at 500 $^\circ$C, and then, the GaN epitaxial layers with thickness 0.35 µm, 1.05 µm and 1.4 µm were grown at 1050 $^\circ$C.

For SH structure, the 110 nm-thick InGaN layer were grown at 800 $^\circ$C on the 0.4 µm-thick GaN. Assuming Vegard's law and neglecting strain effects, the In composition was determined by adopting the x-ray diffraction technique. The 0.1 µm-thick GaN layer was grown at 1050 $^\circ$C on the SH structure to fabricate DH structure. The thickness of the InGaN layer in the DH was 20 to 110 nm.

The reciprocal space mapping diagrams from (0002) plane of the GaN and the InGaN were

Mat. Res. Soc. Symp. Proc. Vol. 449 © 1997 Materials Research Society

measured by the high resolution X-ray diffractometer (Philips X'Pert-MRD).

RESULTS AND DISCUSSION

<u>GaN grown on sapphire</u>

Figure 1 shows the reciprocal space mapping diagrams of the GaN (0002) peak with the three different thickness. The vertical and horizontal axes correspond to ω and ω–2θ scan, respectively. The contour lines indicate 100, 10, 1 and 0.1% of the GaN (0002) peak intensity. The peak broadening in ω and ω–2θ axes show the crystal mosaicity and the crystal coherency in c-axis (lattice coherent length), respectively.

The peak broadening along ω axis with increasing film thickness is clearly observed. The ω-scan full width at half maximum (FWHM) of the 0.35, 1.05 and 1.4 µm-thick films are 24, 36 and 51 arcsec, respectively. However, the ω–2θ scan FWHM, about 30 arcsec in all cases, does not change with increasing film thickness. This means that the crystal mosaicity increases with increasing GaN thickness.

In the case of the 0.35 µm-thick GaN, its reciprocal space mapping has a cross shape, and its diffraction peak along with the ω–2θ axis at $\Delta\omega = 0$ is particularly broadened while the diffraction peak at $\Delta\omega \neq 0$ is narrow. The ω–2θ scan FWHMs at $\Delta\omega = 0$ and $\Delta\omega = 0.03$ degree are 34 and 25 arcsec, respectively. These results show the followings: The crystal with the c-axis perpendicular to the surface has shorter coherence length than those that have the inclined c-axis. Applying the Scherrer formula, the coherence length is estimated from the FWHM of 2θ–ω mode rocking curves. [6] The calculated coherence length of the 0.35, 1.05 and 1.4 µm-thick GaN along the c-axis at $\Delta\omega = 0$ are 0.355, 0.368 and 0.254 µm, respectively. The coherence length of the 0.35 µm-thick film is the same as the thickness of GaN, while those of the 1.05 and 1.4 µm-thick film are smaller than their film thickness. These results indicate the formation of microscopic grain.

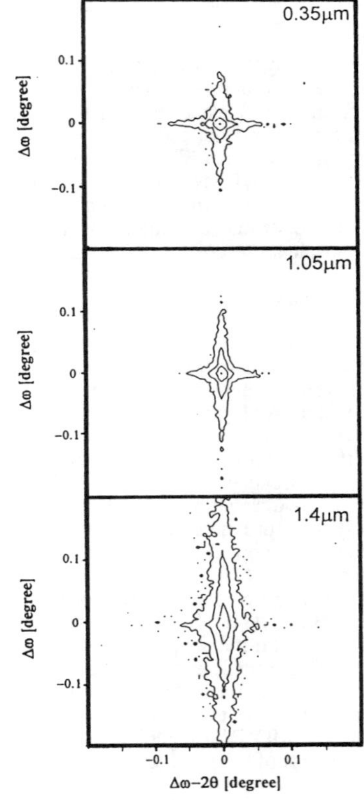

Fig. 1. The reciprocal space mapping from GaN (0002) diffraction with different thickness of 0.35 µm (a), 1.05 µm (b) and 1.4 µm (c).

As for the thick film, the shape of the reciprocal space mapping is nearly oval, and ω–2θ scan FWHMs at $\Delta\omega = 0$ degree (29 arcsec) and $\Delta\omega = 0.03$ degree (31 arcsec) are almost the same. This is interpreted as the results of coalescence of grains formed at the beginning. Small grains with the different tilting angles of the c-plane combine together to form a macroscopic mosaic structure in the thick film.

InGaN/GaN single-heterostructures and GaN/InGaN/GaN double-heterostructures

Figure 2 shows x-ray FWHM from InGaN/GaN single-heterostructures grown on sapphire substrates as a function of In solid composition. The thickness of the InGaN layer is approximately 0.1 μm for all samples. Both ω-mode and ω–2θ mode FWHMs of the InGaN layer rapidly increase when the solid composition exceeds 0.1 μm indicating the degradation in crystal perfection at high indium composition.

Figure 3 is the reciprocal space mapping of the SH and the DH structures. InGaN layers in both structures are grown under the same conditions. It is clear from Fig.3 (a) that both ω-mode and ω–2θ mode FWHMs from InGaN are broaden compared to those from GaN meaning that the mosaicity and lattice coherency in InGaN are inferior to those in the GaN.

Two things are clear by the comparison of Fig. 3 (a) and (b). First is the difference in indium solid composition in SH (indium composition is 0.09) and DH (0.07), although InGaN layers in both structures are grown under the same condition. Evaporation of InN from InGaN may be responsible to this phenomena, since InGaN layer is heated up from InGaN growth temperature of 800 ℃ to that of GaN (1050 ℃) in nitrogen atmosphere.

The second is the broadening in ω-mode FWHM from the GaN in the DH (Fig. 3 (b)). This indicates that the mosaicity in GaN capping layer in the DH is larger than that of the underlying GaN. Since the mosaicity in the InGaN is larger than that of the GaN, the mosaicity of the GaN grown on the InGaN should also be large.

Figure 4 shows the reciprocal space mapping from the (0002) plane of the DH structure which contains the 40 nm and the 110 nm-thick-InGaN layer. In the case of Fig. 4 (b), the ω scan-FWHM of the InGaN (358 arcsec) and that of the GaN (357 arcsec) are almost the same, but the ω–2θ scan-FWHM of the InGaN (275 arcsec) is much larger than that of the GaN (60 arcsec). These results indicate that the coherence length in the InGaN layer is shorter than that in the GaN layer. ω scan-FWHM from the InGaN is very narrow when the InGaN is 40 nm (Fig.4(a)), but it increases at 110 nm (Fig. 4(b)). This indicates that the c-plane of the InGaN is parallel to the substrate at 40 nm, but it becomes tilted at 110 nm. This consideration is consistent with the TEM data shown in Fig. 6(b) in which tilted dislocation is observed in the 110 nm-thick InGaN layer.

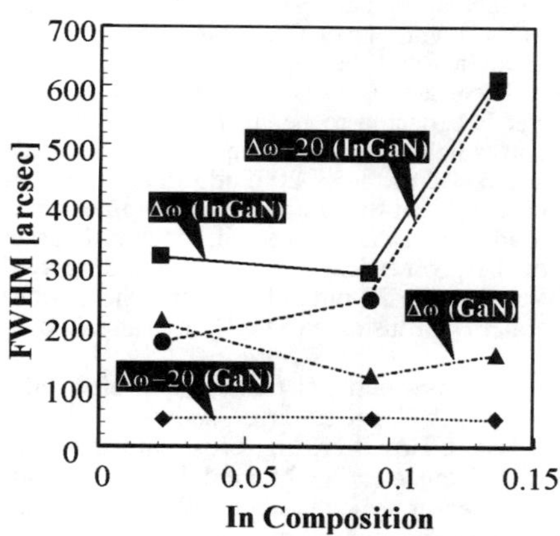

Fig. 2 FWHM as a function of the indium solid composition.

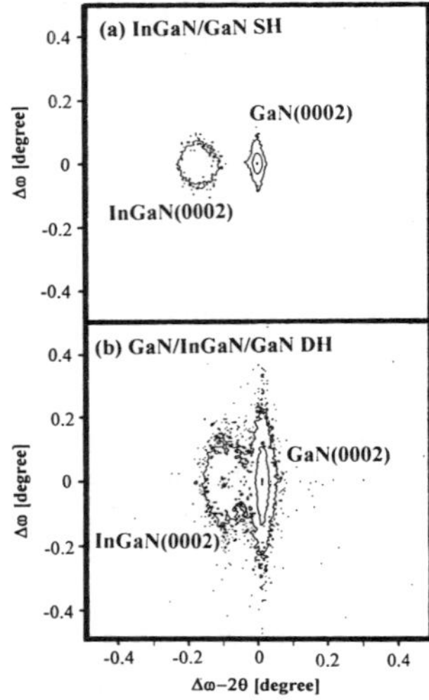

Fig. 3 Comparison of the reciprocal space mapping of the SH and the DH.

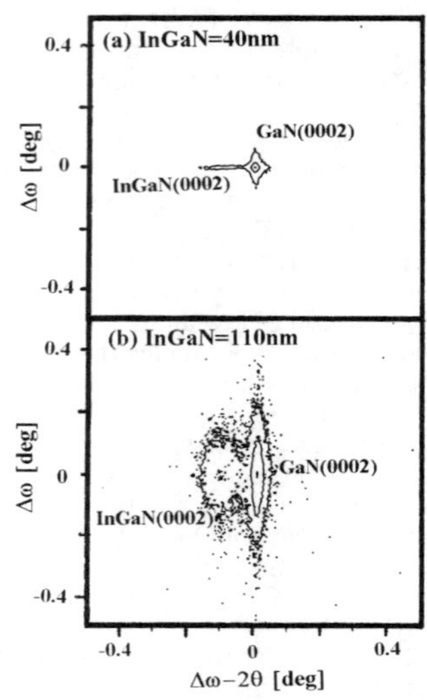

Fig. 4. The reciprocal space mapping from GaN (0002) and InGaN (0002) diffraction for DH structure with different thickness of InGaN layer; (a) 40 nm, (b) 110 nm.

Figure 5 shows the FWHM along both the ω and the ω–2θ axes of (0002) diffraction in the DH structures with different InGaN layer thickness. The FWHMs of the ω mode both from GaN and InGaN increase rapidly as the InGaN layer thickness increases over 40 nm. These findings show that the mosaicity in the InGaN and the capping-GaN layers increase with increasing the thickness of the InGaN layer. There are two reasons for this. First, the X-ray signal from the 0.1 μm-thick capping-GaN layer is expected to be stronger than that from the underlying-GaN layer. Second, this phenomenon is not observed in the case of the single GaN layer. When the InGaN thickness is 20 nm, the FWHM of the ω–2θ mode shows some increase, but at the same time the ω mode FWHM (35 arcsec) and that from the 40 nm-thick InGaN (36 arcsec) remains the same. Since ω mode broadening is not observed, it can be assumed that this FWHM broadening is not the result of the geometrical effect, which is observed in the case of the thin layer. The increase in the FWHM at the 20 nm indicates the short coherence length in InGaN, and which is possibly due to either In diffusion into GaN or to an inhomogeneous lattice strain in the InGaN layer.

Figure 6(a) and (b) show the cross-sectional transmission electron micrograph of the DH structures. Fig. 6(a), with the 40 nm-thick InGaN layer, shows only linear dislocations parallel to the c-axis in GaN and InGaN layers. However, Fig. 6(b), with the 110 nm-thick InGaN layer, tilted dislocations from the c-axis are clearly observed in the 110 nm-thick InGaN and the capping-GaN layers. These observation are consistent with XRD results of Figs. 4 and 5. The inclined dislocations are formed by the increased mosaicity in the InGaN, and they are inherited into the capping-GaN. Therefore, the dislocations shown in Fig. 6(b) are originated from the increased mosaicity in InGaN and capping-GaN layers.

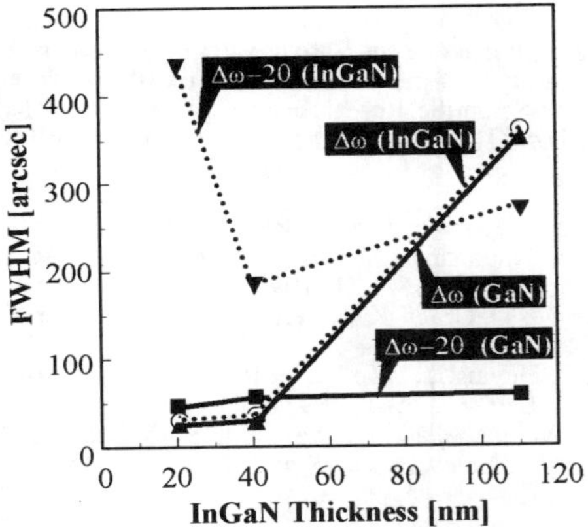

Fig. 5. The FWHM of GaN (0002) and InGaN (0002) diffraction along ω and ω–2θ axis in reciprocal space mapping with different thickness of the InGaN.

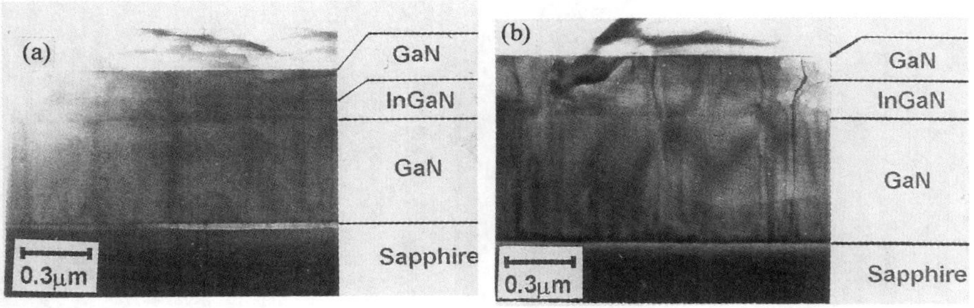

Fig. 6. Cross-sectional TEM micrographs of DH structure with different thickness of InGaN layers; (a) 40 nm, (b) 110 nm.

CONCLUSIONS

The structural properties for the GaN and the InGaN/GaN SH and DH structures grown on the sapphire (0001) substrate using the MOCVD method were investigated adopting the reciprocal space mapping from (0002) plane. For the single GaN layers, the transition of the film structure from grains with relatively independent orientation at 0.35 μm to a uniform film with mosaic structure at 1.4 μm was clearly observed. For the DH structure, the mosaic structure in both InGaN and capping-GaN layer was increased as the increase in the InGaN film thickness from 20 to 110 nm. The dislocations related to this increased mosaic structure in the InGaN and in the GaN grown on it were also observed.

ACKNOWLEDGMENTS

A part of this research was supported by the Satellite Venture Business Laboratory, "Nitride Photonic Semiconductor Laboratory", of the University of Tokushima, by NEDO (project # C-119) and by the grant-in-aid for Scientific Research from the Ministry of Education, Science, Sports and Culture. One of authors (H.S.) was financially supported by JSPS.

REFERENCES
1) S. Nakamura, T. Mukai, and M. Senoh: J. Appl. Phys. **76**, 8189 (1994).
2) S. Nakamura, M. Senoh, S. Nagahama, N. Iwasa, T. Yamada, T. Matsushita, H. Kiyoku, and Y. Sugimoto: Jpn. J. Appl. Phys. **35**, L74 (1996).
3) W. Qian, M. Skowronski, M. DeGraef, K.Doverspike, L. B. Rowland, and D. K. Gaskill: Appl. Phys. Lett. **66**, 1252 (1995).
4) D. Kapolnek , X. H. Wu, B. Heying, , S. Keller, B. P. Keller, U. K. Mishra, S. P. DenBaars, and J. S. Speck: Appl. Phys. Lett. **67**, 1541 (1995).
5) H. Sato, H. Takahashi, A. Watanabe, and H. Ota: Appl. Phys. Lett. **68**, 3617 (1996).
6) For example, B. D. Cullity: *Elements of X-Ray Diffraction* (Addison-Wesley, Reading, Massachusetts, 1978) Chap. 3.

STM OBSERVATION OF NITRIDED-Ga ON Si

Y.NAKADA*, S. MIWA* and H. OKUMURA**, ***
* Joint Research center fotom technology(JRCAT)-Angstrom Technology Partnership, Higashi 1-1-4, Tukuba, Ibaraki, 305, Japan, nnakada@jrcat.or.jp
** JRCAT-National Institute for Advanced Interdisciplinary Research, Higashi 1-1-4, Tukuba, Ibaraki, 305, Japan
*** Electrotechnical Laboratory, Umezono 1-1-4, Tukuba, Ibaraki, 305, Japan

ABSTRACT

To investigate of the initial stage of GaN growth on Si, 0.2 Ga monolayers (ML) on Si (111) was nitrided and then the nitrided surfaces were observed by scanning tunneling microscopy (STM). An aggregation of islands whose longest edges had a direction rotated $15\,^{\circ}$ from Si $[1\bar{1}0]$ direction was observed. The shape of islands looked like a pentagon. Surface roughness was estimated for several nitrided conditions. It was found that surface roughness becomes larger as the nitridation process proceeds.

INTRODUCTION

Recently, much attention has been paid to GaN, which has a large bandgap energy compared with other semiconductor materials, from the viewpoint of realizing blue light emitting diodes (1) and blue lasers (2,3). However, many problems still remain on the growth of GaN single crystals and the epitaxitial growth of GaN thin films. For example, bulk crystals of GaN obtained so far are too small to be used as substrates for homoepitaxial growth, and there are no lattice-matched substrates for heteroepitaxial growth of GaN(4). Therefore, the large differences in lattice constant, crystal structure, and thermal-expansion coefficient between GaN and substrates result. Consequently, the control of nucleation and growth in order to form atomically smooth surface and interface is difficult. Sapphire is the most frequently used substrate material for GaN heteroepitaxial growth, but the lattice mismatch between sapphire and GaN is as large as 12%. Therefore, many defects, e. g., dislocations, are included in GaN epitaxial films on sapphire (5). On the other hand, the lattice mismatch between Si and GaN is as large as 20%. The lattice mismatch between Si and GaN is larger than that between sapphire and GaN. However, both cases are the same considering the larger lattice mismatch. The heteroepitaxitial growths of GaN on Si have the same problems as the heteroepitaxitial growth of GaN on sapphire, but Si would be used as a substrate for GaN as follows. Si is the most developed semiconductor material from the standpoint of cost, size, cleavage and electric conductivity and the realization of optoelectronic devices is well expected for nitride epilayers on Si wafers. Thus, the study of GaN heteroepitaxial growth on Si is desired. However, only a few studies of heteroepitaxial growth of GaN on Si have been reported(6). The investigation of the initial stage of GaN formation on Si is important, as the quality of heteroepitaxial films strongly depends on the initial stage of the growth.

The purpose of this study is to investigate the initial stage of the formation of GaN layers on Si substrates by STM.

EXPERIMENT

The apparatus used for this study is shown in Fig. 1. It consists of three chambers; a STM pre-chamber, a pre-chamber and a lord lock. The pre-chamber is equipped with a knudsen-cell, a gas cracking cell and specimen stages with direct and radiation heating.

P-type Si (111) substrate samples with a conductivity of 0.02 $\Omega \cdot$cm, whose size was 9 mm x 1 mm , were prepared from commercial wafers . Before loading into vacuum, they were cleaned in ultra-pure acetone in an ultra-sonic bath. After loading into vacuum, the samples were degassed at 773 K for 40 hours using the heating stage in the pre-chamber. Subsequently, they were cleaned by flashing several times approximately at 1523 K. The pressure did not exceed 2.5 x 10^{-8} Pa during the flashing cycle. The samples were then slowly cooled down to room temperature (1 $K \cdot s^{-1}$).

Mat. Res. Soc. Symp. Proc. Vol. 449 © 1997 Materials Research Society

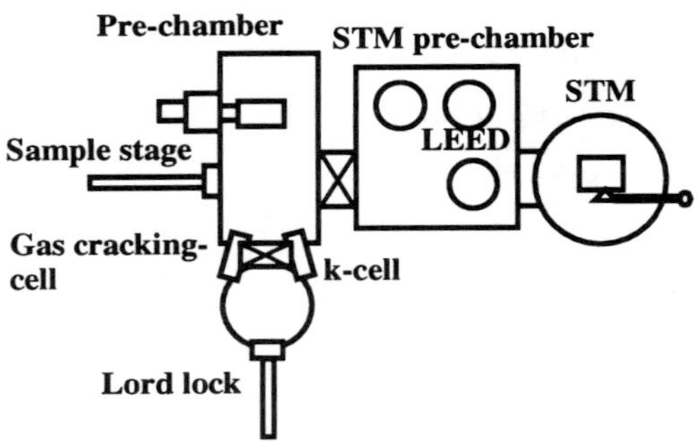

Fig. 1 Schematic diagram of used apparatus.

Ultra pure Ga (7 N) was evaporated onto the Si substrates held at 673 K from a knudsen-cell. 0.2 ML of Ga was deposited on the substrates. The Ga deposited Si substrates were subsequently annealed at 823 K for 5 min and then they were cooled down. Deposited Ga layers on Si substrates were nitrided for 40 min by excited N_2 through a gas-cracking cell. The gas-cracking cell used in this experiment consists of a pipe for injecting N_2 gas and a tungsten filament. The cracking conditions of nitrogen gas were as follows. The temperature of the filament was 2373 K, measured by an optical pyrometer. The chamber pressure was kept at 4×10^{-3} Pa during the nitridation. The purity of nitrogen gas used was 6 N. During the nitridation, the samples were held at 673 K. Then the nitrided Ga on Si was annealed at 933 K or 993 K for 10 min under the exposure to the excited N_2 beam.

All STM imaging was done at room temperature. The STM measurements were not always on the same point accurately but the orientation of the samples was held for all the STM measurements before and after the nitridation of Ga layers in each sample.

RESULTS AND DISCUSSION

1. Islands of GaN

Fig. 2 shows a Si (111) - (7x7) reconstruction image taken at the sample voltage - 1.5 V and the tunneling current 0.3 nA. Fig. 3 shows a STM image of 0.2 ML Ga deposited on Si taken at the sample voltage + 1 V and the tunneling current 0.3 nA.

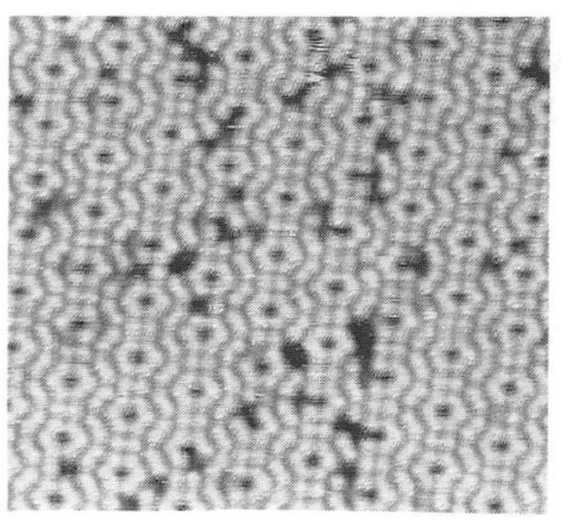

10nm

Fig. 2 A STM image of Si (111) - (7x7) reconstruction image taken at the sample voltage - 1.5 V and the tunneling current 0.3 nA.

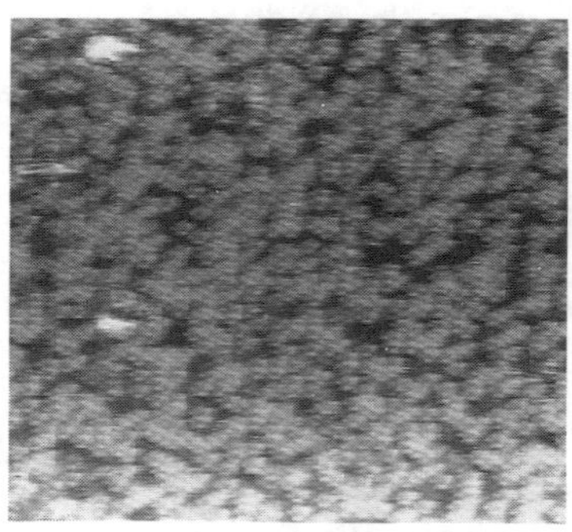

10nm

Fig. 3 A STM image of 0.2 ML Ga deposited on Si taken at the sample voltage + 1 V and the tunneling current 0.3 nA.

100nm

4a

10nm

Fig. 4a, 4b and 4c STM images of nitridation of o.2 ML Ga deposited samples. The sample was nitrided for 40 min. by excited nitrogen through a gas-cracking cell. During the nitridation the samples were held at 673 K. Then the nitrided Ga on Si was annealed at 993 K for 10 min. irradiating the excited N_2. (4a) The sample voltage was - 3 V and the tunnel current was 3 nA. (4b) The sample voltage was - 1.5 V and the tunnel current was 0.3 nA. (4c) The sample voltage was -1.5 V and the tunnel current was 0.3 nA.

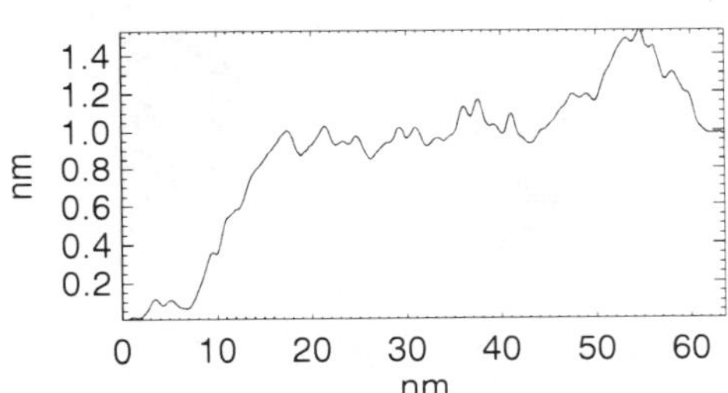

4b

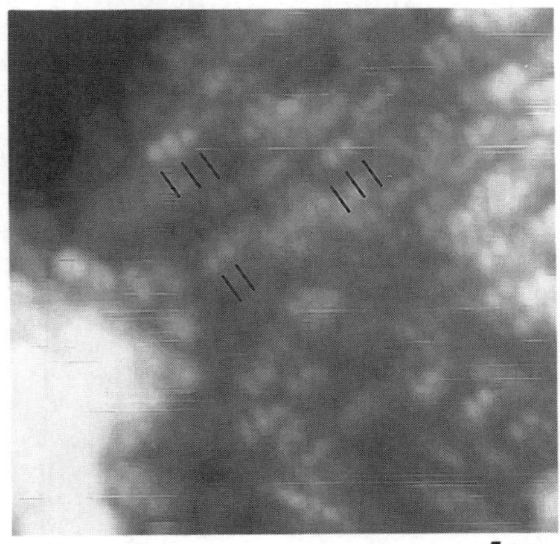

5nm

4c

Fig. 4a, 4b and 4c show nitrided images of 0.2 ML Ga on Si after nitridation process taken at the sample voltage - 2 V, - 1.5 V and -1.5 V and the tunneling current 1 nA, 0.3 nA and 0.3 nA, respectively. In Fig. 4a, an aggregation of islands whose longest edges had a direction rotated 15 ° from Si [1$\bar{1}$0] direction was observed. One of the reasons for this is thought as follows. At first, two cases are thought to the overlapping of lattice points of GaN (0001) lattice and Si (111) lattice. First case, the GaN (0001) lattice is placed on the Si (111) lattice rotated 14° toward the [2$\bar{1}$10] direction from the [1$\bar{1}$0] direction. Second case, the GaN (0001) lattice is placed on the Si (111) lattice overlapped the [2$\bar{1}$10] direction and the [1$\bar{1}$0] direction. The density of overlapping of lattice points of the first case is higher that that of the second case. These relationships are schematically shown in Fig. 5a and 5b. In these figures, circles reveal the overlapped lattice points of two crystals. The shape of islands looks like a pentagon. Fig. 4b is a magnified image of Fig. 4a. In Fig. 4b, the length and the width of this island are 52 nm and 27 nm, respectively.

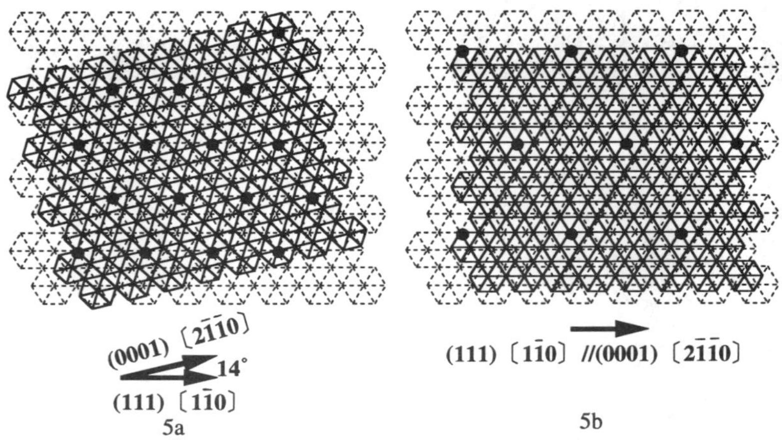

Fig. 5a and 5b Schematic diagrams on the lattice between the lattice of GaN (0001) and Si (111). (5a) The GaN (0001) lattice is placed on the Si (111) lattice rotated 14° toward the [2$\bar{1}$10] direction from the [1$\bar{1}$0] direction. (5b) The GaN (0001) lattice is placed on the Si (111) lattice overlapped the [2$\bar{1}$10] direction and the [1$\bar{1}$0] direction. .The density of overlapping of lattice points of the first case is higher that that of the second case. In these figures, circles reveal the overlapped lattice points of two crystals.

From a line profile on the straight line in the image in Fig. 4b, it was found that the surface in the nitrided region exhibits the features of plateau with a few separated peaks. The height of plateau is 1 nm and the peak is 1.5 nm from the substrate surface. Fig. 6a shows the vertical sectional view of stacking of (0001) basal plane of a lattice structure of wurtzite GaN from seen [$\bar{1}$2$\bar{1}$0] direction . Fig. 6b shows the vertical sectional view of stacking of (111) plane of a lattice structure of zinc-blende GaN from seen [1$\bar{1}$0] direction. From Figs. 6a and 6b, it is found that lattice constant of c-axis of wurtzite GaN and that of zinc-blende GaN corresponding to the c-axis wurtzite GaN is almost the same and it is about 0.52 nm. It is not apparent whether the crystal structure of these islands is wurtzite or zinc-blende structure. However, the height of plateau and the peak is presumed to correspond to 2 ML and 3 ML of GaN regardless of crystal structure of GaN. The amount of Ga is apparently insufficient for the growth of 2 ML thick GaN film in this experiment because the amount of Ga deposited on Si was 0.2 ML. Therefore, Ga must segregate by surface diffusion for forming these islands. As one of the evidence of segregation of Ga, the Si surface of this area is not completely covered by the islands and there are areas where Si surface is exposed here and there among the islands.

The typical periodical spots are shown by arrows in the figure. In Fig. 4b, the periodical spots for the vertical direction in the images were observed and the average of intervals between the spots was 1 nm. That direction well agrees with Si ($\underline{1}$11) [1$\bar{1}$0] direction. In Fig. 4c, the periodical spots for the direction rotated 15 ° from Si (111) [1$\bar{1}$0] direction were observed and the average of these intervals was 0.9 nm. These intervals correspond to 3 times a-axis of wurtzite GaN or lattice point in interval the close-packed planes of zinc-blende GaN.

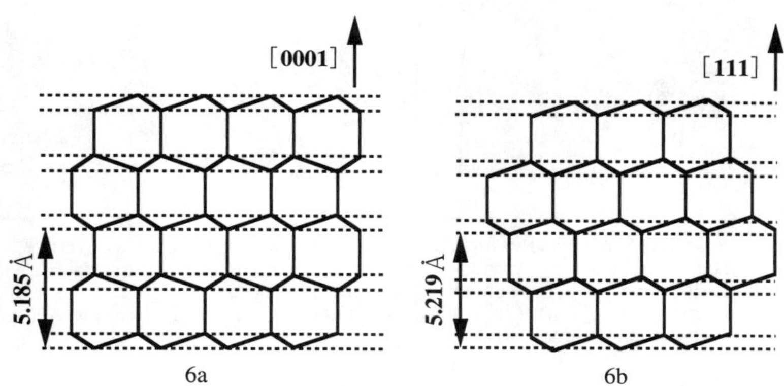

Fig. 6a and 6b (6a) Vertical sectional view of stacking of (0001) basal plane of a lattice structure of wurtzite GaN from seen [$\bar{1}$2$\bar{1}$0] direction . (6b) Vertical sectional view of stacking of (111) plane of a lattice structure of zinc-blende GaN from seen [1$\bar{1}$0] direction .

2. Surface roughness of nitrided Ga

Surface roughness of nitrided Ga layer was evaluated by using the value of root mean square (rms) within the observed STM images. A square of 50 nm x 50 nm was used as an area of the evaluation of surface roughness because STM images of this size could be observed without containing steps on Si (111). Nitridation of Ga layer was done under two different conditions. Ga coverage and annealing temperature during the nitridation was 0.2 ML and 933 K (a) and 0.2 ML and 993 K (b), respectively. Figs. 7a and 7b show typical STM images after nitridation of the Ga layer. Before nitridation, the roughness of Si (111)-(7x7) and the 0.2 ML Ga-deposited surface on Si (111) were estimated to be 0.02 nm and 0.06 nm, respectively. After nitridation of the Ga layer, surface roughness was observed to be increased strongly. Surface roughness of nitrided 0.2 ML Ga annealed at 933 K and that of nitrided 0.2 ML Ga annealed at 993 K were estimated to be 0.09 and 0.17 nm, respectively. It was found that the surface roughness of nitrided Ga layers become larger as the annealing temperature after nitridation of deposited Ga on Si is raised.

From above results, it is concluded as follows. As the lattice mismatch is large between them, a perfect epitaxitial relationship between GaN and Si was not obtained. For example, the observed islands look like a pentagon and the surface roughening of nitrided Ga layers on Si occurs. However, the height of plateau and peak of the island are 2 and 3 ML and these values equal an integral multiple of the lattice constant of GaN. In addition, the intervals of periodical spots equal an integral multiple of the lattice constant of GaN. From these results, the close-packed layer of GaN is suggested to grow on Si (111) surface keeping the coherency to a certain extent between the epitaxitial film and the substrate at the initial stage of GaN growth on Si.

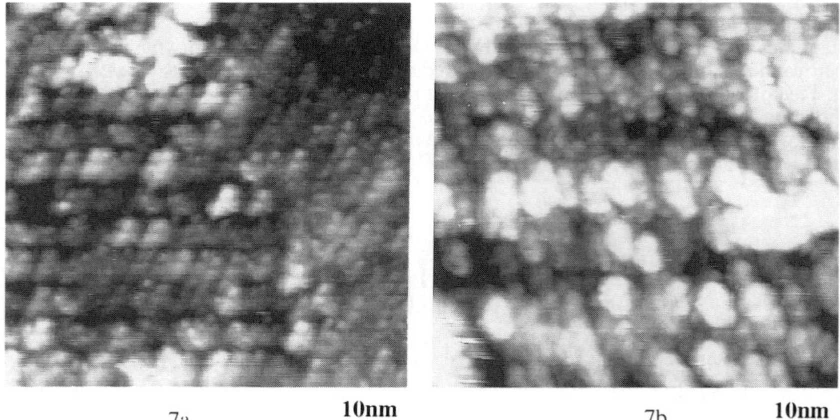

7a   **10nm**   7b   **10nm**

Fig. 7a and 7b STM images after nitridation of Ga deposited samples. Ga coverage and annealing temperature during the nitridation were 0.2 ML and 933 K (7a) amd 0.2 ML and 993 K (7b). (7a) Surface roughness is 0.09 nm. The sample voltage was 3.5 V and the tunnel current was 0.3 nA. (7b) Surface roughness is 0.17 nm. The sample voltage was 5 V and the tunnel current was 0.3 nA.

CONCLUSIONS

The initial stage of the nitridation of Ga layer deposited on Si (111) surface was observed by using STM. It was found that the close-packed layers of GaN grow on Si (111) keeping the coherency to a certain extent between the epitaxitial film and the substrate. It is thought that the growth of crystalline layers of GaN on Si is possible.

ACKNOWLEDGMENTS

This work is supported by the New Energy and Industrial Technology Development Organization(NEDO).

REFERENCES

1.S. Nakamura, T.Mukai and M. Senoh, Appl. Phy. Lett. 64, 1687 (1994)

2. I. Akasaki, H. Amano, S. Sota, H. Sakai, T. Tanaka and M. Koike, Jpn. J. Appl. Phy. 34, L1517 (1995)

3. S. Nakamura, M. Senoh, S. Nagahara, N. Iwase, T. Yamada, T. Matsushita, H. Kiyoku and Y. Sugimoto, Jpn. J. Appl. Phys. 35, L74 (1996)

4. T. Sasaki,J. Cryst. Growth, 129, 81(1993)

5. R. C. Powell et al., Appl. Phys. Lett., 63, 973(1993)

6. T. Lei, T. D. Moustakas, R. J. Graham, Y. He and S. J. Berkowitz, J.appl. Phys. 71, 4993 (1992)

COMPARISON OF THE MICROSTRUCTURE OF AlN FILMS GROWN BY MOCVD AND BY PLD ON SAPPHIRE SUBSTRATES

Yun-Xin Li*, Lourdes Salamanca-Riba*, V. Talyan**, T. Venkatesan**, C. Wongchigul***, P. Zhou***, X. Tang***, and M.G. Spencer***

*Materials and Nuclear Engineering Department, University of Maryland, College Park, MD 20742

**Center for Superconductivity Research, Department of Physics, University of Maryland, College Park, MD 20742

***Materials Science Research Center of Excellence, Howard University, Washington, DC 20059

ABSTRACT

(0001) aluminium nitride thin films were grown epitaxially on (0001) Sapphire substrates by MOCVD at 1200° C and PLD at 800° C. Both films have the same epitaxial growth relationship: $(0001)_{AlN}//(0001)_{Sap}$, and the same in-plane relationship which shows a 30° rotation between AlN and Sapphire: $[\bar{1}2\bar{1}0]_{AlN}//[0\bar{1}10]_{Sap}$ and $[10\bar{1}0]_{AlN}//[\bar{2}110]_{Sap}$. The full width at half maximum (FWHM) of x-ray rocking curve of the MOCVD AlN film was 0.16° and PLD AlN film was 0.2°. Films grown by both MOCVD and PLD showed high crystalline quality. HRTEM images showed that these films are single crystalline with very low density of defects. Dislocations in the film parallel to the film / substrate interface were observed in both AlN films. Atomic force microscopy images showed that the MOCVD films have flatter and larger terraces than the PLD films. The PLD technique for AlN growth needs to be improved further. But both films have a surface roughness of approximately 100nm.

INTRODUCTION

AlN, together with GaN and InN, are wide bandgap semiconductors and are being considered as promising materials for optical devices and high temperature electronic devices. MOCVD has been the most common technique for high quality AlN growth and continues to be very successful [1,2,3]. With the development of pulsed laser deposition and its production of high quality oxides and superconducting films, some researchers have turned to grown AlN on sapphire by PLD for its important advantages in purity. Recently, high quality crystalline AlN films were obtained by PLD [4,5]. Both MOCVD and PLD techniques have their own characteristic advantages and drawbacks[6,7], hence the microstructure of films will naturally be affected by the different growth techniques. The main differences which might affect the microstructure of films grown by MOCVD and PLD are: (1) the growth temperature (for MOCVD, the temperature range is 1050°C - 1250°C; for PLD, 750°C or lower). (2) the purity (gases used in MOCVD reactors can introduce contaminants in the semiconductor films. PLD uses AlN as target, thus, the film is more likely to have the right stoichiometry). In this work, we compare the microstructure of AlN films grown on sapphire by MOCVD and PLD.

Mat. Res. Soc. Symp. Proc. Vol. 449 © 1997 Materials Research Society

EXPERIMENTAL

AlN thin films were grown on sapphire (0001) by MOCVD and PLD. For MOCVD, the growth temperature was 1200°C, the pressure was 10 Torr, and Trimethylauminum (TMA), ammonia (NH_3) and hydrogen (H_2) were used as precursors. For PLD, a KrF excimer UV laser was used to ablate a nominally stoichiometric AlN target with an energy density of 3~4J/cm^2. The base pressure was 10^{-7}~10^{-6} Torr. The films were characterized by x-ray diffraction (θ-2θ scan, ψ scan and rocking curve scan), high resolution TEM and atomic force microscopy (AFM).

RESULTS AND DISCUSSION

Figure 1 shows a XRD θ-2θ spectrum from the AlN/Sapphire grown by MOCVD, d_{002}=2.5068Å. The spectra from the films grown by PLD showed very similar θ-2θ scans with d_{002}=2.4983Å. Thus, the lattice of the film by MOCVD is essentially the same as the one by PLD. Both films are single crystalline. The growth direction of the films is (0001)$_{AlN}$ //(0001)$_{Sap}$. The full width at half maximum of the (002) AlN peak by MOCVD is 0.16° , and the one by PLD is 0.2°, indicating that the crystalline quality of both films is good. Similar ψ scans were also obtained from both films. The ψ scan for the(11 $\bar{2}$2)$_{AlN}$ and (11 $\bar{2}$3)$_{Sap}$ reflections presented in figure 2 for the film grown by PLD shows that the film has very good crystal symmetry and registry with the substrate. The ψ scan in figure 2 also shows a 30° rotation in the basal plane between AlN and Sapphire.

Figures 3 and 4 are the cross-sectional HRTEM images of AlN films grown by MOCVD and PLD, respectively. They both have the same growth direction (0001)$_{AlN}$//(0001)$_{Sap}$. The diffraction patterns indicate that the in-plane epitaxial relationships of AlN are [$\bar{1}$2 $\bar{1}$0]$_{AlN}$//[0 $\bar{1}$10] $_{Sap}$ and [10 $\bar{1}$0]$_{AlN}$ //[$\bar{2}$110]$_{Sap}$, indicating a 30° rotation of the AlN film with respect to the sapphire substrate in the basal plane , as it has previously observed[3, 4].

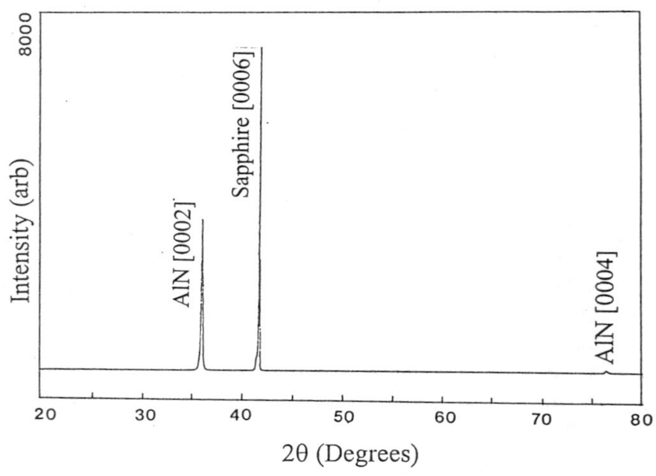

Figure 1 θ-2θ scan of AlN/Sapphire grown by MOCVD

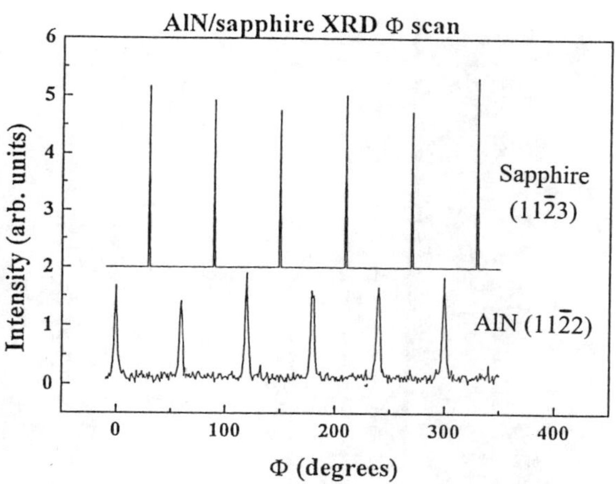

Fig.2 ψ scan for the (11 $\bar{2}$2)$_{AlN}$ and (11 $\bar{2}$3)$_{Sap}$ reflections of the AlN film grown on Sapphire by PLD

Table I   Structural information of AlN and Sapphire and lattice misfit

	AlN	Sapphire
Structure	Wurtzite	Corundum
Crystal system	Hexagonal	Trigonal
Space group	P6₃mc (No.186)	R$\bar{3}$c(No. 167)
Lattice parameter	a=3.112Å, c=4.982Å	a=4.758Å, c=12.991Å
Atom and coordinate	N^{3-} (0,0,0.375)	O^{2-} (0.306,0,0.25)
	Al^{3+} (0,0,0)	Al^{3+} (0,0,0.352)
Lattice misfit	$d_{(10\,\bar{1}0)}$ (Al atoms)=3.112Å	$d_{(10\,\bar{1}0)}$ (O atoms)=4.758Å
(Al-O bonding)	$d_{(10\,\bar{1}0)}$ (Al atoms)=3.112Å	$d_{(\bar{2}110)}$ (O atoms)=2.73 Å

This rotation is due to the big mismatch between AlN and Sapphire. Table I lists the crystallographic data for Sapphire and AlN. [10 $\bar{1}$0]$_{AlN}$ has a very large mismatch of 34.6% with [10 $\bar{1}$0]$_{Sap}$. However, after a 30° rotation about the [0001] c axis , we have [10 $\bar{1}$0]$_{AlN}$ //[$\bar{2}$110]$_{Sap}$. In this case, the lattice mismatch is reduced to 13.99%. The HRTEM images shown in Figures 3 and 4 show that the interfaces in both films are very sharp, and also demonstrate contrast modulations associated with strains in the films near the interface. The films have low density of dislocation. Some dislocations parallel to the interface are marked by arrows. Figure 5 shows that AFM images from the surface of the AlN films grown by MOCVD and PLD. These images show that the MOCVD films have hexagonal facets which are very flat and large compared to the PLD films, thus, the PLD technique for AlN growth needs to be improved further. But both films have a surface roughness of approximately 100nm. The films have good crystalline quality all through the thickness of the film.

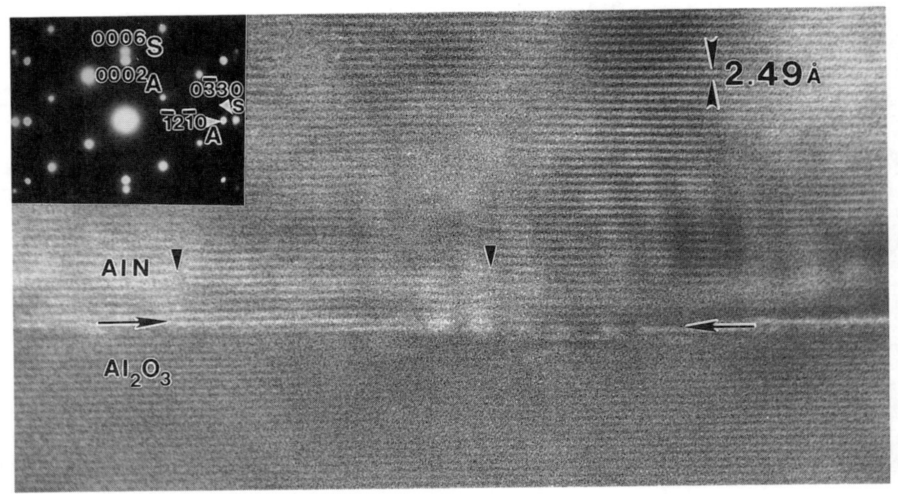

Figure 3 Cross-sectional HRTEM image from AlN/Sapphire grown by MOCVD. The inset is
the corresponding SAD pattern with [$\bar{1}2\ \bar{1}0$]$_{AlN}$//[0 $\bar{1}$10] $_{Sap}$ and [10 $\bar{1}0$]$_{AlN}$ //[$\bar{2}$110]$_{Sap}$

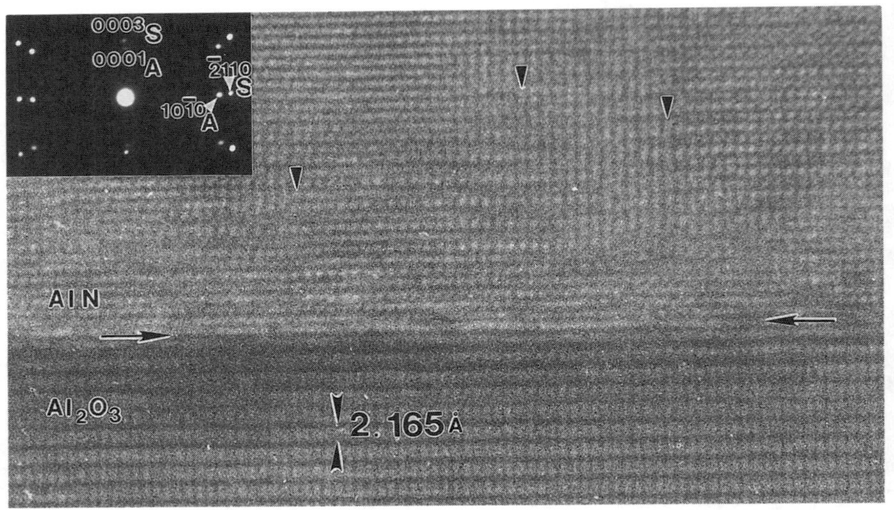

Figure 4 Cross-sectional HRTEM image from AlN/Sapphire by PLD. The inset is the
corresponding SAD pattern with [$\bar{1}2\ \bar{1}0$]$_{AlN}$//[0 $\bar{1}$10] $_{Sap}$ and [10 $\bar{1}0$]$_{AlN}$ //[$\bar{2}$110]$_{Sap}$

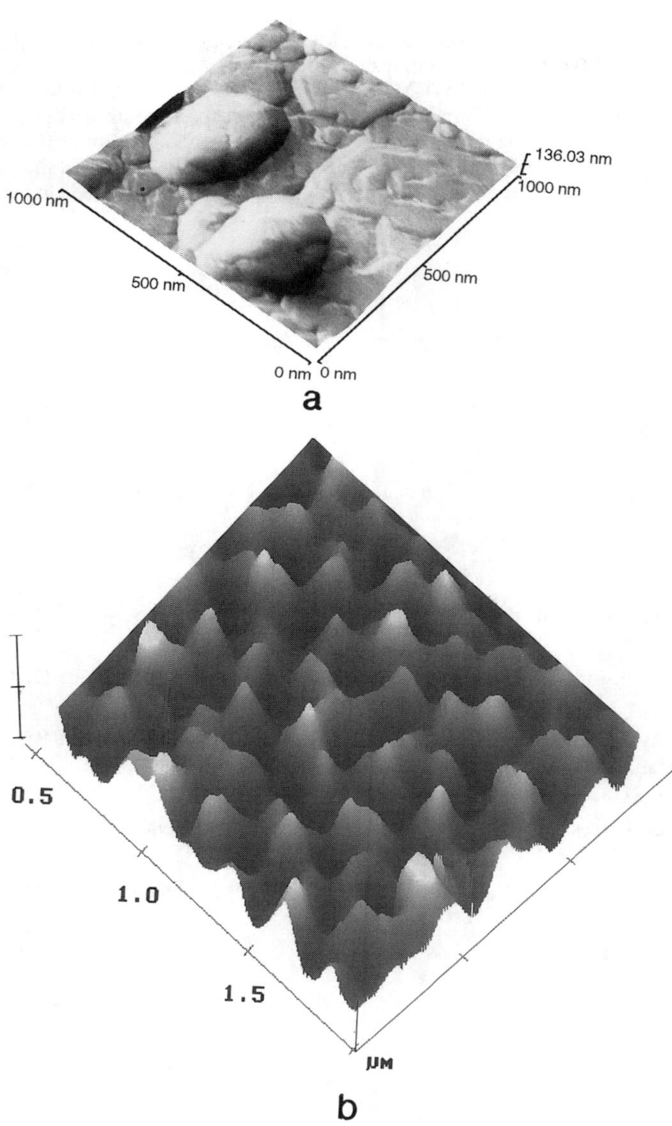

Fig.5   AFM image from AlN films grown by (a) MOCVD
and (b) PLD, x=0.500μm/div, z=50.000nm/div

CONCLUSIONS

High quality AlN films were obtained by both MOCVD and PLD on (0001) sapphire substrates. Both films have the same epitaxial relationships:(0001)AlN//(0001)Sap, [12 10]AlN//[0 110] Sap and [10 10]AlN //[2110]Sap. The FWHM of AlN by MOCVD is 0.16°, that by PLD is 0.2°. The interfaces in the films are very sharp. Contrast modulation caused by strains due to the big misfit and dislocations parallel to the interface were observed in AlN films. Both MOCVD films and PLD films have a surface roughness of approximately 100nm. The surface of the MOCVD films presented clear hexagonal facets. The PLD technique for AlN growth needs to be improved further.

ACKNOWLEDGMENTS

The authors acknowledge support from an MRCP Army grant No. DAAL 0195 23530 and by NSF grant No. DMR 9321957.

REFERENCES

1. S.Yoshida, S.Misawa, Y.Fujii, S.Takada, S.Gonda and A.Itoh, J. Appl.Phys. **53**, 6844(1982).

2. S.Strite and H.Morkoc, J.Vac.Sci.Technol. **B10**, 1237 (1992) and references therein.

3. K. Dovidenko, S.Oktyabrsky, and J.Narayan, J. Appl. Phys. **79** (5), 2439(1996).

4. R.D. Vispute, J.Narayan, Hong Wu, and K.Jagannadham, Appl.Phys. Lett. **77**, 4724(1995).

5. V. Talyansky, R.P.Sharma, S.Chapoon, M.Downes, Y.X. Li, L. Salamanca-Riba, X.Tang, M.G.Spenser and T.Venkatesan, Materials Research Society, 1996, Boston, MA, Fall Meeting.

6. K.L.Saenger, Processing of Advanced Materials, vol **3**(1), 1(1993).

7. M. Razeghi, The MOCVD challenge (Institute of Physics Publishing, Bristle and Philadelphia, 1995), P 22.

N-K-EDGE EXAFS STUDY OF EPITAXIAL GaN FILMS

M. Katsikini[a,b] , E. C. Paloura[a] , M. Fieber-Erdmann[b], T. D. Moustakas[c] , H. Amano[d] and I. Akasaki[d].

[a]Aristotle Univ. of Thessaloniki, Dept. of Physics, 54006 Thessaloniki, Greece.
[b]Hahn-Meitner Institute (A.S), Glienicker Str. 100, D-14109 Berlin, Germany
[c]Boston University, College of Engineering, Boston, MA 02215, USA.
[d]Meijo University, Dept. of Electrical & Electronic Engineering, Nagoya 468, Japan.

Abstract

X-ray absorption measurements at the N and O-K-edges are used to study the local microstructure in cubic and hexagonal GaN films grown by ECR-MBE and HVPE. A distortion in the local microstructure is identified in the 1st nearest neighbor (nn) shell, consisting of Ga atoms, in both the cubic and hexagonal samples. Two N-Ga distances are identified, R_1 and R_2, where R_1 is the expected distance of 1.95Å while $R_2=R_1+0.25$Å. The same distortion is detected in the next nn shell containing Ga atoms, where the two distances are 3.7Å and 4.1Å. All the reported distance variations are larger than the error-bar. The nitrogen 2nd nn neighbor is found at the expected distance of 3.12Å while N deficiency is not detected. Finally, the O-Ga distance is found equal to 1.60Å and therefore it can be proposed that the oxygen atom occupies interstitial positions.

I.   Introduction

Gallium nitride (GaN) has been a subject of intensive study because it finds applications in visible-UV light emitters and detectors as well as high-frequency, temperature and power devices [1 ,2]. GaN exists as the cubic or the stable hexagonal polytype. Device applications require epitaxial layers of high crystalline quality and good control of their electrical conductivity. However, because of the different thermal expansion coefficients and the lattice mismatch between GaN and the available substrates, phase stabilization can be difficult (growth of mixed phase GaN crystals) while crack generation and difficulties in conductivity control often occur[3 ,4 ,5]. Even though the problems related to the mismatch can be surpassed by the use of buffer layers[5], the problems arising due to the different thermal expansion coefficients (bending and cracking) still remain unsolved .

The lattice parameters in GaN can be affected by a number of factors such as free electrons, point (Ga_N, V_N) and extended defects[6]. Furthermore, the lattice constants can be deformed by the thermal strain. GaN films grown on Si or Al_2O_3 are subjected to tensile or compressive stress upon cooling from the growth temperature[7 ,8]. The biaxial stress causes a volume conserving distortion of the unit cell and can be relaxed via formation of microcracks[9].

Here we present for the first time extended x-ray absorption fine structure (EXAFS) characterization results at the N and O-K-edges. EXAFS spectroscopy measures the X-ray absorption coefficient as a function of the photon energy above the threshold of an absorption edge. EXAFS has been established as a powerful structural probe[10 ,11] because it can determine the short range order (bond lengths, bond angles and coordination numbers) in both the amorphous and crystalline states of the matter. One of the attractive features of EXAFS is its atom-specific character, i.e. in

Mat. Res. Soc. Symp. Proc. Vol. 449 © 1997 Materials Research Society

multicomponent systems, it can determine the local structure around each specific atom independently. Previously reported Ga-K-edge EXAFS data were recorded over a short energy range (400eV above the edge) and had a limited precision in the determination of the of the nearest neighbor Ga-N shell. Therefore, the authors reported only on the 2nd nearest neighbohr (nn) shell (Ga-Ga distances)[12].

II.   Experimental Details

The samples used for this study are named GaN57, GaN67, GaN179 and GaN-A4. Their growth conditions are listed in Table I while details on growth and their properties have been reported previously [5,13 ,14 , 15 ,16].

The EXAFS spectra were recorded at room temperature and at 160 K, at the N and O-K-edges (370-1100eV), using the SX-700-I plane grating monochromator at the electron storage ring BESSY in Berlin. The spectra were recorded in the fluorescence yield (FLY) mode using a high purity Ge detector positioned along the electric field vector. Due to the high absorption cross section of the radiation in the soft x-ray region, the experiment was performed in an ultrahigh vacuum chamber (UHV) with a base pressure $\leq 7\text{x}10^{-10}$ mbar. In order to improve the signal-to-noise ratio more than 20 spectra were collected for each sample (data acquisition time 1 s/point). The EXAFS spectra were normalized to the primary photon flux by division with the total electron yield spectra from a clean (non-absorbing) Si wafer which yields a good measure of the monochromator transmission function over the range of interest (370-1100eV).

Table I : Growth conditions. The sample GaN-A4 was grown using Ga solid source, HCl and ammonia (NH_3).

sample name	growth technique	T_{growth} ($^{\circ}$C)	substrate	symmetry	thickness (μm)
GaN57	ECR-MBE	600	p-Si(001)	cubic	1.60
GaN67	ECR-MBE	600	n-Si(111)	mixed	0.80
GaN179	ECR-MBE	600	Al_2O_3 (0001)	hexagonal	1.67
GaN-A4	HVPE	1030	Al_2O_3 (0001)	hexagonal	10.0

III.   Results and Discussion

The EXAFS spectra were recorded at an angle of incidence $\vartheta=55^{\circ}$ (ϑ is defined between the sample surface and the incident beam), while the angle of detection was 35°. The information depth in the used geometry is 0.6μm. The EXAFS spectra, weighted by k^3, were fitted using the FEFFIT 3.23 code. The models used for the fitting and the single scattering paths were constructed using the programs ATOM and FEFF6[17], respectively. The backscattering amplitudes and phase shifts were calculated within the FEFF6 program. The initial values of the Debye-Waller (DW) factors A_i were calculated using the correlated Debye model[18] and iterations were done for R_i, N_i and A_i.

The EXAFS spectra from hexagonal samples depend on the angle of incidence ϑ because the contribution of each bond depends on the angle between the bond and the electric field vector of the incident beam. To elevate this ϑ dependence the measurements were done at $\vartheta=55^{\circ}$ which is equal to the "magic angle" , i.e. the characteristic ϑ for which the EXAFS spectra are independent of the bond orientation, for linearly polarized light. Therefore, the results from the hexagonal samples can be compared to those from the cubic samples. The EXAFS spectra do not suffer from self-absorption effects as shown by the analysis of the angular dependence of the NEXAFS spectra presented previously [19].

The $\chi(k)$ versus k and the corresponding Fourier transforms (FT) of the under study samples are shown in Fig.1 (right and left panels, respectively). The experimental curves and the fitting are shown in thick and thin lines respectively. As shown in the figure in the FT of the experimental spectrum the 1st nn peak splits in two, at distances $R_1=1.95$Å and $R_2=R_1+0.25$Å. This splitting is not predicted by the FEFF model for either the cubic or the hexagonal structures. Fourier filtering of the FT transformed data shows that the envelope functions of both peaks, at R_1 and R_2, correspond to Ga atoms that are displaced with respect to each other.

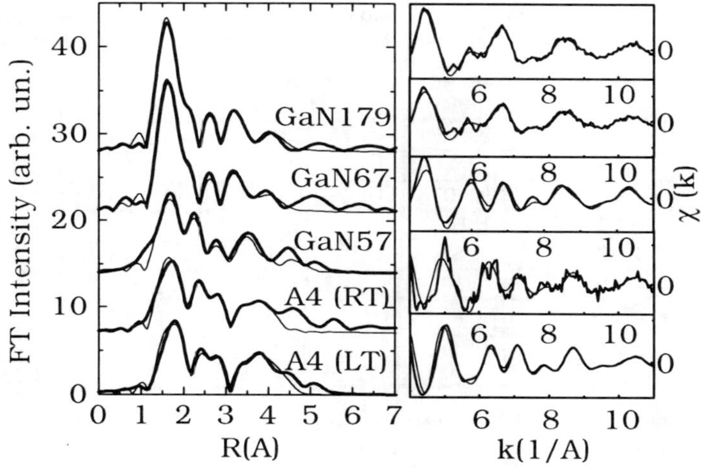

Fig. 1 : The $\chi(k)$ versus k and the FT of the EXAFS spectra at 300K and 160K
(indicated as RT and LT, respectively).

The results of the EXAFS analysis in the 3 nn shells are listed in Table II. A similar distortion is observed in the 4th nn shell where the Ga atoms are found at distances 3.7 and 4.1Å. Using the published value of a=4.50Å for the cubic sample GaN57[14], the theoretically predicted value of R_1 is 1.95Å, i.e. in good agreement with the EXAFS results.

The central N atom is 4-fold coordinated ($N_1+N_2=4$), however, the ratio N_1/N_2 of the Ga atoms at the distances R_1 and R_2 is equal to 1 for GaN57 and GaN-A4, while it takes the value 3 for GaN67 and GaN179. Furthermore, the samples with $N_1/N_2=3$ have smaller values of R_1 and R_2 than those with $N_1/N_2=1$.

The 3[d] nn shell consisting on N atoms is found at the expected distance of 3.12±0.05Å with the expected coordination of 12.5±1.4 atoms. Therefore, N deficiency is not detected, at least within the ±15% accuracy of EXAFS in the determination of the coordination in the N shell.

As shown in Table II the ECR-MBE samples were fitted with similar values of the DW factors in the 3 nn shells. However, those used for the fitting of the GaN-A4 sample are slightly smaller at 300K. This difference in the values of the DW factors could be related to the significant O-contamination in GaN-A4, as discussed in the following.

Among the examined samples, GaN-A4 contains significant amount of O-contamination, the concentration of which, taking into account the relative height of the N and O edge jumps and the differences in the FL yield, is estimated to be about 6% at. In this sample, the N- and O-K-edge EXAFS spectra were recorded simultaneously in two different energy windows and the O-signal was isolated by subtraction of the two channels using appropriate weight factors[20].   The samples grown with ECR-MBE contained detectable traces of O-contamination however the edge jump at the O-K-edge is very small and thus the O-signal could be easily eliminated from the N spectra by a reduction of the width of the detection window.

Table II : Results of EXAFS analysis.   The 1st and 2nd nn shells consist of Ga atoms while the 3^{d} consists of N atoms. N_i, R_i and A_i stand for coordination number, neighbor distance and Debye-Waller factor, respectively. The amplitude reduction factor is 0.93.

	GaN57	GaN-A4 (300K)	GaN-A4 (160K)	GaN67	GaN179
$N_1 \pm 10\%$	2.05	2.09	2.36	2.83	2.90
$R_1 \pm 0.01$Å	1.95	1.96	1.94	1.91	1.91
A_1 (Å^2)	4.9×10^{-3}	3.9×10^{-3}	7.9×10^{-3}	4.9×10^{-3}	5.0×10^{-3}
$N_2 \pm 10\%$	1.94	1.91	1.64	1.66	1.10
$R_2 \pm 0.01$Å	2.20	2.19	2.20	2.15	2.13
A_2 (Å^2)	5.2×10^{-3}	4.1×10^{-3}	8.3×10^{-3}	5.3×10^{-3}	5.3×10^{-3}
$N_3 \pm 15\%$	14.15	11.42	12.14	9.19	10.37
R_3 (Å)	3.11	3.07	3.07	3.14	3.14
A_3 (Å^2)	1.0×10^{-2}	6.7×10^{-3}	8.5×10^{-3}	1.1×10^{-2}	1.1×10^{-2}

The analysis of the O-K-edge spectra was done at the 1st nn shell of the Fourier filtered data following the same procedure used at the N-K-edge. The O-Ga distance is found equal to 1.60Å at 300K and 1.56Å at 160K. This low value of the O-Ga distance indicates that O cannot be a substitutional impurity but it rather occupies interstitial positions.   The interstitial position is further supported by the fact   that the backscattering amplitude has a peak at $k=6$Å^{-1} while that of Ga and N peak at 8Å^{-1} and 2.5Å^{-1}, respectively. Therefore, the nn shell of the oxygen atom consists of both Ga and N atoms, i.e. as it would be expected in an interstitial position. However, since Ga has a much higher backscattering amplitude than N, the fitting was done assuming only Ga atoms as nn. This approximation will affect the calculated coordination number and the value of the DW factor, which were found equal to 6 and 0.018Å^2, respectively, but it will not introduce any significant error in the calculated Ga-O distance. At this point it should be pointed out that the behavior of O in GaN is different than in AlN where O was found to occupy substitutional sites in the N sublattice[21].

The vibrational part (due to thermal disorder) of the DW factor which affects the EXAFS signal is given by $\sigma^2_R = <[U_R-U_o)R]^2>$, where U_o and U_R are the displacement vectors of the central atom and the atom at the lattice point R.   Therefore, the DW factor depends on the maximum amplitudes of the thermal vibrations of the central and neighboring atoms. Since the calculated value of the O-Ga distance is very small, the large value of the corresponding DW factor at 300K indicates a large difference in the thermal vibration amplitudes of the O and Ga atoms, i.e, is consistent with the hypothesis that O is interstitial. The O atom can be accommodated in the hexagonal structure which contains big enough voids as shown in Fig.2. In this figure the radii of the N and Ga atoms are analogous to their ionic radii and the radius of the void is nearly equal to that of the O ion. Contrary to the expected behavior, the DW factors in GaN-A4 further increase at low temperatures (160K). Given that the vibrational part of the DW factor decreases at lower temperatures, the observed increase could be attributed to increased static disorder due to the thermal strain[22].

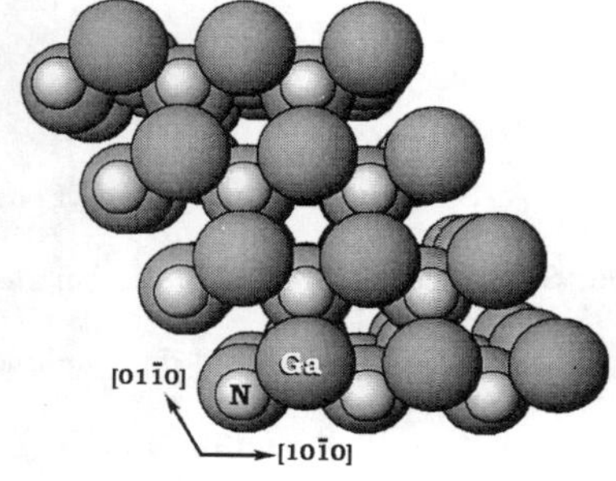

Fig. 2 : Projection of the hexagonal structure along the c-axis. The oxygen atom can be accommodated in the voids present in the structure.

IV.   Conclusions

The EXAFS measurements at the N-K-edge reveal that the local microstructure around the N atom is distorted due to the existence of two different N-Ga distances, R_1 and R_2, where $R_2=R_1+0.25$Å. This distortion appears to be universal in all the examined samples, despite the differences in the growth method, chemical stoichiometry and structural perfection. This distortion could originate from a tendency of the N atom to be 3-fold coordinated (sp^2 hybridization), with the remaining 2 electrons occupying a lone pair orbital, instead of the sp^3 occurring in GaN. Such a distortion was reported in the past for the incorporation of N in the diamond lattice[23]. However, a distortion due to the inhomogeneous strain, induced by the extended defects, and the thermal strain cannot be excluded.

A direct comparison of the numerical results (absolute values of R_1, R_2 and the N_1/N_2 ratio) is not possible since the film properties might vary significantly due to the differences in the film thickness, thermal expansion coefficient of the substrate, growth method and structural perfection. For example GaN57 is reported to be highly faulted[15] with an average grain size of the order of 500Å, while the grain size in GaN179 and GaN67 is $\approx$1500Å. The differences in the grain size could modify the strain and the (N_1/N_2) ratio. The information depth in the used geometry is 0.6μm. Therefore in GaN67 we probe the bulk of the film while in the other ECR-MBE samples we probe 30% of the film thickness and in the MOVPE film we probe only 2% of the film thickness. Another significant difference between the ECR-MBE and the MOVPE samples might arise from the the different film thicknesses. According to Hiramatsu et al [9], in GaN grown on Al_2O_3 partial strain relaxation occurs via microcrack formation in films with a thickness in the range 4-20μm (GaN-A4) while in thinner films (ECR-MBE films) the main relaxation mechanism is via a distortion of the unit cell.

Analysis at the O-K-edge shows that at an oxygen concentration of about 6% a.t. the oxygen atom is found in interstitial positions. The interstitial oxygen could act as a donor and thus can contribute to the "native" n-type conductivity in GaN films, as observed experimentally[24]. Finally, within the experimental error of EXAFS (±15% in the 2nd nn shell consisting of N atoms), we do not detect N deficiency, at least in the local microstructure probed by EXAFS. Due to the large difference in the backscattering amplitudes of the Ga and N atoms and the differences in their FL yields, a more accurate identification of a possible N deficiency (in stoichiometric films as those studied here) using EXAFS will be difficult.

Acknowledgments: This work was realized with partial financial support from the EC-HCM (CHGE-CT93-0027) program. The authors M.K. and E.C.P. wish to thank Dr. D. Arvanitis and Prof. J. Rehr for fruitful discussions.

References

1 . S. D. Lester, F. A. Ponce, M. G. Craford and D. A. Steigerwald, Appl. Phys. Lett. **66**, 1250 (1995).

2 . H. Sakai, T. Koide, H. Suzuki, M. Yamaguchi, S. Yamasaki, M. Koike, H. Amano and I. Akasaki, Jpn. J. Appl. Phys. **34**, L1429 (1995).

3 . T. Lei, M. Fanciulli, R. J. Molnar, T. D. Moustakas, R. J. Graham and J. Scanlon, Appl. Phys. Lett. **59**, 944 (1991).

4 . T. S. Cheng, L. C. Jenkins, S. E. Hooper, C. T. Foxon, J. W. Orton and D. E. Lacklison, Appl. Phys. Lett. **66**, 1509 (1995).

5 . T.Detchprohm, K.Hiramatsu, K.Itoh, I.Akasaki, Jpn. J. Appl. Phys.,**31**, L1454(1992).

6 . M. Leszczynski, H. Teisseyre, T. Suski, I. Gregory, M. Bockowski, J. Jan, S. Porowski, K. Pakula, J. M. Baranowski, C.T. Foxon and T. S. Cheng, Appl. Phys. Lett., **69**, 73, 1996.

7 . A. Watanabe, T. Takenchi, K. Hirosawa, H. Amano, K. Hiramatso and I. Akasaki, J. Cryst. Growth **128**, 391 (1993).

8 . T. Kozawa, T. Kachi, H. Kans, H. Nagase, N. Koide and K. Manabe, J. Appl. Phys. **77**, 4389 (1995).

9 . K. Hiramatsu, T. Detchprohm and I. Akasaki, J. J. Appl. Phys. 32, 1528 (1993).

10 . D. Konigsberger and R. Prins X-ray absorption : Principles, Techniques and Applications of EXAFS and XANES (John Wiley & Sons, New York, 1988).

11 .P.A.Lee, P.H.Citrin, P.Eisenberger and B.M.Kincaid, Rev. Mod. Phys.53, 79 (1981).

12 . P. Perlin, C. Jauberthie-Carillon, J. P. Itie, A. San Miguel, I. Grzegory, A. Polian, Phys. Rev. B. **45**, 83 (1992).

13 . T. D. Moustakas, T. Lei and R. J. Molnar, Physica B **185**, 36 (1993).

14 . T. Lei, T. D. Moustakas, R. J. Graham, Y. He, S. J. Berkowitz, J. Appl. Phys. **71**, 4933 (1992).

15 . S. N. Basu, T. Lei and T. D. Moustakas, J. Mater. Res. **9**, 2370 (1994).

16 . T. Lei, K. F. Ludwig, T. D. Moustakas, J. Appl. Phys. **74**, 4430 (1993).

17 .J.M. de Leon, J. J. Rehr, R. C. Albers, S. I. Zabinsky, Phys. Rev. B **44**, 3937 (1992).

18 . E. Sevillano, H. Meuth, J. J. Rehr, Phys. Rev. B **20**, 4908 (1979).

19 . M. Katsikini, E. C. Paloura, T. D. Moustakas, Appl. Phys. Lett. (in press).

20 . K. M. Behrens, E. D. Klinkenberg, J. Finster, W. Frentrup, K. Holldack, BESSY Jahresbericht 1992, p. 520.

21 . M. Katskini, E. C. Paloura, T. Cheng and C. T. Foxon, Proc. 9th Int. Conf. on X-ray absorption fine structure, Grenoble, August 1996 (to be published in J. de Physique).

22 .B.K.Teo, EXAFS:Basic Principles and Data Analysis (Springer Verlag, Berlin, 1986).

23 . S. Jin and T. D. Moustakas, Appl. Phys. Lett. **65**, 403 (1994).

24 . B. C. Chung, M. Gershenzon, J. Appl. Phys. **72**, 651 (1992).

STUDIES OF GROUP III-NITRIDE GROWTH ON SILICON

A V Blant[+], T S Cheng[+], C T Foxon[+], J C Bussey[O], S V Novikov[*] and V V Tret'yakov[*].
[+]Department of Physics, University of Nottingham, Nottingham, England NG7 2RD. [O] Department of Materials Engineering & Materials Design, University of Nottingham, Nottingham, England, NG7 2RD. [*] Ioffe Physical-Technical Institute, St.Petersburg, 194021, Russia.

Abstract

The growth of Group III-Nitrides on Si substrates offers the possibility of combining optoelectronics with Si technology. We have been studying the growth of Group III-Nitrides on clean and oxidised Si surfaces with a view to local area epitaxy. X-ray data indicates that alloys of $(AlGa)N$ and $(InGa)N$ of controlled composition can be grown on Si using a plasma enhanced Molecular Beam Epitaxy method over the entire composition range from InN to AlN. Films grown on uncleaned, oxidised surfaces of Si are polycrystalline/amorphous in contrast to growth on chemically cleaned Si substrates which show the usual columnar structure common in Group III-Nitrides. XPS studies indicate that there is little tendency for spinodal decomposition, but the In peaks in $(InGa)N$ alloys show that more than one chemical environment is present. The composition of the alloys deduced from electron probe microanalysis studies agree well with those from X-ray measurements, assuming Vegard's law is valid for both alloy systems.

Introduction

The Group III-Nitrides show great promise for both electronic and optoelectronic applications[1]. Nitride-based blue light-emitting diodes (LEDs) are now commercially available from several companies and Nitride-based blue laser diodes (LDs) have just been reported[2].

One of the possible applications for Group III-Nitrides is for inter-chip communication in Si VLSI circuits. For this purpose it will be necessary to develop LEDs or LDs which can be grown locally on already processed Si wafers. As an initial step towards this long-term goal we have been studying the growth of $(AlGaIn)N$ alloys on oxidised and clean Si surfaces, with the intention that the material grown on the oxidised surface can then be removed by differential wet chemical techniques.

$(AlGa)N$ films have been grown previously on Si(111) by both MBE[3] and MOVPE[4], but so far there have been no reports of $(InGa)N$ films grown on Si by either technique. In the films grown by MOVPE the authors report cracking of the epitaxial layers due to differential strain, but no similar observation is reported for the MBE films.

The purpose of this present work is to study the feasibility of the growth of alloys over the whole composition range from InN to AlN. We report on the properties of the films using X-ray diffraction, electron probe microanalysis (EPMA) and X-ray photoelectron spectroscopy (XPS).

Experimental Technique

The samples used in this study were grown using a modified MBE technique in a Varian MOD-II system, equipped with an Oxford Applied Research CARS25 RF activated plasma source to provide the active source of nitrogen. Conventional elemental sources of Al, Ga and In were used for other species. Details of the experimental arrangement have been published elsewhere[5-7]. The growth rate of GaN films used in this study ($0.3\mu m$/hour), was determined by the Group III arrival rate, since excess atomic nitrogen was provided. For the alloy films the composition was set using the in-situ beam

Mat. Res. Soc. Symp. Proc. Vol. 449 © 1997 Materials Research Society

monitoring ion gauge assuming the relative sensitivities for Al and In relative to Ga were 0.5 and 1.4 respectively. Growth temperatures were in the range from 400 to 750°C.

For growth on oxidised surfaces the Si wafers were thermally cleaned in-situ, but growth was initiated at low temperature to avoid any possible oxide desorption.

For growth on clean surfaces the Si wafer was first dipped in 48% aqueous HF for 20 secs and washed with high purity de-ionised water and dried using filtered dry nitrogen. This process is know to result in a H-terminated surface[8-10] which is stable for several minutes in air. Samples were therefore introduced into the vacuum system immediately following the chemical cleaning. They were then heated to 200°C, at which temperature we observed desorption of H with the in-situ mass spectrometer and at the same time the appearence of a RHEED pattern indicative of an oxide-free surface. The temperature was then raised to 750°C for the growth of GaN.

X-ray studies were carried out using a Philips Xpert powder diffractometer and both θ-2θ and ω scans were performed for all samples. Quantitative chemical analysis of the composition of the films was obtained using Electron Probe Micro-analysis (EPMA). Because all the samples are thin it was necessary to employ the thin-film EPMA-method[11] to calculate the thickness and composition of the layers. The samples were investigated at 10 kV primary beam voltages. The depths of EPMA analysis are estimated to be about 0.32 μm for N and 0.50 μm for Ga at 10 keV. XPS studies were carried out in a ESCA LAB 5 MKII using standard methods and Mg k_α X-rays.

Results

Figure 1 shows the θ-2θ data for the whole range of alloys grown on an oxidised Si substrate. This shows, as expected, a progressive variation in peak position from InN to AlN. The general trend is for the material to be more amorphous as we go from pure InN to AlN. In all cases, for the alloys, only a single peak is observed indicating that only one composition is obtained for the crystalline fraction of the material. The ω scans for this material are very broad indicating that growth is highly disordered.

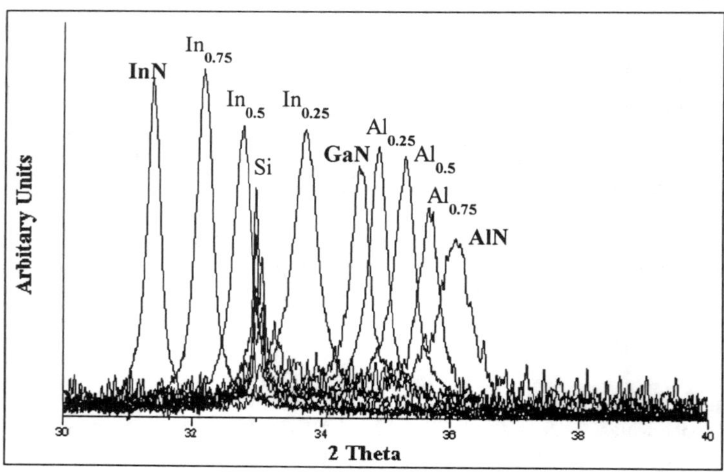

FIGURE 1 : X-ray Peak Positions of the $(Al/In)_x Ga_{(1-x)}N$ Alloys Grown on Oxidised Si (100).

Figure 2 shows the X-ray data from GaN grown on oxidised Si (100) and cleaned Si (111). Gallium Nitride grown on oxidised Si (100) appears to be amorphous with only a small number of crystallites, while growth on cleaned Si (111) shows highly ordered growth with the [0001] direction in the wurtzite GaN parallel to the Si [111] of the substrate. In this case we have epitaxial GaN on the Si substrate in contrast to the situation for growth on the oxidised Si.

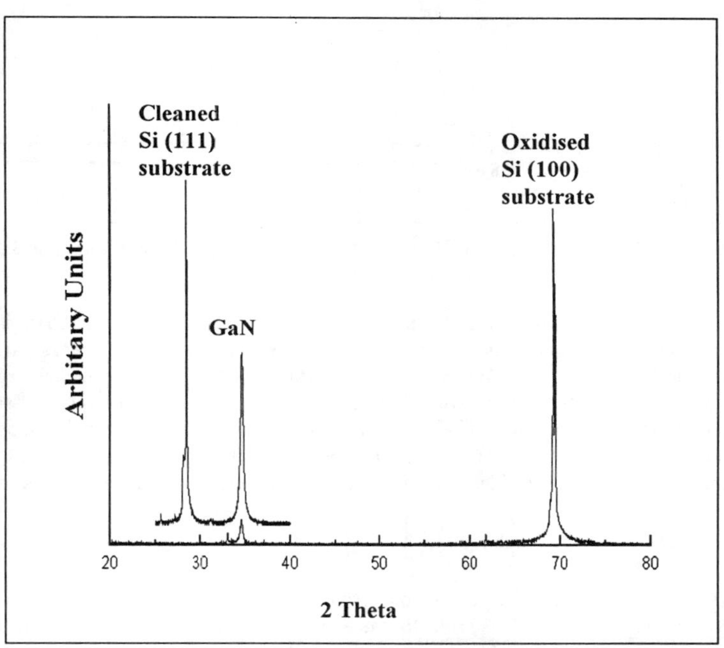

FIGURE 2 : X-ray Spectra of GaN Grown on Oxidised Si (100) and on Cleaned Si (111) Substrates.

Figure 3 shows the XPS data for the (InGa)N alloys in this series. In all cases we see double peaks for the In $3d^{5/2}$ and the $3d^{3/2}$ spectra. We see a progressive change in both the ratio of the peak areas and peak separation as a function of In content. This indicates that for the alloy samples two distinct binding states exist for the In with the difference in energy being larger as the In content increases. A similar doublet is observed for the N 1s peak in the same series of alloys, but the peak shapes of the Ga peaks are very broad and cannot at this stage be deconvoluted. Studies of InN and In_2O_3 are underway to help ascertain the nature of this effect. A key point, however, is that this data again suggests that alloy films of a single composition are being prepared under all conditions. This is contrary to recent suggestions[12] that phase separation occurs in the InGaN alloy system with >20% In. Our films, however, are grown by MBE at a much lower temperature and this may kinetically hinder any such process. Further experiments are being undertaken to elucidate this.

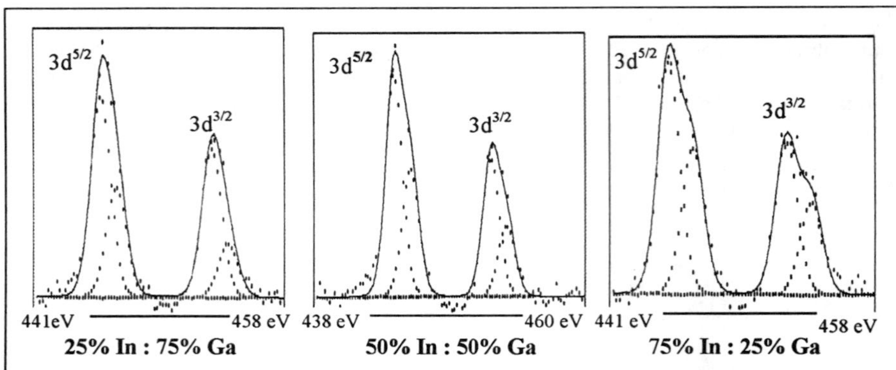

FIGURE 3 : XPS Spectra of In $3d^{5/2}$ and $3d^{3/2}$ in the $In_xGa_{(1-x)}N$ Alloys (Dotted peaks are deconvoluted).

Table (i) shows a comparison between the nominal composition, calculated from the in-situ ion gauge readings, the mole faction deduced from X-ray measurements, assuming Vegard's Law, and the EPMA results for the samples grown on an oxidised Si surface. Within the experimental error there is very good agreement for all the samples which implies that the crystalline and amorphous fractions of the material have similar composition and also that there is no evidence for spinodal decomposition. The X-ray peak positions for AlN and GaN are very close to the theoretical values, but the InN sample shows a considerable discrepancy which might indicate a significantly strained material has been obtained.

Table (i) : A Comparison of the Target Composition ,the Mole Fraction Deduced from X-ray Measurements and the EPMA Results for the Samples Grown on an Oxidised Si Surface.

Nominal Composition	Peep Position [a] (2θ)	Calc. Composition [b] (%)	EPMA Composition (%)
AlN	36.042° (Theoretical 36.056°)	-	-
$Al_{(0.75)}Ga_{(0.25)}N$	35.645°	Al 73% : Ga 27%	Al 74% : Ga 26%
$Al_{(0.5)}Ga_{(0.5)}N$	35.268°	Al 48% : Ga 52%	Al 46% : Ga 54%
$Al_{(0.25)}Ga_{(0.75)}N$	34.854°	Al 20% : Ga 80%	Al 22% : Ga 78%
GaN	34.576° (Theoretical 34.570°)	-	-
$In_{(0.25)}Ga_{(0.75)}N$	33.727°	In 22% : Ga 78%	In 24% : Ga 76%
$In_{(0.5)}Ga_{(0.5)}N$	32.754°	In 48% : Ga 52%	In 50% : Ga 50%
$In_{(0.75)}Ga_{(0.25)}N$	32.160°	In 65% : Ga 35%	In 71% : Ga 29%
InN	31.379° (Theoretical 31.027°)	-	-

(a) Peak profiles fitted to give the peak position.   (b) Calculated assuming Vegard's Law.

Conclusions

We have demonstrated that we can grow (AlGaIn)N samples ranging from InN to AlN of controlled composition, using a modified MBE method, on oxidised Si substrates. Using the in-situ ion gauge readings the nominal and actual composition agree to within ± 3%.

We see considerable differences in the structural quality of the material grown by MBE on oxidised and clean Si surfaces. This in turn suggests that there is the possibility of achieving local area growth on Si samples and that differential etching techniques may selectively remove material deposited on the oxide. Analysis of our data suggests that for alloy samples more than one binding energy exists for the In and N atoms in the surface. At present the situation for Ga is not resolved. Ho and Stringfellow[12] have recently suggested that there is a solid phase immiscibility in (InGa)N alloys for which we find no evidence in this study.

References

1. S Strite and H Morkoc, J Vac Sci Technol, **B10**, 1237, (1992).

2. S Nakamura, M Senoh, S-I Nagahama, N Iwasa, T Yamada, T Matsushita, H Kiyoku and Y Sugimoto, Jpn J Appl Phys, **35**, L74, (1996).

3. S Yoshida, S Misawa and S Gonda, J Appl Phys, **53**, 6844, (1982).

4. K Hirosawa, K Hiramatsu, N Sawaki and I Akasaki, Jpn J Appl Phys, **32**, L1039 (1993).

5. T S Cheng, L C Jenkins, S E Hooper, C T Foxon, J W Orton and D E Lacklison, Appl Phys Lett, **66**, 1509, (1995).

6. J W Orton, D E Lacklison, N Baba-Ali, C T Foxon, T S Cheng, S V Novikov, D F C Johnston, S E Hooper, L C Jenkins, L J Challis and T L Tansley, J.Electronic Materials, **24**, 263, (1995).

7. S E Hooper, C T Foxon, T S Cheng, L C Jenkins, D E Lacklison, J W Orton, T Bestwick, A Kean, M Dawson and G Duggan, J Cryst Growth, **150**, 892, (1995).

8. J M C Thornton and R H Williams, Semi Sci Technol, **4**, 847, (1989).

9. B S Meyerson, F J Himpsel and K J Uram, Appl Phys Lett, **57**, 1034, (1990).

10. D J Eaglesham, G S Higashi and M Cerullo, Appl Phys Lett, **59**, 685, (1991).

11. S V Kazakov, S G Konnikov and V V Tret'yakov, Izv.Akad.Nauk SSSR, **55**, 1627, (1991).

12. I-hsiu Ho and G B Stringfellow, Appl Phys Lett, **69**, 2701, (1996).

HEXAGONAL GROWTH HILLOCKS IN GaN EPILAYERS

P. G. MIDDLETON, C. TRAGER-COWAN, A. MOHAMMED, K. P. O'DONNELL*
W. VAN DER STRICHT, I. MOERMAN AND P. DEMEESTER**
*Dept. Physics and Applied Physics, University of Strathclyde, Glasgow G4 0NG, Scotland,
United Kingdom.
**IMEC-INTEC, University of Gent, Gent 9000, Belgium.

ABSTRACT

We describe a study of the hexagonal growth hillocks commonly present in gallium nitride films. The MOVPE-grown epilayers of the present work exhibit a predominantly smooth morphology but small groups of hexagonal hillocks were found to populate the surface, particularly at the sample edges.

Scanning electron (SE) micrographs were taken of several groups of hillocks. At the maximum beam energy of 25 keV, two types of hexagonal hillock are visible. Hillocks in the first group are terminated by an apex (ie. they are pyramidal in form), while the other, flat-topped, hillocks terminate on (0001)-facets. As one lowers the electron beam energy, thereby reducing beam penetration, some of the flat-topped hillocks disappear from the image. From this we tentatively deduce that these hillocks are buried. The result of further investigations, using an atomic force microscope, are consistent with the presence of sub-surface features.

The relationship between the luminescence and morphological properties of a pyramidal hillock is studied via cathodoluminescence imaging. The band-edge emission originates from the full hexagonal structure, except for the central region, where only the defect-related yellow luminescence is apparent. We suggest this might be explained by defects associated with inversion domain boundaries at the hillock centre.

INTRODUCTION

Considerable progress has been made over the last few years in the development of optical devices based on GaN and its alloys. Such devices are extremely attractive due to their suitability for true-colour displays, underwater communication, and high-density optical data storage applications. Blue light emitting diodes based on Zn-doped InGaN double heterostructures[1] have been on the market for some time, while LEDs based on single quantum wells - exhibiting greatly improved performance[2] - have recently become commercially available. Meanwhile, the demonstration of a working prototype by Nichia Chemical Industries has established gallium nitride as the best candidate for commercial laser diodes emitting in the blue spectral region and beyond.[3]

Sapphire has remained the favoured substrate for the majority of GaN epilayer growth, despite its poor structural and thermal match to both GaN and InN. Many studies of the possible alternatives have been made.[4,5] The introduction of a two-step process, in which a low-temperature AlN or GaN nucleation layer is deposited prior to the main growth sequence, has allowed layers of reasonable quality to be produced.[6] However, much remains to be understood

Mat. Res. Soc. Symp. Proc. Vol. 449 © 1997 Materials Research Society

regarding the mechanisms responsible for growth, the nature of the defects in the resulting material, and their relationship to the optical properties of layers (particularly the "yellow band" luminescence).

In this paper we report optical, atomic force and electron-beam microscopy studies of the hexagonal hillocks commonly observed in gallium nitride epilayers, and discuss their structure and contribution to the luminescence bands of GaN.

EXPERIMENT

GaN films were grown by metalorganic chemical vapour deposition on sapphire (0001) substrates, in a vertical rotating disk reactor. The epilayers discussed in this work comprise a low-temperature GaN buffer layer of thickness 20 nm deposited at 475°C, followed by an epilayer of 1 μm at 1050°C. A full account of the growth procedure has been presented elsewhere.[7]

Optical micrographs of the epilayers were taken using an Olympus microscope equipped with video camera. Photoluminescence measurements were carried out using a He-Cd laser providing several milliwatts of excitation at 325 nm. A Cambridge Instruments scanning electron microscope (SEM) and a Burleigh Personal atomic force microscope were used to examine the surface morphology. A modification of the SEM system permits light (i.e. cathodoluminescence) to be detected as the electron beam is scanned over the surface of a sample. Comparison of SE and CL micrographs of a given sample area allow relationships between the optical and physical morphologies of the sample to be identified.

RESULTS

Optical Microscopy and Photoluminescence

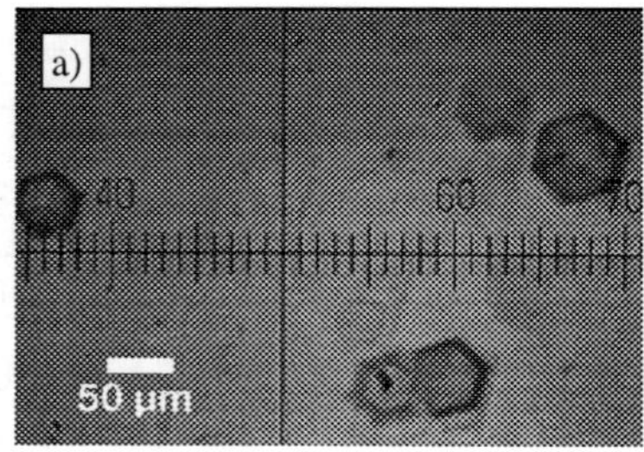

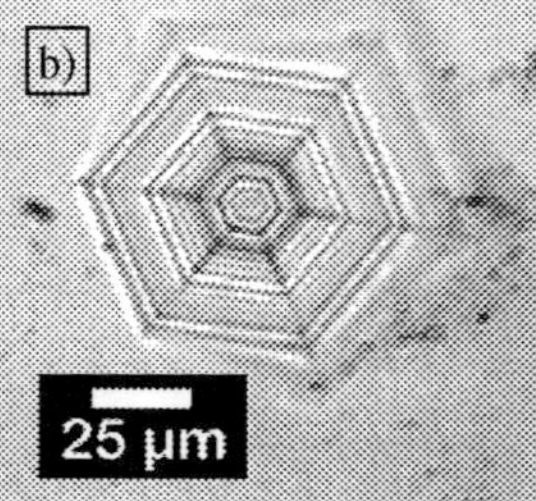

Figure 1. Optical micrograph of a) a typical epilayer surface, b) the stepped structure of a hexagonal pyramid.

The GaN epilayers exhibit a predominantly smooth morphology. However, small groups of hexagonal hillocks (Figure 1a) are found to populate the surface. These range from truncated and plate-like formations to complete hexagonal pyramids. Diameters of around 50 μm are typical. Closer inspection reveals that the sides of the pyramids are not smooth facets, but instead consist of a series of steps, rising to an apex (Figure 1b).

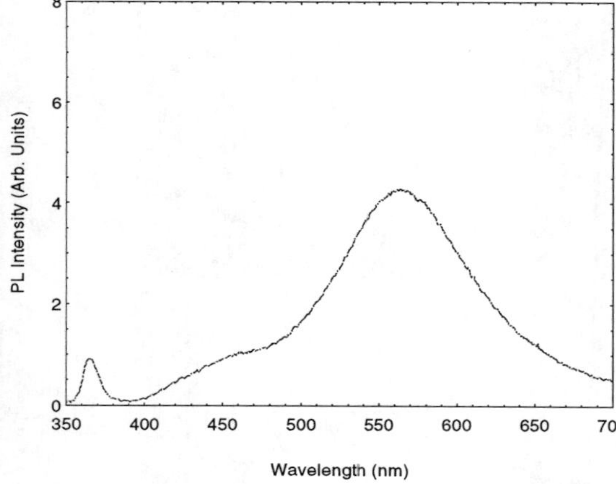

Figure 2 shows a typical photoluminescence spectrum of the epilayer, taken at room temperature. A laser spot diameter of 1 mm was used, exciting both smooth and pyramidal surface regions simultaneously. A strong band edge peak is apparent, along with two broad defect-related bands at 450 nm and 550 nm respectively.

Figure 2. Photoluminescence spectrum of a GaN epilayer with hexagonal hillocks.

SEM and AFM of Hexagonal Pyramids

A series of SEM micrographs were taken of a particular group of hexagons at various electron beam energies. The sample was normal to the incident beam. An optical micrograph at the same orientation is provided for reference (Figure 3). As the beam energy is raised, the depth of penetration of the incident electrons increases. Table 1 lists the depth of maximum energy deposition, and the maximum ranges (Bethe ranges) corresponding to each of the beam energies used. At an energy of 1.5 kV, most of the electrons penetrate only the first few nanometers of material, and surface features dominate the image (see Figure 4). Pyramidal features A and B are visible in these low-energy images. However, several neighbouring features (C and D) do not become apparent until energies of 15 or 25 kV (maximum energy depositions at 270 and 650 nm respectively). At this point the beam penetrates the full depth of the sample. The blurring of the perimeters of features C and D results from the enhanced diffusion of the high energy primary electrons at higher beam energies.

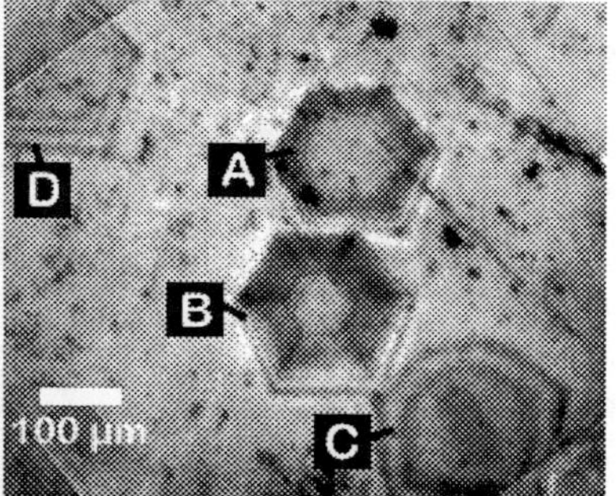

Figure 3. Optical micrograph of hillock region.

Energy (keV)	Depth of Maximum Energy Deposition (nm)	Bethe Range (nm)
1.5	5.5	29
7.5	84	370
15	270	1300
25	650	3000

Table 1. Energy deposition depths for various electron beam energies.

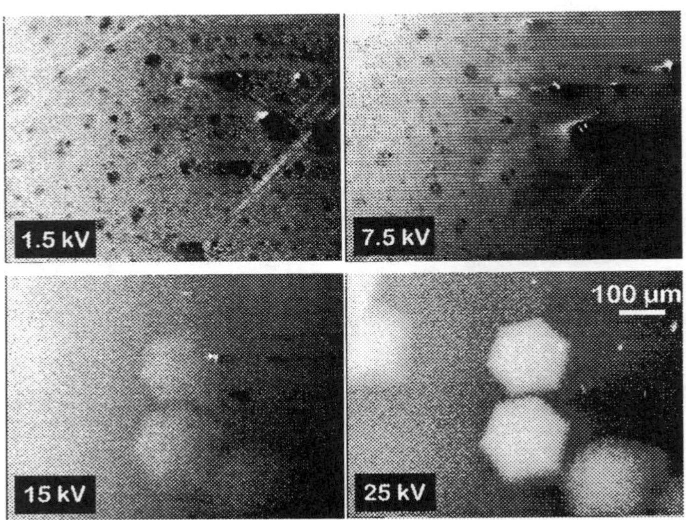

Figure 4. SEM images taken at different electron beam energies.

Hexagon C was mapped using an atomic force microscope, in a series of 70 µm x 70 µm scans. An approximately hexagonal outline can be discerned in these images, in the form of a depression some 100 nm in depth. The only other feature is a slightly raised area at one corner of the hexagon (see figure 5). However, none of the internal structure visible in the optical micrograph is detectable.

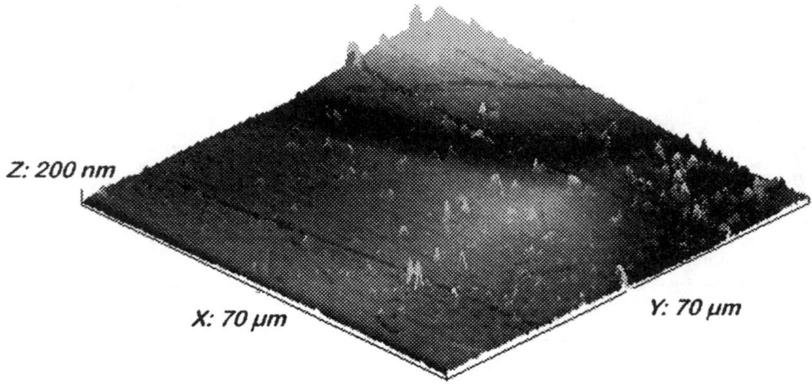

Figure 5. AFM images showing the perimeter trough and elevated region at one corner of hexagon C.

The combination of SEM and AFM data suggests that some of the hexagons observed in GaN epilayers may in fact be sub-surface features, with an additional layer of several nanometers deposited on top. However, further investigation is necessary before any definite conclusions can be drawn.

CL of Hexagonal Pyramids

Finally, we investigated the relationship between the luminescence and morphological properties of a single hexagonal pyramid. By inserting a monochromator into the luminescence exit beam of the SEM, individual bands of the epilayer cathodoluminescence emission can be selected, and the luminescence spatially resolved. An electron beam energy of 25 keV was employed throughout. The wavelengths 373 nm, 450 nm and 550 nm were imaged, corresponding to the UV band-edge, the blue defect band and the yellow band respectively (Figure 6).

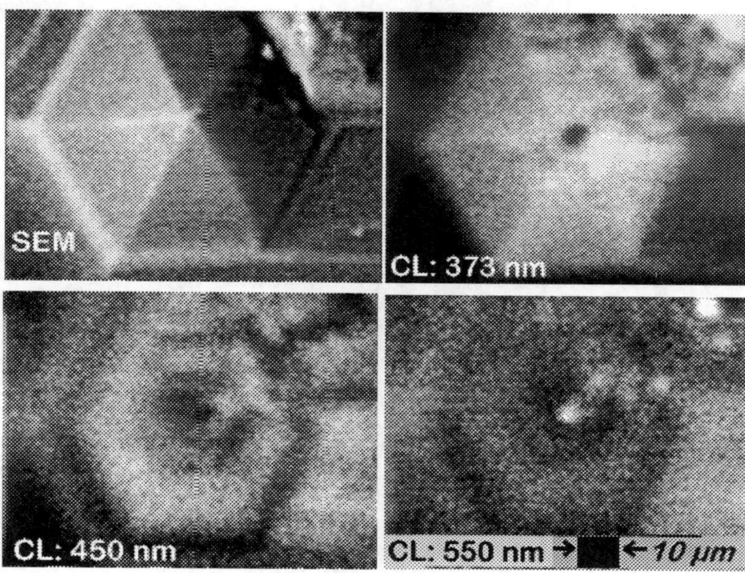

Figure 6. SEM and cathodoluminescence micrographs of a full hexagonal pyramid.

UV luminescence is found to originate from the full hexagon structure, except for a central region of the image approximately 5 μm across where no significant band-edge emission is detected. The 450 nm band also appears to be associated with the main structure. In contrast, the yellow band emission is found primarily at the centre of the hexagonal pyramid. The additional luminescence centres in the upper right of Figure 6 (550 nm) belong to neighbouring crystallites not clearly resolved in this image. Of particular interest is the apparent spatial banding of the luminescence observed when the defect-related wavelengths are selected. This spatial localisation of the luminescence bands suggests that different areas of the hexagon consist of material of quite different character. Rouviere *et al.*, in their study of the polarity of gallium nitride films, report features within hexagonal pyramids structures which may offer an explanation.[8-10] Using convergent beam electron diffraction techniques, films showing pyramidal morphology were found to exhibit both Ga- and N-polarity. The material is predominantly N-polar - however, tiny

Ga-polar domains, which grow more rapidly than the surrounding N-polarity material, dictate the final surface morphology. The resultant hexagonal pyramids are found to contain numerous tiny Ga-polar inversion domains. Importantly, an inversion domain is always found at the apex of the hexagonal pyramids. We suggest that the yellow luminescence observed at the centre of our pyramids originates from defects associated with such an inversion domain boundary.

CONCLUSIONS

In summary, the properties of hexagonal hillocks on MOVPE-grown GaN epilayers have been investigated. SEM and AFM studies suggest the presence of sub-surface hexagonal structures in these epilayers. The luminescence associated with hexagonal hillocks is strongly spatially dependent: band-edge luminescence is emitted across the whole hillock, except at the hillock centre, whereas the yellow band emission is is found almost exclusively in the central region. The latter may be a result of defects associated with inversion domain boundaries in the hillocks.

REFERENCES

[1] S. Nakamura, J. Cryst. Growth **145** (1994) 911.

[2] S. Nakamura, Mat. Res. Soc. Symp. Proc. **395** (1996) 879.

[3] S. Nakamura, M. Senoh, S. Nagahama, N. Iwasa, T. Yamada, T. Matsushita, H. Kiyoku and Y. Sugimoto, Jpn. J. Appl. Phys., **35** (1996) L74-L76.

[4] J. F. H. Nicholls, H. Gallagher, B. Henderson, C. Trager-Cowan, P. G. Middleton, K. P. O'Donnell, T. S. Cheng, C. T. Foxon and B. H. T. Chai, Mat. Res. Soc. Symp. Proc. **395** (1996) 231.

[5] A. Kuramata, K. Horino, K. Domen, K. Shinohara and T. Tanahashi, Appl. Phys. Lett. **67** (1995) 2521.

[6] S. Nakamura, Jpn. J. Appl. Phys **30** (1991) L1705-L1707.

[7] W. Van Der Stricht, I. Moerman, P. Demeester, J. A. Crawley, E. J. Thrush, P. G. Middleton, C. Trager-Cowan, K. P. O'Donnell, Mat. Res. Soc. Symp. Proc. **395** (1996) 535.

[8] J. L. Rouviere, M. Arlery, A. Bourret, R. Niebuhr and K. H. Bachem, Mat. Res. Soc. Symp. Proc. **395** (1996) 393.

[9] J. L. Rouviere, M. Arlery, R. Niebuhr, K. H. Bachem and O. Briot, MRS Internet J. Nitride Semicond. Res. **1** 33 (1996).

[10] B. Daudin, J. L. Rouviere and M. Arlery, Appl. Phys. Lett. **69** (1996) 2480.

HIGH RESOLUTION X-RAY DIFFRACTION OF
GaN GROWN ON SAPPHIRE SUBSTRATES

A. SAXLER*[#], M.A. CAPANO[#], W.C. MITCHEL[#], P. KUNG*, X. ZHANG*, D. WALKER*, AND M. RAZEGHI*

*ECE Department, Northwestern University, Evanston, IL 60208, razeghi@ece.nwu.edu
[#]Wright Laboratory, Materials Directorate, Wright-Patterson AFB, OH, 45433-7707

ABSTRACT

X-ray rocking curves are frequently used to assess the structural quality of GaN thin films. In order to understand the information given by the line shape, we need to know the primary mechanism by which the curves are broadened. The GaN films used in this study were grown by low pressure metalorganic chemical vapor deposition (MOCVD) on (00•1) sapphire substrates. GaN films with both broad and very narrow (open detector linewidth of 40 arcseconds for the (00•2) GaN reflection) rocking curves are examined in this work. Reciprocal space maps of both symmetric and asymmetric reciprocal lattice points are used to determine that the cause of the broadening of GaN rocking curves is a limited in-plane coherence length.

INTRODUCTION

GaN based material is ideal for the fabrication of short wavelength optoelectronic devices. The wide direct bandgap of GaN (3.4 eV) combined with its strong bond strength make it an excellent candidate for short wavelength lasers. A major problem with the growth of GaN has been the absence of a high quality lattice matched substrate. Most GaN growth has been performed on basal plane sapphire substrates resulting in films with many defects in the crystal lattice. Nucleation layers of AlN or low temperature GaN are used to promote quasi-two dimensional growth, dramatically improving the properties of the GaN epilayers [1]. The full width at half maximum (FWHM) of the x-ray rocking curve is often used to assess the crystalline quality of GaN epilayers because of the relatively simple nature of this technique. Recently, very narrow x-ray rocking curves of epitaxial GaN grown on (00•1) sapphire with a FWHM of under 40 arcseconds have been obtained by several researchers [2-5]. A full understanding of the rocking curves is needed if they are to be used as a reliable indicator of epitaxial quality. A better understanding of the defects in GaN has recently been established through x-ray diffraction and transmission electron microscopy (TEM) studies [5,6]. The primary defects are threading dislocations running along the c axis, with the majority being edge type with a Burger's vector of 1/3<11•0> although screw and mixed type dislocations have also been seen. In highly dislocated materials, low angle grain boundaries form. In this paper, we report the analysis of symmetrical and asymmetrical reciprocal space maps to determine what types of imperfections are responsible for the broadening of x-ray rocking curves in GaN.

EXPERIMENT

The GaN samples used in this study were grown on sapphire substrates with a thin AlN buffer by low pressure MOCVD. The sources used were ammonia, trimethylgallium, and trimethylaluminum in a hydrogen carrier gas. The growth conditions for similar samples were discussed in more detail previously [2]. Samples with a wide range of growth conditions were

477

chosen to establish the generality of the type of broadening of x-ray rocking curves in GaN. The growth conditions are presented in Table I.

The x-ray diffraction experiments were performed using a Philips (MRD) high-resolution triple axis diffractometer [7,8].

TEM samples were examined using a Hitachi HF-2000 with a field emission gun operated at 200 keV.

Sample	Sapphire substrate orientation	Buffer layer, thickness (nm), and growth T(C)	Growth pressure (mbar)	Thickness (µm)	Growth Temperature (C)
A (201)	(00.1)	AlN, 16, 850 C	100	1.2	1000
B (114)	(00.1)	AlN, 34, 1000 C	10	1.1	1000
C (23)	(01.2)	GaN, 23, 600 C	10	1.1	1000

Table I. Growth conditions for the GaN films used in this study.

RESULTS AND DISCUSSION

In order to determine the nature of the x-ray broadening in epitaxial GaN, several reciprocal lattice points were mapped. Figure 1(a) shows a reciprocal space map of the (00•2) GaN peak. The most significant broadening is in the ω direction, but from this symmetric peak, we cannot distinguish between broadening due to random tilt or curvature and a finite in-plane coherence length. Pendellösung fringes are clearly visible in the $\omega/2\theta$ direction due to the finite thickness and high structural quality of the epilayer. The spacing corresponds to a thickness of 1.2 µm, in good agreement with the thickness measured by ball polishing and optical transmission interference fringes.

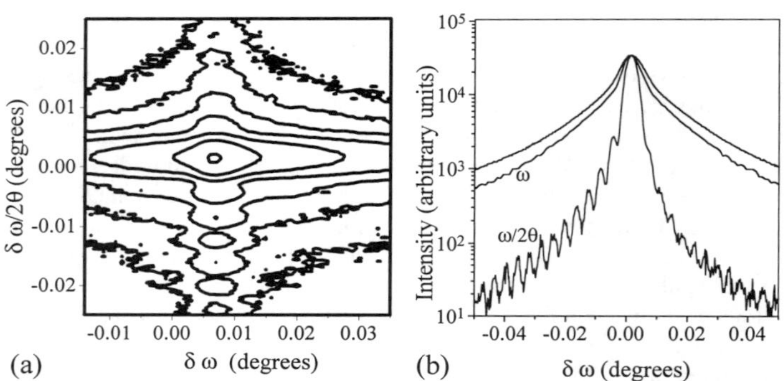

FIG. 1. (a) Reciprocal space map of the (00•2) GaN peak of Sample A. Interference fringes due to the finite film thickness are clearly visible in the $\omega/2\theta$ direction. Isointensity contours are logarithmically spaced by a factor of $10^{0.5}$. (b) Open detector rocking curve, triple axis ω scan, and triple axis $\omega/2\theta$ scan of the (00•2) GaN peak of Sample A.

A conventional rocking curve of the (00•2) GaN peak yielded a FWHM of 40 arcseconds as illustrated in Figure 1(b). For comparison, the normalized cross sections of the reciprocal space map are also shown in Figure 1(b). (The ω curve and the $\omega/2\theta$ curve are the traces through the peak in Figure 1(a) along the ω and $\omega/2\theta$ directions respectively.) The ω curve has a FWHM of 29 arcseconds, while the $\omega/2\theta$ curve has a FWHM of 16 arcseconds. The ω curve makes the dominant contribution to the broadening of the conventional rocking curve, especially in its broad tails, which obscure the fringes. From this information, we can determine that the broadening is primarily due to either a limited coherence length parallel to the film surface or from random tilt or curvature of the film.

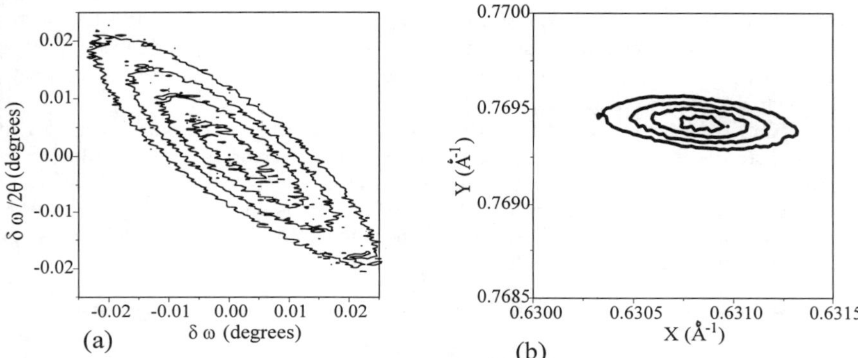

FIG. 2. Reciprocal space map of the (11•4) GaN peak of Sample A plotted in degrees (a) and reciprocal lattice units (b). This shows that the primary cause of broadening is due to a limited in-plane coherence length rather than tilt or curvature.

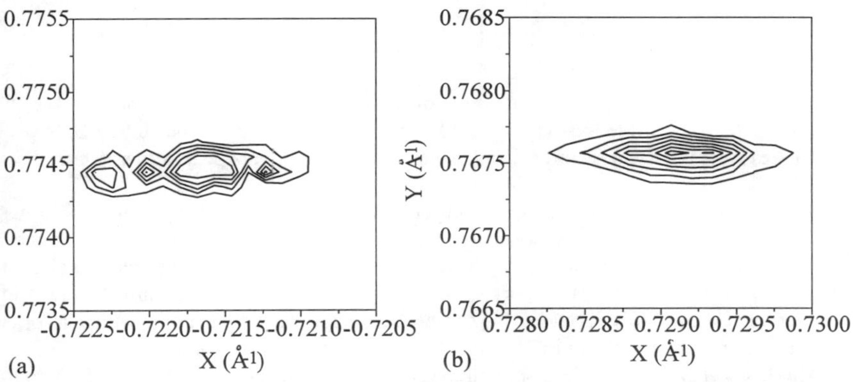

FIG. 3. Reciprocal space maps of the (a) ($\overline{2}$0•4) and (b) (20•4) GaN peaks of Sample A showing that there is a small curvature or tilt component rather than dislocations slightly off normal to the surface.

Investigation of a reciprocal space map of an asymmetrical peak can indicate if the broadening is due to random tilt or curvature or if it is due to a limited in plane coherence length. The (11•4)

GaN peak of Sample A is shown in Figure 2. The (11•4) planes are approximately 39 degrees from the (00•1) planes, allowing the different types of broadening to be easily distinguished. From Figure 2(a), the peak is clearly not oriented along ω, which would be expected if curvature or random tilt caused the broadening. When the same peak is plotted in reciprocal space in Figure 2(b), the broadening is nearly entirely along X, indicative of a limited in-plane coherence length. A small tilt from the horizontal may be due to a contribution of broadening in ω. Two peaks on opposite sides of the origin in reciprocal space were used to determine whether the small tilt from the horizontal is due to broadening in ω or due to the discontinuities not being normal to the surface. The tilts are in opposite directions (Figure 3), consistent with a small contribution of ω broadening. A separate measurement of the peak position as a function of the wafer position indicated a small curvature of 20 m, which contributes to the ω broadening in addition to any random tilts.

The TEM results show that edge threading dislocations dominate with a density of approximately 5×10^9 cm^{-2}. In a plan view image with a **g** vector of 11•0 the dislocations appear to be randomly distributed rather than aligned to form low angle grain boundaries. Both edge and screw dislocations are present in the sample as seen by cross sectional images taken using different **g** vectors. The role of these dislocations in the x-ray line broadening is to disrupt the in-plane coherence length in the GaN film.

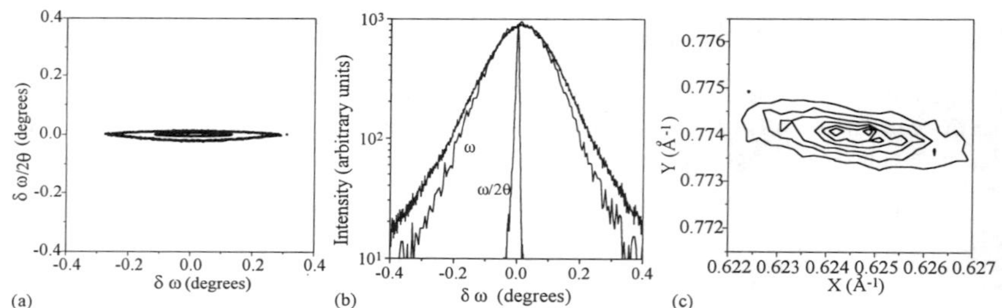

FIG. 4. (a) Symmetrical reciprocal space map of the (00•2) GaN peak of Sample B. (b) Open detector rocking curve, triple axis ω scan, and triple axis ω/2θ scan of the (00•2) GaN peak of Sample B. (c) Asymmetrical reciprocal space map of the (11•4) GaN peak of Sample B.

The rocking curve linewidth was much broader for the second sample of GaN grown on (00.1) sapphire, however, the primary cause for broadening was the same. The symmetrical reciprocal space map is shown in Figure 4(a) and traces through the map are plotted along with the open detector rocking curve in Figure 4(b). Only a small fraction of the broadening is in the ω/2θ direction. Figure 4(c) shows an asymmetrical reciprocal lattice point which is oriented along X in reciprocal space, again indicating a limited in-plane coherence length.

The GaN grown on a (01•2) sapphire substrate was also studied. Figures 5(a) and 5(b) show the symmetrical reciprocal lattice point which again is primarily broadened along ω. The asymmetrical reciprocal lattice point is shown in Figure 5(c). The alignment along X is once again indicative of a limited in-plane coherence length.

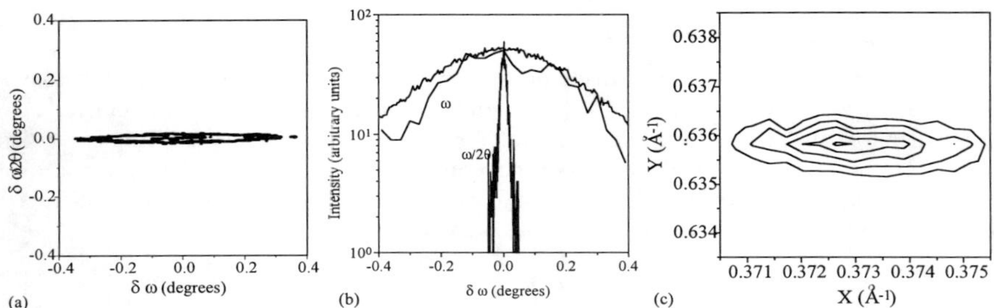

FIG. 5. (a) Symmetrical reciprocal space map of the (11•0) GaN peak of Sample C. (b) Open detector rocking curve, triple axis ω scan, and triple axis ω/2θ scan of the (11•0) GaN peak of Sample C. (c) Asymmetrical reciprocal space map of the (11•2) GaN peak of Sample C.

The full widths at half maxima for the samples are shown in Table II. The primary broadening for all of the samples studied is along X, rather than Y, ω, or ω/2θ. This is indicative of a limited in-plane coherence length. The in-plane coherence length can be estimated by a simple equation for symmetric reflections: $d = 0.89 \lambda / 2 \Delta\omega \sin\theta$, λ where is the x-ray wavelength, $\Delta\omega$ is the FWHM of the rocking curve, and θ is the Bragg angle. However, for randomly sized domains, the coherence length obtained will reflect the longer coherence lengths present in the sample. This type of broadening is consistent with the wide tails commonly present in rocking curves of GaN.

GaN peak	ω (arcsec)	ω/2θ (arcsec)	X (Å⁻¹ x 10⁻⁴)	Y (Å⁻¹ x 10⁻⁴)
Sample A (201)				
(00•2)	29	16	0.539	0.973
(00•4)	56	21	2.02	1.05
(00•6)	62	51	3.5	1.5
(11•4)	72	56	7.1	2.1
(20•4)	50	58	9.5	2.2
($\overline{2}$0•4)	64	76	10	3.0
Sample B (114)				
(00•2)	667	32	12.5	1.8
(00•4)	684	41	25.5	2.1
(11•4)	220	140	10	5
(10•4)	220	130	20	4
Sample C (23)				
(11•0)	1620	61	49	3.6
(11•2)	430	110	47	8.4
(02•0)	360	270	50	7.5

Table II. FWHM of various GaN reflections.

CONCLUSIONS

In summary, reciprocal space mapping of both symmetric and asymmetric reflections was used to determine the primary cause for broadening of the epitaxial GaN x-ray rocking curves. The broadening was shown to be due to a limited in-plane coherence length. This is consistent with TEM studies that show threading dislocations originating at the interface, presumably during the initial stages of growth and island coalescence. These dislocations disrupt the coherence length, resulting in broadening of the x-ray rocking curve.

ACKNOWLEDGMENTS

This work was funded in part by the Ballistic Missile Defense Organization through ONR Grant N00014-93-1-0235. The authors would like to thank Max Yoder and Yoon-Soo Park of the Office of Naval Research for their support and encouragement. The TEM measurements are courtesy of W. Qian and V. P. Dravid at Northwestern University.

REFERENCES

1. K. Hiramatsu, S. Itoh, H. Amano, I. Akasaki, N. Kuwano, T. Shiraishi, and K. Oki, *J. Cryst. Growth* **115**, 628 (1991).

2. P. Kung, A. Saxler, X. Zhang, D. Walker, T. C. Wang, I. Ferguson, and M. Razeghi, *Appl. Phys. Lett.* **66**, 2958 (1995).

3. K. G. Fertitta, A. L. Holmes, J. G. Neff, F. J. Ciuba, and R. D. Dupuis, *Appl. Phys. Lett.* **65**, 1823 (1994).

4. W. E. Plano, J. S. Major, Jr., D. F. Welch and J. Speirs, *Electron. Lett.* **30**, 2079 (1994).

5. B. Heying, X. H. Wu, S. Keller, Y. Li, D. Kapolnek, B. P. Keller, S. P. DenBaars, and J. S. Speck, *Appl. Phys. Lett.* **68**, 643 (1996).

6. W. Qian, M. Skowronski, M. De Graef, K. Doverspike, L. B. Rowland, and D. K. Gaskill, *Appl. Phys. Lett.* **66**, 1252 (1995).

7. P. F. Fewster, *J. Appl. Cryst.* **22**, 64 (1989).

8. P. F. Fewster, *Appl. Phys. A* **58**, 121 (1994).

HIGH RESOLUTION X-RAY DIFFRACTION FROM EPITAXIAL GALLIUM NITRIDE FILMS

T LAFFORD, N. LOXLEY AND B K TANNER*

Bede Scientific, Bowburn, Durham, DH6 5PF, U.K

* *also Department of Physics, Durham University, South Road, Durham, DH1 3LE, U.K.*

ABSTRACT

The width of double axis X-ray rocking curves of epitaxial GaN layers is shown to be critically dependent on the width of the detector aperture. We show that triple axis diffraction measurements using a crystal analyser before the detector enables the instrument function to be defined and the tilt and dilation distributions separated. All GaN samples examined showed a mosaic structure of misoriented sub-grains with little dilation within the mosaic blocks. In reciprocal space maps this was revealed as a wide distribution of intensity in a direction perpendicular to the reciprocal lattice vector.

INTRODUCTION

High resolution X-ray diffraction [1] is extensively used for the characterisation of epitaxial compound semiconductor films used for opto-electronic devices. The excitement caused by the announcement of a room temperature blue laser based on GaN grown on sapphire [2] has led to much work on such materials. As part of attempts to improve epitaxial layer quality, to grow on other substrates and to understand why device lifetimes are relatively high despite extremely high numbers of misfit dislocations, a number of high resolution diffraction studies have been reported [3-6]. However, as with the low band gap II-VI materials such as HgCdTe, there is considerable confusion in the community over the interpretation of the diffraction peak widths in double axis X-ray rocking curves. In this paper, we show that triple axis diffraction provides a much more reliable tool for characterisation of GaN, because of an unambiguous instrument function and the ability to separate tilts and dilations.

X-RAY DOUBLE AXIS DIFFRACTION

Double axis diffraction has a major advantage in simplicity, which has led to it becoming a standard quality control tool on many III-V compound semiconductor production lines. In such quality control instruments, [7,8] there is a single reflection from the beam conditioner which results in a beam of low angular divergence but relatively high wavelength dispersion reaching the specimen. Provided that the specimen and beam conditioner crystals have almost the same Bragg plane spacing, the spread in wavelengths does not affect the rocking curve, which is narrow and the mathematical correlation of the plane wave reflecting curves of the specimen and beam conditioner. In current research instruments, the X-ray beam is conditioned by several Bragg reflections in both a beam conditioner and monochromator resulting in an beam of low angular divergence and low wavelength dispersion striking the specimen. The design of monochromating beam conditioners to satisfy different applications is well documented in the literature [9-11]. It is important in the context of this paper to note that although there may be many reflections from several crystals in the monochromating beam conditioner, there are effectively still only two axes. In both geometries the X-ray beam scattered from the specimen is received by an open area detector.

Mat. Res. Soc. Symp. Proc. Vol. 449 © 1997 Materials Research Society

Although a scan of the specimen gives a direct measure of the difference in Bragg angle between epitaxial layer and substrate, the width of the Bragg peaks is a function of both the distribution of tilts and the distribution of dilations in the sample. Thus in a double axis experiment with an open detector, it is not possible to distinguish between broadening due to lattice tilts and variations in stoichiometry. An example of the effect of restricting the detector aperture is given in Fig 1. This shows a pair of rocking curves taken from a GaN epitaxial layer grown on a (111) GaAs substrate with and without a 0.5 mm slit in front of the detector.

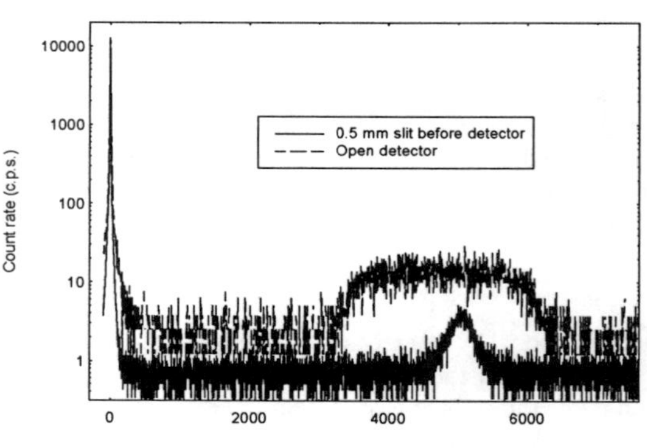

Fig 1 Double axis θ-2θ rocking curves from a GaN epitaxial layer on (111) orientation GaAs

While the diffraction peaks are, except in the tails of the profiles, almost identical for the GaAs substrate, the widths and shapes of the Bragg peak from the GaN are dramatically different. With an open detector, the full width at half height maximum (FWHM) is 2500 arc seconds, while with the 0.5 mm slit in front the FWHM is reduced to 300 arc seconds. Thus it is meaningless to report FWHM values of double axis rocking curves unless the detector geometry is precisely specified. The FWHM depends critically on the detector aperture.

X-RAY TRIPLE AXIS DIFFRACTION

Triple axis X-ray diffraction [1] involves the use of a crystal analyser on a third axis to select X-ray beams that are scattered from the sample within a limited angular range. Despite the implication that the technique is only applicable to very high perfection material, we show in this paper that it is highly appropriate to relatively poor epitaxy materials such as GaN. The importance is that the analyser removes the ambiguity about the detector aperture and enables the contributions of tilts and dilations to be separated. Fig 2 shows a diagram of the triple axis arrangement. Scanning the sample only maps out the distribution of tilts, as only regions of the crystal with the same Bragg plane spacing reach the detector for any one setting of beam conditioner and analyser. Scanning the specimen and analyser in a one-to-two ratio maps out the lattice dilations. From a combination of the two types of scan the intensity of the X-rays scattered around the reciprocal lattice point can be mapped, the so-called reciprocal space map.

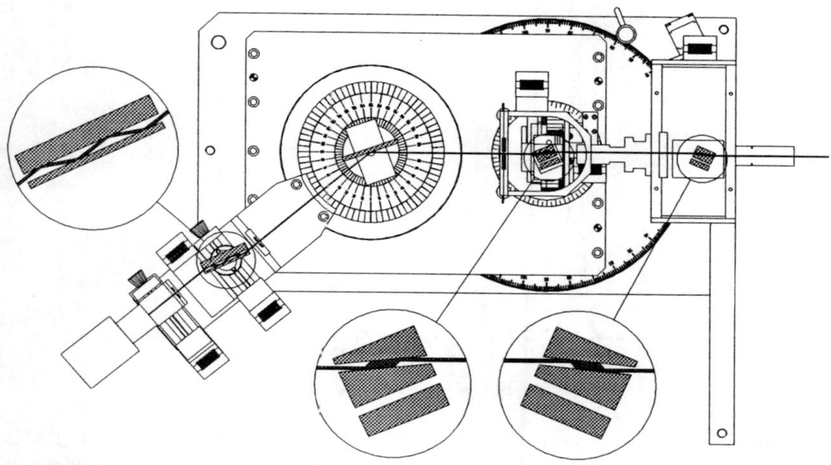

Fig 2 Diagram of the triple axis diffraction arrangement

We show in Fig 3(a) double and triple axis rocking curves for a thicker sample of GaN grown on sapphire. Note that both of these scans are of the specimen only, the double axis having an open detector, while the triple axis scan has a fixed analyser in place. In both cases the beam conditioners were a monochromating duMond pair of Si crystals cut at 17.65° to the (022) planes [11]. Except for an intensity scaling factor, these curves are almost identical, indicating that the rocking curve width (FWHM 608 ± 5 arc seconds) is totally dominated by the tilt distribution. As with low band-gap II-VI compounds [12], the tilt distribution approximates quite well to a Gaussian.

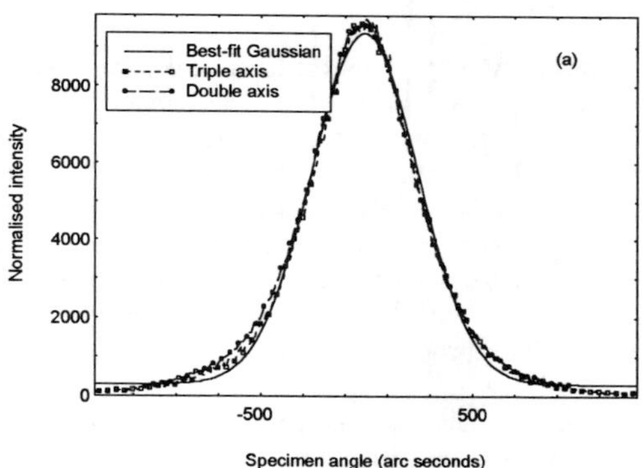

Fig 3 (a) Double and triple axis rocking curves (specimen scans) of a GaN epitaxial layer. Substrate (0001) oriented sapphire. (0002) reflection. Cu Kα₁ radiation. FWHM 608 ± 5 arc seconds

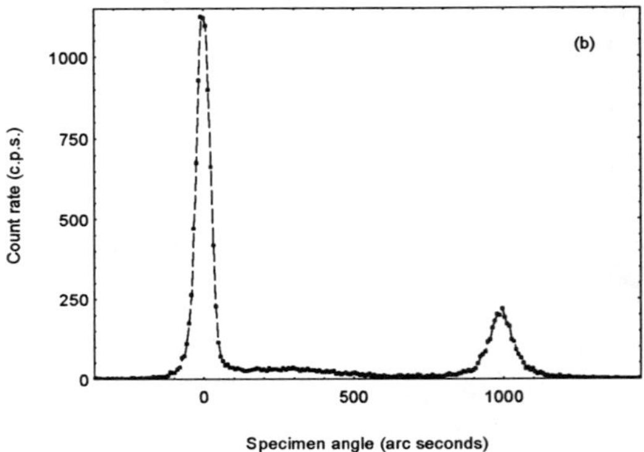

Fig 3(b) Triple axis coupled specimen-analyser scan in ratio 1:2. GaN (0002) reflection. (022) (with 17.65º asymmetric cut) Si duMond beam conditioners, (111) symmetric Si analyser

The triple axis coupled specimen and analyser scan (the so-called θ-2θ scan) in Fig 3(b) shows much narrower peaks, indicating that the strains in the lattice are small, although two sub-grains with slightly different lattice parameters are present. The principal deformation in the layer is therefore in the form of tilts, and the crystal is a mosaic of misoriented sub-grains with little strain or non-stoichiometry within each sub-grain.

RECIPROCAL SPACE MAPPING

The full map of the scattered intensity around the reciprocal lattice point is known as a reciprocal space map and this is achieved by straightforward transformation of the angular co-ordinates into reciprocal lattice units via the X-ray wavelength. The full reciprocal space map of the sample shown in Fig 3 is given in Fig 4.

This map shows how the isointensity contours of the layer peaks are extended in reciprocal space in the direction normal to the reciprocal lattice vector. This supports the assertion that the layer has a mosaic structure with little strain within the sub-grains. The two layer peaks do not lie exactly above one another. Such a feature can be interpreted as arising from two grains, misoriented with respect to one another. It also shows that the single coupled θ-2θ scan in Fig 3(b) does not give a true representation of the relative intensities of the peaks. This scan, although correctly representing the peak from one grain, cuts only the shoulder of the peak from the other grain, giving a false measure of the full width at half height maximum and integrated intensity. The importance of a full reciprocal space map in such a context is apparent.

All GaN layers which we have examined have shown the same characteristic of a relatively high mosaic spread but small amounts of strain within individual sub-grains. Fig 5 shows another GaN layer, again grown on sapphire, this time showing a much larger mosaic spread but with a comparable strain distribution to the sample from which the data of Fig 4 were taken.

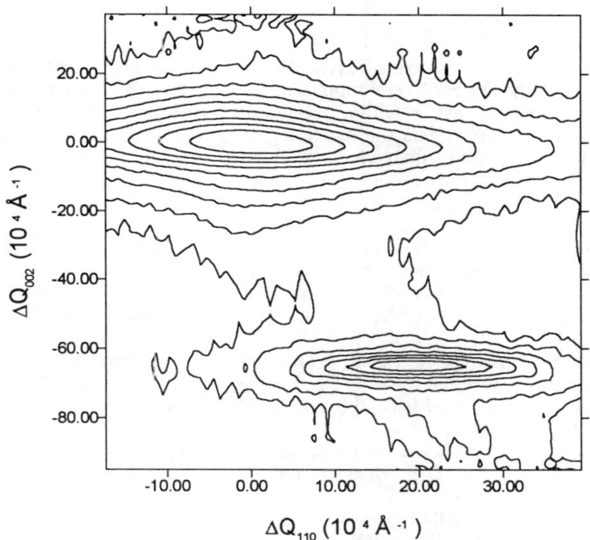

Fig 4 Reciprocal space map of the (0002) reflection from the GaN epitaxial film shown in Fig 3. Logarithmic scale contours. The two features indicate misoriented sub-grains, strained relative to one another. Step size 20" in θ-2θ, specimen (θ) step size 40", counting 4s per point, run overnight. Asymmetric (022) channel cut analyser with 17.65° between Bragg planes and surface.

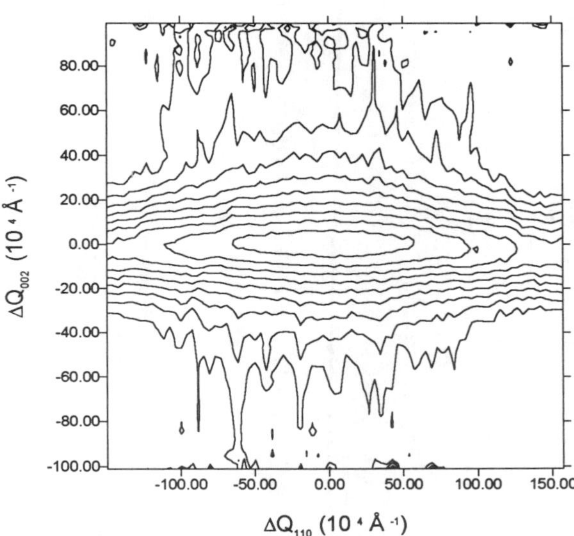

Fig 5 Reciprocal space map around the (0002) reciprocal lattice point of a GaN layer. The GaN was grown on (0001) sapphire. Cu Kα₁ radiation. Logarithmic contours. Step size of 10" in θ-2θ, specimen (θ) step size of 40" and count time 5s per point.

CONCLUSION

High resolution double and triple axis X-ray diffraction and reciprocal space mapping have been successfully applied to the study of GaN films. The GaN layers we have studied have shown quite high mosaic spread but little strain within sub-grains. To establish the cause of rocking curve broadening, it is important to define the scattering angle, and full reciprocal space mapping can give important information not available from double axis measurements.

REFERENCES

1. B.K. Tanner and D.K. Bowen, J. Crystal Growth **126** 1 (1993)
2. S. Nakamura, Japan. J. Appl. Phys. **35** L74 (1996)
3. Z.Q. He, X.M. Ding, X.Y. Hou and X. Wang, Appl. Phys. Lett **64** 315 (1994)
4. K.G. Fertitta, A.L. Holmes, J.G. Neff, F.J. Ciuba and R.D. Dupuis, Appl. Phys. Lett. **65** 1823 (1994)
5. M. Leszczynski, H. Teissevre, T. Suski, I. Grzergory, M. Bockowski, J. Jun, S. Porowski, K. Pakula, J.M. Baranowski, C.T. Foxon and T.S. Cheng, Appl. Phys. Lett. **69** 73 (1996)
6. M. Leszczynski, I. Grzergory, M. Bockowski, J. Jun, S. Porowski and J.M. Baranowski, Acta Phys. Polonica A **88** 799 (1995)
7. D.K. Bowen, N. Loxley and B.K. Tanner, Mater. Res. Soc. Symp. Proc. **208** 119 (1991)
8. S. Cockerton, M.L. Cooke, D.K. Bowen and B.K. Tanner, Mater. Res. Soc. Symp. Proc. (1996)
9. J.W. duMond Phys. Rev. **52** 872 (1937)
10. W.J. Bartels J. Vacuum Sci. Tech. **B1** 338 (1983)
11. N. Loxley, D.K. Bowen and B.K. Tanner, J. Appl. Cryst. **28** 314 (1995)
12. B.K. Tanner, in: Semiconductor Characterisation; Present Status and Future Needs eds. W.M. Bullis, D.G. Seiler and A.C. Diebold (Am. Inst. Phys., Woodbury, New York) p.263 (1995)

HIGH RESOLUTION X-RAY DIFFRACTION ANALYSIS OF GAN-BASED HETEROSTRUCTURES GROWN BY OMVPE

M.S. GOORSKY*, A.Y. POLYAKOV[†], M. SKOWRONSKI[†], M. SHIN[‡], and D.W. GREVE[‡]
*UCLA, Department of Materials Science and Engineering, Los Angeles, CA 90095-1595
[†]Carnegie Mellon University, Department of Materials Science and Engineering, Pittsburgh, PA 15213-3890
[‡]Carnegie Mellon University, Department of Electrical and Computer Engineering, Pittsburgh, PA 15213-3890

ABSTRACT

We demonstrate the use of triple axis diffraction measurements, including Φ scans (in which the sample is rotated about an axis perpendicular to its surface) to assess the crystal perfection of wurtzite GaN layers on sapphire grown using different pre-nitridation growth treatments by organometallic vapor phase epitaxy. The Φ scans determine the in-plane misorientation angles between the crystallites and hence provide information on the edge dislocation density. Using glancing incidence $(10\underline{1}4)$ and $(10\underline{1}5)$ reflections, we determined that the misorientation among the GaN crystallites decreases with increasing layer thickness and that the pre-nitridation conditions control the initial level of misorientation. Triple axis ω and ω-2θ scans around the (0002) reflection did not show a systematic trend with increasing layer thickness. However, layers grown without a pre-nitridation step tended to exhibit higher values of both mosaic spread and strain. The appropriate asymmetric reflections for GaN-based Φ scan measurements are determined using structure factor calculations, which are presented here.

INTRODUCTION

The characterization of GaN epitaxial layers is an extremely important component of understanding the roles which different growth steps play in controlling the presence of and types of defects in the layers. The lack of a lattice-matched substrate for GaN-based materials (including alloys with InN and AlN) results in high defect densities in GaN, AlGaN, and InGaN heterostructures grown on typical substrates such as sapphire or silicon carbide. Several growth optimization steps are typically employed to foster higher quality growth of GaN on these highly mismatched substrates. One standard procedure is to initiate growth using a low temperature GaN or AlN buffer layer [1,2] (typically deposited around 500 °C) which forms islands. Upon ramping to high temperatures (≈ 1100 °C), the islands coarsen and act as nucleation sites for subsequent two-dimensional growth.[3] In addition, exposing the sapphire substrate to ammonia prior to growth can strongly affect the types of dislocations that develop in subsequently grown material.[4] In nearly all cases, the GaN tends to grow in columnar structures of relatively high perfection material bounded by dislocation arrays. These sub-grains are typically on the order of 50-500 nm in diameter with an in-plane misorientation among the subgrains on the order of a few degrees.[5]

Triple axis diffraction techniques have proven to be very valuable in understanding the relationship between crystal growth and defect nucleation in GaN-based epitaxial films. For example, initial reports [6,7] showed that GaN layers exhibited a double axis rocking curve full width at half maximum (FWHM) of 37 or 30 arcsec, respectively, for the (0002) reflection, while others exhibited FWHM of greater than 200 arcsec in GaN used for device structures.[1] However, the dislocation densities were greater than 10^7 cm^{-2} in all cases. Heying, et al., [8] explained this discrepancy by showing that layers with narrow (0002) FWHM contain perfect edge

Mat. Res. Soc. Symp. Proc. Vol. 449 © 1997 Materials Research Society

dislocations (1/3<11$\underline{2}$0>) which lie along the (0001) growth direction. These dislocations do not introduce a tilt component that would cause mosaic broadening in a symmetric (0002) reflection. By using asymmetric reflections ((10$\underline{1}$2) in that case) which are broadened by the edge dislocations, they demonstrated that the nature of the dislocation network present in the GaN layers (pure edge vs. mixed edge and screw) can be determined.

In this paper, we demonstrate that other useful parameters can be gathered from triple axis diffraction measurements. First, we briefly examine the use of a few different asymmetric reflections for GaN characterization. Next, we address the issue of in plane sub-grain misorientation. Although misorientation has been measured using electron microscopy [5], we show that triple axis diffraction Φ scans - which employ asymmetric reflections - to non-destructively determine the range of in-plane misorientation among the GaN subgrains. Using these Φ scans, we assess the evolution of sub-grain misorientation as a function of GaN layer thickness and sapphire pre-nitridation time.

EXPERIMENT

The analysis of different asymmetric reflections for GaN grown on (0001) sapphire employed data from the International Tables for X-Ray Crystallography.[10] The GaN epitaxial layers on (0001) sapphire substrates were grown in a horizontal two-inlet OMVPE system at 76 torr, with TMGa, TMAl and ammonia as precursors and hydrogen as a carrier gas. GaN buffers were grown at 500 °C either without prior high temperature (1050 °C) nitridation of sapphire substrates or with 10 min. nitridation. After buffer growth, the substrate was ramped up to 1025 °C where the GaN layers of 1, 2, and 4 μm were grown with a growth rate of 2 μm/h. The triple axis diffraction measurements used a Bede D3 diffractometer with a four bounce (111) Si analyzer (third) crystal. Three distinct measurements were performed in the triple axis configuration (i.e., with the analyzer or third crystal rediffracting the x-ray beam diffracted from the sample). ω scans measure the variations in lattice tilt (step size = 20 arcsec) by rotating around the sample axis; $\Theta/2\Theta$ scans provide a measurement of the presence of strain in the crystal (step size = 2 arcsec); and Φ scans were performed by rotating the sample about an axis perpendicular to the sample surface (step size = 0.02°). In the triple axis configuration, rotating the sample around an axis which is perpendicular to the sample surface (Φ scan) while diffracting from an asymmetric reflection provides a measurement of sub-grains with a common mosaic tilt direction but different in-plane orientation. The instrumental broadening of a triple axis diffraction Φ scan due to vertical beam divergence was measured using the (115) reflection from an n$^+$ GaAs substrate with an etch pit density of < 100 cm^{-2}.

RESULTS

<u>Asymmetric reflections for hexagonal GaN</u>

A compilation of possible asymmetric reflections for hexagonal (0001) GaN epitaxial layers is generated to complement the much more well established database for cubic semiconductor crystals. Heying, et al.,[8] employed the (10$\underline{1}$2) reflection to determine the presence of the 1/3<11-20> edge dislocations and we tabulate the relevant parameters for this reflection as well as two other reflections [(10$\underline{1}$4) and (10$\underline{1}$5)] with the same zone axis. Table I   lists the reflection (with the (0002) reflection for comparison), the Bragg angle for the reflection using Cu kα radiation, the angle between the plane and the basal plane, the structure factor, and whether one can access the reflection with the plane of diffraction perpendicular to the zone axis for these reflections with either a glancing incidence (g.i.) or a glancing exit (g.e.) geometry or with the zone axis parallel to the plane of diffraction and with the sample rotated along the zone axis by the magnitude of the interplanar angle between the plane in question and the basal plane (skew).

Table I. Parameters for asymmetric reflections

Reflection	Θ_B (Cu Kα)	Interplanar angle w/ basal plane	F (structure factor)	skew/g.e./g.i
(0002)	17.278°	-	47.98	-
(10$\bar{1}$2)	24.044°	39.12°	21.02	skew
(10$\bar{1}$4)	41.013°	22.13°	13.71	all
(10$\bar{1}$5)	52.482°	18.02°	27.01	all

Note that the (0002) reflection possesses the strongest intensity (related to magnitude of the structure factor). Of the asymmetric reflections, the (10$\bar{1}$5) is the strongest. The (10$\bar{1}$2) reflection [8] represents the planes with the greatest interplanar angle with the basal plane. The (10$\bar{1}$2) possesses the greatest "a"-axis component of these reflections and thus the greatest sensitivity to tilts introduced by the pure edge dislocations. However, the use of this reflection is limited by the fact that the interplanar angle is greater than the Bragg angle. This limits the measurement of this reflection compared to the others listed here which can be accessed from any direction by a combination of skew and glancing incident / exit geometries. The (10$\bar{1}$2) reflection, on the other hand, can only be accessed by rotating the sample around an axis that is parallel to the plane of diffraction by $\approx 40°$. This rotation is beyond the range of many diffractometers, but the reflection can be accessed if the sample is placed on a 40° wedge prior to mounting on the diffractometer.

The data in Table I suggest that the (10$\bar{1}$5) is the most convenient for measuring the in-plane misorientation among the GaN subgrains given its high intensity and accessibility from both glancing and skew diffraction geometries. In this case, we accessed this reflection using the glancing incidence geometry (with the angle between the x-ray beam and the sample (0001) surface set to 52.48°-18.02° = 34.46°).

One possible limitation to measuring the in-plane misorientation with a high resolution diffractometer is that the x-ray beam is typically only collimated in one direction. For most diffractometers with the sample mounted vertically, there is much greater vertical divergence (on the order of a degree) than horizontal divergence (on the order of several arcsecs). To assess the broadening in the Φ scan due to the vertical divergence, we performed a Φ scan with the GaAs wafer. The (115) glancing incidence asymmetric reflection was chosen as the angle between the x-ray beam and the sample surface (29.28°) and detector position (90.1°) were similar to those values for the (10$\bar{1}$5) GaN reflection (34.45° and 105.0°, respectively). The Φ FWHM of the GaAs (115) reflection was 0.12°.

Triple Axis Diffraction Measurements of GaN Epitaxial Layers

Figure 1 shows the FWHM of the ω scan measurements from the (0002) reflection for the two series of samples described in the experiment section. The samples that did not undergo a pre-nitridation step possessed a significantly more narrow peak than the samples which had undergone the pre-nitridation step. These results agree with the results of Keller, et al., [4] who demonstrated that longer nitridation times produce layers with more narrow ω FWHM which was attributed to an increase in the pure edge dislocation density and a decrease in the mixed edge/screw dislocations. For comparable GaN layer thicknesses, their FWHM is about a factor of three more narrow, however. In these samples, there is not a systematic FWHM reduction with increasing layer thickness as had been observed in previous studies.[10]

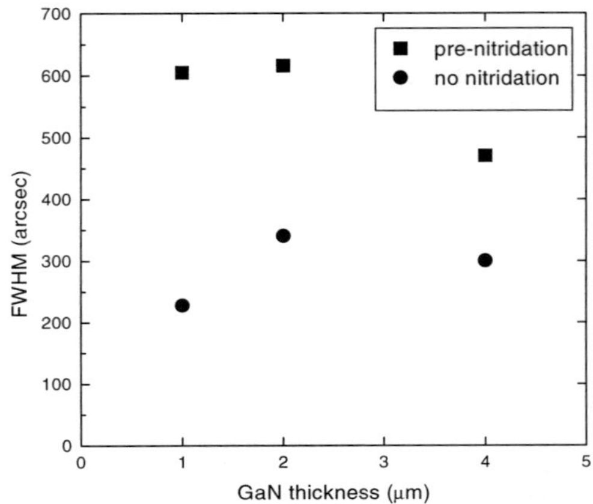

Figure 1. (0002) ω scan FWHM as a function of layer thickness.

Figure 2 includes the Θ/2Θ FWHM from the GaN samples. In this case, there is a pronounced decrease in the FWHM for the 4 μm sample. Except for the thinnest (1 μm) layers, the FWHM are comparable for the different pre-nitridation treatment. The overall decrease in FWHM most likely stems from the thickness of the sub-grains that act to coherently diffract the x-rays. Based on dynamical diffraction theory, a FWHM of 25 arcsec represents a layer thickness of about 0.7 μm and a FWHM of 14 arcsec represents a layer thickness of 2-3 μm. The thickness range for the latter value stems from the nearly intrinsic FWHM ($\approx$ 12 arcsec for (111) Si which is at a similar Bragg angle) for the thickest samples and demonstrates that there is negligible strain-induced broadening in the thickest of these samples. The similar FWHM for a given thickness suggests that the average thickness of the diffracting subgrains is similar in both cases. We are investigating the anomalous increase in FWHM for the samples without nitridation as the layer thickness increases from 1 to 2 μm.

Figure 3 shows the Φ scan FWHM around the (10$\underline{1}$5) reflection for the same samples. In this case, there is a clear trend of decreasing FWHM as the layer thickness increases for both sets of samples. The layers that were exposed to the pre-nitridation step exhibit a much greater in-plane sub-grain misorientation than the layers that were not subjected to a pre-growth nitridation step. The decrease in misorientation with increased thickness most likely stems from the growth of grains which are aligned with each other at the expense of those which show a greater mis-alignment with respect to the average value. The TAD Φ scan results are supported by atomic force microscopy (AFM) measurements of the GaN layers. AFM shows that the layers grown without a pre-nitridation step display a step-flow growth mode, whereas on pre-nitrided samples,

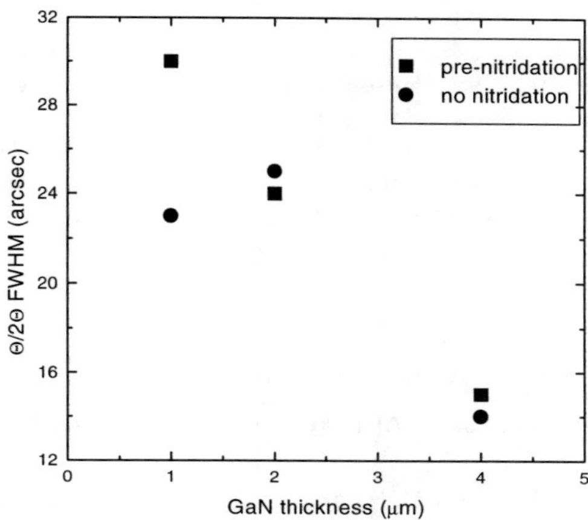

Figure 2. (0002) Θ/2Θ FWHM as a function of layer thickness.

a high concentration of relatively large "holes" of 0.1-1 μm characteristic dimensions exists.[11] These features in the pre-nitrided samples may be evidence of highly misoriented isolated islands which do not coalesce significantly during growth and hence exhibit a high degree of in-plane misorientation.

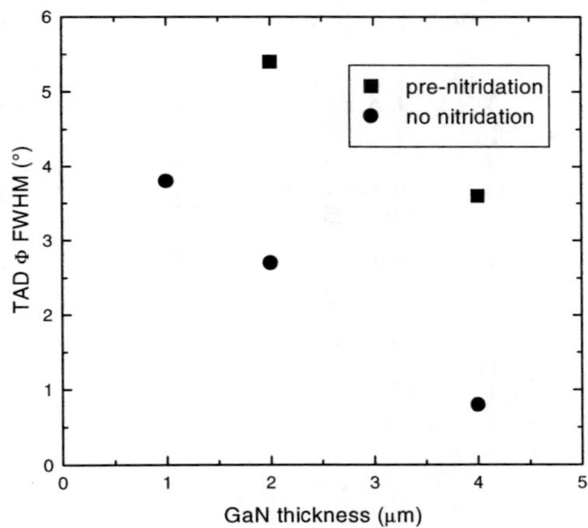

Figure 3. (10$\underline{1}$5) TAD Φ scans as a function of layer thickness.

CONCLUSIONS

We have shown that triple axis x-ray diffraction measurements, in particular Φ scans around an asymmetric reflection, provides useful information about the evolution of microstructure in GaN films. In combination with ω and $\Theta/2\Theta$ symmetric scans, the Φ scans can be used to provide useful non-destructive evaluation of GaN-based heterostructures. The relative merits of different asymmetric reflections are described and the high intensity and relatively easy access to the (10$\underline{1}$5) reflection suggests that it may be very useful for high resolution x-ray diffraction measurements of GaN.

REFERENCES

[1]. S. Nakamura, Jpn. J. Appl. Phys. **30** L1705 (1991)

[2]. K. Hiramatsu, S. Itoh, H. Amano, I. Akasaki, N. Kuwano, T. Shiraishi, and K. Oki, J. Cryst. Growth **115** 628 (1991).

[3] X.H. Wu, D. Kapolnek, E.J. Tarsa, B. Heying, S. Keller, B.P. Keller, U.K. Mishra, S.P. DenBaars, and J.S. Speck, Appl. Phys. Lett. **68** 1371 (1996).

[4]. S. Keller, B.P. Keller, Y.-F. Wu, B. Heying, D. Kapolnek, J.S. Speck, U.K. Mishra, and S.P. DenBaars, Appl. Phys. Lett. **68** 1525 (1996).

[5] W. Qian, M. Skowronski, M. De Graef, K. Doverspike, L.B. Rowland, and D.K. Gaskill, Appl. Phys. Lett. **66** 1252 (1995).

[6] K.G. Fertitta, A.L. Holmes, J.G. Neff, F.J. Ciuba, and R.D. Dupuis, Appl. Phys. Lett. **65** 1823 (1994)

[7] P. Kung, A. Saxler, X. Zhang, D. Walker, T.C. Wang, I. Ferguson, and M. Razeghi, Appl. Phys. Lett. **66** 2958 (1995))

[8] B. Heying, X.H. Wu, S. Keller, Y. Li., D. Kapolnek, B.P. Keller, S.P. DenBaars, and J.S. Speck, Appl. Phys. Lett. **68** 643 (1996).

[9] International Tables for X-Ray Crystallography, C.H. MacGillavry, G.D. Rieck, and K. Lonsdale, eds. Kynoch Press, Bimingham, England 1962 Vol.III

[10] S.D. Hersee, J. Ramer, K. Zheng, C. Kranenberg, K. Malloy, M. Banas, and M. Goorsky, J. Electron. Mater. **24** 1519 (1995).

[11] A. Polyakov, unpublished.

Part V

Electronic Properties

FREE AND BOUND EXCITONS IN GaN EPITAXIAL FILMS

B.K. MEYER

I. Physics Institute, Justus Liebig University Giessen, D-35392 Giessen, Germany

ABSTRACT

We report on photoluminescence experiments on hexagonal GaN epitaxial films grown on 6H-SiC and sapphire substrates by organo-metallic and hydide vapor phase epitaxy. At low temperatures we observe free and neutral donor bound exciton transitions which allow to establish properties of the free excitons and localisation energies of the bound excitons involving different shallow donors. From temperature dependent luminescence experiments thermal activation energies are determined which measure the exciton localization and donor binding energies. The localization energies of the excitons scale with the respective donor binding energies (Haynes rule). The inter-impurity transitions of neutral donors are observed in fourier transform infra-red absorption. Three shallow donors with binding energies of 34.7 meV, 55 meV and 58 meV can be seen. We present evidence for the chemical nature of the shallow impurities.

INTRODUCTION

The technological importance of a semiconductor depends on its ability of controlled, bipolar doping. It is therefore not astonishing that the III-V semiconductor GaN has gained increased interest in the recent years when a major breakthrough was obtained by achieving p-type conduction using Mg as a shallow acceptor dopant [1]. GaN based light emitting devices have been realized and are now commercially available. Despite this tremendous technological success there are still fundamental and important materials parameters which have to be addressed: *i*) the strain modification of the valence bands, *ii*) the Kohn-Luttinger parameters of the valence bands (i.e. the hole masses and g-values) and *iii*) determination of the binding energies of shallow n-type dopants only to mention some. Undoped GaN films are commonly n-type conductive. The free carrier concentrations can vary over a wide range from 10^{17} up to 10^{19} cm^{-3} [1]. There is an ongoing discussion which defects either of intrinsic or extrinsic origin are the source of the high free carrier concentrations. Some theoretical calculations favor intrinsic defects [2] whereas others [3] present evidence for extrinsic impurites being the source of n-type conductivity. Silicon (Si) and oxygen (O) could be prime candidates, and indeed intentional doping of GaN can easily be realized by incorporating Si on Gallium sites. However conclusive data on the binding energies of the extrinsic shallow donors are still not available. Götz et al [4] and Wickenden et al [5] reported on Hall effect measurements on Si doped GaN films. Two donor electron activation energies are found ranging from 12 to 17 meV and 32 to 37 meV. Götz et al [4] attribute the donor with the lower activation energy to Si and results from secondary ion mass spectroscopy seem to support this conclusion.

In high quality undoped epitaxial films the free carrier concentrations are well below 10^{17} cm^{-3} and with the help of optical spectroscopy (photoluminescence (PL), reflectance etc) free and

Mat. Res. Soc. Symp. Proc. Vol. 449 © 1997 Materials Research Society

defect bound excitons could be separated. In the wurtzite crystal structure we deal with three valence bands. The free exciton binding energy with the A-valence band has been determined with high precision [6-11], but there is still no general consensus what the respective values for the B- and C-excitons are. The analysis is complicated due to the fact that the energetic positions of the PL lines are severely influenced by residual strain in the layers. By temperature and time resolved PL experiments free and bound exciton recombinations are distinguishable and precise values for the localisation energies of bound excitons were obtained [6,11]. The localisation energy is measured with respect to the free exciton line position and in many semiconductors there is a constant ratio between the localisation energy at a donor or accceptor and the respective binding energies of the donor or acceptor (Haynes rule) [12]. At present it is not established whether Haynes rule for neutral donor bound excitons is valid in GaN. It holds for neutral acceptor bound excitons and Kaufmann et al reported that the proportionality constant is close to 0.1 [7].

In the following report we use high resolution photoluminescence (PL) to distinguish between different free and neutral donor bound exciton lines. From the luminescence intensity as a function of temperature we obtain activation energies which measure free exciton binding energies, exciton localisation as well as donor binding energies. We compare the PL results with more FTIR absorption measurements on a 400 μm thick substrate free GaN film, where the interimpurity transitions of the shallow donors are seen. A tentative assignement as to the chemical nature of the donors is given.

EXPERIMENTAL DETAILS

The samples used were grown by Hydride Vapor Phase Epitxy (HVPE) (film thickness 400 μm without buffer layer) and by Organo-Metallic VPE (OMVPE) (thickness 3 μm, on a 35 nm AlN buffer) on sapphire or 6H-SiC substrates. For the HVPE film the substrate was removed. The luminescence was excited with the 325 nm line of the HeCd laser, the sample was placed in a temperature variable cryostat. The temperature could be varied in the range from 1.5 K to room temperature. A high resolution monochromator connected to an especially selected photocounting photomultiplier served for spectral resolution better than 1 meV at 3.52 eV. The FTIR experiments employed a BRUKER IFS 113v spectrometer at variable temperatures between 10 and 150 K. A liquid He cooled bolometer served for enhanced signal to noise ratios in the far infrared range (180 to 700 cm^{-1}).

EXPERIMENTAL RESULTS

In Fig.1 we show the luminescence of a 3 μm thick GaN film grown on 6H-SiC at different temperatures. At 7 K the luminescence is dominated by the neutral donor bound exciton line at 3.460 eV. At higher energies free exciton transitions involving the A-valence band (3.467 eV) and the C-valence band (3.4865 eV) are seen (see also below). However, excited states of the A-exciton are also likely to occur in the same energy range which make the analysis not straightforward. The B-exciton is seen only at higher temperatures in the high energy flank of the A-exciton. The transition at 3.4827 eV arises from the A-exciton n=2 excited state. The donor bound exciton is 6.2 meV below the A-exciton. The transition at 3.4378 eV is attributed to a neutral acceptor bound exciton. Its localisation energy is 30±1 meV which is in the range reported for excitons bound to the neutral acceptor Zinc [7]. Note that the positions are red-shifted compared to the homoepitaxial GaN (see Fig.2) which is attributed to the tensile component of the strain in the layer (blue shift for GaN on sapphire, see Fig.2). The free A-exciton and D^{O}X transitions are repeated at 92 meV lower in energy (LO phonon replica). The electron phonon coupling for both recombinations is obviously different since for the one LO-replica the intensity ratio has reversed (for details see ref. [11]).

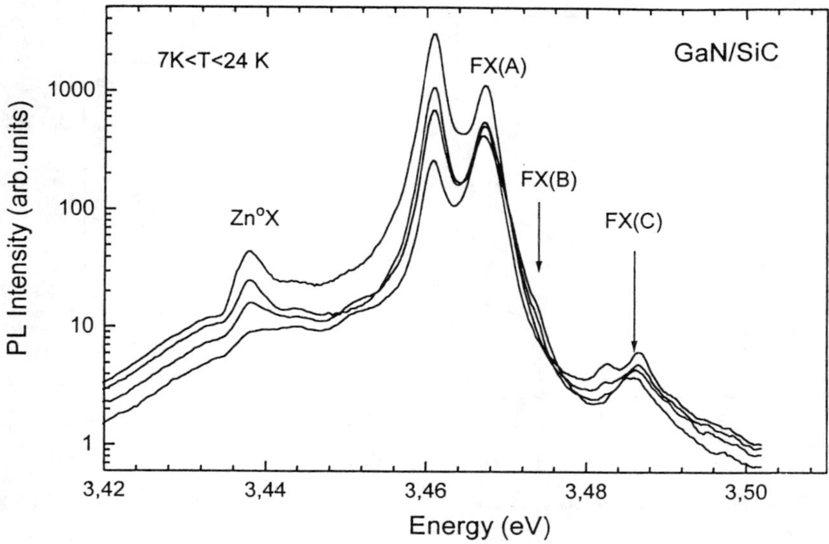

Fig.1: Luminescence spectrum of GaN grown by OMVPE on 6H-SiC measured at different temperatures. Free and bound exciton transitons are observed.

The assignment of the lines becomes clear when comparing with a measurement at elevated temperatures. The free exciton binding energy is of the order of 25 meV, hence free exciton luminescence can be observed up to room temperature. Bound excitons have only small localisation energies and rapidly loose intensity with increasing temperatures. This effect is most pronounced for the acceptor bound exciton at 3.4378 eV. The free exciton as well as its phonon replicas shows considerable broadening on the high energy site of the line with a Maxwellian distribution [11]. Note that donor bound excitons show a symmetrical broadening of the line with increasing temperature. A detailed analysis of the free exciton line shape can be found in ref.[11].

In order to clarify the assignment of the free exciton transitions we plot the transition energies of the B- and C-excitons versus the transition energy of the free A-exciton (Fig.2). For completeness we also include the data from refs.[7,10]. One notes that the B-exciton approaches the A-exciton for films on 6H-SiC which are under tensile stress. Data point drawn with full circles correspond to the n=2 excited state of the free A-exciton. From its energetic distance which is 3/4 of the effective Rydberg its binding energy can be calculated. It is 26 meV for compressively strained films, but decreases to 21 meV for films under tensile strain. We think this is related to the strain modification of the valence bands. More details will be given elsewhere [13].

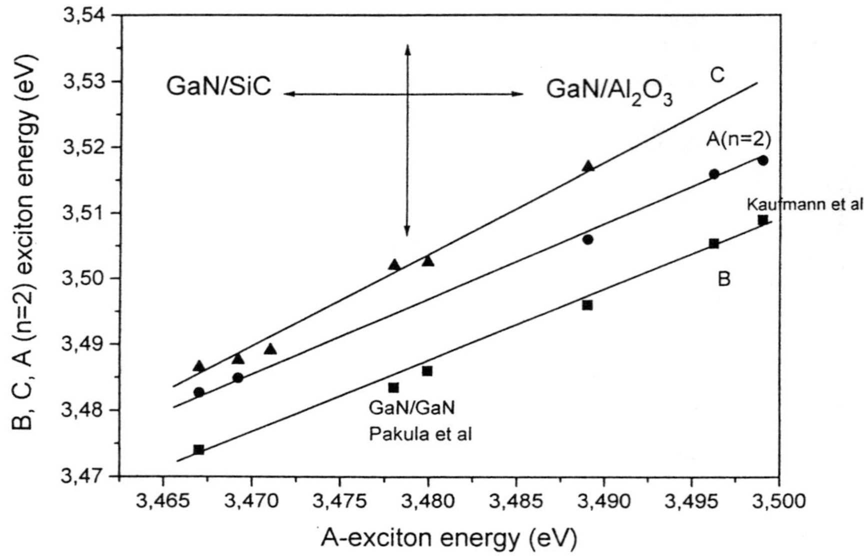

Fig.2: Transition energies of the A,B and A(n=2) excitons as a function of the A-exciton transition energy.

The HVPE grown film on sapphire had a layer thickness of 400 μm and hence corresponds to the strain free case. The exciton line width was as narrow as 0.8 meV at 1.5 K and made it possible to resolve even more details on the free exciton recombinations. Based on a comparison with calorimetric absorption and reflection measurements the transitions of the A-, B- and C-excitons were attributed (for details see refs.[14]). However, the origin of the some of the lines remained unclear. Under high resolution and using polarisation spectroscopy additonal features become visible (see Fig.3). In the following we will only address the localisation energies of the neutral donor bound excitons. As seen in Fig.3 the dominant transition occurs at 7-8 meV below the free A-exciton. It is a doublet with an energy separation of 0.8 meV. We think that this splitting originates from the remaining strain in the film otherwise it would indicate the presence of two shallow donors. There is an additional transition more pronounced for the polarisation vector parallel to the c-axis (Fig.3.b). It is 3.5-4.5 meV below the free A-exciton and mentioned also in a report of Merz et al [15]. From its behaviour upon temperature and electrical field we presented strong evidence that it is a donor bound exciton [6]. With respect to the B-exciton at 3.484 eV it has the same localisation energy i.e. between 7 and 8 meV. We therefore share the conclusion reached in ref. [10] that it is the same donor bound exciton but bound to the B-valence band. A third transition appears at 3.4673 eV and has an energy separation with respect to the A-exciton of 12.1±0.2 meV. It is assigned by others to an acceptor bound exciton [6]. We believe however as outlined below it also due to a neutral donor bound exciton.

The bound exciton lines rapidly decrease in intensity whereas the free exciton emission is observable up to room temperature. This behaviour reflects that the bound exciton lines have small localisation energies. It was possible to follow the intensity decrease of the principal D^oX line in a selected sample over six decades (see Fig.4). We analysed the data using the following expression:

$$I(0)/I(K)= 1/(1+c_1\exp(-\Delta E_1/kT) + c_2\exp(-\Delta E_2/kT)) \tag{1}$$

where c_1 and c_2 are prefactors and ΔE_1 and ΔE_2 are the activation energies, respectively.

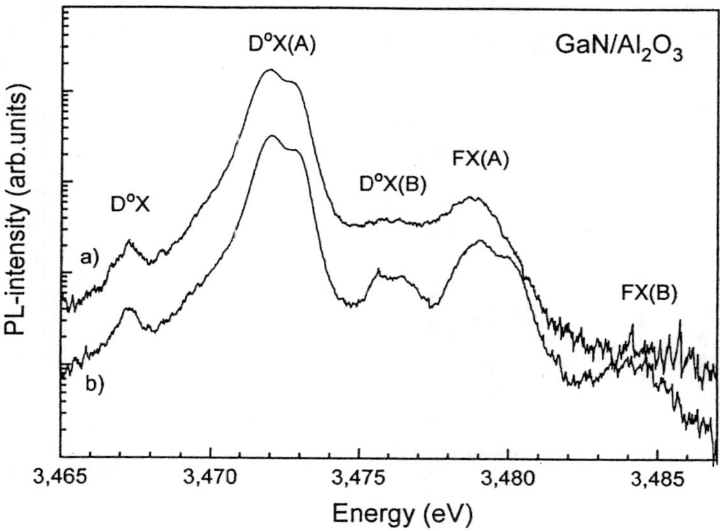

Fig.3: Luminescence spectrum of a 400 μm thick HVPE GaN film on sapphire showing free exciton recombinations with the A- and B-valence bands. Additional fine structure is due to donor bound exciton states.

The thermally activated dissociation of bound excitons should involve two activation energies. By above band gap excitation we create free excitons. They try to localize at the neutral donors (localisation energy). At elevated temperatures we have an interplay between the localisation at and ionization of the neutral donor, reducing the number of available neutral donors therefore also the donor binding energy should be involved.

The two activation energies which govern the thermal decay of the bound exciton were: 4.5±1 meV and 32±2 meV using equ.1. For a precise evaluation of the activation energy in the high temperature range it was essential that the intensity decease was over six orders of magnitude in a rather narrow temperature range. Within this narrow temperature interval the density of states in the conduction band does not vary significantly and hence allows for a precise estimate of the donor binding energy. For the deeper bound exciton which at low temperatures was in intensity lower by a factor of fifty only a measure of the localisation energy was possible. We obtained 7.5 ±1 meV.

The shallow defects also give rise to Rydberg excited states commonly seen in far-infrared conductivity or absorption experiments. From the transition series the ground state binding energy can be calculated within the effective mass theory (EMT) provided the electron effective mass m* is small, the dielectric constant is large and we are dealing with a direct band gap semiconductor. We have recently reported on fourier transform infra-red (FTIR) absorption [16] on GaN thick films where a single absorption line of one residual donor was observed. We assigned it to the 1s to 2p transition and derived a binding energy of 35.6 meV. There is an additional report that residual donors have binding energies of 54 and 57 meV based on the analysis of distant donor acceptor pair recombination lines [7].

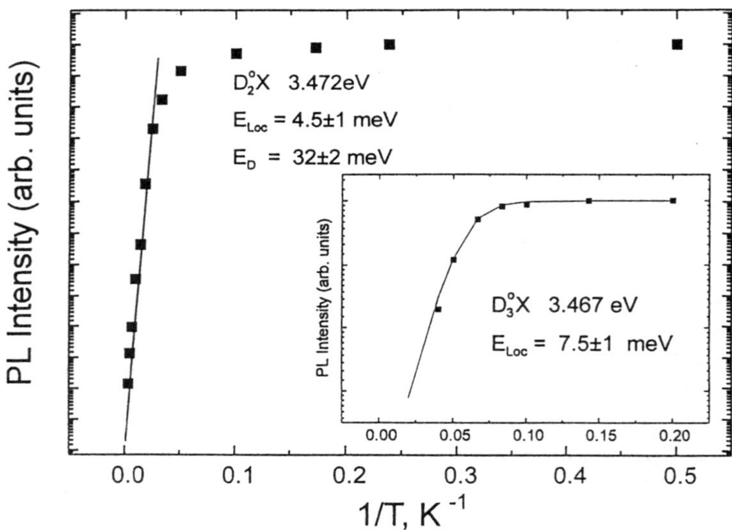

Fig.4: Arrhenius plot for the intensity decrease of the donor bound exciton of an OMVPE grown film GaN on sapphire and a fit with two activation energies. The inset shows a similar analysis for a deeper bound exciton (for details see text).

On a substrate-free HVPE GaN film on which also the PL results have been obtained we performed FTIR absorption experiments in the temperature range between 10 and 90 K. We identify five temperature dependent and hence electronic absorption bands at 209 cm^{-1}, between 240 and 280 cm^{-1}, at 333 and 351 cm^{-1} and a hot line at 422 cm^{-1} (see Fig.5). The line at 209 cm^{-1} has been reported before [16] in experiments on similar films however with the sapphire substrate on. This prominent line is now at slightly lower energies, the shift of 6 wavenumbers might be caused by the release of strain due to the removal of the substrate. The second broader band around 250 cm^{-1} is lower in intensity, the small modulations on top of it are due to interference fringes. From an analysis of the data using the effective mass approach and attributing the line at 209 cm^{-1} with the 1s ground state to 2p excited state transition of a shallow donor we obtained from the series limit an binding energy of 34.7 meV. Hence the 1s to 3p transition should be at 8/9 * 34.7 meV (247.7 cm^{-1}) and the 1s to 4p is expected to show

502

up at 261 cm^{-1}. This range of transition energies is marked in Fig.5 with an arrow. The transitions to higher excited states are expected to be considerably lower in intensity and the broad absorption band between 240 and 280 cm^{-1} can be understood as the overlap of these transitions.

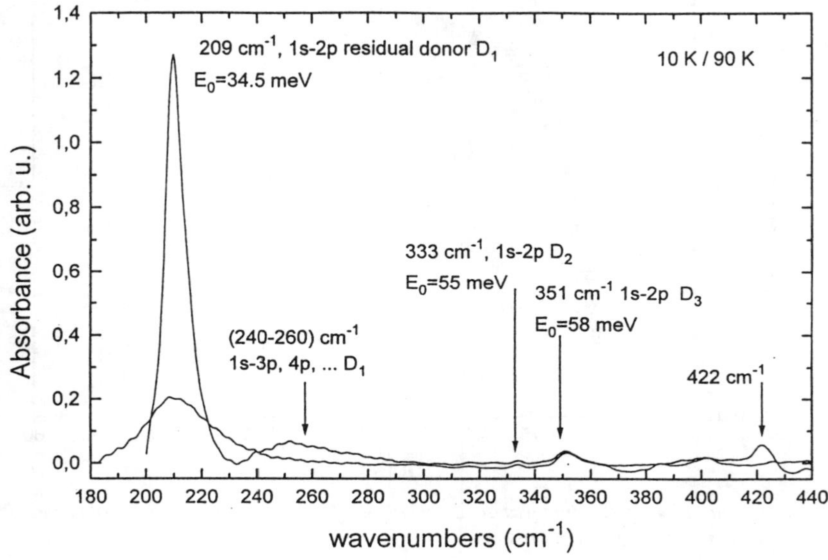

Fig.5: *Fourier transform infrared absorption spectrum of a free standing GaN film grown by HVPE at two different temperatures. Transition energies and assignments are given. Note the hot line at 422 cm^{-1}.*

We observe two additional lines at 333 cm^{-1} and 351 cm^{-1} which are also of electronic origin as demonstrated by temperature dependent measurements. Using the same approach, i.e. attribute them to 1s to 2p transitions the corresponding binding energies come out with 55 and 58 meV. This is in remarkable agreement with the binding energies deduced from luminescence experiments by Kaufmann et al. [7] (54 and 57 meV) and also serves for an independent support of the analysis. Comparing the integrated areas of the absorption bands and assuming equal oscillator strength the concentrations of the deeper donors are lower by one (58 meV) and two (55 meV) orders of magnitude. These are however only rough estimates since the oscillator strengths for hydogenic-like transitions of a EMT donor in GaN are not known. The concentration of neutral donors also depends on temperature. At low temperatures the electrons are frozen in. With increasing temperatures due to ionisation the concentration of neutral donors is reduced and hence also the corresponding absorption bands decrease. This process is thermally activated, the activation energy is a measure of the binding energy. In Fig.6 we show the change in absorbance of the 209 cm^{-1} line as a function of the reciprocal temperature (Arrhenius plot). It was possible to follow the signal decrease over two orders of magnitude. In the high temperature range a single activation energy of 33±1 meV gave the best fit (see Fig.6) and this value is in very good agreement with the results from photoluminescence. The transitions at 333 and 351 cm^{-1} were too weak to allow for such an analysis, also transitions to higher excited states could not be observed.

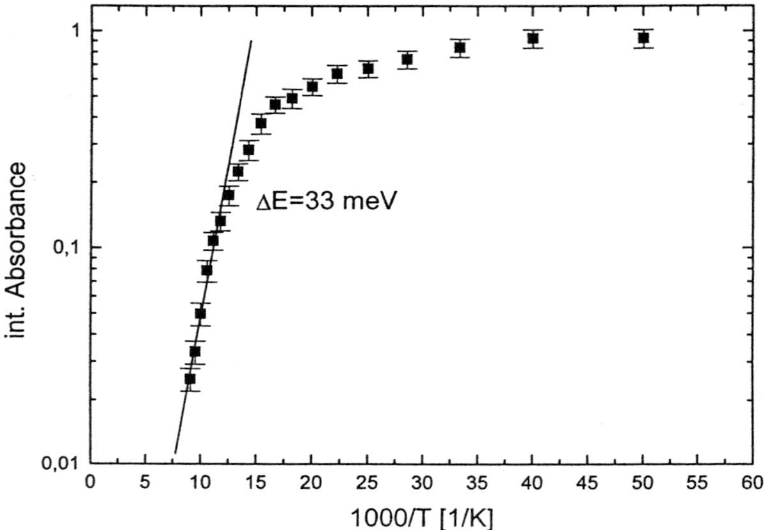

Fig.6: Arrhenius plot of the integrated absorbance of the 209 cm^{-1} line as obtained from fourier transform infrared measurements.

An estimate of the shallow donor binding energy E_D can also be made by using the effective mass theory provided the electron effective mass is known with high precision. We have performed cyclotron resonance experiments [17] on a high quality GaN epitaxial film and obtained a precision value for the electron effective mass (polaron mass), m*=0.22±0.005 m$_0$.

Using 9.7 for the dielectric constant we calculate for E_D=13.58 m*/ε^2 (eV) = 31.7 meV.

Unexpected was the behavior of the transition at 422 cm^{-1} (see Fig.5). It is obviously a hot line, since it is absent a low temperatures and gains in intensity with increasing temperature. A model to explain this behavior could be a split ground state known to occur for shallow donors in the indirect semiconductors Si and SiC (valley-orbit splitting). In such cases in addition to the transitions from the lowest ground state (cold line) hot lines appear at higher temperatures when the first excited state of the ground state manifold is thermally populated. The activation energy involved is the valley-orbit splitting. We analysed the data within a two level system for the ground state in order to get an estimate for the activation energy. It turned out to be 100±5 meV (=806 cm^{-1}). Thus the corresponding cold line should appear at around 1230 cm^{-1} but no transition in this energy range could be detected. At present no explanation can be offered nor are there any further hints as to the origin of this line.

DISCUSSION

Although Si is used as an efficient n-type dopant its binding energy is still not established. In undoped und Si doped GaN films Hall effect measurments derived with the conclusions that two donors are electrically active: the one with a binding energy around 17 meV is attributed to Si whereas for the second one with a binding energy around 32 meV no identification was possible. The conclusions from Götz et al [4] and Wickenden et al [5] are, however, in conflict with recent magneto-optical studies on Si doped GaN [18]. Wang et al [18] monitored the behavior of the 1s to $2p^+$ transition in the magnetic field range from 15 to 27 T. From a linear fit they deduced a transition energy at zero magnetic field of 21.7 meV (3/4 of E_D) and hence calculated for the binding energy of Si a value of 29 meV. This experiment certainly gives at present the best value for the Si donor binding energy. It remains to be investigated why the Hall data are not consistent with the magneto-optical data.

Merz et al [7] reported on the neutral acceptor bound exciton lines in Mg and Zn doped GaN. They concluded that there is a proportionality between the localisation energy and the respective binding energy (Haynes rule). The proportionality constant is close to 0.1. Dependent on the film investigated the localisation energy of the dominant donor bound excitons is between 6-7 meV. On one sample from the infrared measurements a binding energy of 34.7 meV was obtained, whereas from the temperature dependent luminescence measurements an activation energy of 32 meV resulted. Thus the proportionality constant between the localisation and the binding energy is 0.2±0.01 (see Fig.7).

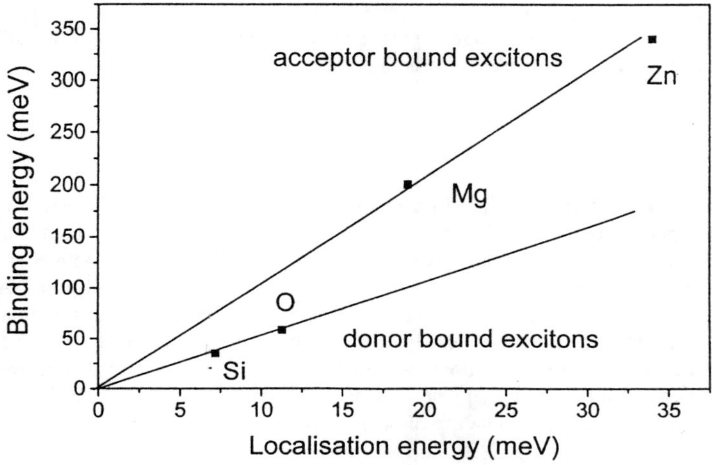

Fig.7: Localisation energy as a function of binding energy for acceptor and donor bound excitons (Haynes rule).

The exciton line at 12.1 meV below the free exciton line is according to ref. [10] an acceptor bound exciton. Additional experimental facts to support this identification were not presented. Using Haynes rule for the acceptors the acceptor binding energy would be 120±10 meV. Such a shallow acceptor would certainly be of great benefit for GaN and an hopefully an efficient p-type dopant. There is currently no experimental evidence for such a shallow acceptor level. In terms of an donor bound exciton from Haynes rule one obtains 59 ± 1 meV which compares very well with the value from the FTIR experiments of 58 meV.

To summarize, the PL and FTIR experiments show consistency, and give strong evidence for the following facts: *i*) Haynes rule is certainly valid for donors in GaN *ii*) the dominant shallow effective mass type donors range between 32 and 35 meV and *iii*) deeper donors are also present in the films.

It is difficult to make a decisive identification as to the chemical nature of the donors. However we try on the knowledge we have from other III-V semiconductors to make reasonable guesses. Substituting for Gallium sites and being donors are possibly the elements from the fourth group of the periodic table, i.e. C, Si, Ge etc. Carbon is believed to act as an acceptor, and some experiments indeed provide evidence for p-type conduction due to Carbon [19]. Si is a very effective n-type dopant in GaN so a prime candidate. On Nitrogen sites the group VII elements can act as donors i.e. O, S, Se, Te etc. GaN films have high oxygen concentrations but the role of oxygen in other III-V semiconductors such as GaAs and GaP is instead of introducing a shallow level oxygen acts as a deep recombination center. Due to the more ionic bond this might not be true in GaN. Oxygen doping experiments concluded that O in GaN introduces a deeper donor state with a binding energy of around 80 meV. HVPE uses quartz reactors and the reactivity of NH_3 and H-radicals can easily introduce Si into the films. In OMVPE films using AlN buffer layers O and Si (the high affinity of oxygen to Al is known and Si is a contaminant of Al-related metal-organica) are most likely to be incorporated. In the absence of controlled doping experiments and without having a clear fingerprint of the residual donors (e.g. local modes and isotope effects) a decisive identification can not be offered, but still the best guess for the 34 meV donor is Si_{Ga}.

CONCLUSIONS

We reported on temperature dependent photoluminescence and fourier transform infrared absorption experiments which allowed to identify free and bound exciton recombinations and the interimpurity transitions of residual shallow donors. From the thermally activated decay we obtain activation energies which are in very close agreement with the localization energies of the excitons and are a measure the donor binding energies. Haynes rule for donor bound excitons is valid in GaN and gives for the proportionality constant between localization and binding energy of the respective donor $\alpha=0.20\pm0.01$. The data show a convergent trend towards established properities of bound excitons in other III-V semiconductors, but are still not conclusive as to the chemical nature of the residual donors in GaN.

ACKNOWLEDGMENT: We thank the DFG for financial support. We express our special thanks to H. Hardtdegen, Jülich for enlightening discussions. The author is thankful to D. Kovalev, D.Volm, D.M. Hofmann, A. Hoffmann, L. Eckey and U. Kaufmann for stimulating discussions.

REFERENCES

[1] S. Strite and H. Morkoc, J. Vac. Sc. Technol. B 10, 1237 (1992)

[2] P. Boguslawski, E. Briggs, T.A. White, M.G. Wensell, and J. Bernholc, Mat. Res. Soc. Symp. Proc. 339, 693 (1994); P. Boguslawski, E. Briggs and J. Bernholc, Phys. Rev. B 51, 17255 (1995)

[3] J. Neugebauer and C.G. Van de Walle, Phys. Rev. B 50, 8067 (1994)

[4] W. Götz, N.M. Johnson, C. Chen, C. Kuo and W. Imler, Appl. Phys. Lett. 68, 3144 (1996)

[5] A.E. Wickenden, D.K. Gaskill, D.D. Koleske, K. Doverspike, D.S. Simons and P.H. Chi, Mat. Res. Proc. Vol. 395, 679 (1996)

[6] D. Volm, K. Oettinger, T. Streib, M. Ben-Chorin, J. Diener, B.K. Meyer, J. Majweski, L. Eckey, A. Hoffmann, H. Amano, I. Akasaki, K. Hiramatsu and T. Detchprohm, Phys. Rev. B 53, 16543 (1996)

[7] C. Merz, M. Kunzer, U. Kaufmann, H. Amano and I. Akasaki, Semicond Science Technol. 11, 712 (1996); U. Kaufmann, M. Kunzer, C. Merz, I. Akasaki and H. Amano, Mat. Res. Proc. Vol. 395, 633 (1996)

[8] B. Gil, O. Briot and R.-L. Aulombard, Phys. Rev. B 52, 17028 (1995)

[9] S. Chichibu, A. Shikanai, T. Azuhata, T. Sota, A. Kuramata, K. Horino and S. Nakamura, Appl. Phys. Lett. 68, 3766 (1996)

[10] K. Pakula, A. Wysmolek, K.P. Korona, J.M. Baranowski, R. Stepniewski, I. Grzegory, M. Bockowski, J. Jun, S. Krukowski, M. Wroblewski and S. Porowski, Solid State Commun. 97, 919 (1996)

[11] D. Kovalev, B. Averboukh, D. Volm, B.K. Meyer, H. Amano and I. Akasaki, Phys. Rev. B 54, 2518 (1996)

[12] J. R. Haynes, Phys. Rev. Lett. 4, 351 (1960)

[13] B.K. Meyer, D. Kovalev (unpublished)

[14] L. Eckey, I. Podlowski, A. Göldner, A. Hoffmann, I. Broser, B.K. Meyer, D. Volm, T. Streibl, K. Hiramatsu, T. Detchprohm, H. Amano and I. Akasaki, Inst. Phys. Conf. Series, 142, 943 (1996)

[15] C. Merz, M. Kunzer, B. Santic, U. Kaufmann, I. Akasaki and H. Amano, Proc. of the E-MRS, Strasbourgh, France, 1996 (in press)

[16] B.K. Meyer, D. Volm, A. Graber, H.C. Alt, T. Detchprohm, H. Amano and I. Akasaki, Solid State Commun. 95, 597 (1995)

[17] M. Drechsler, B.K. Meyer, D. Detchprohm, H. Amano and I. Akasaki, Jap. J. Appl. Phys. L1296 (1995)

[18] Y.J. Wang, H.K. Ng, K. Doverspike, D.K. Gaskill, T. Ikedo, I. Akasaki and H. Amano, J. Appl. Phys. 79, 8007 (1996)

[19] S.J. Pearton, C.R. Abernathy and F. Ren, Electronic Letters, 30, 527 (1994)

CHARACTERIZATION OF OMVPE-GROWN AlGaInN HETEROSTRUCTURES

D. P. Bour, H. F. Chung, W. Götz*, L. Romano, B. S. Krusor, D. Hofstetter, S. Rudaz*, C. P. Kuo*, F. A. Ponce, N. M. Johnson, M. G. Craford*, and R. D. Bringans
XEROX Palo Alto Research Center, Electronic Materials Laboratory, 3333 Coyote Hill Road, Palo Alto, CA 94304, bour@parc.xerox.com
*Hewlett-Packard Optoelectronics Divison, 370 West Trimble Road, San Jose, CA 95131

ABSTRACT

We report on the OMVPE growth and characterization of AlGaInN and its heterostructures, including measurements of electrical properties (Hall), optical properties (photo- and cathodo- luminescence), structural characteristics (x-ray diffraction and TEM); and also the emission of InGaN/AlGaN heterostructures subject to optical and electrical pumping.

INTRODUCTION

Gallium nitride and its alloys AlGaInN are being developed for visible and UV light emitters, detectors, and high-temperature electronic devices. High quality GaN films and AlGaInN heterostructures have been obtained by epitaxial deposition on sapphire (Al_2O_3) substrates, leading to efficient blue and green light emitting diodes; and more recently they have been incorporated into heterojunction laser diodes (1-8). Here we describe our efforts toward the OMVPE growth and characterization of AlGaInN and its heterostructures, including measurements of electrical properties (Hall), optical properties (photo- and cathodo-luminescence), and structural characteristics (x-ray diffraction and TEM); and we have also examined the emission of InGaN/AlGaN heterostructures subject to optical and electrical pumping.

EXPERIMENT

OMVPE Growth and Properties of GaN

Samples were prepared by OMVPE, in a manner similar to that described in the literature (1,2). Our best undoped GaN films, of thickness ~ 4 μm, had an n-type background of mid- 10^{16} cm^{-3}, and a room temperature electron mobility of 626 cm^2/V-sec. This is shown in Figure 1, along with the temperature dependence of the electron mobility. The peak mobility is 1473 cm^2/V-sec, at a temperature of 130 K.

The angular spread of the x-ray diffraction (XRD) peak of such films was typically 4 - 6 arc-minutes (FWHM). However, both the XRD width and the electron mobility were inferior in thinner films ($\leq$ 1 μm), which is most likely a consequence of the highly dislocated "faulted zone" at the sapphire/GaN interface. The defect structure of GaN films has also been investigated by transmission electron microscopy (TEM). These examinations indicate a dislocation density of 10^9 - 10^{10} cm^{-2}.

Mat. Res. Soc. Symp. Proc. Vol. 449 © 1997 Materials Research Society

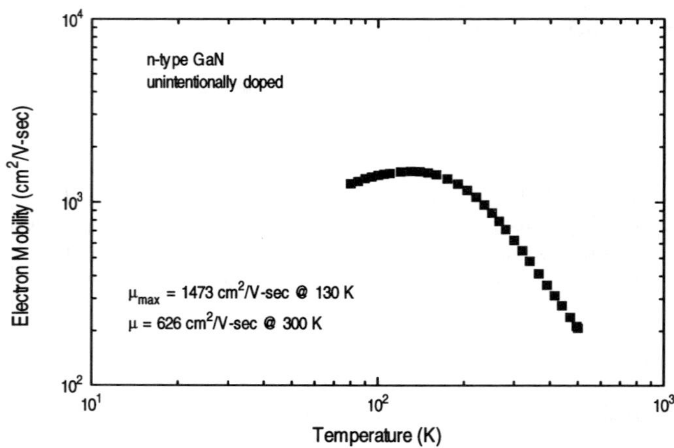

Figure 1: temperature dependence of electron mobility, for undoped 4 μm GaN film.

P-type Doping with Magnesium

The difficulties presented by magnesium p-type doping of GaN and AlGaN are well documented, and largely arise from the high activation energy (~ 170 meV) of the magnesium acceptor (1,2,9). We have grown p-type GaN films with room-temperature hole concentration as high as 5 x 10^{17} cm^{-3}, which is marginally sufficient, but suboptimal, for LEDs and laser diodes. Nevertheless, the activation kinetics of the Mg-acceptor dictate that an active magnesium concentration of mid-10^{19} cm^{-3} is necessary to achieve this hole concentration. Achieving higher hole concentrations is especially important for laser diodes, to permit the formation of a low-resistance ohmic p-contact.

Under our growth conditions, a magnesium concentration of ~ mid 10^{19} cm^{-3} appears to represent a solid-solubility limit. This is illustrated in Figure 2, the hole concentration's dependence upon the magnesium concentration (measured by secondary ion mass spectroscopy) in GaN. The hole concentration reaches a maximum value when the magnesium concentration is approximately mid-10^{19} cm^{-3}.

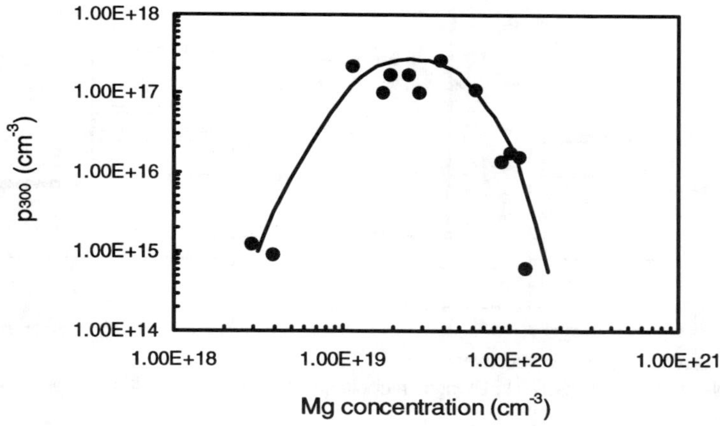

Figure 2: room-temperature hole concentration (p_{300}): dependence upon magnesium concentration in GaN:Mg film

<u>Nitride Device Structures</u>

Typical high-efficiency AlGaInN light emitting diodes incorporate an active region of a single, thin (< 100 Å) layer of InGaN, surrounded by higher-bandgap GaN or AlGaN for the confinement of injected carriers (2). This is in contrast to early devices, where blue or green emission was achieved from an optical transition involving a deep acceptor (zinc) in a thicker (~1000 Å) but higher-bandgap InGaN active region (containing less indium) (3). In these cases of a thick, low [In] active region, the emission spectra were relatively broad, and therefore not ideally suited for display applications. On the other hand, the band-edge or near-band-edge emission of the more recent QW emitters is much more spectrally narrow than the deep-state recombination characteristic of the earlier DH LEDs. This makes them more suitable for full color displays, since purer primary colors (red, green, blue) may be mixed to form a greater pallette of intermediate colors.

From our materials, we have fabricated the two types of device structures illustrated in Figure 3. Single quantum well (QW) InGaN/GaN heterostructures, with a thin InGaN QW at the center of a p-n junction in GaN, form efficient LEDs, where the wavelength depends upon the InGaN alloy composition and QW thickness. Likewise, more sophisticated InGaN/AlGaN multiple quantum well structures are also being developed, for laser diodes. These heterostructures include multiple QWs and AlGaN cladding layers, for creating sufficient optical gain and transverse waveguiding.

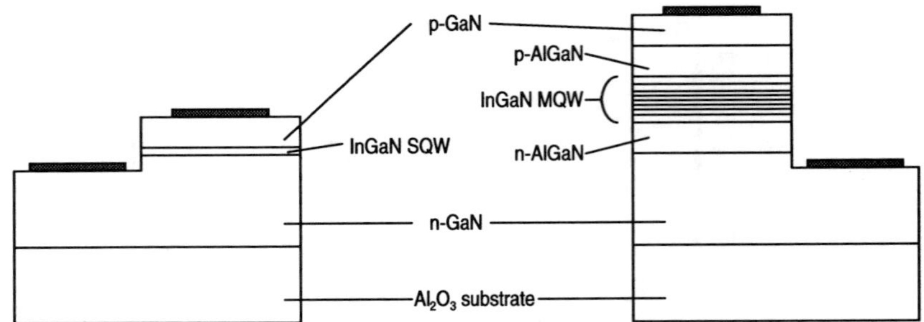

Figure 3: examples of device structures:
(left) single quantum well (SQW) LED; (right) multiple quantum well InGaN/AlGaN heterostructure

Single QW LEDs

The emission wavelength of InGaN SQW LEDs is determined by the InGaN QW composition and thickness. This is indicated by the emission spectra of two single QW InGaN/GaN LEDs with different QW parameters, which are shown in Figure 4. These spectra are measured on unpackaged devices, through the sapphire substrate. The diodes are simple round dots defined by Ar-ion milling, approximately 500 μm in diameter; and the injection current is 100 mA (corresponding to a current density of ~ 50 A/cm^2).

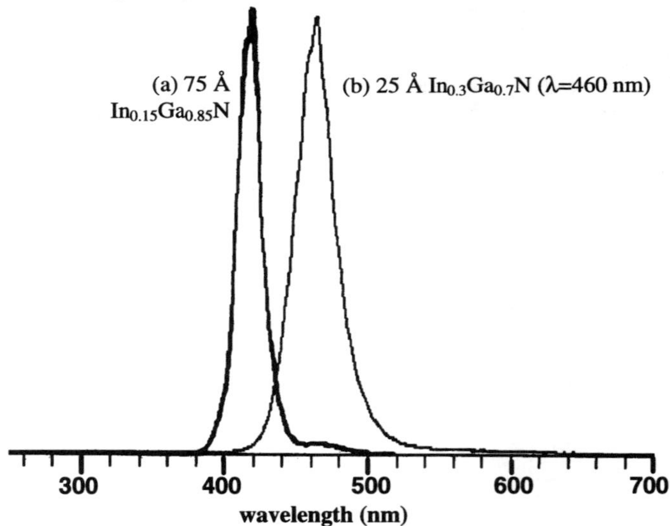

Figure 4: emission spectra of single QW InGaN/GaN LEDs:
(a) 75 Å In$_{0.15}$Ga$_{0.85}$N (λ=408 nm); (b) 25 Å In$_{0.3}$Ga$_{0.7}$N (λ=460 nm)

For one device, a violet LED with a 75 Å $In_{0.15}Ga_{0.85}N$ active region, the spectrum is centered at 408 nm. For the other emitter, a blue LED containing a 30 Å $In_{0.3}Ga_{0.7}N$ active region, the spectrum is centered at 460 nm. Generally, the spectral width is also greater for longer-wavelength LEDs, increasing from 12-15 nm for violet and blue LEDs, to 25 nm for green LEDs. The spectral broadening may be a consequence of segregation of the InGaN alloy active region, whereby the alloy composition becomes nonuniform in the QW plane. Segregation is especially problematic with high indium alloy content films, limiting their thickness and composition uniformity.

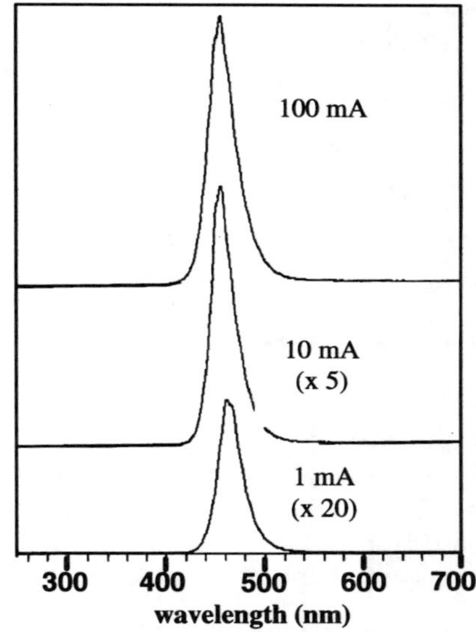

Figure 5: emission spectra of 25 Å $In_{0.3}Ga_{0.7}N$ single QW at three currents

For the 460 nm device, the spectra are also shown for 3 different injection currents in Figure 5. Even at 1 mA (~ 0.5 A/cm^2), the emission is a pure blue, with no indication of any yellow or other deep, defect-related emission. Likewise, Figure 6 shows the emission power as a function of drive current, for a 450 nm blue InGaN SQW LED. The overall quantum efficiency is 2.6% at 20 mA, while the slope efficiency reaches 3.2 %, corresponding to about 4 mW output at 50 mA current. The spectral width of this device's emission was 20 nm.

<u>Multiple Quantum Well InGaN/AlGaN Structures</u>

Although simple, single QW InGaN/(Al)GaN structures can be used to realize efficient blue and green LEDs, laser diodes require somewhat more sophisticated heterostructures. This is so because they must provide for optical waveguiding and the generation of sufficient optical gain. Indeed, the heterostructure design may be especially critical, since the threshold current density is expected to be relatively high in nitride lasers. This is a direct consequence of the high carrier effective masses, which translate into a high

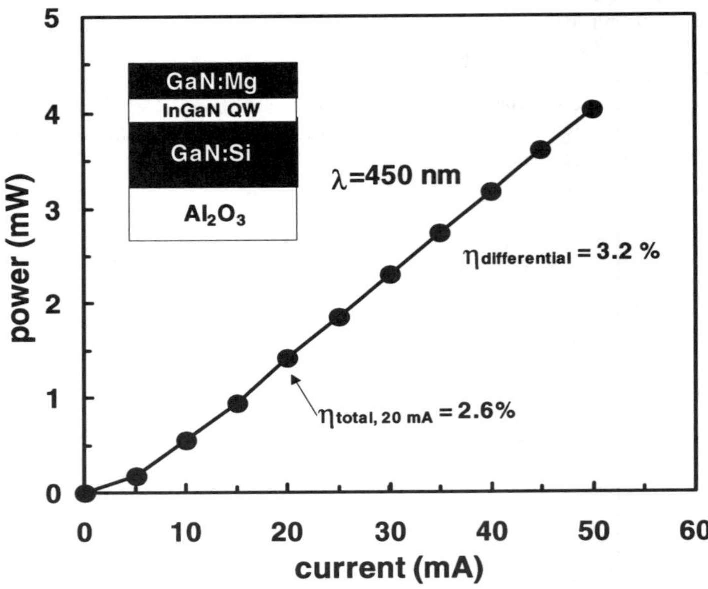

Figure 6: power-current characteristic of 450 nm InGaN/GaN SQW LED.

density of carrier states. As a result, the carrier density at threshold is expected to be quite high, ~ 10^{19} cm^{-3} (corresponding to a current density of several kA/cm^2) compared low 10^{18} cm^{-3} for red or IR devices (10).

In order to produce sufficient optical gain, multiple InGaN QWs are typically required in the active region. Accordingly, the first AlGaInN laser diode of Nakamura *et al.* contained an unusually large number of very thin InGaN QWs (4). Presumably, a single QW produced insufficient optical gain or optical confinement. Since then, Nakamura has continued using multiple QWs, athough their number has decreased significantly from 26 (4), to 3-7 (6-8).

Optical confinement is likewise important for maximizing modal gain in laser diodes. Consequently, it is evident that the incorporation of AlGaN cladding layers would be very beneficial, as they serve to create a stronger waveguide, to increase the spatial overlap between the optical mode and QW gain

Even without AlGaN cladding layers, however, a large number of QWs in the active region can be used to create waveguide. On the other hand, with a large number of QWs, it may become difficult to achieve good spatial overlap between the injected electron and hole distributions, since they are injected from opposite sides of the QW stack.

Overall, optimal laser diode heterostructures differ from LED structures in their requirement for multiple InGaN QWs in the active region, and the additional benefits derived from AlGaN cladding layers. Figure 7 shows the x-ray diffraction spectrum of such a multiple QW active region, incorporated in an InGaN/GaN LED structure containing ten 25 Å

$In_{0.2}Ga_{0.8}N$ QWs, separated by 50 Å GaN barriers. Evidence of the layer uniformity is indicated by coherent reflections from the periodic multilayers comprising the active region, which give rise to visible (-)1st- and ($\pm$)2nd-order satellite peaks in the XRD spectrum (the +1-order peak lies within the GaN peak). The spacing of the satellite peaks indicates the period of the superlattice active region to be 72 Å. Likewise, the absolute position of the 0th-order peak indicates the average InGaN composition to be ~ $In_{0.11}Ga_{0.89}N$.

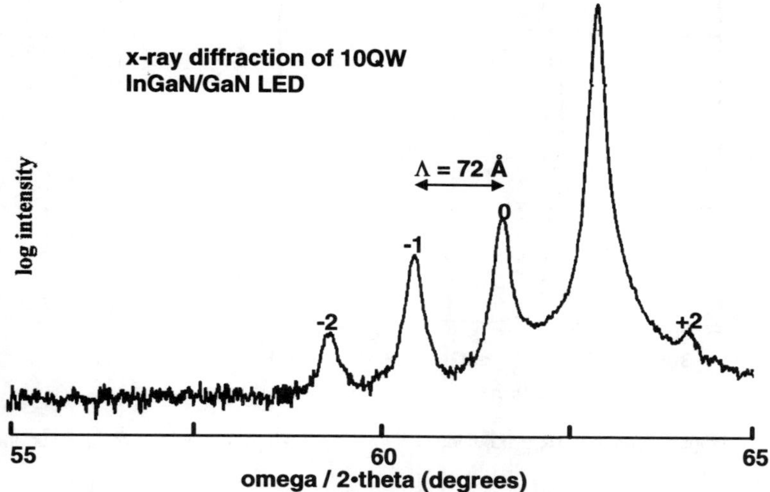

Figure 7: x-ray diffraction spectra of ten-QW InGaN/GaN LED, showing satellite reflections from MQW active region.

The spontaneous emission of an LED with a similar active region (ten 25 Å $In_{0.2}Ga_{0.8}N$ QWs), but also incorporating $Al_{0.1}Ga_{0.9}N$ cladding layers, is shown in Figure 8, for 3 different injection currents. Again, the LEDs are simple ~ 500 μm dots operated under DC bias, probed while lying on a quartz wafer, so that the emission may be detected through the substrate. The spectrum is centered at 422 nm, and the full width at half-maximum (FWHM) of the spectrum is ~ 12 nm, corresponding to ~ 85 meV.

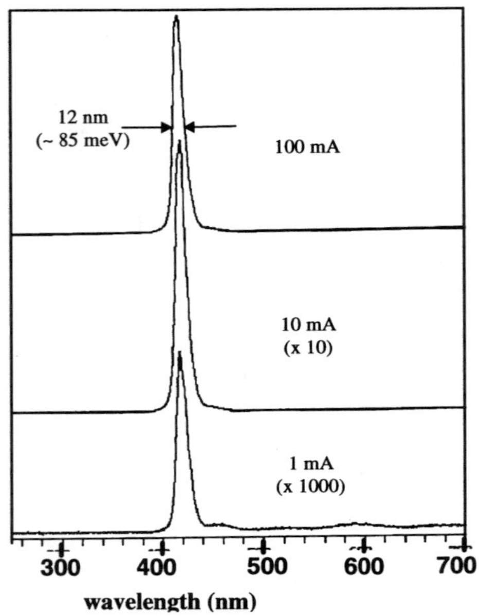

Figure 8: emission spectra of ten-QW $In_{0.2}Ga_{0.8}N$ (25 Å)/AlGaN LED, at three currents

<u>Optical Pumping</u>

Finally, we have observed stimulated emission from optically pumped InGaN/AlGaN multiple QW heterostructures. The sample structure was similar to that shown in Figure (3b), although with a thinner (only 50 nm) AlGaN upper cladding layer, and no GaN p-cap layer. Optical pumping was accomplished by focussing a pulsed, high-power nitrogen laser (λ=337 nm) as a stripe (width of a few hundred μm) onto a bar whose facets were sawed. The excitation power was subsequently adjusted by introducing a stack of glass slides into the pump beam. The intensity and linewidth of the edge-emission is recorded in Figure 9, as a function of the pump power. Evident in the emission intensity is a threshold, which is coincident with a reduction in the emission's spectral linewidth, thus suggesting laser oscillation. The linewidth reaches a value of ~25 Å, compared to an 11 Å resolution limit of the spectrometer (individual Fabry-Perot modes could not be resolved). Furthermore, Figure 10 shows the emission of these structures to be predominantly TE-polarized, as would be expected of a nitride laser. Structures with either 2, 5, or 10 QWs in the active regions, and $Al_{0.1}Ga_{0.9}N$ cladding layers were operated in this manner, and exhibited similar behavior.

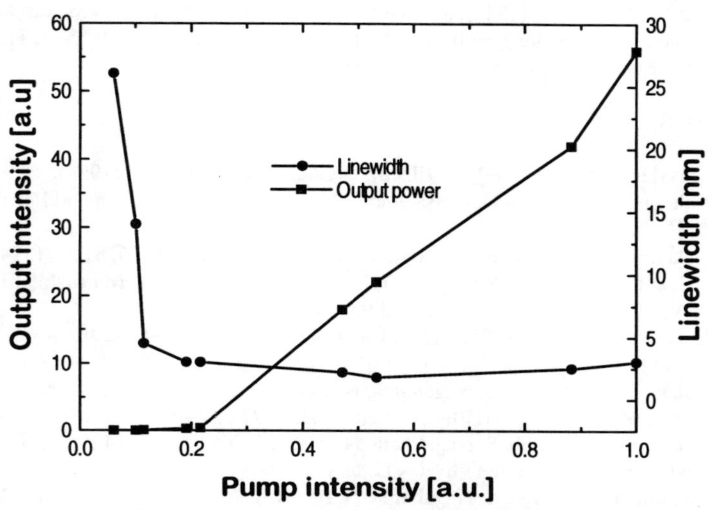

Figure 9: pump-power dependence of edge-emission intensity and spectral width, for optical pumping of a 5-QW $In_{0.2}Ga_{0.8}N/Al_{0.1}Ga_{0.9}N$ structure (λ = 412 nm)

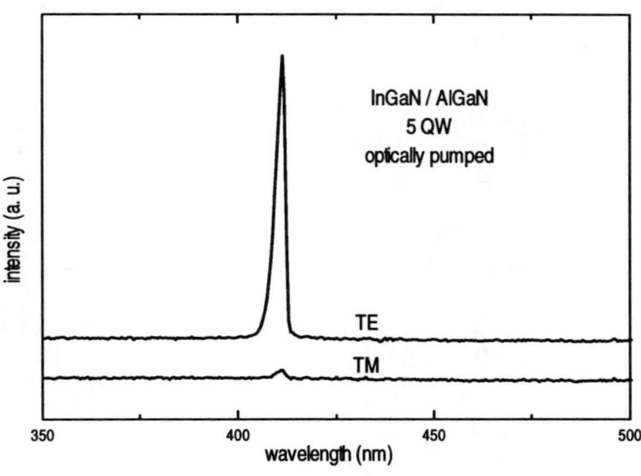

Figure 10: polarization-resolved spectra of optically-pumped InGaN/AlGaN 5-QW heterostructure.

ACKNOWLEDGEMENTS

For support, the authors thank the United States Department of Commerce (DoC contract no. ATP-70NANB2H1241), and the Defense Advanced Research Projects Agency (DARPA). In addition, we are grateful for technical assistance from Jack Walker, Fred Endicott, Steve Ready, David Treat, and Greg Anderson.

REFERENCES

1.   I. Akasaki and H. Amano, Jour. Electrochemical Soc., **141**, 2266 (1994).
2.   S. Nakamura, M. Senoh, N. Iwasa, and S. Nagahama, Jpn. Jour. Applied Physics, **34**, L797 (1995).
3.   S. Nakamura, T. Mukai, and M. Senoh, Jour. Applied Physics, **76**, 8189 (1994).
4.   S. Nakamura, M. Senoh, S. Nagahama, N. Iwasa, T. Yamada, T. Matsushita, H. Kiyoku, and Y. Sugimoto, Jpn. Jour. Applied Physics, **35**, L74 (1996).
5.   I. Akasaki, H. Amano, S. Sota, H. Sakai, T. Tanaka, and M. Koike, Jpn. Jour. Applied Physics, **34**, L1517 (1995).
6.   S. Nakamura, M. Senoh, S. Nagahama, N. Iwasa, T. Yamada, T. Matsushita, Y. Sugimoto, and H. Kiyoku, Applied Physics Letters **69**, 1477 (1996).
7.   S. Nakamura, M. Senoh, S. Nagahama, N. Iwasa, T. Yamada, T. Matsushita, Y. Sugimoto, and H. Kiyoku, Applied Physics Letters **69**, 1568 (1996).
8.   S. Nakamura, M. Senoh, S. Nagahama, N. Iwasa, T. Yamada, T. Matsushita, Y. Sugimoto, and H. Kiyoku, to be published in Applied Physics Letters **69** (1996).
9.   W. Götz, N. M. Johnson, J. Walker, D. P. Bour, and R. A. Street, Applied Physics Letts., **68**, 667 (1996).
10.   P. Rees, C. Cooper, P. M. Smowton, P. Blood, and J. Hegarty, IEEE Photonics Technology Letters, **8**, 197 (1996).

Spatial distribution of electron concentration and strain in bulk GaN single crystals - relation to growth mechanism

Piotr Perlin [1,a], Tadeusz Suski [2], Alain Polian [3], Jean Claude Chervin [3], Elzbieta Litwin-Staszewska [2], Izabella Grzegory [2], Sylwester Porowski [2], J.W. Erickson [4]

1. Center for High Technology Materials, University of New Mexico, Albuquerque, New Mexico 87131-6081
2. High Pressure Research Center, Polish Academy of Sciences, Sokolowska 29,01-142 Warszawa, Poland
3. Physique des Milieux Condensés, Université Pierre et Marie Curie, F-75251 Paris, France
4. Charles Evans and Assoc., Redwood City, CA USA

ABSTRACT

Micro-infrared reflectivity and micro-Raman scattering have been used to determine free electron density and residual strain distribution in bulk GaN crystals. As-grown samples exhibit significant variation of electron concentration and strain along their surfaces. Increase of electron concentration correlates with the growth direction. The observed inhomogeneities in the properties of bulk GaN crystals can be eliminated by mechanical polishing of the crystal and removing the surface layer or growth figures formed during the sample cooling, which follows its high-pressure, high-temperature growth. Properties of that surface layer can differ from the bulk due to different growth mechanisms and impurity distribution. We suggest that oxygen is a likely candidate for the donor impurity supplying high amount of electrons to the conduction band of GaN.

INTRODUCTION

It is well known that heteroepitaxial films of GaN, used as active region material in optoelectronics and electronic devices, are characterized by high density of dislocations (10^8-10^{10} cm^{-2}) [1]. This value exceeds by many orders of magnitude a typical concentration of dislocations in state of art GaAs material. This problem is related to the lack of a substrate lattice matched to GaN. The lattice mismatch between the commonly used sapphire and GaN is as large as 14%. Surprisingly, GaN based optical devices seem to be highly tolerant to the large concentration of extended defects and they show outstanding performance [1]. Recent development of blue and green light emitting diodes (LEDs) by Nichia Chemical Industries [2] represents a convincing argument that epitaxial layers of GaN, InGaN, and AlGaN grown on sapphire can be successfully used in the optoelectronic technology. However, the main technological challenge consists in the fabrication of laser diode, and this requires much better material. Indeed, recent demonstration of a pulsed laser diode based on InGaN shows that although it is already possible to fabricate this kind of device using standard epilayers on sapphire [3], still there is an urgent need for improvement of material quality. An important problem related to development of continuous wave GaN laser is the reduction of still large threshold current. This can be achieved by improving the contact resistance, decreasing the

---

[a] On leave from High Pressure Research Center, Warsaw, Poland

internal losses, and improving the cavity-mirror quality. Two latter goals can be achieved using homoepitaxy of GaN on GaN. Homoepitaxial material should be characterized by a much lower density of dislocations (decrease of scattering losses in the crystal) and good cleaving (mirror quality improvement). The homoepitaxy of GaN was rarely used in the past because of the lack of suitable single crystals of GaN. However, progress in high-pressure, high-temperature growth of GaN [4] makes it possible to obtain single crystals with lateral dimension of 1 cm, which are large enough to be used in homoepitaxy. Indeed, it was recently demonstrated that both crystallographic quality [5] and optical properties [6] of homoepitaxial layers are by far superior compared to standard epilayers on sapphire.

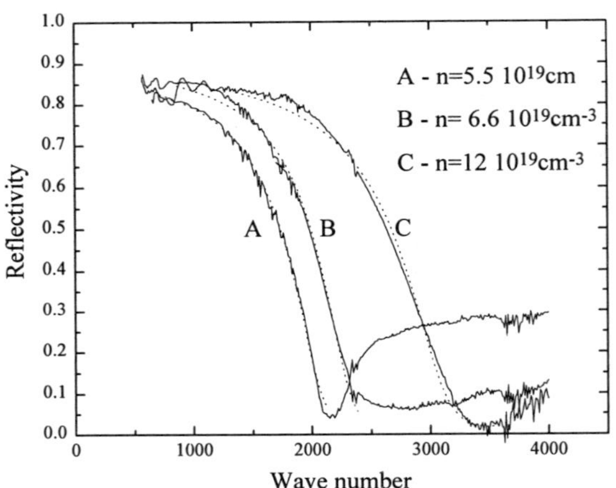

Figure 1 Reflectivity spectra measured for the sample E2ELF in three different spots. Calculated electron concentration are shown on the picture. Dotted line show best fit of Eq. 2 to the experimental data.

In the present paper, we study physical properties and homogeneity of bulk single crystals of GaN obtained by the high-pressure, high-temperature method [4]. Crystals were grown in the temperature gradient from the solution of nitrogen in gallium, at $T \approx 1500$ K and $p \approx 1.5$ GPa. As grown, nonintentionally doped bulk crystals of GaN are characterized by n-type conductivity and high free electron concentration of the order of $5 \cdot 10^{19}$cm^{-3}. The high free carrier concentration was attributed previously to the large concentration of nitrogen vacancies which should form a resonant donor state in the conduction band of GaN [7]. However, very recent state of the art SIMS measurements indicate that surface and subsurface regions of as-grown bulk GaN crystals contain high concentrations of oxygen and carbon. SIMS measurements indicate the following concentration of impurities: hydrogen $< 4 \cdot 10^{17}$ atoms/cm^{-3}, carbon $3 \cdot 10^{19}$ atoms/cm^{-3}, oxygen $2 \cdot 10^{20}$ atoms/cm^{-3}, silicon $< 1 \cdot 10^{17}$ atoms/cm^{-3}. Both carbon [8] and oxygen [9] can form donor states in GaN. Also, very recent experiments [10] suggest that oxygen is responsible for carriers freeze-out observed at high pressure [7] and hence it is more likely the main donor in bulk GaN. In this work we show that the donor distribution in bulk GaN crystals can be related to the growth mechanism of this compound.

SAMPLES

In this paper we present the results obtained from the set of three samples. M2ELF and E2ELF are samples of dimensions approximately 4x5 mm, and the sample Klara is the newest generation 8x6 mm sample. Sample Klara was mechanically polished from one side.

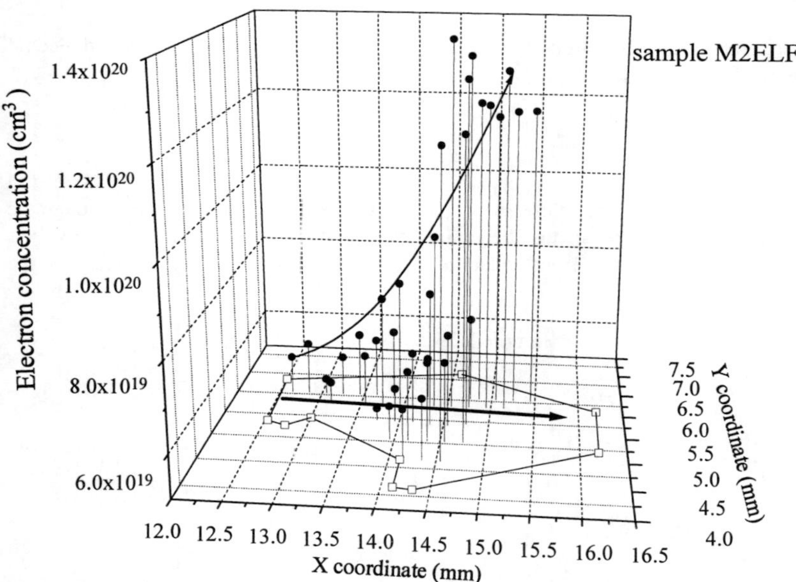

Figure 2. Spatial distribution of the electron concentration in the non-polished bulk GaN sample. Arrow shows the growth direction

EXPERIMENT AND APPLIED MODEL

In order to determine the local concentration of electrons at various points on the GaN bulk crystals we used micro-reflectivity setup consisting of Perkin-Elmer Series 1600 Fourier spectrometer equipped with infrared microscope. The Raman scattering experiments were performed with the use of triple Dilor XY spectrometer, with custom-made micro Raman system. This system can collect the light reflected from a spot of diameter around 50 μm. Assuming the dielectric constant in the form:

$$\varepsilon(\omega) = \varepsilon_\infty \left(1 - \frac{\omega_p^2}{\omega(\omega + i\gamma)} \right) \qquad (1)$$

where ω_p is the plasma frequency and γ is the electron damping constant, we can calculate the reflectivity of the medium using the following equation:

$$R = \frac{(n-1)^2 + k^2}{(n+1)^2 + k^2} \qquad (2)$$

where n is the real part of the refractive index and k is the extinction coefficient. Using equations 1 and 2 we can obtain from the experimental spectra the value of plasma frequency ω_p. The free electron concentration N can be related to the plasma frequency by the relation below:

$$\omega_p^2 = \frac{Ne^2}{m^* \varepsilon_\infty \varepsilon_0} \qquad (3)$$

We take the effective mass of electrons as 0.22 m_0 [11] and we assume the electronic dielectric constant to be equal to 5.2 [12].

RESULTS

Fig. 1 shows three reflectivity spectra measured at three different points. It is clear that there are large differences in plasma edge position depending on the position on the sample. The values of electron concentration obtained from equations 1-3 are shown on Fig. 2. It is clear that the

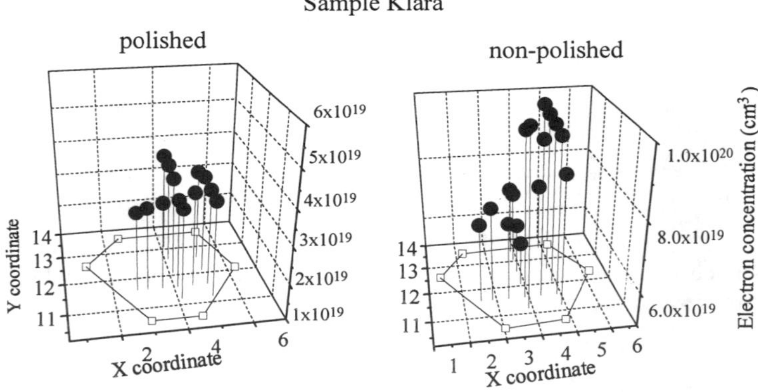

Figure 3. Spatial distribution of electrons for polished and non-polished faces of the sample.

concentration of electrons strongly increases along the direction of growth from around $6 \cdot 10^{19}$ cm^{-3} up to $1.2 \cdot 10^{20}$ cm^{-3}. It means that there are almost macroscopic fluctuations of the conduction band minimum across the sample. The growth of the crystal initiates on the polycrystalline GaN crust formed on the surface of liquid gallium, and then the direction of growth is mostly perpendicular to the gallium surface. The same kind of experiment was performed on a sample in which one side was mechanically polished. During polishing, about 20 µm of GaN was removed from the surface. Fig. 3 shows that while the unpolished face of this surface suffers from large inhomogeneity of electron concentration, the polished face has much lower concentration and is practically homogeneous. This result shows that close to the crystal surface, concentration of dopant is high and not uniform.

To evaluate the influence of the donor concentration on the strain in the crystal we performed a Raman scattering measurement in the same spots on the sample where the infrared reflectivity spectra were measured. In Fig 4 we show the comparison between the E_2 mode peak position for sample with polished and non-polished surfaces. It is clear again that while polished sample does not have any measurable stress at all, the unpolished one has the build in stress as high as 0.3 GPa. It is worthwhile to notice that the phonon frequencies for the non-polished sample lie consistently lower indicating lack of stress.

DISCUSSION

Results of very recent experiments [13,14] indicate that oxygen is the likely source of electrons in GaN semiconductor. Concerning bulk GaN crystals, even small amount of oxygen present in nitrogen gas (used for crystallization) can become quite active due to the compression of the

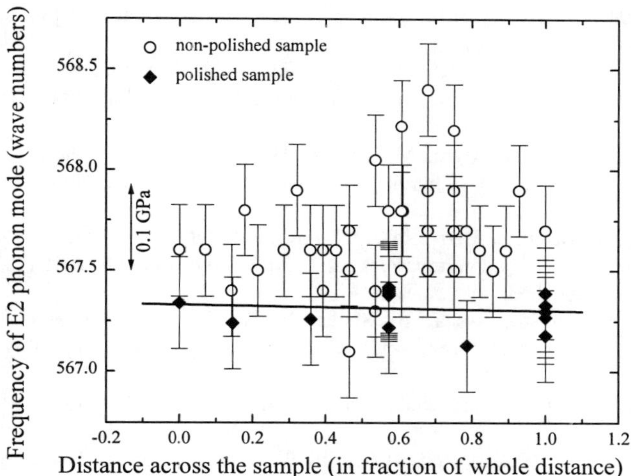

Figure 4. E_2 phonon mode frequency measured across polished and non-polished sample.

gas mixture up to very high pressures. In this case, oxygen can dissolve in liquid gallium and subsequently incorporate into the GaN crystal as an impurity. Results of the present work demonstrate that the donor concentration is the highest near surfaces of the GaN crystals. Moreover the surface is usually covered by a thin layer of slightly different morphology or by separate crystallites which grow during the cooling of the system after the high pressure growth. Concentration of oxygen in these regions (surface and subsurface) can be significantly higher than in the interior of the crystals. For example, different solubilities and segregation coefficient at temperatures lower than the temperature of GaN growth can lead to the described phenomena.

Crystal polishing turns out to be a necessary part of the procedure leading to preparation of highly homogeneous epi-ready GaN substrates. It helps to remove regions of higher (and usually fluctuating) concentration of donors as well as residual strain present sometimes in the surface layer of the bulk crystals. One of the possible ways to reduce free electron concentration can be an additional purification of the system and gettering oxygen generated during high pressure growth of GaN crystals.

REFERENCES

1. S.D. Lester, F.A. Ponce, M.G. Craford, and D.A. Steigerwald., Appl. Phys. Lett. **66**, 1249
2. S. Nakamura, M. Senoh, N. Iwasa, S. Nagahama, T. Yamada, T. Mukai, Jpn. J. Appl. Phys. **34**, L1335, (1995)
3. S. Nakamura, M.Senoh, S. Nagahama, N. Iwasa, T. Yamada, T. Matsushita, H. Kiyoku and Y. Sugimotu, Jpn. J. Appl. Phys.**35**, L74 (1996)
4. I. Grzegory, J. Jun, M. Bockowski, S. Krukowski, M.. Wroblewski; B. Lucznik, S. Porowski, J. Phys. Chem. Solids,.**56**, 639, (1995)
5. F.A. Ponce, D.P. Bour, W.Gotz, N.M. Johnson, H.I. Helava, I. Grzegory, J.Jun, S.Porowski, Appl. Phys. Lett., **68**, 917, (1996)
6. K. Pakula, A. Wysmolek, K.P. Korona, J.M. Baranowski, R. Stepniewski, I. Grzegory, M. Bockowski, J.Jun, S. Krukowski, M. Wroblewski, S.Porowski, Solid State Commun.**97**, 919 (1996)
7. P. Perlin, T. Suski, H. Teisseyre, M. Leszczynski, I. Grzegory, J. Jun, S. Porowski, P. Boguslawski, J. Bernholc, J.C. Chervin, A. Polian, T.D. Moustakas, Phys. Rev. Lett. **75**, 296 (1995)
8. P. Boguslawski, E.L. Briggs, J. Bernholc, Appl. Phys. Lett. **69**, 233, (1996)
9. J.C. Zolper, R.G. Wilson, S. J. Pearton, R. A. Stall, Appl. Phys. Lett. **68**, 1945 (1996)
10. C. Wetzel et al. unpublished.
11. P. Perlin, E. Litwin-Staszewska, B. Suchanek, W. Knap, J. Camassel, T. Suski, R. Piotrzkowski, I. Grzegory, S. Porowski, E. Kaminska, J.C. Chervin, Appl. Phys. Lett. **68**, 1114, (1996)
12. H. Sobotta, H. Neumann, R. Franzheld and W. Seifert, Phys. Status Solidi B **174**, K57, (1992)
13. P. Perlin ,T. Suski, A. Polian, J. C. Chervin, W. Knap, J. Camassel, I. Grzegory , S. Porowski , J.W. Erickson, this Conference.
14. C. Wetzel et al. unpublished

THICKNESS DEPENDENCE OF ELECTRONIC PROPERTIES OF GaN EPI-LAYERS

W. GÖTZ*, J. WALKER, L.T. ROMANO, AND N.M. JOHNSON
Xerox Palo Alto Research Center, Palo Alto, California 94304, USA
R.J. MOLNAR
Massachusetts Institute of Technology, Lincoln Laboratory, 244 Wood Street, Lexington, Massachusetts 02173, USA

ABSTRACT

The electronic properties of heteroepitaxial GaN were investigated for unintentionally doped, n-type films grown by hydride vapor phase epitaxy on sapphire substrates. The GaN layers were characterized by variable temperature Hall-effect measurement, capacitance-voltage (C-V) measurements, and deep level transient spectroscopy (DLTS). The measurements were performed on as-grown, 13 μm thick films and repeated after thinning by mechanical polishing to 7 μm and 1.2 μm. The room temperature electron concentrations as determined by the Hall-effect measurements were found to increase from ~10^{17} cm^{-3} (13 μm) to ~10^{20} cm^{-3} (1.2 μm) with decreasing film thickness. However, the C-V and DLTS measurements revealed that the ionized, effective donor and deep level concentrations, respectively, remained unchanged in regions close to the top surface of the films. These findings are consistent with the presence of a thin, highly conductive near interface layer which acts as a parasitic, parallel conduction path. Possible sources of such a shunt near the GaN/sapphire interface include oxygen contamination from the sapphire substrate or a structurally highly defective, 300 nm thick interface layer.

INTRODUCTION

Hydride vapor phase epitaxy (HVPE) has been revitalized as a promising GaN growth technique to overcome the apparent lack of GaN substrates for the growth of electronic and light emitting III-V nitride devices by metalorganic chemical vapor deposition (MOCVD) or molecular beam epitaxy (MBE) [1,2]. It has been demonstrated that HVPE is capable of growing thick GaN films on sapphire substrates with greatly improved structural and electronic properties as compared to heteroepitaxial GaN films grown by MOCVD or MBE [3,4]. Dislocation densities as low as ~5×10^7 cm^{-2} [5] and room temperature electron mobilities of ~900 cm^2/Vs [6] have been accomplished for ~74 μm thick, unintentionally doped GaN films with an electron background concentration of ~8×10^{16} cm^{-3} (300 K). Characterization of Schottky diodes formed on HVPE-grown GaN films by deep level transient spectroscopy (DLTS) revealed deep level concentrations below 10^{16} cm^{-3} [6]. However, in the same films the presence of deep levels with thermal ionization energies in the range from ~100 to ~200 meV and concentrations as high as ~5×10^{18} cm^{-3} were derived from the temperature dependence of the electron concentration by Hall-effect measurements [5].

The above contradictory results motivated the present study. For the analysis of the Hall-effect data, uniformity of the electronic properties throughout the film thickness is assumed. To investigate the electronic properties as a function of film thickness (d) a 13 μm thick HVPE-grown GaN layer was characterized by variable-T Hall effect, C-V measurements and DLTS. Subsequently, the film thickness was reduced to 7 and to 1.2 μm by mechanical polishing and characterized by the same techniques at each of the thickness steps. Characterization of the HVPE film was complemented by secondary ion mass spectrometry (SIMS).

---

*Permanent address: Hewlett-Packard Company, 370 West Trimble Road, San Jose, CA 95131, USA

Mat. Res. Soc. Symp. Proc. Vol. 449 © 1997 Materials Research Society

EXPERIMENTAL

The GaN material used in this study was grown in a vertical HVPE reactor which is described in Ref. [7]. The nucleation of the GaN on c-plane sapphire substrates was enhanced by a GaCl pretreatment [6,7]. The film was grown at 1050°C to a thickness of 13 μm with a growth rate of 13 μm/h. X-ray diffractometry of the as-grown film revealed a FWHM of the (002) rocking curves of ~5 min. The tilt component of the film was measured by the (002) reflection to be 4.5 arcmin and the twist component was measured by the (112) asymmetric reflection to be 11.5 arcmin [5].

The Hall-effect measurements were conducted in the temperature range between 80 and 500 K with a magnetic field of 17.4 kG. For the measurements, samples of 5 $\times$ 5 mm^2 size were cut from the wafers and Ohmic metal contacts were deposited in the Van der Pauw geometry. For the analysis a temperature independent Hall scattering factor of unity value was assumed.

Conventional capacitance-voltage (C-V) measurements were conducted to investigate the depth profile of the dopants below the surface of the GaN films. Schottky diodes were fabricated by evaporating Au through a shadow mask onto the surface of the GaN samples. Large area Ohmic contacts were also deposited on the surface. The measurements were conducted with a 1 MHz, 10 mV test signal up to a reverse bias of ~20 V.

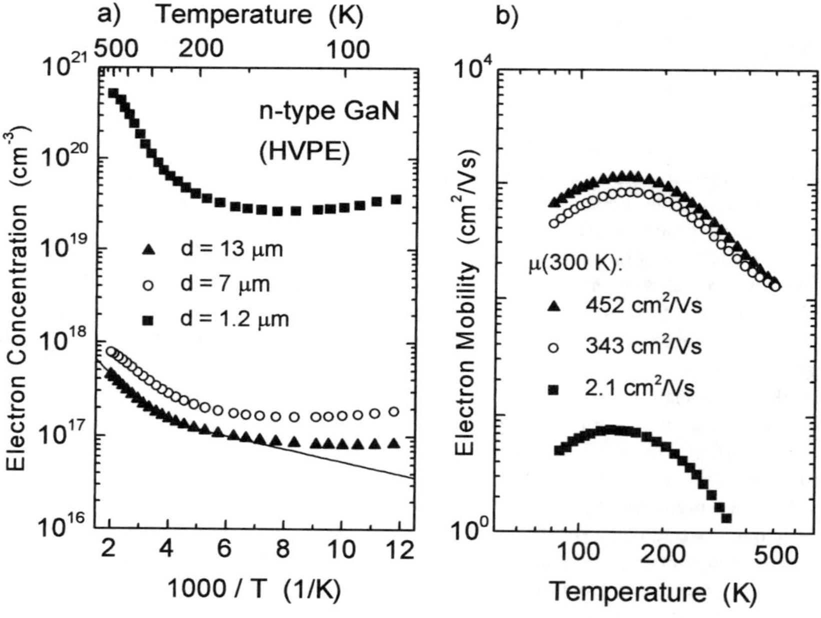

Fig. 1. Electron concentration vs reciprocal temperature (a) and electron mobility vs temperature (b) for three different film thicknesses as determined from variable-T Hall-effect measurements.
The solid line in Fig. 1a results from a fit of the charge neutrality condition to the high temperature branch of the experimental Hall data. The fit yields parameters for two independent donors which are given in the text.

The Schottky diodes were also utilized to perform DLTS measurements in the temperature range between 75 and 475 K. The DLTS system employed in this study is described in Ref. [8].

The film thinning was accomplished by mechanical polishing. The final polish achieved an rms surface roughness of <10 Å.

SIMS depth profiles for Si and O were measured using a Cs^+ primary ion beam in a CAMECA IMS 4F system with GaN implantation standards.

HALL-EFFECT RESULTS

Results from variable-T Hall-effect measurements are shown in Fig. 1 (symbols). Figure 1a displays electron concentrations as a function of temperature for the as-grown film (d = 13 μm) and after thinning by mechanical polishing to thicknesses of 7 μm and 1.2 μm. The electron concentrations significantly increase with decreasing film thickness. Figure 1b shows the electron mobility as a function of temperature for the three different film thicknesses. The electron mobility decreases with decreasing film thickness.

The experimental data for the original film thickness was analyzed using the charge neutrality condition [9] assuming two independent donors and acceptor compensation (solid line in Fig. 1a). The analysis yields a shallow donor with an activation energy (thermal ionization energy) of ~18 meV and a concentration of ~2×10^{17} cm^{-3}. The presence of a second donor is required to explain the high-temperature portion of the experimental Hall-effect data. The parameters for the second donor are ~180 meV and ~5×10^{17} cm^{-3} for the activation energy and concentration,

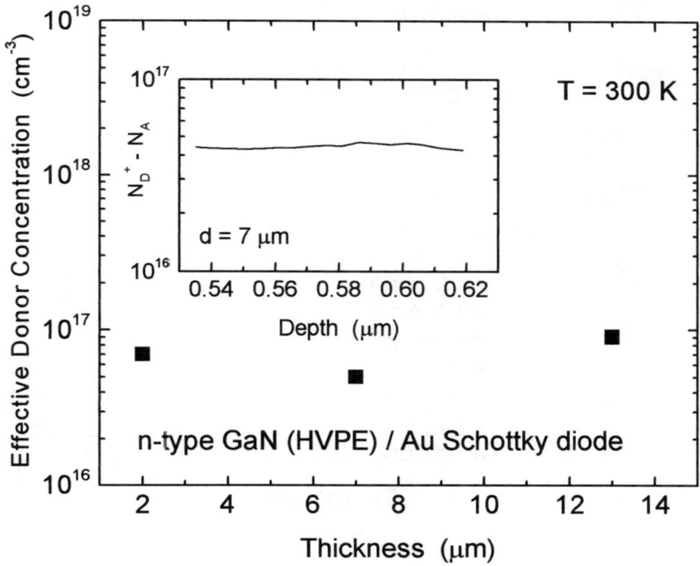

Fig. 2: Net donor concentration (N_D^+-N_A) for HVPE-grown GaN as determined by C-V measurements on Schottky diodes for three different film thicknesses. The inset demonstrates the net donor depth profile for a film thickness of 7 μm.

respectively. The analysis assumes that these donors are uniformly distributed throughout the thickness of the HVPE film.

C-V MEASUREMENTS, DLTS AND SIMS

Results from C-V measurements as a function of film thickness are shown in Fig. 2. The effective donor $(N_D^+-N_A)$ concentrations depicted in Fig. 2 are average concentrations derived from the depth profiles. As an example, the depth profile of $N_D^+-N_A$ determined for a film thickness of 7 μm is shown in the inset of Fig. 2. The C-V results indicate that $N_D^+-N_A$ stays approximately constant from the original film thickness to a depth of ~ 1 μm away from the GaN / sapphire interface.

Results from DLTS are shown in Fig. 3. Displayed is a DLTS spectrum measured for our HVPE-grown GaN material at the original film thickness (13 μm). The spectrum which was recorded for an instrumental emission rate of 46.2 s^{-1} reveals the presence of four discrete deep levels. They are labeled DLN_1, DLN_2, DLN_3, and DLN_4. For this particular sample, the deep level DLN_2 appears only as a shoulder and, therefore, was not considered for analysis. The measurement was repeated after each polishing step; however, the spectra are not shown in Fig. 3. The DLTS spectra were analyzed assuming a temperature independent capture cross section. The defect parameters for DLN_1, DLN_3, and DLN_4 are depicted in Fig. 3 as functions of film thickness. Activation energies for electron emission to the conduction band for DLN_1,

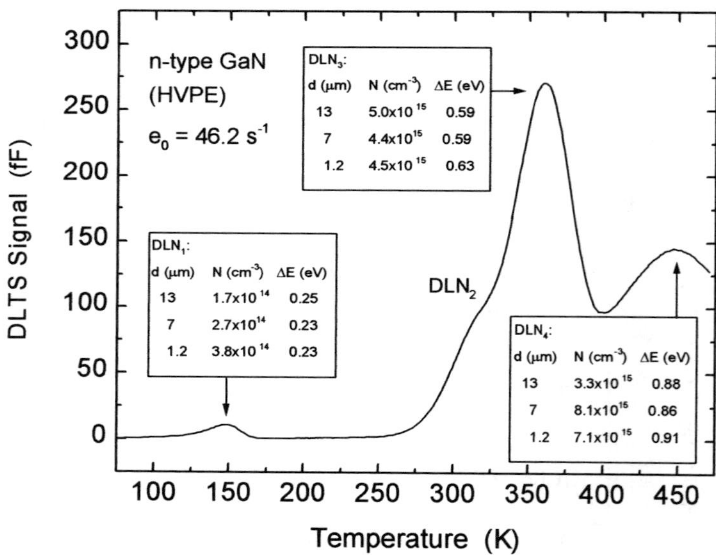

Fig. 3: DLTS spectrum for HVPE-grown GaN (original film thickness). Peaks in the spectrum indicate the presence of discrete deep levels. Parameters (concentration, N, and activation energy for electron emission to the conduction band, ΔE) for three deep levels are depicted for three different film thicknesses (d).

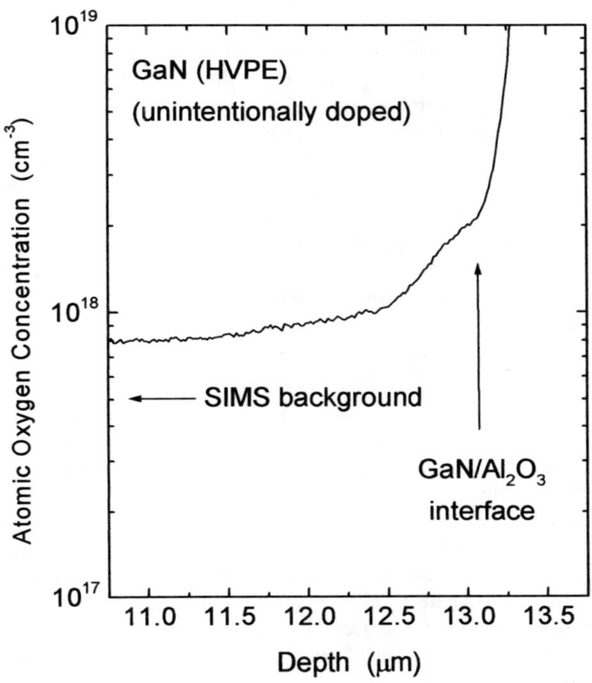

Fig. 4: Depth profile of the oxygen concentration near the GaN / sapphire interface. The oxygen background of the SIMS measurement is also indicated.

DLN_2, and DLN_4 were determined to range from 0.23 to 0.25 eV, 0.59 to 0.63 eV and from 0.86 to 0.91 eV, respectively. The activation energies for the three deep levels only vary within the experimental uncertainties for the measurements at different sample thicknesses indicating that each DLTS measurement detected the same deep levels. The concentrations of DLN_1 and DLN_3 also vary only within experimental uncertainties showing that these deep levels are almost uniformly distributed within the investigated thickness range. An exception is DLN_4, the concentration of which is about three times higher at a film thickness of 1.2 μm than at the original film thickness.

The oxygen concentration profile near the GaN / sapphire interface as determined by SIMS is shown in Fig. 4. The profile from the sample surface to a depth of ~11.5 μm is almost flat at an oxygen level of ~8×10^{17} cm^{-3} (not shown in Fig. 4). At a depth of 11.5 μm the oxygen concentration slightly increases and rises between 12.5 μm and the GaN/sapphire interface from ~9×10^{17} cm^{-3} to ~2×10^{18} cm^{-3}. The Si concentration was also monitored but found not to exceed the background level of the SIMS measurement (~1×10^{17} cm^{-3}) throughout the film thickness (not shown).

CONCLUSIONS

The experimental data presented in this study demonstrate that shallow as well as deep levels are homogeneously distributed in the major portion of the 13 μm film. This is evident from C-V (Fig. 2) and DLTS measurements (Fig. 3), respectively, which were conducted for three different film thicknesses. The DLTS measurements also show that the concentration of detected deep levels do not exceed concentrations of $\sim 10^{16}$ cm^{-3}. However, Hall measurements indicate a significant increase in electron concentration with decreasing film thickness as well as the presence of a deep donor level in concentrations above $\sim 10^{18}$ cm^{-3}. This behavior may be explained by the presence of a thin, highly conductive GaN layer close to the GaN/sapphire interface. For depth inhomogeneities the Hall effect measurements yield an effective areal density of free electrons $n_{s,eff}$ and an effective Hall mobility μ_{eff} [10]. For a two-layer model the product $n_{s,eff}\,\mu_{eff}$ is given by

$$n_{s,eff}\mu_{eff} = n_i\mu_i d_i + n\mu(d - d_i), \tag{1}$$

where d_i is the thickness of the interface layer and n, μ and n_i, μ_i are the electron concentrations, mobilities of the GaN film (without interface layer) and the interface layer, respectively. The presence of a ~ 300 nm thick, highly defective interface layer was detected from transmission electron microscopy for the HVPE film investigated in the present study [5]. Under the assumption that this interface layer is responsible for the observed electrical phenomena a lower limit ($\mu_i = \mu$) of the shallow donor and the deep level concentration in the interface layer can be estimated with Eq. (1). With $d_i \sim 300$ nm the shallow donor concentration in the interface layer becomes $>5\times10^{19}$ cm^{-3} and the deep level concentration becomes $>2\times10^{20}$ cm^{-3}. The atomic concentration of potential donors (O, Si) in the interface layer (Fig. 4) cannot account for the estimated shallow donor concentration. Thus for both shallow donors and deep levels the responsibility of native defects is implied.

Further study is needed to decide whether a two-layer model for the electrical conductivity in heteroepitaxialy-grown GaN films is generally applicable.

ACKNOWLEDGMENT

The work at Xerox was supported by DARPA, agreement # MDA972-95-3-008 and the work at Lincoln Laboratory was supported by the Department of the Air Force and by DARPA. Opinions, interpretations, conclusions, and recommendations are those of the authors and not necessarily endorsed by the United States Air Force.

REFERENCES

1 T. Detchprohm, K. Hiramatsu, N. Sawaki, and I. Akasaki, J. Cryst. Growth **137**, 170 (1994)

2 R.J. Molnar, W. Götz, L.T. Romano, and N.M. Johnson, Electrochem. Soc. Proc. **96-11**, 212 (1996)

3 S.N. Mohammad, A.A. Salvador, and H. Morkoç, Proc. IEEE **83**, 1306 (1995)

4 S.D. Lester, F.A. Ponce, M.G. Craford, and D.A. Steigerwald, Appl. Phys. Lett. **66**, 1249 (1995)

5 W. Götz, L.T. Romano, B.S. Krusor, and N.M. Johnson, Appl. Phys. Lett. **69**, 242 (1996)

6 R.J. Molnar, to be published in J. Cryst. Growth

7 R.J. Molnar, P. Maki, R. Aggarwal, Z.L. Liau, E.R. Brown, I. Melngailis, W. Götz, L.T. Romano, and N.M. Johnson, Mater. Res. Soc. Symp. Proc. **423**, 221 (1996)

8 W. Götz, Mater. Res. Soc. Symp. Proc. **378**, 491 (1995)

9 W. Götz, N.M. Johnson, D. P. Bour, C. Chen, H. Liu, C. Kuo, and W. Imler, Mater. Res. Soc. Symp. Proc. **395**, 443 (1996)

10 N.M. Johnson, in *Hydrogen in Semiconductors*, edited by J.L. Pankove and N.M. Johnson (Academic, San Diego, CA, 1991), p. 118

PERSISTENT PHOTOCONDUCTIVITY IN N-TYPE GaN

A.E. WICKENDEN*, G. BEADIE**, D.D. KOLESKE*, W.S. RABINOVICH**, and J.A. FREITAS, Jr.*

Naval Research Laboratory, Washington, D.C. 20375, wickenden@estd.nrl.navy.mil
* Electronic Science & Technology Division, Code 6800
** Optical Sciences Division, Code 5642

ABSTRACT

Persistent photoconductivity has been observed in n-type Si-doped GaN grown on sapphire substrates by metalorganic vapor phase epitaxy. The effect has been seen both in films which are of electrically high quality, based on low temperature photoluminescence (PL) and Hall analysis, and in films which either appear to be compensated or which exhibit strong donor-acceptor pair recombination. The photoconductivity has been observed to persist for several days at room temperature (300K). Modeling of the resistance recovery by a stretched exponential treatment may suggest a distribution of deep level defect centers in the Si-doped GaN films.

INTRODUCTION

Gallium nitride is a wide bandgap semiconductor which has shown promise in short wavelength light emitting devices and in high power and high temperature electronic applications. Due to the constraint of heteroepitaxial growth, the formation of defects in GaN films is well known. The defects include threading dislocations, present at typical concentrations of 10^8 - 10^{10} cm^{-2}, grain boundaries, and point defects. The concentration of defects may be influenced by growth parameters. Their true origin and range of effects, however, are still not well understood.

Persistent photoconductivity (PPC) is a physical effect which is associated with deep level recombination centers in semiconductor materials. The metastable states indicated by PPC may be associated with local configurational changes in the crystal lattice which act as energy barriers to relaxation for carriers excited out of their ground state by light [1]. PPC may therefore be studied in an attempt to understand the nature of some of the defects found in heteroepitaxial GaN films, as has recently been done for the case of PPC in p-type magnesium (Mg)-doped GaN films [2,3]. PPC has also been observed in unintentionally doped n-type GaN films. [4] Recent reports which note the influence of Si impurity atoms on the structure of the GaN crystal lattice [5,6] are of interest as they may serve to justify the interpretation of PPC in Si-doped GaN films.

EXPERIMENTAL

High quality GaN films are grown in a vertical, water-cooled, RF-heated metalorganic vapor phase epitaxy (MOVPE) reactor. Triethylaluminum (TEA) and ammonia (NH$_3$) are used in a V/III ratio of 74,000 for the deposition of an AlN nucleation layer (NL) at 450°C. Trimethylgallium (TMG) and NH$_3$ are used for GaN growth, in a V/III ratio of 1,900 at a growth temperature of 1040°C. Hydrogen (H$_2$) is used as carrier gas, and a dilute mixture of disilane in H$_2$ is used to effect silicon (Si) doping of the GaN films. The reactor pressure during growth is

Mat. Res. Soc. Symp. Proc. Vol. 449 © 1997 Materials Research Society

58 torr. Growth is typically undertaken on a-plane (11.0) sapphire substrates, but growth on c-plane (00.1) sapphire is also performed. Hall electron transport measurements are made at room temperature using the van der Pauw geometry, with indium contacts and a magnetic field intensity of 0.2 T. During the measurements, the sample is shielded from ambient light. Photoluminescence (PL) spectra of the films have been taken at 6K, using the 325 nm HeCd laser line at a power density of approximately 12 W/cm^2.

To facilitate the optical experiments in this study, ohmic contact strips (Ti/Al) are patterned on the samples. Each sample is placed in a light-tight oven with pressure-contact electrical connections to the contact pads, and a thermocouple probe in contact with the brass sample mounting stage. The resistance of the GaN film is monitored using a digital multimeter. In darkness, the sample is heated to 100°C for 2 hours to thermally release electrons from trap states, then cooled to room temperature and a baseline dark resistance value, R_{sat}, is measured. To excite metastable trap states, the sample is then illuminated with broad spectrum light (unfocused halogen lamp) through a window in the side of the oven until the measured resistance decreases to a steady state level, typically within a five minute period. The light is then turned off, and the increase in resistance is monitored until it approaches the dark resistance level.

RESULTS AND DISCUSSION

The MOVPE growth process which we have developed for electronic-device quality GaN repeatably yields GaN films which are highly resistive (HR) when grown without added dopant, and which under the same growth conditions can be controllably doped with Si to form uniformly-doped GaN:Si films or 2000Å thick doped channel layers on a thick (typically 3 μm) HR GaN buffer layer. [7] These doped GaN films typically exhibit mobilities in the range of 400-600 cm^2/V·s at dopant levels of 2×10^{17} cm^{-3}, and have dopant activation efficiency between 70-100%, and apparently low levels of compensation, as measured by PL and variable temperature Hall. [8,9] The low temperature photoluminescence (PL) spectrum of a device quality ($\mu = 400$ cm^2/V·s at $n = 2 \times 10^{17}$ cm^{-3}) uniformly Si-doped GaN film which was investigated for PPC is shown in Figure 1a. The PL spectrum is seen to have strong near bandedge emission due to recombination of an exciton bound to a neutral Si donor, D°X, very

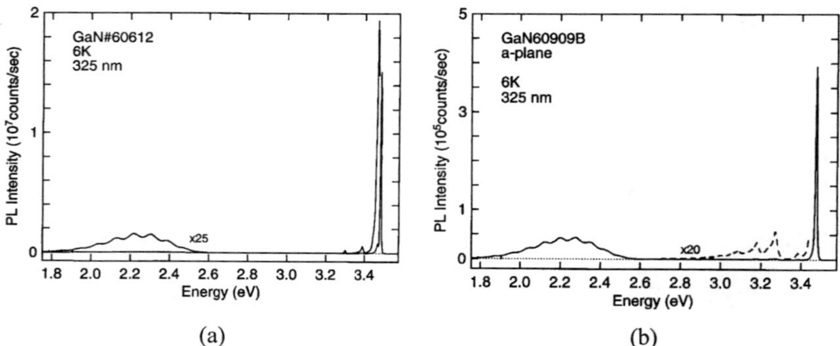

(a)   (b)

Figure 1. Low temperature PL spectra taken of a) a device quality Si-doped GaN film (2.2 eV emission intensity magnified 25X) and b) a Si-doped GaN film grown while increasing the growth temperature (D/A recombination emission intensity near 3.2 eV magnified 20X).

weak 2.2 eV emission, and no visible emission attributable to donor/acceptor pair recombination (D/A) or other near bandedge radiative emission. [10] The magnified peaks seen in the PL spectrum at energies just below the bandedge emission are phonon replicas of the D°X emission peak. This spectrum is typical of Si-doped GaN films, both uniformly doped as well as channel layers on HR-GaN buffer layers, grown on a-plane sapphire in our laboratory.

On occasion, e.g. due to equipment modifications, the growth process must be reoptimized. During such periods of non-optimized growth, films are sometimes produced which exhibit low electron mobility and lower than expected electron concentrations in Hall effect measurements. Since it appeared that compensation centers may exist in these films, they were among the first samples considered as candidates for PPC studies. The PL spectra of these films is identical to that of the high quality film shown in Figure 1a, possibly suggesting a non-radiative compensating center. Both uniformly doped and channel layer structures of this type were investigated for PPC. Uniformly Si-doped GaN films grown on both a- and c-axis sapphire, having a 1000C thick GaN NL and grown while varying the growth temperature from 450°C to 1040°C, [11] were also studied. These films generally exhibited the expected electron concentrations but very low Hall mobilities. Figure 1b illustrates the PL spectrum of such a film grown on a-axis sapphire. The film shown exhibits a much more intense 2.2 eV band emission relative to the near bandedge (NBE) emission, and weak donor/acceptor (D/A) pair recombination emission not typically seen in our device quality GaN films. The PL spectrum of a film grown simultaneously on c-axis sapphire exhibits much stronger D/A emission. Excess carbon impurities or structural defects may be incorporated in these films while growing at low temperatures, providing acceptor sites for Si donor recombination. These films were grown as candidates for PPC studies because of the presence of similar PL features seen in earlier samples grown under these non-optimum conditions.

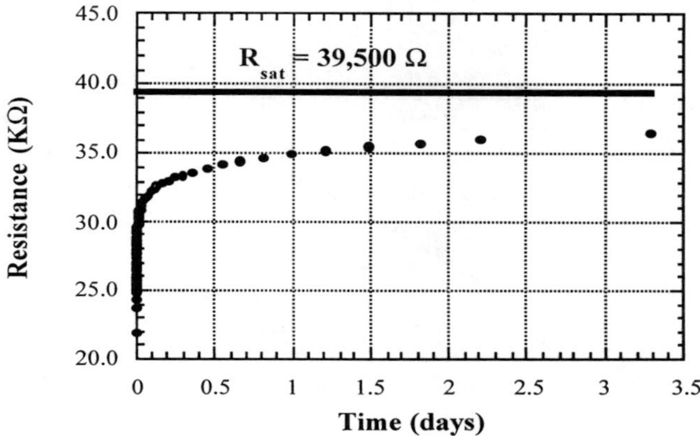

Figure 2. Resistance recovery in a 2000Å GaN:Si channel layer on a 3μm HR-GaN buffer layer. The data was taken at room temperature, and the persistent photoconductivity effect is seen to last for several days.

In addition to the GaN:Si films outlined above, a device quality 2000 Å channel layer on a 3 μm HR-GaN buffer layer, which exhibited a Hall mobility of 410 cm^2/V·s at n = 2 x 10^{17} cm^{-3}, was investigated for PPC. In all of the films studied, the target Si doping concentration was 2 x 10^{17} cm^{-3}. Persistent photoconductivity was seen in all samples with the exception of the uniformly doped device quality GaN:Si film. A typical profile of resistance recovery is illustrated in Figure 2, where the PPC was seen to exist for several days. The PPC is seen to be a small effect (approximately 10%) in the GaN:Si films, similar to PPC results seen at 30K in MOVPE-grown p-type GaN:Mg films. [2] It is worth noting, however, that results shown here are for room temperature experiments. One experiment done at increased temperature showed a substantially reduced PPC response, presumably due to thermal release of trapped electrons, and which suggests that low temperature experiments may show a stronger PPC response. Further experiments are planned to study PPC at varying temperatures. The fact that the effect is seen in films with engineered defects is not too surprising. Of more interest is the fact that PPC is not seen in a high-mobility uniformly doped GaN:Si film, but is seen in films which have an interface between a high mobility GaN:Si channel layer and a HR-GaN layer. As the experiment investigates the bulk of the GaN epitaxial film, these results may suggest that the PPC effect is simply masked by an increased donor population in the uniformly doped film. Alternatively, the results may suggest that trap states exist in the HR-GaN film into which electrons from the Si-doped channel layer may drift, or that localized configurational defects exist at the interface between Si-doped and HR-GaN films caused by the addition of the Si impurity to the GaN lattice. [5,6] An impurity-associated lattice relaxation mechanism has been suggested for PPC seen in p-type GaN:Mg films grown by reactive MBE, [3] and is consistent with our experimental results. Experiments are ongoing to determine the character of the metastable state resulting in PPC.

In light of the nonexponential recovery of our samples, the data was analyzed using a stretched-exponential model. Persistent photoconductivity has been observed to follow this type

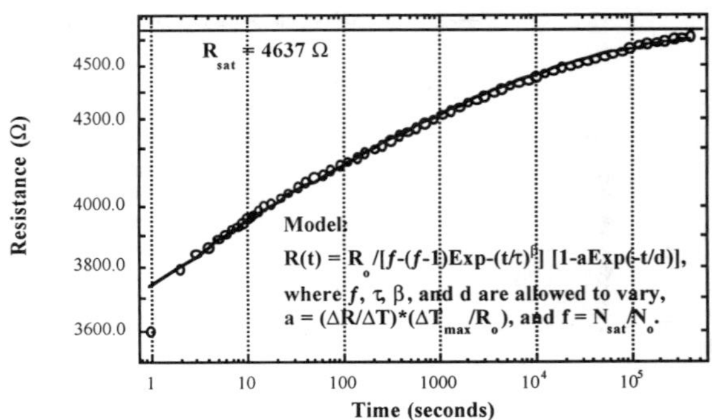

Figure 3. Stretched exponential curve fit (solid line) applied to the resistance data of a uniformly doped GaN:Si film grown at varying temperature (open circles). The fitting parameters were constrained to R$_o$ = 3017Ω, ΔR/ΔT = 8.1 Ω/°C, and ΔT$_{max}$= 5°C.

of recovery in Mg-doped GaN films just mentioned as well as in other systems, [12,13,14]
consistent with the view that the carriers behave according to:

$$N(t) = N_{sat} - (N_{sat} - N_o)\exp[-(t/\tau)^\beta],$$

where N_{sat} is the dark value of the carrier density, N_o is the carrier density value after
photoexcitation, τ is a constant which sets the time scale of the process, and β (between 0 and 1)
is the "stretch" exponent, which determines the deviation from exponential recovery. In the limit
that β equals 1, the decay is exponential, while in the limit that β equals 0, the population
remains a constant. Microscopic models which predict this functional form have been proposed,
but as they are very composition-dependent and not universally accepted, such models will not
be addressed here. The resistance data, assumed to be proportional to $1/N$, was fitted according
to:

$$R(t) = R_O \frac{1}{N(t)}[1 - a\exp(-t/d)],$$

where $N(t)$ is given above, and with fitting parameters R_o, N_o, N_{sat}, τ, β, $a =$
$(\Delta R/\Delta T)(T_{max}/R_o)$ and d, a thermal decay constant. The use of the exponential term in the
resistance equation was found to be necessary to model the initial heating of the sample by the
light source, which was assumed to follow an exponential decay. Figure 3 illustrates the fit of the
above model to the PPC data for a uniformly doped GaN:Si film grown at varying temperature.
In this fit R_o, $\Delta R/\Delta T$, and ΔT_{max} were constrained to 3017Ω, $8.1\ \Omega/°C$, and $5°C$, respectively,
and the fit was obtained with $N_{sat}/N_o = 0.65$, $\tau = 3.8$ sec, $\beta = 0.12$, and $d = 4.4$ sec. The
stretched-exponential fit models the data well for t > 2 seconds. The stretched-exponential form
may suggest a distribution of energy levels for the defect associated with the PPC. Further
investigation at varying temperatures may serve to determine energy levels or distributions
associated with the PPC effect.. The fact that PL data taken over the range of 1.8 eV to 3.5 eV
does not show emission associated with defects responsible for PPC may suggest either that the
process is a non-radiative one (e.g., a multi-phonon process), or a radiative process with lower
energy. Infrared PL experiments are planned to investigate this question further.

CONCLUSIONS

Metastable behavior evidenced as persistent photoconductivity has been observed at
room temperature in Si-doped GaN films grown by MOVPE on sapphire substrates Resistance
recovery to a dark level was incomplete even after several days. This study may suggest that the
PPC effect in the films which we have studied is due to localized lattice distortion at the interface
of the Si-doped and HR-GaN layers, or that trap states (possibly nonradiative in nature) exist in
our unintentionally doped GaN films which cause the highly resistive behavior. The PPC data
has been modeled using a stretched-exponential function, which has previously been applied to
decay kinetics of DX centers in AlGaAs.

ACKNOWLEDGEMENTS

The authors would like to thank S.C. Binari for helpful discussions and for ohmic contact
metallization, and R.J. Gorman for Hall transport measurements. GB gratefully acknowledges
the support of the National Research Council/NRL Research Associateship Program. This work
was sponsored in part by the Office of Naval Research.

REFERENCES

1. See, for example, D. Renfield and R.H. Bube, *Photoinduced Defects in Semiconductors*, Cambridge University Press (1996).
2. C. Johnson, J.Y. Lin, H.X. Jiang, M.A. Khan and C.J. Sun, Appl. Phys. Lett. **68** (13), p. 1808 (1996).
3. J.Z. Li, J.Y. Lin, H.X. Jiang, A.Salvador, A. Botchkarev and H. Morkoc, Appl. Phys. Lett. **69**, p. 1474 (1996).
4. F. Binet, J.Y. Duboz, E. Rosencher, F. Scholz and V. Harle, Appl. Phys. Lett. **69** (9), p. 1202 (1996).
5. S. Ruvimov, Z. Liliental-Weber, T. Suski, J.W. Ager III, J. Washburn, J. Krueger, C. Kisielowski, E.R. Weber, H. Amano and I. Akasaki, Appl. Phys. Lett. **69** (7), p. 990 (1996).
6. A.E. Wickenden, W.B. Alexander, D.D. Koleske and J.A. Freitas, Jr., J. Crystal Growth, in press (1996).
7. K. Doverspike, A.E. Wickenden, S.C. Binari, D.K. Gaskill and J.A. Freitas, Jr., Mat. Res. Soc. Symp. Proc. Vol. 395, p. 897 (1996).
8. A.E. Wickenden, L.B. Rowland, K. Doverspike, D.K. Gaskill, J.A. Freitas, Jr., D.S. Simons and P.H. Chi, J. Electron. Matl. **24** (11), p. 1547 (1995).
9. A.E. Wickenden, D.K. Gaskill, D.D. Koleske, K. Doverspike, D.S. Simons and P.H. Chi, Mat. Res. Soc. Symp. Proc. Vol. 395, p. 679 (1996).
10. For a complete discussion, see J.A. Freitas, Jr. A.E. Wickenden, D.D. Koleske and M.A. Khan, in press.
11. A technique developed for GaN growth on scandium layers; see D.D. Koleske, A.E. Wickenden, J.A. Freitas, Jr., R. Kaplan and S.M. Prokes, this volume.
12. A. Campbell and B. Streetman, Appl. Phys. Lett. 54, p. 445 (1989).
13. H. X. Jiang and J. Y. Lin, Phys. Rev. Lett. 64, p. 2547 (1990).
14. A. S. Dissanayake, M. Elahi, H. X. Jiang and J. Y. Lin, Phys. Rev. B 45, p. 13996 (1992).

PERSISTENT PHOTOCONDUCTIVITY IN P-TYPE GaN EPILAYERS AND N-TYPE AlGaN/GaN HETEROSTRUCTURES

J.Z. LI, J.Y. LIN, H.X. JIANG
Department of Physics, Kansas State University, Manhattan, KS 66506-2601

M.A. KHAN, Q. CHEN
APA Optics Inc., 2950 N. E. 84th Lane, Blaine, Minnesota 55449

A. SALVADOR, A. BOTCHKAREV, H. MORKOC
Materials Research Laboratory and Coordinated Science Laboratory,
University of Illinois at Urbana-Champaign, Urbana, Illinois 61801

ABSTRACT

Persistent photoconductivity (PPC) effect has been observed in p-type GaN epilayers grown both by metal-organic chemical vapor deposition (MOCVD) and reactive molecular beam epitaxy (MBE) as well as in a two-dimensional electron gas (2DEG) system formed by an AlGaN/GaN heterostructure grown by MOCVD. Its properties have been investigated at different conditions.

INTRODUCTION

GaN based devices offer great potential for applications such as high power electronics, UV-blue lasers, and solar-blind UV detectors. Researchers in this field have made extremely rapid progress toward materials growth as well as device fabrication.[1] An important aspect remains to be understood and improved is the p-type doping of GaN. Understanding and control of impurity properties and p-type conduction in these materials remain one of the foremost obstacles hindering device efforts.

High electron mobility transistors (HEMT) and field effect transistors (FET), based on AlGaN/GaN heterostructures, hold promise for high frequency microwave as well as for high-power and high-temperature electronic device applications and offer the advantage of high carrier mobilities due to the formation of two-dimensional electron gas (2DEG) by a heterojunction.[2,3] However, practical operation of these devices still requires detailed material and device characterization and optimization. In this paper, we report the observation of persistent photoconductivity (PPC) effect in p-type GaN epilayers grown both by MOCVD and MBE as well as in a 2DEG system formed by an AlGaN/GaN heterostructure grown by MOCVD. Its behavior has been studied at different conditions.

Mat. Res. Soc. Symp. Proc. Vol. 449 © 1997 Materials Research Society

SAMPLES

Samples used here were grown either by MOCVD or reactive MBE. For p-type GaN epilayers grown by MOCVD, triethylgallium, trimethylindium, and ammonia were used as the precursor. Prior to deposition of the GaN epilayer a thin 50 nm AlN buffer layer was grown on the sapphire (Al_2O_3) substrate. The thickness of GaN epitaxial layer was about 0.2 μm. Mg doping was provided by transporting bismethylcyclopentadienyl magnesium (MCp_2Mg) into the growth chamber with ammonia during the growth. Post-growth annealing at 750 °C in nitrogen gas ambient at 76 torr for about 20 minutes resulted in p-type conduction. MBE grown p-type GaN epilayers were deposited on sapphire substrates and doped with Mg. Prior to the GaN growth a AlN buffer layer with a thickness of about 65 nm was used.

The 2DEG of AlGaN/GaN heterostructure sample consisted of a 2 μm highly insulating GaN epilayer followed by a 25 nm thick n-type GaN conducting channel, again followed by a 25 nm thick n-type $Al_{0.1}Ga_{0.9}N$ epilayer. The structure was deposited over a basal plane sapphire substrate using MOCVD. The typical room temperature carrier concentrations were respectively 1 x 10^{17} cm^{-3} and 5 x 10^{17} cm^{-3} for n-GaN and n-$Al_{0.1}Ga_{0.9}N$ epilayers grown under similar conditions.[3] The formation of 2DEG at the AlGaN/GaN interface has been confirmed previously.[2,3]

MEASUREMENTS

Ohmic contacts for p-type GaN epilayers samples were formed by depositing four Au and followed by In spots with diameters of about 1 mm. For n-type 2DEG AlGaN/GaN heterostructure sample, four gold wires were soldered directly onto the sample using In. The carrier concentrations and mobilities were measured by the variable-temperature Hall-effect technique. For the PPC measurements, illumination of the sample was achieved using a mercury lamp (hν > E_g of GaN) or a neon lamp (hν < E_g of GaN).

Low temperature photoluminescence (PL) spectra were measured by using a picosecond laser spectroscopy system with an average power of about 20 mW, a tunable photon energy up to 4.5 eV, and a spectral width of about 0.2 meV.[4]

RESULTS AND DISCUSSIONS

Fig. 1 shows persistent photoconductivity (PPC) behavior in a p-type GaN epilayer grown by MOCVD measured at two representative temperatures (a) T=200 K and (b) T=240 K. It can be seen that the light-enhanced conductivity persists for a very long period of time and that the PPC relaxation time, τ_d, increases with a decrease of temperature. Since the hole mobility, μ_h, does not change with time at a fixed temperature, the decay of PPC reflects the capture of

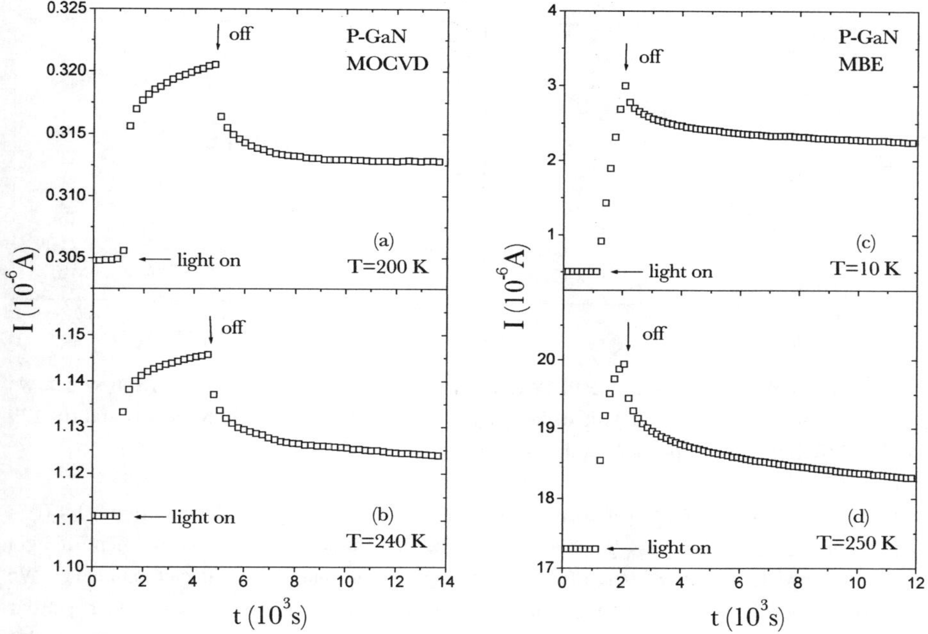

Fig. 1. PPC build-up and decay characteristics in p-type GaN epilayers grown by MOCVD measured at (a) T=200 K and (b) T=240 K and by MBE measured at (c) T=10 K and (d) T=250 K.

photoexcited free holes by the ionized Mg impurities. A typical PPC behavior in a p-type GaN epilayer grown by MBE is illustrated in Fig. 1 (c) T=10 K and (d) T=250 K, which show that the conductivity increases by more than one order of magnitude after exposure to light and that the PPC effect seen in MBE grown layers is much more pronounced than that in the MOCVD grown p-type GaN epilayers.

We have measured the PPC decay time constants, τ_d, in both p-type samples at different temperatures. Fig. 2 shows the Arrhenius plot of τ_d (ln τ_d vs $1/T$) for (a) MOCVD and (b) MBE grown p-type GaN epilayers. Although the magnitudes of PPC in MOCVD and MBE p-type GaN epilayers are quite different, τ_d measured in both MOCVD and MBE epilayers shows a thermally activated behavior as indicated by the solid lines in Fig. 2, which gives an energy barrier for the hole capture E_c to be about 55 meV in MOCVD and 129 meV in MBE grown p-type epilayers, respectively.[5,6]

PPC effect has also been observed in a 2DEG system formed by an AlGaN/GaN heterostructure. This is illustrated in Fig. 3, where the PPC build-up and decay kinetics are shown for two representative temperatures, (a) T=40 K and (b) T=300 K. The decay kinetics

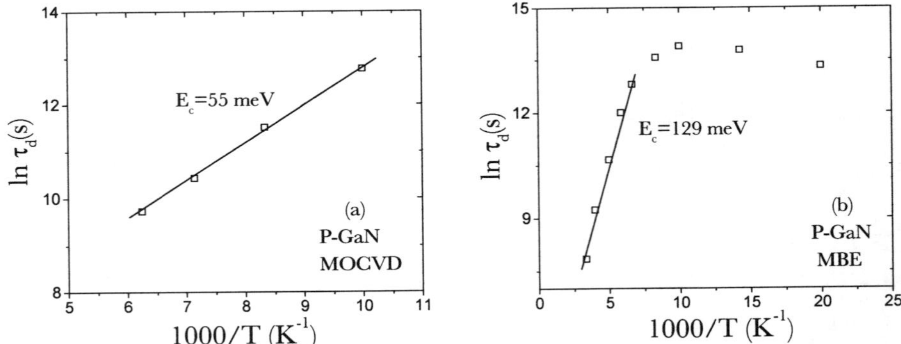

Fig. 2. The Arrhenius plots of the PPC decay time constant τ_d (ln τ_d *vs* *1/T*), from which we obtain an energy barrier for the hole capture E_C to be about (a) 55 meV in MOCVD and (b) 129 meV in MBE grown p-type GaN epilayers.

of PPC seen in AlGaN/GaN heterostructures follow stretched-exponential functions, $I_{PPC}(t) = I_{PPC}(0)\exp[-(t/\tau)^{\beta}]$, $(\beta < 1)$. $I_{PPC}(0)$ is defined as the PPC buildup level at the moment of light excitation being removed, τ is the PPC decay time constant, and β is the decay exponent. We have performed a comparison experiment on n-type GaN epilayers grown under similar conditions and found that PPC effect is absent in n-type GaN epilayers. Furthermore, we also found that illumination by both mercury lamp ($h\nu > E_g$ of GaN) and neon lamp ($h\nu < E_g$ of GaN) give similar PPC results in the AlGaN/GaN heterostructure investigated here. It has been demonstrated previously that PPC,[7] caused by spatial separation of photogenerated electrons and holes by electric field from macroscopic barrier due to band bending, decays logarithmically in time. In fact, the observed PPC decay kinetics shown in Fig. 3(a) and (b) are identical to those seen in Si doped $Al_{0.3}Ga_{0.7}As$ materials,[8] in which the deep level impurities (or DX centers) are a well known cause of PPC. Thus we feel that the PPC effect seen in 2DEG system here is most likely associated with the transfer of photoexcited electrons from deep level impurities in the AlGaN barrier material. The PPC decay time constants, τ, are very long, especially at low temperatures. At temperatures T > 200 K, τ is thermally activated, from which we obtain an energy barrier for the electron capture to be about 230 meV for the heterojunction investigated here.

PL properties of the AlGaN/GaN heterostructure have also been studied. Fig. 3(c) shows an emission spectrum measured at T=10 K, in which three emission lines are clearly resolved. The emission lines can be fit quite well by Lorentzian functions,

$$I(E) = \sum_{i=1}^{3} \frac{2A_i}{\pi} \frac{\Gamma_i}{4(E-E_i)^2 + \Gamma_i^2} \tag{1}$$

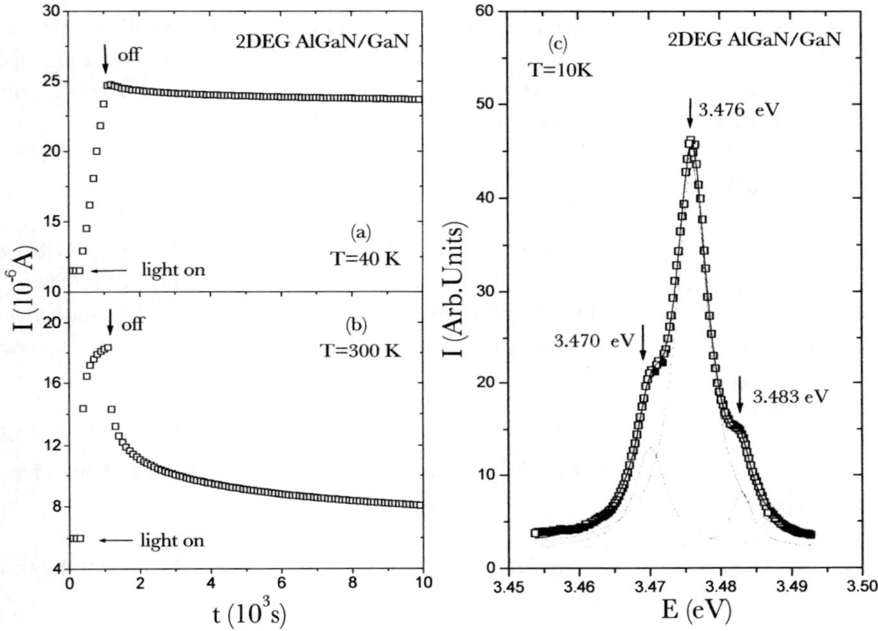

Fig. 3. PPC build-up and decay characteristics in a 2DEG system formed by an AlGaN/GaN heterostructure measured at (a) T=40 K and (b) T=300 K; (c) PL spectrum measured at T=10 K. The solid curve in (c) is the least square fit of data with Eq. (1).

as shown as the solid curve in Fig. 3(c). The dotted lines are the fitted three individual peaks of the emission lines. In Eq. (1), A_i, Γ_i, and E_i (i=1, 2 and 3) denote the total luminescence intensities, the full widths at half maximum (FWHM), and the peak positions of the observed three transition lines, respectively. The fitted values are E_1=3.470 eV, E_2=3.476 eV, E_3=3.483 eV, Γ_1=5.1 meV, Γ_2=5.7 meV, and Γ_3=3.5 meV from the T=10 K PL spectrum.

We have attempted to measured the recombination lifetimes of these emission lines. However, we found that these transitions decay faster than our system response (20 ps). This seems to indicate that non-radiative recombination may be a dominant process in the 2DEG sample investigated here.

CONCLUSION

PPC effect has been observed in p-type GaN epilayers grown by MOCVD and reactive MBE as well as in a 2D electron gas systems formed by an AlGaN/GaN heterostructure and has been systematically investigated. Our studies seem to indicate that PPC is a rather common property associated with impurities in wide bandgap semiconductors.

ACKNOWLEDGEMENTS

The research at Kansas State University is supported by ARO and ONR/BMDO (monitored by Dr. John Zavada and Dr. Yoon S. Park), and DOE (96ER45604/A000). The research at the University of Illinois is supported by ONR, AFOSR and BMDO and monitored Max Yoder, Yoon S. Park, C. Wood, G. L. Witt, K. P. Wu, J. Johnson, and S. Hammonds.

REFERENCES

1. H. Morkoc, S. Strite, G. B. Gao, M. E. Lin, B. Sverdlov, and M. Burns, J. Appl. Phys. **76**, 1363 (1994).

2. M. Asif Khan, M. S. Shur, J. N. Kuznia, Q. Chen, J Burn, and W. Schaff, Appl. Phys. Lett. **66**, 1083 (1995).

3. M. Asif Khan, Q. Chen, C. J. Sun, J. W. Yang, M. S. Shur, and H. Park, Appl. Phys. Lett. **68** 514 (1996).

4. M. Smith, G. D. Chen, J. Y. Lin, H. X. Jiang, A. Salvador, W. K. Kim, O. Aktas, A. Botchkarev, and H. Morkoc, Appl. Phys. Lett. **67**, 3387 (1995).

5. C. Johnson, J. Y. Lin, H. X. Jiang, M. A. Khan, and C. J. Sun, Appl. Phys. Lett. **68**, 1808 (1996).

6. J. Z. Li, J. Y. Lin, H. X. Jiang, A. Salvador, A. Botchkarev, and H. Morkoc, Appl. Phys. Lett. **69**, 1474 (1996).

7. H. J. Quieser and D. E. Theodorou, Phys. Rev. **B33**, 4027 (1986).

8. A. Dissanayake, M. Elahi, H. X. Jiang, and J. Y. Lin, Phys. Rev. **B45**, 13996 (1992); J. Y. Lin, A. Dissanayake, G. Brown, and H. X. Jiang, Phys. Rev. **B42**, 5855 (1990).

MAGNETIC RESONANCE STUDIES OF HIGH-RESISTIVITY GaN FILMS GROWN ON Al_2O_3

E.R. GLASER, T.A. KENNEDY, A.E. WICKENDEN, D.D. KOLESKE, and J.A. FREITAS, Jr.
Naval Research Laboratory, Washington, D.C. 20375-5347, glaser@bloch.nrl.navy.mil

ABSTRACT

We have made an attempt to obtain electronic and atomic structure information on the residual defects that exist in high-resitivity GaN epitaxial layers from optically-detected magnetic resonance (ODMR) experiments performed on the broad 3.0 and 2.2 eV photoluminescence (PL) bands observed from these films. The first observation of magnetic resonance on this 3.0 eV band reveals two ODMR signals. The first resonance is sharp (FWHM ~ 3.5 mT) with g ~ 1.950 and is assigned to effective-mass (EM) donors based on previous studies. The second feature is much broader (FWHM ~ 18 mT) with a donor-like g-value of ~ 1.977. This new resonance may be associated with partially EM-like donor states located ~ 54 - 57 meV below the conduction band edge proposed recently to be involved in this 3.0 eV PL band.

INTRODUCTION

There has been much progress recently in the growth of high-resistivity (HR) GaN layers for which the numbers of residual donors and acceptors is much lower than those typically found for as-grown, undoped films. There have now been reports of HR GaN films grown by both organo-metallic chemical vapor deposition (OMCVD) [1] and molecular beam epitaxy (MBE) [2] techniques. Most noteworthy, these HR GaN films are important building blocks for high-power microwave and millimeter wave electronic devices fabricated from this material system, such as heterojunction field-effect transistors (HFETs) [3,4]. However, the identity of the defect(s) responsible for the high-resistivity character is still not known.

Recent low-temperature PL studies [5] of HR GaN films grown at NRL have revealed strong, free excitonic recombination at 3.487 eV and a weak, broad emission band with peak energy at 2.2 eV (the so-called "yellow" emission band). In addition, these films exhibit a second weak and broad PL band with peak energy at 3.0 eV, not typically observed from undoped, n-type GaN layers. Two other groups [6,7] have also observed the broad 3.0 eV band from OMCVD-grown GaN layers that exhibit both donor-bound and free excitonic recombination at low temperatures.

In this work we have made a preliminary attempt to obtain electronic and atomic structure information on the residual defects that exist in the HR GaN films grown at NRL from optically-detected magnetic resonance (ODMR) experiments performed on these emission bands. ODMR and related optical/magnetic resonance techniques will be particularly useful if the defects that lead to the high-resistivity electrical behavior also participate (either directly or indirectly) in the recombination processes observed from these layers. Previous ODMR work by our group [8,9] and several others [6,10,11] on emission bands from undoped (n-type), Mg-, and Zn-doped GaN epitaxial layers has revealed evidence for shallow, effective-mass like and deep states of donor or acceptor character.

We will discuss in this paper the first ODMR results obtained on the broad 3.0 eV PL band. The ODMR is compared with that obtained on the 2.2 eV band found from the same layer. Two luminescence-increasing signals are found on the 3.0 eV emission. The first resonance has g ~ 1.950 and is assigned to effective-mass (EM) donors based on previous studies [6,8-11]. The

Mat. Res. Soc. Symp. Proc. Vol. 449 © 1997 Materials Research Society

second feature also has a donor-like g-value of ~ 1.977. This feature may be associated with partially EM-like donor states located ~ 55 meV below the GaN conduction band edge proposed recently by the Freiburg group [6] to be involved in the 3.0 eV emission band. Within this model, residual acceptors with binding energy of ~ 116 meV (not observed in these ODMR experiments) are the recombination partners in this emission and may play an important role as compensation centers and, hence, with the high-resistivity character of these GaN films.

EXPERIMENTAL BACKGROUND

The PL and ODMR experiments reported in this work were performed on a 2.5 μm - thick, HR GaN film grown on a-plane (1120) sapphire by OMCVD. Similar PL results as discussed below were obtained on a second HR GaN film grown in the same reactor ~ 14 months apart from the sample that is the focus of this paper. The growth of these HR GaN layers was achieved through an optimum combination of nucleation layer temperature, nucleation layer thickness, NH_3/TMG ratio, and GaN growth temperature [1]. The resistivity of the layer at 300 K was estimated to be ~ 10^{10} Ω-cm from I-V measurements made with large area In contacts on samples from the same wafer.

The PL from the HR GaN layers was excited with above-band-gap radiation provided by the 325 nm line of a HeCd laser at a power density of ~ 40 W/cm^2. The emission from 1.8 - 3.6 eV was analyzed by a 0.85-m double-grating (1800 lines/mm) spectrometer and detected with a UV-enhanced GaAs photomultiplier tube.

The ODMR was detected as the change in the total intensity of PL which was coherent with the on/off amplitude modulation (35 Hz - 10 kHz) of 50 mW of microwave power. The GaN film was studied in a K-band (24 GHz) spectrometer. For the ODMR, the PL was continuously generated by the 351.1 nm line from an Ar$^+$-ion laser with typical power densities near 1 W/cm^2. The emission was detected by a room-temperature, UV-enhanced Si photodiode. A combination of visible cut-off and bandpass filters were placed in front of the detector to separately study the emission bands observed from the HR GaN layer. In addition, the PL below 3.4 eV was analyzed at 1.6 K for these excitation conditions by a 0.25-m double-grating (1200 lines/mm) spectrometer and detected by the same Si photodiode employed for the ODMR studies.

RESULTS AND DISCUSSION

The photoluminescence obtained between 1.8 and 3.6 eV at 6K from the HR GaN film reported in this work is shown in Fig. 1. Three recombination bands are revealed.

The dominant emission observed from this layer is a sharp peak (FWHM ~ 5 meV) at 3.487 eV (labeled FX). This band has been assigned to free excitonic recombination (associated with the n=1 ground state of the Γ_9 valence band) from temperature dependent PL studies [5]. The strongest near band-edge PL emission typically observed from undoped GaN layers with residual room temperature electron concentrations $\geq 3 \times 10^{16}$ cm^{-3} studied at NRL and reported by many other groups is attributed to excitons bound to neutral donors. This suggests that the residual numbers of shallow donors and acceptors in this HR GaN film are less than 10^{16} cm^{-3}. Also, the reduced numbers of both residual donors with E_d ~ 35 meV and acceptors with E_a ~ 200 meV in this layer is supported by the absence of a significant shallow donor - shallow acceptor recombination band with zero-phonon line (ZPL) at ~ 3.27 eV [12].

Two broad emission bands with similar Gaussian lineshapes and linewidths (~ 400 meV) are also found from this HR GaN film with peak energies at ~ 2.2 and 3.0 eV, respectively. These bands are much weaker in amplitude than the FX recombination observed from this film. In ad-

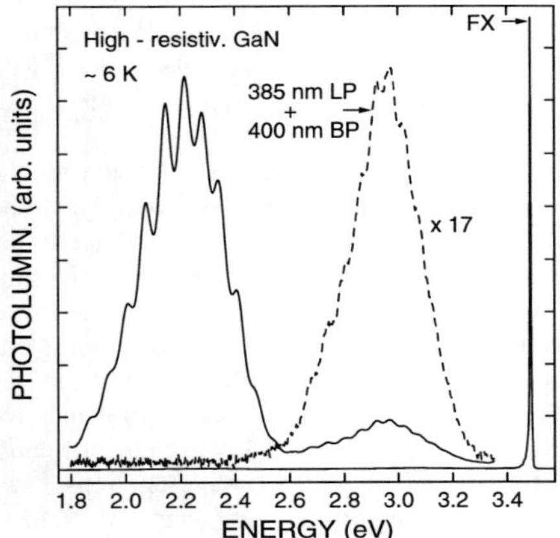

Fig. 1. PL spectra obtained from the high-resistivity GaN film at ~ 6K. The curves displaced vertically show the PL < 3.4 eV under the same excitation conditions as employed for the ODMR studies (see text). The periodic structure on the 2.2 and 3.0 eV bands is due to Fabry-Perot interference.

dition, the intensity of the 2.2 eV emission band is about six times as strong compared to the 3.0 eV PL band under the excitation power conditions employed for the ODMR studies to be discussed below. We note that the Gaussian lineshapes and broad linewidths suggest strong electron (or hole) - lattice coupling interactions for these bands, as is typically observed for deep emission bands from a variety of semiconductors involving a center with a very localized wavefunction.

The 2.2 eV PL band observed from this HR film has been invariably found to some degree of strength from most undoped, n-type GaN layers grown at NRL and from many other laboratories. The microscopic origin of this band is still under discussion. We have not found evidence for the 3.0 eV band from n-type GaN films grown at NRL with carrier concentrations $\geq 3 \times 10^{16}$ cm^{-3}.

The ODMR spectra obtained at 24 GHz on the 2.2 and 3.0 eV emission bands from this HR GaN film with **B** $\perp$ c are shown in Fig. 2. The combination of a 385 nm long-pass filter and a 400 nm band-pass filter was ideal to spectrally isolate the 3.0 eV band from all other emission bands (see dashed curve in Fig. 1). Three distinct resonances are revealed in these spectra. A summary of the magnetic resonance parameters for the three defect states observed on these emission bands is given in Table I.

The g-tensors and linewidths obtained for the resonances (labeled EM and (DD)$_2$) detected on the 2.2 eV band (bottom half of Fig. 2) are identical within experimental error with those found for the two features observed on similar emission from undoped, n-type GaN layers reported previously by our group [8,9] and by several other workers [6,10,11]. We have assigned these features to effective-mass (EM) and deep donor (DD) states, respectively, based on these magnetic resonance parameters [8].

Two luminescence-increasing ODMR signals are also found on the 3.0 eV PL band (top half of Fig. 2). The first ODMR feature has nearly the same magnetic resonance parameters as the line assigned to EM donors observed on the 2.2 eV PL band. Thus, we also assign this sharp resonance to electrons bound at effective-mass donors.

The second ODMR signal (labeled (DD)$_1$) obtained on this band is a broad resonance with g ~ 1.977(2). The full-width at half-maximum amplitude of this new feature is ~ 18 mT. Preliminary angular rotation studies at 24 GHz indicate that the resonance, within experimental error, is isotropic. In addition to the different g-tensors found for the two broad ODMR features ((DD)$_1$ and (DD)$_2$) observed from this HR GaN layer, it is also seen that the resonance labeled (DD)$_1$ can be fitted with a Lorentzian lineshape (dashed curve in Fig. 2) in contrast to the Gaussian lineshape (dotted curve in Fig. 2) typically observed for the ODMR signal labeled (DD)$_2$ detected on the 2.2 eV emission band from this HR GaN layer and on similar emission from the usual undoped, n-type GaN films. The linewidths determined from these fits are given

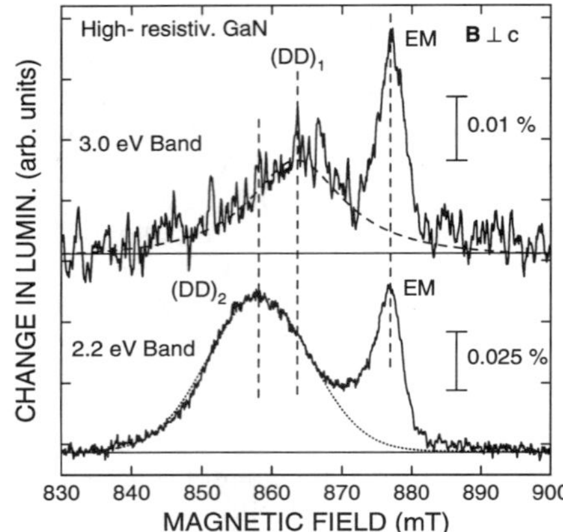

Fig. 2. ODMR spectra found at 24 GHz on the 3.0 and 2.2 eV emission bands from the HR GaN layer with **B** $\perp$ c. The dashed and dotted curves are lineshape fits (see text) of the broad resonances labeled $(DD)_1$ and $(DD)_2$, respectively.

in Table I. We also note that no evidence was found from wider magnetic field scans for a resonance that could be associated with acceptors of shallow or deep character in this film.

Following the arguments [8] made for the assignments of the resonances labeled EM and $(DD)_2$, the negative g-shift with respect to the free electron g-value of 2.0023 found for this new resonance ($(DD)_1$ with g ~ 1.977 suggests an assignment of this line to a third donor-like state in this HR GaN film. Based on the decreasing shift of the g-values from the free electron g-value of 2.0023 expected for donors with more localized wavefunctions, we tentatively assign the new resonance ($(DD)_1$) to an intermediate donor state located below the conduction band (CB) edge between the EM shallow donors with g ~ 1.950 and the deeper donors ($(DD)_2$) with g ~ 1.993.

High-resolution, temperature-dependent PL studies suggest that the broad 3.0 eV emission band is due to strongly phonon-coupled distant DAP recombination (with very weak ZPLs observed near 3.37 eV) involving three residual donor states (with E_d ~ 34, 54, and 57 meV, respectively) and residual acceptor centers (with E_a ~ 116 meV) [6]. Thus, it is plausible that the new ODMR feature with g ~1.977 may be associated with the deeper donor states (partially EM-like) located ~ 54 - 57 meV below the CB edge proposed to be involved in this broad 3.0 eV PL band. Within this model, the residual acceptors with binding energy of ~ 116 meV (not found in these ODMR experiments) participating in this emission may play an important role as compensation centers of residual shallow donors and, hence, with the high-resistivity character of these GaN films.

Table I. Magnetic resonance parameters of the three defect states found from the high-resistivity GaN film discussed in this work and, for comparison, of the acceptor-related states from heavily Mg- and Zn-doped GaN layers reported previously.

Sample	Defect	PL Band	g-values	FWHM	Ref.
HR GaN	EM-Donor	3.0 eV, 2.2 eV	{$g_{||} = 1.9525(5)$, $g_\perp = 1.9485(5)$}	3.5 mT, 4.8 mT	This Work
	$(DD)_1$	3.0 eV	g = 1.977(2); ≈ isotropic	16.6 mT	
	$(DD)_2$	2.2 eV	$g_{||} = 1.992(2)$, $g_\perp = 1.995(2)$	17.7 mT	
GaN:Mg	Mg-related acceptor	~ 3.0 eV	$g_{||}$~ 2.085, $g_\perp$~ 2.000	20-30 mT	[6,8]
GaN:Zn	Zn-related acceptor	~ 2.9 eV	$g_{||} = 1.997(2)$, $g_\perp = 1.992(2)$	7 mT	[6]

In addition to the 3.0 eV PL band, the Freiburg group [6] observed a free-to-bound transition involving the same residual acceptors with $E_a \sim 116$ meV. This band is characterized by a ZPL at ~ 3.41 eV and a series of weaker LO-phonon replicas ($E_{LO} \sim 92$ meV) at lower energies. This suggests effective-mass character for these residual acceptors. Thus, the deeper donors with $E_d \sim 55$ meV described above may be the source of the strong electron-phonon coupling interaction that give rise to the 3.0 eV band. This is supported by the larger degree of wavefunction localization inferred from the g-value (~ 1.977) associated with the new resonance detected on this band compared to the g-value (~ 1.950) found for the effective-mass donors with $E_d \sim 34$ meV. We note that the origin of the Lorenztian lineshape and the broad linewidth found for this new feature (see Fig. 2) are not understood at this time.

The results of microwave modulation frequency studies of the ODMR found on the 2.2 and 3.0 eV PL bands from this HR GaN layer reveal evidence for different lifetimes and/or recombination processes associated with these emission bands. In particular, the ODMR features on the 3.0 eV band were only strongly observed for a very limited range of microwave modulation frequencies ($\nu \sim 3 - 5$ kHz). However, the ODMR on the 2.2 eV PL band was found for ν between 35 Hz and 10 kHz. More details of these frequency studies will be discussed in a future publication.

Finally, an interesting comparison can be made between the ODMR found on the 3.0 eV band from this HR GaN film and the ODMR previously reported [6,8] on PL bands with similar peak energies ($\sim 2.9 - 3.1$ eV) and linewidths from GaN layers doped heavily with either Mg or Zn. One common feature is the EM donor resonance that is found on this emission for all three cases. However, the g- tensor of the broad ODMR feature ($(DD)_1$) observed on the 3.0 eV band from the present HR GaN film is quite different compared to the g-tensors (see Table I) found for the perturbed, optically-active Mg and Zn acceptor states involved (along with the EM donors) in the ~ 3.0 eV DAP recombination bands from these heavily-doped GaN layers [6,8].

SUMMARY

Optically-detected magnetic resonance experiments have been performed on the broad 3.0 and 2.2 eV PL bands from high-resistivity GaN layers grown on a-plane sapphire substrates by OMCVD. The ODMR on the 2.2 eV band revealed two resonances with $g \sim 1.950$ and $g \sim 1.993$ that were observed previously on similar emission from undoped, n-type GaN layers [6,8-11]. These features are attributed to effective-mass (EM) and deep donor (DD) states, respectively [8]. The first ODMR results reported for this 3.0 eV band revealed the same EM donor state and a new donor-like resonance with $g \sim 1.977$. The new feature may be associated with partially EM-like donor states located $\sim 54 - 57$ meV below the GaN conduction band edge proposed to be involved in this 3.0 eV emission band [6]. Within this model for the broad 3.0 eV PL band, the residual acceptors with $E_a \sim 116$ meV (not observed in these ODMR experiments) may play an important role as compensation centers and, hence, with the high-resistivity character of these GaN films.

Further experiments are planned to provide more detailed information on the atomic and electronic structure of the residual defects revealed from these ODMR studies. These include ODMR experiments at 35 GHz in order to better resolve the degree and sense of g-anisotropy associated with the new broad ODMR signal detected on the 3.0 eV band from this HR GaN layer. In addition, optically-detected electron-nuclear double resonance (ODENDOR) experiments will be performed on the 2.2 and 3.0 eV bands which can, in principle, provide chemical information towards the identity of the residual point defects in these layers through electron-nuclear hyperfine interactions. The first ODENDOR study of undoped, n-type GaN films has been recently reported [13].

ACKNOWLEDGMENTS

This work was supported by the Office of Naval Research.

REFERENCES

1. K. Doverspike, A.E. Wickenden, S.C. Binari, D.K. Gaskill, and J.A. Freitas, Jr., Mat. Res. Soc. Symp. Proc. Vol. **395**, 897 (1996) and references therein.
2. M. Smith, G.D. Chen, J.Z. Li, J.Y. Lin, H.X. Jiang, A. Salvador, W.K. Kim, O. Aktas, A. Botchkarev, and H. Morkoç, Appl. Phys. Lett. **67**, 3387 (1995).
3. S.C. Binari, Electrochem. Soc. Proc. **95-21**, 137 (1995).
4. M.A. Khan, Q. Chen, J.W. Yang, M.S. Shur, B.T. Dermott, and J. Higgins, IEEE Electron Device Lett. **17**, 325 (1996).
5. J.A. Freitas, Jr., K. Doverspike, and A. Wickenden, Mat. Res. Soc. Symp. Proc. Vol. **395**, 485 (1996).
6. U. Kaufmann, M. Kunzer, C. Merz, I. Akasaki, and H. Amano, Mat. Res. Soc. Proc. Vol. **395**, 633 (1996) and references therein.
7. W. Shan, X.C. Xie, J.J. Song, and B. Goldenberg, Appl. Phys. Lett. **67**, 2512 (1995).
8. E.R. Glaser, T.A. Kennedy, K. Doverspike, L.B. Rowland, D.K. Gaskill, J.A. Freitas, Jr., M. Asif Khan, D.T. Olson, J.N. Kuznia, and D.K. Wickenden, Phys. Rev. B **51**, 13326 (1995) and references therein.
9. E.R. Glaser, T.A. Kennedy, S.W. Brown, J.A. Freitas, Jr., W.G. Perry, M.D. Bremser, T.W. Weeks, and R.F. Davis, Mat. Res. Soc. Proc. Vol. **395**, 667 (1996).
10. D.M. Hofmann, D. Kovalev, G. Steude, D. Volm, B.K. Meyer, C. Xavier, T. Monteiro, E. Pereira, E.N. Mokov, H. Amano, and I. Akasaki, Mat. Res. Soc. Proc. Vol. **395**, 619 (1996).
11. P.W. Mason, G.D. Watkins, and A. Doernen, in Proceedings of the 1996 Fall Meeting of the Materials Research Society, III-V Nitrides, to be published.
12. R. Dingle and M. Ilegems, Solid State. Commun. **9**, 175 (1971).
13. F.K. Koschnick, K. Michael, J.-M. Spaeth, B. Beaumont, and P. Gibart, Phys. Rev. B **54**, R11042 (1996).

OBSERVATION OF MID-GAP STATES IN GaN WITH OPTICAL-ISOTHERMAL CAPACITANCE TRANSIENT SPECTROSCOPY

P. HACKE, H. MIYOSHI, K. HIRAMATSU[1], H. OKUMURA, S. YOSHIDA, H. OKUSHI
Electrotechnical Laboratory, 1-1-4 Umezono, Tsukuba-shi 305 Japan
[1] Nagoya University Dept. of Electronics Eng. Furo-cho, Chikusa-ku 464-01 Japan

ABSTRACT

Optical-isothermal capacitance transient spectroscopy (O-ICTS) was used to distinguish the deep levels which occur in unintentionally doped n-type GaN by means of their characteristic optical cross section. GaN grown by metalorganic vapor phase epitaxy (MOVPE) and hydride vapor phase epitaxy (HVPE) were compared. Correspondence between optical and thermal emission characteristics of previously discovered levels, E2 ($\sim E_c$–0.55 eV) and E4 ($\sim E_c$–1.0 eV), were clearly determined by observing their sequential appearance in the ICTS spectra. Whether by thermal or optical stimulation, the emission from E4 was found to be broad in nature; it is consequently believed to involve a defect. The total measured concentration of deep levels, including a prominent level which photoionizes in the range 2.5 to 3.0 eV below the conduction band, is greater in the GaN grown by MOVPE than by HVPE that was tested.

INTRODUCTION

GaN has been shown to be a good material for the fabrication of light emitting diodes and laser diodes operating in the blue to UV region and transistors; however, carrier traps existing within the band gap of a semiconductor can alter luminescent recombinations, act as non-radiative recombination sites, and compensate carriers. For high efficiency devices, it is desirable to characterize the deep levels for the ultimate goal of controlling their concentration.

In unintentionally doped n-type GaN, a series of deep levels below the conduction band have already been characterized by deep level transient spectroscopy (DLTS). E1 (E_c–0.2 eV) exists in GaN grown by MOVPE[1] as well as HVPE[2] in concentrations of $\sim 1 \times 10^{14}$ cm^{-3} or below and may originate from some impurity existing in low concentration. E2 (E_c–0.55 eV) usually exists in GaN grown by both MOVPE[1] and HVPE[2] with concentration in the mid 10^{14} cm^{-3} range. It has been shown that the concentrations of both E1 and E2 can be minimized so that they are inconsequential.[3,4] Level E3 has been found in HVPE-GaN with energy (E_c–0.66 eV) and concentration in the mid 10^{14} cm^{-3} range or lower.[2] A level near this energy has been found in N implanted MOVPE-GaN; it was therefore suggested that it could be an energy state caused by the N interstitial.[5] Levels E2 and E3 were previously characterized by isothermal capacitance transient spectroscopy (ICTS) showing that the thermal emission from these traps are purely exponential in nature with a unique time constant and energy.[2] Deeper levels have been observed with energies between 0.8 eV[6] and 1.6 eV[4] by DLTS and 1.0 eV[7] by ICTS; however, high temperature required for the thermal stimulation of carriers from these deeper levels involves greater measurement uncertainty.

For the characterization of mid gap states below 1.0 eV from the conduction band edge, it is useful to use optical stimulation. Steady state photocapacitance results indicated the existence of a deep level with threshold energy 0.87 eV followed by a series of somewhat less resolvable levels which begin to photoionize at about 0.97, 1.25 and 1.45 eV.[8] With higher energy excitation, photoemission capacitance revealed a strong peak with full ionization at 2.5 eV and a relationship between this signal and the intensity of the yellow band photoluminescence emission was shown.[9]

The yellow band emission has been attributed to a number of causes. A model involving a deep donor to shallow acceptor recombination was proposed by Glaser, Kennedy and coworkers.[10] On the other hand, numerous studies indicate a transition from near the conduction band to a deep level in the lower region of the band gap.[9,11-13] Proposed origins of this deep state include the gallium vacancy[14], impurities (such as C)-Ga vacancy complex.[11], and the nitrogen antisite.[15] Other studies have found that the yellow emission is stronger with defects from ion-implantation[16] or at dislocations[17].

Capacitance transient methods using photonic stimulation to detect trapped carriers in GaN exist as described above; however, the above work typically involves measurement of only the change in capacitance some period after a trap filling voltage pulse while the Schottky capacitor is under illumination. With these methods, information about the optical cross-section of the traps which controls the rate at which carriers are stimulated is discarded. In this work, we use optical-isothermal

Mat. Res. Soc. Symp. Proc. Vol. 449 © 1997 Materials Research Society

capacitance transient spectroscopy to simultaneously observe thermal and optical emission processes from deep levels in n-type GaN in a spectroscopic manner so that the deep levels can be distinguished by their characteristic thermal or optical emission time constant. This spectroscopic method yields information about the correspondence of the levels observed by optical and thermal stimulation means, and the various mid gap-states can be more clearly characterized by the signature of their optical cross-section.

Secondly, we compare the O-ICTS spectrum of GaN fabricated by MOVPE and HVPE. Crystals grown by these two methods are different in two ways. First, the HVPE film is ~ 0.1 mm thick whereas the MOVPE film is ~ 6 μm thick. The HVPE-GaN is believed to have a relaxed lattice;[18] the thermal mismatch is accommodated in part by cracks which originate at the sapphire interface – sometimes up to the surfaces.[19] Dislocations have been observed to extend to the surface;[19,20] however, the concentration is lower than in MOVPE-grown GaN films grown on sapphire, which exhibit a rotated cell structure with numerous dislocations to accommodate the thermal mismatch.[21] Secondly, the HVPE growth uses inorganic precursors and growth is carried out at high temperature (1090°C) which is found to yield good quality GaN.[22]

EXPERIMENT

The MOVPE-grown GaN sample was deposited on sapphire using an AlN buffer layer.[23] The 24°C Hall carrier concentration and mobility are respectively 7.8×10^{16} cm^{-3} and 500 cm^2/V·s. C-V measurements yielded N_D-N_A=5.9×10^{16} cm^{-3} at -130°C (at which temperature most of this O-ICTS work was done). The HVPE sample displays bulk Hall carrier concentration 4.8×10^{16} cm^{-3} and mobility 200 cm^2/V·s at room temperature; however, it must be noted that the quality is likely to be higher near the crystal surface than the bulk and sapphire interface region.[19] C-V measurements yielded a net donor concentration N_D-N_A=1.2×10^{16} cm^{-3} at -130 °C.

Capacitance measurements were done using Au Schottky contacts of area $A = 0.0769$ mm^2 with larger area Ga ohmic contacts on the crystal face. A 100 W quartz tungsten halogen lamp and a 675 μm grating monochromator with appropriate filters was used to provide photons in a narrow band centered at energy E in the range 0.83 to 3.26 eV for photoionization of trapped charges in the depletion region. The monochromator output was focused onto the Schottky contact region with a quartz lens. The Schottky barrier was placed in reverse bias V = –4 V with 100 ms trap filling pulses. The reverse bias leak current was less than 0.1 μA at –10 V.

Details of the ICTS method using thermal and optical excitation of carriers from deep levels have been previously published.[23-25] The characteristic time constant for the capacitance transient due to stimulation of trapped carriers from a level i in the depletion region after a trap filling pulse is given by

$$\tau_i = (e^t_{ni} + e^t_{pi} + e^o_{ni} + e^o_{pi})^{-1}. \tag{1}$$

Here, e is the emission rate (s^{-1}) of carriers from the trap; the process may be controlled by thermal or optical stimulation as denoted by the superscripts t and o, respectively. The subscripts n and p respectively denote the rates for majority and minority carriers for this n-type material. The thermal emission rate follows the well known Arrhenius function, while the optical emission rates are given by $e^o_{ni} = \sigma^o_{ni}\Phi$ and $e^o_{pi} = \sigma^o_{pi}\Phi$ where Φ is the flux of the incident photons and σ^o_{ni} and σ^o_{pi} are the photoionization cross sections of electrons and holes, respectively. The photoionization cross section is dependent on the incident photon energy and temperature. Trapped carriers will ionize faster with higher photon flux. In the following analysis, we neglect interaction of carriers from the valence band; however, it is noted that with optical excitation $E>E_{gap}/2$, filling of the deep levels by excited valence band electrons during capacitance transient measurements can cause the trap concentration to be underestimated.

The general relation between capacitance $C(t)$ and the change in net ionized impurity concentration $N_I(t)$ at time t after the application of a gap state-filling pulse is given by

$$C(t)^2 = q\varepsilon\varepsilon_0 A^2 N_I(t)/2(V_0 - V). \tag{2}$$

Where V_0 is the contact potential. From ICTS theory, the ICTS signal $S(t)$ is given by

$$t\left\{\frac{d\left[C^2(t)-C^2_\infty\right]}{dt}\right\} = \frac{q\varepsilon\varepsilon_0 A^2}{2(V_0 - V)}\sum_i N_o^i\left(\frac{-t}{\tau^i}\right)\exp\left(\frac{-t}{\tau^i}\right) \tag{3}$$

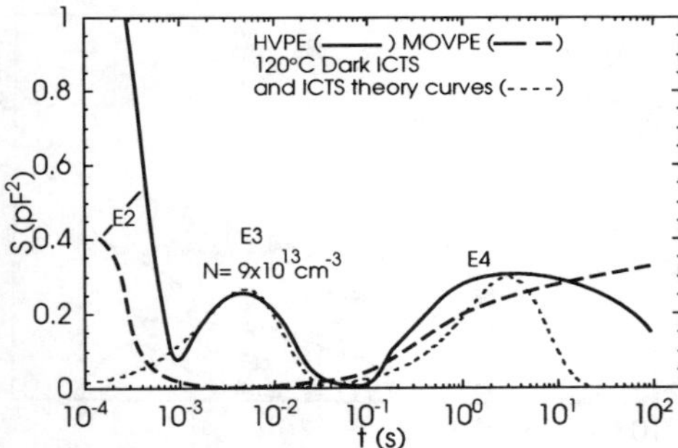

Fig. 1. The 120°C dark ICTS spectrum of the HVPE and MOVPE samples. E4 is found not to be well fitted by the ICTS theory curve. The curves have been smoothed to show significant features.

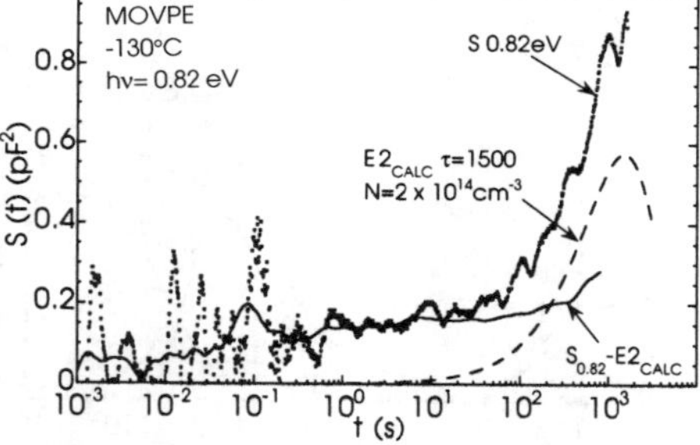

Fig. 2. O-ICTS spectrum of MOVPE sample under 0.82eV illumination is shown deconvoluted by subtracting a calculated signal corresponding to level E2.

where C_∞ is the capacitance of the junction under quiescent reverse bias V and N_o^i is the concentration of trapped carriers at $t=0$ in the ith level. The capacitance transients were transformed into the ICTS signal $S(t)$ using the left side of Eq. 3, whereas deconvolutions of $S(t)$ for each level i were accomplished, where possible, using the formula on the right. The total ionized concentration was also determined using Eq. 2. It should be noted that it is convenient to use the magnitude of an ICTS peak as an indicator of the photoionized charge since N_o^i is proportional to the ICTS signal $S(t)$.

RESULTS

Fig. 1 shows the 120°C dark (thermoionization) ICTS signal for both the HVPE and the MOVPE sample. Within the observation window, we can see the later part of the peak corresponding to E2, E3 (in the case of the HVPE sample only), and a portion of E4. It is seen that E4 is very broad in comparison to the ICTS peak of level E1 (discussed below), E2 and E3.[2] Unlike these shallower levels, E4 is not well fit by an ICTS theory curve as demonstrated in Fig. 1. In fact, the level is not completely ionized within the time frame of the transient capture; this is more true for the MOVPE sample. Using the peaks of the ICTS curves which can be distinguished in the case of the HVPE sample, we previously estimated the thermal activation energy of this peak to be about 1.0 eV.[7] Broad ICTS spectra of this nature are characteristic of continuously distributed gap states observed in polycrystalline diamond,[26] and amorphous Si:H, in which case they have been attributed to dangling bond states.[25]

A series of O-ICTS tests were done at -130°C using various excitation energies with some representative results shown in Figs. 2-6. The spectrum for the MOVPE sample under 0.82 eV illumination is shown in Fig. 2. The main peak of emission from E2 develops over a low base line signal due to other minor deep levels and possibly some carriers in the low energy tail of E4. A calculated curve corresponding to E2, the concentration $N = 2.0 \times 10^{14}$ obtained from standard room temperature ICTS testing (not shown), was fitted under the peak and the difference was calculated. These results (and those that follow) clearly show how the level E2, well known to have thermal activation energy ~0.55 eV, is photoionized. These results are in good accord with the level OL1 described by Götz and coworkers in MOVPE-grown GaN,[8] and supports their conclusion that OL1 is the same as the DL2 level observed in AlGaN.[27] The measured photoionization "threshold energy" will vary with measurement temperature, transient observation time window, photon flux, and will be sensitive to illumination bandwidth.

Figs. 3 and 4 respectively show the 1.24 and 1.55 eV spectra for the HVPE sample. E1 is excited thermally, thus the emission time constant is independent of the illumination; it is controlled by e_{n1}^i. The good fit to ICTS theory indicates ionization from a discrete energy state. For 1.24 eV illumination of the HVPE sample in Fig. 3, E2 is fully photoionized. Deconvolution of the spectrum is demonstrated by subtracting calculated curves for E2 and E3, known by thermal excitation ICTS to have concentration 6.0×10^{14} and 9.0×10^{13} cm^{-3}, respectively. It is seen that E4 is only partially photoionized. Deconvolution of the 1.55 eV curve is done similarly (neglecting the relatively minor E3 level). The remaining broad curve, characteristic of excitation from an extended state, is E4. The

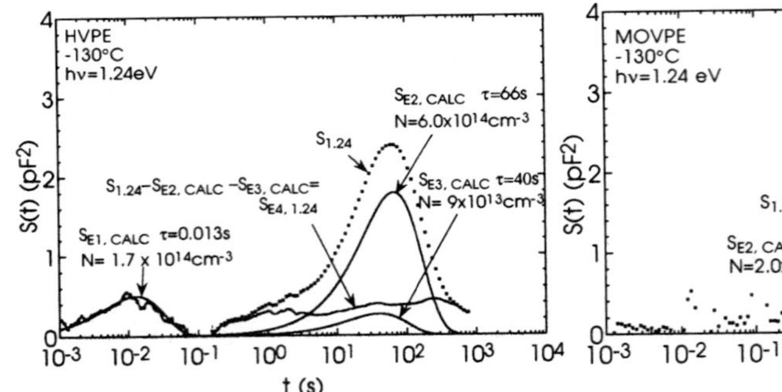
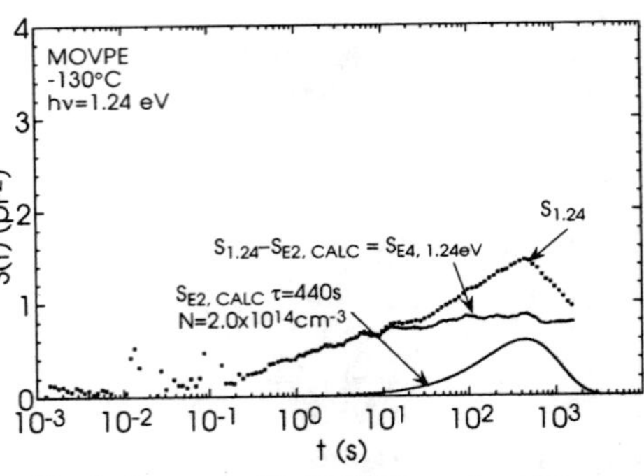

Fig. 3. Spectrum of the HVPE sample under 1.24 eV illumination. E1 is thermally ionized and E2 is photoionized–the curve is shown as components of levels E2, E3, and a portion of E4.

Fig. 5. Spectrum of the MOVPE sample under 1.24 eV illumination. E1 is of negligible concentration (not seen). E2 is fully photoionized. E4 is partially photoionized, yet of higher concentration than the HVPE sample at this point.

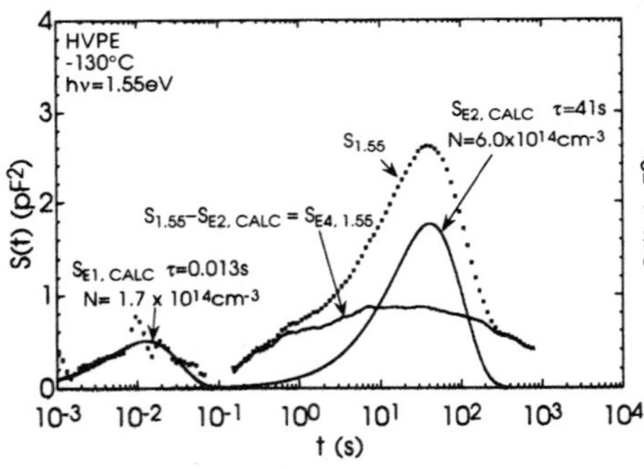
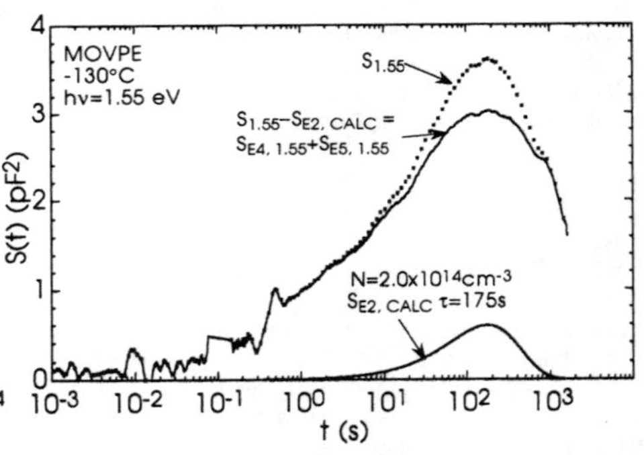

Fig. 4. Spectrum of the HVPE sample under 1.55 eV illumination. E1, as in Fig. 3 is unchanged. The photoionized portion of the curve is shown as components of E2 and E4, which is seen to be a very broad level. E3 is weak and therefore neglected here.

Fig. 6. Spectrum of the MOVPE sample under 1.55 eV illumination. E2 is shown deconvoluted; however, the pronounced form of the ICTS curve suggests the photoionization of a deeper level, E5.

integrated photoionized charge is 7.2×10^{14} cm^{-3}. Considering the tail end of the curve beyond 10^3 s, the concentration is estimated to be 8.0×10^{14} cm^{-3} by assuming a symmetric curve. Subtracting the concentration of level E3, the net concentration of carriers from E4 becomes 7.1×10^{14} cm^{-3} at 1.55 eV illumination.

Results for the MOVPE sample under 1.24 eV illumination in Fig. 5 also show E2 superimposed on a partially photoionized broad E4 curve. After subtracting a calculated curve corresponding to the E2 level, known to have concentration 2×10^{14} cm^{-3}, the broad curve representing the partially photoionized E4 is seen. With 1.24 eV illumination, the integrated photoionized charge is 7.4×10^{14} cm^{-3}. Considering the tail end of the curve beyond 10^3 s, the total concentration is estimated to be 1.1×10^{15} cm^{-3} by assuming a symmetric curve. It should be noted that the concentration of carriers swept out of the broad E4 level is already higher in the MOVPE sample with lower energy (1.24 eV) illumination than in the HVPE sample at higher (1.55 eV) illumination; thus, E4 is seen to be of significantly higher concentration in the MOVPE sample.

At 1.55 eV illumination, the volume under the ICTS curve is greatly expanded as seen in Fig. 6. It is seen that the form of the curve develops into a shape more characteristic of emission from a discrete level. Deconvolution into E2 and E4 alone is not found to be successful as in Figs. 4 and 5. This suggests another level E5, which we can begin to detect around 1.38 eV, becomes a component of the O-ICTS curve. It is possible that this level corresponds to OL3 reported by Götz and coworkers and a level at ~1.4 eV detected by Sánchez and coworkers. Interestingly, E5 is not detected in the HVPE sample with illumination up to 1.55 eV as shown in Fig. 4.

The total integrated deep level concentration for all O-ICTS curves is given in Fig. 7. It is seen that the HVPE sample, because it has higher E2 concentration which is fully ionized at 1.0 eV, yields a higher detected deep level concentration at the y axis. Total measured deep level concentration from the MOVPE sample, because of the much higher concentration of levels E4 and E5, quickly increases as the photon energy is increased to

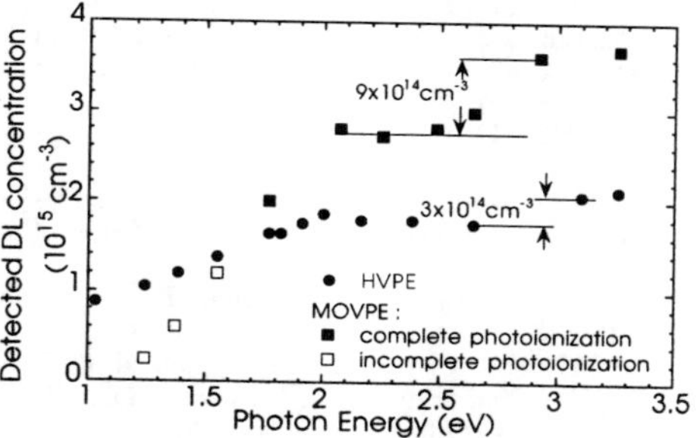

Fig. 7. The integrated concentration of the ICTS signal for both samples. The MOVPE sample is seen to have higher total concentration of mid-gap states, including a prominent level which begins to photoionize around 2.5 eV.

~2 eV. It should be noted that the very broad nature of E4 in the MOVPE sample makes it difficult to obtain complete photoionization at lower energies, therefore the total deep level concentration is not measured in this area. There is a fair amount of inconsistency between reports of deep levels observed by DLTS,[4,6] photoemission capacitance,[6,8] and photocapacitance studies[9,28] in the range around 1 eV below the conduction band. The characteristic non-exponential emission of carriers from level E4 may be partially responsible for these inconsistencies. Further, the broad nature of the peak implies that it occurs from a trapping center with a non-unique photoionization cross section or broad energy band as can be observed from dangling bonds or grain boundaries, and is less likely to be from an impurity or a discrete point defect. Threading dislocations, which are found to exist in high concentrations in thin GaN films grown by MOVPE on sapphire, are found in lower concentrations in thicker HVPE grown sapphire.[20] The very high mobilities of 0.3 mm thick HVPE-grown GaN using a ZnO buffer layer are certainly consistent with this.[22] The relationship between crystallographic defects and E4 warrants consideration.

A significant rise in the total detected concentration is observed in the range of 2.5 to 3.0 eV in both samples (Fig. 7); the estimated concentration is 3 x 10^14 cm^-3 and 9 x 10^14 cm^-3 in the HVPE and the MOVPE sample, respectively. Sánchez and coworkers[9] as well as Yi and Wessels[28] detected a level having relatively high concentration which photoionized in the range 2.0 to 2.5 eV using photocapacitance. The former attributed the commonly observed yellow luminescence from n-type GaN to this level. O-ICTS uses a trap filling pulse before the measurement of deep levels at each photon energy, whereas photocapacitance uses a single pulse before the measurement of deep levels from the entire energy spectrum of the band gap. The occupation statistics of the deep levels are therefore likely to be different, especially with photoexcitation energy greater than $E_{gap}/2$. Because of the prominence of this level in this study as well as those using photocapitance, it is possible that they are the same. A number of theoretical studies show that the Ga vacancy can produce a deep level defect in this region[14,15]. Indeed, the HVPE-grown GaN shows weak yellow luminescence,[20,29] but we cannot conclude a relationship based on the present results. The HVPE sample displays lower total deep level concentration than the MOVPE sample that was tested. Possible reasons include that the HVPE sample is thicker with a lower dislocation concentration, the use of non-organic precursors which produces a purer crystal, and a relatively high growth temperature which may produce fewer point defects.

CONCLUSIONS

ICTS was used to characterize levels in n-type GaN to simultaneously observe carriers ionized from deep levels by thermal and optical excitation. Clear relationships between the thermal and optical ionization characteristics of levels E2 and E4 could be established. While carriers trapped in shallower deep levels E1, E2, and E3 exist with discrete ionization energies, E4 is found to be broad in nature, with significantly higher concentration in the MOVPE sample that was tested. The broad and inconsistent nature of E4 indicates that it is probably not from an impurity or a discrete point defect; therefore, crystallographic defects such as dislocations are suspected. The total concentration of measured deep levels, especially E4, E5 and a prominent level with ~2.6 eV photoionization energy below the conduction band edge, was greater in the MOVPE-grown sample that was tested.

REFERENCES

1 W. Götz, N. M. Johnson, H. Amano and I. Akasaki, Appl. Phys. Lett. 65, 463 (1994).
2 P. Hacke, T. Detchprohm, K. Hiramatsu, N. Sawaki, K. Tadatomo and K. Miyake, J. Appl. Phys. 76, 304 (1994).
3 P. Hacke, A. Maekawa, N. Koide, K. Hiramatsu, and N. Sawaki, Jpn. J. Appl. Phys. 33, 6443 (1994).
4 W. I. Lee, T. C. Huang, J. D. Guo, and M. S. Feng, Appl. Phys. Lett. 67, 1721 (1995).
5 D. Haase, M. Schmid, W. Kürner, A. Dörnen, V. Härle, F. Scholz, M. Burkard, and H. Schweizer, Appl. Phys. Lett. 69, 2525 (1996).
6 W. Götz, N. M. Johnson, D. P. Bour, C. Chen, H. Liu, C. Kuo, and W. Imler in GaN and Related Mat (Mater. Res. Soc. Proc. 395).
7 P. Hacke, H. Nakayama, T, Detchprohm, K. Hiramatsu, and N. Sawaki in Proceedings of the International Symposium on Blue Laser and Light Emitting Diodes, (Ohmsha, Tokyo, 1996) p.184-187.
8 W. Götz, N. M. Johnson, R. A. Street, H. Amano, and I. Akasaki, Appl. Phys. Lett. 66, 1340 (1995).
9 F. J. Sánchez, et. al. MRS Internet Journal Nitride Semiconductor Research 1 (7) 1996.
10 T. A. Kennedy, E. R. Glaser, J. A. Freitas, Jr, W.E. Carlos, M. Asif Khanand and D. K. Wickenden, J. Electronic Mat. 24, 219 (1995).
11 T. Ogino and M. Aoki, Jpn. J. Appl. Phys. 19, 2395 (1980).
12 P. Perlin, T. Suski, H. Tiesseyre, M. Leszczynski, I Grzegory, J. Jun, S. Porowski, P. Boguslawski, J. Bernholc, J. C. Chervin, A. Polian and T. D. Moustakas, Phys. Rev. Lett. 75, 296 (1995).
13 D. M. Hofmann, D. Kovalev, G. Steude, B. K. Meyer, A. Hoffmann, L. Eckey, R. Heitz, T. Detchprohm, H. Amano and I. Akasaki. Phys. Rev. B, 52, 16702 (1995).
14 J. Neugebauer and C. G. Van de Walle, Phys. Rev. B 50, 8067 (1994)
15 T. Mattila, A. P. Seitsonen and R. M. Nieminen, Phys. Rev. B. 54, 1474 (1996).
16 J. I. Pancove and J. A Hutchby, J. Appl. Phys, 47, 5387 (1976).
17 F. A. Ponce, D. P. Bour, W. Götz and P. J. Wright, Appl. Phys. Lett. 68, 57 (1996).
18 B. Monemar, J. P. Bergman, H. Amano, I. Akasaki, T. Detchprohm, K. Hiramatsu, and N. Sawaki, in Proceedings of the International Symposium on Blue Laser and Light Emitting Diodes, (Ohmsha, Tokyo, 1996) pp. 135-140.
19 K. Hiramatsu, T. Detchprohm, I. Akasaki, Jpn. J. Appl. Phys. 32, 1528 (1993).
20 W. Götz, L. T. Romano, B. S. Krusor, N. M. Johnson, and R. J. Molnar, Appl. Phys. Lett. 69, 242 (1996).
21 K. G. Fertitta, A. L. Holmes, F. J. Ciuba,R. D. Dupuis, and F. A. Ponce, J. Electronics Materials 24, 257 (1995).
22 T. Detchprohm, K. Hiramatsu, H. Amano and I. Akasaki, Appl. Phys. Lett. 61, 2688 (1992)
23 H. Okushi and Y. Tokumaru, Jpn. J. Appl. Phys. 20 (suppl. 20-1) 261 (1981).
24 H. Okushi and Y. Tokumaru and H. Naka, Semicond. Sci. Techno. 7, A196 (1992).
25 H. Okushi, Phil. Magazine B, 52, 33 (1985).
26 H. Kiyota, H. Okushi, K. Okano, Y. Akiba, T. Kurosu and M. Iida, Appl. Phys. Lett. 61, 1808 (1992).
27 W. Götz, N. M. Johnson, M. D. Bremster and R. F. Davis, Appl. Phys. Lett. 69, 2379 (1996).
28 G. C. Yi and B. W. Wessels, Appl. Phys. Lett. 86,3769 (1996).
29 T. Kuroda, Master Thesis (Nagoya University, 1995) in Japanese.

STRUCTURAL CHARACTERISTICS OF MOCVD GROWN AlN FILMS WITH DIFFERENT CARBON CONCENTRATION

Yun-Xin Li*, Lourdes Salamanca-Riba*, M.G. Spencer**, K. Wongchigul**, P.Zhou**, X.Tang**, V. Talyansky*** and T. Venkatesan***

*Materials and Nuclear Engineering Department, University of Maryland, College Park, MD 20742

**Materials Science Research Center of Excellence, Howard University, Washington, DC 20059

***Center for Superconductivity Research, Department of Physics, University of Maryland, College Park, MD 20742

ABSTRACT

The structural characteristics of MOCVD AlN films with different carbon doping concentrations grown on sapphire were investigated by XRD (θ–2θ scans, φ scans and rocking curves), HRTEM and Auger spectroscopy. The AlN:C films have very high crystalline quality and low resistivity. The epitaxial relationship between AlN and Sapphire is: $(0001)_{AlN}$ //$(0001)_{Sap}$, $[1\overline{2}10]_{AlN}$//$[0\overline{1}10]_{Sap}$ and $[10\overline{1}0]_{AlN}$ //$[\overline{2}110]_{Sap}$. With increasing carbon concentration, the AlN films have higher carrier concentrations, and lower resistivities even though they have higher defect density. The resistivity decreases by 8 orders of magnitude with C doping. When the carbon concentration reaches 11%, an interfacial layer of ~5nm was observed in HRTEM images. This layer suggests that some C in the film is diffusing into the sapphire substrate. However, optical diffractograms obtained from the negatives of the HRTEM showed no appreciable change in the structure of the interfacial layer compared to the pure substrate.

INTRODUCTION

AlN is a wide band-gap semiconductor (Eg=6.2eV at 300K) with a good combination of physical properties such as high temperature stability, high thermal conductivity, high elastic stiffness, varying electrical properties from semiconducting to insulating. AlN, along with other nitrides, has long been considered as a futuristic material for semiconductor device applications in the blue and UV wavelengths [1-3]. However, because of its insulating properties and the lack of a lattice matched substrate, the type of devices fabricated from this materials system is very limited at the present time. Sapphire is the most commonly used substrate for AlN, it gives the best AlN crystalline quality. Very low resistivity of AlN films grown on sapphire has been recently reported[4]. It is known that because of the large lattice mismatch between AlN and sapphire the films usually have a density of dislocations of 10^{10} cm^{-2} without doping. A high concentration of carbon dopants must affect the microstructure of this material. In this work, we present the structural characteristics of AlN:C films grown on sapphire with different carbon concentrations.

EXPERIMENTAL

Aluminum nitride films were grown in a low pressure metalorganic chemical vapor deposition (MOCVD) vertical reactor on sapphire (0001). Trimethylaluminum (TMA), ammonia (NH$_3$) and hydrogen (H$_2$) were used as precursors. Propane (C$_3$H$_3$) was utilized as an independent carbon (C)

Mat. Res. Soc. Symp. Proc. Vol. 449 © 1997 Materials Research Society

source. The propane flow rate was 2-8 sccm. The growth temperature and pressure were 1200 °C and 10 Torr, respectively. The structure of the AlN films with different carbon concentrations was characterized by X-ray diffraction (XRD) and high resolution TEM. The carbon concentration was measured by Auger spectroscopy. The carrier concentration and type were determined by Van der Pauw Hall measurements.

RESULTS AND DISCUSSION

Table I lists the samples investigated in this work along with their carbon concentrations, carrier type and concentration, the AlN (002) interplanar distance obtained from the θ-2θ scan, the full width at half maximum (FWHM) of the rocking curve of film and substrate, and their resistivities. When the carbon concentration increases, the carrier concentration also increases and the resistivity decreases. The p-type carrier concentration is much smaller than the carbon concentration. It demonstrates that only a small portion of the carbon atoms contributes to the p-type doping and/or that a large portion of the dopants are not active or that there is a large compensation. In any case C doping gives rise to a decrease in the resistivity of ~8 orders of magnitude. We know that only when carbon substitutes N , it is an acceptor. Thus, some carbon atoms must be replacing Al to compensate for the doping or sitting at interstitial sites. The p-type doping should be the net difference between the number of C on N and Al sites. Thus, a large number of carbon atoms are on Al sites. Also, the strain produced by C atoms sitting on N sites can be reduced if some C atoms sit on the Al sites. When the C concentration increases, the crystalline quality decreases as seen by the increase in the FWHM. However, even in the sample with the highest C concentration, the film still has a fairly good crystalline quality. Figure 1 is the (10 $\bar{1}$1) AlN x-ray φ scan from the sample A5. The figure shows that the film has a good symmetry and a good crystalline quality even though it is somewhat worse than the one in the undoped sample. The d of AlN also shows no significant difference in the lattice structure among the three doped samples. There is also a small difference between these and the undoped 002 interplanar spacing. This is probable due to carbon atoms substituting both Al and N. Thus, part of the strain and distortion in the lattice is relaxed.

Table I   Electrical and x-ray data for AlN films doped with different C concentrations

samples	C (at%)	carrier type	carrier concentration (Na-Nd) ($\times 10^{18}$ /cm^3)	FWHM(°) AlN Sap	AlN,d$_{002}$ (Å)	resistivity (Ω-cm)
A1	4.9	p	4.6	0.38 0.06	2.5058	< 1
A3	8.3	p	8.8			< 1
A5	11.0	p	38	0.67 0.06	2.5058	< 1
unintentionally doped				0.16 0.06	2.4983	10^8

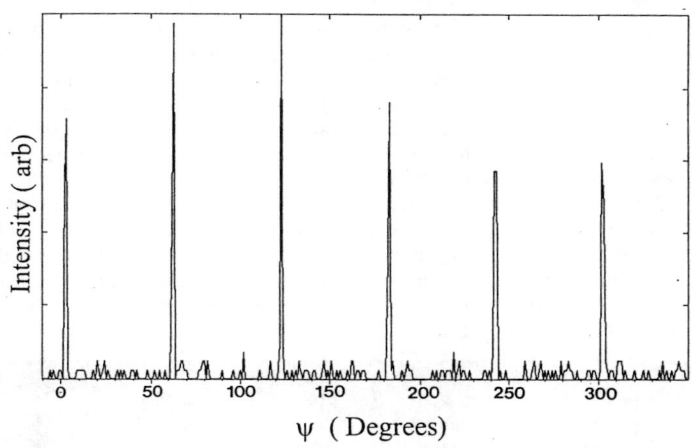

Figure 1   φ scan from the AlN film with 11% carbon

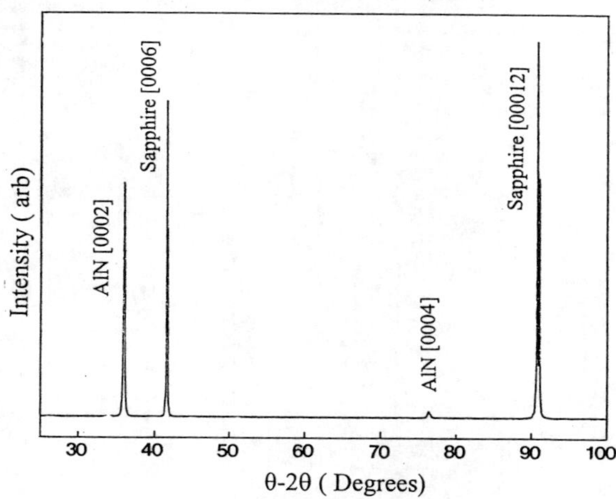

Figure 2   θ-2θ scan of AlN/Sapphire with 4.9% carbon

Figure 2 shows an XRD θ–2θ scan of a (0001) AlN film grown on (0001) Sapphire (sample A1). This spectrum shows a single orientation of the film with a (0001) growth direction. All the AlN films with different C concentrations showed the same epitaxial relationship. Figures 3 and 4 are the HRTEM images and corresponding selected-area diffraction patterns from the undoped and C doped samples, respectively. In the direction of growth, (0001) AlN//(0001)Sap. The in-plane epitaxial relationship is [$\bar{1}2\bar{1}0$]AlN//[0 $\bar{1}$10] Sap and [10 $\bar{1}$0]AlN //[$\bar{2}$110]Sap. This corresponds to a 30° in-plane rotation between AlN and sapphire[5,6]. The HRTEM images show that the interfaces are clean and fairly sharp without indication of second phases in the film. These results suggest that C incorporates fairly well into the AlN lattice. Figures 3 and 4 also show contrast modulations due to the strain near the interface, which are indicated by arrow heads. The density of dislocations in the film in Figure 4 is slightly larger than that in Figure 3, some dislocations are labeled. This is because the dopants cause more strain in the film with higher carbon concentration and produce more defects. The low magnification image shown as inset in Figure 4 also shows an interfacial layer of ~5nm (labeled I) in the substrate. The interfacial layer is also seen in the HRTEM image shown in Fig. 4 which does not show any appreciable difference (other than the contrast) with the image of the substrate. We believe that at the growth temperature of 1200°C, and for the high doping concentrations, carbon diffused into the substrate. A few dislocations are also observed around the interfacial layer in the TEM image labeled. Optical diffractograms obtained from the negative of the image at the interfacial layer and at the substrate did not show any difference in the interplanar spacings within the resolution of our measurement. Thus, the interfacial layer still has the same lattice structure as sapphire, indicating that carbon did not cause appreciable distortion of the lattice.

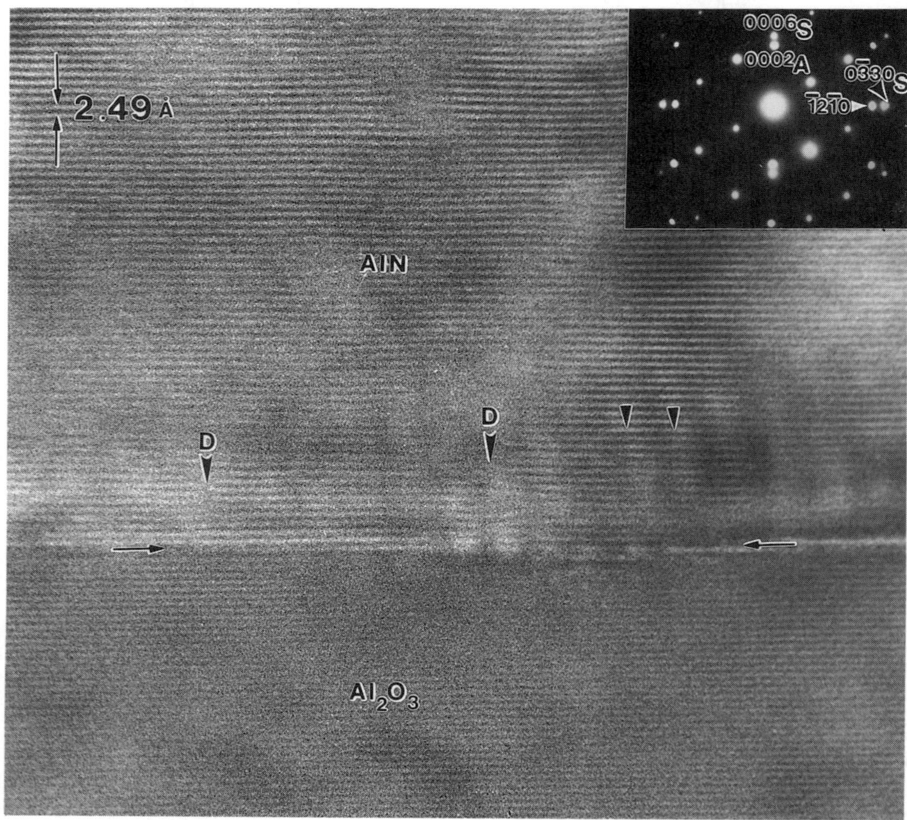

Figure 3 Cross sectional HRTEM image from AlN/Sapphire without carbon doping, the inset shows the corresponding SAD pattern with [$\overline{1}2\overline{1}0$]AlN//[0 $\overline{1}$10] Sap and [10 $\overline{1}$0]AlN //[$\overline{2}$110]Sap

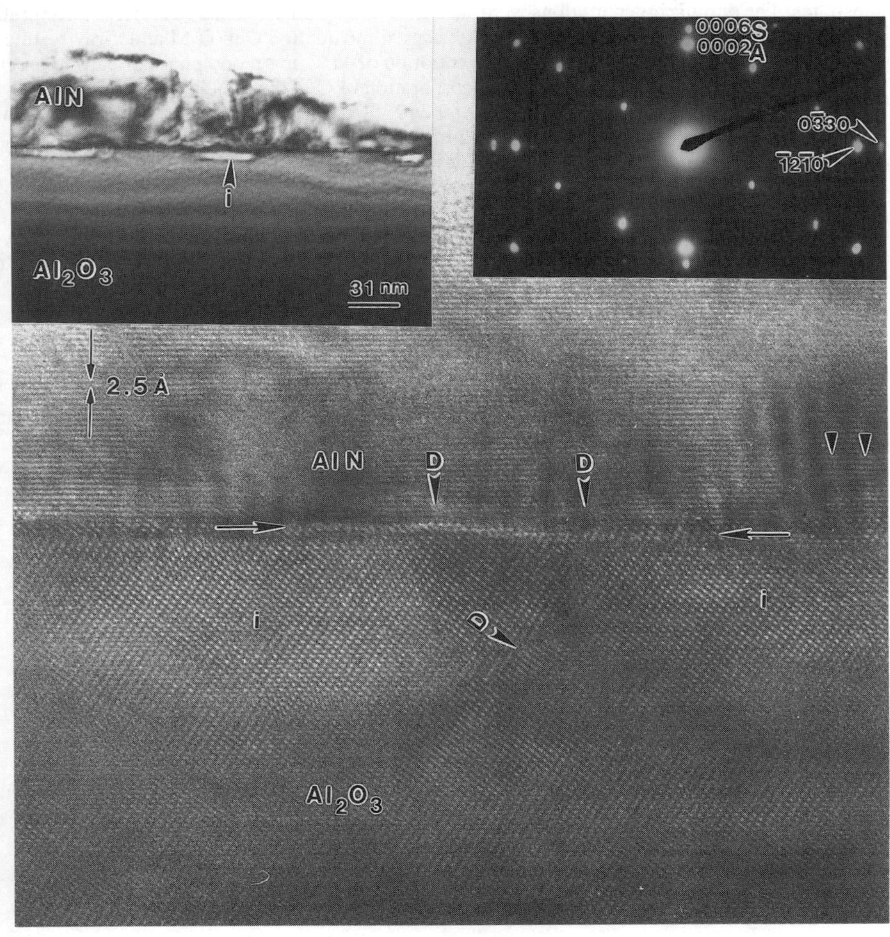

Figure 4 HRTEM cross sectional image from AlN/Sapphire with 11% carbon. The insets are left: low magnification bright field image and right: the corresponding SAD pattern with [$\bar{1}2\,\bar{1}0$]AlN//[0$\bar{1}$10] Sap and [10$\bar{1}$0]AlN //[$\bar{2}$110]Sap. The interdiffusion layer is labeled I.

CONCLUSIONS

Carbon doped AlN films with low resistivity (<1 Ω-cm) were grown by MOCVD on sapphire substrates. The doping concentrations were as high as 11% with a net charge carrier concentration as high as 3.8×10^{19}/cm^3. The films show high crystalline quality. Our TEM and x-ray results show that even for the higher C doping concentration of 11% no second phase or carbon precipitates were observed. This suggests a high incorporation of C in the AlN lattice. The epitaxial relationship is : $(0001)_{AlN}//(0001)_{Sap},$ [12 $\overline{1}$0]$_{AlN}$//[0 $\overline{1}$10] $_{Sap}$ and [10 $\overline{1}$0]$_{AlN}$ //[$\overline{2}$110]$_{Sap}$ which shows a 30° in-plane rotation between film and substrate.   With increasing carbon concentration, the p-type carrier concentration increases, and the resistivity decreased even though the density of dislocations increases. When the carbon concentration reaches 11%, an interfacial layer was observed in the substrate. This suggests interdiffusion of carbon into sapphire during growth . The layer still has the same structure as sapphire. Our results clearly suggest that the defects introduced by the dopant C do not act as effective traps and recombination centers for carries.

ACKNOWLEDGMENTS

The authors acknowledge support from an MRCP Army grant No. DAAL 019523539 and an NSF grant No. DMR 9321957.

REFERENCES

1. T.L.Tanslay and R.J.Egan, Mater.Res.Soc.Symp.Proc., 242 (1992).

2. R.F.Davis, Proc. IEEE, 702(1991)

3. S.Strite and H.Morkoc, J.Vac.Sci.Technol. B10, 1237 (1992) and references therein.

4. K. Wongchotigul, N.Chen, D.P.Zhang, X.Tang, M.G. Spencer, Materials Letters, 26(3) , 223(1996).

5. K.Dovidenko, S.Oktyabrsky, and J.Narayan, J.Appl.Phys., 79(5) 2439(1996).

6. Yun-Xin Li, Lourdes Salamanca-Riba, V. Talyan, T. Venkatensan M.G. Spenser, C. Wongchigul, P.Zhou, X.Tang, Materials Research Society, 1996, Boston, MA, Fall Meeting.

DEEP LEVELS IN GaN STUDIED BY EXTRINSIC PHOTOCONDUCTIVITY MEASUREMENT

Rong Zhang[a,d]*, Zhenchun Huang[b], Bo Guo[a], J.C. Chen[a], Li Yan[a], Youdou Zheng[c] and T.F. Kuech[d]

[a] Department of Electrical Engineering, University of Maryland Baltimore County, Baltimore, MD 21228

[b] NASA Goddard Space Flight Center, Code 718, Greenbelt, MD 20771

[c] Department of Physics, Nanjing University, Nanjing 210093, P.R. China

[d] Department of Chemical Engineering & Materials Science Program, University of Wisconsin - Madison, Madison, WI 53706

* rzhang@cae.wisc.edu

ABSTRACT

Modulation Extrinsic photoconductivity spectra between 1.44eV and 1.75eV of unintentionally n-doped high resistance GaN film grown by MOCVD are measured at room temperature by using wavelength adjustable Ti:Sapphire laser. We find that there are two major deep levels in the GaN material in the used photon energy range. The relaxation time of excess carriers controlled by those levels are in the order of 10^{-4}sec. The concentration of localized states are determined as $1.8 \times 10^8 cm^{-3}$ and $2.5 \times 10^9 cm^{-3}$, respectively. A physical model is developed to explain the results and process the data. Using a new method we have determined the optical absorption cross section of deep levels are $1.5 \times 10^{-17} cm^2$ and $2.7 \times 10^{-18} cm^2$, respectively.

INTRODUCTION

Recently GaN-based III-V nitride semiconductors attract extensive attentions for their device application[1,2]. The direct wide bandgap structure makes them very useful for fabricating blue, green and UV light-emitter due to their efficient radiative recombination. The attractive physical properties, together with the outstanding thermal and chemical stability of those materials also make it ideally suitable for developing solar blind UV detector, high power, high temperature transistors and other novel devices. For most applications, generally speaking, the deep levels in the material is the key factor to influence the features of devices. Many studies have been performed on the deep levels in the GaN by using PL[3-4], DLTS[5], TSC[6] and optical absorption[7] techniques. But most of those studies were focused on 2.2eV related deep traps that is called "yellow band"[8] now, and very little is known about other deep levels in the bandgap, especially near the center of the bandgap which should have significant effects on recombination dynamics of excess carriers. The photoconductivity spectroscopy is a versitile sensitive probe to detect electronic structures in semiconductors. It holds both advantages of high sensitivity of electric measurements and high resolution of optical measurements, so it is popularly used in weak optical absorption measurements. Both intrinsic and extrinsic photoconductivity can offer information of deep traps. But compared to intrinsic photoconductivity (IPC), by analyzing the transient and steady-state or modulation photoconductivity at different wavelength, one could directly get information about the energy position, capture cross section, density of deep traps and relaxation time of excess carriers from extrinsic photoconductivity (EPC). Another advantage of EPC is a long penetrating length due to its relatively small absorption coefficient.

Mat. Res. Soc. Symp. Proc. Vol. 449 © 1997 Materials Research Society

Because absorption through the sample in EPC is much more uniform than in IPC, the effect of surface recombination can be ignored in the case of EPC. Pankove and Berkeyheiser [9] were first authors employing the PC spectroscopy to study GaN sample which were Zn-doped films grown by the chloride transport method. Recently C.H. Qiu et al[10] reported PC measurement of GaN samples of unintentionally heavily n-type doped and p-type Mg doped films grown by metalorganic chemical vapor deposition (MOCVD). Wide deep state distributions were observed, and no discrete deep levels was found in their results. The sensitivity of PC measurement depends on resistance of the sample. The higher resistance the sample has, the higher the sensitivity will be. In this letter we report an EPC measurement of a unintentionally high resistance n-type GaN film in the energy range between 1.44 and 1.75 eV through a metal-semiconductor-metal (M-S-M) system. Two deep levels at 1.44 and 1.56 eV are found and studied.

EXPERIMENT AND RESULTS

The sample used in this study was a coplanar M-S-M structure of unintentionally n-doped high-resistance GaN on (0001) sapphire substrate grown by LP-MOCVD. The M-S-M finger-type patterns were fabricated using a lift-off technique. The fingers were 3μm wide and 5.2mm long and the spacings between the fingers were 8μm. The Ohmic contacts were formed by evaporating Al/Au on the GaN front surface, and alloying at 450°C for 5 min. The I-V curve of the sample measured in the dark was linear, indicating a good Ohmic contact.

The exciting light was from a Coherent 3900S CW Ti:Sapphire laser pumped by a 5W Ar$^+$ laser. The output light wavelength of the Ti:Sapphire laser was adjustable between 710nm and 860nm. The output laser light was focused on the surface of the sample. The focused light spot totally covered the measured M-S-M pattern. The steady-state photoconductivity spectrum was measured and we found that the photoconductivity is much lower than the dark conductivity, i.e., the concentration of excess carrier caused by illumination is much lower than thermal equilibrium carrier concentration. Then the modulation photoconductivity measurement was performed. A DC voltage was supplied to the M-S-M structure through a series sampling resistor whose resistance was much smaller than the measured sample. The output voltage across the sampling resistor was then detected through a lock-in amplifier. All measurements were operated at room temperature.

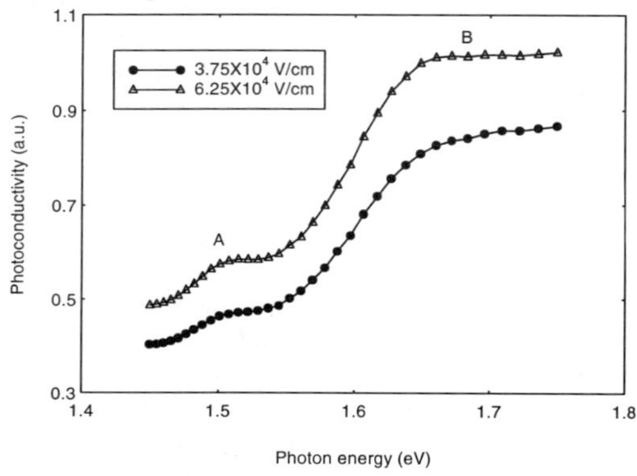

Fig.1 EPC spectra of GaN sample under different electrical field strength. Chopping frequency: 39.4Hz.

Fig. 1 are typical EPC spectra of the measured GaN sample. Prior to EPC measurement, the sample was kept at room temperature under dark for a long time to let it arrive thermal equilibrium. During EPC measurement the chopper frequency was set at 39.4Hz, and the DC electrical field strength were 3.75. The spectra have been normalized to equal incident photon density. From Fig.1, we see clearly two absorption band as marked by A and B in both spectra. According to the classical method for determining depth of the trap levels from EPC, we can obtain the ionization energy of two deep states from the long-wavelength edge of the bands as 1.44eV and 1.56eV, respectively. Since the GaN film is unintentionally n-type doped, both levels are electron-occupied under thermal equilibrium. That means the EPC signal comes from the transition from deep levels to the conduction band. We can conclude that there are two levels with energy of 1.44eV and 1.56eV under the conduction band. As seen, we also find that the shapes of EPC peaks are weakly dependent on the applied electrical field strength, which will be discussed elsewhere[11].

The relation between PC signal and incident photon intensity is very useful to analyze the injection condition. Fig. 2 shows the relation of responsibility to incident optical power intensity of both level A and B. The curve B is almost linear, while the curve A behaves slight nonlinearly. The linearity of curve B indicates that the stimulating optical power intensity we use is due to small-injection and the relaxation time of excess carriers is independent on incident optical power intensity. But in the case of large optical power we use the nonlinearity of curve A shows deviation from small-injection condition and the relaxation time of excess carriers weakly depends on incident optical power intensity.

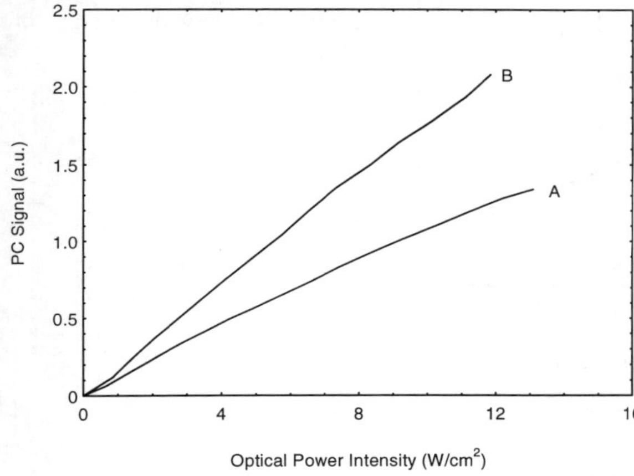

Fig. 2 Relation of PC signal to incident optical power density at 1.50eV (A) and 1.65eV (B).

The dependence of PC signal on chopping frequency of incident light stimulation is informative of relaxation time of excess carriers. Those relations for level A and B are displayed in Fig. 3 and Fig. 4, respectively. Both curves can be well fitted by[12]

$$\Delta\sigma = \Delta\sigma_0 \tanh\frac{1}{4\tau f} \qquad (1)$$

where $\Delta\sigma_0$ is the steady-sate photoconductivity, τ is the relaxation time of excess carriers and f is the chopping frequency. From the equation we determine the relaxation time of excess carriers stimulated from level A and B as 2.5×10^{-4} sec and 1.8×10^{-4} sec, respectively.

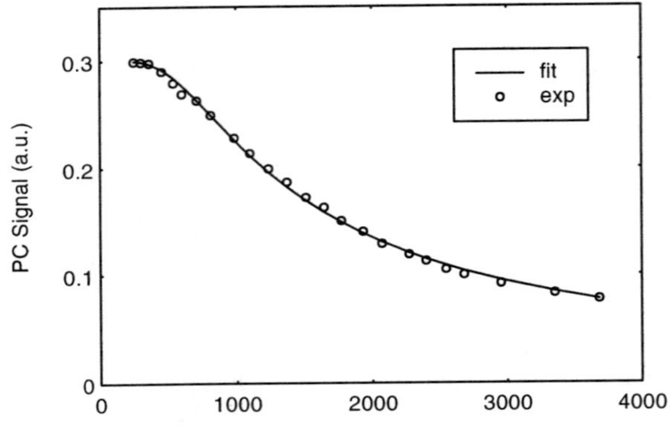

Fig. 3 Dependence of PC signal on chopping frequency of incident lightwith energy of 1.50eV.

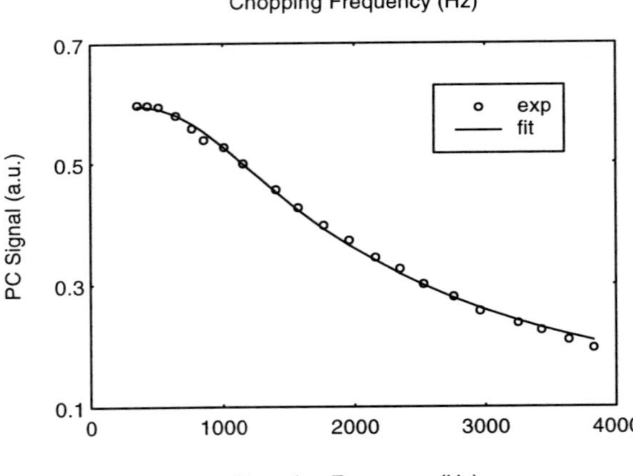

Fig. 4 Dependence of PC signal on chopping frequency of incident light with energy of 1.65eV.

DISCUSSION

For a single level extrinsic photoabsorption in an n-type semiconductor, usually one has the rate equation under illumination by a monochromatic light with energy larger than ionization energy but less than bandgap as

$$\frac{d\Delta n}{dt} = qJ(m_0 - \Delta n) - \alpha\Delta n(n_0 + N_I - m_0 + n' + \Delta n) \tag{2}$$

where n_0 is the thermal equlibrium electron concentration, N_I is the total concentration of the localized level studied, m_0 of which are occupied by electrons in the dark, α is the electron capture coefficient of the level from the conduction band, q is the cross section for optical absorption by bound electrons, J is the incident photon flux density, n'=$N_c$exp[-(E_c-E_t)/kT] with E_C, the bottom energy of the conduction band and E_t, the energy level of the localized state. Here we assume quantum efficiency for photon absorption is 1. Solving the Eqn (2) leads to an expression for the steady-state photoconductivity, $\Delta\sigma_0$ under small-injection condition as

$$\Delta\sigma_0 = \frac{qJm_0 e\mu}{qJ + \alpha(n_0 + N_I - m_0 + n')} \tag{3}$$

Here, e is the charge of electron and m is the electron mobility of the GaN sample. From Eqn (3) we can calculate the optical absorption cross section of bound electrons, q and densities of electrons of electrons occupying on both levels. All parameters of level A and B are summarized in Table I. There we ignore the difference between N_I and m_0 because of level depth as large as 1.44 and 1.56eV, and use 100cm2/Vs as an evaluated value for electron mobility of the GaN sample.

TABLE I Parameter of level A and B

Level	E_C-E_T(eV)	relaxation time(sec)	Density(cm^{-3})	optical cross-section(cm^2)
A	1.44	2.5×10^{-4}	1.8×10^8	1.5×10^{-17}
B	1.56	1.8×10^{-4}	2.5×10^9	2.7×10^{-18}

SUMMARY

We have measured EPC of LP-MOCVD grown high-resistance GaN film. The results indicate that there are two major deep levels locate at 1.44eV and 1.56eV. Two principal phenomenological parameters, cross section for optical absorption by bound electrons and relaxation time corresponding to transitions between deep levels and the conduction band have been obtained together with the densities and energy positions of deep levels.

REFERENCES

1. S. Strite, H. Morkoc, J. Vac. Sci. Technol., **B10**, p. 1237(1992)

2. H. Morkoc, S. Strite, G.B. Gao, M.E. Lin, B. Sverdlov, M. Burns, J. Appl. Phys., **76**, p. 1363(1994)

3. J.C. Zolper, M. Magerott Crawford, A.J. Howard, J. Ramer, S.D. Hersee, Appl. Phys. Lett., **68**(2), p.200(1996)

4. J. Baur, U. Kaufmann, M. Kunzer, J. Schneider, H. Amano, I. Akasaki, T. Detchprohm, K. Hiramatsu, Appl. Phys. Lett., **67**(8), p.1140(1995)

5. G.C. Yi, B.W. Wessels, Appl. Phys. Lett., **68**(26), p.3769(1996)

6. D.C. Look, Z.Q. Fang, W. Kim, O. Aktas, A. Botchkarev, A. Salvador, H. Morkoc, Appl. Phys. Lett., **68**(26), p.3775(1996)

7. L. Balagurov, P.J. Vhong, Appl. Phys. Lett., **68**(1), p.43(1996)

8. J. Neugebauer, C.G. Van de Walle, Appl. Phys. Lett., **69**(4), p.503(1996)

9. J.I. Pankove, J.E. Berkeyheiser, J. Appl. Phys., **45**, p.3892(1974)

10. C.H. Qiu, C. Hoggatt, W. Molton, M.W. Leksono, J.I. Pankove, Appl. Phys. Lett., **66**(20), p.2712(1995)

11. Rong Zhang, Z.C. Huang, Bo Guo, J.C. Chen, Li Yan, Youdou Zheng, T.F. Kuech, *in preparation*

12. Ryvkin, <u>Photoelectric Effects in Semiconductors</u>, Consultants Bureau, New York, 1964, pp35-72

ELECTRON - PHONON SCATTERING IN Si DOPED GaN

C. WETZEL, W. WALUKIEWICZ, and J.W. AGER III
Lawrence Berkeley National Laboratory, Materials Science Divison, MS 2-200,
1 Cyclotron Road, Berkeley, CA 94720, USA, C_Wetzel@LBL.GOV

ABSTRACT

Phonon-plasmon scattering in non-resonant Raman spectroscopy is used to determine the free electron concentration in Si doped GaN films. For various doping concentration and variable temperature the correlation with magneto-transport data is established. The freeze-out of the carrier concentration at low temperature is thus observed in a purely optical detection scheme. We observe a very long transient time of several hours for the carrier concentration as a reaction to temperature variation. This indicates an indirect capture and emission process with a very small cross section. The value of the Faust-Henry coefficient is determined.

INTRODUCTION

Assessment of electronic transport properties in group-III nitrides are a basic necessity for both revealing the fundamental physics and as a standard characterization of process parameters. Magneto-transport measurements are typically performed to derive free electron concentration and electron mobilities in Van-de-Pauw or Hall bar electrical contact schemes [1-3]. A convenient purely optical method is to study the interaction of the free electron plasma with the phonons of the lattice in Raman spectroscopy [4-6]. Within a wide range of device relevant doping levels, i.e. free electron concentrations this technique provides a fast contactless method to both determine the parameters on an area defined only by the focal size of the used laser beam or imaging of wafer size areas. Due to the spectroscopic character of the technique further information of the system under investigation is obtained simultaneously.

Established in other binary compound semiconductors, i.e. GaAs the method has been applied to GaN to derive the coupling parameters of the system [5] and determine the carrier concentration in samples under large hydrostatic pressure in diamond anvil cells [6]. In this paper we provide additional information on the sign of the Faust-Henry coefficient and compare data obtained in Raman spectroscopy with magneto-transport results. We investigate a series of Si doped samples at various doping levels and also present the temperature dependence.

EXPERIMENTAL

A series of nominally undoped and Si doped GaN films were grown by metalorganic vapor phase epitaxy (MOVPE) on (0001) sapphire in two different reactors [3]. Doping and magneto-transport data of the samples are presented in Table I. Raman spectroscopy was performed using 120 mW of the 476.5 nm line of an Ar ion laser and matching narrow band dielectric band blocking filters. Spectra at variable temperature and various polarization were taken in backscattering geometry.

RESULTS

The coupling of the lattice phonon and the free electron plasmon is described in an oscillator model of the dielectric function $\varepsilon(\omega)$.

$$\frac{\varepsilon(\omega)}{\varepsilon_0} = 1 + \frac{\omega_L^2 - \omega_T^2}{\omega_L^2 - \omega^2 + i\omega\Gamma} - \frac{\omega_P^2}{\omega(\omega - i\gamma)}, \tag{1}$$

where ω is the photon frequency, ω_L and ω_T the longitudinal and transverse optical phonon frequencies, respectively. Γ is the phonon damping coefficient. $\omega_P^2 = e^2 N / (\varepsilon_0 \varepsilon_r m^*)$ is the square of the plasma frequency connecting to the free electron concentration N. The carrier mobility is related to the electron scattering frequency γ.

Raman scattering in two distinct configurations $z(x,\bar{x})\bar{z}$ and $x(z,\bar{z})\bar{x}$ (Cartesian coordinates) geometry is shown in Fig. 1. The selection rules (Fig. 1) [7] are well obeyed and allow for an interpretation of the relative intensities I_L and I_T of the A_1(LO) and A_1(TO) modes. The ratio can be related to the Faust-Henry coefficient C by [8,4]:

$$\frac{I_{LO}}{I_{TO}} = \left(\frac{\omega_1 + \omega_L}{\omega_1 + \omega_T}\right)^4 \frac{\omega_T}{\omega_L}\left(1 + \frac{\omega_T^2 - \omega_L^2}{C\omega_T^2}\right)^2, \tag{2}$$

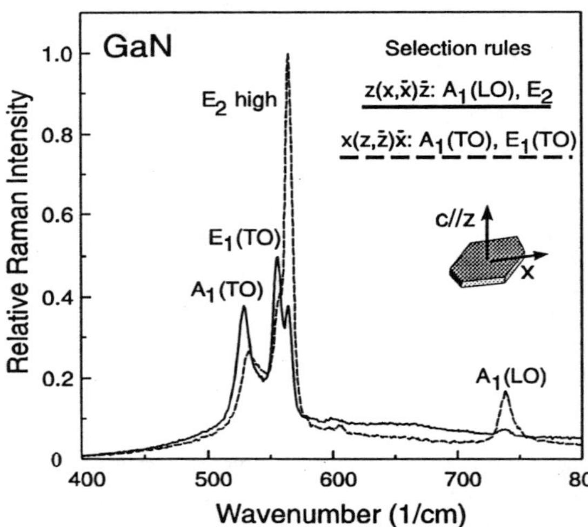

Fig. 1: Non-resonant Raman scattering in backscattering geometry of a low doped GaN/sapphire film in two different orientations along with the selection rules. In this sample the sapphire mode is very weak.

where ω_1 is the laser frequency. Solving for C we find two roots $C \in \{0.4, -5.2\}$. In a previous work only C=0.4 was found [5].

In order to determine the sign of C we compare the measured intensities of the A_1(LO) with the calculated ones [6,5] for samples with various electron concentrations. In Fig. 2 room temperature $z(x,-)\bar{z}$ spectra scaled to the intensity of the electron concentration independent E_2 mode are presented along with N_{Hall} from magneto-transport [3]. We find that for increasing concentration the LO mode *decreases* in peak height. From a simulation of the lineshape and the absolute intensity compared to the unaffected E_2 mode we find that this corresponds to the negative value of $C = -5.2$.

Values N_{Raman} from the interpretation of the lineshape are also collected in Table I. The carrier mobility was assumed to be identical to the value from the transport measurement μ_{Hall}. We find good agreement for the derived carrier concentrations. The assumption of a constant mobility in both experiments is typically good in the limit of high mobility [5]. For low mobility values typically μ_{Hall} shows higher values than optically derived mobilities.

This originates in the different weighing of the scattering events. Similar effects have been described comparing cyclotron resonance, optically detected cyclotron resonance, Shubnikov-de-Haas effect with Hall-effect in GaInAs quantum well systems [9]. In principle for the macroscopic property only the projection of the scattering event along the drift direction matters, whereas for the microscopic detection any scattering direction destroys the particle of the coupled mode. A detailed analysis is currently under way.

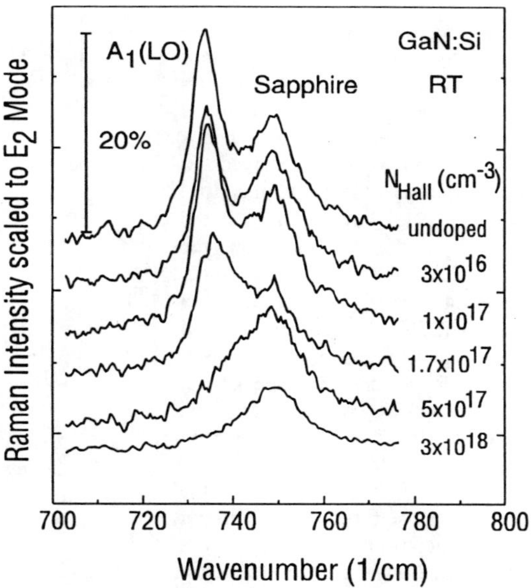

Fig. 2: Raman spectra of GaN films with various Si doping levels scaled to the intensity of the E_2 mode. The lineshape and intensity of the $A_1(LO)$ -plasmon coupled mode is sensitive to the free electron concentration. From comparison with calculations we find $C=-5.2$. In this material an additional sapphire mode at 750 cm^{-1} is seen.

In the next step we correlate the phonon-plasmon coupled mode with transport data as a function of temperature. Fig. 3 gives the center of the E_2 mode in the range of 16 K -- 95 K. A total shift of 1.72 cm^{-1} is found. The linewidth 3.0 ± 0.4 cm^{-1} varies very little. A similar shift of the $A_1(LO)$ mode and the sapphire mode is found for the little doped sample $N_{Hall}= 8.6 \times 10^{16}$ cm^{-3}. The self energy effects will be considered elsewhere.

A full set of spectra at various temperatures were taken on the higher Si doped sample

Table I: Data for Si doped MOVPE GaN/sapphire films. N_{Hall} and μ_{Hall} refer to free electron concentration and mobility from magneto-transport, N_{Raman} and μ_{Raman} refer to data from the Raman study. All data represent room temperature values.

Sample	A	B	C	D	E	F	G
N_{Hall} (10^{16}cm^{-3})	'undoped'	3.5	8.6	10	17	50	300
N_{Raman} (10^{16}cm^{-3})	<5	7	9	4	10	35	--
μ_{Hall} (cm^2/Vs)	--	--	410	240	190	355	230

N_{Hall}= 5x10^{17} cm^{-3} (Fig. 4). The whole temperature cycle consists of a 1.5 h cool down phase, a 17 h settling time and a 3 h warm up phase. The spectra at the various extreme points are displayed in Fig. 5. Starting at room temperature only a weak contribution of the A_1(LO) mode is found on the background of the sapphire mode. Lowering the temperature a monotonically growing and narrowing LO mode is observed. From a fit to the lineshape of the coupled mode we derive a total decrease in the free electron concentration from 3.5x10^{17} cm^{-3} to ~1x10^{17} cm^{-3}. During the subsequent settling time without laser irradiation the temperature

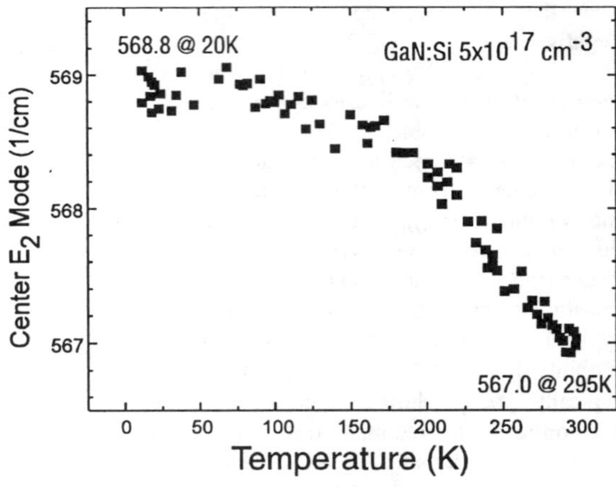

Fig. 3: *Center of the E_2 mode as a function of temperature for N_{Hall}=5x10^{17} cm^{-3}. The absolute position is sample dependent.*

varies little however, there is a considerable change observable in the coupled mode. We derive an electron concentration as low as ≤ 5x10^{16} cm^{-3}. During the long warm up phase under laser irradiation a similar retardation with a time constant of 2 -- 3 h in the change in the LO mode is observed. Reaching a temperature of 265 K the mode is still very well pronounced superseding

the mode right after cool-down. This indicates that the carrier concentration has not yet reached its equilibrium room temperature concentration. During the whole temperature cycle the correct sample temperature is directly monitored from the position of the E_2 mode.

DISCUSSION

The tempera-ture induced freeze-out of the free electron concentration is

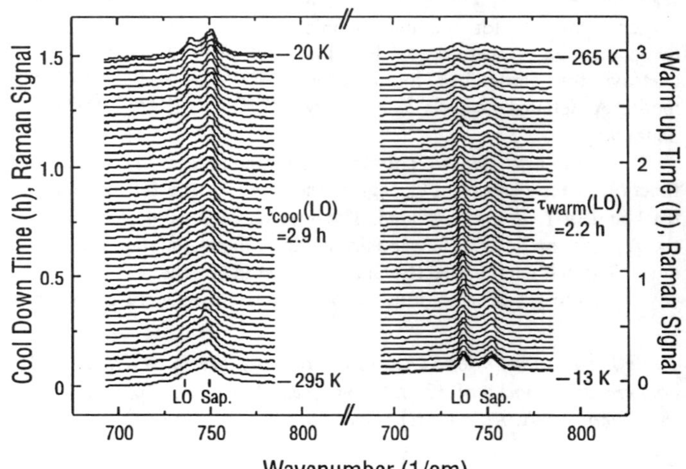

Fig. 4: *The A1(LO) phonon range at various temperatures. A significant transient time for the carrier concentration, i.e. lineshape is observed. The sample temperature is monitored in parallel by the E_2 mode.*

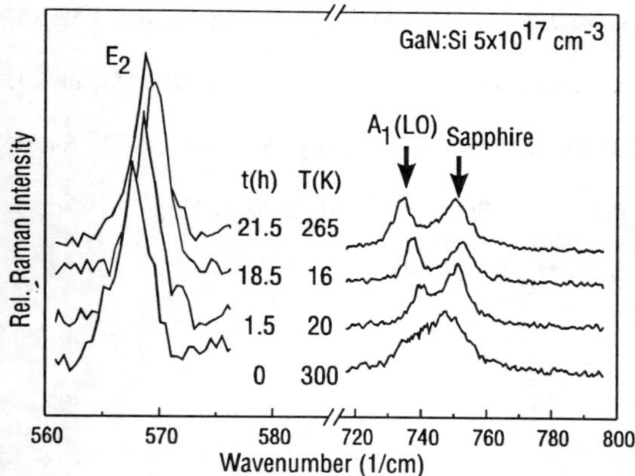

Fig. 5: Raman spectra at the several stages of the temperature cycle (N_{Hall}=5x10^{17} cm^{-3}). The appearance of the A_1(LO) mode corresponds to the freeze-out of the carriers.

directly observed in the lineshape and position of the A1(LO) mode. The total variation of the derived concentration is equivalent to the freeze-out behavior observed in the magneto-transport experiment presented before (Sample No 4 in Ref. [3]). This allows for a direct correlation of the two methods to derive free electron concentration in a contactless optical method.

The strong variation of the mode shape during the very long settling time indicates a very slow electron capture and emission rate from the donor site. Comparing with the settling time during the temperature variation we find time constants for both emission and capture in the order of 2 -- 3 h. Considering the small binding energy of the Si donor these long time scales indicate the participation of intermediate steps and possibly reacting partners. Likely candidates are participating acceptors or mobile compensating defects. The typically highly defective crystal structure of epitaxial GaN films offers a high density of possible impurity and complex partners.

These findings provide important hints for a dynamical compensation mechanism active in GaN thin films and possibly also in devices.

ACKNOWLEDGMENTS

We gratefully acknowledge helpful discussions with Professor E.E. Haller. Part of the work was supported by the Director, Office of Energy Research, Office of Basic Energy Sciences, Materials Science Division of the U.S. Department of Energy under Contract No. DE-AC03-76SF00098.

REFERENCES

[1] N. Koide, H. Kato, M. Sassa, S. Yamasaki, K. Manabe, M. Hashimoto, H. Amano, K. Hiramatsu, I. Akasaki, J. Cryst. Growth, **115**, 639 (1991).

[2] P. Hacke, A. Maekawa, N. Koide, K. Hiramatsu, N. Sawaki, Jpn. J. Appl. Phys. 1, **33**, 6443 (1994).

[3] W. Götz, N.M. Johnson, C. Chen, H. Liu, C. Kuo, W. Imler, Appl. Phys. Lett. **68**, 3144 (1996).

[4]   M.V. Klein in *Light Scattering in Solids I*. Ed. M. Cardona , Topics in Applied Physics, Vol. 8 (Springer Berlin, 1983). p. 159ff

[5]   T. Kozawa, T. Kachi, H. Kano, Y. Taga, M. Hashimoto, N. Koide, K. Manabe, J. Appl. Phys. **75**, 1098 (1994).

[6]   C. Wetzel, W. Walukiewicz, E.E. Haller, J. Ager, I. Grzegory, S. Porowski, T. Suski, Phys. Rev. B **53**, 1322 (1996).

[7]   C.A. Arguello, D.L. Rousseau, S.P.S. Porto, Phys. Rev. **181**, 1351 (1969).

[8]   D.T. Hon, W.L. Faust, Appl. Phys, **1**, 241 (1973).

[9]   C. Wetzel, *Thesis*, Technical University Munich (1994).

INTRINSIC MOBILITY LIMITS OF A TWO-DIMENSIONAL ELECTRON GAS IN AlGaN/GaN HETEROSTRUCTURES

W. WALUKIEWICZ[1], L. HSU[1,2], J. M. REDWING[3]
[1]Lawrence Berkeley National Laboratory, Berkeley, CA 94720
[2]Dept. of Physics, University of California-Berkeley, Berkeley, CA 94720
[3]Advanced Technology Materials Inc., 7 Commerce Dr., Danbury, CT 06810

ABSTRACT

We present the results of a theoretical study of the 2D electron gas mobility at a $Al_xGa_{1-x}N/GaN$ heterointerface. All standard mechanisms, including scattering by acoustic and optical phonons, and remote and background (residual) impurities have been included in our calculation of theoretical mobility limits in a $Al_xGa_{1-x}N/GaN$ structure. Comparison of calculations with experimental mobilities obtained from high quality MOCVD grown $Al_{0.15}Ga_{0.85}N/GaN$ heterostructures shows that the low temperature mobility in these samples is dominated by scattering from ionized impurities, with a smaller contribution from acoustic phonons.

INTRODUCTION

Because of its large tunable direct bandgap, ranging from 3.4 to 6.2 eV, and high saturation drift velocities for electrons, $Al_xGa_{1-x}N$ is ideally suited for high power, high temperature electronic devices. Recent research[1,2,3,4] on $Al_xGa_{1-x}N/GaN$ heterostructures has confirmed the possibility of creating a two-dimensional electron gas (2DEG) at the heterointerface, with mobilities greater than 7000 cm^2/V s and sheet carrier concentrations greater than 6×10^{12} cm^{-2}. In addition, the fabrication of high electron mobility transistors (HEMTs) and heterostructure field effect transistors (HFETs) with operation frequencies in the GHz range has demonstrated the potential for developing devices for high power microwave applications[5,6]. In the future, as the quality of these films increases and a better control over the doping is achieved, improvements in device characteristics and in the electron mobilities in these films can be expected. However, despite the recent experimental work, there have been no published calculations of the highest electron mobilities that might potentially be obtained from these heterostructures, taking into account the two dimensional nature of the electron gas. Thus far, all estimates of the mobilities have used a three dimensional approximation[7] which, though accurate at room temperature, is not suitable for low temperatures and cannot properly describe the scattering from remote impurities, which is the most important characteristic of modulation doped structures.

ELECTRONIC STRUCTURE OF THE CONFINED ELECTRON GAS

The physical configuration of a $Al_xGa_{1-x}N/GaN$ heterostructure is shown in figure 1. During growth, which occurs by metalorganic vapor phase epitaxy (MOVPE), neither film is intentionally doped. However, such nominally undoped films have been shown to have carrier concentrations ranging from 10^{16} to 10^{18} cm^{-3}. Due to the GaN's higher affinity for electrons than $Al_xGa_{1-x}N$, electrons from the $Al_xGa_{1-x}N$ are transferred to the GaN, where they accumulate near the interface[8]. The concentration of charges in that vicinity along with the ionized donors remaining in the $Al_xGa_{1-x}N$ produces a strong electric field perpendicular to the interface, bending the conduction bands. This confines the electron gas and leads to a quantization of the energy band structure, as shown in the figure[9]. Calculating the energy levels and wavefunctions for the 2DEG involves solving the Poisson and Schrodinger equations self-consistently. Previous work[8,10] has shown that

$$(1) \qquad \psi_0(x,y,z) = \phi_{x,y} \sqrt{\frac{2}{b^3}} \, z \exp\left(\frac{-bz}{2}\right)$$

Mat. Res. Soc. Symp. Proc. Vol. 449 © 1997 Materials Research Society

provides a reasonably good approximation to the wavefunction for the ground state, where $\phi_{x,y}$ is the two-dimensional plane wave and b is a variational parameter which depends on the electric field strength. We have found that for our system, this expression gives a ground state energy which is only a few percent higher than that obtained by numerical means. Of course, this wavefunction assumes that there is negligible penetration of the electron density into the alloy region.

Using expression (1) for the wavefunction of electrons in the lowest energy band, we have calculated the energies of the lowest two bands as a function of the two-dimensional electron concentration for an aluminum fraction of 15%. The results are shown in figure 2 and indicate that for electron concentrations greater than 2×10^{12} cm^{-2}, more than one band is filled. For sheet electron concentrations greater than this value, the mobility is lowered by interband scattering. In addition, if many bands are filled, the gas becomes less confined and loses its two-dimensional character if its width becomes greater than either the DeBroglie wavelength or the mean free path of the electrons.

One of the advantages of intentionally doped modulation heterostructures is the ability to separate the 2 DEG from the donors by a spacer, which in this case would consist of an undoped $Al_xGa_{1-x}N$ layer grown in between the doped $Al_xGa_{1-x}N$ and the pure GaN. The additional distance between the two layers reduces the effectiveness of the Coulomb scattering of the electrons in the GaN by the remote donors in the doped $Al_xGa_{1-x}N$, increasing the mobility of the 2 DEG. However, the introduction of the spacer also has the detrimental effect of reducing the concentration of electrons transferred from those remote donors to the GaN well. In figure 3, the sheet electron concentration as a function of the concentration of donors in the $Al_xGa_{1-x}N$ layer is shown for a number of different spacer widths. To facilitate comparison with 3D structures, we note that the range of sheet carrier concentrations shown in the figure corresponds to bulk electron concentrations from 5×10^{16} cm^{-3} to 3×10^{19} cm^{-3} within the well.

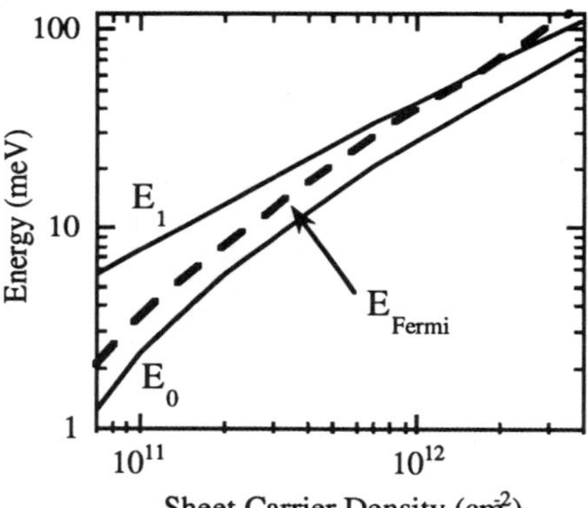

Figure 2. Energy of lowest two quantized energy levels and Fermi level as a function of 2D carrier concentration.

Figure 1. Schematic representation of a $Al_xGa_{1-x}N$/GaN heterostructure, showing ionized donors (o), neutral donors (●), lowest two quantized energy levels, and Fermi level.

MOBILITY CALCULATIONS

Our calculations of the electron mobility in $Al_xGa_{1-x}N$/GaN heterostructures are based on an approach used previously to calculate mobilities in $Al_xGa_{1-x}As$/GaAs heterostructures [11]. We incorporate all of the standard scattering mechanisms, including contributions from acoustic phonons (through deformation potential and piezoelectric mode scattering), optical phonons, the remote ionized donors found in the $Al_xGa_{1-x}N$ alloy, and any residual charged centers in the GaN.

In our calculations, we have assumed an aluminum fraction of 15% and that only the lowest band is occupied by electrons so that interband scattering can be neglected at low temperatures. The contributions from all mechanisms except for optical phonons were calculated using the wavefunction above. Since optical phonons in GaN have such a high energy (90.5 meV) the scattering rate due to this mechanism will be the sum of many highly inelastic interband and intraband scattering processes. Such scattering ruins the confinement of the electron gas by smearing out the electron density. Because many bands over a wide range of energies must be considered, it is justifiable to use a three

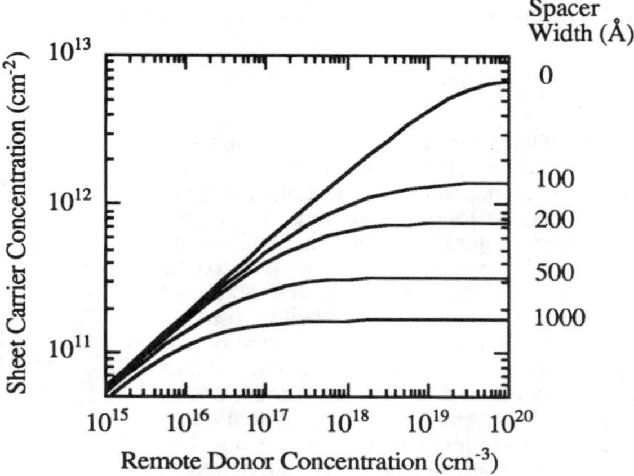

Figure 3. 2D electron concentration plotted as a function of concentration of donors in the $Al_{0.15}Ga_{0.85}N$ layer.

dimensional approach to calculate the optical phonon contribution to the scattering. One mechanism that has been neglected in our calculations is alloy disorder scattering, as our wavefunction assumes that there is no carrier density in the $Al_xGa_{1-x}N$. However, in the case of $Al_xGa_{1-x}As/GaAs$, it has been shown that this mechanism is important only in very high purity, high mobility heterostructures.

Table 1

Parameter	Value
density (g/cm^3)	6.1
ε_0	9.9
ε_∞	5.35
LO phonon energy (meV)	90.5
m^*/m_0	0.21
lattice parameter a_0	4.52
acoustic phonon velocity (cm/s)	6.6 x 10^5
piezoelectric constant (h_{14}) (V/cm)	4.28 x 10^7 [Ref. 1]
deformation potential (eV)	8.5

The materials parameters used in the calculation are shown in table 1 above. All values were taken from the literature except for the deformation potential. In order to estimate this parameter, we took advantage of the fact that the best data on 2DEG in the $Al_xGa_{1-x}N/GaN$ system show mobilities as high as 7500 cm^2/V s at low temperature, as well as a clear temperature dependence. The low temperature mobility in systems with high electron concentrations (such as the carrier concentration of 6 x 10^{12} cm^{-2} found in this sample) is dominated by ionized impurity scattering and scattering by acoustic phonons through the deformation potential. Since only the acoustic phonon scattering is temperature dependent, we were able to estimate the deformation potential by matching the slope of calculated mobilities to that of the experimental mobilities, using the deformation potential as the fitting parameter. As the carrier concentration of the sample was 6 x 10^{12} cm^{-2}, indicating that many energy bands were filled, we used a 3-D approximation to calculate the mobilities. Though the exact width of the electron gas was unknown, we obtained values from 8.0 eV to 9.5 eV, with most between 8.0 eV and 8.5 eV, using a range of reasonable widths.

RESULTS

Mobilities for 2D and 3D electron gases do not differ much in terms of their temperature dependence. In both cases, in the absence of ionized scattering centers, the high temperature mobility is limited by optical phonon scattering, with acoustic phonon scattering becoming the dominant mechanism at temperatures below roughly 140 K. Figure 4 shows a mobility vs. temperature curve for a remote donor concentration of 5.6×10^{17} cm^{-3} and a 200 Å spacer. Depending on the concentration of ionized impurities, Coulomb scattering may be dominant at only very low temperatures (in the case of relatively pure samples), or at all temperatures (in the case of highly doped samples).

At room temperature, the electron mobility in GaN is dominated by optical phonon scattering and thus there is no advantage of a 2 DEG over bulk GaN.

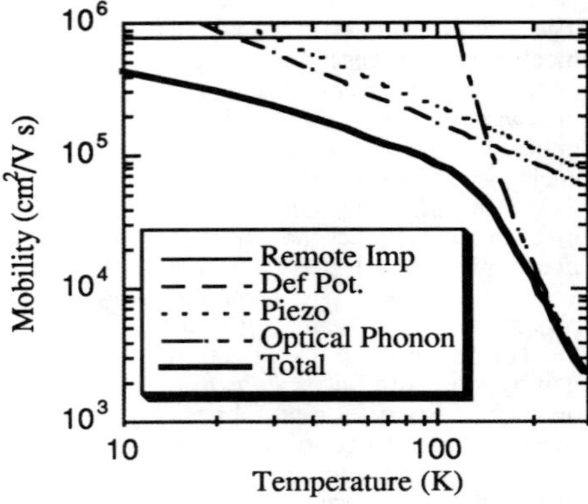

Figure 4. Plot of mobility v. temperature for a 2 DEG at a Al$_{0.15}$Ga$_{0.85}$N/GaN interface.

However, at low temperatures, the difference is dramatic, as can be seen in figures 5A and 5B. In both cases, the total mobility is almost completely determined by scattering from the ionized donors which provide the carriers. However, in the heterostructure, where these impurities are some distance removed from the carriers, the effect of Coulombic scattering is much less and the highest achievable mobilities are one to two orders of magnitude greater than in the bulk case.

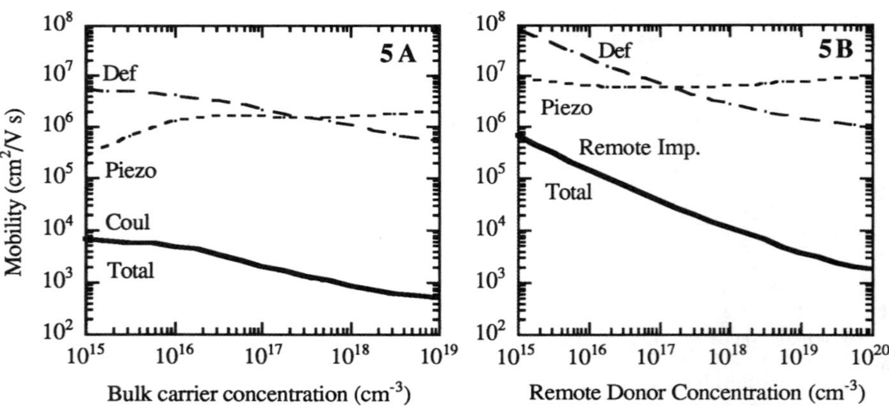

Figures 5A and 5B. Calculated electron mobilities for bulk GaN (A) and for a 2 DEG (B) at a Al$_{0.15}$Ga$_{0.85}$N/GaN heterointerface at 4 K. Contributions from acoustic phonons (Def, Piezo) and ionized impurities (Coul, Remote Imp) are shown, as well as the total mobilities. The ionized impurity curves lie within the linewidth of the total mobility curves.

In addition, unlike the 3D case in which it is impossible to remove Coulombic scattering centers without reducing the electron concentration in the same proportion, in the 2 DEG case, one can reduce the effect of ionized impurity scattering by simply growing a spacer, as discussed above. Figure 6 shows the effect on the low temperature mobility of growing an undoped $Al_{0.15}Ga_{0.85}N$ layer of various thicknesses in between the GaN and doped $Al_{0.15}Ga_{0.85}N$ layers. As the spacer width increases, the electron mobility increases to the point where scattering from acoustic phonons rather than remote donors becomes the dominant mechanism.

Of course, as the spacer width increases, the transfer of electrons from the $Al_xGa_{1-x}N$ to the GaN becomes less efficient as well.

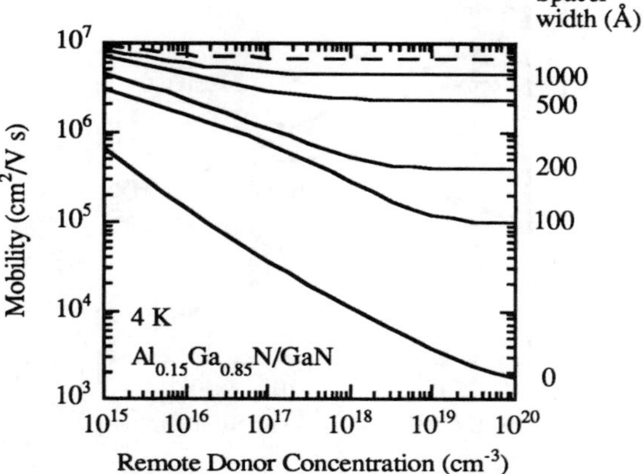

Figure 6. Mobilities as a function of remote donor concentration for a variety of spacer widths. The dotted line is the limit set by acoustic phonon scattering.

For many device applications, such as transistors, the conductivity of the material, which is proportional to the product of the carrier concentration and the mobility, is as important as the mobility. In figure 7, the maximum conductivity that can be achieved by adjusting the remote doping level in the range 10^{15} cm^{-3} to 10^{20} cm^{-3} is shown for three different temperatures as a function of spacer width. At 300K, where optical phonons are the limiting factor in the mobility, the optimum spacer width is small, as it is more important to transfer as many electrons as possible to the GaN to give a large sheet carrier concentration. At low temperatures, however, a larger spacer width is desired, since this decreases the scattering due to remote ionized dopants, which is the dominant scattering mechanism at low temperatures.

Up to now, we have assumed the GaN layer to be completely free of ionized impurities. This will be true if the only defects in the GaN are donor-like. However, there will always be some residual acceptor-like impurities which will be compensated by the excess of electrons and thus contribute to ionized impurity scattering in a way which is very similar to the 3D case. Figure 8 shows the mobilities at 4K of two heterostructures as a function of remote donor concentration. A background charged impurity concentration of 10^{15} cm^{-3} was assumed for the GaN layer. In order to obtain the maximum mobility, the $Al_{0.15}Ga_{0.85}N$ layer of the structure with the 100 Å spacer should be doped more heavily than the one without a spacer. Although the scattering due to remote dopants is increased, the increased electron concentration reduces the scattering

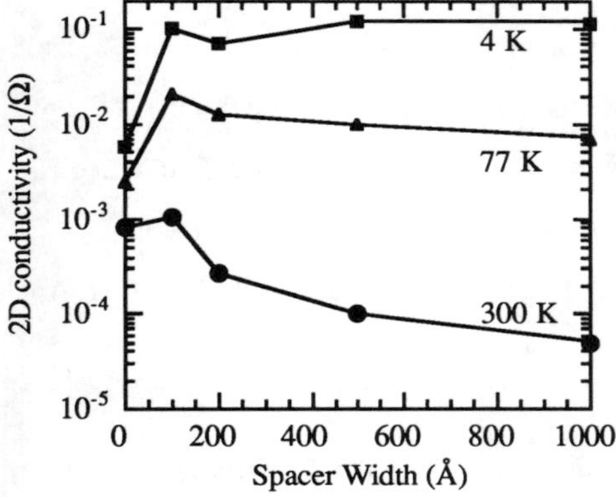

Figure 7. Maximum 2D conductivities for remote doping concentrations in the range 10^{15} cm^{-3} to 10^{20} cm^{-3} at three temperatures.

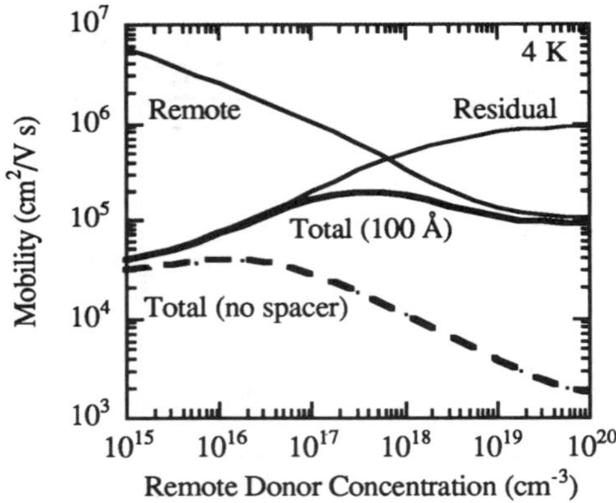

Figure 8. Mobilities in an $Al_{0.15}Ga_{0.85}N/GaN$ heterostructure as a function of remote donor concentration for two different spacer widths and a background impurity concentration of 10^{15} cm^{-3}. The thin lines show contributions from remote and residual impurities for the 100 Å spacer case.

due to the residual impurities in the GaN, increasing the overall mobility. In order to approach the low temperature phonon mobility limit, the residual donor concentration must be reduced below 10^{13} cm^{-3}.

CONCLUSIONS

We have presented the results of theoretical calculations of the electron mobility for a 2 DEG at a $Al_xGa_{1-x}N/GaN$ heterointerface, showing that with the proper structure and material purity, low temperature mobilities of several million should be attainable. We have also determined the deformation potential parameter in GaN by comparing theoretically calculated mobilities with experimentally measured mobilities from high quality $Al_{0.15}Ga_{0.85}N/GaN$ heterostructures.

ACKNOWLEDGMENTS

The authors would like to thank E. E. Haller for his support and encouragement. This work has been supported in part by the Director, Office of Energy Research, Office of Basic Energy Sciences, Materials Science Division of the U. S. Department of Energy under contract number DE-AC03-76SF00098.

REFERENCES

1. J. M. Redwing, M. A. Tischler, J. S. Flynn, S. Elhamri, M. Ahoujja, R. S. Newrock, and W. C. Mitchel, Appl. Phys. Lett. **69**, 963 (1996).
2. M. Asif Khan, J. M. Van Hove, J. N. Kuznia, and D. T. Olson, Appl. Phys. Lett. **58**, 2408 (1991).
3. M. Asif Khan, J. N. Kuznia, J. M. Van Hove, N. Pan, and J. Carter, Appl. Phys. Lett. **60**, 3027 (1992).
4. M. Asif Khan, Q. Chen, C. J. Sun, M. Shur, and B. Gelmont, Appl. Phys. Lett. **67**, 1429 (1995).
5. M. Asif Khan, A. Bhattarai, J. N. Kuznia, and D. T. Olson, Appl. Phys. Lett. **63**, 1214 (1993).
6. M. Asif Khan, J. N. Kuznia, D. T. Olson, W. J. Schaff, J. W. Burm, and M. S. Shur, Appl. Phys. Lett. **65**, 1121 (1994).
7. M. Shur, B. Gelmont, M. Asif Khan, J. Elect. Mat. **25**, 777 (1996).
8. T. Ando, A. B. Fowler, and F. Stern, Rev. Mod. Phys. **54**, 437 (1982).
9. T. Ando, J. Phys. Soc. Jap. **51**, 3893 (1982).
10. F. Stern and W. E. Howard, Phys. Rev. **163**, 816 (1967).
11. W. Walukiewicz, H. E. Ruda, J. Lagowski, and H. C. Gatos, Phys. Rev. B **30**, 4571 (1994).

SPIN RESONANCE INVESTIGATIONS OF GaN AND AlGaN

N. M. REINACHER, H. ANGERER, O. AMBACHER, M. S. BRANDT, M. STUTZMANN
Walter Schottky Institute, Technical University of Munich, Am Coulombwall, D–85748
Garching, email: nmr@wsi.tu-muenchen.de

ABSTRACT

Electron Spin Resonance (ESR) and related methods have been used to study defects
in GaN and in $Al_xGa_{1-x}N$ ternary alloys. In particular, Light-induced ESR of a thick
MOCVD grown layer of GaN shows that the ESR signature of the deep defect appears for
excitation energies > 2.6 eV and saturates above 2.7 eV. This information provides a direct
measure of the energetic level position of the diamagnetic to paramagnetic transition of
this center. Furthermore, standard ESR investigations of MBE-grown layers of $Al_xGa_{1-x}N$
alloys were performed with emphasis on the effective-mass donor resonance. Composition
and temperature-dependent measurements of the resonance position are presented.

INTRODUCTION

Two major centers present in almost all GaN samples independent of fabrication method
are of special interest: the effective-mass donor giving rise to the n-type conductivity of
nominally undoped GaN layers, and the deep defect which is related to the yellow luminescence band at 2.2 eV being a concurrent recombination path to the excitonic recombination.
Electron spin resonance (ESR) in combination with optical excitation (Light-induced ESR)
can not only provide information about defect specific quantities such as number of spins
but furthermore about energetic levels of diamagnetic to paramagnetic transitions of centers
located in the bandgap and about excitation dynamics.

EXPERIMENTAL

In this paper we report on ESR and LESR studies on MOCVD GaN and MBE AlGaN
layers. The 10 μm thick nominally undoped GaN epitaxial layer was grown by metal-organic
chemical vapor deposition (MOCVD) on a sapphire substrate at a rather high growth rate of
about 15 μm/h. The sample exhibits a particularly strong 2.2 eV luminescence band at 4 K,
as shown in Fig. 1(a). This indicates that a large number of defects is present in the sample,
as necessary for ESR measurements. The $Al_xGa_{1-x}N$ films were grown by plasma-induced
MBE (PIMBE) on sapphire substrate with thicknesses varying between .5 and 3 μm. The
Al concentrations were determined by elastic recoil detection analysis (ERDA) and/or the
c_0 lattice constant from X-ray deflection (XRD) measurements. For samples with no ERDA
data availlable a cross-calibration was used to correct XRD data for strain and additional
shifts. Al mole fractions ranged from 0 to 80 %.

The ESR measurements were performed in a standard X-band spectrometer with microwave powers of the order of 20 to 430 mW. With an Oxford liquid helium cryostat,
temperatures in the range from 4–70 K were achieved. In order to evaluate the g tensor
components angle-dependent measurements were made, although all wavelength-resolved

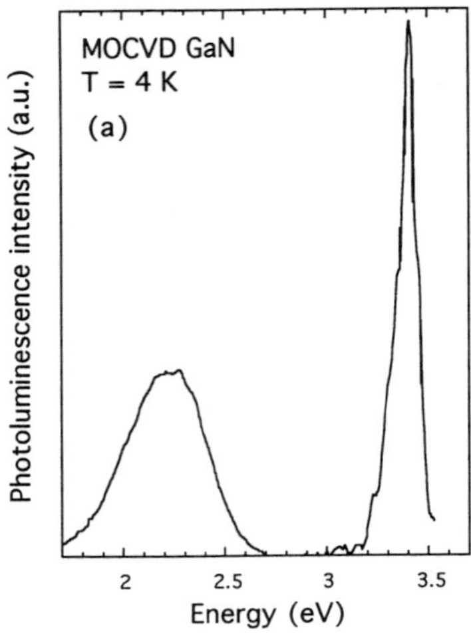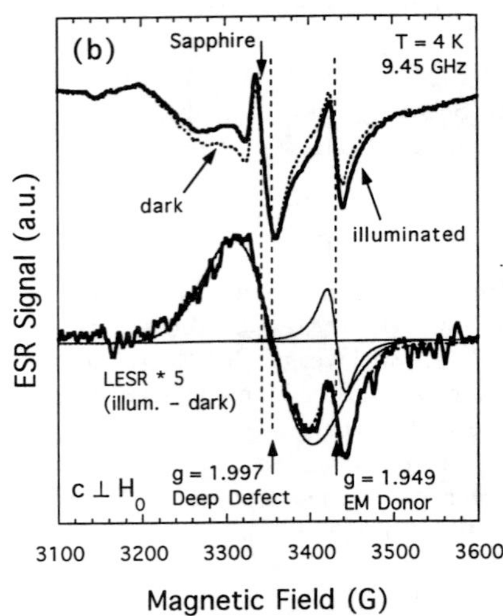

Figure 1: (a) PL spectrum of the MOCVD GaN layer. (b) top: ESR spectra of the sample at 4 K in the dark and under illumination; bottom: numerical difference (LESR) spectrum. The line shape is fitted with two lines: a 95 G broad gaussian at $g = 1.997$ and a narrow lorentzian at $g = 1.949$, indicated by the broken line.

spectra were recorded with the sample's c-axis oriented perpendicular to the external magnetic field H_0. Background ESR lines due to the sapphire substrate were eliminated by special orientations of the samples. The LESR was excited with either single Ar^+ ion laser-lines or the light of a 200 W Hg(Xe) lamp, where different wavelengths from 360 nm to 2000 nm were selected with the help of interference filters. The absolute photon fluxes onto the sample under different illumination conditions were determined using a UV-sensitive calibrated silicon photo-diode.

LEVEL POSITION OF THE DEEP DEFECT

Typical ESR and LESR spectra with $\mathbf{c} \perp \mathbf{H}_0$ are shown in Fig. 1(b). The LESR signal is dominated by a broad resonance at $g = 2.00$ and a smaller signal at $g = 1.95$. The lines are well fitted by two derivative-like structures: a $\Delta H_{\mathrm{pp}} = 95$ G broad Gaussian at $g = 1.997$ and a narrow Lorentzian at $g = 1.949$. In terms of resonance parameters the broad line is identical to the yet unidentified deep-defect (DD) resonance observed in ODMR investigations [1, 2]. The g value of 1.95 and the Lorentzian lineshape of the narrow line lead to the assignment to effective-mass (EM) donors known from ESR experiments [3].

The dependence of the LESR signal intensity of the DD resonance on the energy of the exciting photons is shown in Fig. 2(a). For excitation energies ≥ 2.6 eV the LESR signal shown in Fig. 1(b) is observable, with the signal intensity saturating for energies > 2.7 eV. The solid line in Fig. 2(a) represents a fit of the LESR signal intensity based on the assumption of a Gaussian density-of-states distribution centered around $E_\mathrm{x} = 2.65 \pm 0.05$ eV and having a full width at half maximum of $\Delta E = 0.05$ eV. This threshold energy is

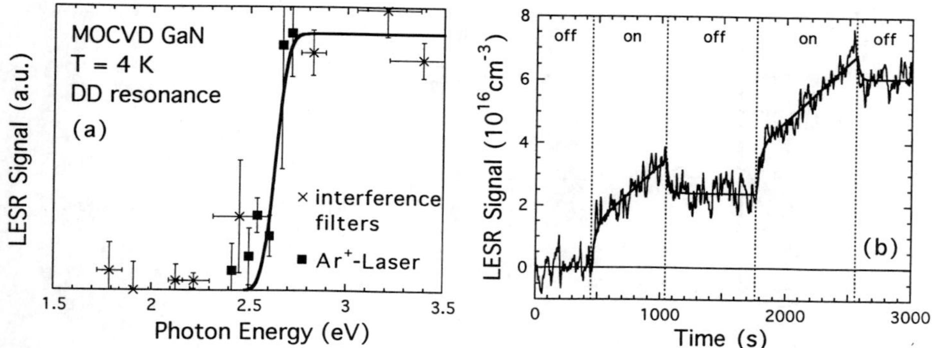

Figure 2: (a) Excitation-energy dependence of the LESR signal amplitude of the DD resonance. The data is well described by an excitation from a level 2.65 eV below the conduction band having a FWHM of 50 meV. (b) Time-dependence of the same signal under alternating white-light illumination. Two distinct processes can be well seperated: a fast increase due to instant ionisation of defects, and a slow rise due to the evolving global equilibrium.

similar to the onset of the PLE as reported by Ref. [4]. However, the LESR saturates at 2.7 eV while the PLE continues to increase.

The straight-forward interpretation of this observation is the optical excitation of electrons from the deep defect level at $E_{\mathrm{C}} - E_{\mathrm{DD}} = 2.65$ eV, as indicated on the left side of Fig. 3. Possible transitions are an intra-center excitation of the deep defect as well as a transfer of an electron to an EM donor or the conduction band. An intra-center excitation would lead to a triplet ESR signal which is not observed. The change in the intensity of the EM resonance indicates that only about 10 % of the electrons are transfered from the DD to this state. The remainder most probably is conduction band electrons which cannot be observed in conventional ESR due to short spin relaxation times. Due to a large electron phonon coupling, the excited electrons partially thermalize to a lower energy level. After radiative recombination giving rise to the yellow luminescence, the defect relaxes back to its initial ground state.

Time-dependent measurements of the LESR signal intensity provide further quantitative information about the excitation process. Fig. 2(b) shows the time-dependence of the DD signal intensity in the dark and under illumination with white light. One can clearly observe two distinct processes: a relatively fast exponential increase with a time constant of about 30 s saturating at an LESR spin density of 1×10^{16} cm⁻³, and a superimposed slow rise of the LESR spin density at an almost constant rate of 4×10^{13}cm⁻³s⁻¹, only present under UV illumination. Only after several hours of illumination saturation of the slow component was observable, allowing an estimate of the saturation spin density of 8×10^{18} cm⁻³ for the slow process. When switching off the light source, the fast component relaxes back to its initial value with a rate faster than the experimental resolution of 5 s, whereas the signal intensity gained by the slow process remains constant at low temperatures. Only warming up the sample to above room temperature restores the initial ESR signal. Such

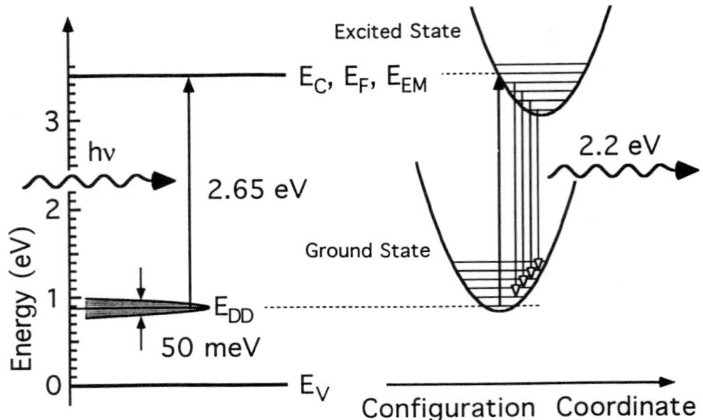

Figure 3: Configuration diagram of the excitation and relaxation process leading to the yellow luminescence. Optically excited carriers quickly thermalize into the potential minimum of the excited state emitting phonons. Concurrently, they recombine radiatively to the ground state, where they further thermalize.

a fast excitation process is expected for electronic transitions from the strongly localized deep defect to a state with an extended electronic wave function. On the other hand, upon illumination, electrons have to be transferred across grain boundaries for a global steady state, giving rise to a slow component. Accordingly, when the exciting source is removed, a part of the carriers recombines instantly, whereas in inhomogeneous material in regions with a low concentration of defects a surplus of electrons remains mobile. Their back-diffusion is prevented by potential fluctuations, traps or grain boundaries, so that a total equilibrium can only evolve very slowly.

g-VALUE SHIFT OF THE EM RESONANCE FOR VARYING Al CONTENT

From previous ESR investigations it is known that all n-type GaN layers exhibit a strong, lorentzian-shaped resonance around $g = 1.95$ with a slight anisotropy of 0.003 [3]. This line is due to extended effective-mass-like states of a donor band close to the conduction band. Therefore, the line position reflects the conduction band-edge g-value, since the wave functions of these states are dominated by band edge states. A simple five-band $\mathbf{k} \cdot \mathbf{p}$ perturbation approach accounts for the observed value of g [3].

Recent work on $Al_xGa_{1-x}N$ alloys with Al mole fractions of up to .26 by the same author show similar resonances at slightly shifted g values [5]. This shift is due to composition dependent contributions of the relevant phenomena, especially acting on the fundamental bandgap E_0 and the spin-orbit coupling term of the second conduction band Δ_0'. Carlos used the perturbation approach to reasonably explain an almost linear shift for lower Al content.

Our data is shown in Fig. 4(a). Besides a good agreement with the data of Carlos for lower Al concentrations, the experimentally determined g values for higher Al fractions seem to be lower than expected from the linear fit, although the difference being smaller than the experimental uncertainty. It has to be noted that we have experimental evidence for a similar

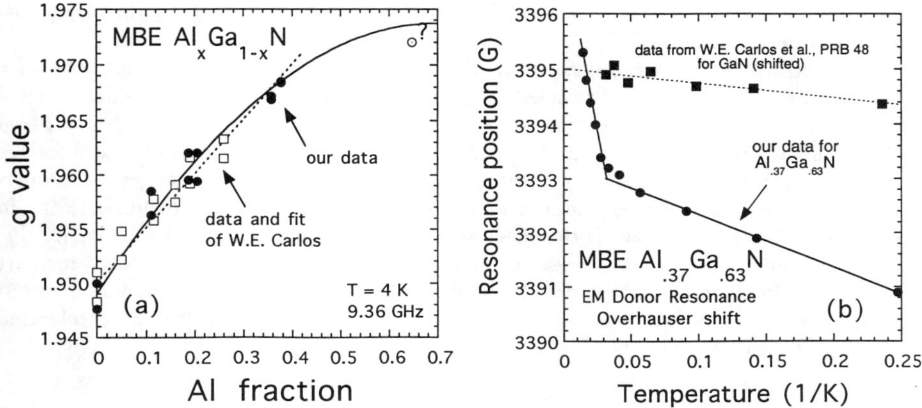

Figure 4: (a) g-value shift of the EM resonance. The circle indicates a similar defect in terms of resonance parameters. (b) Temperature-dependence of the resonance position in Al$_{.37}$Ga$_{.63}$N. A more pronounced Overhauser shift as compared to the data of Carlos for GaN are observed. Additionally, above 35 K the depolarizing effect is enhanced.

resonance in terms of linewidth, saturation behaviour and relaxation times in the Al$_{.64}$Ga$_{.36}$N sample at a g value of 1.974, indicated by the circle mark in Fig. 4(a). Unfortunately, we were not able to detect any EM resonance signal for Al concentrations higher than that. These findings suggest a possible non-linear contribution to the g-value shift. These could result in second-order corrections to the relevant parameters in the $\mathbf{k} \cdot \mathbf{p}$ approach and a non-zero Δ'_0 in AlN which Carlos did not take into account. Further corrections could be introduced by a change in the EM-type character of the defect for higher Al content, or an even stronger effect due to the appearance of an additional defect being hidden in the conduction band for low Al fractions. This is supported by recent calculations of defect scattering in AlGaN alloys [6].

Additional changes occur at higher Al concentrations concerning the anisotropy $g_{\parallel} - g_{\perp}$. Our measurements indicate a decreasing anisotropy for increasing Al fractions, whereas in the Al$_{.37}$Ga$_{.63}$N sample no anisotropy could be resolved experimentally. This could be explained by a reduced spin-orbit coupling at the Al atoms consistent with the reduced g shift, although some anisotropy due to crystal field splitting should remain. Furthermore, the admixture of more localized contributions from an additional defect could reduce the amount of anisotropy.

OVERHAUSER SHIFT IN Al$_{.37}$Ga$_{.63}$N

Further phenomena effect the magnetic-field position of this line. Previous investigations have shown the existence of an Overhauser shift for this resonance in GaN [3, 7]. This B-field shift results from an averaging of spatially extended electrons over the nuclear spin system yielding a modified hyperfine-interaction term $\Delta E = A \cdot \langle I_z \rangle$, where A is the hyperfine interaction parameter and

$$\langle I_z \rangle = I(I+1)\frac{\mu_n g_n}{3k_B T}B \tag{1}$$

is the ensemble average of the nuclear spin polarization. This shift in energy directly transforms into a magnetic field shift.

The experimental findings are shown in Fig. 4(b) in comparison with the data of [3] for GaN which have been shifted and scaled in magnetic field to account for the different resonance position in GaN. One can clearly observe a much stronger effect in AlGaN compared to GaN. This is probably due to the larger Al nuclear spin of $I = \frac{5}{2}$ compared to $\frac{3}{2}$ for Ga nuclei and a larger electronegativity. Furthermore, the slope drastically changes for temperatures above 35 K. Possible explanations concern different parameters to the B shift. In case of different nuclei dominating the effect above 35 K, g_n and I are affected. Additionally, a localization effect due to an increase in the electron concentration could account for part of the change in slope. Furthermore, thermal emission or detrapping of mobile electrons with faster relaxation times (e. g. conduction-band electrons) would result in an accelerated nuclear depolarization and thus affect $\langle I_z \rangle$.

CONCLUSIONS

We have performed electron spin resonance investigations on GaN and AlGaN layers with Al concentrations of up to 80 %. LESR measurements on MOCVD grown GaN supply a direct measure of the excitation energy of the DD signature and lead to the determination of the energetic position of this defect as $E_C - E_{DD} = 2.65 \pm 0.05$ eV. Time-dependent LESR measurements reveal further information about the excitation and relaxation processes involved. Additional ESR studies on the EM donor resonance position in AlGaN alloys including Al mole-fraction dependent and temperature-dependent measurements exhibit a strong influence of the Al admixture in terms of cation spin-orbit splitting and nuclear spin resulting in a decrease of the g-value shift and a more pronounced Overhauser shift.

ACKNOWLEDGEMENTS

The authors would like to thank the Deutsche Forschungsgemeinschaft for financial support.

REFERENCES

1. E. R. Glaser, T. A. Kennedy, H. C. Crookham, J. A. Freitas, Jr., M. Asif Khan, D. T. Olson, and J. N. Kuznia, Appl. Phys. Lett. **63**, 2673 (1993).

2. M. Kunzer, U. Kaufmann, K. Maier, J. Schneider, N. Herres, I. Akasaki, and H. Amano, Mater. Sci. Forum **143–147**, 87 (1994).

3. W. E. Carlos, J. A. Freitas, Jr., M. Asif Khan, D. T. Olson, and J. N. Kuznia, Phys. Rev. B **48**, 17878 (1993).

4. T. Ogino and M. Aoki, Jpn. J. Appl. Phys. **19**, 2395 (1980).

5. W. E. Carlos, Proc. of the Shallow Donor Conf. 1996, Amsterdam (*in print*).

6. L. Hsu, W. Walukiewicz, in Proc. of the 23rd Int. Conf. Phys. Semicond. 1996, p. 2913.

7. G. Denninger, R. Beerhalter, D. Reiser, K. Maier, J. Schneider, T. Detchprohm, K. Hiramatsu, Sol. State Comm. **99**, 347 (1996).

EFFECTS OF x-RAY and γ-ray IRRADIATION ON GaN

C. H. Qiu*, M. W. Leksono*, J. I. Pankove*, C. Rossington**, and E. E. Haller**
* Astralux, Inc., 2500 Central Ave., Boulder, CO 80301, qiu@schof.colorado.edu
** Lawrence Berkeley National Laboratory, 1 Cyclotron Road, Berkeley, CA 94720

ABSTRACT

As part of the feasibility study of using III-V nitride semiconductors for x-ray and γ-ray detection, the irradiation effects on GaN were investigated. GaN films with very different electrical resistivity and electron concentration were used in the study. The electron mobility, photoconductivity spectra, and photo-luminescence spectra were measured before and after irradiation. An enhanced $\eta\mu\tau$ product for undoped GaN films and an enhanced blue luminescence for a Zn-doped sample were observed after irradiation.

INTRODUCTION

The ability to detect x-rays and gamma rays is of great importance since it makes possible a variety of analysis and imaging techniques. However, the popular use of these techniques has been limited by the need to operate the radiation detectors at cryogenic temperatures. Over the years, many semiconductor materials, such as CdZnTe, HgI_2, and diamond have been developed for uncooled x-ray detection[1,2]. CdZnTe and HgI_2 are very promising materials, however, there are problems associated with the intrinsic softness and chemical activity. In comparison, InGaN materials are unique in that they exhibit superior hardness and chemical inertness than CdZnTe and HgI_2, an x-ray absorption coefficient comparable to Ge, and a bandgap energy more favorable than that of diamond. Compared to other III-V semiconductors, the $\eta\mu\tau$ product of InGaN might be much larger since many defects in III-V nitrides (such as dislocations) do not contribute to undesired recombination[3,4]. In this paper, experimental results of irradiation effects on GaN thin films are presented. An enhanced $\eta\mu\tau$ product was found after irradiation, potentially benefiting x-ray detection.

EXPERIMENTAL DETAILS

Four samples were irradiated with a synchrotron x-ray beam at 4 keV for 48 hours. Three samples were irradiated with a Co-60 source for 17 days. Before and after the irradiation, each sample was characterized by photoconductivity and photoluminescence (PL) spectroscopy. The dosage of each sample is listed in table I.

Film GNZN is a Zn-doped GaN grown by chloride transport method[5], with a resistivity of 4.5×10^6 Ω-cm at room temperature. Other samples are undoped n-type. Sample 06203 and 67639, also grown by chloride transport method, had Hall bars and were used to study irradiation-induced changes in electron concentration and

Mat. Res. Soc. Symp. Proc. Vol. 449 © 1997 Materials Research Society

mobility. The other four samples were grown by metalorganic CVD, and the electron mobility at room temperature was in the range of 73 to 360 cm^2/V-sec.

Table I. A list of the irradiation dosage and radiation energy for the samples studied. The electron concentration of the undoped n-type samples spans over two orders of magnitudes.

Sample #	Thickness (μm)	Dosage (Rads)	Radiation Energy	n (cm^{-3})
GNZN	100	1.8×10^9	4 keV	---
06203	30	6.1×10^9	4 keV	1.1×10^{19}
67639	4.0	4.5×10^6	1.2 MeV	1.3×10^{19}
ML190	0.9	4.5×10^6	1.2 MeV	9.6×10^{16}
CQ1107	1.0	4.5×10^6	1.2 MeV	5.3×10^{17}
CQ511	3.0	3.1×10^{10}	4 keV	1.4×10^{18}
CQ918	1.5	2.6×10^{10}	4 keV	1.3×10^{19}

RESULTS AND DISCUSSIONS

After x-ray irradiation, the conductivity of sample 06203 decreased by 2.4%, while the electron mobility increased from 114.5 cm^2/V-sec to 123 cm^2/V-sec. After γ-ray irradiation, the conductivity of sample 67639 increased by 6.4%, while the electron mobility decreased from 80 to 79.1 cm^2/V-sec. Thus, it seems that the radiation-induced changes in electron mobility and concentration are very small.

The photoconductivity was measured using lock-in technique. The light intensity at 3.4 eV was 0.2 μW/cm^2. Other measurement details were the same as described elsewhere[6]. To ascertain irradiation-induced changes in the ημτ product, the same excitation light source, a Halogen lamp operated at the same voltage, was used for photoconductivity measurements before and after irradiation. Furthermore, an un-irradiated GaN sample was used to calibrate the light spectrum at the exit slit of the monochromator. The ημτ product in this work was defined as $(ημτ)_e + (ημτ)_h$ to include the contribution of both electrons and holes to photoconductivity, and was calculated from the photoconductivity data. After irradiation, all the samples exhibited higher ημτ product, as illustrated in Fig. 1. PC data were not collected from sample GNZN, because it became conducting as a result of a spark (a high voltage of >1500 V due to power supply malfunction was accidentally applied across the 0.9 mm gap on GNZN).

The ημτ product in Fig. 1 was calculated from the photoconductivity response of GaN films in the UV wavelengths. If there were some increase in the density of gap states which acted as hole traps, the electron lifetime would become longer. The photoconductivity spectra were normalized at 3.6 eV to reveal possible changes in the subband gap wavelength region. Fig. 2 shows typical cases of normalized

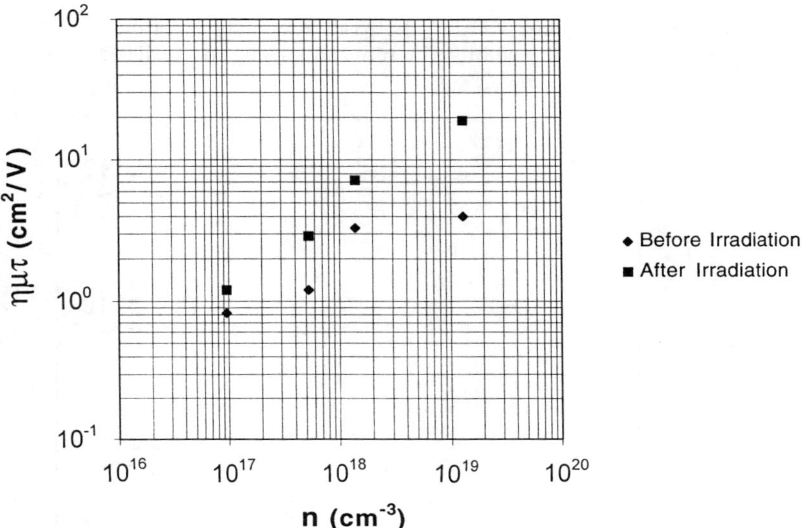

Fig. 1. Both x-ray irradiation and γ-ray irradiation improve the $\eta\mu\tau$ product, and the radiation-induced change seems to be smaller for more resistive samples. The sample number can be looked up in Table I using the electron concentration data.

photoconductivity spectra before and after irradiation. For sample ML190, the sample with the least $\eta\mu\tau$ increase in Fig. 1, the two normalized spectra were almost identical, i.e., the increase in photoconductivity was the same at all the wavelengths. For sample CQ918, the sample with the largest $\eta\mu\tau$ increase in Fig. 1, Fig. 2 indicates a bigger increase in photoconductivity in the subband gap wavelength region than the UV region after irradiation. Similar changes in the normalized spectra were observed for sample CQ511. In the subband gap region, the normalized photoconductivity is given by[6]:

$$Y(h\nu) = r_e \alpha d \tag{1},$$

where α is the optical absorption coefficient, d is the film thickness, and r_e is:

$$r_e = [(\eta\mu\tau)_e]_{@h\nu} / [(\eta\mu\tau)_e + (\eta\mu\tau)_h]_{@3.6\,eV} \tag{2}.$$

Thus, an increase in Y indicated an increase in r_e and/or α. However, film CQ1107 exhibited a different behavior: the photoconductivity in the subband gap wavelength region is increased only slightly after irradiation, while increased much more in the UV region. As a result, the normalized spectra exhibited a decrease in the subband

optical absorption, is increased after the irradiation. The lifetime τ may depend on the photon energy [7], and exhibits different changes upon irradiation.

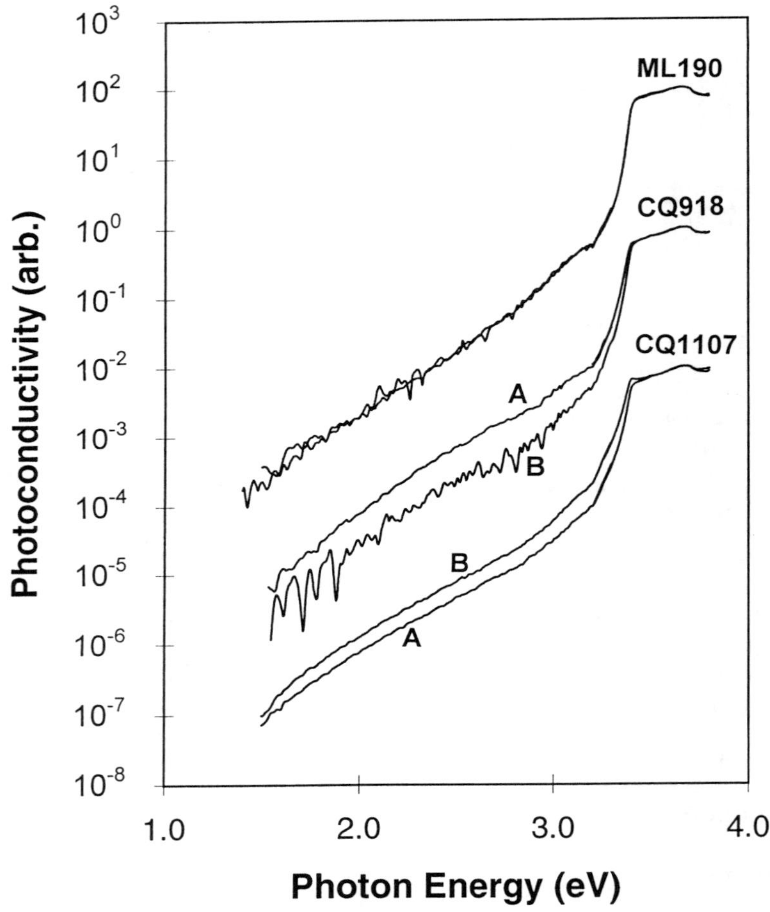

Fig. 2. Normalized photoconductivity spectra before and after irradiation for samples ML190, CQ918, and CQ1107. For clarity, the data for different sample was displaced by a factor of 100.

PL spectra of the samples were measured before and after irradiation both at room temperature and at 3.4 K. The change in PL spectra due to irradiation is very small as shown in Fig. 3 for several samples. In both PL and photoconductivity measurement, the biggest change was observed from sample CQ918, the sample with the highest electron concentration. The yellow luminescence peak became slightly weaker. Slightly stronger yellow luminescence was observed for sample CQ1107 and ML190. The effect of x-ray radiation on Zn-doped sample GNZN seemed to be similar to the effect of electron beam irradiation, i. e., the blue luminescence is enhanced by irradiation as originally observed by Akasaki's group[8]. Sample CQ511 also exhibited a slightly enhanced blue luminescence at room temperature. Below 160 K, this sample had strong D-A pair emission, indicating the presence of acceptor impurities.

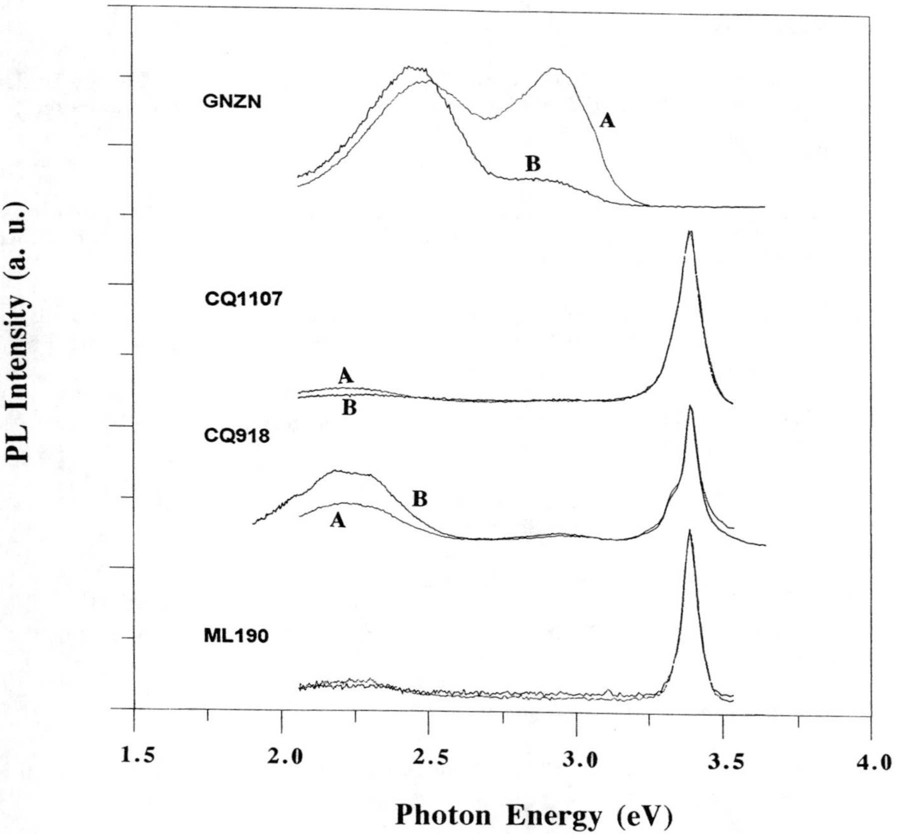

Fig. 3. Photoluminescence spectra before and after irradiation measured at room temperature under identical conditions.

SUMMARY

Enhanced $\eta\mu\tau$ product in undoped n-type GaN films was observed after irradiation, and the irradiation-induced change is smaller for more resistive undoped materials. This effect may make the detector more sensitive during usage. After irradiation, the yellow luminescence intensity is slightly changed, and the biggest change was observed for a sample with the biggest change in photoconductivity. Enhanced blue luminescence was observed in a Zn-doped GaN after x-ray irradiation. Since x-ray detectors are normally made of resistive materials, the latter effect deserves further study.

ACKNOWLEDGMENTS

This work was supported in part by the US Department of Energy under contract number DE-FG03-95ER86024, and monitored by Dr. Richard Rinkinberger.

REFERENCES

1. <u>Semiconductors for Room Temperature Nuclear Detector Applications,</u> Semiconductors and Semimetals, Vol. **43**, ed. by T. E. Schlesinger and R. B. James (Academic Press, San Diego, 1995).

2. <u>Semiconductors for Room-Temperature Radiation Detector Applications,</u> ed. by R. B. James, T. E. Schlesinger, Paul Siffert, and Larry Franks (Mat. Res. Soc. Symp. Proc. **302**, Pittsburgh, PA 1993).

3. S. D. Lester, F. A. Ponce, M. G. Craford, and D. A. Steigerwald, Appl. Phys. Lett. **66**, 1249 (1995).

4. C. H. Qiu, W. Melton, M. W. Leksono, J. I. Pankove, B. P. Keller, and S. P. DenBaars, Appl. Phys. Lett. **69**, 1282 (1996).

5. H. P. Maruska and J. J. Tietjen, Appl. Phys. Lett. **15**, 327 (1969).

6. C. H. Qiu, C. Hoggatt, W. Melton, M. W. Leksono, and J. I. Pankove, Appl. Phys. Lett. **66**, 2712 (1995).

7. T. D. Moustakas, private communication.

8. H. Amano, I. Akasaki, T. Kozawa, K. Hiramatsu, N. Sawaki, K. Ikeda, and Y. Ishii, J. Lumines. **40/41**, 121(1988).

Studies of electrically and recombination active centers in undoped GaN grown by OMVPE

A.Y. Polyakov, A.V. Govorkov*, N.B. Smirnov*, M. Shin, M. Skowronski, D.W. Greve**

Department of Materials Science and Engineering, Carnegie Mellon University, Pittsburgh, PA 15213
* Institute of Rare Metals, B. Tolmachevsky, 5, Moscow, 109017, Russia
** Department of Electrical and Computer Engineering, Carnegie Mellon University, Pittsburgh, PA 15213

ABSTRACT

Deep centers were studied in GaN samples grown by organometallic vapor phase epitaxy (OMVPE). Electron traps 0.2 eV and 0.5 eV below conduction band edge and 0.25 eV and 0.5-0.85 eV above the valence band edge were detected by means of deep levels transient spectroscopy (DLTS), photoelectron relaxation spectroscopy (PERS) and thermally stimulated current spectroscopy (TSC). The photoconductivity at low temperature is shown to be persistent and the magnitude of photosensitivity is dependent on the way the samples are grown. Microcathodoluminescence (MCL) and electron beam induced current (EBIC) measurements indicate that the density of deep recombination centers near the dislocation walls between the misoriented GaN domains is lower than inside the domains. Spatially resolved PERS measurements show that the concentration of the 0.85 eV level is higher in the low angle grain boundary regions that produce bright contrast in EBIC and MCL.

INTRODUCTION

GaN is a wide band gap III-V material with potentially important applications in short wave length optoelectronics and in high power/ high frequency/ high temperature electronics [1]. For most devices now under development for such applications the detailed knowledge of deep centers spectra and of the uniformity of deep centers distribution is of vital importance. Some information regarding these issues can be found in the literature [2-5,7,8]. DLTS [2,3] and photocapacitance spectroscopy [4,5] studies have been reported by several groups and revealed the presence of several electron traps with energies ranging from 0.2 to 2.5 eV and with concentrations in the 10^{13}-10^{15} cm^{-3} range. Admittance spectroscopy [6] measurements carried out on p-GaN samples allowed to determine the activation energy of the dominant shallow acceptors in this material [7]. MCL [8] studies indicate a certain correlation between surface morphology features such as hillocks and the intensity of the band edge and yellow luminescence [9] bands. Still, more has to be done in applying such methods to GaN. In this paper we present the results of capacitance-voltage (C-V), current-voltage (I-V), capacitance-frequency (C-f), DLTS and conductance (at various frequencies) versus temperature (G-f-T) measurements. These measurements are complemented by photoelectron relaxation spectroscopy (PERS) [10]) measurements with electron beam excitation and by thermally stimulated current (TSC) measurements. We also report preliminary results on the distribution of recombination-active defects using MCL at 100 K and electron beam induced current (EBIC) measurements at 300 K.

EXPERIMENTAL

GaN samples studied in this paper were grown by OMVPE using a horizontal two-inlet reactor, with trimethylgallium and ammonia as precursors. Growth was done on sapphire substrates, with low temperature GaN buffers either with or without a relatively long (3-10 min) pre-nitridation step at 1100 °C during the substrate cleaning. The GaN layers themselves were grown at 1025 °C. The growth procedure will be described in detail elsewhere. Resistivities, carrier concentrations and mobilities of the samples were measured using van der Pauw method. Capacitance and AC conductance measurements were made using an HP 4192A impedance analyzer. These measurements were performed in the frequency range of 5 Hz-13 MHz, in the temperature range from 80K to 400K. Current measurements were done with the help of HP4140A picoammeter.

Mat. Res. Soc. Symp. Proc. Vol. 449 © 1997 Materials Research Society

For DLTS spectra measurements we used HP 4280A C-V/C-t meter with an external pulse source. PERS measurements were perfomed with electron beam excitation in our scanning electron microscope. TSC measurements were done by cooling the samples down in the dark to 80 K, shining light from an ultraviolet lamp for half an hour and heating the samples up at heating rates varying from 0.05 to 0.2 K/s. The dark current was monitored during the cooling down stage and the thermally stimulated current was measured during heating up. MCL and EBIC measurements were carried out with electron beam accelerating voltage of 25 kV. Capacitance, admittance, PERS and EBIC measurements were made on Au Schottky diodes of approximately 1 mm in diameter deposited by vacuum evaporation. TSC measurements were carried out on samples with ohmic contacts. Details of experimental set-ups can be found in [10-12].

RESULTS AND DISCUSSION

Most of the measurements discussed in this paper were made on two GaN samples: sample #1 with electron concentration of $8\,10^{17}$ cm^{-3} and sample #2 with electron concentration of $3\,10^{15}$ cm^{-3} at room temperature. Room temperature mobilities for both samples were about 300 cm^2/V·s. Both samples were grown on pre-nitrided (3 min) sapphire substrates.
I-V characteristics of corresponding Schottky diodes are shown in Fig. 1. The ideality factor for forward I-V curves in Fig. 1 is 1.2 for sample #1 and 1.1 for sample #2. Understandably the reverse current is much lower and the series resistance much higher for sample #2. This is reflected in C-f curves in Fig.2 which show that the measured capacitance starts to decrease due to the effect of series resistance for frequencies higher than 1 MHz for sample #1 and higher than 3 kHz for sample #2. If the measurements frequency is chosen well below these threshold frequencies the $1/C^2$ versus voltage plots are linear, with the built-in voltage of about 0.85 V in both cases, in good agreement with the reported results of Au Schottky barrier height measurements on GaN [13]. The donor concentrations yielded by these plots are $8\,10^{17}$ cm^{-3} for sample #1 and about 10^{16} cm^{-3} for sample #2, in reasonable agreement with van der Pauw measurements.

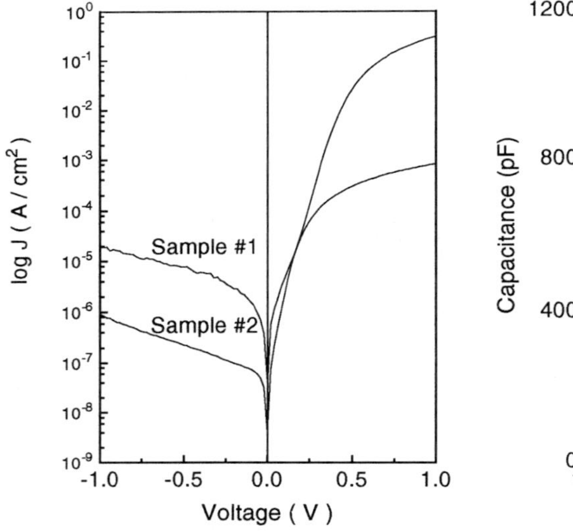

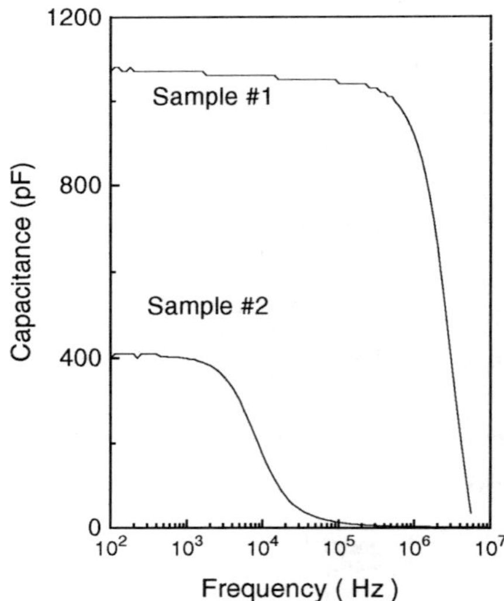

Fig.1. Current-voltage characteristics for samples #1 and #2.

Fig. 2. C-f characteristics for samples #1 and #2.

For sample #1 DLTS measurements were marginally possible and corresponding spectrum for reverse bias of -1 V, excitation pulse of 0 V with duration of 200 ms, and for time windows of 0.2 ms and 200 ms is shown in Fig. 3. Two electron traps with levels at about 0.2 and 0.5 eV below the conduction band and concentration of about 10^{15} cm^{-3} could be detected. These traps are similar to the ones reported previously in [2,3].

No DLTS measurements could be done on sample #2 because of the high series resistance. Hence we tried to do G-f-T, PERS and TSC measurements to derive information on deep traps in this sample. In G-f-T measurements (or admittance spectroscopy) one is monitoring the temperature dependence of conductance G at a chosen frequency f versus temperature or of capacitance versus temperature. When the condition $2\pi f = e_n$ (e_n being the electron emission rate from the trap in question) is satisfied the G value passes through a maximum and a step is observed in capacitance ([6]; the above condition corresponds to the center of the step, to be precise). If such measurements are performed for several frequencies the activation energy and capture cross-section of the traps in question can be calculated. Such G-T curves for frequencies of 3 kHz, 1 kHz and 500 Hz are shown in Fig. 4. From several such curves the activation energy of the trap was estimated to be 0.035 eV with the electron cross section of 10^{-18} cm^2.

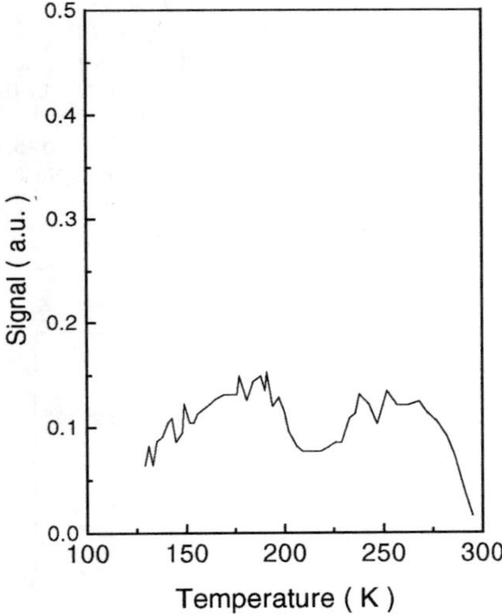

Fig. 3. DLTS spectrum for sample #1.

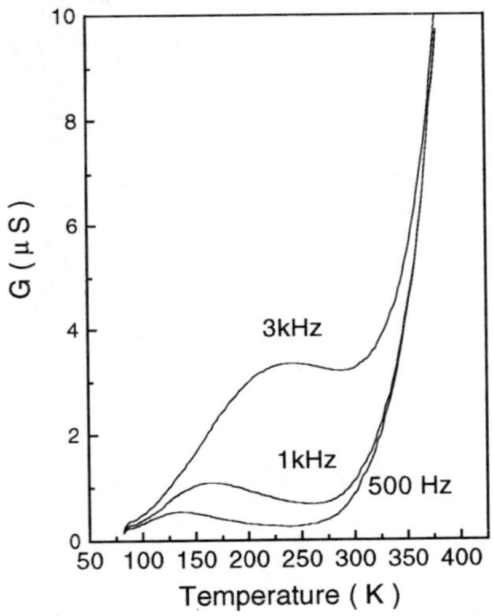

Fig.4. G-T curves for sample #2 measured at 0 V for frequencies of 3 kHz, 1 kHz and 500Hz

Deeper traps were studied by PERS and TSC. In PERS a reverse biased Schottky diode was excited with a pulse of high energy electrons of the electron beam of scanning electron microscope, and the current transients were measured. After a sharp decrease of current due to recombination, "tails" in current transients that come from detrapping of electrons or holes from the deep centers recharged during the excitation pulse were observed. If the transient current values were measured at time windows t_1 and t_2 ($t_2 \gg t_1$) and the difference was plotted against temperature the value of this current difference passed through a maximum each time the condition $1/t_1 = e_n$ was satisfied for one of the traps. By measuring the peak position for different time windows t_1 one can then calculate the activation energy and capture cross section for the traps. It was also argued that the amplitude of the peak is proportional to the concentration of the traps [10] and therefore the uniformity of their distribution can be studied. This is particularly advantageous if the excitation is performed by the electron beam of a scanning electron microscope since it allows to study the distribution of defects on a microscopic scale.

Fig. 5 presents the dominant PERS peak in sample #2 measured for three different sets of t_1/t_2 time windows (corresponding settings are indicated near each curve). From these measurements the activation energy is deduced to be 0.85 eV. A trap similar to ours and giving rise to a PERS peak at about 350 K was observed in [5] and assigned to an acceptor center about 0.9 eV above the valence band. The authors argued that this could be the trap responsible for the yellow luminescence band in GaN. There is unfortunately no easy way to find out whether the transition we observe in PERS is related to the conduction band or the valence band. If we assume that it is the same trap as detected by PERS and photocapacitance in [5], the 0.85 eV activation energy measured above should correspond to hole emission into the valence band and the capture cross section for this trap is $1.4\,10^{-14}$ cm^2. TSC spectrum of sample #2 is shown in Fig. 6. Two overlapping peaks from traps with activation energies of about 0.25 eV and 0.5 eV, as deduced from the dependence of the peak position on the heating rate [14], can be seen. Since the measurements were done on n-type samples with the Fermi level only slightly below the conduction band edge only hole emission from the recharged traps to the valence band should be seen under normal conditions. The temperature of the second peak in TSC somewhat varies from sample to sample changing from about 130K to about 170K probably indicating that the energy of traps responsible for the peak is also slightly different. One very interesting feature of the TSC results is that photoconductivity at low temperature is persistent: at temperatures below 200 K the photocurrent, after initial decay following switching off the light, remains at the same level measurably above the dark current for many hours. There are several possible mechanisms that could account for such persistent photoconductivity (PPC), one of them being trapping of holes at relatively shallow centers that do not participate in recombination. Other mechanisms involve electrical nonuniformity of the samples or the presence of traps with a high barrier for capture of electrons.

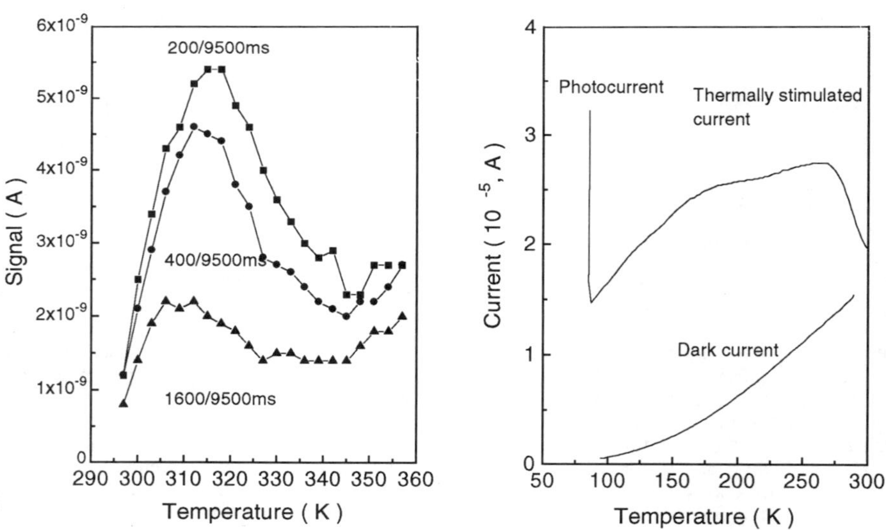

Fig. 5. PERS spectra for sample #2 for three sets of t1/t2 time windows settings.

Fig. 6. TSC for sample #2.

We do not have space in this paper to discuss in detail the results of studies of persistent photoconductivity (these results will be published elsewhere; it seems that electrical nonuniformity of the samples is mainly responsible for the effect), but would like to note here that we have also observed GaN samples without PPC. These were characterized by very low (orders of magnitude lower than for sample #2) photosensitivity in the above band gap region and in such samples only a very weak 270 K trap signal was observed in TSC. The difference in photosensitivity could be most naturally attributed to strong trapping of holes by defects giving rise to TSC peaks at 170 K and 270 K. At the same time, for the samples without PPC, we could not detect any measurable contrast in band edge MCL or EBIC images, as opposed to samples showing strong PPC for which the contrast was very pronounced (see below). The main difference between the two types of the samples is the absence of pre-nitridation step in the samples with low photosensitivity and no PPC.

It should also be said that the TSC current continues to rise and remains higher than the dark current for the highest temperatures used in these experiments (400K) indicating that there are other deeper centers present in the material.

Finally it was of interest to see how uniform is the distribution of deep traps and can any of the traps be associated with certain microdefects. Fig. 7 shows the secondary electron image of the surface of sample #2. The surface is generally very smooth, but displays a pattern of small pits or depressions that seem to define a cellular structure attributable to coalescence of slightly misoriented GaN islands formed during growth and separated by low angle grain boundaries. EBIC image in Fig. 8 demonstrates that the regions around pits are much brighter than the matrix and reveals the presence of the cellular structure with low angle boundary regions being again more bright than the matrix. This shows that the photocurrent is higher in these regions and hence the density of recombination centers is lower. Similar contrast is observed in band edge MCL, the domain walls and the regions around the pits showing substantially higher band edge MCL intensity. Linear PERS scans across several pits and across domain walls whose positions were preliminarily established by EBIC imaging show that the density of 0.85 eV PERS traps is about three times higher in the vicinity of the pits or the walls than in the matrix.

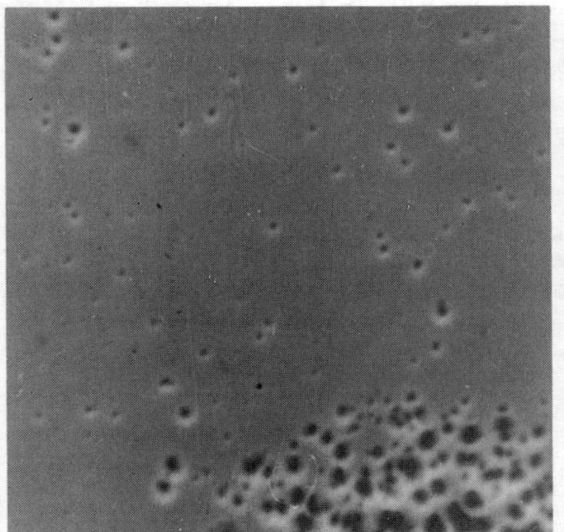 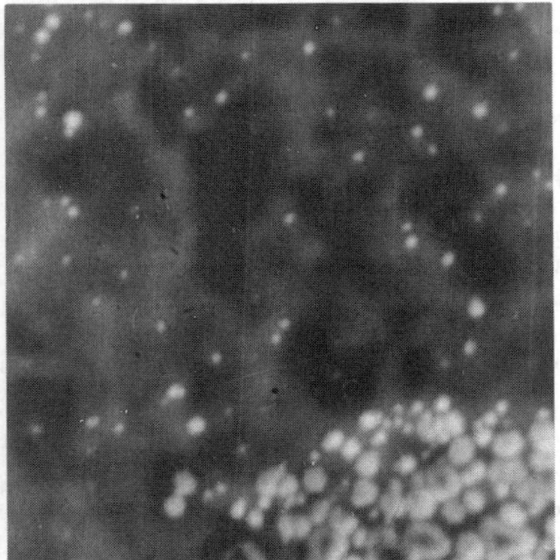

Fig. 7. Secondary electron image of the surface of sample #2, magnification 3000, the scale is 3 μm/cm

Fig. 8. EBIC image of the same region as in Fig.7.

If the 0.85 eV traps are indeed related to the yellow luminescence band their increased concentration in the regions of low angle boundaries is in agreement with the enhanced intensity of the yellow luminescence band near the low angle boundaries, as reported in [8]. At the same time it shows that the 0.85 eV traps cannot be the dominant recombination centers. Otherwise enhanced band edge luminescence and higher electron induced current would not be observed on the low angle boundaries in our samples.

The morphology of sample #1 is quite different from that of sample #2. The surface is very smooth, with occasional low hillocks. These hillocks give a much higher band edge MCL intensity than the matrix around them which is in agreement with the previously published results of Ponce et al. [8]. The hillocks also produce a much higher photocurrent intensity in EBIC, showing, in agreement with MCL, that these are the regions with lower concentration of deep recombination-active defects. And again, as for sample #2, MCL and EBIC measurements on regions free of hillocks reveal the cellular structure with the density of recombination active defects much reduced in the vicinity of the walls.

CONCLUSIONS

Our measurements show that traps with energies about 0.2 eV and 0.5 eV from conduction band edge and 0.25 eV and 0.5-0.85 eV from the valence band edge exist in GaN. The presence of traps at 0.25 eV and 0.5 eV from the valence band edge correlates with increased above band gap photosensitivity in GaN samples, presumably because they efficiently trap holes, but do not participate in recombination. The results described above also strongly suggest that the regions adjacent to low angle boundaries between slightly misoriented GaN domains have lower concentration of some recombination centers but higher concentration of 0.85 eV traps detected in PERS. That means that these traps are not major recombination centers in GaN.

ACKNOWLEDGMENTS

The work at CMU was supported in part by AFOSR Grant F49520-10087. The work at IRM was supported in part by Grant #1200 from the US Civilian Research and Development Foundation.

REFERENCES

1. H. Morkoc, S. Strite, G.B. Gao, M.E. Lin, B. Sverdlov and M. Burns, J. Appl. Phys., **76**, 1363 (1994)
2. W. Gotz, N.M. Johnson, H. Amano and I. Akasaki, Appl. Phys. Lett., **65**, 463 (1994)
3. W.I. Lee, T.C. Huang, J.D. Guo and M.S. Feng, Appl. Phys. Lett., **67**, 1721 (1995)
4. W. Gotz, N.M. Johnson, R.A. Street, H. Amano and I. Akasaki, Appl. Phys. Lett., **66**, 1340 (1995)
5. F.J. Sanchez, D. Basak, M.A. Sanchez-Garcia, E. Calleja, E. Munoz, I. Izpura, F. Calle, J.M.G. Tijero, B. Beaumont, P. Lorenzini, P. Gibart, T.S. Cheng, C.T. Foxon and J.W. Orton, MIJ-NSR, **1**, Article 7 (1996)
6. D.L. Losee, J. Appl. Phys., **46**, 2204 (1975)
7. J.W. Huang, T.F. Kuech, H. Lu and I. Bhat, Appl. Phys. Lett., **68**, 2329 (1996)
8. F.A. Ponce, D.P. Bour, W. Gotz and P.J. Wright, Appl. Phys. Lett., **68**, 57 (1996)
9. T. Ogino and M. Aoki, Jap. J. Appl. Phys., **19**, 2335 (1980)
10. A.V. Govorkov, E.M. Omeljanovsky, A.Y. Polyakov, V.I. Raihstein and V.A. Fridman, in Semi-Insulating III-V Materials 5, eds. G. Grossman and L. Ledebo (Adam Hilger, London and Bristol, 1988) pp 145-148
11. V.A. Gorbylev, A.A. Chelniy, A.Y. Polyakov, S.J. Pearton, R.G. Wilson, A.G. Milnes and A.V. Govorkov, J. Appl. Phys., **76**, 7390 (1994)
12. A.Y. Polyakov, A.A. Chelniy, N.B. Smirnov, A.V. Govorkov and A.G. Milnes, Mat. Sci. & Eng., **B38**, 36 (1996)
13. J.D. Guo, F.M. Pan, M.S. Feng, R.J. Guo, P.F. Chou and C.Y. Chang, J. Appl. Phys., **80**, 1623 (1996)
14. A.G. Milnes, Deep Impurities in Semiconductors (Wiley and Sons, New York, 1972)

Photoconducting Properties of Ultraviolet Detectors based on GaN and $Al_{1-x}Ga_xN$ films grown by ECR-MBE

M. MISRA, D. KORAKAKIS, R. SINGH, A. SAMPATH, T.D. MOUSTAKAS
Molecular Beam Epitaxy Laboratory, Department of Electrical and Computer
Engineering and Center for Photonics Research, Boston University,
44 Cummington Street, Boston, MA 02215.

ABSTRACT

GaN and $Al_{1-x}Ga_xN$ films were grown by the method of ECR-MBE. Absorption constants as a function of wavelength were determined from transmission measurements. Photoconducting detectors were fabricated from these films and characterized in terms of their spectral response and photoconductive gain. Mobility-lifetime products were determined from the measurement of photoconductive gain. The resistivity and mobility-lifetime products of the films were varied from 10-10^9ohm-cm and 10^{-3} -10^{-8} cm^2/V respectively by changing the microwave power in the ECR discharge from 20-60 watts. The change in the mobility-lifetime product is attributed to change in the lifetimes of the photogenerated carriers. This assumption is supported by direct measurement of detector response times. Finally, we report for the first time, the detection of alpha particles using GaN detectors.

INTRODUCTION

Wide bandgap semiconductors, predominantly III-V nitrides, SiC and diamond, are viewed as promising candidates for production of blue-UV detectors because the large bandgap energies promise low noise and 'visible-blind' detection. The III-V nitrides are rapidly becoming the materials of choice because they form a continuous alloy system whose direct bandgaps range from 1.9eV (InN) to 3.4eV (GaN) to 6.2eV (AlN). Thus they offer the potential of fabricating optoelectronic devices which are sensitive over the entire range from red to ultraviolet.

The two techniques that have emerged most successful in growing good quality GaN thin films are metal-organic vapor phase epitaxy (MOVPE)[1,2] and molecular beam epitaxy (MBE).[3,4] Advances in film growth techniques have resulted in the fabrication of several devices. Researchers using the MOCVD technique have reported lasers[5], LEDs[6], detectors[7] and transistors[8]. Similarly, using MBE based techniques, researchers have demonstrated LEDs[9], ultraviolet (UV) detectors[10] and transistors [11].

In this paper, we report on the performance of photoconducting UV detectors made from autodoped n-type GaN and $Al_{1-x}Ga_xN$ films grown by electron cyclotron resonance molecular beam epitaxy (ECR-MBE). We have found that the resistivity and photoconductivity of the material can be varied by controlling the power in the microwave discharge during film growth. In addition, we have investigated the potential for the use of these devices for detection of higher energy radiation and as single photon counters. We report for the first time the detection of alpha particles using GaN detectors. Results on detection of x-rays using GaN detectors have been published earlier. [10]

EXPERIMENTAL PROCEDURES

The GaN films used in this study were grown by the method of microwave plasma-assisted electron-cyclotron resonance molecular beam epitaxy (ECR-MBE). Details of the growth apparatus and procedures have been described earlier.[3] Here, we present a short description of the process. The deposition system consists of a Varian GenII MBE unit with an ASTeX compact ECR source. The films were deposited on (0001) sapphire substrates whose surface, after chemical cleaning and thermal outgassing, was converted to AlN by exposing it to an ECR nitrogen plasma. The films were deposited by a two-step growth process in which a buffer, about 300A thick, was deposited at 550°C and the GaN or $Al_{1-x}Ga_xN$ film was deposited at 800°C. The concentration of Al in the $Al_{1-x}Ga_xN$ films was varied by controlling the Ga to Al flux ratio during growth. The microwave power in the ECR discharge was varied from 20-60 watts, keeping the Ga, Al flux constant, for growth of different samples.

Standard photolithography and lift-off techniques were used to pattern interdigitated electrodes on the films. The electrode width and inter-electrode spacing were 20 μm. Ohmic contacts were formed by depositing a thin film of Ti (200A thick) followed by a film of Al (2000A thick), by electron beam evaporation. The total area of the device was estimated to be $0.27mm^2$.

The spectral response and light transmission through the films as a function of wavelength was measured using a monochromator, illuminated with a xenon lamp. Dark resistivity of the films was calculated by measuring the dark current through the device as a function of bias voltage. Photoconductivity measurements consisted of measuring the photocurrent induced in the detector in response to UV radiation from a He-Cd laser emitting at 325nm. The optical power of the laser beam was determined by measuring the photocurrent from a calibrated UV enhanced silicon photodiode illuminated under identical conditions as our photodetector. The photon flux incident on the detector was estimated to be about $5x10^{15}$ photons/sec. The photoresponse was calculated from measurements of photocurrent produced by the GaN or $Al_{1-x}Ga_xN$ detector as a function of wavelength and normalized to that produced by the calibrated UV-enhanced silicon photodiode of equal active area. Timing response was measured by recording on a digital oscilloscope the response of the detectors to N_2 laser pulses (337nm) of 10ns duration. Alpha particle detection was performed by exposing the detectors to 5.5MeV alpha particles from an Am-241 source. The detector output was amplified prior to pulse shaping and analyzed using a multi-channel analyzer.

EXPERIMENTAL RESULTS AND DISCUSSION

The dark current-voltage (I-V) characteristics of the devices were measured and found to be linear for bias voltages up to 50V. Beyond this value, the I-V characteristics showed some non-linearity, indicating the onset of space-charge effects. From these measurements, we calculated the resistivities of the GaN films at room temperature. We found that the resistivity of the films could be varied from $10-10^9$ ohm-cm by controlling the power in the ECR discharge during film growth.

From transmission measurements, we calculated the absorption coefficients as a function of wavelength for the $Al_{1-x}Ga_xN$ and GaN films.

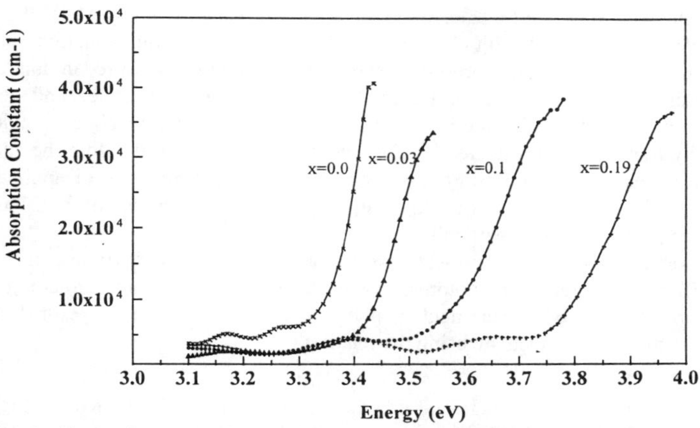

Figure 1: Absorption constants for GaN and $Al_{1-x}Ga_xN$ films, determined from transmission measurements.

Figure 1 shows the change in the bandgap energy as the Al concentration in the films is increased from x=0 to x=0.2.

Spectral response, which corresponds to the gain-quantum efficiency product, of the GaN and $Al_{0.075}Ga_{0.925}N$ detectors was measured and the results are shown in figure 2. For each film the photoresponse drops by several orders of magnitude at the wavelength corresponding to its bandgap energy. For energies less than the bandgap, the quantum efficiency is dominated by the absorption constant, and at energies greater than the bandgap, we assume it remains equal to one.

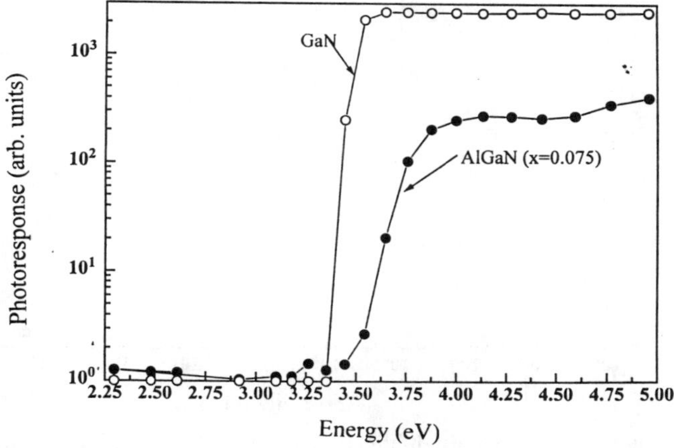

Figure 2: Spectral response of GaN and $Al_{0.075}Ga_{0.925}N$ films.

Similar results were reported by Khan and co-workers for GaN grown by MOVPE.[7] This result should be contrasted with the spectral dependence of semiconductors such as GaAs, where the response falls abruptly at shorter wavelengths, due to recombination in surface states. Thus, the results of figure 2 indicate that surface states and surface recombination are not significant in the case of $Al_{1-x}Ga_xN$ and GaN, as previously concluded by Foresi and Moustakas [12]. The fact that the gain (and therefore the spectral response) remains constant for short wavelengths is important for detector applications because it implies that detectors with high responsivity to wavelengths up to 200nm or less, may be made from these materials.

Photoconductivity measurements were made in order to determine the gain-quantum efficiency product and photoresponse of the detectors in the ultraviolet region of the spectrum. From the measurement of photoconductive gain (G), we estimated the mobility-lifetime ($\mu\tau$) products by using the relation

$$G = \mu\tau V/d^2,$$

where V is the applied bias voltage and d is the effective width of the device[13]. Figure 3 shows the relation between the dark resistivity and $\mu\tau$-product, as observed for the detectors from these measurements. We attribute the dramatic change in the $\mu\tau$-product to change in the lifetimes of the photo-generated excess carriers. Direct measurement of response time supports this assumption. Figure 4 shows the response of two detectors of different resistivities to optical pulses from a N_2 laser. Response times of both detectors show a sharp rise time of about 2ns, a fast component with a decay time constant of 20ns, followed by a slow component whose time constant varies with the resistivity of the film. The detector of figure 4a had a resistivity of 3.3×10^3 ohm-cm and the detector of figure 4b had a resistivity of 2×10^6 ohm-cm.

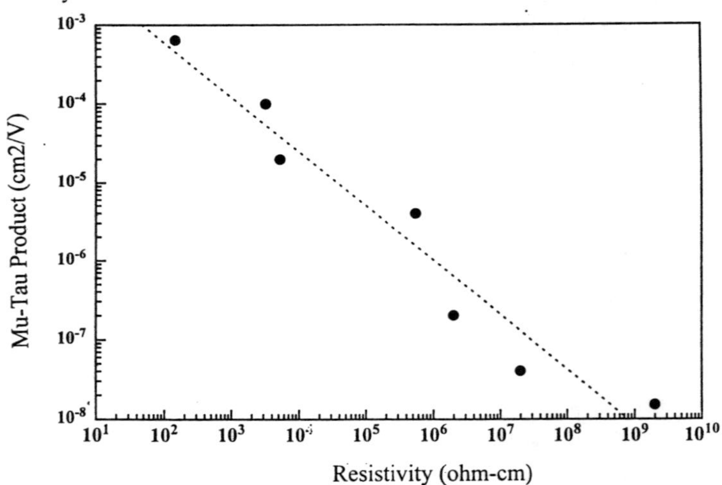

Figure 3: Variation of the mobility-lifetime product with film resistivity, as determined from measurements of photoconductive gain.

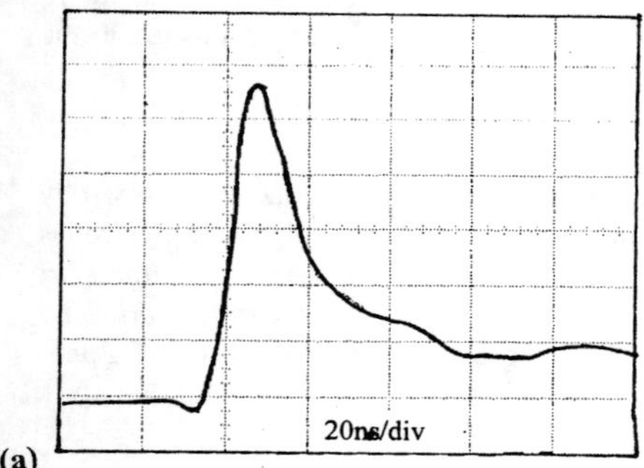

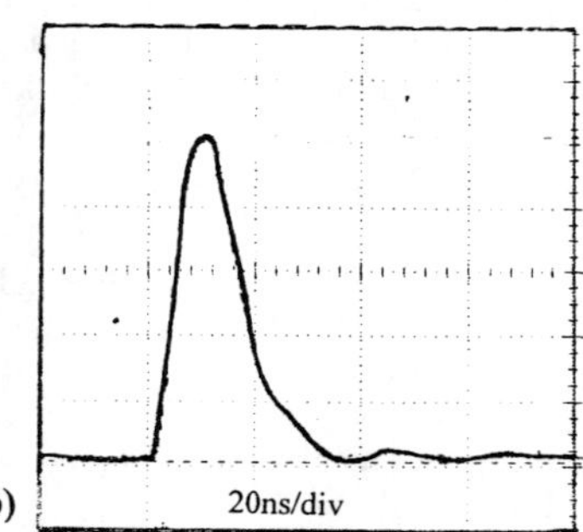

Figure 4: Response to N_2 laser pulses of GaN detectors of resistivity (a) 3.3×10^3 ohm-cm and (b) 2×10^6 ohm-cm.

Detection of alpha particles gives an estimate of the performance of the detector as a single particle radiation detector. In the case of GaN detectors, alpha particle detection was complicated by the fact that the thickness of the films was considerably smaller than the range of alpha particles in the material.

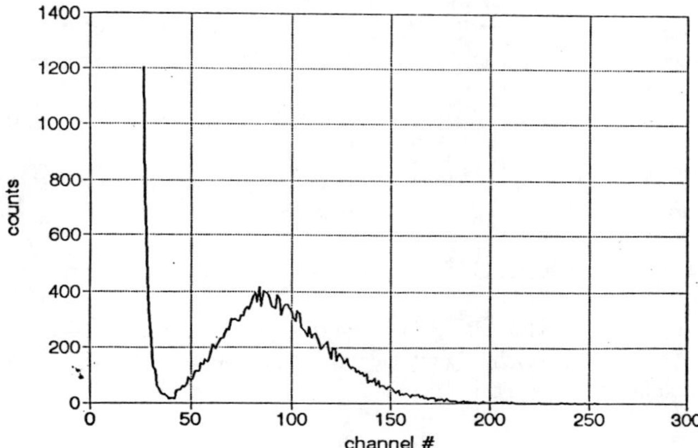

Figure 5: Detection of alpha-particles from Am-241 source using a $2\mu m$ thick GaN detector. The energy deposited in the detector was estimated to be 697.2keV.

Consequently, the alpha particles deposited only a fraction of their total energy in the detector. Figure 5 shows the response of a 2µm GaN photoconducting detector to 5.5MeV alpha particles from an Am-241 source. The fraction of energy deposited in the film was estimated to be 697.2keV.

CONCLUSIONS

$Al_{1-x}Ga_xN$ films with Al concentration ranging from x=0 to x=0.2 were grown by the method of ECR-MBE and photoconducting detectors with interdigitated electrodes were fabricated. For films grown by the MBE method, the electrical and transport properties of the films like resistivity and mobility-lifetime products can be varied by controlling the power in the ECR discharge during growth. The detection of alpha-particles using GaN detectors has been reported for the first time.

ACKNOWLEDGEMENTS

The work was partially supported by the Center for Photonics Research at Boston University.

REFERENCES

(1) S. Nakamura, Y. Harada and M. Seno, Appl. Phys. Lett., 58, 2021, (1991)

(2) M.A. Khan, J.N. Kuznia, J.M. VanHove and D.T. Olson, Appl. Phys. Lett., 58, 526, (1991)

(3) T.D. Moustakas, R.J. Molnar, Mat. Res. Soc. Symp. Proc. vol. 281, 753, (1993)

(4) C. Wang and R.F. Davies, Appl. Phys. Lett., 63, 990, 1993

(5) S. Nakamura, M. Senoh, S. Nagahama, N. Iwasa, T. Yamada, T. Matsushita, Y. Sugimoto, H. Kiyoku, App. Phys. Lett., 69, 3034, (1996)

(6) S. Nakamura, T. Mukai, M. Senoh, J. Appl. Phys., 76, 8189, (1994)

(7) M.A Khan, J.N. Kuznia, D.T. Olsen, J.M. VanHove, M. Blasingame, Appl. Phys. Lett., 60, 2917, (1992)

(8) M.A. Khan, J.N. Kuznia, A.R. Bhattarai, D.T. Olson, Appl. Phys. Lett., 62, 1786, (1993)

(9) R.J. Molnar, R. Singh, T.D. Moustakas, Appl. Phys. Lett., 66, 268, (1995)

(10) M. Misra, T.D. Moustakas, R.P. Vaudo, R. Singh, K.S. Shah, SPIE vol. 2519, 78, (1994)

(11) J. Pankove, S.S. Chang, H.C. Lee, R.J. Molnar, T.D. Moustakas, IEDM-94, 398, (1994)

(12) J.S. Foresi and T.D. Moustakas, Appl. Phys. Lett., 62, 2859, (1993)

(13) B.E.A. Saleh, M.C. Teich, "Fundamentals of Photonics," chapter 17, 644-692, John Wiley and Sons, New York, (1991)

ELECTRICAL CHARACTERIZATION OF Al-AlN(PSMBE GROWN)-Si MIS STRUCTURES

Regina Y. Krupitskaya, Gregory W. Auner, Tom E. Daley
Department of Electrical and Computer Engineering, Wayne State University, Detroit, MI.

ABSTRACT

AlN films were grown by plasma source Molecular Beam Epitaxy (PSMBE) on n-type Si(111) substrates under various growth parameters. I-V and C-V-f dependencies of Al-AlN-Si(111) MIS structures were measured. Electrical characterization of MIS structures with PSMBE grown textured AlN thin films as insulator shows that low current leakage can be achieved, although some samples have rectifying I-V dependencies. C-V-f characteristics reveal evidence of the presence of different types of traps. Additional investigations are needed in order to determine the nature of these traps.

INTRODUCTION

AlN is a wide bandgap semiconductor of the III-V Nitrides Group. This material and its solid solutions with GaN and InN are of current interest for optoelectronics, high temperature and high power device applications, and for Surface Acoustic Wave sensors. As grown high quality AlN is highly insulative, and possibly can be used to fabricate devices such as MIS diodes and MISFET's. Of particular interest is the growth of AlN on Si substrates which are inexpensive and can allow for future integration in Si technology.

In this report we are presenting an investigation of Al-AlN-Si MIS structures. The AlN layer was grown on n-type Si(111) substrates by Plasma Source Molecular Beam Epitaxy (PSMBE). This new technology allows deposition of high quality thin films under ultra-high vacuum conditions.

EXPERIMENT

AlN was deposited by PSMBE. The deposition system consists of a UHV chamber pumped by a 500 l/s ion pump for a base UHV and a throttled CTI cryopump for use under dynamic gas flow conditions. The system also contains effusion cells for evaporation and optical ports for laser plasma studies and ellipsometry. Ultra high purity process gases are controlled by mass flow controllers and additional gas purification is performed by a titanium and copper gettering furnace for argon and nitrogen, respectively. Gas compositions are monitored and controlled by a mass spectrometer connected for feedback control to a mass flow controller. A differentially pumped 35 keV RHEED system is used for in-situ characterization. This experimental setup enables precise control of deposition parameters (such as gas flow and partial pressures, substrate temperatures, plasma densities, r.f. target and substrate bias) for synthesis of almost any thin film material.

This system also has a new and innovative deposition source- a hollow cathode plasma source. This deposition technique uses a magnetically enhanced hollow cathode lined with the target material. An argon/nitrogen plasma is formed within the hollow cathode. A low energy flux of sputtered atoms stream out of the cathode to the substrate. This method allows very

Mat. Res. Soc. Symp. Proc. Vol. 449 © 1997 Materials Research Society

wide ranging parameter control such as the flux energy of the depositing species. The flux rate of each material can be independently controlled, allowing precise composition control. This deposition technique also allows very high target utilization efficiency compared to other deposition methods. The substrate is rotated during deposition to insure very uniform thickness. Large substrates may be deposited in this manner. Detail descriptions of this technology can be found in references [1,2].

For the present investigation, n-type, phosphorus-doped Si(111) substrates, with resistivity of 4-6 Ω-cm were used.. Before deposition the substrates were cleaned in acetone, methanol, and then etched in Hf to remove oxides and hydrogen terminate the Si dangling bonds. Different concentrations and times of etching were tried. In some of the deposition runs, Si substrates were annealed inside of the chamber before AlN deposition in order to dissociate surface hydrogen and reconstruct the Si surface. Surface treatments of substrates and AlN growth parameters are given in Table I.

Table I. Pre-deposition treatments of Si substrates and AlN deposition parameters.

Sample number	Substrate treatment	Substrate temperature, C	Dynamic pressure, mTorr	Substrate bias, V
A1	10%HF, 30 c	450	2	-12
A3	10%HF, 30 c, heating at 850 1 hour	First 200 A-400, then 600.	1	-1
A4	10%HF, 30 c, heating at 850 40 min.	First 200 A-400, then 600.	1	-15
A51 A52 A53 A54 A55 A56	15%HF, 30 c 15%HF, 60 c 10%HF, 30 c 10%HF, 60 c 5%HF, 30 c 5%HF, 60 c	First 200 A-400, then 600.	1	-15

RF power-250 W, argon flow - 40 sccm, and nitrogen flow - 20 sccm.

The base vacuum prior to deposition was typically in the 10^{-9} - 10^{-10} Torr range. The thickness of AlN samples presented in this work was 1000-1300 Å. The crystal properties of the AlN thin films were checked by RHEED, performed in-situ after growth, and by conventional θ–2θ x-ray Diffraction. The samples were shown to be textured with a C-axis AlN growth direction.

For MIS structures Al was used as a metal, and Al circular contacts of 3 different diameters were made. Different techniques of contact fabrication were used, such as direct deposition of Al dots in thermal evaporation chamber; lift-off photolithography with thermal evaporated Al deposition; metal deposition directly after AlN growth in PSMBE high vacuum chamber or thermal evaporation of Al layer with subsequent photolithography.   Current-voltage and capacitance-voltage-frequency measurements of MIS structures were performed, using computer interfaced HP4140b pA meter/ DC voltage Source and HP4192 LF Impedance Analyzer devices correspondingly.

RESULTS AND DISCUSSION

Most of the samples showed similar results. No dependence of electrical characteristics on contact fabrication technique were observed. Typical I-V characteristics of the low leaking MIS structures are shown in Fig. 1. The current is proportional to the contact area and corresponds to an AlN resistivity of 10^{13} - 10^{14} Ω-cm. It was impossible to approximate these I-V plots to any specific current conduction mechanism (for example Frenkel-Poole [3,4]). This an indicator that perhaps mixture of different mechanisms were involved. MIS structures of sample A1, as well as some MIS structures of other samples, showed rectifying characteristics with high forward current (Fig.2).

C-V-f plots for samples A3, and A4 are shown on Fig. 3 and 4, respectively. It is typical that capacitance-voltage characteristics strongly depend on frequency. Here we see that there is not only shifts along the voltage axis, but the capacitance value increases with decreased frequency. At some frequencies a hysteresis is also noticeable. The direction of the hysteresis loop varies from sample to sample. Saturation values of capacitance much lower than ideal theoretical ones for an accumulation regime of MIS structure (insulator capacitance) are observed. C-V characteristics for samples A51-56 at 1 MHz showed capacitance not varying with voltage. Furthermore, annealing of these samples at 250^0C failed to improve the C-V characteristics. The same effects were observed by other groups on AlN grown by different techniques [3,5,6]. All these observations are evidence of the existence of different kinds of states, probably varying from sample to sample. To find out the nature of these states from the present results is difficult-they can be connected with surface states of Si, deep traps in AlN itself (including those which probably exist in an AlN layer near the interface with Si).   The piezoelectric properties of AlN and its large lattice mismatch with Si(111), which causes strain and dislocation formation in the film, makes the situation even more complicated. Additional data is needed to determine a corresponding correlation between substrate surface preparation and growth parameters with AlN electrical properties.

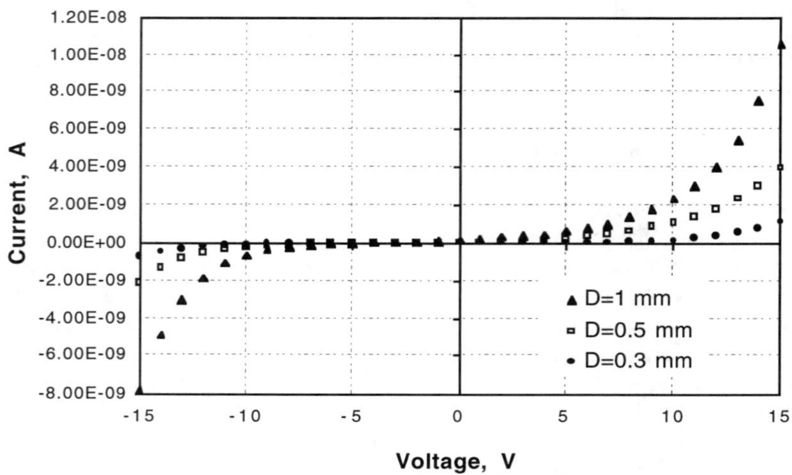

Fig. 1 - I-V characteristics of MIS structures with different contact diameters (D) made from sample A51 . Current leakage is low   and proportional to contact size.

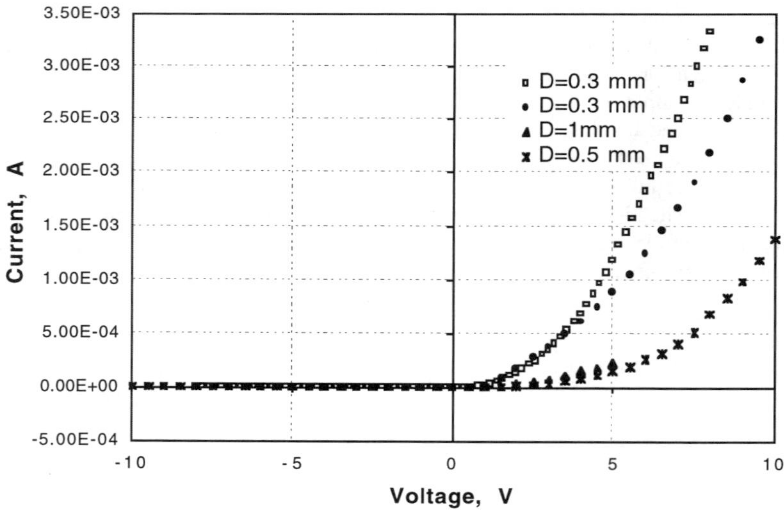

Fig. 2 - I-V characteristics of MIS structures with different contact diameters (D) made from sample A1 showing current rectification.

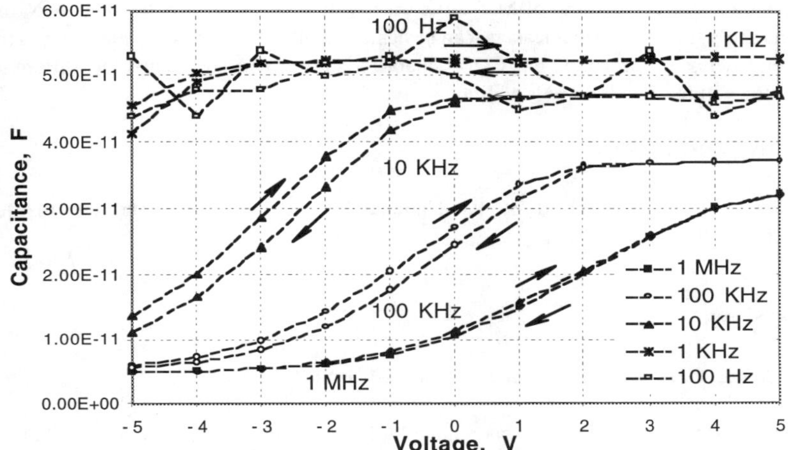

Fig. 3- Typical C-V-f plot for MIS structures made from the sample A3 with contact diameter 0.3 mm. Arrows show scan direction

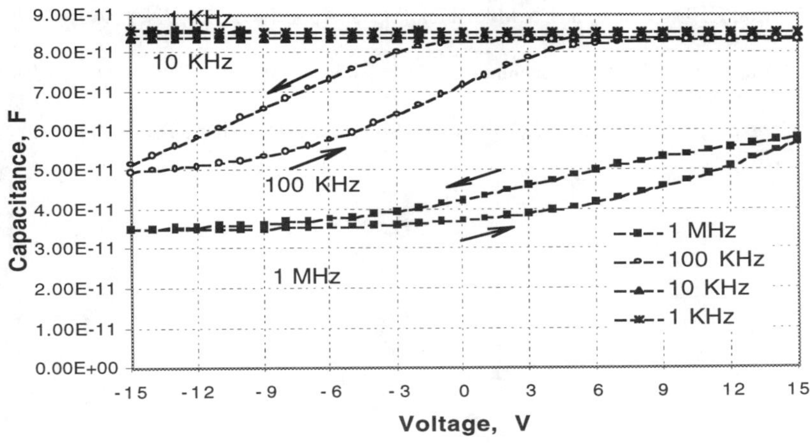

Fig. 4- Typical C-V-f plot for MIS structures made from the sample A4 with contact diameter 1 mm. Arrows show scan direction.

CONCLUSIONS

Electrical characterization of MIS structures with PSMBE grown textured AlN insulator thin films show that low current leakage can be achieved, although some samples show rectifying I-V dependencies. At the same time, C-V-f characteristics provides evidence for different types of traps- the nature of which needs additional investigation.

ACKNOWLEDGMENTS

This work was supported by the National Science Foundation and the Institute for Manufacturing Research (WSU).

REFERENCES

1. G.W. Auner, P.K. Kuo, Y.S. Lu, and Z.L. Wu, IEEE SPIE **V. 248**

2. "Epitaxial Growth of AlN by Plasma Source Molecular Beam Epitaxy", G.W.   Auner, T. Lenane, F. Ahmad, R.Naik, P.K. Kuo and Z.L. Wu, M.A. Prelas   (Eds) Wide Bandgap Electronic Materials, 329-324,(1995), Kluver Publishers.

3. A.H. Khan, J.M. Meese, T. Stacy, E.M. Charlson, E.J. Charlson, G. Zhao, G. Popovici, and M.A. Prelas, Mat. Res. Soc. Proc. Vol. **339**, 637(1994).

4. M. Morita, K. Tsubouchi, and N.Mikoshiba, Jap. J. Appl. Phys., **21**, 728 (1982).

5. M. Morita, S. Isogai, K. Tsubouchi, and N.Mikoshiba, Appl. Phys. Lett., **38(1)**, 50 (1981).

6. J. J. Hantzpergue, Y. Pauleau . J.C. Remy, D. Roptin, and M. Cailler, Thin Solid Films **75**, 167 (1981).

ANALYTICAL THEORY OF ELECTRON MOBILITY AND DRIFT VELOCITY IN GaN

B. L. GELMONT *, M. S. SHUR **, AND M. STROSCIO ***

* EE Department, University of Virginia, Charlottesville, VA 22903, gb7k@virginia.edu
** ECSE Department, Rensselaer Polytechnic Institute, Troy , NY 12180
*** U. S. Army Research Office, P. O. Box 12211, Research Triangle Park, NC, 27709-2211

ABSTRACT

We derive balance transport equations for the electron mobility and drift velocity, which are applicable at any degeneracy of the electron gas. These equations account for the polar optical phonon scattering and ionized impurity scattering and include the effects of screening. These equations are valid only for very high concentrations (above 10^{19} cm^{-3} for GaN). However, the comparison with the results of Monte Carlo simulations shows that they fairly accurately reproduce the field-velocity curves in GaN in moderate electric fields (up to 100 kV/cm). The comparison with the electron mobility calculated using the two-step model [1] shows a much larger difference but allows us to illustrate the trends in mobility dependencies caused by electron-electron collisions. We also derive the balance transport equations accounting for the polar optical phonon scattering in a two-dimensional electron gas. The calculations based on these equations, show that the unscreened polar optical scattering mobility is smaller in the two-dimensional gas than in the bulk intrinsic semiconductor and that the mobility decreases with the decrease of the quantum well thickness.

1. INTRODUCTION.

The polar optical phonon scattering and impurity scattering are often two dominant scattering mechanisms in polar semiconductors, and the theory of electron transport limited by these two mechanisms is a classical problem of semiconductor physics. We recently developed a two-step model [1]-[3], which describes the polar optical scattering for temperatures smaller than the polar optical phonon energy. In this model, the polar optical phonon scattering event is treated as one combined elastic process that includes two successive steps - the absorption and almost immediate emission of a polar optical phonon. We also proposed a new interpolation for the ionized impurity scattering, which is valid for an arbitrary degree of degeneracy [2].

In this paper, we report on the theory based on the balance transport equations, which account for electron-electron collisions and optical polar phonon screening, and apply this theory to GaN.

A large direct band gap, large breakdown field, a relatively large thermal conductivity, and large electron drift velocity make GaN suitable for high-power and high-speed electronic and optoelectronic applications. As follows from the Monte-Carlo simulation [1], [4-6] of the drift velocity-field characteristics, the intervalley transitions in GaN become essential only in very high electric fields (higher than 100 kV/cm). In low electric fields, polar optical phonon scattering and ionized impurity scattering dominate [2]. The polar optical phonon scattering is also the most important mechanism of the electron energy loss. Hence, such an approach gives a fairly complete idea of the velocity-field characteristics in GaN.

The mobility of GaN was analytically calculated in [2] where we demonstrated that the two-step approximation for the polar optical phonon scattering yielded a good agreement with the Monte-Carlo simulations [1] and experimental data.

However, the two-step model does not account for electron-electron collisions and polar optical phonon screening. Hence, this model is valid for relatively low electron concentrations. In the opposite limiting case, when electron-electron collisions are more frequent that all other scattering events, the electron distribution function for a non-degenerate electron gas is the drifted Maxwellian distribution function [7,8]. In this case, the balance transport equations can be applied for the calculation of the electron drift velocity. Such an approach was used for the

Mat. Res. Soc. Symp. Proc. Vol. 449 © 1997 Materials Research Society

calculation of the electron drift velocity in GaN [9]. However, as we show in this paper, the balance equations are valid at electron concentrations which are so high that the electron gas becomes degenerate, and, hence, the degeneracy and polar optical phonon screening effects have to be accounted for.

In Section 2, we derive the balance equations which are applicable for any degree of degeneracy of the electron gas and which account for the polar optical screening and for the ionized impurity scattering. We also estimate the low bound for the electron concentrations for which the balance equation approach is valid. We then calculate the velocity-field characteristics for GaN and compare with the results of the Monte Carlo simulations [10].

In Section 3, we derive the balance equations for the two-dimensional electron gas and discuss the predicted trends for the optical polar scattering mobility in the two dimensional electron gas.

2. THE BALANCE TRANSPORT EQUATIONS

The integration of the Boltzmann equation assuming the shifted Fermi-Dirac distribution function with an effective electron temperature T_e yields the following balance transport equations for the electron momentum and energy:

$$enF = R_{imp} + R_{opt} \tag{1a}$$

$$enFv_d = P_{opt} \tag{1b}$$

where e is the electronic charge, F is the electric field, n is the electron concentration, R_{opt} is the rate of the momentum loss due to the phonon scattering, R_{imp} is the rate of the momentum loss due to the impurity scattering, and P_{opt} is the power loss due to the polar optical scattering.

$$R_{opt} = \frac{\kappa_o m_n a_{Bn} k_o^4 E_{Bn} v_d}{6\pi^2 \kappa^* \hbar} \int_{-\infty}^{\infty} \frac{dt \exp(6t)}{(\exp(2t) + q_o^2 / k_o^2)^2} \{2N_o(T_e)[1 + N_o(T_e)]\ln(1 + \frac{f(\omega_o \hbar \cosh^2(t))}{N_o(T_e)}) +$$

$$[N_o(T) - N_o(T_e)][(1 + \exp(-2t))f(\omega_o \hbar \cosh^2(t)) + (1 - \exp(-2t))f(\omega_o \hbar \sinh^2(t))]\}, \tag{2}$$

$$R_{imp} = \frac{4N_d m_n E_{Bn} v_d}{3\pi\hbar} \int_0^{\infty} dx \Phi(\eta_T x)\exp(\eta_F - x) / [1 + \exp(\eta_F - x)]^2, \tag{3}$$

$$P_{opt} = \frac{\kappa_o E_{Bn} a_{Bn} k_o^4 k_B T_e}{2\pi^2 \kappa^* \hbar}[N_o(T_e) - N_o(T)] \int_{-\infty}^{\infty} \frac{dt \exp(4t)}{(\exp(2t) + q_o^2 / k_o^2)^2}\ln(1 + \frac{f(\hbar\omega_o \cosh^2(t))}{N_o(T_e)}). \tag{4}$$

Here k_B is the Boltzmann constant, T is the lattice temperature, q_o is the screening wave vector, E_{Bn} is the electron Bohr energy, a_{Bn} is the electron Bohr radius, m_n is the electron effective mass, $E_F = k_B T \eta_F$ is the Fermi energy, N_o is the Planck function, κ_o is the static dielectric permittivity, κ_∞ is the high frequency dielectric permittivity, $\omega_o = \hbar k_o^2 / 2m_n$ is the phonon frequency

$$\eta_T = \frac{8k_B T_e m_n}{q_o^2 \hbar^2} \ , \qquad \frac{1}{\kappa^*} = \frac{1}{\kappa_\infty} - \frac{1}{\kappa_o}, \qquad \Phi(\eta) = \ln(1 + \eta) - \frac{\eta}{1 + \eta} .$$

The balance equation for the electron momentum accounts for the two most important scattering processes responsible for the momentum loss: phonon and impurity scattering. The balance equation for the electron energy accounts for the energy loss caused by the polar optical phonon scattering. Both equations account for screening of the polar optical phonon scattering.

At small concentrations, electrons are non-degenerate, and these balance equations can be simplified:

$$R_{opt} = \frac{\kappa_o n m_n x_e^{3/2} k_o a_{Bn} E_{Bn} v_d}{3\pi^{1/2}\kappa^* \hbar} N_o(T)\exp(\frac{x_e}{2})\{ \int_{-\infty}^{\infty} \frac{dt\exp(-3t - x_e\cosh(t)/2)}{(\exp(-t) + q_o^2/k_o^2)^2}[1+$$

$$\exp(x_o - x_e)] + [\exp(x_o - x_e) - 1]\int_{-\infty}^{\infty} \frac{dt\exp(-2t - x_e\cosh(t)/2)}{(\exp(-t) + q_o^2/k_o^2)^2}\}, \tag{5}$$

$$\mathbf{R}_{3i} = \frac{4E_{Bn} n N_d m_n}{3\pi\hbar N_c} \mathbf{v}_d \int_0^{\infty} dx\Phi(\eta_T x)\exp(-x) , \tag{6}$$

$$eFv_d = \frac{2\kappa_o x_e^{1/2} k_o a_{Bn} E_{Bn}\omega_o}{\pi^{1/2}\kappa^*}\sinh(\frac{x_e}{2})[N_o(T_e) - N_o(T)]\int_{-\infty}^{\infty} \frac{dt\exp(-2t - x_e\cosh(t)/2)}{(\exp(-t) + q_o^2/k_o^2)^2} \tag{7}$$

where N_c is the effective density of states,

$$x_o = \frac{\hbar\omega_o}{k_B T}, \qquad x_e = \frac{\hbar\omega_o}{k_B T_e} .$$

If, in addition, we neglect screening of the electron-phonon interaction eqs. (5)-(7) reduce to the balance transport equations derived in [7,8].

As discussed above, strictly speaking, the balance transport equations are valid when electron-electron scattering is faster than the rate of phonon scattering. An approximate low bound for the electron concentration at which the balance equations are valid can be obtained by equating these two scattering rates in the vicinity of the phonon energy [11]

$$n_c = \frac{\kappa_o k_o^4 a_{Bn}\sqrt{x_e}}{2\pi\kappa^*\Phi(\eta_T x_e)} . \tag{8}$$

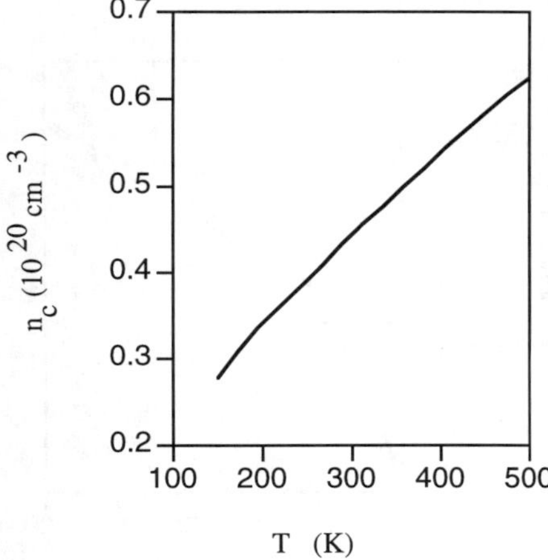

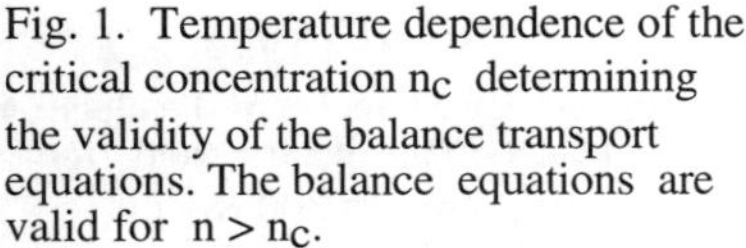

Fig. 1. Temperature dependence of the critical concentration n_c determining the validity of the balance transport equations. The balance equations are valid for $n > n_c$.

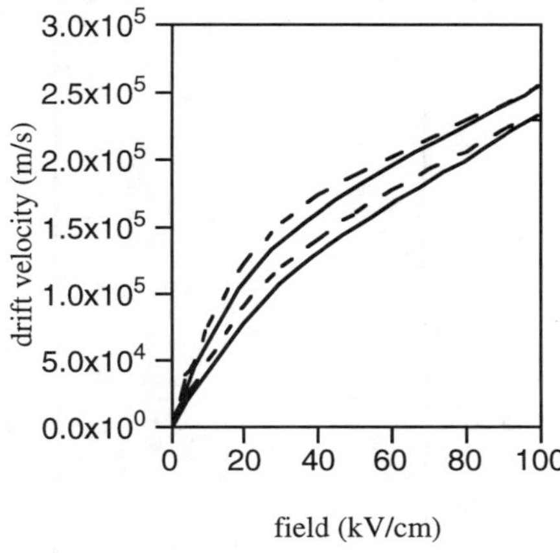

Fig. 2. Electron drift velocity in GaN versus the electric field for electron densities 10^{17} cm^{-3} (two top curves) and 10^{18} cm^{-3} (two bottom (two bottom curves) at 300 K. Solid lines - analytical theory, dashed lines - Monte Carlo simulation [10].

The dependence of n_c on lattice temperature, T, for GaN is shown in Fig. 1. (In all the calculations performed in this paper, we used the parameter values for GaN given in [2].) As seen from Fig. 1, the value of n_c is very high. This confirms that we must account for the degeneracy of the electron gas and for the screening effects in the balance transport equations.

Fig. 2 compares the velocity-field characteristics for GaN calculated using the balance transport equations derived above with the results of the recent Monte-Carlo simulation [10] at 300 K for electron concentrations of 10^{17} cm^{-3} and 10^{18} cm^{-3}. We chose these particular Monte Carlo results for comparison, since the parameter values used in [10] are very similar to the parameter values used in [2] and in the present calculation. The agreement in high electric fields is quite good, in spite of the fact these concentrations are much smaller than n_c. However, the differences for low electric fields are substantial. This is illustrated by Fig. 3, which compares the temperature dependencies of the electron mobility calculated using the two-step model and the balance transport equations. The two-step model predicts a larger value of the mobility. Moreover, the two-step model and the balance transport equation model predict different temperature dependencies of the electron mobility. In GaN, the two step approximation is valid in wider region of electron concentrations than in GaAs, since the phonon energy is much larger (92 meV compared to 35 meV). At 300 K, n_c =4x10^{19} cm^{-3} and 6x10^{17} cm^{-3} for GaN and GaAs, respectively . The two-step approximation [1,2,3,11] should be used for n < n_c.

At the present time, the two-step model has been only developed for low electric fields. However, the energy balance can be incorporated into the two-step using the analytical approximation developed in [12].

Fig. 4 shows the calculated dependencies of the electron drift velocity in GaN on the electric field for the electron concentration of 10^{17} cm^{-3} at 300 K with and without accounting for the ionized impurity scattering. As seen, the contribution of the ionized impurity scattering is less than 5 % at 100 kV/cm. This shows that the electron transport GaN Field Effect Transistors with deep submicron gates (where electric fields are very high) should not be affected too much when the channel is doped. This conclusion is in agreement with very high cutoff frequencies and maximum frequencies of oscillations recently measured for deep submicron doped channel GaN FETs [13]

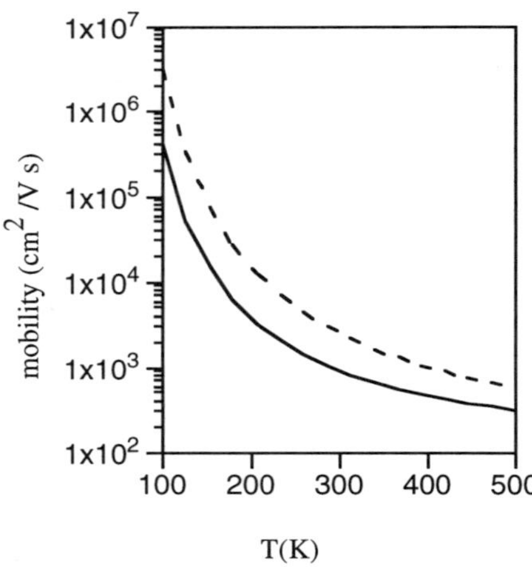

T(K)

Fig. 3. Temperature dependencies of the electron mobility in GaN calculated using the two-step model (dashed line) and the balance transport equations (solid line).

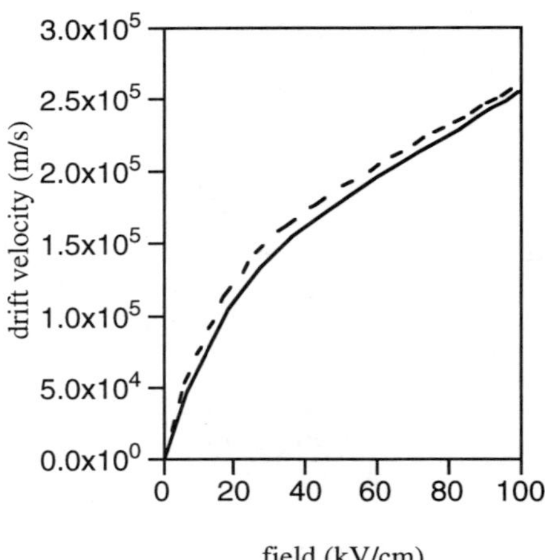

field (kV/cm)

Fig. 4. Calculated dependencies of the electron drift velocity in GaN on the electric field   for electron density 10^{17} cm^{-3} at 300 K with (dashed line) and without (solid line) impurity scattering.

3. BALANCE EQUATIONS FOR TWO-DIMENSIONAL ELECTRON GAS.

The analytical solution for the two-dimensional (2D) electron gas is more difficult to obtain because one may have to account for more than one subband, depending on the 2D gas density and temperature. Here we consider only the limiting case when the difference between the first excited level and the ground state level is larger than the phonon energy. In this case, only intrasubband scattering processes between the states of the lowest subband are essential, and the balance transport equations become

$$eFv_d = \frac{2\kappa_o\omega_o E_{Bn}k_o a_{Bn}(\pi x_e)^{1/2}}{\kappa^*}[N_o(T_e) - N_o(T)]\sinh(\frac{x_e}{2})\int_0^\infty \frac{dt}{t}G(k_o t)\exp(-\frac{x_e(t^2 + t^{-2})}{4}), \quad (9)$$

$$eF = \frac{\pi^{1/2}\kappa_o E_{Bn}k_o a_{Bn}x_e^{3/2}m_n v_d}{\kappa^*\hbar}N_o(T)\int_0^\infty dtG(k_o t)\exp(\frac{x_o}{2} - \frac{x_e(t^2 + 1/t^2)}{4})$$

$$\{t\sinh(\frac{x_o - x_e}{2}) + t^{-1}\cosh(\frac{x_o - x_e}{2})\}, \qquad (10)$$

where

$$G(Q) = \int dz_1 dz_2 \psi^2(z_1)\psi^2(z_2)\exp(-Q|z_1 - z_2|) \qquad (11)$$

is the form factor, $\psi(z)$ is the electron wave function. For a quantum well of width a, the form factor is given by [14]

$$G = \frac{2}{u} + \frac{u}{u^2 + 4\pi^2} - \frac{2(4\pi^2)^2}{u^2(u^2 + 4\pi^2)^2}[1 - \exp(-u)] \qquad (12)$$

where u=Qa.

For narrow quantum wells where $k_o a$ <1, these balance equations can be simplified

$$eFv_d = \frac{2\kappa_o\omega_o k_o a_{Bn}E_{Bn}\sqrt{\pi x_e}}{\kappa^*}[N_o(T_e) - N_o(T)]\sinh(\frac{x_e}{2})K_o(\frac{x_e}{2}), \qquad (13)$$

$$eF = \frac{\pi^{1/2}\kappa_o nm_n E_{Bn}k_o a_{Bn}x_e^{3/2}v_d}{2\hbar\kappa^*}N_o(T)\exp(\frac{x_o}{2})\{K_o(\frac{x_e}{2})\sinh(\frac{x_o - x_e}{2}) +$$

$$K_1(\frac{x_e}{2})\cosh(\frac{x_o - x_e}{2})\}. \qquad (14)$$

The above equations do not account for screening and/or impurity scattering. Also, the assumption of a large subband separation may not be valid in GaN. Hence, we only use the above equations for qualitative comparison with the electron transport in bulk GaN. Our calculations based on eqs. (13) - (14) show that the polar optical phonon mobility and the drift velocity of the two-dimensional electron gas are smaller than those for the three-dimensional gas. The heating effects in the 2D-gas become important in higher electric fields than in a bulk semiconductor. The mobility decreases with a decrease in the quantum well thickness. This behavior can be linked to the step-like density of states of the two-dimensional electron gas.

Equations (13) - (14) have been derived for electron interactions with bulk phonons. The Monte-Carlo simulation for an AlGaAs/GaAs single heterostructure [15] shows that the results obtained with half-space and interface phonons do not differ much, since the sum of scattering rates is practically the same as the scattering rate for bulk phonons. For the low field transport, the situation is likely to be the same for AlGaN/GaN heterostructures. However, a much larger difference of phonon frequencies in AlN and GaN may lead to new features in the shape of the velocity-field characteristics of the two-dimensional electron gas in high electric fields, since the polar optical phonons in AlN and GaN will correspond to two distinct thresholds for the polar

optical phonon emission. Hence, for this materials system, this problem merits further
investigation.

4. CONCLUSIONS.

The analytical theory of the electron transport developed in this paper is valid for high carrier
concentrations (higher than approximately $n_c =4 \times 10^{19}$ cm^{-3} and $n_c =6 \times 10^{17}$ cm^{-3} for GaN and
GaAs, respectively, at 300K). Hence, this theory complements the two-step model valid for the
electron concentrations less than n_c. Moreover, our calculations show that, for higher electric
fields, the balance transport equations may be extrapolated into the range of the electron
concentrations substantially smaller than n_c and still yield a reasonable agreement with Monte Carlo
simulations.

The recent results for doped channel AlGaN/GaN heterostructures showed that, in these
structures, the sheet carrier concentration is nearly an order of magnitude higher than for the
AlGaAs/GaAs materials system. [16] Hence, the electron-electron collisions may be far more
important in the 2D electron gas in AlGaN/GaN, and the approach developed in this paper
will be very useful for the development of the analytical theory of the high-field electron transport
for the 2D electrons in AlGaN/GaN heterostructures.

REFERENCES.

1. B. Gelmont, K. Kim, and M. S. Shur, J. Appl. Phys. , 74 , 1818 (1993).
2. M. Shur, B. Gelmont, and M. A. Khan, J. of Electronic Materials, 25, 5,777 (1996).
3. B. Gelmont, M. S. Shur, and M. Stroscio, J. Appl. Phys., 77, 657 (1995)
4. M. S. Shur, B. Gelmont, K. Saavedra-Munos, and G. Kelner, Inst. of Phys. Conf. Ser. ,137,
p.465 (1994)
5. N. S. Mansour , K. W. Kim, and M. A. Littlejohn, J. Appl. Phys., 77 , 2834 (1995).
6. J. Kolnik, I. H. Oguzman, K. F. Brennan, R. Wang, P. Ruden, and Y. Wang, J.Appl. Phys.,
78 , 1033 (1995).
7. R. Stratton, Proc. Roy. Soc., A246, 406 (1958)
8. E. M. Conwell, <u>High Field Transport in Semiconductors</u>, Solid St. Phys.,Suppl. 9 (1967)
9. D. K. Ferry, Phys. Rev., B12, 2361 (1975)
10. U. Bhapkar and M. S. Shur (unpublished)
11. V. F. Gantmakher and I. B. Levinson,<u> Carrier Scattering in Metals and Semiconductors</u>
(Wiley, New York, 1987).
12. B. L. Gelmont, R. I. Lyagushenko, and I. N. Yassievich, Sov. Phys. Solid State, 14, 445
(1972).
13. M. Asif Khan, Q. Chen, J. Yang, M. Z. Anwar, M. Blasingame, M. S. Shur, J. Burm, and
L. F. Eastman, Recent Advances in III-V Nitride Electron Devices, IEDM-96 Technical Digest,
invited
14. P. J. Price, Phys. Rev., B30,2234 (1984)
15. P. Bordone and P. Lugli, Phys. Rev., B49, 8178 (1994)
16. M. A. Khan, M. S. Shur, and Q. Chen, Appl. Phys. Lett., 68 , p. 3022, (1996)

P-TYPE CONDUCTIVITY WITH A HIGH HOLE MOBILITY IN CUBIC GaN/GaAs EPILAYERS

D.J. AS, A. RÜTHER, M. LÜBBERS, J. MIMKES, K. LISCHKA and
D. SCHIKORA
Universität Paderborn, FB-6 Physik, Warburger Straße 100, D-33095
Paderborn, Germany, d.as@uni-paderborn.de

ABSTRACT

Temperature dependent Hall-Effect-measurements on unintentionally doped cubic GaN epilayers grown by molecular beam epitaxy (MBE) are reported. The cubic GaN layers have been deposited on semiinsulating (001) GaAs-substrates under N-stabilized growth conditions which were controlled by in-situ reflection high energy electron diffraction (RHEED) measurements. GaN-layers, which were fabricated under N-stabilized conditions have a (2x2) surface reconstruction during growth and show p-type conductivity. At room temperature the measured hole concentrations and mobilities are $p = 9.7 * 10^{12}$ cm^{-3}, $\mu_p \cong 350$ cm^2/Vs, respectively. Temperature dependent measurements of the carrier concentration yield an acceptor activation energy of $E_A = 0.445$ eV. The nature of these defects will be discussed in view of intrinsic defects proposed by theoretical calculations already published in literature. The temperature dependence of the mobility is dominated by polar optical phonon scattering in the investigated temperature range.

INTRODUCTION

The metastable configuration of GaN which has cubic crystal structure has recently attracted some interest, since epitaxially grown layers of c-GaN lend themselves to the production of cleaved laser cavities. This offers an interesting alternative to dry etched cavities as already realized with hexagonal GaN based devices [1]. For this application, however, not only the optical properties but also the electrical properties of the layers are of importance. So far, only scarce data on the electrical properties of MBE grown c-GaN layers have been published [2-6]. One major challange developing suitable blue emitters form wide-band gap semiconductor is the reduction of the high residual background electron concentration of about $1 * 10^{17}$ cm^{-3}, which is believed to be due to the incorporation of large amounts of nitrogen vacancies. This high residual background doping has in part made it difficult to achieve p-type nitride material.

Recent developments in heteroexpitaxy resulted in significant advances of the molecular beam epitaxy (MBE) growth of cubic GaN (c-GaN) thin films on various substrates with cubic crystal structure [7-10]. The use of the surface reconstruction phase diagram of c-GaN [8], allows to adjust the stoichiometry

Mat. Res. Soc. Symp. Proc. Vol. 449 © 1997 Materials Research Society

during MBE using in-situ RHEED and enables growth of stoichiometric cubic GaN under N-stabilized and Ga-stabilized conditions. In this contribution we report on production of nearly "intrinsic" cubic p-type material without intentional doping and without post growth treatments. The highest room temperature hole mobilitiy ($\cong 350$ cm^2/Vs) and lowest carrier concentration ($p = 9.7*10^{12}$ cm^{-3}) is achieved under N-stabilized growth conditions.

EXPERIMENTAL

Cubic GaN films are grown on semi-insulating GaAs (001) substrates by plasma assisted molecular beam epitaxy (MBE) at a substrate temperature of 720°C. The nucleation process and the layer growth is controlled by RHEED. Details of crystal growth are reported in Ref.8. The thickness of all layers investigated variied between 0.3 μm and 1.4 μm. The phase purity of our zinc-blende layers has been checked by Raman scattering and has been found to be better than 98 % [11]. Photoluminescence measurements at 2 K show intensive band-edge luminescence at 3.26 eV and 3.15 eV [12], the intensity at 2.2 eV (yellow luminescence) is weaker by more than a factor of 10^3.

Van der Pauw geometry is used for temperature dependent Hall-effect measurements with Au wire soldered In contacts on the GaN layers. The ohmic characteristic of these contacts is checked by measuring current-voltage curves, exhibiting linear behavior over the whole temperature range used. The experimental apparatus is equipped with a high impedance current source and a Keithley 7065 Hall-effect card. This enables measurements up to an input resistance of 100 TΩ, and allows the investigation of high resistivity and semi-insulating samples. The measurements at variable temperature are performed in a cryostate between 240 K to 380 K. Beneath 240 K the resistivity of the samples is to high for accurate measurements and above 380 K the In-contacts degradate. All measurements are performed at a magnetic field of 0.3 T and with the sample in the dark.

RESULTS AND DISCUSSION

Figure 1 shows the carrier concentration as a function of the inverse temperature for the cubic GaN layers GaN66 (open squares) and GaN85 (open triangles), which have been grown under N-stabilized conditions ((2x2) RHEED-pattern). The hole concentration p is obtained from the experimental Hall constant R_H by $p = r_H/qR_H$ (q = electronic charge); the Hall scattering factor r_H is assumed to be isotropic, temperature independent, and one. In the whole temperature range investigated the samples are p-type with a hole carrier concentration between $2.2*10^{11}$ cm^{-3} and $2.8*10^{14}$ cm^{-3}. Under the assumption

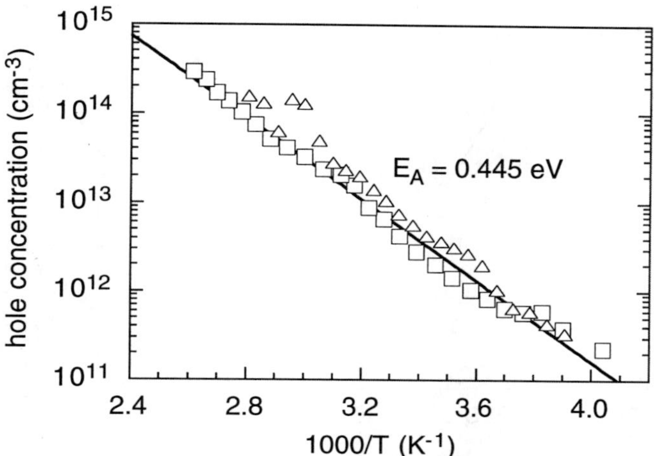

Fig.1: Hole concentration of two cubic GaN epilayers (open triangles GaN85, openl squares GaN66) as a function of 1000/T, layers were grown under N-stabilized conditions ((2 x 2) - RHEED-pattern). The solid line is a least square fits to experimental data.

of compensation, which is very likely at such low carrier concentrations, the temperature dependence of p yields an acceptor activiation energy of E_A= 0.445±0.015 eV.

The hole mobility vs. temperature of sample GaN66 is shown in Fig.2. At room temperature a mobility of 350 cm^2/Vs is measured, which is comparable with the best value reported for h-GaN by Rubin et al [4]. The temperature dependence of the hole mobility is clearly dominated by polar optical phonon scattering and can well be fitted by an expression given by Look [13].

$$\mu_{po} = \frac{2^{\frac{9}{2}} \cdot \pi^{\frac{1}{2}}}{3} \cdot \frac{h^2 (k_B T)^{\frac{1}{2}}}{q \cdot (k_B T_{LO}) \cdot m^{*\frac{3}{2}} \cdot (\varepsilon_\infty^{-1} - \varepsilon_0^{-1})} \cdot \left(e^{\frac{T_{LO}}{T}} - 1 \right) \cdot \chi\left(\frac{T_{LO}}{T}\right) \qquad (1)$$

T_{LO} is the LO phonon temperature and χ (T_{LO}/T) is a slowly varying function of T. The LO-phonon frequency of c-GaN has been measured by Raman scattering [11] to be 740 cm^{-1} corresponding to a T_{LO} value of 1065 K. ε_∞ and ε_0 are 5.35 and 9.5 respectively [2]. The solid line in Fig.2 is a least square fit of the experimental data by Eq.(1). From this fit a lowest effective hole mass of m_h = 0.8*m_0 can be estimated.

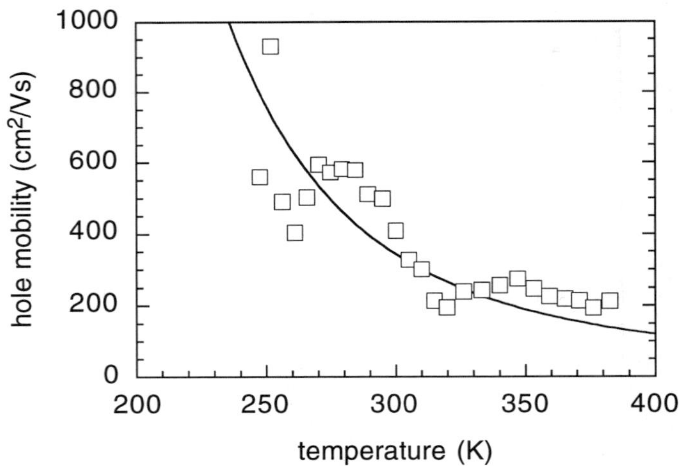

Fig.2: The hole mobility in cubic GaN epilayers (sample GaN66) as a function of temperature, layers were grown under N-stabilized conditions ((2 x 2) - RHEED-pattern). The solid line is a least square fitof experimental mobility data by Eq.1 by polar optical phonon scattering .

In Table I the room temperature values of the hole concentration and mobility of three samples grown under N-stabilized conditions are summarized. All samples are p-type with hole concentrations between $5*10^{13}$ cm^{-3} and $8*10^{15}$ cm^{-3}. In agreement with standard scattering theory the mobility decreases with increasing hole concentration. The high mobility values further demonstrate the excellent quality of our cubic GaN layers.

Table I: Room temperature values of the hole concentrations and mobilities of three samples grown under N-stabilized growth conditions ((2 x 2) surface reconstruction))

hole concentration (cm^{-3})	mobility (cm^2/Vs)	thickness (μm)	sample number
8.10^{15}	50	1.4	GaN 96
1.10^{14}	150	0.35	GaN 85
5.10^{13}	350	0.3	GaN 66

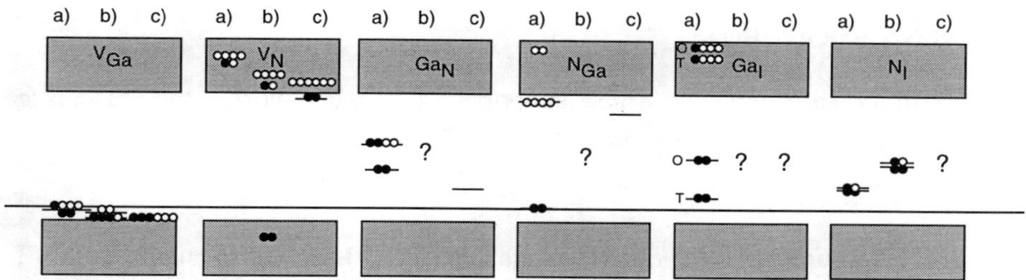

Fig.3: Energy levels for intrinsic defects calculated for hexagonal GaN after Ref. 15 (a), Ref. 16 (b), and Ref.17 (c). The solid line represents the observed activation energy of our p-type cubic GaN epilayers.

We have no evidence for the chemical nature of the acceptor in our GaN-layers. However, it is likely that it may be due to an intrinsic defect, since we are able to grow p-type layers by just decreasing the Ga/N flux ratio within the stoichiometric growth range. In agreement with Brandt et al. [14] a transition of the surface reconstruction from c(2x2) for Ga-stabilized to (2x2) reconstruction for N-stabilized growth is observed. Since the covalent radius of Ga is much larger than that of N, the dimers are assumed to be formed by Ga atoms only. Therefore, the (2x2) reconstruction is due to Ga dimer rows along the [110] direction which are separated by missing dimer rows. Increasing the Ga/N ratio at the surface results in filling up the missing dimer rows. At full monolayer coverage the pattern corresponds to the c(2x2) reconstruction. For that reason we think that either the Ga vacancy V_{Ga} or the nitrogen antisite N_{Ga} may be responsible for the observed defect level. This interpretation is supported by calculations of the energy levels of intrinsic defects by the ab-initio [15, 16] and tight binding method [17]. In Fig.3 the theoretically proposed energy levels are depicted. Due to the lack of data for cubic GaN, data for hexagonal GaN are used. Within the accuracy of calculations and experimental errors only the energy levels of the V_{Ga} or the N_{Ga} are in agreement with the measured activation energy E_A= 0.445 eV.

CONCLUSIONS

We have shown that stoichiometric cubic GaN with a low carrier concentration can be grown by plasma assisted MBE. Using in-situ RHEED to control the surface reconstruction it is possible to grow N-stabilized layers resulting in p-type GaN epilayers. At room temperature hole concentrations and mobilities are about p = $1*10^{13}$ cm^{-3}, μ_p= 350 cm^2/Vs, respectively. From the temperature dependence of the hole concentration an acceptor activation energy of E_A= 0.445 eV is estimated. Since no extrinsic doping is used and the cubic GaN layers are grown under stoichiometric conditions we suggest that these activation energy may correspond to an energy level of an intrinsic defect probably either V_{Ga} or N_{Ga}.

ACKNOWLEDGEMENTS

The authors acknowledge the support of their work by "Deutsche Forschungsgemeinschaft , project number SCHI 345/7-1.

REFERENCES

1. S. Nakamura, M. Senoh, S. Nagahama, N. Iisa, T. Yamaha, and T. Matsushita, H. Kiyoku, Y. Sugimoto, Jpn. J. Appl. Phys. **35**, L74 (1996)
2. S. Strite and H. Morkoc, J. Vac. Sci. Technol. **B 10**, 1237 (1992)
3. R.F. Davis, Proc. IEEE **79**, 701 (1991)
4. M. Rubin, N. Newman, J.S. Chan, T.C. Fu, and J.T. Ross, Appl. Phys. Lett. **64**, 64 (1994)
5. M.E. Lin, G. Xue, G.L. Zhou, J.E. Greene, and H. Morkoc, Appl. Phys. Lett **63**, 932 (1993)
6. D.J. As, D. Schikora, A. Greiner, M. Lübbers, J. Minkes, and K. Lischka, Phys. Rev. **B 54**, R11118 (1996)
7. S.E. Hooper, C.T. Foxon, T.S. Cheng, L.C. Jenkins, D.E. Lacklison, J.W. Orton, T. Bestwick, A. Kean, M. Dawson, G. Duggan, J. Crystal Growth **155**, 157 (1995)
8. D. Schikora, M. Hankeln, D.J. As, K. Lischka, T. Litz, A. Waag, T. Buhrow, and F. Henneberger, Phys. Rev. **B 54**, R8381 (1996)
9. R.C. Powell, N.E. Lee, Y.W. Kim, and J.E. Greene, J. Appl. Phys. **73**, 189 (1993)
10. H. Yang, O. Brandt, and K.H. Ploog, Phys. Stat. Sol. (b) **194**, 1 (1996)
11. H. Siegle, L. Eckey, A. Hoffmann, C. Thomsen, B.K. Meyer, D. Schikora, M. Hankeln, and K. Lischka, Solid State Communications **96**, 943 (1995)
12. D.J. As, F. Schmilgus, C. Wang, B. Schöttker, D. Schikora, and K. Lischka, submitted to Appl. Phys. Lett.
13. D.C. Look, <u>Electrical Characterization of GaAs Materials and Devices</u>, Wiley&Sons, Chichester, 1989, p.83
14. O. Brandt, H. Yang, B. Jenichen, Y. Suzuki, L. Däweritz, and K.H. Ploog, Phys. Rev. **B 52**, R2253 (1995)
15. P. Boguslawski, E.L. Briggs, and J. Bernholc, Phys. Rev. **B 51**, 17255 (1995)
16. J. Neugebauer and C. Van de Walle, Phys. Rev. **B 50**, 8067 (1994)
17. D.W. Jenkins and J.D. Dow, Phys. Rev. **B 39**, 3317 (1989)

MULTIPHOTON EXCITATION STUDIES ON GAN WITH PS PULSES

I. H. LIBON*, C. VOELKMANN*, D. KIM**, V. PETROVA-KOCH*, Y. R. SHEN**
* Technische Universität Munich, Physik-Department E16, D-85747 Garching, Germany
**University of California, Department of Physics, Berkeley, CA 94720, USA

ABSTRACT

We report the observation of UV photoluminescence (PL) from wurtzite GaN by multiphoton excitation. The dependence of the PL intensity on excitation intensity as well as PL excitation measurements with sub-bandgap photon energies indicate the existence of deep defect states centered at about 1.0 eV above the top of the valence band. This result was confirmed by a sum-frequency excitation spectrum. We correlate these measurements with the omnipresent yellow luminescence in GaN. Our two-photon PL excitation spectrum yields a two-photon absorption coefficient that agrees very well with theoretical predictions.

INTRODUCTION

Although much progress has been made in device-oriented applications with the wide bandgap semiconductor GaN, very little information on their nonlinear optical properties has yet been obtained. In fact the optical nonlinearity of those materials provides a source for noninvasive analytical investigations of the electronic and crystalline structure. From the technological viewpoint there are numerous nonlinear optical applications, e.g. optical communications and frequency conversion, that rely on second and higher order optical nonlinearities.

In this report we present some of the results of our nonlinear optical studies on wurtzite GaN. Besides sub-bandgap PLE measurements and sum-frequency excitation

In our studies of the nonlinear properties of GaN we employed samples grown by MOCVD on the basal plane of sapphire. A 500 Å buffer layer of AlN was first grown on the sapphire surface heated to 400 °C. The subsequent growth of the 3.4 µm GaN layer on top of the AlN buffer layer occurred at a substrate temperature of 1100 °C. The precursors used were trimethylaluminum (TMA), trimethylgallium (TMG), and ammonia (NH_3). The flow rates were 88 µmol/min for TMA, 44 µmol/min for TMG, and 10 l/min for NH_3. The n-type conductive samples with mobility 200 cm^2/Vs were shown to have a room-temperature carrier concentration of 10^{17} cm^3 probably due to N vacancies (V_N), or clusters of V_N, and possible Si impurities.

The excitation source in our experiments was a widely tunable picosecond optical parametric generator/amplifier pumped by the third harmonic of a picosecond pulsed Nd:YAG laser [1]. The 18 ps pulses of the system ranging from 400 nm to 9 µm reached a pulse energy of up to 300 µJ. The weakly focused laser beam reached the samples along their optical (hexagonal c) axis with an intensity of 0.1 GW/cm^2 to prevent laser damage. The PL from the sample was detected via a monochromator by a photomultiplier and was recorded with a gated integrator.

Mat. Res. Soc. Symp. Proc. Vol. 449 © 1997 Materials Research Society

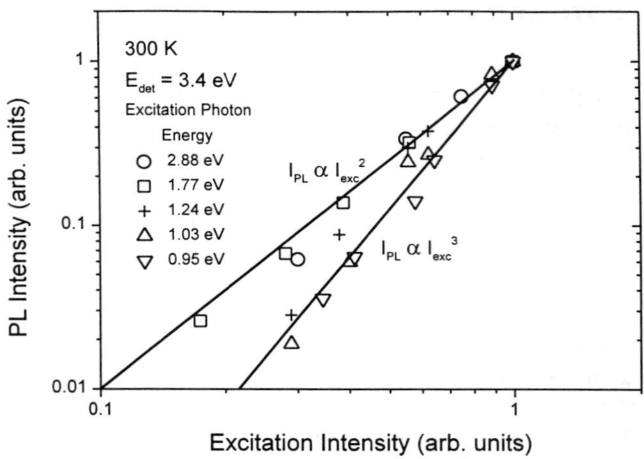

Fig. 1. GaN PL intensity vs. excitation intensity for different excitation photon energies.

PL INTENSITY DEPENDENCE

Ps-Excitation of the sapphire-grown sample with different below-bandgap photon energies provided by the OPG/OPA system yields the well-known GaN PL spectra with their peak at 3.4 eV.

The recorded spectra show the same behavior, irrespective of the applied excitation photon energy. This is a clear indication for the multiphoton excitation.

We have measured the dependence of the 3.4 eV PL on the excitation intensity at 300 K. The results are given in Fig. 1.

The above-bandgap excitation is strictly linear with excitation intensity. Below half bandgap excitation we expect to see a change from I_{exc}^{2} to I_{exc}^{3} at $E_g/2$ = 1.7 eV; we actually detect this behavior at 1.6 eV, which means that we observe a red-shift in respect to the excitation photon energy. A similar behavior is seen at 1.1 eV, where still an intensity dependence of I_{exc}^{3} was observed. Therefore we deduce from the intensity dependence measurements that one of the states involved in the absorption process of the photons is saturated. We suggest that the saturation process may result from deep defect electronic states in our sample, resulting in a nonvanishing electronic density of states in the bandgap.

PL EXCITATION SPECTRA

In Fig. 2 we plot the PL excitation (PLE) spectrum of our sample at 300 K, which clearly reveals features in the energy region shown.

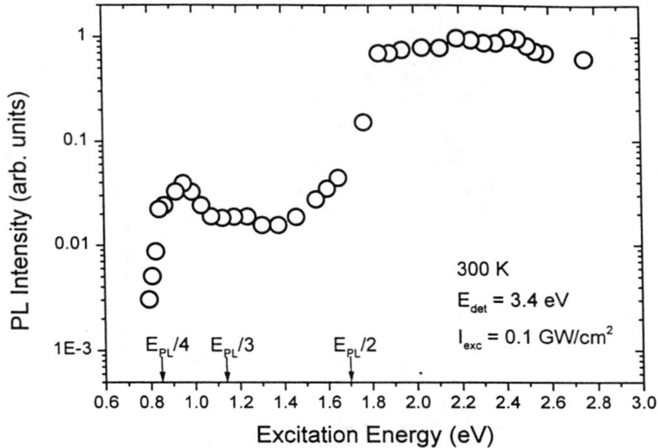

Fig. 2. Normalized PLE spectrum of a wurtzite GaN sample.

We observe a drop at half of the bandgap. In contrast to our expectation we do not observe the next drop in PL intensity at 1.1 eV (= $E_g/3$), but at 0.8 eV. This indicates resonant enhancement of the three-photon excitation due to the existence of real defect states in the bandgap. We take a closer look at this PLE spectrum later on in this paper when we discuss two-photon absorption in GaN.

SUM-FREQUENCY EXCITATION

To further confirm the existence of deep-level intermediate states around 1 eV in our sample we employed a two-color, two-photon excitation scheme instead of the above four-photon excitation at 0.8 eV: sum-frequency excitation (SFE). The optical parametric generator/amplifier (OPG/OPA) works by splitting the pump photon into two less energetic photons whose energy sum is equal to the pump photon energy. The more energetic one is called signal photon and the other is denoted idler photon. Since the pump photon energy (3.5 eV, 355 nm) lies slightly above the bandgap of GaN, if we send in the two beams from the OPG/OPA together into the GaN sample, the combined action of signal and idler photons should cause a two-photon above-bandgap excitation. In case the energy of one of the two beams is resonant with an intermediate real electronic state, we can expect to see resonant enhancement of the PL as the laser is tuned to that level. Fig. 3 is the SFE spectrum obtained by tuning the laser and keeping the energetic sum of the signal and idler beam fixed at 3.5 eV, which lies slightly above the GaN bandgap.

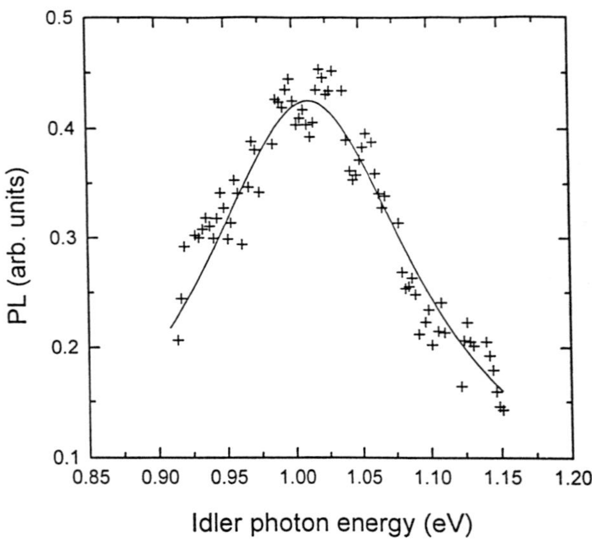

Fig. 3. SFE spectrum of GaN with two-photon excitation. The energetic sum of the two-photon pulses is 3.5 eV (355 nm); only the photon energy of one of the beams is displayed on the x-axis

The pump intensity was about 0.5 GW/cm^2 for the signal beam and 0.1 GW/cm^2 for the idler beam. If the beams were sent separately, the PL from the idler beam was only slightly above noise-level and the PL from the signal beam was less than 1/3 of the PL at the peak value, without any peak or prominent features throughout the tuning range. When they coincided at the sample, however, a pronounced peak was seen at the idler photon energy of 1.0 eV (signal photon energy of 2.5 eV) with a FWHM of 190 meV. This peak at 1.0 eV matches well with the value from Fig. 3 or the proposed deep-level position based on the yellow-luminescence measurement. Also the peak width of 190 meV is reasonable by comparing this with the width of the 2.3 eV yellow luminescence.

In conclusion, the PLE spectrum around 1 eV shows a peak and this peak in consideration with the two-photon, two-color PLE spectrum, discussed in this section, can be related to the known deep level in the bandgap of GaN.

TWO-PHOTON ABSORPTION

A very powerful method to study the optical and electronical properties of GaN is two-photon absorption (TPA) spectroscopy. Because TPA follows other selection rules, TPA provides complementary information to one-photon absorption [2]. From the complete electronic band structure it is possible to obtain via an *ab initio* calculation the two-photon absorption (TPA) coefficient. Murayama *et al.* derived it for the three semiconductors Si, ZnSe and GaAs [3]. We scaled the obtained values of ZnSe and GaAs by the refractive index and the bandgap for wurtzite GaN according to a parabolic two-band model [4], obtaining the TPA coefficient shown in Fig. 4.

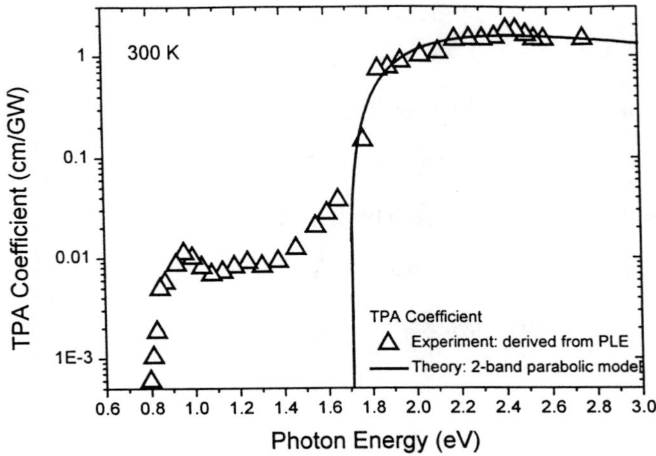

Fig. 4. Two-photon absorption coefficient of wurtzite GaN calculated with a parabolic two-band model. The filled circles denote the values obtained by PL excitation spectroscopy.

For the Kane momentum parameter P [5] we used the value $2P^2m/\hbar^2 = 17.9$ eV [6], which is close to $2P^2m/\hbar^2 = 21$ eV, the value of almost all studied group IV semiconductors and zincblende as well as wurtzite III-V and II-VI compound semiconductors.

Also shown in Fig. 4 are the values for the TPA coefficient we obtained by PL excitation spectroscopy, using the same experimental setup as in the PLE measurement of these samples. The fact that our measured lineshape very well matches with the theoretically derived curve verifies both the validity of our measuring procedure as well as the quality of our GaN samples.

CONCLUSION

In conclusion, we have observed UV PL from wurtzite GaN with multiphoton excitation. The PL intensity dependence on excitation intensity becomes more nonlinear as the pump photon energy is decreased. The sub-bandgap PLE spectrum displays a resonant enhancement at 1.0 eV; we correlate this feature to the presence of deep defect states that are also responsible for the yellow luminescence. The peak appears in a sum-frequency excitation spectrum as well. With a two-photon SFE spectrum we were able to confirm the theoretically predicted lineshape of the GaN two-photon absorption coefficient.

ACKNOWLEDGMENT

V.P.K. thanks the Deutsche Forschungsgemeinschaft (DFG) for the Habitilationsstipendium.

REFERENCES

1. J. Y. Zhang, J. Y. Huang, Y. R. Shen, and C. Chen, J. Opt. Soc. Am. B **10**, 1758 (1993).

2. Y.R. Shen, <u>The Principles of Nonlinear Optics</u>, (Wiley, New York, 1984).

3. M. Murayama and T. Nakayama, Phys. Rev. B **52**, 4986 (1995).

4. B.S. Wherett, J. Opt. Soc. Am. B **1**, 67 (1984).

5. E.O. Kane, J. Phys. Chem. Solids **1**, 249 (1957).

6. B. Segall (private communication).

OBSERVATION OF A TWO-DIMENSIONAL ELECTRON GAS IN THE AlGaN/GaN ON SiC SUBSTRATES

G.C. Chi,
Department of Physics, Natioinal Central University, Chungli, Taiwan.

C. F. Lin, H. C. Cheng and J. A. Huang,
Department of Electronics Engineering and Institute of Electronics, National Chiao Tung University, Hsinchu 30049, Taiwan.

M. S. Feng, and J. D. Guo,
Institute of Materials Science and Engineering, National Chiao Tung University, Hsinchu 30049, Taiwan.

Abstract

GaN/Al$_{0.08}$Ga$_{0.92}$N heterostructure has been grown on(0001) 6H-SiC substrates using low pressure metalorganic chemical vapor deposition(LP-MOCVD). Triethylgallium (TEGa), trimethylaluminum(TMA) and ammonia(NH$_3$) were used as the Ga, Al and N sources, respectively. The carrier gas is hydrogen(H$_2$)and the growth pressure is kept at 76 torr. Five pairs of GaN/Al$_{0.08}$ Ga$_{0.92}$N(100Å/100Å)were used as buffer layer, then follow by an 1.3 μm GaN film. The 500Å AlGaN bulk structure was grown on the GaN and Finally an 100Å GaN cap layer to prevent the oxidation of AlGaN layer. The full width half maxima(FWHM) of x-ray peak is 115 arc second for the thick layer of GaN (1.3 μm), this value is smaller than we reported for the GaN on sapphire substrate (It was about 300 arc second). The Hall mobility and sheet carrier concentration are 887 cm^2/Vs, 1.0×10^{16}cm^{-2} at 300K and 4661 cm^2/Vs, 7.2 $\times$ 10^{12}cm^{-2} at 77K. This high electron mobility is an indication of a two-dimensional electron gas(2DEG) formed at the GaN/AlGaN interface.

Introduction

Gallium nitride is a wide band gap semiconductor. It has attracted considerable interests in applications for blue, green, and ultraviolet light-emitting diodes, detectors and blue lasers. Although a lattice-matched substrate is difficult to obtain, C-face (0001) sapphire has been successfully used as the substrate to grow GaN film by MOCVD growth method. Because of the large differences in lattice constant (lattice mismatch about 13.8%) and thermal expansion coefficient between the sapphire substrate and GaN film, the performance of GaN-based devices depends critically on the initial grown layer that affects the growth of the misfit dislocations and polarity related defects. Substrates that have smaller mismatches would be desirable. 6H-silicon carbide (SiC) has a lattice constant closer to the GaN and the lattice mismatch as small as 3.4%. This substrate is considered more suitable for GaN epitaxial layer growth. A two-dimensional electron gas (2DEG) has been observed at the AlGaN/GaN heterointerface on sapphire[1,2,3] and on SiC[4]. Our result also confirm the existence of 2DEG in AlGaN/GaN heterostructures deposited on 6H-SiC substrates. This paper will describe the observed 2DEG phenomenon.

Mat. Res. Soc. Symp. Proc. Vol. 449 © 1997 Materials Research Society

Experiments

The AlGaN/GaN films were grown by LP-MOCVD on 6H-SiC substrates, from sublimation-grown boules[5]. The wafers were cut along the basal plane orientation. We chose the (0001) silicon face, for which each Si atom has a single dangling bond at the surface, C dangling bounds will occur only at the step sites. Sasaki's results indicated that GaN epitaxial layers on $(0001)_{si}$ and $(0001)_c$ SiC are terminated with nitrogen and gallium [6]respectively. The $(0001)_{si}$ SiC were degreased in sequential ultrasonic baths of acetone for 30 min and rinsed in deionized water. Then cleaned by RCA cleaning procedure[7]. Heavy metals were removed using a heated $(80^{\circ}C)$ H_2SO_4:HNO_3 solution. Next, a surface oxide was grown in a $60^{\circ}C$ HCl:H_2O_2:H_2O (5:3:3) solution[8] to passivate the Si dangling bonds, in a manner analogous to that for Si substrates[9]. This grown oxide was then stripped with a diluted (10:1) H_2O:HF solution. After final HF dip, the SiC substrates were blown dry with N_2 prior to the growth. During the growth the SiC substrate was placed on a graphite susceptor in a horizontal type reactor with a RF heater. Triethylgallium (TEGa), trimethylaluminum (TMA) and ammonia (NH_3) were used as the Ga, Al and N sources respectively. The carrier gas is hydrogen (H_2) and the growth pressure was kept at 76 torr. Before growing GaN Films, the SiC substrates were treated by thermal baking at $1100^{\circ}C$, to clean the contamination on the surface. Then the temperature was decreased to $1025^{\circ}C$ to grow five pairs of GaN/$Al_{0.08}Ga_{0.92}N$ with layer thickness of 100Å/100Å. These buffer layers were used as strain relief purpose prior to the growth of an 1.3 μm GaN epitaxial layer. AlGaN layer (500Å) was then grown on top of the GaN and finally an 100Å GaN cap layer to prevent the oxidation of AlGaN layer.

Results and Discussions

The grown material structure is shown in Fig.1. The heterostructure, as illustrated in Fig 1, consists of GaN (undoped $1.3 \times 10^{17} cm^{-3}$, 1.3 μm thick), grown on 6H-SiC substrate substrate with five pairs of GaN/AlGaN strain layer(total thickness upto 1000Å), undoped-$Al_{0.08}Ga_{0.92}N$ layer ($9.0 \times 10^{19} cm^{-3}$, 500Å thick), and 100Å GaN cap layer on top of $Al_{0.08}Ga_{0.92}N$ layer to prevent the oxidation of $Al_{0.08}Ga_{0.92}N$ layer. The FWHM of x-ray is 1.9 min for the thick layer of GaN (1.3μm). This value is smaller than we previously reported for GaN (5.2 min) epitaxial layer grown on sapphire substrate[10]. The x-ray rocking curves for both samples are shown in Fig 2. We also study various buffer layers to understand the influence of lattice mismatch and thermal expansion coefficient between GaN and SiC substrates. Different buffer layer such as 500Å AlN, 120Å low-temperature GaN and 500Å $Al_xGa_{1-x}N$ were studied.[11] Finally, we found that a 5-period of GaN/$Al_{0.08}Ga_{0.92}N$ thin superlattice (100Å/100Å) will produce good quality GaN epitaxial layer. The mobility and carrier concentration are 612 cm^2/Vs and 1.3×10^{17} cm^{-3} (at 300K) for the GaN epitaxial layer. The AlGaN was then grown on top of this GaN to form the 2DEG structure.

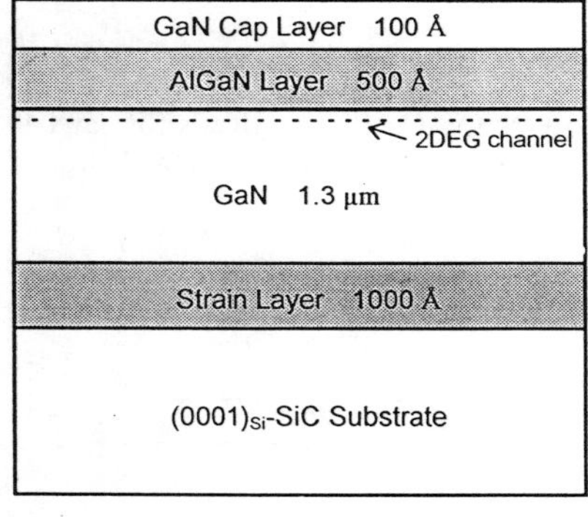

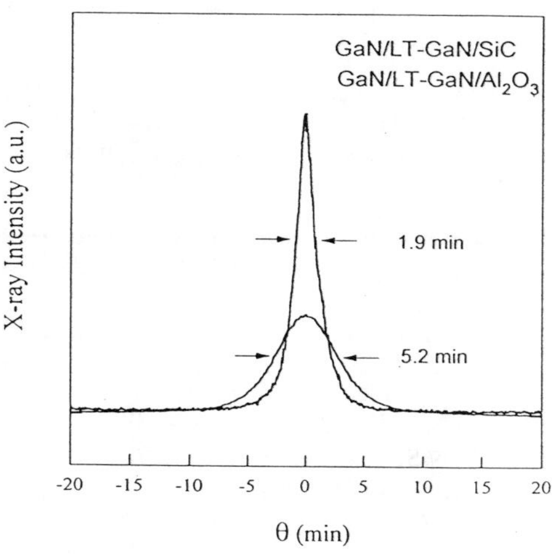

Figure 1. The schematic diagram of the 2DEG structure containing GaN/AlGaN heterostructure on (0001)$_{si}$-SiC substrate grown by LP-MOCVD.

Figure 2. Double-Crystal X-Ray rocking curve linewidths for GaN on SiC and on Al$_2$O$_3$ substrates.

The sheet carrier density and the electron mobility were measured at 300K and 77K using the Van der pauw technique of Hall effect masurement at 0.5T. Hall measurement results were also listed in Table 1. Results of the 2DEG samples and GaN/Al$_2$O$_3$, GaN/SiC and AlGaN/SiC epitaxial layers were shown for comparison. The carrier concentration in AlGaN/SiC is $9.0 \times 10^{19} cm^{-3}$ and GaN/SiC epitaxial layer is $1.3 \times 10^{17} cm^{-3}$ at 300K. We noted that sheet carrier density for AlGaN epitaxial layer is higher and the mobility is lower than GaN epitaxial layer. These results consistent with the observation that the total carriers in AlGaN layers varies with different composition of Al content. In general the mobility increases and concentration decreases when the content of Al are decrease in AlGaN layer. Also, if the carrier density decreases in the 2DEG channel the mobility in 2DEG structures is enhanced. For Al$_{0.08}$Ga$_{0.92}$N 2DEG-bulk structure, the measured Hall mobilities are 887 cm^2/Vs and 4661 cm^2/Vs at 300K and 77K. This 2DEG structure is similar to the structures grown on sapphire substrate reported by Asif Khan et al. (Hall mobilities are 834 cm^2/Vs and 2626 cm^2/Vs at 77K). Fig.3 shows the temperature dependence of mobility and sheet carrier concentration for this 2DEG-bulk structure.

Table

Table1. Sheet carrier density and mobility for GaN/Al$_2$O$_3$, GaN/SiC, AlGaN/SiC and the 2DEG structures measured by Hall measurement of Van der Pauw technique at 300K and 77K.

Structure	Sheet carrier density (cm^{-2})		Carrier mobility (cm^{2}V^{-1}s^{-1})	
	300K	77K	300K	77K
GaN/Al$_2$O$_3$ 1.3 μm	2.2×10^{13}	3.9×10^{12}	435	583
GaN/SiC 1.3 μm	1.7×10^{13}	4.3×10^{12}	612	1064
AlGaN/SiC 1.3 μm	1.1×10^{16}	1.6×10^{13}	403	258
2DEG-bulk	1.0×10^{16}	7.2×10^{12}	887	4661

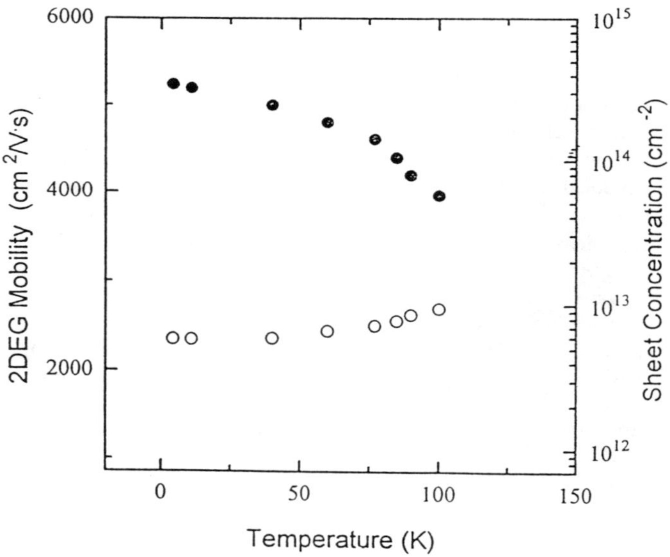

Figure 3. Electron mobility and sheet carrier concentration as a function of temperature from 1.3K to 100K for 2DEG-bulk structure.

Summary

In summary, we describe the growth and measurement of 2DEG mobility in $Al_{0.08}Ga_{0.92}N$/GaN heterostructures on 6H-SiC substrates. The high mobility of 4661 cm^2/Vs at 77K is an indication of 2DEG phenomenon at the $Al_{0.08}Ga_{0.92}N$/GaN interface. The higher electron mobility is attributed to its lower impurities scattering scenters at AlGaN/GaN interface and good GaN epitaxial layer quality on SiC substrates. This study may suggest a choice of heterostructures for field-effect transistors for high-power applications.

ACKNOWLEDGMENTS

The authors wish to thank for the support of the National Sciences Council NSC 85-2221-E008-039 and NSC 85-2221-E009-078.

Reference

1. M Asif Khan, J. M. Van Hove, J. N. Kuznia, and D.T. Olson, Appl. Phys. Lett 58,2048 (1991).
2. M. Asif Khan, A. Bhattarai, J. N. Kuznia, and D. T. Olson, Appl. Phys. Lett. 63, 1214 (1993).
3. M. Aaif Khan, Q. Chen. and C. J. Sun, Appl. Phys. Lett. 67,1429(1995).
4. J. M. Redwing, M. A. Tischler, J. S. Flynn, S. Elhamri, M. Ahoujja, R. S. Newrock and W. C. Mitchel, Appl. Phys. Lett. 69,963 (1996).
5. Cree Research Inc, 2810 Meridian Pkwy, Durham, NC 27713.
6. T. Sasaki and T. Matsuoka, J. APP. Phys. 64,4531 (1988).
7. J. Sumakeris, Z. Sitar, K. S. Ailey-Trent, K. L. More and R. F. Davis, Thin Solid Films 225,244 (1993).
8. M. E.Lin, S. Strite, A. Sgarwal, A. Salvador. G. L. Zhou, N. Teraguchi, A. Rockett and H. Morkoc, Appl. Phy. Lett. 62,702 (1993).
9. B. S. Meyerson, F. J. Himpsel, and K. J.Uram, Appl. Phys. Lett. 57,1034(1990).
10. C.F.Lin, G. C. Chi, M. S. Feng, J. D. Guo, J. S. Tsang, and J. Minghuang Hong, Appl. Phys. Lett. 68,3758 (1996).
11. C. F. Lin and G. C. Chi (to be published elswhere).

STUDY OF TRAPS IN GaN BY THERMALLY STIMULATED CURRENT

Rong Zhang[a,d]*, Zhenchun Huang[b], J.C. Chen[a], Youdou Zheng[c] and T.F. Kuech[d]
[a] Department of Electrical Engineering, University of Maryland Baltimore County, Baltimore, MD 21228
[b] NASA Goddard Space Flight Center, code 718, Greenbelt, MD 20771
[c] Department of Physics, Nanjing University, Nanjing 210093, P.R. China
[d] Department of Chemical Engineering & Materials Science Program, University of Wisconsin - Madison, Madison, WI 53706
* rzhang@cae.wisc.edu

ABSTRACT

In this paper we employed the TSC method to investigate the traps in GaN. The measured sample was a M-S-M UV-detector of high-resistance GaN on sapphire grown by LP-MOCVD. The relation of dark conductance to temperature clearly showed three major donor levels at 0.019, 0.13 and 0.74eV respectively. TSC measurements from 60 to 380K indicated that there were at least 11 traps in the GaN material. The active energy of those traps were 0.15, 0.19, 0.25, 0.28, 0.33, 0.39, 0.47, 0.55, 0.60, 0.63 and 0.67eV. The range of trap density is from $6\times10^{14}cm^{-3}$ to $2\times10^{18}cm^{-3}$. By comparing TSC spectrum to dark current, we consider there are at least 4 hole traps in the measured range with energy of 0.25, 0.28, 0.33 and 0.39eV. The illumination time effect was studied and discussed.

INTRODUCTION

GaN ultraviolet (UV) photodetectors have received more attention recently[1-3] due to their important applications in aerospace, automotive, petroleum, engine monitoring, flame detection, environment monitoring, and solar detection. The high mobility, sharp cutoff absorption edge, wide and useful response wavelength region, excellent response linearity, and high quantum efficiency of GaN make it one of the most promising materials to fabricate UV detectors and other novel devices. The best performance of photoconductive UV detectors require high-resistance semiconductors. In those materials and device applications, traps play an important role in their properties, such as compensation to unintentionally doped impurities, the response time for incident photons, and the sensitivity of detectors. In some cases, traps can significantly degrade the device performance. The image of traps in GaN high-resistance materials have not yet been clear although it is very critical for further development of GaN UV detectors. Few results of traps in high-resistance GaN materials have been reported by employing measuring techniques of photoconductivity[4,5], photoluminescence (PL)[6], Thermally Stimulated Current (TSC)[3,7]. Among them TSC is a traditional technique to characterize traps in semi-insulating materials and has been proved to be very suitable for trap studies in GaN. In this paper, we use TSC to investigate traps in LP-MOCVD grown GaN M-S-M detector structure. TSC measurements from 60 to 380K indicated that there were at least 11 traps in the GaN material. The active energy of those traps were 0.15, 0.19, 0.25, 0.28, 0.33, 0.39, 0.47, 0.55, 0.60, 0.63 and 0.67eV. By comparing those levels to three major donors found in dark current,

Mat. Res. Soc. Symp. Proc. Vol. 449 © 1997 Materials Research Society

we consider there are at least 4 hole traps in the measured range with energy of 0.25, 0.28, 0.33 and 0.39eV. Illumination time effect was studied and discussed.

EXPERIMENT

The sample used in this study was a coplanar M-S-M structure of unintentionally n-doped high-resistance GaN on (0001) sapphire substrate grown by low pressure MOCVD. Fig. 1 is a double crystal x-ray rocking curve of GaN (0004) diffraction and shows a good crystal quality. The M-S-M finger-type patterns were fabricated using a lift-off technique. The fingers were 3μm wide and 5.2mm long and the spacings between the fingers were 8μm. The Ohmic contacts were formed by evaporating Al/Au on the GaN front surface, and alloying at 450°C for 5 min. The I-V curve of the sample measured in the dark was linear even at low temperature, indicating a good Ohmic contact.

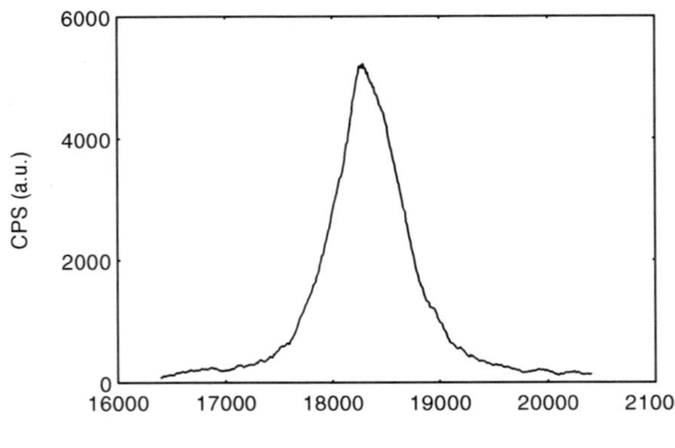

Fig. 1 The double-crystal rocking curve of the GaN sample.

After the sample was cooled down to low temperature, a 10W white-light neon bulb installed in the cryostat was turned on to illuminate the sample. The typical illumination time was more than 10 minutes. Then the sample was allowed to be heated up slowly. A bias was applied to the sample through a series resistor for protecting the sample from burning out. The thermally stimulated current was collected by Keithley Model 617 electrometer. The thermal scan was made at an average heating rate of 0.35K/s in the measured temperature range.

RESULTS AND DISCUSSION

Fig. 2 shows dark current of the sample at the temperature range between 60K to 380K as a solid curve. The x-axis is the reciprocal of temperature. The y-axis is the dark current. The room temperature (300K) is marked at the top of the figure. The measurement was operated under dark during the sample was cooled down from 380K to 60K. Before measurement, the sample was kept at 380K under dark for a long time to let it arrive thermal equilibrium. It is obvious that there are three exponential-featured donor thermally ionization regions between 60K and 380K. All three ionization regions can be well fitted by exponential curve shown in the figure as dash lines. Under 100K the first ionization region has an active energy of 0.0095eV. Between

130K and 200K the second ionization region holds an active energy of 0.063eV. Above 280K, the third ionization region shows an active energy of 0.37eV. Because there is compensation in the sample, energy levels locate at 0.019eV, 0.13eV and 0.74eV, respectively. A plateau between the second and third regions comes from the large interval between those two levels and exhaust of electrons in shallow levels.

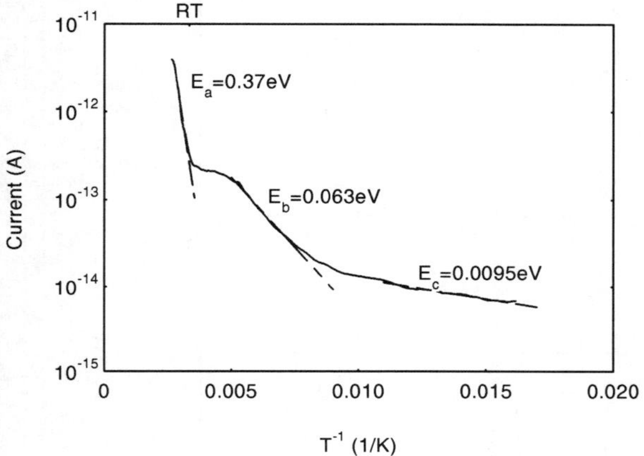

Fig. 2 Dark current of the GaN sample between 60K to 380K.

Fig. 3 are TSC spectra of the used sample with linear and logarithm scale of y-axis. The illumination time at low temperature is 15 min and the heating rate is 0.35K/sec. The bias is 80V. As seen from (a), 5 major TSC peaks appear in the temperature range between 60K to 360K. The shape of the largest peak around 300K indicates that there are more than two peaks involved in the peak. In order to find all minor peaks, we change the scale of y-axis from linear to logarithm as shown in (b). It is obvious that there are totally 11 peaks in the measured temperature range. Every peak indicates a trap level. We can calculate the energy position of those levels in the bandgap, E_t by

$$\ln(T_m^2 / \beta) = E_t / kT_m - \ln(N v \sigma k / E_t) \qquad (1)$$

where E_t is the trap level; T_m is the peak temperature, β is the heating rate, N is the effective density of states of the conduction band, v is the thermal velocity of carriers, and σ is the thermal capture cross section of the trap. Under the condition of $T_m > 100K$ and $\sigma < 10^{-15}$ cm^2, Eqn (1) can be simplified to

$$E_t = kT_m \ln(T_m^4 / \beta). \qquad (2)$$

All parameters of 11 peaks are listed in Table I. The energy level are calculated by Eqn (2), and the densities of traps are determined approximately from the charge emitted, given by the area (current × time) of the emission hump divided by the effective volume of the crystal that is assumed as having completely filled traps.

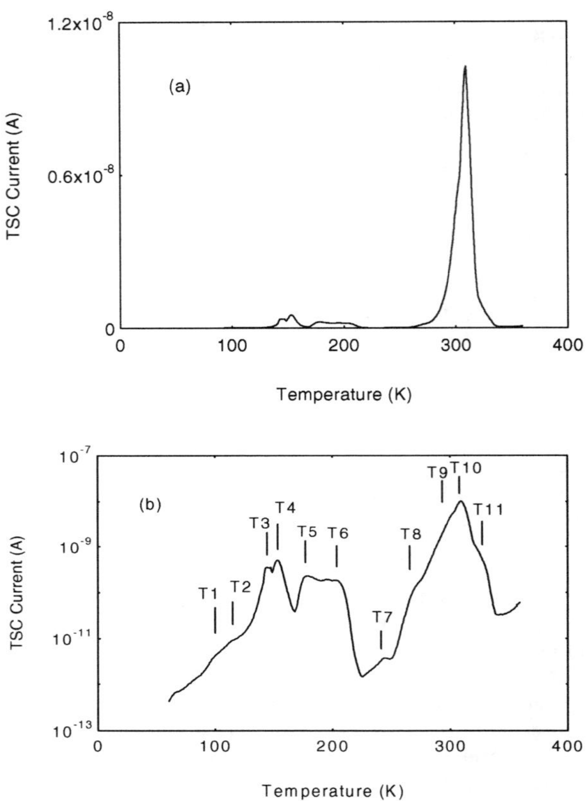

Fig. 3 TSC spectrum of the measured sample with
linear scale (a) and logarithm scale (b) of y-axis.

TABLE I Parameters of measured traps.

	T1	T2	T3	T4	T5	T6	T7	T8	T9	T10	T11
E_t(eV)	0.15	0.19	0.25	0.28	0.33	0.39	0.47	0.55	0.60	0.63	0.67
N_t(cm^{-3})	1e15	3e15	2e17	4e17	1e17	8e16	6e14	2e16	8e17	2e18	5e16

T3/T4, T5/T6, T8 and T10 have been observed in Ref[3]. Here we found there are fine structures in those traps. T10 is the largest peak in the temperature range and controls optical and electrical properties of the material at room temperature.

Compared to dark current shown in Fig. 2, we find the TSC background is much larger than the former as shown in Fig. 4. We believe that comes from persistent photocurrent[7]. From Fig. 4, we can also get information about the type of traps. Because only three major donors shown in the dark current, we consider that at least the peak groups in the medium temperature, including T3, T4, T5 and T6, are hole traps.

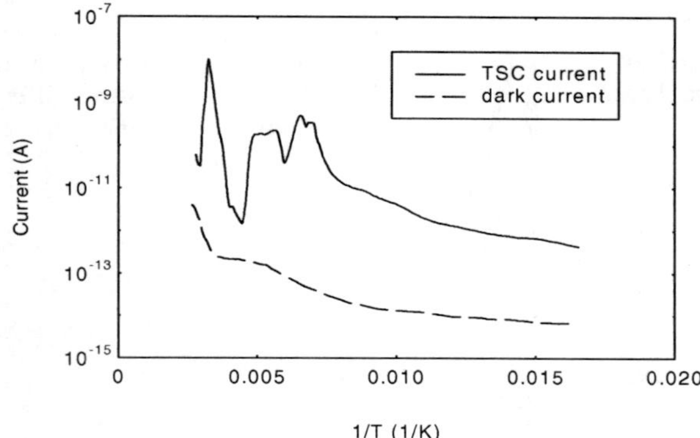

Fig. 4 Comparison of the TSC current to the dark current of the measured sample.

We also measured TSC spectra for different illumination time. Because there are multi-trap and retrapping effects in the sample and we use a white light source, the relation of the TSC current to the illumination time is very complicated. Fig. 5 shows two TSC spectra with illumination time of 10 and 15 min, respectively. It is clear that the shape of the spectrum changes significantly, which indicates that there are different types of trap (both electron and hole traps) and different response time for different traps. We found experimentally that after thermal cycling of illumination at low temperature, heating up under dark, cooling down under dark, the net photocurrent showed a negative value at the beginning of reillumination. Since we use neon bulb as light source, the major output photons distribute in visible range with energy less than the bandgap of GaN. On the other side, traps deeper than 0.75eV do not emit carriers in the thermal cycle. At low temperature when the white light source is on, holes are excited from hole traps to the valence band and recombine with electrons in the conduction band. Thus a negative photocurrent is observed. We can conclude from that phenomenon that there are hole traps located around the center of the gap and they have shorter emission time than electron traps.

CONCLUSION

Dark current and TSC spectra of high-resistance GaN M-S-M structure were measured in the temperature range of 60 to 380K. Dark current clearly showed three major donors at 0.019, 0.13 and 0.74eV, respectively. Totally 11 traps were found in the measured range. The traps have energy of 0.15eV, 0.19eV, 0.25eV, 0.28eV, 0.33eV, 0.39eV, 0.47eV, 0.55eV, 0.60eV, 0.63eV and 0.67eV, respectively. The range of trap density is from $6\times10^{14}cm^{-3}$ to $2\times10^{18}cm^{-3}$. By comparing TSC spectrum to dark current, we consider there are at least 4 hole traps in the measured range with energy of 0.25, 0.28, 0.33 and 0.39eV. Illumination time experiment indicated a very complicated mechanism of retrapping and interaction between different traps.

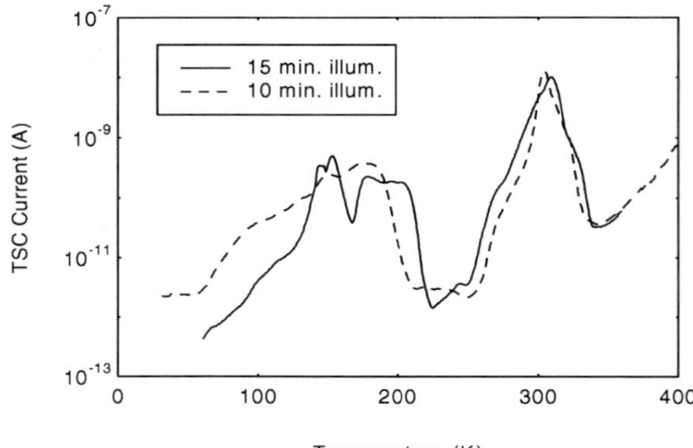

Fig. 5 Comparison of two TSC spectra with different illumination time at low temperature.

REFERENCES

1. M. Asif Khan, J.N. Kuznia, D.T. Olson, J.M. Van Hove, M. Blasingame, L.F. Reitz, Appl. Phys. Lett., **60**(23), p2917(1992)

2. D. Walker, X. Zhang, P. Kung, A. Saxler, S. Javadpour, J. Xu, M. Razeghi, Appl. Phys. Lett., **68**(15), p2100(1996)

3. Z.C. Huang, J.C. Chen, B.D. Mott, Mater. Res. Soc. Symp. Proc. **395**

4. C.H. Qiu, C. Hoggatt, W. Molton, M.W. Leksono, J.I. Pankove, Appl. Phys. Lett., **66**(20), p. 2712(1995)

5. Rong Zhang, Z.C. Huang, Bo, Guo, J.C. Chen, Li Yan, Youdou Zheng, T.F. Kuech, *these proceedings.*

6. J.C. Zolper, M. Magerott Crawford, A.J. Howard, J. Ramer, S.D. Hersee, Appl. Phys. Lett., **68**(2), p.200(1996)

7. D.C. Look, Z.Q. Fang, W. Kim, O. Aktas, A. Botchkarev, A. Salvador, H. Morkoc, Appl. Phys. Lett., **68**(26), p3775(1996).

Part VI

Luminescence and Recombination

SPONTANEOUS AND STIMULATED RECOMBINATION IN THE NITRIDES

A. HANGLEITER, F. SCHOLZ, V. HÄRLE, J. S. IM, AND G. FRANKOWSKY
4. Physikalisches Institut, Universität Stuttgart D-70550 Stuttgart, Germany
E-mail: a.hangleiter@physik.uni-stuttgart.de

ABSTRACT

Both spontaneous and stimulated emission processes are essential ingredients for constructing a laser from the nitrides. Based on our picosecond time-resolved photoluminescence studies we show that spontaneous radiative recombination is strongly influenced by excitonic effects, both in bulk GaN and in quantum wells. Particularly in quantum wells, localization of excitons plays an important role. We have studied the optical gain spectra in GaInN/GaN and GaN/AlGaN double heterostructures and quantum wells, grown by LP-MOVPE, using the stripe excitation method. Both room temperature and low temperature measurements were performed. Based on our results, we discuss the physical mechanism of optical gain in the nitrides as well as consequences for laser operation. We show that localization or, equivalently, the formation of quantum dot like structures, governs the optical gain mechanism in the nitrides.

INTRODUCTION

The group-III nitrides GaN, AlN, and InN have recently attracted much interest due to their application for blue light emitters. Both light emitting diodes (LED's) covering the blue, green, and yellow spectral range [1, 2, 3] and violet injection lasers [4, 5, 6] have been demonstrated.

Particularly for the laser, both spontaneous and stimulated recombination of injected charge carriers is vitally important. Stimulated emission provides the laser gain process, whereas spontaneous recombination is responsible for carrier losses. For conventional III-V semiconductors free-carrier band-to-band transitions are responsible for the gain, whereas for II-VI-based quantum wells excitons are suspected to make substantial contributions at least at low temperatures. Since the nitrides are wide-gap semiconductors as well, this makes the issue of the gain mechanism particularly interesting for the nitrides.

In this paper, we will first discuss the basic mechanisms of spontaneous and stimulated recombination in semiconductors. Based on our experimental data on recombination in GaInN/GaN quantum wells, we will then show that localization is an important concept for the nitrides. The issue of stimulated emission and optical gain will then be elucidated, showing that the experimentally observed optical gain is due to localized states. We will then discuss the similarity between strong localization and artificially prepared quantum dots and highlight the consequences for the nitrides.

Mat. Res. Soc. Symp. Proc. Vol. 449 © 1997 Materials Research Society

MECHANISMS OF SPONTANEOUS AND STIMULATED RECOMBINATION

Spontaneous processes

Excess charge carriers occupying extended band states in a semiconductor (including low-dimensional structures) can recombine either radiatively or non-radiatively. Radiative band-to-band recombination of free carriers is governed by the k conservation rule, leading to the recombination rate being proportional to the electron and hole densities. The temperature dependence of the radiative recombination coefficient follows from k conservation and is determined by the effective combined density of states (i.e. $\propto T^{-3/2}$ for bulk semiconductors, $\propto T^{-1}$ for quantum wells, and $\propto T^{-1/2}$ for quantum wires). For free excitons, i.e. bound electron-hole pairs, similar considerations apply. In general, however, a mixed system of more or less correlated electrons and holes has to be considered [7].

Nonradiative recombination can be either intrinsic or due to defects. Since Auger recombination [8] as an intrinsic mechanism should be negligible for wide-gap materials like the nitrides, we will not discuss it further. More importantly, nonradiative recombination via defect states in the forbidden gap will be active in the nitrides, particularly in the light of the large defect densities in todays epitaxial material. Such recombination follows the statistics of Shockley and Read [9] and Hall [10] and involves the emission of many phonons during the capture of the electron or the hole into the defect state. Due to the nature of the capture processes, the recombination becomes more effective at higher temperatures, i.e. is thermally activated [11].

Localized states, on the other hand, exhibit fixed radiative or nonradiative recombination probabilities, depending on their electronic structure. As long as no excited states are populated, the radiative lifetime of an excitation is independent of temperature. Only nonradiative processes introduce a strong temperature dependence.

Stimulated recombination

Stimulated recombination of excited carriers - or optical gain - is the single most important property of laser-active materials. In a free-carrier two-band picture of a semiconductor, optical gain is achieved if the separation of the electron and hole quasi Fermi energies exceeds the bandgap energy [12]. Due to the extended nature of the states involved, the spectra of the optical gain tend to be rather broad. Even though a large improvement is expected for low-dimensional structures due to their narrow density of states, inhomogeneous broadening due to size fluctuations diminishes the potential improvement.

For III-V semiconductors like GaInAs/InP, GaAs/AlGaAs, or GaInP/GaAs and quantum wells from these materials, the optical gain is well understood on the basis of pure band-to-band transitions. Even for strained quantum wells, model calculations of the gain based on 6-band $k \cdot p$ valence band structures yield excellent quantitative agreement between measured and calculated gain spectra [13]. The most important property of the optical gain in our context is the shape of the gain spectra, as shown in Fig. 1. In case of inversion, optical gain is expected between the (renormalized) band edge and the chemical potential (i.e. the difference of the quasi Fermi energies). At higher energies, a steep absorption edge with up to $\approx 3 \times 10^4 \mathrm{cm}^{-1}$ should be observed.

On the other hand, for II-VI semiconductors, there has been a lively discussion in the literature as to whether the optical gain in such structures is (partly) due to excitonic

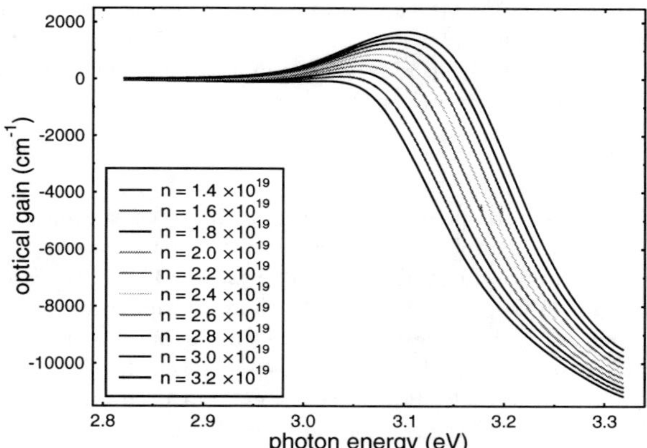

Figure 1: Calculated optical gain spectra for band-to-band transitions in bulk GaInN. Due to the inversion, there is gain close to the band edge, but a strong absorption at photon energies above the chemical potential.

processes [14, 15, 16, 17]. This idea is related to the fact that the excitonic binding is rather strong in wide-gap semiconductors, due to the large effective masses of electrons and holes. If excitons survive at sufficiently high carrier densities allowing for optical gain, an excitonic enhancement of the gain or even purely excitonic gain processes might show up. Recently, it has been shown that at low temperatures the gain in ZnSe-based quantum wells may in fact be due to localized biexcitons [16].

As this will become important later on, we will also briefly discuss optical gain in semiconductor quantum dots. For quantum dot structures, the optical gain has been studied mostly from the theoretical point of view up to now [18, 19]. As in the case of strained quantum wells, it is most difficult to correctly include the complicated electronic structure of such low-dimensional structures when calculating the gain. Since most existing quantum dot structures exhibit quite a number of excited states, the resulting spectra are rather complex. Experimentally, only very few studies have been performed. Due to the inhomogeneous broadening due to size fluctuations, the gain spectra are fairly broad and the excited-state structure is only observed at low temperature [20, 21].

EXPERIMENTAL DETAILS

Samples

Our GaN samples were grown on 0001-oriented sapphire substrates using low pressure metal-organic vapor phase epitaxy (LP-MOVPE) and employing an AlN nucleation layer. The layer thickness was in the range $1 \ldots 3 \ \mu$m. The samples exhibited X-ray diffraction linewidths ((0002), open detector mode) of $50 \ldots 80$ arcsecs and low-temperature photoluminescence linewidths of around $2 \ldots 4$ meV. The net donor concentration of nominally undoped GaN was less than $5 \cdot 10^{16}$cm^{-3}.

Both GaInN/GaN and GaN/AlGaN double heterostructures and quantum wells were grown by adding either In or Al to the appropriate layers. Typically, the structures consisted of a ≈ 0.5 μm GaN or AlGaN buffer layer, a GaInN or GaN active layer, and a ≈ 100 nm GaN or AlGaN top confinement layer. The maximum In-content in the GaInN layers was about 15 %. The AlGaN barriers had an Al-content of about 12 %.

<u>Measurement techniques</u>

The carrier dynamics were investigated using a picosecond time-resolved photoluminescence setup, where the samples were excited with 5 ps pulses from a cavity-dumped frequency-doubled synchronously mode-locked dye laser. The luminescence was detected with a Hamamatsu R3809-U microchannel-plate photomultiplier and processed using time-correlated single-photon-counting electronics. By employing suitable deconvolution techniques, an overall time resolution of less than 20 ps was reached. The samples were mounted in a variable-temperature cryostat, allowing for a temperature range from 2 K up to 400 K.

We have used the stripe excitation method to measure the optical gain in our samples [22]. The samples were optically excited by 6 ... 15 ns pulses of an excimer laser (operating either with N_2 (337 nm) or with XeCl (308 nm)) at a repetition rate of 100 Hz and the amplified spontaneous emission (ASE) emitted from the sample edge was dispersed by a 0.5 m monochromator and detected by a cooled intensified silicon diode array detector. The length of the excited stripe was varied between 0 and typically 200 μm by a computer-controlled stepper motor. The samples were kept at room temperature during most of the measurements. The ASE spectra were analyzed by fitting the relation

$$I_{ASE} = I_0(e^{g \cdot L} - 1), \tag{1}$$

where g is the optical gain coefficient and L is the length of the excited stripe, to the measured dependence of the ASE intensity on stripe length for every wavelength. The actual power density incident on our samples was determined from a careful analysis of the beam profile.

SPONTANEOUS RECOMBINATION

<u>Bulk GaN</u>

The spontaneous recombination was studied both in bulk GaN and in GaInN/GaN and GaN/AlGaN quantum wells using picosecond time-resolved optical spectroscopy. For the present purpose, we concentrate on radiative recombination. Even though the recombination is mainly nonradiative for all structures at elevated temperatures, we have been able to conclude on radiative lifetimes from our measurements. This was accomplished by combining the measured decay times and the corresponding integrated photoluminescence intensities, both as a function of temperature. Since the integrated intensity of the band-edge emission provides at least a relative measure of the quantum efficiency η, we can derive the radiative lifetime τ_{rad} using

$$\tau_{rad} = \frac{\tau_{eff}}{\eta} \tag{2}$$

from the measured decay time τ_{eff}.

For bulk GaN we find (Fig. 2) that the radiative lifetime initially increases proportional to $T^{3/2}$, as expected for a bulk semiconductor [23], due to the dispersion of the electrons and holes in 3 dimensions. In fact, this is an exciton radiative lifetime (polariton effects are

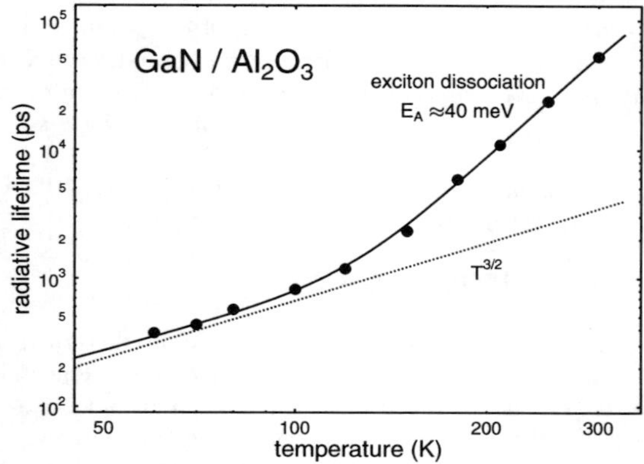

Figure 2: Temperature dependence of the radiative free exciton lifetime in bulk GaN, derived from the measured decay time and the quantum efficiency.

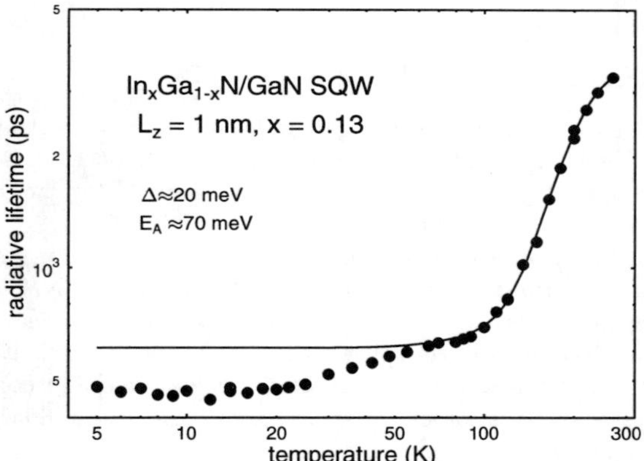

Figure 3: Radiative exciton lifetime in a 1 nm GaInN/GaN quantum well as a function of temperature. The fairly constant lifetime at temperatures below 100 K indicates a strong influence of localization.

negligible as long as the layer thicknesses do not exceed the absorption length), since excitons are stable at these temperatures. At higher temperatures, the increase of the radiative lifetime becomes more pronounced. This is due to dissociation of excitons into free electrons and holes, with an activation energy of about 40 meV. The full line in Fig. 2 represents a fit based on these considerations [24].

GaInN/GaN quantum wells

For GaInN/GaN quantum wells the result is fairly different, as shown in Fig. 3. At low temperatures, below about 100 K, the radiative lifetime remains constant and increases only at higher temperatures. The increase of the lifetime at high temperatures is readily understood on the basis of the exciton dispersion and exciton dissociation. At low temperature, however, the temperature independent radiative lifetime is only understood if we assume a strong localization of the excitons [25]. Localization in this context means binding of the excitons at local potential fluctuations due to alloy fluctuations or due to well width fluctuations. The radiative lifetime of these localized excitons is independent of temperature since there is no energy dispersion of the corresponding states. Moreover, the radiative lifetime is longer than that of free excitons at the same temperature, since the wavefunction of the localized exciton is spread in k-space, leading to a reduced probability of emitting a $k \approx 0$ photon.

Qualitatively, we find the same behavior for GaN/AlGaN quantum wells, even though one would expect far less localization for quantum wells with a binary active layer.

STIMULATED RECOMBINATION - OPTICAL GAIN

We have studied the optical gain in GaInN/GaN and GaN/AlGaN double heterostructures and quantum wells using the stripe excitation method described earlier. The most important advantage of this method is that it is far more sensitive to small optical gain amplitudes (as in single quantum well structures) than pump-probe type transmission experiments. On the other hand, this method works only for spectral regions, where the spontaneous intensity is sufficiently large. However, this condition is satisfied for most of the structures investigated here.

GaInN/GaN double heterostructures and quantum wells

One of the most basic observations regarding the optical gain in the nitrides is the strong polarization anisotropy of the gain: Due to the band structure of the nitrides, i.e. the crystal-field splitting of the valence bands, optical gain is almost exclusively observed for the TE mode (polarization in c-plane), but not for the TM mode [26].

When studying the gain spectra in detail, a surprising behavior is found. Fig. 4 shows a series of room-temperature gain spectra at various pump power levels for a 15 nm GaInN/GaN DH structure. For the lowest pump power level, we only observe absorption, whereas at higher power levels positive net gain is obtained in the low-energy part of the spectra. It is interesting to note that there is fairly little absorption (< 20 cm^{-1}) below the band edge, indicating rather small waveguide losses. This is somewhat surprising, considering the large density of defects present in the epitaxial material.

These data have been analyzed using a band-to-band model and the model of the discussion section. The dashed line represents a fit of the low-energy part of the spectrum at the highest pump power based on the band-to-band model discussed earlier [26]. We note that a good fit is obtained for the low-energy part, but the model is unable to account for decreasing absorption at higher energies. The full line fit drawn in conjunction with the other measured data is based on the model described in "Discussion and Model".

Qualitatively the same behavior is obtained for a 2.5 nm GaInN/GaN quantum well (Fig. 5). At the lowest pump power, only absorption is observed. At higher power levels, positive gain is obtained around 3.1 eV, even though the net waveguide gain remains negative. Again, the absorption at higher energies is limited to the spectral range between about 3.25 and 3.4 eV.

GaN/AlGaN structures

In order to elucidate the optical gain mechanism in the nitrides in more detail, we have also studied GaN/AlGaN structures. Fig. 6 shows gain spectra for GaN/AlGaN structures

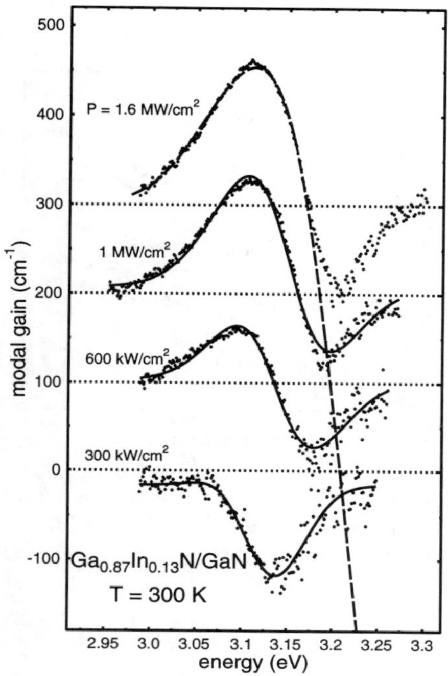

Figure 4: Optical gain spectra of a GaInN/GaN DH structure ($L_z = 15$ nm) at various pump power levels. The dashed and the full line are fits explained in the text. The spectra are vertically displaced for clarity.

with active layer thicknesses between 100 nm and 2.5 nm. The spectra were obtained using non-resonant excitation at 308 nm (4.02 eV). The spectra show an unexpectedly rich structure and we note several remarkable features. First of all, we note the steep onset of absorption due to the AlGaN barriers at about 3.75 eV. Moreover, we note structures in the gain spectra at energies between 3.3 and 3.6 eV similar to the GaInN/GaN structures. Most importantly, we observe <u>no</u> absorption in the energy range between 3.5 and 3.7 eV, i.e. between the "quantum well" emission and the AlGaN barrier energy. For a "normal" quantum well structure, an absorption of a few hundred up to a few thousand inverse centimeters, depending on the optical confinement factor of the waveguide, due to band-to-band transitions would be expected.

In order to clarify the gain mechanism further, we performed optical gain measurements at low temperature (20 K). In addition to the non-resonant pumping as above, experiments with resonant pumping (337 nm $\equiv$ 3.67 eV) of the active layer were carried out. The measurements showed a strong variation of the results, depending on the position of the excited stripe on the sample. A typical gain spectrum, together with an ASE spectrum, is shown in Fig. 7. The ASE spectrum exhibits a considerable number of relatively narrow lines, which find their more or less pronounced counterpart in the gain spectrum. The individual structures in the gain spectrum are similar to those known from the GaInN/GaN samples.

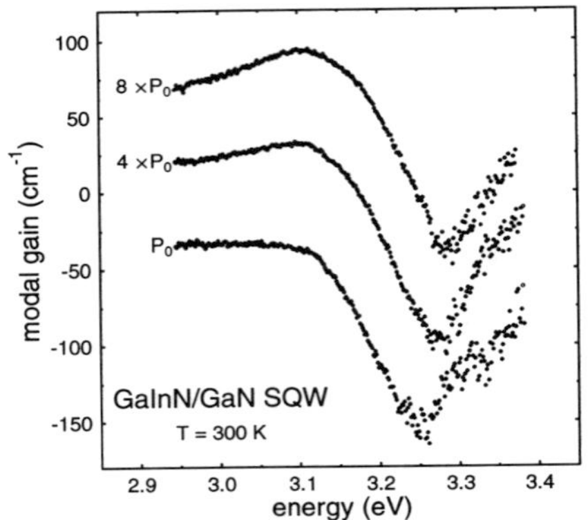

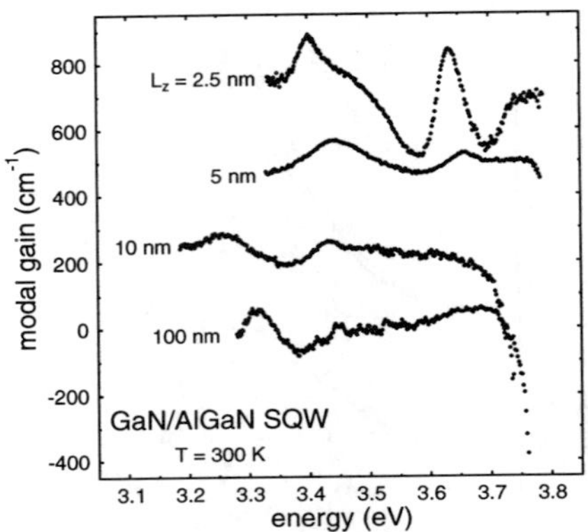

Figure 5: Optical gain spectra of a GaInN/GaN SQW structure ($L_z = 2.5$ nm) at various pump power levels. The spectra are vertically displaced for clarity.

Figure 6: Optical gain spectra for GaN/-AlGaN SQW structures of various widths at room temperature. The spectra are vertically displaced for clarity.

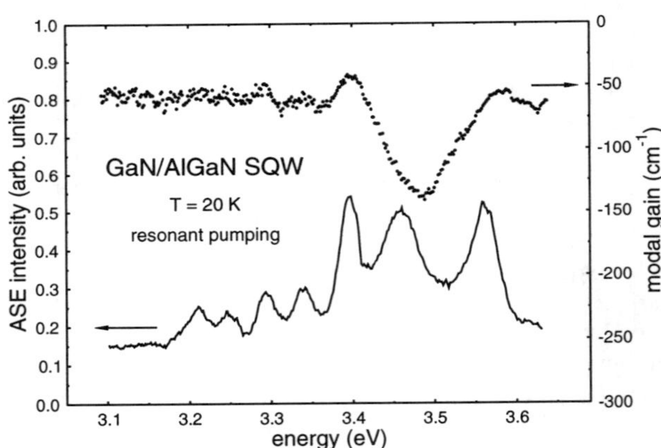

Figure 7: Low-temperature spectra for a GaN/AlGaN structure obtained under resonant excitation. The full line represents an ASE spectrum, whereas the dots give the optical gain.

DISCUSSION AND MODEL

Before drawing further conclusions from our results, let us reiterate the major findings:

- The optical gain spectra both for GaInN/GaN and for GaN/AlGaN structures exhibit a consistent lineshape pattern with the absorption on the high energy side decreasing towards zero.

- Particularly for the GaN/AlGaN structures, absolutely no absorption is observed between the "quantum well" emission and the barrier energy level.

- Resonant pumping experiments at low temperature result in a fairly large number of narrow emission lines and a gain spectrum with rich structure.

- The radiative lifetime of excitons in GaInN/GaN "quantum wells" is constant over a fairly large range of temperatures, indicating a strong influence of localization.

The most important starting point for further considerations is the absence of absorption above the "band edge" of the quantum wells. This is only possible, if the states involved in the optical transitions and in the optical gain are neither 3-dimensional nor 2-dimensional in nature, but rather 0-dimensional, i.e. are localized states.

Nakamura recently pointed out [27] that in his GaInN/GaN multiple quantum well structures, TEM pictures reveal strong local fluctuations of the In content, reminiscent of artificial "self-organized" quantum dot structures. In fact, similar fluctuations of the In content were observed for our GaInN/GaN structures [28].

As a matter of fact, the physical concepts of 3-dimensional size quantization and of localization are strongly interrelated. The term localization is used in a situation, where carriers may be trapped in shallow local potential fluctuations. On the other hand, one typically thinks of quantum dots when there are well defined energetically deep 3-dimensional potential wells.

Actually, it is well known for quantum dot structures that the radiative lifetime is independent of temperature [29], as observed for our GaInN/GaN structures. Only at elevated temperatures, excitons may get thermally excited to barrier states, leading to a lifetime approaching that of the barrier states.

Now let us consider the optical gain from quantum dot structures. The models proposed in the literature are either too simple [18] or too complex [19] to provide a basis for our discussion. In the following, we therefore develop a model adapted to the complexity of our problem.

Consider a single quantum dot small enough to allow only for a single confined electron state, but possibly a larger number of hole states. Due to the Pauli exclusion principle, such a dot can accommodate at most two electrons and two holes (as long as charge neutrality is maintained), or, in other words, two excitons. There are only two possible optical transitions in this system: from the ground state to the one-exciton state and from the one-exciton state to the two-exciton state. These transitions may be assumed to be approximately degenerate. If the quantum dot is in the ground state, the only allowed transition is absorption of a photon to the one-exciton state. If the quantum dot is in the two-exciton state, only stimulated (or spontaneous) emission towards the one-exciton state is possible. Finally, if the dot is in the one-exciton state, both absorption and stimulated emission are possible, with approximately equal probability.

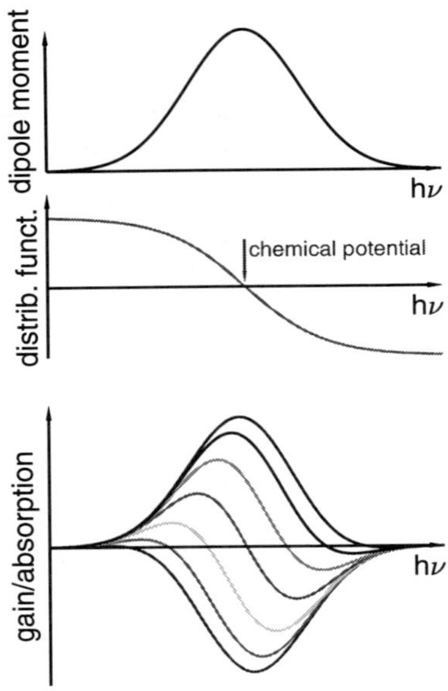

Figure 8: Simple model for optical gain from quantum dots. Top: Inhomogeneously broadened spectral distribution of the dipole moment; middle: Fermi distribution of state filling; bottom: resulting gain lineshapes. There is no absorption at higher energies.

In a real world system, the quantum dots have an inhomogeneous size distribution and therefore an inhomogeneous broadening on the energy scale. Assuming the existence of a mechanism establishing thermal quasi-equilibrium between the dots, the occupation of these states is governed by a Fermi distribution, where those dots at the lowest energy are in the two-exciton state, the ones around the chemical potential are in the one-exciton state, and those at the highest energies are in the ground state.

The consequences for the optical spectra are illustrated in Fig. 8. The upper part of the figure gives the inhomogeneous distribution of the dipole moment, i.e. the number of dots at that energy. In the middle, we have plotted the inversion factor, i.e. $2f(E) - 1$, and the bottom part shows the resulting optical spectra. As can be seen from the figure, there is only absorption for low pumping levels (i.e. chemical potential), but this absorption occurs in a limited spectral range. Optical gain develops with increasing chemical potential, starting from the low-energy side of the spectrum. Finally, dots of all sizes are inverted so that only gain is observed, but no absorption anymore.

The line-shape pattern produced by this model is quite similar to that observed in our experiments. In fact, the model outlined above was used to fit the gain spectra in Fig. 4. We find excellent agreement between the model and the experiments. Moreover, it is most important that this model naturally accounts more the absence of absorption at intermediate energies in Fig. 6. Finally, even the multitude of lines in Fig. 7 finds its natural explanation by a resonant selective excitation of quantum dots of some particular size.

The microscopic origin of these quantum-dot-like structures is not completely clear to date. For GaInN/GaN structures, there are two obvious mechanisms. Firstly, a Stranski-

Krastanow-like growth [30] may take place, where dots are formed due to elastic strain. Secondly, there is some evidence for a miscibility gap in the ternary GaInN [31], which might promote the formation of In-rich and In-poor regions during epitaxial growth. For GaN/AlGaN structures, the microscopic mechanism is less obvious. Since the active layer is a binary compound in this case, only a Stranski-Krastanow-like growth mode might explain the observations.

CONSEQUENCES FOR LASER OPERATION

The nitride based laser structures presented up to now [4, 5, 6] exhibit room-temperature threshold current densities in excess of 3 kA/cm^2. If band-to-band transitions were responsible for the optical gain, this might already be close to the theoretical limit in this material system [32], due to the very large effective masses. However, if a quantum dot-like behavior of nitride structures is involved in the gain mechanism, it is essential to learn to optimize those structures and to reduce size fluctuations in order to optimize the laser properties. If so, the nitride based laser might become the first quantum dot laser to be commercialized.

ACKNOWLEDGEMENTS

Financial support of this work by the Deutsche Forschungsgemeinschaft is gratefully acknowledged.

REFERENCES

1. S. Nakamura, T. Mukai, and M. Senoh, Appl. Phys. Lett. **64**, 1687 (1994).

2. S. Nakamura, M. Senoh, N. Iwasa, and S.-I. Nagahama, Appl. Phys. Lett. **67**, 1868 (1995).

3. S. Nakamura, M. Senoh, N. Iwasa, S.-I. Nagahama, T. Yamada, and T. Mukai, Jpn. J. Appl. Phys. **34**, L1332 (1995).

4. S. Nakamura, M. Senoh, S.-I. Nagahama, N. Iwasa, T. Yamada, T. Matsushita, H. Kiyoku, and Y. Sugimoto, Jpn. J. Appl. Phys. **35**, L74 (1996).

5. S. Nakamura, M. Senoh, S.-I. Nagahama, N. Iwasa, T. Yamada, T. Matsushita, H. Kiyoku, and Y. Sugimoto, Jpn. J. Appl. Phys. **35**, L217 (1996).

6. S. Nakamura, M. Senoh, S. Nagahama, N. Iwasa, T. Yamada, T. Matsushita, H. Kiyoku, and Y. Sugimoto, Appl. Phys. Lett. **68**, 2105 (1996).

7. A. Hangleiter, Phys. Rev. B. **48**, 9146 (1993).

8. A. Beattie and P. Landsberg, Proc. Royal Soc. A **249**, 16 (1958).

9. W. Shockley and W. Read, Phys. Rev. **87**, 835 (1952).

10. R. Hall, Phys. Rev. **87**, 387 (1952).

11. C. Henry and D. Lang, Phys. Rev. B **15**, 989 (1977).

12. M. G. A. Bernard and G. Duraffourg, phys. stat. sol. **1**, 699 (1961).

13. A. Moritz and A. Hangleiter, Appl. Phys. Lett. **66**, 3340 (1995).

14. J. Ding, H. Leon, T. Ishihara, M. Hagerott, A. V. Nurmikko, H. Luo, N. Samarth, and J. Furdyna, Phys. Rev. Lett. **69**, 1707 (1992).

15. R. Cingolani, R. Rinaldi, L. Calcagnile, P. Prete, P. Sciacovelle, L. Tapfer, L. Vanzetti, G. Mula, F. Bassani, L. Sorba, and A. Franciosi, Phys. Rev. B **49**, 16769 (1994).

16. M. Lowisch, F. Kreller, J. Puls, and F. Henneberger, phys. stat. sol. (b) **188**, 165 (1995).

17. J. Gutowsky, A. Diessel, U. Neukirch, D. Weckendrup, T. Behr, B. Jobst, and D. Hommel, phys. stat. sol. (b) **187**, 423 (1995).

18. Y. Arakawa and H. Sakaki, Appl. Phys. Lett. **40**, 939 (1982).

19. Y. Z. Hu, H. Gießen, N. Peyghambarian, and S. W. Koch, Phys. Rev. B **55**, 4814 (1996).

20. A. Moritz, R. Wirth, A. Hangleiter, A. Kurtenbach, and K. Eberl, Appl. Phys. Lett. **69**, 212 (1996).

21. A. Moritz, R. Wirth, A. Hangleiter, A. Kurtenbach, and K. Eberl, in *Proceedings of the 23rd International Conference on the Physics of Semiconductors*, edited by M. Scheffler and R. Zimmermann (World Scientific, London, 1996), p. 1867.

22. P. S. Cross and W. G. Oldham, IEEE J. Quantum Electron. **QE-11**, 190 (1975).

23. G. Lasher and F. Stern, Phys. Rev. **133**, A553 (1964).

24. J. S. Im, A. Moritz, F. Steuber, V. Härle, F. Scholz, and A. Hangleiter, Appl. Phys. Lett. **70**, February 3, 1997.

25. J. S. Im, V. Härle, F. Scholz, and A. Hangleiter, MRS Internet J. Nitride Semicond. Res. **1**, 37 (1996).

26. G. Frankowsky, V. Härle, F. Scholz, and A. Hangleiter, Appl. Phys. Lett. **68**, 3746 (1996).

27. S. Nakamura, in *Proceedings of the 23rd International Conference on The Physics of Semiconductors*, edited by M. Scheffler and R. Zimmermann (World Scientific, Singapore, 1996), p. 11.

28. F. Scholz, V. Härle, F. Steuber, A. Sohmer, H. Bolay, V. Syganow, A. Dörnen, J. S. Im, A. Hangleiter, J. Y. Duboz, P. Galtier, E. Rosencher, O. Ambacher, D. Brunner, and H. Lakner, Metalorganic vapor phase epitaxial growth of GaInN/GaN heterostructures and quantum wells, these proceedings.

29. A. Kurtenbach, W. W. Rühle, and K. Eberl, Solid State Commun. **96**, 265 (1995).

30. L. Goldstein, F. Glas, J. Y. Marzin, M. N. Charasse, and G. Leroux, Appl. Phys. Lett. **47**, 1099 (1985).

31. I. hsiu Ho and G. B. Stringfellow, Appl. Phys. Lett. **69**, 2701 (1996).

32. A. Hangleiter, G. Frankowsky, V. Härle, F. Steuber, and F. Scholz, in *Technical Digest of the 15th IEEE International Semiconductor Laser Conference* (IEEE, Piscataway, 1996), p. 151.

RECOMBINATION OF LOCALIZED EXCITONS IN InGaN SINGLE- AND MULTIQUANTUM WELL STRUCTURES

S. CHICHIBU*, T. AZUHATA**, T. SOTA**, S. NAKAMURA***
*Faculty of Science and Technology, Science University of Tokyo, 2641 Yamazaki, Noda, Chiba 278, Japan
**Department of Electrical, Electronics, and Computer Engineering, Waseda University, 3-4-1 Ohkubo, Shinjuku, Tokyo 169, Japan
***Department of Research and Development, Nichia Chemical Industries Ltd., 491 Oka, Kaminaka, Anan, Tokushima 774, Japan

ABSTRACT

Spontaneous emission mechanisms of InGaN single quantum well (SQW) blue and green light emitting diodes (LEDs) and multiquantum well (MQW) laser diode (LD) structures were investigated. Their static electroluminescence (EL) peak was assigned to the recombination of excitons localized at certain potential minima in the quantum well (QW). The transmission electron micrographs (TEM) indicated fluctuation of In molar fraction in the QWs. The blueshift of the EL peak caused by the increase of the driving current was explained by combined effects of the quantum-confinement Stark effect and band filling of the localized states by excitons.

INTRODUCTION

GaN-based nitrides [1] are attracting much attention as suitable materials for optoelectronic applications in the visible and ultraviolet energy regions. The recent rapid progress of the research on nitrides has realized superbright blue and green InGaN SQW LEDs [2] and room-temperature (RT) pulsed [3-5] and cw [6] oscillation of InGaN MQW [3,5,6] and SQW [4] LDs. It is necessary to investigate the emission mechanisms of InGaN-based QW structures for further improvement of the device performances. Since 3D wurzite GaN exhibits an excitonic photoluminescence (PL) peak even at RT [7], it is also important to investigate the contribution of excitons on the EL from InGaN QW structures.

This report presents the experimental results on optical and structural properties of the InGaN SQW and MQW devices. We investigated mainly their spontaneous emission mechanisms as a first step to clarify the lasing mechanisms of InGaN MQW LDs. The RT spontaneous emission from them are assigned to the recombination of localized excitons in the QWs.

EXPERIMENT

Samples used in this study were grown on a sapphire (0001) substrate by metalorganic vapor phase epitaxy [8]. The SQW LEDs have a 3-nm-thick undoped $In_xGa_{1-x}N$ QW, where x is 0.45 and 0.3 for green (510 nm) and blue (450 nm) LEDs, respectively. The MQW LD structure has ten periods of 2.5-nm-thick $In_{0.2}Ga_{0.8}N$ wells and 7.5-nm-thick $In_{0.05}Ga_{0.95}N$ barriers. Its lasing wavelength is 410 nm , and the spontaneous EL peak is 399 nm (3.11 eV) at RT. We call this device hereafter the MQW LED. The structure of these devices is described in Refs. [2,3,6]. EL and photovoltage (PV) spectra were measured using the LED devices. PL and modulated-electroabsorption (EA) spectra were measured using the device wafers between 10 K and RT. For comparison, we also characterized GaN epilayers alone by the photoreflectance (PR) measurements. Both the PR and EA spectra were analyzed using the Lorentzian lineshape functional form [9].

RESULTS AND DISCUSSION

Excitonic Structures in GaN

As a background of this work, we would like to emphasize here that GaN-based widegap nitrides exhibit excitonic features in optical spectra even at RT. In Fig.1, we summarize the RT optical absorption (OA), PR, and PL spectra of GaN epilayer. For the PL excitation and PR modulation, the 325.0 nm line of a cw He-Cd

Mat. Res. Soc. Symp. Proc. Vol. 449 © 1997 Materials Research Society

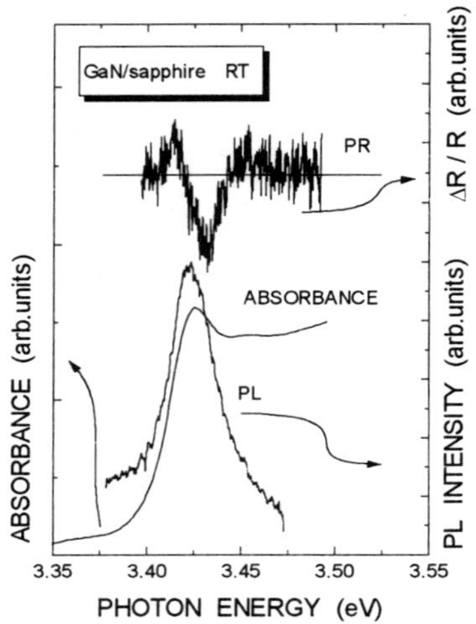

Fig.1 Optical spectra of h-GaN at RT

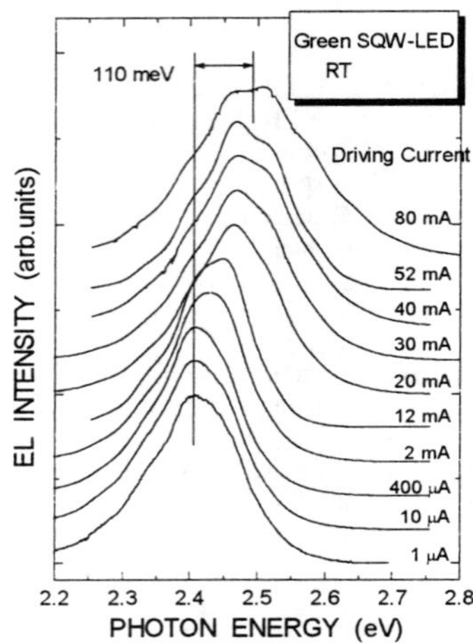

Fig.2 EL spectra of InGaN green SQW LED as a funcrtion of curent

laser was used. It is shown that clear excitonic resonance is found in both OA and PR spectra. Indeed, the structures consist of free excitonic (FE) resonances related to A and B transitions [FE(A) and FE(B), respectively]. The PL peak is also a convolution of FE(A) and FE(B) emissions [7]. It is natural for 3D GaN to exhibit FE emissions even at RT since the binding energy of excitons (E_{ex}) is 26 meV [7,10], the dielectric constant ε is 8.2 [10], and the exciton Bohr radius (R_B) is 3.4 nm [10], leading to the charge density to screen free excitons of about 1×10^{18} cm^{-3} according to Debye-Huckel screening.

EL Property of InGaN SQW LED and Confined Level Energy in InGaN QWs

The EL peak of the green SQW LED shifts to higher energy by 110 meV with increasing the driving current from 1 μ A to 80 mA, as shown in Fig.2. A similar blueshift is found in both the blue SQW and MQW LEDs. The emission intensity increases approximately linearly by increasing the driving current. Similar results are observed for PL spectra of all the structures. The emission intensity is nearly constant from 10 K to RT. Note that the shoulder-like distinct fringes in the EL spectra are due to internal multiple reflections at the surface and n-GaN/sapphire interface.

To investigate the origins of these characteristic emissions from InGaN QWs, we first study the confined energy levels in the QWs. For this purpose we

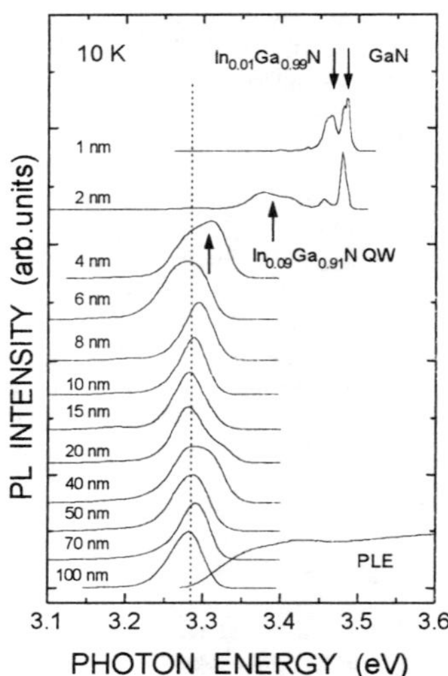

Fig.3 PL spectra of InGaN QWs at 10 K

prepared a series of undoped $In_xGa_{1-x}N$ top layers with various x and thicknesses on a n-GaN:Si layer. Figure 3 shows the PL spectra at 10 K of $In_{0.09}Ga_{0.91}N$ layers with various thicknesses. The blueshift of the confined level becomes significant for wells smaller than 4 nm, which agrees with the calculated result; in the calculation, the confined energy level of GaN wells with $Al_{0.1}Ga_{0.9}N$ barriers is estimated. Thus the emission from the QWs is confirmed to originate from the confined levels.

In Fig.3, the photoluminescence excitation (PLE) spectrum of the 100-nm-thick 3D $In_{0.09}Ga_{0.91}N$ layer is also shown. Comparison of PL and PLE spectra clearly indicates a large Stokes shift up to 80 meV; i.e., the PL peak is located in the lower energy tail of the broadened excitonic absorption band.

<u>Optical Spectra of InGaN QWs</u>

Figure 4 summarizes the spontaneous EL, PV, and EA spectra of SQW and MQW LEDs at RT. In contrast to the EL and PV spectra, the EA spectra do not exhibit reflection fringes since the signal was obtained as Δ I/I, where I is the dc intensity of the transmitted light and Δ I is the change in I due to external electromodulation. The EA measurements monitors FE resonance rather than band-to-band transition even at RT provided that the excitons are stable.

Since the QWs are very thin (3 nm) and both n-GaN and p-$Al_{0.2}Ga_{0.8}N$ barriers are highly-doped, a strong electric field exists across the QW plane. The field strength is as high as 8.5×10^5 V/cm, which is estimated from the values of electron and hole concentrations in n- and p-type barriers (5×10^{18} cm^{-3} and 1×10^{18} cm^{-3}, respectively). The field strength to dissociate excitons in 3D GaN is estimated to be 7.6×10^4 V/cm using the values of E_{ex} and ε. It is known that excitons can survive up to RT in QWs because of the increase of E_{ex} due to the confinement of the wavefunctions. We have calculated E_{ex} in 3-nm-thick GaN QWs by a variational approach [11]. The value obtained is about 1.8 times larger than that in 3D case. Thus the electric field to dissociate free excitons in the QWs is estimated to be as high as 6.0×10^5 V/cm. The EA spectra shown in Fig.4 were measured using a rectanglar modulation bias of -2V to +1.95V (the field strength was 1.2 $\times 10^6$ and 1.1×10^5 V/cm, respectively) in order not to exceed the critical field strength. Therefore the structures observed in the EA spectra are due to FE resonances in the QWs.

PV spectra were taken using monochromated light, and the open-circuit voltage was measured spectroscopically. Each PV peak at 3.21, 2.91, and 2.93 eV for x=0.2, 0.3, and 0.45, respectively, corresponds to FE absorption in the QW since the energies agree with those in the EA spectra. The result that the PV spectra exhibit a peak-like line shape also supports above assignment. The PV peak energy decreases from 3.21 to 2.91 eV with increasing x from 0.2 to 0.3. However, the peak energy is nearly unchanged for x=0.3 and x=0.45. A remarkable difference between the two spectra is that the full width at half maximum (FWHM) for x=0.45 is larger than that for x=0.3. This result implies that InGaN does not form perfect alloys, but has a potential fluctuation due to the compositional inhomogeneity of In. Such a compositional

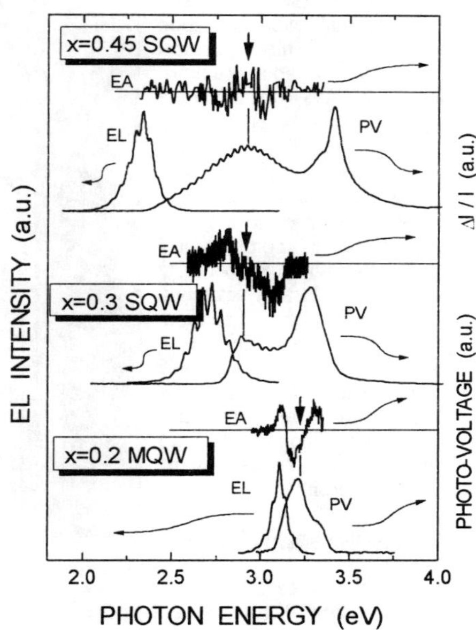

Fig.4 EL,PV, and EA spectra of InGaN QWs measured at RT

tailing in the QW plane can produce certain two-dimensional potential minima.

The EL peak energy is smaller by 100, 215, and 570 meV than the FE energy for QWs with x=0.2, 0.3, and 0.45, respectively, as shown in Fig.4. This behavior was observed in all of the EL measurements; i.e., the EL peak always appeared in the lower energy tail of the broad excitonic resonance. The decay time, τ, of these EL emissions is typically 2-4 ns. This may come from the fact that it takes a longer time for excitons to localize into the potential minima through energy and spacial relaxation, since the peak energy shifted to lower energy with the time after excitation. Note that the FE emission lifetime in GaN is reported to be very short (<100 ps) [12]. From these considerations, the EL emissions observed are considered to be the result of recombinations of localized excitons in the QWs. The carrier density in the SQW for dc operation is estimated to be 4.6×10^{18} cm^{-3} (I=80 mA, τ=3 ns). This value is larger than the charge density to screen FEs in 3D GaN. Thus the observation of the excitonic emission seems to be possible due to a reduction of the long-range Coulomb screening effect in the potential minima. Here we should mention that 3D InGaN also exhibits localized emissions, as shown in the bottom traces of Fig.3. The improvement of the emission intensity in QWs compared to the 3D case [13] may be attributable to the increased E$_{ex}$ and oscillator strength of localized excitons in the QWs.

TEM Observations

A cross-sectional transmission electron micrograph (TEM) of the MQW structure (10 periods) is shown in Fig.5. A number of distinct dark spots is found especially in the In$_{0.2}$Ga$_{0.8}$N wells, and each structure is about 4~5 nm in lateral size. A remarkable lattice distortion is also recognized around each dark spot. The structure represents a potential fluctuation of the compositional disorder in the QWs, which can act as a quantum dot or a quantum mesodot if the potential gap is large enough to confine particles laterally. Recently Kisielowski and Liliental-Weber [14] have observed a similar dot-like nanoscale compositional disorder in the same SQW LED wafers using the electron scattering potential mapping method. Narukawa et al. also reported a compositional inhomogeneity in the 6 periods of the same MQW LD wafer [15]. Therefore the structures observed are considered to correspond to the potential minima where excitons localize.

Quantum Confinement Stark Effect

Figure 6 shows the RT-PL spectra of an In$_{0.45}$Ga$_{0.55}$N SQW structure as a function of external bias. The PL was excited by the 457.9 nm (2.71 eV) line of a cw Ar$^+$ laser (50 mW), which

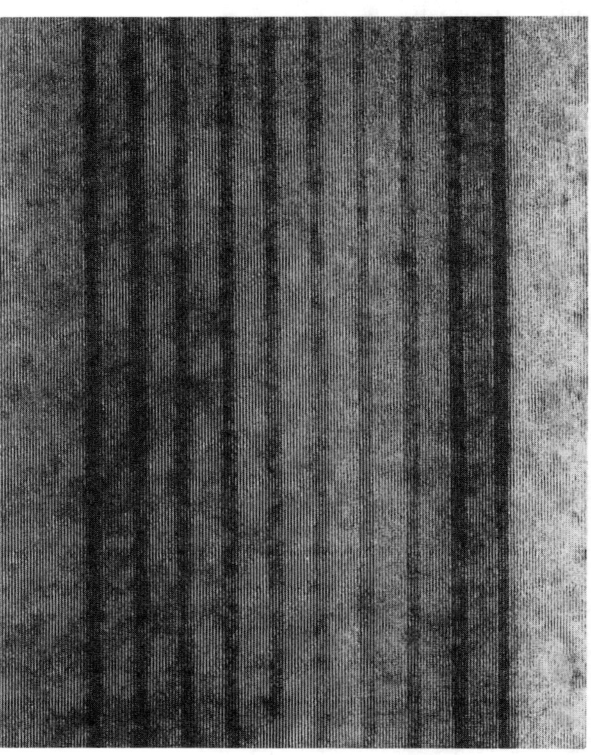

Fig.5 Cross-sectional TEM of the ten periods of InGaN/InGaN MQW LD wafer. The MQW consists of 2.5-nm-thick In$_{0.2}$Ga$_{0.8}$N wells and 7.5-nm-thick In$_{0.05}$Ga$_{0.95}$N barriers

excites carriers only in the SQW. Though the PL spectra involve the interfering fringes, the peak intensity decreases with increasing reverse bias. The PL spectrum for V=+1.991 V corresponds to that taken under open-circuit condition. By applying -2 V reverse bias, the PL intensity decreases to one-third of that for the +2 V bias. The electric field strength is 5.7×10^4 V/cm and 1.2×10^6 V/cm for +2 V and -2 V, respectively. The emission vanishes for a reverse bias of -10 V. The corresponding field strength (2.1×10^6 V/cm) is three times larger than the critical field strength for FEs in the QW. Therefore the quenching of the PL peak is considered to be due to the quantum-confinement Stark effect. Note that the emission is due to the recombination of localized excitons, and it still can be observed for a field ($\sim 1.2 \times 10^6$ V/cm) higher than the critical value.

Next we must consider effects of piezoelectric fields in the (0001)-oriented strained QWs [16]. The piezoelectric field strength is 4×10^6 V/cm for the case of strain induced by 1% mismatch between the barrier and the well, leading to the Stark shift of 274 meV in 3-nm-thick QW. On the other hand, the Stark shift is about 27 meV for the field of 1×10^6 V/cm [17]. Neither estimated value can explain the experimental results. Therefore the shift in the PL peak energy shown in Fig.6 may be due to combined effects of the quantum-confinement Stark shift, screening of the piezoelectric field, and band-filling of the localized states by excitons.

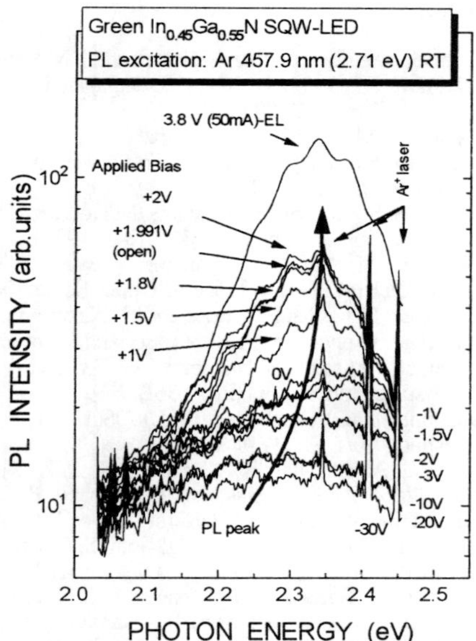

Fig.6 PL spectra of a 3-nm-thick In$_{0.45}$Ga$_{0.55}$N SQW structure as a function of external bias measured at RT

CONCLUSION

The emission mechanisms of InGaN SQW and MQW structures were investigated. The spontaneous EL from InGaN QWs was assigned to the recombination of excitons localized at certain potential minima in the QW plane. The InGaN QWs were shown to have large Stokes shift. The blueshift of the EL caused by an increase of the driving current was explained by combined effects of the quantum-confinement Stark effect, screening of the piezoelectric field, and band filling of the localized states by excitons.

It has been argued that the LD was lasing with the electron-hole plasma mechanism since the threshold carrier density for the cw operation was as high as 2×10^{20} cm^{-3} [3,6]. The difference between the energy gap of the In$_{0.05}$Ga$_{0.95}$N barriers and the EL peak energy in the In$_{0.2}$Ga$_{0.8}$N QWs was as large as 190 meV. Therefore, there is a possibility of the localized exciton related lasing of the InGaN LDs if we could produce appropriate size and density of their quantum dots [18].

ACKNOWLEDGMENTS

The authors are grateful to Dr. M. Sugawara and Professor K. Suzuki for stimulating discussions. The authors wish to thank T. Mitani for TEM observations and T. Shioda for help in the experiments. The authors would like to thank Dr. S. Shirakata for the instructions of PR measurements. They are also grateful to Dr. H. Okumura and Dr. S. Yoshida for stimulating discussions. Professors H. Nakanishi and H. Ikoma are acknowledged for continuous encouragements. This work was supported in part by a Sasagawa Scientific research grant from the Japan Science Society.

REFERENCES

1. For a review see, for example, S. Strite and H. Morkoc, J. Vac. Sci. Technol. **B10**,1237(1992).
2. S. Nakamura, M. Senoh, N. Iwasa, S. Nagahama, T. Yamada and T. Mukai, Jpn. J. Appl. Phys. **34**,L1332(1995).
3. S. Nakamura, M. Senoh, S. Nagahama, N. Iwasa, T. Yamada, T. Matsushita, H. Kiyoku and Y. Sugimoto, Jpn. J. Appl. Phys. **35**,L74(1996), Jpn. J. Appl. Phys. **35**,L217(1996), Appl. Phys. Lett. **68**,3269(1996).
4. I. Akasaki, S. Sota, H.Sakai, T.Tanaka, M.Koike and H. Amano, Electron. Lett. **32**,1105(1996).
5. K. Itaya *et al.*, Jpn. J. Appl. Phys. **35**,L1315(1996).
6. S. Nakamura, M. Senoh, S. Nagahama, N. Iwasa, T. Yamada, T. Matsushita, Y. Sugimoto and H. Kiyoku, IEEE Lasers and Electro-Optics Society Meeting (LEOS'96), **PD1.1**, Nov.20,1996.
7. S. Chichibu, T. Azuhata, T. Sota and S. Nakamura, J. Appl. Phys. **79**,2784(1996), Proceedings of the International Symposium on Blue Laser and Light Emitting Diodes (Ohmsha, Tokyo, 1996), pp.202, S. Chichibu, H. Okumura, S. Nakamura, G. Feuillet, T. Azuhata, T. Sota and S. Yoshida, to be published in Mar.15 issue of Jpn. J. Appl. Phys. **36B**(1997).
8. S. Nakamura, Jpn. J. Appl. Phys. **30**,L1705(1991).
9. D. E. Aspnes, Surf. Sci. **37**,418(1973).
10. S. Chichibu, A. Shikanai, T. Azuhata, T. Sota, A. Kuramata, K. Horino, and S. Nakamura, Appl. Phys. Lett. **68**,3766(1996), A. Shikanai, T. Azuhata, T. Sota, S. Chichibu, A. Kuramata, K. Horino, and S. Nakamura, [to be published in Jan. 1st issue of J. Appl. Phys. (1997)].
11. H. Tsutsui, T. Igarashi, T. Azuhata, T. Sota, S. Chichibu, and S. Nakamura, (submitted to   Phys. Rev. B; unpublished). The E_{ex} value in GaN/Al$_{0.1}$Ga$_{0.9}$N QW was calculated by the   variational method according to G. Bastard, E. E. Mendez, L. L. Chang, and L. Esaki [Phys. Rev.   B 26,1974(1982)]. In the calculation, we started from the Hamiltonian suitable to QWs including   mass and dielectric constant anisotropy, assuming an infinite barrier height to simplify the   calculation. After variable transformations for Z coordinates (perpendicular to the QW plane) which formally remove the mass anisotropy in the Hamiltonian, E_{ex} was calculated using the trial function for the excitonic relative motion given by $\exp\{-[\rho^{2}+(Z_e\text{-}Z_h)^2]^{1/2}/\lambda\}$, considering the fact that the 3D exciton Bohr radius is as small as 3.4 nm. Here λ is the variational parameter, ρ is the absolute value of the relative position of electron and hole in the QW plane, and $Z_e(Z_h)$ is the transformed Z coordinate of the electron (hole).
12. C. I. Harris, B. Monemar, H. Amano, and I. Akasaki, Appl. Phys. Lett. **67**,840(1995).
13. H. Amano and I. Akasaki, Extended Abstracts of Int.Conf. Solid State Mater. & Dev. (1995).
14. C. Kisielowski and Z. Liliental-Weber (private communication 1996).
15. Y. Narukawa, Y. Kawakami, Sz. Fujita, Sg. Fujita, and S. Nakamura, Phys. Rev. B rapid communication (to be published).
16. D. L.Smith and C. Mailhiot, Phys. Rev. Lett. **58**,1264(1987).
17. The piezoelectric field was calculated with the values of piezoelectric constants of GaN (see Ref.1) according to M. P. Halsall, J. E. Nicholls, J. J. Davies, B. Cockayne, and P. J. Wright [J. Appl. Phys. 71,907(1992)], and the Stark shift due to the electric field was calculated by the variational method neglecting exciton binding energy. Calculations are based on works by D. A. Miller, D. S. Chemla, T. C. Damen, A. C. Gross, W. Wiegmann, T. H. Wood, and C. A. Burrus [Phys. Rev. Lett. **53**,2173(1981) and Phys. Rev. B **32**,1043(1985)].
18. M. Sugawara, Jpn. J. Appl. Phys. **35**,124(1996).

GAIN SPECTRA AND STIMULATED EMISSION IN EPITAXIAL (IN,AL) GAN THIN FILMS

D. WIESMANN *, I. BRENER *, L. N. PFEIFFER *, M. A. KHAN **, C. J. SUN **, C. S. CHANG ***, W. FANG***, AND S. L. CHUANG ***

*Bell Laboratories, Lucent Technologies, Murray Hill, 700 Mountain Ave., NJ 07974, igal@bell-labs.com
**APA Optics Inc., 2950 N. E. 84th Lane, Blaine, MN 55449
***University of Illinois at Urbana-Champaign, Department of Electrical & Computer Engineering,1406 W. Green Street, Urbana, IL 61801

ABSTRACT

We measured the emission of (In,Al) GaN films under high intensity optical pumping both in the direction parallel and perpendicular to the film growth. In the edge emission geometry we determine the gain magnitude from the variable stripe length (VSL) method. We use the spontaneous emission collected perpendicular to the layer plane to calculate the spectral dependence of the gain. Theoretical calculations are in good agreement with those experimentally determined gain spectra. We also show that the observation of a stimulated emission peak perpendicular to the film is predominantly due to scattering of the in-plane stimulated emission but without ruling out contributions from microstructures in the films.

INTRODUCTION

GaN and related alloys have drawn a lot of attention for their possible application in blue and near UV light emitting devices. Recent progress has indeed led to the realization of a semiconductor laser structure.[1] High optical excitation emission measurements are widely used to test the quality of III-Nitride samples. These spectra are collected either from the edge of the sample or perpendicular to the layer plane and they consist of a broad spontaneous emission peak and a second narrow peak at slightly lower energies. The narrow peak is usually observed after a certain threshold power density and is due to stimulated emission that can be well-understood as single pass amplification of the spontaneous emission for the edge emission geometry.[2] On the other hand, several mechanisms have been proposed in order to explain the surface emitted stimulated emission.[3,4] In this paper we present calibrated gain spectra for different nitride alloy films and show that the observation of a stimulated emission peak in the nitride films in a direction perpendicular to the film growth is predominantly due to scattering of the in-plane stimulated emission or stimulated emission from some microstructure in the films.

EXPERIMENTAL

An MOCVD system was used to grow the GaN, $Al_{0.05}Ga_{0.95}N$ and $In_xGa_{1-x}N$ samples on basal plane sapphire.[5,6] The GaN and the $Al_xGa_{1-x}N$ layers have a thickness of ~1 μm and are typically n-type with electron concentrations of the order of 10^{16} cm^{-3} and of 10^{18} cm^{-3}, respectively. The $In_xGa_{1-x}N$ layers (0.13<x<0.18) are a few tenths of a micrometer thick and have higher electron concentrations than the other samples. We also grew by MBE a reference sample in another III-V system, namely GaAs/AlGaAs. A stack of alternating thin GaAs and AlAs layers was grown on a GaAs substrate forming an effective AlGaAs alloy with 70% Aluminum. On top of the 0.44 μm thick stack an 8 μm thick GaAs layer was deposited. This sample was used to check the different hypothesis for vertical cavity stimulated emission in a known system with excellent interface quality. This sample was chosen to mimic the environment of the nitride layers used in

Mat. Res. Soc. Symp. Proc. Vol. 449 © 1997 Materials Research Society

these experiments, namely, a thin active layer grown on a lower refractive index substrate but much less scattering due to interface quality or dislocations.

The high excitation density measurements were performed at room temperature with a pulsed Nitrogen laser with an emission wavelength of 337 nm and a pulse width of 2.5 ns. The maximum achievable power density was about 10 MW/cm^2. The exciting light was either focused to a 0.4 mm^2 spot for collecting the front emission or into a narrow stripe with varaible length for the edge emission experiments. The emitted light was dispersed by a 0.27 m spectrometer and detected with a UV-enhanced CCD array. We take advantage of the imaging capabilities of this detector in the edge emission experiments in order to position the sample perpendicular to the collection optics and reject any scattering light from other areas of the sample.

RESULTS AND DISCUSSION

We first present the results for the edge emission measurements where we used the VSL method of Shaklee [2] to determine the gain. Fig. 1(a)-(c) show the dependence of the stimulated emission peak intensity on the excitation stripe length for a single detection wavelength and an excitation power density of 0.8 MW/cm^2. For small excitation lengths the intensity depends exponentially on the stripe length. For longer excitation lengths the increase in intensity saturates strongly. The critical length at which the saturation becomes dominant depends both on the sample and on the wavelength position. This saturation is shown to be due to a carrier depopulation by the stimulated emission. The resulting dependence of the chemical potential on the excitation stripe length is also the reason for the observed red shift of the stimulated emission peak with length.[7,8] The gain is determined from the exponential section of these curves and similar ones for other wavelengths as described in [2].The same has been done for the AlGaN and the InGaN samples. Panes (d), (e) and (f) show the gain as a function of wavelength.

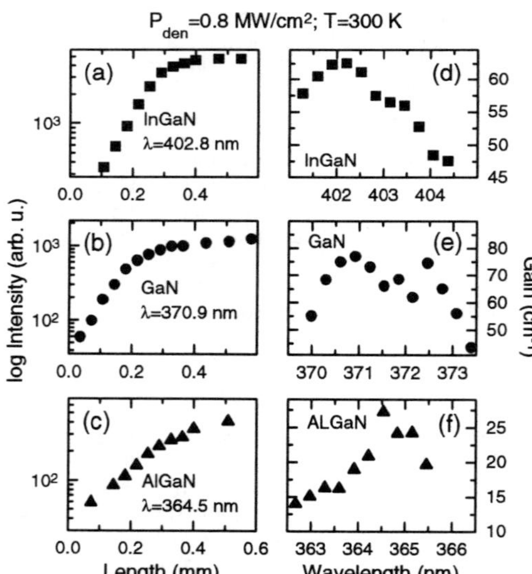

Figure 1: Results of the variable stripe length measurements (excitation density 0.8 MW/cm^2, T=300K). (a), (b), (c) Stimulated emission intensity as a function of stripe length for a single wavelength for InGaN, GaN, and AlGaN, respectively (logarithmic scale). The gain is determined from the exponentiell section. (d), (e), (f) Corresponding gain spectra for InGaN, GaN, and AlGaN.

Due to the limited signal to noise (S/N) ratio we can obtain the gain spectra only for a narrow spectral span. We determined gain values of the order of a few tens cm^{-1}. From all three samples AlGaN shows the lowest gain. Based on the gain values which have been measured lately in (In,Al) GaN quantum wells [9,10] and taking into account that the confinement factor in our experiments is virtually one we expected to find higher gain values. We explain the difference by the higher electron hole plasma density in quantum wells and the worse crystal quality especially close to the sample surface of our samples.[11]

The VSL technique cannot yield the full spectral dependence of the gain. Also nonlinear effects like the gain saturation are included in the obtained gain curves which makes comparisons with theory difficult. We can determine the full gain curves from the spontaneous emission spectra at high excitation intensities, provided that (a) the carriers are in a quasi equilibrium state and that the spontaneous emission spectra are not distorted by (b) reabsorption or (c) a stimulated emission peak. Since the carriers are excited by a 2.5 ns long laser pulse the created eh pairs can be considered to be in a steady state. The second condition is fulfilled best when collecting the emission perpendicular to the layer plane because then the light has traveled the shortest path in the sample. To account for the third limitation we only use spectra for excitation densities not exceeding 1 MW/cm^2 for GaN and 1.8 MW/cm^2 for AlGaN. Those spectra are shown in Fig. 2 (a) and (b) on a logarithmic scale for GaN and AlGaN, respectively. The absorption coefficient α and the gain coefficient g are related to the intensity of spontaneous emission I by the van-Roosbroeck-Shockley relation [12,13]

$$\alpha(h \cdot v, \Delta E_F) = C \cdot \frac{I(h \cdot v, \Delta E_F)}{(h \cdot v)^2} \cdot \left[\exp\left\{ \frac{h \cdot v - \Delta E_F}{k \cdot T} \right\} \right]$$

$$g(h \cdot v, \Delta E_F) = C \cdot \frac{I(h \cdot v, \Delta E_F)}{(h \cdot v)^2} \cdot \left[1 - \exp\left\{ \frac{h \cdot v - \Delta E_F}{k \cdot T} \right\} \right] \tag{1}$$

where C is a proportionality constant that does not depend on hv and ΔE_F is the separation of the quasi Fermi levels for holes and electrons which depends on the excitation density.

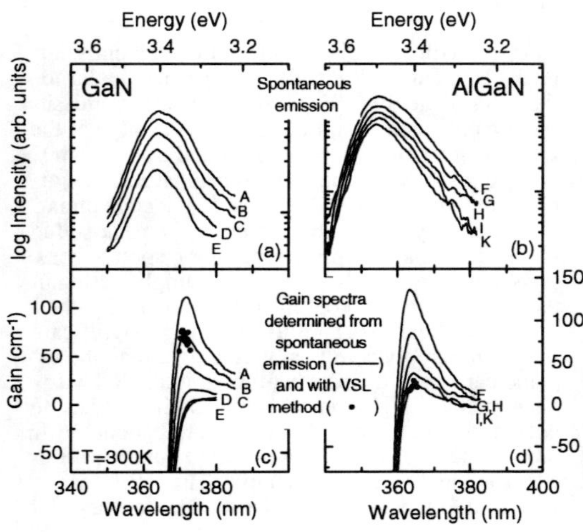

Figure 2: (a), (b) Spontaneous emission spectra collected perpendicular to the layer plane for excitation densities that preclude the observation of stimulated emission for GaN and AlGaN, respectively. The excitation densities are 1 MW/cm^2 (A), 0.8 MW/cm^2 (B), 0.6 MW/cm^2 (C), 0.4 MW/cm^2 (D), 0.3 MW/cm^2 (E) and 1.8 MW/cm^2 (F), 1.3 MW/cm^2 (G), 1.1 MW/cm^2 (H), 0.9 MW/cm^2 (I), 0.8 MW/cm^2 (K). (c) ,(d) Corresponding calculated gain spectra (solid curves) and gain spectra obtained with VSL method for 0.8 MW/cm^2 (solid dots).

For the determination of the separation of quasi Fermi levels we note that the gain spectrum must have its maximum at the energy of the stimulated emission peak. We therefore fit the maximum of the gain curve calculated from one spontaneous emission spectrum for a single excitation density to the peak energy of the stimulated emission by varying the separation of the quasi Fermi levels. Once the separation of quasi Fermi levels ΔE_{F0} for one excitation density is known, ΔE_{Fi} for other excitation densities i can be determined by

$$\frac{I(h \cdot v, \Delta E_{Fi})}{I(h \cdot v, \Delta E_{F0})} = \exp\left\{\frac{\Delta E_{Fi} - \Delta E_{F0}}{k \cdot T}\right\} \tag{2}$$

This is valid for sufficiently high emission energies where the absorption does not depend on the excitation density.[13] When this condition is fulfilled the spectra differ only by a constant factor (high energy tails of measured spectra in Fig. 2(a) and (b)). Using the separation of quasi Fermi levels we calculated the corresponding gain spectra as displayed as solid lines in Fig. 2(c) and (d) for GaN and AlGaN, respectively. This procedure cannot be applied to InGaN films due to the high extrinsic luminescence even at room temperature and high excitation densities.

We then calibrated these qualitative gain spectra with the magnitude of gain we got from the VSL measurements for an excitation density of 0.8 MW/cm^2 (solid dots in Fig. 2(c) and (d)). Whereas the gain determined with the VSL method is the effective gain (gain-losses) the gain calculated from spontaneous emission includes no loss. We assumed the losses to be in the order of 10 cm^{-1}.[10] With this value and the maximum gain value of 77 cm^{-1} and 27 cm^{-1} for GaN and AlGaN, respectively, we calibrated the calculated qualitative gain spectra for the same excitation density of 0.8 MW/cm^2. We used the same calibration factor for the gain spectra according to other excitation densities.

For sufficiently high power densities (>0.8 MW/cm^2 for GaN), the gain peak shifts to longer wavelengths which is in agreement with the red shift of the stimulated emission peak (see Fig. 3). We have not observed this shift in the gain maxima in AlGaN.

We will now discuss the stimulated emission peak observed in all nitride films perpendicular to the layer plane. Figs 3(a) and (b) show the emission spectra collected from the front and the edge of the sample under similar excitation conditions. Both sets of spectra show the same sharp peak at 3.25 eV at comparable excitation densities. In the edge emission geometry this stimulated emission peak is due to the single pass amplification of the spontaneous emission. There is also single pass amplification in the vertical direction but the ratio of optical paths is ~10^3. If vertical pass amplification were significant that would imply a vertical gain bigger than the in-plane gain by many orders of magnitude. Such a phenomena is highly improbable and in fact has not been observed in any other III-V semiconductor. In order to clarify this point we carried out the same experiments in the GaAs sample. The results are shown in Fig. 3(c) and (d): the spectra detected per-

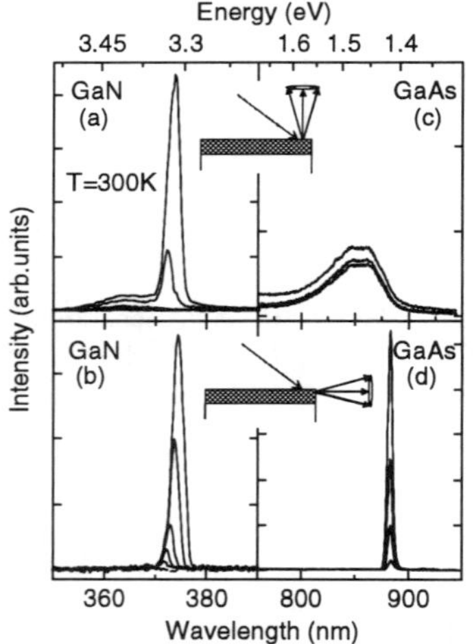

Figure 3: High excitation emission spectra (a) for GaN collected from the front surface. The excitation densities are 0.2 MW/cm^2, 0.5 MW/cm^2, 1.0 MW/cm^2, 2.2 MW/cm^2, and 3.7 MW/cm^2. The laser spot size was 0.4 mm^2, (b) for GaN collected from the edge, the excitation densities are 0.25 MW/cm^2, 0.3 MW/cm^2, 0.35 MW/cm^2, 0.5 MW/cm^2, 0.6 MW/cm^2, and 0.7 MW/cm^2. The excitation stripe length was 2 mm, (c) for GaAs collected from the surface, excitation densities comparable to (a), (d) for GaAs collected from the edge, excitation densities comparable to (b).

pendicular to the layer consist of a broad spontaneous emission band, whereas the edge emission spectra consist mainly of a sharp stimulated emission peak (FWHM of 8 meV). These results agree with a multitude of similar measurements carried out in past years on several III-V semi-conductors (see for instance [8]). We therefore conclude that the stimulated emission peak for the nitride layers is observable in the vertical direction because the crystal quality of the nitride films causes a high degree of scattering. The difference in the excitation densities for the spectra collected from the front and the edge of the GaN sample can be explained by the different in-plane amplification length. For the spectra collected from the front the amplification length is about 0.6 mm (laser spot size) whereas it is in the order of 2 mm (stripe length) for the spectra collected from the edge. One final remark is that our results do not rule out stimulated emission from some microstructure in the nitride films (grains or pillars). Recent results show that the long pillars or grains found in many nitride films show a high degree of crystallinity and lum i-nescence efficiency. [14]

THEORY

The spontaneous emission spectra were calculated theoretically and compared with the experimental data for GaN. Our theoretical model is based on [15]. The model has been used to compare the gains of GaN and GaAs [16]. Three bands, conduction, heavy-hole (HH), and light-hole (LH) bands, are included. The two valence bands are assumed to have parabolic shapes. The proper symmetry of the wurtize crystal is taken into account by including an offset energy between bandedges of HH and LH bands and the optical matrix elements [17]. The band gap renormalization (BGR) is included to account for the bandgap shrinkage due to the carrier injection. A universal formula for the bulk semiconductor is used to model the BGR. A Lorentzian line-shape is used to model the linewidth broadening. The calculated spontaneous emission spectra are then used to obtain the material gain spectra based on the principle of detailed balance [15].

The quasi-Fermi level seperation was first determined from the peak wavelengths of both the experimental spontaneous emission and gain spectra. The carrier density is then determined based on the Fermi-Dirac integral. A good agreement to the experimental data is seen in Fig. 4 although there is still an ambiguity in the determination of the exact gain magnitude.

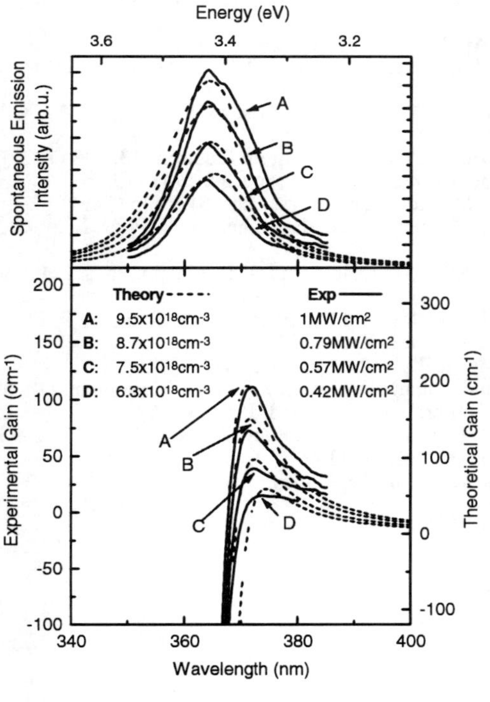

Figure 4: Theoretical fit to the experimental data of Fig. 2 for GaN. See text for details.

CONCLUSION

We have determined the magnitude of the gain for GaN, InGaN, and AlGaN for an excitation density of 0.8 MW/cm^2 with the variable stripe length method. We find the peak gain values to be in the order of a few tens per cm and lowest for AlGaN. We have calculated gain spectra from the spontaneous emission we collected perpendicular to the sample plane for GaN and AlGaN at high excitation densities. We calibrated these spectra with the results from the variable stripe length method and an assumed scattering loss value of 10 cm^{-1}. Our theoretical calculations show a reasonable agreement with the experiment. Finally we have explained the high excitation density stimulated emission peak detected perpendicular to the nitride layer plane as scattered in-plane stimulated emission but without ruling out contributions from microstructures in the films.

ACKNOWLEDGEMENTS

The work at the University of Illinois was supported by ONR under Grant N00014-96-1-0303.

REFERENCES

[1]   S. Nakamura, M. Senoh, S. Nagahama, N. Iwasa, T. Yamada, T. Matsushita, H. Kiyoku, and Y. Sugimoto, Jpn. J. Appl. Phys. 35, L74 (1996)

[2]   K. L. Shaklee, R. E. Nahory, and R.F. Leheny, J. Luminesc. 7, 284 (1973)

[3]   M. A. Khan, D. T. Olson, J. M. Van Hove, and J. N. Kuznia, Appl. Phys. Lett. 58, 1515 (1991)

[4]   S. T. Kim H. Amano, and I. Akasaki, Appl. Phys. Lett. 67, 267 (1995)

[5]   M. A. Khan, S. Krishnankutty, R. A. Skogman, J. N. Kuznia, D. T. Olson, and T. George, Appl. Phys. Lett. 65, 520 (1994)

[6]   M. A. Khan, J. N. Kuznia, J. M. Van Hove, D. T. Olson, S. Krishnankutty, and R. M. Kolbas, Appl. Phys. Lett. 58, 526 (1991)

[7]   E. O. Goebel, O. Hildebrand, and K. Löhnert, IEEE J. Quantum Electron. 13, 848 (1977)

[8]   G. Bongiovanni, J. Butty, and J.-L. Staehli, Optical Engineering 34, 1941 (1995)

[9]   S. T. Kim, H. Amano, I. Akasaki, and N. Koide, Appl. Phys. Lett. 64, 1535 (1994)

[10] G. Frankowsky, F. Steuber, V. Härle, F. Scholz, and A. Hangleiter, Appl. Phys. Lett. 68, 3746 (1996)

[11] S. Borenstain, D. Fekete, M. Vofsi, R. Sarfaty, E. Cohen, and A. Ron, App. Phys. Lett. 50, 442 (1987)

[12] H. B. Bebb and E. W. Williams 'Semiconductors and Semimetals', vol. 8, ed. by R. K. Willardson and A. C. Beer, 181 (Academic, New York 1972)

[13] C. H. Henry, R. A. Logan, and F. R. Merrit, J. Appl. Phys. 51, 3042 (1980)

[14] L.-L. Chao, G. S. Cargill III, C. Kothandaraman, D. Cyr, G. Flynn, E. S. Hellman, D. Wiesmann, D. N. E. Buchanan and I. Brener, to be published, JMR (1996)

[15] S. L. Chuang, J. O'Gorman, and A. F. J. Levi, IEEE J. Quantum Electron. 29, 1631 (1993)

[16] W. Fang and S. L. Chuang, App. Phys. Lett. 67, 751 (1995)

[17] S. L. Chuang and C. S. Chang, Phys. Rev. B, July 15, (1996).

Emission Mechanism of the InGaN MQW grown by MOCVD

Yukio Narukawa, Yoichi Kawakami, Shizuo Fujita and Shigeo Fujita

Department of Electronic Science and Engineering, Kyoto University, Kyoto 606-01,
Japan

Shuji Nakamura

Department of Research and Development, Nichia Chemical Industries Ltd, 491 Oka,
Kaminaka, Anan, Tokushima 774, Japan

ABSTRACT

Dynamical behavior of radiative recombination has been assessed in the $In_{0.20}Ga_{0.80}N$ (3nm)/ $In_{0.05}Ga_{0.95}N$ (6 nm) multiple quantum well (MQW) structure by means of transmittance (TR), electroreflectance (ER), photoluminescence excitation (PLE) and time-resolved photoluminescence (TRPL) spectroscopy. The PL at 20 K was mainly composed of two emission bands whose peaks are located at 2.920 eV and 3.155 eV. The ER and PLE revealed that the transition at 3.155 eV is due to the excitons at quantized level between n=1 conduction and n=1 A(Γ_{9v}) valence bands, while the main PL peak at 2.920 eV is attributed to the excitons localized at the trap centers within the well. The TRPL features were well understood as the effect of localization where photo-generated excitons are transferred from the n=1 band to the localized centers, and then are localized further to the tail state. The origin of the localized centers were attributed to the In-rich region in the wells acting as quantum dots which could be observed by transmission electron microscopy (TEM) and energy-dispersive X-ray microanalysis (EDX).

INTRODUCTION

GaN-based semiconductors are attracting much interest because of both the realization of incandescent blue, green and yellow light emitting diodes (LEDs),[1] and the operation of purplish blue laser diodes (LDs) at room temperature (RT) under the pulsed mode. [2, 3, 4, 5] Although the achievement of the high quantum efficiency of these devices is owing to the use of InGaN ternary alloys as the active layer, little has been known about the optical properties of InGaN quantum wells (QWs).

An important feature of these materials is the role of excitons on the emission mechanism. Since the binding energy of excitons (E_{ex}) in hexagonal GaN (h-GaN) is 28 meV which is larger than the thermal energy of RT, excitonic emissions have been observed up to RT in high-quality h-GaN epilayers.[6] Recently, it has been reported that biexciton (excitonic molecule) binding energy (E_b) in h-GaN is about 6 meV.[10, 11] It is expected that values of both E_{ex} and E_b are enhanced in quasi two-dimensional QW system.

Preliminary PL measurements at low temperature of the InGaN LD structure has shown that the linewidth of the emission is as large as about 80 meV, indicating that the energy levels formed in the QWs are broadened inhomogeneously by the disorder such as a fluctuation of well width and (or) alloy composition. This is interesting because Sugawara has predicted theoretically that excitons or biexcitons localized at deep potential minima (more than about 100 meV) contribute to the optical gain even at RT.[12] In fact, very recently, Chichibu *et al.* has proposed such a model for the lasing mechanism in the InGaN multiple QW (MQW) structure.[13, 14] However, the detailed nature for the exciton localization has not been clearly understood. Therefore, the assessment is motivated on the dynamical behavior of localized excitons in InGaN QWs.

In this paper, the mechanism of radiative recombination has been studied in the $In_{0.20}Ga_{0.80}N$ / $In_{0.05}Ga_{0.95}N$ MQW structure by employing transmittance (TR), electroreflectance (ER), photoluminescence excitation (PLE) and TRPL spectroscopy[7] and the microstructure has been investigated by means of cross sectional observation using transmission electron microscopy (TEM) and energy-dispersive X-ray microanalysis (EDX).[8]

Mat. Res. Soc. Symp. Proc. Vol. 449 © 1997 Materials Research Society

EXPERIMENTAL PROCEDURE

The sample used in this study was grown on a (0001) oriented sapphire (Al_2O_3) substrate by a two-flow metalorganic chemical vapor deposition (TF-MOCVD) technique.[9] The layer consists of the separate confinement heterostructure where the undoped $In_{0.20}Ga_{0.80}N$ (3 nm) / $In_{0.05}Ga_{0.95}N$ (6 nm) MQW with 6 periods is sandwiched between GaN waveguiding layers (0.1 μm in each) and $Al_{0.15}Ga_{0.85}N$ cladding layers (0.4 μm in each). The top of the $Al_{0.15}Ga_{0.85}N$ clad and the GaN waveguide are Mg-doped p-type layers, while the bottom of the clad and the waveguide are Si-doped n-type layers. It is noted that the LD has been operated at 420 nm from this sample under pulsed mode at RT.

In order to make assignment of emissions under low excitation, the cw PL was measured by the excitation of a He-Cd laser (325 nm) and it was compared with TR, ER and PLE spectra. PL detections were carried out using a cooled-charge coupled device (CCD) and a 50 cm monochromator with a 150 lines/mm grating. The light source used for TR, ER and PLE was obtained by passing the Xe-lamp through a 25 cm monochromator. The TR and ER signals were detected by a Si-photodiode and amplified by a lockin amplifier.

The TRPL measurement were performed with a fast scan streak camera in conjunction with a 25 cm monochromator using a 100 lines/mm grating. Pulsed excitation was provided by the frequency doubled beam of a mode-locked Al_2O_3:Ti laser which was pumped by Ar^+ laser. In order to avoid the multi-excitation, repetition rate of the source (80.0 MHz) was selected to 4.0 MHz by the acoustic optic (AO) modulator. The wavelength and the pulse width were 357 nm and 1.5 ps, respectively. The spectral resolution of all measurements were about 1 nm which is well below the linewidth of the PL. Whole measurement have been done at 20K.

The specimen for cross-sectional micro-analysis was prepared by mechanical thinning and ion-milling with Ar^+. Bright field images were observed using a TEM (JEM-2000GX) operated at 200 kV. EDX measurements were performed with a field-emission scanning TEM (STEM) (VG Microscopes-HB501) operated at 100 kV with a probe diameter of 1 nm.

RESULTS AND DISCUSSION

Fig.1 (i) shows cw PL spectrum obtained under a He-Cd laser excitation. The PL was composed of a few (or several) emission bands. Among them, two bands are clearly observed as peaks where the main peak was located at 2.920 eV, while another peak at 3.155 eV was weak in intensity. If the alloy composition of InGaN well layers is randomly distributed, alloy broadening of excitons is calculated to be about 10 meV.[15] The PL linewidth of the main peak was about 80 meV which is much larger than the value estimated above. Consequently, it follows that the energy levels within the well are distributed by the disorder such as the fluctuation of well width and (or) the separation of In composition.

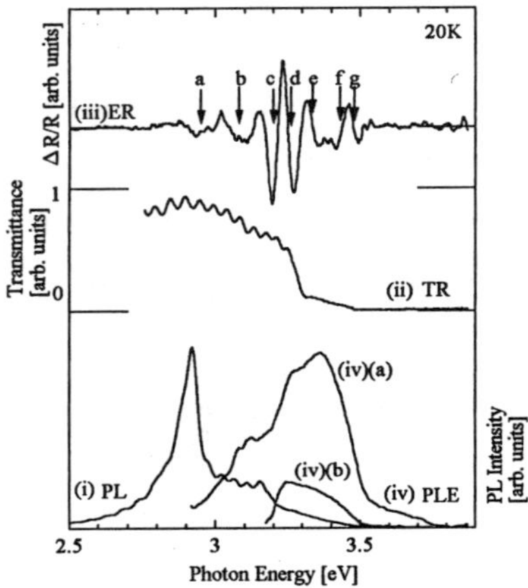

Fig.1 (i) Photoluminescence taken under a He-Cd laser excitation (4 W/cm^2). (ii) transmittance spectrum. Oscillations below about 3.25 eV are due to the interference effect. (iii) Electroreflectance spectrum. Energy positions labeled a - g are theoretical ones fitted. (iv) Photoluminescence excitation spectra monitored at (a) 2.920 eV and (b) 3.155 eV.

The TR spectrum is depicted in Fig. 1 (ii). The dips were observed at about 3.32 eV and about 3.49 eV which can be ascribed to the absorption edge of $In_{0.05}Ga_{0.95}N$ barrier and GaN waveguiding layers, respectively. Although the energy levels within the well are distributed in the range between 3.0 eV to 3.3 eV, no clear structure could be observed due to the periodic oscillations which arise from the interference multi-reflection.

In order to avoid such effects, the ER spectrum (Fig. 1 (iii)) was measured by applying the reverse bias to the *pn* junction. The optimum fitting by theoretical equation [16] was made assuming the seven energy levels labeled (a) to (g). According to the X-ray diffraction measurement, it was found that GaN layers are under compressive stress by about 0.3 % due to the difference of thermal expansion coefficient between epilayers and the substrate. Lattice mismatch between GaN and $In_{0.20}Ga_{0.80}N$ well is 2.2 %. Since the whole layer was almost coherently grown, the well layers are compressively strained by about 2.5 %. Valence bands of h-GaN based semiconductors consist of three bands labeled $A(\Gamma_{9v})$, $B(\Gamma_{7uv})$ and $C(\Gamma_{7lv})$.[17] The transition (g) located at 3.488 eV is attributed to A exciton in the GaN layers. This energy position is blue-shifted by about 10 meV compared to that in bulk GaN due to strain.[18] The B exciton, which should be observed at about 3.496 eV, could not be observed. This may be due to the limitation of spectral resolution. The transition (f), whose oscillator strength is small, is located at 3.430 eV. It is difficult to make an assignment of this transition at the moment. The transition (e) at 3.326 eV is ascribed to the A exciton in the $In_{0.05}Ga_{0.95}N$ barrier layers.

The transitions (a) to (d) are energy levels within the quantum wells. The transitions of (c) and (d) are strong in intensity, while those of (a) and (b) are much weaker and broader. If the deformation potential of InGaN is assumed to be same as that of GaN, [18] blue-shift energy of bandgaps in $In_{0.05}Ga_{0.95}N$ barriers and $In_{0.20}Ga_{0.80}N$ wells are calculated to be approximately 30 meV and 80 meV, respectively. Thus, the bandgap energies of $In_{0.05}Ga_{0.95}N$ barriers and $In_{0.20}Ga_{0.80}N$ wells are estimated to be about 3.40 eV and 3.09 eV, respectively. The energy difference of bandgaps between barriers and wells are about 310 meV. Even if this energy is equal to the conduction band offset, only n=1 quantized level is formed in the conduction band in the case of the thin well layer thickness of 3 nm. For this estimation, the electron effective mass in the $In_{0.20}Ga_{0.80}N$ wells is approximated to be $0.18m_0$ which is obtained by linear extrapolation of effective masses between GaN ($0.20m_0$) and InN ($0.11m_0$). Therefore, the allowed transitions in the well are between n=1 conduction to n=1 A, B and C valence bands. Among them, the excitonic transitions between n=1 conduction to n=1 A valence band (E_{ex1A}), and n=1 conduction to n=1 B valence band (E_{ex1B}) would be major transitions considering their oscillator strength.

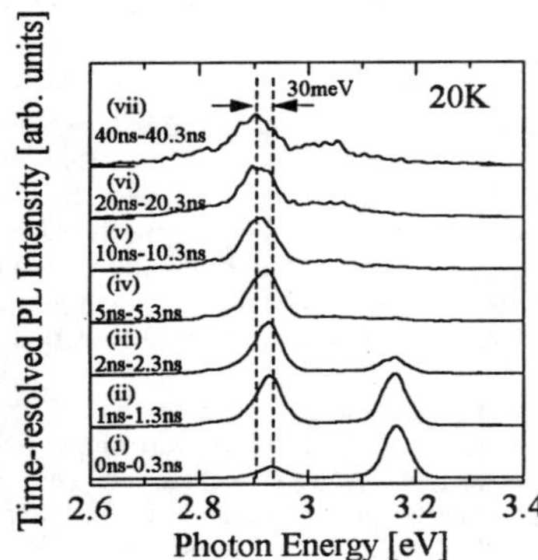

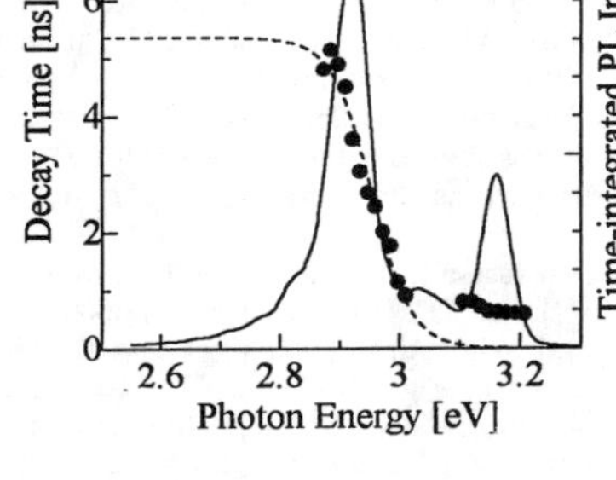

Fig.2 Time-resolved PL spectra monitored at various time-interval after pulsed excitation. Excitation energy density is 210 nJ/cm^2. Each spectra are normalized in intensity.

Fig.3 Time-integrated PL taken under 210 nJ/cm^2 as well as PL decay times monitored at various emission energies. Dotted curve is the theoretical one fitted by the equation (1).

Consequently, the transitions of (c) at 3.203 eV and (d) at 3.255 eV are most probably attributed to the E_{ex1A} and E_{ex1B}, respectively. The propriety of these assignments will be proved if unknown parameters such as deformation potentials, band offsets, effective masses of holes are clarified. It is concluded that the PL peak at 3.155 eV arises from the lowest quantized energy level (E_{ex1A}) whose absorption band corresponds to the transition (c). The transitions (a) and (b) are attributed to the localized trap centers whose density of states are lower than the quantized energy levels ((c) and (d)) because they are located at lower-energy side compared to the estimated bandgap of $In_{0.20}Ga_{0.80}N$ wells (3.09 eV). The transition (b) at 3.071 eV may be related to the weak emission component which are located between two PL peaks. The transition (a) at 2.953 eV is located at higher photon energy by 33 meV compared with the main PL peak. This energy difference corresponds to the Stokes shift of the PL.

Fig.1 (iv) shows PLE spectra monitored at 2.920 eV ((a)) and 3.155 eV ((b)). PLE spectrum shown in Fig.1 (iv) (a) is composed of a number of broad bands whose shoulder almost corresponds to the transition energies ((a) to (g)) described in Fig. 1 (iii). The main PL peak is located at the tail of the absorption edge.

Dynamics of radiative recombinations are studied by the TRPL spectroscopy. Fig.2 shows TRPL spectra monitored at various time after pulsed excitation. The whole spectra ((i) to (vii)) are integrated during the same time-interval (300 ps), and are normalized in intensity. Just after the excitation ((i) 0 ns - 0.3 ns), the spectrum is dominated by the emission band at 3.16 eV. The rise time of this emission band was about 100 ps, while the main PL band reaches maximum at about 300 ps. Moreover, emission bands at 3.16 eV quench more rapidly than the main emission bands. These features indicate that photo-generated excitons are transferred from the n=1 quantized level (E_{ex1A}) to the localized centers. The main PL peak shifts towards lower photon energy with increasing time. The energy difference of the peak between (i) (0 ns - 0.3 ns) and (vii) (40 ns - 40.3 ns) is about 30 meV. This behavior suggests that density of states of localized centers are distributed to some extent, and that localized excitons are transferred further to the lower lying energy levels.

Fig.3 shows PL decay times as a function of monitored emission energies whose range is selected to two major emission bands. Time-integrated PL is also inserted in the figure. Decay times monitored in the vicinity of 3.16 eV are almost constant at about 600 ps. Decay times in the main PL band decreases with increasing monitored photon energy. This is because the decay of localized exciton is not only due to radiative recombination but also due to the transfer process to the tail state. If the density of tail state is approximated as $\exp(-E/E_0)$, and if the radiative recombination lifetime (τ_r) does not change with emission energy, observed lifetime ($\tau(E)$) can be expressed by the following equation, [19, 20]

$$\tau(E) = \frac{\tau_r}{1 + \exp(E - E_{me})/E_0} \tag{1}$$

where E_0 represents the degree of the depth in the tail state and E_{me} is the characteristic energy which is analogous to the mobility edge. The best fit could be obtained using τ_r=5.34 ns, E_0=33.3 meV and E_{me}=2.953 eV. The energy position of E_{me} is same as the transition (a) in the ER spectrum.

ER and TRPL spectroscopy have revealed that the main PL peak is located below the lowest n=1 quantized level by about 250 meV. Moreover, the large increase in recombination time as the energy decreases across the PL band is in favor of the localized exciton model, whose trap centers originate from the disorder such as the fluctuation of the well width and/or the fluctuation of In composition within the wells.

In order to assess the origin of the localization, the structural analysis has been performed by means of cross sectional TEM. Fig.4 shows the 0002 bright field image of the LD structure taken from the [0$\bar{1}$10] crystal axis. Almost flat interfaces are observed between $Al_{0.15}Ga_{0.85}N$ clads and GaN waveguides as shown in Fig.4 (i). The $Al_{0.20}Ga_{0.80}N$ layer with the 20 nm thickness is formed above the $In_{0.20}Ga_{0.80}N/In_{0.05}Ga_{0.95}N$ MQW to avoid the diffusion of In atom to the upper layer. From this figure, the well and barrier layer thicknesses are estimated to be about 3 nm and 6 nm, respectively. However, if the detailed structure of the MQW (Fig.4 (ii)) is viewed, a lot of dark spots are observed in the wells. The diameter of these dark spots distributes from 2 nm to 5 nm and many of them are observed to have the diameter of around 3 nm. Although TEM images have been observed by rotating the specimen in the plane perpendicular to the growth direction ((0002) c-axis) in the range from [0$\bar{1}$10] to [1$\bar{2}$10] [21], no remarkable change has been observed in the shape of dark spots. Therefore, it is suggested that the dark spots originate from the isotropic dot-like structures which are self-formed in the wells.

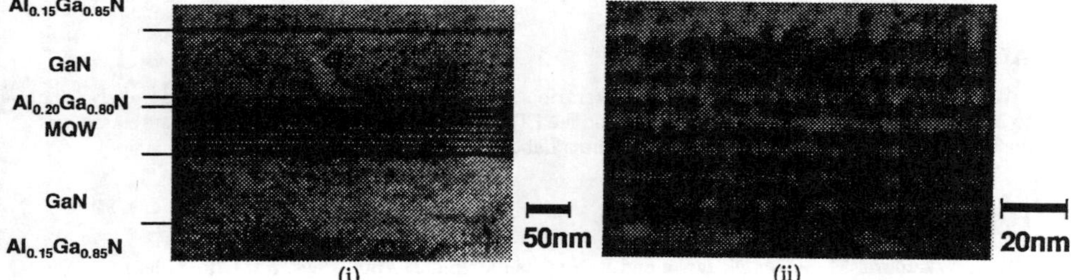

Fig.4 (i) Cross sectional TEM image of the purple LD structure consisting of Al$_{0.15}$Ga$_{0.85}$N clads, GaN waveguides and the In$_{0.20}$Ga$_{0.80}$N/In$_{0.05}$Ga$_{0.95}$N MQW. The Al$_{0.20}$Ga$_{0.80}$N layer above the MQW is made to prevent the diffusion of In to the upper layer. (ii) Magnified TEM image showing the MQW region. Dark contrast which looks like quantum dots is observed in the wells.

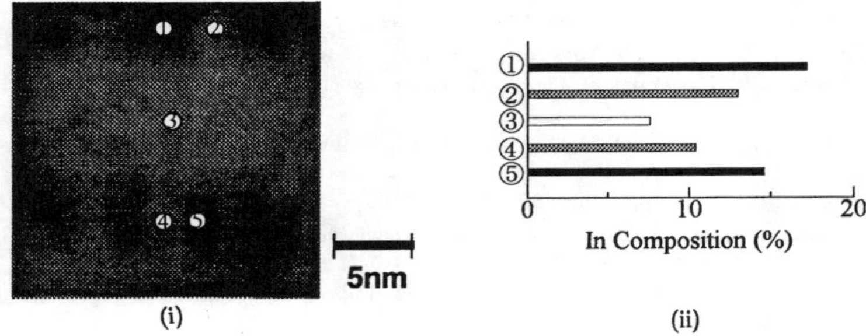

Fig.5 (i) Cross sectional STEM image of the MQW. Probed positions by the EDX are marked by (1)to (5). The In$_x$Ga$_{1-x}$N quantum dots like regions correspond to (1) and (5). (ii) Indium composition at positions (1) to (5) determined by the EDX data.

The spatial distribution of In composition in the MQW was evaluated by EDX using STEM apparatus.[22] In compositions monitored at various positions (labeled (1) to (5) in Fig. 5 (i)) are shown in Fig.5 (ii). In composition in the region (1) corresponding to the InGaN barrier layer was estimated to be 7.63 % which is larger than the designed value (5 %). This is probably because the X-ray signal is generated also from the upper and lower wells due to the spread of incident electron beam within the specimen. In fact, In compositions monitored at wells are smaller than the designed value (20 %). Important finding is that In compositions monitored at quantum dot like regions are always larger than those at neighboring well regions, as can be seen by the comparison of the data between (1) and (2), and also between (4) and (5). This is the direct evidence of the formation of In-rich quantum dots in the InGaN QWs. Since the thickness of the specimen is about 30 nm, the areal density of the dots are estimated to be approximately 5×10^{11} cm^{-2} assuming no overlapping of the quantum dots observed in the TEM image. Consequently, the volume ratio of the quantum dots within the wells is roughly 2 %. Therefore, the origin of localization is likely attributed to the region of In-rich composition which act as quantum dot centers. Such mechanism may result from the intrinsic nature of InGaN alloys [23] because it is reported that randomly mixing of alloy composition is hardly achieved in this system.[24]

CONCLUSIONS

It has been found by means of PL, TR, ER, PLE and TRPL spectroscopy that PL main peak observed in the In$_{0.20}$Ga$_{0.80}$N/In$_{0.05}$Ga$_{0.95}$N MQW structure is ascribed to excitons localized at trap centers which are located below the lowest n=1 quantized level by about 250 meV. The origin of the localization center is possibly ascribed to self-formed In-rich region which may act as a quantum dot. It is interesting to note that the depth of localization was so large that localized excitons have been observed even at RT.

ACKNOWLEDGEMENTS

Authors are grateful to M. Funato for his contribution to the TEM observation. Thanks are also due to Y. Ohtsuka at TORAY Research Center for the EDX measurement. A part of this measurement was done using the facility at the " Venture Business Laboratory (VBL)" in Kyoto University.

References

[1] S. Nakamura, M. Senoh, N. Iwasa and S. Nagahama, Jpn. J. Appl. Phys. **34**, L797 (1995).

[2] S. Nakamura, M. Senoh, S. Nagahama, N. Iwasa, T. Yamada, T. Matsushita, H. Kiyoku and Y. Sugimoto, Jpn. J. Appl. Phys. **35**, L74 (1996).

[3] S. Nakamura, M. Senoh, S. Nagahama, N. Iwasa, T. Yamada, T. Matsushita, H. Kiyoku and Y. Sugimoto, Jpn. J. Appl. Phys. **35**, L217 (1996).

[4] I. Akasaki, S. Sota, H. Sakai, T. Tanaka, M. Koike, and H. Amano, Electron. Lett. **32**, 1105 (1996).

[5] K. Itaya *et al.*, Jpn. J. Appl. Phys. **35**, L1315 (1996).

[6] S. Chichibu, T. Azuhata, T. Sota and S. Nakamura, J. Appl. Phys. **79**,2784 (1996).

[7] Y. Narukawa, Y. Kawakami, M. Funato, Sz. Fujita, Sg. Fujita and S. Nakamura, in press Phys. Rev. B

[8] Y. Narukawa, Y. Kawakami, M. Funato, Sz. Fujita, Sg. Fujita and S. Nakamura, submitted to Appl. Phys. Lett.

[9] S. Nakamura, Jpn. J. Appl. Phys. **30**, L1705 (1991).

[10] K. Okada, Y. Yamada, T. Taguchi, F. Sakaki, S. Kobayashi, T. Tani, S. Nakamura and G. Shinomiya, Jpn. J. Appl. Phys. **35**, L787 (1996).

[11] Y. Kawakami, Z. G. Peng, Y. Narukawa, Sz. Fujita, Sg. Fujita and S. Nakamura, Appl. Phys. Lett. **69**, 1414 (1996).

[12] M. Sugawara, Jpn. J. Appl. Phys. **35**, 124 (1996).

[13] S. Chichibu, T. Azuhata, T. Sota and S. Nakamura, presented at 30th Electronic Materials Conference, W-10, June 26-28, Santa Barbara, (1996).

[14] S. Chichibu, T. Azuhata, T. Sota and S. Nakamura, in press Appl. Phys. Lett.

[15] The value of alloy broadening can be estimated by taking into account the standard deviation of alloy composition within the volume of excitons. See for example : R. Zimmerman, J. Cryst. Growth **101**, 346 (1990).

[16] D. E. Aspnes and J. E. Rowe, Phys. Rev. Lett. **27**, 188 (1971).

[17] R. Dingle, D. D. Sell, S. E. Stokowski and M. Ilegems, Phys. Rev. **B4**, 1211 (1971).

[18] B. Gil, O. Briot and R. Aulombard, Phys. Rev. **B52**, R17028 (1995).

[19] C. Gourdon and P. Lavallard, Phys. Stat. Solidi, (b) **153**, 641 (1989).

[20] S. Yamaguchi, Y. Kawakami, Sz. Fujita, Sg. Fujita, Y. Yamada, T. Mishina and Y. Masumoto, Phys. Rev. B **54**, 2629 (1996).

[21] This range corresponds to the angle of 30 °.

[22] This method was reported to be effective to characterize the self-formed $In_{0.5}Ga_{0.5}As$ quantum dots. See for example, K. Mukai, N. Ohtsuka, M. Sugawara and S. Yamazaki, Jpn. J. Appl. Phys. **33**, L1710 (1994).

[23] Large localization of excitons is also reported in the InGaN single epilayer.: T. Taguchi, T. Maeda, Y. Yamada, S. Nakamura and G. Shinomiya, Proc. of Intern. Symp. on Blue Laser and Light Emitting Diodes, Chiba Univ., Japan, March 5-7, 372 (1996).

[24] I-hisu Ho and G. B. Stringfellow, Appl. Phys. Lett. **69**, 2701 (1996).

DEFECT TRANSITIONS IN GaN BETWEEN 3.0 AND 3.4 eV

W. RIEGER, O. AMBACHER, E. ROHRER, H. ANGERER, M. STUTZMANN
Walter Schottky Institute, D-85748 Garching, Germany

ABSTRACT

We have studied optical transitions (absorption and luminescence) in nominally undoped and Mg-doped GaN deposited by MOCVD and MBE. In the range between 3.0 and 3.4 eV, a variety of well known low-intensity luminescence lines are observed, whose origin is discussed. In particular, by comparing excitation with subgap versus above-gap laser lines as well as by combining optical subgap absorption with spectrally resolved photoconductivity, we identify localized optical transitions occuring in isolated cubic inclusions in the otherwise hexagonal GaN epitaxial layers. Implications of these strucural defects for photocurrent transients are also presented.

INTRODUCTION

The usefulness of GaN for optoelectronic and other devices such as transistors depends crucially on the density of shallow and deep defects in the heteroepitaxial layers grown on sapphire or SiC substrates. To date, mostly photoluminescence (PL) and related techniques have been used to study these defects and to correlate them with structural features such as impurities, native point defects, dislocations, etc. [1-3]. However, a definitive and generally accepted assignment of the various radiative transitions has not emerged yet. In particular, a systematic comparison of otherwise well characterized samples prepared by different methods (and, thus, containing different concentrations of defects and impurities) with the help of additional optical defect spectroscopy methods is still needed. Here, we present results dealing with cubic inclusions in wurtzite GaN by a combination of PL and absorption measurements.

EXPERIMENT

All samples investigated in this study were deposited on c-plane sapphire either by MOCVD using triethylgallium and ammonia, or by plasma-enhanced MBE equipped with an rf plasma source (Oxford CARS25). The background pressure in the MBE system was 10^{-10} mbar, and the N_2 gas was purified prior to feeding into the plasma source. During deposition, the pressure in the MBE system was $4x10^{-5}$ mbar. Luminescence was excited with an Ar^+ ion laser and measured with a double monochromator and a photomultiplier. Optical subgap absorption measurements were performed with photothermal deflection spectroscopy (PDS) and the constant photocurrent method (CPM). Details of these latter two techniques are described elsewhere [4, 5]

RESULTS AND DISCUSSION

We begin our discussion by presenting PL spectra obtained under identical conditions on a number of non-intentionally doped GaN samples prepared with different deposition parameters and methods. Fig. 1 shows samples prepared in the same MOCVD reactor at

Mat. Res. Soc. Symp. Proc. Vol. 449 © 1997 Materials Research Society

950°C with different buffer layers (low temperature GaN or AlN grown at different temperatures. It can be seen from this figure that the type of buffer has a large influence especially on the donor-acceptor-pair recombination (DAP) at 3.26 eV and the corresponding LO-phonon replica.

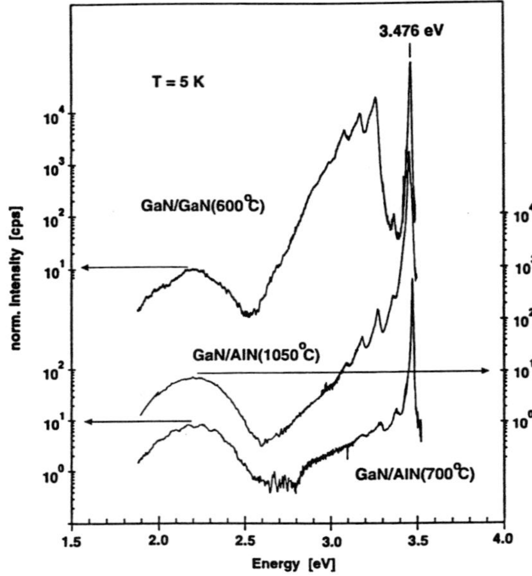

Fig.1: Undoped GaN samples grown by MOCVD on different buffer layers. Type and growth temperature of the buffers are indicated in the figure. Note the large variations in the energy range between 3.0 and 3.4 eV. The sharp line at 3.476 eV is the donor bound exciton radiation.

In a recent publication [3], this DAP recombination was tentatively ascribed to carbon as a likely impurity in GaN, forming a deep acceptor about 230 meV above the valence band edge. This assignment is somewhat problematic in view of Fig. 1, since the GaN layers grown on different buffers have a similar C content of about 10^{18} cm^{-3} as determined by elastic recoil detection analysis, whereas the DAP transitions essentially disappear for the GaN grown on a 700°C AlN buffer.

To emphasize this point, we show in Fig. 2 the PL spectra of one MOCVD and one MBE sample both grown on a low-temperature GaN buffer. Although the impurity content in general and the carbon content in particular should be significantly lower in the MBE sample than in the MOCVD material, the PL spectra are almost identical, except for a much lower deep defect luminescence around 2.2 eV in the MBE sample. It rather appears that the DAP luminescence is connected to an intrinsic defect whose density can be influenced by the buffer, e.g. by the dislocation density as observed in TEM investigations.

An intrinsic point defect which according to recent theoretical investigations should have a rather low formation enthalpy in n-type GaN is the Gallium vacancy, V_{ga} [6]. Under MBE growth conditions, the tendency for the formation of V_{ga} can be influenced via a variation of the Ga flux from the effusion cell. This indeed has pronounced effects on the PL spectra as evidenced by Fig. 3. With a balanced N and Ga flux to the substrate, MBE samples show very little Pl due to shallow or deep defects. The main spectral feature is the D°X line at 3.467 eV, accompanied by two LO-phonon replica at 3.375 and 3.285 eV. There is very little

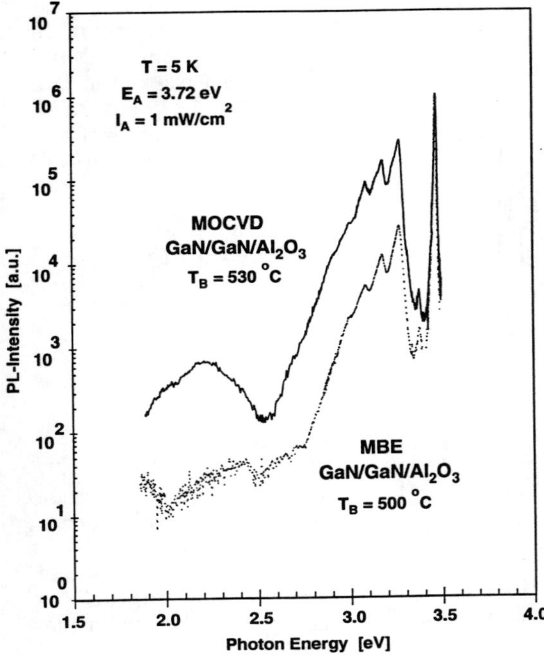

Fig.2: Comparison of the PL spectra of GaN deposited by MOCVD and MBE using a similar low-temperature GaN buffer. The only significant difference between the two samples occurs in the region of the yellow luminescence band around 2.2 eV. The level of impurity contamination in the two samples is expected to be drastically different.

yellow luminescence and only a weak additional feature close to the bandgap of cubic GaN (3.31 eV at 5 K. A more detailed discussion of cubic inclusions follows below.) A sample grown under otherwise identical conditions but with a reduced Ga-flux exhibits much stronger luminescence in the region between 3.0 and 3.4 eV as well as around 2.2 eV. Although at present we have no microscopic explanation for this trend, it emphasizes the importance of controlled variations of deposition conditions for a better understanding of radiative centers in GaN.

We now discuss in more detail the luminescence properties in the region between 3.0 and 3.4 eV observed for a rather typical undoped MOCVD sample showing a strong $D^{\circ}X$ transition together with a noticeable DAP recombination. This type of PL spectra has also been observed in a number of cases discussed in the literature [e.g. 6,7]. As shown on the left hand side of Fig.4, a new set of PL lines labelled L_1 and L_2 appears upon strong illumination of such a sample at 3.369 and 3.313 eV, respectively. Again the assignment of these lines is still uncertain, so that a general description in terms of strongly localized excitons has been used [8]. Characteristic for these additional lines is that they can also be excited with subgap light (3.41 eV in Fig. 4), too low in energy for band-to-band transitions in wurtzite GaN at 5 K. Similar results have been reported by Dai et al. [9]. As shown on the right hand side of Fig. 4, the temperature dependence of the L_2 peak position follows closely the variation of the fundamental gap of cubic GaN deduced from photoreflectance data [10], indicating that these additional radiative processes are connected to cubic inclusions in hexagonal GaN. This would explain why the $L_{1,2}$ transitions can be effectively excited by a laser energy below the bandgap of hexagonal GaN, but above the gap of cubic GaN.

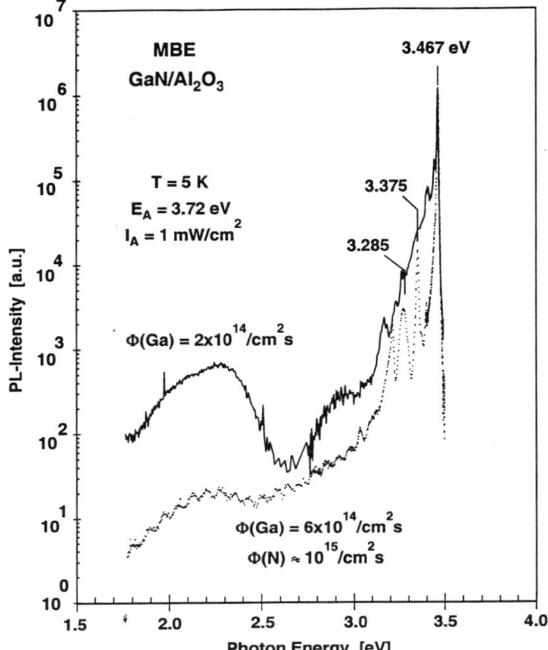

Fig.3: Luminescence spectra of GaN samples grown by MBE with different fluxes of Ga. The peaks labelled 3.375 and 3.285 eV are phonon replica of the donor bound exciton at 3.467 eV

Fig. 4: (below)
Left: PL spectra of a MOCVD sample at different excitation intensities and two different excitation energies.
Right: Temperature dependence of the L_1 and L_2 PL peaks in comparison to the temperature dependence of the fundamental gap of cubic GaN (solid dots).

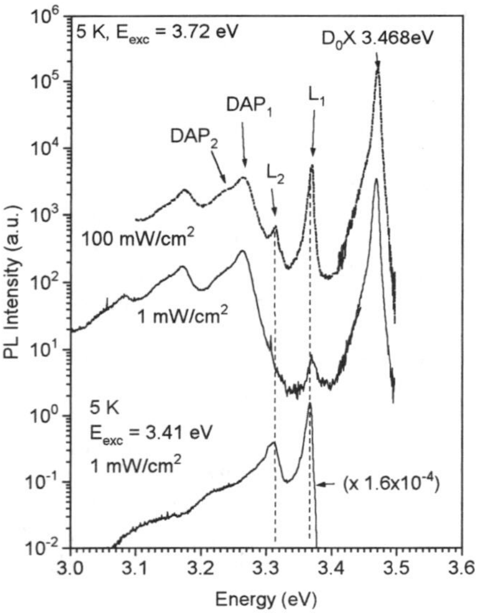

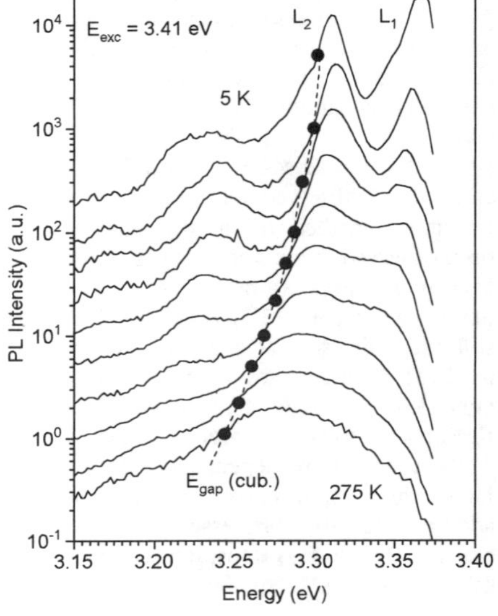

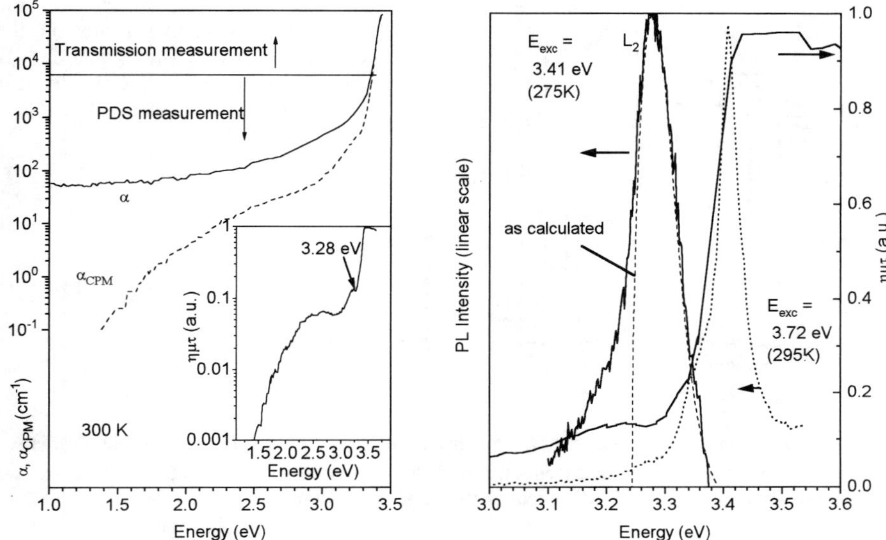

Fig. 5: left hand side: comparison between subgap absorption spectra of the sample in Fig.4 obtained by PDS versus CPM. The inset shows the mobility-lifetime product of photo-generated carriers obtained as the ratio between the two measurements. On the right hand side, this $\mu\tau$-product is shown together with the L_2 transition and the band edge luminescence of wurtzite GaN at room temperature.

We now compare the radiative transitions in Fig 4. with the spectral features observed in subgap absorption measurements. To this end we compare in Fig. 5 PDS and CPM spectra. In PDS, the net absorption coefficient at a given photon energy is determined, whereas in a CPM measurement only those optical transitions giving rise to mobile majority carriers (i.e. electrons) are detected. It can be shown [11] that the ratio of the CPM and PDS spectra is proportional to the mobility-lifetime product of photogenerated carriers (inset in Fig. 5). For the purpose of the present discussion we note that a local minimum of the $\mu\tau$–product appears at 3.28 eV, coinciding with the position of the L_2 PL transition. This indicates that electrons excited at this particular energy remain localized, not contributing to a macroscopic photocurrent before they recombine radiatively. This is in contrast to electrons excited across the bandgap of wurtzite GaN, which according to Fig. 5 have a much higher $\mu\tau$-product.

These experimental results lead us to the following model concerning the influence of cubic inclusions in wurtzite GaN (Fig. 6). It is known from TEM investigations that wurtzite GaN contains a large number of stacking faults which can serve as nucleation centers for cubic regions. Because of the smaller bandgap of cubic GaN, there will be an offset of about 200meV in the conduction band, sufficiently large to cause a net transfer of electrons from the surrounding hexagonal n-type matrix. This gives rise to a local band bending as indicated in Fig. 6. Then, optical transitions below the bandgap of wurtzite GaN occur preferentially in

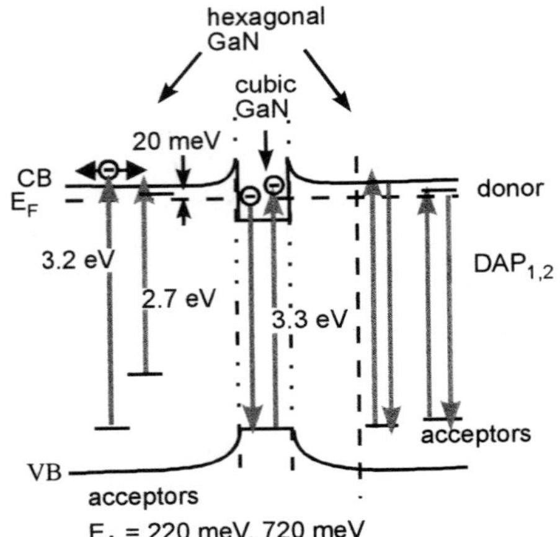

Fig. 6: Schematic model for the influence of cubic inclusions on the optical properties of hexagonal GaN in the enrgy range 3.0 to 3.4 eV. See text for details.

these cubic inclusions, giving rise to localized electron hole pairs with a high radiative recombination rate (L_2 luminescence) and a low $\mu\tau$-product. In addition, other optical transitions (bound excitons, DAP) occur in the same energy range.

CONCLUSIONS

A combination of different optical defect spectroscopies together with a systematic variation of sample properties will be necessary to fully understand the complex defect structure of GaN and the influence of intrinsic and extrinsic defects on carrier recombination. As an example, we have presented experimental evidence for the influence of cubic inclusions on the PL properties of state-of-the-art wurtzite GaN.

REFERENCES

1. T. Matsumoto and M. Aoki, Japan. J. Appl. Phys. **13**, 1804 (1974)
2. W.J. Choyke and I. Linkov, Inst. Phys. Conf. Ser. **137**, 141 (IOP Publishing Ltd. , London, 1994) and references therein.
3. S. Fischer, C. Wetzel, B.K. Meyer and E.E. Haller, Appl. Phys. Lett. **67**, 1298 (1995)
4. O. Ambacher, W. Rieger, P. Ansmann, H. Angerer, T.D. Moustakas and M. Stutzmann, Solid State Commun. **97**, 365 (1996)
5. M. Vanecek, A. Abraham, O. Stika, J. Stuchlik and J. Kocka, phys. stat. sol. (a) **83**, 617 (1984)
6. J. Neugebauer and C.G. Van de Walle, Phys. Rev. **B50**, 8067 (1994)
7. L. Eckey et al., Appl. Phys. Lett. **68**, 415 (1996)
8. C. Wetzel et al., Appl. Phys. Lett. **68**, 2556 (1996)
9. R. Dai, S. Fu, J. Xie, G. Fan, G. Hu, H. Schrey and C. Klingshirn, J. Phys. **C15**, 393 (1982)
10. G. Ramirez-Flores, H. Navarro-Contreras, A. Lastras-Martinez, R.C. Powell and J.E. Greene, Phys. Rev. **B50**, 8433 (1994)
11. W. Rieger, R. Dimitrov, D. Brunner, E. Rohrer, O. Ambacher and M. Stutzmann, Phys. Rev. B, in print

DEPTH-PROFILE OF THE EXCITONIC LUMINESCENCE IN GALLIUM-NITRIDE LAYERS

H. SIEGLE*, A. HOFFMANN*, L. ECKEY*, AND C. THOMSEN*,
T. DETCHPROHM**, K. HIRAMATSU**, T. DAVIS***, J. W. STEEDS***
*Institut für Festkörperphysik, TU Berlin, Hardenbergstraße 36, 10623 Berlin, Germany
**Department of Electronics, Nagoya University, Nagoya 468-01, Japan
***H. H. Wills Physics Laboratory, University of Bristol, Bristol BS8 1TL, UK

ABSTRACT

We present results of spatially-resolved photoluminescence and Raman measurements on a 200 µm thick GaN layer grown on sapphire by hydride vapor phase epitaxy. Our micro-photoluminescence measurements reveal that the peak position of the excitonic and donor-acceptor-pair transitions strongly depends on the distance to the substrate interface. We observed a strong blue shift near the interface and discuss the influence of strain, which we quantified by micro-Raman experiments.

INTRODUCTION

A major problem in growing GaN epitaxially is the large mismatch of lattice constants and thermal expansion coefficients between layer and common substrates as e.g. sapphire or GaAs [1]. Consequently most GaN epilayers are highly strained. Another problem is the inhomogeneous distribution of the photoluminescence in these layers [2,3]. In order to handle thin-film heterostructures and devices based on GaN a knowledge about the influence of strain on the optical properties as well as a control of the homogeneity is necessary.

We performed spatially-resolved photoluminescence measurements at low temperatures on GaN grown on sapphire in order to map the luminescence distribution in the layer on a micrometer scale. Micro-Raman-scattering of the E_2-mode of hexagonal GaN in the same layer regions provides us with independent information about the internal stress in the layer and allows us to correlate structural and optical properties [4].

After presenting the experimental details we first discuss the main features of the near-bandgap luminescence of the layer. We then describe the results of our spatially-resolved measurements and discuss the observations.

EXPERIMENT

The sample under study was an undoped 200 µm thick hexagonal GaN layer grown on [0001] sapphire using hydride vapor phase epitaxy (HVPE) with a free carrier concentration below $1 \cdot 10^{17}\,\mathrm{cm}^{-3}$ as determined by room-temperature Hall measurements.

Micro-photoluminescence measurements were performed using a single-grating Renishaw spectrometer equipped with an UV notch filter and an UV-enhanced CCD detector. The sample was excited parallel to the substrate surface using the 325 nm line of a He-Cd laser. By passing the laser through a microscope objictce (x40, x27) the laser beam was focused to a point spot with a diameter of about 2 µm. With this micro-photoluminescence setup we reached a spatial resolution of about 5 µm.

Mat. Res. Soc. Symp. Proc. Vol. 449 © 1997 Materials Research Society

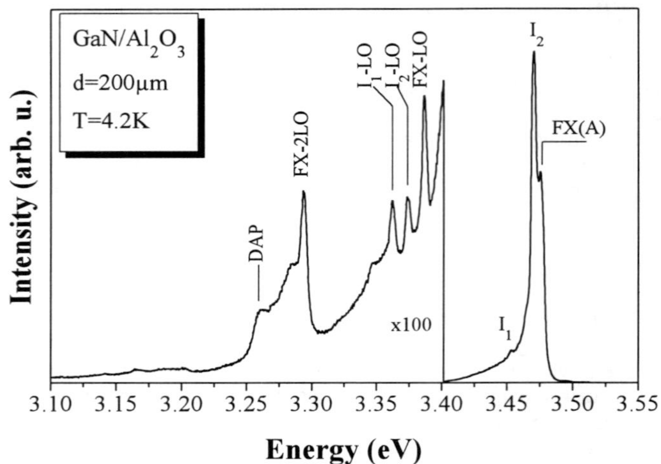

Fig. 1: Low-temperature micro-photoluminescence spectrum taken near the surface. Excitation wavelength was at 325 nm.

Micro-Raman measurements were carried out with a triple-grating Dilor spectrometer and the 514.5 nm line of an Ar$^+$ laser for excitation. The spatial resolution was better than 1 μm, and we were able to detect Raman shifts smaller than 0.1 cm^{-1}.

Both, photoluminescence and Raman measurements were performed at low temperatures (4.2 K) using a microscope cryostat.

RESULTS AND DISCUSSION

Photoluminescence overview

Figure 1 shows as an overview a low-temperature micro-photoluminescence spectrum taken near the surface in a distance d = 200 μm away from the substrate interface where we assume bulk-like conditions. The spectrum is dominated by the free-exciton emission (FX(A)) located at 3.477 eV and the neutral-donor-bound-exciton emission (I$_2$) at 3.471 eV. The weak structure near the I$_2$ at 3.450 eV is due to the annihilation of excitons at neutral shallow acceptors (I$_1$) [5]. The low-energy side of the spectrum exhibits the phonon sidebands of these excitons. The strength of the second LO replica of the free excitons and the weakness of the donor-acceptor-pair luminescence indicate the excellent quality of the layer near the surface.

Spatially-resolved photoluminesence measurements

The energetic positions of all photoluminescence transitions described above depend on the distance to the substrate surface as can be seen, e.g., for the excitonic transitions from the linescan shown in Fig. 2. In this linescan which was taken in 5 μm steps across the entire GaN cross section we normalized the spectra to their maximum intensity.

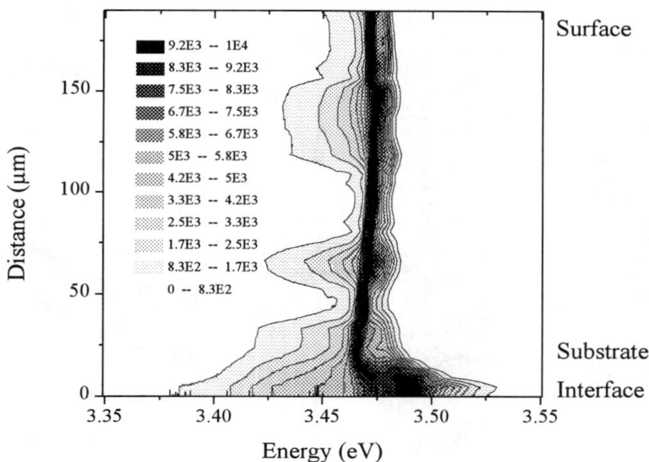

Fig. 2: Depth-profile of the near-bandgap luminescence in a 200 µm thick GaN layer grown on sapphire taken at low temperatures (4.2K) after excitation at 325 nm.

Starting from the surface of the layer (distance d = 200µm) and proceeding in the direction of the substrate interface (d = 0 µm) one can see that the peak position of the excitonic photoluminescence remains nearly constant during the first 100 µm. From d ≈ 100 µm to d ≈ 30 µm the spectra exhibit a small red shift. Proceeding further in the direction of the substrate interface the red shift gives way to a significant blue shift. The corresponding spectra are shown separately in Fig. 3. For clarity the peak positions of the I_2 and of the free-exciton emission are plotted as a function of the distance to the substrate (Fig. 3, right). For the I_2 emission we observed a red shift of 8 meV in total and a blue shift near the substrate interface of about 5 meV. The free-exciton emission follows the I_2 up to a distance of d ≈ 50µm, indicating a constant binding energy of the neutral-donor-bound excitons. For distances smaller than 50 µm this emission shifts strongly to higher energy reaching a maximum blue shift of 21 meV. Consequently the energy difference between both near-bandgap transitions increases.

<u>Spatially-resolved Raman measurements</u>

In order to scrutinize wether this strong blue shift is caused by stress in the GaN layer due to the mismatch of the lattice constants and the thermal expansion coefficients between layer and substrate, we performed micro-Raman measurements of the same spatial region in the layer investigated by photoluminescence measurements. Our measurements reveal that the non-polar E_2 Raman mode exhibits a shift in this region of 0.8 cm^{-1}. This is displayed in Fig. 4 where we plotted the Raman shift of this mode as a function of distance to the substrate interface. We found that the biaxial compressive stress in the layer reduces exponentially with increasing distance from the substrate. The layer is already largely relaxed at a distance d ≈ 30µm away from the substrate. For distances d > 100µm the layer is fully relaxed. This observation is in

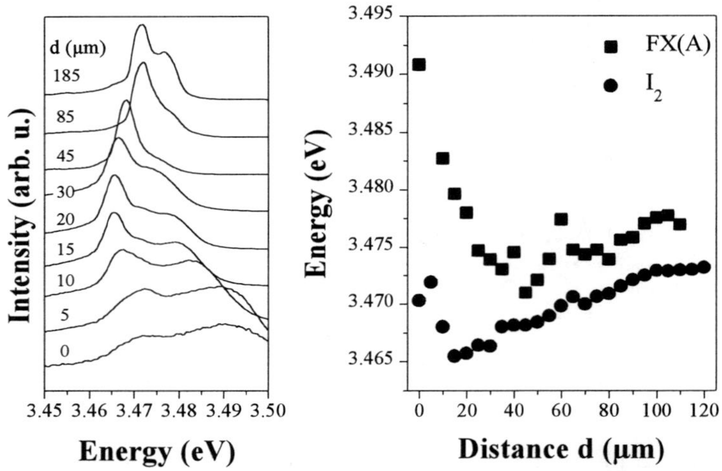

Fig. 3: Low-temperature photoluminescence spectra taken at different distances from the substrate interface (left). Energy positions of the excitonic transitions as a function of distance from the substrate (right)

good agreement with X-ray diffraction measurements on a series of GaN layers also grown on sapphire with varying thicknesses [6].

The observed blue-shift of the near-bandgap luminescence cannot totally be explained by a strain-induced change of the bandgap. According to the relation $\sigma = \Delta\omega/6.2\,\text{cm}^{-1}$ reported by Kozawa et al. [7], where σ is the biaxial compressive stress in GPa and $\Delta\omega$ is the Raman shift in cm^{-1}, we can quantify the strain in our sample. Rieger et al. [8] investigated the luminescence shift due to biaxial compressive stress and determine $dE_{PL}/d\sigma$ to be between 21 and 27 meV/GPa. Using both relations one would expect a maximum blue shift in our sample of about 3.5 meV.

In fact this change is in the order of the shift we observed for the neutral-donor-bound exciton emission (I_2). However we point out that the free-exciton emission band shift with 21 meV is much too strong to be explained solely by stress in the GaN layer.

This emission band cannot be due to the recombination of free excitons; otherwise the

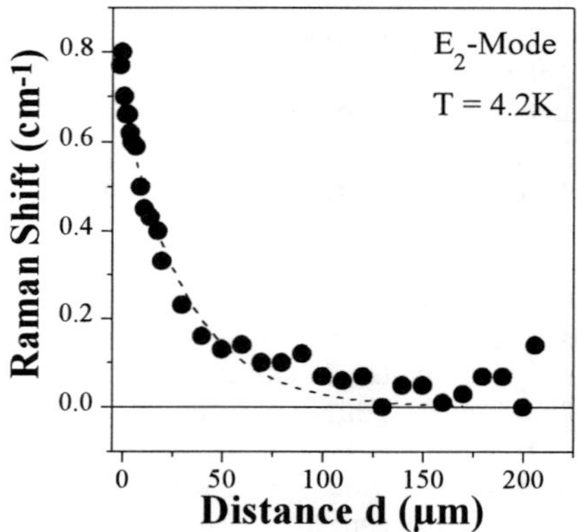

Fig. 4: Shift of the E2-Raman mode as a function of the distance from the substrate interface.

binding energy of the donor-bound-excitons would grow extremely near the interface. We rather think that from the distance at which both, the peak positions of the I_2 and those of the free-exciton emission diverge band-to-band transitions appear. These transitions appear whith increasing doping level and they are known from doped GaN layers or layers of poorer quality [5]. This can be confirmed when considering the intensity of the donor-acceptor-pair luminescence in this layer. As can be seen from Fig. 5 its intensity increases strongly for distances smaller than d < 100μm, indicating that with decreasing distance to the substrate more and more impurities are built in which yields sufficient concentration of impurities to a blue shift of the emissions. The broadening which is also found in the spectra is an additional confirmation.

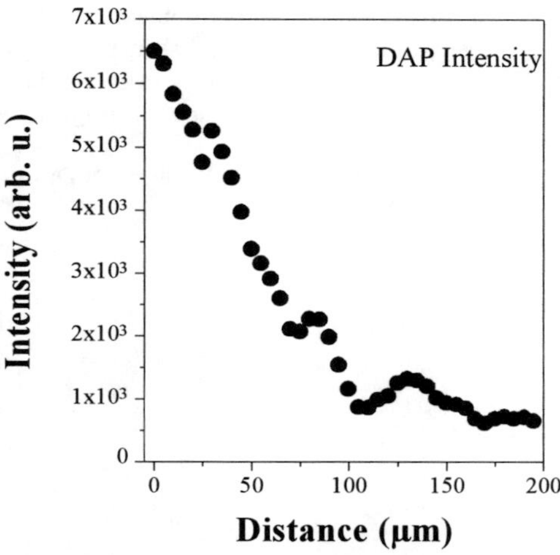

Fig. 5: Peak intensity of the donor-acceptor-pair (DAP) luminescence as a function of distance to the substrate (left).

SUMMARY

We have shown that the photoluminescence of GaN depends strongly on the distance to the substrate interface. Near the substrate we observed a significant blue shift of the near-bandgap emission. Although our micro-Raman measurements reveal that in the same spatial region the biaxial compressive stress in the layer reduces exponentially with increasing distance from the substrate, the blue shift of the photoluminescence with 21 meV is much too strong to be explained solely by stress. We found that with decreasing distance from the substrate the free exciton emission gives way to band-band-transitions, which appear because of the increasing impurity concentration near the substrate.

ACKNOWLEDGMENTS

The authors thank D. Pitt and co-workers of Renishaw plc for using their micro-photoluminescence system.

REFERENCES

1. For a review see for example: S. Strite, M. E. Lin, and H. Morkoç, Thin Solid Films **231**, 197 (1993)
2. H. Siegle, P. Thurian, L. Eckey, A. Hoffmann, C. Thomsen, B. K. Meyer, H. Amano, I. Akasaki, T. Detchprohm, K. Hiramatsu, Appl. Phys. Lett. **68**, 1265 (1996)
3. F. A. Ponce, D. P. Bour, W. Götz, P. J. Wright, Appl. Phys. Lett. **68**, 57 (1996)
4. For a survey of Raman scattering in GaN see for example: H. Siegle, L. Eckey, A. Hoffmann, C. Thomsen, B. K. Meyer, D. Schikora, M. Hankeln, K. Lischka, Solid State Commun. **96**, 943 (1995)

5. A. Hoffmann, in: Festkörperprobleme, ed. by R. Helbig, Advances in Solid State Physics **35**, (Vieweg, Braunschweig/Wiesbaden 1996), in print

6. K. Hiramatsu, T. Detchprohm, I. Akasaki, Jpn. J. Appl. Phys. **32**, 1528 (1993)

7. T. Kozawa, T. Kachi, H. Kano, H. Nagase, N. Koide, K. Manabe, J. Appl. Phys. **77**, 4389 (1995)

8. W. Rieger, T. Metzger, H. Angerer, R. Dimitrov, O. Ambacher, M. Stutzmann, Appl. Phys. Lett. **68**, 970 (1996)

DISLOCATION LUMINESCENCE IN WURTZITE GaN

Y.G. SHRETER[*], Y.T .REBANE[*], T.J. DAVIS[**], J. BARNARD[**], M. DARBYSHIRE[**], J.W. STEEDS[**], W.G. PERRY[***], M.D. BREMSER[***] and R.F. DAVIS[***]

[*]A.F.Ioffe Physico-Technical Institute, Russian Academy of Sciences, Polytechnicheskaya 26, St.Petersburg 194021, Russia, e-mail: Y.Shreter @ cmc.ioffe.rssi.ru

[**]University of Bristol, H.H.Wills Laboratory, Tyndall Avenue, Bristol , BS8 1TL, UK

[***]Departement of Materials Science and Engineering, North Carolina State University, Raleigh, North Carolina 27695

ABSTRACT

The 364 nm PL-system in GaN is attributed to the formation of dislocation excitons and charged dislocation excitons on c-axis screw dislocations.   The binding energy for the dislocation exciton, charged dislocation exciton and a hole on the screw dislocation were determined as 35 meV, 7 meV and 65 meV respectively, in accord with experiment.

INTRODUCTION

Dislocations are the dominant structural defect in GaN. It has been shown recently that the dislocation-related photoluminescence (PL) in Ge, Si and ZnSe crystals can be explained on the basis of one-dimensional dislocation bands, split off by dislocation deformation fields from the edges of the normal energy bands. The strong confinement of the electrons and holes transverse to the dislocation line by the deformation field gives rise to rather deep PL lines, $h\nu \approx E_g - (0.1 - 0.3$ eV $)$, whose energy depends on the value of the edge component of the Burgers vector. The carriers' free movement along the dislocation line determines their weak interaction with the lattice   and   small value of the Huang-Rhys factor. This model was successfully applied for interpretation of   the low temperature PL spectrum of dislocated crystals by dislocation excitons (DEX), for a classification of the dislocation-related PL lines in Si, for explanation of the line splitting in Ge and Si crystals with sets of differently split 60^0-dislocations, for calculations of the g-factor for the electrons on the 60^0-dislocations and for a symmetry analysis of the EDSR lines in Ge and Si [1,2,3].

Dislocation-related PL in GaN has been preliminarily attributed to two PL systems: the so called yellow PL (hv $\approx$520 nm) [4] and a PL line at 364 nm [5]. The results presented here show that the 364 nm PL system is the most probable candidate for dislocation-related PL.

EXPERIMENTAL PROCEDURE

Photoluminescence and TEM investigations were carried out on monocrystalline wurtzite GaN (0001) thin films, $\approx$ 0.6 μm thick, grown   at 1050^0 C on high temperature, $\approx$100 nm thick, monocrystalline AlN(0001) buffer layers predeposited at 1100^0 C concurrently on off- and on-axis α(6H)-SiC(0001)$_{Si}$ substrates via OMVPE [6].   Comparing by atomic force microscopy on-axis with off-axis-substrate-grown films shows rougher film surfaces on the latter, therefore we would expect a higher density of structural defects in films grown on off-axis SiC. The density of threading dislocations was estimated by TEM as ~10^9 cm^{-2} and ~10^{10}

Mat. Res. Soc. Symp. Proc. Vol. 449 © 1997 Materials Research Society

cm^{-2} in the samples grown on on- and off-axis substrates respectively. Inversion domain boundary concentration was much higher $\sim 10^{9}$ cm^{-2} in the samples grown on off-axis SiC.

A second set of scanning PL measurements was made to investigate the variation of the PL intensity, position and width of the bound exciton line (358 nm) and PL intensity of the 364 nm line across the thickness of the film. For this purpose "a wedge" was prepared by polishing a GaN film of thickness ~2.7 μm, at an angle of 0.5° to the substrate, grown on off- and on-axis SiC. In this paper we will present PL scanning results obtained for the films grown on on-axis SiC. Photoluminescence was excited by a He-Cd (325 nm) laser. Renishaw Raman and PL Imaging Microscope System 2000 was used for PL mapping of the samples at 6K and 300K. The spatial resolution of the system was ~2 μm.

RESULTS AND DISCUSSION

<u>Thin Films of GaN Grown on On- and Off-Axis SiC Substrates.</u>

The GaN film grown on a SiC substrate vicinal surface shows a very intense bound exciton line (BE) and 364.4 nm line with TO-phonon replica at 372 nm, Fig.1. The D-A systems also differ; the on-axis-grown sample has additional BE lines despite identical growth conditions. Here the 364 nm line is considered, because of its similarity to dislocation-related lines in cubic semiconductors. This line shows (a) sublinear dependence of intensity as a function of excitation, (b) very rapid disappearance with increase in

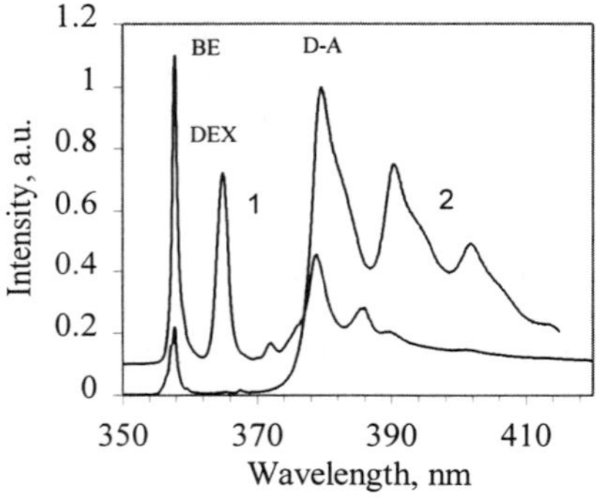

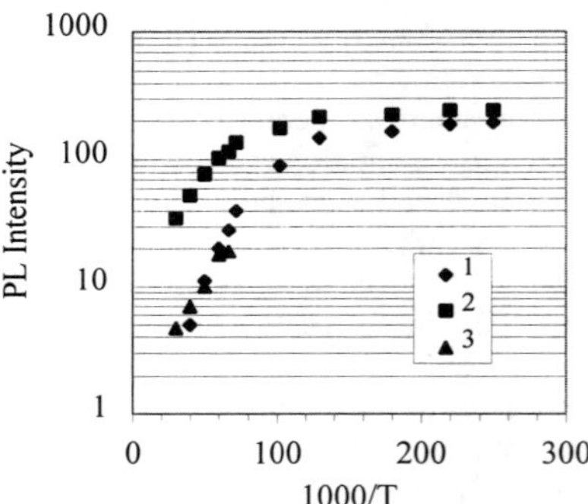

Fig.1.Photoluminescence spectra of GaN grown on off-(1) and on-(2) axis SiC substrates, T=4 K.

Fig.2. Temperature dependence of the DEX - (1), BE - (2) and $(DEX)^{+}$ - (3) PL lines. Level of excitation ~1 W/cm^{2}.

temperature, Fig.2, and (c) a fine structure of the line that develops, in the range 15 - 40 K, as a new line with energy ~7 meV higher, Fig.3. This structure can be seen at low temperature and low excitation level, Fig.4.

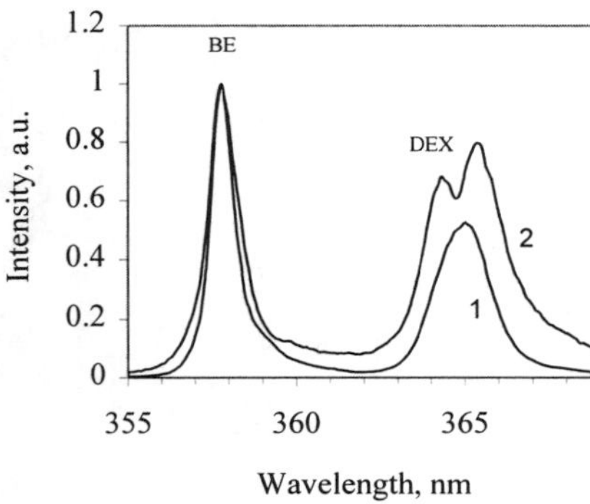

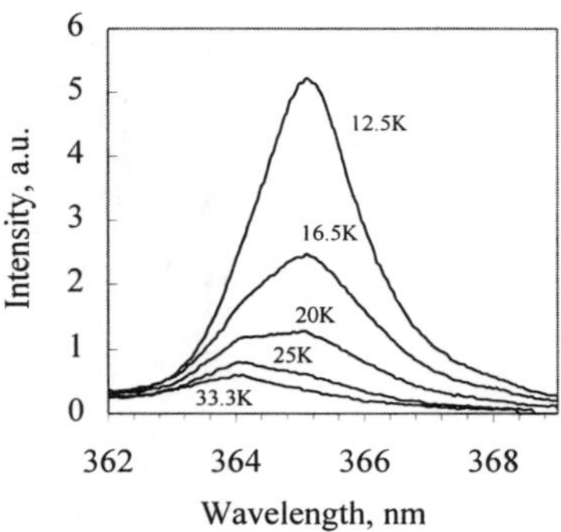

Fig.3. PL of GaN grown on off-axis SiC. 1- 4 K, ~1 W/cm^2; 2- 2 K, ~1 mW/cm^2.

Fig.4. Temperature dependence of the 364 nm PL system, ~1 W/cm^2.

Thick Film of GaN Grown on On-Axis SiC
========

The PL map of the BE line (Fig.5), in the thick sample (~2.7 µm), shows a large fluctuation in the PL intensity near the surface of the GaN film. With decreasing thickness the PL intensity becomes more steady and rapidly decreases near the interface, because of the reduction in the excited volume of the GaN. The line scan in Fig.6a illustrates the variation of the intensity as a function of the position of the laser beam on the wedge of the GaN film. The position of the BE exciton line shows a small variation with thickness of the film, shifting to higher energies near the interface, Fig.6b. There is a more marked variation in the half width half maximum (HWHM) of the PL emission near the energy gap, across the film , Fig.6c. Near the surface of the film the big variation of the HWHM reflects the fact that there are at least three BE lines varying in intensity independently of each other. In the middle of the sample only one BE line

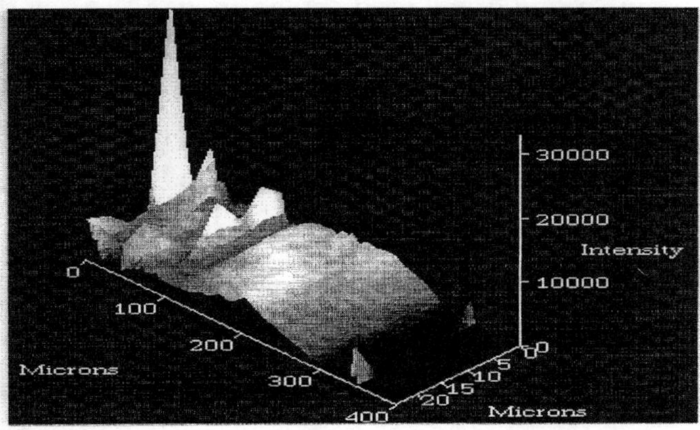

Fig.5. PL map of BE intensity

Fig. 6a

Fig. 6b

Fig. 6c

Fig.6d

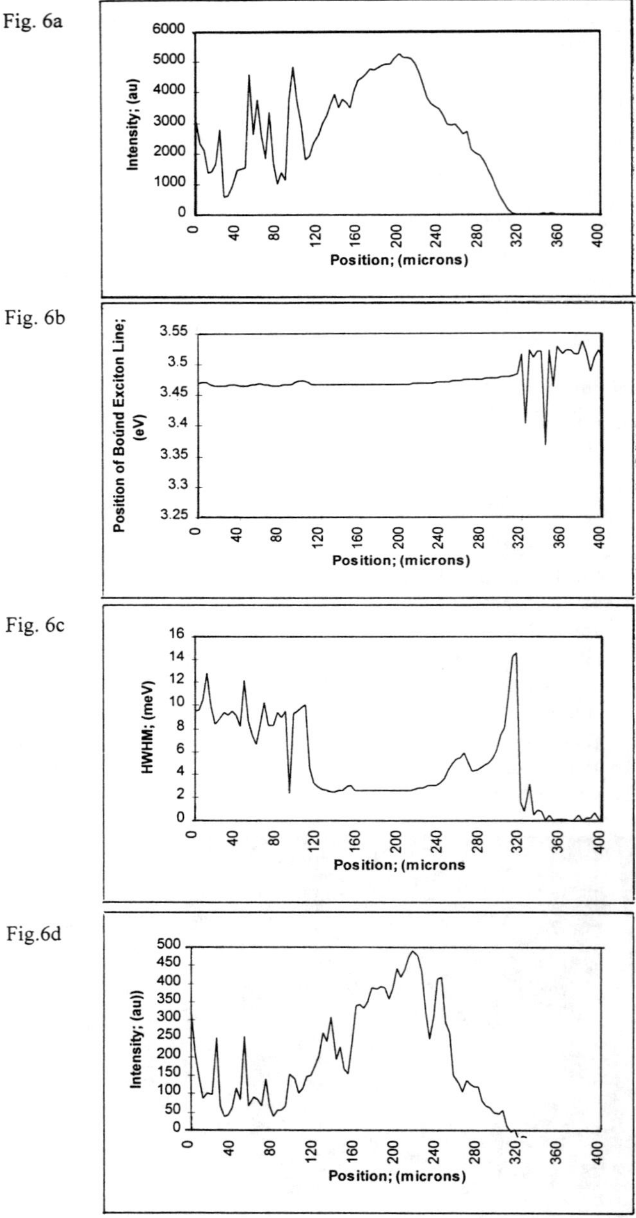

Fig.6 Line scans of GaN interface, a; intensity, b;position, c; HWHM, of BE line and d; intensity of 364 nm PL band

is strong with a FWHM of ~6 meV, Fig.7. The dislocation related line (364nm) intensity is very weak in this sample and shows a very low intensity near the surface of the film. There is a tendency for the intensity of this line to increase towards the substrate. The sharp reduction in the intensity of this line near to the interface is due to the reduction in the excited volume of GaN, Fig.6d.

Our evidence confirms that the 364 nm line is connected with structural defects. This line is very intense in thin GaN films grown on off-axis SiC. Also in the thick sample the intensity of this line increases towards the interface, where there is a higher density of extended defects. But what type of structural defects could be responsible for this PL line? Despite this high density of edge dislocations in GaN they are not radiatively active owing to the inherent charge, which destroys the captured excitons [3]. Only stacking faults, inversion domains and screw dislocations are neutral, and they can be, in principle, responsible for this line. We attribute the 364 nm line to the formation of the DEX on c-axis screw dislocations, because of the high density of screw dislocations near the interface seen in the similar system - GaN/ Al$_2$O$_3$ [7].

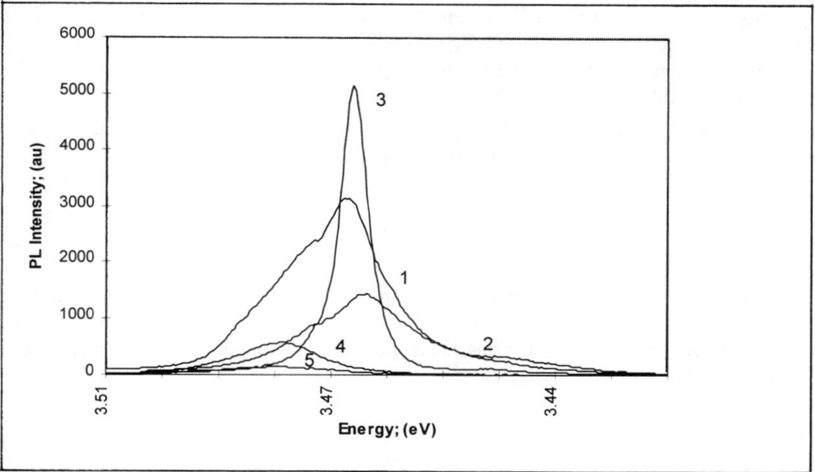

Fig.7. Comparison of spectra taken at different points along the line scan. Curve 1, 0 μm from start of scan, 2, 16 μm, 3, 200 μm, 4, 300 μm, 5, 308 μm.

Dislocation Excitons on Screw Dislocations in Direct-Gap Semiconductors.

The main peculiarity of direct material is that symmetry of the conduction band minimum at Γ-point prevents the deformation-potential interaction of electrons at this valley with the strain field of the screw dislocation. We have two possibilities for an electron to be bound to the dislocation in this case. One possibility is that electron is at a closely higher valley with lower symmetry (for GaN candidates are U and M valleys). However, estimations have shown that this possibility should be ruled out for GaN since the topological interaction [8] strongly reduces the electron binding energy in this case. Another possibility is that an electron is bound by Coulomb field to a hole that in its turn is trapped by the strain field of the screw dislocation. The corresponding dislocation exciton possesses slightly anisotropic reduced mass since the electron mass m_c=0.22 m_0, is much smaller than hole mass, m_h = 0.82 m_0. Thus, its binding energy can be estimated using perturbation theory giving E_{DEX}=e^4($2m_c^2$+$3m_c m$)/$\hbar^2 \varepsilon^2$(m_c + m_h), where ε is dielectric constant. For ε =9.8 we get E_{DEX}= 35 meV.

The main shift of the DEX line from the band edge is produced by the hole binding energy E_h in the strain field of the dislocation. At the distances that mainly contribute to the hole binding energy the deformation field decreases as $1/r$ and the corresponding Shrödinger equation is similar to one for 2d-hydrogen atom with an effective Rydberg R_y^* = 3 m_h b^2 b'^2 / 32 $\pi^2 \hbar^2$, where b is the Burgers vector and b' is the deformation potential constant [9]. For 2d-hydrogen atom the binding energy is 4 R_y^*. The case of a screw dislocation is more complicated since the corresponding strain field as well as kinetic energy operator is spin dependent. This needs to be taken into account by the Berry topological phase related to the spin rotation. This reduces the binding energy to less than R_y^* [9]. Thus, we can estimate $E_h \leq R_y^*$. For GaN, R_y^* = 100 meV for parameters m_h =0.82 m_0, b = 5.19 Å, b' =2 eV. This is in good agreement with $E_h \approx$ 65 meV found from the DEX-line position 3.40 eV with respect to the energy gap 3.50 eV. The DEX-line splitting at low temperatures and excitations, Fig.3, can be attributed to the formation of the DEX$^+$ complex consisting of two holes bound to the screw dislocation and one electron bound to these holes by the Coulomb field (charged DEX). The corresponding binding energy can be found from calculations of adiabatic potential curves for the molecular hydrogen ion [10] E_{DEX+} = e^4 m_c/10 $\hbar^2 \varepsilon^2$ that gives E_{DEX+} = 7 meV in agreement with the experimental splitting of the DEX line.

CONCLUSIONS

PL imaging in taper sections of GaN on SiC (0001) substrates has been used to relate 364 nm emission to extended defects in the GaN. Additional splitting of this line at low temperatures and excitation levels has been interpreted with the aid of a recently developed theory of dislocation excitons.

ACKNOWLEDGEMENTS

We are thankful to P.Pirouz for useful discussion. For financial support, Y.G.S. thanks EPSRC and Y.T.R. thanks the Russian Fund for Fundamental Studies (grant no.96-02-175825-a).

REFERENCES

[1] Y.T.Rebane and Y.G.Shreter in *Polycrystalline Semiconductors II*, Eds.
     J.H.Werner and H.P.Strunk. Springer - Verlag, pp.28-39 (1991).
[2] Y.G.Shreter, Y.T.Rebane and A.R.Peaker. Phys.Stat.Sol..(a), **138**, 681 (1993).
[3] Y.G.Shreter et al, J.Crystal Growth, **159**, 883 (1996).
[4] F.A.Ponce, D.P.Bour, W.Gotz and P.J.Wright, Appl.Phys.Lett., **68**, 57 (1996).
[5] C.Wetzel et al, Appl.Phys.Lett., **68**, 2556 (1996).
[6] T.W.Weeks, Jr. et al, J.Mater.Res., **11**, 1011 (1996).
[7] X.J.Ning et al, J.Mater.Res., **11**, 580 (1996).
[8] Y.T.Rebane, Phys.Rev., **B52**, 1590 (1995).
[9] Y.T.Rebane and J.W.Steeds, Phys.Rev. **B48**, 14963 (1993).
[10] D.M.Bishop, J.Chem.Phys., **53**, 1541 (1970).

Coexistence of shallow and localized donor centers in bulk GaN crystals studied by high pressure Raman spectroscopy.

P. Perlin [*,#], T. Suski[+], A. Polian [**], J. C. Chervin[**], W. Knap [***], J. Camassel [***], I. Grzegory [+], S. Porowski [+], J.W. Erickson [++]

* Center for High Technology Materials, University of New Mexico, EECE Building, Albuquerque, NM 87131-6081 USA
+ High Pressure Research Center "Unipress", Sokolowska 29/37, 01-142 Warszawa, Poland
** Physique Des Milieux Condenses, Université Pierre et Marie Curie, France, Place Jussieu, F-75251 Paris, France
*** Groupe d'Etudes des Semiconducteurs, Université Montpellier 2 CNRS, place Eugene Bataillon F-34095 Montepellier, France.
++ Charls Evans and Associates, Redwood City, CA.

ABSTRACT

Character of the metal-insulator transition which occurs at about 23 GPa in bulk GaN crystals has been studied by means of high pressure Raman spectroscopy. The related freeze-out of electrons is caused by the localized donor state formed by most likely oxygen and emerging at high pressures to the band gap of GaN. As a result, the electron concentration drops from its initial value of $5 \cdot 10^{19}$ cm^{-3} to about $3 \cdot 10^{18}$ cm^{-3}. These remaining electrons originate likely from another donor center with effective mass character, probably carbon. The obtained results raise a question whether the nitrogen vacancy is abundant enough to be observed in bulk GaN crystals.

INTRODUCTION

One of the oldest problems of GaN technology is related to the n-type conductivity of this material. As grown GaN crystals, if undoped, are always characterized by a high background electron concentration of the order 10^{19}-10^{20} cm^{-3} for the bulk crystals and 10^{17}-10^{18} cm^{-3} for the epitaxial layers grown by the molecular beam epitaxy (MBE) or organometallic chemical vapor deposition (MOCVD) method. Maruska and Tietjen [3] and Ilegems and Montgomery [4] proposed that the autodoping is due to native defects, since the concentration of contaminants is lower by a few orders of magnitude than the highest observed electron concentration. This residual donor was tentatively identified as a nitrogen vacancy [3-6]. It was experimentally observed that $Ga_{1-x}Al_xN$ undergoes a metal-insulator transition when x approaches 20-40 % .[7] This situation is similar to that observed in GaAs where donor impurities like S, Si, Sn,Ge form both shallow and localized donor states [8]. These localized states in GaAs are resonant with the conduction band but when the energy gap gets larger with increasing content of Al in GaAlAs alloys they appear in the band gap and can trap the electrons. This similarity suggests that the dominant donor in GaN can have similar properties i.e. can form a localized state degenerate with the conduction band. The recent theoretical calculations of Boguslawski et. al [9] indicated that the nitrogen vacancy can form a resonant donor level as high as 800 meV above the conduction band minimum. Moreover an interstitial Ga forms a resonant donor level in the GaN conduction band. It has been also suggested that unintentionally introduced oxygen or silicon can lead to a highly n-type character of the undoped GaN material [10]. Only very recently, high

689

resolutions SIMS measurements were performed on bulk GaN crystals. They revealed high content of oxygen and carbon approaching 10^{20} cm^{-3} and 10^{18} cm^{-3} respectively. It has been shown that if GaN crystal are submitted to the external pressure exceeding 20 GPa the sudden decrease of electron concentration can be observed. [11] This effect can not be explained by the structural phase transition which occurs much higher in pressures (47 GPa) [12] neither by direct-indirect band gap transition. Fig 1. shows the various energy gaps of GaN [13] as a function of pressure. It is clear that within large error margin the GaN remains direct semiconductor until the phase transition to the rocksalt phase occurs. In that situation the only likely explanation of the observed decrease of electron concentration is the appearance of the localized state in the gap which traps the electron.

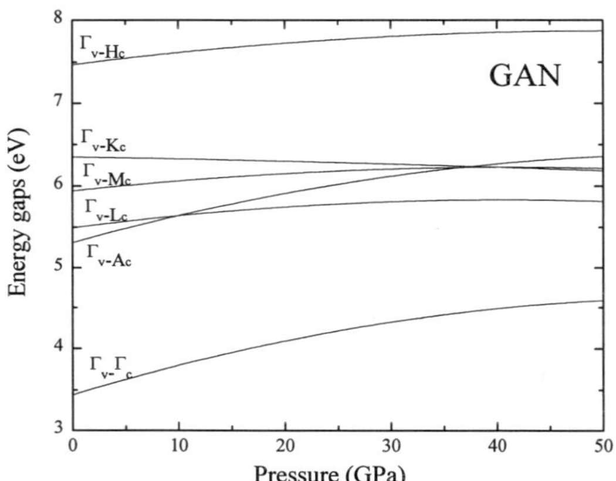

Figure 1. Different energy gaps of GaN in the function of hydrostatic pressure (from Ref. 13). The lowest curve corresponds to the pressure evolution of the main energy gap, i.e. a distance between valence and conduction band maximum and minimum, respectively (at the Γ point of the Brillouin zone)

Although this experimental result was in agreement with the theoretical model of the nitrogen vacancy [9,11] it did not exclude such a native donors like nitrogen interstitial. In order to get more quantitative information about the high pressure freeze-out we decided to use a fact that Raman scattering spectra of GaN are sensitive to the electron concentration via efficient LO phonon - plasmon coupling. It was shown before [14] that in bulk GaN sample of electron concentration $5 \cdot 10^{19}$ cm^{-3} instead of LO modes two coupled plasmon-phonon modes denoted LPP^{-} and LPP^{+} can be observed (Fig. 2 and Fig. 3). One should notice that the frequency and character of the LPP^{-} mode are very similar to the TO mode and these of LPP^{+} to the plasmon. The frequencies of the coupled modes can be expressed in simplified form [14] by the equation:

$$\omega^{\pm} = \frac{1}{2}\left\{ \omega_L^2 + \omega_p^2 \pm \left(\left(\omega_L^2 + \omega_T^2 \right)^2 - 4\omega_p^2 \omega_T^2 \right)^{1/2} \right\} \tag{1}$$

where ω_L and ω_T are LO and TO phonons frequencies and ω_p is a plasmon frequency.

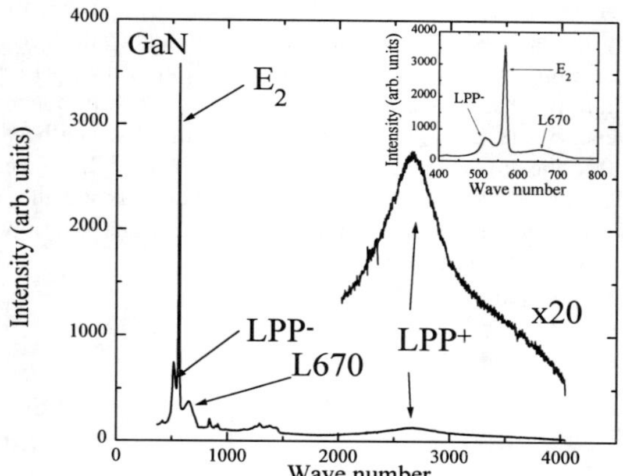

Figure 2. LO phonon - plasmon coupled modes in Raman spectrum of highly conductive GaN bulk sample. Inset shows enlarged spectrum around E_2 phonon mode.

It is demonstrated on Fig. 3 that frequencies of both coupled modes depend on electron concentration and thus can be used to monitor this concentration.

EXPERIMENT

The experiment was performed in a diamond anvil cell (membrane cell [15]). Argon was used as the pressure transmitting medium. The pressure was calibrated by the ruby luminescence method using the power-five relation [16]. The signal was analyzed using a Jobin Yvon 64000 spectrometer equipped with a nitrogen-cooled charge device (CCD) camera and a Notch filter.

RESULTS

Although it is clear from the Fig. 3 that at high electron concentration n_H, the position of LPP^+ mode is more sensitive to n_H changes, the position of LPP^- mode was used here because the LPP^+ mode was covered by strong two-phonon Raman scattering of diamonds which consist of the main part of the high-pressure cell. Fig. 4 shows the pressure evolution of the Raman spectra of bulk GaN. As the pressure is increased two effects can be observed: modes shift to higher frequencies and a few new bands appear above 20 GPa. This effect has been interpreted as a metal-insulator transition related to capture of electrons onto the localized states of a dominant donor. These states emerge to the GaN band gap at p>20 GPa. The band which appears close to 860 cm^{-1} at 23 GPa can be identified as a LO mode or more strictly speaking as a high frequency coupled LO phonon - plasmon mode $(LO'+L_A)$. If one looks at the upper curve in Fig. 3 it is easy to notice that the

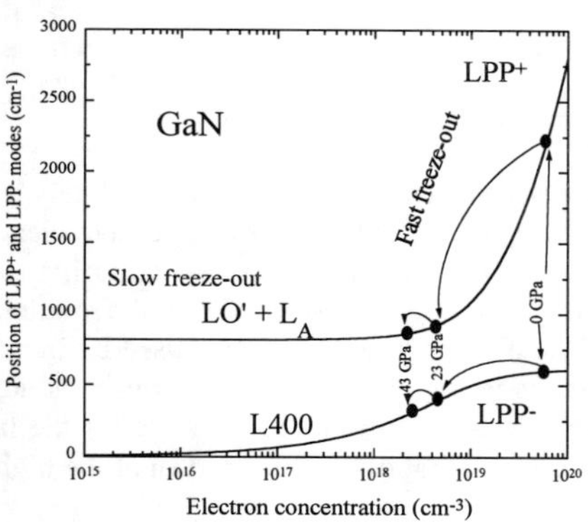

Figure 3. Energy of LO phonon - plasmon coupled modes as a function of electron concentration with shown process of freeze-out of electrons described further in the text.

appearance of this mode in this position means that the concentration of electrons dropped down from the initial $5 \cdot 10^{19} \text{cm}^{-3}$ to few times 10^{18}cm^{-3}. However, the flatness of the LPP^+ curve in this concentration region (or equivalently the applied pressure magnitude) does not allow for the better determination of electron concentration. Instead one can use an evolution of the position of LPP^- peak. At lower electron concentration the LPP^- which has initially very close frequency to $A_1(TO)$ mode should move to lower energies (see Fig. 3). Figure 5 shows the distance between the E_2 - phonon mode of wurtzite GaN the lower energy neighbor, which at higher electron concentration (lower pressure) is LPP^- mode. The sudden change of the inter-peak distance can be interpreted as a disappearance of the LPP^- mode and appearance of the transverse optical $A_1(TO)$ mode which has energy about 30 cm^{-1} lower than the E_2 mode.

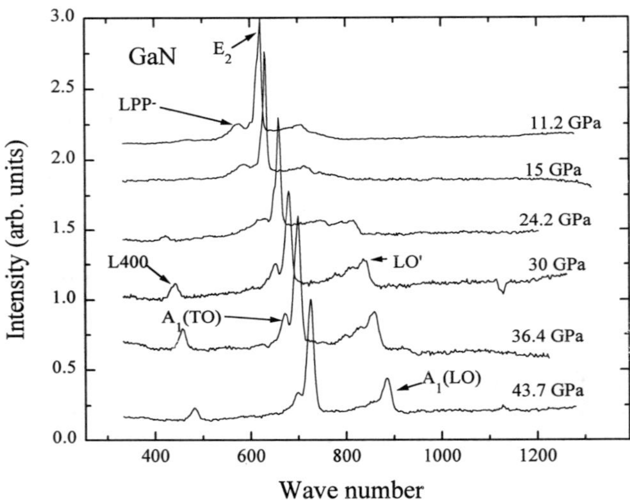

Figure 4. Raman spectra of bulk GaN sample measured at different pressures

Simultaneous appearance of the L400 mode (see Fig. 4) suggests that actually this is likely LPP^- mode observed in a new position. Using the Fig. 3 we can thus evaluate the electron concentration after the freeze-out as being close to $3 \cdot 10^{18}$ cm^{-3} which is still quite high and indicates presence of the other donor center ionized even at high pressures. Gradual evolution of the $\text{LO}^{'}$ line can be related to the slow decrease of the electron concentration caused by increase of donor binding energy. We can conclude this part suggesting that there are two different donor centers which represent active dopant in bulk GaN crystals. First one responsible for the high pressure freeze-out and the second one which preserves its shallow character even at the highest pressures.

DISCUSSION

In order to explain the observed phenomena we have to consider at first the donor which should have localized state which appears in the gap at the pressure close to 20 GPa and secondly find a candidate for shallow donor state which does not form localized levels. The candidates for localized donor states could be nitrogen vacancy [9], gallium interstitial [9] and oxygen [17]. Although nitrogen vacancy is an attractive candidates for a donor in GaN it was predicted by the theoretical calculations [10] that the concentration of this defect is too small to explain high carrier concentration of bulk GaN crystals. Even less likely from the point of view of formation energy is the presence of large number of gallium interstitial [10]. Recent theoretical calculation of oxygen impurity in GaN [17] show that it can form a localized states in the gap in AlGaN alloys. Also similar behavior was predicted for oxygen in GaN subjected to hydrostatic pressure [18]. Formation energy of oxygen in GaN is much lower than that of nitrogen vacancy [17] Very recent experiments of Wetzel et al. [19] showed also that epitaxial samples highly doped with oxygen undergo similar metal-insulator transition like this observed in bulk GaN crystals. Thus taking into account large concentration of oxygen measured by SIMS in bulk GaN crystals we have very strong indication that oxygen in GaN is responsible for the high free electron concentration in this material, as well as for metal-insulator transition observed at 20 GPa. However, we have to point out that there is no definite experimental proof that nitrogen vacancy does not contribute in some extent to the high electron concentration.

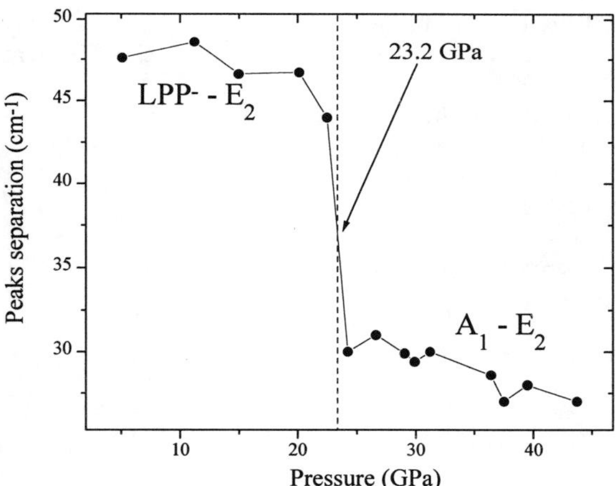

Figure 5. Pressure dependence of the energetic distance between E$_2$ mode and its low energy neighbor.

One can speculate on a possible origin of the second donor (hydrogenic-like) responsible for the presence of electrons in the conduction band of bulk GaN at pressures higher than 23 GPa. Carbon represents a likely candidate. Large concentration of this dopant has been found in bulk GaN crystals. Moreover, first principle calculations of Boguslawski et. al [20] has shown that in GaN C forms a shallow donor state.

REFERENCES

On live from High Pressure Research Center, "Unipress" Warsaw, Poland

1. S. Nakamura, J. Cryst. Growth **145**, 911 (1994)
2. S. Nakamura, M. Senoh, S. Nagahama, N. Iwasa, T. Yamada, T. Matsushita, H. Kiyoku, Y. Sugimoto, Jpn. J. Appl. Phys. **35**, L74, (1996).
3. H.P. Maruska and J.J. Tietjen, Appl. Phys. Lett. **15**, 327 (1992)
4. M. Ilegems and M. C. Montgomery, J. Phys. Chem. Solids, **34**, 885 (1969)
5. B. Monemar, O. Lagerstedt, J. Appl. Phys. **50**, 6480, (1979)
6. T.L. Tansley, R.J. Egan, Phys.Rev. B. **45,** 10942, (1992)
7. D. W. Jenkins, J. Dow, M. Tsai, J. Appl. Phys. **72**, 4131, (1992)
8. T. Suski, Materials Science Forum, **143-147**, 975 (1994)
9. P. Boguslawski, E. Briggs, J. Bernholc, Phys. Rev. B. **51**, 17255, (1995)
10 J. Neugebauer and C.G. Van de Walle, in Proc. 22nd Int. Conf. on the Physics of Semiconductors, Vancouver, 1994 (World Scientific Publishing Co. Pte. Ltd., Singapore) p.2327
11. P. Perlin,T. Suski, H. Teisseyre, M. Leszczynski, I. Grzegory, J.Jun, S. Porowski, P.Boguslawski, J. Bernholc, J.C. Chervin, A. Polian, T.D. Moustakas, Phys. Rev. Lett. **75**, 296 (1995); C. Wetzel, W. Walukiewicz, E.E. Haller, J. W. Ager III, I. Grzegory, S. Porowski and T. Suski, Phys. Rev. B **53**, 1322 (19960
12 P. Perlin, C. Jauberthie-Cariilon, Jean Paul Itie, A. San Miguel, I. Grzegory, A. Polian, Phys. Rev. B **45**, 83, (1992)
13. N.E. Christensen, I. Gorczyca, Phys. Rev. B, **50**, 4397 (1994)
14 P. Perlin, J. Camassel, W. Knap, T. Taliercio, J. C. Chervin, T. Suski, I. Grzegory, S. Porowski, Appl. Phys. Lett. **67**, 2524, (1995)
15. R. Le Toullec, J. P. Pinceaux, P. Loubeyre, High Pressure Res. **1**, 77 (1988)
16. H.K. Mao, P.M. Bell, J.W. Shaner, D. J. Steinberg, J. Appl. Phys. **49**, 3276 (1978)
17. T. Mattila, R.M. Nieminen, Phys. Rev. B in print
18. J. Neugebauer, private communication.
19 C. Wetzel et al. unpublished.
20. P. Boguslawski, E.L. Briggs, J. Bernholc, Appl. Phys. Lett. **69**, 233 (1996)

CHARACTERIZATION OF NEAR EDGE OPTICAL TRANSITIONS IN UNDOPED AND DOPED GaN/SAPPHIRE GROWN BY MOVPE, HVPE AND GSMBE

M.LEROUX, B.BEAUMONT, N.GRANDJEAN, J.MASSIES, P.GIBART.
Centre de Recherches sur l' Hétéro-Epitaxie et ses Applications-CNRS,
Rue B. Grégory, 06560 Valbonne (France), email: mleroux@crhea1.unice.fr

ABSTRACT

We report a comparative study by luminescence and reflectivity of GaN grown on sapphire by MOVPE, HVPE and GSMBE. Whatever the growth technique, undoped GaN shows well resolved reflectivity spectra, allowing hetero-epitaxial strain to be taken into account in the interpretation of luminescence spectra. MOVPE provides the highest purity samples so far. From the luminescence study of n- and p-type doped samples, activation energies of 265 ± 10 meV and 34 ± 2 meV are deduced for Mg acceptors and Si donors respectively. From the luminescence spectra of HVPE grown samples, the presence of a perturbed interfacial n^+ layer is evidenced.

INTRODUCTION

This paper describes the results of an optical characterization of hetero-epitaxial GaN layers on (0001) sapphire grown using three different techniques: metal-organics vapor phase epitaxy (MOVPE), halide vapor phase epitaxy (HVPE) and gas source molecular beam epitaxy (GSMBE). MOVPE is today the most widely used technique for GaN growth, illustrated by the achievement of injection laser diodes [1]. HVPE, allowing high growth rates (~ 30 µm/h) is a promising technique for the obtention of pseudo-substrates for GaN homo-epitaxy. MBE techniques allow low growth temperatures, and also provide access to *in situ* characterization tools, invaluable for the understanding of some basic aspects of the hetero-epitaxial growth [2].

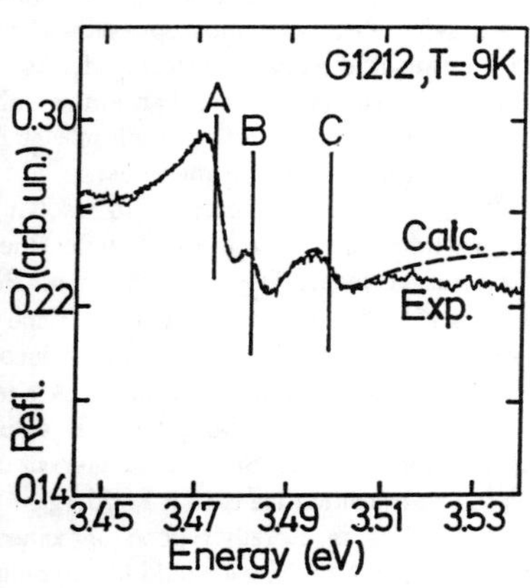

Figure 1: Near edge reflectivity spectrum of undoped MOVPE grown GaN/Al₂O₃ at 9 K, and corresponding calculated spectrum.

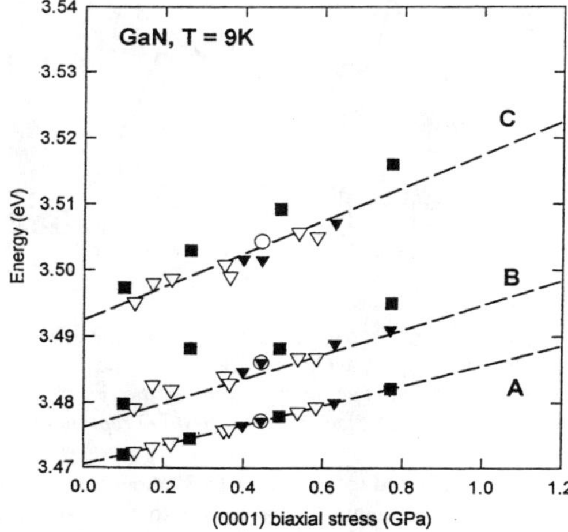

Figure 2: A, B and C exciton energies in GaN at 9 K, A being aligned to the (0001) biaxial strain scale of [3]. Triangles, squares and circles correspond to MOVPE, HVPE and GSMBE grown samples respectively.

Mat. Res. Soc. Symp. Proc. Vol. 449 © 1997 Materials Research Society

RESULTS AND DISCUSSION

Generalities

For the three growth techniques investigated, undoped GaN displays well resolved reflectivity spectra at low temperature (9 K). As an example, figure 1 shows such a spectrum together with the corresponding calculated one. The fitting is performed by adding three Lorentzian oscillators, corresponding to the band edge free excitons (noted A, B and C) of the wurtzite band structure to the background dielectric constant. Such a fitting, systematically performed, allows to obtain the positions of A, B and C excitons with a precision better than 1 meV. Figure 2 illustrates how important this knowledge is: the energies of the free excitons of various GaN/Al$_2$O$_3$ samples are plotted, the A energy being aligned with the biaxial strain scale of Gil *et al.* [3]. Variations in the ground state energies exceeding 10 meV are recorded, important to be known for the interpretation of the photoluminescence (PL) spectra.

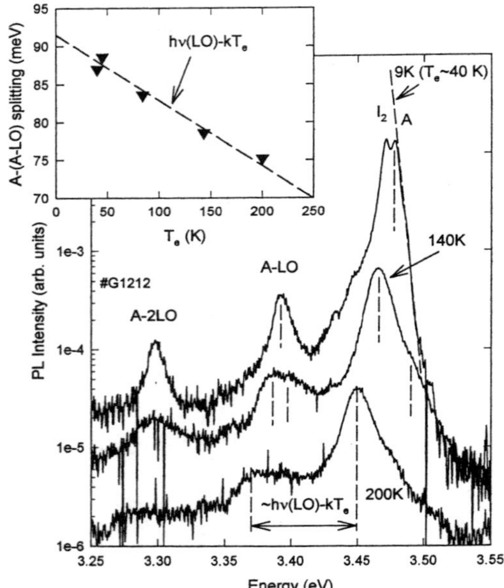

Figure 3: Temperature dependence of the near edge PL spectra of undoped, MOVPE-grown GaN. The inset shows the variation with temperature of the splitting between the free A exciton and its first LO repliqua.

Some additionnal remarks have to be done in the light of figure 2. First, whereas MOVPE (triangles) and GSMBE (circles) data are in good agreement with the (0001) biaxial strain scale of reference 3, it is not the case for HVPE ones (squares). The thickness of HVPE samples ranges from 20 to 100 µm, and their strain state is different from that of thin layers. In particular, their surface is often composed of large (~ 15 µm) 3-dimensionnal crystals, which can be supposed not to be under pure biaxial compression. Another remark is that MOVPE samples grown on an AlN buffer layer (dark triangles) are in the average more compressed than those grown on a GaN buffer (open triangles). Actually, transmission electron microscopy shows that the interfacial density of stacking faults is lower in the AlN buffer case than in the GaN buffer one [4], leading to less strain release in the former case than in the latter one.

It is, in some cases, possible to resolve a fourth weak transition in the GaN reflectance spectrum, corresponding also to weak PL features [5]. It is separated by 19 ± 1 meV from A, and its oscillator strength, deduced from reflectance, is about 1/10 of the A one, *i.e.* not very different from the 1/8 value expected for a 2s state. Such an assignment of this fourth transition leads to a Rydberg of $\approx$ 25 meV. This is in rather good agreement with both early and recent reports [6-8], though controversies still exist [9]. Finally, we would like to point that the data of figure 2 correspond to nominally undoped samples. When doping n or p-type, the reflectivity spectra broaden and become eventually unresolved. They are anyhow still useful by providing a check of the approximate band gap, through the damping of interference oscillations (see for instance figure 5

Undoped MOVPE-grown GaN

Figure 3 displays some PL spectra of a nominally undoped MOVPE-grown GaN layer as a function of temperature. The residual doping lies in the low 10^{16} cm^{-3} range. The spectra are dominated at low temperatures by two transitions coresponding to the recombination of free A excitons and to residual donor bound excitons (I$_2$) [6,8,9]. The low temperature PL energy of A is generally 1-2 meV higher from that deduced from reflectance fitting [5]. This can be due to the high electronic temperature (T$_e$) of 40 K induced by using HeCd laser excitation. At lower energies are observed the 1 and 2 LO phonon repliqua of A. As the inset of figure 3 shows, the A-(A-LO) splitting is the LO phonon energy minus kT$_e$, since the Fröhlich interaction is forbidden at $\mathbf{k} = \mathbf{0}$ for excitons [10]. Note also the absence of acceptor-related PL in the 3.27 eV range. Combination of PL and reflectivity shows that GaN luminescence is dominated by free A excitons at room temperature (RT) [9,11].

Si-doped MOVPE-grown GaN

The low residual doping of the layers, and the purity of the PL spectra as those shown on figure 3, allows to study the optical properties of doped samples. n-type doping can be achieved up to $\sim 10^{20}$ cm^{-3} using silane.

Figure 4 displays the PL spectra at 9 K of four Si doped GaN samples, with RT electron densities ranging from 4×10^{17} cm^{-3} to 6×10^{19} cm^{-3}. The spectrum of the lower doped sample is dominated by a transition at 3.461 eV, whose energy is independent of the excitation intensity. With increasing temperature, this band is progressively replaced by a second one, 9 meV higher in energy, corresponding to A exciton recombination (this is confirmed by reflectivity). We assign the 3.461 eV band to free hole-neutral donor (D^{0}h) recombinations. Actually, we can discard its assignment to the I$_2$ transition, since I$_2$ is only 6 meV lower in energy than A [5, 7, 9], and is thermally quenched at 50 K, whereas the D^{0}h band is dominant up to 90 K. Using the previously quoted value of the A Rydberg, a depth of 34 $\pm$ 2 meV is obtained for Si donors.

It should be stressed that this is in the range quoted for the activation energy of the residual donor in GaN [5, 12, 13]. It

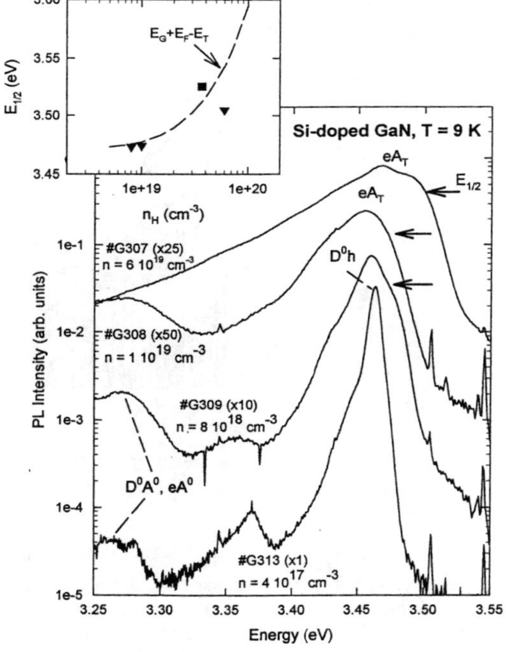

Figure 4: PL spectra (9 K) of Si-doped MOVPE grown GaN for various RT electron concentrations n_H. The inset shows the high energy cutoff of the PL band of highly n-doped GaN as a fonction of n_H (triangles: MOVPE, squares: HVPE).

corresponds also rather well to the depth of an hydrogenic donor, using an electron mass of 0.22 m$_0$ and a static dielectric constant of $\sim$ 9.

Figure 4 shows that the spectra of the highly doped (n > 10^{18} cm^{-3}) samples are asymmetric, with a broad low energy tail and a rather sharp high energy cutoff. In particular, the PL cutoff of sample #G307, doped to 6×10^{19} cm^{-3}, is higher than the GaN band gap. Such spectra are typical of direct gap n-type semiconductors well above degeneracy. Actually, we have determined the donor Mott transition to occur in GaN for n$_c$ = 10^{18} cm^{-3}, in agreement with theory, and it becomes meaningless to talk of donor states above this value. The inset of figure 4 compares the energy of the high energy cutoff of the PL spectra ($E_{1/2}$, defined on figure 4) to that calculated by a semi-empirical formula approximately valid for highly n-doped GaAs [14]. Though the agreement is not excellent, the overall trends are well accounted for, and we assign the broad PL band recorded in highly n-doped GaN to indirect (in **k** space) transitions between free electrons up to the Fermi level E_F and holes in acceptor-like tail states of average energy E_T (hence the labeling as eA_T of the PL bands in figure 4).

<u>Mg-doped MOVPE-grown GaN</u>

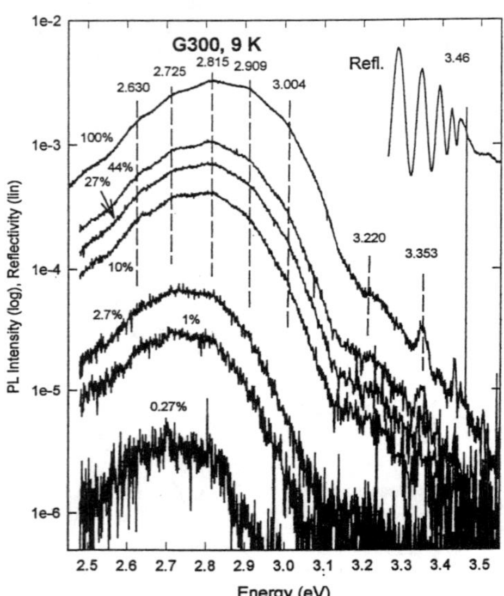

Figure 5: Excitation intensity dependence of the PL spectra at 9 K of highly Mg-doped MOVPE-grown GaN (p $\approx 10^{18}$ cm^{-3} at 300K).

p-type doping up to a RT hole concentration of several 10^{18} cm^{-3} (corresponding to $\sim 10^{20}$ Mg atoms cm^{-3}) using bis-methylcyclopentadienyl magnesium can be achieved [15]. The evolution of the low temperature PL spectra with increasing Mg doping can be found in [5,13]. For weak Mg doping, *i.e.* n-type compensated samples or p $\sim 10^{16}$ cm^{-3}, the spectra become dominated by 'shallow' donor-acceptor pair (DAP) bands. The analysis of the evolution of the DAP peak energy with excitation intensity gives acceptor depths of 220 $\pm$ 5 meV for the residual acceptor (observable without Mg doping) and 265 $\pm$ 10 meV for a deeper acceptor, involved in a PL band appearing with Mg doping, and that we assign to Mg acceptors [5,13]. However, with increasing doping, the PL bands broaden and deepen, being dominated by a 'blue band' in the 2.7-2.9 eV range, depending on doping and excitation intensity. The occurrence of this Mg blue band is very often reported in the litterature [13, 16, 17] and is independent of the GaN growth technique.

Figure 5 shows a typical low temperature PL (and reflectivity) spectrum of highly Mg-doped GaN, and its dependence with excitation intensity I_{exc}. Besides the blue band, structures at 3.22 and 3.27 eV can be seen on the high excitation spectra. They correspond to DAP bands involving shallow acceptors and donors [13]. The blue band shows structures separated by 95 meV in the average, that may correspond to LO phonons. Its maximum shifts of 90 meV for two decades of I_{exc} variation, much higher than what is expected for classical DAP recombinations. Actually, a possible interpretation of the spectra shown on figure 5 is that they involve two overlapping bands, one peaking near 2.7 eV, saturating rapidly with I_{exc}, and a second one peaking near 2.9 eV, observable under high excitation. Remembering that PL involves minority electrons, and noticing that the 'shallow' DAP bands are also observable under high excitation intensity (figure 5) suggests that the occurrence of this deep Mg blue band is related to the formation of deep electron levels rather than deep hole ones, in agreement with Smith *et al.* [17]. This hypothesis agrees with the general observation that the Hall depth of Mg acceptors is roughly independent of the doping level (though Hall and optical depths desagree). Also to be remembered is that large DAP shifts can be recorded in GaAs, but only in the case of strong compensation [18]. In the GaAs case, this strong shifts are interpreted by band structure distorsions by inversely charged impurity atoms and concentration fluctuations [18]. On another hand, the reflectivity spectrum displayed in figure 5 shows that the absorption edge of our sample is in the 3.46-3.47 eV range, suggesting that band tailing is not strong. As a support for our hypothesis, we mention that the formation of deep donor levels induced by Mg doping has been theoretically predicted [19]. Clearly, a definite understanding of the exact nature of PL in highly Mg doped GaN needs further experiments.

HVPE and GSMBE-grown GaN.

HVPE grown samples are much thicker than MOVPE ones, and as mentionned previously, their strain state is different from that of thin layers. Well defined PL and reflectivity spectra are obtained from HVPE samples [5, 13]. A residual doping in the 10^{18} cm^{-3} range is measured, though well resolved bound and free exciton transitions are recorded. A possible explanation of this fact can be the presence of a perturbed n+ interfacial layer, deduced from the comparison of front and substrate side PL spectra [13], that could influence Hall measurements. To get a better insight on this layer, cathodoluminescence (CL) have been performed on the cleaved edge of a 100 μm thick HVPE sample.

The results are shown in figure 6, displaying CL spectra at 120 K recorded as a function of depth in the sample. The spectra correspond to points separated by about 10 μm from each other. As figure 6 shows, the perturbed region, that we identify from a spectrum broadened towards both high and low energies, extends up to about 50 μm from the interface. The existence of a

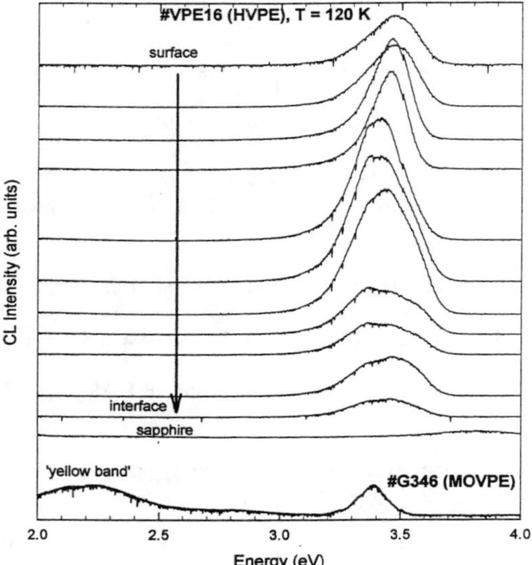

Figure 6: CL spectra at 120 K of the cleaved face of a nominally undoped HVPE-grown GaN sample (100 μm thick) as a function of depth. The CL spectrum of a MOVPE-grown sample is also shown for comparison.

perturbed interfacial region extending a few tenths of μm has already been reported by Siegle *et al.* [20] from Raman studies, and ascribed to a disoriented layer. Finally, figure 6 shows also for comparison the CL spectrum of a MOVPE-grown sample, emphazising the absence of the 'yellow band' in HVPE grown GaN.

GSMBE allows to grow at temperatures as low as 800 °C GaN with good optoelectronic properties. The results are reported elsewhere in this volume [21].

ACKNOWLEDGMENTS

D.Le Si Dang and Y.Genuist (CENG-CNRS, Grenoble) are highly acknowledged for the CL studies. Numerous discussions with B.Gil (CNRS, Montpellier) and R.Planel (CNRS, Bagneux) have been of invaluable help. We thank P.Lorenzini, G.Nataf and C.Golivet for their hand, and J.P. Faurie for continuous support. This work is supported in part by an EU contract ESPRIT-LTR LAQUANI n° 20968.

REFERENCES

1. S.Nakamura, M.Senoh, S.Nagahama, N.Iwasa, T.Yamada, T.Matsushita, H.Kiyoku and Y.Sugimoto, Japan. J. Appl. Phys. **35**, L217 (1996).

2. N.Grandjean, J.Massies and M.Leroux, Appl. Phys. Lett. **69**, 2071 (1996).

3. B.Gil, O.Briot and R.L.Aulombard, Phys. Rev. B **25**, R17028 (1995).

4. P.Vennègues, B.Beaumont, M.Vaille and P.Gibart, J. Cryst. Growth, to be published.

5. M.Leroux, B.Beaumont, N.Grandjean, C.Golivet, P.Gibart, J.Massies, J.Leymarie, A.Vasson and A.M.Vasson, Mat. Sci. Eng. B in press.

6. B.Monemar, Phys. Rev. B **10**, 676 (1974).

7. D.Volm, K.Oettinger, T.Streibl, D.Kovalev, M.Ben-Chorin, J.Diener, B.K.Meyer, J.Majewski, L.Eckey, A.Hoffmann, H.Amano, I.Akasaki, K.Hiramatsu and D.Detchprohm, Phys. Rev. B **53**, 16543 (1996).

8. R.Dingle, D.D.Sell, S.E.Stokowski and M.Ilegems, Phys. Rev. B **4**, 1211 (1971).

9. M.Tchounkeu, O.Briot, B.Gil, J.P.Alexis and R.L.Aulombard, J. Appl. Phys. **80**, 5352 (1996).

10. D.Kovalev, B.Averboukh, D.Volm, B.K.Meyer, H.Amano and I.Akasaki, Phys. Rev. B **54**, 2518 (1996).

11. B.Beaumont and P.Gibart, <u>Semiconductor Heteroepitaxy</u>, ed. by B.Gil and R.L.Aulombard (World Scientific, Singapore 1995), p 258

12. S.Strite and H.Morkoç, J. Vac. Sci. Technol. B **10**, 1237 (1992).

13. M.Leroux, B.Beaumont, N.Grandjean, P.Gibart, J.Massies and J.P.Faurie, MRS Internet J. Nitride Semicond. Res. **1**, 25 (1996).

14. J.De-Sheng, Y.Makita, K.Ploog and H.J.Queisser, J. Appl. Phys. **53**, 999 (1982).

15. B.Beaumont, M.Vaille, P.Lorenzini, P.Gibart, T.Boufaden and B.El Jani, MRS Internet J. Nitride Semicond. Res. **1**, 17 (1996).

16. W.Gotz, N.M.Johnson, J.Walker, D.P.Bour and R.A.Street, Appl. Phys. Lett. **68**, 667 (1996).

17. M.Smith, G.D.Chen, J.Y.Lin, H.X.Jiang, A.Salvador, B.N.Sverdlov, A.Botchkarev, H.Morkoç and B.Goldenberg, Appl. Phys. Lett. **68**, 415 (1996).

18. Phil Won Yu, J. Appl. Phys. **48**, 5043 (1977).

19. J.Neugebauer, <u>Proc. 1st European GaN Workshop</u> (Rigi, Switzerland 1996) to be published.

20. H.Siegle, P.Thurian, L.Eckey, C.Thomsen, B.K.Meyer, H.Amano, I.Akasaki, T.Detchprohm and K.Hiramatsu, Appl. Phys. Lett. **68**, 1265 (1996).

21. N.Grandjean, J.Massies, P.Vennégues, S.Laügt and M.Leroux, <u>this volume</u>.

PHOTOLUMINESCENCE EXCITATION STUDIES OF THE OPTICAL TRANSITIONS IN GaN

D. KOVALEV*, B. AVERBOUKH*, B.K. MEYER**, D. VOLM**,
H.AMANO***, I.AKASAKI***
*Technische Universität München,Physik-Department E16, D-85747 Garching, Germany
**1. Physics Institute, University Giessen, Heinrich-Buff-Ring 16, D-35392
Giessen, Germany
***Department of Electrical and Electronical Engineering, Meiji University,1-501
Shiogamaguchi, Tempaku-ku, Nagoya 468, Japan

ABSTRACT

We present a detailed photoluminescence excitation study of the optical transitions in GaN. This technique is employed to distinguish between band-to-band excitation and exciton contribution to the formation of the free exciton, bound exciton, violet and yellow photoluminescence bands. We show the dominant role of the Fröhlich polar intraband scattering in the formation of the free exciton states. We demonstrate that bound exciton states in a large extent are created by the capture of the free excitons by shallow impurities as well as by phonon-assisted resonant excitation of the bound exciton states. The capture of the free carriers excited in the band continuum is a main excitation source for the violet and yellow bands. However, distinct A- and C-exciton resonances are detected in the excitation spectra of the violet and yellow emission bands.

INTRODUCTION

By the moment the photoluminescence of wurzite-type GaN through the shallow electronic states (such as free and bound excitons) [1-4] as well as through the shallow or deep donor-acceptor pair electronic states [1,5] has been well documented. However, there is a lack of knowledge on the kinetic properties of the electronic states contributing to the PL. The relative strength of the optical transitions cannot be investigated correctly without information about mechanism of the formation of these states under optical excitation. One of the principal features of the optical processes in the polar semiconductors is the intensive interaction of the electronic excitations with LO phonons. This effect affects both the excitation and recombination processes but cannot be seen in the absorption. In our experiments the photoluminescence excitation (PLE) technique has been employed to identify the excitation channels of the free exciton (FEx), bound exciton (BEx), violet and yellow emission bands. Using this technique we distinguish between the contribution to the excitation process via electron-hole pairs and indirect phonon-assisted exciton formation.

EXPERIMENTAL RESULTS AND DISCUSSION

Hexagonal GaN films are prepared by hydride and organometallic vapour phase epitaxy on Al_2O_3 or 6H-SiC with the c axis perpendicular to the surface plane. Photoluminescence is excited by normally incident 3.8 eV line of the He-Cd laser and detected with a single monochromator and lock-in technique. The PLE technique is used to monitor the efficiency of the optical excitation for the different optical transitions. The light from a 1000 W Xe arc lamp is dispersed

Mat. Res. Soc. Symp. Proc. Vol. 449 © 1997 Materials Research Society

by a single monochromator with a spectral resolution of 3 Å and focused on the sample using quartz lenses. The emission is spectrally resolved with a second single monochromator. The intensity of the emission is normalised simultaneously on the intensity of excitation by inserting

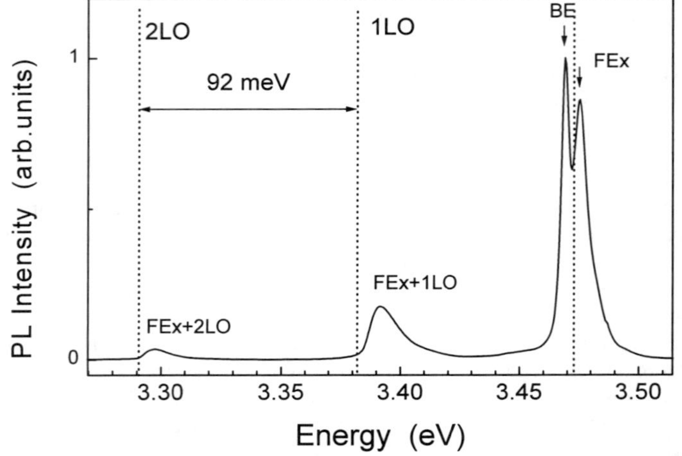

Figure 1. Photoluminescence spectrum of GaN on SiC at T=2K. $E_{ex.}$=3.8eV.
Grids indicate the positions of LO-phonons with respect to the bottom of FEx band.

the beam splitter in the path of the exciting light. A pirodetector with a spectrally independent response is used to measure the flux of the exciting light.

Fig.1 shows the PL spectrum in the region of the exciton emission . One could distinguish the strong emission lines related to the FEx (3.473 eV) and BEx (3.468 eV) transitions as well as 1LO- and 2-LO FEx replicas. Vertical grids indicate the bottom of the exciton band and the positions of the LO phonons with respect to this point. The identification of the nature of these lines was done in [4] using temperature dependent line shape and activation energy analysis. In the low energy region the violet and yellow bands are detected as well.

Optically active electronic excitations could be created through the absorption in the continuum of the density of e-h states as well as by the direct or indirect phonon-assisted resonant formation. The powerful way to study these phenomena is a PLE technique. By separating the optical transition of interest and varying the energy of the excitation the resonances related to the direct or indirect formation of electronic state could be monitored.

Figure 2 shows the PLE spectra measured at the detection energies corresponding to the spectral position of the FEx and FEx+1LO emission lines. Measurements are done at 70K to avoid spectral overlapping of the FEx and BEx lines seen at low temperatures (Fig.1). For the both lines up to 6 resonances with apparent periodicity of 92 meV (energy of LO phonon in GaN) are seen. For the FEx-1LO line the resonance related to the direct FEx formation is detected as well. The amplitude of these phonon-related oscillations increases toward the lower energies and the intensity of structureless background is of the order of magnitude lower. This assures that the indirect free exciton formation with simultaneous emission of several LO phonons or followed by hot exciton cascade [6] rather than e-h interband transitions with subsequent bounding of *e* and *h* into the FEx state takes place in GaN. These hot excitons relax quickly toward the bottom of the exciton band (luminescing zero-phonon FEx state) with emission of LO or acoustical phonons. If the initial kinetic energy of the excitons is equal to the integer number of LO phonons energy, the FEx's thermalization process is completed within 10^{-13}-10^{-12} s. Otherwise, the relaxation with the emission of the acoustical phonon has to be taken into consideration. This process is much slower and the time of relaxation can be longer than the nonradiative one. Therefore, if the thermal equilibrium is not established the significant enhancement of the FEx quantum yield has to be seen at energies exactly equal to the energy of LO phonon with respect to the luminescing (k_{FE}=0) FEx state. As could be seen from Fig.2 this is obviously not the case.

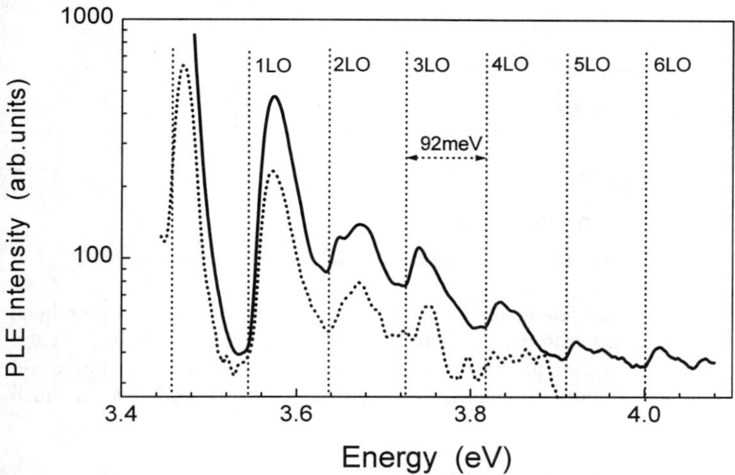

Figure 2. Photoluminescence excitation spectra of GaN on SiC at T=70K. Dotted line - detected at FEx+1LO-phonon line, solid line - FEx line.

The width of these features is comparable with the LO-phonon energy and cannot be explained by the dispersion of the LO-phonon branch. Furthermore, the PLE spectrum measured at the energy of FEx+1LO-phonon transition exhibits all the features mentioned above. The phonon-assisted FEx emission has no forbidden (by **k**- conservation rule) recombination since the LO phonon could always compensate the **k** vector of FEx. The similarity of both PLE spectra implies the very efficient energy relaxation of FEx's via emission of acoustical phonons and the thermal equilibrium of the exciton subsystem with the lattice [6].

The bound exciton states could appear from the subsequent capture of the free *e - h* or FEx on the neutral impurity or via resonant excitation with emission of the host lattice or local phonons. Impurities introduced into semiconductor host matrix give rise to the local vibration modes. For the substitution impurity atoms the frequency depends on their masses and local forces [7]. All these factors have to be considered under analysis of the BEx PL excitation spectra. Fig.3 shows the PLE spectra of free (T=30K) and bound exciton (T=2K) transitions in the vicinity of the exciton band. Both spectra exhibit C-exciton peak as well as the resonant feature related to the LO-phonon assisted free exciton formation. That proofs that the bound exciton states in GaN are created via capture of the free excitons by the neutral donors rather than by subsequent capture of e and h. In addition, the strong background signal with ill-

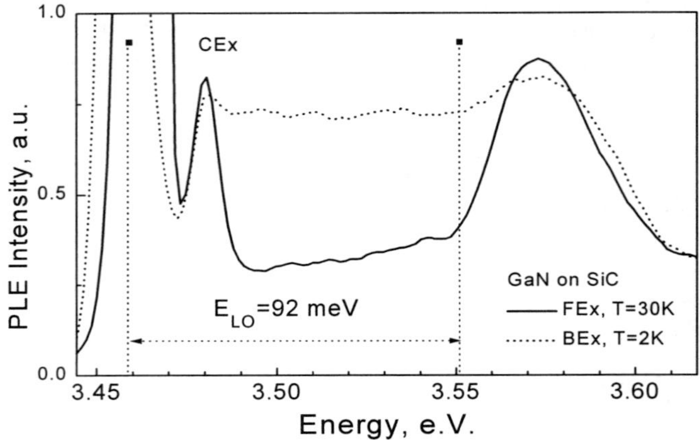

Figure 3. Photoluminescence excitation spectra of the free and bound exciton states.

defined reproducible structure in the intermediate region of energies is seen. Since its amplitude is of the order of magnitude larger than that for nonresonant excitation (far above the exciton resonance), we believe that it is due to the phonon-assisted BEx excitation. The structure could appear from the local vibrational modes (the energy of the BEx excited states has to be much smaller).

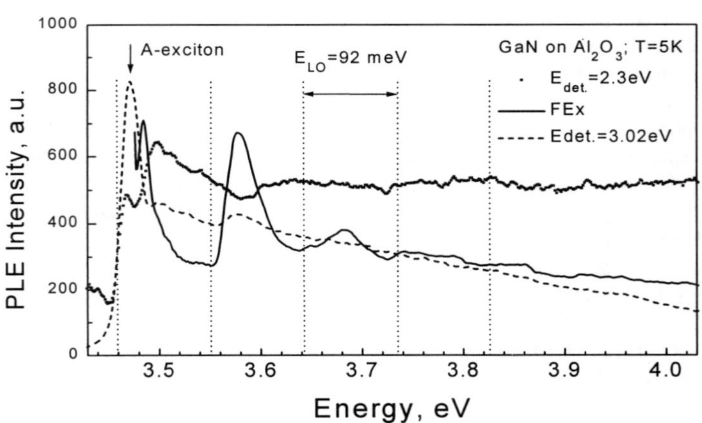

Figure 4. Photoluminescence excitation spectra of FEx, violet and yellow bands.

The PLE spectra of the violet and yellow bands (detection energies 3.02 and 2.3 eV respectively) are shown on Fig.4. The PLE spectrum of the violet band, in general, follows the FEx formation efficiency with relatively weak intensity of the resonance features. The PLE behavior of the yellow one exhibits opposite tendency: maxima in the PLE spectrum of FEx transitions coincide with minima in the PLE spectrum of the yellow one. Furthermore, the excitation efficiency of the yellow band remains constant at energies higher than the energy of the first exciton resonance (E_{FE}+1LO-phonon). Therefore, the electron and hole excited in the band continuum are essential for the yellow emission band. In the energy regions where the most efficient FEx formation takes place and , therefore, there is a lack of the *e-h* excitations the local minima in the PLE signal are seen.

CONCLUSIONS

We show that the formation of the free exciton states in GaN occurs via the indirect LO-phonon assisted process. We demonstrate that bound exciton states in a large extent are created by capture of free excitons on shallow impurities. Violet and yellow bands are excited due to band-to-band transitions.

D.Kovalev is grateful to the Alexander von Humboldt Foundation for support.

REFERENCES

1.R.Dingle, D.D.Sell, S.E.Stokowski, M.Illegems, Phys.Rev.B, v.**4,** 1211 (1971)

2.H.Amano, K.Hiramatsu, I.Akasaki, Jpn. J. Appl. Phys., **v.**27, L1384 (1988)

3.M.Smith, G.D.Chen, J.Z.Li, J.Y.Lin, H.X.Jiang, A.Salvador, W.K.Kim, O.Autas, A.Botchkarev, H.Morkoc, Appl.Phys.Lett., v.67, 3387 (1995)

4.D.Kovalev, B.Averboukh, D.Volm, B.Meyer, H.Amano, I.Akasaki, Phys.Rev.B, **v.**54, 2518 (1996)

5.M.I.Ilegnis, R.Dingle, J.Appl.Phys., v.44, 4234 (1975)

6.S.Permogorov, in *Modern problems in Condensed Matter Sciences*, ed. By E.I Rashba and M.D.Sturge (North-Holland, Amsterdam, 1982), Vol.2, 177

7. M. Vandevyver, P.Plumelle, Phys.Rev.B, v.17, 675 (1978)

PHOTOLUMINESCENCE OF Fe-COMPLEXES IN GaN

P. THURIAN, A. HOFFMANN, L. ECKEY, P. MAXIM, R. HEITZ, I. BROSER,
K. PRESSEL*, B.-K. MEYER**, J. SCHNEIDER***, J. BAUR***, M. KUNZER***
TU-Berlin, Institut für Festkörperphysik, Hardenbergstr. 36, 10623 Berlin, Germany
*Institut für Halbleiterphysik, P. O. Box 409, 15204 Frankfurt (Oder), Germany
**Justus-Liebig-Universität Giessen, Heinrich-Buff-Ring 16, D-35392 Giessen Germany
***Fraunhofer-Institut, Tullastr. 72, D-79108 Freiburg, Germany

ABSTRACT

We report a photoluminescence (PL) and photoluminescence excitation (PLE) investigation of the deep iron acceptor in hexagonal GaN. Codoped samples give rise to a new zero-phonon line (ZPL) at 1.268 eV. The spectral shape of its phonon side band is very similar to that of the Fe^{3+}-spectrum. A lifetime of 2.6 ms was measured for this ZPL which indicates a spin-forbidden transition. In contrast to isolated Fe^{3+}, no Zeeman-splitting is observed. PLE spectra of the Fe^{3+} ($^4T_1$-$^6A_1$) zero-phonon line at 1.299 eV in semi-insulating GaN samples reveal intracenter excitation processes via excited states of the Fe^{3+} center. Three excited Fe^{3+} crystal field states at 1.299 eV ($^4T_1$), 2.01 eV ($^4T_2$), and 2.731 eV (^4E) above the $^6A_1$ ground state were identified. PLE spectroscopy for the 1.268 eV zero-phonon line reveals resonances at 2.3 eV and 2.65 eV. The 1.268 eV ZPL is tentatively attributed to a Fe-complex with a nearby donor.

INTRODUCTION

GaN and related alloy systems are very promising for applications as widegap LEDs and injection lasers[1], as well as for high-temperature electronics. The investigation of impurity, native-point or extended-defect properties is essential to optimize the device quality. However, only few studies of transition metals (TMs) exist. A luminescence band with a ZPL at 1.299 eV was unambiguously assigned to the Fe^{3+} ($^4T_1$-$^6A_1$) transition.[2,3] Recently, detailed PLE data of isolated Fe^{3+} in GaN was presented and new excited states of the d^5 configuration were identified.[4,5] Additionally, a new zero-phonon line (ZPL) at 1.268 eV was observed and tentatively attributed to Fe-related defects. The aim of this paper is to compare the properties of the isolated Fe^{3+} center and iron-related defect centers in GaN by means of PL and PLE spectroscopy. Special emphasis is put on PL studies in magnetic fields up to 15 T for identifing the electronic structure.

EXPERIMENTAL RESULTS

We investigated a series of hexagonal GaN samples epitaxially grown on (0001) sapphire. Results presented in this paper are obtained from a 400 μm thick n-type sample (crystal 1) and a 38 μm thick semi-insulating sample doped with Fe (crystal 2). Luminescence was excited by various lines of an Ar^+-, Kr- or a HeCd-laser. The Zeeman PL experiments are carried out using a superconducting 15 T magnet built in split-coil configuration. Depending on the spectral region the excitation source in PLE measurements was either a xenon lamp or a tungsten-halogen lamp, spectrally dispersed by a 0.35 double-grating monochromator. The luminescence was detected by a cooled Ge photodiode. We employed a double-prism monochromator to control the detection window in the PLE experiments. High-resolution PLE experiments were performed using a dye laser as tunable excitation source.

Mat. Res. Soc. Symp. Proc. Vol. 449 © 1997 Materials Research Society

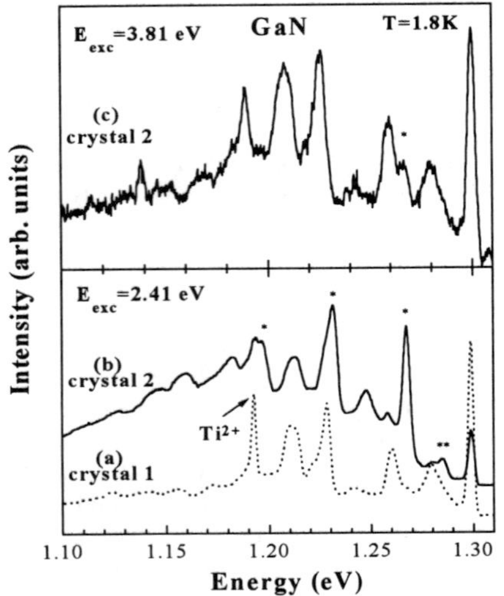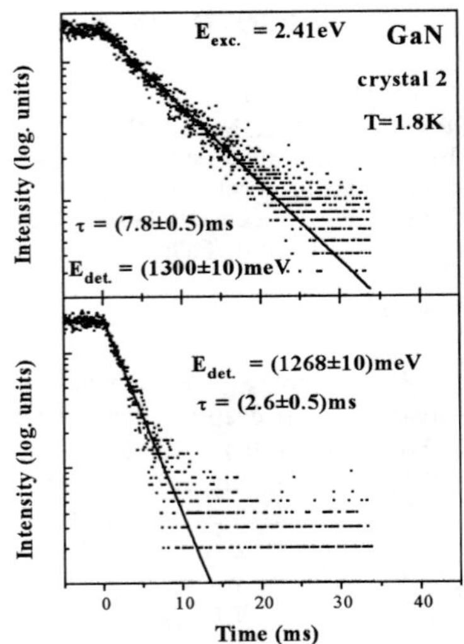

Figure 1: (left) Low-temperature PL of the Fe^{3+} ($^4T_1$-$^6A_1$) center. Shown are the n-type crystal 1 (a) and the semiinsulating sample 2 (b,c). Spectra (a) and (b) were recorded using the green Ar^+-laser line at 2.412 eV for excitation. New ZPLs at 1.268 eV and 1.286 eV are observed in the semiinsulating sample and tentatively attributed Fe-related defect complexes. These lines and the corresponding phonon replica are labeled by * and **, respectively. (right): Decay of the 1.299 eV ZPL and of the 1.268 eV ZPL.

PL Results

Figure 1 represents typical PL spectra of the Fe^{3+} ($^4T_1$-$^6A_1$) luminescence for different excitation energies. Exciting the sample at 2.412 eV new ZPLs at 1.268 eV and 1.286 eV are observed (indicated by * and **), superimposed to the luminescence of isolated Fe^{3+} with its ZPL at 1.299 eV. The full width at half maximum (FWHM) of the 1.268 eV ZPL is 3 meV whereas for the 1.299 eV line the FWHM is only 1 meV. A lifetime of 7.8±0.5 ms and 2.6±0.5 ms is detected for the 1.299 eV and 1.268 eV ZPL, respectively. These lifetimes in the ms range are typical for spin-forbidden transitions. Temperature dependent PL experiments reveal the detection of additional ZPLs on the high-energy side of the respective ZPL due to the thermal population of higher excited fine-structure states. For the $^4T_1$ state of isolated Fe^{3+} these three states are 1.8, 2.6 and 3.8 meV above the lowest component at 1.2988 eV. For the 1.2687 eV ZPL these states are 2.0 and 3.5 meV above the lowest component.

In order to obtain insight into the spectral shape of both superimposed luminescence bands we have to separate the PL of the new ZPLs from the PL of isolated Fe^{3+}. Therefore we normalize the intensity of the 1.299 eV ZPL of isolated Fe^{3+} in the spectra c and b of figure 1 and subtract both normalized spectra. The result is shown in curve a of figure 2. The vibronic structure of the 1.268 eV ZPL (spectrum a) looks quite similar to that of Fe^{3+} luminescence band (spectrum b)

ZPL. However, the relative intensity and the energy difference of the vibronic peaks to each ZPL vary slightly. Thus, different local vibrational modes are involved for both defects.[6]

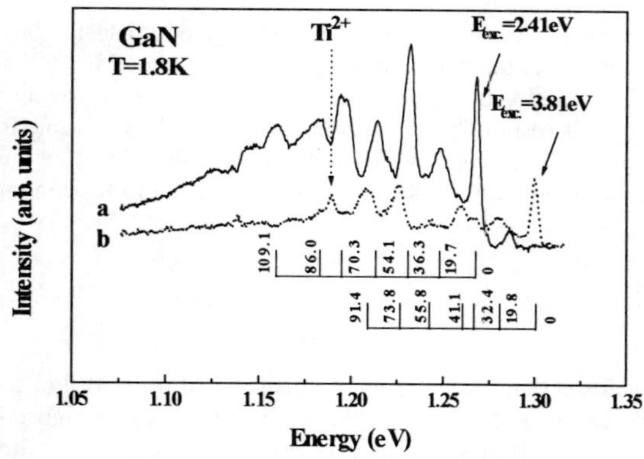

Figure 2: PL-spectrum of the 1.268 eV luminescence (a) in comparison to the Fe^{3+} lumines-cence (b) of the semiinsulating GaN sample 2. The energy differences of the vibronic peaks to the respective ZPL are given in meV.

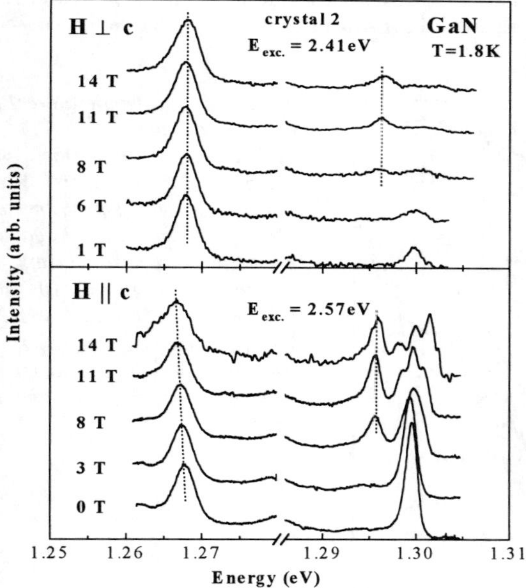

Figure 3: Zeeman spectra of the ZPL-region in the configuation H||c and H⊥c for T = 1.8 K. For H||c the crystal is excited at 2.57 eV, whereas for H⊥c the crystal is excited at 2.41 eV. No Zeeman-splitting is observed for the 1.268 eV ZPL. With increasing magnetic field, a new ZPL at 1.296 eV is observed for both configurations. The relative intensity change between the 1.299 eV ZPL and the 1.268 eV ZPL in both Zeeman configurations is due to the different excitation behavior of the ZPLs.

In order to get insight in the electronic structure of the defect, we performed Zeeman spectroscopy of the ZPL region. For H||c, the three components[3] of the Fe^{3+} ZPL are well resolved above B=11 T. In contrast to the Fe^{3+} behavior, no Zeeman-splitting is observed for the 1.268 eV ZPL. Only a slight shift to lower energies of 61±5 μeV/T is observed in this configuration. For H⊥c the sixfold splitting[3] of the Fe^{3+} ZPL is not resolved, but a broadening with increasing magnetic field is observed. The 1.268 eV ZPL does not show any shift or broadening in this configuration. This indicates that the electronic structure of the Fe-related defect is strongly modified by the nearby impurity atom. The relative intensity change between the 1.299 eV ZPL and the 1.268 eV ZPL in both Zeeman configurations is due to the different excitation behavior of the ZPLs.

Additionally, a new ZPL at 1.296 eV is observed with increasing magnetic field for both Zeeman configurations. The excitation behavior of this line is similar to that of isolated Fe^{3+}. This line is also specific for the Fe-doped crystal 2 and was not observed for GaN epilayers grown on SiC.[3] This might indicate that the Fe-doping is responsible for this line. If we explain this line by ferromagnetic coupling between two nearby iron atoms[7], we should expect a fast decay within the ground state, because the spin-selection rule is lifted. More investigations have to be done to clarify the origin of this new ZPL.

PLE Results

The presence of PL from Fe complexes renders the spectral position and the width of the detection window critical for the reliability of the PLE experiments. Spectrum (a) in Figure 4 presents the PLE of isolated Fe^{3+} detecting only luminescence in a 10 meV window around 1.299 eV. The excitation behavior of the Fe^{3+} luminescence depends critically on the stable charge state of iron. The spectra of semi-insulating samples show structured absorption bands not observed for the n-type samples (c).

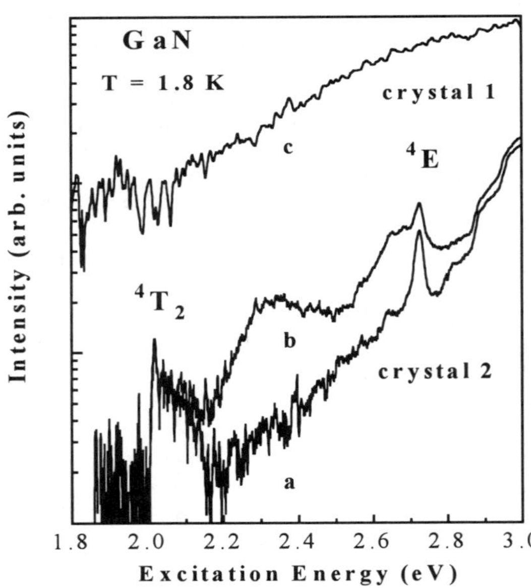

Figure 4: Low-temperature PLE spectra of the Fe^{3+} ($^4T_1$-$^6A_1$) luminescence for the two samples. Crystal 1 is n-type containing Fe^{2+} (spectrum c). Crystal 2 contains practically only Fe^{3+} (spectra a and b). Luminescence in a 10 meV window around the ZPL at 1.299 eV was detected for spectra a and c, whereas the detection energy is 1.268 eV for the spectrum b.

For the semi-insulating sample, fine structure is resolved around 2.0 eV and 2.8 eV and a broad excitation band appears in the uv spectral region (Fig. 4 (a)). The occurrence of sharp ZPLs in the PLE spectrum of the Fe^{3+} luminescence together with the Fe^{3+} EPR signal in the crystal directly demonstrates the presence of the neutral 3+ charge state in the unexcited samples.

Using a dye laser as tunable excitation source at least four ZPLs (at 2.0091 eV, 2.0152 eV, 2.0170 eV, and 2.0188 eV) are resolved with FWHM down to 1.2 meV. Transitions into higher excited quartet states are known to lead to structured bands in PLE spectra of the $^4T_1$-$^6A_1$ luminescence. This was experimentally verified for Mn^{2+} in ZnS.[8] Thus, we attribute the 2.01 eV absorption band to the $^6A_1$-$^4T_2$ transition of Fe^{3+}. The structure at 2.8 eV starts with a single ZPL at 2.731 eV. The energy separation between this ZPL and the first step amounts to 157 meV and, thus, corresponds neither to the step period of 75 meV nor to typical phonon modes of hexagonal GaN. Therefore, we treat the 2.731 eV absorption as a separate feature and tentatively attribute it to an intracenter transition involving the next excited quartet state, the $^6A_1$-^4E transition. From our results we conclude that the crystal field splitting of the ^4G multiplet of Fe^{3+} in GaN is approximately three times larger than that of Mn^{2+} in II-VI compounds. This strong crystal field may be the result of the comparatively small lattice constant of GaN, of the 3+ charge state of iron, and of the high electronegativity of nitrogen.

The spectral appearance of the step-like PLE structure showing at least four replica with increasing intensity is unusual for an intracenter transition of a TM. However, it is rather typical for deeply bound excitons or deeply bound electron-hole pairs.[9,10] The steps are located on the low-energy onset of the broad UV excitation band, which we assign to the $Fe^{3+/2+}$ charge-transfer transition. We therefore attribute this structure to the formation of a shallow bound state at the Fe center. The exact location of the $Fe^{3+/2+}$ acceptor level in GaN is not clear yet. The observation of the $Fe^{3+/2+}$ charge-transfer band in PLE should provide this information. However, the superposition of shallow bound states of TMs to the low-energy slope of a charge-transfer band makes a fit of the ionization band difficult in order to determine the onset energy. But it seems reasonable to identify the energy (3.17 ± 0.10 eV) at which the step-like structure vanishes and the broad structureless charge-transfer band begins to dominate with the energy position of the $Fe^{3+/2+}$ acceptor level. This assignment yields a binding energy of (280 ± 100) meV for the shallow bound state of the isolated Fe center in GaN. This complex is also an excited state of Fe^{3+} and relaxes nonradiatively to the $^4T_1$ state. Due to the position of the deep $Fe^{3+/2+}$ acceptor level close to the conduction band, hybridization of the core states of Fe is more important than in the other III-V compounds. Together with the high binding energy, these effects add up to a strong exchange interaction between the core hole and the bound hole. Therefore, a deeply bound electron-hole complex (Fe^{3+},e,h) is the appropriate description of this state. A very similar situation was demonstrated for shallow bound states of Ni comparing cubic ZnS and hexagonal CdS.[9] Recent calculations[10] show that the hybridization of the TM ground state with band-states is favored in hexagonal host crystals. Indeed, all known shallow states of TMs in wurtzite crystals have the character of a (TM,e,h) complex.

Compared to the Fe^{3+} PLE in n-type and semiinsulating samples, two additional features peaking at 2.3 eV and 2.65 eV are observed in the PLE spectrum of the 1.268 eV ZPL. (spectrum b). The 2.65 eV resonance is rather broad in comparison with the 2.73 eV PLE resonance of Fe^{3+}. This broadening can be caused by the larger energy separation within the ^4E state of the Fe^{3+} complex due two the larger crystal field. A similar observation was made for Mn^{2+} centers in cubic ZnS.[11] Here, additional axial crystal fields due to polytype effects or nearby impurity atoms cause a larger energy separation within the ^4E-state of the d^5 configuration of Mn^{2+}. The other additional feature is a broad PLE structure peaking at 2.3 eV. The peak energy of 2.3 eV agrees well with the expected $Fe^{2+/3+}$ low energy threshold of 2.36 eV within the $^4T_2$ state of isolated Fe^{3+}.[5] This indicates that the other atom within the Fe complex might by a nearby donor, resulting in a (Fe^{2+}, D$^+$) complex in the unexcited crystal. The 1.268 eV PL of the Fe-complex is excited via a charge-transfer process of the Fe^{2+}

$$(Fe^{2+}, D^+) + h\nu \ (2.3 \ eV) \longrightarrow (Fe^{3+*}, D^+) + e_{CB}$$

followed by the radiative relaxation of the excited (Fe^{3+*}, D$^+$) center causing the 1.268 eV luminescence and the (very slow) capture of the electron.

CONCLUSION

The similar spectral shape of the vibronic sideband and the energy position together with the decay time indicate the close correlation between the isolated Fe^{3+} ($^4T_1$-$^6A_1$) transition at 1.299 eV and the 1.268 eV defect. Additionally, the PLE results can be explained in the Fe term-scheme. On the basis of these observations, the 1.268 eV is tentatively attributed to radiative recombination within a (Fe^{3+}, D$^+$) defect. However, the Zeeman-behavior of the 1.268 eV ZPL do not show the clear fingerprint of the d^5-configuration. This indicates, that the electronic structure of the 1.268 eV transition is strongly modified by the nearby impurity atom.

REFERENCES

1.   S. Nakamura, Jpn. J. of Appl. Phys. 35 (1996) L74
2.   K. Maier M. Kunzer, U. Kaufmann, J. Schneider, H. Amano, I. Akasaki,
        T. Detchprohm, K. Hiramatsu, Mat. Science Forum 143-147 (1994) 93
3.   R. Heitz, P. Thurian, I. Loa, L. Eckey, A. Hoffmann, I. Broser, K. Pressel,
        B. K. Meyer, E. N. Mokhov, Appl. Phys. Lett. 67 (19) (1995) 2822
4.   P. Thurian, R. Heitz, L. Eckey, P. Maxim, V. Kutzer, A. Hoffmann, I. Broser,
        K. Pressel, B.-K. Meyer, Proc. of the 23th ICPS 1996, Berlin, Ed. M. Scheffler,
        R. Zimmermann, World Scientific   pp. 2897-2901
5.   R. Heitz, P. Maxim, L. Eckey, P. Thurian, A. Hoffmann, I. Broser, K. Pressel,
        B.-K. Meyer, Phys. Rev. B, accepted
6.   P. Thurian, G. Kaczmarczyk, H. Siegle, R. Heitz, A. Hoffmann, I. Broser,
        B.-K. Meyer, R. Hoffbauer, U. Scherz
        Mat. Science Forum, Vol. 196-201 (1995) 1571
7.   J. Kreissl, W. Gehlhoff, Phys. Status Solidi A 81 (1984) 701
8.   R. Parrot, A. Geoffroy, C. Naud, W. Busse, H.-E. Gumlich,
        Phys. Rev. B 23 (1981) 5288
9.   R. Heitz, A. Hoffmann, I. Broser, Phys. Rev. B 48 (1993) 8672
10.   P. Dahan, V. Fleurov, K. A. Kikoin, Mat. Science Forum 196-201 (1995) 755
11.   U. W. Pohl, H.-E. Gumlich, Phys. Rev. B, 40 (1989) 1194

PHOTOLUMINESCENCE, REFLECTANCE, AND MAGNETOSPECTROSCOPY OF SHALLOW EXCITONS IN GaN

B.J. SKROMME[a], H. ZHAO[a], B. GOLDENBERG[b], H.S. KONG[c], M.T. LEONARD[c], G.E. BULMAN[c], C.R. ABERNATHY[d], S.J. PEARTON[d]
[a]Department of Electrical Engineering and Center for Solid State Electronics Research, Arizona State University, Tempe, AZ 85287-5706, skromme@asu.edu
[b]Honeywell Technology Center, Plymouth, MN 55420
[c]Cree Research, Inc., Durham, NC 22713
[d]Department of Materials Science and Engineering, University of Florida, Gainesville, FL 32611

ABSTRACT

We report several new aspects of the excitonic properties of heteroepitaxial GaN grown on sapphire or 6H-SiC. In particular, we observed the $n = 2$ free exciton associated with both A and B excitons (which are distinct from the $n = 1$ C exciton) using reflectance and 1.7 K photoluminescence. We also studied the behavior of the $n = 2$ A-exciton using magnetoluminescence in fields up to 12 T. The large diamagnetic shift and splitting positively confirm the identification, yielding an exciton binding energy of about 26.4 meV. Several previous identifications of the $n = 2$ free exciton yielding a smaller exciton binding energy are probably in error, based on our results. We have also detected the two-electron replica of the neutral donor-bound exciton for the first time in GaN and observed its splitting pattern in magnetic fields up to 12 T. This feature is 22 meV below the principal neutral donor-bound exciton peak, independently of strain shifts in the overall spectrum. It yields a precise donor binding energy of 29 meV for the shallow residual donor in material grown by metalorganic chemical vapor deposition and gas-source molecular beam epitaxy, considerably smaller than that of the residual donor reported earlier in hydride vapor phase epitaxial material (about 35.5 meV).

INTRODUCTION

Gallium nitride and related materials are of intense current interest for short wavelength optoelectronic applications as well as for high power, high temperature electronic devices. Many fundamental materials parameters have yet to be accurately determined in this system, however, and the properties of shallow impurities and excitonic states in GaN are only beginning to be understood. We have therefore undertaken a fundamental study of the excitonic luminescence and reflectance properties to deduce some of these parameters. In particular, we employ high field magnetospectroscopy to provide positive identification of spectral features involving free and bound exciton states and thereby permit determination of the shallow donor and free exciton binding energies.

EXPERIMENTAL TECHNIQUES

The samples in this study were grown in several different laboratories using metalorganic chemical vapor deposition (MOCVD) on (0001) sapphire or 6H-SiC substrates, or using metalorganic molecular beam epitaxy (MOMBE) with trimethylgallium and NH_3 sources on (0001) sapphire. Conventional low temperature AlN buffer layers were employed. Layer thicknesses ranged from 3 to 7 μm and none of the samples were intentionally doped. Low temperature photoluminescence (PL) and reflectance measurements were performed with the samples suspended strain-free in flowing He vapor or superfluid He. Magnetospectroscopy was performed in Faraday configuration with **B** parallel to the c-axis of the layers in magnetic fields up to 12 T, using a system described elsewhere.[1] The polarization of the PL emission was not analyzed. The pump source was either an Ar^+ laser operating at 351 nm or a He-Cd laser operating at 325 nm. Simple reflectance measurements were performed using a tungsten-halogen or Xe arc lamp and a Si photodiode or photomultiplier tube as a detector.

Mat. Res. Soc. Symp. Proc. Vol. 449 ©1997 Materials Research Society

RESULTS AND DISCUSSION

Strain Effects and Identification of Intrinsic and Extrinsic Features

A series of 1.7 K PL spectra are shown in Fig. 1 for four representative samples grown on sapphire and 6H-SiC substrates. This figure illustrates the profound effects of strain on the spectrum of heteroepitaxial GaN, due to the large lattice and thermal mismatch between GaN and the commonly employed substrate materials.[2-13] The neutral donor-bound exciton (D^0,X) peak shifts from 3.459 eV in the lower spectrum to 3.4852 eV in the upper one, as a consequence of the varying strains in these layers. Most of the large lattice mismatch is expected to relax at the growth temperature, but differences in thermal expansion coefficients result in substantial residual strains at low temperature. Room temperature measurements using surface profilometry similar to those reported by Kozawa et al.[7] show that at 300 K, the lower sample is bowed concave upward, whereas the upper three samples show increasing amounts of convex bowing. These observations imply that the latter three samples are under biaxial compression at 300 K in contrast to the lower sample, which is in biaxial tension at 300 K. The tensile strain in the lower sample is confirmed by its tendency to show cracking in the epilayer after repeated thermal cycling, whereas none of the other samples did. The results suggest that the one layer on SiC is actually in biaxial compression, which is the opposite that would be expected from the differences in thermal expansion coefficients.[10,11,13] Such behavior could be due to residual lattice mismatch compressive stress, and further work is in progress to confirm this conclusion.

Given the enormous variations in strain that exist in GaN, even for samples grown on the same type of substrate, excitonic peak positions alone are clearly of no use for peak identification. It is first necessary to establish the positions of the free exciton peaks in each sample using a method that is sensitive to intrinsic rather than extrinsic transitions. We have employed simple reflectance for this purpose, as it yields well resolved structures corresponding to each of the three ground state free exciton states in this wurtzite material (conventionally denoted A, B, and C). A first example is shown for the sample under tensile strain in Fig. 2, where we compare the PL spectrum to reflectance. Aside from the low energy Fabry-Perot interference fringes, two prominent structures appear in reflectance corresponding to a superposition of the A and B excitons and to the C exciton, respectively. These two structures match the energies of corresponding PL peaks perfectly, confirming the intrinsic nature of the latter. A weaker reflectance feature corresponds to the $n = 2$ excited state of the A/B free excitons, which will be discussed more in

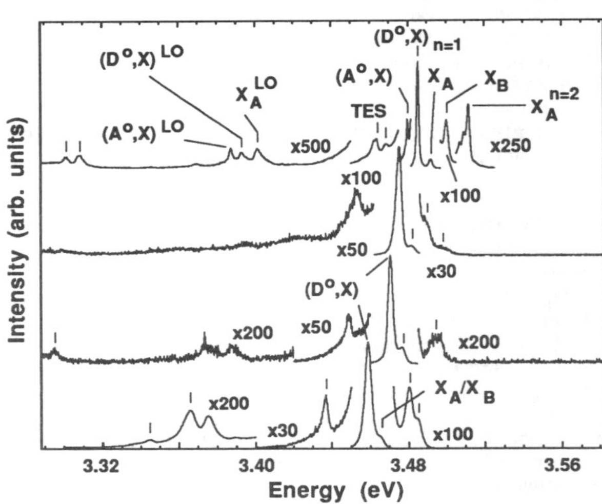

Fig. 1. 1.7 K PL spectra of four GaN layers grown on sapphire (upper two) and 6H-SiC (lower two) substrates.

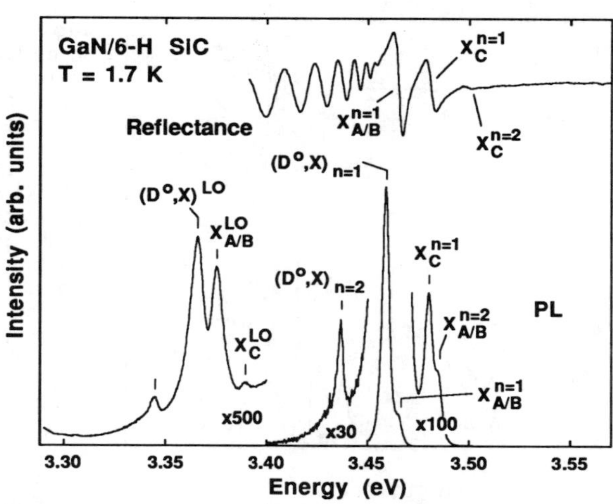

Fig. 2. A comparison of PL and reflectance spectra for the sample of Fig. 1 that is under biaxial tensile strain.

connection with another sample below. The free exciton peaks exhibit characteristically stronger coupling to LO phonons than the bound excitons. Temperature-dependent measurements (not shown) also verify the free/bound exciton assignments by the thermal quenching of the latter in favor of the former. Further, measurements as a function of excitation intensity show saturation of the bound excitons with respect to free excitons at high pumping levels, as expected.

The (D^o,X) peak is identified in Fig. 2 from its known localization energy (6 meV).[14] We also observe a weaker peak about 22 meV below it in energy, which we assign as a two-electron replica of the (D^o,X) peak in which the donor is left in its $n = 2$ excited state after recombination of the exciton. This assignment implies a $1s$-$2s$ splitting of 22 meV, for a donor binding energy of 29 meV if we assume a simple hydrogenic model for the donor. This assignment is strongly supported by our magnetospectroscopy results discussed below.

In Fig. 3 we display PL and reflectance for the sample under the greatest biaxial compression, which also exhibits the sharpest bound exciton peaks. Reflectance structures corresponding to the A, B, and C $n = 1$ free excitons are observed, as well as weak structures we assign to the $n = 2$ states of the A and B free excitons. Corresponding PL peaks are observed for each of these peaks (except the C exciton). A detailed theoretical analysis of the free exciton splitting as a function of strain is in progress and will be reported elsewhere. The excited-state exciton peaks are particularly valuable as they allow us to deduce the free exciton binding energy. Again assuming a simple hydrogenic model (neglecting exchange and crystal field splitting) we obtain a free exciton binding energy of about 26.4 meV for the A and B excitons (which are the same within experimental error).

Considerable controversy has existed in the literature about the free exciton binding energy in GaN. Chichibu et al.[12] found 25.3 meV, Volm et al.[11] reported 26.7 meV, and Merz et al.[15] reported 26.1 meV, all in reasonable agreement with our value. On the other hand, Smith et al. claimed to observe free exciton-dominated PL spectra with an excited exciton state yielding a binding energy of 20 meV[16] or 18.3 meV.[17] However, these authors did not consider the possibility of strain shifts in their spectra, which we have shown can be very important. Moreover, the data in Ref. [17] show the "B free exciton" peak becoming stronger than the "A free exciton" peak at high temperature, which is impossible as the oscillator strength is larger for the A transition.[18] A similar value of 20 meV was quoted by Reynolds et al.,[19] but their reflectance spectrum although distorted actually supports assignment of their "A free exciton" peak as the (D^o,X) peak. It appears that the workers obtaining values around 20 meV misassigned the peaks in their spectra and those samples were actually dominated by bound excitons, not free excitons.

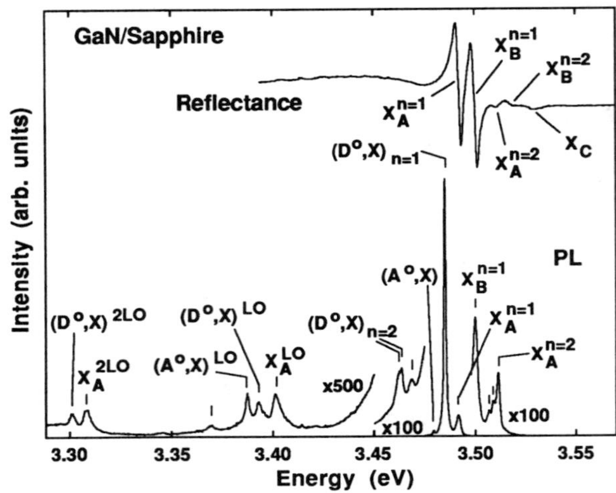

Fig. 3. As Fig. 2, but for the sample of Fig. 1 under the greatest biaxial compression.

Most importantly, our identifications of the excited states are positively confirmed by magnetospectroscopy as discussed below.

Below the principal (D^o,X) peak in Fig. 3, we again observe a two-electron replica with a small doublet splitting (probably related to crystal field splitting of the $n = 2$ state). The strongest component lies 21.6 meV below the main (D^o,X) peak, again implying a donor binding energy around 29 meV. We have seen the same value consistently (within experimental error) in all samples studied to date that exhibit a clear two electron satellite. This value is in agreement with a recent infrared determination of the binding energy of Si donors (29.0 meV),[20] but is much lower than the 35.5 meV value found by infrared spectroscopy on residual donors in material grown by hydride vapor phase epitaxy (VPE).[21] It appears that the residual donor in MOCVD and MOMBE material is different from that in hydride VPE material.

A second weak peak appears above the two-electron replica of the (D^o,X) peak in Fig. 3. It has the same energy separation from the A free exciton peak as the $(D^o,X)_{n=2}$ peak has from the $(D^o,X)_{n=1}$ peak, and it exhibits an analogous magnetic field splitting behavior (see below). We therefore assign it as inelastic scattering of A free excitons from neutral donors, exciting the donors from their ground states to their first excited states. A weak peak tentatively denoted (A^o,X) in Fig. 3 lies 5.6 meV below (D^o,X); the identification is based on its LO phonon coupling strength. Further magnetic field experiments are needed to confirm this identification.

<u>Magnetospectroscopy</u>

PL spectra for the same sample are shown in Fig. 4 as a function of magnetic field up to 12 T. The principal (D^o,X) and A and B $n = 1$ free exciton peaks broaden slightly at high field due to unresolved spin splittings but show no gross splitting, as would be expected and has been reported previously in the (D^o,X) case.[22] The peak identified as the $n = 2$ A free exciton exhibits much more pronounced behavior, including a strong diamagnetic shift and a clear orbital splitting. The diamagnetic shift is characteristic of a highly extended wave function, and the splitting is assigned as Zeeman splitting of the $n = 2$ state in the magnetic field. The lower energy shoulders on this peak may involve $n = 2$ excitons bound to impurities but require further study. The two electron satellites and the analogous scattering replica of the A free exciton display large shifts and splittings also, which can only be characteristic of orbital (Zeeman) splittings. The separations between the observed two electron peak positions and the principal $(D^o,X)_{n=1}$ peak are plotted as a function of field in Fig. 5. The symbols represent the experimental data points and the lines represent a very simple hydrogenic model for the data, based on the calculations of Makado and McGill for a H atom in a magnetic field.[23] This theory neglects the axial splitting of the donor excited states due to the anisotropy of the wurtzite crystal in electron effective mass and/or dielectric constant.[24] Also, a fixed electron effective mass of 0.20 m_o and a dielectric constant of 9.5 were assumed, without any attempt to fit the data. Nonetheless, the theory gives a good qualitative explanation of the splittings, particularly in view of the limited resolution in the experimental data. A more detailed

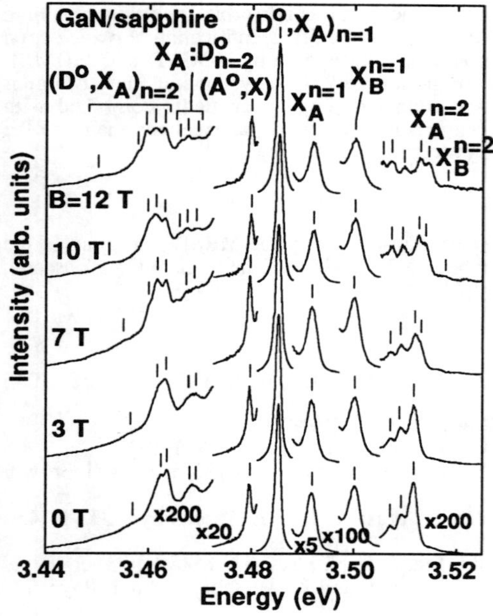

Fig. 4. Magnetic field-dependent PL spectra for the sample of Fig. 3 in Faraday configuration.

analysis incorporating the effects of anisotropy is in progress, which is necessary to explain the zero-field splitting in particular, but the validity of our identification is well established by the present results.

A quantitative analysis of the $n = 2$ free exciton splitting also requires the inclusion of anisotropy, but the large diamagnetic shift observed for the average of the two resolved components (1.5 × 10^{-5} eV/T^2) clearly implies an excited state. (In the low field limit, the various components of the $n = 2$ free exciton shift at 7, 3, or 6 × γ^2, where γ is the dimensionless magnetic field strength, compared to a shift of $\gamma^2/2$ for the $1s$ ground state.[24]) The experimental resolution (limited by inhomogeneous strain and Stark effect broadening) is insufficient for a highly detailed analysis, but the qualitatively correct nature of the identification is clearly established.

SUMMARY

In conclusion, we have employed reflectance and magnetospectroscopy to study the excitonic properties of GaN epilayers grown on sapphire and 6H-SiC substrates. We positively identify the $n =$

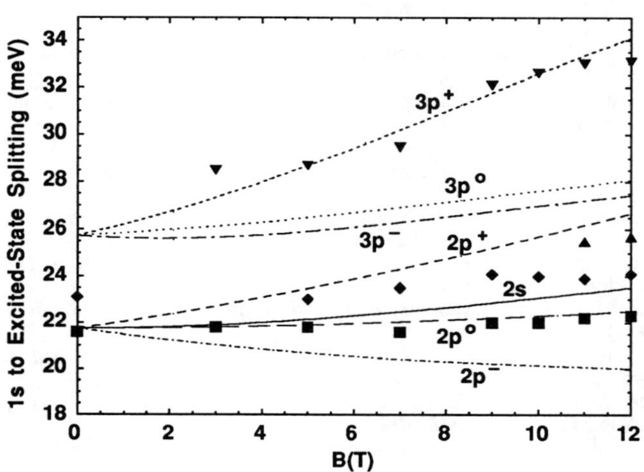

Fig. 5. Separations of the $n = 1$ and $n = 2$ final state (D^o,X) peaks in the spectra of Fig. 4 as a function of magnetic field (symbols). Lines are a rough theoretical calculation (see text).

2 state of the free excitons on the basis of its diamagnetic shift rate and orbital splitting, obtaining the free exciton binding energy as 26.4 meV. We also report the first identification of two-electron transitions in GaN, which yield the residual donor binding energy in MOCVD and MOMBE material as 29 meV. These observations lay the groundwork for a systematic study of donor dopants and their properties in GaN, using intentional doping experiments combined with magnetospectroscopy. The observations have been made possible by the excellent quality of the epitaxial material and its narrow luminescence linewidths.

ACKNOWLEDGMENT

The ASU portion of this work was supported primarily by the MRSEC program of the National Science Foundation under Award Number DMR-9632635. The work at Cree Research was supported by DARPA under Contract MDA972-95-C-0016.

REFERENCES

1. B.J. Skromme, R. Bhat, H.M. Cox, and E. Colas, IEEE J. Quantum Electron. **25**, 1035 (1989).
2. T. Matsumoto and M. Aoki, Jpn. J. Appl. Phys. **13**, 1583 (1974).
3. H. Amano, K. Hiramatsu, and I. Akasaki, Jpn. J. Appl. Phys. **27**, L1384 (1988).
4. K. Naniwae, S. Itoh, H. Amano, K. Itoh, K. Hiramatsu, and I. Akasaki, J. Crystal Growth **99**, 381 (1990).
5. T. Detchprohm, K. Hiramatsu, K. Itoh, and I. Akasaki, Jpn. J. Appl. Phys. **31**, L1454 (1992).
6. K. Hiramatsu, T. Detchprohm, and I. Akasaki, Jpn. J. Appl. Phys. **32**, 1528 (1993).
7. T. Kozawa, T. Kachi, H. Kano, H. Nagase, N. Koide, and K. Manabe, J. Appl. Phys. **77**, 4389 (1995).
8. B. Gil, O. Briot, and R.-L. Aulombard, Phys. Rev. B **52**, R17028 (1995).
9. W. Rieger, T. Metzger, H. Angerer, R. Dimitrov, O. Ambacher, and M. Stutzmann, Appl. Phys. Lett. **68**, 970 (1996).
10. W. Li and W.-X. Ni, Appl. Phys. Lett. **68**, 2705 (1996).
11. D. Volm, K. Oettinger, T. Streibl, D. Kovalev, M. Ben-Chorin, J. Diener, B.K. Meyer, J. Majewski, L. Eckey, A. Hoffmann, H. Amano, I. Akasaki, K. Hiramatsu, and T. Detchprohm, Phys. Rev. B **53**, 16543 (1996).
12. S. Chichibu, A. Shikanai, T. Azuhata, T. Sota, A. Kuramata, K. Horino, and S. Nakamura, Appl. Phys. Lett. **68**, 3766 (1996).
13. W. Shan, A.J. Fischer, J.J. Song, G.E. Bulman, H.S. Kong, M.T. Leonard, W.G. Perry, M.D. Bremser, and R.F. Davis, Appl. Phys. Lett. **69**, 740 (1996).
14. R. Dingle, D.D. Sell, S.E. Stokowski, and M. Ilegems, Phys. Rev. B **4**, 1211 (1971).
15. C. Merz, M. Kunzer, U. Kaufmann, I. Akasaki, and H. Amano, Semicond. Sci. Technol. **11**, 712 (1996).
16. M. Smith, G.D. Chen, J.Z. Li, J.Y. Lin, H.X. Jiang, A. Salvador, W.K. Kim, O. Aktas, A. Botchkarev, and H. Morkoç, Appl. Phys. Lett. **67**, 3387 (1995).
17. M. Smith, G.D. Chen, J.Y. Lin, H.X. Jiang, M.A. Khan, C.J. Sun, Q. Chen, and J.W. Yang, J. Appl. Phys. **79**, 7001 (1996).
18. G.L. Bir and G.E. Pikus, *Symmetry and Strain-Induced Effects in Semiconductors* (Wiley, New York, 1974).
19. D.C. Reynolds, D.C. Look, W. Kim, O. Aktas, A. Botchkarev, A. Salvador, H. Morkoç, and D.N. Talwar, J. Appl. Phys. **80**, 594 (1996).
20. Y.J. Wang, R. Kaplan, H.K. Ng, K. Doverspike, D.K. Gaskill, T. Ikedo, I. Akasaki, and H. Amano, J. Appl. Phys. **79**, 8007 (1996).
21. B.K. Meyer, D. Volm, A. Graber, H.C. Alt, T. Detchprohm, A. Amano, and I. Akasaki, Solid State Commun. **95**, 597 (1995).
22. D. Volm, T. Streibl, B.K. Meyer, T. Detchprohm, H. Amano, and I. Akasaki, Solid State Commun. **96**, 53 (1995).
23. P.C. Makado and N.C. McGill, J. Phys. C.: Solid State Phys. **19**, 873 (1986).
24. J.C. Miklosz and R.G. Wheeler, Phys. Rev. **153**, 913 (1967).

CHARACTERIZATION OF GaN FILMS ON SAPPHIRE BY CATHODOLUMINESCENCE

L.-L. CHAO, G. S. CARGILL III and C. KOTHANDARAMAN
Department of Chemical Engineering, Materials Science, and Mining Engineering, Columbia University, New York, NY 10027

ABSTRACT

Cathodoluminescence (CL) spectroscopy and microscopy were used to study the luminescent properties of a variety of GaN films, both Si-doped and unintentionally-doped, grown on sapphire substrates. A narrow and intense near band-edge emission was found in the CL spectrum of each film examined, and deep-level emission was also observed for some of the films. The luminescence efficiency of near band-edge emission increased with a faster rate than that of deep-level emission when the pumping current was increased. Spatial nonuniformities of luminescence were observed in monochromatic CL microscopy, and microstructures were observed in scanning electron microscopy. No correlations between luminescence features and microstructural features were seen. Degradation of near band-edge luminescence was observed, accompanied by growth of deep-level emission.

INTRODUCTION

Due to their superior physical properties, GaN and related compounds have been considered as the promising materials for electronic devices operating under high temperature and high power conditions and, especially, for light emitting devices operating in the blue and ultraviolet wavelength range. Remarkable, rapid success has been achieved in developing devices based on III-V nitrides.[1] Light emitting diodes fabricated from InGaN/AlGaN double-heterostructure layers on sapphire substrates are commercially available.[2] A nitride-based laser diode operating at 417 nm under pulsed conditions has also been demonstrated.[3]

In this paper, we report the results of cathodoluminescence (CL) studies of GaN films grown on sapphire substrates by molecular beam epitaxy (MBE), metalorganic chemical vapor deposition (MOCVD) and hydride vapor phase epitaxy (HVPE). The films were characterized in terms of spectral features, spatial nonuniformity in luminescence and surface morphology, and degradation behavior. The competition between near band-edge and deep-level emission was also investigated.

EXPERIMENTAL

Sample description

Four samples were used in this study. The growth methods, doping and film thicknesses are given below:

Sample No.	Growth Method	Doping	Thickness (µm)
1	MBE	Si ($n \approx 10^{18}/cm^3$)	1.5
2	MBE	undoped ($n \approx 10^{16}/cm^3$)	1.5
3	MOCVD	undoped (high resistivity, $\rho \geq 10^5 \, \Omega\text{-cm}$)	3
4	HVPE	undoped ($n \approx 10^{18}/cm^3$)	10

Mat. Res. Soc. Symp. Proc. Vol. 449 © 1997 Materials Research Society

<u>Cathodoluminescence system</u>

The instrument used for CL measurements is based on a JEOL JSM-6400 scanning electron microscope equipped with an Oxford Instrument CF302 system. More detailed description of the CL system is given in a previous publication.[4] In the present study, CL measurements were made at room temperature with beam voltages of 5-25 kV and beam currents of 1 nA to 400 nA.

RESULTS and DISCUSSION
<u>Spectroscopy</u>

Fig. 1 shows CL spectra for each sample taken with a beam voltage of 10 kV and a beam current of 20 nA. All spectra were dominated by an intense near band-edge emission which peaked at 362 nm. The Si-doped MBE sample (#1) had the highest intensity of near band-edge emission, and the undoped MBE sample (#2) had the lowest intensity, as shown by scaling factors given in Fig. 1. Besides near band-edge emission, deep-level emission bands related to impurities or structural defects also appeared in CL spectra for sample #1 at 423 nm and for sample #3 at 430 nm and 555 nm. The 555 nm "yellow luminescence"[5] only appeared in sample #3.

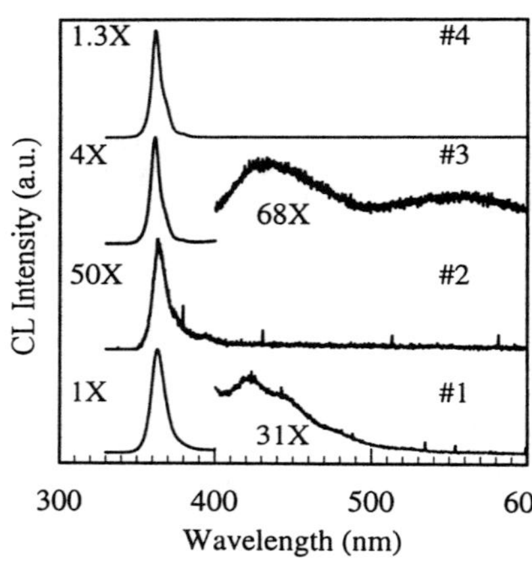

Fig. 1 CL spectra taken with 10 kV beam voltage and 20 nA beam current.

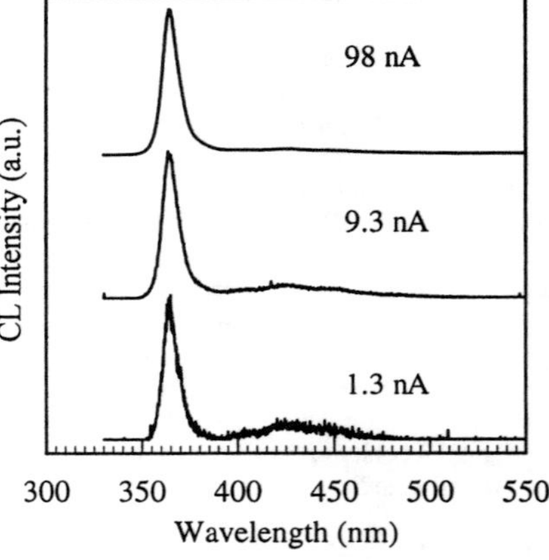

Fig. 2 CL spectra of sample #1 taken with 15 kV beam voltage and with three different beam currents, showing competition between near band-edge and deep-level emissions.

It was found that the CL intensity of near band-edge emission increased with a faster rate than that of deep-level emission when the beam current was increased, as shown in Fig. 2 for sample #1. More quantitative analysis showed, in the case of sample #1, that the intensity of near band-edge emission increased linearly with beam current, but that it increased superlinearly (by a power of 1.5) with beam current in the case of sample #3. The intensity of deep-level emission increased sublinearly (by a power of 0.5) with beam current for both samples.

A model has been developed by Grieshaber *et al.*[6] to describe the intensities of these two types of radiative transitions as a function of excitation density. Based on bimolecular rate equations and taking into account shallow impurities, deep level and continuum states, the model predicted that dependencies of the two luminescence channels should follow power laws with exponents which depend upon excitation density and doping. In the case of high excitation density, as for CL, the dependence of intensity of band-edge emission on excitation density followed a power law with an exponent of 1 for both highly doped and intrinsic samples. The dependence of intensity of deep-level emission followed a power law with an exponent of 0.5 for both highly doped and intrinsic samples. Our results from the Si-doped sample (#1) agree with their model and PL results, while the results from the undoped sample (#3) exhibits a stronger dependence for band-edge emission.

In contrast to the beam current effect, increased beam voltage increased the intensity of deep-level emission more than of near band-edge emission of each sample, as shown in Fig. 3 for sample #1. These results suggest that the film quality is better for the material closer to the film surface, since the higher the beam voltage, the deeper the electrons penetrate into solid samples.

As also shown in Fig 3, the peak position of near band-edge emission red-shifted as the beam voltage increased. Similar CL results were reported by Trager-Cowan et al.[7], who attributed the red shift to variation of strain with depth. However, the PL results of Rieger et al.[8] indicate that a blue shift, rather than a red shift, should result from increased biaxial compressive strain. Increasing beam voltage would excite material closer to the GaN-Al$_2$O$_3$ interface, where the GaN is expected to have higher biaxial compressive stress.

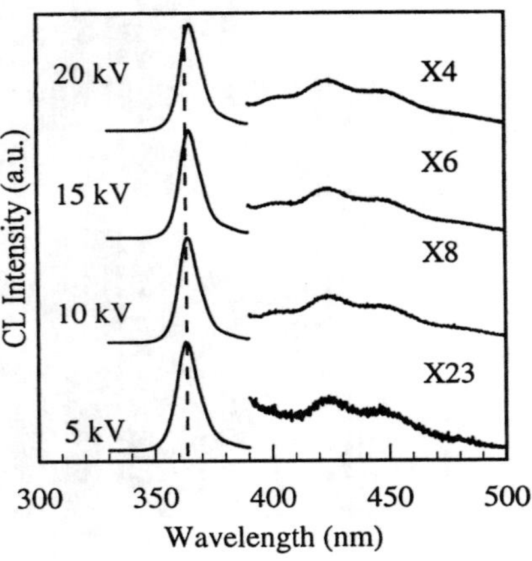

Fig 3 CL spectra of sample # 1 taken with 20 nA beam current and with different beam voltages, showing red shifts accompanying increased voltage.

Microscopy

The scanning electron (SE) micrographs in Fig. 4(a)-(d) for sample #1 through #4 show striking differences in surface morphology between the MBE-grown and non-MBE-grown samples. Submicron surface features with irregular shapes were observed for the MBE-grown samples, while the non-MBE-grown samples had much smoother surfaces. The surface of the MOCVD-grown sample was almost featureless, and the surface morphology of the HVPE-grown sample showed granular features with characteristic diameters of ~10 µm. Fig. 5(a)-(d) show the monochromatic CL, at 362 nm, taken at the same areas as in Fig. 4(a)-(d), which did not appear to be related to the physical features shown in the corresponding SE micrographs. Even for samples showing smooth surfaces, CL was spatially very nonuniform. For all the samples studied, CL micrographs showed micron-scale features. Spatially-resolved CL of deep-level emission was also obtained, as shown in Fig. 6(a) for sample #1 at 423 nm and in Fig. 6(b)-(c) for sample #3 at 430 nm and 555 nm respectively. In the case of sample #1, CL of deep-level emission was more uniform than that of near band-edge emission, which suggests that the two radiative transitions may occur independently. In the case of sample #3, CL images at 430 nm and 555 nm, with weaker contrasts between bright and dark patches, were similar to that at 362 nm, which suggests a common species involved in both deep-level transitions and near band-edge transitions. Similar observations have been reported in the literature which correlated microstructures obtained by TEM.[9]

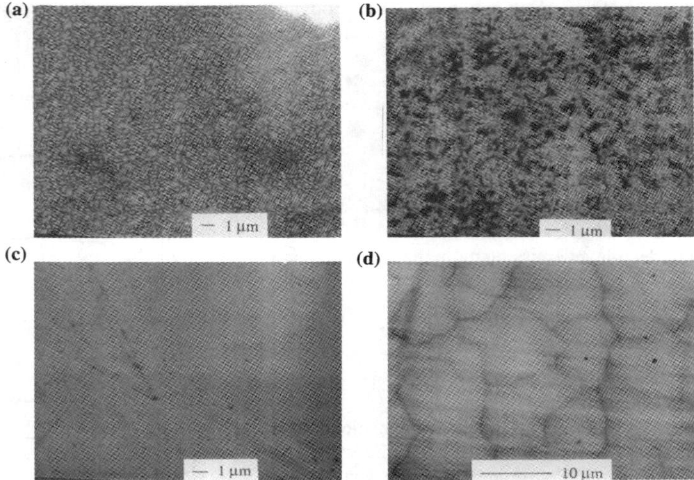

Fig. 4 SEM images of (a) sample #1, (b) sample #2, (c) sample #3 and (d) sample #4.

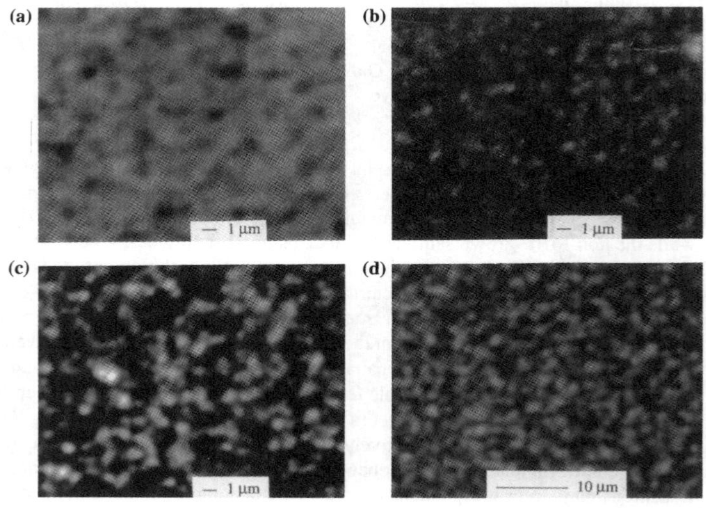

Fig. 5 Monochromatic CL images at 362 nm of (a) sample #1, (b) sample #2, (c) sample #3 and (d) sample #4, taken at the same areas used in Fig. 4.

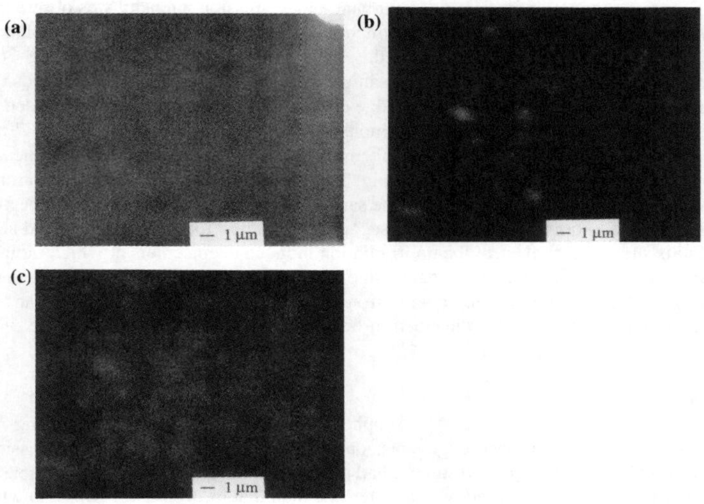

Fig. 6 Monochromatic CL images of (a) sample #1 at 423 nm. (b) sample #3 at 430 nm and (c) sample #3 at 555 nm.

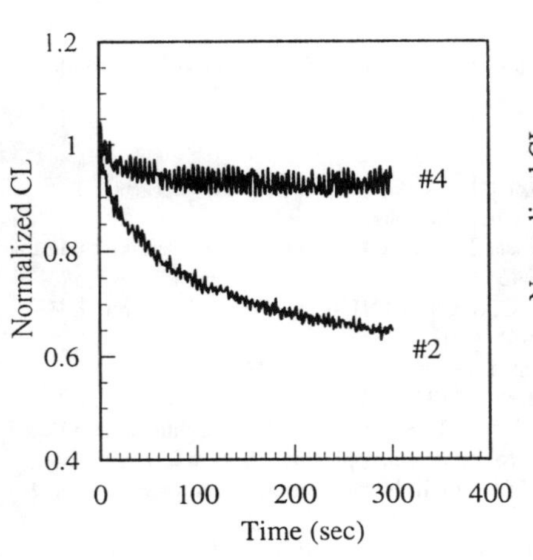

Fig. 7(a) Degradation of near band-edge emission of samples #2 and #4 at 362 nm.

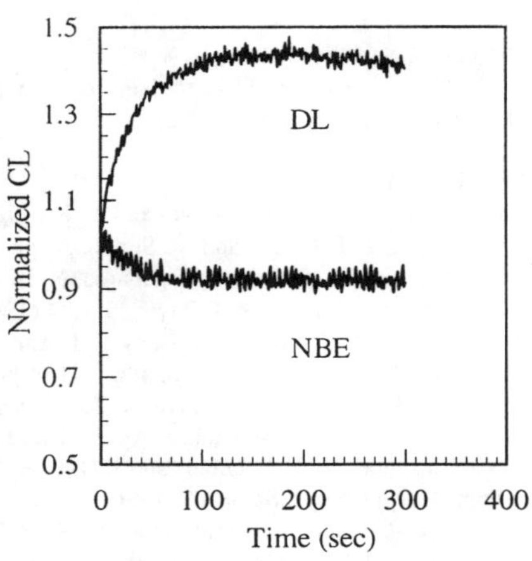

Fig. 7(b) Growth of deep-level (DL) emission at 430 nm and degradation of near band - edge (NBE) emission at 362 nm of sample #3.

<u>Degradation</u>

Some permanent degradation of near band-edge emission intensity was observed for all samples. Sample #2, MBE-grown and showing the weakest luminescence strength among the samples studied, degraded much faster than the rest of the samples. As shown in Fig. 7(a), the CL intensity of near band-edge emission of sample #2 dropped by 35% in 300 seconds for 50 nA beam current. CL intensities of near band-edge emission of the other samples degraded between 8% and 10% under the same conditions. Sample #4 degraded least, as shown in Fig. 7(a).

Growth of CL intensity of deep-level emission accompanied degradation of near band-edge emission for sample #3. As shown in Fig. 7(b), the luminescence intensity of sample #3 at 430 nm grew by 45% in 150 seconds with the same beam current used in Fig. 7(a). A much higher growth rate, more than 300% in 300 seconds, was found for the emission band at 555 nm. These results suggest that electron-beam irradiation induced deep centers at which radiative transitions occurred at the expense of band-edge transitions. However, in contrast to the case of sample #3, the deep-level emission (at 423 nm) of sample #1 degraded just as the near band-edge emission did, which suggests that the electron-beam irradiation induced nonradiative centers in this case.

SUMMARY

A variety of GaN films grown on sapphire substrates were characterized using cathodoluminescence. Near band-edge emissions and, in some cases, deep-level emissions were observed. A MBE-grown Si-doped sample had the highest CL intensity. The CL intensity of near band-edge emission increased with a faster rate than that of deep-level emission when the beam current was increased. No correlations were observed between luminescence features shown in monochromatic CL images and physical features shown in SE images. Degradation of near band-edge emission and growth of deep-level emission were observed under electron irradiation.

ACKNOWLEDGMENT

The authors would like to thank Dr. S. Guha and Dr. J. Redwing for providing samples and for fruitful discussions.

REFERENCE
[1] T. Matsuoka, A. Ohki, T. Ohno and Y. Kawaguchi, J. Cryst. Growth **138**, 727 (1994).
[2] S. Nakamura, T. Mukai and M. Senoh, Appli. Phys. Lett. **64**, 28 (1994).
[3] S. Nakamura, M. Senoh, S. Nagahama, N. Iwasa, T. Yamada, T. Matsushita, H. Kiyoku and Y. Sugimoto, Jpn. J. Appl. Phys. **35**, L74 (1996).
[4] L.-L. Chao, M. B. Freiler, M. Levy, J.-L. Lin, G. S. Cargill III, R. M. Osgood, Jr. and G. F. McLane, Mat. Res. Soc. Symp. **406**, 541 (1996).
[5] T. Suski, P. Perlin, H. Teisseyre, M. Leszczynski, I. Grzegory, J. Jun, M. Bockowski, S. Porowski and T. D. Moustakas, Appl. Phys. Lett. **67**, 2188 (1995).
[6] W. Grieshaber, E. F. Schubert and I. D. Goepfert, R. F. Karlicek, Jr., M. J. Schurman and C. Tran, J. Appl. Phys. **80**, 4615 (1996), see also Singh *et al.* Appl. Phys. Lett. **64**, 336 (1994).
[7] C. Trager-Cowan, P. G. Middleton and K. P. O'Donnell, Nitride Semiconductor Research, **1**, Article 6 (1996), see http://nsr.mij.mrs.org/.
[8] W. Rieger, T. Metzger, H. Angerer, R. Dimitrov, O. Ambacher and M. Stutzmann, Appl. Phys. Lett. **68**, 970 (1996).
[9] F. A. Ponce, D. P. Bour, W. Gotz and P. J. Wright, Appl. Phys. Lett. **68**, 57 (1996).

Raman Analysis of Electron-Phonon Interactions in GaN Films

L. Bergman*, M.D. Bremser**, J.A. Christman*, S.W. King**, R.F. Davis**, R.J. Nemanich*
*Physics Department, **Department of Material Science and Engineering,
North Carolina State University, Raleigh, NC 27695-8202

ABSTRACT

Raman analysis of film quality, phonon-plasmon coupling, and phonon-exciton interaction in GaN films with varying Si doping levels is presented. The films exhibit a small stress component ~0.1 GPa, calculated from the frequency shift of the E2 mode. No correlation between the stress and the doping concentration was found. No forbidden Raman lines were detected in the spectra, implying high quality oriented films. The Raman lineshape of the E2 mode is a Lorentzian with similar linewidths at room temperature and at 10K indicating a homogeneous lifetime broadening mechanism which is not significantly affected by the change in temperature. The linewidth is also independent of Si concentration. The phonon-plasmon mixed frequency modes were calculated to be at $\omega_-=86$ cm^{-1} and at $\omega_+=741$ cm^{-1}. The modes are not present in the spectra and the only effect of the plasmons is a change of the A1(LO) lineshape. Analysis of the A1(LO) line indicated a uniform spatial doping distribution. A resonance effect was observed for the symmetry-allowed A1(LO) mode at T=10K with sub bandgap excitation light. The resonance interaction is consistent with the free exciton model.

INTRODUCTION

GaN films grown on 6H-SiC with varying doping Si concentrations were investigated via Raman spectroscopy. The analysis focused on the determination of film quality and on the phonon interactions with other excitations, specifically plasmon waves and excitons.

The Raman modes investigated are the E2 at frequency 568 cm^{-1} and the A1(LO) at 735 cm^{-1}. The A1(LO) mode possesses an electric field which interacts with the electric field produced by the free carrier waves known as plasmons [1]. A study of the A1(LO) Raman line was carried out to determine the effect of the free carrier concentration, introduced via Si doping, on the LO-plasmon coupled modes. The E2 non-polar mode, due to the relatively large Raman cross section and its non-interacting nature with plasmons, has a relatively intense Raman signal. The E2 Raman line was selected for these reasons in order to analyzed and determine film quality. The enhancement of Raman intensity of the A1(LO) via the resonance effect [2] was also studied. At resonance frequencies the intensity of the LO optical modes may be enhanced via coupling to the intermediate states of the crystal. This effect can occur via Frohlich interaction for which symmetry-forbidden LO modes appear in the spectra [2,3] as well as via the coupling of the symmetry-allowed LO modes to the intermediate states [2]. The present study focuses on the investigation of the resonance enhancement of the symmetry-allowed A1(LO) mode. It was found that the resonance is due to the interaction of the Raman phonons with the free exciton states.

EXPERIMENT

The GaN films were grown on AlN buffer layer on 6H-SiC(0001) substrates via organometallic vapor phase epitaxy [4]. The films are 2 - 3.7 μm thick and contain varying Si doping concentrations which were measured utilizing capacitance-voltage (CV) technique. The room temperature (RT) micro-Raman spectra were acquired in a back scattering geometry from the c-axis. The allowed modes from this scattering geometry are the E2 and the A1(LO) modes. The laser spot size and spectral resolution were ~4μm diameter and 0.2 cm^{-1} respectively.

The 10K macro-Raman experiments were acquired in a 45° scattering geometry from the c-face of the samples. The polarization of the incoming incident light was perpendicular to the c-axis and the scattered light was unpolarized. The geometry allows the A1(LO), A1(TO), and the E2 modes. The spot size and spectral resolution were ~100μmx2mm diameter and 0.2 cm^{-1} respectively.

Mat. Res. Soc. Symp. Proc. Vol. 449 © 1997 Materials Research Society

RESULTS AND DISCUSSION

Figure 1a shows the RT Raman spectra of the E2 and the A1(LO) modes. Figure 1b presents the high resolution RT spectra of the E2 mode for undoped and Si-doped GaN films ~ 2μm thick. The undoped film with a free carrier concentration ~ 10^{16} cm^{-3} exhibits the E2 frequency at ~ 567.7 cm^{-1}, while the $8x^{16}$ cm^{-3} and the $3x10^{17}$ cm^{-3} Si-doped samples exhibit the peak frequencies at 568.7 cm^{-1} and 568.3 cm^{-1} respectively. The stress, P, in the films may be approximated by [5]

$$P \text{ (GPa)} \sim (\omega - \omega_0) / 4.2 \tag{1}$$

where ω is the mode frequency and $\omega_0 = 568$ cm^{-1} is the unstressed value. Thus the undoped film is under tensile stress ~ 0.07 GPa, and the films of low and high Si concentration are under compressive stress ~ 0.15 GPa and 0.07 GPa respectively. The values of the stress are small and do not correlate with the free carrier concentration. As is shown in Figure 1, the Raman signal of the forbidden E1(TO) mode at 599 cm^{-1} is not present, indicating the film is well oriented along the c-axis and does not contain a high concentration of structural defects. If the film were to be missaligned or a high concentration of structural defects were present, the Raman selection rules would relax and forbidden modes would appear in the spectra.

The full width at half maximum (FWHM) of all the Raman lines is ~3 cm^{-1} independent of Si concentration. Moreover, the Raman line is a Lorentzian (see the inset to the figure) implying an homogeneous lifetime broadening mechanism. Thus it may be concluded that the Si impurities do not contribute significantly to the life-shortening mechanism of the E2 phonons. To further investigate the line broadening mechanism, Raman spectra at 10K were acquired from a thicker sample (2.7 μm). Figure 2 presents the Raman spectra at RT and at 10K. As is shown in the figure, the RT frequency of the E2 line is at 566.6 cm^{-1} while that of 10K is at 567.8 cm^{-1}. The ~ 1.2 wavenumber shift is the same as that expected from a high quality free standing (1mm thick) GaN crystallite [6]. Thus in that respect, the phonon response to temperature of the film is similar to that of the bulk crystallites. Moreover, the linewidth at 10 K did not change significantly from its RT-value implying that the lifetime of the E2 phonons is not strongly temperature dependent.

The interaction of the A1(LO) mode with plasmons is investigated. In polar semiconductors the free carrier waves, plasmons, can interact with the LO optical modes via their macroscopic electric fields. The interaction results in the coupled modes-frequencies [7,8]

$$2\omega_{\pm}^2 = \omega_L^2 + \omega_P^2 \pm \left[\left(\omega_L^2 + \omega_P^2\right) - 4\omega_P^2\omega_T^2\right]^{1/2} \tag{2}$$

where ω_L, ω_T, and ω_P are the frequencies of the LO, TO, and the plasmons respectively. For GaN the frequencies of the A1(TO) and A1(LO) are 534 cm^{-1} and 736 cm^{-1} respectively [9], and the plasmons frequency is 119 cm^{-1} [10]. The coupled modes calculated from Eq.2 are: $\omega_- = 86$ cm^{-1} and $\omega_+ = 740$ cm^{-1}. Figure 3a shows the Raman spectra from the ~ $3x10^{17}$ Si/cm^3 doped sample. This doping level introduces a significant free carrier concentration which should result in the coupling mechanism. However, Fig.3a indicates that the coupled mode-frequencies are not present in the Raman spectra. Similar behavior has been previously observed in nitrogen-doped SiC [7] and also discussed in the case of GaN [10]. The undetectable signal of the coupled modes was argued to be due to the high collision rate and thus the short lifetime of the plasmons with an outcome of very broad and low intensity Raman lines. Fig.3b shows the Raman spectra of the A1(LO) mode, normalized to the intensity of E2 as a function of Si doping. For Si concentration ~$8x10^{16}$ cm^{-3}, the intensity of the Raman line is smaller, and the line frequency and broadening is shifted toward the frequency of ω_+. The inset to the figure shows the Raman intensity as a function of the Si concentration.

The exact functional behavior of the intensity and frequency versus the free carrier concentration is left to a future investigation; such knowledge may be beneficial to the estimation of doping levels in very small samples for which electrical measurements of the doping level is not possible. Moreover, the intensity of the A1(LO)/E2 was measured at several locations across the film and no significant difference was observed, indicting a uniform doping distribution. A feature

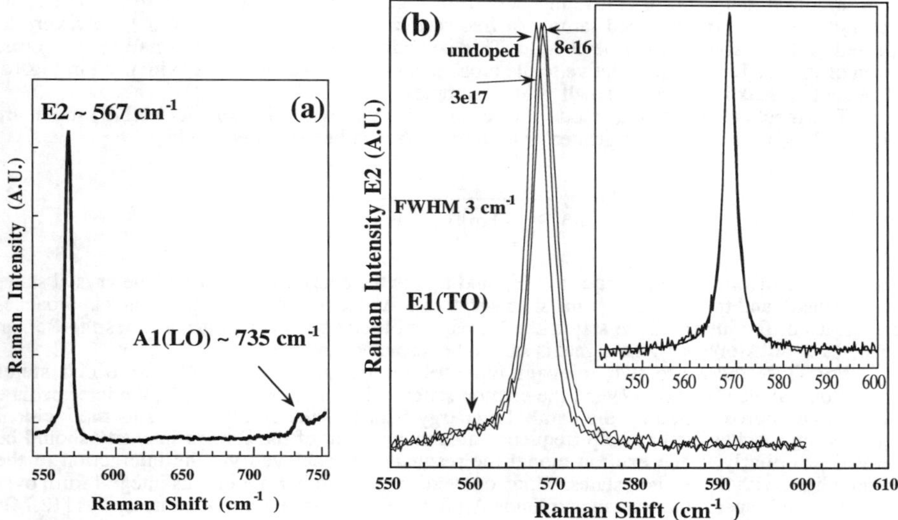

Figure 1. The Raman spectra of 2 μm GaN films (a) The E2 and the A1(LO) Raman lines and (b) The E2 line for varying free-carrier concentrations. The inset shows the Lorentzian line-fit to the Raman data.

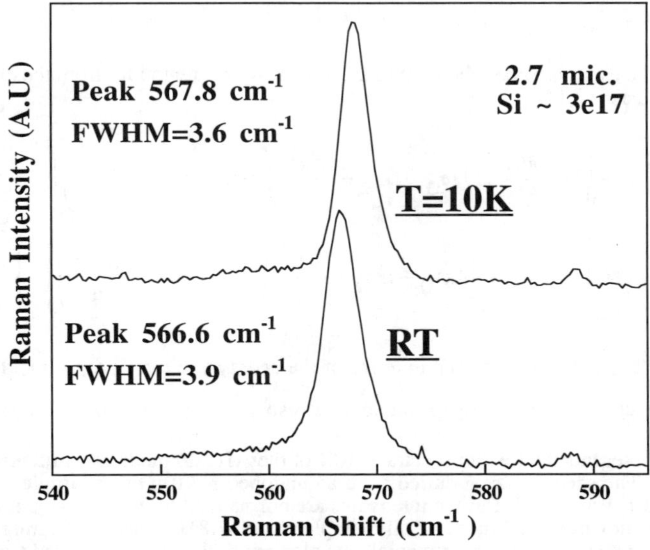

Figure 2. The spectra of the E2 Raman line at RT and at T=10K of 2.7 μm GaN film .

of the sample thickness bears close inspection: in Fig.3b the dashed line represents the Raman spectra of the ~ 0.7 μm undoped sample of free carrier concentration $\sim 10^{16}$ cm^{-3}; however, the expected A1(LO) signal is not detectable. This behavior is attributed to the small Raman cross section of the A1(LO) mode (relative to the cross section of the E2 mode, as evidenced in Fig.1a) and hence the weak intensity for small volume samples.

The mechanisms giving rise to the resonant Raman scattering of the A1(LO) mode are analyzed. In general the Raman scattering amplitude, A, can be expressed as [11]

$$A(\omega) \propto \sum_{\alpha\beta} \frac{1}{\left(E_\alpha - \hbar\omega\right)\left(E_\beta - \hbar\omega\right)}$$

(3)

where ω is the incoming laser frequency, E_α and E_β are the energies of intermediate crystal states (to be defined), and the sum is over the state spectrum. As the incident laser frequency approaches the energies of the intermediate states, the Raman amplitude becomes larger as does the Raman intensity, $I \propto \omega^4 |A(\omega)|^2$, and the signal is said to be resonance-enhanced.

For GaN there are two relevant types of intermediate states [9]: the Bloch states (conduction-valance bands) [12] and the exciton states [13]. In the Bloch model the intermediate states can be approximated by the parabolic energy bands of the crystal. Thus as the incident frequency approaches the bandgap frequency an enhancement of the Raman intensity should be observed. Similarly in the exciton model the resonance is achieved via the interaction of the incoming light with the exciton states. Upon converting the sum in Eq.3 into its integral form over the wavevector k-space, the Raman amplitude A_B in the Bloch model may be expressed as [12,14]

$$A_B = \int_0^k dk \left\{ \left[\left(\hbar\omega_G - \hbar\omega_2 + \frac{h^2 k^2}{2\mu} \right)\left(\hbar\omega_G - \hbar\omega_1 + \frac{h^2 k^2}{2\mu} \right) \right]^{-1} + \right.$$
$$\left. + \left[\left(\hbar\omega_G + \hbar\omega_2 + \frac{h^2 k^2}{2\mu} \right)\left(\hbar\omega_G + \hbar\omega_1 + \frac{h^2 k^2}{2\mu} \right) \right]^{-1} \right\}$$

(4)

Accordingly, in the exciton model, the Raman amplitude A_E in a region far from the exciton energy may be expressed as [13,14]

$$A_E = \frac{1}{(2\pi)^3} \int_0^k dk \frac{\pi\theta \exp(\pi\theta)}{\sinh(\pi\theta)} \left\{ \left[\left(\hbar\omega_G - \hbar\omega_2 + \frac{h^2 k^2}{2\mu} \right)\left(\hbar\omega_G - \hbar\omega_1 + \frac{h^2 k^2}{2\mu} \right) \right]^{-1} + \right.$$
$$\left. + \left[\left(\hbar\omega_G + \hbar\omega_2 + \frac{h^2 k^2}{2\mu} \right)\left(\hbar\omega_G + \hbar\omega_1 + \frac{h^2 k^2}{2\mu} \right) \right]^{-1} \right\}$$

(5)

where $\theta \equiv \left| R / \left(h^2 k^2 / 2\mu \right) \right|^{\frac{1}{2}}$ is expressed in terms of the energy exciton R; ω_1, ω_2 and $\omega_G = 3.5$ eV, are the incident, scattered and bandgap frequencies respectively; and $\mu = 0.174 m_0$ is the reduced effective mass.

Figure 4a presents the Raman spectra at 10K of the A1(LO) mode as a function of the laser incident energy. The spectra was acquired from an undoped 3.7 μm thick sample of free carrier concentration $\sim 10^{16}$ cm^{-3}. All Raman intensities are normalized to the ω^4 spectral dependence. The Raman frequency of the E2 mode of this sample is at 567.8$\pm$0.2 cm^{-1} indicating a negligible stress state. Figure 4b shows the experimental data along with the plots calculated from the Bloch and exciton models of Eqs. 4 and 5 respectively. The results indicate that although the incident

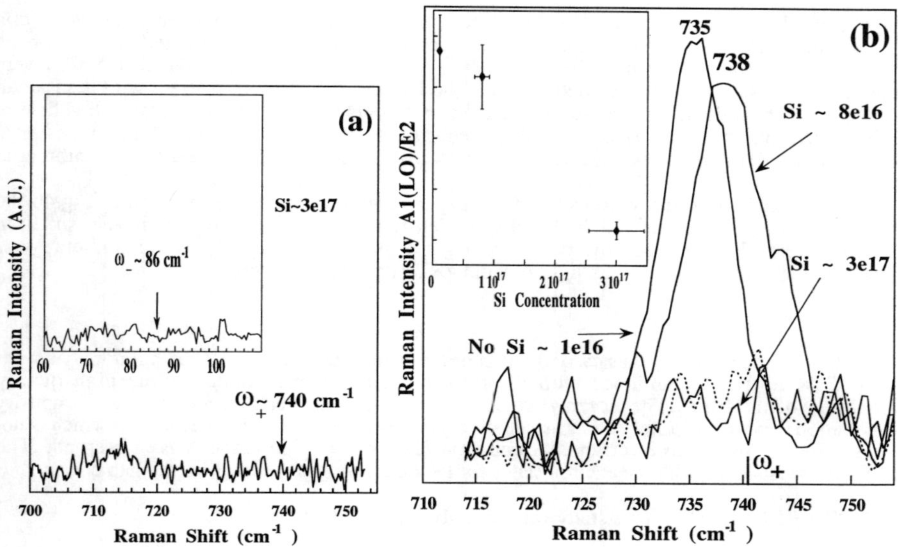

Figure 3. The phonon-plasmon mixed modes: (a) The spectra in the Raman energy range of the mixed modes and (b) The A1(LO) Raman line as a function of Si concentration. The inset presents the Raman intensity vs Si concentration.

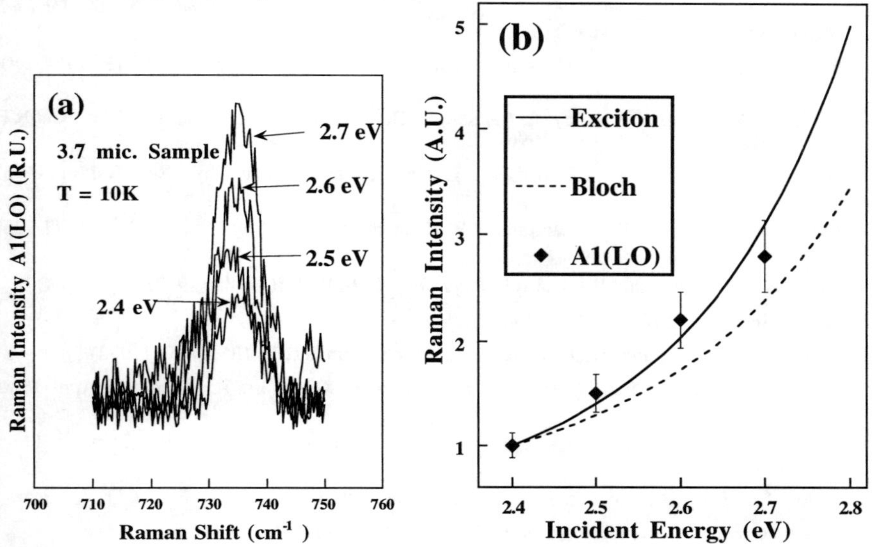

Figure 4. The resonance of the A1(LO) mode at T=10K: (a) The Raman line vs laser energy and (b) The exciton and Bloch models (lines), Raman data (dotes).

729

energy is at sub-bandgap level the intensity of the A1(LO) line is enhanced via the mode interaction with the exciton states of R ~ 28±2 meV. Results complementary to that presented here have been reported by Lemos et al [9], where the E2, E1(TO), and A1(TO) modes of ~ 20 μm GaN film were shown to be at resonance via the exciton state. The fact that the excitons interact with the Raman phonons even at sub-bandgap energies indicate their stability in the very thin films studied herein. The inference concerning the exciton stability is consistent with the free exciton photoluminescence of high quality films which was detected up to room-temperature as reported by Chichibu et al [15].

A note has to be made concerning the experimental setup: as can be seen in Fig.4b the difference between the Bloch and the Exciton models in the sub-bandgap regime is very small; in order to differentiate between the two and to get a meaningful signal the Raman setup was configured to acquire the maximum light without losing spectral resolution.

CONCLUSIONS

Raman Spectroscopy was carried out in order to analyze film quality and phonon dynamics of thin GaN films. Compliance with the Raman selection rules indicates the high quality crystallinity of the films. The Lorentzian E2 lineshape and its weak temperature and impurity concentration dependence imply a thermal stability and lifetime broadening mechanism which is not strongly affected by the free carriers. No phonon-plasmon coupled modes were observed. The A1(LO) is at resonance with the exciton states for incident energies well bellow the gap energy.

ACKNOWLEDGMENTS. This study was partially supported by the ONR.

REFERENCES

1. M.V. Klein in Light Scattering in Solids I Topics in Applied Physics, Edited by M. Cardona (Springer-Verlag, New York, 1983), p. 147-204.

2. M. Cardona in Light Scattering in Solids II Topics in Applied Physics, Edited by M. Cardona and G. Guntherodt (Springer-Verlag, New York, 1982), p. 19-179.

3. D. Behr, J. Wagner, J. Schneider, H. Amano, and I. Akasaki, Appl. Phys. Lett. 68, P. 2404-2406 (1996).

4. T.W. Weeks, M.D. Bremser, S. Ailey, E. Carlson, W.G. Perry, E.L. Piner, N.A. El-Masry, and R.F. Davis, J. Mater. Res. 11, P. 1081-1088 (1996).

5. P. Perlin, C.J. Carillon, J.P.Itie, A.S. Miguel, I. Grzegory, and A. Polian, Phys. Rev. B 45, p. 83-89 (1992).

6. D.D. Manchon, A.S. Baker, P.J. Dean, and R.B. Zetterstrom, Solid State Comm. 8, P. 1227-1230 (1970).

7. M.V. Klein, B.N. Ganguly, and P.J. Colwell, Phys. Rev. B, 6, P. 2380-2388 (1972).

8. B.B. Varga, Phys. Rev. 137, P. A1896-1902 (1965).

9. V. Lemos, C.A. Arguello, and R.C.C. Leite, Solid State Comm. 11, P. 1351-1353 (1972).

10. T. Kozawa, T. Kachi, H. Kano, Y. Taga, M. Hashimoto, N. Koide, and K. Manabe, J. Appl. Phys. 75, P. 1098-1101 (1994).

11. J.L. Birman in Light Scattering in Solids, Edited by M. Balkanski (Flammarion Sciences, Paris, 1971), P. 15-19.

12. R. Loudon, Prc. Roy. Soc. London A 275, P. 218-232 (1963).

13. A.K. Ganguly and J.L. Birman, Phys. Rev. 162, P. 806-816 (1967).

14. B. Bendow, J.L. Birman, A.K. Ganguly, T.C. Damen, R.C.C. Leite, and J.F. Scott, Optical Commu. 1, P. 267-270 (1970).

15. S. Chichibu, T. Azuhata, T. Sota, and S. Nakamura, J. Appl. Phys. 79, P. 2784-2786 (1996).

MAPPING OF DONOR IMPURITIES IN GaN BY RAMAN IMAGING

F. A. PONCE*, J. W. STEEDS°, C. D. DYER˝, and G. D. PITT˝
* Xerox Palo Alto Research Center, Palo Alto, CA 94304
° University of Bristol, H. H. Wills Physics Laboratory, Bristol BS8 1TL, UK
˝ Renishaw, Old Town, Wotton-under-Edge, Gloucestershire, GL12 7DH, UK

ABSTRACT

Raman scattering experiments on silicon-doped GaN show that donor impurities quench the A1(LO) Raman line at 735 cm^{-1}. This is due to interaction between lattice vibrations and the free carrier plasma. The spatial variation of the A1(LO) signal has been imaged directly using newly developed instrumentation. Features with dimension under on micron are observed in faceted GaN crystallites. The variation in free carrier concentration is attributed to preferential incorporation of donor impurities during growth.

INTRODUCTION

Probing the lattice vibrational modes by Raman spectroscopy provides insight into the properties of crystalline materials. The crystal quality can be judged from the peak shape and the adherence to selection rules [1]. It has been recently shown that Raman scattering is a very sensitive and straightforward method for distinguishing the hexagonal and cubic phases in GaN [2]. Unexpected peaks in the Raman spectra may be the consequence of local modes or electronic transitions of dopants or impurity atoms in the material [3,4]. Crystal strain may be determined from shifts of the peaks of the Raman spectra [5, 6] and in the case of the ternary AlGaN such shifts may be used to determine the alloy composition [7]. Phonon-plasmon modes have been observed in n-type GaN [8], GaAs [9], and ZnSe [10] and a close correlation between these modes and the free carrier concentration has been observed. It has also been shown that it is possible to use Raman scattering in the determination of the surface space-charge layer [11].

The longitudinal optical (LO) phonons interact strongly with a free carrier plasma when the plasma frequency is close to the LO phonon frequency. The coupled LO phonon plasma modes thereby created, diminish the intensity of the LO phonon peak in a systematic fashion that can be used to determine the donor concentration in Si-doped GaN [12]. Recent improvements in Raman spectroscopic equipment have produced faster and more efficient means to acquire the spectra and have made feasible Raman spectroscopic imaging [13-15]. The preferred method is to tune by tilting a multilayer dielectric band pass filter, transmitting a narrow spectral range (10 - 20 cm^{-1}), and to image the sample directly to a two-dimensional array detector.

In this paper we show that n-type dopants like silicon produce a reduction in the A1(LO) Raman signal. We next use this signal to directly image the spatial variation of the plasmon-phonon coupling in epitaxial layers of GaN. The images thus produced have resolutions of under 1 micron, and yield a mapping of the donor-impurity distribution in the material. It is observed that the impurity incorporation appear to depend strongly on the sense of the growth (polarity).

Mat. Res. Soc. Symp. Proc. Vol. 449 © 1997 Materials Research Society

EXPERIMENT

The specimens used in this study were grown by metalorganic chemical vapor deposition (MOCVD) on (0001) sapphire [16, 17] and on $(000\bar{1})$ GaN bulk single crystal [18] substrates. The epitaxial growth direction was [0001] GaN/sapphire and $[000\bar{1}]$ for homoepitaxial GaN/GaN [19]. Raman spectra and images were collected on a Renishaw Raman Imaging Microscope System 2000. The measurements were performed in backscattering $z(xx)z'$ or $z(xy)z'$ geometry, with the z direction parallel to the c axis of the GaN wurtzite structure. In this notation, xx means that the polarization of the incident light and the of the polarizer are parallel to each other; xy means that they are perpendicular. A 25 mW, 514.5 nm, Argon ion (Ar^+) laser was used as light source, typically with 4 to 8 mW at the sample. A 50X (NA 0.8) objective was employed, giving an illuminated spot of 1 to 1.5 microns in diameter for spectra acquisition and 20 microns diameter for images. A thermoelectrically-cooled (-70°C) front-illuminated charge coupled device (CCD) array was used as the detector. Two high-performance holographic notch filters (HNFs) were used to reject the scattered and reflected light at the laser wavelength; the first of these filters was also employed as the microscope beamsplitter.

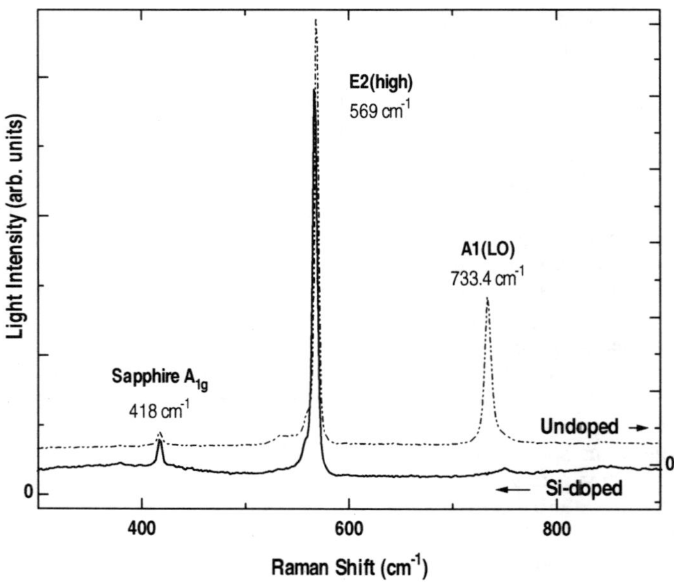

Figure 1. Raman spectra of undoped, and silicon-doped GaN/sapphire thin films. The spectra were taken in the z(xx)z' configuration, with incident and analyzer polarizations parallel. Note the absence of the A1(LO) phonon line in the silicon-doped film.

Raman features were easily determined without any attenuation to 100 cm^{-1}, and increasing attenuation to within 50 cm^{-1} of the laser line. Spectral analysis was performed using the standard 250 mm focal length spectrograph, fitted with a 1800 g/mm diffraction grating, giving a spectral resolution of 2 cm^{-1} and a spectral window of 950-1000 cm^{-1} when centered at 570 cm^{-1} shift. Spectral accuracy was determined to be $\pm$ 1 cm^{-1} versus a neon (Ne) lamp, and precision an order of magnitude higher, since the diffraction grating angle and room temperature were held constant throughout the series of experiments. Typical acquisition time for spectra was 5-30 seconds.

RESULTS

Raman spectra corresponding to undoped and silicon-doped GaN epitaxial layers on (0001) sapphire are shown in Fig. 1. The free carrier concentration was <5x10^{16} cm^{-3} and ~ 2x10^{17} cm^{-3} at room temperature, respectively. Note that the main difference in the Raman spectra is that the strong A1(LO) peak present in the undoped film is absent in the Si-doped case. As already mentioned, quenching of the A1(LO) phonon has been observed previously [12], and it has been explained in terms of phonon-plasmon coupled modes associated with the free electrons introduced by the silicon dopant.

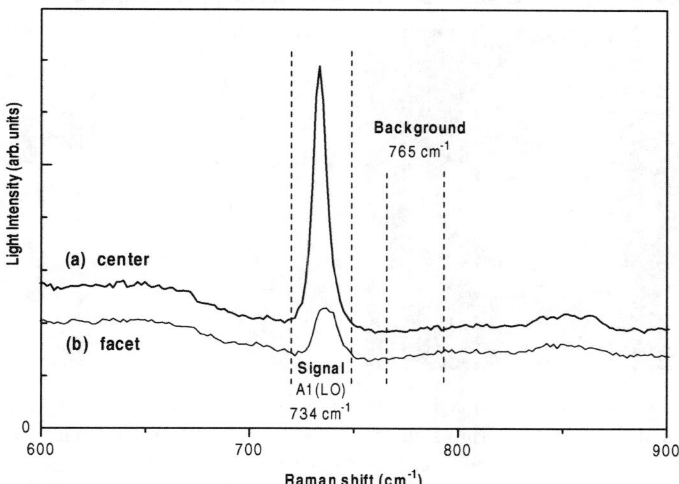

Figure 2. Raman spectra taken from a hexagonal crystallite in homoepitaxial GaN. It shows a significant decrease in the A1(LO) line away from the center of the hexagon, and in particular at the edges between the triangular facets.

A hexagonal crystallite, typical of GaN homoepitaxy on the [000$\bar{1}$] nitrogen-terminated surface of GaN bulk single crystals is shown in Fig. 3a. These hexagonal hillocks have typical dimensions of the order of 20 to 50 mm, and are characterized by a faceted dome with six triangular facets and a well-pronounced tip in the center of the hexagon. The feature at the center of the hillock has been characterized by TEM and it is identified as an inversion domain column with opposite polarity to the matrix [20]. Raman spectra taken at the center of the hexagon and in the middle of the facets are shown in Fig. 2. It is clear that the center has a strong A1(LO) peak, similar to the undoped case in Fig. 1, whereas the facets show weak peaks. Spectra taken at the vertex of the facets show no A1(LO) peak, in a very similar fashion as the silicon-doped case in Fig. 1.

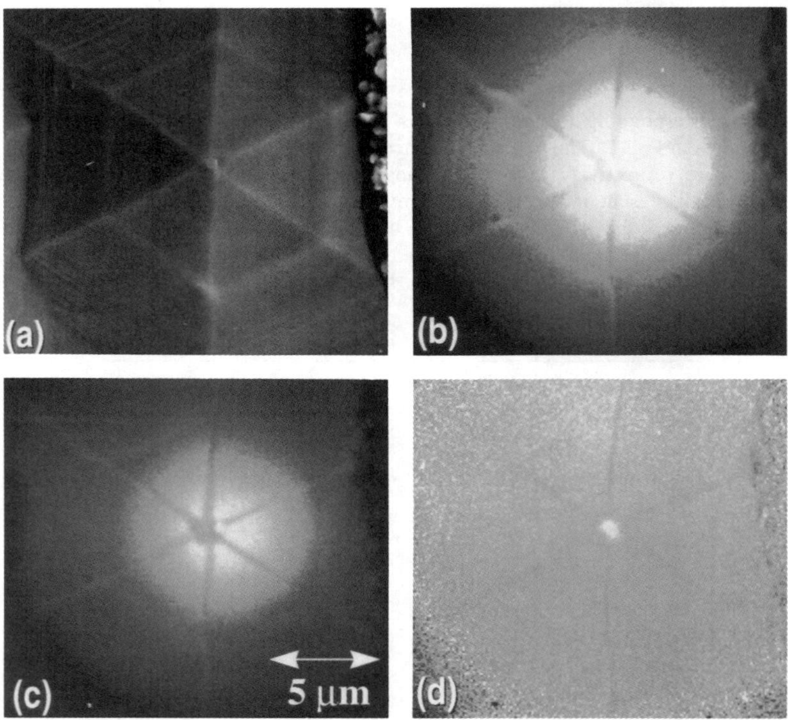

Figure 3. Raman scattering images of hexagonal feature in homoepitaxial GaN/GaN grown on the [000$\bar{1}$] N-terminated side: (a) Secondary electron image taken at 30kV; (b) scattering at 735cm^{-1}; (c) background reference at 765 cm^{-1}; and (d) flat-field image obtained by dividing the images at 735 and 765 cm^{-1}.

The area about the A1(LO) was used to form an image, shown in Fig. 3b. Direct imaging was performed by illuminating the desired area on the sample and passing the collected light through dielectric multilayer bandpass filters onto the 2-D CCD array. This gives a high-throughput (60-80% transmission) technique for the determination of scattered intensity as a function of position on the sample for a specific Raman band at high spatial resolution (one micron). The images have been normalized against a reference image, to account for any residual nonuniformity in illumination and for sample topography. The normalization file was generated at 765 cm^{-1}, so that the background was similar in intensity to the background at 735 cm^{-1}, where the A1(LO) phonon was imaged, and is shown in Fig. 3c. Spectral accuracy through the filter lightpath was estimated at ± 2 cm^{-1}.

The normalized image in Fig. 3d shows a symmetric distribution of the 735 cm^{-1} Raman signal. The center of the hexagon shows a strong signal, indicative of low free-electron densities (or absence of silicon and other n-type dopants). The edge between the triangular facets appear dark over a linear thickness of about 500nm, indicating high donor concentration. Full Raman spectra taken at various sample positions indicate that only the A1(LO) peak has a spatial variation. This fact makes unlikely the possibility of the geometric factors such as reflections or waveguiding as being responsible for the image contrast. This leads us to conclude that the variation of the image is due to local variations in the free electron concentration, likely produced by preferential incorporation of donor impurities during growth. The spatial variation in the coupling of the LO phonon to overdamped plasmons is responsible for the image characteristics of the A1(LO) line in GaN shown in Fig. 3.

DISCUSSION

The reduction in the LO phonon line, associated with an increase in carrier concentration, is accompanied by broadening and a shift towards the high frequency side. This has already been observed previously by Kozawa et al. [8], who concluded that the dominant scattering mechanisms in GaN involve deformation potential and electro-optic mechanisms. Absence of the 735 cm^{-1} A1(LO) line may not be limited solely to silicon in GaN. It may be a property of donor impurities in general. Incorporation and segregation of oxygen are suspected in GaN and could be in part responsible for the observations reported here.

It is interesting to note that the growth sense (polarity) seems to be associated with different levels of impurity distribution. The inversion domain in the center of the hillock has the same polarity as growth on sapphire substrates shown in Fig. 1. The implication is that the incorporation of (donor) impurities such as silicon and oxygen is sensitive to the growth direction. MOCVD epitaxy is usually carried out under nitrogen rich conditions, with the growth rate limited by the availability of Ga. In the [0001] case, gallium atoms are located on the top position of the basal planes, whereas in the [000$\bar{1}$] case, nitrogen atoms are located at the top position of the basal planes [19]. From our observations, it is clear that the Ga-on-top case is less reactive to the incorporation of this type of impurities than the N-on-top case.

The techniques required to reduce recording and display times to reasonable levels are only now being mastered, and our results show that it has become technically possible to produce direct images of Raman signals with high spatial resolution. This is also the first case that we know where variations caused by plasmons coupling are used to map out dopant distribution in a specimen at high spatial resolution. The technique is non-destructive and yields images with resolutions better than 1 micron. No vacuum methods or cryogenic techniques have been used, while the Raman data is gathered very rapidly - typically 1 to 10 sec per spectrum and between 10 to 20 minutes per image.

CONCLUSIONS

In summary, we have observed quenching of the A1(LO) Raman line by silicon in GaN. We have used Raman scattered light to directly image the spatial variation of this effect, and thereby of the donor distribution, with a spatial resolution of under 1 micron. These results demonstrate that Raman imaging can be used for directly mapping the local variations of donor impurities in semiconductors, a technique that could eventually prove critical for device technology advancement.

ACKNOWLEDGEMENTS

The authors are grateful to I. Grzegory, S. Porowski, and H. Helava who provided the single crystal GaN substrate, and to D. Bour who grew the epitaxial GaN reported in Ref. 18. This work was partially supported by Department of Commerce Advanced Technology Program (70NANB2H1241) and by ARPA (Agreement # MDA972-95-3-0008).

REFERENCES

1. C. A. Argüello, D. L. Rousseau, and S. P. S. Porto, Phys. Rev. **181**, 1351 (1969).
2. H. Siegle, L. Eckey, A. Hoffmann, C. Thomsen, B. K. Meyer, B. Schikora, M. Hankein, K. Lischka, Sol. State Comm. **96**, 943 (1995).
3. A. S. Barker, Jr. and M. Ilegems, Phys. Rev. B **7**, 743 (1973).
4. M. V. Klein, in *"Light Scattering in Solids I"*, editted by M. Cardona (Springer Verlag, Berlin 1975), p. 147.
5. W. Rieger, T. Metzger, H. Angerer, R. Dimitrov, O. Ambacher, and M. Stutzmann, Appl. Phys. Lett. **68**, 970 (1996).
6. T. Kozawa, T. Kachi, H. Kano, H. Nagase, N. Koide, and K. Manabe, J. Appl. Phys. **77**, 4389 (1994).
7. K. Hayashi, K. Itoh, N. Sawaki, and I. Akasaki, Sol. State Comm. **77**, 115 (1991).
8. R. Ruppin and J. Nahum, J. Phys. Chem. Solids **35**, 1311 (1974).
9. D. Olego and M. Cardona, Phys. Rev. B **24**, 7217 (1981).
10. H. Shen, F. H. Pollak, and R. N. Sacks, Appl. Phys. Lett. **47**, 891 (1985).
11. J. Kraus and D. Hommel, Semicond. Sci. Technol. **10**, 785 (1995).
12. T. Kozawa, T. Kachi, H. Kano, Y. Taga, M. Hashimoto, N. Koide, and K. Manabe, J. Appl. Phys. **75**, 1098 (1994).
13. I. P. Hayward, K. J. Baldwin, D. M. Hunter, D. N. Batchelder and G. D. Pitt, Diamond and Rel. Mat. **4**, 617 (1995).
14. N. C. Burton, J. W. Steeds, G. M. Meader, Y. G. Shreter, and J. E. Butler, Diamond and Rel. Mat. **4**, 1222 (1995).
15. F. A. Ponce, J. W. Steeds, C. D. Dyer, and G. D. Pitt, Appl. Phys. Lett. **69**, 2650 (1996).
16. F. A. Ponce, J. S. Major, Jr., W. E. Plano, and D. F. Welch, Appl. Phys. Lett., **65**, 2303 (1994).
17. F. A. Ponce, D. P. Bour, W. Götz, and P. J. Wright, Appl. Phys. Lett. **68**, 57 (1996).
18. F. A. Ponce, D. P. Bour, W. Götz, N. M. Johnson, H. I. Helava, I. Grzegory, J. Jun, and S. Porowski, Appl. Phys. Lett. **68**, 917 (1996).
19. F. A. Ponce, D. P. Bour, W. T. Young, M. Saunders, and J. W. Steeds, Appl. Phys. Lett. **69**, 337 (1996).
20. J.-L. Rouviere, M. Arlery, A. Bourret, R. Niebuhr, and K.-H. Bachem, Mat. Res. Soc. Proc. **395**, 393 (1996).

SUBPICOSECOND TIME-RESOLVED RAMAN STUDIES OF NONEQUILIBRIUM EXCITATIONS IN WURTZITE GAN

K.T. TSEN*, R.P.JOSHI**, D.K. FERRY***, A.BOTCHKAREV****,
B.SVERDLOV****, A. SALVADOR****, H. MORKOC****
*Department of Physics and Astronomy, Arizona State University, Tempe, AZ 85287
**Department of Electrical Enigineering, Odl Dominion University, Norfolk, VA 23529
***Department of Electrical Engineering, Arizona State University, Tempe, AZ 85287
****Coordinated Science Laboratory, University of Illinois, Urbana, IL 61801

ABSTRACT

Non-equilibrium electron distributions as well as phonon dynamics in wurtzite GaN have been measured by subpicosecond time-resolved Raman spectroscopy. Our experimental results have demonstrated that for electron densities $n \geq 5 \times 10^{17} \mathrm{cm}^{-3}$, the non-equilibrium electron distributions in wurtzite GaN can be very well described by Fermi-Dirac distribution functions with the temperature of electrons substantially higher than that of the lattice. The population relaxation time of longitudinal optical phonons was directly measured to be $\tau \cong 5 \pm 1$ ps at T = 25 K. The experimental results on the temperature dependence of the lifetime of longitudinal optical phonons suggest that the primary decay channels for these phonons are the decay into (1) one transverse optical phonon and one high energy, longitudinal or transverse acoustical phonons; and (2) one transverse optical phonon and one E_2 phonon.

INTRODUCTION

Due to great potential for many device applications, wide bandgap semiconductors such as GaN have recently attracted much attention[1]. We note that although much progress has been made in the device-oriented applications with these wide bandgap semiconductors, very little information concerning their dynamical properties has yet been known. Knowledge of carrier as well as phonon dynamics is very crucial for device engineers to design better and faster devices. In this paper, we have studied non-equilibrium electron distributions and phonon dynamics in wurtzite GaN with time-resolved sub-picosecond Raman spectroscopy. Our experimental results show that for electron densities $n \geq 5 \times 10^{17} \mathrm{cm}^{-3}$, non-equilibrium electron distributions in wurtzite GaN can be very well described by Fermi-Dirac distribution functions with the temperature of electrons much higher than that of the lattice, as a result of large momentum randomization. In addition, the population relaxation time of longitudinal optical phonons was determined to be $\tau \cong 5 \pm 1$ ps. From the analysis of the temperature dependence of the population relaxation time of longitudinal optical phonons, we conclude that the primary decay channels for these phonons are the decay into (1) one transverse optical phonon and one high energy, longitudinal or transverse acoustical phonon; and (2) one transverse optical phonon and one E_2 phonon.

SAMPLES AND EXPERIMENTAL TECHNIQUE

The sample investigated in this work was a wurtzite structure, undoped GaN (with electron density $n \cong 5 \times 10^{16} \mathrm{cm}^{-3}$) grown by molecular beam epitaxy on a (0001)-oriented sapphire substrate. The z-axis of this wurtzite structure GaN is perpendicular to the sapphire substrate plane. The thickness of the GaN layer was about 2 μm.

Very short ultraviolet laser pulses were generated by the frequency-doubling of a dual-jet R6G dye laser, which was synchronously pumped by the second harmonic of a cw mode-locked YAlG laser operated at a repetition rate of $\cong 76$ MHz. The pulse width of the dye laser was about 800 fs as determined from the autocorrelation measurements. This dye laser, after

Mat. Res. Soc. Symp. Proc. Vol. 449 © 1997 Materials Research Society

frequency-doubling, provides a train of ultrashort, ultraviolet pulses of $\cong 600$ fs in duration. The photon energy was chosen to be $\hbar\omega_L \cong 4.36$ eV so that electron-hole pairs could be photoexcited with an excess energy of $\cong 1$ eV. For the study of electron distributions, the same laser pulse was used to both excite and probe non-equilibrium electron distributions, therefore, the experimental results represent an average over the laser pulse width. On the other hand, in the investigation of phonon dynamics, the laser was split into two equally intense but perpendicularly polarized beams. One is used to excite electron-hole pairs and the other to probe non-equilibrium LO phonon populations through Raman spectroscopy. The density of photoexcited electron-hole pairs was determined by the fitting of time-integrated luminescence spectrum[2]. The sample was kept in contact with a cold finger tip of a closed-cycled refrigerator. The temperature of the sample was estimated to be T $\cong 25$ K. The backward scattered Raman signal was collected and analysed with a standard Raman setup with a photomultiplier tube (for electrons) as well as a CCD (for phonons) detection systems.

EXPERIMENTAL RESULTS AND DISCUSSIONS

Figs. 1(a) and 1(b) show two transient single-particle scattering (SPS) spectra of a GaN sample taken at T$\cong$ 25 K and for electron densities n $\cong 3$x10^{18} and 5x10^{17}cm^{-3}, respectively. Here, the scattering geometry of z(x,y)$\bar{z}$ was used, which ensures that the SPS signal comes from light scattering from spin-density fluctuations[3,4]. One intriguing respect of the Raman scattering technique is that since Raman scattering cross section is inversely proportional to the square of the effective mass, it effectively probes electron distributions even if holes are simultaneously present. We have used the theory originally developed by Hamilton and McWhorter[5] (but appropriately modified for our experimental conditions) that has been demonstrated to be able to explain the SPS spectrum very well in the equilibrium case[6] to interpret these SPS spectra.

We note that in fitting the experimental results with the theory, three parameters were used: Γ(the damping constant involved in the Raman scattering process), τ (electron collision time) and T_{eff} (the effective electron temperature). In addition, beacuse the photon energy used in the experiments is sufficiently far from any relevant energy gaps, the fitting processes are not sensitive to the detailed band structure of GaN.

The solid curves in Figs. 1(a) and 1(b) are theoretical calculations based on the theory mentioned above with Fermi-Dirac distribution functions and parameter sets that best fit the data: T_{eff} = 800 K, τ= 15 fs, Γ= 20 meV for n $\cong 3$x10^{18}cm^{-3}; and T_{eff}= 500 K, τ= 20 fs, Γ= 20 meV for n $\cong 5$x10^{17}cm^{-3}. The effective temperature of electrons have been found to be much higher than that of the lattice, as expected. We note that the damping constant Γ= 20 meV involved in the Raman scattering process is very close to the value ($\cong$ 13meV) obtained for GaAs from an analysis of resonance Raman profile under the equilibrium conditions[7]. From the quality of the fit, we conclude that for electron densities n ≥ 5x10^{17}cm^{-3}, due to efficient momentum randomization the electron distribution functions in wurtzite GaN can be very well described by Fermi-Dirac distributions with effective temperature of the electrons much higher than that of the lattice.

Fig. 2(a) shows a transient polarized Stokes Raman spectrum of a GaN sample taken at T= 25 K and with a frequency-doubled, cavity-dumped single-jet R6G dye laser (which has a pulse width of about 2 ps and a photon energy of $\hbar\omega_L$ = 4.36 eV), in z(x,x)$\bar{z}$ scattering geometry; where x= (100) and z= (001). Similar Raman spectrum was reported by Azuhata et al. at T= 300 K[8]. The sharp peak around 757 cm^{-1} comes from scattering of light by the E_g phonon mode of sapphire; whereas the shoulder close to 741 cm^{-1} belongs to the A_1(LO) phonon mode. Here, we note that E_1(LO) phonon mode, which has a frequency very close to 750 cm^{-1}, does not contribute to the Raman spectrum in Fig. 2(a) due to the Raman scattering

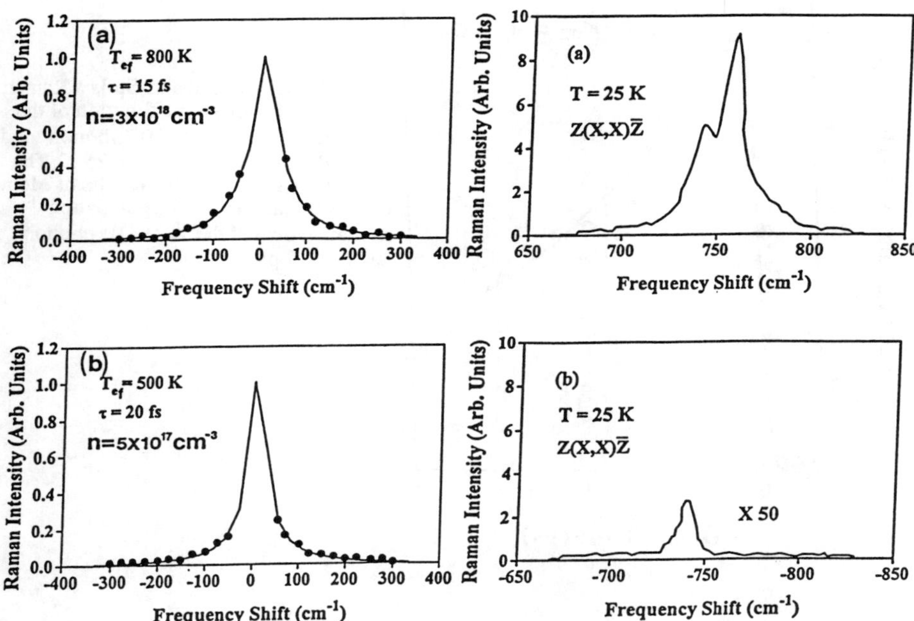

Fig. 1: Single-particle scattering spectra of a wurtzite structure GaN sample taken at T = 25 K and with an ultraviolet ($\hbar\omega_L$ = 4.36 eV), ultrashort ($\cong$ 600 fs) laser for electron densities (a) 3×10^{18} cm^{-3} and (b) 5×10^{17} cm^{-3}, respectively. The spectra were fit by the theory in Ref. 6 with electron effective temperature T_{eff} and electron collision time τ as fitting parameters.

Fig. 2: Transient polarized Stokes spectrum (a) and anti-Stokes Raman spectrum; (b) for a GaN sample taken at T = 25 K and with an ultraviolet ($\hbar\omega_L$ = 4.36 eV), veryshort ($\cong$ 2 ps) laser. For clarity, the anti-Stokes Raman signal has been magnified by a factor of 50.

selection rule[8]. The anti-Stokes Raman spectrum corresponding to Fig. 2(a) was shown in Fig. 2(b). Since at very low temperatures, thermal occupations of phonons are vanishingly small, any Raman signal observed in the anti-Stokes Raman spectrum must arise from the

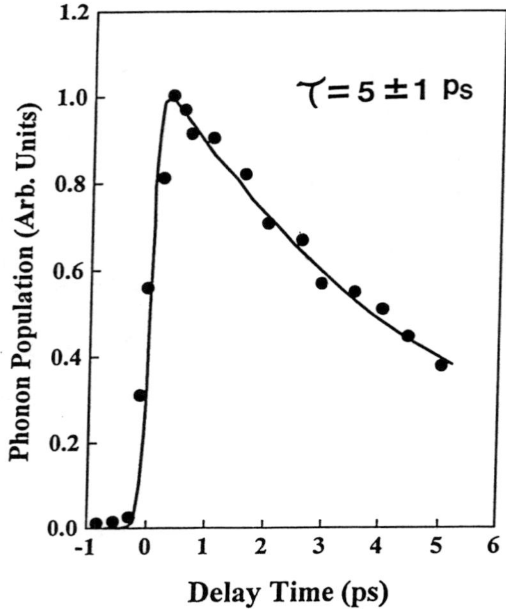

Fig. 3: Time-resolved anti-Stokes Raman signal (solid circles) as a function of the time delay for the $A_1(LO)$ phonon mode in a GaN sample taken at T = 25 K. The solid curve represents the results of Monte Carlo simulation with the population relaxation time of the $A_1(LO)$ phonon mode τ= 5 ps.

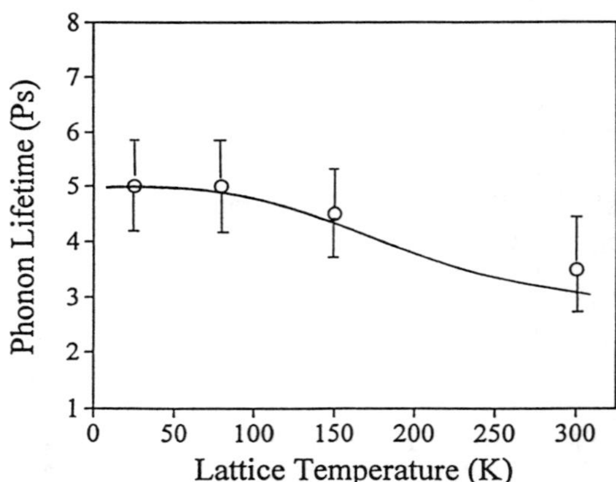

Fig. 4: Temperature dependence of the lifetime of $A_1(LO)$ phonon mode in a GaN sample. The solid curve is based on the theoretical calculations of Eq. (1) with the possible decay channels discussed in the text.

presence of non-equilibrium phonon modes[9,10]. With the help of Stokes Raman spectrum in Fig. 2(a), we can make the following identification: The broad structure centered around -741 cm^{-1} comes from Raman scattering of light from non-equilibrium $A_1(LO)$ phonon mode. We note that within our experimental uncertainty there is no detectable non-equilibrium phonons for the E_g mode of sapphire, which is consistent with the fact that no electron-hole pairs are photoexcited in the sapphire layer under our experimental conditions.

Fig. 3 shows the time-resolved non-equilibrium population of $A_1(LO)$ phonon mode taken at T = 25 K and with an ultrashort ($\cong 600$ fs), ultraviolet ($\hbar\omega_L \cong 4.36$ eV) laser as described before. The density of electron-hole pairs was $n \cong 10^{17} cm^{-3}$. The rise and the decay of the signal correspond to the generation and decay of non-equilibrium $A_1(LO)$ phonons[11,12]. To obtain a better insight onto these processes, we have used a standard Monte Carlo scheme[13] to simulate the transient dynamics of the coupled carrier-phonon system. A single phonon decay constant of 5 ps was used here under the relaxation time approximation to obtain the best fit to the experimental data.

Fig 4 shows the measured $A_1(LO)$ phonon lifetime as a function of lattice temperature. Because of the requirements of momentum and energy conservation in the interaction of the three phonons, an optical phonon at q=0 decays into two phonons of opposite momenta with energies which sum to the optical phonon energy. Examination of the phonon dispersion curve for GaN reveals that there are two possible decay channels. One channel is decay into one transverse optical phonon and a high energy, longitudinal or transverse acoustical phonon. The other channel involves the creation of a TO phonon and a E_2 phonon. Klemens[14] used perturbation theory to show that the decay of the LO phonon can be express as

$$dn \, / \, dt = -n\left[1 + n_1(T) + n_2(T)\right] / \, \tau_0 \qquad (1)$$

where τ_0 the decay time at T = 0 and $n_1(T)$ and $n_2(T)$ are the decayed lower energy phonon occupation numbers at the lattice temperature T. The derivation of the above equation assumes that the two lower energy phonons into which the LO phonon decay are always in thermal equilibrium, which is trus for the present experimental conditions. As shown in Fig. 4, we have found that both of the possible decay channels described above are consistent with our experimental results. Here, the energies of the lower phonon branches for these two possible decay channels are : TO(540 cm^{-1}), LA or TA(300 cm^{-1}); and TO(540 cm^{-1}), E_2(300 cm^{-1}), respectively.

CONCLUSION

In conclusion, an ultraviolet, ultrashort laser has been used to study non-equilibrium electron distributions and phonon dynamics in wurtzite GaN. It has been demonstrated that for electron densities $n \geq 5x10^{17} cm^{-3}$, the electron distributions can be very well described by Fermi-Dirac distribution functions with the temperature of electrons substantially higher than that of the lattice. In addition, the population relaxation time of logitudinal optical phonons was directly measured to be $\tau \cong 5 \pm 1$ ps. Our experimental results on the temperature dependence of the lifetime of longitudinal optical phonons suggest that the primary decay channels for these phonons are the decay into (1) one transverse optical phonon and one high energy, longitudinal or transverse acoustical phonons; and (2) one transverse optical phonon and one E_2 phonon.

ACKNOWLEDGEMENTS
This work was supported by the National Science Foundation under Grant No. DMR-9301100.

REFERENCES

1. For a revew, see H. Morkoc, S. Strite, G.B. Gao, M.E. Lin, B. Sverdlov and M. Burns, J. Appl. Phys. Rev. Vol. 76, No. 3, 1363 (1994).

2. D.S. Kim and P.Y. Yu in "Light Scattering in Semiconductor Structures and Superlattices", edited by D.J. Lockwood and J.F. Young (plenum Press, New York, 1991), p.383.

3. M.V. Klein, in "Light Scattering in Solids I", edited by M. Cardona and G. Guntherodt, Topics in Appl. Phys. Vol. 8 (Springer, New York, 1983), p.151.

4. G. Abstreiter, M. Cardona, and A. Pinczuk, in "Light Scattering in Solids III", edited by M. Cardona and G. Guntherodt, Topics in Appl. Phys. Vol. 51 (Springer, New York, 1983), p.5.

5. D.C. Hamilton and A.L. McWhorter, in "Light Scattering Spectra of Solids", edited by G.B. Wright (Springer, New York, 1969), p. 309.

6. C. Chia, O.F. Sankey and K.T. Tsen, Modern Phys. Lett. B7, 331 (1993).

7. A. Pinczuk, G.A. Abstreiter, R. Trommer and M. Cardona, Solid State Commun. 30, 429 (1979).

8. T. Azuhata, T. Sota, K. Suzuki, S. Nakamura, J. of Phys. Condensed Matter 7, L129 (1995).

9. K.T. Tsen, K.R. Wald, T. Ruf, P.Y. Yu and H. Morkoc, Phys. Rev. Lett. 67, 2557 (1991).

10. K.T. Tsen. R.P. Joshi D.K. Ferry and H. Morkoc, Phys. Rev. B39, 1446 (1989).

11. J.A. Kash, J.C. Tsang and J.M. Hvam, Phys. Rev. Lett. 54, 2151 (1985).

12. K.T. Tsen and H. Morkoc, Phys. Rev. B38, 5615 (1988).

13. R.P. Joshi, J. Appl. Phys. 74, 4434 (1994); R.P. Joshi, A.N. Dharamsi and J. Mcadoo, Appl. Phys. Lett. 64, 3611 (1994).

14. P.G. Klemens, Phys. Rev. 148, 845 (1966).

Part VII

Optical Properties

EFFECTS OF STRAIN FIELDS ON EXCITONS AND PHONONS IN WURTZITE GAN EPILAYERS

B.GIL*, O.BRIOT*, R.L.AULOMBARD*, J.F.DEMANGEOT**, J.FRANDON**, AND M.RENUCCI*,

*CNRS - GES , Université de Montpellier II -Case Courrier 074, 34095 Montpellier - France
gil@ges.univ-montp2.fr
**Laboratoire de Physique des Solides - Université Paul Sabatier, 31062 Toulouse France

ABSTRACT

The electronic structure of a set of several tens of a few micrometer-thick GaN epilayers grown by MOVPE on (0001)-oriented sapphire is investigated by means of photoluminescence and reflectance spectroscopy. The strain-fields effects are quantitatively interpreted using a group theory analysis which predicts seven pertinent parameters: the crystal field splitting, two spin-orbit interaction parameters and four deformation potentials. From a theoretical point of view, we also examine the exciton problem. We establish the selection rules which control the strength of optical transitions for Γ_5 and Γ_1 excitons, and estimate the electron hole exchange interaction and the longitudinal transverse splitting of A, B, C excitons. Raman spectroscopy measurements are performed at various temperatures to observe the strain-induced shift of Raman frequencies. We determine four phonon deformation potentials in w-GaN.

INTRODUCTION

Gallium Nitride and its alloys with InN and AlN have recently emerged as important semiconductor materials with applications to yellow, green, blue and ultraviolet portions of the spectrum as emitters and detectors and high temperature electronics. The difficulties to realize single crystals with prehensile sizes have long prevented large scale investigations and have compensated potentialities of GaN linked to its robustness and to the large value of its direct fundamental bandgap. Modern deposition techniques have recently revealed the possibility to synthesize micrometer thick GaN layers in epitaxial relation with a substrate, which display very good structural and purity properties, which can be doped n and p type. GaN heteroepitaxies, where the thin layer is deposited on lattice mismatched substrates like sapphire or 6H-SiC, or ZnO, or zeolites constitute a prototype case for modern semiconductor studies[1]. Very recently, pulsed room temperature operation of 410 nm semiconductor lasers with important implications to digital storage, the shortest wavelength ever from a semiconductor, has been reported[2]. Such lasers are based on the epitaxial growth of several nitrides layers in sophisticated architectures in order to produce both carrier and photon confinements , and efficient high current injection. The alloys of interest, namely AlGaN and InGaN, display strong lattice mismatch with sapphire and 6H-SiC, the substrates currently used. From extensive investigations in strained epilayers, it has been clearly demonstrated that residual strain often creates defects, which can have severe and deleterious influence on device performances in general, and on lifetime in particular. Control of these residual strains

Mat. Res. Soc. Symp. Proc. Vol. 449 © 1997 Materials Research Society

constitutes an actual issue in the race towards production of advanced devices, since for instance, advantage of them can be taken to reduce lasing thresholds in quantum wells[3].

Optical spectroscopy techniques have long been recognized as extremely efficient methods to investigate semiconductor bulk compounds as well as heterostructures. If excitons where early found to be well adapted probes, phonons which are straightforward signatures of bonds can be used to sample strains fields, interface morphologies, phase mixing....In this article we address, using group theory, the modeling of strain field effects on excitons and phonons in w-GaN. The predictions of this approach quantitatively account for experimental results taken on epilayers grown by low pressure Metal Organic Vapor Phase Epitaxy on c-plane sapphire which are found to experience in-plane biaxial compression. A quantitative agreement if found between the shifts of the exciton lines and the hardening of the Raman modes. The prediction of our modeling are extended to optical properties of w-GaN grown on 6H-SiC and we show that in that case, the crystal field splitting may be offset by the biaxial tension in the epilayer.

BAND STRUCTURE PROPERTIES IN THE WURTZITE PHASE

In a spin-less description, the properties of the top most of the valence band of the III-V semiconductors can be qualitatively related to the properties of p-like orbitals. In case of cubic crystal we have a threefold degeneracy which will be lifted by symmetry breaking. Strain fields applied along one highly symmetric orientation like (001) or (111) are efficient symmetry-breaking perturbations giving a valence band splitting between one single state and a doublet one, when the symmetry of the crystal is lowered from T_d towards D_{2d} or C_{3v} . Similarly, in the wurtzite C_{6v} simple group , the threefold triplet of p states transform like $\Gamma_1 + \Gamma_5$, with eigenvectors being p_z and (p_x, p_y) respectively. The Γ_1 - Γ_5 splitting is named crystal field splitting and noted Δ_1. A convenient way to write the related hamiltonian is $H_{cr} = \Delta_1\, l_z^2$, where l_z is the z projection of the hole (spinless) angular momentum. Group theory predicts dipole allowed transitions towards the s-like conduction band, which are submitted to selection rules in σ and π polarization as shown on figure 1. The crystal field parameter Δ_1 is a very sensitive probe of the value of the c/a ratio in the wurtzite crystal. c/a is coupled to the internal   structural parameter   u, the relative displacement of the anion versus the cation sublattice along the c axis. This makes Δ_1 extremely difficult to be computed accurately. A more convenient way is to measure it experimentally, what will be done below. In T_d symmetry, when the valence electron is attributed its spin, a fourfold Γ_8 level is distinguished from a deeper Γ_7 valence band doublet. The transitions between the conduction band and all these levels are dipole allowed. Any lowering of cubic symmetry of the kinds mentioned above leads to a splitting of the Γ_8 level. When the perturbation is sufficiently symmetric, one state remains uncoupled with rest of the valence band while the second one is coupled by the perturbation with the spin-orbit split-off state. In the C_{6v} double group, the symmetries of the valence band states are labeled Γ_9, Γ_7 and Γ_7. In our parametrization, we prescript the spin-orbit hamiltonian to be invariant under symmetry operations in both spin and ordinary spaces. This gives: $H_{SO} = 1/3\ \Delta_{SO}\ (l.\sigma)$ and $H_{SO} = \Delta_2\ l_z\sigma_z + \Delta_3\ (l_x\sigma_x + l_y\sigma_y)$ in cubic and wurtzite symmetry respectively. There is a general trend in the III-V and II-VI semiconductors: the splitting of the top most valence band due to the spin orbit interaction increases with the atomic number of the anion. In the case of GaN, and in the framework of the tight binding description of the valence band, we expect the spin orbit interaction to be close to the value computed for the 2p state of the free nitrogen atom (13.6 meV)[4].

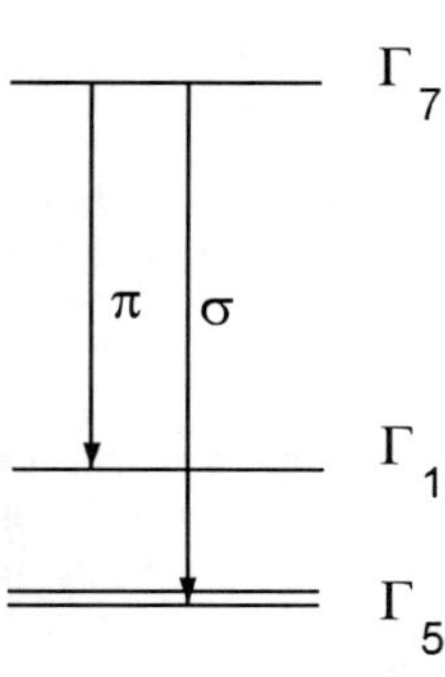

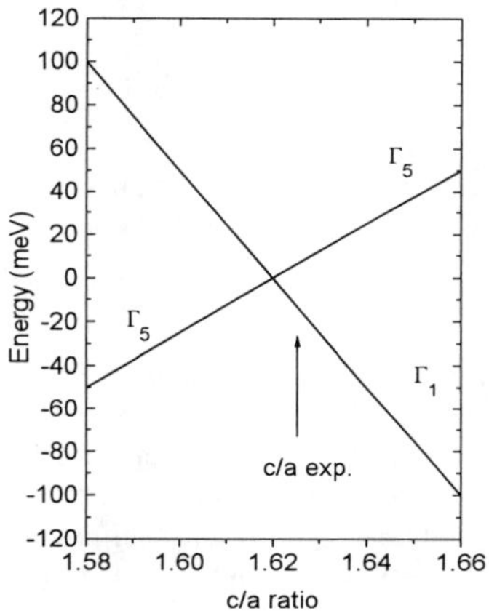

Figure 1: Sketch of the $\Gamma = 0$ band gap transitions in a spinless description of the wurtzite GaN crystal. π and σ polarizations correspond to electric field parallel and perpendicular to the c axis respectively.

Figure 2: Crystal field splitting in wurtzite GaN calculated as a function of the c/a ratio with u relaxed at each point. The result is extracted from the theoretical calculation of ref.[4].

In the unstrained crystal the valence band Hamiltonian writes $H_{val} = H_{cr} + H_{so}$ to which we have to add a third contribution, H_{strain} to account for the strain-fields, if any. In this paper, following arguments we earlier exposed in [5], we will limit this hamiltonian to most *spin-independent* deformation potentials. For the convenience of any future discussion, in this paper, we choose to label the deformation potentials following Sandomirskii[6]. The strain hamiltonian writes:

$$H_{strain} = (C_1 + C_3 L_z^2)e_{zz} + (C_2 + C_4 L_z^2)e_\perp + C_5(L_-^2 e_+ + L_+^2 e_-) + C_6 \{[L_z L_+] e_{+z} + [L_z L_-]e_{-z}\}$$

where

$$L_\pm = 1/\sqrt{2}(L_x \pm iL_y), \quad 2[L_i L_j] = L_i L_j + L_j L_i$$

$$e_\perp = e_{xx} + e_{yy}, \quad e_\pm = e_{xx} \pm 2ie_{xy} - e_{yy}, \quad e_{\pm z} = e_{xz} \pm ie_{yz}$$

In our case, making the assumption that the residual stress X is biaxial in the (0001) plane, and using the compliance tensor of the s_{ij} quantities in wurtzite symmetry, we obtain two non-vanishing components of the strain $e_{xx} = (s_{11} + s_{12})X$, $e_{zz} = (s_{13} + s_{12})X[5]$ Thus, in our case, the evolution of the valence band extrema is ruled by four deformation potentials: C_i with i running from 1 to 4. Analytical expressions are obtained which give the evolution of the three valence band levels.

$$E(\Gamma_{9v}) = E_0 + \Delta_1 + \Delta_2 + C_1 e_{zz} + C_2(e_{xx} + e_{yy}) + C_3 e_{zz} + C_4(e_{xx} + e_{yy})$$

$$E(\Gamma_{7v}^{\pm}) = E_0 + \frac{\Delta_1 - \Delta_2}{2} + C_1 e_{zz} + C_2(e_{xx} + e_{yy}) + \frac{C_3 e_{zz} + C_4(e_{xx} + e_{yy})}{2}$$

$$\pm \frac{\sqrt{[\Delta_1 - \Delta_2 + C_3 e_{zz} + C_4(e_{xx} + e_{yy})]^2 + 8\Delta_3^2}}{2}$$

We have earlier shown in [5] that, making a linear approximation of the stress behavior for the two Γ_7 levels can give a relatively good estimation of δ_1 / δ_2 (where $\delta_1 = C_1 e_{zz} + C_2(e_{xx} + e_{yy})$, $\delta_2 = C_3 e_{zz} + C_4(e_{xx} + e_{yy})$), and $\eta = \dfrac{\Delta_1 - \Delta_2}{\sqrt{(\Delta_1 - \Delta_2)^2 + 8\Delta_3^2}}$.

To go further, the secret of the method we detailed in our original work in ref.[5] requires to plot energies of A,B,C excitons against the energy of the fundamental one A. In such a dimensionless diagram, from data collected in the literature of the time, we got $\delta_1 = 1.86$ meV/kbar, $\delta_2 = - 0.86$ meV/kbar, and $\eta \approx 0.22$. This led us to propose the following values for the crystal field splitting and spin-orbit interaction parameters $\Delta_1 = 10 \pm 0.1$ meV, $\Delta_2 = 6.2 \pm 0.1$ meV , $\Delta_3 = 5.5 \pm 0.1$ meV.

EPITAXY AND OPTICAL CHARACTERIZATION OF GaN EPILAYERS ON C-PLANE Al$_2$O$_3$

The GaN layers of which we study the properties here have been grown by low pressure MOVPE (76 Torrs) in an ASM OMR 12 MOCVD equipment. The precursors we used are triethylgallium (TEGa), triethylaluminum, (TEAl) and ammonia (NH$_3$). The substrates were (0001) sapphire. Both sides were polished by the manufacturer so that accurate transmission experiments could be easily performed to calibrate growth rates of the thin low temperature-grown GaN or AlN buffer layers and, when further deposited , growth rates of the crystalline epilayers. This is of crucial importance since a lot of parameters had to be varied so that we could change the magnitude of the residual strain in the epilayers. At the initial steps of this work we prepared the sapphire substrate as follows: after cleaning with organic solvents and etching using a H$_3$PO$_4$: 3 H$_2$SO$_4$ hot solution, it underwent a surface nitridation step by heating at 1070°C under a NH$_3$ flow. We investigated samples grown at temperatures ranging between 950 and 1040°C. Last but not least, the influence of the V/III molar ratio was studied by keeping the molar flow rate of TEGa constant (65 sccm at 20°C), but scaling the NH$_3$ flow rate in the range 300 - 4000 sccm. Extensive details were given in refs. [7,8].

Figure 3(bottom) displays a typical 2K sharp near band edge photoluminescence spectrum together with its corresponding reflectivity data (top). The 2K photoluminescence spectrum is dominated by free exciton contribution as evidenced by the absence of *Stokes-shift* with respect to the reflectance data. At lower energy we observe the weak contribution of a residual donor as an evidence of the exceptional purity of the sample. Obtention of so nice photoluminescence data is not trivial and most of our samples display photoluminescence dominated by bound exciton recombination at 2K. However their 300K luminescence is intrinsic in nature. This we illustrate more concretely in figure 4. The top of the figure shows reflectance and photoluminescence spectra collected at 2K. In the 2K reflectance data, the signatures of the three A,B,C excitons are clearly observed, which give the position of the bandgaps, as well as undulations at low energy in the region where the crystal is transparent to the light. These contrasted undulations which were fitted in order to calibrate growth rates, are again evidence of the remarkable control of epilayer thickness. At the bottom part of the figure straightforward comparison of the 300K photoluminescence spectrum with the photoreflectance spectrum taken at the same temperature is nice evidence of the intrinsic nature of the PL in our samples at this temperature.

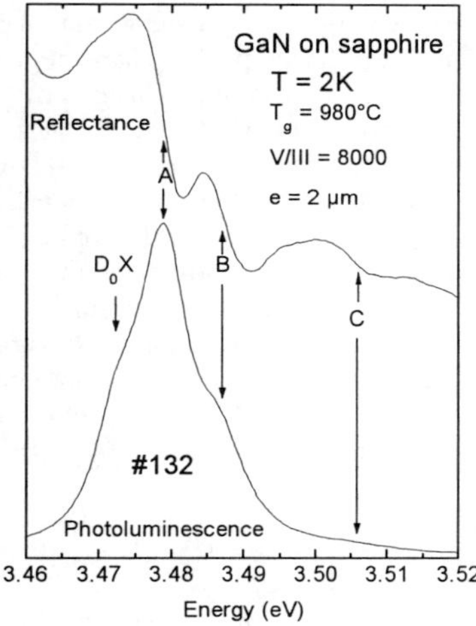

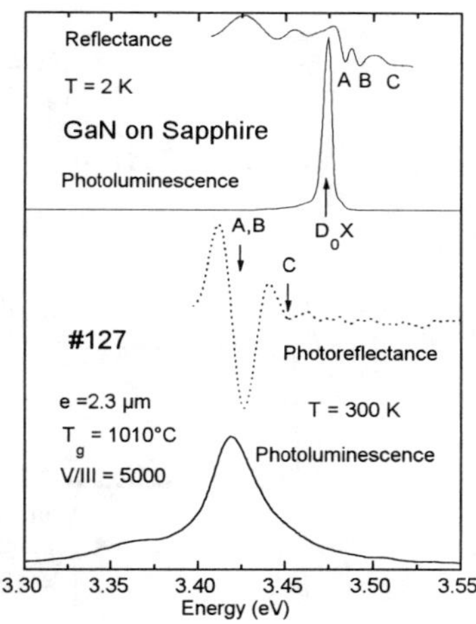

Figure 3: 2K intrinsic photoluminescence in near band edge region of a thin epilayer and associated reflectivity (top) with three marked signatures of A, B, C excitons.

Figure 4: Photoluminescence at 2K and 300K for unoptimized GaN epilayers. The intrinsic nature of the RT photoluminescence is obvious.

Growing the GaN samples under slightly different conditions may give epilayers still having nice photoluminescence and reflectance features. We however observed that the reflectance (or photoluminescence) lines are energy shifted in relation with these growth conditions. The V/III molar ratio is for instance found to have significant impact on both residual electron incorporation, and residual strain fields. Using a GaN buffer layer and scaling the V/III molar ratio between 2,000 and 10,000 epilayers with residual electron concentrations ranging from 10^{20} cm^{-3} down to 10^{16} cm^{-3} were grown. Decreasing this molar ratio through this range of values, we could simultaneously and continuously correlate it to a *blue-shift* of the reflectance and photoluminescence lines reaching up to 20 meV. Using an AlN buffer layer and further depositing a GaN epilayer in a given experimental context led to epilayers generally more strained than if using a GaN buffer layer. If the growth process is not varied except for the deposition time, the thicker the layer, the smaller the residual strain. Last, we have found that the recrystallization process of the buffer prior to deposition of the w-GaN layer is found to be a relevant parameter. The influence of this is not fully understood to date, but work is currently in hands in our laboratory to address this issue.

A careful line shape analysis of the reflectivity data was made on a set of samples grown under a large variety of experimental conditions. This analysis was made in the context of a non local description of the dielectric constant and using three oscillators to estimate the energies of A, B and C excitons[7]. Simultaneously with the change in photoluminescence energy from sample to sample, we observe, as shown in figure 5, a significant change in the reflectivity line shape together with a remarkable change of the exciton energies. These changes we attribute to effects of residual strain fields in the epilayers. Monemar et al.[9,10] also reported the effect, and in addition, qualitatively showed that the compressive strain field characteristics of sapphire substrates could be switched to dilatation if grown onto SiC. Their reflectance structures are dominated by a strong

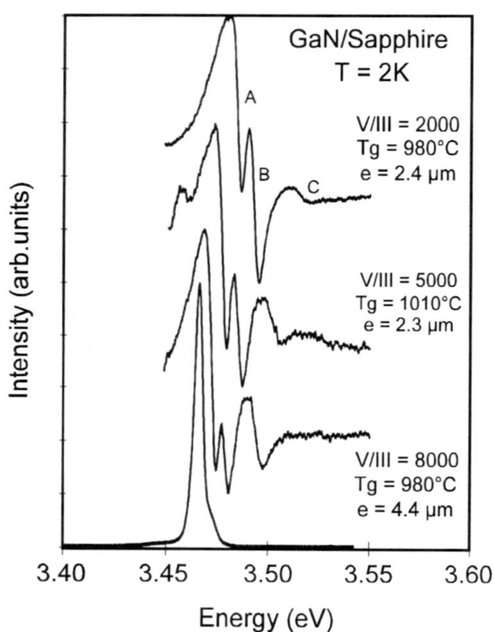

Figure 5 : some typical 2K Photoluminescence and reflectance features observed in our samples.

3.473 meV structure and a significantly intense one at 3.489 meV[11]. They suggested to interpret their data classically, in terms of effects of differences between thermal expansion coefficients of the substrates and epilayers. We now wish to evidence strain-related effects on oscillator strengths. The experimental data were plotted in figure 5 by scaling them in such a way that amplitudes of A lines are identical from sample to sample. *Simultaneously with the blue-shift of the three transitions observed when the biaxial compression experienced by the GaN epilayer increases, we observe a collapse of the intensity of C*

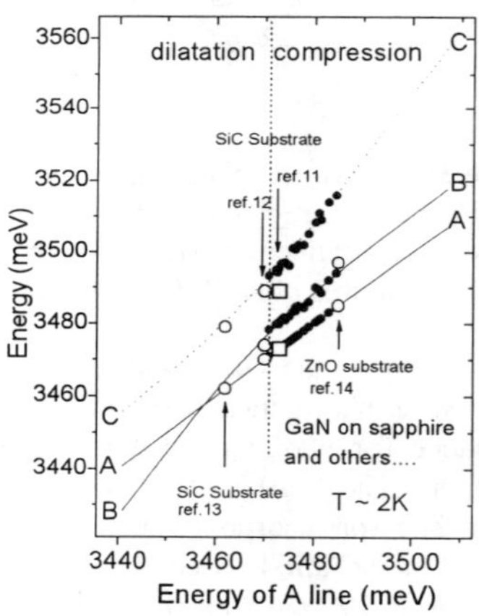

Figure 6: Fit to the data using our set of valence band parameters given above in the paper.

and an increase of the strength of B. The fit to the data is addressed in figure 6. Full and dotted lines correspond to the result of the calculation using the valence band parameters given above and obtained from the whole literature[5]. The full circles correspond to some selected points taken on our own epilayer (they all match to the fitting). Added to the plot are other results corresponding to samples grown on 6H-SiC (open circles)[12,13] and ZnO[14] . We remark from ref.[13], that the crystal field splitting can be offset under biaxial dilatation as predicted theoretically. Last we conclude this section by emphasizing the disagreement between the data of refs. [10,11] and our modeling which seems to indicate that some different relaxation phenomena may occur in some samples. In the case of ref. 10 (open squares), it is obvious that data interpretation requires to superimpose to the biaxial stress an additional component which behaves like a hydrostatic pressure. There is however a general trend which is observed by all authors: the stress induced variation of the excitonic oscillator strengths.

EXCITON STATES AND EXCHANGE INTERACTION

The exciton hamiltonian Ξ_{exc} is written[15]:

$$\Xi_{exc} = H_{c=0} + H_{cstrain} + H_{v=0} + H_{vstrain} + H_{exc}$$

where $H_{v=0}$ ($H_{c=0}$) are the strain-free valence (conduction) band hamiltonians and $H_{vstrain}$ ($H_{cstrain}$) are their respective, which account for strain-related effects on the evolution of band extrema. The last operator in the right hand part of equation above, we write it :

$$H_{exc} = R^* + \tfrac{1}{2}\, \gamma\, \sigma_h . \sigma_c$$

where R^* is the customary exciton binding energy and the last term, γ is the crystalline exchange interaction. Operators σ_h and σ_c operate on valence hole and conduction electron spin functions. In the configuration the experiments are made, when the photon is propagating along the z direction, the radiative exciton states are those having Γ_5 symmetry . Their eigen energies are obtained from solution of the following matrix:

$$
\begin{array}{ccc}
|\Gamma_5> & |\Gamma_5>^{\prime} & |\Gamma_5>^{\prime\prime} \\
\Delta_1 + \Delta_2 + \delta_1 + \delta_2 - 1/2\gamma & -\gamma & 0 \\
-\gamma & \Delta_1 - \Delta_2 + \delta_1 + \delta_2 - 1/2\gamma & \sqrt{2}\Delta_3 \\
0 & \sqrt{2}\Delta_3 & \delta_1 + 1/2\gamma
\end{array}
$$

The excitons eigen vectors are written $\Psi_{exc\Gamma5} = \upsilon |\Gamma_5> + \omega\, |\Gamma_5>^{\prime} + \varpi\, |\Gamma_5>^{\prime\prime}$, where these vectors express as a function of the following linear combinations: $p_+ = -(p_x + ip_y)/\sqrt{2}$, and $p_- = (p_x - ip_y)/\sqrt{2}$, where spin components of the missing valence electron are $\uparrow$ and $\downarrow$, and where α and β represent the spin components of the conduction electron :
$|\Gamma_5> = \{p_+ \uparrow\alpha\,, p_-\downarrow\beta\}$, $|\Gamma_5>^{\prime} = \{p_+ \downarrow\beta\,, p_-\uparrow\alpha\}$, and $|\Gamma_5>^{\prime\prime} = \{p_z \downarrow\alpha\,, p_z\uparrow\beta\}$.

Their oscillator strengths (σ-polarization) are $|\Omega_{exc\Gamma5}|^2 \sim (\upsilon^2 + \omega^2)/2$.

We have calculated γ, extending to the GaN case the two-body calculation developed in ref.16 for a few II-VI compounds. The pertinent parameter of the model is the product p of the exciton radius with an *had hoc* radius k_m of an effective Brillouin zone (BZ hereafter). This k_m is introduced to replace the tedious integration in k space through the whole *actual* BZ by an integration over a spherical space having the volume of this BZ. It has been shown in ref. 16 that k_m can be connected to the lattice parameter a using a well adapted scaling argument. For GaN, taking a = 3.19Å gives k_m = 0.873 Å^{-1}. To calculate p , we need the exciton Bohr radius in GaN. Recent determination of exciton masses[17] suggest a_B = 34 ± 2 Å. This gives p = 30 ± 2, which corresponds to an exchange-influenced 1s exciton binding energy of about 85% of R^* ($1/\rho_1^2$ = 0.851± 0.006 in the model of ref 16). The exciton exchange energy 2γ is thus estimated to be

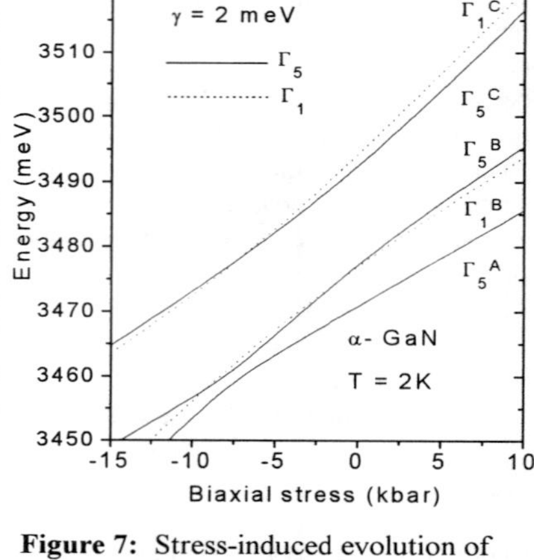

Figure 7: Stress-induced evolution of the radiative exciton levels in w-GaN

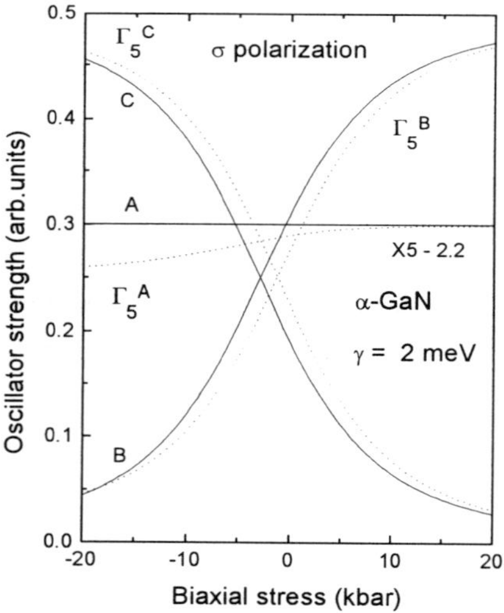

Figure 8: Stress-induced evolution of the excitonic (dashed lines) and band to band (full lines) oscillator strengths in the case of σ-polarization.

some 15% of the binding energy. Further using a Rydberg energy of 26.7 meV[18], the magnitude of the corresponding γ we get (~ 2 meV), is far from being negligible. Figure 7 illustrates the evolution of these Γ_5 excitons versus biaxial stress. We have also included the two Γ_1 excitons which are allowed in π polarization. A detailed treatment of this can be found in ref. 15.We now plot the excitonic (dashed lines) and band to band (full lines) oscillator strengths in σ-polarization as a function of the biaxial stress in figure 8.

We remark that inclusion of the exchange interaction does not significantly alter the predictions of the band to band calculation. Band to band transitions between Γ_7^c and Γ_9^v keep a constant oscillator strength (0.5 in our notation) whatever the stress is. Band to band transitions we note that B and exchange their oscillator strength when the nature of the residual stress varies from biaxial tension to biaxial compression. We also remark that values expected in a cubic environment (1/6 and 1/3 respectively) are computed when the crystal field splitting is offset by the biaxial tension. *An exact matching to these values would be obtained by using an isotropic spin-orbit interaction.*

LONGITUDINAL-TRANSVERSE SPLITTING OF THE EXCITON

When fitting the lineshape of the three imbricate reflectance structures, the dielectric constant was modeled using a set of three oscillators, as shown in the equation below[19]:

$$\varepsilon(k,\omega) = \varepsilon_b + \sum_{j=1}^{3} \frac{4\pi\alpha_{0j}\omega_{0j}^2}{\omega_{0j}^2 - (\hbar^2 k^2 / m_j^*)\omega_{0j} - \omega^2 - i\Gamma_j\omega}$$

where ε_b is the background dielectric constant, ω_{0j} is the transverse frequency related to the exciton j with an effective mass m_j^*, $A_j = 4\pi\alpha_{0j}$ is the polarizability of the exciton resonance at $\omega = 0$ and $k = 0$, and Γ_j is the damping parameter used to account the interactions of the excitons with the phonons and extrinsic defects. For each exciton, the longitudinal transverse splitting is given by : $\omega_{LT} = 2\pi\alpha\omega/\varepsilon_b$. For excitons A, B, and C we found $4\pi\alpha$ to be $17\ 10^{-3}$, $6.5\ 10^{-3}$, and $7.8\ 10^{-3}$ respectively for a sample giving these resonances at 3477 meV, 3.486 meV, and 3.506 meV at 2K. From the relation above, this simplified model enables us to estimate in the simplest approach the following values for the longitudinal splittings of these Γ_5 excitons: $\omega_{LT}(A) = 3$ meV, $\omega_{LT}(B) = 1.2$ meV, $\omega_{LT}(C) = 1.4$ meV. Monemar [11] reports observation of a doublet structure at 3.477 eV and 3.479 eV and thus suggests $\omega_{LT}(A) = 2$ meV. Although these determinations are very preliminary, they are consistent with each other. It is obvious that these calculated values are only qualitative ones, and we believe that better determination requires detailed investigations of the exciton polariton problem in w-GaN layers. The lack of accurate information concerning the six mass parameters , the equivalent of Lüttinger parameters in zinc blende crystal currently prevents to address this issue: full treatment requires to know exciton dispersion relations in various strain situations. The determination of the oscillator strength of C exciton is further complicated by its interaction with either the 2S excited state of A or its corresponding continuum, depending on strain. This could explain why although the fitted oscillator strength for A and B exciton follow the calculated trend on figure 8, the oscillator strength we find for C is larger than expected theoretically.

PHONON DEFORMATION POTENTIALS

The $\mathbf{q} = \mathbf{0}$ phonon frequencies have been measured by Raman spectroscopy in a series of layers grown under different V/III ratio conditions, in which stress calibration has been carried out through reflectance measurements[20]. A group theory analysis of $\mathbf{q} = 0$ lattice vibrations predicts six optical modes which decompose as follows into the following representations of the C_{6v} point group: $\Gamma_{opt} = A_1 + E_1 + 2E_2 + 2B_1$. The polar modes A_1 and E_1 -both infra-red and Raman active- are polarized along the z optical axis and in the basal (x,y) plane respectively (x,y,z represent the   crystal principal axes). Concerning the other -non polar- lattice vibrations, the E_2 modes are Raman active whilst the B_1 modes are silent. Obviously, the anisotropic structure of GaN together with the partially ionic character of crystal bonding have strong influence on the dynamical properties. The anisotropy in short-range atomic forces is thus responsible for the A_1-E_1 splitting while the long-range Coulomb field is at the origin of the longitudinal-transverse (LO-TO) splitting of polar modes. This results in direction-dependent frequencies and polarization properties when approaching the limiting $\mathbf{q} = \mathbf{0}$ value along or perpendicularly to the optical axis. The changes in frequency of $\mathbf{q} = \mathbf{0}$ phonons under stress are obtained in first order perturbation theory in terms of deformation potential constants. Under hydrostatic or biaxial (0001) oriented stress, the shifts of the phonon frequencies can be expressed in the following way:

$$\Delta v_J = 2a'_J\, \sigma_{xx} + b'_J\, \sigma_{zz}\,, \qquad J = A_1,\ E_1,\ \text{or}\ E_2$$

The σ_{ii}'s are the diagonal components of the stress tensor, with $\sigma_{xx} = \sigma_{zz}$ for the hydrostatic pressure and $\sigma_{zz} = 0$ for biaxial (0001) stress. The coefficients a'_J and b'_J are expressed in terms of elastic constants and deformation potential constants as:

$$a'_J = a_J\,(S_{11} + S_{12}) + b_J\, S_{13} \quad \text{and} \quad b'_J = 2a_J\, S_{13} + b_J S_{33}$$

where a_J represents the change in frequency of phonon J per unit relative change of length along z, and b_J is its counterpart along x.

Values of the biaxial stress coefficient $K^B_J = dv_j/d\sigma$ where $\sigma = -\sigma_{xx}$, ($\sigma > 0$ for compression) were estimated from Raman data on the strained layers. A first relation between the a_J's and b_J 's can be established:

$$2a_J\,(S_{11} + S_{12}) + 2b_J\, S_{13} = -K^B_J$$

One more relation is needed for the determination of the { a_J, b_J } set of deformation potential constants. It can be obtained from hydrostatic pressure Raman measurements carried out by Perlin et al.[21] on bulk GaN crystals since:

$$2(S_{11} + S_{12} + S_{13})a_J + (2S_{13} + S_{33})\, b_J = -K^H_J$$

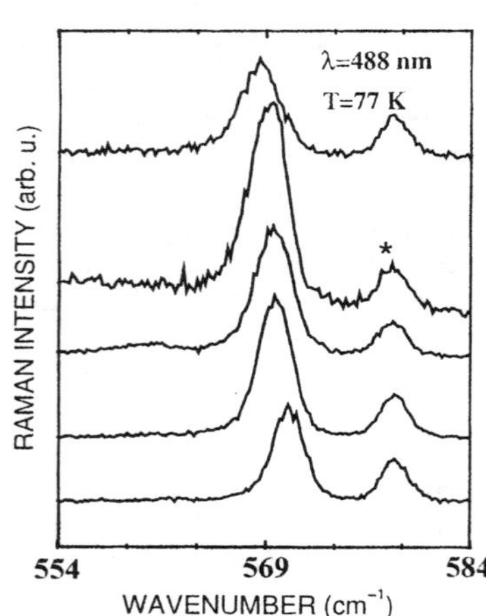

Figure 9: Unanalyzed Raman Spectra, for E_2 mode in different GaN layers.

The Raman spectra were excited in backscattering geometry along <0001>, using the 488 nm radiation of an Ar^+ laser. The scattered light was analyzed with a T800 triple spectrometer equipped with a conventional photon counting system. Although the actual scattering geometry allows the observation of the A_1 (LO) extraordinary phonon and that of the two E_2 modes, the low frequency E_2 mode at 144 cm^{-1} was hardly detected due to far-off resonant conditions with respect to the yellow photoluminescence band. No reliable conclusions could then be drawn concerning this mode. Figure 9 shows the non-analyzed Raman spectra of the heterostructures in the frequency range of the highest E_2 mode. The asterisk marks one of the sapphire lines at 578 cm^{-1}. A shift of the E_2 mode frequency is clearly observed with respect to the sapphire line at 578 cm^{-1}, corresponding to mode hardening with decreasing V/III ratio. To a lesser extent, the same trend is noted for the A_1 (LO) mode. Phonon frequency as a function of biaxial stress in GaN is deduced from our Raman

measurements for the E_2 and A_1 (LO) modes. Details cand be found in ref.[20]. Least square fit gives for the slopes the following values K_{E2}^B = 2.9±0.3 cm^{-1}/GPa and $K_{A1(LO)}^B$ = 0.8±0.4 cm^{-1}/GPa, taking for elastic compliance constants values, results of a calculation obtained from the elastic stiffness coefficients C_{11} = 296 GPa, C_{12} = 130 GPa, C_{13} = 158 GPa, C_{33} = 267 GPa [22]. Despite the large imprecision in the $K_{A1(LO)}^B$ pressure coefficient, due to rather weak Raman efficiency of the mode, its lower value undoubtfully reflects the weak influence of the in-plane biaxial stress to the A_1(LO) compared to the E_2 mode. The hydrostatic pressure coefficients of the modes were taken from the measurements of Perlin et al[21], performed apparently on a heavily doped sample as they failed to observe the A_1(LO) mode in backscattering geometry along <0001>. We adopt the pressure coefficient of their A_1 (TO) mode, in fact the phonon-like component of the A_1(LO) plasmon coupled mode[23]. For K_{E2}^H = 4.17 cm^{-1}/GPa and $K_{A1(LO)}^H$ = 4.06 cm^{-1}/GPa, we obtain the two sets of deformation potential constants for each mode. $a(E_2)$ = -818 ± 14cm^{-1}/unit strain, $b(E_2)$ = -797 ± 60 cm^{-1}/unit strain, $a(A_1^{LO})$ = -685 ± 38 cm^{-1}/unit strain, $b(A_1^{LO})$ = -997 ± 70 cm^{-1}/unit strain.
As expected, they are two orders of magnitude smaller than those describing strain effects on the electronic band extrema[19] which we found to be : C_1 = - 8.16 eV, C_2 = - 816 eV, C_3 = 1.87 eV, and C_4 = - 0.94 eV. These values are in agreement with the values used by Chuang and Chang [24], but are smaller than the one obtained theoretically by Suzuki and Uenoyama using a full-potential linearized augmented plane wave method[3,25].

CONCLUSION

The influence of residual strain fields on band structure, excitons and Raman active phonons has been studied in detail for wurtzite GaN deposited on sapphire along the <0001> direction. We could derive important parameters valence band parameters like the crystal field splitting and the two components of the spin orbit interaction. Four band deformation potentials were determined as well as their analogues to account of the strain-induced variation of the phonon frequencies. An estimation of the excitonic exchange interaction was done and we addressed the problem of the longitudinal-transverse splitting of the excitons. Last, we computed the variation of the strengths of optical transition with the built-in strain and compared our finding with experimental data.

REFERENCES

1- « Gallium Nitride and Related Materials » Materials Research Soc. Symposium Proceedings volume **395**, edited by F.A.Ponce, R.D.Dupuis, S.Nakamura, and J.A.Edmond (1996)
2 - S. Nakamura, M. Senoh, Shin-ichi Nagahama, N.Iwasa, T.Yamada, T. Matsushita, H. Kiyoku and Y. Sugimoto, Jpn. J.Appl. Phys. **35**, L74 (1996)
3 - M.Suzuki and T.Uenoyama, Jpn. J.Appl. Phys. **35**, 1420, (1996); J.Appl. Phys, 80, xxxx, (1996); T.Uenoyama and M.Suzuki, Mat. Res. Soc. Symp. Proc. vol. **395**, edited by F.A.Ponce, R.D.Dupuis, S.Nakamura, and J.A.Edmond, 925, (1996); S.L.Chuang and C.S. Chang Appl. Phys. Lett. **68**, 1657, (1996); Yu. M.Sirenko, J.B.Jeon, K.W.Kim, M.A.Littlejohn and M.A.Stroscio, Appl. Phys. Lett. **69**, 2504, (1996);
4 - W.R.L.Lambrecht, K.Kim, S.Rashkeev, and B.Segall, Materials Research Soc. Symposium Proc. vol. **395**, edited by F.A.Ponce, R.D.Dupuis, S.Nakamura, and J.A.Edmond, 455, (1996)
5 - B.Gil, O.Briot, and R.L.Aulombard, Phys. Rev. B **52**, R 17 028, (1995)
6 - V.B. Sandomirskii, Fisika Tverdogo Tela **6**, 324, (1964), [Sov.Phys. Sol. St.**6**,261,(1964)]

7 -O. Briot, B.Gil, S.Sanchez, and R.L.Aulombard,Inst. Physics Conf. Ser. **142**, 891, (1996)

8 - O.Briot, J.P.Alexis, B.Gil, and R.L.Aulombard Materials Research Society Symposium Proceedings volume **395** Edited by F.A.Ponce, R.D.Dupuis, S.Nakamura, and J.A.Edmond pages 411-415 (1996)

9 -J.P.Bergman, B.Monemar, H.Amano, I.Akasaki, K.Hiramatsu, N.Sawaki, and T.Detchprohm , Inst. Physics Conf. Ser. **142**, 931, (1996)

10 - I.A.Buyanova, J.P.Bergman, B.Monemar, H.Amano, I.Akasaki, Appl. Phys. Lett. **69**, 1255, (1996)

11- B. Monemar (private communication)

12 - D.K.Nelson, Yu. V. Melnik, A.V.Selkin, M.A.Yacobson, V.A.Dmiriev, K.Irvine, and C.H.Carter Jr. Fiz. Tverd. Tela **38**,651, (1996) [Soviet Physics Solid State , 38, 455, (1996)]

13 - S.N.Mohammad and H. Morkoç, in Progress in Quantum Electronics october 1995 edited by Marek Osinski, Elsevier Science Publishers

14 - B.Gil, F.Hamdani, and H.Morkoç, Phys. Rev. B, **54**, 7678, (1996)

15 B.Gil, and O.Briot, Phys. Rev. B (in press)

16 - P.G.Rohner, Phys. Rev. B. **3**, 433, (1971)

17 - M.Suzuki, and T.Uenoyama, Jpn. J.Appl. Phys. **34**, 3442, (1995), M.Susuki, T.Uenoyama and A.Yanase, Phys. Rev. B **52**, 8132, (1995), M.Drechsler, D.M.Hoffmann, B.K.Meyer, T.Detchprohm, H.Amano, and I.Akasaki, Jpn. J.Appl. Phys **34**, L1178, (1995),W.Knap, H.Alause, J.M.Bluet, J.Camassel, J.Young, M.Asif Khan, Q.Chen, S.Huant, and M.Shur, Solid State Communications **99**, 195, (1996), C.Merz, M.Kunzer, and U.Kaufmann Semicond. Science Technol. **11**, 712, (1996), D.Kovalev, B.Averboukh, D.Volm, B.K.Meyer, H.Amano and I.Akasaki, Phys. Rev B **54**, 2518, (1996))

18 - D.Volm, K.Oettingen, T.Streibl, D.Kovalev, M.Ben-Chorin, J.Diener, B.K.Meyer, J.Majewski, L.Eckey, A.Hoffmann, H.Amano, I.Akasaki, K.Hiramatsu and T.Detchprohm, Phys. Rev B. **53**, 16543, (1996)

19 - M. Tchounkeu, O.Briot, B.Gil, and R.L.Aulombard Journal Appl. Phys. **80**, 5352, (1996)

20 - F.Demangeot, J.Frandon, M.Renucci and O.Briot, B.Gil and R.L.Aulombard Solid State Communications, **100**, 207, (1996)

21 - P.Perlin, C.Jauberthie-Carillon, J.P.Itie, A.San Miguel, I.Grzegory and A.Polian, Phys. Rev. B, **45**, 83, (1992)

22 -Data in Science and Technology - Semiconductors: Group IV Elements and III-V Compounds Edited by O.Madelung Springer Verlag Berlin Heidelberg (1991)

23 - P.Perlin, J.Camassel, W.Knap, T.Talierco, J.C.Chervin, T.Suski, I.Grzegory, and S.Porowski, Appl. Phys. Lett.Lett. **67**, 2524, (1995)

24 - S.L.Chuang and C.S. Chang Phys. Rev B **54**, 2491, (1996)

25 - M.Suzuki and T.Uenoyama Jpn. J.Appl. Phys. **35**, L935, (1996)

MAGNETIC RESONANCE STUDIES OF GaN-BASED SINGLE QUANTUM WELL LEDS

W.E. Carlos[a], E.R. Glaser[a], T.A. Kennedy[a] and Shuji Nakamura[b]
[a]Naval Research Laboratory, Washington, D.C. 20375
[b]Nichia Chemical Industries, Ltd.,491 Oka, Kaminaka, Anan, Tokushima 774, Japan

ABSTRACT

We report electrically-detected magnetic resonance (EDMR) and electroluminescence-detected magnetic resonance (ELDMR) results on InGaN/AlGaN single-quantum-well light emitting diodes. The dominant feature detected by either technique is a broad resonance ($\Delta B \approx 13$ mT) at $g \approx 2.01$ which is enhanced by high current stressing. Our ELDMR measurements show that, depending on bias, this defect is predominately associated with either an increase or a decrease in electroluminescence at resonance while our EDMR measurements show that this resonance is associated with an increase in current at resonance before stressing and a decrease after stressing. We suggest that this is associated with a nonradiative recombination path, in parallel with the radiative recombination path and with recombination in the depletion region of a contact. A second resonance, more prominent before stressing, with $g \approx 1.99$ and $\Delta B \sim 7$ mT is very similar to the deep donor trap, previously observed in double heterostructure diodes and is associated with a decrease in both the current and electroluminescence at resonance.

INTRODUCTION

GaN-based devices show great promise in electro-optic applications.[1] Advances in materials growth[2] have made devices such as efficient blue and green light emitting diodes (LEDs) possible.[3,4] The first of these diodes were double heterostructures (DHs) with an optically active layer of InGaN co-doped with Zn and Si and a bright blue luminescence due to donor-acceptor recombination. Recently, GaN-based LEDs whose electroluminescence (EL) is due to bandedge emission from a single quantum well (SQW) of undoped InGaN have been introduced.[5] These diodes have more intense, sharper electroluminesence than the DHs. The brightness is especially impressive given the high densities of stacking faults, dislocation and other defects in these devices. These structures are the subject of this investigation. Even more recently, multiquantum well laser diodes have been introduced.[6] Spectroscopic probes have provided valuable insights into the basic materials properties and defects in the nitrides as well as other semiconductors.[7-12] A critical question is one of the relevance of spectroscopic results to actual device performance. In this work we discuss the application of magnetic resonance to these SQW LEDs.

We use electrically detected magnetic resonance (EDMR) and electroluminescence detected magnetic resonance (ELDMR) to study both point defects which are present in unstressed InGaN/AlGaN SQW LEDs and those which are introduced by high current stressing. These techniques are especially well suited to the problem of defects in device structures in that they are much more sensitive than conventional paramagnetic resonance and are selectively responsive to those defects involved in the electrical and/or optical properties of the structure. A study of the bias and modulation frequency dependence of the two techniques and the relationship between the two results allows us to make some arguments about the location of the defects. The studies of current-stressing effects in these diodes are relevant not only to failure mechanisms in these structures but also to degradation of laser diodes which operate at much higher currents.[6]

Magnetic resonance techniques based on spin-dependent processes in semiconductor devices are finding increasing application in a number of materials systems. ELDMR has been applied to organic semiconductor-based[13] and amorphous Si-based [14] LEDs and EDMR has been applied to

757

Mat. Res. Soc. Symp. Proc. Vol. 449 © 1997 Materials Research Society

a number of devices and materials.[15,16] In previous work on double heterostructure devices[17] we resolved two resonances, one of which we attributed to a zinc acceptor and another which we attributed to a deep donor. ESR and ODMR (optically detected magnetic resonance in which changes in photoluminescence are monitored) have been applied to doped and undoped wurtzite GaN thin films. Several resonances are observed,[8,9,10] including an effective mass (EM) donor, a deep double donor and Zn- and Mg-related acceptor states (in doped films).

The structure and manufacture of the single quantum well AlGaN/InGaN LEDs used in this work have been discussed in detail in previous publications.[5] These devices are now available commercially in both blue (λ_{peak}=450 nm) and green (λ_{peak}=520 nm). In this report we will focus our discussion on the green diodes for which we have a more complete set of data and make only a few comparisons with results on the blue diodes. The optically active region is a 2 nm thick layer of undoped $In_xGa_{1-x}N$ (x=0.43 for the green LEDs). Samples were examined before and after stressing at DC currents of ~100-150 mA. (i.e., ~3-5 times the recommended maximum current) Lower currents do not produce appreciable degradation in a reasonable time while higher DC currents rapidly destroy the devices. Samples were typically stressed at sufficient currents to reduce the EL intensity by about an order of magnitude in 1-2 weeks. The magnetic resonance apparatus is based on a Varian E9 X-band (9.25 GHz) ESR spectrometer. Samples were cooled using a flow cryostat with either liquid helium or liquid nitrogen as a cryogen and current was applied from a constant current source through a ~1 mm diameter coaxial cable. The c-axis of the thin film LED was always perpendicular to the magnetic field and light was extracted using a quartz light pipe and detected by a UV-enhanced Si photodiode. Electroluminescence (EL) spectra were measured using a 0.22 m single grating spectrometer. The EDMR and ELDMR results presented here were obtained by detecting the AC voltage across the sample and the EL, respectively, in phase with magnetic field modulation. (Modulation frequencies were 10 Hz - 10 kHz)

RESULTS FOR UNSTRESSED SAMPLES

In this section we restrict our discussion to results for samples which have not been subjected to high current stressing. In Figure 1 we show the DC characteristics for a typical green diode at T=4.2K. As most of our results are for liquid helium temperatures, it is necessary to examine the characteristics at such temperatures and their differences with room temperature characteristics. One important difference is the dependence of the electroluminescence intensity on current. At higher temperatures there is a much stronger bias dependence, particularly at higher currents (I>100 μA) where the approximately linear bias dependence is maintained up to about 10 mA. At low temperatures the voltage across the sample is significantly higher, presumably due to the increase in contact resistance. The spectral dependence of the EL is shown in the inset. Other than a slight shift to higher energy the peak is very

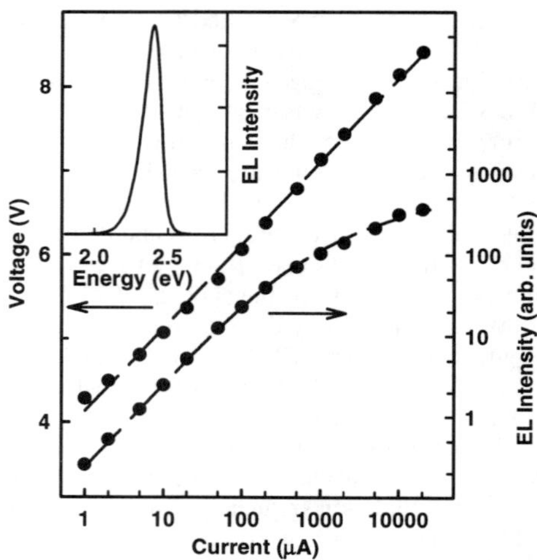

Figure 1. The voltage-current and EL-current characteristics of a green unstressed SQW diode at T=4.2 K. The spectral dependence of the EL is shown in the inset.

similar in width and intensity to that observed at room temperature.

In Figure 2 the EDMR spectra at several biases is plotted. In this and subsequent figures a derivative signal which is first positive then negative (i.e., integrates to a positive signal) is due to a current-enhancing resonance and one which is negative and then positive is due to a current-quenching resonance. The measurements are taken at constant current so the relative change is related to the measured change in voltage across the sample by;

$$\frac{\Delta I}{I} = - \Delta V \frac{d(\ln I)}{dV}. \tag{1}$$

As can be seen from Figure 1, the derivative term is approximately constant (≈ 2.3 V^{-1}) and so the

relative change is approximately proportional to the measured AC voltage signal. Examples of the ELDMR spectra are shown in Figure 3. Here the relative signal is simply the ratio of the peak to peak AC signal to the DC electroluminescence, i.e., ΔEL/EL. In Figure 1 we see that the EL intensity is a relatively strong function of bias current, i.e., approximately linear at low currents. (Thus, the relative signals are not even approximately proportional to the measured AC signals with those at low biases being enhanced relative to those at higher currents.) Before discussing differences in the two measurements, we should first notice that the lineshapes (ΔB$\approx$13 mT), positions (g$\approx$2.01) and amplitudes ($\sim 10^{-5}$ - 10^{-4}) are quite similar for the two measurements. In both the spectra are dominated by a broad line at g$\approx$2.01 over most of this bias range. At higher biases a weak second line begins to emerge[18] and will be discussed below. Perhaps the most obvious difference between the two spectra is in the phase; at low currents the EDMR shows a

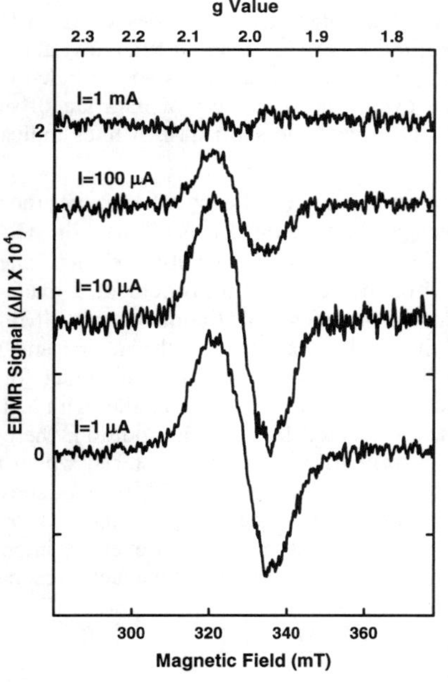

Figure 2. The EDMR spectra at selected biases. All spectra were taken at T=4.2 K and at a microwave power level of 500 mW.

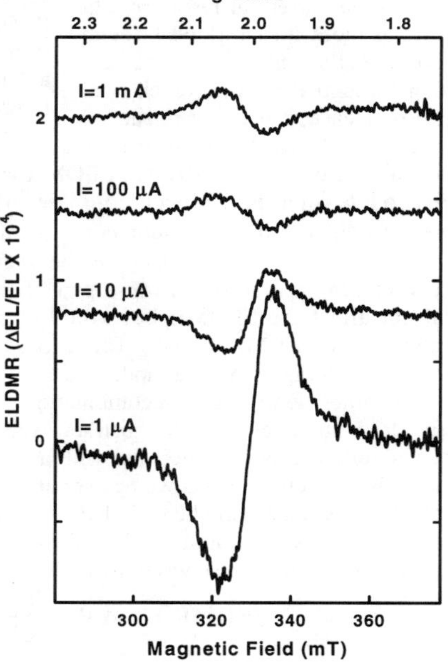

Figure 3. The ELDMR spectra taken under the same conditions as the EDMR spectra in Figure 2.

current enhancement at resonance while the ELDMR shows a quenching of the EL at resonance. Furthermore we see that the ELDMR goes from an EL-quenching signal at low biases to an EL-enhancing signal at higher biases. The integrated intensities of the EDMR and ELDMR spectra are shown in Figure 4. Note that while the two signals have approximately equal magnitudes at I=1 μA, the magnitude of the EDMR signal remains relatively constant up to about 20 μA but the magnitude of the ELDMR is nearly zero at that bias.

The current-enhancing signal seen in EDMR under the same conditions as an EL-quenching signal is observed is understood as follows. We believe that the resonance is due to an increase in the non-radiative recombination rate. The EDMR measures the total $\Delta I/I$ and hence shows an increase in recombination current at resonance. In contrast, as more current flows through the non-radiative channel the current flowing through the radiative channel decreases. Hence, the EL is quenched as seen in the ELDMR. However, as

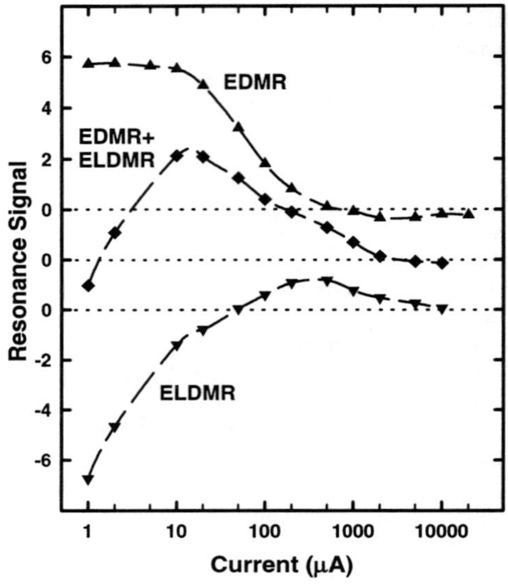

Figure 4. The integrated intensities of the EDMR (triangles), the ELDMR (inverted triangles) and their sum (diamonds). The dashed lines are included as an aid to the eye. The three sets of data are offset to improve clarity with the short dashed lines indicating the zeros for the three curves.

we see in Figure 4, the EDMR and ELDMR are not mirror images of each other, rather the sum of the two intensities is significant for biases between $\sim 10\,\mu$A and ~ 1 mA. This indicates that we are detecting this defect in another region of the device, such as the depletion width of one the contacts, in series with the optically active region of the device. The measurement is made at a constant current so a resonance in a part of the device in series with the quantum well region would not affect the ELDMR but would change the total voltage across the device and thus contribute to the EDMR signal. These ideas are illustrated in a simple circuit model for the device shown in Figure 5. We model the LED structure as an ideal diode in parallel with leakage paths and other nonradiative recombination paths, represented by $R_{\parallel}$. In addition, there are voltage drops in series with the quantum well region primarily due to the contacts, which may have a significant capacitance as well as a lossy component (represented by C_C and R_C, respectively). In our analysis we neglect any voltage drops due to the film resistance. This and other terms would naturally be included in a more realistic circuit model. However our intention here is to include a minimum number of elements necessary to interpret our magnetic resonance data. From equation (1) we write the EDMR signal as:

$$\text{EDMR} = -\Delta V_{R_c}\frac{d(\ln I)}{dV} - \Delta V_{R_{\parallel}}\frac{d(\ln I)}{dV} \quad . \tag{2}$$

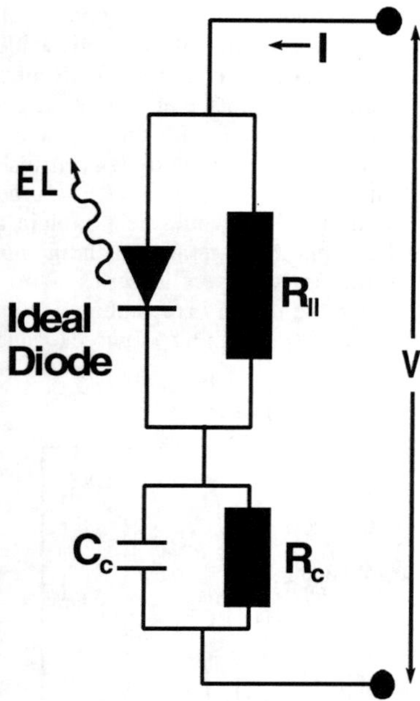

Figure 5. A simple circuit model for the radiative and non-radiative recombination processes in the SQW diode. R$_C$ and C$_C$ represent the contact resistance and capacitance, respectively. The leakage paths through the QW region are modeled by R$_{||}$.

Similarly, the ELDMR signal is approximately the fractional change in the EL which at low currents is approximately proportional to the fractional change in the current through the ideal diode; i.e., the signal is written as;

$$\text{ELDMR} = \frac{\Delta \text{EL}}{\text{EL}} = \frac{\Delta I_D}{I_D} = \Delta V_{R_{||}}\frac{d(\ln I_D)}{dV_D}. \quad (3)$$

In these equations the AC component of the voltage across a circuit element is given by ΔV, the current change is given by ΔI (with the appropriate subscript) and the electroluminescence is given by ΔEL. If we make the approximation that the two logarithmic derivatives are equal we can then combine equations (2) and (3) to obtain;

$$\text{EDMR} + \text{ELDMR} \approx -\Delta V_{RC}\frac{d(\ln I)}{dV}. \quad (4)$$

That is, at low currents the sum of the two measurements gives an approximate measure of the spin dependent recombination in the contact region. As we see in Figure 4 this quantity is quite significant for currents less than a few hundred microamps. At very low currents (I≤2 μA) the EL is very weak and therefore it is difficult to say whether the magnitude of the ratio ΔEL/EL continues to increase at even lower biases. However, for the modulation frequency (100 Hz) used in these measurements, at biases

less than ~10 μA we observe a significant signal in quadrature with the field modulation in ELDMR down to ~1 μA and in EDMR down to ~10 nA. We attribute this signal in quadrature to the effects of the capacitance of the contact depletion layers. In EDMR the quadrature signal increases monotonically with the modulation frequency as would be expected from a capacitance effect. It is more important in the EDMR measurements, partly because they remain sensitive down to very low currents where the dynamic resistance of the diode is high. At higher modulation frequencies, the contact capacitance should not effect the ELDMR signal because the measurements are all performed at a constant DC current. The fact that some quadrature signal is seen in the ELDMR, at all, suggests that a more complicated circuit model may be necessary to fully explain all of the data, incorporating leakage paths in parallel with the contact capacitance, for example. At higher currents (I>100 μA) the ELDMR changes sign and the EDMR drops to near zero as seen in Figures 2-4. In this bias range various aspects of the above semi-quantitative analysis are no longer valid. For example, as seen in Figure 1, the EL is not proportional to the current in this bias range. We suggest that the change in sign in the ELDMR may be due to a change from diffusion limited currents to drift limited currents, as was suggested by Lips and Fuhs[19] to explain their EDMR results in a-Si p-i-n diodes. The basic idea is that at low biases the current is primarily diffusion limited and therefore the spin dependent recombination enhances the current by decreasing the distance a carrier must diffuse before recombining. At higher biases both electrons and holes are effectively injected into the region of the recombination center and the current is limited by drift. In this regime the spin dependent recombination decreases the current by removing carriers. In our case the effect is to take current from the leakage paths and divert it to the radiative recombination path, thereby increasing the EL. The distribution of the electric fields within the device is certainly not uniform and there are probably still regions, perhaps around one or both contacts, in which the current is diffusion limited, especially for these low temperatures at which most carriers are frozen out. From these regions we would still have a current enhancement at resonance. This would help explain the near zero EDMR signal, i.e., the current-enhancing signal cancels the current-quenching signal from the leakage paths. The ELDMR is not affected by the device layers in series with the optically active region, only by the current paths in parallel with the radiative recombination path and therefore we only detect the reflection of the current-quenching spin dependent recombination in the leakage paths.

At high bias currents we observed a second sharper ($\Delta B \approx 7$ mT) resonance at $g \approx 1.99$ in both techniques as illustrated in Figure 6. The relative intensity of the resonance is somewhat enhanced at lower temperatures and high magnetic field modulation frequencies. (1 kHz was used for the ELDMR spectrum and 10 kHz was used for the EDMR spectrum.) The g value and linewidth are identical to the resonance which we observed in double heterostructure LEDs[17,18] and attributed to a deep donor. The g value is also similar to a resonance observed in ODMR[9] on GaN films and also attributed to a deep donor, although the linewidth in ODMR is larger than that observed with either technique in either type of diode. Based on their electrical measurements, Götz and co-workers[20] have suggested that there are several important deep levels and very recently Glaser and co-workers[21] have reported ODMR evidence for a second deep donor in some GaN films. So, the donor resonance observed in the diodes may be more closely related to these

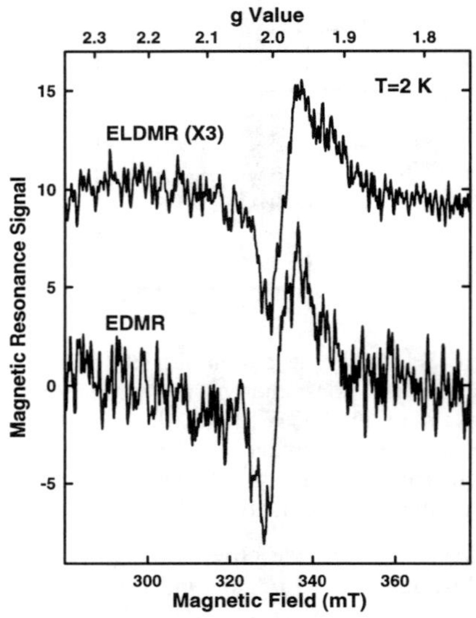

Figure 6. The resonance due to the donor at $g \approx 1.99$ as observed in EDMR (I=1 mA) and in ELDMR (I=20 mA).

recently reported centers than to the deep donor reported in the original ODMR work.[9,10] We note that both EDMR and ELDMR give a quenching resonance and speculate that this may be the result of recombination in either the lightly In-doped layer on which the InGaN quantum well is grown or conceivably in an AlGaN barrier layer. This would take carriers away from the EL process and result in a quenching of the EL as well as quenching of the current. We observe a very similar resonance in the blue SQW diodes, which may argue that this defect is located in the low x $In_xGa_{1-x}N$ layer (x=0.02 for the blue diode and x=0.04 for the green diode). There is no observable difference in the resonance for the two diodes contrary to what might be expected for a defect in another layer where the alloy concentrations are significantly different for the blue and green diodes.[4] In the blue SQW diodes,[22] and the double heterostructures,[17,18] we did not obtain a lineshape for this donor as well resolved from a broad resonance as the spectra shown in Figure 6.

Finally we note that the lineshape for the ELDMR at low biases has somewhat more intensity in the wings than the corresponding EDMR signal. (e.g., compare the I=1 μA spectra in Figures 2 and 3.) This could be the result of a subtle difference in the physical mechanism for the two techniques or it could be due to an underlying broad resonance at $g \approx 2$. We have not been able to enhance this feature by manipulating the various operating parameters.

RESULTS FOR STRESSED SAMPLES

We have also studied samples which had been subjected to high current stressing. The diodes showed significant degradation after stressing at currents of 100-150 mA. There was some sample-to-sample variation in the sensitivity to current stressing however, we were always able to observe changes in the EL on the order of 50% in 1-2 days. The changes in the optical and electrical properties are illustrated in Figure 7 which shows the room temperature characteristics (note that the voltages are much lower than at T=4.2 K.) of a diode and the effects of the current stressing (I_{Stress}=150 mA). The EL is reduced by over an order of magnitude after approximately 400 hrs. The fractional change in the EL varies only slightly over a wide range of bias currents. As seen in the V-I characteristics, this is not true for the electrical characteristics. At high bias the voltage across the diode is increased with stress, presumably due to increased contact resistances, while at low bias currents the voltage is decreased as a result of increased

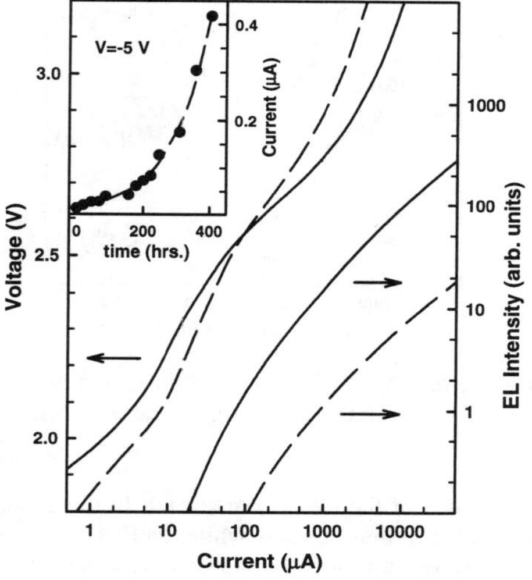

Figure 7. The DC V-I and EL-I characteristics of a green SQW diode before (solid lines) and after (dashed lines) 400 hours of high current stress. The reverse bias leakage current at -5V is shown as a function of stress time in the inset.

leakage currents. The increase in leakage current is further illustrated in the inset which shows the reverse bias leakage current at V=-5V as a function of stress time.

In Figures 8 and 9, respectively, we show the EDMR and ELDMR spectra for a diode which had been stressed at 100 mA for 200 hrs. As before, all measurements were taken at T=4.2 K, with a magnetic field modulation frequency of 100 Hz. We follow the same procedures as outlined in the previous section to obtain $\Delta I/I$ and $\Delta EL/EL$ for the EDMR and ELDMR measurements, respectively. As in the case of an unstressed diode, shown in Figures 2 and 3, we see that over a wide range of biases the spectra are dominated by a broad line at $g \approx 2.01$ for both techniques. We also see that the magnitudes of the resonance are quite similar for the two techniques, although they are about an order of magnitude smaller than those observed for the unstressed diodes. The lineshape and position of the resonance signal are the same as seen in the unstressed diodes indicating that the stress does not introduce a new defect but rather, changes the absolute and relative concentrations of the same defect at different locations in the device. After stress we do not observe the deep donor resonance under any operating conditions. Again,

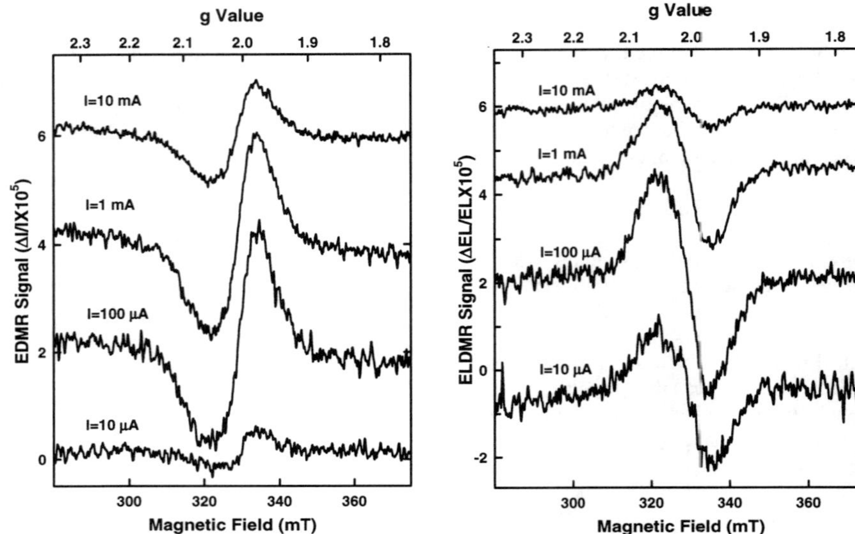

Figure 8. The EDMR spectra measured at the indicated currents for a diode after current stressing.

Figure 9. The ELDMR spectra for the same diode used for the EDMR in Figure 8.

the phases of the two signals differ. In this case, the EDMR signal is current-quenching over the full range of bias currents while the ELDMR signal is EL-enhancing at all bias currents.

The variation of the intensity with bias current is summarized in Figure 10 which gives the integrated intensities of the resonance signals as a function of bias current. Comparing Figure 10 to the intensities for the unstressed diode shown in Figure 4, we note that in addition to the signs of the signals being opposite, the general trends are very different. In the unstressed diode, the intensity of the EDMR signal is relatively constant at low currents and then falls off at high currents, while in the stressed diode the intensity is weak at low as well as high currents and goes through a maximum at $I \approx 200$ μA. The ELDMR signal is also a maximum for $I \approx 100\text{-}200$ μA (and very similar in magnitude to the EDMR signal). Following the basic model outlined in the previous section, the EL enhancement at resonance is due to a quenching of the drift-limited current through the leakage paths, thereby increasing the current through the radiative recombination channel and hence, the EL. This results in an increase in voltage across the device and a negative EDMR signal. Again we do not observe a simple one-to-one correspondence between the two signals and we see that the sum of the two signals peaks at about 10-20 μA as in the case of the unstressed diode. Following our previous arguments, we assign this portion of the signal to recombination in the contact region. The sum becomes negative for $I \approx 50\text{-}100$ μA and it is tempting to suggest that this indicates that the current in the contact region has gone from a diffusion limited regime to a drift limited regime. However, as in the unstressed diode, our simple analysis breaks down at high current. In particular, the EL of the stressed diode is an even more sublinear function of current in this range than the EL of the unstressed diode and so $\Delta EL/EL$ is probably an underestimate of $\Delta I_D/I_D$ which means this sum underestimates the recombination in the contacts. It is then prudent to wait for more sophisticated device models to quantitatively analyze the resonance data in this bias range.

In these measurements the primary difference between the stressed and unstressed diodes results from the nature of the current through the leakage paths. This is, of course, most important at low bias currents. In the unstressed diode this current is diffusion-limited, giving a current-enhancing and EL-quenching resonance. In the stressed diodes this current is larger and is drift-limited. The same resonance is then seen as current-quenching and EL-enhancing. The contact resistance is certainly increased by the current-stressing; however, we are not yet able to say whether this results in a change in the bias dependence of the resonance signal. We note that pulsed high current stressing of double heterostructure diodes (~ 1 kA/ cm^2) is thought to give rise to degradation by metal diffusion through the grain boundaries and not through the incorporation of new defects.[23] In our work the current density is about an order of magnitude smaller and may not be adequate to give rise to such effects.

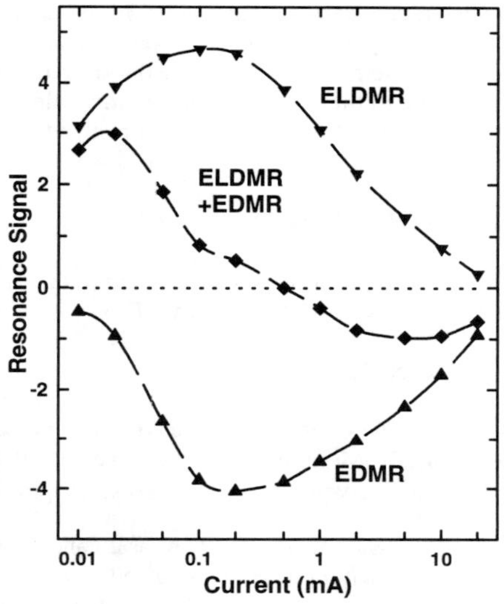

Figure 10. The integrated intensities (in arbitrary units) of the EDMR (triangles) and the ELDMR (inverted triangles) and their sum (diamonds) for a stressed diode. The dashed lines are included as aids to the eye.

We believe that the degradation of the diode is primarily due to the heat generated by the high current rather than directly by the current. To test this hypothesis, we subjected a diode to a high current (150 mA) in liquid nitrogen and did not observe any degradation of the EL. This indicates that the lack of effective heat sinking at room temperature permits the device temperature to rise sufficiently to degrade the device, while in liquid nitrogen the heat is dissipated more readily, thus inhibiting degradation.

SUMMARY AND CONCLUSIONS

We have presented results of our electrically detected and electroluminescence detected magnetic resonance measurements on GaN-based single quantum well LEDs. The spectra are dominated by a broad resonance at g$\approx$2.01 which is apparently due to a deep defect present in both the non-radiative leakage paths through the optically active region of the diode and in at least one of the contact depletion widths. We have used a simple circuit model to give a semi-quantitative understanding of the role of this defect in different bias ranges. We have presented results for both samples which were unstressed and ones which have been subjected to high-current stress. The high-current stress increases the leakage current through the diode and we observe that this current changes from being diffusion limited before stressing to being drift-limited after stressing. This is observed as a change in the sign of the resonance signal. We are also able to observe a second defect in unstressed samples which we attribute to a deep donor. While the bias dependencies of the EDMR and ELDMR signals and a comparison of their amplitudes provides some understanding of the physical location of these defects, we have not been able to obtain information on their chemical identity. Hopefully, with the success demonstrated with the application of paramagnetic resonance to these devices we and others will be able to apply even more powerful spectroscopic probes such as double resonance techniques to the problem of understanding the structure of defects which limit GaN device performance and contribute to their degradation. In this vein we are encouraged by the first successful application of electrically detected double resonance to defects in Si[24] and the first double resonance results in GaN.[25]

ACKNOWLEDGMENTS

The authors wish to thank J.A. Freitas, Jr. for many helpful discussions regarding GaN and the optical aspects of this study. This work was funded by the Office of Naval Research.

REFERENCES

1.   For reviews of work in the Group III nitrides see; R.F. Davis, Proc. IEEE **79**, 702 (1991); S. Strite and H. Morkoç, J. Vac. Sci. Technol. B **10**, 1237 (1992); H. Morkoç, S. Strite, G.B. Gao, M.E. Lin, B. Sverdlov and M. Burns, J. Appl. Phys. **76**, 1363 (1994).

2.   H. Amano, N. Sawaki, I. Akasaki and Y. Toyoda, Appl. Phys. Lett. **48**, 353 (1986); H. Amano, M. Kitoh, K. Hiramatsu and I. Akasaki, J. Electochem. Soc. 137, 1639 (1990); S. Nakamura, M. Senoh and T. Mukai, Jpn. J. Appl. Phys. **30**, L1708 (1991).

3.   Shuji Nakamura, Takashi Mukai and Masayuki Senoh, Jpn. J. Appl. Phys. **30**, L1998 (1991).

4.   Shuji Nakamura, Takashi Mukai and Masayuki Senoh, Appl. Phys. Lett. **64**, 1687 (1994).

5.   Shuji Nakamura, Masayuki Senoh, Naruhito Iwasa and Shin-ichi Nagahama, Jpn. J. Appl. Phys. **34**, 797 (1995); Appl. Phys. Lett. **67**, 1868 (1995).

6.   Shuji Nakamura, M. Senoh, S. Nagahama, N. Iwasa, T. Yamada, T. Matsushita, H. Kiyoku and Y. Sugimoto, Jpn. J. Appl. Phys.. **35**, L74 (1996); Shuji Nakamura, Masayuki Senoh, Shin-ichi Nagahama, Naruhito Iwasa, Takao Yamada, Toshio Matsushita, Yasunobu Sugimoto, and Hiroyuki Kiyoku, Appl. Phys. Lett. **69**, 3034 (1996).

7.   For a review of recent optical and magnetic resonance results see; U. Kaufmann, M. Kunzer, C. Merz, I. Akasaki and H. Amano, Gallium Nitride and Related Materials, ed. by F.A. Ponce, R.D. Dupuis, S. Nakamura and J.A. Edmond (Materials Research Society, Pittsburgh, 1996), p. 633.

8.   W.E. Carlos, J.A. Freitas, Jr., M. Asif Khan, D.T. Olson and J.N. Kuznia, Phys. Rev. B**48**, 17878 (1993).

9.   E.R. Glaser, T.A. Kennedy, K. Doverspike, L.B. Rowland, D.K. Gaskill, J.A. Freitas, Jr., M. Asif Khan, D.T. Olson, J.N. Kuznia, and D.K. Wickenden, Phys. Rev. B**51**, 13326 (1995).

10. M. Kunzer, U. Kaufmann, K. Maier, J. Schneider, N. Herres, I. Akasaki, and H. Amano, Mat. Sci. Forum **143-147**, 87 (1994); K. Maier, M. Kunzer, U. Kaufmann, J. Schneider, B. Monemar, I. Akaski and H. Amano, Mat. Sci. Forum **143-147**, 93 (1994).

11. D.H. Hofmann, D. Kovalev, G. Steude, B. K. Meyer, A. Hoffmann, L. Eckey, R. Heitz, T. Detchprom, H. Amano and I. Akasaki, Phys. Rev. B52, 16702 (1995).

12. M. Ramsteiner, J. Menniger, O. Brandt, H. Yang and K.H. Ploog, Appl. Phys. Lett. 69, 1276 (1996).

13. L.S. Swanson, J. Shinar, A.R. Brown, D.D.C. Bradley, R.H. Friend, P.L. Burn, A. Kraft and A.B. Holmes, Phys. Rev. B46, 15072 (1992).

14. K.P. Homewood, B.C. Cavenett, I.G. Austin, T.M. Searle, W.E. Spear and P.G. Lecomber J. Phys. C17, L103 (1984).

15. D.J. Lepiné, Phys. Rev. B6, 436 (1972); I. Solomon, Solid State Commun. 20, 215 (1976); F.C. Rong, G.J. Gerardi, W.R. Buchwald, E.H. Poindexter, M.T. Umlor, D.J. Keeble and W.L. Warren, Appl. Phys. Lett. 60, 610 (1992); B. Stich, S. Greulich-Weber and J.-M. Spaeth, J. Appl. Phys. 77, 1546 (1995).

16. N.M. Reinacher, M.S. Brandt and M. Stutzmann, Materials Science Forum 196-201, 1915 (1995).

17. W.E. Carlos, E.R. Glaser, T.A. Kennedy and S. Nakamura, Appl. Phys. Lett. 67, 2376 (1995); Materials Science Forum 196-201, 25 (1995); J. Electron. Mater. 25, 851 (1996).

18. W.E. Carlos, E.R. Glaser, T.A. Kennedy and S. Nakamura, in The Physics of Semiconductors, ed. by M. Scheffler and R. Zimmermann (World Scientific, Singapore, 1996) p. 2921.

19. Klaus Lips and Walter Fuhs, J. Appl. Phys. 74, 3993 (1993).

20. W. Götz, N.M. Johnson, H. Amano and I. Akasaki, Appl. Phys. Lett. 65, 463 (1994); W. Götz, N.M. Johnson, R.A. Street, H. Amano and I. Akasaki, Appl. Phys. Lett. 66, 1340 (1995).

21. E.R. Glaser, T.A. Kennedy, A.E. Wickenden, D.D. Koleske and J.A. Freitas, Jr., this conference.

22. W.E. Carlos, Proceedings of Seventh International Conference on Shallow Level Centers in Semiconductors, ed. by C.A.J. Ammerlaan and B. Pajot (World Scientific, Singapore, 1996) in press.

23. Marek Osinski, Joachim Zeller, Per-Chih Chiu, B. Scott Phillips and Daniel L. Barton, Appl. Phys. Lett. 69, 898 (1996).

24. S. Stich, S. Greulich-Weber and J.-M. Spaeth, Appl. Phys. Lett. 68, 1102 (1996).

25. G. Denninger, R. Beerhalter, D. Reiser, K. Maier, J. Schneider, T. Detchprohm, K. Hiramatsu, Solid State Commun. 99, 347 (1996); F.K. Koshnick, K. Michael, J.-M. Spaeth, B. Beumont and P. Gibart, Phys. Rev. B54, 11042 (1996).

STRUCTURAL AND OPTICAL PROPERTIES OF AlGaN/GaN QUANTUM WELL STRUCTURES GROWN BY MOCVD ON SAPPHIRE

R. NIEBUHR, K.H. BACHEM, D. BEHR, C. HOFFMANN, U. KAUFMANN, Y. LU,
B. SANTIC, J. WAGNER *
M. ARLERY, J.L. ROUVIERE **
H. JÜRGENSEN ***
* Fraunhofer-Institut für Angewandte Festkörperphysik, Tullastrasse 72,
  D-79108 Freiburg, Germany
** CEA-Grenoble, Departement de Recherche Fondamentale sur la Matiere
   Condensee, 17, Rue de Martyrs, 38054 Grenoble Cedex 9, France
*** AIXTRON Semiconductor Technologies, Kackertstr. 16-17,
   D-52072 Aachen, Germany

ABSTRACT

AlGaN/GaN single quantum wells (QW) have been grown on 2" sapphire substrates (c-plane) by metal-organic chemical vapor deposition (MOCVD). The well width was varied between 20 and 40 Å for barriers containing 4 % and 16 % of aluminium. Cathodoluminescence (CL) and Photoluminescence (PL) spectra of the samples show, as expected, a shift of the quantum well emission to higher energies with decreasing well width, whereas the barrier luminescence stays at constant energy. Examination of the QWs by resonant Raman spectroscopy tuned to the gap of the well, clearly shows the GaN A_1(LO) phonon besides the AlGaN A_1(LO) phonon from the barrier. For a well width of 20 Å we observe a shift of the A_1(LO) GaN phonon indicating a certain degree of intermixing at the GaN/AlGaN interface. Atomic Force Microscopy (AFM) reveals that the layers are growing in a 2-dimensional step flow growth mode with step heights of 3 and 6 Å corresponding to mono- and biatomic steps. High Resolution Transmission Electron Microscopy (HRTEM) micrographs of the 40 Å well show a very low interface roughness of 1-2 atomic layers.

INTRODUCTION

Up to now only a few groups have investigated the properties of AlGaN/GaN QWs [1], [2], [3]. In this paper we report on the growth and the properties of group-III nitride QWs focusing on AlGaN/GaN structures. Because of the relatively large masses of electrons and holes in GaN [4] quantum confinement effects are expected only for rather narrow QWs as compared to more conventional III-V semiconductors. To prove the quantum effect the aim was to establish a relationship between the well width and the luminescence peak energies of the wells. The wells studied here are made up from binary GaN between ternary AlGaN. In this case for a given barrier height the energy of the QW states is determined by the well width only and not by its composition neglecting a possible misplacement of aluminium into the well. Using resonant Raman spectroscopy the binary composition of the well could be confirmed, and information on the degree of interdiffusion at the GaN/AlGaN interface was obtained. Two series of QWs with widths between 20 and 40 Å and Al contents x of the barriers of 0.04 and

Mat. Res. Soc. Symp. Proc. Vol. 449 © 1997 Materials Research Society

0.16, respectively, have been grown. For such QW layers with only a few atomic layers in width it is important to control the growth mode of the AlGaN and GaN and to induce a 2-dimensional growth. Introducing an AlN buffer layer deposition occurs in a step flow mode and atomic steps can be observed on the surface by AFM. Likewise, HRTEM studies reveal high quality interfaces. In both QW series PL and CL spectra confirm quantum confinement effects.

GROWTH CONDITIONS AND CHARACTERISATION TECHNIQUES

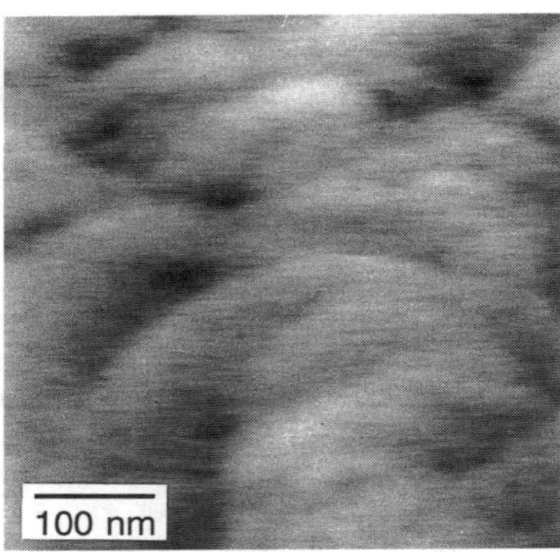

Fig. 1: AFM micrograph of the surface of an $Al_{0.04}Ga_{0.96}N$/GaN QW. The contrast is due to mono- and biatomic steps on the surface.

The AlGaN/GaN QWs were grown by low pressure (LP) MOCVD using trimethylgallium (TMG) and trimethylaluminium (TMA) as metal-organic and ultra high purity ammonia as nitrogen precursor. The layers were deposited on c-plane sapphire (2") using nitrogen as carrier gas. After inserting the wafer into the reactor a nitridation step was performed. For this purpose the wafer was annealed for 30 minutes under ammonia. After nitridation a 300 Å thick AlN buffer layer was grown at 875 °C. The buffer layer was annealed for 1 minute at 1100 °C before the film was deposited at 1025 °C. The growth rate for the quantum wells amounted to 0.5 µm/h. With this rate the growth time for the wells was between 15 and 30 seconds. The well width calculated from the macroscopic growth rate was checked by Secondary Ion Mass Spectroscopy and HRTEM of the samples. The total gas flow was adjusted to lead to a mean gas velocity of about 1m/s. This was considered to be high enough to minimize misplacement of the barrier material into the wells. The growth started immediately with the ternary barrier material. The quantum well was placed 1.5 µm above the sapphire-interface and 0.1 µm below the sample surface. The wells were grown without growth interruption.

The surface morphology of the samples was examined by a commercial AFM (Park Scientific, Autoprobe cp). Some of the $Al_{0.16}Ga_{0.84}N$/GaN QWs were studied by HRTEM. The sample preparation technique and the observation mode are described elsewhere [5]. For $Al_{0.04}Ga_{0.96}N$/GaN QWs the contrast between well and barrier is too low to be observable by HRTEM. The $Al_{0.04}Ga_{0.96}N$/GaN quantum wells were characterized by low temperature photoluminescence using the 325 nm line of a He-Cd laser for excitation. In order to observe luminescence from the well and the barrier the $Al_{0.16}Ga_{0.84}N$/GaN

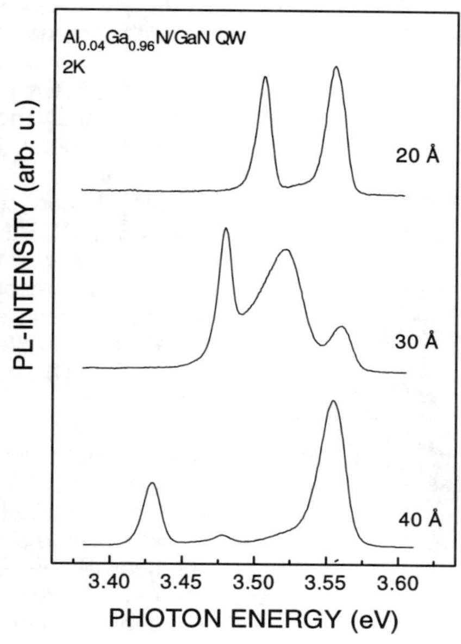

Fig. 2: PL spectra of 20, 30 and 40 Å wide Al$_{0.04}$Ga$_{0.96}$N/GaN QWs

QWs were examined by CL at 20 K . The Raman spectra were recorded in backscattering geometry. The incident light was polarized perpendicular to the c-axis, the scattered light was not analyzed for its polarization. For optical excitation either an Ar-ion or a Kr-ion laser was used. Off-resonance spectra were taken with laser lines of 2.57 or 2.60 eV. For resonant excitation of the GaN QWs a photon energy of 3.54 eV was chosen.

Apart from the quantum structures we have grown bulk layers of AlGaN reported elsewhere [6]. The aluminium concentration was calibrated by microprobe analysis (EDX) .

RESULTS AND DISCUSSION

Fig. 1 shows an AFM picture (0.5 µm x 0.5 µm) of the surface of the upper barrier of an Al$_{0.04}$Ga$_{0.96}$N/GaN QW. The contrast in the picture is caused by mono- and biatomic steps. The terrace width is between 50 and 100 nm. The mean roughness of the portion shown amounts to 3.3 Å. We take the surface structure of the sample as an indication of the quality of the interfaces which are assumed to be of comparable roughness. In Fig. 2 PL spectra at 2 K of three Al$_{0.04}$Ga$_{0.96}$N/GaN QWs are shown. The peak at highest energy stems from the barrier. The peak position compares well with that found in bulk Al$_{0.04}$Ga$_{0.96}$N reference layers. The shift of the barrier peak from sample to sample is about ± 5 meV while the FWHM amounts to about 15 - 20 meV. In addition, all three samples in Fig. 2 reveal one or two luminescence lines which shift to higher energy with decreasing well width. These lines are assigned to emissions from quantum confined excitons. A more accurate analysis in terms of heavy and light hole excitons as well as strain effects is beyond the scope of the present paper.

Fig. 3 presents a HRTEM micrograph of a 40 Å Al$_{0.16}$Ga$_{0.84}$N/GaN QW. The roughness of the interfaces is about 1-2 atomic layers. Within the well small contrast variations are observed which are possibly due to TEM sample preparation. In Fig. 4 the CL spectra of 20, 30 and 40 Å wide Al$_{0.16}$Ga$_{0.84}$N/GaN QWs are shown. Two peaks are observed. The peak higher in energy results from excitonic recombination in the barrier material and was also observed at the same position in bulk samples. The lower peak shows a clear shift to higher energies with decreasing well width. This peak is attributed to the quantum well. The emission of the 40 Å quantum well is at 3.444 eV and is too low in energy to originate from free exciton recombination. Presumably this emission is due to bound excitons or confined defect states as already observed in the $Al_{0.04}Ga_{0.96}N/GaN$ QWs. Remarkably, the intensity of the QW emission decreases with increasing well width relative to that of the barrier peak.

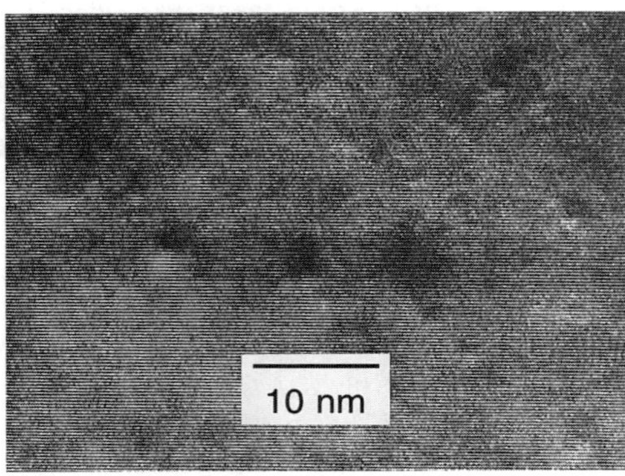

Fig. 3: HRTEM micrograph of an $Al_{0.16}Ga_{0.84}N/GaN$ QW

The $Al_{0.16}Ga_{0.84}N/GaN$ QWs were also analysed by Raman spectroscopy. Fig. 5 shows three room-temperature Raman spectra of an $Al_{0.16}Ga_{0.84}N/GaN$ QW excited off-resonantly (middle), and resonantly (above), respectively, and of a bulk GaN layer excited off-resonant (below). Excited off-resonantly the $Al_{0.16}Ga_{0.84}N/GaN$ QW spectrum is dominated by Raman scattering from the barriers. For excitation at 3.54 eV we observe a strong enhancement of the Raman scattering by the GaN $A_1(LO)$ phonon from the QWs due to a resonant Raman effect. The excitation energy of 3.54 eV is close to the calculated energy of 3.55 eV for interband transitions between the topmost valence band state and the lowest quantized electron level in the quantum well [7]. In Fig. 6 Raman spectra of three different QWs of 20, 30 and 40 Å width are shown. The spectra were recorded under resonant conditions with an excitation energy of 3.54 eV. For the 30 and 40 Å quantum wells the GaN $A_1(LO)$ phonon is observed at 732 cm^{-1} which is very close to the GaN bulk value (see Fig. 5). For the 20 Å quantum well the $A_1(LO)$ mode broadens and shifts to higher energies by 7 cm^{-1} towards the $A_1(LO)$ phonon mode of the AlGaN barriers. This shift is likely due to misplaced aluminium within the well and a resulting interface width comparable to the well width. However, for the 30 and 40 Å quantum well the Raman spectra show that the wells consist of binary GaN.

Fig. 4: (right) CL spectra of 20, 30 and 40 Å wide $Al_{0.16}Ga_{0.84}N$/GaN QWs

Fig. 5: (below,left) Off-resonant Raman spectra of bulk GaN (below) and of a $Al_{0.16}Ga_{0.84}N$ /GaN QW excited off-resonantly (middle) and resonantly (above)

Fig. 6: (below, right) Raman spectra of 20, 30 and 40 Å wide $Al_{0.16}Ga_{0.84}N$/GaN QW under resonant excitation

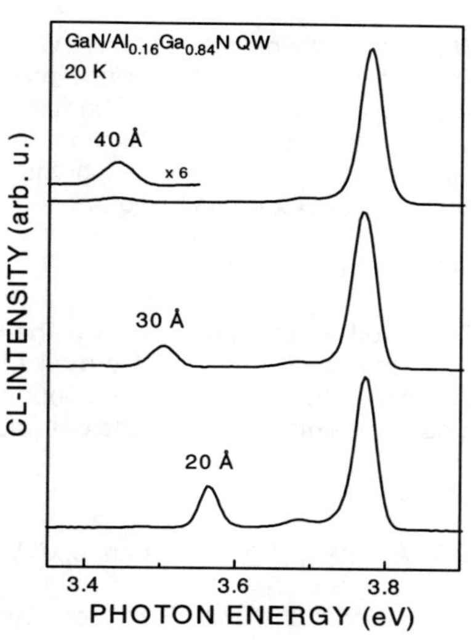

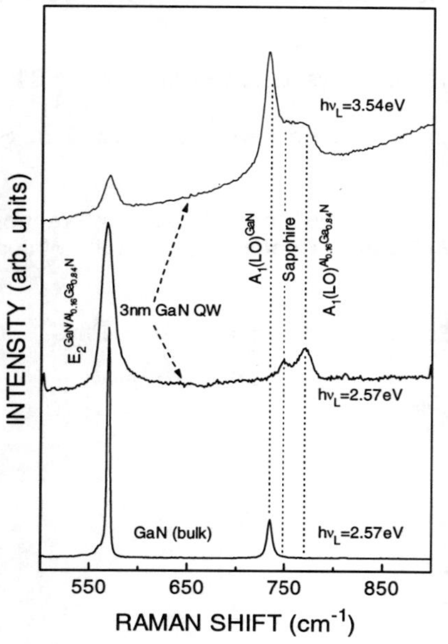

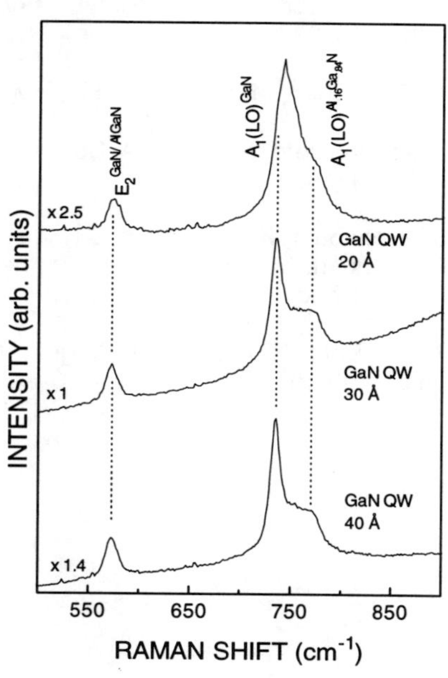

CONCLUSIONS

We have grown AlGaN/GaN QW structures with well widths between 20 and 40 Å.
They exhibit excellent interface properties as proved by HRTEM and resonant Raman
scattering measurements. The luminescence properties have been studied by PL and
CL at low temperatures. Excitonic recombination within the barrier material as well as
recombination of confined excitons in the well is observed. Further work is required to
elucidate the exact nature of these excitons.

ACKNOWLEDGEMENT

This work was supported by the European Community under the Brite-Euram II
Programme and by the German Ministery for Education, Research and Technology.
We thank W. Rothemund for support with the CL measurements, M. Maier for the SIMS
measurements and J. Schneider for useful discussions.

REFERENCES

1. M.A. Khan, R.A. Skogman, J.M. Van Hove, Appl. Phys. Lett 56 (13) 1990, 1257

2. J.F. Schetzina, Mat. Res. Soc. Symp. Proc. Vol 395 (1996), 123

3. A. Salvador, G. Liu, W. Kim, Ö. Aktas, A. Botchkarev, H. Morkoc, Appl. Phys. Lett. 67
(22) 1995, 3322

4. G.D. Chen, M. Smith, J.Y. Lin, H.X. Jiang, Appl. Phys. Lett. 68 (20) 1996, 2784

5. J.L Rouviere, M. Arlery, A. Bourret, R. Niebuhr, K.H. Bachem, Inst. Phys. Conf. Ser.,
Vol. 146 (1995) 285

6. R. Niebuhr, K.H. Bachem, C. Merz, S. Santic, U. Kaufmann, D. Behr, J. Wagner, W.
Rothemund, Y. Lu, presented at the 1996 OMVPE Conference, Cardiff, Wales UK
(unpublished)

7. D. Behr, R. Niebuhr, J. Wagner, K.H. Bachem, U. Kaufmann, Appl. Phys. Lett. (to
be published)

INTRINSIC AND THERMAL STRESS IN GALLIUM NITRIDE EPITAXIAL FILMS

J. W. AGER III*, T. SUSKI**, S. RUVIMOV*, J. KRUEGER*, G. CONTI*, E. R. WEBER*, M. D. BREMSER***, R. DAVIS***, AND C. P. KUO****
*Materials Sciences Division, Lawrence Berkeley National Laboratory, Berkeley, CA 94720
**UNIPRESS, Polish Academy of Sciences, Warsaw 01-142 POLAND
***MS&E Department, North Carolina State University, Raleigh, NC 27695
****Hewlett Packard, Optoelectronics Division, San Jose, CA 95131

ABSTRACT

Strain in GaN epitaxial layers at room temperature is measured with three complementary methods: Raman spectroscopy (via shifts of phonon frequencies), low temperature photoluminescence (via shifts of band-edge luminescence), and X-ray diffraction (via shifts in lattice spacings). GaN films grown on the c-plane of sapphire tend to be in compression. Increasing the Si-dopant concentration (up to 10^{19} cm^{-3}) is observed to add compressive strain to the layer. Axially resolved measurements obtained by micro-Raman in 4 µm thick Si-doped films reveal strain relaxation toward the sample surface at Si concentrations above 10^{18} cm^{-3}. Mg- and Si-doped GaN films on SiC substrates are found to be in tension. An experimental methodology is presented that separates two contributions to the room temperature residual stress in GaN epilayers: (1) the thermal stress due to differences in the thermal expansion coefficients of the epilayer and substrate and (2) the intrinsic stress, which is influenced by the growth conditions. We measure stress as a function of temperature up to 325 C, about one-third of the growth temperature, by monitoring the frequency of the E_2 phonon mode by Raman spectroscopy. A high-quality bulk single crystal of GaN is used as a strain-free standard. Over this temperature range, most layers behave elastically; the observed stress trends are well-fit by a thermal expansion model using previous reported values of the thermal expansion coefficients of GaN and the substrates. The intrinsic stress states at the growth temperature for films grown on sapphire and SiC are predicted to be tensile and compressive, respectively, in agreement with the a-plane lattice coefficient mismatch.

INTRODUCTION

GaN epitaxial films have a high density of extended defects and considerable residual stress due in part to the large lattice and thermal expansion coefficients mismatches between GaN and most commonly used substrates [1]. Consideration of the relative a-plane lattice constants, SiC (3.08 Å) < GaN (3.19 Å) < Al$_2$O$_3$ (4.76 Å) would predict tensile and compressive stress for GaN grown on sapphire and SiC, respectively. Consideration of the average thermal expansion coefficients, SiC (4.2×10^{-6}/K) < GaN (5.6×10^{-6}/K) < Al$_2$O$_3$ (7.5×10^{-6}/K), predicts the opposite, compressive stress for alumina and tensile stress for SiC. There is increasing evidence that impurity concentration can also influence the residual stress state. For example, Ruvimov *et al.* [2] found that incorporation of moderate levels of Si doping decreased the compressive residual stress (and increased the layer quality) in MOCVD films grown on the A-face of sapphire.

We present here data showing the influence of Si doping on residual stress in GaN grown by MOVPE on the c-plane of sapphire. Both the average stress in the layer and the axial gradient along the growth direction are measured. Average stress values are obtained for

Mat. Res. Soc. Symp. Proc. Vol. 449 © 1997 Materials Research Society

doped GaN films grown by MOVPE on SiC. The thermal stress contribution to the observed residual stress is evaluated by performing Raman measurements at elevated temperature.

EXPERIMENTAL

Si-doped films, 4 µm thick, were grown by metal organic vapor phase epitaxy on (0001) sapphire, using a GaN buffer layer [3]. Mg- and Si-doped films were grown on SiC substrates using MOVPE [4]. Bulk single crystals of GaN grown by a high pressure, high temperature process were used as the standard of strain-free material [5].

Raman spectroscopy was used as a local stress probe. A 100x microscope objective was used to focus 476.5 nm light from an Ar ion laser to 1 µm spot on the surface of the samples. Shifts of the Raman-allowed E_2 phonon from its stress-free value in the bulk single crystal were used to measure local stress. The precision of the Lorentzian line-fitting procedure was ±0.1 cm^{-1}. Raman spectra were taken in two measurement modes. (1) Stress values averaged through the epitaxial layer are obtained with the beam normal to the GaN surface. (2) Axially resolved stress data is obtained moving the sample in 0.5 - 1 µm steps with the laser beam incident on a cleaved surface normal to the growth direction. The spatial resolution of 1 µm in this case was confirmed by fitting of the observed E_2 Raman intensity as a function of the measurement position as the beam was scanned over the substrate/GaN interface. Raman spectra of GaN films and single crystals up to 325 C in air were obtained by fastening the samples with mechanical pressure to a ceramic block which was heated by cartridge heaters. The sample temperature was monitored by a thermocouple in contact with the back of the sample.

Two previous measurements of the shift of the E_2 phonon under hydrostatic pressure are in good agreement: 4.17 cm^{-1}/GPa by Perlin *et al.* [6] and 3.75 cm^{-1}/GPa by Wetzel *et al.*[7]. Using recent measurements of the GaN elastic constants [8], a value of

$$\Delta E_2 = -4.1 \text{ cm}^{-1}/\text{GPa} \tag{1}$$

for the shift of the Raman phonon under biaxial stress in the *a*-plane is obtained [9]. We know of two other determinations of this parameter in the literature. Kozawa *et al.* [10] obtained -6.2 cm^{-1}/GPa by correlating Raman measurements with substrate curvature, and Demangeot *et al.* [11] obtained -2.9±0.3 cm^{-1}/GPa from an analysis of Raman and reflectance data. The former value was used in the study of Reiger *et al.* [12] in which the effect of AlN buffer thickness on the Raman and PL spectra of a series of heteroepitaxial layers was studied.

RESULTS AND DISCUSSION

Figure 1 illustrates the trend in compressive stress caused by Si-doped for material grown on the *c*-plane of sapphire. The overall compressive stress increases as the Si concentration is raised from 8.6×10^{16} to 1×10^{19} cm^{-3} from about 120 MPa to 320 MPa. The Raman measurements of stress are in excellent agreement with photoluminescence and X-ray diffraction data obtained from the same samples [13]. Interestingly, a gradient is observed in the stress in the axial direction for the 3×10^{18} and 1×10^{19} cm^{-3} Si-doped samples, with the maximum located at the GaN/sapphire interface. It should be noted that due to the 1 µm size of the laser spot size (and the effects of the depth of focus of the beam in the transparent layer) the observed maximum in stress near the interface is a lower limit. The origin of the

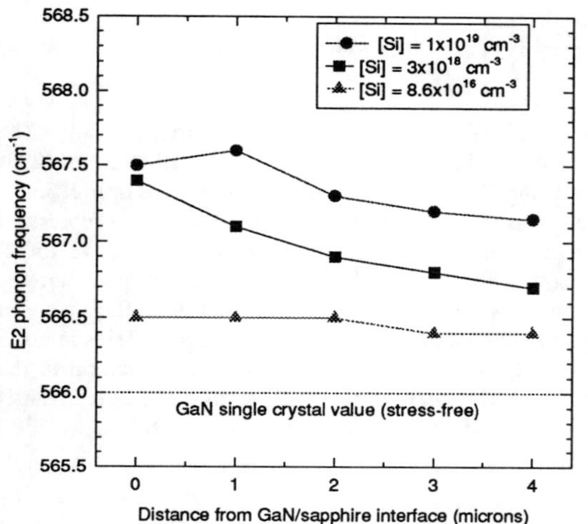

Figure 1. Axially resolved micro-Raman measurements in 4-micron-thick, Si-doped GaN grown on sapphire. The laser was incident on the (110) cleaved edge of sample.

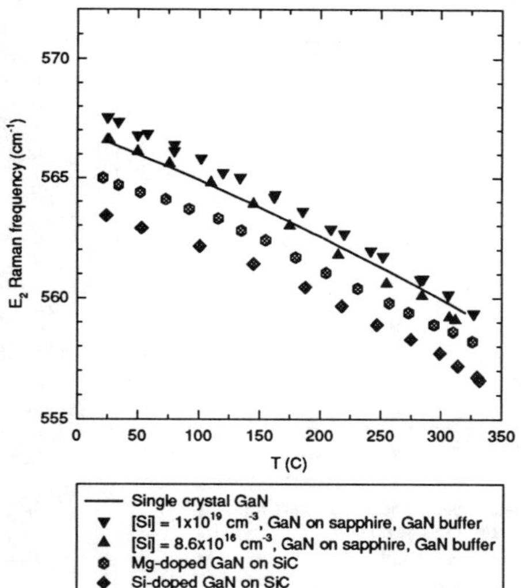

Fig. 2 E_2 phonon frequency vs. temperature. The solid line is a fit of the single crystal data (not shown) to a 2nd-order polynomial.

stress concentration at the interface is unclear at present, although it may be related to the influence of Si on the rate of formation of threading dislocations [2]. The existence of this stress gradient might be responsible for reports of epilayer cracking in heavily Si-doped material. As shown in Fig. 2 at room temperature, the residual stress in GaN grown on SiC is influenced by the dopant species. Mg doped material (ca. 10^{18} cm^{-3}) is observed to be 270 MPa in tension while Si doped material at a similar concentration is 660 MPa in tension.

Figure 2 illustrates the E_2 phonon frequency vs. temperature for the samples considered in this study. The separation of the shifts to lower frequency into thermal expansion and self-energy effects [14] will be considered elsewhere. The bulk single crystal data was fit to a parabolic temperature dependence which is shown as a solid line in Fig. 2. Figure 3 shows the difference in the E_2 phonon frequency for the epitaxial layers relative to the single-crystal value as a function of temperature. We expect the temperature trend of the E_2 phonon frequency of the epitaxial films to differ from that of the single crystal due to the different thermal expansion coefficients of the GaN films and the substrates. The change in stress state as the epilayers are heated is:

$$\sigma(T) = \frac{E}{1-v} \int_{20C}^{T} (\alpha_{GaN} - \alpha_{substrate}) dT \,, \tag{2}$$

where E is the a-plane Young's modulus, 324 GPa, and v is the Poisson's ratio, 0.25 [8]. The thermal expansion values are obtained from the literature [1,15]. The thermal stress in (2) is converted to shift of the Raman phonon using (1) and graphed as the dotted lines in Fig. 3.

Both Si-doped samples follow the stress model very well. The observed compressive stress observed at room temperature relaxes and becomes tensile at temperatures above 150 C for the lightly doped sample and above 300 C for the more heavily doped sample. Extrapolation of the thermal model to the growth temperature predicts that these films were in 0.5 - 1.0 GPa of tensile stress at the growth temperature. The thermal stress model is in good agreement with the data in Fig. 3 for the Mg-doped layer on SiC. The extrapolation to the growth temperature predicts a slight (<100 MPa) amount of tensile stress. The agreement is poor for the Mg-doped sample; the stress state does not change with temperature below 300

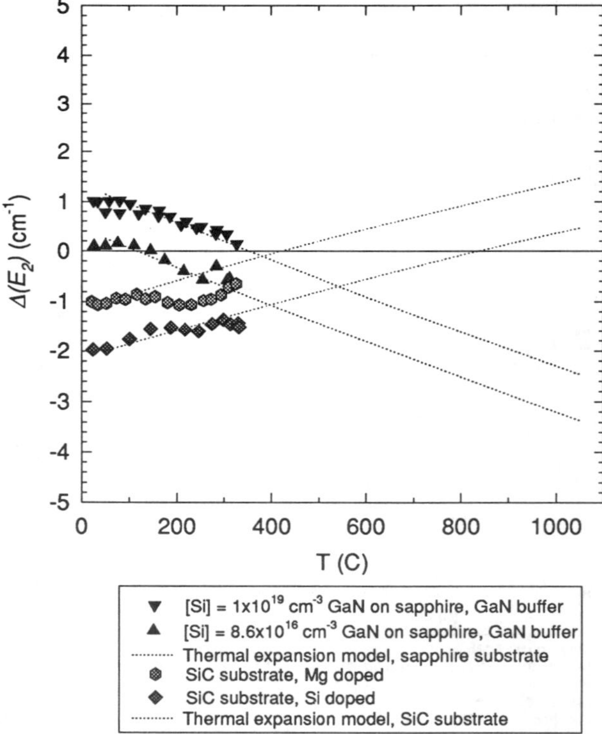

Fig. 3 Stress in epitaxial films as a function of temperature measured relative to GaN bulk single crystal. A positive shift corresponds to compressive stress (4.1 cm^{-1}/GPa biaxial stress).

C (i.e. it behaves like bulk GaN). This sample had been previously exposed to high temperature annealing to attempt to activate the Mg acceptor. One possible explanation for the discrepancy is that the thermal cycling may have damaged the GaN/SiC interface.

Intrinsic stress is believed to be driven by lattice mismatch and is modified by production of stress-relieving point and extended defects. GaN differs from other III-Vs in that the stress-relieving dislocations are less mobile than in, for example, GaAs. The predictions of tensile stress at the growth temperature for GaN on sapphire and slight compression for Si-doped GaN on SiC are consistent with the lattice mismatches: $a(Al_2O_3) >> a(GaN) > a(SiC)$. Our results imply that this "intrinsic" lattice mismatch stress is not completely relieved by defect formation at the growth temperature. In the case of Si-doped GaN, high concentrations of Si appear to aid the formation of stress-relieving defects, leading to a lower tensile stress at the growth temperature.

CONCLUSIONS

In addition to the substrate, the concentrations of impurities and dopants influence the residual stress in GaN epilayers. In heavily Si-doped material stress gradients are found that may explain mechanical failure in these films. Stress measurements at elevated temperature obtained with Raman spectroscopy are shown to be a practical way to separate the thermal and intrinsic components of the observed stress. Extrapolation of the stress trends to the growth temperature allows an estimate the amount of mismatch strain relieved at or close to the growth temperature. It is found that the estimated stress states at the growth temperature are in agreement with the lattice mismatches for GaN grown on sapphire and SiC, implying that lattice mismatch strains are not completely relieved by defect formation at the growth temperature.

ACKNOWLEDGMENTS

The work at LBNL was supported by the Director, Office of Energy Research, Office of Basic Energy Sciences, Division of Materials Sciences, of the U.S. Department of Energy, under contract No. DE-AC03-76SF00098.

REFERENCES

1. H. Morkoç, S. Strite, G. B. Gao, M. E. Lin, B. Sverdlov, M. Burns, J. Appl. Phys. **76**, 1363 (1994).

2. S. Ruvimov, Z. Liliental-Weber, T. Suski, J. W. Ager III, J. Washburn, J. Krueger, C. Kisielowski, E. R. Weber, H. Amano, and I. Akasaki, Appl. Phys. Lett. **69**, 990 (1996).

3. T. Suski, J. W. Ager III, G. Conti, Z. Liliental-Weber, J. Krueger, S. Ruvimov, C. Kisielowski, E. R. Weber, C.-P. Kuo, I. Grzegory, and S. Porowski, Proceedings of the 23rd International Conference on the Physics of Semiconductors (World Scientific, Singapore, 1996).

4. T. W. Weeks Jr., M. D. Bremser, K. S. Ailey, K. S. Carlson, and R. F. Davis, Appl. Phys. Lett. **67**, 401 (1995).

5. P. Perlin, I. Gorczyca, N. E. Christensen, I. Grzegory, H. Teisseyre, and T. Suski, Phys. Rev. B **45**, 13307 (1992).

6. P. Perlin, C. Jauberthie-Carillon, J. P. Itie, A. San Miguel, I. Grzegory, A. Polian, Phys. Rev. B **45**, 83 (1992).

7. C. Wetzel, W. Walukiewicz, E. E. Haller, J. W. Ager III, I. Grzegory, S. Porowski, and T. Suski, Phys. Rev. B, **53**, 1322 (1996).

8. A. Polian, M. Grimsditch, and I. Grzegory, J. Appl. Phys. **79**, 3343 (1996).

9. C. Kisielowski, J. Krueger, S. Ruvimov, T. Suski, J. W. Ager III, E. Jones, Z. Liliental-Weber, M. Rubin, E. R. Weber, M. D. Bremser, and R. F. Davis, Phys. Rev. B, in press.

10. T. Kozawa, T. Kachi, H. Kano, H. Nagase, N. Koide, and K. Manabe, J. Appl. Phys. **77**, 4389 (1995).

11. F. Demangeot, J. Frandon, M. A. Renucci, O. Briot, B. Gil, and R. L. Aulombard, Solid State Commun. **100**, 207 (1996).

12. W. Rieger, T. Metzger, H. Angerer, R. Dimitrov, O. Ambacher, and M. Stutzmann, Appl. Phys. Lett. **68**, 970 (1996).

13. J. Krueger, C. Kisielowski, T. Suski, S. Ruvimov, Z. Liliental-Weber, J. W. Ager III, M. Rubin, and E. R. Weber, Proceedings of the IEEE-SIMC, Toulouse, France, in press.

14. J. Menendez and M. Cardona, Phys. Rev. B **29**, 2051 (1984).

15. Thermophysical Properties of High Temperature Solid Materials, ed. Y. S. Touloukian (Macmillan, New York, 1967). Vol. 4, pp. 22-25 and Vol. 5, pp. 129-130.

VARIATION OF GaN VALENCE BANDS WITH BIAXIAL STRESS: QUANTIFICATION OF RESIDUAL STRESS AND IMPACT ON FUNDAMENTAL BAND PARAMETERS

N.V. Edwards[a], S.D. Yoo[a], M.D. Bremser[a], M.N. Horton[b], N.R. Perkins[b], T.W. Weeks, Jr.[a], H. Liu[c,*] R.A. Stall[c], T.F. Kuech[b], R.F. Davis[a], and D.E. Aspnes,[a]; [a]NCSU, Raleigh, NC; [b]U. of Wisc., Madison, WI; [c]EMCORE, Somerset, NJ; *now at Hewlett-Packard, San Jose, CA.

ABSTRACT

We provide the widest estimate thus far of the range of tensile and compressive stress (-3.8 to 3.5 kbar) that GaN epitaxial material can withstand before relaxation occurs, and an unambiguous determination of the spin-orbit splitting $\Delta_{SO} = 17.0 \pm 1$ meV for the material. These are achieved by analyzing 10K reflectance data for the energy separation of transitions between the uppermost valence bands and the lowest conduction band of wurtzitic GaN as a function of biaxial stress for a series of GaN films grown on both Al_2O_3 and 6H-SiC substrates. Our data explicitly show the nonlinear behavior of the excitonic energy splittings B-A and C-A vs. the energy position of the A exciton, which stands in contrast to the linear approximations used by previous workers analyzing material grown only on Al_2O_3 substrates. Further, the lineshape ambiguities present in GaN reflectance spectra that hindered the accurate determination of such excitonic energies have also been resolved by analyzing these data in reciprocal space, where critical point energies are determined by phase effects to an accuracy of ±0.5 meV.

INTRODUCTION

With the achievement of the first multilayer nitride laser structure,[1] the task of quantifying residual strain in GaN films becomes more pressing. However, to do this by optical methods requires the identification and analysis of A, B, and C excitonic structures when the current low temperature reflectance lineshape theory is plagued by ambiguities. Lineshape ambiguities appear to be of diminished seriousness when only a narrow range of energy is analyzed; however, in this laboratory we studied a variety of GaN layers with thicknesses, buffer layers, and substrates representing a wide range of residual in-plane strain. In each case the customary comparisons to historical lineshapes[2] obtained from 100-200 μm thick GaN films grown on Al_2O_3 were not useful; observed variations in excitonic energy splittings and oscillator strengths rendered ambiguous any assignment based on lineshape alone, in addition to raising the more fundamental issue concerning the correct positions of the excitonic critical point energies in the individual spectra. We solve the latter problem by analyzing our lineshapes in reciprocal space,[3] where critical point energies are determined by phase effects independent of baseline artifacts and lineshape interpretation to an accuracy of ±0.5 meV provided that the features are at least 2 meV apart.

With the energy positions E_A, E_B, and E_C of the A, B, and C excitons known to this precision over a wide range of in-plane strain, parameters such as the spin-orbit splitting V_{SO} can be calculated with increased confidence. For example, we find $V_{SO} = 17.0 \pm 1$ meV, a value significantly larger than most values reported by previous workers [2,4,5] though similar to that obtained by Gil, *et al.*[6] on the basis of a linear approximation to the correct variation of the energies with strain. By examining a fuller range of samples grown on both Al_2O_3 and 6H-SiC, our data unambiguously show the nonlinear dependence of the excitonic energy splittings ΔE_{BA} and ΔE_{BC} on the energy E_A of the A exciton, confirming our earlier application of the Hopfield

Mat. Res. Soc. Symp. Proc. Vol. 449 © 1997 Materials Research Society

model[7], where the nonlinear variation of the valence band energies with crystal field potential and biaxial strain is incorporated explicitly.[8]

EXPERIMENT

Data were obtained with a single-beam low-temperature reflectometer consisting of a Xe arc lamp, a Cary 14 monochromator with a resolution better than 1 meV at 3.4 eV, and an EMI 9558 photomultiplier. We initially used a more complex double-beam arrangement, but the contribution of the optical instrumentation to the optical structure in the spectral region of interest was negligible. Data reported here were obtained on samples cooled to 10K with an Air Products cryotip, although other temperatures were also used to confirm assignments. The GaN layers examined were grown by (1) metallorganic chemical vapor deposition (MOCVD) on Al_2O_3 substrates with 250Å GaN buffer layers; (2) Hydride Vapor Phase Epitaxy (HVPE) on Al_2O_3 substrates without buffer layers, and (3) MOCVD on 6H-SiC substrates with 1000Å AlN buffer layers. These shall be referred to as Category 1, 2, and 3 samples, respectively. Details of crystal growth are given elsewhere.[9,10,11]

RESULTS

A selection of reflectance spectra is shown in Fig. 1 in order of descending $\Delta E_{BA} = E_B - E_A$ excitonic energy separation. Critical point energies determined by reciprocal-space analysis are indicated by the points; the historical lineshape of Dingle, *et al.*[2] is shown for the sake of comparison in (d), with the original excitonic assignments indicated by the arrows. In this case the lineshape trends revealed by reciprocal-space analysis show that the excitonic energy splittings ΔE_{BA} and ΔE_{CA} are closer to 5 and 23 meV, respectively, rather than the 6.0 and 20.5 meV reported in Ref. 2. Note that we observe excitonic splittings both significantly larger and smaller than those reported in Ref. 2.

Figure 1(a) is typical of reflectance spectra of Category (1) films. The feature farthest to the left on the low energy end of the spectrum is an interference oscillation that is unremarkable for our purposes; the excitonic features of interest lie to the right. For this sample ΔE_{BA} and ΔE_{CA} are 9.2 and 27.8 meV, respectively. Interestingly, real-space assignments for the same sample yield ΔE_{BA} and ΔE_{CA} values of 11 and 34 meV. This disparity between real- and reciprocal-space results does not appear to be sample-specific, at least for this category. Indeed, reciprocal-space values obtained for other Category (1) samples of similar thicknesses are listed in Table I. Clearly for small differences in thickness in this group there is a small but non-negligible variation in ΔE_{BA}, though ΔE_{CA} varies more widely. In contrast, as a result of baseline ambiguity, the real-space values for ΔE_{BA} for these samples were all ~11 meV, though the ΔE_{CA} values still showed a relatively wide variation. Perhaps it has been this type of real-space analysis that has led other workers,[4] analyzing data obtained on films grown solely on Al_2O_3, to believe that ΔE_{AB} is essentially invariant. Here, we observe a non-negligible variation of ΔE_{BA} for a sample set varying by only 0.5μm in thickness, though this variation is small compared to the range of ΔE_{CA} values obtainable from seemingly similar samples.

We observe a similar pattern in a sample set within Category (3), a 1.32 μm representative of which is shown in Fig. 1(g). These samples all are in the 1.03 to 1.45 μm thickness range, grown on 6H-SiC with 1000Å AlN buffer layers, and have $\Delta E_{BA} < 2$ meV with ΔE_{CA} values ranging from 14.2 to 20.4 meV. For this set the growth temperature of each sample is given in Table I, providing a rationale for the observed range of ΔE_{CA}. Reciprocal-space analysis

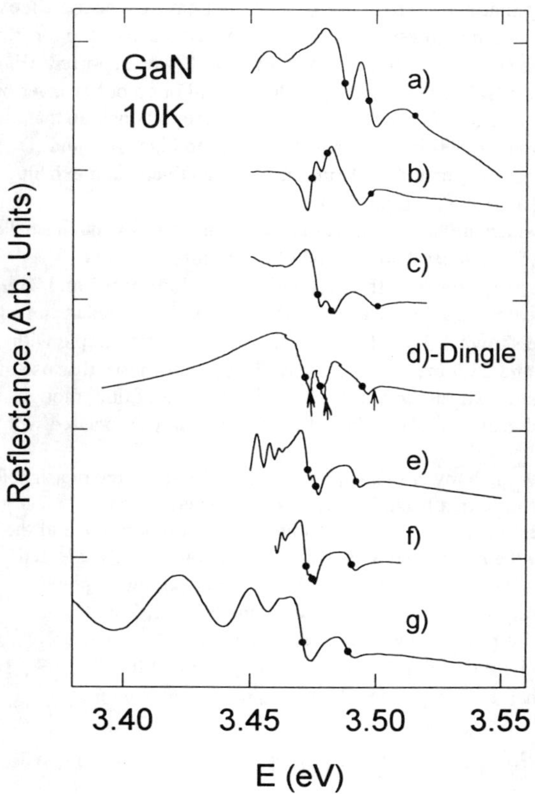

Figure 1. Representative GaN lineshapes. Reciprocal space assignments of threshold energies are given by points. From top, after Table I: (a) Category 1, sample 4; (b) Category 2, sample 5; (c) Category 3, sample 8; (d) Dingle reference lineshape[2]; (e) Category 3, sample 7; (f) Category 3, sample 6; (g) Category 3, sample 12.

Table I. Summary of GaN samples and corresponding growth parameters.

Sample	Category	Growth Technique	Substrate Material	Thickness (μm)		Film Growth Temp. (°C)	E_A (eV)	ΔE_{BA} (meV)	ΔE_{CA} (meV)
1.	I	MOCVD	Al$_2$O$_3$				3.4857 ± 0.5	8.6±1	26.5±1
2.	"	"	"	2.7 + .025 GaN buffer			3.4900	7.1	34.7
3.	"	"	"	2.6 +	"		3.4890	8.3	30.9
4.	"	"	"	2.2 +	"		3.4876	9.2	27.8
5.	II	HVPE	Al$_2$O$_3$	250			3.4743	6.7	23.5
6.	III	MOCVD	6H-SiC	3.7 + .1 AlN buffer		1050	3.4724	2.5	18.1
7.	"	"	"	3.1 +	"	1050	3.4729	3.4	19.7
8.	"	"	"	2.0 +	"	1050	3.4767	6.2	23.9
9.	"	"	"	1.5 +	"	975	3.4726	<2	20.4
10.	"	"	"	1.4 +	"	1025	3.4651	<2	14.2
11.	"	"	"	1.4 +	"	975	3.4680	<2	16.3
12.	"	"	"	1.3 +	"	975	3.4709	<2	18.9
13.	"	"	"	1.3 +	"	1075	3.4668	<2	14.8
14.	"	"	"	1.0 +	"	975	3.4668	<2	18.8

confirmed these assignments by showing that in each case $|\Delta E_{BA}|$ could not have exceeded 2 meV and that ΔE_{CA} is as given in Table I. Our temperature-dependent reflectance data also support this interpretation. These data are also consistent with those of Buyanova *et al.*[12], who reported ΔE_{BA} < 5 meV and ΔE_{CA} = 17 meV for a 2.0µm GaN film grown by MOCVD without a buffer layer on 6H-SiC. However, as in the case of the Category (1) samples, it is premature to conclude that ΔE_{BA} has a single value for material grown on 6H-SiC. The spectra shown in Figs. 1(e) and (f) were obtained from Category (3) samples 3.10µm and 3.71 µm thick, respectively, and exhibit ΔE_{BA}, ΔE_{CA} values of 3.4, 19.7 and 2.5, 18.1 meV, respectively.

Though Category (3) samples, which are all grown on 6H-SiC, tend to show much smaller ΔE_{BA} splittings than those grown on Al_2O_3, a general conclusion of this nature would be premature. Figure 1(c) shows a Category (3) sample with ΔE_{BA} and ΔE_{CA} values of 6.2 and 23.9 meV, respectively. Such values are typical for GaN grown on Al_2O_3[2,6,13-17] and are similar, in fact, to the 250µm thick Category (2) sample shown in Fig. 1(b). For this HVPE grown sample with no buffer layer, ΔE_{BA} and ΔE_{BA} are 6.7 and 23.5 meV, respectively. This result shows that use of a particular type of substrate does not automatically guarantee a particular type of film, though unresolvable ΔE_{BA} splittings for material grown on Al_2O_3 have been traditionally associated with poor material quality.[2]

To interpret these data we follow the conventional approach, i.e., we apply the quasicubic model introduced by Hopfield.[7] We first note that biaxial in-plane stress generates both hydrostatic and shear strains, which affect the mean valence-conduction band separation and the valence-band splittings, respectively. To the extent that anisotropic relaxation can be neglected (not necessarily a good assumption), we can determine the relationship between the in-plane stress $\sigma_{xx} = \sigma_{yy} = \sigma_{11}$ and the in-plane ($\varepsilon_{xx} = \varepsilon_{yy} = \varepsilon_{11} = \varepsilon_{22}$), out-of-plane ($\varepsilon_{zz} = \varepsilon_{33}$), and hydrostatic ($\varepsilon_H = \varepsilon_{11} + \varepsilon_{22} + \varepsilon_{33}$) strains from the known elastic constants of GaN[18]. After Nye,[19] we have $\{C\}^{-1} \equiv \{S\}$ and $\varepsilon_{ij} = S_{ijkl}\,\sigma_{kl}$, where C_{11}, C_{12}, C_{13}, C_{33}, C_{44}, C_{66} = 296, 130, 158, 267, 24 and 83 GPa[18], respectively. Assuming that $\sigma_{11} = \sigma_{22}$ and that all remaining $\sigma_{ij} = 0$ we find

$$\varepsilon_{11} = \left[C_{11} + C_{12} - 2\left(C_{13}^2/C_{33}\right)\right]^{-1}\sigma_{11} = \left(4.18\times10^{-12}\right)\sigma_{11} \tag{1a}$$

$$\varepsilon_{33} = \left(-4.93\times10^{-12}\right)\sigma_{11} \tag{1b}$$

The hydrostatic, and therefore the in-plane strain, can be obtained within an additive constant E_{A0} from the measured gap energy E_A and the deformation potential a according to the empirical expression

$$E_A = E_{A0} + a\,\varepsilon_H \tag{2}$$

where E_{A0} = 3.477 eV,[20] and $a \approx$ -10 eV. The quasicubic model therefore allows the B-A and C-A splittings to be plotted vs. E_A as shown in Fig. 2. Using Eqs. (1) and (2), we find for the data of Fig. 2 that σ_{11} ranges from -3.80 kbar compressive at the high energy limit of the data to 3.47 kbar tensile at the low energy limit. These values (1) provide an estimate of the amount of stress the material can withstand before relaxation occurs and (2) suggest, not surprisingly, that the epitaxial material can withstand roughly equal amounts of tensile and compressive stress. Recalculation of ranges of stress reported by previous workers[6] according to Eqs. (1) - (2) using the reported energy position of the A exciton shows that the previously reported range of -12.8 to -0.8 kbar (compressive) actually ranges from -3.78 kbar (compressive) to 1.16 kbar (tensile).

Given ε_{xx}, and following the notation of ref. 5, ΔE_{BA} and ΔE_{CA} can be written in the quasicubic model as

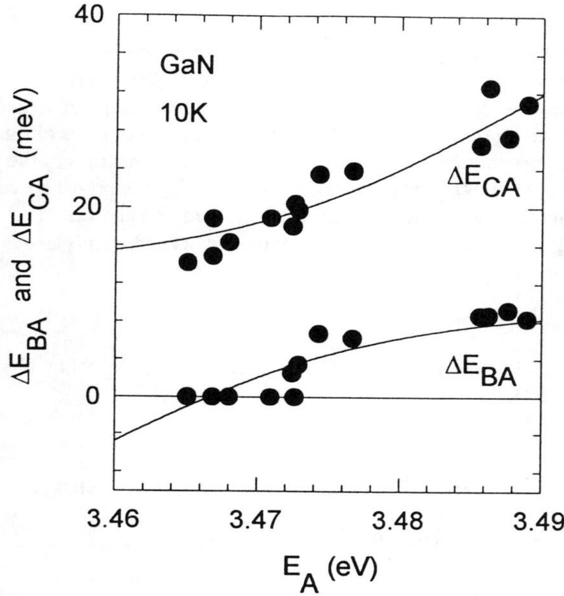

Figure 2. Excitonic energy splittings ΔE_{BA} and ΔE_{CA} vs. energy position of the A exciton; data shown by points. Least squares fit shown by straight lines. Note the nonlinear character of the fit.

$$\Delta E_{BA}, \Delta E_{CA} = (E_B - E_A) = 2\Delta_2 + \frac{1}{2}(\Delta_1 - \Delta_2 + \Theta_\varepsilon)$$

$$\pm \sqrt{\left(\frac{\Delta_1 - \Delta_2 + \Theta_\varepsilon}{2}\right)^2 + 2\Delta_3^2} \tag{3}$$

where $\Theta_\varepsilon = k\varepsilon_{xx}$ is the shear term, Δ_1 is the crystal-field potential, and $\Delta_2 = \Delta_3 = 1/3\Delta_{SO}$, where Δ_{SO} is the conventional spin-orbit splitting parameter. Note that Δ_{so} is the A-C splitting where $E_A=E_B$, i.e., where the shear and crystal-field terms cancel. Since ΔE_{AB} and ΔE_{AC} are determined experimentally, and since the independent variable Θ_ε can be expressed in terms of ε_{xx}, then $\Delta_1 = \Delta_{cr}$ and $\Delta_2 = \Delta_3 = 1/3\ \Delta_{so}$ are simply adjustable parameters for the least-squares fit shown in Fig. 2 by the solid lines. The data of Fig. 2 clearly cannot be represented by a straight line. The calculated variation is approximately parabolic, in contrast to linearization formulas offered by previous workers[6] which had been previously used to obtain the relationships between GaN deformation potentials. This has recently also been recognized by Gil *et al.*[21] in a revision of their earlier work on this topic. Although our wider range of data show more scatter than the more narrow range of previous workers,[6,21] we believe (1) that this scatter is due to partial violation of the 1:1 connection assumed between hydrostatic and shear strain and (2) that it is logical to assume that any partial, anisotropic relaxation would occur differently for each sample represented in a set of dissimilar samples.

CONCLUSION

We have presented a range of ΔE_{AB} and ΔE_{AC} splitting data for a variety of GaN films grown on both Al_2O_3 and 6H-SiC, all nominally varying with respect to film thickness and buffer layer composition. Some of these splittings are heretofore unreported. In fact, the range of values observed supports the conclusion that no a priori correlation between splittings, lineshapes of the excitons, and type of substrate exists. Unambiguous determination of the energy positions of the A, B, and C excitons in these spectra to within 0.5 meV has been achieved using reciprocal-space analysis, which has in turn enabled a more accurate determination of V_{so}. Additionally, the correlation between E_A and ΔE_{AB} and ΔE_{AC} has enabled the calculation of the residual in-plane stress in the films.

ACKNOWLEDGEMENTS

It is a pleasure to acknowledge the financial support of the Office of Naval Research (ONR) under contract number N-00014-93-1-0255.

REFERENCES

[1] S. Nakamura, M. Senoh, S. Nagahama, N. Iwasa, T. Yamada, T. Matsushita, H. Kiyoku, and Y. Sugimoto, Jpn. J. Appl. Phys. **35**, L74 (1996).

[2] R. Dingle, D.D. Sell, S.E. Stokowski, and M. Ilegems, Phys. Rev. B **4**, 1211 (1971).

[3] D.E. Aspnes, Surf. Sci. **135**, 284 (1983).

[4] J.W. Orton, Semicond. Sci. Technol. **11**, 1026 (1996).

[5] S.L. Chuang and C.S. Chang, Phys. Rev. B **54**, 2491 (1996).

[6] B. Gil, O. Briot, and R.L. Aulombard, Phys. Rev. B **52**, R17 028 (1995).

[7] J.J. Hopfield, J. Phys. Chem. Solids **15**, 97 (1960).

[8] N.V. Edwards, M.D. Bremser, T.W. Weeks, Jr., R.S. Kern, H. Liu, R.A. Stall, A.E. Wickenden, K. Doverspike, D.K. Gaskill, J.A. Freitas, Jr., U. Rossow, R.F. Davis, and D.E. Aspnes, Proc. Mat. Res. Soc. **395**, Boston, Ma., 1995.

[9] T.W. Weeks, Jr., M.D. Bremser, K.S. Ailey, E. Carlson, W.G. Perry, and R.F. Davis, Appl. Phys. Lett. **67**, 401 (1995).

[10] EMCORE Corporation, Somerset, NJ 08873.

[11] N.R. Perkins, M.N. Horton, and T.F. Keuch, Proc. Mat. Res. Soc. **395**, Boston, Ma, 1995.

[12] I.A. Buyanova, J.P. Bergman, B. Monemar, H. Amano, and I. Akasaki, Appl. Phys. Lett. **69**, 1255 (1996).

[13] B. Monemar, Phys. Rev. B **10**, 676 (1974).

[14] C.I. Harris, B. Monemar, H. Amano, and I. Akasaki, Appl. Phys. Lett. **67**, 840 (1995).

[15] W. Shan, T.J. Schmidt, X.H. Yang, S.J. Hwang, J.J. Song, and B. Goldenberg, Appl. Phys. Lett. **66**, 985 (1995).

[16] S.N. Mohammad, A.A. Salvador, and H. Morkoç, Proc. IEEE **83**, 1306 (1995).

[17] D.C. Reynolds, D.C. Look, W. Kim, Ö. Aktas, A. Botchkarev, A. Salvador, H. Morkoç, and D.N. Talwar, J. Appl. Phys. **80**, 594 (1996).

[18] *Properties of Group III Nitrides*, edited by J.H. Edgar (INSPEC, IEEE, London, 1994).

[19] J.F. Nye, *Physical Properties of Crystals* (Clarendon Press, Oxford, 1985).

[20] K. Pakula, A. Wysmolek, K.P. Korona, J.M. Baranowski, R. Stepniewski, I. Gregory, M. Bockowski, J. Jun, S. Krukowski, M. Wroblewski, and S. Porowski, Solid State Commun. **97**, 919 (1996).

[21] B. Gil, F. Hadani and H. Morkoç, Phys. Rev. B **54**, 7678 (1996).

BULK AND SURFACE ELECTRONIC STRUCTURE OF GAN MEASURED USING ANGLE RESOLVED PHOTOEMISSION, SOFT X-RAY EMISSION AND SOFT X-RAY ABSORPTION

KEVIN E. SMITH,*[†] SARNJEET S. DHESI,* LAURENT -C. DUDA,* CRISTIAN B. STAGARESCU,*[‡] J.H. GUO,** JOSEPH NORDGREN**, RAJ SINGH*** and THEODORE D. MOUSTAKAS***
* Department of Physics, Boston University, Boston, MA 02215
** Department of Physics, Uppsala University, Uppsala, Sweden
*** Department of Electrical and Computer Engineering, Boston University, Boston, MA 02215.

ABSTRACT

The electronic structure of thin film wurtzite GaN has been studied using a combination of angle resolved photoemission, soft x-ray absorption and soft x-ray emission spectroscopies. We have measured the bulk valence *and* conduction band partial density of states by recording Ga L- and N K- x-ray emission and absorption spectra. We compare the x-ray spectra to a recent *ab initio* calculation and find good overall agreement. The x-ray emission spectra reveal that the top of the valence band is dominated by N $2p$ states, while the x-ray absorption spectra show the bottom of the conduction band as a mixture of Ga $4s$ and N $2p$ states, again in good agreement with theory. However, due to strong dipole selection rules we can also identify weak hybridization between Ga $4s$- and N $2p$-states in the valence band. Furthermore, a component to the N K-emission appears at approximately 19.5 eV below the valence band maximum and can be identified as due to hybridization between N $2p$ and Ga $3d$ states. We report preliminary results of a study of the full dispersion of the bulk valence band states along high symmetry directions of the bulk Brillouin zone as measured using angle resolved photoemission. Finally, we tentatively identify a non-dispersive state at the top of the valence band in parts of the Brillouin zone as a surface state.

INTRODUCTION

Despite the technological importance of refractory III-V nitrides as materials for wide band gap semiconductor devices, there is a significant lack of experimental data concerning the basic electronic structure of such films [1]. Only recently have the intrinsic band structure, the density of states, and the electronic properties of surfaces come under scrutiny. A fundamental understanding of the electronic structure is required if these materials are to achieve their full technological potential. We present here preliminary results of a comprehensive study of the electronic structure of wurtzite (hexagonal) GaN thin films using a combination of angle resolved photoemission, soft x-ray emission, and soft x-ray absorption spectroscopies. This combination of spectroscopies allow the partial density of states (PDOS) of both the valence and conduction bands, *and* the dispersion of the valence band states to be directly measured. Such measurements can be directly compared to electronic structure calculations.

SPECTROSCOPIC TECHNIQUES

Three complimentary techniques were used in this work: angle resolved photoemission (ARP), soft x-ray emission (SXE) and soft x-ray absorption (SXA). Each reveals a different aspect of the GaN electronic structure. In ARP the sample is illuminated with monochromatic ultraviolet radiation, which leads to the photoemission of electrons. The kinetic energy and momentum of these

Mat. Res. Soc. Symp. Proc. Vol. 449 © 1997 Materials Research Society

electrons is then measured. The kinetic energy can be related to the binding energy of the electron inside the solid, while the electron momentum in vacuum can be related to the band momentum (k vector) of the electron inside the solid. Given the short mean free path of electrons in solids at the kinetic energies typically encountered in ARP, this spectroscopy is inherently surface sensitive, with a sampling depth of on the order of 10 Å. Since bulk wavefunctions extend continuously to the surface in a perfect crystal, ARP can nevertheless measure bulk electronic states. However, great care has to be taken to prepare and maintain clean, ordered and stoichiometric surfaces if information on the bulk electronic structure is sought. Naturally, the high degree of surface sensitivity of ARP enables the surface electronic structure to be measured easily. In so-called normal emission, only electrons emitted normal to the surface are measured as the photon energy is varied. These electrons have no momentum parallel to the surface, and thus states along a direction in the bulk Brillouin zone perpendicular to the surface are probed. States along other directions in the zone can be measured by coordinated variation of off normal detection angles and photon energies. Many reviews of the use of ARP can be found in the literature [2].

In the SXE experiment the sample is illuminated with monochromatic synchrotron radiation in the 100 - 1000 eV range. For a fixed photon energy, core holes are selectively created and some of these de-excite by the radiative decay of a valence electron into the hole [3]. By use of a suitable high resolution spectrometer, the spectrum of emitted photons can be measured [4]. The transition is governed by dipole selection rules, and assuming a non dispersive core hole, the resulting spectrum of emitted photons reflects the occupied valence band PDOS [4]. The sampling depth for SXE in this energy range is on the order of 1000 Å, and so it is the bulk electronic structure that is being probed.

Finally, in the SXA experiment, the sample is again illuminated with monochromatic radiation in the 100 - 1000 eV range. However, in this case the photon energy is swept through an x-ray absorption edge and the current through the sample is measured. The photons excite core electrons into the conduction band, and again assuming a non-dispersive core state, the current is proportional to the PDOS of the unoccupied conduction band [5]. Note that the resolution in SXE is primarily determined by the resolution of the emission spectrometer, while the resolution in SXA is determined by that of the synchrotron radiation monochrometer.

EXPERIMENTAL DETAILS

The wurtzite GaN(001) films were grown using electron cyclotron resonance assisted molecular beam epitaxy on sapphire substrates; the growth procedure has been reported elsewhere [6]. Samples were auto doped n-type and of high quality as determined from resistivity, mobility, carrier concentration and photoluminescence measurements. SXA/SXE measurements were performed at the HASYLAB synchrotron, DESY, Hamburg, on undulator beamline BW3, which is equipped with a modified SX-700 monochrometer. Absorption spectra were recorded in the total electron yield mode by measuring the sample drain-current and were taken with energy resolutions of 0.13 eV at 400 eV (in the vicinity of the N $1s$ edge) and 1.0 eV at 1000.0 eV (in the vicinity of the Ga $2p$ edge). Emission spectra were recorded using a high resolution grazing-incidence grating spectrometer [7]. A 5 m spherical grating with 1200 lines/mm was used for wavelength dispersion resulting in resolutions of 1.1 eV at photon energies of 400.0 eV and 1.7 eV at photon energies of 1000 eV. The base pressure in the SXE/SXA chamber was 1.0×10^{-8} Torr. This vacuum is sufficient for these bulk probes. Sample surfaces were not atomically cleaned for SXA and SXE.

ARP experiments were performed at the National Synchrotron Light Source (NSLS), on the Boston University/North Carolina State University/NSLS bending magnet beamline U4A, which is equipped with a 6 m toroidal grating monochrometer and a custom designed hemispherical electron analyzer [8]. Typical energy and full angular resolution were 100 meV and 1°, respectively. Samples were transported in air and mounted in the analyzer chamber which has a base pressure $< 2 \times 10^{-10}$ Torr. Clean GaN(0001) surfaces were prepared following the procedure of Bermudez *et al* [9]. A Ga layer is evaporated onto the surface, then removed by heating to 850°C. This results in the simultaneous removal of much of the surface O as revealed by Auger electron spectroscopy (AES). The sample is then subjected to repeated cycles of sputtering with 1000 eV N^+ ions and annealing in ultra-high vacuum at 850°C. Following this procedure AES shows no C and minimal O contamination, and low energy electron diffraction (LEED) displays a sharp hexagonal pattern with no evidence of surface reconstruction. All measurements were performed at room temperature. Binding energies are referenced to E_F determined from an atomically clean Ta foil in electrical contact with the sample, and to the valence band maximum (VBM) as determined from the spectra.

RESULTS AND DISCUSSION

<u>N Partial Density of States</u>

Figure 1A shows the SXE spectrum for the radiative decay of the valence band into a N 1s hole (N *K* emission), and the SXA spectrum for excitation of N 1s electrons into the empty conduction band (N *K* absorption) [10]. In the SXE spectrum, the N 1s core state is excited by 430 eV photons.

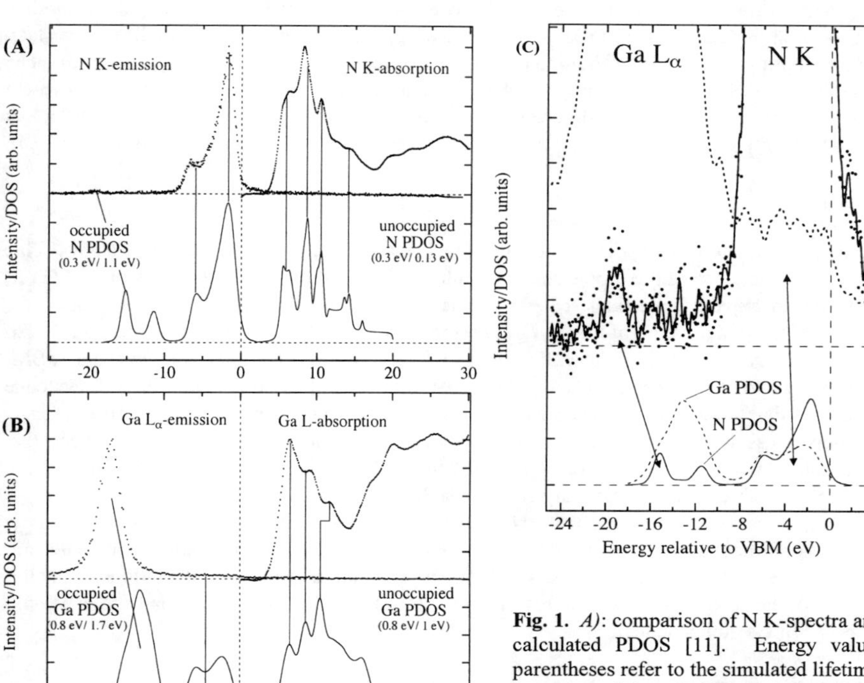

Fig. 1. *A)*: comparison of N K-spectra and the calculated PDOS [11]. Energy values in parentheses refer to the simulated lifetime and instrumental broadening of the PDOS. *B)*: similar comparison for Ga L-spectra and the calculated PDOS. *C)*: same emission data on an expanded vertical scale. See text for details.

Details of how the emission and absorption spectra are placed on a common binding energy scale relative to the VBM can be found in Ref. 10. Note that dipole selection rules imply that the emission spectrum is the N 2p PDOS for the valence band. Emission from Ga 4s states is dipole forbidden, while the crosssection for inter-atomic radiative decays is much smaller than for an intra-atomic decay, resulting in a negligable contibution from Ga 4p decays. Also shown in Fig. 1A are the results of an LDA calculation of the PDOS for wurtzite GaN [11]. Here we have convoluted the theoretical DOS by Lorentzians to simulate the core-hole lifetime broadening and then convoluted the theoretical DOS by Gaussians to simulate the instrumental broadening. The agreement between the calculated and measured PDOS for the valence and conduction bands is very good. Structure in the emission spectra at binding energies below the valence band minimum will be discussed below. Note that Lambrect *et al* have also recently reported a study of electronic structure in GaN using x-ray absorption and reflectivity [12]. Our absorption data are in good agreement with that study for similar polarizations of the incident synchrotron radiation.

Ga Partial Density of States

Figure 1B shows the SXE spectrum for the radiative decay of the states near and in the valence band into a Ga $2p_{3/2}$ hole (Ga L_α emission), and the SXA spectrum for excitation of Ga $2p_{3/2}$ electrons into the empty conduction band (Ga L absorption) [10]. As before, details concerning the common binding energy scale can be found in Ref. 10. Here, emission from N 2p states is dipole forbidden, and the spectrum reflects the Ga PDOS of the valence band, including the Ga 3d shallow core state, with the exception that Ga 4p emission is dipole forbidden and missing from the spectrum. The calculated Ga PDOS, with broadening as before, is also shown in Fig. 1B [11]. Again, agreement between the calculated and measured PDOS for both the valence and conduction bands is very good. The binding energy of the Ga 3d is similar to that observed in photoemission experiments [12,13].

Hybridization

A unique characteristic of SXE is the ability to directly observe hybrid electronic states. This is simply a consequence of the dipole selection rules. Fig. 1C shows the N K emission and Ga L_α emission on an expanded vertical scale; the calculated N and Ga PDOS are also plotted for comparison [11]. Examination of the Ga L_α emission shows that it is dominated by the Ga 3d → 2p transition at about 18 eV below the VBM. However, emission is clearly visible up to the VBM, where it ceases abruptly. Since the Ga 4p → 2p transition is forbidden, this emission in the valence band region is due to the Ga 4s → 2p transition. Note that this spectrum indicates a uniform Ga 4s contribution across the entire valence band. Likewise, examination of the N K emission reveals that it is dominated by the N 2p → 1s transition, giving the N PDOS for the valence band. However, a weak signal is visible in the energy range of the Ga 3d states. Again, dipole selection rules imply that this signal can only originate from N 2p states hybridized into the Ga 3d states. This is conclusive experimental evidence of the (weak) valence state nature of this shallow core state, and indicates the need to explicitly treat this hybridization in band structure calculations. Finally, comparison of the absorption spectra in Figs 1A and 1B indicate that the conduction band is heavily mixed between Ga and N states.

Bulk Band Dispersion

Bulk band dispersion in GaN can be studied using ARP. Ding *et al* have measured the band dispersion along selected directions in cubic GaN films grown on GaAs by using normal emission

ARP [15]. We have very recently measured the band structure of wurtzite GaN along all the high symmetry directions [14]. Figure 2 shows a series of normal emission spectra from GaN where the photon energy is varied from 29 to 100 eV. Since the surface of the film indexes as (0001), the bulk Brillouin zone direction probed in this experiment is $\Gamma\Delta A$. Clear dispersion of the valence band states is visible, and variation of the band width with location along Δ can also be determined. By measuring in an off-normal mode, dispersion along $\Gamma\Sigma M$, $\Gamma T K$, and from K to M has also been measured. Preliminary analysis of the data indicates very good agreement for the magnitude of the dispersion with the band structure calculations of Rubio *et al* [16]. The theoretical and experimental binding energies can be brought into reasonable agreement by a rigid shift of the calculated bands to 1 eV higher energy. Full analysis of the results of our experiment will be reported elsewhere [14].

Surface State

One glaring disagreement between our measurements and band structure calculations is our observation a state at a binding energy equivalent to the valence band maximum throughout much of the bulk Brillouin zone. The calculated band structures all indicate that the bands forming the top of the valence band disperse to higher binding energies by as much as 2 to 2.5 eV away from the Γ point. In

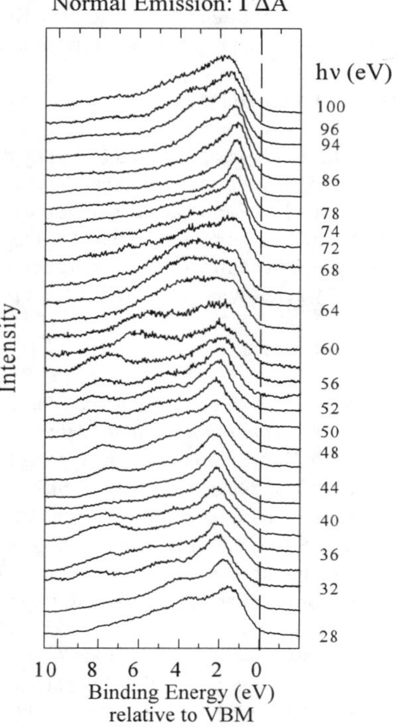

Fig. 2: ARP normal emission spectra from GaN(0001)

contrast, we observe strong, non-dispersive emission at the VBM throughout the surface Brillouin zone. In order to test if the origin of this emission is a surface state, we dosed the surface with O_2. The $\bar{K}$ point emission in the surface Brillouin zone at the VBM was found to be sensitive to the presence of adsorbed oxygen, consistent with identification as a surface state. Bermudez *et al* have studied the adsorption of O_2 on GaN(0001) using angle integrated photoemission, and have reported evidence for the existence of a surface state at the VBM [17]. However, angle integrated photoemission is unable to measure band dispersion, nor the location of states in the surface Brillouin zone. Our preliminary measurements indicate that the state is non-dispersive, and thus has a high density of states at the VBM, and is located throughout the zone. Further measurements are underway to identify the symmetry properties of this state.

CONCLUSIONS

We have reported a study of the electronic structure of wurtzite GaN using angle resolved photoemission, soft x-ray absorption and soft x-ray emission spectroscopies. We have measured the bulk valence and conduction band PDOS and find good agreement with calculations. We have also measured the hybridization between Ga *4s*- and N *2p*-states in the valence band, and between N *2p* and Ga *3d* states. We also reported preliminary results of our ARP study of band dispersion along high symmetry directions of the bulk Brillouin zone, and have tentatively identified an observed non-dispersive state at the top of the valence band throughout the Brillouin zone as a surface state.

ACKNOWLEDGMENTS

This work was supported in part by the National Science Foundation under DMR-9504948 and INT-9515370. Part of the research was also supported by the Swedish Natural Research Council (NFR) and the Göran Gustafsson Foundation. Photoemission experiments were undertaken at the NSLS which is supported by the U.S. Department of Energy, Divisions of Materials and Chemical Sciences. X-ray emission and absorption experiments were performed at HASYLAB/DESY. L.C.D gratefully acknowledges the NFR for a postdoctoral fellowship at Boston University.

REFERENCES

†: *author to whom correspondence should be addressed. Electronic mail: ksmith@bu.edu*
‡: *On leave from the Institute of Microtechnology, Bucharest, Romania.*
1. **Wide Band Gap Semiconductors**, MRS Symp. Proc. **242**, Ed. T.D. Moustakas, J.I. Pankove, and Y. Hamakawa, (1992); S. Strite and H. Morkoç H, J. Vac. Sci. Technol. B **10**, 1237 (1992)
2. See for example: **Angle Resolved Photoemission**, Ed. S.D. Kevan, Elsevier, Amsterdam, 1991; K.E. Smith and S.D. Kevan, Prog. Solid State Chem. **21**, 49 (1991).
3. L.G. Paratt, Rev. Mod. Phys. **31**, 616 (1959).
4. J. Nordgren and N. Wassdahl, Phys. Scr. **T31**, 103 (1989); J. Nordgren, J. Physique **C9**, 693 (1987). T.A. Callcott, C.H. Zhang, D.L. Ederer, D.R. Mueller, J.E. Rubensson and E.T. Arakawa, Nuc. Inst. Methods **A291**, 13 (1990).
5. See for example **X-Ray Absorption: Principles, Applications, Techniques of EXAFS, SEXAFS, and XANES,** Ed. by D.C. Koningsberger, R. Prins (Wiley, New York, 1988).
6. T. Lei, M. Fanciulli, R.J. Molnar, T.D. Moustakas, R.J. Graham, and J. Scanlon, App. Phy. Lett. **59**, 944 (1991).
7. J. Nordgren, G. Bray, S. Cramm, R. Nyholm, J-E. Rubensson, and N. Wassdahl, Rev. Sci. Instrum. **60**, 1690 (1989); J. Nordgren and R. Nyholm, Nuc. Inst. Methods, **A246**, 242 (1986)
8. S.D. Kevan, Rev. Sci. Instrum. **54**, 1441 (1983).
9. V.M. Bermudez, R. Kaplan, M.A. Khan and J.N. Kuznia, Phys. Rev. B **48**, 2436 (1993)
10. C. B. Stagarescu, L.-C. Duda, K.E. Smith, J.H. Guo, J. Nordgren, R. Singh and T.D. Moustakas, Phys. Rev. B **54** (1996) (in press).
11. Y-N. Xu and W.Y. Ching, Phys. Rev. **B48**, 4335 (1993).
12. W. R. L. Lambrecht, S. N. Rashkeev, B. Segall, K. Lawniczak-Jablonska, T. Suski, E. M. Gullikson, J. H. Underwood, R. C. C. Perera, J. C. Rife, I. Grzegory, S. Porowski, and D. K. Wickenden, Phys. Rev. B (in press).
13. W.R.L. Lambrecht, B. Segall, S. Strite, G. Martin, A. Agarwal, H. Morkoc, and A. Rockett, Phys. Rev. B **50**, 14155 (1994)
14. S.S. Dhesi, C.B. Stagarescu, K.E. Smith, R. Singh and T.D. Moustakas (unpublished).
15. S.A. Ding, G. Neuhold, J.H. Weaver, P. Häberle, K. Horn, O. Brandt, H. Yang, and K. Ploog, J. Vac. Sci. Technol. A **13**, 819 (1996).
16. A. Rubio, J.L. Corkill, M.L. Cohen, E.L. Shirley, and S.G. Louie, Phys. Rev. B **48**, 11810 (1993).
17. V.M. Bermudez, J. Appl. Phys. 1996.

YELLOW LUMINESCENCE AND ASSOCIATED ODMR IN MOVPE GaN: A COMPARISON OF DEFECT MODELS

P.W. MASON[*], A. DÖRNEN[**], V. HAERLE[**], F. SCHOLZ[**], and G.D.WATKINS[*]
[*]Dept. of Physics, Lehigh University, Bethlehem, PA 18015
[**]IV Physikalisches Institut, Universität Stuttgart, D-70550 Stuttgart, Germany

ABSTRACT

Two positive ODMR resonances are commonly observed on a luminescence band in GaN at 2.2 eV, one identified as a shallow donor, the other currently unidentified. We here report a study of their dependencies on a variety of experimental parameters, including microwave modulation frequency, microwave power, photoexcitation power and photoexcitation energy. ODMR simulations using two theoretical models are compared to experimental results which are consistent with spin-dependent recombination between the two defects, assuming the donor has a spin-lattice relaxation time shorter than the spin-dependent recombination lifetime. The photoexcitation energy dependence suggests that the spin-dependent recombination associated with the 2.2 eV band is not the same recombination that is responsible for the luminescence. This supports the two stage model put forth by Glaser et al. for the luminescence process.

INTRODUCTION

The III-V nitrides have a variety of technologically important applications including use in blue and UV LED's and lasers and are promising candidates for high temperature, frequency, and power electronic devices. Under photoexcitation, GaN commonly emits a broad yellow luminescence band centered near 2.2 eV[1], but the precise nature of the defects responsible have not been clearly identified. Two popular models found in the literature offer different recombination routes for the production of this luminescence. The Ogino model, based upon luminescence studies, suggests radiative transfer of an electron from a shallow donor to a deep acceptor (distant d-a pair recombination).[1] The Glaser model, based primarily on Optically Detected Magnetic Resonance (ODMR) studies, suggests a two step decay process.[2] Specifically, an initial spin-dependent (SD) non-radiative electron transfer between a shallow effective mass (EM) donor and a deep double-donor (DD) is followed by radiative electron transfer from the deep DD to a shallow acceptor. The main arguments supporting the Glaser model concern g-value analysis and luminescence energy considerations. Recently, several additional ODMR studies[3,4,5] and hydrostatic pressure measurements[6] have further investigated the recombination nature of the 2.2 eV band, but with yet no consensus as to the correct model.

Here we study a variety of aspects of the ODMR signals to probe whether or not they are sensitive to the recombination processes involved. Microwave modulation frequency, microwave power, and photoexcitation (PE) intensity dependencies, as well as effects from varying the PE energy are considered. Numerical simulations of the SD transfer process and how it shows up in the ODMR for both the Ogino and Glaser models are presented and compared to the experimental findings.

EXPERIMENTAL RESULTS

The samples were grown using the MOVPE technique on a sapphire substrate with a thin AlN buffer layer. They have a wurtzite structure (bandgap of 3.5 eV) and are not intentionally doped. Seven samples were studied, but the majority of the data to be discussed was measured using one of the highest quality samples. All photoluminescence measurements were taken using

Mat. Res. Soc. Symp. Proc. Vol. 449 © 1997 Materials Research Society

an attenuated beam from a 20 mW HeCd laser for PE (325 nm), unless otherwise specified. The measurements were performed at pumped liquid helium temperatures (T≅1.7 K) in a 35 GHz ODMR spectrometer which has been described previously.[7]

The epilayers were typically found to emit three luminescence bands that are commonly observed for high quality GaN[2]: near bandgap (related to excitonic decay), shallow d-a pair recombination (zero-phonon line at 3.27 eV) and the broad 2.2 eV luminescence. The intensity of the near bandgap luminescence was found to scale linearly with PE power over three orders of magnitude, while the 2.2 eV band scaled roughly as the square root of power over the same range. Such a sub-linear dependence is expected for easily saturated distant pair recombination. Further, both recombination models mentioned in the introduction predict a spectral energy shift vs. PE power associated with d-a pair luminescence. However, in our samples a broad background existed in this spectral region making the necessary comparisons difficult. Others have also addressed this issue[3,5], with similar uncertainties.

The remaining experimental results in this paper refer only to the 2.2 eV band. Shown in Figure 1 are two ODMR spectra taken on this band using relatively low PE intensity (~ 20 μW), at two microwave modulation frequencies. The g-values measured for these two resonances agree within experimental error with those reported by Glaser et al.[2] The high and low field line g-values are approximately 1.95 and 1.99, respectively (referred to hereafter as the EM and g=1.99 resonances) and were attributed by Glaser et al. to the shallow EM donor[8] and a deeper DD. The spectral dependencies of both ODMR resonances follow the shape of the 2.2 eV band. Both resonances broaden as the modulation frequency is increased, and a very similar behavior is observed if the PE power is increased.

The amplitudes of the EM and g=1.99 resonances behave very similarly vs. PE power. Both increase approximately with the square root of power until they level off at ~ 0.5 mW power. The luminescence intensity was already noted to increase roughly with the square root of PE power, and thus the ODMR normalized to the 2.2 eV luminescence intensity is approximately constant at low PE power, but decreases above ~0.5 mW.

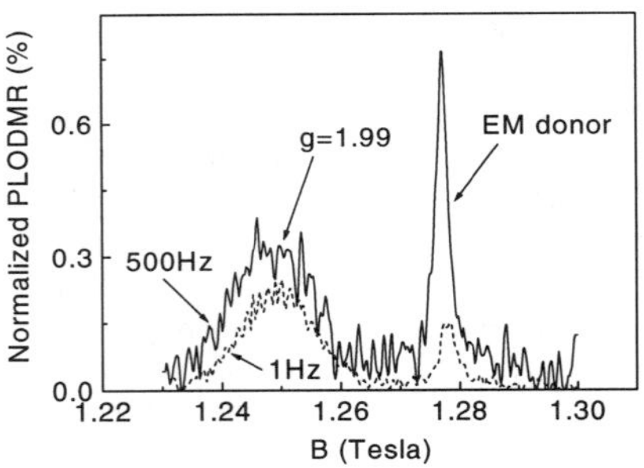

Figure 1. ODMR spectra of the 2.2 eV luminescence band at two microwave modulation frequencies, 1 Hz and 1 kHz; with low intensity (20 μW) PE at 325 nm. A slowly increasing background signal vs. magnetic field has been subtracted.

Also at low PE power (~ 20 μW) and low modulation frequency (1 Hz), a microwave power dependence over two orders of magnitude has been measured. The two resonances behave differently as shown in Figure 2. First the integrated intensity of the g=1.99 resonance increases with the square root of the microwave power. This dependence is consistent with an inhomogeneously broadened resonance where only the individual spin packets near resonance are saturated, the ODMR signal increasing linearly with microwave field amplitude (square root of the microwave power) as it progressively widens the field range of saturated spin packets. The inhomogeneously broadened character of this resonance has been commented on before.[2] On the other hand, the linear dependence observed for the EM resonance in Figure 2 is consistent with a homogeneously broadened

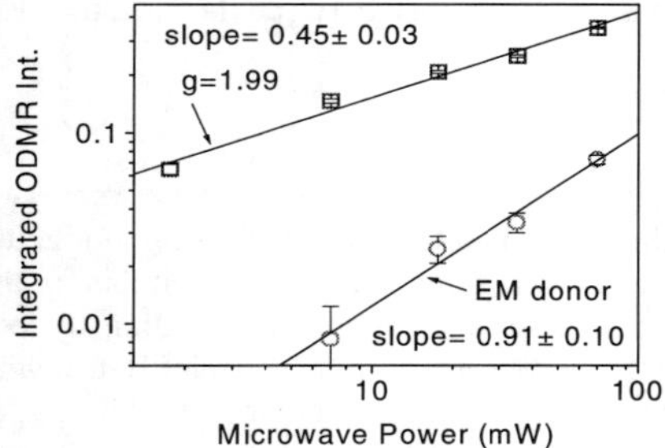

Figure 2. ODMR integrated intensities for the EM and g=1.99 resonances vs. microwave power taken at relatively low (~20 μW) PE power and low modulation frequency (1 Hz). The lines are least squares linear fits.

resonance that is not saturated. EPR measurements have confirmed that this resonance is difficult to saturate.[8] In the literature a wide variety of intensity ratios for the EM and g=1.99 resonances have been reported. The operating conditions may vary greatly from one experimental setup to another, and this different saturation behavior may partially explain this discrepancy. However, we also note differences in the ratios of the two signals amongst our different samples taken under otherwise identical conditions, which appears to be the case also for some of the samples described in reference 2. This suggests the possible importance of competing processes involving the EM donor, as well.

A study of the luminescence and ODMR intensities as a function of the PE energy has been performed using some of the prominent Hg lines of a Hg-Xe arc lamp passed through a monochromator. We find that the 2.2 eV luminescence band can be excited effectively using either above (3.50 eV) or below (3.07 or 2.84 eV) bandgap PE energies. A drastic change is found, however, when we compare their respective ODMR intensities. When using below bandgap PE energy, the ODMR signals normalized to the total photoluminescence intensity ($\Delta I/I \sim 3\times10^{-4}$ at 3.07 eV, $\sim 1\times10^{-4}$ at 2.84 eV) are at least an order of magnitude weaker than those using above bandgap excitation ($\Delta I/I \sim 3\times10^{-3}$). This result implies that there is a way to excite the luminescence that does not involve the SD process (perhaps direct excitation of a pair). This is strong evidence that the SD and luminescence recombinations are separate processes, supporting that part of the Glaser model.

Modulation frequency dependencies of the ODMR signals, at both high (~2 mW) and low (~20 μW) PE power have also been measured and are shown in Figure 3. The relative intensities of the two resonances vary dramatically, and the EM signal peaks at a higher frequency than the g=1.99 signal. These differences in response are not observed, for example, in the well studied electron transfer d-a pair system comprised of zinc-vacancy--zinc-interstitial Frenkel pairs of different separations in ZnSe.[9,10] The question naturally arises therefore as

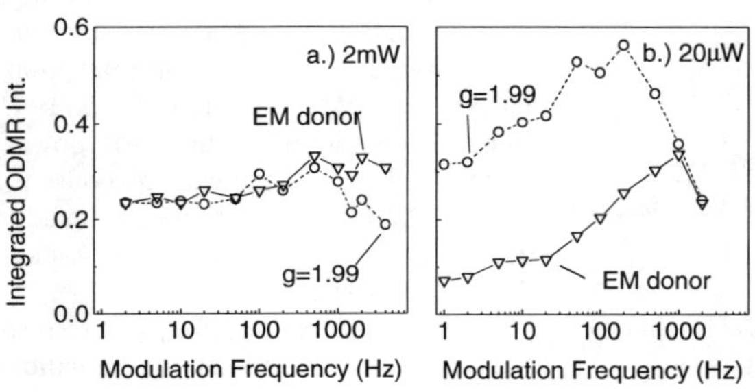

Figure 3. Integrated ODMR intensities for the EM and g=1.99 resonances vs. microwave modulation frequency taken at two PE powers: (a) high (~2 mW), and (b) low (~20μW).

to whether they are consistent with such a recombination, as proposed for the SD process in both the Ogino and Glaser models, and if, perhaps, they serve to distinguish between them. In the next section, we attempt to answer this question.

NUMERICAL SIMULATION

Here we describe the results of a simple model for SD recombination between two defects, and its effect on ODMR detection for both the Ogino and Glaser models. For simplicity, we have considered an isolated pair of defects, each (d, a) with an unpaired s = ½ spin before the recombination. There are four combinations for the relative spin alignments, leading to four excited states. Two states, with both spins up or down together, are pure S =1 triplet states, with a slower recombination rate (R_{slow}). The other two are mixtures of singlet and triplet with a faster recombination rate (R_{fast}). The PE rates were set equal (P) for all the excited states. Arbitrary spin-lattice relaxation rates for the two spins (T_{1d}^{-1}, T_{1a}^{-1}) were included, as well as the microwave-induced transition rates, which were assumed to be fast compared to the other rates. This is sufficient for the Ogino model, where the SD transfer is the radiative transition. In the Glaser model, a second luminescence decay rate (R_{lum}) was included after the SD transfer. For each model appropriate rate equations were developed, following an approach previously outlined.[11] Square wave modulation of the microwave power at a particular frequency was added, giving the rate equations an explicit time dependence and the steady-state time dependent populations of the system were solved for numerically. The time-dependent luminescence intensity was then calculated and convoluted with both in-phase and quadrature sine functions of the same frequency as the microwave amplitude modulation (simulating the lock-in amplifier). The values obtained were the simulated ODMR signals.

1.Ogino model. It is instructive to first consider the limit where neither defect spin system is thermally relaxed ($R_{eff} = R_{slow} + R_{fast} >> T_{1d}^{-1}, T_{1a}^{-1}$). The two resonances exhibit identical frequency responses since their rate equations are symmetric. From Figure 4, however, we see that this is no longer the case when one of the defects is relaxed. In the case shown, the donor defect (d) is relaxed but not the partner (a) ($T_{1d}^{-1} >> R_{eff} >> T_{1a}^{-1}$). The donor's frequency response for both low and high PE power extends to a higher frequency than the partner signal's response. The relaxation also reduces the ODMR amplitude for both. Under saturation conditions at high PE power and low modulation frequency the response is non-zero and flat for both because the steady-state equilibrium luminescence intensity increases when the microwaves are on.

The theoretical frequency response effect produced by relaxation is qualitatively similar to what is observed experimentally in Figure 3. The low power data appears to fall somewhere in between the two theoretical curves, but the samples actually contain, of course, a distribution of d-a pair separations,

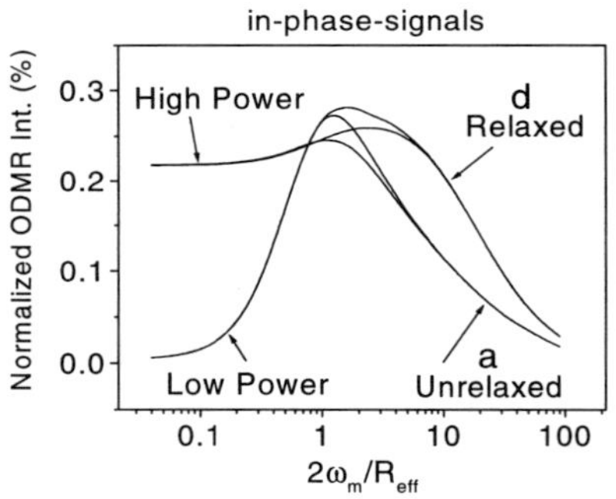

Figure 4. Simulated modulation frequency (ω_m) dependence of the ODMR intensities for distant d-a pair recombination for $T_{1d}^{-1}=10R_{eff}=100T_{1a}^{-1}$. Two PE powers are considered: low ($P=0.001R_{eff}$) and high ($P=0.1R_{eff}$).

and therefore pair lifetimes, so convolution of the theoretical response with a suitably weighted distribution of recombination lifetimes would be required before quantitative comparisons could be made.

2. Glaser Model. This model differs from the Ogino model in that the excited state undergoes a SD decay to an intermediate state before the luminescent transition occurs. Since the system is linear (no back-transfer processes) we find that the results of the numerical simulation are fully equivalent to the intermediate state acting as an effective low pass filter on the output of the SD recombination. Thus, if the luminescence decay rate is much faster than the SD rate ($R_{lum} >> R_{eff}$), this model behaves identically to the Ogino model, including the relaxation effects observed in Figure 4. If the reverse case holds ($R_{eff} >> R_{lum}$) then the system cannot respond above the "cut-off frequency" associated with the luminescence decay. In that case, the donor and its partner (the DD becomes a "net acceptor") exhibit similar frequency responses that peak at a frequency governed by the luminescent state's decay lifetime. If only one of the defects is relaxed, a difference at higher frequencies still exists but is drastically reduced due to the filtering effect.

Summarizing, we see that a short spin-lattice relaxation time for the EM donor could account for the different frequency responses observed for the EM and g=1.99 resonances. This is always true for the Ogino model, but also true for the Glaser model if the radiative rate of the second luminescing system is rapid. The kinetic response of the ODMR signals does not appear to serve by itself therefore to distinguish between the models.

DISCUSSION AND CONCLUSIONS

Several groups have reported ODMR work on the 2.2 eV luminescence band in GaN.[2,3,4,5] The two resonances commonly observed have generally been interpreted to arise from a SD recombination process between a shallow donor and a deeper defect. In this work on MOVPE grown samples a variety of findings also point to the same conclusion: partial saturation (vs. PE power) of the luminescence over many decades has been observed. The amplitudes of both resonances were found to scale similarly with PE power, both saturating at higher power. Also the spectral dependencies of each follow the 2.2 eV band. In addition, broadening of the ODMR resonances with increasing PE power and modulation frequency was observed. This is an expected characteristic for distant pair recombination, the closer pairs having the shorter recombination lifetimes and the larger exchange broadening. These findings all indicate that distant pair, SD recombination is involved. However, the Ogino and Glaser models, give different interpretations of the relationship between the SD and luminescence processes.

Other aspects of the ODMR resonances have therefore been investigated to search for distinguishing features, including microwave modulation frequency and microwave power dependencies. The two resonances typically exhibit different modulation frequency responses. Numerically simulated ODMR signals using the Ogino and Glaser models have been used to try to understand this result. They predict that if the donor has a spin-lattice relaxation time shorter than the pair's luminescence lifetime, it will respond at a higher modulation frequency than the partner, which is consistent with the data. The numerical results are unable to distinguish between the two models, but if the Glaser model is correct, they indicate that the luminescence process is much faster than the initial SD process.

Microwave power dependence studies show that neither resonance is fully saturated, the linear dependence of the donor being consistent with an unsaturated, homogenously broadened line, while the square root dependence of the g=1.99 resonance agrees with an only partially saturated, inhomogeneously broadened line (saturated spin-packets). Thus, the integrated intensities of the donor and g=1.99 resonances are not expected to be identical (only at large microwave power where they are both fully saturated should this occur). This possibly explains some of the differences between the relative amplitudes of the two ODMR signals reported in the literature. However, sample-related differences also appear to exist, suggesting the additional role of competing recombination effects involving the EM donor.

Common to the specific models of Ogino and Glaser is the further assumption that the luminescence itself results from donor to *acceptor* recombination. This aspect of the models has not yet been satisfactorily tested due to the failure so far to detect a Coulomb shift of the broad band vs. PE power.

Finally, in our samples, we observe a large reduction in the ODMR intensity (normalized to the total luminescence intensity) when using below bandgap PE energy as compared with above bandgap energy. This finding gives significant evidence that the spin-dependent and luminescent processes do not occur simultaneously. It, and it alone of the results presented here, provides a strong argument for a two step process as incorporated in the Glaser model.

ACKNOWLEDGMENTS

This work was supported jointly by the National Science Foundation under Grant No. DMR-9204114, and the Office of Naval Research under Grant No. N00014-94-1-0117. The participation at Lehigh by A. Dörnen was partially supported by a fellowship from the Deutsche Forschunggemeinschaft. The crystal growers, V. Haerle and F. Scholz, were also supported by Deutsche Forschungsgemeinschaft.

References

1. T. Ogino and M. Aoki, Jap. Jour. Appl. Phys. **19**, 2395 (1980).
2. E.R. Glaser, T.A. Kennedy, K. Doverspike, L.B. Rowland, D.K. Gaskill, J.A. Frietas, Jr., M. Asif Khan, D.T. Olson, J.N. Kuznia, and D.K. Wickenden, Phys. Rev. B **51** 13,326 (1995).
3. F.K. Koschnick, J.M. Spaeth, E.R. Glaser, K. Doverspike, L.B. Rowland, D.K. Gaskill, and D.K. Wickenden, Mat. Sci. Forum 196-201, **37** (1995).
4. U. Kaufmann, M. Kunzer, C. Merz, I. Akasaki, and H. Amano, in *Gallium Nitride and Related Materials*, ed. By F.A. Ponce, R.D. Dupuis, S. Nakamura and J.A. Edmond (MRS Symposium Vol. 395, Pittsburgh 1996), p. 633.
5. D.M. Hofmann, D. Kovalev, G. Steude, B.K. Meyer, A. Hoffmann, L.Eckey, R. Heitz, T. Detchprom, H. Amano, and I. Akasaki, Phys. Rev. B **52**, 16,702 (1995).
6. T. Suski, P. Perlin, H. Teisseyre, M. Leszczynski, I. Grzegory, J. Jun, M. Bockowski, S. Porowski, and T.D. Moustakas, Appl. Phys. Lett. **67**, 2188 (1995).
7. M.H. Nazare, P.W. Mason, G.D. Watkins, and H. Kanda, Phys. Rev. B **51**, 16,741 (1995).
8. W.E. Carlos, J.A. Freitas, Jr., M. Asif. Khan, D.T. Olson, and J.N. Kunzia, Phys. Rev. B **48**, 17,878 (1993).
9. F.C. Rong, W.A. Barry, J.F. Donegan, and G.D. Watkins, Phys. Rev. B **54**, 7779 (1996).
10. W.A. Barry and G.D. Watkins, Phys. Rev. B **54**, 7789 (1996).
11 D.J. Dunstan and J.J. Davies, J. Phys. C **12**, 2927 (1979).

OPTICAL QUENCHING OF PHOTOCONDUCTIVITY IN GaN PHOTO-CONDUCTORS

Z.C. HUANG[+], D.B. MOTT[+], P.K. SHU[+], R. ZHANG[*], J.C. CHEN[*] AND D.K. WICKENDEN[**]
[+] NASA Goddard Space Flight Center, Solid State Device Development Branch, Code 718, Greenbelt, MD 20771
[*] Department of Computer Science and Electrical Engineering, University of Maryland Baltimore County, Baltimore, MD 21228
[**] Applied Physics Laboratory, The Johns Hopkins University, Laurel, MD 20723

ABSTRACT

We report the first observation of optical quenching of photoconductivity in GaN photoconductors at room temperature. Three prominent quenching bands were found at $E_V+1.44$, 1.58 and 2.20 eV, respectively. These levels are related to the three hole traps in GaN materials based on a hole trap model to interpret the quenching mechanism. The responsivity was reduced about 12% with an additional He-Ne laser shining on the detector.

INTRODUCTION

There has been extensive studies on the wide bandgap III-nitrides due to their applications in light-emitting devices and ultraviolet detectors. Some common electrical and optical properties in other III-V materials have also been observed in GaN, such as persistent photoconductivity (PPC)[1] and transferred electron effect (TEE)[2], etc. Deep levels in GaN have been studied by several groups using both electrical and optical methods.[3-7] Despite many efforts on these materials, the electrical and optical properties of the impurities and native defects in these materials are sfill far from well understood. In this letter, we report the observation of optical quenching of photoconductivity in GaN photoconductors. The responsivity of GaN photoconductor in ultraviolet region was found to decrease about 12% when the detector was illuminated with a He-Ne laser. We attribute this optical quenching to the existence of acceptor levels in the material.

Optical quenching of photoconductivity has been reported for CdS[8], ZnSe[9] and Ge[10], ...etc. In principle, a given photocurrent, excited by light with its energy greater than the semiconductor bandgap, can be considerably reduced or "quenched" by the addition of light with energy lower than the bandgap. Assume that there are two deep centers, recombination centers I and hole trap II, within the bandgap of an n-type semiconductor. When the photoconductor is illuminated with a strong larger energy light, both centers are filled with electrons and holes (Fig.1 a). With the addition of sub-bandgap energy light, the photons excite valence band electrons to hole trap levels II, and recombine with holes trapped in centers II, then the resultant excess free holes are captured by centers I (Fig.1 b) or recombine with electrons in the conduction band, causing the decrease of the free electrons in the conduction band, i.e., optical quenching of photoconductivity occurs. This photon energy could then be related to the activation energy of the hole trap in the material, providing an effective way to determine the deep levels in photoconductors.

799

EXPERIMENT

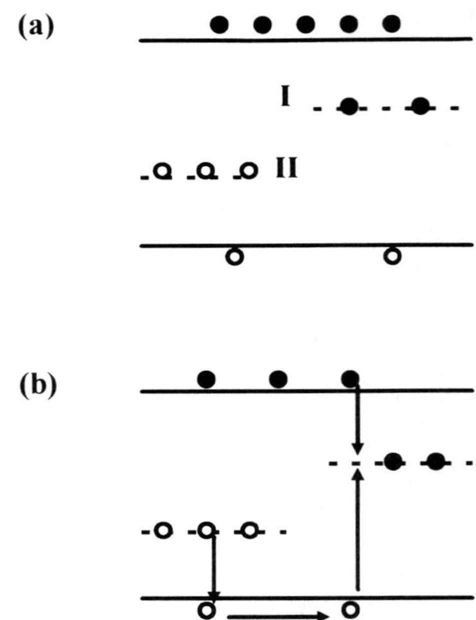

Figure 1. Model for optical quenching: (a) Photoconductor illuminated with strongly absorbed light. (b) Addition of low energy light to (a).

The GaN material used in this study was deposited on (0001) oriented sapphire by metalorganic chemical vapor deposition (MOCVD) using trimethyl-gallium (TMG) and ammonia (NH_3) as reactants and nitrogen as the carrier gas in a vertical spinning disc reactor operating at 700 torr[11]. GaN nucleation layers were deposited at 540°C using a TMG flow of 40 μmol/min., an NH_3 flow of 500 sccm and an N_2 flow of 4000 sccm. The single crystal overlayers were grown at 1050°C using NH_3 flow rate of 2000 sccm. Metal-semiconductor-metal (M-S-M) photoconductor with 10 fingers on each side was made by the conventional lift-off technique[2]. The fingers are 8 μm wide, 100 μm long and 8 μm in spacing. Ohmic contacts were made by evaporating 300A Al followed by 1200A Au, and annealing at 450°C for 5 mins. After fabrication, the detectors were packaged for testing. For the photocurrent measurement, we used a Xenon lamp as light source. The light beam was split into two paths: one is for providing UV source through a 365 nm filter, and the other is for low energy illumination through a monochromator. The data were taken automatically by a computer.

RESULTS AND DISCUSSIONS

In Fig.2, we present the normalized transient photocurrent from GaN photoconductor when illuminated by different wavelength light pulses. The background photocurrent was the steady response from the 365nm light (5.6μW/ cm^2) with no additional light added. For comparison, all three wavelength lights were adjusted to have the same incident power of 2.0 μW/cm^2 during the measurement. When additional light is added, the 360nm light caused an increase in photocurrent (a), which is reasonable because the 360nm light excites more carriers to the conduction and valence band. Whereas the 560nm and 780nm lights caused a decrease in photocurrent (b and c), showing an optical quenching effect. The time constants of transient quenched photocurrent caused by 560nm light and 780nm light are different. The quenching by 560nm light is slow, it took about 4s to be stable. However, the 780nm light causes a fast drop first, and then increases to a stable level within 2s, indicating that different deep centers may be involved in these two cases. After the quenching light is turned off, the recovery processes of photocurrent to the normal response level (with 365nm light) for both 560nm and 780nm lights are similar (b and c), and gradually (~60s) reach to the current level before quenching.

Figure 3 demonstrates the photocurrent of GaN photoconductor as a function of incident light energy. During the measurement, the 365nm light constantly illuminated the photoconductor with its response as background level. Notice that in Fig.2 (b and c) the quenched photocurrent levels need about 4s to be stable. Therefore for each measurement point in Fig.3, after each wavelength changed, we waited 5s before reading the current value. The photocurrent in Fig.3 was also normalized by the incident light intensity although for Xenon lamp the output power in 360-1000nm region is relatively stable. Three quenching peaks were observed with their energies at E_V+1.44, 1.58 and 2.20 eV, respectively. The peaks at 1.44 and 1.58 eV are very sharp, while the 2.20 eV peak is relatively broad. The photocurrent was quenched by the light with energy from 1.2 to 2.6 eV. The low energy region was limited by our system (grating limit). However, it seems that the quenching also happens in the lower energy region. When the light energy is greater than 2.6 eV, the photocurrent increases to above the background level, and increases rapidly when the bandgap transition occurs.

Based on the model described to interpret the optical quenching effect by Bube[12] and Rose[13], the observed three quenching bands are related to three deep levels in GaN with the same energies. With the 365nm light shinning on the photoconductor, we have found that the GaN material used in this study is n-type. Therefore, these three levels at 1.44, 1.58 and 2.20 eV are hole traps because only hole traps in n-type material can cause quenching of photocurrent. When a light with the same energy as hole trap level illuminates the photoconductor, the excess electrons are excited to the hole trap level, and the holes are freed and may recombine with electrons either in conduction band or in other recombination centers (Fig.1). Either of these two processes would cause a decrease of free electron density in the conduction band, resulting in a quenching of photoconductivity. Deep levels in GaN have been reported by several groups using deep level transient spectroscopy[3,4] and thermally stimulated current[5]. Most of those reported levels were electron traps. The hole traps we observed in this study are the first reported in GaN materials.

Finally, we would like to stress that the observed optical quenching of photoconductivity in GaN photoconductors is harmful for the UV detection applications. In general, the detected sources contain lights with different wavelengths. With the existence of other lights which have energies below the band gap, the responsivity of GaN photoconductor may be reduced due to quenching effect. This is illustrated in Fig.4, showing that the responsivity of GaN photoconductor decreased in the entire UV region when a He-Ne laser illuminated the detector, especially in the near band gap region (up to 12%). This would

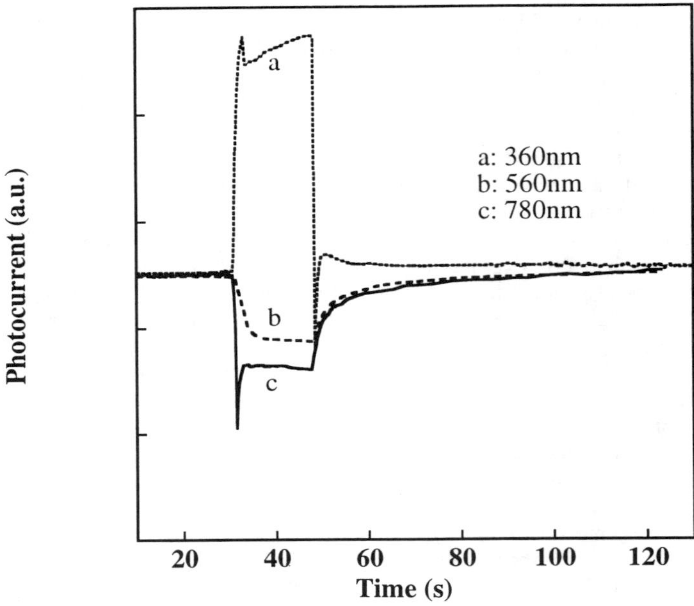

Figure 2. Transient photocurrent showing optical quenching by 560 nm and 780 nm lights. The background is the photocurrent of the detector with 365 nm light.

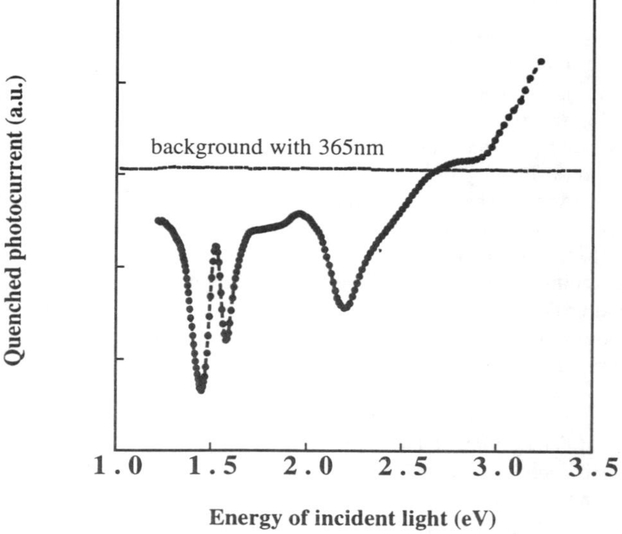

Figure 3. Optical quenching spectrum for GaN photoconductor.

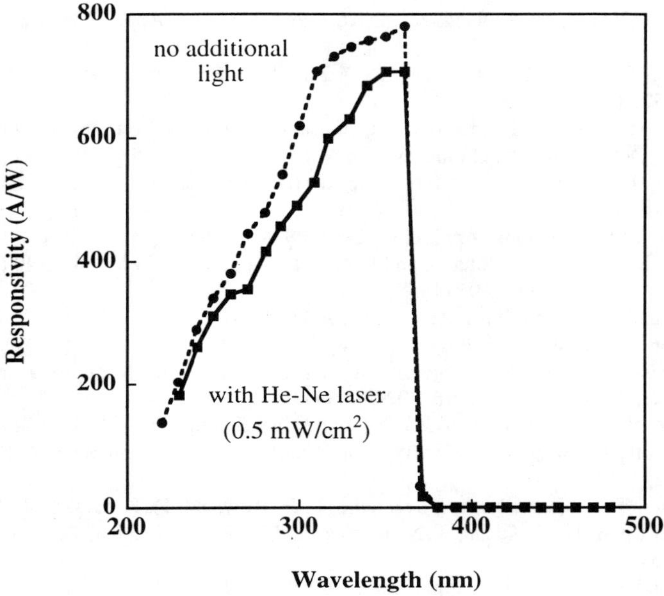

Figure 4. Responsivity spectra for GaN photoconductor (a) without quenching light, (b) with He-Ne laser as quenching light.

confuse the output signal from the photoconductors. However, the problem can be solved by either removing or saturating these deep centers. Further investigation is needed to identify the origin of these levels.

CONCLUSION

In conclusion, we have observed the optical quenching of photoconductivity in GaN photoconductors. Three prominent quenching bands were found at $E_V+1.44$, 1.58 and 2.20 eV, respectively. These levels are corresponded to the three hole traps in GaN materials based on a hole trap model to interpret the quenching mechanism.

ACKNOWLEDGMENTS

One of the author (Dr. J.C. Chen) acknowledges the support from Army Research Lab. (Dr. F. Semendy) and Army Research Office (Dr. J. Zavada and the AASERT Program).

REFERENCES

1. C. Johnson, J. Y. Lin, H. X. Jiang, M. A. Khan and C. J. Sun, Appl. Phys. Lett., **68** 1808 (1996)
2. Z. C. Huang, R. Goldberg, J. C. Chen, Y. D. Zheng, D. B. Mott and P. Shu, Appl. Phys. Lett., **67**, 2528 (1995)
3. W. Gotz, N. M. Johnson, H. Amano, and I. Akasaki, Appl. Phys. Lett., **65**, 463 (1994)
4. P. Hacke, T. Detchprohm, K. Hiramatsu, and N. Sawaki, J. Appl. Phys., **76**, 305 (1994)
5. Z.C. Huang, J. C. Chen, D. B. Mott, Proc. of 1st. Symposium on Nitrides and Related Devices, Boston (1995)
6. L. Balagurov and P. J. Chong, Appl. Phys. Lett., **68**, 43 (1996)
7. W. Gotz, N. M. Johnson, H. Amano, and I. Akasaki, Appl. Phys. Lett., **66**, 1340 (1995)
8. R. H. Bube, Phys. Rev., **99**, 1105 (1955)
9. R. H. Bube and E. L. Lind, Phys. Rev., **110**, 1040 (1958)
10. R. Newman and W. W. Tyler, Phys. Rev., **96**, 882 (1954)
11. D. K. Wickenden, C. B. Bargeron, W. A. Bryden, J. A. Miragliotta, and T. J. Kistenmacher, Appl. Phys. Lett., **65**, 2024 (1994)
12. R. H. Bube, *Photoconductivity of Solids*, (John Wiley & Sons, New York), p. 173, 1960
13. A. Rose, Concepts in *Photoconductivity and Allied Problems*, (John Wiley & Sons, New York), p. 52, 1963
14. E. R. Glaser, T. A. Kennedy, K. Doverspike, L. B. Rowland, D. K. Gaskill, J. A. Freitas, Jr., M. Asif Khan, D. T. Olson, J. N. Kuznia, and D. K. Wickenden, Phys. Rev. **B 51**, 13326 (1995)
15. T. Ogino and M. Aoki, Jpn. J. Appl. Phys., **19**, 2395 (1980)
16. T. Suski, P. Perlin, H. Teisseyre, M. Leszczynski, I. Grzegory, J. Jun, M. Bockowski, S. Porowski, and T. D. Moustakas, Appl. Phys. Lett., **67**, 2188 (1995)

OPTICAL DIELECTRIC RESPONSE OF GALLIUM NITRIDE STUDIED BY VARIABLE ANGLE SPECTROSCOPIC ELLIPSOMETRY

H. YAO[a], C. H. YAN[a], H. A. JENKINSON[b], J. M. ZAVADA[c], J. S. SPECK[d], S. P. DENBAARS[d]

a) University of Nebraska, Center for Microelectronic and Optical Material Research, and Department of Electrical Engineering, Lincoln, NE 68588, hyao@ unl.edu
b) US Army ARDEC, Picatinny Arsenal, NJ 07806
c) US Army Research Office, Research Triangle Park, NC 27709
d) University of California, Santa Barbara, CA 93106

ABSTRACT

Variable angle spectroscopic ellipsometry (VASE) and transmission measurements have been employed to study the dielectric response of gallium nitride (GaN) thin films — an important material for light emitting diodes (LEDs) and laser diodes applications. The GaN films were grown by atomsphere pressure metal organic chemical vapor deposition (MOCVD) on c-plane sapphire substrates (α-Al$_2$O$_3$). Room temperature VASE measurements were made, in the range of 0.75 to 5.5eV, at the angle of incidence of 73, 75, and 77 degree, respectively. Evidence of anisotropy is observed especially in the spectral range under the band gap (~3.4 eV), reflecting the nature of wurtzite crystal structure of GaN. The ordinary dielectric function $\varepsilon_\perp(\omega)$ of GaN were obtained through the analysis of transmission and VASE data in the range below and above the band gap. The thickness of these GaN films is also determined via the analysis.

INTRODUCTION

The wide-band-gap semiconductors like GaN, with their excellent thermal conductivity, large breakdown field, and resistance to chemical attack, have a very promising application potential for both high temperature electronic devices and short wave-length optical emitters [1,2]. The latest development of high-brightness, blue and green, light emitting diodes (LEDs) [3] and room temperature, pulsed quantum well lasers [4] has greatly encouraged researchers to continue the work on this material. Most of high quality GaN films were grown by MOCVD and MBE techniques on sapphire and silicon substrates. GaN on sapphire usually has the wurtzite structure (α-GaN), while GaN films on Si or GaAs substrates may have zinc-blende structure (β-GaN), and their fundamental band gaps are about 3.4 eV and 3.3 eV respectively [5]. As the growth techniques are improving, much higher quality GaN material have been made in recent years (lower defect density, lower background electron concentration, and higher mobility). More accurate measurements on electrical and optical properties are needed for these improved materials for further device fabrication and research activities. In this paper, we report the ordinary dielectric function of GaN determined by VASE and transmission measurements via a multiple-sample, multiple-model analysis in the range of 0.75 to 5.5 eV.

THEORY

The standard ellipsometry parameters ψ and Δ are related to the complex ratio of reflection coefficients for light polarized parallel (p) and perpendicular (s) to the plane of incidence [6]. This ratio is defined as:

805

Mat. Res. Soc. Symp. Proc. Vol. 449 © 1997 Materials Research Society

$$\rho = \frac{r_p}{r_s} = \tan(\psi)e^{i\Delta}. \tag{1}$$

The electric-field reflection coefficient at an incident angle of ϕ is defined as r_p (r_s) for p (s) - polarized light. The pseudodielectric function of the sample $\langle\varepsilon\rangle = \langle\varepsilon_1\rangle + i\langle\varepsilon_2\rangle$ is obtained from the ellipsometrically measured values of ρ in a two phase model (ambient/substrate) regardless of the possible existence of surface overlayers [6]. To determine optical constants of a thin film on the substrate, VASE data must be analyzed using a parametric model that is adjusted to fit the measured data. A regression analysis is used to vary the model parameters (e.g., optical constants or layer thickness, etc.) until the calculated and measured values match as closely as possible. This is done by minimizing the mean square error (MSE) function, defined as:

$$MSE = \sqrt{\frac{1}{N-M}\sum_{i,j}\left\{\left[\frac{\psi\left(hv_i,\phi_j\right)-\psi^c\left(hv_i,\phi_j\right)}{\sigma_{i,j\psi}}\right]^2 + \left[\frac{\Delta\left(hv_i,\phi_j\right)-\Delta^c\left(hv_i,\phi_j\right)}{\sigma_{i,j\Delta}}\right]^2\right\}}, \tag{2}$$

where N is the total number of experimental observations (for ellipsometry measurements, there are two observations Ψ and Δ, for each data point), M is the number of fit parameters, σ is the measured standard deviation of the measurement, ϕ is the external angle of incidence, hv is the photon energy, and i and j are used to sum over all the photon energies and external angles of incidence, respectively.

EXPERIMENTAL

The GaN films were grown by atmospheric pressure MOCVD on c-plane sapphire substrates (α-Al$_2$O$_3$). The substrates were cleaned in solvents, and then subjected to an *in situ* 1050°C pre-treatment prior to growth. A nucleation layer of GaN, approximately 0.019μm thick, was deposited at 600°C at a growth rate of 3.7×10^{-3}μm/sec. This layer contains a large number of crystal defects and the resulting GaN crystals have zinc-blende symmetry. Following this nucleation layer, the growth temperature was then increased to 1050°C for film growth at a rate of 2.5μm/hour. The resulting films are single crystal with wurtzite symmetry. Stacking faults, at a density of approximately 5×10^8cm^{-2}, are present in the films and extend from the nucleation layer to the film surface. The films studied in these experimental had a nominal thickness of 1-2μm. Our Micro-probe Raman spectra indicate that these GaN films have very good single crystal quality.

Optical transmission intensity measurements were made from 0.5 eV to 6.5 eV on a Perkin Elmer Lambda 9 spectrophotometer. This instrument is a dual-beam, ratio-recording model which compares the optical energy passing through the sample compartment to that passing through the reference compartment. Identical apertures were placed in each compartment and a background correction was performed to balance the instrument to 100% transmission across the spectral range of interest. The transmission data were obtain from a double-side polished GaN/sapphire sample at normal incidence.

Another one-side polished GaN/sapphire sample was measured in photon energies from 0.75 to 5.5 eV at room temperature by using a Woollam Co. VASE system. The photon energy increment was set at 0.01 eV in order to resolve the possible fine features in the spectrum. A beam-chopped, rotating-analyzer system and an auto-retarder were used for achieving more accurate results. Three angles of incidence 73°, 75°, and 77° were used in the VASE measurements. Even though multiple-angle can not bring more independent information for samples with thin overlayer, it does help average out measurement noise, and it does ensure that for each measured wavelength at least one pair of ψ and Δ will be near the optimum regime for the type of ellipsometer being used (e.g., a rotating-analyzer ellipsometer is most accurate for Δ ~ 90°) [6, 7].

RESULTS AND ANALYSIS

Ordinary part of dielectric function $\varepsilon_\perp(\omega)$ of GaN films grown on sapphire substrate has been determined via a multiple-sample, multi-model analysis of the VASE and transmission data. The optical dielectric response of the two GaN samples with different thickness (one-side polished for VASE measurement and double-side polished for transmission) are assumed identical in the data analysis. Fig. 1 shows both the real part (ε_1) and the imaginary part (ε_2) of the ordinary dielectric function of GaN. For the reason of simplicity, we will use ε for the ordinary dielectric function in the rest of the paper. The rapid increase of ε_2 near 3.4 eV indicates the fundamental absorption edge. This is consistent to the reported band gap value of GaN in references[1] and [5].

In order to determine the GaN optical constants as accurate as possible, optical constants, shown in Fig. 2, of the sapphire substrate was measured and determined by VASE analysis via the same approach. A detailed description about sample measurement and data analysis will be reported in a separated paper.

Under the GaN band gap value of 3.4 eV, the GaN film and GaN/sapphire structure are all transparent. The film thickness can be fully determined through the fitting in this region. Since GaN film is not a chemically reactive material, it is reasonable to assume that there is no oxide overlayer formed on top of the GaN films. Therefore, a three-phase model: ambient/GaN/sapphire, was assumed for the data fitting. A user-defined dispersion model was used to calculate the index of reflection (n) and the extinction coefficient (k) of the ordinary ray which are described as below:

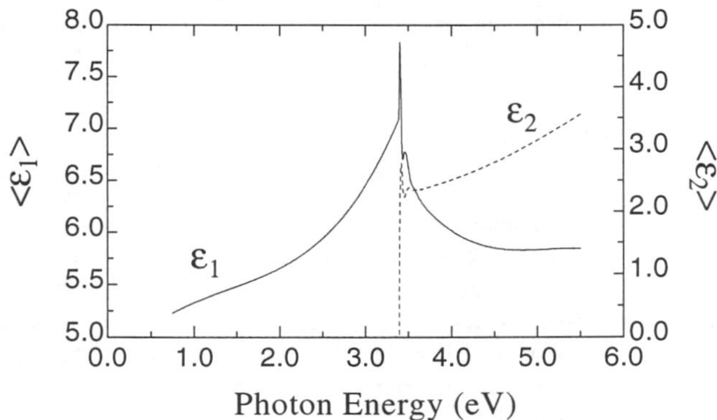

Fig. 1. Ordinary dielectric function of GaN.

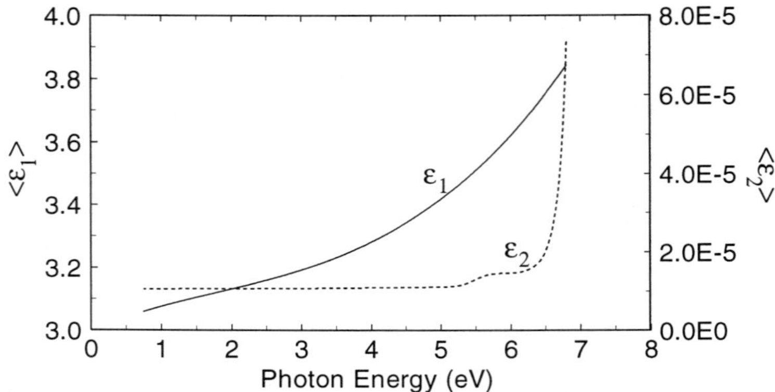

Fig. 2. Dielectric function of sapphire substrate.

$$n(\lambda)=A+B/\lambda+C/\lambda^2+D/\lambda^4,\tag{3}$$

$$k(\lambda)=P*EXP[-(\lambda-Q)/R]^3.\tag{4}$$

In equation (3) and (4), coefficients A, B, C, D, and P, Q, R, are fitting parameters along with the thin film layer thicknesses of the two samples. Using the dispersion functions to describe the dielectric responses greatly reduced the number of fitting parameters in the determination of dielectric functions of the interested material. Assisted by the transmission data, the VASE analysis system is over determined (more measured values than fitting parameters). Thus, it is insured to minimize the possible correlation between the fitting parameters, such as the optical constants of GaN and the thicknesses of the films.

Above the band gap, the samples are no longer transparent. The dispersion relation becomes more complicated, especially at the near band gap region. We employed Lorentz oscillator model to describe the dielectric response in this region [8, 9]. It is a collection of absorption peaks which is also Kramers-Kronig consistent. The dielectric function can be expressed as:

$$\varepsilon(h\nu) \equiv \varepsilon_1 + i\varepsilon_2 = \varepsilon_{1\infty} + \sum_k \frac{A_k}{E_k^2 - (h\nu)^2 - iB_k h\nu}.\tag{5}$$

For the kth oscillator, A_k is the amplitude, E_k is the center energy, B_k is the broadening of each oscillator. $h\nu$ is the photon energy in eV. $\varepsilon_{1\infty}$ is an additional offset term defined in the model. By using the results of sample structures determined by the VASE analysis from the region below the band gap, the dielectric response of GaN in the above band gap region can be described by a seven-oscillator model. Each oscillator's parameters, i.e., A_k, E_k, and B_k along with the value of $\varepsilon_{1\infty}$ in Eq. 5 were determined via the regression analysis. Combining the dielectric functions of the two regions, a complete set of dielectric function of GaN for ordinary ray is therefore obtained in the range of 0.75 to 5.5 eV, as shown in Fig. 1. The best fit results from the multiple-sample, multiple-model analysis are shown in Fig. 3. Figure 3 (a) exhibits the VASE data and its best fit, in which the interference fringes are ideally fit, and the fitting goes very well in the near band gap and above the gap region. Figure 3 (b) shows the transmission data and its fitting. All the oscillations in the transparent region are fit very well from which the layer thickness and nonuniformity were accurately determined. The non-perfect fitting of the amplitudes below the band gap indicates that there is a room to improve our current model. Fig. 4 shows the pseudo dielectric function of the GaN/sapphire structure. It is obvious that there is a clear dispersion due to the different angle of incidence. This indicates that the material has an anisotropic nature which mainly result from the wurtzite symmetry of GaN crystal. In order to determine the extraordinary dielectric function of GaN, we need VASE measurements on a-plane GaN samples and set up an uniaxial model for analysis.

SUMMARY

We have presented the ordinary part of dielectric function ε of GaN Crystal, as determined

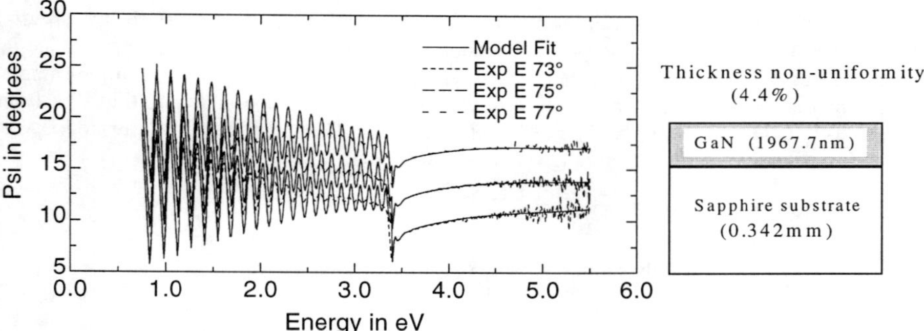

Fig. 3 (a). Best fit of VASE data with a three-phase model.

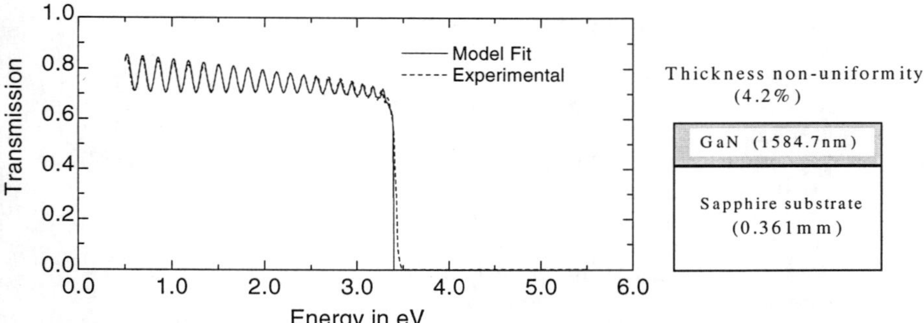

Fig. 3 (b). Best fit of transmission data with a three-phase model.

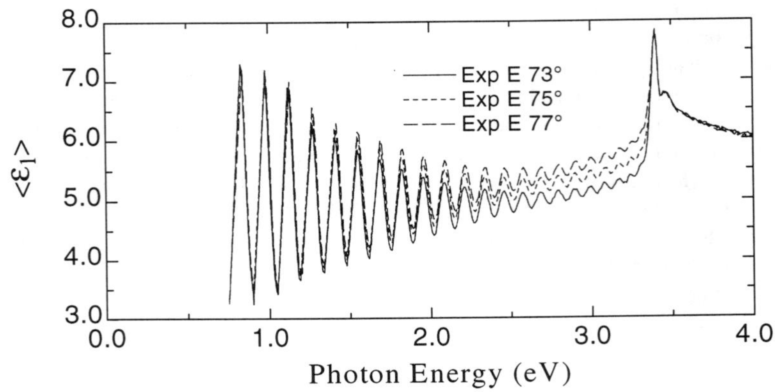

Fig. 4. Pseudo dielectric function $<\varepsilon_1>$ of GaN/sapphire at different angles of incidence.

by VASE and transmission measurements from c-plane GaN/sapphire samples, in the spectral range of 0.75 to 5.5 eV. Both of the transmission data and the dielectric functions indicate an absorption edge of ~ 3.4 eV, which is consistent to other measurements. Multiple-sample, multiple-model analyses were employed to determined the optical constants and thin film layer thicknesses nondestructively. Evidence of anisotropy of GaN crystal optical response was observed, especially in the spectral range under the band gap.

ACKNOWLEDGEMENTS

This work was partially supported by US Army Research Office.

REFERENCES

1.   H. Morkoc, S. Strite, G. B. Gao, M. E. Lin, B. Sverdlov, and M, Burns,
        J. Appl. Plys. **76**, 1363 (1994).
2.   S. Nakamura, T, Mukai, and M. Senoh, Appl. Phys. Lett. **64**, 1687 (1994).
3.   S. Nakamura, M. Senoh, N. Iwasa, and S. Nagahama, Jpn. J. Appl. Phys. **34**,
        L797 (1995).
4.   S. Nakamura, M. Senoh, S. Nagahama, N. Iwasa, T. Yamada, T. Matsushita,
        H. Kiyyoku and Y. Sugimoto, Jpn. J. Appl. Phys. **35**, L74 (1996).
5.   S. Logothetidis, J. Petalas, M. Cardona, and T. D. Moustakas, Physical Review B,
        50, 18017 (1994).
6.   R. M. A. Azzam and N. M. Bashara, *Ellipsometry and Polarized Light*,
        (North-Holland, Amsterdam, 1977).
7.   D. E. Aspnes, in Handbook of *Optical Conatants of Solids*, edited by E. D. Palik,
        (Academic Press, New York, 1985), p. 89.
8.   H. Yao, P. G. Snyder, and J. A. Woollam, J. Appl. Phys. **70**, 3261 (1991).
9.   H. Yao, P. G. Snyder, K. Stair, T. Bird, Mat. Res. Soc. Symp. Proc. **242**, 481 (1992).

CHANGES IN OPTICAL TRANSMITTANCE OF ALUMINUM NITRIDE THIN FILMS EXPOSED TO AIR

Yoshifumi Sakuragi, Yoshihisa Watanabe, Yoshikazu Nakamura and
Yoshiki Amamoto
Department of Materials Science and Engineering, National Defense
Academy, 1-10-20 Hashirimizu, Yokosuka, Kanagawa 239, Japan

ABSTRACT

Aluminum nitride (AlN) thin films have been synthesized by ion-beam
assisted deposition method and the influence of air-exposure on the optical
transmittance has been studied. The kinetic energy of nitrogen ion beam was
kept at 0.1 or 1.5 keV under the constant current density. Synthesized films
have been exposed to controlled air (23 °C and RH; 50%) and optical
transmission spectrum from 190 to 2200 nm has been measured by UV-
visible spectrometer every week. Surface morphology of the films has been
observed with an optical microscope (OM). The optical transmittance has not
changed drastically up to one year. Observations by OM show that round
features of some microns were produced on the surface after about 25 weeks
exposure. These substances seem to be reaction products between AlN and
water in air.

INTRODUCTION

Aluminum nitride (AlN) has many unique properties, such as high
electrical resistivity, high thermal conductivity and high hardness. In
addition, AlN thin films are transparent in the visible and near infrared
regions. Therefore, AlN thin films are a potential material for protection of
optical storage media, and thus optical properties of AlN films have been
intensively studied [1-10].

For practical application, corrosion resistivity of films in atmosphere is
of importance. Although corrosion behavior of bulk AlN samples has been
reported under various corrosive environments in the literature [11], there is
limited information about corrosion resistivity of AlN films [12]. Especially,
very little is known about the influence of long-time air-exposure on the
optical transmittance of AlN films.

The present authors have synthesized AlN thin films by ion beam assisted
deposition (IBAD) method and have studied the dependence of optical
properties of AlN films on the ion beam energy [13]. We reported that the
refractive index of AlN films depends on the nitrogen ion beam energy and
seems to be related to film microstructure. Therefore, it is an exciting
objective to study the dependence of corrosion resistivity of AlN films as a
function of nitrogen ion beam energy.

In the present paper, using the IBAD method, AlN thin films are

Mat. Res. Soc. Symp. Proc. Vol. 449 © 1997 Materials Research Society

synthesized with changing the ion beam energy and the influence of exposure to controlled air on the optical transmittance of films is studied.

EXPERIMENTAL

Nitrogen gas (99.999% pure) and aluminum (99.99% pure) were used as ion source and target, respectively. Film synthesis was carried out on the fused silica glass substrate. After evacuating the vacuum chamber to about 2.7×10^{-4} Pa, pure nitrogen gas was introduced to the ionization chamber and nitrogen ions were generated by an arc discharge. Then a nitrogen ion beam was obtained with electric field lenses for focusing and accelerating. The kinetic energy of the nitrogen ions is kept at 0.1 or 1.5 keV and the deposition rate was kept at 0.07 nm/s. The current density was approximately 70 μA/cm^2. Aluminum was evaporated by electron bombardment from a 10 kW electron gun and the evaporation rate was monitored by a quartz sensor. The substrate temperature was kept at room temperature. Films were exposed to controlled air (temperature 23 °C and relative humidity 50%)

The thickness of the as-deposited films was measured with a stylus device (Rank Taylor Hobson Ltd.). Optical transmission was measured in the wavelength region from 190 to 2200 nm using a UV-visible spectrometer (JASCO U-best 570) every week. The surface morphology was observed with an optical microscope (OM).

RESULTS AND DISCUSSION

Figure 1 shows typical transmission spectra of as-deposited AlN films, of 300 nm in thickness, synthesized with the ion beam energy of 0.1 keV and 1.5 keV. Both films are transparent in the visible and near infrared region, and the average transmittance in the visible region is slightly higher in the film synthesized with 1.5 keV nitrogen ion beam than in the film synthesized with 0.1 keV. The wavy patterns in the measured spectra are attributed to interference, and the wavy pattern is more emphasized in the film synthesized with 0.1 keV nitrogen ion beam than in the film synthesized with 1.5 keV, showing the good agreement with the previous results [13].

Figure 2 shows typical transmission spectra of AlN films after exposure to controlled air for 56 weeks. From both Figs. 1 and 2, it can be seen that the transmission spectra are not affected by exposure to controlled air for 56 weeks in both films synthesized with the ion beam energy of 0.1 keV and 1.5 keV. In other words, the AlN films show excellent stability in controlled air.

The averaged transmittance over the wavelength from 400 to 800 nm, viz. the visible region, is plotted as a function of the exposure time and shown in Fig. 3. This figure shows that the average transmittances of both films are not varied within the experimental error and the corrosion resistivity against air-exposure is independent of the nitrogen ion beam energy. These results support the applicability of AlN films as an optical coating layer.

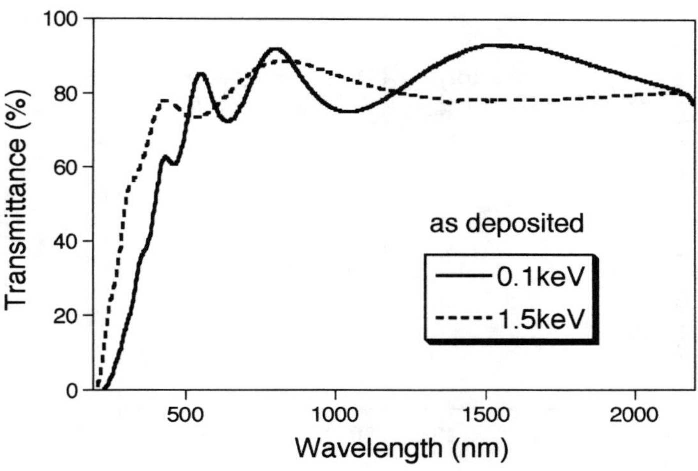

Figure 1 Optical transmission spectra of as-deposited AlN thin films. Nitrogen ion beam energy was kept at 0.1 keV or 1.5 keV during synthesis.

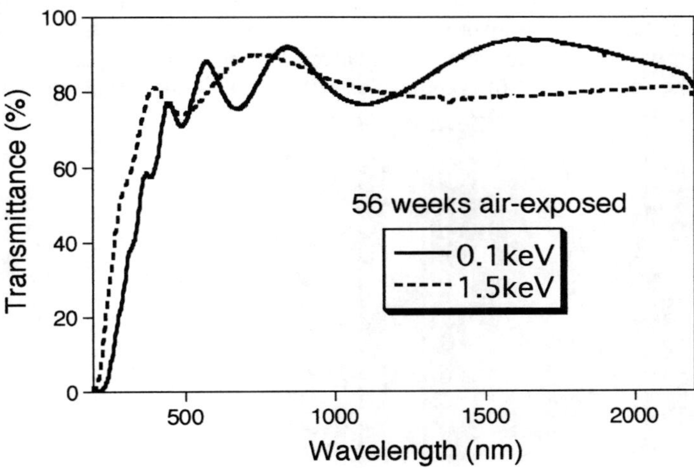

Figure 2 Optical transmission spectra of AlN thin films after exposure to controlled air for 56 weeks.

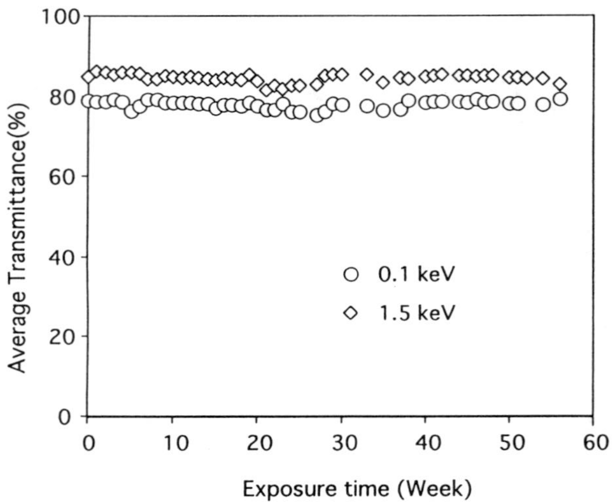

Figure 3 Relation between the average optical transmittance of AlN thin films and air exposure time. The transmittance is averaged over the wavelength region from 400 to 800 nm.

Optical microscope observations reveal that the surface of as-deposited films is smooth and featureless. However, after exposure to air for 25 weeks, many bubble-like features appeared. Furthermore, after exposure to air for 50 weeks, although the transmission spectrum was not changed, bubble-like features grew up to round shaped substances of some microns, as shown in Fig. 4 in the case of the film synthesized with 1.5 keV nitrogen ion beam.

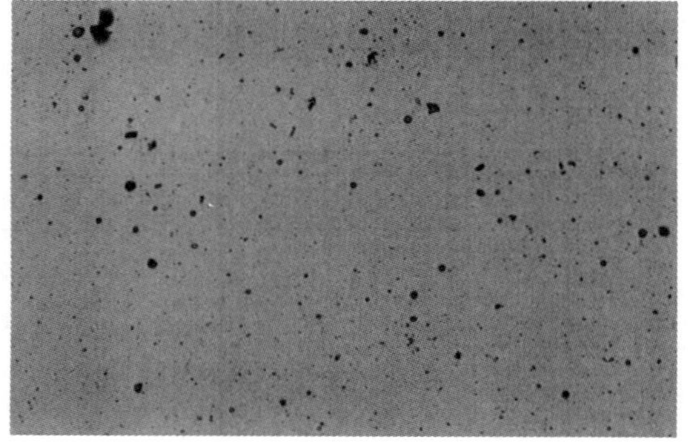

Figure 4 Typical OM image of the surface of AlN films after exposure to air for 50 weeks. Nitrogen ion beam energy was kept at 1.5 keV during synthesis.

According to Young *et al*, bulk AlN corrodes in water, yielding alumina and aluminum hydroxide [11]. The observed round shaped substances on the surface of the films after exposure to air for 50 weeks seem to be reaction products between AlN and water in air. In order to confirm the reaction products, AlN film of 300 nm in thickness synthesized similarly to the specimen for optical measurements was immersed into pure water and then the immersed film was analyzed by thin film X ray diffraction, where the angle between the incident X ray beam and the sample surface is kept at 0.3 degree so as to detect substances on the surface. A typical XRD pattern of the films after immersion into water for 36 weeks is shown in Fig. 5, and it is found that aluminum hydroxide (Bayerite) is produced on the surface.

In addition, ammonia was detected in the water, into which the AlN film was immersed, by Nessler's reagent. This fact supports that AlN films reacted with water, resulting in aluminum hydroxide (Bayerite) and ammonia.

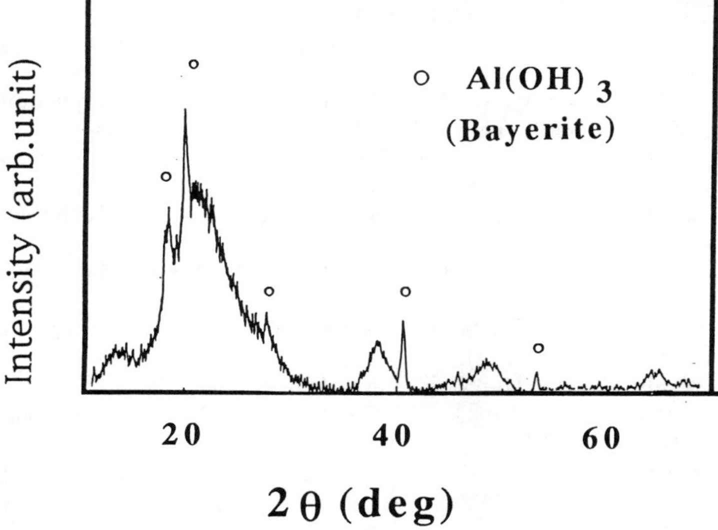

Figure 5 X-ray diffraction pattern of the AlN film immersed into pure water for 36 weeks. The film was synthesized on the fused silica substrate.

The reason for the fact that the optical transmittance of the films remains unchanged in spite of the reaction products on the surface is considered to be relatively small amount of the reaction products occupied in the measurement area of optical transmission, 2 mm in diameter. Furthermore, it is necessary to clarify the optical properties of the reaction products so as to estimate the absorption by them.

SUMMARY

Aluminum nitride thin films have been synthesized by the IBAD method
and the effect of long-time air-exposure on the optical transmittance has been
studied. It is found that the optical transmittance has not changed drastically
up to one year and round shaped features of some microns were produced on
the surface after about 25 weeks exposure. These materials are considered to
be reaction products between AlN and water in air, perhaps aluminum
hydroxide. The large difference in corrosion resistivity is not found between
the films synthesized with the nitrogen ion beam energy of 0.1 and 1.5 keV.
It is proposed that the AlN films synthesized by the IBAD have the potential
to be used for protection of optical media.

<u>References</u>

[1]   W. M. Yim, E. J. Stofko, P. J. Zanzucchi, J. I. Pankove, M. Ettenberg,
       and S. L. Gilbert, J. Appl. Phys. **44**, p. 292 (1973).
[2]   R. G. Gordon, U. Riaz, and D. M. Hoffman, J. Mater. Res. **7**, p. 1679
       (1992).
[3]   S. V. Krishnaswamy, W. A. Hester, J. R. Szedon, M. H. Francombe,
       and M. M. Driscoll, Thin Solid Films **125**, p. 291 (1985).
[4]   M. M. Ramos, J. B. Almeida, M. I. C. Ferreira, and M. P. Dos Santos,
       Thin Solid Films **176**, p. 219 (1989).
[5]   P. B. Legrand, J. P. Dauchot, and M. Hecq, J. Vac. Sci. Technol. A **10**,
       p. 45 (1992).
[6]   R. Zarwasch, E. Rille, and H. K. Pulker, J. Appl. Phys. **71**, p. 5275
       (1992).
[7]   K. Seki, X. Xu, H. Okabe, J. M. Frye, and J. B. Halpern, Appl. Phys.
       Lett. **60**, p. 2234 (1992).
[8]   H. Windischmann, Thin Solid Films **154**, p. 159 (1987).
[9]   X-D. Wang, K. W. Hipps, J. T. Dickinson, and U. Mazur, J. Mater.
       Res. **9**, p. 1449 (1994).
[10] P. Martin, R. Netterfield, T. Kinder and A. Bendavid, Appl.
       Optics **31**, p. 6732 (1992)
[11] C.-D Young and J.-G. Duh, J. Mater. Sci. **30**, p. 185 (1995).
[12] T. K. Hatwar, S. C. Shin and D. G. Stinson, IEEE Tras,
       Magnetics **MAG-22**, p. 946 (1986).
[13] Y. Watanabe, Y. Nakamura, S. Hirayama, and Y. Naota in
       <u>Polycrystalline Thin Films: Structure, Texture, Properties and
       Applications II</u>. edited by H. J. Frost, M. A. Parker, C. A. Ross
       and E. A. Holm (Mater. Res. Soc. Proc. 403, Pittsburgh, PA,
       1996), p.539-544.

AN OPTICAL WAVEGUIDE FORMED BY
ALUMINUM NITRIDE THIN FILM ON SAPPHIRE

XIAO TANG, YIFANG YUAN*, K. WONGCHOTIGUL, MICHAEL G. SPENCER
Materials Science Research Center of Excellence, School of Engineering, Howard University,
2300 6th Street, NW, Washington, DC 20059, spencer@msrce.howard.edu

ABSTRACT

We have investigated an optical waveguide formed by aluminum nitride (AlN) thin film on
sapphire. A good quality AlN thin film on sapphire substrate was prepared by metal organic
chemical vapor deposition (MOCVD) in this laboratory. A rutile prism coupler was employed to
display the waveguide modes (N-lines) with wavelengths of 632.8, 532.1, 514.5 and 488.0 nm.
The refractive index and thickness of the waveguide material is obtained by prism-coupler
measurement. The dispersion curve of the AlN film is given and the dispersion equation is
derived. The attenuation in the waveguide is evaluated by scattering loss measurements using a
fiber probe. The attenuation coefficient alpha (α) is 1.5- 2.1 cm^{-1} depending on the sample and the
different modes of waveguide. The accuracy of the measurement is discussed.

INTRODUCTION

Aluminum nitride is a wide band gap semiconductor. The insulating properties can be used in
the fabrication of GaAs and InP based electronic device structures. The outstanding physical
properties of hardness, high thermal conductivity, and its resistance to high temperature have
attracted much interest. Optoelectronic integration is a new technological revolution in the field
of electronics and optics. The III-V nitrides devices are the most promising competitors.
Aluminum nitride is now not only a insulator but also a good conductive material which leads to
many new applications[1]. It is hopeful that AlN devices can be used in the blue and ultraviolet
wavelength. Actually, some III-V nitride devices have been available recently such as laser
diodes[2] and detectors[3], which are disparate elements. The integration of these elements together
on the same substrate is significant for future applications. In this paper, study of an AlN
waveguide on sapphire is presented. A high quality AlN waveguide sample on a sapphire
substrate were prepared by MOCVD in this laboratory. With the method of prism coupler
waveguide measurement, the refractive index of AlN thin films are obtained from a set of samples.
The dispersion equation at room temperature was determined by curve fitting with the measured
data at different wavelengths. The results were compared with those issued in a previous work[4] .
Waveguide loss, an important parameter, is also evaluated in this work. To our knowledge, this is
the first demonstration of an optical waveguide of AlN film on sapphire substrate.

EXPERIMENT AND RESULTS

Material

The AlN samples used in this work were grown in a low pressure metal organic chemical
vapor deposition (MOCVD) vertical reactor on (0001) orientated sapphire substrates in our
laboratory. Sapphire is an inexpensive material commonly employed for CVD growth of III-

Mat. Res. Soc. Symp. Proc. Vol. 449 © 1997 Materials Research Society

nitrides. The preference towards sapphire substrates can be attributed to its wide availability, hexagonal symmetry, its easy of handling and pregrowth cleaning, and its stability at high temperature (1000°C). All these advantages are beneficial to device integration.

The reactor in the growing system has two separated chambers joined by a common load lock. One of the chambers is equipped with a reflection high energy electron diffraction (RHEED) system which allows a quick determination of film crystallinity after growth. The common load lock provides a way of transporting samples between chambers without exposing them to the ambient air. The growth temperature and pressure were 1200°C and 10 torr, respectively. Trimethylaluminum (TMA), ammonia (NH_3) and hydrogen (H_2) were used as precursors. The TMA was supplied by passing H_2 through a TMA bubbler which was kept at 35°C. Samples were grown in the nitride growth chamber (typical growth rate was 2000 Å/10 minutes) and then transported under vacuum into a second chamber where RHEED analysis was performed. After growth, surface composition analysis was performed by auger spectroscopy or secondary ion mass spectroscopy.

<u>Measurement of thickness and refractive index</u>

In various application the main parameters characterizing thin film Opticalwaveguides are the refractive index and the film thickness. There are several methods available for the determination of these parameters. One method that is particularly well adapted to this problem is the prism coupling technique which has several advantages when compared with other methods. Some of these advantages are: high precision measurements are easily obtained, waveguide mode information is readily available for other applications, and measurement of both thickness and index without prior knowledge of either parameter are possible.

A rutile prism (45°- 90°- 45°) coupler is employed to display the waveguide modes (N-lines). The detailed arrangement of measurement and measuring process can be found in many references[4]. The experiment is carried out by using different light sources with wavelengths of 632.8, 532.1, 514.5 and 488.0nm. The initial data such as refractive index and angles of rutile prism, and refractive index of substrate are carefully evaluated.

Using the precisely measured value of angles between the incident beam and the direction normal to the prism surface synchronous to the propagating modes m, where m is the order of the modes (taken from zero to m-1), the refractive index for each mode is obtained from the classical relation[6].

$$N_m = \sin \alpha_m \cos \varepsilon + (n_p^2 - \sin \alpha_m)^{1/2} \sin \varepsilon \qquad (1)$$

Where ε is the prism angle.

The unknown quantities for a given film are the refractive index (n) and thickness (d). In principle, two modes are sufficient to calculate these two values. The sample, which is single crystal thin film and optically anisotropic, makes the measurement complicated. The dispersion equation for a planar waveguide is given

$$Kd (n^2 - N_m^2)^{1/2} = W_m (n, N_m) \qquad (2)$$

Where

$$W_m(n, N_m) = m \pi + \Phi_0 (n, N_m) + \Phi_2 (n, N_m)$$

and

$$\Phi_i (n, N_m) = \arctan ((n / n_i)^{2\rho}(N_m^2 - n_i^2) / (n^2 - N_m^2))^{1/2} \qquad (3)$$

with

$$i = 0, 2,$$

where

$$\rho = 0 \quad \text{for TE mode}$$
$$\rho = 1 \quad \text{for TM mode}$$

With the measured data and above equations, the results for TE modes are shown in Table 1. The dispersion equation at room temperature can be expressed as

$$n^2 = A \lambda^2 / (\lambda^2 - B^2) \tag{4}$$

Table 1. Measured data and the results for TE mode

		λ_1 632.8 (nm)	λ_2 532.1 (nm)	λ_3 514.5 (nm)	λ_4 488.0 (nm)
N_m	0	2.06133	2.08495	2.08883	2.08999
	1	1.96219	2.00889	2.01828	2.01839
	2	1.80679	1.88380	1.90307	1.90441
n		2.094	2.105	2.1075	2.10875
d(μm)		0.72			

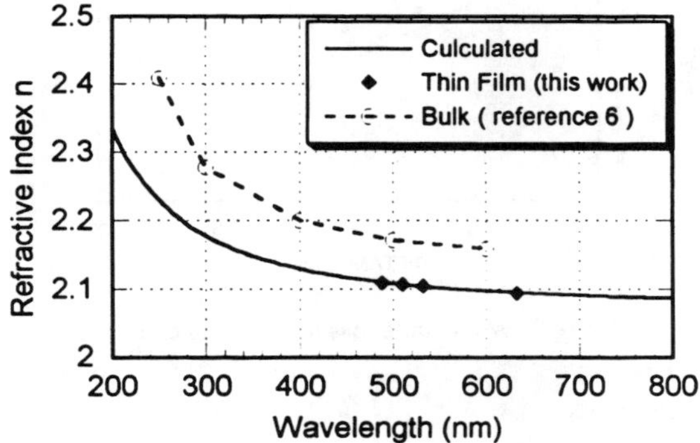

Figure 1. Dispersion curve for AlN

where A = 4.2943, B = 92.369 are constants obtained by fitting the data at different wavelength for the AlN film. The dispersion curve is given in Fig.1. Our results compare reasonably well with those given in reference[5,6].

<u>Losses in AlN waveguides</u>

The attenuation or loss is the important parameter of a waveguide when the light wave travels through the guiding layer. There are two main loss mechanism in semiconductor materials: scattering and absorption. Surface scattering loss in our case is the dominant mechanism. There are two different ways to collect the surface scattered light: fiber probe and lens probe. The prism coupler and the optical fiber probe are employed in our experiment for measuring the scattering loss of an AlN waveguide. A He-Ne laser beam is coupled into the waveguide with a rutile prism-coupler. An optical fiber with diameter of 1 mm is used as a probe to collect light scattered from the surface of the waveguide. One end of the fiber is held at right angle to the waveguide and scanned along the light path using a precision translation stage. The other end of the fiber is connected to a semiconductor detector. The incident light beam is chopped by a beam chopper. A lock-in amplifier and an oscilloscope were connected with the detector for measuring the scattered light intensity and monitoring the process. Also, we have employed a lens as a probe which scans the surface of the waveguide as a fiber probe. Fig. 2 shows a relative scattered light intensity versus length of the light path. The loss per unit length (α) is determined from the slope of the curve in Fig. 2 and is 1.96cm^{-1}. In this technique of loss measurement, the scattering centers are assumed uniformly distributed and the intensity of the scattered light in the transverse direction is proportional to the number of scattering centers. In this case, it is not necessary to collect all of the scattered light. The only requirement is that the detector aperture be constant.

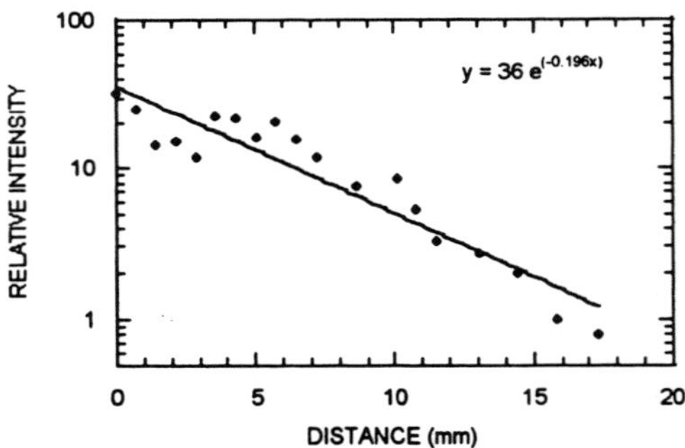

Figure 2. Waveguide loss for single mode

DISCUSSION AND CONCLUSION

The refractive indices from different samples show a small variation which can be attributed to film quality. It was believed that the smaller refractive index could be caused by nitrogen vacancies in the crystal[5]. Several TEM studies of our AlN samples grown on sapphire have shown the material columnar in nature with low angle grain boundary seperating individual grains. We suggest that the low angle grains (which are planar defects that can be decomposed into arrays of dislocation) are responsible for reducing the refractive index.

The experiments prove that the transverse optical fiber probe method is the most useful tool to measure the loss in waveguide with relatively large scattering. The loss coefficient α of the waveguide is very much dependent on the surface preparation of the film. Surface topography from the atomic force microscope shows the surface roughness with small crystal nucleations. The loss is different for different samples. The higher order modes show higher losses. The curve in Fig. 2 (a straight line variation) implies the accuracy of the measurements. It may be caused by randomly located large scattering centers or non-uniform scattering.

A good quality AlN waveguide can be produced if the growing condition is appropriate. The surface of the film can be polished if the thickness of the film is thick enough. Therefore, the surface quality will be improved and loss reduced. This AlN waveguide could be used in applications of surface acoustic devices and other optoeletronic devices.

ACKNOWLEDGMENTS

The authors acknowledge support from the National Science Foundation (NSF) and the US Army Research Office. The authors wish to thank Dr. Peizhen Zhou for surface analysis and topography measurements.

REFERENCES

1. K. Wongchotigul N. Chen, D. P. Zhang, X. Tang and M. G. Spencer, "Materials Letters," **26**, 223-226 (1996)
2. S. Nakamura, M. Senoh, S. Nagahama, N. Iwasa, T. Yamada, T. Matasushita, H. Kiyoku and Y. Sugimoto, "Jpn J. Appl. Phys". **35**, L74, (1996)
3. M.A. Khan, et al., "Appl. Phys. Lett.", **60**, 2917, (1992)
4. R.Ulrich and R.Torge, "Applied Optics" **12**, 12, 2901 (1973)
5. J. Bauer, L. Biste, and D. Bolze, "Phys. Stat. Sol. (a)", **39**, 173, (1977)
6. L. Roskovcova, J. Pastrnak, and R. Babuskova, Physica Status Solidi, **20**, k29 (1967)

*Permanent address: East China University of Technology, 516 Jun Gong Road, Shanghai 200093 China

OPTICAL STUDIES OF MOCVD $In_xGa_{1-x}N$ ALLOYS

B.D. Little*, W. Shan*, J.J. Song*, Z.C. Feng**, M. Schurman**, and R.A. Stall**
*Center for Laser and Photonics Research and Department of Physics, Oklahoma State
University, Stillwater, OK 74078
**EMCORE Corporation, 394 Elizabeth Avenue, Somerest, NJ 08873

ABSTRACT

We present the results of optical studies of $In_xGa_{1-x}N$ alloys ($0<x<0.2$) grown by metalorganic chemical vapor deposition on top of thick GaN epitaxial layers with sapphire as substrates. Photoluminescence (PL) and photoreflectance (PR) measurements were performed at various temperatures to determine the band gap and its variation as a function of temperature for samples with different indium concentrations. Carrier recombination dynamics in the alloy samples were studied using time-resolved luminescence spectroscopy. While the measured decay time for the alloy near-band-edge PL emissions was observed to be generally around a few hundred picoseconds at 10 K, it was found that the decay time decreased rapidly as the sample temperatures increased. This indicates a strong influence of temperature on the processes of trapping and recombination of excited carriers at impurities and defects in the InGaN alloys.

INTRODUCTION

Wide band-gap III-V nitrides have recently attracted much attention as the most promising material for the applications of light emitting devices operating in blue and ultraviolet (UV) spectral regions. Recently commercialized superbright high-efficiency blue light emitting diodes and demonstrated current-injection laser diodes are all based on $GaN/In_xGa_{1-x}N$ heterostructures using $In_xGa_{1-x}N$ layers as the active light emitting medium. The well recognized importance of $In_xGa_{1-x}N$ alloys due to its direct band gap covering a wide spectral range from UV (~365 nm for the GaN band gap) to red (~650 nm for the InN band gap) has promoted a great deal of interest in the study of this alloy system to improve its material quality and understand the physics involved. In this report, we present the results of a study of the optical properties of $In_xGa_{1-x}N$ alloys ($0<x<0.2$) grown on top of thick GaN epilayers by metalorganic chemical vapor deposition (MOCVD). Photoluminescence (PL) measurements were performed to assess the optical properties of samples with different alloy compositions. Photomodulation spectroscopy was used to determine the energy gap of the samples and to examine the effect of temperature on the band gap. Transient luminescence measurements were carried out to study photoluminescence decay processes in the alloy samples.

EXPERIMENTS

The InGaN alloy samples used in this work were nominally undoped single crystal epifilms grown by MOCVD. Before the depositions of alloys, thick GaN layers were grown on sapphire substrates at a temperature of ~1050°C with 20-nm GaN buffers. The alloy layers were deposited at a temperature around 800°C. The thicknesses of the $In_xGa_{1-x}N$ epitaxial layers were typically around a few thousand Ångstroms. Optical measurements in this work were carried out on the InGaN samples over a temperature range from 10 K up to room temperature (295 K). Samples were mounted onto the cold finger of a closed cycle refrigerator and cooled down to desired temperatures for the measurements. Photoluminescence spectra were measured using an experimental setup

Mat. Res. Soc. Symp. Proc. Vol. 449 © 1997 Materials Research Society

consisting of a HeCd laser as the excitation source and a 1-M double-grating monochromator connected to a photon-counting system. For photomodulation reflectivity measurements, quasi-monochromatic light dispersed by a 1/2-M monochromator from a xenon lamp and a chopped HeCd laser modulating beam were focused on the samples and the reflected signals were detected by a UV-enhanced photomultiplier tube connected to a lock-in amplification and data acquisition system. For transient luminescence measurements, a frequency-doubled pulsed dye laser synchronously pumped by a frequency-doubled modelocked Nd:YAG laser was used as the primary excitation source (2 ps, 76 MHz). The luminescence signals were dispersed by a 1/4 M monochromator and detected by a synchroscan streak camera with a temporal resolution of two picoseconds. The overall time resolution of the system is less than 15 ps.

RESULTS AND DISCUSSIONS

Shown in Fig.1 is a comparison of 10-K PL and photoreflectance (PR) spectra taken from an $In_{0.14}Ga_{0.86}N$ sample. The PL spectrum exhibits two dominant spectral features: a sharp, strong emission line at higher energy arising from the near-band-edge excitonic transitions in GaN, and a relatively broad strong luminescence structure related to the alloy. The weak spectral structures observed in between were mainly luminescence signatures involving donor-acceptor-pair transitions in the GaN layer. The spectral feature with derivative-like line shapes on the lower-energy side of the PR spectrum corresponds to the optical transition associated with the alloy band gap, and the differential spectral structures at high energy are free-exciton transitions from the edges of different valence bands to that of the conduction band of wurtzite GaN. Photoreflectance is a spectroscopic method utilizing modulation of the built-in electric field in the samples. While a PR spectrum exhibits sharp derivative-like structures on a featureless background due to the optical modulation, the nature of the derivative is not immediately clear. Therefore, it is necessary to fit the PR curve to different line-shape functional forms[1,2] in order to determine the energy positions associated with the optical transitions. We found that using a third derivative Gaussian line-shape functional

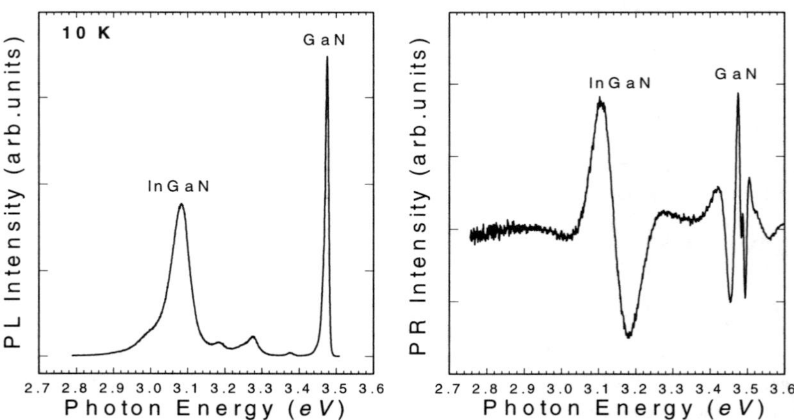

Fig.1. Photoreflectance (right curve) and photoluminescence (left curve) spectra of a $In_{0.14}Ga_{0.86}N$ sample at 10 K are shown for comparison.

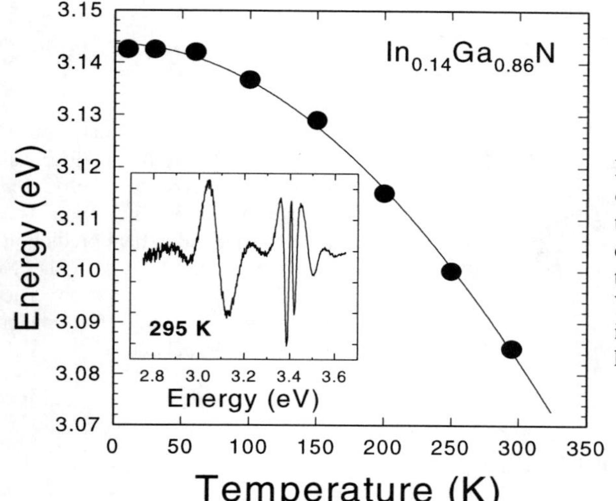

Fig.2. Temperature dependence of interband transition energy of the In$_{0.14}$Ga$_{0.86}$N sample. The solid curve is the least-squares fits to the experimental data using Varshni empirical equation. The inset shows a room-temperature PR curve of the sample.

form which is appropriate for describing band-to-band transitions results in a much better fit to the line positions and widths of the PR spectra than using a Lorentzian functional form[5,6]. This suggests that the optical transition observed at even the lowest temperature is of the nature of band-to-band transition rather than excitonic transition, presumably because of the strong inhomogeneous broadening due to alloying effects. The best fits to the 10-K PR spectral features yield an energy of 3.143 eV for the transition in In$_{0.14}$Ga$_{0.86}$N, indicating a large Stokes shift (~60 meV) between the PL peak position and the actual band-to-band transition energy.

With increasing temperature, the PL signal from the alloy layer quickly lost its intensity and spectral signature due to the increasingly enhanced nonradiative recombination processes and thermal broadening of the emission structures, whereas PR spectral structures associated with InGaN band gap could be well resolved up to room temperature for all alloy samples used. Fig. 2 shows the shift of optical transition energy as a function of temperature for the In$_{0.14}$Ga$_{0.86}$N sample. The solid line in the figure represents the best fit to the Varshni empirical equation[7]

$$E_0(T)=E_0(0)-\alpha T^2/(\beta+T). \tag{1}$$

Where $E_0(0)$ is the transition energy at 0 K, and α and β are constants referred to as Varshni thermal coefficients. The best fits yield $E_0(0)$=3.1438 eV, α=1.0x10^{-3} eV/K, and β=1196 K for the sample. The much better spectral resolution of PR spectroscopy compared to PL measurement enables us to determine the actual energy gaps for the InGaN alloy samples used in this work. In Fig.3, we plot the measured energy gaps of various samples at 10 K and 295 K against their alloy compositions, with 10-K PL results for comparison. Within the alloy composition range studied in this work, the PR results are in reasonably good agreement with the theoretical prediction for the dependence of In$_x$Ga$_{1-x}$N band gap on In concentration including the bowing effect[8]

$$E(x)=3.5-2.63x+1.02x^2 \ eV. \tag{2}$$

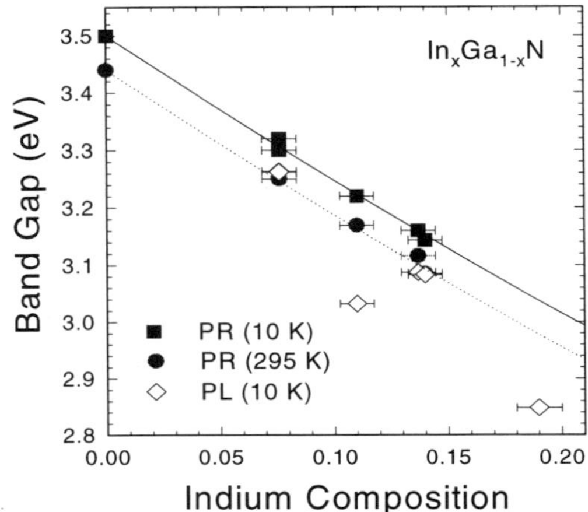

Fig.3. Energy gaps of various samples measured by PR at 10 K and 295 K vs. their alloy compositions. 10-K PL results are shown for comparison. The solid line is the theoretical prediction for the dependence of the In$_x$Ga$_{1-x}$N band gap on In concentration. The dotted line is that for room temperature (295 K) data.

On the other hand, the scattered PL data, as indicated by the figure, are apparently not a reliable way to map energy gap evolution for the alloy system because the PL signals are most likely associated with impurities and the effects of alloy disorder broadening.

We have also performed time-resolved PL measurements to study carrier recombination dynamics in the In$_x$Ga$_{1-x}$N samples. Fig. 4 shows the temporal evolution of spectrally integrated luminescence associated with an In$_{0.08}$Ga$_{0.92}$N sample at selected temperatures. The time evolution of the luminescence is dominated by exponential decay with a measured effective lifetime around 340 ps at 10 K. The temporal profile of the luminescence gradually evolved into an increasingly nonexponential decay with a drastic decrease of the effective lifetime for the main decay process as the sample temperature increased. At temperatures around 150 K, the deduced lifetime for the PL decay has already reached the limit of our instrumental resolution of 15 ps. Similar results were obtained for all samples used in this work: the measured decay time for near-band-edge PL emissions was generally around a few hundred picoseconds at 10 K, and it decreased rapidly as sample temperatures increased. The measured decay time was examined as a function of energy positions of the 10-K PL peak shown in the inset of Fig.4. The reason for doing so is to verify whether the PL structure resulted from recombination of localized excitons or emission directly involving impurity states and alloy potential fluctuations[9,10]. If the PL emission originated from localized excitons in the InGaN alloys, the lifetime is predicted to follow the so-called power-law dependence[11]:

$$\tau \propto E_{BX}^{3/2}, \tag{3}$$

where E_{BX} is the exciton localization energy. That means the measured lifetimes increase with increasing exciton localization energy which corresponds to a decrease in electron-hole wavefunction overlap. Our experimental results did not exhibit such a three-halves dependence of measured effective lifetime upon the energy position, suggesting impurity states and alloy potential fluctuations are most likely responsible for the alloy PL signals.

Although the radiative recombination processes involved in PL emissions in the InGaN samples are not of excitonic nature as aforementioned, it is still worthy to make a comparison of the results obtained in this work with the excitonic emission decay processes observed in pure GaN[12]. While

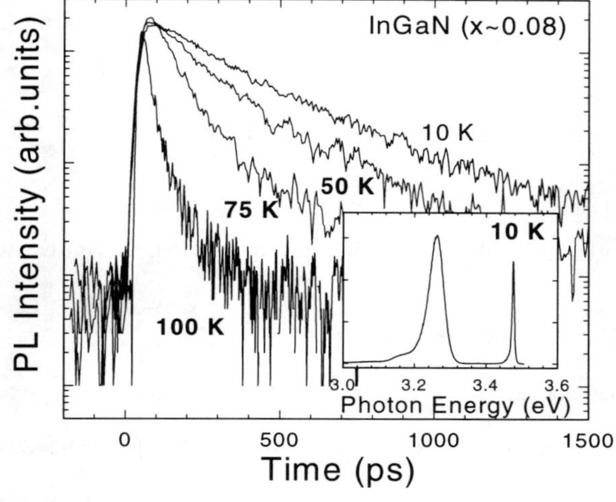

Fig.4. Temporal variation of the alloy PL peak for a In$_{0.08}$Ga$_{0.92}$N sample at selected temperatures. The inset shows the 10-K PL spectrum over a broad spectral range including both emissions from InGaN and GaN layers.

the measured effective lifetime for PL decay in InGaN samples was found to be an order of magnitude longer than that in GaN, the dependence of the PL decay process on temperature in InGaN alloys are much stronger than that in GaN[13]. Such a strong temperature influence suggests that although recombination from carriers bound to extrinsic states such as defects or impurities can be very efficient at low temperatures, the measured decay time is still determined by detailed decay kinetics in the alloy samples. The observed decrease of PL decay time along with the increasingly nonexponential transient characteristics with temperature indicates that incrementally stronger nonradiative relaxations occur as the temperature increases, resulting in continued faster decay of the photogenerated carrier population. The decrease of PL in both intensity and observed effective lifetime with temperature suggests that nonradiative processes including trapping and recombination at energy levels associated with impurities and defect centers at or near the mid gap (Schockley-Read-Hall recombination), surface recombination, and Auger recombination prevail in the InGaN alloys at relatively high temperatures. It is well known that the lifetime of photogenerated carriers is strongly dependent on the sample quality for semiconductor alloy systems[10]. The observations reported here indicate that the impurities and defects incorporated during the InGaN epitaxial growth generate PL quenching centers in the layers which adversely affect the carrier radiative recombination.

CONCLUSIONS

We have studied optical transition processes in In$_x$Ga$_{1-x}$N alloys using a few different approaches including both conventional and time-resolved PL measurements as well as PR spectroscopy. The highly sensitive PR spectroscopy allows us to unambiguously determine the band gap energy for the alloy samples within the alloy composition range (0<x0.2) studied. We found that the low-temperature PL emission from alloy layers are primarily recombinations directly involving impurity states and alloy potential fluctuations. Strong dependence of PL decay on temperature observed in the various alloy samples indicates that the trapping and recombination of photogenerated carriers

at impurities and defect centers are dominant channels in determining the carrier population decay process.

ACKNOWLEDGEMENTS

This work at OSU was supported by AFOSR, ARO, DARPA, and ONR.

REFERENCES

1. D.E. Aspnes, in <u>Optical Properties of Solids</u>, ed. M. Balkanski (North-Holland, Amsterdam, 1980), Chap. 4A.

2. F.H. Pollak and O.J. Glembocki, SPIE Proc. **946**, p.2(1988).

3. O.J. Glembocki and B.V. Shanabrook, Superlattices and Microstructure, **5**, p.235(1987).

4. H. Shen, S.H. Pan, F.H. Pollak, M. Dutta, and T.R. AuCoin, Phys. Rev. **B36**, p.9,384(1987).

5. O.J. Glembocki and B.V. Shanabrook, in <u>Semiconductors and Semimetals</u>, **vol.36**, ed. D.G. Seiler and CL. Little, (Academic Press, 1992), Chap. 4.

6. F.H. Pollak and H. Shen, Mater. Sci. Eng. **R10**, p.275(1993).

7. Y.P. Varshni, Physica, **34**, p.149(1967)

8. A.F. Wright and J.S. Nelson, Appl. Phys. Lett. **66**, p.3,051(1995).

9. M. Voos, R.F. Leheny, and J. Shah, in <u>Optical Properties of Solids</u>, edited by M. Balkanski, (North-Holland, Amsterdam, 1980), Chap. 6.

10. L. Pavesi and M. Guzzi, J. Appl. Phys. **75**, p.4,779(1994).

11. E.I. Rashba and G.E. Gurgenishvili, Sov. Phys. Solid State, **4**, p.759(1962).

12. W. Shan, X.C. Xie, J.J. Song, and B. Goldenberg, Appl. Phys. Lett. **67**, p.2,512(1995).

13. J.J. Song, W. Shan, T. Schmidt, X.H. Yang, A. Fischer, S.J. Hwang, B. Taheri, B. Goldenberg, R. Horning, A. Salvador, W. Kim, Ö. Aktas, A. Botchkarev, and H. Morkoç, in <u>Physics and Simulation of Optoelectronic Devices IV</u>, ed. W.W. Chow and M. Osiński, (SPIE Proceedings Series, 1996), pp.86-96.

OPTICAL TRANSITIONS AND RECOMBINATION LIFETIMES IN GaN AND InGaN EPILAYERS, AND InGaN/GaN AND GaN/AlGaN MULTIPLE QUANTUM WELLS

M. SMITH, J. Y. LIN, H. X. JIANG
Department of Physics, Kansas State University, Manhattan, KS 66506-2601

A. KHAN, Q. CHEN
APA Optics Inc, Blaine, MN 55449

A. SALVADOR, A. BOTCHKAREV, H. MORKOC
Coordinated Science Laboratory, University of Illinois at Urbana-Champaign, Urbana, IL 61801

ABSTRACT

Time-resolved photoluminescence (PL) has been employed to study the optical transitions and their dynamical processes in GaN and $In_xGa_{1-x}N$ epilayers, and $In_xGa_{1-x}N$/GaN and GaN/$Al_xGa_{1-x}N$ multiple quantum wells (MQW). We compare the results from both metal-organic chemical vapor deposition (MOCVD) and molecular beam epitaxy (MBE) grown samples. In addition, results are also compared with GaAs/$Al_xGa_{1-x}As$ MQW. It was found for all samples that the low temperature emission lines were dominated by radiative recombination transitions of either localized or free excitons, which demonstrates the high quality and purity of these III-nitride materials.

INTRODUCTION

GaN and its allied alloys have been recognized recently as among the most important semiconductors due to their potential applications for many optoelectronic devices which are active in the UV-blue wavelength region.[1] The possibilities of GaN based devices such as UV-blue lasers and light emitting diodes (LEDs), solar-blind UV detectors, and high-power electronics have fueled recent research in these materials. However, the understanding of their fundamental optical properties is still incomplete, in part to material quality in the past. In this work we present results from both epilayers and MQW samples grown by both MOCVD and MBE techniques. Results from different epilayers and MQW are compared with each other to extrapolate the mechanisms of optical transitions in these materials. Recombination lifetimes are compared to further our understanding of these materials.

Mat. Res. Soc. Symp. Proc. Vol. 449 © 1997 Materials Research Society

EXPERIMENT

Three of the samples considered here were grown by low pressure metal-organic chemical vapor deposition (MOCVD). The first is a nominally undoped 3.8 μm n-type (n = 5x10^{16} cm^{-3}) GaN epilayer, a second is a nominally undoped 0.25 μm n-type In$_x$Ga$_{1-x}$N (x ≈ 0.12) epilayer, and the third is an In$_x$Ga$_{1-x}$N/GaN MQW (x ≈ 0.15) composed of 45 periods of alternating 25 Å In$_x$Ga$_{1-x}$N wells and 25 Å GaN barriers. The others were grown by molecular beam epitaxy (MBE). These are a nominally undoped, insulating 4μm GaN epilayer, and a GaN/Al$_x$Ga$_{1-x}$N MQW (x ≈ 0.07) composed of 10 periods of alternating 25 Å GaN wells and 50 Å Al$_x$Ga$_{1-x}$N barriers. The time resolved PL spectra were measured using a picosecond laser spectroscopy system with an average output power of about 20 mW, a tunable photon energy up to 4.5 eV, and a spectral resolution of about 0.2 meV. The time resolution of the detection system is about 20 ps.[2,3]

RESULTS

In Fig. 1, we plotted the continuous-wave (CW) PL spectra of the MOCVD grown In$_x$Ga$_{1-x}$N/GaN MQW sample (open circles) obtained at (a) T = 10 K and (c) T = 300 K. For comparison, the PL spectra of an MOCVD grown InGaN epilayer is also shown. We have also plotted the CW PL spectra of the MBE grown GaN/Al$_x$Ga$_{1-x}$N MQW sample (open triangles) obtained at (b) T = 10 K and (d) T = 300 K. The PL spectra of an MBE grown GaN epilayer is also shown for comparison. In the In$_x$Ga$_{1-x}$N epilayer, the dominant transition line at low temperature is from the recombination of localized excitons.[2] As can be seen, the transition peak position in the In$_x$Ga$_{1-x}$N/GaN MQW is blue shifted. This blue shift is smaller then we expected from the confinement of the electrons and holes, possibly due to the slight differences in the In composition. The measured blue shift is approximately 16±2 meV. In the GaN epilayer, the dominating transition line at T = 10 K is due to the recombination of the ground-state of the A exciton.[3,4] The blue shift of the emission line from the MQW at room temperature (79 meV) is what we expected for our MQW structure with a 67% (33%) conduction (valence) band offset. One of the interesting features shown in Fig. 1 is that the blue shift observed at 10 K is only 54 meV for the GaN/Al$_x$Ga$_{1-x}$N MQW, or about 25 meV less than the shift at room temperature. We attribute this difference to the fact that the PL emissions in the GaN/Al$_x$Ga$_{1-x}$N MQW at low and room temperatures result from the recombinations of localized and free excitons, respectively. The exciton localization at low temperatures is caused by interface roughness of the MQW. The 25 meV difference then measures the localization energy, indicating a well thickness fluctuation of about ±4 Å.

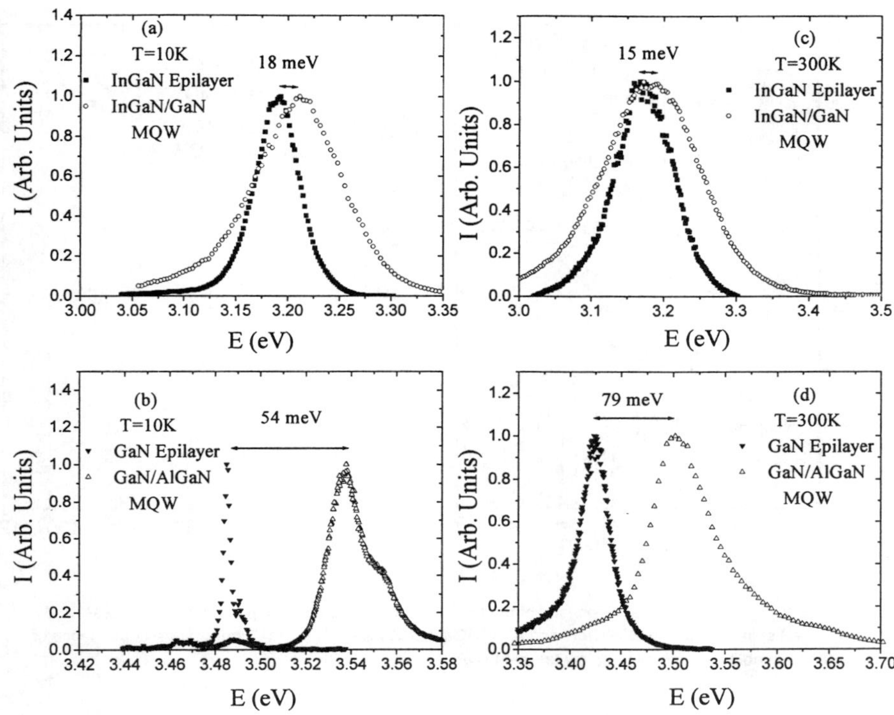

FIG. 1. CW PL spectra of an InGaN epilayer (filled squares) and In_xGa_{1-x}/GaN MQW (open circles) measured at (a) T=10 K and (c) T=300 K, and of a GaN epilayer (filled triangles) and GaN/Al_xGa_{1-x}N MQW (open triangles) measured at (b) T=10 K and (d) T=300 K.

The recombination dynamics of the PL emissions have also been studied. Fig. 2(a) shows the temporal responses of the GaN epilayer, GaN/Al_xGa_{1-x}N MQW, In_xGa_{1-x}N epilayer, and the In_xGa_{1-x}N/GaN MQW measured at the spectral peak positions of the respective samples at T = 10 K. As can be seen from Fig. 2(a), PL decay in all samples but the In_xGa_{1-x}N/GaN MQW can be well fit by a single exponential decay. The PL decay in the In_xGa_{1-x}N/GaN MQW was fit using a two-exponential function with the fast component contributing over 85% of the PL signal. The temperature dependencies of the recombination lifetime τ are shown in Fig. 2(b). In the In_xGa_{1-x}N epilayer (filled triangle), the localized exciton lifetime is about 0.53 ns at T < 40 K and increases almost linearly with temperature from 40 K up to 100 K. Comparing this with the In_xGa_{1-x}N/GaN MQW (open triangle), where only the recombination lifetime of the fast component is plotted, only a slight temperature dependency of the recombination lifetime is seen, up to 100 K. Considering the GaN epilayer (filled square), the recombination lifetime of the

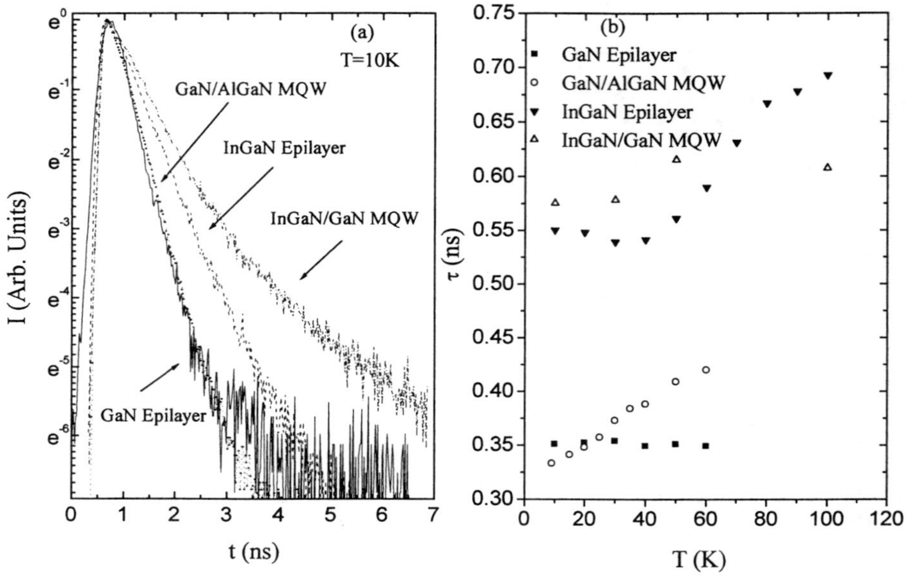

FIG. 2. (a) The temporal responses of the PL measured at the spectral peak positions in GaN and InGaN epilayers, and InGaN/GaN and AlGaN/GaN MQW at T=10K. (b) Temperature dependence of the recombination lifetimes at low temperatures.

A-exciton ($\sim$ 0.35 ns) is nearly temperature independent below 60 K. In the GaN/Al$_x$Ga$_{1-x}$N MQW (open circle), however, τ increases linearly with temperature up to 60 K.

All samples are seen to have recombination lifetimes which are either temperature independent, or which show a linear increase with temperature at low temperatures. A similar linear behavior at low temperatures has been observed previously in GaAs/Al$_x$Ga$_{1-x}$As MQW and is now regarded as a unique property of radiative excitonic recombination in MQW.[5] Thus, we see that radiative recombinations are the dominant process in all samples at low temperatures. A summary of these results can be seen in Table I, which also includes results from GaAs/Al$_x$Ga$_{1-x}$As MQW for comparison. Comparing the behavior between the three MQW of comparable well size, the behavior of the In$_x$Ga$_{1-x}$N/GaN MQW stands out. It shows no clear linear dependence, unlike the GaN/Al$_x$Ga$_{1-x}$N MQW and the GaAs/Al$_x$Ga$_{1-x}$As MQW.[5] The slope of τ vs temperature T, dτ/dT, for the GaN/Al$_x$Ga$_{1-x}$N MQW is about 1.8×10^{-3} ns/K while that of the GaAs/AlGaAs of the same well width is about 1.7×10^{-2} ns/K. The high quality In$_x$Ga$_{1-x}$N epilayer, on the other hand, does show this linear dependence, with a slope dτ/dT $\sim$ 4×10^{-3} ns/K, unlike the GaN

TABLE I. Recombination lifetime behavior of epilayers and MQW

	MOCVD InGaN/GaN MQW	MOCVD InGaN Epilayer	MOCVD GaN Epilayer	MBE[3] GaN Epilayer	MBE GaN/AlGaN MQW	MBE[5] GaAs/AlGaAS MQW		
Well (Å)	25	—	—	—	25	25	40	55
Temperature (K)	100	$40 \lesssim T \leq 100$	≤ 60	≤ 30	≤ 60	≤ 30	≤ 25	≤ 20
$d\tau/dT$ (10^{-3} ns/K)	—	4	~0	~0	1.8	17.5	58.8	33.8
τ_o (ns)	0.57	0.54	0.35	0.22	0.32	0.32	0.44	0.82

epilayer samples which have a constant recombination lifetime at low temperatures. By extrapolating the plots in Fig. 2(b) to T=0 K, the radiative recombination lifetime, τ_o, of the samples can be determined. The lifetime of the localized exciton in the $In_xGa_{1-x}N$ epilayer is slightly less then that of the $In_xGa_{1-x}N$/GaN MQW, 0.54 ns compared to 0.57 ns. For the GaN epilayer and the GaN/$Al_xGa_{1-x}N$ MQW, the behavior was different.

CONCLUSIONS

In conclusion, time-resolved PL has been employed to study the optical transitions as well as recombination lifetimes in GaN and $In_xGa_{1-x}N$ epilayers, and $In_xGa_{1-x}N$/GaN and GaN/$Al_xGa_{1-x}N$ MQW. The results have been compared with GaAs/Al_xGa_{1-x}As MQW. Our results have revealed (i) the PL emissions in $In_xGa_{1-x}N$ epilayers result primarily from localized exciton recombination, (ii) the quantum confinement effect of excitons in GaN/$Al_xGa_{1-x}N$ and $In_xGa_{1-x}N$/GaN MQW, (iii) recombination lifetimes either increase with temperature ($In_xGa_{1-x}N$ epilayers, $In_xGa_{1-x}N$/GaN MQW, GaN/$Al_xGa_{1-x}N$ MQW) or is independent of it (GaN epilayer) , (iv) dominance of the radiative recombination for all samples at low temperatures, indicative of the high quality and purity of the samples used here. These findings are expected to have important implications on optoelectronic designs employing GaN-based MQW.

ACKNOWLEDGMENTS

The research at Kansas State University is supported by ONR/ BMDO, ARO, and NSF (DMR-9528266) and monitored by Dr. Yoon S. Park and Dr. John Zavada. The research at the University of Illinois is supported by ONR, AFOSR, and BMDO and monitored by Dr. Yoon S. Park, Dr. C. E. Wood, Dr. G. L. Witt, Dr. K. Wu, Dr. J. Johnson, and Dr. S. Hammonds.

REFERENCES

1. H. Morkoc, S. Strite, G. B. Gao, M. E. Lin, and M. Burns, J. Appl. Phys. **76**, 1363 (1993); S. N. Mohammad, A. Salvador, and H. Morkoc, Proc. IEEE **83**, 1306 (1995).

2. M. Smith, G.D. Chen, J. Y. Lin, H. X. Jiang, M. Asif Khan, and Q. Chen, Appl. Phys. Lett. **69**, 2837 (1996).

3. M. Smith, G.D. Chen, J. Z. Li, J. Y. Lin, H. X. Jiang, A. Salvador, W. K. Kim, O. Aktas, A. Botchkarev, and H. Morkoc, Appl. Phys. Lett. **67**, 3387 (1995).

4. D. C. Reynolds, D. C. Look, A. Salvador, A. Botchkarev, and H. Morkoc, J. Appl. Phys. **80**, 3387 (1995).

5. J. Feldmann, G. Peter, E. O. Gobel, P. Dawson, K. Moore, C. Foxon, and R.J. Elliot, Phys. Rev. Lett, **59**, 2337 (1987).

IN-PLANE OPTICAL ANISOTROPIES OF $Al_xGa_{1-x}N$ FILMS
IN THEIR REGIONS OF TRANSPARENCY

U. ROSSOW*†, N. V. EDWARDS#, M. D. BREMSER#, R. S. KERN#, H. LIU**, R. F. DAVIS#, and D. E. ASPNES*#§
*Department of Physics, North Carolina State University, Raleigh, NC 27695-8202
†Now at the Department of Physics, University of Ilmenau, Ilmenau, Germany
#Department of Materials Science and Engineering, North Carolina State University, Raleigh, NC 27695-7907

##Emcore, 35 Elizabeth St., Somerset, NJ 08873. Now at Hewlett-Packard, San Jose, CA 95131
§Corresponding author: aspnes@unity.ncsu.edu

ABSTRACT

GaN, $Al_xGa_{1-x}N$, and AlN layers exhibit interference oscillations and bandgap-related features in their reflectance-difference (-anisotropy) (RD/RA) spectra. We concentrate on the interpretation of interference-related data, providing a general expression for these optical anisotropies and discussing mechanisms that originate in the layers themselves. These include anisotropic strain in the plane of the layer, a tilt of the c axis with respect to the surface normal, and non-normal-incidence illumination. We estimate the magnitudes of these contributions, and show that they are consistent with those observed. In principle these contributions can be separated by their different azimuthal dependences. The complex pattern of the data for $Al_xGa_{1-x}N$ and AlN indicate that contributions from several layers are present.

INTRODUCTION

Wide-bandgap semiconductors have attracted much recent attention with the realization of blue light emitting and laser diodes [1,2]. These materials are well suited for optical characterization. Below-bandgap radiation can access buried layers and interfaces, while above-bandgap radiation returns information about the surface and near-surface region of the top layer. Here, we report and discuss data on GaN, $Al_xGa_{1-x}N$, and AlN layers obtained by reflectance-difference (-anisotropy) spectroscopy (RDS/RAS), a technique that capitalizes on reduced symmetry of surfaces, films, and interfaces with respect to the underlying material to provide information about them [3,4]. In particular, our data exhibit bandgap-related features, which in principle allow compositions to be determined over the entire alloy range, and below-bandgap interference oscillations, which we describe in terms of propagation in a weakly optically anisotropic layer. We formulate the problem in general using virtual-interface (V-I) theory [5], and discuss three mechanisms specific to the film in its region of transparency: (1) anisotropic strain in the plane of the layer; (2) a tilt of c axis of one or more layers with respect to the surface normal; and (3) non-normal incidence illumination by the optical beam. The estimated magnitudes of these mechanisms are consistent with our data, and in principle the individual contributions can be separated by their energy and azimuthal dependences.

EXPERIMENTAL

Optical anisotropy data were obtained from 1.5 to 5.5 eV at an angle of incidence of 0.6° using a standard RD configuration [6]. For the AlN sample the signal was strong enough to allow measurements to be extended to somewhat higher photon energies. Samples were mounted on a rotation stage to determine the azimuthal dependences of the signals, to find their

Mat. Res. Soc. Symp. Proc. Vol. 449 © 1997 Materials Research Society

minimum and maximum amplitudes, and to establish an unambiguous baseline. To minimize instrumentation artifacts, the spectra presented here were calculated as the average of the difference between spectra measured for linear polarizations along the two different principal axes in the plane of the surface, which are separated by an azimuth angle of 90°.

Layers were grown by organometallic chemical vapor deposition (OMCVD) on on-axis sapphire and offcut 6H-SiC(0001) substrates as described elsewhere [7]. The layers were between 1 and 3 μm thick except for the AlN layer, which was 0.250 μm thick. Samples were measured as-grown after roughening the reverse side, and also after chemical treatment to remove overlayers [8].

RESULTS

Representative RD spectra of GaN, $Al_xGa_{1-x}N$, and AlN layers are shown at the bottom, middle, and top, respectively, of Fig. 1. The thicknesses of these particular layers, as determined from the spacing of the interference oscillations, are 1.3, 1.7, and 0.29 μm, respectively. The thickness determined for the AlN film is within the experimental uncertainty of the nominal 0.25 μm thickness of this layer. These spectra are typical, exhibiting interference oscillations below the bandgap, structure at the bandgap, and an absence of structure above the bandgap within our spectral range. The oscillations are nearly independent of the position of the light beam on the sample, which indicates that the layer thicknesses are homogeneous. The below-bandgap oscillations typically do not vanish completely at any single azimuth, as expected for interference-related anisotropies as discussed below. On the other hand, for the bandgap-related features an azimuth can always be found where the associated RD structure vanishes, allowing a positive identification of bandgap features as shown for AlN in Fig. 2. As expected, the bandgap-related features shift to higher energies with increasing Al content.

While attention in RD measurements is usually directed toward surface or interface effects, we concentrate here on the below-bandgap oscillations, which are clearly related to transmission of light within the outermost and/or underlying layers. We can describe these features completely generally in terms of the V-I expression [5]

$$\tilde{r}_{voa} = \frac{\tilde{r}_{oa} + Z\tilde{r}_v}{1 + Z\tilde{r}_v \tilde{r}_{oa}} \tag{1}$$

where $\tilde{r}_{voa}$ and $\tilde{r}_{oa}$ are the complex reflectances of the sample and the overlayer/ambient interface, $\tilde{r}_v$ is the virtual reflectance at the virtual interface located a distance d beneath the surface (here taken to be the same as the overlayer thickness), and $Z = \exp(2ik_o d)$, where k_o is the propagation vector in the overlayer. We note that Eq. (1) is formally identical to the three-phase-model representation recently used by Yasuda et al. [9] to discuss surface and interface anisotropies of heteroepitaxial ZnSe on (001) GaAs, except that $\tilde{r}_v$ replaces the term $\tilde{r}_{so}$ describing reflection at the substrate-overlayer boundary. The difference is primarily one of interpretation: V-I theory shows that $\tilde{r}_v$ (or for that matter $\tilde{r}_{so}$ in the three-phase model) incorporates all optical effects below the outermost layer, defined as the outermost region where the dielectric response is independent of d. Further discussions can be found in refs. [5] and [10].

In general anisotropy effects can arise from the r_v term if the substrate, underlying layers, or interfaces are optically anisotropic, or from the k_o term if the overlayer itself is optically anisotropic. The appearance of optical anisotropy related to the overlayer is not surprising since hexagonal materials are intrinsically birefringent, with the component of the dielectric tensor along

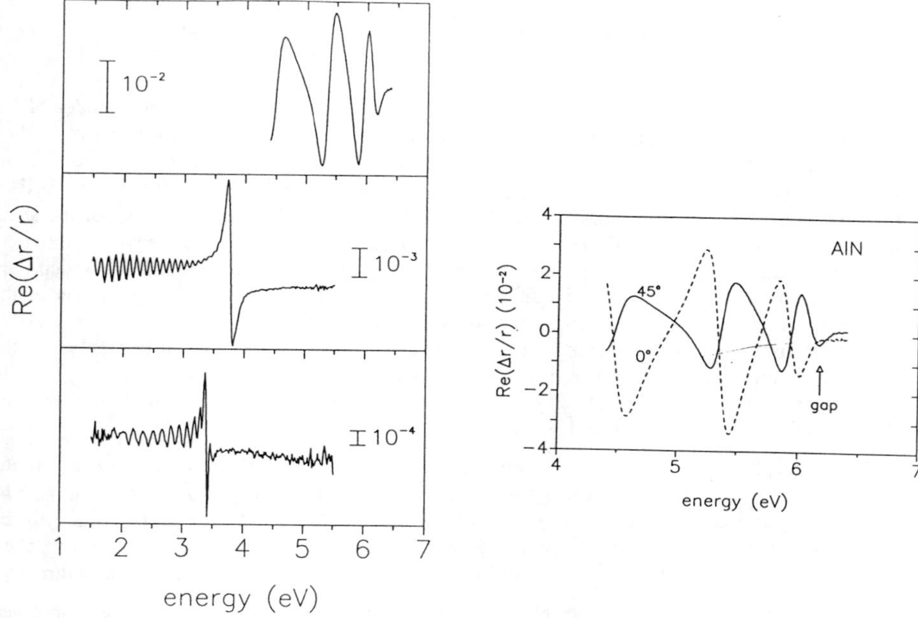

Fig. 1. RD spectra for layers of GaN (bottom), $Al_xGa_{1-x}N$ (middle), and AlN (top), all grown on 6H-SiC.

Fig. 2. RD spectra of an AlN film for two sample azimuths differing by 45°. While the bandgap-related features vanish at some azimuth, such cancellation does not occur for the below-bandgap features.

the c axis being different from that perpendicular to the c axis. However, the sixfold rotational symmetry about the c axis in these materials ensures that this anisotropy could not be observed unless (1) the sixfold rotational symmetry is broken by the presence of an inhomogeneous strain in the plane of the layer; or (2) $\mathbf{k}_o$ is not entirely parallel to the c axis, i.e., $\mathbf{k}_o$ and the c axis are not collinear. The latter situation can arise if (2a) the c axis is perpendicular to the surface but the probe beam is not incident normal to the surface, or (2b) the c axis of the overlayer is inclined to the surface normal. For any of these three conditions the two orthogonal propagation modes in any given layer become nondegenerate, exhibiting different amplitudes of the propagation vector. In principle these mechanisms can be distinguished by their azimuthal dependences: a tilted c axis and in-plane strain should exhibit one- and twofold rotational symmetries, respectively, while non-normal-incidence illumination should yield no azimuthal dependence, and thus be undetectable when spectra are calculated by taking differences as done here.

The relative anisotropy $\Delta\tilde{r}/\tilde{r} = (\tilde{r}_\alpha - \tilde{r}_\beta)/[2(\tilde{r}_\alpha + \tilde{r}_\beta)]$ for either case can be determined directly from Eq. (1) by taking appropriate differentials, and we find:

$$\frac{\Delta \tilde{r}_{voa}}{\tilde{r}_{voa}} = (\Delta \tilde{r}_v + \frac{4\pi i\, d\, E\, \Delta\varepsilon}{\bar{n}\, h\, c})\, \frac{1 - \tilde{r}_{oa}^2}{(\tilde{r}_{oa} + Z\tilde{r}_v)(1 + Z\tilde{r}_{so}\tilde{r}_{oa})}\, e^{2ikd}, \tag{2}$$

where $\Delta \tilde{r}_v$ is the anisotropy of $\tilde{r}_v$, $\Delta\varepsilon$ expresses the difference between the amplitudes of the wavevectors of the two propagating modes in the overlayer as a difference between their effective dielectric functions, $\bar{n}$ is the average refractive index of the two modes, and E is the photon energy. In Eq. (2) both $\Delta \tilde{r}_v$ and the accumulated phase difference $\Delta k\, d = \pi\, E\, \Delta\varepsilon\, d /(\bar{n}\, h\, c)$ in the propagating modes are treated in first order, and the effect of the small difference of $\bar{n}$ on interface reflectances is ignored. The fact that $\exp(2ikd)$ is common to both virtual-reflectance and propagation terms shows that interference oscillations in RD spectra are a natural consequence of the optical anisotropy of the layer or any underlying portion of the sample, even though such anisotropy may be small. In particular, Eq. (2) shows that, to the extent that dispersion in the relevant dielectric functions of the overlayer can be neglected, the amplitude of the oscillations arising from the propagation term should increase linearly with E, in agreement with the GaN data of Fig. (1). This provides a means of discriminating layer from interface anisotropies, which would be expected to show dispersion.

Since the physical origin of the effects described by the propagation term of Eq. (2) is the slight difference in optical thickness nd of the two modes, and since this difference can also be simulated by a slight change of energy, this term of Eq. (2) should be closely related to the numerical derivative of $\Delta \tilde{r}/\tilde{r}$ with respect to energy. In fact, this derivative is obtained by replacing the combination $E\,\Delta\varepsilon/\bar{n}$ in Eq. (2) with $2(n + E\, d\bar{n}/dE)$, illustrating the connection directly.

To assess the magnitude of the relative phase difference necessary to obtain the observed values of $\Delta \tilde{r}/\tilde{r}\,|_{pp}$ by means of the propagation term in Eq. (2), we consider this term in detail. For GaN where $\varepsilon_a = 5.53 + i0.00$ at 2.5 eV [11] we have $|\tilde{r}_{so}| \approx 0.08 << 1$ and $|\tilde{r}_{oa}| \approx 0.5$ at 2.5 eV, so $\Delta k\, d$ can be written approximately as $\Delta k\, d \approx \Delta \tilde{r}/\tilde{r}\,|_{pp}/6$, where $\Delta \tilde{r}/\tilde{r}\,|_{pp}$ is the peak-to-peak value of the interference envelope at 2.5eV. From the GaN data of Fig. 1 we have $\Delta \tilde{r}/\tilde{r}\,|_{pp} \approx 1.0 \times 10^{-4}$, whence $\Delta k\, d \approx 0.001°$ and $\Delta k \approx 0.0007°/\mu m$. Thus the assumption that the accumulated phase difference is small is clearly satisfied for this example. In addition, the result shows that RDS is sensitive to extremely small differences of the accumulated phase difference.

We now consider microscopic mechanisms due to the overlayer, beginning with inhomogeneous relaxation of stress. In the quasicubic model [12], isotropic stress in the plane of the layer is treated simply as an additive term to the crystal potential. However, for anisotropic stress relaxation the problem is much more difficult. To estimate the magnitude of the effect we bypass some of these difficulties by assuming that the transitions that contribute most strongly to $Re(\varepsilon)$ near 2.5 eV are those associated with the E_1 critical points of the equivalent cubic material, i.e., the critical points along the (111), ($\bar{1}$11), (1$\bar{1}$1), and (11$\bar{1}$) directions in reciprocal space. This model is clearly incomplete since other, lower-symmetry critical points will also contribute, but we can expect it to yield an estimate of the correct order of magnitude. Representing the dielectric response in the transparent region in Sellmeier form and using the appropriate polarization selection rules we find

$$\Delta\varepsilon(E) \approx \frac{2E_p^2\,\Delta_\sigma^2\,(4E_1^4 - (E_1^2 + E^2)^2)}{9(E_1^2 - E^2)^4}, \tag{3}$$

where E_p is the plasma energy, E_1 is the unperturbed threshold energy of the E_1 transition, and Δ_σ is the splitting between the ($\bar{1}$11) and (1$\bar{1}$1) levels for a (110) in-plane strain [for this case the (111) and ($\bar{1}11$) critical points do not contribute]. Using representative values $E_p = 22$ eV and $E_1 = 9$ eV, and considering an extreme upper limit of 0.1 eV for Δ_σ, we find that $\Delta\varepsilon \approx 0.0006$, which at 2.5 eV leads to an accumulated phase difference per unit length of $\Delta k \approx 0.07°/\mu m$. Thus this mechanism can easily provide anisotropy signals on the scale of Fig. 1.

We now consider the situation where $\mathbf{k}$ and the c axis are not parallel. If the components of the dielectric tensor parallel and perpendicular to c are ε_c and ε_a, respectively, then for a small angle θ_{kc} between $\mathbf{k}$ and the c axis the difference $\Delta\varepsilon$ in the dielectric functions of the two modes is given by $\Delta\varepsilon = \varepsilon_a \theta_{kc}^2 (1 - \varepsilon_a / \varepsilon_c)$. For GaN at 2.5 eV where $\varepsilon_c = 5.98 + i0.0$ [11] and ε_a is as given above, we have $\Delta k \approx 1 \times (\theta/°)^2 \, °/\mu m$. Thus a value of $\Delta r/r |_{pp} \approx 10^{-4}$ would correspond to a value of θ_{kc} of approximately 0.4°. In our present configuration the beam is incident at an angle of 0.6°, which is reduced by refraction to an internal angle of 0.3°. If the c axis were parallel to the surface normal, this would lead to signals of the order of 10^{-4}, although as pointed out above this contribution exhibits no azimuthal dependence and hence cannot appear in the difference spectra of the type shown in Fig. 1. On the other hand, if the c axis were inclined by this amount from the surface normal, the signal strengths would be the same and would be observable in these spectra.

For the $Al_x Ga_{1-x} N$ sample the accumulated phase difference per unit length is $0.025°/\mu m$, an order of magnitude larger than that of the GaN sample. In this case the situation is more complicated, as evidenced by the decrease of oscillation envelope with increasing energy. This behavior can be reproduced numerically by adding a second, relatively thin anisotropic layer below the first, which in Eq. (2) enters through r_v. A contribution such as this could arise from the AlN buffer layer.

For the AlN sample the accumulated phase difference per unit length is still larger. Here, as shown in Fig. 2, it was impossible to obtain a null signal for any azimuth at any energy except for the bandgap feature near 6.1eV. Again, numerical simulations show that this is the type of behavior expected if another layer with a different symmetry axis is involved. In general, our simulations show that two anisotropic layers on an isotropic substrate can exhibit rather complicated behavior depending on the properties of the layers, for example beat patterns similar to that suggested by the $Al_x Ga_{1-x} N$ spectrum in the middle of Fig. 1. These results will be discussed in more detail elsewhere.

We comment finally on the bandgap-related features, which contain infomration about alloy composition. The interesting aspect here is the bandgap features are large enough to allow them to be followed to the AlN limit, although in view of the mecanisms discussed above this capability will clearly be sample-dependent. We have analyzed these by differentiating the $\Delta\tilde{r}/\tilde{r}$ spectra with respect to energy and fitting a Lorentz oscillator lineshape to the resulting real and imaginary parts. Estimated accuracies are about 0.05eV for AlN and 0.005eV in the other cases. These aspects will also be discussed elsewhere.

CONCLUSIONS

We have presented a general expression for describing optical anisotropy in the region of transparency of an overlayer, as seen in the data of Fig. 1. We have also discussed several mechanisms by which this anisotropy can occur within the overlayer: (1) anisotropic strain in the plane of the layer, (2) a tilt of the c axis with respect to the surface normal, and (3) non-normal-incidence illumination. The magnitudes of these contributions are consistent with data and in principle can be separated by their different azimuthal dependences. The complex pattern of the data for $Al_x Ga_{1-x} N$ and AlN indicate contributions from several mechanisms, including contributions from buffer layers.

ACKNOWLEDGMENTS

It is a pleasure to acknowledge financial support from the Alexander von Humboldt Foundation and the Office of Naval Research (ONR) under contract N-00014-93-1-0255.

REFERENCES

1. S. Nakamura, M. Senoh, N. iwasa, and S. Nagahama, J. Appl. Phys. Jpn. **34**, L797 (1995).

2. V. A. Dmittiev, K. Irvine, C. H. Carter, Jr., A. S. Zubrilov, and D. V. Tsvetkov, Appl. Phys. Lett. **67**, 115 (1995).

3. W. Richter, Philos. Trans. Roy. Soc. ser. A **34**, 453 (1993) and references therein.

4. D. E. Aspnes, Mater. Sci. and Engineering **B30**, 109 (1995) and references therein.

5. D. E. Aspnes, J. Opt. Soc. Am. **A10**, 974 (1993).

6. D. E. Aspnes, J. P. Harbison, A. A. Studna, L. T. Florez, and M. K. Kelly, J. Vac. Sci. Technol. **A6**, 1327 (1988).

7. T. W. Weeks, Jr., M. D. Bremser, K. S. Ailey, E. Carlson, W. G. Perry, and R. F. Davis, Appl. Phys. Lett. **67**, 401 (1995).

8. N. V. Edwards, M. D. Bremser, T. W. Weeks, Jr., R. S. Kern, R. F. Davis, and D. E. Aspnes, Appl. Phys. Lett. **69**, 2065 (1996).

9. T. Yasuda, K. Kimura, S. Miwa, L. H. Kuo, C. G. Jin, K. Tanaka, and T. Yao, Phys. Rev. Lett. **77**, 326 (1996).

10. D. E. Aspnes, J. Vac. Sci. Technol. **A14**, 960 (1996).

11. Landolt-Bornstein tables, data for GaN.

12. J. J. Hopfield, J. Phys. Chem. Solids **15**, 97 (1960).

STRAIN EFFECTS ON EXCITONIC TRANSITIONS IN GaN

W. Shan*, R.J. Hauenstein*, A.J. Fischer*, J.J. Song*, W.G. Perry**, M.D. Bremser**, R.F. Davis** and B. Goldenberg***
*Center for Laser and Photonics Research, Oklahoma State University, Stillwater, OK 74078
**Department of Materials Science and Engineering, North Carolina State University, Raleigh, NC 27695
***Honeywell Technology Center, Plymouth, MN 55441

ABSTRACT

We present the results of experimental studies of the strain effects on the excitonic transitions in GaN epitaxial layers on sapphire and SiC substrates. Photoluminescence and reflectance spectroscopies were performed to measure the energy positions of exciton transitions and X-ray diffraction measurements were conducted to examine the lattice parameters of GaN epitaxial layers grown on different substrates. Residual strain induced by the mismatch of lattice constants and thermal-expansion between GaN epitaxial layers and substrates was found to have a strong influence in determining the energies of excitonic transitions. The overall effects of the strain generated in GaN is compressive for GaN grown on sapphire and tensile for GaN on SiC substrate. The uniaxial and hydrostatic deformation potentials of wurtzite GaN were derived from the experimental results. Our results yield the uniaxial deformation potentials $b_1 \approx -5.3$ eV and $b_2 \approx 2.7$ eV, as well as the hydrostatic components $a_1 \approx -6.5$ eV and $a_2 \approx -11.8$ eV.

INTRODUCTION

Great efforts have recently been devoted to the preparation of high quality GaN crystals and epitaxial films, the characterization of GaN crystals and films, and the development of devices using GaN based materials. In the course of these studies, a number of investigations on the optical properties of GaN have revealed that there is a relatively large difference in the reported values of observed energy positions of excitonic transitions in GaN[1-8]. It is found that the energy positions vary from sample to sample depending on the epitaxial layer thickness, growth techniques and substrate materials. Such discrepancy has been attributed to the effects of residual strain in the epilayers due to the mismatch of lattice parameters and coefficients of thermal expansion between GaN and the substrate materials[5,8,9]. In this report, we present the results of experimental studies of the strain effects on the excitonic transitions in GaN epitaxial layers on sapphire and SiC substrates. The residual strain built in GaN epitaxial films induced by the mismatch of lattice constants and thermal-expansion between GaN epitaxial layers and substrates was found to play an important role in determining the energies of excitonic transitions. The overall effects of the strain generated in GaN is compressive for GaN grown on sapphire and tensile for GaN on SiC substrate. The uniaxial and hydrostatic deformation potentials were derived from the experimental results.

EXPERIMENTS

The GaN samples used in this work were nominally undoped single-crystal films grown by metalorganic chemical vapor deposition on (0001) 6H-SiC or basal-plane sapphire substrates. AlN buffer layers were deposited on the substrates at about 775°C before the growth of the GaN epilayers. Photoluminescence (PL) measurements were conducted with a cw HeCd laser (325 nm) as the

841

Mat. Res. Soc. Symp. Proc. Vol. 449 © 1997 Materials Research Society

excitation source and a 1-M double-grating monochromator connected to a photon-counting system. For reflectance measurements, the quasi-monochromatic light dispersed by a ½-M monochromator from a xenon lamp was focused on samples, and the reflectance signals were detected using a lock-in amplification system. To determine the lattice parameters of GaN epitaxial films, four-crystal X-ray rocking curves were measured. Absolute lattice parameter measurements were performed in the triple-axis mode of a Philips high-resolution diffractometer with four-bounce Ge (220) incident beam optics and three-bounce Ge (220) diffracted beam optics. The resolution limit of this configuration is ~10 Arcsec. The data were collected using 2θ-ω scans, with corrections made for refraction. The accuracy of lattice parameters measured using this system is predicted to be better than 0.0002 Å. The high resolution of the triple-bounce analyzer crystal and the steps taken to eliminate inaccuracies due to the 2θ zero error as well as sample centering account for this. The lattice parameter c was measured using the (002) reflection and a was measured using both the (002) and (015) reflections.

RESULTS AND DISCUSSIONS

<u>Effects of Residual Strain on Excitonic Transitions</u>

The GaN based samples studied in this work all exhibit strong near-band-edge exciton luminescence. Results of PL measurements from a 3.7-μm GaN epilayer on SiC and a 4.2-μm sample with sapphire substrate are shown in Fig. 1. The intensity of the strongest emission line marked by BX in Fig.1 is the emission from the radiative recombination of excitons bound to neutral donors.[4] The second strongest luminescence structure, together with the weak emission feature on the higher energy side, are radiative recombination of the intrinsic free exciton. As clearly illustrated by the figure, the values of PL transition energies obtained here from GaN on SiC substrate are lower than those from GaN on sapphire. More indicative results are presented in Fig.2, where three reflection spectra taken from two GaN epilayer grown on sapphire and a GaN epilayer on SiC at 10 K are given. Three exciton resonances associated with the transitions referred to as the A, B, and

C exciton transitions[4-7] between the bottom of the conduction band (Γ_7^C) and three topmost valence band edges ($\Gamma_9^V+\Gamma_7^V+\Gamma_7^V$) are indicated by vertical arrows. The energy positions of these transitions are listed in Table I. The differences in the measured exciton transition energies can be attributed to the effects of residual strain in the epilayers due to the mismatch of lattice parameters and coefficients of thermal expansion between GaN and the substrate materials[5,8,9]. The effects of strain can be further evidenced by the variations of the lattice parameters of GaN relative to that of the virtually strain-free bulk GaN as shown in the inset of the figure, where the lattice constants of the a-axis measured from two GaN epilayers grown on sapphire and one GaN epilayer on SiC were plotted against those of the c-axis. It clearly indicates that GaN epilayers grown on sapphire substrates are under biaxial compression and those on SiC substrates exhibit basal tensile strain. Therefore, the results

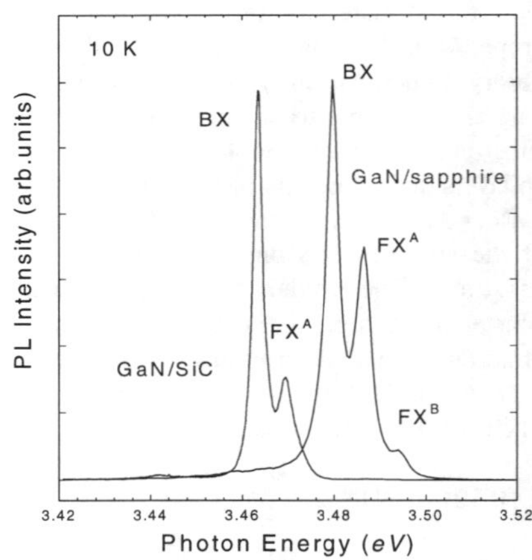

Fig.1. Exciton luminescence spectra taken from a 3.7-μm GaN epilayer on SiC and a 4.2-μm GaN epilayer on sapphire at 10 K.

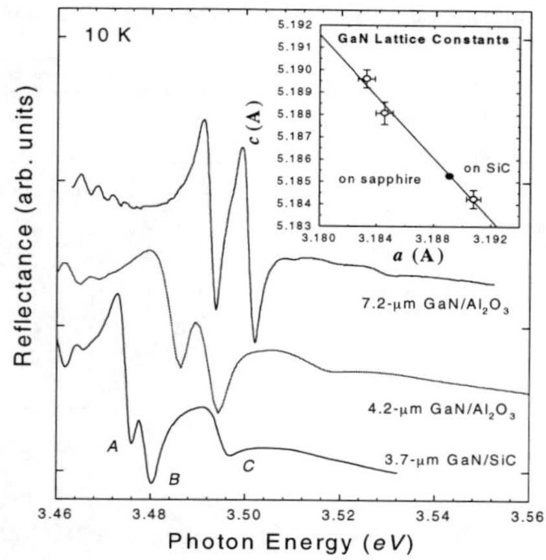

Fig. 2. Reflection spectra of near-band-edge excitonic transitions in a GaN/SiC and two GaN/sapphire samples at 10 K. The curves are vertically displaced for clarity. The inset shows measured in-plane GaN lattice constants (*a*-axis) versus the lattice constants along the growth direction (*c*-axis). The solid circle represents the lattice constants of strain-free GaN.

presented here lead to the conclusion that residual strain induced by thermal-expansion mismatch in GaN based epitaxial layers has the prevailing influence on the energy variations of exciton transitions, since lattice-mismatch induced strain is of the opposite sign and hence would have the opposite effect on the variation of GaN band gap from that observed.

Principal Deformation Potentials

The observed strain shifts in excitonic transition energies permit a direct estimate of the deformation potentials, including both hydrostatic and uniaxial components for wurtzite GaN. Under the assumption of a strain-independent and isotropic spin-orbit interaction, the energies of the three free excitons A, B, and C can be described as[10,11]

$$E_A = E_A(0) + a_1 \epsilon_{zz} + a_2(\epsilon_{xx} + \epsilon_{yy}) + b_1 \epsilon_{zz} + b_2(\epsilon_{xx} + \epsilon_{yy}), \tag{1}$$

$$E_B = E_B(0) + a_1 \epsilon_{zz} + a_2(\epsilon_{xx} + \epsilon_{yy}) + \Delta_+[b_1 \epsilon_{zz} + b_2(\epsilon_{xx} + \epsilon_{yy})], \tag{2}$$

$$E_C = E_C(0) + a_1 \epsilon_{zz} + a_2(\epsilon_{xx} + \epsilon_{yy}) + \Delta_-[b_1 \epsilon_{zz} + b_2(\epsilon_{xx} + \epsilon_{yy})], \tag{3}$$

where the $E_i(0)$ represent strain-free exciton transition energies, a and b are deformation potentials, and the ϵ_{ii} are components of the strain tensor for the GaN film. Since the energy variation given in Eqs.(1)-(3) by a_1 and a_2 is analogous to the hydrostatic shift of a cubic semiconductor, a_1 and a_2 are combined hydrostatic deformation potentials for transitions between the conduction and the valence bands, while b_1 and b_2 are uniaxial deformation potentials characterizing the further splitting of the three topmost valence band edges for tension or compression along and perpendicular to (0001), respectively. Since the growth direction of our epilayers is the z-axis, the strain components are described by

$$\epsilon_{xx} = \epsilon_{yy} = \epsilon_{\parallel} = (a_s - a_o)/a_o; \quad \epsilon_{zz} = \epsilon_{\perp} = (c_s - c_o)/c_o, \tag{4}$$

where a_o and c_o are lattice parameters for strain-free bulk GaN, and a_s and c_s are those for the

Table I. Values of measured excitonic transition energies and built-in residual strain for GaN on SiC and sapphire, together with those for strain-free bulk GaN.

	GaN/sapphire		GaN/bulk	GaN/SiC
Thickness (μm)	4.2	7.2	>100[a]	3.7
A-exciton (eV)	3.485	3.491	3.4735[a]	3.470
B-exciton (eV)	3.493	3.499	3.4800[a]	3.474
C-exciton (eV)	3.518	3.528	3.4993[a]	3.491
$\epsilon_{\parallel}$ (10^{-4})	-13.5	-18.2	0	2.8
$\epsilon_{\perp}$ (10^{-4})	4.5	7.9	0	-1.5

a)   Refs.2,3 and 5

strained GaN epilayer. Under biaxial-stress conditions, the components of ϵ_{xx}, ϵ_{yy}, and ϵ_{zz} are related through the elastic stiffness coefficients as $\epsilon_{\perp}=-2C_{13}/C_{33}\epsilon_{\parallel}$. The coefficients $\Delta_{\pm}$ represent the mixing of valence-band orbital states by the spin-orbit interaction and are given by[10,11]

$$\Delta_{\pm}=\tfrac{1}{2}\{1\pm[1+8(\Delta_3/(\Delta_1-\Delta_2))^2]^{-1/2}\}, \tag{5}$$

where Δ_1 is the crystal-field splitting of the Γ_9 and Γ_7 orbital states, Δ_2 and Δ_3 are parameters which describe the spin-orbit coupling. In principle, one has to know the values of these three band-structure parameters, in addition to the deformation potentials to predict strain shifts from Eqs.(1)-(3) above. This approach is complicated by the lack of consistent numerical values for the parameters, Δ_i[1,5,12-14]. Fortunately, Eqs.(2) and (3) only require a knowledge of the *ratio*, $\gamma\equiv\Delta_3/(\Delta_1-\Delta_2)$, rather than the individual numerical values for these parameters. By plotting the observed excitonic transition energies against the residual strain in Fig.3, we were able to obtain a value of ~0.531 for γ from the slopes of the solid lines in the figure.

The observed shifts in excitonic transition energies relative to the values of strain-free GaN result from an overall effect of strain on the band gap which includes contributions from both hydrostatic and uniaxial components of the stress. To determine the respective values for the uniaxial and hydrostatic potentials of wurtzite GaN, one has to separate their contributions to the energy variations in observed exciton transition energies under strain compared to the strain-free case. By taking the differences between the experimentally obtained values of each individual excitonic transition according to Eqs.(1)-(3), the uniaxial component of strain induced energy shift of the conduction-band edge relative to the valence-band edges can be readily separated from the total energy shift. With linear fits to the whole set of data listed in Table I using least-squares fitting, our results yield the relationship of the uniaxial deformation potentials b_1 and b_2 as well as the combined hydrostatic deformation potentials a_1 and a_2:

$$b_1-C_{33}/C_{13}b_2=-15.2 \text{ eV}, \tag{6}$$
$$a_1-C_{33}/C_{13}a_2=37.9 \text{ eV}. \tag{7}$$

Based on the facts that the strain caused total energy shift relative to the excitonic transition energy is very small and the elastic properties of GaN are of quasi-cubic nature ($C_{11}\approx C_{33}$)[15,16], we found that it is appropriate to estimate the numerical values, within the range of linear dependence on

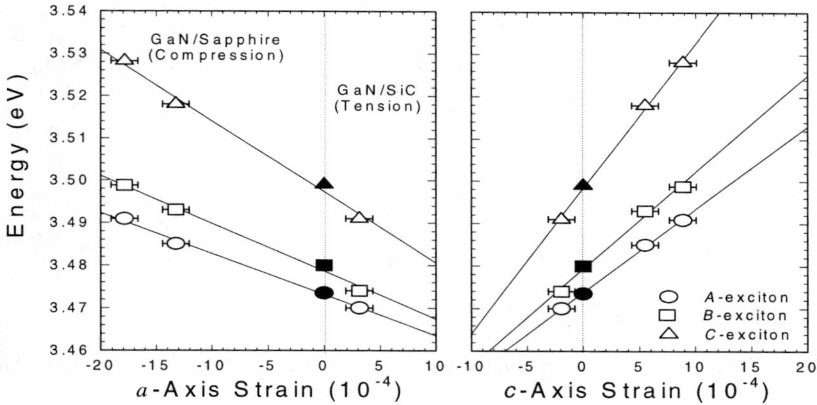

Fig. 3. The measured excitonic transition energies from GaN samples used in this work as a function of relative in-plane (biaxial) strain (left portion) as well as the relative strain along c-axis (right portion). The exciton transition energies of strain-free GaN were included in the figure for reference. The solid lines are the best linear fits to the experimental data.

strain, for the uniaxial and hydrostatic deformation potentials using quasi-cubic approximation[17] with $b_1 \approx -2b_2$, and $a_1-a_2 \approx 2b_2$. The deformation potentials are readily estimated to be $b_1 \approx -5.3 \ eV$ and $b_2 \approx 2.7 \ eV$, as well as $a_1 \approx -6.5 \ eV$ and $a_2 \approx -11.8 \ eV$ using the values $C_{13}=106$ and $C_{33}=398$ GPa[16]. The uncertainty of the above estimates is ~15%, originating primarily from experimental error in the precise determination of lattice parameters.

CONCLUSIONS

We have studied the effects of residual strain in GaN epitaxial films using spectroscopic methods combined with X-ray diffraction measurements, with the emphasis on determination of deformation potentials. Strong, sharp spectral structures associated with exciton transitions in GaN epitaxial layers grown on sapphire and 6H-SiC substrates by MOCVD were observed in photoluminescence and reflectance spectra. The observation of exciton transitions with *lower* energies in GaN grown on SiC, and with *higher* energies for the same transitions for GaN grown on sapphire, in comparison to those obtained from strain-free bulk GaN, suggests that residual strain in GaN epilayers resulting from lattice-parameter and thermal-expansion mismatch plays an important role in determining the precise exciton transition energies. X-ray diffraction measurements were performed to determine the variations in the lattice parameters of GaN epilayers on SiC and sapphire, respectively. Our results clearly indicate that GaN epilayers grown on SiC exhibit basal tensile strain, while those on sapphire substrates are under biaxial compression. Based on these results, the values of the four principal deformation potentials of wurtzite GaN have been determined.

ACKNOWLEDGEMENTS

This work at OSU was supported by AFOSR, DARPA, ONR and ARO. The work at NCSU was supported by ONR under contract No. N00014-92-J-1477.

REFERENCES

1. R. Dingle and M. Ilegeme, Solid State Commun. **9**, p.175(1971).

2. R. Dingle, D.D. Sell, S.E. Stokowski, and M. Illegems, Phys. Rev. **B4**, p.1,211(1971).

3. B. Monemar, Phys. Rev. **B10**, p.676(1974).

4. W. Shan, T.J. Schmidt, X.H. Yang, S.J. Hwang, J.J. Song, and B. Goldenberg, Appl. Phys. Lett. **66**, p.985(1995).

5. B. Gil, O. Briot, and R.-L. Aulombard, Phys. Rev. **B52**, p.R17,028(1995).

6. M. Smith, G.D. Chen, J.Y. Lin, H.X. Jiang, M. Asif Khan, C.J. Sun, Q. Chen, and J.W. Yang, J. Appl. Phys. **76**, p.7,001(1996).

7. G.D. Chen, M. Smith, J.Y. Lin, and H.X. Jiang, S.H. Wei, M. Asif Kahn, and C.J. Sun, Appl. Phys. Lett. **68**, p.2,784(1996).

8. W. Shan, A.J. Fischer, J.J. Song, G.E. Bulman, H.S. Kong, M.T. Leonard, W.G. Perry, M.D. Bremser, and R.F. Davis, Appl. Phys. Lett. **69**, p.740(1996).

9. W. Rieger, T. Metzger, H. Angerer, R. Dimitrov, O. Ambacher, and M. Stutzmann, Appl. Phys. Lett. **68**, p.970(1996).

10. V.B. Sandomirskii, Soviet Phys. -Solid State **6**, p.261(1964).

11. A. Gavini and M. Cardona, Phys. Rev. **B1**, p.672(1970).

12. M. Suzuki, T. Uenoyama, and A. Yanase, Phys. Rev. **B52**, p.8132(1995).

13. Yu. M. Sirenko, J.B. Jeon, K.W. Kim, M.A. Littlejohn, and M.A. Stroscio, Phys. Rev. **B53**, p.1997(1996).

14. S. Chichibu, A. Shikanai, T. Azuhata, T. Sota, A. Kuramata, K. Horino, and S. Nakamura, Appl. Phys. Lett. **68**, p.3,766(1996).

15. V.A. Savastenko and A.U. Sheleg, Phys. Stat. Sol. (a)**48**, p.K135(1978).

16. A. Polian, M. Grimsditch, and I. Grzegory, J. Appl. Phys. **79**, p.3,343(1996).

17. G.L. Bir and G.E. Pikus, in <u>Symmetry and Strain-induced Effects in Semiconductors</u>, John Wiley & Sons, New York, 1974), Chapt. V.

BOUND EXCITON ENERGIES, BIAXIAL STRAINS, AND DEFECT MICROSTRUCTURES IN GAN/ALN/6H-SIC(0001) HETEROSTRUCTURES

W.G. Perry*, T. Zheleva*, K.J. Linthicum*, M.D. Bremser*, R.F. Davis*, W. Shan** and J. J. Song**

*Department of Materials Science and Engineering, North Carolina State University, Box 7907, Raleigh, NC 27695-7907
**Center for Laser Research, Oklahoma State University, Stillwater, OK 74078

ABSTRACT

Biaxial strains resulting from mismatches in thermal expansion coefficients and lattice parameters in 22 GaN films grown on AlN buffer layers previously deposited on vicinal and on-axis 6H-SiC(0001) substrates were measured via changes in the c-axis lattice parameter (c). Six of the films were in compression, indicating the residual strain due to lattice mismatch was not relieved. A Poisson's ratio of $v=0.18$ was calculated. The bound exciton energy (E_{BX}) was a linear function of these strains. The shift in E_{BX} with film stress was 23 meV/GPa. The role of the SiC off-axis tilt was investigated for GaN films grown concurrently on the vicinal and on-axis 6H-SiC substrates. Marked variations in E_{BX} and c were observed, with a maximum shift of ΔE_{BX} = 15 meV and $\Delta c = 0.0042$ Å. Threading dislocations densities of $\sim 10^{10}/cm^2$ and $\sim 10^8/cm^2$ were determined for GaN films grown on vicinal and on-axis SiC, respectively. A 0.9% residual compressive strain at the GaN/AlN interface was observed by high resolution transmission electron microscopy (HRTEM). It is proposed that the on-axis SiC substrate does not offer a sufficient density of steps for defect formation to relieve the lattice mismatch between GaN and AlN and AlN and SiC.

INTRODUCTION

Mismatches in the a-axis and c-axis lattice parameters (denoted hereafter as a and c) and/or the coefficients of thermal expansion (α) exist between heteroepitaxial GaN films and all of the presently used substrates. These mismatches result in interfacial biaxial strain which leads to misfit dislocations and associated threading defects that degrade film quality. The most commonly used substrate for GaN film growth is sapphire (0001). Buffer layers of either AlN or GaN have been employed to improve GaN film quality, although a significant lattice mismatch exists between each material and sapphire ($\Delta a/a_o$ =13.6 and 16.1%, respectively). Recent work[1] has used a 6H-SiC(0001) substrate/AlN(0001) buffer layer combination. The relatively smaller lattice mismatch between AlN and SiC (~1%) have resulted in a measurable reduction in the dislocation density within the first micron of film growth. For GaN films grown on both substrates the mismatch strain is compressive and alters both a (decreases) and c (increases). This strain is thought to be relieved after several nanometers of growth[2] by the formation of dislocations at the film/substrate interface. Upon cooling, the difference in the thermal expansion coefficients results in an additional stress contribution. For growth on sapphire the stress is compressive ($\alpha_{sap.} > \alpha_{AlN} > \alpha_{GaN}$), while for SiC it is tensile ($\alpha_{GaN} > \alpha_{SiC} \approx \alpha_{AlN}$). These stresses are often assumed to be wholly responsible for the observed residual strain present in GaN films grown on both substrates.[3] However, HRTEM analysis of the GaN/AlN interface on sapphire[4] suggests that the lattice mismatch stress is not fully relieved.

One consequence of film strain is a variation of band gap energy (E_g). Changes in E_g may be detected by the shift of excitonic features via low temperature photoluminescence (PL), which for high-quality GaN reveals intense near-band edge emission attributed to the recombination of both free excitons and/or excitons bound to shallow neutral donors.[5] The bound exciton feature is often the dominant feature due to the nature of GaN, which is always n-type for undoped films. Reported E_{BX} values range from 3.467 to 3.493 eV for GaN on sapphire,[6] and from 3.463 to 3.472 eV for GaN on SiC.[1,7] The following reports the relationship between E_{BX} and lattice parameters (and hence strain) observed for 22 GaN films grown on AlN buffer layers previously deposited on 6H-SiC(0001). Poisson's ratio was calculated from lattice parameter measurements and used to determine the shift of E_{BX} with biaxial stress (σ_a). The roles of the AlN buffer layer

Mat. Res. Soc. Symp. Proc. Vol. 449 © 1997 Materials Research Society

and the off-axis tilt of SiC on film stress were also determined. Transmission electron microscopy (TEM) was employed to compare and contrast the crystal structure of GaN films grown on the vicinal (off-axis) and on-axis SiC wafers. The GaN/AlN interface was analyzed via high resolution (HR)TEM to determine whether or not residual lattice mismatch stress occurred.

EXPERIMENTAL PROCEDURES

The GaN/AlN films were grown[1] via organometallic vapor phase epitaxy (OMVPE) on both on-axis and vicinal (2-4° off-axis) 6H-SiC(0001) substrates. For each sample a ~0.03-0.1 μm AlN buffer layer was grown at 1100°C. The GaN films were grown at 950-1100°C and ranged in thickness from 0.3-3.7 μm. All of the latter films were unintentionally doped, with n-type carrier concentrations ranging from $<1 \times 10^{16} - 1 \times 10^{17}/cm^3$.

Absolute values of the lattice constants were measured using a Philips X'Pert MRD X-ray diffractometer in the triple-axis mode using a technique proposed by Fewster.[8] The data were collected using 2θ-ω scans, with corrections made for refraction. The accuracy of the lattice parameters measured using this system is predicted to be better than 0.0004 Å. Values of c were measured using the (002) reflection; values of a were obtained using both the (002) and (015) reflections. The low count rate for the (015) reflections limited the number of samples for which a could measured accurately. Photoluminescence (PL) measurements of the GaN films on SiC were made at 4.2 K using a He-Cd laser (λ=325 nm) as the excitation source, unless otherwise noted. Microscopy was performed using a TOPCON EM - 002B HRTEM, operated at 200 kV with a point-to-point resolution of 0.18 nm. The cross-sectional samples in the [11$\bar{2}$0] orientation were prepared by conventional techniques using mechanical grinding, polishing, and dimpling, as well as Ar^+ ion-milling at a low angle in the final stage.

RESULTS AND DISCUSSION

Low temperature PL of the GaN(0001) films exhibited strong near-band-edge emission due to the recombination of both free and donor-bound excitons. The bound exciton peak was always the dominant feature in our films with the free exciton peak most often a shoulder of this emission. The bound exciton energies (E_{BX}) determined at 4.2 K as a function of c for 22 GaN films on SiC are shown in Fig. 1. It is expected that the tensile strain in the GaN films due to thermal expansion mismatch increased as the temperature was decreased from room temperature to 4.2 K, but lattice parameter measurements were not possible over this temperature range. A linear relationship between E_{BX} and c was observed. Any change in c was in response to a strain (ε_a) along the a-axis (a), $\Delta a/a_o$, which resulted from biaxial stress produced in the plane of the film by residual stresses. Thus a decrease in c indicated that the biaxial tensile strain in the film increased, which caused E_g to decrease and E_{BX} to shift to a lower energy.

Six films had values of c greater than that for relaxed GaN (5.1855 Å[9]), which indicated an overall compressive residual strain in these films. This suggested that the lattice mismatch strain had not been fully relieved by defect formation in these samples. The highest value of c observed was

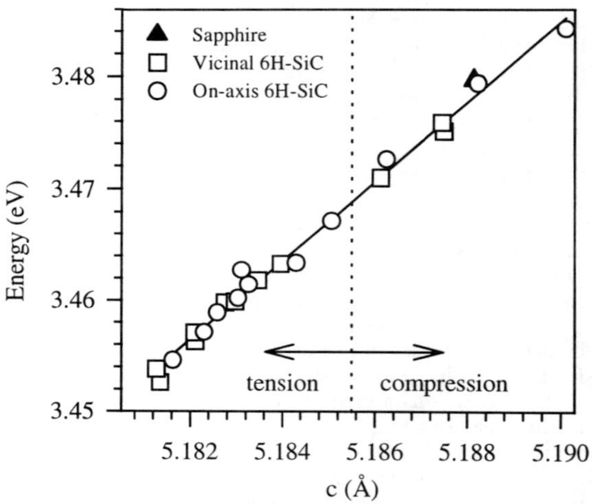

Figure 1. E_{BX} vs. c vs. for the GaN/AlN/SiC heterostructure. The dashed line indicates c_o (5.1855 Å) for fully relaxed GaN. Included is a point for a 4.2 μm GaN film on sapphire.

5.190 Å, with E_{BX} = 3.484 eV. The highest E_{BX} value previously reported for GaN grown on SiC was 3.472 eV.[7] Values of E_{BX} can be calculated from the equation

$$E_{BX}(eV) = (3.546*c(\text{Å}) -14.92)eV, \qquad (1)$$

obtained via a linear regression of the data in Fig 1. The value, E_{BXo}, for relaxed GaN was predicted from the previous expression to be 3.469 eV. This compares favorably to reported values of 3.467[10], 3.469[5] and 3.472[11] eV. Included in Fig. 2 is a data point for a 4.7 μm GaN film grown on sapphire. The excellent fit of this data with that determined in this research indicates that E_{BX} can be predicted for GaN grown on any substrate using values of c and the relationship determined in this study.

For biaxially strained films the relationship between c and a is given by the strain ratio[12]

$$\varepsilon_c/\varepsilon_a = (\Delta c/c_o)/(\Delta a/a_o) = - 2\nu/(1-\nu), \qquad (2)$$

where ν is Poisson's ratio and a_o and c_o are the relaxed lattice parameters. For GaN, values of υ have been calculated using anisotropic elastic constants and range from 0.372[13] to 0.20.[14] A recent[15] survey of x-ray data in the literature determined $\nu = 0.23 \pm 0.06$ for samples dominated by biaxial strain. In this study a and c were measured simultaneously for 23 GaN films on SiC, including 10 of those shown in Fig. 1. The results are shown in Fig. 2. Average strain ratios of $\varepsilon_c/\varepsilon_a$ = -0.48, -0.33, -0.48, and -0.73 were determined from the same data using various published values for a_o and c_o.[9,15-17] The disparities of these average strain ratios was due to individual values of c and a that were close to the various values for a_o and c_o. This resulted in abnormal values in the nominator and/or denominator of equation (2) for that sample's strain ratio.

Two methods were used to obtain a more reasonable estimation of the strain ratio and the resulting Poisson's ratio. The first eliminated those data points (4 total) that had adverse effects on the average strain ratio of the data set. An average strain ratio of $\varepsilon_c/\varepsilon_a$ = -0.420 was then calculated using the four sets of published values for a_o and c_o. The second method fit a line to the data in Fig. 2, and then used the experimentally determined values of c to calculate values of a that fit this line. The strain ratio for each point was then calculated using values of c_o = 5.1850 and a_o = 3.1892, where the latter value was determined from the line fit to the data in Fig. 2. The average strain ratio using this method was $\varepsilon_c/\varepsilon_a$ = -0.446 ± .03. Poisson's ratio for each of the strain ratios was ν=0.18 ± .02, which compared favorably with values determined from anisotropic elastic constants[14] (0.20) and x-ray data[15] (0.23 ± .06).

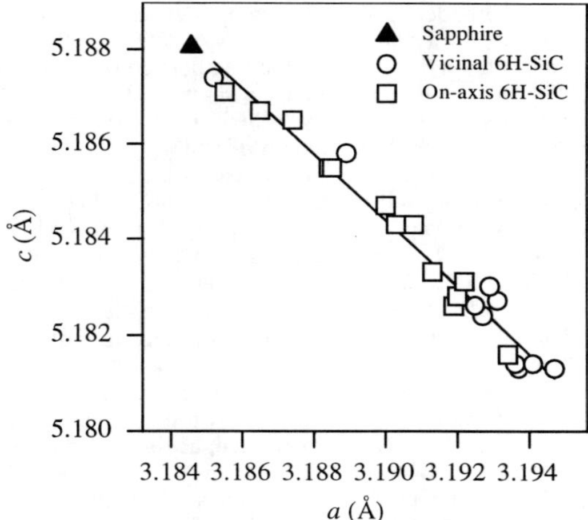

Figure 2: Values of a vs. c for GaN/AlN/SiC thin film/wafer heterostructures. Poisson's ratio was ν = 0.18 ± .02. Included is a GaN film on sapphire.

The shift of E_{BX} as a function of the biaxial film stress ($\Delta E_{BX}/\Delta\sigma_a$) was estimated from the data in Fig. 1 using the relationship[12]

$$\sigma_a = Y/(1-\nu)\varepsilon_a = -Y\varepsilon_c/2\nu, \qquad (3)$$

where Y is the Young's modulus[15] (290 ± 20 GPa). The first step was to calculate $\Delta E_{BX}/\Delta\varepsilon_c$ from the data in Fig. 1, where the relaxed values of E_{BXo} and c_o were assumed. A linear regression of the data yielded $\Delta E_{BX}/\Delta \varepsilon_c = 18.1$ eV. Using this expression and the previously determined Poisson's ratio of $v = 0.18$ a value of $dE_{BX}/d\sigma_\alpha = 23$ meV/GPa was calculated. This value is lower than that reported[18] for hydrostatic pressure experiments (39meV/GPa), but it compares favorably to results for GaN films where biaxial stress was assumed[15] (27 meV/GPa).

The role of SiC off-axis tilt on GaN film stress was investigated using films grown concurrently on vicinal and on-axis SiC(0001) wafers. The PL spectra of the bound exciton emission from two separate GaN film sets (a and b) are displayed in Fig. 3. The film thickness was ~ 0.4-0.5 μm for each sample, and the PL measurements were performed at 10 K. The bound exciton peak was strain shifted to a higher energy for each film on the on-axis SiC compared to its counterpart on vicinal SiC. This trend was repeated for 8 different sample sets, with a maximum strain shift of $\Delta E_{BX} = 15$ meV. A corresponding strain shift in c was observed, with a maximum value of $\Delta c = 0.0042$ Å.

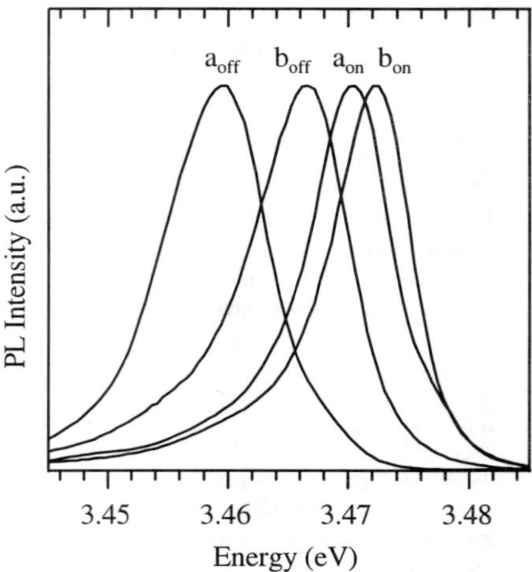

Figure 3. PL of two sets of GaN films (a, b) grown concurrently on vicinal and on-axis SiC.

To understand these results it is necessary to initially examine the defect microstructures of the AlN buffer layers. The primary defects in these layers grown on the off-axis SiC substrates were inversion domains and their associated boundaries (IDBs), as shown in the TEM micrograph in Fig. 4(a). Recent studies[19] showed that the steps on the vicinal 6H-SiC (0001) surface provided sites for the growth of inversion domains as a result of the mismatch in the stacking sequence of the Si/C and Al/N bilayers in the 6H-SiC(0001) and 2H-AlN(0001)(wurtzite structure), respectively. These defects introduce a marked amount of strain in the AlN grown on the off-axis SiC substrate. There was also a high density of threading dislocations in the buffer layer. By comparison, the crystal quality of AlN films grown on the on-axis SiC substrates is of improved crystalline quality, as shown in Fig. 4(b). Most significantly, there was a reduced density of inversion domain boundaries due to the reduction in the density of SiC steps. The primary defects were stacking faults parallel to the AlN/SiC interface and associated partial dislocations. Threading dislocations running from the top to the bottom of the film are also present.

The crystal quality of the GaN is directly influenced by the AlN buffer. The inferior quality of the AlN buffer layer grown on vicinal SiC resulted in a high dislocation density($\sim 10^9$-10^{10}/cm^2) at the AlN/GaN interface, as determined from plan-view TEM analysis by counting the number of dislocations per unit area. The dislocation density decreased markedly from this interface. The predominant defects were threading dislocations and threading segments that persisted throughout the film. Stacking faults and dislocation loops were observed close to the interface. Domain boundaries that originated at the interface were also visible. The improved quality of the AlN buffer layers on the on-axis SiC also carried over into the GaN, as shown in Fig. 4(b). A dislocation density in the order of $\sim 10^8$ /cm^2 in the GaN layer was observed in plan-view TEM near the GaN/AlN interface. Most notably there were no planar defects such as stacking faults or domain boundaries in the GaN. The dominant defects in the on-axis GaN films were edge-type misfit dislocations with Burgers vectors $\mathbf{b} = 1/3\langle 11\bar{2}0 \rangle$ and $\mathbf{b} = 1/3\langle 1\bar{1}00 \rangle$ which formed at the GaN/AlN interface. This TEM analysis, as well as that of others[20] with samples grown in this laboratory, revealed that the lattice mismatch stress in GaN films grown on the on-axis SiC substrate is accommodated mostly by misfit dislocations. However, our PL and XRD studies suggest that these defects do not provide full stress relief.

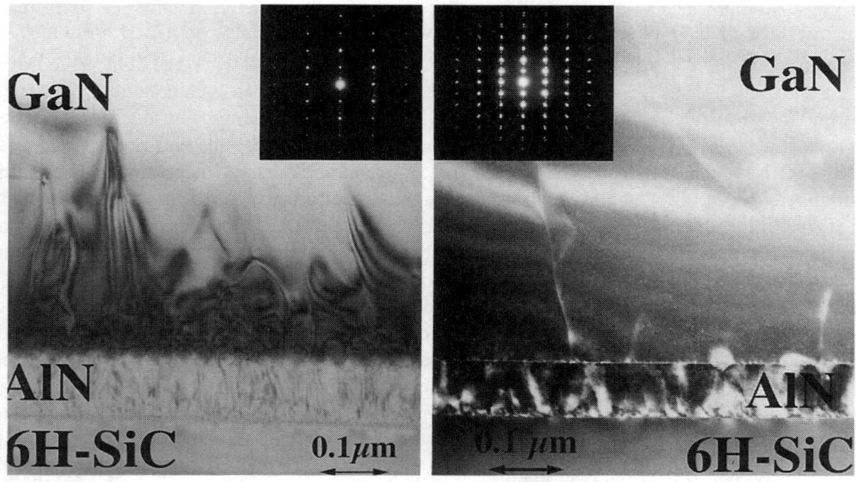

Figure 4: (a) Low magnification TEM micrographs from: (a) GaN/AlN/vicinal 6H-SiC(0001) heterostructure and (b) GaN/AlN/on-axis 6H-SiC(0001) heterostructure.

In order to further assess the residual strain in the GaN films, HRTEM studies were performed in the $[11\bar{2}0]$ orientation at the GaN/AlN interface grown on the on-axis SiC . Using the results, the residual strain at the interface can be calculated for planes that are perpendicular to the interface and parallel to each other. Ning, et $al.$,[4] used this method to calculate a ~1% residual strain in GaN film at the GaN/sapphire interface. In this research the average lattice mismatch was calculated along the GaN/AlN interface from the HRTEM micrographs by counting the number of $(1\bar{1}00)$ GaN and AlN planes along $[1\bar{1}00]$ direction that was bounded on each end by commensurate GaN and AlN planes. The experimental mismatch is equal to $N_{GaN} - N_{AlN}/N_{AlN}$, where N is the number of $(1\bar{1}00)$ planes perpendicular to the interface. The average experimental mismatch was 0.9% lower than the theoretical misfit value of 2.5% for GaN and AlN for ~20 different locations along the interface. Thus there is a 0.9% residual compressive strain in the on-axis GaN film, which supports our PL and XRD results.

CONCLUSIONS

Compressive residual strain in GaN films grown on AlN buffer layers previously deposited on 6H-SiC(0001) wafers is reported for the first time This residual strain was due both to mismatches in the thermal expansion coefficients and the lattice parameters. A linear dependence between E_{BX} and c was observed and expressed by $E_{BX} = -14.92 + 3.546*c$ eV. Poisson's ratio for GaN was determined to be $\nu = 0.18$. The shift of E_{BX} with biaxial film stress was estimated to be 23 meV/GPa. Marked variations in E_{BX} and c were observed for GaN films grown concurrently on vicinal and on-axis SiC wafers. Results from TEM showed that the higher density of steps on the vicinal SiC wafers acted as formation sites for IDBs. Threading dislocation densities of $\sim10^{10}/cm^2$ and $\sim10^8/cm^2$ were observed for GaN grown on vicinal and on-axis SiC, respectively. The on-axis SiC substrate did not contain sufficient steps for defect formation to fully relieve the lattice mismatch via defect formation at the GaN/AlN and AlN/SiC interfaces. This resulted in residual compressive stresses that counteracted the tensile stresses that formed upon cooling due to the thermal expansion mismatch. A 0.9% residual compressive strain at the GaN/AlN interface was calculated by comparing the average experimental lattice mismatch at the interface (1.6%) observed via HRTEM with that theoretically predicted for GaN/AlN (2.5%). It is proposed that this strain was accommodated elastically and resulted in biaxially compressive stresses in the films.

ACKNOWLEDGMENTS

The authors acknowledge the support of the Office of Naval Research via contract #N00014-92-J-1477, Cree Research, Inc. for the SiC wafers, Barbara Goldenberg of Honeywell, Inc. for the GaN/sapphire sample, and Dr. S. K. Streiffer of NCSU and Drs. Jaime Freitas and Evan Glaser of the Naval Research Laboratory for helpful discussions.

REFERENCES

1   T. W. Weeks, M. D. Bremser, K. S. Ailey, E. Carlson, W. G. Perry, E. L. Piner, N. A. El-Masry, and J. M. R. R. F. Davis, J. Mater. Res. **11**, 1011 (1996).
2   S. Tanaka, PhD Dissertation, North Carolina State University, 1993.
3   I. A. Buyanova, J. P. Bergman, B. Monemar, H. Amano, and I. Akasaki, Appl. Phys. Lett. **69**, 1255 (1996).
4   X. J. Ning, F. R. Cein, P. Pirouz, J. W. Yang, and M. A. Khan, J. Mater. Res. **11**, 580 (1996).
5   B. Monemar, Phys. Rev. B **10**, 676 (1974).
6   C. Merz, M. Kunzer, U. Kaufmann, I. Akasaki, and H. Amano, Semicond. Sci. Technol. **11**, 712 (1996).
7   L. Eckey, J. C. Holst, P. Maxim, R. Heitz, A. Hoffmann, I. Broser, B. K. Meyer, C. Wetzel, E. N. Mokhoc, and P. G. Baranov, Appl. Phys. Lett. **68**, 415 (1996).
8   P. F. Fewster and N. L. Andrew, J. Appl. Cryst. **28**, 451 (1995).
9   C. M. Balkas, C. Basceri, and R. F. Davis, Powder Diffraction **10**, 266 (1995).
10   A. Gassmann, N. N. T. Suski, C. Kisielowski, E. Jones, E. R. Weber, A. Liliental-Weber, M. D. Rubin, H. I. Heleva, I. Grzegory, M. Bockowski, J. Jun, and S. Porowski, J. Appl. Phys. **80**, 2195 (1996).
11   K. Pakula, A. Wysmolek, K. P. Korona, J. M. Baranowski, R. Stepniewski, I. Grzegory, M. Bockowski, J. Jun, S. Krukowski, M. Wroblewski, and S. Porowski, Solid State Commun. **97**, 919 (1996).
12   K. N. Tu, J. W. Mayer, and L. C. Feldman, *Electronic Thin Film Science*, Vol. 84 (Macmillan, New York, 1992).
13   V. A. Savastenko and A. U. Sheleg, Phys. Status Solidi A **48**, 135 (1978).
14   T. Azuhata, T. Sota, and K. Suzuki, J. Phys.: Condens. Matter **8**, 3111 (1996).
15   C. Kisielowski, J. Krüger, S. Rumimov, T. Suski, J. W. A. III, E. Jones, Z. Liliental-Weber, M. Rubin, E. R. Weber, M. D. Bremser, and R. F. Davis, submitted to Phys. Rev. B (1996).
16   T. Detchprohm, K. Hiramatsu, K. Itoh, and J. J. A. P. I. Akasaki, Jpn. J. Appl. Phys. **10B**, L1454 (1992).
17   *Landolt-Börnstein*; *Vol. 17*, edited by O. Madelung (Springer, New York, 1982).
18   W. Shan, T. J. Schmidt, R. J. Hauenstien, J. J. Song, and B. Goldenberg, Appl. Phys. Lett. **66**, 3492 (1995).
19   P. Pirouz, presented at the Nitride Workshop on Wide Bandgap Nitrides, St. Louis (1996)
20   F. Chien, X. Jing, P. Pirouz, M. D. Bremser, and R. F. Davis, Appl. Phys. Lett. **68**, 2678 (1996).

INVESTIGATION OF VACANCIES IN GaN BY POSITRON ANNIHILATION

L. V. JØRGENSEN [*], A. C. KRUSEMAN [*], H. SCHUT [*], A. VAN VEEN [*],
M. FANCIULLI [**] AND T. D. MOUSTAKAS [***]
[*] Interfaculty Reactor Institute, Delft University of Technology, 2629 JB Delft, Netherlands
[**] Institute of Physics and Astronomy, University of Aarhus, 8000 Aarhus C, Denmark
[***] Department of ECS, Boston University, Boston, MA 02215

ABSTRACT

Positron beam analysis has been performed on autodoped n-type, semi-insulating and Mg doped p-type epitaxially grown layers of GaN on sapphire. Doppler Broadening measurements clearly indicate the presence of vacancies in the intrinsically autodoped n-type GaN by an increase in the annihilation Doppler lineshape S-parameter of 1.04 relative to the value for the high resistivity sample. This value is typical for vacancy-type defects in compound semiconductors such as GaAs. Results of experiments with higher sensitivity to core-electrons are also presented. These two detector coincidence measurements yield information on the chemical environment surrounding the vacancies. The results are consistent with the presence of Ga vacancies in the autodoped n-type sample.

INTRODUCTION

GaN has attracted much attention in the last years due to its physical and electronic properties and above all its recently demonstrated qualities as a high efficiency material for optical devices operating in the blue and ultraviolet (UV) regions of the optical spectrum [1,2]. Despite this increased interest little is known about the nature of the intrinsic defects in GaN. As-grown samples of GaN are typically heavily autodoped n-type with carrier concentrations as high as 10^{20} cm^{-3}. This intrinsic autodoping has long been associated with the N-vacancy [3-5]. However, recent *ab initio* calculations seem to rule out this vacancy as the source of the autodoping [6] and generally predict a much lower formation energy for the Ga-vacancy in the n-type case than for any other point defect, including the N-vacancy [6,7].

While there have been many theoretical investigations on the nature of the intrinsic defects little experimental work has been done. However, all the experimental data so far either directly or indirectly ascribes the autodoping to N-vacancies [8,9]. It should be noted that none of the experimental data so far have been able to conclusively pinpoint the nature of the autodoping.

The use of positrons as a non-destructive probe to study defects in materials have been shown to be a useful tool [10,11]. The technique is based on the propensity of positrons to be trapped at defects such as vacancies and voids [12] making them a sensitive probe for studying these kinds of defects. After being trapped the positron will annihilate with an electron from the near vicinity of the defect emitting two 511 keV γ-photons. These two photons carry with them information on the electron momentum density of the annihilation site. Measuring the Doppler broadening of the annihilation γ-photons yields information about the electron momentum distribution in the direction of the detector. Furthermore, if a beam of mono-energetic positrons is used, the measurements will also yield information on the depth distribution of the defects.

EXPERIMENT AND METHOD

The measurements were performed on three different GaN samples, an autodoped n-type, a Mg-doped p-type and a semi-insulating (SI) sample. The samples were grown by the ECR-MBE method on sapphire substrates [13]. The autodoped n-type sample has a thickness of 1.28 μm with a buffer layer of about 30 nm grown at 550 °C. The rest of the sample was grown at 800 °C. The GaN layer has a carrier concentration of 2×10^{18} cm^{-3} and an electron mobility of 80 cm^{2}V^{-1}s^{-1}. The semi-insulating sample (SI) has a thickness of 1.38 μm with a buffer layer of about 30 nm grown at 600

Mat. Res. Soc. Symp. Proc. Vol. 449 © 1997 Materials Research Society

°C. The rest of the film was in this case grown at 900 °C. The p-type sample is 1.09 μm thick with a slight Mg doping and a resistivity of 6×10^3 Ωcm. The samples were grown at a slight N_2 overpressure.

The experiments were performed using the variable energy positron beamline (VEP) at Delft. This magnetically guided beam delivers 1×10^5 positrons per second in a spot with a diameter of 7 mm. The energy of the positrons is tunable in the range from 0 to 25 keV [14]. The annihilation radiation was measured using a high purity Ge γ-counter placed perpendicular to the beam axis at a distance of about 45 mm from the center of the samples.

Traditionally in Doppler broadening measurements the lineshape parameter, S, is measured as a function of incident positron energy. The S parameter measures the integrated area in the central region of the peak compared to the total area of the peak. Since this corresponds to annihilations with low-momentum electrons this parameter is especially sensitive to the density of valence electrons. Another lineshape parameter that has become increasingly more used in recent years is the W or wing parameter [15]. This measures changes in the far wings of the annihilation peak and thus monitors annihilations with high momentum, i.e. core electrons.

The experimental data was analyzed using the VEPFIT fitting and modelling program [16,17]. This program is based on a layered structure model and includes implantation and interlayer diffusion. The implantation was modeled by a Gaussian derivative profile with $2^{1/2} \sigma = z_0 = $ $<z>/\Gamma[(1/m)+1]$ where Γ is the gamma function and $<z>=(\alpha/\rho)E^n$ is the mean implantation depth for incident positron energy E in a material with density ρ. The values used for the two material independent constants α and n were 3.6 μg cm^{-2}keV^{-n} and 1.62, respectively. The studied system is assumed to consist of a stack of different positron trapping layers. The program assigns to each layer a value for the Doppler lineshape parameter S as well as a value for the positron diffusion length in the layer. The S value for a given implantation energy E is then given by

$$S(E) = \sum_{i-1}^{n} f_i(E)S_i + f_{surface}(E)S_{surface} , \tag{1}$$

where f_i (E) is the fraction of positrons implanted at energy E which after thermalization and diffusion annihilate in layer i. A similar equation can be set up for W and by fitting the two sets of experimental data information about the layer specific S and W values, the positron diffusion length in the layers and the layer thicknesses can be obtained. In combination the two lineshape parameters can therefore yield a good characterization of the involved materials and their defects. By plotting the measured S and W data as a trajectory in the S - W plane, using the implantation energy as the running parameter, a direct interpretation of the experimental data in terms of positron trapping layers with distinct S and W values can be obtained [14]. Since absolute S and W values can vary with the specific experimental setup while relative values are reproducible, all reported S and W values are relative to the value for SI-GaN.

In a conventional Doppler broadening measurement the sensitivity to annihilations with core electrons is hampered by the background at high momenta. Based on the ideas of Lynn *et al.* [18] we have used a two detector coincidence setup to reduce this background. For these measurements a second high purity Ge γ-counter was placed directly opposite to the other Ge detector and at the same distance to the samples. By using both timing coincidence and energy conservation considerations in a manner similar to that described by Asoka-Kumar *et al.* [19] we were able to improve the peak to background ratio by a factor of almost 1000 compared to a single detector system. At the same time, by using this setup, the resolution of the system was improved from 1.3 keV (FWHM) at the 511 keV photopeak for a single detector to 0.93 keV with the two detector system. This increased sensitivity to core electrons comes at the price of much lower count rates.

RESULTS

<u>One-detector measurements</u>

The results of the Doppler Broadening measurement is shown in Fig. 1. It shows a striking difference between the results for the semi-insulating and p-type samples on one side and the autodoped n-type on the other. The increase in the S parameter of about 4 % for the n-type in comparison to the semi-insulating and p-type samples is typical of vacancy-type defects in compound semiconductors such as GaAs and strongly implies the presence of such defects in the autodoped n-type sample. This is further confirmed by the decrease in the W value for the same sample. The Doppler broadening data was analyzed using the VEPFIT program and the fits are shown as solid lines in fig. 1. The plot in fig. 1(c) shows the trajectory traversed in the S-W plane for the cases of the n-type and semi-insulating samples using the positron energy as a running parameter. The open circles show the fitted values for each layer. The plot illustrates how the positrons first probe the surface, then the GaN layer and finally just begins to probe the Al_2O_3. Note also that for the case of the n-type there is an interface layer between the GaN and sapphire layers with a high S value and a low W value indicating the presence of vacancy-clusters or microvoids. This layer is quite thin but acts as a very efficient trap for positrons. The trajectory for the p-type sample is not shown in fig. 1(c) since it would essentially run on top of the SI trajectory, as can be seen from fig. 1(a) and (b).

Two detector measurements

Figure 2(a) shows the result of the two-detector measurements at an incident positron energy of 7 keV for the semi-insulating sample and the autodoped n-type sample. This energy was chosen because it yields a maximum fraction of the incident positrons annihilating in the GaN layer (see fig. 1.) Figure 2(b) shows the calculated momentum distribution for defect-free GaN as well as with

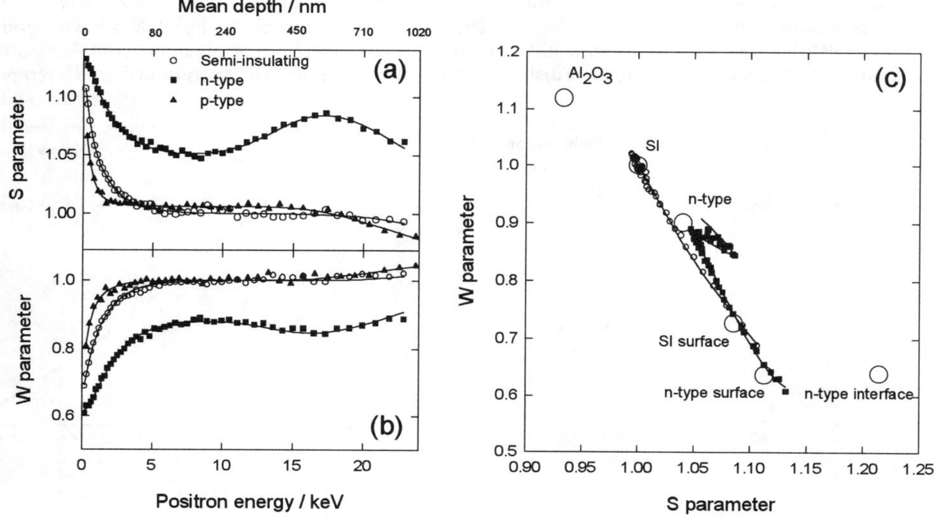

Fig. 1.(a) and (b) S and W parameters as a function of incident positron energy for the three samples studied. (c) S-W plot for the autodoped n-type and semi-insulating samples showing the probing of the different layers. All lines in the figures are fitted values obtained from the VEPFIT fitting program.

N vacancies, GaN-V_N, and Ga vacancies, GaN-V_{Ga}. For comparison the calculated values for defect-free GaN is also shown in fig. 2(a).

The program used for the calculations solves the Dirac equation to obtain relativistic electron wave functions and charge densities. A supercell is constructed by placing the atoms at the appropriate lattice positions. Periodic boundary conditions are imposed at the supercell boundaries. The electron density of the solid is approximated by the superposition of charge densities of free atoms using the, non-self-consistent, atomic superposition (AT-SUP) method of Puska and Nieminen [20]. The electron density and the positron potential are calculated at the node points of a three-dimensional mesh. The Schrödinger equation is then solved to obtain the positron energy and the positron wave function at the node points.

It has been shown that the momentum distribution of the annihilating positron-electron pair can be approximated by [21]:

$$\rho_j(p) = \pi r_0^2 c u_j^2(0) \left| \int dr\, e^{-ip\cdot r} \psi_+(r) \psi_j(r) \right|^2, \tag{2}$$

where $\psi_+(r)$ and $\psi_j(r)$ are the single particle wave functions for the positron and electron respectively, p is the total momentum of the annihilating pair, r_0 is the classical electron radius, c is the speed of light and $u_j^2(0) = \lambda_j^{GGA} / \lambda_j^{IPM}$ the state-dependent enhancement factors. Here, λ_j^{IPM} is the annihilation rate with the electrons in the state j, calculated within the independent particle model (IPM) and thus neglecting all the correlations between the electrons and the positron. Correspondingly, λ_j^{GGA} is the annihilation rate for the same electron state calculated with the generalized gradient approximation (GGA) for the electron-positron correlations [22]. The generalized gradient approximation has been shown to give good results for both positron lifetimes [22] and the momentum distribution [21].

In the calculation of the momentum distribution the wavefunctions are assumed to be spherically symmetric. This assumption is not true for valence electrons, so only core-electrons are included in the calculation of the momentum distribution. The restriction to core-electrons inhibits the program to calculate accurately the momentum distribution between 0 and 20 mrad since the contribution of valance electrons to the momentum distribution will dominate the contribution of core electrons.

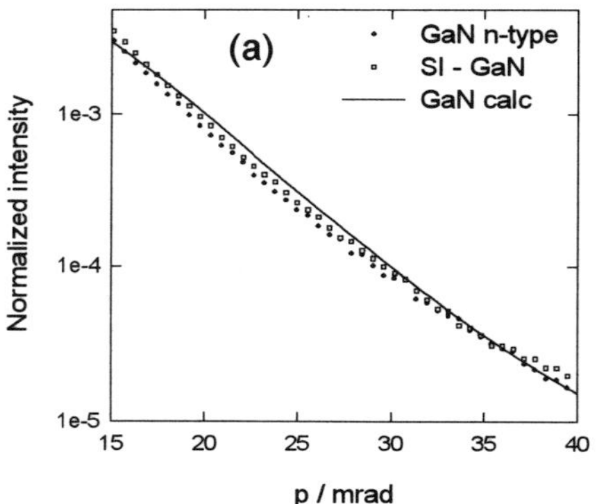
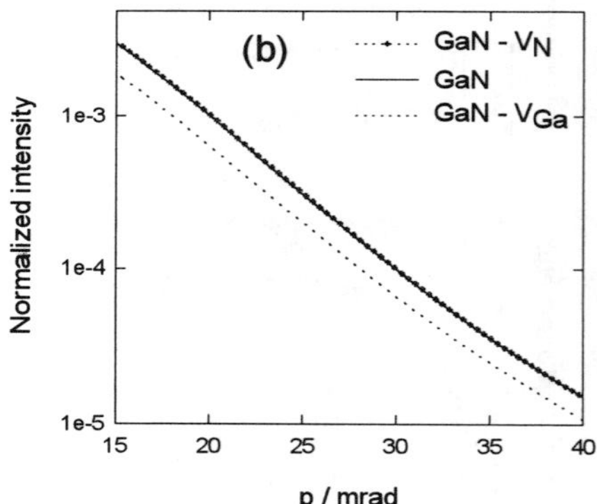

Fig. 2. (a) Results of two-detector coincidence measurements. At 40 mrad there were 100-200 counts per point, making the difference between the two measurements at lower momentum statistically significant. (b) Results of calculations of positron annihilations with core-electrons for defect free GaN as well as at N-vacancies and Ga-vacancies in GaN. Note that the calculated profiles for defect-free GaN and GaN - V_N are virtually identical. All experimental curves are normalized to unit area. The calculated curves are scaled to $\sum_{j=core} \lambda_j / \sum_{j=all} \lambda_j$

This approach is essentially the same as done by Alatalo *et al.* [20] except that we use, as in ref. [23], the AT-SUP positron wave functions (directionally averaged) in the calculation of the momentum densities.

The calculation is done for a $90\times90\times90$ node points using a supercell of 64 atoms for the perfect GaN and 63 atoms for GaN with a vacancy defect. No relaxation is taken into account. The calculated distribution is convoluted with a gaussian with a FWHM of 0.93 keV to account for the experimental resolution. Up till the Ga-3d electrons are included in the calculation. Ref. [21] showed that, although the 3d electrons of Ga cannot be considered core-electrons, the calculated Doppler-spectra show the same trends as the experimental spectra when including the d-electrons.

<u>Discussion</u>

The observed increase in the S parameter value of 4 % and the accompanied decrease in the W parameter are clear evidence of the presence of vacancy-type defects. In GaAs increases in the S parameter of 2-3 % have been associated with the presence of mono-vacancies and increases of 4-5 % with divacancies [24]. The two-detector measurement on SI-GaN showed good agreement with the calculated momentum distribution for defect-free GaN and yielded values for the n-type GaN that were consistently below those for SI-GaN. The calculated momentum distributions showed very little difference between defect-free GaN and GaN with N vacancies (GaN-V_N). This is an indication that such vacancies are difficult to observe with positron techniques. A further indication of this can be obtained by calculating the positron lifetimes involved. The calculated lifetime for defect-free wurtzite GaN is 168 ps. The presence of a N vacancy only slightly increases the lifetime to 170 ps. Ga-vacancies on the other hand show an increase of over 30 ps in the lifetime and a clear difference in the momentum distribution. It is not possible to perform conventional lifetime measurements to compare the calculated lifetimes since this is a bulk method and the GaN layer was only ~ 1 μm thick. Thus it is unlikely that positrons trap at N vacancies as can be seen by the virtually identical high momentum distributions and the small increase in the lifetime. Also recent *ab-initio* calculations predict a positively charged state of the N vacancy [6,7], thus making it cease completely to be a trapping site for positrons. Therefore it is unlikely that the observed change in the momentum distribution is caused by N vacancies. Ga vacancies are more likely, although the calculated Doppler curve predicts a bigger difference between the defect-free material and the Ga vacancy. The discrepancy between the calculated Doppler spectrum of a Ga vacancy and the experimental spectrum of n-type GaN can be caused by relaxation of atoms nearby the defect site or the formation of defect complexes. It should be noted that this does not solve the mystery of the origen of intrinsic autodoping since the most likely candidate for the observed vacancies, Ga vacancies, acts as an acceptor.

CONCLUSIONS

Positron beam experiments have been performed on autodoped n-type, semi-insulating and Mg doped p-type GaN. The results from Doppler broadening yield clear evidence of the presence of vacancy-type defects in the autodoped n-type sample compared to the two other samples. Calculations and charge state considerations exclude that the observed vacancies can be attributed to the N-vacancy. Core-electron sensitive two-detector measurements show that the most likely candidate ascribable to these vacancies is the Ga-vacancy, though vacancy-complexes or di-vacancies can not be ruled out.

ACKNOWLEDGEMENTS

M. Alatalo and M. J. Puska are gratefully acknowledged for the use of the atomic superposition program. LVJ and ACK would like to thank the Netherlands Organisation for Scientific Research (NWO) for providing travel grants.

REFERENCES

[1] S. Nakamura, M. Senoh, N. Iwasa, and S. Nagahama, Appl. Phys. Lett. **67** (1995) 1868
[2] S. Nakamura, M. Senoh, S. Nagahama, N. Iwasa, T. Yamada, T. Matsushita, H. Kiyoku, and Y. Sugimoto, Jpn. J. Appl. Phys. **35** (1996) L74
[3] H. P. Maruska and J. J. Tietjen, Appl. Phys. Lett. **15** (1969) 327
[4] D. W. Jenkins and J. D. Dow, Phys. Rev. B **39** (1989) 3317
[5] P. Bogusławski, E. L. Briggs, and J. Bernholc, Phys. Rev. B **51** (1995) 17255
[6] J. Neugebauer and C.G. van de Walle, Phys. Rev. B **50** (1994) 8067
[7] T. Mattila and R.M. Nieminen, submitted to Phys. Rev. B
[8] P. Perlin, T. Suski, H. Teisseyre, M. Leszczynski, I. Grzegory, J. Jun, S. Porowski. P. Bogusławski, J. Bernholc, J. C. Chervin, A. Polian, and T. D. Moustakas, Phys. Rev. Lett. **75** (1995) 296
[9] T. L. Tansley and R. J. Egan, Phys. Rev. B **45** (1992) 10942
[10] P. J. Schultz and K. G. Lynn, Rev. Mod. Phys. **60** (1988) 701
[11] P. Asoka-Kumar, K. G. Lynn, and D. O. Welch, J. Appl. Phys. **76** (1995) 4935
[12] M. J. Puska and R. M. Nieminen, Rev. Mod. Phys. **66** (1994) 814
[13] T. D. Moustakas and R. J. Molnar, Mat. Res. Soc. Proc. Vol. 281 (1993) 753
[14] A. van Veen, J. Trace Microprobe Techn. **8** (1990) 1
[15] M. Clement, J.M.M. de Nijs, P. Balk, H. Schut, and A. van Veen, J. Appl. Phys. **79**   (1996) 9029
[16] A. van Veen, H. Schut, J. de Vries, R.A. Hakvoort and M.R. Ijpma in: Positron Beams for Solids and Surfaces, AIP Conf. Proc., Vol. 218, eds. P. J. Schultz, G. R. Massoumi and P. J. Simpson (AIP, New York, 1990) p. 171
[17] A. van Veen, H. Schut, M. Clement, J.M.M. de Nijs, A. Kruseman and M.R. Ijpma, Appl. Surf. Sci. **85** (1995) 216
[18] K.G. Lynn, J.R. MacDonald, R.A. Boie, L.C. Feldman, J.D. Gabbe, M.F. Robbins, E. Bonderup and J. Golovchenko, Phys. Rev. Lett. **38** (1977) 241
[19] P. Asoka-Kumar, M. Alatalo, V.J. Ghosh, A.C. Kruseman, B.Nielsen and K.G. Lynn, Phys. Rev. Lett. **77** (1996) 2097
[20] M.J. Puska and R.M. Nieminen, J. Phys. F:Metal Phys. **13** (1983) 333
[21] M. Alatalo, B. Barbiellini, M. Hakala, H. Kauppinen, T. Korhonen, M.J. Puska, K. Saarinen, P. Hautojärvi and R.M. Nieminen, Phys. Rev. B **54** (1996) 2397
[22] B.Barbiellini, M. J. Puska, T. Torsti, and R. M. Nieminen, Phys. Rev. B **51** (1995) 7341
[23] V. Ghosh, M. Alatalo, P. Asoka-Kumar, K. G. Lynn, and A. C. Kruseman, submitted to Appl. Surf. Sci.
[24] P. Hautojärvi, J. de Physique IV (1995) C1-3

Part VIII

Point Defects

THEORY OF POINT DEFECTS AND INTERFACES

CHRIS G. VAN DE WALLE* and JÖRG NEUGEBAUER**
*Xerox Palo Alto Research Center, 3333 Coyote Hill Road, Palo Alto, California 94304;
vandewalle@parc.xerox.com
**Fritz-Haber-Institut der Max-Planck-Gesellschaft, Abt. Theorie, Faradayweg 4–6, D-14195 Berlin, Germany

ABSTRACT

First-principles theoretical results can predict and explain a variety of materials properties of the nitride semiconductors. For n-type GaN, we summarize the current understanding about incorporation of unintentional donor impurities, as opposed to nitrogen vacancies. For p-type GaN, we discuss the cause of the limited doping levels, and the role of hydrogen. We describe the role of gallium vacancies in the yellow luminescence, and the interaction between these vacancies and donor impurities. Finally, we discuss our first-principles investigations of the atomic and electronic structure of heterojunction interfaces between the III-nitrides, and provide values for natural band lineups.

INTRODUCTION

Tremendous progress has recently been made in the growth and fabrication of GaN-based electronic and optoelectronic devices. A number of problems still exist, however, which may hamper further progress. One such problem is doping. In 1994 we countered the conventional wisdom by suggesting that nitrogen vacancies were *not* responsible for the commonly observed n-type conductivity in GaN [1]. Instead we proposed that donor impurities are unintentionally incorporated, with oxygen and silicon the main candidates for donors in GaN [2]. Our proposals have recently been confirmed in a number of experimental investigations, showing that oxygen and silicon concentrations in well characterized samples are high enough to explain the observed electron concentrations. Our current understanding will be discussed in the section on "N-TYPE DOPING."

With regard to p-type doping, the doping levels are still lower than desirable for low-resistance cladding layers and ohmic contacts. Achieving higher hole concentrations with Mg as the dopant has proved difficult; various explanations have been proposed for this limitation. Our investigations of compensation mechanisms [3] have revealed that the determining factor is the solubility of Mg in GaN, which is limited by competition between incorporation of Mg acceptors and formation of Mg_3N_2. Incorporation of Mg on interstitial or substitutional nitrogen sites was found to be unfavorable. Some compensation by native defects may occur, in particular by nitrogen vacancies; however, such compensation is significantly suppressed in the presence of hydrogen. We addressed the role of hydrogen during p-type doping and subsequent anneals in Refs. [4] and [5]. One may wonder whether the limitations encountered with Mg could be overcome by using other acceptor impurities; we will report on these issues in the section on "P-TYPE DOPING.'

A major concern for optoelectronic devices is the presence of alternate recombination channels, such as the "yellow luminescence" (YL) in GaN. Our investigations of native defects have revealed that gallium vacancies are the most likely source of this YL [6], and an increasing number of experiments support this assignment. An assessment of the current understanding follows in the section on "YELLOW LUMINESCENCE".

Mat. Res. Soc. Symp. Proc. Vol. 449 ©1997 Materials Research Society

Finally, we address another issue of importance for device design, namely the atomic and electronic structure of heterojunction interfaces. The III-nitride semiconductors exhibit large differences in lattice constant, and the resulting lattice mismatch complicates an assessment of the heterojunction band discontinuties. We will report on our results in the section on "INTERFACES".

METHODS

Our calculations are based on first-principles density-functional theory [7], using a supercell geometry and soft Troullier-Martins pseudopotentials [8]. The effect of d electrons in GaN and InN was taken into account either through the so-called non-linear core correction or by explicit inclusion of the d electrons as valence electrons; the latter proved to be necessary for obtaining accurate results in certain cases [9]. Our results should apply to both the wurtzite and zincblende phases of the nitride semiconductors; indeed, in Ref. [9] we reported that the wurtzite and the cubic phase show nearly equivalent formation energies and electronic structure for defects. Further details of the computational approach can be found elsewhere [1, 10, 11].

The key to describing doping issues is the calculation of the equilibrium concentrations of impurities and native defects:

$$c = N_{\text{sites}} \exp^{-E^f/k_B T} \tag{1}$$

where N_{sites} is the number of sites the defect or impurity can be incorporated on, k_B the Boltzmann constant, T the temperature, and E^f the formation energy. Equation (1) shows that defects with a *high* formation energy will occur in *low* concentrations.

The formation energy is not a constant but depends on the various growth parameters. For example, the formation energy of an oxygen donor is determined by the relative abundance of O, Ga, and N atoms, as expressed by the chemical potentials μ_O, μ_{Ga} and μ_N. If the O donor is charged (as is expected when it has donated its electron), the formation energy depends further on the Fermi level (E_F), which acts as a reservoir for electrons. Forming a substitutional O donor requires the removal of one nitrogen atom and the addition of one O atom; the formation energy is therefore:

$$E^f(\text{GaN:O}_N^q) = E_{\text{tot}}(\text{GaN:O}_N^q) - \mu_O + \mu_N + qE_F \tag{2}$$

where $E_{\text{tot}}(\text{GaN:O}_N^q)$ is the total energy derived from a calculation for substitutional O, and q is the charge state of the O donor. E_F is the Fermi level, Similar expressions apply to other impurities and to the various native defects. We refer to Refs. [1] and [12] for a more complete discussion of formation energies and their dependence on chemical potentials.

The Fermi level E_F is not an independent parameter, but is determined by the condition of charge neutrality. However, it is informative to plot formation energies as a function of E_F in order to examine the behavior of defects and impurities when the doping level changes. As for the chemical potentials, these are variables which depend on the details of the growth conditions. For ease of presentation, we set these chemical potentials to fixed values in the figures shown below; however, a general case can always be addressed by referring back to Eq. (2). The fixed values we have chosen correspond to Ga-rich conditions ($\mu_{Ga} = \mu_{Ga(bulk)}$), and to maximum incorporation of the various impurities, with solubilities determined by equilibrium with Ga_2O_3, Si_3N_4, and Mg_2N_3.

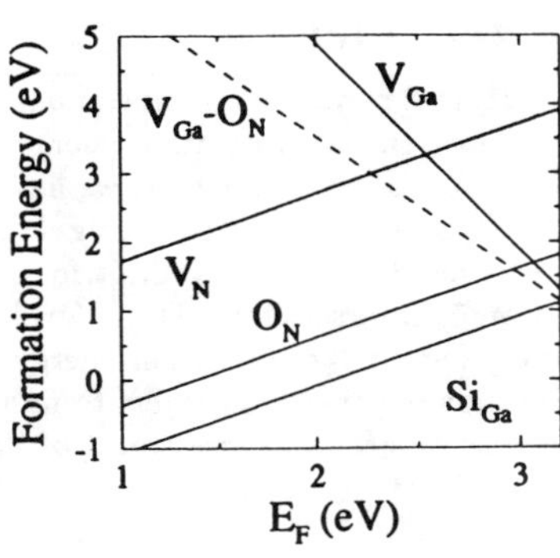

Figure 1: Formation energy *vs.* Fermi energy for native defects (nitrogen and gallium vacancies), donors (oxygen and silicon) and the V_{Ga}-O_N complex. The zero of Fermi energy is located at the top of the valence band.

N-TYPE DOPING

Figure 1 summarizes our first-principles results for native defects and impurities relevant for *n*-type doping. We observe that nitrogen vacancies (V_N) are high-energy defects in GaN, and are thus unlikely to occur in significant concentrations. We have also found that self-interstitial and antisite defects are high-energy defects [1]. Similar results for native defects were obtained by Boguslawski *et al.* [13]. These findings allow us to conclude that nitrogen vacancies are not responsible for *n*-type conductivity in GaN. In contrast, Fig. 1 shows that oxygen and silicon have relatively low formation energies in *n*-type GaN, and can thus be readily incorporated. Both oxygen and silicon form shallow donors in GaN. The slope of the lines in Fig. 1 indicates the charge state of the defect or impurity: Si_{Ga}, O_N, and V_N all appear with slope +1, indicating single donors.

The possibility that oxygen could be responsible for *n*-type conductivity in GaN was recognized by Seifert *et al.* [14] and by Chung and Gershenzon [15]. Still, the prevailing conventional wisdom, attributing the *n*-type behavior to nitrogen vacancies, proved hard to overcome. Recent experiments have confirmed that unintentionally doped *n*-type GaN samples contain silicon or oxygen concentrations high enough to explain the electron concentrations. Götz *et al.* [16] reported electrical characterization of intentionally Si-doped as well as unintentionally doped samples, and concluded that the *n*-type conductivity in the latter was due to silicon. They also found evidence of another shallow donor with a slightly higher activation energy, which was attributed to oxygen. Götz *et al.* have also recently carried out SIMS (secondary-ion mass spectroscopy) and electrical measurements on hydride vapor phase epitaxy (HVPE) material, finding levels of oxygen or silicon in agreement with the electron concentration [17].

High levels of *n*-type conductivity have traditionally been found in GaN bulk crystals grown at high temperature and high pressure [18]. It has recently been established that the characteristics of these samples (obtained from high-pressure studies) are very similar to epitaxial films which are intentionally doped with oxygen [19]. The *n*-type conductivity of bulk GaN can therefore be attributed to unintentional oxygen incorporation.

Finally, we note in Fig. 1 that gallium vacancies (V_{Ga}^{3-}) have relatively low formation energies in highly doped *n*-type material (E_F high in the gap); they could therefore act as compensating centers. Yi and Wessels [20] have found evidence of compensation by a triply charged defect in Se-doped GaN.

Figure 2 summarizes some of our results for acceptor doping in GaN. The Mg acceptor has a low enough formation energy to be incorporated in large concentrations in GaN. For the purposes of the plot, we have assumed Ga-rich conditions (which are actually the least favorable for incorporating Mg on Ga sites), and equilibrium with Mg_2N_3, which determines the solubility limit for Mg. We note that the formation energies for Mg_{Ga}^0 and Mg_{Ga}^- intersect for a Fermi level position around 250 meV; this transition level would correspond to the ionization energy of the Mg acceptor. However, since our calculated formation energies are subject to numerical error bars of ±0.1 eV, this value should not be taken as an accurate assessment of the ionization energy.

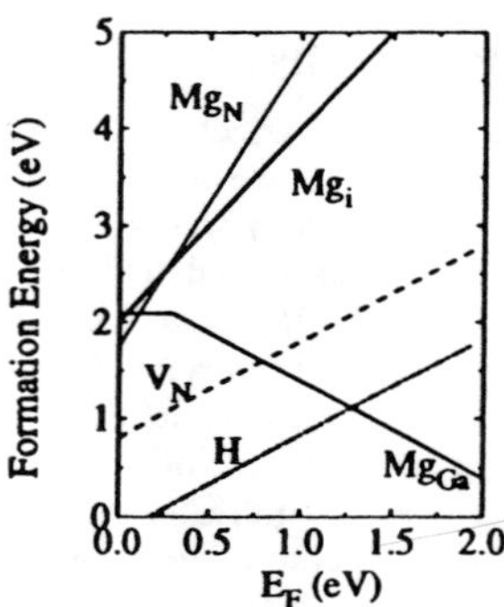

Figure 2: Formation energy as a function of Fermi level for Mg in different configurations (Ga-substitutional, N-substitutional, and interstitial configuration). Also included are the native defects and interstitial H.

We have investigated other positions of Mg in the lattice, such as on substitutional N sites (Mg_N) and on interstitial sites (Mg_i), always finding much larger formation energies. We therefore conclude that Mg overwhelmingly prefers the Ga site in GaN, the main competition being with formation of Mg_2N_3, which is the solubility-limiting phase. It would be interesting to investigate experimentally whether traces of Mg_2N_3 can be found in highly Mg-doped GaN.

Other potential sources of compensation are also illustrated in Fig. 2. The nitrogen vacancy, which had a high formation energy in *n*-type GaN (see Fig. 1) has a significantly lower formation energy in *p*-type material, and could potentially act as a compensating center. However, we also note that hydrogen, when present, has a formation energy much lower than that of the nitrogen vacancy. In growth situations where hydrogen is present [such as metal-organic chemical vapor deposition (MOCVD) or HVPE] Mg-doped material will preferentially be compensated by hydrogen, and compensation by nitrogen vacancies will be suppressed. The presence of hydrogen is therefore beneficial – at the expense, of course, of obtaining material which is heavily compensated by hydrogen! Fortunately, the hydrogen can be removed from the active region by post-growth treatments, such as low-energy electron-beam irradiation [21] or thermal annealing [22]. A more complete discussion of the role of hydrogen in GaN is given in Refs. [4] and [5].

For Mg, we thus conclude that achievable doping levels are mainly limited by the solubility of Mg in GaN. We have investigated other candidate acceptor impurities, and evaluated them in terms of solubility, shallow *vs.* deep character, and potential compensation due to incorporation on other sites. Some of these results are summarized in Ref. [23]. None of the candidate impurities exhibited characteristics superior to Mg.

Finally, we note the importance of avoiding any type of contamination during growth of *p*-type GaN. For instance, the oxygen formation energy shown in Fig. 1 clearly extrapolates to very low values in *p*-type GaN. Any oxygen present in the growth system will therefore be readily incorporated during *p*-type growth. In addition, complex formation between oxygen and magnesium can make oxygen incorporation even more favorable [24].

YELLOW LUMINESCENCE

The yellow luminescence (YL) in GaN is a broad luminescence band centered around 2.2 eV. Its origins have been extensively debated; we have recently proposed that gallium vacancies are the source of the YL [6]. Here we summarize the arguments in favor of this assignment, and discuss recent experimental results.

n-type *vs.* *p*-type

The gallium vacancy is an acceptor-type defect, and hence its formation energy decreases with increasing Fermi level (see Fig. 1). Gallium vacancies are therefore more likely to occur in *n*-type than in *p*-type GaN. The correlation of the YL with Ga vacancies is therefore consistent with experimental observations indicating suppression of the YL in *p*-type material [25, 26, 27]. Conversely, an increase in *n*-type doping increases the intensity of the YL [28]. Additional systematic studies of the YL as a function of doping level are desirable; however, care should be taken in the analysis of such experiments, since the ratio of the magnitude of the YL to the band-gap luminescence depends on the excitation intensity [29, 30].

Ga-rich *vs.* N-rich

It is obvious that the concentration of gallium vacancies will be lower in Ga-rich material. The YL was indeed found to be suppressed in MOCVD samples grown with higher TMGa flow rates [25, 31]. Singh *et al.* [29] observed that the YL was stronger in samples grown at higher microwave power in ECR (electron-cyclotron resonance) assisted MBE (molecular beam epitaxy); this could be consistent with higher V_{Ga} concentrations when the growth is more N-rich, provided no plasma-induced damage is involved. It has also been observed that the YL is weak in material grown by HVPE [32], which could be consistent with growth conditions in HVPE being more Ga-rich than in MOCVD.

Role of carbon

A number of authors have related the YL to the presence of carbon in the material [33, 34]. However, our extensive investigations of carbon on different sites and as a component of various complexes did not produce any defect with properties consistent with the known facts about the YL [6]. We have concluded that the presence of carbon in material that exhibits YL is merely coincidental; indeed, an increase in *n*-type doping facilitates the incorporation of C (an acceptor) at the same time as enhancing the YL.

Recombination mechanism

Our calculations indicate that the Ga vacancy has a deep level (the 2-/3- transition level) about 1.1 eV above the valence band. Transitions between the conduction band (or shallow donors) and this deep level therefore exhibit the correct energy to explain the YL. Various experiments have indeed linked the YL with a deep level, located about 1 eV above the valence band [35, 36]. In addition, our calculated pressure dependence of this level is also consistent with experiment [35].

Complexing with donor impurities

Gallium vacancies can form complexes with donor impurities in GaN. The V_{Ga}-Si_{Ga} complex has a rather small binding energy, due to its components being only second-nearest neighbors. The V_{Ga}-O_N complex, on the other hand, has a large binding energy (1.8 eV), and can therefore play a role in enhancing the concentration of Ga vacancies (see Fig. 1). The electronic structure of this complex is very similar to that of the isolated gallium vacancy, giving rise to a deep level again about 1.1 eV above the valence band.

Ion implantation

The YL has been observed after ion implantation: Pankove and Hutchby [37] found that implantation with a variety of elements produced a broad luminescence band around 2.15 eV [37]. Implantation damage is likely to result in preferential creation of Ga-site defects; indeed, the displacement energy threshold in III-V compounds tends to be lower for the cation site [38]. Formation of Ga vacancies is thus likely during implantation, once again consistently explaining the increase in the YL.

Similarity with SA centers in II-VI compounds

Finally, we point out the similarity between the YL in GaN and the so-called self-activated (SA) luminescence in II-VI compounds. Metal vacancies and their complexes with donor impurities are well known in II-VI compounds (e.g., ZnS, ZnSe). The metal vacancy complexes (the so called *SA* centers) exhibit features which are strikingly similar to the YL: recombination between a shallow donor-like state and a deep acceptor state, and a broad luminescence band of Gaussian shape [39, 40].

INTERFACES

Most nitride-based devices incorporate heterojunctions between GaN, AlN, or InN, or their alloys. The most important parameters characterizing such heterojunctions are the band discontinuities in the conduction and valence bands. It is well known that these band offsets sensitively depend on the strain condition of the materials joined at the interface; however, these strains have not always been properly taken into account in the analysis of experimental data or computational results.

Lattice constants and band gaps for the nitrides (in the zincblende phase) are listed in Table 1. The lattice mismatch between AlN and GaN is about 3%; between GaN and InN, the mismatch is over 11%. The band offsets are only well defined in the case of a pseudomorphic interface, in which the materials on either side of the junction are strained in order to match a common in-plane lattice constant. For instance, when InN is grown on a thick layer of GaN, the InN should be compressed in the plane of the interface to match the GaN lattice constant, and expanded in the perpendicular direction (by Poisson's ratio). The critical layer thickness (beyond which dislocation formation sets in and the interface is no longer pseudomorphic) for growth of pure InN on GaN is probably vanishingly small, but the same logic applies for growth of InGaN alloys on GaN. Results for alloys can usually be obtained by linear interpolation.

Table 1: Experimental lattice constants (a, in Å) and room-temperature band gaps (E_g, in eV) for zincblende AlN, GaN, and InN.

	AlN	GaN	InN
a	4.37	4.50	4.98
E_g	6.20	3.39	1.89

We have focused on computations for (110) interfaces between the nitrides in the zincblende phase. We expect very little difference for the wurtzite phase, which differs from zincblende only in the atomic arrangements beyond third nearest neighbors. We also expect only minor changes for other interface orientations.

Our calculations are performed in a superlattice geometry, for various values of the in-plane lattice constant. The materials are strained according to their elastic constants, and relaxation of the atoms around the interface is explicitly allowed. The superlattice calculation yields the lineup of average electrostatic potentials across the interface; the position of the valence-band maximum with respect to the average electrostatic potential is obtained from bulk calculations [41].

Instead or re-calculating the bulk electronic structure for every strain situation, we have derived *deformation potentials* describing the changes in band edges due to various strain components. We have also calculated the *absolute* deformation potentials for the valence-band maximum. Using the band lineups at strained interfaces together with the deformation potentials allows us to derive a so-called *natural* band lineup between unstrained materials. This natural band lineup can be used as a starting point to calculate offsets at an arbitrary interface, by using information about the strains and the deformation potentials. Complete information about deformation potentials will be published elsewhere [42].

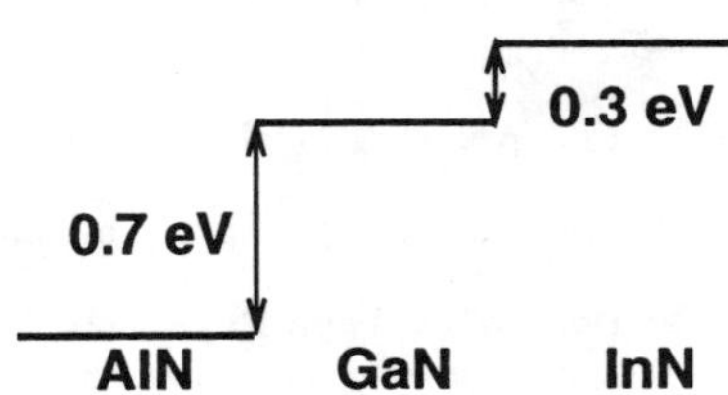

Figure 3: Natural valence-band lineups between AlN, GaN, and InN, obtained from first-principles calculations for zincblende (110) interfaces.

The natural band lineups for the nitrides are illustrated in Fig. 3. the valence-band offset between AlN and GaN (for which the lattice mismatch is relatively modest) is about

0.7 eV, consistent with other recent determinations [43, 44, 45]. For GaN/InN we find a surprisingly small offset, 0.3 eV. Atomic relaxations play an important role at this interface, driving the valence-band offset toward lower values.

CONCLUSIONS

We have presented a variety of results obtained from theoretical and computational investigations of the III-nitride semiconductors. Strong evidence now exists attributing the residual n-type conductivity of GaN to unintentional donor impurities. Additional computational work is in progress to address the behavior of these donors in GaN under pressure, and in alloys. For p-type material, we attribute the limitation in doping levels to solubility constraints, rather than compensation. More work is needed to clarify the behavior of acceptors other than Mg. It is very clear, however, that oxygen contamination is detrimental to p-type conductivity. Regarding the yellow luminescence, we have presented our arguments for linking this luminescence with gallium vacancies. Finally, we have reported results for natural band lineups at interfaces between the III-nitrides.

ACKNOWLEDGMENTS

We gratefully acknowledge stimulating conversations with W. Götz, N. Johnson, and T. Suski. This work was supported in part by ARPA under agreement no. MDA972-95-3-0008.

References

[1] J. Neugebauer and C. G. Van de Walle, Phys. Rev. B **50**, 8067 (1994).

[2] J. Neugebauer and C. G. Van de Walle, in *Proceedings of the 22th International Conference on the Physics of Semiconductors*, Vancouver, 1994, edited by D. J. Lockwood (World Scientific Publishing Co Pte Ltd., Singapore), p. 2327.

[3] J. Neugebauer and C. G. Van de Walle, in *Gallium Nitride and Related Materials*, edited by R. D. Dupuis, J. A. Edmond, F. A. Ponce, and S. Nakamura, Materials Research Society Symposia Proceedings, Vol. 395 (Materials Research Society, Pittsburgh, Pennsylvania), p. 645.

[4] J. Neugebauer and C. G. Van de Walle, Phys. Rev. Lett. **75**, 4452 1995).

[5] J. Neugebauer and C. G. Van de Walle, Appl. Phys. Lett. **68**, 1829 1996).

[6] J. Neugebauer and C. G. Van de Walle, Appl. Phys. Lett. **69**, 503 (1996).

[7] P. Hohenberg and W. Kohn, Phys. Rev. **136**, B864 (1964); W. Kohn and L. J. Sham, *ibid.* **140**, A1133 (1965).

[8] N. Troullier and J. L. Martins, Phys. Rev. B **43**, 1993 (1991).

[9] J. Neugebauer and C. G. Van de Walle, Proc. Mater. Res. Soc. Symp. **339**, 687 (1994).

[10] R. Stumpf and M. Scheffler, Comp. Phys. Commun. **79**, 447 (1994).

[11] J. Neugebauer and C. G. Van de Walle, Proc. Mater. Res. Soc. Symp. **408** (1996).

[12] C. G. Van de Walle, D. B. Laks, G. F. Neumark, and S. T. Pantelides, Phys. Rev. B **47**, 9425 (1993).

[13] P. Boguslawski, E. L. Briggs, and J. Bernholc, Phys. Rev. B **51**, 17 255 (1995).

[14] W. Seifert, R. Franzheld, E. Butter, H. Sobotta, and V. Riede, Cryst. Res. & Technol. **18**, 383 (1983).

[15] B.-C. Chung and M. Gershenzon, J. Appl. Phys. **72**, 651 (1992).

[16] W. Götz, N. M. Johnson, C. Chen, H. Liu, C. Kuo, and W. Imler, Appl. Phys. Lett. **68**, 3114 (1996).

[17] W. Götz *et al.*, these proceedings.

[18] P. Perlin, T. Suski, H. Teisseyre, M. Leszczyński, I. Grzegory, J. Jun, S. Porowski, P. Boguslawski, J. Bernholc, J. C. Chervin, A. Polian, and T. D. Moustakas, Phys. Rev. Lett. **75**, 296 (1995).

[19] C. Wetzel, T. Suski, J. W. Ager III, W. Walukiewicz, S. Fisher, B. K. Meyer, I. Grzegory, and S. Porowski, Proc. ICPS-23 (World Scientific, Singapore, 1996), p. 2929.

[20] G.-C. Yi and B. W. Wessels, Appl. Phys. Lett. **69**, 3028 (1996).

[21] H. Amano, M. Kito, K. Hiramatsu, and I. Akasaki, Jpn. J. Appl. Phys. **28**, L2112 (1989).

[22] S. Nakamura, N. Iwasa, M. Senoh, and T. Mukai, Jpn. J. Appl. Phys. **31**, 1258 (1992).

[23] J. Neugebauer and C. G. Van de Walle, in *Proceedings of the 23rd International Conference on the Physics of Semiconductors*, Berlin, 1996, edited by M. Scheffler and R. Zimmermann (World Scientific Publishing Co Pte Ltd., Singapore, 1996). p. 2849.

[24] C. G. Van de Walle (unpublished).

[25] X. Zhang, P. Kung, and M. Razeghi, in *Gallium Nitride and Related Materials*, edited by R. D. Dupuis, F. A. Ponce, J. A. Edmond, and S. Nakamura (MRS Symposia Proceedings, World Scientific, Singapore, 1995), Vol. 395.

[26] W. Götz, N. Johnson, J. Walker, D. P. Bour, H. Amano, and I. Akasaki, in *Proceedings of the 6th International Conference on SiC and Related Materials*, Kyoto, Japan, Sept. 18-21, 1995, edited by S. Nakashima, H. Matsunami, S. Yoshida, and H. Harima, Inst. Phys. Conf. Ser. No 142 (IOP Publishing, Bristol, 1996), 1031.

[27] W. Kim, A. Salvador, A. E. Botchkarev, O. Aktas, S. N. Mohammad, and H. Morkoç, Appl. Phys. Lett. **69**, 559 (1996).

[28] N. Kaneda, T. Detchprohm, K. Hiramatsu, and N. Sawaki, Jpn. J. Appl. Phys. **35**, L468 (1996).

[29] R. Singh, R. J. Molnar, M. S. Ünlü, and T. D. Moustakas, Appl. Phys. Lett. **64**, 336 (1994).

[30] W. Grieshaber, E. F. Schubert, I. D. Goepfert, R. F. Karlicek, Jr., M. J. Schurman, and C. Tran, J. Appl. Phys. **80**, 4615 (1996).

[31] X. Zhang, P. Kung, A. Saxler, D. Walker, T. Wang, and M. Razeghi, Acta Phys. Polonica A **88**, 601 (1995).

[32] W. Götz, L. T. Romano, B. S. Krusor, N. M. Johnson, and R. J. Molnar, Appl. Phys. Lett. **69**, 242 (1996).

[33] T. Ogino and M. Aoki, Jpn. J. Appl. Phys. **19**, 2395 (1980).

[34] R. Niebuhr, K. Bachem, K. Dombrowski, M. Maier, W. Pletschen, and U. Kaufmann, J. Electron. Mater. **24**, 1531 (1995).

[35] T. Suski, P. Perlin, H. Teisseyre, M. Leszczyński, I. Grzegory, J. Jun, M. Boćkowski, and S. Porowski, Appl. Phys. Lett. **67**, 2188 (1995).

[36] F. J. Sánchez, D. Basak, M. A. Sánchez-García. E. Calleja, E. Muñoz, I. Izpura, F. Calle, J. M. G. Tijero, B. Beaumont, P. Lorenzini, P. Gibart, T. S. Cheng, C. T. Fozon, and J. W. Orton, MRS Internet J. Nitride Semicond. Res. **1**, 7 (1996).

[37] J. I. Pankove and J. A. Hutchby, J. Appl. Phys. **47**, 5387 (1976).

[38] K. W. Böer, *Survey of Semiconductor Physics* (Van Nostrand Reinhold, New York, 1990), p. 629.

[39] *Point Defects in Crystals*, edited by R. K. Watts (John Wiley & Sons, New York, 1977), p. 248ff.

[40] P. J. Dean, Phys. Stat. Sol. (a) **81**, 625 (1984).

[41] A. Franciosi and C. G. Van de Walle, Surface Science Reports Vol. 25, Nos. 1-4, pp. 1-140 (1996).

[42] C. G. Van de Walle and J. Neugebauer (unpublished).

[43] G. Martin, A. Botchkarev, A. Rockett, and H. Morkoç, Appl. Phys. Lett. **68**, 2541 (1996).

[44] E. A. Albanesi, W. R. L. Lambrecht, and B. Segall, J. Vac. Sci. Technol. B **12**, 2470 (1994).

[45] X. Chen, X. Hua, J. Hu, J.-M. Langlois, and W. A. Goddard III, Phys. Rev. B **53**, 1377 (1996).

INCOMPLETE SOLUBILITY IN NITRIDE ALLOYS

I.H. Ho and G.B. Stringfellow
Department of Materials Science and Engineering
University of Utah, Salt Lake City, Utah 84112.

ABSTRACT

A model based on the valence-force-field (VFF) model has been developed specifically for the calculation of the miscibility gaps in III-V nitride alloys. In the dilute limit, this model allows the relaxation of the atoms on both sublattices. It was found that the energy due to bond stretching and bond bending was lowered and the solubility limit was increased substantially when both sublattices were allowed to relax to distances as large as the sixth nearest neighbor positions. Using this model, the equilibrium mole fraction of N in GaP was calculated to be 6×10^{-7} at 700°C. This is slightly higher than the calculated results from the semi-empirical delta lattice parameter (DLP) model. Both the temperature dependence and the absolute values of the calculated solubility agree closely with the experimental data. The solubility is more than three orders of magnitude larger than the result obtained using the VFF model with the group V atom positions given by the virtual crystal approximation, i.e., with relaxation of only the first neighbor bonds. Other nitride systems, such as GaAsN, AlPN, AlAsN, InPN, and InAsN were investigated as well. The equilibrium mole fractions of nitrogen in InP and InAs are the highest, which agrees well with recent experimental data where high N concentrations have been produced in InAsN alloys. Calculations were also performed for the alloy systems with mixing on the group III sublattice that are so important for device applications. Allowing relaxation to the 3rd nearest neighbor gives an In solubility in GaN at 800°C of less than 6%. Again, this is in agreement with the results of the DLP model calculation. This result may partially explain the difficulties experienced with the growth of these alloys. Indeed, evidence of solid immiscibility has recently been reported. A significant miscibility gap was also calculated for the AlInN system, but the AlGaN system is completely miscible.

INTRODUCTION

The III/V nitrides and their alloys have recently become important materials for the fabrication of blue and green light emitting diodes (LEDs)[1]. This is due to breakthroughs in the organometallic vapor phase epitaxial (OMVPE) growth of GaN on sapphire substrates. The structures giving the highest performance LEDs also contain thin GaInN layers in double heterostructures. Similar structures have also recently been used for fabrication of violet and blue injection lasers (LDs)[2]. Since the bandgap energy of GaInN can be varied from 2.0 to 3.5 eV by increasing the GaN concentration[3], the operating wavelengths for light emitting devices with GaInN active layers can, in principle, cover nearly the entire visible spectral range. Al containing nitrides are also important as the confining layers in light emitting devices. The AlGaN/GaN system may also be important for high electron mobility transistors[4].

The growth of GaInN has proven to be extremely challenging, mostly due to the trade-off between the epilayer quality and the amount of InN incorporation into the alloy as the growth temperature is changed.   Growth using high temperatures of approximately 800 °C typically results in high crystalline quality but the amount of InN in the solid is limited to low values, partly because of the high volatility of N over InN[5]. Katsui et. al.[5] found that, using organometallic vapor phase epitaxy (OMVPE), lowering the growth temperature from 800 to 500 °C results in an increase in the allowable InN concentrations, but at the expense of reduced crystalline quality. Attempts to increase the InN concentration in the solid by raising the In pressure in the vapor results in In droplets on the surface.[6]   Some evidence of phase separation was demonstrated in early annealing experiments, where GaInN samples were treated in an argon ambient at various temperature below 700 °C.[7] More recent results indicate evidence of phase separation in GaInN

Mat. Res. Soc. Symp. Proc. Vol. 449 ©1997 Materials Research Society

alloys grown by both OMVPE[8] and molecular beam epitaxy (MBE)[9]. The thermodynamic stability of the GaInN system has not been discussed in detail until now, a somewhat surprising situation in light of the importance of these alloys.

Other nitrogen containing III-V semiconductors are potentially useful for wide-bandgap optoelectronic applications. Vegard's law indicates that several nitride alloys will be lattice-matched to silicon, for example, $GaP_{0.98}N_{0.02}$ and $GaAs_{0.80}N_{0.20}$, have been considered for the integration of III-V semiconductors to silicon technology[10]. The dearth of papers discussing nitrides with mixing on the group V sublattice is mainly due to the expectation that such alloys will have large miscibility gaps[11,12]. For example, the theoretical value of the solid solubility of N in GaP is about $3 \times 10^{-7} cm^{-3}$, i.e., about 0.3 ppm, at 700 °C[12], a number that is in good agreement with the available experimental data[13]. This value, calculated using the delta lattice parameter (DLP) model, is much less than the desired lattice-matched concentration of 0.02.   However, it has been demonstrated that epitaxial growth techniques operating with a large supersaturaton, such as OMVPE, are able to produce metastable alloys lying inside the miscibility gap[14]. Thus, values of N concentration as large as 0.16 have been reported for layers grown by chemical beam epitaxy (CBE) at temperatures between 500 and 650°C[15]. The solubility of N in GaAs is expected to be even less than in GaP due to the larger size difference[11,12]. Nevertheless, N concentrations of 1.5% have been reported for layers grown by OMVPE[16-18]. Even larger values of 14.5% and 20% have been reported for layers grown by CBE[15] and MBE[19], respectively. A correlation is generally observed between the magnitude of the miscibility gap and the difficulty of growing the alloys by OMVPE[14]. Thus, it is worthwhile to have an estimate of the solid solubility when considering nitrides with mixing on the group V lattice.

In this work, the solubility of nitrogen in GaP, GaAs, and other III/V compounds was calculated using the delta-lattice parameter (DLP) model[12] and the valence-force-field (VFF) model[20]. It was found that the VFF model significantly under-estimates the solubility when only one sublattice is allowed to relax. This is unfortunate, since the calculation can be performed relatively simply, without relying on large machine calculations involving the relaxation of hundereds of atoms. Thus, a modification of the VFF model was developed, for application to dilute alloys, which allows the relaxation of both sublattices in a shell of several nearest neighbor atoms surrounding each N atom in III-V-N alloys. This is still a relatively simple calculation requiring only modest computer resources. Using this model, the microscopic strain energy was found to be reduced significantly until the lattice deformation around that nitrogen was allowed to extend to the sixth nearest neighbor. The nitrogen solubility limits were calculated using this model, referred to as VFF(6), for several alloys with mixing on the group V sublattice: GaPN, GaAsN, AlPN, AlAsN, InPN, and InAsN. Comparison of the VFF(6) and DLP results suggests that the DLP model is quite accurate for these N-containing III/V alloys. Both the DLP and VFF models have also been used to calculate the extent of the miscibility gap for the GaInN and AlInN systems. The calculated equilibrium InN solubility in GaN at 800 °C is less than 6%. The results of these calculations suggests that the GaInN and AlInN alloys are unstable over nearly the entire range of solid composition useful for photonic and electronic devices.

CALCULATION OF THE ENTHALPY OF MIXING OF III-V ALLOYS

The calculation of the enthalpy of mixing or the interaction parameter in III/V systems has been a topic of interest for decades. In 1972 Stringfellow[11-12] developed the semi-empirical delta-lattice-parameter (DLP) model which is found to yield surprisingly accurate interaction parameters for a wide range of III/V alloys knowing only the lattice constants of the binary constituents. The DLP model was developed based on the dielectric theory of electronegativity[21,22]. A regular solution interaction parameter, Ω, is calculated by forcing the mixing enthalpy to take the regular solution form[12]:

$$\Delta H^M(x) = \Omega \ x(1-x) \qquad (1)$$

In the DLP model, the interaction parameter can be approximated as[12]

$$\Omega = 4.375 \; K \cdot (\Delta a)^2 / \overline{a}^{\,4.5} \qquad (2)$$

where $\overline{a}$ is the average of the lattice constants, and Δa is the difference in lattice constants for the two binary compounds AC and BC.   Note that Ω is a function of Δa , hence the name delta-lattice-parameter model. The constant K can be obtained by fitting eq. (2) to available experimental data, hence the model is semi-empirical. The form of equation (2) suggests that the main term in the enthalpy of mixing is the strain energy.

The entropy of mixing is generally considered to be equal to the ideal configurational term for a random system. In a system where the enthalpy of mixing is large compared to the entropy term in the free energy (TΔS), i.e., for $\Delta H^M >> T\Delta S$, the solubility may be approximated as[11]

$$x = \exp(-\Omega /RT) \qquad (3)$$

where T is the temperature and R is the ideal gas constant. The DLP model gives a very accurate prediction of nitrogen solubility in GaP[13,14].

The DLP model always gives a positive enthalpy of mixing. The Δa^2 dependence in Eqn (2) suggests that the main effect in the DLP model is the strain energy. In recent years, the major component of the enthalpy of mixing in III/V alloys has been determined to be the microscopic strain energy associated with deforming the bonds in the alloy. Thus, the VFF model, developed by Keating[20], provides an attractive alternative not containing the adjustable constant, K. In the VFF model, the short-range energy due to stretching and bending of the bonds constitutes the enthalpy of mixing.   In the case of a ternary alloy $A_xB_{1-x}C$ the simplest form of the VFF calculation[23] assumes that the lattice is composed of five types of tetrahedra. Each has an atom C at the center and the apexes are occupied by a combination of A and B atoms. The atoms on the mixed sublattice (A and B) are fixed at positions determined by the virtual-crystal approximation (VCA) while the atoms of the common element (C) are allowed to relax in each tetrahedron to minimize the total strain energy. The strain energy (E_m) can be written for each of the 5 tetrahedra,

$$E_m = \frac{3}{8} \sum_{i=1}^{4} \alpha_i \frac{[d_i^2 - d_{io}^2]^2}{d_{io}^2} + \frac{3}{8} \sum_{S} \sum_{i=1}^{4} \sum_{j=i+1}^{4} \frac{\beta_i + \beta_j}{2} \cdot \frac{[\vec{d}_i \cdot \vec{d}_j + d_{io}d_{jo}/3]^2}{d_{io}d_{jo}} \qquad (4)$$

where d_i is the distance between the center atom and a corner atom in the tetrahedron, while d_{i0} is the equilibrium length of this bond in the binary compound AC or BC. α and β are the bond stretching and the bond bending force constants, respectively.   The mixing enthalpy is the summation of E_m over the 5 types of tetrahedra weighted by the distribution probability P_m.

$$\Delta H^M = \Sigma E_m P_m \qquad (5)$$

Note that ΔH^M, E_m, and P_m are all implicit functions of x. In the general case, E_m and P_m are coupled and must be solved simultaneously[23]. Assuming a random alloy uncouples E_m and P_m and simplifies the calculation[23]. The interaction parameter can be obtained by substituting ΔH^M of eq. (5) into eq.(1). The solubility limit is again calculated using eq. (3).

There are two major drawbacks of this simple form of the VFF model: 1) One of the sublattices is not relaxed, causing an overestimate of total strain energy. 2) The difference in energy between the several tetrahedra types is much greater than kT for many III/V alloys. This, of course, gives a nonrandom distribution of the five types of tetrahedra. Taking into account the effects of the resulting short range order (SRO) makes solving E_m and P_m in eq. (5) difficult, since it couples these two factors.[23] The first difficulty can be surmounted by considering a large ensemble of several hundred atoms with the positions of each allowed to relax[24]. However, this involves large-scale calculations and does not address problem 2. This dilemma is easily resolved while maintaining a relatively simple calculation by considering only the dilute limit, where the effect of the SRO is negligible. With this limitation, a full relaxation of the entire lattice can be considered without undue complexity, although the calculation must be done numerically.

At the dilute limit ($x \ll 1$), equation (2) can be simplified to

$$\Delta H^M = \Omega \ x. \tag{6}$$

To calculate Ω, the N atom is fixed in space and its neighbors are allowed to relax to the nth neighbor. If the N atoms are sufficiently far apart, interactions between the clusters can be neglected. Fortunately, the lattice distortion is extremely small outside a shell with a radius of a few nearest neighbor spacings.

The concentration satisfying the constraints for the dilute limit depends, of course, on the size of the relaxed region surrounding the N atom. For n=6, the dilute limit will extend to concentrations as high as 2×10^{20} cm^{-3} or approximately 1%. For n=3 the dilute limit extends to approximately 7%.

The input parameters for both the DLP and VFF models are listed in table I. The zinc-blende lattice constants of several nitrides are listed in ref.25.   Since the bond-stretching and -bending force constants for the nitrides have not been determined experimentally, they are approximated as[25,27]

$$\alpha \ a^3 \sim \text{constant} \tag{7}$$
$$\beta / \alpha \sim 0.3 \ (1 - f_i) \tag{8}$$

where f_i is the Phillip's ionicity.[28]

For AlN, GaN, and InN a second set of values of the force constants from first-principles calculations[29] are also included.

RESULTS AND DISCUSSION

Figure 1 shows the strain energy associated with deformation of the bonds versus the number of nearest neighbors allowed to relax around N dissolved in GaP. As discussed below, this system was chosen because experimental solubility data is available for comparison with the calculated results. The strain energy (or equivalently, Ω) is the highest if all atoms are fixed at their virtual crystal approximation (VCA) sites (no relaxation of either sublattice).   As the first nearest-neighbors are relaxed, the strain energy (Ω(SVFF) in the figure) is lowered dramatically. When the number of relaxed shells extends to the sixth nearest neighbor (the 7th neighbors take the VCA positions), the strain energy approaches a constant value corresponding to $\Omega = 27.4$ kcal/mole. This value is nearly the same as the value of Ω calculated using the DLP model (the dashed line labeled Ω(DLP). Relaxation of the lattice beyond the sixth shell has little effect. In addition, as discussed above, expanding the size of the relaxed volume limits the concentration range satisfying the dilute limit. Thus, relaxation to the 6th shell was used for subsequent calculations for systems where the solubility is extremely small, such as N in GaP, GaAs, etc. Obviously, there is a small error, resulting in an underestimate of the solubility, in allowing relaxation to extend only to the 3rd shell. However, this is necessary for systems with larger N concentrations such as the nitride alloys with mixing on the group III sublattice. An alternate approach would be to use the DLP model, but it appears to significantly overestimate the enthalpy of mixing of these alloys, as indicated by the data in Table II.

The temperature dependence of the solubility of N in *GaP* is plotted in figure 2. The experimental data are compared with the results obtained using the three models discussed. The results from the DLP model and the VFF model with relaxation to the 6th neighbor are in close agreement, as expected from the results plotted in Fig. 1. Both match the experimental data rather

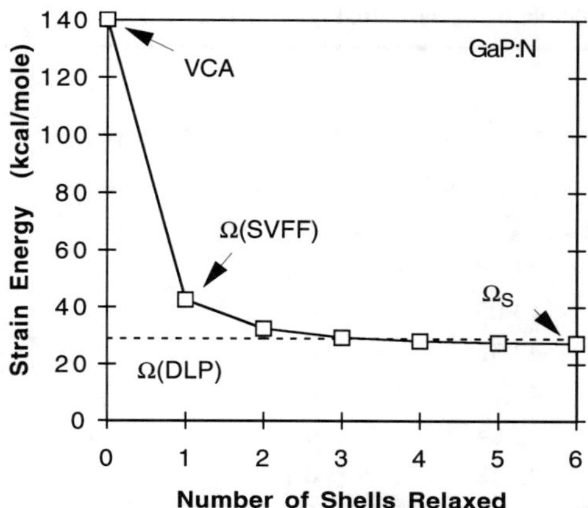

Fig. 1: Strain energy of the nitrogen-centered cluster vs. number of neighboring shells relaxed in the GaPN system. The solid line is simply drawn to connect the data points. The horizontal dashed line labelled Ω(DLP) is the interaction parameter calculated using the DLP model.

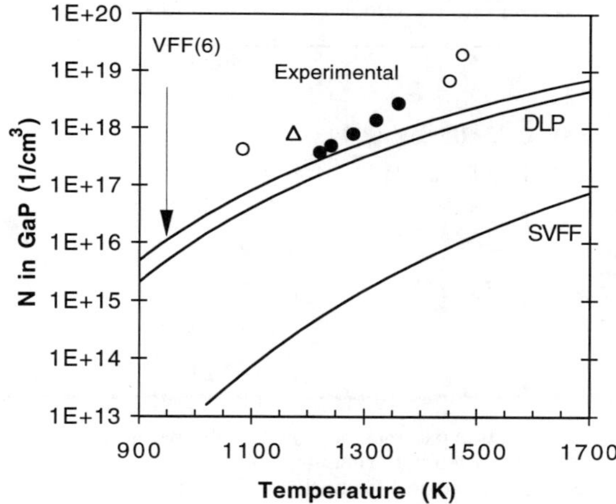

Fig. 2: The solid solubility of nitrogen in GaP vs. temperature calculated using several models. The lines (——)are the result of calculation and the points are the experimental data from references [13,30, and 31].

Table I. Input parameters for calculation of the enthalpy of mixing

System	$a_{0\,(cubic)}$ (Å)	α (N/m)	β / α	fi
AlN	4.380[a]	86.53[c], 98.00[d]	0.1653[c], 0.153[d]	0.45
GaN	4.520[a]	81.09[c], 96.30[d]	0.1500[c], 0.154[d]	0.50
InN	4.980[a]	63.58[c], 79.20[d]	0.1266[c], 0.090[d]	0.58
AlP[b]	5.466	47.29	0.192	-
AlAs[b]	5.660	43.05	0.229	-
GaP[b]	5.450	47.32	0.221	-
GaAs[b]	5.653	41.19	0.217	-
InP[b]	5.868	43.04	0.145	-
InAs[b]	6.055	35.18	0.156	-

[a] ref. 32
[b] ref. 33
[c] calculated using eqs.7 and 8
[d] ref. 29

Table II. The interaction parameter estimated for various III-V nitrides using the VFF model. The value given is for the alloy dilute in N except where indicated.

System	$\Omega - \text{VFF(6)}$ (kcal/mole)	$\Omega - \text{DLP}$ (kcal/mole)
AlPN	36.56	45.53
AlAsN	53.42	57.93
AlGaN	0.88* (0.85, 0.90),1.00**	1.19
AlInN	11.45* (10.52, 12.37),11.80**	17.45
GaPN	27.38	28.90
GaAsN	36.84	42.78
GaInN	5.98* (5.63, 6.32), 6.57**	9.60
InPN	16.33	19.68
InAsN	21.61	26.71

*Average of the two dilute interaction parameters shown in the parentheses. The first value is for the alloy dilute in the first element listed in the chemical formula given.
**Average value calculated using input elastic constants from ref. 29.

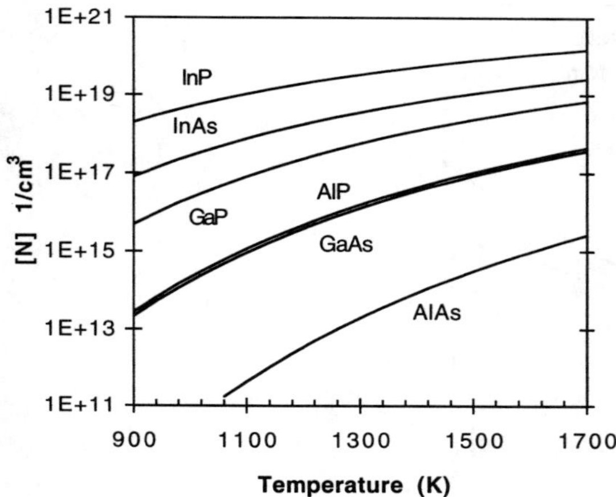

Fig. 3: Nitrogen solubility vs. temperature in GaAs, GaP, AlP, AlAs, InP, and InAs calculated using the VFF model with relaxation to the 6th neighbor.

closely. The curve calculated using the simple VFF model is approximately 3 orders of magnitude too small. Clearly this model over-estimates the total strain energy. As discussed above, this is due to the neglect of the SRO and, mainly, to the restriction of lattice relaxation to the first nearest neighbors. Thus, in what follows only the VFF model with relaxation to several neighbors will be used.

The input parameters used for the several III/V compounds are tabulated in Table I. The interaction parameters calculated using the VFF model for several alloys of interest here are tabulated in table II. The results are compared with the values of interaction parameter obtained using the DLP model. In each case, the values are similar. This indicates the general utility of the simpler DLP model. However, in what follows, only the results obtained using the dilute VFF model will be presented, since this model contains no adjustable parameters and is more physical, it may be more reliable for systems for which no experimental data exist.

The calculated nitrogen solubilities in GaAs, AlP, AlAs, InP, and InAs are shown in figure 3. Clearly, the N solubility in InP and InAs are predicted to be the highest for these III/V compounds. The values are 5 orders of magnitude larger than for GaAs. This is due to the lower force constants for the InAs, InP, and InN bonds. This is thought to represent a real and very significant difference between the addition of N to GaAs and InP or InAs. Indeed, recent experimental data obtained for the growth of InAsN by chemical beam epitaxy support this conclusion[34]. Data for InPN are limited, but the maximum solubility was reported to be approximately 1%[15,35].

Again, for the GaN-InN system the calculated interaction parameters for both GaN-rich and InN-rich solids decreases dramatically as the number of neighbors allowed to relax increases from 1 (equivalent to the VFF model) to approximately 3. For relaxation to the 5th and 6th nearest neighbors the effect saturates. The small increase in accuracy obtained by allowing relaxation to the 5th or 6th nearest neighbor is more than offset by the necessity to limit the calculation to very dilute alloys in order to prevent the interaction of neighboring clusters. For n=3, the calculation is

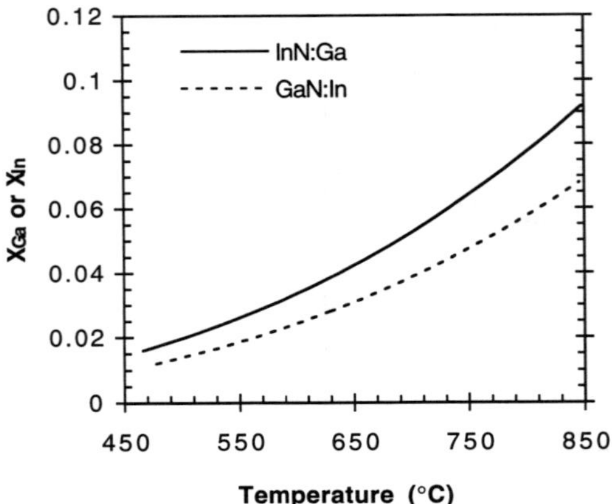

Fig. 4: Solubility of GaN in InN (InN:Ga) and InN in GaN (GaN:In) calculated using the VFF model with bond relaxation to the 3rd neighbor.

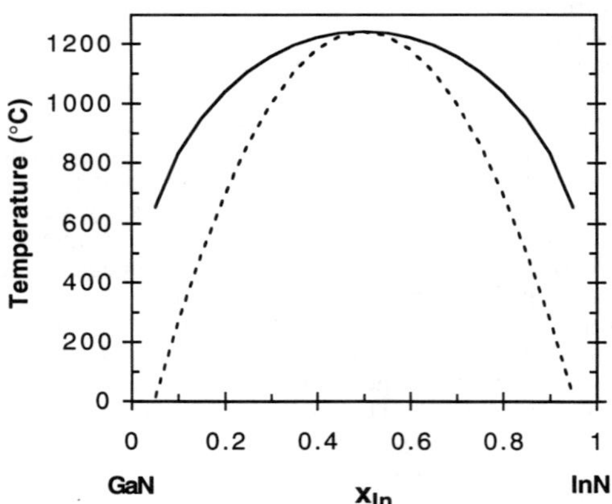

Fig. 5: Binodal (solid) and spinodal (dashed) curves for the $Ga_{1-x}In_xN$ system, calculated assuming a constant, average value for the solid phase interaction parameter.

valid to a concentration of 7%. The limiting values of Ω are approximately 6.32 kcal/mole for the GaN-rich solid and 5.63 kcal/mole for the InN-rich solid using the values of α and β from Eqns. 7 and 8. The interaction parameters are slightly higher using the larger elastic constants from ref. 29. The average value, 5.98 kcal/mole, is about 38% smaller than that calculated by the delta-lattice parameter model (9.60 kcal/model), as seen in Table II.

As seen in Fig. 4, the solubility of In in GaN is less than 6% at 800°C. The solubility of Ga in InN is slightly larger. Again, the larger elastic constants from ref. 29 would give a small reduction in the solubility limits. For example, the solubility of In in GaN at 850°C is decreased from 8.6% to 4.1%. The binodal (solid line) and spinodal (broken line) curves were calculated versus temperature for the entire GaInN system using the average value of Ω=5.98 kcal/mole and are shown in Fig. 5. The critical temperature, above which the miscibility gap disappears, is found to be near 1250 °C which exceeds the melting point of InN.

The above results indicate that, in general, the $Ga_{1-x}In_xN$ alloys, which are highly desirable for both photonic and electronic devices, are unstable at the temperatures commonly used for epitaxialgrowth. This is supported by the results of annealing experiments performed by Osamura et. al.[7] where phase separation in GaInN with InN concentrations exceeding 10% was observed after annealing in argon at 600 and 700°C and the more recent data showing evidence of phase separation in GaInN alloys grown by OMVPE[8] and MBE[9].

SUMMARY

A version of the VFF model has been developed for the calculation of the regions of solid immiscibility in III/V nitrides. In this model the lattice is allowed to relax around an atom on the sublattice where mixing occurs. In the dilute limit, the surrounding lattice can be relaxed in a shell extending to several neighbors. This results in a significant reduction in the strain energy and a concomitant increase in the solubility. The effect is seen to be small for relaxation beyond the sixth nearest-neighbor shells. Thus, for alloys with mixing on the group V sublattice, where the solubility is small, a value of n=6 was used. For the GaPN system, where experimental data are available, this VFF model calculation gives the correct magnitude and temperature dependence of N solubility. The result is similar to that obtained using the semi-empirical DLP model. The simplified VFF model, where only the nearest neighbor bonds are allowed to relax, underestimates the N solubility in GaP by approximately 3 orders of magnitude. The interaction parameters and nitrogen solubility limits are also calculated for GaAs, AlAs, AlP, InAs, and InP. The solubility limit for nitrogen in InP and InAs were found to be 4 or 5 orders of magnitude larger than for GaAs. This suggests that the InPN and InAsN alloys may be easier to grow. For the alloys GaInN and AlInN, with mixing on the group III sublattice, the solubilities are larger, so a value of n=3 was used. At the maximum temperature typically used for the epitaxial growth of $Ga_{1-x}In_xN$ (800°C), the solubility of InN is less than 6%. An average interaction parameter of 5.98 kcal/mole was calculated, yielding a critical temperature for phase separation of 1250°C. These results indicate that $Ga_{1-x}In_xN$ alloys are unstable over most of the composition range at normal growth temperatures. This conclusion is supported qualitatively by the results of published experimental investigations that report evidence of phase separation in the GaInN system. The large region of solid immiscibility may partially explain the difficulties reported in the epitaxial growth of these alloys.

ACKNOWLEDGMENTS

This work was financially supported by the Department of Energy and Texas Instruments Inc.

REFERENCES

1. S. Nakamura, M. Senoh, N. Iwasa and S. Nagahama, Jpn. J. Appl. Phys. **34**, L798 (1995).
2. S. Nakamura, M. Senoh, S. Nagahama, N. Iwasa, T. Yamada, T. Masushita, H. Kiyoku, and Y. Sugimoto, Jpn. J. Appl. Phys. **35** L74 (1996).
3. S. Nakamura, M. Senoh, N. Iwasa and S. Nagahama, Jpn. J. Appl. Phys. **34** L797 (1995).
4. M. Redwing, M.A. Tischler, J.S. Flynn, S. Elhamri, M. Ahoujja, R.S. Newrock, and W.C. Mitchel, Appl. Phys. Lett. **69** 963 (1996).
5. T. Matsuoka, N. Yoshimoto, T. Sasaki, and A. Katsui, J. Electron. Mater. **21** 157 (1992).
6. M. Shimizu, K. Hiramatsu, and N. Sawaki, J. Cryst. Growth **145** 209 (1994).
7. K. Osamura, S. Naka, and Y. Murakami, J. Appl. Phys. **46** 3432 (1975).
8. M. Funato (private communication).
9. R. Singh and T.D. Moustakas, Mat. Res. Soc. Proc. Vol. 395, 163 (1996); R. Singh and T.D. Moustakas, Electrochemical Society Proceedings, Vol. 96-11, 186 (1996).
10. S. Nakamura, M. Senoh, S. Nagahama, N. Iwasa, T. Yamada, T. Masushita, H. Kiyoku, and Y. Sugimoto, Jpn. J. Appl. Phys. **35** L74 (1996).
11. G. B. Stringfellow, J. Electrochem. Soc. **119** 1780 (1972).
12. G. B. Stringfellow, J. Crystal Growth **27** 21 (1974).
13. J. Karpinski, J. Jun, I . Grzegory, and M. Bugajski, J. Cryst. Growth **72** 711 (1985).
14. G. B. Stringfellow, "Organometallic Vapor-phase Epitaxy: Theory and Practice" (Academic press, New York, 1989) Chapter 3.
15. W.G. Bi and C. W. Tu, Materials Research Society Fall Meeting, December 1996, Paper N3.41.
16. M. Kondow, K. Uomi, K. Hosomi and T. Mozume. Jpn. J. Appl. Phys. **33** L1056 (1994).
17. M. Weyers and M. Sato, Appl. Phys. Lett. **62** 1396 (1993).
18. M. Weyers. M. Sato and H. Ando, Jpn. J. Appl. Phys. **31** L853 (1993).
19. C. T. Foxon, T. S. Cheng, S. V. Novikov, D. E. Lacklison, L. C. Jenkins, D. Johnston, J. W. Orton, S. E. Hooper, N. Baba-Ali, T. L. Tansley, V. V. Tretyakov, J. Cryst. Growth **150** 892 (1995); S. V. Novikov, C. T. Foxon, T. S. Cheng, T. L. Tansley, J. W. Orton, D. E. Lacklison, D. Johnston, N. Baba-Ali, S. E. Hooper, L. C. Jenkins, L. Eaves, J. Cryst. Growth **146** 340 (1995).
20. P. N. Keating, Phys. Rev. **145** 637 (1966).
21. J. C. Phillips and J. A. Van Vechten, Phys. Rev. B **2** 2147 (1970).
22. J. C. Phillips, Phys. Rev. Lett. **20** 550 (1968).
23. M. Ichimura, and A. Sasaki, J. Appl. Phys. **60** 3850 (1986).
24. M.C. Schabel and J.L. Martins, Phys. Rev. B **43**, 11873 (1991).
25. D. Chandrasekhar, D. J. Smith, S. Strite, M. E. Lin, and H. Morkoc, J. Crystal Growth **152** 135 (1995).
26. T. N. Morgan and M. Maier, Phys. Rev. Lett. **27** 1200 (1971).
27. R. M. Martin, Phys. Rev. B **1** 4005 (1970).
28. J. C. Phillips, "Bonds and Bands in Semiconductors" (Academic Press, New York, 1973) p. 42
29. K. Kim, W.R.L. Lambrecht, and B. Segall, Phys. Rev. B **53**, 16310 (1996).
30. V. Thierry-Mieg, A Marbeuf, J. Chevallier, H. Mariette. M. Bugajski, and K. Kazmierski, J. Appl. Phys. **54** 5358 (1983).
31. T. J. Hayes, A. Mottram, and A. R. Peaker, J. Cryst. Growth **45** 59 (1978).
32. S. Strite and H. Morkoc, J. Vac. Sci. Tech. B **10**, 1237 (1992).
33. J.L. Martins and A. Zunger, Phys. Rev. B **30**, 6217 (1984).
34. Y. C. Kao, T. P. E. Broekaert, H. Y. Lin, S. Tang, I. H. Ho, and G. B. Stringfellow, Paper E12.4, Presented at the Spring MRS Meeting, April, 1996, San Francisco, CA.
35. W.G. Bi and C.W. Tu, J. Appl. Phys. **80**, 1934 (1996).

X-RAY ABSORPTION AND REFLECTION AS PROBES OF THE GaN CONDUCTION BANDS: THEORY AND EXPERIMENT OF THE N K-EDGE AND Ga $M_{2,3}$ EDGES

W. R. L. Lambrecht, S. N. Rashkeev[‡] and B. Segall
Department of Physics, Case Western Reserve University, Cleveland, OH 44106-7079
K. Lawniczak-Jablonska,[§] T. Suski,[¶] E. M. Gullikson, J. H. Underwood and R. C. C. Perera
Lawrence Berkeley National Laboratory, University of California, Berkeley, CA 94720
and J. C. Rife
Naval Research Laboratory, Washington, D. C. 20375

ABSTRACT

X-ray absorption and glancing angle reflectivity measurements in the energy range of the Nitrogen K-edge and Gallium $M_{2,3}$ edges are reported. Linear muffin-tin orbital band-structure and spectral function calculations are used to interpret the data. Polarization effects are evidenced for the N-K-edge spectra by comparing X-ray reflectivity in s- and p-polarized light.

INTRODUCTION

X-ray absorption measurements are a well-known probe of the unoccupied states in a material. The same information can also be obtained by using glancing angle X-ray reflectivity. In spite of several existing band structure calculations of the group-III nitrides [1] and previous optical studies in the UV range [2, 3, 4], a direct probe of their conduction band densities of states is of interest. Here, we present a joint experimental and theoretical investigation using both of these experimental techniques for wurtzite GaN. A more complete account of this work can be found in a forthcoming paper [5].

EXPERIMENT

X-ray absorption measurements and variable angle X-ray reflection measurements were carried out at the bending magnet 6.3.2. beamline of the Advanced Light Source (ALS) at LBL (Lawrence Berkeley Laboratory). Details of the experimental set-up can be found in Ref. [6]. The investigated samples were stuck into a conducting indium foil which was isolated from the sample holder in the ultra high vacuum reflectometer. The X-ray absorption curves were obtained by measuring the direct photocurrent from the sample, positioned perpendicularly to the X-ray beam as a function of the incoming radiation energy. The X-ray reflectivity was measured with a GaAsP photodiode positioned at the angle 2θ (where θ is the angle between sample surface and incident beam). These measurements used s-polarized light, i.e. $\mathbf{E}$ perpendicular to the plane containing the incident light ray and the normal of the sample surface and were thus for $\mathbf{E} \perp \mathbf{c}$ for any incident angle when using basal plane oriented epitaxial films or single crystals [1]. Both of these as

---

[‡]Permanent address: P. N. Lebedev Physical Institute, Russian Academy of Sciences, 117924 Moscow, Russia

[§]Permanent address: Institute of Physics, Polish Academy of Sciences, 02-668 Warszawa, Poland

[¶]Permanent address: UNIPRESS, Polish Academy of Sciences, 01-142 Warszawa, Poland

[1]Obtained from I. Grzegory and S. Porowski, UNIPRESS, Polish Academy of Sciences, Warszawa, Poland

Mat. Res. Soc. Symp. Proc. Vol. 449 © 1997 Materials Research Society

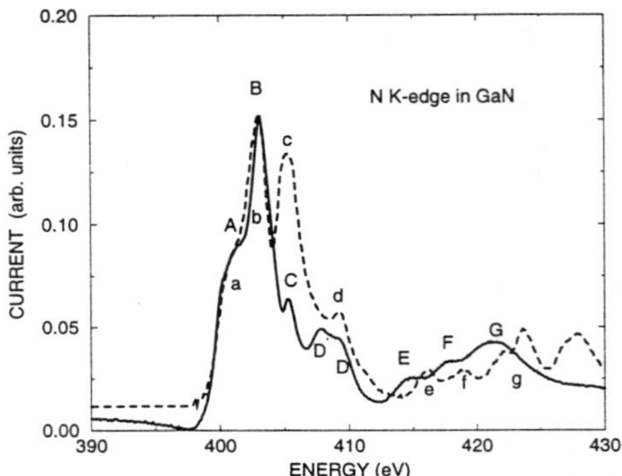

Figure 1: Nitrogen K-edge X-ray absorption spectrum of GaN as measured by total photocurrent (solid line) compared to theory (dashed line).

well as polycrystalline materials were used as samples. Reflection measurements were done for θ varying between 2.3° and 8.3 °. Additional measurments of the X-ray reflectivity were performed on different samples[2] at the Naval Research Laboratory beamline X24C at the National Synchrotron Light Source (NSLS) in Brookhaven. These measurements used both a p-polarization and s-polarization set-up and hence allowed us to detect polarization effects. Details of this experimental set-up can be found in Refs. [7, 8]. Kramers-Kronig analysis of these data was performed using the method of Veal and Paulikas [9] modified to accommodate glancing angle reflectance measurements using the method of Roessler [10].

COMPUTATIONAL METHOD

The theoretical interpretation of the data is based on local density functional theory [11] using linear muffin-tin orbital band-structure calculations [12]. Using the muffin-tin orbital basis set and averaging over polarization directions, the imaginary part of the dielectric function contribution from the transitions of the relevant core level (N1s for K-edge spectra, Ga $3p_{1/2}$, $3p_{3/2}$ for $M_{2,3}$ edges) to the conduction band can be calculated as a sum of the $l \pm 1$ partial densities of states weighted by momentum matrix elements, where l is the angular momentum of the core state. From it, the real part of the optical dielectric function is obtained by a Kramers-Kronig transormation. We then obtain either directly the reflectivity $|r(\omega)|^2$ using the Fresnel equations [13],

$$r(\omega) = \frac{\epsilon \sin \theta - \sqrt{\epsilon - (\cos \theta)^2}}{\epsilon \sin \theta + \sqrt{\epsilon - (\cos \theta)^2}},$$

(1)

or the absorption coefficient $\alpha(\omega) = 4\pi k(\omega)/\lambda$, the two quantities that are directly measured.

We note that in the N K-edge region, $\epsilon_2(\omega)$ and $\epsilon_1(\omega) - 1$ are of order 10^{-2}. For the chosen angles near 5°, we thus have $|\epsilon - 1|$ of the same order as θ^2, the first term in the Taylor expansion of $(\cos \theta)^2 - 1$ in the above equation. In the limit where $\theta^2 \gg |\epsilon - 1|$, but

---

[2]Obtained from D. K. Wickenden at Applied Physics Laboratory, Johns Hopkins University

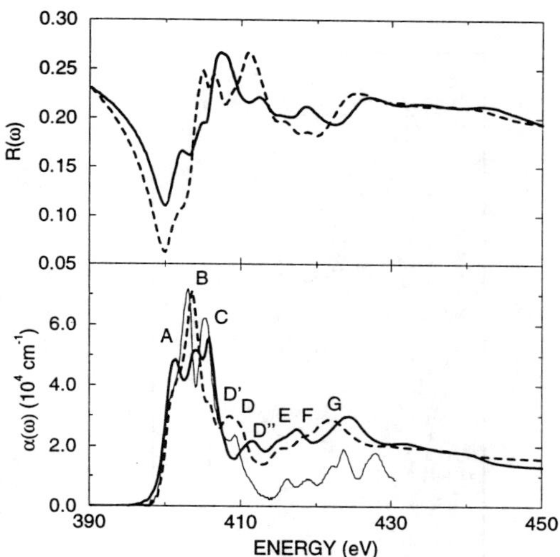

Figure 2: Nitrogen K-edge X-ray reflection spectrum (top) and Kramers-Kronig deduced absorption spectrum (bottom) for p-polarization (solid line) and s-polarization (dotted line) and polarization averaged calculated spectrum (thin solid line).

still $\theta^2 \ll 1$ we obtain by Taylor expansion that reflectivity $r(\omega) \propto |\epsilon - 1|$ and thus starts resembling the absorption function since both are dominated by the imaginary part ϵ_2. On the other hand, the absolute reflectivity drops dramatically with angle approximately as θ^4 and therefore high photon flux synchrotron radiation is required to measure it.

RESULTS

Fig. 1 shows the X-ray absorption coefficient of GaN as measured by means of the total photocurrent method compared with the calculated spectrum. The absolute energy of the theoretical spectrum, which starts at the conduction-band minimum, was determined by using the position of N1s core level with respect to the valence-band maximum as determined by X-ray photoelectron spectroscopy [14] and adding the experimental gap. Using this single adjusted parameter, we see that the various peak positions are in good agreement between theory and experiment up to at least 10 eV above the edge. The peaks E–G appear to be overestimated by the theory. The differences in intensity and some apparent shifts are primarily due to the fact that the calculated spectrum is an average over polarization directions while the measurement is for $\mathbf{E} \perp \mathbf{c}$.

This becomes clear when considering the absorption spectra obtained by Kramers-Kronig analysis of the X-ray reflectivity measurements in Fig. 2 for both s- and p-polarization. One can see that the polarization dependence is reponsible for the suppression of peaks A and C with respect to B in $\mathbf{E} \perp \mathbf{c}$ (s-polarization). Peaks D and D' are suppressed in $\mathbf{E} \parallel \mathbf{c}$ (p-polarization) and another peak D" at slightly higher energy appears. Peaks E and F are more pronounced in p-polarization and peak G is shifted to higher energy. Similar polarization effects on the X-ray absorption were reported by Katsikini et al. [17]. By comparing to the average polarization theoretical curve one can see that these apparent shifts are really due to changes in relative intensity. Similar relative shifts in intensity of peaks A–C were also found in X-ray absorption measurements on a powder

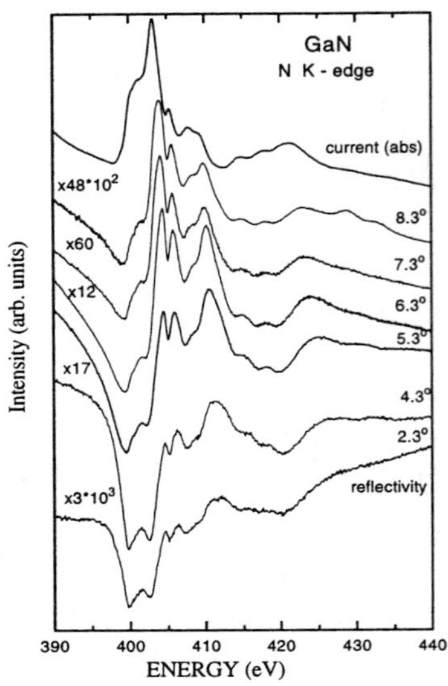

Figure 3: Normalized reflectivity spectra of the Nitrogen K-edge as function of angle. Normalization factors are indicated for each curve, except for the 5.3° case where it is one. The spectra are arbitrarily shifted vertically for convenience. The topmost curve gives the photoyield measurement of the absorption spectrum for comparison.

sample, which is consistent with the fact that randomly oriented crystals exhibit a mixture of the $\mathbf{E} \parallel \mathbf{c}$ and $\mathbf{E} \perp \mathbf{c}$ spectra.

The good agreement in peak positions indicates that the conduction band density of states is well described by the local density approximation (LDA). This means that self-energy corrections beyond LDA of the unoccupied states are approximately constant and not systemetically increasing with energy above the conduction-band minimum as suggested by GW calculations [15, 16]. (G and W stand for the exact one-electron Green's function and screened Coulomb interaction respectively.) This conclusion is consistent with analysis of UV reflectivity measurements [2]. The peaks in partial density of states can further be associated with speficic regions of the Brillouin zone [5].

The angular dependence of the s-polarized reflectivity spectrum is shown in Fig. 3. As discussed above, it shows that for increasing angle the spectrum starts to resemble the absorption spectrum. The factor indicated along with each curve indicates the normalization factor applied and show that the intensitiy is dramatically increasing for angles below 5.3° as expected from the Fresnel theory. The decrease of signal amplitude for angles below 5.3° is an indication of the rapid decrease of the penetration depth into the bulk. Therefore, for small angles, the oxygen and carbon contamination layers of the surface may overwhelm the GaN signal and only an increase of the background signal can be seen. In [5] we show that the relative intensity changes of the peaks in reflectivity is well reproduced by our calculations.

Fig. 4 shows the Ga $M_{2,3}$ absorption spectra obtained by direct photocurrent measurement (under s-polarization) as well as by Kramers-Kronig analysis of X-ray reflectivity

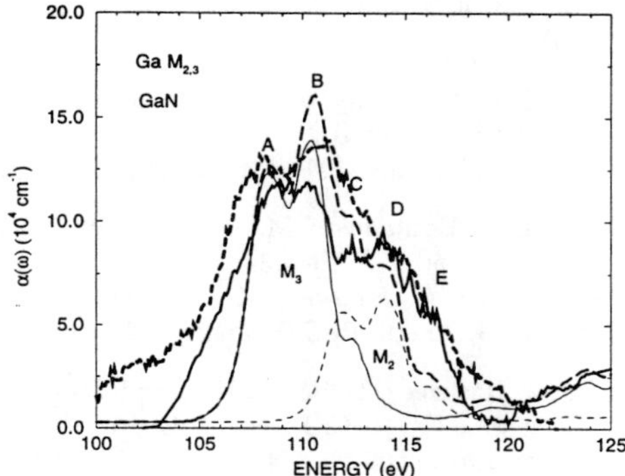

Figure 4: Gallium $M_{2,3}$-edge of X-ray absorption spectrum of GaN as measured by total photocurrent (thick short-dashed line), and from KK analysis of reflectivity (thick solid line) compared to theory (long dashed line). Separate calculated contributions of $3p_{1/2}$ and $3p_{3/2}$ are indicated by thin dashed and thin solid lines respectively.

(under p-polarization) compared to the theory. Again, some changes in intensity due to the different polarizations are apparent but due to the overlap of the M_2 and M_3 components it becomes more cumbersome to analyze them. It is important to note that again a good agreement is obtained as far as peak positions are concerned with the LDA partial density of states. Hence these measurements further support the notion of a constant rather than energy dependent shift of the conduction band with respect to its LDA position, at least up to about 10 eV above the minimum gap.

CONCLUSIONS

X-ray reflectivity and absorption measurements on GaN were reported. Good agreement was obtained between the two experimental techniques and with theoretical calculations. A polarization dependence of the spectra was evidenced by using both s- and p-poarized light in the reflectivity measurements. First-principles calculations indicate good agreement for peak positions with LDA theory and hence support an approximately constant rather than energy dendent self-energy correction in the first 10 eV of the conduction band.

The work at CWRU was funded by the National Science Foundation under grant No. DMR 95-29376. The measurements were performed at the National Synchrotron Light Source and the Advanced Light Source which are both sponsored by the U.S. Department of Energy. K.L-J and T.S. kindly acknowledge the financial support of the Fulbright Foundation.

REFERENCES

1. W. R. L. Lambrecht and B. Segall, in *Properties of Group III Nitrides*, edited by J. H. Edgar, Electronic Materials Information Service (EMIS) Datareviews Series (Institution of Electrical Engineers, London 1994), Chapt. 4.

2. W. R. L. Lambrecht, B. Segall, J. Rife, W. R. Hunter, and D. K. Wickenden, *Phys. Rev. B* **51**, 13516 (1995).

3. W. R. L. Lambrecht, K. Kim, S. N. Rashkeev, and B. Segall, in *Gallium Nitride and Related Materials*, edited by R. D. Dupuis, F. Ponce, S. Nakamura, and J. A. Edmond, Mater. Res. Soc. Symp. Proc. Vol. 395, (MRS, Pittsburgh 1996) p. 455

4. J. Petalas, S. Logothetidis, S. Boutadakis, M. Alouani, and J. M. Wills, *Phys. Rev. B* **52**, 8082 (1995); S. Logothetidis, J. Petalas, M. Cardona, and T. D. Moustakas, *Phys. Rev. B* **50**, 18017 (1994); C. Janowitz, M. Cardona, R. L. Johnson, T. Cheng, T. Foxon, O. Günther, and G. Jungk, *BESSY Jahresbericht* (1994) p. 230.

5. W. R. L. Lambrecht, S. N. Rashkeev, B. Segall, K. Lawniczak-Jablonska, T. Suski, E. M. Gullikson, J. H. Underwood, R. C. C. Perera, J. C. Rife, I. Grzegory, S. Porowski and D. K. Wickenden, *Phys. Rev. B* **54**, (1996), to be published.

6. J.H. Underwood, M.K. Gullikson, M. Koike, P.J. Batson, P.E. Denham, K.D. Franck, R.E. Tackaberry, and W.F. Steele, *Rev. Sci. Instrum.*, in press.

7. J. C. Rife, H. R. Sadeghi, and W. R. Hunter, *Rev. Sci. Instrum.* **60**, 2064 (1989).

8. W. R. Hunter and J. C. Rife, *Nucl. Instrum. Meth.* **A246**, 465 (1986).

9. B. W. Veal and A. P. Paulikas, *Phys. Rev. B* **10**, 1280 (1974).

10. D. M. Roessler, *Brit. J. Appl. Phys.* **16**, 1359 (1965).

11. P. Hohenberg and W. Kohn, Phys. Rev. **136**, B864 (1964); W. Kohn and L. J. Sham, *ibid.* **140**, A1133 (1965).

12. O. K. Andersen, O. Jepsen, and M. Šob, in *Electronic Band Structure and its Applications*, edited by M. Yussouff, (Springer, Heidelberg, 1987), p.1.

13. J. D. Jackson, *Classical Electrodynamics*, Second Edition, (John Wiley & Sons, New York 1975), Chapter 7.

14. G. Martin, S. Strite, A. Botchkarev, A. Agarwal, A. Rockett, H. Morkoç, W. R. L. Lambrecht, and B. Segall, *Appl. Phys. Lett.* **65**, 610 (1994).

15. A. Rubio, J. L. Corkill, M. L. Cohen, E. L. Shirley, and S. G. Louie, *Phys. Rev. B* **48**, 11810 (1993).

16. M. Palummo, L. Reining, R. W. Godby, C. M. Bertoni, and N. Börnsen, *Europhys. Lett.*, **26**, 607 (1994); and in *Proc. 21st Int. Conf. on the Physics of Semiconductors*, Eds. Ping Jiang, Hou-Zhi Zheng (World Scientific Press, Singapore 1993), p. 89.

17. M. Katsikini, E. C. Paloura, J. Kalomiros, P. Bressler, and T. Moustakas, *Proceedings of the 23rd International Conference on The Physics of Semiconductors*, ed. by M. Scheffler and R. Zimmermann (World Scientific, Singapore 1996), p. 573.

ELECTRONIC STRUCTURE OF BIAXIALLY STRAINED WURTZITE CRYSTALS GaN and AlN

J. A. MAJEWSKI, M. STÄDELE, and P. VOGL
Walter Schottky Institute, Technical University of Munich, D-85748 Garching, Germany

ABSTRACT

We present first-principles studies of the effect of biaxial (0001)-strain on the electronic structure of wurtzite GaN, and AlN. We provide accurate predictions of the valence band splittings as a function of strain, which may facilitate the interpretation of data from strained samples. The conduction and valence band effective mass tensors for AlN and GaN are also presented. The computed crystal-field and spin-orbit splittings in unstrained materials as well as the computed deformation potentials are in accord with available experimental data. We show that the numerically computed band energies can be excellently represented in terms of a 6-band $k \cdot p$ model. The present calculations are based on the first-principles pseudopotential method within the local-density formalism and include the spin-orbit interactions non-perturbatively.

INTRODUCTION

The group-III nitrides AlN, GaN, and InN have recently attracted much attention as candidates for short-wavelength optical devices [1]. The stable structure of bulk nitrides is the wurtzite structure. For these direct gap nitrides, the detailed knowledge of the carrier spectra in the vicinity of the center of the Brillouin zone is necessary for understanding and design of optoelectronic devices. Unfortunately, only little is known about the rather complex structure of the hole spectra. Additionally, the effect of strain on the electronic structure is appreciable due to the large lattice mismatch in nitride heterostructures.

In this paper, we provide quantitative *ab-initio* predictions of the electronic structure of group-III nitrides as a function of biaxial strain. The band structure near the band gap and the major optical transitions across the energy gap are calculated. We have determined the electron and hole effective mass tensors and the deformation potentials from our ab-initio calculations. These results permit us to determine the band parameters in terms of a 6x6 k·p Hamiltonian for the valence bands in the wurtzite structure.

The band structures of bulk AlN, GaN, and InN in the wurtzite and zincblende phase have been studied before extensively [2]. However, few reports have dealt with the details of the valence band edge and strain effects. First-principles calculations of the effect of biaxial strain on the electronic structure of the valence band edge in wurtzite nitrides have been recently presented [3, 4, 5]. Non-relativistic *ab-initio* calculations of the effective masses in AlN and GaN have been also performed [4, 6].

In the wurtzite structure, the top of the valence band at the Γ point consists of a doubly degenerate Γ_9 and two doubly degenerate Γ_7 states. In GaN, the separations between these states are of the order of 10 meV reflecting the comparable strength of the crystal field and spin-orbit interactions. The exciton energies corresponding to these split valence edge states have been recently measured in strained GaN films [7].

Employing the relativistic local density functional pseudopotential method and elasticity theory, we have analyzed the electronic band energies of three different wurtzite structure

Mat. Res. Soc. Symp. Proc. Vol. 449 © 1997 Materials Research Society

nitrides as a function of biaxial and hydrostatic strain. It turned out to be crucial to fully optimize internal structural parameters that are not determined by the symmetry, in order to ensure reliable and truly quantitative results.

THEORY

First-principles calculations

Our calculations are based on the first-principles total-energy pseudopotential method within the local-density-functional formalism [8]. We have used norm-conserving separable pseudopotentials [9, 10] and a preconditioned conjugate gradient algorithm [11] for minimizing the total crystal energy with respect to the electronic as well as the ionic degrees of freedom. These pseudopotentials are highly transferable, yet sufficiently soft so that a kinetic energy cutoff of 62 Ry suffices to yield converged total energies. We have used 14 special points [12] for the k-space integrations. The semicore Ga 3d-electrons are treated as part of the frozen core, but their considerable overlap with the valence electrons is accounted for by including the nonlinear core exchange-correlation correction [13]. This procedure yields lattice constants, atomic positions, and bulk moduli in very good agreement with experiment [5]. In order to realistically account for the interplay between strain and the spin-orbit interaction, we have taken into account relativistic effects nonperturbatively by using relativistic pseudopotentials. This method has been shown to predict spin-orbit splittings very reliably in other III-V compounds [14].

We have calculated the unstrained electronic band structures with the experimental lattice constants a_0 and c_0. This is known to yield valence band splittings in GaN that are in better agreement with experiment than those calculated at the self-consistently determined theoretical lattice constants [4]. The changes in lattice constants with strain are determined according to elasticity theory with experimental elastic constants [15, 16]. For a given in-plane lattice constant a, corresponding to the in-plane strain $\epsilon_{xx} = \epsilon_{yy} = a/a_0 - 1$, the lattice constant along the hexagonal axis is given by $c = c_0(1 + \alpha\epsilon_{xx})$, where $\alpha = -2c_{13}/c_{33}$ for biaxial strain, and $\alpha = (c_{11} + c_{12} - 2c_{13})/(c_{33} - c_{13})$ for hydrostatic pressure. The ϵ_{zz} strain component equals $\alpha\epsilon_{xx}$. Once the lattice constants are given, we have performed a full optimization of the atomic positions within the unit cell by calculating the Hellmann-Feynman forces. The latter determine the distance uc between cation and anion along the c-axis. This optimization of u has a significant influence on the valence band splittings [4, 5, 6].

Model 6×6 k $\cdot$ p Hamiltonian

The valence-band spectrum in the neigborhood of the band edge of wurtzite type materials can be reproduced by a 6×6 matrix $k \cdot p$ Hamiltonian that includes one Γ_9 and two Γ_7 band states [17, 18, 19]. This matrix also includes terms that are linear in the strain tensor. For biaxial or hydrostatic strain, these strain dependent matrix elements are proportional to four deformation potentials D_1, D_2, D_3, D_4. Three parameters Δ_1, Δ_2, and Δ_3 describe the splittings of the valence band maximum ($\mathbf{k} = 0$) and seven parameters $A_1 \ldots A_7$ determine the dispersions along different directions in the Brillouin Zone. In the present paper we use the definition of these valence band parameters and deformation potentials given in Ref. [19].

In the absence of the spin-orbit coupling, the Hamiltonian can be diagonalized analytically. For zero strain, the doubly degenerate energies of the valence band maximum are conventionally written as

$$E(\Gamma_9) = \frac{1}{3}\Delta_1 + \Delta_2 , \tag{1}$$

$$E(\Gamma_{7+}) = -\frac{1}{2}\left(\frac{\Delta_1}{3} + \Delta_2\right) + \sqrt{\left(\frac{\Delta_1 - \Delta_2}{2}\right)^2 + 2\Delta_3^2} , \tag{2}$$

$$E(\Gamma_{7-}) = -\frac{1}{2}\left(\frac{\Delta_1}{3} + \Delta_2\right) - \sqrt{\left(\frac{\Delta_1 - \Delta_2}{2}\right)^2 + 2\Delta_3^2} . \tag{3}$$

The constants Δ_2 and Δ_3 are spin-orbit Hamiltonian matrix elements, whereas Δ_1 is conventionally termed crystal-field splitting. When $\Delta_1 > 0$, as is the case in GaN and InN [5, 18], the ordering of these band states is $\Gamma_{7-} < \Gamma_{7+} < \Gamma_9$. For $\Delta_1 < 0$, which is the case for AlN [5, 18], the states form the sequence $\Gamma_{7-} < \Gamma_9 < \Gamma_{7+}$. The Γ_9 state is always termed heavy-hole (hh) state and the energy below forms the light-hole state (lh). This is Γ_{7+} for GaN, and InN and Γ_{7-} for AlN. The third eigenvalue defines the crystal field split-off state (cs). These definitions are kept for $\mathbf{k} \neq 0$.

The shift of the conduction band edge at Γ-point with strain and the band dispersion in the vicinity of this point is governed by two deformation potentials D_{1c} , and D_{2c} and two effective masses ($m_e^{\parallel}$ and $m_e^{\perp}$), respectively,

$$E_c(\mathbf{k}) = E_c(0) + \frac{\hbar^2 k_z^2}{2m_e^{\parallel}} + \frac{\hbar^2(k_x^2 + k_y^2)}{2m_e^{\perp}} + D_{1c}\epsilon_{zz} + D_{2c}(\epsilon_{xx} + \epsilon_{xx}) . \tag{4}$$

We measure this strain dependence relative to the center of gravity of the valence band.

RESULTS

Effective masses and valence band splittings

For unstrained wurtzite GaN, the present ab-initio calculations predict $E_{hh}(\Gamma_9)$ - $E_{lh}(\Gamma_{7+})$ = 6.8 meV (6 meV [20], [21], [22]) and $E_{hh}(\Gamma_9)$ - $E_{cs}(\Gamma_{7-})$ = 33.7 meV (28meV [21], 25meV [22], 23meV [20]) in fair agreement with experimental data (given in parenthesis) of bulk GaN and epitaxial GaN films that are believed to be strain free. In InN, the $hh - lh$ and $hh - cs$ splittings are predicted to be 5 meV and 34 meV, respectively. In AlN, $E_{cs}(\Gamma_{7+})$ forms the top of the valence band, and the calculations yield the $cs - hh$ and $hh - lh$ splitting to be equal to 212 meV and 14 meV, respectively.

The calculated components of the effective mass tensor at Γ for electrons and holes in AlN and GaN are given in Table I.

Table I. Electron and hole effective masses (in units of m_0)

	$m_e^{\parallel}$	$m_e^{\perp}$	$m_{hh}^{\parallel}$	$m_{hh}^{\perp}$	$m_{lh}^{\parallel}$	$m_{lh}^{\perp}$	$m_{cs}^{\parallel}$	$m_{cs}^{\perp}$
AlN	0.32	0.33	3.52	0.73	3.44	0.73	0.26	4.49
GaN	0.18	0.20	2.09	0.37	0.74	0.39	0.18	0.94

The masses were calculated by numerical differentiation of the dispersion relations obtained from the present relativistic *ab-initio* calculations. The calculated conduction band masses for GaN are in good agreement with recent optically detected cyclotron resonance data [23] that givea polaron mass of $0.22m_0$. The corresponding bare mass is estimated to be $0.20m_0$ [23]. We are not aware of experimental data on the valence band effective masses in nitrides. The predicted hole masses show a strong anisotropy. We have excluded the calculated masses of InN from Table I since the LDA calculations grossly underestimate the energy gap of wurtzite InN (it is only 0.1 eV) and are therefore unlikely to yield reliable masses.

<u>Biaxial strain effects</u>

Very recently, free exciton lines at Γ have been measured in epitaxial GaN films. In the same samples, the amount of strain could be directly measured by X-rays [7]. These data allow us to directly compare theory with experiment, as depicted in Fig. 1. We have increased all valence to conduction band transition energies by a rigid, strain-independent self energy of 1.209 eV. In this way, the calculated hh to conduction-band transition energy for zero pressure coincides with the observed exciton transition energy of 3.4745 eV measured for bulk GaN [21]. Fig. 1 shows excellent agreement between theory and experiment for all other transition energies. This may help to determine the actual strain in epitaxial films from the measured optical excitonic spectra. We notice that for a tensile strain of about 0.2% the light-hole band becomes the top of the valence band.

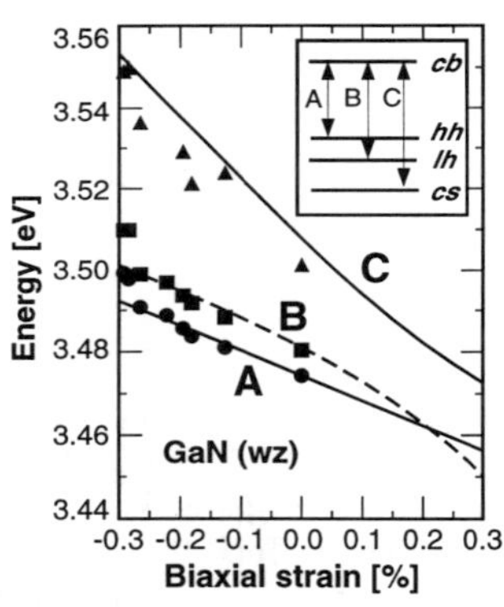

Figure 1: Comparison of experimental free exciton energies of strained GaN from Ref. 7, and unstrained GaN [21] (full dots, squares, and triangles denote energies of A, B, and C exciton lines, respectively) with present theoretical calculations (lines). The theoretically calculated energies were rigidly shifted by 1.209 eV. This energy shift includes the self-energy correction to the LDA gaps and the exciton binding energy.

Valence band parameters and deformation potentials

From the calculated band structures, we have determined all parameters in the 6×6 $k \cdot p$ Hamiltonian, following the notation of Ref.[19]. In order to determine Δ_1 independently from Δ_2 and Δ_3, we have also performed nonrelativistic calculations. From the relativistic bands, we can deduce Δ_2 and Δ_3. The parameters $A_1 \ldots A_5$ can easily be determined from the effective masses parallel and transverse to the hexagonal axis. Since the masses are independent of A_6, it is difficult to determine this parameter from the eigenvalues at small $\mathbf{k}$. Therefore, we have used the quasicubic model that gives $A_6 = (A_3 + 4A_5)/\sqrt{2}$. Finally, A_7 can be determined from the spin splittings of the bands along the direction orthogonal to the hexagonal axis. In order to determine the valence (D_3 and D_4) and conduction deformation potentials (D_{1c} and D_{2c}), we have computed the derivative $(dE_i/d\epsilon_{xx})|_{\epsilon=0}$, $i =$ valence and conduction band, for two different strain tensors.

The complete sets of $k \cdot p$ parameters for AlN and GaN are summarized in Tables II. The conduction band deformation potentials D_{1c} and D_{2c} are equal to -10.23 eV and -9.65 eV for AlN, and -9.47 eV, -7.17 eV for GaN, respectively.

Table II. Valence band parameters of the 6×6 k $\cdot$ p model for AlN and GaN. Δ_1, Δ_2, and Δ_3 are given in meV, all other parameters in eV.

	Δ_1	Δ_2	Δ_3	A_1	A_2	A_3	A_4	A_5	A_6	A_7	D_3	D_4
AlN	-219	6.6	6.7	-3.82	-0.22	3.54	-1.16	-1.33	-1.25	0.00	9.02	-3.99
GaN	24	5.4	6.8	-6.40	-0.80	5.93	-1.96	-2.32	-3.02	-0.026	6.26	-3.29

In Fig. 2, we depict the calculated valence band edge structure of wurtzite GaN perpendicular to the hexagonal axis ($\Gamma \to M$) for three different biaxial strains. The discrepancies between the full diagonalization of the Hamiltonian and the 6-band $k \cdot p$ model are negligible up to wave vectors of the order of 0.15 Å^{-1}.

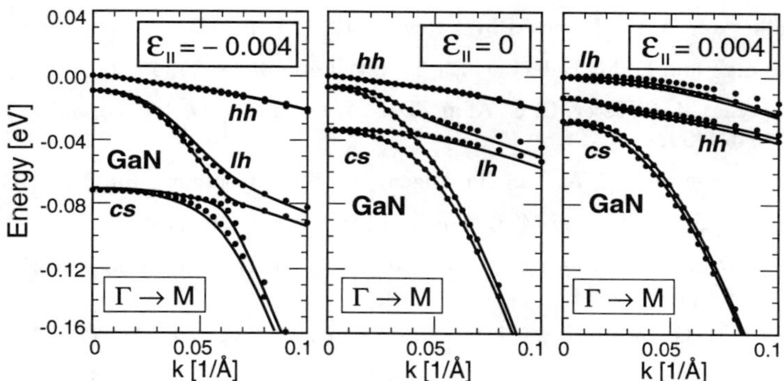

Figure 2: Valence-band dispersions along the transverse axes for biaxially strained and unstrained wurtzite GaN. The full dots denote the present *ab-initio* results, whereas the lines represent the solutions of the 6×6 $k \cdot p$ model Hamiltonian with the parameters from Table II.

CONCLUSIONS

In summary, we have predicted the hole effective mass tensor for AlN and GaN. We have determined the parameters of 6-band model Hamiltonian that very accurately reproduces the hole spectra near the valence band edge in biaxially strained AlN and GaN.

ACKNOWLEDGMENTS

This work was supported by the Bayerische Forschungsverbund FOROPTO and the Deutsche Forschungsgemeinschaft, project SFB 348.

REFERENCES

1. S. Strite and H. Morkoç, J. Vac. Sci. Technol. B **10**, 1237 (1992).

2. For a review see, J. H. Edgar editor, *Properties of Group III Nitrides* (Electronic Materials Information Service (EMIS), London, 1994).

3. K. Kim, W. R. L. Lambrecht, and B. Segall, Phys. Rev. B **53**, 16310 (1996).

4. W. R. L. Lambrecht, K. Kim, S. N. Rashkeev, and B. Segall, Mater. Res. Soc. Symp. Proc. **395**, 455 (1996).

5. J. A. Majewski, M. Städele, and P. Vogl, MRS Internet J. Nitride Semicond. Res. **1**, 30 (1996).

6. M. Suzuki, T. Uenoyama, and A. Yanase, Phys. Rev. B **52**, 8132 (1995).

7. S. Chichibu, A. Shikanai, T. Azuhata, T. Sota, A. Kuramata, K. Horino, and S. Nakamura, Appl. Phys. Lett. **68**, 3766 (1996).

8. W. E. Pickett, Comp. Phys. Rep. **9** 115 (1989).

9. N. Troullier and J. L. Martins, Phys. Rev. B **43**, 1993 (1991).

10. L. Kleinman and D. M. Bylander, Phys. Rev. Lett. **48**, 1425 (1982).

11. M. C. Payne, M. P. Teter, D. C. Allan, T. A. Arias, and J. D. Joannopoulos, Rev. Mod. Phys. **64**, 1045 (1992).

12. P. J. H. Denteneer and W. Van Haeringen, Sol. State Commun. **59**, 829 (1986).

13. S. G. Louie, S. Froyen, and M. L. Cohen, Phys. Rev. B **26**, 1738 (1982).

14. J. A. Majewski, in *The Physics of Semiconductors*, Ed. D. J. Lockwood (World Scientific, Singapore, 1995) pp. 711-714.

15. A. Polian, M. Grimsditch, and I. Grzegory, J. Appl. Phys. **79**, 3343 (1996).

16. L. E. Neil, M. Grimsditch, and R. H. French, J. Am. Ceram. Soc. **76**, 1132 (1993).

17. G. L. Bir and G. E. Pikus, *Symmetry and strain-induced effects in Semiconductors* (John Wiley & Sons, New York, 1974), p. 1111.

18. Y. M. Sirenko, J.-B. Jeon, K. W. Kim, M. A. Littlejohn, and M. A. Stroscio, Phys. Rev. B **53**, 1997 (1996).

19. S. L. Chuang and C. S. Chang, Phys. Rev. B **54**, 2491 (1996).

20. D. Volm, K. Oettinger, T. Streibl, D. Kovalev, M. Ben-Chorin, J. Diener, B. K. Meyer, J. Majewski, L. Eckey, A. Hoffmann, H. Amano, K. Hiramatsu, and D. Detchprohm, Phys. Rev. B **53**, 16543 (1996).

21. R. Dingle, D. D. Sell, S. E. Stokowski, and M. Ilegems, Phys. Rev. B **4**, 1211 (1971).

22. B. Monemar, Phys. Rev. B **10**, 676 (1974).

23. M. Drechsler, B. K. Meyer, D. M. Hoffmann, D. Detchprohm, H. Amano, and I. Akasaki, Jpn. J. Appl. Phys. **34**, L1178 (1995).

THEORY OF INTERFACES IN WIDE-GAP NITRIDES

M. BUOGIORNO NARDELLI, K. RAPCEWICZ, E.L. BRIGGS, C. BUNGARO,
J. BERNHOLC

Department of Physics, North Carolina State University
Raleigh, NC 27695, nardelli@quinn.physics.ncsu.edu

ABSTRACT

The results of theoretical studies of the bulk and interface properties of nitrides are presented. As a test the bulk properties, including phonons of GaN at the Γ-point, are calculated and found to be in excellent agreement with the experimental data. At interfaces, the strain effects on the band offsets range from 20% to 40%, depending on the substrate. The AlN/GaN/InN interfaces are all of type I, while the $Al_{0.5}Ga_{0.5}N$ on AlN zinc-blende (001) interface is of type II. Further, an interface similar to those used in the recent blue laser diodes is of type I and does not have any electronically active interface states. The valence band-offset in the (0001) GaN on AlN interface is -0.57 eV and the conduction band-offset is 1.87 eV.

INTRODUCTION

Interest in wide band-gap nitrides is stems from possible applications in blue/UV light-emitting diodes and lasers, and in high-temperature electronics [1, 2]. The recent demonstration of stimulated emission in the blue region has served to highlight the potential of nitride-based devices.[3]

In this paper we describe several of our recent results [4] concerning the bulk properties of the III-V nitrides and their interfaces.

Using novel pseudopotentials that permit the treatment of the Ga 3-d and In 4-d electrons as core electrons, we have investigated the bulk properties of the nitrides and also phonons in GaN. With these pseudopotentials, we have also studied interfaces of AlN/GaN/InN [4]. The calculated values of the valence-band offset for AlN on GaN is -0.44 eV, while for GaN on AlN it is -0.73 eV, indicating that the effects of strain on the valence-band offset are significant. The band offsets between AlN, GaN and InN were computed using the AlN in-plane lattice constant and including strain effects. They are all of type I and the transitivity rule is satisfied. We have also studied the zinc-blende (001) $Al_{0.5}Ga_{0.5}N$ on AlN and $Al_{0.2}Ga_{0.8}N$ on $In_{0.1}Ga_{0.9}N$ interfaces in the virtual crystal approximation. The latter is based upon the laser diode of Nakumura et al. [3].

The standard *ab initio* plane-wave pseudopotential method was employed in the calculations. For the phonons, density functional linear response theory [5] was used. Further details can be found in Ref. [4]. An efficient multigrid-based method that uses a real-space grid as a basis was also used in connection with a large cell in order to simulate an interface based on the recently demonstrated blue laser diode [6].

BULK PROPERTIES

The calculated bulk properties are presented in Table 1. In general, agreement with experiment is excellent [4].

Mat. Res. Soc. Symp. Proc. Vol. 449 © 1997 Materials Research Society

	AlN	GaN	InN
zinc-blende			
a_0 (Å)	4.37 (4.38)	4.51 (4.50)	5.01 (4.98)
B_0 (MBar)	2.02 (2.02)	1.92 (1.90)	1.58 (1.37)
E_Γ (eV)	4.09	2.15 (3.45)	0.16
ΔE_{vbw} (eV)	14.86	15.73	14.01
wurtzite			
a (Å)	3.09 (3.11)	3.20 (3.19)	3.55 (3.54)
c/a	1.62 (1.60)	1.63 (1.63)	1.63 (1.61)
u (units of c)	0.378 (0.382)	0.376 (0.377)	0.375
B_0 (MBar)	1.99 (2.02)	1.91 (1.95, 2.37)	1.62 (1.26, 1.39)
E_Γ (eV)	4.44 (6.28)	2.29 (3.50)	0.16 (1.89)
ΔE_{vbw} (eV)	14.89	15.60	14.00

Table 1: Calculated bulk properties of zinc-blende and wurtzite nitride semiconductors. The values of the gap at the Γ-pt (E_Γ) and of the valence-band width (ΔE_{vbw}) are the LDA results. Note that the LDA indirect gap in zinc-blende AlN is 3.2 eV. Experimental values are in brackets and follow Ref. [7].

For comparison, the bulk properties of GaN were calculated with the $3d$-electrons of gallium as valence electrons after Ref. [7]. With a cut-off of 240 Ry, we find $a_0 = 4.46$ Å and $B_0 = 2.14$ Mbar. The quality of agreement of this result with experiment is of the same order as our calculations employing pseudopotentials with the d-electrons in core.

	c_{11} (10^{11} dyne/cm^2)	c_{12} (10^{11} dyne/cm^2)
GaN	26.62 (26.4)	15.49 (15.3)
AlN	24.85	13.37
InN	20.24	12.96

Table 2: Elastic constants calculated for zinc-blende AlN, GaN and InN (in units of MBar). The experimental values, in brackets, are from Ref. [8].

The calculated elastic constants are shown in Table 2. These elastic constants were used to determine the strain present in interfaces. In addition, we have calculated the band-gap deformation potential for zinc-blende GaN and find it to be -8.9 eV. Experimental measurements are available only for wurtzite and they range from -7.8 to -9.8 eV. Since in a previous theoretical calculation of this quantity, the values for wurtzite and zinc-blende were found to be very close, our results are in good agreement with the existing experimental data and we expect that our treatment of strain effects on the band structure is accurate. The frequencies of the phonon modes at the Γ-point are shown in Table 3. The agreement with experiment is satisfactory.

	Present Theory	Experiments
E_2^1	138	144
B_1^1	334	–
A_1	556	533
E_1	568	560
E_2^2	574	568
B_1^2	694	–

Table 3: Phonon dispersion at Γ in wurtzite GaN. Experimental data are from Ref. [8].

INTERFACES OF WIDE-GAP NITRIDES

Superlattices based upon the nitrides will be strained as a result of the lattice-mismatch between AlN and GaN (2.5%) and between InN and AlN (12.1%). The interface energy, defined as the excess or deficit energy due to the presence of the interface, is extremely small, namely ~ 1 meV/atom, which is of the order of the precision of the calculations. The interfaces therefore show similar bonding characteristics.

The band offsets of the strained heterojunctions have also been studied. As the elastic energy of GaN on AlN is lower than the elastic energy of AlN on GaN, AlN is taken to be the substrate in all of the cases presented below. The valence-band offsets shown in Table 4 indicate that the effect of strain on the value of the valence-band offset in GaN on AlN is significant. In each case, $\mathrm{VBO_{ave}}$ was computed from the average of the split valence-band manifold.

	ΔV	VBO	$\mathrm{VBO_{ave}}$
AlN lattice constant	1.05	-0.73	-0.55
Average lattice constant	0.33	-0.58	-0.53
GaN lattice constant	0.74	-0.44	-0.53

Table 4: Electrostatic potential line-up ΔV and valence-band offsets, VBO and $\mathrm{VBO_{ave}}$. All quantities are in eV and are quoted with respect to AlN.

We also investigated the band offsets of the (001) GaN on AlN, InN on AlN and InN on GaN strained heterojunctions. The band offsets are shown in Figure 1. In determining the conduction-band offsets, the conduction band minima were shifted to their experimental values using the so-called scissors operator. In each case, the interface is of type I and, overall, the transitivity rule is satisfied. Further, there are no interfaces states in the gap of the superlattice and the states at the top of the valence band are largely confined to the epilayer with the smaller gap.

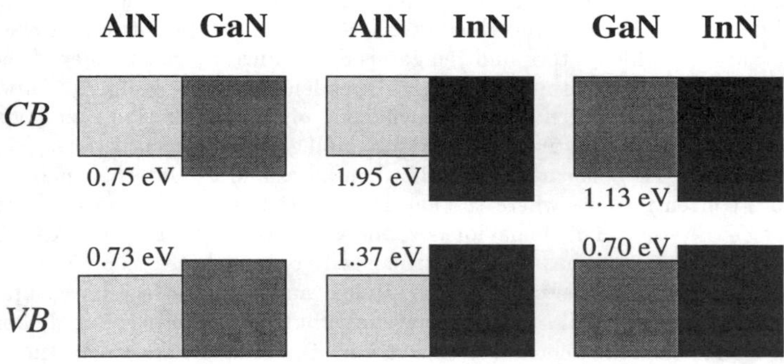

Figure 1: Calculated band offsets for the three interfaces described in the text (an AlN substrate is assumed).

The case of the wurtzite (0001) GaN on AlN interface was examined. As for the (001) interfaces, strain effects were included using macroscopic elasticity theory [10] and a total energy optimization. The calculated valence-band offset is estimated to be -0.57 eV. This value is smaller than the result for the strained non-polar (001) GaN on AlN interface.

The ratio of the conduction-band to valence-band offset is 65:20. These values agree well with the experimental measurements of the (0001) wurtzite interface [11].

The low symmetry of the wurtzite system means that pyroelectric and piezoelectric behavior may be present[12]. In accordance with the observations of Satta and coworkers [15], we find that there is a substantial electric field in (0001) strained GaN on AlN. We have calculated the spontaneous bulk polarization of unstrained AlN and the strain induced polarization for the GaN epilayer [16] in order to distinguish the bulk pyroelectric and piezoelectric contributions to this field from that induced by the interface. The spontaneous polarization (P_3) of AlN and GaN in equilibrium are -1.227 $\mu C/cm^2$ and -0.448 $\mu C/cm^2$ respectively; the polarization of the strained GaN is -0.454 $\mu C/cm^2$. These values are comparable to the computed bulk polarization in BeO [16]. The estimated contribution of the interface dipole, which includes the response of one epilayer to the field of the other, is only 0.057 $\mu C/cm^2$. It is an order of magnitude smaller and has the opposite sign to the bulk polarizations. The computed value of the polarization in the multi-quantum-well agrees with that estimated from experiment by Martin *et al.* [11].

The band offsets can be tuned both through changes to the strain in the system and changes to the electronic structure accomplished by alloying. The virtual crystal approximation, in which the pseudopotentials of the constituent species are averaged according to their fractional composition [17], is the theoretical tool by which this was achieved. Strain effects were explicitly included using macroscopic elasticity theory (using the theoretically determined elastic constants) and a total energy minimization. The values for scissors corrections of the band gap of the alloys were estimated using a Vegard-type rule, namely the gap of the alloy was taken to be the average of the band gap of the pure phase weighted by its fractional composition.

We first analyzed the behavior of the zinc-blende (001) $Al_{0.5}Ga_{0.5}N$ on AlN interface. As opposed to the situation in the pure nitride interfaces, this particular alloy interface is a staggered type II interface and a change in the character of the interface is to be expected upon variation in the Ga concentration. In pure zinc-blende AlN, the gap is indirect ($\Gamma_1^v \rightarrow X_1^c$), while the gap in pure zinc-blende GaN is direct. In superlattices grown along the (001) direction X is folded into Γ and the gap is determined by the smaller of the two. In the alloy, the gap at Γ is much more strongly dependent upon the gallium composition than is the gap at X. Consequently, the dependence of the conduction band offset on gallium concentration has two distinct regimes: for small gallium concentrations, ($\leq 50\%$), where the gap is at X, ΔE_{gap} is small, the VBO dominates and the interface is of type II; for large gallium concentrations, where the gap is at Γ, the larger difference in the gaps make the interface of type I. This behavior is common to other systems, like $Al_{1-x}Ga_xAs$ on AlAs, where the same effect has been experimentally observed [18].

In the same theoretical framework, we also studied an alloy interface based upon the nitride-based multi-quantum-well structure that Nakumura et al. [3] used to demonstate stimulated emission in the blue region of the spectrum. The band offsets are for the valence and conduction band -0.26 eV and 0.70 eV respectively. The interface is of type I and there are no interface states in the gap. To verify this latter conclusion, we have simulated the multi-quantum-well using a 192 atom cell in which we have not made the virtual crystal approximation but, rather, have used a number of atoms exactly corresponding to the concentration of each epilayer, namely $Al_{0.2}Ga_{0.8}N$ and $In_{0.1}Ga_{0.9}N$. This calculation confirmed the absence of electrically active interface states.

The omission of the anion p- and cation d-state repulsion [9], spin-orbit effects and the well-known neglect of many-body effects in the LDA constitute the major sources of systematic error. Spin-orbit effects in both AlN and GaN have been shown to be of the order of 20 meV, so that their difference is indeed negligible [19]. In GaN on AlN interfaces, the inclusion of the $3d$-electrons as valence electrons results in a constant shift of 0.2 eV, which is less than the experimental error [11], and does not change the character of the interface. Incorporating this shift gives results in agreement with previous estimates using a d-valence pseudopotential [20] and an all-electron calculation [21]. The importance of many-body effects on the band offsets is not known.

REFERENCES

[1] R. F. Davis, Physica B **185**, 1 (1993).

[2] H. Morkoc, S. Strite, G. B. Gao, M. E. Lin, B. Sverdlov, and M. Burns, J. Appl. Phys. **76**, 1363 (1994).

[3] S. Nakumura et al., Jpn. J. Appl. Phys. **35**, L74 (1996).

[4] M. Buongiorno Nardelli, K. Rapcewicz and J. Bernholc, submitted to Phys. Rev. B (1996).

[5] S. Baroni, P. Giannozzi, and A. Testa, Phys. Rev. Lett. **58**, 1861 (1987); P. Giannozzi, S. de Gironcoli, P. Pavone, and S. Baroni, Phys. Rev. B, **43**, 7231 (1991).

[6] E. L. Briggs, D. J. Sullivan and J. Bernholc, Phys. Rev. B **52**, R5471 (1995).

[7] A. F. Wright and J. S. Nelson, Phys. Rev. B **50**, 2159 (1994).

[8] K. Kim, W.R.L. Lambrecht and B. Segall, Phys. Rev. B, 53, 16310 (1996).

[9] S. Wei and A. Zunger, Phys. Rev. Lett. **59**, 144 (1987).

[10] The starting point of the total energy minimization was determined using the experimental ratio [22], $c_{13}/c_{33} = 0.59$.

[11] G. Martin et al., Appl. Phys. Lett. **65**, 610 (1994).

[12] N. W. Ashcroft and N. D. Mermin, *Solid State Physics* (Saunder College, Philadelphia 1976). Ch. 27.

[13] D. Smith, Solid State Commun. **57**, 919 (1986).

[14] A. Bykhovski, B. Gelmont and M. Shur, Appl. Phys. Lett. **63**, 2243 (1993); J. Appl. Phys. **74**, 6734 (1993).

[15] A. Satta, V. Fiorentini, A. Bosin, F. Meloni and D. Vanderbilt, MRS Proceedings **395**, 515 (1996).

[16] M. Posternak, A. Baldereschi, A. Catellani and R. Resta, Phys. Rev. Lett. **64**, 1777 (1990).

[17] M. Peressi, S. Baroni, A. Baldereschi and R. Resta, Phys. Rev. B **41**, 12106 (1990).

[18] B.A. Wilson, P. Dawson, C.W. Tu and R.C. Miller, J. Vac. Sci. Technol. B **4**, 1037 (1986).

[19] M. Suzuki, T. Uenoyama and A. Yanase, Phys. Rev. B **52** 8132 (1995).

[20] V. Fiorentini, M. Methfessel and M. Scheffler, Phys. Rev. B **47**, 13353 (1993).

[21] E. Albanesi, W. Lambrecht and B. Segall, J. Vac. Sci. Technol. B **12**, 2470 (1994).

[22] *Landolt-Börnstein: Numerical Data and Functional Relationships in Science and Technology*, ed. O. Madelung (Springer, New York, 1982), vol. 17.

Energetics of AlN epitaxial wetting layers on SiC(0001)

R. DI FELICE*, J. E. NORTHRUP*, and J. NEUGEBAUER**

* Xerox PARC, 3333 Coyote Hill Road, Palo Alto, CA 94304
** Fritz-Haber-Institut, Faradayweg 4-6, D-14195 Berlin, Germany

ABSTRACT

We present a first-principles characterization of the initial stages of formation of AlN films on c-plane SiC substrates. Studying the competition between two-dimensional films and three-dimensional islands as a function of Al and N abundances, we find that a two-dimensional film can wet the surface in N-rich conditions. Ordered layer-by-layer growth can proceed to some extent on this wetting layer, and is improved by the formation of an atomically mixed interface which eliminates interface charge accumulation. Our results indicate that the stable AlN films grow in the (0001) orientation on the Si-terminated SiC(0001) substrate.

INTRODUCTION

The group III-nitride compounds are wide band gap materials having a strong ionic component in the atomic bonds. Because of their potential for applications in optoelectronics and high-temperature electronics, they have recently attracted a great deal of interest. The difficulty of growing high quality epitaxial nitride films on the most commonly available substrate, sapphire, has been a central challenge in the development of nitride-based devices. Silicon carbide is currently studied as an alternative substrate [1-5].

Because of its reasonably close lattice match (~ 1 %) to SiC, AlN has been successfully employed as a buffer layer for growth of nitride films on SiC. It has been demonstrated that AlN undergoes two-dimensional (2D) growth on SiC(0001) up to a thickness of ~ 15 Å [1,2]. An issue which has arisen in connection with the growth of AlN on SiC(0001) is that of the preferred orientation of the epitaxial film. From a simple analysis of Transmission Electron Microscopy (TEM) images, Ponce et al. [3] argued that the polarity of AlN on Si-terminated SiC(0001) would correspond to the Al-terminated surface. This conclusion is opposite to a previous experimental determination of the polarity of GaN on Si-terminated SiC(0001) made by Sasaki and Matsuoka [4] on the basis of X-ray photoemission spectroscopy.

In this paper we present a microscopic characterization of the initial stages of formation of the AlN/SiC(0001) interface based on first principles total energy calculations. We address the question of the polarity of the epitaxial film as well as the energetics of 2D vs. three-dimensional (3D) growth. We argue that the initial 2D growth mode is a consequence of AlN having a lower surface energy than SiC. We have analyzed many possible epitaxial structures having between 1 and 4 monolayers (MLs) of AlN. These films span a wide range of Al and N stoichiometries and exhibit a variety of different surface reconstructions. For all the structures we have studied, we find that a (0001)-oriented film is preferred energetically with respect to its (000$\bar{1}$)-oriented counterpart: the preferred orientation corresponds to the formation of an Al-terminated AlN film on Si-terminated SiC. We have also identified certain 2D films which are stable with respect to 3D cluster formation, suggesting that an initial layer-by-layer growth of AlN on SiC is energetically favorable. This result is

Mat. Res. Soc. Symp. Proc. Vol. 449 © 1997 Materials Research Society

consistent with recent experimental indications [1] of good wetting behavior of AlN on SiC.

METHOD

We have simulated by repeated supercells AlN/SiC structures having 1×1, $\sqrt{3} \times \sqrt{3}$ and 2×2 surface periodicities. For all the structures considered, the equilibrium geometries have been calculated in a dynamical approach according to Car and Parrinello [6-8], in which the electronic structure is updated along with the atomic positions. Details of the technical parameters used in our calculations can be found elsewhere [9].

Since we have studied geometries including different numbers of N and Al atoms, in describing their energetical competition we have taken into account the chemical potentials of Al and N. The relative energy of each phase with respect to a given reference phase is expressed as

$$\Delta E = \Delta E_0 - (\Delta n_{Al} - \Delta n_N)(\mu_{Al} - \mu_{Al(bulk)}).$$

In writing this expression we have assumed that the structures are in equilibrium with large AlN islands and have therefore enforced the constraint $\mu_{Al} + \mu_N = \mu_{AlN(bulk)}$. μ_{Al} and μ_N are the chemical potentials of Al and N respectively, n_{Al} and n_N are the numbers of Al and N atoms in the structures, $\mu_{Al(bulk)}$ is the Al chemical potential in fcc Al, and ΔE_0 is the relative energy in Al-rich conditions. The Al chemical potential varies between a maximum allowed value of $\mu_{Al(bulk)}$ and a minimum allowed value of $\mu_{Al(bulk)} - \Delta H$, where ΔH is the heat of formation of bulk AlN.

RESULTS AND DISCUSSION

If the surface energy of material A is much less than that of material B, then one expects that a 2D film comprised of material A will wet material B. The criterion for the stability of the film is

$$\sigma_{int} + \sigma_{film} \leq \sigma_{sub},$$

where σ_{int} is the interface energy, and σ_{film} and σ_{sub} are the surface energies of the film and of the substrate respectively. σ_{int} is usually small in comparison to the surface energies, hence it is useful to calculate the surface energies of the substrate and of the overlayer in order to predict whether wetting will occur. We have calculated the surface energies for the non-polar surfaces [10,11] of AlN, GaN, and SiC, employing stoichiometric ideal topology structures, and the results are reported in Table 1. It is shown there that the surface energies of the nitrides are lower than those of SiC for both the $(10\bar{1}0)$ and the $(11\bar{2}0)$ orientations. In particular, $\sigma_{AlN(10\bar{1}0)} - \sigma_{SiC(10\bar{1}0)} \simeq -0.27$ eV/(1×1). By calculating the formation energy of the 1×1 structure AlN/SiC$(10\bar{1}0)$, obtained by replacing Si with Al and C with N in the top SiC$(10\bar{1}0)$ layer, we actually verified that the interface energy is very small, ~ 0.05 eV/(1×1), and that the stability of the film is a consequence of the lower surface energy. Hence, on the basis of the values of the surface energies, one can predict that AlN and GaN will wet the non-polar surfaces of SiC.

Turning to the polar (0001) direction, we find that the addition of an Al-terminated 1×1 bi-layer of AlN to the 1×1 ideal surface of SiC lowers the surface energy by 0.33 eV/(1×1). In this case we are faced with the complication that stoichiometric ideal topology structures are not necessarily appropriate for the interface. The 1×1 ideal surface of SiC is not a particularly good model for the SiC surface which, after all, exhibits a complex reconstruction [12]. Hence, from the analysis of only the 1×1 models, it is not possible

Table I: Surface energies of non-polar surfaces of AlN, GaN, and SiC, and formation energies for nitride overlayers on SiC.

	SiC eV/(1 × 1)	AlN eV/(1 × 1)	GaN eV/(1 × 1)	AlN/SiC eV/(1 × 1)
$(10\bar{1}0)$	2.64	2.37	1.95	−0.22
$(11\bar{2}0)$	5.00	4.25	3.50	

to conclude that $\sigma_{SiC} > \sigma_{AlN}$ for the (0001) direction. A more appropriate model for the SiC(0001) surface is the Si-adatom $\sqrt{3} \times \sqrt{3}$ reconstruction [12] (denoted CSi-Si$_{1/3}$). Below we discuss the relative surface energies of this structure and thin films of AlN on SiC(0001).

$\underline{1 \times 1 \text{ and } \sqrt{3} \times \sqrt{3} \text{ structures.}}$ We have considered initially structures with 1×1 and $\sqrt{3} \times \sqrt{3}$ periodicities (see Fig. 1), modeled by $\sqrt{3} \times \sqrt{3}$ 2D supercells. In addition to the structures illustrated in Fig. 1, we have also considered the Al-adatom and the Al-monolayer configurations on the bare SiC(0001) substrate (CSi-Al$_{1/3}$ and CSi-Al respectively); moreover, for each type of surface reconstruction (vacancy, adatom, monolayer) we have considered both polarities of the aluminum nitride film. In the following discussion we denote an Al-terminated bi-layer by NAl and a N-terminated bi-layer by AlN. For example, CSi-NAl is an ideal Al-terminated bi-layer of AlN on Si-terminated SiC (see Fig. 1(d)).

Thick AlN films ($\sim$ 30 nm) grown heteroepitaxially on SiC(0001) have shown a 2×2 reconstruction [13]. Nevertheless, it is not clear whether the 2×2 structure will form in the earliest stages of growth. In fact, for very thin films it is possible to satisfy the electron counting rule [14] with a $\sqrt{3} \times \sqrt{3}$ reconstruction. For thicker films the requirement of local charge neutrality necessitates atomic mixing at the interface, and the smallest cell which can describe the mixed interface is a 2×2 cell. The 2×2 reconstruction will be discussed later on in the paper.

In Fig. 2(a) we have plotted the energies of the 1×1 and $\sqrt{3} \times \sqrt{3}$ structures as a function of the Al chemical potential, under the assumption that the films have equilibrated with large AlN islands. The shaded area represents the energy of the Si-adatom-terminated SiC(0001) surface, which depends on the Si chemical potential. The low energy border of this region corresponds to Si-rich conditions, and the high energy border corresponds to C-rich conditions. The results shown in Fig. 2(a) suggest that under Al-rich conditions the equilibrium structure would be comprised of AlN islands surrounded by an Al-adatom-terminated SiC(0001) surface. Such a morphology is clearly undesirable. One may stabilize a nitrogen-containing film either by growing under more N-rich conditions, in which case the CSi-NAl$_{2/3}$ structure is stable, or by growing with an Al supersaturation, $\mu_{Al} > \mu_{Al(bulk)}$, in which case the Al-adlayer structure CSi-NAl-Al would become stable. The Si-adatom-terminated substrate is stable only in a very small range of the Al chemical potential, and for Si-rich conditions. This suggests that it is more favorable to have Al dangling bonds on the surface, rather than Si dangling bonds. More generally, the surface energy of the nitride is lower than that of SiC, and wetting of the SiC(0001) surface is predicted in a large range of μ_{Al}.

As seen in Fig. 2(a) structures comprised of NAl bi-layers are more stable than the

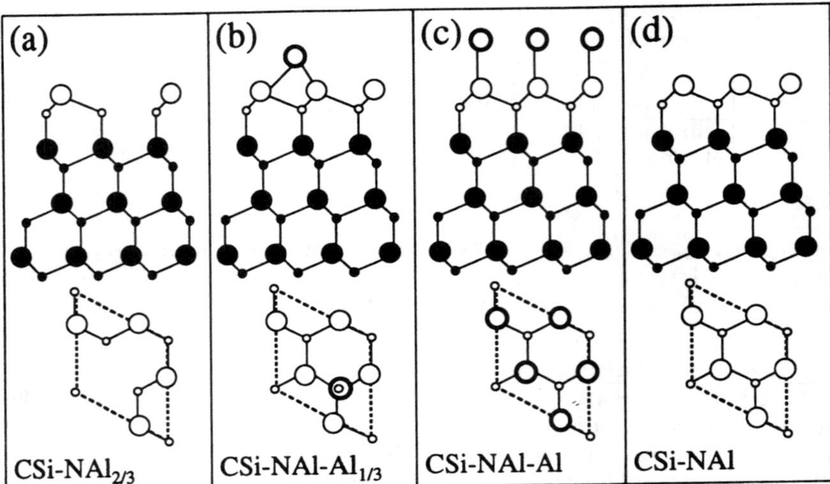

Figure 1: Possible structures which could form in the initial stages of growth of AlN on SiC(0001). (a) $\sqrt{3} \times \sqrt{3}$ Al-vacancy; (b) $\sqrt{3} \times \sqrt{3}$ Al-adatom; (c) 1×1 Al-monolayer; (d) 1×1 ideal. Each panel illustrates both a side view (top) and a top view (bottom) of the elementary cell. Small dots (circles) represent C (N) atoms. Large dots (circles) represent Si (Al) atoms. Large double circles represent Al adatoms. The structures are labelled according to their stoichiometries.

corresponding structures comprised of AlN bi-layers: the CSi-NAl structure is lower in energy than the CSi-AlN structure by 1 eV/(1×1), the CSi-NAl-Al$_{1/3}$ structure is lower in energy than the CSi-AlN-Al$_{1/3}$ structure by 0.7 eV/(1×1), and the CSi-NAl-Al structure is lower in energy than the CSi-AlN-Al structure by 0.3 eV/(1×1). Thus, given that the polarity of the film becomes frozen in during the initial stages of growth, we may conclude that the AlN film would exhibit the (0001) orientation on a (0001)-oriented Si-terminated SiC substrate. We have also verified that our conclusion regarding the polarity of the film does not depend on any assumptions regarding the chemical potentials. It is consistent with the interpretation of the TEM images for the AlN/SiC(0001) interface [15], and with the result of *ab initio* total energy calculations for the 1×1 GaN/SiC(0001) interface [16].

Fig. 2(a) also shows that incorporation of additional AlN onto the Al-vacancy-stabilized wetting layer CSi-NAl$_{2/3}$ is endothermic with respect to incorporation of the material into islands. The energy cost of the reaction, in which CSi-NAl$_{2/3}$ is transformed into CSi-NAl-NAl$_{2/3}$, is ~ 0.21 eV/(1×1). The energy increases because electrons must be transferred from the Si-N bonds at the interface to the N-derived vacancy states at the surface. This charge transfer sets up a dipole in the potential which, along with the total energy, would diverge as the film thickness increased. This increase of energy is a driving force towards atomic mixing at the interface, which leads to local charge neutrality and to the elimination of the interface dipole.

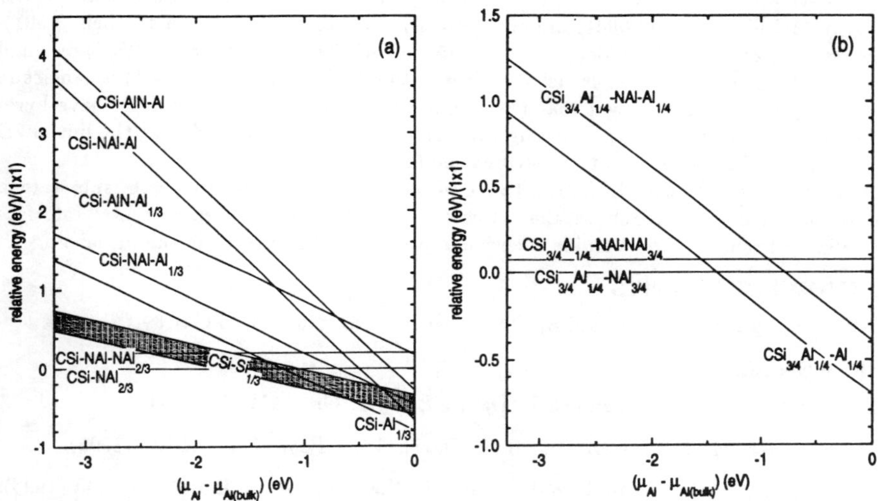

Figure 2: (a) Relative energy of the $\sqrt{3} \times \sqrt{3}$ structures, with respect to the Al-vacancy structure CSi-NAl$_{2/3}$, as a function of the Al chemical potential. (b) Relative energy of the 2×2 structures, with respect to the Al-vacancy structure CSi$_{3/4}$Al$_{1/4}$-NAl$_{3/4}$, as a function of the Al chemical potential. Equilibrium with bulk AlN is assumed in both cases.

2×2 **structures.** 2×2 structures exhibiting a neutral interface have been studied in order to determine the extent to which the elimination of the interface dipole reduces the energy cost of adding bi-layers of AlN to an initial 2D film. The neutral interface that we have considered is obtained from the abrupt interface by substitution of $\frac{1}{4}$ ML of Si atoms by Al atoms in the top substrate layer. We focus here on the low-energy structures obtained in our study of the $\sqrt{3} \times \sqrt{3}$ geometries, i.e., the Al-vacancy and the Al-adatom on the Al-terminated NAl bi-layer, as well as the Al-adatom on the SiC substrate. The energetical comparison of the different configurations that have been studied is illustrated in Fig. 2(b). The qualitative behavior is quite similar to that of the $\sqrt{3} \times \sqrt{3}$ structures: an Al-terminated film with $\frac{1}{4}$ ML of Al vacancies wets the substrate in the initial stages of growth in N-rich conditions, while in Al-rich conditions bulk AlN islands are formed, with the excess Al terminating the SiC surface in the form of adatoms. The most important quantitative difference is that the addition of a second NAl bi-layer to the initial 2D wetting layer costs less energy, only 0.075 eV/(1 $\times$ 1). Although the deposition of additional bi-layers is still endothermic with respect to island formation, only islands larger than ~ 60 Å are stable with respect to the 2×2 2D films containing Al vacancies [9]. Intuitively, one expects that growth will occur most readily in regions where the energy cost of adding AlN layers is least. Thus we think that the initial 2×2 wetting layer should be more conducive to layer-by-layer growth than the $\sqrt{3} \times \sqrt{3}$ wetting layer.

CONCLUSIONS

We have performed first-principles total energy calculations which indicate that (0001)-oriented AlN films on SiC(0001) are energetically favorable with respect to their $(000\bar{1})$-oriented counterparts. Thus, deposition of AlN on SiC(0001) results in an (0001)-oriented AlN film. If growth occurs in N-rich conditions, the AlN surface contains Al vacancies in the top layer. We have shown that the non-polar surfaces of AlN and GaN have lower surface energies than the corresponding SiC surfaces, and have argued that the initial 2D growth observed for AlN on SiC is attributable to this fact.

Both 2×2 and $\sqrt{3}\times\sqrt{3}$ AlN films with Al vacancies on top are able to wet the substrate in N-rich conditions. We have shown that atomic mixing at the (2×2)-reconstructed interface neutralizes the charge dipole and is effective in promoting 2D growth in the initial stages.

ACKNOWLEDGMENTS

This work has been supported by ONR Contract No. N00014-95-C-0169.

REFERENCES

[1] Z. Sitar, L. L. Smith, and R. F. Davis, *J. Cryst. Gr.* **141**, 11 (1994).

[2] S. Tanaka, R. Scott Kern, and R. F. Davis, *Appl. Phys. Lett.* **66**, 37 (1995).

[3] F. A. Ponce, C. G. Van de Walle, and J. E. Northrup, *Phys. Rev. B* **53**, 4743 (1996).

[4] T. Sasaki and T. Matsuoka, *J. Appl. Phys.* **64**, 4531 (1988).

[5] S. Stemmer, P. Pirouz, Y. Ikuhara, and R. F. Davis, *Phys. Rev. Lett.* **77**, 1797 (1996).

[6] R. Car and M. Parrinello, *Phys. Rev. Lett.* **55**, 2471 (1985).

[7] R. Stumpf and M. Scheffler, *Comp. Phys. Commun.* **79**, 447 (1994).

[8] J. Neugebauer and C. Van de Walle, to be published in: *Materials Theory, Simulations, and Parallel Algorithms, MRS Symposia Proceedings* **408**, edited by E. Kaxiras, J. Joannopoulos, P. Vashista, and R. K. Kalia (MRS, Pittsburgh, Pennsylavania, 1995).

[9] R. Di Felice, J. E. Northrup, and J. Neugebauer, *Phys. Rev. B* **54** (1996).

[10] J. E. Northrup and J. Neugebauer, *Phys. Rev. B* **53**, 10477 (1996).

[11] R. Di Felice, J. E. Northrup, and J. Neugebauer, *unpublished.*

[12] J. E. Northrup and J. Neugebauer, *Phys. Rev. B* **52**, 17001 (1995).

[13] M. E. Lin, S. Strite, A. Agarwal, A. Salvador, G. L. Zhou, N. Teraguchi, A. Rockett, and H. Morkoç, *Appl. Phys. Lett.* **62**, 702 (1993).

[14] M. D. Pashley, *Phys. Rev. B* **40**, 10481 (1989).

[15] F. A. Ponce, B. S. Krusor, J. S. Major Jr., W. E. Plano, and D. F. Welch, *Appl. Phys. Lett.* **67**, 410 (1995).

[16] R. B. Capaz, H. Lim, and J. D. Joannopoulos, *Phys. Rev. B* **51**, 17755 (1995).

BAND STRUCTURE AND CATION ORDERING IN LiGaO$_2$

Sukit Limpijumnong, Walter R. L. Lambrecht, Benjamin Segall, and Kwiseon Kim
Department of Physics, Case Western Reserve University, Cleveland, OH 44106-7079

ABSTRACT

Full-potential linear muffin-tin orbital calculations were performed for LiGaO$_2$ in different crystal structures in order to investigate the nature and origin of the cation ordering, structural relaxation and their effects on the band structure. It is found that the most important factor for the bonding is the exclusive occurrence of Li$_2$Ga$_2$ tetrahedra surrounding oxygen. Structures including LiGa$_3$ and Li$_3$Ga tetrahedra have significantly higher total energies and smaller bandgaps. The band-offset between GaN and LiGaO$_2$ is estimated using the dielectric midgap approach.

INTRODUCTION

One of the main difficulties in developing GaN as electronic materials is the lack of a closely lattice-matched substrate. While most efforts have concentrated on sapphire and SiC as substrates, several alternative substrates have been suggested recently and initial growth experiments on them have been carried out [1]. One of these materials is LiGaO$_2$ [2]. This material which has a wurtzite derived structure is closely lattice matched to GaN, having a mismatch of less than 1 % in the basal plane lattice constant a. Furthermore, large LiGaO$_2$ crystals can be grown by the Czochralsky method. Very little is presently known about the electronic properties of this material. This motivated us to perform an investigation of the electronic and related properties of this material. The present work is the first step towards an investigation of LiGaO$_2$/GaN interfaces. Our present results already allow us to make an estimate of the band-offset at this interface using the dielectric midgap approach [3].

One may think of this material as being derived from ZnO by replacing the two Zn (group-II) atoms by a pair of Li (group-I) and Ga (group-III) atoms, thus preserving the average valence of the cations. This kind of chemical substitution is well known in zincblende materials and leads in that case to the chalcopyrite structure. The latter constitutes a particular ordering of cations on the zincblende lattice which is accompanied by relaxation of the structure. The chalcopyrite structure is also of interest for III-V alloys. Among the various possible orderings of a 50 % ABC$_2$ alloy, the chalcopyrite structure appears to be particularly stable for large size mismatch between the constituent A and B atoms, for example in InGaAs$_2$ and Ga$_2$AsSb [4]. The structure is also a good representative of random alloy behavior, being one of the so-called special quasirandom structures (SQS) [5]. This suggests that the ordering of cations in LiGaO$_2$ in the naturally occurring structure may play a similar role in wurtzite alloys as chalcopyrite does for zincblende alloys. Thus, it is important to understand the nature and origin of the cation ordering in natural LiGaO$_2$. This information can subsequently be used to investigate cation ordering in III-N alloys. The latter are the subject of a separate investigation presented elsewhere in these proceedings [6].

In this paper, we present a first-principles electronic structure study of LiGaO$_2$ in various structures and discuss the above issues.

Mat. Res. Soc. Symp. Proc. Vol. 449 © 1997 Materials Research Society

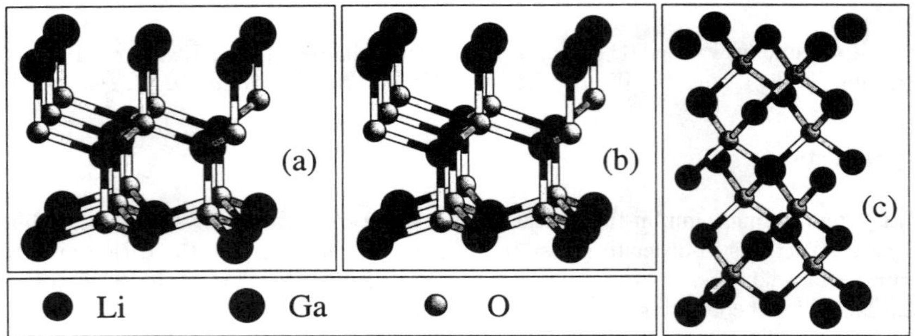

Figure 1: Crystal structures of LiGaO$_2$ investigated in the present work: (a) natural structure, (b) $1+1$ (0001) superlattice, (c) chalcopyrite.

CRYSTAL STRUCTURE

The crystal structure of natural LiGaO$_2$ as determined by Marezio [7] is shown in Fig. 1. The structure is derived from wurtzite with a particular ordering of the Li and Ga cations accompanied by structural relaxation. A full description of the structural parameters is found in [7]. In particular, we note that all oxygens are surrounded by tetrahedra containing two Li and two Ga atoms and that different Li-O and Ga-O bond lengths occur in the structure. Each of the basal planes contains both Ga and Li atoms in a 2×2 supercell of the wurtzite unit cell.

In order to gain insight into the structural relaxation and its effects on electronic structure, we performed calculations for several other crystal structures. The first is an idealized LiGaO$_2$ structure in which bond lengths are all equal but the cation ordering is the same as the natural structure. Secondly, we considered a partially relaxed structure in which only the positions of the oxygens inside their Li$_2$Ga$_2$ tetrahedra are relaxed. This constitutes a nearest neighbor only relaxation model. Third, we considered an alternative ordering of cation in the wurtzite structure, consisting of alternating layers of LiO and GaO along the c-direction($1+1$ (0001) superlattice). In this structure, there are alternatingly LiGa$_3$ and Li$_3$Ga tetrahedra surrounding oxygen. This structure was relaxed by adjusting both the Li-Ga interplanar spacings and the positions of oxygens inside the tetrahedra. We also considered the chalcopyrite structure. While we did not fully relax the latter (c/a of the tetragonal structure was kept ideal) we did relax the internal structural parameter determining the oxygen position inside its Li$_2$Ga$_2$ tetrahedra. As in the natural structure, this structure has only Li$_2$Ga$_2$ tetrahedra but the overall structure is derived from the cubic zincblende lattice instead of the hexagonal wurtzite lattice. Finally, we also performed calculations for the parent compound ZnO in the wurtzite structure. In all cases, the total energy was minimized with respect to volume or in-plane a lattice constant after internal parameter relaxation at each volume.

The above set of structures allows us to investigate the relative importance of local neighborhoods (chemical types of tetrahedra), the structural relaxation and the overall cubic vs. hexagonal stacking of layers.

METHOD OF CALCULATION

We use the full-potential linearized muffin-tin orbital method [8] in conjunction with the local density functional theory [9] with Hedin-Lundqvist parametrization [10]. Triple κ basis sets (*ddp* on Ga, *dpp* on O, *ddp* on Li) were employed. Some initial convergence test calculations used larger basis sets. The tails of the muffin-tin orbitals were expanded in empty sphere as well as atom centered muffin-tin augmentation spheres with an angular momentum cut-off of $l_{max} = 4$. The choice of empty spheres in wurtzite and zincblende derived structures is the same as in previous work on nitrides [11]. Brillouin zone summations were carried out using well-converged symmetry-reduced k-point sets on a regular $(4 \times 4 \times 4)$ mesh in the reciprocal unit cell, displaced to the center of the microzones. Structural relaxations were performed in the manner described above.

RESULTS

Table I gives our structural and binding energy results. Focusing first on the last column which gives the cohesive energy, we notice that the natural structure has a significantly higher binding energy (by 0.72 eV/formula unit) than the $1 + 1$ (0001) superlattice (SL), but is very close in energy to the chalcopyrite structure. In view of the discussion of the structural elements given above these results indicate that the chemical arrangement of the local tetrahedral neighborhood is the dominant factor in the total energy. Both the natural and the chalcopyrite structure have only Li_2Ga_2 tetrahedra, while the SL has a mixture of $LiGa_3$ and Li_3Ga tetrahedra. This preference for a Li_2Ga_2 tetrahedron is, of course, closely related to maintaining local charge neutrality and average valence of the first neighbor coordination shell of each oxygen atom.

Considering the effect of relaxations, we can see in Table I that the relaxation energies are 0.13 eV for the natural structure and 0.29 eV for the (0001) SL, which are at least a factor 2 smaller than the energy difference between the SL and the natural structure itself. The above indicates that the relative stability of the natural structure is not primarily related to a particularly effective structural relaxation but is already present even in the ideal structure. In principle, of course, other cation ordering schemes need to be considered before claiming a rigorous conclusion.

We also note that the nearest neighbor relaxation only model appears to be quite satisfactory. The natural structure with only the oxygen position relaxed within each Li_2Ga_2 tetrahedron differs by only 0.01 eV/formula unit from that of the experimental structure. Furthermore, the Li-O and Ga-O bond lengths obtained from the nearest neighbor only relxation model are close to the average Li-O and Ga-O bond lengths of the experimental structure. The latter has four different Li-O and Ga-O bond lengths indicating that a secondary relaxation of the Li-Ga sublattice is taking place. The bond lengths in the chalcopyrite structure are also close to these averages.

It is notewhorthy that (in agreement with the experimental data on the crystal structure), the Li-O bond lengths are increased with respect to the idealized crystal structure while Ga-O bond lengths are decreased. This indicates a predominance of electrostatic over atomic size effects. In fact, it suggests rather weaker Li-O bonds and stronger Ga-O bonds than Zn-O bonds. The average bond length in ZnO is larger than that in $LiGaO_2$. This interpretation is confirmed by the analysis of the electronic structure given below, which indicates very little Li component in the occupied bands. That means that the LiO bond is very ionic and that Li essentially donates its electrons to oxygen.

Fig. 2 shows the band structure of $LiGaO_2$ in the natural crystal structure (i.e cation ordering) with experimental crystal structure parameters. The character of the bands

Table I: Structure, binding energy and LDA band gap results for $LiGaO_2$ and ZnO.

structure	a (Å)	c/a	V/atom ($Å^3$)	bond lengths (Å) Li-O	Ga-O	E_g (eV)	E_{coh}^a (eV)
			$LiGaO_2$				
natural (expt. param.)	3.170	1.605	10.61	1.991	1.826	3.415	21.61
				1.939	1.851		
				1.987	1.843		
				1.986	1.835		
		average		1.976	1.839		
natural (ideal)	3.056	1.633	10.09	1.872	1.872	3.227	21.47
natural (relaxed)	3.098	1.633	10.51	1.954	1.844	3.399	21.60
chalcopyrite (ideal)	3.047^b	1.633	10.00	1.865	1.865	3.234	21.60
chalcopyrite (relaxed)	3.086	1.633	10.40	1.939	1.844	3.377	21.60
$1+1$ (0001) (ideal)	3.066	1.633	10.19	1.877	1.877	-0.053	20.59
$1+1$ (0001) (relaxed)	3.103	1.559	10.09	1.874	1.754	0.680	20.88
				1.877	1.906		
		average		1.876	1.868		
			ZnO				
				Zn-O			
wurtzite (ideal)	3.187	1.633	11.45	1.952		0.804	17.79
wurtzite (relaxed)	3.206	1.580	11.27	1.943		0.845	17.98
				1.941			
		average		1.943			
experiment [16]	3.250	1.601	11.91	1.992		3.44	
				1.976			
		average		1.988			

^{a} Normalized per formula unit, not including spin-polarization of free atoms or zero-point motion corrections.

^{b} This is $a_{tetragonal}/\sqrt{2}$, so as to be consistent with lattice-parameter in wurtzite basal plane of other cases.

is O2s like for the lowest band centered at -17 eV, Ga3d for the band at -12 eV and primarily O2p like for the bands between -6 eV and the valence band maximum (chosen as energy reference). The conduction band minimum is predominantly of Ga4s character. The bandgap is direct at Γ. The gap obtained from our calculations is 3.415 eV. We expect that the LDA underestimates the band gap and that the correction should be intermediate between those in ZnO (about 2.4 eV) and GaN (about 1 eV) but closer to GaN for reasons explained below. Thus $LiGaO_2$ itself may be a very interesting wide band gap semiconductor with an estimated direct gap of 4.4–5.8 eV. Self-energy corrections are also expected to be important for the Ga3d bands [12]. We estimate that these would shift the Ga3d bands down by about 4 eV which would then make them overlap with the O2s bands.

In comparing the $LiGaO_2$ bands to those for the parent compound ZnO, we note that in ZnO the Zn3d bands are higher up and, in fact, in the LDA overlap and hybridize with the bottom of the O2p like bands. While self-energy effects would again shift these levels down by a few eV they will still remain significantly closer to the valence-band maximum than in $LiGaO_2$. This tends to push the valence band maximum up in ZnO and is partly responsible for the reason why the gap in ZnO is smaller than that in $LiGaO_2$. This effect

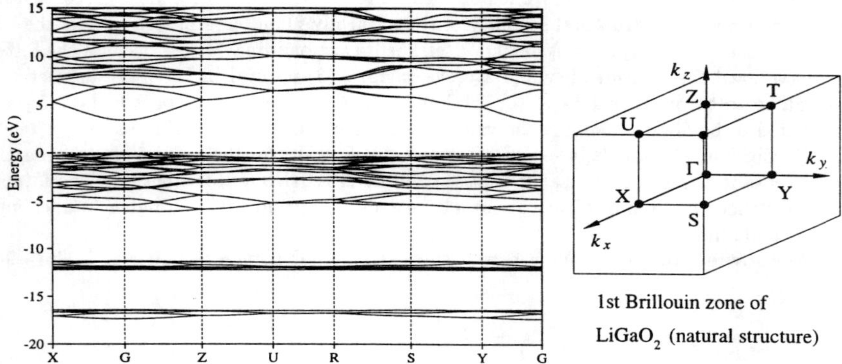

Figure 2: Band structures of LiGaO$_2$ in the experimental crystal structure.

of the Zn3d bands is also responsible for the larger self-energy correction of the gap in ZnO. Hence, in our above estimate of the gap correction of LiGaO$_2$ we expect the self-energy correction to be closer to that of GaN, i.e. the gap is expected to be $\sim$4.4 eV. No experimental value for the gap of LiGaO$_2$ has been published to our knowledge.

The nearest neighbor cation ordering also has a marked influence on the gap. As can be seen in Table I, we find that the $1+1$ (0001) SL has a much smaller gap than the natural structure while the chalcopyrite structure has a gap which is much closer. The structural relaxation effects on the gap are also significant but are smaller than those due to the chemical arrangement.

The band-offset between GaN and LiGaO$_2$ is an important parameter because the primary interest in LiGaO$_2$ is as a substrate for GaN growth. A simple estimate can be made using the dielectric midgap approach [3]. In this method, one assumes that the line-up of bands is in such a way as to preserve local charge neutrality [13]. Secondly, one assumes that there is a charge neutrality point (CNP) intrinsic to each semiconductor. This is a point in the gap such that the states above it must remain empty and below it must be filled when such states arise at the interface. This concept can be justified in various ways [3, 14, 15]. In practice, one identifies the charge neutrality point with a Brillouin zone (BZ) averaged midpoint of the gap. We used here the same averaging scheme as we did for our BZ summations of the charge density in the self-consistent calculations. We obtain the dielectric midgap points measured from the valence band maximum in LiGaO$_2$, GaN and ZnO to be 2.44 eV, 1.81 eV, and 1.99 eV, respectively, leading to valence-band offset estimates of 0.63 eV for GaN/LiGaO$_2$ and 0.18 eV for GaN/ZnO, respectively. For ZnO, this is in good agreement with our previous estimate using zincblende calculations [1].

Since the total gap difference between LiGaO$_2$ and GaN is $4.4 - 3.5 = 0.9$ eV, the offset is predicted to be about 2/3 in the valence bands and 1/3 in the conduction bands. To our knowledge no experimental results on this offset are available.

CONCLUSIONS

The most important conclusions of the present first-principles electronic structure investigation of LiGaO$_2$ are: (1) the bandgap and structural stability in LiGaO$_2$ are primarily affected by the cation ordering while bond length relaxations play a secondary role; (2) the main reason for the structural stability of the naturally found crystal structure is the occurrence of only Li_2Ga_2 nearest neighbor tetrahedra surrounding oxygen because the latter preserve local charge neutrality and average valence; (3) natural $LiGaO_2$ has a direct gap estimated to be of order 4.4 eV; (4) the band-offsets of GaN with respect to $LiGaO_2$ are predicted to be about 0.6 eV in the valence and 0.3 eV in the conduction band.

Conclusion (2) above is based on a comparison of results for the natural cation ordering with those in the $1 + 1$ (0001) superlattice and in the chalcopyrite structure. The former crystal structure is found to have about a 2.5 eV lower gap and a 0.7 eV higher total energy per formula unit.

Acknowledgments: This work was supported by ONR under grant No. N00014-94-1-100.

REFERENCES

1. See e.g. *Gallium Nitride and Related Materials,* edited by F. A. Ponce, R. D. Dupuis, S. Nakamura and J. A. Edmond, (Mater. Res. Soc. Proc. **395,**Pittsburgh, PA, 1996), p. 27, p. 51, p. 55, p. 61, p.67, and p. 535.

2. J. F. H. Nicholls, H. Gallagher, B. Henderson, C. Trager-Cowan, P. G. Middleton, K. P. O'Donnell, T. S. Cheng, C. T. Foxon, and B. H. T. Chai, in Ref. [1], p. 535.

3. M. Cardona and N. E. Christensen, Phys. Rev. B **35**, 6182 (1987).

4. S.-H. Wei, L. G. Ferreira, and A. Zunger, Phys. Rev. B **41**, 8240 (1990).

5. A. Zunger, S.-H. Wei, L. G. Ferreira, and J. E. Bernard, Phys. Rev. Lett. **65**, 353 (1990).

6. K. Kim, S. Limpijumnong, W. R. L. Lambrecht and B. Segall, in this proceeding.

7. M. Marezio, Acta Cryst. **18**, 481-4 (1965).

8. M. Methfessel, Phys. Rev. B **38**, 1537 (1988).

9. P. Hohenberg and W. Kohn, Phys. Rev. **136**, B864 (1964); W. Kohn and L. J. Sham, ibid. **140**, A1133 (1965).

10. L. Hedin and B. I. Lundqvist, J. Phys. C **4**, 2064 (1971).

11. K. Kim, W. R. L. Lambrecht and B. Segall, Phys. Rev. B **53**, 16310 (1996).

12. W. R. L. Lambrecht, B. Segall, S. Strite, G. Martin, A. Agarwal, H. Morkoç, and A. Rockett, Phys. Rev. B **50**, 14155 (1994).

13. F. Flores and C. Tejedor, J. Phys. C **20**, 145 (1987).

14. J. Tersoff, Phys. Rev. Lett. **30**, 3874 (1984); W. A. Harrison and J. Tersoff, J. Vac. Sci. Technol. B **4**, 1068 (1986).

15. W. R. L. Lambrecht and B. Segall, Phys. Rev. Lett. **61**, 1764 (1988); Phys. Rev. B **41**, 2832 (1990).

16. O. Madelung, M. Schulz, H. Weiss, *Numerical Data and Functional Relationships in Science and Technology, Group III Vol 17(b),* (Springer-Verlag, Berlin, 1982), pp. 35-44.

OPTICAL SIGNATURE OF THE GaN (10$\bar{1}$0) SURFACE

C. NOGUEZ*, R. ESQUIVEL-SIRVENT*, D.R. ALFONSO**, S.E. ULLOA***, and D.A. DRABOLD***
*Instituto de Física, UNAM, Apartado Postal 20-364, México D.F. 01000
**International Center for Theoretical Physics, Condensed Matter Section, Trieste, Italy
***Department of Physics and Astronomy, Ohio University, Athens OH 45701

ABSTRACT

We present a theoretical study of the optical properties of the GaN (10$\bar{1}$0) surface. We employed a semi-empirical tight-binding method to calculate the surface electronic structure. The parameters were adjusted to reproduce the correct band structure of the bulk wurtzite GaN. These parameters were interpolated to the surface using Harrison's rule. From the surface electronic structure the surface dielectric response was obtained. The dielectric response is analized in terms of surface-surface, and surface-bulk electronic transitions.

INTRODUCTION

The study of the optical and electronic properties of Gallium Nitride (GaN) has gained importance due to its potential application in near-ultraviolet opto-electronic devices [1,2]. GaN has a direct gap of 3.4 eV and at ambient conditions it crystalizes in the wurtzite structure, or in some cases, like thin films of GaN, the zinc-blende structure is obtained [3]. The bulk properties (electronic and optical) have been known for some time now. Bloom et al. [4] measured the reflectivity and calculated the band structure and reflectivity using an empirical pseudopotential method. It is not until very recently that systematic studies of the surface reconstruction of GaN were done. Jaffe et al. [5] studied the anomalous surface relaxation of GaN (10$\bar{1}$0) and (110), using an *ab initio* Hartree-Fock method. The equilibrium geometries for the principal nonpolar cleavage faces of GaN, are characterized by very small surface bond rotations and large surface and back bond contractions ($\sim$ -7%). Here, we use these results for the atomic structure of the GaN (10$\bar{1}$0) surface to study its optical properties.

MODELS AND METHOD OF CALCULATION

The GaN (10$\bar{1}$0) surface was modeled using a slab of 8 bi-layers, yielding a free reconstructed surface on each face of the slab. The thickness of the slab is large enough to decouple the surface states at the top and bottom surfaces of the slab. Periodic boundary conditions were employed parallel to the surface of the slab to effectively model a two dimensional crystal system. The x direction on the surface plane corresponds to the minor axis of the bulk wurzite unit cell, since the y axis corresponds to the long axis of the bulk unit cell. The calculations are done taking 32 atoms in the slab. The atomic coordinates of the two first layers were obtained from Reference [5], that utilizes an *ab initio* Hartree-Fock approximation, using linear combinations of Gaussian orbitals. The surface reconstruction is then characterized by no surface bond rotation and large surface and back-bond contraction.

To calculate the optical properties of the system, we generate the electronic level structure of the slab using a well known parametrized tight-binding approach. This method employs a sp^3s^* atomic-like basis, that provides a good description of the conduction band of semiconductors. The parameters for the GaN crystal with a wurzite unit cell were fitted to reproduce

Mat. Res. Soc. Symp. Proc. Vol. 449 © 1997 Materials Research Society

Table 1: Tight-binding interaction parameters for GaN

V_{ss}	-1.93		
V_{sp}	2.516	V_{ps}	-3.34
$V_{pp\sigma}$	3.99	$V_{pp\pi}$	-1.0925
V_{s^*p}	1.08	V_{ps^*}	-1.33

the bulk band structure reported on Refs. [4,6,7]. The values of the orbital energies for Ga are $\epsilon_s = -10.7806$eV, $\epsilon_p = 0.897$eV, $\epsilon_{s^*} = 11.5150$eV, and for N are $\epsilon_s = -0.5494$eV, $\epsilon_p = 5.6019$eV, and $\epsilon_{s^*} = 9.1850$eV. The parameters of tight-binding interactions are listed in Table 1. For the surface, we interpolated the parameters using Harrison's rule of $1/d^2$, where d is the bond length of any two first-neighbor atoms.

The optical properties of the surface region are determined by its dielectric function $\epsilon_s(\omega)$. We calculate the imaginary part of the average slab polarizability, which is related to the transition probability between eigenstates induced by an external radiation field. The surface dielectric response was obtained substracting the bulk dielectric response from the slab dielectric function. Here, the thickness of the surface layer of about 2.5 Å, was cosidered. Only two additional parameters to those of the tight-binding Hamiltonian were needed in order to reproduce the bulk dielectric function. These parameters are the so-called intra-atomic sp and s^*p dipoles, with best fitted values of 0.255 Å and 0.878 Å, respectively. The details of the calculation are fully explained in Ref. [8].

RESULTS AND DISCUSSION

Bulk properties

The bulk band electronic structure of GaN is shown in Figure 1. The tight- binding parameters were adjusted to fit the band structure of Refs. [4,6,7]. The direct gap at the Γ point has a value of 3.6 eV, which is similar to the value found theoretically and experimentally [4,6,7]. We observed that the calculated electronic structure is very similar for the valance states to those reported in Refs. [4,6,7], while some descrepancies are found for the conduction states. This is expected since previous theoretical formalisms had not taken into account an extended basis to describe correctly the conduction states. Notice that large cutoff energies are necessary in order to obtain a complete basis set using plane waves.

The anisotropic dielectric function of the bulk GaN was obtained using the calculated electron levels at 5808 points on the irreducible Brillouin zone. Then, the reflectivity transversal and parallel to the long axis of the wurzite unit cell is calculated. We show in figure 2 the reflectivity as a function of the frequency. These calculations are in agreement with those presented by Bloom et al. [4] using an empirical pseudopotential method. Compared to the experimental results [4], the reflectivity spectra of Figure 2 is in agreement in regard to the energies and relative shape of the peaks. The agreement holds up to an energy of ~ 8 eV due to experimental limitations reported by the authors [4]. Furthermore, our method of calculation for the optical properties was tested by comparing the zero frequency dielectric function with experimental values. The theoretical prediction gives a value of $\epsilon(0) = 5.4$ while the experimental values are in the range of 5.2 – 5.8 [4].

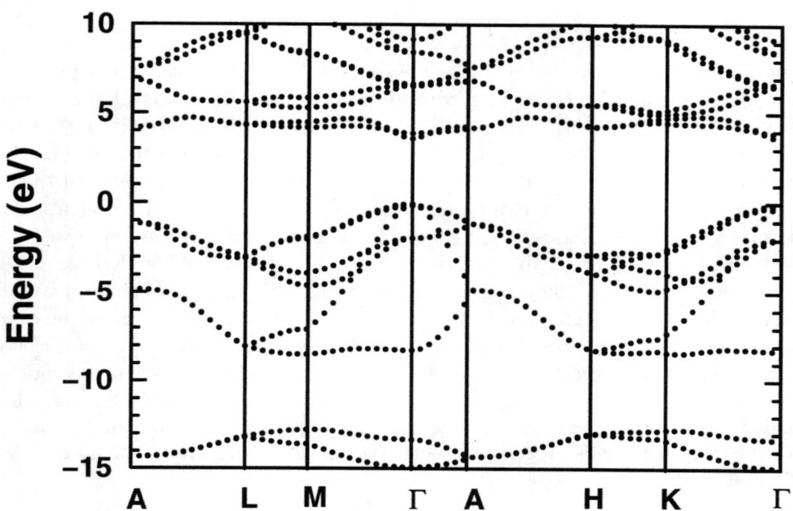

Figure 1: The bulk band electronic structure of wurzite GaN.

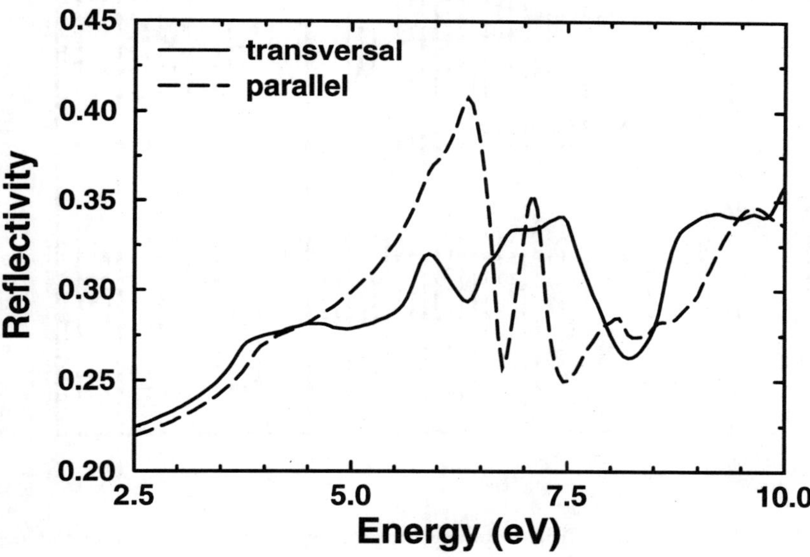

Figure 2: Reflectivity of bulk GaN for an external field transversal to the long axis (solid line) and for an external field parallel to the long axis (dashed line).

The electronic band structure of the reconstructed GaN (10$\bar{1}$0) surface is presented in Figure 3. The electronic structure is shown along the main directions of the irreducible surface Brillouin zone, from Γ to J (x direction), and from J to K (y direction). The surface states are represented by stars, while dots correspond to the projected bulk states. The top of the bulk valence band is set at 0 eV. The calculated Fermi level is in between the top of the valence band (0 eV), and the empty surface states in the gap at 2.7 eV. These surface states in the gap are due to the dangling bonds and backbonds on the first two layers of the reconstructed surface. We can observe that along $J - K$ direction they do not show disperssion, since along $\Gamma - J$ they split in two, showing a disperssion of about 1 eV. There are also some other surface states within the valence band along the $J - K$ direction at about -1.3 eV and -2 eV, also with a backbond character. In Fig. 4, we present the total electronic density of states and its projection on the first, second and third layers. In this figure it is easier to observe that the surface states at 2.7 eV are due to the backbonds between the first and second layers, while the extended states from 2.7 eV to 3.7 eV are on the first layer and are due to the dangling bonds. We can also observe surface states at -1.3 eV and -2 eV that are on the first layer.

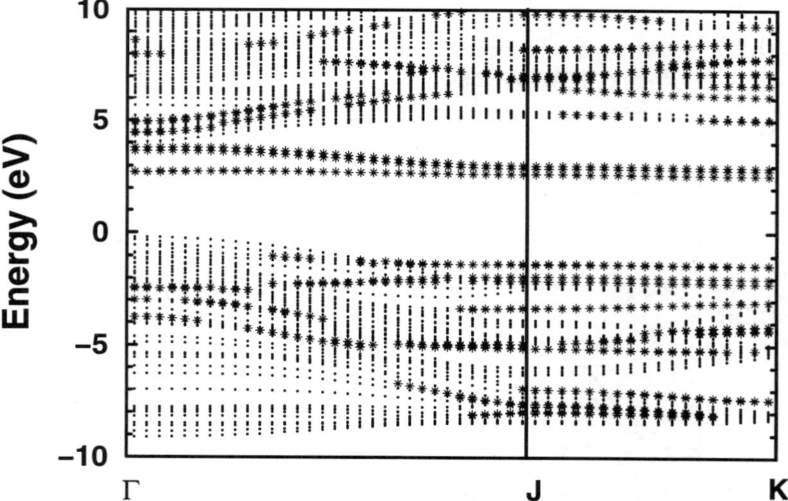

Figure 3: Surface electronic structure along the main symmetry directions of the surface unit cell. Dots correspond to the projected bulk states, while stars represent surface states. Resonance states embedded in projected bulk bands are also represented by stars.

The imaginary part of the average polarizability of the slab was obtained using the electron levels calculated at 4900 points distributed homogeneously on the irreducible surface

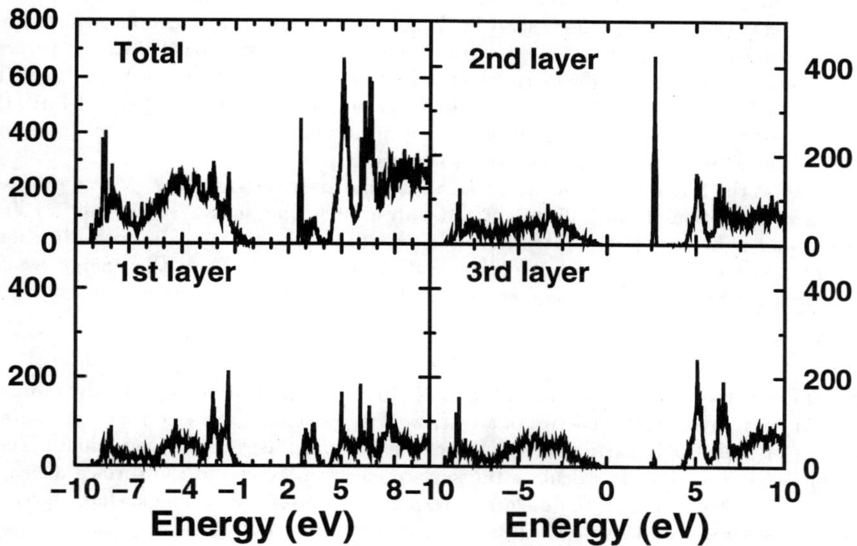

Figure 4: Calculated density of states for (a) total, (b) projected on first layer, (c) on second layer, and (d) on third layer.

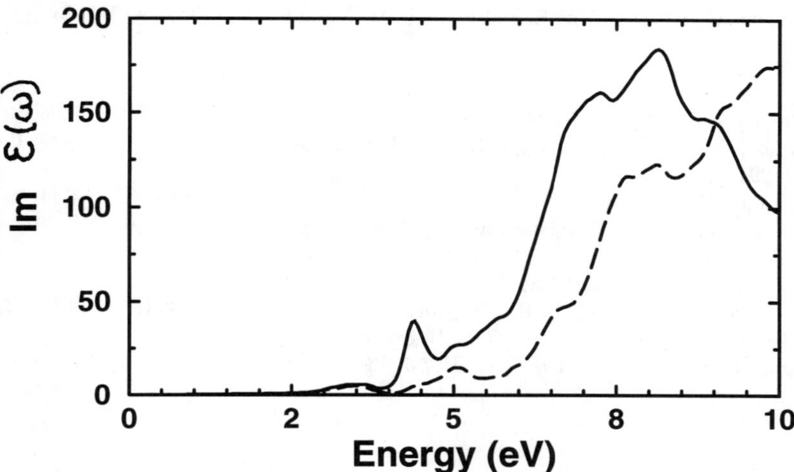

Figure 5: Imaginary part of the surface dielectric response. Solid line corresponds to light polarized along the x axis, while dashed line corresponds to light polarized along the y direction.

Brillouin zone. Figure 5 shows the imaginary part of the surface dielectric function, for both transverse (along x direction, solid line) and parallel (along y direction, dashed line) polarizations. For both polarizations, electron transitions start at about 2.7 eV. In general, the line shape of both surface dielectric functions is very similar up to about 8.5 eV, but the response to an external field along the x direction is more intense. At an energy of 4 eV, the transveral polarizability (along x), shows a sharp peak attributed to the optical transition from the valence bulk band to the empty surface states at 4 eV at Γ point.

Looking at the origin of the electron transitions one finds that from 2.7 eV to 4 eV there are electron transitions from bulk to surface states, while the peak at 4 eV only for light polarized along the x axis is due to transition between surface states. In both directions, electron transitions from surface to bulk states start at about 5.5 eV. At this energy, we can observe that for both polarizations the surface dielectric function is more intense.

SUMMARY

We studied the optical properties of the GaN ($10\bar{1}0$) surface using a tight-binding formalism based on a sp^3s^* orbital basis. As a test, we first calculated the electronic and optical properties of bulk GaN that agree with experimental and theoretical observations. Then, we employed the method to calculate the surface electron levels and the surface dielectric response. We observed that the dielectric response is anisotropic. The surface dielectric function shows a gap of 2.7 eV, which is 1 eV lower than the bulk gap.

ACKNOWLEDGMENTS

This work has been supported in part by Consejo Nacional de Ciencia y Tecnología through grant CONACyT-0075P-E, and ONR-BMDO grant No. N00014-96-1-1183.

REFERENCES

[1] J.H. Edgar, J. Mater. Res. **7**, 235 (1992).
[2] H. Morkoc, S. Strite, G. B. Gao, M.E. Lin, B. Sverdlov, and M. Burns, J. Appl. Phys. **76**, 1363 (1994).
[3] G. Martin, S. Strite, J. Thornton, and H. Morkoc, Appl. Phys. Lett. **58**, 2375 (1991).
[4] S. Bloom, G. Harbeke, E. meiner, and I.B. Ortenburger, phy. stat. sol. (b) **66**, 161 (1974).
[5] J.E. Jaffe, R. Pandey, and P. Zapol, Phys. Rev. B **53**, R4209 (1996).
[6] M. Palummo, C.M. Bertoni, L. Reining, and F. Finocchi, Physica B, **185**, 404 (1993).
[7] K. Miwa and A. Fukumoto, Phys. Rev. B **48**, 7897 (1993).
[8] C. Noguez and S.E. Ulloa, Phys. Rev. B, **53**, 13138 (1996); and references therein.

STABILITY AND BAND OFFSETS OF SiC/GaN, SiC/AlN, AND AlN/GaN HETEROSTRUCTURES

J. A. MAJEWSKI, M. STÄDELE, and P. VOGL
Walter Schottky Institute, Technical University of Munich, D-85748 Garching, Germany

ABSTRACT

We present first-principles calculations of structural and electronic properties of heterovalent SiC/GaN, SiC/AlN, and isovalent AlN/GaN heterostructures that are grown pseudomorphically on (001) or (110) SiC substrates. For the polar interfaces, we have investigated reconstructed stoichiometric interfaces consisting of one and two mixed layers with lateral c(2 x 2), 2 x 1, 1 x 2, and 2 x 2 arrangements. The preferred bonding configurations of the reconstructed interfaces are found to be Si-N and Ga-C. With respect to vacuum, the valence band maximum is found to be highest in SiC and lowest in AlN. In these systems, the valence band offsets deviate substantially from the transitivity rule and depend sensitively on the microscopic details of the interface geometry. The SiC/AlN and AlN/GaN heterostructures are predicted to be of type I, whereas SiC/GaN heterostructure can be of type I or II.

INTRODUCTION

Thin nitride films are currently attracting much interest because of their potential applications in opto- and microelectronics. Recently, cubic nitride films grown on cubic SiC have obtained considerable attention [1, 2, 3]. The band discontinuities of the heterojunctions are vital for assessing the degree of carrier confinement, and therefore the potential for optoelectronic device operation. However, no definitive measurements for the band offsets between cubic SiC and GaN (AlN) or between AlN and GaN have been reported. The valence-band offsets for nonpolar interfaces in the zincblende structure, namely (110) SiC/AlN [4] and (110) GaN/AlN [5], have been calculated previously.

In this work, we present first-principles pseudopotential studies of the band offsets as well as the formation enthalpies of the polar [001] nitride heterostructure interfaces of AlN/SiC, GaN/SiC, and GaN/AlN. Our calculations are based on the first-principles total-energy pseudopotential method within the local-density-functional formalism [6]. We have used norm-conserving separable pseudopotentials [7] and a preconditioned conjugate gradient algorithm [8] for minimizing the total crystal energy with respect to both the electronic and ionic degrees of freedom. The semicore Ga 3d-electrons are treated as part of the frozen core, but their considerable overlap with the valence electrons is accounted for by including the nonlinear core exchange-correlation correction [9]. This procedure increases the transferability of the pseudopotentials and yields lattice constants that agree very well with experiment [10].

INTERFACE RECONSTRUCTIONS AND FORMATION ENTHALPIES

Interface configurations

In heterovalent SiC/AlN and SiC/GaN heterostructures without a common anion or cation, the "ideal" abrupt (001) interface contains tetrahedral bonds with more than 2

Mat. Res. Soc. Symp. Proc. Vol. 449 © 1997 Materials Research Society

or less than 2 electrons per bond, leading to a macroscopically charged interface that is energetically unstable [11]. Local charge neutrality can be restored by forming chemically intermixed interfaces. We have studied the electronic and structural properties of charge compensated, pseudomorphic (001) interfaces with one and two mixed interface layers. We have taken cubic SiC as the unstrained substrate that fixes the lateral lattice constant. There are four abrupt SiC/GaN (and SiC/AlN) interfaces, two of which lead to energetically favorable heterostructures with a single mixed interface layer, namely N-Si and Ga-C. In the Ga-C interface, the mixed layer is formed by either replacing 50% of the Ga atoms by Si atoms (Si/Ga interface) or, alternatively, 50% of the C atoms by N (C/N interface). Other structures can be eliminated because they lead to unfavorable anion-anion (C-N) type and cation-cation (Si-Ga) type bonds with large dipole moments. In addition, we have considered three different lateral atomic arrangements of the atoms, namely c(2×2), 2×1, and 1×2 [10]. The other cases that we considered in detail are configurations with two mixed layers in the interface. Here, it turns out that the energetically favorable reconstructions contain two types of atoms A, B with $A : B = 1 : 3$ in each mixed layer, and a 2×2 lateral arrangement. Out of the 32 different interface structures that we have studied, the energetically most favorable ones contain only cations (Ga (or Al) and Si) within one layer and anions (C and N) in the adjacent layer. In contrast to these cases, the (110) SiC/AlN and SiC/GaN as well as the homovalent (001) AlN/GaN heterostructures do not possess 'wrong' acceptor and donor bonds. Consequently, intermixing is less important for these interfaces and we have assumed them to be abrupt. In the case of (001) AlN/GaN heterostructures, we have performed calculations for substrates with a lateral lattice constant equal to (i) $a_\| = a_{bulk}^{AlN}$, (ii) $a_\| = (a_{bulk}^{GaN} + a_{bulk}^{AlN})/2$ (corresponds to the Al$_{0.5}$Ga$_{0.5}$N alloy as substrate), and (iii) $a_\| = a_{bulk}^{GaN}$, respectively.

These interfaces have been modeled by supercells with up to 16 monolayers. All atomic positions in the unit cell have been optimized since the valence band offsets (VBO's) depend sensitively on them. If we assume, for example, the interface lattice constant of the AlN/GaN (001) heterostructure to be equal to the arithmetic average of the bulk lattice constants (giving an interface spacing of 1.126 Å), we obtain a VBO of 1.11 eV. The relaxed interface height, on the other hand, is 1.174 Å and yields a VBO equal to 0.76 eV. The details of the atomic relaxations of the (001) AlN/GaN heterostructure (grown on AlN) are depicted in Fig. 1.

One can easily see that the distance between the N and Ga layers right at the interface layers is significantly larger than an estimation based on averaged bulk lattice constants. Such an estimation works well in many heterostructures, but the present calculations indicate that this assumption is not applicable for nitrides.

Formation enthalpies

We have calculated the formation enthalpy δH of all heterostructures discussed above. With the exception of the SiC/GaN (110) and AlN/GaN (001) heterostructures, they represent metastable configurations with $\delta H > 0$. In the case of the heterovalent (001) interfaces, δH of the Al(Ga)/C and Si/N interfaces is one order of magnitude larger than that of the Al(Ga)/Si and C/N interfaces (which yield $\delta H \approx 0.04 - 0.16$). Indeed, the former contain predominantly cation-cation or anion-anion (Ga-Si or C-N) bonds, whereas the more stable Ga(Al)/Si and C/N interfaces contain only cation-anion bonds. Analogous conclusions can be drawn for the interfaces with two mixed layers but they are found to be energetically

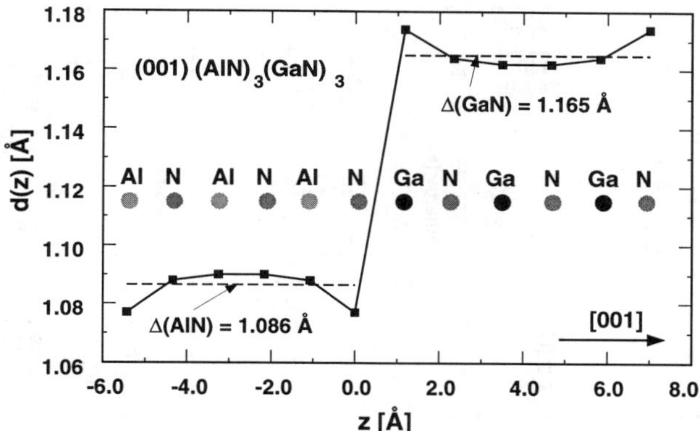

Figure 1: Relaxation of the atomic layers near the abrupt (001) interface between AlN and GaN, grown on AlN substrate. The horizontal axis labels the position z of the atomic layers along the (001) growth direction and the vertical axis gives the distance $d(z)$ to the preceding adjacent layer. Δ(AlN) and Δ(GaN) denote the distance between the atomic layers in unstrained bulk AlN, and strained bulk GaN, respectively. The full line only guides the eye. The interface was modelled by a (AlN)$_3$(GaN)$_3$ superlattice.

less favorable ($\delta H \geq 0.23$). The lateral arrangement of the atoms in the mixed layers has only little effect on δH. For nonpolar SiC/AlN and SiC/GaN (110) heterostructures, we obtain $\delta H = 0.07$ eV and -0.02 eV, respectively, and the isovalent AlN/GaN (001) and (110) heterostructures lead to $\delta H = 0$ within an accuracy of 1 meV. For AlN/GaN (001) heterostructures, we find the formation enthalpy to be practically independent on the substrate lattice constant.

BAND OFFSETS

The valence band offset ΔE_V at an interface between two semiconductors can be conveniently split up into two terms [12]

$$\Delta E_V = \Delta \overline{V} + \Delta E_{BS}, \tag{1}$$

where $\Delta \overline{V}$ is the asymptotic difference between the laterally and vertically averaged electrostatic (Hartree plus ionic) potential $V(\mathbf{r})$ in the superlattice far from the interface. For neutral interfaces, $\Delta \overline{V}$ equals the dipole moment of the electronic and ionic charge density across the interface. It is nonzero due to the rearrangement of the ions and electrons near the interface and thus depends on the detailed interface geometry. The second term in Eq. 1, $\Delta E_{BS} = E_v(GaN) - E_v(SiC)$, is the difference between the eigenvalues of the top of the valence band in the two bulk materials, measured with respect to the average electrostatic potential $\overline{V}$ (that is arbitrary and may be set to zero for a bulk solid). ΔE_{BS} is, by definition, a bulk property of the two constituent solids, and can be obtained from standard bulk band-structure calculations for SiC and strained GaN, respectively. The spin-orbit coupling influences ΔE_{BS} and consequently the band offsets by less than 0.01 eV in

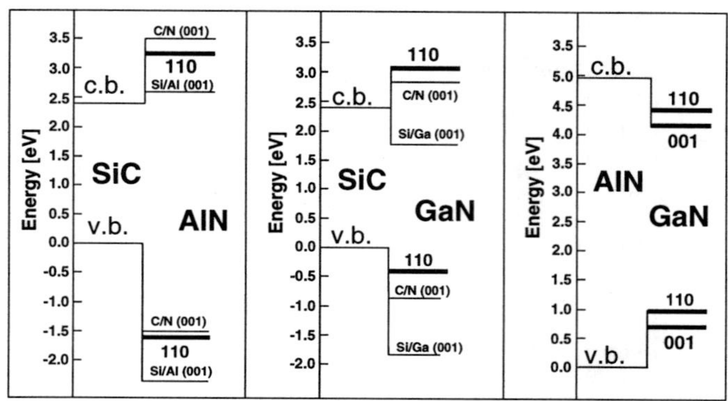

Figure 2: Band offsets of SiC/AlN, SiC/GaN, and AlN/GaN heterostructures for (001) and (110) substrate surface orientation. C/N and Si/Al(Ga) indicate the reconstructed interfaces with a single mixed layer containing C and N, or Si and Al(Ga) atoms, respectively. The zero of energy is chosen at the top of the valence band of the substrate which is the left material in each panel.

the presently studied systems. Relativistic effects can therefore be neglected.

In Fig. 2, we depict the band offsets predicted by the present theory. On an absolute energy scale, the calculations give the highest valence band maximum in SiC and the lowest in AlN. For the (110) heterostructures, we observe that their VBO's deviate from the transitivity rule by roughly 0.3 eV.

The calculated valence band offsets depend sensitively on microscopic details of the interface geometry. Consider, for example, the c(2 x 2) SiC/GaN (001) heterostructure with one mixed layer. By replacing the Si/Ga interface by C/N, i.e. by swapping cations and anions, we just reverse the polarity of the interface and obtain nearly the same formation enthalpy. However, the VBO of the Si/Ga and the C/N interface differs by 1.0 eV. Furthermore, we find that the VBO's of the reconstructed interfaces are largely determined by the chemical composition of the interface but they depend very little on the *lateral* atomic arrangement. Altogether, we predict the VBO's of the SiC/GaN and SiC/AlN (001) heterostructures to lie in the range 0.8 - 1.8 eV, and 1.5 - 2.4 eV, respectively. The VBO's of the AlN/GaN interfaces grown on AlN and SiC substrates, respectively, are found to differ by only 49 meV. The calculated VBO's of the nonpolar SiC/GaN and SiC/AlN (110) interfaces equal 0.4 eV and 1.6 for SiC/GaN and SiC/AlN, respectively. Thus, they differ markedly from the average of the (001) VBO's, in contrast to the situation in the GaAs/ZnSe system [13].

In the case of (001) AlN/GaN heterostructures, we find the VBO to be equal to 0.74, 0.58, and 0.41 eV for $a_\| = a_{bulk}^{AlN}$, $a_\| = (a_{bulk}^{GaN} + a_{bulk}^{AlN})/2$, and $a_\| = a_{bulk}^{GaN}$, respectively. The top of the valence band is lower in AlN. We see that the VBO of a AlN/GaN junction can be taylored quite significantly by using an appropriate substrate AlGaN alloy. The change of the VBO with substrate lattice constant comes mainly from the bulk contribution to the VBO and is therefore determined by the strain induced by the pseudomorphic growth; ΔE_{BS} decreases with increasing $a_\|$ and equals -0.15, -0.33, and -0.55 eV for the three cases mentioned above, respectively. The dipole contribution, on the other hand, is practically independent of the substrate lattice constant for these isovalent heterojunctions.

For cubic systems, no experimental data on VBO's are yet available. However, recent photoemission measurements of the VBO for hexagonal AlN/GaN structures [14] yield 0.60 ± 0.24 eV and 0.57 ± 0.22 eV for GaN on AlN, and AlN on GaN, respectively. These data indicate a rather weak dependence of the offset on the strain set by $a_{||}$. However, the dense network of threading dislocations in these samples may help to reduce the strain near the interface [14]. It is not yet clear to what extent these measurements reflect the situation of pseudomorphic heterostructures. Earlier photoemission measurements for a thin GaN film grown on hexagonal AlN gave the valence-band offset to be equal to $\Delta E_V = 0.8 \pm 0.3$ eV [15]. Using the energetic position of the iron impurity level in both materials as a reference, one obtains a rather small value of ΔE_V 0.5 eV [16].

CONCLUSIONS

We have discussed the stability of various reconstructed (001) SiC/GaN and (001) SiC/AlN interface geometries with one and two mixed layers at the interface. The preferred bonding configurations across the interface are found to be Si-N and Ga-C, which corresponds to cation-anion bonding. We found that the atomic relaxation near and right at the interface plays a crucial role for the formation enthalpies and band discontinuities. We predict that the VBO's in pseudomorphic AlN/GaN heterostructures can be altered considerably by growing the structure on various substrates because the isovalent VBO's are largely determined by the strain at the interface. Finally, our studies show that the SiC/AlN and AlN/GaN heterostructures are always of type I, whereas SiC/GaN heterostructure can be of type I or II depending on the orientation and chemical composition of the interface.

ACKNOWLEDGEMENTS

This work has been supported by the Deutsche Forschungsgemeinschaft (Project SFB 348) and Bayerischer Forschungsverbund.

REFERENCES

1. J. Paisley, Z. Sitar, J.B. Posthill, and R. F. Davis, J. Vac. Sci. Technol. A7, 701 (1989).

2. H. Liu, C. Frenkel, J. G. Kim and R. M. Park, J. Appl. Phys. **74**, 6124 (1993).

3. A. Barski, U. Rössner, J. L. Rouvière and M. Arlery, MRS Internet. J. Nitride Semicond. Res. **1**, 21 (1996).

4. W. R. L. Lambrecht, and B. Segall, Phys. Rev. B **43**, 7070 (1991).

5. E. A. Albanesi, W. R. L. Lambrecht, and B. Segall, J. Vac. Sci.Technol. B **12**, 2470 (1994).

6. W. E. Pickett, Computer Physics Reports **9**, 115 (1989).

7. N. Troullier and J. L. Martins, Phys. Rev. B **43**, 1993 (1991); L. Kleinman and D. M. Bylander, Phys. Rev. Lett. **48**, 1425 (1982).

8. M. C. Payne, M. P. Teter, D. C. Allan, T. A. Arias, and J. D. Joannopoulos, Rev. Mod. Phys. **64**, 1045 (1992).

9. S. G. Louie, S. Froyen, and M. L. Cohen, Phys. Rev. B **26**, 1738 (1982).

10. M. Städele, J. A. Majewski, and P. Vogl, Acta Phys. Polon. **A88**, 917 (1995).

11. R. M. Martin, J. Vac. Sci. Technol. **17**, 978 (1980).

12. L. Colombo, R. Resta, and S. Baroni, Phys. Rev. B **44**, 5572 (1991).

13. G. Bratina, L. Vanzetti, L. Sorba, G. Biasiol, A. Franciosi, M. Peressi, and S. Baroni, Phys. Rev. B **50**, 11723 (1994).

14. G. Martin, A. Botchkarev, A. Rockett, and H. Morkoç, Appl. Phys. Lett. **68**, 2541 (1996).

15. G. Martin, S. Strite, A. Botchkarev, A. Agarwal, A. Rockett, H. Morkoç, W. R. L. Lambrecht, and B. Segall, Appl. Phys. Lett. **65**, 610 (1994).

16. J. Baur, K. Maier, M. Kunzer, U. Kaufman, and J. Schneider, Appl. Phys. Lett. **65**, 2211 (1994).

OFFSETS AND POLARIZATION AT STRAINED AlN/GaN POLAR INTERFACES

Fabio Bernardini,* Vincenzo Fiorentini*, and David Vanderbilt**

*INFM – Dipartimento di Scienze Fisiche, Università di Cagliari, I-09124 Cagliari, Italy
** Department of Physics and Astronomy, Rutgers University, Piscataway, NJ, U.S.A.

ABSTRACT

The strain induced by lattice mismatch at the interface is responsible for the different value of the band discontinuities observed recently for the AlN/GaN (AlN on GaN) and the GaN/AlN (GaN on AlN) polar (0001) interface. We present a first-principles calculation of valence band offsets, interface dipoles, strain-induced piezoelectric fields, relaxed geometric structure, and formation energies. Our results confirm the existence of a large forward-backward asymmetry for this interface.

INTRODUCTION

A reliable determination of the valence-band offset (VBO) at the (0001) polar interface between wurtzite AlN and GaN is still missing. The few experimental investigations available [1, 2] are in mutual disagreement, and theoretical studies refer either to zincblende interfaces [3], or artificially lattice-matched wurtzite interfaces [4]. The latter approximation leads to a less accurate determination for the VBO, and cannot pick up the possible forward-backward asymmetry characteristic of lattice-mismatched interfaces. In the case of the AlN/GaN interface, lattice mismatch amounts to 2.5 %, and may cause a very large asymmetry. This asymmetry has not yet been clearly determined experimentally (it was not even found in early experimental work [5]), being hidden by the large uncertainties in the measured data. Even the best experimental investigations available face two kinds of problem: (i) the determination of the core level alignment with the valence-band maximum (VBM) is obtained indirectly using theoretical estimates of the VBM position, which (as underlined by Vogel *et al.* [6]) is affected by large systematic errors; (ii) the existence of strong polarization fields in both the substrate and the overlayer tends to modify the apparent value of the VBO deduced from the core-level shift measurements.

The present *ab-initio* investigation includes all strain and relaxation-induced effects, and overcomes the difficulty in VBO determination due to polarization fields by the use of a novel charge-decomposition technique. An estimate of the formation energy of the interfaces studied is also given.

BULK PROPERTIES

Valence-band offset calculations at lattice-mismatched interfaces require the evaluation of the band structure energies for the bulk crystals in equilibrium, and subjected to biaxial strain. The calculations are done using density-functional theory in the local-density approximation (LDA) to describe the exchange-correlation energy, and ultrasoft pseudopotentials [7] for the electron-ion interaction. Plane-wave basis sets up to 25 Ry, and 24 special k-points are found to give fully converged values for the bulk properties. Since the properties of GaN are affected by Ga $3d$ states [8], our Ga pseudopotential includes $3d$ electrons in the valence. This yields very good structural bulk parameters.

Mat. Res. Soc. Symp. Proc. Vol. 449 © 1997 Materials Research Society

Table I: Predicted structural parameters and valence band maxima for equilibrium and strained AlN and GaN.

Material	AlN	AlN	GaN	GaN
Substrate	—	GaN	—	AlN
a	5.814	6.04	6.04	5.814
c/a	1.619	1.51	1.6336	1.73
u	0.38	0.3927	0.3761	0.3653
E_{strain} (eV)		+0.179		+0.155
E_{VBM} (eV)	−0.16	0.09	−4.90	−4.69

In wurtzite crystals, the determination of the atomic structure at a given lattice constant a implies the calculation of the c/a and u parameters. The equilibrium c as been determined by fitting with a polynomial the total energy computed for six different values of c, with u being determined for each value of c/a via minimization of the Hellmann-Feynman forces, with a threshold of 10^{-4} Hartree/bohr. The calculated structural parameters are given in Table I. The structural parameters of AlN and GaN behave similarly under strain $[\Delta(c/a)/(c/a) \sim 7\%, \Delta u/u \sim 2\%]$ with similar total energy variations. Instead, the effect of strain on the valence-band edge is very different. A rationale for this difference is that the AlN (GaN) band edge is a singlet (doublet) formed by the hybridization along the c-axis (in the a-plane) of N $2s$ orbitals with Al p_z (Ga p_{xy}) states, so that biaxial compression pushes the edge upward in GaN and downward in AlN.

BAND OFFSET

As pointed out by Baldereschi *et al.* [9], the valence-band offset ΔE_v may be split in two terms:

$$\Delta E_v = \Delta E_{\text{VBM}} + \Delta V_{el}.$$

The first contribution ΔE_{VBM} is the difference between the valence-band edge energy in the two bulk materials, each edge being referred to the average bulk electrostatic potential. The second contribution, the potential lineup ΔV_{el}, is the drop of the macroscopic average of the electrostatic potential across the interface. The latter term requires a selfconsistent calculation of the electronic density distribution for the real interface system. Our interface has been modeled using a $(GaN)_4/(AlN)_4(0001)$ superlattice (see below), both ideal and fully relaxed. The material being grown epitaxially on the chosen substrate, has been pre-strained to have the same a lattice constant as the substrate.

The lineup term is customarily obtained by solving the Poisson equation for the macroscopic average of the charge density, neutralized by a suitable distribution of gaussian charges centered on the ion sites. The potential drop across the interface is usually calculated as the difference of potential values in bulk-like regions inside the two interfaced materials. This turns out to be non-trivial for a system such as the present one, in which the existence of polarization fields in the equilibrium bulk makes it impossible to define asymptotic values for the electrostatic potentials. The existence of such fields, moreover, limits the maximum length of our slab. Indeed, beyond a certain critical thickness the drop of the potential inside each slab would make the system metallic, with a related transfer of charge, which would spoil the exact determination of the lineup term. Our choice of a

16-atom supercell is a compromise between the need to have an insulating system, and at the same time to avoid a spurious coupling of the interfaces at the sides of the slabs. Tests were performed in supercells of up to 40 atoms. In Fig. 1, we show the macroscopic average of the charge density and of the electrostatic potential. The potential drop is inextricably linked to the polarization fields within the AlN and GaN bulks. We have circumvented

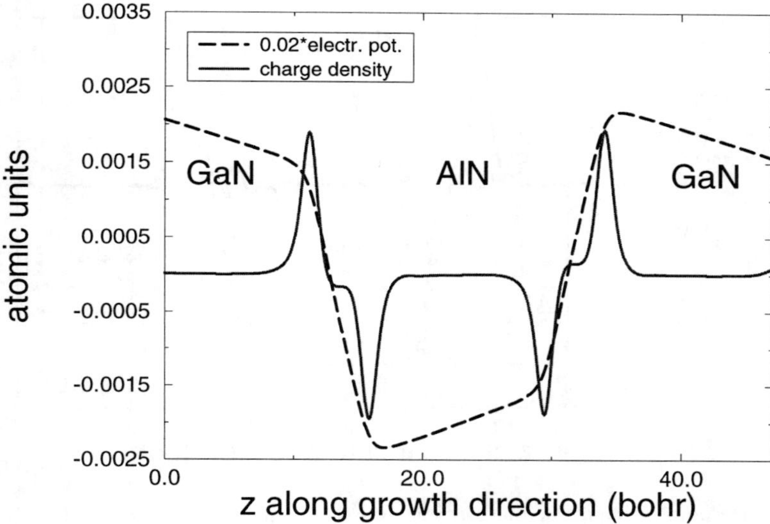

Figure 1: Supercell electron density and electrostatic potential. Electron density has been compensated in the two bulks by a distribution of gaussians placed at the atomic sites.

this problem by employing a new method. The basic idea is that at the polar AlN/GaN interfaces, the existence of polarization fields reveals itself by an accumulation of charge in the form of a *monopole* distribution whose density is proportional to the difference between the polarizations inside the two interfaced bulks. On top of this monopole term, we have the traditional *dipole* term representing the local charge transfer across the interface. This dipole term is the quantity we are interested in, as the band offset is by definition related to the dipolar part of the potential drop. Since the monopole contributions are related to the polarization fields, they must be equal and opposite for the two (geometrically inequivalent) interfaces in our AlN/GaN superlattice. To filter out the monopole term we superimpose the two interface distributions by folding them around a plane placed halfway between the two junctions. We define the dipole term $\bar{\bar{n}}^{dip}$ as the average of the superimposed charges,

$$\bar{\bar{n}}^{dip}(z - z_0) = \frac{1}{2}\left[\bar{\bar{n}}(z - z_0) + \bar{\bar{n}}(z_0 - z)\right],$$

where z is a coordinate along the c-axis, z_0 the plane position and $\bar{\bar{n}}$ the macroscopic average for the charge density. The monopole term $\bar{\bar{n}}^{mono}$ is just the difference between the dipole term and the total macroscopic charge:

$$\bar{\bar{n}}^{mono}(z) = \bar{\bar{n}}(z) - \bar{\bar{n}}^{dip}(z).$$

Such a decomposition allows a determination of the polarization charges and dipole terms which is nearly independent of the position of the folding plane. Fig. 2 shows the decomposition for the AlN/GaN interface. The decomposition reveals the origin of the asymmetry

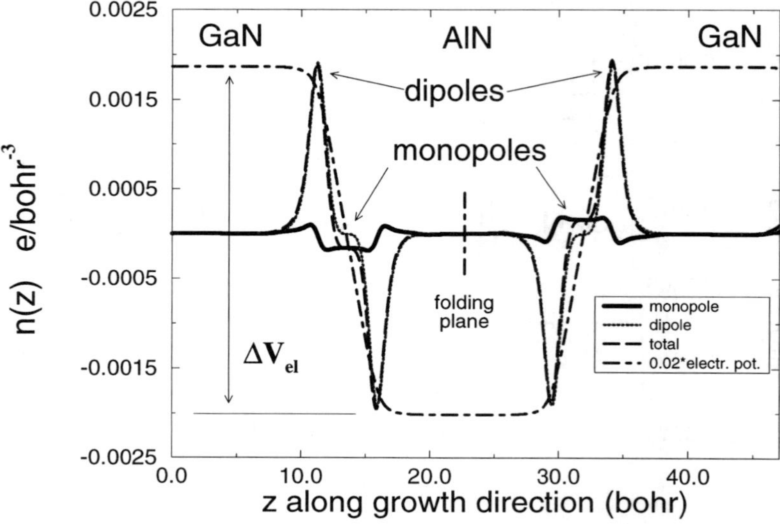

Figure 2: Decomposition of the macroscopic average of the electronic density (dotted line) into monopole (solid) and dipole (dashed) terms. Such a decomposition allows the determination of the lineup term ΔV_{el} from the solution of the Poisson equation (dot-dashed) of the dipole term.

in the total charge distribution, and at the same time it enables us to evaluate the lineup term. In Table II we report the values for the VBO obtained via this decomposition. There

Table II: Valence-band offset ΔE_v, potential lineup ΔV_{el}, relaxation energies E_{rel}, monopole charge densities σ_{int} and electric fields $\vec{E}$ in the ideal and relaxed the AlN/GaN (0001) interface.

Interface	AlN/GaN		GaN/AlN		units
structure	ideal	relaxed	ideal	relaxed	
ΔE_v	0.29	0.20	1.00	0.85	eV
ΔV_{el}	5.28	5.18	5.52	5.36	eV
σ_{int}	0.029	0.014	0.022	0.011	C/m^2
$\vec{E}$	32.7	15.6	24.4	12.9	10^8 V/m

is a very large forward-backward asymmetry of 0.65 eV between AlN/GaN and GaN/AlN interface VBOs. This is only marginally due to the lineup term (contributing 0.18 eV), its main component being the band structure term (0.47 eV). The relaxation is responsible for comparatively small deviations of ~ 0.1 eV from the ideal-interface values. The relaxation pattern is characterized in both cases by a contraction of the Al-N axial interface bond ($\sim$ −0.04 a.u.) and an expansion of the axial Ga-N bond ($\sim$ +0.02 a.u.).

POLARIZATION

Supercell calculations are not the only way to obtain the interface charge density σ_{int}. As shown in Ref. [10], given the polarization P_1 and P_2 and the dielectric constants ε_1 and ε_2 of the component materials, σ_{int} is given by

$$\sigma_{int} = \pm\, 2\,(P_2 - P_1)/(\varepsilon_1 + \varepsilon_2). \tag{1}$$

in periodic boundary conditions. We have calculated the macroscopic polarization for equilibrium and strained GaN and AlN via the Berry phase technique of Ref. [11]. The (high-frequency) dielectric constants of AlN and GaN have been calculated using the relation

$$\Delta P_{\mathrm{T}} = \varepsilon_\infty \Delta P_{\mathrm{L}},$$

where ΔP_{T} is the (so-called transverse) polarization change induced by a small cation sublattice displacement in the bulk in zero field, and ΔP_{L} is the (so-called longitudinal) polarization change due to a uniform displacement of few cation planes in a periodic bulk supercell. In the latter, a depolarizing field is present due to the periodic boundary conditions. As a by-product of our calculations, we obtained the Born effective charges for AlN and GaN which, as expected for highly polar semiconductors, are quite close to the nominal ionicity. The results are shown in Table III. Substituting P_{tot} and ε in Eq.1 we

Table III: Polarization in GaN and AlN: electronic P_{el} and total P_{tot} polarization, derivative of the latter with respect to u, Born effective charge, and dielectric constant are shown.

System	a	P_{el}	P_{tot}	$\partial P/\partial u$	Z^*	ε_∞
——	bohr	C/m^2	C/m^2	C/m^2	e	——
AlN	5.814	−0.178	−0.0812	10.51	2.69	4.59
AlN	6.04	−0.492	−0.1712	9.95	2.74	4.64
GaN	5.814	+0.223	+0.0343	10.69	2.74	5.27
GaN	6.04	−0.0511	−0.0308	9.88	2.72	5.52

obtain for the AlN/GaN (GaN/AlN) interface a monopole density of 0.028 (0.023) C/m^2 in embarrassing agreement with the outcomes of the supercell calculations. This proves directly that the sources of the internal fields in the interface system are the charges accumulated at the interface by the bulk polarization effects. Also, the latter finding provides an a posteriori justification of our somewhat ad-hoc charge decomposition procedure.

FORMATION ENERGIES

An important issue for the present selfconsistent calculation is the evaluation of the formation energy for the AlN/GaN interfaces. Contrary to the case of non-polar interfaces, for the wurtzite (0001) system it is impossible to build a superlattice with symmetric interfaces. This means that only the average value of the formation energy for the two interfaces can be obtained from a total-energy calculation. We define the average formation energy for the AlN/GaN interface as

$$E_f = \frac{1}{2A} \left[E^{\text{tot}} - n^{\text{Ga}} \mu^{\text{GaN}} - n^{\text{Al}} \mu^{\text{AlN}} \right],$$

where μ^X are the total energies per pair of GaN and AlN, n^X the number of Ga and Al atoms, E^{tot} the supercell total energy and A its cross-sectional area. A reliable determination of E_f requires equivalent k-point sampling for bulk and interface calculations. This is easily accomplished if the interface is lattice-matched. In the present case, the supercell length is not simply an integer multiple of the bulk unit cell of either constituent material. This means that an exact equivalence between k-point meshes cannot be achieved. A good approximation for E_f can however be obtained by defining, for each component material, an auxiliary bulk cell having the same lattice constant a, and an axial length $\bar{c}$ being a sub-multiple of the supercell length l. This value in the present case is just the average of c_{AlN} and c_{GaN}. It is then possible to downfold exactly the supercell mesh into the auxiliary bulk cell. The next step is to uniformly scale the k-points coordinates to adapt the mesh to the real value of c. We should point out that the accuracy of this procedure (compared with an exact computation of the energy integral over the IBZ) increases with the number of points in the mesh. It is therefore possible to find a suitable mesh to accomplish any required accuracy. The results for the formation energies reported in Table IV have been

Table IV: Average formation energy for the AlN/GaN (0001) interfaces.

Interface	AlN/GaN		GaN/AlN		units
	ideal	relaxed	ideal	relaxed	
E_f	−3.4	−16.4	+11.7	−6.2	meV

obtained using a 6-point Chadi-Cohen mesh [12] in the supercell, which when downfolded in the auxiliary cell produces 24 special points. We estimate the k-point sampling error in the formation energies to be $\sim$ 10 meV. It should be noted that such low formation energies are not surprising when compared with the results obtained by Chetty *et al.* [13] for GaAs/AlAs(111) interfaces.

REFERENCES

1. J. W. Waldrop and R. W. Grant, Appl. Phys. Lett. **68**, 2879 (1996).
2. G. Martin, A. Botchkarev, A. Rockett, and H. Morkoc, Appl. Phys. Lett. **68**, 2541 (1996).
3. E. Albanesi, W. R. L. Lambrecht, and B. Segall, J. Vac. Sci. Tech. **B 12**, 2470 (1994).
4. S. Wei and A. Zunger, to be published.
5. G. Martin *et al.*, Appl. Phys. Lett. **65**, 610 (1994).
6. D. Vogel, P. Krüger and J. Pollmann, to be published
7. D. Vanderbilt, Phys. Rev. **41**, 7892 (1990).
8. V. Fiorentini *et al.*, Phys. Rev. B **47**, 13353 (1993); V. Fiorentini *et al.*, Proc. ICPS-22, D. J. Lockwood ed. (World Scientific, Singapore 1995), p.137; A. Satta *et al.*, MRS Proc. **395**, 515 (1996).
9. A. Baldereschi, S. Baroni, and R. Resta, Phys. Rev. Lett. **61**, 734 (1988).
10. D. Vanderbilt and R. D. King-Smith, Phys. Rev. B **48**, 4442 (1993).
11. R. D. King-Smith and D. Vanderbilt, Phys. Rev. B **47**, 1651 (1992).
12. D. J. Chadi and Marvin L. Cohen, Phys. Rev. B **8**, 5747 (1973).
13. N. Chetty and R. M. Martin, Phys. Rev. B **45**, 6089 (1992).

THEORETICAL STUDY OF GROUP-III NITRIDE ALLOYS

KWISEON KIM, SUKIT LIMPIJUMNONG, WALTER R. L. LAMBRECHT and B. SEGALL
Department of Physics, Case Western Reserve University, Cleveland, OH 44106-7079

ABSTRACT

Band gap bowing, structural relaxations, and energies of formation were calculated for the three pseudobinary nitride zincblende alloy systems Al-Ga, In-Ga and In-Al using the full-potential linearized muffin-tin orbital method. The cluster expansion and Connolly-Williams approaches were used to relate calculated band structures and energies of formation of ordered compounds to the behavior of disordered alloys. Effects of bond length and volume variation on those properties are discussed. An interpolation formula for the gap of the full pseudoternary $Al_xGa_yIn_zN$ system is proposed and tested by separate calculations. Extension of the results to the wurtzite alloys is discussed.

INTRODUCTION

Alloying among group-III nitrides in principle allows one to change the band gap from 1.9 eV in InN, to 6.3 eV in AlN. GaN intermediate in the series has a bandgap of 3.5 eV. Previous theoretical studies and most experimental work (see e.g. [1]) have focused on the pseudobinary[1] alloy combinations, AlGaN and InGaN [2, 3, 4]. However, for each concentration in those alloy systems, both the gap and the lattice constant are determined. Pseudoternary systems allow more freedom. For any desired band gap (except those for the extremes), there are several possible alloy combinations and hence lattice constants to choose from. Since it is desirable to have different layers in a heterostructure lattice-matched, it is useful to employ the full pseudoternary $Al_xGa_yIn_zN$ system ($x + y + z = 1$). To guide such an alloy design, an accurate knowledge of the band gap and the lattice constant as functions of x, y and z is required.

A second important concern for the alloy design is the issue of phase separation. While it is widely believed that the nitrides are fully miscible, there is, in reality, a miscibility gap. In a closely lattice-matched $Al_xGa_{1-x}N$ alloy system, the miscibility gap temperature was shown [2] to be rather low and hence at a typical growth temperature one may expect to have a true solid solution. In $In_xGa_{1-x}N$, however, phase separation problems have been reported experimentally [5, 6].

In this paper, we present the first systematic study of the full psudoternary system using throughout the same first-principles computational scheme and including an accurate treatment of bond length relaxations.

METHOD OF CALCULATION

Our alloy model is based on the cluster expansion approach [7]. While the natural crystal structure for the nitrides is the wurtzite (wz), zincblende (zb) structures have also been stabilized. Since there is considerably more experience with alloy modeling in cubic

---

[1] We call and $A_xB_{1-x}N$ alloy pseudobinary and an $A_xB_yC_zN$ alloy pseudoternary to emphasize that the composition variation and disorder pertain only to the cation sublattice. The commonly used terminology is ternary for the former and quaternary for the latter.

929

Mat. Res. Soc. Symp. Proc. Vol. 449 © 1997 Materials Research Society

based materials than in hexagonal based materials, we have started our alloy investigation for the zincblende-based structures. Nevertheless, our results can easily be adapted to wz alloys because the electronic structure in both crystal structures are closely related and the relations between the two are well understood [8]. The wz/zb total energy differences are much smaller than the typical alloy energy differences. Direct calculations for wz alloys to verify these predictions are in progress.

As in previous work,[2, 4] we use the Connolly-Williams method [9] to obtain average properties of disordered alloys,

$$\overline{\Omega}(x) = \sum_n P_n(x, T)\Omega_n, \tag{1}$$

where Ω is the property of interest, e.g. band gap or energy of formation, $\overline{\Omega}(x)$ is its ensemble averaged value for alloy concentration x, Ω_n the contribution from "cluster" n and $P_n(x, T)$ the frequency of occurrence of cluster n in a disordered alloy of concentration x and at temperature T. Specifically, we use the nearest neighbor tetrahedron approximation, in which case $n = 0 \ldots 4$ labels the number of B atoms in the $A_{4-n}B_n$ tetrahedron surrounding each anion C. Because thermodynamic equilibrium is not guaranteed under typical growth conditions for nitrides, we assume truly random probabilities $P_n(x, T) = \binom{4}{n}x^n(1 - x)^{4-n}$ for simplicity. The values associated with each cluster n are obtained by using a similar expansion for known ordered compounds, for which we choose the usual fcc based structures, $L1_0$, $L1_2$ and pure fcc. Some additional ordered structures, such as chalcopyrite and $L1_1$, are calculated to investigate the sensitivity of the properties to particular ordering.

The main difference from our previous work is the way in which we treat bond-length relaxations. In a previous report on $In_xGa_{1-x}N$ by one of the authors [4], a rather high energy of formation was found and it was suggested that deviations from Vegard's law might be important in this system. However, as was cautioned in that work, the structural relaxation was not fully taken into account. In the ideal structure, the volume was relaxed first. The internal parameter, the relative position of nitrogen atom to cations, was then relaxed at the fixed volume. We will show here that a significantly smaller deviation from Vegard's law is found when the minimum energy is calculated with relaxed bond lengths at each volume. This also has important consequences for the band-gap bowings and the energies of formation which determine the miscibility gap. We find, in fact, that the band gap bowing is predominantly determined by the lattice constant since the various calculated results fall on a single curve when the band gap is plotted as a function of lattice constant (see Fig. 1(b)) instead of as a function of concentration. The present work assumes that full relaxation of the second nearest neighbor distance can take place. This may not be completely the case in a disordered alloy because of the residual strain [10] and may lead to a slight underestimate of the present energies of formation and bowing coefficients. Still, this residual strain would result from frustration due only to disorder in the cation lattice and not because of the unrealistic strain induced by keeping nearest neighbor cation-anion bond lengths unrelaxed, which in the previous work [4], yielded a significant overestimate of energies of formation and band-gap bowings.

We use the full-potential linearized muffin-tin orbital method [11] in conjunction with the local density functional theory (LDA) [12] with Hedin-Lundqvist parametrization [13] for all band structure and total energy calculations. Details of the method as applied to the nitrides are given elsewhere [14].

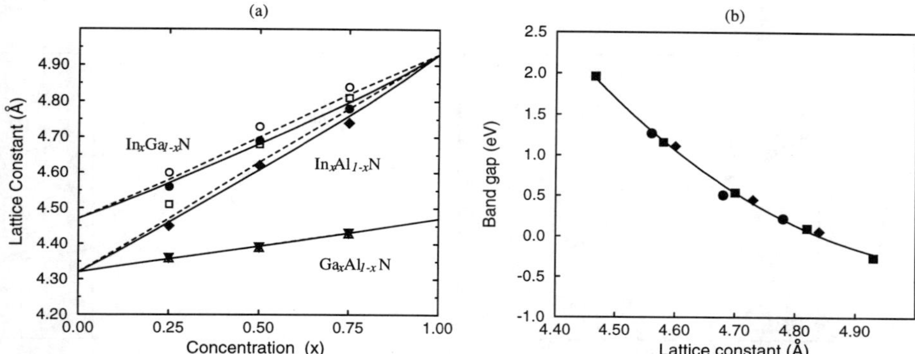

Figure 1: (a) Lattice constants of III-N alloy systems vs. concentration: dashed lines, Vegard's law; open symbols, volume relaxation only; filled symbols, full relaxation of bond lengths; solid lines, random alloy deduced from full relaxation. (b) Band gap (LDA) of $In_xGa_{1-x}N$ ordered compounds as a function of lattice constant: squares, Vegard's law; diamonds, volume relaxation only; circles, full relaxation; solid line, best fit parabola.

RESULTS AND DISCUSSION

Fig. 1(a) shows the lattice constants of the three pseudobinary nitride alloy systems as functions of concentration, once with only volume relaxation and ideal (i.e. equal bond length) structures, and once with full relaxation of the internal parameters before volume relaxation. It can be seen that the second set of results are much closer to Vegard's law, particularly for the $In_xAl_{1-x}N$ alloys where the lattice mismatch is the largest.

Fig. 1(b) shows the band gap as a function of lattice constant for $In_xGa_{1-x}N$ using the two relaxation approaches mentioned above. One can see that all results fall on the same curve, showing that the lattice constant is the primary factor in determining the band-gap bowing.

Fig. 2(a) shows the band gaps of $In_xGa_{1-x}N$ in our final fully relaxed calculation as a function of concentration for the various ordered compounds and the ensemble averaged value for the disordered compound deduced from it. One may note that the value of chalcopyrite lies on the disorder averaged line. This confirms the fact that the relaxations in chalcopyrite and the cation ordering are good representatives of average random behavior. This is the main idea behind the use of so-called special quasirandom structures [15]. We also see that particular ordered structures such as the $L1_1$ may have significantly lower gaps than the average. This can, in this case, be traced back to the occurrence of only A_3B and AB_3 tetrahedra as opposed to the A_2B_2 tetrahedra which predominate in the random alloy.

Since the wz and zb gaps are both direct at Γ in InN and GaN and differ only by a constant of about 0.3 eV in each case, we expect that the zb bowing coefficients should also apply to wz. The nearest neighbor ordering in a $1 + 1$ (0001) superlattice of wz structure is the same as in the $L1_1$ structure. It is thus of interest to note that the calculated gaps for the superlattice, 3.14 eV, 1.26 eV, and 0.51 eV for $AlGaN_2$, $AlInN_2$, and $GaInN_2$ repectively, are indeed close to the values of $L1_1$, 3.04 eV, 1.13 eV, and 0.38 eV after including the constant shift between wz and zb gaps.

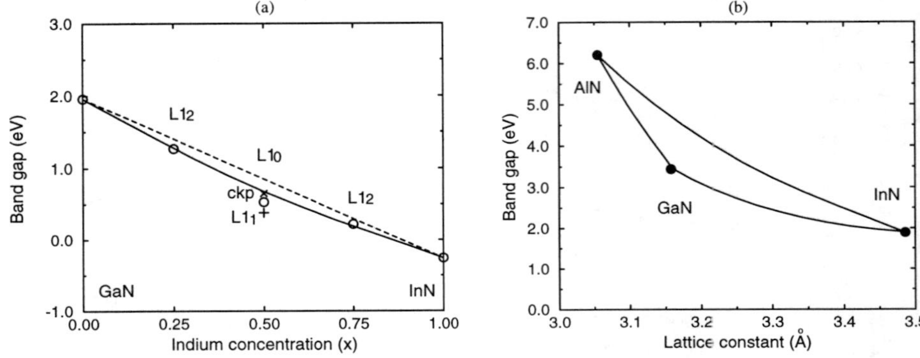

Figure 2: (a) Band gap (LDA) of $In_xGa_{1-x}N$ in ordered compounds and configurational average (solid line). Deviations from the linear variation (shown by the dashed line) correspond to the bowing. (b) Band gap (including estimated self-energy corrections, see text) of wurtzite III-nitride alloys as a function of wurtzite lattice constant.

We note that here we have shown the LDA gaps which are well known to be too small by a self-energy correction of about 1 eV in GaN, 2 eV in AlN and InN. Assuming that the correction varies linearly with concentration, it will not affect the bowing coefficients. In Fig. 2(b) we show the configurationally averaged band gaps of the three pseudobinary wurtzite alloy systems as a function of the wurtzite lattice constant. These were obtained by assuming that the wz gaps differ from the zb direct gaps by a constant shift. Self-energy corrections were linearly interpolated between those for the end point pure compounds, which in turn were chosen so as to yield the experimental gaps. We see that all of the curves can be well approximated by parabolas.

For the full pseudoternary system $Al_xGa_yIn_zN$, we propose the following interpolation scheme for the gap:

$$\begin{aligned} E_g(x, y, z) &= xE_g(\text{AlN}) + yE_g(\text{GaN}) + zE_g(\text{InN}) \\ &\quad - b(\text{AlGaN})xy - b(\text{InGaN})yz - b(\text{InAlN})xz, \end{aligned} \tag{2}$$

with $b(\text{AlGaN})$ the bowing coefficient of $Al_xGa_{1-x}N$, etc. Values for the bowing parameters are given in Table I. To test this equation we have directly calculated the gaps of several pseudoternary compounds. Full structural details on those calculations will be given elsewhere. They are obtained by a substitution in the $L1_2$ structures. The comparison between

Table I: Bowing parameters of of III-Nitrides in eV.

compound	b
AlGaN	0.27
AlInN	1.67
GaInN	0.69

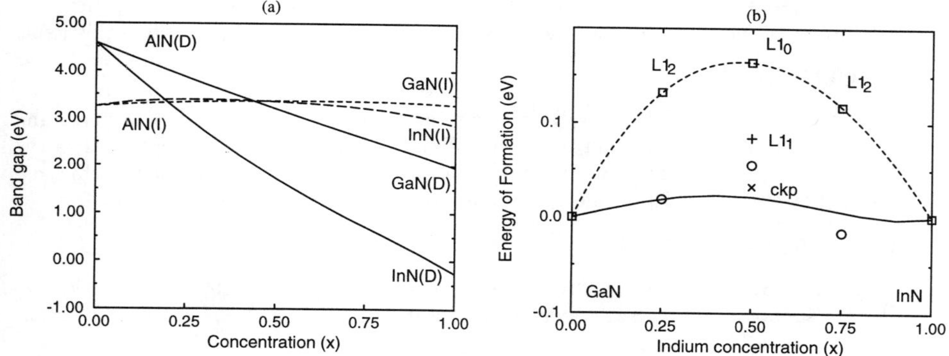

Figure 3: (a) Direct(solid line) and indirect(short dashed and long dashed lines) band gaps in $Ga_xAl_{1-x}N$ and $In_xAl_{1-x}N$ alloys. (b) Energy of formation of $In_xGa_{1-x}N$ alloys: squares, Vegard's law; dashed line, fit; circles, full bond length relaxation; solid line, random alloy deduced from it.

the results predicted by Eq. 2 and the directly calculated values is shown in Table II. The agreement is reasonably good.

Since AlN has an indirect gap at X in zb, it is also of interest to consider the X gaps as functions of concentration in InAlN and GaAlN. These are shown along with the direct gaps in Fig. 3(a). The indirect gaps are seen to vary much less with concentration than do the direct gaps. This is because the p-like states involved in the X-gap are less sensitive to the nuclear charge than are the s-like states involved in the minimum at Γ. The curves indicate that the crossover between the direct and the indirect gaps takes place at $x = 0.44$ and 0.19 for $Ga_xAl_{1-x}N$ and $In_xAl_{1-x}N$ respectively.

Fig. 3(b) shows the energy of formation of $In_xGa_{1-x}N$ as a function of concentration in various relaxation steps. It is evident that the relaxation of bond lengths has a strong and asymmetric effect on the energy of formation. The formation energies at 50 % give a rough indication of the tendency towards phase separation [2] because in a simple regular solution model it is proportional to the miscibility gap temperature T_{MG}. They are 20 meV/atom, 25 meV/atom and 12 meV/atom respectively for $In_xGa_{1-x}N$, $In_xAl_{1-x}N$ and $Ga_xAl_{1-x}N$. While a reliable calculation of T_{MG} would need a more advanced theory in view of the asymmetry of the energy of formation curve, the relative values indicate that T_{MG} for the In-containing systems would be about twice those of the $Al_xGa_{1-x}N$ system.

Table II: Band gaps (LDA) of pseudoternary III-Nitrides in eV.

compound	calculated value	interpolation
$Al_{0.25}Ga_{0.25}In_{0.5}N$	1.05	1.12
$Al_{0.25}Ga_{0.5}In_{0.25}N$	1.64	1.79
$Al_{0.5}Ga_{0.25}In_{0.25}N$	2.14	2.39

This is consistent with the experimental indications that higher growth temperatures are needed to incorporate In into the alloys. Kinetic effects may also play a role in this issue.

CONCLUSIONS

We have presented FP-LMTO results for the three pseudobinary alloy systems and the full pseudoternary $Al_xGa_yIn_zN$ system. Bond-length relaxation strongly affects the equilibrium lattice constant, the value of which is close to the Vegard's law prediction. Band gap bowing coefficients and energies of formation were obtained.

ACKNOWLEDGMENTS

This work was supported by ONR under grant No. N00014-94-1-100. K. Kim acknowledges the support from Korean Ministry of Education.

REFERENCES

1. <u>Gallium Nitride and Related Materials</u>, edited by F.A. Ponce, R.D. Dupuis, S. Nakamura, and J.A. Edmond (Mater. Res. Soc. Proc. **395**, Pittsburgh, PA, 1996)

2. E.A. Albanesi, W.R.L. Lambrecht, and B. Segall, Phys. Rev. B **48**, 17841 (1993).

3. A.F. Wright and J.S. Nelson, Appl. Phys. Lett. **66**, 3051 (1995); ibid., **66**, 3465 (1995).

4. W.R.L. Lambrecht, Solid State Electronics, Special Issue (Topical Workshop on Nitrides) (1996), in press.

5. K. Osamura, S. Naka, and Y. Murakami, J. Appl. Phys. **46**, 3432 (1975).

6. R. Singh and T.D. Moustakas, in [1], p. 163.

7. J.M. Sanchez, F. Ducastelle, and D. Gratias, Physica A **128**, 334 (1984).

8. W. R. L. Lambrecht in <u>Diamond, SiC and Nitride Wide Bandgap Semiconductors</u>, edited by C.H. Carter, Jr., G. Gildenblat, S. Nakamura, and R.J. Nemanich (Mater. Res. Soc. Proc. **339**, Pittsburgh, PA, 1994) p. 565.

9. J.W.D. Connolly and W. R. Williams, Phys. Rev. B **27**, 5169 (1983).

10. C. Amador, W.R.L. Lambrecht, M. van Schilfgaarde, and B. Segall, Phys. Rev. B **47**, 15276 (1993).

11. M. Methfessel, Phys. Rev. B **38**, 1537 (1988).

12. P. Hohenberg and W. Kohn, Phys. Rev. **136**, B864 (1964); W. Kohn and L.J. Sham, ibid. **140**, A1133 (1965).

13. L. Hedin and B.I. Lundqvist, J. Phys. C **4**, 2064 (1971).

14. K. Kim, W.R.L. Lambrecht, and B. Segall, Phys. Rev. B **53**, 16310 (1996).

15. A. Zunger, S.-H. Wei, L.G. Ferreira, and J.E. Bernard, Phys. Rev. Lett. **65**, 353 (1990).

SHALLOW IMPURITY STATES IN WURTZITE AND ZINCBLENDE STRUCTURE GaN

R. Wang*, P.P. Ruden*, J. Kolnik**, I. Oguzman**, and K.F. Brennan**
*Dept. of EE, Univ. of Minnesota, Minneapolis, MN 55455, rpwang@ee.umn.edu
**School of ECE, Georgia Tech., Atlanta, GA 30332

ABSTRACT

Calculations of shallow donor states for wurtzite and zincblende structure GaN, and acceptor states for zincblende GaN in an effective mass approximation are presented. Band parameters were taken from experiment or determined from band structure calculations. The effect of wavevector dependent screening is examined, based on dielectric functions calculated from empirical pseudopotential band structures. For donor states, the effects of electron-phonon coupling and free carrier screening in the Thomas-Fermi and Debye approximations are discussed briefly.

INTRODUCTION

Gallium nitride has become a material of considerable interest for optoelectronic and electronic device applications. The material has been grown epitaxially in both wurtzite and zincblende modifications. Although the wurtzite phase has been much more thoroughly explored, some properties of zincblende structure GaN appear to be very favorable for device applications [1]. The development of GaN materials technology is currently very rapid, in particular in the area of controled n- and p-type doping. Much of the recent theoretical work in this area has concentrated on first-priciples calculations [2]. While these calculations provide considerable insight they are computationally rather intensive. It therefore is desirable, in view of the development of device models, to explore simple phenomenological models specifically suited for the shallow donors and acceptors that are of importance to virtually all device applications. The theory of shallow impurity states in semiconductors has been reviewed thoroughly [3]. Effective mass theory (EMT) establishes the basic framework of this paper. While it has to be borne in mind that the applicabilty of EMT is limited to substitutional impurity levels that are sufficiently shallow, it appears that some insights may indeed be obtained from this simple treatment even in the case of GaN.

MODELS

In effective mass theory for direct bandgap semiconductors with a single, isotropic, spin degenerate valley at the Γ point, the shallow donor states are completely determined by the dispersion of the conduction band near the band edge and by the relevant screening mechanisms. The effective mass Schroedinger equation for the donor system is

$$H\psi = [-\frac{\hbar^2}{2m^*}\nabla^2 + V]\psi = E\psi \tag{1}$$

where m^* is the effective mass at the band edge. The potential, $V(\vec{r})$, represents a screened Coulomb interaction. If the dielectric screening is simply described by a constant, equation (1) is just the hydrogen atom problem, and the solution is well known. However, in general

935

the screening is a function of the wavevector, q, or, equivalently, of the relative coordinates. This is negligible if the donor bound states extend over many lattice constants. For the case of q-dependent screening, we may solve the problem by a variational technique. A further modification of the hydrogenic problem relevant to polar materials concerns the inclusion of the (Fröhlich) electron-phonon interaction. It has been shown [4] for impurity levels that are shallow on the scale of the LO phonon energy, $\hbar\omega$, that in the weak coupling limit the ground state energy, E_0, is enhanced by a factor $(1 + \frac{\alpha_F}{6} + \frac{1}{24}\frac{\alpha_F E_0}{\hbar\omega})$, where α_F is the Fröhlich coupling constant.

The acceptor problem is more challenging because of the more complex band structure at the valence band edge. We here do not consider the case of wurtzite structure material but rather focus on the simpler case of the zincblende structure. For GaN the spin-orbit splitting is very small and the top six valence bands are almost degenerate at the Γ point. The bands are not isotropic, but show considerable warping. Employing a spherical model, Baldereschi and Lipari analyzed two extreme situations, namely strong and weak spin-orbit coupling, and recast the problem in a new EMT frame suitable for the calculation of shallow acceptor states [5]. In their theory the Hamiltonian consists of a spherical term and a cubic correction, and in most semiconductor materials the spherical term dominates and the cubic correction can be treated as a small perturbation. We apply the weak spin-orbit coupling limit of this theory to the case of GaN.

Baldereschi and Lipari introduced two parameters:

$$\mu = \frac{6\gamma_3 + 4\gamma_2}{5\gamma_1}, \qquad \delta = \frac{\gamma_3 - \gamma_2}{\gamma_1} \tag{2}$$

where γ_1, γ_2 and γ_3 are three Luttinger band parameters. The parameters μ and δ measure the strength of the spherical and cubic terms of the dispersion, respectively.

In reference [5], the dielectric screening was treated as a constant. This treatment can produce satisfactory accuracy for very shallow acceptor states, such as those found in Ge, because in that case the average bound state orbital radius is much larger than the lattice constant and the short range interaction is negligible. However, in order to obtain good results for acceptor ground states in most semiconductors, a wavevector or distance dependent dielectric function has to be used. Baldereschi and Lipari later studied Si and Ge with wave vector dependent dielectric functions [6].

In the present paper, we use calculated static dielectric functions, $\epsilon_\infty(\vec{q})$, for GaN as obtained from pseudopotential band structure calculations. The method of determining the form factors and a comparison with other band structures has been presented elsewhere [7]. These "electronic" dielectric functions do not account for the lattice polarizability. Figures 1(a) and (b) show $\epsilon_\infty(\vec{q})$ for wurtzite and zincblende GaN, respectively.

Neglecting the anisotropy, the static dielectric functions can be approximated by an analytic form:

$$\frac{1}{\epsilon_\infty(\vec{q})} = \frac{Aq^2}{q^2 + \alpha^2} + \frac{(1-A)q^2}{q^2 + \beta^2} + \frac{[1/\epsilon_\infty]\gamma^2}{q^2 + \gamma^2} \tag{3}$$

where ϵ_∞ is the optical dielectric constant. The parameters α, β, and γ are all positive and determined by fitting the calculated data. This expression can be easily transformed into a real space, r-dependent dielectric function and consequently the point charge potential,

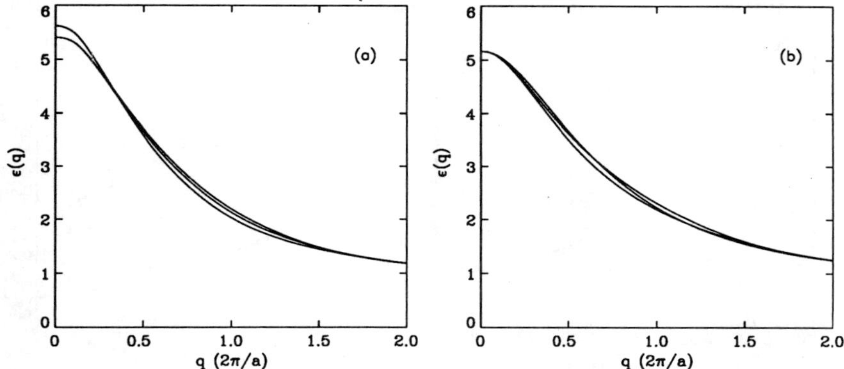

Figure 1: (a) Static dielectric functions, $\epsilon_\infty(\vec{q})$, for $\vec{q}$ along (100), (010) and (001) directions in wurtzite GaN and (b) along the (100), (110) and (111) directions in zincblende GaN. The zincblende GaN lattice constant, a, is the normalization length.

$V(\vec{r})$, is written as:

$$V(\vec{r}) \equiv -\frac{e^2}{\epsilon_\infty r} + V_s(\vec{r}) = -\frac{e^2}{\epsilon_\infty r}[1 + A\epsilon_\infty e^{-\alpha r} + (1 - A)\epsilon_\infty e^{-\beta r} - e^{-\gamma r}] \tag{4}$$

where $V_s(\vec{r})$ is the short-range part of the interaction. For strongly polar materials the lattice polarizability makes a large contribution to the overall dielectric screening. Based on the generalized, q-dependent LST relationship, we could, in principle, calculate the screening function including lattice polarization. To our knowledge however, there is no simple model for calculating the phonon dispersion frequency in strongly polar compounds. Neither does there appear to be enough experimental data on GaN for empirical models. To account for the lattice polarization in GaN, we therefore use the scaling method of Bernholc and Pantelides [8]. With this scaling the boundary conditions are satisfied, i.e. $\epsilon_0(\vec{q}=0)$ equals the static dielectric constant (9.5 for wurtzite GaN, the same value is used for zincblende GaN) and $\epsilon_0(\vec{q}=\infty)$ equals 1. Figure 2(a) shows the static dielectric function.

<u>Shallow Donor Levels for Wurtzite and Zincblende GaN</u>

Recent experiments have yielded a conduction band effective mass value for wurtzite GaN of $0.20m_0$ with negligible anisotropy [9]. For zincblende GaN a value of $0.15m_0$ has been determined [10]. With these effective masses and constant screening, characterized by $\epsilon_0 = 9.5$, we obtain a binding energy for the donor state of 30.1meV for wurtzite GaN, and 22.6meV for zincblende GaN. Using wavevector dependent screening as discussed above the corresponding values are 31.6meV and 23.7meV, i.e. enhancements of less than 5%. Figure 2(b) shows the donor ground state radial wavefunctions for q-dependent and constant screening. We also evaluated the polaron correction discussed above and found that for a coupling constant value $\alpha_F = 0.4$ and an LO phonon energy of 92meV the binding energy is enhanced relative to the hydrogenic value by a factor of 1.07.

More dramatic screening effects are associated with free electrons. To analyze this phe-

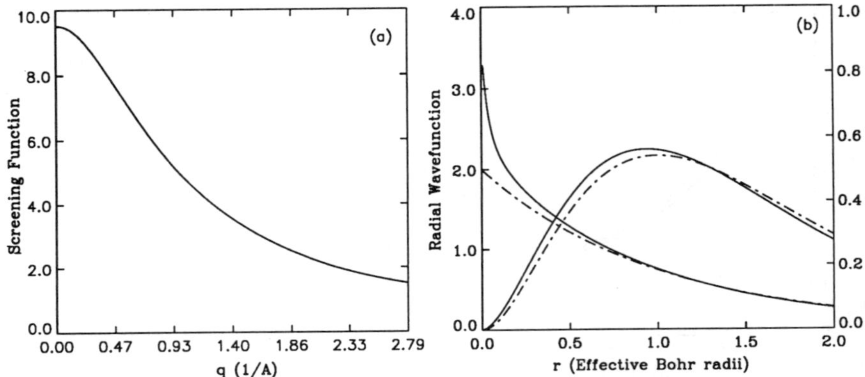

Figure 2: (a) Static dielectric function, $\epsilon_0(\vec{q} = 0)$, of wurtzite GaN; (b) The radial donor envelope wavefunction for constant screening (dashed line) and that for the q-dependent screening (solid line) and the corresponding probability distribution

nomenon we neglect the small effects associated with the q-dependence of the lattice dielectric function and with the electron-phonon interaction and allow for free carrier screening within the Thomas-Fermi approximation. In this framework the Coulomb potential is replaced by:

$$V_s(\vec{r}) = \frac{e^2}{\epsilon_0 r}(1 - e^{-k_{TF}r}) \tag{5}$$

where $k_{TF} = \frac{k_0}{\sqrt{\epsilon_0}}$, and k_0 is taken as either the Thomas Fermi (T=0) or the Debye (T=300K) screening parameter for electrons of concentration, n. Figures 3(a) and 3(b) show the donor binding energies in wurtzite and zincblende GaN, respectively, as a function of the free electron density. From these two diagrams we find that the bound donor states disappear for electron densities near $10^{19} cm^{-3}$. Free carriers affect the shallow impurity donor states severely in a range which covers typical densities of practical devices.

EMT Acceptor States for Zincblende GaN

Because of the current uncertainty of the valence band parameters of zincblende GaN, we used three different sets of Luttinger parameters. One was obtained from our pseudopotential calculation [7], another was published by Meney and O'Reilly [11], based on calculations in the framework of $\vec{k} \cdot \vec{p}$ theory, and the third was derived from first-principles band structure calculations by Kim et al. [12]. In table (1) we list the relevant deduced quantities.

Table (1): Band parameters from [7], [11], and [12] and the corresponding acceptor binding energies in zincblende GaN in the spherical approximation. E_b', constant screening (ϵ_0); E_b, q-dependent screening ($\epsilon_0(q)$).

	γ_1	γ_2	γ_3	$R^*(meV)$	$a^*(\AA)$	μ	δ	$E_b'(meV)$	$E_b(meV)$
Meney et al.	3.06	0.91	1.03	49.3	15.38	0.64	0.04	95	130
Kim et al.	2.40	0.62	0.95	62.8	12.07	0.68	0.14	130	200
Present	2.94	0.89	1.24	51.3	14.78	0.75	0.12	135	350

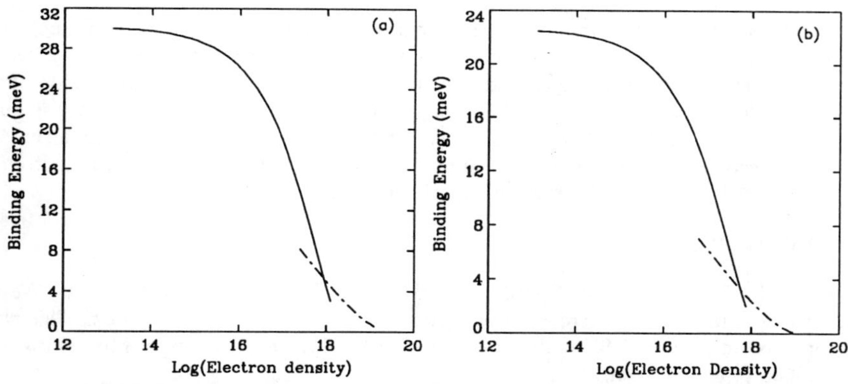

Figure 3: (a) Shallow donor binding energy as a function of electron density for wurtzite GaN. The solid line is for non-degenerate case and dash-dotted line is for the degenerate case; (b) Results for zincblende GaN.

DISCUSSION

It is to be expected that the EMT gives a reasonable description of simple donors in GaN, even in its most basic, hydrogenic form. Corrections due to the wavevector dependence of the dielectric function and those due to the electron-phonon interaction are relatively small. Free carrier screening, on the other hand, may change the donor binding energy significantly. We would expect our simple treatment to be best applicable to isocoric donors. Most experimental work currently focuses on Si donors. However, we find reasonable agreement between our results and, e.g. the data of Götz et al. [13]. In particular, we may attribute the relatively low activation energies measured (12meV...17meV) to the effect of free carrier screening which is significant for samples with the carrier concentrations of those investigated. However, we would expect the donor activation energy to decrease more rapidly with increasing carrier concentration than was observed in the experiments [13]. Our results do not explain the additional deeper level found experimentally [13].

Concerning the calculation of the acceptor level, we note that even the EMT values are relatively deep and the ground state is quite localized, hence raising questions about the fruitfulness of the effective mass concept. The experimental binding energy of Zn (substituting Ga in wurtzite GaN) is about 340 meV that of C (substituting N) about 230meV [14]. The calculated acceptor binding energy is very sensitive to the large q screening. We also find that the binding energy is crucially dependent on the band parameters. Unfortunately, values for these parameters are currently not very well established. In the present work we used three sets of band parameters and find the discrepancies to be remarkably large. We also expect the polaron correction to be relatively large for acceptors, however, a quantitative discussion is beyond the scope of this work. The free carrier screening is of less importance for the acceptor ground state because of its strong localization.

CONCLUSION

We have calculated the binding energies of shallow donors in wurtzite and zincblende GaN in EMT, and estimated corrections associated with wavevector dependent dielectric screening, electron-phonon coupling, and free electron screening. Our brief discussion of the acceptor problem in zincblende GaN indicates that even the EMT acceptor state is quite deep, implying a rather localized ground state. We also note that the EMT in the form used here cannot account for differences in the binding energies of isocoric donors or acceptors on the different lattice sites, as observed in experiment.

ACKNOWLEDGMENTS

RW thanks Profs. Socrates Pantelides and David Fox for helpful comments. We acknowledge the Minnesota Supercomputer Institute (MSI) for technical support. This work was supported in part by NSF under contract ECS-9408479 (University of Minnesota) and ECS-9313635 (Georgia Tech.).

REFERENCES

1. J. Kolnik, I. H. Oguzman, K. F. Brennan, R. Wang, and P. P. Ruden, *J. Appl. Phys.*, **78**, p.1033, (1995).

2. J. Neugebauer and C. v.d. Walle, *Mat. Res. Soc. Symp. Proc.*, **Vol.395**, p.645, (1996).

3. S. T. Pantelides, *Rev. Mod. Phys.*, **50**, p.797, (1978).

4. J. Sak, *Phys. Rev.*, **B3**, p.3356, (1971).

5. A. Baldereschi and N. O. Lipari, *Phys. Rev.*, **B8**, p.2697, (1973).

6. N. O. Lipari and A. Baldereschi, *Solid State Comm.*, **25**, p.665, (1978).

7. R. Wang, P. P. Ruden, J. Kolnik, I. H. Oguzman and K. F. Brennan, *J. Phys. Chem. Sol.*, (in press), 1996.

8. J. Bernholc and S. T. Pantelides, *Phys. Rev.*, **B15**, p.4935, (1977).

9. M. Drechsler, D. M. Hofmann, B. K. Meyers, T. Detchprohm, H. Amano, and Akasaki, *Jpn. J. Appl. Phys.*, **34**, p.1178, (1995).

10. M. Fanciulli, T. Lei, and T. D. Moustakas, *Phys. Rev.*, **B48**, p.15144, (1993).

11. A. T. Meney and E. P. O'Reilly, *Appl. Phys. Lett.*, **67**, p.3013, (1995).

12. K. Kim, W. Lambrecht, and B. Segall, private communication.

13. W. Götz, N. M. Johnson, C. Chen, H. Liu, C. Kuo, and W. Imler, *Appl. Phys. Lett.*, **68**, p.3144, (1996).

14. S. Fischer, C. Wetzel, E. E. Haller, and B. K. Meyer, *Appl. Phys. Lett.*, **67**, p.1298, (1995).

STRUCTURE, ELECTRONIC PROPERTIES AND DEFECTS OF GaN USING A SELF-CONSISTENT MOLECULAR DYNAMICS METHOD

PETRA STUMM AND D. A. DRABOLD*
*Department of Physics and Astronomy, Ohio University, Athens, OH 45701
stumm@helios.phy.ohiou.edu

ABSTRACT

Molecular dynamics simulations are employed to study defects in GaN. We use local basis density functional theory within the local density approximation where charge transfer between the ions is included in an approximate fashion. We find good agreement for the band structure of wurtzite and zincblende GaN compared to other recent calculations, suggesting the suitability of our method to describe GaN. A 96 atom GaN supercell is used to study the relaxations and electronic properties of common defects in the crystal structure, including Ga and N vacancies and antisites. We analyze the electronic signatures of these defects.

INTRODUCTION

III-V nitride semiconducting materials are some of the most promising wide-bandgap materials, because of their suitability for optoelectronic devices. Recent progress in fabricating a blue light-emitting diode based on silicon doped GaN has led to increased interest in understanding its fundamental structural and electronic properties. Some calculations on the band structure and intrinsic defects of wurtzite GaN have been performed with tight binding models [1] and first-principles methods [2, 3, 4]. Only one theoretical study of vacancies in a 32 atom supercell in the zincblende phase is available [2]. The main aim of this paper is to describe the structure and relaxation of intrinsic defects in zincblende GaN in large supercells. We have determined the role of native defects in zincblende GaN and identified their structural and electronic consequences.

METHOD

Our method is based on an approximate first-principles electronic-structure approach, first introduced by Sankey and coworkers in 1989 [5]. Demkov, Sankey, Ortega and Grumbach [6] generalized this non-self-consistent local basis Harris functional LDA scheme to an approximate self-consistent form, "Fireball96".

In this approach, Demkov and coworkers [6] exploited the original idea of the Harris functional which allowed *non neutral* input charge densities. Spherical *charged* atom

Mat. Res. Soc. Symp. Proc. Vol. 449 © 1997 Materials Research Society

densities are used as Harris input charge, and these charges are then self-consistently determined. The method is very efficient, combining the advantages of charge transfer with a fixed atom-centered basis (and therefore efficient look-ups for matrix elements).

Basis functions of four pseudoatomic orbitals per site are used, with a confinement [5] radius of $r_C = 3.8a_B$ and $r_C = 5.4a_B$ for nitrogen and gallium, respectively. Studies comparing the equilibrium lattice constants and structural relaxation around vacancies with and without explicitly treating the Ga 3d states have been performed by others [2]. These calculations show that the main effect of handling the Ga 3d states as part of the core is to make the lattice constant too small. In this work the Ga 3d electrons are treated as core electrons.

APPLICATION TO WURTZITE AND ZINCBLENDE GaN

The electronic band structures for wurtzite and zincblende GaN were computed at the theoretical equilibrium lattice constant (6% smaller than the experimental value, due to taking the Ga 3d states as part of the core) and the energy dispersion along the high-symmetry lines in the first Brillouin zone is shown in Fig. 1 and Fig. 2. The slightly excited four orbital basis set used for these calculations typically overestimates the bandgap by a factor of two.

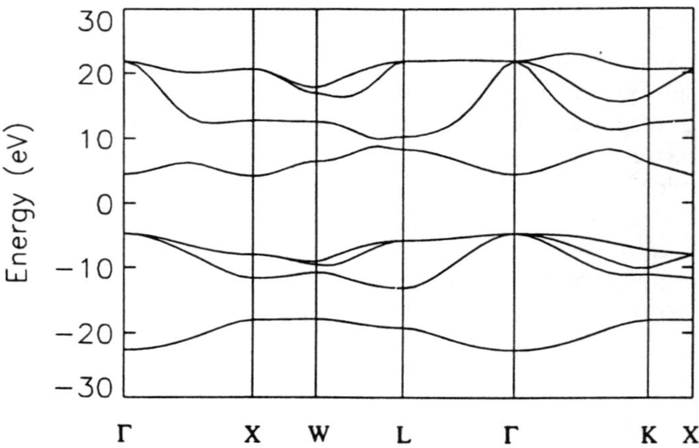

Figure 1: Energy band structure of zincblende GaN.

Comparing these band structures to other *ab initio* calculations by Lambrecht et al [7], we find good agreement between the bands. Even the lower conduction band states are represented rather well.

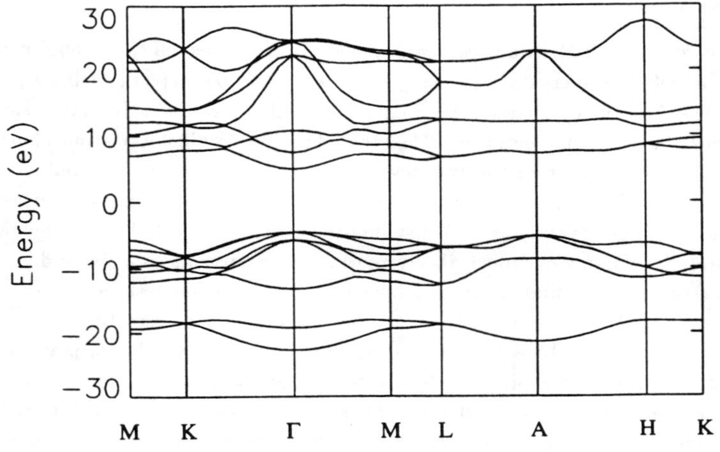

Figure 2: Energy band structure of wurtzite GaN.

Finite Size Effects

The influence of finite size effects on structural relaxations and positions of gap states was investigated. We performed calculations at different k-points for the N vacancy in the zincblende phase using three different cell sizes. To check the dispersion dependency of the k-points on the energy eigenvalues the N vacancy in a 216 atom and a 96 atom cell was investigated using 4 k-points and just the Γ point. Further a N vacancy was relaxed in a 32 atom cell using 14 special k-points. All cells were relaxed for 100 steps.

The structural relaxations in the 216, 96 and 32 atom cells are identical within 0.02 Å, justifying the use of 32 atom cells to describe relaxation of intrinsic defects [2]. There is no variation (less than 0.003 eV) in the 96 atom and 216 atom cells for a given energy eigenvalue around the gap for the four different k-points in the Brillouin zone. Γ point eigenvalues in the 96 atom cell differ by less than 0.25 eV from the eigenvalues at different k-points.

For the 32 atom cell we find a large variation of the energy eigenvalues of a given state for different k-points. Typically energies at different k-points for the same state vary *by about 1 eV* . Further, the energy difference between the N donor state and the conduction band edge (averaged over all 14 k-points) is smaller by 0.5 eV, compared to the two larger cells. This is a sign that a larger cell is needed to completely characterize the electron states. There is no reason to expect this result to be peculiar to a specific method.

While structural changes seem to be described well in the 32 atom cell, our results show that larger cells are needed to adequately describe the electronic structure. Our calculations indicate that a 96 atom cell, using only the Γ point for Brillouin zone sampling, does give the energy eigenvalues without significant k dispersion.

Vacancies in Wurtzite GaN

Calculations for Ga and N vacancies were carried out in a supercell that would contain 96 atoms for the defect free crystal. The cell was relaxed for 100 steps in each case using the Γ point. The N vacancy shows only small structural relaxations. In-plane vacancy nearest neighbors have no change in bond lengths. The inequivalent Ga atom along the z-axis increases its distance from the vacancy by 0.04 Å. A singlet and a doublet state are formed at the conduction band edge. Scaling our calculated bandgap of the perfect crystal to the experimental value, we find that the singlet state is located 0.7 eV below the conduction band. It is singly occupied, therefore the N vacancy acts as a single donor.

For the Ga vacancy we find a larger relaxation. In plane nearest neighbors move to a distance of 2.02 Å, while the distance in the z-direction increases to 2.10 Å. This asymmetrical relaxation can be attributed to the different symmetry for in plane versus out of plane atoms. The N dangling bonds formed by the vacancy exhibit a quasi-triplet state 0.17 eV above the valence band top. This quasi triplet is split by 0.1 eV with strongly localized wave functions. Our results for the Ga vacancy are in good agreement with previous calculations [2, 3].

Vacancies in Zincblende GaN

For each vacancy and antisite calculation in the zincblende structure we used two different supercell sizes, a 32 atom cell and a 96 atom cell. Calculations were performed using 14 k-points in the smaller cell and the Γ point in the 96 atom cell.

Distortion of the lattice around the N vacancy in both cells is limited to nearest neighbors, which move outward by less than 0.05 Å. The N vacancy is a single donor, with the donor state positioned 0.7 eV below the conduction band edge in the 96 atom cell and 0.2 eV (averaged over 14 k-points) below the conduction band edge in the 32 atom cell. The vibrational density of states is nearly identical to that of the defect free crystal except for two additional peaks at 110 and 540 cm^{-1} due to vacancy nearest and next nearest neighbors, respectively. The exact location of these vibrational peaks is probably shifted compared to vibrational spectra from experiments, due to the 6 % lower lattice used in these calculations.

Relaxation effects for the Ga vacancy are more pronounced than for the N vacancy. The Ga vacancy nearest neighbor N atoms move outward by 0.22 Å to a distance of 2.05 Å from the vacancy for both cell sizes. This result compares well to that of Reference [2], who use a plane wave basis set and investigated the Ga vacancy in a 32 atom cell. The Ga vacancy is a triple acceptor and the two highest valence band states are strongly localized on the four 3-fold coordinated N atoms surrounding the vacancy.

Double Vacancy, Gallium and Nitrogen Antisites

A double vacancy in the 96 atom zincblende GaN cell was studied, by removing a neighboring Ga and N atom and relaxing the structure for 100 steps. The three neighboring atoms of each vacancy relax differently than for the isolated vacancies. The nearest neighbor to vacancy distance increases symmetrically to 2.0 Å for the Ga vacancy and 2.17 Å for the N vacancy. An occupied triplet has formed above the valence band top, localized on the three N atoms neighboring the Ga vacancy. The N vacancy single donor state is now unoccupied and is positioned close to midgap, not near the conduction band edge, as for the isolated N vacancy.

We also investigated N and Ga antisites in the 96 atom zincblende cell. The relaxation around N_{Ga} is symmetric as the nearest neighbor N shell relaxes outward by 0.17 Å. The N antisite introduces a doubly occupied singlet 0.5 eV above the valence band top and an unoccupied triplet in the middle of the bandgap. The N antisite is the only defect that leads to a qualitative difference between the electronic structure of the wurtzite and zincblende structure. N_{Ga} in the wurtzite structure introduces a doubly occupied singlet and an empty doublet [3]. This difference is due to the lower symmetry of the wurtzite compared to the zincblende structure, which allows for different structural relaxations and the corresponding electronic structure changes. The vibrational density of states for the N antisite shows three strongly localized peaks associated with the antisite, located at 440, 566 and 840 cm^{-1}.

The largest relaxation effect of the lattice for all single defect investigations in the zincblende structure was observed for the Ga antisite. The Ga-Ga bond lengths increases by 0.33 Å. Three gap states have formed close to the middle of the bandgap, that are all localized on the antisite Ga atom.

CONCLUSIONS

In summary we have reported the first large scale, detailed *ab initio* treatment of vacancies and native defects in zincblende GaN. We find that the electronic consequences of defects in wurtzite and zincblende GaN are similar except for the N antisite, where the lower symmetry of the wurtzite structure leads to a different bonding and electronic environment. Finite size effects, and their influence on structural and electronic properties are addressed. Structural relaxations are described well even in cells as small as 32 atoms, while k dispersion of the energy eigenvalues in these cells introduces a large uncertainty in the placement of defect states.

ACKNOWLEDGEMENTS

This work was supported in part by the Ballistic Missile Defense Organization through the Office of Naval Research Grant No.: N00014-96-1-1183. We thank Dr. Alex Demkov and Prof. Otto Sankey for the use of Fireball96 and for helpful discussions.

REFERENCES

[1] D. W. Jenkins and J. D. Dow, Phys. Rev. B **39**, 3317 (1989).

[2] J. Neugebauer and C. G. Van de Walle, Phys. Rev. B **50**, 8067 (1994); Mat. Res. Soc. Symp. Proc. **339**, 687 (1994).

[3] P. Boguslawski, E. L. Briggs, T. A. White, M. G. Wensell and J. Bernholc, Mat. Res. Soc. Symp. Proc. **339**, 693 (1994).

[4] P. Boguslawski, E. L. Briggs and J. Bernholc, Phys. Rev. B **51**, 17255 (1995).

[5] O.F. Sankey and D.J. Niklewski, Phys. Rev. B **40**, 3979 (1989); O. F. Sankey, D. A. Drabold, G. B. Adams, Bull. Am. Phys. Soc. **36**, 924 (1991).

[6] A. A. Demkov, J. Ortega, O. F. Sankey and M Grumbach, Phys. Rev. B **52**, 1618 (1995).

[7] W. R. L. Lambrecht and B. Segall, in Properties of III-Nitrides, edited by J. H. Edgar, Ed., INSPEC, London, 1994.

A FIRST MODEL OF AMORPHOUS GaN FROM AB INITIO MOLECULAR DYNAMICS

D. A. DRABOLD, PETRA STUMM*

*Department of Physics and Astronomy, Ohio University, Athens, OH 45701

drabold@roma.phy.ohiou.edu

ABSTRACT

A probable byproduct of growth of crystalline GaN is an amorphous phase of the material. In this paper, we propose a structural model of amorphous GaN obtained from *ab initio* molecular dynamics. The radial distribution function, local bonding, electronic density of states and vibrational spectra are described. The network we obtain is highly disordered but exhibits a large state-free optical gap, and has no wrong (homopolar) bonds. These predictions are intended to elucidate the experimental signatures of amorphous GaN.

INTRODUCTION

The dramatic success of GaN as an electronic and optical material has motivated an enormous effort to probe all aspects of the material and seek improved means of growth. In this contribution, we describe a first model of *amorphous* GaN, and characterize it as completely as possible. Our motive is to determine experimental signatures of the amorphous state, which is almost certainly encountered in some methods of growth.

In this paper we report the properties of a first model of a-GaN. The calculation is exploratory, in the sense that we have not yet tried to optimize the density of the model or attempted different simulated annealing/cooling rates. Thus the "details" could differ somewhat from the laboratory material. Nevertheless, this calculation provides new insight into the character of bonding, defects and their potentially measurable experimental signatures. It is of particular interest that our calculation leads to *no wrong (homopolar) bonds*, a remarkable property of the model given the short time scales one is compelled to employ in *ab initio* simulations of any kind.

Mat. Res. Soc. Symp. Proc. Vol. 449 © 1997 Materials Research Society

LOCAL BASIS AB INITIO MOLECULAR DYNAMICS

For a-GaN, we used thousands of time steps of local basis *ab initio* molecular dynamics to form a small (64 atom) model. This would be a *very* challenging calculation in a plane wave approach (see however the recent plane wave *tour de force* of Sarnthein and coworkers for the comparably difficult problem of silica.[1]). We use the methods of Demkov, Sankey, Ortega and Grumbach[2], "Fireball96", who generalized the non-self-consistent local basis Harris functional LDA scheme of Sankey and coworkers[3] to an approximate self-consistent form. In this approach, Demkov and coworkers exploited the original idea of the Harris functional which allowed *non neutral* input charge densities in the language of density functional theory. Spherical *charged* atom densities are used as Harris input fragments, and these charges are self-consistently determined. They could (in principle) be determined from the Harris *maximum* principle, applicable to this class of input fragment densities[2]. The method is very efficient, combining the advantages of charge transfer with a fixed atom-centered basis (and therefore efficient look-ups for matrix elements). The long range Coulomb effects are handled in the conventional way. The basis functions are slightly excited pseudoatomic orbitals with confinement radii[3] of $5.4a_B$ and $4.0a_B$ for Ga and N, respectively.

In zincblende GaN, Fireball96 produces a band structure in rather close agreement with Lambrecht and Segall[4], even in the structure of the conduction states (which is initially surprising with a minimal basis). The value of the gap is *overestimated* in zincblende GaN E_g=6.5 eV, (vs 3.2-3.5eV for experiment). LDA with a complete basis tends to underestimate the gap; the slightly excited sp^3 basis leads to an overestimate. See the work of Stumm elsewhere in this volume for details on the band structure.

From studies of crystalline phases of GaN (reported elsewhere in this volume) we have found that we closely reproduce the essentially exact (within LDA) result of Neugebauer and Van de Walle[5] for intrinsic defect relaxations, when the Ga 3-d states are taken to be part of the core. The Ga 3-d states have a significant effect on the lattice constant and energies, as shown by these workers[5], but cannot be expected to lead to topologically different results in the study reported here. The main effect on our work is to make the lattice constant 6% too small, which probably also introduces a moderate shift in the phonon energies. Accurate *ab initio* simulations on the crystal and certain defects have also been reported by the North Carolina State group[6].

Since the density of a-GaN is unknown, but is probably somewhat less dense than crystalline GaN, we performed our calculations at constant volume (fixed to the experimental volume), which roughly means that we are modeling a form of a-GaN with about 83% of the bulk density. This choice clearly requires further investigation in future modeling work, a first step should involve a study of energy vs. density for thoroughly annealed cells at a range of densities.

CONSTRUCTION OF THE NETWORK

The structural model of this paper was formed by simulated quenching of liquid GaN. We began with a 64 atom (zincblende) cubic supercell of c-GaN, and randomized the atomic positions by performing a 400 fs MD run starting at 10^4K. Inspection of the network showed that it was substantially disordered, and a typical interatomic force was 5 eV/Å, implying that the structure was far from equilibrium. Then, Harris functional-LDA MD was used to slowly cool the network over 7 ps to 300K (a Berendsen[7] thermostat was used). The resulting network was then fully relaxed to equilibrium with the self-consistent method of Ref. [2]. We found that the self-consistent relaxation was essential to obtain reasonable network structure, an unsurprising result for such an ionic material.

PROPERTIES OF THE NETWORK

While the final network has no wrong (N-N or Ga-Ga) bonds, the coordination of N and Ga is variable, with 34 three-fold sites and 30 four-fold sites. Probably there are many different forms of a-GaN possible with varying "sp^3/sp^2" fractions (an imprecise concept here since the bonding is far from ideally covalent) as in a-C. The striking feature of the network topology is that even with the extremely fast quench rates required by *ab initio* molecular dynamics, there are no wrong bonds, a result likely to be hold up in a wide range of densities for this ideal stochiometry. It also suggests that atomic segregation is unlikely except for stochiometries significantly different than 1:1.

We briefly summarize our model in several figures. Fig. 1 is the total pair correlation function g(r) and shows a sharply defined nearest neighbor peak and a rapid loss of interatomic correlations.

As we show in Fig. 2, the electronic density of states shows a Γ point gap of 3.60 eV. When we roughly compensate for the general overestimate of the gap, *this suggests a predicted experimental gap of roughly half the predicted gap or 1.8eV*. It is quite revealing that there are *no midgap states*, as one would certainly find in amorphous Si or a-Ge, for example. By analysis of the electronic eigenvectors, we find that the band tail states (on either side of the Fermi level) are significantly localized, with the valence-tail state mostly

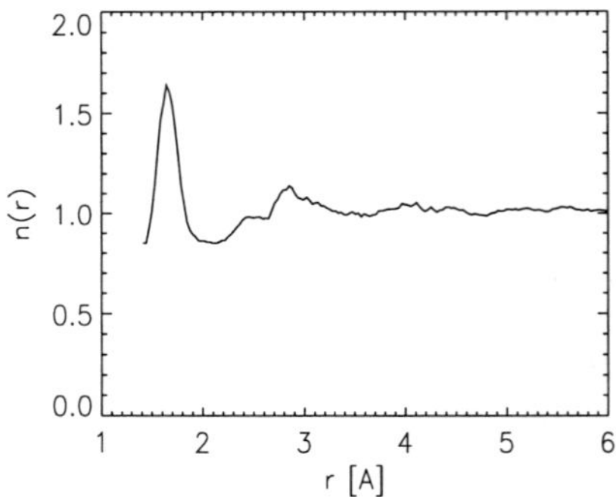

Figure 1: Radial distribution function g(r) for 64 atom model of a-GaN. Note the rapid loss of order after the first neighbor peak.

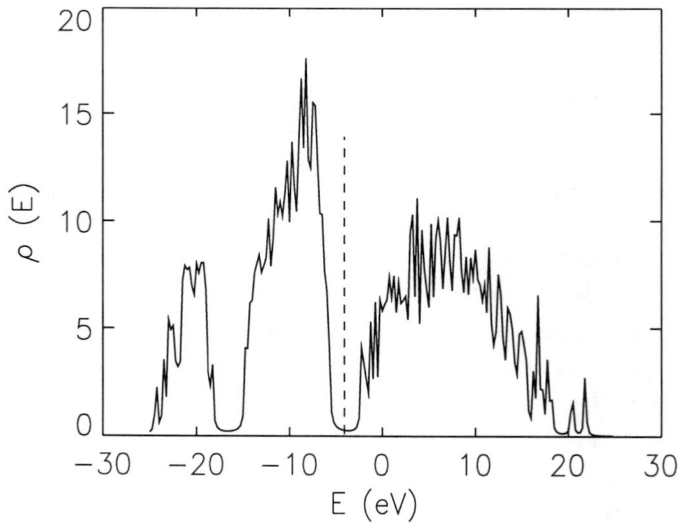

Figure 2: Electronic density of states. The vertical dashed line is the Fermi level. We can roughly predict an experimental gap of 1.8eV in this highly disordered material (see text).

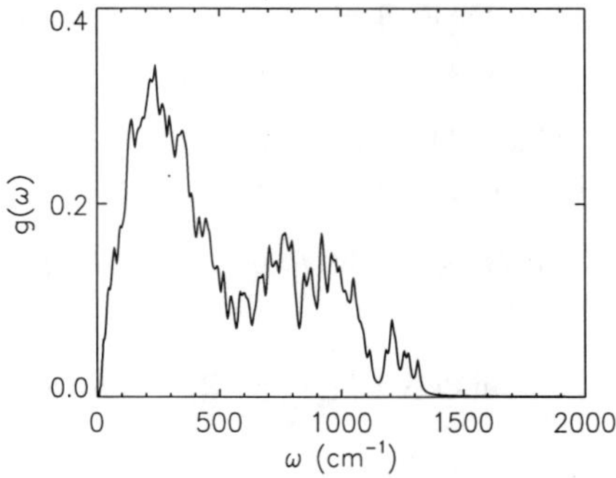

Figure 3: Vibrational power spectrum predicted for a-GaN.

localized on three-fold N sites and the conduction tail states largely localized on three-fold Ga sites. This again bears some interesting resemblance to the case for a-C, where the sp^2 states form π and π^* bands near the Fermi levels (exactly closing the gap in graphite). *It is possible that a-GaN could be a useful electronic material in its own right, as a highly defective model possesses no gap states and therefore no deep carrier traps – without defect passivating H.* This conjecture requires *much* more work to justify however: for example it is likely that the carrier mobility would be quite low, as seen in tetrahedral amorphous carbon (ta-C), for example[8]; this is connected to the localization properties of the band tails.

Finally we present the vibrational density of states, reproduced in Fig. 3. The structure is unremarkable, except for a weakly resolved splitting between acoustic and optic features, a signature of a large degree of topological disorder.

ACKNOWLEDGEMENTS

This work was supported by the Ballistic Missile Defense Organization through the Office of Naval Research Grant No.: N00014-96-1-1183. We thank Dr. Alex Demkov and Prof. Otto Sankey for the use of Fireball96 and for helpful discussions. We also acknowledge helpful discussions with S. Ulloa, M. Kordesch, C. Noguez and D. R. Alfonso.

References

[1] J. Sarnthein, A. Pasquarello and R. Car, Phys. Rev. Lett. **74** 4682 (1996).

[2] A. A. Demkov, J. Ortega, O. F. Sankey and M Grumbach, Phys. Rev. B **52** 1618 (1995).

[3] O.F. Sankey and D.J. Niklewski, Phys. Rev. B **40**, 3979 (1989); O. F. Sankey, D. A. Drabold, G. B. Adams, Bull. Am. Phys. Soc. **36**, 924 (1991).

[4] W. R. L. Lambrecht and B. Segall in *Properties of III-Nitrides*, J. H. Edgar, Ed., INSPEC, London, 1994.

[5] J. Neugebauer and C. G. Van de Walle, Phys. Rev. B **50** 8067 (1994).

[6] P. Boguslawski, E. Briggs, T. A. White, M. G. Wensell and J. Bernholc, Mat. Res. Soc. Symp. Proc. **339** 693 (1994).

[7] H. J. C Berendsen *et al* J. Chem. Phys. **81** 3684 (1984).

[8] D. A. Drabold, P. Stumm and P. A. Fedders, Phys. Rev. B **49** 16415 (1994).

RELAXATIONS AT GAN($10\bar{1}0$) AND (110) SURFACES

Alessio Filippetti, Manuela Menchi and Andrea Bosin

INFM and Dipartimento di Scienze Fisiche, Università di Cagliari, Italy

Giancarlo Cappellini

INFM and Istituto di Fisica, Facoltà di Medicina e Chirurgia, Università di Cagliari, Cagliari, Italy

ABSTRACT

We present an *ab-initio* calculation of GaN wurtzite ($10\bar{1}0$) and zinc-blende (110) surface structures and formation energies. Our method employs ultrasoft pseudopotentials and plane-wave basis. These features enable us to obtain accurate results using small energy cut-off and large supercells. The (110) surface shows a Ga-N surface dimer rotation of $\sim 14°$, i.e. about one half that of the ordinary III-V non-nitride compounds, and a 5% contraction of the surface bond-length (more than the double that occurring in GaAs). For the ($10\bar{1}0$) surface, a layer rotation angle of about $11°$ and a bond-length contraction of 6% has been found. Zinc-blende GaAs (110) and wurtzite ZnO ($10\bar{1}0$) surfaces have been studied as well, for the sake of comparison. GaAs results are in good agreement with the experimental findings. For ZnO a large bond contraction and a rotation angle of around 11% result. Thus, our findings place GaN closer in behaviour to the highly ionic II-VI compounds than to the non-nitride III-V semiconductors.

Introduction

A great effort is devoted today to the study of gallium nitride (GaN) properties, due to its large range of applications in the field of high-temperature electronics and near-ultraviolet electro-optics[1]. GaN is a wide band-gap semiconductor crystallizing at room environment in the wurtzite structure. At room temperature it shows a direct band gap of 3.4 eV. Although the natural structure is wurtzite, GaN thin films having zinc-blende structure have been epitaxially grown on various substrates[2]. For this reasons, detailed studies of GaN surfaces in both wurtzite and zinc-blende structures are now whortwhile. In this paper we report results concerning the structure of non-polar ($10\bar{1}0$) wurtzite and (110) zinc-blend surfaces and their formation energies. Their relaxations, as well as that of all II-VI and III-V compounds, are characterized by the contraction and the rotation (with respect to the surface plane) of the anion-cation planar bonds. The chemical picture of what happens at surface, at least for the non-nitride compounds, is well understood[3]. The symmetry breaking due to the surface formation causes a partial broken bond dehybridization. In case of GaAs (one of the best known III-V semiconductors) the (110) surface atoms present a dehybridized dangling bond (mainly s-like for As and surface normal p-like for Ga) and rehybridized backbonds (sp^2-like for Ga and p-like for As). To this end a dimer rotation (with As moving outwards and Ga inwards with respect to the substrate) is needed. It gives rise to an angle between the ideal surface plane and the plane formed by the surface dimer chains (*tilt angle*). For GaAs the tilt angle is quite large (around $30°$). Also a (small for GaAs) charge transfer from the cation to the anion and a resulting contraction of

953

Mat. Res. Soc. Symp. Proc. Vol. 449 © 1997 Materials Research Society

	GaN(110)		GaAs(110)		
	This work	ref.[4]	This work	ref.[4]	exp.
θ	14.3°	2.06°	30.1°	24.3°	31.1°
ω	7.3°	1.00°	16.5°	13.4°	16.7°
BC	4.9	6.5	0.9	1.3	2

Table I: Layer rotation angle (θ), bond rotation angle (ω) and bond contraction (BC) in percentage of its ideal value for the zinc-blende (110) GaN and GaAs surfaces

bond length occurs. GaAs can be considered as prototype of sp^3-coordinated weakly ionic semiconductor, while ZnO is generally taken as an example of highly ionic structure. Its strong ionicity causes the surface dimers to have a smaller rotation and a considerably larger contraction. Thus, the surface relaxations can be understood as a competition between ionicity (leading to large contraction and small bond rotation) and dehybridization forces (causing strong rotations and weak contractions). It seems clear that surface relaxations faithfully reflect the ionicity degree of the structure. Then, our calculations will address to the question of how GaN can be placed into this ionicity trend. To our knowledge, only one previous theoretical calculation[4] (performed within a quite different method) about GaN (110) surface is present in the literature, whereas two calculations about GaN ($10\bar{1}0$) surface have been reported[4, 5]. At present, no experimental results exist for GaN surfaces. For the sake of completeness, we also determined formation energies and relaxed structures of GaAs (110) and ZnO ($10\bar{1}0$) surfaces. In particular, for the latter, theoretical and experimental results are not as reliable as for the former, therefore a further accurate calculation seemed to be needed for ZnO ($10\bar{1}0$). Our *ab-initio* calculations are performed by means of plane-wave basis and ultrasoft (Vanderbilt)[6] pseudopotentials. Ultrasoft pseudopotentials, a key tool of our calculations, allow to employ a very reasonable cut-off energy (25 Ryd), overcoming the well known difficulties related to the strongly localized nature of nitrogen p electrons. For both Ga and Zn, $3d$ electrons are taken into account as valence states. The theoretically determinated bulk structure parameters result in close agreement with experiments and have been reported elsewhere[7]. Large size cells (8 atomic layers for wurtzite and 9 for zinc-blende surfaces) have been used. Relaxations have been performed by allowing all slab layers to relax within a force threshold of 1 mRyd/a.u., k-point grids have been downfolded by bulk sets to allow an accurate evaluation of formation energies. The paper is organized as follows. We show in section I results for (110) surface relaxations, in section II the relaxations related to the ($10\bar{1}0$) surfaces, and in section III the surface formation energies of all the investigated compounds.

I) Structure of Zinc-Blende GaN(110) Surface

The (110) surface relaxations are characterized by a rotation of the planes formed by the [$1\bar{1}0$]-oriented atomic chains with respect to the surface plane (see Fig.1). Also, a contraction of the dimer bond length occurs. We call layer rotation angle (θ) the angle formed by the (110) plane and the chain plane, and bond rotation angle (ω) that one between the (110) plane and the line connecting surface first neighbours. The bond contraction (BC) corresponds to the first neighbour distance change, in percentage of its ideal value. In Table I we show our calculated values of θ, ω and BC for GaN (in comparison with that obtained by Jaffe et $al.$[4]) and GaAs (compared with ref.[4] and LEED experiments). Our results, on some extent, confirm the anomalous behavior of GaN with respect to GaAs and to the

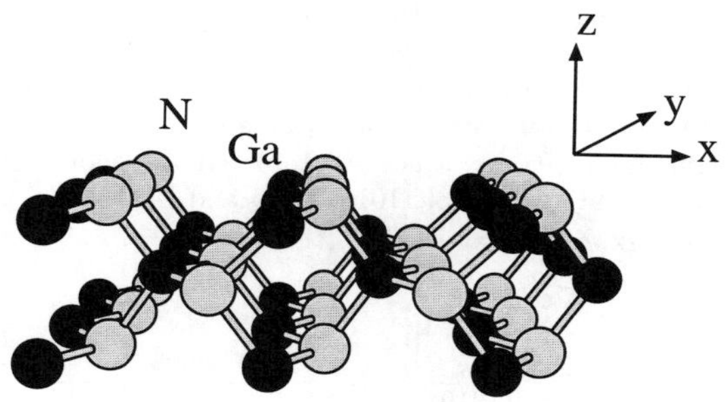

Figure 1: Side view of zinc-blende GaN (110) relaxed surface.

	Δx	Δz
A_1	−0.04	0.05
C_1	0.17	−0.18
A_2	−0.05	0.02
C_2	−0.03	0.07

Table II: Displacements of first layer (A_1 and C_1) and second layer (A_2 and C_2) atoms from their ideal positions (in Å) for the zinc-blende GaN (110) surface. A and C indicate anion and cation, respectively.

other III-V non-nitride compounds (rotation angles of GaN are near half that one of GaAs and the bond contraction is strongly increased), although our values are far away from that of ref.[4] (obtained by means of an Hartree-Fock method). We believe such a discrepancy can be ascribed to the approximations (small size cells and top-layer only relaxations) used in the calculations presented in ref.[4]. On the other hand we tested our scheme in the case of GaAs obtaining very good agreement between our results and experiments. Thus, GaN shows the characteristic of an highly ionic compound, with relaxations similar to that typical of II-VI semicondutors. This features will be better appreciated in next section, where we compare results for the wurtzite ($10\bar{1}0$) surface of GaN and ZnO. Anyway, if for

GaAs the driving force to relaxation is the surface bond dehybridization, for GaN one has to consider the role of ionicity also. Our preliminary results about the other III-V nitrides (i.e. AlN and InN) suggest that this is a common feature of all the nitride compounds.

In Table II the atomic displacements (in Å) with respect to the ideal positions are reported for the first two layer atoms. (see also Fig.1). As expected, the anion goes up, but the largest displacement is that of the first layer cation, moving deeply inside the surface. Also, surface anions and cations come one to the other along x.

II)Structure of Wurtzite GaN($10\bar{1}0$) Surface

As pointed out before, the wurtzite structure is the natural phase in which GaN crystallizes. The ($10\bar{1}0$) surface relaxations consist on a contraction and rotation of the [0001]-oriented first-neighbour bonds at surface. In Table III we report our results for BC and ω (in such a case coincident with θ) of ($10\bar{1}0$) GaN and ZnO surfaces compared to that of other theoretical and experimental works.

	GaN($10\bar{1}0$)		ZnO(($10\bar{1}0$)		
	This work	ref.[5]	This work	ref.[8]	exp.
ω	$11.5°$	$7°$	$11.5°$	$3.6°$	$11.47° \pm 5°$
BC	6.0	6	6.0	8	-0.9

Table III: Bond rotation angle (ω) and bond contraction (BC) in percentage of the ideal value for the ($10\bar{1}0$) GaN and ZnO surfaces.

	GaN				ZnO	
	This work		ref.[5]		This work	
	Δx	Δz	Δx	Δz	Δx	Δz
A_1	0.04	0.08	0.01	0.02	0.02	-0.13
C_1	-0.15	-0.28	-0.11	-0.20	-0.14	-0.50
A_2	0.04	-0.02	0.05	0.05	-0.02	-0.09
C_2	0.05	0.15	0.05	0.05	0.03	-0.09

Table IV: Displacements of first layer (A_1 and C_1) and second layer (A_2 and C_2) atoms from their ideal positions (in Å) for the wurtzite GaN and ZnO ($10\bar{1}0$) surfaces. A and C indicate anion and cation, respectively.

For GaN we obtain larger size relaxations with respect to that of ref.[5] (a recent pseudopotential and plane-wave calculation). However, the bond contraction is in perfect agreement and substantially similar to that we found for the (110) GaN surface, as we should expect. Thus, the picture designing GaN as a strongly ionic compound is clearly confirmed.

In Table IV we point out the atomic displacements of the two first layers with respect to the ideal positions, compared to that of ref.[5] (see Figure 2). The first layer N atoms move sligthly up, the Ga atoms down. The global vertical displacement between surface N and Ga atoms is 0.36 Å, against the 0.22 Å of ref.[5]. Such a difference causes our rotation angle to be around 40% greater. Also for the second layer atoms we found larger displacements than that of ref.[5], with Ga atoms moving considerably outwards.

The main feature resulting from our calculations is the close agreement between ZnO and GaN. If compared with previous *ab-initio* calculations [8, 9], we obtain larger layer

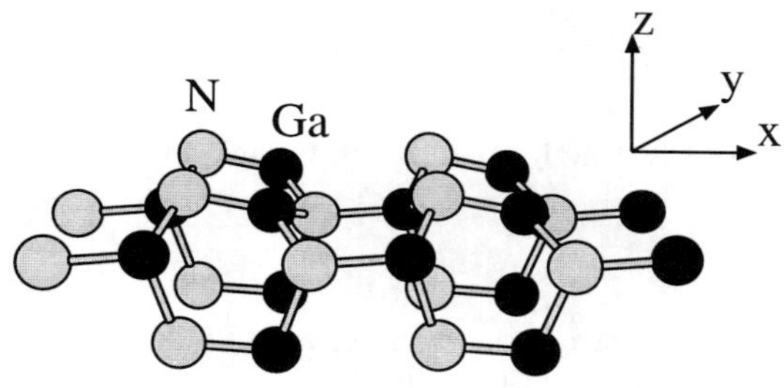

Figure 2: Side view of wurtzite GaN (10$\bar{1}$0) relaxed surface.

rotation and smaller bond contraction. It suggests an interplay of ionic and rehybridization forces in ZnO, and places ZnO and GaN close one to each other. Our value for ω of ZnO is supported by the only given experimental measure[10]. However, it should be noted that the experimental data reported in Table IV (obtained from LEED analysis) suffer of large uncertainties[9]. From our preliminar calculations about AlN and InN surface relaxations, we argue that the whole series of III-V nitrides presents structural features close to that of the highly ionic II-VI compounds. The appealing hypotesis that such a similarity also regards the electronic properties arises. It is well known that the large relaxations occurring in GaAs (110) pull the surface states out from the fundamental band gap, whereas, for ZnO (see ref.[8]) a filled surface state in the gap, nearly unaffected by relaxation, results. Calculations to verify the occurrence of the same feature for GaN surfaces are timely. Indeed a recent surface band study of GaN (10$\bar{1}$0)[5] found the filled surface state slightly below the valence band edge. Some suggestions about the key elements differentiating GaN and GaAs surface relaxations have just been formulated[4]. We can exclude that such a difference resides in the hybridization of Ga $3d$ and N $2s$ states, because, for AlN, relaxations of the same order of magnitude than that of GaN have been found.

III) Surface Formation Energies

In table V formation energies of GaN and GaAs (110), and GaN and ZnO ($10\bar{1}0$) surfaces are reported. σ is the formation energy per atom (i.e. one half the two-atom unit cell formation energy), and refers to the fully relaxed surfaces, whereas $\Delta\sigma$ is the relaxation energy (i.e. the difference between relaxed and ideal surface energy). Our results for σ look equal (to within one percent of eV) to previous calculations for GaN ($10\bar{1}0$)[5] and GaAs (110)[11]. Formation energies can be understood in terms of the energy needed to break

	(110)		($10\bar{1}0$)	
	GaN	GaAs	GaN	ZnO
σ	0.97	0.60	0.99	0.85
$\Delta\sigma$	0.22	0.34	0.39	0.37
$E_{coh}/4$	1.09	0.81	1.09	0.94

Table V: Surface formation energies (σ), relaxation energies ($\Delta\sigma$) and bond cutting energy (i.e. 1/4 the cohesive energy). All results are in eV/atom.

a bond (and, in turn, in terms of cohesive energy). In Tab.V we also show, for all the considered compounds, the bond cutting energies. (i.e. 1/4 the cohesive energies). This are in fair agreement with σ (within 0.1∼0.2 eV, a discrepancy noticeably smaller than the surface relaxation energies), and delineates the same trend: the surface energy increases with the ionic character that clearly enhances the bond strength. Also, notice that, although the GaN (110) and ($10\bar{1}0$) values of σ are nearly the same, a quite larger relaxation energy is found for the latter.

Summary

We calculated structures and formation energies of GaN ($10\bar{1}0$) wurtzite and (110) zinc-blende surfaces. The resulting features pose GaN nearer to II-VI compounds than to the non-nitride III-V semiconductors. The relaxation mechanism comes out to be an interplay between hybridization and ionic forces.

Acknowledgement

We wish to thank Riccardo Valente for providing us with the parallel code. IBM-SP2 computer time was provided by the Center for Advanced Studies, Research and Development in Sardinia (CRS4, Cagliari, Italy).

REFERENCES

1. S. Nakamura, T. Mukai, M. Senoh, Appl. Phys. Lett. **64**, 1687 (1994)

2. *Diamond, Silicon Carbide and Related Wide Bandgap Semiconductors*, edited by J. T. Glass, R. Messier, and N. Fujimori, MRS Symposia Proceedings No. 162 (Materials Research Society, Pittsburgh, Pa, 1990); M.J. Paisley, Z. Sitar, J.B. Posthill, and R.F. Davis, J. Vac. Sci. Technol. A. **7**, 701 (1989); G. Martin, S. Strite, J. Thornton, and H. Morkoc, Appl. Phys. Lett. **58**, 2375 (1991).

3. F. Bechstedt and R. Enderlein, *Semiconductor Surfaces and Interfaces* (Akademie-Verlag, Berlin, 1988)

4. J.E. Jaffe, R. Pandey, P. Zapol, Phys. Rev. B **53**,4209 (1996)

5. J.E. Northrup, J. Neugebauer, Phys. Rev. B **53**, 10477 (1996)

6. K. Laasonen, A. Pasquarello, R. Car, Changyol Lee, D. Vanderbilt, Phys. Rev. B **47**, 10142 (1993)

7. A. Satta, V. Fiorentini, A. Bosin, F. Meloni and D. Vanderbilt, *Gallium Nitride and related compounds* MRS Proceedings Vol.**395**, R.D. Dupuis, J.A. Edmond, F. Ponce, and S. Nakamura, eds. (Material Research Society, Pittsburgh, PA, 1996), p.503

8. P. Schröer, P. Krüger, J. Pollman, Phys. Rev. B **49**, 17092 (1994)

9. J.E. Jaffe, N.M. Harrison, A.C. Hess, Phys. Rev.**49**, 11153 (1994)

10. C. B. Duke, R. J. Meyer, A. Paton and P. Mark, Phys. Rev. B **18**, 4225 (1978)

11. G. X. Qian, R. M. Martin and D. J. Chadi, Phys. Rev. B **37**, 1303 (1988); **38** 7649 (1988).

FIRST MICROSCOPIC OBSERVATION OF CADMIUM-HYDROGEN PAIRS IN GaN

A. BURCHARD [1], M. DEICHER [1], D. FORKEL-WIRTH [2], E.E. HALLER [3], R. MAGERLE [1], A. PROSPERO [1], R. STÖTZLER [1], AND THE ISOLDE-COLLABORATION [2]

[1] Fakultät für Physik, Universität Konstanz, D-78434 Konstanz, Germany
[2] CERN / PPE, CH-1211 Geneva 23, Switzerland
[3] University of California, Dept. of Materials Science, Berkeley CA94270, USA

ABSTRACT

The formation and properties of acceptor-hydrogen pairs in GaN have been studied using radioactive ^{111m}Cd acceptors and the perturbed $\gamma\gamma$ angular correlation spectroscopy (PAC). After H-loading by low energy implantation (100 eV) at temperatures between 295 K and 473 K, the formation of two Cd-H complexes involving about 30% of the Cd-acceptors is observed. The complexes have been identified as single hydrogen atoms bound to the Cd acceptor in two different configurations. The dissociation enthalpies of these configurations have been determined as 1.1(1) eV and 1.8(1) eV, respectively.

INTRODUCTION

The passivation of shallow acceptors, donors and of deep defects in III-V semiconductors by hydrogen is a generally observed phenomenon. Due to the fact, that H is present during almost all processing steps from the fabrication of the III-V compounds up to the processing steps necessary to produce a device, the understanding of the passivating mechanisms and the knowledge on the thermal stability of this passivation is essential. The properties of hydrogen in the "classic" III-V materials like GaAs and InP are widely understood and have been summarized in several reviews [1,2]. Much less is known about the group III nitrides like GaN, which is a very promising material for optoelectronic devices like LEDs and laser diodes. A common procedure to grow GaN layers on a substrate like sapphire or SiC is MOCVD which involves precursors containing hydrogen. As-grown GaN is usually n-type and contains many structural defects due to the lattice mismatch between substrate and GaN layer. P-type conductivity is more difficult to obtain. Experiments reported in the literature for Mg-doped GaN show that Mg acceptors can be activated by low-energy electron irradiation [3] or by thermal annealing [4]. The concentration of electrically active acceptors can be reduced by annealing in a hydrogen atmosphere [4]. These observations lead to the conclusion, that hydrogen is involved in the problems to achieve p-type GaN.

In this paper, we report on the microscopic observation of the formation and the break-up of acceptor-hydrogen pairs formed after low energy (100 eV) hydrogen implantation using radioactive ^{111m}Cd acceptors as probe atoms for perturbed $\gamma\gamma$ angular correlation spectroscopy (PAC) [5].

EXPERIMENT

N-type GaN grown on sapphire has been implanted with ^{111m}Cd ions at the online separator ISOLDE at CERN with an energy of 60 keV and doses of typically 5×10^{11} cm^{-2} corresponding to a Gaussian shaped implantation profile centered at 190 Å below the surface with a peak concentration of 3×10^{17} cm^{-3}. The implantation induced damage has been removed by annealing the samples filled with N_2 for 600 s at temperatures between 1100 K and 1300 K in a closed quartz

Mat. Res. Soc. Symp. Proc. Vol. 449 © 1997 Materials Research Society

ampoule. The implanted and annealed samples were loaded with hydrogen by using a low energy implanter that provides a mass separated H^+ or D^+ beam in the energy range between 100 eV and 1 keV and doses up to 1×10^{16} cm^{-2}. This procedure allows a quantitative control of the introduced amount of H and avoids the contamination by other impurity atoms and, for the case of the 100 eV implantation, the creation of native lattice defects. To identify the effect of hydrogen, additional experiments with low energy implantations of N^+ and Li^+ have been performed.

The immediate neighborhood of the ^{111m}Cd acceptors and their interactions with hydrogen was monitored with the perturbed $\gamma\gamma$ angular correlation spectroscopy (PAC) [5]. Here the electric field gradient tensor (EFG) at the site of the radioactive probe atom is measured. The EFG causes a three-fold hyperfine splitting of an excited state of the ^{111}Cd nuclei created by the γ decay of ^{111m}Cd. This splitting is measured by PAC and is characteristic for the presence of a defect in the immediate neighborhood of the acceptor. The EFG is described by the quadrupole coupling constant $v_Q = eQV_{zz}/h$ (Q denotes the nuclear quadrupole moment and V_{zz} the largest component of the diagonalized EFG tensor) and the asymmetry parameter η. These quantities are unique for specific defects and both, the symmetry of a formed probe-atom-defect complex and the fraction of probe atoms involved in this complex, can be determined from the characteristic modulation of the PAC spectrum R(t) generated by this hyperfine interaction. The short half-life of ^{111m}Cd (49 min) allows to record only two PAC spectra per implanted sample.

RESULTS AND DISCUSSION

Figure 1a shows a PAC spectrum after ^{111m}Cd implantation and annealing at 1100 K without any intentional introduction of hydrogen. The observed reduction of the R(t) signal with increasing time corresponds to a slow modulation caused by an axially symmetric EFG with

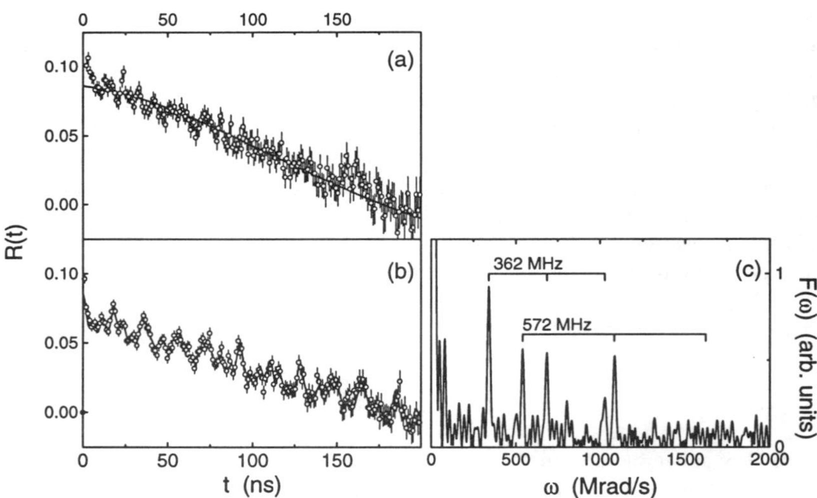

Fig. 1: a) PAC spectrum of a GaN layer implanted with ^{111m}Cd (60 keV, 5×10^{11} cm^{-2}) and annealed at 1100 K; b) PAC spectrum along with its Fourier transform (c) after H implantation (100 eV, 1×10^{15} cm^{-2}) at 323 K.

$\nu_Q = 7.0(5)$ MHz. The same EFG has been observed after implantation of ^{111}In and annealing between 1000 K and 1300 K [6]. [111]In is an impurity isoelectronic to Ga and should occupy Ga sites in GaN and it should have no attractive interaction with other defects present in the lattice. That means the observed EFG after [111m]Cd doping corresponds to the intrinsic EFG at a Ga site created by the hexagonal wurtzite structure of GaN.

In Fig. 1b, a PAC spectrum after H implantation (1×10^{15} cm^{-2}) with 100 eV at 323 K is shown. The details of the observed modulation can be seen in the Fourier transform of the same spectrum (Fig. 1c). The two observed frequency triplets correspond to two defect complexes formed at Cd which both possess axial symmetry ($\eta = 0.0(1)$) and are characterized by $\nu_Q = 360(2)$ MHz and $\nu_Q = 572(2)$ MHz, respectively. The observed amplitudes correspond to about 30% of the Cd acceptors involved in these defects. None of these EFGs have been observed after implantations of Li and N with energies of 100 eV or 1 keV. On the other hands, implanting deuterium creates the same defects. This clearly shows, that H has to be involved in the observed defects. Varying the implanted hydrogen dose between 3×10^{14} cm^{-2} and 1×10^{16} cm^{-2} always creates the same defects and no other complexes are formed. This strongly indicates that the observed defects involve only one hydrogen atom, i.e., electrically neutral Cd-H pairs are formed. In none of our experiments we have been able to decorate more than about 30% of all Cd acceptors with hydrogen. This is in contrast to GaAs and InP where about 80% of the Cd acceptors form Cd-H pair under similar conditions [7]. Because the acceptor-hydrogen pair formation is driven by the Coulombic attraction between the two species, the independence of the observed fraction of Cd acceptors forming Cd-H pairs in GaN from the total amount of hydrogen incorporated favors the conclusion that only about 30% of the implanted Cd acceptors act as trap for H$^+$.

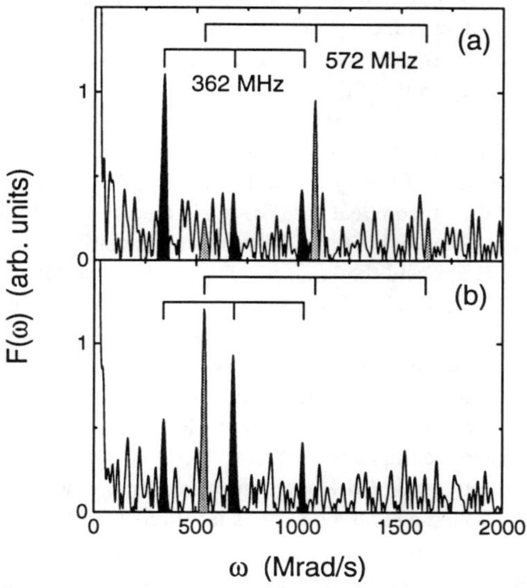

Fig. 2: Fourier transforms of PAC spectra recorded for H loaded GaN with the c-axis of the crystal oriented 45° between two γ-detectors (a) and perpendicular to the detector plane (b).

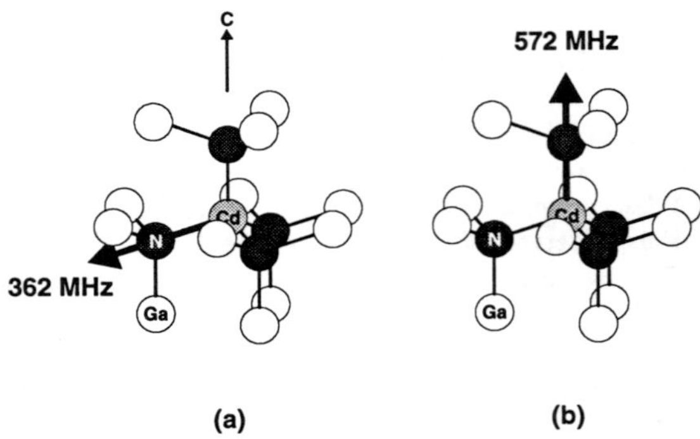

Fig. 3: Symmetry axes of the observed H-related electric field gradients in the hexagonal Wurtzite lattice of GaN.

Figure 2 shows the Fourier transforms of two PAC spectra recorded after H loading where the crystallographic c-axis has been oriented in the detector plane at an angle of 45 degrees between the γ detectors (Fig. 2a) and perpendicular to the detector plane (Fig. 2b). From the amplitude ratios of the frequency triplet characterizing each EFG, the symmetry axis of the respective defect can be extracted. Each of the two observed Cd-H pair configurations possess a different symmetry axis. Fig. 3 illustrates the two possible symmetry axes which are in agreement with the measurements within the angular resolution of the γ detectors (±10 degrees). The hydrogen atom has to be located on these axis near to the Cd acceptor. Due to the lack of a sufficiently precise theory on the strength of defect induced EFGs, the exact position of the hydrogen can not be determined by PAC measurements alone. Both a "bond-center" or an "antibonding" site is compatible with our results. Similar findings have been reported by Brandt et al. [8] who observed by infrared absorption different localized vibrational modes for Mg doped GaN which the authors assigned to two different Mg-H pairs. For the III-V semiconductors not containing nitrogen, the "bond-center" position of the hydrogen is well established, both by experiments and by theory [2]. In contrast to this, theoretical studies by Neugebauer et al. [9,10] for Mg-H pairs in GaN reveal the nitrogen antibonding site as the most stable configuration. The authors attribute this to the strongly ionic nature and the large bond strength of the Ga-N bond. In this work, the hexagonal distortion of the real GaN was not taken into account which may be the explanation for the two different configurations observed experimentally. The same authors [9] also calculated a migration barrier for H$^+$ diffusion in GaN of 0.7 eV. This is in perfect agreement with our experiments which show the formation of Cd-H pairs already at room temperature. That means the hydrogen ions deposited just below the surface by the low energy (100 eV) implantation have to migrate about 200 Å to reach the implanted Cd ions.

To determine the stability of the Cd-H pairs, for each sample two PAC spectra have been recorded: first after hydrogen loading and then after annealing. The observed number of Cd-H pairs observed after annealing has been normalized to the number observed directly after hydrogen loading. Fig. 4 shows this normalized fraction as function of the annealing temperature for both Cd-H pair configurations. The configuration characterized by $\nu_Q = 572$ MHz is only stable up to 323 K and has totally vanished at 373 K. On the assumption, that the dissociation of the pairs occurs by a single jump of the hydrogen and the retrapping probability can be neglected, this corre-

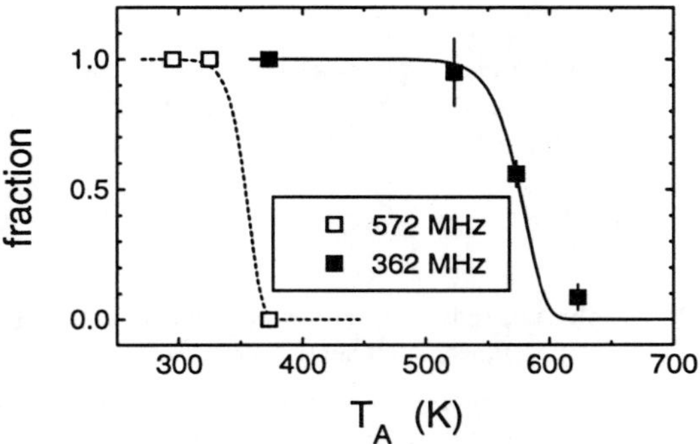

Fig. 4: Normalized fractions of the observed Cd-H pairs in GaN as function of the annealing temperature. The dashed and full lines correspond to a dissociation energy of the Cd-H pairs of 1.1(1) eV and 1.8(1) eV, respectively.

sponds to a dissociation enthalpy of 1.1(1) eV (assuming an attempt frequency of 10^{13} s^{-1}). The second configuration ($v_Q = 362$ MHz) shows a significantly larger stability corresponding to a dissociation enthalpy of 1.8(1) eV. Theoretical calculations by Neugebauer et al. [9] revealed a dissociation enthalpy of 1.5 eV for the Mg-H pair which is very similar to the values for Cd-H pairs reported here and what probably reflects the Coulombic attraction between the negatively charged acceptor and H$^+$. The microscopically observed break-up of Cd-H described here does not implicate a macroscopically reactivation of acceptors observable by electrical measurements or that hydrogen leaves the GaN layer. Nakamura et al. [4] and Pearton et al. [11] have observed that annealing temperatures above 970 K are necessary to reduce the resistivity of Mg doped hydrogenated GaN films. For the case of Ca doped GaN, the observed reactivation of the Ca acceptors after plasma hydrogenation requires annealing at about 720 K [12]. SIMS measurements performed by Zavada et al. [13] at GaN deuterated either by plasma or implantation with 40 keV show no distribution of hydrogen up to annealing temperatures of 1070 K and temperatures of 1170 K are necessary to observe an out-diffusion of the hydrogen. The apparent discrepancy between these findings and our results which are supported by theory shows clearly, that the interaction of hydrogen not only with intentionally introduced acceptors but with other defects present in GaN dominates the electrical properties of the material.

CONCLUSIONS

The formation of two differently configured Cd-H pairs in GaN has been shown on a microscopical scale and the geometrical arrangements of these pairs within the lattice have been proposed. The dissociation enthalpies for Cd-H pairs in GaN have been determined as 1.1(1) eV and 1.8(1) eV, respectively. Both, the formation and the stability of the Cd-H pairs are in good agreement with theoretical calculations.

ACKNOWLEDGEMENT

We acknowledge the Hewlett-Packard Optoelectronics Division for supplying the GaN samples. This work has been supported by the Bundesminister für Bildung, Wissenschaft, Forschung und Technologie under Grant No. 03-RE4KO1-5.

REFERENCES

1. J.I. Pankove and N.M. Johnsons (eds.), *Hydrogen in Semiconductors*, (Semiconductors and Semimetals Vol.34, Academic Press, San Diego, 1991).

2. S.J. Pearton (ed.), *Hydrogen in Compound Semiconductors*, (Materials Science Forum Vol. 148-149, Trans Tech Publications, Aedermannsdorf, 1994).

3. H. Amano, M. Kito, K. Hiramatsu, and I. Akasaki, Jpn. J. Appl. Phys. **28**, L2112 (1989).

4. S. Nakamura, N. Iwasa, M. Senoh, and T. Mukia, Jpn. J. Appl. Phys. **31**, 1258 (1992).

5. Th. Wichert, N. Achtziger, H. Metzner, and R. Sielemann, in: *Hyperfine Interactions of Defects in Semiconductors*, ed. G. Langouche (Elsevier, Amsterdam 1992) p. 77.

6. A. Burchard, R. Magerle and M. Deicher, to be published.

7. M. Deicher and W. Pfeiffer, in Ref. 2, p. 481.

8. M.S. Brandt, J.W. Ager III, W. Götz, N.M. Johnson, J.S. Harris, R.J. Molnar, and T.D. Moustakas, Phys. Rev. B **49**, 14758 (1994).

9. J. Neugebauer and C.G. Van de Walle, Phys. Rev. Lett. **75**, 4452 (1995).

10. J. Neugebauer and C.G. Van de Walle, Appl. Phys. Lett. **68**, 1829 (1996).

11. S.J. Pearton, S. Bendi, K.S. Jones, V. Krishnamoorthy, R.G. Wilson, F. Ren, R.F. Karlicek Jr., and R.A. Stall, Appl. Phys. Lett. **69**, 1879 (1996).

12. J.W. Lee, S.J. Pearton, J.C. Zolper, and R.A. Stall, Appl. Phys. Lett. **68**, 2102 (1996).

13. J.M. Zavada, R.G. Wilson, C.R. Abernathy, and S.J. Pearton, Appl. Phys. Lett. **64**, 2724 (1994).

Part IX

Etching, Hydrogenation and Other Material Processes

CHLORINE-BASED PLASMA ETCHING OF GaN

R. J. Shul,[a] R. D. Briggs,[a] S. J. Pearton,[b] C. B. Vartuli,[b] C. R. Abernathy,[b] J. W. Lee,[b]
C. Constantine,[c] and C. Barratt[c]

[a]Sandia National Laboratories, Albuquerque, NM 87185-0603
[b]University of Florida, Department of Materials Science and Engineering, Gainesville, FL 32611
[c]Plasma-Therm, Inc., St. Petersburg, FL 33716

ABSTRACT

The wide band gap group-III nitride materials continue to generate interest in the semiconductor community with the fabrication of green, blue, and ultraviolet light emitting diodes (LEDs), blue lasers, and high temperature transistors. Realization of more advanced devices requires pattern transfer processes which are well controlled, smooth, highly anisotropic and have etch rates exceeding 0.5 μm/min. The utilization of high-density chlorine-based plasmas including electron cyclotron resonance (ECR) and inductively coupled plasma (ICP) systems has resulted in improved etch quality of the group-III nitrides over more conventional reactive ion etch (RIE) systems.

INTRODUCTION

Interest in GaN and related group-III nitride materials continues to grow as demonstrations of blue, green, and UV LEDs, blue lasers, and high temperature transistors are reported.[1-9] Commercially available LEDs and advances in device fabrication may be attributed to recent progress in material growth technology. Although further improvements in material properties can be expected, enhanced device performance can only be obtained with improved process capabilities including dry etching. Laser facet fabrication is especially dependent upon dry etch pattern transfer since the majority of epitaxially grown group-III nitrides is on sapphire substrates which inhibits cleaving the sample with reasonable yield.

The importance of dry etch development for the group-III nitrides is further accentuated by the fact that they resist etching in standard, room temperature wet chemical etchants. These materials are chemically inert and have strong bond energies as compared to other compound semiconductors. GaN has a bond energy of 8.92 eV/atom, InN 7.72 eV/atom, and AlN 11.52 eV/atom as compared to GaAs which has a bond energy of 6.52 eV/atom. Therefore, essentially all device patterning has been accomplished using dry etching technology. For example, commercially available LEDs from Nichia are fabricated using a Cl_2-based reactive ion etch (RIE)[3] while the first GaN-based laser diode was also fabricated using RIE to form the laser facets.[4] The etched sidewalls were somewhat rough with vertical striations which may have contributed to the scattering loss and high lasing threshold, thus demonstrating the need for improved dry etch processes.

With the recent emphasis placed on the development of dry etch processes for the group-III nitrides, perhaps the most crucial advancement has been the utilization of high-density plasmas. Plasma etching of GaN has been reported using several dry etch techniques including RIE, electron cyclotron resonance (ECR), inductively coupled plasma (ICP), magnetron reactive ion etch (MRIE), and chemically assisted ion beam etching (CAIBE). Using RIE, GaN etch rates as high as 650 Å/min have been reported at dc-biases of -400 V.[10-13] Under similar dc-bias conditions, high-density plasmas typically yield higher etch rates than RIE due to ion densities which are 2 to 4 orders of magnitude greater. Etch profiles also tend to be more anisotropic due to lower process pressures which results in less collisional scattering of the plasma species. GaN etch rates have been reported up to 1.3 μm/min at -150 V dc-bias in an ECR,[14-22] 3500 Å/min at -100 V dc-bias in a MRIE,[23] 2100 Å/min at -500 V in a CAIBE,[24] and 6875 Å/min at -280 V dc-bias in an ICP.[25, 26]

Mat. Res. Soc. Symp. Proc. Vol. 449 © 1997 Materials Research Society

GaN has also been etched using low energy electron enhanced etching (LE4) at ~2500 Å/min in Cl_2 at 100°C and 0 V dc-bias.[27] In this paper, we compare ECR and ICP etch results for GaN in Cl_2- and BCl_3-based plasmas. GaN etch rates obtained in an ECR and ICP will also be compared to RIE results in several plasma chemistries.

EXPERIMENT

The GaN films etched in this study were grown by one of three techniques; metal organic-molecular beam epitaxy (MO-MBE), radio-frequency-MBE (rf-MBE), or metal organic chemical vapor deposition (MOCVD). The MO-MBE GaN films were grown at 925°C on either GaAs or Al_2O_3 substrates in an Intevac Gen II system described previously.[28] The group-III source was triethylgallium and the atomic nitrogen was formed in an ECR Wavemat source operating at 200 W forward power. The rf-MBE GaN film was grown in a commercial MBE system equipped with a conventional Ga effusion cell and a rf plasma source to supply atomic nitrogen during growth. The epitaxial layers were grown on a n^+ GaAs substrate. After deposition of a 5000 Å Sn-doped GaAs buffer layer, growth was interrupted while the substrate temperature was stabilized at 620°C. The GaN epi-layer was comprised primarily of the cubic phase. The MOCVD GaN film was approximately 1.8 μm thick and was grown on a c-plane sapphire substrate in a multiwafer rotating disk reactor at 1040°C with a 20 nm GaN buffer layer grown at 530°C.[29]

The ECR plasma reactor used in this study was a load-locked Plasma-Therm SLR 770 etch system with a low profile Astex 4400 ECR source in which the upper magnet was operated at 165 A. Energetic ion bombardment was provided by superimposing an rf-bias (13.56 MHz) on the sample. Etch gases were introduced through an annular ring into the chamber just below the quartz window. To minimize field divergence and to optimize plasma uniformity and ion density across the chamber, an external secondary collimating magnet was located on the same plane as the sample and was run at 25 A. Plasma uniformity was further enhanced by a series of external permanent rare-earth magnets located between the microwave cavity and the sample.

ICP etching offers an attractive alternative high-density dry etching technique. The general belief is that ICP sources are easier to scale-up than ECR sources and are more economical in terms of cost and power requirements. ICP does not require the electromagnets or waveguiding technology necessary in ECR. Additionally, automatic tuning technology is much more advanced for rf plasmas than for microwave discharges. ICP plasmas are formed in a dielectric vessel encircled by an inductive coil into which rf-power is applied. A strong magnetic field is induced in the center of the chamber which generates a high-density plasma due to the circular region of the electric field that exists concentric to the coil. At low pressures ($\leq$ 10 mTorr), the plasma diffuses from the generation region and drifts to the substrate at relatively low ion energy. Thus, ICP etching is expected to produce low damage while achieving high etch rates. Anisotropic profiles are obtained by superimposing a rf-bias on the sample to independently control ion energy. In order to study ICP, the ECR source and chamber were removed from the SLR 770 etch system and replaced with a Plasma-Therm ICP source. The ICP reactor used in this study was a cylindrical coil configuration with an alumina vessel encircled by a three-turn inductive coil into which 2 MHz rf-power was applied. Identical to the ECR, energetic ion bombardment was provided by superimposing an rf-bias (13.56 MHz) on the sample and etch gases were introduced through an annular region at the top of the chamber. The RIE plasma used in this study was generated by a 13.56 MHz rf-power supply in the ECR chamber configuration discussed above with the ECR source power turned off.

All samples were mounted using vacuum grease on an anodized Al carrier that was clamped to the cathode and cooled with He gas. Samples were patterned using AZ 4330 photoresist. Etch rates were calculated from the depth of etched features measured with a Dektak stylus profilometer after the photoresist was removed with an acetone spray. Each sample was approximately 1 cm^2 and depth measurements were taken at a minimum of three positions. Standard deviation of the etch depth across the sample was nominally less than ± 10% with run-to-run variation less than ± 10%. The gas phase chemistry for several plasmas was studied using a quadrupole mass spectrometer (QMS) or an optical emission spectrometer (OES). Surface morphology, anisotropy, and sidewall undercutting were evaluated with a scanning electron microscope (SEM). The root-mean-square (rms) surface roughness was quantified using a Digital Instruments Dimension 3000 atomic force microscope (AFM) system operating in tapping mode with Si tips. Auger electron spectroscopy (AES) was used to investigate the near-surface stoichiometry of GaN before and after exposure to several plasmas.

RESULTS AND DISCUSSIONS

Chlorine-based plasmas have been the dominant etch chemistries used for GaN and compound semiconductors in general due to the higher volatility of the group-III chlorides as compared to the other halogen-based plasmas. Table I shows possible etch products and their boiling points for GaN etched in a variety of halogen- and hydrocarbon-based plasma chemistries. In a Cl-based plasma, the high volatility of $GaCl_3$ and the nitrogen based etch products implies that the etch rates are not limited by desorption of the etch products. However, due to the strong bond energies of the group-III nitrides, the initial bond breaking of the GaN which must precede the etch product formation may be the rate limiting step.[30] Faster GaN etch rates obtained in high-density plasma etch systems (ECR, ICP, and MRIE) as compared to RIE may be attributed to a two step process directly related to the plasma flux. Initially the high-density plasmas increase the bond breaking mechanism allowing the etch products to form and then produce efficient sputter desorption of the etch products. This can be seen in Figure 1 where GaN etch rates are plotted as a function of rf-power for ECR, ICP, and RIE discharges. Plasma etch conditions were; 1 mTorr pressure, 10 sccm Cl_2, 15 sccm H_2, 3 sccm CH_4, 10 sccm Ar, 1000 W ECR power, and 500 W ICP power. The lowest rf-power required to generate a stable RIE plasma was 50 W, whereas both ECR and ICP plasmas were stable at 1 W rf-power. Independent of etch technique, the GaN etch rate increased as the rf-power or ion energy increased due to improved sputter desorption of the etch products. The ECR and ICP GaN etch rates were approximately 5 to 10 times faster than those obtained in the RIE mode due to higher plasma densities resulting in more efficient GaN bond breaking followed by enhanced sputter desorption of the etch products. The etched surface morphology was evaluated and quantified using AFM as a function of plasma etch technique. The rms roughness measured for samples exposed to the plasmas were normalized to the rms roughness for the as-grown samples since different GaN samples were used in the etch matrix. The normalized rms roughness remained relatively constant and smooth (< ~3 nm) independent of ion energy and etch platform.

Etch Products	Boiling Point (°C)
$GaCl_3$	201
GaF_3	~1000
$GaBr_3$	279
GaI_3	sub 345
$(CH_3)_3Ga$	55.7
NCl_3	<71
NF_3	-129
NBr_3	
NI_3	explodes
NH_3	-33
N_2	-196
$(CH_3)_3N$	-33

Table I. Boiling points of possible GaN etch products for halogen and hydrocarbon based plasma chemistries.

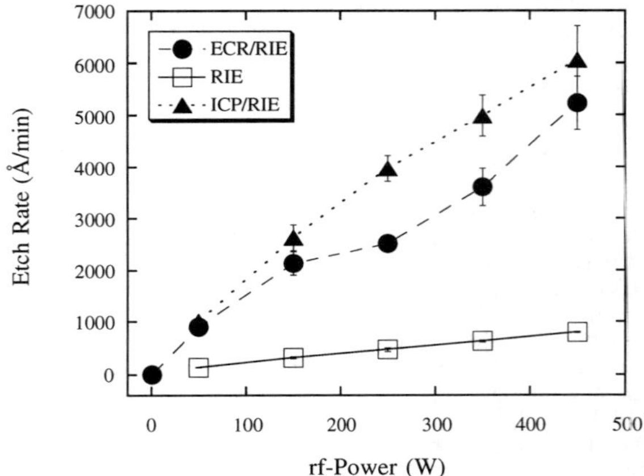

Figure 1. GaN etch rates as a function of rf-power as etched in ECR-, ICP-, and RIE-generated $Cl_2/H_2/CH_4/Ar$ plasmas.

Chlorine-based discharges usually produce fast etch rates and smooth surface morphologies for compound semiconductors whereas CH_4/H_2-based plasma chemistries typical result in smooth etch morphologies at much slower rates. This is unexpected based on the information in Table 1 where the volatility of the $Ga(CH_3)_3$ etch product is much higher than that for $GaCl_3$. This demonstrates the complexity of the etch process where redeposition, polymer formation, or gas-phase kinetics can have a significant impact on the etch rates. An example is shown in Figure 2 where GaN etch rates are plotted as a function of rf-power for Cl_2/Ar and $CH_4/H_2/Ar$ ECR- and RIE-generated

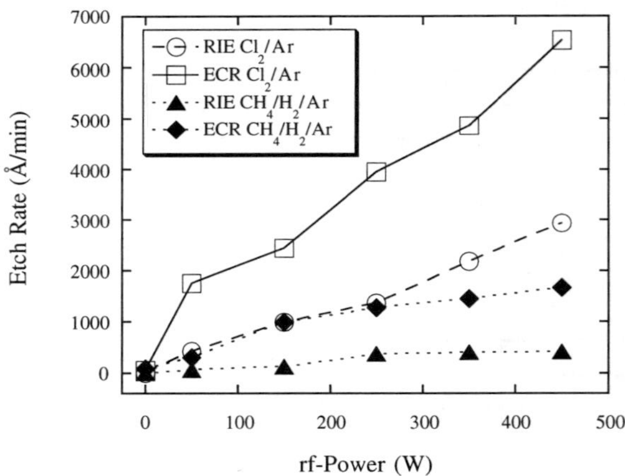

Figure 2. GaN etch rates as a function of rf-Power for ECR- and RIE-generated Cl_2/Ar and $CH_4/H_2/Ar$ plasmas.

plasmas. These experiments were performed in the same chamber with either 1000 or 0 W of applied ECR power. The pressure was held constant at 1.5 mTorr and the plasma chemistry was 5 sccm Cl_2/10 sccm Ar or 5 sccm CH_4/15 sccm H_2/10 sccm Ar. GaN etch rates were significantly faster in Cl_2/Ar possibly due to more efficient sputter desorption of the $GaCl_3$ etch products as compared to $Ga(CH_3)_3$. CH_4/H_2 plasmas may also initiate redeposition or polymer formation on the etched surfaces thereby reducing the etch rates. Faster GaN etch rates in the high-density ECR were attributed to enhanced GaN bond breaking and sputter desorption of the etch products.

Although fast GaN etch rates have been observed in chlorine-based plasmas, the source of reactive Cl as well as the use of additive gases have not been discussed. Unless otherwise noted, the following plasma conditions remained constant: ECR; 1 mTorr, 850 W ECR power, 150 W rf-Power with corresponding dc-biases of -170 to -210 V, and 30 sccm total flow with 5 sccm Ar and ICP; 2 mTorr, 500 W ICP power, 95 to 115 W rf-Power with a constant dc-bias of -250 $\pm$ 10 V, and 30 sccm total flow with 5 sccm Ar. Samples etched in the ECR were grown by MO-MBE whereas samples etched in the ICP were grown by MOCVD. In Figure 3, GaN etch rates are shown as a function of $\%H_2$ concentration for ECR- and ICP-generated Cl_2/H_2/Ar plasmas. GaN etch rates in the ECR and ICP increased slightly as H_2 was initially added to the Cl_2/Ar plasma (10% H_2) implying a reactant limited regime. Using quadrupole mass spectrometry (QMS) in the ECR discharge, the Cl concentration (indicated by m/e = 35 peak intensity) remained relatively constant at 10% H_2. As the H_2 concentration was increased further, the Cl concentration decreased and the HCl concentration increased as the GaN etch rates decreased in both plasmas, presumably due to the consumption of reactive Cl by hydrogen. In Figure 4, the rms roughness is plotted as a function of $\%H_2$ for the ECR and ICP plasmas. The rms roughness for the as-grown GaN samples etched in the ECR was 6.4 $\pm$ 0.5 nm and 3.5 $\pm$ 0.2 nm for samples etched in the ICP. The rms roughness for GaN etched in the ECR increased as the $\%H_2$ increased from 0 to 10 and then decreased as the H_2 concentration was increased further. The roughest surface was observed at 10% H_2 where the etch rate was greatest. The rms-roughness for samples etched in the ICP were less than 1 nm except for the pure Cl_2 plasma where the rms roughness was 41 nm.

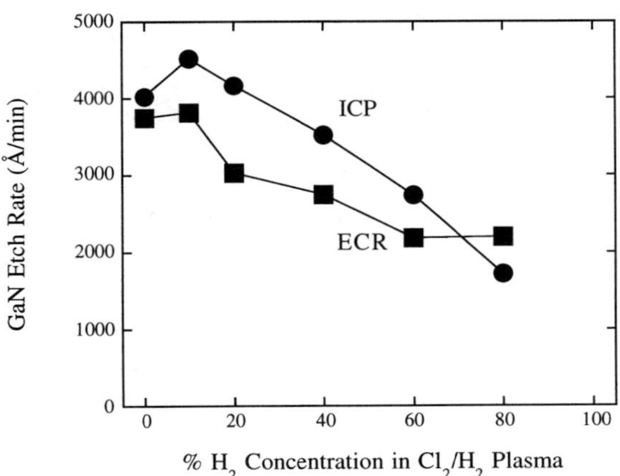

Figure 3. GaN etch rates as a function of $\%H_2$ for ECR- and ICP-generated Cl_2/H_2/Ar plasmas.

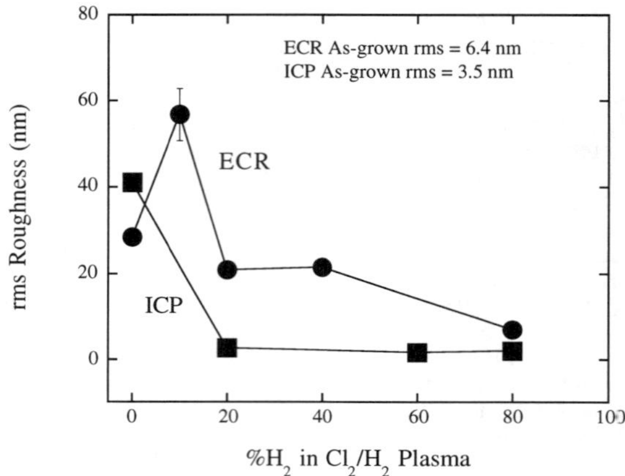

$\%H_2$ in Cl_2/H_2 Plasma

Figure 4. GaN rms-roughness as a function of $\%H_2$ for ECR- and ICP-generated $Cl_2/H_2/Ar$ plasmas.

GaN etch profiles showed a strong dependence on the $\%H_2$ in the $Cl_2/H_2/Ar$ ECR plasma (Figures 5a - c). The etched surface was quite rough (Figure 5a) in the Cl_2/Ar plasma possibly due to preferential removal of the $GaCl_3$ etch products or micromasking of the surface. The foot observed at the edge of the etched feature may be attributed to mask-edge erosion due to the aggressive attack of photoresist by reactive Cl. As the H_2 concentration was increased to 20%, the etch became smooth and very anisotropic (Figure 5b). However, the SEM micrograph showed a lower density of surface roughness near the etched feature than AFM images scanned in open 10 x 10 μm areas. This may be attributed to a proximity effect of the etch where redeposition or micromasking was worse in open areas. At 60% H_2, the etch remained smooth and anisotropic with a slight foot at the base of the feature (Figure 5c). For GaN samples etched in the $Cl_2/H_2/Ar$ ICP-generated plasma (Figures 5d-f), the features were anisotropic and smooth independent of $\%H_2$. Smooth GaN etched surfaces were observed in a pure Cl_2 ICP plasma (Figure 5d) where the rms-roughness was ~ 40 nm which may again be explained by proximity effects.

In Figure 6, BCl_3 was substituted for Cl_2 and was used to etch GaN in both the ECR and ICP reactor. The increase in etch rate observed at 10% H_2 concentration in the ECR-generated BCl_3 plasma correlated with an increase in the reactive Cl concentration as observed by QMS. As the H_2 concentration was increased further, the Cl concentration decreased, the HCl concentration increased, and the GaN etch rates decreased due to the consumption of reactive Cl by hydrogen. In the ICP reactor the GaN etch rates were quite slow and decreased as hydrogen was added to the plasma up to 80% H_2 where a slight increase was observed. The GaN etch rates were consistently faster in the Cl_2-based plasmas as compared to BCl_3 due to the generation of higher concentrations of reactive Cl.

In Figure 7, GaN etch rates are shown for ECR-generated $Cl_2/SF_6/Ar$ and $BCl_3/SF_6/Ar$ plasmas as a function of $\%SF_6$. With the substitution of SF_6 for H_2 in the Cl_2-based plasma, the GaN etch rates were typically a factor of 2 slower. As the concentration of SF_6 was increased, the etch rate decreased up to 30% SF_6 followed by a slight increase at 40%. As the $\%SF_6$ was increased from 0 to 20, the Cl concentration (m/e = 35) decreased but remained significant; faster GaN etching at 20% SF_6 might be expected based on the Cl concentration alone. However, formation of SCl (m/e = 67) was observed at 20% SF_6 which may be responsible for the reduced GaN etch rate due to consumption of the reactive Cl by S. At 30 and 40% SF_6, the Cl

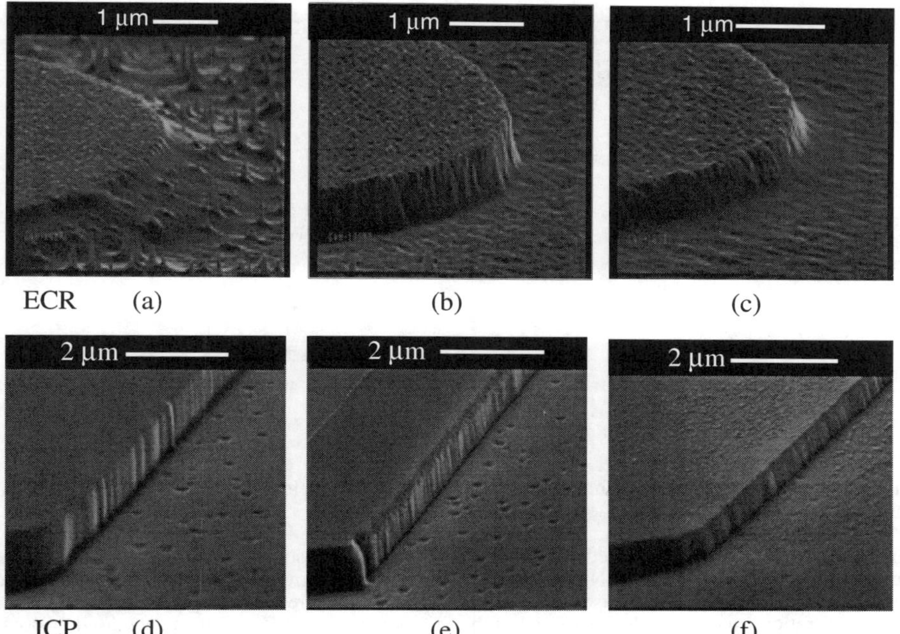

Figure 5. SEM micrographs of GaN samples etched in either an ECR or ICP $Cl_2/H_2/Ar$ plasma at (a, d) 0% H_2 , (b, e) 20% H_2, and (c, f) 60% H_2. The photoresist mask has been removed.

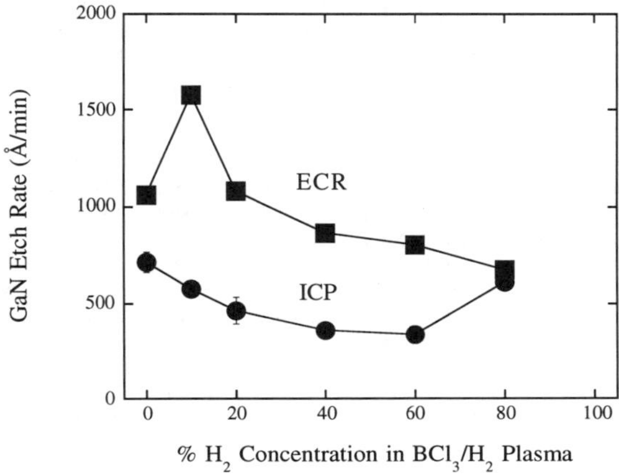

Figure 6. GaN etch rates as a function of %H_2 for ECR- and ICP-generated $BCl_3/H_2/Ar$ plasmas.

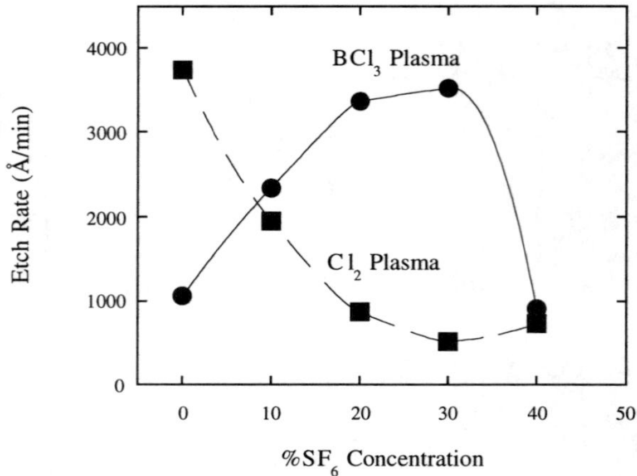

Figure 7. GaN etch rates as a function of %SF$_6$ concentration for Cl$_2$/SF$_6$/Ar and BCl$_3$/SF$_6$/Ar plasmas.

concentration was greatly reduced and slow GaN etch rates resulted. The opposite trend was observed for BCl$_3$, where the GaN etch rates were significantly greater when SF$_6$ was substituted for H$_2$. The GaN etch rate increased up to 30% SF$_6$ and then decreased sharply at 40% SF$_6$. The Cl concentration (m/e = 35) also increased as the SF$_6$ increased to 30% and then decreased at 40%. As with the Cl$_2$-based plasma, there appeared to be a competitive reaction of sulfur with chlorine as the SCl concentration increased above 30% SF$_6$. Under most etch conditions, the trend of the Cl concentration correlated with the trends observed for the GaN etch rates.

Auger spectra were taken to determine the near-surface stoichiometry of GaN following exposure to ECR-generated Cl$_2$/H$_2$/Ar, Cl$_2$/SF$_6$/Ar, BCl$_3$/H$_2$/Ar, and BCl$_3$/SF$_6$/Ar plasmas. The Auger spectrum for the as-grown GaN showed a Ga:N ratio of 1.5 with normal amounts of adventitious carbon and native oxide on the GaN surface. In general, the Ga:N ratio increased as the %H$_2$ or %SF$_6$ concentration increased in either BCl$_3$ or Cl$_2$. These trends implied that the GaN film was being depleted of N perhaps due to preferential chemical etching of the N atoms with the addition of H$_2$ or SF$_6$ to the plasma.

In the ICP reactor, GaN etch rates were obtained for Cl$_2$/N$_2$/Ar and BCl$_3$/N$_2$/Ar plasmas. Figure 8 shows the GaN etch rates as a function of %N$_2$ concentration in both Cl$_2$ and BCl$_3$. As the %N$_2$ concentration increased in the Cl$_2$ plasma, the GaN etch rates decreased due to less available reactive Cl. However, in the BCl$_3$ plasma the GaN etch rates increased significantly up to 40% N$_2$ and then decreased as more N$_2$ was added. This trend was similar to that observed in ECR and ICP etching of GaAs, GaP, and In-containing materials.[31, 32] Ren *et al.* observed peak etch rates for In-containing materials in an ECR plasma at 75% BCl$_3$- 25% N$_2$. As N$_2$ was added to the BCl$_3$ plasma, Ren observed maximum emission intensity for atomic and molecular Cl at 75% BCl$_3$ using OES. Correspondingly, the BCl$_3$ intensity decreased and a BN emission line appeared. It was suggested that at 75% BCl$_3$, N$_2$ enhanced the dissociation of BCl$_3$ resulting in higher concentrations of reactive Cl and Cl ions and higher etch rates. This may explain the faster GaN etch rates observed from 20 to 60% N$_2$ in BCl$_3$/Ar in this study, however higher concentrations of reactive Cl and BN emission were not observed using OES in the ICP.

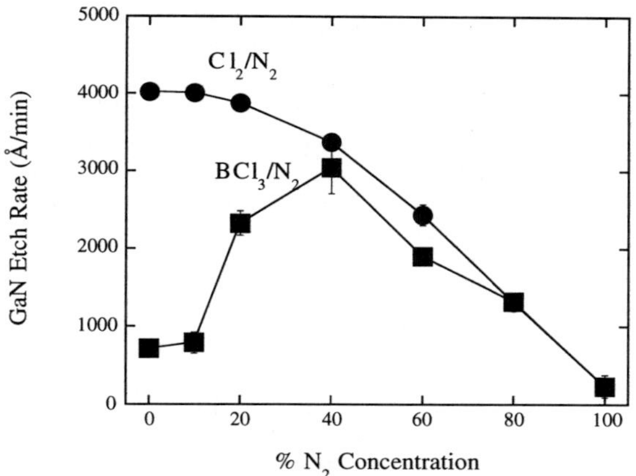

Figure 8. GaN etch rates as a function of %N$_2$ concentration for Cl$_2$/N$_2$/Ar and BCl$_3$/N$_2$/Ar ICP-generated plasmas.

Additional halogen-containing plasmas have been used to etch GaN including ICl/Ar and IBr/Ar. In Figure 9, GaN etch rates are shown as a function of rf-Power in an ECR plasma. Etch conditions were; 1.5 mTorr pressure, 1000 W ECR power, 4 sccm ICl/IBr, and 4 sccm Ar. The GaN etch rates increased with increasing rf-Power (sputter desorption) at similar rates up to 150 W, however at 250 W rf-power, the GaN etch rate in ICl/Ar increased to > 1.3 μm/min whereas the IBr etch rate decreased slightly. This is the fastest etch rate reported to date for GaN. Near-surface Auger electron spectroscopy (AES) showed no loss of N in the Ga:N ratio at low

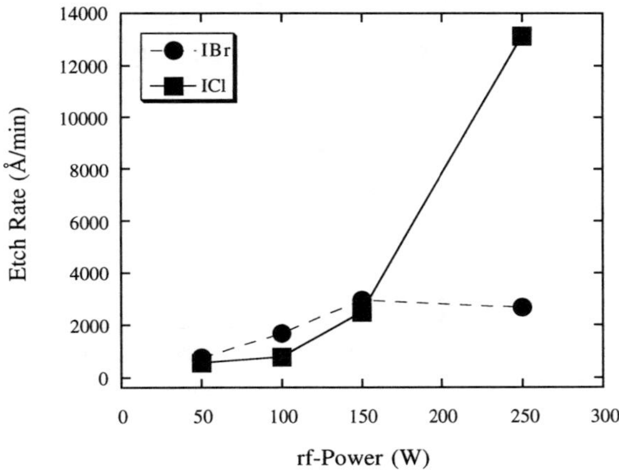

Figure 9. GaN etch rate as a function of rf-Power for ECR- and RIE-generated ICl/Ar and IBr/Ar plasmas.

rf-powers. The high volatility of the GaI_3 etch products may have increased the chemical etch mechanism of the Ga and thereby minimized preferential loss of the lighter N atoms maintaining the stoichiometry of the as-grown film.[14, 15]

Since several GaN samples were used in this study, it is important to identify any etch dependence on growth technique. In Figure 10, GaN ECR etch rates are shown as a function of rf- power for samples grown by MO-MBE, rf-MBE, and MOCVD. The ECR plasma conditions were; 2 mTorr pressure, 22.5 sccm Cl_2, 2.5 sccm H_2, 5 sccm Ar, 30°C electrode temperature, 1000 W microwave power, and rf-powers ranging from 1 to 450 W with a corresponding dc-bias range of -25 to -275 ± 25 V. GaN samples were etched simultaneously. As the rf-power was increased the GaN etch rates increased due to higher ion energies and improved sputter desorption efficiency, independent of growth technique. Etch rates approaching 9000 Å/min were obtained for the MO-MBE and rf-MBE GaN samples at 450 W rf-power (~ -275 V dc-bias). We observed a trend where the MO-MBE GaN etched faster than the rf-MBE GaN which was faster than the MOCVD GaN. Faster etch rates correlated with higher rms roughness for the as-grown GaN samples. The rms-roughness for the as-grown GaN samples were 19.38 ± 0.44 nm for MO-MBE, 3.12 ± 0.84 nm for rf-MBE, and 1.76 ± 0.29 nm for MOCVD.

Figure 11 shows a SEM micrograph of MOCVD grown GaN etched in a $Cl_2/H_2/Ar$ ICP-generated plasma. The GaN was overetched by approximately 15%. The plasma conditions were; 5 mTorr pressure, 500 W ICP power, 22.5 sccm Cl_2, 2.5 sccm H_2, 5 sccm Ar, 25°C electrode temperature, and 150 W rf-power with a corresponding dc-bias of -280 ± 10V. Under these conditions the GaN etch rate was ~6880 Å/min with highly anisotropic, smooth sidewalls. The vertical striations observed in the sidewall were due to striations in the photoresist mask which were transferred into the GaN feature during the etch. The sapphire substrate was exposed during the overetch period and showed significant pitting possibly due to defects in the substrate or growth process. With optimization of the masking process, these etch parameters may yield profiles and sidewall smoothness which improve etched facet laser performance.

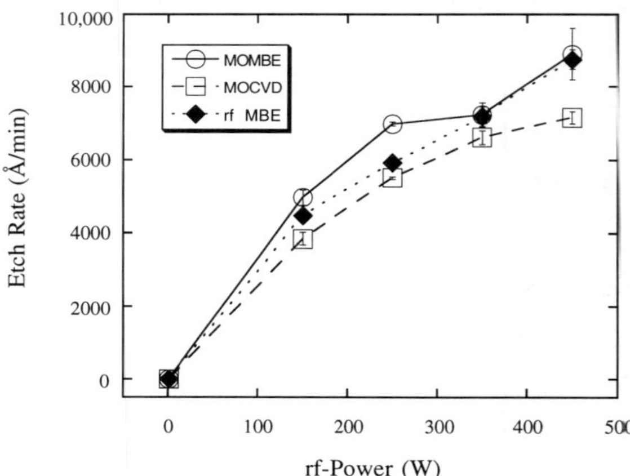

Figure 10. GaN etch rates as a function of rf-Power for MO-MBE, MOCVD, and rf-MBE grown GaN in a $Cl_2/H_2/Ar$ ECR-generated plasma.

Figure 11. SEM micrograph of MOCVD GaN etched in an ICP-generated $Cl_2/H_2/Ar$ plasma.

CONCLUSIONS

In summary, the utilization of high-density ECR and ICP chlorine-based plasmas has resulted in high rate (> 1 μm/min), smooth, anisotropic etching of GaN. The source of reactive Cl (Cl_2, BCl_3, ICl, etc.) and the use of additive gases (H_2, N_2, SF_6) have several effects on the etch characteristics of GaN. Using Cl_2-based plasmas typically resulted in relatively high concentrations of reactive Cl which increased the GaN etch rate. The addition of H_2, N_2, or SF_6 to Cl_2- or BCl_3-based plasmas appeared to effect the chemical removal the N atoms from the GaN as well as the concentration of reactive Cl in the plasma which directly correlated to the GaN etch rate. Very smooth pattern transfer was obtained for a wide range of plasma chemistries and conditions. The mechanism of breaking the GaN bonds appears to be critical and perhaps the rate limiting step in the etch mechanism. The use of high-density plasmas resulted in improved GaN etch results possibly due to a two step process directly related to the plasma flux. Initially the high-density plasmas increase the bond breaking mechanism allowing the etch products to form and then produce efficient sputter desorption of the etch products. ICP etching of the group-III nitrides in Cl_2/H_2 plasmas resulted in etch profiles and sidewall smoothness which may improve the yield and performance of etched facet lasers. Identifying plasma conditions which maintain the stoichiometry of the as-grown films and minimize plasma-induced-damage are critical to the fabrication of high performance group-III nitride devices and must be addressed.

ACKNOWLEDGMENTS

The authors would like to thank P. L. Glarborg and L. Griego for their technical support. This work was performed at Sandia National Laboratories supported by the U.S. Department of Energy under contract # DE-AC04-94AL85000. Sandia is a multiprogram laboratory operated by Sandia Corporation, a Lockheed Martin Company, for the United States Department of Energy.

REFERENCES

1. S. Nakamura, T. Mukai, M. Seno, and N. Iwasu, Jpn. J. Appl. Phys. **31**, L139 (1992).
2. H. Amano, M. Kito, K. Hiramatsu, and I. Akasuki, Jpn. J. Appl. Phys. **28**, L2112 (1989).
3. S. Nakamura, T. Mukai, and M. Senoh, Appl. Phys. Lett. **64**, 1687 (9194).
4. S. Nakamura, M. Senoh, S. Nagahama, N. Iwasa, T. Yamada, T. Matsushito, H. Kiyoku, and U. Sugimoto, Jap. J. Appl. Phys. **35**, L74 (1996).
5. J. C. Zolper, R. J. Shul, A. G. Baca, R. G. Wilson, S. J. Pearton, R. A. Stall, Appl. Phys. Letts. **68**, 2273 (1996).
6. S. Strite and H. Morkoc, J. Vac. Sci. Technol. B **10**, 1237 (1992).
7. M. A. Kahn, J. N. Kuzina, J. M. Van Hove, D. T. Olson, S. Krishnankutty, and R. M. Kolbas, Appl. Phys. Lett. **58**, 526 (1991).
8. T. Matsuoka, T. Sasaki, and A. Katsui, Optoelectronic Devices and Technologies **5**, 53 (1990).
9. M. A. Khan, A. Bhattarai, J. N. Kuznia, and D. T. Olson, Appl. Phys. Lett. **63**, 1214 (1993).
10. I. Adesida, A. Mahajan, E. Andideh, M. Asif Khan, D. T. Olsen, and J. N. Kuznia, Appl. Phys. Lett. **63**, 2777.
11. M. E. Lin, Z. F. Zan, Z. Ma. L. H. Allen, and H. Morkoc, Appl. Phys. Lett. **64**, 887 (1994).
12. A. T. Ping, I. Adesida, M. Asif Khan, and J. N. Kuznia, Electron. Lett. **30**, 1895 (1994).
13. H. Lee, D. B. Oberman, and J. S. Harris, Jr., J. Electron. Mat. **25**, 835 (1996).
14. C. B. Vartuli, S. J. Pearton, J. W. Lee, J. Hong, J. D. MacKenzie, C. R. Abernathy, and R. J. Shul, Appl. Phys. Lett. **69**, 1426 (1996).
15. C. B. Vartuli, S. J. Pearton, J. W. Lee, J. D. MacKenzie, C. R. Abernathy, and R. J. Shul, J. Vac. Sci. and Technol A, in press, (1997).
16. C. B. Vartuli, J. D. MacKenzie, J. W. Lee, C. R. Abernathy, S. J. Pearton, and R. J. Shul, J. Appl. Phys. **80**, 3705 (1996).
17. S. J. Pearton, C. R. Abernathy, F. Ren, J. R. Lothian, P. W. Wisk, A. Katz, and C. Constantine, Semicond. Sci. Technol. **8**, 310 (1993).
18. S. J. Pearton, C. R. Abernathy, and F. Ren, Appl. Phys. Lett. **64**, 2294 (1994).
19. S. J. Pearton, C. R. Abernathy, and F. Ren, Appl. Phys. Lett. **64**, 3643 (1994).
20. R. J. Shul, S. P Kilcoyne, M. Hagerott Crawford, J. E. Parmeter, C. B. Vartuli, C. R. Abernathy, and S. J. Pearton, Appl. Phys. Lett. **66**, 1761 (1995).
21. L. Zhang, J. Ramer, J. Brown, K. Zheng, L. F. Lester, and S. D. Hersee, Appl. Phys. Lett. **68**, 367 (1996).
22. R. J. Shul, A. J. Howard, S. J. Pearton, C. R. Abernathy, C. B. Vartuli, P. A. Barnes, and M. J. Bozack, J. Vac. Sci. Technol. **B13**, 2016 (1995).
23. G. F. McLane, L. Casas, S. J. Pearton, and C. R. Abernathy, Appl. Phys. Lett. **66**, 3328 (1995).
24. I. Adesida, A. T. Ping, C. Youtsey, T. Dow, M. Asif Khan, D. T. Olson, and J. N. Kuzina, Appl. Phys. Lett **65**, 889 (1994).
25. R. J. Shul, G. B. McClellan, S. J. Pearton, C. R. Abernathy, C. Constantine, and C. Barratt, Electron. Lett. **32**, 1408 (1996).
26. R. J. Shul, G. B. McClellan, S. A. Casalnuovo, D. J. Rieger, S. J. Pearton, C. Constantine, C. Barratt, R. F. Karlicek, Jr., C. Tran, and M. Schurman, Appl. Phys. Lett. **69**, 1119 (1996).
27. H. P. Gillis, D. A. Choutov, and K. P. Marlin, JOM, 50 (1996).
28. C. R. Abernathy, J. Vac. Sci. Technol. A **11**, 869 (1993).
29. C. Yuan, T. Salagaj, A. Gurary, P. Zawadzki, C. S. Chern, W. Kroll, R. A. Stall, Y. Li, M. Schurman, C.-Y. Hwang, W. E. Mayo, Y. Lu, S. J. Pearton, S. Krishnankutty, and R. M. Kolbas, J. Electrochem. Soc. **142**, L163 (1995).
30. S. J. Pearton and R. J. Shul, "III-Nitrides", Academic Press, in press.
31. R. J. Shul, G. B. McClellan, R. D. Briggs, D. J. Rieger, S. J. Pearton, C. R. Abernathy, J. W. Lee, C. Constantine, and C. Barratt, J. Vac. Sci. and Technol A, in press, (1996).
32. F. Ren, J. R. Lothian, J. M. Kuo, W. S. Hobson, J. Lopata, J. A. Caballero, S. J. Pearton, and M. W. Cole, J. Vac. Sci. Technol. **B14**, 1 (1995).

ION IMPLANTATION AND ANNEALING STUDIES IN III-V NITRIDES

J. C. ZOLPER,[a] S. J. PEARTON,[b] J. S. WILLIAMS,[c] H. H. TAN,[c]
R. J. KARLICEK,[d] JR., R. A. STALL[d]

[a] Sandia National Laboratories, Albuquerque, NM 87185-0603,
[b] University of Florida, Department of Materials Science and Engineering,
   Gainesville, FL 32611,
[c] Dept. of Electronic Materials Engineering, Australian National University,
   Canberra, 0200, Australia,
[d] Emcore Corp., Somerset, NJ 08873

ABSTRACT

Ion implantation doping and isolation is expected to play an enabling role for the realization of advanced III-Nitride based devices. In fact, implantation has already been used to demonstrate n- and p-type doping of GaN with Si and Mg or Ca, respectively, as well as to fabricate the first GaN junction field effect transistor.[1-4] Although these initial implantation studies demonstrated the feasibility of this technique for the III-Nitride materials, further work is needed to realize its full potential.

After reviewing some of the initial studies in this field, we present new results for improved annealing sequences and defect studies in GaN. First, sputtered AlN is shown by electrical characterization of Schottky and Ohmic contacts to be an effective encapsulant of GaN during the 1100 °C implant activation anneal. The AlN suppresses N-loss from the GaN surface and the formation of a degenerate n^+-surface region that would prohibit Schottky barrier formation after the implant activation anneal. Second, we examine the nature of the defect generation and annealing sequence following implantation using both Rutherford Backscattering (RBS) and Hall characterization. We show that for a Si-dose of 1×10^{16} cm^{-2} 50% electrical donor activation is achieved despite a significant amount of residual implantation-induced damage in the material.

INTRODUCTION

The group III-Nitride semiconductors (GaN, AlN, and InN) have become the focus of intense research following the demonstration of high brightness light emitting diodes based first on GaN p/n junctions and more recently based on InGaN quantum wells.[5,6] In addition III-Nitride based laser diodes have also been reported.[7] The impressive photonic device results where achieved in epitaxial III-nitride material grown on sapphire substrates with a lattice mismatch in excess of 12% and with a resulting defect density on the order of 10^{10} cm^{-2}.[8]

In addition to photonic device demonstrations, these materials are also very attractive for electronic devices that operate at high power or high temperature. Already Heterostructure Field Effect Transistors, Junction Field Effect Transistors, and

Mat. Res. Soc. Symp. Proc. Vol. 449 © 1997 Materials Research Society

Heterojunction Bipolar Transistors have been demonstrated with impressive properties considering the highly defective nature of these materials.[4,9-14] While continued improvements in epitaxial material quality can be expected to yield improved device results, advances are also needed in device processing technology.

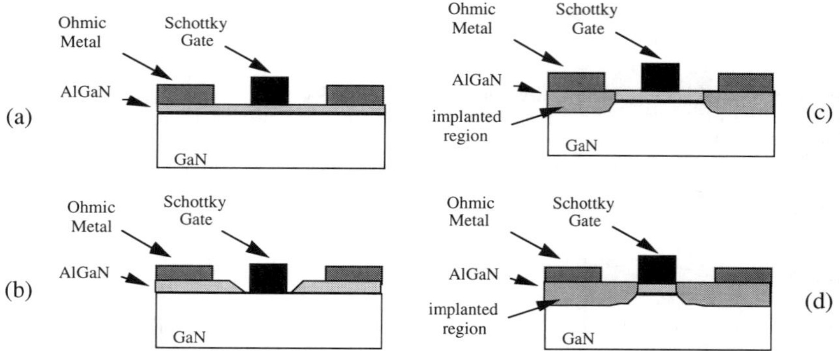

Fig 1. Schematics of typical FET structures a) fully planar with uniform lateral doping, b) recessed gate, c) non-self-aligned, implanted, d) self-aligned, implanted.

One processing technique of particular utility for electronic devices is ion implantation that can be employed for selective area doping or isolation. This is illustrated in Fig 1 with four transistor structures. Fig 1.a shows the fully planar (ignoring possible mesa etching for interdevice isolation) structure that has been employed for the majority of AlGaN/GaN heterostructure field effect transistors (HFETs) reported to date.[9-13] This structure suffers from a high access resistance since the doping is uniform in both the channel and ohmic regions and must therefore be a compromise between the two. An improved structure is the recessed gate approach shown in Fig 1.b that is widely used in other compound semiconductor materials.[15] The recess gate design reduces the access resistance by having a highly doped capping layer that is removed in the gate region, typically by a wet etch process. For the III-Nitride materials, however, no appropriate wet etch chemistry has been identified thus making the use of this structure problematic. Although some success has been reported with dry recess etch processes, control of the plasma damage becomes critical and has limited this approach so far.[16,17] Figure 1c show an alternative approach to realizing higher doping in the ohmic contact regions without impacting the channel. In this case, the contact regions are defined by a photolithographic process and selectively implanted with the appropriate doping species. In this way, no etching of the semiconductor is required. This approach will require the GaN surface to maintain its stiochiometry during the implant activation anneal so that a Schottky contact can be formed for the gate electrode after the activation anneal. Finally, Fig 1d shows a more optimum approach with the source and drain implants self-aligned to the gate Schottky contact metal. This final structure will require the Schottky metal to be in place during the implant activation anneal and to maintain its rectifying properties following the thermal processing. The development of such a high temperature metallization scheme will be challenging due to the high implant activation annealing temperature (~1100 °C) required for GaN.

In this paper, we review the progress in ion implantation doping of GaN such as would be needed to realize the FET structures of Fig 1c,d. This includes the successful demonstration of n-type doping with Si and p-type doping with Mg and Ca. We address the implant activation annealing sequence and in particular the use of an AlN encapsulation layer to maintain the surface stoichiometry to allow Schottky contact formation. The nature of implantation-induced damage will then be discussed. Finally, we outline areas which need further work if ion implantation is to become a viable technology for application to the group III-Nitride materials.

IMPLANTATION DOPING OF GaN

Implantation has been used to achieve n-type doping in GaN with both Si and O.[1,2] For Si-implantation in Ref 1, although the unimplanted sample showed a reduction in sheet resistance that may be due to the formation of N-vacancies or the depassivation of unintentional impurities, the Si-implanted samples demonstrated a 3 order-of-magnitude decrease in resistance after annealing out to 1100 °C where the Si is becoming an active donor. This result was the first to demonstrate that the activation temperature for implanted dopants in GaN was > 1000 °C. The implant activation temperature of other common compound semiconductors are summarized in Table I and compared to the melting point of the material. For Si, GaSb, InP, and GaAs the implant activation temperature is seen to be roughly 70% for the melting point while for SiC and GaN it is closer to 50%. This suggests that the optimum implant activation temperature for GaN may be closer to 1700 °C. This hypothesis is support by damage studies presented later in this paper.

Table I
Comparison of the melting point and implant activation temperature for the materials listed.

material	T_{mp} (°C)	T_{act} (°C)	T_{act} (°C)/T_{mp} (°C)
Si	1415	950	0.67
GaSb	707	550	0.77
InP	1057	750	0.71
GaAs	1237	850	0.69
SiC	2797	1300-1600	0.46-0.57
GaN	2518	1100	0.44

The nature of the Si-activation process in GaN is thought to be dominated by a substitutional diffusion process of the host elements with a activation energy of 6.7 eV.[3] This means that the Si-ions are substitutional on Ga-sites at relatively low temperatures but only become electrically active when the anti-site defects (N_{Ga}, Ga_N) are annihilated.

A primary limitation of the III-Nitride materials, and all wide bandgap semiconductors in general, has been the high ionization energy of acceptor species which significantly limits the number of ionized free holes at room temperature. To date, the shallowest acceptor in GaN is Mg with an ionization energy level of 150 to 170 meV.[18] Therefore, the search for an alternative acceptor species that has a lower ionization energy is of particular interest. In the regard, implantation has the flexibility to introduce a wide array of elements into the semiconductor lattice to study their properties. Once p-type implantation doping of GaN was demonstrated with Mg with the use of a P co-implantation scheme,[1] other potential acceptor species were studied. The first example of this was Ca-implant doping of GaN where Ca was shown to be an acceptor in GaN. Ca was studied since it had been suggested to be a shallow acceptor in GaN.[19] Both Ca-only and Ca+P samples were examined with both seen to convert from n-type to p-type after an 1100 °C anneal. From an Arrhenius plot of the sheet hole concentration for the Ca-implanted samples annealed at 1150 °C, the ionization energy of Ca in GaN is estimated to be 169 meV or equivalent to that of Mg.[1] Further work to use implantation to identify alternative acceptor species for GaN is on going. To date, two other attractive elements, Be and C, have been investigated by implantation without p-type conductivity being achieved either with or with a co-implantation scheme. It may be that these lighter elements will require a different ratio of co-implant ion to acceptor ion dose than that used for Mg and Ca of 1:1. Alternatively, there may be other defect or impurity interactions, such as complex formation (Be:O in Refs. 20,21 or C:N in Ref. 22), that compensate these elements but do not occur for implanted Mg or Ca. This areas of alternative dopant species requires further study.

AlN ENCAPSULATION DURING ANNEALING

As discussed above, to realize the FET structure shown in Fig. 1c, the surface stiochiometry of GaN must be maintained during the high temperature activation anneal. To address this problem in other compound semiconductors encapsulation of the surface with a dielectric film is often employed.[23] For this approach to be effective for GaN, the encapsulating film must maintain its integrity (i.e. not crack or blister) and suppress out-diffusion at a temperature of ~1100 °C. Although SiO_2 and SiN are typically used for encapsulating other compound semiconductors, they are not viable at these high temperatures due to cracking or hydrogen desorption.[24] AlN, on the other hand, has excellent thermal and mechanical stability even at these high temperatures and can be readily deposited by sputtering Al in a nitrogen ambient. Therefore, sputtered AlN was investigated as an annealing cap for GaN.[25]

Two sets of GaN samples were processed into Pt/Au Schottky contacts following a 1100 °C, 15 s rapid thermal anneal. One set of samples (A-samples) was n-type as-grown (n ~ 1×10^{17} cm^{-3}) while the second set (B-samples) were initially semi-insulating and was implanted with Si-ions at an energy of 100 keV and a dose of 5×10^{13} cm^{-2} to simulate a MESFET channel implant. One sample from each set was encapsulated with 120 nm of reactivity sputtered AlN prior to annealing. Following annealing, the AlN was removed in a selective KOH-based etch (AZ400K developer) at 60-70 °C.[26] This etch has been shown to etch AlN at rates of 60 to 10,000 Å/min, depending on the film quality, while under the same conditions no measurable etching of GaN was observed.[27] The etch rate of the sputter deposited AlN film was found to increase as a function of post-deposition annealing temperature.[28] Ti/Al ohmic contacts were deposited and defined by conventional lift-off techniques on all samples and annealed at 500 °C for 15 s. Pt/Au Schottky contacts were deposited and defined by lift-off within a circular opening in the ohmic metal. Electrical characterization was performed on a HP4145 at room temperature on 48 μm diameter diodes. Samples prepared in the same way, except without any metallization, were analyzed with Auger Electron Spectrometry (AES) surface and depth profiles. The surface morphology was also characterized by atomic force microscopy (AFM) before and after annealing.

AlN ENCAPSULATION: RESULTS AND DISCUSSION

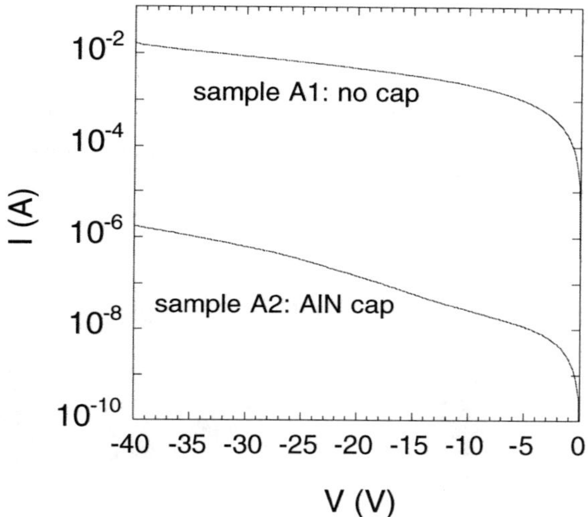

Fig 2. Reverse current/voltage characteristics for 48 μm diameter Pt/Au Schottky contacts on GaN annealed at 1100 °C, 15 s with (sample A2) and without (sample A1) an AlN cap.

Figure 2 shows the reverse current/voltage characteristics of 48 μm diameter Pt/Au Schottky diodes on the initially n-type samples. The sample annealed without the AlN cap has over 3 orders-of-magnitude higher reverse leakage current than the capped sample. The implanted samples showed a similar offset in the reverse leakage current between the capped and uncapped samples. The reason for the difference in reverse bias characteristics is discussed below.

Figure 3 shows the current/voltage characteristics between two adjacent Ti/Al ohmic contacts on the same two samples as in Fig 2. Here the sample annealed without the AlN cap displays dramatically lower resistance than the AlN-capped sample. This is consistent with the Schottky characteristics and suggests the creation of an n$^+$-surface layer formed on the sample annealed without the AlN encapsulant.

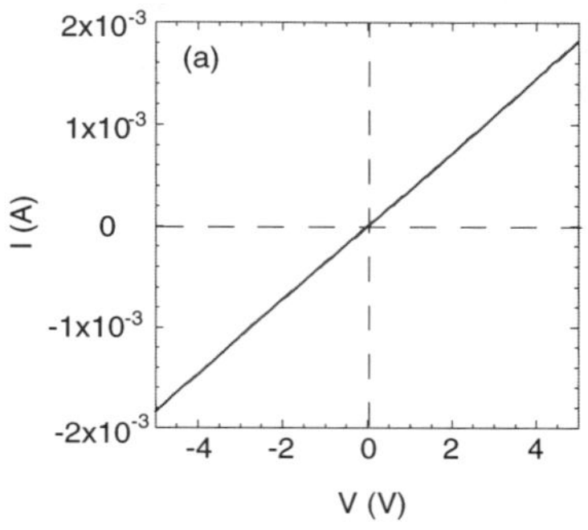 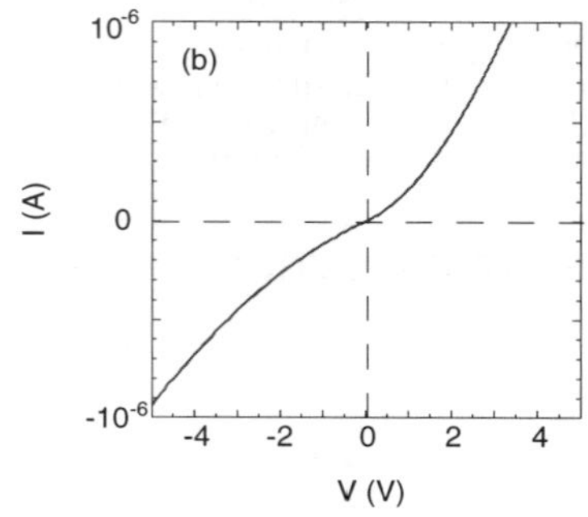

Fig 3. Current/voltage characteristics between two adjacent Ti/Al ohmic contacts on GaN annealed at 1100 °C, 15 s (a) without and (b) with an AlN cap. Note the change in current scale for the two samples.

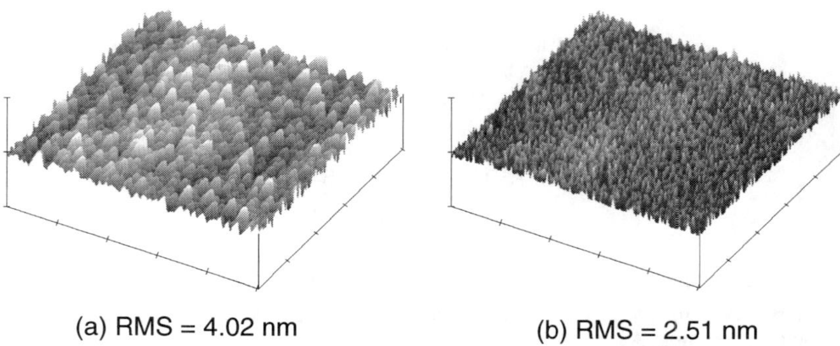

(a) RMS = 4.02 nm   (b) RMS = 2.51 nm

Fig 4. Atomic force microscope images of GaN after an 1100 °C, 15 s anneal either a) uncapped or b) capped with reactively sputtered AlN. The AlN film was removed in a selective KOH-based etch (AZ400K developer) at 60 - 70 °C. On both images, the vertical scale is 50 nm per division and the horizontal scale is 2 μm per division.

Further evidence of the difference between the two annealing processes is seen in the atomic force microscopy images shown in Fig 4. The sample annealed without the AlN cap is markedly rougher than the capped sample, again suggesting some degree of surface decomposition. In an attempt to quantify the change in the GaN surface resulting from annealing with and without the AlN cap, AES surface and depth profiles were performed on annealed samples (capped and uncapped) after removal of the AlN encapsulant. Unfortunately, it is difficult to compare the absolute Ga and N concentrations from the AES data. However, when comparing the Ga/N ratio for each case we do see an increase for the sample annealed without the AlN cap (Ga/N ratio = 2.34) as compared to the as-grown sample (Ga/N ratio = 1.73). This can be understood by N-loss from the GaN during the annealing process.[29] For the sample annealed with the AlN cap, the AES spectrum had a strong signal from C contamination at the surface that masked the absolute N-concentration. AES depth profiles of the uncapped and annealed sample suggests that the N-loss is occurring in the very near surface region (~ 50 Å).

N-loss and the formation of N-vacancies during the high-temperature anneal is proposed as the key mechanism involved in changing the electrical properties of the Schottky and ohmic contacts. Since N-vacancies are thought to contribute to the background n-type conductivity in GaN, an excess of N-vacancies at the surface should result in a n^+-region (possibly a degenerate region) at the surface.[30] This region would then contribute to tunneling under reverse bias for the Schottky diode and explain the increase in the reverse leakage current in the uncapped samples. Similarly, a n^+-region at the surface would improve the ohmic contact behavior as seen for the uncapped samples.[31] The effectiveness of the AlN cap during the anneal to suppress N-loss is readily understood by the inert nature of AlN and its extreme thermal stability thereby acting as an effective diffusion barrier for N from the GaN substrate.

IMPLANTATION DAMAGE AND DOSE EFFECTS

To minimize the access resistance in a transistor structure employing ion implantation such as in Fig 1c,d the implantation dose should be selected at some maximum level determined by dopant solubility and the ability to remove implantation damage. Therefore, it is important to study the activation properties of Si-implanted GaN versus dose.

Experimental Approach

The GaN layers used in the experiments were 1.5 to 2.0 μm thick grown on c-plane sapphire substrates by metalorganic chemical vapor deposition (MOCVD) in a multiwafer rotating disk reactor at 1040 °C with a ~20 nm GaN buffer layer grown at 530 °C.[32] The GaN layers were unintentionally doped, with background n-type carrier concentrations $\leq 1 \times 10^{16}$ cm^{-3} as determined by room temperature Hall measurements. When annealed at 1100 °C for 15 s the material maintained its high resistivity. The as-grown layers had featureless surfaces and were transparent with a strong bandedge luminescence at 356 nm at 4 K. Additional luminescence peaks were observed near 378 and 388 nm. We speculate that these second peaks are due to carbon contamination in the film from the graphite heater in the growth reactor. Si ions were implanted at 100 keV at doses from 5×10^{13} cm^{-2} to 1×10^{16} cm^{-2}. Ar-ions were implanted at 140 keV and over the same dose range to place its peak range at the equivalent position as the Si. All samples were annealed at 1100 °C for 15 s in flowing N_2 with the samples in a SiC-coated graphite susceptor. This annealing sequence has previously been shown to yield activated implanted dopants in GaN.[1,2] Following annealing the samples were characterized at room temperature by the Hall technique by evaporating Ti/Au ohmic contacts at the corners of each sample. Structural analysis of select Si-implanted GaN samples was performed with channeling Rutherford backscatter (C-RBS) and cross-sectional transmission electron microscopy (XTEM).

Results and Discussion

Figure 5 shows the sheet electron concentrations versus implant dose for the Si and Ar implanted samples after the 1100 °C anneal. For the Si-implanted samples, there appears to be no significant donor activation until a dose of 5×10^{15} cm^{-2} is achieved. This is in contrast to earlier results at a dose of 5×10^{14} cm^{-2} where roughly 10% of the implanted Si-ions were ionized at room temperature, corresponding to 94% of the implanted Si forming active donors on the Ga-sublattice assuming a Si-donor ionization energy of 62 meV.[1] The lack of free electrons for the lower dose Si-implanted samples in this study may be due to compensation by background carbon in the as-grown GaN as was postulated to exist based on the photoluminescence data. For the two highest dose Si-samples, 5 and 10×10^{15} cm^{-2}, 35% and 50% , respectively, of the implanted Si-ions created ionized donors at room temperature. The possibility that implant damage alone is generating the free electrons can be ruled out by comparing the Ar-implanted samples at the same dose with the Si-samples which have over a factor of 100 times more free electrons. If the implantation damage were responsible for the carrier generation then the Ar-samples, which will have more damage than the Si-samples as a results of Ar's heavier mass, would demonstrate at least as high a concentration of free electrons as the Si-samples. Since this is not the case, implant damage can not be the cause of the enhanced conduction and the implanted Si must be activated as donors. The significant activation of the implanted Si in the high dose samples and not the lower dose samples is understood based on the need for the Si-concentration to exceed the background carbon concentration that is thought to be compensating the lower dose Si-samples.

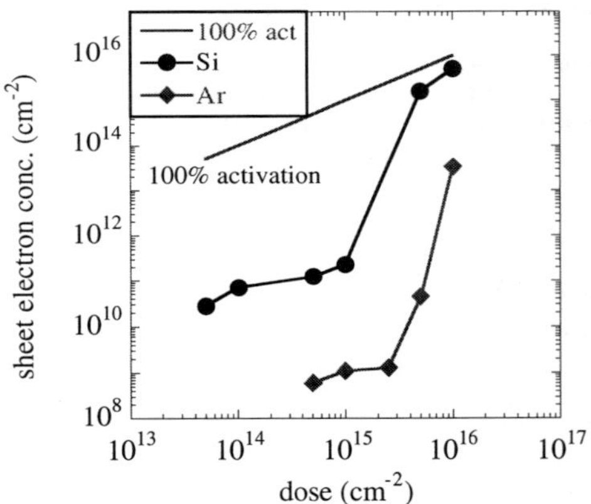

Fig 5. Sheet electron concentration versus implant dose for Si and Ar implanted GaN.

Figure 6 shows the C-RBS spectra for as-implanted (Si: 100 keV, 6×10^{15} cm^{-2}) and annealed (1100 °C, 30 s) GaN. These figures demonstrate the difficulty in removing the implantation-induced damage from the GaN layer. Although the C-RBS spectra in Fig. 6 displays improved channeling after annealing as seen by the decrease in the backscattered signal, when changes in dechanneling due to the reduced surface peak are accounted for, no significant change in the buried defect concentration is evident.[33] This has been confirmed by the cross-sectional transmission microscopy where the annealed sample still has significant damage apparently consisting of clusters, loops, and planar defects.[33,34] The fact that significant damage remains for an implantation and annealing sequence that produces significant electrical activation of Si-donors as seen in Fig. 5 suggests that complete defect removal is not required to activate implanted Si-donors in GaN. This is in stark contrast to the case of Si implanted GaAs, where damage removal and dopant activation are sequential processes.[35] However, since the defects in GaN will most likely act as scattering centers and degrade transport properties of the implanted layers, it will be desirable to develop an annealing procedure which more effectively restores the initial crystal quality. This will most likely require using a higher annealing temperature.

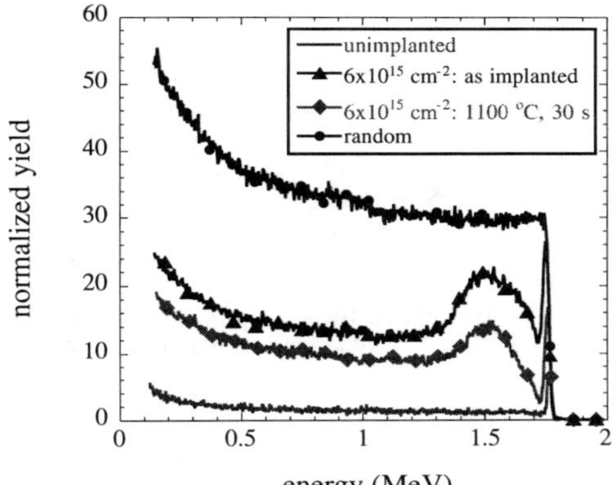

Fig 6. Channeling Rutherford Backscatering spectra of as-implanted (Si: 100 keV, 6x10^{15}) and annealed at 1100 °C, 30 s.

CONCLUSION

In conclusion, ion implantation is expected to play an enabling role in the realization of many high performance devices in the III-Nitrides as it has in other mature semiconductor material systems. In particular, ion implantation doping is effective for reducing the access resistance of transistors. Therefore, we have review the progress made in ion implantation doping of GaN with the successful demonstration of n- and p-type doping. The utility of an AlN encapsulating layer was also reported and shown to be effective at maintaining the surface stoichiometry of GaN during the activation anneal and thereby allow the formation of Schottky contact after annealing. It was further shown that high dose Si-implants can be activated up to 50% after a 1100 °C anneal. Finally, C-RBS data was reported that demonstrates the difficulty in removing implantation-induced damage in GaN. However, it appears the complete damage removal is not required to achieve activated implanted Si-donors in GaN. This is sharp contrast to GaAs where donor activation and damage removal are serial processes.

FUTURE WORK

While significant progress has been made in implantation doping and isolation of GaN, further work remains. In particular, additional work is needed to understand the nature of implantation-induced damage and in developing an annealing sequence to restore the material closer to its as-grown condition. This may require annealing temperatures well in excess of 1100 °C. Other areas of study include implant doping of AlGaN as used in HFET's and continued exploration of alternative p-type dopants.

Acknowledgments: The authors gratefully acknowledge the technical support of J. A. Avery at Sandia. The portion of this work performed at Sandia was supported by the United

States Department of Energy under contract #DE-ACO4-94AL85000. Sandia is a multiprogram laboratory operated by Sandia Corporation, a Lockheed Martin Company, for the United States Department of Energy. The work at UF is partially supported by a National Science Foundation grant (DMR-9421109) and a University Research Initiative grant from ONR (N00014-92-5-1895). Additional support for the work at Sandia, UF, and EMCORE was provided from DARPA (A. Husain) and administered by AFOSR (G. L. Witt).

References

[1] S. J. Pearton, C. R. Abernathy, C. B. Vartuli, J. C. Zolper, C. Yuan, R. A. Stall, Appl. Phys. Lett. **67**, 1435 (1995).

[2] J. C. Zolper, R. G. Wilson, S. J. Pearton, and R. A. Stall, Appl. Phys. Lett. **68** 1945 (1996).

[3] J. C. Zolper, M. Hagerott Crawford, S. J. Pearton, C. R. Abernathy, C. B. Vartuli, C. Yuan, and R. A. Stall, J. Electron. Mat. **25** 839 (1996).

[4] J. C. Zolper, R. J. Shul, A. G. Baca, R. G. Wilson, S. J. Pearton, and R. A. Stall, Appl. Phys. Lett. **68** 2273 (1996).

[5] S. Nakamura, T. Mukai, M. Senoh, and N. Iwasa, Jap. J. Appl. Phys. **30** L1998 (1991).

[6] S. Nakamura, T. Mukai, M. Senoh, N. Iwasa, and S. Nagahama, Appl. Phys. Lett. **67** 1868 (1995).

[7] S. Nakamura, M. Senoh, S. Nagahama, T. Yamada, T. Matsushita, H. Kiyoku, and Y. Sugimoto, Jap. J. Appl. Phys. **35**, L74 (1996).

[8] S. D. Lester, F. A. Ponce, M. G. Craford, and D. A. Steigerwald, Appl. Phys. Lett. **66** 1249 (1995).

[9] M. A. Khan, A. Bhattarai, J. N. Kuznia, and D. T. Olson, Appl. Phys. Lett. **63**, 1214 (1993).

[10] S. C. Binari, L. B. Rowland, W. Kruppa, G. Kelner, K. Doverspike, and D. K. Gaskill, Elect. Lett. **30**, 1248 (1994).

[11] J. Burm, W. J. Schaff, L. F. Eastman, H. Amano, and I. Akasaki, Appl. Phys. Lett. **68**, 2849 (1996).

[12] Z. Fan, S. N. Mohammad, O. Aktas, A. E. Botchkarev, A. Salador, and H. Morkoc, Appl. Phys. Lett. **69**, 1229 (1996).

[13] Y-F. Wu, B. P. Keller, D. Kapolnek, S. P. Denbaars, and U. K. Mishra, Appl. Phys. Lett. **69**, 1438 (1996).

[14] J. I. Pankove, Tech Digest Inter. Elec. Dev. Meeting, San Francisco, CA, Dec. 11-14, 1994, p. 389.

[15] Gallium Arsenide: Materials Devices, and Circuits: ed. M. J. Howes and D. V. Morgan (John Wiley & Sons, New York, 1985) chapter 10.

[16] S. C. Binari, Proc. of the Sym. Wide Bnadgap Semiconducotrs and Devices, Electrochemical Society, October 8-13, Chicago, IL **vol 95-21**, 136 (1995).

[17] F. Ren, et al. J. Vac Sci. Tech, B, (in press).

[18] I. Akasaki, H. Amano, M. Kito, and K. Hiramatsu, J. Lumin. **48/49** 666 (1991).

[19] S. Strite, Jpn. J. Appl. Phys. **33**, L699 (1994).

[20] A. E. Von Neida, S. J. Pearton, W. S. Hobson, and C. R. Abernathy, Appl. Phys. Lett. **54**, 1540 (1989).

[21] J. C. Zolper, A. G. Baca, and S. A. Chalmers, Appl. Phys. Lett. **62**, 2536 (1993).

[22] J. C. Zolper, M. E. Sherwin, A. G. Baca, and R. P. Schneider, Jr., J. Elec. Mat. **24**, 21 1995).

[23] S. J. Pearton, Inter. J. Modern Physics B, **7**, 4686 (1993).

[24] S. Strite and P. W. Epperlein, Conf. Proc. MRS, Fall 1995, symposium AAA (Material Research Society, Pittsburgh PA, 1996) p. 795.

[25] J. C. Zolper, D. J. Rieger, A. G. Baca, S. J. Pearton, J. W. Lee, and R. A. Stall, "Sputtered AlN Encapsulant for High-Temperature Annealing of GaN," Appl. Phys. Lett. **69**, 538 (1996).

[26] J. R. Mileham, S. J. Pearton, C. R. Abernathy, J. D. MacKenzie, R. J. Shul, and S. P. Kilcoyne, Appl. Phys. Lett. **67**, 1119 (1995).

[27] J. R. Mileham, S. J. Pearton, C. R. Abernathy, J. D. MacKenzie, R. J. Shul, and S. P. Kilcoyne, J. Vac. Sci. Tech. B (in press).

[28] C. B. Vartuli, J. W. Lee, J. D. MacKenzie, S. J. Pearton, C. R. Abernathy, J. C. Zolper, R. J. Shul, F. Ren, Materials Research Society, Boston MA, Dec. 2-6, 1996, in press.

[29] C. D. Thurmond and R. A. Logan, J. Electrochem. Soc. **119**, 622 (1972).

[30] H. P. Maruska and J. J. Tietjen, Appl. Phys. Lett. **15**, 327 (1969).

[31] L. F. Lester, J. M. Brown, J. C. Ramer, L. Zhang, S. D. Hersee, and J. C. Zolper, Appl. Phys. Lett. **69**, 2737 (1996).

[32] C. Yuan, T. Salagaj, A. Gurary, P. Zawadzki, C. S. Chern, W. Kroll, R. A. Stall, Y. Li, M. Schurman, C.-Y. Hwang, W. E. Mayo, Y. Lu, S. J. Pearton, S. Krishnankutty, and R. M. Kolbas, J. Electrochem. Soc. **142**, L163 (1995).

[33] H. H.Tan, J. S. Williams, J. Zou, D. J. H. Cockayne, S. J. Pearton, and C. Yuan, Proc. 1st Symp. on III-V Nitride Materials and Processes, Electrochemcial Society, vol. 96-11, 142 (1996).

[34] J. C. Zolper, H. H. Tan, J. S. Williams, J. Zhou, D. J. H. Cockayne, S. J. Pearton, M. H. Crawford, and R. F. Karlicek, Jr., Appl. Phys. Lett. (submitted).

[35] Ion Implantation and Beam Processing ed. J. S. Williams and J. M. Poate (Academic Press, Sydney, 1984).

REACTIVATION OF ACCEPTORS AND TRAPPING OF HYDROGEN IN GaN/InGaN DOUBLE HETEROSTRUCTURES

S.J. PEARTON, S. BENDI, K.S. JONES, V. KRISHNAMOORTHY*, R.G. WILSON**, F. REN***, R.F. KARLICEK, JR. AND R.A. STALL****
*University of Florida, Gainesville, FL 32611
**Hughes Research Laboratories, Malibu, CA 90265
*** Bell Laboratories, Lucent Technologies, Murray Hill, NJ 07974
****EMCORE Corporation, Somerset, NJ 08873

ABSTRACT

The apparent thermal stability of hydrogen passivated Mg acceptors in GaN is a function of the annealing ambient employed, with H_2 leading to a reactivation temperature approximately 150°C higher than N_2. The dissociation of Mg-H complexes and the loss of hydrogen from GaN are sequential processes, with reactivation occurring at ≤ 700°C for annealing under N_2, while significant concentrations of hydrogen remain in the crystal even at 900°C in implanted samples. The hydrogen is gettered to regions of highest defect density such as the InGaN layer in a GaN/InGaN double heterostructure. The addition of an accelerating potential for $^2H^+$ ions in the plasma did not greatly affect the deuterium profiles.

INTRODUCTION

Hydrogen plays an important role in GaN and related alloys, since it is part of the growth environment for Metal Organic Chemical Vapor Deposition (MOCVD), which employs $(CH_3)_3Ga$ and NH_3 as its standard precursors.[1] Hydrogen remaining in Mg-doped GaN layers has been identified as the cause of the high resistivity of these films, since it appears that the hydrogen forms stable neutral complexes $(Mg\text{-}H)^0$, with the Mg acceptors.[1] Post-growth annealing at ~700°C[1,2], or exposure to a low energy electron beam[3] reactivates the Mg acceptors and produces conductivities 5-6 orders of magnitude higher than in the as-grown material. Hydrogen has been shown to be injected into GaN during a number of processing steps, including dry etching in CH_4/H_2 plasma chemistries, boiling in water, wet etching in KOH solutions and chemical vapor deposition of dielectrics using SiH_4.[4]

The behavior of hydrogen in device structures is likely to be more complicated. For example light-emitting diodes or laser diodes contain both n-and p-type GaN cladding layers with one or more InGaN active regions. The first laser diode reported by Nakamura et al.[5] contained 26 InGaN quantum wells. In other III-V semiconductors the diffusivity of atomic hydrogen is a strong function of conductivity type and doping level since trapping by acceptors is usually more thermally stable and more efficient than trapping of hydrogen by donor impurities.[6-14] Moreover hydrogen is attracted to any region of strain within multilayer structures and has been shown to pile-up at heterointerfaces in the GaAs/Si[15,16], GaAs/InP[14,15] and GaAs/AlAs[6] materials systems. Therefore it is of interest to investigate the reactivation of acceptors and trapping of hydrogen in double heterostructure GaN/InGaN samples, since these are the basis for optical emitters. We find that the reactivation of passivated Mg acceptors also depends on the annealing ambient, with an apparently higher stability for annealing under H_2 rather than N_2. Hydrogen is found to redistribute to the regions of highest defect density within the structure.

Mat. Res. Soc. Symp. Proc. Vol. 449 © 1997 Materials Research Society

EXPERIMENTAL

The sample was grown by MOCVD in a rotating disc reactor on c-plane Al_2O_3.[2] The sapphire substrate was rinsed in H_2SO_4, methanol and acetone prior to loading into the growth chamber, where it was first baked at 1100°C under H_2. A low temperature (~510°C) GaN buffer (~300Å thick) was followed by 3·3μm of n[+] GaN (n = 10^{18}cm^{-3}, Si-doped), 0·1μm InGaN (undoped) and 0·5μm thick p[+]GaN (p = $3X10^{17}$cm^{-3}, Mg-doped). The growth temperature was 1040°C for the GaN and ~800°C for the InGaN. Cross-sectional transmission electron microscope (XTEM) analysis was carried out on the MOCVD grown InGaN/GaN double-heterostructure. An XTEM bright-field image, obtained using two-beam diffraction conditions with $\mathbf{g}$=(2$\underline{1}$10) along the [0$\underline{1}$10]$_{GaN}$ zone axis of the DH-LED structure is shown in Figure 1 (top). The interface between the various layers appears to be abrupt with no indication of interfacial phases. Selected-area-diffraction (SAD) and high resolution electron microscopy revealed that the entire DH-LED structure grew epitaxially on the substrate.

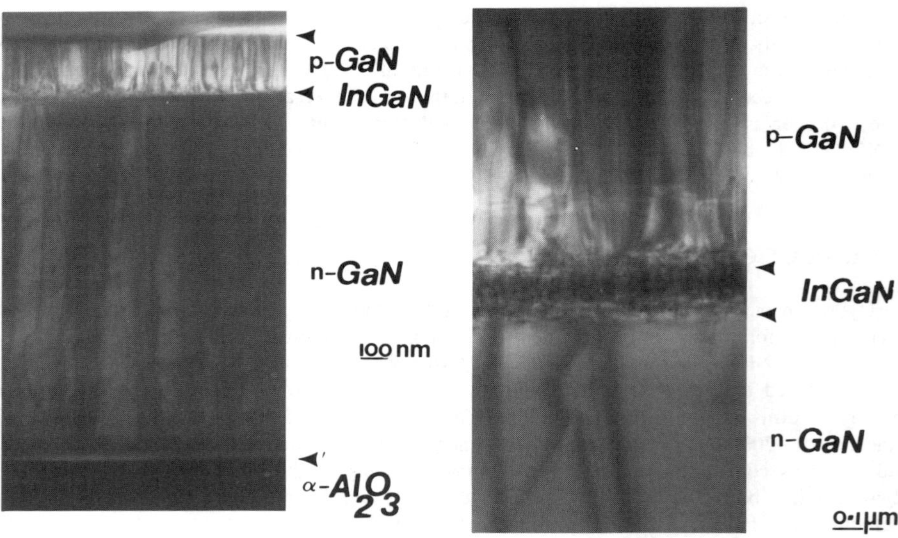

Figure 1. (Left) Bright-field XTEM image showing the defect distribution along the device obtained using two-beam diffraction condition with $\mathbf{g}$=(2$\underline{1}$10) along [0$\underline{1}$10]$_{GaN}$ zone axis. (Right) Bright-field XTEM image of the top region of the device including the double heterostructure active region showing the increase in threading dislocation density after the growth of the active layer. The image was obtained using two-beam condition with $\mathbf{g}$=(2$\underline{1}$10) along the [0$\underline{1}$10]$_{GaN}$ zone axis.

RESULTS AND DISCUSSION

In the immediate vicinity of the n-GaN/Al_2O_3 interface, the defect density was high but was reduced with increasing film thickness. However, after the growth of the active layer (InGaN) the defect density of the threading dislocations increased as shown at the right of Figure 1. A possible reason could be due to the different growth conditions used for growing the active layer and the GaN layers. The growth mechanism for p-GaN on InGaN in the DH-LED structure could be similar to that proposed by Hiramatsu.[17] During the growth of the subsequent p-GaN layer the underlying active layer may be undergoing solid-phase-epitaxy. Hence, the quality of p-GaN grown on top of the active layer depends on the amount of epitaxy undergone by the active layer, and in this structure the thermal degradation of the InGaN upon raising the growth temperature for the p-GaN leads to a higher defect density in this overlayer.

XTEM of the DH-LED showed dislocations as dark lines propagating in the direction normal to the substrate. Most of the dislocations appeared to traverse the entire double heterostructure, while some appeared to bend and follow the interface for a short distance before threading out to the surface. The nature of the threading dislocations was studied by conventional XTEM using the $\mathbf{g}.\mathbf{b}$=0 criteria. The dislocation will be invisible when $\mathbf{b}$ lies in the reflecting plane.

Some of the dislocations were invisible both in $\mathbf{g2}$=(0002) and $\mathbf{g5}$=(1$\bar{1}$01) and, because $\mathbf{b}$ was common to both reflections, $\mathbf{b}$ was found to be 1/3[11$\bar{2}$0]. Assuming that the growth is the same as the translation vector of the dislocation, these defects would be pure edge type in nature. The average threading dislocation density was also found along the plane normal to the growth direction. The dislocation density was found to be ~8×10^{10}/cm^2.

The double-heterostructure sample was exposed to an Electron Cyclotron Resonance (ECR) plasma (500W of microwave power, 10 mTorr pressure) for 30 min at 200°C. The hole concentration in the p-GaN layer was reduced from 3×10^{17}cm^{-3} to ~2×10^{16}cm^{-3} by this treatment, as measured by capacitance voltage (C-V) at 300K. Sections from this material were then annealed for 20 mins at temperatures from 500-900°C under an ambient of either N$_2$ or H$_2$ in a Heatpulse 410T furnace. Figure 2 shows the percentage of passivated Mg remaining after annealing at different temperatures in these two ambients. In the case of N$_2$ ambients the Mg-H complexes show a lower apparent thermal stability (by ~150°C) than with H$_2$ ambients. This has been reported previously for Si donors in InGaP and AlInP, and Be and Zn acceptors in InGaP and AlInP, respectively[18], and most likely is due to indiffusion of hydrogen from the H$_2$ ambients, causing a competition between passivation and reactivation. Therefore an inert atmosphere is clearly preferred for the post-growth reactivation anneal of p-GaN to avoid any ambiguity as to when the acceptors are completely active. Previous experimental results by Brandt et.al.[19] and total energy calculations by Neugebauer and Van de Walle[20] suggest that considerable diffusion of hydrogen in GaN might be expected at ≤600°C.

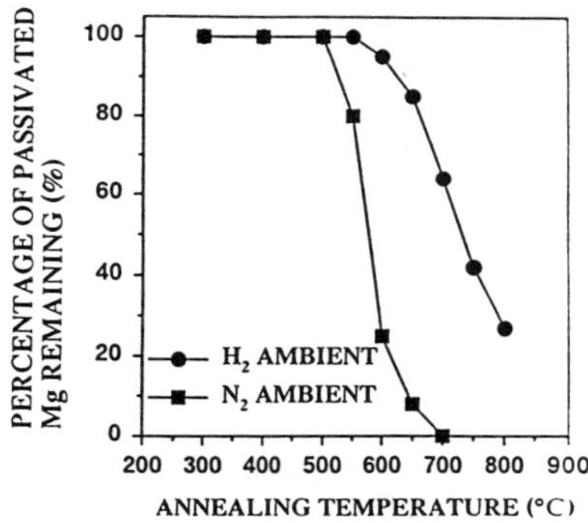

Figure 2. Fraction of passivated Mg acceptors remaining in hydrogenated p-type GaN after annealing for 20 min at various temperatures in either N$_2$ or H$_2$ ambients.

Other sections of the double-heterostructure material were implanted with $^2H^+$ ions (50keV, $2 \times 10^{15} cm^{-2}$) through a SiN_x cap in order to place the peak of the implant distribution within the p^+GaN layer. Some of these samples were annealed at 900°C for 20 min under N_2. As shown in the Secondary Ion Mass Spectrometry (SIMS) profiles of Figure 3, the ^2H diffuses out of the p^+GaN layer and piles-up in the defective InGaN layer, which we saw from the TEM results suffers from thermal degradation during growth of the p^+GaN. Note that there is still sufficient ^2H in the p^+GaN ($\sim 10^{19} cm^{-3}$) to passivate all of the acceptors present, but electrical measurements show that the p-doping level was at its maximum value of $\sim 3 \times 10^{17} cm^{-3}$. These results confirm that as in other III-V semiconductors, hydrogen can exist in a number of different states, including being bound at dopant atoms or in an electrically inactive form that is quite thermally stable. We expect that after annealing above 700°C all of the Mg-H complexes have dissociated, and the electrical measurements show that they have not re-formed. In other III-V's the hydrogen in p-type material is in a bond centered position forming a strong bond with a nearby N atom, leaving the acceptor 3-fold coordinated.[1] Annealing breaks this bond and allows the hydrogen to make a short-range diffusion away from the acceptor, where it probably meets up with other hydrogen atoms, forming molecules or larger clusters that are relatively immobile and electrically inactive. This seems like a plausible explanation for the results of Figures 2 and 3, where the Mg electrical activity is restored by 700°C, but hydrogen is still present in the layer at 900°C. In material hydrogenated by implantation, there is almost certainly a contribution to the apparently high thermal stability by the hydrogen being trapped at residual implant damage as is evident by the fact that the ^2H profile retains a Pearson IV type distribution even after 900°C annealing. The other important point from Figure 3 is that as in other defective crystal systems hydrogen is attracted to regions of strain, in this case the InGaN sandwiched between the adjoining GaN layers.

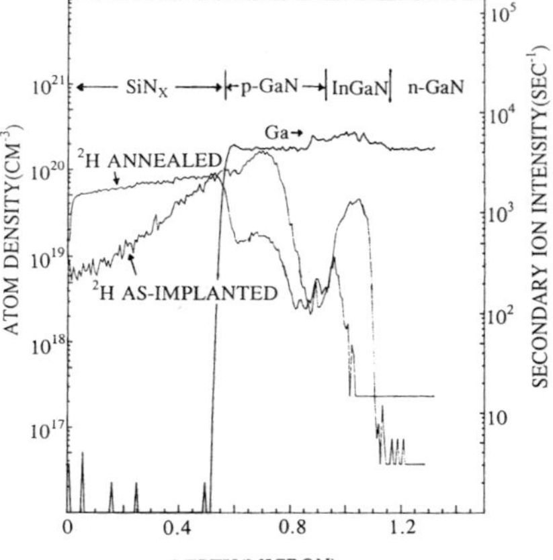

Figure 3. SIMS profiles of ^2H in an implanted (50keV, $2 \times 10^{15} cm^{-2}$ through a SiN_x cap) double-heterostructure sample, before and after annealing at 900°C for 20 min.

Figure 4 shows SIMS profiles of ^2H in InAlN epilayers treated in a deuterium plasma at 200°C for 30 min and subsequently annealed at 500 or 700°C. A high concentration of ^2H has permeated the entire layer structure under these conditions and the plateau shape of the profiles suggests the deuterium is trapped at defects or impurities. Similar results were obtained for GaN and AlN.[6] The addition of a dc self-bias to the sample to accelerate the $^2H^+$ ions did not greatly influence the depth of incorporation, in contrast to other semiconductors.

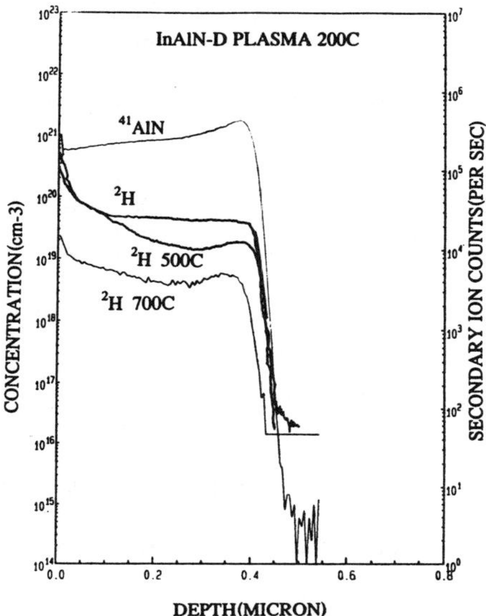

Figure 4.   SIMS profiles of ^2H in InAlN treated in deuterium plasma at 200°C, 30 min and then annealed at 500 or 700°C.

CONCLUSIONS

In conclusion, the apparent thermal stability of hydrogen-passivated Mg acceptors in p-GaN is dependent on the annealing ambient, as it is in other compound semiconductors. While the acceptors are reactivated at ≤700°C for annealing under N_2, hydrogen remains in the material until much higher temperatures and can accumulate in defective regions of double-heterostructure samples grown on Al_2O_3. It will be interesting to compare the redistribution and thermal stability of hydrogen in homoepitaxial GaN in order to assess the role of the extended defects present in the currently available heteroepitaxial material.

ACKNOWLEDGMENTS

The work at UF is partially supported by a National Science Foundation grant through the Division of Materials Research (DMR-9421109). The work at HRL is partially supported by a grant from ARM (J.M. Zavada), while a grant from ARPA (A. Husain) administrated by AFOSR (G.L. Witt) partially supports the work at UF and EMCORE.

REFERENCES

1. S. Nakamura, N. Iwasa, M. Senoh and T. Mukai, Jap. J. Appl. Phys. <u>31</u>, 1258 (1992).

2. C. Yuan, T. Salagaj, A. Gurary, P. Zawadzki, C.S. Chen, W. Kroll, R.A. Stall, W. Yi, M. Schurman, C.Y. Hwang, W.E. Mayo, Y. Lu, S.J. Pearton, S. Krisnakutty and R.M. Kolbas, J. Electrochem. Soc. <u>142</u>, 1163 (1995).

3. H. Amano, M. Kito, K. Hiramatsu and I. Akaski, J. Appl. Phys. 28, L112 (1989).

4. S.J. Pearton, R.J. Shul, R.G. Wilson, F. Ren. J.M. Zavada, C.R. Abernathy, C..B. Vartuli, J.W. Lee, J.R. Mileham and J.D. MacKenzie, J. Electron. Mater. 25, 845 (1996).

5. S. Nakamura, M. Senoh, S. Nagahama, N. Iwasa, T. Yannada, T. Matsushita, H. Kyoku and Y. Sugimoto, Jap. J. Appl. Phys. 35, L74 (1996).

6. J.M. Zavada and R.G. Wilson, Mat. Sci. Tor. 148/149, 189 (1994).

7. S.K. Estreicher, Mat. Sci. Eng. Dep. R 14, 319 (1995).

8.. J.I. Pankove and N.M. Johnson (eds.), Semiconductors and Semimetals Vol. 34 (Academic Press, San Diego, 1991).

9. S.J. Pearton, J.W. Corbett and M. Stavola, Hydrogen in Crystalline Semicond. (Springer-Verlag, Heidelber, 1992).

10. N.M. Johnson, C. Doland, F. Ponce, J. Walker and G. Anderson, Physica B170, 3 (1991).

11. J. Chevallier and M. Aucouturier, Ann. Rev. Mater. Sci. 18, 219 (1988).

12. D. Mathiot, Phys. Rev. B 40, 5869 (1989).

13. A.Y. Polyakov, M. Stam, A.Z. Li and A.G. Milnes, Inst. Phys. Conf. Ser. 120, 83 (1992).

14. S.A. Ringel and B. Chalberhjee (in press); B. Chatterjee, S.A. Ringal, R. Sieg, R. Hoffman and I. Weinberg, Appl. Phys. Lett. 65, 58 (1994).

15. U.K. Chabrabarti, S.J. Pearton, W.S. Hobson, J. Loparta and V. Swaminathan, Appl. Phys. Lett. 57, 887 (1990).

16. S.J. Pearton, C.S. Wu, M. Stavola, F. Ren, J. Loparta and W.C. Dautremont-Smith, Appl. Phys. Lett. 51, 496 (1987).

17. K. Hiramatsu, S. Itoh, H. Amano, I. Akaski, N. Kuwano, T. Shiraishi and K. Oki, J. Cryst. Growth 115, 628 (1991).

18. S.J. Pearton, Int. J. Mod. Phys. B 8, 1247 (1994).

19. M.S. Brandt, J.W. Ager, W. Gotz, N.M. Johnson, J.S. Harris, R.J. Molnar and T.D. Moustakas, Phys. Rev. B 49, 14758 (1994).

20. J. Neugebauer and C.G. Van der Walle, Phys. Rev. Lett. 75, 4452 (1995).

THE INFLUENCE OF HYDROGEN ADDITION ON
THE CHEMICAL PROPERTIES OF HYDROGENATED
ALUMINUM NITRIDE FILMS PREPARED BY RF REACTIVE SPUTTERING

Jai-Young Lee, Yoon-Joong Yong
Department of Materials Science and Engineering, Korea Advanced Institute of Science and
Technology, Kusong-dong 373-1,Yusung-gu, Taejon, 305-701, South Korea

ABSTRACT

Hydrogenated aluminum nitride (AlN:H) films have been deposited on the (100) silicon wafers
by the RF reactive magnetron sputtering method with H_2 gas in addition to an Ar-N_2 gas
mixture. Stoichiometric AlN films without oxygen impurities can be prepared by adding 10 %
H_2 to reactive gas, which is proven by Rutherford Backscattering Spectrometry (RBS). The
bonding aspects of Al, N, O and H atoms in AlN:H films have been examined by X-ray
Photoelectron Spectroscopy (XPS) and Fourier Transform Infrared (FTIR) to understand the
effects of H_2 addition. The chemical shift of the binding energies of Al, N and O atoms in AlN:H
films from XPS analysis and the change of N-H bonding in FTIR with respect to different partial
pressures of H_2 gas have been confirmed. The role of H atoms is suggested to facilitate bonding
with unbound N atoms in AlN:H films and hinder N-O bonding, thus, reducing oxygen
concentration in AlN:H films. Also, the activation energy for the evolution of H_2 gas from
AlN:H film has been determined to be 0.11 eV/atom through a Kissinger-type analysis by a
thermal desorption test using Gas Chromatograph(GC). This result implies that the hydrogen
atom in film forms the hydrogen bond.

INTRODUCTION

Recently, aluminum nitride (AlN) films have drawn attention as a piezoelectric material for
high-frequency surface acoustic wave (SAW) devices due to its high phase velocity[1]. The
stoichiometry of AlN and lower concentrations of impurities such as oxygen are important for
the use of SAW filters because the non-stoichiometry or impurity in films creates a leakage
current between the input and output interdigital transducer (IDT) of SAW filters, which acts as
noise in filtering characteristics.

Nowadays, several notable results have been reported from experiments using the addition of
hydrogen gas to reactive gas during the deposition of AlN film. Hasegawa et al. used chemical
vapor deposition to produce AlN films and found that with the addition of hydrogen to the
reactive gas flow, a lower oxygen content determines by Auger Electron Spectroscopy (AES)[2].
From our previous X-ray Photoelectron Spectroscopy (XPS) study about the deposition of AlN
film by RF reactive sputtering, it was also found that the concentration of aluminum and
nitrogen atoms in AlN film increased and that of oxygen atoms decreased by increasing the
partial pressure of hydrogen gas to Ar-N_2 reactive gas[3]. In the previous papers, only relative
concentrations could be known because quantitative composition analysis was impossible when
applying AES and XPS without a standard sample. So it is necessary to confirm the absolute
film composition by Rutherford Backscattering Spectrometry (RBS), from which the
quantification of the composition is possible without a standard sample[4]. Also, Wang et al.
reported that the injection of hydrogen into the feed gas eliminated stress failures, and atomically
smooth coatings could be produced on virtually any substrate at room temperature[5,6].

Mat. Res. Soc. Symp. Proc. Vol. 449 ©1997 Materials Research Society

Summarizing the effects of H_2, it is known that the film surface becomes smoother and oxygen impurities decrease. These are positive effects of increasing the electromechanical coupling coefficient and signal-to-noise ratio of SAW filters. In order to understand these phenomena and develop a high performance SAW filter using AlN:H thin film, the change of the bonding nature of atoms and the bonding form of H atom in films made by H_2 gas injection should be observed. In this paper, hydrogenated AlN (AlN:H) films have been deposited on silicon wafers by the RF reactive magnetron sputtering method with H_2 gas mixed to an Ar-N_2 gas mixture. The quantitative composition analysis using RBS has been performed. The bonding aspects of Al, N, O and H atoms in AlN:H films have been examined by XPS and Fourier Transform Infrared (FTIR), and the roles of hydrogen atoms in AlN:H films have been discussed. Also, the activation energy for the evolution of H_2 gas from AlN:H films has been determined through a Kissinger plot[7] using Gas Chromatograph (GC) to discover the bonding form of the H atom.

EXPERIMENTAL

The AlN:H films were fabricated by the RF magnetron sputtering of an aluminum target in a mixture of argon, N_2 and H_2 gases. The purity of these gases were 99.9999 % each. The aluminum target was 5 cm in diameter and the purity was 99.999 %. The substrates used in this experiment were Si (100) wafers (2.0×2.0 cm^2). Degreasing of substrates was carried out in ultrasonic baths of acetone, ethanol and de-ionized water, successively. Native oxide on the Si wafer was removed by etching in the diluted HF solution. The substrates were blown dry with N_2 gas and immediately inserted into a vacuum chamber. The base pressure was lower than 6.7×10^{-5} Pa (5×10^{-7} Torr). The experimental conditions of the AlN films are listed in Table 1, which was selected to obtain c-axis oriented AlN film with a reasonable deposition rate[3]. The concentration of H_2 in the gas mixture varied from 0 to 20 volume %. The chemical shifts of Al, N, O atoms in films were measured by XPS (VG ESCALAB210) with monochromatic Mg K_a X-rays. For XPS analysis, AlN and AlN:H films with thickness of 2500 Å were deposited and the spectra were obtained after sputtering the film with Ar ions sufficiently to exclude the surface contamination. Also, the flood gun emitting electrons of 8 V and 0.1 mA has been used to avoid the charge effect which is due to the insulating property of AlN films. The quantitative analysis of the absolute composition of Al, N and O atoms in the films was performed by RBS (NEC 3SDH) using 2.236 MeV He atoms and the film thickness was 2500 Å to ease the separation of peaks from different atoms. FTIR (BOMEN, Michelson FTIR) was used to measure the change of bondings in AlN:H films with different H_2 partial pressures. The thickness of the film for FTIR analysis was 1.5 mm to acquire large peak intensities. For the gas evolution experiments, the samples were inserted into a quartz tube under Ar flow, and heated to 1073 K at uniform heating rates of 5 ~ 30 K/min. High-purity (99.999%) Ar gas was used to carry the evolved hydrogen into a GC apparatus (Hewlett-Packard, 5890 II). The amount of continuously evolved gas, mainly H_2, was measured by GC with a thermal conductivity detector (TCD) which distinguishes the species of gas from the difference in the thermal conductivity between them, such as H_2 and Ar. The details concerning this apparatus were presented in a previous paper [8]. All the samples were stored in a desiccator until the time of measurement to minimize the reaction of films with water vapor in air.

Table 1. Sputtering conditions of AlN:H films

Target	Aluminum (99.999 %, 5 cm diameter)
Substrate	Si (100) wafer
Base pressure (Pa)	$< 6.7 \times 10^{-5}$ (5×10^{-7} Torr)
Target-substrate distance (cm)	8.0
R.F.power (W)	250
Sputtering pressure (Pa)	1.1 (8 mTorr)
Ar gas flow rate (sccm)	6.0
N_2 gas flow rate (sccm)	6.0
Amount of H_2 addition (%)	$0 \sim 20$

RESULTS AND DISCUSSION

Figure 1 shows the XPS spectra of AlN:H films with different amounts of H_2. Comparing the heights of Al, N and O peaks, we can find that the concentration of nitrogen atom in the film increases and that of oxygen atom decreases with increasing amounts of H_2.

To know the absolute concentration, AlN:H film with 10% H_2 has been examined by RBS and Fig. 2(a) shows its spectra. From this, the absolute atomic concentration is calculated as shown in Fig. 2(b) in which the error range is less than 5%. Beneath 1000 Å from the surface, the concentrations of Al and N atoms are both 50% and that of O atom is 0%. Taking into account the error range of analysis, near stoichiometric AlN films without oxygen impurity have been obtained by adding 10% H_2 to reactive gas.

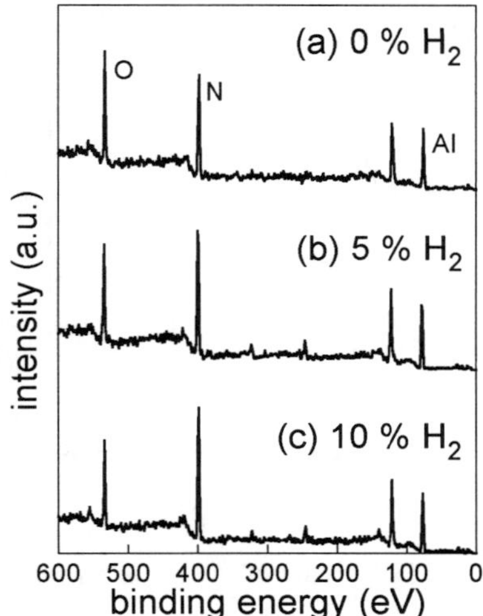

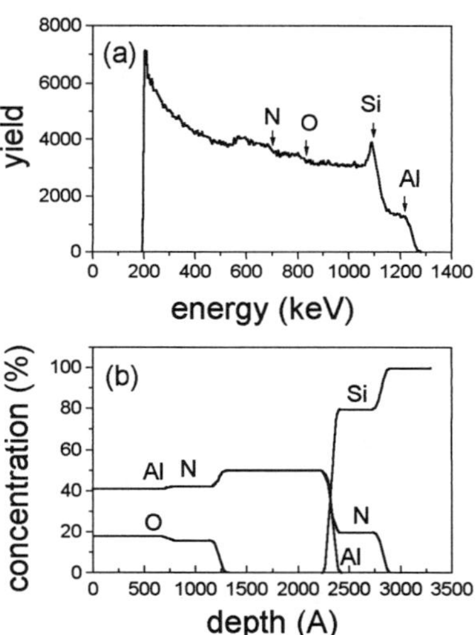

Fig. 1. XPS spectra of AlN:H films with different amount of hydrogen addition; (a) 0 %, (b) 5 %, and (c) 10 %.

Fig. 2. (a) RBS spectra of AlN:H film with 10% H_2 addition and (b) absolute atomic concentration of Al, N and O calculated from (a).

Fig. 3 shows the binding energies of Al 2p, N 1s and O 1s in AlN:H films with different amounts of H_2 analyzed by XPS. The binding energies have been calibrated by taking the C 1s peak (285 eV) as a reference. As the amount of H_2 increases, all the binding energies shift to lower energy sides. It is the evident that Al, N and O atoms in AlN:H films form bonding with H atoms because the electronegativity of hydrogen is low, so the large screening effect makes the binding energy lower. Also, it is known that most H atoms interact better with N and O atoms than with Al atoms because the changes of binding energies in N and O atoms are larger than those of Al atoms as shown in Fig. 3.

In order to analyze the bonding related to hydrogen atoms, FTIR spectra of AlN:H films with different amounts of H_2 have been taken, as shown in Fig. 4. Because many peaks are superimposed, peak deconvolution is performed to understand the changes of bonding nature. Each of the peaks are positioned and assigned by referring to other works[9,10] as revealed in Table 2. The primary N-H bond represents the N-H bond in $AlNH_3^+$ or NH_4^+ molecules and the secondary N-H bond represents that in $AlNH_2$ or NH_3 molecules. The peaks appearing at 3376 and 3532 cm^{-1} are deduced to be free ones, while the peaks raised at 3074 and 3230 cm^{-1} are found to be bounded with water molecules, so the wave number moves to a lower value.

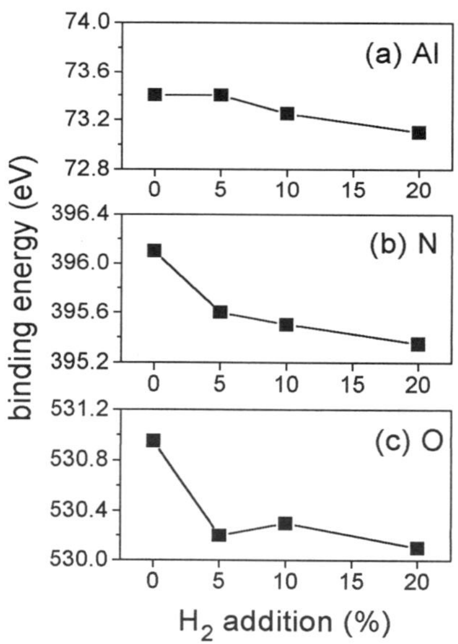

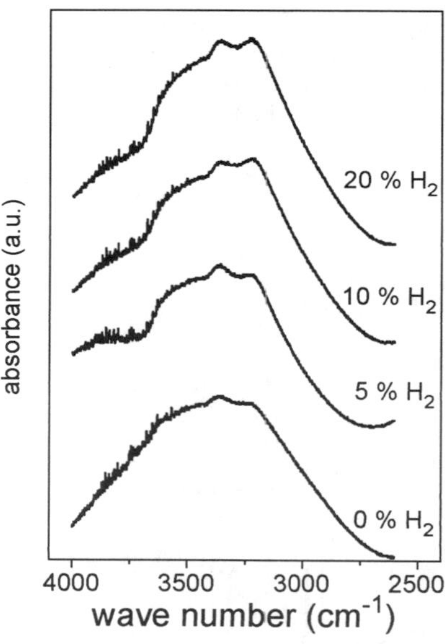

Fig. 3. The binding energies of (a) Al 2p, (b) N 1s and (c) O 1s of AlN:H films analyzed by XPS with different amount of H_2 addition.

Fig. 4. FTIR spectra of AlN:H films with different amount of H_2 addition.

Table 2. Peak position and its assignments of FTIR spectra.

peak position (wave number ; cm^{-1})	assignments
3074	secondary-bound N-H
3230	primary-bound N-H
3376	secondary-free N-H
3532	primary-free N-H

Fig. 5 shows the amplitudes of the four peaks assigned in Table 2 with different amounts of H_2. When H_2 gas is added by as little as 5% to reactive gas, the amplitudes of the secondary-bound and secondary-free N-H bonds abruptly increase and saturate. As the partial pressure of H_2 increases, primary-bound and primary-free N-H bonds increase linearly. From these phenomena, one can see that hydrogen atoms form secondary N-H bonds at low H_2 concentrations, and that primary N-H bonding increases with increasing H_2 content. That is to say, hydrogen atoms make bonding with unbound nitrogen atoms in films and hinder N-O bonding. It is the role of hydrogen to reduce the oxygen concentration in films. The reason for the high amplitudes of primary N-H bonds of AlN films without H_2 gas is that the nitrogen atoms are bound with the hydrogen atoms of water molecules in the atmosphere.

To confirm the bonding nature of H atom, the kinetics of hydrogen for AlN:H films have been studied by applying a thermal analysis technique, and the activation energies for hydrogen evolution were obtained by Kissinger's method[7].

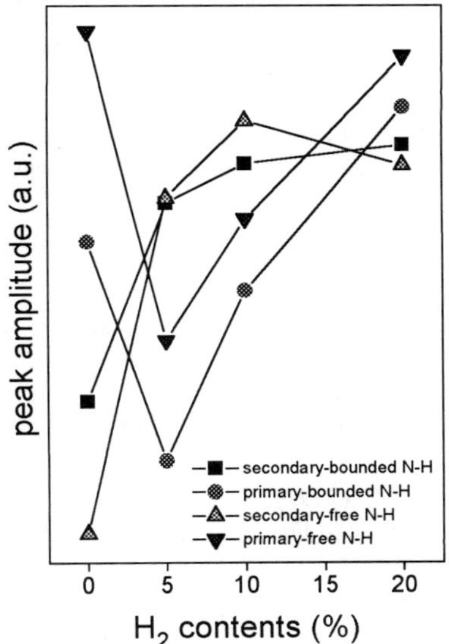

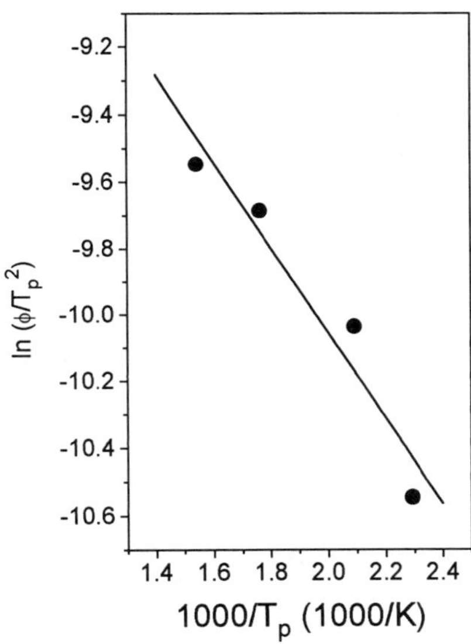

Fig. 5. Amplitudes of four peaks assigned in Table 2 with different amount of H_2 addition.

Fig. 6. Kissinger plot for the calculation of the activation energy for the evolution of H_2 gas.

Fig. 6 shows its Kissinger plot where ϕ is the heating rate, T_p is the peak temperature and from the slope of the line, the activation energy for the evolution of H_2 gas is found to be 0.11 ± 0.02 eV/atom. This value is similar to the bonding energy of the hydrogen bond (0.1 eV), being formed only between the most electronegative atoms, particularly F, O, and N[11]. So it is confirmed that the bonding form of H atoms in AlN:H film is the hydrogen bond with the large electronegative atom such as nitrogen, and this activation energy corresponds to that which is required to break the hydrogen bond between H and N atoms in the film.

CONCLUSION

Hydrogenated aluminum nitride (AlN:H) films have been deposited on silicon wafers by the RF reactive magnetron sputtering method with H_2 gas addition to Ar-N_2 gas mixture. The experimental results can be summarized as follows.
(1) As the amount of H_2 increases, the surface of film becomes smooth, the stress in the film is relieved, the concentration of nitrogen atom increases and that of oxygen atom decreases. And the stoichiometric AlN films without oxygen impurities can be prepared by adding 10 % H_2 to reactive gas.
(2) The role of H atoms is suggested to facilitate bonding with unbound N atoms in AlN:H films and hinder N-O bonding, thus, reducing oxygen concentration in AlN:H films. Also, the activation energy for the evolution of H_2 gas from AlN:H film has been determined to be 0.11 eV/atom and this result implies that the hydrogen atom in film forms the hydrogen bond.

ACKNOWLEDGMENTS

The authors gratefully acknowledge the financial support of the Ministry of Information and Telecommunication, Korean Government on this work, and would like to thank Prof. H.-J. Kang of the Dept. of Physics, Chungbuk National Univ. for XPS analysis, Electronics and Telecommunication Research Institute (ETRI) for RBS analysis, and Mr. C.-H. Kim of the Dept. of Chemical Engineering, KAIST for FTIR analysis.

REFERENCES

1. T. Shiosaki, T. Yamamoto, T. Oda and A. Kawabata, Appl. Phys. Lett. **36**, p. 643 (1980).
2. F. Hasegawa, T. Takahashi, K. Kubo and Y. Nannichi, Jpn. J. Appl. Phys. **9**, p. 1,555 (1987).
3. Y.- J. Yong and J.- Y. Lee, J. Vac. Sc. Technol. submitted (1996)
4. W. A. Grant in Method of Surface Analysis, edited by J. M. Walls (Cambridge Univ. Press, Cambridge, 1987), p. 299-337.
5. X. D. Wang, K. W. Hipps and U. Mazur, Langmuir **8**, p. 1,347 (1992).
6. X. D. Wang, W. Jiang, M. G. Norton and K. W. Hipps, Thin Solid Films **251**, p. 121 (1994).
7. H. E. Kissinger, Anal. Chem. **29**, p. 1,702 (1957).
8. W. Y. Choo and J. Y. Lee, Metall. Trans. **A13**, p. 135 (1982).
9. U. Mazur, Langmuir **6**, p. 1,331(1990).
10. L. J. Bellamy, The Infra-red Spectra of Complex Molecules, Chapman and Hall, London, 1975, pp. 233.
11. C. Kittel, Introduction to Solid State Physics, John Wiley & Sons, New York, 1986, pp. 74.

Monitoring of indium x-ray peak to optimize In$_x$Ga$_{1-x}$N layer grown by metalorganic chemical vapor deposition

Hongqiang Lu, Malathi Thothathiri,* Ziming Wu,* Ishwara Bhat

Electrical, Computer and System Engineering Department, Rensselaer Polytechnic Institute, Troy, NY12180

* Material Engineering Department, Rensselaer Polytechnic Institute, Troy, NY12180

Abstract

Indium droplet formation during the epitaxial growth of In$_x$Ga$_{1-x}$N films is a serious problem for achieving high quality films with high indium mole fraction. In this paper, we studied the formation of indium droplets on the In$_x$Ga$_{1-x}$N films grown by metalorganic chemical vapor deposition (MOCVD) using single crystal x-ray diffraction. It is found that the indium (101) peak in the x-ray diffraction spectra can be utilized as a quantitative measure to determine the amounts of indium droplets on the film. It is shown by monitoring the indium diffraction peak that the density of indium droplets increases at lower growth temperature. To suppress these indium droplets, a modulation growth technique is used. Indium droplet formation in the modulation growth is investigated and it is revealed in our study that the indium droplets problem has been partially relieved by the modulation growth technique.

Introduction

In$_x$Ga$_{1-x}$N ternary alloys, which have a direct bandgap from 1.9 eV to 3.4 eV, are an important wide bandgap semiconductor for optoelectronic applications. Recently, using In$_x$Ga$_{1-x}$N as an active layer, the first blue edge-emitting laser-diodes [1], as well as the high-efficiency green and blue light-emitting diodes (LEDs) have been demonstrated [2, 3]. However, despite these encouraging achievements in device fabrication, there are still few reports on the growth and characterization of In$_x$Ga$_{1-x}$N films, especially with respect to In droplets formation. Nagatomo *et. al* [4] first reported the successful growth of In$_x$Ga$_{1-x}$N at 500 $^\circ$C by MOCVD technique, but the crystal quality was poor at this growth temperature and no photoluminescence(PL) could be detected.   Yoshimoto and Matsuoka obtained better quality In$_x$Ga$_{1-x}$N films by raising the growth temperature up to 800 $^\circ$C [5, 6]. To get indium incorporated into the film at this temperature, they had to apply very high nitrogen over pressure (V/III ratio is 20,000) and high indium source flow rate. Very recently, there were several reports on the successful growth of In$_x$Ga$_{1-x}$N films [7-9], using high V/III (> 10,000) ratio. But none of them has reported detailed information about the effect of growth conditions on the indium droplets formation which is a common problem in In$_x$Ga$_{1-x}$N growth, especially when thick In$_x$Ga$_{1-x}$N films with higher indium mole fraction are grown. The understanding of the effects of the growth conditions on the In droplets formation is essential to grow high quality In$_x$Ga$_{1-x}$N films. In our study, we found that the indium (101) peak in the single crystal x-ray diffraction spectra is very sensitive to the density of indium droplets formed on the In$_x$Ga$_{1-x}$N film and could be utilized as a quantitative means to characterize the amount of indium droplets. Therefore, using this technique, we investigated the formation of indium droplets on In$_x$Ga$_{1-x}$N grown by metalorganic chemical vapor deposition (MOCVD), and also the possible ways to suppress the indium droplet formation during the growth.

Mat. Res. Soc. Symp. Proc. Vol. 449 © 1997 Materials Research Society

Experimental

In$_x$Ga$_{1-x}$N films were grown on c-plane sapphire substrates in a low-pressure (100 Torr), horizontal MOCVD system. Trimethylgallium (TMG), trimethylindium (TMI) and ammonia (NH$_3$) were used as the precursors for Ga, In and N, respectively. Hydrogen was employed as the carrier gas. Prior to the In$_x$Ga$_{1-x}$N growth, the sapphire substrate was first annealed at 1100 °C for 10 minutes in a mixture of H$_2$ and NH$_3$, which is believed to form a thin layer of AlN on the substrate. Then, the substrate temperature was lowered down to 600 °C to grow a 100 nm thick GaN buffer layer. After that the growth temperature was raised to 1050 °C to grow ~0.6 μm thick undoped GaN layer. The TMG and NH$_3$ flow rates were 20 μmol/min and 2 l/min, respectively. Finally, the In$_x$Ga$_{1-x}$N film was grown at 700 ~ 800 °C on this GaN layer. The TMG and NH$_3$ were introduced at a flow rate of 11 μmol/min and 2.5 l/min, respectively, with a total flow rate of 3.5 l/min. The TMI flow rate was varied from 0 to 44 μmol/min. For the In$_x$Ga$_{1-x}$N modulation growth discussed below, the growth sequence was as shown in Fig.1. For every cycle, In$_x$Ga$_{1-x}$N was grown for x seconds followed by y seconds annealing in NH$_3$ ambient at the same growth temperature. The growth time was adjusted such that the total In$_x$Ga$_{1-x}$N thickness was around 0.5 μm. Single crystal x-ray diffraction, as well as optical micrographs were utilized to monitor the indium droplets.

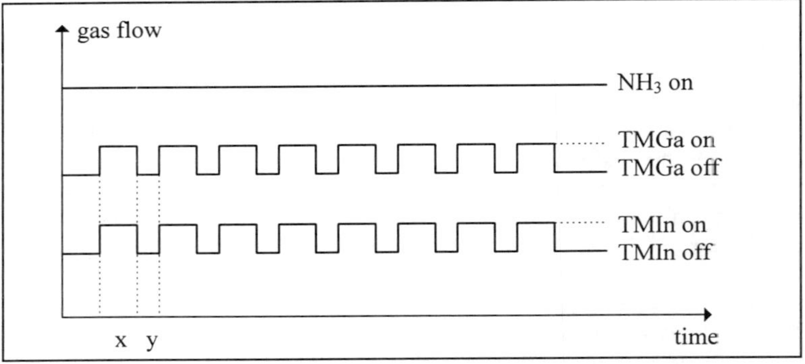

Fig. 1 Control sequence of gas flow in the modulation growth of In$_x$G$_{1-x}$N. NH$_3$ gas is always on, while TMG and TMI repeat the cycle in which gas flow is turned on for **x** seconds and then turned off for **y** seconds.

Results and Discussion

Fig.2 shows the morphology of 0.5 μm thick In$_x$Ga$_{1-x}$N films grown continuously on GaN. At low TMI flow rate (such as 6 μmol/min), indium droplets are not present on the surface though the film morphology is not smooth, as shown in Fig.2a. Single crystal x-ray diffraction measurements only show (0001) group peaks for the GaN and In$_x$Ga$_{1-x}$N on this sample, indicating that the film is probably single crystal. The indium mole fraction x of this film is about 6%. Indium droplets begin to appear when the TMI/TMG flow rate ratio is raised to above 1.5. Both the size and the number of these In droplets increase with TMI/TMG flow ratio. Fig.2b shows the morphology of a In$_x$Ga$_{1-x}$N film grown under a TMI/TMG ratio of 3, on which indium droplets are clearly observed. Single crystal x-ray diffraction spectra of this film, as shown in Fig.3, reveals an extra peak located at 33° compared to that of the sample shown in Fig.2a. This peak disappears if the indium droplets are etched off in a diluted HCl solution (also shown in Fig.3). The intensity of this peak increases whenever the amount of indium droplets on the surface increases. Hence we believe that this peak corresponds to the indium (101) diffraction from the In droplets present on the surface. In fact, this x-ray peak is more sensitive to the density of In droplets than the optical microscope. Therefore it can be utilized as a quantitative means to characterize the amounts of indium accumulated on the $In_xGa_{1-x}N$ surface and to arrive at a growth condition to eliminate them.

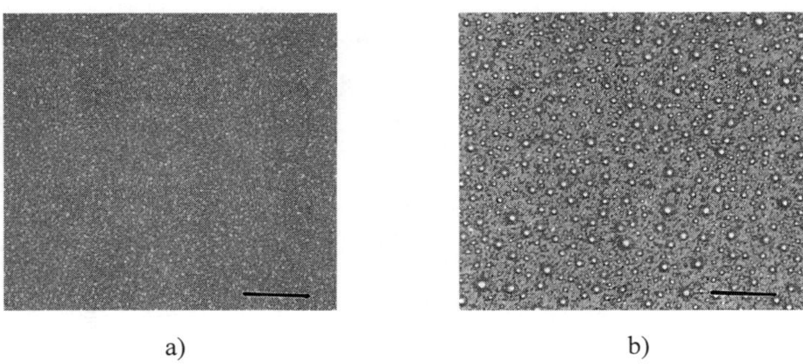

a)   b)

Fig. 2 Morphology of 0.5 μm thick $In_xGa_{1-x}N$ films which were grown continuously with a TMI/TMG flow rate ratio of: a) 0.5; b) 3. Marker represents 25 μm.

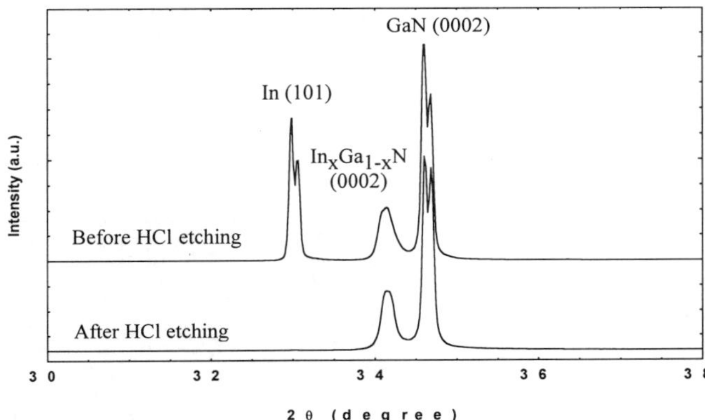

Fig. 3 X-ray diffraction spectra of the $In_xGa_{1-x}N$ film before and after HCl treatment. The film was grown continuously with a TMI/TMG flow rate ratio of 3.

Growth temperature is a very important and sensitive parameter in the growth of $In_xGa_{1-x}N$ film. Figure 4 shows the indium mole fraction in the film as well as the intensity of the indium

(101) x-ray peak as a function of the $In_xGa_{1-x}N$ growth temperature. These films are grown continuously, and the TMI and TMG flow rates are kept constant for all these films with the ratio of 1.5. It is seen that the indium incorporation in the film increases rapidly with the decrease of the growth temperature. This is usually explained by the reduction of indium species evaporation rate at the low temperature [6, 7]. Though reducing the growth temperature can increase the incorporation of indium into the film, the quality of the films is also poorer at lower growth temperature. This latter fact may be partially related to the formation of In droplets on the surface. As shown in the Fig.4, the amount of these indium droplets (represented by the indium (101) x-ray intensity) increases approximately exponentially with the reduction of the growth temperature.

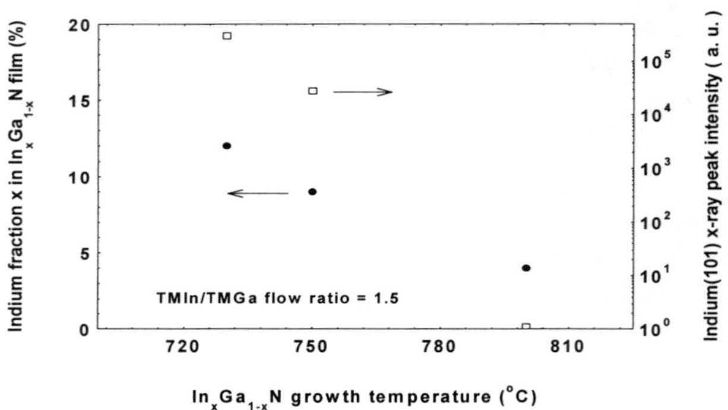

Fig.4   Indium mole fraction in the $In_xGa_{1-x}N$ film and the intensity of x-ray diffraction peak from indium droplets on the film as a function of growth temperature.

The problem of the In droplets formation becomes more severe whenever we attempt to increase the indium mole fraction in the film during the continuous growth. It is postulated that during the growth, the TMI molecules which adsorb onto the substrate would decompose completely at the high substrate temperature. Some of these indium adatoms would react with N species from NH_3 to form $In_xGa_{1-x}N$, and others travel along the surface until they either reevaporate to the ambient or meet some In nuclei to form indium droplets. At the initial growth stage of a few monolayer in thickness, most indium adatoms are still sparsely distributed and the size of the In nuclei is still smaller than the critical size for stable nucleation. If we stop the flow of the In and Ga precursors while supplying the NH_3 at this moment, these nuclei would then decompose and indium adatoms would either reevaporate or react with N provided by the NH_3 gas. If a

Fig. 5 Morphology of $In_xGa_{1-x}N$ film grown by modulation growth technique at 800 ^{o}C with a TMI/TMG flow ratio of 3. Both the growth and annealing time is 30 seconds. Marker represents 25 μm.

proper growth time/annealing time sequence is chosen, it may be possible to grow smooth $In_xGa_{1-x}N$ films. This (growth + annealing) cycle can be repeated several times to obtain thick films without indium droplets.

Fig.5 shows the morphology of $In_xGa_{1-x}N$ films grown by the modulation growth technique suggested above. The TMG and TMI flow rate were 11 μmol/min and 33 μmol/min, respectively. Every cycle includes 30 seconds of $In_xGa_{1-x}N$ growth followed by 30 seconds of annealing in NH_3 ambient. The process is repeated 30 times. The surface morphology shown in Fig.5 is free from In droplets and is better than those shown in Fig.2, and the x-ray diffraction does not show the presence of In droplets either. However, non-optimized growth/annealing sequence do show the presence of In droplets as identified by the x-ray diffraction.

The indium (101) peak in the x-ray diffraction spectra is again utilized here as a quantitative measure of the amount of In droplets formed during the modulation growth. Fig.6 shows variation of the In x-ray intensity with the TMI/TMG flow ratio for both continuous and modulation growth. The TMG flow rate is kept at 11 μmol/min and the growth temperature is 800 °C. In modulation growth, each growth cycle includes 30 seconds of $In_xGa_{1-x}N$ growth plus 30 seconds of annealing process. The indium peak arises at smaller value of TMI/TMG ratio for continuous growth and its intensity increases much faster. On the other hand, this peak does not appear until the TMI/TMG ratio reaches 3 in the case of modulation growth. Even under this condition, the indium droplets could not be easily observed using the optical microscope. For both growth technique, the indium mole fraction obtained in the film is almost the same if the TMI and TMG flow ratio is kept same. Even though the growth parameters for the modulation growth are not optimized in our case, the modulation growth technique has already demonstrated effective reduction of indium droplets and the improvement of the film morphology.

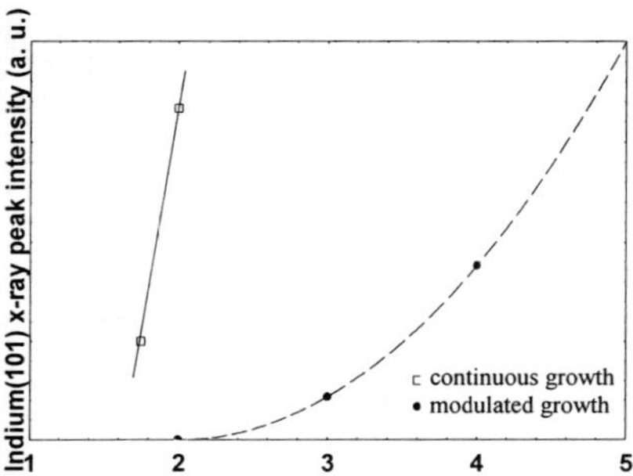

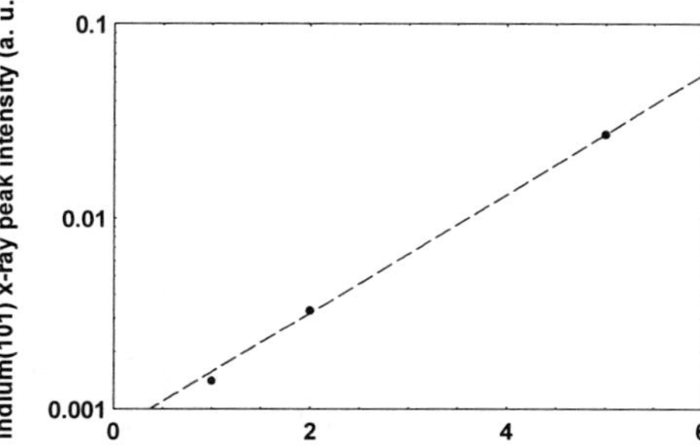

Fig. 6 Intensity of the x-ray diffraction peak of indium droplets as a function of TMI/TMG flow ratio.

Fig. 7 Intensity of x-ray diffraction peak of indium droplets as a function of growth/annealing time ratio.

Fig.7 shows the influence of the growth and the annealing time on the formation of indium droplets in the modulation growth. The annealing time is set at 2 minutes while the growth time varied from 2 minutes to 10 minutes. The total $In_xGa_{1-x}N$ growth time is kept at 20 minutes. It is noticed that the intensity of this indium peak increases exponentially with the growth time, implying that the process of indium droplets generation and growth will accelerate with the time after the initial formation of indium droplets. Also it seems that short time annealing right after the growth could not remove this indium as long as they form droplets of some critical size.

Therefore the growth time in each growth cycle should be greatly reduced to prevent the formation of initial indium nuclei.

Conclusions

In summary, we have studied the indium droplet formation on the $In_xGa_{1-x}N$ films grown by MOCVD. It is found that the indium (101) peak in the x-ray diffraction spectra originated from the indium droplets can be used as a quantitative means to study the amount of indium droplets on the film surface. By this means, it is found that the indium mole fraction, as well as the amount of indiums droplets, increase with the reduction of the growth temperature. It is also demonstrated using the x-ray peaks that the modulation growth can suppress the indium droplets formation on the $In_xGa_{1-x}N$ films. It is expected that by optimizing the growth and annealing conditions such as reducing the growth time and the growth rate, indium droplet free $In_xGa_{1-x}N$ films could be achieved by this modulation growth technique.

Acknowledgment

This work is partially supported by DARPA (subcontracted through Texas Instruments through Johnson Matthey Prime Contract No. MDA972-91-C-0046). We also acknowledge the support through a grant from Philips Laboratory, Briarcliff Manor, NY. We also thank J. Barthel for technical assistance.

Reference

1. S. Nakamura, M. Senoh, S. Nagahama, N. Iwasa, T. Yamada, T. Matsushita, H. Kiyohu, and Y. Sguimoto, Jpn. J. Appl. Phys. **35**, L74 (1996)
2. S. Nakamura, M. Senoh, N. Iwasa and S. Nagahama, Jpn. J. Appl. Phys. **34**, L797 (1995)
3. S. Nakamura, M. Senoh, N. Iwasa and S. Nagahama, T. Yamada and T. Mukai, Jpn. J. Appl. Phys. **34**, L1332 (1995)
4. T. Nagatomo, T. Kuboyama, H. Minamino, and O. Omoto, Jpn. J. Appl. Phys. **28**, L1334 (1989)
5. N. Yoshimoto, T. Matsuoka, T. Sasaki, and A. Katsui, Appl. Phys. Lett. **59**, 2251 (1991)
6. T. Matsuoka, N. Yoshimoto, T. Sasaki, and A. Katsui, J. Electron. Mat. **21**, 157 (1992)
7. S. Nakamura and T. Mukai, Jpn. J. Appl. Phys. **31**, L1457 (1992)
8. K. S. Boutros, F. G. McIntosh, J. C. Roberts, S.M. Bedair, E. L. Piner, and N. A. EI-Masry, Appl. Phys. Lett. **67**, 1856 (1995)

PULSED EXCIMER LASER PROCESSING OF AlN/GaN THIN FILMS

W.S. Wong*, L.F. Schloss*, Sudhir G.S.*, B.P. Linder**, K.-M. Yu***, E.R. Weber*,
T. Sands*, N.W. Cheung**

* Department of Materials Science and Mineral Engineering, University of California, Berkeley, CA 94720
** Department of Electrical Engineering and Computer Science, University of California, Berkeley, CA 94720
*** Lawrence Berkeley National Laboratory, Materials Science Division, Berkeley, CA 94720

Abstract

A KrF (248 nm) excimer laser with a 38 ns pulse width was used to study pulsed laser annealing of AlN/GaN bi-layers and dopant activation of Mg-implanted GaN thin films. For the AlN/GaN bi-layers, cathodoluminescence (CL) showed an increase in the intensity of the GaN band-edge peak at 3.47 eV after pulsed laser annealing at an energy density of 2000 mJ/cm^2. Rutherford backscattering spectrometry of a Mg-implanted AlN (75 nm thick)/GaN (1.0 μm thick) thin-film heterostructure showed a 20% reduction of the ^{4}He$^+$ backscattering yield after laser annealing at an energy density of 400 mJ/cm^2. CL measurements revealed a 410 nm emission peak indicating the incorporation of Mg after laser processing.

Introduction

Gallium nitride (GaN) and its solid-solution alloys with InN and AlN comprise an optoelectronic materials system spanning the infra-red to the ultra-violet regime. Developments in GaN growth and doping [1-4] have resulted in the fabrication of blue light emitting diodes [5-8] and more recently a blue laser diode [9]. As a result, there is significant research interest in the III-V nitrides materials system for optoelectronic devices.

Advances in the area of GaN growth have significantly improved the materials quality in recent years while inherent difficulties in p-type doping and alloying with Al, to form Al$_x$Ga$_{1-x}$N, still prevent full utilization of the III-V nitrides. P-type material has been realized by thermal annealing of metal-organic chemical vapor deposition (MOCVD)-grown Mg-doped GaN [10]. N- and p-type doping of GaN using Si$^+$ and Mg$^+$/P$^+$ implantation, respectively, in conjunction with a post implant rapid thermal anneal (RTA) at ~1100 °C has also been reported [11]. Recently, high sheet electron concentration and high Hall mobility were reported for a doped channel AlGaN/GaN heterostructure [12].

In response to the doping and alloying issues, we have investigated UV pulsed laser annealing (PLA) as an alternative approach to non-equilibrium processing of the Group III nitrides. The nanosecond-scale pulse lengths and the materials selectivity afforded by the UV wavelength are potential advantages over more conventional rapid thermal processing for alloying, dopant activation and selective etching of the III-V nitrides. A KrF (248 nm) excimer laser may be utilized to selectively process AlN/GaN bi-layers or activate dopants in AlN/GaN heterostructures. AlN, with a band-gap energy of 6.2 eV, is transparent to the 248 nm UV light which is absorbed primarily by the underlying GaN film. This selectivity allows for non-equilibrium annealing of the AlN/GaN interface to form metastable (Al,Ga)N alloy layers. As a demonstration of PLA for the Group III nitrides, this laser processing technique was used to selectively anneal AlN capped GaN and Mg-implanted AlN/GaN heterostructures.

Mat. Res. Soc. Symp. Proc. Vol. 449 © 1997 Materials Research Society

Experimental Conditions

PLA was performed on bare GaN, AlN capped GaN, Mg-implanted GaN and Mg-implanted AlN/GaN heterostructures. The GaN thin films, of 1.0 μm nominal thickness grown on c-plane sapphire, were obtained from CREE Research Inc. A Lambda Physik Lextra 200 KrF pulsed excimer laser (38 ns pulse width) was used for pulsed laser deposition (PLD) of the AlN cap layers. The AlN was deposited in a cryo-pumped high vacuum chamber using a N_2 ambient ($P_{N_2} = 5 \times 10^{-5}$ Torr) at a substrate temperature of 400 °C. The incident laser energy, focused on a 99.98% pure AlN target, was nominally set at 500 mJ with a pulse repetition rate of 5 Hz. Deposition times were varied between 15 minutes and 1 hour to achieve cap layer thicknesses between 75 and 300 nm.

Mg-implanted GaN was accomplished by conventional ion implantation. In this study, the Mg implant profile was tailored with the following implant conditions: (1) energy: 40 keV, dose: 1.0×10^{15} cm^{-2}; (2) energy: 100 keV, dose: 2.5×10^{15} cm^{-2}; (3) energy: 150 keV, dose: 3.0×10^{15} cm^{-2}. These parameters were used to target a flat profile to a depth of ~150 nm below the GaN film surface.

PLA was done in air using the same KrF laser described above but utilizing a separate optical setup. Variation of the energy density was accomplished by defocusing the laser light using a fused silica plano convex lens with a 350 mm focal length. In this way, the laser spot size could be varied between 0.25 and 4 cm^2 with energy densities between 200 and 2000 mJ/cm^2. Characterization by scanning electron microscopy (SEM), cathodoluminescence (CL), Rutherford backscattering (RBS) and energy dispersive x-ray (EDX) was performed.

Results and Discussion

PLA of the GaN thin films was examined using SEM to determine ablation thresholds (Figure 1). At energy densities up to 350 mJ/cm^2, the bare GaN surface morphology remained smooth. At a fluence of 600 mJ/cm^2, the surface morphology began to roughen and surface features appeared indicating a maximum energy threshold had been reached. For the case of the Mg implanted GaN, SEM characterization found similar features on the GaN surface at a lower energy density of 350 mJ/cm^2. The lower threshold energy of the Mg-implanted films suggest that the implant altered the optical absorption and reflectivity of the GaN. In an attempt to preserve the surface morphology and prevent the decomposition of the GaN surface, an AlN cap layer was deposited onto the GaN. A 300 nm AlN cap layer effectively prevented any roughening of the GaN surface up to laser fluences of 600 mJ/cm^2 for both unimplanted and Mg-implanted GaN.

At an energy density of 2000 mJ/cm^2, micro-crack features appeared on the surface of the AlN/GaN structure (Figure 2a). The micro-crack formation indicates that the tensile stress wave resulting from the difference in the thermal coefficient of expansion between the AlN, GaN and sapphire exceeded the tensile strength of the AlN cap. After a second pulse under the same energy conditions the AlN film appears to have been lifted off the GaN (Figure 2b). This observation suggests that the tensile cracks in the AlN film resulting from the first pulse allow ablation of the GaN film during the second pulse, ejecting fragments of the AlN cap.

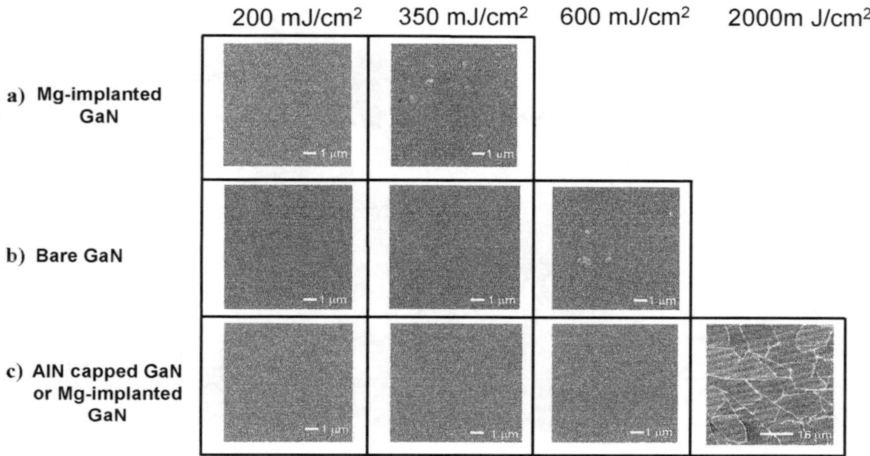

Figure 1. SEM characterization of surface morphology comparing: a) Mg-implanted GaN, b) bare GaN and c) AlN capped GaN or Mg-implanted GaN for laser fluences between 200-2000 mJ/cm². The AlN cap is observed to preserve smooth surface morphology for fluences up to 600 mJ/cm².

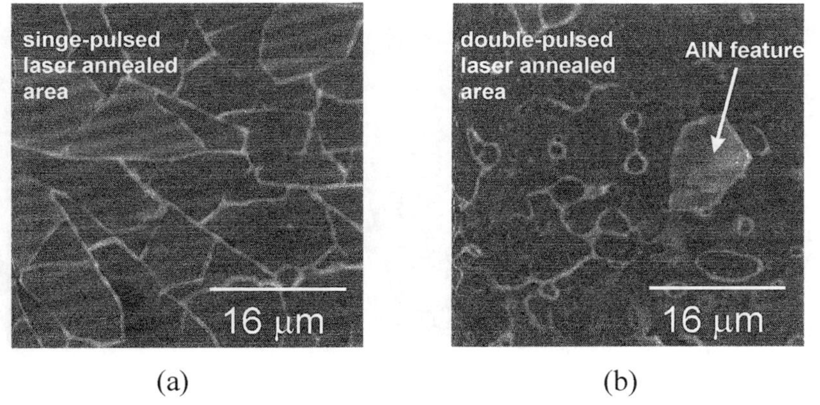

(a)   (b)

Figure 2.   (a) SEM micrograph showing micro-crack formation on AlN capped surface after a 2000 mJ/cm² single laser pulse.
(b) SEM micrograph showing lift-off of the AlN cap layer after second annealing pulse.

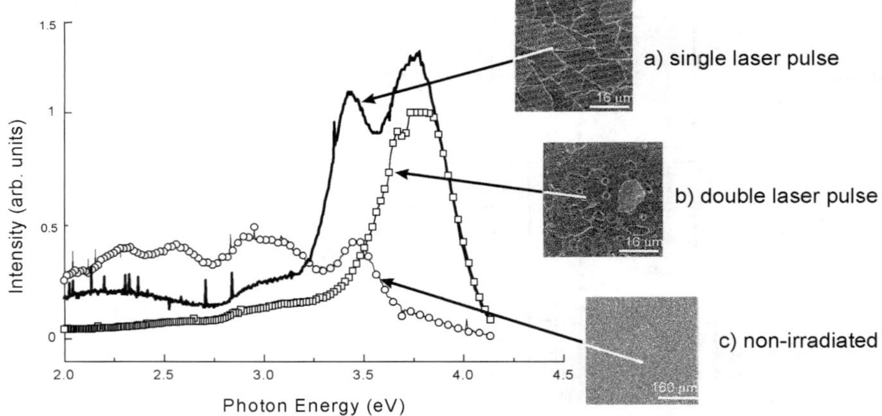

Figure 3. CL characterizarion comparing: a) single-pulsed laser annealed AlN capped GaN at 2000 mJ/cm^2 (solid line), b) double-pulsed laser annealed AlN capped GaN (squares) and c) non-irradiated AlN capped GaN (circles). The associated surface morphology is shown for each PLA condition. The band-edge peak at 3.47 eV is observed to increase by a factor of five after a single laser anneal pulse. Lift off of the AlN/GaN film occurs after a second anneal pulse indicated by the absence of the GaN band-edge peak. The 3.77 eV peak is due to the sapphire substrate.

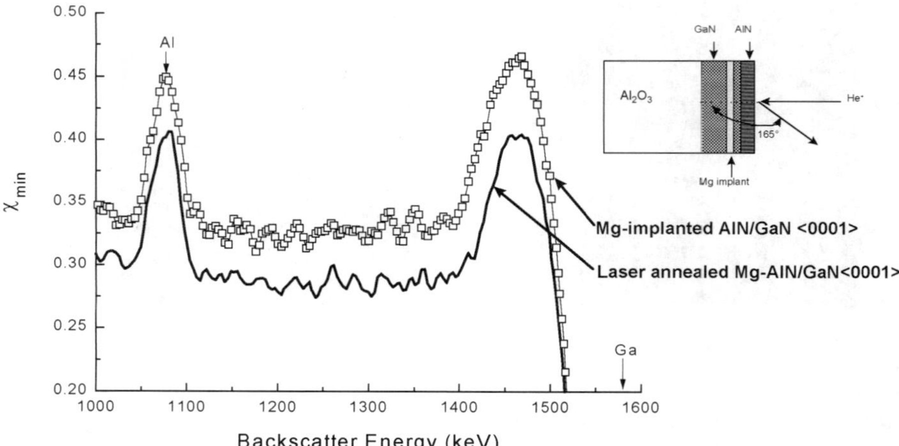

Figure 4. RBS channeling of Mg-implanted AlN/GaN heterostructures. The decrease in χ_{min} after laser annealing at   400 mJ/cm^2 indicates partial recovery from the ion implantation damage.

CL characterization was used to confirm the surface morphology observations. Figure 3 shows a CL spectrum comparing three different process conditions. Improvement in the luminescence properties of the GaN film were observed after an initial laser energy pulse of 2000 mJ/cm^2. The characteristic GaN band-edge peak at 3.47 eV is seen to increase by a factor of five after initial laser processing indicating improved GaN film quality. The low energy deep-level emissions at ~2.2 eV associated with impurity recombination centers in GaN [13,14] have decreased after PLA, suggesting a reduction of the deep level impurities. CL measurement of the double-pulsed region reveal the GaN band-edge peak has been extinguished leaving only a bare sapphire peak at 3.77 eV. We can conclude, therefore, that the implementation of the AlN cap preserves the GaN film and prevents GaN decomposition. PLA at a fluence higher than the ablation threshold improved the GaN film quality but once the mechanical integrity of the AlN cap layer was diminished, ablation of the GaN occurred upon subsequent laser irradiation.

Rutherford backscattering spectroscopy, using ^{4}He$^+$ ions at an energy of 1.96 MeV, was implemented to examine Mg-implanted AlN (75 nm thick)/GaN (1.0 µm thick) heterostructures. Figure 4 shows the RBS channeling spectra comparing a laser annealed and an unannealed Mg-implanted GaN film. For the implant conditions used, the GaN film remained crystalline before and after annealing. More significantly, a decrease of χ_{min} by as much as 20%, compared to the as-implanted GaN, was measured following PLA at an energy density of 400 mJ/cm^2. The backscatter reduction in the annealed sample indicates partial removal of the ion implant damage in the GaN and improved crystallinity after PLA.

Examination of multiple-pulse annealing was performed on uncapped Mg-implanted GaN by SEM, CL and EDX. For multiple laser pulses at 400 mJ/cm^2, a Mg related peak at 410 nm [15] was found with CL characterization at 95 K (Figure 5). The intensity of the peak was observed to increase with the number of laser pulses indicating incorporation of Mg into the GaN (Figure 6). SEM analysis of the surface morphology showed droplet formations. These droplets were determined by EDX to be Ga rich suggesting some decomposition of the GaN surface at these annealing conditions. AlN cap layers are currently being investigated to suppress the GaN surface decomposition.

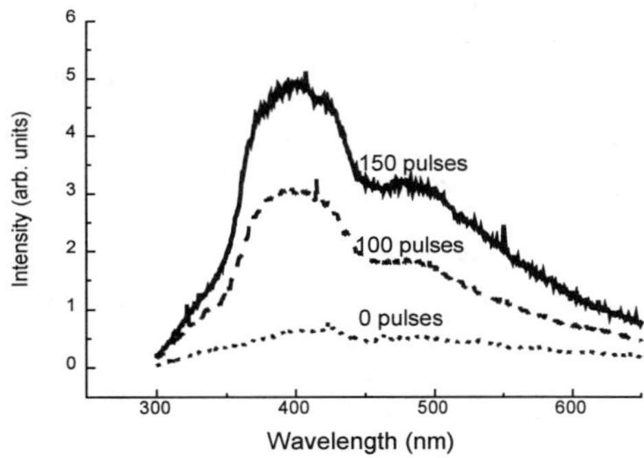

Figure 5. CL of Mg-implanted GaN. Luminescence peak intensity at 410 nm, indicating incorporation of Mg, increases with the number of pulses at a fluence of 400 mJ/cm2.

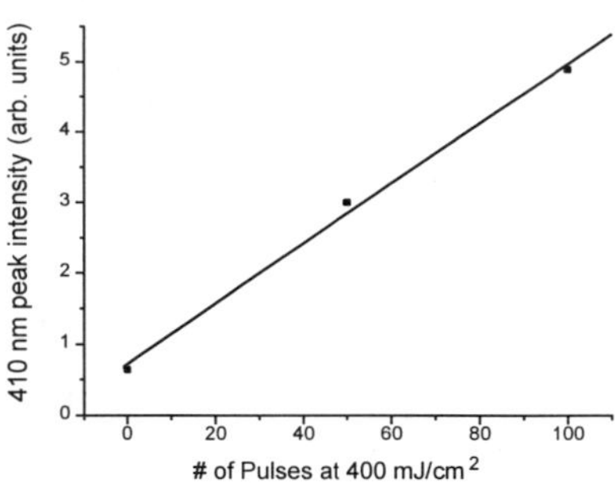

Figure 6. 410 nm CL peak intensity as a function of laser annealing pulses.

Summary

Demonstration and feasibility of pulsed excimer laser processing of GaN was presented. From SEM analysis, an AlN cap layer was found to be an effective barrier in preventing surface decomposition of GaN during pulse laser annealing. CL measurements of AlN capped GaN showed improved GaN film quality after PLA. AlN cap layers were also effective in preventing surface decomposition of Mg-implanted GaN at the same laser energy densities. RBS channeling showed improved crystalline properties of Mg-implanted AlN/GaN heterostructures following PLA. Mg incorporation in implanted GaN was also observed by CL after multiple laser annealing pulses. The results presented demonstrate that PLA is a viable alternative processing technique that offers non-equilibrium and bandgap selective processing of the Group III nitrides.

Acknowledgments

This work benefited from the use of the UC-Berkeley Integrated Materials Laboratory which is sponsored by the National Science Foundation. This research is supported in part by the Air Force Office of Scientific Research (AFOSR/JSEP) under contract # F49620-94-C-0038, and by the U.S. Department of Energy under contract # DE-AC03-76SF00098.

References

1. S. Yoshida, S. Misawa and S. Gonda, Appl. Phys. Lett. **42**, 427 (1983).
2. H. Amano, N. Sawaki, and I. Akasaki, Appl. Phys. Lett. **48**, 353 (1986).
3. S. Nakamura, Jpn. J. Appl. Phys. **30**, L1705 (1991).
4. Z. Sitar, L.L. Smith and R.F. Davis, J. Crystal Growth **141**, 11 (1994).
5. S. Nakamura, M. Senoh, N. Iwasa, and S. Nagahama, Jpn. J. Appl. Phys. Lett. **34**, L797 (1995).
6. S. Nakamura, M. Senoh, N. Iwasa, and S. Nagahama, T. Yamada and T. Mukai, J. Appl. Phys. Lett. **34**, L797 (1995).
7. S. Nakamura, M. Senoh, and T. Mukai, Jpn. J. Appl. Phys. **30**, L1708 (1991).
8. S. Nakamura, T. Mukai, and M. Sonah, Jpn. J. Appl. Phys. **30**, L1998 (1991).
9. S. Nakamura, M. Senoh, S. Nagahama, N. Iwasa, T. Yamda, T Matsushita, H. Kiyoku and Y. Sugimoto, Jpn. J. Appl. Phys. **35**, L74 (1996).
10. S. Nakamura, N. Iwasa, M. Senoh, and T. Mukai, Jpn. J. Appl. Phys. **31**, 1258 (1992).
11. S.J. Pearton, C.B. Vartuli, J.C. Zolper, C. Yuan and R.A. Stall, Appl. Phys. Lett. **67**, 1435 (1995).
12. M.A. Khan, M.S. Shur, and Q. Chen, Appl. Phys. Lett., **68**, 3022 (1996).
13. W. Götz, N.M. Johnson, R.A. Street, H. Amano, and I. Akasaki, Appl. Phys. Lett. **66**, 1340 (1995).
14. T.A. Kennedy, E.R. Glaser, J.A. Freitas, Jr., W.E. Carlas, M.A. Khan and D.K. Wickenden, J. Electron. Mater. **24**, 219 (1995).
15. S.S. Liu, T.R. Cass, and D.A. Stevenson, J. Electron. Mat. **6**, 237 (1977).

CHEMICAL ETCHING OF AlN AND InAlN IN KOH SOLUTIONS

C.B. Vartuli,* J.W. Lee,* J.D. MacKenzie,* S.J. Pearton,* C.R. Abernathy,* J.C. Zolper,** R.J. Shul** and F. Ren***
*University of Florida, Gainesville FL 32611, **Sandia National Laboratories, Albuquerque NM 87185, ***Lucent Technologies, Bell Laboratories, Murray Hill NJ 07974

ABSTRACT

Wet chemical etching of AlN and $In_xAl_{1-x}N$ was investigated in KOH-based solutions as a function of etch temperature, and material quality. The etch rates for both materials increased with increasing etch temperatures, which was varied from 20 °C to 80 °C. The crystal quality of AlN prepared by reactive sputtering was improved by rapid thermal annealing at temperatures to 1100 °C with a decreased wet etch rate of the material measured with increasing anneal temperature. The etch rate decreased approximately an order of magnitude at 80 °C etch temperature after a 1100 °C anneal . The etch rate for $In_{0.19}Al_{0.81}N$ grown by Metal Organic Molecular Beam Epitaxy was approximately three times higher for material on Si than on GaAs. This corresponds to the superior crystalline quality of the material grown on GaAs. Etching of $In_xAl_{1-x}N$ was also examined as a function of In composition. The etch rate initially increased as the In composition changed from 0 to 36%, and then decreased to 0 Å/min for InN. The activation energy for these etches is very low, 2.0 ± 0.5 kcal•mol^{-1} for the sputtered AlN. The activation energies for InAlN were dependent on In composition and were in the range 2-6 kcal·mol^{-1}. GaN and InN layers did not show any etching in KOH at temperatures up to 80 °C.

INTRODUCTION

Much progress has recently been made in the areas of growth, dry etching and implant isolation and doping of the III-V nitrides and their ternary alloys. This has resulted in nitride-based blue/UV light emitting and electronic devices.[1-9]. There has been less success in developing wet etch solutions for these materials due to their excellent chemical stability. High etch rates have been achieved in dry etch chemistries, but damage may be introduced by ion bombardment, and controlled undercutting is difficult to attain. In addition, since dry etching has a physical component to the etch, selectivities between different materials is generally reduced.

Amorphous AlN has been reported to etch in 100 °C HF/H_2O,[10-12] HF/HNO_3,[13] and $NaOH$,[14] and polycrystalline AlN in hot ($\leq$ 85 °C) H_3PO_4 at rates less than 500 Å/min.[15,16] Mileham et al.[17] reported the etching of AlN defective single crystals in KOH based solutions at etch temperatures ranging from 23- 80 °C. They reported decreased etch rates with increasing crystal quality, as the reactions occurred favorably at grain boundaries and defect sites. InN in aqueous KOH solutions was reported to etch at a few hundred angstroms per minute at 60 °C.[18]

Zolper et. al.[19] observed that the luminescence and surface morphologies of GaN annealed in flowing N_2 actually improved for RTA temperatures up to 1100 °C. Similar results were obtained at lower temperatures by Lin et. al.[20] This is a common situation for lattice-mismatched systems (e.g. GaAs/Si), where post-growth or even in-situ annealing is generally found to improve structural and optical properties, provided that group V loss from the surface can be suppressed and that external impurities do not diffuse in during the anneal. Thus we might expect that annealing of group III-nitrides would affect their wet etching characteristics.

In this paper, we report an examination of wet etching of AlN and $In_xAl_{1-x}N$ in KOH solutions as a function of crystal quality, etch temperature, and film composition. AlN samples prepared by reactive sputtering on Si substrates at ~ 200 °C were annealed at temperatures from

Mat. Res. Soc. Symp. Proc. Vol. 449 © **1997 Materials Research Society**

400 to 1100 °C, and as expected, the etch rate decreased with anneal temperature, indicating improved crystal quality. We found that InAlN on Si substrates had higher wet etch rates. Both AlN and InAlN samples had an increase in etch rate with etch temperature. The etch rate for the InAlN increased as the In composition increased from 0 to 36%, and then decreased to zero for InN. Finally the effect of doping concentration in InAlN samples of similar In concentration (~ 3%) was examined- much higher etch rates were observed for the heavily doped material at solution temperatures above 60 °C.

EXPERIMENTAL

The AlN was reactively sputter deposited on a Si substrate to a thickness of ~ 1200 Å using a N_2 discharge and a pure Al target. This type of AlN film has been shown to be an effective annealing cap for GaN at a temperature of 1100 °C.[21] The InAlN samples were grown using Metal Organic Molecular Beam Epitaxy (MO-MBE) on semi-insulating (100) GaAs substrates or p-type (1 $\Omega \cdot$cm) Si substrates in an Intevac Gen II system as described previously.[22,23] The group-III sources were triethylgallium, trimethylamine alane and trimethylindium, respectively, and the atomic nitrogen was derived from an ECR Wavemat source operating at 200 W forward power. The layers were single crystal with a high density $(10^{11} - 10^{12}$ cm$^{-2})$ of stacking faults and microtwins. InAlN samples were found to contain both hexagonal and cubic forms. The $In_xAl_{1-x}N$ were either conducting n-type as grown (~ 10^{18} cm^{-3}) for x ≥ 0.03 due to residual autodoping by native defects or fully depleted for x < 0.03. The compositions examined were 100, 75, 36, 29, 19, 3.1, 2.6 and 0 % In.

The AlN samples were annealed in a rapid thermal anneal (RTA) system (AG 410T) face down on a GaAs substrate for 10 s at temperatures between 500 - 1150 °C in a N_2 atmosphere. For wet etching studies, all samples were masked with Apiezon wax patterns. Etch depths were obtained by Dektak stylus profilometry after the removal of the mask, with an approximate 5% error. Scanning electron microscopy (SEM) was used to examine the undercutting on the etched samples. AZ400K developer solution, with an active ingredient of KOH,[17] was used for the etch, at etch temperatures between 20 and 80 °C.

RESULTS AND DISCUSSION

(a) AlN

Figure 1 shows the etch rate of the sputtered AlN as a function of etch temperature for samples as deposited or annealed at 500, 700, 900, 1000 and 1100 °C. The etch rates of both the as deposited and 500 °C annealed sample increase sharply as the etch temperature increases from 20 to 50 °C, and then level off; the rate drops by approximately 10 % with a 500 °C anneal. The samples annealed at 700, 900 and 1000 °C also show similar trends, with a monotomic decrease in rate for higher anneal temperatures. The crystal quality appears to increase significantly with anneal temperature, as the etch rate drops accordingly. The etch rate continues to drop by ~ 10 % with each successive anneal, up to 1000 °C. After the 1100 °C the etch rate drops and is less temperature dependent. Overall there is an ~ 90 % reduction in etch rate from the as deposited AlN to that annealed at 1100 °C for etching at 80 °C.

The activation energy for an etch solution can be determined from an Arrhenius plot, and is shown in Fig. 2. The activation energies for all samples was the same within experimental error, 2.0 ± 0.5 kcal•mol^{-1}. This is indicative of a diffusion limited reaction. This is much lower than the activation energy of 15.45 kcal•mol^{-1} reported by Mileham et al,[26] for AlN grown by metal organic molecular beam epitaxy. The quality of the material in the current experiment is much lower though, and the etch may be proceeding at such a rapid rate that the solution is becoming depleted of reactants near the materials surface.

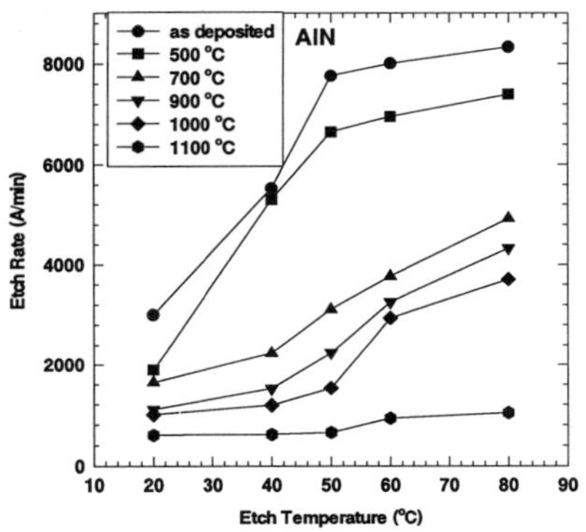

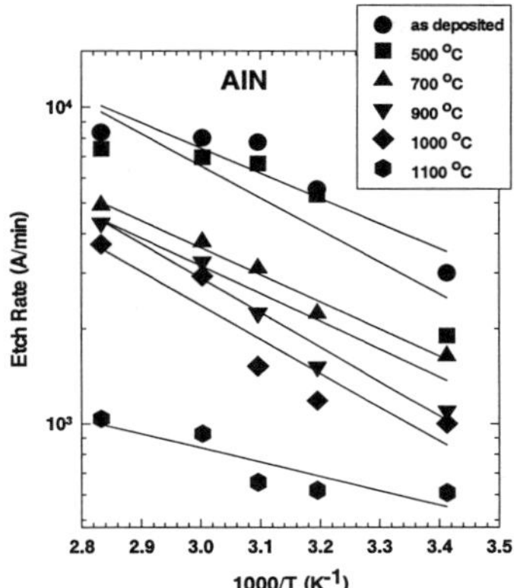

Figure 1 Etch rate of AlN as a function of etch temperature for samples as deposited or annealed at 500, 700, 900, 1000 and 1100 °C.

Figure 2 Arrhenius plots of etch rates for as-deposited or annealed AlN as a function of reciprocal etch temperature.

(b) $In_xAl_{1-x}N$

The etch rate as a function of solution temperature for $In_xAl_{1-x}N$ grown on either GaAs or Si, for 19 % In is shown in Fig. 3. At 20 °C etch temperature there is no difference in etch rate. The etch rates for both materials increase with etch temperature, with the differential in etch rates also increasing with temperature. As was mentioned previously, the InAlN grown on Si has a greater concentration of crystalline defects as evident from x-ray diffraction and absorption measurements. At 80 °C the etch rate for the film on the Si substrate is approximately three times faster than for the film grown on GaAs. This is another clear indication of the dependence of wet etch rate on material quality, and emphasizes why it has proven very difficult to find etch solutions for high quality single-crystal nitrides.

Etch rates for $In_xAl_{1-x}N$ grown on GaAs for $0 \leq x \leq 1$ are shown in Fig. 4, for etch temperatures between 20 and 80 °C. Up to 40 °C, the etch rates are very low and show little dependence on In composition. The AlN etches much faster at these temperatures than any composition of the ternary alloy InAlN. As the etch temperature increases to 60 °C, the etch rates increase, showing a peak for 36% In. This is presumably due to tradeoff between the reduction in average bond strength for InAlN relative to the pure binary AlN, and the fact that the chemical sensitivity falls off at higher In concentrations. Thus the etch rates initially increase for increasing In, but then decrease at higher concentrations because there is no chemical driving force for etching to occur. InN did not etch in this solution at any temperature, but was occasionally lifted off during long etches because of the defective interfacial region between InN and GaAs being attacked by the KOH.

Arrhenius plots of etch rates for $In_xAl_{1-x}N$ for $0 \leq x \leq 1$ giving activation energies for the etches are shown in Fig. 5. There is substantial scatter in the data, but the activation energies are all in the range 2-6 kcal•mol^{-1}, which again is consistent with diffusion-controlled etching. This is not desirable for device fabrication processes because the rates are then dependent on solution agitation and the etched surface morphology are generally rougher than for reaction-controlled solutions.

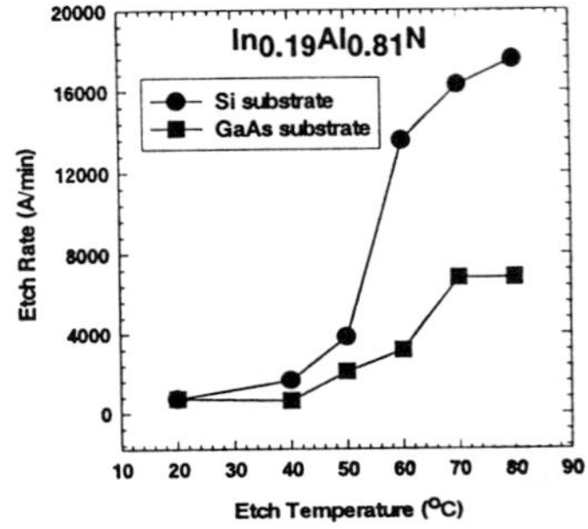

Figure 3 Etch rates as a function of etch temperature for $In_xAl_{1-x}N$ grown on GaAs and Si, for 19 % In.

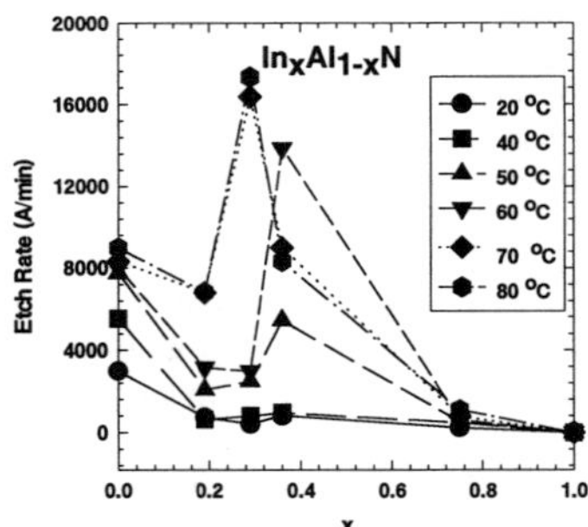

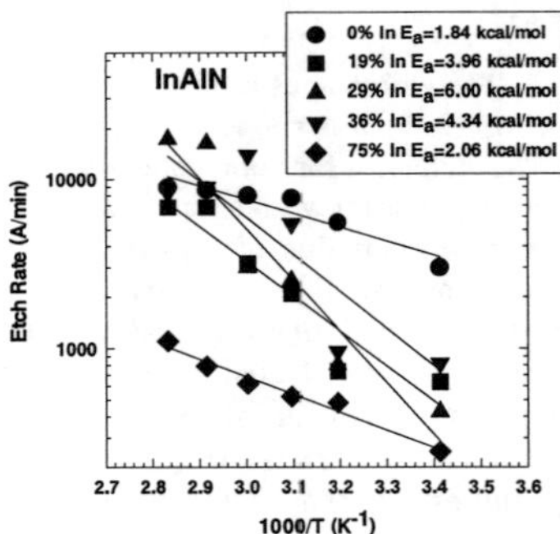

Figure 4. Etch rate for $In_xAl_{1-x}N$ for $0 \leq x \leq 1$ grown on GaAs at solution temperatures between 20 and 80 °C.

Figure 5 Arrhenius plots of etch rates for $In_xAl_{1-x}N$ for $0 \leq x \leq 1$ as function of reciprocal etch temperature, giving activation energy for etch.

CONCLUSIONS AND SUMMARY

Annealing of sputtered AlN improved the crystal quality of the film, decreasing the chemical etch rate in KOH-based solutions. InAlN etch rates also increased with decreasing crystalline quality. Both AlN and InAlN samples had activation energies for etching in KOH $\leq$ 6 kcal•mol^{-1}, which is typical of a diffusion-controlled etch mechanism. The etch rate for the InAlN initially increased as the In composition increased from 0 to 36%, and then decreased to zero for pure InN.

ACKNOWLEDGMENTS

The authors would like to thank the staff of the Microfabritech Facility for their help with this work. The work at the University of Florida is supported by NSF (DMR-9421109), an AASERT grant through ARO (Dr. J. M. Zavada), a DARPA grant (A. Husain) administered by AFOSR (G.L. Witt), and a University Research Initiative grant #N00014-92-J-1895 administered by ONR. The work at Sandia is supported by DOE contract DE-AC04-94AL85000.

REFERENCES

1. S. Nakamura, M. Senoh, and T. Mukai, Jpn. J. Appl. Phys. **30**, L1708 (1991).

2. S.C. Binari, L.B. Rowland, W. Kruppa, G. Kelner, K. Doverspike, and D.K. Gaskill, Electron. Lett. **30**, 1248 (1994).

3. M.A. Khan, M.S. Shur, and Q. Chen, Electron. Lett. **31**, 2130 (1995).

4. M.A. Khan, J.N. Kuznia, A.R. Bhattarai, and D.T. Olson, Appl. Phys. Lett. **62**, 1248 (1993).

5. S. Nakamura, M. Senoh, and T. Mukai, Appl. Phys. Lett. **62** 2390 (1993).

6. I. Akasaki, H. Amano, M. Kito, and K. Kiramatsu, J. Lumin. **48/49**, 666 (1991).

7. S. Nakamura, M. Senoh, N. Iwasa, and S. Nagahama, Appl. Phys. Lett. **67**, 1868 (1995).

8. J.C. Zolper, A.G. Baca, R.J. Shul, R.G. Wilson, S.J. Pearton and R.A. Stall, Appl. Phys. Lett. **68**, 166 (1996).

9. S. Nakamura, M. Senoh, S. Nagahama, N. Iwasa, T. Yamada, T. Matsushita, H. Kiyoku and Y. Sugimoto, Jap. J. Appl. Phys. **35** L74 (1996).

10. K.M. Taylor and C. Lenie, J. Electrochem. Soc. **107** 308 (1960).

11. G. Long and L.M. Foster, J. Am. Ceram. Soc. **42** 53 (1959).

12. N.J. Barrett, J.D. Grange, B.J. Sealy and K.G. Stephens, J. Appl. Phys. **57** 5470 (1985).

13. C.R. Aita and C.J. Gawlak, J. Vac. Sci. Technol. A 1 403 (1983).

14. G.R. Kline and K.M. Lakin, Appl. Phys. Lett. **43** 750 (1983).

15. T. Pauleau, J. Electrochem. Soc. **129** 1045 (1982).

16. T.Y. Sheng, Z.Q. Yu and G.J. Collins, Appl. Phys. Lett. **52** 576 (1988).

17. R.J. Mileham, S.J. Pearton, C.R. Abernathy, J.D. MacKenzie, R.J. Shul and S.P. Kilkoyne, Appl. Phys. Lett. **67**, 1119 (1995).

18. Q.X. Guo, O. Kato and Y. Yoshida, J. Electrochem. Soc. **139** 2008 (1992).

19. J.C. Zolper, M. Hagerott-Crawford, A.J. Howard, J. Rainer and S.D. Hersee, Appl. Phys. Lett. **68**, 200 (1996).

20. M.E. Lin, B.N. Sverdlov and H. Morkoc, Appl. Phys. Lett. **63**, 3625 (1993).

21. J.C. Zolper, D.J. Reiger, A.G. Baca, S.J. Pearton, J.W. Lee, R.A. Stall, Appl. Phys. Lett. (in press).

22. C.R. Abernathy, J. Vac. Sci. Technol. A **11** 869 (1993).

23. C.R. Abernathy, Mat. Sci. Eng. Rep. **14**, 203 (1995).

COMPARISON OF ICl AND IBr FOR DRY ETCHING OF III-NITRIDES

C.B. Vartuli,* J.W. Lee,* J.D. MacKenzie,* S.J. Pearton,* C.R. Abernathy* and R.J. Shul**
*Department of Materials Science and Engineering, University of Florida, Gainesville FL 32611,
**Sandia National Laboratories, Albuquerque NM 87185

ABSTRACT

ICl/Ar ECR discharges provide the fastest dry etch rates reported for GaN, 1.3 µm/min. These rates are much higher than with Cl_2/Ar, CH_4/H_2/Ar or other plasma chemistries. InN etch rates up to 1.15 µm/min and 0.7 µm/min for $In_{0.5}Ga_{0.5}N$ are obtained, with selectivities up to 5 with no preferential loss of N at low rf powers and no significant residues remaining. The rates are much lower with IBr/Ar, ranging from 0.15 µm/min for GaN to 0.3 µm/min for InN. There is little dependence on microwave power for either chemistry because of the weakly bound nature of ICl and IBr. In all cases the etch rates are limited by the initial bond breaking that must precede etch product formation and there is a good correlation between materials bond energy and etch rate. The fact that low microwave power can be employed is beneficial from the viewpoint that photoresist masks are stable under these conditions, and there is no need for use of silicon nitride or silicon dioxide. Selectivities for GaN over AlN with ICl and IBr are still lower than with Cl_2- only.

INTRODUCTION

Dry etching of GaN and related compounds in various chemistries have been investigated in both reactive ion etching, (RIE),[1-4] and electron cyclotron resonance (ECR) modes.[5-8] ECR etching has proven much more efficient than RIE for the nitrides due to higher ion and excited neutral densities and more effective bond-breaking, allowing the etch by-products to form more readily.[9,10] Another concern is preferential loss of the group V element from the surface due to its higher volatility, which can result in rough morphologies during etching of compound semiconductors. This is of particular concern for the nitrides. The dry etching of III-V materials is often rate limited by removal of the group III etch product, particularly for In-containing materials. Various alternative plasma chemistries have been explored in the search for a fast, smooth, anisotropic etch for nitrides, such as Br_2 and I_2.[11,12] Pearton et al. investigated the ECR etching of InP in HI/H_2/Ar[13] and found etch rates > 1 µm/min, smooth anisotropic etching with no residue after etch. This chemistry was also used to etch GaAs, InAs, InSb, InP, and GaSb,[14] reporting features that were anisotropic, smooth, with no deposition during etch, and an order of magnitude faster rates than with the CH_4/H_2/Ar chemistry. Less work has been done with Br_2-based plasma discharges. GaN has been etched in HBr, HBr/Ar, and HBr/H_2 under reactive ion etch conditions, with etch rates around > 600 Å/min at 400 V dc.[12] Somewhat faster rates were achieved under ECR conditions in HBr/H_2[11] (~900 Å/min at -150 V dc). Plasma chemistries based on ICl and IBr are of interest as plasma dissociation produces active chlorine or bromine and iodine and should provide efficient etching of the nitrides. A review of nitride etching results has recently appeared.[15] In this paper we report experiments comparing ICl/Ar and IBr/Ar ECR plasma etching of GaN, InN, InAlN, AlN and InGaN.

EXPERIMENTAL

The GaN, AlN, InN, $In_{0.36}Al_{0.64}N$ and $In_{0.5}Ga_{0.5}N$ samples were grown by Metal Organic Molecular Beam Epitaxy (MO-MBE) on semi-insulating, (100) GaAs and Si substrates or on Al_2O_3 substrates in an Intevac Gen II system as described previously.[16,17] The group-III sources were triethylgallium, trimethylamine alane and trimethylindium, respectively, and the atomic nitrogen was derived from an ECR Wavemat source operating at 200 W forward power. The layers were single crystal with a high density of stacking faults and microtwins. The GaN and AlN were resistive as-grown, and the InN was highly autodoped n-type (>10^{20} cm^{-3}) due to the presence of native defects. InAlN and InGaN

Mat. Res. Soc. Symp. Proc. Vol. 449 © 1997 Materials Research Society

were found to contain both hexagonal and cubic forms. The $In_{0.36}Al_{0.64}N$ and $In_{0.5}Ga_{0.5}N$ compositions we employed were conducting n-type as grown ($\sim 10^{19}$ cm^{-3}) due to residual autodoping by native defects.

The samples were patterned with either a carbon-based mask or photoresist, and were etched in a Plasma-Therm SLR 770 reactor with an Astex 4400 low profile ECR source. The ICl and IBr are crystalline solids with melting temperatures of ~ 23 and 50 °C respectively.[18] We loaded ~ 50 g of ICl or IBr into a stainless steel vacuum vessel directly connected to a mass flow controller which injected the vapor into the ECR source. The vacuum vessel was wrapped in Al-foil and heated to ~ 45 °C. We obtained flow rates up to 12 standard cubic centimeters per minute (sccm). The process pressure was held constant at 1.5 mTorr and the temperature of the He back-side cooled chuck was held at 23 °C. The rf power (13.56 MHz) was varied between 50 and 250 W and the microwave power between 400 and 1000 W. The plasma chemistries used were ICl/Ar or IBr / Ar with respective flows of 4 sccm/4 sccm, 2 sccm/ 6 sccm, 1 sccm/ 7 sccm and 8 sccm/ 0 sccm. Step heights were obtained from Dektak stylus profilometry measurements after the removal of the carbon mask with acetone, and used to calculate the etch rates. The error in these measurements is approximately ± 5 %. The surface morphology of selected GaN samples were examined with Atomic Force Microscope (AFM) using a Si tip in tapping mode and Scanning Electron Microscopy (SEM), while near-surface composition was measured by Auger Electron Spectroscopy (AES).

RESULTS AND DISCUSSION

The etch rates as a function of plasma composition for GaN, InN, InAlN, AlN and InGaN are shown in Fig. 1. Microwave power was held at 1000 W and rf power at 150 W (corresponding to a dc self bias of -170 V at the sample position). For the ICl based etch (Fig. 1, top), the GaN and InGaN etch rates rise as the amount of ICl in the etch increases from 12.5 to 50%, and then level off. Above 50% ICl there appears to be a competition between the formation of $GaCl_3$ with has a boiling point of 201 °C, with that of GaI_3 which sublimes at 345[19]. The $GaCl_3$ may form preferentially at some plasma compositions. The InN shows a sharp increase in etch rate above 25 % ICl. This suggests that at 25 % ICl the InI_3, which is much more volatile ($InCl_3$ boils at 600 °C, InI_3 at 210 °C), can form easily. The InAlN and AlN are not greatly effected by changes in the composition of the plasma in ICl (or IBr, Fig. 1, bottom), perhaps because both Al containing etch products have similar volatility ($AlCl_3$ boils at 183 °C, $AlBr_3$ at 263 °C and AlI_3 at 191 °C) and because the etch rate is probably limited by the initial bond breaking in the Al-containing materials. We expect that AlN and InAlN will be difficult to etch because of their high average bond energies (11.52 eV/atom for AlN, 7.72 eV/atom for InN, compared to 6.52 eV/atom for GaAs).[19] The N containing etch products are much more volatile than the group-III etch products, with NCl_3 boiling at < 71 °C while NI_3 is explosive.

The etch rates for InN and InGaN increased as the amount of IBr in the etch increased from 12.5 to 25%, and remained constant at higher percentages (Fig. 1, bottom). This suggests that above 25 % IBr the etching is no longer reaction-limited. The $InBr_3$ etch product is much less volatile than InI_3, as mentioned earlier. Above that composition however, there may have been competition between the formation of $InBr_3$ and InI_3, which slowed the etch, or the etch may have been limited by the removal of the reactants from the surface. GaN etch rates showed little change with IBr composition to 50 % IBr plasma composition, but at 100 % IBr the etch rates increased sharply. There may not have been enough reactants at the etch surface at 50 % IBr, or at these lower percents of IBr there may be a competition between the formation of $GaBr_3$ and GaI_3.

In Figure 2 the etch rate as a function of microwave power for GaN, InN, InAlN, AlN and InGaN is shown for values between 400 and 1000 W for ICl/Ar plasmas (top) and IBr/Ar (bottom). The rf power was held at 150 W, and 4 sccm ICl or IBr / 4 sccm Ar gas flows were used. Both InAlN and AlN have low etch rates in ICl/Ar, and show no significant change in etch rate with increasing microwave power. This indicates that they are not reaction limited in this chemistry, since increasing the microwave power results in a higher concentration of reactive species which enhances the chemical component of the etch mechanism. GaN and InGaN showed a slight increase in etch rate from 400 W microwave power to 600 W. Thereafter the GaN etch rate dropped gradually with increasing microwave power, while the InGaN etch rate dropped sharply at 800 W and then remains constant at 1000 W. This would indicate either a diffusion-limited etch, where the number of reactants becoming available exceeds the rate at which the iodine and/or chlorine etch products can be removed, or competition between reactants occurs above 600 W microwave power. The InN had a maximum in etch rate at 800 W ECR power. This might result from the large difference in volatilities of the etch products for this material, leading to a strong sensitivity to reactant density. We expect that below that density the etch rate is reaction-limited and above it there is competition between the reactants that limits the etch rate. A similar trend is observed for InN in the IBr/Ar mixtures although the peak is not as distinct.

The etch rates for InAlN and AlN were again quite low in IBr based plasmas (Fig. 2, bottom). GaN had constant etch rates for powers between 400 W and 800 W in the IBr chemistry, and then increased sharply at 1000 W. The InGaN etch rate again decreased with increasing microwave power. As the InGaN etch rate increased monotonically with increasing rf power (as will be seen shortly), the removal of the etch products would seem to be limiting the etch rates for this material.

Figure 3 shows the etch rate as a function of rf power for GaN, InN, InAlN, AlN and InGaN in ICl/Ar (top) and IBr/Ar (bottom) plasmas for chuck powers between 50 and 250 W. Microwave power was held at 1000 W and the flow was held at 4 sccm ICl or IBr / 4 sccm Ar. The AlN and InAlN rates were affected very little by increasing rf power in either chemistry. GaN, InN and InGaN all have large increases in etch rate as the rf is increased from 150 to 250 W in the ICl chemistry. This could mean that the bombarding ions have enough energy to remove the less volatile etch product at this power or to more efficiently break bonds, allowing the etch to proceed with both I- and Cl- etch products. GaN etch rates in IBr/Ar increased with increased rf power to 150 W, and then decreased slightly at 250 W, where

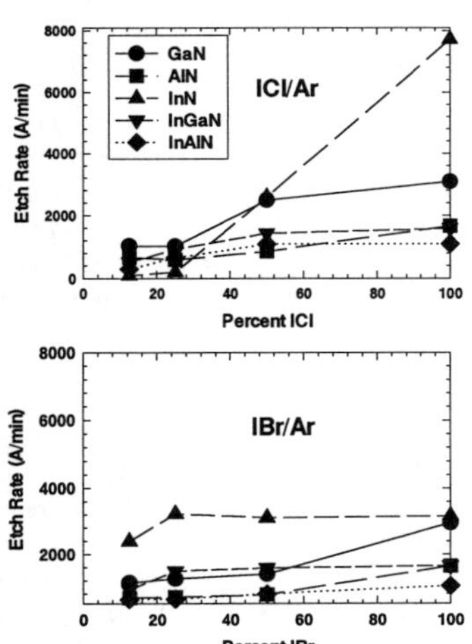

Figure 1 Etch rate as a function of percent ICl (top) or IBr (bottom) for GaN, InN, InAlN, AlN and InGaN in 1000 W (ECR), 150 W rf, 1.5 mTorr discharges.

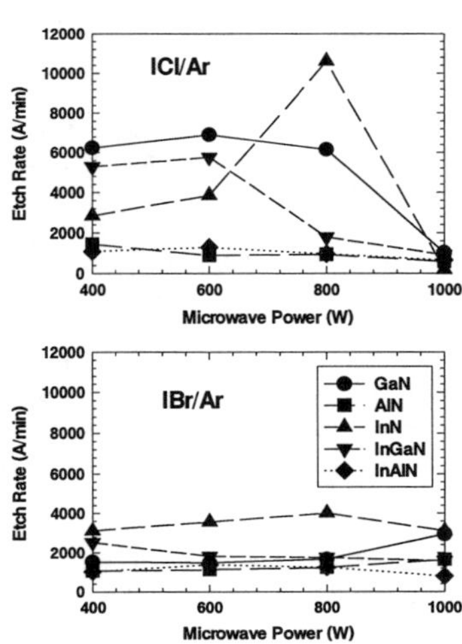

Figure 2. Etch rate as a function of microwave power for GaN, InN, InAlN, AlN and InGaN in 4 ICl/4 Ar (top) or 4 IBr/4 Ar (bottom) plasmas (150 W rf, 1.5 mTorr).

the sputtering ions may have removed reactants before the etch could proceed. InN and InGaN in IBr/Ar plasmas had large increases in etch rate as the rf was increased from 150 to 250 W, similar to the results in ICl/Ar.

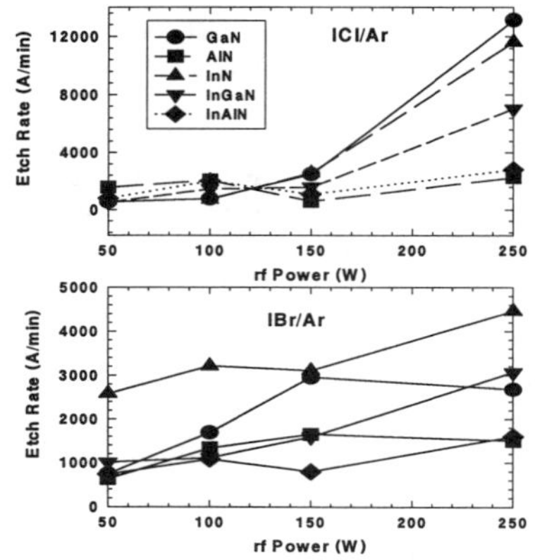

Figure 3 Etch rate as a function of rf power for GaN, InN, InAlN, AlN and InGaN in 4 ICl/4 Ar (top) or 4 IBr/4 Ar (bottom) plasmas (1000 W ECR, 1.5 mTorr).

Figure 4. RMS roughness for GaN as a function of rf power in 4 ICl/4 Ar 1000 W ECR, 1.5 mTorr discharges plasmas.

In Fig. 4 the RMS roughness for GaN etched in ICl/Ar as a function of rf power is shown. The RMS roughness for the as grown sample is shown for reference. These samples were unpatterned to avoid roughness caused by mask material being redeposited. The etched surfaces were found to be significantly smoother than that of the as grown sample indicating that surface features are removed predominantly by sputtering. Sharp features will be removed by ion milling faster than flat features because of the angular dependence of milling rate, and as long as preferential sputtering of N does not occur, this will lead to a degree of smoothing of the surface.[20]. Above 150 W the surface roughness begins increasing again, probably due to the onset of preferential sputtering.

AES depth profiles of GaN as-grown and etched in ICl/Ar at 50 W or 100 W rf power are shown in Fig. 5. At these powers we find no reduction in the N/ Ga ratio. This means that there is little preferential loss of N during the etching at these powers. Adventitious C and O from native oxide are also observed on the surface of these samples. I and Cl are found in the top 25 Å of the etched samples. Similar results were found for GaN etched in IBr/Ar plasma, though no Br was detected on the surface. At high rf powers one would expect preferential N-loss, as reported previously by Shul et. al.[7]

Figure 6 shows the etch selectivity of GaN over InN, InAlN, InGaN or AlN under ICl/Ar conditions as a function of rf power (top), percent ICl (middle) and microwave power (bottom). The selectivity of GaN over the other nitrides rose with increasing rf power, with GaN/AlN reaching ~ 6 and GaN/InAlN almost 5 at 250 W rf. The volatility of the InCl$_3$ was lower than that of GaCl$_3$, and as the percent ICl in the etch increased, so did the selectivity for GaN/InN, reaching ~ 10 at 100% ICl. With both GaCl$_3$ and GaI$_3$ having high volatilities, with increasing reactant concentration, GaN was etched faster than the In-containing compounds, which may still be limited by removal of InCl$_3$. GaN etched much faster than AlN and InAlN as well for most microwave powers (Figure 6, bottom), achieving selectivities of ~8 and 5 respectively at 600 W microwave power. In IBr/Ar chemistries the selectivities were low, never going above 4 for any set of rf or microwave powers of plasma compositions. We assume this is due to the similar volatilities of iodide and bromide etch products.

CONCLUSION

The etch rates for GaN, InN, InAlN, AlN and InGaN were measured in ICl/Ar and IBr/Ar plasmas. The sensitivity to changes in plasma chemistry, microwave power and rf power appears to be directly influenced by the volatility of the group-III -I and Cl- or Br-etch products. InN, with the largest difference between volatility of etch products, proved to be the most sensitive to the plasma composition and ion density in ICl/Ar plasma chemistries. Very fast etch rates were achieved for GaN, InN and InGaN in ICl/Ar chemistries. At 250 W rf power AlN and InAlN had slow etch rates in this mixture and were affected very little by changes in etch conditions. GaN and AlN etched in IBr/Ar showed a sharp increase in etch rate as the IBr composition increased from 50 to 100 %, while the etch rates for the other materials stayed relatively constant above 25% IBr. All the materials showed a general increase in etch rate with increasing rf power in both chemistries. The etched surface of GaN under both plasma chemistries was found to be extremely smooth with little preferential loss of N from the surface at low rf powers. There was no detectable residue from the etch after IBr/Ar plasma, and only slight Cl residue in ICl/Ar chemistry.

ACKNOWLEDGMENTS

The authors would like to thank the staff of the Microfabritech Facility for their help with this work. The work at the University of Florida is supported by NSF (DMR-9421109), an AASERT grant

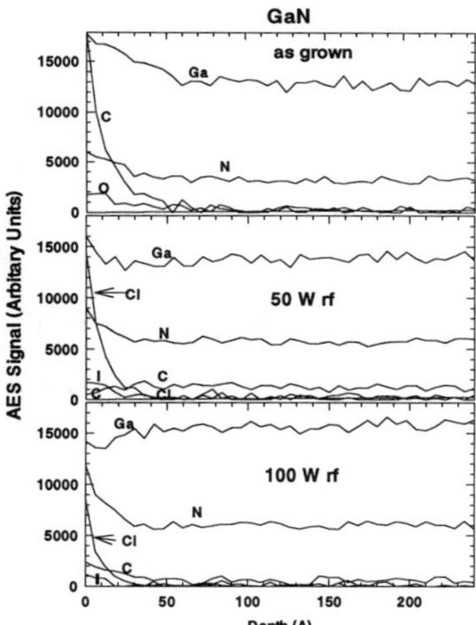

Figure 5 AES depth profiles of GaN as-grown (top), and etched in 4 ICl/ 4 Ar at 50 W rf (middle) and at 100 W rf (bottom) power. The ECR source power was 1000 W and the pressure 1.5 mTorr

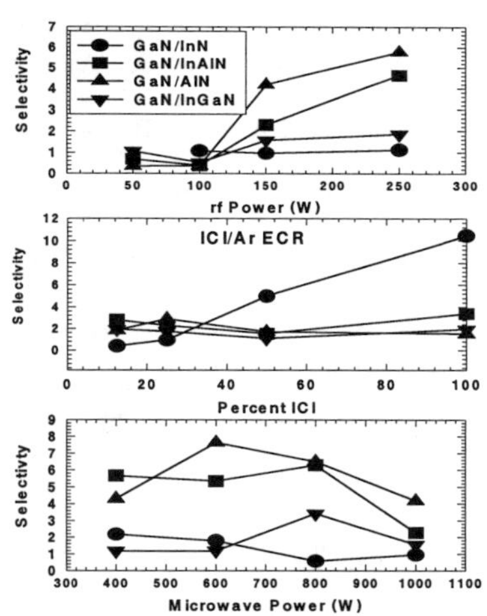

Figure 6.Selectivity of GaN over InN, InAlN, InGaN or AlN in ICl/Ar plasmas as a function of rf power (top), percent ICl (middle) and microwave power (bottom). The ECR power was 1000 W for the top two plots, the rf power 150 W for the bottom two plots and the plasma composition 4 ICl/4 Ar for the top and bottom plots.

through ARO (Dr. J. M. Zavada), and a University Research Initiative grant #N00014-92-J-1895 administered by ONR, and a DARPA grant (A. Husain) administered by AFOSR (G. Witt). The work at Sandia was supported by DOE under contract no. DE-AC04-94AL85000.

REFERENCES

1. M.E. Lin, Z.F. Zan, Z. Ma, L.H. Allen and H. Morkoc, Appl. Phys. Lett. **64** 887 (1994).

2. A.T. Ping, I Adesida, M. Asif Khan and J.N. Kuznia, Electron. Lett. **30** 1895 (1994).

3. H. Lee, D.B. Oberman and J.S. Harris, Jr., Appl. Phys. Lett. **67** 1754 (1995).

4. S.J. Pearton, C.R. Abernathy, F. Ren, J.R. Lothian, P.W. Wisk, A. Katz and C. Constantine, Semicond. Sci. Technol. **8** 310 (1993).

5. S.J. Pearton, C.R. Abernathy and F. Ren, Appl. Phys. Lett. **64** 2294 (1994).

6. L. Zhang, J. Ramer, K. Zheng, L.F. Lester and S.D. Hersee, MRS Fall Meeting, Boston MA, 1995).

7. R.J. Shul, S.P. Kilcoyne, M. Hagerott Crawford, J.E. Parmeter, C.B. Vartuli, C.R. Abernathy and S.J. Pearton, Appl. Phys. Lett. **66** 1761 (1995).

8. R.J. Shul, presented at 189th ECS meeting, Los Angeles CA, May 1996.

9. C.B. Vartuli, S.J. Pearton, C.R. Abernathy, R.J. Shul, A.J. Howard, S.P. Kilcoyne, J.E. Parmeter and M. Hagerott-Crawford, J. Vac. Sci. Technol. A **14** 1011 (1996).

10. C.W. Krueger, C.A. Wang, D.Hsieh and M. Flytzani-Stepanopoulos, J. Cryst.Growth **153** 81 (1995).

11. C.W. Krueger, C.A. Wang and M. Flytzani-Stepanopoulos, Appl. Phys. Lett. **60** 1459 (1992).

12. D.C. Flanders, L.D. Pressman and G. Pinelli, J. Vac. Sci. Technol. B **8** 1990 (1990).

13. S.J. Pearton, U.K. Chakrabarti, A. Katz, F. Ren and T.R. Fullowan, Appl. Phys. Lett. **60** 838 (1992).

14. S.J. Pearton, U.K. Chakrabarti, W.S. Hobson, C.R. Abernathy, A. Katz, F. Ren, T.R. Fullowan, and A.P. Perley, J. Electrochem. Soc. **139** 1763 (1992).

15. H.P. Gillis, D.A. Choutov and K.P. Martin, JOM, August 1996 pp 50-55.

16. C.R. Abernathy, J. Vac. Sci. Technol. A **11** 869 (1993).

17. C.R. Abernathy, Mat. Sci. Eng. Rep. **14**, 203 (1995).

18. CRC Handbook of Chemistry and Physics (CRC Press, Boca Raton, FL 1990).

19. W.A. Harrison, Electronic Structures and Properties of Solids (Freeman, San Francisco, 1980).

20. S.J. Pearton, C.R. Abernathy, F. Ren,and J.R. Lothian, J. Appl. Phys. **76** 1210 (1994).

REACTIVE ION BEAM ETCHING OF GaN GROWN BY MOVPE

K. SAOTOME, A. MATSUTANI, T. SHIRASAWA, M. MORI,
T. HONDA, T. SAKAGUCHI, F. KOYAMA and K. IGA
Precision and Intelligence Laboratory, Tokyo Institute of Technology
4259 Nagatsuta, Midori-ku, Yokohama 226, Japan

ABSTRACT

A dry etch technique using Cl_2 based reactive ion beam etching (RIBE) has been developd for GaN-based semiconductor lasers. The etching rate of 350 - 1000 Å/min was obtained. This is applicable for micro fabrication of GaN based materials in the same way as used for other III-V group semiconductors. Furthermore, it is found that the surface damage of GaN layers induced by the RIBE-etch can be removed using ultra-violet assisted wet-etching using alkali solution. The PL intensity of damaged GaN layers is increased after the post-process wet-etching.

INTRODUCTION

Gallium nitride and related compounds are attracting considerable attention for light emitting devices operating in blue and ultraviolet spectral regions. In order to realize ultra low-threshold GaN based vertical cavity surface emitting lasers (VCSELs), a microfabrication process is required to minimize the volume of an active layer. We have investigated a reactive ion beam etching (RIBE) technique to fabricate a microresonator for VCSELs [1,2]. The RIBE is recognized as one of the most effective techniques among various dry-etching techniques for conventional III-V compound semiconductor lasers[3-6]. This technique also enables the formation of smooth and vertical facets . In the case of applying this dry-etch technique to fabricate microstructures such as VCSELs, the etched surface must be smooth and vertical to avoid light scattering[7, 8] and surface recombination. Up to now, the reactive ion etching (RIE)[9,10], chemically assisted ion beam etching (CAIBE)[11], reactive fast-atom-beam etching (FAB)[12] and electron cyclotron resonance etching (ECR)[13,14] techniques using Cl_2 layers have been reported for GaN compounds. However, the RIBE etching technique for GaN based materials has not been fully clarified yet.

When we reduce the VCSEL volume, the dry-etched microstructure may suffer from surface recombination on etched surfaces which increases the threshold current[15]. VCSELs are especially sensitive to surface recombination when their size is reduced to several μm or lower, and even a low damage induced by the RIBE is unacceptable. Thus, the wet-etching technique is necessary to remove an induced damage layer. Recently, for GaN-based materials,   Watanabe *et. al.* and Minsky *et al.* reported that the photo-assisted wet-etching using alkali solution may be effective to fabricate device GaN structure[16,17].

In this paper, etching characteristics of undoped-GaN layers are studied, in particular, the dependence of the etch rate on the ion extraction voltage and the substrate temperature is discussed. Also, we report the removal of RIBE-induced damage in GaN layers by using a photo-assisted wet-etching for the first time.

RIBE OF GaN

We used a high-vacuum RIBE system (ANELVA UHV-ECR etching system) [18]. This system has an electron cyclotron resonance (ECR) ion source, and the ECR plasma is generated

Mat. Res. Soc. Symp. Proc. Vol. 449 ©1997 Materials Research Society

by introducing a 2.45GHz microwave into a horizontal discharge chamber on which a magnetic field is superimposed. The etching gas is pure chlorine with a fixed chamber gas pressure of 8×10^{-4} Torr. The dry etch can be performed with a sample temperature ranging from room temperature to 250°C. The ion extraction voltage was changed from 300V to 500V. The etching time is 5 min for all samples. The etching mask material was a photo resist (OFPR 8600). The depth of etched surface was measured by a DEKTAK stylus profilometer.

Undoped GaN layers were grown on (0001) sapphire substrates by atmospheric pressure metal organic vapor phase epitaxy (MOVPE). After thermal cleaning in H_2 ambient at 1150 °C for 15 min, an AlN buffer layer [19,20] was grown at 800 °C for 10 minutes, followed by the growth of a GaN layer was grown at 1150 °C for one hour. Trimethylgallium (TMGa) and NH_3 were used as a Ga and N source, respectively. The V/III ratio was about 500. The thickness of the GaN layer as about 0.8-1.0 μm for each growth. The GaN (0002) peak is clearly observed in the double-crystal X-ray diffraction measurement(2θ-ω mode) . The FWHM of 2θ-ω and rocking-curve associated with GaN (0002) were 55 arcsec and 450 arcsec, respectively. In 10K photo-luminescence spectra of GaN layers grown by MOVPE, the near-band-edge emission can be clearly observed.

The dependence of the etch rate on the substrate temperature is shown in Fig.1. The temperature dependence of the etch rate of the GaN is smaller than that of the GaAs. This result indicates that the activation energy in etching GaN is lower than GaAs due to a large binding energy between Ga and N atoms.

The dependence of the etched depth on the ion extraction voltage is also shown in Fig.2. The etching rate is almost proportional to the ion extraction voltage. The etching rate of the GaN increases about 3 times between 300V and 500V. This result indicates that an

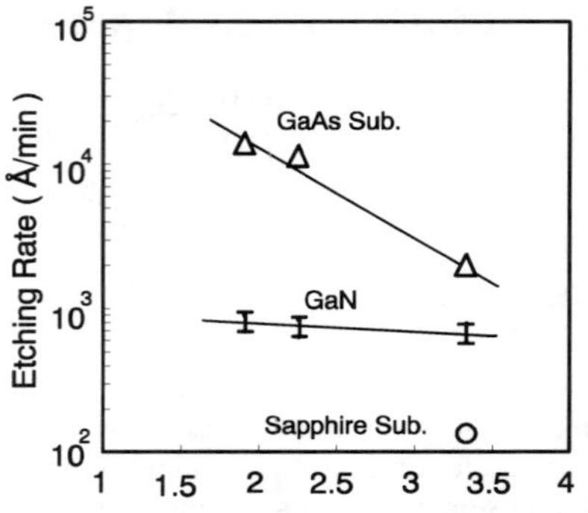

Fig.1 Etching rate dependence on substrate temperature.

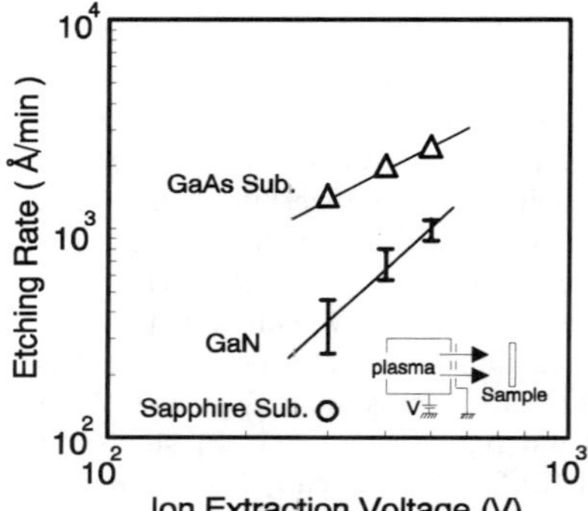

Fig. 2 Etching rate dependence on ion extraction voltage.

Fig.3 Scanning electron microscope of etched GaN on sapphire :Cl_2-RIBE,8×10^4Torr,RT, Ion.Ext.Volt.: 400V, 30min.

effect of an ion milling is bigger than chemical etch, and that the generative reaction of the etch products is thermally difficult. We think that it will be necessary to increase an ion current density to increase the etching rate of GaN.

Figure 3 shows a scanning electron microscope (SEM) of a GaN etched by RIBE. The etched depth was about 1μm. The ion extraction voltage was 400V. The etching time was 30min. You can see that the etched surface is very smooth in this figure.

REMOVAL OF DAMAGED LAYER BY PHOTO-ASISSTED WET-ETCHNG

A photo-assisted wet-etching setup with three low-pressure mercury lamps as excitation light sources were used in this experiment. The samples were located 10 cm away from the center of the excitation light sources. The power density was estimated to be below $5mW/cm^3$. The unetched region was covered with wax. We used $85\%KOH:H_2O$ (1:3) as an alkali solution. The etched depth was measured by a DEKTAK stylus profilometer.

We used samples grown by MBE and MOVPE. The carrier concentration of GaN layers grown by MOVPE was below that of GaN layer grown by MBE was measured around 1-$3\times10^{16}/cm^3$.

Figure 4 shows the relationship between the etch depth and the etching time. The etched depth of the MBE-sample and MOVPE-sample were about 6000Å and 400Å per 2 hours, respectively. This result shows that, similarly to other III-V materials[21], the photo-assisted wet-etching of GaN strongly depends on the background carrier density . Figure 5 shows an SEM photograph of an etched surface after 5 hours of photo-assisted wet-etching . The GaN layer was completely etched and the sapphire substrate was exposed. The etched bottom has several protuberances with a hexangular pyramidal shape in almost all the samples.   However, after the etching condition is optimized, this wet-etching technique would be expected as an effective method for GaN-based device fabrication.

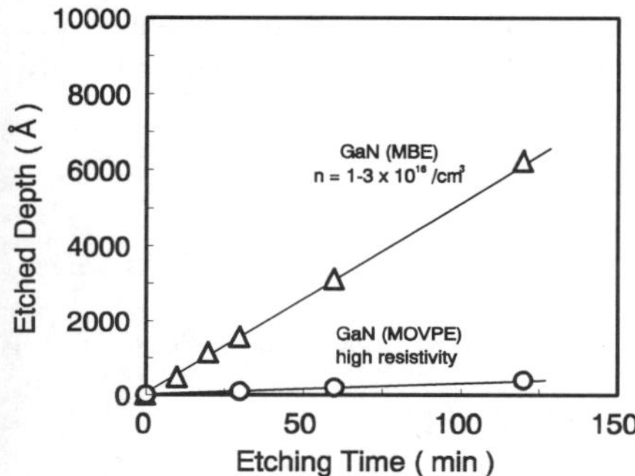

Fig.4 Etched depth as a function of etch time.

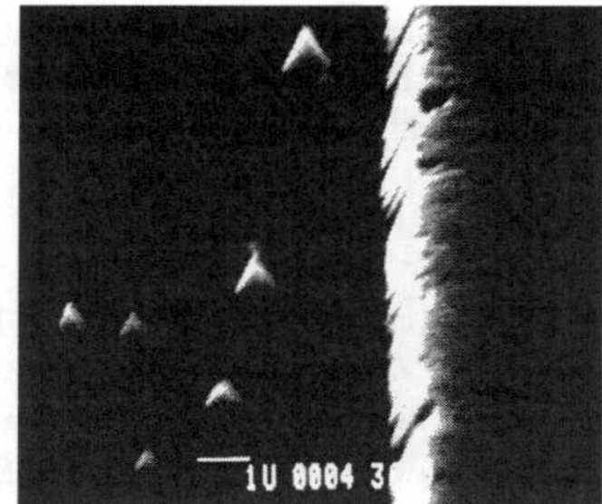

Fig.5 Scanning electron microscope of photo-assisted   wet-etched GaN grown by MBE on sapphire.

Figure 6 shows the room temperature photoluminescence (PL) spectra of RIBE processed samples before and after photo-assisted wet-etching. A He-Cd (325nm) laser was used as an excitation light source. The GaN layer was etched to a depth of about 400Å by Cl_2-RIBE. Then, some of the samples were etched by photo-assisted wet-etching for 2 hours. The PL intensity near-band-edge emission was decreased after etching by RIBE. This result indicates that the GaN surface was damaged due to the ion irradiation. On the other hand, after the treatment of a photo-assisted wet-etching, the PL intensity was recovered. This result indicates that the induced damage by RIBE was removed by photo-assisted wet-etching. Figure 7 shows the room temperature PL spectra near band edge of this sample. After the photo-assisted wet-etching, the peak was shifted by about 0.73nm (7meV) toward a short wavelength side. We think that this blue shift may be caused by decreasing of some defects or stress relaxation.

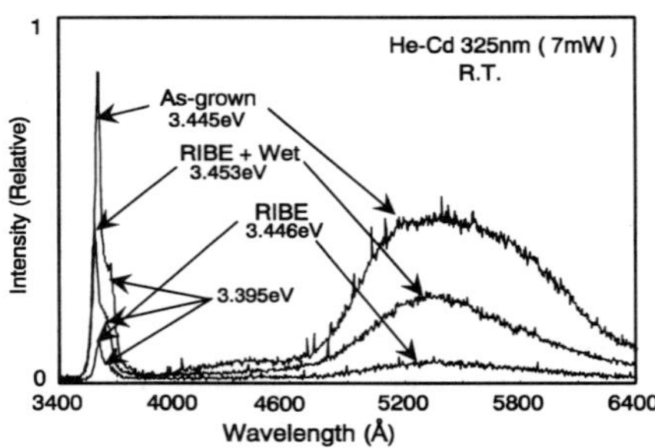

Fig.6 PL spectra of as-grown, Cl$_2$-RIBE and Cl$_2$-RIBE + photo-assisted wet-etch samples.

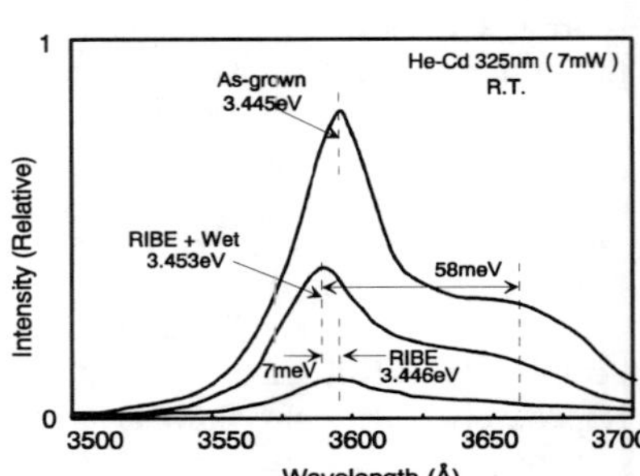

Fig.7 PL spectra near band edge of as-grown, Cl2-RIBE and Cl2-RIBE + photo-assisted wet-etch samples.

SUMMARY

In summary, we have studied the Cl$_2$-RIBE etching characteristic of GaN. Also, the photo-assisted wet-etching was shown to be useful in removing RIBE induced damages. We believe that this technique can be effective for the reduction of a surface recombination. The developed technique can be applied to fabricate GaN-based VCSELs.

REFERENCES

1. A. Matsutani, F, Koyama, and K. Iga, *Spring Meet. of Appl. Phys. Soc.* 27p-M-19, Asaka, p. 1227 (1996). (in Japanese)
2. S. Nunoue, M. Suzuki, M. Yamamoto, and M. Onomura, *Fall Meeting of Appl. Phys. Soc.* 9p-C-8, Fukuoka, p. 1125 (1996). (in Japanese)
3. K. Asakawa, and S. Sugata, J. Vac. Sci. & Technol., **B3**, 402, (1985).
4. K. Iga, F. Koyama and S. Kinoshita, IEEE J. Quantum Electron., **24,** 1845 (1988)
5. Jewell, J. L., Scherer, A., A., McCall, S.L., Lee, Y.H., Waiker, S.J., Harbison, J. P. and Florez, L.T., Electron. Lett., **25,** 1123, (1989).
6. A. Matsutani, F. Koyama and K. Iga, Jpn. J. Appl. Phys., **30**, 389, (1990).
7. H. Saito, Y. Noguchi and H. Nagai, Jpn. J. Appl. Phys., **28**, 1836, (1989).
8. K.Saotome, T. Honda, A. Matsutani, F. Koyama, and K. Iga: *International Symposium on Blue Laser and Light Emitting Diodes* (Chiba Univ., 1996) L506-509.
9. M. E. Lin, Z. F. Fan, Z. Ma, L. H. Allen and H. Morkoc, Appl. Phys. Lett., **64**, 887, (1994) .

10. S. Nakamura, M. Senoh, S. Nagashima, N. Iwasa, T. Yamada, T. Matsushita, H. Kiyoku and Y. Sugimoto, Jpn. J. Appl. Phys., **35**, 14, (1996).

11. I. Adesida, A. T. Ping, and M. Asif Khan, *Topical Workshop on III- V Nitrides* (*TWN'95*), **D-3**, Nagoya (1995).

12. H. Tanaka, A. Nakadaira, and T. Matsuoka, *Topical Workshop on III- V Nitrides* (*TWN'95*), **D-2**, Nagoya (1995)

13. S. J. Pearton, C. R. Abernathy, F. Ren, Appl. Phys. Lett., **64**, 2294, (1994).

14. R. J. Shul, S. P. Kilcoyne, M. Hagerott Crawford, J.E. Parmter, C. B. Vartuli, C. R. Abernathy, and S.J. Pearton, Appl. Phys. Lett., **66**, 1761, (1995).

15. T. Tamamuki, F. Koyama and K. Iga, Jpn. J. Appl. Phys., **31**, 3292, (1992).

16. A. Watanabe, K. Uchida and S. Minagawa, Fall Meeting of Appl. Phys. Soc. ,27a-ZE-16 , Kanazawa, p. 246, (1995). (in Japanese)

17. M. S. Minsky, M. White, and E. L., Hu, Appl. Phys. Lett., **68**, 1531, (1996).

18. H. Amano, N. sawai, I. Akasaki, and Y. Toyoda, Appl. Phys. Lett., **48**, 353, (1986).

19. I. Akasaki, H. Amano, Y. Koide, K. Hiramatsu, and N. Sawaki, J. Cryst. Growth, **98**, 209, (1989).

20. T. Tadokoro, F. Koyama, and K. Iga, Jpn. J. Appl. Phys., **27**, 389, (1988).

21. F. Kuhn-Kuhnenfeld, J. Electrochem. Soc., **119**, 1063, (1972).

Photo-assisted Anodic Etching of Gallium Nitride Grown by MOCVD

Hongqiang Lu*, Ziming Wu, Ishwara Bhat
Department of Electrical, Computer and System Engineering, Rensselaer Polytechnic Institute,
Troy, NewYork 12180
* luh@rpi.edu

ABSTRACT

In this paper, the first study of photo-assisted anodic etching of unintentionally doped n-GaN at room temperature is reported. The electrolyte used is a mixture of buffered aqueous solution of tartaric acid and ethylene glycol. The etching rate varies from ~20 Å/min to as high as 1600 Å/min. A systematic study shows that i) the etch rate, as well as the surface roughness, increases with the current density; ii) the etching rate is the highest when the pH of the electrolyte is around 7; iii) the etching is faster when there is more ethylene glycol in the electrolyte solution.

INTRODUCTION

GaN based III-V nitrides are important candidates for applications in blue-light emitters and lasers, high voltage and high temperature electronics. However, to realize these applications, etching techniques for defining the geometry of the electronic and optical devices must be established. GaN however, is chemically stable and very insoluble in most common etchants at room temperature. The most successful approach, to date, has been based on the dry etching technique [1-5]. Etching rates as high as 3500 Å/min were reported by McLane *et al.* [2] using reactive ion etching in BCl_3 plasmas. In addition, Nakamura *et al.* [6] recently fabricated InGaN multi-quantum-well structure laser diodes using a dry etching technique. Nevertheless it is still important to further explore viability of wet etching techniques for GaN and its alloys. Historically in the mid 1970's, Pankove [7] reported electrolytic etching of GaN using a NaOH solution, but no particular etching rate was given. At the same time, Shintani *et al.* [8] published a study on wet etching of GaN using phosphoric acid. The etch rate reported was 0.2 - 1.0 μm/min in the temperature region of 50° - 200°C. This work is inconclusive due to the quality of GaN at that time. Recently Pearton *et al.* [9] reported wet etching of GaN using 30% - 50% $NaOH/H_2O$ which gave an etching rate of ~ 20 Å/min at room temperature. To date, the only encouraging work in GaN wet etching was presented by Minsky *et al.* [10] using a photo-electrochemical etching technique. This work reported an etch rate of 4000 Å/min in KOH solution at room temperature, but the etching was only localized.

Anodic etching is another important wet etching method, involving the electrochemical process. For GaAs and InP, anodic etching has been successfully employed in depth profiling and to planarize the surface [11, 12]. For GaN, on which common chemical etching methods have not been successful, this technique becomes more attractive. In this paper, we report photo-assisted anodic etching of GaN using a mixed solution of glycol and water (AGW). Anodic oxidation in this AGW solution has commonly been used as a reproducible means to form an anodic oxide layer on GaAs and InP. We have found that it can be used to etch GaN effectively by choosing a proper pH value and water/glycol ratio. Ultra-violet light was employed here to generate electron-hole pairs in GaN to aid the etching process. An etch rate of 1600 Å/min was realized at room temperature in these preliminary studies.

Mat. Res. Soc. Symp. Proc. Vol. 449 © 1997 Materials Research Society

EXPERIMENT

GaN layers were grown on (0001)-oriented sapphire substrates in a low pressure metalorganic chemical vapor deposition system (LP-MOCVD). The samples were undoped with a residual electron concentration $\leq 5 \times 10^{17} \, cm^{-3}$. The thickness of the epitaxial GaN layers varied from 2 to 4 μm and the typical x-ray rocking curve linewidth of these samples was around 6 arc-minutes. The AGW electrolyte solution used in these studies was a mixture of a buffered aqueous solution of tartaric acid and ethylene glycol [13]. The concentration of tartaric acid was 3% by weight, and NH_4OH was used to buffer the acid to a specific pH value. This solution was then mixed with glycol in the desired ratio. Ethylene glycol has been used previously for the anodic oxidation of GaAs, but the purpose of the glycol was to make the process more reproducible. In our study it appeared that the presence of glycol is essential to GaN etching. The experimental

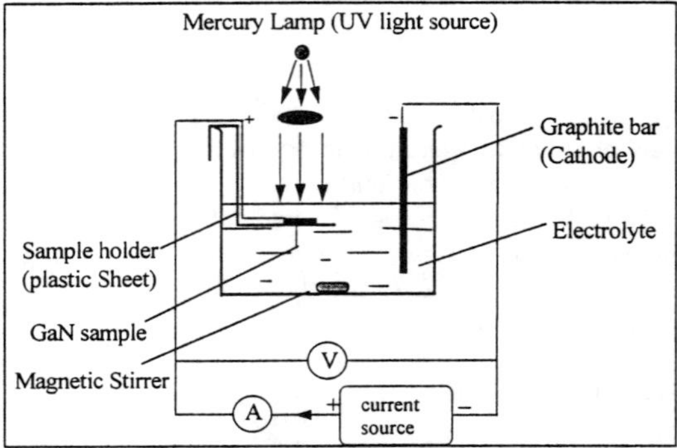

Fig. 1 Schematic diagram of the experimental setup for photo-assisted anodic etching of GaN

setup used for this photo-assisted anodic etching is shown in Fig.1. Samples were patterned with wax to provide an exposed area of 0.4 x 0.5 cm². They were attached to the polypropylene plastic sheet, and an indium contact was made on the corner or side of the GaN film and protected by wax. The samples were immersed in the solution and positioned such that the wafer surface could be exposed to UV light. The UV source was the mercury lamp, which emitted at 365 nm and 405 nm, with an intensity approximately 60 mW/cm² and 150 mW/cm², respectively. The electrolyte solution was stirred magnetically to keep the liquid flowing over the sample. The etching depth was determined by a Tencor Alpha Step stylus profilometer.

RESULTS AND DISCUSSION

Fig.2 shows the morphology of GaN sample after 1 hour anodic etching. The electrolyte used in this experiment had a pH value of 8 and glycol/water ratio of 2. A current density of 1 mA/cm² was employed and the sample was illuminated uniformly with the UV light. The measured etch depth was ~1.2 μm. This corresponded to an etch rate of ~200 Å/min. The surface became rough after etching. Atomic force microscopy of several etched samples was carried out to determine the surface roughness. For example, the sample shown in Fig. 2 had surface roughness of 815 Å.

To confirm that the etching is due to the photo-electrochemical effect, a corresponding 'dark' etching experiment was conducted. Initially, the GaN sample was immersed into the same AGW solution used above without current and UV illumination. There was no measurable etching detected after immersing in the solution for 24 hours. Subsequently, the same experiment was repeated with UV light, but without any current passed through the sample. No etching was observed after 2 hours. Finally, the sample was immersed in the electrolyte with a current passing through, but with no UV illumination. We found that the voltage across the sample gradually increased while the current was maintained at 2 mA/cm². After the voltage increased to near the maximum output of the constant current source employed (~160V) in 4 minutes, the current began to drop dramatically. This experiment also did not show any observable etching even after the sample was left in the solution for 1 hour. However, the atomic force microscopy (AFM) revealed several very deep holes on the GaN surface (shown in Fig.3). It is possible that these holes are related to dislocations and non-stoichometric regions in the GaN film. As a result, these regions offer low breakdown voltage for the GaN-electrolyte barrier and get preferentially etched.

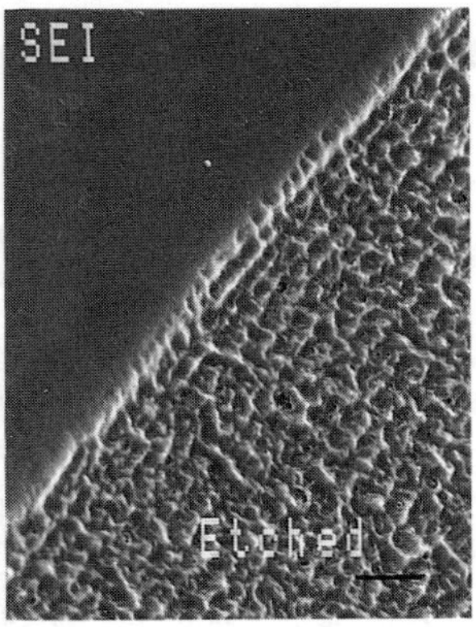

Fig. 2 SEM picture of the surface morphology of GaN after anodic etching. The unetched potion was covered with wax during etching. Marker represnets 2 μm.

Fig. 3 AFM micrograph of GaN surface after one hour anodization without UV light illumination. The dark dots on the picture are deep holes, which were not present before etching

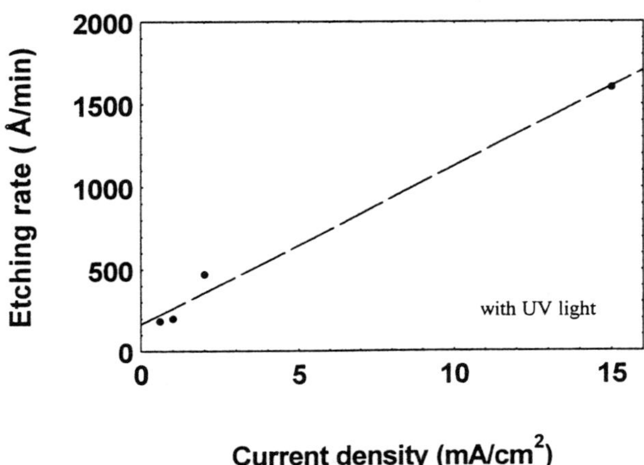

Fig. 4 Effect of current density on the anodic etching rate. The pH value of the electrolyte is 8 and the glycol/water ratio in the electrolyte is 2.

The electrochemical nature of the process dictates that the etching rate should be closely related to the current according to Faraday's law of electrolysis. Fig.4 shows the etching rate as a function of the current density (J_a). The etch rate was ~180 Å/min for J_a = 0.6 mA/cm^2, and gradually increased with current density to ~1600 Å/min at J_a = 15 mA/cm^2. At this high current density, however, the surface of the sample became very rough and non-uniform. The current efficiency, defined as the fraction of GaN etched per electron passed, is less than 0.5% in this study. It should be noted that current efficiency is about 95% for GaAs in AGW solution and about 1-3% for Si.

The process found in this study can be understood in terms of the anodic etching of GaAs. First, the GaAs surface is oxidized to form Ga_2O_3 and As_2O_3 at the GaAs/electrolyte interface in the presence of OH$^-$ ions and holes. Holes are important here to convert atoms in GaAs to higher oxidation states. In n-GaAs, these holes can either photo- or thermally generated. In the second step, the oxides dissolve in the solution thus exposing the fresh GaAs surface beneath. Both Ga_2O_3 and As_2O_3 can easily be dissolved in acidic or basic solutions and therefore, an electrolyte with a pH value far away from 7 is preferred for such anodic etching purposes. On the other hand, Hasegawa *et al.* [13] have used a AGW solution with a pH of 7 to grow oxide layers on GaAs. Since the GaAs oxide does not dissolve in a solution with pH of 7, the process is self-limiting with the final oxide layer thickness dependent on the applied voltage. The presence of glycol seems to make the process more reproducible. In case of GaAs, the oxides have been found to have a higher As_2O_3 content than Ga_2O_3, indicating that Ga_2O_3 preferentially dissolves in the solution. However, to GaN, only Ga_2O_3 would be grown on the surface, and we found that the presence of glycol was required to dissolve the oxide, and the pH value of the electrolyte should be close to 7. Fig.5 shows the anodic etching rate as a function of the pH value of the electrolyte (pH was measured before the solution was mixed with ethylene glycol). Here the ratio of glycol to water was 2 and the current density was 2 mA/cm^2. Interestingly, a maximum etch rate of 570 Å/min was obtained for pH values around 7. In the acidic solution, the etching process decreased to ~290 Å/min and the surface appeared more porous after etching. Conversely, if the solution was buffered to strongly basic, the etching process almost stopped even though the solution has higher concentration of (OH)$^-$ ions. Even though only three data points are shown in Fig.5, the experiments were repeated several times with similar results, viz, higher etch rate for solutions with the pH around 7 .

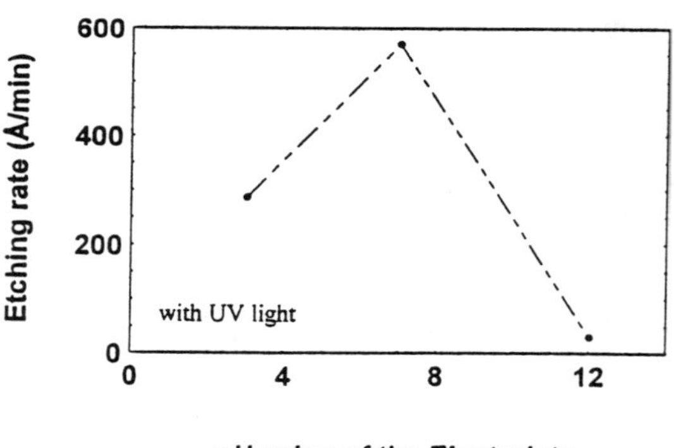

Fig. 5 Effect of the pH value of the electrolyte on the anodic etchig rate. The current density is 2 mA/cm2. The glycol to water ratio is 2.

The primary goal of the UV light is to generate electron-hole pairs in the GaN. In addition, it may also play a role in the etching of gallium oxide. A GaAs sample with previously grown oxide layer was kept in the AGW solution. This oxide layer could not be etched in the AGW solution with a pH of 7, but could be etched, albeit slowly, in the same solution when the sample was exposed to UV light. It is possible that in the presence of UV, the glycol may react with the oxide forming a complex which then will dissolve in the aqueous solution.

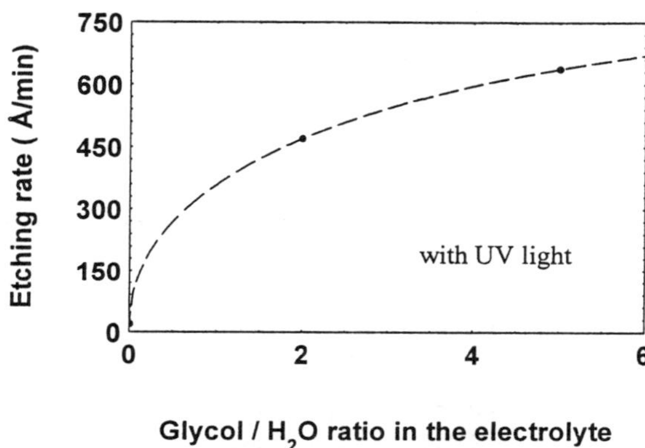

Glycol / H$_2$O ratio in the electrolyte

Fig. 6   Effect of the glycol to water ratio on the anodic etching rate. The current density is 2 mA/cm^2 and the pH of the electrolyte is 8 .

The effect of glycol was then studied and the result was shown in Fig.6. The pH value was maintained around 8 and the current density at 2 mA/cm^2. It was found that without glycol the etching rate was negligible (20 Å/min). The addition of glycol to the solution increased the etch rate dramatically. For a glycol /water ratio of 5, an etching rate of 610 Å/min was obtained. However, if pure glycol was used as the electrolyte, no current passes through the solution resulting in no etching. Thus, in case of GaN, glycol plays a significant role in etching. We speculate that with the assistance of UV light ethylene glycol acts as a "chelating agent" for Ga [14], resulting in the dissolution of Ga$_2$O$_3$. However, more work is needed to clearly understand the role of ethylene glycol and the pH value in the etching process. In the study of etching of p-GaN, we found that oxide layer can be grown without UV illumination, since holes are available without illumination. However, with UV illumination, the process is found to etch p-GaN. These results also supports the above arguments on the role of ethylene glycol in the etching process.

CONCLUSIONS

In summary, UV-light-assisted anodic etching of unintentionally doped n-GaN at room temperature is described. The preferred electrolyte was a mixture of buffered aqueous solution and ethylene glycol. Etching rate ranges from ~20 Å/min to as high as 1600 Å/min. It was observed that the etching rate increases with current density and with the ratio of glycol to water in the electrolyte. In addition, the etching rate was influenced by the pH of the electrolyte and showed a maximum value when the pH was around 7. We postulate, the UV light generated e-h pairs in GaN to aid the formation of oxide and also appeared to help the dissolution of the oxide. The detailed mechanism of the etching process is not clear at present and more work is needed.

ACKNOWLEDGMENT

This work is partially supported by DARPA (subcontracted through Texas Instruments through Johnson Marthey Prime Contract No. MDA972-91-C-0046). We also acknowledge the support through a grant from Philips Laboratory, Briarcliff Manor, NY. We also thank Y. Zhao for AFM measurements and J. Barthel for technical assistance.

REFERENCES

1.   S. J. Pearton and C. R. Abernathy, Appl. Phys. Lett. **64**, 2294 (1994)
2   G. F. McLane, L. Casas, S. J. Pearton, and C. R. Abernathy, Appl. Phys. Lett. **66**, 3328 (1995)
3.   I. Adesida, A. Mahajan, E. Andideh, M. A. Khan, D. T. Olson, and J. N. Kuznia, Appl. Phys. Lett **63**, 2777 (1993)
4.   M. E. Lin, Z. F. Fan, Z. Ma, L. H. Allen, and H. Morkoc, Appl. Phys. Lett **64**,   887 (1994)
5.   A. T. Ping, I. Adesida, and M. asif. Khan, Appl. Phys. Lett **67**, 1250 (1995)
6.   S. Nakamura, M. Senoh, S. Nagahama, N. Iwasa, T. Yamada, T. Matsushita, H. Kiyoku, and Y. Sugimoto, Jpn. J. Appl. Phys. **35**, L74 (1996)
7.   Pankove, J. Electrochem. Soc. **119**, 1118 (1972)
8.   A. Shintani, and S. Minagawa, J. Electrochem. Soc. **123**, 706 (1976)
9.   S. J. Pearton, C. R. Abernathy, F. Ren, J. R. Lothian, P. W. Wisk, and A. Katz, *J.* Vac. Sci. Technol. A **11**, 1772 (1993)
10.   M. S. Minsky, M. White, and E. L. Hu, Appl. Phys. Lett **68**, 1531 (1996)
11.   H. Muller, F. H. Eisen, and J. W. Mayer, J. Electrochem. Soc. **122**, 651 (1975)
12.   H. D. Barber, H. B. Lo, and J. E. Jones, J. Electrochem. Soc. **123**, 1404 (1976)
13.   H. Hasegawa and H. L. Hartnagel, J. Electrochem. Soc. **123**, 713 (1976)
14.   R. M. Finne, and D. L. Klein, J. Electrochem. Soc. **114**, 965 (1967)

PATTERNING OF LiGaO$_2$ AND LiAlO$_2$ BY WET AND DRY ETCHING

J. W. Lee,[1] S. J. Pearton,[1] C. R. Abernathy[1], R. G. Wilson[2], B. L. Chai[3], F. Ren[4] and J. M. Zavada[5]

[1]University of Florida, Gainesville FL 32611
[2]Hughes Research Laboratories, Malibu CA 90265
[3]CREOL, University of Central Florida, Orlando FL 32816
[4] Lucent Technologies, Bell Laboratories, Murray Hill NJ 07974
[5]US Army Research Office, Research Triangle Park NC 27709

ABSTRACT

LiGaO$_2$ and LiAlO$_2$ have similar lattice constants to GaN, and may prove useful as substrates for III-nitride epitaxy. We have found that these materials may be wet chemically etched in a number of acid solutions, including HF, at rates between 150-40,000 Å/min. Dry etching with SF$_6$/Ar plasmas provides faster rates than Cl$_2$/Ar or CH$_4$/H$_2$/Ar under Electron Cyclotron Resonance conditions, indicating the fluoride etch products are more volatile that their chloride or metalorganic/hydride counterparts. Dry etch rates are low (< 2, 000 Å/min), providing high selectivity (>5) over the nitrides. The incorporation of hydrogen in these materials is also of interest because this could provide a reservoir of hydrogen that may passivate dopants in overlying nitride films. In ^{2}H implanted samples, 50 % of the deuterium is lost by evolution from the surface by annealing at 400 °C for 20 min and all of the deuterium is gone at 700°C. The diffusivity of ^{2}H is ~10^{-13} cm^2/s at 250°C in LiAlO$_2$, approximately two orders of magnitude higher than in LiGaO$_2$.

INTRODUCTION

GaN and related alloys are generally grown on (0001) Al$_2$O$_3$ substrates in spite of the large lattice mismatch (~14% for GaN), due to the lack of commercially available bulk single crystal nitrides.[1-4] The Al$_2$O$_3$ is relatively inexpensive, available in large diameters, stable at the nitride growth temperatures and simple to clean. Growth of AlGaN/InGaN heterostructures on Al$_2$O$_3$ has led to the high brightness light-emitting diodes[5] and lasers[6] reported by Nakamura et al. Alternative substrates that have been employed include SiC (3.1% lattice mismatch to GaN for the 6H polytype), which is attractive because it is cleavable and available in doped form, but is expensive and difficult to clean,[7-12] MgAl$_2$O$_4$ which has a 10% mismatch with GaN and is cleavable,[13,14] ZnO, MgO, Si and III-V materials such as GaAs and InP.[4] Thick GaN layers grown by vapor phase epitaxy on Al$_2$O$_3$ substrates have been employed as templates for GaN homoepitaxy by removing the Al$_2$O$_3$,[15] and the first report of growth on small bulk GaN crystals recently appeared.[16] The high dissociation pressure of nitrogen (15 kbar) at the growth temperature (1,600°C) makes bulk GaN crystals difficult to produce on a large scale.[17,18]

Two promising substrates for nitride epitaxy are the wurtzite crystals LiGaO$_2$ and LiAlO$_2$, which have a-axis mismatches of +0.19% and -1.45%, respectively, to GaN.[19] These are grown by the Czochralski technique in diameters currently up to 1.5 inches, and have bandgaps of 6.5eV (LiAlO$_2$) and 5.6eV (LiGaO$_2$). Both were (100) oriented. Dissociation of these materials at the typical Metal Organic Chemical Vapor Deposition growth temperature of ~1040°C may restrict their use to the Molecular Beam Epitaxy techniques, but promising x-ray and photoluminescence results have already been reported for AlGaN/GaN structures grown on LiGaO$_2$.[20] Growth on patterned substrates, or creation of through-wafer via holes for microwave power electronics, requires that etching processes be developed for both materials. In this paper we report several wet etch solutions for LiGaO$_2$ and LiAlO$_2$, and compare three different dry etch chemistries in terms of rate, surface morphology and anisotropy.

Mat. Res. Soc. Symp. Proc. Vol. 449 © 1997 Materials Research Society

EXPERIMENT

The polished faces of both materials were patterned either with AZ5209E photoresist or a carbon-based mask in a resolution test array with feature sizes from 0.5-200 μm. Dry etching was performed in a Plasma Therm SLR 770 system utilizing an Astex 4400 Electron Cyclotron Resonance (ECR) source operating at 2.45 GHz and at powers up to 1000W. The Si carrier wafer is mechnically clamped to a He backside cooled, rf powered chuck. The process pressure was varied from 1-20 mTorr, but was held at 1.5mTorr for most runs. Three different plasma chemistries, Cl_2/Ar, $CH_4/H_2/Ar$ and SF_6/Ar, were investigated since they enable us to determine which of the typical types of etch products (i.e. chlorides, metalorganic/hydride, or fluoride, respectively) are most volatile for the oxide substrates. Etch rates were determined by stylus profilometry after removal of the mask materials and surface morphology was examined by scanning electron microscopy and atomic force microscopy.

RESULTS AND DISCUSSION

Figure 1 shows etch depth versus time for $LiAlO_2$ and $LiGaO_2$ in various wet etch solutions at 25°C. The etch rates were very high (~3-4μm/min) for $LiAlO_2$ with HF, and for $LiGaO_2$ with HCl and HF. More controlled rates (150-350Å/min) were obtained for $LiAlO_2$ with H_3PO_4, HNO_3 and H_2SO_4. Note that HCl is highly selective for $LiGaO_2$ over $LiAlO_2$. Most of the solutions provided linear etch depth with time, an absence of etch rate dependence on agitation, and smooth surface morphologies. These are all characteristics of reaction-limited etching where the rate-limiting step is formation and dissociation of the etch products. The exception was HF, which for both materials was found to have a square root dependence of etch depth on solution immersion time, and whose rate was a strong function of agitation. The morphology of the etched surface was also considerably rougher than those obtained with the other solutions. These are all hallmarks of diffusion-controlled etching, where the rate is controlled by diffusion of the reactant to the gallate or aluminate surface. This is a less attractive situation for reproducible etch processing because it is more difficult to achieve the same degree of agitation in a mixture than to control its temperature, which is the key parameter in reaction-limited etching. As mentioned above, the etched surface morphology for both $LiGaO_2$ and $LiAlO_2$ in HF solutions was worse (typically by a factor of approximately five in terms of root mean square roughness) than with the other etchants investigated. We did not study in detail the crystallographic dependence of the wet etching.

Figure 2 shows the etch rate of $LiGaO_2$ in ECR plasmas of $5CH_4/15H_2/25Ar$, $5Cl_2/10Ar$ or $5SF_6/10Ar$ (where the numbers refer to the individual gas flow rates in standard cubic centimeters per minute), as a function of rf power. The microwave power was held constant at 1000 W, and the process pressure was 1.5 mTorr. These etch conditions are typical of those we employed for etching III-V materials, including GaN ($CH_4/H_2/Ar$) or dielectrics (SF_6/Ar). We obtained similar rates for $LiAlO_2$ in all three mixtures, and show only those for SF_6/Ar. The fluoride etch products appear to be the most volatile, and for each plasma chemistry the etch rates continue to increase with rf power due to two reasons. Firstly, the increased average ion energy at higher rf powers is more efficient at initially breaking the bonds in the substrate material, which must precede etch product formation, and secondly, there is more efficient sputter-enhanced removal of the products once they form. The first mechanism has been found to be the rate-limiting step in dry etching of GaN and AlN, which have volatile chloride etch products but very stable bonding that must be overcome before the products can form. In addition, the LiF_X, GaF_X and AlF_X species that form in SF_6/Ar plasmas have quite low volatilites and must be removed by sputter assistance. The etched surface is quite smooth (similar RMS roughness for an unetched control). The high ion density under these conditions produces severe distortion of the photoresist mask, causing the resist to reticulate and leading to rough sidewalls. For these slow etching materials, dielectrics or metal masks must be employed because of their superior etch resistance compared to photoresist.

The dependence of $LiGaO_2$ and $LiAlO_2$ etch rate on microwave power at fixed pressure (1.5 mTorr) and rf power (250 W) is shown in Figure 3. There is only a slight increase in rate for Cl_2/Ar and $CH_4/H_2/Ar$ at higher microwave powers, confirming that the etching is not reactant-limited since the active neutral species density increases rapidly under these conditions.[21] By contrast the rates for both materials approximately double in the SF_6/Ar mixture between 400 and 1000 W microwave power, which is a more typical result for ECR reactors. Since the fluoride etch products are more volatile than those from the Cl_2/Ar and $CH_4/H_2/Ar$, an increase in ion density and fluoride radical density at high microwave powers will enhance the etch rates.

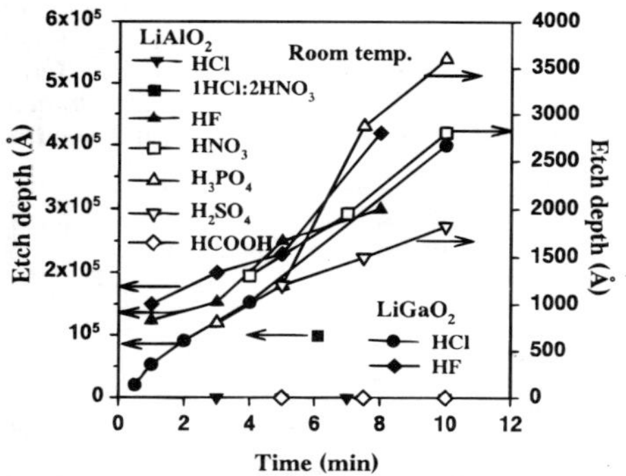

Figure 1. Etch depth versus time for $LiAlO_2$ and $LiGaO_2$ in various acid solutions at 25°C. No etching was observed for $LiAlO_2$ with HCl or with HCOOH.

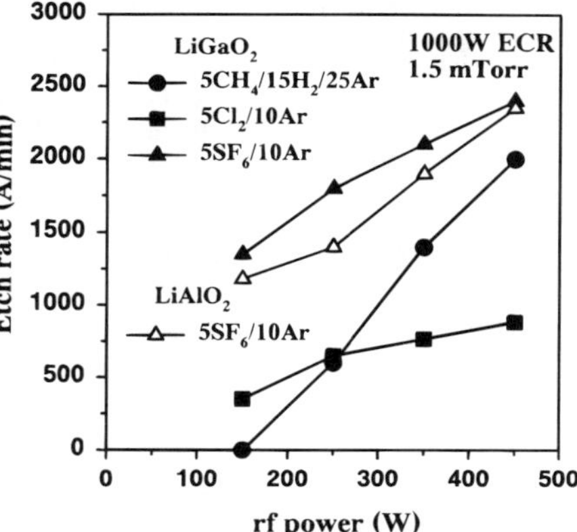

Figure 2. Etch rates of $LiGaO_2$ and $LiAlO_2$ in 1000W microwave, 1.5mTorr discharges as a function of rf power.

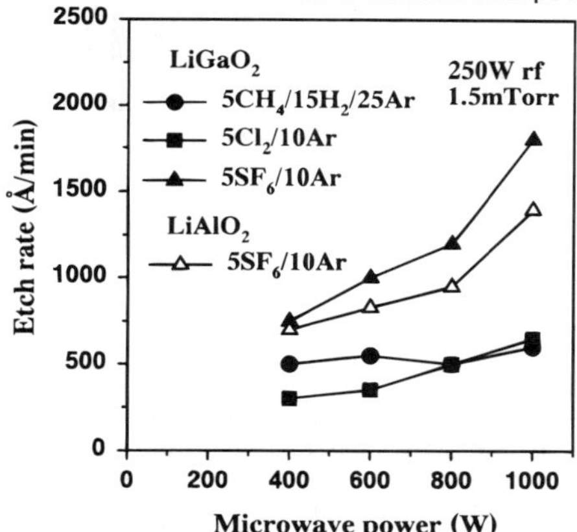

Figure 3. Etch rates of $LiGaO_2$ and $LiAlO_2$ in 250W rf, 1.5mTorr discharges as a function of microwave.

To facilitate measurements of the hydrogen incorporation and stability we introduced deuterium (an isotope of ^{1}H which can be detected with much better sensitivity by Secondary Ion Mass Spectrometry) by two different methods. In the first, $^2H^+$ ions were implanted at an energy of 100 keV to a dose of 2×10^{15} cm^{-2}. Different sections of these samples were subsequently annealed up to 700°C for 20 mins, with SIMS measurements performed for each temperature. In the second method the substrates were exposed to Electron Cyclotron Resonance (ECR) ^2H plasmas for 1 hour at 250 °C. The microwave power was 750 W, and the process pressure 10 mTorr. SIMS measurements were again performed before and after a subsequent anneal at 500 °C for 1 min under N_2.

Figure 4 shows the deuterium depth profiles for as-implanted and annealed $LiAlO_2$ (left) and $LiGaO_2$ (right) samples. Annealing at $\geq$ 400 °C produces a decrease in the peak concentration of deuterium in both materials. Note that the implanted profiles do not show diffusional broading as is seen in doped semiconductors where the hydrogen can attach to acceptor or donor dopant impurities to form stable neutral complexes. It appears that in the LiAlO$_2$ and LiGaO$_2$ the hydrogen is basically lost from the crystal upon annealing, with the remaining hydrogen at each temperature decorating the residual implant damage. This is an area of potential concern because residual hydrogen in the oxide substrate could be readily transferred into overlying GaN epilayers during device process steps, such as contact alloying, implant activation, or implant isolation annealing. However as seen below, it may be less of an issue than in some other potential substrates.

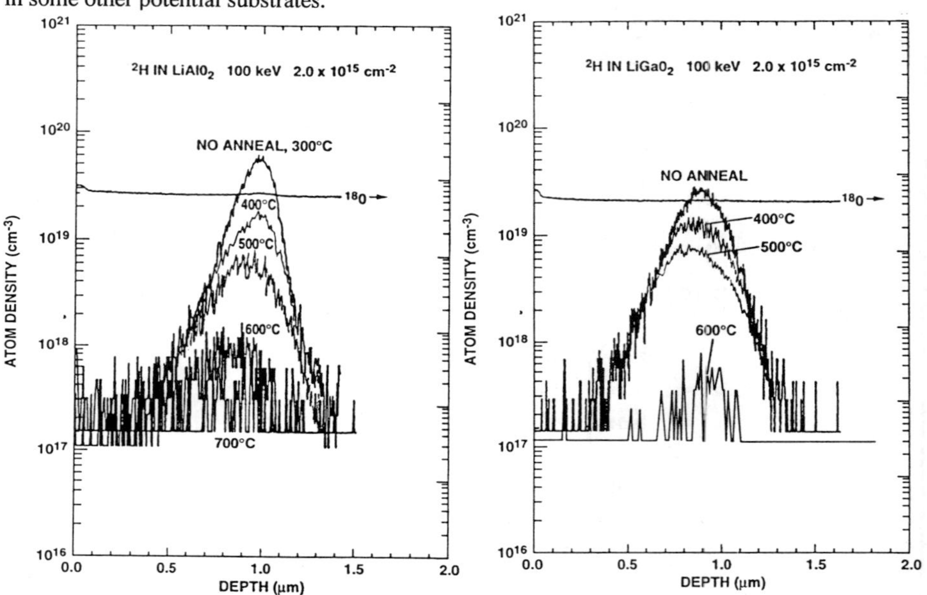

Figure 4. SIMS profiles of implanted ^{2}H in LiAlO$_2$ (left) and LiGaO$_2$ (right) before and after annealing.

The implanted deuterium is much less thermally stable in LiAlO$_2$ and LiGaO$_2$ relative to Al$_2$O$_3$. Figure 5 shows ^{2}H profiles before and after annealing up to 950 °C for 20 min in Al$_2$O$_3$. There is no measurable change under these conditions. In Figure 6 we have plotted the percentage of implanted deuterium that remains in LiGaO$_2$, LiAlO$_2$, Al$_2$O$_3$, SiC, GaN, AlN and GaAs after various annealing treatments, based on the current data and that reported earlier for the nitrides and GaAs. These materials are all potential substrates for nitride epitaxy. Clearly the LiAlO$_2$ and LiGaO$_2$ fall at the lower end of the stability range for hydrogen retention, which is an advantage because there will be lower levels of residual hydrogen after a high temperature growth step. The deuterium is presumably in random positions after implantation and in semiconductors it moves to regions of strain (defects or impurities) upon annealing. The fact that hydrogen is not retained to high temperatures in the oxide materials is at least an indication that they are basically free of large densities of defects or impurities, or one would expect to see plateaus forming in the deuterium profiles at certain annealing temperatures.

The fact that deuterium is readily mobile in both materials at relatively low temperatures is obvious from the data in Figure 7, which shows the ^{2}H profiles after exposure to a plasma for 1 hour at 250 °C. Based on a simple $(4Dt)^{1/2}$ calculation for the diffusion distance, one can estimate ^{2}H diffusivities of ~1.1 x 10^{-13} cm/s in LiAlO$_2$ and 1.7 x 10^{-15} cm^2/s in LiGaO$_2$ at 250°C. These values are approximately one and three orders of magnitude, respectively, lower than in GaAs at the same temperature. Note that annealing at 500 °C again reduces the deuterium density in each material, without significant diffusional broadening of the profiles. The deuterium appears to be lost by evolution from the surface.

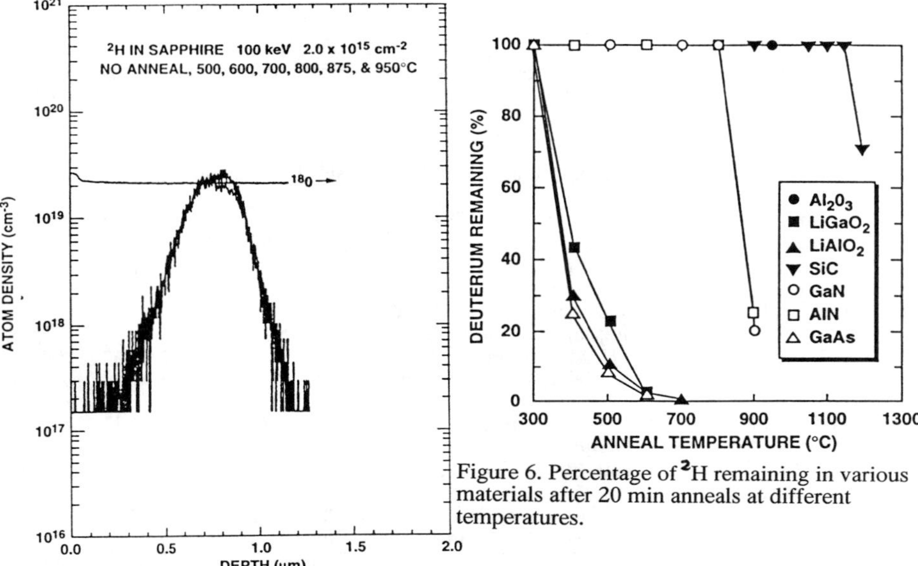

Figure 6. Percentage of ^{2}H remaining in various materials after 20 min anneals at different temperatures.

Figure 5. SIMS profiles of implanted ^{2}H in Al$_2$O$_3$, before and after annealing.

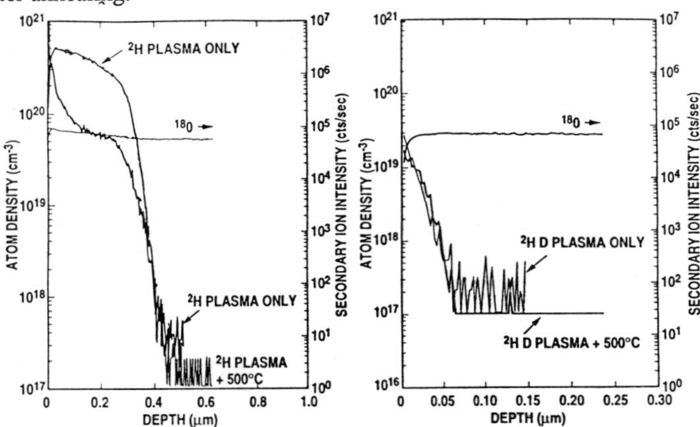

Figure 7. SIMS profiles of ^{2}H in plasma exposed LiAlO$_2$ (left) and LiGaO$_2$ (right) before and after annealing for 1 min at 500 °C.

CONCLUSIONS

We have found that lithium gallate and lithium aluminate bulk wafers can be wet chemically etched in many common acid solutions with rates ranging from 150-40,000 Å/min. Chlorine, fluorine and methane-hydrogen plasma chemistries can all be employed for dry etching of both materials, with SF$_6$/Ar providing the fastest rates for similar plasma parameters. These processes should prove useful for patterning $LiGaO_2$ and $LiAlO_2$ substrates prior to growth of GaN and related alloys.

ACKNOWLEDGMENTS

The work at UF is partially supported by grants from ARO-AASERT and DARPA (contract monitor Dr. G. R. Witt., AFOSR). The work is performed in the Microfabritech facility.

REFERENCES

1. I. Akasaki, H. Amano, Y. Koide, K. Hiramatsu and N. Sawaki, J. Cryst. Growth **98**, 209 (1989).
2. S. Nakamura, Jap. J. Appl. Phys. **30**, L1705 (1991).
3. J. N. Kuznia, M. Asif Khan and D. T. Olson, J. Appl. Phys. **73**, 4700 (1993).
4. S. Strite and H. Morkoc, J. Vac. Sci. Technol. B **10**, 1237 (1992) and references theirs.
5. S. Nakamura, T. Mukai and M. Senoh, Appl. Phys. Lett. **64**, 1687 (1994).
6. S. Nakamura, M. Senoh, S. Nagahama, N. Iwasa, T. Yanada, T. Matshusita, H. Kiyokuand and Y. Sugimoto, Jap. J. Appl. Phys. **35**, L74 (1996).
7. B. N. Sverdlov, G. A. Martin, H. Morkoc and D. J. Smith, Appl. Phys. Lett. **67**, 2063 (1995).
8. F. A. Ponce, B. S. Krusor, J. S. Major, W. E. Plano and D. F. Welch, Appl. Phys. Lett. **67**, 410 (1995).
9. A. S. Zubrilov, V. I. Nikolaev, V. A. Dimitriev, K. G. Irvine, J. A. Edmond and C. H. Carter, Inst. Phys. Conf. Ser. **141**, 525 (1994).
10. Z. Sitar, M. J. Paisley, B. Yuan, J. Ruan, W. J. Choyke and R. F. Davis, J. Vac. Sci. Technol. B **8**, 316 (1990).
11. M. E. Lin, B. Sverdlov, G. L. Zhou and H. Morkoc, Appl. Phys. Lett. **62**, 3479 (1993).
12. H. Liu, A. C. Frenhel, J. G. Kim and R. M. Park, J. Appl. Phys. **74**, 6124 (1993).
13. A. Kuramata, K. Horino, K. Domen, K. Shinohara and T. Tanahashi, Appl. Phys. Lett. **67**, 2521 (1995).
14. T. George, E. Jacobson, W. T. Pike, P. Chang-Chien, M. A. Khan, J. W. Yang and S. Mahajan, Appl. Phys. Lett. **68**, 337 (1996).
15. T. Detchprohm, H. Amano, K. Hiromatsu and I. Akasaki, Appl. Phys. Lett. **61**, 2688 (1992).
16. F. A. Ponce, D. P. Bour, W. Gotz, N. M. Johnson, H. I. Helava, I. Grzegory, J. Jan and S. Porowski, Appl. Phys. Lett. **68**, (1996).
17. S. Porowski, I. Gregory and J. Jun, High Pressure Chemical Synthesis, ed. J. Jurczak and B. Baranowski (Elsevier, Amsterdam 1989).
18. M. Leszczynski, I. Gregory and M. Bockowsk, J. Cryst. Growth **126**, 601 (1993)
19. B. L. Chai J. Cryst. Growth (to be published)
20. J. F. Schetzina et al (to be published).
21 S. J. Pearton, T. Nakano and R. A. Gottscho, J. Appl. Phys. **69**, 4206 (1991).

Part X

Metal Contacts and Surfaces

THERMAL STABILITY OF Pt, Pd, and Ni on GaN

K.J. Duxstad[1,2], E.E. Haller[1,2], K.M. Yu[2], M.T. Hirsch[1,2], W. R. Imler[3], D.A. Steigerwald[3], F.A. Ponce[4], and L.T. Romano[4]
[1] Dept. of Materials Science and Mineral Engineering, University of California, Berkeley, CA 94720
[2] Lawrence Berkeley National Laboratory, Berkeley, CA 94720
[3] Hewlett-Packard Company, San Jose, CA
[4] Xerox Palo Alto Research Center, Palo Alto, CA

ABSTRACT

The development of reliable Ohmic and Schottky contacts on GaN will require an understanding of the thermal stability and metallurgy of metal-GaN contact structures. We investigated the behavior of Pt, Pd, and Ni on GaN as a function of annealing temperature. Rutherford backscattering spectrometry, x-ray diffraction, and scanning electron microscopy were used to characterize the interface and surface behavior. 800Å thin metal films were deposited by sputtering on 2μm GaN films grown by metal organic chemical vapor deposition on sapphire substrates. No reactions occur in the Pt, Pd, and Ni systems with annealing up to 800°C. The Pt film begins to form submicron spheres and islands after annealing above 600°C. The Pd and Ni films begin to island with annealing above 700°C. Below these temperatures no structural changes were observed.

INTRODUCTION

Research into the properties of GaN has increased dramatically since the first announcement of the development of blue light emitting diodes (LEDs) using this material.[1] GaN has a wide direct band gap of 3.4 eV which is useful for blue and ultra-violet light emission. Moreover, the materials properties of GaN also make it suitable for high temperature electronics. In order to optimize the electrical behavior of these devices it is crucial to develop stable and reliable Ohmic and Schottky contacts.

Pt, Pd, and Ni have large work functions which makes them ideal for use as Schottky contacts on most n-type semiconductors, including GaN. They are also useful for contacts because of their resistance to oxidation and corrosion. The barrier height of these contacts on n-type GaN has been measured by several groups. The barrier height of Pt lies between 1.03 and 1.27 eV.[2-4] The barrier height of Pd ranges from 0.91 to 1.24 eV,[2,4] while the barrier height of Ni ranges between 0.95 and 1.15 eV.[5,6] The large variation in the measured barrier heights is partially due to differences between results of current-voltage (I-V) and capacitance-voltage (C-V) measurements used in these studies.

Ternary phase diagrams for many metal(M)-GaN systems have recently been computed.[7] Under thermodynamic equilibrium conditions it is predicted that Pt and Pd will react to form M-Ga phases, even at room temperature. Ni is predicted to form M-Ga phases at higher temperatures.

EXPERIMENT

The GaN films used in this study were grown by low pressure metal-organic chemical vapor deposition. They had a thickness of 2μm and were Si doped yielding a free electron concentration

Mat. Res. Soc. Symp. Proc. Vol. 449 © 1997 Materials Research Society

of 1×10^{18} cm^{-3}. The GaN surface was prepared by cleaning for three minutes each in a sequence of hot xylenes, acetone, and methanol. The samples were then subjected to a 30 sec. HF dip etch, followed by a rinse in distilled water. They were blown dry with nitrogen. The samples were immediately placed into a model 2400 Perkin Elmer RF sputtering system which was pumped down to a pressure of ~3×10^{-6} Torr. Prior to metal deposition, the surface was backsputtered with Ar$^+$ ions at a power of 100W. The metals were then deposited by RF sputtering at a power of 300W to a nominal thickness of 800Å. During sputtering the samples were kept at room temperature by water cooling. Thermal annealing was performed in an open tube furnace in flowing N_2.

Changes in composition with depth were determined with Rutherford backscattering spectrometry (RBS) using a 1.95 MeV He$^+$ beam and a backscatter angle of 165°. X-ray diffraction (XRD) spectra were obtained on a Siemens D5000 diffractometer in the Bragg-Brentano (θ-2θ) geometry to observe the formation of new phases and solid solutions. Scanning electron microscopy (SEM) was used to observe surface morphology before and after annealing. Auger electron spectroscopy (AES) provided information regarding the composition of the surface.

RESULTS

<u>Platinum</u>

No change in the structure of the Pt/GaN interface is observed by RBS after annealing for 30 min. below 700°C. Fig. 1 shows RBS spectra of as-deposited and annealed Pt films on GaN. Between 700°C and 800°C a significant change is observed. There is an increase in signal between the Pt signal tail and the Ga edge (1.4-1.6 MeV) in the RBS spectrum. This could be interpreted

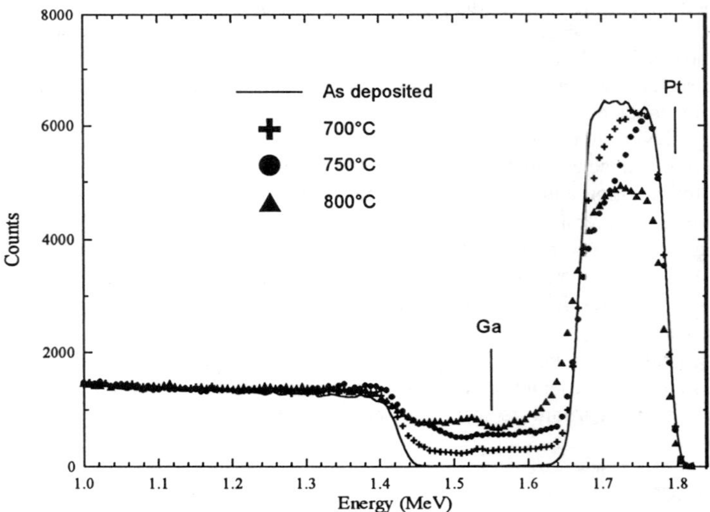

Fig. 1: RBS spectra of Pt on GaN as deposited and annealed for 30 min. at various temperatures.

as Pt indiffusion taking place, however the flat signal in this energy range indicates a uniform distribution of Pt. A normal diffusion profile would follow an error function distribution. Therefore, the RBS spectra show Pt coalescing on the surface of the film through lateral surface diffusion. The formation of islands leads to much thicker metal coverage in some areas, resulting in the low signal extending beyond the main Pt peak, as well as decreasing the height of the main Pt peak. More of the film agglomerates as the temperature increases; after annealing at 800°C approximately 14% of the surface is covered by particles. No reaction is observed by XRD for annealing up to 800°C. The only new peaks observed correspond to different orientations of Pt. In the as-deposited case Pt is highly (111) oriented. During annealing, grain growth and reorientation occurs due to the agglomeration, which leads to the observation of x-ray peaks from many different planes. A more detailed explanation of this process is given in ref. 8.

The morphology of the surface after annealing was determined using SEM. The growth of small Pt particles was observed for samples annealed at temperatures as low as 600°C. Fig. 2 shows a sequence of micrographs showing the evolution of the morphology. The as-deposited surface is smooth and featureless. After annealing at 600°C, small particles of Pt begin to form on

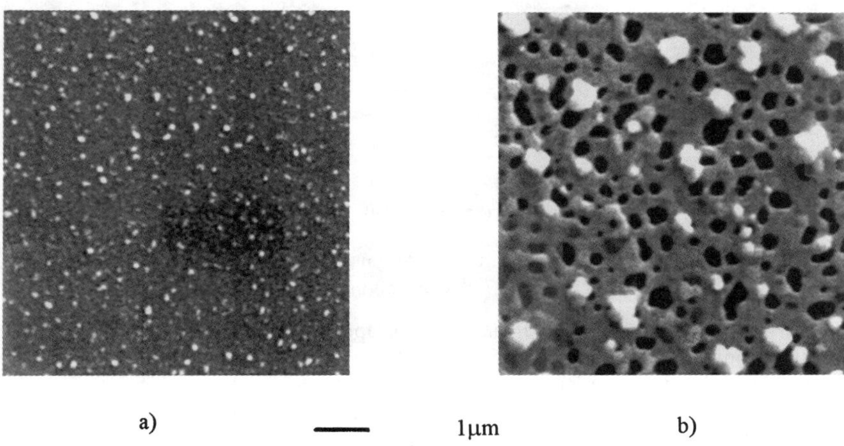

a)   ————   1μm   b)

Fig. 2: SEM of Pt on GaN annealed at a)600°C for 30 min., b)800°C 60 min.

the surface. With further annealing (90 min.) at 600°C, the surface morphology does not change significantly. The particles increase slightly in size. Increasing the temperature to 725°C results in an increase in particle size, as well as coalescence of the remaining film underneath. This behavior becomes more pronounced after annealing at 800°C. Energy dispersive spectroscopy shows clearly that the bright particles are Pt and the black areas are exposed GaN. The gray film in between may be a thin layer of Pt over GaN or a solid solution of Pt and Ga.

The phenomena of island formation is typically observed for interfaces between thin metal films and ceramics.[9,10] The dewetting occurs because the ceramic has a low surface energy, while the metal has a high surface energy. Therefore the metal agglomerates to minimize the total interfacial energy of the system. Since dewetting is not observed in the Pt/Si system, we conclude from this behavior that GaN has a lower surface energy than Si.

<u>Palladium and Nickel</u>

In the cases of Pd and Ni, we again observe behavior which is usually seen in metal/ceramic systems. No change in the structure of the Pd/GaN interface is detected by RBS for annealing below 700°C. Above 700°C, islanding behavior is observed by RBS. No new phases are observed by XRD for annealing up to 800°C. Fig. 3a shows a micrograph of the Pd surface annealed at 725°C for 60 min. The as-deposited metal film is smooth and featureless. Fig. 3a

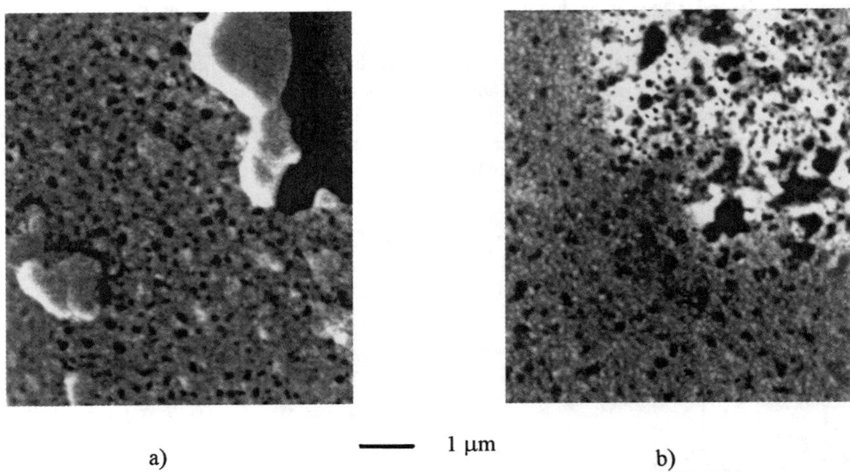

a)   —— 1 μm   b)

Fig.3: SEM micrographs of Ni and Pd surfaces after annealing. a) Pd film after annealing at 725° for 60 min., b) Ni film after annealing at 750°C for 60 min.

shows that the film has begun to agglomerate and approximately 1μm areas of the film have delaminated from the GaN surface. AES clearly shows that there is no GaN on the backside of the delaminated Pd and no Pd remains on the exposed GaN surface.

In the Ni system, no changes are observed by RBS after annealing up to 700°C as long as the sample is capped with a1000 Å layer of SiO_2. Without the capping layer the Ni film will oxidize at these temperatures. After annealing, the SiO_2 layer was etched off with HF and AES studies verified that no reaction occurred between the SiO_2 and Ni. At 700°C the film begins to island. Fig. 3b shows the agglomeration that has occurred after annealing at 750°C for 1 hour. In this case compressive stress is also generated as the thermal expansion coefficient is 13.3×10^{-6} K^{-1}. The presence of the capping layer, however, maintains the planarity of the film and suppresses delamination.

This type of behavior in thin metal film/ceramic systems is attributed to compressive stress generated in the metal film due to differences in thermal expansion coefficients of the metal and substrate.[11] Indeed, the thermal expansion coefficient of Pd is almost twice that of GaN (11.7×10^{-6} K^{-1} vs. 6×10^{-6} K^{-1}). The change between room temperature and the annealing temperature generates a compressive stress which is over twice the ultimate tensile strength of Pd, thus causing delamination failure. The thermal expansion coefficient for Pt is only 9×10^{-6} K^{-1}, therefore less compressive stress is generated during annealing and no delamination is seen.

CONCLUSIONS

From a practical standpoint, Pt, Pd, and Ni are suitable for Schottky contacts on n-type GaN as long as temperatures remain below 600° (Pt) and 700°C (Pd and Ni). Above this temperature, the formation of islands and delamination failure limit the use of these metals as contacts. The lack of lateral homogeneity after annealing will degrade the electrical properties. For example, Ping et al.[12] observed that a Pd/Al contact had an increased contact resistance after annealing at 650°C and above. This was likely due to islanding and delamination similar to that observed in this study. No solid phase reaction has been observed for Pt, Pd, or Ni below 800°C.

The results obtained in this study indicate that the surface of GaN behaves more like a ceramic than a typical semiconductor. The surface energy and linear thermal expansion coefficient are low compared to most metals. This behavior must be considered when determining what metals may act as good contacts on GaN. To obtain stable, non-reactive contacts which withstand high temperatures it may be necessary to use metals with low surface energies and low thermal expansion coefficients. To form an alloyed contact, metals which form stable nitrides, such as Ti, must be used. When engineering metal contacts to GaN a different approach must be taken than has been typically used for metal-semiconductor contacts.

ACKNOWLEDGMENTS

The authors wish to thank the Hewlett-Packard Optoelectronics Division for providing the GaN samples and R. Wilson at UC-Berkeley for his assistance with SEM. KJD is supported by a Robert Noyce Intel Foundation Fellowship. This work was supported by the Director, Office of Energy Research, Office of Basic Energy Sciences, Materials Science Division of the U.S. Department of Energy under contract No. DE-AC03-76SF00098.

REFERENCES

1   S. Nakamura, T. Mukai, and M. Senoh, Jpn. J. Appl. Phys. **30,** L1998 (1991).

2   J. D. Guo, M. S. Feng, R. J. Guo, F. M. Pan, and C. Y. Chang, Appl. Phys. Lett. **67,** 2657 (1995).

3   S. N. Mohammad, Z. Fan, A. E. Botchkarev, W. Kim, O. Aktas, A. Salavador, and H. Morkoc, Electron. Lett. **32,** 598 (1996).

4   L. Wang, M. I. Nathan, T.-H. Lim, M. A. Khan, and Q. Chen, Appl. Phys. Lett. **68,** 1267 (1996).

5   E. V. Kalinina, N. I. Kuznetsov, V. A. Dmitriev, K. G. Irvine, and J. C.H. Carter, Journal of Electronic Materials **25,** 831 (1996).

6   A. C. Schmitz, A. T. Ping, M. A. Khan, and I. Adesida, in *Schottky barrier heights of Ni, Pt, Pd, and Au on n-type GaN*, Boston, 1995 (MRS), p. 831.

7   S. E. Mohney and X. Lin, Journal of Electronic Materials **25,** 811 (1996).

8   K. J. Duxstad, E. E. Haller, and K. M. Yu, submitted to Journal of Applied Physics (1996).

9   J. D. Cawley, in *Metal-Ceramic Joining*, edited by P. Kumar and V. A. Greenhut (TMS, 1991), p. 3.

10   W. C. Maskell, N. M. Sammes, and B. C. H. Steele, Journal of Physics D: Applied Physics **20,** 99 (1987).

11   R. Peddada, K. Sengupta, I. M. Robertson, and H. K. Birnbaum, in *Metal-Ceramic Interfaces*, *Vol. 4*, edited by M. Ruhle, A. G. Evans, M. F. Ashby, and J. P. Hirth (Pergamon Press, New York, 1990), p. 115.

12   A. T. Ping, M. A. Khan, and I. Adesida, Journal of Electronic Materials **25,** 819 (1996).

OHMIC CONTACT TO n-GaN WITH TiN DIFFUSION BARRIER

E.KAMIŃSKA*, A.PIOTROWSKA* M.GUZIEWICZ*, S.KASJANIUK*, A. BARCZ*,
E.DYNOWSKA**, M.D.BREMSER***, O.H.NAM***, AND R.F.DAVIS***
*Institute of Electron Technology, Al.Lotników 46, Warszawa, Poland, eliana@ite.waw.pl
**Institute of Physics, PAS, Al.Lotników 46, Warszawa, Poland
***Department of Materials Science and Engineering, North Carolina State University, Raleigh,
NC   27695-7907

ABSTRACT

The formation of n-GaN/Ti ohmic contacts with TiN diffusion barriers has been investigated by
electrical measurements, x-ray diffraction and SIMS. It has been shown that the onset of the
ohmic behaviour is associated with the thermally induced phase transformation of Ti into TiN at
the GaN/Ti interface. It is suggested that the process is accompanied by an increase in the
doping level in the semiconductor subcontact region. The presence of a TiN barrier is found to
inhibit excessive decomposition of GaN and to confine the reaction between n-GaN and Ti.

INTRODUCTION

The most successful approach to fabricate ohmic contacts to n-type GaN involves the use
of Ti-based metallization, such as TiAl [1], TiAu [2], TiAg [3], and recently also Ti/Al with
Ni/Au overlayer [4]. It is believed that during annealing of GaN/Ti interfacial nitrogen, extracted
from GaN, reacts with Ti to form TiN. The outdiffusing nitrogen atoms create a substantial
concentration of nitrogen vacancies, and thus a heavily doped semiconductor subcontact region
[1]. Although Ti-based contacts may yield excellent contact resistivity, they do not satisfy the
requirement of thermal stability [5].

In recent years there has been significant progress in the development of thermally stable
metallization schemes for Si VLSI and III-V semiconductor devices. The use of diffusion
barriers in contact structures enabled stable and reliable contacts and became widely accepted as
a standard technology. One of the most attractive materials used for barrier layer is TiN due to
its thermodynamic stability, its high melting point and low resistivity. In electrical contacts the
effectiveness of TiN as a diffusion barrier between gold, aluminum, or copper overlayers for
wiring purposes and underlying films has been shown by several authors [6-9].

In this paper we present an approach for improving the reliability of Ti-based metallization
to n-GaN by using a reactively sputtered TiN diffusion barrier to prevent the intermixing of GaN
and contact metallization with overlayers. We correlated the electrical properties and the
microstructures of n-GaN/Ti contacts, encapsulated with TiN layers. In an attempt to clarify the
particular role of Ti and TiN during the formation of the ohmic contact, we compared the
interfacial microstructure of pure Ti, pure TiN, and Ti/TiN contacts both as-deposited and
annealed.

EXPERIMENTAL PROCEDURE

N-type, Si doped GaN films for this study were grown by OMVPE on 6H-SiC at 1050^0C
[10]. Since the current transport across the potential barrier at the metal/semiconductor interface
depends essentially upon the doping level of the semiconductor, to follow this

Mat. Res. Soc. Symp. Proc. Vol. 449 © 1997 Materials Research Society

dependence the experiments were performed on four sets of samples with various doping concentrations N_D namely: $3.7*10^{16}$cm^{-3} (μ=342cm^2/Vs), $3.7*10^{17}$cm^{-3} (μ=411cm^2/Vs), $2.0*10^{18}$cm^{-3} (μ=257cm^2/Vs), $1.1*10^{19}$cm^{-3} (μ=170cm^2/Vs).

Special attention was paid to the surface preparation prior to metal deposition. First, the surface of GaN was cleaned by boiling in hot organic solvents and by plasma ashering. Next, the samples were soaked in buffered HF for 5 min., rinsed in H_2O DI, dipped in NH_4OH:H_2O 1:10 for 15 s. and immediately loaded into the deposition chamber. The cleaning of the substrates was completed in the deposition chamber by heat treatment at 550^{0}C for 10 min.

Metallic films were deposited by rf magnetron sputtering at a base pressure of $1*10^{-7}$ Torr. Before starting the deposition of metallization the substrates were cooled down to 220^{0}C. Ti films of the thickness 15-100 nm and 80 nm thick TiN layers were sequentially deposited, without breaking the vacuum, using Ti target, by Ar and Ar/N_2 reactive sputtering, respectively. Ti films were deposited at a pressure p=$3.3*10^{-3}$ Torr, at a rate 1.4 nm/s. The parameters of the deposition of stoichiometric TiN films were as follows: a gas flow ratio of N_2/Ar = 0.18, p=$3.7*10^{-3}$ Torr, the substrate bias voltage U_B = -60V. The deposition rate was 0.2 nm/s. The resistivities of as deposited Ti and TiN layers were 71.5 $\mu\Omega$cm and 38 $\mu\Omega$cm, respectively.

A complementary study of Ti contacts encapsulated with Si_3N_4 film has been performed. Ti/Si_3N_4 (100 nm/200 nm) double layers were deposited by Ar rf magnetron sputtering from Ti and Si_3N_4 targets, respectively.

Heat treatments were performed using rapid thermal annealing (RTA) in N_2 flow at temperatures up to 900^{0}C for 30s. During RTA the samples were protected by a piece of oxidized Si as a proximity cap.

The electrical characterization involved the measurements of I-V characteristics and contact resistivity r_c. Samples for I-V measurements were patterned of dots with diameter varying from 120 to 350 μm and one large area contact. I-V measurements were taken between one of the dots and the large area contact. The contact resistance test pattern had 100x100 μm contact pads with pad separations 2, 4, 6, 10 and 12 μm. Mesa structures, required for r_c measurements by transmission line method (TLM), were patterned with SiO_2 mask and dry etched using CCl_4/H_2 plasma. The RIE process pressure and the gas flow ratio were $1.1*10^{-1}$ Torr and 9.3 sccm/12 sccm, respectively. The etching rate was 16.7 nm/min. at a rf power (13.56 MHz) of 0.4W/cm^2.

The thermally induced transformations in the contact region were investigated by the complementary use of x-ray diffraction (XRD) and secondary ion mass spectrometry (SIMS).

RESULTS

Electrical properties

As-deposited n-GaN/Ti/TiN contacts were ohmic with r_c $5.3*10^{-3}\Omega$cm^2, $1.9*10^{-3}\Omega$cm^2, and $2*10^{-5}\Omega$cm^2 for dopant concentration $3.7*10^{17}$ cm^{-3}, $2.0*10^{18}$ cm^{-3}, and $1.1*10^{19}$cm^{-3}, respectively. Contacts deposited on material doped to $3.7*10^{16}$ cm^{-3} exhibited non-ohmic behavior. Figure 1 shows the I-V characteristics of contacts formed on lightly doped n-GaN, before heat treatment and after annealing at temperatures up to 900^{0}C. Slightly non-linear after deposition, the behaviour of Ti/TiN contacts annealed up to 400^{0}C remained unchanged. However, as a result of annealing at 500^{0}C the contacts became rectifying, with a ideality factor n=1.19 and the barrier height $\Phi_B\approx$0.71 eV, as inferred from the forward biased I-V characteristics. After annealing at higher temperatures, contacts gradually lost their rectifying properties. After the 900^{0}C anneal they were ohmic with r_c=$6*10^{-2}\Omega$cm^2. In contrast, pure TiN contacts deposited on the similarly doped substrates were ohmic in the as-deposited state and remain such after heat treatment up to 400^0C, with $r_c=8.8*10^{-2}\Omega cm^2$. Anneals at higher temperatures resulted in progressive degradation of their linear characteristics and in increase of the contact resistivity. The I-V characteristics of n-GaN/TiN contacts are plotted in Fig. 2.

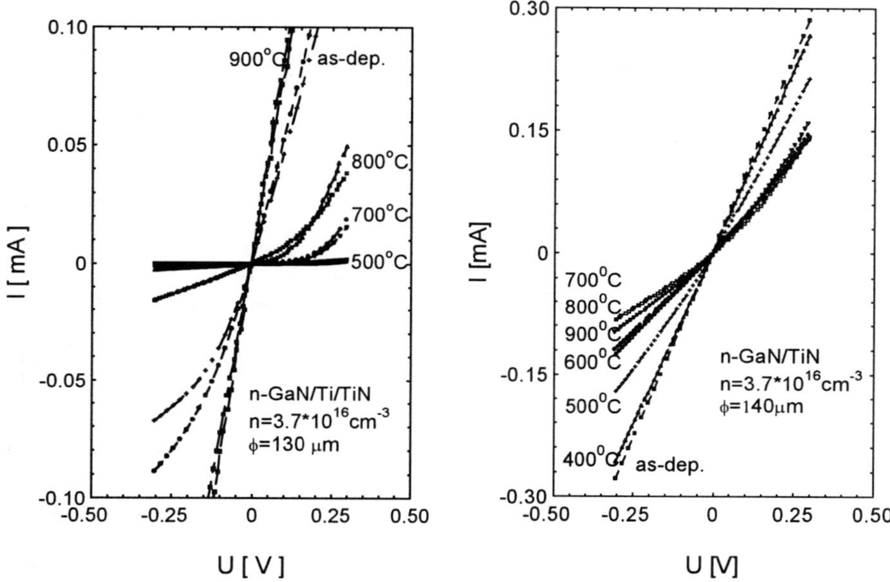

Fig.1. I-V characteristics of as-deposited and annealed n-GaN/Ti/TiN contacts.

Fig.2. I-V characteristics of as-deposited and annealed n-GaN/TiN contacts.

The resistivity of Ti/TiN contacts heat treated at 900^0C as a function of doping level of n-type GaN is shown in Fig.3. For dopant concentration up to $2*10^{18}$ cm^{-3}, an important decrease of r_c as a result of annealing is observed, and the contact resistivity exhibits an inverse proportionality to N_D. For doping level of $1.1*10^{19}$cm^{-3} the contact resistivity practically does not change upon annealing.

<u>Thermally induced interactions at the n-GaN/Ti interface</u>

Since the ohmic behaviour of n-GaN/Ti contacts may be attributed to the annealing induced growth of TiN phase at the interface, special attention was paid to properly distinguish between the deliberately deposited TiN barrier layer and the newly formed TiN. To facilitate conclusive phase identification, a complementary study of Ti contacts annealed under Si_3N_4 cap was performed. The results of XRD analysis of n-GaN/Ti(15nm)/TiN(80 nm) contacts are presented in Fig. 4. The as-deposited Ti/TiN contact consisted of two separate layers: Ti and TiN (fcc) phases. Hexagonal (cph) Ti was found to have a <001> preferred orientation, and a lattice parameter of c=4.726 Å in the direction perpendicular to the surface (c=4.684 Å for bulk Ti). Cubic (fcc) TiN film exhibited a <111> preferred orientation, with a lattice parameter a=4.299 Å (a=4.24 Å for bulk TiN). The sequence of thermally activated phase transformations in TiN- and Si_3N_4-encapsulated contacts was similar. Up to 700^0C the lattice parameter of Ti progressively increased, while the lattice parameter of TiN decreased. During 800^0C anneal, the phase transformation Ti→TiN took place.

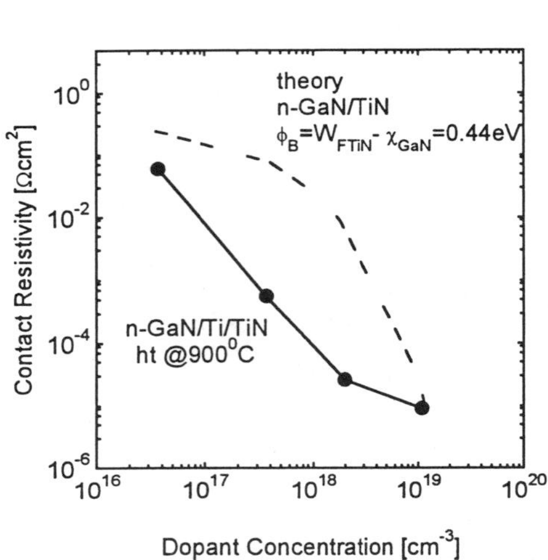

Fig.3. The dependence of r_c of Ti/TiN contacts on the doping level of n-GaN; dots -experimental data points, dashed line - theoretical prediction for Φ_B=0.44 eV.

Fig.4. X-ray diffractograms of as-deposited and annealed GaN/Ti (15 nm)/TiN(80nm) contacts.

The observed phase transformations can be interpreted as follows. As indicated by the values of lattice constants, both the as-deposited Ti and TiN films are strained. Upon annealing, the nitrogen released from neighboring GaN at one side, and from TiN at the other, spreads throughout the Ti film. At lower temperatures nitrogen atoms enter the Ti lattice, filling the interstitial sites. As they are larger than interstitial sites, they cause further distortion of the Ti lattice. When the concentration of nitrogen exceeds the solubility limit, the structure transformation from hcp to fcc takes place [11]. The stress in TiN film was relaxed during annealing.

The degree of decomposition of GaN during the formation of Ti-based ohmic contacts was studied with SIMS. Annealing in a N_2 flow of the as-deposited GaN/Ti(50nm) /TiN(80nm) structure leads to the limited dissolution of the substrate which manifests in a migration of Ga atoms into the initial Ti layer with pronounced accumulation at the Ti/TiN interface (Fig.5.). The Ga content in the Ti layer can be estimated to be a few at.%, but is undetectable in the initial TiN cap. This suggests that the main supply of nitrogen required for the observed transformation of the Ti film into TiN originates from the ambience rather than from the substrate. This is confirmed by a similar measurement of a pure Ti(100nm) deposited on GaN (Fig.6.) Gradual decrease of the Ga concentration towards the surface seems associated with the inhibiting of the dissolution of GaN in the course of annealing and subsequent formation of TiN. Thermal stability of the GaN/TiN system is demonstrated in Fig.7. where the respective changes in the N, Ga and Ti profiles before and after heating to 900^0C are insignificant.

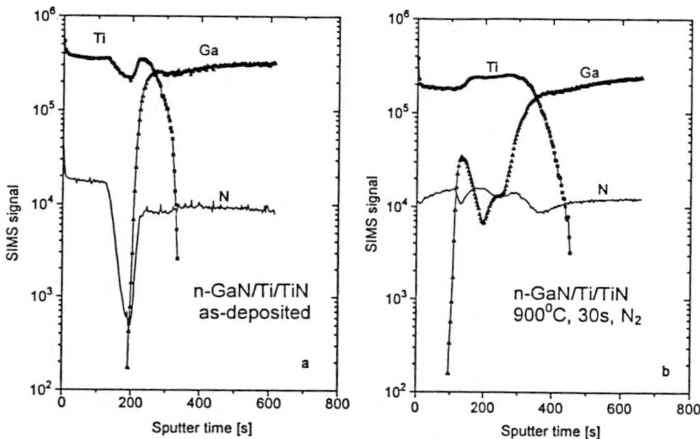

Fig.5. SIMS in-depth profiles for n-GaN/Ti/TiN contacts: a) as deposited, b) annealed at 900^0C.

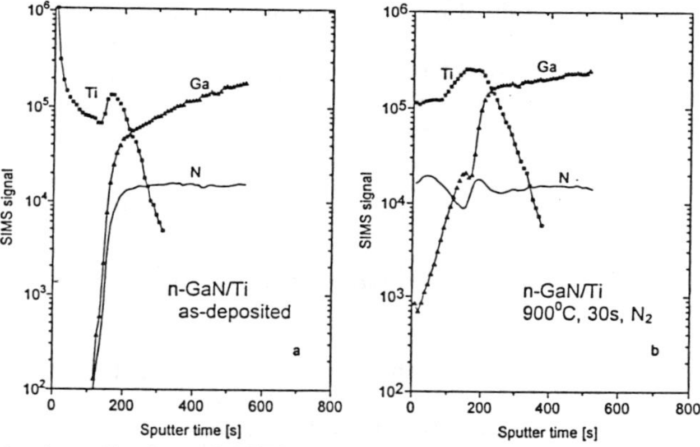

Fig.6. SIMS in-depth profiles for n-GaN/Ti contacts: a) as deposited, b) annealed at 900^0C.

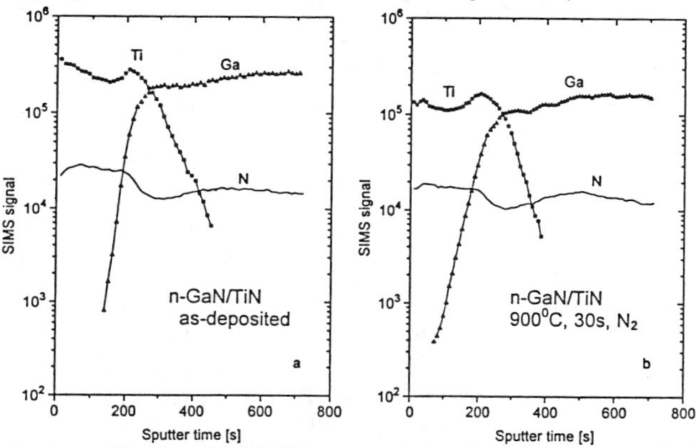

Fig.7. SIMS in-depth profiles for n-GaN/TiN contacts: a) as deposited, b) annealed at 900^0C.

DISCUSSION AND CONCLUSIONS

It is for a number of reasons that Ti plays a key role in producing low resistivity contacts to n-GaN. Ti is known to reduce the native Ga_2O_3 on GaAs surface [12], thus we can expect a similar behaviour for the native oxide on GaN. The improvement of the diode characteristics of n-GaN/Ti/TiN contacts after annealing at 500^0C can be interpreted as the reduction of the residual oxide and/or the annealing out of defects at the metal/semiconductor interface. In contacts annealed at higher temperatures, the barrier height is dominated by the interfacial reactions, leading to the formation of TiN at the metal/semiconductor interface. From the simple Schottky barrier theory, the barrier height at n-GaN/TiN interface should be equal to the difference between the work function of TiN (3.74 eV [13]) and the electron affinity of GaN (3.3 eV [14]). In Fig.3., together with the experimental data points, we have plotted a theoretical dependence for $r_c=f(N_D)$, calculated from the simple thermionic and thermionic-field emission theory [15], for $\phi_B=0.44$ eV. The discrepancy between the theoretical prediction and experimental results can be explained by the increase of the doping concentration in the semiconductor subcontact region. The assumption that the N_D had increased up to $\sim 1.0*10^{19}$ cm^{-3} would explain the unchanged resistivity of contacts deposited on highly doped GaN, and the decrease of r_c for $N_D=2.0*10^{18}$ cm^{-3}. For contacts formed on lightly doped substrates, the contact resistivity would be limited by the barrier between highly doped subcontact region and low-doped bulk, the latter being characterized by an inverse proportionality to N_D [16]. TiN barrier inhibits excessive decomposition of GaN and thus confines the reaction between n-GaN and Ti.

REFERENCES

1. M.E.Lin, Z.Ma, F.Y.Huang, Z.F.Fan, L.H.Allen, H.Morkoc, Appl.Phys.Lett., **64**, 1003 (1994).
2. M.A.Khan, J.N.Kuznia, A.R.Bhattarai, D.T.Olson, Appl.Phys.Lett., **62**, 1786 (1993).
3. J.D.Guo, C.I.Lin, M.S.Feng, F.M.Pan, G.C.Chi, C.T.Lee, Appl.Phys.Lett., **68**, 235 (1996).
4. Z.Fan, S.N.Mohammad, W.Kim, O.Aktas, A.E.Botchkarev, H.Morkoc, Appl.Phys.Lett., **68**, 1672 (1996).
5. H.Morkoc, Proc.Int.Symp.Blue Laser and LEDs, Chiba Univ.,Japan, 23 (1996).
6. N.Kumar, J.T.McGinn, K.Pourezaei, B.Lee, J.Vac.Sci.Technol., **A6**, 1602 (1988).
7. J.Schulte, S.B.Brodsky, T.Lin, R.V.Joshi, Tungsten and other refractory metals for VLSI applications, Mat.Res.Soc. 1987 Workshop Proc. 367 (1988)
8. Jian Li, P.F.Chapman, F.Goodwin, Mat.Res.Soc.Conf.Proc. **ULSI-VIII**, 75 (1993).
9. A.Piotrowska, E.Kamińska, M.Guziewicz, S.Kwiatkowski, A.Turos, Mat.Res.Soc.Symp. Proc. **300**, 219 (1993).
10. T.W.Weaks, Jr., M.D.Bremser, K.S.Ailey, E.Carlson, W.G.Perry, E.L.Piner, N.A.El-Masry, R.F.Davis, J.Mat.Res., 10, 1011 (1996).
11. M.Wittmer, J.Vac.Sci.Technol., **A3**, 1797 (1985).
12. S.P.Kowalczyk, J.R.Waldrop, R.W.Grant, Appl.Phys.Lett., **38**, 167 (1981).
13. V.S.Fomenko, <u>Emission Properties of Materials</u>, Naukova Dumka, Kiev, 1981.
14. R.J.Nemanich, M.C.Benjamin, S.P.Bozeman, M.D.Bremser, S.W.King, B.L.Ward, R.F.Davis, B.Chen, Z.Zhang, J.Bernholc, Mat.Res.Soc.Symp. Proc. **395**, 777 (1996).
15. A.Y.C.Yu, Solid-State Electronics, **13**, 239 (1970).
16. W.Dingfen, W.Dening, K.Heime, Solid-State Electronics, **29**, 489 (1986).

Cr/Ni/Au ohmic contacts to the moderately doped p- and n-GaN

Taek Kim, Myung C. Yoo, and Taeil Kim
Photonics Semiconductor Lab., Samsung Advanced Institute of Technology, P.O. Box 111,
Suwon 440-600, Korea, taek@saitgw.sait.samsung.co.kr

ABSTRACT

We report new Cr/Ni/Au and Ni/Cr/Au tri-layer metallization schemes for achieving low resistance ohmic contacts to moderately doped p- ($\sim$1 x 10^{17}/cm^3), and n-GaN ($\sim$1 x 10^{18}/cm^3) respectively. The metallizations were thermally evaporated on 2 μm-thick GaN layers grown on c-plane sapphire substrates by metalorganic chemical vapor deposition (MOCVD). Comparisons with bi-layer metallizations such as Ni/Au and Cr/Au were also made. The Cr/Ni/Au contacts showed a low specific contact resistivity of 9.1 x 10^{-5} $\Omega \cdot$cm^2 to n-GaN while that of Ni/Cr/Au to p-GaN was 8.3 x 10^{-2} $\Omega \cdot$cm^2. The Ni/Cr/Au contacts also showed a low specific contact resistivity of 2.6 x 10^{-4} $\Omega \cdot$cm^2 to n-GaN. The Ni/Cr/Au metallization could made reasonable ohmic contacts to p- and n-GaN simultaneously

INTRODUCTION

III-V nitrides are very attractive materials for blue and green light emitting diodes (LEDs) [1], metal-semiconductor field effect transistors (MESFETs) [2], and modulation-doped field effect transistors (MODFETs) [3]. Low resistance and thermally stable ohmic contacts are imperative in successful implementation of these devices. A very low resistance n-type ohmic contact using Ti/Al/Ni/Au was demonstrated by Z. Fan *et al.* [4]. In the case of laser diodes (LD) and light emitting diodes (LED), Ti/Al and Ni/Au contacts are widely used for n-, and p-GaN ohmic contacts, respectively [5][6]. However, there has been no detailed reports for p-type GaN ohmic contacts. In fact, the high contact resistance of p-GaN is one of major obstacles for realizing continuous wave operation of GaN based laser diodes at room temperature

In this paper, we report our initial investigation of new ohmic contact schemes to moderately doped p-, and n-GaN. We have tried the Cr/Ni/Au and the Ni/Cr/Au and compared the results to those of the Ni/Au and the Cr/Au contacts.

EXPERIMENT & RESULTS

The GaN specimens used in this study were grown by MOCVD on c-plane sapphire substrates. Low temperature GaN buffer layers were grown first. The thickness of n- and p-GaN layers was 2 μm. The carrier concentration of the n-GaN epilayer was found to be 1 x 10^{18}/cm^3 by Hall measurements method. The p-doping concentration ranged from 1 x 10^{16}/cm^3 to 1 x 10^{17}/cm^3. The GaN samples were patterned and then etched to make mesa areas for TLM measurements. The mesas were isolated by chemically-assisted ion beam etching (CAIBE) using Cl$_2$. The dimension of the rectangular pads was 200 μm wide and 100 μm long. The separations between the contact pads were increased from 5 to 30 μm in 5 μm increments. The metallization patterns were defined using lift-off.

All metal contacts were deposited by thermal evaporation. The deposition sequences and the layer thickness were as follows: (1) Cr - 15 nm / Ni - 15 nm / Au - 500 nm ; (2) Ni - 15 nm / Cr - 15 nm / Au - 500 nm ; (3) Ni - 20 nm / Au - 500 nm and (4) Cr - 20 nm / Au - 500 nm. The

Mat. Res. Soc. Symp. Proc. Vol. 449 © 1997 Materials Research Society

Cr/Ni/Au, Ni/Cr/Au, and Cr/Au contacts were deposited on both n-, and p-GaN epilayers. The Ni/Au contacts were applied to p-GaN. All epitaxial samples of the same polarity were taken from the same wafer. After sample preparation, the metal patterns were inspected using optical microscope and defective ones were excluded from the measurements. The I-V characteristics of the metal contacts were measured by a HP 4145B analyzer. The n-type I-V characterizations were carried out by current injecting up to ± 20 mA and measuring the voltage across the contacts. The p-type I-V curve exhibited linear characteristics for small current level and the measurements were performed within ± 10 μA range. The measurements were performed on both as-deposited and annealed contacts at 500 °C for 30 seconds in N_2 ambient.

<u>n-GaN Ohmic Contacts</u>

Fig. 1 shows the I-V characteristics of various n-type contact metals. All data, except that of Ni/Cr/Au before annealing, show linear I-V characteristics. The Cr/Au contacts exhibited the lowest resistance after annealing. However, to compare the ohmic characteristics quantitatively, specific contact resisitivties were calculated using TLM. The gradients and the intersection with the resistance axis were found using the least-square fit method. The plotted data of the three contact materials are shown in Fig. 2. The specific contact resistivity of the Cr/Ni/Au contacts was calculated to be 9.1 x 10^{-5} $\Omega \cdot cm^2$ before annealing and slightly increased to 2.7 x 10^{-4} $\Omega \cdot cm^2$ after annealing. It should be noted that the specific contact resistivities of the Cr/Ni/Au contacts was lower than that of the Cr/Au contacts (ρ = 3.1 x 10^{-4} $\Omega \cdot cm^2$). In the case of the Cr/Au contacts, no significant difference in specific contact resistivity was observed between as-deposited and annealed samples. In terms of work function, Ni should make a better contacts to p-GaN. Thus, we assume that the Ni plays an certain role in the Cr/Ni/Au system as a intermediate layer lowering contact resistance. On the other hand, the I-V curve of the Ni/Cr/Au contacts showed significant rectifying characteristics before annealing.

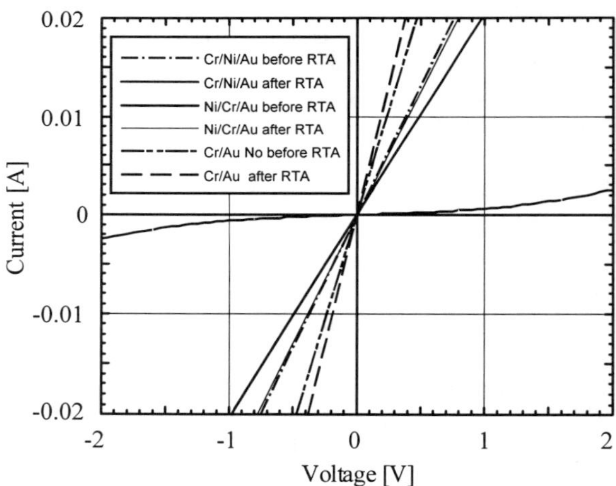

Figure 1. Current (I) - voltage (V) characteristics of
various contact metals on n-GaN.

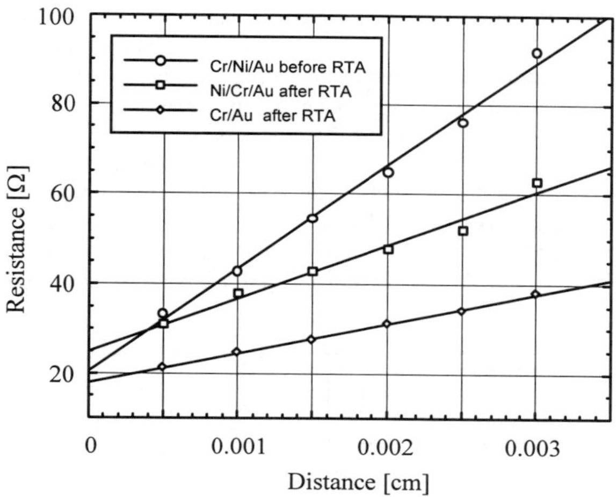

Figure 2. Least-squares regression of ohmic contacts on n-GaN

However, a linear I-V curve was obtained after annealing and the specific contact resistivity was calculated to be 2.6 x 10^{-4} $\Omega \cdot cm^2$. We attribute this result to the diffusion of Cr into the GaN.

<u>p-GaN Ohmic Contacts</u>

The same tri-layer schemes (ie, the Cr/Ni/Au and the Ni/Cr/Au) were also applied on p-GaN. Figure 3 shows I-V characteristics of the Cr/Ni/Au, the Ni/Cr/Au, the Cr/Au, and the Ni/Au contacts on p-GaN. Non-linear I-V characteristics were shown in the Ni/Au contacts before

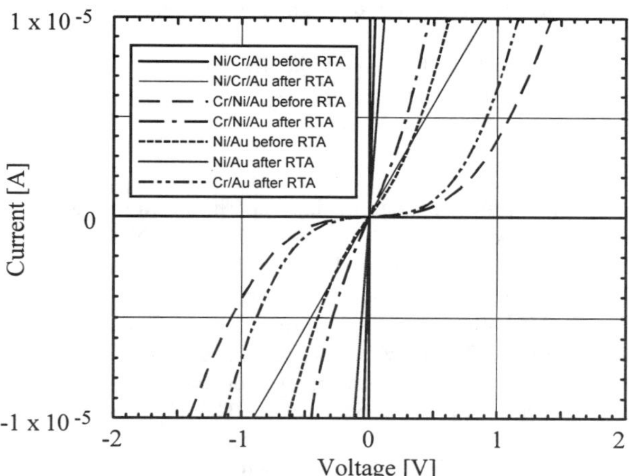

Figure 3. Current (I) - voltage (V) characteristics of
various contact metals on p-GaN.

annealing. The ohmic characteristics were obtained after annealing and the specific contact resistivity was calculated to be 3.4 x 10^{-1} $\Omega \cdot cm^2$ as shown in Fig.4. On the contrary, the Ni/Cr/Au contacts showed linear I-V characteristics before annealing and the specific contact resistivity was 8.3 x 10^{-2} $\Omega \cdot cm^2$ which was the lowest value in our experiment. After annealing, the linearity was maintained but specific contact resistivity was increased by about one order of magnitude. We attribute this phenomenon to that the Cr plays a role of diffusion barrier and/or wetting agent when confined in the contact metallization lowering contact resistance, but deteriorates the p-ohmic characteristics if diffused into GaN. The Cr/Ni/Au contacts showed better contact resistance after annealing although they did not show perfectly linear characteristics. The Cr/Au contacts showed clear rectifying characteristics whether annealed or not.

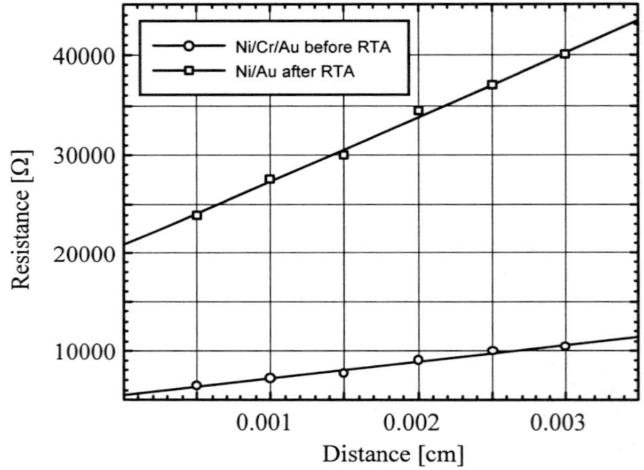

Figure 4. Least-squares regression of ohmic contacts on p-GaN

These results also supported the hypothesis that the Cr does not directly produce ohmic properties but helps Ni improve contact resistance.

<u>InGaN / GaN DH LED</u>

The best p-, and n-GaN ohmic contacts in this experiment were the Ni/Cr/Au and the Cr/Ni/Au contacts respectively. However, we found that the alloyed Ni/Cr/Au contacts could be used as p-, and n-ohmic contacts simultaneously. In this case, the specific contact resistivities of p-, and n-GaN were calculated to be 4.5 x 10^{-1} $\Omega \cdot cm^2$ and 2.6 x 10^{-4} $\Omega \cdot cm^2$ respectively. We fabricated $In_{0.15}Ga_{0.85}N$ / GaN double hetero-structure (DH) LEDs with the Ni/Cr/Au ohmic metallization for both p-, and n-GaN. Individually metallized LEDs were also fabricated for comparisons. As shown in Fig. 5, the best I-V curve was obtained with the LED which was p-metallized with Ni/Cr/Au and n-metallized with Cr/Ni/Au. The I-V curve of the simultaneously metallized LEDs showed higher a turn-on voltage. However, the I-V characteristics was comparable to those of the LEDs individually metallized with Ni/Au and Ti/Al.

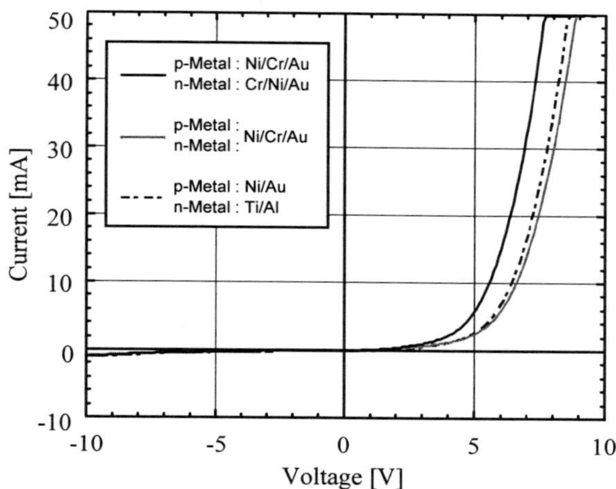

Figure 5. I-V Characteristics of InGaN/GaN Double Heterostructure LED with various Ohmic metals

CONCLUSIONS

In conclusion, we have developed new tri-layer ohmic contact scheme for moderately doped n-, and p-GaN. The low specific contact resistivitiy of 9.1×10^{-5} $\Omega \cdot cm^2$ for n-GaN and 8.3×10^{-2} $\Omega \cdot cm^2$ for p-GaN was obtained using the Cr/Ni/Au and the Ni/Cr/Au contacts, respectively. We attribute these low values to the intermediate layer which plays a certain role lowering contact resistance. Further analytical study is required to verify our hypothesis. The Ni/Cr/Au contacts also show low resistant ohmic characteristics for n-GaN as well as p-GaN suggesting the possibility of simultaneous metallization for p-, and n-type contacts. We also fabricated InGaN / GaN DH LEDs using this simultaneous metallization scheme which provides cost and time effective process.

REFERENCES

1. Shuji Nakamura, Masayuki Senoh, Naruhito Iwasa and Shin-ichi Nagahama, Jpn. J. Appl.Phys. Vol. 34 (1995) pp.L797 - L799
2. M. A. Khan, J. N. Kuznia, A. R. Bhattarai, and D. T. Oslon, Appl. Phys. Lett. **82**, 1786 (1993)
3. M. A. Khan, A. R. Bhattarai, J. N. Kuznia, and D. T. Oslon, Appl. Phys. Lett. **63**, 1214 (1993)
4. Zhifang Fan, S. Noor Mohammad, Wook Kim, Ozgur Aktas, Andrei E. Botchkarev, and Hadis Morkoc, Appl. Phys. Lett. **68**, 1672 (1996)
5. Georg Mohs, Brian Fluegel, Harald Giessen, Habib Tajalli, and Nasser Peyghambarian, Appl. Phys. Lett. **67**, 1515 (1995)
6. Shuji Nakamura, Masayuki Senoh, Shin-ichi Nagahama, Naruhito Iwasa Takao Yamada, Toshio Matsushuta, Hiroyuki Kiyoku, and Yasunobu Sugimoto, Appl. Phys. Lett. **68**, 3269 (1996)

InN-BASED OHMIC CONTACTS TO AlInN

*S. M. Donovan, *F. Ren, J. D. MacKenzie, C. R. Abernathy, S. J. Pearton and +K. Jones,* Department of Materials Science and Engineering, University of Florida, Gainesville, *Bell Laboratories, Lucent Technologies, Murray Hill, NJ, +U. S. Army Research Lab, Ft. Monmouth, NJ.

ABSTRACT

In order to maximize the performance of III-Nitride devices, it is necessary to develop thermally stable low resistance Ohmic contacts to III-N based electronic structures. This paper reports on the utility of InN as an aid to contact formation on widegap materials such as InAlN. For n-type materials, several questions relating to the growth conditions have been explored. Specifically, the impact of substrate type (GaAs vs. Sapphire), cap layer growth temperature and V/III ratio on contact resistance has been investigated. It was found that the use of sapphire substrates combined with high growth temperatures (575°C) and high V/III ratios produced acceptable contact resistances ($\sim 10^{-6}$Ohm-cm^2) to InAlN.

INTRODUCTION

III-V nitrides are becoming increasingly important for the development of high temperature electronic and optical devices due to their wide bandgaps and high saturation velocities[1]. Before the performance of devices such as metal enhanced semiconductor field effect transistors (MESFETs) or heterojunction bipolar transistors (HBTs) can be optimized, however, several key processing issues must be addressed. Among the most important of these is the ability to fabricate low resistance, thermally stable Ohmic contacts to n-type material. This is a critical technology for III-V nitrides.

In electronics applications, narrow bandgaps and/or high doping levels are usually needed in order to minimize the resistances arising from Ohmic contacts. In GaAs-based devices, for example, heavily n-doped InGaAs layers on n-GaAs can be used to reduce parasitic resistances as much as one order of magnitude relative to contacts directly on GaAs[2]. The higher doping and smaller bandgap which can be achieved in InGaAs result in contact resistances of $\sim 2 - 6 \times 10^{-7}$ Ω-cm^2 vs. $\sim 1 \times 10^{-6}$ Ω-cm^2. In this paper we present a similar approach to contact formation to InAlN using InN as an aid to contact formation. In the past, InN has been deposited primarily using various sputtering[3-8] or electron beam[9] techniques which are not compatible with *in situ* deposition of III-V device structures. Growth by metalorganic chemical vapor deposition[10] (MOCVD) and electron cyclotron resonance molecular-beam epitaxy[11] (ECR MBE) have been demonstrated as well. Here the advantages of both techniques are utilized by electron cyclotron resonance metalorganic molecular-beam epitaxy (ECR MOMBE).

EXPERIMENTAL

The samples were grown on either (0001) sapphire or semi-insulating, (100) GaAs substrates in an Intevac Gas Source Gen II. The group III precursors, dimethylethylamine alane (DMEAA) and trimethylindium (TMI) were transported by a He carrier gas in order to avoid possible hydrogen passivation effects. An Electron Cyclotron Resonance plasma source (Wavemat MPDR 610) operating at 2.45 GHz and 200 W forward power was used to provide the

Mat. Res. Soc. Symp. Proc. Vol. 449 © **1997 Materials Research Society**

nitrogen flux. N_2 flows of 5 or 20 sccm were used. Growth temperatures for In-containing layers ranged from 500°C to 575°C, while AlN layers were grown at 700°C. The growth sequence on sapphire consisted of a nitridation step of 5 minutes at 700°C, followed by a low temperature (425°C) AlN nucleation layer. The film compositions were varied by altering the relative group III gas flow rates. Electrical transport properties were obtained from Van der Pauw geometry Hall measurements at 300 K using alloyed (400° C, 2 min) HgIn ohmic contacts. The compositions of the calibration samples were determined by electron microprobe analysis using a 6 kV beam, and by powder x-ray diffraction assuming Vegard's law. Surface morphology was examined by scanning electron microscopy (SEM) and atomic force microscopy (AFM). XRD was used to distinguish between the wurtzite and zincblende phases. Contact resistances were measured using the TLM method on WSi_x contacts. Fabricated by sputtering, WSi_x was chosen for its refractory properties. Two types of contact structures were evaluated, those with InN contacting regions, and those without. The layer structures for both are shown schematically in Figure 1. The graded regions between the InAlN and the InN were used to avoid band discontinuities at the InN and InAlN interface.

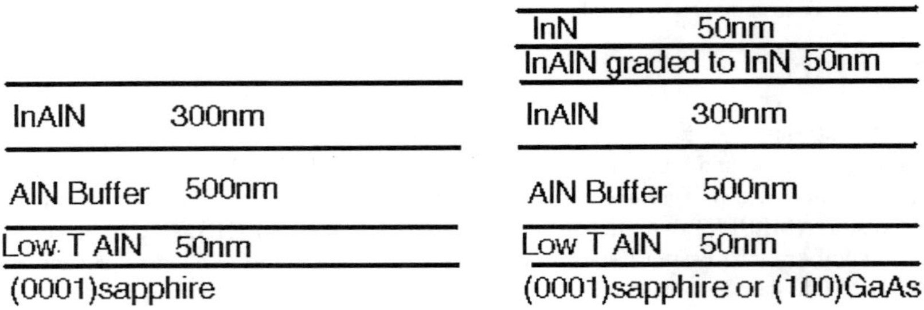

<table>
<tr><td></td><td colspan="2">InN   50nm</td></tr>
<tr><td></td><td colspan="2">InAlN graded to InN 50nm</td></tr>
<tr><td>InAlN   300nm</td><td colspan="2">InAlN   300nm</td></tr>
<tr><td>AlN Buffer 500nm</td><td colspan="2">AlN Buffer 500nm</td></tr>
<tr><td>Low T AlN 50nm</td><td colspan="2">Low T AlN 50nm</td></tr>
<tr><td>(0001)sapphire</td><td colspan="2">(0001)sapphire or (100)GaAs</td></tr>
</table>

Figure 1. Schematic of layer structures used for this study.

RESULTS AND DISCUSSION

The initial experiments have been conducted on InAlN structures, such as would be used in lattice matched InAlN/InGaN HEMT devices, using $TiWSi_x$ contacts. Contact structures based on InAlN were grown with and without InN connected with graded interlayers. Both GaAs and sapphire substrates were investigated. As shown in Figure 2, the GaAs substrates produce InAlN surface morphologies comparable to those obtained on sapphire, in spite of the larger lattice mismatch between the GaAs and the epilayer. As expected the addition of the graded region and the InN contact layer roughens the surface considerably, particularly for the structures grown on sapphire. This is also observed for InAs growth on GaAs-based structures. This roughening typically becomes worse with increasing In-containing layer thickness, thus limiting the thickness of the contacting layer which can be used. For the structures used in this study, with 500A InN layers, the surface morphology is still acceptable for metallization. Hall measurements of InAlN layers gave mobilities of ~5-15 $cm^2/V \cdot s$ while InN layers were ~25-50 $cm^2/V \cdot s$. In all cases the InN electron concentration was measured to be ~ 1 - 3 x 10^{20} cm^{-3} with InAlN layers measuring an order of magnitude less.

Given the similarity in electrical behavior and surface morphology, it was expected that the contact resistances would also be similar in structures grown on the two substrates. However, as shown in Table 1, the structures grown on sapphire produce significantly lower

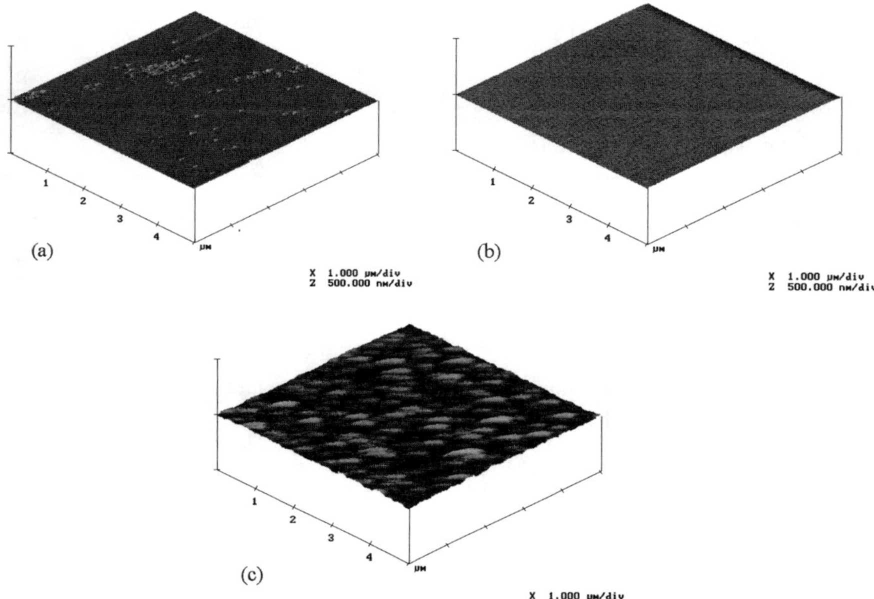

Figure 2. AFM scans of InAlN on either GaAs (a) or sapphire (b), and the InN/InAlN contact structure on GaAs (c).

contact resistances. This difference does not appear to be related to doping level as the sheet resistances of the InN cap layers do not appear to be substantially different between the two substrates. As shown in Table 2, x-ray diffraction studies suggest that the sapphire grown material has slightly better crystallinity as determined by the width of the XRD peaks. This may account for the improved contact. Clearly it appears that from the point of view of parasitic resistances, electronic devices fabricated on sapphire will most likely provide superior performance.

Contrary to what we have observed for GaN, where even relatively low In mole fractions ($X_{In} \sim 0.25$) produce significant improvements in contact resistance[12], it was found that InAlN contact layers produce unacceptably high contact resistances even at low aluminum concentrations ($X_{Al} \sim 0.25$). As shown in Table 1, while the addition of InN cap layers does reduce the resistance relative to InAlN/GaAs structures, it still does not reach the level which can be obtained with sapphire and InAlN alone. Interestingly the contact resistance obtained with InN layers on sapphire based structures, while still better than comparable layers on GaAs, was worse than for InAlN alone on sapphire. Presumably the poorer contact behavior on sapphire was due to the rougher surface morphologies. It was found that by increasing the nitrogen flow from 5 to 20 sccm through the ECR plasma source during growth of the InN, the

Table 1. Contact resistance and rms surface roughness of films grown as a function of nitrogen flow and growth temperature.

STRUCTURE	CONTACT RESISTANCE (Ω-cm^2)	AFM RMS ROUGHNESS (nm)	N$_2$ FLOW DURING InN GROWTH (sccm)	InN GROWTH TEMP. (°C)
InAlN/AlN/Sapphire	1.9×10^{-5}	1.4	5	525
InAlN/AlN/GaAs	1.4×10^{-3}	1.6	5	525
InN/InAlN/AlN/Sapphire	4.8×10^{-4}	68.7	5	525
InN/InAlN/AlN/GaAs	$8 - 9 \times 10^{-4}$	7.0	5	525
InN/InAlN/AlN/Sapphire	3.6×10^{-5}	20.5	20	525
InN/InAlN/AlN/Sapphire	3.5×10^{-6}	54.7	20	575

Table 2. Primary XRD peaks for contacts grown on GaAs and sapphire.

STRUCTURE	N$_2$ FLOW DURING InN GROWTH (sccm)	InN GROWTH TEMP. (°C)	PEAK POS. (degrees)	PEAK ID	FWHM (arcsec)
InN/InAlN/AlN /GaAs	5	525	31.4	InN(002)	3320
			36.1	AlN(002)	2650
InN/InAlN/AlN /sapphire	5	525	31.4	InN(002)	1895
			33.5	InAlN	12300
			36.1	AlN(002)	1420
InN/InAlN/AlN /sapphire	20	525	35.9	AlN(002)	1910
			64.7	InN(004)	1800
			76.2	AlN(004)	3700
InN/InAlN/AlN /sapphire	20	575	31.4	InN(002)	1950
			33.5	InAlN	17000
			36.1	AlN(002)	1890

contact resistance and to a lesser extent the surface roughness could be reduced to values closer to those of InAlN alone. This is reflected in the SEM micrographs shown in Figure 3. The rough surface at low nitrogen flow may indicate that growth has occurred under nitrogen deficient conditions, leading to the formation of In-rich regions in which 2D growth does not occur. Such behavior is also observed in other In-containing materials such as InAs.

Further improvement in the contact resistance was obtained by increasing the growth temperature from 525°C to 575°C. While this actually degraded the surface roughness slightly, as evidenced by the AFM mean surface roughness shown in Table 1, increasing the growth temperature reduced the as-deposited resistance by an order of magnitude. Presumably this is

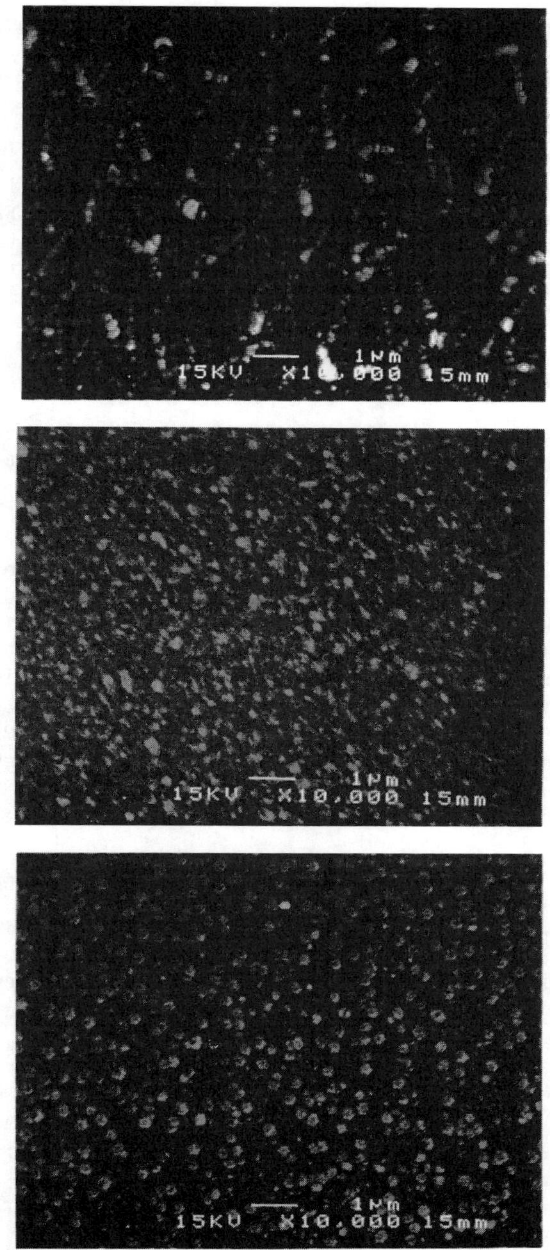

Figure 3. SEM micrographs of contact structure grown on sapphire at 525C and 5 sccm N_2 (top), 525C and 20 Sccm N_2 (middle) and 575C at 20 sccm N_2 (bottom).

the cause of the reduced contact resistance obtained using the higher growth temperature. The reason for this drop in resistivity is currently under investigation. The contact resistances obtained using this growth sequence is acceptable for either digital or analog operation.

CONCLUSION

The ability to realize the potential of wide bandgap III-V nitrides for high performance electronic and optical devices lies partly with the formation of useful Ohmic contacts. This can be accomplished most easily through the use of In-containing materials due to their smaller bandgaps. In this study it was shown that by optimizing the growth temperature, the N_2 flow rate, and the crystallinity (through the choice of substrate) acceptable contact resistances ($\sim 10^{-6}$ Ohm-cm^2) to InAlN could be achieved.

ACKNOWLEDGEMENTS

The authors acknowledge the staff at the Microfabritech facility for their assistance. This work was partially supported by grant #ECS-9522887 from the National Science Foundation.

REFERENCES

1. S. Strite and H. Morkoc, J. Vac. Sci. Technol. **B10** 1237 (1992) and references therein.
2. F. Ren, S. N. G. Chu, C. R. Abernathy, T. R. Fullowan, J. Lothian and S. J. Pearton, Semicond. Sci Tech. **7** 793 (1992).
3. H. J. Hovel and J. J. Cuomo, Appl. Phys. Lett. **20**, 71 (1972).
4. K. Kubota, Y. Kobayashi, and K. Fujimoto, J. Appl. Phys. **66**, 2984 (1989).
5. T. L. Tansley and R. J. Egan, Mater. Res. Soc. Symp. Proc. **242**, 395 (1992).
6. B. T. Sullivan, R. R. Parsons, K. L. Westra, and M. J. Brett, J. Appl. Phys. **64**, 4144 (1988).
7. W. A. Bryden, S. A. Ecelberger, J. S. Morgan, T. O. Poehler, and T. J. Kistenmacher, Mater. Res. Soc. Symp. Proc. **242**, 409 (1992).
8. T. J. Kistenmacher, S. A. Ecelberger, and W. A. Bryden, Mater. Res. Soc. Symp. Proc. **242**, 441 (1992).
9. K. Osamura, S. Naka, and Y. Mukakami, J. Appl. Phys. **46**, 3432 (1975).
10. T. Matsuoka, J. Cryst. Growth **124**, 433 (1992).
11. W. E. Hoke, P. J. Lemonias, and D. G. Weir, J. Cryst. Growth **111**, 1024 (1991).
12. F. Ren, C. B. Vartuli, S. J. Pearton, S. M. Donovan, J. D. Mackenzie, R. J. Shul, J. C. Zolper, M. L. Lovejoy, A. G. Baca, M. Hagerott-Crawford, and K. A. Jones, J. Vac. Sci. Technol. **A15** (1997).

Characterization of Metal/$Al_xIn_{1-x}N$ Interface
Thermal Stability and Electrical Properties

Guohua Qiu, Fen Chen, and J.O.Olowolafe
Department of Electrical Engineering, University of Delaware, Newark, DE 19716

C.P.Swann and K.M.Unruh,
Department of Physics and Astronomy, University of Delaware, Newark, DE 19716

D.S.Holmes
Lake Shore Cysotronics, Inc., 64 East Walnut Street, Westerville, Ohio 43081-2399

Abstract

The interfaces between metals and semiconductors are very crucial to the performance and reliability of solid-state devices. At the moment information on the interfaces between metals and group III-nitride semiconductors are very rare. In this study, linear I-V characteristics of titanium and aluminum to $Al_xIn_{1-x}N$ of three different composition (x=0.18, 0.50,0.85) were obtained exhibiting ohmic characteristics. Specific contact resistance of these metals to $Al_{.18}In_{.82}N$ and $Al_{.5}In_{.5}N$ was measured by transmission line measurement.   Interdiffusion between the metals and the semiconductors, induced by annealing in N_2 ambient, was determined using RBS and thermal stability was evaluated.

Introduction

The group III-nitrides are promising materials for semiconductor device applications in the blue and ultra-violet (UV) wave region, particularly in the optical emission and data storage applications[1]. This is primarily because of their large and direct band-gaps, which range from 1.9ev to 6.2ev. Thus there has been considerable amount of reports on the growth of these materials. However, there has been few reports on metal contacts to nitrides and metal/nitride interactions[2,3,4]. The ohmic contact between metals and semiconductors are of great importance and an essential part of all solid-state device performance. However, the development of adequate and reliable ohmic contacts to compound semiconductors has met a number of challenges. In addition, thermal stability of metal contacts plays important roles in device performance. In this study, the contact resistance of titanium (Ti) and aluminum (Al) to $Al_xIn_{1-x}N$ (x=0.18, 0.50, 0.85) was measured and the thermal stability was evaluated.

Experimental

$Al_xIn_{1-x}N$ films for this study were deposited on p-type Si substrate by magnetron sputtering at room temperature, using Denton Vacuum Discovery 18 sputtering system.

Mat. Res. Soc. Symp. Proc. Vol. 449 © 1997 Materials Research Society

The detail of the film preparing is in elsewhere[5]. Hall effect measurement was carried out by using a standard Van der Pauw structure with indium contacts under a magnetic field of 0.5T. Electrical characterization of the contacts was done using standard transmission line measurements (TLM)[6]. The samples were patterned by standard photolithography and then etched down to the Si substrate in hot H_2PO_4 to generate mesa structures for the TLM measurement. The contact area was then defined by lift-off technology. Metal pads of dimensions 200 μm wide and 100 μm long were deposited by sputtering with Ar gas. The base pressure was $2*10^{-7}$ Torr. The purity of Ti and Al target is 99.95% and 99.999%, respectively. I-V measurement was carried out by injecting up to 100mA with a current source and measuring the voltage across the contact with an electrometer. A four point probe arrangement was used to eliminate the probe contact resistance. Samples, before and after annealing, were characterized by 2MeV $^4He^+$ Rutherford backscattering spectroscopy (RBS) to profile the interdiffusion/interface reaction between the metals and the nitrides. X-ray diffraction was also used to evaluate the interface diffusion and reaction products.

Results

Three $Al_xIn_{1-x}N$ alloy films with different Al/In ratio were used in this study. Carrier concentration, measured by Hall effect measurement, as well as composition and film thickness of these $Al_xIn_{1-x}N$ alloys were listed in table 1.

Table 1 Composition, film thickness and carrier concentration of $Al_xIn_{1-x}N$

sample No.	composition (x)	film thickness (μm)	carrier concentration (cm^{-3})
1	0.18	1.14	$3.64*10^{21}$
3	0.50	1.23	$6.55*10^{20}$
5	0.85	1.29	$4.64*10^{16}$

Electrical properties of Ti and Al contacts on $Al_xIn_{1-x}N$ were evaluated using standard transmission line measurement. The metal contact size is 200μm×100μm and the thickness is about 2000Å. The I-V characteristics of Ti and Al contacts on $Al_{.18}In_{.82}N$ was shown in figure 1. The I-V curves are linear both for Ti and Al contacts indicative of Ohmic behavior between the metals and the $Al_xIn_{1-x}N$ films.. Figure 1 (b) showed a plot of the total resistance R_T measured between metal contacts as a function of the contact spacing between them. Three parameters, sheet resistance (ρ_c), contact resistance (R_c), and transfer length (L_T), were extracted from a least-squares interpolation line of these data. The values of specific contact resistance were then derived from these parameters using:

$$\rho_c \equiv R_c ZL_T \times \tanh(\frac{L}{L_T})$$

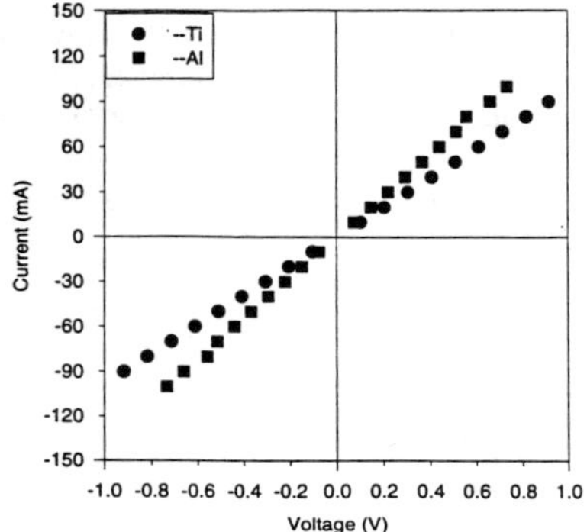

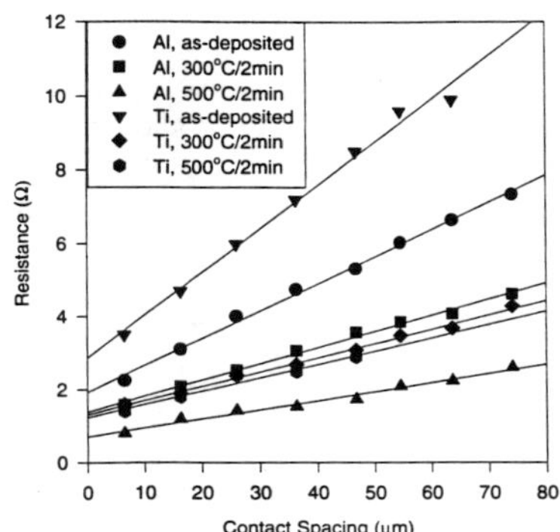

Figure 1. I-v characteristics of Ti or Al/ Al$_{.18}$In$_{.82}$N contacts.

Figure 2. Total resistance R$_T$ measured between Ti or Al contacts pads in the TLM structure as a function of the contact separation for Al$_{.18}$In$_{.82}$N.

The contact resistance of Ti and Al to Al$_{.18}$In$_{.82}$N were both pretty low, about $10^{-3}\Omega cm^2$. These values decreased after annealing at 300°C and 500°C for 2 minutes, which were listed in Table 2, indicative of interface modification.

Table 2. Specific contact resistance of Ti and Al to Al$_{.18}$In$_{.82}$N

	Ti	*Al*
as-deposited	$3.5*10^{-3}$	$2.38*10^{-3}$
300°C/2min	$1.95*10^{-3}$	$2.19*10^{-3}$
500°C/2min	$1.82*10^{-3}$	$9.66*10^{-4}$

Interdiffusion at the metal/nitride interface was mornitored using 2 MeV He+ ion backscattering spectrometry (RBS). The measurements were performed on these samples before and after annealing. Figure 2 showed the RBS spectrum of Ti/Al$_{.18}$In$_{.82}$N sample. It was clearly shown in the figure that in the as-deposited condition the In leading edge is depressed from 1.74 ev to 1.48ev by the top Ti layer. No In signal was detected beyond 1.48 ev before annealing. After 300°C annealing, however, signals from In were showing beyond 1.48 ev, indicating In atoms from the Al$_{.18}$In$_{.82}$N layer diffused into the Ti top layer. After annealing at 500°C for 2 minutes a complete In peak was formed with leading edge at 1.74ev, which means the In atoms diffused very fast, reaching the top surface of 2000Å thick Ti layer in a very short time. RBS detected the same for the Al/Al$_{.18}$In$_{.82}$N sample: no In signal was detected beyond 1.64ev, which is the In leading edge depressed by the Al top layer. After annealing at 300°C for 2 minutes, In diffused into Al metal layers, showing In signals beyond 1.64 ev in RBS. Further annealing at 500°C for 2 minutes showed significant out diffused of In segregation to the surface. No extra peak was detected by XRD after annealing 2 minutes suggesting no reaction product was formed.

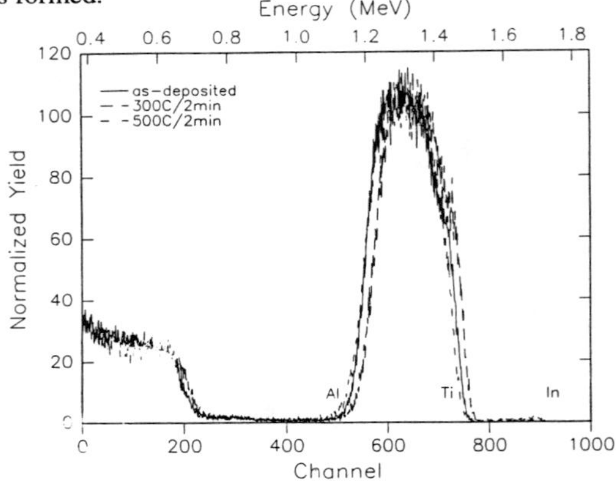

Figure 3. RBS-spectra of the Ti/Al$_{.18}$In$_{.82}$N contact before and after annealing at 300°C and 500°C for 2 minutes.

I-V curves of Ti and Al contacts to Al$_{.5}$In$_{.5}$N were also linear as those to Al$_{.18}$In$_{.82}$N, except the values of resistance are higher. Table 3 listed the specific contact resistance calculated using TLM method. These values for Ti/ Al$_{.5}$In$_{.5}$N and Al/ Al$_{.5}$In$_{.5}$N contacts are very close to each other and are two orders of magnitude higher than those of Ti/Al$_{.18}$In$_{.82}$N and Al/Al$_{.18}$In$_{.82}$N contacts. After annealing at 300°C and 500°C the contact resistance decreased. RBS measurements were also performed on these samples. No In out diffusion into the top metal layer were detected in these samples until annealed at 500°C, but the amount was much less than those in Ti/Al$_{.18}$In$_{.82}$N and Al/Al$_{.18}$In$_{.82}$N.

Table 3 Specific contact resistance of Ti and Al to Al$_{.5}$In$_{.5}$N

	Ti	Al
as-deposited	$1.15*10^{-1}$	$1.17*10^{-1}$
300°C/2min	$6.11*10^{-2}$	$4.32*10^{-2}$
500°C/2min	$9.78*10^{-3}$	$2.1*10^{-2}$

For the Ti/Al$_{.85}$In$_{.15}$N and Al/ Al$_{.85}$In$_{.15}$N contacts the I-V curves measured are linear before and after annealing, indicating ohmic characteristics. However, the total resistance is not sensitive to the changing of contact spacing. The R$_T$ versus contact spacing was plotted as a flat horizontal line, thus make it impossible to calculate the specific contact resistance using the TLM method. No change was detected by RBS in those samples after annealing up to 500°C for 2 minutes.

Conclusion

Titanium and aluminum contacts to three Al$_x$In$_{1-x}$N alloy films were evaluated. The I-V characteristics of both metals are linear, indicating that Ti or Al contact to the Al$_x$In$_{1-x}$N is ohmic. The specific contact resistance values of Ti to Al$_{.18}$In$_{.82}$N or Al$_{.18}$In$_{.82}$N were similar to those of Al contacts. Those values were reduced upon heat treatment at 300°C and 500°C. Significant indium out diffusion occured in Ti or Al/Al$_{.18}$In$_{.82}$N contacts after annealing, indicative of unstable interfaces. Segregation of indium, out of the Al$_{.5}$In$_{.5}$N , into the Ti or Al metal layer were also observed by RBS. The contact resistance was much higher for the Ti or Al/ Al$_{.85}$In$_{.15}$N contact than those observed for other compositions. No evidence of diffusion in Ti or Al/Al$_{.85}$In$_{.15}$N samples after annealing up to 500°C for 2 minutes.

Reference

1. S.Strite and H.Morkoc, J.Vac.Sci.Technol., **B10**(1992), 1237
2. J.S.Foresi and T.D.Moustakas, Appl.Phys.Lett., **62**(1993), 2859
3. M.E.Lin, Z.Ma, F.Y.Huang, Z.F.Fan, L.H.Allen, and H.Morkoc, Appl.Phys.Lett., **64**(1994), 1003
4. J.S.Chan, T.C.Fu, N.W.Cheung, N.Newman, X.Liu, J.T.Ross. M.D.Rubin, and P.Chu, Mat.Res.Soc.Symp.Proc., **Vol.339**(1994), 223
5. G.Qiu, O.Olowolafe, K.Unruh, T.Peng, C.Swann, and J.Piprek, in this proceeding
6. H.H.Berger, Solid-State Electron. 15(1972), 145

ON THE EPITAXY OF METAL FILMS ON GAN

Q.Z. LIU [*], K.V. SMITH [*], E.T. YU [*], S.S. LAU [*], N.R. PERKINS [**], T.F. KUECH [**]
[*] Department of Electrical and Computer Engineering, University of California, San Diego, La Jolla, CA 92093
[**] Department of Chemical Engineering, University of Wisconsin, Madison, WI 53706-1691

ABSTRACT

A variety of metal films deposited at room temperature have been found to grow epitaxially under conventional vacuum conditions on GaN grown by metalorganic vapor phase epitaxy on sapphire substrates. The metal films have been characterized by X-ray diffraction using a thin-film Read camera and by MeV ion channeling measurements. Lattice mismatch between the epitaxial metals and the GaN basal planes ranges from ~ 0.2% to ~ 22%, and does not appear to be the determining factor in the epitaxy reported here. Surface structure of the epitaxial metal films has been studied by atomic force microscopy and found to differ considerably from that of nonepitaxial metal films grown on similar GaN substrates.

INTRODUCTION

GaN-based materials have been the subject of intensive research recently for blue and ultraviolet light emission [1] and high temperature/high power electronic devices [2-4]. For device applications, control of metal/semiconductor interface quality is important; ideally, metal/semiconductor interfaces should be inert, epitaxial, oxide and defect-free, and atomically smooth. While the epitaxy of metal films on GaAs and other III-V semiconductors has been studied extensively [5], studies of metal epitaxy on GaN have been rare. For GaAs and other common III-V semiconductors, a thin oxide layer (several to tens of Å) grows rapidly on the surface when exposed to air [6], necessitating in-situ cleaning of GaAs surfaces under ultra high vacuum (UHV) conditions for the epitaxial growth of metal films [5]. For GaN, an in-situ cleaning method under UHV has been developed [7] and used in the growth of thin Al and Ni films on GaN (0001) substrates [8, 9]. Epitaxy of Sc films on GaN with a thin ScN intermediate layer formed at high temperatures (645 °C and above) has been reported [10]. Recently, we have found that Pd, Ni, and Pt can be deposited epitaxially on GaN substrates grown by metalorganic vapor phase epitaxy (MOVPE) [11]. The metal films were deposited on GaN substrates cleaned only by acid-etching, and in a conventional e-beam evaporation system without in-situ cleaning under UHV.

In this paper, we report studies of the epitaxy of a wide range of metals on GaN. While most of the metal films we have investigated so far can be grown epitaxially on GaN, two metals - Ag and Cr - consistently grow nonepitaxially on GaN. The epitaxial quality of the metal films was characterized by X-ray diffraction using a thin-film Read camera and MeV ion channeling measurements. The surface structure of the epitaxial metal films was studied by atomic force microscopy (AFM) and compared with that of the nonepitaxial metal films.

EXPERIMENTAL

GaN/AlN/sapphire structures used in this study were produced in a horizontal metalorganic vapor phase epitaxy (MOVPE) system. The layer structures and growth system have been described in detail elsewhere [12]. The thickness of the GaN layers was about 3 μm, and full width at half maximum of the double-crystal X-ray rocking curve was about 330 arcsec. For deposition of metal films, GaN samples were first cleaned using organic solvents. Acid etching was then performed to clean the GaN surfaces, followed by loading in a conventional evaporation chamber equipped with dry pumping, yielding a base pressure of ~ 2×10^{-8} Torr. Metal films other than Al were e-beam deposited on GaN surfaces at an evaporation rate of about 2 Å/sec in a vacuum of ~1×10^{-7} Torr during evaporation. The thickness of the metal films

Mat. Res. Soc. Symp. Proc. Vol. 449 © 1997 Materials Research Society

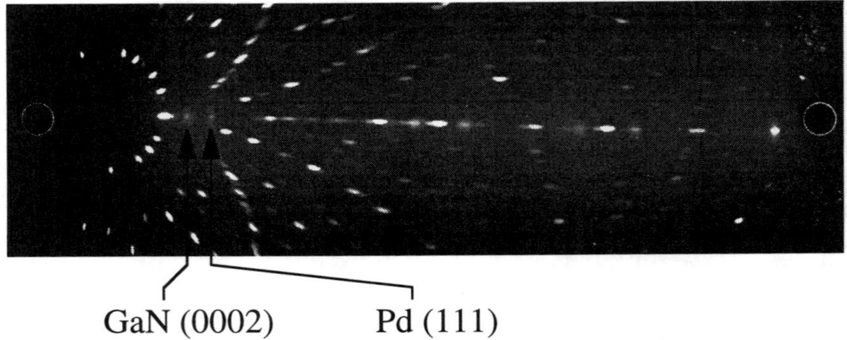

Figure 1. X-ray diffraction pattern from a Read camera of a Pd/GaN sample where the GaN substrate was etched in HCl:H$_2$O. The X-ray diffraction spots form GaN (0002) and Pd (111) planes are marked by the arrows, respectively. The other diffraction spots are due to the sapphire substrate.

was ~ 1000 Å. Al films were deposited on GaN with a deposition rate of 0.26 Å/sec in an MBE system, in which the base pressure was ~ 1x10^{-10} Torr. All metal films were deposited at room temperature. MeV ^{4}He$^+$ backscattering and channeling techniques in conjunction with an X-ray thin-film Read camera were used to evaluate the deposited metal films on GaN. AFM in the tapping mode was used to study the nanoscale surface structure of the GaN layers and the metal films.

RESULTS AND DISCUSSION

Figure 1 shows the X-ray diffraction pattern from a Read camera at a glancing incidence angle of 12° for a Pd/GaN sample where the GaN surface was cleaned in HCl:H$_2$O prior to metal deposition. The epitaxial nature of the Pd film is demonstrated vividly by a single extra spot from the Pd (111) planes shown in Fig. 1

Epitaxy of the Pd film has also been confirmed by MeV ion channeling measurements. Figure 2 shows the random and <111> aligned backscattering spectra of the same Pd/GaN sample used in Fig. 1. The epitaxial nature of the Pd film can be seen from the reduced channeling yield compared to the random yield of the sample. The epitaxial quality of the film improves from the film/GaN interface to the film surface with the lowest channeling yield, χ_{min}, of ~ 19% occurring near the surface; this improvement in epitaxy is commonly observed in lattice-mismatched systems [13]. For Pd/GaN samples where the GaN surfaces were cleaned in HF: H$_2$O, χ_{min} is also ~ 19%. It is worth noting that the epitaxy of Pd films was achieved without an in-situ cleaning procedure, indicating the uniqueness of the GaN surface compared with other III-V semiconductor compounds. This seems to suggest that after the GaN surface has been cleaned by acid-etching, the presence of oxides and/or other contaminants should be minimal, and that native oxides do not grow rapidly on the GaN surface during the time (less than 30 minutes) required to load the samples in the vacuum deposition chamber [11]. Auger studies have shown that acid-etching can remove part of the oxides and carbon contaminants on unetched GaN surfaces [14]. However, these studies also revealed the presence of oxygen and carbon on acid-etched GaN surfaces [14], i.e., the acid-cleaned GaN surfaces are not atomically clean. Despite this, metal epitaxy on GaN is achieved under conventional vacuum evaporation conditions.

Auger studies have shown that HCl:H$_2$O is almost as effective as HF:H$_2$O in removing the oxides and carbon contaminants [14]. It is interesting to note that the epitaxial quality of our Pd films deposited on GaN cleaned by HCl:H$_2$O is almost the same, as determined by channeling yield, as for Pd deposited on GaN cleaned by HF:H$_2$O. We have determined that boiling aqua

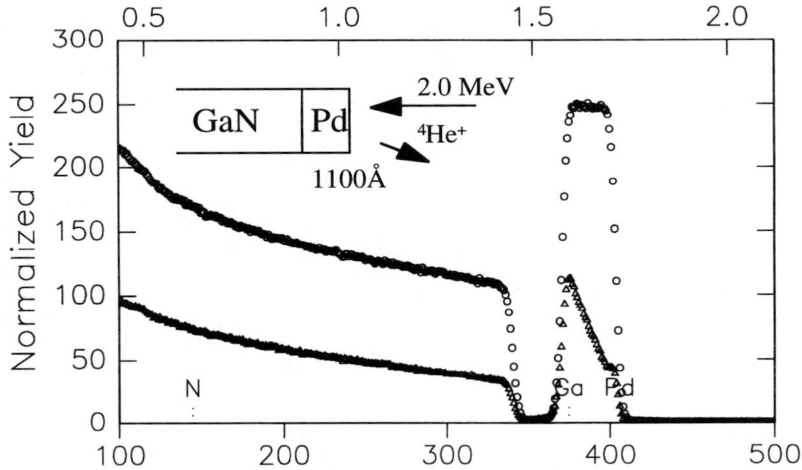

Figure 2. Random and <111> (Pd) aligned spectra of the Pd/GaN sample used in figure 1. The circles represent the random spectrum, and the triangles depict the channeling spectrum. The incident beam energy of $^4He^+$ ions was 2.0 MeV with a detector angle at 168°.

regia is more effective than either HCl:H_2O or HF: H_2O in cleaning the GaN surfaces [11], and this has become our standard cleaning solution for GaN in our laboratory whenever possible.

<u>Epitaxy of Metal Films on GaN - Lattice Mismatch</u>

A variety of other metal films have also been investigated for epitaxy on GaN. Table 1 shows a summary of our results. Epitaxy is achieved for Au, Al, Pt, Pd, Ni, Hf, and Ti deposited on GaN, but not for Ag and Cr. Au, Pt, Pd and Ni have a large barrier height (~ 1 eV) on n-GaN, and are candidates for Schottky contacts; Al, Ti, and Hf have low barrier height on n-GaN, and are candidates for ohmic contacts [15]. The epitaxy of these metals on GaN under conventional vacuum conditions suggests that a relatively clean interface between metal contacts and GaN can be easily obtained, resulting in better-controlled contact properties, which should have significant impact on nitride contact technology. For the metals with face-center-cubic (FCC) structure such as Au, Al, Pt, Pd and Ni, the (111) planes of the metal films match the basal planes of GaN; for the metals with hexagonal structure such as Hf and Ti, the basal planes of the metal films match the basal planes of GaN. In heteroepitaxy, lattice mismatch is usually a very important factor. Small lattice mismatch generally leads to improved epitaxial

Table 1. A survey of metal epitaxy on GaN

Metal	Crystal Structure	Mismatch to GaN basal plane	Epitaxial
Ag	FCC	9.41%	No
Au	FCC	9.57%	Yes
Al	FCC	10.21%	Yes
Pt	FCC	13.00%	Yes
Pd	FCC	13.74%	Yes
Ni	FCC	21.86%	Yes
Hf	hexagonal	-0.18%	Yes
Ti	hexagonal	7.48%	Yes
Cr	BCC	13.86%	No

quality. As shown in Table 1, the lattice mismatch between the metal films and the basal plane of GaN ranges from as small as ~ 0.2% for Hf to as large as ~ 22% for Ni. However, the epitaxial quality of these metal films does not appear to correlate with the degree of lattice mismatch.

<u>Surface Morphology of Epitaxial and Nonepitaxial Metal Films</u>

AFM studies have provided insight into the role of initial GaN surface quality on film growth and detailed information concerning the surface structure of epitaxial metal films. We have found that epitaxial metal growth occurs only on microscopically flat GaN surfaces on which atomically flat terraces, typically hundreds of Å in width, are discernible by AFM [16]. Furthermore, the surface morphologies of epitaxial and nonepitaxial metal films are considerably different. AFM images of a) Pd, b) Au, c) Al, and d) Ag films on GaN are shown in Figure 3. The Pd, Au, and Al films on GaN are epitaxial, but the Ag film is not. In the case of Al, epitaxial islands with a characteristic hexagonal geometry are present, with the island-like epitaxial structures separated by shallow trenches (~40 Å in depth) with an overall RMS surface roughness of ~ 12Å. Somewhat similar structure is observed for Pd and Au: epitaxial islands

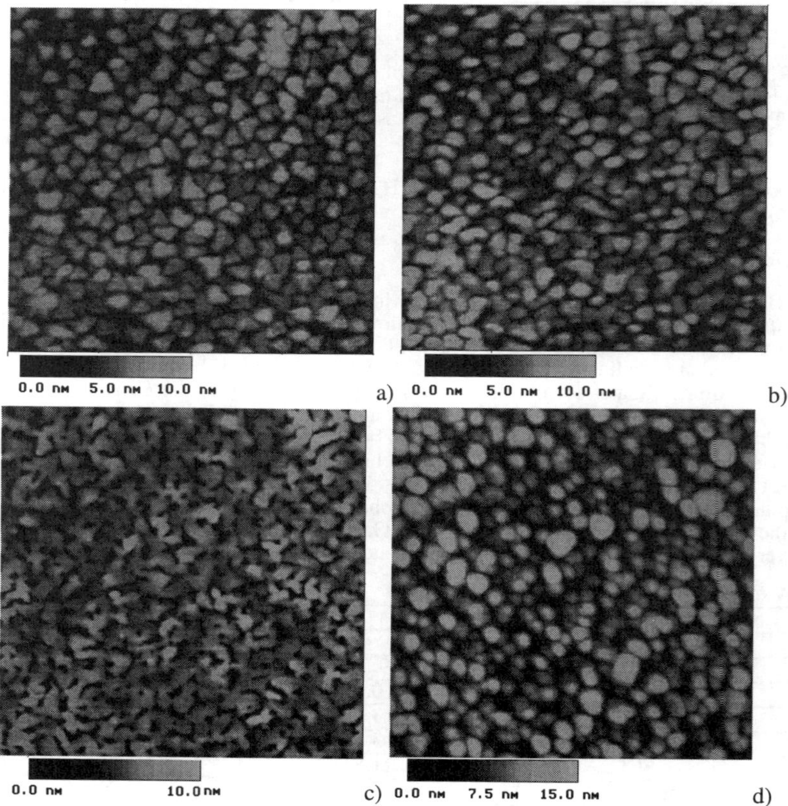

Figure 3. 1μmx1μm AFM images of metal films on GaN: a) epitaxial Pd films, b) epitaxial Au films, c) epitaxial Al films, and d) non-epitaxial Ag films.

with a more limited degree of hexagonal symmetry, separated by shallow trenches, are still observed, but the islands are more granular; the overall RMS surface roughness is ~ 12Å, essentially the same as for the Al films. In the case of Ag, there is no apparent grain structure, and the surface is much rougher (with an overall RMS surface roughness of 22 Å) than that of the epitaxial metal films. It should be noted that the surface morphology of the GaN substrates used in this study is very uniform; the larger surface roughness of the Ag film compared with the other epitaxial films therefore is associated entirely with the metal.

SUMMARY

A variety of metal films deposited at room temperature have been found to grow epitaxially on certain GaN layers under conventional vacuum conditions. These epitaxial films include candidates suitable for both Schottky contacts and ohmic contacts on n-GaN, and could have a significant impact on nitride contact technology. Lattice mismatch between the epitaxial metals and the GaN basal planes ranges from less than 1% to ~ 22%, and does not seem to be the determining factor in the epitaxy reported here. The surfaces of the epitaxial metal films typically have island- or mesa-like structure with readily apparent hexagonal symmetry, as determined by AFM, and are smoother than those of the nonepitaxial metal films deposited on the same GaN substrates.

ACKNOWLEDGMENT

The work at UCSD is supported by BMDO (Dr. Kepi Wu) monitored by the US Army Space and Strategic Defense Command. The University of Wisconsin acknowledges financial support from the Naval Research Laboratory and the ARPA URI on Visible Light Emitters.

REFERENCES

1.   See for example, S. Nakamura, M. Senoh, S. Nagahama, N. Iwasa, T. Yamada, T. Matsushita, Y. Sugimoto, and H. Kiyoku, Appl. Phys. Lett. **69**, 3034 (1996).

2.   Zhifang Fan, S. N. Mohammad, O. Aktas, A. E. Botchkarev, A. Salvador, and H. Morkoc, Appl. Phys. Lett. **69**, 1229 (1996).

3.   Q. Chen, M. A. Khan, J. W. Wang, C. J. Sun, M. S. Shur, H. Park, Appl. Phys. Lett. **69**, 794 (1996).

4.   J. C. Zopler, R. J. Shul, A. G. Baca, R. G. Wilson, S. J. Pearton, and R. A. Stall, Appl Phys. Lett. **68**, 2273 (1996).

5.   T. Sands, C. J. Palmstrom, J. P. Harbison, V. G. Keramidas, N. Tabatabaie, T. L. Cheeks, R. Ramesh, and Y. Siberberg, Materials Science Reports, **5** (3), 99 (1990), and references therein.

6.   See for example, C. W. Wilmsen, in "Physics and chemistry of III-V compound semiconductor interfaces," edited by C. W. Wilmsen (Plenum Press, New York, 1985), chap. 7.

7.   M. A. Khan, J. N. Kuznia, D. T. Olson, and R. Kaplan, J. App. Phys. **73**, 3108 (1993).

8.   V. M. Bermudez, R. Kaplan, M. A. Khan, and J. N. Kuznia, Phys. Rev. B **48**, 2458 (1993).

9.   V. M. Bermudez, T. M. Jung, K. Doverspike, and A. E. Wickenden, J. Appl. Phys. **79**, 110 (1996).

10. R. Kaplan, S. M. Prokes, S. C. Binari, and G. Kelner, Appl. Phys. Lett. **68**, 3248 (1996).

11. Q. Z. Liu, S. S. Lau, N. R. Perkins, and T. F. Kuech, Appl. Phys. Lett. **69**, 1722 (1996).

12. N. R. Perkins, M. N. Horton, D. Zhi, Z. Z. Bandic, T. C. McGill, and T. F. Kuech, presented at Spring 1996 Meeting of the MRS, April, 1996, San Francisco, CA.

13. See for example, L. C. Feldman, J. W. Mayer, and S. T. Picraux, Materials Analysis By Ion Channeling, (Academic Press, New York, 1982), chap 8.

14. L. L. Smith, S. W. King, R. J. Nemanich, and R. F. Davis, J. Electron. Materials, **25**, 805 (1996).

15. See for example, J. S. Foresi, and T. D. Moustakas, Appl. Phys. Lett. **62**, 2859 (1993); S. N. Mohammad, A. A. Salvador, and H. Morkoc, Proc. of the IEEE, vol. **83**, 1306 (1995).

16. Q. Z. Liu, L. Shen, K. V. Smith, C. W. Tu, E. T. Yu, S. S. Lau, N. R. Perkins, and T. F. Kuech, submitted to Appl. Phys. Lett.

INFLUENCE OF SURFACE DEFECTS ON THE CHARACTERISTICS OF GaN SCHOTTKY DIODES

J.-Y. DUBOZ, F. BINET, N. LAURENT, E. ROSENCHER, F. SCHOLZ*, V. HARLE*, O. BRIOT**, B. GIL**, R.L. AULOMBARD**
Laboratoire Central de Recherches, Thomson-CSF, 91404 ORSAY Cedex, France
*4.Physikalisches Institut, Universität Stuttgart, D-70550 STUTTGART, Germany
**GES, Université Montpellier II, 34095 MONTPELLIER, France

ABSTRACT

We have studied the fabrication of Pt/Au Schottky diodes on n-type GaN. We show that the electrical characteristics of the diodes are strongly dependent on the surface chemical treatment before the metal deposition. Lowest leakage currents were obtained by the use of a HCl solution. We also show that annealing the diode at a moderate temperature (400°C) leads to reduced reverse currents. In order to explain these results, we measured the density of deep levels in the Schottky diode depletion region before and after the annealing process. We did not observe any significant difference in the bulk density of defects due to the anneal. We also studied the temperature dependence of the reverse currents and found a low activation energy. Our results are interpreted in terms of electrical defects at the metal-GaN surface.

INTRODUCTION

GaN related materials have already demonstrated a very high potential for opto-electronic applications. Very efficient blue LED's[1] and even blue LASER's[2] have been fabricated. High power and high temperature field effect transistors[3,4] have also been shown to be competitive with devices based on other materials. However, some physical properties in GaN still remain unclear. Among them, the electrical properties of GaN surfaces and interfaces with metals have not yet been fully investigated. We present here a study of Schottky diodes on n-type GaN and we show the importance of surface related defects for the electrical characteristics of the diodes.

SURFACE PREPARATION

The GaN layers used here are grown by metal organic vapor phase epitaxy[5,6]. They are around 2 μm thick. Although undoped, they show an n-type conductivity with electron densities in the range of $10^{17} cm^{-3}$. The X-ray diffraction on these layers shows intense peaks with a width of 300-400 arcs. The low temperature photoluminescence is dominated by a bound exciton with linewidths in the range of 5 to 8 meV. The dislocation density is between 10^9 and 10^{10} cm^{-2}. On the surface, very few morphological defects (nanopipes...) can be observed and the roughness is of the order of 1 nm.
Devices were fabricated by the following process: First, the ohmic contact was defined by optical lithography. Ti/Al was evaporated and annealed. Second, the Schottky diode was defined by lithography to form 100x100 μm squares surrounded by the ohmic contact. The surface was prepared by different chemical treatments. We report here the effect of NH_3 (10%): H_2O (90%) and HCl (50%): H_2O (50%) solutions. Although aqua regia is known to efficiently clean the GaN surface[7], it was not used here as it also etches the photoresist. The sample was dipped in the solution for roughly 15 s and then rinsed in deionized water. Then, the Schottky diode (1000 Å Pt /1000 Å Au) was evaporated in a vacuum system. The pressure before the evaporation was in the range of 10^{-6} Torr. Finally, some diodes were annealed at various temperatures under a N_2 flux. Figure 1 shows the electrical characteristics of the Schottky diode with and without chemical treatment of the surface. Let us mention that all these diodes were annealed at 400°C. We clearly observe the reduction of reverse currents when going from the untreated to the NH_3 and HCl treated samples. The forward current curve is shifted towards larger voltages indicating a larger effective barrier height. The linear range (in log scale) in forward bias is also increased in these samples up to 6 orders of magnitude in the case

1085

of the NH_3 treatment and would be even larger for the HCl surface preparation if we could measure the current below 1 pA . The ideality is around 1.25 for the three diodes. In this latter sample, the reverse current is as low as 5 pA (5×10^{-8} Acm^{-2}) at -2V and 8 nA (8×10^{-5} Acm^{-2}) at -40V. The saturation current extrapolated from the linear part (in log scale) of the forward current was measured as a function of temperature. We found an exponential dependence on the inverse temperature. From the value of the activation energy and taking the ideality factor into account, we calculated Schottky barrier heights equal to 0.69, 0.81 and 1.01 eV for the untreated, the NH_3 and the HCl treated diodes respectively. The capacitance C as a function of voltage was also measured on these diodes. The electrostatic barrier height deduced from the extrapolation of $1/C^2$ was in the range of 1 eV for all diodes, in agreement with values found in the literature[8]. Hence, in the HCl cleaned diode, the barrier height deduced from transport measurements agrees with the electrostatic value, while in the other diodes it is much lower than the electrostatic value. This has to be related to the improved characteristics in the HCl cleaned sample and will be interpreted later on.

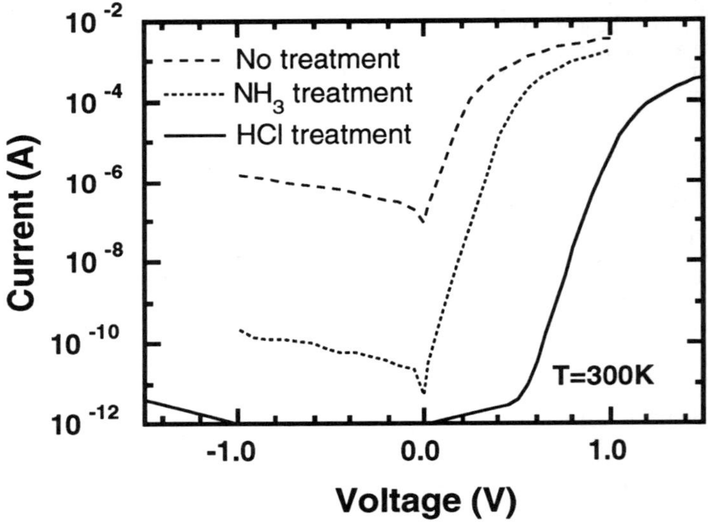

Figure 1: Current-voltage characteristics of Schottky diodes on n-type GaN for different surface preparations. All diodes were annealed at 400°C for 10 min.

EFFECT OF ANNEALING ON THE ELECTRICAL CHARACTERISTICS

The Schottky diodes were annealed at various temperatures from 200°C to 600°C for 10 min. The electrical characteristics were measured and we found that the reverse current was strongly reduced for annealing temperatures up to 400°C while a degradation was observed for higher temperatures. The forward current curve was shifted to larger voltages, which can be interpreted as an increase of the barrier height. The improvement due to the annealing was observed in all samples. However, the effect was larger in samples with larger reverse leakage currents before the annealing. Figure 2 shows the case for the NH_3 treatment. The voltage shift in forward current is of the order of 0.1 eV. The electrostatic Schottky barrier height deduced from C(V) measurements was found to increase by roughly 0.15 eV on this sample. One thus deduces that the Schottky barrier height increases by 0.1-0.15 eV.This increase accounts for part of the reduction of the reverse current. However, a factor of 5 to 10 of the reverse current decrease cannot be accounted for by the barrier increase and is likely to be related to a reduction in the density of electrical defects at the interface. We must also keep in mind that the Fermi level at metal/semiconductor interfaces may be pined by defects and a modification of their density may change the pinning thus resulting in a Schottky barrier height change. Let us mention that we obtained similar results with other metals such as Cr which indicates that the nature of the metal does not play a major role in this characteristics improvement by thermal annealing. Moreover, a decrease of reverse currents related to an increase of the barrier height with annealing has also been reported in Ti on n-GaN Schottky diodes[9].

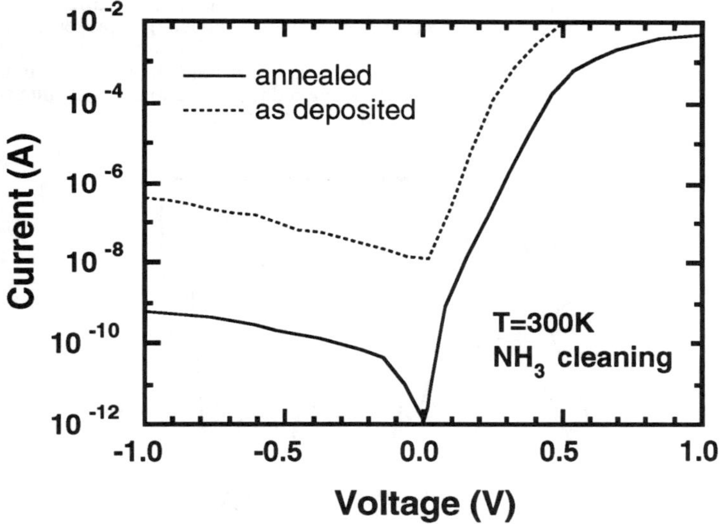

Figure 2: Current-voltage characteristics of a Schottky diode on n-type GaN before and after the annealing. The GaN surface was cleaned in an NH_3 solution before metallization.

In the case of HCl treatment, the effect of annealing was to reduce the reverse currents with almost no voltage shift of the forward current. This shows that the reverse current is not fully controlled by the barrier height but rather due to parasitic conduction paths related to interface defects. In addition, the ideality factor of the diode is reduced from 1.84 before annealing to 1.25 after annealing. This evolution is in agreement with conduction dominated by defects (even in forward bias) before annealing and a more intrinsic thermionic emission after annealing.

ELECTRICAL DEFECTS IN THE BULK

The presence of defects in the vincinity of the Schottky barrier can be studied by different means. Deep Level Transient Spectroscopy in n-type Schottky diodes allows investigation of electron traps lying in the first 0.5 eV below the conduction band. On the NH_3 treated sample, we found two levels at 0.18 and 0.46 eV with concentration in the 10^{-15} cm^{-3} range before or after the annealing. In order to measure deeper levels, we had to use optical excitation. Therefore, we performed Deep Level Optical Spectroscopy on these samples. A forward bias (0.5 V) pulse is applied for a short time (1s). Traps in the diode depletion region are thus filled with electrons. Then, a reverse bias (-1 V) is applied while the sample is illuminated with a monochromatic light (Xe lamp filtered by a monochromator). The capacitance is measured as a function of time and is found to increase indicating that electrons are emitted from traps. As the incident optical power is low, the capacitance transient occured on a long time scale (1 min).

The measurement was done at low temperature (77 K) in order to reduced the thermal emission and verified that the capacitance transient in the dark after 1 min was not measurable. Thus, the capacitance transient with illumination is proportional to the density of all the states that can be emptied by the optical excitation. It is important to recall that for photon energies larger than half the band gap energy, electron emission from traps toward the conduction band coexists with hole emission toward the valence band so that traps cannot be completely emptied anymore. As a result, the apparent integrated density of states may differ from the real one for photon energies larger than $E_g/2$, i.e. 1.75 eV. In order to investigate a possible change in the defect density with annealing, we measured the NH_3 cleaned sample before and after annealing. In this diode, the effect of annealing was a reduction of the reverse current by a factor 500. The comparison between DLOS spectra before and after annealing is presented in Fig. 3. We observe that the spectra are very similar to each other which indicates that the density of defects in the bulk below the GaN-metal interface is not changed by the annealing. Thus, both the DLTS and DLOS show that the reverse current reduction due to the annealing is not related to a reduction of the density of electrical defects in the volume.

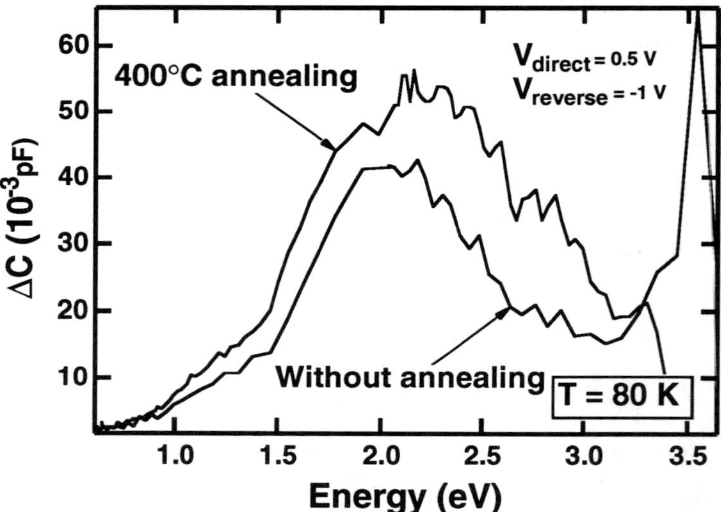

Figure 3: DLOS spectra in Schottly diode on n-GaN before and after the annealing. The surface of this diode was cleaned in an NH_3 solution.

INTERFACIAL DEFECTS

The reverse current in the NH_3 cleaned sample was studied as a funtion of temperature. Before annealing, the curent was found to be exponentially dependent on the inverse temperature as shown in Fig. 4 for V= -0.5 V. The activation energy below 240 K is 0.16 eV while it is 0.23 eV between 240 and 300 K. Thus, at room temperature, the transport is dominated by a parasitic path with a low activation energy E_a. This is particularly true for reverse bias ($E_a = 0.23$ eV) but also for forward bias as the Schottky barrier height deduced from the forward current ($E_a = 0.69$ eV) is much smaller than the electrostatic value. These low activation energies are not intrinsic but are due to defect levels in the band gap. Their density is reduced by the chemical surface treatment and by the annealing. All our results suggest that the most important defects for the Schottky diode are interface defects. We cannot rule out that these defects are the same as in the bulk. However, they can be efficiently removed from the surface by chemical cleaning or annealing while bulk defects cannot. It has been reported that the presence of oxygen on the surface may lead to the formation of a thin oxide[10]. Different groups are mentioning a chemical treatment of the GaN surface in the Schottky diode fabrication in order to remove a possible oxide[11,12]. Our results are in agreement with the presence of such an oxide on the GaN surface and its removal by HCl.

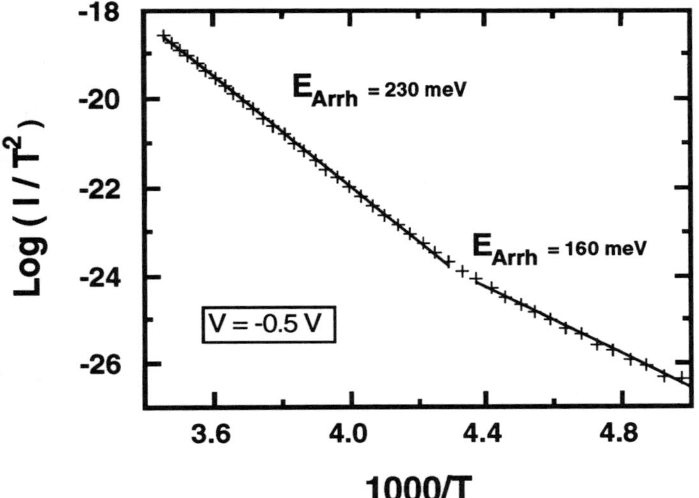

Figure 4: Reverse current (V= -0.5 V) in a Schottky diode as a function of the inverse temperature. The sample was NH_3 cleaned and was not annealed.

CONCLUSION

Schottky diodes have been fabricated and measured. The electrical characteristics have been found to be strongly dependent on the surface preparation before the metal deposition. Cleaning the GaN surface with HCl led to the lowest leakage currents, in the range of 10^{-8} Acm^{-2}. An annealing process allows also a reduction of the reverse dark current by one to three orders of magnitude depending on the surface chemical treatment. These results are interpreted in terms of surface defects inducing a parasitic conduction path with a lower activation energy. With an appropriate cleaning of the surface, good Schottky characteristics can be obtained, with an ideality factor of 1.25, barrier heights around 1 eV and very low leakage currents.

REFERENCES

1 S. Nakamura, M. Senoh, N. Iwasa, S.-I. Nagahama, T. Yamada, and T. Mukai, Jpn. J. Appl. Phys. **34**, 1332 (1995).

2 S. Nakamura, M. Senoh, S. I. Nagahama, N. Iwasa, and T. Yamada, Jpn. J. Appl. Phys. **35**, 74 (1996).

3 M. Asif Khan, Q. Chen, J. W. Yang, M. S. Shur, B. T. Dermott, and J. A. Higgins, Elect. Dev. Lett. **17**, 325 (1996).

4 Y.-F. Wu, B. P. Keller, S. Keller, D. Kapolnek, S. P. Denbaars, and U. K. Mishra, IEEE Elec. Dev. Lett. **17**, 455 (1996).

5 O. Briot, J. P. Alexis, M. Tchounkeu, and R. L. Aulombard, Proc. Mat. Sc. Eng. to be published (1996).

6 F. Scholz, V. Haerle, H. Bolay, F. Steuber, B. Kaufmann, G. Reyher, A. Drnen, O. Gfroerer, S.-J. Im, and A. Hangleiter, Solid State Electronics **accepted for publication, in print,** (1996).

7 Q. Z. Liu, S. S. Lau, N. R. Perkins, and T. F. Kuech, Appl. Phys. Lett. **69**, 1722 (1996).

8 K. Suzue, S. N. Mohammad, Z. F. Fan, W. Kim, O. Aktas, A. E. Botchkarev, and H. Morkoç, J. Appl. Phys. **80**, 4467 (1996).

9 M. Hirsch, K. J. Duxstad, and E. E. Haller. *Evolution of Ti Schottky barrier heights on n-type GaN with annealing* (These proceedings).

10 K. Prabhakaran, T. G. Andersson, and K. Nozawa, Appl. Phys. Lett. **69**, 3212 (1996).

11 L. Wang, M. I. Nathan, T. H. Lim, M. A. Khan, and Q. Chen, Appl. Phys. Lett. **68**, 1267 (1996).

12 A. C. Schmitz, A. T. Ping, M. Asif Khan, Q. Chen, J. W. Yang, and I. Adesida, Semicond. Sci. Technol. **11**, 1464 (1996).

COMPARISON OF Ni/Au, Pd/Au, AND Cr/Au METALLIZATIONS FOR OHMIC CONTACTS TO p-GaN

J.T. Trexler*, S.J. Pearton*, P.H. Holloway*, M.G. Mier**, K.R. Evans**, and R.F. Karlicek***
*Department of Materials Science and Engineering, University of Florida, Gainesville, FL
32611-6400. **Solid State Electronics Directorate, Wright Laboratory, Wright-Patterson Air
Force Base, OH 45433-7323. ***EMCORE Corporation, Somerset, NJ 08873.

ABSTRACT

Reactions between electron beam evaporated thin films of Ni/Au, Pd/Au, and Cr/Au on
p-GaN with a carrier concentration of 9.8×10^{16} cm^{-3} were investigated in terms of their structural
and electronic properties both as-deposited and following heat treatments up to 600°C (furnace
anneals) and 900°C (RTA) in a flowing N_2 ambient. Auger electron spectroscopy (AES) depth
profiles were used to study the interfacial reactions between the contact metals and the p-GaN.
The electrical properties were studied using room temperature current-voltage (I-V)
measurements and the predominant conduction mechanisms in each contact scheme were
determined from temperature dependent I-V measurements. The metallization schemes consisted
of a 500 Å interfacial layer of Ni, Pd, or Cr followed by a 1000 Å capping layer of Au. All
schemes were shown to be rectifying as-deposited with increased ohmic character upon heat
treatment. The Cr/Au contacts became ohmic upon heating to 900°C for 15 seconds while the
other schemes remained rectifying with lower breakdown voltages following heat treatment. The
specific contact resistance of the Cr/Au contact was measured to be 4.3×10^{-1} Ωcm^2. Both Ni and
Cr have been shown to react with the underlying GaN above 400°C while no evidence of a
Pd:GaN reaction was seen. Pd forms a solid solution with the Au capping layer while both Ni
and Cr tend to diffuse through the capping layer to the surface. All contacts were shown to have
a combination of thermionic emission and thermionic field emission as their dominant
conduction mechanism, depending on the magnitude of the applied reverse bias.

INTRODUCTION

The group III nitrides, in particular GaN, are promising materials for semiconductor
device applications in the blue and ultra-violet (UV) wavelength region. InN, GaN, and AlN
with direct band gaps of 1.9, 3.4, and 6.2 eV respectively, can be grown in the wurtzite crystal
structure to form a continuous alloy system with the ability to grade the wavelength of the
emitted light over the entire visible spectrum. In recent years, various groups have produced
light emitting diodes (LED's)[1,2] based on group III-Nitrides and also the first pulsed lasers have
been fabricated.[3]

For these semiconductor diodes and lasers there is a need to make electrical contact to
both the n- and p-type layers through the use of an ohmic contact. An ohmic contact is defined
as a metal-semiconductor contact that has a negligible contact resistance relative to the bulk or
spreading resistance of the semiconductor.[4] In metal/semiconductor contacts there are three
mechanisms of current conduction, thermionic emission (TE), thermionic field emission (TFE),
and field emission (FE), also known as tunneling. For a rectifying or Schottky contact,
thermionic emission dominates while in an ohmic contact tunneling is preferred. The dominant
conduction mechanism will depend on the height and thickness of the potential barrier formed at
the metal/semiconductor interface. The thickness of this barrier is related to the depletion width

1091

Mat. Res. Soc. Symp. Proc. Vol. 449 © 1997 Materials Research Society

in the semiconductor. This depletion width is in turn inversely proportional to the square root of the carrier concentration in the near surface region of the semiconductor. When there is a wide barrier thermionic emission over the potential barrier dominates. Field emission can occur when the potential barrier formed at the metal-semiconductor interface is very narrow, increasing the tunneling probability. The third mechanism, thermionic field emission occurs when the potential barrier has sufficiently thinned at the top so the carriers can be excited to a level where tunneling through the thinned barrier is possible

Temperature dependent I-V data can be used to determine the dominant conduction mechanism in a reverse biased contact. A model for thermionic emission over a barrier was presented by Wagner[5] and Schottky and Spence.[6] The following equation can be used to describe contacts dominated by thermionic emission over the valence band offset

$$\ln(I_o/T^2)=\ln(SA^*)+(\Delta\phi_B-\phi_B)/kT, \qquad (1)$$

where I_o is the saturation current, S is the diode area, A^* is Richardson's constant, $\Delta\phi_B-\phi_B$ is the effective barrier height and T is the absolute temperature. From this equation, a plot of $\ln(I_o/T^2)$ vs. 1/T would produce a straight line if thermionic emission was the only mechanism resulting in conduction. For the case of field emission, currents in reversed biased contacts were modeled by Padovani and Stratton,[7] who developed the following relationship:

$$I=SJ_s\exp(-qV_R/E'), \qquad (2)$$

by taking the natural log of each side of this equation, it can be rewritten:

$$\ln(I)=\ln(SJ_s)+\ln(-qV_R/E'). \qquad (3)$$

Based on this equation, a plot of the natural log of current vs. voltage should produce a straight line if tunneling is dominating conduction.

For GaN, low resistance contacts have been reported to n-type material using Ti/Al.[8] For contacts to p-GaN early research and devices used thin films of Au,[9,10] while commercially available LED's use a Ni/Au contact scheme.[11,12] This contact scheme is also presently being used in the experimentally reported LD's.[13] Neither of these contact schemes have been proven to provide low resistance contacts to p-GaN and thus contact to the p- material has been indicated as one of the obstacles in the ability to achieve CW operation.

Formation of low resistance ohmic contacts to p-GaN is complicated due to the nature of the semiconductor. In general there are two methods of forming an ohmic contact to a p-type semiconductor. The semiconductor can be degenerately doped to decrease the depletion distance and allow tunneling to occur or a large work function metal can be used to decrease the potential offset at the metal/semiconductor interface. In GaN the Fermi level is believed to be unpinned leading to the potential barrier being work function dependent. However, in p-GaN neither of these methods are presently viable. GaN tends to auto-dope n-type and has only been reliably doped to the mid 10^{18} cm^{-3} p-type with Mg. This is due to problems with activation of the deep Mg acceptor and the formation of H-Mg complexes decreasing the number of active carriers. Concerning the second approach, GaN, with a band gap of E_g=3.4 eV and an electron affinity of χ=4.1 eV, requires a metal with a work function of $\approx$7.5 eV to provide no offset at the interface. Unfortunately, metal work functions are never much larger than 5 eV[14] so there will still be a noticeable energy barrier to conduction. The purpose of this experiment is to determine the reactions which take place between the commonly used Ni/Au contact and p-GaN, and compare these to Pd/Au and Cr/Au contacts.

EXPERIMENTAL

All metals were deposited on MOCVD grown GaN doped p-type with Mg acceptors to a carrier concentration of 9.8 x 10^{16} cm^{-3}. The contact systems studied included Ni/Au, Pd/Au and Cr/Au (500Å/1000Å) thin film metallizations deposited in an Airco Temescal electron beam evaporator with a base pressure of ~6 x 10^{-7} Torr for all depositions. Before insertion into the vacuum chamber, the samples were degreased and then any residual oxide was etched for 5 minutes in a 10:1 DI:HCl solution. In all cases, dot contacts (diameter=0.5 mm) were patterned during deposition using a stainless steel shadow mask. All samples were heat treated using a Applied Test Systems conventional tube furnace at 200, 400, and 600°C for 5, 15, and 30 minutes in a flowing N_2 atmosphere and the Pd/Au and Cr/Au contacts were heated to 900°C for 15 seconds in a rapid thermal annealing (RTA) furnace. Following each heat treatment, current-voltage (I-V) data were measured over a range of -5 to +5 V using a Tektronix 577 Curve Tracer with a 177 Standard Text Fixture. Compound formation and elemental concentrations at the surface and metal/p-GaN interface were evaluated using a Phi 660 Scanning Auger Microprobe to perform surface analysis and depth profiles with an accelerating voltage of 5 kV and a beam diameter of ~1 μm. Current transport mechanisms were determined using temperature dependent I-V measurements. The samples were mounted and Au wire bonded onto TO5 headers. The I-V characteristics were measured using a HP 6111A DC power supply with a HP-3455A digital voltmeter, the current was measured across a 1kΩ resistor. Temperature was measured using a Pt resistance thermometer calibrated to 0.01 degrees and ranged from 80-400 K. The temperature was controlled by a Lake Shore Cryogenics DRC 82C temperature controller. All measurements were taken at a pressure of 10 mTorr.

RESULTS

The contact schemes investigated were chosen on the basis of their large work functions and their ability to react with GaN to provide the opportunity for doping of the near surface region which would increase the probability for tunneling to occur. The work functions are Au:5.1 eV, Ni:5.15 eV, Pd:5.12 eV and Cr:4.5 eV. Bermudez et al.[15] have shown that Ni films on GaN will decompose the GaN film upon heat treatment, and Mohney et al.[16] have predicted that Ni and Pd will react with GaN.

<u>Ni/Au</u>

For the Ni/Au contacts the I-V curves were rectifying as-deposited and remained so upon heat treatment. There was a decrease in breakdown voltage at 400°C which is attributed to Ni breaking up some of the interfacial contamination between the metal and semiconductor allowing for more current to be transported across the interface. The Ni however does not begin to dissociate the GaN until temperatures ≥600°C have been reached. In the AES depth profile (Figure 1) it can be seen that upon heating to 600°C there is large scale diffusion of Ni through the Au capping layer to the surface of the contact where it oxidized. There is a slight Ga tail into the Au region which is indicative of GaN dissociation. As-deposited there was no indication of Ni diffusion either to the surface of the contact or into the GaN. These results are different from previous experiments in which Ga diffused through the entire contact layer to the surface and near linear I-V curves were obtained.[17] It is postulated that the loss of nitrogen to the environment and the subsequent formation of N vacancies may be the cause of the lower current transport in this study.

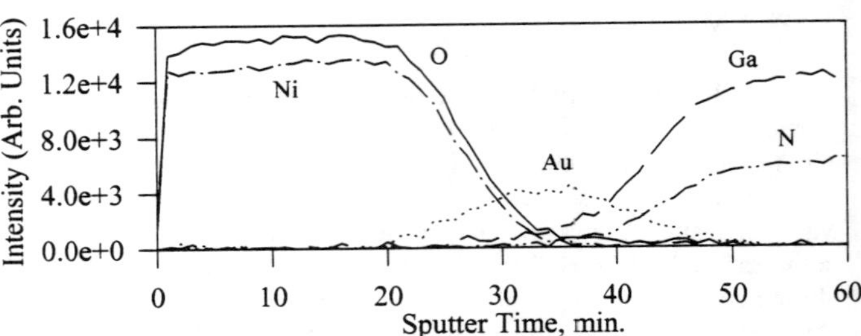

Fig 1. AES depth profile for Au/Ni/p-GaN 600°C, 5 min.

Pd/Au

For the Pd/Au contacts the I-V was rectifying as-deposited and remained so upon all heat treatments. The lowest breakdown voltage came from the sample which was rapid thermally annealed (RTA) to 900°C for 15 seconds. Figure 2 shows the AES depth profiles for this sample. In this sample the reactions are very different than in the Ni/Au contact. For a Pd/Au contact, Pd began to diffuse into the Au capping layer, but not through it like Ni. The Pd formed a Au:Pd solid solution that encompassed the entire contact region. There was also evidence of a slight incorporation and diffusion of Pd into the GaN from the tail of the Pd signal in Figure 2. It is postulated that this diffusion of Pd into the GaN causes lower breakdown voltage upon heating to 900°C. The Pd is though to act as an acceptor in the GaN matrix causing the near surface region to be highly doped leading to increased conduction. The formation of N vacancies may be compensating these Pd acceptors which would lead to the rectifying nature of the contacts.

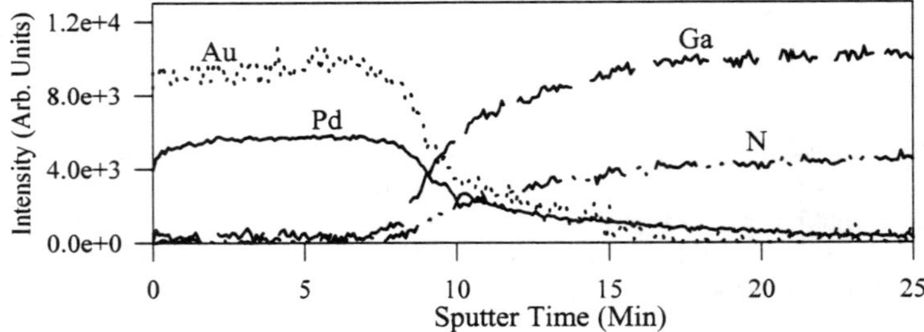

Fig 2. AES depth profile for Au/Pd/p-GaN 900°C, 15 seconds.

Cr/Au

For the Cr/Au samples, the I-V was rectifying as-deposited and remained so upon heating up to 600°C. However, upon a RTA of 900°C for 15 seconds, the I-V curve became linear, indicative of an ohmic contact. The specific contact resistance of this Cr/Au contact has been calculated to be $\leq 4.3 \times 10^{-1}$ Ωcm^2 assuming a bulk resistance of zero for the underlying GaN. Figure 3 shows the best case I-V curves for all of the contact schemes investigated. From AES depth profiles, at lower temperatures Cr behaves like Ni in that it diffused to the surface through the Au capping layer but did not form a Cr:Au solid solution as the Pd did. However, at 900°C, there is evidence of dissociation of GaN by Cr. (Figure 4) It addition to Cr diffusing to the surface, a Au:Ga phase formed with the excess N released from the GaN being incorporated into what appears to be a CrN compound at the metal/semiconductor interface. The formation of these phases prevented the loss of nitrogen and the formation of N vacancies which are detrimental to contact performance . It is also suggested that the CrN phase may have a higher

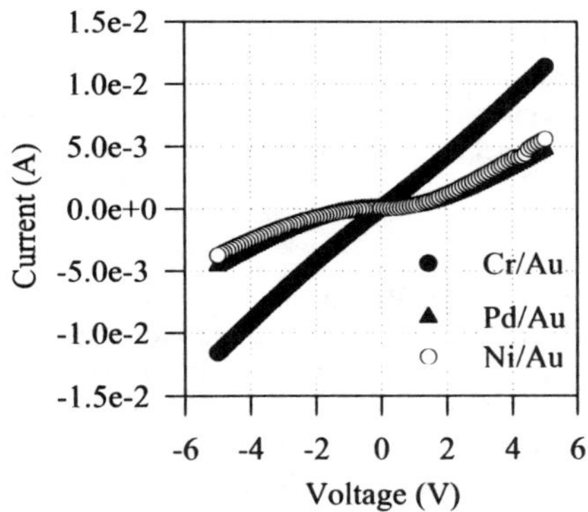

Fig 3. I-V Data for Ni/Au, Pd/Au, and Cr/Au on p-GaN.

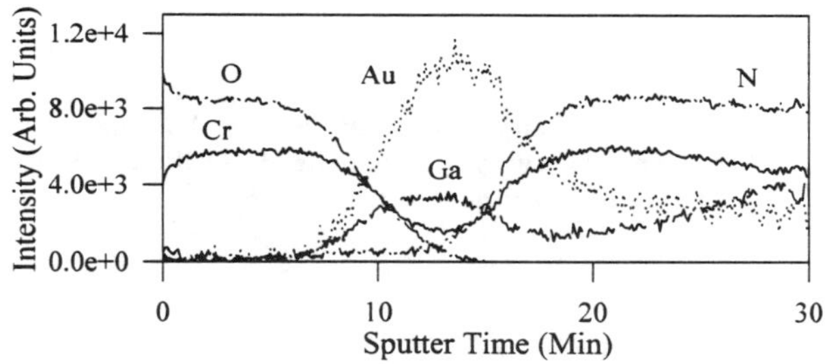

Fig 4. AES depth profile for Au/Cr/p-GaN, 900°C, 15 seconds.

work function than Cr metal, which would lead to a reduced potential offset at the metal/semiconductor interface. More investigation of this CrN phase needs to be performed to determine the properties of this phase that are leading to the ohmic behavior of the contacts.

<u>Temperature Dependent I-V</u>

Temperature dependent I-V was performed on a Ni/Au sample heated at 600°C for 5 min, a Pd/Au sample heated at 600°C for 30 min and a Cr/Au sample heated at 200°C for 5 min. The data showed two trends. For the Ni/Au and Pd/Au the current increased substantially with an increase in measurement temperature, while for the Cr/Au contacts the current was relatively stable over the entire temperature range. This is indicative of the current in the Cr/Au sample being dominated by field emission due to the lack of temperature dependence in the I-V measurements. However, upon plotting the data as $\ln (I_o/T^2)$ vs $1/T$ for all the contact schemes, it appeared that thermionic emission is dominating over a portion of the temperature and reverse bias range studied. Over the entire range of 80-400 K the curves were non-linear but became more linear as the temperature was raised. From the $\ln(I)$ vs. V curve for field emission it was determined that in each of the contact schemes, thermionic field emission is the dominant conduction mechanism above a reverse bias of $\approx$2.5-3 V. Below this threshold value thermionic emission dominates current conduction in each of the schemes studied. More work need to be done on these contact schemes to exact regions in which each conduction process dominates.

CONCLUSIONS

Interfacial reactions between Ni/Au, Pd/Au, and Cr/Au thin films and p-GaN have been investigated. Both Ni/Au and Pd/Au have been shown to develop rectifying contacts even upon heat treatment while Cr/Au forms an ohmic contact with a specific contact resistance of $\leq 4.3 \times 10^{-1}$ Ωcm^2 following a 900°C, 15 second RTA. Both Ni and Cr appear to dissociate the underlying GaN and also diffuse through the Au capping layer to the surface of the contact. The reaction products in the Cr/Au contact are a Au:Ga phase and a Cr:N phase. It is postulated that this reaction leads to the ohmic behavior of the contact due to the prevention of the formation of N vacancies in the GaN. The Pd/Au contact has been shown to form a Pd:Au solid solution and the Pd diffuses into the GaN where it may act as an acceptor. The temperature dependent I-V characteristics for all the contacts exhibited thermionic emission below $\approx$2.5 V and thermionic field emission above 2.5 V.

ACKNOWLEDGEMENT

The authors would like to thank Larry Callahan of TSSI for all his help mounting the temperature dependent I-V samples. This work was supported by ONR/ARPA Grant N00014-92-J-1895 and the Air Force Office of Scientific Research.

REFERENCES

[1] H.P. Maruska and J.J. Tietjen, Appl. Phys. Lett. **15**, 327 (1969).

[2] J. Pankove, J. of Lumin. **7**, 114 (1973).

[3] D.W. Jenkins and J.D. Dow, Phys. Rev. B **39**, 3317, (1989).

[4] S.M. Sze, <u>Physics of Semiconductor Devices</u>, 2nd ed. (John Wiley + Sons, Inc. Publishers, New York, 1981) p. 304.

[5] C. Wagner, Phys. A. **32**, 641 (1931).

[6] W. Schottky and E. Spence, Wiss. Veroff. Siemens-Werken, **18**, 225 (1939).

[7] F. Padovani and R. Stratton, Solid St. Electron., **9**, 695 (1966).

[8] M.E. Lin, Z. Ma, F.Y. Huang and H. Morkoc, Appl. Phys. Lett. **64**, 2557 (1994).

[9] S. Nakamura, T. Mukai, and M. Senoh, Jpn. J. Appl. Phys. **30**, L1998 (1991).

[10] S. Nakamura, M. Senoh, and T. Mukai, Jpn. J. Appl. Phys. **32**, L8 (1993).

[11] S. Nakamura, M. Senoh, and T. Mukai, Appl. Phys. Lett. **62**, 2390 (1993).

[12] S. Nakamura, T. Mukai, M. Senoh, Appl. Phys. Lett. **64**, 2557 (1994).

[13] S. Nakamura, M. Senoh, S. Nagahama, N. Iwasa, T. Yamada, T. Matsushita, H. Kiyoku and Y. Sugimoto, Jpn. J. Appl. Phys. **35**, L74 (1996).

[14] S.M. Sze, <u>Physics of Semiconductor Devices</u>, 2nd ed. (John Wiley + Sons Inc. Publishers, New York, 1981) p. 251.

[15] V. M. Bermudez, R. Kaplan, M.A. Khan and J.N. Kuznia, Phys. Rev. B. **48**, 2436 (1993).

[16] S.E. Mohney, B.P. Luther and T.N. Jackson, Mat. Res. Soc. Proc. **395**, 843 (1996).

INVESTIGATION OF ALUMINUM AND TITANIUM/ALUMINUM CONTACTS TO N-TYPE GALLIUM NITRIDE

B. P. LUTHER *, S. E. MOHNEY **, T. N. JACKSON *, M. ASIF KHAN ***, Q. CHEN ***, AND J. W. YANG ***
*Department of Electrical Engineering, The Pennsylvania State University, University Park, PA 16802
**Department of Materials Science and Engineering, The Pennsylvania State University, University Park, PA 16802, mohney@ems.psu.edu
***APA Optics, Inc., 2950 N.E. 84th Lane, Blaine, MN 55449

ABSTRACT

We report on a study of Al and Ti/Al contacts to n-type GaN. Al contacts on n-GaN $(7\times10^{17}cm^{-3})$ annealed in forming gas at 600°C reached a minimum contact resistivity of $8\times10^{-6}\Omega cm^{2}$ and had much better thermal stability than reported by previous researchers. Ti/Al (35nm/115nm) contacts on n-GaN $(5\times10^{17}cm^{-3})$ had resistivities of $7\times10^{-6}\Omega cm^{2}$ and $5\times10^{-6}\Omega cm^{2}$ after annealing in Ar at 400°C for 5min and 600°C for 15sec, respectively. Depth profiles of Ti/Al contacts annealed at 400°C showed that low contact resistance was only achieved after Al diffused to the GaN interface. We propose that the mechanism for ohmic contact formation in Ti/Al contacts annealed in the 400–600°C range includes reduction of the native oxide on GaN by Ti and formation of an Al-Ti intermetallic phase in intimate contact with the GaN. Contacts with different Ti/Al layer thicknesses were investigated and those with 50nm/100nm layers had the same low resistance and better stability than 25nm/125nm contacts.

INTRODUCTION

GaN has received great interest in the past few years for applications in blue/ultraviolet lasers, light emitting diodes, photodetectors, and high temperature/high power electronic devices. High quality ohmic contacts are critical to these applications. Although many researchers have made ohmic contacts to n-GaN with low contact resistivities, the mechanism for ohmic contact formation has not been explained fully. The purpose of this investigation was to determine the mechanism for Al-based ohmic contacts on n-GaN formed by annealing in the temperature range of 400–600°C.

Contacts containing Ti/Al layers have displayed the lowest contact resistivity to n-type GaN.[1,2] TiN has been observed[3] at the interface of Ti/Al and Ti/Al/Ni/Au contacts annealed at 900°C. Researchers have speculated[1-3] that due to TiN formation when Ti reacts with GaN, a high concentration of nitrogen vacancies is created near the interface, causing the GaN to be heavily doped n-type. A thin layer of TiN has also been observed at the interface of unannealed samples that were etched by RIE before metal deposition. Without RIE prior to metal deposition, good ohmic contacts have been attained by annealing Ti/Al films on n-GaN between 550 and 750°C for 20sec,[4] however, there are no reported investigations of Ti/Al contacts annealed in this temperature range that would reveal the mechanism for ohmic contact formation.

Mat. Res. Soc. Symp. Proc. Vol. 449 © 1997 Materials Research Society

EXPERIMENT

N-type GaN layers on Al_2O_3 substrates were supplied by APA Optics for electrical measurements. Samples were etched in dilute HCl and blown dry with nitrogen before they were loaded into the vacuum system. Al (150nm) was deposited on n-GaN ($7\times10^{17}cm^{-3}$) and Ti/Al (35nm/115nm) layers were deposited on n-GaN ($5\times10^{17}cm^{-3}$) by dc magnetron sputtering. Al layers were patterned by etching and Ti/Al layers by liftoff to define circular structures for measuring specific contact resistance by the transfer length method. The inner circular contact had a radius of 100µm and gaps between inner and outer contacts ranged from 2–50µm. Specific contact resistance was calculated from the circular contacts taking into account the circular geometry.[5] Four probes were used to eliminate the effects of resistance between the probes and metal contacts. Al contacts were annealed in a quartz lamp furnace in an Ar/4%H_2 atmosphere. Ti/Al contacts were annealed in an AG 610 rapid thermal annealing furnace in Ar that first flowed through a titanium getter furnace to remove oxygen and water vapor. X-ray photoelectron spectroscopy (XPS) depth profiles were performed using a Kratos Analytical XSAM 800pci system.

RESULTS

As deposited Al-only contacts ranged from ohmic (as low as $1\times10^{-4}\Omega cm^2$) to rectifying contacts. After being annealed at 600°C in Ar/4% H_2, all Al contacts were ohmic with a best contact resistivity of $8\times10^{-6}\Omega cm^2$. Two samples prepared on different occasions were annealed together at 600°C in Ar/4% H_2 and their contact resistivities were measured and plotted in Fig. 1 as a function of annealing time. Both had nonlinear I-V curves as deposited but became low resistance ohmic contacts after annealing. Sample A reached its minimum resistivity after 1min, while sample B was annealed for 8min before it reached its minimum.

The different annealing times needed for formation of Al-only ohmic contacts may be the result of different initial thicknesses of GaN native oxide, as exposure to humidity after HCl etching varied between samples prepared on different days. Al is known to reduce Ga_2O_3 on GaAs[6] and would be expected to do the same on GaN. Thus, we expect that upon annealing the Al reduces the native oxide on GaN and improves the electrical properties of the contact. To test

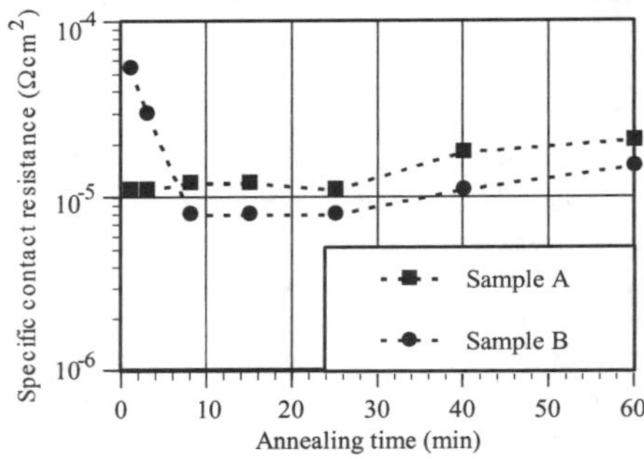

FIG. 1. Specific contact resistance as a function of annealing time in Ar/4%H_2 of two n-GaN samples with 150nm Al contacts.

this hypothesis, we deposited Al on GaN that had not been dipped in dilute HCl to remove the native oxide prior to Al deposition. As expected, these contacts were rectifying as deposited but became ohmic after annealing, with lowest contact resistance higher than for etched samples.

Al has a work function of 4.08eV,[7] which is near the electron affinity of GaN (4.1eV[8]) and may result in a low energy barrier between Al and n-GaN[9] when they are in intimate contact. Other researchers have reported Al contacts on n-GaN to be ohmic as deposited but have seen these contacts degrade rapidly upon annealing.[2,4,10] By annealing in forming gas we minimized the oxidation of Al, which may be the cause of the increased resistance in earlier reports. In our experiments, the sheet resistance of the Al was observed to increase upon annealing at 600°C for 1hr and is likely the cause of the slight increase in the measured contact resistivity, as shown in Fig. 1. Although the samples were annealed below the Al melting point of 660°C, some degradation of the Al surface morphology was noticeable after 1hr at 600°C.

Ti/Al (35nm/1150nm) contacts were annealed in Ar at 400°C and 600°C, and their contact resistivities are plotted as a function of annealing time in Fig. 2. The sample annealed at 600°C was ohmic with contact resistivity $5\times10^{-6}\Omega cm^2$ after 15sec. Further annealing caused the resistivity to increase slightly, possibly due to oxidation. The sample annealed at 400°C had high resistance after being annealed for 45sec but became ohmic after 3min and had a contact resistivity of $7\times10^{-6}\Omega cm^2$ after 5min. We know of no reports in the literature of contacts to n-type GaN with such a low contact resistance following an anneal at a temperature this low, except when there was intentional RIE damage to the GaN surface.[1]

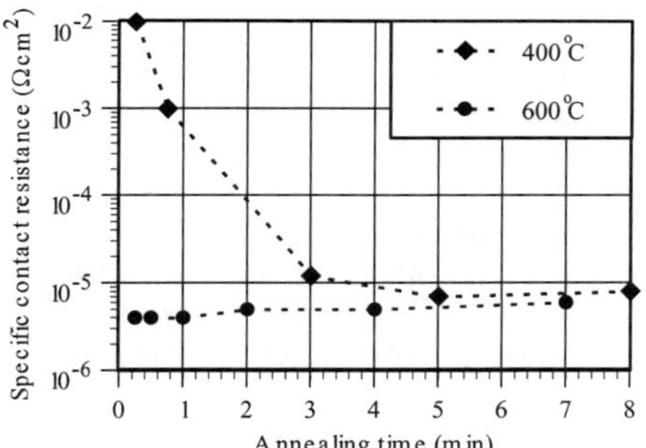

FIG 2. Specific contact resistance as a function of annealing time in Ar of Ti/Al (35nm/115nm) contacts on n-GaN annealed at 400°C and 600°C.

XPS depth profiles of Ti/Al layers (35nm/1150nm) on GaN as deposited and annealed at 400°C for 45sec and 3min are shown in Fig. 3a-c. The separate layers of Al and Ti on GaN can be seen in the as deposited profile. Annealing for 45sec results in intermixing of the Ti and Al layers, with no Al at the Ti/GaN interface. The depth profile of the sample annealed for 3min clearly shows that Al has diffused through the Ti and reached the GaN surface. Thus, we observe that Ti/Al becomes a low resistance ohmic contact to n-GaN at this low annealing temperature only after Al comes into contact with the GaN surface. The metal/GaN interface does not appear any less abrupt in the sample annealed for 3min than in the as deposited sample, indicating that no

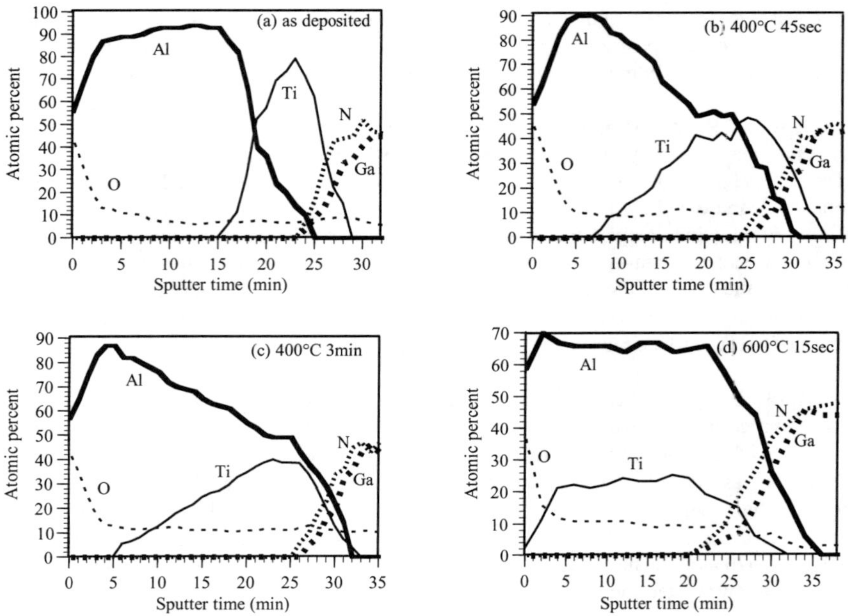

FIG. 3. XPS depth profiles of Ti/Al (35nm/115nm) contacts on GaN as deposited (a) and annealed at 400°C for 45sec (b) and 3min (c) and at 600°C for 15sec (d). Atomic percents of Al, Ti, O, Ga, and N are shown as a function of sputter time.

detectable reaction occurred between the metal contact and GaN layer. Fig. 3d. shows an XPS depth profile of Ti/Al layers (35nm/1150nm) on GaN annealed at 600°C for 15sec, which resulted in a contact resistivity of $5 \times 10^{-6} \Omega cm^2$. Almost uniform mixing of Al and Ti is evident, and, as in the ohmic contacts annealed at 400°C, Al is present at the metal/GaN interface.

Reports of Ti-only contacts to GaN[10,11] further suggest that the mechanism for ohmic contact formation in the low-temperature Ti/Al contacts is not reaction between Ti and GaN. Ti-only contacts become good ohmic contacts only after being annealed at 900°C and above.[10] On the other hand, Ti is important in the Ti/Al contacts since the Ti/Al contacts had slightly lower contact resistivity and became ohmic after shorter annealing times than did Al-only contacts. Gallium oxide will be reduced by Ti,[6] and for small amounts of oxygen in Ti, the most stable phase is α-Ti with oxygen in solid solution. Therefore, we would expect the native oxide on GaN to be reduced by the Ti with the oxygen dissolved into the Ti film, leaving no insulating oxide at the metal/GaN interface.

Based on our experimental observations we propose the following mechanism for ohmic contact formation in Ti/Al contacts to n-GaN: upon annealing the contact, Ti reduces the native gallium oxide on the GaN surface and Al diffuses through the Ti to make contact with the surface of the GaN in the form of an Al-Ti intermetallic phase that has a low work function, similar to that of Al. The intimate contact between the low work function intermetallic and n-GaN results in a low barrier height, allowing current to flow in either direction by thermionic or thermionic-field emission.

Contacts with Ti/Al layers of 25nm/125nm, 50nm/100nm, and 75nm/75nm were made to investigate the effect of layer thicknesses. The contacts were annealed together in Ar at 600°C and their specific contact resistances are plotted as a function of annealing time in Fig. 4. These contacts were all rectifying as deposited and became ohmic after being annealed for 15sec. The 25nm/125nm and 50nm/100nm samples had resistivities of $4\times10^{-6}\Omega cm^2$ after 15sec and the 75nm/75nm sample had a resistivity of $1.2\times10^{-5}\Omega cm^2$. After subsequent anneals, the 25nm/125nm contact showed the most degradation, reaching a resistivity of $1.0\times10^{-5}\Omega cm^2$ after 4min. The 50nm/100nm and 75nm/75nm contacts only degraded slightly, increasing to $5\times10^{-6}\Omega cm^2$ and $1.3\times10^{-5}\Omega cm^2$, respectively, after 4min. From these measurements we see that a thicker Ti layer in the contact results in better stability, while thinner Ti gives lower contact resistivity. Ti/Al layers of approximately 50nm/100nm display both low contact resistance and better stability under these conditions.

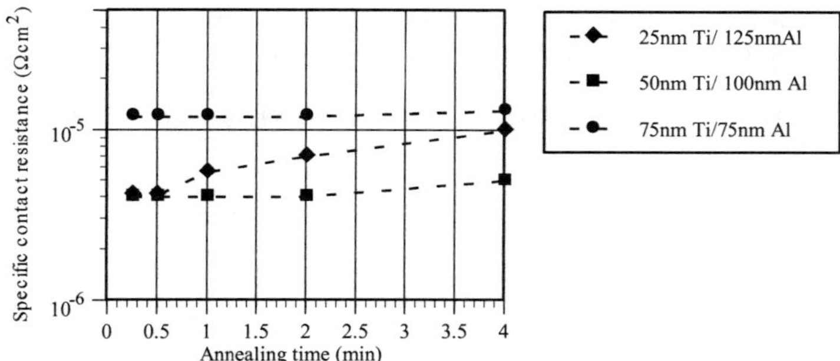

FIG. 4. Specific contact resistance as a function of annealing time in Ar at 600°C of contacts to n-GaN with Ti/Al layer thicknesses of 25nm/125nm, 50nm/100nm, and 75nm/75nm.

CONCLUSION

Al (150nm) and Ti/Al (35nm/115nm) contacts to n-GaN were investigated to determine the mechanism for ohmic contact formation in contacts annealed between 400 and 600°C. Al contacts improved after being annealed at 600°C in Ar/4% H_2. Improvement in Al contacts after annealing has not been reported elsewhere and is probably due to oxidation being inhibited by the forming gas annealing environment. Ti/Al contacts became ohmic after annealing for 3min at 400°C, and the following mechanism was proposed: Ti reduces the native oxide on GaN and Al diffuses through the Ti layer, resulting in an ohmic contact when a low work function Al-Ti intermetallic phase comes into contact with the GaN surface. Contacts with different Ti/Al layer thicknesses were investigated. Those with 50nm/100nm Ti/Al layers had the same low resistance and better stability than contacts with a thinner Ti layer.

ACKNOWLEDGMENTS

The authors gratefully acknowledge DARPA for supporting this work through AFOSR grant F49620-95-1-0516.

REFERENCES

1. Z.-F. Fan, S. N. Mohammad, W. Kim, O. Aktas, A. E. Botchkarev and H. Morkoc, Appl. Phys. Lett. **68**, 1672 (1996).
2. M. E. Lin, Z. Ma, F. Y. Huang, Z. F. Fan, L. H. Allen, and H. Morkoc, Appl. Phys. Lett. **64**, 1003 (1994).
3. S. Ruvimov, Z. Liliental-Weber, J. Washburn, K. J. Duxstad, E. E. Haller, Z.-F. Fan, S. N. Mohammad, W. Kim, A. E. Botchkarev and H. Morkoc, Appl. Phys. Lett. **69**, 1556 (1996).
4. A. T. Ping, M. Asif Khan, and I. Adesida, J. Electron. Mat. **25**, 819 (1996).
5. G. S. Marlow and M. B. Das, Solid St. Electron. **25**, 91 (1982).
6. S. P. Kowalczyk, J. R. Waldrop, and R. W. Grant, J. Vac. Sci. Technol. **19**, 611 (1981).
7. K. W. Boer, *Survey of Semiconductor Physics* (Van Nostrand, New York, 1990).
8. J. I. Pankove and H. Schade, Appl. Phys. Lett. **25**, 53 (1971).
9. J. S. Foresi and T. D. Moustakas, Appl. Phys. Lett. **62**, 2859 (1993).
10. L. L. Smith, R. F. Davis, M. J. Kim, R. W. Carpenter and Y. Huang, J. Mater. Res. **11**, 2257 (1996).
11. S. C. Binari, H. B. Dietrich, G. Kelner, L. B. Rowland, K. Doverspike, and D. K. Gaskill, Electron. Lett. **30**, 909 (1994).

PtIn₂ ohmic contacts to n-GaN

D. B. Ingerly and Y. A. Chang
Department of Materials Science and Engineering, University of Wisconsin, Madison, Wisconsin
53706-1595

N. R. Perkins and T. F. Kuech
Department of Chemical Engineering, University of Wisconsin, Madison, Wisconsin 53706-1691

ABSTRACT

A new metallization scheme has been developed to form ohmic contacts to n-GaN. Contacts were fabricated by sputtering the intermetallic compound, $PtIn_2$ on n-GaN ($n \sim 5 \times 10^{17}$ cm^{-3}) grown by metalorganic vapor phase epitaxy (MOVPE) with some of the contacts subjected to rapid thermal annealing. Contacts in the as-deposited state exhibited nearly ohmic behavior with a specific contact resistance of 1.2×10^{-2} Ω cm^2. Contacts subjected to rapid thermal annealing at 800 °C for 1 minute exhibited linear current-voltage characteristics and a decrease in contact resistance. To rationalize the reaction at the contact interface the phase diagram of the $PtIn_2$-$PtGa_2$-InN-GaN reciprocal system at 800 °C was calculated based on known and estimated thermodynamic values. This phase diagram suggests that there will be a solid state exchange of In and Ga atoms at the $PtIn_2$/GaN contact interface producing $(In_xGa_{1-x})N$ and $Pt(In,Ga)_2$. Results from Auger depth profiling used to examine the $PtIn_2$/n-GaN contacts are consistent with this estimated phase diagram information. It is proposed that the formation of $(In_xGa_{1-x})N$ at the contact interface is responsible for the ohmic behavior of $PtIn_2$ contacts.

INTRODUCTION

GaN is a III-V compound semiconductor with a wurtzite crystal structure having a 3.4 eV direct energy band gap at room temperature. When alloyed with the other Group III-Nitrides, GaN can form a continuous alloy system whose room temperature band gaps range from 1.9 eV (InN) to 6.2 eV (AlN).[1] This makes GaN an excellent material for optoelectronic devices, such as light emitting diodes (LEDs), performing in the blue and ultraviolet regions. In addition, GaN's high thermal conductivity, hardness and thermal stability make it a candidate material for the fabrication of high temperature and high power devices.[2]

Despite its promise, GaN could not be used for high efficiency optical devices until success in obtaining p-type GaN films lead to a renewed and intense interest in GaN.[2] Blue and blue-green LEDs are commercially available and Nichia Chemical Industries has also demonstrated the operation of a nitride laser operating at 416 nm.[3] In addition to optical devices GaN has also been used for metal semiconductor field effect transistors (MESFETs)[4] and high electron mobility transistors (HEMTs).[5]

Ohmic contacts are an important element in the fabrication of devices. High contact resistance can substantially reduce the performance of GaN optical and electrical devices; therefore, to obtain optimum performance the contact resistance should be minimized. The present technology for the formation of ohmic contacts to n-GaN usually involves the use of Ti or Al metallization schemes.[6,7] In this study we report a different type of metallization scheme,

1103

Mat. Res. Soc. Symp. Proc. Vol. 449 © 1997 Materials Research Society

where PtIn$_2$ is utilized to form ohmic contacts to n-GaN. PtIn$_2$ is an intermetallic compound with a CaF$_2$ (C1) crystal structure, good chemical stability and a peritectic melting point at 1039 °C.[8]

EXPERIMENTAL PROCEDURE

The following procedures were used to prepare all samples examined in this study, with the lithography being omitted on substrates not used for electrical measurements. The n-GaN substrates used in this study are thin layers (2-3 μm) of single crystal GaN grown by metalorganic vapor phase epitaxy grown (MOVPE) on sapphire (0001) with a 10-20 nm AlN buffer layer. The n-GaN epilayer is typically unintentionally doped to 5 x 10^{17} cm^{-3} with a sheet resistance of 1000 Ω/□. Prior to the lithography, the substrates where ultrasonically degreased with trichloroethylene, acetone, and methanol for 5 min each. The substrates were then dipped into a H$_2$SO$_4$:H$_3$PO$_4$ (1:1) solution and rinsed in H$_2$O. They were then patterned using standard photolithographic techniques. For the I-V measurements, equally spaced circular dots 400 μm in diameter and 550 μm apart were used. For specific contact resistance (ρ_c) measurements a circular transmission line model (TLM) pattern[9] was used. The patterned substrates were then placed in a BOE solution for 5 min, rinsed in DI water and immediately loaded into a vacuum chamber with the background pressure less than 4 x 10^{-7} torr.

Nominally 150 nm PtIn$_2$ films were deposited onto the substrates by DC magnetron sputtering from a single alloy target. The composition of the sputtered film was analyzed with a Cameca electron microprobe to be 31.5 at% Pt and 68.5 at% In. X-ray diffraction was used to confirm that the sputtered film's crystal structure matched that of PtIn$_2$. After sputter deposition, the photoresist was lifted off in an acetone bath leaving the patterned metallization on the wafers. Following lift-off, samples were annealed in an AG Associated MiniPulse rapid thermal annealing (RTA) system with a flowing high purity Ar atmosphere.

The electrical properties of the contacts were measured with a Keithley Model 236 electrometer which was employed as a current source and voltage meter. To characterize the interfacial reactions of the contacts, Auger depth profiling was used. The Auger depth profiling was done using a Perkin-Elmer scanning electron microprobe.

PHASE EQUILIBRIA

In this study we utilized the exchange mechanism for the selection of a suitable metal for the formation of ohmic contacts to n-GaN. The exchange mechanism is a solid state reaction that has been identified as a systematic approach for tailoring metal/semiconductor contact properties.[10] During the exchange reaction one element in the metal exchanges with a second element in the compound semiconductor. Therefore, at the contact interface the composition of each phase is changed; in this way the contact properties can be altered. In order to select suitable metals and metal compounds for participating in the exchange reaction Jan[11] and Chang[12] derived criteria based on thermodynamic and kinetic considerations.

Based on these criteria PtIn$_2$ was selected as a potential ohmic contact to n-GaN. For reasons that will be discussed below, the annealing of PtIn$_2$/GaN contacts should lead to an exchange of In and Ga atoms at the interface producing (In$_x$Ga$_{1-x}$)N and Pt(In,Ga)$_2$. (In$_x$Ga$_{1-x}$)N has a lower band gap energy than GaN and low resistance ohmic contacts to InN have demonstrated.[13] Based on these observations the authors propose that the formation of (In$_x$Ga$_{1-x}$)N at the contact interface is responsible for the ohmic behavior of PtIn$_2$/n-GaN contacts.

PtIn$_2$, PtGa$_2$, InN and GaN should form a reciprocal system; this statement can be made based on the following information. It is known that GaN and InN are thermodynamically very stable compounds (large negative values for Gibbs energy of formation) and that they form a continuous solid solution. Swenson[14] has shown that PtIn$_2$ and PtGa$_2$ also form a continuous solid solution. From binary phase diagram data, ternary phase diagrams of the Pt-In-N and Pt-Ga-N were estimated. These phase diagrams show that a tie line connects GaN to PtGa$_2$ and InN to PtIn$_2$. If the end members of continuous solid solution phases are in equilibrium it can be expected that tie lines connect the two solution phases at all compositions. This would make PtIn$_2$-PtGa$_2$-InN-GaN a reciprocal system, simplifying what could otherwise be a very complex four component system.

Figure 1 shows the calculated PtIn$_2$-PtGa$_2$-InN-GaN reciprocal system at 800 °C. This calculation was done treating (In$_x$Ga$_{1-x}$)N as an ideal solution and using the reported Gibbs free energy of formation of the binary end members and thermodynamic data of Pt(In,Ga)$_2$ solid solution.[14,15] This phase diagram represents the thermodynamic equilibrium between these two solid solutions, with calculated tie-lines connecting two compositions at equilibrium. According to this phase diagram there should be an In-Ga exchange at the PtIn$_2$/GaN contact interface at 800 °C and the resulting (In$_x$Ga$_{1-x}$)N should be In rich, a beneficial result for the proposed mechanism of ohmic contact formation.

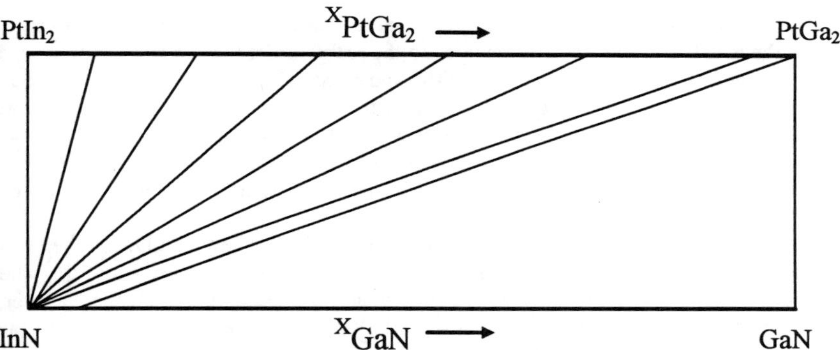

Figure 1. Calculated PtIn$_2$-PtGa$_2$-InN-GaN reciprocal system at 800 °C

RESULTS

Figure 2 shows the I-V characteristics of the PtIn$_2$/n-GaN contacts for three different annealing conditions. These data show that the contacts do not exhibit rectifying behavior even in the as-deposited state. When the contacts are annealed at 550 °C for 1 min they exhibit a more linear I-V behavior than the as-deposited contacts, but the contact resistance increases. In contrast, the contact resistance is lower for the contacts annealed at 800 °C for 1 min. Also, the contacts subjected to this annealing condition exhibit nearly linear I-V characteristics.

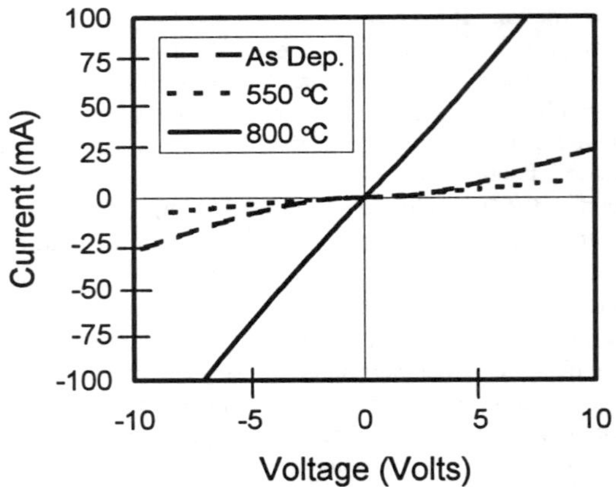

Figure 2. Current-Voltage characteristics of $PtIn_2$ contacts on n-GaN with varied annealing conditions. All contacts were annealed for 1 min.

In this study a circular TLM was utilized to measure the ρ_c.[9] This method requires the resistance of the metal layer to be small compared to the resistance of the substrate. However, due to the relatively high electrical resistivity of the $PtIn_2$ sputtered film ($\sim 1.5 \times 10^{-4}\ \Omega$ cm) it was not possible to measure a resistance that would be valid for the TLM. To solve this problem a more conductive second layer, 100 nm of Au, was thermally evaporated on top of the $PtIn_2$ metallization. For as-deposited contacts, a ρ_c of $1.2 \times 10^{-2}\ \Omega$ cm^2 was measured using TLM. This is as low as as-deposited Al and Ti based metallization schemes on n-GaN films that were not subjected to reactive ion etching.[7]

While the use of Au as a conductive layer was a very effective way of attaining the as-deposited measurement, Auger depth profiles showed that Au reacts extensively with $PtIn_2$ when annealed at 800 °C for 1 min. Co and W were also examined as metals for the conductive layer, but Auger depth profiles found that they also react with $PtIn_2$ when annealed at 800 °C for 1 min. Since the reaction between the two metal layers could affect the In-Ga exchange reaction, the ρ_c of the annealed contacts was not measured.

To characterize the interfacial reaction of the contacts, Auger depth profiles were done on $PtIn_2$/n-GaN both in the as-deposited state and after two different annealing conditions: 550 and 800 °C both for 1 min. The Auger depth profile of the as-deposited contact, Fig. 3a, shows a relatively sharp interface between $PtIn_2$ and GaN. Figure 3b shows the effect of annealing at 550°C for one min. This demonstrates clearly that there is little or no change in the contact interface when compared to the as-deposited sample. If there is any reaction, it is on a scale too small for this technique to resolve. In contrast, Fig. 3c does not show a sharp interface. It has a broad region where In and Ga are both present and the Pt level is decreasing while the N level is increasing, suggesting the formation of $(In_xGa_{1-x})N$. The Auger depth profiles are thus consistent with the proposed reaction which predicts the formation of interfacial $(In_xGa_{1-x})N$ during annealing.

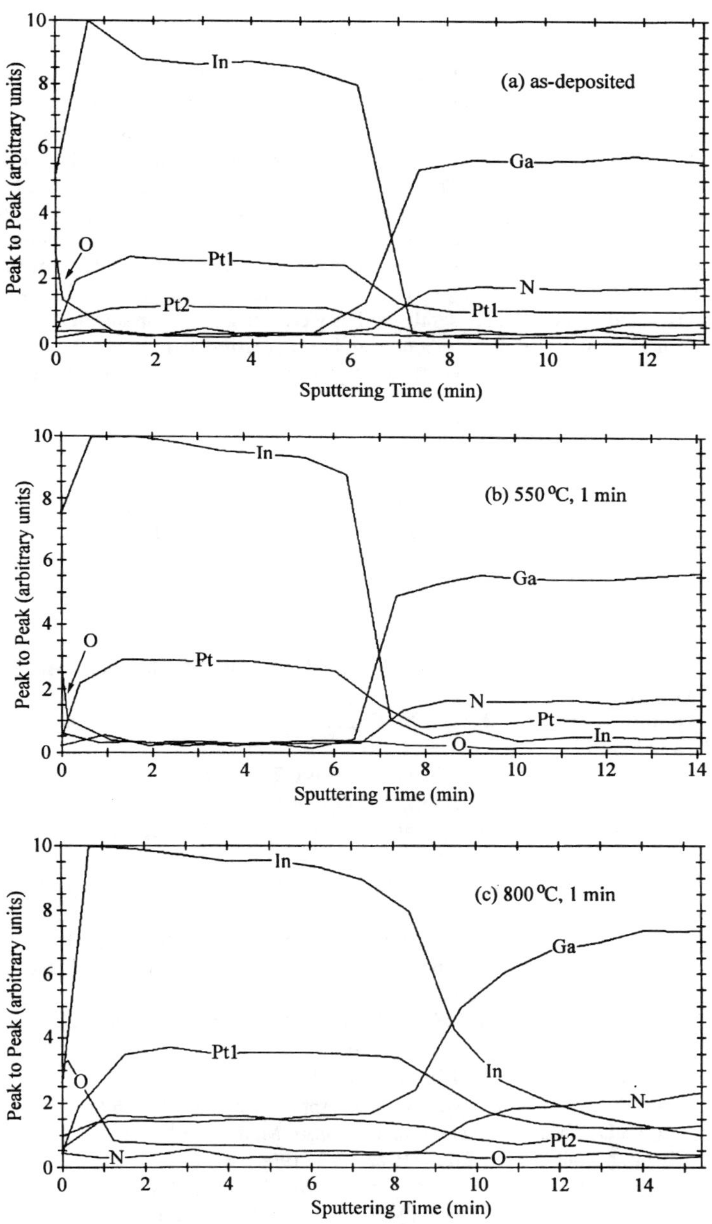

Figure 3. Auger depth profiles of PtIn$_2$/n-GaN (a) as-deposited state (b) annealed at 550 °C for 1 min and (c) annealed at 800 °C for 1 min.

CONCLUSIONS

Using sputter deposited $PtIn_2$ we have fabricated nonalloyed ohmic contacts to n-GaN (n ~ 5 x10^{17} cm^{-3}). The as-deposited contacts exhibited nearly linear I-V characteristics with a measured ρ_c of 1.2 x 10^{-2} Ω cm^2. Contacts annealed at 800 °C for 1 min showed a decrease in contact resistance and had linear I-V characteristics. The authors propose the formation of $(In_xGa_{1-x})N$ at the contact interface during annealing is responsible for the ohmic behavior of $PtIn_2$/GaN contacts. To rationalize the reaction that occurs between $PtIn_2$ and GaN the phase equilibria of the $PtIn_2$-$PtGa_2$-InN-GaN reciprocal system at 800°C were estimated from thermodynamic data. According to these equilibria there will be an exchange of In and Ga atoms at the contact interface producing $(In_xGa_{1-x})N$ and $Pt(In,Ga)_2$. To examine the interfacial reactions between $PtIn_2$ and n-GaN, Auger depth profiles were used. For contacts in the as-deposited state and after annealing at 550°C for 1 min no significant reaction was observed. The contacts annealed at 800°C for 1 min showed Auger depth profiles consistent with the proposed reaction mechanism.

ACKNOWLEDGMENTS

The authors would like to thank the National Science Foundation for its support of this project through grant number NSF-DMR-94-24478 as well as ARPA though the URI on visible light emitters.

REFERENCES

1. A. Botchkarev, A. Salvador, B. Sverdlov, J. Myoung and H. Morkoc, J. Appl. Phys. 77(9), 4455 (1995)
2. H. Morkoc, S. Strite, G. B. Gao, M. E. Lin, B. Sverdlov, and M. Burns, J. Appl. Phys. 76(3), 1363 (1994)
3. S. Nakamura, M. Senoh, S. Naghama, N. Iwasa, T. Yamada, T. Matsushita, H. Kiyoku and Y. Sugimoto, Jpn. J. Appl. Phys. 35, L217 (1996)
4. S. C. Banari, L. B. Rowland, W. Kruppa, G. Kelner, K. Doverspike, and D. K. Gaskill, Electron. Lett. 30(15), 1248 (1994)
5. M. Asif Khan, A. R. Bhattarai, J. N. Kuznia and D. T. Olson, Appl. Phys. Lett 63, 1214 (1993)
6. J. S. Foresi and T. D. Moustakas, Appl. Phys. Lett. 62(22), 2859 (1993)
7. Z. Fan, S. Noor Mohmmad, W. Kim, O. Aktas, A. E. Botchkarev and H. Morkoc, Appl. Phys. Lett. 68(12), 1672 (1996)
8. H. Okamoto, in T. B. Massalski (ed.), *Binary Alloy Phase Diagrams*, ASM International, Materials Park, OH, 2276 (1990) 2nd edition
9. G. K. Reeves, Solid-State Electron. 23, 487 (1980)
10. D. Y. Chen, Y. A. Chang and D. Swenson, Appl. Phys. Lett. 68, 96 (1996)
11. C.-H. Jan, Ph.D. Thesis, University of Wisconsin-Madison, (1991)
12. Y. A. Chang, Metall. Mater. Trans. B 25, 789 (1994)
13. F. Ren, C. R. Abernathy, S. J. Pearton, and P. W. Wisk, Appl. Phys. Lett. 64, 1508 (1994)
14. D. Swenson, Ph.D. Thesis, University of Wisconsin-Madison, (1994)
15. O. Kubaschewski, C. B. Alcock and P. J. Spencer, Materials Thermochemistry, Pergamon Press, Oxford, UK (1993) 6th edition

WIDE BAND-GAP SEMICONDUCTORS FOR COLD CATHODES: A THEORETICAL ANALYSIS

PETER LERNER*, P. H. CUTLER , N. M. MISKOVSKY
Penn State University, Department of Physics, University Park, PA 16802
*pbl@hbar.phys.psu.edu

ABSTRACT

In this paper we describe the field emission from wide band-gap semiconductor thin film electron sources as a three-step process. Internal field emission is the mechanism for electron injection at the metal-semiconductor cathode interface. Under an internal field, electrons injected into the conduction band can propagate quasi-ballistically through the thin semiconductor film. At the vacuum interface, they are field emitted across a PEA or NEA surface. Consistent with the electron injection mechanism we have done molecular dynamics simulations for GaN films with an initial energy distribution corresponding to a Fowler-Nordheim (FN) spectrum. Results demonstrate quasi-ballistic propagation and approximate preservation of the FN energy distribution. Furthermore, high levels of n-doping in GaN ($\sim 10^{17}$cm^{-3}) do not inhibit transport in thin films (<0.1 μm).

INTRODUCTION

The use of wide band-gap semiconductors (e.g. diamond, AlN, GaN, BN, TiN, etc.) as copious electron sources is currently being studied [1, 2, 3]. In addition to their well-known physical and electrical properties, these materials have the unique advantage of exhibiting small positive electron affinities (PEA) or negative electron affinities (NEA) on some crystalline surfaces. If a mechanism can be found to inject or supply electrons into the conduction band of these wide band-gap materials, then high current emission at low power is attainable. In this paper we describe a three step model for emission from such composite metal-semiconductor devices: internal field emission at the Schottky barrier formed at the metal-semiconductor interface, field-assisted transport to the surface of the film and the electron emission at the vacuum-semiconductor surface, the exact nature of which depends upon the electron affinity at the surface.

Some differences between diamond and the nitrides are the following: For diamond, the NEA property of its (111) surface is well established (χ=-0.7eV), whereas NEA is surmised for AlN and BN but the precise numerical values for χ are unknown. Some nitrides, such as GaN and AlN, can be n-doped,with concentrations on the order of 10^{17}cm^{-3}, but similar doping is problematic for diamond [4].

INTERNAL FIELD EMISSION MODEL

One of the methods recently proposed by Geis *et al.* ([5], hereafter GTL; see also Ref. [6]) to create significant concentrations of conduction-band electrons is *internal field emission*. To utilize this mechanism for electron injection, GTL fabricated a Schottky barrier at a metal diamond cathode interface by heavily doping the semiconductor with substitutional nitrogen (N). The difference in valency between the nitrogen and carbon atoms is the source of the negative charge transferred to the metal, by tunneling, to achieve thermodynamic equilibrium of the Fermi levels. According to the simple model of Schottky barrier formation [7], the localized N-ions in the diamond give rise to the steep potential near the interface, i.e., the Schottky barrier. As of now we do not know, how much of the achieved doping levels (up to 0.3 % in experiments of Ref. [8]) is truly substitutional, i.e., preserves the crystalline symmetry of diamond. Some, or most of it, can be related to the charged defects in diamond. In the latter case, what we are referring to as "donor

Mat. Res. Soc. Symp. Proc. Vol. 449 © 1997 Materials Research Society

levels" may be the states inside the sub-band(s) produced by the defects due to the nitrogen implantation into diamond. The presence of large concentrations (in excess of 10^{19} cm^{-3}) of nitrogen doping bends the energy barrier upwards in the vicinity of the interface making electron injection into the conduction band possible, as described earlier [6]. GTL obtained low power levels (below 8 eV) for the electron emission of a diamond-based field emitting device with a roughened metal cathode. Currently, experiments are under way, which exploit Si-diamond junctions [8]. One can also employ other NEA wide band gap semiconductors, like the nitrides, for which true n-doping is possible (GaN, AlN, BN, etc.) [1-3].

Once an electron is in the conduction band and has traversed the width of the diamond film, it can "slip" into vacuum, without further acceleration by the field, if there is a negative electron affinity (NEA) surface. Quasi-ballistic electron transport through thin diamond films has been predicted by the authors [9], suggesting that the injection current (and its energy distribution) is sustained during transport to the vacuum interface. Field emission from the device can be further enhanced despite the high dielectric constant (ε= 5.7, 8.9, 8.5 for diamond, GaN and AlN, respectively), by creating artificial asperities on the metal which protrude into the film. The resulting concentration of the field lines around these "tips" enhances further the tunneling from the metal into the conduction band by barrier reduction and thinning. In principle, the barrier may be thinned to such an extent that it is possible to form an ohmic contact. In our earlier analysis, the current density of the whole device was found to be roughly proportional to the current density through the junction. Assuming that the injected distribution is not significantly changed by the transport process, we can use an expression identical to the conventional FN equation for the current through the interface [6,9].

The FN characteristics of the device as a whole are predicated on the following:
1)   A modified Fowler-Nordheim equation is a valid description of the tunneling through the Schottky barrier at the interface;
2)   Quasi-ballistic propagation so that every electron supplied from the metal into the conduction band of the film is presumed to be emitted into vacuum. Hence   the current transport through the metal-semiconductor interface is thus equated with the I-V characteristic of the whole device;
3)   Near zero-field conditions in the semiconductor outside of the depletion layer, so that little or no acceleration of electrons occurs on its path to the vacuum-semiconductor interface. The charge carriers in the conduction band "slip" out of the film because of the negative electron affinity at the surface.

We exhibit in Fig. 1 calculated results of internal field emission (in FN coordinates) from metal-diamond and metal-GaN thin film junctions. The somewhat higher threshold for internal field

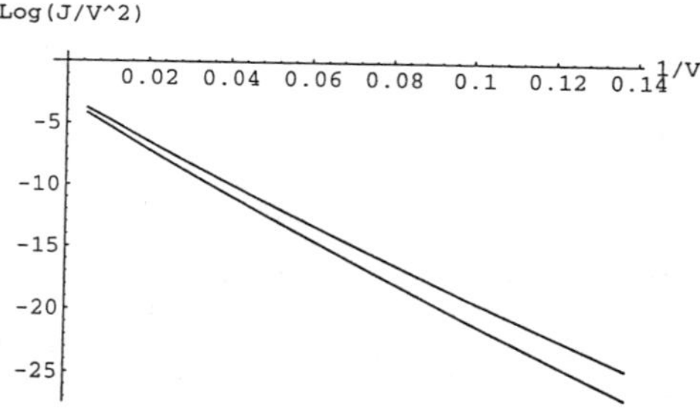

Fig. 1 FN plot for diamond (upper curve) and GaN (lower curve) for radius of curvature, R=10nm, N-doping concentration, N_d=10^{19} cm^{-3}, and Schottky barrier height, ϕ_{bn}=1.9 eV. The zero of energy corresponds to the bottom of the conduction band.

emission for GaN is due to its higher dielectric constant than that of the diamond. Existing experimental results for the undoped samples seem to confirm this tendency [2].

The threshold bias voltage for a current emission density of $j \sim 10^{-8}$ A/cm^2 from N-doped diamond is on the order of 8V. A comparable GaN device (eg, same doping, radius of curvature, etc.) would have a threshold bias voltage $V_b \sim 16V$. In the case of n-doping in GaN, the Fermi level is pinned to the donor level. When the applied bias is greater than the threshold, the injected electrons with energies much above the bottom of the conduction band in the bulk have an average energy, $< E >= e(V_b - V_a)$, where V_a, a contact potential on the internal interface, is less than 1 V. The final position of the peak in the energy distribution is shifted by a value of $\chi \leq 1 eV$. Although the field F in the depletion layer is $\geq$ 1000V/μm, the field in the bulk is much weaker, i.e., most of the potential drop (and electron acceleration) occurs in the vicinity of the metal-diamond interface[5].

MONTE CARLO SIMULATIONS OF TRANSPORT IN GaN

After the electron is in the conduction band, it propagates through the bulk of the film to the vacuum interface. We study this propagation using a Monte Carlo method, previously described in [10]. An approximate Fowler-Nordheim energy spectrum is used to describe the injected electrons:

$$P(\varepsilon) = J_0 \frac{\exp(c_0 \varepsilon)}{1 + \exp(\frac{\varepsilon}{k_B T})} \quad ,$$

where $\varepsilon = E - V_b$ and J_0 and c_0 are Fowler-Nordheim constants [11]. Assuming the field is in the (111) direction, as a first approximation, the electron momentum is assumed randomized around this direction. Simulations were performed for GaN with conduction electron density equal to 10^{17}cm^{-3} corresponding to the levels of unintentional n-doping.

Since electrons are injected into the film with energies of tens of eV above the bottom of the conduction band, their effective mass is different from the electron's effective mass at the bottom of the conduction band [10]. However, for high effective mass, i.e., $m* = m_e$, the free electron mass, the scattering rates for high-energy electrons grow inordinately high if one ignores the wave vector dependence of the deformation potential. Since there is, as yet, no consistent treatment of transport for such high energy electrons, we performed the Monte Carlo simulations for GaN both with the band-bottom effective mass (option I) and with free electron mass (option II). In the latter case we assumed optical phonon rate saturation for wave vectors larger than the size of an elementary cell in k-space: $k > k_{cr} = 2\pi/a_{cell}$ (option II). The resulting spectra for option II are shown in Fig. 2. We see that most of the electron energies are concentrated in a narrow range around the energy of injection, though there is a substantial broadening, which is much more pronounced in the case of "lighter" electrons (option I). For low fields (F~1V/μm), the electron spectrum is skewed to the lower energies (as in the initial FN spectrum) . In this regime, the energy gained by the electrons from the field is approximately equal to the scattering losses. For high fields there is a net gain of energy and a shift to higher average energies (F~100V/μm). We observe from Fig. 3 that, as expected, transport becomes more ballistic for the higher fields in accordance with the results of Ref. [12]. Emission of acoustic phonons dominates the scattering picture; the importance of intravalley optical phonons decreases with higher fields. This is confirmed by Fig. 4 where we plot the mean energy in (a) and the standard deviation (STD) in (b). In the case of option II, the spectra exhibit a monotonic decrease in the STD with increase of the field. This behavior persists at least for film thicknesses of 0.1μm and for the electron density $n_e \sim 10^{17}$cm^3 and confirms the quasi-ballistic transport.

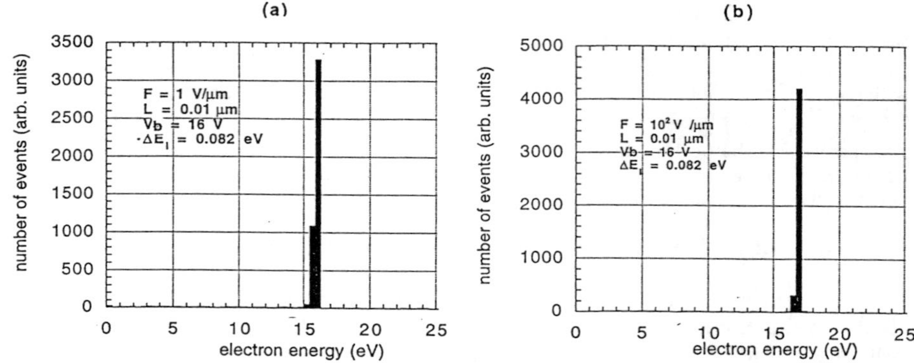

Fig. 2 Electron energy spectra for conduction-band transport of a FN distribution of electrons: (a) low field; (b) high field. The density of conduction band electrons is 10^{17} cm^{-3}.

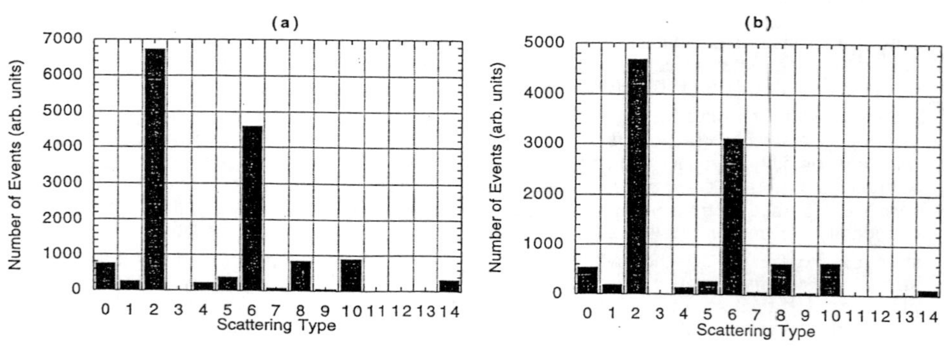

Fig. 3 The scattering statistics for conduction-band electrons. The conditions correspond to those in Figs. 2(a) and 2(b). Unless stated, all even(odd) processes refer to phonon emission(absorption). Process 0, elastic acoustic; 1 and 2 inelastic acoustic; 3 and 4 polar acoustic; 5 and 6 polar optical; 7-10 intervalley optical; 11, e-e scattering; 12, e-h scattering; 13 and 14 e-pl.

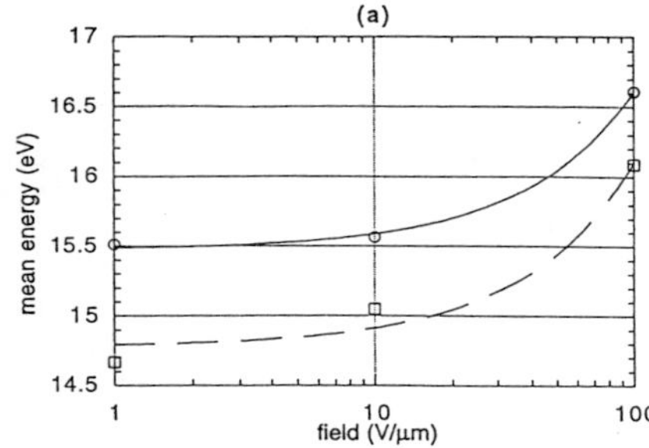
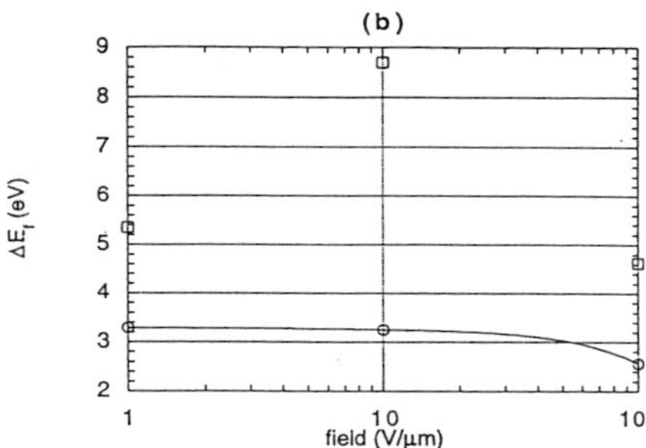

Fig. 4 (a) Mean energy and (b) standard deviation of energy as a function of electric field inside the semiconductor: option I (circles), option II(squares).

For the NEA surface all particles transported to the film-vacuum interface will be emitted with a small energy shift $\sim\chi$. Thus we can identify the spectra in Fig. 2 with the spectra of the field-emitted electrons from the whole device. We note that the vacuum energy spectrum still retains a narrow-line shape typical of the field emission from the surface of a metal. In conclusion, our simulations show that thin nitride films ($< 0.1\mu m$) sustain unimpeded transport of electrons through the film.

CONCLUSIONS

In this paper we describe a three step process for electron emission from wide-band gap semiconducting thin films. This process, originally suggested by Geis et al. [5] and developed by the authors [6,9], consists of: Electron injection via internal field emission at the metal semiconducting interface; Electron transport through the semiconducting film; Electron emission at the NEA or PEA semiconducting-vacuum interface. For diamond and GaN, the internal field emission current can be described by a FN type equation. Using a molecular dynamic approach (with a Monte Carlo algorithm) to model the transport, it was found that scattering reduces the average energy of the electrons by about 0.5 eV for very thin films (L=0.01 µm) which is compensated by the energy gained from the field during its transit to the surface. In addition, the narrow FN energy distributions characteristic of the internal field emission process are sustained for films with L$\leq$0.1µm and doping densities $\leq 10^{17}$ cm^{-3}.

ACKNOWLEDGEMENT

The work was supported in part by the Ballistic Missile Defense Organization administered by the Office of Naval Research through the ONR grant No.00014-95-0905.

REFERENCES

[1] R. W. Pryor, Mat. Res. Soc. Proc., 416, 425 (1996).

[2] V. V. Zhirnov, G. J. Wojak, W. B. Choi, J. J. Cuomo, J. J. Hren, NCSU preprint (1996).

[3] M. Endo, H. Nakane, H. Adachi, J. Vac. Sci. Technol. B14, 2114 (1996).

[4] G. Popovici, M. A. Prelas, in: M. A. Prelas, P. Gielisse, G. Popovici, B. V. Spitsyn, T. Stacy (eds.) *Wide band gap electronic materials,* (Kluwer, Dordrecht, 1995), p.1.

[5] M. W. Geis, J. W. Twichell, and T. M. Lyszczarz, J. Vac. Sci. Technol. B 14, 2060 (1996).

[6] P. Lerner, P. H. Cutler, and N.M. Miskovsky, submitted to JVST **B** (Nov. 1996)

[7] S. M. Sze, *Physics of Semiconductor Electronic Devices,* (John Wiley and Sons, New York, 1981).

[8] K. Okano, S. Koizumi, S. Ravi, P. Silva and G.A. Amaratunga, Nature **381**, 140(1996).

[9] P. Lerner, P. H. Cutler and N. M. Miskovsky, Journ. de Physique (in press).

[10] P. H. Cutler, Z.-H. Huang, N. M. Miskovsky, P. D' Ambrosio and M. Chung, J. Vac. Sci. Technol. **B14**, 2020(1996).

[11] A. Modinos, *Field, Thermionic and Secondary Electron Emission Spectroscopy,* (Plenum, New York, 1983).

[12] H. J. Fitting and A. Van Czarnovski, Phys. Stat. Sol. **A 93**, 385(1986).

Evolution of Ti Schottky barrier heights on n-type GaN with annealing

Michèle T. HIRSCH, Kristin J. DUXSTAD, and E. E. HALLER
Department of Materials Science and Mineral Engineering, University of California at Berkeley and Lawrence Berkeley National Laboratory, Berkeley, California 94720, USA

ABSTRACT

We report the effect of mild annealing on Ti Schottky diodes on n-type GaN. The Ti films were deposited by electron beam evaporation on n-type GaN grown by metal organic vapor deposition. We determine the effective barrier height Φ_{b0} by current–voltage measurements as a function of temperature. The as–deposited Ti contacts show rectifying behavior with low barrier heights $\Phi_{b0} \leq 200$meV. At annealing temperatures as low as 60°C we observe an increase of the barrier height to values of 250meV. After annealing at 230°C and above a stable barrier height of 450meV is measured. The increase in barrier height is not due to any macroscopic interfacial reaction. The origin of the observed changes are discussed in terms of the Schottky–Mott model and possible microscopic interfacial reactions.

INTRODUCTION

Extensive research and development of GaN have recently made possible the realization of new GaN based opto–electronic devices and high temperature electronics. Metal contacts are typically an essential ingredient in the technology of such devices. The formation of both Ohmic [1, 2] and rectifying Schottky contacts [3, 4, 5, 6] are, therefore, central issues in current GaN research. Titanium is of special interest since it has been observed to act as an Ohmic contact on 10^{17}cm^{-3} n-type GaN after high temperature rapid thermal annealing (RTA) [1] or when the surface is exposed to reactive ion etching prior to metal deposition [2]. However, Ti deposited by electron-beam evaporation on chemically prepared n-type GaN shows rectifying behavior [2, 3]. Binari et al. reported a flat band barrier height of $\Phi_{bf} = 0.58$eV for as deposited Ti contacts on 3×10^{17}cm^{-3} n-type GaN, restricting their measurements to room temperature.

Interfacial reactions have been proposed to be responsible for these observed changes from rectifying to low resistivity ohmic contacts [1, 2]. To obtain a better understanding of the relevant processes, it is of interest to study the evolution of barrier height with low temperature annealing. Schmitz et al. reported an increase in Φ_{b0} with annealing of Pd/GaN contacts above 300°C [7]. Since Ti is a highly reactive metal and a strong nitride former, annealing effects can be expected to be of importance for the Ti/GaN system.

From a more fundamental point of view, Ti is also of interest as a representative of low work function metals for the study of the correlation of metal work function and barrier height. This correlation is expected to be strong for the ionic semiconductor GaN [8]. In this paper we report on the Schottky barrier behavior and the thermal stability of Ti on n-type GaN.

Mat. Res. Soc. Symp. Proc. Vol. 449 © 1997 Materials Research Society

EXPERIMENTAL

The Ti contacts in this study were deposited on unintentionally doped n-type GaN epi–layers on sapphire provided by Hewlett-Packard Opto–Electronic Division and Honeywell Inc. All GaN layers were grown by metal-organic vapor phase epitaxy (MOVPE) on a sapphire substrate. The layer thickness was 1–2 μm. Hall effect measurements indicate doping levels in the mid $10^{17}cm^{-3}$ range and room-temperature mobilities between $240cm^2/Vs$ and $355cm^2/Vs$. Prior to metallization, the samples were degreased in hot xylenes, acetone and methanol. Subsequently, they were etched for 5 min. in buffered oxide etch (BOE— $NH_3F:HF, 6:1$), diluted with H_2O (BOE:H_2O, 1:10), rinsed in DI water and finally blown dry with N_2. A Ti layer of 1500Åwas deposited by electron beam evaporation through a shadow mask in a vacuum better than 5×10^{-6}Torr. The contact areas ranged from 0.92 to 2.92 $\times10^{-3}cm^2$. Some Ti contacts were covered by a Au capping layer without breaking vacuum to prevent oxygen incorporation into the Ti films. Large area Ohmic contacts were formed by evaporation of Al. Annealing of the samples was performed either in a tube furnace in flowing forming gas $N_2:H_2$ (24:1) or in a vacuum of 60mTorr. The vacuum annealing allowed in situ current–voltage measurements. Annealing temperatures ranged from 60 to 350°C and times varied between 20 minutes and 3 hours.

RESULTS

The characteristics of the resulting diodes were investigated by current–voltage (I-V) measurements. The as-deposited Ti yields rather leaky diodes with reverse bias currents in the range of mA for voltages of -2V. Series resistance effects were clearly visible under forward biasing. Series resistance effects and high conductance have significant influence on capacitance voltage (C-V) measurements even under reverse bias condition. They do not allow a reliable determination of the built-in voltage and hence the barrier heights of the diodes from the measured C-V spectra. This led us to restrict ourselves to current–voltage experiments.

For a succesful study of the forward I-V characteristic of our diodes, the determination of the series resistance R_s is crucial. We used the method of Werner [9] to determine R_s and used the series resistance corrected voltage $V' = V - IR_s$ for further investigations. The diode characteristic according to the thermionic emission theory is then described by

$$I = I_0(\exp(\frac{qV'}{nkT}) - 1) \quad \text{with} \quad I_0 = AA^{**}T^2\exp(-\frac{\Phi_{b0}}{kT}) \tag{1}$$

where n is the ideality factor of the diode, T the temperature, k the Boltzmann factor, A the device area, A^{**} the effective Richardson constant, I_0 the saturation current, and Φ_{b0} the effective barrier height.

The rectifying behavior of the Ti contacts improved significantly upon annealing. Room-temperature I-V curves after sequential annealing steps are depicted in Fig. 1. For the evaluation of I_0 and n we employed a plot of $\ln(I/\{1 - \exp(-qV'/kT)\})$ vs. V' which allows us to include biases as small as $3kT/q$ [10].

Temperature dependent I-V measurements (I-V-T) were used to determine Φ_{b0} and A^{**} of the contacts for each annealing step. Samples were annealed at 60°C, 120°C, and 230°C. I-V-T measurements were performed up to the respective annealing temperature. Figure 2 shows an activation energy plot of $\ln(I_0/T^2)$ vs. $1/T$ for as-deposited and annealed contacts. The corresponding effective barrier heights are determined from the slope of these plots

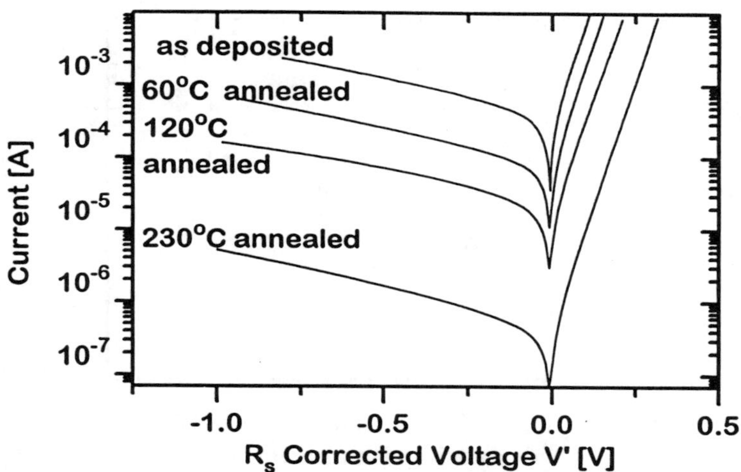

Figure 1: Room temperature I-V characteristics for Ti contacts ($A = 2.45 \times 10^{-3}$ cm^2) as deposited and after annealing at 60°C , 120°C , and 230°C shift to lower currents with increasing annealing temperature. The saturation currents decrease by a factor of ~ 1000, whereas the ideality factors remain constant at a value of $n \sim 1.29$: $I_0 = 3.4 \times 10^{-4}$A, $n = 1.29$ (as-deposited), $I_0 = 9.57 \times 10^{-5}$A, $n = 1.25$ (60°C annealed), $I_0 = 1.96 \times 10^{-5}$A, $n = 1.32$ (120°C annealed), and $I_0 = 4.87 \times 10^{-7}$A, $n = 1.29$ (230°C annealed).

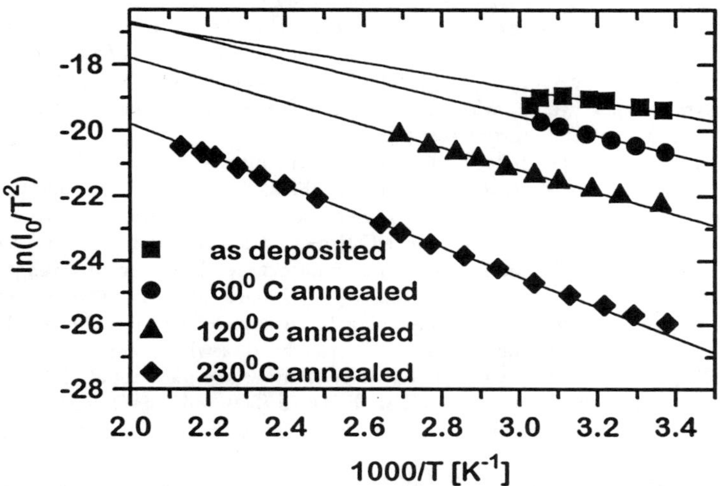

Figure 2: Activation energy plot for Ti contacts shows an increase in slope (i.e., increasing barrier height) but a nearly constant y-axis intercept (i.e., constant area × Richardson constant) after annealing treatments.

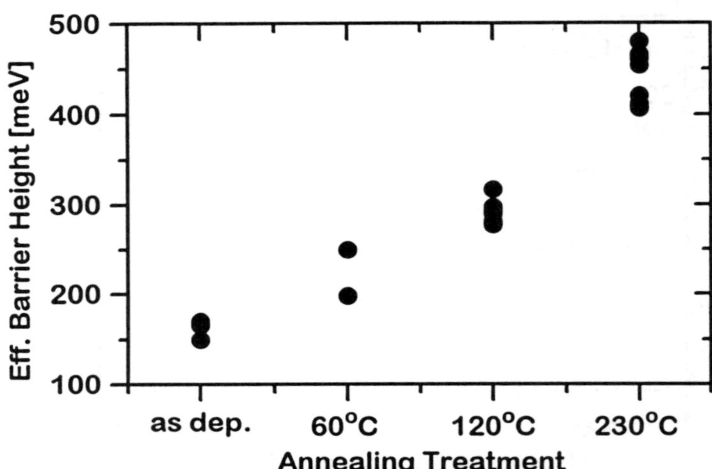

Figure 3: Evolution of Ti barrier heights with different annealing steps. Each filled circle corresponds to the barrier height determined from I-V-T measurements of one specific sample after the respective annealing step. The Ti Schottky diodes were formed on HP and on Honeywell GaN films.

where the ideality factor becomes close to unity (n $\leq$ 1.15) and constant with temperature, i.e., where thermionic emission should be the predominant transport mechanism. For Φ_{b0} we find 250meV (60°C), 294meV (120°C) and 407meV (230°C). The corresponding effective Richardson constants times the diode area $A^{**}A$ remain nearly constant and A^{**} lies around a value of $1\times10^{-2}\mathrm{Acm}^{-2}\mathrm{K}^{-2}$ for all three annealing steps. This value is small compared to the theoretical value of $26\mathrm{Acm}^{-2}\mathrm{K}^{-2}$ [4], but of the same order of magnitude as those reported by several authors [4, 5], who attribute this discrepancy to an additional thin barrier through which electrons must tunnel [4].

The observed changes in our measured I-V-curves are mainly due to a change in the effective barrier height for the electron transport. For the as-deposited Ti contacts there is significant uncertainty in the extrapolation of I_0 because of a pronounced series resistance influence. Additionally, the temperature induced changes of the Schottky barriers at temperatures as low as 60°C prevent measurements over an extended temperature range. We can only estimate the barrier height of as-deposited Ti to lie near 200meV. The determined barrier heights versus the corresponding annealing temperatures for several contacts are shown in Fig. 3. We observe an increase of the barrier heights with increasing annealing temperatures. The barrier height, averaged for more than 10 samples, equals 447meV for the 230°C annealing step. Higher annealing temperatures (300 and 350°C) resulted in no further increase in the barrier height but in some cases led to an increase of the ideality factors at room temperature.

Performing a rapid thermal anealing step of 30 seconds at 900 °C in nitrogen atmosphere similar to the treatment described by Lin et al. [1] results in nearly linear, i.e., ohmic I-V behavior of our contacts. Sputtered Ti contacts also exhibit nearly ohmic behavior after deposition. Mild annealing of these sputtered Ti contacts led in most cases again to more rectifying I-V charactersistics. In some cases, however, the I-V curves did not undergo significant annealing induced changes. We attribute this observed behavior to stronger surface damage due to the sputtering than during e–beam evaporation.

DISCUSSION

The change in barrier height with annealing suggests that structural changes occur at the interface. Rutherford backscattering and x-ray diffraction experiments, however, indicate no *macroscopic* interfacial reaction, i.e. phase formation on a scale larger than 50 Å, between the Ti film and GaN after annealing up to 230°C. Changes during annealing at such low temperatures may happen on a more microscopic scale.

An interfacial layer of a few monolayers of oxide and/or water at the Ti/GaN interface may be consumed by the metal, leading to a more intimate and therefore possibly more ideal contact. For those contacts on GaN one may expect the Schottky–Mott model to be valid, where the difference between the Ti work function (4.3eV [10]) and the GaN electron affinity ($\sim$ 4.06eV [6]) determine the barrier height. The observed barrier heights after the 60 and 120 °C annealing step of 220 to 300meV may support the Schottky–Mott theory, which is expected to be applicable because of the high ionicity of GaN [8, 11].

However, the barrier heights increase with higher annealing temperatures and, therefore, deviate from that simple relationship. It is also possible that a reaction takes place on a microscopic scale resulting in the formation of new phases at the interface. The most likely candidate as a reaction product at the interface is expected to be TiN. The formation of a cubic TiN phase has been observed in Ti/Al bilayer and Ti/Al/Ni/Au multilayer ohmic contacts treated with RIE before and RTA after deposition [12]. The observed ohmic behavior of these contacts is explained in terms of surface damage due to the RIE and/or the possible accumulation of N vacancies in the GaN due to the outdiffusion of nitrogen from the GaN both leading to a heavily doped n–type interface layer [1, 2]. If the surface damage is the predominant mechanism leading to ohmic contact behavior, which is backed by our results for the sputtered Ti contacts, TiN may, however, also be responsible for the observed behavior of our Schottky contacts. If a very thin layer of TiN is formed at our contacts, which should have a different work function than Ti, it may be responsible for the increase of the barrier heights to values higher than those predicted by the Schottky–Mott model.

Furthermore, we consider the possible formation of bandgap states at the interface, e.g. metal induced gap states (MIGS), during annealing when, for example, an interfacial layer between the Ti and the GaN is consumed by the Ti. Bandgap states at the interface influence the position of the Fermi level depending on the neutrality level of the gap state density and may again lead to a barrier higher than expected from the pure Schottky–Mott rule [13].

Further studies are necessary to lead to full understanding of the underlying mechanisms involved in the observed changes in the Schottky barrier heights. High resolution transmission electron microscopy studies may detect an interfacial layer. Such studies are in progress. Measurement of barrier heights between GaN and a wider variety of metals, especially nitride forming metals, should also help to understand which model of barrier height formation might be appropriate.

CONCLUSIONS

In conclusion, the influence of annealing on the electrical characteristics of Ti contacts to n-type GaN has been investigated. We find an increase of the barrier height with increasing annealing temperatures. The lower $\Phi_{b0} \approx$250–300meV is described by the simple Schottky–Mott model of Φ_{b0} being equal to the difference in the work function and electron affinity of Ti and GaN. Annealing increases Φ_{b0} to values of $\sim$450meV, as measured by the I-V-T-method. The reason for this increase is not known at this time, but is not due to any macroscopic interfacial reaction between the Ti and the GaN. The formation of a microscopic layer of TiN at the interface or the introduction of states in the GaN near the Ti and GaN interface are possible candidates for an explanation of our results.

ACKNOWLEDGMENTS

We wish to acknowledge W. R. Imler and D. A. Steigerwald at Hewlett-Packard Opto-Electronic Division and W. Yang at Honeywell for providing the GaN samples. M. T. H. thanks the Deutsche Forschungsgemeinschaft for a research grant. This work was supported by the Director, Office of Energy Research, Office of Basic Energy Sciences, Materials Sciences Division of the U. S. Department of Energy under Contract No. DE-AC03-76SF00098.

REFERENCES

[1] M. E. Lin, Z. Ma, F. Y. Huang, Z. F.Fan, L. H. Allen and H. Morkoc, Appl. Phys. Lett. **64**, 1003 (1994).

[2] Z. Fan, S. N. Mohammad, W. Kim, Ö. Aktas, A. E. Botchkarev, and H. Morkoc, Appl. Phys. Lett. **68**, 1672 (1996).

[3] S. C. Binari, H. B. Dietrich, G. Kelner, L. B. Rowland, K. Doverspike, and D. K. Gaskill, Electron. Lett. **30**, 909 (1994).

[4] P. Hacke, T. Detchprohm, K. Hiramatsu, and N. Sawaki, Appl. Phys. Lett. **63**, 2676 (1993).

[5] L. Wang, M. I. Nathan, T.-H. Lim, M. A. Khan and Q. Chen, Appl. Phys. Lett. **68**, 1267 (1996)

[6] E. V. Kalinina, N. Kuzenetsov, V. A. Dmitriev, and K. G. Irvine and C. H. Carter, JR., Journ. of Electron. Mat. **25** 831 (1996).

[7] A. C. Schmitz, A. T. Ping, M. A. Khan, Q. Chen, J. W. Yang, I. Adesida, Electron. Lett. (accepted for publication).

[8] J. S. Foresi and T. D. Moustakas, Appl. Phys. Lett. **62**, 2859 (1993).

[9] J. H. Werner, Appl. Phys. A **47**, 291-294 (1988).

[10] E. H. Rhoderick and R. H. Williams in *Metal-Semiconductor Contacts*, (Clarendon, Oxford, 1988).

[11] S. Kurtin, T. C. McGill and C. A. Mead, Phys. Rev. Lett. **22**, 1433, (1963).

[12] S. Ruvimov, Z. Liliental–Weber, J. Washburn, K. J. Duxstad, E. E. Haller, Z.-F. Fan, S. N. Mohammad, W. Kim, A. E. Botchkarev, and H. Morkoc, Appl. Phys. Lett. **69**, 1556, (1996).

[13] W. Mönch in *Semiconductor Surfaces and Interfaces*, pp. 74, (Springer, Berlin, 1993).

NITRIDE BASED THIN FILM COLD CATHODE EMITTERS

JAMES A. CHRISTMAN*, ANDREW T. SOWERS*, MICHAEL D.
BREMSER**, BRANDON L. WARD*, ROBERT F. DAVIS**,
AND ROBERT J. NEMANICH*
*Department of Physics, North Carolina State University, Raleigh, NC 27695
**Department of Materials Science and Engineering, North Carolina State
University, Raleigh, NC 27695

ABSTRACT

Cold cathodes have been fabricated using two different nitride structures as a thin film emitting layer. The AlN and graded AlGaN structures are prepared by metalorganic chemical vapor deposition (MOCVD) on an n-type 6H-SiC substrate. Individual aluminum grids are perforated with an array of either 1, 3, or 5μm holes through which the emitting surface is exposed. After device fabrication, a hydrogen plasma exposure was found to be necessary to activate the cathode. The devices have displayed a limited lifetime and a small percentage of the devices operate, although half of the devices with 5μm holes functioned. The highest measured collector currents are 0.1μA for AlN and 10nA for AlGaN at grid voltages of 110V and 20V, respectively. The grid currents are typically 10 to 10^4 times the collector currents.

INTRODUCTION

The electrical properties of wide bandgap semiconductors in combination with their chemical stability make them candidates for use in flat panel displays and high power, high frequency devices. Wide bandgap semiconductors such as diamond [1], AlN [2], and $Al_xGa_{1-x}N$ for $x \geq .75$ [3] show promise for use as cold cathode materials since they each exhibit a negative electron affinity (NEA). At the surface of an NEA material, the vacuum level lies below the conduction band minimum; therefore, electrons in the conduction band and near the semiconductor surface can be emitted into vacuum.

Two different nitride emitting layers were used in these experiments. The first is a 1000Å AlN layer and the second is graded from n+ GaN to $Al_{0.9}Ga_{0.1}N$. In each of these films, a thin (~1000Å) AlN buffer layer lies between the conducting SiC substrate and the emitting layer. The simplified band diagrams for these two structures are shown in Figure 1. A major device issue is how to populate the conduction band at the surface with electrons. These structures employ two similar approaches to this problem. For the AlN structure, if the AlN layer is thin enough, the application of an electric field will increase electron tunneling across the AlN and into vacuum. In the graded AlGaN structure, electrons must first tunnel through the AlN barrier (the buffer layer) from the conducting SiC to the GaN. Then the electrons move into the AlGaN layer and some are swept into vacuum because of the applied grid voltage. An additional barrier would exist at the surface of a positive electron affinity material.

Most wide bandgap field emission research has been dedicated to depositing diamond on silicon field emitting tips [4, 5] or studying diamond films themselves [6, 7]. Our approach is to use a planar nitride surface as the emitter, rather than a structure deliberately exploiting field enhancement at a sharp projection. This approach is similar to recently fabricated diamond cathodes[8]. The nitride cold cathode design is shown in Figure 2. An aluminum grid lies on an

Mat. Res. Soc. Symp. Proc. Vol. 449 © 1997 Materials Research Society

SiO$_2$ layer which separates the grid from the nitride emission layer. An array of square emission holes is etched through the aluminum and SiO$_2$ to the nitride layer. At the emission hole bottom, there is a vacuum-nitride interface across which electrons tunnel with the application of the grid voltage (V$_g$) between the aluminum pad and the backside contact.

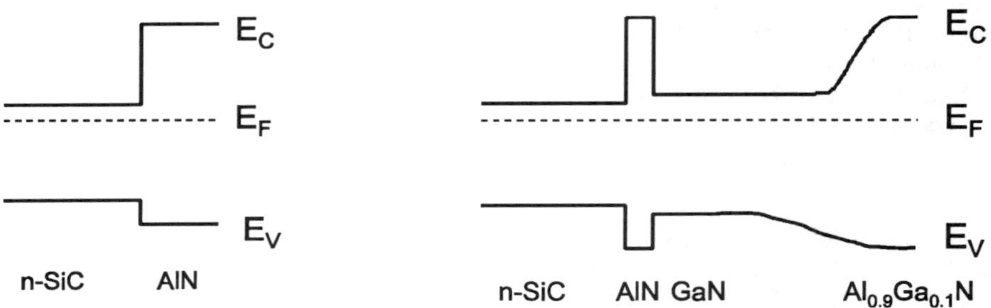

Figure 1. Band diagrams of AlN/SiC and graded AlGaN/SiC structures.

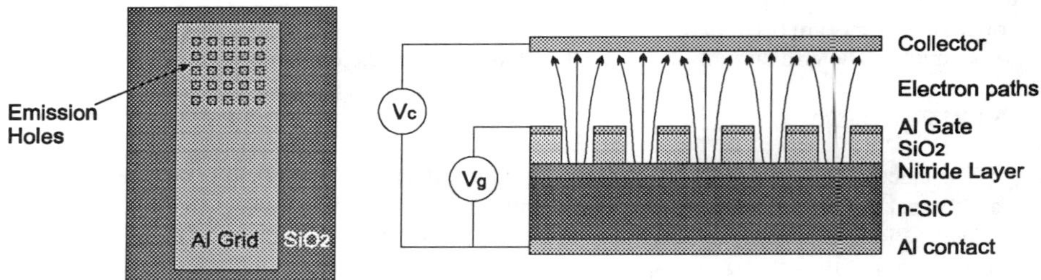

Figure 2. Cold cathode design. The top view (left) shows a 5 x 5 array of emission holes. The Al grid is ½ x 2mm and the emission holes are either 1, 3, or 5µm. The nitride layer is exposed through the holes. The cross section (right) is across the emission holes.

EXPERIMENT

<u>Cold Cathode Fabrication</u>

Fabrication of the cold cathodes is accomplished with a two mask process. The first mask creates the emission holes and the second forms the aluminum grid pads for electrical probing. Prior to patterning, the nitride emitting layer is deposited on a 6H-SiC substrate using the MOCVD system described elsewhere [9, 10]. Two different nitride emitting layers were used in these experiments. The first is a 1000Å AlN layer and the second is graded from n+ GaN:Si (n > 1 x 10^{19} cm^{-3}) to Al$_{0.9}$Ga$_{0.1}$N:Si. As shown by atomic force microscopy, the nitride films are smooth with root mean square roughnesses of ~20Å on a 1 x 1µm scan.

SiO$_2$ is used as the dielectric separating the aluminum grid from the emitting layer. A one micron SiO$_2$ layer is deposited on the nitride layer at 400°C. The deposition system is a horizontal, low pressure CVD system using diethylsilane and oxygen precursors. Finally, 200-300nm of aluminum is thermally evaporated onto the oxide.

The first mask step creates the emission holes which are either 1, 3, or 5μm squares. The aluminum is patterned with a standard aluminum etch and the oxide is reactive ion etched (RIE) to achieve high aspect ratio features. A mixture of SF_6 and O_2 at 30mTorr is used because of the high etch rate. Since the RIE environment may damage the emitting surface, a wet oxide etch is used to etch the last tenth micron of oxide down to the nitride layer. The second mask step defines the ½ x 2mm metal pads that form the grids which will be electrically probed. Standard oxygen descum procedures to remove residual photoresist are not performed since this may damage the nitride emitting layer. The last processing step is thermal evaporation of 200-300nm of aluminum on the backside of the SiC for an electrical contact.

Cold Cathode Electrical Testing System

The electrical testing system is a stainless steel 4-1/2" six-way ultra-high vacuum cross with two electrical probes. One probe is used to make contact to the grid and the other probe is used to collect the emission current ~1mm above the holes. The device wafer is held down onto a stainless steel platform which forms the backside electrical contact. The electrical probes are connected to wobblestick feedthroughs that can be finely controlled with two external XYZ stages. With the aid of a microscope, individual devices on a wafer are probed in vacuum. Two computer controlled Keithley source measure units (SMUs) apply the grid and collection voltages to a device. The grid voltage can be varied from 0-110V and the collector voltage from 0-1100V. The pumping system is oil free and testing is performed at $< 5 \times 10^{-7}$ Torr as measured by a cold cathode gauge.

RESULTS

Cathode Testing

The nitride cathodes were electrically tested directly after processing and no collector current above the noise level was measured. It was determined that a post processing clean is necessary to activate emission from the nitride cathodes. Since the aluminum grids are exposed, we are limited to cleans which will not etch away the grids. The samples were cleaned ultrasonically in methanol for 10 min. and then subjected to a remote hydrogen plasma clean at 25mTorr and 450°C for 10 min. A hydrogen plasma exposure with these parameters has been shown to remove hydrocarbons from AlN and GaN surfaces [11]. Also, a hydrogen plasma will remove some residual photoresist. After plasma exposure, the sample is immediately transported in air to the electrical testing system.

Several AlN and AlGaN cathodes have had collector currents well above the current background level. None of the cathodes with 1 and 3μm holes functioned. Only cathodes with 5μm emission holes operated and the cathode lifetimes varied from minutes to half an hour. For all measurements shown in this paper, the grid voltage was held constant and the collection voltage was varied. For most of the data shown, the changing collector voltage is not as important as the passage of time. At a constant grid voltage the varying collector current is due to cathode instability rather than to the changing collector voltage.

Emitter structures without emission holes were also fabricated to test the SiO_2 properties. These test structures are on the same wafer as the cathodes, and therefore undergo the exact same processing and plasma treatments. The oxide breakdown voltage for these structures was found to be >800V which is much higher than the grid voltages of 0-110V used during normal cathode testing. No collector currents above the noise level have been measured for the structures without emission holes.

Figure 3 shows the data for an AlN cathode that operated for about 30min. This cathode is a 5 x 5 array of 5µm square holes (see Figure 1). The center-to-center spacing between holes is 75µm. Collector current data is shown for four different grid voltages. Each curve represents a constant grid voltage, and as expected, the collector current increases with increasing grid voltage. The V_g=0 and 110V measurements were repeated several times and all data was consistent with the data shown. Also, when the grid voltage was turned off in the middle of a scan, the collector current would drop to the V_g=0 level. The ratio of the grid current (I_g) to the collector current (I_c) varied from 1 - 100 for this cathode.

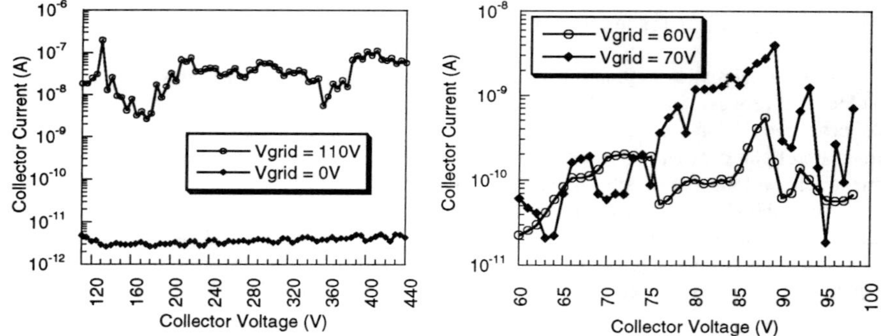

Figure 3. Collector current measurements for an AlN cathode that is a 5 x 5 array of 5µm square holes spaced by 75µm. The grid currents for V_g = 60, 70, 110V are I_g = 2.3 - 2.8nA, 4 - 13nA, 0.08 - 0.8µA, respectively.

Figure 4 shows collector current data for an AlGaN cathode that is also a 5 x 5 array of 5µm square holes spaced by 50µm. During the 30V measurement, the cathode failed. This cathode had the largest collector current of the AlGaN cathodes measured ($I_c \cong$ 10nA). Other AlGaN cathodes had longer lifetimes and withstood testing at higher grid voltages. Grid currents for the AlGaN devices were 10^3 to 10^4 times the collector current. Figure 5 shows data for a different AlN cathode that is a 5 x 5 array of 5µm square holes spaced by 75µm. Below V_c = 137V, when the device was operating, the grid current was low (~1µA). The grid current increased to 10mA when the device turned off (V_c > 137V). This plot also demonstrates the sometimes sudden extinction of emission current.

<u>Current Density Calculations</u>

The current densities (J) can be estimated assuming that all cathode pixels are activated and emitting uniformly. Using the AlN V_g = 110V data from Figure 3 gives J $\cong$ 0.01 Acm^{-2}. The V_g = 20V AlGaN data from Figure 4 yields J $\cong$ 0.001 Acm^{-2}. These current densities are minimum values because the emission area is probably smaller than the exposed nitride area.

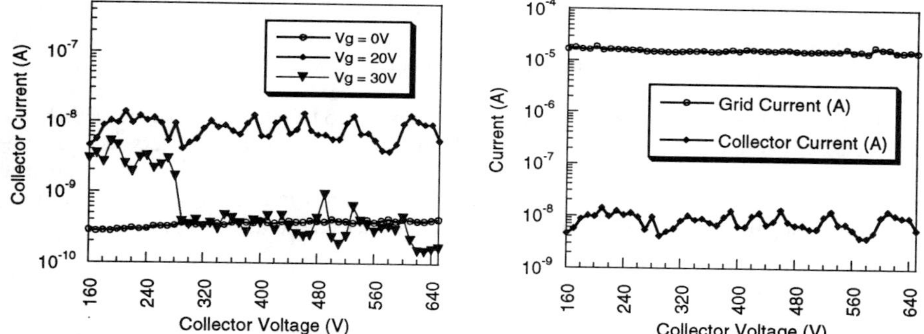

Figure 4. Current measurements for a graded AlGaN cathode that is a 5 x 5 array of 5μm square holes spaced by 50μm. The right plot shows the grid and collector currents for V_g = 20V. The cathode failed during the V_g = 30V measurement.

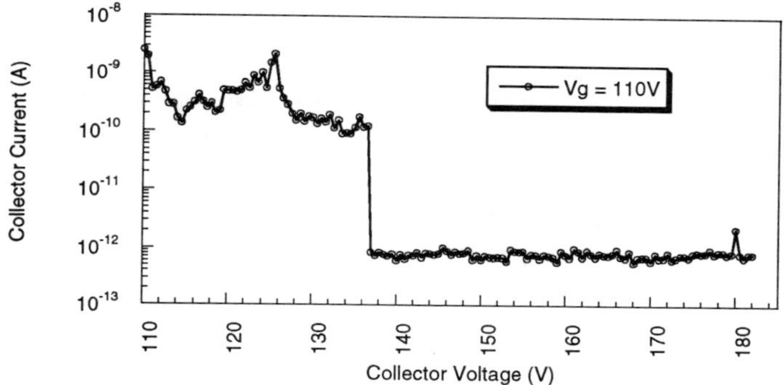

Figure 5. Collector current measurements for an AlN cathode that is a 5 x 5 array of 5μm square holes spaced by 75μm.

DISCUSSION

The cathodes that functioned followed identical patterns during testing. The grid current would be high, either 10 or 100mA, depending on the SMU compliance value. Then the grid current would drop and a collector current could be measured from several minutes to half an hour. Observing the cathodes after testing revealed melting of the aluminum grid. Melting was observed at the point of contact between the aluminum pad and the grid probe. Also, the emission holes would be enlarged and rounded, and sometimes more severely damaged. Current is either flowing along the sidewalls of the emission holes, or electrons emitted from the nitride surface are colliding with the oxide and aluminum grid. We suspect that after the many device processing steps, a conducting residue coats the sides of the emission holes. This creates a short between the grid and the nitride layer. During device testing, a high current flows through the residue and eventually it decomposes and also melts areas of the grid. After this decomposition occurs, an electric field can build up at the nitride surface and electrons are emitted. We also suspect that the devices with 1 and 3μm holes do not function because of the many processing steps. There may be residual photoresist in these smaller holes since oxygen descums are not performed.

In an effort to solve these problems, in the next devices to be fabricated, the emission holes will be etched after the ½ x 2mm aluminum pads are created. In this new process sequence, the emission holes are never exposed to photoresist. These new devices will also have top contact to the n+ GaN layer; thus circumventing the resistive AlN buffer layer. Also, conducting buffer layers are being developed which will be used with the backside contact design.

CONCLUSIONS

Cold cathodes have been successfully fabricated using nitride structures as the emitting surface. Although most cathodes did not function, a high percentage of cathodes with 5μm holes operated from several minutes to half an hour. The grid current is much higher than the collector current for both the AlN and graded AlGaN cathodes. The grid current to collector current ratio for the AlN cathodes is 1-100, and for the AlGaN cathodes it is 10^3 - 10^4. We suspect that many of the limitations in this study are related to residuals after the fabrication process.

ACKNOWLEDGEMENTS

We acknowledge the Advanced Electronic Materials Processing (AEMP) laboratory staff for help with the lithography and processing necessary to fabricate the cathodes. Field emission measurements performed by Bill Partlow of Northrup-Grumman were helpful. We also acknowledge Chris Hattfield and Griff Bilbro for device simulations. This research is supported by the Office of Naval Research and the Northrup-Grumman Science and Technology Center.

REFERENCES

1. J. van der Weide, R. J. Nemanich, *Phys. Rev. B* **49**, 13629 (1994).
2. M. C. Benjamin, C. Wang, R. F. Davis, R. J. Nemanich, *Appl. Phys. Lett* **64**, 3288 (1994).
3. M. C. Benjamin, M. D. Bremser, J. T. W. Weeks, S. W. King, R. F. Davis, R. J. Nemanich, *to be published in Surf. Sci. Reports ICFSCI-5 Proceedings* (1996).
4. E. I. Givargizov, *J. Vac. Sci. Technol. B* **13**, 414-417 (1995).
5. J. Liu, V. V. Zhirnov, G. J. Wojak, A. F. Myers, W. B. Choi, J. J. Hren, S. D. Wolter, M. T. McClure, B. R. Stoner, J. T. Glass, *Appl. Phys. Lett.* **65**, 2842-2844 (1994).
6. N. S. Xu, Y. Tzeng, R. V. Latham, *J. Phys. D: Appl. Phys.* **27**, 1988-1990 (1994).
7. M. W. Geis, J. C. Twichell, *Appl. Phys. Lett.* **67**, 1-4 (1995).
8. M. W. Geis, J. C. Twichell, N. N. Efremow, K. E. Krohn, C. Marchi, T. M. Lyszczarz, *Proceedings of the 8th International Vacuum Microelectronics Conference,* **p. 277** *(1995)*.
9. M. D. Bremser, W. G. Perry, N. V. Edwards, T. Zheleva, N. Parikh, D. E. Aspnes, R. F. Davis, *Mater. Res. Cos. Symp. Proc.* **395**, 195 (1996).
10. M. D. Bremser, W. G. Perry, T. Zheleva, N. V. Edwards, O. H. Nam, N. Parikh, D. E. Aspnes, R. F. Davis, *MRS Internet J. Nitride Semicond. Res.* **1**, 8 (1996).
11. S. W. King, L. L. Smith, J. P. Barnak, J. Ku, J. A. Christman, M. C. Benjamin, M. D. Bremser, R. J. Nemanich, and R. F. Davis, in <u>GaN and Related Materials</u> edited by F. A. Ponce, R. D. Dupuis, S. Nakamura, and J. A. Edmond (Mater. Res. Soc. Proc. 395, Pittsburgh, PA, 1996) pp. 739-744.

ELECTRON EMISSION FROM COLD CATHODES

R. W. PRYOR*, LIHUA LI* AND H. H. BUSTA[†]
*Institute for Manufacturing Research, Wayne State University, Detroit , Michigan 48201
rpryor@hal.physics.wayne.edu
[†]David Sarnoff Research Center, Princeton NJ, 08543
hbusta@sarnoff.com

ABSTRACT

New observations are presented on the emission of electrons from n-type polycrystalline diamond and n-type boron nitride (BN) cold cathode films, both as synthesized and after post-synthesis annealing. The films have been observed to show an increase in electron emission after annealing by one to several orders of magnitude, depending upon the type of emitter and the specific surface treatment. Observations from both plasma and laser annealing treatments will be presented.

The annealed BN cold cathodes have been observed to yield stable emission currents as high as 2 A cm^{-2} (~4 mA total current) at fields as low as ~30 V μm^{-1}. This is believed to be the highest reported current density yet observed from a planar film cold cathode emitter.

INTRODUCTION

The electron emission studies of solids have generated an extensive literature. The presentation of typical examples of studies relating to that area and the associated principles can be found elsewhere [1,2]. Those studies include observations in the areas of thermionic, field and secondary electron emission. Typical electron emitting materials used as cathodes require relatively high temperatures [3] (from ~1000 °K to ~2600 °K) and/or relatively high applied fields (~10^3 to 10^4 V / μm) to obtain adequate electron emission per unit area for use in modern high power device structures. The development goal for this cold cathode materials research is to obtain the highest room temperature electron emission density, with the least energy expenditure and with a minimum of post-synthesis processing.

The initial results reported [4] earlier on the electron emission from these new nitride-based cold cathode materials and n-type polycrystalline diamond indicated the potential that these materials have as low temperature electron emitters. This paper presents further observations on the emission of electrons from the n-type boron nitride-based (BN) group of cold cathode materials [5] and also presents some results from n-type polycrystalline diamond for comparison. The nitride-based cold cathode materials were synthesized by two different processes; a reactive laser ablation process [6] and a reactive magnetron sputtering process [7]. The n-type polycrystalline diamond was synthesized in the Diamond Laboratory at Wayne State University using a microwave plasma enhanced chemical vapor deposition (MPECVD) technique [8].

Mat. Res. Soc. Symp. Proc. Vol. 449 © 1997 Materials Research Society

EXPERIMENT

<u>Cold Cathode Film Synthesis: Reactive Laser Ablation (RLA)</u>

The first method used to synthesize the polycrystalline BN films on MPECVD diamond on Si was a reactive, biased, laser ablation technique (Fig. 1), discussed in detail elsewhere [9]. The n-type diamond coated substrates were loaded into the deposition system as shown below. Once the synthesis chamber was evacuated and the substrates heated to temperature (~450 C), the substrates were exposed to an atomic hydrogen plasma etch (~10 min.) to remove any residue that may have formed during the loading / pump-down / heat-up process. Exposure to the atomic hydrogen plasma removed any residue and simultaneously hydrogenated the diamond surface.

The source material for the growth of the BN cold cathode film was a split rotating disc target. The split target was composed of one half hBN and the other graphite. The cold cathode material was ablated from the split target by a KrF (248 nm) excimer laser in the presence of an Ar-N atmosphere (~80% Ar, 20% N). The cold cathode layer was deposited on a heated (~450°C) diamond coated (~24 µm) Si substrate under biased conditions (~350-450V) in the argon-nitrogen atmosphere. The plume excitation was enhanced by an electrical discharge into the plume from the bias ring. The total energy available under those conditions was as much as 11.5 J/pulse (~1.5 J_{uv}, ~10.0 J_e^- (electrical)). Typically, the deposition conditions were ~4 J/pulse (~0.8 J_{uv}, ~3.2 J_e^-).

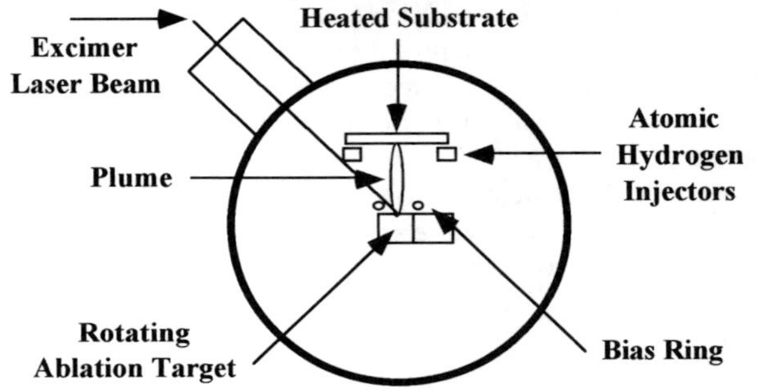

Figure 1. Reactive Laser Ablation System

<u>Cold Cathode Film Synthesis: Reactive Magnetron Sputtering (RMS)</u>

The second method used to synthesize the polycrystalline BN films on MPECVD diamond on Si was the Reactive Magnetron Sputtering technique. The n-type diamond coated substrates were loaded into the deposition system as shown here. Once the synthesis chamber was evacuated and the substrates heated to temperature (~450 C), the substrates were exposed to an atomic hydrogen plasma etch (~10 min.) to remove

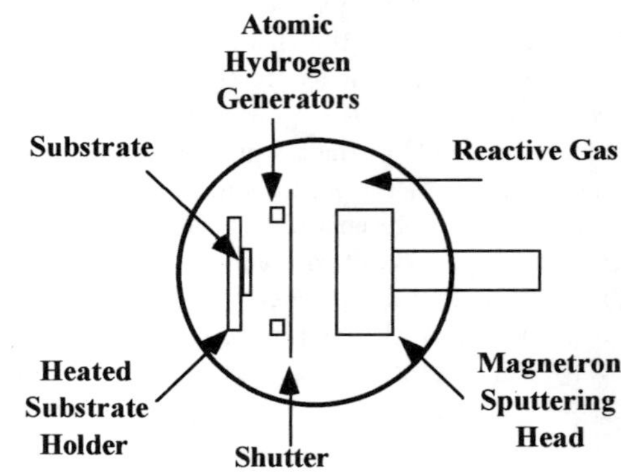

Figure 2. Reactive Magnetron Sputtering

any residue that may have formed during the loading / pump-down / heat-up process. Exposure to the atomic hydrogen plasma removed the residue and simultaneously hydrogenated the diamond surface. This left that surface prepared for the subsequent BN cold cathode deposition.

The source material for the growth of the BN cold cathode film was a rectangular racetrack magnetron hBN target. The carbon doping for the BN cold cathode was supplied by the reactive gas (CH_4). The cold cathode material was sputtered from the target through this reactive technique, by an RF energy (250 W) in the presence of an Ar-CH_4 atmosphere (~99% Ar, 1% CH_4). Films made by this process were synthesized directly on Mo, W, (100)Si and on n-type diamond on (100)Si.

The dopant for the nitride-based films was C. The dopant for the polycrystalline diamond was phosphorus (P). The doped BN films typically had a measured resistance on the order of 8×10^4 Ω. Values of doped BN film resistance as low as 6×10^3 Ω were measured. Undoped films of BN read values in excess of $2\times10^7\Omega$ (the limit of the measuring equipment). The carrier type in each film was independently determined through use of the absolute thermoelectric power measurement [10, 11]. No determination has yet been made of the dopant concentration or the dopant energy levels. The exact nature of the sites contributing the carriers for the n-type conductivity (i.e. substitutional, defect, etc.) has not yet been determined and is the subject of on-going work.

Figure 3. Movable Test Probe Fixture

<u>Electron Emission Measurement Techniques</u>

Two different test fixtures were used to measure the characteristics of the electron emission from the exposed surface of the cold cathode substrates under vacuum conditions (~10^{-4} Pa or less). The Movable Test Probe Fixture (MTP) (Fig. 3, 4) [12] allows the scanning of a large number of positions and anode to cathode separations on a given test substrate. The exact anode (~0.51 mm) to cathode separation for the MTP was less certain for small separations, under 300µm, than for the Fixed Test Probe Fixture (FTP), since the FTP spacing was measured to an accuracy of less than 1 µm by laser techniques. The emission data measured on each system are clearly identified, as such, when presented.

In the Fixed Test Probe Fixture (FTP) (Fig. 5), several wires (~0.51mm dia.) were mounted through a dielectric block. The surface of the block, with the wires in place, was then machined flat. The cold cathode substrate was mounted on an equally flat dielectric block with the emitting surface facing the anode block. The metal strip spacers (~ 12 to 40 µm) defined the separation between the anode and the cathode surfaces.

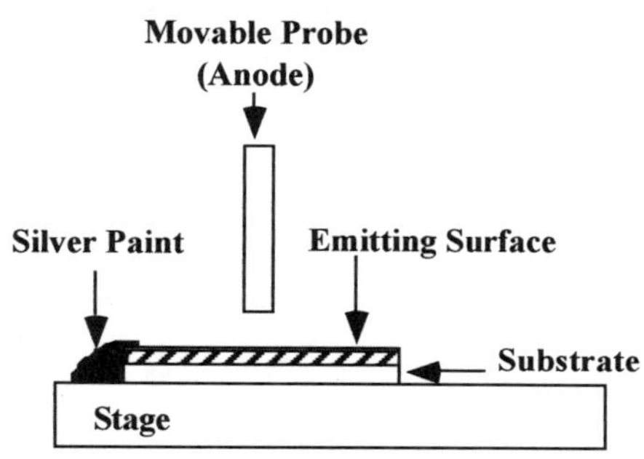

Figure 4. Movable Test Probe Fixture

Electrical contact to the emitting surface was made by the thin metal spacers under pressure. It was found to be especially important to apply adequate pressure to those spacers, since the spacer-substrate interface pressure defined the quality of the electrical contact ("ohmic", etc.) in this fixture. The spacers were then contacted by spring loaded wires passing through clear holes in the anode block. After mounting, the sample in the FTP was inserted into a chamber and a vacuum established.

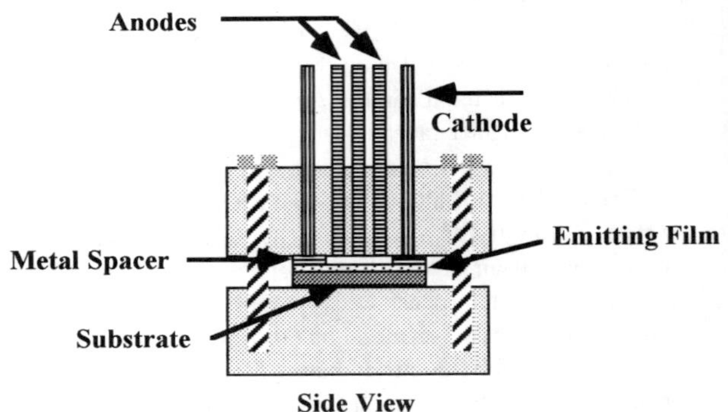

Figure 5.Fixed Probe Test Fixture

In the case of both the MTP and the FTP systems, once an adequate level of vacuum was achieved, a voltage was applied between the anode (+) and ground(-) to measure the extent of electron emission from the sample surface by measuring the current collected at the anode. The current flow through the device was calculated by measuring the voltage drop across an external series cathode resistor.

Electron Emission Measurements

For the cold cathodes reported herein, the process started with the synthesis of an n-type diamond layer (~24 µm) on an n-type (100)Si substrate. For the diamond emission experiments, the MPECVD n-type diamond cold cathodes were annealed in a microwave generated atomic hydrogen plasma for ~30 min. The results of a typical annealed sample are shown here. Similar results were first reported earlier [4]. A significant post-anneal increase in the current-density (J) / extraction field (F) curve was observed (Fig. 6). This shift in emission current showed the importance of the emission surface barrier layer in the extraction of electrons from the conduction band of wide bandgap semiconductor cold cathodes [13].

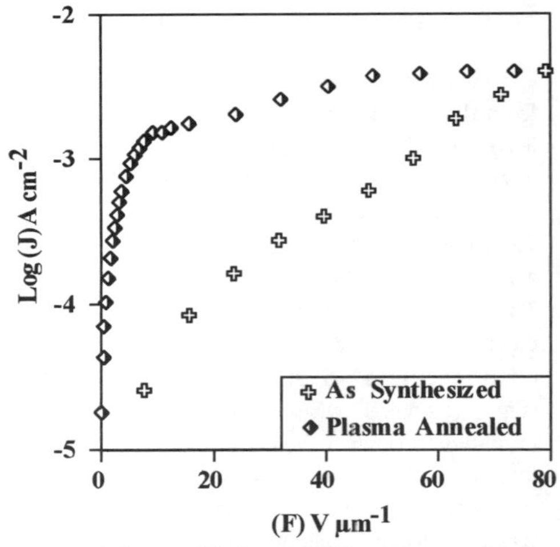

Figure 6.Polycrystalline Diamond Cold Cathode

Two different processes (RLA, RMS) have been used to synthesize n-type carbon doped BN cold cathodes. The n-type BN cold cathode films (~150 nm) synthesized over n-type diamond (~24 μm) on a (100)Si substrate have been the most extensively tested. Annealed RLA films of that type have shown better emission to date than the RMS films (Fig. 7). Films of that type have been tested in both the MTP and the FTP electron emission vacuum test systems (Fig. 8). The results of testing the same cold cathode in both the FTP and the MTP are shown below.

In the case of the FTP, the separation between the anode and the cathode were measured by laser diffraction to be 20 μm. The separation between the anode and the cathode of the MTP were estimated to be 75 μm, based on mechanical considerations. Additionally, the cathode contact electrode area was significantly smaller in the MTP system, potentially adding internal series resistance, which would shift the resulting I-F curve at the higher current levels (Fig. 8). It can be seen that in both the MTP and the FTP cases, currents on the order of 1 mA or greater were measured.

Conclusions

Measurements have been presented on both n-type diamond and n-type BN cold cathode films that show that post-processing annealing plays a significant role in the emission of electrons from the cold cathode surface. This paper also presents evidence of currents measured in excess of 1 mA for 0.51 mm diameter anode probes by two different laboratories on two different electron emission testing systems on the same sample.

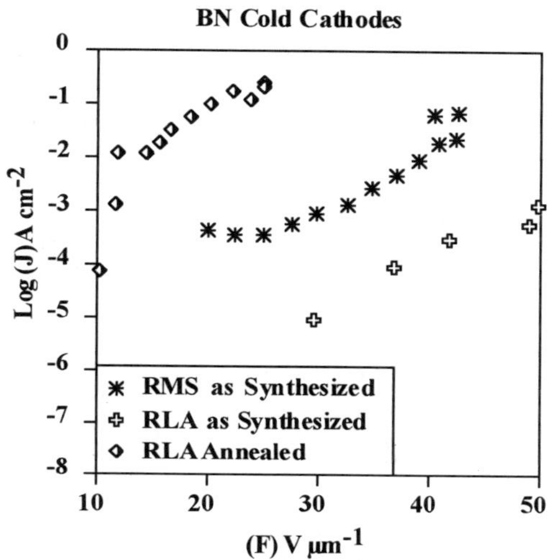

Figure 7. RLA and RMS Cold Cathodes

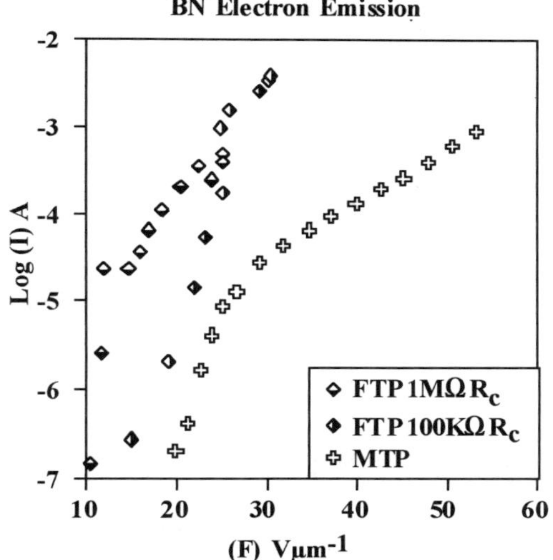

Figure 8. MTP and FTP Measurements Compared for the same BN on Diamond on Silicon Cold Cathode

ACKNOWLEDGMENTS

The authors would like to thank D. Durisin for his invaluable assistance. This work was supported in part by the Michigan Research Excellence Fund and BMDO/IST through ONR Grant N00014-93-1-0254 and a Subcontract under ONR Contract N00014-95-C-0170.

REFERENCES

1. E.U. Condon and H. Odishaw, <u>Handbook of Physics,</u> (McGraw-Hill, New York, 1958) p 8-74.

2. A. Modinos, <u>Field, Thermionic, and Secondary Electron Emission Spectroscopy,</u> (Plenum Press, New York, 1984).

3. R. W. Landee, D. C. Davis and A. P. Albrecht,, <u>Electronic Designer's Handbook,</u> (McGraw-Hill, New York, 1957) p 2-9.

4. R.W. Pryor, "Polycrystalline Diamond, Boron Nitride and Carbon Nitride Thin Film Cold Cathodes", Mat. Res. Soc. Symp. Proc., Vol. 416, (Materials Research Society 1996) p425

5. R.W. Pryor, Patent Pending.

6. R.W. Pryor, Patent Pending.

7. R.W. Pryor, Publication Pending.

8. R.W. Pryor, K.R. Padmanabhan, K. Chawla, "Diamond on Heteroepitaxial cBN on Si(100) ", Diamond and Related Materials **4**, 128 (1995).

9. R.W. Pryor, Z.L. Wu, K.R. Padmanabhan, S. Villanueva, R.L. Thomas, "Characterization of laser-ablated boron nitride thin films on silicon", Thin Solid Films **253**, 243(1994).

10. C. Kittel, <u>Introduction to Solid State Physics</u> , 5th Edition (John Wiley & Sons, Inc., New York, 1976) p 238.

11. R. A. Smith, <u>Semiconductors</u> , (Cambridge University Press, Cambridge, 1968) p 170.

12. H. H. Busta, David Sarnoff Research Center, Princeton NJ, 08543

13. R.E. Thomas, T.P. Humphreys, C. Pettenkofer, D.P. Malta, J.P. Posthill, M.J. Mantini, R.A. Rudder, G.C. Hudson and R.J. Markunas, "Influence of Surface Terminating Species on Electron Emission from Diamond Surfaces", Mat. Res. Soc. Symp. Proc., Vol. 416, (Materials Research Society 1996) p263

Part XI

Devices

CHRACTERISTICS OF InGaN MULTI-QUANTUM-WELL-STRUCTURE LASER DIODES

SHUJI NAKAMURA
Department of Research and Development, Nichia Chemical Industries, Ltd.,
491 Oka, Kaminaka, Anan, Tokushima 774, Japan, shuji@nichia.co.jp

ABSTRACT

The continuous-wave (CW) operation of InGaN multi-quantum-well-structure laser diodes (LDs) was demonstrated at room temperature (RT) with a lifetime of 35 hours. The threshold current and the voltage of the LDs were 80 mA and 5.5 V, respectively. The threshold current density was 3.6 kA/cm^2. Longitudinal modes with a mode separation of 0.042 nm were observed under CW operation at RT. When the temperature of the LDs was varied, large mode hopping of the emission wavelength was observed. The carrier lifetime and the threshold carrier density were estimated to be 10 ns and 2 x 10^{20}/cm^3, respectively. The beam full width at half-power values for the parallel and the perpendicular near-field patterns were 1.6 μm and 0.8 μm, respectively. Those of the far-field patterns were 6.8° and 33.6°, respectively.

INTRODUCTION

Short-wavelength-emitting devices, such as blue laser diodes (LDs), are currently required for a number of applications, including full-color electroluminescent displays, laser printers, read-write laser sources for high-density information storage on magnetic and optical media, and sources for undersea optical communications. Major developments in wide-gap III-V nitride semiconductors have recently led to the commercial production of high-brightness blue/green light-emitting diodes (LEDs) [1] and to the demonstration of room-temperature (RT) violet laser light emission in InGaN/GaN/AlGaN-based heterostructures under pulsed currents [2-9]. These developments are a result of the realization of high-quality crystals of AlGaN and InGaN, and p-type conduction in AlGaN [10-13]. Also, the recombination of localized excitons has been proposed as an emission mechanism for these InGaN quantum-well-structure LEDs [14]. Currently, the main focus of research is the realization of a current-injected LD capable of continuous-wave (CW) operation at RT. Here, we describe the RT CW operation of InGaN multi-quantum-well (MQW)-structure LDs with a lifetime of 35 hours .

EXPERIMENT

III-V nitride films were grown by the two-flow metalorganic chemical vapor deposition (MOCVD) method. Details of two-flow MOCVD have been described elsewhere [15]. The growth was conducted at atmospheric pressure, and (0001) C-face sapphire was used as the substrate. The InGaN MQW LD device consisted of a 300-Å-thick GaN buffer layer grown at a low temperature of 550 °C, a 3-μm-thick layer of n-type GaN:Si, a 0.1-μm-thick layer of n-type In$_{0.05}$Ga$_{0.95}$N:Si, a 0.5-μm-thick layer of n-type Al$_{0.08}$Ga$_{0.92}$N:Si, a 0.1-μm-thick layer of n-type GaN:Si, an In$_{0.15}$Ga$_{0.85}$N/In$_{0.02}$Ga$_{0.98}$N MQW structure consisting of four 35-Å-thick Si-doped In$_{0.15}$Ga$_{0.85}$N well layers forming a gain medium separated by 70-Å-thick Si-doped In$_{0.02}$Ga$_{0.98}$N barrier layers, a 200-Å-thick layer of p-type Al$_{0.2}$Ga$_{0.8}$N:Mg, a 0.1-μm-thick layer of p-type GaN:Mg, a 0.5-μm-thick layer of p-type Al$_{0.08}$Ga$_{0.92}$N:Mg, and a 0.5-μm-thick layer

Mat. Res. Soc. Symp. Proc. Vol. 449 © 1997 Materials Research Society

of p-type GaN:Mg. The 0.1-μm-thick n-type and p-type GaN layers were light-guiding layers. The 0.5-μm-thick n-type and p-type $Al_{0.08}Ga_{0.92}N$ layers acted as cladding layers for confinement of the carriers and the light emitted from the active region of the InGaN MQW structure. The structure of the ridge-geometry InGaN MQW LD was almost the same as that described previously [6].

First, the surface of the p-type GaN layer was partially etched until the n-type GaN layer and the p-type $Al_{0.08}Ga_{0.92}N$ cladding layer were exposed, in order to form a ridge-geometry LD [6]. A mirror facet was also formed by dry etching, as reported previously [2]. The area of the ridge-geometry LD was 4 μm x 550 μm. High-reflection facet coatings (30 %) consisting of 2 pairs of quarter-wave TiO_2/SiO_2 dielectric multilayers were used to reduce the threshold current. A Ni/Au contact was evaporated onto the p-type GaN layer, and a Ti/Al contact was evaporated onto the n-type GaN layer. The electrical characteristics of the LDs fabricated in this way were measured under a direct current (DC). The structure of the InGaN MQW LDs is shown in Fig. 1.

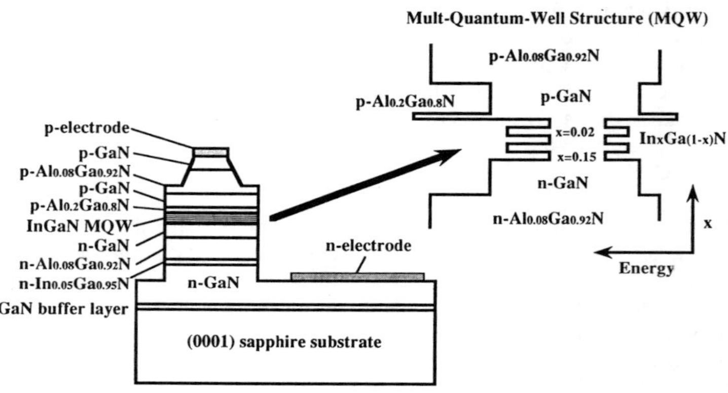

Fig. 1. The structure of the InGaN MQW LDs.

RESULTS AND DISCUSSION

Figure 2 shows typical voltage-current (V-I) characteristics and the light output power per coated facet of the LD as a function of the forward DC current (L-I) at RT. No stimulated emission was observed up to a threshold current of 80 mA, which corresponded to a threshold current density of 3.6 kA/cm^2, as shown in Fig. 2. The operating voltage at the threshold current was 5.5 V. We were able to reduce the operating voltage significantly in comparison with values obtained previously (about 20-30 V) by adjusting the growth, Ohmic contact and doping profile conditions [2-7].

Figure 3 shows the results of a lifetime test of CW-operated LDs carried out at RT, in which the operating current is shown as a function of time under a constant output power of 1.5 mW per facet controlled using an autopower controller (APC). The operating current gradually increases due to the increase in the threshold current from the initial stage and sharply increases after 35 hours. This short lifetime is probably due to the large heat generation resulting from the high operating currents and voltages. Breakdown of the LDs occurred after a period of more than 35 hours due to the formation of a short circuit in the LDs.

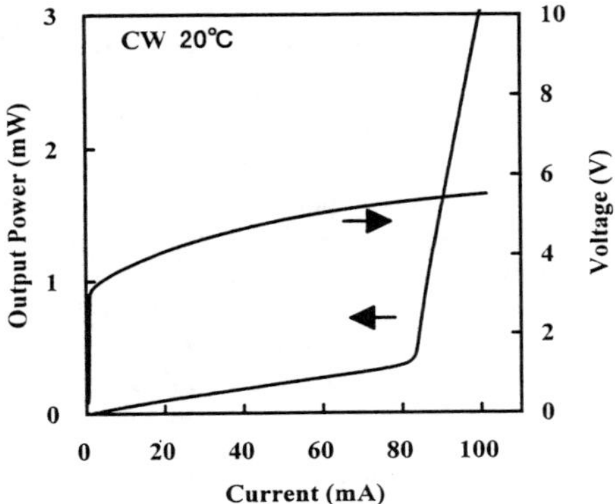

Fig. 2. Typical L-I and V-I characteristics of InGaN MQW LDs measured under CW operation at RT.

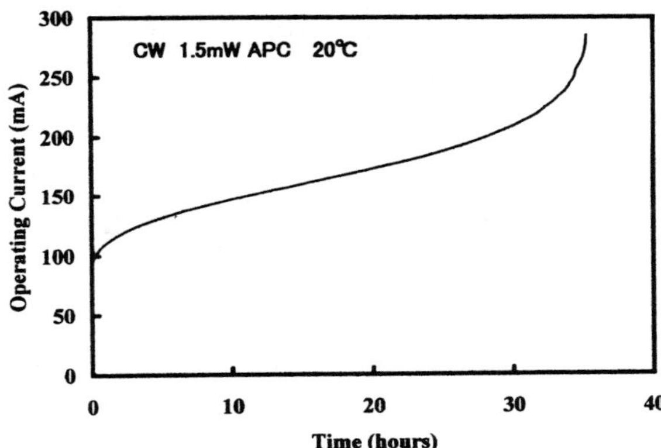

Fig. 3. Operating current as a function of time under a constant output power of 1.5 mW per facet controlled using an autopower controller. The LD was operated under DC at RT.

Next, the emission spectra of the LDs were measured under RT CW operation at an output power of 1 mW. An optical spectrum analyzer (ADVANTEST Q8347) which utilized the Fourier-transform spectroscopy method by means of a Michelson interferometer was used to measure the spectra of the LDs with a resolution of 0.001 nm. At $J = 1.0J_{th}$, where J is the current density and J_{th} is the threshold current density, longitudinal modes with many sharp peaks with a peak separation of 0.042 nm ($\Delta E=0.3$ meV, where ΔE was the mode separation energy) were observed, as shown in Fig. 4(a). If these peaks arise from the longitudinal modes of the LD, then the mode separation $\Delta\lambda$ is given by

$$\Delta\lambda=\lambda_0{}^2/2/L/n_{eff}, \tag{1}$$

1137

where n_{eff} is the effective refractive index and λ_0 is the emission wavelength (405.83 nm). L is 0.055 cm. Thus, n_{eff} is calculated as 3.6, which is relatively large due to the wavelength dependence of the refractive indices of GaN and InGaN. Also, other periodic subband emissions are observed with a peak separation of 0.25-0.29 nm ($\Delta E=1.8–2.1$ meV). The origin of these subband emissions has not yet been clarified. However, it is possible that these emissions result from transitions between quantum well or quantum dot subband energy levels as mentioned previously [3,6,7]. Several peaks with a different peak separation (energy separation of 1-5 meV) from that of the longitudinal mode appeared under pulsed current operation as described in our previous reports [3,5-7]. These subband emissions with an energy separation of 1-5 meV are probably caused by mode hopping between adjacent quantum well or quantum dot subband energy levels due to the temperature fluctuation in the active region under pulsed current operation, as shown in Figs. 5 and 6. At $J = 1.2J_{th}$, the main peak at 405.83 nm becomes dominant, as shown in Fig. 4(b).

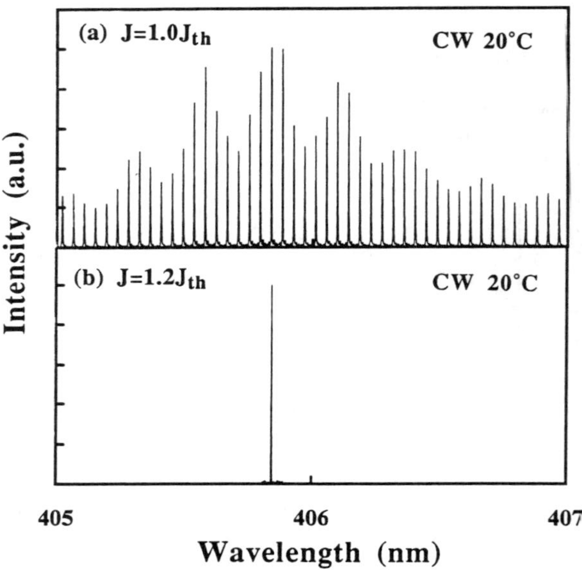

Fig. 4. Laser emission spectra measured under RT CW operation with current densities of (a) $J = 1.0J_{th}$ (b) $J = 1.2J_{th}$.

The temperature dependence of the emission spectra was measured between 20 °C and 60 °C under CW operation with a constant output power of 1mW, as shown in Fig. 5. Large mode hopping of the peak emission wavelength with an energy step of 1-7 meV is observed, which results from the temperature dependence of the gain profile. Mode hopping is probably a result of the transitions between adjacent quantum well or quantum dot subbands, as shown in Fig. 4. The change in the actual emission spectra with temperature between 47 °C and 48 °C is shown in Fig. 6. When the temperature is increased from 47 °C to 48 °C, the peak wavelength varies from 407.428 nm to 408.523 nm (with an energy difference of 7 meV) due to the change in the gain profile.

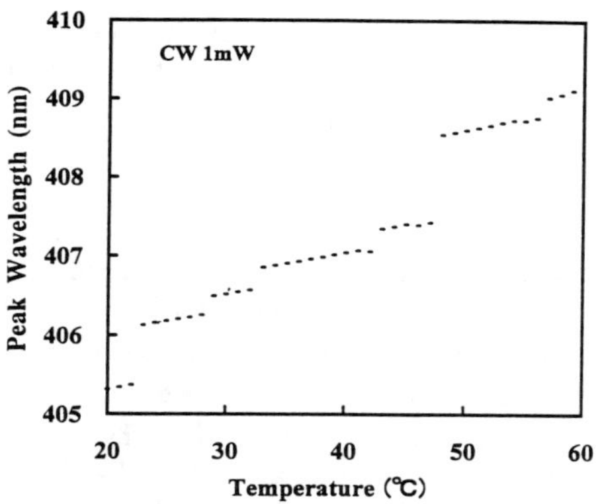

Fig. 5. Temperature dependence of the peak emission wavelengths of InGaN MQW LDs under CW operation with a constant output power of 1 mW.

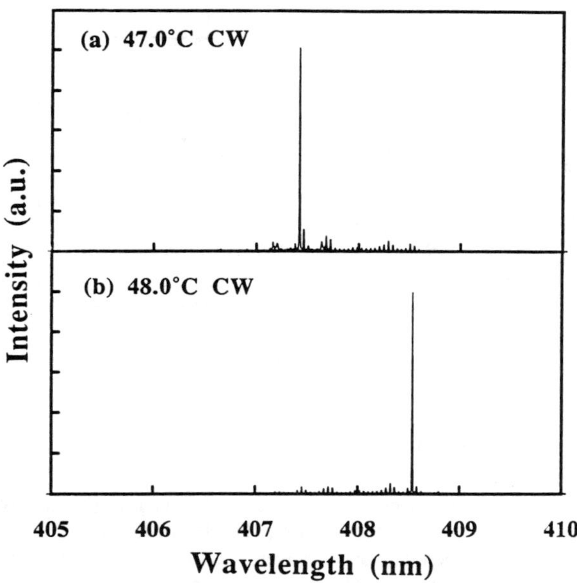

Fig. 6. Optical spectra of InGaN MQW LDs measured under CW operation at temperatures of (a) 47 °C and (b) 48 °C. The intensity scales for these two spectra are in arbitrary units, and each one is different.

Next, the delay time of the laser emission of the LDs as a function of the operating current was measured under pulsed current modulation using the method described in ref. 7 in order to estimate the carrier lifetime (τ_s). The delay time t_d is given by

$$t_d = \tau_s \ln(I/(I-I_{th})), \tag{2}$$

where τ_s is the minority carrier lifetime, I is the pumping current, and I_{th} is the threshold current. Figure 7 shows the delay time t_d of the laser emission as a function of $\ln(I/(I-I_{th}))$. From this figure, τ_s was estimated to be 10 ns, which was relatively large in comparison with the previous value of 3.2 ns [7]. The threshold carrier density (n_{th}) was estimated to be 2 x 10^{20}/cm^3 for a threshold current density of 3.6 kA/cm^2, a carrier lifetime of 10 ns, and an active layer thickness of 150 Å [7]. The thickness of the active layer was determined as 140 Å assuming that the injected carriers were confined in the InGaN well layers in the active layer. Typical values are τ_s = 3 ns, J_{th} = 1 kA/cm and n_{th} = 2 x 10^{18}/cm^3 for AlGaAs lasers and n_{th} = 1 x 10^{18}/cm^3 for InGaAsP lasers. In comparison with these values for conventional lasers, n_{th} for our structure is relatively large (two orders of magnitude higher), probably due to the large density of states of carriers resulting from their large effective masses [7].

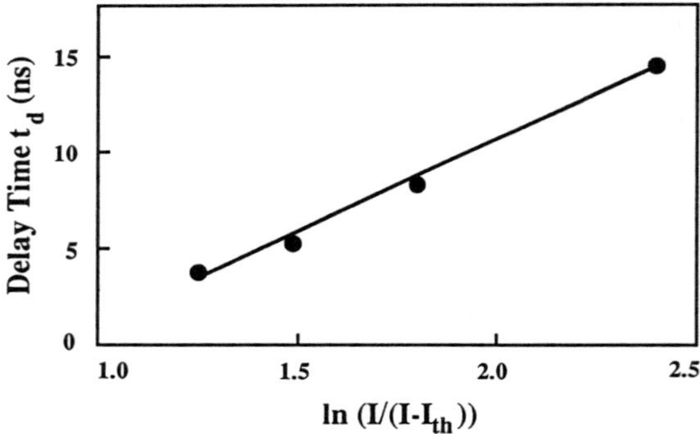

Fig. 7. The delay time t_d of the laser emission as a function of $\ln(I/(I-I_{th}))$. I is the pumping current and I_{th} is the threshold current.

Figure 8 shows typical near-field radiation patterns (NFP) for the InGaN MQW LDs in the planes parallel and perpendicular to the junction. The beam full width at half-power (FWHP) values for the parallel and perpendicular NFPs were 1.6 µm and 0.8 µm, respectively. The beam width, shown by 1/e^2, was 3.3 µm which was almost the same as the ridge width (4 µm). The transverse optical confinement resulting from the ridge geometry is relatively good using this ridge waveguide. The astigmatism of the laser diodes depends on the optical mode profile, which in turn is determined by the laser structure. A typical value of the astigmatism was 2 µm.

Typical far-field radiation patterns (FFP) of the InGaN MQW LDs in the planes parallel and perpendicular to the junction are shown in Fig. 9. The FWHP values for the parallel and perpendicular FFPs are 6.8° and 33.6°, respectively. From Figs. 8 and 9, only fundamental transverse mode operation with an output power of 2 mW was observed in the laser emission. When the output power was changed, the NFP and FFP were almost the same as those in Figs. 8 and 9.

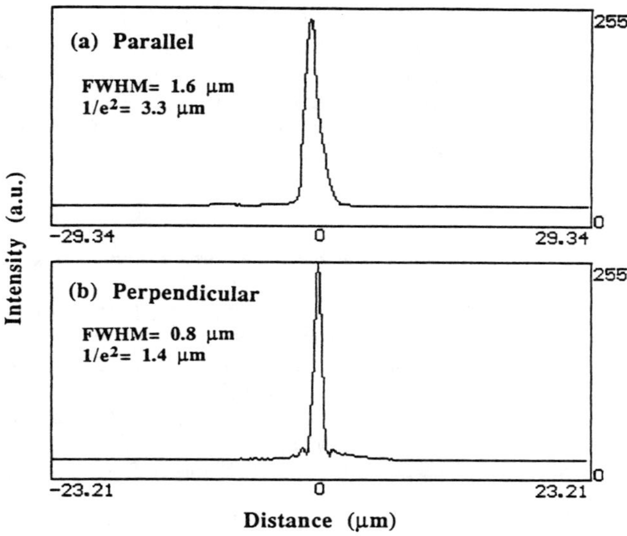

Fig. 8. Typical NFP of the InGaN MQW LDs in the planes parallel and perpendicular to the junction under RT CW operation with an output power of 2 mW.

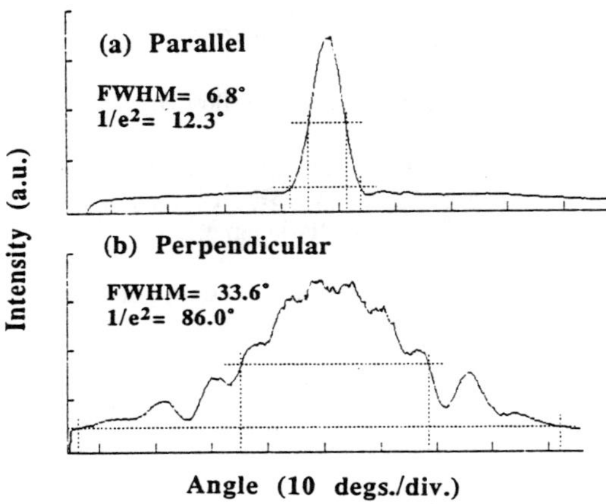

Fig. 9. Typical FFP of the InGaN MQW LDs in the planes parallel and perpendicular to the junction under RT CW operation with an output power of 2 mW.

SUMMARY

In summary, the RT CW operation of InGaN MQW LDs was demonstrated with a lifetime of 35 hours. The laser emission was fundamental single mode emission with a peak wavelength of 405.83 nm. The carrier lifetime and the threshold carrier density were estimated to be 10 ns and 2 x 10^{20}/cm^3, respectively. The beam full width at half-power values for the parallel and the perpendicular near-field patterns were 1.6 μm and 0.8 μm, respectively. Those of the far-field patterns were 6.8° and 33.6°, respectively. Further improvement in the lifetime of the LDs can be obtained by reducing the threshold current and voltage.

REFERENCES

1. S. Nakamura, M. Senoh, N. Iwasa, S. Nagahama, T. Yamada, and T. Mukai, Jpn. J. Appl. Phys. Lett. **34**, L1332 (1995).
2. S. Nakamura, M. Senoh, S. Nagahama, N. Iwasa, T. Yamada, T. Matsushita, H. Kiyoku, and Y. Sugimoto, Jpn. J. Appl. Phys. **35**, L74 (1996).
3. S. Nakamura, M. Senoh, S. Nagahama, N. Iwasa, T. Yamada, T. Matsushita, H. Kiyoku, and Y. Sugimoto, Jpn. J. Appl. Phys. **35**, L217 (1996).
4. S. Nakamura, M. Senoh, S. Nagahama, N. Iwasa, T. Yamada, T. Matsushita, H. Kiyoku, and Y. Sugimoto, Appl. Phys. Lett. **68**, 2105 (1996).
5. S. Nakamura, M. Senoh, S. Nagahama, N. Iwasa, T. Yamada, T. Matsushita, H. Kiyoku, and Y. Sugimoto, Appl. Phys. Lett. **68**, 3269 (1996).
6. S. Nakamura, M. Senoh, S. Nagahama, N. Iwasa, T. Yamada, T. Matsushita, Y. Sugimoto, and H. Kiyoku, Appl. Phys. Lett. **69**, 1477 (1996).
7. S. Nakamura, M. Senoh, S. Nagahama, N. Iwasa, T. Yamada, T. Matsushita, Y. Sugimoto, and H. Kiyoku, Appl. Phys. Lett. **69**, 1568 (1996).
8. I. Akasaki, S. Sota, H. Sakai, T. Tanaka, M. Koike, and H. Amano, Electron. Lett. **32**, 1105 (1996).
9. K. Itaya, M. Onomura, J. Nishino, L. Sugiura, S. Saito, M. Suzuki, J. Rennie, S. Nunoue, M. Yamamoto, H. Fujimoto, Y. Kokubun, Y. Ohba, G. Hatakoshi, and M. Ishikawa, Jpn. J. Appl. Phys. **35**, L1315 (1996).
10. H. Morkoç, S. Strite, G. B. Gao, M. E. Lin, B. Sverdlov, and M. Burns, J. Appl. Phys. **76**, 1363 (1994).
11. H. Amano, M. Kito, K. Hiramatsu, and I. Akasaki, Jpn. J. Appl. Phys. **28**, L2112 (1989).
12. S. Nakamura and T. Mukai, Jpn. J. Appl. Phys. **31**, L1457 (1992).
13. M. A. Khan, J. N. Kuznia, D. T. Olson, M. Blasingame, and A. R. Bhattarai, Appl. Phys. Lett. **63**, 2455 (1993).
14. S. Chichibu, T. Azuhata, T. Sota, and S. Nakamura, presented at 38th Electronic Material Conference, W-10, June 26-28, Santa Barbara, (1996).
15. S. Nakamura, Jpn. J. Appl. Phys. **30**, 1620 (1991).

STRUCTURAL AND OPTICAL PROPERTIES OF NITRIDE BASED HETEROSTRUCTURE AND QUANTUM WELL STRUCTURE

H.AMANO, T.TAKEUCHI, S.SOTA, H.SAKAI and I.AKASAKI
Department of Electrical and Electronic Engineering, Meijo University, 1-501 Shiogamaguchi, Tempaku-ku, Nagoya 468, Japan, amano@meijo-u.ac.jp

ABSTRACT

Structural and optical properties of nitride based heterostructure and quantum well structure were investigated. Both AlGaN and GaInN ternary alloys are found to grow coherently on the underlying GaN layer. Compressive strain of GaInN is found to cause quantum confined Stark effect, thus affects the luminescence properties of nitride-based quantum wells.

INTRODUCTION

The use of group III nitrides were prohibited for a long time due to the difficulty in growing large size bulk crystals, the lack of substrate materials with lattice constant and thermal expansion coefficient close to those of nitrides, and the difficulty in obtaining p-type films. Use of low-temperature-deposited thin buffer layer on the growth of nitrides by MOVPE[1,2] and MBE[3] drastically changed the situation and enabled the growth of high-quality nitrides on sapphire substrates.

P-type GaN was realized by low-energy electron beam irradiation treatment[4] or thermal treatment[5] of such high-quality MOVPE-grown films doped with Mg. In case of MBE, no special treatment is necessary to obtain highly conductive p-type GaN. Progress in the nitrides technologies has brought about today's availability of commercial bright blue, blue-green and green light emitting diodes[6-8]. Recently, stimulated emission by current injection[9] and laser action have been successfully achieved with GaInN multi-quantum well[10,11] or single quantum well[12] active layer sandwiched with AlGaN cladding layer.

For the fabrication of laser diodes (LDs), heteroepitaxial growth of ternary alloys and quantum well structures are indispensable. Recently, we found that both AlGaN and GaInN ternary alloys can be grown coherently on binary GaN[13], which means that AlGaN is under tensile stress while GaInN is under compressive stress. Therefore, in order to understand the mechanism of lasing in nitride based laser diode, it is important to understand the fundamental structural and optical properties of strained ternary alloys and quantum wells. Optical properties of bulk or thick ternary alloys have been already reported by many groups[14-18]. However, those of strained ternary alloys and quantum well structures are still controversial.

In this paper, firstly, structural and optical properties of AlGaN and GaInN single layer grown on thick GaN is reported. Secondly, influence of piezoelectric fields on optical properties of GaInN strained MQWs is reported.

EXPERIMENT

All the samples used in this study were grown on sapphire (0001) substrate by atmospheric pressure MOVPE. Single heterostructure shown in fig.1 were used for the characterization of structural and optical property. All the layers were nominally undoped. The thickness of underlying GaN was fixed at 2 μm. The thickness of $Al_xGa_{1-x}N$ was somewhere

Mat. Res. Soc. Symp. Proc. Vol. 449 © 1997 Materials Research Society

in between 0.35-0.65μm, and its AlN molar fraction ranged from 0 to 0.25.

In case of Ga$_{1-x}$In$_x$N, the thickness was fixed around 40 nm, and the InN molar fraction ranged from 0 to 0.2. At first, strain and relaxation of ternary heterostructured alloy layers were analyzed by reciprocal space mapping (RSM) of X-ray diffraction intensity around the (20$\bar{2}$4) asymmetrical diffraction spot of nitrides. We also measured grazing incidence X-ray diffraction to characterize the lattice constant parallel to the plane near the surface.

In order to estimate precise composition of strained ternary alloys grown on GaN, we measured 2θ-scan and ω/2θ-scan profiles of X-ray symmetrical (0004) diffraction using detectors having analyzer crystals to collimate the diffracted X-rays, and then calculated alloy composition using elastic stiffness constants[19-21]. Elastic stiffness constants of ternary compounds were linearly interpolated. Alloy composition x was calculated using the following formula.

$$\frac{\Delta c_0}{c_0} = -2 \frac{c_{13}}{c_{33}} \frac{\Delta a_0}{a_0}$$

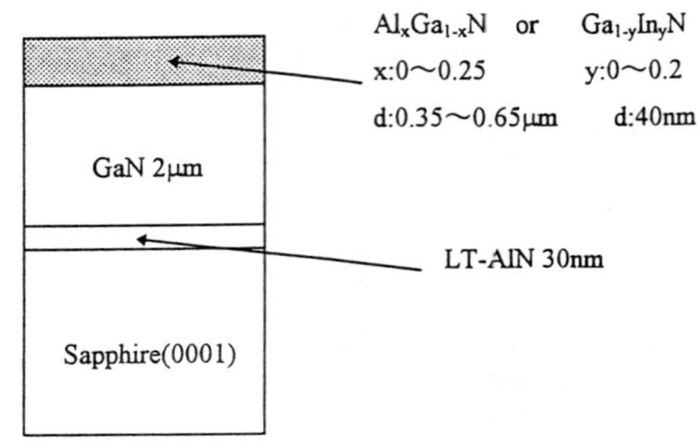

Fig.1(a)   Single heterostructure used for characterization of effect of strain and relaxation.

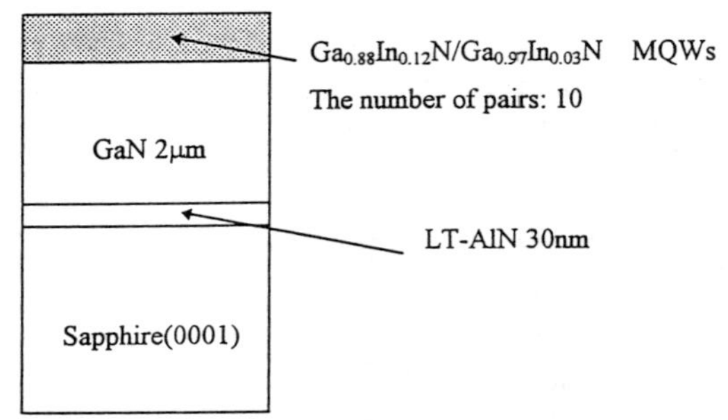

Fig.1(b) Structure of MQWs used for characterization of piezoelectric induced quantum confined Stark effect.

where, Δc_0:difference of the lattice constant c of alloy between fully relaxed one and measured one. Δa_0:difference of the lattice constant a between fully relaxed one and that of GaN. c_0:fully relaxed lattice constant c of alloy. a_0:fully relaxed lattice constant a of alloy. c_{ij}:elastic stiffness constants

For the characterization of bandgap of AlGaN, we used spectroscopic ellipsometry. The bandgap energy was determined by fitting calculated dielectric function to measured one[22]. In this fitting, contributions of joint density of states and three dimensional exciton of E_0 critical-point were considered.   In the case of GaInN, PL spectra were measured at room temperature (RT).   A pulsed N$_2$ laser and a He-Cd laser were used for extremely high(200 kW/cm^2) and relatively low($\sim$2 W/cm^2) intensity excitation, respectively.

In order to clarify the effect of piezoelectricity on the luminescence properties,   the structure consisting of 30 nm low-temperature-deposited AlN buffer layer, 2μm GaN layer and $Ga_{0.88}In_{0.12}N/Ga_{0.97}In_{0.03}N$ multiple QWs, which were nominally undoped, was fabricated.   For all the samples, well layers have InN mole fractions of 0.12, barrier layers have that of 0.03, respectively. Compositions and the thickness of these layers were determined by X-ray diffraction measurement.

RESULTS

AlGaN/GaN and GaInN/GaN single heterostructure

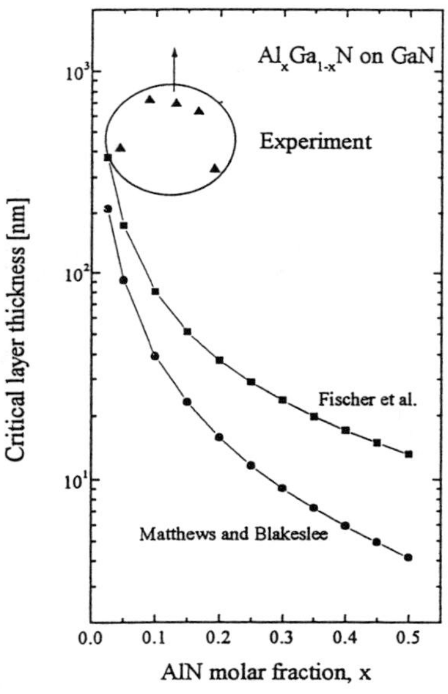

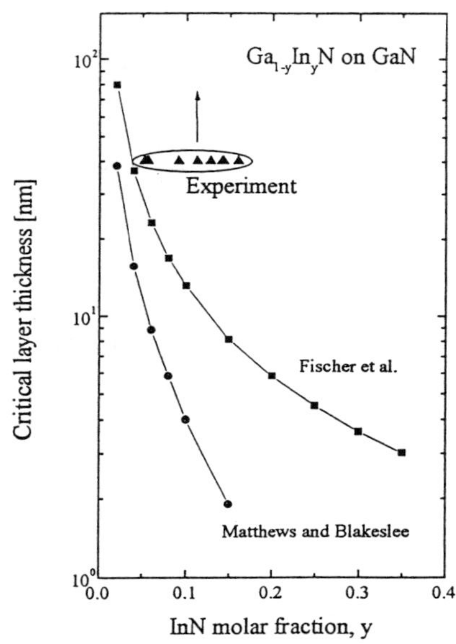

Fig.2(a) Critical layer thickness of Al$_x$Ga$_{1-x}$N grown on GaN.

Fig.2(b) Critical layer thickness of Ga$_{1-y}$In$_y$N grown on GaN.

Solid circle shows critical layer thickness estimated from Matthews and Blakeslee model[25], and solid square shows that estimated from Fischer model[26]. Solid triangle shows experimental result. It should be noted that real critical layer thickness might be thicker than the experimental result.

Reciprocal space mapping of the X-ray diffraction intensity around ($20\overline{2}4$) diffraction spot from AlGaN layer and from GaInN layer showed that lattice constant *a* of the diffraction spot from AlGaN layer and GaInN layer are almost perfectly aligned to that of GaN. The difference of the lattice constant *a* between that of ternary layers and that of GaN is less than or equal to

0.0001nm. This clearly shows that both AlGaN layer and GaInN layer are coherently grown on GaN. Therefore, AlGaN layer is under tensile stress, while GaInN layer is under compressive stress. All the AlGaN layers and GaInN layers inspected in this study showed similar results.

Calculated ω/2θ-scan profiles coincide very well with experimental profiles. Therefore, macroscopic fluctuation of alloy composition or that of strain relaxation in AlGaN layers and in GaInN layers is negligibly small. For all the samples in this experiment, the lattice constant *a* of thick GaN and that of fully-strained ternary alloys were both 0.3182 ± 0.0001nm, which is slightly different from that of bulk value (0.3188 ± 0.0001nm)[23,24] due to the thermal stress originating from the difference of the thermal expansion coefficient between GaN and sapphire.

Figures 2(a) and 2(b) summarizes the estimated critical layer thickness of AlGaN(fig.2(a)) and GaInN(fig.2(b)) on GaN based on Matthews and Blakeslee model (solid circle)[25] and Fischer model (solid square) [26]. Even though thickness of ternary layers on GaN exceeds critical layer thickness estimated from two models, the ternary layer showed coherent growth. This study suggests that the actual critical layer thickness is thicker than that estimated from these models.

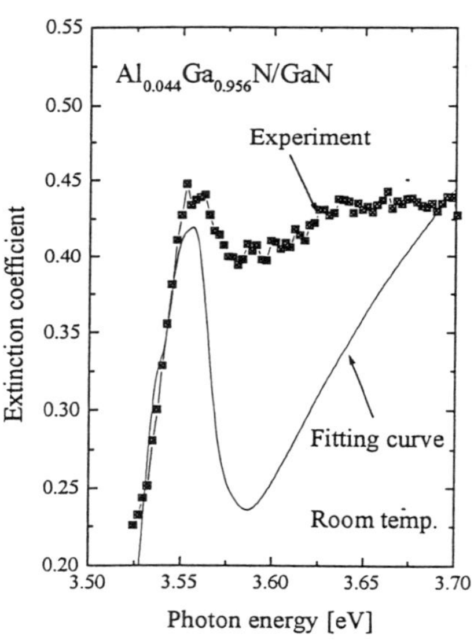

Fig.3(a) Spectra of extinction coefficient of Al$_{0.044}$Ga$_{0.956}$N/GaN. Solid square is the measured one, and the solid line is the fitting curve to obtain the bandgap.

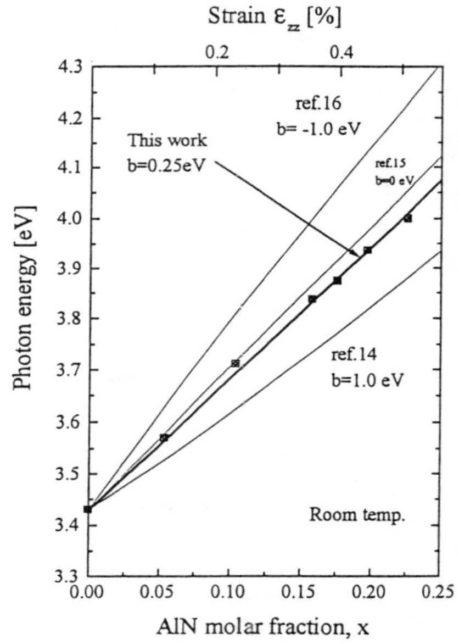

Fig.3(b) Compositional dependence of the bandgap of strained AlGaN coherently grown on GaN. Results ever reported are also shown for comparison.

Figures 3(a) and 3(b) represent comparison of the spectrum of extinction coefficient of Al$_{0.044}$Ga$_{0.956}$N between measured one and calculated one(fig.3(a)) and compositional dependencies of the bandgap energy of the fully-strained AlGaN layer(fig.3(b)). In this calculation, contribution of joint density of state and three dimensional exciton of E$_0$ specific point were considered, and that from higher energy specific point is neglected. Therefore fitted absorption spectrum is slightly different from that of measured one especially at higher energy side. Nevertheless, since we can clearly observe the contribution of both joint density of state and three dimensional exciton close to the bandgap energy, we can determine the bandgap of AlGaN by fitting within a error of less than 10meV. The tensile strain of AlGaN layer is less than 0.5%, so that a change in transition energy due to the deformation potential is expected to be small. Therefore, as shown in fig.3(b), the bandgap of strained AlGaN layer determined in this study is close to bandgap of unstrained AlGaN, especially for to the value reported by Wickenden et al[15], who grew AlGaN by MOVPE on the sapphire substrate using AlN buffer layer technique. In this experiment, a bowing parameter of strained AlGaN is calculated to be 0.25 eV.

Figure 4 shows the compositional dependence of the PL peak energy of strained thick(40nm) GaInN. In case of thick strained layer, red-shift of PL peak energy due to piezoelectric field is negligibly small because overlap of wavefunctions of separate electrons an holes are smaller than that of confined case. Therefore, PL peak energy of strained thick GaInN is much closer to the bandgap of unstrained GaInN than that of MQW structure having very thin well layer. PL peak energy of strained thick GaInN layer showed very large red-shift from reported one which is determined from 100nm thick GaInN on GaN[18]. In this experiment, a bowing parameter of strained GaInN is calculated to be 3.2 eV, which is not so astonishing because the difference of the lattice constant between GaN and InN is as large as about 11%. It should be in mind, however, that the GaInN layer is largely strained by biaxial compressive stress, which causes blue-shift of bandgap from that of unstrained GaInN. Therefore, bandgap of unstrained GaInN is expected to be smaller than that determined in this study.

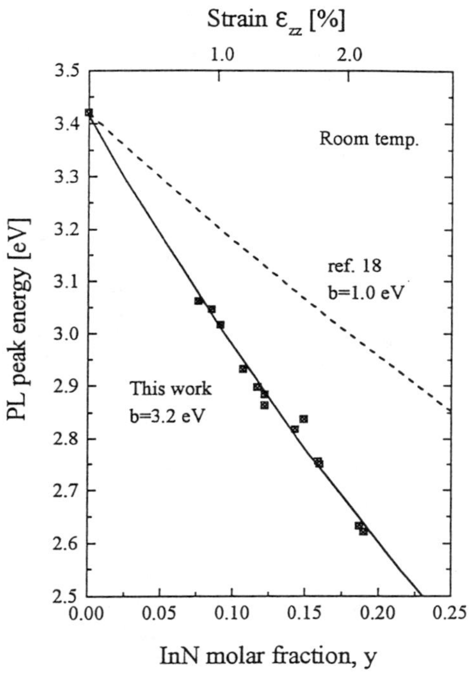

Fig.4 Compositional dependence of the PL peak energy of strained GaInN grown coherently on GaN.

Quantum confined Stark effect in MQW

Figure 5 shows the RT PL peak energies of the $Ga_{0.88}In_{0.12}N$ strained QWs as a function of well width. The calculation results are also shown for comparison. Solid squares represent the measured PL peak energies under the high excitation with N_2 laser, and open squares under the low excitation with the He-Cd laser. The electric field along [0001] in the strained layer due to piezoelectric effect are given by

$$E_z = -\frac{P_z}{\varepsilon \varepsilon_0} = -\frac{1}{\varepsilon_r \varepsilon_0}(e_{33} - \frac{c_{33}}{c_{13}}e_{31})\varepsilon_{zz}$$

where ε_r and ε_o is the dielectric constant and the permitivity in vacuum, e_{13} and e_{33} are the piezoelectric constant, c_{13} and c_{33} is the elastic stiffness constant, and ε_{zz} is the strain parallel to c-direction. We use the piezoelectric constants of GaN[27] even in case of ternary alloys. Piezoelectric field up to 1.0 MV/cm can be calculated along [0001] direction when the InN molar fraction is 0.12. A solid line is the calculated transition energies assuming such a high electric field is induced in the $Ga_{0.88}In_{0.12}N$ strained quantum well layer. Transition energy without electric field is also shown by a dashed line. Very thin well layer of less than 3.5 nm showed blue shift due to quantum size effect (QSE). Under low excitation, transition energy of the well layer is smaller than that of thick GaInN single layer especially at the well layer thicker than about 3.5 nm because of the so-called quantum confined Stark effect (QCSE).

However, as the excitation intensity increase, transition energy of the well layer becomes close to that of strained thick GaInN single layer, which is caused by the Coulomb screening of the piezoelectric induced electric field by plenty of electrons and holes generated in the well layer. It should be emphasized that if the well layer becomes too thick,

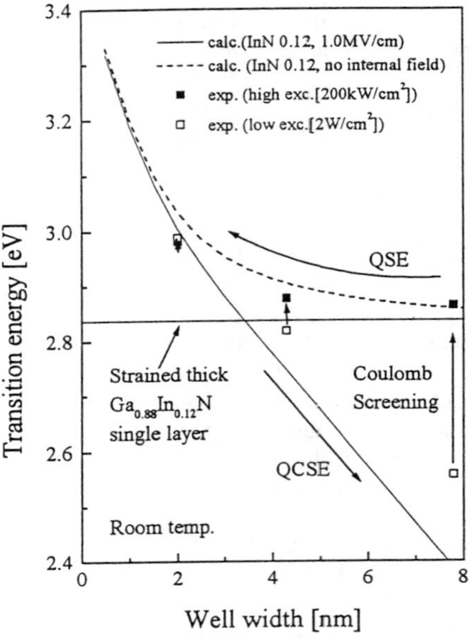

Fig.5 PL peak energy of 10 periods of $Ga_{0.88}In_{0.12}N/Ga_{0.97}In_{0.03}N$ MQWs as a function of well width. Solid and open squares represent the measured PL peak energies at room temperature under high and low excitation, respectively. Solid and dashed lines represent the calculation results with piezoelectric field of 1.0MV/cm and without any field, respectively. The measured PL peak energy of $Ga_{0.88}In_{0.12}N$ single layer is also shown.

even though the well layer is strained, PL peak energy becomes close to that of thick single layer. This is because overlap of wavefunction of separate electrons and holes becomes smaller as the well layer thickness increase. Figure 6 summarizes the compositional dependence of the lowest transition energy of strained GaInN as a function of well width. Bandgap of MQWs having rather thick wells is expected to show large red-shift by QCSE.

CONCLUSION

Structural and optical properties of AlGaN and GaInN ternary layers grown on thick GaN were obtained. It is found that both AlGaN and GaInN layer are coherently grown on GaN. Bandgap of strained AlGaN is found to almost linearly increase with increase of AlN molar fraction, while that of GaInN is largely red-shifted. Optical properties of strained ternary compounds were clarified. Quantum confined Stark effect cause large red-shift of the lowest transition energy especially in nitride-based MQWs when a growth direction parallel to [0001]. These results are very important for the design and fabrication of future nitride-based heterostructured devices.

ACKNOWLEDGEMENT

This work was partly supported by the Ministry of Education, Science, Sports and Culture of Japan, (contract nos. 06452114, #07505012,#07650025 and High-Tech Research Center Project), Japan Society for the Promotion of Science (Research for the Future) and Daiko Foundation.

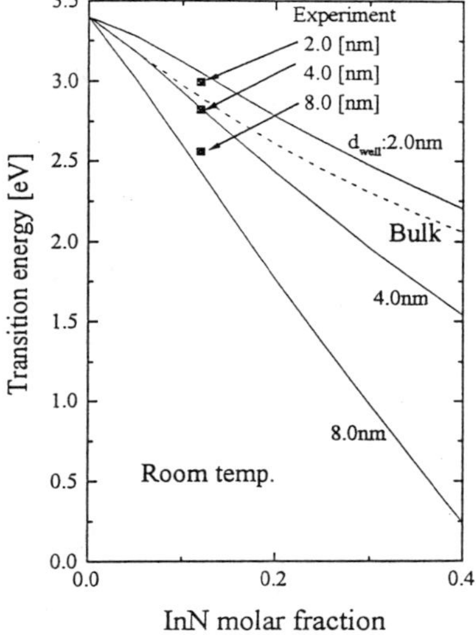

Fig.6 Compositional dependence of the transition energy of strained MQW as a function of well width. That of bulk GaInN is also shown for comparison.

REFERENCES

1. H. Amano, N. Sawaki, I. Akasaki and Y. Toyoda, Appl. Phys. Lett. **48**(1986)353.
2. S. Nakamura, T. Mukai and M. Senoh, J. Appl. Phys. **71**(1992)5543.
3. T. D. Moustakas, R. J. Molnar, T. Rei, G. Menon and C. R. Eddy Jr., Mat. Res. Soc. Symp. Proc., **242**(1992)427.
4. H. Amano, M. Kitoh, K. Hiramatsu and I. Akasaki, Jpn. J. Appl. Phys. **28**(1989)L2112 .
5. S. Nakamura, M. Senoh and T. Mukai, Jpn. J. Appl. Phys. **30**(1991)L1708.
6. S. Nakamura, T. Mukai and M. Senoh, Appl. Phys. Lett. **64**(1994)1687.
7. S. Nakamura, M. Senoh, N. Iwasa, and S. Nagahata, Jpn. J. Appl. Phys. **34**(1995)L797.
8. M. Koike, N. Shibata, H. Kato, S. Yamasaki, N. Koide, H.Amano and I. Akasaki, Proc.1st

Int.Sym.GaN and Related Materials, Boston, 1995, **395**(1995)889.

9. I. Akasaki, H. Amano, S. Sota, H. Sakai, T. Tanaka and M. Koike, Jpn. J. Appl. Phys. **34**(1995)L1517.

10. S. Nakamura, M. Senoh, S. Nagahata, N. Iwasa, T.Yamada, T.Matsushita, H. Kiyoku and Y. Sugimoto, Jpn. J. Appl. Phys. 35(1996)L74.

11. K. Itaya, M. Onomura, J. Nishio, L. Sugiura, S.Saito, M. Suzuki, J. Rennie, S. Nunoue, M. Yamamoto, H. Fujimoto, Y. Kokubun, Y. Ohba, G. Hatakoshi and M. Ishikawa, Jpn. J. Appl. Phys., 35(1996)L1315.

12. I. Akasaki, S. Sota, H. Sakai, T. Tanaka, M. Koike and H.Amano, Electron. Lett. 32(1996)1105.

13. S. Sota, H. Takeuchi, T. Takeuchi, H. Sakai, H.Amano and I. Akasaki: unpublished.

14. Y. Koide, H. Itoh, M. R. H. Khan, K. Hiramatsu, N. Sawaki and I. Akasaki, J. Appl. Phys. **61**(1987)4540.

15. D. K. Wickenden, C. B. Bargeron, W. A. Bryden, J. Miragliova and T. J. Kistenmacher, Appl. Phys. Lett. **65**(1994)2024.

16. S. Yoshida, S. Misawa and S. Gonda, J. Appl. Phys. **53**(1982)6844.

17. K. Osamura, K. Nakajima, Y. Murakami, P. H. Shingu and A. Otsuki, Solid State Commun. **11**(1972)617.

18. S. Nakamura and T. Mukai, J. Vac. Sci. Technol. **A13**(1995)705.

19. K. Tsubouchi, K. Sugai and N. Mikoshiba, 1981 Ultrasonic Proc. (IEEE, New York, 1981) 375.

20. M. Yamaguchi, T. Yagi, T. Azuhata, T. Sota, K. Suzuki, S. Chichibu and S. Nakamura, to be published in J.Phys.Condens.Matter.

21. G. Martin, A. Botchkarev, A. Rockett and H. Morkoc, Appl. Phys. Lett. **68**(1996)2541.

22. S. Ninomiya and S. Adachi, J. Appl. Phys. **78**(1995)1183.

23. L. Leszczynski, H. Teisseyre, T. Suski, I. Grzegory, M. Bockowski, J. Jun, S. Porowski, K. Pakula, J. M. Baranowski, C. T. Foxon and T. S. Chueng, Appl. Phys. Lett. **69**(1996)73.

24. T. Detchprohm, K. Hiramatsu, K. Itoh and I. Akasaki, Jpn. J. Appl. Phys. **31**(1992)L1454.

25. J. W. Matthews and A. E. Blakeslee, J. Crystal Growth **32**(1974)265.

26. A. Fischer, H. Kuhne and H. Richter, Phys. Rev. Lett., **73**(1994)2712.

27. A. D. Bykhovski, V. V. Kaminski, M. S. Sur, Q. C. Chen and M. A. Khan, Appl. Phys. Lett., **68**(1996)818

DESIGN CONSIDERATION OF GaN-BASED SURFACE EMITTING LASERS

T. Honda, F. Koyama and K. Iga
Precision and Intelligence Laboratory, Tokyo Institute of Technology,
4259 Nagatsuta, Midori-ku, Yokohama 226, Japan
thonda@pi.titech.ac.jp

ABSTRACT

The threshold current density of GaN-based vertical cavity surface emitting lasers (VCSELs) has been estimated. It is clarified that the introduction of a quantum well structure as an active layer is very effective for a low threshold operation and that high reflective mirrors are required for low threshold GaN-based VCSELs. Also, attempts on micro-fabrication process of GaN is presented.

INTRODUCTION

A GaN-based vertical cavity surface emitting laser (VCSEL) operating from ultraviolet to blue spectral regions is recognized as an important device for full-color display panels and high-density optical recording, because of its low threshold potentiality and two-dimensional array configuration. Some groups have been challenging to fabricate those VCSELs [1-5]. It is substantial to estimate the threshold current for a suitable design of GaN-based VCSELs. In this paper, a design concept for low threshold GaN-based QW surface emitting lasers is presented. Also, some key issues including high reflective mirrors, dry etching techniques and current confinement structures for VCSELs are discussed.

BASIC CONSIDERATION

Optical Gain of Threshold Current Density of GaN-Based VCSELs

The volume of an active region in VCSEL is smaller than that in a conventional stripe laser. Therefore, it is possible to fabricate devices not only with low threshold current, but also with long lifetime because a defect-free epilayer with a small active region can be used for device fabrication. If the GaN-based material is used for VCSELs operating with such a short wavelength, the threshold current estimation should be crucial to realize GaN-based VCSELs with low threshold operation. The threshold current depends on the volume of an active region. Thus, it is required, for the design of low-threshold GaN-based VCSELs, to obtain the relationship between a threshold current and a volume of the active layer. Especially, the quantum well (QW) layer as an active region is very interesting for the realization of a low-threshold GaN-based VCSEL.

The optical gain is one of the important parameters to estimate the threshold current density of GaN-based VCSELs. The estimation of linear gain for $GaN/Al_{0.1}Ga_{0.9}N$ quantum well is carried out using the density-matrix theory with intraband broadening [6]. We have considered TE mode and employed the same method as Asada carried out for the strong polarization dependence of QWs [7]. This model has already been established for the GaAs-based and InP-based semiconductor lasers and its result shows good agreement with experiments. In this

Mat. Res. Soc. Symp. Proc. Vol. 449 ©1997 Materials Research Society

estimation, the square of dipole moment is an important parameter and we assume that this is dependent on band-gap energy of an active region, reported previously [1]. Also we assumed the dipole moment as a constant, which is independent on the transition energy.

In the case of QWs, the optical gain can be written as:

$$\alpha(\omega) = \omega \sqrt{\frac{\mu}{\varepsilon}} \left\{ \frac{1}{\pi \hbar d} \left(\frac{m_c m_h}{m_c + m_h} \right) \right\}$$

$$\times \sum_{n=0}^{M-1} \int_{Ecn+Ehn+Eg}^{\infty} \langle R_{ch}^2 \rangle_n (f_c - f_v) L(E_{ch} - \hbar\omega) dE_{ch}$$

(1)

where ω denotes optical angular frequency, μ_0 is magnetic permeability in vacuum, ε_0 is dielectric constant in vacuum, n is reflective index of an active region, f_c and f_v are the Fermi-Dirac distribution functions at the conduction band and valence band, and τ_{in} is intraband relaxation time. The Lorentzian function is assumed as a line shape function, $L(E_{ch}-\hbar\omega)$.

$$L(E_{ch} - \hbar\omega) = \frac{\hbar / \tau_{in}}{(E_{ch} - \hbar\omega)^2 + (\hbar / \tau_{in})^2}$$

(2)

The square of the dipole moment $\langle R_{ch}^2 \rangle$ for the TE mode of a quantum well is given by

$$\langle R_{ch}^2 \rangle \approx \frac{3}{2} \cdot \left(\frac{1 + \frac{E_{cn}}{\varepsilon_{cn}}}{2} \right) \left(e \cdot \frac{4.4}{E_g[eV]} \right) \qquad \left(C^2 \cdot \overset{\circ}{A}{}^2 \right)$$

(3)

In this equation, ε_{cn} denotes the transition energy of photon. We treat that $(E_{cn}/\varepsilon_{cn})$ is unity because the transition energy is corresponding to the energy near band edge emission.

The maximum gain is plotted as a function of an injected carrier density as shown in Fig. 1 [8]. The material parameters used for the estimation are listed in Table I. In this figure, solid lines and dashed lines denote the maximum gain of GaN/Al$_{0.1}$Ga$_{0.9}$N QWs and bulk crystals, respectively. At the present stage, the intraband relaxation time τ_{in} is a variable parameter ranging from 5.0×10^{14} to $> 1.0 \times 10^{-11}$ s. The transparent carrier density of GaN is higher than other III-V materials like as GaAs, originating from its heavy electron and hole mass. Generally, the effective masses of an electron and a hole depend on the bandgap energy. Thus it seems that the wide-bandgap semiconductors require higher transparent carrier densities than narrow-bandgap materials. This tendency resulted in a heavy effective mass can be seen in the estimations about the optical gain reported by other authors [9-11]. This result indicates that the GaN/Al$_{0.1}$Ga$_{0.9}$N QW having high material gain is useful for the low threshold operation a VCSELs. It is also useful for a conventional edge emitting lasers.

Table I   Material parameters used for the calculation.

GaN

	Bandgap energy	E_g	3.4 eV	ref. 12
	Effective mass	m_c	$0.2m_0$	ref. 13
		m_0	$0.8m_0$	ref. 14
	Reflective index	n	2.8	ref. 15

$Al_{0.1}Ga_{0.9}N$

	Bandgap energy	3.7 eV	ref. 15

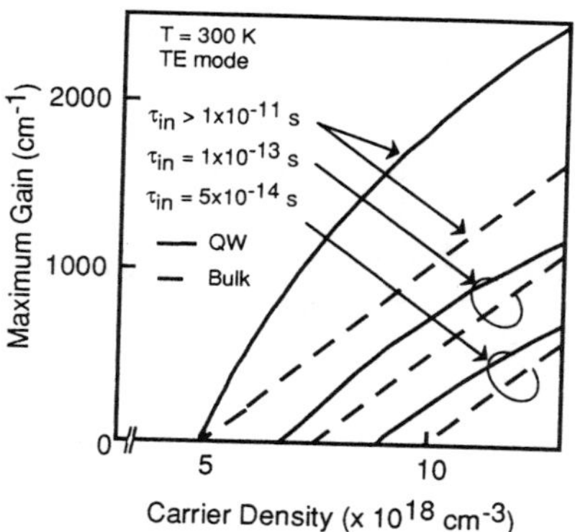

Fig.1   Maximum gain of GaN as a function of injected carrier density.

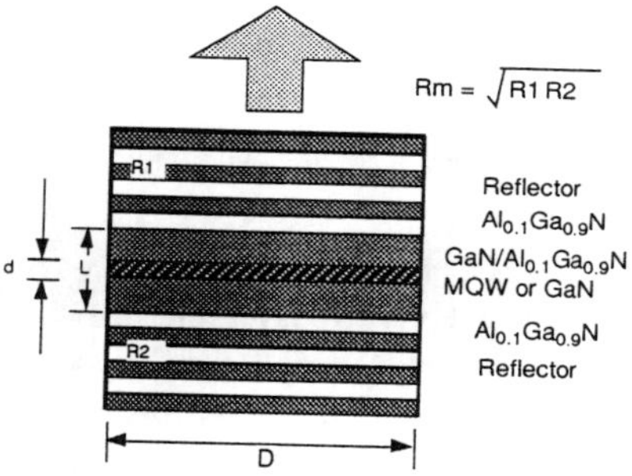

Fig. 2 Estimated structure in this study.

Estimation of Threshold Current Density of GaN-Based VCSELs

We have estimated the threshold current density of GaN-based VCSELs with the quantum well structure as an active region shown in Fig. 2 based on the estimation of the maximum gain. We calculated the spontaneous lifetime by the same way reported by Yamada and Ishiguro [16]. We also considered the GaN/Al$_{0.1}$Ga$_{0.9}$N single quantum well layer as an active region and the optical standing wave in the cavity [17]. The threshold current density as a function of the active layer thickness is shown in Fig. 3. If we can fabricate the thickness of the GaN active layer less than 10 nm, the threshold current density of GaN-based VCSEL could be 1 kA/cm^2 or lower. These results indicate that the low threshold GaN-based VCSEL can be expected, if high quality GaN-based layers and high reflective mirrors higher than 99.9 % are realized.

The device diameter defined by the radius of an active region is a very important parameter for low threshold devices. We have to decrease the diameter to realize a low threshold. In Fig. 4, the relation of a device diameter and threshold current is shown. If we can fabricate the device with a small diameter less than 10 μm, the GaN-based VCSEL with sub milliampere low threshold currents can be excepted.

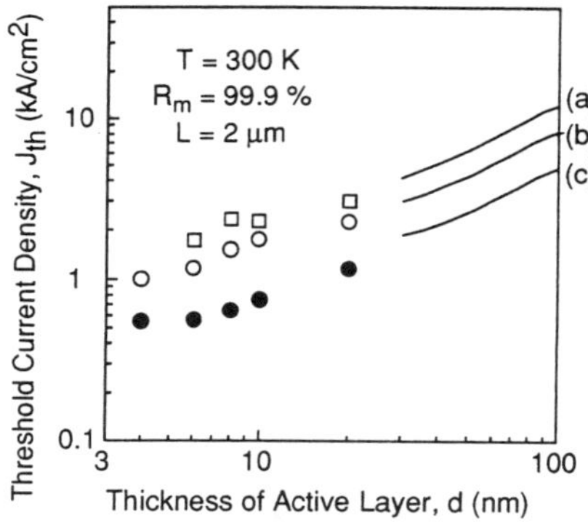

Fig. 3 Threshold current density as a function of active layer thickness. (a), (b) and (c) are intraband relaxation times, τ_{in}, of 5×10^{-14} s, 1×10^{-13} s and more than 1×10^{-11} s, respectively.

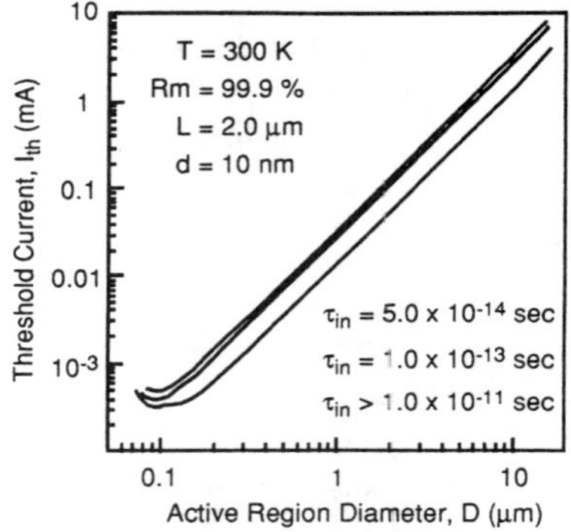

Fig. 4 Threshold current as a function of active region diameter.

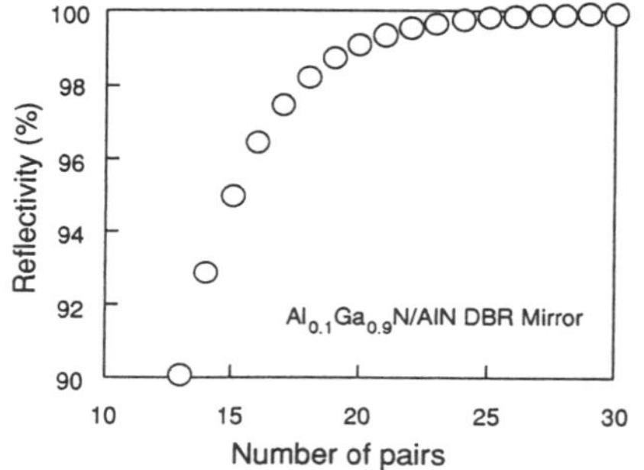

Fig. 5 Reflectivity dependence on number of pairs.

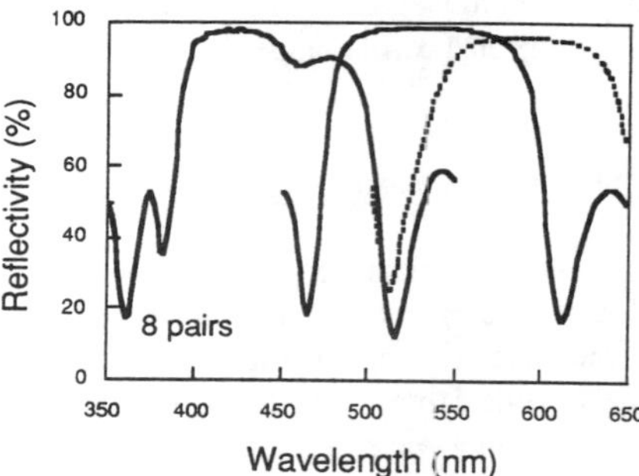

Fig. 6. Spectral reflectivity of SiO_2/ZrO_2 multilayer DBR.

FABRICATION PROCESS

<u>Highly Reflective Mirror</u>

The fabrication of a highly reflective mirror is required for low-threshold operations of VCSELs. A semiconductor DBR mirror can be fabricated monolithically by epitaxial growth. However, this mirror requires a large number of pairs to obtain high-reflectivity owing to the small refractive index difference of the two semiconductor materials. Since the band-gap of AlN is larger than that of GaN, it can be considered that the absorption of AlGaN can be negligible at the bandgap wavelength of GaN. It was reported that the refractive indices of AlN and GaN are 2.2 and 2.67, respectively [18-21]. The refractive index difference of $AlN/Al_{0.1}Ga_{0.9}N$ is comparable to that of AlAs and GaAs. Therefore, an $AlN/Al_{0.1}Ga_{0.9}N$ DBR is considered to be a suitable mirror for GaN-based VCSELs operating in the ultraviolet spectral regions. The theoretical reflectivity of the AlN/AlGaN DBR mirror is calculated and shown as a function of the number of pairs in Fig. 5. In this calculation, we have applied a transfer matrix method and the absorption of $AlN/Al_{0.1}Ga_{0.9}N$ layers is not considered. In the case of 20 pairs of $AlN/Al_{0.1}Ga_{0.9}N$ layers, the reflectivity at the GaN band-gap energy is higher than 99%. From this result we found that the AlN/AlGaN DBR structure is a good candidate for a high-reflectivity mirror.

On the other hand, dielectric multilayer stacks such as SiO_2/MgO, SiO_2/ZrO_2 and SiO_2/Y_2O_3 and so on are also the candidates for high reflective mirrors. However, if we adopt these mirror as highly reflective mirrors, it is required to establish the technique of removing a substrate. In any way, dielectric multilayer stacks can be used for a top side mirror. SiO_2/TiO_2 stacks are used for VCSELs operating in infrared spectral region, however, those can not be used VCSELs operating in ultraviolet spectral region because of its large absorption. Because the absorption edges of MgO, ZrO_2 and Y_2O_3 are in ultraviolet, those can be used for the low index materials for multilayer reflectors in VCSELs. Figure 6 shows some reflective spectra of SiO_2/ZrO_2 Bragg reflectors. Those are deposited by electron beam evaporation with a thickness monitor. During the deposition, oxygen gas was introduced into the chamber to prevent a lack of oxygen in the stacks. It is concluded that SiO_2/ZrO_2 pairs are useful for the application of GaN-based and others VCSELs operating between red to ultraviolet spectral regions.

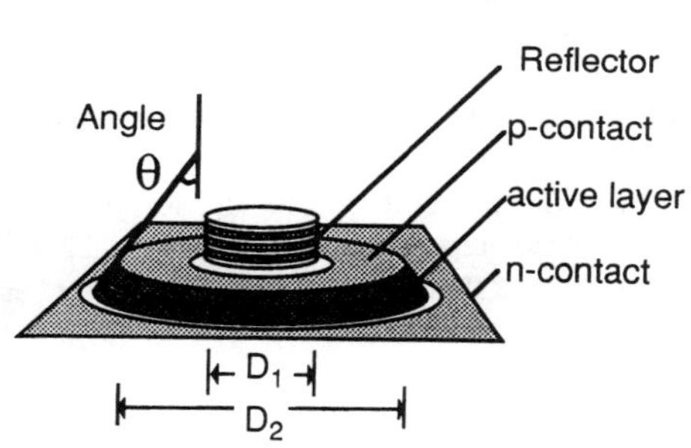

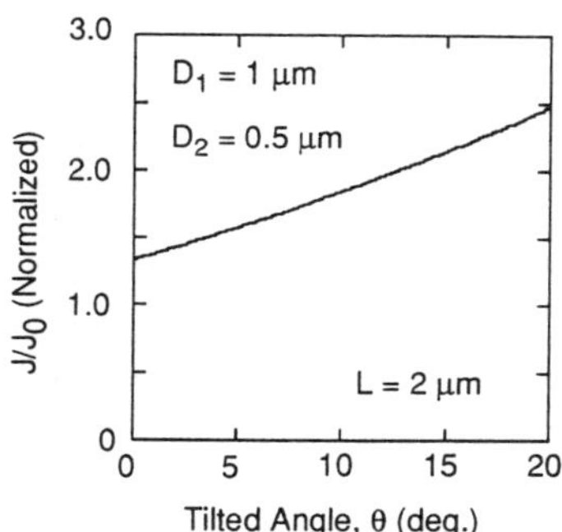

Fig. 7 Schematic diagram of etched structure.

Fig. 8 Normalized threshold current density as a function of tiled angle.

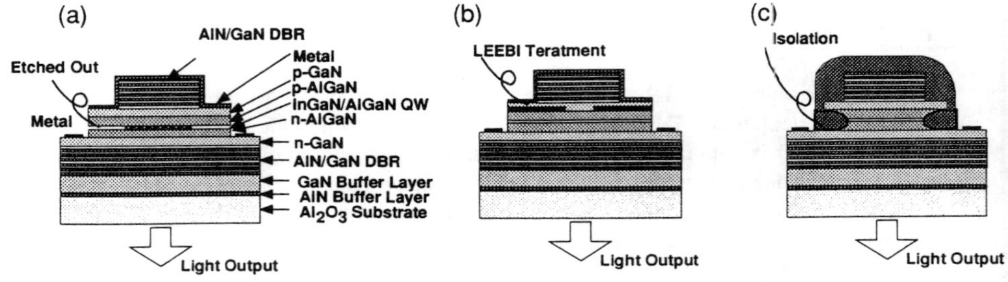

Fig. 9 Schematic diagram of current confinement structure. (a), (b) and (c) are confinement structures using side-etch-out of InGaN active layer , LEEBI treatment, and isolation, respectively.

<u>Dry Etch of GaN Layers for VCSELs</u>

GaN and related compounds grown on sapphire substrates have hexagonal structure, which have been applied to light emitting diodes [22 - 25]. If we want to develop the GaN-based lasers, we have to fabricate a laser cavity including the high quality mirrors with smooth surface and vertical walls. In the case of GaN-based VCSELs, it is one of the crucial issues to fabricate smooth and vertical side-walls. A dry etching technique is one of the effective method to fabricate a high quality cavity. The required quality of etched side-walls for GaN-based VCSELs are discussed.

The mesa structure of a GaN VCSEL is shown in Fig. 7, and a uniform injection is assumed. In this figure, J and J_0 are defined as the threshold current density on the p-side contact and the ideal threshold current density on an active layer, respectively. In the case of VCSELs with a large diameter, a tilted side wall does not affect the threshold. However, if we consider the VCSELs with a 10 μm diameter, the tilted angle of a side wall under 10 degree is required to suppress the increase of threshold as shown in Fig. 8. However, if we can adopted the current confinement structure, the threshold current density on the p-side contact layer will be suppressed. On the other hand, the surface recombination is crucial issue for the micro devices. We developed a wet etch technique with ultraviolet light irradiation in an alkali solution for removing a damage layer on the etched surface [26].

For GaN-based semiconductor lasers with a stripe shape, it is necessary to fabricate a smooth and vertical facet for the resonator, which have been reported in Saotome *et al.* [27] and Itaya *et al.* [28]. We found a reactive ion beam etching (RIBE) technique is one of the best solution to fabricate the laser cavity.

<u>Current Confinement Structure</u>

The current confinement is another critical issue to realize GaN-based VCSELs with low threshold operation. Furthermore, the contact resistivity at the p-side boundary between a metal and a GaN-based layer do not show a good ohmic characteristic at the present stage. Good ohmic contacts are required to suppress Joule heat and power consumption. Thus, the wide contact area on the p-side have to be required to reduce the total resistivity in the p-side contact. If we adopt the DBR with high aluminum composition layers, we have to consider an intra-cavity current injection. The side-etch-out of InGaN active layers for the current confinement have been presented by DenBaars *et al.* [4].

The LEEBI technique [29] is useful to fabricate the confinement structure. That can be fabricated to change the exposure time of LEEBI treatment. In these cases, we have to consider the effects of a surface recombination in the active layer and a distribution of vertical carrier profiles in the current confinement layer. The thermal anneal [30] and proton implantation techniques are also possible ways for the fabrication of a current confinement structure as shown in Fig. 9. High resistive GaN can be formed from the side of mesas by a H_2 annealing treatment.

THRESHOLD CRITERION

When we face a new and/or ultra-low threshold device to check the lasing operation, the nonlinear relation in the injection current vs light output characteristics is an easy way for threshold criterion. However sometimes this is observed even at non-lasing samples owing to "filtering effect" and electron-hole plasma emission as well. The methods to definitely confirm the laser operation of vertical cavity for example are shown as follows [31];

1) Narrow line width < 1 Å.
2) Difference of near field patterns (NFP) and far field patterns (FFP) below and above threshold.
3) Linearly polarized light of the emission above the threshold.

CONCLUSIONS

In summary, the optical gain of quantum well GaN was estimated based on the density-matrix theory. The introduction of GaN quantum wells as an active region in GaN-based surface emitting laser can reduce the threshold current lower than that of bulk GaN layer. Sub micro ampere operation can be expected by reducing the diameter of devices. Some fabrication process such as high reflective mirrors and dry-etch techniques are also discussed.

There may be some unknown substantial problems to realize GaN-based VCSELs in addition some technical issues which should be developed. Further engineering studies are required to realize high performance GaN VCSELs.

ACKNOWLEDGMENTS

The author would like to thank T. Sakaguchi, A. Katsube, K. Saotome, T. Shirasawa, M. Mori, N. Mochida and A. Inoue for discussions. This study was partly supported by Grant-in-Aid for COE Research from the Ministry of Education, Science, Sports and Culture (#07CE2003, "Ultra-parallel Optoelectronics").

REFERENCES

1. T. Honda, A. Katsube, T. Sakaguchi, F. Koyama and K. Iga, Jpn. J. Appl. Phys. **34**, 3527 (1995).

2. M. A. Khan, S. Krishnankutty, R. A. Skogman, J. N. Kuznia, T. Olson and T. George, Appl. Phys. Lett., **65**, 520 (1994).

3. J. M. Redwing, D. A. S. Loeber, N. G. Anderson, M. A. Tischler and J. S. Flynn, Appl. Lett. **69**, 1 (1996).

4. S. P. DenBaars, *High Speed Optoelectronics Conference for Communications*, Snowbird, UT, 1996 (unpublished).

5. M. Oshinski, V. A. Smagley and P. G. Eliseev, *High Speed Optoelectronics Conference for Communications*, Snowbird, UT, 1996 (unpublished).

6. M. Asada, A. Kameyama and Y. Suematsu, IEEE J. Quantum Electron. **20**, 745 (1984).

7. M. Asada and Y. Suematsu, IEEE J. Quantum Electron. **21**, 434 (1985).

8. A. Katsube, T. Honda, T. Sakaguchi, F. Koyama and K. Iga, *Topical Workshop on III-V Nitrides*, P-8, Nagoya, Japan, 1995.

9. S. Kamiyama, K. Ohnaka, M. Suzuki and T. Uenoyama, Jpn. J. Appl. Phys. **34**, L821 (1995).

10. K. Domen, K. Horino, A. Kuramata and T. Tanahashi, *15th IEEE International Semiconductor Laser Conference*, W2.7, Haifa, Israel, 1996.

11. S. L. Chung, IEEE J. Quantum Electorn., **32**, 1791 (1996).

12. H. P. Maruska and J. J. Tietjen: Appl. Phys. Lett. **15**, 327 (1969).

13. A. S. Barker, M. Ilegems: Phys. Rev. **B7** (1973) 743.

14. J. I. Pankove, S. Bloom, G. Harbeke: RCA Rev. **36**, 163 (1975).

15. H. Amano, N. Watanabe, N. Koide and I. Akasaki: Jpn. J. Appl. Phys. **32**, L1000 (1993).

16. M. Yamada and H. Ishiguro, Jpn. J. Appl. Phys. **19**, 135 (1980).

17. T. Tamanuki, F. Koyama and K. Iga, Jpn. J. Appl. Phys. **30**, L593 (1991).

18. H. Amano, N. Watanabe, N. Koide and I. Akasaki, Jpn. J. Appl. Phys. **32** , L1000 (1993).

19. E. Ejder, Phys. Status Solidi a **6**, K39 (1971).

20. A. T. Collins, E. C. Lightowlers and P. J. Dean, Phys. Rev. **158**, 833 (1967).

21. M. A. Khan, J. N. Kuznia, J. M. Van Hove and D. T. Olson, Appl. Phys. Lett. **59**, 1449 (1991).

22. T. Detchprohm, K. Hiramatsu, N. Sawaki and I. Akasaki, J. Cryst. Growth **145**, 192 (1994).

23. S. Nakamura, T. Mukai and M. Senoh, Appl. Phys. Lett., **64**, 1687 (1994).

24. I. Akasaki, H. Amano, S. Sota, H. Sakai, T. Tanaka, M. Koike, Jpn. J. Appl. Phys. **34**, L1517 (1995).

25. S. Nakamura, M. Senoh, S. Nagahama, N. Iwasa, T. Yamada, T. Matsushita, H. Kiyoku and Y. Sugimoto, Jpn. J. Appl. Phys. **35**, L74 (1996).

26. K. Saotome, A. Matsutani, T. Shirasawa, M. Mori, T. Honda, T. Sakaguchi, F. Koyama and K. Iga, will be presented at the 1996 MRS Fall Meeting, Boston, MA, 1996.

27. K. Saotome, T. Honda, A. Matsutani, F. Koyama and K. Iga in *Proceeedings of International Symposium on Blue Laser and Light Emitting Diodes* (Ohmsha, Tokyo, 1996) pp. 506 -509.

28. K. Itaya, M. Onomura, J. Nishio, L. Sugiura, S. Saito, M. Suzuki, J. Rennie, S. Nunoue, M. Yamamoto, H. Fijimoto, Y. Kokubun, Y. Ohba, G. Hatagoshi and M. Ishikawa, Jpn. J. Appl. Phys., **35**, L1315 (1996).

29. H. Amano, M. Kito, K. Hiramatsu and I. Akasaki, Jpn. J. Appl. Phys. **28** , L2112 (1989).

30. S. Nakamura, N. Iwasa, M. Senoh and T. Mukai, Jpn. J. Appl. Phys. **31**, 1258 (1992).

31. K. Iga in *Proceeedings of International Symposium on Blue Laser and Light Emitting Diodes* (Ohmsha, Tokyo, 1996) pp. 263 -266.

Stacked InGaN/AlGaN Double Heterostructures

J. C. Roberts[*], F. G. McIntosh[*], M. E. Aumer[*], E. L. Piner[**], V. A. Joshkin[*], S. Liu[**], N. A. El-Masry[**], S. M. Bedair[*]
[*]ECE Dept., North Carolina State University, Raleigh, North Carolina 27695-7911.
[**]MSE Dept., North Carolina State University, Raleigh, North Carolina 27695-7916.

ABSTRACT

InGaN ternary alloys can be the basis for light emission from the near UV to the red region of the electromagnetic spectrum. When InGaN/AlGaN double heterostructures emitting different colors are stacked in a single structure, simultaneous emission of different wavelengths will be achieved. If the color and the intensity of emission for each well are adjusted properly, tailored emission spectra, including white light, will be feasible. We demonstrate this concept with two wells emitting at different wavelengths that are stacked between AlGaN barrier layers. The emitted PL spectra for the stacked structure is found to be the superposition of the emission from the individual double heterostructures that were grown separately.

INTRODUCTION

The AlGaInN materials system is unique in that its direct energy bandgap can be theoretically engineered over a range extending through the entire mid ultraviolet to the red region of the electromagnetic spectrum. This wide spectral range is extremely attractive for light emitting devices used in full color displays, and has provided an impetus toward the rapid development of this materials system. In the past several years, dramatic advances in the epitaxial growth, doping and processing of the III-nitride alloys have led to the development of commercially available high brightness blue and green light emitting diodes (LEDs)[1] and the realization of violet emitting laser diodes.[2] $In_xGa_{1-x}N$ based double heterostructures (DHs) and quantum wells (QWs) form the active layers of these optoelectronic devices, with the emission wavelength being determined primarily by the value of x.

To date, optical devices based on the III-nitride materials system are designed to emit one color. There are many applications, however, where several colors, or even white light may be desired from a single device. Examples of such applications include full color indoor/outdoor displays and spectrally tailored white light sources that are bright, compact, light weight, long lived, and efficient. The potential demand for a solid state white light source as a viable replacement for conventional incandescent or fluorescent light bulbs is by itself enormous. In this paper, we present preliminary results of a device structure designed for multicolor emission that is based on stacked $Al_yGa_{1-y}N/In_xGa_{1-x}N$ DHs. By changing the growth conditions of the $In_xGa_{1-x}N$ active layers in this stacked structure, two distinct colors are emitted, one color from each active layer, as observed in photoluminescence (PL) spectra. The colors emitted from each active layer in the stacked structure match the colors emitted from conventional DHs having $In_xGa_{1-x}N$ active layers grown under the same

Mat. Res. Soc. Symp. Proc. Vol. 449 © 1997 Materials Research Society

conditions as those of the stacked structure.

EXPERIMENT

Epitaxial growth of these $In_xGa_{1-x}N$ and $Al_yGa_{1-y}N$ layers was performed in a hybrid ALE/MOCVD growth system that has been previously described[3]. This hybrid ALE/MOCVD system was previously used for the epitaxial growth of $In_xGa_{1-x}N$ (0 < x < 0.27) by ALE[4] and of AlInGaN quaternary alloys by MOCVD[5]. Source gases used were, trimethylgallium (TMG, -10 °C), trimethylaluminum (TMA, +18 °C), ethyldimethylindium (EdMIn, +10 °C) and NH_3; N_2 was used as the carrier gas. Basal plane sapphire substrates were solvent cleaned and annealed in N_2 and NH_3 for 15 minutes and 1 minute respectively at 1050 °C prior to epitaxial growth. A schematic of the stacked DH is shown in Figure 1. ALE growth of the AlN buffer layer[6] was performed at 700 °C while the AlGaN (~7 - 10% AlN) cladding layers were grown at 950 °C. The lower and upper InGaN active layers, grown at temperatures of 780 and 750 °C respectively, are separated by an AlGaN spacer layer which was grown at 950 °C. The active layers were deposited on InGaN prelayers (not shown in the figure) whose growth temperatures were ramped during growth from 790 °C down to the active layer growth temperature of 780 or 750 °C. The InGaN prelayers are believed to provide a graded InGaN buffer that seems to improve the optical properties of the active layer.

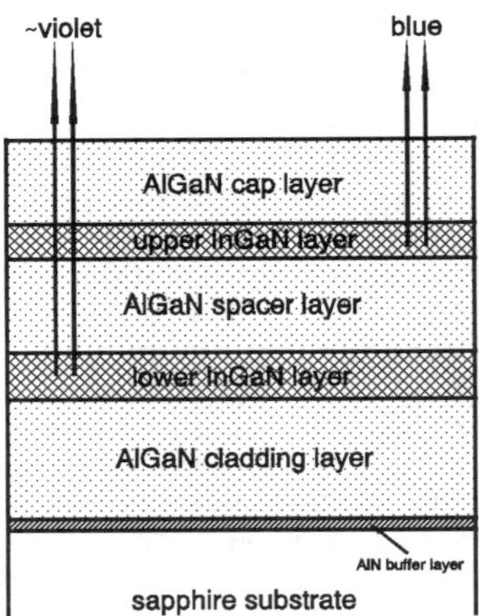

Figure 1. Schematic of stacked double heterostructure with two InGaN active layers.

In order to further study the emission from the stacked structure, two conventional DHs were deposited having InGaN active layers grown at 780 and 750 °C respectively. These active layers were sandwiched in between AlGaN cladding layers that were grown at 950 °C. It should be noted that the growth conditions of the active layers and cladding layers of these conventional DHs were the same as those used for the stacked DH.

RESULTS

PL spectra of the two conventional DHs with the InGaN active layers grown at 780 °C and 750 °C are shown in Figures 2 and 3, respectively. The excitation source is a HeCd laser operating at 325 nm with cw power of approximately 25 mW. All spectral measurements were performed at room temperature. The DH grown at 780 °C exhibits peak emission near 390 nm with a full width at half maximum (FWHM) of ~15 nm, while the DH deposited at 750 °C has peak emission near 470 nm and a FWHM of ~38 nm.

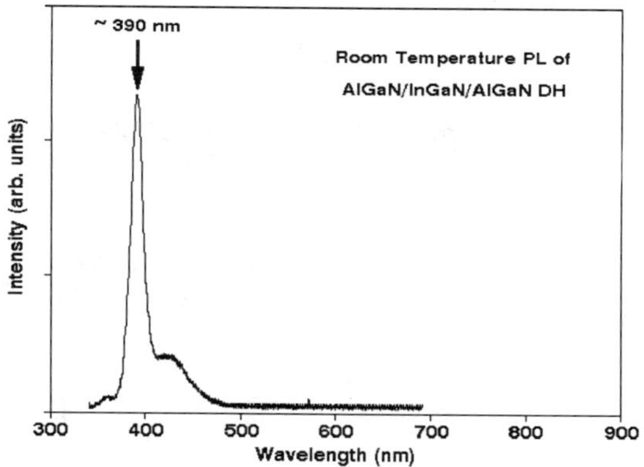

Figure 2. Room temperature PL of a conventional DH with an InGaN active layer grown the same as the lower InGaN layer of the stacked DH.

The PL spectrum of the stacked DH, shown in Figure 4, exhibits two main peaks at 384 nm and ~470 nm respectively. The peak at 384 nm has a FWHM of ~17 nm, while the peak near 470 nm has a FWHM of ~55 nm. There is good correspondence between the PL spectra from the two conventional DHs to the two colors emitted from the stacked DH. In comparing the spectra between the conventional and stacked structures, the FWHM of the peak corresponding to 390 nm has nearly the same value while the FWHM of the peak corresponding to 470 nm is broader in the stacked DH than in the conventional DH. This can be attributed to the upper InGaN active layer being grown on a potentially rougher surface. The AlGaN spacer layer, grown at 950 °C on top of the first InGaN active layer, can suffer from poor surface morphology and can create subsequent nucleation problems for overlying layers. Maintaining good surface morphology throughout the entire structure will

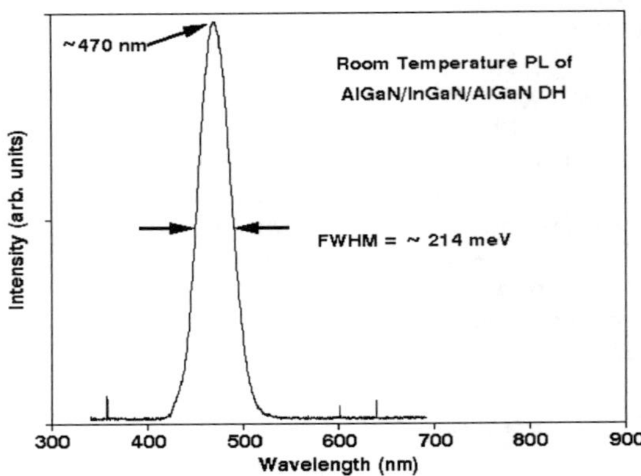

Figure 3. Room temperature PL of a conventional DH with an InGaN active layer grown the same as the upper InGaN layer of the stacked DH.

depend on optimizing the growth conditions during the transition between the InGaN active layers grown at relatively lower temperatures and the AlGaN layers that are grown at higher temperatures.

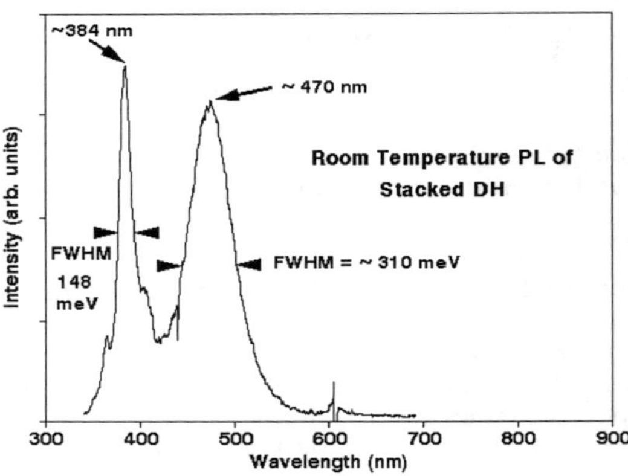

Figure 4. Room temperature PL of stacked DH showing two color emission from upper and lower InGaN active layers, respectively.

CONCLUSIONS

A novel stacked AlGaN/InGaN DH has been epitaxially grown on a sapphire substrate. The room temperature PL spectrum from this structure exhibits two color emission at 390 nm ($\sim$violet) and 470 nm (blue). These emission wavelengths match the emission wavelengths of conventional DHs with active layers grown under the same conditions as the stacked DH. This demonstrates the potential for the realization of multicolor optical devices based on III-nitride stacked DHs. Further work needs to be done to expand the range of multiple color emission so that colors may be mixed to obtain white light emitters, and further refinements in the growth process need to be made to obtain stacked structures that are uniform and have smooth surface morphologies.

ACKNOWLEDGEMENTS

This work is supported by the Office of Naval Research (ONR), University Research Initiative (URI) program.

REFERENCES

1. S. Nakamura, M. Senoh, N. Iwasa, and S. Nagahama, Jpn. J. Appl. Phys. **34**, L797 (1995).

2. S. Nakamura, M. Senoh, S. Nagahama, N. Iwasa, T. Yamada, T. Matsushita, Y. Sugimoto, and H. Kiyoku, Appl. Phys. Lett., **69**, 4056 (1996).

3. J. C. Roberts, F. G. McIntosh, K. S. Boutros, S. M. Bedair, M. Moussa, E. L. Piner, Y. He and N. A. El-Masry, Mat. Res. Soc. Proc. **395**, 273 (1996).

4. K. S. Boutros, F. G. McIntosh, J. C. Roberts, S. M. Bedair, E. L. Piner and N. A. El-Masry, Appl. Phys. Lett. **67**, 1856 (1995).

5. F. G. McIntosh, K. S. Boutros, J. C. Roberts, S. M. Bedair, E. L. Piner and N. A. El-Masry, Appl. Phys. Lett. **68**, 40 (1996).

6. E. L. Piner, Y. W. He, K. S. Boutros, F. G. McIntosh, J. C. Roberts, S. M. Bedair, and N. A. El-Masry, Mat. Res. Soc. Symp. Proc. **395**, 307 (1996).

TUNNEL EFFECTS IN LUMINESCENCE SPECTRA OF INGAN/ALGAN/GAN LIGHT-EMITTING DIODES

A.E.YUNOVICH*, A.N.KOVALEV**, V.E.KUDRYASHOV*, F.I.MANYACHIN**,
A.N.TURKIN*, K.G.ZOLINA*.
*Moscow State Lomonosov University, Department of Physics,
**Moscow Institute of Steel and Alloys, 119235, Leninsky Prospect, 4.
119899 Moscow, Russia. E-mail: yunovich@scon175.phys.msu.su

ABSTRACT

Tunnel effects in luminescence spectra and electrical properties of blue InGaN/AlGaN/GaN LEDs were studied. The tunnel radiation in a spectral region of 2.1-2.4 eV predominates at low currents (J<0.2 mA). The role of tunnel effects grows as the maximum of the main blue line in LEDs is shifted to short wavelengths. The position of the tunnel maximum $\hbar\omega_{max}$ is approximatly proportional to the voltage eU. The spectral band is described by the theory of tunnel radiative recombination. Current-voltage characteristics have a tunnel component at low direct and reverse currents. The distribution of charged impurities was received from dynamic capacitance measurements. There are charged layers at heterointerfaces and adjacent compensated layers in the structures. There is a high electric field in the active layer. The energy diagram is analysed.

INTRODUCTION

We have shown in references [1-3] that a tunnel radiation spectral band dominates at low currents in luminescence spectra of superbright blue light-emitting diodes (LEDs) based on heterostructures of InGaN/AlGaN/GaN. Current-voltage characteristics of LEDs had a tunnel component. The difference between blue and green LEDs (with different content of In in the active InGaN layer) was described by a lower width of the space-charge layer and a higher electric field in blue LEDs [2]. It was pointed out [1,3] that the models of tunnel radiative recombination elaborated for other $A^{III}B^{V}$ compounds [4,5] can be applicated to an analysis of the new results.

Tunnel spectra and electrical properties of blue LEDs described in [1-3] are thoroughly studied in this paper. Dynamic capacitance measurements and charge distribution in the structures are analyzed. It is shown that the tunnel effects dominate when the electric field in the active 2D-layer is sufficiently high. An energy diagram is proposed which takes into account the experimental results. Spectra are described using theories of diagonal tunneling. The role of potential fluctuations caused by alloy inhomogenities and heterointerfaces is discussed.

EXPERIMENTAL

We studied LEDs made from the structures InGaN/AlGaN/GaN grown by MOCVD [6]. A layer of n-GaN:Si ($\approx$ 5 μm) on the sapphire substate is a base; an active layer $In_xGa_{1-x}N$ (x=0.2-0.43) has a thickness 2-3 nm; the In-content and a thickness vary to change spectra from violet-blue to green. A barrier layer of $p-Al_{0.1}Ga_{0.9}N$:Mg ($\approx$ 100 nm) is a barrier layer and a cap layer is p-GaN:Mg ($\approx$ 0.5 μm); see also [1-3, 6].

Spectra were studied at currents J = 0.02-0.2 mA. The forward and reverse J(V) characteristics were measured in the interval J = 10^{-7}-3$\cdot10^{-1}$ A. Dynamic capacitance measurements gave it possible to determine the nonlinear part of the capacitance.

Mat. Res. Soc. Symp. Proc. Vol. 449 © 1997 Materials Research Society

EXPERIMENTAL RESULTS

Tunnel radiation spectra

The luminescence spectra of a blue LED at room temperature and J = 0.03-0.1 mA are given in Figure 1. A spectral band with spectral maxima changing with the voltage V in the interval $\hbar\omega_{max}$= 2.12÷2.30 eV (green) is attributed to the tunnel radiative recombination. The main blue line dominates at higher currents, V > 2.4 V. The relative role of the tunnel band in different diodes is growing (at a fixed V) as the main blue peak shifts to higher energies (Figure 2).

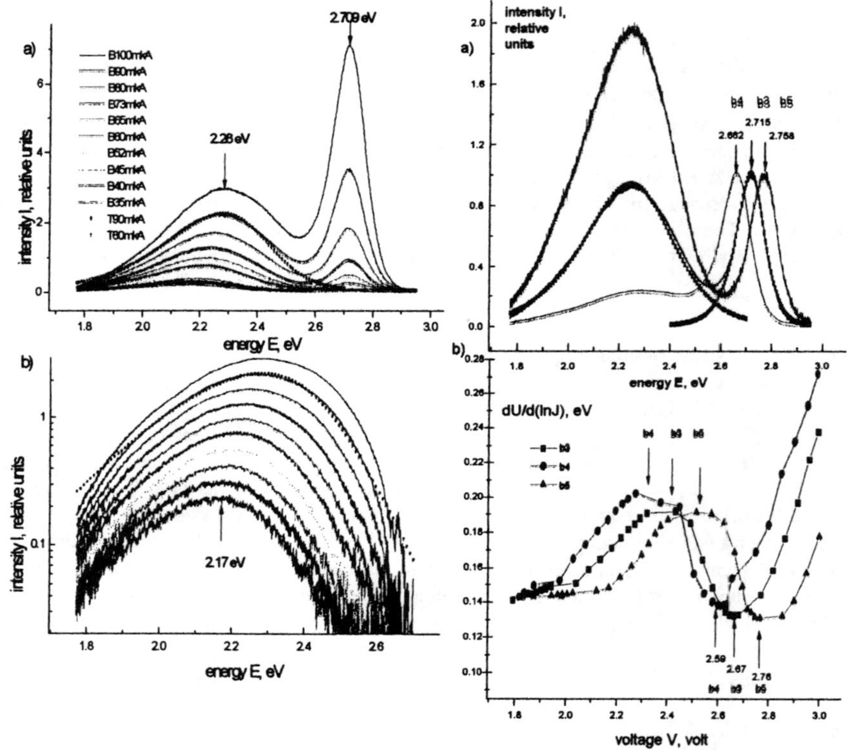

Fig.1 (left). Tunnel spectra; blue LED N3, room T, currents 35-100 μA; V = 2.16-2.34 V.
a) Linear scale; b) logarithmic scale, blue line is subtracted.
Fig. 2 (right). a) Spectra of three blue LEDs at the same voltage, V =2.304 V. Spectra are normalized at the maximum of the blue line. Dots - theoretical fit to the main (blue) and tunnel lines. b) Logarithmic derivative of J(V) characteristics, dV/d(lnJ), versus V for the same LEDs.

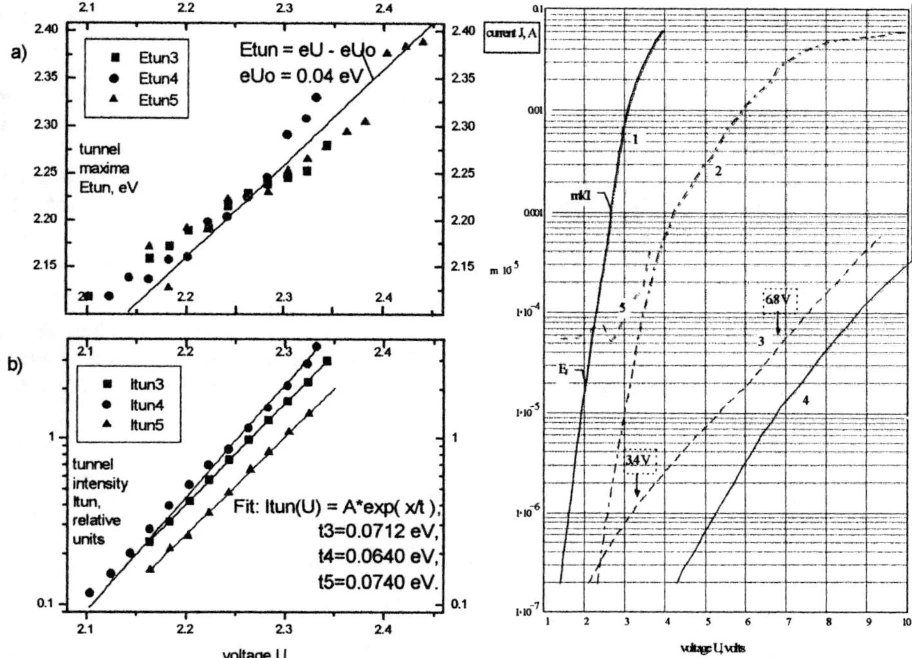

Fig.3. Tunnel maxima (a) and intensity (b) versus voltage. Three blue LEDs (see Figs. 1, 2).

Fig.4. Current-voltage characteristics of blue LED N3. 1,2- forward voltage; 3,4- reverse V; 1,3 - T=300 K; 2,4 - T=77 K; 5- dV/d(lnJ).

The dependences of $\hbar\omega_{max}$ and intensity I_{max} versus $eU = e(V-JR_s)$ (R_s - series resistance) are shown in Figure 3. The value of $\hbar\omega_{max}$ depends on eU approximately linearly: $\hbar\omega_{max} = eU - \Delta$. This is consistent with a model of diagonal tunneling - radiative transitions are going between states near quasi-Fermi levels F_n, F_p from both sides of the junction. The value of I_{max} depends on eU exponentially, with an energy in the exponent E_I of 70-80 meV: $I_{max} \sim \exp(eU/E_I)$.

Current-voltage characteristics

The current-voltage characteristics are shown in Figure 4. The exponential part at low direct currents ($J \sim \exp(eU/E_J)$) has an exponent E_J = 130-140 meV (V = 1.5 - 2.5 V) (see also [7]). This value slightly depends on T (E_J = 155-165 meV at T = 77 K). This part corresponds to the tunnel component of the current. In this interval of U the intensity of the tunnel radiation depends approximately quadratically on the current: $I_{max} \sim J^2$. At higher voltages, $V > 2.6 - 2.7$ V, an injection current component predominates: $J \sim \exp(eU/mkT)$; $U = V - JR_s$.

It is to be noted, that in an interval $1.8 < U < 3.0$ V the logarithmic derivative $E_J = dV/d(lnJ)$ has a maximum near V = 2.3 - 2.5 eV, and a minimum near V = 2.6 - 2.7 V, that is in the region of a change in spectra from tunnel band to the main line (see Figure 2). The shift of these extrema from one LED to the other corresponds to the shift of the maxima of the blue line. The higher is

$\hbar\omega_{max}$, the lower is a minimum of E_J, the lower is R_s. This is evidence that the spectral change is due to a change from tunneling to the injection into the active layer.

The reverse current-voltage curves are also shown on in Figure 4. They have two exponential parts, $J \sim \exp(-eV/E_J)$, with $E_J \approx 0.59$-0.60 eV at voltages $3.4 < -V < 6.8$ V and $E_J \approx 0.96$-0.98 eV at $6.8 < -V < 10$ V. The values of E_J did not depend on T. These components of the current can be described as a tunnel breakdown.

At voltages $-eV > 10$ eV $\approx 3Eg(GaN)$, avalanche breakdown began; radiative recombination caused by electron-hole plasma was detected. Reverse currents and spectra of avalanche breakdown luminescence will be described elsewere.

<u>Capacitance-voltage measurements and the potential disribution</u>

Measurements of capacitance versus volage C(V) have shown that C has two parts: one almost independent and the second dependent on V (see also [3]). The dependent part is determined mostly by the p-side (AlGaN:Mg) of the structure; a part of a contact potential ΔV_p, determined from C(V) curves, was $\Delta V_p \approx 0.5$-0.6 V.

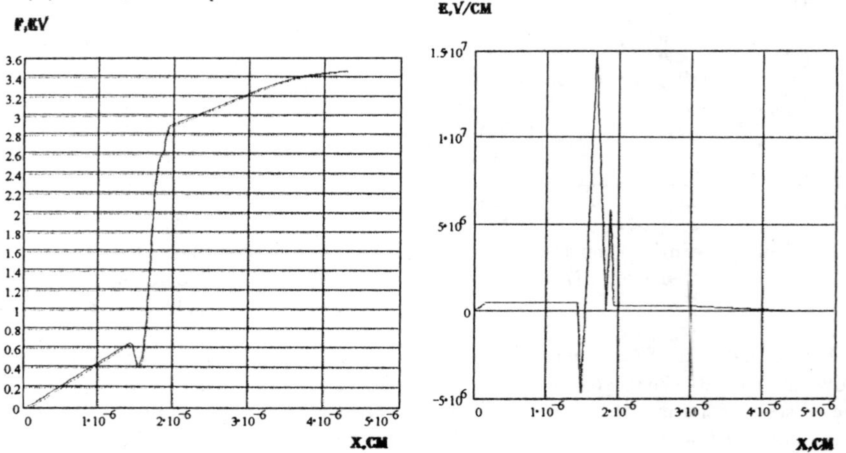

Fig.5. A model potential (a, left) and an electric field (b, right) distribution.

A part of the contact potential on the n-side was $\Delta V_n \approx 0.4$-0.5 V; $\Delta V_p + \Delta V_n \approx 0.9$-$1.0$ V. The part of the capacitance independent of V is determined by compensated layers (with a thickness of 10-15 nm) adjacent to the active layer. Part of the contact potential ΔV_{qn} drops on these quasi-neutral regions, but about $\Delta Vi = 2.0$-2.4 eV drops on the thin (2.5-4 nm) active layer (quantum well). The potential distribution of such a type is formed by charged walls on the interfaces (p-AlGaN/InGaN and InGaN/n-GaN). A model for charged walls in other case was discussed in [6].

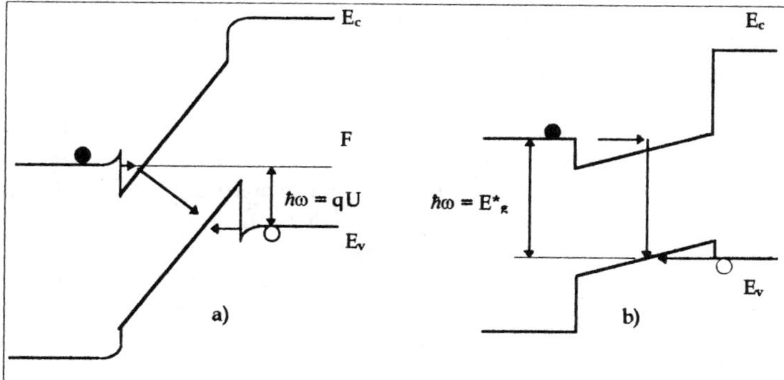

Fig.6. Energy diagram at direct voltages; a) tunneling; b) injection.

DISCUSSION

Energy diagram.

The model potential and electric field distributions deduced from the analysis of experimental results are shown in Figure 5. It is to be noted that there is a high electric field up to $E \approx 10^7$ V/cm in the active layer. More quantitative analysis is in progress. The tunnel component of J dominates at V corresponding to Figure 6a (V $\approx \Delta V_i$), the injection one to 6b (V $\approx \Delta V_i + \Delta V_{qn}$).

Tunnel radiative recombination.

The theory of tunnel radiative recombination [4,5] was elaborated for homogeneous degenerate p-n-junctions. We have used it with some modifications in our complicated case of a heterojunction. A spectrum of tunnel radiation can be described by the equation

$$I(\hbar\omega) \sim [\hbar\omega/(E_g - \hbar\omega)][(\hbar\omega - eU)/(\exp((\hbar\omega - eU)/mkT) - 1)][\exp(-(4/3)((E_g - \hbar\omega)/E_0)^{3/2}], \quad (1)$$

where E_g -is the effective energy gap, m - parameter depends on effective masses ratio, and

$$E_0 = [(\hbar/(2m^*_{cv})^{1/2})eE]^{2/3} \quad (2)$$

is an exponent of the theory of the Franz-Keldysh effect, m^*_{cv} is a reduced effective mass,
E - electric field assumed constant near intersection of electron and hole wave functions. A fit of the experimental tunnel radiation spectra by the equation (1) is shown in Figures 1,2. A blue line was subtracted from spectra using the theoretical fit described in [1-3]. A fit has shown that the voltage eU in (1) is equal to the measured voltage on the structure within the limits of error; an effective gap needed for a good fit was E_g = 2.8-2.9 eV, parameter m = 1.6.

The essential result was that the parameter E_0 needed for a good fit was changing in the limits E_0 = 0.35 - 0.42 eV. An evaluation of the electric field E needed for the tunnel radiation effects gives values E = (4-6)·10^6 V/cm which is consistent with the analysis of capacitance measurements and charge distributions (effective masses in GaN $m^*_C = 0.20m_0$ and $m^*_V = 0.54m_0$ [9] were used).

<u>Tunnel effects and potential fluctuations</u>

The model of radiative recombination for the main (blue) line in these structures [1-3] took into account quantum-size effects and potential fluctuations due to alloy inhomogenities, interface roughness and impurities. All these factors influence tunnel efects also. A high electric field in the heterojunction really is accompanied by high electric fields of fluctuations.

We do not discuss here questions of quantum-size levels which determine band-edges in the structures with a high electric field in the well. Out of scope in this publication is a role of impurity levels (Mg acceptors and Si donors) in the tunnel radiative recombination. We have to take into account the study of the nature of 2.2 eV GaN luminescence band in [10]. These problems are for the next study.

CONCLUSIONS

Spectral bands observed at low currents (J < 0.2 mA) at $\hbar\omega = 2.1\text{-}2.5$ eV in blue LEDs made from the heterostructures InGaN/AlGaN/GaN with a thin space-charge region are caused by radiative recombination of electrons and holes tunneling from large-band-gap layers to the active quantum well layer.

The charge distribution in these structures includes not only space-charge regions but also compensated quasi-neutral layers and charged walls at the interfaces. This charge distribution cause a high electric field in the quantum well, up to $(4\text{-}6)\cdot10^6$ V/cm.

The model of diagonal tunneling describes qualitatively radiative recombination spectra and the change of spectral maxima with the voltage.

The proposed energy diagram describes luminescent and electrical properties of superbright blue InGaN/AlGaN/GaN LEDs.

ACKNOWLEDGMENTS

The authors thank Dr. S.Nakamura for sending LEDs to Moscow University and S.S.Shumilov for help with computer problems.

REFERENCES

1. K.G.Zolina, V.E.Kudryashov, A.N.Turkin, A.E.Yunovich, S.Nakamura. MRS Intern. Journ. of Nitride Semiconductor Research, 1/11; http:\\nsr.mij.mrs.org\1\11.
2. A.N.Kovalev, V.E.Kudryashov, F.I.Manyachin, A.N.Turkin, K.G.Zolina, A.E.Yunovich. Abstr. Abstr. of 23rd Int. Symp. on Semic. Comp., ISCS-23, S.-Peterburg, Sept. 1996, Abst.03.P3.04.
3. K.G.Zolina, V.E.Kudryashov, A.N.Turkin, A.E.Yunovich. FTP, 1996, to be published.
4. A.E.Yunovich, A.B.Ormont. ZhETP, 1966, v. 51, N 11, pp.1292-1305.
5. T.N.Morgan. Phys. Rev., 1966, Vol.148, N 2, pp.890-903.
6. S.Nakamura, M.Senoh, N. Iwasa, S.Nagahama, T.Yamada, T.Mukai. Jpn. J. Appl. Phys. Vol.34 (Oct. 1995), Part 2, N 10b, pp.L1332-1335.
7. J.Zeller, P.G.Eliseev, P.Sartori, P.Perlin, M.Osinski. MRS Symp. Proc. Vol.395, pp.937-942.
8. A.Satta, V.Fiorentini, A.Bosin, F.Meloni. MRS Symp. Proc. Vol.395, pp.515-520.
9. U.Kaufmann, M.Kunser, C.Mers, I.Akasaki, H.Amano. MRS Symp.Proc. Vol.395, p.633-643.
10. D.M.Hoffmann, D.Kovalev, G.Steude, D.Volm, B.K.Meyer, C.Xavier, T.Monteiro, E.Pereira, E.N.Mokov, H.Amano, I.Akasaki. MRS Symp. Proc. Vol.395, pp.619-624.

Optical and electrical characteristics of single-quantum-well InGaN light emitting diodes

Piotr Perlin[a], Marek Osiński, and Petr G. Eliseev[b],
Center for High Technology Materials, University of New Mexico,
Albuquerque, New Mexico 87131-6081

ABSTRACT

We have studied the electroluminescence and photoluminescence of Nichia single-quantum-well $Al_{0.2}Ga_{0.8}N/In_{0.45}Ga_{0.55}N/GaN$ green light-emitting diodes over a broad range of temperatures (15-300 K) and currents (0.2 μA - 2 A). The most striking behavior is an anomalous temperature shift of both photo- and electroluminescence, with the emission peak moving towards higher energies with increasing temperature. This blue shift is opposite to that of the energy gap of the active layer, which practically excludes interband transitions as responsible for the observed optical transitions. We suggest that population effects within the band tails can account for the observed anomaly. We also determined that the current flowing through the *p-n* junction is dominated by carrier tunneling, the omnipresent effect in the GaN-based optoelectronic devices.

INTRODUCTION

Recently, Nichia Chemical Industries introduced new commercial blue- and green-light emitting diodes (LEDs) based on single-quantum-well (SQW) AlGaN/InGaN/GaN structures [1,2]. These new devices have absolutely outstanding performance in comparison with all existing LEDs. Luminosity as high as 12 cd has been achieved for the green LEDs, and external quantum efficiency is as high as 6.3% and 9.1% for the green and blue LEDs, respectively. Very similar device structure was also applied in the newly developed blue laser diodes [3]. In the present paper, we focus on properties of NSPG-500 green SQW LEDs [2], with an undoped active layer consisting of a 30-Å thick $In_{0.45}Ga_{0.55}N$ layer sandwiched between *n*-GaN and *p*-$Al_{0.2}Ga_{0.8}N$ claddings. One should notice that there is a large lattice mismatch (~6%) between the quantum well and barrier materials. However, it is at this stage an open question whether the structure has a pseudomorphic character or the strain is relaxed by a large (~10^{10} cm^{-2}) concentration of dislocations [4] always present in Nichia materials. Another puzzling property of these diodes is that the energy of emitted photons (2.28 - 2.33 eV) is much lower than the bandgap energy of $In_{0.45}Ga_{0.55}N$ (2.50 - 2.68 eV) [5]. It was suggested by Nakamura [3] that this discrepancy could be caused by tensile strain in the quantum well, induced by differences in thermal expansion coefficients of the quantum well and barrier materials. Alternative explanations involve unidentified localized states [6] or localized excitons [7]. In any case, no defi-

---

[a] On leave from High Pressure Research Center, Warsaw, Poland
[b] On leave from P. N. Lebedev Physics Institute, Moscow, Russia

Mat. Res. Soc. Symp. Proc. Vol. 449 © 1997 Materials Research Society

nite answer has been given yet as to the nature of recombination in Nichia SQW LEDs. Here, we report on extensive electroluminescence (EL) measurements over a broad current and temperature range performed in order to shed more light on the nature of radiative transitions responsible for electroluminescence in these devices.

EXPERIMENTAL

Emission spectra were recorded using a CVI double-grating spectrometer model DK 242 D and a cooled GaAs photomultiplier. High-pulsed-current measurements (100 ns pulsewidth and the duty cycle of 10^{-4}) were performed using an Avtech AV-1010-C pulse generator. I-V characteristics were measured with the use of HP 4140 B picoamperometer. For low temperature measurements, the LEDs were fixed on the cold finger of closed-circuit refrigerator (CTI Cryogenics, model 22).

For photoluminescence (PL) measurements, the diodes were de-encapsulated, the chip was removed from the package, and fixed directly on the cold finger of the cryostat. PL was excited with a 20 mW 442 nm blue line of He-Cd laser. On should notice that a thick cap layer of p-GaN makes it impossible to use laser lines with photon energies higher than GaN bandgap.

RESULTS

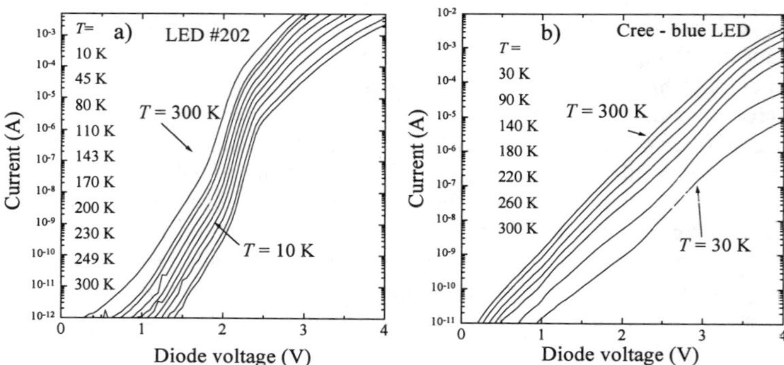

Figure 1. Temperature dependence of I-V characteristics of: (a) Nichia NSPG-500 green SQW LEDs, and (b) Cree C430-DH85 blue LEDs.

<u>Electrical characteristics</u>

Fig. 1(a) shows clearly that the slope of the forward part of *I-V* characteristics of NSPG-500 LEDs is practically independent of temperature. In diodes where the usual Shockley mechanism of thermal diffusion dominates transport properties, the slope should be inversely proportional to temperature. Insensitivity of *I*-V characteristic to temperature is usually interpreted as a domination of carrier tunneling over thermal diffusion. Previously, we observed a similar behavior in Nichia NLPB-500 double heterostructure blue LEDs [8,9] and also in GaN/SiC LEDs commercially manufactured by Cree, as illustrated in Fig. 1(b). This omnipresence of tunneling transport in GaN- and InGaN-based optoelectronic devices seems to be related to a large density of electronic states within the energy gap of these materials.

We can distinguish three different regions in the forward *I-V* characteristics of NSPG-500 LEDs. First, at voltages *V* lower than 2 V, tunneling seems to be related to nonradiative processes, as no light emission takes place. The second region, at 2 < *V* < 2.3 V, is related to tunneling of carriers which can subsequently recombine radiatively. Third, at voltages higher than 2.3 V, the *I-V* curves are dominated by the potential drop on the series resistance of the device and the slope gradually decreases. We should, however, point out that we have not observed any evidence of the so-called photon-assisted tunneling, contrary to the situation in double-heterostructure Nichia LEDs [9].

<u>Optical characteristics</u>

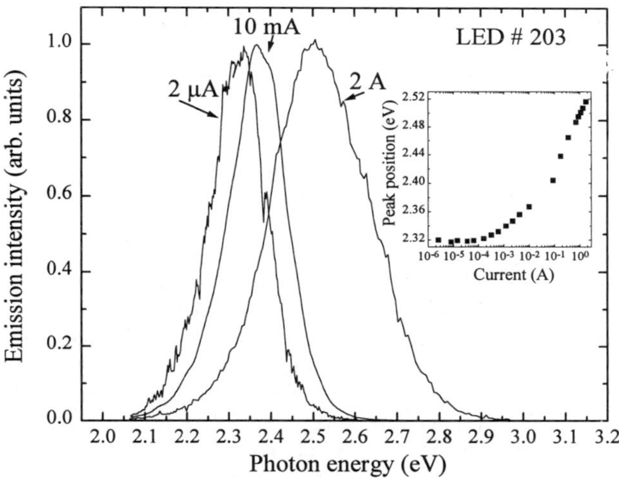

Figure 2. NSPG-500 LED electroluminescence spectra measured at different values of pulsed current. Insert shows the emission peak position as a function of current.

It is shown in Fig. 2 that the EL peak position moves quickly to higher energies once the pumping current exceeds 0.1 mA. This shift is about two orders of magnitude faster than what we would expect from the filling of conduction band states, thus much lower density of states is required to account for the magnitude of the observed effect. Consequently, we interpret our results in terms of *filling of band tail states* rather than the bands themselves. Interestingly enough, the energy of emitted photons at the largest applied currents is quite close to estimated separation between the confined states in a 3-nm thick $In_{0.45}Ga_{0.55}N$ quantum well.

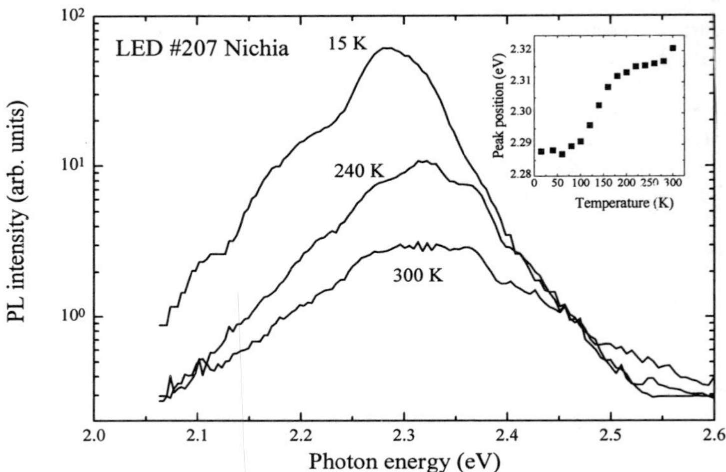

Figure 3. Photoluminescence spectra of NSPG-500 LEDs measured at various temperatures. Insert shows the temperature shift of the emission peak position.

Fig. 3 shows the PL spectra measured at different temperatures. It is clear that the PL peak shifts towards higher energy when the temperature is increased. At the same time, the bandgap of InGaN active layer should decrease by approximately 50 meV between 15 K and 300 K [5]. Fig. 4 shows complementary data obtained from EL measurements performed at various temperatures. It can be seen that at a fixed current lower than 0.1 mA (*i.e.* before the band tail filling effect begins to be important), the EL undergoes blue shift with increasing temperature. Between 15 and 300 K, this shift can be as large as 70 meV for the lowest applied currents and it is of the opposite sign than the shrinkage of InGaN bandgap. As the temperature is increased, the low-current plateau also shifts towards higher energies. Furthermore, the emission bandwidth ($\approx$ 130 meV at 1 mA) remains practically unchanged over the entire temperature range.

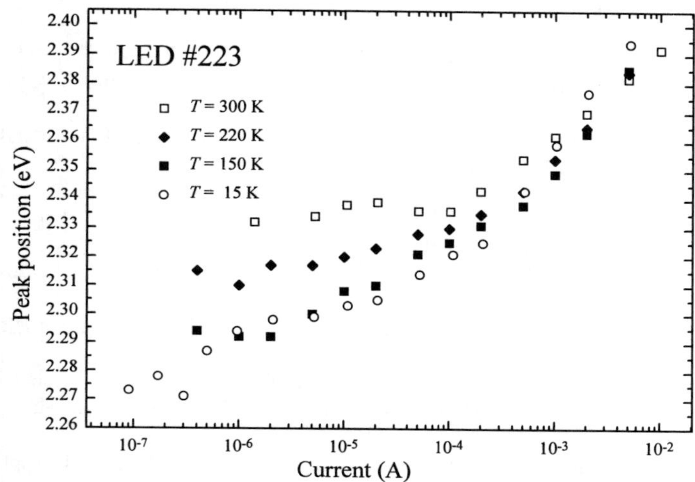

Figure 4. Position of EL emission peak as a function of current for NSPG-500 LEDs, measured at various temperatures.

DISCUSSION

The anomalous temperature behavior of both electro- and photoluminescence in Nichia NSPG-500 SQW green-light emitting diodes, in conjunction with the fact that this emission occurs below the $In_{0.45}Ga_{0.55}N$ bandgap, strongly suggests that the radiative recombination does not directly involve the conduction and valence band quantum well subbands. It was proposed by Chichibu *et al.* [7] that radiative recombination in these devices is related to presence of highly localized excitons, localized on fluctuations of indium contents in the active layer material. However, it is not clear whether the exciton-related recombination should be important for electroluminescence, especially when it is measured at high currents. Also, our observations of rapid blue shift of electroluminescence with current indicate that a continuous density of states may be present within the energy gap of the active layer material. In such case, the anomalous temperature behavior can be explained in terms of population effects. In a future publication, we will show that we can provide a qualitative picture of the observed phenomena within the framework of band-tail theory [10].

ACKNOWLEDGMENT

This work was supported by DARPA under the Optoelectronic Materials Center program.

REFERENCES

1. S. Nakamura, M. Senoh, N. Iwasa, S. Nagahama, Jpn. J. Appl. Phys. **34**, L797 (1995).
2. S. Nakamura, M. Senoh, N. Iwasa, S. Nagahama, T. Yamada, and T. Mukai, Jpn. J. Appl. Phys. **34**, p. L1332 (1995).
3. S. Nakamura, M. Senoh, S. Nagahama, N. Iwasa, T. Yamada, T. Matsushita, H. Kiyoku and Y. Sugimoto, Jpn. J. Appl. Phys. **35**, p. L74 (1996).
4. S. D. Lester, F. A. Ponce, M. G. Craford, and D. A. Steigerwald, Appl. Phys. Lett. **66**, p. 1249 (1995).
5. K. Osamura, S. Naka, and Y. Murakami, J. Appl. Phys. **46**, p. 3422 (1975).
6. K. G. Zolina, V. E. Kudryashov, A. N. Turkin, A. E. Yunovich, and S. Nakamura, MRS Internet J. Nitride Semicond. Res. **1**, p. 11 (1996).
7. S. Chichibu, T. Azuhata, T. Sota, and S. Nakamura, in <u>Proc. of 38th Electronic Materials Conf.</u>, UC Santa Barbara, June 1996.
8. J. Zeller, P. G. Eliseev, P. Sartori, P. Perlin, and M. Osiński, <u>Gallium Nitride and Related Materials</u>, edited by F. A. Ponce, R. D. Dupuis, S. Nakamura, and J. A. Edmond (Mater. Res. Soc. Proc. 395, Pittsburgh, PA 1996), p. 937-942.
9. P. Perlin, M. Osiński, P. G. Eliseev, V. A. Smagley, J. Mu, M. Banas, and P. Sartori, Appl. Phys. Lett. **69**, p. 1680 (1996).
10. P. G. Eliseev, P. Perlin, and M. Osiński, to be published.

DEGRADATION OF SINGLE-QUANTUM WELL InGaN GREEN LIGHT EMITTING DIODES UNDER HIGH ELECTRICAL STRESS

Marek Osiński*, Piotr Perlin*, Ptr G. Eliseev*, Gungtan Liu*, and Daniel L. Barton**

* Center for High Technology Materials, University of New Mexico, Albuquerque NM 7131-6081, USA, osinski@chtm.eece.unm.edu
** Sandia National Laboratories, Albuquerque NM, 87185-1081, USA, bartondl@sandia.gov

ABSTRACT

We performed a degradation study of high-brightness Nichia single-quantum well AlGaN/InGaN/GaN green light-emitting diodes (LEDs). The devices were subjected to high current electrical stress with current pulse amplitudes between 1 A and 7 A and voltages between 10 V and 70 V with a pulse length of 100 ns and a repetition rate of 1 kHz. The study showed that when the current amplitude was increased to the 6 A - 7.5 A range, a fast (about 1 s) degradation occurred, with a visible discharge between the p and n-type electrodes. Subsequent failure analysis revealed severe damage to metal contacts which lead to the formation of shorts in the surface plane of diode. For currents smaller than 6 A, a slow degradation was observed as a decrease in optical power and an increase in the reverse current leakage. After between 24 and 100 hours however, a rapid degradation occurred which was similar to the rapid degradation observed at higher currents. Failure analysis results suggest that carbonization of the plastic encapsulation material on the diode surface leads to the discharge which destroys the diode.

INTRODUCTION

Single quantum well light emitting diodes manufactured by Nichia Chemical Industries with GaN/GaInN/GaAlN active layer [1,2] are some of the best light emitters ever fabricated. Not only they are the best in the short-wavelength part of the spectrum, but their efficiency is close to that of the best devices manufactured in the GaAs/ GaAlAs system. Luminosity as high as 12 cd has been achieved for the green LEDs and quantum efficiencies as high as 6.3 % for the green LEDs and 9.1 % for the blue LEDs have been reported [1]. Very similar structures were also used for the newly developed blue laser diode [3]. In the present paper we will focus on the properties of the green light emitting single quantum well (SQW) LED. The active layer of this device consists of 30 Å thick $In_{0.45}Ga_{0.55}N$ layer sandwiched between n-GaN and p-$Al_{0.2}Ga_{0.8}N$ claddings. These diodes emit light with a center wavelength of about 530 nm. It was been previously shown that devices based on group-III nitrides are characterized by very high longevity and are difficult to degrade by high current stress [4]. It is most likely related to the very high hardness of these materials and consequently very low mobility of dislocations. However, the lifetime of CW laser diodes based on this structure is still very short (about 2 s) because of the high threshold currents these devices currently have. Because of these types of problems, it is very important to understand high current degradation mechanisms in these devices.

EXPERIMENT

Diodes under study were biased with the use of Avtech AV-1010-C pulse generator with a pulse width of 100 ns and repetition rate of 1kHz. The light output was monitored by a

Mat. Res. Soc. Symp. Proc. Vol. 449 © 1997 Materials Research Society

Newport model 1830-C power meter with a silicon diode detector. The I-V characteristics were measured with an HP4140B pico-amp meter/voltage source.

RESULTS

<u>Optical characteristics</u>

It was determined early in this study that the LEDs would degrade immediately if they were subjected to current pulses with amplitudes between 6 A and 7.5 A. The degradation takes about 1s to occur and is accompanied by a visible electrical discharge between the electrodes. At these currents, the voltage drop across the diode is between 70 V and 80 V. In order to investigate the effects of longer term exposure to reduced currents, it was found that using pulsed current amplitudes less than 6 A would cause a very slow degradation allowing the diode to continue working for between 24 and 90 hours. Several LEDs were stressed at a forward current of 5 A with a duty cycle of 10^{-4} %.

In Figure 1, the intensity of light emission from a diode subjected to high current pulses between 4.5 A and 5.8 A is shown as a function of time. Two different regions can be distinguished in this figure. First, for times lower than approximately 5 hours, a fast degradation is observed with the slope -0.17 %/h. This is followed by a second region for times longer than 5 hours where the slope is -0.03 %/h. by using a simple extrapolation from this data, the lifetime of diode should be as long 300 h, but in reality, the diodes degraded rapidly after 24 to 90 hours. Analysis of the damage showed the same kind of damage was found on LEDs which were subjected to currents above 6 A.

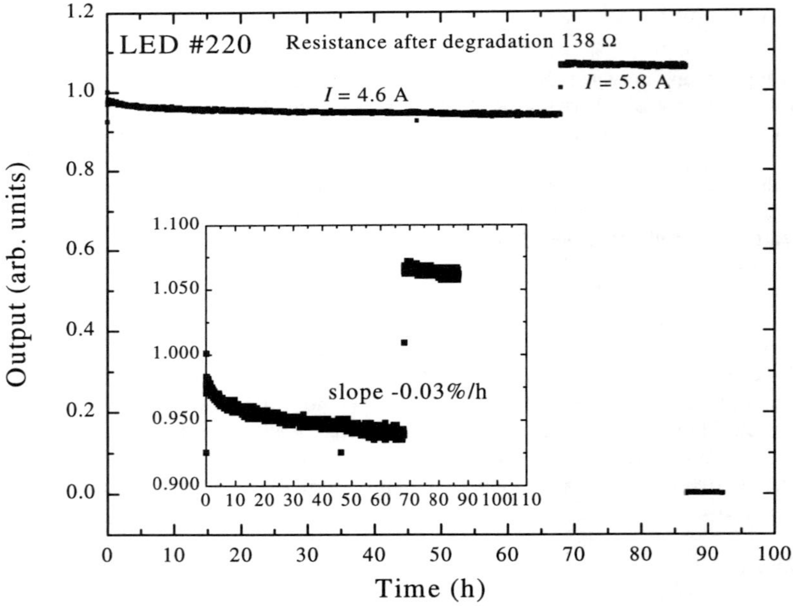

Figure 1. Light output of a stressed LED (#220) as a function of time.

<u>Electrical characteristics</u>

In Figure 2, the *I-V* characteristics of an LED (#219) both before and after being subjected to high current stress are shown. It is characteristic of these LEDs that the reverse current leakage increases fast when first subjected to stress but slows considerably as the stress is continued. This behavior is similar to that that observed in the optical power (faster degradation at the beginning and slower degradation after). Interestingly, the last *I-V* curve shown in Figure 2 was measured just before the diode failed demonstrating that neither the electrical nor the optical measurements could predict that catastrophic degradation was about to occur. This observation was consistent in all of the LEDs subjected to pulsed currents less than 6 A.

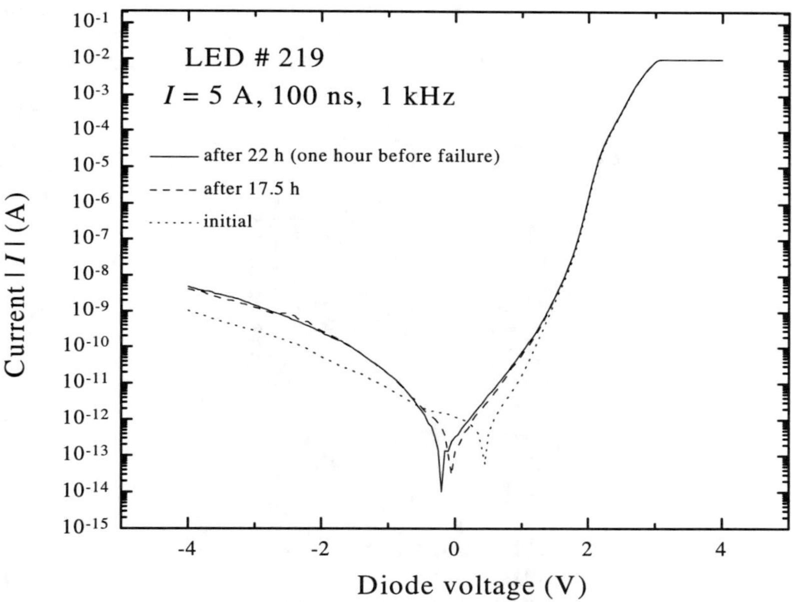

Figure 2. *I-V* characteristic of an LED (#219) after before and after degradation.

FAILURE ANALYSIS OF DEGRADED DIODES

Three of the devices (numbers 202, 219, and 220) which were stressed to the point of failure (identified by a sudden and complete loss in light output) were analyzed to determine the cause of the degradation. These LEDs lasted several hundred hours under stress before failure. The I-V characteristics of the three devices were all linear with slopes indicating that the LEDs had resistive shorts in the 18 Ω to 140 Ω range. A fourth device, #221, was subjected to the same stress, but it was removed from the test prior to failure. The I-V characteristics of #221 showed no significant changes but was decapsulated anyway for comparison to the other three devices. Figure 3 shows an optical micrograph of LED #202 after decapsulation. It is apparent from Figure 3 that the LED has been severely damaged by the pulsed current stress. Further inspection of the device indicated that the observed damage was mostly the plastic encapsulation material whose material properties had been altered by the testing to the point where it was not removed during the decapsulation process. This indicates that most of the damage may be contained in the plastic and not in the LED. These observations were consistent with those made on the other two samples which were stressed to failure, LEDs #219 and 220.

Figure 4 shows an optical micrograph of LED #221 which was not stressed to failure. This image shows several very important details. First, as in the LEDs stressed to failure, some of the plastic packaging material was left over after decapsulation. This means that even though there were no significant changes in light output or in I-V characteristics, there was a change in the plastic composition. And second, there are several black spots in the image, mostly at the edges of the semi-transparent metallization on the p-contact layer. These spots are believed to be damaged plastic areas created when the junction below went into a non-permanent breakdown under the electrical stress. Under continued stress, enough of the damaged plastic sites will form to create a conductive layer which forms a short circuit across the LED. Further studies are in progress to demonstrate the current conduction path in the shorted devices.

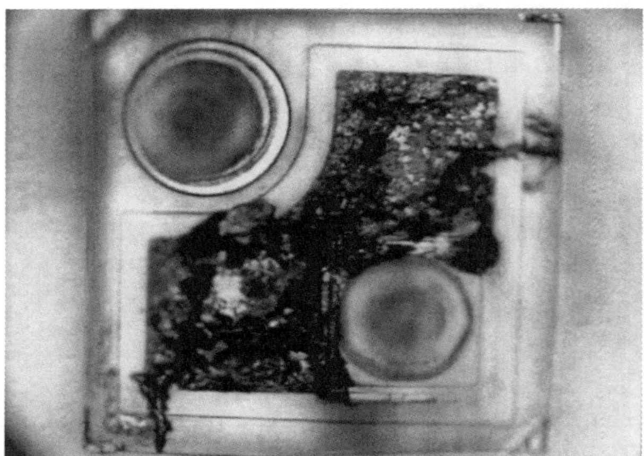

Figure 3. Optical micrograph of LED #202 after plastic removal (the darkened areas over the p-contact area are damaged plastic which could not be removed).

CONCLUSIONS

In order to determine if the same degradation mechanism identified on the earlier double heterostructure LEDs would limit maximum current density in quantum well LEDs, several were stressed at very large pulsed currents. The LEDs were stressed using pulsed currents of approximately 5 A with a 1 kHz repetition rate and a 10^{-4} percent duty cycle to minimize heating. This setup yielded an average power dissipation of only 25 mW.

In light of our experimental results, we can speculate about two very different degradation mechanisms of Nichia single-quantum-well green light-emitting diodes. The first mechanism which is characterized by a relatively slow degradation is probably related to the creation of the internal defects and is yields a device lifetime of about 300 hours. The second mechanism which leads to very fast (1 second) device degradation is related to heat generated by degradation of the plastic packaging material.

The data collected on the earlier double heterojunction LEDs [5] indicated a possible connection between the large number of crystalline defects and a tendency for metal to migrate from the p-contact across the junction and short out the device. The newer generation, quantum well LEDs showed a significant improvement in resistance to this type of stress and revealed that the limitation may be in the plastic packaging material and not in the diode itself.

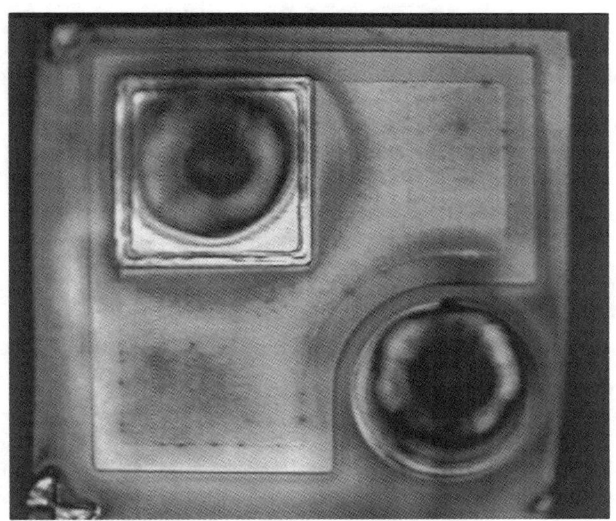

Figure 4. Optical micrograph of damaged plastic areas on LED #221.

ACKNOWLEDGEMENTS

This work was performed at Sandia National Laboratories and is supported by the U.S. Department of Energy under contract DE-AC04-94AL85000.

REFERENCES

1. S. Nakamura, M. Senoh, N,. Iwasa, S. Nagahama, Jpn. J. Appl. Phys. **34**, L797, (1995).

2. S. Nakamura, M. Senoh, N. Iwasa, S. Nagahama, T. Yamada, T. Mukai, Jpn. J. Appl. Phys. **34**, L1332, (1995).

3. S. Nakamura, M.Senoh, S. Nagahama, N. Iwasa, T. Yamada, T. Matsushita, H. Kiyoku and Y. Sugimotu, Jpn. J. Appl. Phys.**35**, L74 (1996).

4. M . Osiński, J. Zeller, P. Chiu and B.S. Phillips, Appl. Phys. Lett. **69**, 898, (1996).

5. D. L. Barton, J. Zeller, B. S. Phillips, P. -C. Chiu, S. Askar, D. -S. Lee, M. Osiński, and K. J. Malloy, (IEEE Int. Rel. Phys. Symp. Proc., Las Vegas, NV 1995), p. 191-199.

Carrier Recombination Dynamics of InGaN/GaN LEDs and its Applications to the Optimization of UV Generation Efficiency

J.P. Basrur, F.S. Choa, P.-L. Liu*, J. Sipior**, G. Rao**,***, G.M. Carter, Y.J. Chen
CSEE Department, University of Maryland Baltimore County, Baltimore, MD 21250
* On Leave to Naval Research Laboratory, Washington DC 20375, from Department of
Electrical and Computer Engineering, SUNY at Buffalo, Amherst, NY 14260
** Department of Chemical and Biochemical Engineering, UMBC
*** The Medical Biotechnology center of the Maryland Biotechnology Institute, University
of Maryland at Baltimore, Baltimore, MD 20201

ABSTRACT

To obtain small size high speed ultraviolet sources, we studied the UV generation process
and efficiency of GaN Blue LEDs. The blue and UV emissions follow a 4-level recombination
model. Depending on a given pump pulse amplitude the UV to blue generation ratio increases
and saturates with the increasing of pump pulse duration. High efficiency, upto 450 μ W
UV power at 380 nm, can be obtained from a 1.2 mW blue LED.

INTRODUCTION

The quest for compact UV light sources has been driven by their potential use in portable
fluorescence measurement sensors, drug testing[1], and other medical applications[2]. Gallium
Nitride (GaN) material is currently being investigated due to its potential for optical devices
operating in the UV and blue–violet region. Intensity dependent photoluminescence in GaN
thin films[3], and time resolved spectra of GaN LEDs have been recently studied[4-6]. The
diodes with either an InGaN active layer (Nichia Blue LED) or a Zn disordered GaN active
layer (CREE Blue LED) normally emit blue light. When pumped with short electrical pulses,
UV light was generated. The UV generation efficiency depends on the characteristics of the
electrical pump pulses. To understand the generation process and obtain good generation
efficiency of UV light with any given electrical pump sources, we did both theoretical and
experimental studies of the generation process. The results are reported in the following.

EXPERIMENTAL RESULTS

The output spectra of the Blue LED were measured under both DC biases and pulsed
operations with a SPEX 1250M monochromator and a photomultiplier tube detector. The
LEDs were soldered to SMA connectors and driven through a bias-T by a DC current source
and an HP 214B pulse generator capable of generating voltage pulses upto 50 V with 10 ns
minimum pulse duration. The Nichia Blue LED is found to produce very weak UV light
when a DC bias of above 30 mA was applied. Much stronger UV light can be generated
from the LED under pulsed operation. The UV to blue ratio varied with the amplitude and
duration of the pump pulse.

Mat. Res. Soc. Symp. Proc. Vol. 449 © 1997 Materials Research Society

Fig.1. shows the output spectra generated with 10 ns pump pulses at various amplitudes. The injected carriers are guidely trapped by the upper states of the blue band due to very short transition time from the band edge to these states. When these states are nearly filled up, the UV recombination becomes more effective. By increasing the pump pulse amplitude more carriers can be provided for the UV generation. Fig.2. shows the output spectra generated with 50 V pulses at various pulse durations. The carrier injection rate remains unchanged since the pump pulse amplitude is constant. With a constant trap number, the longer the pump pulses, the more the number of available carriers for UV recombination and more the amount of UV light generation.

Therefore, to generate more UV light, we can either increase the pump pulse amplitude or the duration. In order to maximise the UV light generation, we need to optimize the pump pulse according to those parameters controlling the recombination process. A simulation has been done to understand the generation and recombination mechanism and estimate these parameters. Experimental work is done to extract information and allow us to examine the model and estimate those parameters.

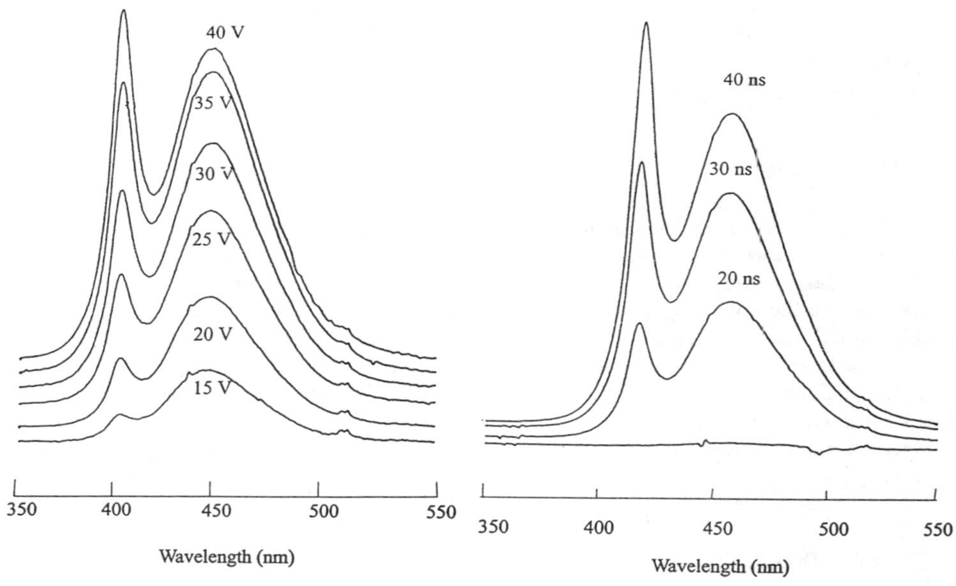

Fig.1 Spectral Response of the Nichia Blue LED with varying pulse amplitudes at a pulse duration of 40 ns.

Fig.2 Spectral output of the Nichia Blue LED for varying pulse duration at 50 V

Fig.3 (a) and (b) show respectively the optical pulse shapes at 380 nm and 450 nm when the LED is pumped with 50 V, 100 ns pulses without DC bias. Time resolved spectral studies and optical outputs of shorter electrical pump pulses can be found in our previous work[6].

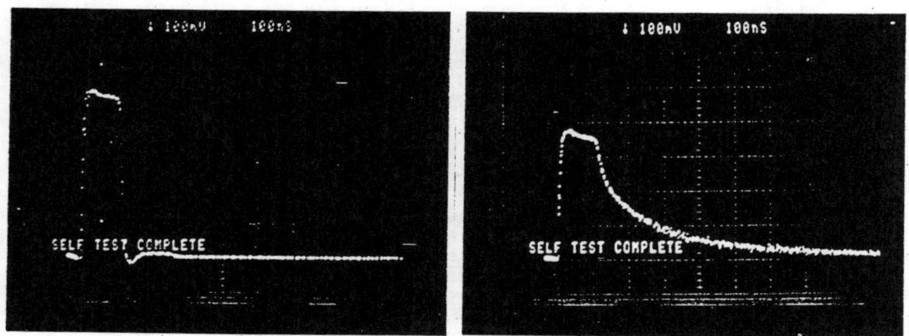

Fig.3 (a) Output Pulse at 380 nm (UV) (b) Output Pulse at 450 nm (Blue)

THEORETICAL MODELING

The UV and blue generation mechanism can be described by a 4-level system as shown in Fig.4. Both the N_1 and N_2 states are created by Zn dopants. The N_1 states are corresponding to the usual acceptor levels of Zn. The N_2 states, which equal in number with N_1 states, are shallow states near the conduction band. The N_2 states have to be included to explain the UV and Blue generation behaviour as described above. Using only a 3-level system, we found that it is impossible to obtain any shapes closer to the experimental results of the time response as shown in Fig. 3 (a) and (b) no matter how we change the parameters like time constants and doping density etc.. It can actually be understood these states are corresponding to the initially dominated carrier capturing of the dopant states.

Therefore the rate equations for the 4-level recombination system are shown below:

$$\frac{dN_3(t)}{dt} = \frac{J(t)}{qd} - \frac{N_3}{\tau_{30}} - \frac{N_3(t)}{\tau_{32}} - \frac{N_3}{\tau_{non3}} \tag{1}$$

$$\frac{dN_2(t)}{dt} = \frac{N_3}{\tau_{30}} + \frac{N_3(t)}{\tau_{32}} - \frac{N_2(t)}{\tau_{21}} - \frac{N_2(t)}{\tau_{non2}} \tag{2}$$

$$\frac{dN_1(t)}{dt} = \frac{N_2(t)}{\tau_{21}} - \frac{N_1(t)}{\tau_{10}} - \frac{N_1(t)}{\tau_{non1}} \tag{3}$$

where J is the current density, d is the thickness of the active layer, τ_{ij} represents the lifetime of the transition from level i to level j, τ_{noni} represents the non-radiative recombination lifetime at level i. While the lifetimes τ_{30} and τ_{10} are assumed to be constant, the lifetimes τ_{32} and τ_{21} are dependent on N_2 and N_1 respectively. More specifically the value of τ_{32} and τ_{21} can be expressed in terms of the empty band recombination lifetime τ_{32}' and τ_{21}' as[3]

$$\tau_{32} = \frac{N_{max}}{N_{max} - N_2(t)}\tau_{32}' \tag{4}$$

$$\tau_{21} = \frac{N_{max}}{N_{max} - N_1(t)}\tau_{21}'$$

(5)

where N_{max} is the dopant density.

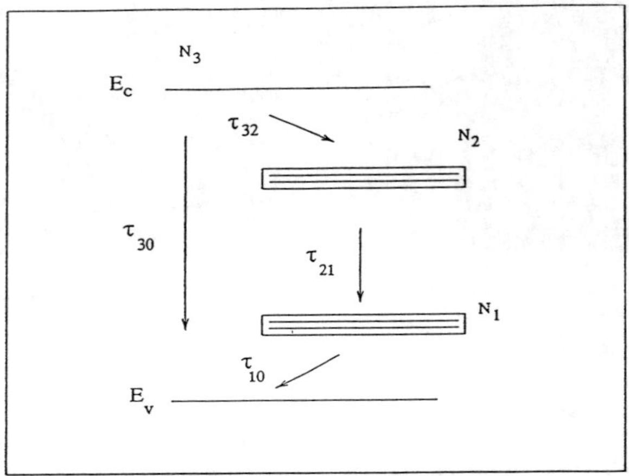

Fig. 4 Schematic diagram of Four level model of the LED

From the above two equations we can infer that τ_{32} and τ_{21} increase as their respective bands get filled up. The transition will become infinitely slow when N_1 or N_2 approaches N_{max}. To simplify the model we have neglected the higher order non-radiative carrier recombinations (Auger and Biomolecular) at each level. Solving equations (1), (2) and (3) we can obtain the time dependent response of N_3, N_2 and N_1 for any given input, $J(t)$. The instantaneous uv and blue light intensity are given by[3]

$$I_{blue}(t)\alpha\frac{hc}{\lambda_{blue}}\frac{N_2(t)}{\tau_{21}}$$

(6)

$$I_{uv}(t)\alpha\frac{hc}{\lambda_{uv}}\frac{N_3(t)}{\tau_{30}}$$

(7)

Using this, the output blue and uv pulses are plotted in Fig.5. By setting the values of the parameters $\tau_{30}, \tau_{32}', \tau_{21}', \tau_{10}$ and the $\tau_{non}s$ along with the doping concentration N_{max}, a best fit to the experimental data was obtained as shown in Table I. The value of τ_{30} is set to be 5ns which is derived from our previous time resolved spectral studies. The value of N_{max} is estimated by the UV generation threshold of 30 mA.

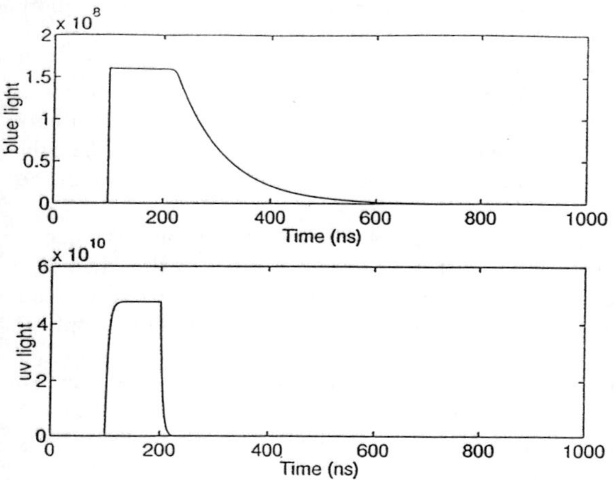

Table I

Parameter	Value
τ_{30}	5 ns
τ_{32}	0.1 ns
τ_{21}	163.5 ns
τ_{10}	3.021 ns
τ_{non3}	55 ns
τ_{non2}	185 ns
τ_{non1}	15 ns
N_{max}	6.0e18

Fig.5 Time Response of the output UV and Blue Pulses

DISCUSSION

Based on the fitted parameters, we can calculate the UV-to-blue ratio as a function of the duration and amplitude of the pump pulse. Under 50V pulse operation, the diode has input impedance of 30 ohm which determines the relationship between the pump voltages and their corresponding currents. Fig.6 shows the results.

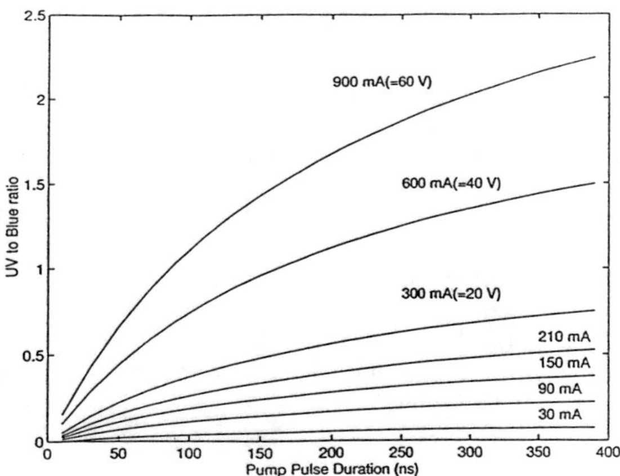

Fig.6 Dependence of UV to Blue ratio on Pulse Duration

The UV to blue ratio increases with both the pump pulse amplitude and duration. Using pulses with larger amplitude has the advantage of generating UV with higher efficiency and larger slope of the UV-to-blue ratio vs duration. That means the ratio will be more effectively improved with long pulses when their amplitude is larger. The slope decreases with the duration and the improvement will gradually saturate. The UV-to-blue ratio will approach to a constant when the pulse is approaching to be like a DC signal. However, the diode can accept 30 mA DC bias and will burn out after 120 mA. We cannot unlimitedly increase the duration. Thermal effect will come in when the pulse duration is long. The light generation efficiency will decrease and the emitting wavelength will be red shifted when the diode temperature is increased.

With a Nichia diode rated at 1.2 mW blue, we can obtain 450μW UV output when it is pumped with a 2.5 MHz, 50 ns, 50 V pulses with no apparent damage to the LED. The center wavelength of the UV band moved from 380 nm to 395 nm under such operation and the generated UV power is limited by this thermal shifting owing to the use of an optical notch filter cutting near 395 nm.

Another very important factor which determines the UV-to-blue ratio is the value of the N_{max} which corresponds to the Zn doping. Provided no other recombination mechanisms come into play, reducing the doping will increase the amount of UV light generation. It would therefore be interesting to see the behaviour of the LED for low doping density.

CONCLUSIONS

In conclusion, to obtain compact size UV sources, we have studied the UV generation efficiency from blue GaN LEDs. Short electrical pulses facilitate the generation of UV light. High efficiency and up to 0.45 mW UV has been generated from an 1.2 mW blue LED. A 4-level recombination model is presented to explain the response to the electrical pulses and conformity with experimental results is obtained. Using the model we can predict the UV generation efficiency dependence on pump pulse characteristics.

G.Rao acknowledges support from NSF Grant No. BES-9413262 and NIH Grant No. RR10955.

References:
[1] F. -S. Choa, M. -H. Shih, A. R. Toppozada, M. Block and M. E. Eldefrawi Analytical Letters, 29(1), 29-33(1996)
[2] J. Sipior, L. Randers-Eichhorn, J. R. Lakowicz, G. Rao, Biotechnol. Prog., **12**, 266 (1996)
[3] R.Singh, R.J. Molnar, M.S.Ünlü, and T.D.Moustakas Appl. Phys. Lett **64** (3), 17 Jan 1994
[4] Tsutomu Araki and Hiroaki Misawa, Rev. Sci. Instrum, **66** (12), Dec 1995
[5] G. Mohs, B. Fluegel, H. Giessen, H. Tajalli, N. Peyghambrian, P-C. Chiu, B. S. Phillips, M. Osinski, Appl. Phys. Lett., **67**, 1515(1995)
[6] F. S. Choa, J. Y. Fan, P.-L. Liu, J. Sipior, G. Rao, G. M. Carter, Y. J. Chen to be published in Appl. Phys. Lett December 9, 1996

DEGRADATION OF InGaN/AlGaN LED ON SAPPHIRE SUBSTRATE GROWN BY MOCVD

T. EGAWA[*], H. ISHIKAWA[**], T. JIMBO[*] and M. UMENO[**]
[*]Research Center for Micro-Structure Devices, Nagoya Institute of Technology, Gokiso-cho, Showa-ku, Nagoya 466, Japan
[**]Department of Electrical and Computer Engineering, Nagoya Institute of Technology, Gokiso-cho, Showa-ku, Nagoya 466, Japan

ABSTRACT

We report an optical degradation of an InGaN/AlGaN double-heterostructure light-emitting diode (LED) on a sapphire substrate grown by metalorganic chemical vapor deposition. The InGaN/AlGaN LED exhibited an optical output power of 0.17 mW, external quantum efficiency of 0.2 %, and the peak emitting spectrum at 437 nm with full width at half-maximum of 63 nm under 30 mA dc operation at 300 K. The InGaN/AlGaN LED showed the optical degradation under high injected current density. Electroluminescence, electron-beam induced current and cathodoluminescence observations show that the degraded InGaN/AlGaN LED exhibits formation and propagation of dark spots and a crescent-shaped dark patch, which act as nonradiative recombination centers. The values of degradation rate under injected current density of 0.1 kA/cm^2 were determined to be 1.1×10^{-3}, 1.9×10^{-3} and 3.9×10^{-3} h^{-1} at ambient temperatures of 30, 50 and 80 °C, respectively. The activation energy of degradation was also determined to be 0.23 eV.

INTRODUCTION

Wide-band-gap III-V nitrides and ZnSe-based II-VI compound semiconductors have attracted much attention because their large direct band gap at room temperature is appropriate for short wavelength light-emitting diodes (LEDs) and laser diodes, which are needed for a wide range of applications such as high-density optical data storage and full-color display. GaN-based LEDs have been commercially available and about 100 times brighter than SiC-based LEDs. The recent study on III-V nitrides has been focussed mainly on room-temperature reliable continuous-wave (CW) operation of laser diode since pulsed and CW operations have been achieved [1, 2]. One of the main problems encountered in epitaxial growth of GaN is a lack of suitable substrate that match with GaN in lattice constant and in thermal expansion coefficient. Sapphire is to date the most commonly used substrate because it allows the epitaxial growth of high-quality GaN layer following the growth of a buffer layer at low temperature.

It is widely recognized that high density of dislocations, which act as nonradiative recombination centers, are introduced in epitaxial layers grown by use of heteroepitaxial growth technique. Dislocations migrate during device operation under high injected current density and ambient temperature, and result in limited stable operation of optical devices. For example, GaAs-based laser diodes on Si substrate, which involve differences of lattice constants and thermal expansion coefficients between GaAs and Si materials, suffer from rapid degradation due to high dislocation density ($>10^6$ cm^{-2}) and large tensile stress ($\sim 10^9$ dyn/cm^2) in the active region. We have shown that rapid degradation in AlGaAs/GaAs single quantum well laser diodes on Si substrates is caused by formation of dark-line defects (DLDs) and degraded current-voltage (I-V) characteristic under high injected current density [3]. Guha et al. reported that major degradation in II-VI blue-green light-emitting device occurred due to microstructural changes such as the formation of dark spots, <100> DLDs and dark patches, acting as nonradiative recombination centers [4]. Hua et al. also reported that formation of dislocation networks in quantum well region by climb motion of dislocations degraded the characteristics of II-VI blue-green laser diode during the current injection, which was suggestive of dislocation network formation in degraded AlGaAs/GaAs DH laser diodes [5]. Thus, the degradation of optical characteristic is caused by the dislocations in the epitaxial layer.

On the other hand, Lester et al. reported that high density of dislocations ($2 \sim 10 \times 10^{10}$ cm^{-2}) in GaN-based LED on sapphire substrate do not act as efficient minority carrier recombination

Mat. Res. Soc. Symp. Proc. Vol. 449 © 1997 Materials Research Society

sites in comparison to other III-V materials [6]. Osinski et al. reported stable operation of GaN-based LED under relative low injected currents ranging from 20~30 mA at 300 K [7]. However, studies under high injected current and ambient temperature are important in order to realize reliable CW operation for GaN-based laser diode at 300 K.

We have confirmed that the degraded characteristics of InGaN/AlGaN LED arise from the deterioration of ohmic electrode and the generation of dark-spot regions. In this study we focus on an optical degradation of InGaN/AlGaN LED on sapphire substrate under high direct current (dc) density and high ambient temperature. We estimate the degradation rate and the activation energy of degradation, and observe the formation and propagation of dark regions.

EXPERIMENTAL

The sample was grown on sapphire substrate with (0001) orientation (c face) by metalorganic chemical vapor deposition (MOCVD) at atmospheric pressure using a modified two step growth technique. After the substrate was heated at 1050 °C in a hydrogen ambient, the InGaN/AlGaN DH was grown. Figure 1 shows schematic cross sectional structure of InGaN/AlGaN LED on sapphire. The structure consists of the following growth sequence: a 25-nm-thick GaN buffer layer at 500 °C, a 4-μm-thick n-GaN layer at 1020 °C, a 150-nm-thick n-$Al_{0.15}Ga_{0.85}N$ layer at 1020 °C, a 50-nm-thick $In_{0.06}Ga_{0.94}N$ layer at 780 °C, a 150-nm-thick p-$Al_{0.15}Ga_{0.85}N$ layer at 1020 °C, and a 350-nm-thick p-type GaN cap layer at 1020 °C. After the growth, the sample was partially etched until the n-GaN layer was exposed. The ohmic electrodes of Ni/Au and Ti/Al were formed by vacuum evaporation on the p- and n-GaN layers, respectively.

Aging tests were performed for five samples from two wafers under various dc densities and ambient temperatures. Optical degradation studies were carried out by electroluminescence (EL), electron-beam induced current (EBIC) and cathodoluminescence (CL) methods. EL imaging, to study the formation and propagation

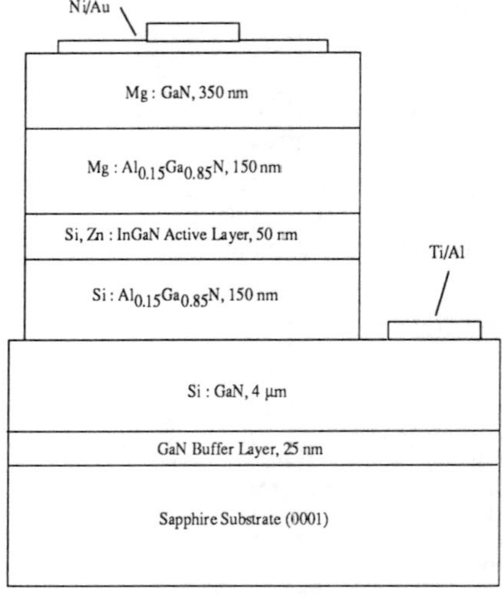

Fig. 1. Schematic cross sectional structure of InGaN/AlGaN LED on sapphire substrate grown by MOCVD.

of nonradiative recombination centers, was carried out by passing the light exiting from the top surface through the thin Ni pad of the InGaN/AlGaN LED. The degraded samples were also studied by EBIC and CL measurements at accelerating voltage of 20 kV.

RESULTS AND DISCUSSION

Figure 2 shows the light output power-current (L-I) characteristic of the InGaN/AlGaN LED on the sapphire substrate under dc operation at 300 K. The light output power increased linearly with increasing current, and saturated at high injected current level. The emission spectra at various currents are shown in Fig. 3. For low injected current, the impurity related broad spontaneous emission was observed around 443 nm with full width at half-maximum (FWHM) of 57 nm. As the injected current increases, the emission intensity of 443 nm increases and a new emission is observed at 380 nm. Fabry-Perot fringes with a spacing of 7 nm are clearly observed as shown in Fig. 3. The fringe spacing agrees with the calculated value based on the total epitaxial layer thickness. The presence of these Fabry-Perot fringes indicates the high quality of interfaces and thickness uniformity of our sample. The InGaN/AlGaN LED exhibited an optical output power of 0.17 mW, external quantum efficiency of 0.2 %, and the peak emission spectrum at about 440 nm with full width at half-maximum of 63 nm at 30 mA (0.06 kA/cm^2).

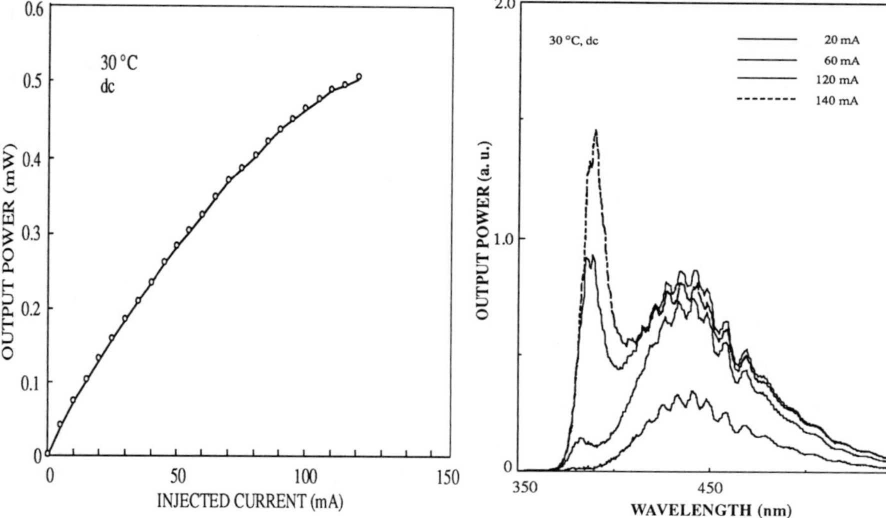

Fig. 2. L-I characteristic of the InGaN/AlGaN LED on the sapphire substrate under dc operation at 300 K.

Fig. 3. Emission spectra at various dc currents for the InGaN/AlGaN LED on the sapphire substrate.

Figure 4 shows the variation of output power as a function of aging time under various injected current densities. Each aging test was performed for 24 h under constant current densities from 0.04 to 0.28 kA/ cm^2 at 30 °C. Although a gradual decrease was observed in the output power for the injected current density of 0.12 kA/cm^2, stable operation was obtained for low injected current densities. However, the output power from the sample tested under high injected current densities decreased rapidly in a few minutes. For an injected current density of 0.28 kA/cm^2, the output power initially decreased rapidly, from 0.18 to 0.13 mW in one minute, and then decreased to 0.07 mW. The L-I characteristics were also measured after each aging test was finished. The output power and the external quantum efficiency measured at 30 mA (0.06 kA/cm^2) were 0.17 mW and 0.2 % at initial stage, and 0.07 mW and 0.08 % after aging at 0.28 kA/cm^2 for 24 h.

To investigate the degradation, accelerated aging tests were carried out under the injected current density of 0.1 kA/cm^2 at ambient temperatures of 30, 50 and 80 °C. Figure 5 shows the variation of relative output power from the InGaN/AlGaN LED as a function of aging time at various temperatures. The half-intensity lifetimes obtained from Fig. 5 were 656.7, 365.7 and 170 h at the ambient temperatures of 30, 50 and 80 °C, respectively. The output power of P can be expressed by [8]

$$P = P_0 \cdot \exp(-\beta t) \tag{1}$$

where P_0, β and t are the initial output power, the degradation rate and operating time, respectively. The degradation rate depends on the device temperature, and is given by [8]

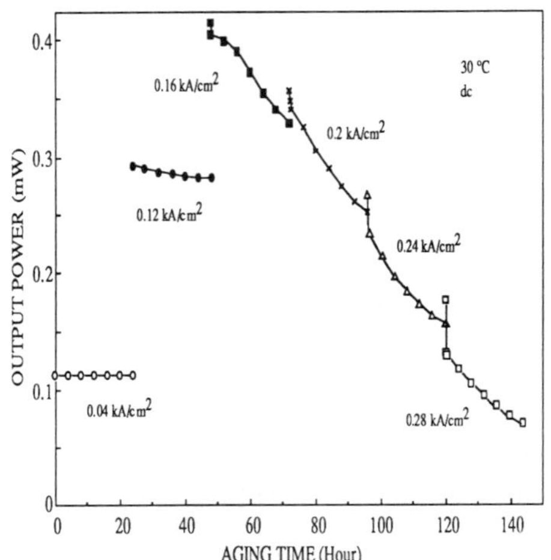

Fig. 4. Variation of output power from InGaN/AlGaN LED as a function of aging time under various injected current densities. Each aging test was performed for 24 h at 30 °C.

Fig. 5. Variation of relative output power from InGaN/AlGaN LED as a function of aging time at ambient temperatures of 30, 50 and 80 °C. The injected current density was 0.1 kA/cm².

$$\beta = \beta_0 \cdot \exp(-E_a/kT) \qquad (2)$$

where β_0, E_a, T and k are a constant, the activation energy of degradation, the device temperature, and Boltzmann's constant, respectively. The values of β were estimated to be 1.1×10^{-3}, 1.9×10^{-3} and 3.9×10^{-3} h^{-1} at 30, 50 and 80 °C, respectively.

Figure 6 shows the comparison of temperature dependence of degradation rate for the InGaN/AlGaN, InGaAsP and AlGaAs LEDs. The activation energy of E_a and the value of β_0 for the InGaN/AlGaN LED were determined to be 0.23 eV and 7 h^{-1}, which were much smaller than the values of 1.0 eV and 1.84×10^7 h^{-1} for InGaAsP LED and 0.57 eV and 93 h^{-1} for AlGaAs LED[8, 9]. The temperature rise due to the operating current was not taken into account because of relatively lower injected current density. Thus, the output power from the InGaN/AlGaN LED decreases during the aging test under higher injected current density and ambient temperature.

In order to study the optical degradation process, EL, EBIC and CL observations were carried out on the InGaN/AlGaN LED. Figure 7 (a), (b) and (c) shows the EL images of the progressive stages of degradation during the aging test under 0.4 kA/cm² at 30 °C. Figure 7 (a) shows the faint dark spots at initial stage, which indicate the pre-existing defects in the structure acting as nonradiative recombination centers. At the first stage of degradation shown in Fig. 7 (b), the faint dark spots become darker and a dark region appears in Fig. 7 (c), the dark spots enlarge individually and the dark region also enlarges. The reason why the dark spots and region were observed in the vicinity of the corner of the left electrode is that the injected current was concentrated at that location. The growth rate of dark spot at 0.4 kA/cm² was estimated to be 0.02 ~ 0.04 μm/h. We also carried out EBIC and CL measurements on the degraded InGaN/AlGaN LED. Figures 8 and 9 show the EBIC and CL images of the degraded LED observed by the EL image as shown in Fig. 7 (c). The EBIC and CL images showed the dark spots and a crescent-shaped dark patch, which indicates nonradiative recombination centers in the active region [4].

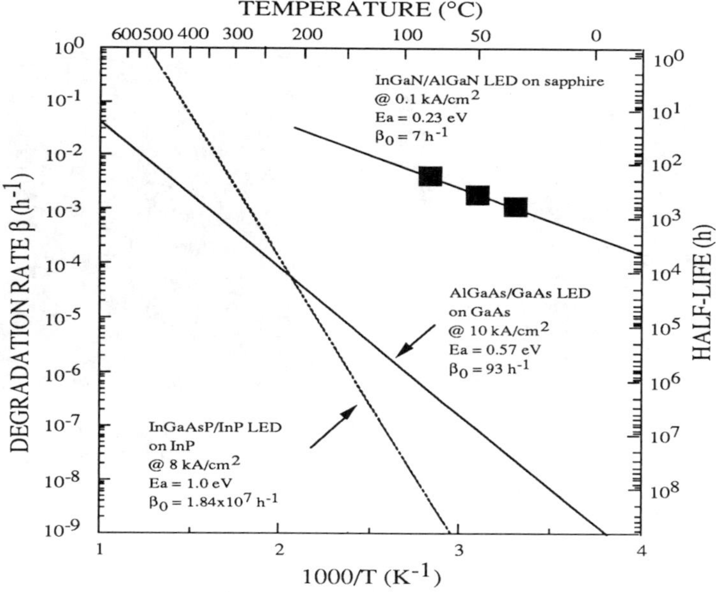

Fig. 6. Comparison of temperature dependence of degradation rate for the InGaN/AlGaN, InGaAsP and AlGaAs LEDs.

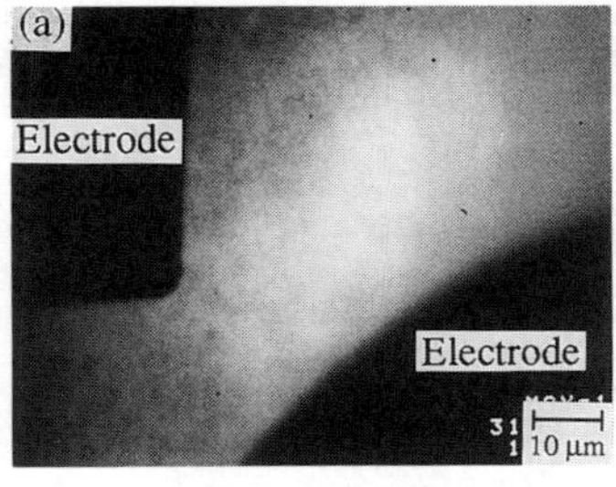

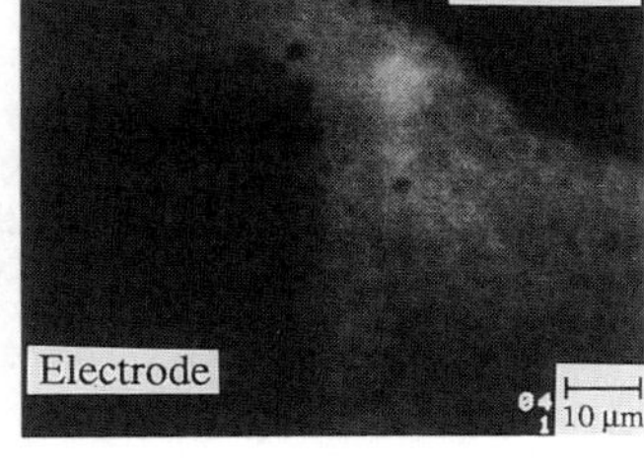

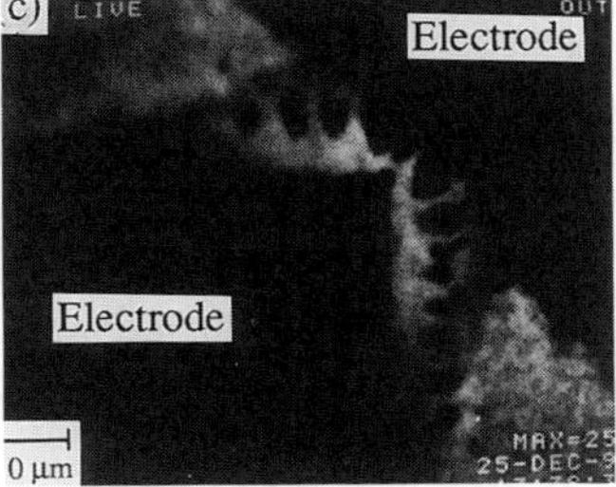

Fig. 7. EL images of progressive stages of degradation for InGaN/AlGaN LED during the aging test under 0.4 kA/cm^2 at 30 °C. (a), (b) and (c) correspond to initial stage and aging of 67 and 310 n, respectively.

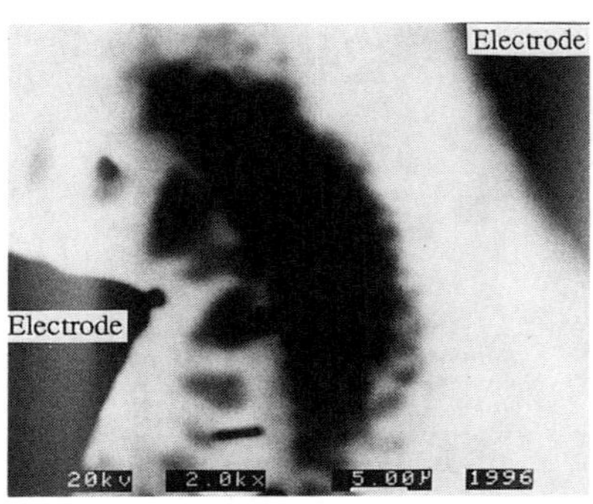

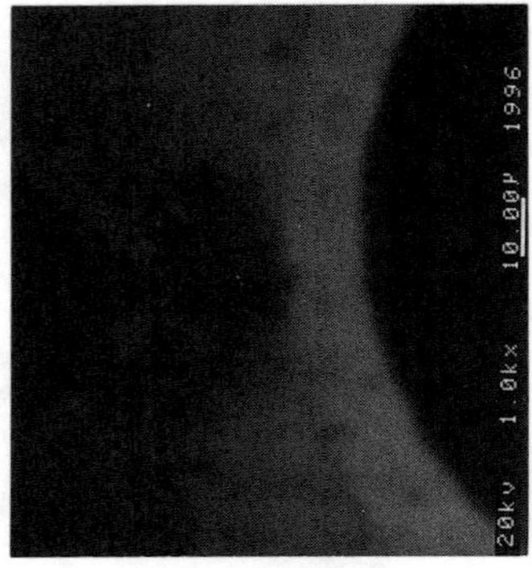

Fig. 8. EBIC image of the degraded InGaN/AlGaN LED shown in Fig. 7 (c).

Fig. 9. CL image of the degraded InGaN/AlGaN LED shown in Fig. 7 (c).

CONCLUSIONS

The InGaN/AlGaN LED on the sapphire substrate grown by MOCVD exhibited the optical output power of 0.17 mW, the external quantum efficiency of 0.2 %, and the peak emitting spectrum at 437 nm with full width at half-maximum of 63 nm under 30 mA dc operation at 300 K. We observed the formation and propagation of dark spots and a crescent-shaped dark patch in the degraded InGaN/AlGaN LED on the sapphire substrate. The decrease in the output power under high injected current density and ambient temperature is thought to be caused by the formation of dark regions, which act as nonradiative recombination centers.

REFERENCES

1. S. Nakamura, M. Senoh, S. Nagahama, N. Iwase, T. Yamada, T. Matsushita, H. Kiyoku and Y. Sugimoto, Jpn. J. Appl. Phys. **35**, L74 (1996).
2. S. Nakamura, M. Senoh, S. Nagahama, N. Iwase, T. Yamada, T. Matsushita, Y. Sugimoto and H. Kiyoku, IEEE Lasers & Electro-Optics Society, Boston, 1996, PDP1.1.
3. T. Egawa, Y. Hasegawa, T. Jimbo and M. Umeno, Appl. Phys. Lett. **67**, 2995 (1995).
4. S. Guha, J. M. DePuydt, M. A. Haase, J. Qiu and H. Cheng, Appl. Phys. Lett. **63**, 3107 (1993).
5. G. C. Hua, N. Otsuka, D. C. Grillo, Y. Fan, J. Han, M. D. Ringle, R. L. Gunshor, M. Hovinen and A. V. Nurmikko, Appl. Phys. Lett. **65**, 1331 (1994).
6. S. D. Lester, F. A. Ponce, M. G. Craford and D. A. Steigerwald, Appl. Phys. Lett. **66**, 1249 (1995).
7. M. Osinski, C. J. Helms, N. Berg, D. L. Barton and B. S. Phillips, Proc. Mat. Res. Soc. **395**, 931 (1996).
8. S. Yamakoshi, M. Abe, O. Wada, S. Komiya and T. Sakurai, IEEE J. Quantum Electron. **QE-17**, 167 (1981).
9. S. Yamakoshi, O. Hasegawa, H. Hamaguchi, M. Abe and T. Yamaoka, Appl. Phys. Lett. **31**, 627 (1977).

REALIZATION AND CHARACTERIZATION OF OPTICALLY PUMPED GaInN-DFB LASERS

R. HOFMANN, V. WAGNER, H-P. GAUGGEL, F. ADLER, P. ERNST, A. SOHMER, H. BOLAY, F. SCHOLZ, H. SCHWEIZER,
4. Physik. Institut, Universität Stuttgart, Pfaffenwaldring 57, D-70550 Stuttgart, Germany

ABSTRACT

We demonstrate room temperature laser activity of optically pumped GaN/GaInN-DFB-lasers. The ridge-waveguide DFB lasers were realized on GaN/GaInN double heterostructures grown by low-pressure metal-organic vapor phase epitaxy. The best laser threshold we achieved is $1.9\,\mathrm{MW/cm^2}$.

By varying the grating period, the laser emission wavelength could be tuned from 399 to 415 nm. This allows to determine the dispersion relation of the effective refractive index $n_{\mathrm{eff}}(\lambda)$ and the spectral dependence of the pump power density at the laser threshold $P_{\mathrm{th}}(\lambda)$ over the whole emission range. Furthermore, the shift of the emission wavelength with temperature of the DFB-lasers is investigated and is found to be small compared to the emission wavelength shift of the gain maximum.

INTRODUCTION

In the last few years, great progress has been achieved in the development of III-V nitride-based optoelectronic devices. Stimulated by the improvement of the crystal quality of these materials, intense investigations led to the commercialization of high-brightness blue/green light-emitting diodes (LEDs) [1,2] and to the demonstration of electrically pumped violet lasers based on InGaN/GaN/AlGaN-based multi-quantum-well-structures [3,4]. Besides further investigations towards electrically pumped laser diodes operating under continuous-wave (cw) conditions, an important area of present III-V nitride research is the development and demonstration of different laser designs. Stimulated emission has been observed for optically pumped GaN [5,6] and GaInN films [7] in surface and edge emission geometries, and on AlGaN/GaInN [8] and GaN/AlGaN double heterostructures [9–11] for edge emitting structures. The achieved pump power densities range from a few hundred $\mathrm{kW/cm^2}$ to a few $\mathrm{MW/cm^2}$. Recently, the realization of a vertical cavity surface emitting laser (VCSEL) with AlGaN Bragg reflectors was reported [12]. Laser activity could also be obtained for distributed feedback (DFB) structures, realized by etching a grating into the top cladding layer of a GaInN/GaN double heterostructure [13]. Due to their wavelength selectivity, DFB lasers are suitable devices to investigate the spectral behavior of laser parameters such as the effective refractive index $n_{\mathrm{eff}}(\lambda)$ and the laser threshold $P_{\mathrm{th}}(\lambda)$. In this paper, we report the realization and characterization of improved GaInN/GaN DFB lasers.

EXPERIMENT

The GaN/GaInN double heterostructures were grown on a (0001)-oriented sapphire

Mat. Res. Soc. Symp. Proc. Vol. 449 © 1997 Materials Research Society

substrate by low-pressure metal-organic vapor phase epitaxy (LP-MOVPE). First, a thin (15 nm) AlN nucleation layer was grown followed by a 100 nm GaN buffer layer, a 10 nm $Ga_{0.88}In_{0.12}N$ active layer, and a 100 nm GaN top cladding layer [14]. Second-order DFB gratings with grating periods between 120 and 175 nm were defined by high resolution electron beam lithography using 120 nm PMMA as resist and were transfered into the top GaN barrier layer by electron cyclotron resonance reactive ion etching (ECR-RIE) using a $1.80{:}0.30/Ar{:}CCl_2F_2$ gas mixture. The etch depth was 50 nm. Lateral confinement of the laser modes was achived by a $6\,\mu m$ wide ridge wave guide, fabricated by electron beam lithography and ion beam etching (IBE). The overall length of the cavities was $1000\,\mu m$.

The DFB resonators and the reference mesa without grating were optically pumped with a pulsed XeCl-excimer laser (10 ns pulse width, 10 Hz repetition rate) which emits at a wavelength of 308 nm. In order to pump the DFB cavities homogeneously, the excimer beam was shaped to a stripe by a cylindrical lense and focused on the sample surface. The intensity of the pump beam was regulated by a variable attenuator placed in the laser beam and measured by a joulemeter. The spontaneous and the stimulated emission were measured perpendicularly to the sample surface with a 0.85 m monochromator and a cooled GaAs photomultiplier tube. For temperature control, the samples were mounted on a Peltier element.

RESULTS

To characterize the GaN/GaInN double heterostructure and it's gain behavior, we measured the luminescence spectra of the GaN/GaInN heterostructure at different pump power densities. As shown in Fig. 1, at low pump power densities the spectrum shows two peaks with peak positions at 362 nm and 395 nm which are due to the spontaneous recombination in GaN and GaInN, respectively. At increased pump power densities, a red shift of the spontaneous emission peaks is observed. This is attributed to heating of the sample during pumping. At pump power densities above $6\,MW/cm^2$, the spectra show a strong, nonlinear

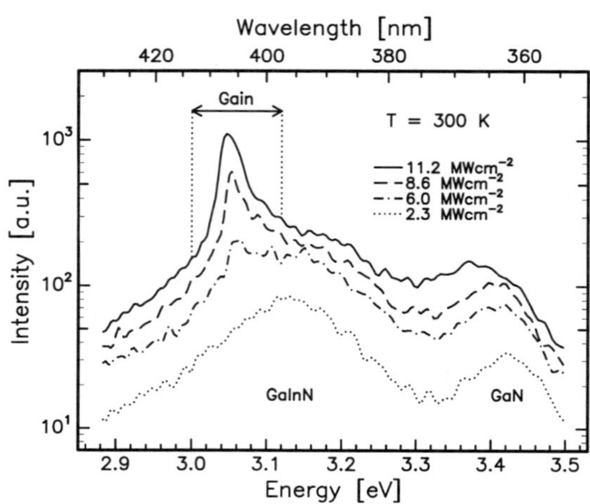

Figure 1: Luminescence spectra of a GaInN/GaN double heterostructure at different pump power densities. The spectum taken at a pump power density of $2.3\,MW/cm^2$ shows the spontaneous emission peaks of GaN and GaInN with peak positions at 362 nm and 395 nm, respectively. For increased pump power densities, amplified spontaneous emission appears on the low energy side of the GaInN peak.

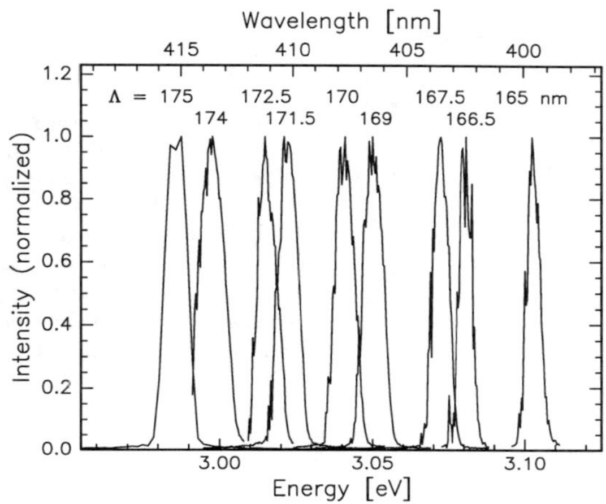

Figure 2: Emission spectra of a series of GaInN/GaN-DFB lasers with different grating periods. The spectra are normalized.

increase of the luminescence intensity on the low energy side of the GaInN peak. This is caused by amplified spontaneous emission and marks the spectral range of optical gain. We therefore expect laser activity only in the spectral range between 400 and 415 nm.

Fig. 2 shows the emission spectra of a series of DFB resonators with different grating periods if pumped above laser threshold. We conclude from the following facts that DFB-laser activity is observed: First, the laser emission wavelength depends linearly on the grating period as expected for DFB lasers according to the Bragg condition $\lambda = 2n_{\text{eff}}\Lambda/m$, where n_{eff} is the effective refractive index of the heterostructure, m is the order of the grating (here $m = 2$), and Λ is the grating period. Second, the linewidths of the laser peaks, which are less than 1 nm, are much smaller than the linewidth of the ASE. This proves that the index grating causes the feedback necessary for laser action. Fig. 2 also shows that the laser emission wavelengths obtained by DFB-lasers cover the whole spectral range of optical gain.

A further comment should be made on the linewidth of the laser emission because it is still rather broad for laser modes. We attributed this spectral broadening mostly to frequency chirping caused by sample heating and dynamic carrier density fluctuations during the excitation pulse. It might also be enhanced by the presence of unresolved transversal modes.

The wavelength selectivity of the DFB lasers allows the precise determination of the effective refractive index $n_{\text{eff}}(\lambda)$ over the entire emission range. The experimentally determined values of the effective refractive index are shown in Fig. 3. They range from 2.42 to 2.36 in the observed spectral range. The dispersion of the effective refractive index was determined by a linear fit. It amounts to $-3.19 \cdot 10^{-3}\,\text{nm}^{-1}$. Also shown are calculated values determined by means of a waveguide model using published data for the refractive index of GaN and GaInN [15,16]. The experimental values are systematically smaller than the calculated values. This effect is attributed to an increased carrier density during pumping resulting in a reduction of the refractive index. A similar reduction of the refractive index was observed for optically and electrically pumped DFB lasers in the material system GaInP/AlGaInP [17,18].

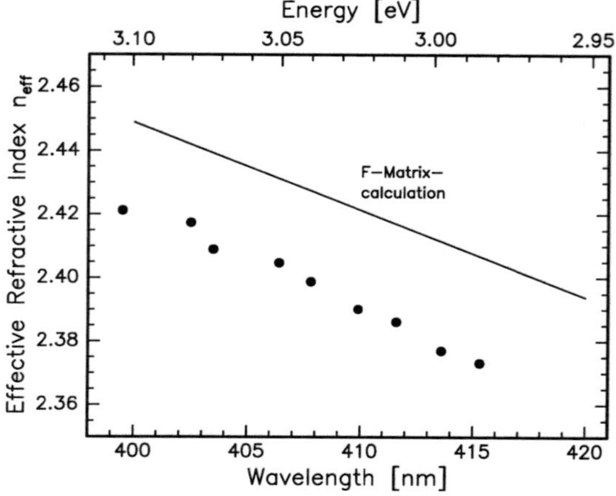

Figure 3: Experimentally determined values of n_{eff} in comparison with calculated values. The dispersion of the effective refractive index was obtained by a linear fit and amounts to $-3.19 \cdot 10^{-3}\,\text{nm}^{-1}$. The experimental values are systematically smaller than the calculated values due to an increased carrier density during pumping.

Fig. 4 shows the emission wavelengths of two DFB lasers as a function of temperature. For comparison, Fig. 4 also shows the wavelength shift of a reference mesa without gratings, which was processed on the same wafer as the DFB lasers, with temperature. In the investigated temperature range from -5 to $60°\,\text{C}$, the emission wavelength shift of the FP cavity was determined to be $0.065\,\text{nm/K}$ ($-0.47\,\text{meV/K}$). This reflects the temperature shift of the gain maximum and is of the same magnitude as the band gap shift of GaN. The published values for the temperature coefficient dE_{g}/dT at room tmperature range from -0.45 to $-0.67\,\text{meV/K}$ [19]. As expected, the emission wavelength of the DFB lasers shifts much less with temperature than the FP emission and the shift is the same for both DFB lasers. This is resonable because the wavelength shift of DFB lasers depends only on the change of the effective refractive index with temperature. For the DFB lasers, the wavelength shift

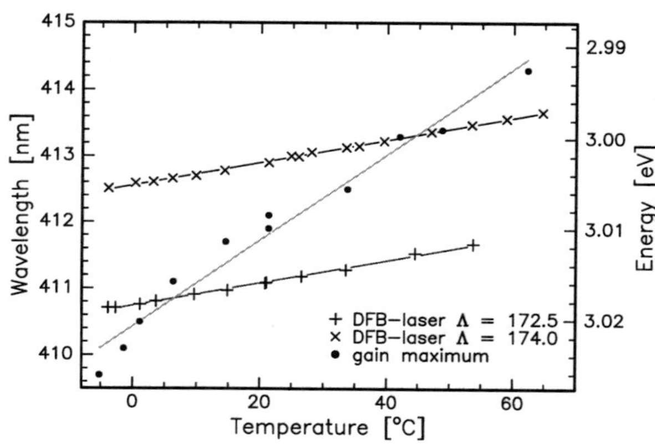

Figure 4: Temperature dependence of the emission wavelength of two DFB lasers in comparison with the temperature dependent emission wavelength shift of a reference mesa without gratings. The wavelength shift is 0.017nm/K for both DFB lasers and is much smaller than the shift (0.065nm/K) for the reference mesa.

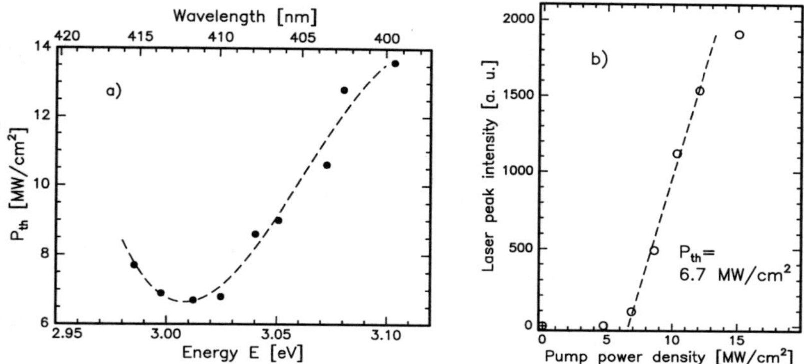

Figure 5: a) Spectral dependence of the laser threshold. It resembles the behavior of the netto gain. The lowest laser threshold achieved was $6.7\,\mathrm{MW/cm^2}$ at an emission wavelength of 412 nm. b) Typical input-output characteristic.

was determined to be $0.017\,\mathrm{nm/K}$ ($-0.124\,\mathrm{meV/K}$).

A typical input-output power characteristic for the DFB-lasers can be seen in Fig. 5b). It shows that the DFB-lasers exhibit a distinct threshold behavior. The dependence of the laser threshold on the emission wavelength is shown in Fig. 5a). It resembles the behavior of the netto gain curve. The lowest laser threshold ($6.7\,\mathrm{MW/cm^2}$) was achieved for a DFB resonator with an emission wavelength of 412 nm corresponding to the maximum position of the material gain spectra. Even lower laser thresholds ($1.9\,\mathrm{MW/cm^2}$) could be realized for DFB-lasers with a thicker GaN buffer layer (450 nm) because of their better crystal quality. But DFB lasers based on this asymetric heterostructure show clearly resolvable higher vertical modes.

CONCLUSION

We demonstrated laser operation of optically pumped GaInN/GaN DFB lasers at room temperature. The second-order DFB ridge waveguide structures were realized by electron beam lithography and dry etching on MOVPE grown symmetric double heterostuctures. DFB laser emission was observed over the whole spectral range of optical gain from 400 to 415 nm with a lowest laser threshold of $6.7\,\mathrm{MW/cm^2}$ at 412 nm. The effective refractive index was experimentally determined to be ≈ 2.40 and compared with calculated values. The emission wavelength shift of DFB lasers with temperature is significantly smaller than the wavelength shift of the gain maximum.

The authors wish to thank H. Gräbeldinger, P. Burkard, and I. Grigoridis for technical support.

[1] S. Nakamura, T. Mukai and M. Senoh. *Appl. Phys. Lett.* **64**, 1687 (1995).

[2] S. Nakamura, M. Senoh, N. Iwasa and S. Nagahama. *Appl. Phys. Lett.* **67**, 1868 (1995).

[3] S. Nakamura, M. Senoh, S. Nagahama, N. Iwasa, T. Yamada, T. Matsuchita, H. Kiyoku and Y. Sugimoto. *Jap. Journ. Appl. Phys.* **35**, L217 (1996).

[4] S. Nakamura, M. Senoh, S. Nagahama, N. Iwasa, T. Yamada, T. Matsuchita, H. Kiyoku and Y. Sugimoto. *Appl. Phys. Lett.* **68**, 3269 (1996).

[5] H. Amano and T. Asahi I. Akasaki. *Jap. Journ. Appl. Phys.* **29** (1990).

[6] M. Asif Khan, D. T. Olson, L. M. Van Hove and J. N. Kuznia. *Appl. Phys. Lett.* **58**, 1515 (1991).

[7] M. Asif Khan, S. Krishnankutty, R. A. Skogman, D. T. Olson and T. George. *Appl. Phys. Lett.* **65**, 520 (1994).

[8] H. Amano, T. Tanaka, Y. Kunii, S. T. Kim and I. Akasaki. *Appl. Phys. Lett.* **64**, 1377 (1994).

[9] R. L. Aggarwal, P. A. Maki, R. J. Molnar, Z.-L. Liau and I. Melngailis. *Appl. Phys. Lett.* **79**, 2148 (1996).

[10] S. T. Kim, H. Amano, I. Akasaki and N. Koide. *Appl. Phys. Lett.* **64**, 1535 (1994).

[11] T. J. Schmidt, X. H. Yang, W. Shan, J. J. Song, A. Salvador, W. Kim, O. Aktas, A. Botchkarev and H. Morkoç. *Appl. Phys. Lett.* **68**, 1820 (1996).

[12] J. M. Redwing, D. A. S. Loeber, N. G. Anderson, M. A. Tischler and J. S. Flynn. *Appl. Phys. Lett.* **69**, 1 (1996).

[13] R. Hofmann, H.-P. Gauggel, U. A. Griesinger, H. Gräbeldinger, F. Adler, P. Ernst, H. Bolay, V. Härle, F. Scholz, H. Schweizer and M. H. Pilkuhn. *Appl. Phys. Lett.* **69**, 2068 (1996).

[14] V. Härle, H. Bolay, F. Steuber, F. Scholz, V. Syganow, G. Frankowsky and A. Hangleiter. In: *Proc. Int. Symp. on Blue Lasers and Light Emitting Diodes, Chiba, Japan* page 62. 1996.

[15] S. N. Mohammad, A. A. Salvador and H. Morkoç. *Proceeding of the IEEE* **83**, 1306 (1995).

[16] E. Ejder. *Phys. stat. sol. (a)* **6**, 445 (1971).

[17] C. Kaden, U. A. Griesinger, H. Schweizer, C. Geng, M. Moser and F. Scholz. *Appl. Phys. Lett.* **63**, 3414 (1993).

[18] H.-P. Gauggel, C. Geng, H. Schweizer, F. Barth, J. Hommel, R. Winterhoff and F. Scholz. *Electron. Lett.* **31**, 367 (1995).

[19] H. Teisseyre, P. Perlin, T. Suski, I. Grzegory, S. Porowski, J. Jun, A. Pietraszko and T. D. Moustakas. *J. Appl. Phys.* **76**, 2429 (1994).

STIMULATED EMISSION FROM SINGLE- AND MULTIPLE-QUANTUM-WELL GaN-AlGaN SEPARATE-CONFINEMENT HETEROSTRUCTURES

D.A.S. LOEBER*, N.G. ANDERSON*, J.M. REDWING**, J.S. FLYNN**, G.M. SMITH**, and M.A. TISCHLER**

* Department of Electrical and Computer Engineering, University of Massachusetts at Amherst, Amherst, MA 01003
** Epitronics, 7 Commerce Drive, Danbury, CT 06810

ABSTRACT

Stimulated emission characteristics are examined for GaN-AlGaN separate-confinement quantum-well heterostructures grown by MOVPE on 4H-SiC substrates. We specifically focus on comparison of structures with different quantum well active region designs. Polarization resolved edge emission spectra and stimulated emission thresholds are obtained under optical pumping using a stripe excitation geometry. Stimulated emission characteristics are studied as a function of the number of quantum wells in the structure, and are correlated with surface photoluminescence properties. We find reduced stimulated emission thresholds and increased surface photoluminescence intensities as the number of quantum wells is reduced, with the best results obtained for a single-quantum-well structure. These results should provide useful information for the design of GaN-based quantum well lasers.

INTRODUCTION

The explosive growth in research on the III-V nitrides over the past several years has yielded rapid progress toward the realization of commercial short-wavelength visible and ultraviolet laser diodes. The first successful demonstrations of current injection lasers have been reported within the last year, including blue InGaN-based diodes fabricated on sapphire [1-3] and spinel [4] at Nichia Chemical Industries. While all of the injection lasers demonstrated so far have been edge-emitting Fabry-Perot devices with quantum well active regions, laser action has also been observed under optical pumping in horizontal-cavity distributed feedback lasers [5] and vertical-cavity surface emitting lasers grown on sapphire [6] and SiC [7].

In this work, we study the surface photoluminescence (PL) and stimulated edge emission from optically excited GaN-AlGaN separate-confinement quantum-well heterostructures (SCQWH's). We specifically compare the emission properties of samples which differ in the number of GaN quantum wells in the active region, but are otherwise identical in all respects including the total thickness of quantum well material. The observed trend in the stimulated emission thresholds is discussed within the context of a simple model incorporating interface recombination at GaN-AlGaN interfaces. Implications of these results for the design of III-V nitride laser diodes are also discussed briefly.

SAMPLE DESCRIPTION, GROWTH, AND CHARACTERIZATION

In order to systematically study the dependence of light emission properties on active region design, three GaN-AlGaN SCQWH's were grown. The three samples, which were grown by metalorganic vapor-phase epitaxy (MOVPE), each consist (from bottom to top) of a 4H-SiC substrate, a thin AlN buffer, a $1\mu m$ $Al_{0.17}Ga_{0.83}N$ lower cladding layer, a 750Å GaN-$Al_{0.07}Ga_{0.93}N$ quantum well active region, and a $0.2\mu m$ $Al_{0.17}Ga_{0.83}N$ upper cladding layer. Only the design of the 750Å quantum well active region differs for the three structures, as discussed below. The samples were grown consecutively under identical conditions in a vertical reactor operating at 100 Torr, with all epilayers grown at 1100°C. Trimethylgallium (TMGa), trimethylaluminum (TMAl), and ammonia were used as precursors, with H_2 as the carrier gas. Note that the aluminium mole fractions specified here (x=0.07 and x=0.17) are target values corresponding to those achieved in growths of bulk layers carried out under the same conditions.

Mat. Res. Soc. Symp. Proc. Vol. 449 © 1997 Materials Research Society

The active region of the first structure consists of a 150Å GaN single quantum well (SQW) sandwiched between 300Å $Al_{0.07}Ga_{0.93}N$ barriers. (This structure is very similar to one grown on a 6H-SiC substrate which has clearly exhibited laser action under optical pumping [8].) The active region of the second structure consists of three 50Å GaN quantum wells separated by 50Å $Al_{0.07}Ga_{0.93}N$ barriers. This 250Å multiple quantum well (MQW) structure is set back from the upper and lower cladding layers by 250Å $Al_{0.07}Ga_{0.93}N$ spacer layers, yielding a 750Å total active region thickness. Finally, the active region of the third structure consists of six 25Å GaN quantum wells separated by 50Å $Al_{0.07}Ga_{0.93}N$ barriers, with the entire 400Å MQW structure set back from the upper and lower cladding layers by 175Å $Al_{0.07}Ga_{0.93}N$ spacers. These three samples are hereafter referred to as the SQW, 3-well MQW, and 6-well MQW, respectively. Note that the active region of each sample consists of a total of 150Å of active GaN quantum well material and 600Å of $Al_{0.07}Ga_{0.93}N$ barrier material.

Surface PL spectra were obtained at 300K using a pulsed nitrogen laser as an excitation source and at 50K using a CW HeCd laser. Edge emission spectra and stimulated emission thresholds were obtained by focusing light from the nitrogen laser to a narrow (70 μm) stripe on the sample surface using a cylindrical lens. The sample is oriented such that the pump light is incident normal to the sample surface and the excitation stripe intersects the cleaved sample edge at a right angle. In this configuration, the excitation stripe is colinear with the input axis of the spectrometer used to collect and disperse the edge emission. Pump power densities at the sample surface were calculated from the measured pulse energy, the stripe width, and the pump pulse duration (0.8 ns).

RESULTS AND DISCUSSION

Surface Photoluminescence Experiments

Surface PL spectra obtained at 300K using the nitrogen laser pump source are shown for the three quantum well samples in Fig. 1(a). These spectra were obtained at a pump intensity of 7.6 MW/cm^2 from regions of the sample surfaces which exhibited maximum PL intensity. Note that the three spectra are quite similar to one another, each exhibiting a single broad (~200 meV) feature peaked near 355 nm with a shoulder near 363 nm. The overall PL intensity is, however, strongest for the SQW and weakest for the 6-well MQW. This is illustrated for a wide range of photoexcitation intensities in Fig. 1(b), which shows the dependence of the integrated PL intensity on pump intensity for the three samples.

Surface PL experiments were also performed at 50K and at low photoexcitation intensities using a HeCd laser as a pump source. Spectra for the three samples are shown in Fig. 2. The spectrum for the SQW (Fig. 2(a)) consists of a feature peaked at 354.8 nm (3.495 eV), which lies 27 meV above the dominant exciton peak observed for bulk GaN on SiC at 50K (data not shown) and is presumably due to the fundamental excitonic transitions in the 150Å quantum well. The spectral full-width-at-half-maximum (FWHM) for this feature is 36 meV. A doublet is observed at slightly higher energy with peaks at 346.6 nm and 349.6 nm (3.578 and 3.547 eV), which we associate with emission from the $Al_{0.07}Ga_{0.93}N$ barrier layers. Note that the energy spacing between the dominant feature and this higher energy doublet is similar to the spacing between the peak and shoulder of the room temperature spectrum for this sample (i.e. ~80 meV; see Fig. 1(a)).

The 50K PL spectrum for the 3-well MQW is shown in Fig. 2(b). This spectrum is quite similar to that of the SQW, with a dominant feature peaked at essentially the same energy lying below a substantial emission band in the 346-350 nm range. We again attribute the peak, which has a FWHM of 50 meV, to the fundamental excitonic transitions in the three 50Å wells. A small shoulder near 362 nm is also present in this spectrum. This feature is enhanced in spectra taken from other regions of the sample surface, where, perhaps significantly, the "barrier" emission is also stronger relative to the main PL peak. This may suggest that the 362 nm feature originates from transitions involving impurity levels in the AlGaN barrier layers. Note that a similar feature appears in the 50K PL spectrum for the 6-well MQW, which is shown in Fig. 2(c). The intensity of this 362 nm feature is even larger than that of the 355 nm feature, which we again attribute to fundamental excitonic transitions in the quantum wells. These results suggest that barrier recombination accounts for a significant fraction of the surface PL in our samples over a wide range of temperatures.

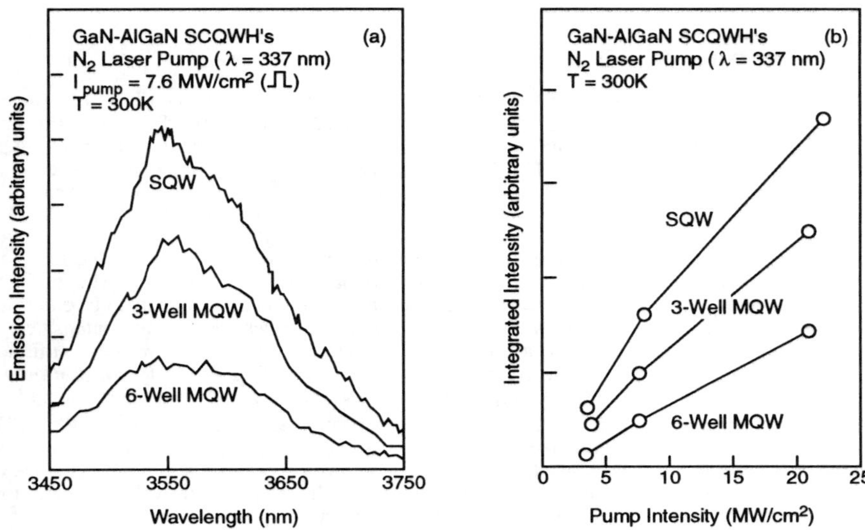

Fig. 1 (a) 300K surface photoluminescence spectra for GaN-Al$_{0.07}$Ga$_{0.93}$N separate confinement quantum well heterostructures (SCQWH's) with SQW, 3-well MQW, and 6-well MQW active regions. The three spectra were taken at the same pump intensity, and are shown on the same scale. (b) Integrated emission intensity as a function of pump intensity for all three samples.

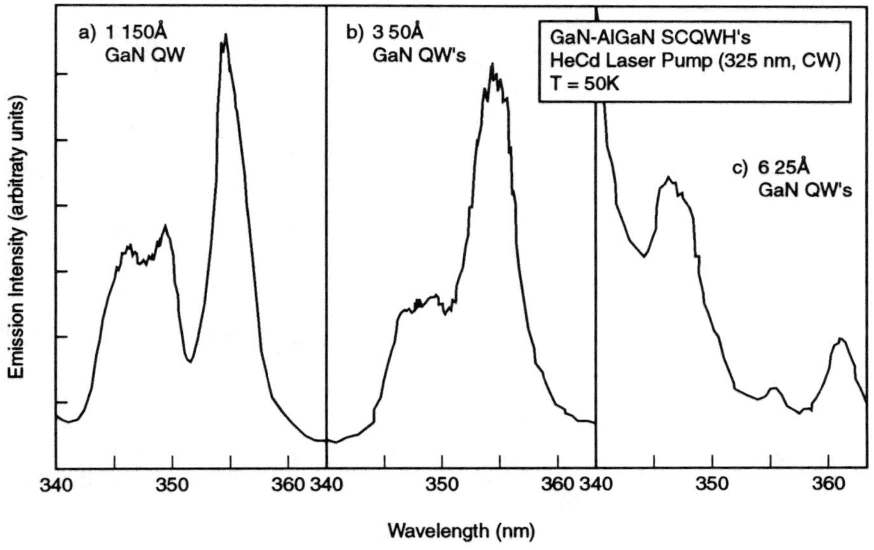

Fig. 2 50K PL spectra for the GaN-Al$_{0.07}$Ga$_{0.93}$N SQW (a), 3-well MQW (b), and 6-well MQW (c) samples obtained at low photoexcitation intensities.

<u>Stimulated Edge Emission</u>

Room temperature polarization-resolved edge emission spectra are shown in Fig. 3 for the SQW (Fig. 3(a)), 3-well MQW (Fig. 3(b)), and 6-well MQW (Fig. 3(c)). These spectra were obtained by photopumping the samples with a nitrogen laser using the stripe-excitation geometry described above. The TE (upper) and TM (lower) polarized components of the edge luminescence are shown for each sample at two excitation intensities near threshold. Note that, for each case, the emission spectrum at lower power levels (thin traces) is relatively broad and only weakly polarized. At sufficiently high pump intensities, however, a sharp and strongly TE-polarized stimulated emission peak emerges from the background luminescence and dominates the spectrum (bold traces). This behavior is generally very similar to that which we have observed for GaN-AlGaN double heterostructures [8], although an unexpected narrow feature appears to be emerging from the TM-polarized emission spectra of these quantum well samples at high pump intensities.

Integrated edge emission intensity, as measured by collecting and focusing the edge emission onto a photodiode, is shown as a function of pump intensity in Fig. 4. Note that the output intensity for each sample exhibits a superlinear dependence on pump intensity, which is characteristic of stimulated emission. Threshold pump intensities are ~1.8 MW/cm^2 for the SQW, ~4.0 MW/cm^2 for the 3-well MQW, and ~7.6 MW/cm^2 for the 6-well MQW. Stimulated emission thresholds thus clearly increase with the number of quantum wells in the structure. This trend, together with the observed *decrease* in surface PL intensity with the number of wells, indicates that the radiative efficiency is reduced as the number of quantum wells in the structure is increased.

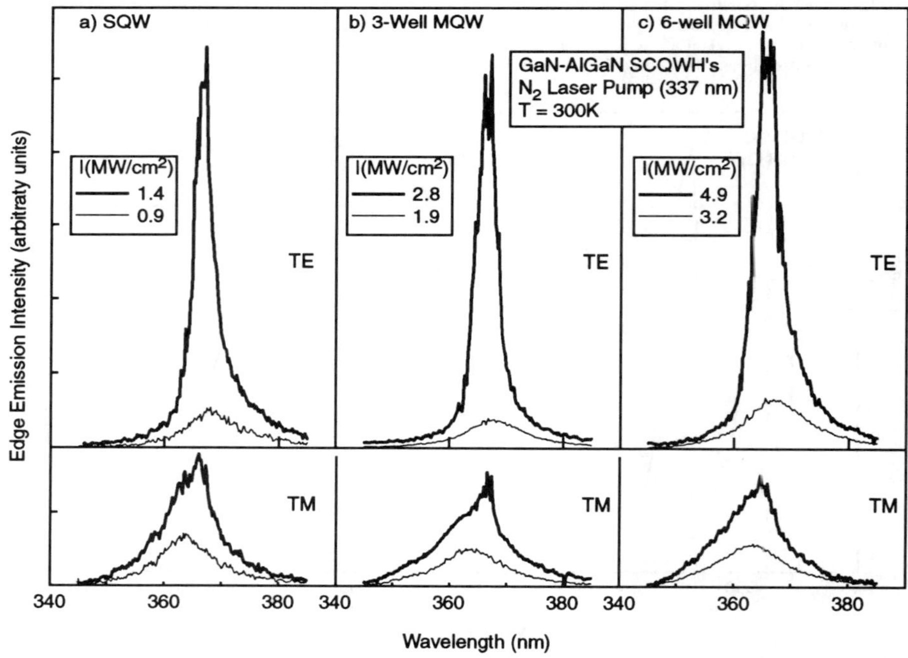

Fig. 3 300K polarization-resolved edge-emission spectra for GaN-Al$_{0.07}$Ga$_{0.93}$N SQW (a), 3-well MQW (b), and 6-well MQW (c) samples photoexcited in a stripe excitation geometry. Results are shown for TE (upper) and TM (lower) polarizations at two excitation intensities near threshold. Results for the two polarizations are shown on the same scale to allow direct comparison.

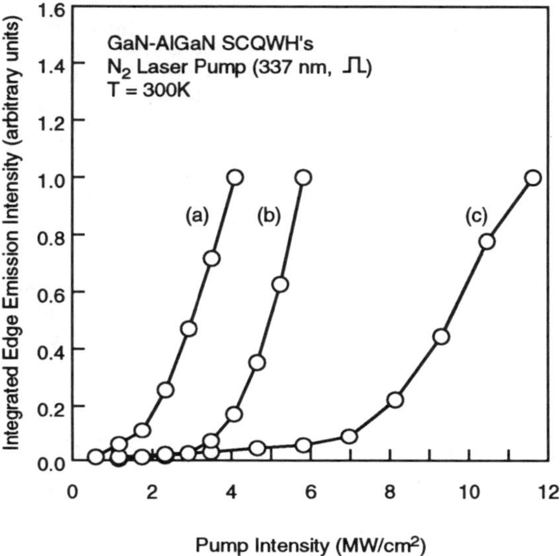

Fig. 4 Integrated edge emission intensity (unpolarized) as a function of pump intensity for the GaN-Al$_{0.07}$Ga$_{0.93}$N SQW (a), 3-well MQW (b), and 6-well MQW (c) samples photoexcited in a stripe excitation geometry. Note that all samples exhibit a superlinear dependence on pump intensity, and that the threshold pump intensity increases with the number of quantum wells in the structure.

Before discussing this trend further, we should note that the stimulated emission threshold intensities quoted above are likely overestimates. We expect that the pump powers quoted above are roughly 20% too high because of incomplete filtering of scattered pump laser light into the photodiode. We should also emphasize that they are incident surface intensities, and are thus perhaps an order of magnitude larger than the power actually contributing to carrier generation. The relatively low efficiency for photon-to-electron-hole-pair conversion in these structures results from a combination of surface reflection and the fact that only the 750Å active region has a low enough band gap to absorb the pump photons.

Discussion

A strong dependence of radiative efficiency and threshold pump intensity on quantum well thickness has obvious implications for laser design. The dependence we have observed in our samples is perhaps unexpected for several reasons. The total thickness of quantum well material is the same in all three structures, so the radiating volume is identical for each structure and the optical confinement factors should be essentially the same. Quantum size effects should be minimal, as expected from the large effective masses for GaN and the absence of significant quantum shifts in the 50K PL spectra, and should thus contribute little to sample-to-sample variations in densities of states and optical matrix elements. Finally, while strain effects are difficult to assess in these structures, the likelihood that relevant critical thicknesses for dislocation formation are exceeded in the GaN layers should increase with well thickness. This would imply an opposite trend in radiative efficiency than the one we have observed.

One possible explanation for the observed increase in threshold intensity with increasing number wells (N) is interface-related recombination. Provided that the optical gain for a MQW with a total active thickness L depends primarily on L and injected carrier density, but only weakly on the widths of the individual wells (L/N), the threshold intensity should increase linearly with the number of interfaces in the structure (2N). To assess the plausibility of interface recombination as a cause for the observed N-dependence of threshold pump intensity, we have analyzed our experimental results in the context of a simple model which includes both quantum well material recombination (radiative and non-radiative) and interface recombination [9]. Assuming a quantum well recombination lifetime of ~0.1 ns [10] and a transparency current density of n_{th}=8x10^{18} cm^{-3} [11], this model describes our experimental results well if we take the effective interface recombination velocity S$_I$ at the GaN-AlGaN heterointerface to be in the low 10^4 cm/s range.

This estimate for S$_I$, while necessarily approximate, seems plausible. The interface recombination velocity is a few orders of magnitude larger than values obtained for high-quality lattice-matched GaAs-AlGaAs MQW lasers, as might be expected given the lattice mismatch present at the GaN-AlGaN heterointerfaces, but is a few orders of magnitude lower than those typical of free semiconductor surfaces. Interestingly, application of the parameters from our analysis to the GaN-Al$_{0.07}$Ga$_{0.93}$N MQW structure studied by Smith and co-workers (N=10, L=250Å) [12], yields a net effective lifetime at room temperature which is very close to their measured value (~0.02 ns). While such close agreement is obviously fortuitous given the simplicity of this analysis and the uncertainties involved, it supports the plausibility of interface recombination coefficients in the 10^4 cm/s range for GaN-Al$_{0.07}$Ga$_{0.93}$N MQW structures.

If interface recombination is nearly as significant as this analysis would suggest, then it can be expected to represent a dominant carrier loss mechanism in GaN-AlGaN MQW's. This would clearly imply that GaN-AlGaN MQW laser structures should incorporate as few quantum wells as possible. While such a design strategy is easily implemented in the GaN-AlGaN system, difficulties with the growth of thicker high-quality InGaN layers [13] may preclude such an approach in InGaN-based laser structures. Large numbers of thin wells may be required to simultaneously ensure acceptable crystalline quality and large optical confinement factors in InGaN-based MQW lasers, even if at the expense of substantial carrier loss through interface recombination.

REFERENCES

1. S. Nakamura, M. Senoh, S. Nagahama, N. Iwasa, T. Yamada, T. Matsushita, H. Kiyoku, and Y. Sugimoto, Jpn. J. Appl. Phys. **35**, L74 (1996).
2. S. Nakamura, M. Senoh, S. Nagahama, N. Iwasa, T. Yamada, T. Matsushita, Y. Sugimoto, and H. Kiyoku, Appl. Phys. Lett. **69**, 1477 (1996).
3. S. Nakamura, M. Senoh, S. Nagahama, N. Iwasa, T. Yamada, T. Matsushita, Y. Sugimoto, and H. Kiyoku, Appl. Phys. Lett. **68**, 2105 (1996).
4. S. Nakamura, M. Senoh, S. Nagahama, N. Iwasa, T. Yamada, T. Matsushita, H. Kiyoku, and Y. Sugimoto, Appl. Phys. Lett. **68**, 2105 (1996).
5. R. Hofmann, H.-P. Gauggel, U.A. Griesinger, G. Grabeldinger, G. Adler, P. Ernst, H. Bolay, V. Harle, F. Scholz, H. Schweizer, and M.H. Pilkuhn, Appl. Phys. Lett. **69**, 2068 (1996).
6. J.M. Redwing, D.A.S. Loeber, M.A. Tischler, N.G. Anderson, and J.S. Flynn, Appl. Phys. Lett. **69**, 1 (1996).
7. J.M. Redwing, D.A.S. Loeber, M.A. Tischler, N.G. Anderson, and J.S. Flynn, in Proceedings of the International Conference on Blue Lasers and Light Emitting Diodes, edited by Y. Yoshikawa, K. Kishino, M. Dobayashi, and T. Yasuda (Ohmsha Ltd., Tokyo, 1996). pp. 267-270.
8. D.A.S. Loeber, J.M. Redwing, N.G. Anderson, and M.A. Tischler in <u>Gallium Nitride and Related Compounds</u>, edited by F.A. Ponce, R.D. Dupius, S. Nakamura, and J.A. Edmond (Mater. Res. Soc. Proc. 395, Pittsburgh, PA 1996), pp. 949-954. Note that, owing to a calibration error, photoexcitation thresholds quoted in this reference were overestimated by a factor of ~2.5.
9. This analysis assumes $I_{th}(N)= \{h\nu/T[1-\exp(\alpha L_{MQW})]\}\{L[1/\tau_r + 1/\tau_{nr}] + 2NS_I\}n_{th}$, where I_{th} is the threshold pump intensity, $h\nu$ is the pump photon energy, T is the transmission coefficient for the air/semiconductor interface, α is the effective MQW absorption coefficient at the pump photon energy, τ_r and τ_{nr} are the radiative and non-radiative recombination coefficients for the quantum wells, S_I is the interface recombination velocity, and n_{th} is the transparency carrier density. Note that the threshold pump intensities of this work are not lasing thresholds: They represent estimates of the pump intensities at which the transparency condition is met.
10. Assuming a 300K radiative lifetime of 0.8 ns, which is obtained from the low temperature data and temperature coefficient given in Ref. 12, a linear fit to our experimental data for $I_{th}(N)$ extrapolated to N=0 implies a nonradiative lifetime for the quantum well material of 0.15 ns. This yields a quantum well recombination lifetime (less surface recombination) of 0.13 ns.
11. This value is consistent with calculated transparency carrier density estimated for bulk GaN (W.W. Chow, A. Knorr, and S.W. Koch, Appl. Phys. Lett. **67**, 754 (1995)) and GaN quantum wells (S. Kamiyama, K. Ohnaka, M. Suzuki, and T. Uenoyama, Jpn. J. Appl. Phys. **31**, L821 (1995); A.T. Meney and E.P. O'Reilly, Appl. Phys. Lett. **67**, 3013 (1995)).
12. M. Smith, J.Y. Lin, H.X. Jiang, A. Salvador, A. Botchkarev, W. Kim, and H. Morkoc, Appl. Phys. Lett. **69**, 2453 (1996).
13. I Akasaki, H. Amano, S. Sota, H. Sakai, T. Tanaka, and M. Koide, Jpn. J. Appl. Phys. **34**, L1577 (1995); M. Koide, S. Yamasaki, S. Nagai, N. Koide, S. Asami, H. Amano, and I. Akasaki, Appl. Phys. Lett. **68**, 1403 (1996).

Portions of this work carried out at the University of Massachusetts were supported by NSF Grant ECS-9414510.

STIMULATED EMISSION AND GAIN MEASUREMENTS FROM InGaN/GaN HETEROSTRUCTURES

I.K. SHMAGIN*, J.F. MUTH*, S. KRISHNANKUTTY*†, R.M. KOLBAS*, S. KELLER**, U.K. MISHRA**, S.P. DenBAARS**
* Electrical and Computer Engineering Department, North Carolina State University, Raleigh, NC 27695-7911, kolbas@eos.ncsu.edu
** Electrical & Computer and Materials Departments, University of California, Santa-Barbara, CA 93106.
*† Presently at Honeywell Technology Center, Plymouth MN 55441.

ABSTRACT

InGaN/GaN Heterostructures were deposited on c-plane sapphire substrates by atmospheric pressure Metalorganic Chemical Vapor Deposition (MOCVD). A frequency tripled modelocked Ti-sapphire laser with a 250 fs pulse operating at 280 nm was used for photoexcitation. Photopumped stimulated emission was observed from InGaN/GaN single heterostructures (SH's) in both edge and surface emitting configurations. A sharp threshold at the onset of stimulated emission and a strong nonlinear dependence of output emission on input power density was observed. Distinct Fabry-Perot modes corresponding to both cavity configurations were also observed. Gain coefficients were measured from an $In_{0.14}Ga_{0.86}N$ film using the method developed by Shaklee and Leheny for the edge emitting configuration.

INTRODUCTION

InGaN/GaN and InGaN/AlGaN heterostructures play a very important role in today's optoelectronics. InGaN has a direct band gap varying with In composition from 3.4 eV (GaN) to 1.9 eV (InN) at room temperature. This offers the opportunity to manufacture nitride based light emitters covering all visible regions of the spectrum.
InGaN multiple quantum wells (MQW's) form the active region in high power blue, green light emitting diodes [1] and nitride based lasers [2]. The optical properties of InGaN/GaN heterostructures were studied extensively over the last decade. Observation of photopumped stimulated emission from a sample provides an effective nondestractive method of optical characterization and gives an indication of the optical quality of the material. Optical gain coefficients are needed for laser diode design.

Mat. Res. Soc. Symp. Proc. Vol. 449 © 1997 Materials Research Society

EXPERIMENT

Samples in the present study were grown on c-plane sapphire substrates by atmospheric pressure MOCVD. Prior to the deposition of the epitaxial layers a thin low-temperature GaN buffer layer was deposited on a c-plane sapphire substrate (300 μm thick). The heterojunctions consisted of a 2.4 μm GaN layer grown at 1080°C followed by a 0.1 μm layer of Si-doped $In_xGa_{1-x}N$ deposited at 700°C. Two different In compositions, x=0.1 and 0.14, were studied. Details of growth can be found elsewhere [3].

An Ar-ion pumped mode-locked femtosecond Ti-sapphire laser with a frequency tripler was used as an excitation source. The tripled output at a wavelength of 280 nm was focused to a spot size of 50-200 μm on the sample. The temporal width of the frequency tripled pulse was 250 fs.

Specimens were cut from the wafer by scribing with the diamond scribe on the substrate side of the wafer. The samples were mounted on an oxygen-free-heat-conductive copper plug and placed on the cold finger of a liquid nitrogen dewar. The measurements were taken at liquid nitrogen temperature (77 K).

The samples were studied in surface and edge emitting configurations. For the case of surface emission the pump was focused on the surface at an angle of approximately 45°, and the output emission exiting the front surface was collected. In the edge emitting configuration the pump beam was normal to the surface (close to the sample edge), and the photoluminescence exiting the edge was collected. The output signal in both cases was collected, collimated and focused on the entrance slits of a 0.32 m spectrometer with a cooled GaAs photomultiplier.

RESULTS AND DISCUSSION

The 77 K pulsed PL emission spectrum from a Si doped $In_{0.1}Ga_{0.9}N$/GaN single heterostructure in the surface emitting configuration is shown in Fig. 1. At input power densities below the threshold for the onset of stimulated emission (P_{th}=10 MW cm^{-2}; 0.66 nJ/pulse) the PL spectrum from the 0.1 μm $In_{0.1}Ga_{0.9}N$ sample has a peak position at 400 nm and full width at half maximum (FWHM) of 20 nm (Fig. 1(b)). The spontaneous emission originated from two radiative transitions: from the Si donor level to the valence band, and from the conduction to the valence. Above the threshold power density (P_{th}=10 MW cm^{-2}; 0.66 nJ/pulse) a very narrow (FWHM=1.0 nm) line of stimulated emission (peak wavelength 392.0 nm) originates from the band to band transitions that dominate the high energy side of the spontaneous emission spectrum, as shown in Fig. 1(a). At threshold (P_{th}=10 MW cm^{-2}; 0.66 nJ/pulse) the broad spontaneous emission spectrum in the surface emitting configuration is modulated by Fabry-Perot mode structure originating from the high optical and structural quality of the epilayers (Fig. 2). The cavity length calculated from

Fig. 1. Output emission spectra from an In$_{0.1}$Ga$_{0.9}$N/GaN SH in the surface emitting configuration (a) below, and (b) above threshold (P_{th}=10 MW cm^{-2}; 0.66 nJ/pulse).

Fig. 2. Output emission spectrum from an In$_{0.1}$Ga$_{0.9}$N/GaN SH in the surface emitting configuration at threshold input power density (10 MW cm^{-2}; 0.66 nJ/pulse).

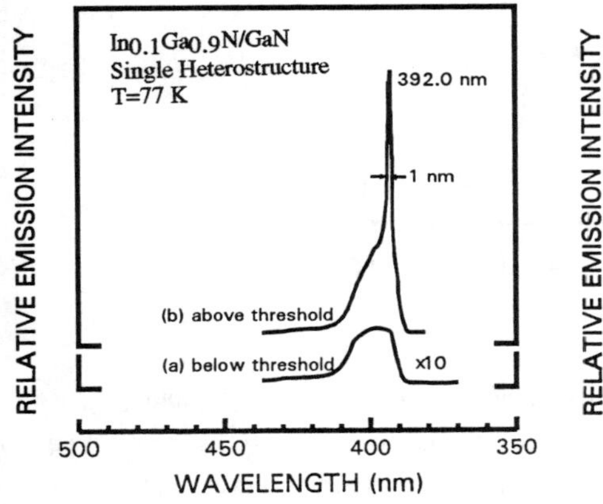

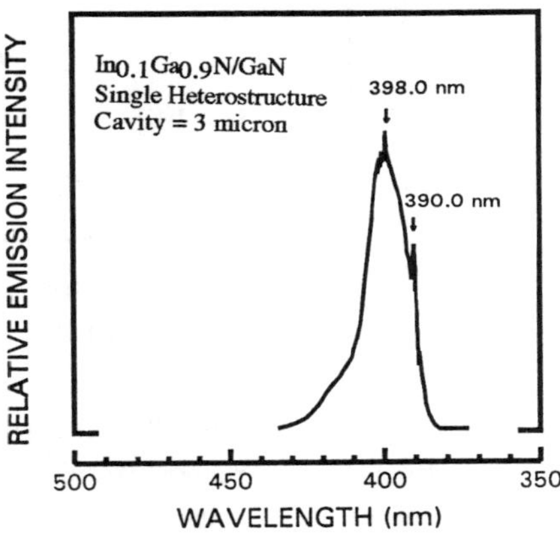

$d=\lambda^2/[2\Delta\lambda n_{eff}]$ [4] is 3 μm, which is approximately equal to the epitaxial layer thickness (2.5 μm), where λ is the emission wavelength; $\Delta\lambda$ is the mode spacing; and n_{eff} is the effective refractive index of the epilayer (n~2.7). Photopumped stimulated emission in the surface emitting configuration was also observed from the In$_{0.14}$Ga$_{0.86}$N/GaN SH with a peak position at 395.0 nm and FWHM of 0.75 nm.

The 77 K pulsed output emission spectrum from an In$_{0.14}$Ga$_{0.86}$N/GaN SH in the edge emitting configuration is shown in Fig. 3. For input power densities above threshold (P_{th}=12 MW cm^{-2}; 0.79 nJ/pulse) the output emission exhibits a multi-mode laser spectrum with the peak position at 396.0 nm (Fig. 3). The mode spacing is 1 nm which corresponds to a 30 μm long cavity. Fabry-Perot modes on the high energy side of Fig. 3 are indicative of feedback from the resonator cavity that is formed by the sample/air interface on one side and by the boundary between the illuminated and nonilluminated parts of the sample on the other. High power densities used for excitation (up to 16 MW cm^{-2}) are sufficient to create a change in the refractive index due to the local increase in temperature and excess carrier density [5]. Reflection from the illuminated/dark area interface and sample/air interface provide the necessary feedback for laser action.

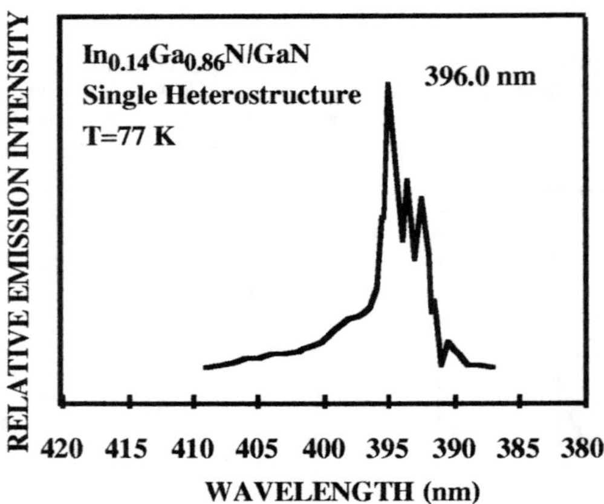

Fig. 3. Output emission spectrum from $In_{0.14}Ga_{0.86}N/GaN$ SH in the edge emitting configuration for input power density above threshold.

The output emission from the $In_{0.14}Ga_{0.86}N/GaN$ sample in the edge emitting configuration exhibits a strong nonlinear dependence on input power density, as shown in Fig. 4. The onset of stimulated emission in the edge emitting configuration occurs at an input power density of about 4 MW cm^{-2} (0.26 nJ/pulse), when a narrow (FWHM=0.75 nm) sharp line of stimulated emission emerges on the high energy shoulder. Above the threshold for

Fig. 4. Output stimulated emission intensity from $In_{0.14}Ga_{0.86}N/GaN$ as a function of input power density.

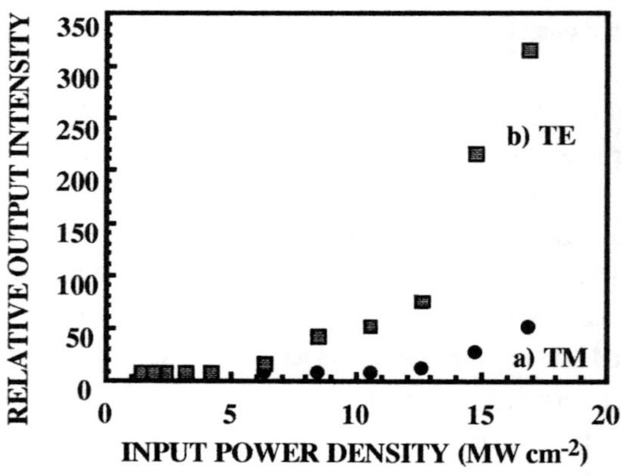

stimulated emission (4 MW cm^{-2}; 0.26 nJ/pulse) the output intensity increases for both the transverse electric (TE) and the transverse magnetic (TM) polarizations up to an input power density of 12 MW cm^{-2} (0.79 nJ/pulse). This is consistent with the characteristics of a mirrorless laser operating on amplified spontaneous emission. Above 12 MW cm^{-2} (0.79 nJ/pulse) the output emission rises sharply in intensity as a function of the input power density and is strongly TE polarized (Fig. 4). Predominance of TE polarized output is expected for an edge emitting laser with feedback.

Gain coefficients from the 0.1 μm In$_{0.14}$Ga$_{0.86}$N sample were measured using the technique developed by K.L. Shaklee [6]. The collimated laser beam is shaped into a rectangular geometry and is then focused near the edge of the sample. The side of the rectangular spot facing the edge is fixed, while the opposite side can be moved to change the length of the excited region (cavity length). The cavity length is controlled by slits with the adjustable width.

The ratio of the stimulated emission to spontaneous emission (I/I_0) intensities is plotted on a Log scale as a function of the cavity length in Fig. 5. The gain coefficient was calculated for the In$_{0.14}$Ga$_{0.86}$N/GaN SH to be 170 cm^{-1} (77 K, 10 MW cm^{-2}) from $I=(I_0/g)\exp(gL-1)$ [6], where I is the stimulated emission intensity; I_0 is the spontaneous luminescence intensity; g is the net optical gain; L is the length of the excited region. This result is somewhat larger, than the gain reported for InGaN/GaN MQW based laser diodes, where g=110 cm^{-1} was measured [7].

Fig. 5. Log of the ratio of the stimulated to spontaneous emission intensities(I/I_o) as a function of cavity length L.

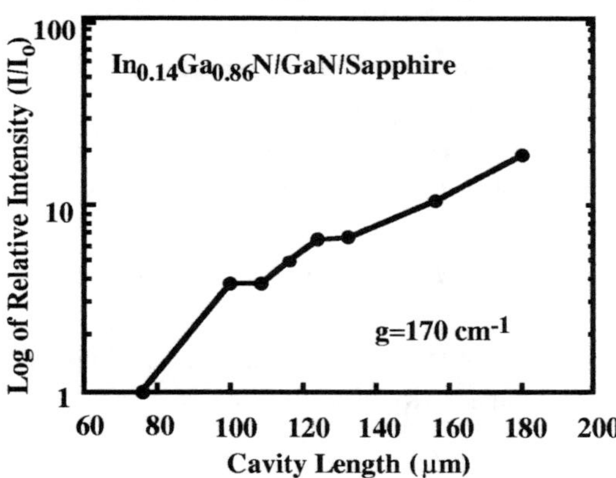

CONCLUSIONS

Photopumped stimulated emission was observed from $In_{0.1}Ga_{0.9}N/GaN$ and $In_{0.14}Ga_{0.86}N/GaN$ single heteroctructures at 77 K in both surface and edge emitting configurations. Observations of photopumped laser action from the $In_{0.14}Ga_{0.86}N/GaN$ SH in the edge emitting configuration was confirmed by a strong nonlinear dependence of the output emission intensity on input power density; dramatic line narrowing (narrows by a factor 27) at threshold (P_{th}=4 MW cm^{-2}; 0.26 nJ/pulse); mode structure indicative of feedback, and; predominance of TE polarized output at input power densities above 12 MW cm^{-2}, which is consistent with the operation of an edge emitting laser with feedback. The net optical gain was measured for an $In_{0.14}Ga_{0.86}N/GaN$ SH to be equal to 170 cm^{-1} at 77 K and 10 MW cm^{-2} input power density. This work was supported by U.S. Army Research Office under contract # DAAH04-93-D-003-4 supervised by Dr. J. Zavada, and ARPA Optoelectronics Technology Center.

REFERENCES

1. S. Nakamura, M. Senoh, N. Iwasa, and S. Nagahama, Appl. Phys. Lett. **67**, 1868 (1995).

2. S. Nakamura, M. Senoh, S. Nagahama, N. Iwasa, T. Yamada, T. Matsushita, H. Kiyoku, and Y. Sugimoto, Jpn. J. Appl. Phys. **35**, L74 (1996).

3. S. Keller, B. P. Keller, D. Kapolnek, A. C. Ambare, H. Masui, L. A. Coldren, U. K. Mushra, and S. P. DenDaars, Appl. Phys. Lett. 68, 3147 (1996).

4. J.I. Pankove, in Optical Processes in Semiconductors, (Dover Publications, Inc., New York, 1971), p. 222.

5. C. D. Poole and E. Garmire, in *Optical Bistability*, edited by C. M. Bowden, H. M. Gibbs, and S. L. McCall (Plenum Press, New York, 1984), v. 2, p. 279.

6. K.L. Shaklee, R.E. Nahory, and R.F. Leheny, J. of Luminescence 7, 284 (1973).

7. S. Nakamura, M. Senoh, S. Nagahama, N. Iwasa, T. Yamada, T. Matsushita, Y. Sugimoto, and H. Kiyoku, Appl. Phys. Lett. 69, 1568 (1996).

HIGH QUALITY PHOTOCONDUCTIVE ULTRAVIOLET GaN/6H-SiC DETECTOR AND ITS PROPERTIES

K. Yang*, R. Zhang*, L. Zang*, B. Shen*, Z.Z. Chen*, Y.D. Zheng*, X.M. Bao*
Z.C. Huang**, and J.C. Chen**
*Department of Physics, Nanjing University, Nanjing 210093, China
**Department of Electrical Engineering, University of Maryland Baltimore County, Baltimore, MD 21228-5398

ABSTRACT

The properties of photoconductive ultraviolet detector based on GaN epilayer grown on 6H-SiC substrate by metalorganic chemical vapor deposition were investigated in this paper. We obtained the detectable energy span of the device up to ultraviolet by photocurrent measurement. The spectral responsivity remained nearly constant for wavelengths from 250 to 365 nm and dropped by three orders of magnitude within 10 nm of the band edge (by 380 nm). The detector was measured to have a responsivity of 133 A/W at a wavelength of 360 nm under a 5-V bias, and the voltage-dependent responsivity was performed. Furthermore, an easy method was developed to determine the response time, and the relationship between response time and bias was obtained.

INTRODUCTION

Gallium Nitride, a direct wide band gap semiconductor, has recently attracted considerable interest because of its great potential in electronic and optoelectronic applications[1,2]. It is being considered as material for efficient blue and ultraviolet (UV) light emitters and for laser diode (LD) application[3,4,5], as well as for use as photodetectors at high energy[6]. With the substantial development of the material and device, an energy span ranging from the blue to near-UV wavelengths, which is at present inaccessible to semiconductor technology, has been opened. GaN has a large direct band gap, high electric breakdown strength, high thermal conductivity, and great physical hardness, which make it ideally suitable for blue-and-ultraviolet emitter devices, high energy photodetectors and high temperature transistors[7,8]. Since the technology for growth of GaN-based material by epitaxial method has been greatly improved, high quality single crystal GaN films have been successfully fabricated by various methods, such as metalorganic chemical vapor deposition (MOCVD)[7,8,9,10], and molecular beam epitaxy (MBE)[11]. Photoconductive ultraviolet sensors[12], blue-light-emitting diodes (LEDs)[8], metal semiconductor field effect transistors (MESFETs)[7,13] and heterostructure field effect transistors (HFETs)[9] have been developed.

In this paper, we report the properties of photoconductive UV detector based on the GaN epilayer grown on 6H-SiC substrate by MOCVD. GaN based detectors, from a theoretical

Mat. Res. Soc. Symp. Proc. Vol. 449 ©1997 Materials Research Society

standpoint, have by far higher responsivities at energy above 3.4 eV than below. This property makes the detectors capable of detecting ultraviolet emissions in the presence of infrared background. The response time was estimated to several ps for a metal-semiconductor-metal GaN detector[14]. Furthermore, the large photoabsorption depth makes GaN-based photodetector free from the problems associated with surface recombination and surface scattering. In our work, photocurrent (PC) measurement exhibited a cutoff at 3.4 eV (365 nm) and a continued photo-response through the ultraviolet region. An easy method was used to determine the response time of the detector, and the kinetics of photo-conductivity in UV region was discussed.

EXPERIMENTS

The sample used in this study is an ultraviolet detector based on single crystal hexagonal GaN film grown on 6H-SiC substrate by MOCVD. The Ga and N source gases were trimethylgallium (TMGa) and ammonia (NH_3), respectively. The epitaxial layer was grown using TMGa and high purity NH_3 as sources and H_2 as carrier gas, and a thin GaN buffer layer was employed. The thickness of the film was determined as ~1.2 μm from the interference features on the optical reflection curve. The sample was unintentionally doped n-type, and the electron concentrations were typically around 1×10^{17} cm^{-3}.

Fig.1 The micrograph of the interdigitated electrode detector.

The photoconductive detector was fabricated by using the GaN epilayer, and using standard photolithography and lift-off procedures to pattern the device. An interdigitated finger structure, as shown in Fig.1, was employed for the photoconductive response measurements. The interdigitated gold electrodes were 5 μm wide, 1 mm long, and had a 10 μm spacing. The area of its photosensitive surface was 1.0 mm^2. The current-voltage characteristics were measured by a semiconductor parameter analyzer. The I-V behavior was linear in both forward- and reverse-bias region, showing ohmic behavior. The resistance of the GaN detector was about 120 Ω.

The spectral response was measured using an Xenon lamp, monochronmator, chopper, and lock-in amplifier (EG&G 5210). The measurement circuit consisted of a constant voltage source, the photoconductor, and a sense resistor connected in series. The photocurrent induced in the device was obtained by measuring the voltage across the sense resistor under dark and illuminated conditions.

RESULTS AND DISCUSSIONS

Photosignal current was measured as a function of the wavelength of the incident light, which was from the high pressure Xenon lamp mechanically chopped at 30 Hz, with a 5-V bias. The photo response was read by the lock-in amplifier, and also normalized to allow for the Xenon lamp emission intensity variation as a function of wavelength. Figure 2 is a plot of the spectral responsivity data for a single detector element. As seen, the detector had response from 380 nm up to ultraviolet, and it reached its peak value at 365 nm (~3.4 eV),   and then remained nearly constant up to 250 nm. The photo response at higher energy was three orders of magnitude larger than that at lower energy. The response below the band gap is obviously come from the energy levels among the forbidden band, just like those in the photoluminescence, which mainly depend on defects and impurities in GaN films.

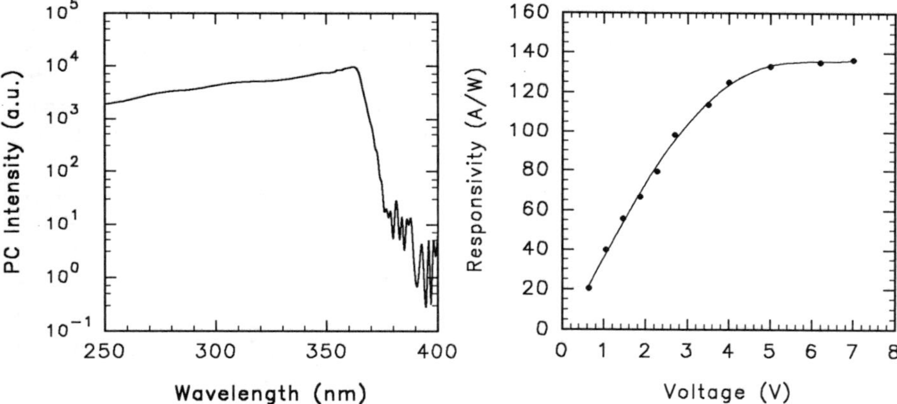

Fig.2 Photocurrent of the photoconductive detector based on the GaN epilayer on 6H-SiC substrate.

Fig.3 The voltage-dependent responsivity of the GaN photoconductor.

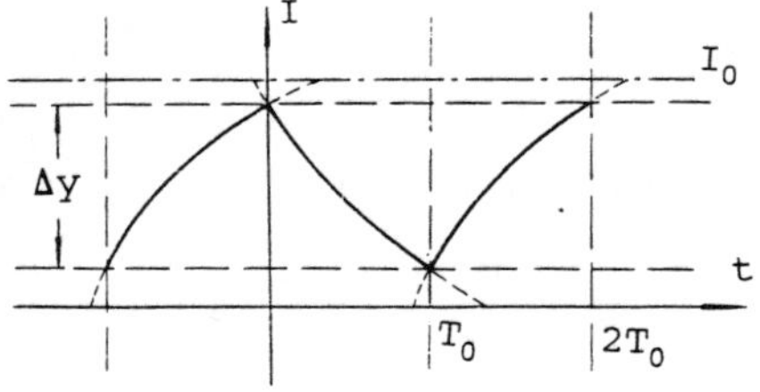

Fig.4 Schematic diagram of the intensity of the photocurrent read by lock-in amplifier.

To quantify the peak responsivity we measured the Xenon lamp intensity using a calibrated silicon UV detector at wavelength of 360 nm. Using the silicon detector area, and 0.25 mm^2 as the exposed area for the GaN detector, we estimate the peak responsivity of the GaN detector to be   around 133 A/W under a 5-V bias. To further characterize the photoconductors, the voltage-dependent responsivity was performed, as shown in Fig.3. Below 6 V, the photoconductor responsivity increased nearly linearly with applied voltage.

Above 6 V the responsivity saturated. Assuming essentially complete saturation at 6 V yield a saturation field of approximately 5000 V/cm. The saturation behavior can be explained by the sweep-out effect[15].

Instead of using a pulse laser, we chose a different way to determine the time constant of the detector. During the measurement of photocurrent, we changed the chopper frequency. As shown in Fig.4, we assumed that the transient behavior of device was exponential with time $I(t) = I_0 \exp(-t/\tau)$. When the light was off, the intensity of photocurrent I decreased as I(t), but was stopped at $t = T_0$, at which time the light was turned on , and then the intensity increased. So the AC signal that the lock-in amplifier detected was $\Delta y(f) = I_0 \tanh(1/(4\tau f))$, where f is the frequency of the chopper and $2f = 1/T_0$. Thus fitting the $\Delta y(f)$ curve to the measured intensity (as shown in Fig.5), we obtained the time constant τ. It consisted two parts: the lifetime of carriers τ_0 and the trapping time τ_t, which was dominant in this case. Though traps can enhance the intensity of the photodetector, the response time will be enlarged simultaneously due to the increasing τ_t.

By using the method, we plotted the response time dependence on the bias of the GaN detector from the experimental data, as shown in Fig.6. The response time of the device under a 5-V bias was determined as 6 ms through fitting the experimental curve.   As seen, the response time decreases with increasing the bias on the detector, agreeing well with the effect of sweep-out field increase.

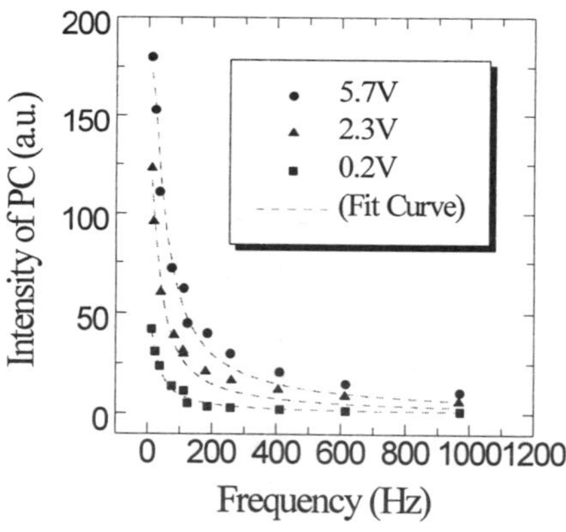

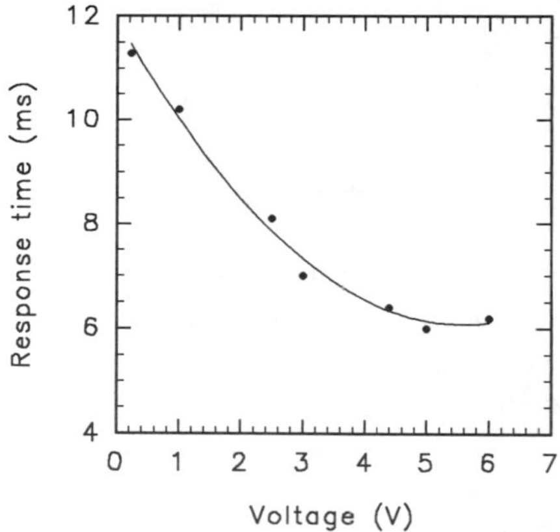

Fig.5 Dependence of the photocurrent on the chopper frequency.

Fig.6 Dependence of the response time on the bias of the GaN detector.

CONCLUSIONS

In summary, we studied the properties of photoconductive UV detector based on GaN epilayer grown on 6H-SiC substrate by MOCVD. We obtained the detectable energy span of the device up to UV by photocurrent measurements. The PC signal exhibited a cutoff at 3.4 eV (~365 nm) and a continued constant photo-response through the UV region. The detector was measured to have a responsivity of 133 A/W at a wavelength of 360 nm under a 5-V bias, and the voltage-dependent responsivity was performed. Furthermore, an easy method was developed to determine the response time, and the relationship between response time and bias was obtained.

REFERENCES

1. S.Strite and H.Morkoc, J.Vac.Sci.Technol. B10(4), 1237(1992)

2. H.Morkoc, S.Strite, G.B.Gao, M.E.Lin, B.Sverdlov and M.Burns, J.Appl.Phys. 76(3), 1363(1994)

3. S.Nakamura, M.Senoh and T.Mukai, Appl. Phys. Lett. 62 , 2390(1993)

4. S.Nakamura, T.Mukai and M.Senoh, J Appl. Phys. 71, 5543 (1992)

5. R.J. Molnar, R. Singh, T.D.Moustakas, Appl. Phys. Lett., 66, 268(1995)

6. M.Misra, T.D. Moustakas, R.P.Vaudo, R.Singh, K.S. Shah, SPIE, vol.78, 2519(1994)

7. M.A.Khan, J.N.Kuznia, A.R.Bhattarai, and D.T.Olson, Appl. Phys. Lett. 62(15),1786(1993)

8. S.Nakamura, T.Mukai and M.Senoh, Appl. Phys. Lett. 64(13), 1687(1994)

9. M.A.Khan, J.N.Kuznia, D.T.Olson, W.J.Schaff, J.W.Burm and M.S.Shur, Appl. Phys. Lett. 65(9), 1121(1994)

10. K.G.Fertitta, A.L.Holmes, J.G.Neff, F.J.Ciuba, and R.D.Dupuis, Appl. Phys.Lett.65 (14), 1823(1994)

11. T.D.Moustakas, R.J.Molnar, Mat. Res. Soc. Symp. Proc. Vol.281, 753(1993)

12. M.A.Khan, J.N.Kuznia, D.T.Olson, J.M.Van Hove, M.Blasingame, L.F.Reitz, Appl. Phys. Lett. 60(23), 2917(1992)

13. J. Pankove, S.S. Chang, H.C. Lee, R.J. Molnar, T.D. Moustakas, IEDM-94, 398(1994)

14. R.P.Joshi, A.N.Dharamsi, and J.McAdoo, Appl. Phys. Lett. 64(26), 3611(1994)

15. D.Walker, X.Zhang, P.Kung, A.Saxier, S.Javadpour, J.Xu, and M.Razeghi, Appl.Phys.Lett. 68(15), 2100(1996)

GaN BASED MSM UV PHOTODETECTORS

S. Liang*, W. Cai*, Y. Li*, Y. Liu*, Y. Lu*, C.A. Tran**, R.F. Karlicek**, and I. Ferguson**
*Department of ECE, Rutgers University, Piscataway, NJ 08855-0909
**EMCORE Corporation, Somerset, NJ 08873

ABSTRACT

Results obtained from interdigital metal-semiconductor-metal (MSM) type of GaN based UV photodetectors are presented. MSM devices were fabricated using two types of GaN; high-resistive GaN and Mg doped GaN. For the high-resistive GaN detector, the lowest dark current is ~0.1 nA and the UV responsivity of the device was about 460 A/W at a DC bias of 30 V. The Mg doped GaN exhibited large gains, 1150 A/W at 2.0 V, but at much higher dark currents, 400 nA. The high gain in this device is not well understood but was attributed to an 'avalanche' effect and is under investigation. It was found that the surface plasma treatment plays an important role in the device performance. After an appropriate plasma treatment the detectors showed higher responsivity with lower dark current.

INTRODUCTION

The recent progress in the area of GaN based materials and devices shows that GaN based materials have great potential applications in the area of optoelectronics in the UV/blue wavelength region and in the area of high temperature/high power electronics. A number of researchers have reported their results in the various types of GaN based UV photodetectors, such as, MSM [1-6], Schottky [7], p-n junction [8-9], and HFET [10]. M. A. Khan et al reported UV photoconductive detectors based on i-GaN layers on sapphire with a responsivity of 2000 A/W and a gain of 6×10^3 at a wavelength of 365nm under a 5.0 V bias, and a bandwidth in excess of 2 kHz [5]. They also reported Schottky barrier photodetectors based on Mg-doped GaN films on sapphire with a responsivity of 0.13 A/W at a wavelength of 365nm under 0 V bias [7]. K. S. Stevens et al reported GaN:Mg on silicon UV photodetectors with a cutoff at 3.3 eV and a responsivity of 12A/W at 4.0 V bias for optical intensities on the order of $1W/m^2$ and below [4]. M. Misra et al reported a photoconductive UV detector based on semi-insulating GaN films grown by molecular beam epitaxy with a responsivity of 125 A/W and gain-quantum efficiency product of 600 at 254nm at 25V, and a response time on the order of 20ns corresponding to a bandwidth of 25MHz [1]. Q. Chen et al reported the UV photodetectors based on GaN p-n junctions with an abrupt long-wavelength cutoff wavelength at ~370nm, responsivity values as high as 0.09 A/W at 360nm, and rise and fall times of 300µs at 325nm were achieved [8]. Recently a 0.2µm gate photodetector based on GaN/AlGaN heterostructure field effect transistor was reported [10]. Its responsivity is as high as 3000A/W for wavelengths from 200 to 365nm, and a response time is of order 0.2ms. In this work preliminary results are reported for interdigital MSM UV photodetectors fabricated using both high-resistive GaN and Mg doped GaN.

EXPERIMENTAL

The GaN films have been grown in an EMCORE D-180 multi-wafer rotating disc low pressure MOCVD system. Detailed description about the system and growth process has been

Mat. Res. Soc. Symp. Proc. Vol. 449 © 1997 Materials Research Society

reported previously [11]. Trimethylgallium (TMG) and ammonia (NH$_3$) were used as the Ga and N sources, respectively, with H$_2$ as carrier gas. Cp$_2$Mg was used as the doping source for p-type carriers. The typical growth sequence included a low temperature 20 nm GaN buffer/nucleation layer, deposited at approximately 500 °C, before the substrate was then raised to the growth temperature of about 1050 °C before the active layer was grown.

UV-detectors were designed and fabricated based on MSM structures. The device material is made up of 2μm thick undoped or Mg doped GaN on Sapphire. The Mg was not activated so the GaN was n-type in character. Ti/Au was deposited by e-beam evaporation (20nm and 240nm, respectively) to form the metal contact. An inter-digit electrode pattern shown in Figure 1 was used for MSM detectors. The electrode width and the inter-digit spacing vary from 2μm to 16μm. The total inter-digit area is 1.0mm^2.

Image reversal photolithography, E-beam metallization and lift-off techniques were used for generation of the inter-digit electrode pattern. Oxygen plasma was used for the surface treatment after lift-off process.

Shown in Figure 2 is the schematic of the measurement setup for measuring dark and photo-illuminated I-V characteristics. A 200Watt tungsten lamp was used as UV light source. One set of laser line filters (10nm bandwidth) was employed instead of monochromator for higher light power output. The light input power to the detectors was calibrated by Newport 1830-C Optical Power Meter. Dark and photo-illuminated I-V characteristics was measured using Tektronix Semiconductor Work Bench 271.

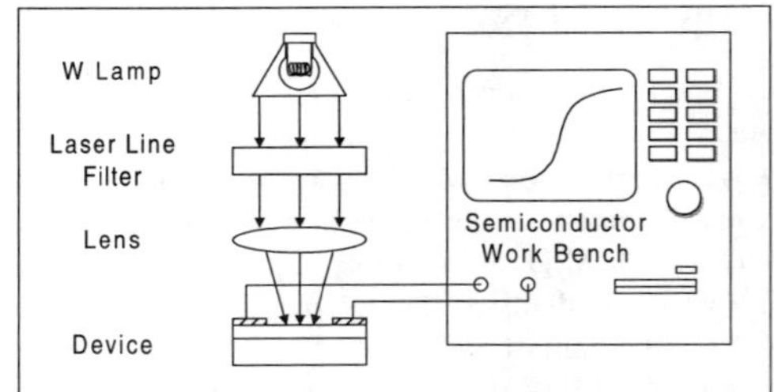

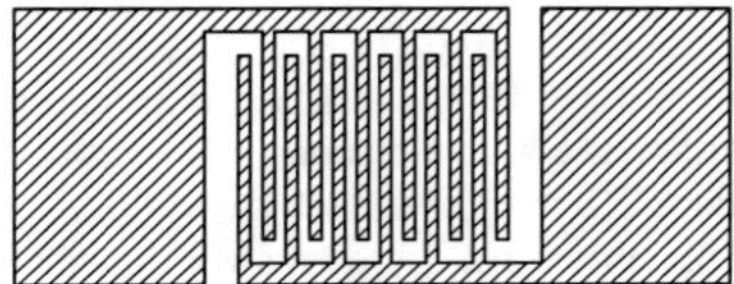

Figure 1:   Contact pattern of the MSM detectors

Figure 2:   Measurement setup for photo-illuminated I-V characteristics

RESULTS AND DISCUSSIONS

The dark and photo-illuminated I-V characteristics was measured using the setup in Figure 2. The test results of a high resistive GaN and a Mg doped GaN MSM photodetector are shown in Figure 3 and 4, respectively. Two laser line filters, 351nm and 308nm were used for the photo-illuminated I-V measurements in Figure 3 and 4. The light power density was 2nW/cm^2. Figure 3 shows symmetric I-V characteristics at forward and reversed bias. The linear I-V relations in low DC bias range (<5V) exhibits 'ohmic-like' metal-semiconductor contacts. The detector was operated in the photoconductive mode. The life-time mobility product ($\tau\mu$) can be estimated using [120]

$$G = \frac{\tau\mu V}{d^2}, \qquad (1)$$

where G = 460 (at V=5V) is photoconductive gain, d ~ 8mm is the effective width of the device. $(\tau\mu)$ = 5.9cm^2/V. if μ ~ 500V/cm^{2}S for undoped GaN, the hole life-time thus can be estimated in the range of 0.1μS. The photoresponse exhibits saturation phenomenon at the high DC bias (>5V) range, which might be due to the saturation of electrons passing through the Schottky barrier via generation-recombination centers. The dark current at bias less than 10V is below 0.1nA which beyond our detection limit.

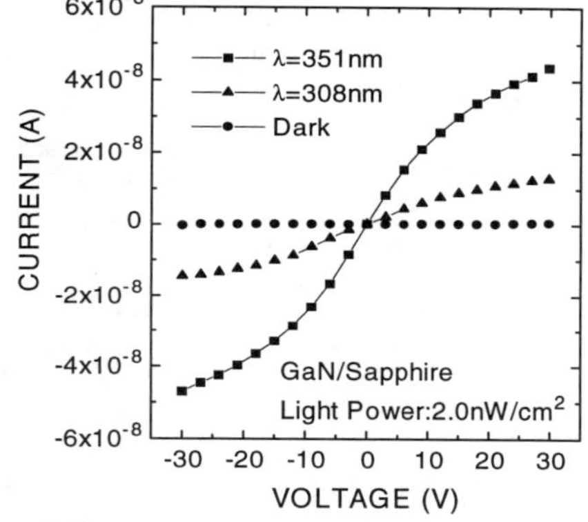

Figure 3: Dark and photo-illuminated I-V characteristics of a high resistive GaN MSM photodetector

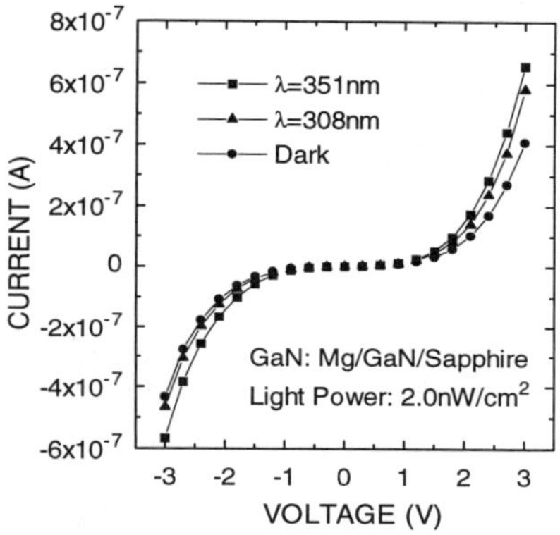

Figure 4: Dark and photo-illuminated I-V characteristics of a Mg doped GaN MSM photodetector

The I-V characteristics in Figure 4 shows different behavior from those in Figure 3. Both dark current and photoresponses are very small at low DC bias (<1.5V) and exhibit a rapid increase with bias and breakdown at 2.5V. The non-linear I-V relations exhibit Schottky mode of operation. Shown in Figure 5 are the calculated photoresponse spectra using the photo-illuminated I-V characteristics under different DC bias. When DC bias increase from 1.0V to 2.0V, the photoresponsivity increases one order. High photoresponsivity (1150A/W at bias of 2.0V and wavelength of 351nm) was achieved, which is attributed to the 'avalanche' breakdown near the reverse-biased junction. Similar phenomenon was observed in GaAs Schottky detector [13]. Further study of the 'avalanche' effect is under progress.

Figure 6 shows I-V characteristics (dark and UV illuminated) for both of non-surface plasma treatment and after surface treatment using oxygen plasma. It can be seen in Figure 6 that after surface plasma treatment, the UV responsivity of the device was increased. This is attributed to the reduction of the surface UV absorbing and surface recombination centers. The dark current was decreased due to the removal of a surface conducting layer or the neutralization of the surface charges by the plasma treatment.

CONCLUSION

MSM UV photodetectors were fabricated on both undoped GaN and Mg doped GaN. Both 'ohmic-like' and 'Schottky avalanche' modes of operation were observed. Low dark current of 0.1nA at 10V bias for the device operating at ohmic-like mode. High responsivity of 1150A/W was obtained at bias of 2.0V and wavelength of 351nm for the device apparently operating in an 'avalanche' mode. The surface treatment using oxygen plasma resulted in reduction of dark current and increase of photoresponsivity.

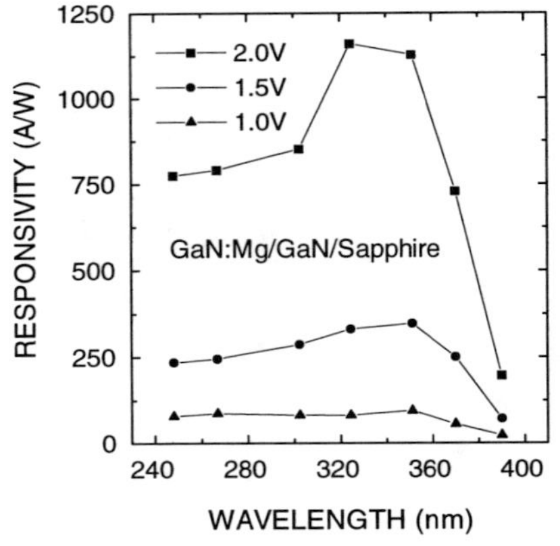

Figure 5: Photoresponse spectra of Mg doped GaN MSM Photodetector

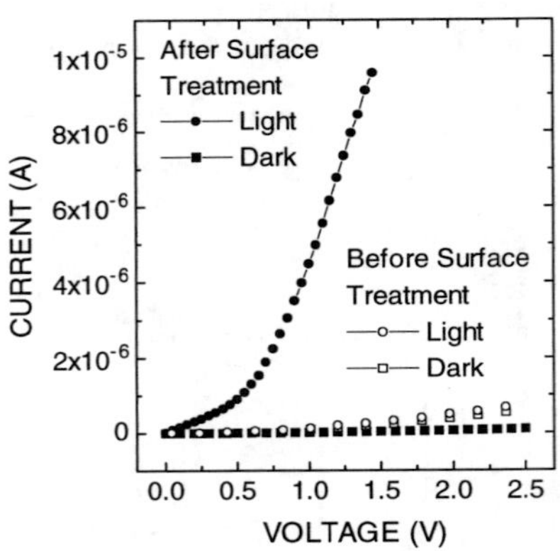

Figure 6: Oxygen plasma passivation effect on MSM type of GaN UV detector

ACKNOWLEDGMENT

This work was sponsored by NASA under Contract # NAS5-33248.

REFERENCES

1. M. Misra, T.D. Moustakas, R.P. Vaudo, Rajminder Singh and K.S. Shah, SPIE, **2519,** p. 78 (1995).

2. Larry F. Reitz, SPIE, **1952,** p. 14 (1993).

3. D. Walker, X. Zhang, P. Kung, A. Saxler, S. Javadpour, J. Xu and M. Razeghi, Appl. Phys. Lett., **68**, p. 2100 (1996).

4. K.S. Stevens, M. Kinniburgh and R. Beresford, Appl. Phys. Lett., **66**, p. 3518 (1995).

5. M. Asif Khan, J.N. Kuznia, D.T. Olson, J.M. Hove and M. Blasingame, Appl. Phys. Lett., **60**, p. 2917 (1992).

6. Z.C. Huang, D.B. Mott and P.K. Shu, Annual Device Research Conference Digest 1996, IEEE, Piscataway, NJ, USA, pp. 188-189.

7. M. Asif Khan, J.N. Kuznia, D.T. Olson, M. Blasingame and A.R. Bhattarai, Appl. Phys. Lett., **63**, p. 2455 (1993).

8. Q. Chen, M. Asif Khan, C.J. Sun And J.W. Yang, Electronics Letters, **31**, p. 1781 (1995).

9. D. Walker, X. Zhang, P. Kung, A. Saxler, J. Xu And M. Razeghi, MRS, **395**, p. 955 (1996).

10. M.A. Khan, M.S. Shur, Q. Chen, J.N. Kuznia and C.J. Sun, Electronics Letters, **31**, p. 398 (1995).

11. C. Yuan, T. Salagaj, R. A. Stall, Y. Li, M. Schurman, C. Y. Hwang, W. E. Mayo, Y. Lu, S. J. Pearton, S. Krishnankutty and R. M. Kolbas, J. Electrochem. Soc., **142**, p. L163 (1995).

12. B.E.A. Saleh and M.C. Teich, <u>Fundamentals of Photonics</u>, John Wiley and Sons, New York, 1991, pp. 644-692.

13. W.T. Lindley, R.J. Phelan, Jr., C.M. Wolfe and A.G. Foyt, Appl. Phys. Lett., **14**, p. 197 (1969)

VISIBLE BLIND UV GaN PHOTOVOLTAIC DETECTOR ARRAYS GROWN BY RF ATOMIC NITROGEN PLASMA MBE

J. M. VAN HOVE, P.P. CHOW, R. HICKMAN, A.M. WOWCHAK, J.J. KLAASSEN, C.J. POLLEY
SVT Associates, 7620 Executive Dr., Eden Prairie, MN 55344, svta@svta.com

ABSTRACT

RF atomic nitrogen plasma molecular beam epitaxy (MBE) was used to deposit gallium nitride (GaN) p-i-n junction photovoltaic detectors on (0001) sapphire. The detectors consisted of a bottom contact layer n-type silicon doped to $5x10^{18}$ cm^{-3}. The intrinsic layer was undoped and possessed an n-type background carrier concentration of $1x10^{16}$ cm^{-3}. The top p-GaN layer was doped with magnesium to give a Hall concentration of $5x10^{17}$ cm^{-3}. The p-type GaN cathodoluminescence (CL) spectra showed a strong 372 nm emission level in contrast to the 430 nm level observed in MOCVD samples. These layers were fabricated into 1x10 element detector arrays using a chlorine-based reactive ion etch (RIE) and refractory metal ohmic contacts. Peak responsivity of 0.11 A/W on detectors without anti-reflection coating were obtained at the GaN bandedge of 360 nm. The ultraviolet (UV) to visible rejection ratio was greater than 10^3-10^4 and was accredited to the reduction of the yellow defect levels in MBE material. Preliminary results on Al$_x$Ga$_{1-x}$N detectors with responsivity peaks at 313 and 343 nm are presented as well.

INTRODUCTION

GaN and its alloys with aluminum and indium are among the most promising of materials for the development of semiconductor photonic devices operating in the blue and UV regions of the spectrum [1-3]. GaN and Al$_x$Ga$_{1-x}$N are also extremely robust materials suitable for high temperature, high power applications. Unlike silicon carbide (SiC), which is a potentially valuable semiconductor for optoelectronic applications in the UV, GaN is a semiconductor with a direct energy gap of 3.39 eV (366 nm). The transparency of high quality GaN at wavelengths longer than the bandgap make GaN an ideal material for the fabrication of photodetectors capable of rejecting near infrared and visible regions of the solar spectrum while retaining near unity quantum efficiency in the UV. UV sensitive photodetectors that do not respond to visible light have numerous applications ranging from simple fast response switches for flame sensing to diode arrays for satellite borne UV spectrometers.

EXPERIMENT

Epitaxial Growth

The GaN p-i-n photodiode layers were grown on (0001) basal-plane 2" sapphire substrates by MBE using an RF atomic nitrogen plasma source [4]. In this system, effusion cells were used to provide flux of the group III element gallium as well as the p-type dopant magnesium. The n-type doping was performed with silicon evaporated from a compact e-beam source.

The growth sequence was initiated with deposition of an AlN buffer at low temperature to accommodate GaN lattice mismatch with sapphire. Following the buffer, the GaN photodiode layers were deposited at 800°C as shown in Figure 1 with growth rates of 0.5 µm/hr. The

Mat. Res. Soc. Symp. Proc. Vol. 449 © 1997 Materials Research Society

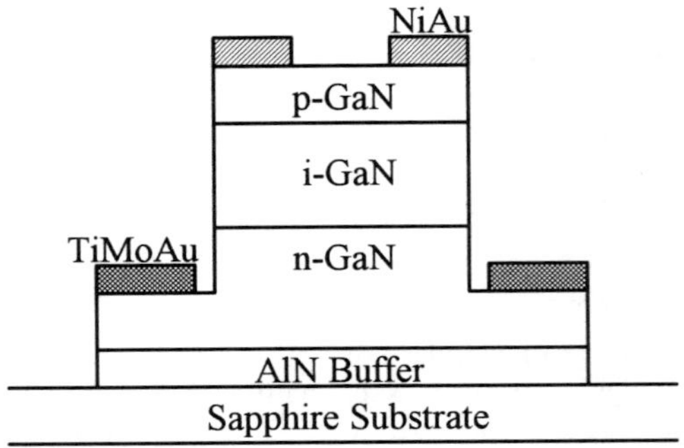

Figure 1: Schematic drawing of the GaN *p-i-n* photovoltaic diode.

structure consisted of a 1 μm *n*-type contact layer doped at 5×10^{18} cm^{-3} followed by an intrinsic region with an unintentional *n*-type doping of 1×10^{16} cm^{-3}. The epitaxial structure was completed with the deposition of a 3000 Å GaN layer Mg-doped *p*-type at 5×10^{17} cm^{-3}. The *p*-type concentration was determined by Hall measurement on prior samples, and no post growth anneal was required to activate the Mg impurities. The photovoltaic diode arrays were fabricated with elements having rectangular and octagonal geometries. The photodiode *p-i-n* junction areas varied from 0.033 mm^2 to 0.20 mm^2.

The MBE grown GaN exhibited good surface morphology and uniformity. The best intrinsic mobility was obtained in films with (0002) X-ray diffraction peaks that had full width at half maximums (FWHMs) of 2-5 arcminutes. The narrowest FWHM measured in the MBE grown epilayers was 39 arcseconds. Cathodoluminescence (CL) was used to monitor the optical quality of the GaN material which had a typical FWHM of 7.4 nm (66 meV) as shown in Figure 2. The "yellow" defect level emission intensity at 550 nm was less than 10^2 times that of the bandedge intensity. In contrast to MOCVD grown *p*-GaN with 430 nm emission [5-6], that grown by RF atomic nitrogen plasma MBE exhibited luminescence spectra centered at 372 nm as seen in Figure 3. CL spectra from *p*-type GaN material were typically characterized by broader FWHM values than *n*-type material. For example, the spectrum of Figure 3 has a FWHM of 14.8 nm (148 eV). Since the *p*-type emission broadening is not gaussian, a more relative CL peak width occurs at approximately 1/3 of the maximum where the peak is 29.3 nm (302 meV) wide.

<u>Device Processing</u>

Device processing of the UV photovoltaic detectors on 2" sapphire substrates began with a recess etch through the *p*-type and intrinsic regions of the structure to the *n*-type contact region. The recess was performed by RIE in a low pressure SiCl$_4$ plasma which yielded etch rates of approximately 125 Å/min. A mesa etch was then performed to electrically isolate the 10-element arrays using the same RIE process. Refractory Ti/Mo/Au contacts were first deposited on the *n*-GaN and annealed at 875°C to yield specific contact resistances which were typically $<3 \times 10^{-5}$ Ωcm^2 [7]. The *p*-GaN contacts followed with deposition of Ni/Au and a lower temperature anneal at 425°C for 30 s in an N$_2$ ambient. All contact metals were patterned by liftoff. The

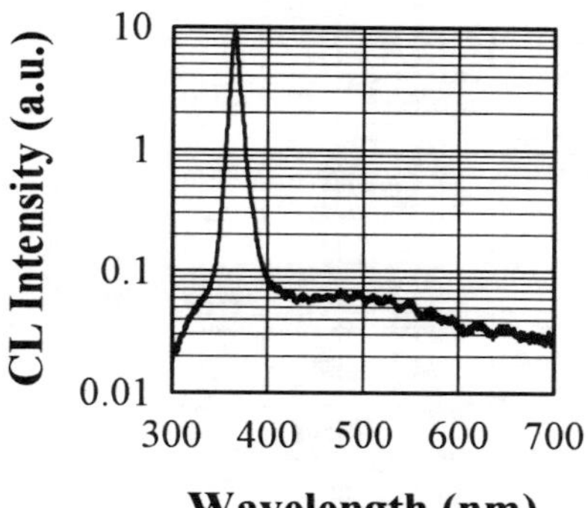

Figure 2: CL spectrum of *n*-GaN material with bandedge emission peak at 365 nm and a FWHM of 7.4 nm (66 eV).

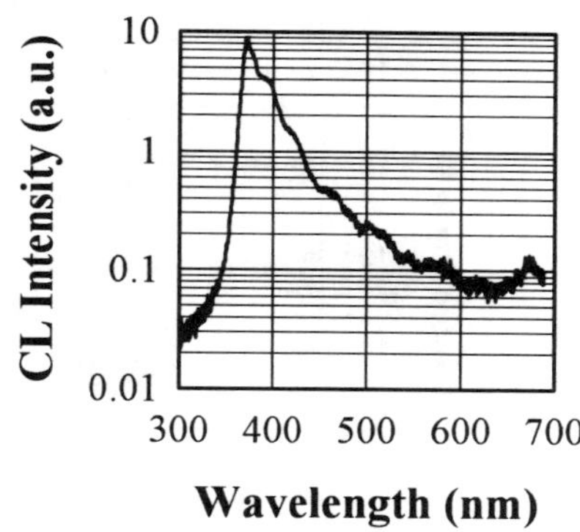

Figure 3: CL spectrum of *p*-GaN material with bandedge emission peak at 372 nm and a FWHM of 14.8 nm (148 eV). At 1/3 maximum, the peak width is 29.3 nm (302 meV).

device processing was completed with electroplating of soft gold and dicing of the 10-element arrays.

Spectral Response Measurement

The spectral responsivities of the UV photodiodes were measured at zero bias voltage using a 75 W xenon arc lamp chopped at 400 Hz and filtered by a 1/8 meter monochromator set to a 5 nm bandpass. The power of the monochromatic light was measured with a calibrated, NIST traceable, silicon photodiode (Newport 1815-C) and then focused onto the GaN *p-i-n* photodiodes which were mounted on a micropositioner stage. The diode photocurrent was amplified, and the power spectral density in a 1 Hz bandwidth at the modulation frequency was monitored with a FFT spectrum analyzer. Measurements of incident power and photodiode current were made in 10 nm intervals from 250 to 800 nm.

$Al_xGa_{1-x}N$ Detectors

The deposition and fabrication of photodetectors with peak response at wavelengths closer to 290 nm for solar blind applications were also investigated. $Al_xGa_{1-x}N$ photovoltaic *p-i-n* detectors with epitaxial doping and thicknesses identical to those of the GaN photodetectors were fabricated. Two epitaxial wafers were grown with ternary $Al_{0.05}Ga_{0.95}N$ and $Al_{0.10}Ga_{0.90}N$ and processed into photodiode arrays for spectral characterization. The fabrication and measurement processes were unchanged for the $Al_xGa_{1-x}N$ photodiodes.

RESULTS

Electrical Characteristics

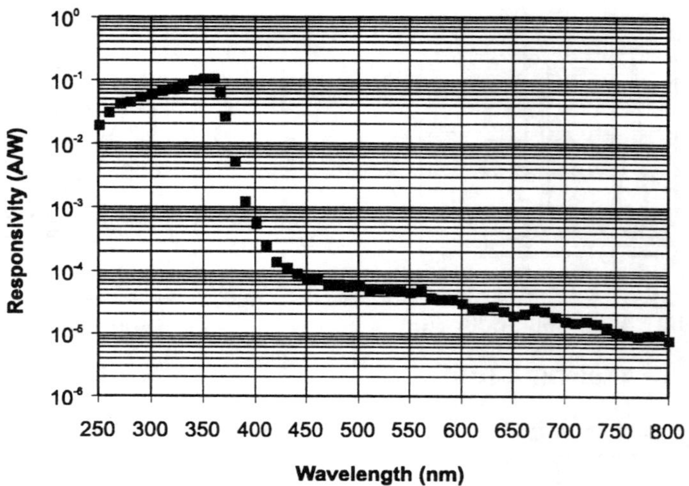

Figure 4:   The spectral response curve of a GaN photovoltaic diode with a peak responsivity at 360 nm of 0.11 A/W and 3 to 4 orders of magnitude rejection of radiation in the visible and near IR.

RESULTS

Electrical Characteristics

The GaN *p-i-n* diode forward turn-on voltages were approximately 3 V, and they exhibited soft current-voltage (IV) characteristics beyond -7 V reverse bias. Typical leakage current densities were 54-171 μA/mm^2 at -3 V bias. Surface damage occurring along the junction mesa from the parallel plate RIE process is suspected to be the primary contributor to the leakage current in the IV characteristics.

GaN Photovoltaic Spectral Response

The photovoltaic GaN UV diodes exhibited peak responsivities of 0.11 A/W at 360 nm as shown in Figure 4. The visible rejection with respect to the peak responsivity at 360 nm was greater than 3 and 4 orders of magnitude at 500 and 800 nm, respectively. The visible rejection was attributed to the absence of yellow defect states in the MBE grown material. The maximum peak responsivity at unity external quantum efficiency was calculated to be 0.23 A/W including a 19% reflection loss at the air-GaN interface; therefore, the internal quantum efficiency deduced from the experimental data was approximately 48%. Excellent linearity of the photodiode response at 360 nm was observed over the incident power range of 5×10^{-10} to 1×10^{-6} W.

Al$_x$Ga$_{1-x}$N Photovoltaic Response

The spectral response for photovoltaic diodes fabricated with nominal 5% and 10% aluminum mole fractions proved the tunability of the Al$_x$Ga$_{1-x}$N materials system. As shown in

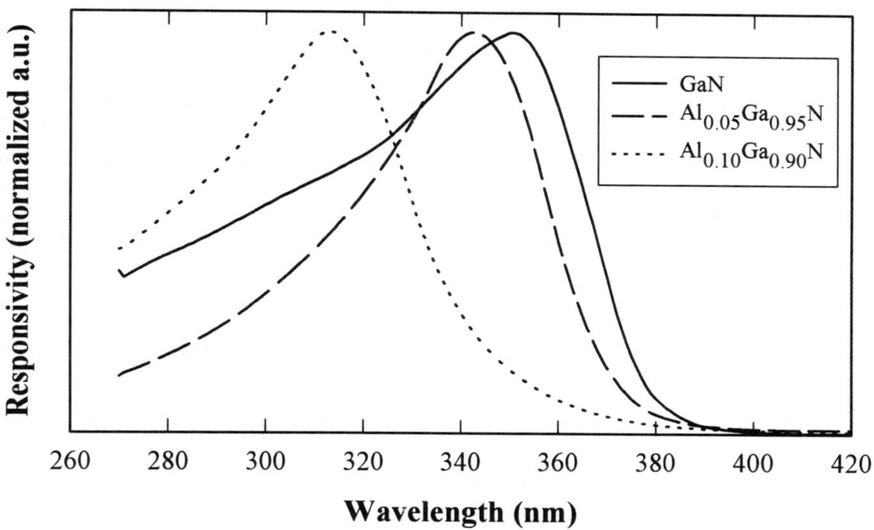

Figure 5: The normalized spectral responsivity for *p-i-n* photovoltaic diodes fabricated from GaN, $Al_{0.05}Ga_{0.95}N$ and $Al_{0.10}Ga_{0.90}N$ epitaxial materials with peak responsivity positions of 360, 343 and 313 nm, respectively.

Figure 5, the responsivity peak of the two *p-i-n* photovoltaic detectors fabricated from $Al_{0.05}Ga_{0.95}N$ and $Al_{0.10}Ga_{0.90}N$ epitaxial materials shifted to 343 nm and 313 nm, respectively, from the peak position of 360 nm for GaN photovoltaics.

CONCLUSIONS

GaN *p-i-n* photovoltaic UV detectors grown by MBE were fabricated with peak responsivities of 0.11 A/W. Spectral characterization was performed over the range of 250 nm in the mid-UV to 800 nm in the red yielding 3 to 4 orders of magnitude visible rejection over the peak responsivity at 360 nm. Spectral data for $Al_{0.05}Ga_{0.95}N$ and $Al_{0.10}Ga_{0.90}N$ *p-i-n* photovoltaic detectors demonstrated the tunability of the GaN-based materials toward the solar blind region of the UV spectrum with peak responsivities at 343 and 313 nm, respectively.

ACKNOWLEDGMENTS

This work was supported by NASA, under contract NAS5-32828, D. B. Mott monitor.

REFERENCES

[1]   M. N. Yoder, IEEE Trans. Electron Dev. **43**, 1633 (1996).

[2]   S.N. Mohammad, A. A. Salvador, and H. Morkoc, Proc. IEEE **83**, 1306 (1995).

[3]   H. Morkoc, S. Strite, G. B. Gao, M.E. Lin, B. Sverdlov, and M. Burns, J. Appl. Phys. **76**, 1363 (1994).

[4]   J. M. Van Hove, G. J. Cosimini, E. Nelson, A. M. Wowchak, and P. P. Chow, J. Cryst. Growth **150**, 908 (1995).

[5]   C. Yuan, T. Slagaj, A. Gurary, A. G. Thompson, W. Kroll, R. A. Stall, C. Y. Hwang, M. Schurman, Y. Li, W. E. Mayo, Y. Lu, S. Krishnankutty, I. K. Shmagin, R. M. Kolbas, and S. J. Pearton, J. Vac. Sci. Technol. B **13**, 2075 (1995).

[6]   S. Nakamura, M. Senoh, and T. Mukai, Jpn. J. Appl. Phys. **31**, L1708 (1991).

[7]   M. A. Smith, V. J. Kapoor, R. Hickman, and J. Van Hove, Electrochem. Soc. Symp. Proc. **96-11**, 133 (1996).

STUDY OF IBAD DEPOSITED AlN FILMS FOR VACUUM DIODE ELECTRON EMISSION

E.W. Forsythe, J.A. Sprague[*], B.A. Khan[**], S. Metha[***], D.A. Smith[***], I.H. Murzin, B. Ahern[****], D.W. Weyburne[****], and G.S. Tompa
Structured Materials Industries, Inc., 120 Centennial Av., Piscataway, NJ 08854
[*]Naval Research Laboratory, Surface Modification Branch, Washington, D.C. 20375
[**]N.A. Philips, Briarcliff Manor, NY
[***]Lehigh University, Bethlehem, PA
[****]ASM International, Hanscom AFB, MA

ABSTRACT

We have demonstrated the basic operation of a vacuum diode based on the negative electron affinity polycrystalline AlN thin film emitters. The AlN films, both undoped and Ge doped, were deposited by ion beam assisted deposition (IBAD). The IBAD process utilizes thermal evaporation from either electron-beam or resistance heated sources with ion bombardment from Kaufman-type ion sources at energies from 50 to 1500 eV. Films were post-annealed by rapid thermal annealing and long-time tube furnace annealing in a N_2 atmosphere to test improvements in crystallinity. The electrical and transport properties of the films were tested by DC I-V measurements. The structure of the AlN films was investigated by TEM, SIMS, optical absorption, and RBS as a function of growth parameters and annealing. The field emission was tested for films with different Ge doping concentrations, film thickness, diode voltage, and post annealing conditions. Field emission was observed for the undoped AlN films with a thickness of approximately 10 nm.

INTRODUCTION

The recent demonstration of high density emission from diamond films has spurred a great deal of interest in the cold cathode applications of diamond films, which offers an opportunity for achieving the goal of stable, efficient and manufacturable vacuum microelectronics devices [1]. Diamond negative electron affinity (NEA) cold cathodes would be a good choice for vacuum microelectronics devices, if diamond could be doped n^+. Otherwise, the parasitic series resistance of the cathode will be too high for high speed applications. However, so far only minimal levels of n-doping have been demonstrated [2]. This may not be the case for AlN, which is a rigid, thermally stable material and was shown to possess a phenomena of NEA [3,4]. The present paper reports the results of using polycrystalline AlN films to make vacuum diodes. To prepare the AlN films, we used the IBAD, since, along with high flexibility and independent control over the main process parameters it also provides deposition of AlN films at substantially lower temperatures (400°C to 600°C) as compared with MOCVD (>800°C), which makes this process simpler.

EXPERIMENT

Film deposition was performed using a custom built Ion Beam Assisted Deposition system (Figure 1). The IBAD system utilizes thermal evaporation from either electron-beam or resistance heated sources with ion bombardment from a Kaufman-type ion source at energies from 50 to 1500 eV [5]. The salient feature of this system is that is allows flexible and independent control of deposition rate, ion energy, and ion

Mat. Res. Soc. Symp. Proc. Vol. 449 © 1997 Materials Research Society

beam current density. For the formation of Ge-doped AlN films, the so-called arrival ratio of N atoms to Al atoms (R-value) was controlled to supply just enough active nitrogen to achieve stoichiometric AlN and co-deposit Ge. Ge/AlN films, in the thickness range of 0.001 to 0.5 μm, were deposited on (001) n- and p-type Si substrates. The N_2^+ ions were obtained by flowing N_2 gas through the ion gun, which raised the background pressure in the chamber to 1×10^{-4} Torr (N_2). Films were grown with an Al deposition rate, controlled by a quartz micro-balance, of 2A/s, with total Al deposition (for most films) equivalent to 2000 A of the metal. Ge deposition rates were controlled at 0, 0.2, 0.4, 0.6, or 1.0 A/s during each deposition run. Since the absolute deposition rate for the Ge was somewhat different from the control set point due to the low deposition rates employed, the total equivalent Ge deposited during each run was reordered. A range of substrate temperatures from 300 to 600°C was used. During deposition, films were bombarded with N_2^+ ions at an energy of 50, 100 (most times) or 200 eV. Ion current densities at the substrates, as measured by three Faraday cups surrounding the substrate holder, were controlled to produce an arrival rate for energetic N atoms to Al atoms of 1.0, 1.25 (most films), or 2.0.

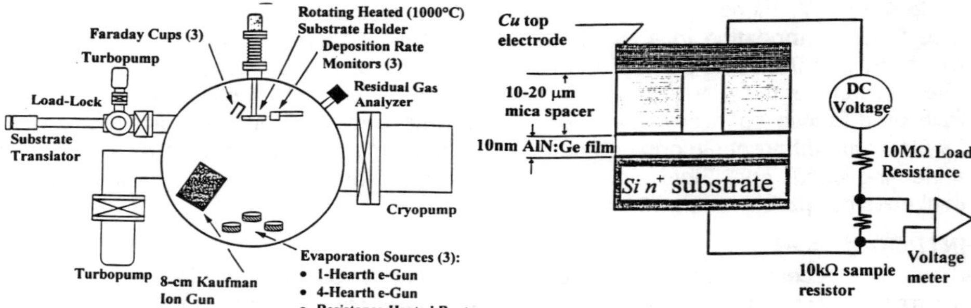

Figure 1. Schematics of IBAD system.   Figure 2. Schematic of vacuum diode device.

For I-V testing, a 3 mm dot of ITO was deposited on top of the AlN film by sputtering the ITO through a physical mask. The ITO was sputtered in a Perkin Elmer Randex sputtering system from a 10" ITO target, in an argon ambient. The argon pressure was set between 6-10 mTorr and the power at 200 W. A DC voltage was applied across the n$^+$ type Si wafer/AlN film/Metal contact structure with a Keithley voltage source model 225 and the corresponding current measured with a Keithley 175 multimeter. The current was sampled with a 1 sec integration time as a function of increasing and decreasing applied voltage.

Table 1. Table of growth conditions for samples selected for field emission testing.

Sample	Al (A/s)	Ge (A/s)	R (N/Al)	T, °C	Thick (A)	Field emission
950707	1	0.001	1.0	600	1000	No
951025a	2	0.2	1.2	615	2660	No
951028	2	0.4	1.0	400	2960	No
951026	2	0.6	1.2	620	3115	No
960215	2	1.0	1.0	600	~100	No
960217	~1	none	~1	565	150	YES

The as-deposited films were annealed using furnace annealing or RTA at various temperatures up to 1200°C in a N_2 ambient. Each sample listed in Table I was divided into four pieces. Two of the pieces were annealed at 900°C and 1000°C for 60 sec, respectively. A third piece was annealed in tube furnace at 1000°C for 30 min. The fourth piece was left as-grown. Several of so-prepared samples that were selected for field emission testing are listed in Table 1.

Figure 2 depicts the field emission device structure. The device consisted of the AlN:Ge film grown on an n+ Si substrate. The substrate was the bottom electrode for the device. Next, a 10-20 μm mica sheet with a 2 mm hole was pressed between the film and the Cu top electrode. A mica spacer was used in place of the SiO_2 layer as this offered a quicker turn around time for testing multiple samples. Before each measurement, the mica and the samples were carefully degreased and cleaned with an acetone wash followed by methanol, which was blown dry with N_2 gas. The samples were immediately loaded into the vacuum chamber to avoid any contamination. The pressure during testing was 8×10^{-8} Torr. First, a forward and then reverse DC voltage was applied with the DC power supply and the field emission current measured as a voltage across a 10 kΩ with a Kiethley multimeter. The voltage across the 10 kΩ resistor was sampled every 2 msec and averaged over 1 sec. The current was averaged over the voltage measurements with error bars for the 1-σ of the average.

The evolution of the structure of the films with temperature was analyzed by observing microstructural changes in the film as a function of annealing with *in-situ* high resolution TEM. A 3 mm disc was cut out for TEM studies from the silicon wafer. These samples were then mounted, substrate side up, on a glass slide covered with molten wax. A small hole was scribed on the sample to allow chemical etching. The sample was then etched using a solution of HF, CH_3COOH and HNO_3 mixed in the ratio of 1:2:6 until the sample perforated. The TEM examination was performed on a Philips CM30 HRTEM at an operating voltage of 300 keV. The in-situ annealing was carried out in the same microscope using a Philips (model # PW6363/00) hot stage holder. A ramp experiment was used to see the microstructural changes accompanying the annealing treatment; the sample was heated in steps of 50°C with an intermediate hold of 5 minutes to allow thermal stabilization of the sample. The maximum temperature attained in the TEM experiment was 850°C.

RESULTS AND DISCUSSION

XRD and TEM results showed that stoichiometric AlN can be formed at temperatures ranging from ambient to 700°C for bombardment with (50 - 200) eV N_2^+ ions, if the arrival rate ratio of N_2^+ to Al is greater than or equal to 0.5 (one energetic N atom per deposited Al atom). As was shown earlier for silicon nitride [6], very little N is incorporated in the deposited Al from the N_2 background gas (P~1×10^{-4} Torr) in the absence of ion bombardment. This is due to the large binding energy of the N_2 molecule. The optical properties of the stoichiometric AlN films deposited at all temperatures and ion energies were good, while the crystalline quality was best for films deposited at temperatures greater than 600 °C with ion bombardment energies less than or equal to 100 eV. The Ge doped AlN films were deposited by simply co-depositing Ge while the Al was being evaporated. The Ge flux was changed to vary the concentration of Ge in the films. Since Ge is also an n-type dopant in AlN, the rate of Ge evaporation was kept substantially higher than the rate required for simple doping of the AlN with Ge. The SIMS depth profiling showed the presence of Al, N, and Ge uniformly distributed through the film. In addition, the results revealed oxygen in the as-grown films, with the highest concentration at the surface. Rutherford backscattering spectroscopy measurements also confirmed the film stoichiometry. The optical absorption was deduced, by measuring the transmission and reflection of the films over the wavelength range 200-2000 nm (0.62 - 6.2 eV photon energy). The absorption edge shifts to longer wavelengths, with increasing Ge content.

The I-V traces have been performed for the thick (~100 nm) as-deposited films only for two growth conditions with a high and low Ge concentration, respectively. The low Ge concentration I-V curves showed a typical diode response, where as the high Ge concentration appeared to be almost resistive (Figure 3).

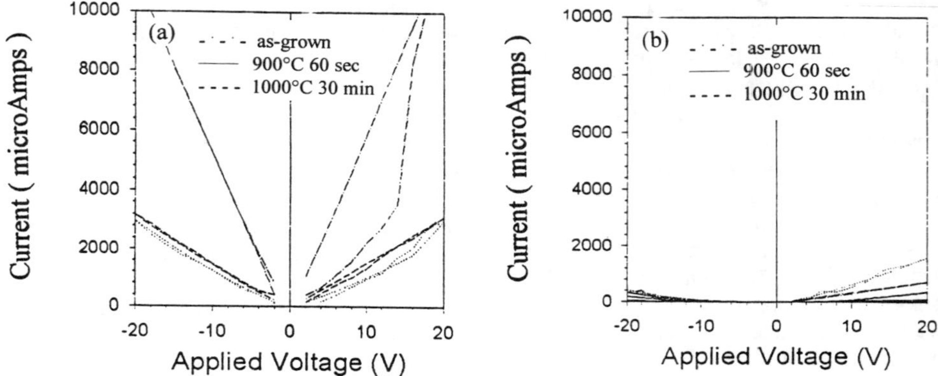

Figure 3. DC I-V traces for sample 951028 doped to 4.9×10^{16} Ge/cm^2 (a) and for sample 960707 doped to at least two orders of magnitude fewer Ge atoms (b).

Figure 4 is a bright field electron micrograph showing the microstructure of as-deposited AlN/Ge film. This representative film was deposited onto (100) Si substrate held at 500°C during deposition with an Al deposition rate of 2Å/s, Ge rate of 0.4 Å /s and N/Al arrival rate ratio of 1.25. Electron diffraction pattern of the as-deposited film shows continuous diffraction rings corresponding to reflections from [0001] zone axis in AlN, indicative of the [0001] fiber texture in the as-deposited film, with [0001] direction of AlN oriented perpendicular to the plane of the film. The high resolution observations of the same film ramp annealed at 800°C and 850°C also confirmed three-fold lattice symmetry of the AlN basal planes. Variation in intensity observed along several rings in the diffraction pattern is due to slight tilt of the sample (Fig. 4). Such variations in ring intensity are commonly observed in textured films. The absence of any "Ge" spots in the diffraction pattern suggests the absence of Ge crystallites in the as-deposited film.

The average AlN crystallite size of the films ramp annealed at 800°C and 850°C obtained from the high resolution images is ~50 nm. The convergent beam electron diffraction pattern from these regions showed diffraction rings corresponding to AlN phase. An electron diffraction pattern taken from the film ramp annealed at 850°C showed extra AlN rings, which were absent in the diffraction pattern of as-deposited sample. This, and the previous observation suggest that possibly new AlN crystallites, with random orientation, nucleated upon annealing the film. Additional tests showed that films with higher Ge concentrations could form Ge_3N_4 and possibly a tetragonal phase of Ge. While not shown, this films also exhibited weak red PL and EL output.

Figure 4. Bright field image of the as-deposited AlN/Ge sample and its electron diffraction pattern.

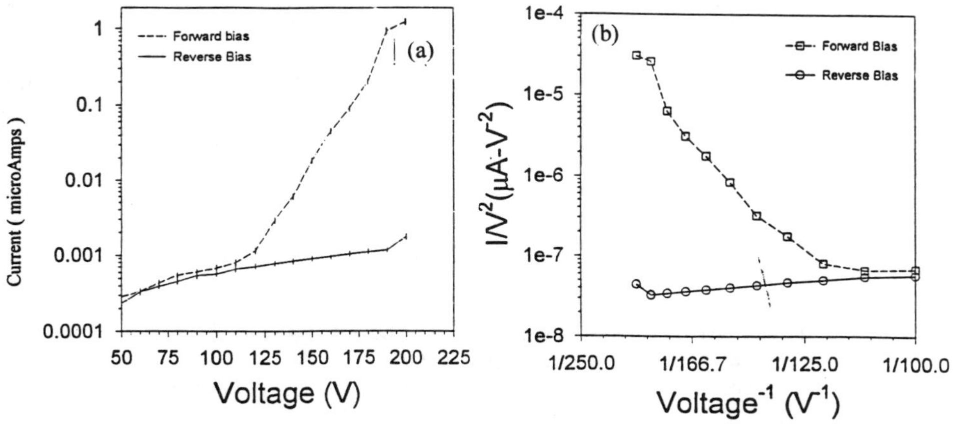

Figure 5. The I-V (a) and I/V^2-V^{-1} curves (b) for the sample 960217 (RTA at 1000°C for 60 sec) demonstrating Fowler-Nordheim current characteristics.

We observed field emission from the AlN films rapid thermal annealed at 900°C and 1000°C for 60 sec (Figure 5a). These films did not contain Ge. The emission currents for the 900°C annealed sample were as high as 1μA with 170 V applied across the diode structure. The field emission current versus forward applied voltage had an exponential dependence. Further, the emission current exhibited a Fowler-Nordheim type dependence, as indicated in Figure 5b, where the data is re-plotted with an I/V^2 versus $1/V$ dependence. The straight line verifies the single barrier characteristics of the materials. This barrier may be the surface barrier due to the work function. The emission current showed an onset voltage at approximately 100 V and a saturation current at approximately 200 V. The AlN films tested in this effort were grown n-type on Si substrates. Thus, the onset voltage may result from the limited electron injection from the substrate into the AlN film due to the band misalignments. The emission measurements were repeated several times. Reverse bias showed no leakage. We speculate the saturation current results from an effective series resistance.

The additional samples listed in Table 1 were also tested for field emission and current was not observed. UPS measurements showed NEA properties on the emitting sample, but not on not emitting sample.

SUMMARY

XRD and TEM results showed that stoichiometric AlN can be formed at temperatures ranging from ambient to 700°C for bombardment with $(50 - 200)$ eV N_2^+ ions, if the arrival rate ratio of N_2^+ to Al is greater than or equal to 0.5. The as-deposited AlN/Ge film was almost completely crystalline and had a strong fiber texture with [0001] axis of hexagonal AlN perpendicular to the plane of the film. Additional measurements showed that films with higher Ge concentrations could form Ge_3N_4 and possibly a higher presence tetragonal phase of Ge. SIMS and RBS showed the presence of Al, N, and Ge uniformly distributed through the film.

The DC I-V curves taken from the as-deposited films exhibited typical diode behavior. We have demonstrated field emission from thin AlN films grown on n-type Si substrates. The films were post annealed at 900°C and 1000°C for 60 sec. The AlN films were approximately 10 nm thick. The emission current exhibited a Fowler-Nordheim type dependence. UPS showed no NEA property on non-electron emitting samples and NEA property on the emitting sample.

ACKNOWLEDGMENT

This work was supported by BMDO/IST as administered by the AF in contract Nu. F19628-95-C-0206. We would also like to thank M.C. Benjamin and R.J. Nemanich for their assistance with UPS NEA measurements.

References

1. "Diamond Films and Coatings: Development, Properties and Applications", Edited by Robert F. Davis, Noyes Publications (1993).
2. K.Okano, H.Kiyota, J.Iwasaki, Y.Nakamura, Y.Akiba, T.Kurosu, M.Iida and J.Nakamura, Appl. Phys. A. 51, 1990, 344.
3. R.F. Davis, 2nd Workshop on Wide Bandgap Nitrides, Oct. 17, 1994, St. Louis, MO.
4. R.J. Nemanich, M.C. Benjamin, S.P. Bozeman, M.D. Bremser, S.W. King, B.L. Ward, R.F. Davis, B. Che, Z. Zhang, J. Bernholc, "Negative Electron Affinity of AlN and AlGaN Alloys", MRS Proc, Symp. AA, 1995.
5. G.K. Hubler, D. Van Vechten, E.P. Donovan, and C.A. Carosella, J. Vac. Sci. Technol. A8 (1990) 831-839.
6. "AlN Negative Electron Affinity Surface", Final Report, Phase I SBIR, Contract Nu. F19628-95-C-0206, 1996.